Microbiology:
Dynamics and Diversity

Jerome J. Perry, Ph.D.
Professor
Department of Microbiology
North Carolina State University

James T. Staley, Ph.D.
Professor
Department of Microbiology
University of Washington

Saunders College Publishing
Harcourt Brace College Publishers

Fort Worth • Philadelphia • San Diego • New York • Orlando • Austin • San Antonio
Toronto • Montreal • London • Sydney • Tokyo

Text Typeface: Baskerville
Compositor: York Graphic Services, Inc.
Executive Editor: Edith Beard Brady
Developmental Editor: Cathleen E. Petree
Art Developmental Editors: Beverly McMillan, Leslie Ernst
Managing Editor: Carol Field
Project Editor: Linda Boyle
Copy Editor: John Beasley
Manager of Art and Design: Carol Bleistine
Senior Art Director: Joan Wendt
Cover Designer: Joan Wendt
Text Artwork: Rolin Graphics
Photo Researcher: Amy Ellis Dunleavy
Vice President of EDP: Tim Frelick
Manager of Production: Joanne Cassetti
Senior Product Manager: Sue Westmoreland
Cover Artwork: This biomass pyramid illustrates three different trophic levels of a four level aquatic food chain. The base of the pyramid shows a primary producer, in this example a cyanobacterium, which serves ultimately as a food source for all other higher trophic levels, called consumers. The primary producer is eaten by small aquatic crustaceans, which are themselves eaten by fish, that are in turn, eaten by animals, including humans at the apex. (Photos courtesy of James T. Staley)

Printed in the United States of America

Microbiology

ISBN: 0-03-053893-9

Library of Congress Catalog Card Number: 96-069306
6789012345 032 10 9 8 7 6 5 4 3 2 1

Dedication

To Elizabeth for her many contributions, including typing my manuscript and to Jerome, Jr., Marianne, and Chuck.

J. J. Perry

To Sonja for her support and patience and to Greg and Wendy.

J. T. Staley

Preface

One of the many reasons we wrote this textbook, *Microbiology: Dynamics and Diversity,* was to portray in a straightforward and lucid manner the exciting developments that are occurring in microbiology today. Just as quickly as new microorganisms are being identified and classified, they are changing the direction of the health, agricultural, and environmental industries. At the same time, microorganisms are important life forms to study for their own sake, not just as research tools toward other ends. Microbial biomass on Earth exceeds that of plants and animals. Therefore, we consider it imperative for general understanding to discuss the basic biochemical and physiological principles underlying microbial life, because these principles direct all life.

In this textbook, emphasis has been placed on the evolution of concepts in microbiology from the observations of van Leeuwenhoek to the contributions of the great scientists in the latter part of the 19th century—Robert Koch, Louis Pasteur, Martinus Beijerinck, and Sergei Winogradsky—which was indeed the Golden Age of Bacteriology. Their discoveries, and the discoveries of many that followed in their footsteps, led to the broad area of science known as microbiology. Some of today's major voices in microbiology appear in this text's unit opening interviews.

A remarkable and fascinating aspect of the microbial world is the staggering diversity of the forms and their importance to humans and the biosphere. This diversity is exemplified by the impact of bacteria on the evolution of all other living organisms, the physiological tolerances and ranges over which microorganisms live and function, the fascinating and unique metabolic pathways utilized by microorganisms that evolved through their four billion years on Earth, and the consequences of ecological activities and geochemical transformations carried out by microorganisms. Microorganisms drive many of the life processes on Earth—macroscopic organisms are utterly and completely dependent on them. Yet, as surprising as it may seem, biologists are only just beginning to appreciate the fascinating diversity of microbial life.

Although the microorganisms are worthy of even more attention than they have been given in this text, we recognize that their lives and activities are of interest, not only in and of themselves, but to the whole of biology. Microbiology is the biological science that has influenced and shaped all other biological sciences. For this reason, we have attempted to provide the student with a broader context for the study of microorganisms, keeping in focus our other basic themes: the evolution of ideas within microbiology; the process of scientific inquiry; the significance of molecular approaches in the study of microbiology. Therefore, we present their cell and molecular biology, genetics, physiology, metabolism, and the importance they have played in developing our understanding of other biological fields. In addition, we explore the microbial roles in plant and animal symbioses, disease, and immunity. We also discuss their larger role in the biosphere and ecosystems. Finally, we cover important microbial activities as they shape other fields including biotechnology and the environmental, food, and health industries.

Student Audience

This text is organized so that students are presented topics in a sequence that allows them to develop their understanding of microbiology in a logical, reinforcing manner. We believe that the microbe is the perfect model system for studying metabolism, genetics, and molecular interactions, and this view has been adopted in the organization and presentation of the text material. We recognize that undergraduate students who use this book will have varied backgrounds. Some students who are microbiology or biology majors may have formally studied other biological areas prior to reading this text. The text is organized to accommodate different groups of students including students with an intermediate level of understanding. For example, it may also serve as supplementary reading in a diversity or taxonomy course.

Organization

The text uses a levels of organization approach beginning with basic biochemistry, structure/function, nutrition, growth, metabolism, physiology, and genetics. These basic concepts are then followed by microbial taxonomy and evolution, a treatment of the diverse microbial groups, microbial and viral disease, microbial ecology, immunology, and industrial and environmental microbiology.

Microbiology: Dynamics and Diversity is presented in eight parts, with two to seven chapters in each part. Part 1 is an overview of the science of microbiology. Chapter 1, "The Scope of Microbiology," introduces the reader to the microbial world. Chapter 2 provides the reader with the development of concepts in the field. For many students, Chapter 3 about chemistry will be a useful review. Part 1 ends with Chapter 4 on the structure and function of Eubacteria and Archaea.

Microbial physiology is the subject of Parts 2 and 3, which illustrate in particular the diverse energy-sustaining strategies adapted by microorganisms for survival. Chapter 5 describes the wide variety of nutritional types of microorganisms. The growth of single-celled organisms is treated kinetically in Chapter 6, followed by the control of microbial growth in Chapter 7. Part 3 covers metabolism, particularly the diversity of metabolic types. Chapter 8 presents the variety of pathways utilized by microorganisms in generating energy chemically, while Chapter 9 addresses the variety of photosynthetic processes used to generate energy by an array of microorganisms. Following these two chapters on energy generation are two chapters on biosynthesis. Chapter 10 discusses the synthesis of major monomers, including amino acids, needed for macromolecular synthesis. The synthesis of protein and peptidoglycan is treated in Chapter 11. The final chapter in this part, Chapter 12, presents information on the biodegradation of selected organic compounds.

Part 4 presents an up-to-date treatment of genetics and virology. Basic genetics principles are discussed in Chapter 13. In order to thoroughly present gene transfer between bacteria through viral transduction, which is covered in Chapter 15, viruses are introduced in the preceding chapter, Chapter 14. The last chapter in this part, Chapter 16, presents recombinant DNA.

In Part 5, we look at microbial evolution and diversity. The authors have attempted to provide as complete coverage of these topics as reasonable for this level of treatment. The taxonomy and phylogeny of bacteria is discussed in Chapter 17, which also presents the universal tree of life, derived from Carl Woese's three Domains model. The taxonomy chapter is followed by one on evolution, Chapter 18, that illustrates the prominent role microorganisms have played in the evolution of all life and in the diversification of the biosphere.

The diversity chapters provide a well-balanced treatment of the various microbial groups. They can be taught in their entirety or used as reference. The chapters are organized phylogenetically according to the latest research developments. Three chapters discuss the various Eubacterial groups: Chapter 19 on gram-negative heterotrophic Eubacteria; Chapter 20 on gram-positive heterotrophic Eubacteria; and Chapter 21 on phototrophic, chemolithotrophic, and methylotrophic Eubacteria. Our treatment of the cyanobacteria in Chapter 21 exemplifies our belief that this important group of bacteria is often neglected, if not overlooked, in most other comparable texts. This chapter is followed by Chapter 22 on the novel group, Archaea, which contains some of the most fascinating, unusual forms of life on Earth. Finally, Chapter 23 closes Part 5 with a discussion of the evolution and remarkable diversity of eukaryotic microorganisms.

The significance of microorganisms and their diversity in ecological processes and geochemical transformations becomes clear in Part 6, "Microbial Ecology." Chapter 24 presents information about the roles microorganisms play in ecosystems. Chapter 25 highlights some fascinating beneficial symbiotic associations between microorganisms and other microorganisms, as well as microorganisms and macroorganisms. Chapter 26 covers the special symbiotic relationship of host/parasite, focusing on nonspecific host resistance.

An especially important area of host/parasite associations impacts on health and disease. Part 7 presents medical microbiology and immunology. Chapters 27 and 28 discuss immunology, the specific strategies of host resistance. This is followed by Chapter 29 which addresses microbial diseases, and Chapter 30 on viral diseases. The final chapter in Part 7, Chapter 31, highlights epidemiology and clinical microbiology.

The last part, Part 8, is entitled "Applied Microbiology," about microbes put to work—microbes as tools. Many new and important careers for microbiologists will continue to be found in these areas. The first of these chapters, Chapter 32, deals with industrial microbiology and biotechnology. The final chapter, Chapter 33, covers the area of environmental microbiology, including water and wastewater treatment and bioremediation.

Features

This text has been written and designed for the student's ease of use and encouragement in learning this vast body of knowledge.

- Each chapter is structured with the following elements: a chapter-opening quote from a past or present scientist or writer gives cultural context; a chapter outline of section headings previews the chapter contents; and the first paragraph of each chapter offers an introductory overview. At the end of each chapter, a narrative summary with boldfaced key terms, suggested readings, and thought questions all help to review the chapter contents.
- Each chapter includes one or more special highlight boxes that fall within three major themes: "Research Highlights" focus on recent or significant re-

search; "Milestones" offer historical or biographical essays on the evolution of important ideas or concepts in microbiology; "Methods and Techniques of Microbiology" discuss the tools used by microbiologists in the course of their work.

- The presentation of diversity in this textbook is phylogenetic in orientation and emphasizes the three Domain classification of organisms. The diversity coverage is more evolutionary in its approach than in other texts because it is based on the newest ribosomal RNA sequencing techniques of molecular biology.
- In addition to the normal tables that appear in all chapters, Chapters 19–23 present uniquely designed diversity tables that help the student categorize and compare the immense variety of microorganisms. The diversity tables are of three types: Reference tables are a means for quick identification; Differential tables list the differences among the taxa; and Descriptive tables present typical common features of particular taxa.
- The text presents microbiology for majors, focusing on the microbe itself rather than on diseases caused by microorganisms. Recurring themes are the evolution of ideas within microbiology; an emphasis on the process of scientific inquiry; the significance of molecular approaches in the study of microbiology.
- A balanced and contemporary chapter on the Archaea explores these extraordinary organisms, which live under some of the most unusual conditions for life on Earth.
- A separate chapter on beneficial symbioses clearly defines the role of the prokaryotes in the evolution and survival of eukaryotic life forms.
- A separate chapter on eukaryotic microorganisms brings the taxonomy of this important group into perspective. It is the first textbook chapter of its kind in which the presentation is organized around the phylogenetic classification of these microorganisms.
- Taxonomy, phylogeny, and evolution appear in separate chapters to give more emphasis to each area. Because microbial taxonomy is becoming more phylogenetic in nature, increased emphasis is given to molecular sequencing and phylogenetic analysis procedures in this text compared with other available texts.
- The microbial ecology chapter provides a basic and comprehensive overview of this important developing area of investigation.
- An important chapter on biodegradation, Chapter 12, is often overlooked in texts. It is offered here as an additional metabolism chapter.

Art Program

Special care was taken to develop an art program that immediately clarifies the main concepts and is accurate, as well as aesthetically appealing. The rendering of the line art was coordinated with the development of text and legends to ensure pedagogical soundness. A specially designed system of phylogenetic tree diagrams appears throughout the diversity chapters to provide a quick relational reference for the microorganisms under discussion.

Interviews

One of the special features of this textbook is the interview entitled "A Conversation With . . ." that opens each of the eight parts. Each of the interviews conveys a different personal view into the career of a microbiologist. We are appreciative to the scientists who took time from their busy schedules to provide their unique life stories and insights into the particular fields in which they work. We believe this special feature of the book will prove to be inspirational to students who are curious about careers and research activities in microbiology.

Part 1, The Scope of Microbiology

A Conversation with R.G.E. Murray

Part 2, Microbial Physiology: Nutrition and Growth

A Conversation with Lucy Shapiro

Part 3, Microbial Physiology: Metabolism

A Conversation with Fred Neidhardt

Part 4, Genetics and Basic Virology

A Conversation with Martha Howe

Part 5, Microbial Evolution and Diversity

A Conversation with Carl Woese

Part 6, Microbial Ecology

A Conversation with Eugene Nester

Part 7, Immunology and Medical Microbiology

A Conversation with Stanley Falkow

Part 8, Applied Microbiology

A Conversation with Arnold Demain

Ancillaries

The supplements package available for *Microbiology: Dynamics and Diversity* includes the following:

Investigating Microbiology: A Laboratory Manual for General Microbiology, by Philip Stukus, Denison University. This is a flexible combination of traditional and investigative experiments, allowing the instructor to choose the extent to which the lab experience will be investigative. This exciting new manual encourages independent thought and experience in microbial analysis, experimental design, critical thinking, and cooperative learning using basic techniques.

Student Study Guide, William Coleman, University of Hartford. This guide will help students focus their study efforts and further develop critical thinking skills. Each chapter features a key concept summary, comprehensive self-test questions, critical thinking questions, a vocabulary review, and text figures for labeling.

Many chapters will include study diagrams requiring students to complete a concept map or to draw their own illustration to encourage active learning.

Instructor's Manual with Test Bank, William Coleman, University of Hartford. The instructor's manual contains chapter-by-chapter guidelines for teaching the course. The test bank comprises questions written by William Coleman with contributions from the authors and professors actively teaching microbiology courses. The bank of 1,000 multiple choice and essay type questions will range in difficulty from simple recall to questions requiring synthesis and application of concepts.

Computerized Test Bank. ExaMaster™ Computerized Test Bank enables instructors to edit, revise, add to, or delete from the printed Test Bank. Available in IBM or Macintosh versions.

Overhead Transparencies. A set of 100 transparencies is available to adopters of the textbook to make classroom instruction more productive.

Saunders College Publishing may provide complimentary instructional aids and supplements or supplement packages to those adopters qualified under our adoption policy. Please contact your sales representative for more information. If as an adopter or potential user you receive supplements you do not need, please return them to your sales representative or send them to

Attn: Returns Department
Troy Warehouse
465 South Lincoln Drive
Troy, MO 63379

Contributors

Microbiology is such a burgeoning field that it would be impossible for just two authors to write with authority about all the areas of the discipline. We are indebted to several "guest" microbiologists who contributed their expertise in writing important chapters in this book. Stephen Lory, University of Washington, wrote three well-received chapters on microbial genetics. Ian Tizard took time from his own books to offer here the two chapters on immunology. We also want to express our sincere thanks to William "Barney" Whitman and John Gunderson, who wrote the fascinating and timely chapters on the Archaea and eukaryotic microorganisms, respectively. Finally, we wish to thank Holger Jannasch for his major contribution to the chapter on symbiotic interactions. Please see the "Book Team" pages to learn more about these scientists.

Acknowledgments

This book is the result of many years of writing and teaching, or as we like to say, "years of rumination interrupted by infrequent bouts of writing." We recognize that it would not have been possible to complete this book without the encouragement and help of many colleagues in our own departments and elsewhere in the United States, some of whom were reviewers of chapters. Some individuals are especially deserving of our gratitude and are singled out; but we are indebted to all microbiologists with whom we interact at meetings, seminars, and at other forums in the hurly-burly world of microbiology.

We wish to thank William G. Bryden, who first talked with us about the book and with whom we signed the original contract at HBJ. Several intervening years elapsed while the publishing industry went through a series of upheavals. We are especially grateful to Cathleen Petree, our freelance editor at SciQ Publishing, who resurrected our own interest in the book. Her unstinting dedication to the book stirred us during the gestation period and her persistence kept us writing until the end (is it really over yet, Cathleen?). Zanae Rodrigo, a freelance editor/science writer made excellent suggestions during the initial development of the chapters.

We also owe the editorial staff at Saunders College Publishing many thanks for their help. We particularly thank Edith Beard Brady, Executive Editor, for her help with the interviews, her novel ideas, her uncommon economic sense, and good humor. Leslie Ernst and Beverly McMillan did marvelous things for the graphics and illustrations as the art developmental editors. Amy Ellis Dunleavy greatly assisted our photo procuring efforts. Others at Saunders have also provided much necessary insight and support. Linda Boyle, Project Editor, provided excellent and timely manuscript proof. We also thank Julie Alexander, former Executive Editor, for her past support at the home office in Philadelphia over the years.

Reviewers

Our heartfelt thanks go to our colleagues who provided reviews of many of the chapters throughout revisions of the manuscript, and who participated in focus groups. They are listed.

John Adams, University of Wyoming
Kevin Anderson, Mississippi State University
Larry Barton, University of New Mexico
Ronald L. Bridges, Kansas State University
Alfred E. Brown, Auburn University
Kim Burnham, Oklahoma State University
William Coleman, University of Hartford
Tim Donohue, University of Wisconsin/Madison
Louis R. Fina, Kansas State University
J. W. Fitzgerald, University of Georgia
William C. Ghiorse, Cornell University
George Hegeman, Indiana University
Ilsa Kaattari, Oregon State University
Robert Kearns, University of Dayton
Timothy Kral, University of Arkansas
Susan Leschine, University of Massachusetts/Amherst
Richard Myers, Southwest Missouri State University
Penelope Padgett, Shippensburg University
Grace Ann Spatafora, Middlebury College
George Stewart, University of South Carolina Medical School
Philip Stukus, Denison University
Timothy Wong, University of Washington

We also wish to acknowledge the many scientists and journal and text publishers that have provided permission for use of figures and tables.

Special Thanks

J. J. Perry wants to express his gratitude to several colleagues at North Carolina State University: C. Lee Campbell, Larry F. Grand, Geraldine H. Luginbuhl, Wesley E. Kloos, Eric S. Miller, Ian Petty, and Thomas J. Schneeweis. Thanks to Elizabeth A. Foley for her invaluable assistance. I am especially indebted to Linda E. Rudd for her library trips and for reading much of the manuscript.

Jim Staley wishes to thank several faculty members at the University of Washington for their advice and support including Eugene W. Nester, John Leigh, Mary Bicknell, Dale Parkhurst, J. C. Lara, Janis Fulton, Denise Anderson, Beth Traxler, Stephen Lory, and Tim Wong. In addition, he appreciates the interest and helpful comments of several graduate students, in particular, John Gosink, Allison Geiselbrecht, Brian Hedlund, and Matt Stoecker. Douglas Vollgraff and Nancy Quense in the departmental office provided much welcome assistance with support services.

Inasmuch as this is the first edition of this textbook, we welcome comments from students and faculty alike on how the book can be modified. Please send us your suggestions for improvement. We hope that you will all enjoy what you read about these microorganisms who could tell us all many more marvelous stories if only they could speak and write.

Dr. J. J. Perry
Department of Microbiology
North Carolina State University
Raleigh, NC 27695
jjp@mbio.ncsu.edu

Dr. Jim Staley
Department of Microbiology
University of Washington
Seattle, WA 98195
jtstaley@u.washington.edu

Book Team

Microbiology: Dynamics and Diversity is written by experienced teachers and expert researchers to offer students an authoritative and accurate voice across the wide range of topics essential in a microbiology course.

Authors & Contributors List

Jerome J. Perry North Carolina State University: Dr. Perry received his Ph.D. from The University of Texas in 1956. He held the position of Research Scientist in the pharmaceutical industry from 1951–1958, working primarily on the isolation of antibiotics. He returned to academia as Research Scientist at The University of Texas 1961–64, then left for North Carolina State University, where he has been teaching general microbiology for more than 20 years. His major research interests are in metabolism of gaseous alkanes, co-oxidation, and bioremediation. Dr. Perry has written extensively on hydrocarbon metabolism in reviewed papers, chapters, and various proceedings. In addition to this textbook, Dr. Perry co-edited *Introduction to Environmental Toxicology,* a major textbook in its field. His major avocation is track and field. He serves as chair of the Track and Field Officials Committee in USATF and has officiated in major meets including the Summer Olympics of 1984 and 1996.

James T. Staley University of Washington: Dr. Staley received his B.A. degree in mathematics from the University of Minnesota, his M.Sc. in bacteriology from The Ohio State University, and his Ph.D. in microbiology from the University of California at Davis. He was a faculty member at Michigan State University and the University of North Carolina before coming to the University of Washington, where he is now Professor of Microbiology. He serves on the editorial boards of several journals and is a trustee of Bergey's Manual Trust. Dr. Staley's research interests include bacterial taxonomy and phylogeny as well as microbial ecology. He has studied the prosthecate bacteria and the planctomycetes group and has conducted ecological studies of freshwater, marine, and desert microorganisms. He is currently investigating polar sea ice bacteria and bioremediation.

Stephen Lory University of Washington: Dr. Lory attended the University of California, Los Angeles, as an undergraduate majoring in bacteriology, where he also did his graduate work in microbiology. For his doctoral thesis project he studied the structural and functional characteristics of bacterial protein toxins. He then did his post-doctoral training at Harvard Medical School, where he studied the mechanism of bacterial protein export. In 1984 he joined the Microbiology Department at the University of Washington, where he is currently a Professor. His research laboratory at the University of Washington is engaged in studies of pro-

tein secretion by gram-negative microorganisms, and genetic regulatory mechanisms controlling bacterial virulence. He serves on the editorial boards of several microbiological journals. He teaches graduate courses in bacterial physiology, bacterial pathogenesis, and participates in a team-taught course for medical and dental students, where he lectures on aspects of bacterial physiology and genetics.

Holger Jannasch Woods Hole Oceanographic Institution: Dr. Holger Windekilde Jannasch studied microbiology, zoology, botany, and organic chemistry at the Universities of Göttingen and Munich obtaining a Ph.D. in 1955 at Göttingen. As an Assistant Scientist of the Max-Planck Institute for Limnology, he spent the years 1957–59 at Scripps Institution of Oceanography at La Jolla, California, Hopkins Marine Station, Pacific Grove, California, and Department of Microbiology at the University of Wisconsin, Madison. On his return to Germany, he joined the Department of Microbiology at the University of Göttingen to complete the habilitation procedure for "Privatdozent" in microbiology. In 1963 he accepted an appointment as Senior Scientist at the Woods Hole Oceanographic Institution, Woods Hole, Massachusetts. He received the Henry Briant Bigelow Medal in Oceanography (1980), the Fisher Scientific Award in Applied and Environmental Microbiology (1982), and the Cody Award in Ocean Sciences (1992). He was elected member of the Göttingen Academy of Sciences (1984), the American Academy of Arts and Sciences, Cambridge (1987), and the National Academy of Sciences, Washington, D.C. (1995).

John Gunderson Tennessee Tech University: Dr. Gunderson received a B.S. degree (Zoology and Russian) from the University of Nebraska and a Ph.D. from the University of California at Berkeley, where he specialized in protozoology. He did post-doctoral research in Mitchell Sogin's laboratories at the National Jewish Center for Immunology and Respiratory Medicine (Denver, Colorado) and at the Center for Molecular Evolution (Marine Biological Laboratory, Woods Hole, Massachusetts), where he used ribosomal RNA sequences to study protistan phylogeny. He is interested in protistan phylogeny and the application of molecular techniques to the study of protistan-bacterial interactions. He is currently an adjunct professor at Tennessee Technological University, where he teaches microbiology.

William Whitman University of Georgia: Dr. Whitman received his B.S. in 1973 from SUNY at Stony Brook and his Ph.D. from The University of Texas in 1978. He is currently teaching at the University of Georgia where he is a Professor of Microbiology, Biochemistry, and Marine Sciences. He has always been interested in unusual microorganisms. His current research focuses on carbon metabolism in the methanogens in the larger picture of trying to understand their highly specialized physiology and growth properties.

Ian Tizard Texas A & M University: Dr. Ian Tizard obtained his BVMS and B.Sc. degrees from the University of Edinburgh and his Ph.D. degree in immunology from the University of Cambridge. He worked at the University of Guelph, Ontario, Canada for ten years and developed a program in parasite immunology focusing on African trypanosomiasis. In 1982 he moved to Texas A & M University to be Professor and head of the Department of Veterinary Microbiology for eight years. At the present time he is Professor in the Department of Veterinary Pathobiology and has an active research program on the use of immunomodulators, focusing on their roles in promoting wound healing and reducing inflammation. He has written the standard text in veterinary immunology which is now in its fifth edition. He has also written a basic immunology text now in its fourth edition, published over 100 scientific papers in refereed journals and has made scientific presentations worldwide. He is actively engaged in consulting on immunological aspects of animal diseases and on animal vaccines.

Contents Overview

Contents

Part 1

The Scope of Microbiology
A Conversation with R. G. E. Murray

Professor **R. G. E. Murray's** leadership accomplishments in microbiology are manifested in many ways. Foremost among them was his service as president of both the Canadian Society of Microbiology as well as the American Society for Microbiology. In addition, he served as chair of *Bergey's Manual* Trust for 14 years. Dr. Murray was also an editor of several journals, including *Bacteriological* (now *Microbiological*) *Reviews* and the *International Journal of Systematic Bacteriology,* and served on the editorial and publication boards of numerous others.

R. G. E. Murray is best known scientifically for his work on bacterial structure–function and taxonomy. He has been a leader in these fields for many years and has made numerous contributions to our understanding of the bacterial cell—in particular, the structure and function of cell envelope layers. Dr. Murray has also made important research contributions in the area of taxonomy. Perhaps the best example of this is the characterization of the *Deinococcus* group of bacteria and discovering that they constitute, along with *Thermus,* a separate phylogenetic group of the Eubacteria.

Professor Murray recently retired from the University of Western Ontario in London, Ontario, where he was professor and chair for many years.

JS: How did you become interested in microbiology as a career?

RM: I had two fortunate circumstances. I chose the right parents. My father was a bacteriologist and interested very generally in biology, aside from his concentration on pathogenic bacteria. Also, I was introduced to an old brass microscope early in life. So, I was exposed to microbes in a Leeuwenhoekian way. I enjoyed watching microbial life, although admittedly early [on] it was mostly protozoa and other larger microorganisms.

I went into science with the intention of going into medicine, not necessarily of practicing medicine, but as a basis for scientific work. I thought I might be interested in pathology—the pathology aspect of disease rather than microbial. It just turned out that my interest got smaller and smaller. Initially, I did an honors degree in pathology at Cambridge University, and pathology included bacteriology. But I had a broad interest, starting in zoology and entering biochemistry, physiology, anatomy, and pharmacology at Cambridge. In 1941 I returned to Canada to finish medical school. Then I did an internship in bacteriology, which was interesting because I had access to every infected patient in a big hospital. So I got all sorts of interesting experience there. I worked at the bench for a year using early versions of *Bergey's Manual* and was indoctrinated in clinical bacteriology.

(Courtesy of Dr. R.G.E. Murray)

JS: You have made many important contributions to the study of prokaryotic structure and function. What would you regard as your most important contributions and why?

RM: I suppose the most useful thing was that I helped describe the structure of bacteria. At that time (1940–1960), people weren't sure what bacteria were or even that you could really describe them as being cellular. I was fortunate in getting interested early on in the cytology of bacteria and having an able cytologist coming to work with me—Carl Robinow, a remarkable man. We studied bacterial structure and how to describe them. Many microbiologists were still talking about bacteria as if they were plants. My father and I had argued together about this and agreed with C. B. van Niel, R. Y. Stanier, and others who thought you had to include cell structure in descriptions. Really, this was the most important thing we did, gradually, along with efforts by many colleagues around the world.

JS: What are the most important contributions that studies of prokaryotic structure function have made to the field of biology?

RM: For one, discovering bacteria as cells; and secondly, enabling their use as model systems for the exploration of cellular biology. There are many examples. In the structure–function area we needed to find out what things were made of, and this required combining electron microscopy with cell fractionation, the analysis of the chemical composition of bacterial cell walls, culminating in the description of peptidoglycan, teichoic acids, and so forth.

JS: What exciting areas do you see developing during the next decade or so in the study of structure and function of bacteria?

RM: I think there is still a lot to be done on the integration of processes and function and the structure that is involved with that function. Many functions are interrelated in very complex ways, requiring sophisticated biochemistry to understand them. That is today's donkey work. My personal bias as to what needs to be done and probably what will have to be done is more in the line of physical chemistry, and that is to understand why so many biological reactions are so remarkably rapid and work so well. These chemical transformations occur in the interstices of structure of one sort or another—in other words, in very thin layers of water, water with all sorts of other things in it. I think there is a lot to be done in understanding how biological functions really progress. Why is it that they're so efficient? And, of course, the work of today is all involved in genes and gene products and the regulation of gene function. Putting the whole thing together requires another map and that is yet to be drawn. It is really not well enough integrated yet. There are all sorts of interesting items, isolated bits and pieces, which will gradually form a picture.

JS: What about higher microorganisms?

RM: Cells are cells are cells. But bacteria are more diverse than anything else as far as energy transformations are concerned and that is what is crucial to the biosphere. It's the reason we operate at all on this Earth.

JS: Bacteria are so surprising because of their great diversity. Can you provide one or two examples of a time when you learned something that you could not have foreseen that really surprised you?

RM: Probably the most exciting surprise that I remember was when we first used a negative stain (which is the use of a heavy metal salt in solution to put around fractions of cells so that the metal will scatter electrons, because cells and organic substrates are relatively nonscattering) in the electron microscope and actually saw the images and contours of macromolecules and how they were arranged. The first time seeing that was really something. That was the basis for understanding macromolecular arrangements and suggested experiments to do to control fractionation of cellular components. Then we learned how they assembled and the rules of assembly and how structures were made. It was exciting.

JS: Can you give one or two examples of how the development of instrumentation has changed or enhanced our view of bacterial structure?

RM: The reason the electron microscope became important is that it offered much higher resolution than was possible with light microscopy. The thing was, you took your microscopy seriously and tried to see what you could see. I tell students if you want to see something in the bush, let's say a deer, you have to have in your head a picture that will allow you to recognize it. Suddenly, maybe you *will* see something. In the same way, when McFarlane Burnett, the great Australian microbiologist, heard about influenza virus causing agglutination of red corpuscles, he said, "I've already seen that—it just didn't gel." We used the chorionic membrane of eggs to cultivate many viruses. When you had an infected egg, you broke this membrane and the blood spilled out. If influenza virus was present, you had agglutination. Virologists saw it all the time, but you had to *recognize* it. You have to be prepared to see things. So it is with instruments, they are really only as good as the person who is using them.

JS: You have been involved in studies of bacterial taxonomy, phylogeny, and diversity. What changes do you foresee in this field in the next decade?

RM: Taxonomy has two faces; one is practical. That is, you need to be able to recognize friend or foe or to recognize that something's there that you've never met before. This means you need to have a record of bacteria you've met before and they have to be findable in that record. Therefore, they have to be put in order. So, taxonomy is a way of looking at life in an ordered way so that it can be used.

The other side of taxonomy is that life forms, we believe, were not all created; each one evolved from some relative that lived before. Maybe one original set that we don't know yet. Then, there must be an order that is determined by the order of evolution. And, are the two sides of taxonomy the same? Now, we have in the last 15 to 20 years developed methods of deciding about the order of evolution. We've spent the last 150 years or so trying to list and order the things we've isolated or looked at in terms that we were able to determine at those times. So now we have two sets of information. One set says we have a lot of living things and some of them are alike and some are not alike and they form patterns of forms, so-called phenotypes, so that if they are sufficiently alike we'll classify them together. On the other hand, you have a phylogeny that says it's fine to put it there, they are related, but also, you may find you have strange bedfellows, which are evolutionarily separate although they have converged in their form, physiology, and structure. Life is a complex of microcosms forming macrocosms of varying degrees of complexity. In the end, taxonomy must reflect the best science.

We can't do without our bacteria. We can't really. We could live in a special environment without bacteria. In fact, we grow gnotobiotic animals in special enclosures with some difficulty and can study them—how immune systems are developed, and what is necessary to stimulate that to happen. But that's not a way of life; that's a test tube. So, you have to take a very general view. Taxonomy is concerned with what living things there are and how one can arrange, recognize, and understand them. The important thing to tell students is that taxonomy takes everything that you know, you think you know, or you partly know about life forms, and puts them together to try to make sense. So at every level of your understanding, taxonomy is going to change slightly, whether you like it or not.

JS: Along that line, one of the surprises that I've had in microbiology related to taxonomy has to do with the genus *Deinococcus*, which was once thought to belong to the genus *Micrococcus*. You worked extensively with this organism, and now we know it's a different critter all together. Can you explain this?

RM: I can say something about *Deinococcus*. It's an interesting story. In the mid-1950s, somebody dug out the drains outside the hospital and there was this great pile of black stuff that had come out of the drain. Carl Robinow took a sample and plated it out. Among all sorts of things that grew were pink and red colonies. At the same time he had other plates around, one of which was contaminated with airborne dust, presumably from the lab; it too, had a very distinctive red colony and he, being Robinow, spread it out. It formed the most beautiful cells, in regular sets of four—it was elegant. He used it for teaching purposes. I grew it and thought it was just lovely, but I didn't know what the hell it was! I studied it intermittently and then, in 1958, I gave a paper on it in Stockholm describing it as a most unusual sort of gram-positive coccus. Someone in the audience said "You know, somebody else has something like that. Anderson in Oregon isolated the same sort of organism from irradiated meats."

I got in touch with Anderson. He and his students were studying what they were calling *Micrococcus radiodurans,* which they isolated from meats they were irradiating to test for food preservation using very high-flux gamma radiation

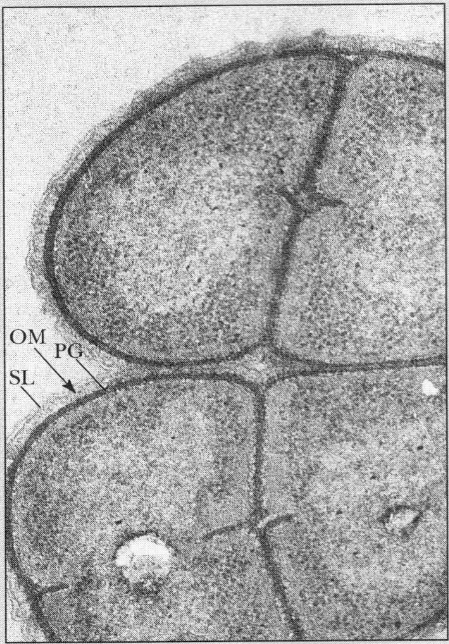

Deinococeus cell envelope layers.
(© Science VU/Canumette)

(one mega rad). They were getting survivors, and these survivors were these red colonies. As far as I could tell, they were essentially the same organism that had fallen out of the air, and had never been irradiated as far as I knew in London, Ontario.

Later, I happened to be busting up this complicated cell and examining cell fractions that had been negatively stained using uranium acetate, and there in the electron microscope a fraction of the cell wall displayed a marvelous moire pattern because of the interference of two layers of macromolecules. This layer looked like a marvelous shirt covering the cell. And the London strain is still known as "sark." A sark is a shirt, an elegant vestment. Anyway, in the end, various other organisms rather like it appeared from sources in Japan and other places, so this is a genus with several species.

In the late 1970s, I had a really good student who helped do this more complex study of the phenotype of the strains. At the same time I knew about Carl Woese. I wrote to him and said this is an extraordinary organism and it is likely to be different from all others. I wanted to know if it is related to the micrococci because biochemically in terms of lipids it is gram negative, although it stains as gram positive. It really is a gram negative pretending to be something it isn't. Stackebrandt and Woese showed that this is very different from the micrococci, in terms of phylogeny, which was determined using sequences of the 16S rRNA. The documents of evolution are written in stable macromolecules like ribosomal RNA. So, then, it was apparent that this was not only a separate kind of microorganism but, after accumulation of a few hundred different bacterial sequences, it became apparent that the deinococci came off the evolutionary stem rather early and they are now considered one of the earliest derivatives of the stem of true bacteria.

JS: You were involved in another interesting study of the *Deinococcus* group. You looked in a field and obtained different strains.

RM: I just came in on the periphery of that. It was really Ian Master's work. He came from Nottingham in England and he was just finishing a Ph.D. at Edinburgh under Mosley on *Deinococcus radiodurans.* He knew that these bacteria are in the environment, broadly distributed. So, when home he collected a number of samples on the banks of a little lake near Nottingham in a rather small area maybe ten square feet and grew about 19 strains. The thing that was interesting is that genotypically each one could be recognized as different from the others [using RFLP patterns of their 16S rDNA]. We just cannot understand this. Phenotypically they were essentially identical. It's odd because it is always in low numbers where it is found, and you would expect the strains to be representative of a single clone.

JS: Is that an example of an environment shaping an organism?

RM: We don't know. It's quite possible. The thing that is amusing is here you have an organism that is quite stable. You can irradiate it with things that should produce mutations or you can use mutagens. And, they don't do much. Nitrosoguanadines produce some mutants, but they are remarkably few.

JS: Is it partly because they have repair strategies?

RM: They have remarkable repair. They don't have SOS repair but they have more direct repair mechanisms that are remarkable. That's true. But that still doesn't say why an organism can survive a megarad. People can't understand what a megarad is. If I put that glass into a radiator and gave it a megarad, it would be turned brown like actinicware. It would be dark. I don't know how any membrane can survive that bombardment. But one thing about *Deinococcus radiodurans,* its membranes aren't made in the same way. They don't have any of the normal phospholipids; the phosphatidyl glycerol, phosphotidyl serine, and phosphotidyl this and that, they're just not there. They have a lot of polar lipids if you run them out as we have done. Bob Anderson at Dalhousie University actually identified three of the lipids, and they are very strange indeed—they're phosphoglycoaminolipids. They are really quite extraordinary. So maybe they're important. Perhaps, like a Michelin self-sealing tire, they can plug the holes! The deinococci are UV resistant, too. Before you start getting a recognizable death rate, they have absorbed an enormous amount of radiation. The survival curve has a tremendous shoulder on it before it drops off, which means they must be able to repair more than DNA. I don't know how they do that.

JS: What's the function of the S layers?

RM: That's also interesting. Because we have done some work on that as well. There are various possibilities. One of them is, certainly, that it is a barrier. It's a sieve made of remarkably nonreactive proteins. And, if God was going to make a sieve, she would do it that way. The S layer may serve as a nonreactive network through which all sorts of molecules could pass, if small enough. Because if you had a reactive layer, it would select what went through it and that might be bad for you. Bacteria take in food that is in solution. One of the functions that is a possibility, a distinct possibility, is that it protects them from predators. Bacteria do suffer predation, even by certain bacteria like *Bdellovibrio,* which is a small rod, swims very fast, and when it runs into a suitable bacterium, it sticks there, drills a hole in the surface, and falls in between the wall and the plasma membrane. It reproduces there using up the goodies in the cell underneath. *Bdellovibrio* grows as a long filament and then scissors up, like sausages, and out they go, hunting again, for another host to run into. Bdellovibrios are very pervasive in soils and sediments, many natural environments, mostly aerobic ones as far as we know. There's no doubt that a complete S layer protects gram negative bacteria from predation by these species and the heavy investment in making a special protein (maybe 10 percent of the cell protein) has survival value. Susan Koval, one of my associates, is doing experiments with isogenic S^+ and S^- strains for feeding ciliates and protozoan predators. There you get different sorts of results. For some protozoa it looks as if the S-layered bacteria are not as easily ingested, and in others there's no difference. So there may be a lot of variation.

A possibility is that S layers provide very few attachment sites for bacteriophages or bacterial viruses. That seems to be true but rare. However, there are a couple of bacterial viruses that can use an S layer, on a *Bacillus* sp. for instance. Some, as on a fish pathogenic *Aeromonas* sp., actually contribute to pathogenicity.

Another possibility is that large enzymes that are destructive to important surface components of bacteria can't get through S layers. Schleifer has shown that is true for some muramidases. It's then a barrier filter so to speak. Obviously S layers must be important because then nakedness would be a hazard. Nakedness is not good for free-living bacteria—a high proportion have S layers.

JS: Do you have any kind of general statement you could make about the state of microbiology education?

RM: I have quite a lot of feelings about that and even concerns, if you're interested. Well, I think the most important one is this: If there are about 10^6 bacteria per cc of all water on the surface of Earth, a little bit of arithmetic will convince you that this is the biggest biomass of anything on this Earth, not by one order but by a number of orders. If that is true and it is also true that the cycles of nature are dependent on microbes, and the remediation of the messes we make depends on the processes of microbes, then microbiology proper is absolutely crucial and must be available to students. It is not just microbiology designed to make a product to combat disease, or to use up this or that chemical, or to live more comfortably. But it gives a deeper perspective to understanding where we live, the environment we live in, and what's happening in the biosphere. If that is true, microbiology is poorly represented in the studies of our young people.

Well, this is a problem and it is a very real one. One of the directors of NSF recently remarked to the Society for Industrial Microbiology that there is now the interesting situation where most of the people who understand bacterial physiology are emeritus professors! Not much of that kind of thing is being taught. That's why I said, as advice to microbiologists, it is important to be as broadly based as possible. Don't forget that there's more than technology involved, and a lot of important microbiology doesn't need high technology. It needs a microscope, a brain and eyes that work, some energy, a few very simple things, and a bench, where you can find out what is involved in a particular environment. What you get from the instruments available to you, depends on you, nobody else, not on the instruments.

It is not just microbiology designed to make a product to combat disease, or to use up this or that chemical, or to live more comfortably. But it gives a deeper perspective to understanding where we live, the environment we live in, and what's happening in the biosphere.

As I said earlier, you see what you are prepared to see. If you don't have the shape of that buck in the bush, you're not going to see it. In teaching students, you have to lead them into seeing with a different perspective. Computers and things like that do not teach experience, really. Do not let anyone abandon lectures, even by people who don't lecture very well, because they do have something to say. I have students come up to me and say, "That was really terrible." And I say to them, "Have you found out what he really knows? Have you tried to find out what you could learn from him?" Each person is different. I've had some really terrible teachers, but they each had a shining light in them.

I am afraid that the experiments you quote, M. Pasteur, will turn against you. . . . The world into which you wish to take us is really too fantastic . . .

French Journalist (1860)[1]

Chapter 1

The Microbial World

Microorganisms, or as they are sometimes called **microbes,** are the smallest living organisms. Although we do not see them, microorganisms are everywhere and play important roles in our daily lives and in the natural activities of Earth. A major objective of the book is to acquaint the reader with both the diversity of these small creatures with whom we share this Earth, and the significance of their activities. In this chapter we introduce the various microbial groups—the bacteria, algae, protozoa, fungi, and viruses—and compare the simple bacterial cell with the more complex cells of other microorganisms and higher plants and animals. Furthermore, we discuss how each microbial group is differentiated from the others, as well as current views on how biologists classify them.

[1]Remark in the French newspaper La Presse (1860) regarding Louis Pasteur's research. From Vallery-Rodot, Rene. 1923. *The Life of Louis Pasteur.* Garden City, NY: Garden City Publishing Co. Inc.

An Introduction to Microbial Life

The microbial world is a fascinating place of living creatures so small we cannot see them without magnifying them about 500 times with a microscope. Miniature though they are, microbes are very important to us and to natural activities on Earth. For example, fully half of the Earth's living material, termed **biomass,** consists of microorganisms, whereas plants constitute only 35 percent and animals a mere 15 percent.

Microbes live almost everywhere on Earth. Incredible as it may seem, some grow at boiling temperatures in hot springs (Figure 1.1**a**) and at over 110°C in undersea volcanic hydrothermal vents. Others live in sea ice at temperatures below freezing (Figure 1.1**b**). Some produce sulfuric acid from sulfur compounds and live at a pH of 1.0, equivalent to 0.1 N H_2SO_4. A few microbes live in saturated salt brine solutions (Figure 1.1**c**), while others live in pristine mountain lakes as pure as distilled water. Mi-

(a)

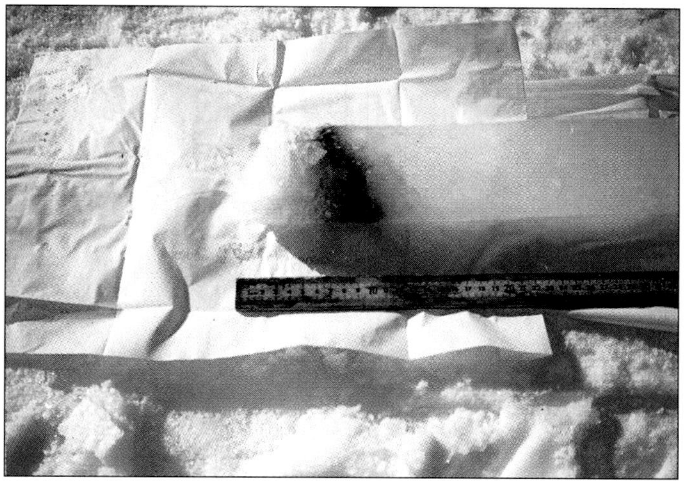

(b)

(c)

(d)

***Figure* 1.1 (a)** Thermal hot springs, such as Mammoth Terrace in Yellowstone National Park, contain thermophilic (heat-loving) bacteria. (Courtesy of J. T. Staley) **(b)** A core taken through the sea ice of Antarctica. The sea ice microbial community appears as the dark red band within the ice layer. Algae and bacteria are the principal inhabitants of this community. (Courtesy of J. T. Staley) **(c)** Salt evaporation ponds, such as these in The Great Salt Lake, are used commercially to convert salt water to table salt. When they become saturated with salt, only halophilic (salt-loving) microorganisms grow in them, and their pigments impart a red coloration to the water. (Courtesy of Fred Post) **(d)** Antarctica has a cold desert area in Victoria Land called the Dry Valleys where microbes are the only living organisms. They reside *inside* the rocks. This sandstone rock has been broken to show a lichen consisting of bands of fungi (upper dark layer) and algae (lower greenish layer) that grow beneath the surface of the rock. (Photo courtesy of R. Vestal)

crobes even live inside rocks in the harsh environment of Antarctica (Figure 1.1**d**).

Microbes preceded humans on Earth by billions of years (see Chapter 17). Thus, we have evolved in their world and, more recently, they in ours. For this reason, it should not seem surprising that microbes live intimately with us, on and in our bodies. Like it or not, microbes live on our skin (Figure 1.2**a**) and in our intestines (Figure 1.2**b**). Without our permission, they inhabit all of our orifices including our mouth, nose, eyes, ears, and anal and genitourinary tracts. However, most of them are harmless symbionts using our bodies as their home and actually pro-

BOX 1.1 RESEARCH HIGHLIGHTS

Some Like it Hot!

The scalding temperatures of hot springs may seem an unlikely place for life of any kind. Yet microbiologists find that the Yellowstone Hot Springs—as well as other thermal habitats throughout the world such as at Rotorua, New Zealand—are teeming with bacterial life. Scientists do not yet know the highest temperature at which some microbes can grow. A century ago bacterial endospores were found to be very resistant to heat and could *survive,* but not grow, at boiling temperatures. However, during the past 20 or so years, extremely thermophilic bacteria have been discovered that not only survive, but actually grow at temperatures higher than the boiling point of water.

A contest is on today among microbiologists who study thermal habitats to find the organism that grows at the highest temperature. Some Archaea that have been isolated in pure culture grow at temperatures between 110° and 120°C. In order to grow these bacteria at temperatures above the boiling point of water, it is necessary to incubate them in an autoclave!

The hot springs at Rotorua, New Zealand. (Courtesy of J.T. Staley)

tecting us from **pathogenic** (disease-causing) species. If we lost the protective natural microflora of the respiratory or intestinal tracts, pathogenic species would readily invade our tissues and body cavities.

Microorganisms cause bad breath, athlete's foot, pimples on our skin, and cavities in our teeth. Human infectious diseases caused by bacteria and viruses have afflicted all of us and include such maladies as the common cold,

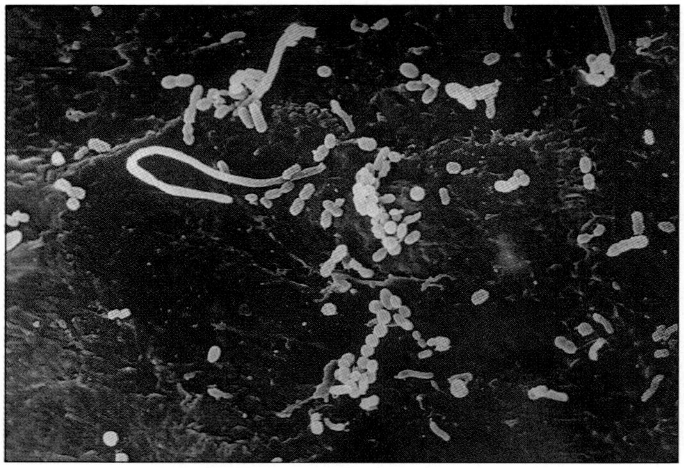

(a)

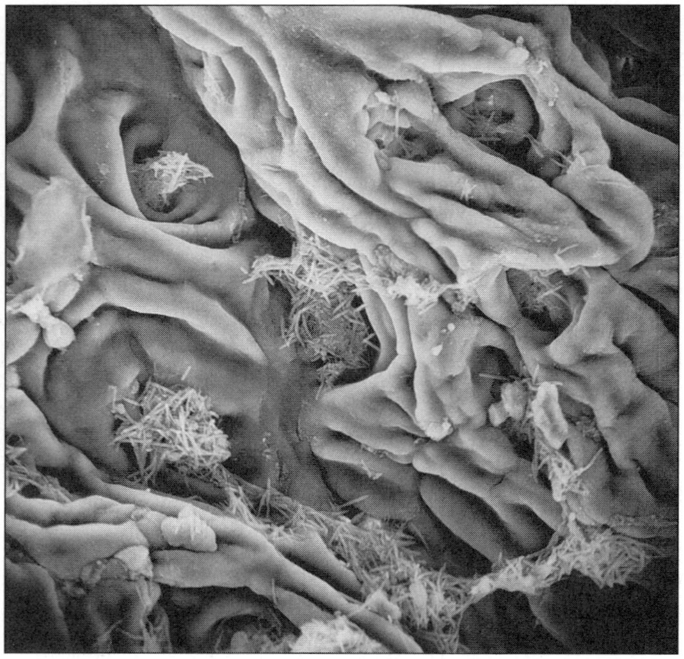

(b)

***Figure* 1.2 (a)** A scanning electron micrograph showing bacteria growing near a sebaceous gland on mammalian skin. (© David M. Phillips / VU) **(b)** An electron micrograph showing bacteria that grow attached to mammalian intestinal tract. (Courtesy of R. Vestal) chicken pox, food "poisoning," and measles. More deadly diseases include typhoid fever, AIDS, tetanus, syphilis, cholera, and bubonic plague, just to name a few.

Our bodies have approximately 10^{13} cells, and it has been estimated that we carry an additional 10^{14} bacterial cells! Many of these bacteria reside in our intestinal tract. The feces we discharge consist largely of bacteria that have grown inside our intestines, helping to degrade the food we eat and providing us with vitamins, amino acids, and other needed nutrients.

One of the major functions of microorganisms is in **geochemical processes,** the natural chemical processes that occur on Earth. The importance of these activities can be illustrated by briefly considering the cycling of just one of the elements—carbon (Figure 1.3). As shown in this simplified carbon cycle, carbon exists in the atmosphere primarily as the gas carbon dioxide. In this form it is available to photosynthetic organisms as a source of carbon for their nutrition. Virtually all energy available for life on Earth is derived ultimately from sunlight. Plants and photosynthetic microorganisms (algae and some bacteria) use the energy from sunlight to convert carbon dioxide into organic chemicals by the process of **photosynthesis.** In photosynthesis carbon dioxide is "fixed," that is, it is converted from a gas to cellular organic material by the following overall chemical reaction:

$$CO_2 + H_2O \longrightarrow (CH_2O)_n + O_2$$

where $(CH_2O)_n$ is organic carbon. Animals are dependent upon plants for organic carbon because they cannot use inorganic forms for their nutrition. Herbivores eat green

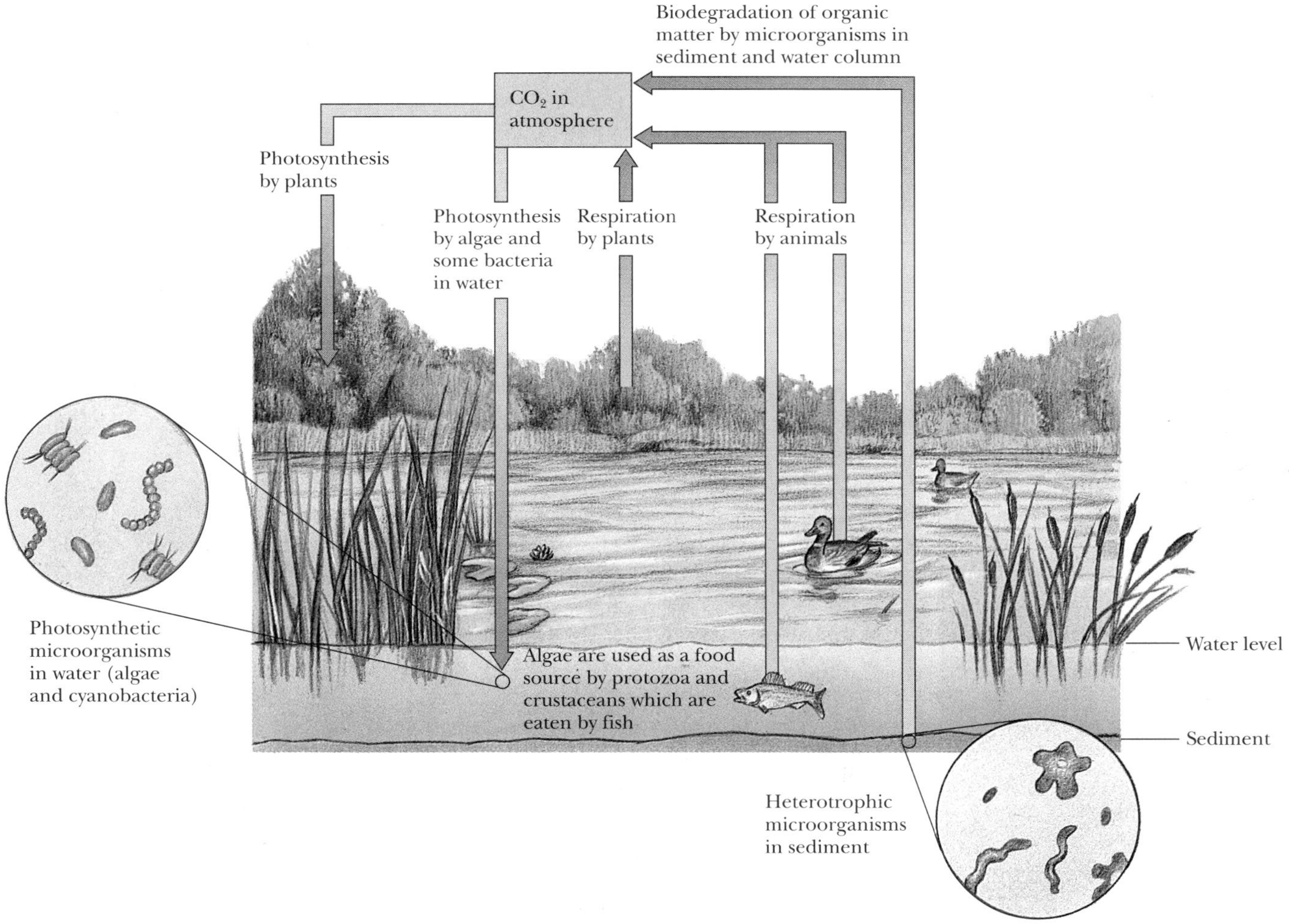

***Figure* 1.3** A diagram of the carbon cycle. Some microorganisms (algae and cyanobacteria) are involved in photosynthesis (light-driven carbon dioxide fixation) and other microorganisms (bacteria, protozoa, and fungi) in the biodegradation of organic materials back to carbon dioxide and other inorganic materials. See text for details.

plants and therefore rely directly on photosynthesis, while carnivores eat herbivores and rely indirectly on the organic carbon fixed by green plants.

The CO_2 in the atmosphere available for photosynthesis (see chemical reaction above) is derived from two primary biological sources. Some of the carbon dioxide (about 5 percent to 10 percent) is released by plants and animals in the process of **respiration.** However, the majority of it (90 percent to 95 percent) is derived from organic material that is degraded to inorganic materials by microorganisms in a process called **biodegradation** (**mineralization** or **decomposition** are other commonly used terms which describe this process; see Chapter 12), in which organic material is used by the microorganisms as an energy and carbon source for growth. Both respiration and mineralization have the same overall formula, which is the reverse of the reaction above (although some bacteria use other oxidizing agents rather than oxygen):

$$(CH_2O)_n + O_2 \longrightarrow CO_2 + H_2O$$

In the global carbon cycle, a balance exists between the overall consumption of carbon dioxide by photosynthesis and its production by mineralization and respiration. This balance is, however, influenced by human activities, primarily the burning of fossil fuels, which is slowly causing a build-up of additional carbon dioxide in the atmosphere and a resulting "greenhouse" effect.

Plant cellulose, the most abundant organic material on earth, can be degraded or mineralized by microorganisms. Cellulose is a rather tough, refractory material that cannot be degraded by animals, with the exception of some invertebrates such as snails. However, it is readily utilized as a carbon and energy source by many types of microorganisms, including certain fungi, protozoa, and bacteria. Some of these microorganisms live in herbivorous animals, such as termites or cattle. These animals' existence depends on the cellulolytic activity of the microorganisms that degrade cellulose to produce utilizable carbon sources for the animals. Other cellulose-degrading microbes live by themselves in soil and aquatic environments. Thus, these microorganisms are responsible for the natural recycling of cellulosic materials in the environment, converting them ultimately to inorganic matter. Likewise, all organic constituents of living and dead plants, animals, and other microorganisms are ultimately converted by microbes to carbon dioxide and inorganic materials. Only small fractions become fossils or are preserved in geological deposits.

Even the most toxic and refractory compounds can be mineralized by microorganisms. For example, chemically synthesized compounds such as polychlorinated biphenyls (PCBs), the pesticides 2,4,5-T, and DDT are not degraded by higher plants and animals and are toxic to humans, other animals, and plants. Fortunately, some bacteria and fungi are known to degrade these toxicants, allowing them to be recycled back into the environment as harmless inorganic compounds. These compounds have often been haphazardly handled and disposed of in chemical dumps and thus have contaminated many areas that now need to be treated to remove them. Microorganisms will be used, at least in part, for this cleanup process (see Chapter 33). Bacteria and other microorganisms play many other roles in the carbon cycle and other elemental cycles, a topic we will consider in more detail, later, in Chapters 12 and 24.

Microorganisms are also important in many industrial processes (see Chapter 32). Since prehistoric times, they have been used unwittingly to produce wine, beer, and other fermented beverages. They are used in cheese processing as well as in the leavening of bread. In addition, they are important for making pickles and sauerkraut, and they are used to produce vinegar. Farmers use certain microbes to make silage, a fermented plant material used for cattle feed. In addition, microbes produce the antibiotics used to treat infectious diseases caused by other microorganisms (see Chapters 29 and 30).

More recently, microbes have been "engineered" genetically to be used in industry. For example, insulin needed by diabetics was previously obtained only by extraction from the pancreas of pigs. The product was both expensive and different from human insulin. Now some insulin is produced by **genetically engineered** (the process by which genes from one organism are inserted into another in the laboratory) bacteria in large fermentation tanks. This is particularly important for diabetics who are allergic to pig insulin.

Bacteria are also employed to carry genetic traits into plants and therefore alter them permanently. This new application of genetic engineering will revolutionize agriculture in many ways. For example, agriculturally significant plants have been engineered to be resistant to specific herbicides that effectively kill nuisance weeds, or to produce insecticides so insects will not be able to eat them.

In the following sections, we discuss the properties of microorganisms and the differences among the various types and their current classification.

Cell Theory: A Definition of Life

Microorganisms can be divided into four groups, on the basis of form and function: bacteria, fungi, algae, and protozoa. Like higher plants and animals, all of these microorganisms consist of one or more cells, and therefore, they must be alive. This is referred to as the **cell theory of life.** Stated in the reverse way, if something is alive, it must be cellular. Viruses are also treated in this book, but they are not cellular and therefore are not regarded as living

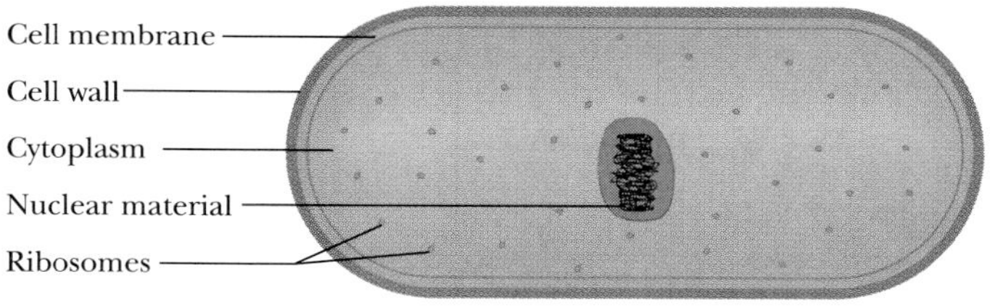

***Figure* 1.4** A diagram of a typical cell showing the cell wall, cell membrane, cytoplasm, and nuclear area.

organisms. Nonetheless, they are important biological agents that develop only as intracellular parasites of organisms, including microorganisms.

The **cell** is defined as a unit with characteristic functional and structural features. These functions include **metabolism,** the chemical and physical activities by which cells obtain energy and carry out synthesis through biochemical reactions. The biochemical reactions are catalyzed by proteins called **enzymes,** which are discussed in Part III. The other function is **reproduction,** the process by which cells duplicate themselves to produce progeny.

The structural components of cells (Figure 1.4) include: (1) **cytoplasm,** the aqueous fluid of the cell in which most of the enzymatic and metabolic activities occur (for example, **ribosomes,** small structures responsible for protein synthesis, are located in the cytoplasm); (2) a central **nuclear area** that contains deoxyribonucleic acid (DNA), the hereditary material that is duplicated during reproduction; and (3) a **cell** or **cytoplasmic membrane,** the boundary between the cell's cytoplasm and its environment. The cell membrane consists of lipids and proteins. Many microorganisms contain a layer external to the cell membrane that is referred to as (4) the **cell wall,** a rigid structure that confers shape to the cell. All fungi have cell walls as well as most algae and bacteria (see Table 1.1). Protozoa lack cell walls. Chapter 4 covers these structures and their functions in greater detail.

Unlike plants and animals, which are all **multicellular** (containing millions of cells), many microorganisms consist of a single cell and are therefore called **unicellular.** Most—but not all—bacteria, protozoa, and algae are unicellular. Only one group of fungi is unicellular—the yeasts. Plants and animals are macroscopic because they consist of many cells organized into tissues and organs, neither of which is found in microorganisms.

Nutritional Types of Microorganisms

Algae and a group of bacteria called the cyanobacteria are **photosynthetic,** that is, as with plants, they obtain their energy from sunlight. Also, as in plants they use carbon dioxide as their principal source of carbon for growth. This type of nutrition, which is based entirely on *inorganic* compounds, is referred to as **autotrophic** (self-nourishing or feeding). Algae are therefore called **photoautotrophic** in that they obtain their energy from sunlight and their carbon source is carbon dioxide.

In contrast to algae, fungi obtain their energy directly from chemical compounds, not sunlight. This type of nutrition is referred to as **chemotrophic** (chemical feeding). Fungi require *organic* chemical compounds as their sources of energy and carbon. Such nutrition is termed **heterotrophic** (other or different feeding as distinguished from autotrophic) or **organotrophic.** Thus, fungi are **chemoheterotrophic,** that is, they use chemical compounds as energy sources and organic compounds as carbon sources.

Like fungi, protozoa are chemoheterotrophic organisms that use organic compounds as sources of carbon and energy for growth. A few protozoa such as the parasitic sporozoa use soluble nutrients and therefore resemble

***Table* 1.1 Microorganisms and their carbon nutritional types**

Microbial Group	Number of Cells per Organism	Cell Walls	Nutritional Type
Algae	Usually one, some filamentous	Yes	Photoautotrophic
Protozoa	One	No	Chemoheterotrophic
Fungi	Filamentous, except yeasts	Yes	Chemoheterotrophic
Bacteria	Usually one, some multicellular	Yes[1]	Photoautotrophic, photoheterotrophic, chemoautotrophic, or chemoheterotrophic

[1]A few bacteria, namely, the mycoplasmas and thermoplasmas, lack cell walls.

fungi in their nutrition. However, typical protozoa, which lack cell walls, engulf bacteria and other microorganisms much like higher animals eat food. Therefore, their source of food is *particulate* organic material, and this type of feeding is called **phagocytosis.** Only *dissolved* organic carbon sources can enter fungal cells because of their cell walls. Thus, fungi are well known for their ability to use simple sugars and other dissolved substances as carbon sources. Some fungi can also degrade particulate organic materials, such as cellulose, by producing and excreting enzymes that solubilize the organic material outside the cell. They then transport the dissolved compounds into the cell.

As a group, bacteria are exceedingly diverse in their nutritional capabilities. Some are similar to algae in being photoautotrophic. Others are **photoheterotrophic,** that is, they can obtain energy from sunlight, but they use organic compounds as carbon sources. *Most bacteria are chemoheterotrophic,* deriving energy from organic compounds and using them also as carbon sources. One especially interesting group of bacteria can obtain energy by the oxidation of inorganic compounds, such as ammonia or hydrogen sulfide, and use carbon dioxide as their principal carbon source. This type of nutrition is termed **chemoautotrophic,** a nutritional category that is found only in these specialized prokaryotic microorganisms. The four principal groups of microbes and the types of nutrition they exhibit are shown in Table 1.1. Microbial nutrition will be discussed in more detail in Chapter 5.

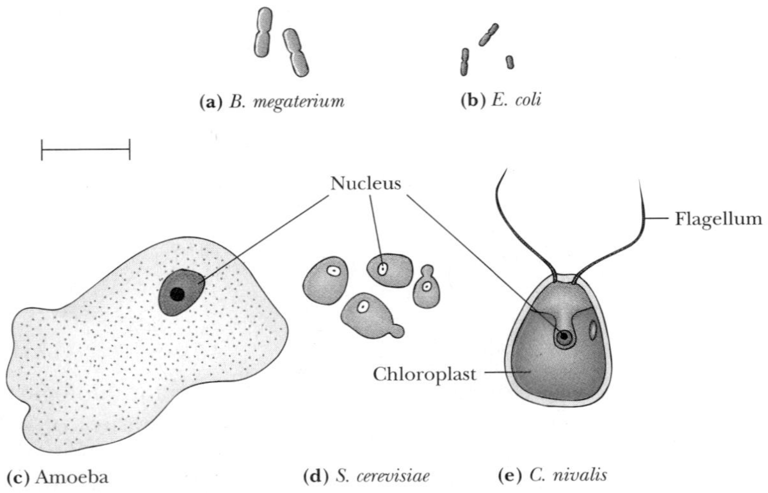

***Figure* 1.5 (a–e)** Drawing depicting representative microorganisms as they appear by light microscopy. Included are two examples of bacteria, a large rod, *Bacillus megaterium,* and a small rod, *Escherichia coli.* The eukaryotic organisms are an amoeba (a protozoan), a yeast (*Saccharomyces cerevisiae*), and an alga (*Chlamydomonas nivalis*). Note that the nucleus is clearly discernible in eukaryotic microorganisms. Also note the cup-shaped chloroplast of *C. nivalis.* Bar is 10.0 μm.

Prokaryotic Versus Eukaryotic Microorganisms

Bacteria appear different from algae, protozoa, and fungi when they are examined under the light microscope (Figure 1.5). Bacterial cells are usually very small and they have no apparent nucleus. In contrast, cells of algae, protozoa, fungi, and higher life forms are typically much larger and have a distinct nucleus.

These differences noted by observations with the light microscope are borne out by more detailed examination using the electron microscope. Microorganisms can be sliced into very thin sections and examined at high magnification with the transmission electron microscope (TEM) (see Chapter 4). When viewed in this manner, the structural differences between bacteria (Figure 1.6) and other microorganisms (Figure 1.7) are striking. Bacteria have a much simpler cell structure and are referred to as **prokaryotic** (from Greek meaning "before nucleus") organisms. The higher microorganisms are called **eukaryotic** (from Greek meaning "good" or "true nucleus"). Table 1.2 lists the major differences between these two basic types of organisms. As the terms imply, the single major difference between these two cell types is related to their nucleus. The nucleus of the cell of a eukaryotic **(Box 1-2)** microorganism (as well as higher plants and animals) is bound by a membrane referred to as a **nuclear membrane** (Figure 1.7). In Eubacteria and Archaea (prokaryotic organisms), the nuclear material, which appears as a central fibrous mass in thin sections, is not bound by a membrane but is in direct contact with the cytoplasm (Figure 1.6).

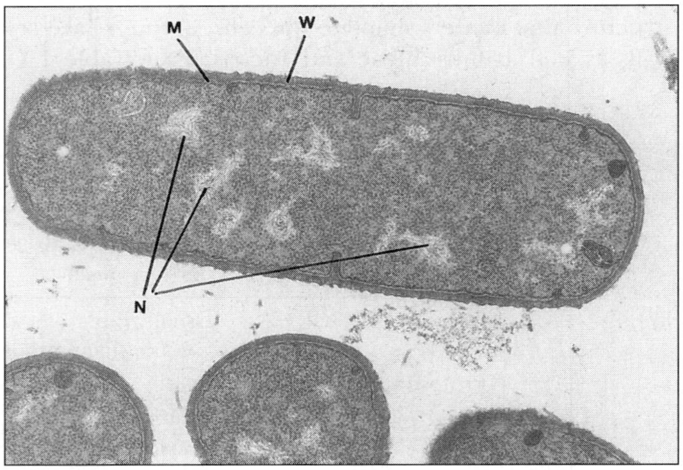

***Figure* 1.6** An electron micrograph of a thin section of the bacterium *Bacillus megaterium.* The cell wall (W), cell membrane (M), and nuclear material (N) can be seen clearly. The nuclear material appears as fibrous patches scattered about the cytoplasm. However, it is all connected together. (Courtesy of Stuart Pankratz)

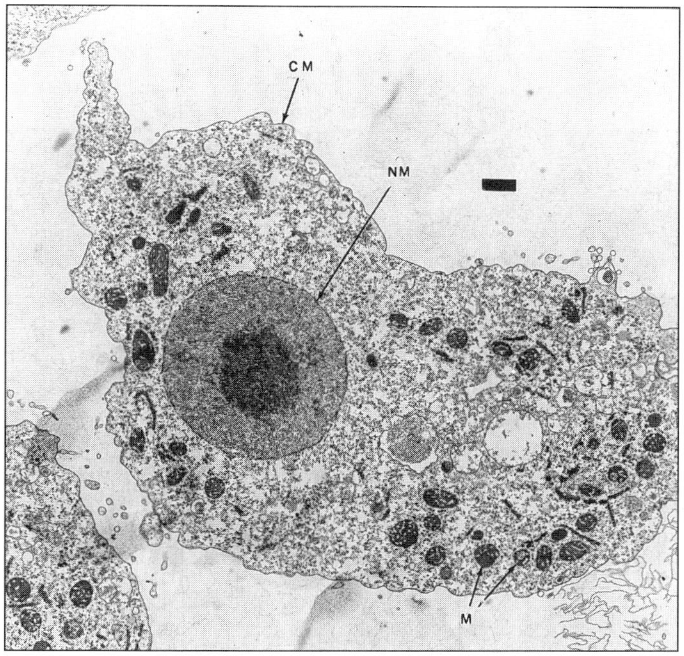

***Figure* 1.7** A thin section of a eukaryotic microorganism, a protozoan in the genus *Acanthamoeba*. Note the nuclear membrane (NM) with pores (P) and the mitochondrion (M). Bar is 10 μm. (Courtesy of T. Fritsche)

Other differences exist between prokaryotic and eukaryotic cells, some structural, others genetic and physiological. The nucleus of prokaryotes contains a single type of DNA molecule. More than one copy of it may be present, depending upon the growth state of the organism. Thus, rapidly growing cells might have two or four copies of the DNA molecule, but all (with the exception of mutations) are identical. In contrast, there are different types of DNA in the nucleus of eukaryotic organisms. Each type is referred to as a **chromosome.** Thus, bacteria can be regarded as having a single chromosome, while eukaryotic microorganisms have more than one chromosome. It is because eukaryotes have multiple chromosomes that they undergo **mitosis** during cell division. In this process, each chromosome aligns along the division axis of the cell and then replicates before asexual cell division occurs (Figure 1.8). This process is an elaborate physiological and morphological orchestration that never occurs in prokaryotes.

Other Structural Differences

Ribosomes are small cytoplasmic or intracellular structures responsible for protein synthesis. These appear as granules (about 5 nm in diameter) in the cytoplasm. Prokaryotic ribosomes are called 70S ribosomes (the S refers to the Svedberg sedimentation unit). Eukaryotes, with rare exceptions in some protozoa, have slightly larger ribosomes called 80S ribosomes.

One of the striking features of eukaryotes is their internal organelles. These are membrane-bound structures found in the cytoplasm which, like the cell membrane, contain both protein and lipid. The most common or-

***Table* 1.2 Characteristics of prokaryotes versus eukaryotes**

Characteristic	Prokaryote	Eukaryote
Nuclear structure and function		
Nucleus with bounding membrane	No	Yes
Chromosomes	One	> 1
Mitosis	No	Yes
Sexual reproduction	Rare; only part of genome exchanged	Common; entire chromosome exchanged
Meiosis	No	Yes
Cytoplasmic structure		
Mitochondria	No	Yes[1]
Chloroplasts	No	Yes (if photosynthetic)
Ribosomes	70S	80S[2]
Typical cell volume	$< 5\ \mu m^3$	$> 5\ \mu m^3$

[1]A few primitive eukaryotic microorganisms lack mitochondria.

[2]Some rare, primitive eukaryotic microorganisms have 70S ribosomes.

BOX 1.2 MILESTONES

"Higher" and "Lower" Forms of Life

Although his views were largely ignored in the 1930s, the French biologist E. Chatton noted the differences in cellular structure among "higher" and "lower" forms of life. He coined the terms eukaryotic and prokaryotic based upon his light microscopic observations of the differences between the cells of higher organisms and bacteria, respectively. It was only after the invention of the electron microscope (late 1930s) and the subsequent development of appropriate procedures to thin-section organisms (1950s to 1960s) that the detailed differences between these two types of cellular organization were confirmed by other biologists (see A Conversation with R. G. E. Murray). In addition to these morphological features, a number of other differences were also discovered that permitted the clear distinction of these two types of cells. The major features that distinguish prokaryotic from eukaryotic cells were eloquently stated in the important publication by Roger Stanier and C. B. van Niel in 1962.

ganelle of this type found in almost all eukaryotic cells is the **mitochondrion** (Figure 1.9). The mitochondrion is the site of respiratory activity in eukaryotes (see Chapter 23). Mitochondria are sufficiently large that they have their own internal DNA, cytoplasm, and ribosomes. One of the exciting facts of cell biology is that the DNA of the mitochondrion is similar to prokaryotic DNA—that is, it is not bound by a membrane. Furthermore, the ribosomes of the mitochondrion are 70S ribosomes, like those of prokaryotes. These features and other lines of evidence (see Chapters 18 and 23) have led biologists to conclude that the mitochondrion evolved from a bacterium that developed a close symbiotic association with another cell over one billion years ago.

The **chloroplast** is the site of photosynthesis in eukaryotes. Like the mitochondrion, the chloroplast is a membrane-bound organelle found in the cytoplasm. It also resembles the mitochondrion in that its DNA is not membrane bound, and its ribosomes are 70S; however, it also contains internal membranes having pigments associated with photosynthesis. The chloroplast is found in algal and plant cells. It, too, is thought to be derived through

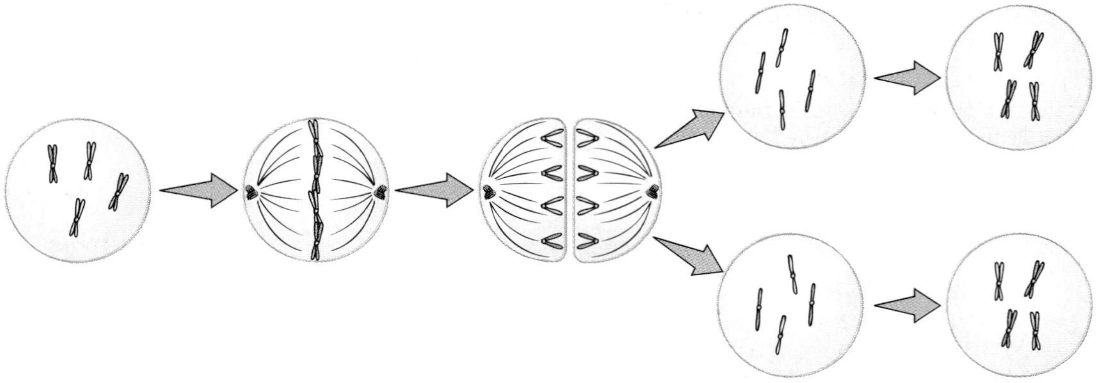

(a) (b) (c) (d) (e)

***Figure* 1.8** The process of mitosis, in which a dividing eukaryotic cell duplicates its chromosomes and distributes one copy to each of the newly forming daughter cells. **(a)** This particular cell has two sets of chromosomes. **(b)** The four chromosomes align along the central axis of the cell during metaphase. **(c)** They separate during anaphase, which is followed by **(d)** Asexual division of the cell, telophase. **(e)** The resulting cells replicate their chromosomes before the process is repeated.

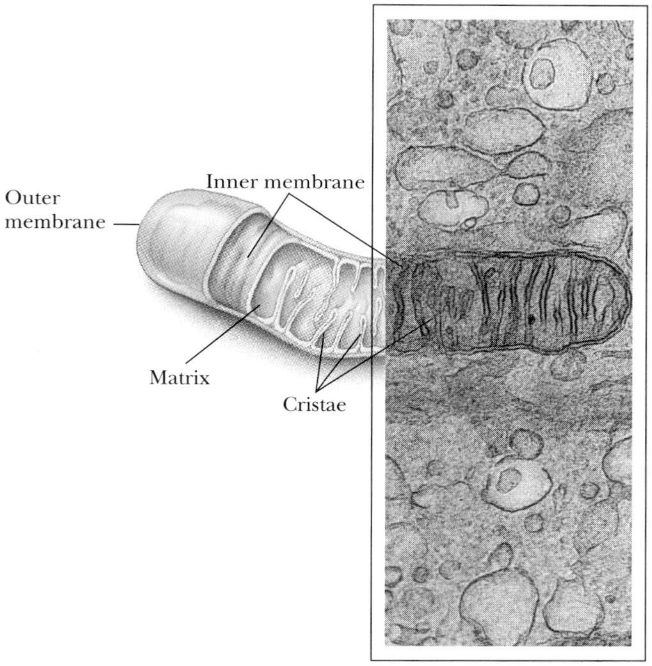

***Figure* 1.9** Electron micrograph of a section of a mitochondrion from a eukaryotic microorganism. Note the outer membrane and the inner, folded membranes (cristae). (Courtesy of artist's depiction on the left; actual electron micrograph on the right) (© Don Fawcett/Photo Researchers Inc.)

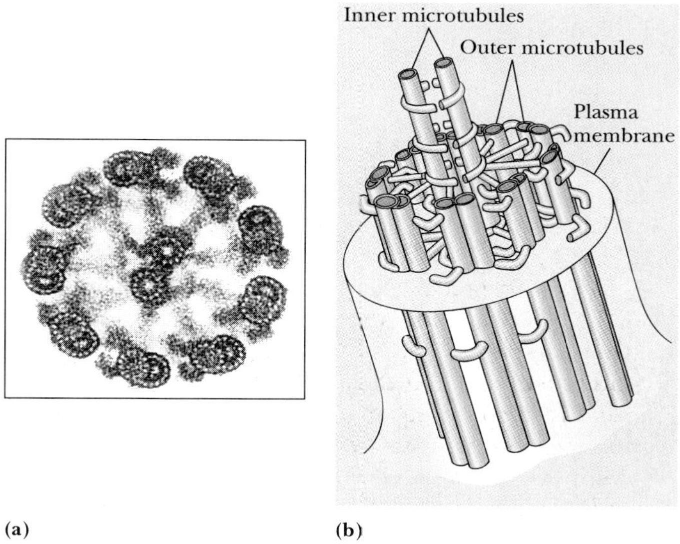

***Figure* 1.10 (a)** An electron micrograph of a cross section through a eukaryotic flagellum showing the characteristic "9 + 2" fibrillar arrangement. (© Don Fawcett/Photo Researchers Inc.) **(b)** Drawing illustrating the structure of the flagellum. In contrast, the bacterial flagellum is a single fibril.

evolution from prokaryotic organisms, specifically the photoautotrophic group called the cyanobacteria (see Chapter 21).

The organelles of motility of eukaryotic cells—the **flagellum** and **cilium**—are also larger and more complex than those of prokaryotes. Their cross section reveals an elaborate fibrillar system called the "9 + 2" arrangement with nine outer doublets of fibrils and an inner pair (Figure 1.10). In contrast, the prokaryotic flagellum has a single fibril when viewed in cross section. It is so fine a thread that a single flagellum cannot be seen when observed by light microscopy. Eukaryotic flagella and cilia, in contrast, are readily observed with the light microscope (see Figure 1.5).

Reproductive Differences

All prokaryotes reproduce by asexual cell division. Cells simply enlarge in size, replicate their DNA (that is, produce a second identical copy of their DNA), and divide to form two new cells, each with a copy of the DNA molecule (Figure 1.11). Thus, prokaryotes have only one type of DNA and are called **haploid.** Sexual reproduction is rare in prokaryotes. Though many bacteria are able to exchange genetic material between mating types, this is not commonly done and is not known to be a universal characteristic. As discussed later (see Chapter 15), this rarely results in the formation of a **diploid** cell, which is a cell with a full copy of the chromosome(s) from each of the mating cells.

In contrast to prokaryotes, most eukaryotes exist as diploid organisms or have major diploid stages in their life

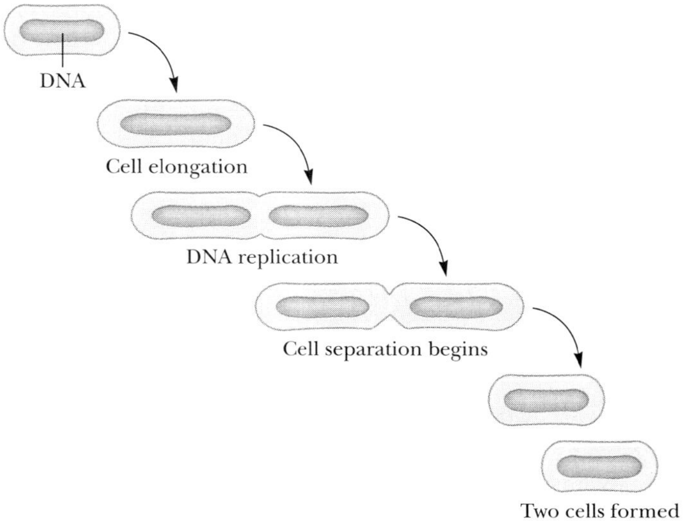

***Figure* 1.11** A diagram of a prokaryotic cell undergoing division. Though this process is analogous to mitosis in eukaryotic organisms (compare with Figure 1.8), none of the structural features of mitosis occur in bacteria. The DNA replicates to produce two identical molecules that are separated into each of the two new daughter cells.

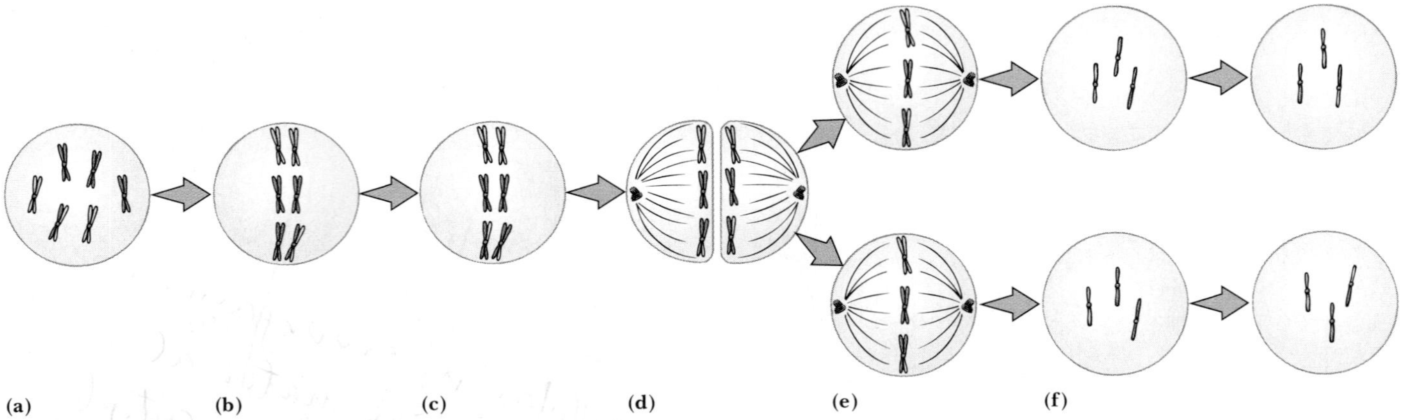

Figure **1.12** The meiotic process, shown for one parent. **(a)** The original three pairs of chromosomes. **(b)** These combine. **(c)** Crossovers may occur, and they align along the axis. **(d)** The chromosomes then separate in the **(e)** Formation of two cells. **(f)** These cells divide again mitotically to yield **(g)** Four haploid gametes, which are either eggs or sperm.

cycles. Thus, their cells have two sets of chromosomes, one set from the "male" and another set from the "female" mating types. For example, human cells have 46 chromosomes. These exist as 23 paired chromosomes. Half (or one set of 23) is derived from the man and the other 23 from the woman parent. **Meiosis** is the process whereby the 46 chromosomes are reduced to 23 in preparation for sexual reproduction. Meiosis results in the formation of special mating cells, called **gametes**—the sperm and the ovum—produced by male and female mating types, respectively (Figure 1.12). During sexual reproduction these cells fuse together in the formation of a **diploid zygote** (the fertilized egg). Therefore, the zygote contains a full genetic complement from each of the parental mating cells. Sexual reproduction is very common among eukaryotic organisms. Except for haploid gametes, the cells of most higher eukaryotes are diploid. Genetic studies of prokaryotes are simplified because of the single copy of each gene (thus, there are no dominant and recessive characteristics). This means that any genetic alteration can be expressed immediately.

Cell Size

As mentioned earlier, cell size is an important characteristic for an organism. It is generally true that most eukaryotic organisms have larger cells than prokaryotic organisms (see Figure 1.5), but there are some exceptions. For example, although typical bacterial cells range in diameter from 0.5 μm to 1.0 μm, some wider than 50 μm have been reported. The cells of typical eukaryotes range in diameter from 5 μm to 20 μm, with most averaging about 20 μm, although some species have larger ones. Specialized cells in multicellular organisms can be much larger. A human neuron can be as long as one meter.

Many bacteria grow and reproduce at very rapid rates. Some can double in size or numbers of cells every 15 minutes under optimum growth conditions. This implies that metabolic processes can be extremely rapid in these organisms. This rapid metabolic rate is due in part to the small size of bacteria, which ensures that the cytoplasm is in close proximity to the surrounding environment from which the bacteria derive their nutrients. The farthest distance between the cytoplasm and the growth environment is only 0.5 μm in a bacterium with a diameter of 1.0 μm, whereas it is 10.0 μm in a eukaryotic organism with a diameter of 20.0 μm.

Another way to consider the close spatial relationship between the cytoplasm of a cell and its environment is to calculate the ratio of its surface area to its volume (Figure 1.13). If one assumes that a bacterial cell is 1.0 μm in diameter, and, although highly unlikely, cubical in shape (actually, one extreme halophilic bacterium is a cube!), then its surface area would be 6.0 μm^2 and its volume would be 1.0 μm^3. Thus, its surface area to volume ratio (SA/V) would be 6.0. In comparison, a hypothetical eukaryotic microorganism of the same shape that is 10.0 μm in diameter would have a SA/V of 0.6. This smaller value for the eukaryote indicates that there is a tenfold greater amount of cytoplasm per unit of cell membrane than in the smaller prokaryote. Since the source of nutrients for growth comes from the external environment, more nutrients are available per unit of cytoplasm in the smaller prokaryote, thereby enabling faster metabolism and growth.

Classification of Microorganisms

The ribosome is a complex intracellular structure found in all organisms. Studies indicate that its structure has

BOX 1.3 RESEARCH HIGHLIGHTS

You Can't Tell a Bacterium by Its Size Alone!

Perhaps the largest bacterium is *Epulopiscium,* which lives in the intestinal tract of tropical marine fish called surgeonfish. *Epulopiscium* has not yet been grown in pure culture. However, based upon analyses of its ribosomal RNA, it is known to be related to the gram positive bacteria. Note how the *Epulopiscium* dwarfs the ciliate protozoan alongside it.

These and other large bacteria illustrate that some bacteria are exceptions to the general rule that prokaryotic cells are smaller than eukaryotic cells. These unusually large bacteria may not grow as rapidly as the smallest bacteria, and they may also have different mechanisms for transport of nutrients. In any event, they have found a habitat in which to successfully live despite their unusually large size.

Like *Epulopiscium,* many other large bacteria, such as *Thioploca* species (see Chapter 21), have not yet been grown in pure culture in the laboratory. It is not known why these large bacteria are so difficult to cultivate.

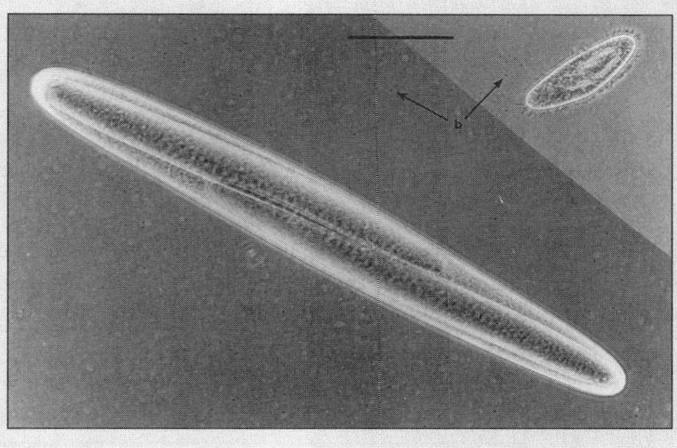

An *Epulopiscium* cell from the intestinal tract of a surgeonfish. Note that the inset, at the same magnification, shows a smaller eukaryotic ciliate protozoan. (Courtesy of Ester Angert and Norman Pace)

changed very slowly during evolution. Its highly conserved nature and universal occurrence have resulted in its being used in the study of the evolutionary relatedness among organisms. As mentioned previously, most eukaryotic organisms have a larger ribosome than do prokaryotes, which alone indicates a difference between these two types of organisms. Furthermore, the 5S and 16S subunits of the 70S ribosomes from a variety of prokaryotes have been analyzed chemically. The differences found from these studies support the classification of eukaryotes as separate from prokaryotes (see Chapter 17).

In addition, analyses of the ribosomal RNA and certain other features of prokaryotic organisms affirm that they should be divided into two groups, called **Domains.** One Domain is called the **Eubacteria,** which contains most of the common bacteria encountered in the environment. However, another separate group of prokaryotes has been recognized that differs from the Eubacteria in ribosomal RNA composition, cell wall structure, and metabolism. This Domain is referred to as the **Archaea** (from Greek meaning "ancient" although there is no definitive evidence that indicates they are older than the Eubacteria). This group of prokaryotes differs from the Eubacteria and both of them differ from the third Domain, the **Eucarya,** which contains all eukaryotic organisms. As a consequence of these studies, three major Domains of organisms are now recognized by microbiologists: the Eubacteria, the Archaea, and the Eucarya (Figure 1.14). The major differentiating characteristics among these organisms are shown in Table 1.3. The eukaryotic microorganisms—the algae,

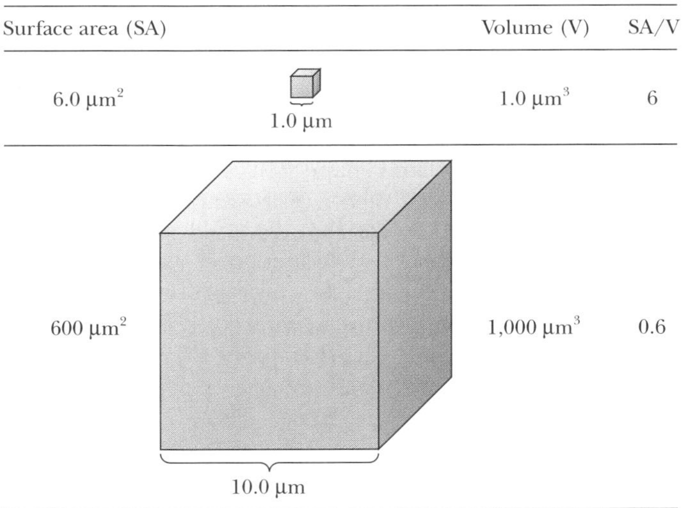

***Figure* 1.13** Hypothetical cubical cells. One is 1.0 μm in diameter and another is 10.0 μm in diameter. The larger cell has a much smaller surface area to volume ratio. See text for implications of this.

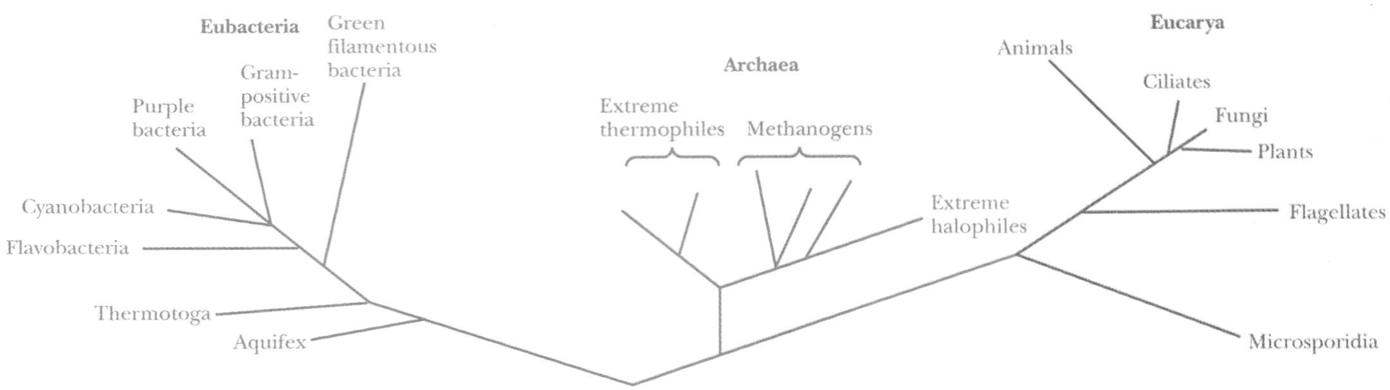

Figure **1.14** The three Domains of living organisms recognized by microbiologists. The two prokaryotic Domains are the Eubacteria and Archaea. All higher organisms, including eukaryotic microorganisms are placed in a separate Domain, the Eucarya. Although enormous morphological differences exist among plants, animals, algae, protozoa, and fungi, they all belong in one Domain, whereas the morphologically simple bacteria belong to two Domains, the Archaea and the Eubacteria. (Adapted from Carl Woess)

***Table* 1.3 Major differential characteristics of the three domains of life**

	Eubacteria	Archaea	Eucarya
Nuclear membrane	No	No	Yes
Mitochondria and chloroplasts	No	No	Yes
Peptidoglycan cell walls	Yes[1]	No	No
Membrane lipids	ester-linked	ether-linked	ester-linked
Ribosome size	70S	70S	80S

[1]Three bacterial groups, the chlamydia, planctomycetes, and mycoplasmas, lack cell wall peptidoglycan (the structure of this material will be discussed in Chapter 4).

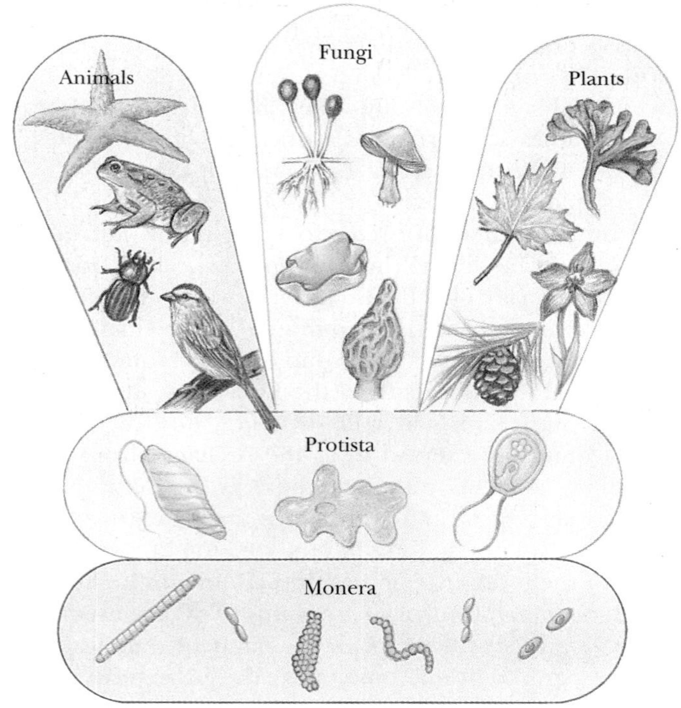

Figure **1.15** The five-kingdom system of classification of living organisms preferred by some biologists.

protozoa, and fungi—along with the plants and animals are all placed in the Domain Eucarya by this classification.

Although most microbiologists now accept this classification of organisms, the five-kingdom classification system proposed in 1969 by the zoologist Robert Whittaker, remains popular with some other biologists (Figure 1.15). In this system, bacteria are placed in the kingdom Monera, protozoa in the kingdom Protista, algae in the kingdom Plantae, Fungi in their own kingdom, and animals in the kingdom Animalia. The five-kingdom system is based primarily on the morphology of organisms, whereas the three-domain system is based on the molecular analyses of organisms, particularly the sequence of bases of highly conserved macromolecules such as ribosomal RNA, and therefore presumably more accurately reflects the true evolution of organisms (see Chapters 17 and 18 and A Conversation with Carl Woese).

Introduction to Eukaryotic Microorganisms

The bacteria will be treated most thoroughly in this book. Chapters 19, 20, 21, and 22 discuss various groups of the Eubacteria and the Archaea. The remainder of this chapter is devoted to an introduction to eukaryotic microorganisms. A more detailed treatment of eukaryotic microorganisms and what is currently known about their relatedness to higher organisms will be presented in a later chapter, Chapter 23. The discussion that follows provides an overview of the various eukaryotic microbial groups as they are traditionally viewed.

Algae

Most algae are microscopic photosynthetic organisms; however, some larger forms such as kelp and other seaweeds are of macroscopic size. Algae reside in aquatic habitats where they carry out photosynthesis. Their type of photosynthesis, which is metabolically identical to that of higher plants and cyanobacteria, is referred to as **oxygenic photosynthesis** because oxygen is evolved in the process (reaction is not balanced):

$$CO_2 + H_2O \longrightarrow (CH_2O)_n + O_2$$

where $(CH_2O)_n$ refers to organic material produced.

All algae contain chlorophyll *a* (Figure 1.16). Some groups have, in addition to chlorophyll *a*, other types of chlorophyll. Many algae are unicellular and reside in the euphotic (light-receiving) zones of various aquatic habitats. The green algae are especially common in freshwater habitats, and brown algae in marine habitats (Table 1.4). Green algae often cause "blooms" in the spring and summer when their numbers become large in lakes. Some major groups of algae are described in Table 1.4.

Algae and cyanobacteria are the principal photosynthetic organisms of aquatic habitats. Most are innocuous organisms that live as **phytoplankton,** that is, photosynthetic organisms that are carried by water currents in aquatic habitats. They are very important in the carbon cycle because they account for about one half of the world's primary production. They are the bottom layer of the food chain in aquatic habitats (Figure 1.17). Therefore, they serve as the principal food source for all organisms that occur at higher levels. Protozoa and other larger zooplankton consume them, and they in turn are

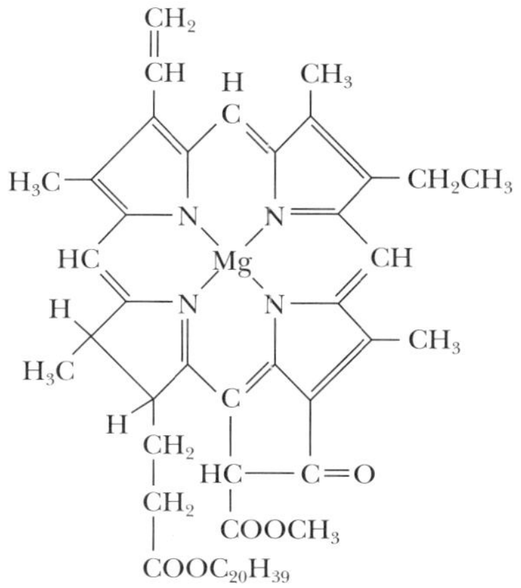

Chlorophyll A

***Figure* 1.16** The chemical structure of chlorophyll *a*.

eaten by larger animals such as fish and ultimately marine mammals such as seals and whales.

Certain dinoflagellates are responsible for "red tides" or algal blooms when high concentrations of them occur. In some cases, depending on the species of dinoflagellate, animal toxins are produced. For example, some species produce a detergent-like toxin that can cause fish kills. Others, which may or may not reach red tide levels of concentration, produce substances that are toxic to humans. These toxins enter into the aquatic food chain through bivalves such as clams and oysters that feed on the dinoflagellates (thereby causing paralytic or diarrhetic shellfish poisoning to humans who eat them) or through crustaceans and into fish such as grouper (causing ciguaterra poisoning to those who eat them).

Diatoms are unusual in that they produce silica cell walls. Diatomaceous earth is formed when large numbers of them die and are deposited in sediments. The organic material of the cells is degraded by bacterial activities and the silicon dioxide remains in the deposit. The silica walls of diatoms have very elaborate and beautiful shapes (Figure 1.18). The shapes are characteristic of each species and can be used for their identification.

Many of the species of the brown algae grow to macroscopic sizes and are commonly known as **seaweeds** or **kelp.**

***Table* 1.4 Important algal groups**

Algal Group	Chlorophyll Type	Cell Wall Polysaccharide
Green algae	*a, b*	Cellulose
Diatoms	*a, c*	Silica
Dinoflagellates	*a, c*	Cellulose
Brown algae	*a, c*	Cellulose and algin

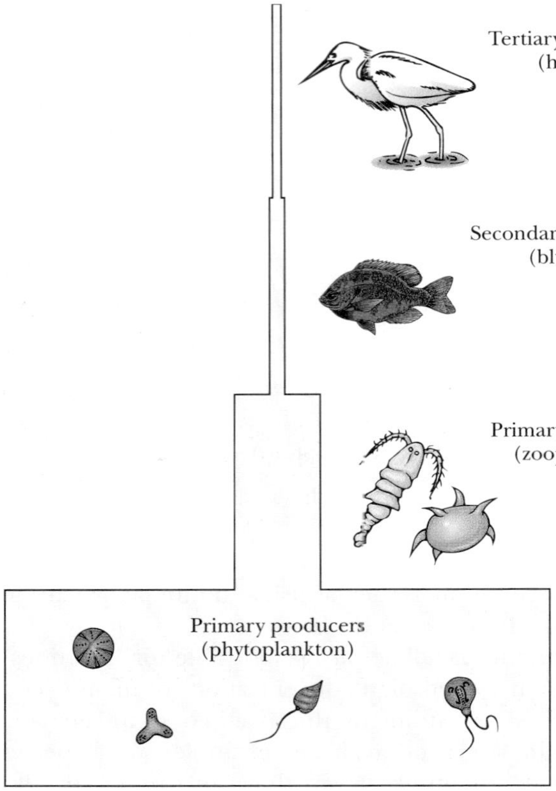

***Figure* 1.17** This is a "biomass" pyramid which illustrates the amount of living biological matter (biomass) at each of the nutrient levels in an aquatic habitat. Thus the bottom layer corresponds to the primary producers (algae or phytoplankton) which serve as the base of the pyramid. The next level corresponds to the biomass of primary consumers, such as the water fleas (daphnia) and other zooplankton that consume the primary producers for food. The biomass of this level is considerably less than that of the primary producers (typically about 10 percent of that of the primary producers). Likewise, similarly lower biomass amounts occur for carnivores (secondary consumers) such as small fish, and even lower levels for tertiary consumers, e.g., birds, mammals, etc. For simplicity, bacteria are not included.

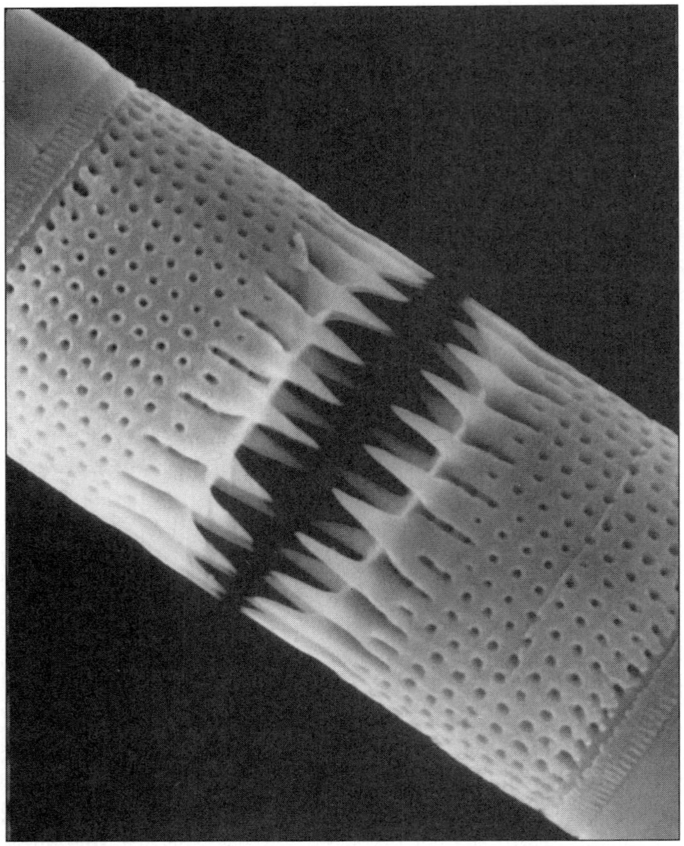

***Figure* 1.18** A scanning electron micrograph of a diatom showing the elaborate structure of its silica shell or frustule. (Courtesy of Barbara Reine)

Some are harvested from near-shore environments for alginate and carageenan, cell wall polysaccharides, which are used in the food industry as stabilizers and thickeners. Agar, used as a food source by some in Asia as well as a solidifying agent for bacteriological media, is also derived from seaweed.

Some of the simplest algae are members of the green algal genus, *Chlorella.* This alga is nonmotile and lives in ponds and lakes all over the world. Another common alga is the green algal genus, *Chlamydomonas.* This organism is larger than *Chlorella* and is motile by two flagella (see Figure 1.5). Its chloroplast is cup-shaped. This genus is also common in freshwater habitats. However, some members of this genus grow in snow fields (Figure 1.19**a**) where they impart a reddish color to the snow (Figure 1.19**b**). The red color is due to carotenoid pigments which mask the green color of the chlorophyll *a.* Another common group of green algae, called the desmids, are found in ponds and lakes (Figure 1.20).

A colonial green alga, called *Volvox,* consists of a spherical colony in which hundreds of individual cells are joined together to form an outer shell. Each cell is motile by two flagella as in *Chlamydomonas.* As reproduction occurs, small "daughter colonies" are formed within the original mother colony (Figure 1.21). The genus *Zygnema* is an example of a filamentous alga (Figure 1.22).

The Fungi

The fungi are nonphotosynthetic microorganisms that utilize a variety of organic compounds as growth substrates. Table 1.5 lists the various groups of fungi based on cell type and reproduction. Almost all of the fungi form a multicellular branching filamentous structure referred to as a **mycelium** (Figure 1.23). A familiar example is bread mold, which grows as a dark filamentous mass.

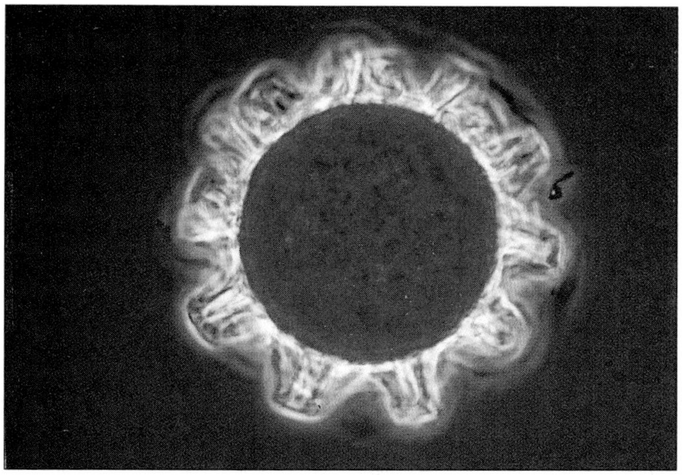

(a)

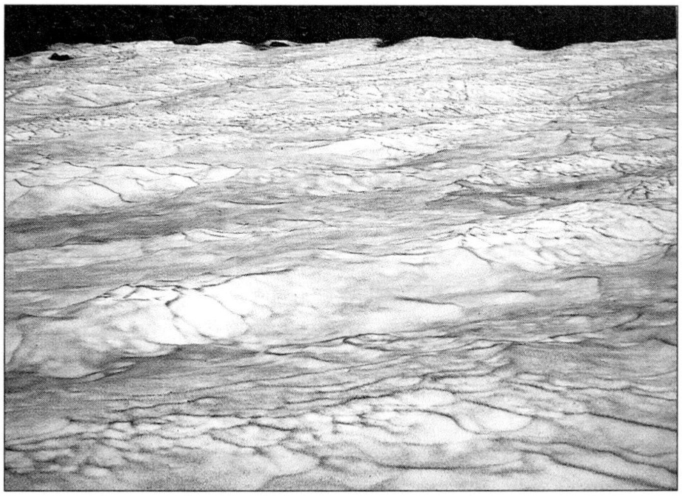

(b)

***Figure* 1.19** **(a)** A phase photomicrograph of the cyst of a snow alga. **(b)** A snow field on Mt. Rainier. The red color is due to snow algae. (Courtesy of J. T. Staley)

Chitridiomycetes

Most fungi occur in terrestrial habitats. However, the chitridiomycetes contain both aquatic and terrestrial forms. An example of an aquatic chitrid is the genus *Allomyces. Allomyces* grows as a coenocytic mycelium. In its haploid state it produces a mycelium bearing spores (a **gametothallus**) containing the male and female mating types (Figure 1.24). When the sporangium matures, motile spores of each mating type are released. These can fuse together to form a zygote cell. The zygote, in time, will develop a diploid mycelium. The diploid mycelium develops its own sporangia, some of which are diploid, and others of which are haploid. The diploid spores can repeat the life cycle of the diploid stage by giving rise to motile diploid spores, and the haploid spores released will produce a haploid mycelium.

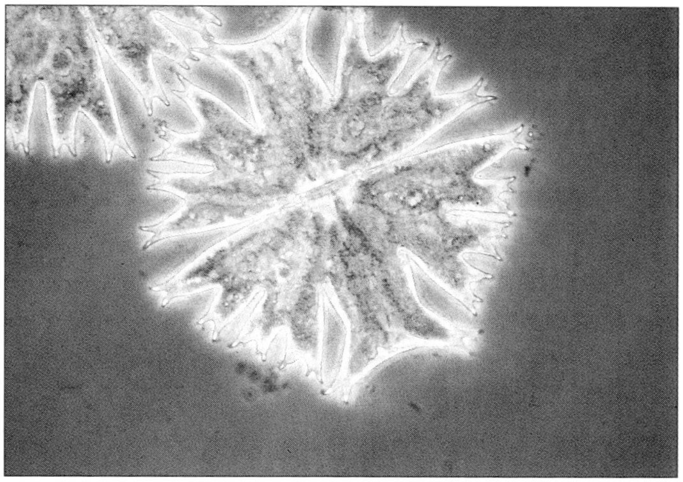

***Figure* 1.20** A desmid of the genus *Micrasterias* commonly found in ponds. (Courtesy of J. T. Staley)

Ascomycetes

Baker's yeast, *Saccharomyces cerevisiae,* is an example of an ascomycete. It divides by bud formation and can do so repeatedly until its entire surface is scarred by bud-formation, at which time it stops dividing. In this relatively simple one-celled form, the organism exists mostly in the diploid stage. Meiosis occurs at some point in the life cycle to produce within the cell four haploid mating cells.

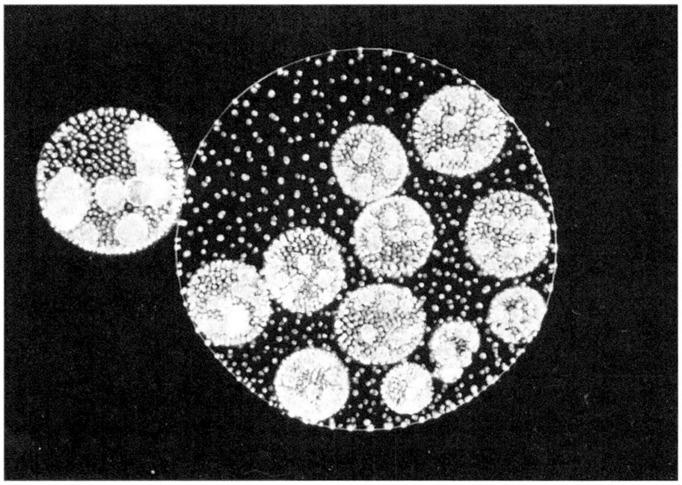

***Figure* 1.21** A phase photomicrograph of *Volvox,* a colonial green alga. Each bright, green dot corresponds to a cell. Note that the two largest colonies contain small daughter colonies that are formed inside the "mother" colony as a means of protection. (Courtesy of Dennis Kunkel)

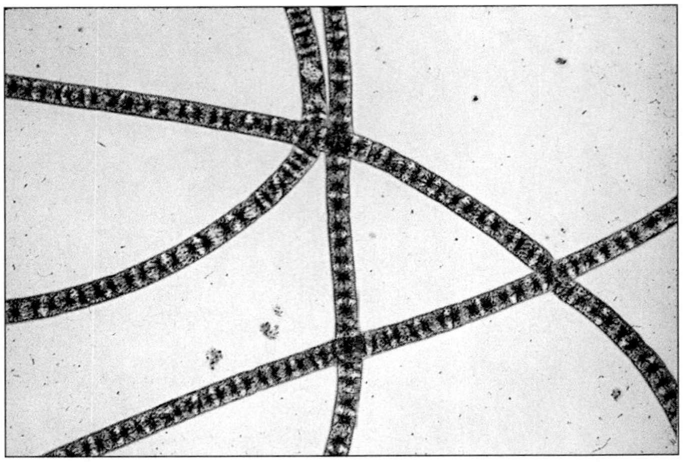

Figure **1.22** *Zygnema,* a genus of filamentous green algae. (Courtesy of J. T. Staley)

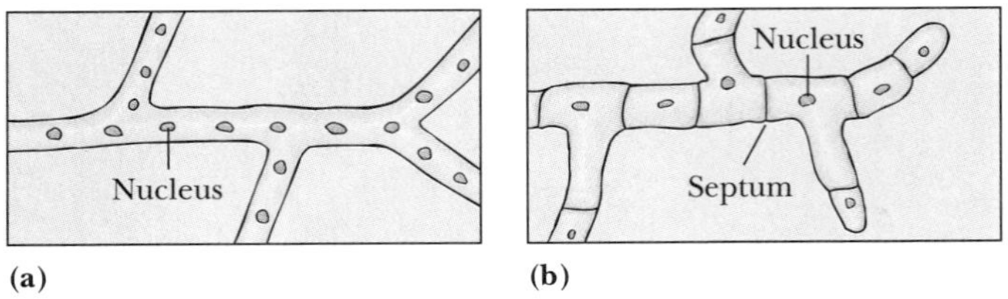

Figure **1.23** The branching filamentous appearance of a mycelium. **(a)** A coenocytic mycelium lacks cross wall septations between cells, whereas **(b)** The septate mycelium has them.

The mother cell therefore becomes the **ascus,** i.e., the structure which bears the spores. Two mating types exist within the mother ascus, the + and − form. In this species zygote formation occurs normally by the fusion of the two mating types upon germination of the ascospores within the mother ascus. This is then followed by budding and continuation of life in the diploid state. Thus, the haploid state is short-lived.

One of the macroscopic forms of ascomycetes is the morel, a delicious edible type of fruiting fungus.

Basidiomycetes

Most of the mushrooms are basidiomycetes. They have a characteristic diploid mycelium which contains + and − nuclei. During replication of DNA a special device in the cell called a "clamp" is used to partition the newly formed nuclei from the initial ones (Figure 1.25). Sexual spores are produced on a structure called a **basidium,** in which the spores are held externally at the tips of a structure called **sterigma** (Figure 1.26). These are released into the environment where they are carried by the wind to a new location.

The life cycle of the common edible mushroom *Agaricus campestris* is shown in Figure 1.27. The mycelium begins by the germination of a basidiospore. Two mating types fuse to produce a binucleate, diploid mycelium which persists during most of its life cycle in the soil. Frequently mushrooms grow in association with tree or other plant roots from which they derive nutrients (some fungi exist in special associations with tree roots called mycorrhizae). When conditions favorable for fruiting occur (in the natural environment this normally occurs in the fall, after the summer growth period has ceased and is followed by a wet period), the mycelium develops beneath the soil and produces the characteristic above-ground fruiting structure distinctive of the species.

There are many species of mushrooms ranging from the simple puffballs to the bracket fungi. Most of them are basidiomycetes; however, the morels and truffles are examples of ascomycetes.

Table **1.5 The groups of fungi**

Group	Special Features	Reproduction	Examples
Chitridiomycetes	Coenocytic[1], asexual spores in sac	Free zygote	*Allomyces*
Ascomycetes	Septate, asexual spores on chains	Sexual spores (zygotes) in sac (ascus)	Most yeasts, morels, truffles
Basidiomycetes	Septate, clamp cells	Sexual spores on stalks (basidium)	Most mushrooms
Fungi Imperfecti	External spores in chain	Sexuality unknown	*Penicillium*

[1]Filamentous organisms in which there is no crosswall or septum between the nuclei.

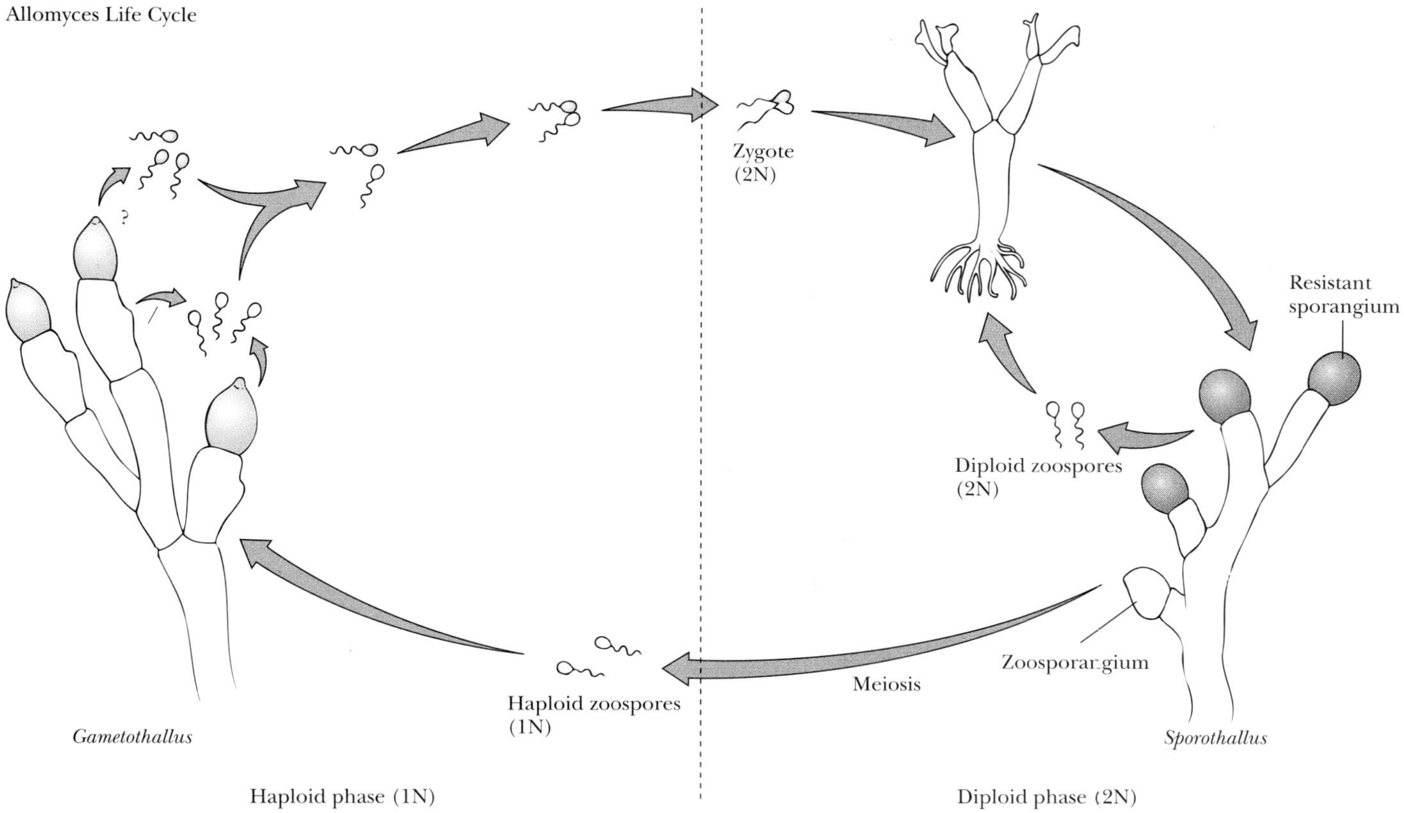

***Figure* 1.24** Alternation of generations as illustrated for the genus *Allomyces* an aquatic fungus. The haploid generation releases male and female gametes which fuse to form a diploid zygote. The diploid zygote germinates to produce a mycelium which forms spores. The meiosporangium produces haploid cells meiotically and these germinate to form the haploid generation.

Fungi imperfecti

This is a group of fungi whose sexual stage has not yet been observed. They can be grown in the laboratory, and many form asexual spores (Figure 1.28) at the tips of aerial mycelia. This feature clearly separates them from the Phycomycetes, but they have not yet been found to produce a diploid stage. Thus, it is not certain whether they are ascomycetes or basidiomycetes.

Protozoa

The protozoa are mostly single-celled organisms. They are nonphotosynthetic and most grow by ingesting particulate food items, a type of nutrition referred to as **phagotrophic.** Although most of the protozoa consist of a single cell, their

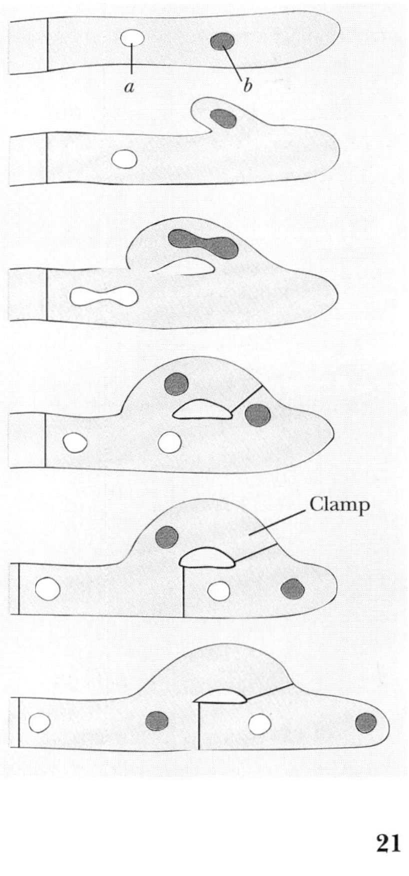

***Figure* 1.25** In basidiomycetes, the diploid mycelium contains two nuclei, *a* and *b,* which makes it diploid. During asexual division of the cells a clamp structure is formed so that each of the resulting daughter cells can receive one set of nuclei.

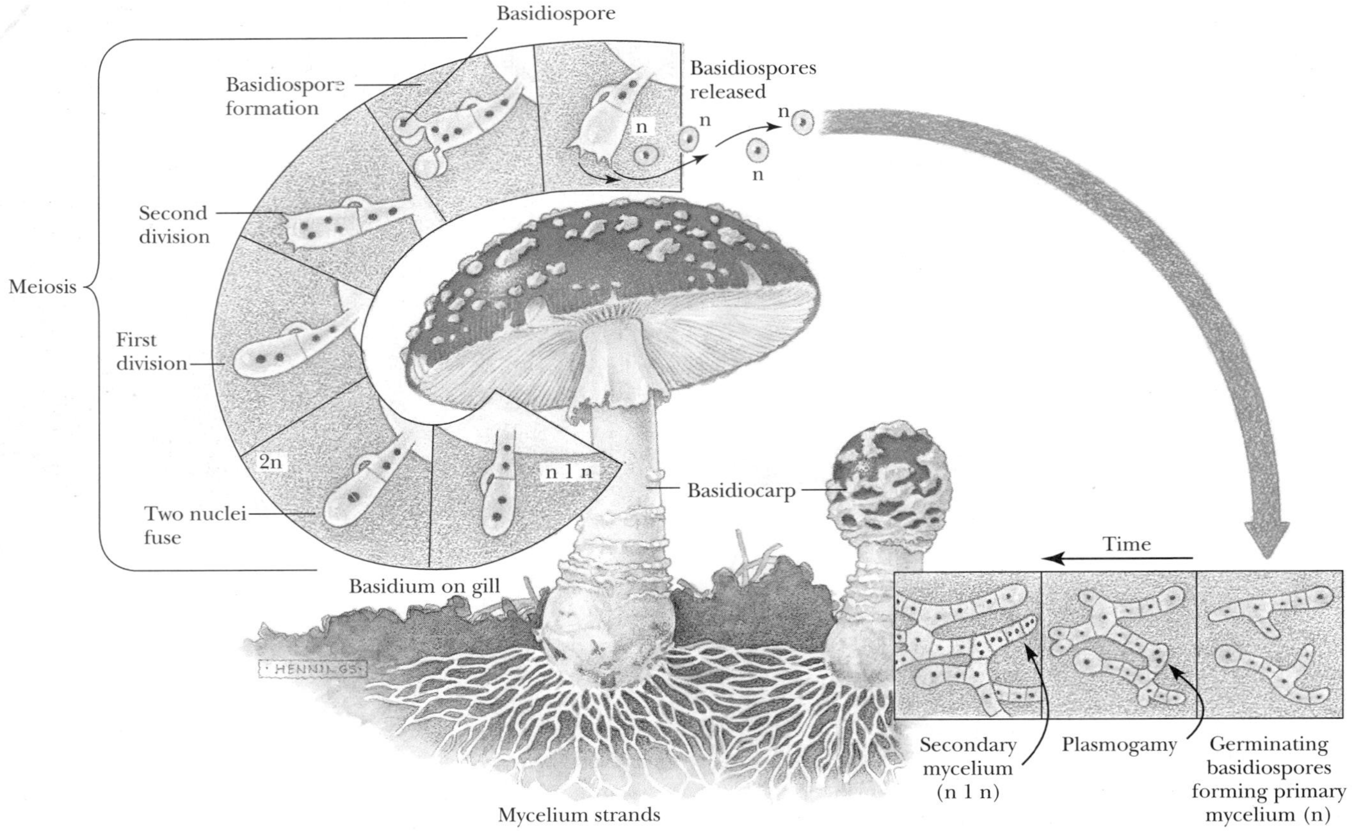

***Figure* 1.26** Life cycle of a typical basidiomycete showing its basidiocarp (mushroom) and sexual spore development. The structure that forms the gametes, or sexual spores, in basidiomycetes is called a basidium, large numbers of which are produced in the gills of the mushroom. The nuclei (n + n) fuse to produce the true diploid nucleus (2n) during basidiospore formation. This is followed by meiosis to produce the four gametes, the nucleus of each of which is incorporated into its own basidiospore. Basidiospores are released to the wind and, when they germinate in soil, result in the formation of 1n mycelium. Later, two different mating types fuse to form the n + n mycelium that predominates in the mycelium that forms the fruiting structure.

***Figure* 1.27** Photo of a fruiting structure of a common mushroom. (©Dick Poe / VU)

cells have elaborate internal structures and functions. For convenience the protozoa are treated here in four separate groups (Table 1.6). The evolutionary grouping of these organisms is treated later in the book (Chapter 23).

Amoebae

The amoebae are noted for their peculiar amoeboid movement, in which the cell projects membrane extensions of the cytoplasm called **pseudopodia** (false feet) into the environment. The cytoplasm then flows into the area of the pseudopodium.

These organisms feed by enclosing small particulate materials such as bacteria and bringing them inside the cells in their engulfment process. The result is the creation of a vacuole about the food, which is then digested enzymatically. Though the amoebae have no definite shape as

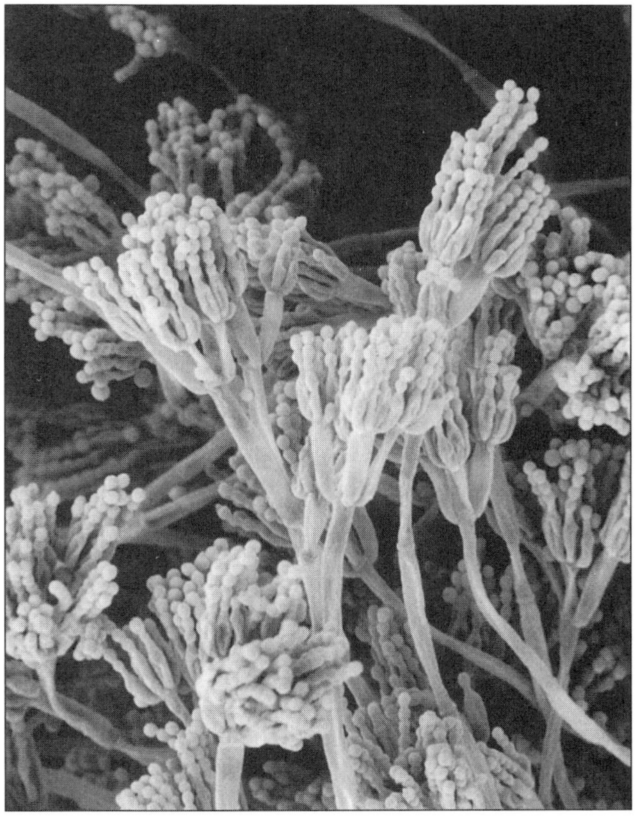

***Figure* 1.28** Photo of the asexual fruiting structure and spores formed by a *Penicillium* species, the genus of fungi that produces penicillin. (Courtesy of Valerie Knowlton)

they lack a cell wall, some members of this group such as the foraminiferans and radiolarians produce and reside in an external shell (Figure 1.29).

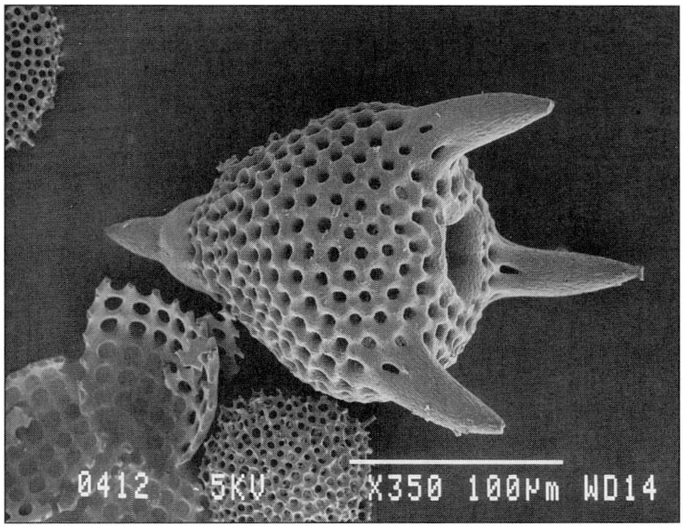

***Figure* 1.29** Scanning electron micrograph of a radiolarian shell. (Courtesy of Barbara Reine)

Flagellates

The flagellate protozoans move by long flagella. Not all of them are phagotrophic; some use soluble nutrients and feed like fungi. Two examples of flagellate protozoans are *Trichomonas vaginalis,* a sexually transmitted protozoan that causes an infection of the vagina, and *Trypanosoma gambiense,* which causes African sleeping sickness.

Ciliates

The ciliate protozoa contain numerous short flagella, called **cilia,** which, by acting in a concerted fashion, move the cell about. They have a gullet, oftentimes lined with special cilia that bring food particles into the gullet for ingestion. The food particles enter the cell in vacuoles and circulate through the cytoplasm by cytoplasmic streaming, during which they are digested.

Some of the ciliate protozoans are anaerobic and grow in the rumen of cattle and other animals.

***Table* 1.6 Distinguishing features among the groups of protozoa**

Group/motility	Examples
Amoebae/amoeboid	Amoebae, e.g., *Entamoeba histolytica,* causes amoebic dysentery; *Naegleria fowleri,* causes encephalitis; foraminiferans, radiolarians
Flagellates/ flagella	*Trypanosoma* which causes sleeping sickness; *Trichomonas* which causes a type of vaginitis; *Giardia* which causes diarrhea
Ciliates/cilia	*Paramecium, Tetrahymena*
Sporozoa/most immotile	Symbiotic forms; mostly parasitic including *Plasmodium,* the malarial parasite

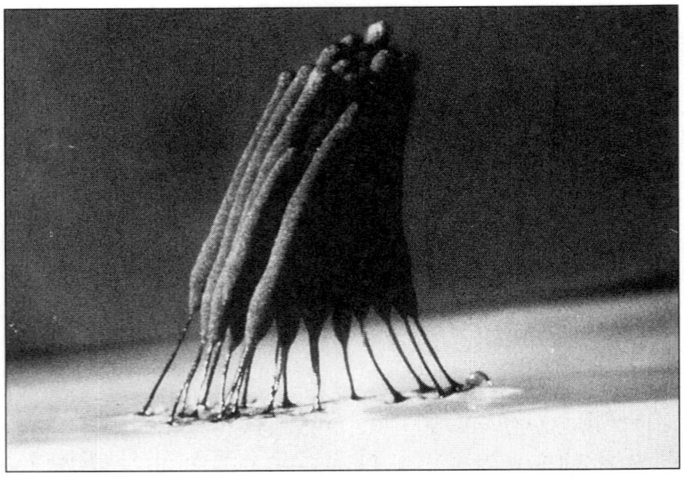

***Figure* 1.30** Fruiting body of a plasmodial slime mold, *Stemonitis.* (Courtesy of Edward Haskins)

Sporozoans

The sporozoans, also called apicomplexans, are mostly nonmotile. They are all, without exception, symbionts, and many are pathogenic. These are saprotrophic in their nutrition and thus feed by taking in dissolved substances from their host.

Slime Molds

Another microbial group is called slime molds. They share some characteristics of protozoa and fungi. They are of special interest to biologists because of their complex life cycles. There are two types of slime molds, the plasmodial or true slime molds and the cellular slime molds. Both are heterotrophic and feed on particulate organic materials.

The **plasmodial slime molds** grow in an organic-rich environment and are multinucleate. They lack cell walls and feed on particulate matter such as bacterial cells and dead organic matter in the same manner as amoebae. As they develop, the cell becomes increasingly large, perhaps up to as much as several hundred grams in weight, with millions of nuclei. This cell mass, which is referred to as a *plasmodium,* moves like a large amoeba and stops when it has run out of food or when conditions become too dry. It then undergoes fruiting body formation (Figure 1.30).

The **cellular slime molds** are quite different. In contrast to the plasmodial types, they remain unicellular through their life cycle, during which they resemble typical unicellular amoebae. However, at some point in their life cycle, when nutrients become limiting or conditions become dry, they aggregate together to form a moving mass that resembles a slug (Figure 1.31). This mass glides slowly along surfaces and eventually undergoes fruiting body formation.

Lichens

Lichens—one of the most fascinating forms of life—actu-

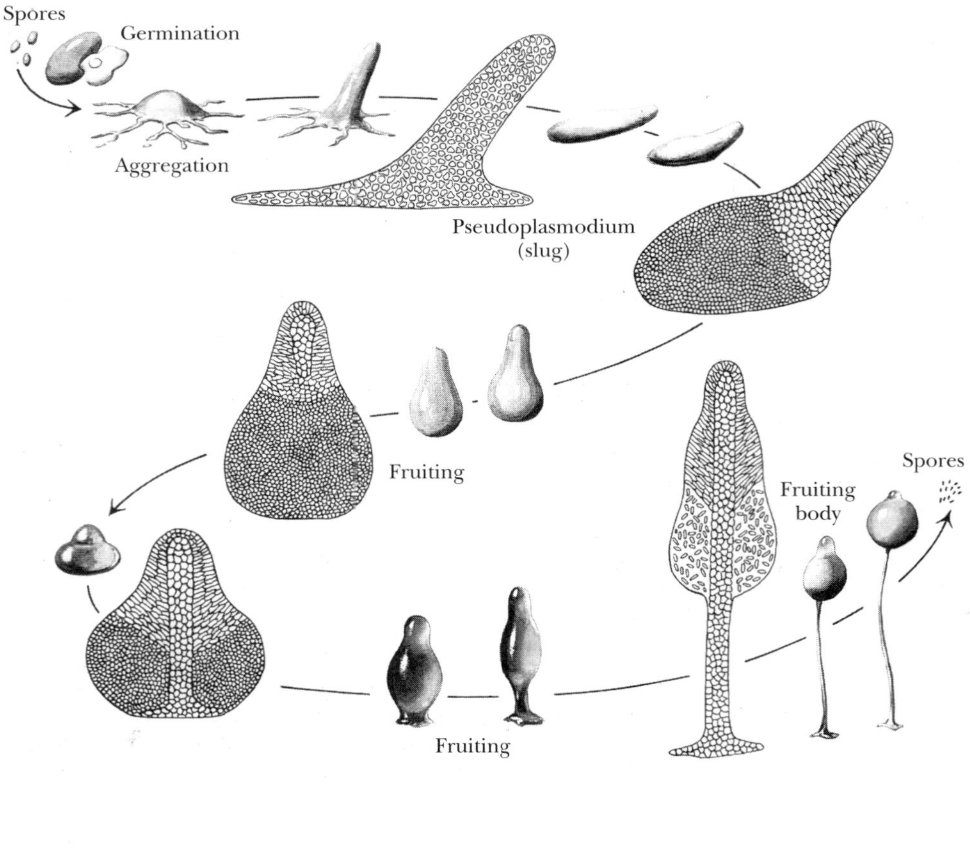

***Figure* 1.31** Life cycle of a cellular slime mold *Dictyostelium discoideum.* Spores germinate to form myxamoebae when nutrients are available. These feed until nutrients become depleted. Then the myxamoebae aggregate to form the pseudoplasmodium—so-called because each cell retains its cell membrane (a true plasmodium is multinucleate). The pseudoplasmodium is also called a "slug" because it resembles a miniature shell-less snail. The slug moves through the environment and undergoes fruiting body formation, resulting in the production of spores at the tip of a stalk. (Drawing by H. C. Lyman; Courtesy of Edward Haskins)

***Figure* 1.32** A lichen, *Cladonia cristatella,* so-called the "British soldier" because of its red top. (© L. West 1985/Photo Researchers, Inc.)

ally live on, and sometimes even inside of, rocks. Some lichen species reside on other surfaces as well, such as tree bark or the surface of soil. Spanish "moss" is actually a type of lichen that is found hanging from trees in certain parts of the Southern United States, and "reindeer moss" is a type of lichen that grows on soils in polar tundra regions. But the most common form of lichen is found on the surface of rocks, where they form greenish, gray, orange, yellow, or black growths that adhere strongly to the rock surface (Figure 1.32).

Lichens are remarkable in that they consist of two types of microorganisms: a fungus and an alga or cyanobacterium. They are examples of a mutualistic symbiosis (life together), meaning that these two forms of life live together for the benefit of both. The algal partner is photosynthetic and therefore fixes carbon dioxide and provides organic matter for the fungus. The fungus provides inorganic nutrients for the alga and protects it from dehydration. Between the two of them they can live where no other forms of life reside.

The surface of a rock is a most inhospitable place for other forms of life, and, because other organisms cannot colonize it successfully, it is available for the lichens. The lichen structure, termed its **thallus,** consists mostly of fungal hyphae within which are enmeshed the algal cells. In cross section the layer of algal cells can be seen to be located near the surface and therefore close to the light (Figure 1.33). It is possible to cultivate the fungal partner and the algal partners of the symbiosis separately; however, no one has yet been able to produce a lichen in the laboratory by combining the two partners together.

Fungi are noted for their ability to grow at low relative humidities (see Chapter 6) such as are often found on rock surfaces. They also produce acids that can dissolve rock minerals. The role of the fungus is to protect the algal cells during dry periods and provide water and nutrients during wet periods. Because rocks are largely devoid of organic materials, these are provided by the algal partner. The lichen metabolizes quickly during wet periods, and survives well during periods of dryness. The distinctive features of both partners allow for the symbiosis to succeed in this harsh environment.

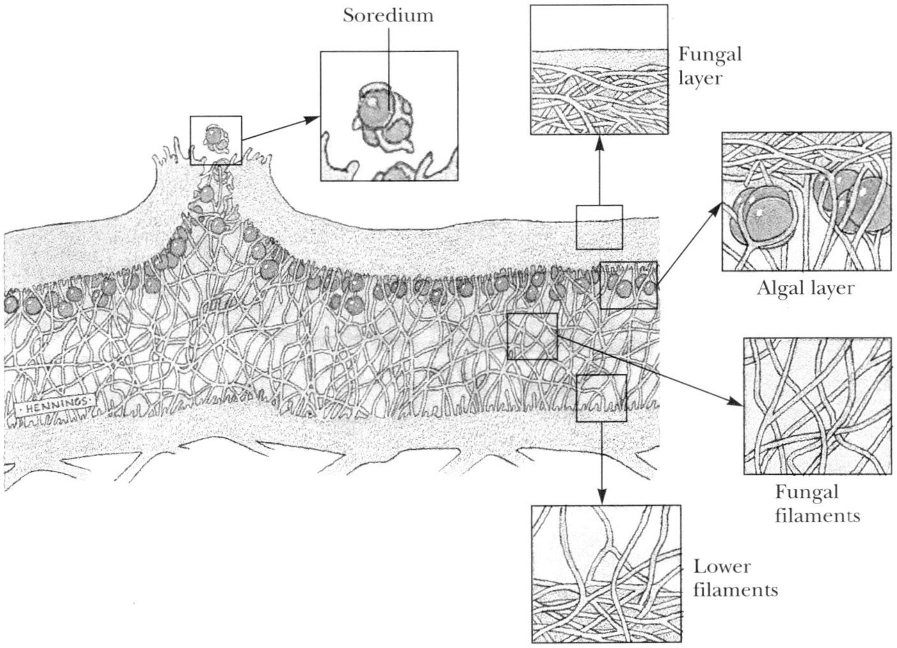

***Figure* 1.33** A cross section through *Cladonia cristatella* showing its mycelial nature and the location of the algal cells.

Summary

- **Microorganisms** are the smallest living creatures; most cannot be seen unless they are magnified with a microscope.
- Microorganisms are ubiquitous in the biosphere; some grow in extreme habitats such as boiling hot springs, saturated brine solutions, and sea ice; others produce sulfuric acid.
- Microorganisms that use light as an energy source and carbon dioxide as a carbon source are called **photoautotrophic.**
- Microorganisms that use light as an energy source and organic carbon as a carbon source are called **photoheterotrophic.**
- Microorganisms that use organic chemical compounds as energy and carbon sources for growth are called **heterotrophic, chemoheterotrophic,** or **chemoorganotrophic.**
- Microorganisms that use inorganic compounds as energy sources and carbon dioxide as a carbon source are called **chemoautotrophic.**
- The term **bacteria** is synonymous with **prokaryotic** organism.
- The two groups of prokaryotic microorganisms are called **Eubacteria** and **Archaea.**
- The three groups of eukaryotic microorganisms are **algae, protozoa,** and **fungi.**
- **Viruses** have RNA or DNA, not both, they lack cytoplasm, and they do not produce cells.
- Prokaryotic organisms lack a **nuclear membrane** and do not have membrane-bound organelles such as mitochondria and chloroplasts.
- Prokaryotic organisms have **70S ribosomes,** whereas eukaryotic organisms have larger **80S ribosomes.**
- **Eukaryotic** organisms have a nuclear membrane, and most possess **mitochondria** and/or **chloroplasts.**
- **Mitosis** is the process by which the chromosomes are duplicated and separated during asexual division of eukaryotic cells.
- **Meiosis** is the process by which diploid chromosomes of eukaryotic cells are separated in the formation of haploid gametes.

Questions for Thought and Review

1. Why do bacteria grow in such unusual habitats?
2. **Gnotobiotic** animals are animals that are kept germ free from birth by placing them in special living chambers. What would be the advantages and disadvantages of a bacterial-free environment to these animals?
3. Cellulose can be degraded by many microorganisms and a few higher animals including termites and ruminant animals (which have intestinal cellulolytic microbes). It can also be burned to form carbon dioxide. Compare wood burning with microbial decay of cellulose. What would happen in a world in which cellulolytic microorganisms did not exist?
4. What are viruses and are they alive?
5. Compare the nutrition of humans with that of a typical chemoheterotrophic bacterium.
6. Some bacteria can reproduce to form a progeny cell in less than 15 minutes. Why is human reproduction so much slower?
7. In the *Exxon Valdez* oil spill in Alaska in 1989, the spill was cleaned by both physical treatment (washing rocks by hand) and microbial treatment (fertilizer was added to enhance microbial growth). Why was microbial degradation preferable?
8. Compare the three-Domain classification of living organisms accepted by bacteriologists with the five-kingdom system. Which kingdoms have prokaryotes and eukaryotes in both classifications?
9. Why don't bacteria undergo meiosis?
10. Compare and contrast the various groups of eukaryotic microorganisms.

Suggested Readings

Chatton, E. 1932. *Titres et Travaux Scientifiques.* Séte: Sottano.

Dubos, René. 1962. *The Unseen World.* New York: The Rockefeller Institute Press.

Gest, H. 1987. *The World of Microbes.* Madison, WI: Science Tech Publishers, Inc., 249 pp.

Lwoff, A. 1957. The concept of virus. *Journal of General Microbiology,* **17:**239–253.

Stanier, R. Y., and C. B. van Niel. 1962. The concept of a bacterium. *Archiv für Mikrobiologie* **42:**17–35.

Woese, C. R., O. Kandler, and M. C. Wheelis. 1990. Towards a natural system of organisms: proposal for the domains Archaea, Bacteria, and Eucarya. *Proceeding of the National Academy of Sciences, USA.* **87:**4576–4579.

The accidents of health had more to do with the march of great events than was ordinarily suspected.

H. A. L. Fisher 1865–1940

Chapter 2

Historical Perspective

The previous chapter was devoted to an outline of that part of the living world termed "microbial." Microbiology is the study of microorganisms, a multifaceted discipline concerned with infectious disease, agricultural practice, sanitation, and industrial production of chemicals. Microorganisms have been and remain important models for studies in nutrition, metabolism, and genetics; they have been instrumental in the development of biotechnology. Microbes also make up a considerable part of the biosphere and are therefore studied for their own sake. Many scientists contributed to the development of microbiology, and this chapter presents that background as well as the contributions made by some of the pioneers in the field.

Effects of Disease on Civilization

The microbe has markedly affected the course of human history. The morale of populations, the strength of armies, and the outcome of battles have often been decided by disease-bearing microorganisms. The mobilization of armies themselves, with the consequent concentration of young soldiers from across the country, created environments ripe for the spread of epidemics. Every soldier was both a potential carrier of infectious disease from his home and a potential victim of disease from lack of previous exposure. In rapidly growing urban areas, the absence of proper sanitation under crowded conditions also contributed to the spread of infectious agents. And as world trade increased, infectious disease was readily carried from region to region. It then retraced its steps, returning to a new generation of susceptible individuals. This section briefly covers some examples of these and other effects of microbial disease on human populations, institutions, and wars throughout the history of our civilization.

The decline of Rome under the Emperor Justinian (A.D. 565) was hastened by epidemics of bubonic plague and smallpox. The inhabitants of Rome were decimated and demoralized by this massive epidemic and, as a result,

BOX 2.1 MILESTONES

The Spread of the Black Death in Europe

The Great Plague of the fourteenth century was also referred to as the "Black Death." The epidemic originated in China in 1331 and moved slowly across Asia, reaching the outskirts of Europe in 1347. The disease spread from caravan stop to caravan stop, moving with the rat-flea-human community. The high rate of fatalities among these hosts in thinly populated areas of Asia was probably responsible for the slow movement of the epidemic across Asia. When the infected flea/rat population reached the Mediterranean area the disease agent *(Yersinia pestis)* spread into the black rat populations. Shipping and commerce carried the black rats and their bubonic plague–infected fleas upward across Europe. The Black Death reached Sweden by late 1350.

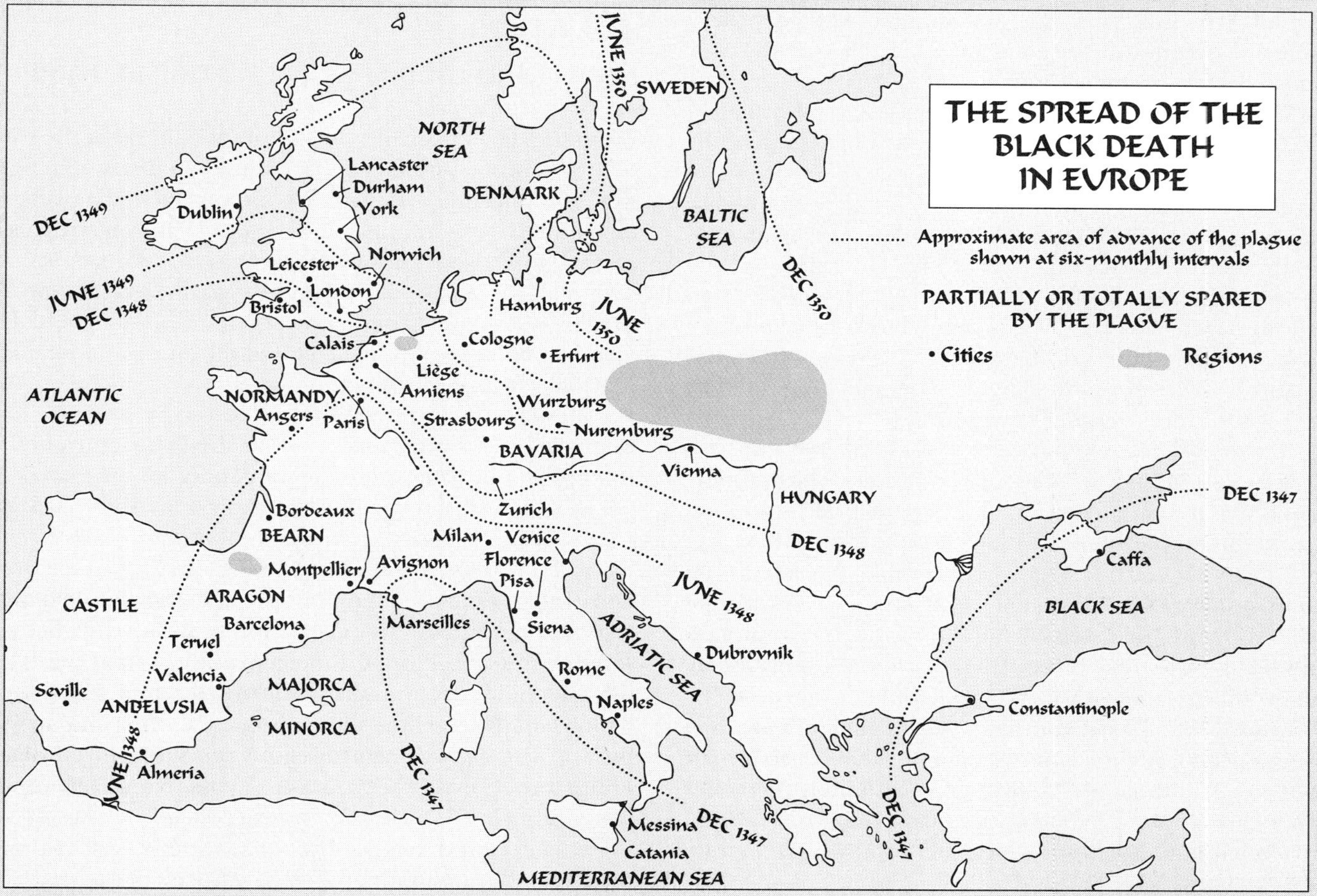

were left powerless against barbarian hordes that destroyed the empire. Through the Middle Ages and beyond, each generation was subject to renewed epidemics. Some of these epidemics spread across the continents; others were more localized. Typhus, plague, smallpox, syphilis, and cholera were some of the infectious diseases that caused suffering and great loss of human life.

The populations of Europe, North Africa, and the Middle East totaled about 100 million when a plague (the "Black Death") struck in 1346. The epidemic came down the "silk road" (the main trade route to China), bringing death to Asia; it then spread throughout Europe, with the loss of 25 million people in a few short years (see **Box 2.1**). Recurrences of plague through the sixteenth and seventeenth centuries kept populations in check. Between 1720 and 1722, one last great epidemic occurred in France. Deaths of people living in these cities were 60 percent in Marseilles, 60 percent in Poulon, 44 percent in Arles, 30 percent in Aix, and 30 percent in Avignon. The most recent major plague pandemic originated in Yunnan, China, in 1892, moved across India, and arrived in Bombay in 1896. This outbreak killed an estimated 6 million people in India alone.

In wars fought prior to World War II, the outcome was generally decided by arms, strategy, and pestilence—with pestilence, more often than not, playing the leading role. Following are a few examples of the role played by disease.

In 1566, Maximilian II of Germany outfitted an army of 80,000 to face the Sultan Soliman of Hungary. This battle was to drive the Eastern Hordes out of Europe. The German Army was encamped at Komorn when a violent and deadly typhus outbreak occurred. The German Army dispersed without engaging the Turks.

The Thirty Years War (1618–1648) was essentially a Protestant revolt against Catholic oppression. This lengthy war had a marked effect on the religion and geography of Europe, and it was also dominated by debilitating disease epidemics. One episode involved the opposing armies before the battle of Nuremberg in 1632. The armies of the Swedish king, Gustavus Adolphus, and of the Roman Catholic Duke Albrecht Wallenstein faced one another, and before they could do battle 18,000 soldiers from the two forces succumbed to typhus. The surviving soldiers from both armies fled the field of battle.

By the time Napoleon began his retreat from Moscow in 1812, most of his army had fallen victim to typhus, pneumonia, dysentery, and other illnesses. Disease, cold, and deprivation all played leading roles in his departure. The following year (1813), the irrepressible Napoleon recruited a new army of 500,000 young soldiers. As with the previous army, their youth and the crowded unsanitary conditions under which they lived rendered them susceptible to infectious disease. When Napoleon faced the allies at Leipzig, the preliminary battles and disease had reduced his army of approximately 500,000 to about 170,000. An estimated 105,000 were casualties of earlier battles, but about 220,000 were incapacitated by illness. Thus, microbial infections were a major factor in the ultimate defeat of Napoleon.

Why Study the History of a Science?

Microbiology, as with any field of endeavor, can be comprehended best if one has a reasonable understanding of the historical development of the field. The ingenious experimentation and insights that led scientists, such as Pasteur and Koch, to a logical explanation for the observable manifestations of disease should be of interest to the modern student in microbiology. Just as a scholar of modern political science must be knowledgeable about the forces and conflicts that shaped the modern state, studies in a scientific discipline should be initiated by gaining an understanding of the historical development of concepts in that field. After all, can one fully comprehend present theories and concepts without understanding the logical steps that led to those ideas?

The remainder of this chapter presents the scientific contributions made by several intellectual giants in microbiology. The list is by no means complete, nor is space sufficient to discuss all those who contributed to the foundations of microbiology. Suggested readings at the end of the chapter will be especially useful for those interested in furthering their knowledge of the historical development of microbiology.

Status of Microbial Science Prior to 1650

From the dawn of civilization until the middle of the nineteenth century, any success in combating disease, in fighting the microbe's destructive power or harnessing its fermentative capabilities, came about by inexact processes of trial and error. Those occurrences now known to have been caused by the microbe were ascribed to a supreme being, "miasmas," spontaneous generation, magic, chemical instabilities, or other factors and interpretations limited only by the individual imagination.

Human thought was influenced by the great philosophers whose writings indicate that they were intrigued by theories supporting **spontaneous generation,** the origin of living organisms from inert organic materials. Aristotle (384–322 B.C.) and others wrote abundantly of the formation of frogs from damp earth and of mice from decaying grain. After the fall of Rome and the decline of civilization, inquiry and acquisition of knowledge became severely limited. Through the centuries of the Dark Ages, disease epidemics and plagues were recorded, but little of scientific consequence was written. The first major writings on the causation of disease were those of **Girolamo Fracastoro** (ca. 1478–1553), who wrote extensively on the contagions involved in the disease process.

It is evident from the literature of that time (prior to the "Age of Enlightenment") that scholars in the sixteenth and seventeenth centuries were seeking logical explanations for the phenomena of nature. Theory and experimentation were one thing, but acceptance of concepts contrary to the dogma of the time was quite another. It was particularly difficult to gain acceptance of biological explanations for natural phenomena when many renowned scientists, particularly chemists, clung to theories of chemical instability and spontaneous generation. Typical of this group was van Helmont (1580–1644), a forerunner of scientific chemistry, who published a recipe for producing mice from soiled clothing and a little wheat.

Microbiology from 1650 to 1850

The eighteenth century was the Age of Enlightenment, a time when people questioned traditional doctrines. Science, reason, and individualism replaced strict obedience to accepted dogma. However, during the seventeenth century there had been individuals of a scientific bent whose contributions gave impetus to the later awakening of the human spirit. Biology benefitted significantly from such developments as the microscope and the realization that living matter was composed of individual cells.

Fabrication of the original microscope is generally attributed to the Dutch spectacle maker, Zacharias Janssen, and his father, Hans, between 1590 and 1610. The first to employ a microscope extensively in the examination of biological material was **Marcello Malpighi** (1628–1694), and he is considered by many the "father of microscopic biology." Malpighi made major contributions in his discovery of capillaries and understanding of embryonic development. **Antony van Leeuwenhoek** developed a solar microscope that led to the first recorded observations of bacteria in 1683. Robert Hooke (1635–1703) examined the structure of cork and, based on these observations, suggested that all living creatures were made up of individual cells. The following section covers Leeuwenhoek and his discovery of bacteria in some detail. About 200 years passed before the true nature of these organisms was generally accepted. Why 200 years? Because humans were not yet ready to disregard completely the prevailing dogma that chemical instabilities and spontaneous generation were responsible for all activities now ascribed to microbes.

Antony van Leeuwenhoek

Antony van Leeuwenhoek was born in Delft in 1632 into a relatively prosperous family (Figure 2.1). At the age of 16 he was sent to Amsterdam to apprentice in a draper's shop and learn a useful trade. He was a bright lad and was appointed cashier in the business before returning to Delft to spend the remaining 70 years of his life. He had not attended a university, but he was learned in mathematics and became a successful businessman, a surveyor, and the official wine gauger for the town of Delft. In the latter capacity, he assayed all of the wines and spirits entering the town, and it was his responsibility to calibrate the vessels in which they were transported. His lack of a university education was inconsequential since, at that time, institutions were mostly devoted to theology, law, and philosophy. Scientific research was done by amateurs as an avocation, by the wealthy, or by individuals under the sponsorship of a patron.

Leeuwenhoek apparently developed microscopes and made extensive observations of bacteria prior to 1673 because, in that year, several of his studies were communicated to the Royal Society in England by his friend Renier de Graaf, a noted Dutch anatomist. Fortuitously, the editor of *Philosophical Transactions* (published by the Royal Society) was Henry Oldenberg, a German-born scientist well versed in several languages, including Dutch. Oldenberg firmly believed that scientific information should have no nationalistic restrictions and carried on an extensive correspondence with scientists throughout Europe. Leeuwenhoek was encouraged to communicate with the Royal Society. His first report, published in the *Philosophical Transactions,* dealt with molds, mouth parts and the eye of the bee, and gross observations on the louse. Due to Oldenberg's contacts and immediate translation of Leeuwenhoek's correspondence into several languages, his work generated considerable interest in the scientific world. Leeuwenhoek continued to send letters to the Royal Society throughout his life, and, in 1680, he was unanimously elected a Fellow of the Royal Society.

***Figure* 2.1** Antony van Leeuwenhoek (1632–1723) developed a microscope and was the first to observe and accurately describe bacteria. (Courtesy of Rijkmuseum, Amsterdam)

Over 200 of Leeuwenhoek's original letters have been preserved in the library of the Royal Society. Upon his death, Leeuwenhoek bequeathed a cabinet containing 26 microscopes to the Royal Society, and these were delivered by his daughter Maria with the handwritten passage, "every one of them ground by myself and mounted in silver and furthermore set in silver . . . that I extracted from the ore . . . and therewithal is writ down what object standeth before each glass." The cabinet and microscopes remained in the Royal Society collection for a century, but, unfor-

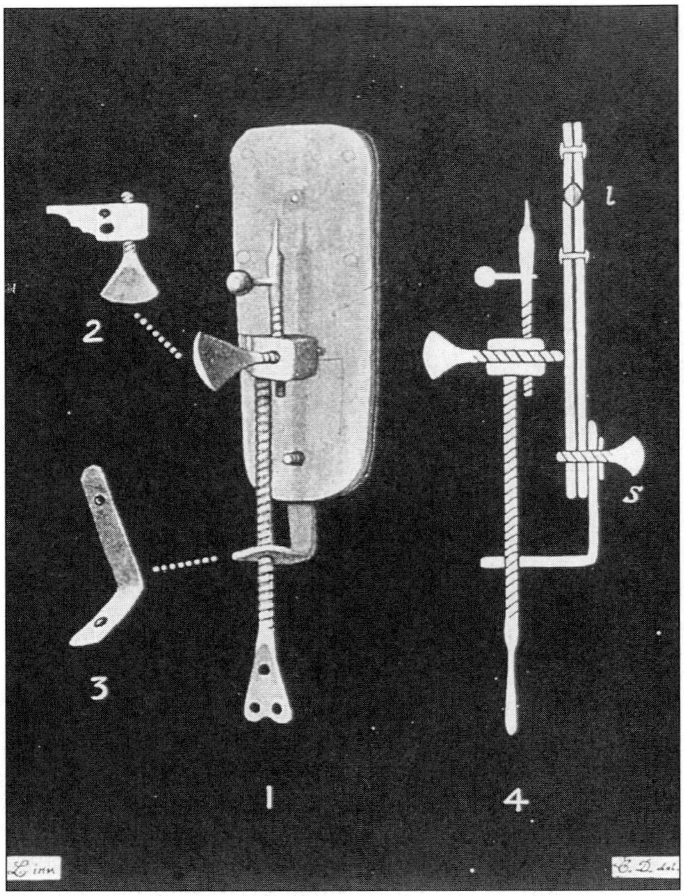

***Figure* 2.2** The microscope developed by Leeuwenhoek. It is very difficult to observe microbes with this crude instrument but Leeuwenhoek held the implement to the light in such a way that cells of both prokaryotes and eukaryotes could be viewed. His drawings confirm the accuracy of his observations. (From *Antony van Leeuwenhoek and His "Little Animals"* edited by Clifford Dobell, Dover Publications, 1960.)

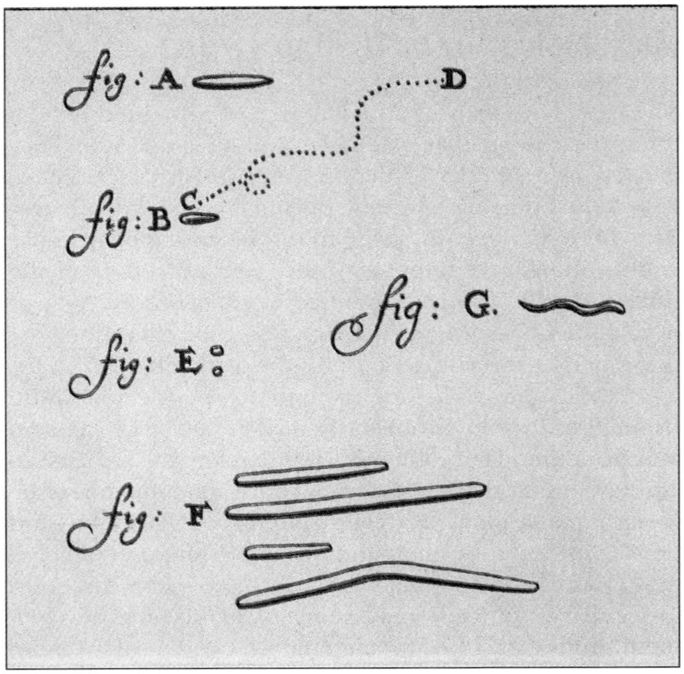

***Figure* 2.3** Drawings made by Antony van Leeuwenhoek and their probable identification: **(a)** *Bacillus;* **(b)**, **(c)**, **(d)** motile *Selenomonas;* **(e)** *Micrococcus;* **(f)** *Leptothrix;* and **(g)** a spirochete. (From *Antony van Leeuwenhoek and His "Little Animals"* edited by Clifford Dobell, Dover Publications, 1960.)

tunately, were lost and never recovered. The microscopes were rather simple in design, and Leeuwenhoek left no description of how they were operated (Figure 2.2). The exactness of his drawings suggests that he had a keen eye, but most certainly he also had an unexplained method for examination of bacteria. Since his microscopes could magnify by only 300 diameters (less than one third of what modern light microscopes can do), it is probable that he achieved his detailed observations of bacteria by careful focusing of solar light.

It is evident from his descriptions of what he called **"animalcules"** and from size estimates by comparison with grains of sand or human red blood corpuscles (which he measured fairly accurately at 1/300 inch) that Leeuwenhoek indeed closely observed the living forms he depicted. His writings describe some of the major bacterial forms now known—spheres, rods, and spirals—and he described motility in rod- and spiral-shaped cells (Figure 2.3). Leeuwenhoek's descriptions of bacteria were superior to those made by Louis Joblot (1645–1723) and Robert Hooke (1635–1703), who used slightly more sophisticated compound microscopes. Leeuwenhoek's extensive letters described an array of protozoa, spermatozoa, and red blood cells, and he is generally considered to be the founder of protozoology and histology.

Leeuwenhoek emphasized the abundance of protozoa, yeast, bacteria, and algae in environments as diverse as the mouth and seawater. Although these observations by Leeuwenhoek were widely known, apparently no one attributed biological processes such as fermentation and decay to these "animalcules." He was firm in his belief that his animalcules arose from preexisting organisms of the same kind and did not arise by spontaneous generation.

From Leeuwenhoek to Pasteur

A number of experimentalists (now much ignored) worked between 1725 and 1850, and their studies laid the foundation for the great advances made in the latter half of the nineteenth century. These scientists did much to discredit the theories of spontaneous generation, provide a background for studies on the immune response, and suggest rational classification schemes for bacteria. A major factor in affirming the role of microbes in nature was in disproving the generally held concept of spontaneous generation.

Francesco Redi (1626–1697), an Italian physician, published a book in 1688 attacking the doctrine of spontaneous generation, of which the presence of maggots in decaying meat was considered a prime example. Redi placed meat in beaker-like containers and covered some with fine muslin and left others exposed to invasion by blow flies. Although eggs were deposited on the muslin over the covered jars, maggots did not develop. Extensive growth of maggots, however, did occur in the meat that was left uncovered. When Redi placed the eggs deposited on the muslin onto the surface of the meat, maggots quickly appeared. Studies of this type led to careful experimentation and clear evidence that animals and insects, discernible by the eye, could not arise spontaneously from decaying matter. In response, the proponents of spontaneous generation turned to phenomena whose causes were not so apparent. The inability to identify the factor responsible for fermentation and putrefaction in infusions gave their theory continued life.

Lazzaro Spallanzani (1729–1799), an Italian naturalist and priest, was familiar with the studies of Redi and did a series of experiments to confirm and extend Redi's earlier work. Spallanzani hermetically sealed (airtight) meat broth in glass flasks and reported that 1 to 2 hours of heating the enclosed infusions was sufficient to render the contents incapable of supporting growth. These studies were attacked by John Needham, an English cleric and proponent of spontaneous generation. Needham proposed that a "vegetative force" was responsible for spontaneous generation and that this vital force was destroyed by hermetically sealing and heating the flasks. Spallanzani then did a series of experiments which he considered to be conclusive and from which he affirmed that, to render a broth sterile, it was necessary to seal the flask and not allow unsterile air to enter. Spallanzani also concluded that most organisms were destroyed by boiling for a few minutes, but that others, now known to be spore formers, withstood boiling for a half hour. Spallanzani's work was quite advanced for his time, and his deductions and conclusions were similar to those reported later by Pasteur.

During the latter part of the eighteenth century, Antoine Lavoisier and others demonstrated the indispensability of oxygen to animals. This led to the assumption that oxygen was the mysterious element (vegetative force?) necessary for spontaneous generation and that it was this element that had been excluded in Spallanzani's experiments. In 1836, **Franz Schulze** (1815–1873), a German chemist, performed a crucial experiment indicating that oxygen depletion was not the sole reason for sterility in Spallanzani's flasks. Schulze took a flask half filled with vegetable infusion and closed it with a cork through which two bent glass tubes were fitted (Figure 2.4). He thoroughly boiled the infusion in a sand bath, and while steam was being emitted, he attached an absorption bulb to each of the two glass tubes. One bulb contained concentrated sulfuric acid and the other a solution of potassium hydrate. Every day for several months, Schulze drew air out through the potash bulb and passed the air drawn in through the acid. The flask remained sterile, while a control flask without acid sterilization of incoming air had visible mold growth after a few days. When the sterile infusion was opened and exposed to the atmosphere, growth followed in a few days. His experiment demonstrated that infusions could be sterilized (no viable microbes) and that microbes could be introduced from the air. Many modifications of the Schulze experiment were done by John Tyndall, and further refinements were made by Theodor Schwann; these finally led to refutation of the doctrine of spontaneous generation.

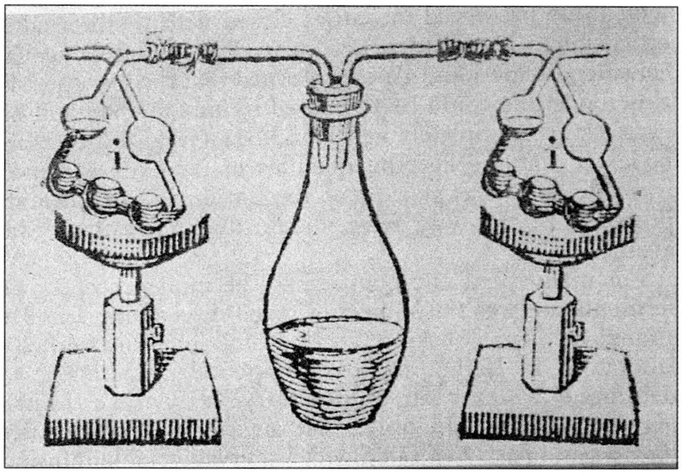

***Figure* 2.4** The experiment of Franz Schulze that did much to confirm that air does not contain a "vital force." (From *The History of Bacteriology* by William Bulloch, M.D., Oxford University Press, 1960.)

At the same time, other scientists were helping to disprove spontaneous generation by examining the involvement of microbes in fermentation. The first clear account of the yeast cell and its role in fermentation of beer and wine was given in an 1836 report by **Charles Cagniard-Latour** (1777–1859). He stated that **yeast** were nonmotile organized globules capable of reproduction by budding and that they probably belonged to the vegetable kingdom. In 1837, Cagniard-Latour suggested that the vital activity of the yeast cell was responsible for converting a sugar solution to carbonic acid and alcohol. **Theodor Schwann** (1810–1882), a German physiologist, independently discovered and described the yeast cell in 1837. Although his report also concerned the doctrine of spontaneous generation, he wrote that beer yeast consisted of granules arranged in rows and that they resembled fungi. He believed them to be plants and observed their reproduction by budding. The relationship between growth of yeast and the process of fermentation was clear to Schwann, and he called the organisms "zuckerpilz" (sugar fungus), from which the term *Saccharomyces* (a genus of common yeast) was derived. He also noted the indispensable requirement for nitrogenous compounds in the fermentation process.

A third independent worker who contributed to the discovery of the **fermentation** process was **Friedrich Kützing** (1807–1893), a German naturalist, and his major publication was also dated 1837. He described the nucleus of the cell and developed the concept that all fermentation is caused by living organisms. He was among the first to suggest that different types of fermentation were brought about by physiologically distinct organisms.

The role ascribed to the yeast cell in fermentation was contemptuously attacked by the chemists of that time. J. J. Berzelius, F. Wöhler, and J. von Leibig (all influential chemists) attributed a chemical character to every vital process and suggested that chemical instabilities were responsible for fermentation. Although much rhetoric ensued, the criticisms of the chemists led to more definitive experimentation and a clearer understanding of the role of yeast and fungi in many different types of fermentation.

Classification

The earliest classification scheme that included the bacteria was formulated by Linnaeus (1707–1778) in *Systema Naturale* (1743), in which he described the animalcules (Leeuwenhoek) as "infusoria." The infusoria defined by Linnaeus, which included the bacteria, were placed in the genus *Chaos.* A major shortcoming of the Linnaean classification scheme was the emphasis that he placed on a few selected characteristics. This emphasis has permeated classification schemes until recent times; for example, we often use terms such as *rod* or *coccus* without considering the metabolic capabilities of the microbe. The **Adansonian classification** scheme proposed by Michael Adanson in 1730 gives equal weight to all characteristics of a species and is more amenable to modern computerized classification systems.

The Danish naturalist **Otto F. Muller** (1730–1784) presented several works that described, arranged, and named a number of animalcular forms. His major study, *Animalcula infusoria et Marina,* published posthumously in 1786, was 367 pages in length and described 379 species of bacteria. In his scheme, two of five genera described contained bacterial forms. The genera were named *Monas* and *Vibrio.* The term *Vibrio* has been retained to the present day.

The first attempt to separate bacterial forms from more complex organisms was made by Christian G. Ehrenberg (1795–1876) in 1838. His treatise, entitled *Die Infusionsthierchen als Volkommene Organismen (The Infusoria as Complete Organisms),* used some of Leeuwenhoek's drawings of microbial cells. It was 547 pages in length and recognized three families that comprised forms we now recognize as bacteria. These families—*Monadina, Crystomonadina,* and *Vibrionia*—encompassed several genera: *Bacterium, Vibrio, Spirochaeta,* and *Spirillum.* All of these terms are still in common usage.

Medical Microbiology and Immunology

Several scientists merit mention for their contributions in the area of causation of disease and immunology prior to 1850. In 1822, the Italian **Enrico Acerbi** postulated that parasites existed that were capable of entering the body and their multiplication caused typhus fever. This theory was advanced by the 1835 work of **Agostino Bassi** (1773–1856), who made classic observations on the diseases of silkworms. A disease that caused the death of the worms was rampant at that time; the dead silkworms were covered with a hard, white, limy substance. At the time, the limy coat was considered to arise spontaneously from unknown factors. Bassi demonstrated that aseptic (sterile) transfer of subcutaneous material from sick living worms to healthy worms resulted in disease. He suggested that the disease was caused by a fungus, ultimately renamed in his honor *Botrytis bassiana.* In later life, although nearly blind, he developed his theory that contagion in such diseases as cholera, gangrene, and plague resulted from living parasites.

During 1798, **Edward Jenner** (1749–1823) published studies on the immunization of humans against smallpox (Figure 2.5). Jenner and others observed that individuals

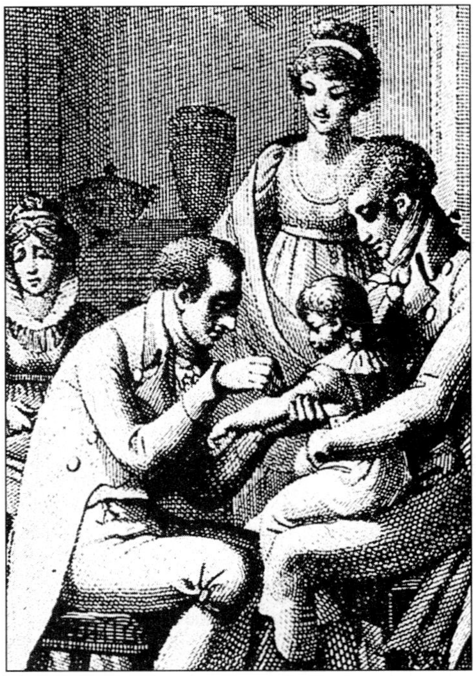

***Figure* 2.5** An early depiction of Edward Jenner vaccinating a child against smallpox. (From *The Eradication of Smallpox from India,* by Basw, Jezek, and Ward, a World Health Organization publication, 1979.)

routinely exposed to cows often developed pustules on their hands and arms that were similar to those caused by the dreaded smallpox. This "cowpox" was not fatal in humans, and, apparently, those infected with the cowpox did not contract the smallpox. Jenner inoculated an eight-year-old boy (James Phipps) with material from the infected cowpox pustules on the hands of a milkmaid. The boy became ill and had pustules at the site of inoculation but quickly recovered. A challenge dose later from an active case of smallpox was without effect. Jenner then inoculated several healthy individuals with cowpox exudate and observed that they were made immune to smallpox. Jenner has often been criticized for the empirical nature of his experimentation, but, given the context of the time, he deserves much credit for his contributions.

Microbiology After 1850: The Beginning of Modern Microbiology

During the latter part of the nineteenth century, a number of gifted scientists working independently established the disciplines that are now encompassed by the term *microbiology*. Much was accomplished during this half century, and the foundations for immunology, medical microbiology, protozoology, systematics, fermentation, and mycology were all set in place. However, up until 1860, the doctrine of spontaneous generation of microbes remained a generally accepted concept. This was largely due to the widespread acceptance that chemists were infallible; also, they tended to be more dogmatic than naturalists. They would soon meet their match in a strong-willed Frenchman named Louis Pasteur.

A notable publication appeared in 1861 by Pasteur entitled *Mémoire sur les corpuscules organisés qui existent dans l'atmosphere: Examen de la doctrine de générátions spontanées (Report on the organized bodies that live in the atmosphere: Examination of the doctrine of spontaneous generation)*. It marked the beginning of a new epoch in bacteriology.

Louis Pasteur (1822–1895) was a chemist, and his approach to the study of microorganisms was markedly influenced by his background (Figure 2.6). Pasteur was a genius in his thinking, argument, and experimentation, but also among his unique qualities was the ability to communicate and convince scientists and laypersons alike of his views. He laid to rest forever the idea of spontaneous generation, established immunology as a science, developed the concepts of fermentation and anaerobiosis, and developed many microbiological techniques. In short, he contributed to every phase of microbiology.

The Pasteur School

Louis Pasteur entered the controversy surrounding spontaneous generation at a time when the dogma of *heterogenesis* was being used to explain the origin of living matter. This theory was avidly expounded by Felix-Archimede Pouchet (1800–1872), a noted French physician and naturalist and honored member of many learned societies in France. Pouchet believed that life can spring *de novo* from a fortuitous collection of molecules and that the "vital force" comes from preexisting living matter. In contrast Pasteur's studies on fermentation indicated that "ferments" were actually organic living beings that reproduce and, by their vital activities, generate the observed chemical changes. In a series of brilliant experiments, Pasteur showed that "germs" present in the air were the cause of ferments and that such organisms were widely distributed in nature. Pasteur's "germs" were actually what we call bacteria and fungi today.

***Figure* 2.6** Louis Pasteur (1822–1895), an outstanding researcher and the developer of the rabies vaccine. Pasteur is credited with disproving "spontaneous generation." (From *Life of Pasteur* by Rene Vallery-Radot, Doubleday Page, Garden City, NY, 1923.)

Pasteur dealt with the problem of microbes in air, in 1859, by designing an aspirator filter system to recover them. Pasteur was aware that H. G. F. Schröder and T. von Dusch had found spun cotton-wool to be an effective filter for airborne microbes. Pasteur drew copious quantities of air through spun cotton and then dissolved the cotton in a mixture of alcohol and ether. Microscopic examination of the resulting sediment revealed a considerable number of small, round or oval bodies indistinguishable from "germs" previously described. Pasteur noted that the number of these organisms varied with the temperature, moisture, and movement of the air.

During this era (in the 1870s), the English physicist

John Tyndall (1820–1893), also opposed to the dogma of spontaneous generation, did a series of experiments that supported the work of Pasteur. Tyndall used optics to demonstrate that microbes were present in air and that heated infusions placed in optically clear chambers remained sterile, whereas infusions placed under an ordinary atmosphere exhibited growth. A major contribution was his empirical observation that some bacteria have phases: one is a *thermolabile* (unstable when heated) phase during which time the bacteria are destroyed at 100°C, and the other is a thermoresistant phase that renders some microbes incredibly resistant to heat. Tyndall's suppositions were confirmed by Ferdinand Cohn (1828–1898) with the demonstration that hay bacilli could form heat-resistant bodies we now call **endospores.**

To dispel the theories of spontaneous generation, Pasteur initiated a series of experiments with long-necked flasks of various types (Figure 2.7) to improve upon experiments by earlier workers such as Schwann and Spallanzani. He fashioned one flask with a horizontal neck and placed in it distilled water containing 10 percent sugar, 0.2 percent to 0.7 percent albuminoid (soluble proteins), and the mineral matter from beer yeast. The flask was boiled for several minutes, and the neck was attached to a platinum tube that was maintained at a red-hot temperature as the flask cooled down. The air drawn into the flask was sterilized by passage through this heated tube. Pasteur noted that flasks containing various infusions treated in this manner remained clear and free of microbial growth. He also showed that swan-necked flasks, which were long and bent down in a way that excluded passage of dust upon cooling, also remained sterile. Microbes rapidly grew in all of the flasks if the neck was broken off or if its infusion was spilled into the neck and allowed to drain back into the flask.

Much heated controversy continued during the period from 1860 to 1880, and the heterogenesists continued to attack the experiments and writings of Pasteur, Tyndall, Lister, Cohn, and others. Despite this, the voice of the opposition was weakening. The Academy of Science in Paris and scientists everywhere were convinced of the correctness of Pasteur's experimental work. Pasteur gave a public lecture in 1864 about his experiments and demonstrated some infusions that had remained unspoiled for four years. In his address, he stated:

> *And, therefore, gentlemen, I could point to that liquid and say to you, I have taken my drop of water from the immensity of creation, and I have taken it full of the elements appropriated to the development of inferior beings. And I wait, I watch, I question it, begging it to recommence for me the spectacle of the first creation. But it is dumb, dumb since these experiments were begun several years ago; it is dumb because I have kept it from the only thing that man cannot produce, from the germs that float in the air, from life, for life is a germ and a germ is life. Never will the doctrine of spontaneous generation recover from the mortal blow of this simple experiment.*
>
> (From Vallery-Radot, Rene. 1920. *The Life of Pasteur.* Garden City, NY: Garden City Publishing Company, Inc., pp. 108–109.)

The work of Cagniard-Latour, Schwann, and Kützing (1837) reported that the fermentation of sugar to an alcohol was due to the biological activities of a viable organism. Pasteur initiated studies on fermentation about 1857 and published an extensive paper three years later that established a number of properties of alcohol fermentation: First it is caused by a yeast; second, nitrogen is required by the organism; and third, alcohol and carbon dioxide are not the sole products of sugar fermentation because some of the sugar is utilized by the yeast to synthesize cell protein, carbohydrate, and fat.

In 1865, Napoleon III asked Pasteur to investigate the causative agent of bad wine. Pasteur reported that it was caused by one specific organism that was different from the agent of good wine. He suggested that the juices of grapes be heated, or **pasteurized,** to destroy the resident populations and that the resultant material be inoculated with a proven producer of good wine. In addition to working with wine, Pasteur studied the fermentation of beer over several years. He felt that it was his patriotic duty to make French beer superior to the brew produced elsewhere. Whether he laid the foundation for the ultimate production of a superior brew is left to the taste of the connoisseurs. His research resulted in the publication of a book on beer fermentation in 1876.

Pasteur was very successful in his experimentation of butyric acid fermentation (conversion of sugar to butyric acid and other products) because it was during this work that he confirmed that anaerobic forms of life do exist. While microscopically examining fluid from a butyric acid fermentation, he noted that organisms near the edge of

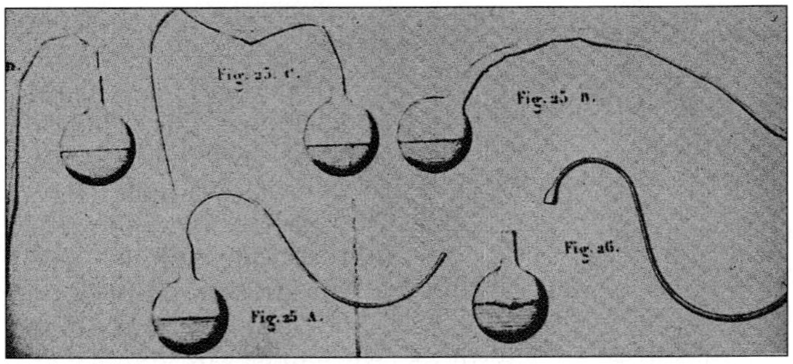

***Figure* 2.7** Pasteur's drawings of his bent neck flasks that were employed in disproving spontaneous generation. (From *The History of Bacteriology* by William Bulloch, M.D., Oxford University Press, 1960.)

the cover glass placed on the slide ceased movement while those in the middle swam about vigorously. He wondered whether oxygen in the air might be harmful to the organisms. To test his theory, he passed a stream of oxygen through a group of cells that were actively fermenting. He found that butyric acid production was halted and the organism died. He was the first (1863) to use the terms *aerobe* and *anaerobe* to describe the effects of air on microorganisms.

Knowledge of the extraordinary sterilizing effects of superheated steam came from studies in Pasteur's laboratory. He and his coworker Charles Chamberland noted that when hermetically sealed flasks were placed in a bath of calcium chloride and heated above 100°C, the solutions in the flasks were free from viable organisms. Experimental work resulted in an improvement in this process so that equally effective sterilization resulted if the flasks were plugged with wool and heated in a closed container. The modern apparatus used for this purpose is the **autoclave,** which was originally manufactured in 1884 by a Parisian engineering firm under the name Chamberland's Autoclave.

In 1865, Pasteur's chemistry professor from his earlier days at the Sorbonne (who was in later life a senator from the South of France) asked his former pupil to investigate a devastating problem in silkworms. Pasteur was reluctant to go because, as a chemist, he knew nothing of silkworms. Eventually he took his entourage to Alais, France, to initiate his studies on the disease. Most of his remaining years were dedicated to the understanding and prevention of infectious diseases in animals and humans. It took Pasteur five years and generous help from his able assistant, M. Gernez, to show clearly that the agent that was destroying the extensive silkworm industry in France was a transmissible microbe—specifically, the protozoan *Nosema.* Pasteur outlined a course for eradication of the disease based on isolating healthy worms, retaining the eggs produced, and examining the progenitor for a period of time. If the parent worm remained healthy, the eggs were allowed to hatch. These robust progeny were free of the disease, and the French silk industry was restored to its former glory.

Pasteur studied the disease anthrax in farm animals and demonstrated that immunization against this scourge was possible. He also devised methods whereby cholera in chickens could be prevented by immunization of the animals. Pasteur obtained the chicken cholera organism and grew it in culture. Inoculation of chickens with laboratory cultures of this organism resulted in a mild illness from which they recovered. At a later date, a virulent (infectious) strain of the cholera organism was used to inoculate these chickens, but they were completely resistant to the disease. Apparently, growth of the culture in the laboratory yielded a less virulent strain. Pasteur realized the implications and potentials of this research and then turned to a study that was to be the crowning achievement of his career—the use of a weakened virus to prevent hydrophobia in humans bitten by rabid dogs.

During Pasteur's later years, rabies was a serious health problem in France, and he decided to devote his efforts to the eradication of this dreadful disease. Perhaps he was intrigued by the always fatal consequence of a bite by a rabid dog and sensed the profound effect that a cure would have on the scientific world. Pasteur recognized that the infection settled in the brain and nervous system of animals. All previous efforts to isolate a microscopically visible agent of this disease had been unsuccessful. He outlined an empirical method to develop a vaccine that would counteract the fearful effects of rabies. In 1885, he published a method for protecting dogs after they had been exposed to rabies. A dog was protected by inoculation with an emulsion prepared from the dried spinal cord of a rabbit that had succumbed to the disease. Because Pasteur was able to prevent symptoms of rabies in animals exposed to other rabid animals through this procedure, he was convinced that the method was also applicable to unfortunate humans bitten by rabid animals.

A nine-year-old boy, Joseph Meister, was brought to Pasteur in July 1885 after he had been severely bitten by a dog that was certainly rabid. Meister was injected over several days with the emulsions prepared from animal spinal cord material. After 2 weeks, the boy was given an injection of virus that had maximal virulence when tested in a rabbit. The boy survived, as did thousands of others treated by the same procedure, and Pasteur received worldwide acclaim.

In 1886, a commission was appointed within the Academy of Sciences in Paris to erect a scientific institute in honor of the man who had contributed so much to world health. More than 2.5 million francs were collected from throughout the world, and the Institute Pasteur was established. Although Pasteur had many noble accomplishments during his lifetime, he may actually have erred in some of his judgments and experimentation. However, this should not detract from his great contributions to society. The well-known microbiologist A. T. Henrici summed it up best:

> *It has been hinted that Pasteur was not always willing to give credit to those who had preceded him that his ideas and experiments were not always strictly original with him. It is one thing to discover a truth, another to get it established as an accepted fact. Whatever criticism may be directed towards Pasteur as regards the originality of his ideas, nothing can be said to belittle his ability to put them across. His genius for quick and accurate thinking, for keen argument, and for obtaining publicity was not less important to the development of microbiology than his ingenious experiments. He "sold" the science to the public.*
>
> (From Henrici, A. T. 1939. *The Biology of Bacteria,* 2nd ed. New York: D.C. Heath & Co., p. 10.)

***Figure* 2.8** Robert Koch (1843–1910), the great medical microbiologist. Koch confirmed the "germ theory" of disease. (From *The History of Bacteriology* by William Bulloch, M.D., Oxford University Press, 1960.)

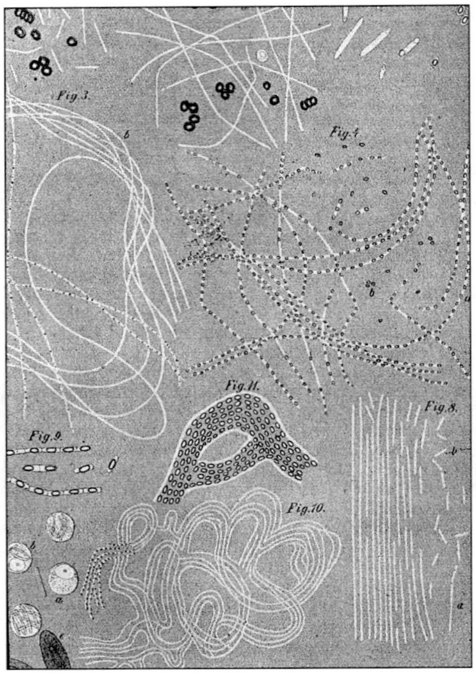

***Figure* 2.9** Koch's drawings of *Bacillus anthracis,* which he recognized as the causative agent of anthrax in cattle. The drawing in the lower right indicates why he described the bacterium as "sticklike." (From *The History of Bacteriology* by William Bulloch, M.D., Oxford University Press, 1960.)

The Koch School

This school of thought concentrated on the isolation of pure cultures of pathogenic (disease causing) and saprophytic (live on dead or decaying matter) bacteria. The members of Koch's laboratory were the originators of pure culture methods and made significant contributions to many specialized branches of bacteriology.

Robert Koch (1843–1910) was trained as a physician and became a country doctor in Wollstein, East Prussia (Figure 2.8). His wife, seeking to allay his restless curiosity, gave him a microscope as a gift, little knowing the far-reaching consequences this small instrument would have in alleviating human misery. Koch examined many specimens, among them the blood of an ox that had succumbed to **anthrax**—an infectious disease of warm-blooded animals. He noted the constant presence of sticklike bodies in diseased animal blood, which were absent in blood taken from healthy animals (Figure 2.9). He found that the disease symptoms could be transmitted by inoculating a healthy mouse with blood from an animal that had died from anthrax. His tool for inoculation was a fire-sterilized splinter. Koch surmised that these long, cylindrical bodies might be the viable causative agent of the disease and proceeded to culture the organism in fluid obtained from the eye of an ox. The organism, transferred in several passages of the fluid medium, would again cause anthrax when injected into the tail of a mouse. This was the first clear experimental evidence that a bacterium was an agent of disease (see Figure 2.9).

From these experiments, Koch formulated his theories on the causal relationships between microbes and disease. From this came what we now know as **Koch's postulates:** (1) that a specific microorganism is present in all cases of a disease; (2) that the organism can be obtained in pure culture outside the host; (3) that the organism will, when inoculated into a susceptible host, bring about symptoms equivalent to those observed in the host from which it was isolated; and (4) that the organism may be isolated in pure culture from the experimentally infected host.

Koch observed that the causative agent of anthrax has a life cycle involving a dormant spore, and he theorized correctly that these were the resistant bodies responsible for survival in soil. In 1876, Koch wrote to Ferdinand Cohn, a noted scientist in Breslau, Germany, telling him of his investigations, and he was invited there to demonstrate his work. He was received with enthusiasm, and his discoveries were acknowledged with acclaim. Later, he was invited to Berlin to set up a laboratory and devote his energies to studying microbes.

The origin of pure culture methods can be traced to Koch's observation that individual bacterial colonies growing on potato slices often differed in appearance (Figure 2.10). Microscopic examination of stained cells revealed

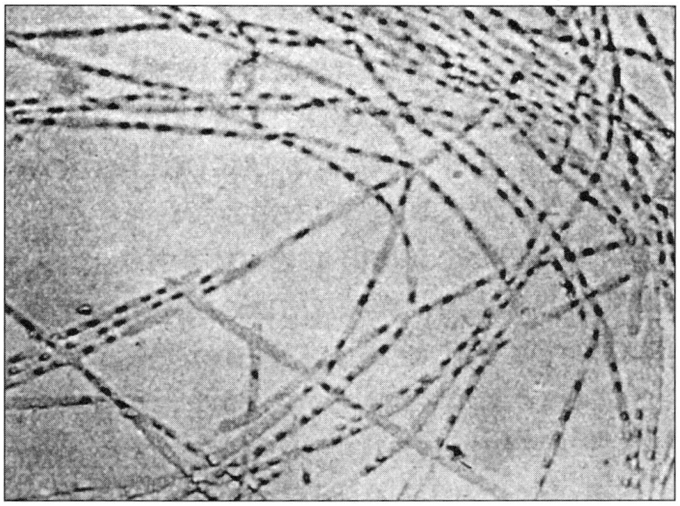

***Figure* 2.10** Koch's photomicrograph of bacteria from a single colony confirming that it was composed of one morphological type. (From *The History of Bacteriology* by William Bulloch, M.D., Oxford University Press, 1960.)

that the organisms within a colony were similar, but were often unlike organisms in other colonies. Koch theorized that a colony arose from a single cell, and he developed streaking methods using a platinum loop (his invention) that enabled him to isolate organisms in pure culture. **Gelatin** was the solidifying agent for culture media during the early years of Koch's career (see **Box 2.2**). Gelatin had major shortcomings, and Fannie Hesse, the American-born wife of Walter Hesse, a coworker in Koch's laboratory, suggested that **agar** could be used as a solidifying agent. At that time (1882), agar was added as a jellying agent in fruit, and that discovery was a considerable aid in the isolation of pathogenic microorganisms. An assistant in Koch's laboratory, R. J. Petri, in 1887 developed the dish (or plate)—named in his honor—that is still used in culturing bacteria. The design of the **petri dish** is virtually unchanged from his original except that glass has been replaced with plastic.

Paul Ehrlich (1854–1915) was another coworker in Koch's laboratory and made far-reaching discoveries in immunology and chemotherapy. In devising staining methods for infectious microbes in tissue, he observed that bacterial cells often absorbed selected dyes to a markedly greater extent than did surrounding tissue. He reasoned that a toxic dye might destroy the bacterium without significant damage to the host tissue because so much more would adhere to the invading microbe. Ehrlich theorized that organic arsenicals might be synthesized that would be harmless to animals but toxic to invading parasites. **Arsenicals** are organic derivatives of arsenic, a toxic element. In Ehrlich's laboratory, scores of arsenicals were synthesized and tested on trypanosomal infections in mice. **Trypanosomes** are protozoan parasites that infect the blood of vertebrates. The 606th compound (Salvarsan) that Ehrlich synthesized proved to be effective in curing these protozoan infections.

At the time, the spirochete that caused syphilis was reported by its discoverer, Fritz Schaudinn, to be related to the trypanosomes. Ehrlich proceeded to test 606 as a cure for syphilis. It was remarkably successful, and thus the "magic bullet" against the dreaded disease syphilis was discovered.

The magnitude of the discoveries in Koch's laboratory can be appreciated by considering that he developed his methods in 1881, when only anthrax was suspected to be a causative agent of disease. During the next 20 years, Koch and his coworkers confirmed that microbes were the etiological (causative) agents of significant numbers of human and animal diseases. These diseases and their discoverers are listed in Table 2.1.

Chemical antiseptics originated with **Joseph Lister** (1827–1912), an English physician who employed carbolic acid for antisepsis during surgery. Koch extended the work of Lister and devised a method for comparing the efficiency of chemical antiseptics. He dried cultures of bacteria, generally anthrax spores, on small pieces of silk thread, which were then immersed in the antiseptic solution. At intervals, a thread was removed from the antiseptic, washed in sterile water, and placed in a growth medium to determine whether the organism remained viable. Koch found that carbolic acid was relatively weak in its disinfecting (killing) power, and of all the substances he tested, perchloride of mercury was most effective. It destroyed bacterial spores at a high dilution and in the shortest period of time. Mercuric compounds remain widely used as disinfectants, although we are now concerned about the fate of toxic mercury in the environment.

Microbes in Nature

The adverse effects that microbes had on human beings were the major considerations of the Koch and Pasteur schools. The causative agents of diseases such as tuberculosis or rabies, which caused human suffering, would obviously have been a challenge to the curious scholar of scientific bent. During the latter years of the nineteenth century, knowledge of disease expanded rapidly along with a broad acceptance of the germ theory of disease. Another area that caught attention during the latter 1800s was the possible role of microbes in the continuation of viable systems on Earth. This leads us to the next conceptual development in microbiology—the role of microbes in nature.

Ferdinand Cohn suggested in 1872 that microbes were involved in the cycling of all matter and that the activities of organisms in the biosphere allowed for the reutilization of cellular constituents. Our knowledge of the indispens-

BOX 2.2 MILESTONES

Discovery of Agar as a Solidifying Agent

Agar has long been the universal solidifying agent used in preparing media for growing microbes. The first solid medium was employed by Robert Koch and it was an aseptically cut slice of potato. He was able to isolate colonies on this substrate and developed his theory that all organisms in a colony were of one species. Koch assumed correctly that all organisms in a discrete colony grew from a single cell. These studies were the origin of pure culture methods. The next solidifying agent employed by Koch was gelatin. Gelatin is a protein and was unsatisfactory for two principal reasons: (1) it melts at 37°C, the favored incubation temperature for most pathogens, and (2) many bacteria can digest gelatin. In 1882, Fannie Hesse, the American-born (New Jersey) wife of Walter Hesse, suggested that a jellying agent used in making fruit preserves might be a replacement for gelatin in bacterial media. Walter Hesse (1846–1911) was an associate in Koch's laboratory and probably discussed the problems involving gelatin with his wife. Fannie Hesse's suggestion led to the adoption of agar as the choice solidifying agent.

The Dutch apparently brought agar from their East Indian colonies, where it was used to improve the setting quality of jam. Agar is an extract of algae that thrive in the Pacific and Indian Oceans and Japan Sea. It is a complex polysaccharide containing sulfated sugars, is of poor nutritional value, and is digested by only a few microorganisms, mostly of marine origin. Laboratory-grade agar is inhibitor free, and virtually all organisms grow well in its presence.

Agar melts at 100°C and remains in the liquid state down to about 45°C. The high melting point makes agar useful for growing all organisms including thermophiles at temperatures up to near 100°C. Cultures growing at higher temperatures should be placed in closed containers to prevent dehydration. The low solidifying temperature permits one to add bacteria to melted agar (45°C). They can be distributed by mixing, and the isolated colonies can be counted or obtained in pure cultures.

Thus, an observation by Fannie Hesse on the qualities of a simple kitchen commodity became an object of worldwide utility.

Fannie Hess. (Courtesy of National History of Medicine Library, NIH)

able role that diverse microorganisms play in recycling constituents of living cells has expanded greatly since this pronouncement by Cohn.

Martinus Beijerinck

A major technique developed by the great botanist **Martinus Beijerinck** (1851–1931) provided much of the foundation for the elucidation of the various functions of microbes in the cycles of matter (Figure 2.11). He introduced the principles of enrichment culture, which gave clarity and rationality to microbial ecology. **Enrichment culture** is a means by which an organism evolved to exist under any chosen conditions of temperature, pH, salinity, osmolarity, and so on can be isolated. The sole limitation on isolation is the existence of the desired organism in the inoculum.

Combined with the studies of Sergei Winogradsky (dis-

Table **2.1 Diseases Whose Causative Agents Were Discovered by Koch and His Coworkers**

Date	Discoverer	Disease
1882	Koch	Tuberculosis
	Loeffler and Schutz	Glanders
1884	Koch	Asiatic cholera
	Loeffler	Diphtheria
	Gaffky	Typhoid fever
	Rosenbach	Staphylococcal and streptococcal infections
1885	Bumm	Gonorrhea
1886	Fraenkel	Pneumonia
1887	Bruce	Malta fever
	Weichselbaum	Meningococcal infections
1889	Kitasato	Tetanus
1891	Wolff and Israel	Actinomycosis
1894	Kitasato and Yersin	Plague
1897	van Ermengen	Botulism
1898	Shiga	Acute dysentery

cussed in the next section), the enrichment culture technique provided a means for the isolation of various physiological types of microorganisms that exist in natural environments. An enrichment medium is prepared with a defined chemical composition and inoculated with soil or water rich in microbes; only those microbes capable of growth on that particular medium will come to the fore. The predominant microbes will be those best equipped by heredity to survive on that medium at the temperature, pH, and other conditions chosen. With this technique, Beijerinck and his followers readily obtained microbes with differing physiological capabilities, and they were able to assess the potential role of that microbe under natural conditions.

Figure **2.11** Martinus Beijerinck (1851–1931), a major contributor to our understanding of the role of microbes in nature. (From Van Iterson, Den Dooren De Jong and Kluyver, *Martinus Willem Beijerinck: His Life and His Work,* published by Martinus Nijholt, The Hague 1940.)

Beijerinck discovered free-living, nitrogen-fixing bacteria by the application of enrichment, using a medium devoid of nitrogenous compounds. The aerobic microbe that he obtained was given the genus name *Azotobacterium.* He also published extensively on the symbiotic nitrogen-fixing organisms *(Rhizobium)* that form nodules in the roots of legumes such as peanuts.

Beijerinck discovered and described many major groups of bacteria: the luminous organisms *(Photobacterium),* the sulfate reducers *(Desulfovibrio),* the methane-generating bacteria, and *Thiobacillus denitrificans,* an organism involved with denitrification (removal of organic nitrogen to the atmosphere). Beijerinck did much of the early work on lactic acid bacteria and proposed the genus name *Lactobacillus.* He recognized the existence of "soluble" living germs, which he called "contagium vivum fluidum." This is generally accepted as the initial description of a virus (specifically, the tobacco mosaic virus). Beijerinck's contributions to microbiology are legion, and his perceptions of the great role of microbes established the foundation of modern approaches to microbial physiology and ecology.

***Figure* 2.12** Sergei Winogradsky (1856–1953), a Russian-born microbiologist. Winogradsky was the father of autotrophy. He lived from the days of Pasteur and Koch to the modern era of microbiology. (From Selman A. Waksman, *Sergei N. Winogradsky: His Life and Work,* Copyright © 1953 by the Trustees of Rutgers College. Reprinted by permission of Rutgers University Press.)

Sergei Winogradsky

During the era of Koch, Pasteur, and Beijerinck, there arose another major figure in the field of general bacteriology (Figure 2.12). **Sergei Winogradsky** (1856–1953) was born in Russia, and during his long and fruitful life he witnessed the origins of the science of microbiology and survived to see the Age of Antibiotics. It is noteworthy that both Winogradsky and Beijerinck spent time in the laboratory of the great mycologist and plant pathologist, **Anton De Bary** (1831–1888), at Strassburg, Germany. Winogradsky, who arrived in Strassburg shortly after Beijerinck left, initiated studies on sulfur-oxidizing bacteria. He concluded from his studies on the bacterium *Beggiatoa* that the organism could utilize inorganic H_2S as a source of energy and atmospheric CO_2 for carbon in the synthesis of cellular material. He named these organisms "orgoxydants" and thus opened up the entire concept of **autotrophy,** which is the ability of bacteria to manufacture their cells from CO_2 and use inorganics or light as a source of energy (see Chapter 9). Previous to Winogradsky's study, only chlorophyll-containing plants were believed to use CO_2 as the sole carbon source.

Following the death of De Bary, Winogradsky went to Zurich, where he isolated and clarified the role of the nitrifying (converting ammonia to nitrate) autotrophic bacteria. During this period, he also showed that green and purple bacteria could oxidize hydrogen sulfide to sulfate, but he was uncertain whether or not his was a photosynthetic process. In addition, Winogradsky isolated the nitrogen-fixing anaerobe *Clostridium pastorianum.*

About this time, the noted scientist and Nobel Laureate **Elie Metchnikoff** (1845–1916) carried a personal letter from Pasteur to Zurich inviting Winogradsky to work at the Pasteur Institute. Metchnikoff was the discoverer of phagocytosis and cellular immunity. Winogradsky refused and elected to return to Russia, thus ending the first half of his illustrious scientific career. He did go to Paris in 1892 to represent Russia at the seventieth birthday celebration for Pasteur and met Beijerinck for the first and only time. The Revolution in Russia led Winogradsky to emigrate, and in 1922 he did go to the Pasteur Institute, where he spent the remainder of his life studying the broad aspects of soil bacteriology. Winogradsky compiled his life's work and published it in a monumental treatise entitled *Microbiologie du Sol.*

Microbes and Plant Disease

Among the first to recognize that microbes might be directly involved in plant diseases was Anton De Bary, who in 1853 suggested that brandpilze (plant rust) was caused by a parasitic fungus. He later proved experimentally that a fungus, *Phytophthora infestans,* caused late blight of the potato. This disease had caused the crop failure that brought widespread famine to Ireland in the 1890s. The famine resulted in thousands of deaths and led to the immigration of 1.5 million Irish to the United States.

Although the causal relationship between bacteria and animal disease had gained widespread acceptance by 1900, few botanists were willing to believe that bacteria could cause disease in plants. The fungi were generally acknowledged to be the agents of plant disease. **J. H. Wakker,** working with De Bary in 1881, isolated and identified the bacterium *Xanthomonas hyacinthia* as the pathogen causing widespread damage to hyacinth bulbs in Holland. **Erwin Smith,** in 1895, demonstrated that a bacterium caused a wilt in cucurbits (cucumbers and other members of the gourd family), and in 1898, Beijerinck concluded that tobacco mosaic was caused by a virus.

Despite these and other reports, there remained considerable controversy about the role of bacteria in plant disease. It took two American scientists, Thomas Burrill and Erwin Smith, to convince the scientific world that bacteria did indeed cause plant disease. Burrill was the first to discover and demonstrate a bacterial disease of plants when he reported, in 1877, that pear blight was caused by *Micrococcus (Erwinia) amylophorus.* Burrill investigated a number of diseases of corn, potatoes, and fruit, but he is probably best known as the first person in America to offer a laboratory course in bacteriology.

During the period from 1895 to 1900, Erwin Smith

implicated bacteria in diseases of tomatoes, cabbage, and beans and in the wilt of maize. By applying the postulates of Koch (a hero to Smith), his group at the U.S. Department of Agriculture found definitive proof that strains of *Xanthomonas, Pseudomonas, Erwinia,* and *Corynebacterium* were the causative agents of disease in agricultural crops. Smith also demonstrated that crown gall, a plant tumor, resulted from an *Agrobacterium tumefaciens* infection.

Transition to the Modern Era

Louis Pasteur died in 1895, Koch in 1910. Beijerinck retired in 1921, and the first part of Winogradsky's scientific career ended at the turn of the century when he returned to Russia. Thus ended what is considered to be the **"Golden Era of Microbiology,"** for during the 50-year period from 1870 to 1920, the discipline of microbiology was firmly established.

Albert Jan Kluyver (1888–1956) replaced Beijerinck as Professor of Microbiology at the Technical University at Delft in 1922. Knowledge of metabolic processes in microbes and in living cells in general was then quite limited. Most questioned whether metabolic reactions in microbes and in higher forms of life could occur in any equivalent manner. Knowledge of biochemical activities in microbes was mostly restricted to a few unrelated transformations brought about by individual organisms. During the decade following his arrival at Delft, Kluyver introduced order to chaos by presenting a simple coordinated model for metabolic events in all living cells. His experimental approach led to our modern understanding that unity exists in biochemical reactions, a concept termed **comparative biochemistry.** This concept arose from Kluyver's proposal that the basic feature of virtually all metabolic processes is a transfer of hydrogen. This transfer is universal and occurs whether the organism is aerobic or anaerobic, autotrophic or heterotrophic. Kluyver also believed that biosynthetic and biodegradative pathways in cells are highly coordinated and relatively few in number. He proposed that metabolic processes are functionally equivalent in all living cells.

Kluyver and Cornelius B. van Niel, one of his students, proposed that aerobic and anaerobic respiration can be illustrated by the following simple formula:

$$AH_2 + B \longrightarrow A + BH_2$$

in which A represents a more reduced element (NH_4^+, CH_4) and B represents oxygen in aerobic respiration or some less reduced metabolic intermediate in anaerobic respiration.

They also suggested that the general formula for all photosynthetic reactions would be as follows:

$$CO_2 + 2\,H_2A \xrightarrow{\text{light}} CH_2O \text{ (cell material)} + H_2O + 2\,A$$

In plant photosynthesis, A would represent oxygen because the hydrogen atom for CO_2 reduction is donated by H_2O. Molecular oxygen (O_2) is released in this process. Prior to the studies of Kluyver and van Niel, it was widely considered that light was used in photosynthesis to decompose carbonic acid and generate oxygen. For photosynthesis to occur in the anaerobic photosynthetic bacteria, A could be hydrogen, sulfide, or a reduced organic compound. As a consequence of the reduction of CO_2, an oxidized product such as sulfate would be generated.

One should not underestimate the contribution that the comparative biochemistry concept made to advances in the area of cellular metabolism. The bacteria became a major model system for study only with the general acceptance of this unity by all biologists. During the decade following World War II, a fundamental understanding of the role and structure of DNA (as the genetic material double helix), metabolic regulatory mechanisms, and microbial genetics was attained. These discoveries led to a broad, integrated study on cellular growth that we now call **molecular biology,** a term coined by William Astbury in 1945.

Many characteristics of Eubacteria and Archaea—rapid growth, simple nutritional requirements, growth under harsh conditions, and ease in handling—make them attractive tools for physiological studies. Since researchers generally choose to progress from simpler systems to those of greater complexity, the Eubacteria and Archaea have provided an attractive model system for studying animal and plant processes. The broad array of bacterial and archaeal types available—from those that use light as an energy source to the myxobacteria, which form fruiting bodies—makes them a system of choice for investigations on regulation, biosynthesis, gene expression, and molecular interactions. Studies of their ability to grow under extreme conditions of pH, temperature, salinity, and anaerobiosis (and combinations of these extremes) have yielded many clues to the elucidation of life strategies. Other unique characteristics that allow study at the molecular level include nitrogen fixation, antibiotic synthesis and action, parasitism (for bacteria and eukaryotes), directed motility, and development of bacteriophages (bacteria viruses).

Much might be written on the growth of microbiology as a scientific discipline over the last 50 years. The following chapters give ample evidence that much has been accomplished. Each is an abbreviated chronicle of the developments in that area of microbiology. A list of Nobel prize winners that contributed to our knowledge of microbiology is listed in Table 2.2.

Table **2.2 Nobel Prizes Given for Work Closely Related to Microbiology**

Year	Laureate(s)	Work
1901	Emil von Behring–German	Antitoxin—principles of serotherapy.
1905	Robert Koch–German	Pure culture methods. Causation of disease.
1908	Paul Ehrlich–German	Chemotherapy immunology.
	Elie Metchnikoff–Russian	Phagocytosis theory of immunity.
1913	Jules Bordet–Belgian	Immunity.
1927	Julius Wagner-Jauregg–Austrian	Therapeutic effect of malaria infections in syphilis treatment.
1928	Charles Nicolle–French	Typhus fever transmitted by lice.
1930	Karl Landsteiner–American	Human blood groups—immunology.
1939	Gerhard Domagk–German	Antibacterial effect of prontosil (sulfonamide chemotherapy).
1945	Sir Alexander Fleming–British Ernst Boris Chain–British Citizen Sir Howard Florey–English	Discovery and development of penicillin.
1946	Wendell M. Stanley–American John H. Northrop–American	Production of purified enzymes and virus protein.
1951	Max Theiler–South African	Discoveries concerning yellow fever.
1952	Selman S. Waksman–American	Discovery of streptomycin.
1953	Fritz Albert Lipman–German-American	Discovery of coenzyme A.
1954	John F. Enders–American Thomas H. Weller–American Frederich C. Robbins–American	Discovery that poliomyelitis viruses multiply in tissue cultures.
1958	George W. Beadle–American Joshua Lederberg–American Edward L. Tatum–American	Discovery that genes act by regulating chemical processes. Neurospora genetics.
1959	Severo Ochoa–American Arthur Kornberg–American	Discoveries of the mechanism in the biological synthesis of ribonucleic and deoxyribonucleic acids.
1960	Frank MacFarland Burnet–Australian Peter Brian Medawar–British	Discovery of acquired immunological tolerance to tissue transplants.
1961	Melvin Calvin–American	Elucidating chemical steps that occur during photosynthesis.

Summary

- **Epidemic diseases** played a major role in battles throughout recorded history. Napoleon was driven from Russia in 1812 by disease and deprivation and ultimately lost at Waterloo because about one half of his army was incapacitated by illness.

- The first significant writings on contagious disease were those of **Fracastoro** in the first half of the sixteenth century.

- Antony van Leeuwenhoek observed and made extensive drawings of bacteria in 1683, but it was 200 years before "spontaneous generation" was disproved and the "germ theory" of disease was accepted.

- There were a number of scientists who preceded Louis Pasteur who did much careful experimentation that disproved **spontaneous generation.** Among these were: Redi, Spallanzani, Schulze, Cagniard-Latour, Schwann, and Kützing.

- Edward Jenner developed an effective **vaccination** that protected humans against smallpox long before the germ theory of disease was established.

- When it became clear that animals could not spring from inanimate material, the proponents of spontaneous generation moved to changes caused by decay or putrefaction as proof that **chemical instabilities** were the basis for the observable changes in organic matter.

- By 1864, Pasteur had clearly proved that **sponta-**

Table 2.2 ***continued***

1962	James Watson–American
	Francis Crick–British
	M. Wilkins–British
	DNA structure.
1964	Konrad E. Block–American
	F. Lynen–German
	Fatty acid metabolism.
1965	J. Jacob–French
	A. Lwoff–French
	J. Monod–French
	Regulation in cells.
1966	C. R. Huggins–American
	F. Peyton Rous–American
	Viruses in cancer.
1968	M. Nirenberg–American
	G. Khorana–American
	R. W. Holley–American
	Genetic code.
1969	S. E. Luria–American
	M. Delbruck–American
	A. Hershey–American
	Bacteriophages in molecular biology. Virus infection in cells.
1972	Gerald M. Edelman–American
	Rodney R. Porter–British
	Chemical structure and nature of antibodies.
1975	David Baltimore–American
	Howard M. Temin–American
	Renato Dulbecco–American
	Interaction of tumor viruses and genetic material in the cell.
1976	Baruch S. Blumberg–American
	D. Carleton Gajdusek–American
	Mechanisms for origin and dissemination of infectious diseases.
1978	Daniel Nathans–American
	Hamilton Smith–American
	Werner Arber–Swiss
	Restriction enzymes and their application to problems of molecular genetics.
1980	B. Benacerraf–American
	George D. Snell–American
	Jean Dausset–French
	Structure of cells and relationship to disease and organ transplants.
1988	Gertrude Elion–American
	George Hitchings–American
	James Black–British
	Chemotherapy.
	Hartmut Michel–German
	Discoveries on the three-dimensional structure of closely linked proteins that are essential for photosynthesis.

neous generation was a myth and his ideas on fermentation were generally accepted.

- Pasteur and his colleague Chamberland developed the **autoclave.**

- Pasteur was the first to observe and report that **anaerobic bacteria** existed. He also recognized that immunity to disease was possible and developed an effective rabies vaccine.

- The Academy of Sciences in Paris collected funds from around the world and established the renowned **Pasteur Institute** in honor of Louis Pasteur.

- Robert Koch was the first to clearly demonstrate that bacteria were a causative agent of infectious disease.

- Koch proposed a set of postulates, known as **Koch's postulates,** that could be employed to demonstrate that a culturable organism was the causative agent of disease symptoms.

- Koch developed **pure culture** methods. His laboratory developed the loop, petri dish, and was the first to use agar as a solidifying agent.

- Paul Ehrlich, an associate of Koch, is considered the founder of **chemotherapy.**

- Martinus Beijerinck was a pioneer in assessing the role of microorganisms in the **cycles of matter** (carbon, nitrogen, and sulfur) in nature.

- Beijerinck discovered the free-living nitrogen fixing *Azotobacter* sp and the symbiotic nitrogen fixing

Rhizobium sp. He was also the codiscoverer (with Dmitrii Ivanowsky of Russia) of viruses.

- Sergei Winogradsky was the first to recognize that **autotrophic** bacteria were of widespread occurrence.
- Winogradsky spent the last 34 years of his life at the Pasteur Institute studying the activities of bacteria in soil.
- The role of bacteria in plant disease was elucidated by the American Scientists Thomas Burrill and Erwin Smith.
- Albert Jan Kluyver, a Dutch bacteriologist, proposed that metabolic reactions in all living cells were equivalent, a concept known as **comparative biochemistry.**
- Eubacteria and Archaea have proven to be an excellent model system for studies in metabolism, genetics, and molecular interactions.

Questions for Thought and Review

1. What are some of the influences the microbe has had on human history? Discuss in terms of war, food, population control, and other factors.
2. It took 200 years from Antony van Leeuwenhoek's observations and descriptions of bacteria (1673) until the latter part of the nineteenth century for the role of bacteria to be elucidated. Why did it take so long?
3. What is "spontaneous generation" and why did this concept have so many proponents?
4. Spallanzani and Redi made significant contributions to disproving "spontaneous generation." What were some of these and why were their experiments not accepted as definitive proof? What did Franz Schulze add that was crucial?
5. Koch and Pasteur were interested in the effects of microbes on humans, Winogradsky and Beijerinck in the role of microbes in nature. Give examples of the differences in their research.
6. How did Louis Pasteur contribute to the conquering of rabies? What experiment(s) led him to conclude that a treatment was possible?
7. Most of the basic techniques used in microbiological studies originated in Robert Koch's laboratory. What are some of these?
8. What are Koch's postulates? How did he use these in discovering the causative agents of disease? What are some of the common diseases for which Koch's laboratory discovered the causative agent?
9. Do bacteria play a role in plant disease? How did our understanding of the role of fungi and bacteria in plant disease evolve? What important historic development resulted from plant disease?
10. How did the philosophies of Albert Kluyver and C. B. van Niel influence modern science?

Suggested Readings

Brock, T. D., ed. and trans. 1975. *Milestones in Microbiology.* Washington, DC: American Society for Microbiology.

Bulloch, W. 1938. *The History of Bacteriology.* London: Oxford University Press.

Dobell, C., ed. and trans. 1932. *Antony van Leeuwenhoek and His "Little Animals."* New York: Harcourt, Brace and Company.

Kamp, A. F., J. W. M. La Rivière, and W. Verhoeven, ed. 1959. *Albert Jan Kluyver: His Life and His Work.* Amsterdam: North-Holland Publishing Co.

Karlen, A., 1995. *Man and Microbes: Diseases and Plagues in History and Modern Times.* New York: G.P. Putnam Publ.

Lechevalier, H. A., and M. Solotorovsky, 1965. *Three Centuries of Microbiology.* New York: McGraw Hill.

Vallery-Radot, R. 1920. *The Life of Pasteur* (translated by R. L. Devonshire). Garden City, NY: Garden City Publishing Co.

Van Iterson, G., L. E. Den Dooren De Jong, and A. J. Kluyver. 1940. *Martinus Willem Beijerinck: His Life and His Work.* The Hague, Netherlands: Martinus Nijhoff.

Waksman, S. A. 1953. *Sergei N. Winogradsky: His Life and Work.* New Brunswick, NJ: Rutgers University Press.

Life, atom of that Infinite Space that stretcheth, twixt the Here and There

Sir Richard Francis Burton
1821–1890

Chapter 3

Fundamental Chemical Principles

The basic physical principles that govern the chemical activities of all elements apply equally to living and nonliving systems. The living cell differs in that it has the capacity to utilize these physical properties of component atoms to generate energy, to grow, and to reproduce. To understand viable systems, we must have a fundamental knowledge of chemical principles and how these principles relate to the structure and function of a cell. For example, we know that all matter is composed of molecules that are made up of atoms. We also know that matter has characteristics that are determined by the manner in which these atoms are joined to form molecules. These molecules, in turn, join to form compounds (monomers). Compounds are interconnected to form macromolecules that, working in concert, produce life. This chapter is devoted to an overview of chemical principles underlying the formation of the molecules that collectively make up a living cell.

Atoms

The **atom** is the smallest unit that has all of the characteristics of an element. There are 92 different naturally occurring elements. An atom can exist as a single unit or in combination with other atoms. The smallest atom is that of hydrogen, which consists of one **proton** and one **electron.** The proton is located in the dense core of the atom called the **nucleus,** and the electron orbits the nucleus at great speed. The nuclei of all elements except hydrogen also contain **neutrons.** A proton has a positive charge, an electron has a negative charge, and a neutron has no charge. The number of protons in the nucleus of an atom is equal to the number of electrons orbiting around it, which effectively renders an atom electrically neutral. The attraction between the positively charged nucleus and the negatively charged electrons keeps the atom intact (Figure 3.1**a,b**).

(a) (b)

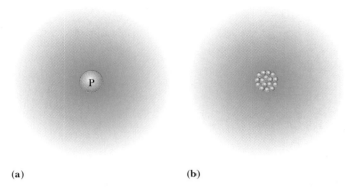

***Figure* 3.1 (a)** An atom of the simplest element hydrogen consists of a proton (P) at the center of a spherical electron cloud of negative charge. **(b)** The dense core of an atom with an equal number of protons and neutrons. The electrons are equal in number to the protons and form the negative cloud around the nucleus. The nucleus contains 99.9 of the mass but occupies only one hundred trillionth (10^{-14}) of the volume of the atom.

The number of protons in the nucleus of any element, called the **atomic number,** is constant. The number of neutrons and electrons in an atom may vary. There is close to the same number of protons and neutrons in the nucleus, and the sum of these is the **atomic weight.** Electrons are inconsequential in the atomic weight of an atom since an electron is only 1/1836 the relative mass of a proton or neutron. The structure of an element that has an equal (or close to equal) number of protons and neutrons is considered the **elemental** form. The elemental form is most abundant in nature.

Isotopes are atoms that have a greater number of neutrons than protons. Some isotopes of biological interest are listed in Table 3.1. The atomic number of an isotope does not differ from the natural element because the number of protons is invariant. The atomic weight is, of course, different because of the greater number of neutrons.

Some isotopes are stable, while others spontaneously decay. Those that decay with the release of subatomic particles (radioactivity) are termed **radioisotopes.** Radioisotopes are generally obtained by neutron bombardment. Thermal neutrons that are activated during nuclear fission may enter the nucleus of an atom, creating an isotope.

The three naturally occurring isotopes of oxygen are ^{16}O, ^{17}O, and ^{18}O, and their relative abundance in the environment is 99.76 percent, 0.04 percent, and 0.01 percent. The naturally occurring ^{16}O obviously constitutes the bulk of the oxygen in the environment. The radioisotopes of carbon (^{14}C), phosphorous (^{32}P), and sulfur (^{35}S), and the radioisotope of hydrogen (^{3}H)—commonly called tritium—are widely employed in research. The natural form of each of these elements is ^{12}C, ^{31}P, ^{32}S, and ^{1}H. The use of a radioisotope in studying metabolic pathways is illustrated in **Box 3.1.** The stable isotopes of carbon (^{13}C), nitrogen (^{15}N) and oxygen (^{18}O) are employed in biological research, and their presence is determined by mass spectrometry.

***Table* 3.1 Isotopes of sulfur, oxygen, carbon, and phosphorous**

	Protons	Neutrons		Type[1]
Sulfur	16	16	^{32}S	Elemental
	16	18	^{34}S	Stable isotope
	16	19	^{35}S	Radioisotope
Carbon	6	6	^{12}C	Elemental
	6	7	^{13}C	Stable isotope
	6	8	^{14}C	Radioisotope
Oxygen	8	8	^{16}O	Elemental
	8	9	^{17}O	Stable isotope
	8	10	^{18}O	Stable isotope
Phosphorous	15	16	^{31}P	Elemental
	15	17	^{32}P	Radioisotope
Nitrogen	7	7	^{14}N	Elemental
	7	8	^{15}N	Stable isotope
Hydrogen	1	0	^{1}H	Elemental
	1	2	^{3}H	Radioisotope

[1]The elemental form is most abundant in nature and has virtually the same number of protons and neutrons in the nucleus. Stable isotopes have a greater number of neutrons but the excess neutrons do not decay. The excess neutrons in radioisotopes spontaneously decay with the release of subatomic particles.

BOX 3.1 MILESTONES

CO₂ Fixation and the Nuclear Age

In 1796, Jan Ingen-Housz, a Dutch scientist, wrote that plants obtain their cellular carbon by assimilating carbon dioxide from the environment. Subsequently, others discovered that selected bacteria can also obtain their cell carbon by "CO_2 fixation." Elucidating the pathway whereby CO_2 was incorporated into a cell proved to be very difficult. From a biochemical standpoint, autotrophs (CO_2 as sole carbon source) were indistinguishable from heterotrophs (organics as carbon source). Data available indicated that CO_2 entered into metabolic sequences much as did sugars and other substrates. But how? The answer to this intriguing mystery came only when better tools were available to solve it. Ironically, the research that provided these tools came indirectly in a search for weapons of mass destruction.

The atom bomb was developed in the United States during World War II (1941 to 1945). It was a weapon of horrible consequence and remains a threat to human populations. Research that led to atom bombs brought us into a period called the nuclear age. The lessons learned in bomb development have been extended to provide us with nuclear energy for electrical power generation, nuclear medicine, and important research tools. In a way, the problem of CO_2 fixation ran headlong into the nuclear age.

Under proper conditions, a nuclear reactor can be manipulated to introduce extra neutrons into the nuclei of selected elements. Among the elements that can be so manipulated are several that are of biological interest, including carbon, phosphorous, sulfur, and iron. These elements are key components of living cells. The extra neutrons in the nucleus of these elements tend to decay with the release of radioactivity (subatomic particles). This radioactivity is readily detectable and permits us to locate the site of a radioactive compound in a cell. A radioactive compound is also called a **tracer** because we can "trace" the movement of the radioactive compound in a biological system. With the knowledge available in 1947, it was possible to make radiolabeled carbon dioxide and follow its incorporation into an organism during photosynthesis. Radiolabeled carbon dioxide is abbreviated $^{14}CO_2$, indicating that there are 14 neutrons present rather than the normal 12.

In 1949, Melvin Calvin and Andrew Benson initiated a series of experiments using $^{14}CO_2$, and their studies led to an understanding of the biochemistry of CO_2 fixation. They exposed photosynthesizing algae to

Calvin's Experiment. (Melvin Calvin, University of California, Berkley)

$^{14}CO_2$ for varying lengths of time. They then isolated and identified the radioactive compounds in the algae cells. They reasoned that the compounds that became radioactive after brief exposure were those in which the CO_2 was initially fixed. Those compounds involved in succeeding reactions would become radioactive at a later time. Analysis of the pattern of $^{14}CO_2$ fixation into compounds and the distribution of the radiolabel within compounds ultimately answered the question of how a cell can fix carbon from the atmosphere. CO_2 fixation occurs in most autotrophs as follows:

$$^{14}CO_2 + \begin{array}{c} H \\ | \\ H-C-O-PO_3H_2 \\ | \\ C=O \\ | \\ H-C-OH \\ | \\ H-COH \\ | \\ H-C-OPO_3H_2 \\ | \\ H \end{array} \longrightarrow \left[\begin{array}{c} H \\ | \\ H-C-O-\textcircled{P} \\ | \\ HO-C-^{14}CO^- \\ | \\ C=O \\ | \\ H-C-OH \\ | \\ H-C-O\textcircled{P} \\ | \\ H \end{array} \right] \longrightarrow \begin{array}{c} H \\ | \\ H-C-O\textcircled{P} \\ | \\ H-C-OH \\ | \\ ^{14}COOH \\ + \\ COOH \\ | \\ H-C-OH \\ | \\ CH_2O\textcircled{P} \end{array}$$

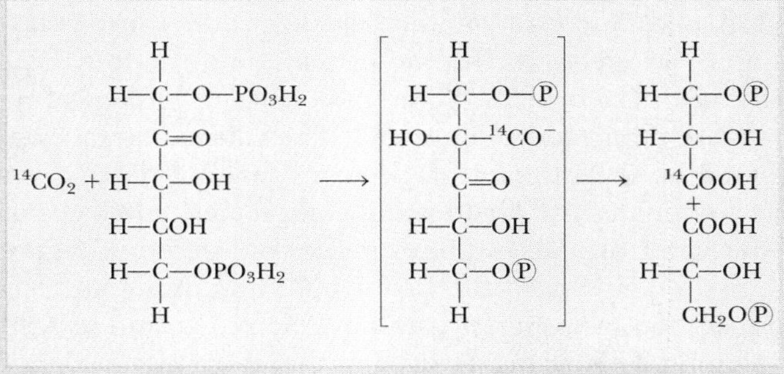

Thus, a seemingly unrelated event, the development of the atomic bomb, presented the unlikely solution to a major biological question. We now know that there are other CO_2 fixation reactions in autotrophic bacteria. These reactions will be discussed in Chapter 10.

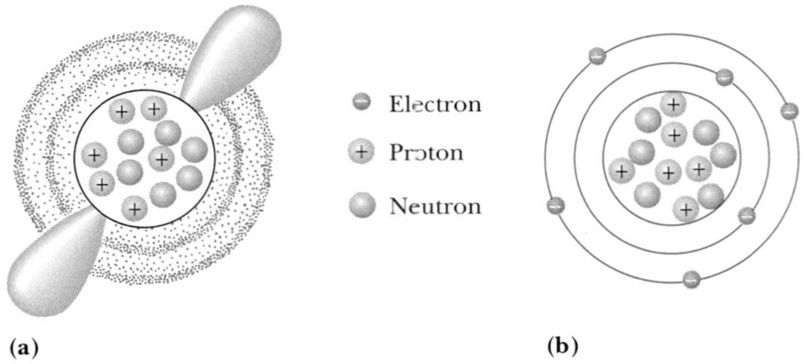

Figure **3.2** **(a)** A model of a carbon atom with spherical and dumbbell-shaped orbitals represented. The dots represent the probability that the electrons will be in that location at any moment. **(b)** The Bohr model (named for Niels Bohr, a noted Danish physicist) is less accurate but commonly used for convenience.

Electronic Configurations of Molecules

The electrons that spin around the nucleus assume definite patterns, termed **electronic shells** or **orbitals.** The electrons located in the shells farthest from the nucleus travel the fastest and are at the highest energy level. The first orbital (1*s*) is spherical in shape and has the lowest energy level. The other orbitals farther from the nucleus are termed the *p*, *d*, and *f* orbitals. While the 1*s* orbital is spherical, the other orbitals can be either spherical or dumbbell-shaped. The electrons can be depicted as an electron cloud that surrounds the nucleus (Figure 3.1) or can be visualized as presented in Figure 3.2. The arrangement of electrons around the nucleus may also be referred to as the **electron configuration.** An electron will always fill an orbital nearest the nucleus, and the total number of electrons that can occupy the first orbital is 2; the second shell has four orbitals (one spherical and three dumbbell shaped) and can contain 8 electrons; the third a maximum of 18 electrons in nine orbitals; and the fourth 32 electrons in 16 orbitals. Some molecules of biological interest are depicted in Figure 3.3. The **valence** of an atom is based on the number of electrons that are needed (or may be donated) to create a full complement of electrons in the outer orbital. In the examples presented in Figure 3.3, hydrogen can fill the first orbital by gaining an electron, carbon by losing or gaining 4 electrons, and oxygen by gaining 2 electrons. Hydrogen therefore has a valence of +1, carbon +4, and oxygen −2. There is a tendency for atoms to interact so that the outer shell will contain the maximum allowable number of electrons. They therefore donate, accept, or share electrons to attain this configuration. When the outer orbit is filled, a chemically stable state is attained. Gases such as argon, neon, and helium have filled outer shells and are, therefore, mostly inert. They are unreactive.

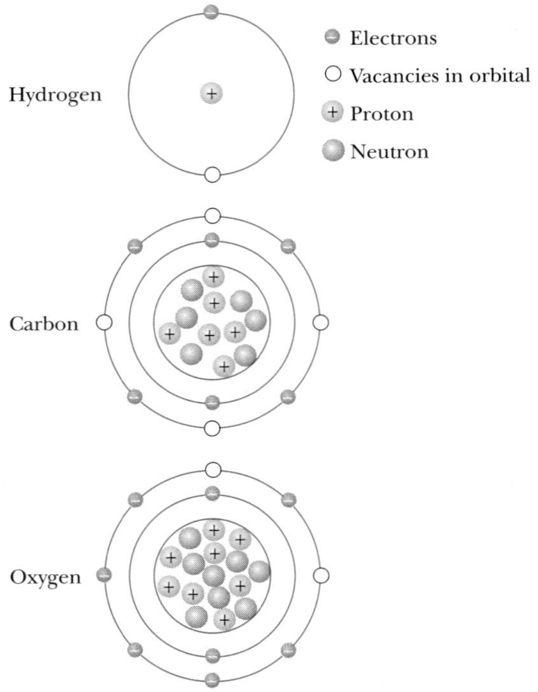

Figure **3.3** Bohr models of three important biological atoms. The mass of the electrons is very small but they carry a negative charge that markedly affects the chemical properties of the atom.

Molecules

An atom may exist separately; exist in combination with dissimilar atoms; or, in some cases, combine with like atoms. For example, 2 atoms of hydrogen may share electrons (one from each atom) to form H_2. Oxygen and nitrogen atoms are often paired, and the earthly atmosphere contains N_2 and O_2. Atoms most often pair with unlike atoms, and this results in the formation of **compounds.** A compound formed by combinations of atoms other than carbon is considered an **inorganic** compound, and those that contain carbon are generally termed **organic** compounds. The configurations of some simple compounds are illustrated in Figure 3.4. For example, oxygen shares a pair of electrons to form double bonded O_2, methane results from the sharing of 4 electrons in the second orbital with 4 atoms of hydrogen. Oxygen picks up 1 electron from each of 2 hydrogen atoms to fill its outer orbital and form water.

When we wish to denote the chemical composition of a molecule we use a **chemical formula.** H_2O, CH_4, and CO_2 are the formulas for water, methane, and carbon dioxide. The bonds between these atoms are discussed below.

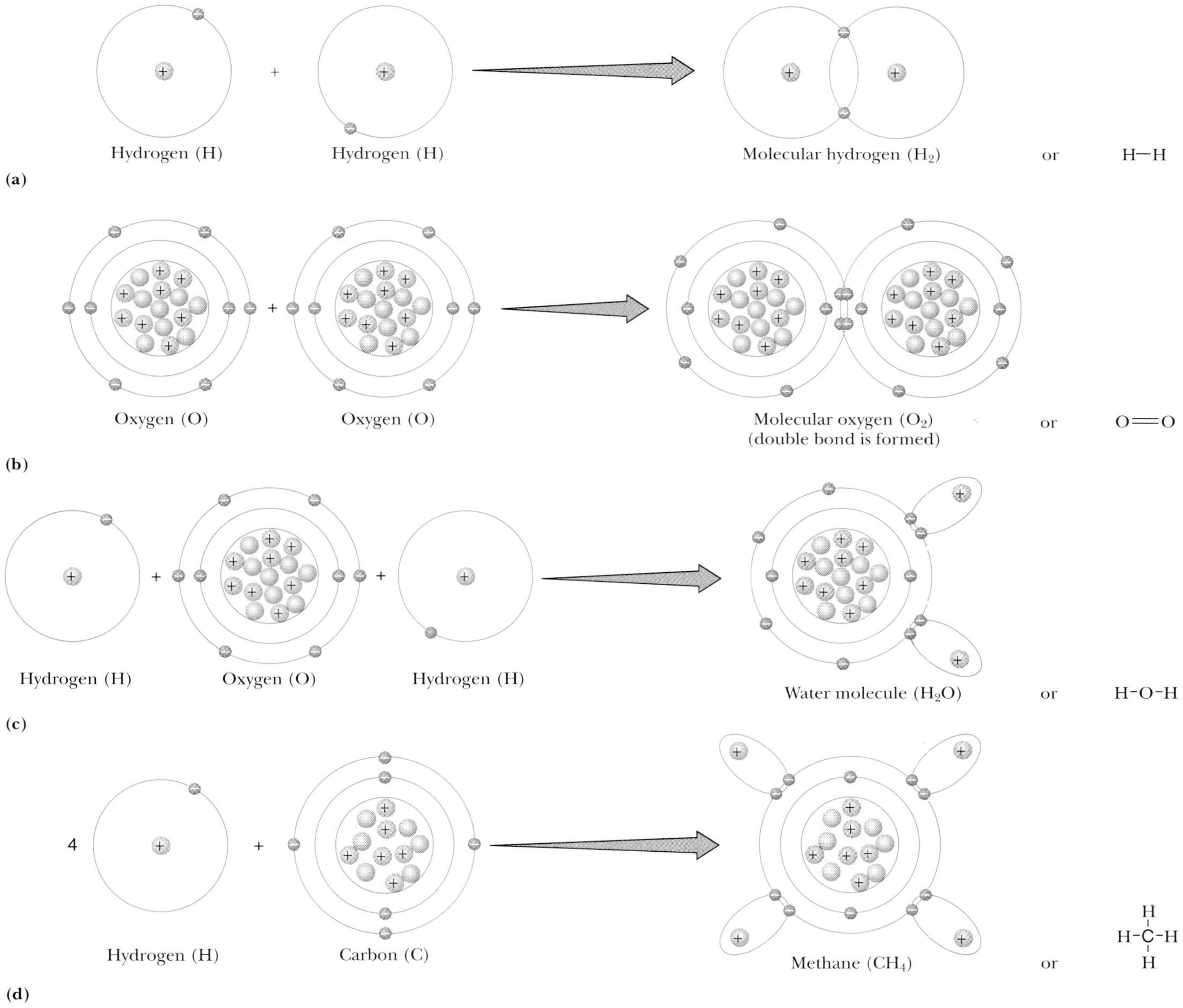

***Figure* 3.4** The formation of compounds by sharing electrons. **(a)** Hydrogen shares two electrons to form a more stable dihydrogen. The structural formula on the right is molecular hydrogen with a single covalent bond between the atoms. **(b)** Two molecules of oxygen share two pairs of electrons with a double bond representing the electron pairs. **(c)** Two molecules of hydrogen share electrons with an atom of oxygen. **(d)** Carbon shares an electron with four atoms of hydrogen to form methane.

Chemical Bonds

The forces that hold molecules together are called **chemical bonds.** Energy is required for the formation of chemical bonds, and, as a result, each bond has some potential chemical energy. There are three main types of bond that join atoms or molecules together: **ionic bonds, covalent bonds,** and **hydrogen bonds.** These bonds are formed because atoms seek a full complement of electrons in the outer shell or tend to achieve neutrality through interactions between polar (charged) molecules. A discussion of these bonds follows.

Ionic Bonds

Two atoms achieve a full complement of electrons in their outer orbital when one of the atoms donates an appropriate number of electrons and the other atom accepts these electrons. The donor has a **positive charge (cationic),** and the acceptor a **negative charge (anionic).** The loss of electron(s) in the donor results in an excess of protons relative to electrons, and the gain of an electron(s) results in an excess of electrons over protons. An element with an excess number of protons is referred to as positive (+). Some examples include Na^+, K^+, Ca^{2+}, Mg^{2+}, and Mn^{2+}. An element with an excess of electrons is negatively (−) charged and examples include Cl^-, I^-, and S^{2-}. An element with a positive charge can bond with an element with a negative charge, and the bond that joins them is considered an **ionic bond.**

The classical example of the formation of a compound by ionic bonding is the reaction between Na^+ and Cl^- to form sodium chloride—NaCl (table salt). The overall process is depicted in Figure 3.5. Elemental sodium donates an electron and achieves a full complement of electrons (8) in the outermost orbital. It is an ion (Na^+) because it carries an electrical charge due to the excess of one proton relative to electrons. Chlorine picks up one electron, resulting in a full outer orbital, but now has one more electron than protons in the nucleus (Cl^-). The attraction between Na^+ and Cl^- results in the formation of a salt.

Ionic bond formation has limited applicability to biological systems. However, it is important in the neutralization of DNA in a bacterial cell. DNA is an acidic molecule. Divalent cations such as Mg^{2+} form ionic bonds that neutralize the anionic phosphate PO_4^{-2} in bacterial DNA. Neutralization of DNA in eukaryotes occurs by interacting with basic (positively charged) proteins such as histones.

Covalent Bonds

The **covalent bond** is more relevant to biological systems because it is the major bond that joins the component elements that form the molecules of the cell. Carbon to carbon bonds are examples of covalent bonds and are the basic backbone in cellular lipids, proteins, nucleic acids, and carbohydrates. Covalent bonds that are important in cell structures are presented in Table 3.2. A covalent bond is formed by the sharing of one or more pairs of electrons. The simplest example of shared electrons is the hydrogen molecule, where two protons share two electrons (Figure 3.4). These electrons orbit around both nuclei. This is an example of a **single covalent bond** because one pair of electrons is shared. Other examples of the sharing of a single pair of electrons are methane and water. A **double covalent bond** is formed when two pairs of electrons are shared. An example of this is carbon dioxide (Figure 3.6). Carbon dioxide is a stable molecule because each oxygen molecule has a valence of −2 and carbon has a valence of +4. The 4 electrons in the outer shell of the carbon atom are shared with 2 electrons from each of 2 oxygen atoms, resulting in a full complement of electrons in the outer orbital of both atoms. Double bonds, the sharing of two electron pairs, occur frequently between adjacent carbons (C=C) in many of the important compounds in cells.

In some cases, a triple bond occurs because 3 pairs of electrons are shared. A biomolecule of this type is N≡N (N_2). Eubacteria and some of the Archaea can cleave triple bonds as some species utilize molecular nitrogen gas from the atmosphere as a source of organic nitrogen. This cleaving of N_2 by these organisms requires considerable expenditure of energy as the N≡N triple bond is one of the strongest of all chemical bonds (Table 3.3).

Hydrogen Bonds

When hydrogen forms a covalent bond with an electronegative atom such as nitrogen or oxygen there is an uneven sharing of the electrons. The relatively large nucleus of the nitrogen or oxygen attracts the electron from

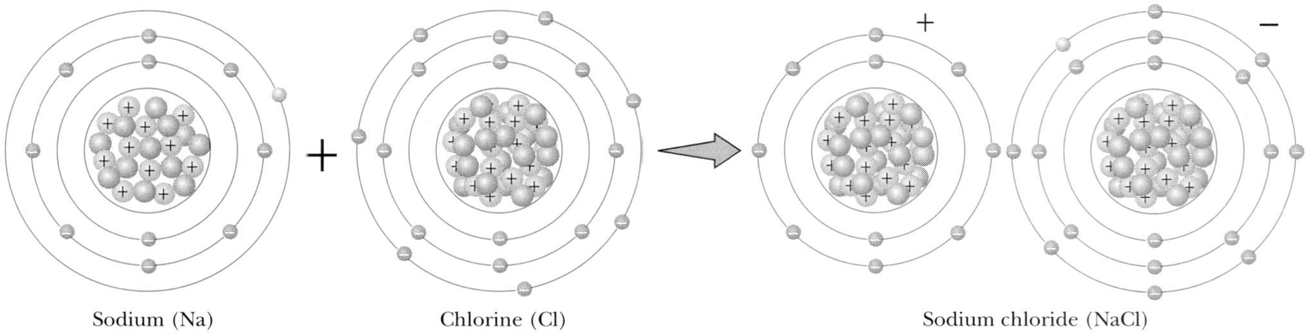

***Figure* 3.5** Formation of a compound by ionic bonding. Sodium donates its single valence electron to chlorine, which has seven electrons in its outer shell. With this addition the chlorine has a complete outer energy level but 17 protons to 18 electrons, which creates a negative charge. Sodium after donating an electron to chlorine has 11 protons and 10 electrons or a net positive charge. The atoms are attracted to one another by this unlike electrical charge forming the ionic compound sodium chloride.

Table 3.2 Types of covalent bonds that are an integral part of biomolecules. These are the major functional groups present in living cells.

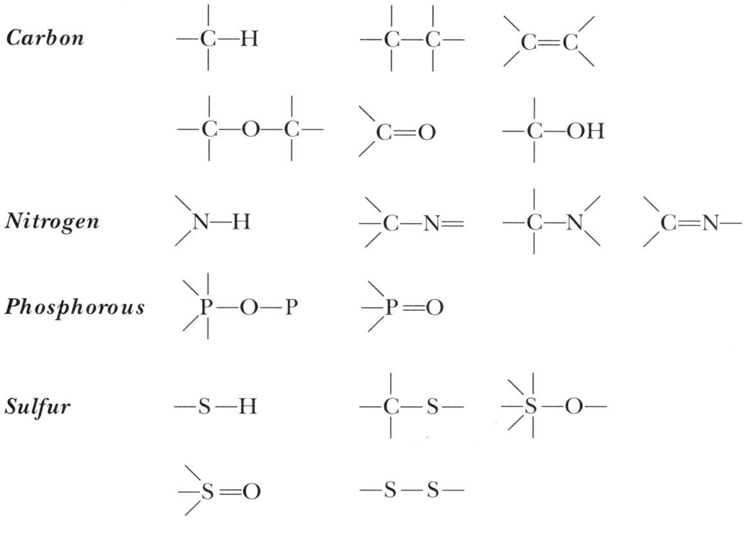

hydrogen more strongly than does the small single proton in the hydrogen nucleus. In a molecule of water, the total electrons orbiting the two nuclei at any given moment are closer to the oxygen nucleus than to the hydrogen nucleus. This generates a greater electronegativity surrounding the oxygen nucleus and greater positivity around the hydrogen nucleus. This results in an attraction between the positive part of a water molecule and the negative part of another electronegative atom, as depicted in Figure 3.7. The hydrogen of the water molecule is attracted to the nitrogen atom, and the resultant bond between the two molecules is called a **hydrogen bond.** A hydrogen bond has about 5 percent of the strength of a covalent bond.

Hydrogen bonds are of considerable importance in biological systems. The double helix in nuclear DNA and conformation of cellular proteins are dependent on hydrogen bonding. The role of hydrogen bonding in the

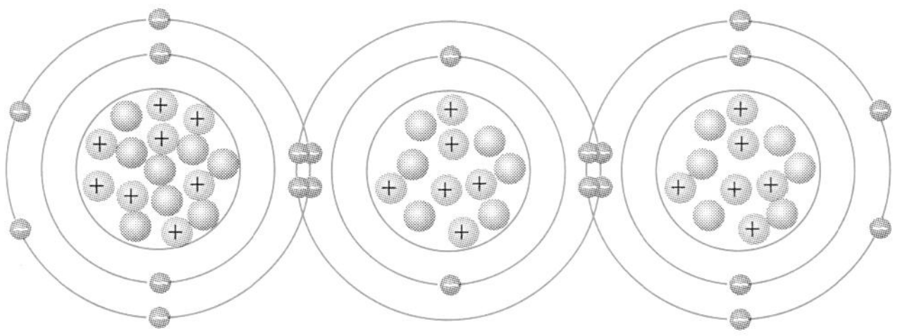

O=C=O

***Figure* 3.6** The sharing of electrons between an atom of carbon and two atoms of oxygen. This results in the formation of two sets of double bonds.

Table 3.3 Bond energies between atoms of biological importance—Kcal/mole consumed in breaking bonds

Single Bonds	*Kcal/mole*
C—C	82
O—O	34
S—S	51
C—H	99
N—H	94
O—H	110
Multiple Bonds	
C=C	147
O=O	96
C=N	147
C=O	167
N≡N	226

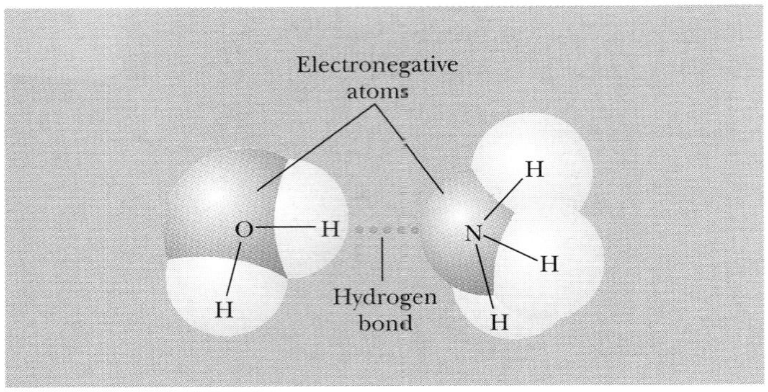

Figure **3.7** A hydrogen bond. The nitrogen atom in a molecule of ammonia (NH_3) is bonded to a hydrogen atom of a molecule of water (H_2O). A hydrogen bond results when a hydrogen atom is covalently bonded to a larger electronegative atom (oxygen) and has a weak electrical attraction to another electronegative atom (nitrogen).

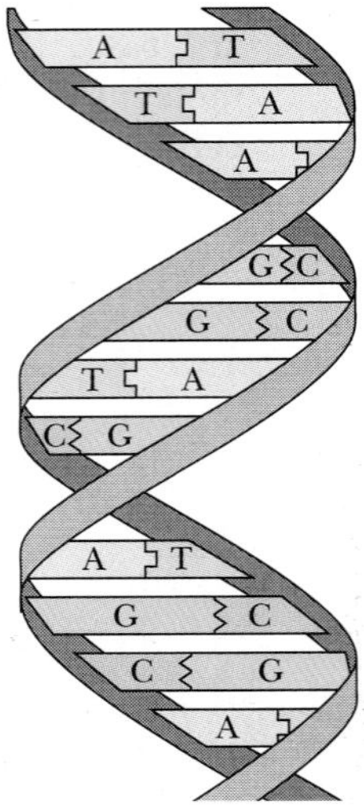

Figure **3.9** The double helical structure of DNA. This open model illustrates the hydrogen bonding of base pairs at the center of the structure.

double helix conformation of DNA is illustrated in Figure 3.8, where hydrogen bonding is involved in the planar arrangement of base pairs. Adenine and thymine are always bonded via two hydrogen bonds. Guanine is always paired with cytosine through three hydrogen bonds. The helical core is also maintained by **stacking interactions** between the planes of adjacent base pairs. These stacking interactions involve dipole-dipole and van der Waals forces (discussed below) that are equal in magnitude to the stabilizing energy of the hydrogen bonds between base pairs. The open model of the DNA double helix is presented in Figure 3.9. This figure illustrates the interior arrangement of the molecule, and the atomic model is presented in Figure 3.10. The base pairs are arranged on the interior of the spiral, and sugar phosphates are on the outside.

Hydrogen bonding is also important in the conformation of functional proteins. Bonding can occur between polar molecules on separate protein chains or hydrogen bonds can form between polar areas within a molecule. Polypeptide chains with internal bonding can assume a helical shape as shown in Figure 3.11. In this case, the hydrogen bonds between three amino acids form a loop (helix). Two individual polypeptide chains may form sheets by bonding between chains as depicted in Figure 3.12.

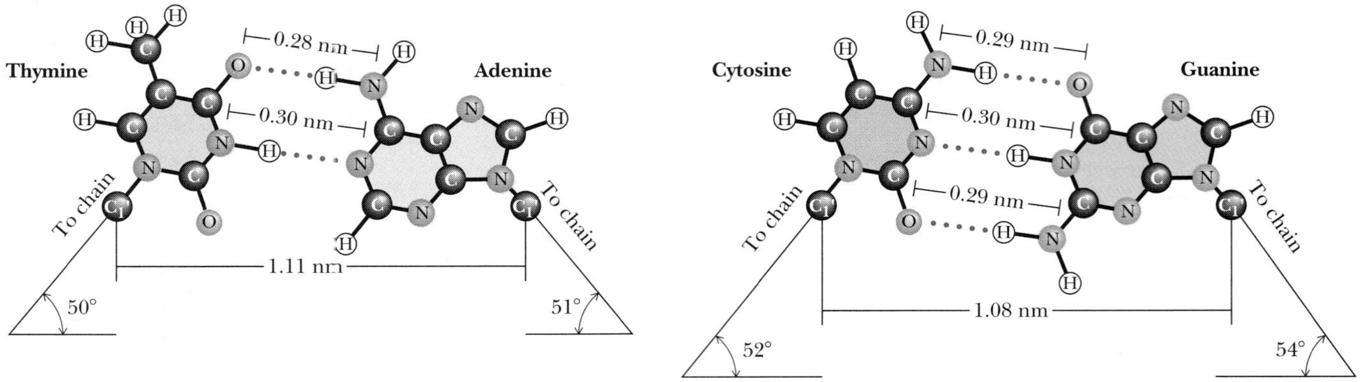

Figure **3.8** Hydrogen bonding retains the planar arrangement of purines and pyrimidines in the double helix of DNA. The dimensions and hydrogen bonds formed: **(a)** adenine-thymine and **(b)** guanine-cytosine. Both A—T and G—C are equivalent in dimension, and this is instrumental in the compact helical structure of DNA. Other interactions at the periphery maintain the "stacked" arrangement of base pairs one above the other.

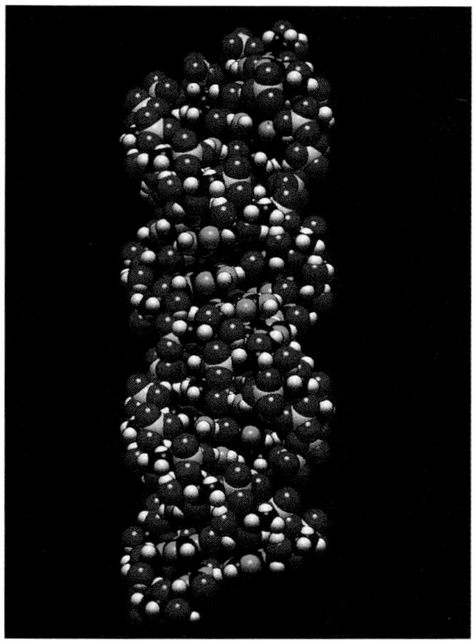

Figure **3.10** A computer-generated simulation of a colored plastic, space-filling model of DNA: red–oxygen, blue–nitrogen, dark blue–carbon, yellow–phosphorus, and white–hydrogen. (© Account Phototake/Phototake NYC)

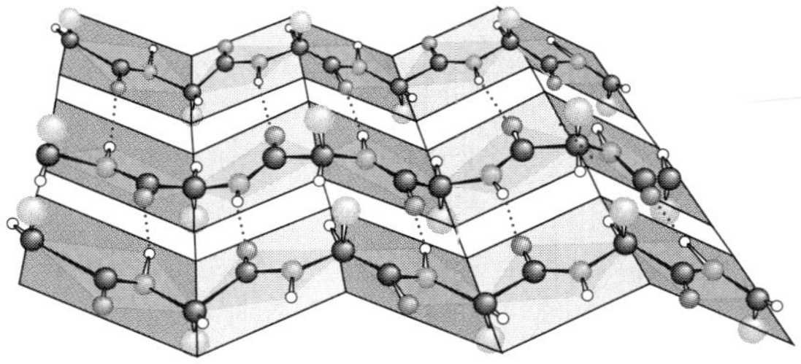

Figure **3.12** Several polypeptide chains may run parallel to one another, forming a pleated sheet structure held together by hydrogen bonds.

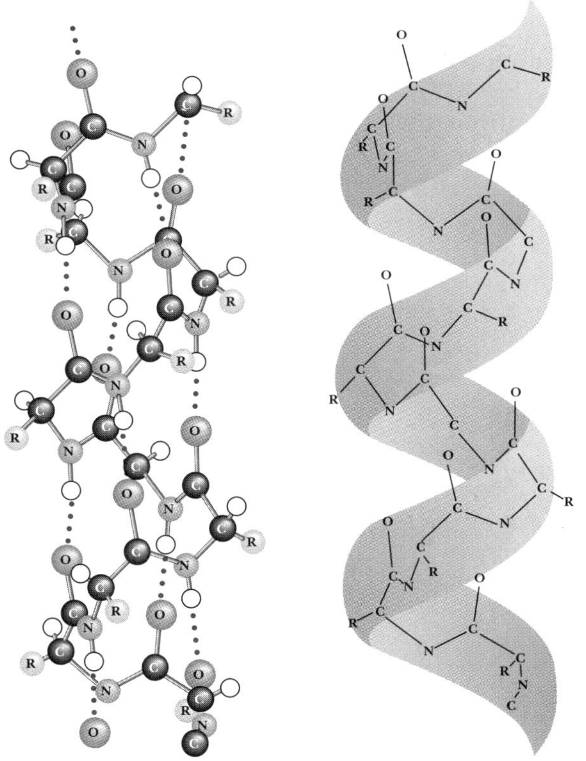

Figure **3.11** Model of an α helix as it would occur in a protein. There are three amino acids in one hydrogen-bonded loop. All of the C=O and $^{-}NH_2$ groups in the series of amino acids depicted form a hydrogen bond.

van der Waals Forces

van der Waals forces are interactions that occur between adjacent molecules that are not due to ionic, covalent, or hydrogen bonds. These interactions can be based on either an attraction or a repulsion. The **attractions** are due to a short-lived fluctuation in the electron charge densities that surround adjacent nonbonded atoms. This fluctuation is called a **dipole moment.** A dipole occurs when a molecule that is electrically neutral becomes polar. This polarity results when the center of negative charge (the electrons) does not center around the site of positive charge (Figure 3.13). As a result, there is a momentary polarity in the otherwise neutral molecule, which can attract an opposite charge in a neighboring molecule. Some hydrophobic molecules such as hexane are liquids rather than gases because the molecules are held together by the weak van der Waals forces. These van der Waals forces also play a role in enzyme substrate binding and protein interactions with nucleic acid.

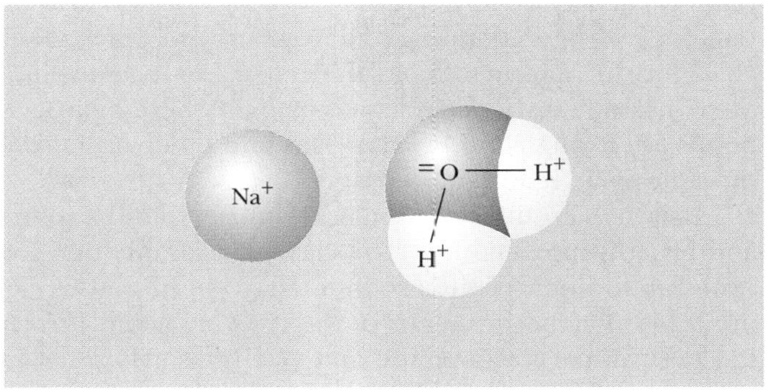

Figure **3.13** An ion dipole between a sodium ion (Na^+) and the polar molecule, water. The electrons orbiting about the water molecule may stray and may not be centered over the nucleus. As a result, the water becomes a dipolar molecule. Water molecules orient around the ion, a process called hydration.

Repulsion occurs when atoms that are not covalently bonded come too close together and repel one another. The electrons surrounding one molecule overlap the space occupied by the electrons of another, and the negatives tend to push the atoms apart.

Hydrophobic Forces

Hydrophobic forces are significant in biological systems but they are not well understood. These forces essentially prevent a solute from interacting with a solvent. Hexane is virtually insoluble in water because the water (solvent) withdraws in the area of contact with the apolar hydrophobic molecule (solute). The water molecules form a rigid hydrogen bonded network among themselves. This network effectively restricts the orientation of water molecules at the interface. The molecules of the hydrophobic compound cannot interact with the tightly oriented water molecules. This is why oil floats on water and lipids in membranes are effective in creating an impenetrable wall. Hydrophobicity is also involved in protein/nucleic acid interactions that preserve the tertiary structure of DNA.

Water

Water is the basic requirement of all living cells. Life evolved in water, and without it there is no life. If moisture is present in a microenvironment it is probable that microorganisms (bacteria, cyanobacteria, or fungi) will be present as well. Water is the solvent in living cells and a cell is 60 percent to 95 percent water. Even inert spores or seeds are 10 percent to 20 percent water by weight.

Water is essentially a neutral compound. However, the two constituent elements, oxygen and hydrogen, differ in electronegativity, and the distribution of charge around these two elements is asymmetrical. As a result, water is actually polar, as was depicted in Figure 3.7. This polarity allows for hydrogen bonding and an affinity of water molecules for one another. Water molecules are dynamic and are constantly involved in arrangements where hydrogen bonds are formed and broken. This and other properties of water make it an ideal solvent.

Polarity permits water molecules to surround a solute and bring it into solution. The classical example of a solvent/solute interaction is sodium chloride in water (Figure 3.14). The negative part of the H_2O molecule attracts the positive parts of a solute and vice versa. These interactions permit the sodium and chloride ions to solubilize. Organic acids and amino acids are also ionizable, and this is a major factor in their water solubility. Sugars also contain many polar hydroxyl (OH) groups and are generally quite soluble in water.

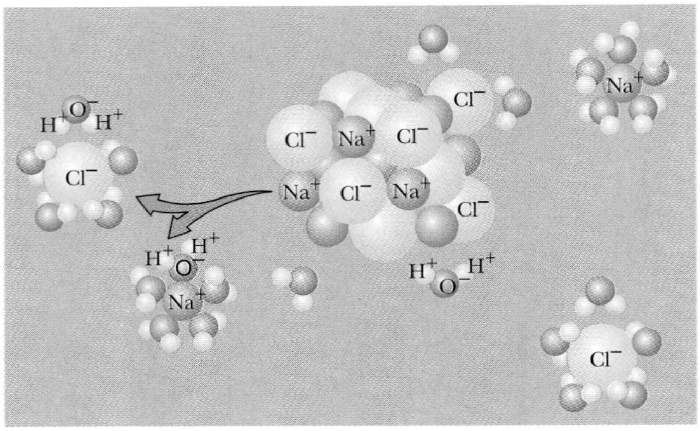

***Figure* 3.14** Ionization of a solute permits hydration of the component parts. When NaCl is added to water the weakly negative ends of the water molecules (O^-) are attracted to the sodium ions (Na^+) and draw them away from the chlorine ions. The positive (H^+) ends of the water molecule are attracted to the negative chloride ions (Cl^-). When the NaCl is dissolved, every ion is surrounded by water molecules.

Rapid temperature changes do not occur when water is exposed to heat because of the extensive hydrogen bonding between water molecules (Figure 3.15). Generally, heat absorption by molecules increases their kinetic energy (both the rate of motion of molecules and their reactivity). Absorption of heat by water, however, results in the breaking of hydrogen bonds rather than increasing rates of motion of the atoms. Much more heat is required to raise the temperature of water than would be required for a nonhydrogen-bonded liquid. This makes water an ex-

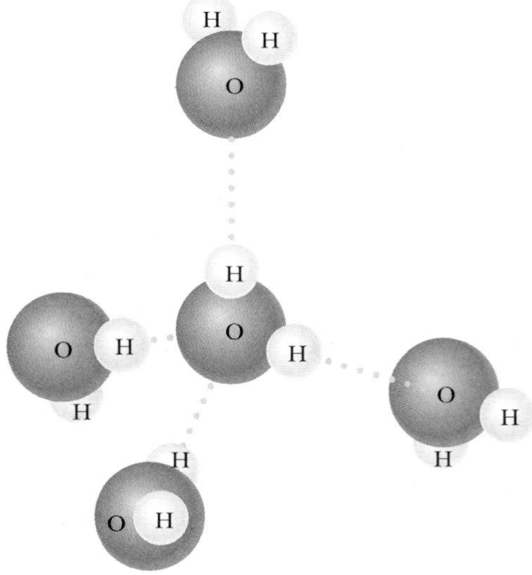

***Figure* 3.15** Hydrogen bonding between water molecules. These hydrogen bonds cause water molecules to stick to one another. They are cohesive, and water molecules stick to other kinds of substances by adhesive forces. These forces cause surface tension and result in capillary action of water.

cellent insulator because it is slow to heat and slow to cool. This property is important to warm-blooded animals, but it is of somewhat limited consequence to microbes that assume the temperature of their environment. A microbe can only exist in an environment that generally is at a tolerable temperature for the survival of that particular species.

Water can form thin layers on surfaces because of its hydrogen bonds. This phenomenon is termed **surface tension** and is due to the attraction of water molecules for one another (Figure 3.15). Surface tension occurs on the surfaces of bodies of water and insects can "walk" on water because of the adhesion between water molecules.

Biological membranes are protected by a thin film of water. Water atoms actually bond with oppositely charged atoms at or near the surface of the membrane. As a result of this surface layer of water, a membrane retains fluidity and flexibility and is slow to dry out.

Most nutrients for microbes must be water soluble because microbes are unable to ingest particles. When a microbe attacks a large molecule such as protein, lipid, or starch it must digest the substrate to low molecular weight soluble monomers extracellularly before that compound can be transferred into the cell.

Water interacts with hydrophobic molecules such as lipids, and, as a result, the lipids aggregate into compact molecular arrangements. This is an essential property in biological membranes where the lipid portion of the cellular membrane lines up to form a tight barrier to the flow of polar molecules in and out of the cell. Even protons cannot penetrate through a cell membrane. This role of membranes will be covered in Chapter 4.

Acids, Bases, and Buffers

Acids are substances that ionize in water, and this ionization liberates hydrogen ions. A strong acid is one that ionizes readily in water, even in dilute solution. Among the strong acids are hydrochloric acid, nitric acid, and sulfuric acid. Organic acids such as acetic acid are much weaker because they ionize poorly and are only partly ionized even when in dilute solution.

A **base** is a compound that, when ionized, releases one or more cations and negatively charged ions (OH^-). These negative ions can accept protons from solution. Thus the OH^- radicals remove H^+ from solution, leading to an excess of hydroxyl groups. A solution with an excess of OH^- is basic. A strong base such as sodium hydroxide ionizes completely in dilute solution while ammonium hydroxide, a weaker base, does not.

Water has a slight tendency to ionize—that is to dissociate into hydrogen ions (H^+) and hydroxide ions (OH^-).

$$HOH \rightleftharpoons H^+ + OH^-$$

In pure water the hydrogen ion concentration in solution is 0.000000 1M or 10^{-7}. The term pH is defined as the logarithm of the reciprocal of the hydrogen ion concentration. Thus a neutral solution would have a pH of 7. When the concentration of H^+ is greater than 10^{-7} the solution is acidic, and if less than 10^{-7} the solution is basic. The pH scale is logarithmic, so a solution of pH of 6 has a hydrogen ion concentration 10× greater than a solution with a pH of 7. The pHs of some often encountered materials are presented in Figure 3.16.

Microbial growth in laboratory media can result in the production of organic acids. These acids can lower the pH of the medium, which can result in cessation of growth or death of the cell. To prevent a significant change in pH, a **buffer** is generally added to the growth medium. A buffering system often employed in growth media is composed of potassium dihydrogen phosphate (KH_2PO_4) and the salt dipotassium hydrogen phosphate (K_2HPO_4). This buffering system is effective around the neutral range (pH 6 to 8) because it can accept protons when the medium becomes acidic and donate protons if the medium becomes basic.

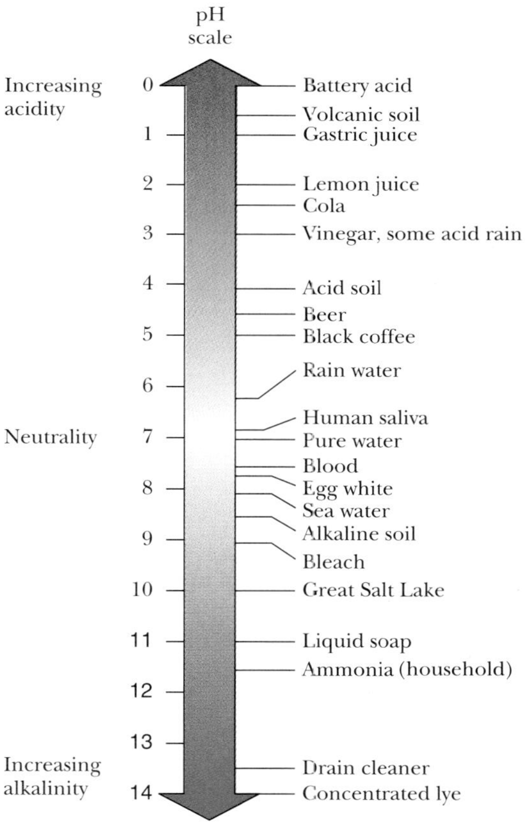

***Figure* 3.16** The pH scale and the pH of some familiar materials.

Chemical Constituents of the Cell

The number of reactions that are possible in living cells is limited by the restricted range of elements present in cells. The most abundant element in cells is hydrogen, followed by carbon, oxygen, nitrogen, phosphorous, and sulfur. Cations that are present in virtually all cells are sodium, potassium, magnesium, calcium, zinc, and iron. Some organisms require trace elements (required in small amounts) such as cobalt, molybdenum, copper, vanadium, or manganese.

There are a limited number of functional groups present in microbial cells, and the major groups were presented in Figure 3.2. Carbon to carbon bonds are the backbone of cell macromolecules. Subunits (monomers) of these macromolecules are generally joined together through an oxygen molecule. Phosphorous is the medium of energy exchange and also a structural component of nucleic acids. Nitrogen is a part of many cellular components including proteins and the nucleic acid bases. Sulfur is generally a component of proteins and some of the B vitamins.

Major Monomers

The discussion here is limited to the structure of typical monomeric chemicals that make up a cell. The pathways involved in the synthesis of the major monomers (building blocks) that are present in a cell will be covered in Chapter 10. The polymerization of these monomers to form macromolecules is presented in Chapter 11.

Carbohydrates (sugar) are a primary source of energy in many microbial cells. They are also a constituent of many structural units in the microbe, including nucleic acids, cell walls, and ATP. The general formula for a sugar is $(CH_2O)n$. The commonly occurring sugars contain 3 to 7 carbons. A simple sugar, and the most abundant product of living cells, is glucose (Figure 3.17). Glucose is also called blood sugar and is depicted along with fructose (fruit sugar). Other monosaccharides of biological interest are presented in Figure 3.18. A common disaccharide (2-sugar) is sucrose or cane sugar. It is composed of one molecule of glucose and one of fructose. Lactose (milk sugar) is also a disaccharide and is made up of glucose and galactose. The structures of these two important disaccharides are presented in Figure 3.19. The disaccharide lactose has a glycosidic linkage between 1 and 4 carbons of the sugars, and this configuration is called beta (β). Lactose is galactose-β-1,4-glucose. The glycosidic linkage in sucrose is between the 1 and 2 carbons and in the alpha (α) configuration. Sucrose is glucose-α-1,2-fructose. Glycogen, starch, and cellulose are composed of glucose units polymerized into large molecules, called polysaccharides. Glycogen and starch are important carbon and energy reserves in bacteria, animals, and plants. Cellulose is present in fungi and plants as a structural component of the cell wall and is also present in a limited number of bacteria.

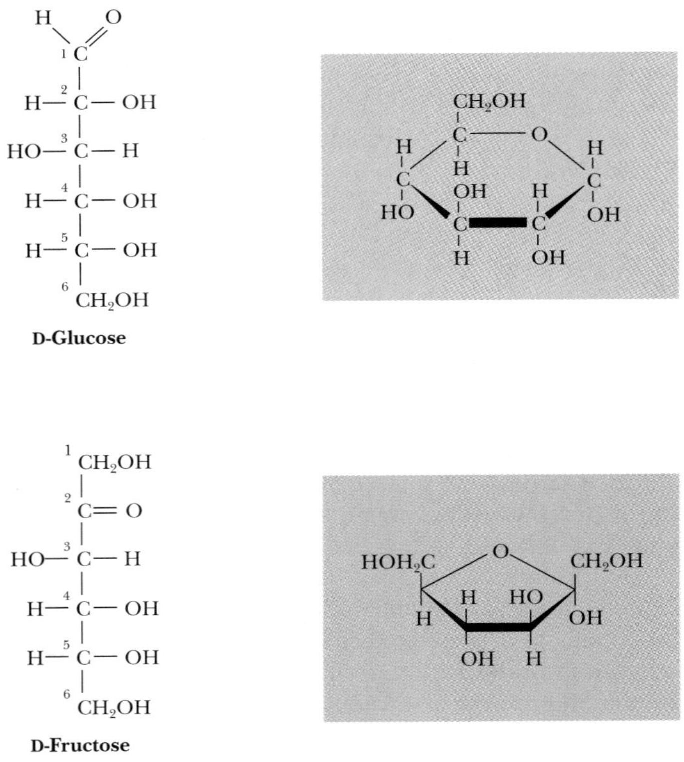

Figure **3.17** The structure of glucose and fructose. The figures on the left are stick formulas and are a convenient way to depict sugars. A more accurate depiction of the ring structure is on the right. The ring presentation is called a Haworth projection formula. Glucose and fructose are structural isomers, as they have the same chemical formula ($C_6H_{12}O_6$).

Lipids are composed of fatty acids and a sugar molecule, generally glycerol. The fatty acids in bacteria are

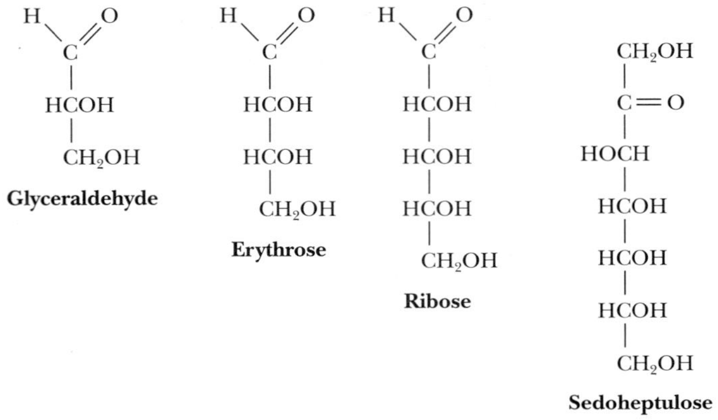

Figure **3.18** Examples of sugars that have three to seven carbons.

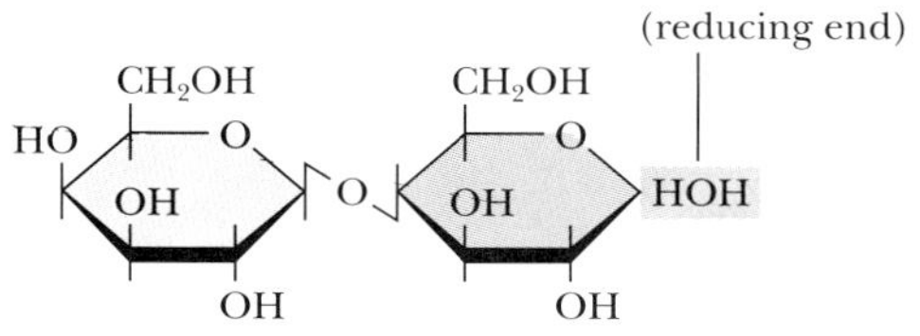

Lactose (galactose-β-1,4-glucose)

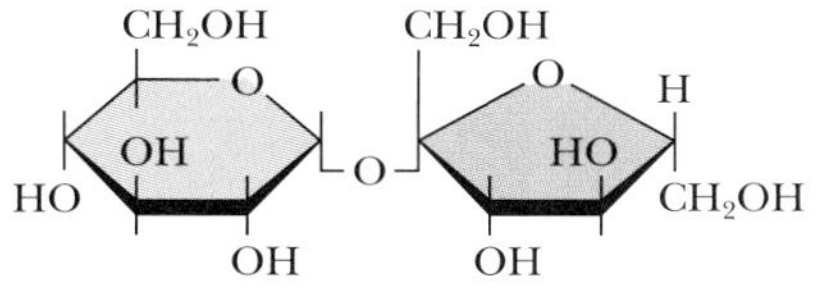

Sucrose (glucose-α-1,2-fructose)

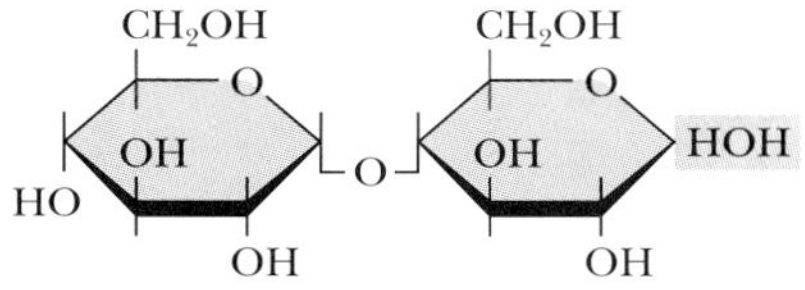

Maltose (glucose-α-1,4-glucose)

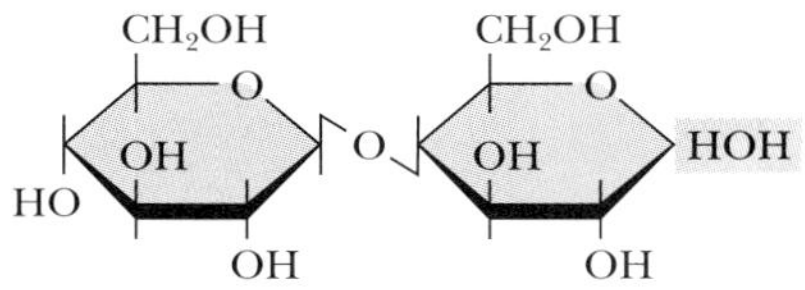

Cellobiose (glucose-β-1,4-glucose)

***Figure* 3.19** Common linkages that occur in disaccharides and in polysaccharides. There are specific enzymes that cleave these linkages.

mostly saturated (hexadecanoic) or mono-unsaturated fatty acids 16 or 18 carbons in length (Figure 3.20**a**). A mono-unsaturated fatty acid has a double bond between two adjacent carbons, most often 9 carbons from the carboxyl end of the long chain fatty acid. The lipids in the Archaea are isoprenoid and hydroisoprenoid hydrocarbons linked to glycerol through an ether linkage (Figure 3.20**b**). Higher organisms (plants and animals) have more than one set of double bonds in their fatty acids.

Proteins are a major component of cells and function as the workhorse of the cell. They are involved in synthetic reactions, energy production, and serve as structural components. About 50 percent (dry weight) of a cell is protein and a bacterium, such as *Escherichia coli,* can synthesize well over 1000 different proteins. The chemical components of a protein are amino acids. An amino acid has one or more amino groups ($-NH_2$) and one or more carboxylic acid groups ($-COOH$). The general structure of an amino acid is as follows:

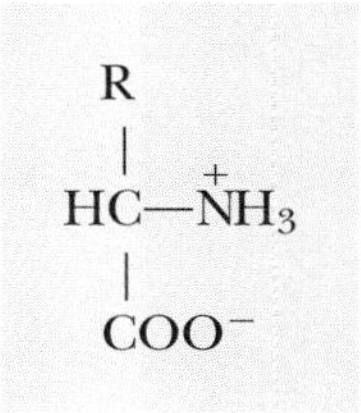

The R-group differs among the 20 amino acids. In water, the amine gains a proton and has a positive (+) charge while the carboxyl loses a proton and has a negative (−) charge.

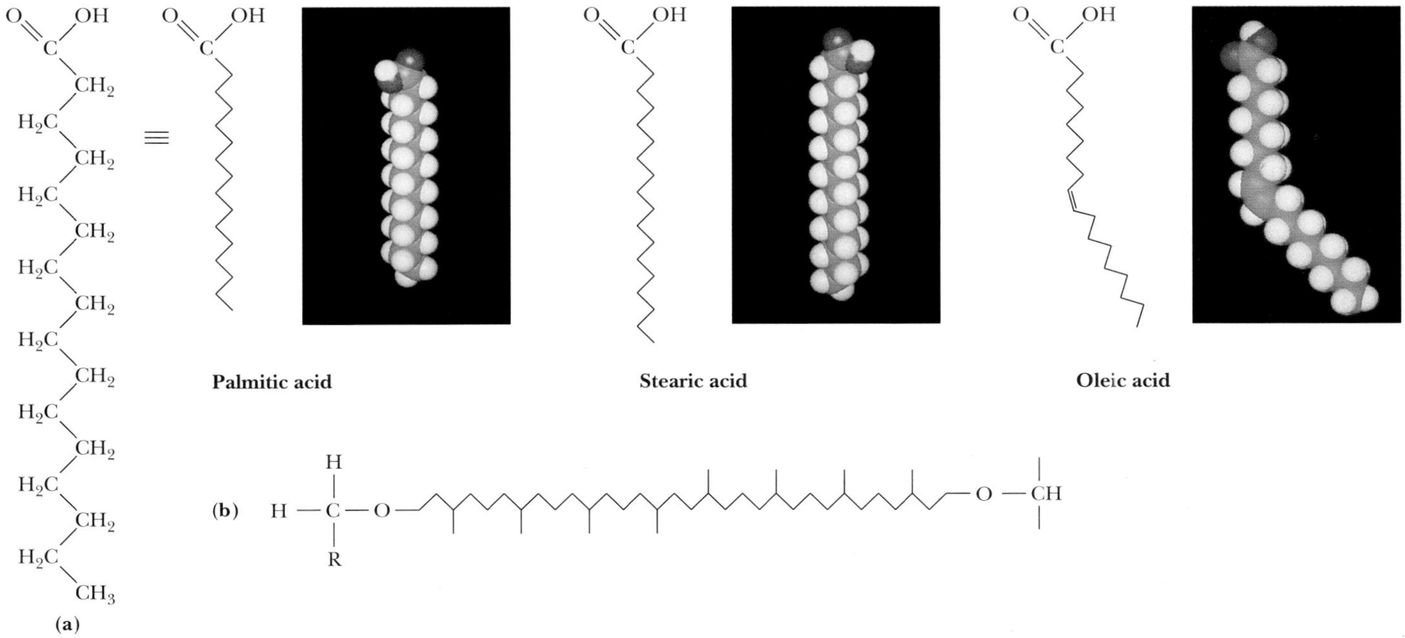

***Figure* 3.20** The structures of some fatty acids commonly present in the phospholipids of bacteria **(a)**. An archael glycerol ether lipid **(b)**.

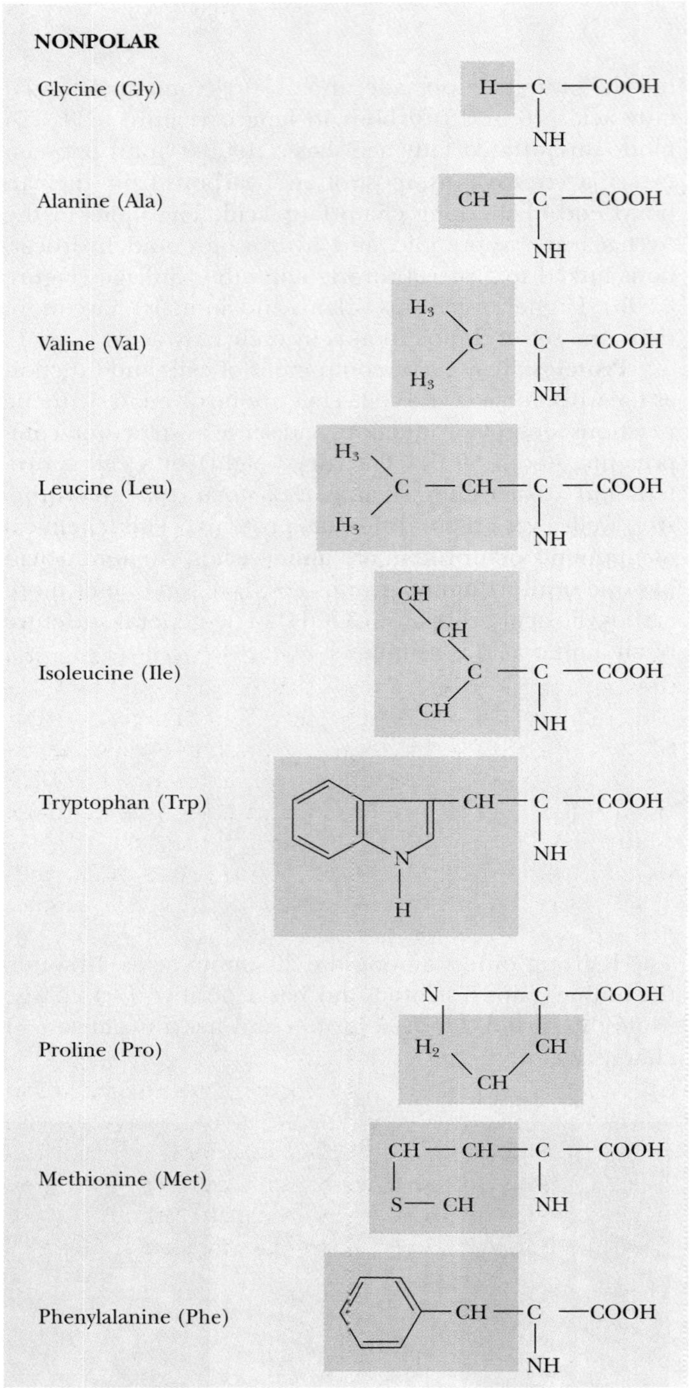

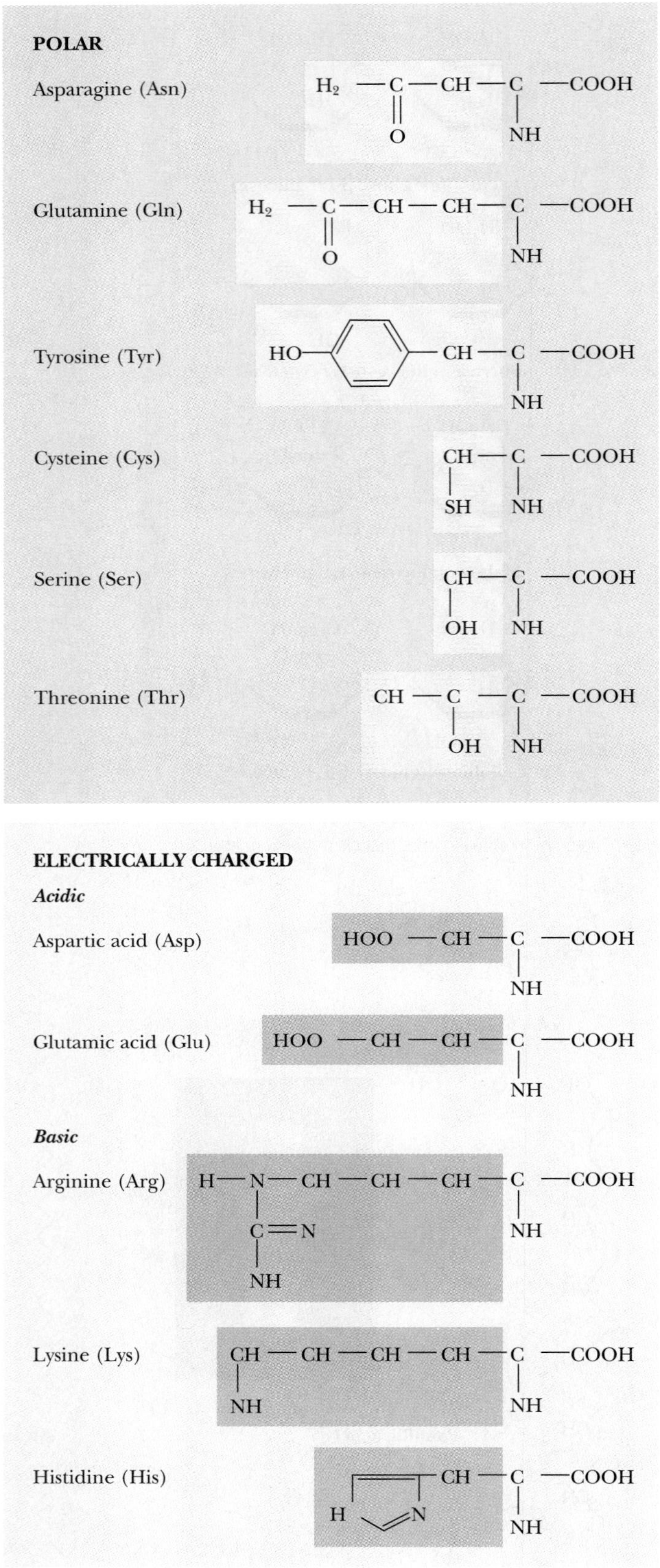

***Figure* 3.21** The structure of the 20 amino acids most commonly present in proteins. It is the sequence of these amino acids in a protein that determines its character. All amino acids except proline exist in the zwitterion form.

This **zwitterionic** form of the alpha amino acids occurs at physiological pH values. Molecules that bear charged groups of opposite polarity are termed zwitterions or **dipolar ions.** At near neutral pH both the carboxylic acid and amino groups are completely ionized—an amino acid can act as either an acid or a base. The structures of the 20 major amino acids are presented in Figure 3.21. Amino acids are joined together to make an amino acid chain called a **peptide.** The bond between amino acids in

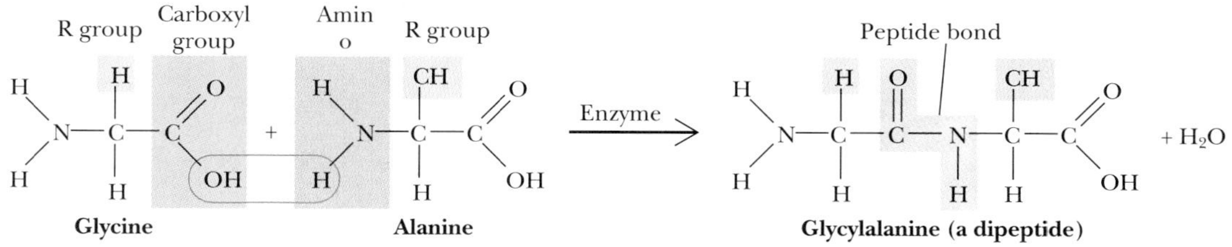

***Figure* 3.22** The peptide bond. Two amino acids combine via a covalent bond to form a dipeptide. Water is the by-product of this reaction.

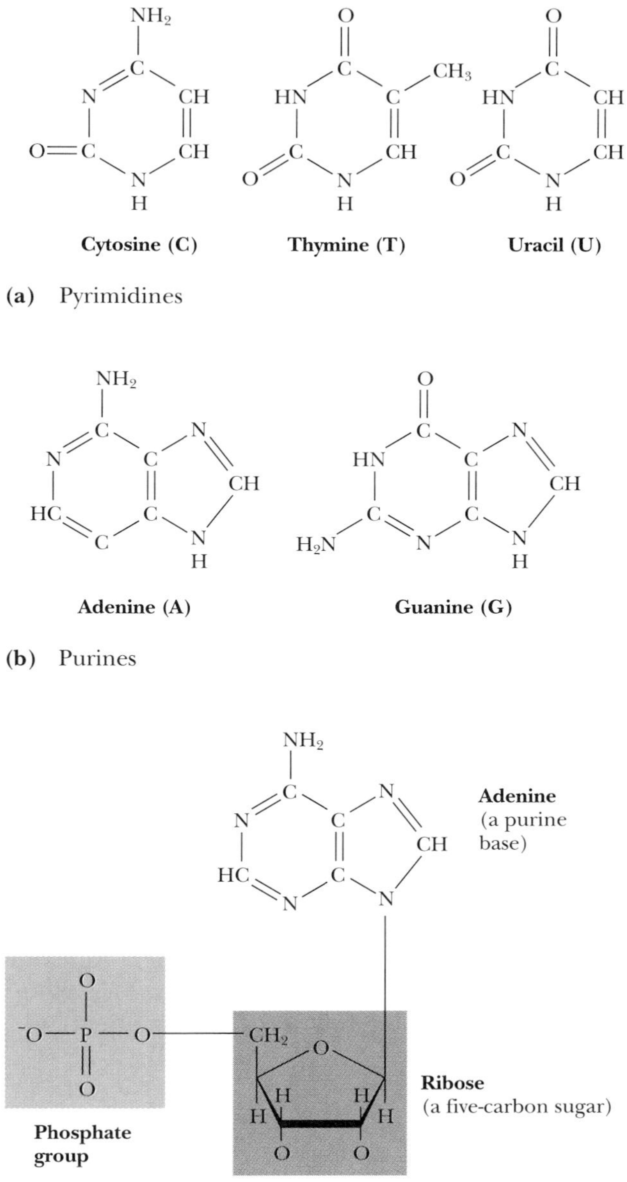

***Figure* 3.23** The basic structures present in nucleic acid: **(a)** pyrimidines, **(b)** purines, and **(c)** a nucleotide adenosine monophosphate. The nucleotide consists of one of the nitrogenous bases, a 5-carbon sugar, and a phosphate molecule.

a peptide chain is called a **peptide bond** (Figure 3.22).

There are two major nucleic acids in all cells: **ribonucleic acid (RNA)** and **deoxyribonucleic acid (DNA),** the nucleic acid that carries the blueprint of the cell. RNA is responsible for the translation of information inherent in DNA into functional proteins. Just as amino acids are the structural units of a protein, nucleotides are the structural units of nucleic acids. The overall structure of DNA was presented in Figures 3.8 through 3.10. There are four bases present in DNA and these are adenine, guanine, cytosine, and thymine. Thymine is replaced by uracil in RNA (Figure 3.23). The 5-carbon sugar in DNA is deoxyribose and in RNA it is ribose. The order of bases in DNA determines the sequence of amino acids in a protein, and, as mentioned previously, the characteristics of a protein depend on this amino acid sequence.

Isomeric Compounds

Isomers are compounds that have identical molecular formulas but differ in the arrangement of the constituent atoms. Such isomers are termed **structural** isomers. Isomers composed of six carbons and fourteen hydrogens (C_6H_{14}) are illustrated in Figure 3.24. Although these three compounds have identical molecular weights

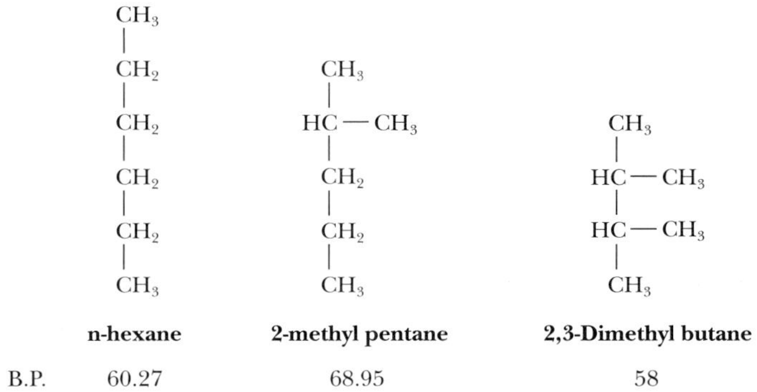

***Figure* 3.24** Structural isomers of a 6-carbon saturated hydrocarbon (C_6H_{14}). All have an equivalent molecular weight (86.18) but differ in physical properties as evident from the boiling point (b.p.) of each hydrocarbon.

BOX 3.2 MILESTONES

Louis Pasteur, the Crystallographer

Louis Pasteur began his scientific career as a chemist. He became intrigued with the observation that a chemical such as sulfur would crystallize in two distinct forms. Earlier workers had reported that tartaric acid from grapes deflected the plane of polarized light, but sodium ammonium tartrate was optically neutral. When examining crystals of the tartrate salt, Pasteur noted that some crystals had "faces" that inclined to the right, and those of others inclined to the left. He meticulously separated the two crystal types, dissolved each in water, and placed the respective solutions in the polarizing apparatus. Pasteur found that the dissolved crystals with faces on one side rotated polarized light in one direction, and the others rotated light in the opposite direction. An equal mixture of the two types was optically neutral. This was a clear proof that there can be a relationship between optical activity and molecular structure. That living things selectively synthesized only one optical form was intriguing to Pasteur and ignited a lifelong interest in biological studies.

Louis Pasteur in his laboratory.

(86.18) and molecular formulas they differ in physical properties as indicated by their boiling point. The structure would also influence biodegradability as hexane would be the most readily degraded and 2,3 dimethyl butane the least.

Stereoisomers are isomers where all of the bonds in the compounds are the same but the spatial arrangements of the atoms differ. Essentially stereoisomers are mirror images of one another (Figure 3.25). Stereoisomers are also called **enantiomers** and are optical isomers. In the one configuration a stereoisomer will rotate a plane of polarized light passing through it in one direction and the mirror image isomer will rotate light in the opposite direction. If the light is rotated to the right the isomer is dextrorotatory (D), and rotated to the left the isomer is levorotatory (L). The D,L system of nomenclature is based on the optical properties of D-glyceraldehyde and L-glyceraldehyde, respectively. The absolute configurations of all other carbon-based molecules are referenced to D- and L-glyceraldehyde.

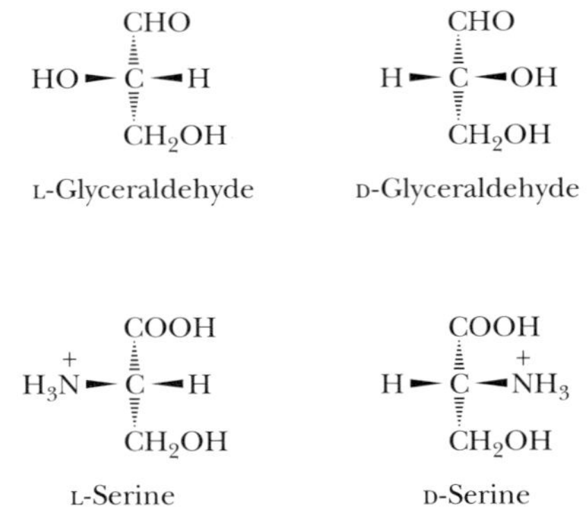

***Figure* 3.25** The configuration of the common amino acid L-serine can be related to the configuration of L-glyceraldehyde. The horizontal lines (dark) indicate bonds coming outward from the central carbon, and the vertical lines (stippled) indicate bonds going behind the central carbon.

Proteins are normally composed of L-amino acids. The cell wall peptidoglycan contains D-amino acids, and the D form is also present in selected antibiotics. Sugars are generally present in nature in the D configuration. There are enzymes present in bacteria (racemases) that can convert the D-amino acids to the L form. Racemases are also present in bacteria that interconvert the L configuration of sugars to the D form.

Summary

- Essentially an atom is electrically neutral because it has an equal number of negatively charged electrons in orbit about the nucleus to the positively charged **protons** present within the nucleus. **Neutrons** are present in all elements except hydrogen and they *have no charge.*
- The atomic number is equal to the number of pro-

tons in the nucleus of an element. The sum of the number of neutrons and protons present is the **atomic weight.**

- The three major bonds that join atoms or molecules together are **ionic, covalent,** and **hydrogen bonds.** Carbon to carbon bonds are covalent bonds and vital in the formation of biological molecules.

- Hydrogen bonds are important in the conformation of nucleic acids and proteins.

- **Water** is essential to life and a cell is 60 percent to 95 percent water. It is actually a polar compound and polymerized at room temperature.

- **Sugars** are essential components of living cells and are present in nucleic acids, cell walls, and ATP. They are also important sources of energy. Glucose is considered the most abundant product of living cells.

- Bacterial membranes contain **fatty acids,** that are generally 16 or 18 carbons in length, linked to glycerol via an ester linkage. They may be saturated fatty acids or mono-unsaturated. Archaeal membrane lipids are composed of **isoprenoid** chains that are ether linked to glycerol.

- One-half of the dry weight of a **cell** is protein, and the constituent parts of a protein are the amino acids. The bond that joins one amino acid to another is the peptide bond.

- The major nucleic acids in cells are deoxyribonucleic acid (DNA) and ribonucleic acid (RNA). **DNA** is the **blueprint** of the cell and **RNA** is involved in **translating** the information present in DNA to functional proteins. Purines and pyrimidines are the bases present in nucleic acids.

- Structural isomers have equivalent molecular formulas but a different arrangement of the constituent atoms. In stereoisomers the bonds between atoms are in the same place but the spatial arrangement differs. They are mirror images of one another.

Questions for Thought and Review

1. Draw an atom and label the parts. What retains the structure of an atom?

2. How does atomic weight differ from atomic number?

3. Define isotope. How would you design an isotope and experimentation to determine how methane is assimilated?

4. What structural feature is involved in the reactivity of an atom? How is this related to valence?

5. There are three types of bonds that occur between molecules. What are they and how do they differ? Which are most prevalent in a bacterium?

6. How are hydrogen bonds involved in the basic configuration of proteins? Nucleic acids?

7. What are van der Waals forces? Where might they play a role in biological systems?

8. Give several reasons why water is an effective solvent in biological systems.

9. How does an acid differ from a base? Why is water neutral?

10. Why are buffers necessary in bacterial culture media?

11. What are the five most abundant elements in a cell? What sort of function would they have in macromolecules?

12. What are the major monomers in polysaccharides, lipids, proteins, and nucleic acids?

13. How are the major monomers attached to one another?

14. Basically, how does one protein differ from another?

Suggested Readings

Brown, T. L., H. E. LeMay, and B. E. Bursten. 1991. *Chemistry: The Central Science.* Englewood Cliffs, NJ: Prentice Hall.

Garrett, R. H., and C. M. Grisham. 1995. *Biochemistry.* Philadelphia: Saunders College Publishing.

Hand, C. W. 1994. *General Chemistry.* Philadelphia: Saunders College Publishing.

Zubay, G. 1993. *Biochemistry.* 3rd ed. Dubuque, IA: W. C. Brown.

C'est un grand progres, monsieur.

Louis Pasteur (1881)[1]

Chapter 4

Structure and Function of Eubacteria and Archaea

Chapter 1 compared and contrasted prokaryotes with eukaryotes and affirmed that prokaryotes have simpler shapes and structures than eukaryotic organisms. Nonetheless, prokaryotes have all the necessary structural features to enable them to perform essential life functions, including metabolism, growth, and reproduction. More complex prokaryotic organisms have additional structures and life cycles not found in the simpler species. This chapter introduces the various unicellular and multicellular shapes of prokaryotes and the types of cell division found among them. It also discusses the structure, chemical composition, and function of common morphological features found in prokaryotic organisms.

However, before we discuss the properties of these microorganisms in greater detail, it is important to recognize that special instruments, techniques, and procedures are needed to study them. Microbiology, in fact, did not become a scientific discipline until appropriate instruments and techniques were developed to enable people to observe and study these small organisms (Chapter 2). Of foremost importance was the development of quality microscopes. Indeed, it is not surprising that microorganisms were discovered by a lensmaker, Antony van Leeuwenhoek, rather than by a biologist—an early illustration of the importance of instrumentation in microbiology. In this chapter we will initially focus on microscopy, because it is routinely used by microbiologists, and follow with a discussion of the structure and function of prokaryotic organisms.

Microscopy

The human eye has some ability to magnify objects. Natural magnification is simply achieved by bringing the object closer to the eye. The closer the object, the larger it appears, because the image on the retina is larger (Figure

[1] "That is important progress, sir." Remark made by Louis Pasteur to Robert Koch at the International Medical Congress in London, 1881. Made after Koch demonstrated his pure culture methods.

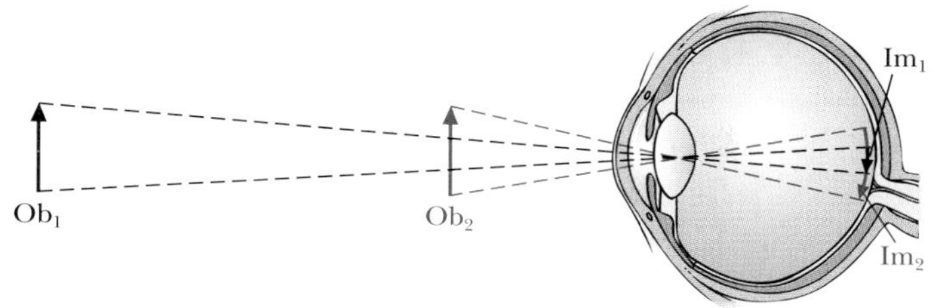

***Figure* 4.1** Diagram illustrating the magnifying capacity of the human eye. The object (Ob) is shown at two locations, Ob_1 and Ob_2. When the object is farther from the eye, the image formed on the retina (Im_1) is smaller than when the object is closer (Im_2).

4.1). Maximum enlargement depends on how close the object can be brought to the eye and still remain in focus. This **near point** is typically 250 mm for an adult human, and the distance increases as people age. Objects smaller than about 0.1 mm (the "eye" of a needle is about 1.0 mm wide) cannot be seen distinctly because their image does not occupy a sufficiently large area on the retinal surface.

Therefore, the unaided human eye cannot see small organisms such as bacteria, which typically have a cell diameter of only 0.001 mm (1.0 μm or 10^{-6} m). However, microscopes have been developed to aid the eye by increasing its ability to magnify. In the next sections we discuss several different types of microscopes commonly used by microbiologists.

Simple Microscope

The simplest optical device that can be used to assist the eye in enlarging objects is appropriately called the **simple microscope,** which in principle is a magnifying glass **(Box 4.1).** Leeuwenhoek ground his own specially constructed simple microscopes, which he used in his observations of microorganisms (see Figure 2.2).

Magnifying glasses are not particularly powerful. They usually magnify objects about fivefold. It is a tribute to Leeuwenhoek that his simple microscopes could not only magnify 200- to 300-fold, but did so with great clarity. Indeed, the images Leeuwenhoek produced were comparable to those obtained with the light microscopes of similar magnification that are used today.

Compound Microscope

Another Dutchman, Zacharias Janssen (late sixteenth, early seventeenth century), is generally given credit for the development of the compound light microscope, although the origins of this instrument are somewhat obscure. It is interesting to note that compound microscopes were available at the time Leeuwenhoek made his momentous discoveries. However, the early compound microscopes were inferior to Leeuwenhoek's simple microscope. Thus, it was not possible for Robert Hooke, the English scientist who coined the term *cell,* to confirm Leeuwenhoek's reports of microorganisms with the early compound microscopes available to him.

The **compound light microscope** is named for the two lenses that separate the object from the eye. The **objective lens** is placed next to the object or specimen to be viewed, whereas the **eyepiece** or **ocular lens** is located next to the eye. The object to be viewed is normally placed on a glass slide and illuminated with a light source.

Image formation is achieved by "focusing" the specimen—moving the objective lens and ocular (eyepiece) lens together relative to the specimen, until the image is clear. When the specimen has been properly focused, the objective lens produces a **real image** (one that can be displayed on a screen) within the dark body tube of the microscope. This real image is formed within the F_1 of the eyepiece lens. When the viewer looks through the ocular lens a **virtual image,** which cannot be displayed on a screen, is perceived (Figure 4.2**a**).

The magnification of a compound light microscope is determined by multiplying the magnification of the objective lens by that of the eyepiece lens. Usually the eyepiece magnification is 10×. Several separate objective lenses are usually mounted on a rotating nosepiece and typically give magnifications of 10× (low power), 60× (high, dry power), and 100× (oil immersion). The resulting magnifications attainable from this microscope would be 100×, 600×, and 1000×, respectively.

The typical compound microscope (Figure 4.2**b**) also has a third lens system. Light from a lamp or other source is focused on the specimen by a **condenser lens.** The condenser lens, which is not directly involved in image formation, is necessary to provide high-intensity light because the lack of brightness of the object becomes a limiting factor as the specimen is enlarged. The light intensity is adjusted by opening or closing a diaphragm called the **iris diaphragm,** located in the condenser.

It would appear that one could increase the magnification indefinitely using the compound light microscope simply by constructing more and more powerful objective and eyepiece lenses. In reality, the light microscope has a useful magnification maximum of only 1000- to 2000-fold. Although magnification beyond that level is attainable, it is referred to as **empty magnification** because it is not possible to see greater detail of the specimen.

Improving Image Formation in Light Microscopy

The ability of an optical system to produce a detailed image is termed its **resolving power.** The resolving power, *d*, is defined as the distance between two closely spaced

Optics of the Simple Microscope

The simple microscope consists of a convex lens made out of glass that is mounted on a bracket. An object placed an infinite distance in front of a convex lens will form an image on the opposite side of the lens (Figure **a**). This image, which is termed a **real image** because it can be visualized on a screen, is located one focal length (**F**) from the center of the lens. The $\mathbf{F_1}$ of the lens is defined as the position on the object side of a lens that is located one focal length from the lens, whereas the $\mathbf{F_2}$ of the lens is located one focal length from the lens on the opposite side (where the real image is formed).

If an object is placed *beyond* the F_1 of the simple microscope, a real image is produced on the opposite side of the lens (Figure **b**). This principle is used in the construction of slide projectors. In this example, the slide is the object and the real image is that which is projected on the screen. The size of the real image is determined by the distance between the screen and the projector (the greater the distance, the larger the image) and by the distance of the slide from the lens (this is varied by moving the lens during focusing).

To use the simple microscope, the object to be viewed is placed *within* the F_1 of the lens, that is, between the F_1 of the lens and the lens (Figure **(c)**). In such an arrangement, a real image cannot be formed. However, if the eye peers through the opposite side of the lens, then one sees a **virtual** image of the object. Unlike the real image, the virtual image cannot be projected. Nonetheless, the eye perceives an enlarged image of the object, the size being determined by the near point of the eye and the curvature of the lens.

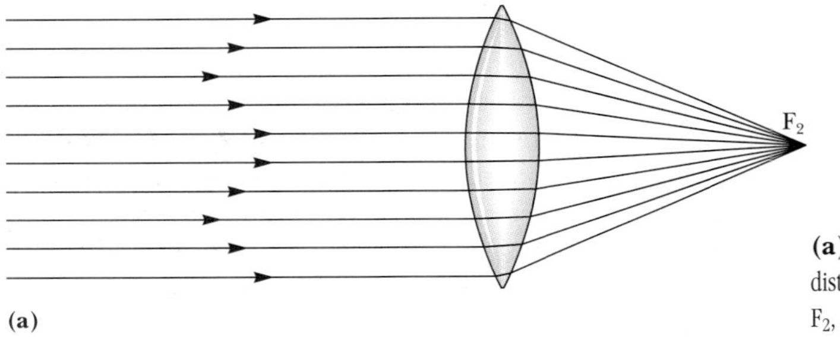

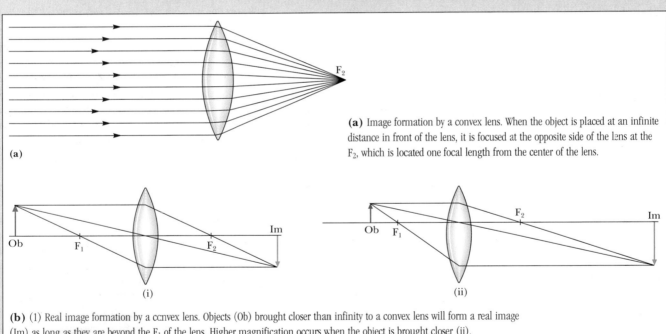

(a) Image formation by a convex lens. When the object is placed at an infinite distance in front of the lens, it is focused at the opposite side of the lens at the F_2, which is located one focal length from the center of the lens.

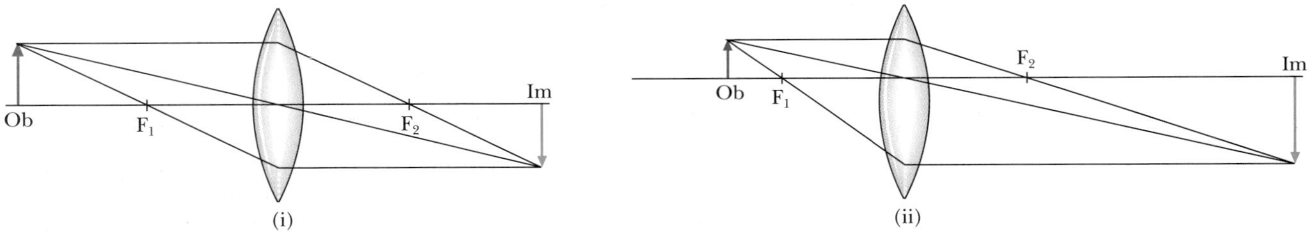

(b) (1) Real image formation by a convex lens. Objects (Ob) brought closer than infinity to a convex lens will form a real image (Im) as long as they are beyond the F_1 of the lens. Higher magnification occurs when the object is brought closer (ii).

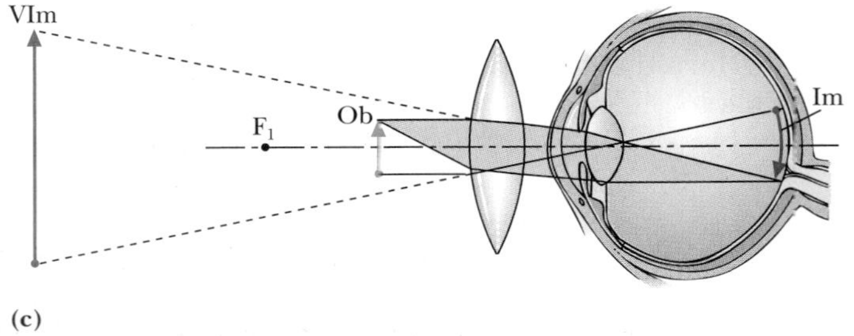

(c) The use of a simple microscope with the human eye. When the object is placed within one focal length, F_1, of the lens, the rays from the object are divergent and will not form a real image. They are, however, converged by the eye lens to form a real image on the retina, which perceives this image (called a virtual image) as an object about 25 cm in distance from the eye.

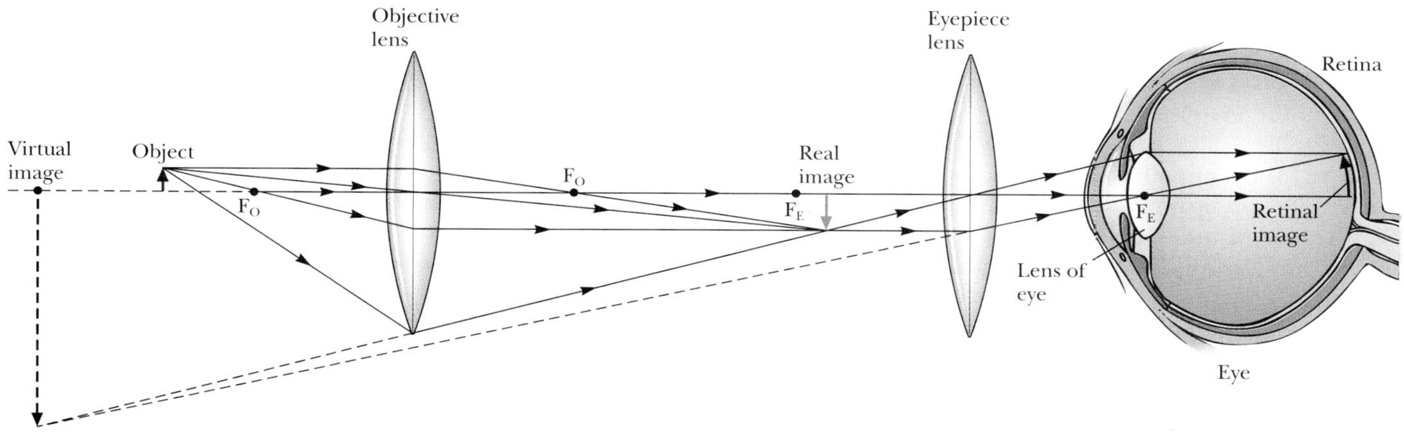

(a)

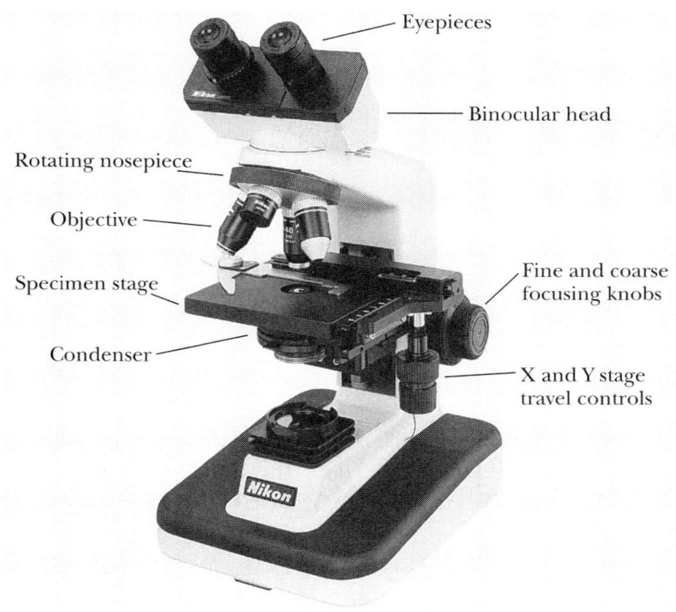

(b)

***Figure* 4.2** **(a)** Image formation by a compound microscope. The specimen is placed beyond the F_1 of the objective lens and therefore forms a real image in the ocular diaphragm of the microscope. This real image is located within the F_1 of the ocular or eyepiece lens, so the eye perceives an enlarged virtual image of the specimen. **(b)** A typical compound light microscope with binoculars, a rotating nosepiece with objectives, and a condenser lens system with a built-in light source. (Courtesy of Nikon Corporation)

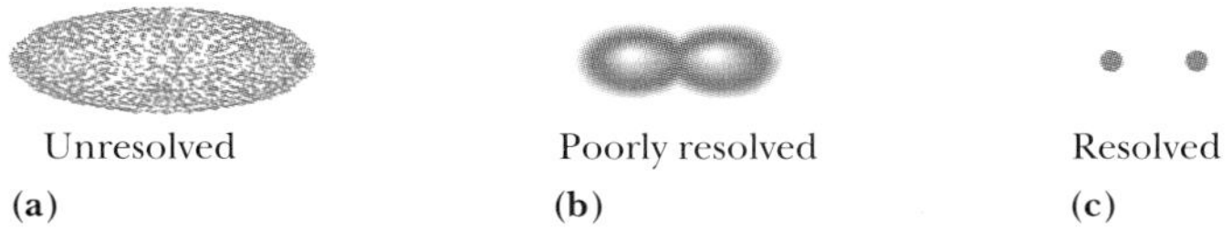

***Figure* 4.3** Two separate points are **(a)** unresolved, **(b)** partially resolved, and **(c)** resolved. Increased resolution is due to improved lenses.

points in the object that can be separated by the lens in the formation of the image (Figure 4.3). Its equation is:

$$d = \frac{0.61\ \lambda}{n \sin\ \theta} \qquad [4.1]$$

Therefore, for greater resolving power or **resolution** (lower values of d), one must use short wavelength (λ) light, a high refractive index (n) for the suspending medium for the specimen, and a large half-angle of the aperture or opening ($\sin\ \theta$) of the objective lens (Figure 4.4). The light microscope cannot be operated with wavelengths of light less than 400 to 500 nm, as the eye cannot see shorter wavelengths than the violet range. Thus, for

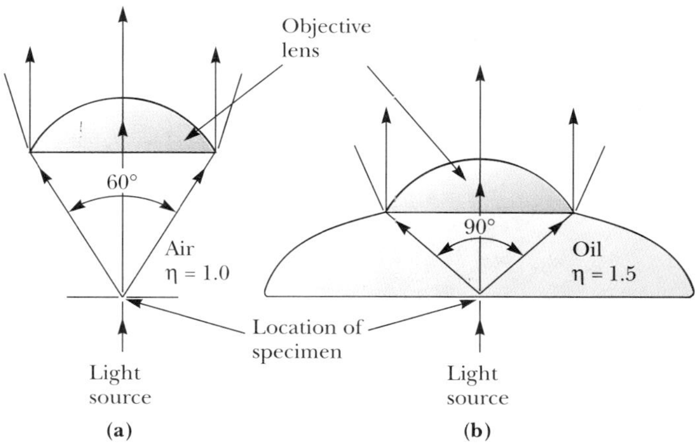

***Figure* 4.4** A comparison of **(a)** nonhomogeneous immersion (specimen in air) to **(b)** homogeneous immersion (specimen in oil). When the specimen is immersed in oil, *homogeneous immersion* occurs (the refractive index between the lens and specimen is 1.5). In contrast, if air is used, the index is only 1.0, with a corresponding decrease in resolution. Also, immersion in oil enables the specimen to be brought closer to the lens, thereby increasing the angular aperture of the lens and, hence, the resolving power.

maximum resolution, the shortest wavelength that can be used is about 500 nm (or 0.5 μm).

The **refractive index** of a material is a measure of the ability of the material to bend light rays. Crown glass used in lenses has a high refractive index. The refractive index of air is 1.0 (essentially identical to that of a vacuum), water is 1.33, and crown glass is 1.5. To permit a continuous high refractive index between the condenser and the objective, immersion oils of high refractive index are placed on the specimen and on the condenser lens system. Such **homogeneous immersion** systems not only permit a high value for the refractive index but also enable the objective to be brought closer to the specimen, thereby increasing the angular aperture (Figure 4.4).

The final factor affecting resolution is the aperture or opening of the objective itself. Because the objective lens has a finite aperture, a point on the specimen is not imaged as a point on the image, but as a disc (called an *Airy disc*) having alternate dark and light rings. The size of this disc can be decreased by increasing the aperture of the lens, but there is a limit to this. As a result, two closely spaced points on the object will be seen as fuzzy discs on the image, and if they are too close, may actually overlap and not be separated (Figure 4.5**a** and 4.5**b**). The theoretical maximum value for the angular aperture is 1.0. When these ideal values of the shortest possible wavelength (0.5 μm), homogeneous oil immersion (1.5), and the angular aperture of 1.0 are substituted into the equation for resolving power, then $d = 0.61(0.5\ \mu\text{m})/1.5$, which is equal to about 0.2 μm. Since typical bacterial cells are about 1.0 μm in diameter, the compound light microscope has a satisfactory resolution for their observation. It is not well suited, however, for observing internal cell structures that are much smaller.

Original versions of the compound microscope were plagued by lens aberrations. Two types of lens aberrations occur: chromatic aberration and spherical aberration (**Box 4.2**). Correction of these aberrations is made possible by use of multicomponent lens systems (Figure 4.5) in the objective lens. These developments and the introduction of the condenser transpired over a period of about two centuries, until the compound light microscope was finally perfected in the late nineteenth century by opticians such as Ernst Abbey in Germany (1840–1905).

Because the refractive index of bacteria is similar to that of water, bacteria are almost invisible when they are viewed with an ordinary compound light microscope (Figure 4.8**b**). Two approaches have been taken to overcome their transparency. In one approach, the bacterial cells are stained by dyes so they exhibit higher contrast—as discussed immediately below. An alternative approach is to use modifications of the ordinary compound microscope such as the darkfield microscope or the phase contrast microscope, which will be discussed in sections to follow.

Dyes and Stains

Bacteria observed in the light microscope are often stained with dyes because dyes increase the contrast between the cells and their environment. Most dyes used to stain microorganisms are aniline dyes, which are intensely pigmented organic salts derived from coal tar. They are called **basic dyes** if the **chromophore** (pigmented portion) of the molecule is positively charged. For example, crystal violet and methylene blue are basic dyes (Figure 4.6). Other basic dyes commonly used to stain bacteria are basic fuchsin, malachite green, and safranin. Under normal growth conditions for most bacteria, their internal pH is near neutrality (pH 7.0) and the cell's surface charge is negative. Therefore, basic dyes with positively charged chro-

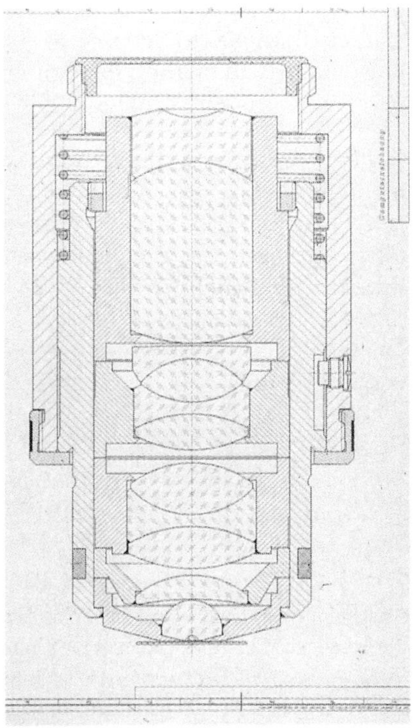

***Figure* 4.5** Cross-sectional view of a microscope objective lens showing its multicomponent lens system. (Courtesy of Carl Zeiss, Inc.)

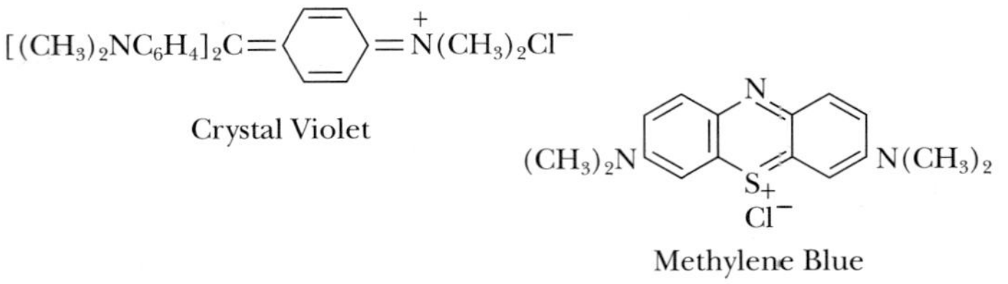

***Figure* 4.6** Chemical structures of crystal violet and methylene blue, two basic dyes that are chloride salts.

BOX 4.2 METHODS & TECHNIQUES

Lens Aberrations

Chromatic aberration (Figure **a**) is due to the inherent properties of lenses to separate light into its various wavelengths to produce a spectrum. Another problem, called **spherical aberration,** is the tendency of light rays from a point on the object to be bent or refracted differently, depending upon their distance from the center of the lens (Figure **b**).

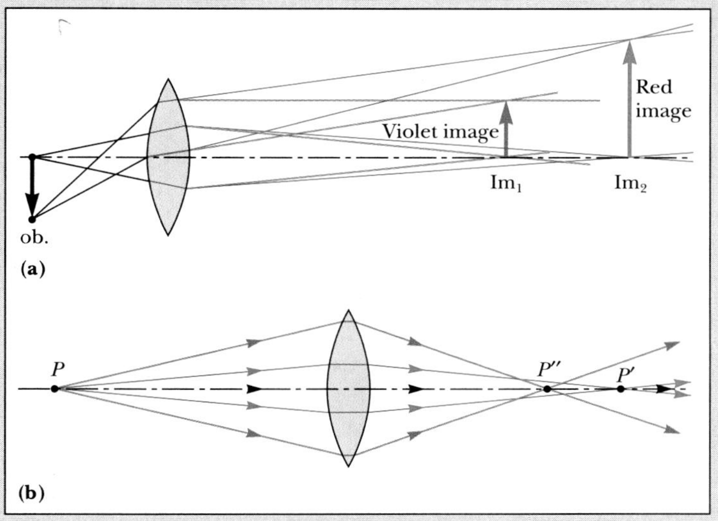

(a) Chromatic aberration occurs because lenses refract and thus focus different wavelengths of light at different focal planes. The result is a blurred, fuzzy image of various sizes and colors.

(b) Spherical aberration occurs because rays passing through the center of the lens are focused differently from rays entering the perimeter of the lens.

mophores are generally the most effective staining agents. Acid dyes such as nigrosin, Congo red, eosin, and acid fuchsin have negatively charged chromophores and are useful in staining positively charged cell components such as protein.

Simple stains of bacteria are made by spreading a suspension of the organism on a glass slide, allowing it to dry, and then gently heating it to **fix** it to the slide (a **fixative,** in this case heat, allows it to adhere—similar to frying an egg in a pan without oil). Such a preparation is called a **smear.** The stain is added, and, after a brief period of exposure, the excess dye is removed by gently rinsing and the slide is then viewed with the microscope.

Differential stains, such as the Gram stain (see **Box 4.7**), distinguish one microbial group from another—in this instance, gram-positive bacteria from gram-negative bacteria. Likewise, **acid fast** bacteria such as those in the genus *Mycobacterium* retain the color of a dye when the stained preparation is rinsed in a solution of ethanol containing hydrochloric acid at a final concentration of 3 percent.

Some staining procedures allow the identification of structures in the cell. Thus, specific stains exist for bacterial endospores, flagella, capsules, and other cell structures. Lipophilic dyes, such as Sudan Black, can be used to specifically stain lipid inclusions such as poly-β-hydroxybutyric acid (see below).

Fluorescence Microscope

Certain dyes used for staining microorganisms are called **fluorescent dyes,** because, when they are illuminated by short wavelength light, they emit light of a longer wavelength. One of the most commonly used fluorescent dyes is acridine orange, which specifically stains nucleic acid components of cells. When preparations are illuminated with an ultraviolet or halogen light source, the cells fluoresce green to red in color (Figure 4.7). One special advantage of fluorescence microscopy is that it enables the observation of cells that are located on an opaque surface such as a soil particle. These would be impossible to see

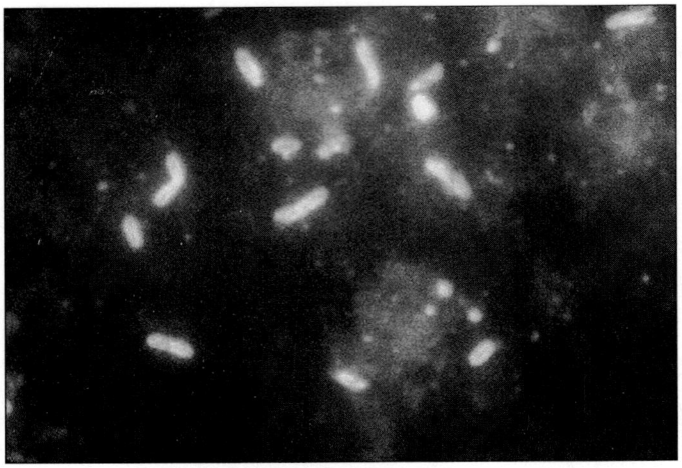

***Figure* 4.7** Bacterial cells that have been stained by a fluorescent dye and illuminated with ultraviolet light appear as green, yellow, or red to the eye. (Courtesy of VU/Soad Tabuqchali)

with an ordinary transmission light microscope in which light is transmitted through the specimen.

In typical fluorescence microscopy short wavelength light is provided by either a mercury lamp (ultraviolet) or a halogen lamp (near ultraviolet). This is passed through the objective lens system onto the specimen to provide **incident** illumination, which is illumination from *above* the specimen (not transmitted through the specimen as in an ordinary light microscope—although some fluorescence microscopes do use transmitted light). The longer wavelength light emitted by the fluorescent dyes is visible when viewed through the ocular. Special *barrier filters* prevent any harmful short wavelength ultraviolet light from reaching the eyes.

Confocal Scanning Microscope

The confocal scanning microscope is especially useful when viewing microorganisms in a three-dimensional space. The preparation, which may be a natural community containing microorganisms, is illuminated with a laser beam that is focused on one point of the specimen using an objective lens mounted between the condenser lens and the specimen. Mirrors are used to pass (scan) the laser beam across the specimen in the X and Y directions. The objective lens used for viewing the specimen magnifies the image, which is free from diffracted light, and the image is reconstructed on a video display screen.

Epifluorescence scanning microscopy uses a laser beam to illuminate the specimen, which has been stained with a fluorescent dye. The laser beam is focused at a particular plane of the preparation. The image viewed is that of a cross section of the preparation in which only those cells that are in the plane of illumination are observed on the display screen (see Figure 24.2).

Darkfield Microscope

An ordinary light microscope is called a **brightfield** microscope because the entire field of view containing the specimen and the background are illuminated and appear bright. The brightfield microscope can be modified to become a **darkfield microscope,** so named because the cells appear bright against a dark background **(Box 4.3).**

Phase-Contrast Microscope

As mentioned earlier, most bacterial cells appear to be colorless, transparent objects when observed by ordinary brightfield microscopy. A slight difference exists, however, between the refractive index of a bacterial cell (n is about 1.35) and its aqueous environment (n is 1.33). The **phase-contrast microscope,** or phase microscope, amplifies this slight difference in refractive index and converts it to a difference in contrast **(Box 4.4).** The result is that the cells appear very dark against a bright background (Figure 4.8**a**). As with the darkfield microscope, cells in wet mount preparations can be observed in their natural living state.

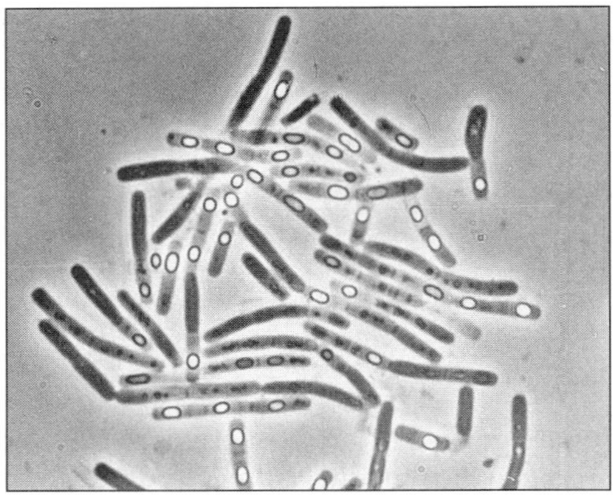

(a)

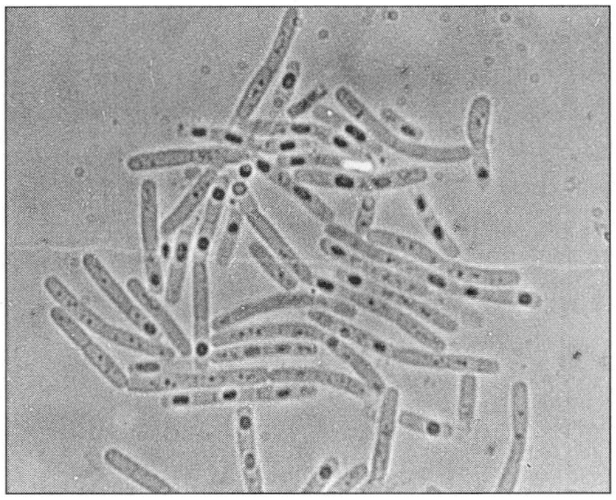

(b)

***Figure* 4.8** Photomicrographs of an unstained bacterium, *Bacillus megaterium,* comparing **(a)** its dark appearance by phase contrast microscopy with **(b)** its appearance by brightfield microscopy. (Courtesy of J. T. Staley)

BOX 4.3 METHODS & TECHNIQUES

The Darkfield Microscope

The darkfield effect is produced by illuminating only the cells and not the background. For this procedure, **wet mount** preparations are made by placing a droplet of cells on a slide and covering the aqueous suspension with a coverslip. Unlike smears, the cells in wet mount preparations remain alive in their normal growth environment while being observed. The darkfield effect is created by using a special condenser that introduces light onto the specimen at such an angle that the only light entering the objective is light that is scattered into the objective lens from the cells (Figure **a**). As a result, the cells appear bright against a dark background (Figure **b**).

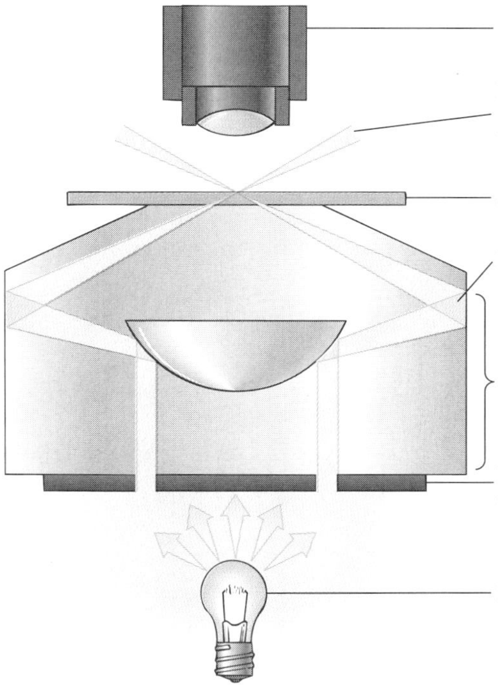

(a)

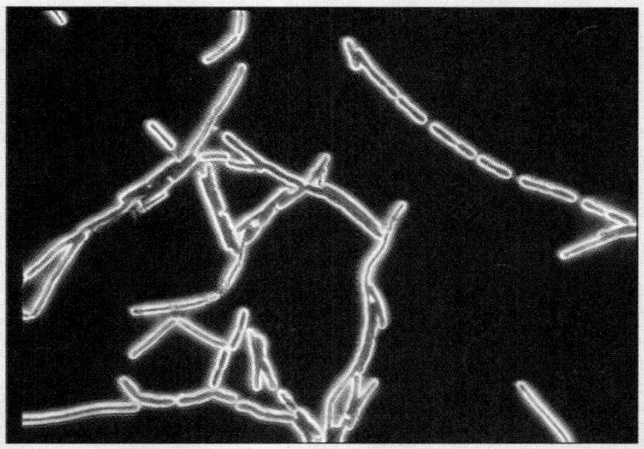
(b)

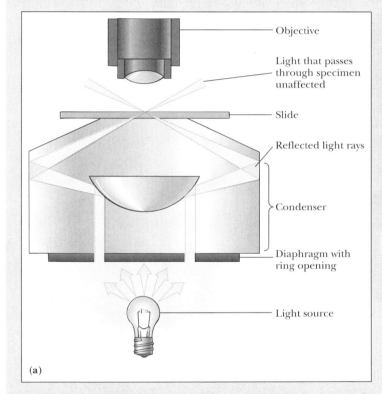

(a) A cross-sectional view illustrating the principle of darkfield illumination. None of the light rays are permitted to pass directly through the specimen and onto the lens. This is prevented by the diaphragm, which allows only peripheral light to enter the condenser lens. This light passes through the specimen but not the objective lens. Therefore, the only light entering the objective lens is light that is refracted or diffracted by the specimen. The result is that the specimen appears bright on a dark background.
(b) Cells of *Bacillus megaterium* as viewed by darkfield microscopy. (© Carolina Biological Supply Co./Phototake)

Transmission Electron Microscope (TEM)

The light microscope has a useful magnification of about 1000× to 2000×. Although greater magnification can be achieved, finer detail will not result, in part due to the properties of the illuminating source itself—light (see Equation 4.1). It is not possible to observe objects well if they are smaller than the wavelength of the illuminating source. An analogy may help illustrate this point. Suppose you wish to make an impression of a starfish on a piece of cardboard. Further, assume that two materials are available with which to make the impression, either fine particles of sand (analogous to a short wavelength) or rocks the size of the starfish (illuminating source of long wavelength). A much more detailed and accurate impression of the starfish could be made with the sand.

Very short wavelengths can be attained in the transmission electron microscope (TEM) by using a beam of electrons as the illuminating source. The wavelength is controlled by the voltage applied to an electron gun, which is the source of electrons. If the accelerating voltage is 60,000 V, the wavelength of the electron beam would be 0.005 nm or 0.000005 μm. This value is 100,000 times shorter than the wavelength of violet light (about 500 nm).

BOX 4.4 METHODS & TECHNIQUES

Optics of the Phase-Contrast Microscope

The phase microscope closely resembles the ordinary compound light microscope. However, two components differ. In place of the ordinary iris diaphragm of the light microscope, a condenser ring called the **condenser annulus** is used on the phase microscope. The other difference is the addition of a **phase plate** which is placed at the F_2 of the objective lens. In the absence of the specimen, light entering the objective lens comes entirely from the condenser annulus and falls only upon an annulus located in the phase plate. When a specimen is viewed, some of the light passing through it is scattered so that it falls outside of the phase annulus onto the complementary area of the phase plate. Therefore, at the phase plate, light scattered from points in the cell is separated from nonscattered light from the cell and the aqueous environment surrounding the cells. An absorptive, transparent material is placed in the phase annulus to decrease the much greater intensity of light passing into this area relative to that which has been scattered onto the complementary area. Because of the slightly different refractive index between the cell and its surroundings, the scattered light from the cell has a slightly retarded wave. This retardation is roughly equivalent to 0.25 wavelength. If these waves are retarded an additional 0.25 wavelength, they become one-half (0.5) wavelength out of phase with the unscattered light and cause maximum interference, and hence, contrast in the formation of the image. This effect is accomplished by incorporating a transparent retarding material in the complementary area where these scattered rays fall. Therefore, the slight difference between the refractive index of the cell and its surroundings has been changed in the formation of the real image into a dramatic difference in contrast.

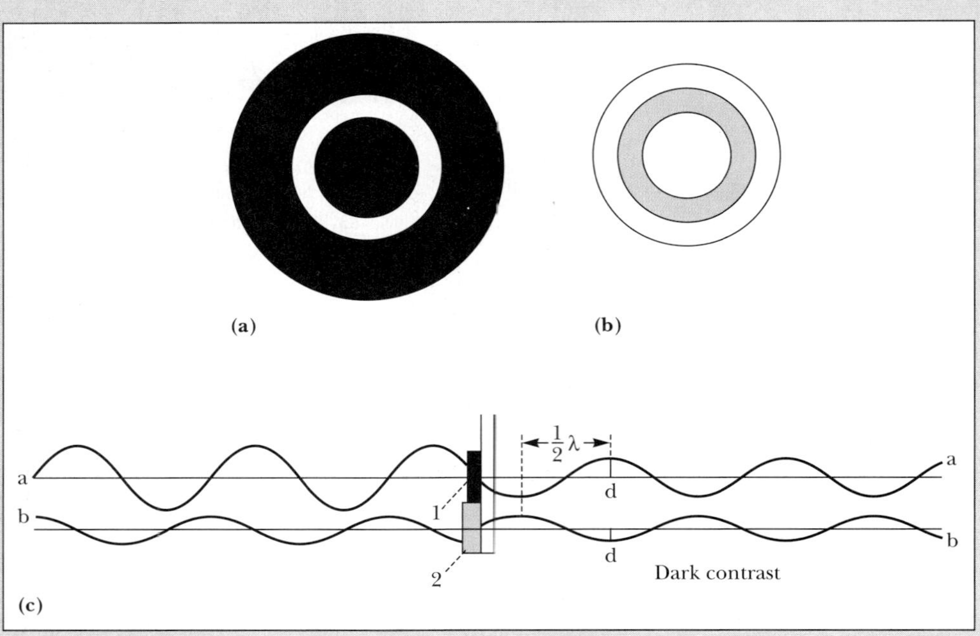

The optical components of the phase contrast microscope. (a) The condenser annulus, which is part of the condenser system, only permits light to pass through the ring or annulus. (b) The phase plate is located in the objective lens. (c) Undiffracted light a from the specimen passes through the phase annulus (1) and its intensity is retarded by 75 percent by use of a coating. Diffracted light b from the specimen is also altered as it passes through the phase plate where it falls on the complementary area (2). This light is retarded a half wavelength so that maximum interference occurs between the diffracted and undiffracted light from each point on the specimen. The result is an enhancement of contrast between the specimen and its background.

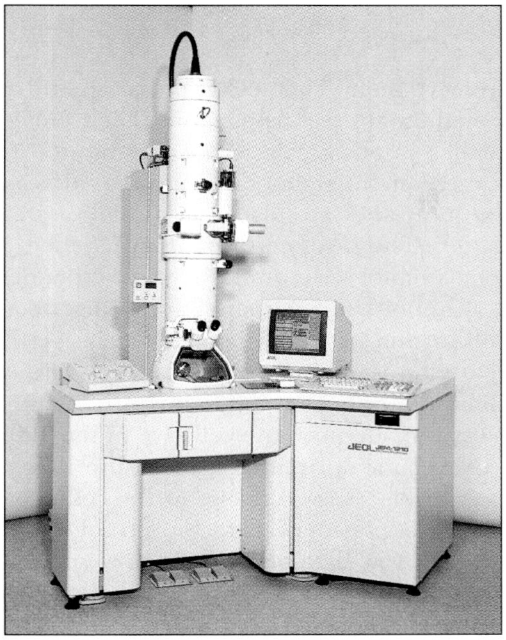

***Figure* 4.9** A transmission electron microscope (TEM). The large size of the instrument is due to the need to contain the entire device in a vacuum and to use electron magnets for lenses. The illuminating source is an electron gun which is located at the top of the microscope. The specimen is placed between the electron gun and the phosphorescent screen which is used to view the image. (Courtesy of JEOL USA, Inc.)

As a result the theoretical resolution of the electron microscope is about 2 Å (1 Å = 10^{-10} m), which is twice the diameter of the hydrogen atom.

In place of optical lenses, the TEM uses electromagnetic lenses to bend the electron beam for focusing. A vacuum is essential in permitting the flow of electrons through the lens system. Thus, the entire electron microscope must have an enclosed chamber and accompanying vacuum pumps. This makes the TEM a much larger instrument than the ordinary light microscope (Figure 4.9). The real image is formed by electrons bombarding a phosphorus screen, and the photographs, called **electron micrographs,** are taken using a camera mounted below the screen containing film sensitive to electron radiation. It is interesting to note that the lenses of the TEM are equivalent to that of the compound light microscope—a condenser lens, an objective lens, and a projector (ocular) lens. The only differences are that electrons are the illuminating source and magnets replace optical lenses (Figure 4.10).

Observation of Organisms Using the TEM

Figure 4.11 compares the appearance of bacterial cells observed in a light microscope and in the TEM. Note the

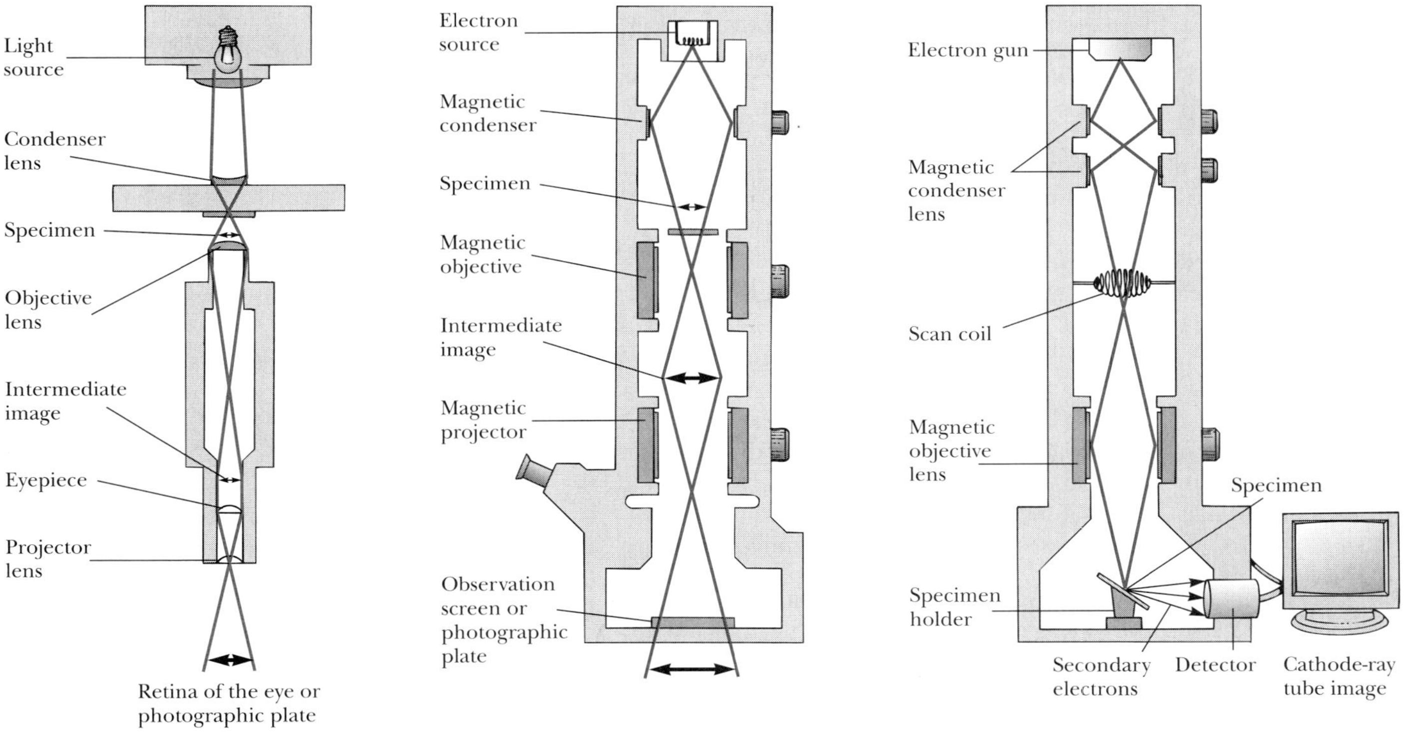

(a) Light microscope

(b) Transmission electron microscope (TEM)

(c) Scanning electron microscope (SEM)

***Figure* 4.10** A comparison of the illuminating paths in the light microscope with the transmission electron microscope (TEM) and the scanning electron microscope (SEM). Note that all three have illuminating sources, condenser lenses, and objective lenses. Since it is not possible to view a virtual image, a projector lens is needed in the TEM or when cells are photographed in the light microscope. The image of the SEM specimen is viewed in a cathode ray tube.

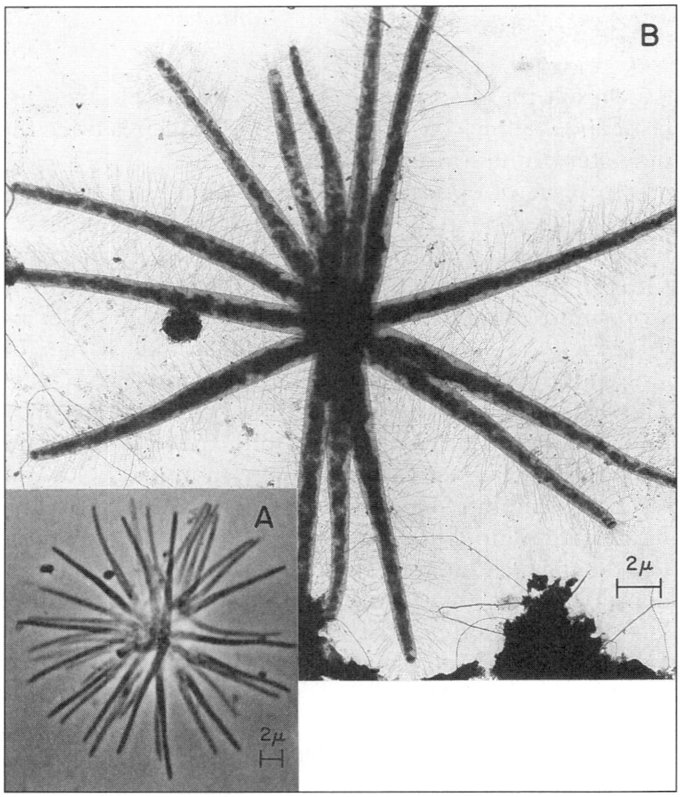

Figure **4.11** A comparison of the appearance of a large unidentified colonial bacterium from a lake by use of **(a)** the light microscope and **(b)** the electron microscope. This particular bacterium has many fine fibrils called fimbriae (see section on cell appendages later in Chapter 4 and Figure 4.55) that can be seen in the electron micrograph but cannot be resolved by the light microscope because they are too fine. (Courtesy of J. T. Staley and Joanne Tusov)

increased detail in the electron micrograph even though the magnification is about the same for both preparations. Since the preparation to be observed in the electron microscope is placed in a vacuum chamber, it is not possible to observe living cells with the TEM. In place of a glass slide, cell preparations are placed on a small screen (3.0 mm diameter, about the size of this "O") called a **grid.** A thin plastic film is placed over the grid to hold the cell preparation. In the simplest method of preparation, cells are placed on the plastic-coated grid and allowed to dry. This whole cell preparation is then stained with a heavy metal stain such as phosphotungstic acid or uranyl acetate, allowed to dry, and then placed in the electron microscope.

A more elaborate procedure is necessary for the observation of the intracellular structure of microorganisms by TEM. This procedure is called **thin sectioning.** Since the preparation will ultimately be observed in a vacuum, some procedure must be used to preserve the structure in the absence of water. This is accomplished through a process called **dehydration** and **embedding,** in which the cells are taken from their aqueous environment and transferred into a plastic resin. Although more complex, it is similar to embedding insects in plastic resins as hobbyists do to preserve them. First, cells are harvested from their growth medium by centrifugation. They are gradually dehydrated by transferring them step by step from the aqueous medium into ethanol solutions of increasing concentration until they are placed in a solution of 100 percent ethanol. The next step is to transfer them through an acetone-ethanol series until they are suspended in 100 percent acetone. Unlike ethanol and water, acetone is a plastic solvent and is miscible in plastic resins. At this point, the preparation is completely dehydrated. The next step is to embed the cells in a plastic mixture. Again, a series of transfers is made until the cells are in the 100 percent plastic resin. Time is permitted to ensure that the resin completely displaces the acetone in the cells. The preparation is **cured** (polymerized to form a solid) by heating at 60°C in an oven. The organism is now embedded.

The embedded cells are then sliced into **thin sections** with an **ultramicrotome.** The ultramicrotome is analogous to a meat slicer except it has a diamond knife and cuts extremely thin sections (about 60 nm thick). The thin sections are stained with heavy metals (lead citrate and uranyl acetate) to increase contrast. They are then placed on TEM grids and examined. A typical thin section is shown in Figure 4.12.

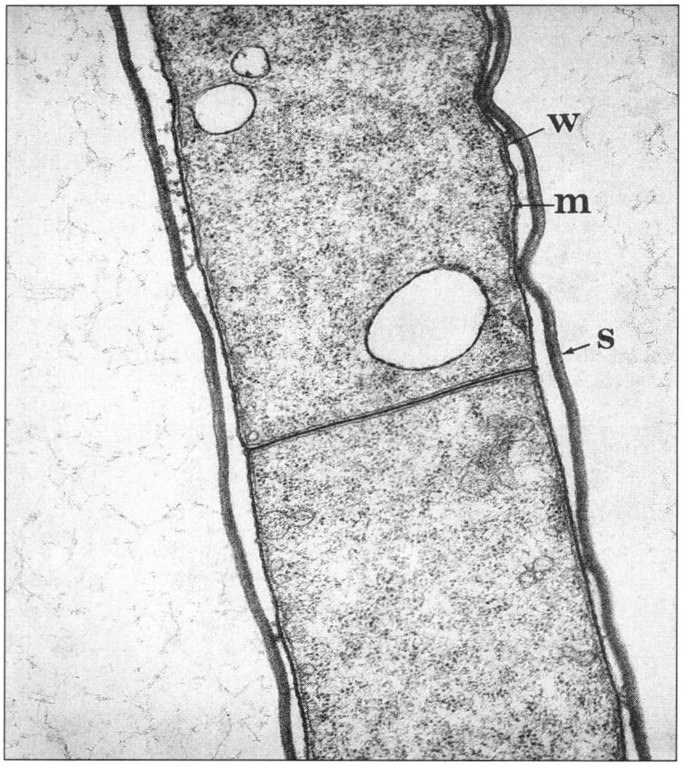

Figure **4.12** A thin section of a gram-negative bacterium, *Thiothrix nivea,* showing its sheath (S) lying outside the cell wall (W) and membrane (M). (Courtesy of Judith Bland and J. T. Staley)

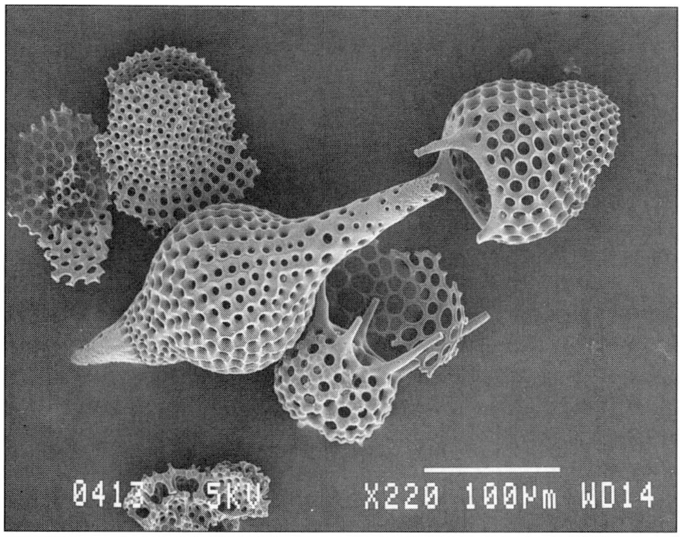

(a)

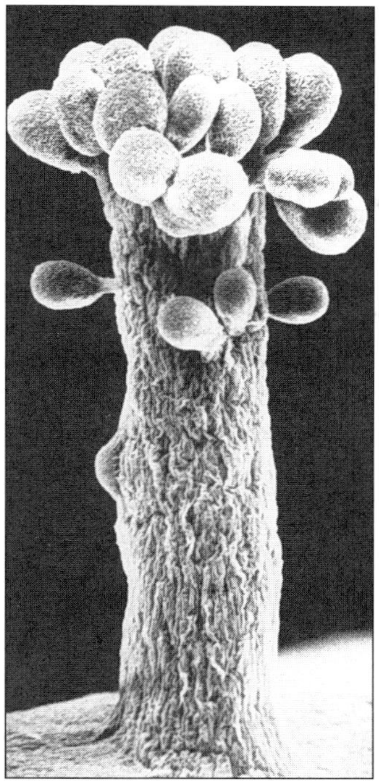

(b)

***Figure* 4.13** **(a)** Scanning electron micrograph (SEM) of a radiolarian, a protozoan with a siliceous shell. Bar is 100 μm. (Courtesy of Barbara Reine) **(b)** SEM of a fruiting structure of the myxobacterium *Chondromyces crocatus*. Fruiting structure is about one mm high. (Courtesy of Patricia Grilione and Jack Pangborn)

Scanning Electron Microscope (SEM)

As in the TEM, electrons are the illuminating source for the **scanning electron microscope** (SEM), but here the electrons are not transmitted *through* the specimen as in transmission electron microscopy. In SEM the image is formed by incident electrons scanned across the specimen that are back-scattered (reflected) from the specimen, as previously described for the fluorescent microscope. The reflected radiation is then observed with the microscope, in the same manner we observe objects illuminated by sunlight. Solid metal **stubs** are used to hold the preparations. Before the organism is viewed on the stub, it is dried by **critical point drying,** which is carefully controlled drying in which water is removed as vapor so that structural damage to cells is minimized. The specimen is then coated with an electron-conducting noble metal such as gold or palladium.

The SEM has a larger **depth of field** than the TEM so there is a greater depth through which the specimen remains in focus. Thus, all parts of even a relatively large specimen, such as a eukaryotic cell, will remain in focus when viewed by the SEM (Figure 4.13**a**). The result is an image that looks three-dimensional. Therefore, SEM offers the best procedure available to examine colonial forms of microorganisms such as the fruiting structures of the myxobacteria (Figure 4.13**b**).

An SEM can be equipped with an X-ray analyzer, enabling one to determine the elemental composition of microorganisms or parts of them. Elements with atomic weights greater than 20 can be assayed in a semiquantitative manner using this instrument (Figure 4.14).

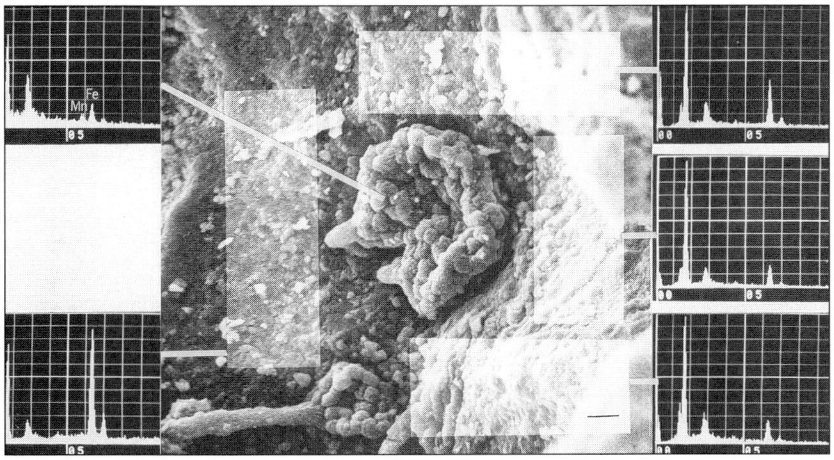

***Figure* 4.14** SEM of a fungal microcolony growing on a desert rock. X-ray analyses (shown as the five graphs) indicate that manganese is found only inside the colony, whereas iron and other elements are found in all sectors bordering the colony as well as in the colony. The data suggest that the colony is accumulating manganese. Bar equals 10 μm. (Courtesy of F. Palmer and J. T. Staley)

The variety of microscopes and associated procedures just described were instrumental in the developing field of microbiology. We now discuss the various morphological attributes of prokaryotes, at both the organismal and subcellular level—their chemical composition and their function for the organism. Keep in mind that most bacteriologists study the behavior of microorganisms in the laboratory, although their ultimate interest lies in understanding what the function of a given structure is in the natural environment in which the organism lives.

Morphology of Eubacteria and Archaea

Bacteria come in a variety of simple shapes. Most are single-celled but some are multicellular forms consisting of numerous cells living together. Their shapes and types of cell division are discussed below.

Unicellular Organisms

The simplest shape for a single-celled bacterium is the sphere (Figure 4.15**a**). Unicellular spherical organisms are called **cocci.** Some cocci grow and divide along only one axis. If they remain attached after cell division, this results in the formation of a **chain** of cells of various lengths. A **diplococcus** is a very short chain of only two cells, whereas a **streptococcus** can contain many cells in its chain (Figure 4.15**b**).

Some cocci divide along two perpendicular axes in a

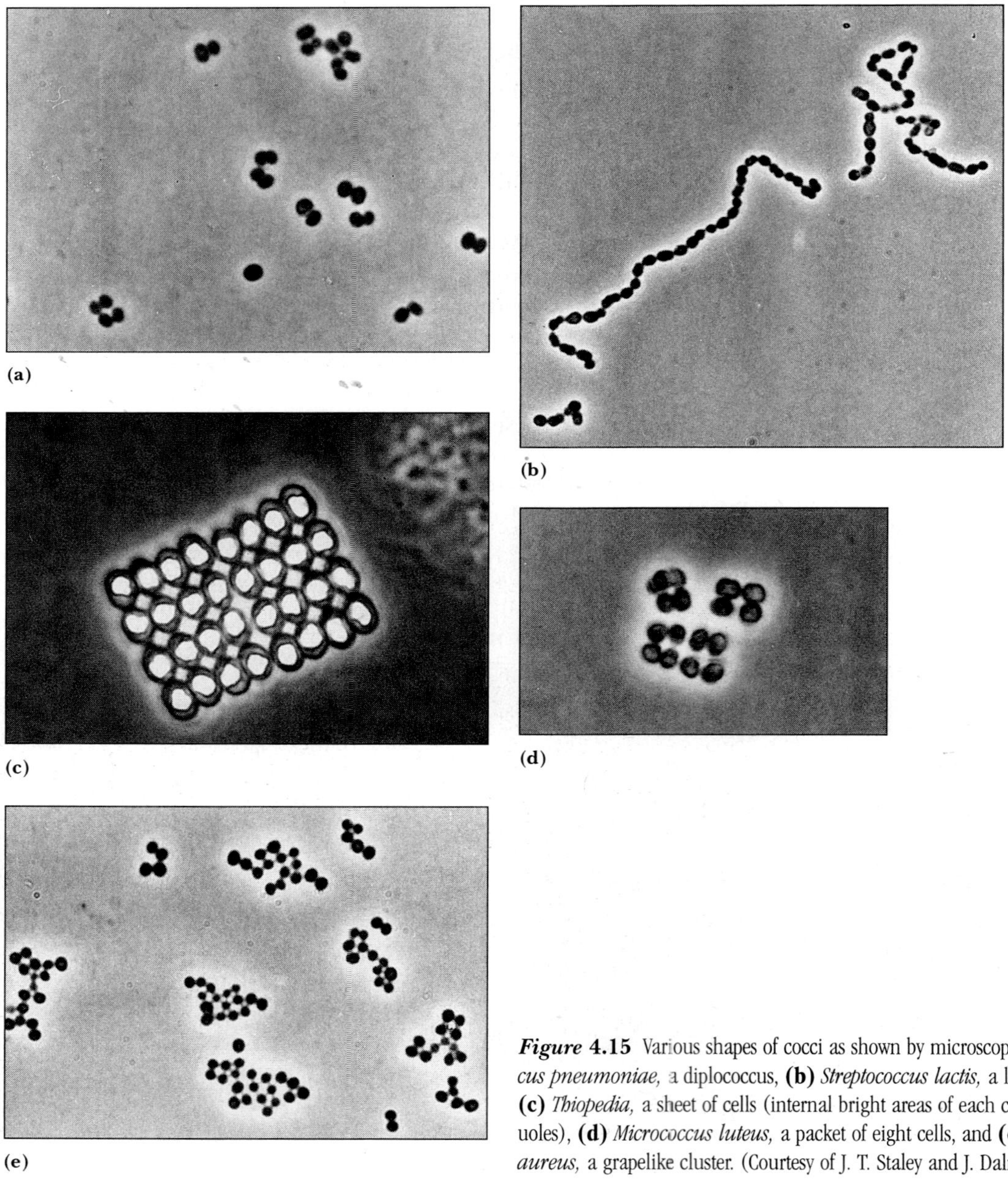

***Figure* 4.15** Various shapes of cocci as shown by microscopy: **(a)** *Streptococcus pneumoniae,* a diplococcus, **(b)** *Streptococcus lactis,* a longer chain, **(c)** *Thiopedia,* a sheet of cells (internal bright areas of each cell are gas vacuoles), **(d)** *Micrococcus luteus,* a packet of eight cells, and **(e)** *Staphylococcus aureus,* a grapelike cluster. (Courtesy of J. T. Staley and J. Dalmasso)

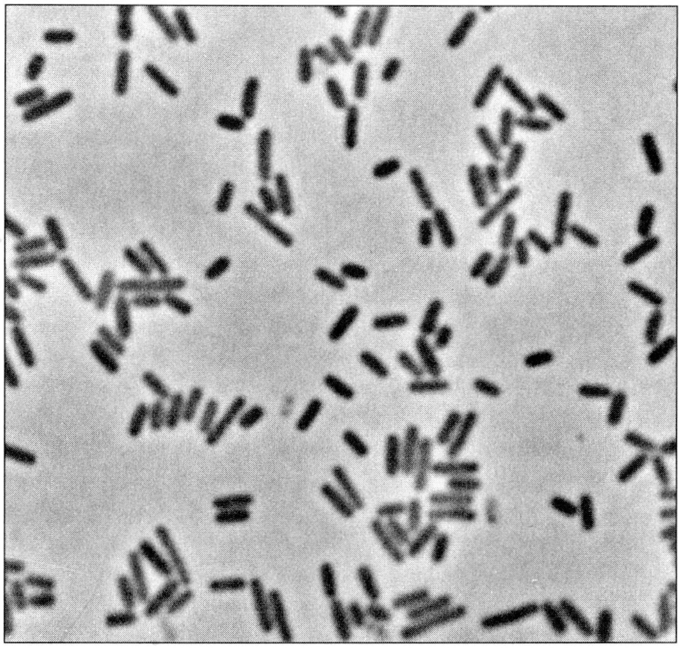

***Figure* 4.16** A unicellular rod, a *Pseudomonas* sp., as shown by phase contrast microscopy. (Courtesy of J. T. Staley)

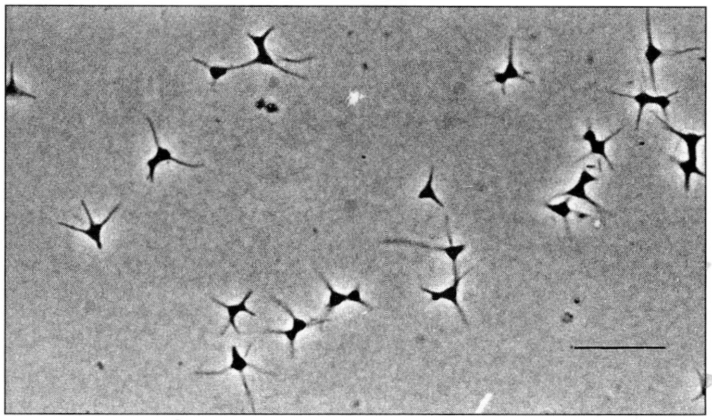

***Figure* 4.18** A star-shaped bacterium, *Ancalomicrobium adetum*. Bar equals 10 μm. (Courtesy of J. T. Staley)

regular fashion, which results in the formation of a **sheet** of cells (Figure 4.15**c**). Other cocci divide along three perpendicular axes, resulting in the formation of a **packet** or **sarcina** of cells (Figure 4.15**d**). Finally, random division of a coccus will produce a grapelike cluster of cells referred to as a **staphylococcus** (Figure 4.15**e**).

The most common shape in the prokaryotic world is not a sphere, however. It is a cylinder with blunt ends referred to as a **rod** or **bacillus** (Figure 4.16). Some rods remain attached to one another after division across the short or transverse axis of the cell, forming a chain **(Box 4.3).**

A less common shape for unicellular bacteria is a helix. A very short helix (less than one wavelength long) is called a **bent rod** or **vibrio** (Figure 4.17**a**). A longer helical cell is a **spirillum** (Figure 4.17**b**) if the cell shape is rigid and unbending, or a **spirochete** if the organism is flexible and changes its shape during movement (Figure 4.17**c**). Variations of these common shapes of unicellular bacteria also exist. For example, some bacteria produce appendages that are actually extensions of the cell, called **prosthecae,** that give them a star-shaped appearance (Figure 4.18). The morphological diversity of Eubacteria and Archaea will be discussed further in Chapters 19 to 22, where individual genera will be treated.

Multicellular Organisms

Numerous prokaryotic organisms exist as multicellular forms. One group exemplifying this is the **streptomycetes** or **actinomycetes.** They produce an extensive **mycelium**

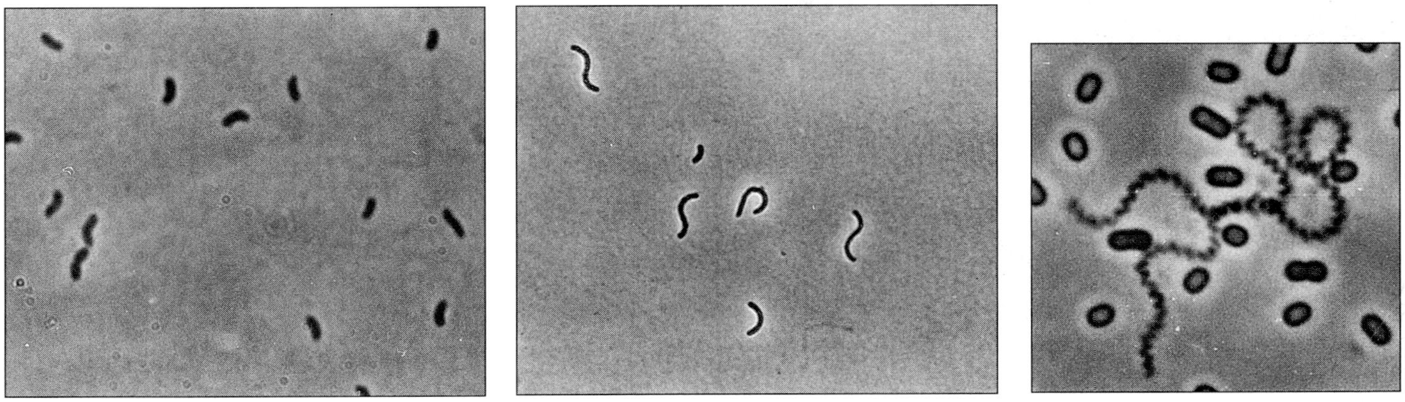

(a) (b) (c)

***Figure* 4.17** Various curved shapes of bacteria as shown by microscopy: **(a)** a *Vibrio* species, **(b)** a *Spirillum* species, and **(c)** a large spirochete from a natural mud sample that also contains many rod-shaped bacteria. (Courtesy of J. T. Staley)

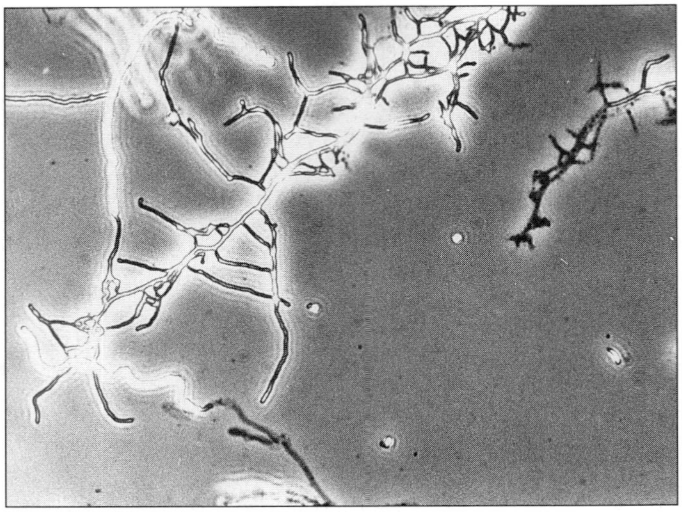

***Figure* 4.19** A *Streptomyces* species illustrating the complex network of filaments called a mycelium. (Courtesy of J. T. Staley and J. Dalmasso)

(network of branching filaments) as they grow (Figure 4.19), much like that of the fungi (see Chapter 1). In fact, prior to 1950, these bacteria were classified with the filamentous fungi.

Another common multicellular shape is the **trichome,** which is frequently encountered in the cyanobacteria (Figure 4.20**a**). Although a trichome superficially resembles a chain, there is a much closer spatial and physiological relationship among adjoining cells in a trichome. Motility and other functions result from the concerted action of all cells of a trichome. And some cells in the trichome may have specialized functions that benefit the entire trichome. For example, the heterocyst (Figure 4.20**b**) is the site of nitrogen fixation in some filamentous cyanobacteria (Chapter 21).

Cell Division of Eubacteria and Archaea

Prokaryotes maintain their shapes during the process of reproduction—an asexual process in which a single organism divides to produce two progeny. For unicellular prokaryotes there are two ways in which this may be accomplished, either by binary transverse fission or by budding.

Binary Transverse Fission

The most common type of bacterial cell division is **binary transverse fission.** In this process, the cell (which may be a coccus, rod, spirillum, or other shape) elongates as growth occurs along its longitudinal axis (Figure 4.21). When a certain length is reached, a **septum** (wall structure) is produced along the transverse axis of the cell midway between the cell ends. When the septum has completely formed, the two cells that result become separate entities. This process is called **binary fission** because two cells are produced by a division or "splitting" of one orig-

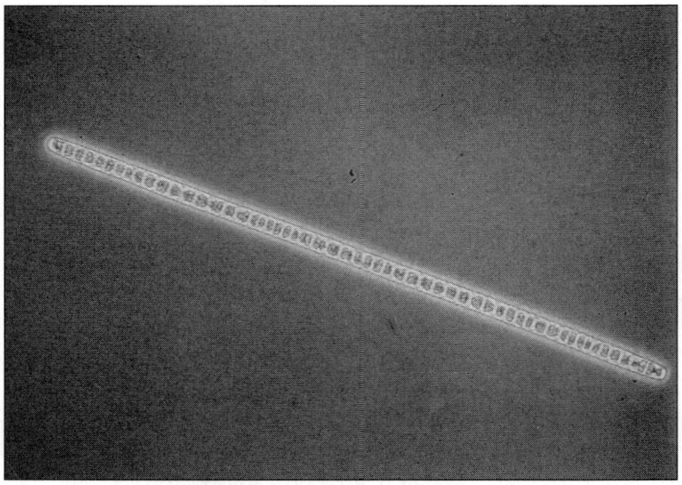

(a)

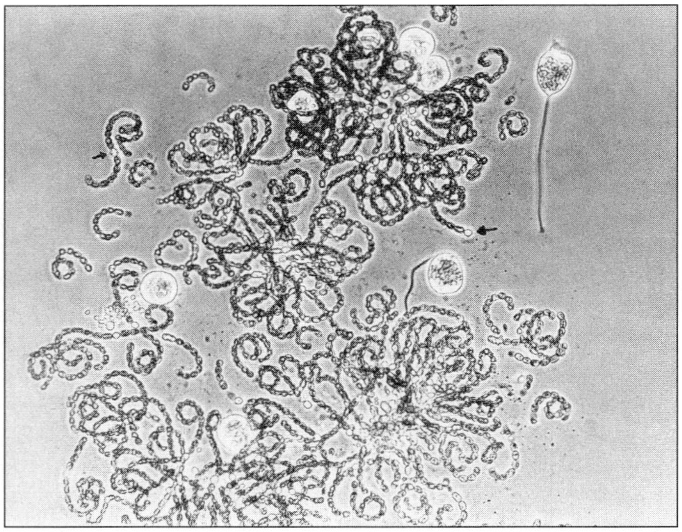

(b)

***Figure* 4.20** **(a)** A multicellular filamentous cyanobacterium, a species of *Oscillatoria* showing the close contact between cells in the trichome, and **(b)** an *Anabaena* sp. from a natural lake sample growing in association with a stalked protozoan. Some of the curved trichomes have a nonpigmented specialized cell, called a heterocyst (see arrows), which is the site of nitrogen fixation. (Courtesy of J. T. Staley)

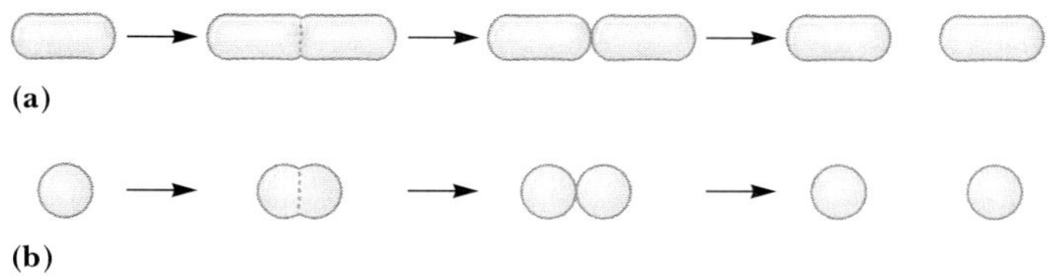

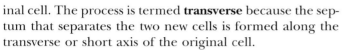

***Figure* 4.21** A diagram illustrating typical binary transverse fission in **(a)** a rod-shaped bacterium and **(b)** a coccus.

inal cell. The process is termed **transverse** because the septum that separates the two new cells is formed along the transverse or short axis of the original cell.

In binary transverse fission, DNA replication precedes septum formation. The two resulting cells are mirror images of one another. Analyses of cell wall components of dividing cells indicate that the chemical constituents of the original "mother" cell wall are equally shared in the cell walls of the two resulting "daughter" cells. In typical filamentous multicellular organisms the cells divide by binary transverse fission and the filament or trichome ultimately separates into two separate filaments.

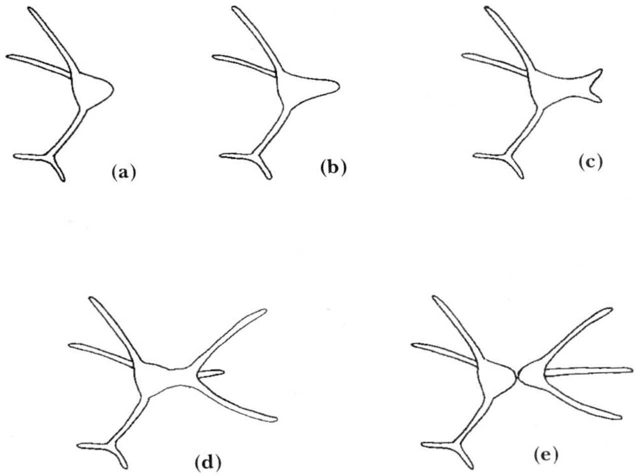

***Figure* 4.22** Bud formation in *Ancalomicrobium adetum.* **(a)–(c)** A small protuberance (bud) on this prosthecate bacterium enlarges as growth proceeds. **(d)** When the bud becomes sufficiently mature, **(e)** it separates from the mother cell. Note that the mother cell maintains its shape throughout the divisional process. It is now able to produce another bud from the same location on the cell surface and repeat the division process again, assuming nutrients are available for growth. (Courtesy of J. T. Staley)

Budding

Budding or bud formation is a less common form of cell division among prokaryotic organisms. As in binary transverse fission, this is an asexual division process that results in the formation of two cells from the original cell. In the budding process, however, a small protuberance, a bud, is formed on the cell surface. The protuberance enlarges as growth proceeds. In time, the bud becomes sufficiently large and mature to separate from the mother cell (Figure 4.22).

A number of differences exist between binary transverse fission and budding. During binary transverse fission, symmetry of the cell with respect to the longitudinal and transverse axes is maintained throughout the entire process of division (Figure 4.21). This results in the mother cell producing two daughter cells and losing its identity in the process. In the budding process, however, symmetry with respect to the transverse axis does not occur during the process of division (Figure 4.22). Also, in contrast to binary transverse fission, most of the new cell wall components are used in the synthesis of the bud instead of being divided equally between the two progeny cells. The result is that, during budding, the mother cell produces one daughter cell while the mother cell retains its identity generation after generation. Thus, a primitive "aging" process occurs in the budding bacteria. It is not yet known whether there is a limit to the number of buds a mother cell might produce during its existence.

Fragmentation

Another type of cell division process occurs with the mycelial bacterial group called the actinomycetes or streptomycetes. These organisms form "multinucleate" filaments that lack septa between cells. Such filaments are referred to as being **coenocytic** and are analogous to that found in some fungi (Chapter 1). Some actinomycete species undergo a **multiple fission** process called **fragmentation** in which the filament develops septations between the nuclear areas, resulting in the simultaneous formation of numerous unicellular rods.

Fine Structure, Composition, and Function of Eubacteria and Archaea

The remainder of this chapter is devoted to the description of various prokaryotic structures, their chemical composition, and their functions. The terms "fine" or "ultra" structure refer to subcellular features that are best observed using the electron microscope. Studies of the fine structure of microbial cells began in the 1950s and 1960s when electron microscopy procedures were perfected. Scientists used a combination of procedures to "break open" or **lyse** cells, followed by centrifugation to separate the various subcellular components. These components were

BOX 4.5 METHODS & TECHNIQUES

Cell Fractionation, Separation, and Biochemical Analyses of Cell Structures

In order to determine the composition of various components of the cell, scientists separate these components from the rest of the cell, purify them, and analyze them biochemically. The initial step is to break open the cells, for which either chemical or physical procedures can be used. For example, chemical procedures include lysis of the cells by enzymes or detergents. Physical methods include ultrasound (called **sonication** by biologists), in which high-frequency sound waves vibrate cells until they break. A sonicator probe is inserted into a cell suspension for this purpose as shown in the illustration. Alternatively cells can be broken by passing them, while frozen in thick suspensions, through a small orifice (French press) at high pressure.

Once the cells have been broken, the various structural fractions are separated from one another, usually by centrifugation. Two types of centrifugation can be used.

In **differential** or **velocity** centrifugation, fractions are separated by the length of time they are centrifuged at different gravitational forces. More dense structures such as unbroken cells or bacterial endospores, cell membranes, or cell walls will sediment at low speeds (15,000 × g for 10 min). The supernatant can then be removed and centrifuged at higher speed to spin out less dense structures. For example, ribosomes will sediment only after centrifugation at higher speeds (100,000 × g for 60 min). The remaining material that does not sediment in the centrifuge tube contains soluble constituents such as cytoplasmic enzymes.

Alternatively, **buoyant density** or **density gradient centrifugation** can be used to separate the various cell fractions. In this procedure a density gradient is set up in the centrifuge using different concentrations of a solute, such as sucrose. The sample is layered on the surface and centrifuged at moderate speed until the cellular fractions equilibrate with the layer in the gradient that has the same buoyant density. They can then be removed with a pipette and studied as purified fractions.

When the cell fraction that is of interest to the microbiologist is separated and purified by the procedures outlined above, then the material can be analyzed chemically. The electron microscope is used to check the identity and purity of the material at each step in the process.

Cells in suspension are disrupted by sonication and the various cell fractions are separated by either differential centrifugation or by buoyant density centrifugation.

purified and then analyzed biochemically. The electron microscope was used at various steps in the procedure to ensure the identity and purity of the structures **(Box 4.5).**

In this section we begin with internal structures found in the cytoplasm and then consider the outer layers of the cell.

Internal Structures

Foremost among the intracellular material of all cells is their DNA, which constitutes the hereditary material of the cell. DNA and other intracellular components commonly found in many different Eubacteria and Archaea are discussed individually below.

DNA

As discussed in Chapter 3, the deoxyribonucleic acid (DNA) of prokaryotes is a circular, double-stranded helical molecule. The two strands are held together by hy-

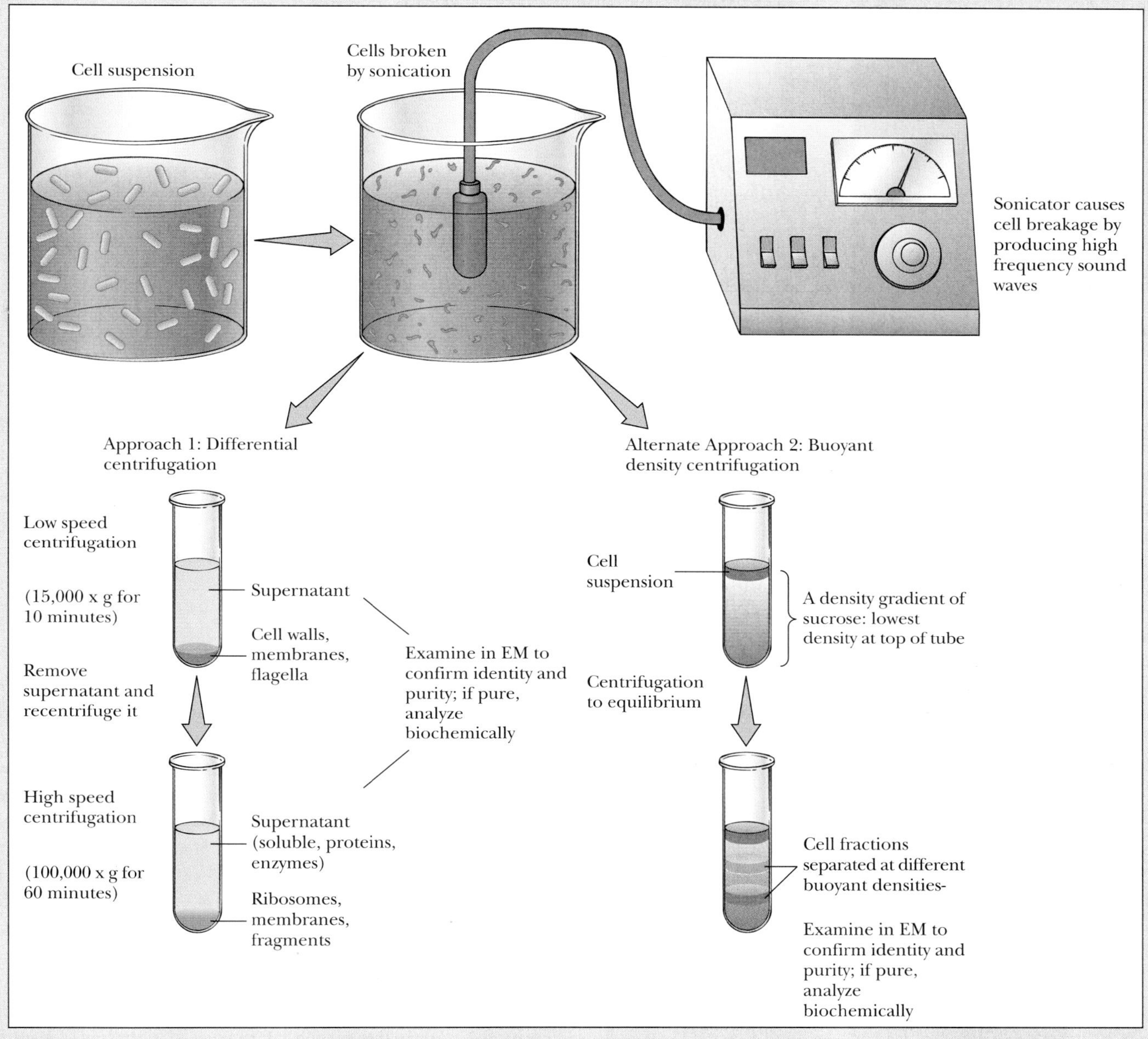

drogen bond formation with the bases on one strand forming hydrogen bonds with bases of the opposite strand. Thus, adenine and thymine pair with one another and cytosine and guanine pair with one another (see Figures 3.7 and 3.8).

The DNA appears as a fibrous material in the cytoplasm when cells are viewed in thin sections (Figure 4.23). As noted in Chapter 1, the DNA of prokaryotes is not bound by any membrane. Therefore it does not appear in a confined area within the cell, but rather as a somewhat diffuse, dispersed fibrous area within the cell. For this reason it is not called a nucleus, but rather a **nucleoid** or **nuclear area.**

If gentle conditions are used to lyse bacterial cells (like eggs, cells can be broken carefully—in such a manner that the cytoplasm can be freed from the cell membrane and wall), the DNA is released and appears as a coiled structure spilled from the cell (Figure 4.24). When stretched out, the length of the DNA molecule is about 1 mm, about a thousand times longer than the 1 to 3 μm length of the

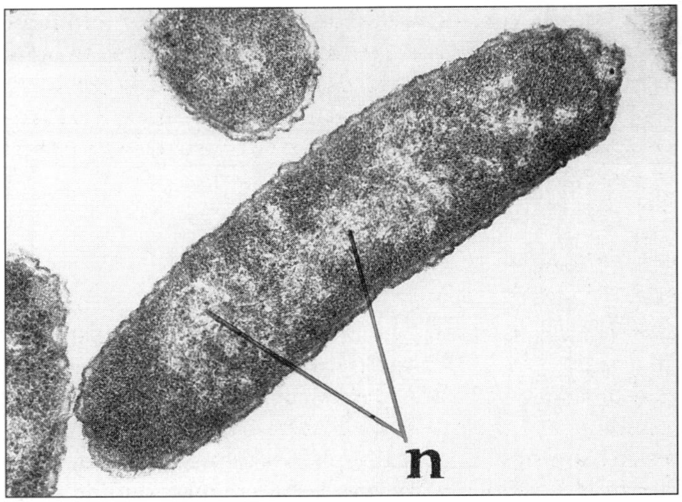

***Figure* 4.23** A thin-section through *Salmonella typhimurium* showing the appearance of the nuclear material (n). (Courtesy of J. Lara)

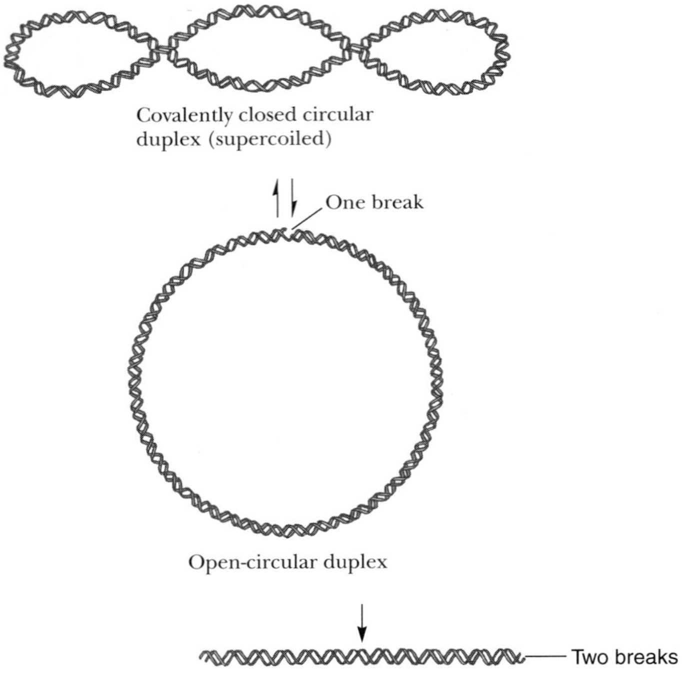

***Figure* 4.25** A diagram illustrating the phenomenon of supercoiling of DNA. A single break in one strand results in complete uncoiling. If both strands are broken, the molecule becomes a linear duplex.

typical bacterial cell! In order to package all of this material within the cell, the DNA molecule is tightly wound in supercoils (Figure 4.25). Special enzymes are responsible for supercoiling and controlling the unwinding of the DNA during DNA synthesis **(DNA replication)** and in **transcription** (production of RNA from a DNA template; see Chapter 13).

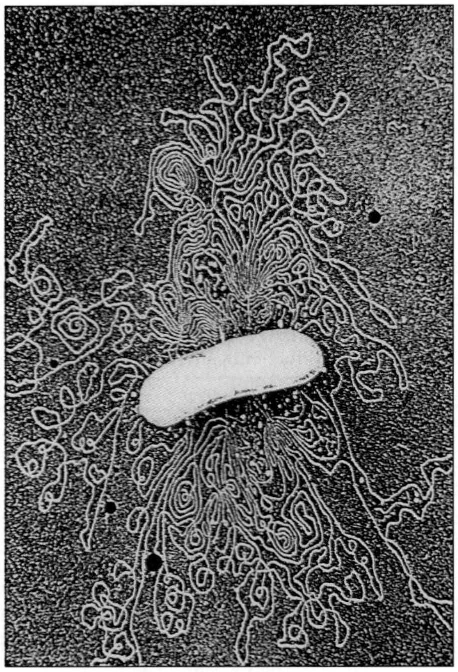

***Figure* 4.24** A photomicrograph showing DNA strands released from a lysed cell. (© VU/K.G. Murti)

The molecular weight of the DNA molecule of prokaryotes ranges from about 10^9 to 10^{10}. The typical bacterium contains about 4×10^6 nucleotide base pairs (4 megabase pairs or 4 mgb). This is considerably smaller than the size of eukaryotic genomes, but larger than those of viruses. Bacterial cells may have more than one copy of the DNA molecule. For example, when the cell is growing and dividing rapidly, two to four more copies, or partial copies, may be present.

In addition to the genomic DNA molecule of the cell, bacteria often contain other, extrachromosomal molecules of DNA called **plasmids.** These, too, are double-stranded DNA molecules (see Chapter 15), but do not carry genetic material that is essential to the organism, although they may contain features that enhance the survivability. For example, some soil bacteria carry a plasmid having a gene that allows them to degrade naphthalene—which is not useful to the bacterium unless naphthalene is in the immediate environment of the organism.

The primary function of the prokaryotic genome is to store hereditary information, which is stored in its genes, as discussed in Chapter 13. Furthermore, genetic material can be transferred from some bacteria to others, under the appropriate conditions. This can be accomplished by three different processes, depending on the prokaryote involved. Thus, **transduction** is a process in which

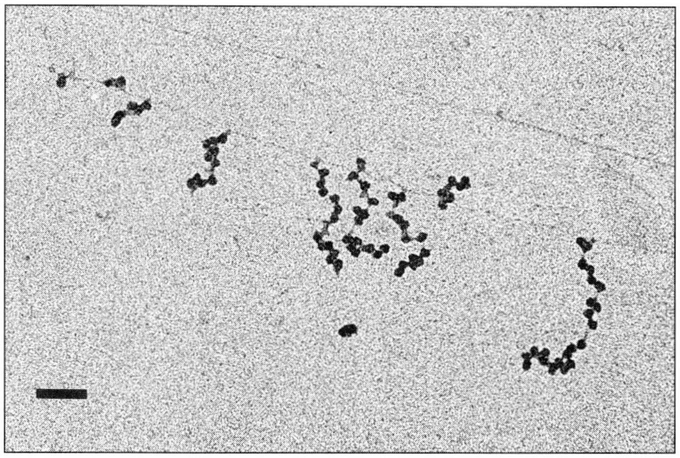

***Figure* 4.26** An electron micrograph showing ribosomes from *E. coli* in action. Filaments are DNA strands. Ribosomes are attached as polyribosomes to mRNA, which they are translating to produce protein. Bar is 0.1 μm. (Courtesy of Oscar L. Miller)

prokaryotic viruses are involved in transferring DNA from one organism to another. **Transformation** occurs when DNA, released by lysis (cell breakage) from one organism into the environment, is taken up by another organism. In some bacteria, DNA can also be directly transferred between two organisms by **conjugation,** in which cell-to-cell contact occurs between two appropriate bacterial strains.

Ribosomes

As mentioned in Chapter 1, ribosomes are small structures that carry out protein synthesis (Figure 4.26), a process referred to as **translation** (Chapter 13), in which the messenger RNA (mRNA) carries the genetic information from the genome to the ribosome.

At high magnification the ribosome of prokaryotes can be seen to comprise two subunits, the small 30S subunit and the larger 50S subunit (Figure 4.27). Please note that the 30S and 50S ribosome subunits comprise a 70S ribosome indicating that the Svedberg coefficient is not directly related to molecular mass, but to the density of particles in ultracentrifugation. Likewise, the eukaryotic 80S ribosome consists of a small 40S subunit and a larger 60S subunit. Ribosomes consist of both protein and ribonucleic acid (RNA). Figures 4.27**b** and **c** show the RNA and protein components of each of these subunits from the 70S and 80S ribosomes, respectively.

Even though they have the same sedimentation coefficient, some differences exist in the structure and composition between eubacterial and archaean ribosomes. As a result, differences are observed when these organisms are exposed to the same **antibiotic** (a substance produced by one organism that inhibits or kills other organisms). For example, certain antibiotics, including chloramphenicol and the aminoglycosides, affect ribosome activity (and therefore inhibit protein synthesis) in Eubacteria, but have no adverse effect on protein synthesis in the Archaea. Likewise, eukaryotic ribosomes are not sensitive to some of the antibiotics that affect Eubacteria.

Gas Vesicles

One of the most unusual structures found in bacteria is produced by some aquatic prokaryotes. These are special membrane-bound structures called **gas vesicles.** They provide buoyancy to many aquatic prokaryotes and are not found in any other life forms. When bacteria that have gas vesicles are observed in the phase microscope, they contain bright, refractile areas with an irregular outline (Figure 4.28**a**). These bright areas were initially called gas vacuoles—the numerous individual gas vesicles that comprise a gas vacuole could not be seen prior to the introduction of the electron microscope. However, when gas vacuolate cells are viewed with the transmission electron microscope, the vacuoles are found to consist of numerous subunits called gas vesicles (Figure 4.28**b**).

Gas vacuoles are found in widely disparate organisms. They are common in photosynthetic groups such as the cyanobacteria, proteobacteria, and green sulfur bacteria. They are also found in heterotrophic bacteria such as the genera *Ancylobacter* as well as the prosthecate genera *Prosthecomicrobium* and *Ancalomicrobium.* The anaerobic gram-positive genus, *Clostridium,* also has gas vacuolate strains. Finally, some Archaea—including members of the genus *Methanosarcina,* a methanogen, and *Halobacterium*—produce gas vacuoles.

Gas vesicles have been isolated from bacteria and studied in the isolated state. They are obtained by gently lysing cells to release the vesicles. The vesicles are then separated from cellular material by low-speed differential centrifugation **(Box 4.5).** Cell material is relatively dense and is spun to the bottom of the centrifuge tube, while the buoyant gas vesicles float to the surface. The purified vesicles have a very distinctive shape. They appear as cylindrical structures with conical endpieces (Figure 4.29). The size varies from about 30 nm in diameter in some species to about 300 nm in others. They can exceed 1000 nm in length, depending upon the organism.

Each vesicle consists of a thin (about 2 nm thick) membrane surrounding a hollow space. The membrane is composed of one predominant protein whose repeating subunits have a molecular mass of about 7500 daltons. Amino acid analyses of this protein from different prokaryotes indicate that the composition of the vesicle is highly uniform from one organism to another. About half of the protein consists of hydrophobic amino acids (such as alanine, valine, leucine, and isoleucine). It is thought that the hydrophobic amino acids are located on the inside of the

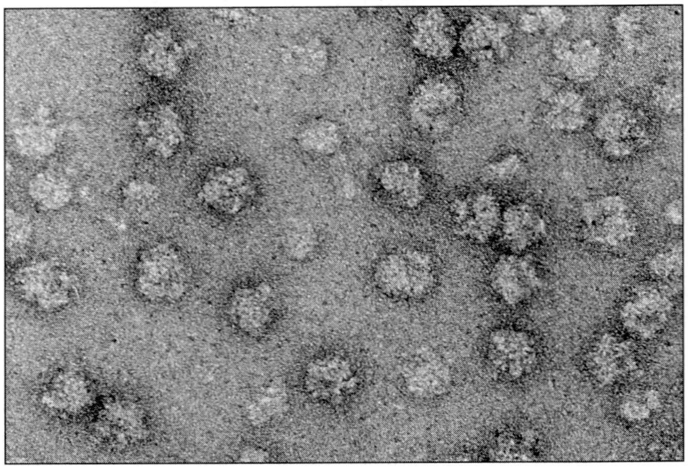

(a)

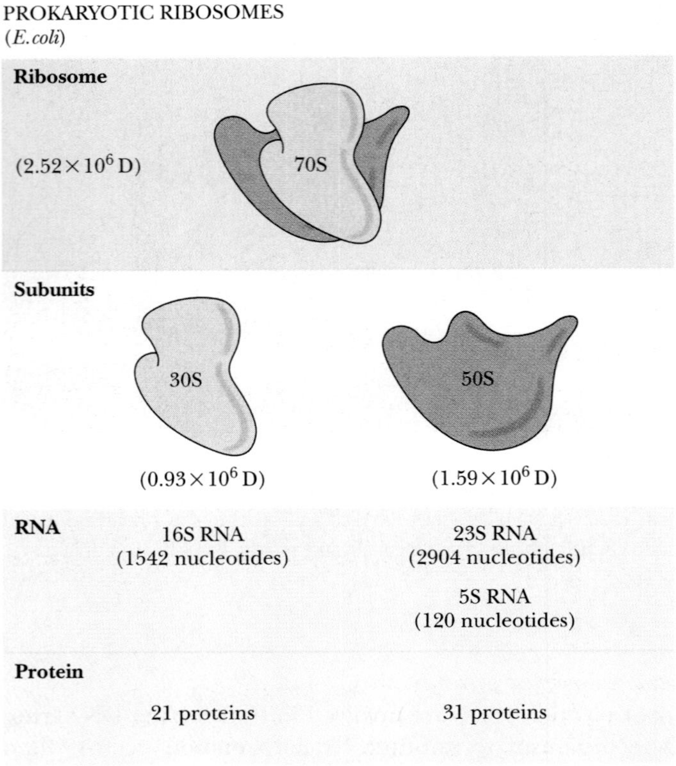

(b)

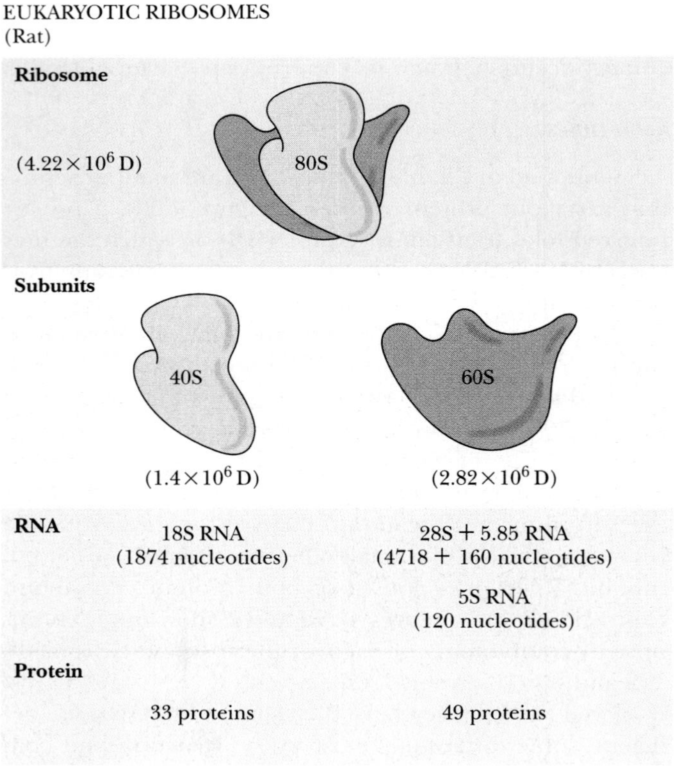

(c)

***Figure* 4.27** **(a)** A high-resolution electron micrograph of 70S ribosomes also showing large (50S) and small (30S) subunits (© James Lake, UCLA); **(b)** the protein and ribosomal RNA composition of the 70S ribosome of *E. coli,* and **(c)** the protein and ribosomal RNA composition of a eukaryotic organism (rat). D refers to the unit of mass called a dalton which is approximately equivalent to that of the hydrogen atom.

membrane and that their presence prevents water from entering the interior of the vesicle. In contrast, gases freely diffuse through the membrane and are thus the sole constituents of the interior.

It is important to recognize that the gas vacuole *cannot* store gases like a balloon, but that gases freely diffuse through the membrane. Therefore, the membrane does not maintain its shape because it is inflated, but instead is a rigid protein framework that is impermeable to water. Because all gases freely diffuse through the gas vesicle membrane, the gases found in the interior of the vesicles are those present in the environment of the organism.

The primary function of gas vesicles is to provide buoyancy for aquatic prokaryotes **(Box 4.6).** The density of the organism is reduced when the cell contains these structures, thus permitting the organism to be buoyant in planktonic habitats. The mechanisms by which bacteria regulate gas vacuole formation in nature are only poorly understood. Some cyanobacteria can descend in the environment by producing more dense storage materials such as polysaccharides, or by actually collapsing weaker vesicles. Depending upon their requirements for light, oxygen, and hydrogen sulfide, various bacterial groups are found in different strata in lakes during summer thermal stratification (Chapter 24). The vertical position of the gas vacuolate species depends on their cell density, which is determined by the proportion of cell volume occupied by their gas vesicles at a particular time.

Intracellular Reserve Materials

A variety of organic and inorganic materials are stored in-

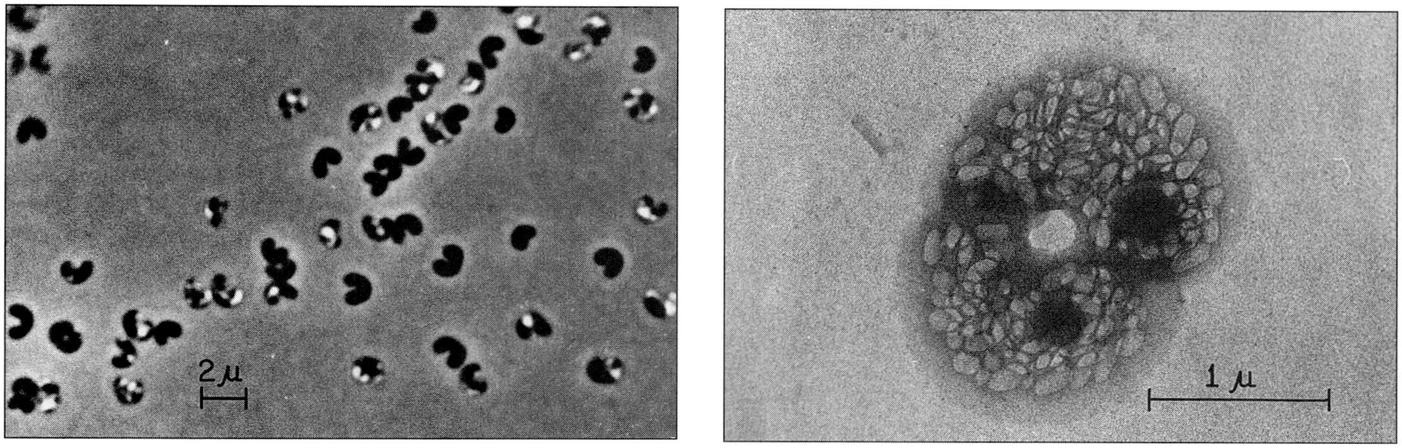

(a) (b)

***Figure* 4.28 (a)** Phase photomicrograph of *Ancylobacter aquaticus,* a vibrioid bacterium with gas vacuoles. Bar equals 2.0 μm. **(b)** Electron micrograph of *Ancylobacter aquaticus* cell during cell division showing the numerous transparent gas vesicles which comprise its gas vacuole. Bar is 1.0 μm. (Courtesy of M. van Ert and J. T. Staley)

tracellularly by prokaryotic organisms as nutrient reserves. Almost all are stored as polymers, thereby maintaining the internal osmotic pressure at a low level (see the section, Cell Walls).

ORGANIC COMPOUNDS Glycogen and starch are often stored in prokaryotic organisms. These are in the form of polymers of glucose linked together primarily by α 1 $\rightarrow$ 4 linkages (Chapter 11). These storage materials cannot be seen using a light microscope, but when observed with the electron microscope, they appear as either small, uniform granules in some cyanobacteria (see Figure 21.17) or as larger spheroidal structures in heterotrophic bacteria. Glycogen and starch are stored when organic carbon is abundant and nitrogen is limiting for growth. These substances can then be degraded as energy and carbon sources when conditions are favorable. Interestingly, glycogen is also formed by animal cells, and starch is formed by eukaryotic algae and higher plants.

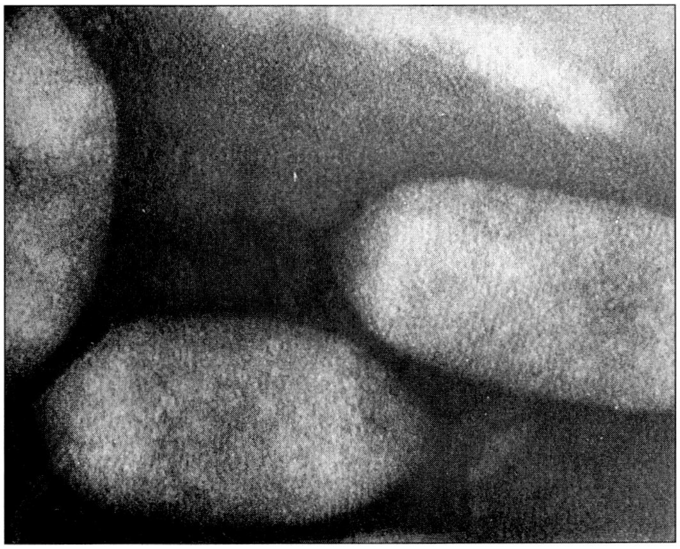

***Figure* 4.29** Gas vesicles isolated from *Ancylobacter aquaticus.* (Courtesy of J. T. Staley and A. E. Konopka)

Unlike glycogen, poly-β-hydroxybutyric acid (PHB) and related polymeric acids are lipids that appear as visible granules in bacteria when viewed with a light microscope. In phase microscopy, they appear as bright, refractile, spherical granules (Figure 4.30). As with glycogen, PHB granules are synthesized usually during periods of limiting nitrogen in environments that have excess utilizable organic carbon. In contrast to glycogen and starch, PHB is not found in eukaryotic organisms.

PHB is a polymer of β-hydroxybutyric acid and is synthesized from acetyl coenzyme A (see Chapter 11). The native granules are enclosed in a nonunit membrane protein, which may be responsible for their synthesis and/or degradation.

The only organic nitrogen polymer stored by prokaryotes is **cyanophycin.** As implied by the name, these granules occur only in cyanobacteria. They consist of a copolymer of aspartic acid and arginine and have a distinctive appearance when viewed by thin section with an electron microscope (see Figure 21.16).

INORGANIC COMPOUNDS Inorganic compounds are also stored by some bacteria. **Volutin,** or **metachromatic granules,** consists of polymers of phosphate covalently linked to form a linear **polyphosphate** molecule. These appear as dark

BOX 4.6 MILESTONES

The Hammer, Cork, and Bottle Experiment

In the early 1900s C. Klebahn conducted an important but simple experiment called the "hammer, cork, and bottle" experiment which provided the first evidence that the gas vacuoles of cyanobacteria actually contained gas. For the experiment, cyanobacteria containing the purported gas vacuoles were taken from a bloom from a lake. Microscopic examination showed that they contained bright areas indicative of gas vacuoles. Samples were placed into two bottles, one to be used as a control for the experiment.

(a) A cork was then placed in one bottle and secured in such a way that no air space was left between the cork and the water (b). Then, a hammer was used to strike a sharp blow on the suspension of cyanobacteria in the corked bottle The blow briefly increased the hydrostatic pressure of the water in the experimental bottle. The cyanobacteria in that bottle soon sank to the bottom, whereas those in the control bottle floated to the top. Furthermore, an air space formed between the water and the cork in the experimental bottle. When the cells from this bottle were subsequently examined in the microscope, no bright areas remained in the cells thus illustrating that the cells had lost their buoyancy and their gas vacuoles at the same time. The air space that had collected in the experimental bottle was due to the gas released from the broken vesicles and contained by the cork.

(a) (b)

(a) The two bottles on the left contain gas vacuolate cyanobacteria that have been collected from a lake. A cork has been placed in the bottle on the left. That bottle has just been struck with a hammer to collapse the gas vesicles which explains its darker appearance (gas vacuolate cells cause much greater refraction of light and therefore the bottle on the right appears more turbid). **(b)** After a few minutes, the cells in the bottle on the left have descended because they have lost their gas vacuoles and hence, their buoyancy, whereas the cells in the bottle on the right have risen to the surface because their gas vacuoles have remained intact. (Courtesy of A. E. Walsby)

This simple experiment showed that, indeed, gas vacuoles do contain gas, and that they are essential in providing buoyancy to the cyanobacteria.

granules when viewed by phase microscopy. They can be stained with methylene blue, which results in a red appearance caused by the metachromatic effect of the dye-polyphosphate complex. They appear as dense areas when viewed by the electron microscope (Figure 4.31) and may actually volatilize under the electron beam, leaving a "hole" in the specimen.

Volutin is synthesized by the stepwise addition of single phosphate units from ATP onto a growing chain of polyphosphate. Although its degradation has not been studied, it may serve as an energy source for the synthesis of ATP from ADP, at least in some bacteria, as well as a reserve material for nucleic acid and phospholipid synthesis.

Volutin is apparently stored by organisms when nutrients are limiting to growth. However, bacteria of the genus *Acinetobacter* are known to store volutin during active growth when there is no nutrient limitation, an effect called luxurious phosphate uptake. Because of this feature, *Acinetobacter* spp. have been studied as possible candidates for the removal of phosphate during tertiary wastewater treatment (see Chapter 33). In sewage treatment it is desirable to remove potential nutrients such as organic carbon, inorganic nitrogen (NH_3 and NO_3^-), and phosphate.

Sulfur is stored as elemental sulfur by certain bacteria involved in the sulfur cycle. Sulfide-oxidizing photosynthetic proteobacteria and colorless filamentous sulfur

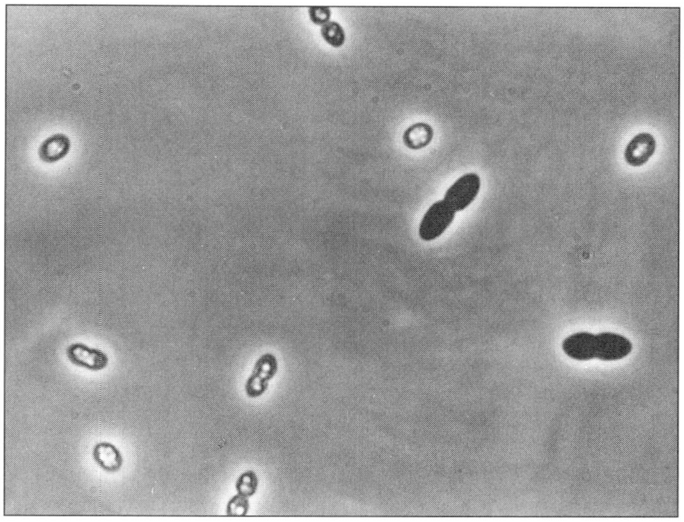

***Figure* 4.30** Poly-β-hydroxybutyrate (PHB) granules appear as bright, refractile areas by phase microscopy as seen in an *Azotobacter* species. (Courtesy of J. T. Staley)

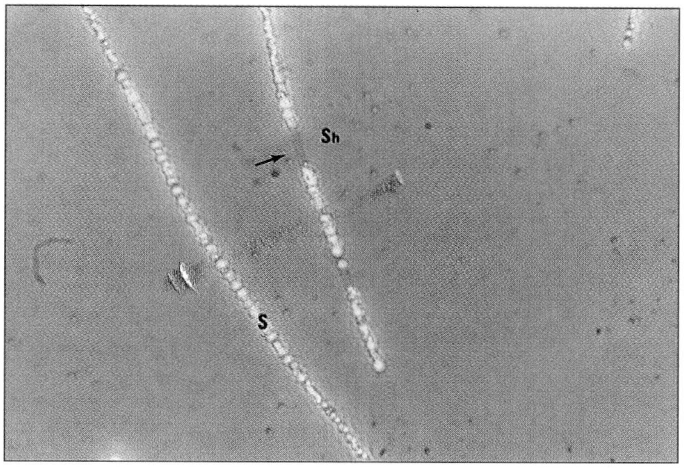

***Figure* 4.32** Phase photomicrograph of a sulfur bacterium *Thiothrix nivea,* showing its many bright yellow and orange sulfur (s) granules. Also, note sheath (sh) in area devoid of cells. (Courtesy of Judith Bland)

bacteria, both of which produce sulfur by the oxidation of sulfide, store the sulfur as granules, which appear as bright, slightly yellow, spherical areas in their cells (Figure 4.32). The sulfur can be further oxidized to sulfate by both groups of organisms. Thus, the granules are transitory and serve as an energy source (when oxidized by filamentous sulfur bacteria, see Chapter 21) or as a source of electrons (for carbon dioxide fixation by photosynthetic bacteria, see Chapter 21).

Cell Membranes of Eubacteria

The **cell** or **cytoplasmic membrane** serves as the boundary for the cell. In prokaryotes, it looks like a typical unit membrane—two thin lines separated by a transparent area—when viewed in thin section by the electron microscope (see Figures 4.12 and 4.23). At the molecular level, a model termed the **fluid mosaic model** best explains the structure and composition of the cell membrane (Figure 4.33). The cell membranes of bacteria are composed of phospholipid and protein. As many as 7 different phospholipids and approximately 200 proteins have been identified in the typical membrane of bacteria such as *Escherichia coli.* One of the phospholipids is phosphatidyl serine (Figure 4.34). This complex lipid consists of a glycerol moiety covalently bonded to two chains of fatty acids by an ester linkage. In addition, phosphoserine is covalently linked by its phosphate group to the third carbon of the glycerol moiety. The result is a bipolar molecule with one hydrophobic end (the hydrocarbon chains of the fatty acids) and a hydrophilic end (the glycerol phosphatidyl serine). Two membrane leaflets are formed by such molecules in the aqueous environment: the hydrophobic portions of the two fatty acid hydrocarbon chains or tails extend into the center of the leaflet and the hydrophilic portion extends away (see Figure 4.33).

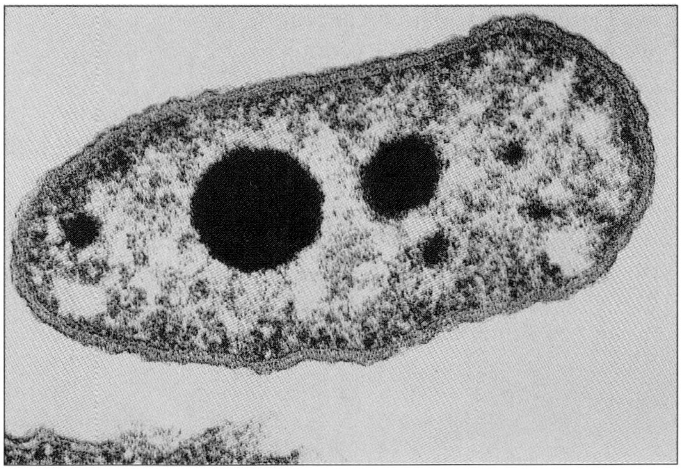

***Figure* 4.31** Polyphosphate granules as observed in bacterial cells. (© VU/ T. J. Beveridge)

The proteins of the cell membrane are embedded in the phospholipid bilayer (Figure 4.33). They are associated with the lipid bilayer of the membrane, particularly with its hydrophobic portion. Many of the proteins serve in transport of substances into the cell. The membrane is stabilized by divalent cations including calcium and magnesium.

Most prokaryotic organisms do not have sterols in their membranes. However, the mycoplasmas, without cell walls, are exceptions and they derive their sterols from the environment in which they live. The only other group of prokaryotes known to contain sterols is the methan-

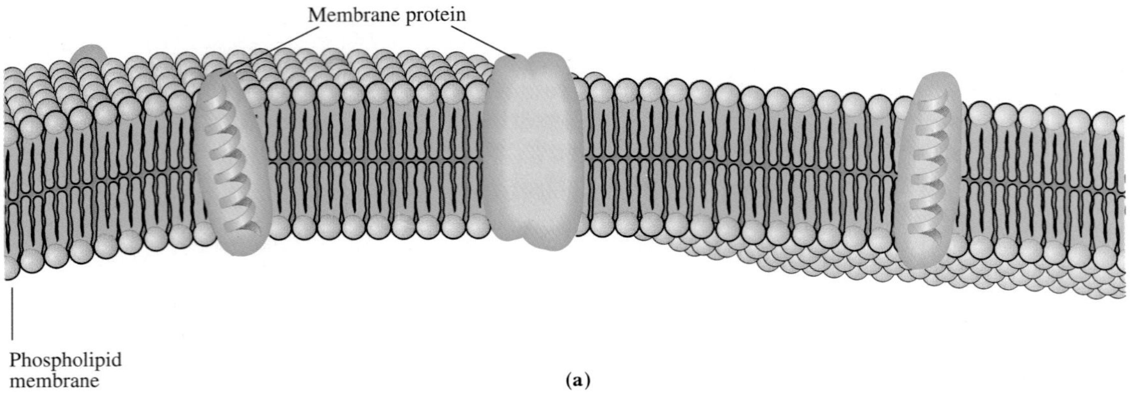

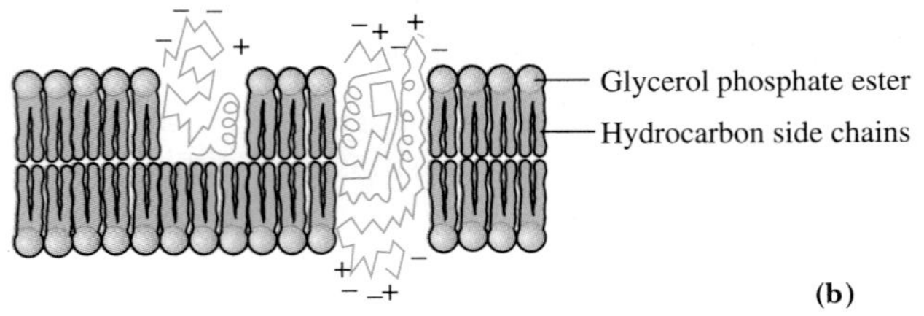

***Figure* 4.33 (a)** Diagram of the cell membrane structure of a gram-negative bacterium. The hydrophobic tail of the two fatty acids (designated by two wavy lines) of each glycerol moiety face inward whereas the hydrophilic glycerol phosphate moiety of each layer (designated by a circle) faces the aqueous environment. **(b)** Proteins are represented as charged, globular structures embedded in the lipid bilayer.

otrophs, which have special membranes involved in the oxidation of ammonia.

The cell membrane is the ultimate physical barrier between the cytoplasm and the external environment of the cell. It is selectively permeable to chemicals; thus the molecular size, charge, polarity, and chemical structure of a substance determines its permeability properties. For example, water and gases diffuse freely through cell membranes, whereas sugars and amino acids do not. These important organic compounds, which serve as substrates for growth and energy metabolism, are concentrated in the cell by special transport systems (see Chapter 5). This selective permeability is also important in cell energetics as discussed in Chapter 8.

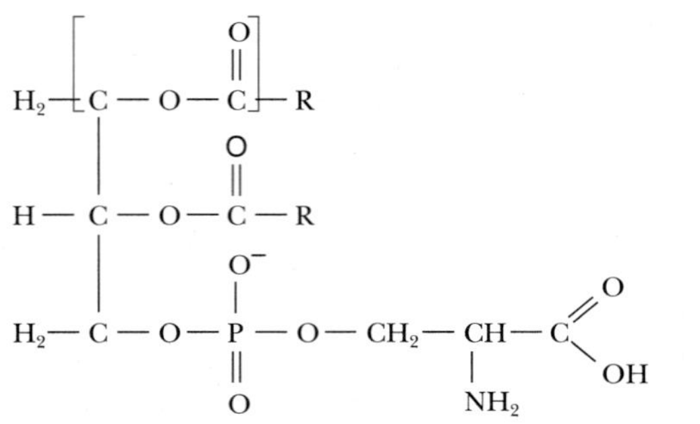

***Figure* 4.34** Chemical structure of phosphatidyl serine, a typical phospholipid. The hydrocarbon side chains (R) of the fatty acids of such molecules typically contain 12 to 18 or more carbon atoms. The ester linkage is shown in the box.

One additional function attributed to the cell membrane is its role in the replication of DNA. Following replication, the nuclear material separates in the cell at the point where it was attached to the membrane. As nuclear growth proceeds, the two new nuclear bodies are physically separated by new cell membrane synthesis. The eventual outcome is the separation of the two new nuclear structures prior to cell septum formation and cell division.

Some bacteria contain membranes that extend into the cell itself. These intracytoplasmic membranes are often internal extensions or invaginations of the cell membrane. **Mesosomes** are tube-shaped cell membrane invaginations that are formed by some gram-positive bacteria (Figure 4.35). These tubular membranes may serve important roles in cell division, although their true function has not been resolved.

Some autotrophic gram-negative bacteria also have intracytoplasmic membranes. For example, phototrophic bacteria have internal membranes that are involved in photosynthesis. Also some chemoautotrophs (also called chemolithotrophs), which oxidize ammonia and nitrite, and methanotrophs, which oxidize methane, possess similar intracytoplasmic membranes. These bacteria are discussed individually in Chapter 21.

Cell Membranes of Archaea

Much less is known about the structure and function of archaeal cell membranes. Although their appearance in thin section is like that of bacterial cell membranes, their chemical composition is different. The major difference

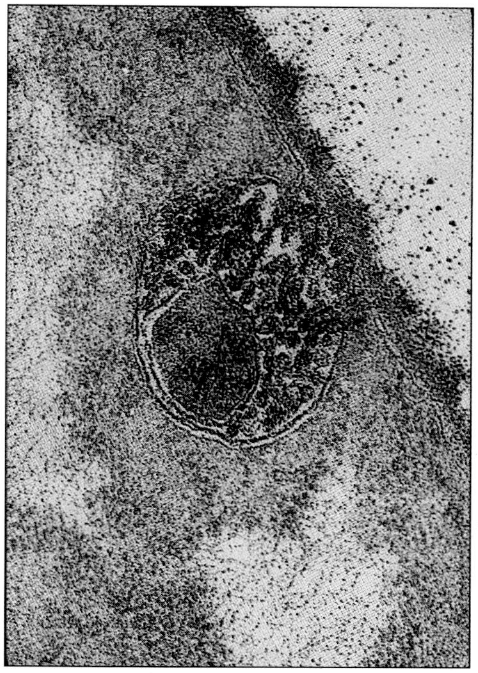

***Figure* 4.35** A thin section of *Bacillus megaterium* showing a mesosome. (© VU/Ralph Slepecky)

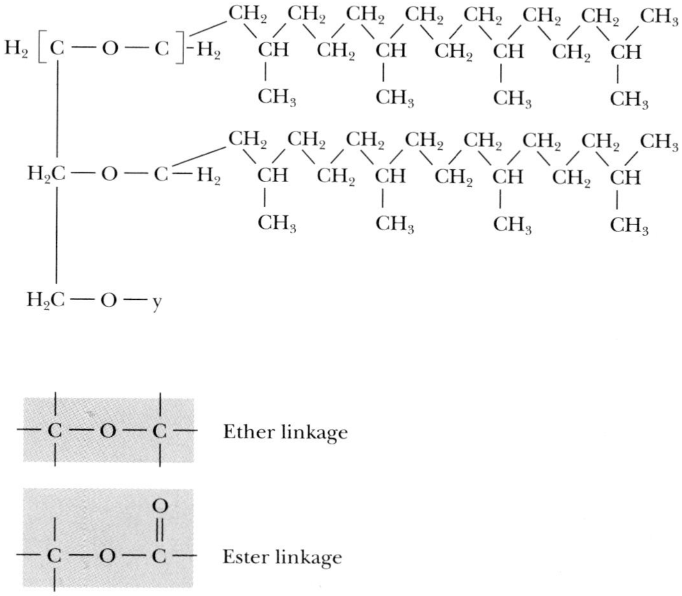

***Figure* 4.36** An example of the ether-linked lipid found in archaean cell membranes. The two hydrocarbon side chains are C-20 phytane groups. The ether bond is shown outlined. The "y" group may be an H atom, a saccharide, or a phosphorylated moiety.

is that the lipid moieties of the membrane are not glycerol-linked esters, but instead are glycerol-linked *ethers* (Figure 4.36). Nonetheless, their membranes need to carry out all of the same functions of transport and energy generation as the Eubacteria.

For the remainder of the chapter, the layers external to the cell are considered, beginning with those *farthest* from the cell and working toward the cell membrane.

Extracellular Layers

These are discrete layers produced by some bacteria that are external to the cell wall. If these layers, called **capsules, sheaths, slime layers,** and **protein jackets,** depending on their structure and function, are removed from the organism, the organism does not lose its viability. Thus, they do not appear to be essential to the bacteria that produce them, at least under some conditions of growth. Depending upon the prokaryotic group and the nature of the extracellular layer, the layer is called a capsule (also glycocalyx), a sheath, slime layer, or S layer (also protein jacket).

Capsules

Capsules constitute barriers that protect the cell from the external environment, no doubt aiding the cell in unknown ways, such as prevention of virus attachment or slowing the desiccation process when cells encounter dry conditions in the environment.

Capsules can be removed from bacteria in the laboratory without affecting the viability of the organism, indicating they are *not always essential* to survival. Nevertheless, several functions have been attributed to capsules, and, under certain conditions, they may be crucial in determining the survival of the organism in their natural environment. For example, the capsules produced by some bacteria are important **virulence factors** (factors that enhance the ability of a bacterium to cause disease). This can be illustrated by considering the species *Streptococcus pneumoniae,* one of the bacteria responsible for bacterial pneumonia. Unencapsulated strains of this bacterium are not pathogenic, that is, they are not capable of causing disease. The reason is that the unencapsulated cells are more readily killed by phagocytes (white blood cells) in the potential victim's circulatory system. Apparently the white blood cells can ingest cells without capsules much more readily than those with capsules. Thus, capsules may provide a means by which these bacteria evade the host defense system in their natural environment.

Capsules are also important in mediating attachment of some bacteria. This is well illustrated by one of the bacteria responsible for dental cavity formation. In the presence of sucrose, *Streptococcus mutans* produces a polysaccharide capsule that permits the bacterium to attach to dental enamel. As *S. mutans* grows on the surface of the

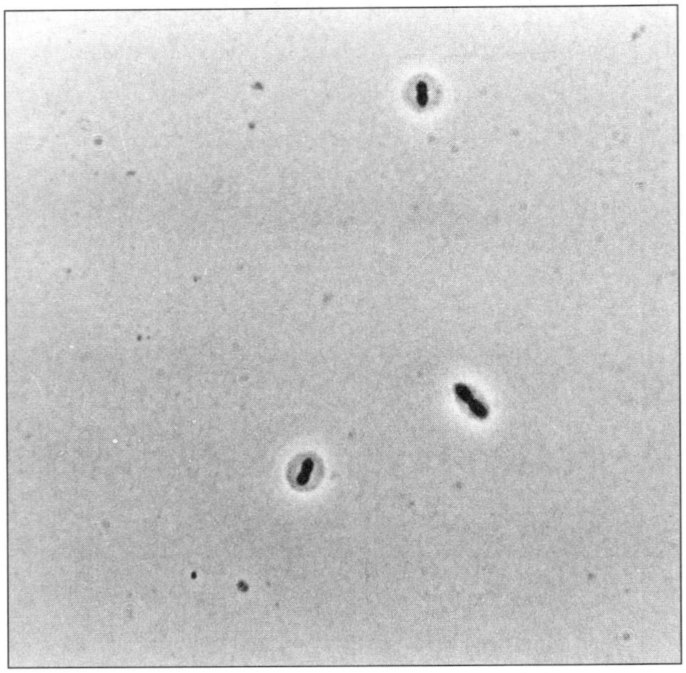

***Figure* 4.37** The Quellung reaction in *Streptococcus pneumoniae.* (Courtesy of J. Dalmasso and J. T. Staley)

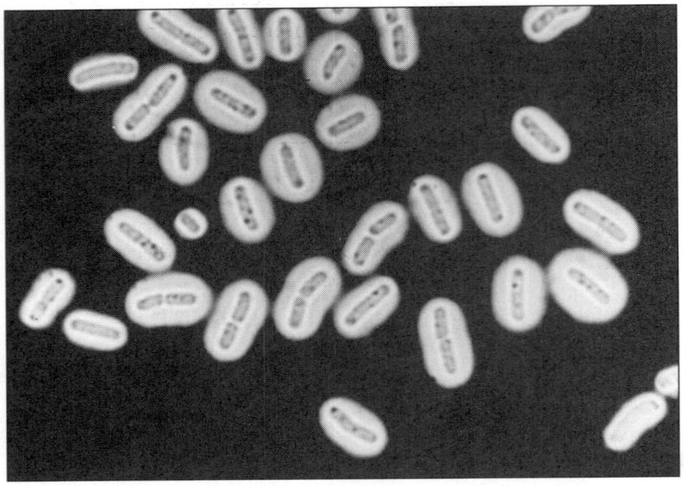

***Figure* 4.38** An India ink negative stain showing the capsule of a *Bacillus* sp. This is called a negative stain because it stains the *background* around the capsule, not the capsule itself. (Courtesy of C. Robinow)

tooth, acid produced by fermentation of sucrose causes etching of the enamel, and a cavity may eventually develop.

Capsules are diffuse structures that are often difficult to visualize without special techniques. This is well exemplified by the capsule of *Streptococcus pneumoniae.* Its capsule is not visible when the organism is observed by ordinary phase microscopy or by simple staining techniques because of the diffuse consistency of the polysaccharide that forms the capsule. A special technique has been developed to permit its visualization called the **Quellung** (German for "swelling") **reaction.** An antiserum (see Chapter 27) is prepared using capsular material from *S. pneumoniae,* and when it is used to stain *S. pneumoniae,* it complexes with the capsular material. After staining, the bacterium appears to have grown much larger or to have swollen. In reality, the bacterium has not swollen, but rather the stained antiserum has complexed with the extensive but previously transparent capsule and made it visible (Figure 4.37).

Negative stains can also be used to stain capsules for light microscopy. These stains consist of insoluble particulate materials such as India ink. The dark particles of the stain do not penetrate the capsule, so the cell appears with a large unstained halo about it (Figure 4.38).

Most capsules are composed of polysaccharides (Table 4.1). **Dextrans** (polymers of glucose) and **levans** (polymers of fructose or levulose) are common homopolymers found in capsules produced by lactic acid bacteria (*Streptococcus* spp. and *Lactobacillus* spp.). However, some bacteria produce heteropolymers containing more than one monomeric subunit. Hyaluronic acid is an example of a heteropolymeric capsule consisting of two subunit molecules, *N*-acetyl glucosamine and glucuronic acid. Another example is the glucose-glucuronic acid polymer, also produced by some *Streptococcus* spp.

Capsules can also be composed of polypeptides. For example, some members of the genus *Bacillus* have capsules made of polymers that are repeating units of D-glutamic acid (Table 4.1). It is interesting to note that the D-stereoisomers of amino acids are found almost exclusively in the capsules and cell walls of some bacteria. Virtually all other biologically produced amino acids are of the L-stereoconfiguration. Thus, all proteins in bacteria (as well as in higher organisms) have L-amino acids. Furthermore, the poly D-glutamic acid of *Bacillus* is also noteworthy because the amino acids are linked together by their gamma-carboxyl group, not the alpha-carboxyl group as is characteristic of proteins.

Sheaths

Some bacteria, called *sheathed bacteria,* produce a much more dense and highly organized external layer termed a *sheath.* Unlike a capsule, this structure can be easily discerned with a light microscope (Figure 4.39). Sheathed bacteria form filamentous chains and grow in flowing aquatic habitats such as rivers and springs. The sheath protects the cells in the chain from being disrupted by turbulence in the water and ultimately from being removed from the area of the habitat that is most favorable for growth. Few chemical analyses of sheaths have been made, but those that have indicate they are more complex than

***Table* 4.1 Chemical composition of capsules from various bacteria**

Bacterium	Name of Capsular Material	Structure of Repeating Unit
Leuconostoc mesenteroides	dextran	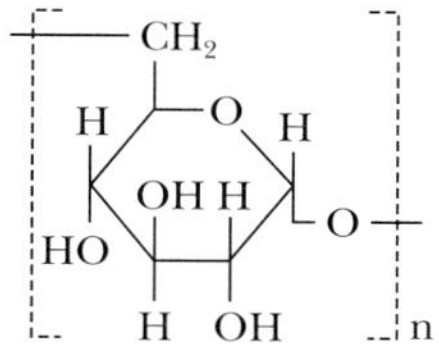 α-1,6-poly D-glucose
Streptococcus pneumoniae III	polyglucose glucuronate	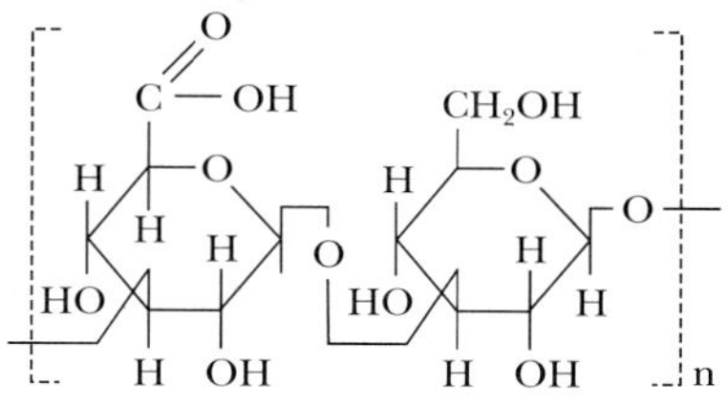 Glucuronic acid, Glucose
Streptococcus spp.	hyaluronic acid	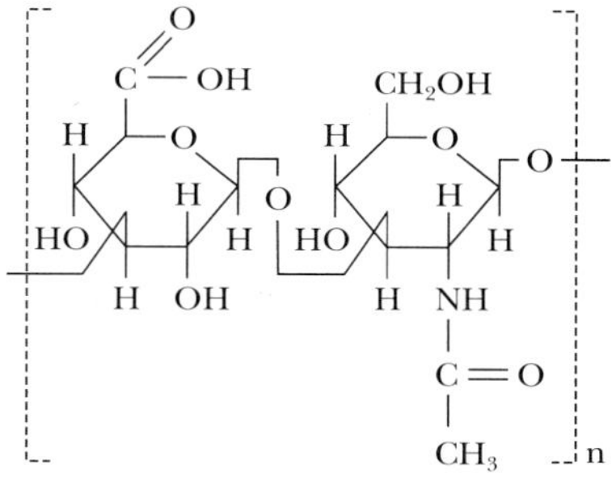 Glucuronic acid, N-acetyl glucosamine
Bacillus anthracis	gamma poly D-glutamic acid	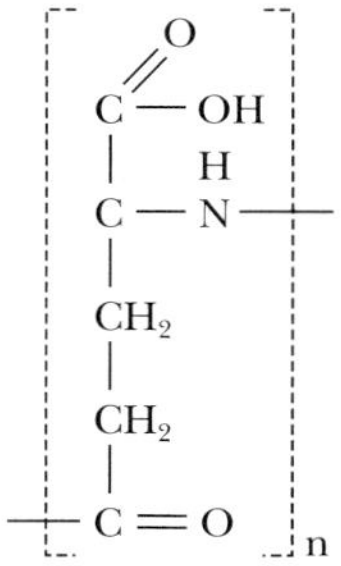 γ D-glutamic acid

capsules, having in addition to polysaccharides amino sugars and amino acids.

Slime Layers

Some bacteria such as *Cytophaga* species move by a type of motility referred to as **gliding.** This type of movement requires that the organism remain in contact with a solid substrate. Gliding bacteria produce a chemical "slime," part of which is lost in the trail left by the bacterium during movement (Figure 4.40). Members of the genus *Cytophaga* are noted for their ability to degrade substances

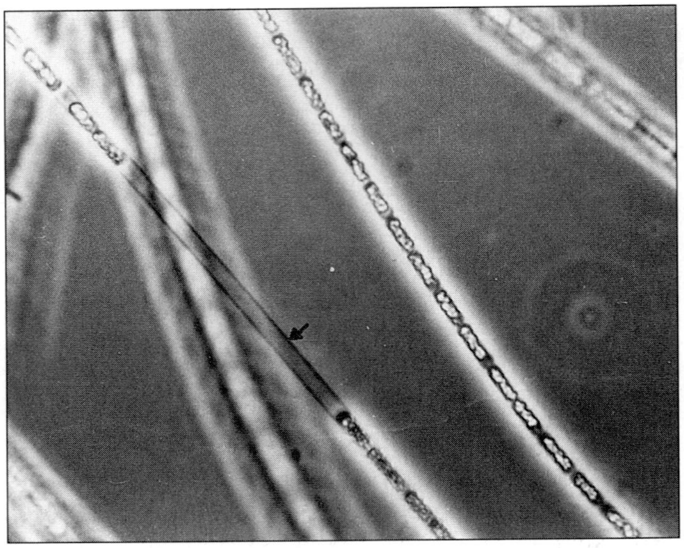

Figure **4.39** The filamentous gram-negative bacterium, *Sphaerotilus natans,* has a sheath (see arrow where section of sheath lacks cells) and grows in flowing aquatic environments. (Courtesy of J. T. Staley)

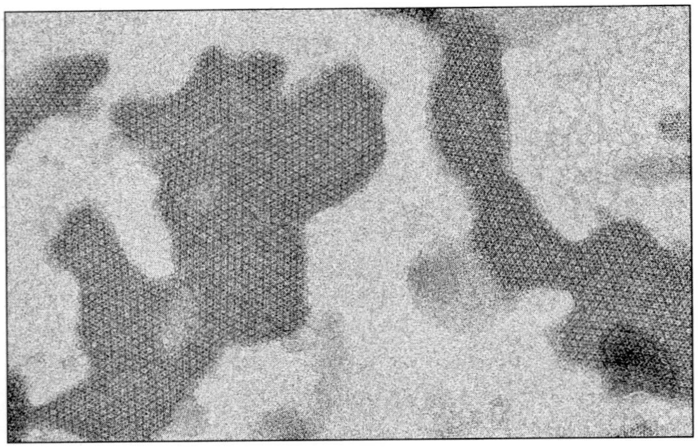

Figure **4.41** Electron micrograph of negatively stained fragments of a highly textured protein jacket, or S layer, from *Aquaspirillum serpens* strain VHA. (Courtesy of S. F. Koval)

such as cellulose and chitin and other high molecular weight polysaccharides. These materials occur in nature in dead plant material (cellulose) and insect or arthropod shells (chitin). There is a perceived advantage for the bacterium that digests these materials to be in physical contact with them. Thus, by gliding on the surface of the material, the cellulase and chitinase enzymes, which are produced extracellularly by these bacteria, can be kept close to the substrate as well as the cells. In this manner the cells can derive nutrients from their enzymes more effectively.

The chemical nature of the slime varies. In some species it is a polysaccharide. For example, *Cytophaga hutchinsonii* produces a heteropolysaccharide of arabinose, glucose, mannose, xylose, and glucuronic acid. The mechanism for gliding motility is unknown, however, recent interesting research on *Cytophaga johnsonae* indicates that a special sulfur-containing lipid (sulfonolipid) found on the surface of the cell is essential for gliding in this organism. Furthermore, it is produced only when the organism is in contact with a surface.

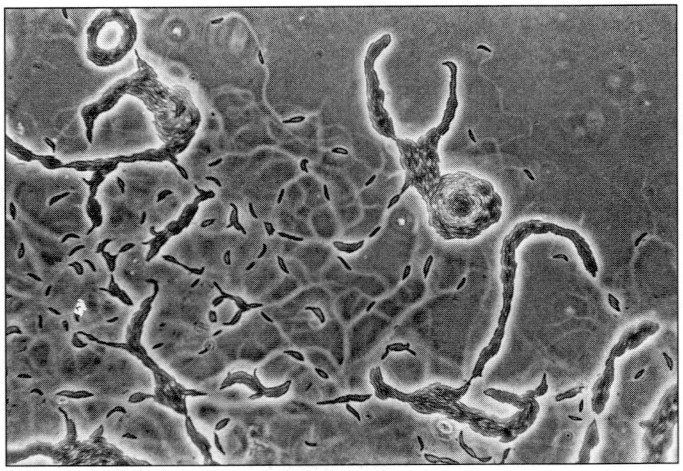

Figure **4.40** The myxobacterium, *Stigmatella aurantiaca,* produces slime trails as the cells glide across an agar surface. Note that even the smallest units comprise at least two cells gliding together. (Courtesy of Hans Reichenbach)

Protein Jackets

External protein layers are produced by certain prokaryotes such as the genus *Spirillum.* These protein jackets, also called **S layers,** are often highly textured surfaces (Figure 4.41). Their function is unknown, although in some Archaea they comprise the only layer external to the cell membrane and, therefore, in these organisms may serve as a cell wall. However, bacteria such as some *Spirillum* species and certain cyanobacteria that have these layers also have a cell wall, so they cannot serve that role in these organisms. For some prokaryotes, the S layer serves as a site on the cell surface to which bacterial viruses adhere.

Cell Walls

The cell wall is the outermost cellular constituent of prokaryotic organisms. In contrast to extracellular layers, they are essential to those microorganisms that produce them. Without this protection, the microorganisms could not survive in their normal habitats because they would lyse (cells would break open and the cytoplasm would be lost due to osmotic reasons; see below). Considerable variation in cell wall composition exists among prokaryotes. One special group of bacteria lack a cell wall entirely, whereas all other bacteria produce them. These bacteria that lack cell walls survive because their cell membranes differ from typical Eubacteria. For example, their membranes contain sterols or other compounds that help with stabilization; they are, in this respect, similar to animal cell membranes, which also lack cell walls and contain sterols.

Two major groups of bacteria are recognized based upon the chemistry of their cell walls: **Eubacteria** and **Archaea.** Almost all Eubacteria have a chemical polymer called **peptidoglycan** in their cell wall structure, while the Archaea do not. Peptidoglycan is a complex polymeric substance containing amino sugars and amino acids as described in detail below.

Composition of the Eubacterial Cell Wall

The peptidoglycan cell wall structure varies from one species of bacteria to another. However, all have the same general chemical composition: two amino sugars, glucosamine and muramic acid, are joined together by β 1 $\rightarrow$ 4 linkages to form a chain or a linear polymer. The chains of these amino sugars are crosslinked together by polypeptides. Figure 4.42 shows the peptidoglycan of the eubacterium, *Staphylococcus aureus.* Note that the specific amino sugars are *N*-acetyl glucosamine and *N*-acetyl muramic acid. The *N*-substituents are acetyl groups in almost all bacteria except for some *Bacillus* species, which have amino groups that are unsubstituted, and in mycobacteria and some of the streptomycetes, which have *N*-glycollyl groups. The amino group of the first amino acid in the peptide crosslink is joined to the carboxyl group of the lactic acid moiety of *N*-acetyl muramic acid by a peptide bond. Some of the amino acids such as alanine and glutamate exist in the *D*-stereoconfiguration.

Note that one of the amino acids, in this example, lysine, is a **diamino** acid, that is, it contains two amino groups (Figure 4.43). It is essential that one of the amino acids in the crosslinking peptide bridge is a diamino acid. This extra amino group is needed because the amino groups of the other amino acids are tied up in the peptide linkages all the way back to the muramic acid. The diamino acid can link up one amino group to one chain and the other to the other peptide chain to crosslink the two peptides and their amino sugar chains to one another. The result of this crosslinking is the formation of a two-

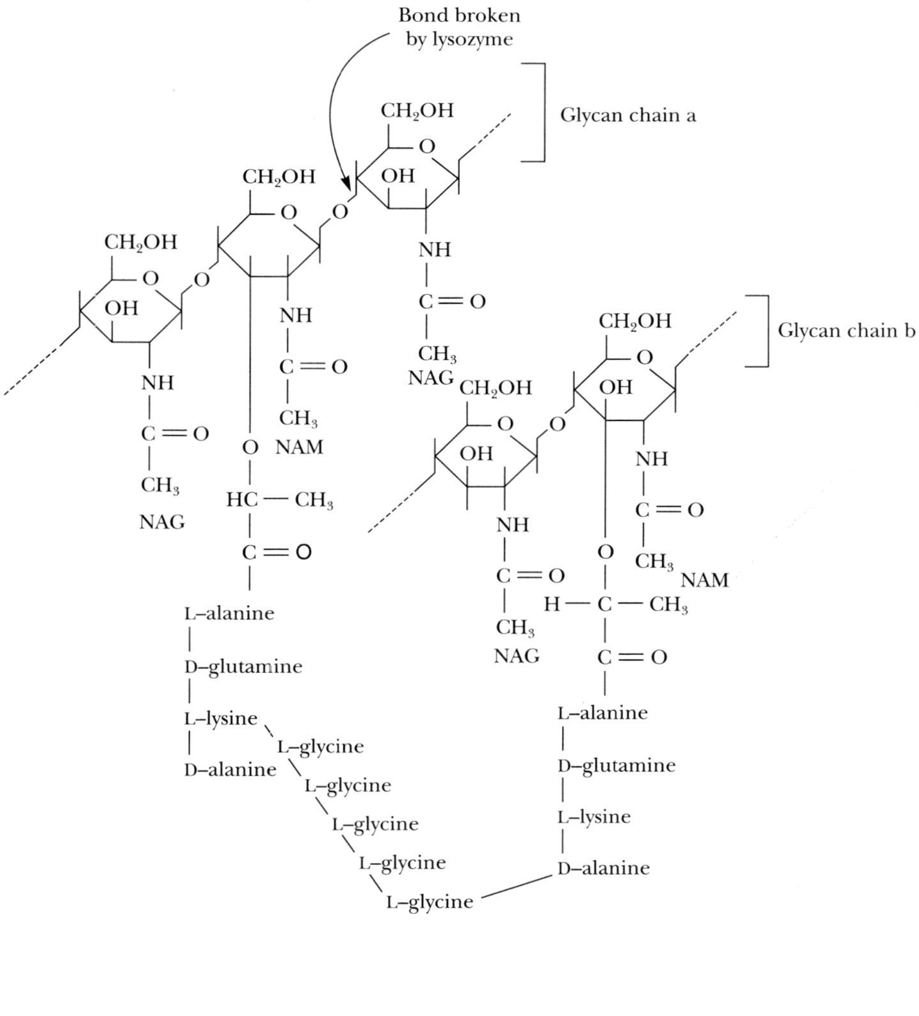

***Figure* 4.42** Chemical structure of the peptidoglycan layer of *Staphylococcus aureus.* The two amino sugars, *N*-acetyl glucosamine (NAG) and *N*-acetyl muramic acid (NAM) are joined together by β 1 $\rightarrow$ 4 glycosidic bonds and alternate to form the backbone of the structure. Chains of these amino sugars are held together by peptide crosslinks that are connected to the carboxyl group of the lactic acid moiety of the muramic acid through a peptide bond with the amino group of the first amino acid in the chain. Note that some amino acids, alanine and glutamic, are in the D-stereoconfiguration. Also note that lysine is a diamino amino acid that serves to interlink the peptide bridges together.

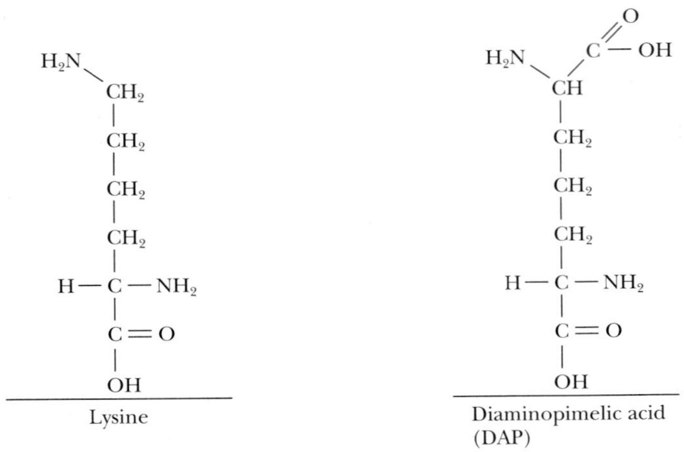

***Figure* 4.43** Structure of two diamino acids, lysine and diaminopimelic acid, which are found in peptidoglycan.

dimensional network around the cell (Figure 4.44). It is this two-dimensional sac that provides the rigidity and strength of the cell wall. Lysine or diaminopimelic acid (Figure 4.43) often serve as the diamino acids responsible for this crosslinking. Diaminopimelic acid is *never* found in proteins; it occurs uniquely in the cell wall structures of some prokaryotic organisms. It is found in the peptidoglycan of virtually all gram-negative bacteria (Figure 4.45).

Considerable variation exists in the amino acids that form the crosslinking peptides of bacteria. Thus, about 100 types of cell wall peptidoglycan structures are known.

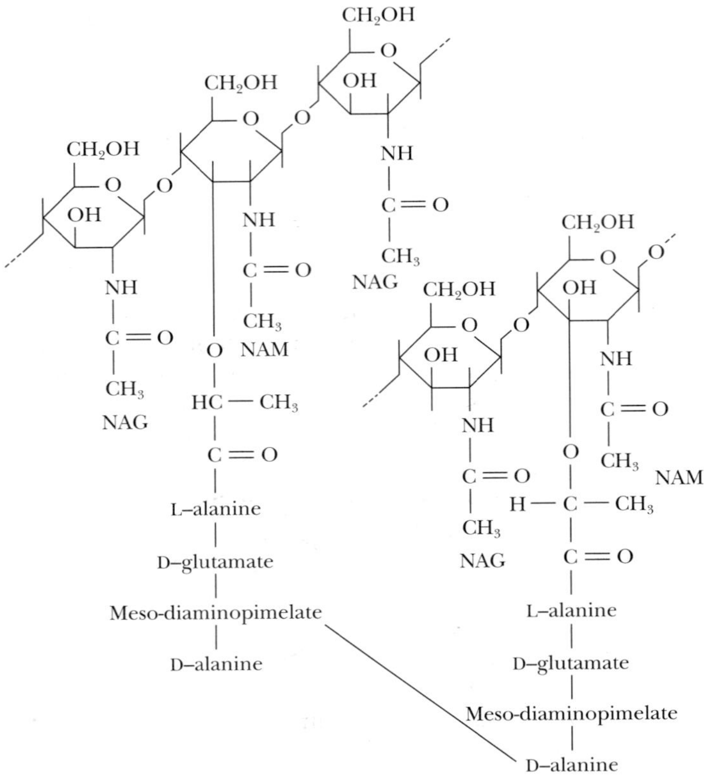

***Figure* 4.45** Chemical structure of the peptidoglycan of a gram-negative bacterium *Pseudomonas aeruginosa.* Note that the diamino acid is diamino pimelic acid (DAP), which is found as the diamino amino acid in all gram-negative bacteria and in some gram-positive bacteria. As in Figure 4.44 the peptidoglycan forms a network around the cell of gram-negative bacteria.

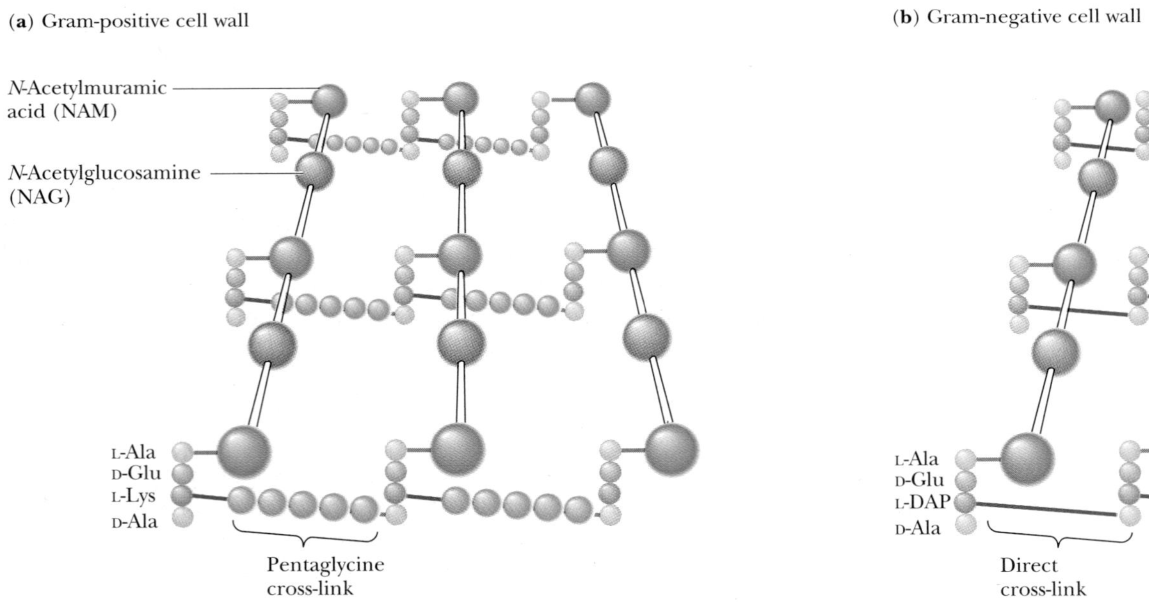

***Figure* 4.44** Diagram of two-dimensional network formed by peptidoglycan as a sac around **(a)** a gram-positive and **(b)** gram-negative cell. This layer is the major structural component of eubacterial cell walls. The *N*-acetylglucosamine (NAG) is joined to the *N*-acetylmuramic acid (NAM) to form the amino sugar backbone. These amino sugar chains are held together by peptide bridges in which amino acids are represented as small balls.

BOX 4.7 MILESTONES

The Gram Stain

The Gram stain procedure was developed unwittingly in 1888 by Christian Gram, a Danish physician studying in Berlin. He was examining lung tissues during the autopsy of individuals who had died of pneumonia. He noted that *Streptococcus pneumoniae* found in the lung tissues retained the primary stain, Bismark Brown (crystal violet is now used), whereas the lung tissue did not. It was subsequently determined that certain bacteria, now termed **gram-positive,** are like *S. pneumoniae* and retain the purple dye when stained by this method, whereas other bacteria, the **gram-negative** ones, do not. The Gram stain is called a **differential stain** because it distinguishes between these two groups of bacteria.

The modern Gram stain procedure is as follows (Figure **a**):

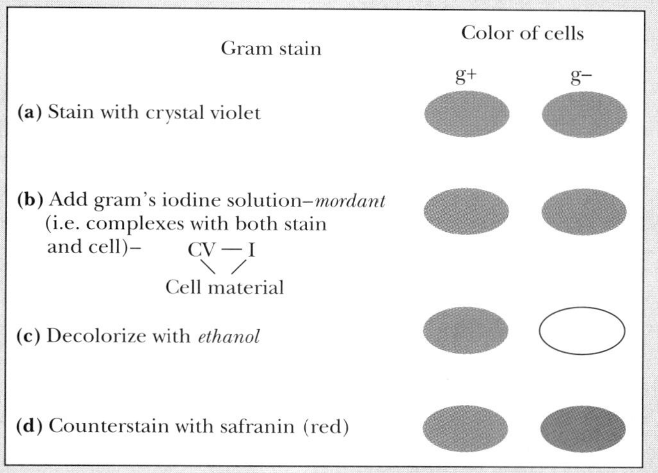

1. A smear of the bacterium is prepared and stained with the **primary stain,** crystal violet.
2. The preparation is then flooded with a solution of potassium iodide and iodine. This solution serves as a **mordant** by complexing with the crystal violet as well as with the cellular material.
3. In the next, most delicate step, called **decolorization,** ethyl alcohol is added dropwise to the stained smear. If the organism is gram-negative, the dye complex that has been used to stain the preparation is readily removed. However, if the organism is gram-positive, the dye is not entirely removed so the organism retains its purple color.
4. The final step is **counterstaining.** The slide preparation is stained with the red dye safranin. Gram-negative organisms, which are colorless at this time, are readily stained by this dye and appear red. Since the gram-positive bacteria are already stained with the intense crystal violet dye, they remain purple in color.

However, certain amino acids are never found in the peptide bridge including the sulfur containing amino acids, the aromatic amino acids, and branched chain amino acids, as well as arginine, proline, and histidine.

The Eubacteria are separated into two groups based upon their reaction to a staining procedure called the **Gram stain (Box 4.7).** The differences between **gram-positive** and **gram-negative** bacteria relate to differences in their cell wall structure and chemical composition. Thin-sections of gram-positive bacteria reveal that their walls are thick and nearly uniformly dense layers (Figure 4.46**a,b**). In contrast, the cell walls of gram-negative bacteria are more complex because they have in addition to a peptidoglycan layer an additional layer called an **outer membrane** (OM) (Figure 4.46**c,d**).

GRAM-POSITIVE EUBACTERIAL CELL WALL The structural differences between the cell walls of gram-positive and gram-negative bacteria reflect differences in biochemical composition. Techniques were developed to permit the separation of cell walls from cytoplasmic constituents so that they could be chemically analyzed. The major constituent found in the cell wall of gram-positive bacteria is peptidoglycan, making up from 40 percent to 80 percent of the dry weight of the wall depending upon the species. Other constituents of gram-positive cell walls include teichoic acids and teichuronic acids.

Teichoic acids are polyol phosphate polymers such as polyglycerol phosphate and polyribitol phosphate (Figure 4.47). Sugars, such as glucose and galactose, and amino sugars, such as glucosamine, and the amino acid, *D*-alanine, are found in some of these compounds. Teichoic acids are attached covalently to the 6-hydroxy position of muramic acid in the peptidoglycan.

Teichuronic acids are polymers of two or more repeating subunits, one of which is always a uronic acid, such

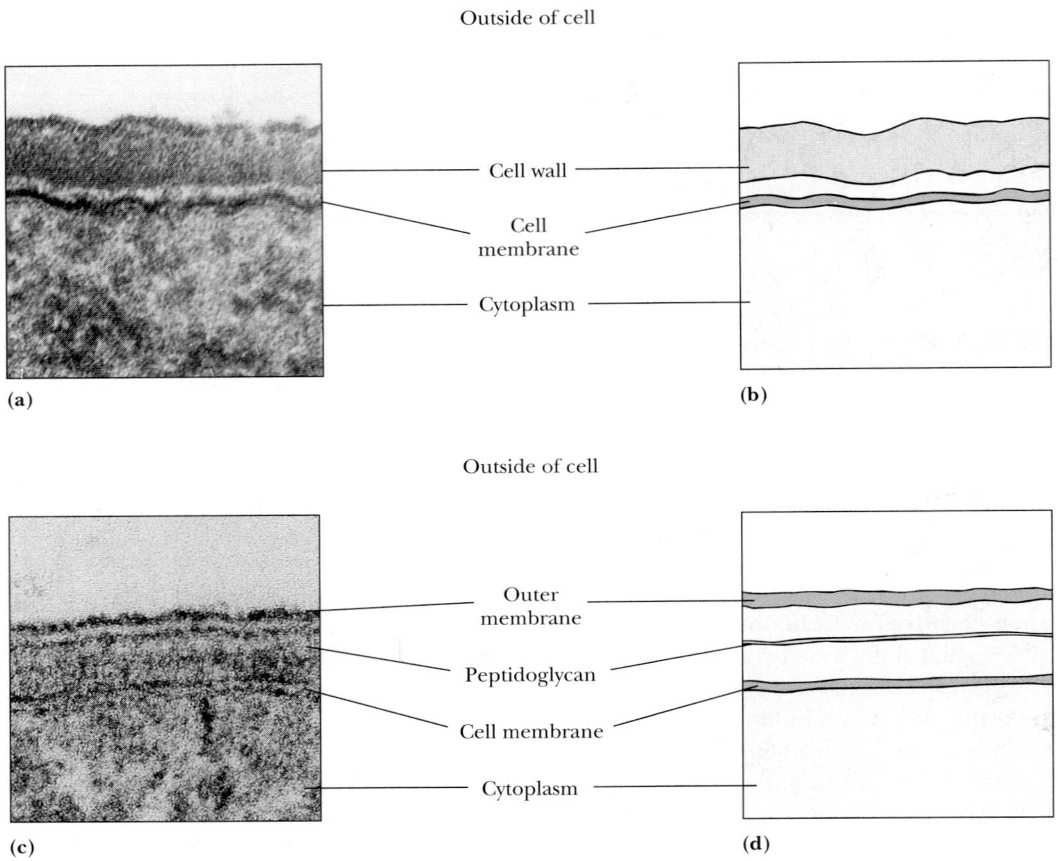

***Figure* 4.46** A comparison of the cell envelope of gram-positive bacteria with gram-negative bacteria. **(a)** A thin section showing the appearance of the cell envelope of a gram-positive bacterium *Bacillus megaterium*. (Courtesy of J. T. Staley) Note the uniformly dense outer layer comprising the cell wall (CW) outside the cell membrane (CM). **(b)** A diagram illustrating the structure of the envelope of a typical gram-positive bacterium. **(c)** A thin section through a gram-negative bacterium, a *Hyphomicrobium* sp., shows the cell membrane (CM) and cell wall with its characteristic outer membrane (OM). The peptidoglycan layer of gram-negative bacteria is very thin and not always evident in electron micrographs. (Courtesy of J. T. Staley) **(d)** As shown in this diagram the cell wall of gram-negative bacteria consists of two layers, an inner peptidoglycan layer (P) and an outer wall membrane (OM).

as glucuronic acid, or the uronic acid of an amino sugar, such as aminoglucuronic acid (Figure 4.48). They are linked covalently to peptidoglycan, but the linkage group in unknown.

GRAM-NEGATIVE EUBACTERIAL CELL WALL Gram-negative bacteria do not have teichoic or teichuronic acids. However, in other respects their cell wall structures are more complex than those of the gram-positive bacteria, as suggested by their multilayered appearance (see Figure 4.46**c,d**). Peptidoglycan is present, but it makes up a much smaller proportion of the cell wall material. In contrast to gram-positive bacteria, which have several layers of peptidoglycan encasing the cell, only a single macromolecular sheet is found in gram-negative organisms. Indeed, it comprises only about 5 percent by weight of the gram-negative cell wall. Despite the reduced amount of this material, it still serves an important role as a rigid barrier outside the cell membrane because it is covalently attached to one of the major proteins in the outer membrane.

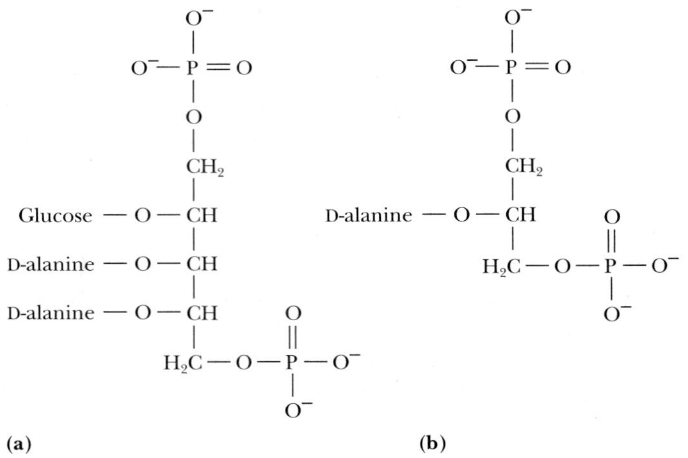

***Figure* 4.47** Examples of chemical structures of two teichoic acids found in gram-positive bacteria. **(a)** The polyribitol teichoic acid monomer structure from *Bacillus subtilis,* and **(b)** the polyglycerol teichoic acid monomer from *Lactobacillus* species. Polymers are joined together by phosphate linkages.

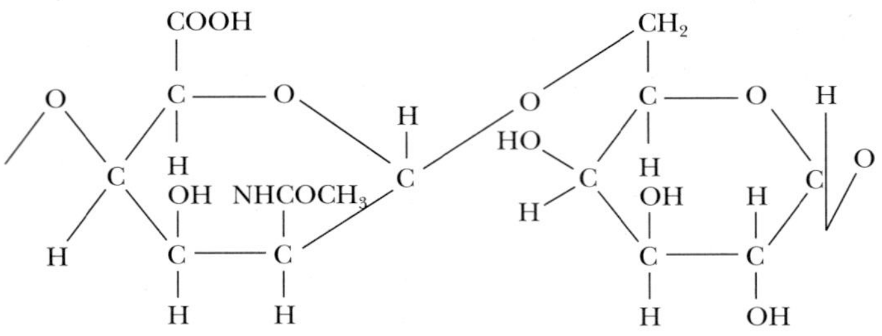

4-*N*-acetyl-D-mannosaminuronosyl-β-(1-6)-glucose

***Figure* 4.48** An example of a teichuronic acid found in the cell wall of *Micrococcus luteus.*

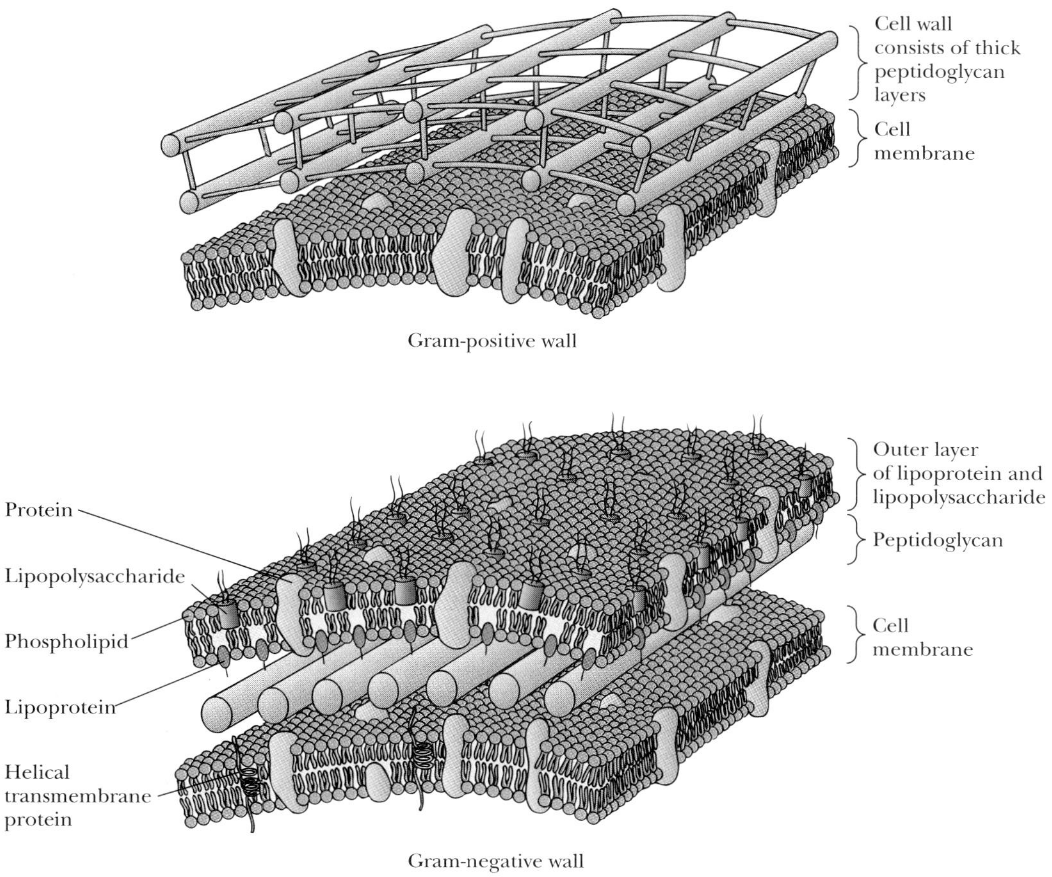

***Figure* 4.49** Diagram of the cell envelope of a gram-negative bacterium showing the outer membrane, peptidoglycan layer, and the cell membrane. The cell membrane complex consists of two lipid leaflets held together by hydrophobic forces. The outer membrane is a similar structure. Lipid A is attached to the outer leaflet. Lipid A is, in turn, bonded covalently to a polysaccharide forming the lipopolysaccharide (LPS) of gram-negative bacteria. The polysaccharide tail extends away from the cell into the external environment. The outer membrane contains porins and other proteins.

The outermost layer, or **outer membrane,** of gram-negative cells appears similar to a cell membrane when viewed by electron microscopy (see Figure 4.46 **c, d**). As in cell membranes, both lipid and protein are present; however, in addition, polysaccharides are found. As in the cell membrane, each of the two leaflets of this membrane consists of molecules of lipid with their hydrophilic moieties (glycerol ester linkages) facing away from the other leaflet and outward toward the aqueous environment. Likewise, the hydrophobic portions (hydrocarbon chains) of the leaflets face inward and hold the two leaflets together in the aqueous environment.

The outer leaflet of the membrane serves as the site of attachment for **lipid A** (Figure 4.50). Bonded covalently to lipid A is a **core** polysaccharide (Figure 4.50) with specific side-chain polysaccharides (called **somatic** or ***O*-polysaccharides**) that vary from one species of gram-negative bacterium to another. This lipid A-polysaccharide complex is referred to as **lipopolysaccharide,** or simply **LPS.** LPS that contains lipid A is called **endotoxin** because it is toxic to a variety of animals including humans. Animals injected with endotoxin may experience fever, tumor necrosis, hemorrhage, shock, miscarriage, and a variety of other symptoms depending upon dose and source. Many gram-negative bacteria including *Salmonella* spp. (which cause enteric diseases including typhoid fever) and *Yersinia pestis* (which causes bubonic plague) owe their toxic properties to this compound. Other gram-negative bacteria with LPS contain a version that is not toxic because they lack specific parts of the lipid A moiety.

The outer membrane of gram-negative bacteria also contains protein; however, the amounts are less than that of the cell membrane. The most prevalent protein found is a small polypeptide (molecular weight of 7200) containing a lipid moiety. This lipoprotein is joined covalently to the diamino acid of the peptidoglycan to form a complex called **peptidoglycan-lipoprotein.** Thus this protein serves to anchor the outer membrane to the cell wall peptidoglycan layer.

Other important proteins serve as transport proteins, which permit the passage of molecules through the outer membrane. For example, the **matrix proteins** (or **porins**) Omp C and Omp F of *Escherichia coli* are trimeric structures that form channels across the bilipid membrane (Figure 4.51). These water-filled pores in the membrane allow the free diffusion of water-soluble nutrients, such as sugars and amino acids, as well as inorganic ions. Finally, some specific proteins are responsible for the entry of certain compounds such as vitamin B_{12}, iron chelates, disacharides, or phosphorylated compounds. Passage of solutes through these porins is quite different from transport across a cell membrane (Chapter 5).

Archaean Cell Walls

Until recently it was believed that all bacteria had a pep-

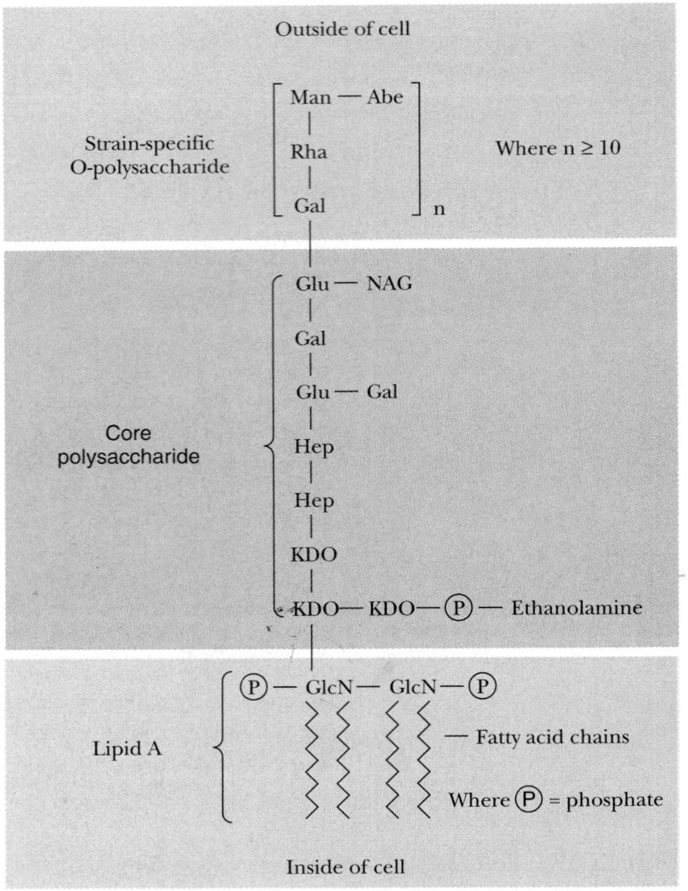

***Figure* 4.50** Chemical composition of the polysaccharide portion of LPS in a *Salmonella* strain. The polysaccharide portion extends outside the cell where it acts as a strain-specific antigen. KDO (ketodeoxyoctanoate); Hep (heptose); Glu (glucose); NAG (*N*-acetyl glucosamine); Man (mannose); Abe (abequose).

tidoglycan layer in their cell walls. Thus, one of the most exciting developments in bacteriology has been the recognition that prokaryotic organisms in the Domain of Archaea lack peptidoglycan (see Chapter 22). This was an important finding because it was consistent with the 16S rRNA analyses showing that the Archaea are different from Eubacteria in more than one way. Three separate groups comprise the Archaea including the extreme thermophilic bacteria, the methane-producing bacteria (methanogens), and the extreme halophiles (salt-loving bacteria; see Chapter 22).

Some of the methanogens have cell walls that are protein in composition. In other methanogens, a material like peptidoglycan, called **pseudopeptidoglycan,** is formed (Figure 4.52). In this substance, the amino sugar chain of the peptidoglycan contains *N*-acetylglucosamine and *N*-acetyltalosaminuronic acid in place of *N*-acetylmuramic acid. Furthermore the two amino sugars are linked in a $1 \rightarrow 3$ configuration rather than a $1 \rightarrow 4$. Moreover, the amino acids of the peptide crosslink lack D-stereoisomers.

The extreme halophilic Archaea, the extreme thermophilic Archaea, and some of the methanogens have glycopeptide cell walls containing a variety of sugars linked covalently to protein (see Chapter 22 for more detail). Differences in the cell walls of the bacteria may explain how they are able to survive in their special environments. For example, halophilic Archaea that live in high osmotic concentrations may not need the rigid peptidoglycan cell walls typical of Eubacteria.

Function of the Cell Wall

Bacteria generally grow in **hypotonic** aquatic environments; that is, the concentration of particles (ions and

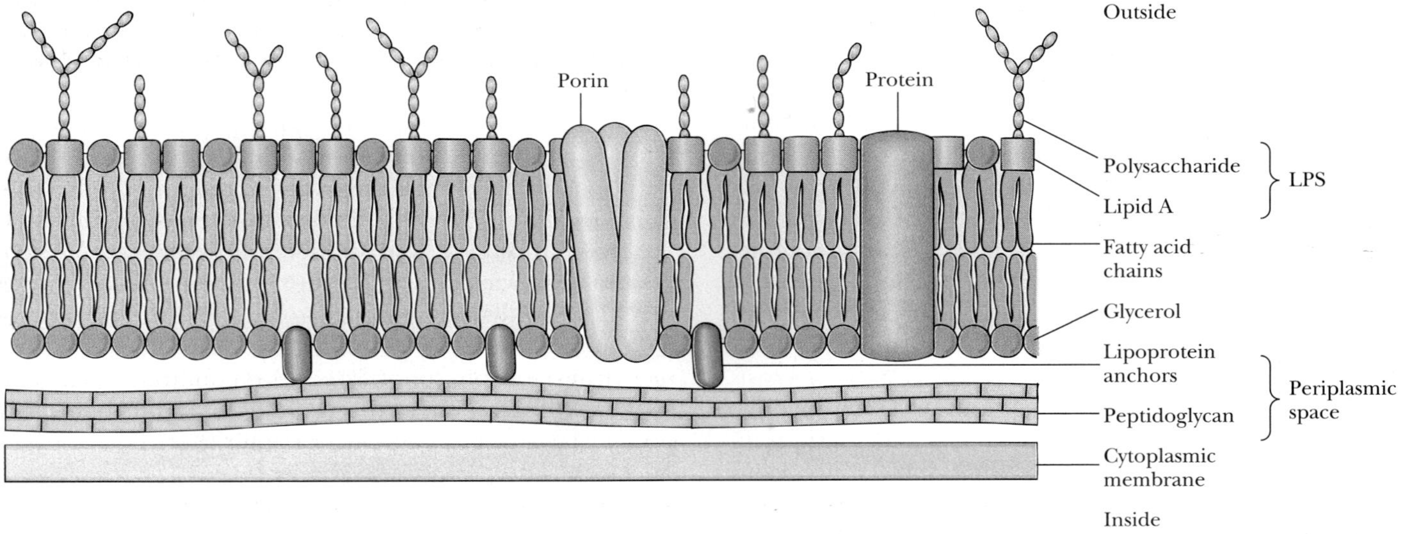

***Figure* 4.51** Diagram of the outer membrane of a gram-negative cell showing transmembrane porin proteins.

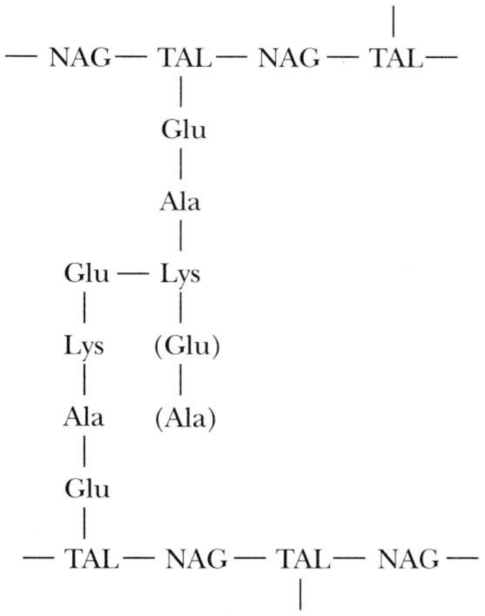

***Figure* 4.52** The pseudopeptidoglycan of *Methanobacterium thermoautotrophicum.* The amino sugar backbone consists of alternating NAG (*N*-acetyl glucosamine) and NAT (*N*-acetyl talosaminuronic acid). The cross-linking amino acids are Glu (glutamate), Ala (alanine), and Lys (lysine).

molecules) inside the cell is greater than outside the cell. The cell membrane, which is freely permeable to water, permits uptake of water by osmosis so that the concentration of particles on both sides of the cell membrane can be equalized. Thus, the cell swells and the membrane distends to accommodate the increase in water. The cell wall prevents this osmotic pressure, called **turgor pressure** (which is the normal consequence of living in hypotonic environments), from stretching the membrane too far. Excessive distension of the membrane would cause the membrane to break and result in **lysis** (rupture) of the cell.

Thus, one function of the cell wall is to protect the cell from turgor cell lysis, or **plasmoptysis.** This function is readily illustrated by gram-positive bacteria such as those of the genus *Bacillus.* The cell wall layer can be removed by treatment of the cell with **lysozyme,** an enzyme that breaks the covalent bonds between the amino sugars (see Figure 4.42) of the peptidoglycan in the cell wall. If this is performed in a normal hypotonic growth environment, the cells simply lyse. However, if the growth environment is rendered isotonic (that is, the osmotic pressure on both sides of the membrane is equalized) by suspending the cells in a 2-*M* sucrose solution, then the wall-less cells do not lyse. Instead, the rod-shaped cells round into spherical units called **protoplasts,** which lack a cell wall (Figure 4.53). A protoplast can persist in an isotonic environment because the turgor pressure of the cell is balanced by the concentration of sucrose molecules in the medium.

Antibiotics and Cell Walls

Some antibiotics exert their activity against Eubacteria by preventing the synthesis of normal cell walls. Antibiotics in the **penicillin** (Figure 4.54) and **cephalosporin** groups, both of which are produced by certain fungi, act by inhibiting the enzymes responsible for biosynthesis of peptidoglycan (Chapter 11). As a result, when bacteria are exposed to them, the bacteria produce a defective cell wall, which results in the ultimate lysis, and destruction, of the bacterial cell. It is noteworthy that penicillin will cause lysis and death of bacteria only under conditions that permit growth and cell wall synthesis. Thus, when antibiotics are used therapeutically, bacterial cells that are not growing and dividing can survive in the body and may begin growing following the treatment. They must be removed by the normal defense activities, such as by phagocytes, before they are completely eliminated.

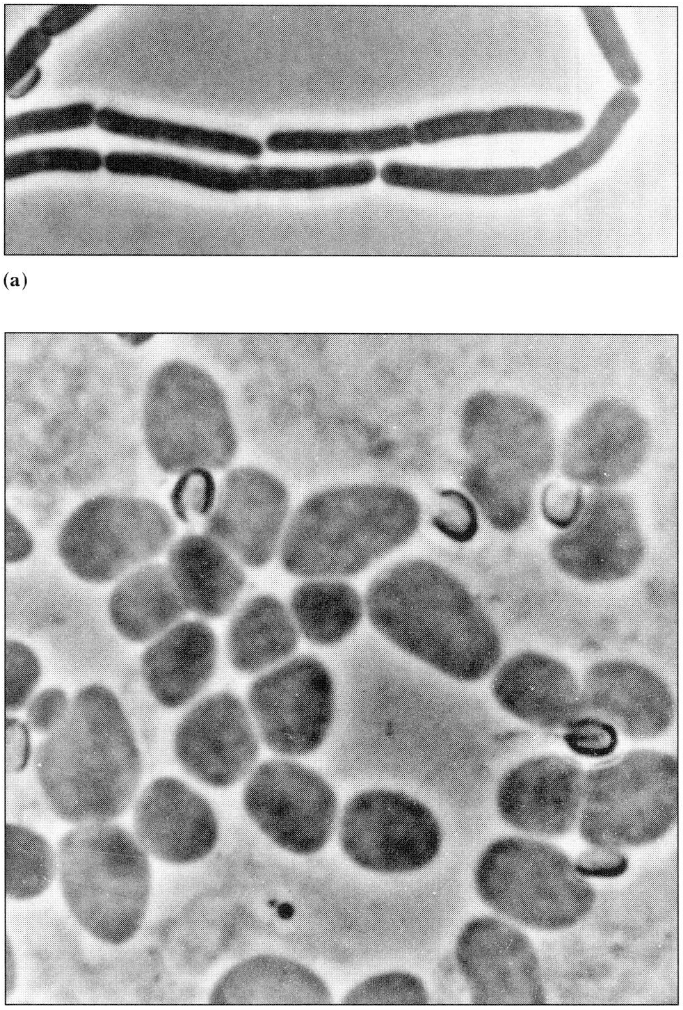

(a)

(b)

***Figure* 4.53** **(a)** Normal cells of *Bacillus megaterium* showing rod-shaped cells in chains. **(b)** Protoplasts of the same *Bacillus megaterium* preparation produced by lysozyme treatment and stabilized by M/7 sucrose. Note that the protoplasts are distended because of low sucrose concentration. Empty spore coats in the preparation are derived from spores used as the inoculum. (Courtesy of Carl Robinow)

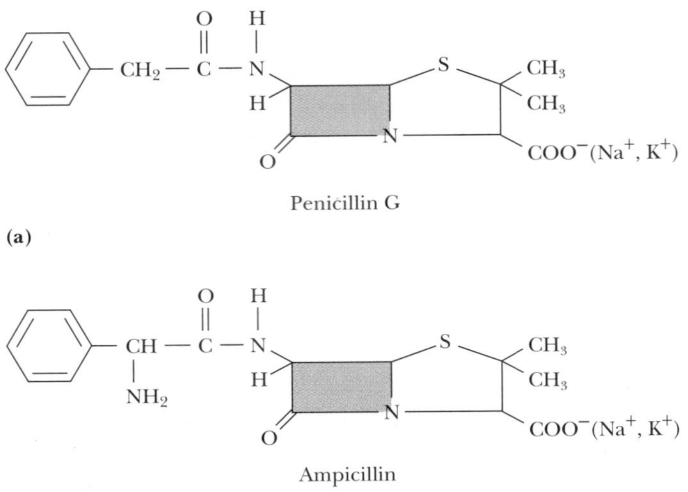

***Figure* 4.54** Chemical structure of **(a)** penicillin G and **(b)** ampicillin. Blue area is characteristic β-lactam ring.

These antibiotics owe their **selective toxicity** to Eubacteria due to the absence of peptidoglycan outside the eubacterial world. Thus, humans treated with these antibiotics are not affected (unless they have an allergy to them) because they do not synthesize peptidoglycan. As one would predict, penicillin is generally much more effective against gram-positive bacteria than against gram-negative bacteria. There are two primary reasons for this. First, the major structural component of gram-positive bacterial cell walls is peptidoglycan. Also, in gram-negative organisms the outer membrane provides additional protection by acting as a permeability barrier to some antibiotics such as penicillin G. For this reason, penicillin and other antibiotics are sometimes modified to permit better solubility in the outer lipid membranes. Ampicillin, for example, is a form with greater hydrophobicity that is used for gram-negative infections (Figure 4.54**b**). Penicillin is effectively used in the treatment of certain gram-negative infections, including syphilis, caused by *Treponema pallidum,* and gonorrhea, caused by *Neisseria gonorrhoeae,* as well as in most infections by gram-positive bacteria.

Treatment of gram-positive bacteria with penicillin results in the formation of protoplasts in much the same manner that they form using lysozyme. Protoplasts are not produced by gram-negative bacteria because their cell walls cannot be completely removed in this way. Instead, removal of the peptidoglycan layer results in the formation of **spheroplasts** which, although lacking in cell wall peptidoglycan, still possess the outer membrane.

In summary, the cell walls of bacteria are rigid structures that not only give cells their shape, but protect the cell from lysis. In addition, they serve a limited role in determining which molecules will pass into the cell membrane, especially for gram-negative bacteria.

Periplasm

The space between the cell wall and the cell membrane is called the **periplasm.** Although in thin sections it appears to be an empty space (see Figure 4.46), it is important to the physiology of the cell. The periplasm serves as an area of considerable enzymatic activity. For example, several steps in the synthesis of the cell wall occur in this area. Also, **chemoreceptors** (see Flagella section below) involved in the chemotactic response are located here. Furthermore, gram-negative bacteria produce proteins called **binding proteins** that are located in the periplasm. These binding proteins combine reversibly with substrate molecules, and, in doing so, concentrate them for membrane carrier proteins. They then release the substrates to the carrier proteins for transport into the cytoplasm.

Some controversy exists among microbiologists as to whether gram-positive bacteria have a periplasm. No doubt this is due to the extensive research that has been done with gram-negative bacteria, which have served as model organisms for the study of the periplasmic area. Furthermore, the periplasm of gram-negative organisms undoubtedly differs from that of the gram-positive bacteria because gram-negatives have the additional permeability layer, the outer membrane.

Bacterial Appendages

Many bacteria produce special appendages that extend from the surface of the cell. The two types that are widely found on prokaryotes include fimbriae and flagella.

Fimbriae, Pili, and Spinae

These are protein appendages that range in diameter from about 3 nm to 5 nm in most bacteria (Figure 4.55**a**) to about 100 nm to 200 nm in some aquatic varieties, a size that is sufficiently large for them to be seen with the light microscope. The three forms are structurally identical in that they are all proteinaceous appendages formed by subunit protein structures. The protein of these appendages is helically wound, leaving a pore through the center. Each consists of a repeating protein subunit molecule. Some also have a glycoprotein tip on them.

The largest appendages of this type are called **spinae** (Figures 4.55**b** and **c**). The spinae of aquatic bacteria serve to increase the surface area and, hence, their drag, thereby enabling them to remain suspended for longer periods of time or to be carried further by currents. Thus, they assist the organism in maintaining its position in the plankton.

The terms **fimbriae** and **pili** are often used interchangeably; however, fimbriae is usually considered a more general term for the thin fibrillar appendages so commonly found on prokaryotes. A single bacterial cell may produce more than one type of fimbria. In fact, some bacterial viruses attach to one type of pilus but do not attach to the cell surface or to other types of pili.

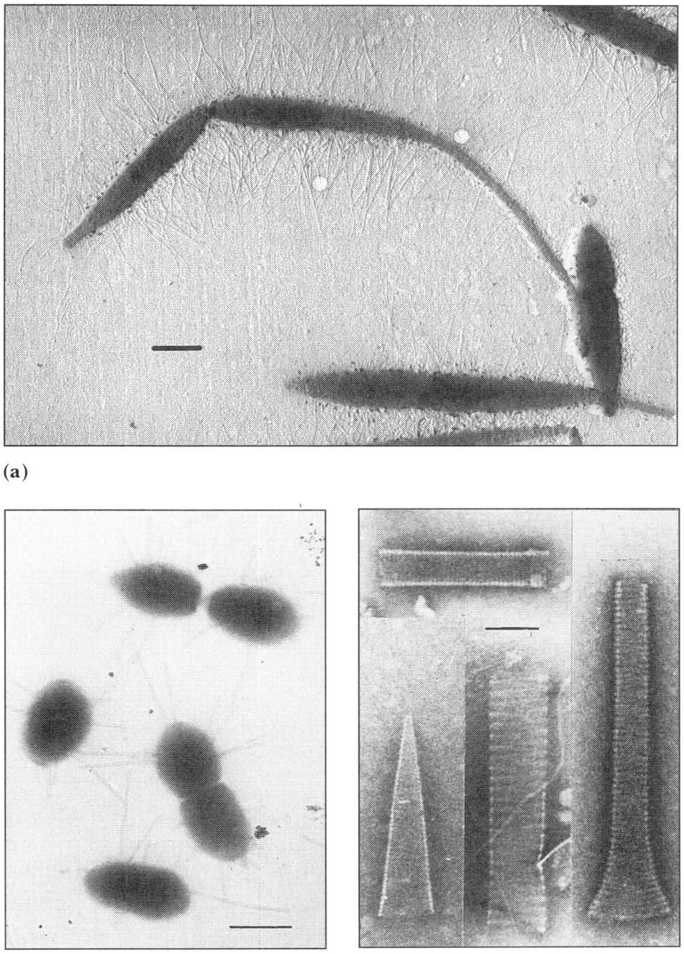

(a) (b) (c)

***Figure* 4.55 (a)** An electron micrograph of a cell of *Prosthecobacter fusiformis* with many fimbriae extending from it; bar equals 0.5 μm. (Courtesy of R. Bigford and J. T. Staley) **(b)** An example of a spinate bacterium with several large spinae; bar is 0.3 μm. **(c)** Electron micrographs of isolated spinae; bar is 0.1 μm. (Courtesy of K. B. Easterbrook)

Fimbriae are known to be important for the attachment of some bacteria. For example, one pilus of *Escherichia coli* mediates the attachment of the two types of mating cells during sexual conjugation and is therefore specifically called the "sex pilus" (see Chapter 15). Likewise, pili produced by *Neisseria gonorrhoeae* are responsible for the attachment of pathogenic strains to endothelial tissue of the genito-urinary tract in humans. Only piliated strains cause gonorrhea.

In some *Pseudomonas* spp. certain fimbriae have been shown to be involved in the export of protein from the cell. Furthermore, a type of motility termed **twitching** in some bacteria may also be caused by fimbriae. Twitching is a somewhat obscure phenomenon whereby cells that seem to be sessile (immotile) suddenly jerk 1 μm or so in distance.

Flagella

Many prokaryotes that are motile produce **flagella,** which are long, flexible appendages resembling "tails." These proteinaceous appendages, generally about 10 nm to 20 nm in diameter, are found on some members of the gram-positive and gram-negative Eubacteria as well as on some Archaea. They act like a propeller—the cell rotates them so the cell can move through an aqueous medium.

The number and location of flagella on the cell surface have been used as important features in taxonomy. Some bacteria have a single polar flagellum extending from one end of the cell (and thus move by what is called **monotrichous polar flagellation**). These single flagella filaments cannot be resolved by the light microscope, but are readily discerned with the electron microscope (Figure 4.56). Other species produce polar tufts or bundles **(lophotrichous flagellation),** which can be seen with light microscopy (Figure 4.57). Still other prokaryotes produce flagella from all locations on the cell surface **(peritrichous flagellation).** Motile enteric bacteria, such as *Proteus vulgaris,* exhibit this latter type of flagellation (Figure 4.58). Finally, a few bacteria are known to move by mixed flagellation—involving one type of polar flagellum and another type of flagellum from other locations on the cell surface.

A flagellum consists of three major sections: a long **filament,** which extends into the surrounding environment; a **hook,** which is a curved section connecting the filament to the cell surface; and a **basal structure,** which anchors the flagellum into the cell wall membrane of the bacterium by special disc-shaped structures called plates or rings. The

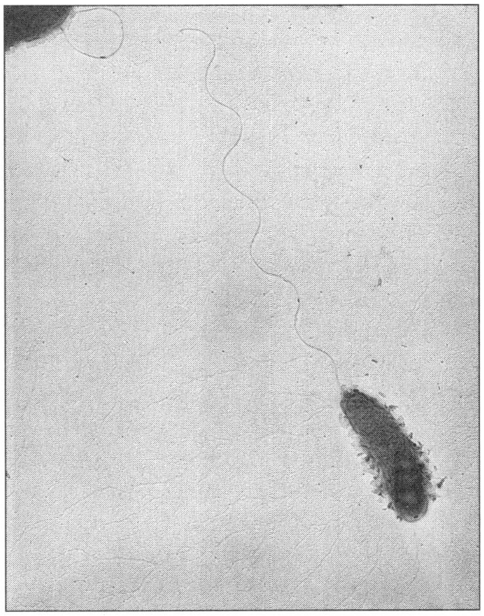

***Figure* 4.56** An electron micrograph of a whole cell of a gram-negative bacterium showing its single polar flagellum. (Courtesy of J. T. Staley)

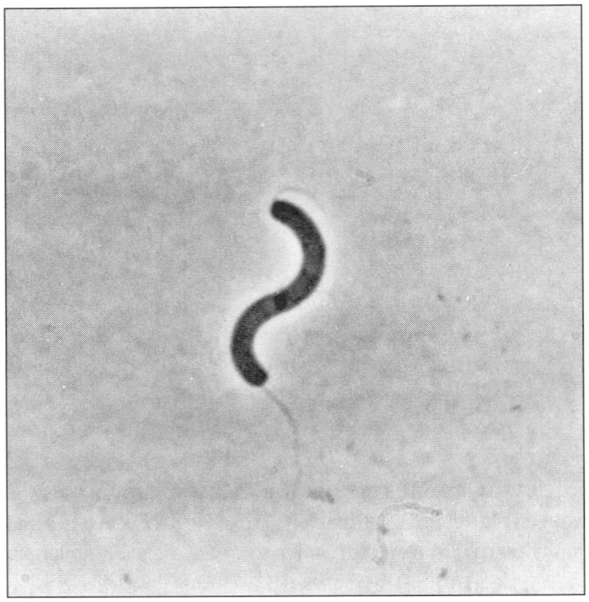

Figure **4.57** A phase microscope image of a spirillum showing the polar tuft of flagella. (Courtesy of J. T. Staley)

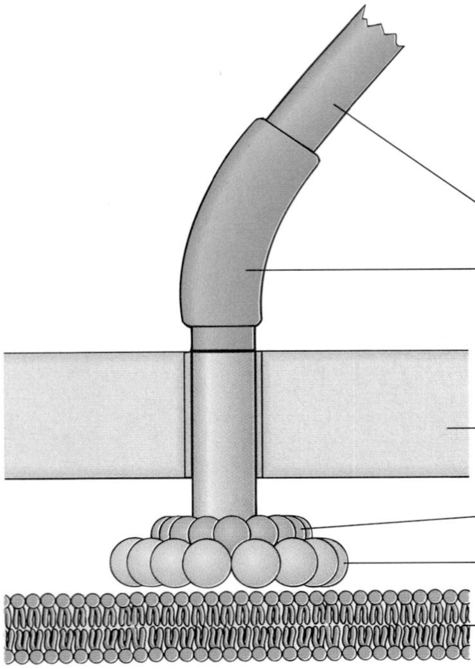

Figure **4.59** Diagram showing the anchoring of the flagellum to the cell envelope of a gram-positive bacterium.

anchoring varies somewhat between gram-positive bacteria (Figure 4.59) and gram-negative bacteria (Figure 4.60) because of the differences in cell wall layers.

The flagellum filament is composed of a protein called **flagellin.** With few exceptions, all bacteria studied thus far have a single type of protein subunit (two exist in *Caulobacter* and some *Vibrio* species). Repeating subunits of flagellin are joined together to form the filament, which is helically wound and has a hollow core. The filament (and thus the flagellum) is about 20 nm in diameter. The filaments are not straight. Each has a characteristic sinusoidal curvature depending upon the bacterial species. During the synthesis of flagella, new protein subunits are added at their distal tips.

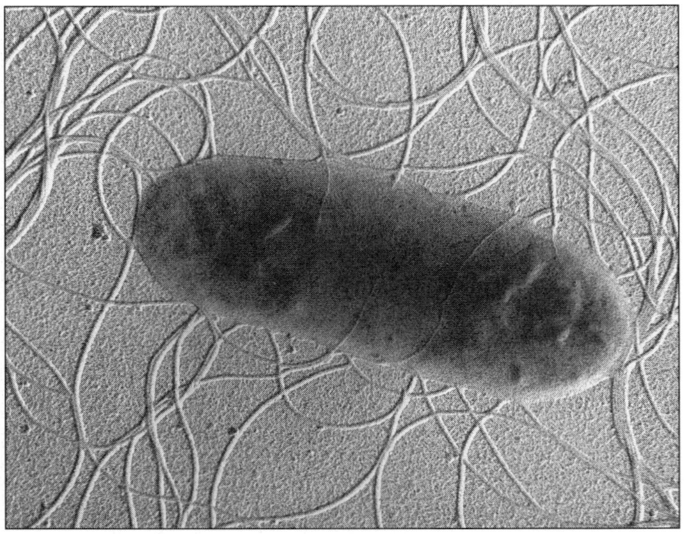

Figure **4.58** An electron micrograph of a peritrichously flagellated bacterium showing many flagella about the cell surface. (© BPS/T. J. Beveridge)

Flagella rotation and hence, cell movement, requires energy that is generated by the proton motive force. This can comprise a major expenditure of energy for the organism. But the result is immensely important allowing the cell to translocate by converting chemical energy into mechanical energy.

Tactic Responses: Chemotaxis and Phototaxis

Heterotrophic bacteria are able to propel themselves by their flagella into areas where organic nutrients are concentrated. This phenomenon, termed **chemotaxis,** can be illustrated with a capillary technique. If a capillary containing an organic energy source such as the amino acid serine is placed in a suspension of heterotrophic bacteria, the bacteria will accumulate at the tip of the capillary from which the nutrient is diffusing (Figure 4.61).

The development of a special microscope called a **tracking microscope** enables researchers to follow single bacterial cells during their movement. Through the use of this instrument, scientists were able to study the chemotactic behavior of individual cells exposed to nutrients. Chemotaxis has been studied most thoroughly in enteric bacteria such as *Escherichia coli.* These peritrichously flagellated bacteria orient their flagella in a polar bundle during motility. The bundle is rotated counterclockwise during forward movement called **running.** Running can be disrupted by **twiddling,** meaning brief periods when the flagella cease to operate in a coordinated bundle. As a result the flagella reverse direction and rotate clockwise, and the cell is seen to stop and somersault or tumble. After a

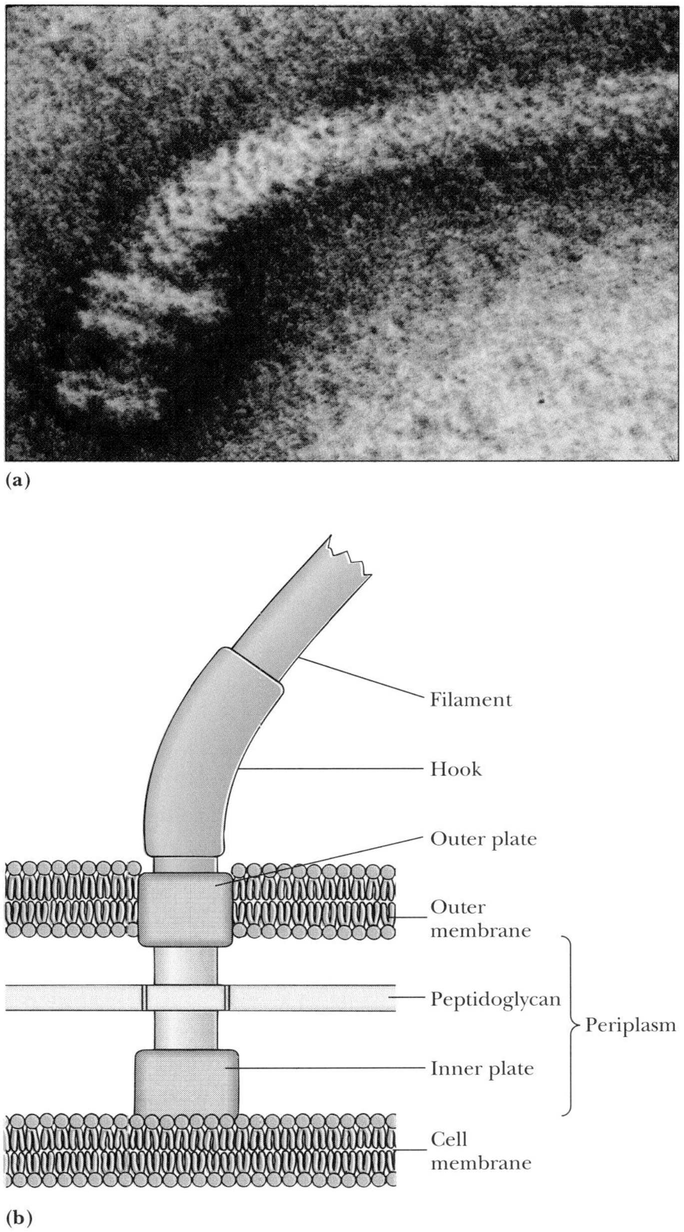

***Figure* 4.60** **(a)** An electron micrograph showing the basal ring structure of a gram-positive flagellum. (© Science VU/Visuals Unlimited) **(b)** Illustration of a cell cross section showing the flagellum attachment site of a gram-negative bacterium.

twiddle, the organism will again begin another run. However, the direction is completely random.

When the organism is in the gradient of a utilizable nutrient, less twiddling occurs, allowing the organism to spend more time in the favorable area (Figure 4.62). This phenomenon, in which a bacterium effectively becomes directed toward a utilizable nutrient, an **attractant,** is termed **positive chemotaxis. Negative chemotaxis** refers to movement *away* from toxic materials, called **repellants,** and is also commonly exhibited by bacteria.

Scientists are just now beginning to unravel the biochemical mechanism of chemotaxis. Many utilizable organic nutrients serve as attractant molecules in chemotaxis and thereby cause positive chemotaxis. These molecules

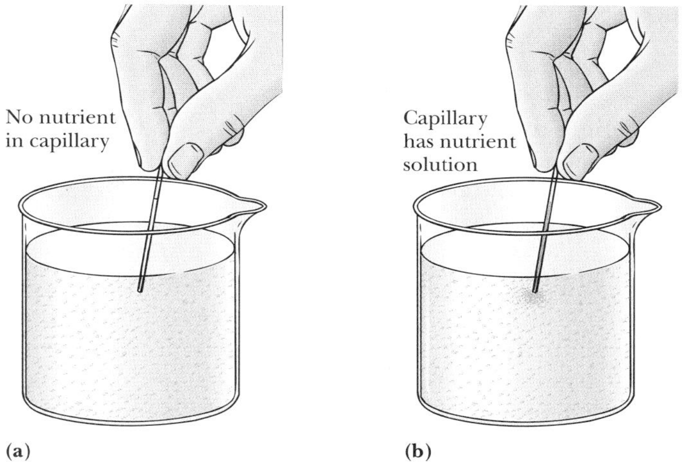

***Figure* 4.61** An illustration of the capillary chemotaxis effect. **(a)** Capillary in suspension has no nutrient. Note the even distribution of cells (represented by dots) in suspension. **(b)** A nutrient gradient has been established in this bacterial suspension by introducing a solution of glucose into the capillary. The cells are attracted to the glucose by positive chemotaxis, and therefore an increased turbidity is observed near the orifice of the capillary.

are bound to specific receptor proteins that are located in either the periplasmic space or the cell membrane of gram-negative bacteria. For some attractants such as the amino acid serine, the binding protein is a cell membrane protein that can be methylated in the presence of a methyltransferase using *S*-adenosylmethionine as a substrate. The process of binding and subsequent methylation of the methyl-accepting chemotaxis protein (MCP) results in a signal that causes the flagellin subunits to form a bundle, exciting the flagella motors into counterclockwise rotation, and thus a run. Some sugars that serve as

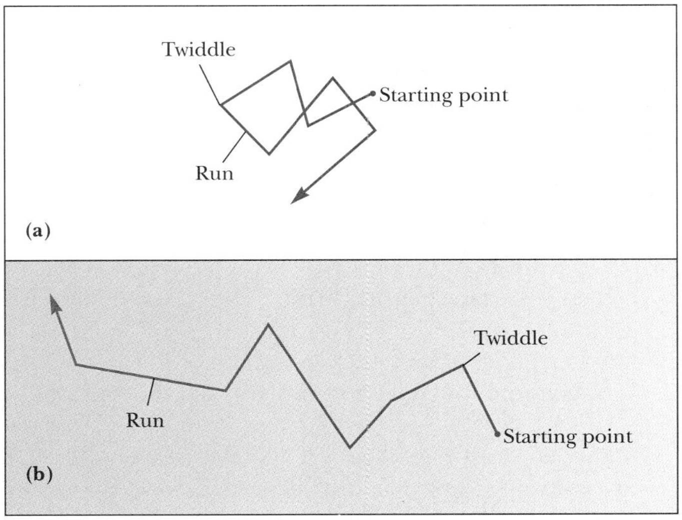

***Figure* 4.62** A diagram illustrating the phenomenon of runs and twiddles in chemotaxis. **(a)** Cells twiddle more frequently and have shorter runs when no nutrient is present. **(b)** When cells are in a nutrient gradient, positive chemotaxis occurs so that runs are longer and twiddles less frequent.

attractant molecules can effect positive chemotaxis at concentrations as low as 10^{-8} M.

It is noteworthy that, in chemotaxis, both spatial gradients and temporal changes in the chemical environment of the organism have an effect on cell movement.

Adaptation of the bacteria to their environment also occurs in the chemotaxis process. In the prolonged presence of the attractant substrate, remethylation of MCP proteins occurs to some extent and twiddles occur, but the rate of twiddling is less than it would be in the absence of the attractant. This may be a mechanism whereby the organism is able to conserve energy that would be needlessly spent on changing direction when it is not needed.

The simple behavior exhibited by bacteria during motility is exciting to biologists because understanding it at the molecular level may lead to a fuller understanding of more complex behavioral responses of higher organisms such as the sense of smell.

Likewise, photosynthetic bacteria use their flagella to move themselves toward specific wavelengths of light which change with depth in a water column. This phenomenon, called **phototaxis,** can be illustrated by projecting a spectrum of visible light onto a glass microscope slide. A suspension of flagellated bacteria placed on the illuminated slide will quickly concentrate at the wavelengths of light that are absorbed by the photosynthetic pigments of the particular organism (Figure 4.63). Thus, the cells can position themselves at specific wavelengths that are used for photosynthesis.

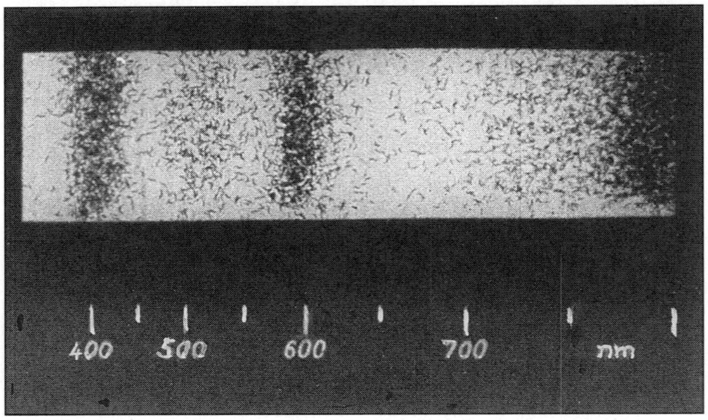

***Figure* 4.63** A spectrum of visible light has been projected on this slide of *Thiospirillum jenense.* These purple sulfur photosynthetic bacteria have migrated and accumulated by positive phototaxis at those wavelengths of light at which their photosynthetic pigments absorb light best. (Courtesy of Norbert Pfennig)

As in chemotaxis, both positive (toward light) and negative (away from light) responses are observed. However, very little is known about the chemical signals and mechanisms underlying phototaxis.

Summary

- A **simple microscope** is analogous to a magnifying glass with a single lens, whereas **compound microscopes** have two lens systems including the **eyepiece** (or ocular lens) and the **objective lens.** In addition a **condenser lens** system is used to concentrate light on the specimen.

- The **resolution** of the light microscope is determined by the **wavelength** of light, the **refractive index** of the suspending medium, and the closeness of the objective to the specimen, which determines the **angular aperture** of the lens.

- Under most conditions of growth, bacteria have a negative surface charge and are therefore stained with **basic dyes,** which have a positive charge. In contrast, **negative stains,** such as India ink, do not stain cell material but stain the aqueous background.

- **Phase contrast microscopy** increases the contrast of bacteria by amplifying the difference in refractive index between the cells and the aqueous environment.

- **Transmission electron microscopes (TEMs)** have high resolution because they use electrons as their source of illumination. **Thin-sections** that reveal intracellular details are prepared for electron microscopy by dehydrating and embedding cells in plastic resins, which are then sliced with ultramicrotomes.

- **Scanning electron microscopes (SEMs)** produce images of whole bacterial cells or colonies.

- Many bacteria are unicellular **rods, cocci, vibrios,** or **spirilla.** Some species form chains, filaments, or packets of cells. The **actinomycetes** are gram-positive bacteria that produce branching filamentous structures called **mycelia.**

- Most bacteria divide by **binary transverse fission** in which a mother cell divides to produce two mirror image sister cells, but a few bacteria divide by **budding** or **fragmentation.**

- The bacterial chromosome is a double-stranded circular molecule of DNA. Many bacteria harbor **plasmids,** which are double-stranded DNA molecules containing genes that are not essential for the livelihood of the host bacterium but may carry important genes for determining their physiological capabilities.

- **Gas vesicles** are protein membrane structures that exclude water and therefore reduce cell density, enabling cells to become buoyant in aquatic habitats.

- Organic storage compounds of bacteria include glycogen, starch, **poly-β-hydroxybutyrate (PHB),** and **cyanophycin.** Inorganic storage compounds of some bacteria include **volutin** (also called polyphosphate) and elemental sulfur.

- The cell membranes of Eubacteria contain proteins embedded in lipid bilayer membranes that contain fatty acids linked to glycerol with **ester linkages.** However, the cell membranes of Archaea have proteins and lipids that contain phytanyl groups linked to glycerol by **ether linkages.**

- **Extracellular layers** are not found on all bacteria but those that have them have either **capsules, sheaths, slime layers,** or **protein jackets,** depending on the species.

- Almost all Eubacteria produce **peptidoglycan** or **murein,** a unique cell wall structure that contains an amino sugar backbone of *N*-acetyl glucosamine and *N*-acetyl muramic acid and a peptide bridge used for crosslinking the amino sugar backbone. Peptidoglycan contains some unique **D-amino acids** such as D-alanine and D-glutamine acid not found in protein. Peptidoglycan contains one diamino amino acid, such as **diaminopimelic acid** or **lysine,** that crosslinks the peptide bridge. Peptidoglycan can be hydrolyzed by **lysozyme.** Normal peptidoglycan synthesis is inhibited by **penicillin** and certain other antibiotics.

- Gram-positive bacteria may also contain **teichoic acids** and **teichuronic acids** which are not found in the cell walls of gram-negative bacteria. In contrast, gram-negative bacteria contain, in addition to peptidoglycan, an **outer membrane** that has lipid, protein, and polysaccharides. The polysaccharides of the outer membranes of gram-negative bacteria are linked to **lipid A.** The lipid A–polysaccharide complex is termed **lipopolysaccharide** (or **LPS**) and is toxic to vertebrates (called **endotoxin**).

- **Archaea** lack peptidoglycan in their cell walls; most have polysaccharide or protein cell walls.

- **Fimbriae** and **pili** are proteinaceous extracellular fibrillar appendages of some bacteria used primarily in attachment.

- The **periplasm** is an area found between the cell membrane and the outer membrane of gram-negative bacteria. A variety of enzymatic activities occur in the periplasmic space including cell wall synthesis, chemotaxis, and transport.

- **Positive chemotaxis** involves moving toward a nutrient. **Negative chemotaxis** occurs when a cell moves away from a toxic substance. Photosynthetic bacteria can respond to light by either positive or negative **phototaxis.**

Questions for Thought and Review

1. Why does the TEM have better resolution than the light microscope?
2. Explain the differences between brightfield and darkfield microscopy. Would bacteria be stained for viewing in darkfield or phase contrast microscopy?
3. What advantages does phase contrast microscopy have over ordinary brightfield?
4. How does fluorescence microscopy work?
5. Which prokaryotes are not sensitive to penicillin? Why?
6. In what ways do the Archaea differ from the Eubacteria?
7. Why is a diamino acid important in peptidoglycan?
8. It has been hypothesized that the gas vacuole is a primitive organelle of prokaryotic motility. What evidence is consistent with this?
9. Distinguish among positive and negative chemotaxis and phototaxis. What is known about the molecular basis for chemotaxis?
10. List the various types of storage granules of prokaryotes under the categories of energy source versus nutrient source.

Suggested Readings

Beveridge, R. L., and L. L. Graham. 1991. Surface layers of bacteria. *Microbiological Reviews* 55:684–705.

Gerhardt, P., R. G. E. Murray, W. A. Wood, and N. R. Krieg. 1993. *Methods for General and Molecular Bacteriology.* Washington, DC: American Society for Microbiology.

Neidhardt, F. C., J. L. Ingraham, and M. Schaechter. 1990. *Physiology of the Bacterial Cell: A Molecular Approach.* Sunderland, MA: Sinauer.

Schatten, G., and J. B. Pawley. 1988. Advances in optical, confocal, and electron microscopic imaging for biomedical researchers. *Science* 239:164.

Part 2

Microbial Physiology: Nutrition and Growth
A Conversation with Dr. Lucy Shapiro

Dr. Lucy Shapiro has devoted her scientific career to understanding the process of cellular differentiation using bacteria. In her research she has focused on the prosthecate or stalked bacterium, *Caulobacter crescentus.* Her seminal work has shown that cells can localize molecules at their poles, not only in this bacterium, but in other typical eubacteria, such as *Escherichia coli.* In addition, her lab has made important contributions to understanding the structure and assembly of the bacterial flagellum. Dr. Shapiro is currently a professor at Stanford University.

JS: What led to your decision to study microbiology?

LS: My training as a graduate student—we won't discuss my undergraduate work in fine arts—was really in physics, chemistry, and physical chemistry. My graduate degree is in molecular biology, but because I had little training in the sciences as an undergraduate, I had to take a lot of very basic science courses as a graduate student. I concentrated on physics and chemistry—the idea being that if I learned to think analytically, I could read biology.

By the time I became an Assistant Professor, I was really interested in a simple question: How do you go from a one-dimensional genetic code to a cell that exists in three-dimensional space? I wanted to solve that. Clearly, the shape of something, how something is built, how something looks in three dimensions is genetically encoded. Well, how do you do that? How do you translate that information? It's inherited, so it's got to be in the genes, but where?

I wanted to get my hands on molecules. I was also fascinated by all that was being learned in those days about phage lambda and the power of genetics. What I did was spend about two months just reading, looking for a system that was akin to lambda, where I could identify all the genes, all their products, all their regulatory feedback loops, and figure out how it was built—but I didn't want a phage. I wanted a cell. I didn't care what kind of cell it was. It had to be a cell that knew which end was up, that built something, and had landmarks at the cell poles, so that I could take the organism apart and understand how to put it together again, and then understand how information was translated from a genetic code to three-dimensional space. While reading, I came upon a review written by Roger Stanier, where he showed a small figure with a diagram of the gram nagative bacterium *Caulobacter crescentus,* in which a swarmer cell becomes a stalked cell, and a stalked cell divides to yield two different progeny: a new swarmer cell and a stalked cell. He said that this was a wonderful organism in which to study cell differentiation. And I said, "Right!"

Then I read more. Some very pretty work had been done by his two students, Jeanne Poindexter and Jean Schmidt. They had put together a beautiful body of work on microbiology of *Caulobacter.* For me it was a good starting place, because what was needed was to build a whole system of genetics, do sophisticated biochemistry, and truly understand how you go from gene to structure. I chose *Caulobacter* because it potentially held the answers to my questions.

I literally started out in 1969–1970, knowing all the questions that I was going to ask 20 years later. And I knew that I had to build this thing pyramid style—there was no shooting from the hip. I couldn't go in and ask exotic questions about how cells build things, until I knew more about how cells are organized and the

(Courtesy of Dr. Lucy Shapiro)

mechanisms that control the cell cycle. A wonderful friend and mentor, Barbara McClintock, at Cold Spring Harbor, would say whenever I would get discouraged, something like: "Know your organism, do not go to any other organism. Stay with *Caulobacter.* Don't follow fads. Stick by your questions and you'll get answers. Don't get discouraged. Find out all you can about the complete physiology of your organism and you will be able to get answers to important questions." She was right. Everything that drove me to microbiology was based on answering important questions in biology.

> **When I started my quest for a system to study I didn't care what the organism was, as long as it was suitable for my questions.**
>
> **I literally started out in 1969–1970, knowing all the questions that I was going to ask 20 years later.**

It turns out that the microbial world is a very good place to get answers about many fundamental questions in biology. So my initial question of how you go from a one-dimensional genetic code to three-dimensional space then got translated into asking how cells become polar! How does a cell know one end from the other? That easily got translated into what makes an asymmetric cell division? One of the most important happenings in human life, in early development, is that as we develop, cell divisions occur that yield two kinds of cells: stem cells that divide to give one cell that is differentiated, and another cell that remains a stem cell. This is a basic paradigm. *Caulobacter* does that in every cell cycle.

JS: Earlier you mentioned your interest in art. How does that dovetail with your interest in biology?

LS: I think the way it dovetails is in structure. I'm interested in how to build things. The combination of learning a lot of chemistry and understanding quantum mechanics and statistical mechanics forces me to think in an analytical, orderly way. The fact that I can visualize things in three dimensions prompted me to address these questions: Where is the epigenetic memory in a cell that lets a cell know which end is up? There have to be interactions among genetic regulatory mechanisms that direct the cell to make specific proteins at specific times in specific places, culminating in complex cell structure, such as the flagellum. This quest ultimately led me to cell biology. So, I believe very strongly that it is important to learn multiple disciplines.

JS: What advice would you give to undergraduate students who are interested in pursuing a career in microbiology?

LS: We have so many tools at our fingertips nowadays (and we need them), that I would advise a very solid grounding in biochemistry and genetics. You really need to know hard core science and you also have to be curious about the biological world. But I would stress hard core training both at an undergraduate and graduate level. The world of microbiology has opened many windows to us, but to get through the window you have to know quantitative biology.

JS: What about careers for women in microbiology?

LS: I've thought about this question. My conclusion is that in any field, there are real hurdles and there are real opportunities. In microbiology, I believe, the hurdles and opportunities are the same for men and women. No matter how hard I try, I can't separate out anything that would be women specific.

You know, there is a great Barbara McClintock story. I think she was in her early eighties. There was a lecture she was going to give at a big university which I will not name, and you will know why in a moment—University X. They had gone to extraordinary lengths to take care of her, to get her out there, to do all this stuff. Maybe it was even after she won the Nobel Prize. I was in her office when she got a call from this University X, three days before the big event, just to finalize things, and Barbara claimed they said, "By the way, the money for the event is being provided to support a lecture for a woman in science." And, I heard Barbara say, "I'm not coming." Then I hear, "I don't believe in women in science. I don't believe in men in science. Science is gender-blind." She didn't go. I agree. It is gender-blind.

However, there is a difference between saying that when you are already in it, and when you are a young woman trying to get into it. I think that the best advice I can give in that area, is you have to be very good at what you do and you have to have a passion for your science. You also have to focus your mind on doing science, not fighting battles. As a young student I gave a talk at a Federation Meeting and Dr. X came up after and said, "I didn't know whether to listen to your talk or look at your legs." My response at that time was, "Your loss." And that was that.

I can recommend by the way, a wonderful biography of Marie Curie that I'm reading right now. It's written by Susan Quinn. The Quinn biography shows the passion that Marie Curie had for her science. And, I must say I don't know a single, successful woman scientist who has not lived this passion. Yes, some of us have kids, some of us don't have kids. We live as many different lives as there are women, but one thing we have in common is that the central core of our existence is our science. Sure there are hurdles, sure there are problems, but they are there for everyone. The successful women scientists I have seen pursued their work through hell and high water because of their passion for science. This is a tough business whether you are a woman or a man. You give up a lot. I don't have a normal life. I'll never have a normal life. But I have an interesting life.

JS: Your laboratory has been instrumental in studying differentiation in *Caulobacter.* Among your contributions, what do you think are most important in this area and why?

LS: I think that there are at least three. The first, which is probably one of the more recent ones, is the discovery that bacteria are not like swimming pools. I had always been told bacteria were bags of enzymes: there was no organization, and the chromosome, proteins and everything else were simply floating around. There was free diffusion. Untrue, untrue! It is not a swimming pool.

Free diffusion is rare, and the bacterial cell is, in fact, compartmentalized. The first indication in *Caulobacter* came from the localization of the chemoreceptors that are involved in chemotaxis. We had done many biochemical experiments

The world of microbiology has opened many windows to us, but to get through the window you have to know quantitative biology.

that told us the chemoreceptors had to be localized at one pole of the cell. When we purified the protein, made antibody to it, and Janine Maddock, a very talented postdoctoral fellow, did immunogold electron microscopy on thin sections, as well as indirect immunofluorescence on these little bacterial cells, we were able to demonstrate chemoreceptors located at the pole of the cell, not only in *Caulobacter,* but in *E. coli* and every other bacterium we looked at. This was a surprise. The paper reporting this result was a lead article in Science, was 99% controls, and generated a lot of interest. I think if we had just shown it in *Caulobacter,* it would not have gotten as much attention as it did.

The second thing we found is that the bacterial cell is quite asymmetric just before cell division. The two newly replicated chromosomes behave differently; different genes are transcribed from the two chromosomes, and only one of the two chromosomes is competent to initiate DNA replication. We now know that protein factors that control differential transcription and differential replication initiation are asymmetrically distributed in the predivisional cell. Because we know that there has to be compartmentalization before cell division, it seems likely to me that bacterial cells have some sort of cytoskeleton. Demonstration of a cytoskeleton equivalent in bacteria is really the next challenge.

Third, a graduate student in my lab, Kim Quon discovered a global transcriptional regulator, an essential response regulator homolog of the two component signal transduction system, that regulates the transcription of genes controlling important landmarks in the cell cycle: from flagellar biogenesis, to DNA methylation, to the differential initiation of DNA replication, and cell division. Thus, the bacterial cell cycle will soon yield its biochemical checkpoints, in a manner not unlike cyclins in eukaryotic cells.

JS: Why is it desirable to study developmental biology in a bacterium compared to a eukaryotic organism?

LS: I don't think it matters. However, the great diversity of microbes make them extremely valuable in studying differentiation, because you can ask questions about such things as cell-cell communication and about how cells build complicated structures, in a way that easily allows you to go from genes to cellular structure and function. The goal is to find out how nature solves a problem and I don't care what organism provides a solution.

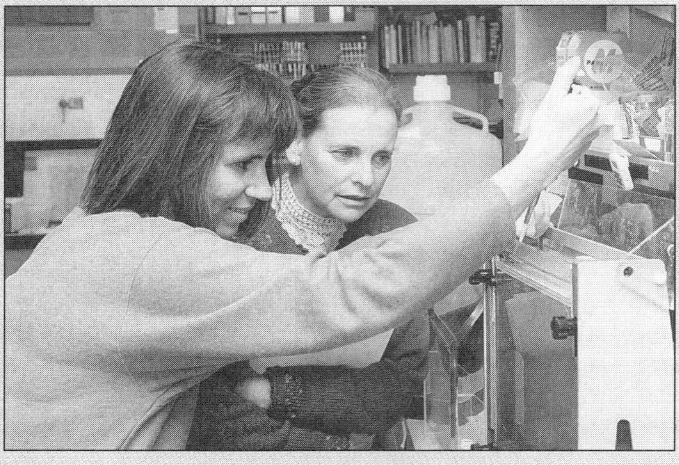
(Courtesy of Dr. Lucy Shapiro)

JS: In some cases is it simpler to study prokaryotes instead of eukaryotes?

LS: Sometimes but not always. The great advantage in bacteria and a few eukaryotic organisms is provided by the "Security Council" of organisms, as Jerry Fink at the Whitehead Institute always calls them. The "Security Council" comprises bacteria, yeast, flies, worms, and mice because you can use sophisticated genetics in these organisms. If you can't do genetics, you are severely limited. When I started working with *Caulobacter,* I knew if I was going to learn anything meaningful, I had to be able to take it completely apart genetically. I've had some wonderful colleagues. Bert Ely, at the University of South Carolina, did the major part of the work on the genetics of *Caulobacter.* The power of prokaryotes, as opposed to mice, for example, is that you can look at the entire system. You can look at the complete physiology of the organism. You can look at all parts of its cell cycle, or life cycle. You can get at every single gene. It's got a smaller genome. It does fewer things. But even the complexity of the single-celled bacterium is awesome.

JS: What future work on cell differentiation in *Caulobacter* needs to be conducted and why?

LS: Since everything we've done up to now tells me that bacteria must have a cytoskeleton, I'm intent on finding it. I want to understand what controls the segregation of the chromosomes, how they get to the right place, what tracks they are riding on, how asymmetry is generated and maintained.

We can't see any septum. We can't see any membrane, but experimentally, we see two things: when we do EM and immunogold labeling for cytoplasmic proteins we see the gold dots all through the cell, but as we approach division a region at mid-cell becomes refractory to immunogold labeling. Doing different kinds of EM staining we can actually see fibrils running across the waist of the cell. We know there's structure there.

In our department here at Stanford we have *Caulobacter* people sitting next to fly embryo people. The *Caulobacter* students say, "Gee, if you can see stripes on your fly embryos, then why can't we see parts of the bacteria. They're only smaller." We've been doing all kinds of cell biology in a way we weren't supposed to be able to do in bacteria. I think it is unproductive to segregate microbiologists from all other biologists. This has been done for too many years and it's artificial. We are limiting our future. By mixing disciplines, mixing geneticists, yeast people, worm people, fly people, everybody talking and listening and learning from one another, you open your horizons.

It comes back to the 'one tool chest' analogy. If fact, what will be done with bacteria is parallel to what is being done on yeast—understanding the role of the cytoskeleton in polarity. Only we don't know the nature of 'cytoskeleton-like' molecules in bacteria. We don't know what the proteins are. But that's the future. How does the cell become compartmentalized? What are the secretory mechanisms like? Are there multiple secretory mechanisms? Do some bacteria selectively send a subset of proteins out at the cell poles? Is there lateral differentiation? How does the cell architecturally do its job? What functions as the contractor? What coordinates the building of the cell? What's the role of proteases? Our recent experiments suggests that the cell is building and rebuilding constantly, by making and getting rid of proteins at the right place and at the right time. Temporal concerns, spatial concerns, turnover concerns. I want to construct a cell. That's the future.

How one organism or living entity chooses to solve a problem is used in various forms by all other kinds of organisms from *Escherichia coli* to *Homo sapiens*. We are only now beginning to understand the extent of the universality of basic mechanisms in biology.

I think what makes microbiology so different is the technology that is feeding into it. There are two very major influences that have happened to change microbiology. One is the sequencing of genomes. We are going to have all the genetic information for a great variety of microbes. As far as I'm concerned, it's done. It may be a few years yet, but it's done.

The second major thing is the end of the era of our antibiotic safety net. We are into an era of drug-resistant organisms and, I think pandemics. The pressure on us to find new targets for antimicrobials is going to escalate. I'm not saying that we should all simply switch to the study of pathogenic organisms. I'm not saying that at all. That's a different issue. What I'm saying is that microbiologists have to understand basic phenomena, fundamental mechanisms about how bacteria carry out their daily life, in order to uncover new mechanisms, new enzymes, new proteins, new things that are happening in the cell to provide targets for antimicrobial agents.

I think we stand now at the beginning of an explosion, a new flowering in microbiology. We have been lulled into complacency by antibiotics. We have been lulled into complacency by genetic engineering which has reduced bacteria to test tubes in which to make plasmids. And now suddenly it has become important that we uncover new targets. It's become very obvious that nature works with one tool kit. How one organism or living entity chooses to solve a problem is used in various forms by all other kinds of organisms from *Escherichia coli* to *Homo sapiens*. We are only now beginning to understand the extent of the universality of basic mechanisms in biology. Just now.

It appears incongruous to delineate metabolism of bacteria from that of fungi, since the two overlap in so many respects and since each group displays, more than any other groups of microorganisms, a kaleidoscopic array of metabolic diversity.

Jackson W. Foster

Chapter 5

Nutrition, Isolation, and Cultivation of Microorganisms

Most all plants are photosynthetic, gaining their energy from sunlight and feeding on atmospheric CO_2. Animals, on the other hand, require preformed organic matter as the source of both food and energy. This preformed organic matter comes either from plants or from an animal that ate plants. A plant or an animal species is differentiated from any other plant or animal species by observable differences in size and component structures. Mites to elephants or moss to sequoias are examples of the extremes in size/structure in the plant and animal kingdoms.

The other major component of the biosphere is the various microorganisms that are not markedly different from one another in either size or structure (for a notable exception see Chapter 8). Microorganisms basically differ from one another in the substrates that they can utilize as food and in their mechanisms for gaining energy. Bacterial and archaeal nutrition (sum of the processes whereby the organism takes in and utilizes food substances) varies from species to species. There is a microorganism in nature that can utilize, as nutrient source, any carbon-containing constituent that is a component part of living cells.

Many other compounds unrelated to living cells can also be utilized as substrate for growth. Among these compounds that microorganisms can actually utilize as carbon/energy source are carbon monoxide, methane, or cyanide. Others readily utilize toxic aromatic compounds such as benzene. These marked nutritional differences are referred to collectively as representing the microbes' "metabolic diversity." This chapter examines the major nutritional differences that distinguish microorganisms—how various nutritional types can be isolated and how they may be cultivated and preserved.

Fundamental Aspects of Diversity

Microorganisms live in virtually every environmental niche on earth. These microbial populations can utilize the naturally occurring organic compounds that fall into these niches as sources of carbon and/or energy (Chapter 12). Many products of chemical synthesis, such as pesticides, are also metabolized by natural microbial populations. However, some chemically derived compounds are resistant to microbial degradation and tend to accumulate in our environment. The evolutionary processes that yielded the bacterial and archaeal types that flourish under the various constraints imposed by nature have resulted in a seemingly endless array of microbial types. These organisms differ from one another in their nutritional capabilities and their adaptation to the physical conditions in which they live. Selected Eubacteria and Archaea are adapted to growth under wide-ranging variations in physical conditions, such as pH ($<$1 to $>$11), temperature ($<$0 to $>$100°C), salinity (fresh to saturated salt water), or availability of oxygen (aerobic to anaerobic) (see Figure 1.1**a,b,c,d**). These physical conditions are further imposed on the nature and availability of the essential substrates for growth of a microorganism (carbon, nitrogen, phosphorous, etc.).

Some bacteria can utilize simple monomeric substrates while others digest complex proteins and carbohydrates. Although some bacteria, such as *Pseudomonas cepacia,* can utilize any of over 100 different carbon-containing compounds as growth substrate, most organisms are capable of growth on a more limited number. There are others that are restricted to growth on one or two substrates. For example, *Methylomonas methanooxidans* will grow only when methane or methanol is provided as growth substrate. The ability of microbial populations to collectively grow on the unending variety of substrates encountered in nature is essential to maintain the level of CO_2 in the atmosphere. It is this CO_2 that sustains the photosynthetic populations that are the basis for all viable systems on this planet.

The Evolution of Metabolic Diversity

The presence of diverse nutritional types, as the quote at the beginning of the chapter affirms, is a consequence of the evolutionary role that microbes have assumed. Much of the microbial population has evolved as recyclers, able to mineralize constituents or end products of all biota. In the course of microbial evolution, processes of mutation and selection operated constantly, thus giving rise to populations able to utilize all available carbonaceous compounds as carbon and/or energy source. A population of microbes also evolved that could utilize virtually all potential energy sources including light, hydrogen, sulfides, reduced iron, and ammonia. Organisms also evolved that could utilize gaseous carbon sources, such as CO, CO_2, methane, ethane, or propane. The ability to "fix" nitrogen from the air gave a distinct advantage to selected Eubacteria and Archaea living in environments where little of this element would be present. Basically, survival advantage has been the key to the four billion years of evolution and a major factor in generating microbial diversity **(Box 5.1).** The ability to utilize a substrate better than others, or a substrate under physical conditions that others could not, afforded survival advantage.

BOX 5.1 MILESTONES

The Unique Character of Eubacteria and Archaea

Bacteria have a significant number of unique capabilities not found anywhere among the eukaryotes. Among those capabilities, bacteria can:

- Fix atmospheric nitrogen (N_2)
- Synthesize Vitamin B_{12}
- Use inorganic energy sources NH_4, H_2S, H_2, Fe^{2+}
- Photosynthesize without chlorophyll
- Utilize inorganic (CO_2, NO_3^-, SO_4^{2-}, S°, Fe^{3+}) terminal electron acceptor as an alternative to O_2
- Employ extensive mechanisms for anaerobic growth
- Use H_2S, H_2, or organics as electron donor in photosynthesis
- Grow at temperatures in excess of 100°C

Nutritional Types

Bacteria can be divided into four major groups, based on their source of carbon and their source of energy. (The nutrition and characteristics of the Archaea are discussed in Chapter 22.) All microorganisms must have carbon, which provides building blocks, and they require a source of energy to drive the reactions involved in cell synthesis. The grouping presented is based on the use of CO_2 or organic compounds as carbon source and light or chemicals as energy source. The ratio of carbon/hydrogen/oxygen in a cell is 1–2–1, respectively, and an organism growing with CO_2 as carbon source would need a source of [H] to **reduce** CO_2 to CH_2O. This source of [H] is called the electron donor as the electron would carry along a proton. An organism that utilizes reduced substrates such as methane (CH_4) generally must have an electron acceptor to **oxidize** the substrate to the CH_2O level. Most microorganisms utilize substrates such as glucose ($C_6H_{12}O_6$) that have a C/H/O ratio equivalent to that in a cell. Although the categories presented here are not all inclusive, they do provide a framework for separating microbes into compatible nutritional units. These groupings are defined as follows:

Photoautotroph (from *photo* meaning "light," *auto* meaning "self," and *troph* meaning "feeding"): Photoautotrophs use light as their source of energy and CO_2 as their source of carbon. An obligate photoautotroph is an organism that will grow only in the presence of both light and CO_2. The obligate photoautotrophs use inorganics such as H_2O, H_2, or H_2S as electron donor to reduce CO_2 to cellular carbon (CH_2O).

Photoheterotroph (from *hetero* meaning "different"): Photoheterotrophs grow by photosynthesis if provided with an electron donor (H_2 or organic) for reductive assimilation of CO_2. Often, organisms in this group require selected growth factors, such as B-vitamins, and many will grow on organic substrates if oxygen is available. They may also utilize light as an energy source while assimilating organic compounds from the environment as growth substrate.

Chemoautotroph (from *chemo* meaning "chemical"): Chemoautotrophs use **reduced** inorganic substrates for both the reductive assimilation of CO_2 and as a source of energy. The major energy sources for these microbes are H_2, NH_3, NO_2^-, H_2S, and Fe^{2+}. Aerobic chemoautotrophs utilize O_2 as terminal electron acceptor and some of the anaerobic Archaea can utilize inorganic sulfur as terminal electron acceptor (see Chapter 22).

Chemoheterotroph: Chemoheterotrophs are microorganisms that assimilate **preformed** organic substrates as source of both carbon and energy, and often this is a single substrate, for example, glucose or succinate. However, the sources of carbon and energy may be different; for example, the sulfate reducers use H_2 for energy but require an organic carbon source for cellular biosynthesis. The vast majority of bacteria that have been studied are chemoheterotrophs.

It is apparent from this brief outline that the various bacterial and archaeal species do have significant differences in their nutritional requirements. Bacteria representing each of these categories are presented in Figure 5.1. The ability of some organisms to fit into more than one of the above groupings is discussed in **Box 5.2.** Growth of microorganisms under laboratory conditions occurs in an artificial environment generally not equivalent to that encountered in nature. A microorganism living in soil, water, or the intestinal tract of a warm-blooded animal sur-

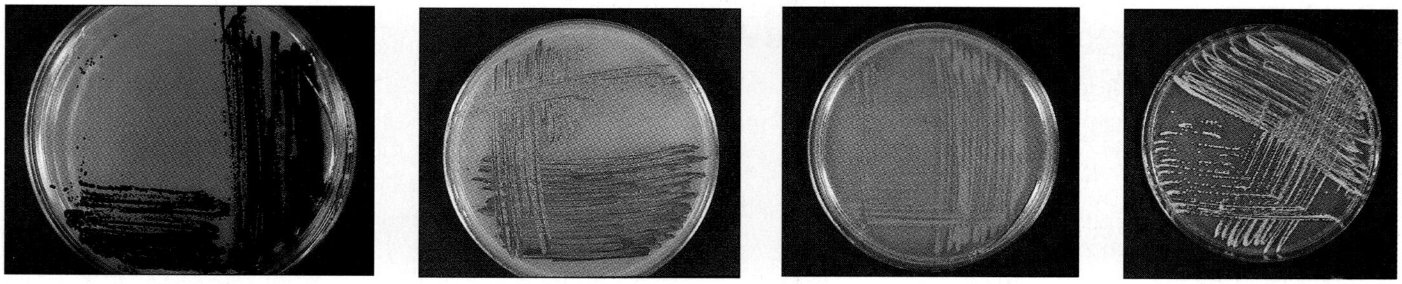

(a) (b) (c) (d)

***Figure* 5.1** Microorganisms that represent the major nutritional groups: **(a)** *Anabaena viribilis* (photoautotroph), **(b)** *Rhodomicrobium* sp. (photoheterotroph), **(c)** *Thiobacillus perometabolis* (chemoautotroph), and **(d)** *Micrococcus luteus* (chemoheterotroph). (**a,** courtesy of Robert Tabita and **b, c, d,** courtesy of J. J. Perry)

BOX 5.2 MILESTONES

Microbial Versatility: Obligate vs. Facultative

A microorganism that will grow only if provided with CO_2 for carbon and light as energy source is an **obligate photoautotroph.** One that grows only with CO_2 and inorganic (NH_4^+, $S^=$) as energy source is an **obligate chemoautotroph.** Restriction to an obligate lifestyle can be found in the microbial world, but versatility is a common attribute among microbes. An obligate autotroph can actually assimilate a limited amount of selected organic substrates while growing with CO_2 as bulk carbon source and light (for photo-) or inorganics (for chemo-) as energy source. However, many of the organisms that will grow as photo- or chemo-autotrophs can grow as chemoheterotrophs. Such organisms are termed **facultative.** The term facultative is also used to describe an organism that can grow aerobically if O_2 is available or anaerobically in the absence of O_2.

vives there because its nutritional requirements are met. These requirements may be simple or they may be complex. If we wish to take a microorganism from an environmental niche and grow it in the laboratory we must somehow mimic the conditions that permitted it to survive in the environment. These would include both physical conditions (pH, temperature, etc.) and substrate availability. A laboratory medium must provide nutrients that are essential for growth. The following section examines in some detail how one would prepare a growth medium.

Substrate Constituents

Biochemical analysis of bacterial and archaeal cell mass affirms that their cellular composition is mostly equivalent regardless of the species or the composition of the medium on which they grew. They are composed of the same major components including the monomers that make up the major macromolecules (proteins, nucleic acids, carbohydrates, and lipids). A bacterium is 70 percent to 80 percent water. The elemental composition is likewise very similar, with little variance from the percentages of elements shown in Table 5.1. Consequently a growth medium must provide a microorganism with these elements in one form or another. Requirements for iron and the trace metals may vary. The macromolecules that are present in all microorganisms are composed of the first six elements in Table 5.1. These are discussed here as component parts of the growth medium.

Carbon

The major elemental component of any living cell is carbon. Carbon is the backbone of functional biological molecules. Microbial diversity reflects the ability of a given organism to synthesize all of its component parts from the carbon source that is available. Marked differences exist among microbes in this capacity. For example, a cyanobacterium can synthesize *de novo* all of its cellular components (protein, nucleic acids, etc.) from CO_2 if mineral salts are available (see Table 5.2). Other organisms may have a very limited synthetic capacity and require a complex medium, and one such is shown in Table 5.3. A medium with all of these components would be necessary for the growth of the fastidious lactic acid–producing bacterium *Streptococcus agalactiae.*

***Table* 5.1 A typical analysis of the elements that may be present in a bacterial or archaeal cell**

Element	Percentage of Dry Weight
Carbon	50
Oxygen	20
Nitrogen	14
Hydrogen	8
Phosphorus	3
Sulfur	1
Sodium	1
Potassium	1
Calcium	0.5
Magnesium	0.5
Chlorine	0.5
Iron	0.25
Cu, Zn, Mo, Bo, Se, Ni, Cr, Co, and Wo	0.25
	100%

***Table* 5.2 A typical mineral salts medium for the isolation and growth of free-living bacteria. It would be necessary to add a carbon source such as glucose at 0.5 to 2.0%**

Constituents	Amount
	(mg/l H_2O)
NH_4Cl	500
$NaNO_3$	500
Na_2HPO_4	210
NaH_2PO_4	90
$MgSO_4 \cdot 7\ H_2O$	200
KCl	40
$CaCl_2$	15
$FeSO_4$	1
Trace elements	µg/l H_2O
$ZnSO_4 \cdot 7\ H_2O$	70
H_3BO_3	10
$MnSO_4 \cdot 5\ H_2O$	10
MoO_3	10
$CoSO_4$	10
$CuSO_4 \cdot 5\ H_2O$	5

Table* 5.3 A typical defined medium for the growth of *Streptococcus agalactiae

Compound	Final Concentration (µg/mL)	Compound	Final Concentration (µg/mL)
L-alanine	250	K_2HPO_4	1,000
L-arginine	320	KH_2PO_4	1,000
L-aspartic acid	500	$NaHCO_3$	500
DL-asparagine	100	$FeSO_4 \cdot 7\ H_2O$	10
L-cystine	600	$MnCl_2$	25
L-cysteine–HCl	600	NaCl	10
L-glutamic acid	400	$ZnSO_4 \cdot 7\ H_2O$	10
L-glycine	200	$MgSO_4 \cdot 7\ H_2O$	80
L-glutamine	100		
L-histidine	320	Glucose	10,000
L-leucine	200		
L-lysine	320		
L-isoleucine	200	Adenine	10
DL-methionine	200	Guanine	10
L-phenylalanine	200	Xanthine (Na salt)	10
L-proline	200	Uracil	10
DL-serine	400		
L-tryptophan	200		
L-tyrosine	200		
L-valine	200		
Nicotinic acid	10		
Ca pantothenate	10		
Pyridoxal HCl	10		
Thiamine HCl	0.6		
Riboflavin	1		
Biotin	0.2		
Folic acid	0.02		

From N. P. Willett and G. E. Morse. Long-chain fatty acid inhibition of the growth of *Streptococcus agalactiae* in a chemically defined medium. *J. Bacterial* 91:2245, 1966.

The carbon source for microbial growth can range from CO or CH_4 to any naturally occurring complex organic compound present in the biosphere (carbohydrates, peptides, organic acids, etc.). A variety of microbes also exist that can, under appropriate conditions, attack many synthetic organic compounds. This is the basis for "bioremediation" of polluted environments such as soil or ground water (see Chapter 33). Bioremediation is the removal of pollutant chemicals from an environment by use of selected microbes that can utilize these pollutants as a carbon/energy source.

Hydrogen

Hydrogen plays a number of roles in the Eubacteria and Archaea—from an atom in the structural organic molecules to a participant in the complex process of energy generation. Protons (H^+) along with electrons are involved in the production of ATP via the ATPase system in the cell membrane (see Chapter 8) of most microorganisms. A source of hydrogen is essential for autotrophic organisms to reduce CO_2 to the level present in the cell (CH_2O). Anaerobes (CH_4 producers, denitrifiers, and sul-

***Table* 5.4 Types of respiration that occur in Eubacteria and Archaea. Examples of organisms that perform this respiration are given**

Aerobic respiration	
Oxygen	
$O_2 \rightarrow H_2O$	*Pseudomonas fluorescens*
Anaerobic respiration	
Iron	
$Fe^{+++} \rightarrow Fe^{++}$	*Shewanella putrefaciens*
Nitrate	
$NO_3^- \rightarrow NO_2^-, N_2O, N_2$	*Thiobacillus denitrificans*
Fumarate	
Fumarate $\rightarrow$ Succinate	*Proteus rettgeri*
Sulfate	
$SO_4^{2} \rightarrow HS^-$	*Desulfovibrio desulfuricans*
Sulfur	
$S^\circ \rightarrow HS^-$	*Desulfurococcus mucosus*
Carbonate	
$CO_2 \rightarrow CH_4$	*Methanosarcina barkeri*
$CO_2 \rightarrow CH_3COO^-$	*Acetobacterium woodii*

fate reducers) gain their energy by transfer of an electron and hydrogen from a substrate to a selected acceptor (Table 5.4). Aerobes transfer electrons (and protons) to O_2 during energy generation.

Nitrogen

Nitrogen is an integral constituent of amino acids, nucleic acids, membranes, cell walls, and most cell macromolecules. A unique property of some Eubacteria and Archaea is the ability to obtain cellular nitrogen by fixation of N_2 from the atmosphere. The nitrogen fixers reduce N_2 to NH_4^+, and assimilate this into the synthetic machinery of the cell. This ability is not limited to a few species, as was previously believed, but occurs in an array of microbial types, as outlined in Table 5.5. Most free-living microorganisms assimilate ammonia from their environment or they can reduce nitrate, and one or both of these are commonly added as a constituent of the growth medium.

A requirement for an organic nitrogen source is generally confined to microorganisms that evolved in richer environments where some of the amino acids, nucleic acids, and B-vitamins were readily available. For example, the medium presented in Table 5.3 is a growth medium for a microorganism with a limited ability to synthesize nitrogen-containing intermediates. The lactic acid bacteria evolved in or on animals or plants where these monomers were available. Growth of most heterotrophic bacteria is stimulated by adding rich nitrogenous material, such as yeast extract at 0.05 percent, to a basic mineral salts medium. Growth stimulation occurs because otherwise considerable cellular energy would be expended to synthesize B-vitamins, amino acids, purines, pyrimidines, and other nitrogen-containing compounds. Yeast extract is the water-soluble portion of autolyzed brewer's yeast.

***Table* 5.5 Some genera of Eubacteria and Archaea that have nitrogen fixation ability**

- ***Eubacteria***
  - Heterotrophs
    - Aerobes
      - *Azotobacter*
      - *Klebsiella*
      - *Beijerinckia*
    - Anaerobes
      - *Clostridium*
      - *Bacillus* (facultative)
  - Photosynthetics
    - Cyanobacteria
      - *Anabaena*
      - *Oscillatoria*
      - *Gloeocapsa*
    - Purple and green bacteria
      - *Chromatium*
      - *Chlorobium*
      - *Rhodospirillum*
  - Symbiotic
    - Legumes
      - Clover + *Rhizobium*
      - Soybeans + *Rhizobium* or *Bradyrhizobium*
      - Bluebonnets + *Rhizobium*
    - Nonlegumes
      - Bayberry + *Actinomycete*
      - Alder + *Frankia*
- ***Archaea***
  - Methanogens
    - Anaerobes
      - *Methanococcus*
      - *Methanosarcina*
      - *Methanobacterium*
      - *Methanothermus*

Sulfur

Sulfur is a constituent part of a few of the amino acids that are constituents of proteins. It is also a part of some of the B-vitamins (biotin and thiamine) and other essential constituents of a cell. Sulfur is usually added to a growth medium as a sulfate salt. $MgSO_4$ added to a medium would serve as a source of both sulfur and magnesium. The capacity to reduce sulfate to the sulfide level (SH), the form present in cellular constituents, is a common attribute of bacteria. If an organism cannot reduce sulfate, the addition of the amino acid cysteine may suffice. Yeast extract or peptone (enzyme digest of protein) in a culture medium meets the need for reduced sulfur compounds in most microorganisms. Reduced inorganic sulfur compounds such as H_2S or pyrite (iron sulfide) can serve as the energy source for a group of bacteria called the thiobacilli. Oxidation of sulfides generates sulfate. Sulfur can serve as a terminal electron acceptor in some of the Archaea, and sulfate serves this purpose in the sulfate reducing Eubacteria.

Phosphorus

This element has played a major role in the evolution of viable systems on earth. Phosphorus is a constituent of high energy compounds, the phospholipids in cell membranes, and nucleic acids. Adenosine triphosphate (ATP) is the principal medium of energy exchange in cellular metabolism and is an indispensable contributor to biosynthetic reactions involved in reproduction and growth. For this reason, phosphates are an integral constituent in culture media. Inorganic phosphates are also an effective buffer at near neutral pHs and at concentrations not generally inhibitory to bacterial growth. Phosphate salts are commonly added to culture media to satisfy the phosphate requirement and to provide a buffer to prevent significant changes in pH during growth.

Oxygen

The total amount of oxygen present as a cellular component is equivalent in aerobes and anaerobes. However, free oxygen (uncombined with other atoms) is toxic to most strictly anaerobic bacteria and some of the Archaea, so they obtain this element in a combined form from the substrate. As a general rule, anaerobes utilize growth substrates that are in an oxidation-reduction state equal to or more oxidized than cellular material (CH_2O). Aerobic bacteria can grow on reduced substrates (such as methane and propane), but generally only when molecular oxygen is available. Aerobic microorganisms use oxygen as a terminal electron acceptor through the electron transport system.

Growth Factors

Growth factors are specific, relatively low molecular weight organic compounds that are required for the growth by some microorganisms. They must be available in their growth medium because the microorganisms cannot synthesize them. The substances that frequently serve as growth factors are selected amino acids, purines, pyrimidines and B-vitamins (microbes do not require the fat-soluble vitamins such as A, C, and D as these are not constituents of microorganisms).

The three types of growth factor most often required in bacterial nutrition are as follows:

1. **Vitamins:** These are special organic compounds that serve as the prosthetic group (nonprotein catalytic part) of a number of enzymes. Small catalytic amounts of vitamins are required as they are present in cells in low quantity (varying from a nanogram of vitamin B_{12} to 250 micrograms/gram dry weight cell of nicotinic acid). The vitamins most frequently required are those less susceptible to destruction by light. Thiamine, biotin, and nicotinic acid are frequently required by microorganisms. The function of the B-vitamins in nutrition is outlined in Table 5.6.
2. **Amino acids:** Proteins are made up from about 20 α-amino acids (Chapter 3), and some Eubacteria and Archaea have a requirement for one or more of these. For example, most strains of *Staphylococcus epidermidis,* a normal inhabitant of human skin, require proline, arginine, valine, tryptophan, histidine, and leucine. The lactic acid bacteria require a greater complement, as indicated in Table 5.3. The level required is proportional to the amount of that amino acid in the cell. This can be calculated as follows: a cell is about 50 percent protein, and the average amount of the aromatic amino acid phenylalanine present in protein is about 5 percent; therefore, to grow one gram of an organism that required this amino acid one should add at least 25 mg phenylalanine to the growth medium (1 gram cell = 500 mg protein, $500 \times 0.05 = 25$ mg).
3. **Purines and pyrimidines:** Requirements for the nucleic bases are most often observed in lactic acid bacteria (Table 5.3) and other fastidious organisms. The need for added nucleic acid bases is rare in free-living soil microbes.

Elements

A number of other elements are present in Eubacteria and Archaea and are generally involved with enzymatic activity or in cell stability. These are mostly cations and include potassium, sodium, magnesium, iron, and cobalt (Table 5.7). The chloride (Cl^-) anion is also required by some microorganisms. Iron is a constituent of electron transport chains and therefore an absolute requirement among aer-

***Table* 5.6 The function of the various B-vitamins in nutrition of Eubacteria and Archaea**

Compound	Function
p-Aminobenzoic acid	Precursor of folic acid, a coenzyme involved in one carbon unit transfer
Folic acid	A coenzyme involved in one carbon unit transfer
Biotin	A prosthetic group for enzymes that act in carboxylation reactions
Nicotinic acid	Precursor of NAD and NADP, which are coenzymes involved with hydrogen transfer
Riboflavin	A component of the flavin mononucleotide (FMN) and dinucleotide (FAD) involved in hydrogen transfer
Pyridoxine	A component of the coenzyme for transaminase and amino acid decarboxylase
Vitamin B_{12}	A coenzyme involved in molecular rearrangements
Thiamin	The prosthetic group for a number of decarboxylases, transaldolases, and transketolases
Pantothenic acid	A functional part of coenzyme A and the acyl carrier proteins
Coenzyme M	A coenzyme in methane-generating bacteria

***Table* 5.7 The function of various elements in eubacterial and archaeal nutrition**

Element	Function
Potassium	Utilized in a number of enzymatic reactions as a cofactor and especially in protein synthesis
Sodium	Involved, along with chloride, in the regulation of osmotic pressure; affects activity of some enzymes; uptake of solutes in some species with Na^+ dependent transport systems
Magnesium	Integral part of chlorophyll; cation required in enzymatic reactions including those involved in ATP synthesis or hydrolysis
Iron	Reactive center of heme-containing proteins (cytochromes, catalase, etc.) and component of other proteins
Cobalt	Constituent of Vitamin B_{12}, complexed to some enzymes
Copper, zinc, molybdenum, nickel, tungsten, and selenium	Essential components of selected enzymes

obes. Copper, zinc, molybdenum, and others are required in small amounts and are therefore called **trace** elements.

Uptake of Nutrients

The cytoplasmic membrane of a cell is a highly selective barrier between the external environment and the cytoplasm. It permits or facilitates the entry of essential nutrients inward and rejects many of those that are not essential. The hydrophobic nature of the lipid bilayer is responsible for the high degree of impermeability inherent in cytoplasmic membranes. Water can pass freely across the membrane while alcohols, fatty acids, and other fat-soluble compounds may also pass through at varying rates. Polar solutes such as amino acids or sugars cannot traverse unaided across the cytoplasmic membrane.

Microorganisms in nature generally live in environments where virtually all nutrients are available at quite low concentrations. Effective balanced growth can occur only if nutrients can be accumulated inside the cell at levels that far exceed those present externally. Microorganisms have evolved with mechanisms that permit them to concentrate nutrients such as sugars or required cations to levels inside the cell that are a thousandfold or more higher than the level in the environment. To accomplish this accumulation of nutrients a microorganism may utilize several distinct transport mechanisms. Among the transport mechanisms most utilized are **facilitated diffusion, active transport,** and **group translocation.** These carrier-mediated processes are necessary because simple diffusion, at best, would raise the internal concentration only to that present externally. Under most environmental conditions this would not support balanced growth. The following is a discussion of the mechanisms that are important in concentration of nutrient solutes inside a bacterial cell.

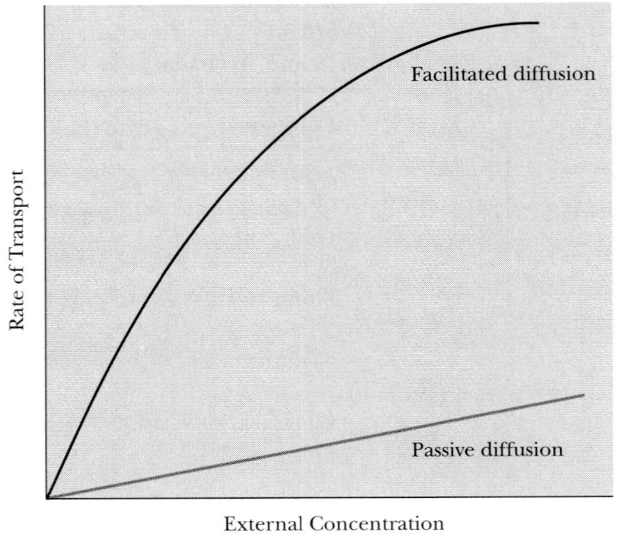

***Figure* 5.2** Passive and facilitated diffusion. In passive diffusion, molecules move from the area of higher concentration to one of lower concentration. Facilitated diffusion occurs across a selectively permeable membrane with the aid of carriers. The plateau indicates that the carrier-mediated transport shows saturation at relatively low external concentrations.

Diffusion

Some solutes such as alcohols can pass through the cytoplasmic membrane by simple diffusion. The concentration inside the membrane will equilibrate with that outside in a process termed **passive** diffusion. Generally, passive diffusion occurs with fat-soluble compounds and is dependent on a relatively high concentration outside the cell. Glycerol is a compound that may enter a cell by passive diffusion or by a process termed **facilitated diffusion** (Figure 5.2). The rate of glycerol uptake in passive diffusion is solely dependent on the external concentration. In facilitated diffusion there is a rapid initial uptake of glycerol that levels off as solute concentration inside reaches a certain level.

Facilitated diffusion is a carrier-mediated transport process. The transmembrane proteins involved are also termed **permeases.** One end of the transport protein protrudes to the outside of the membrane and the other end extends to the interior. Some of these permeases are highly selective in that they transport only a single type molecule, but most are active with classes of substrates such as a group of amino acids or a number of related sugars. Facilitated transport does not require energy but depends on a conformational change in the transport protein. The solute to be transported is bound to the external portion of the transport protein, and then the solute moves inward by a change in conformation that results in the release of the solute inside the cell (Figure 5.3). Although facilitated diffusion is effective in carrying solutes into a cell, the total concentration internally will not exceed the level in the immediate external environment. The process is reversible and can carry solutes out when the internal concentration is higher than the external concentration. Facilitated diffusion is effective because it "speeds up" the diffusion process and generally the transported material is metabolized upon entry. This maintains a low internal concentration and promotes continued uptake. Since facilitated diffusion will not function against a concentration gradient, a microbe in nature needs a better mechanism for concentrating a nutrient that is present at a low external concentration. There are two major mechanisms available to accomplish this, active transport and group translocation.

Active Transport

Energy is required to move a solute from the side of a membrane where the concentration is low to establish a higher concentration on the other side. This in effect is

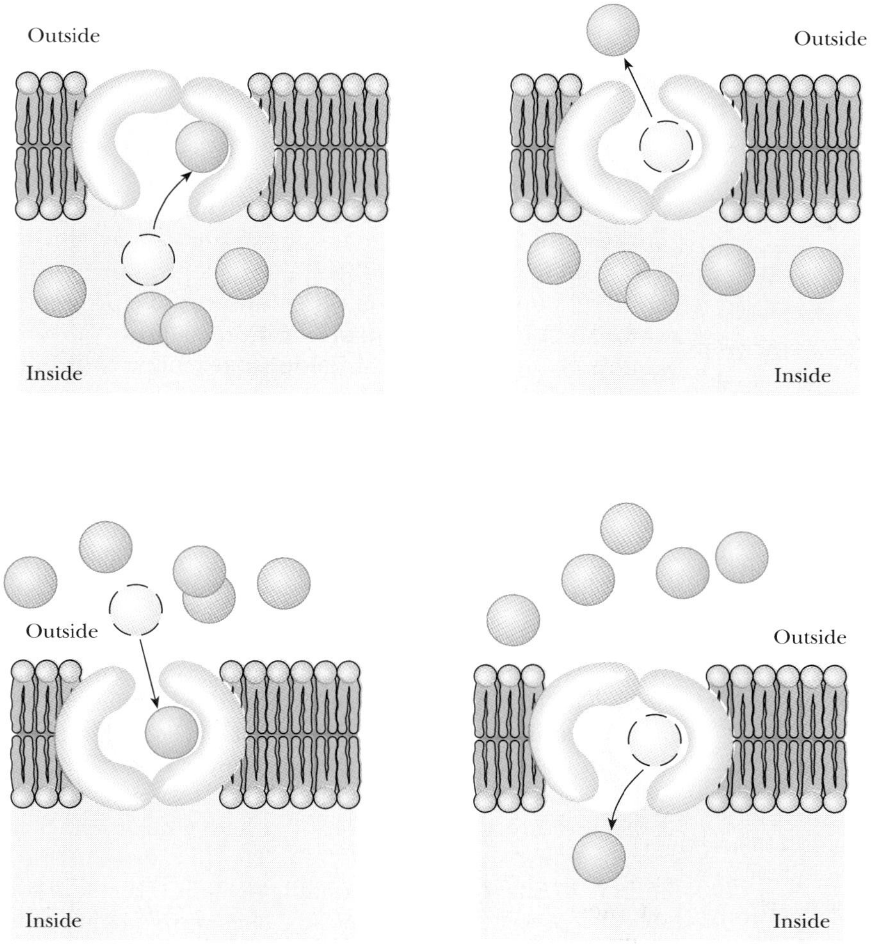

***Figure* 5.3** Facilitated diffusion is promoted by transmembrane proteins. The solute binds to the external part of the protein. Changes in configuration push the solute through and release it into the cytoplasm. If the internal concentration exceeds the external concentration, the process is reversed.

the accumulation of a solute against a concentration gradient. As with facilitated diffusion, transmembrane proteins are involved in active transport. These transmembrane proteins are also specific for the type of solutes they transport. The solute transported is not altered as it passes into the cytoplasm, and an organism may have multiple transport systems for one solute. These diverse transport systems for a single solute may differ in the energy source required and affinity for the solute that is transported. Among the nutrients that are taken into the cell by active transport are some of the sugars, amino acids, organic acids, and inorganic ions. The energy for active transport generally comes from ATP or from a proton gradient established across the membrane. The proton gradient can result from the metabolism of organic substrates or can be generated by light energy (in photosynthetics).

In ATP-driven transport a nutrient is bound to the external portion of a transmembrane protein and is moved inward by configurational changes in the protein (Figure 5.4). The solutes are generally bound to a specific **binding protein** that passes the solute to the transport or carrier protein. These binding proteins accumulate small molecules within the periplasmic space and stimulate transport. Energy is transferred to the carrier protein by a specific **transducer protein** located on the inner surface of the cytoplasmic membrane.

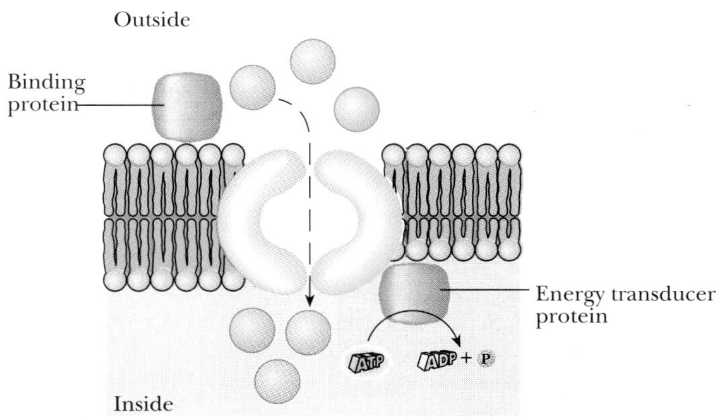

***Figure* 5.4** Active transport involving ATP. The binding proteins accumulate the substrate at the outer surface of the cytoplasmic membrane and pass it along to the carrier protein. Configurational changes in the carrier protein transport the substrate into the cell. The energy required for the configurational changes is transmitted to the carrier protein by a specific transducer protein.

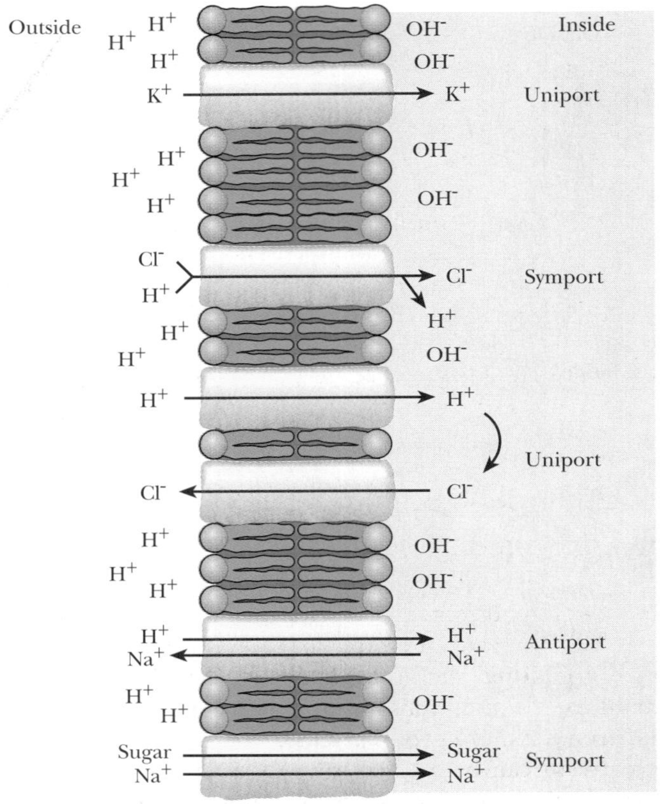

***Figure* 5.5** Active transport mechanisms involving uniport, symport, and antiport. The electrochemical charge on the membrane is generated by the electron transport which serves as a "proton pump" that effects an accumulation of protons on the outside. The charge can carry cations into the cell by uniport, carry anions or solutes inward by symport, or effect an exchange of proton entering for driving out a cation such as Na^+ (antiport).

Bacteria also transport solutes by utilizing the energy inherent in an electrical charge separation across the membrane. This charge separation is established by a proton gradient generated during electron transport and is also termed the proton motive force. This is the same proton force that drives ATP synthesis (see Chapter 8). The electrochemical potential generated by the protons drives uptake of solutes through the transport proteins. **Uniporters** are transport proteins that take in cations such as K^+ through the electrochemical gradient established when there is a high proton ($^+$) concentration on the outside and negativity on the inside (Figure 5.5). Anions would be carried in by a **symport** mechanism when a proton would accompany the anion. An anion may also be driven out in concert with the uptake of a proton, and this would be a uniport reaction. The transport of a proton inward can concomitantly drive a cation such as Na^+ outward, and this would be an **antiport** reaction. A sodium ion gradient on the outside may be utilized to drive the uptake of amino acids or other nutrients by symport. All three of these mechanisms depend on the generation of a proton gradient across the membrane.

Group Translocation

Group translocation is a transport process utilized by bacteria where the transported compound is chemically altered. The best defined of the group translocations is the **phosphotransferase system (PTS)** that is involved in transport of sugars into the cell including glucose, fructose, and β-glucosides. The PTS system is quite complex and involves the direct participation of at least four enzymes to transport one sugar (Figure 5.6). The first two enzymes, Enzyme I (Enz. I) and a heat-stable protein (HPr), are soluble and present in the cytoplasm. These enzymes are nonspecific and function in the PTS system for the transport of an array of PTS substrates. The transmembrane protein (Enz. II) is an integral part of the cytoplasmic membrane and final recipient of the high-energy phosphate, which it passes on to phosphorylate the sugar in the transport process. Enzyme III (Enz. III) is a peripheral protein attached to the inner surface of the membrane at the site of Enz. II. Both Enz. II and Enz. III are quite specific and involved in the transport of a single sugar, such as glucose. The PTS transport system is energy conserving in that the high-energy phosphate from phosphoenolpyruvate ulti-

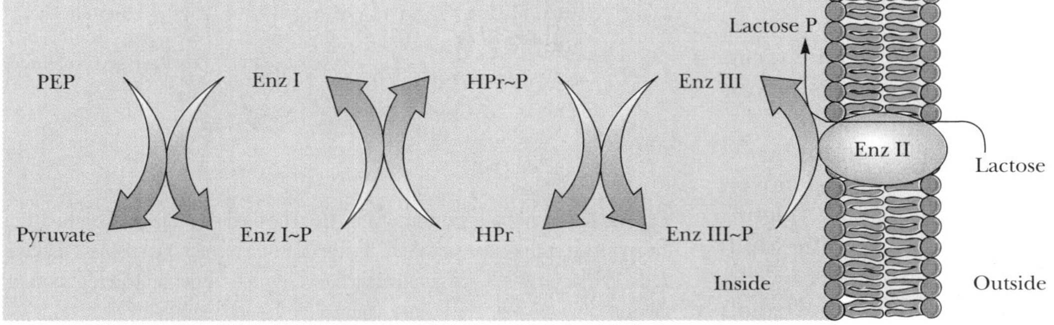

***Figure* 5.6** Transport of lactose via the phosphotransferase system (PTS). The high-energy phosphate bond of phosphoenolpyruvate is transferred consecutively to Enzyme I (Enz. I), to heat-stable protein (HPr), and to Enzyme III (Enz. III). The transmembrane protein (Enz. II) transports the lactose inward as the phosphate is transferred from Enz. III to the sugar.

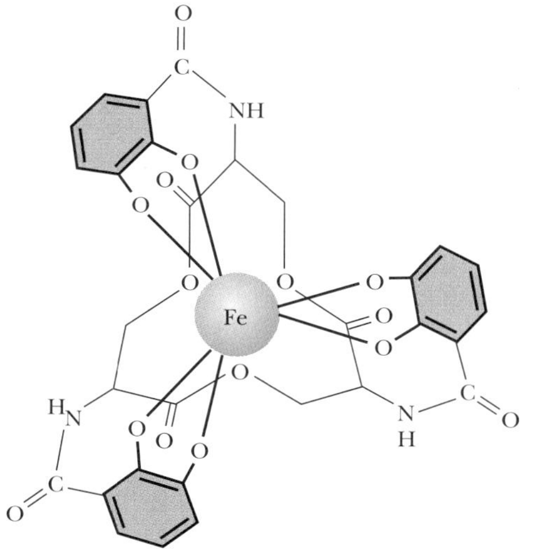

***Figure* 5.7** A siderophore—enterobactin, a catechol derivative that chelates iron through the oxygen groups on the catechol ring.

mately becomes a constituent of the transported sugar. The overall reaction can be drawn as follows:

$$\text{Phosphoenolpyruvate} + \text{HPr} \xrightarrow[\textit{Enz. I}]{} \text{Pyruvate} + \text{Phospho} - \text{HPr}$$

$$\text{Phospho} - \text{HPr} + \text{Sugar} \xrightarrow[\textit{Enz. III}]{} \text{Sugar Phosphate} + \text{HPr}$$

The PTS system is present in obligately anaerobic bacteria such as *Clostridium* and *Fusobacterium.* It is also present in many genera of facultative anaerobes including *Escherichia, Staphylococcus, Vibrio,* and *Salmonella.* It is rare in obligately aerobic genera. Other substrates that may be transported by group translocation include purines, pyrimidines, and fatty acids.

Iron Uptake

Iron is a constituent part of the cytochromes and iron-sulfur proteins that are involved in electron transport. Most iron that is available in nature is in the insoluble oxidized state as the ferric ion (Fe^{3+}) or rust and is not freely available for transport into a cell. Iron may be provided in a culture medium in a complex with a **chelating agent.** A chelating agent is a compound such as ethylene diamine tetraacetic acid that binds to iron to prevent oxidation but can release it to the cell's uptake system.

Many microorganisms can themselves produce chelating agents that solubilize iron salts and make the iron available. These low molecular weight compounds are called **siderophores.** A typical chelating agent (siderophore) produced by *Escherichia coli* is enterobactin. It is a derivative of catechol (Figure 5.7), and chelates iron through the oxygen molecules on the catechol ring. Microbes tend to secrete siderophores under growth conditions where little iron is available from the environment. The siderophore-iron complex is bound by a specific siderophore receptor in the cell envelope, and the iron is transferred through the cytoplasmic membrane by specific transport proteins.

Isolation of Selected Microbes by Enrichment Culture

The basic requirements for the growth of bacteria have been considered in the previous sections. The isolation of microorganisms that have selected nutritional requirements can be accomplished by applying this knowledge to the preparation of growth media. The physical conditions for growth can be adjusted to select for microorganisms that can thrive under specific conditions of temperature, pH, oxygen availability, osmotic pressure, and other parameters. Isolation of specific types by a combination of nutrient and physical conditions is generally termed **enrichment culture.** The isolation of Archaea is discussed in Chapter 22.

Origins of Enrichment Culture Methodology

The first scientists to apply enrichment culture on a broad scale were Martinus Beijerinck and Sergei Winogradsky (see Chapter 2). Their experimentation resulted in the isolation of a broad array of bacterial types and led to a rational approach to microbial ecology. Beijerinck's basic method was to prepare petri dishes with a mineral salts agar medium that had been inoculated with a sample of soil or water taken from diverse environments. Beijerinck then placed selected organic compounds on the surface of the inoculated agar. After a few days at room temperature, he noted that, in virtually every case, colonies developed on the agar surface where soluble compounds were added (Figure 5.8). Beijerinck also noted that equivalent conditions applied to soil or water samples obtained from widely separated habitats generally yielded similar bacterial types.

Thus, Beijerinck essentially demonstrated the Darwinian concept of natural selection in a petri dish. The organism or organisms most capable of development under the specific regimen of pH, temperature, and substrate were the ones that arose in greatest numbers from the soil or water inoculum. The selection of a specific organism(s)

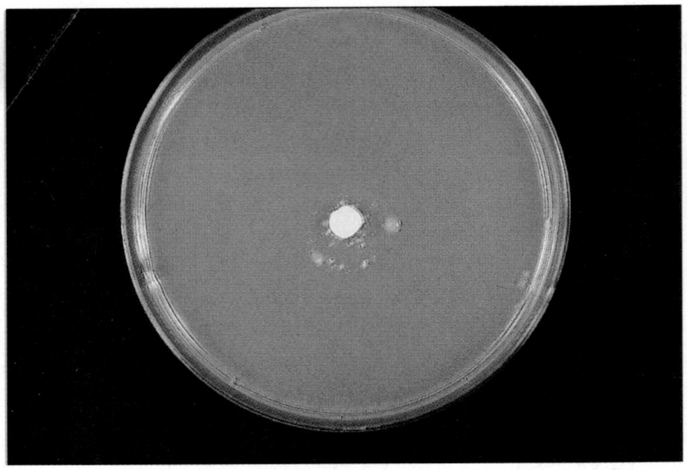

***Figure* 5.8** An example of enrichment culture. A soil inoculum was added to mineral salts/melted agar that had been cooled to 48°C and poured into a petri dish. A disc containing glucose was placed on the agar surface. Note growth of colonies on glucose that diffused outward from the disc. (Courtesy of J. J. Perry)

from all those in the inoculum (1 g of fertile soil generally contains 10^7 to 10^9 microbes) is the essence of enrichment culture.

If one were to set up enrichment cultures under equivalent selected conditions using soil from North America, eastern Asia, or northern Europe, the organisms obtained from these three soils would be physiologically similar. The characteristics of the organisms from the separate continents would probably not differ any more than two cultures obtained from the confines of one acre.

Applications of Enrichment Culture

Enrichment culture isolation of microorganisms that can biodegrade the biochemical constituents of cells has been of considerable utility in metabolic studies. **Biodegradation,** in this sense, refers to the stepwise conversion of an organic compound to $CO_2 + H_2O$. Although there are notable exceptions, biosynthetic pathways in cells generally occur by a reversal of the biodegradative route. **Biosynthetic pathways** are the molecular changes that transform a substrate, such as glucose or CO_2, to the various macromolecular constituents of the cell. It is generally easier to isolate sufficient quantities for structural analysis by following biodegradation of a molecule than to accumulate identifiable quantities of biosynthetic intermediates. Identification of metabolic intermediates was particularly difficult when analytical methods were far less sophisticated than they are today. It should also be noted that during the period when biosynthetic pathways were elucidated the genetic manipulation of bacteria was also in its infancy. Use of bacteria obtained by enrichment was of considerable value in studies that led to the elucidation of both biodegradative and biosynthetic pathways.

Often the goal in isolating microorganisms is to obtain a specific type for practical reasons. For example, the isolation of thousands of actinomycetes has been accomplished over the past 45 years in the quest for those that yield useful antibacterial, antiviral, or antitumor agents. These organisms are commonly obtained from soil, compost, freshwater, and the atmosphere. A number of enrichment substrates have been devised for this purpose. One containing low levels of hydrolyzed casein (animal protein), soytone (plant protein), and yeast extract (microbial digest) (each at 0.1 percent) has proven quite effective. The low level of sugars present in a medium of this composition would curtail overgrowth by fast-growing motile genera, such as *Bacillus* and *Pseudomonas*.

Colonies of microbes that have the enzymatic capacity for digesting insoluble higher molecular weight sugar polymers, lipids, and proteins can easily be selected. To do this the polymer would be incorporated into the agar medium as carbon source. The surface of the agar is then inoculated with soil or water, and digestion of the polymer is indicated by a clear area in the agar around the colony (Figure 5.9).

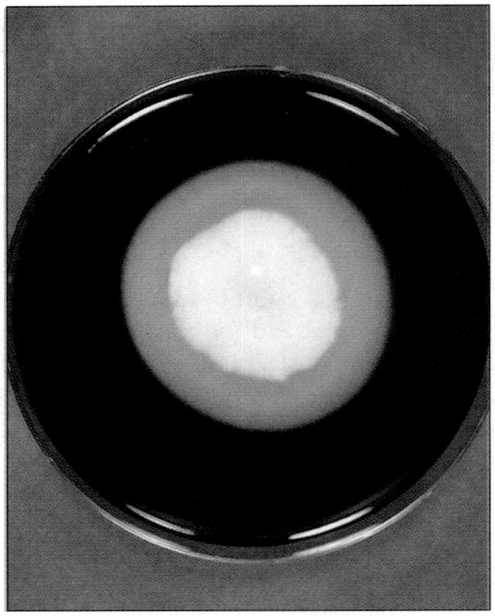

***Figure* 5.9** Clearing in the area around the colony indicates that the microorganism produces an extracellular enzyme. A plate containing starch (1 percent) was inoculated in the center with *Bacillus subtilis*. After three days an alcohol solution of iodine was flooded over the surface and the excess poured off. The clear area is evidence of starch hydrolysis. The outer blue ring stained with iodine because the starch was not hydrolyzed. This basic technique can be effective in enriching for microorganisms that utilize high molecular weight water-insoluble compounds such as protein or lipids. (Courtesy of J. J. Perry)

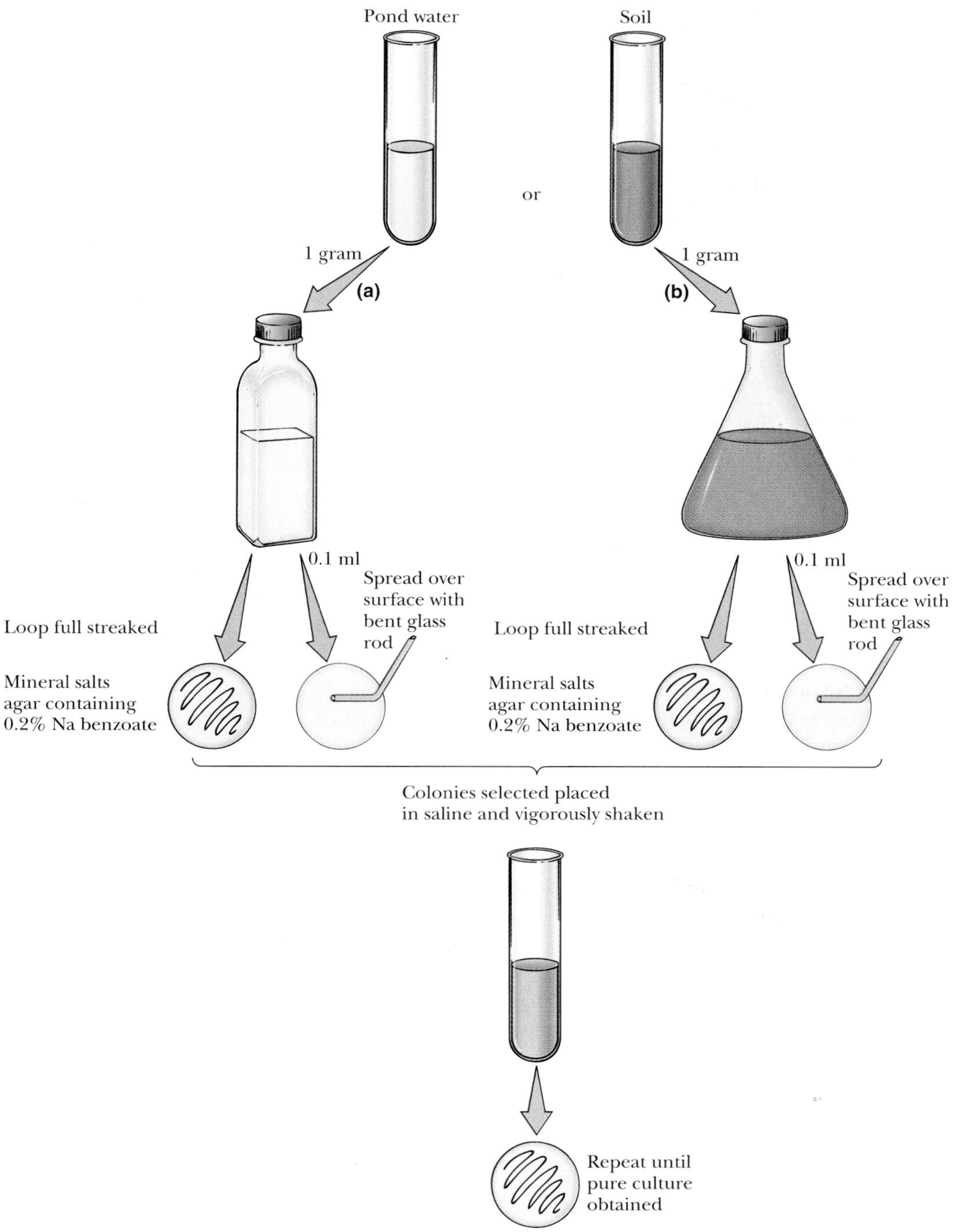

***Figure* 5.10** Isolation of a bacterium that can utilize a specific compound such as benzoic acid as carbon and energy source. **(a)** Solid enrichment. The enrichment on a solid medium is accomplished by streaking or spreading a soil or water sample directly onto an agar plate after diluting in physiological saline. **(b)** In liquid enrichment the soil or pond water is added directly to a liquid medium containing the substrate. The ultimately pure culture is obtained by streaking on an agar medium containing the enrichment substrate.

General Enrichment Methods

Aerobic or anaerobic microorganisms can be isolated readily by enrichment by applying proper methodology. Samples should be obtained from environmental niches that favor the type of organism sought—soil for aerobes, mud for anaerobes. Basic conditions for enrichment would be as follows.

Aerobes

There are two general procedures that can be followed in enriching for aerobic bacteria. These procedures are outlined in Figure 5.10: a Liquid enrichment and an enrichment directly on a solid medium such as agar or silica gel. A streak from a liquid enrichment for an organism that might use a compound such as sodium benzoate would

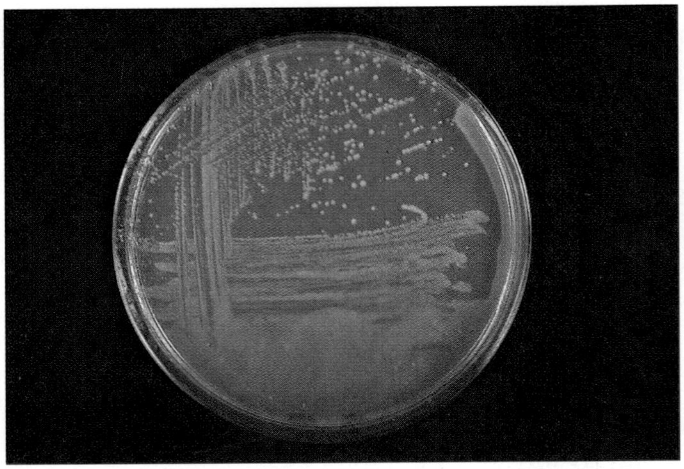

***Figure* 5.11** A plate streaked from a liquid enrichment. The liquid enrichment technique tends to favor rapidly growing organisms. Note that only one type of colony has developed. (Courtesy of J. J. Perry)

appear as shown in Figure 5.11. The direct application of soil or spreading of a water-diluted sample on an agar surface would result in colonies as depicted in Figure 5.12.

A shortcoming of liquid enrichment is a marked tendency for faster growing organisms, such as pseudomonads, to be favored, and these become dominant on continued transfer to a newly prepared medium. Enrichment on a solid medium will generally yield a greater variety of organisms provided that one has the patience to wait for slower growing colonies to appear. Unfortunately, organisms that are present as a small percentage of the total population may be overlooked by enrichment procedures.

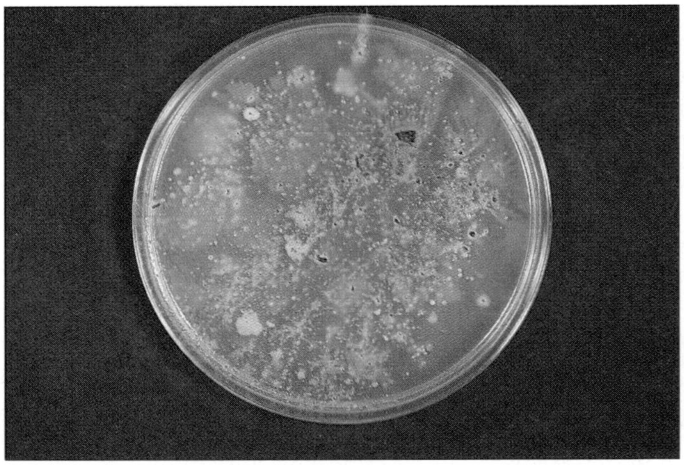

***Figure* 5.12** Sprinkling of soil particles directly on the surface of an agar plate containing sodium succinate (0.2 percent). It is apparent that this method yields a variety of microorganisms. (Courtesy of J. J. Perry)

Anaerobes

Enrichment for anaerobic bacteria is technically more difficult, particularly for strict anaerobes that are killed by exposure to even minute levels of oxygen. Samples for enrichment must be handled carefully to preclude exposure to oxygen prior to isolation procedures. Elaborate methodology has been developed for handling the obligately anaerobic bacteria, including the fabrication of entire rooms from which all traces of oxygen can be removed. A technique now in general use is an oxygen-free **glove box,** which permits one to apply techniques generally used for aerobes to strictly anaerobic bacteria. The isolation of anaerobic organisms less sensitive to brief exposure to air can be accomplished in glass-stoppered bottles and in shake tubes, as illustrated in Figure 5.13. In this example, manipulation to obtain photosynthetic organisms was selected to demonstrate the procedure.

Mud from an environment such as the shallow area of a pond would contain many anaerobic photosynthetic bacteria. Placing a sample of this mud in a glass-stoppered bottle filled to the top with the medium outlined would result in anaerobiosis. Any oxygen present would be removed by aerobes or facultative microorganisms present in the sample. The organism would utilize light as energy source and glycerol as electron donor for CO_2 assimilation. Such an enrichment could be done from a Winogradsky column, as illustrated in **Box 5.3.**

Enrichment for Specific Metabolic Types

The foregoing discussion presented an outline of general methods for isolating microorganisms. Proper manipulation of these conditions can lead to the isolation of an organism that will utilize a selected substrate as the sole nutrient. More restrictive selective conditions can be established that will lead to the isolation of a microorganism of a distinct metabolic type. Some of these selective methods are outlined here.

Chemoheterotrophic Aerobic Bacteria

Conditions are outlined in Table 5.8 for the isolation of a number of chemoheterotrophic aerobic or facultative bacteria. It should be emphasized that a soil or water inoculum will contain a great variety of microorganisms and many of these can grow under primary enrichment conditions. They grow on products of, or in association with, an organism metabolizing the enrichment substrate. Obtaining a pure culture of a desired organism may require careful and continued restreaking on the appropriate medium. The procedures outlined are to familiarize the reader with the broad aspects of enrichment culture. To actually obtain selected microorganisms, one might consult references cited at the end of this chapter.

Many distinctly different bacterial species can fulfill

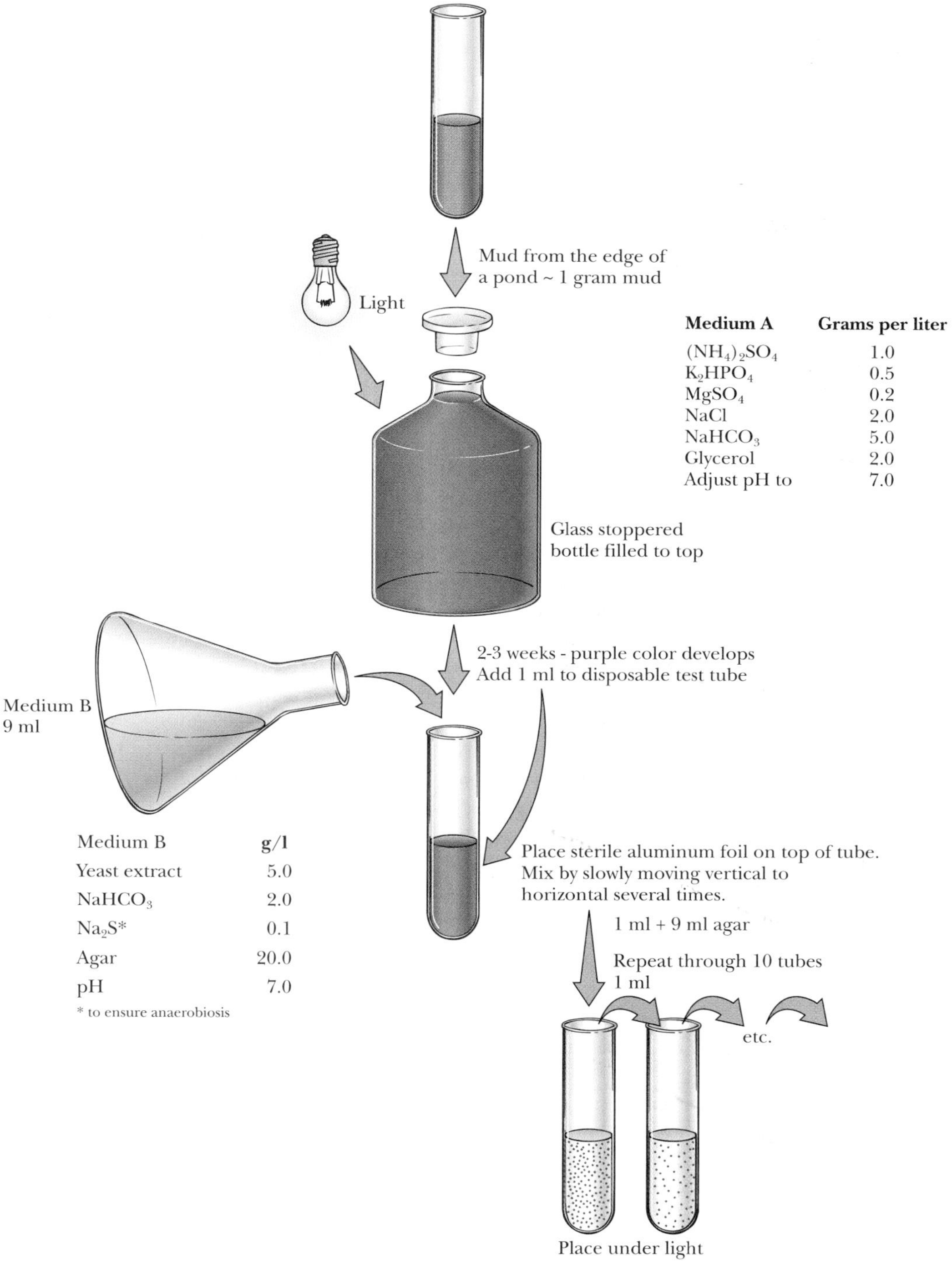

***Figure* 5.13** An isolation technique for photosynthetic microorganisms. This would be effective for other anaerobic organisms that are not sensitive to brief exposure to low levels of oxygen.

their nitrogen requirement by fixing atmospheric N_2 (see Table 5.5). For the isolation of *Azotobacter* or any nitrogen-fixing organisms, a nitrogen source such as NH_4^+ and NO_3^- must be eliminated from the medium. Eliminating a fixed nitrogen source from the medium would select for microorganisms that obtain this key cellular component from the atmosphere. An element that must be present in such enrichments is molybdenum, as this metal is a constituent part of the enzyme nitrogenase that is involved in nitrogen fixation.

Chemoheterotrophic Anaerobic Bacteria

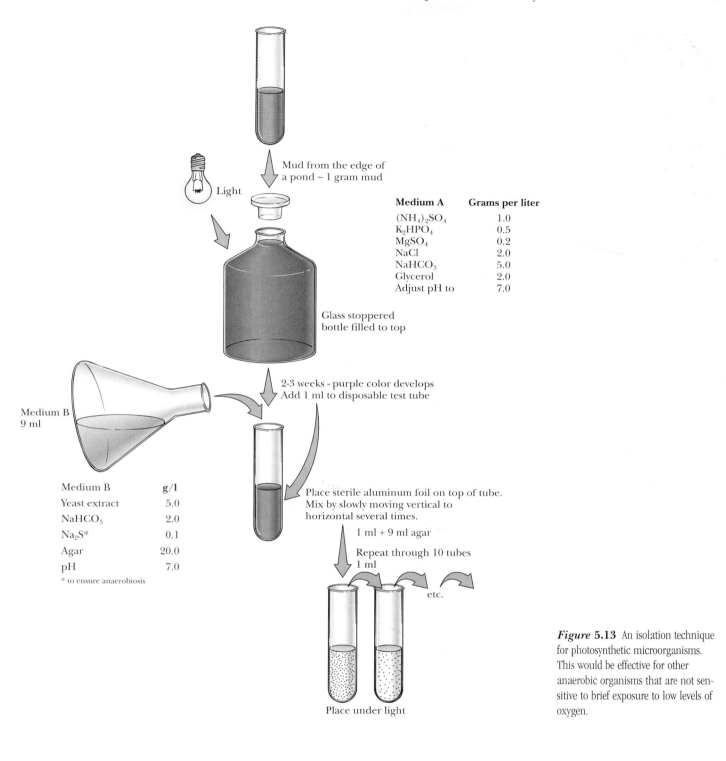

Procedures for the isolation of selected chemoheterotrophic anaerobic bacteria are outlined in Table 5.9. Many anaerobes are killed by the presence of very low levels of

BOX 5.3 METHODS & TECHNIQUES

Special Enrichment—The Winogradsky Column

The Winogradsky Column is an enrichment culture technique developed by Sergei Winogradsky in the latter part of the nineteenth century (see Chapter 2). A typical Winogradsky Column is a long glass tube (1-1/2 × 24 inches), closed at one end, and with about two thirds of the tube filled with rich mud. The column should be placed in a north window where it receives adequate but not intensive light. It is a microcosm where one can follow growth and succession in an anaerobic environment.

Organic-rich mud from a shallow area of a pond is a suitable source of mud for a column. Before placing the mud in the column, a few grams of calcium carbonate and calcium sulfate can be incorporated. In practice it is preferable to mix the calcium sulfate in the mud that will occupy the lower fourth of the column along with some starch, cellulose powder, or shredded filter paper. These carbon sources can be varied depending on the imagination of the individual preparing the column. There should be no air pockets in the mud column, and any that form can be disrupted with a glass rod. The mud column is topped with a layer of pond water.

The appearance of the column after a few weeks would be as illustrated. Anaerobic bacteria ferment the cellulose or starch at the bottom to hydrogen, organic acids, and alcohols. These would be utilized as substrate by sulfate-reducing bacteria, and sulfate would serve as terminal electron acceptor. This would cause the formation of hydrogen sulfide resulting in an H_2S gradient from the bottom of the column upward. An O_2 gradient would occur from the top downward. The bottom area of the column becomes black due to formation of metal sulfides. Restricting calcium sulfate to the lower area curtails an excessive production of black precipitate throughout. The green sulfur bacteria develop immediately above the dark area because of their tolerance for hydrogen sulfide. The purple sulfur bacteria are less tolerant and appear in the area above the green. The purple nonsulfur bacteria grow nearer the top in the absence of sulfide and they may tolerate the low levels of oxygen present. Aerobic microorganisms, including cyanobacteria and algae, grow in the water layer. One would find anaerobic bacteria growing throughout and some sulfide oxidizers such as *Beggiatoa* and *Thiothrix* growing in the upper area. Samples can be removed periodically with a length of glass tubing narrowed at the end. Organisms can be isolated from these samples by aerobic or anaerobic techniques described in this chapter.

oxygen, and these microorganisms require special precautions in their handling.

A medium for the isolation of nitrogen-fixing anaerobes is presented in Table 5.9. With starch as enrichment substrate and pasteurized soil as inoculum, the organism most likely isolated would be *Clostridium pasteurianum*. This anaerobe forms heat-resistant endospores that readily survive the pasteurization process.

Chemoautotrophic Bacteria

The carbon source for the isolation of the chemoautotrophs is limited to carbon dioxide and any of the following inorganic sources of energy: NH_4^+, NO_2^-, H_2S, Fe^{2+}, or H_2 (Table 5.10). Probably the most important factor in the isolation of the nitrifying bacteria (organisms that oxidize $NH_4^+ \longrightarrow NO_2^- \longrightarrow NO_3^-$) is patience, as the appearance of visible colonies on petri plates can take from one to four months. Enrichment and isolation of the nitrifiers is generally carried out by **serial dilution** of rich soil (Figure 5.14). Growth of the ammonia (NH_4^+) oxidizers is measured by chemically analyzing for an increase in nitrite concentration in the medium. Growth of those that oxidize nitrite to nitrate is assessed by adding

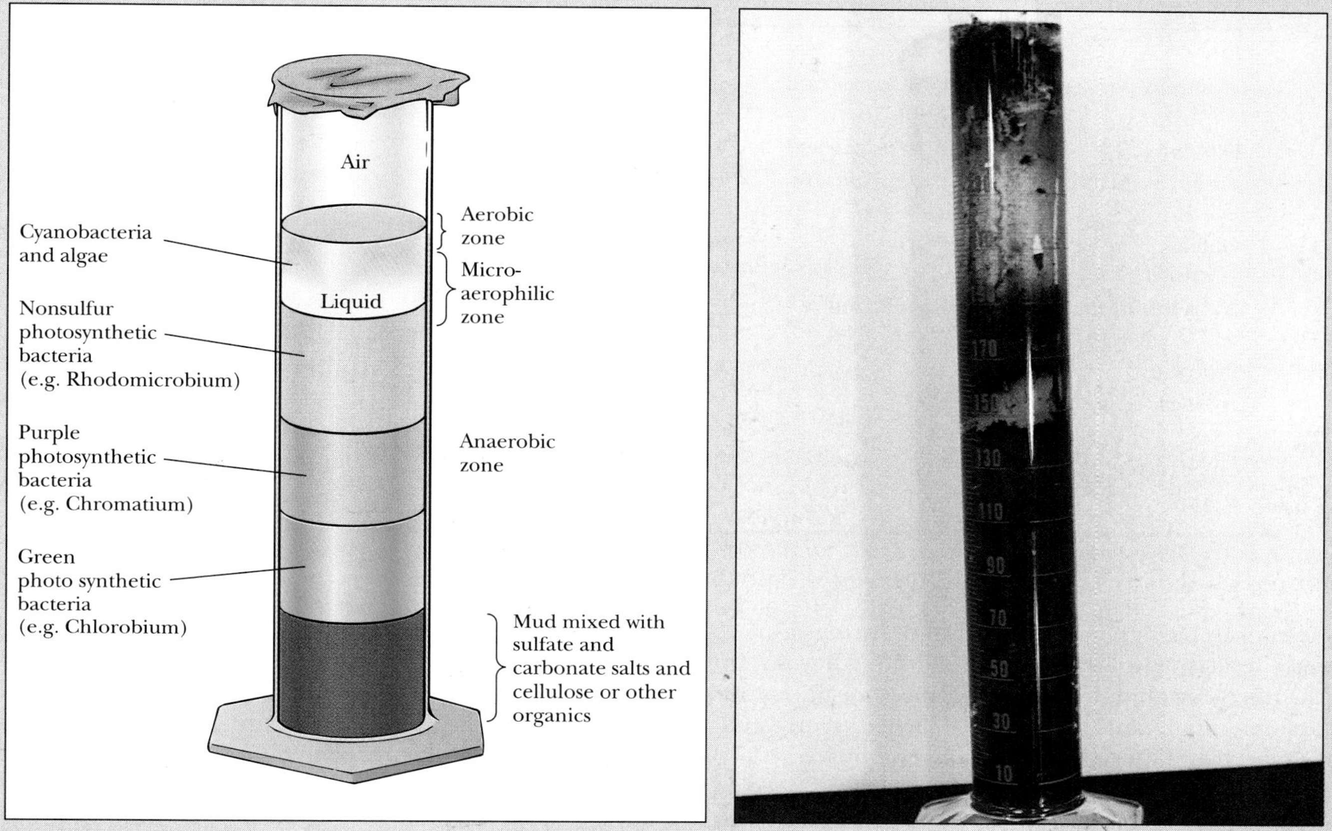

A Winogradsky Column. Anaerobic decomposition of organic matter in the bottom of the cylinder creates an anaerobic sulfide-rich environment. The green and purple sulfur bacteria are present immediately above the dark sulfide area. The purple nonsulfur bacteria would be above the purple sulfur bacteria. Cyanobacteria would grow at the top in the area exposed to air. (Courtesy of J. J. Perry)

measured amounts of nitrite and determining the amount that disappears.

The sulfur-oxidizing autotrophic bacteria play a key role in the conversion of reduced sulfur compounds to sulfate. They are present in a wide range of habitats, from pH 1 to 9, and many species are thermophilic ("heat loving").

The hydrogen-oxidizing bacteria do not form a cohesive taxonomic unit, and the ability to grow using hydrogen as the energy source occurs across many of the prokaryotic phyla. Enrichment is rather straightforward because these organisms grow chemolithotrophically under an atmosphere of hydrogen, carbon dioxide, and oxygen. The hydrogen-oxidizing bacteria can be isolated by sprinkling soil on an agar surface and incubating under the proper CO_2/H_2 atmosphere.

Phototrophic Bacteria

The major requirement for the isolation of phototrophic bacteria is a constant source of light. Other conditions of enrichment are outlined in Table 5.11. Provided light is available, photosynthetic bacteria can flourish under a broad range of conditions. Cyanobacteria, for example,

***Table* 5.8 Enrichment conditions for chemoheterotrophic aerobic or facultative aerobic organisms**

Carbon and Energy Source	Inoculum	Special Conditions	Organism Favored
Ethanol	Soil	N_2 as nitrogen source	*Azotobacter*
Uric acid	Pasteurized soil (80°C for 15 min)		*Bacillus fastidiosus*
Glucose	Pasteurized soil		*Bacillus*
Casein + thiamine	Pasteurized soil	Add urea at pH 9.0	*Bacillus pasteurii*
Glucose + yeast extract	Soil	Incubate at 60°C	*Bacillus stearothermophilus*
Nutrient broth	Pasteurized soil	10% NaCl 0.5% $MgCl_2$	*Sporosarcina halophila*
Propane	Soil	Gas 50/50 in air	*Mycobacterium*
n–Hexadecane	Soil	Incubate at 60°C	*Bacillus thermoleovorans*
Filter paper	Soil		Cytophaga
Chitin	Soil		Actinomycetes

grow in various harsh natural habitats: hot springs, Antarctic lakes, deserts, and areas of high salinity. They are the photosynthetic symbiont in many lichen associations and are common inhabitants of ecological niches in terrestrial, marine, and freshwater habitats. As the cyanobacteria produce oxygen during photosynthesis they are all aerobic. Isolation can be accomplished in vessels exposed to air.

The green and purple bacteria do not produce oxygen during photosynthesis and grow under anaerobic conditions. The isolation of members of the nonsulfur purple bacteria can be accomplished by using mud or water samples from ponds, ditches, and shores of eutrophic lakes (that is, lakes rich in nutrient but low in oxygen). Nonsulfur purple bacteria are organisms that utilize reduced

***Table* 5.9 Enrichment conditions for chemoheterotrophic anaerobic Eubacteria or Archaea**

Carbon and Energy Source	Inoculum	Special Conditions	Type Organism
Sugars + yeast extract	Plant material	pH 5–6	Lactic acid bacteria
Mixed amino acids	Pasteurized soil		*Clostridium*
Starch	Pasteurized soil	N_2 as nitrogen source	*Clostridium pasteurianum*
Uric acid + yeast extract	Pasteurized soil	pH 7.8	*Clostridium acidiurici*
Organic acids	Pond mud	Added SO_4^{2-}	Desulfovibrio
Organic acids	Soil	Added NO_3^-	Denitrifying bacilli and pseudomonads
Organic acids	Rumen fluid	Added CO_2	Methane producers
Lactate + yeast extract	Swiss cheese		*Propionibacterium*

***Table* 5.10 Enrichment for chemoautotrophic Eubacteria or Archaea**[1]

Energy Source	Inoculum	Special Conditions	Organism Obtained
H_2	Soil or water	Aerobic	Hydrogen-utilizing bacteria
NH_4^+	Soil or water	Aerobic	*Nitrosomonas*
NO_2^-	Soil or water	Aerobic	*Nitrobacter*
H_2	Rumen fluid	Anaerobic	Methanogens
$Na_2S_2O_3$	Soil or water	Anaerobic + KNO_3	*Thiobacillus denitrificans*
Fe^{2+}	Estuarine mud	Aerobic, pH 2.5	*Thiobacillus ferrooxidans*

[1]In the presence of CO_2 or added $NaHCO_3$.

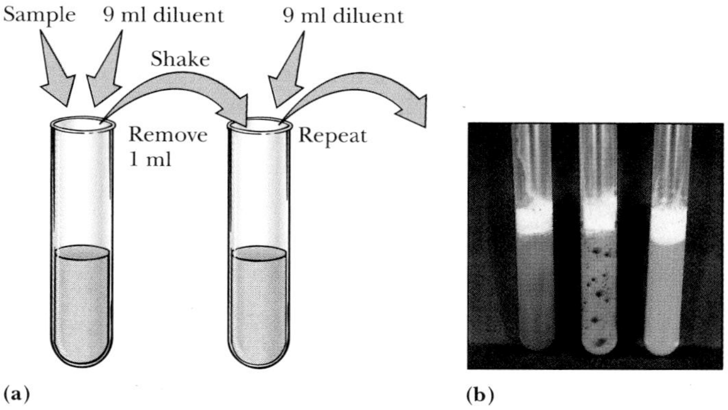

***Figure* 5.14** An example of a serial dilution that may be utilized in isolation of anaerobic microorganisms. Manipulations for anaerobes would be accomplished in a glove box. **(a)** Dilutions are made in melted agar (at 48°C), the tubes are sealed to exclude air. After incubation, colonies can be enumerated or the agar may be removed from the tube and individual colonies selected for further purification. **(b)** Results of a dilution series with the photosynthetic organism *Rhodomicrobium* sp. (**b**, courtesy of J. J. Perry)

***Table* 5.11 Enrichment for photosynthetic microorganisms**

Special Conditions	Source of Inoculum	Organism
Aerobic		
N_2 as nitrogen source	Surface water or soil	Nitrogen-fixing cyanobacteria
NH_4^+ as nitrogen source	Surface water or soil	Cyanobacteria and eukaryotic algae
Anaerobic		
Glycerol	Mud from pond edge	Purple or green nonsulfur bacteria
H_2S (high levels)	Sulfide-rich mud	Green sulfur bacteria
H_2S (low levels)	Mud from pond edge	Purple sulfur bacteria

electron donors other than sulfur compounds. Hydrogen (H_2) would be a typical electron donor for a nonsulfur purple photosynthetic. The mud from the bottom of a shallow eutrophic lake may contain as many as 1 million purple nonsulfur bacteria per gram. The major component of an enrichment medium for these organisms would be hydrogen or a reduced organic compound to serve as electron donor in photosynthetic growth. It is important that yeast extract (0.05 percent) be added to the enrichment because the purple nonsulfur bacteria may require some of the B-vitamins.

The green sulfur bacteria and purple sulfur bacteria can also be obtained by enrichment culture, but are more difficult to isolate than nonsulfur bacteria. They utilize soluble sulfides as electron donor for photosynthetic growth. The sulfide concentration is critical for the enrichment of these bacteria, and a higher population density and species diversity can be obtained by repeated addition of sulfide during enrichment. These sulfur-utilizing photosynthetic organisms are present in aquatic environments, especially where anaerobic conditions are stringent and constant.

One can select for green sulfur bacteria by controlling the amount of light available. A low intensity direct sunlight (north-facing window) or a tungsten light of low intensity (5 to 2000 lux) favors green sulfur bacteria as these intensities are not sufficient to support growth of most other phototrophic bacteria. The purple sulfur bacteria can be enriched by higher light intensities.

Cultivation of Microorganisms

Microbiologists have devised techniques and procedures that enable them to separate a single species of microorganism from a mixed culture. These organisms can then be cultivated in **pure culture.** A pure culture is also called an **axenic culture** (*a* = "without," *xenos* = "strange"). Some of the standard procedures for the growth, maintenance, and preservation of pure cultures are presented here.

A Culture Medium

A culture medium is composed of biologically or chemically derived materials that provide a local environment for the growth of a microorganism. There are hundreds of described media, and a variety of ingredients can be used in their preparation. Among the common media components are extracts of beef heart or brain, whole blood or serum, yeast extract (autolyzed yeast), peptone (protein digest), hydrolyzed casein (milk protein), or soil extract. These ingredients are not chemically defined, and a medium composed of any of these would be termed a **complex medium.** Sugars or organic acids might also be added to a complex medium. When a solid medium is desired one would add the complex polysaccharide, agar, as solidifying agent. Silica gel can be used for the solidifying agent when agar is undesirable.

A **defined** or **synthetic** medium can be employed for growth of many microorganisms. All of the ingredients in a defined medium are chemically characterized, and one such medium was presented in Table 5.2. The defined medium (see Table 5.3) for *Streptococcus agalactiae* is considerably more detailed.

Isolation in Pure Culture

In nature, virtually all microorganisms coexist with countless other microbial species. For example, hundreds of different species live in the intestinal tracts of animals. A gram of fertile soil contains 10^7 to 10^9 bacteria and thousands of species. This natural state where many different species coexist is referred to as a **microbial community** or a **mixed population.** Enrichment or isolation procedures generally result in the selection of a mixture of microbial species. Therefore it is necessary to apply techniques that will permit one to **isolate** individual species from a conglomerate of many species. Isolation involves separation of these small microorganisms from one another. This is accomplished by use of **aseptic** isolation procedures. **Asepsis** means in the absence of microorganisms (its antonym is **sepsis,** which means nonsterile, or in the presence of microorganisms). In this context, aseptic isolation also means that one handles materials in such a way that unwanted microorganisms are not introduced during isolation.

Sterilization of Media

Sterilization is the process of killing or removing *all* living things from a particular container or environment. Heating with live steam under pressure (autoclave) is commonly used for sterilization (see Figure 7.3).

Growth media are autoclaved in covered or cotton-stoppered test tubes or flasks. Agar-containing petri dishes are prepared by autoclaving the medium in a flask and aseptically pouring the liquified agar/nutrient into a sterile petri dish. Plastic petri dishes that have been presterilized by ethylene oxide gas or radiation are generally purchased.

Incubation Conditions

Aerobic organisms require oxygen for growth and should be cultivated on petri plates or slants in **incubators.** An incubator is an insulated chamber used for the growth of microorganisms. Incubators have a set, thermostatically controlled temperature and are usually maintained at 25° to 30°C for common soil organisms or at 37°C (body temperature) for those from human or animal sources.

Aerobic organisms grown in broth culture tend to deplete the oxygen from the medium during active respiration. This will curtail growth of the organism. Thus, **shak-**

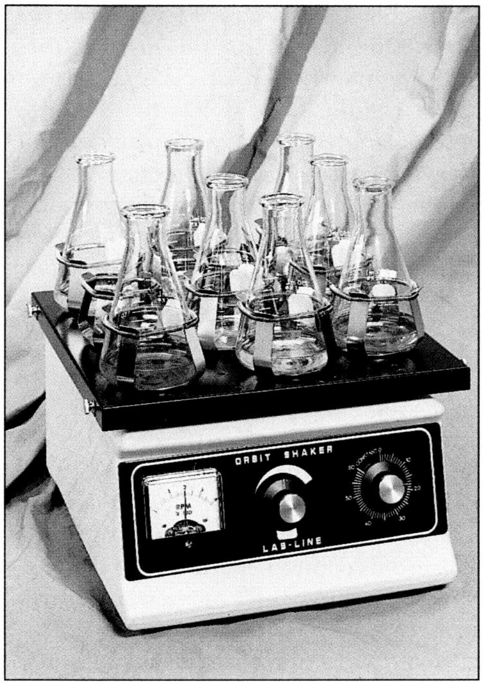

Figure **5.15** A bench top shaker. The platform on top can gyrate at controlled rates from 25 to 500 rpm. These shakers can be operated at room temperature or placed in a refrigerator or incubator at a desired temperature. The cotton stoppers ensure that only filter-sterile air will enter the flasks. (Courtesy of Lab-Line Instruments)

ers are sometimes used to agitate the cultures during growth so more oxygen is provided and at a constant rate. Shakers are mechanical devices that rotate a culture to promote the introduction of oxygen into the medium (Figure 5.15).

Obligately anaerobic organisms, in contrast, cannot be grown in the presence of oxygen. Special media and media-preparation procedures are needed to cultivate strict anaerobes. This can be accomplished in a variety of techniques as will be discussed in Chapter 6.

Streak Plate Procedure

The classic method for isolation of bacteria is the **streak plate procedure** developed in Robert Koch's laboratory (Chapter 2). In this technique, a sample of the natural community is picked up with a sterile **loop,** a metal wire that has a circular loop at its end (Figure 5.16). The wire is first sterilized by heating it directly in a burner flame until red hot. The loop should be cooled by touching on the sterile liquid or solid medium. The bacterium of interest would be collected on the loop by touching a colony or immersing in a liquid medium. A sample may be a droplet of lake water, a colony, a liquid culture, or a mixture of microbes obtained by enrichment. For example, lake water typically contains about 10,000 viable bacteria

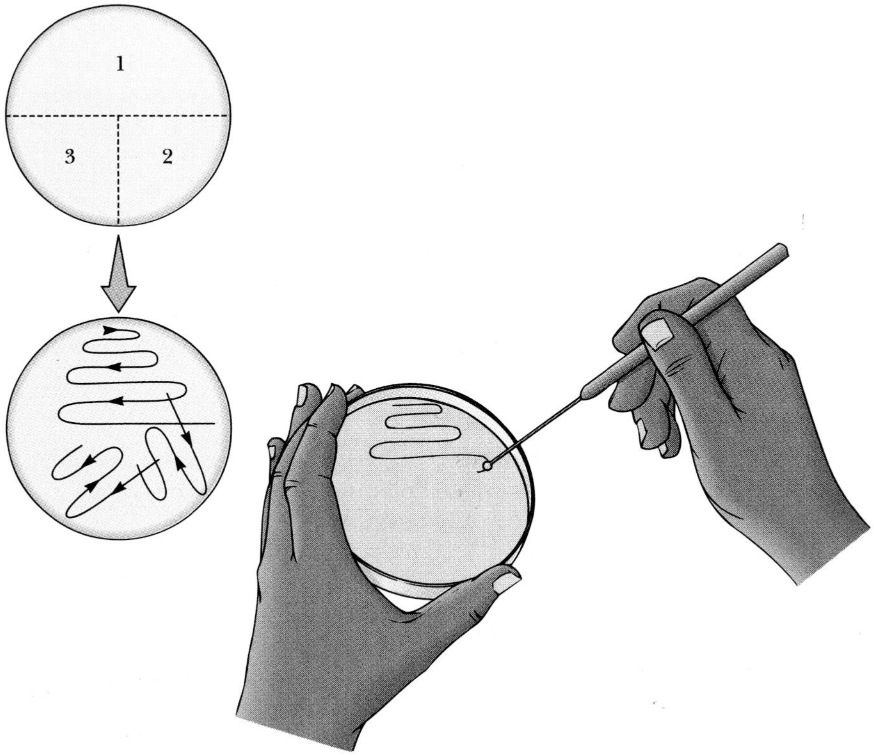

Figure **5.16** The proper streaking procedure for obtaining isolated colonies. First streak is with the sample from an enrichment, soil, water, or a culture. Second streak is accomplished by contacting a small amount from one after sterilizing the loop and the third is a repeat of the second, with contact at the latter part of 2.

per milliliter (ml). The loop, which has an approximate volume of 0.01 ml, would thus pick up 100 or more bacterial cells. To separate the bacteria in the droplet, dilution is necessary.

A common method of "dilution" is by **streaking** the loop's contents on the surface of an **agar plate.** Streaking over the surface is accomplished in three successive steps. In the first streak, a loop bearing the sample is spread back and forth in nonoverlapping lines to cover about one-third of the agar surface. This is followed by flaming and cooling of the loop. The sterile loop would be passed across the first streak to obtain some of the cells, and streaked over another sector of the plate. This process would be repeated on a third sector to further ensure the separation of bacteria (see Figure 5.16). In this streaking procedure, individual microorganisms from the original sample become spaced farther and farther apart. The plate can then be incubated at the appropriate temperature to permit growth of the organisms.

Individual microorganisms that are spaced apart by the streaking process can multiply to produce a **colony** (or **clone**) made up of cells derived from a single parent. The colony represents a **pure culture** or **isolated strain** of the microorganism. To ensure purity of a desired species or colony type, however, it is necessary to repeat the streaking process one or more times on a freshly prepared plate. A pure culture is obtained when only one type of colony grows on the plate (Figure 5.17). This pure culture is referred to as a bacterial **strain.**

Some organisms may not grow well or at all on an agar surface. There may be several reasons for this. For example, the composition or pH of the medium or the incubation conditions may not be satisfactory for their growth. Some bacteria cannot grow as separate colonies, but require closely associated bacteria to provide biochemicals for their growth. These required biochemicals may not be provided in the medium. Organisms that require the presence of other organisms for growth are referred to as **syntrophic bacteria.** They grow very well in **consortia** (cultures containing more than one species), but not as pure cultures.

Pour Plate Procedure

Another method for obtaining pure cultures is through the **pour plate procedure,** in which a soil or water sample is mixed with a molten agar medium *before* the medium solidifies. After a thorough mixing, the molten agar (45–48°C) is poured into a petri dish and allowed to solidify. In this procedure, the colonies are distributed uniformly throughout the medium (Figure 5.18). This procedure has one disadvantage in that the high temperature needed to keep the medium molten may actually kill some species. For example, bacteria from marine samples or temperate zone lakes rarely, if ever, encounter temperatures greater than 25° to 30°C, and may be killed by brief exposures to 45°C. In contrast, most human intestinal bacteria can survive heating to 45°C because they grow at body temperature (37°C).

Spread Plates

Spread plates may also be used for isolation of bacteria. A

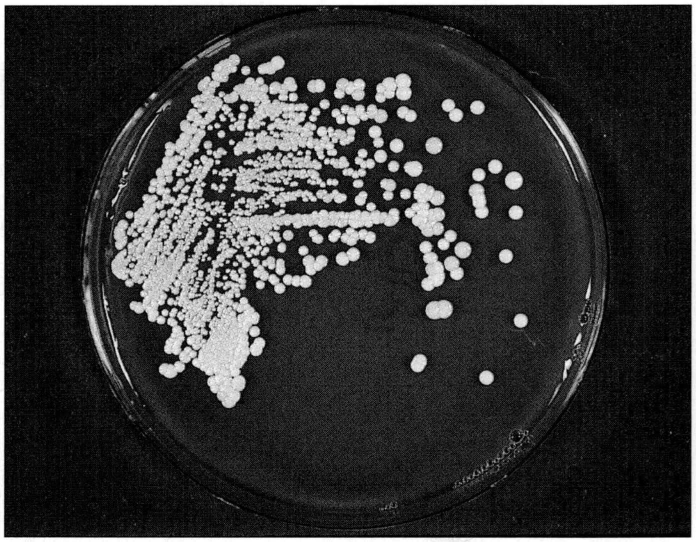

***Figure* 5.17** Results of a three-way streak of four strains of *Micrococcus luteus.* A dark yellow strain and a white mutant. A light yellow strain and its pink mutant. (Courtesy of Wesley Kloos)

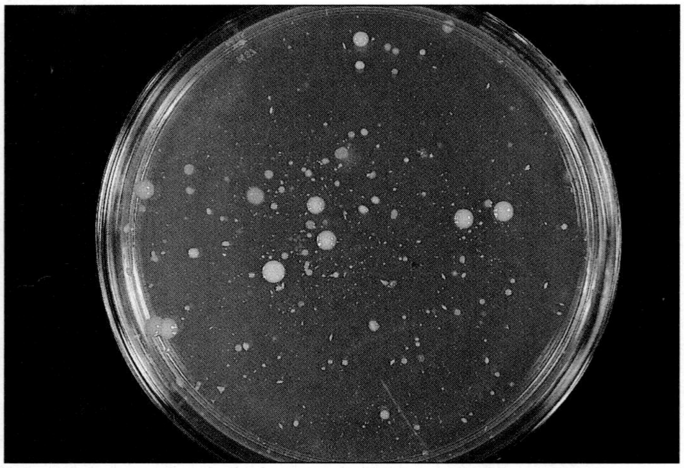

***Figure* 5.18** Isolation of colonies by the pour plate method. A culture or mixture of organisms can be added to melted agar (48°C). After incubation the individual colonies are "trapped" in the agar, resulting in separated colonies. (Courtesy of J. J. Perry)

(a)

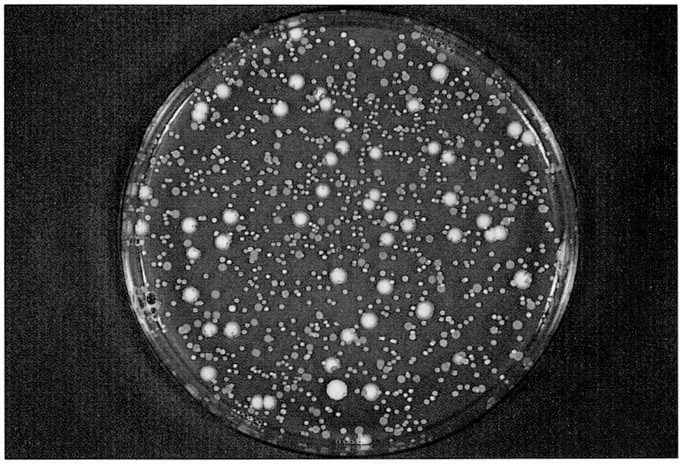

(b)

***Figure* 5.19** Individual colonies can be obtained by spreading a liquid sample containing microorganisms over the agar surface. **(a)** The "hockey stick" is sterilized by placing the lower part in ethanol, it is removed, and the excess is burned off by contact with a bunsen burner flame. The turntable on which the agar plate sits can be rotated while holding the hockey stick in one place. **(b)** Results of a spread plate inoculated with a diluted sample of material obtained from nasal mucus. The small gray colonies are *Staphylococcus epidermidis,* the small white colonies are *S. capitis,* the large white colonies are *S. hemolyticus,* and the single yellow colony is *S. aureus.* (**a**, courtesy of J. J. Perry and **b**, courtesy of Wesley Kloos)

small volume (typically about 0.1 ml) of the sample is placed on the agar surface and then spread with a sterile bent glass rod that resembles a "hockey stick." This process distributes the cells evenly on the surface, so they grow as discrete colonies (Figure 5.19). This procedure is less restrictive than pour plating because the cells are not exposed to the high temperature of the molten agar.

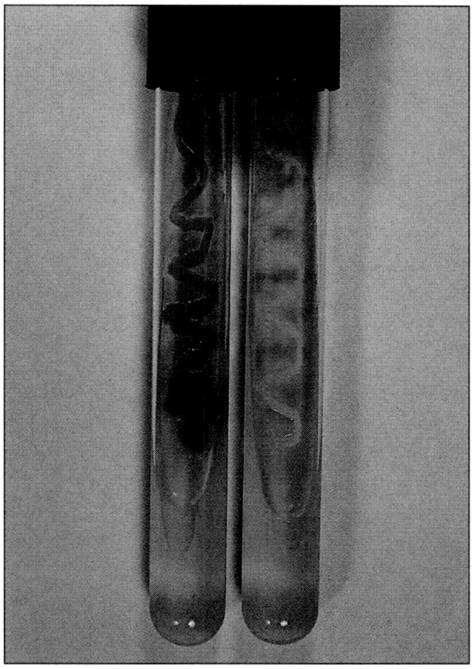

***Figure* 5.20** An agar slant. The organisms are the yellow pigmented organism *Micrococcus luteus* and the red pigmented *Serratia marcescens.* (Courtesy of J. J. Perry)

Maintenance of Cultures

After an axenic culture is obtained it can be transferred to a suitable agar medium and maintained as a **stock culture** by transfer at suitable intervals to a newly prepared medium. It is of utmost importance that a researcher retains such cultures in an axenic state, as faulty research will result from **contaminated cultures** (cultures that contain a mixture of two or more organisms).

Stock cultures are often stored on an agar surface in a test tube. To increase the surface area for growth in the tube, the agar medium is allowed to solidify at an angle resulting in a **slant** or **slope** (Figure 5.20). Cultures transferred to a slant would be placed in an incubator, and after growth they would be placed in a refrigerator (4–5°C). Generally, microbial cultures are viable on refrigeration for at least a month. Survival under these conditions varies from species to species.

Longer term maintenance is accomplished either by **lyophilization** (freeze-drying) or by freezing bacterial or archaeal mass in glycerol at exceedingly low temperatures (−70°C or lower in special freezers or in liquid nitrogen). Lyophilized cultures maintain viability in freezers for long periods of time: most cultures can be stored for 10 years or longer, and therefore require minimum maintenance.

Summary

- **Eubacteria and Archaea differ** from one another based on the compounds they utilize as food and/or energy source.
- The **substrates** utilized by Eubacteria and Archaea as source of carbon include all constituents of living cells and many synthetic chemicals. This microbial biodegradation can occur over a broad range of pH, temperature, salinity, and other physical parameters.
- Microorganisms can be divided into four nutritional groupings: **photoautotrophs, photoheterotrophs, chemoautotrophs,** and **chemoheterotrophs.**
- Bacteria, regardless of species or growth substrate, are composed of the same elements. The monomers from which the macromolecules are synthesized are equivalent. A growth medium must, in one way or another, supply all elements that make up a microbial cell.
- Some microorganisms can synthesize all cellular components from CO_2, whereas others require a source of amino acids, vitamins, or nucleic acid bases.
- Many eubacterial and archaeal species can **"fix" atmospheric N_2** into cellular material. N_2 fixation is not known among the eukaryotes.
- **Phosphate** is present in microorganisms in relatively low amounts (~ 3% dry weight) but is an essential element in nucleic acids and in energy generation.
- The **growth factors** (needed in small amounts) most commonly required by microorganisms are: B-vitamins, amino acids, or nucleic acid bases.
- A **trace element** is one required in very small amounts (traces). Among these are Ca^{++}, Mg^{++}, Na^{+}, and Cl^{-}.
- Microorganisms need specific nutrient uptake systems, as they generally live in environments where many nutrients are present at low levels. The three major systems for nutrient uptake are **facilitated diffusion, active transport,** and **group translocation.**
- **Facilitated diffusion** does not require energy input but cannot concentrate nutrients internally to a concentration greater than the external concentration.
- **Active transport** requires energy that may come from ATP or a proton gradient.
- Proton gradients can effect uptake by three mechanisms—**uniport, symport,** and **antiport.**
- **Group translocation** results in an alteration of the compound transported. Sugars may be transported into the cell via this process. Group translocation is the most elaborate of the transport systems and involves a number of enzymes in uptake of a single solute.
- Iron is required by aerobic organisms, and microorganisms may have specific chealators called **siderophores** for iron uptake.
- **Enrichment culture** is a method for isolating microorganisms that grow on a specific nutrient and/or under selected physical conditions. The organism obtained from an environmental sample by this procedure will have the ability to grow under the specific enrichment conditions employed.
- Most microoganisms are maintained in collections as **pure (axenic) cultures.** All studies with pure cultures must be done under aseptic conditions.
- Pure cultures can be obtained by **streaking, pour plate,** or **spread plate** techniques.
- Microorganisms can be maintained for extended periods of time by **lyophilization** or by placing them in **liquid nitrogen.**

Questions for Thought and Review

1. What is metabolic diversity? How did it come about? How does this relate to the role of microbes in nature?
2. Microbes survive and actually thrive in virtually every environmental niche on earth. How does this relate to the perpetration of life on earth?
3. What are the basic requirements in the development of a growth medium? Why is each added? How does this relate to the composition of a bacterial cell?
4. Define "growth factors." Why do some organisms require them?

5. How would one use enrichment culture to isolate an organism capable of growth with *p*-aminobenzoic acid as sole source of carbon? What would one do to obtain an organism present in primary enrichment in axenic culture?

6. Why have bacteria been utilized in studying nutrition rather than using higher organisms such as rats or other animals? How does this relate to comparative biochemistry?

7. Where would one obtain samples to be used in isolating various photosynthetic bacteria? Would the source for cyanobacteria differ from that for green sulfur photosynthetics? Why?

8. What is a microbial community? Are these prevalent in nature?

9. What is the streak plate method, and how does one employ this to obtain a pure culture?

10. Define axenic culture, consortia, and a bacterial strain.

11. How does a pour plate differ from a streak plate? What are the advantages of each?

12. What are the two major methods for long-term storage of bacteria?

Suggested Readings

Bridson, E. Y., and A. Brecker. 1970. Design and Formulation of Microbial Culture Media. In J. R. Norris and D. W. Ribbons, eds. *Methods in Microbiology.* vol. 3A, pp. 229–295. New York: Academic Press.

Gerhardt, Philipp (Editor-in-Chief). 1993. *Methods for General and Molecular Bacteriology.* Washington, DC: American Society for Microbiology.

Gottschal, J. C., W. Harden, and R. A. Prins. 1992. *Principles of Enrichment, Isolation, Cultivation, and Preservation of Bacteria.* In A. Balows, H. G. Truper, M. Dworkin, W. Harden, and K. H. Schleifer. *The Prokaryotes.* 2d ed. New York: Springer-Verlag.

Ingraham, J. L., O. Maaløe, and F. C. Neidhardt. 1988. *Growth of the Bacterial Cell.* Sunderland, MA: Sinauer Associates Inc.

Norris, J. R., and D. W. Ribbons, eds. 1969. *Methods in Microbiology.* vol. 3B. New York: Academic Press.

Veldkamp, H. 1970. *Enrichment Cultures of Prokaryotic Organisms.* In J. R. Norris and D. W. Ribbons, eds. *Methods in Microbiology.* vol. 3A, pp. 305–361. New York: Academic Press.

Williams, S. T., and T. Cross. 1971. *Actinomycetes.* In Norris, J. R. and D. W. Ribbons, eds. *Methods in Microbiology.* vol. 4, pp. 295–334. New York: Academic Press.

The paramount evolutionary accomplishment of bacteria as a group is rapid, efficient cell growth in many environments. Bacteria grow and divide as rapidly as the environment permits.

J. L. Ingraham
O. Maaløe
F. C. Neidhardt

Chapter 6

Microbial Growth

Population Growth
Growth Curve
Measure of Growth
Effect of Nutrient Concentration on Growth Rate
Effect of Environmental Conditions on Growth

The nutritional requirements for the growth of representative eubacterial and archaeal cultures were presented in the previous chapter. The indispensable need for carbon, nitrogen, sulfur, and various cations and anions was discussed. Methods were also outlined for the isolation of specific autotrophic and heterotrophic microorganisms.

When the nutrients and physical conditions required by a specific organism are met, the organism will grow (see quote). The term *growth* generally refers to an increase in the number of cells in a population, or **population growth.** An individual eubacterial or archaeal cell may increase in size, and growth in this sense would be called **cell growth.** During population growth, a eubacterial or archaeal cell divides to generate two equivalent progeny. The duplicate progeny each receives a genome that is a precise copy of the genetic information in the parent cell (see Chapter 13).

A microorganism is generally considered viable only if it is able to reproduce. In fact, the test most often used for viability is to place the organism under a growth condition, incubate it, and then examine for visible growth after a suitable time. However, we must be aware that most of the microorganisms in nature have not yet been cultivated in the laboratory. These organisms are most assuredly viable, and our inability to grow them is probably due to our lack of understanding of their growth requirements.

Population growth is the culmination of a complex series of biochemical events that are driven by light or chemical energy. This energy is utilized to synthesize or assimilate monomers, and these are in turn assembled into macromolecules. In the course of growth and division, a eubacterial or archaeal cell will synthesize an estimated 1800 different proteins, more than 400 different RNA molecules, a copy of the genomic DNA, cytoplasmic membrane, and, where necessary, sufficient cell wall to surround the newly formed cell.

As discussed in Chapter 5, accessory compounds such as B-vitamins may be required by a microorganism and must be available for growth to occur. Many organisms can grow by utilizing a single substrate, such as glucose, as the sole source of carbon and energy. Autotrophs use light or inorganic sources of energy with atmospheric carbon dioxide serving as the sole carbon source. All the synthetic processes involved in population growth are regulated and integrated to produce a duplicate cell, and this occurs in a short period of time called the **generation time.** Generation time is defined as the time required for a population of cells to double in number.

Population Growth

A bacterial cell generally grows in size as a prelude to cellular division. Thus, this phase of the cell cycle (see **Box 6.1**) is considered cell growth, while cellular division results in an increase in the number of cells (Figure 6.1). When a bacterium is placed in a suitable growth medium, an initial adjustment period often occurs and during this time there is no increase in cell number (see Growth Curve section). This is followed by an exponential increase in the bacterial population. This increase in numbers per unit time occurs at a constant and reproducible rate for a given organism and is called the **growth rate.** Because growth of a population is exponential (that is 1 cell $\longrightarrow$ 2 cells $\longrightarrow$ 4 $\longrightarrow$ 8, etc.), it is generally depicted on a logarithmic scale, as shown in Figure 6.2. The generation time for a species is the time required for the population to double during the exponential phase of growth (Figure 6.3). The generation time (also called doubling time) is usually determined under what would be considered optimal conditions for growth **(Box 6.2).**

Typical generation times for a number of eubacterial and archaeal species are presented in Table 6.1. It is apparent from these values that some organisms divide much more rapidly than others. An organism such as *Escherichia coli* that can reproduce in 20 min could, starting with one cell and under unlimited growth conditions, yield 4.7×10^{21} progeny in 24 hr **(Box 6.3).** In 13 hr, a microbe with a generation time of 20 min can produce 1×10^{12} cells, approximately the number of cells in the entire human body! A single cell of a slower growing soil organism such as *Bacillus subtilis* (generation time 30 min) would generate 1×10^{12} progeny in about 20 hr. These figures would

BOX 6.1 RESEARCH HIGHLIGHTS

The Cell Cycle in Bacteria

The cell cycle is the period of time in which a newly formed bacterium elongates, replicates its DNA, and divides to generate two cells. The replication of DNA and events involved in division to form two cells is under tight regulatory control.

In a newly formed cell there is a regulatory event that initiates the replication of the bacterial genome. The time required to replicate the 4.2×10^6 base pairs in the DNA of various bacteria with a generation time between 20 and 60 minutes is 40 minutes. This is the C phase (see diagram). After completion of DNA replication (termination) there is a 20-minute period before division occurs. This is the D (delay) phase. At termination, regulatory proteins initiate the D phase wherein the replicated DNA is separated into opposite ends of the elongated cell. The cytoplasmic membrane is involved in this process. After separation of the DNA the process of constructing cytoplasmic membrane and cell wall begins at the midpoint of the cell. The division into two distinct daughter cells is called **transverse fission.**

You might ask how a species can have a generation time of 25 minutes when replication of DNA requires 40 minutes. The answer is that a second round of replication begins on the part of the DNA that has already divided before the C phase of the first replication is complete. Thus both daughter cells may receive a genome that has replication forks that are ongoing. This will be explained further in Chapter 13.

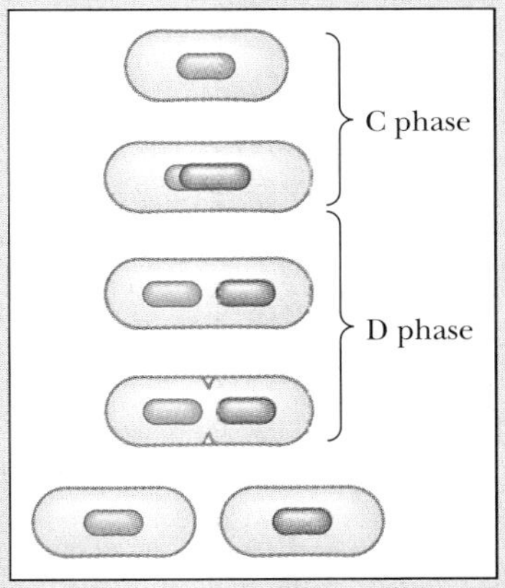

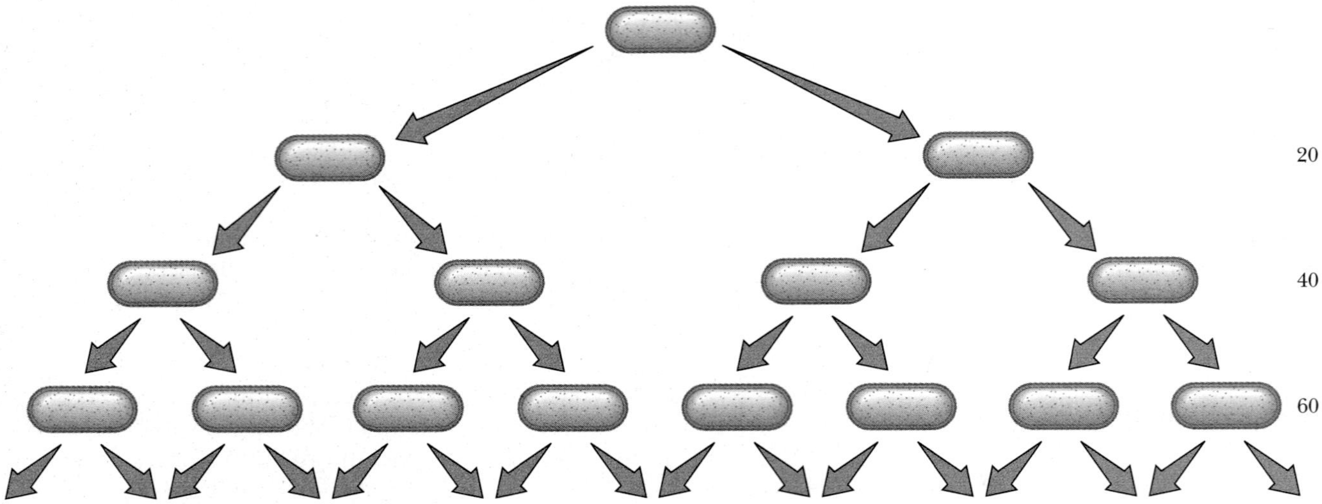

***Figure* 6.1** A depiction of exponential phase population growth. One cell of a microorganism with a 20-minute generation would yield 8 progeny in one hour.

apply only if an unlimited supply of all nutrients were available and if the culture medium was free of excreted inhibitory products.

Growth Curve

Adding a small population of bacteria to a suitable volume of culture medium that can support growth results in a predictable increase in cell numbers. If the number of organisms present is determined at intervals throughout population growth and plotted on semilogarithmic paper, a **growth curve** for that microorganism is obtained. A typical growth curve is shown in Figure 6.4. The numbers on the ordinate and abscissa of the graph vary quantitatively for different bacterial species, but the overall picture does not. The growth curve is divided into four distinct phases, each having a different slope: the lag phase, the expo-

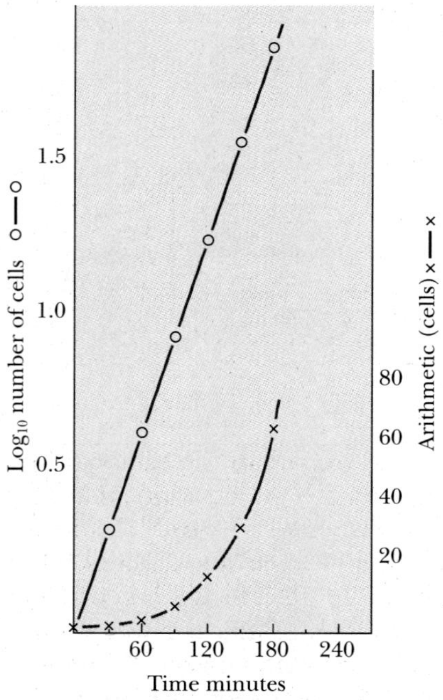

***Figure* 6.2** Exponential growth of a bacterium plotted on a logarithmic and arithmetic scale. The generation time for this microorganism is 30 minutes.

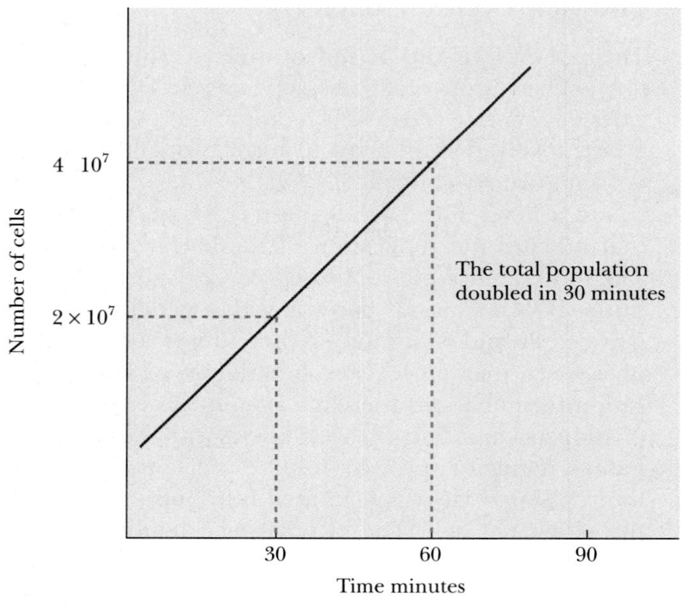

***Figure* 6.3** Calculation of the generation time in an exponentially growing bacterial population. Generation time may be determined from either a logarithmic or an arithmetic plot. It is the time required for the population to double in number.

BOX 6.2 METHODS & TECHNIQUES

Calculating the Generation Time

The generation time for a microorganism that is growing under a determined set of conditions is constant. It can be calculated if three bits of information are available. The information required is as follows:

- N_o The number of bacteria present at an early stage in exponential growth
- N_t The number of bacteria present after a period of exponential growth
- t The time interval between N_o and N_t

Exponential (logarithmic) growth is a geometric progression of the number 2. Exponential growth proceeds as follows: 2 cells $\longrightarrow 4 \longrightarrow 8 \longrightarrow 16 \longrightarrow 32 \ldots$ or 2^n where n = number of generations. Exponential growth can be described mathematically by the following equation:

$$N_t = N_o \times 2^n$$

To solve for n one would take the logarithm of the two sides:

$$\log N_t = \log N_o + n \log 2$$

or solving for n

$$n = \frac{\log N_t - \log N_o}{\log 2}$$

Applying real numbers to the equation one would proceed as follows:

$$\text{Assume } N_o = 6{,}000 \text{ cells}$$
$$N_t = 38{,}000{,}000 \text{ cells}$$

consulting a log table and filling in the equation:

$$n = \frac{7.5798 - 3.7782}{.301}$$

$$n = 12.6 \text{ generations}$$

Assume that the elapsed time between N_o and N_t was 5 hours (300 minutes) the generation time would be:

$$\text{Generation Time} = \frac{300}{12.6} = 23.8 \text{ minutes}$$

nential growth phase, the stationary phase, and the death phase.

Lag Phase

Transfer of a culture into a fresh medium often results in a period of adjustment. During this initial phase there is no increase in cell number and, in many cases, there is actually a decrease. This is called the **lag phase.** The cells, however, may grow in size, particularly if the inoculum is taken from a culture late in the stationary phase (see below). The lag phase is a period of **unbalanced** growth when the various components of individual cells are synthesized to provide coenzymes and metabolites necessary for **balanced** growth and division. During this phase proteins necessary for the uptake and metabolism of available nutrients may also be synthesized. Balanced growth is a steady-state situation where every component of a cell culture increases by the same constant factor per unit time.

The length of the lag phase can vary considerably. It

***Table* 6.1 Approximate generation times for several organisms growing in media optimum for growth**

Species	Generation Time
Escherichia coli	20 min
Bacillus subtilis	28 min
Staphylococcus aureus	30 min
Pseudomonas aeruginosa	35 min
Thermus aquaticus	50 min
Thermoproteus tenax	1 hr 40 min
Rhodobacter sphaeroides	2 hr 20 min
Sulfolobus acidocaldarius	4 hr
Thermoleophilum album	6 hr
Thermofilum pendens	10 hr
Mycobacterium tuberculosis	13 hr 20 min

can be short or zero when the inoculum is obtained from a culture in exponential growth, or it can be quite long if the inoculum is obtained from a rich complex medium and the microorganisms are placed in a defined minimal medium. In the latter case, it would probably be necessary for the organism to synthesize many of the proteins that are involved in generating the cellular components that were readily available in the complex medium.

Exponential Growth Phase

The **exponential growth phase** can also be termed the **logarithmic phase** because the increase in cells is logarithmic and is often plotted on semilogarithmic paper, as was done in Figure 6.2.

During the exponential phase, each cell in the population doubles within a unit time (generation time). If there is one organism present at the beginning of the exponential phase, the total number present after a defined period of time would be 2^n where n would be the number of generations. For example, starting with one organism with a generation of 20 minutes, after 4 hours or 12 generations there would be 2^{12}, or 4,096 progeny. This illustrates the explosive nature of exponential growth. The rate of exponential growth depends on the composition of the medium, with the most rapid growth occurring in a rich medium. A rich medium supplies preformed cellular components, and thus energy is not expended in synthesizing them.

Exponential growth is an important factor in the rapidity of food spoilage or the onset of an infectious disease. For example: consider that a human becomes in-

BOX 6.3 RESEARCH HIGHLIGHTS

Growth—A Matter of Perspective

A population increase where each cell divides during a unit of time is termed exponential growth. If the generation time of an organism is 20 minutes, each cell in the total population will reproduce every 20 minutes. This rate of population growth will continue as long as the organism remains in the exponential phase. When a bacterium is growing in a culture medium, its exponential phase is relatively short. It can be limited by nutrient depletion, oxygen deprivation, the accumulation of inhibitory products, or other factors.

Imagine that we devise a means for sustaining a culture of *Escherichia coli* (generation time 20 minutes) in the exponential phase for 24 hours. What would the number of progeny be?

From one bacterium we would obtain 4,722,366,478,574,681,194,496 cells.

They would weigh 4,722,366,478 grams, which is 10,401,687 pounds or 5,200 tons.

Assuming the average weight of individuals in a crowd to be 160 lbs, this biomass would equal 65,010 people.

Assuming that each cell is 2 μm long, end to end they would stretch 9,444,732,957,149,362 meters or 9,444,732,957,149 kilometers or 5,855,734,433,432 miles.

Placed end to end these bacteria would circle the earth 243,988,935 times or stretch to the moon and back 12,459,009 times.

Utilizing glucose as growth substrate they would consume about 10,000 tons in the 24-hour growth period.

Assuming the biomass to be a satisfactory food source, it would feed the entire population of North Carolina for one day.

Never underestimate the power of a microbe.

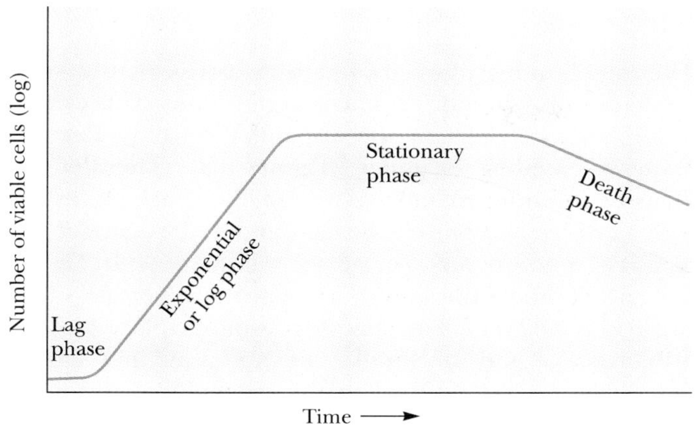

Figure **6.4** Growth curve of a typical bacterium grown in batch culture illustrating the four phases: lag phase, exponential growth phase, stationary phase, and decline or death phase.

fected with 32 disease-causing organisms with a generation time of 30 minutes. Assume further that the infecting organisms are in the exponential phase. During the next 30 minutes each bacterial cell would divide once, adding 32 more organisms for a total of 64. However, should the disease causer continue growing exponentially, after 7.5 hours there would be 1,176,576 progeny and these would double in the next 30 minutes yielding 2,353,152 organisms after 8 hours. This is obviously an important consideration in the early treatment of an infectious disease.

Stationary Phase

Microbial populations, in general, will not maintain exponential growth for extended numbers of generations. Either the substrate or an essential nutrient will be exhausted or inhibitory products of microbial metabolism will accumulate in the medium. These factors combined bring an end to exponential growth. The culture then enters the **stationary phase.** During this phase, some of the cells die and lyse. The lytic products can provide nutrients for other microorganisms and these replace the dead ones. However, most of the population in the stationary phase survives but simply does not proliferate. The cells of microorganisms in this phase differ in certain biochemical components from cells in the exponential phase. Generally, stationary phase cells are more resistant than exponential phase cells to adverse physical conditions such as increased heat, radiation, or change in pH.

The stationary phase is a period of survival and may mimic the conditions in nature when the microorganism is growing slowly or ceases to grow. There are a number of genes that are expressed in microorganisms as they enter the stationary phase that are not expressed during exponential growth. Included among these are the genes for secondary metabolites such as antibiotics (see Chapter 32). Other genes expressed are known as the survival (sur) genes, and these are indispensable to an organism entering the stationary phase. *Escherichia coli* mutants have been obtained that lack or have defective sur genes, and these mutants die rapidly as they enter the stationary phase. Microbes living in nature often face conditions that are unfavorable for growth, and it appears that evolution has given them mechanisms for protecting themselves in times of stress.

Death Phase

If a culture is maintained in stationary phase beyond a certain length of time (depending on the species and the conditions), the microorganisms die. Estimation of cell mass by turbidimetry and direct microscopic counting of cells suggests that the total number of cells remains constant (see Measure of Growth section). Viability counts, however, suggest otherwise. Generally, the **death phase** is an exponential function during which a logarithmic decrease in the number of viable cells occurs with time (Figure 6.4). The death rate depends on the particular organism involved and conditions in the environment.

Measure of Growth

A reasonably accurate estimation of biomass is a necessity in most studies involving the growth of microorganisms. The total biomass vs. time is required to accurately measure growth rate, substrate utilization, effect of inhibitors, and other parameters. Cell mass can be determined by several different methods, some direct and others indirect. The method generally employed to determine mass directly is wet/dry weight of cells; indirect determination is usually by chemical analysis of a specific cellular component (such as nitrogen). Total number of cells can be determined by direct counting of individual cells or by a viability count. Turbidimetry is an indirect method of obtaining a relative estimate of cell mass. Each of these is discussed here.

Total Weight

The **wet weight** of packed centrifuged cells can be ascertained by placing the biomass in a tared weighing pan and determining the actual weight. A tared pan is one of known weight. This method is quite often employed to estimate a cell mass taken for enzyme isolation or lipid analysis.

The total **dry weight** is another reasonably accurate method for determining cell mass. A cell mass is placed in a tared pan and dried to constant weight. One can also filter cells onto a tared membrane filter having a pore size that will capture the desired organism. The filter and cells would then be dried to a constant weight. Drying is usually at 100°C to 105°C for 8 to 12 hours. Unfortunately, the drying process renders cells unsuitable for most biochemical analyses.

Chemical Analysis

The total biomass can also be determined by analytical chemical procedures designed to quantitate the amount of some cellular constituent. The constituent most widely quantitated in laboratory research is total cell nitrogen. Because it is known that cells are about 14 percent nitrogen, an analysis of the total nitrogen present in a sample will give a reasonable approximation of total biomass. In this procedure, cellular organic nitrogen is released and converted to ammonia by sulfuric acid digestion of the carbonaceous material. After digestion is complete, the total NH_3^+ present can be determined colorimetrically.

Total cellular carbon can be estimated by digesting a sample of cells with a strong oxidizing agent such as potassium dichromate in sulfuric acid. The amount of carbon present is proportional to the dichromate reduced.

Direct Count

The total number of microorganisms in a suspension can be determined by a direct microscopic count. This can be accomplished by placing the suspended cells in a **Petroff-Hausser counting chamber** (Figure 6.5). The Petroff-Hausser counting chamber is essentially a thick glass slide with a grid etched onto its surface, and each square of the grid is of a known area (in square millimeters). In Figure 6.5, the depth of the liquid between the cover glass and the grid is 0.02 mm. The chamber would be placed under a microscope and the number of organisms in a square (total volume of 0.02 mm^3 in this example) determined by direct count. This number can be multiplied by a factor to give the total number of cells per milliliter in the original suspension.

A direct counting of microorganisms in suspension can also be accomplished with an electronic instrument called a **Coulter counter.** The cells to be counted are suspended in a conducting fluid (saline solution) that flows through a minute aperture in the instrument. There is an electric current across this aperture. During passage of any nonconducting particle, such as a bacterium, the electri-

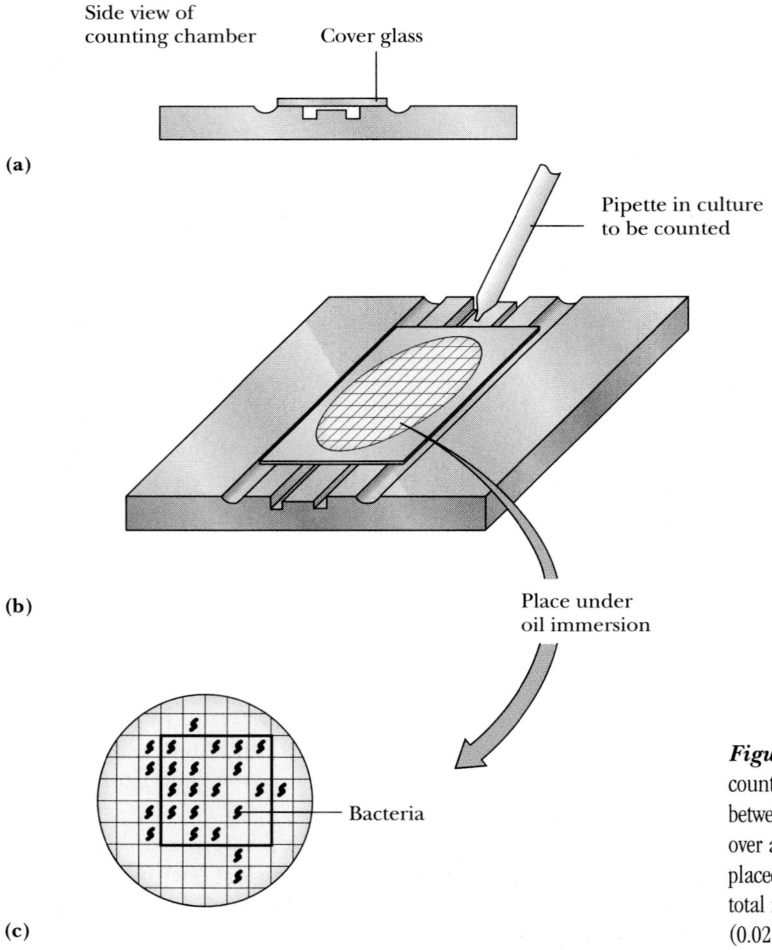

***Figure* 6.5** The Petroff-Hausser counter: **(a)** Side view. The sample to be counted is added to the edge of the cover slip placed over the grid area. The space between the cover slip and the slide is 0.02 mm (1/50 mm). There are 25 squares over a total area of 1 mm. The volume is 0.02 mm^3. **(b)** The sample is then placed under the microscope. Sixteen cells are in a large square. To calculate the total number per ml of original aliquot: 16 cells × 25 squares (mm) × 50 (0.02 mm^3) × 10^3 (cubic cm or ml) = 2×10^7 cells.

cal resistance within the aperture is increased, causing a brief spurt of increased voltage across the aperture. The voltage pulse can be recorded electronically, and the number of pulses per unit volume is directly proportional to the number of particles flowing through the aperture. Since the amplitude of the voltage pulse is proportional to the particle volume, the sizes of the cells can also be determined as they pass through the aperture. The aperture for bacteria is generally 10 to 30 μm in diameter so suspensions of cells to be counted must be diluted to reduce the probability of simultaneous passage of two or more cells.

Viability Count

A viability count is a determination of the number of living cells (able to reproduce) present in a suspension. A determination of the number of aerobic or facultatively aerobic bacteria present in a sample can be accomplished by adaptations of the spread or pour plate methods discussed in Chapter 5. Generally, viable counts are made on samples (soil, ground meat, etc.) that contain significant numbers of bacteria and must be diluted. Dilutions are tenfold and done in physiological saline. Saline protects the microorganisms from the osmotic shock that might occur if distilled or tap water served as diluent.

A dilution series is illustrated in Figure 6.6**a**. The number of colonies that are countable and would be statistically valid is 100 to 200 per plate. All dilutions should be plated in triplicate. The preferred volume of liquid that can be spread on the surface is 0.1 to 0.2 ml (more will flood the plate and result in indistinct colonies). Therefore

(a)

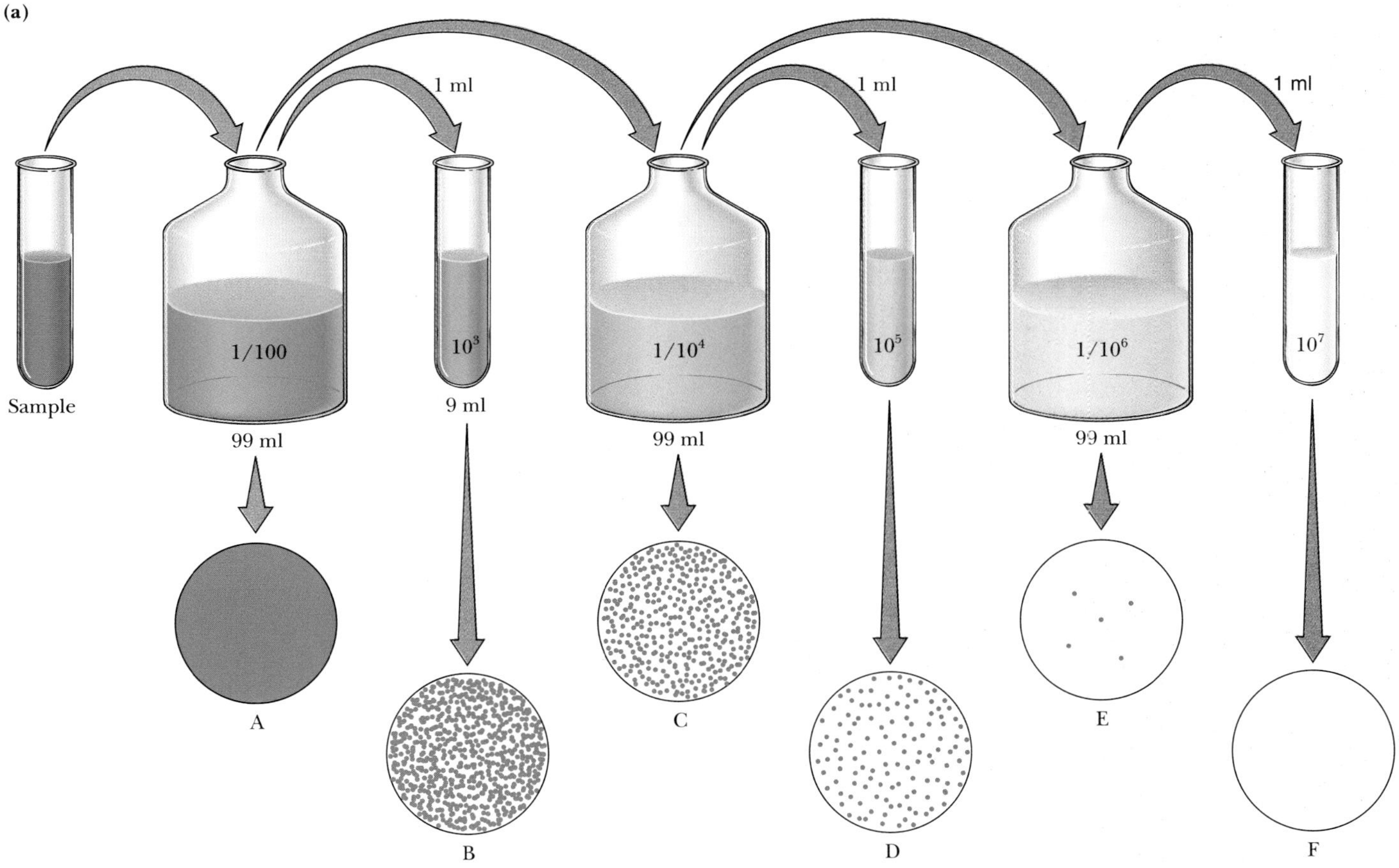

(b)

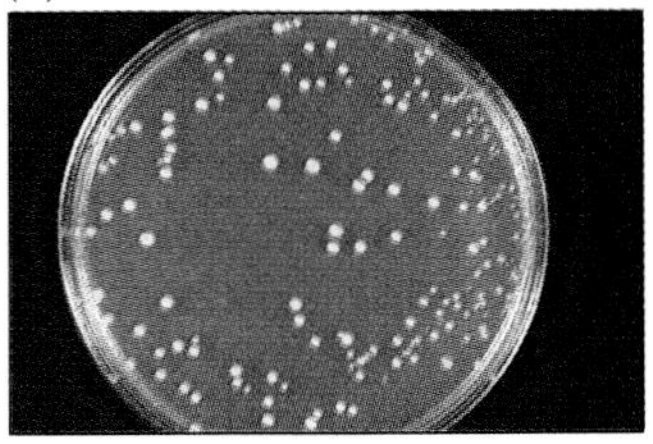

***Figure* 6.6** Determination of viable cells in a sample by serial dilution and plate count **(a).** The aliquots of each dilution would be spread on the surface as outlined in Figure 5.20. The appearance of the plates after incubation might be as follows: confluent growth at the lowest dilution in (A). (B) and (C) would be too numercus to count (TNC). (D) would be countable at 124 colonies per plate. (E) would be too few and (F) would have no colonies. **(b)** is an example of a countable spread plate of *Micrococcus luteus*. (**b**, Courtesy of J. J. Perry)

the dilution that would yield a countable plate should have 500 to 1000 cells per ml. Spreading 0.2 ml of such a dilution would yield 100 to 200 colonies—a number that is readily countable. In the series illustrated (Figure 6.6**a**), the number of cells in the original suspension would be:

$$124 \times 5(0.2 \text{ ml aliquot}) \times 10^5 = 6.2 \times 10^7 \text{ viable organisms per ml of sample}$$

The number of organisms in samples that contain fewer than ten organisms per milliliter, such as some natural waters, can be determined by concentrating the sample rather than diluting it. Concentration can be accomplished by centrifugation of a measured volume of water. A known volume of water may also be filtered through a sterile membrane filter having a pore size that retains microorganisms. The filter may then be placed on the surface of suitable medium in an agar plate, and the number of colonies that develop determined after a suitable incubation period (Figure 6.7). The problem with viable counts is in selecting a medium that will permit the growth of a reasonable number of the microorganisms present in the sample. Determining the viable organisms in soil would not be possible.

The **most probable number (MPN)** method is another viability count and is a statistical method based on the theory of probability. A suspension of cells is placed in an appropriate liquid growth medium and serially diluted in more of the same medium. Dilutions would be generally tenfold (see Figure 6.8). Multiple tubes are used at each dilution level, and the highest *dilution* that becomes turbid (cloudy) is *an estimate* of the number of bacteria in the

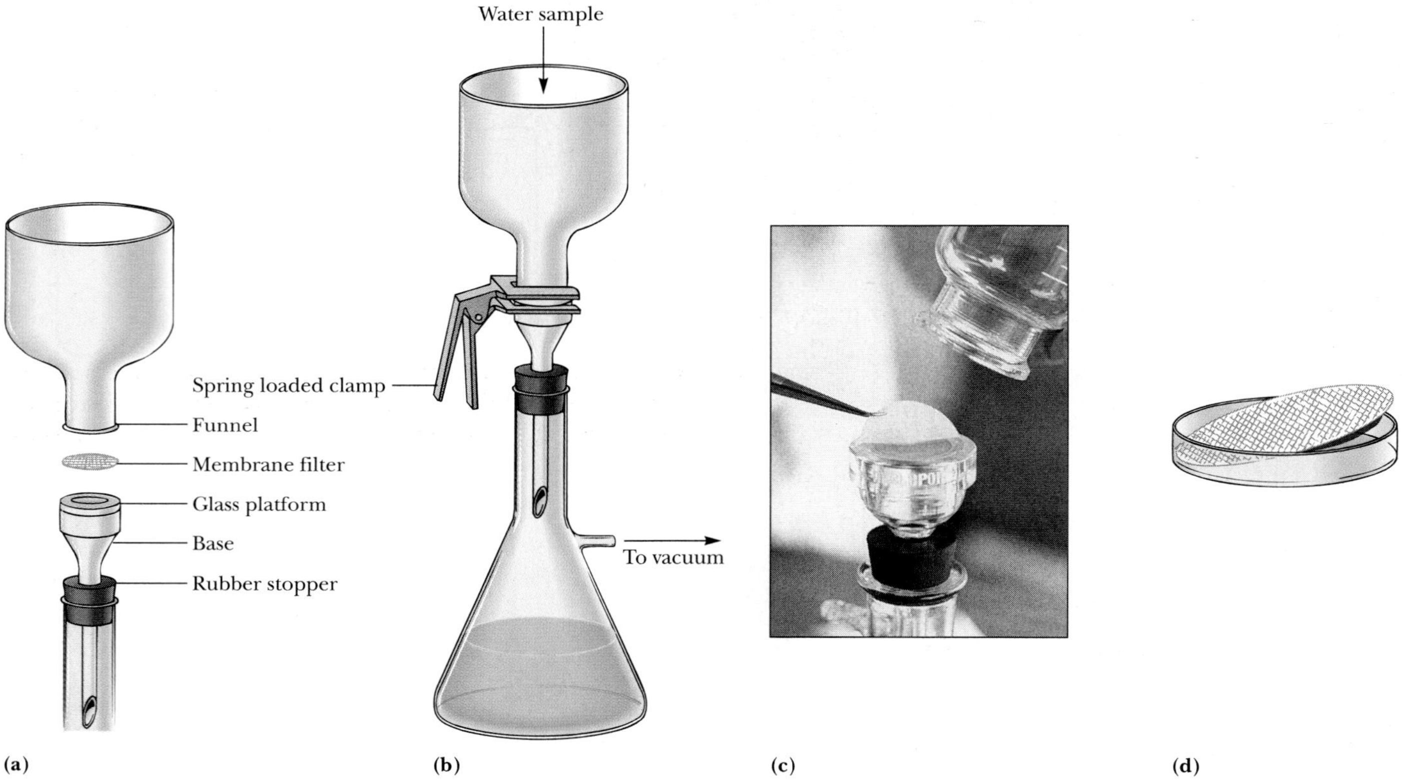

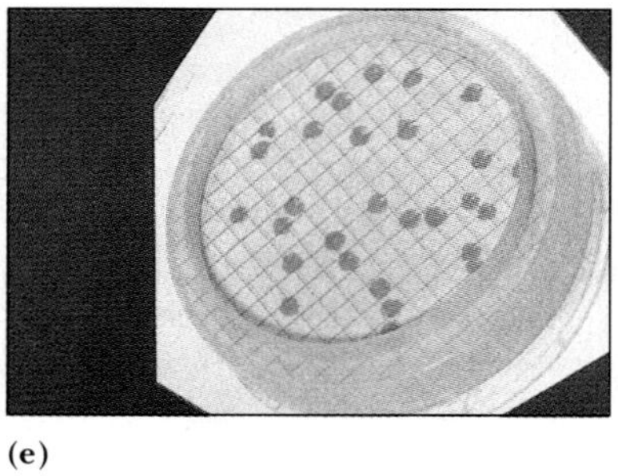

***Figure* 6.7** Recovery of bacteria from liquid samples by membrane filtration. The disassembled apparatus is shown in **(a).** The apparatus can be sterilized by autoclaving after assembly and the liquid sample poured into the funnel **(b).** Liquid is drawn through the membrane filter by applying a vacuum. The filter can be removed aseptically (Courtesy of Corning Costar Corporation / Jim Clements) **(c)** and placed on the surface of an agar medium **(d).** After incubation the colonies would appear as in **(e).** (**e**, Courtesy of Millipore Corporation)

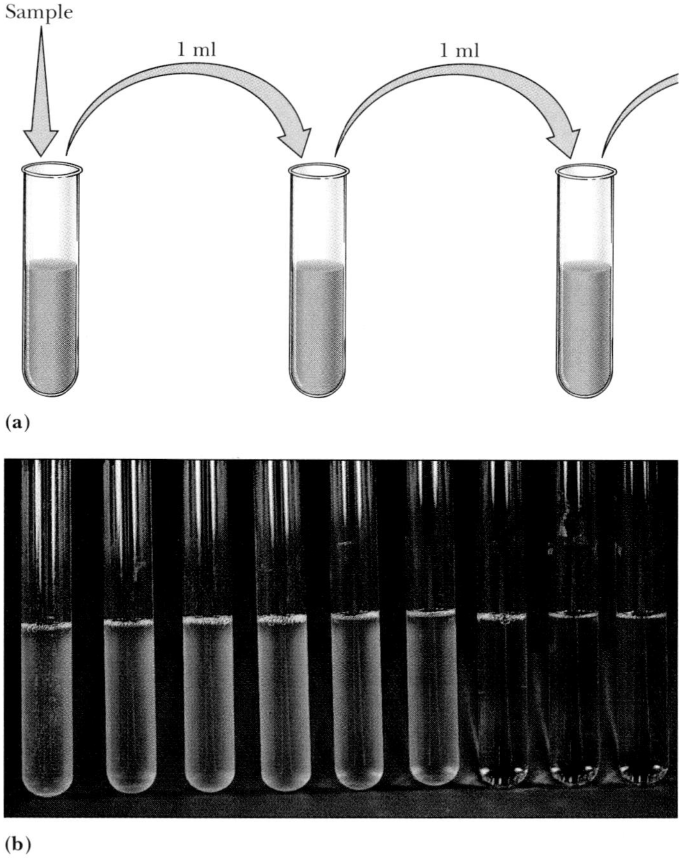

***Figure* 6.8** A most probable number (MPN) series **(a).** The first tube might be inoculated with a sample of drinking water, ground meat, soil, etc. Each tube would contain 9 ml of a selected medium. One ml from tube 1 would be placed in tube 2, mixed, and one ml from this to tube 3, etc. The highest dilution yielding turbidity would indicate total in the original sample as illustrated in **(b).** (Courtesy of J. J. Perry)

original sample. If growth occurs in the one to one million dilution, the original suspension would be considered to have one million cells. Viable count and MPN are not effective for organisms that form chains or clumps.

Turbidimetry

The most convenient and rapid method for measuring cell mass, and consequently the one generally employed in the laboratory, is **turbidimetry.** This method involves the use of a **spectrophotometer.** When a beam of light is passed through a suspension of bacteria, the light is scattered in proportion to the number of cells present. A bacterial suspension appears turbid because each cell scatters light. The more cells present, the more light scattered, the greater the visible turbidity. The optical system employed to quantitate this light scattering is a spectrophotometer. The functional parts of a spectrophotometer are shown in Figure 6.9. The wavelength of light passing through the suspension can be varied over a broad range, but bacterial suspensions are generally analyzed with the wavelength set between 400 and 600 nm.

Readings from the spectrophotometer are in absorbency (A), which is defined as the log of the ratio of incident light (I_O) to the light transmitted (I) through the suspension:

$$A = \log (I_O/I)$$

Because the shape and size of bacteria affect the scattering of light, the absorbency is a relative figure. To estimate the actual cell mass for a particular species of microor-

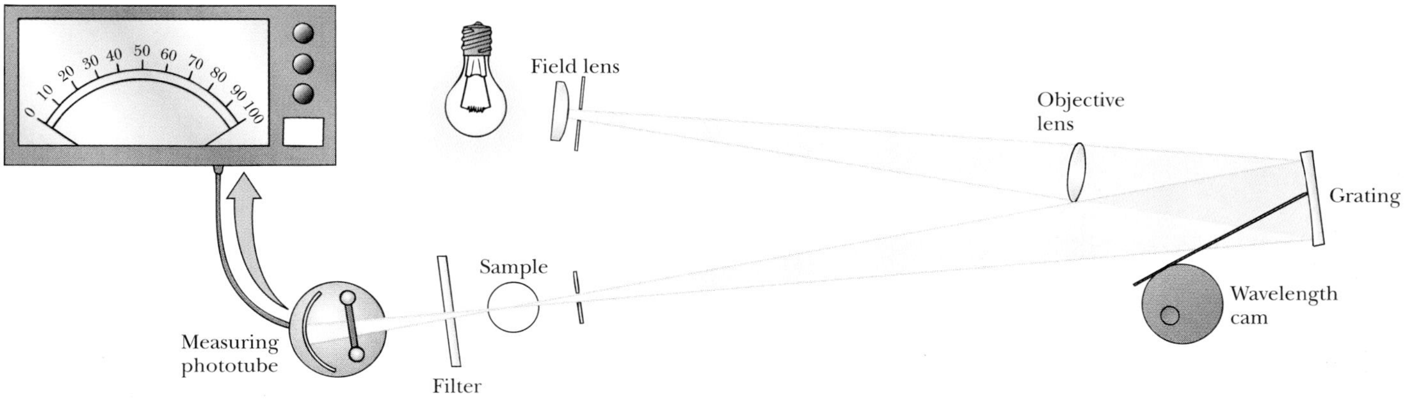

***Figure* 6.9** The basic components of a spectrophotometer for routine measurement of the turbidity of bacterial cultures. The wavelength cam permits one to adjust the grating to vary the wavelength of light that passes through the sample. As the number of microorganisms in the sample increases, more light is scattered, and this decreases the amount of light reaching the measuring phototube. The relative absorbency is indicated by an absorbency meter.

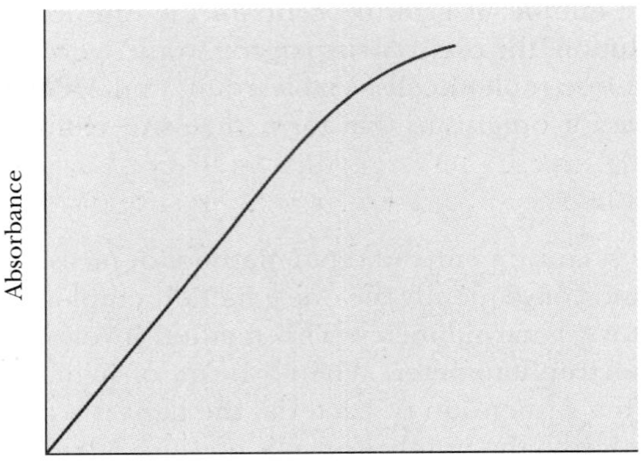

***Figure* 6.10** The relationship between absorbency and cell mass. At high cell densities the absorbency is not proportional to cell numbers. Measurements must be made at cell densities in the straight portion where absorbency is proportional to cell mass.

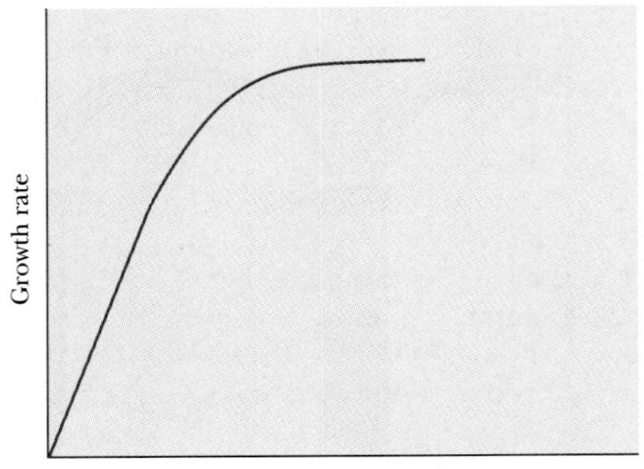

***Figure* 6.11** The relationship between nutrient concentration and growth rate. At low nutrient concentrations an organism cannot attain balanced growth.

ganism, one must relate the absorbency to the actual cell dry weight or cell numbers that give a particular absorbency. This relationship is illustrated in Figure 6.10. Note that, at high densities of cell mass, the absorbency deviates from linearity. As a consequence, one must dilute thicker suspensions to an absorbency that is in the linear range and multiply by the dilution factor to obtain a more accurate measure of cell mass. Turbidity is superior to other methods of determining cell mass because it is quick, readily repeatable, and does not adversely affect the cells in suspension. Turbidimetric measurements are not practical for cells that grow in clumps or for suspensions containing fewer than 10^7 cells per ml (lower limit of visible turbidity). Mycelial organisms such as actinomycetes or fungi do not form uniform suspensions. Consequently, one must resort to other methods for these organisms. Total nitrogen, total carbon, or wet/dry weight would be the methods of choice for any microbe that does not give uniform suspensions.

Effect of Nutrient Concentration on Growth Rate

Nutrient is added to growth media to provide a source of building blocks and energy. The amount generally added is in excess of the level necessary to attain a maximum growth rate. The relationship between nutrient concentration and rate of growth is illustrated in Figure 6.11. Note that at a certain point, the growth rate ceases to increase even though additional nutrient is added. When substrate is added below a threshold level, the growth rate is adversely affected. A decreased growth rate at low nutrient levels results from the inability of an organism to transport substrate at a rate sufficient to support balanced and rapid growth. The intracellular accumulation of building blocks is less than at substrate saturation, and not all of the precursors are available simultaneously to provide for a maximal rate of cell synthesis. The nutrient concentration also affects the total cell mass produced, as shown in Figure 6.12. As nutrient level increases, the yield of cell mass also increases. However, if the concentration of nutrient in a medium is increased beyond a certain threshold, the growth yield does not correspondingly increase.

When the source of energy available to an organism is at a very limited level, growth of the organism will cease. Organisms require a fixed amount of energy, called **maintenance energy,** simply to stay alive and perform basic functions. Maintenance energy is used for DNA repair, accumulation of nutrient inside the cell, motility, turnover of macromolecules, maintenance of a potential across the cytoplasmic membrane, and other basic cellular needs.

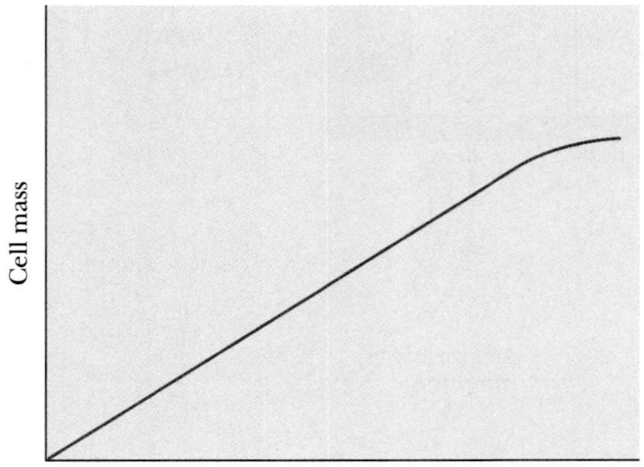

***Figure* 6.12** The relationship between nutrient concentration and total cell mass produced.

Growth or an increase in cell mass can occur only when available energy exceeds these nongrowth-related demands. It should be emphasized that growth rates, total growth, and the concentration of substrate necessary to support growth all vary with species. Some organisms will not grow to heavy suspension even though they are presented with a favored nutrient and what are assumed to be optimal growth conditions. Generally, the cessation of growth is attributed to exhaustion of nutrient or to the accumulation of metabolic products that inhibit further growth. Other unknown factors are also involved.

Synchronous Growth

The sequential events that occur during the cell cycle (see **Box 6.1**) cannot be measured in a single bacterial cell. Techniques are just not available that are sensitive enough to follow DNA replication, protein synthesis, polymerization reactions, and the delay (D) period before division occurs in a single growing cell. The sequence of events during a cell cycle can be quantitated accurately only when we examine a **population** in which all of the cells are at the same stage of division. This means that an equal phasing of all growth processes exists in every cell in the population. A culture of organisms where every cell is in the same stage of division is called a **synchronous culture.** Data obtained from experimentation on such a population may then be extrapolated to a single cell. A growth curve for a synchronous culture is presented in Figure 6.13. Note in the graph that there is a periodicity in cell number versus time. This indicates that all cells in the culture divide at the same time (approximately 60 minutes).

There are a number of techniques that have been developed that permit one to establish a synchronous culture. A synchronous culture can be obtained by selecting organisms of equivalent size from a population. Young cells are generally the ones of choice in synchronous culture studies, and since they tend to be the smallest cells, physical means of obtaining this population have been devised. One effective method is to filter a population of microorganisms through a cellulose nitrate filter having a pore size that retains most of the cells. Many bacterial species will adhere tightly to this type of filter. The filter can be turned over, and liquid passed through to dislodge all microorganisms not tightly attached to the membrane. After removing these unattached cells, a nutrient is poured through to permit the attached cells to divide. The cells released by the nutrient wash are the small progeny of the attached cells. The effluent collected over a short period of time will contain bacteria of equivalent size and in the same stage of the cell cycle.

Synchronous cultures can also be obtained in certain species by light stimulation, alternating the temperature of growth, and nutrient limitation. The technique of passing a culture directly through a membrane filter of a pore size that allows the passage of small cells but retains mature large cells has also been employed.

Bacteria do not divide simultaneously for more than two or three generations. Thus, experimentation involving metabolic activities occurring during the cell cycle should be done in the first or second generation.

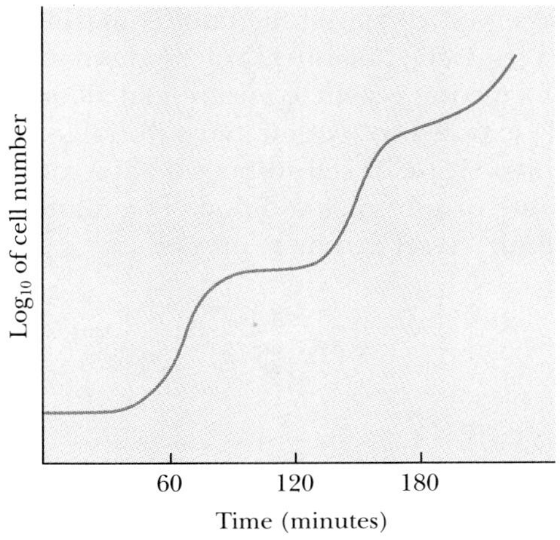

***Figure* 6.13** A typical growth curve for a microorganism in a synchronous culture. Cultures tend to lose synchrony after a few generations. Compare with the straight line obtained in an asynchronous culture (see Figure 6.2).

Growth Yield

The efficiency of an organism in converting the substrate to cell mass is designated the **efficiency of growth.** Efficiency can be measured by comparing the grams biomass produced per mole substrate utilized. The ratio of the cell mass obtained to the substrate consumed can be defined as the **growth yield** and determined as follows:

$$y_{sub} = \frac{x_{max} - x_0}{\text{moles Substrate consumed}}$$

In this case the x_{max} is the cell mass obtained and x_0 the weight of cell in the inoculum. The growth yield can be expressed as y_{max} or y_{sub}. In organisms for which the catabolic pathway for utilization of the substrate is known, the energy can also be calculated. Growth yield can then be expressed as y_{ATP}, which is the weight (in grams) of biomass per mole of ATP available.

Anaerobic bacteria, such as those that produce lactic acid, require a complex growth medium (see Chapter 5), and generally these species utilize glucose solely as a

***Table* 6.2 Growth yields of fermentative bacteria utilizing glucose as the energy source**

	Mol ATP/mol Glucose	y_m (g of cell/mol glucose)	y_{ATP} (g of cell/mol ATP)
Lactobacillus delbruckii[1]	2	21	10.5
Enterococcus faecalis[1]	2	20	10
Zymomonas mobilis[2]	1	9	9

[1]Homolactic fermentation, Embden-Meyerhof pathway.

[2]Alcoholic fermentation, Entner-Doudoroff pathway.

source of energy according to the following reaction:

$$\text{Glucose} + 2\ \text{ADP} + 2\ \text{Pi} \longrightarrow 2\ \text{lactic acid} + 2\ \text{ATP}$$

One can readily determine the mass of cells produced per mole of ATP generated in organisms such as this, that utilize a substrate solely as energy source.

The growth yield, y_{ATP}, is generally constant among bacteria. One mole of ATP will support the growth of about 10 grams of biomass. Consequently, the cell yield on a substrate utilized solely as the energy source is a direct function of the number of ATP molecules produced per mole of energy source utilized (Table 6.2).

The growth yield in a chemoheterotrophic bacterium utilizing a substrate as both energy and carbon source is more difficult to assess in terms of ATP generated. Aerobic organisms utilizing a sugar or other substrate as sources of carbon and energy source vary in their efficiency. Experimentation indicates that the efficiency of substrate conversion to biomass in chemoheterotrophs varies from 15 percent to 50 percent. The maximum conversion of substrate to biomass for a sugar is about 50 percent (one-half is assimilated into cells and one-half released as CO_2 or waste product). An organism growing on methane (CH_4) for which 1 mol of atmospheric oxygen is added per mole of methane assimilated can have a higher growth yield. Cell yields of 70 percent have been reported for organisms growing on methane.

Continuous Culture of Bacteria

Bacteria, yeasts, or filamentous fungi are generally cultured in the laboratory by the **batch** method. A nutrient liquid medium is prepared in an erlenmeyer flask, sterilized, and inoculated. The inoculated flask is then placed on a shaker (if aerobic) at the appropriate growth temperature for a suitable growth period.

Culturing bacteria by the batch method results in a population that is in a **balanced state** of growth for only a few generations. This balanced state occurs during the early to middle exponential growth phase (see Figure 6.4). Bacteria growing in batch culture actually go through an "aging" process as they progress through the growth cycle. This means that cells in the latter stage of exponential growth can be physiologically different from cells present during the middle exponential growth phase. Differences can be detected in enzymatic composition, presence of storage material, capsules, and membrane lipid composition. Techniques have been devised that permit one to maintain a bacterial culture in the exponential growth phase for an extended period of time. This is accomplished by employing a **continuous culture** system in an instrument termed a chemostat.

A typical **chemostat apparatus** is illustrated in Figure 6.14. A reservoir of sterile medium is fed into a culture vessel in a controlled manner. The medium in the culture vessel has a mixing system to ensure that all organisms are constantly exposed to oxygen and substrate. An overflow line permits release of cell suspension at a rate consistent with the rate of substrate addition. The dilution rate (D) in the culture vessel can be expressed as:

$$D = f/v$$

where f is the flow rate of the incoming medium and v is the volume of medium in the culture vessel.

One can control the population density and growth rate in a chemostat by changing the dilution rate. Regulating the supply of some limiting nutrient can also be used to control the cell density. In such a situation, the total

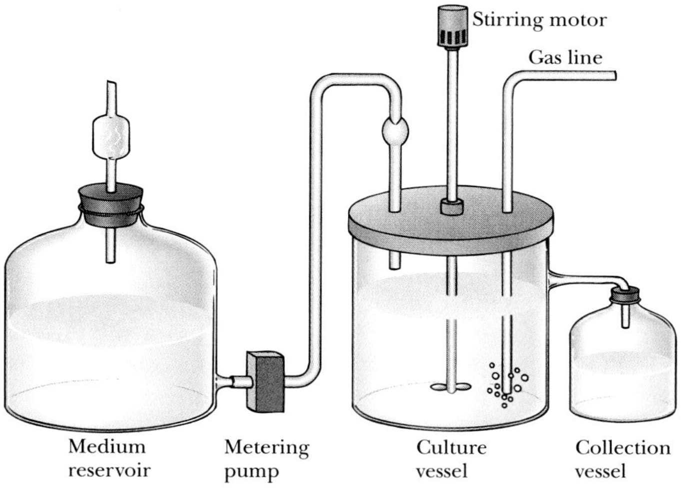

***Figure* 6.14** Model of a chemostat apparatus for the continuous growth of a bacterial culture.

cell mass present will not exceed the number of cells able to grow from the limiting nutrient. When the total cell concentration and the nutrient supply are balanced and in equilibrium, the system is in a **steady state.** The theoretical steady state relationship between substrate concentration and output of bacterial mass is depicted in Figure 6.15. The maximum bacterial **growth yield (*y*)** is the ratio of the cell mass to the substrate consumed and is equal to about 0.5 for virtually all bacterial species:

$$y = \frac{\text{Weight of bacterial mass}}{\text{Substrate consumed}} = 0.5$$

Under ideal conditions, the maximum growth rate constant (μ_{max}) at a saturation level for all substrates is 1.0/hr. The critical dilution rate (d_c) that produces the highest cell yield at a given level of entering substrate (s_r) is calculated as:

$$d_c = \mu_{max}\left(\frac{s_r}{k_s + s_r}\right)$$

where k_s is a saturation constant at which the specific growth rate is one-half the maximum ($0.5\mu_{max}$). If $s_r > k_s$, that is, if the entering substrate is greater than saturation, then $d_c \approx \mu_{max}$. At a dilution rate greater than d_c, the bacteria are washed out of the vessel faster than they can grow. In Figure 6.14, d_{max} is the dilution rate at which the maximum output of exponentially growing cells occurs per unit time. At this point, the doubling time is about 1 hr and less than 100 percent of the incoming substrate is being used. The excess substrate then overflows. The bacterial concentration in the culture vessel remains relatively constant. Dilution by the incoming medium is compensated by an increased doubling rate (shortened doubling time). A flow rate and medium concentration at or below

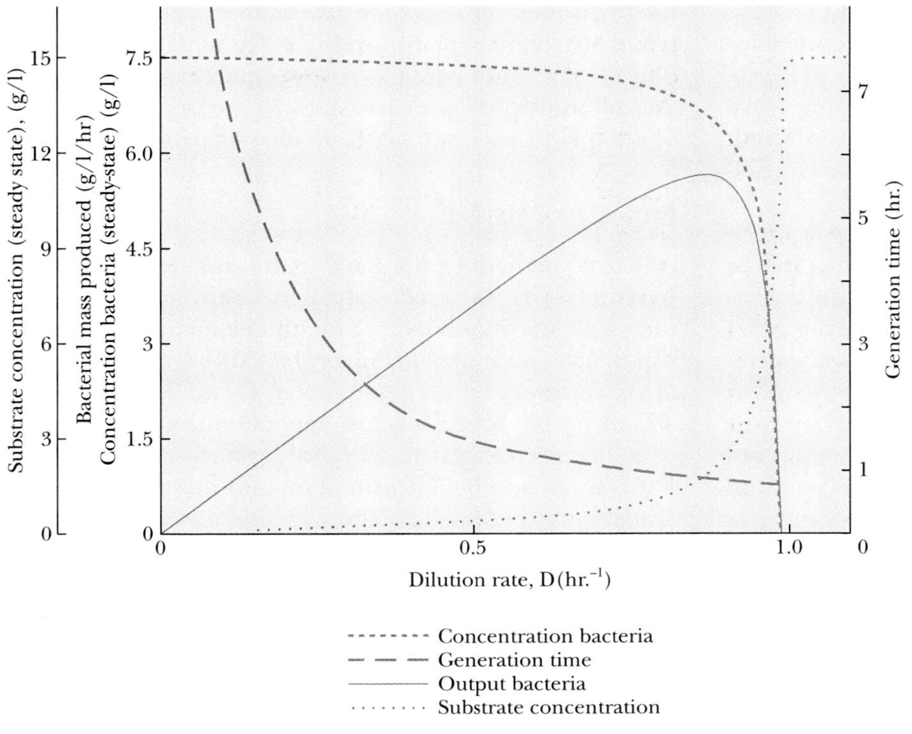

***Figure* 6.15** The relationship between substrate concentration, growth rate, doubling time, and yield of bacterial biomass in a steady state chemostat.

the d_{max} permits the constant production of cell mass in the exponential phase of growth.

A chemostat operates with a regulated flow rate such that the organism can grow at a rate equal to the dilution rate. Another instrument that functions solely by regulation of cell density is called a **turbidostat.** A turbidostat is equipped with a device that monitors the cell density in the growth chamber. If the density exceeds a preset level, the culture medium automatically enters the culture vessel. If the cell density falls below the preset level, no culture medium flows into the vessel.

Effect of Environmental Conditions on Growth

The prokaryotes have evolved with a remarkable capacity for growth under the varied conditions that they encounter in the environment. There are microorganisms that thrive at pHs from near 0 to 11 and with or without available oxygen. Other species grow at temperatures below 0°C (polar ice) while some live at 115°C (volcanic areas under the sea). Some of the mechanisms that have evolved in microorganisms that permit them to survive under the various physical conditions encountered are presented.

Acidity and Alkalinity

The relative hydrogen ion concentration in a solution is expressed as the pH. It is based on the dissociation of water, in which there would be an equal concentration of H^+ and OH^-. This is defined as pH 7 (actually the negative log of the H^+ ion concentration in water). The pH scale is from 0 (the most acidic) to 14 (the most alkaline). Because pH is a log function, a difference of one pH unit represents a tenfold difference in hydrogen ion concentration.

Among the Eubacteria and Archaea there is a range of individual species that grow at a pH of from near 0 to those that can thrive at pH 11. Filamentous fungi grow over a pH range of 2 to 9 with a majority favoring a pH of 4 to 6. Yeasts cover the same range, but most species prefer a slightly higher pH. Most soil bacteria grow best at a pH near neutral.

The pH range for growth of any individual organism is somewhat narrow. Most microorganisms, however, are quite insensitive to small variations in pH and can grow within a range of 2 to 3 pH units (which is actually 100 to 1000 times the difference in H^+ concentration). Growth of a microorganism in a culture medium requires that the pH be adjusted and maintained near that favored by the organism.

In the course of growth, an organism utilizing a fermentable substrate such as glucose may excrete acidic intermediates (acetic, lactic, and formic acids, among others) causing a lowering in the pH of the medium. Conversely, growth on proteins or amino acids results in ammonia production, causing the culture fluid to become alkaline. To prevent significant changes in pH, a chemical **buffer** is routinely incorporated into a culture medium. Buffers retard pH changes by removing excess H^+ or by donating H^+ to balance excess acidity or alkalinity, respectively. The most commonly used buffering system for organisms that grow near a neutral pH is a mixture of KH_2PO_4 and K_2HPO_4. This also provides the organism with phosphate for the synthesis of phosphorylated cellular constituents. Carbonates may be added as a buffer to control pH when excess acid production is anticipated. For example:

$$CH_3COOH + CaCO_3 \longrightarrow (CH_3COO)_2Ca \text{ (insoluble)} + CO_2$$

Microbes that grow at low pH are called **acidophiles.** The thiobacilli that are present in acid mine drainage are acidophiles and grow well at a pH of 1. *Thiobacillus thiooxidans* has a pH optimum of 2.5 and can survive at a pH of nearly 0. An interesting organism is *Thermoplasma acidophilum,* an archaeal species that grows in coal refuse piles and grows optimally at 59°C and a pH of 1. This is remarkable because survivability of bacteria generally decreases at any given temperature as pH is lowered.

In contrast, **alkalophiles** are organisms that grow optimally under very alkaline conditions. The Wadi el Natrum, a lake in Egypt, for example, has a pH ranging from 9 to 11. The water contains a rich population of halophilic and phototrophic bacteria. Many of the organisms in this environment have not yet been characterized.

Presence of Oxygen

About 20 percent of the Earth's atmosphere is molecular oxygen (O_2). As a consequence, many environmental niches are constantly exposed to this element. Swamps and subsurface areas may be anaerobic, and some subsurface areas have very limited amounts of O_2 available. Virtually all eukaryotic organisms are dependent on the presence of molecular oxygen for survival. Eukaryotes evolved on Earth in an aerobic environment, and most of these generate energy by "burning" hydrogen in the presence of O_2 with the concomitant production of water. Among the eukaryotes only a few fungi and protozoa that inhabit environments devoid of molecular oxygen can survive in its absence.

Eubacteria and Archaea exhibit a mixed response to oxygen. The atmosphere of the primordial Earth was completely anaerobic, and for more than one billion years, eubacterial and archaeal populations evolved without O_2-associated respiratory mechanisms for energy generation. Thus, throughout evolution, some eubacterial and archaeal populations have occupied anaerobic niches and have retained lifestyles that are independent of molecular oxygen. Some of these organisms are unaffected by atmospheric levels of oxygen, while others are killed by a brief exposure. Eubacteria and Archaea can be divided into three major groupings based on their response to molecular oxygen:

1. Aerobes: Obligate (strict) and facultative
2. Microaerophiles
3. Anaerobes: Obligate and tolerant

An **obligate aerobe** can grow only in the presence of molecular oxygen. It has no mechanism for energy generation other than a respiratory pathway that is geared to O_2 as a terminal electron acceptor. A **facultative aerobe** follows a respiratory pathway when oxygen is available, but can use alternate anaerobic energy-generating systems when oxygen is not available (see Chapters 8 and 9). In facultative organisms, growth is generally better in the presence of air. **Microaerophiles** are organisms that occupy niches where the atmosphere has limited levels of oxygen. While they must have oxygen for respiration, they grow best at reduced levels that range from 2 percent to 10 percent (dry atmospheric air is 20 percent oxygen). An **obligate anaerobe** cannot grow in the presence of oxygen and may be killed by minute levels. **Tolerant anaerobes** are oblivious to the presence of oxygen and can survive both with or without oxygen. However, they do not use molecular oxygen in energy generation.

Aerobic microorganisms generally respond favorably to increased oxygen availability during growth. Growth of the aerobic actinomycetes, which produce most of the major antibiotics, is greatly augmented by increased aeration. In industrial operations, these microorganisms and aerobes in general are grown in elaborate fermenters where high levels of oxygen can be introduced. Some obligate aerobes, however, fail to grow when oxygen concentrations are greater than atmospheric.

Toxicity of Oxygen

Oxygen is a highly reactive molecule, and metabolic reactions involving molecular oxygen can generate toxic byproducts. Most organisms that live in the presence of air synthesize enzymes that protect them from these toxins. The major toxic products are superoxide, hydrogen peroxide, hydroxyl radical, and singlet oxygen. These products, if not removed, can damage the cell by reacting with proteins, lipids, or nucleic acids. These toxic intermediates are generated as follows:

$$O_2 + e^- \longrightarrow O_2^- \quad \text{Superoxide}$$

$$O_2^- + e^- + 2\,H^+ \longrightarrow H_2O_2 \quad \text{Hydrogen peroxide}$$

$$H_2O_2 + e^- + H^+ \longrightarrow H_2O + OH^\cdot \quad \text{Hydroxyl radical}$$

$$\text{Chlorophyll} + h\nu \longrightarrow \text{chlorophyll*}$$

$$\text{Chl*} + O_2 \longrightarrow {}'O_2 \quad \text{Singlet oxygen}$$

Superoxide is a free radical that has gained an extra unpaired electron. The reduction of oxygen to water requires the addition of four electrons. This stepwise reduction occurs during respiration by single electron transfer. An intermediate product is O_2^- and small amounts of this toxic compound may be released during respiration. Superoxide must be removed as it is formed to protect the cell.

Hydrogen peroxide is the product of two electron additions to form O_2^{2-}. The anion is present as H_2O_2. It is commonly formed by a two-electron reduction of oxygen. Such reactions are mediated by flavoproteins, which are normal electron carriers.

Hydroxyl radical is the most reactive of the toxic products of oxygen respiration and is a very strong oxidizing agent. It is readily formed by ionizing radiation.

Pigments such as chlorophyll are sensitive to light (photosensitizers). Exposure to light can covert these pigments to a triplet state (indicated by the asterisks), which can result in a secondary reaction that produces **singlet state oxygen** (${}'O_2$). Singlet state oxygen is a very powerful oxidant and thus highly lethal to cells.

Enzymes that are effective in destroying toxic oxygen by-products include the following:

Superoxide dismutase (SOD)

$$O_2^- + O_2^- + 2\,H^+ \longrightarrow H_2O_2 + O_2$$

Catalase

$$H_2O_2 + H_2O_2 \longrightarrow 2\,H_2O + O_2$$

Peroxidase

$$H_2O_2 + NADH^+ \longrightarrow 2\,H_2O + NAD$$

Aerobic organisms contain active superoxide dismutase and either catalase or peroxidase to break down hydrogen peroxide. These enzymes are an essential defense mechanism for aerobic microorganisms.

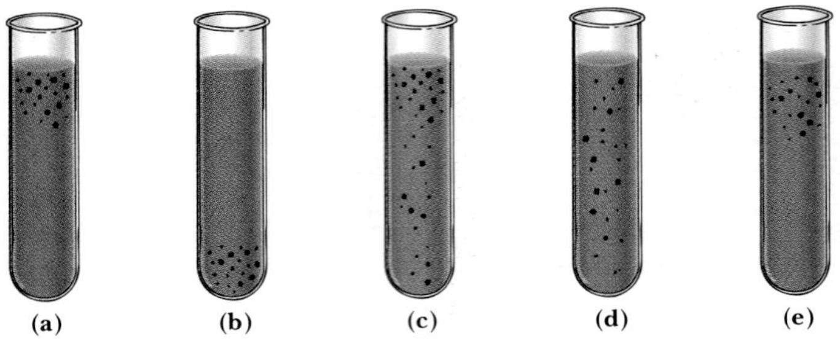

***Figure* 6.16** Patterns of bacterial growth relative to oxygen availability. **(a)** Strict aerobe grows only where oxygen is available. **(b)** Strict anaerobe grows only where all traces of oxygen are removed. **(c)** Facultative anaerobe grows throughout the tube, but best where oxygen is available. **(d)** Aerotolerant grows throughout the tube. **(e)** Microaerophile grows at a reduced level of oxygen.

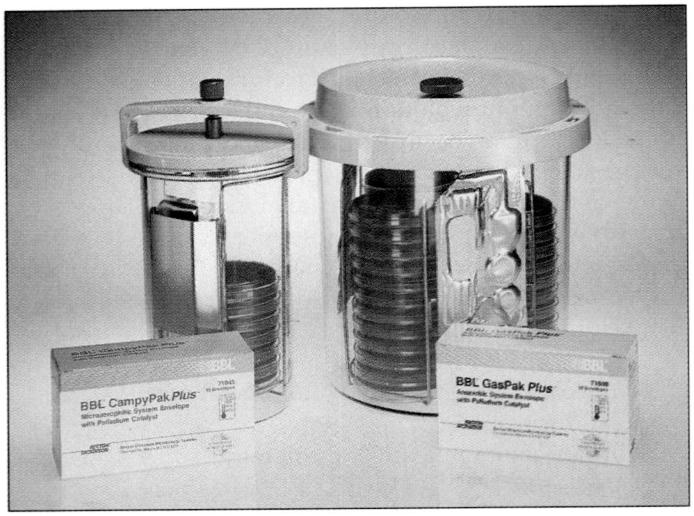

***Figure* 6.17** A Gas-Pak anaerobic jar. Inoculated plates or tubes can be placed in the jar along with a catalyst. After sealing, the catalyst is activated to remove all atmospheric oxygen. Because transfer and manipulation of the cultures occur outside the jar, these cannot be used for strict anaerobes. (Courtesy of Becton Dickinson Microbiology Systems)

Growth of Anaerobic Microorganisms

Growing anaerobic Eubacteria and Archaea in culture requires elaborate procedures and specialized appliances for handling them, as many of these microorganisms are destroyed by traces of oxygen. The major obligate anaerobic bacteria include the endospore-forming clostridia, the sulfate reducers, the methanogens, the bacteroids, and many microorganisms that inhabit the animal intestinal tract. Anaerobic protozoa are also present in the rumen of ruminants.

Anaerobic microorganisms can be grown on media placed in glass tubes to which a chemical compound is added to remove any trace of oxygen. For example, addition of thioglycollate is effective in removing oxygen from a culture medium. If a bacterial population is distributed in an agar culture medium prior to solidification, the growth patterns depicted in Figure 6.16 occur after suitable incubation.

Microorganisms can be grown in specialized chambers where all traces of oxygen can be removed from the atmosphere. An **anaerobic jar** is a convenient apparatus for growing anaerobes in the laboratory. Figure 6.17 depicts the Gas Pak anaerobic jar, which has a gas-tight seal and thick walls that are impermeable to gas. After innoculated tubes or plates are placed in the jar, the jar is sealed and the air inside is replaced with hydrogen gas. A catalyst is also added to the jar to remove all traces of oxygen. These jars are suitable for growth of air-tolerant anaerobes that can be manipulated in the atmosphere but must have anaerobiosis for growth.

Strict anaerobes must be maintained in the absence of O_2 at all times. This has been accomplished by manipulating them under a stream of nitrogen gas, an awkward procedure that is no longer in widespread use. It is much easier to carry out all manipulations of strict anaerobes in sealed chambers (Figure 6.18). The atmosphere in these glove boxes can be replaced with inert gases such as nitrogen. Oxygen scavenger chemicals are placed inside the box to remove all traces of O_2. These glove box anaerobic chambers permit one to perform all operations, including transfer and streaking, inside the box in the complete absence of molecular oxygen.

Temperature

Microorganisms are of widespread occurrence in environments in which the ambient temperature can range

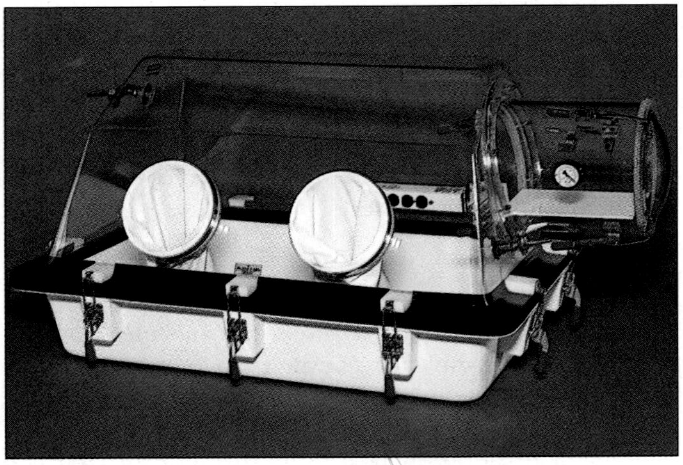

***Figure* 6.18** An anaerobic glove box. The atmosphere in a chamber of this type can be replaced by inert gases and maintained under anaerobiosis for the transfer of cultures. It may be employed in the purification of enzymes or other molecules that are sensitive to molecular O_2. (Courtesy of Plas Labs)

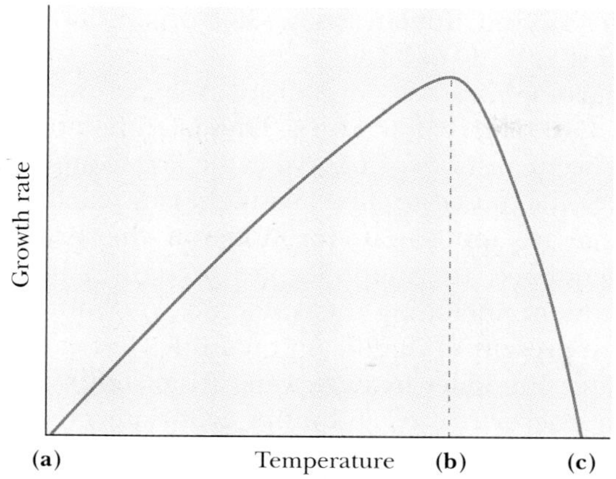

***Figure* 6.19** The cardinal temperatures for the growth of a bacterium: **(a)** minimal temperature, **(b)** optimum temperature, and **(c)** maximum growth temperature.

from less than 0°C to over 100°C. For example, salt water beneath polar ice can reach −10°C and volcanic areas under the sea can exceed 110°C. Microorganisms thrive at these extremes and those that live there have little competition from other life forms. It is a maxim that species diversity tends to decrease as the harshness of the environment increases.

The effect of temperature on the growth rate of a microorganism is depicted in Figure 6.19. Any given species has a minimum temperature, a maximum temperature, and an optimum temperature for growth. These three temperatures are referred to as **cardinal** temperatures. As the temperature rises above the minimum, the metabolic reactions in the cell accelerate and growth becomes more rapid. This increase leads to a physiological state in which all cellular reactions are operating at their maximum or optimum level. A little above this temperature, the cytoplasmic membrane may lose function, proteins or nucleic acids may be denatured (break-down), and the cell ceases to function. The cellular reaction(s) that is least heat stable is primarily responsible for the demise of the microorganism as the temperature rises.

Temperature Range

Microbes are divided into four groups based on the range of temperature at which they can grow (Figure 6.20). These groups include the following: the *psychrophiles,* which grow at temperatures below 20°C; the *mesophiles,* which generally grow between 20°C and 44°C; the moderate *thermophiles,* which grow between 45°C and 70°C; and the *hyperthermophiles,* microorganisms that require growth temperatures from 70°C to 110°C. The optimal growth temperature for microorganisms that fall into each category are also presented. Table 6.3 presents a listing of organisms in each group and their growth temperature range.

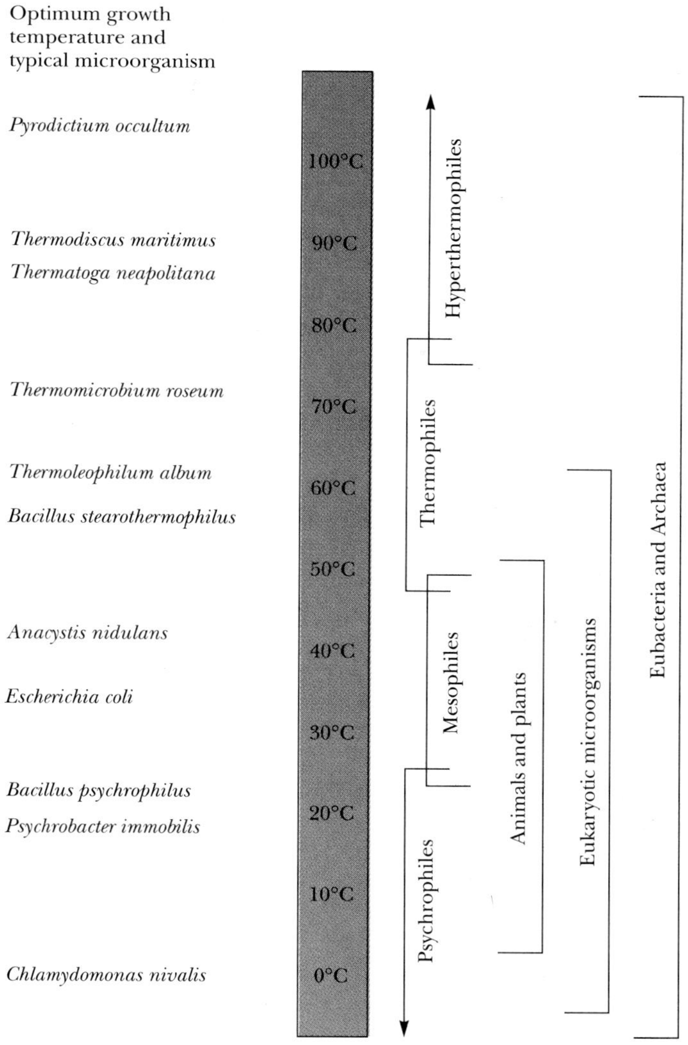

***Figure* 6.20** The growth temperature range for various life forms. The prokaryotes are the most versatile, with the ability to grow from < 0°C to 110°C. The only microorganisms so far described that grow at temperatures over 92°C are *Archaea.*

A **psychrophile** (from *psychro* meaning "cool") is defined as a microorganism that grows at 0°C and has an optimum growth temperature at less than 15°C. An obligate psychrophile cannot grow at a temperature above 20°C. Truly psychrophilic types are present in seawater where the average temperature is 5°C, and these microorganisms are particularly abundant in polar ice and water. The only requirement for survival is that liquid water be available, and this does occur in microenvironments within sea ice.

The term *psychrophile* is somewhat less precise than the other terms used to describe the temperature ranges for growth. Many microorganisms that grow optimally at 25°C to 30°C can also grow very slowly at 0°C. The ability to

Table **6.3 Temperature ranges for growth of Eubacteria and Archaea**

Species	Range (°C)
Psychrophiles	
Cytophaga psychrophila	4–20
Bacillus insolitus	< 0–25
Aquaspirillum psychrophilum	2–26
Mesophiles	
Escherichia coli	10–40
Lactobacillus lactis	18–42
Bacillus subtilis	22–40
Pseudomonas fluorescens	4–40
Thermophiles	
Bacillus thermoleovorans	42–75
Thermoleophilum album	45–70
Thermus aquaticus	40–79
Chloroflexus aurantiacus	45–70
Hyperthermophiles (Archaea)	
Hyperthermus butylicus	85–108
Methanothermus fervidus	65–97
Pyrodictium occultum	80–110
Thermococcus celer	70–95

grow at psychrophilic temperatures is not considered of major significance in classification.

Facultative psychrophiles are sometimes present in the soil and water from temperate climates. These microorganisms grow very slowly at 0°C but grow well at 22°C. They are often present in or on food and can cause spoilage in food that is refrigerated for extended periods of time. These organisms are generally referred to as **psychrotrophs.** Psychrotrophic fungi also cause spoilage of refrigerated foods. Food is best preserved for extended periods of time by drying or freezing because these processes eliminate liquid water.

Mesophiles comprise the vast population of microorganisms that are so ubiquitous in soil and water and are inhabitants of nearly all animals and plants. They generally grow between 20°C and 45°C although some will grow at temperatures as low as 10°C. The mesophiles are mostly responsible for the cycles of matter, disease, fermentation, deterioration of material, and other activities that are important to humans. Consequently, they have been the subject of the most research. Human pathogens such as *Streptococcus pneumoniae* grow optimally at 37°C, the temperature of the human body. Soil organisms generally grow best at 26°C to 30°C.

Thermophiles are organisms that grow optimally above 45°C and below 70°C. These temperatures commonly occur in hot springs, volcanic areas, and compost heaps. Soil exposed to direct sunlight can reach temperatures that are appropriate for moderate thermophiles. As a consequence, thermophiles are present in most every environment, including the Antarctic. Thermophilic bacteria are present in significant quantities in heated industrial water, hot water heaters, water from cooling towers at nuclear energy plants, and other artificially created thermal environments.

Hyperthermophiles are microorganisms that grow at temperatures above 70°C, and many grow at temperatures near or above the boiling point of water. The hyperthermophiles might appropriately be divided into two groups based on the cardinal growth temperature. Those that grow from 70°C to 90°C in one group and those that grow from 80°C to 110°C in the other. Liquid water is a necessity for all life forms, so these microorganisms that grow at temperatures near 100°C or higher are present in habitats that are under sufficient pressure to maintain water as a liquid. These conditions occur at volcanic areas beneath the ocean surface.

Several genera of the Archaea have been obtained in culture that grow at temperatures around 100°C. Not all Archaea are hyperthermophiles, but all microorganisms described so far that grow at temperatures over 90°C are Archaea.

There have been reports of organisms that grow under pressure at considerably higher temperatures (> 200°C), but evidence for growth at these temperatures is scant. Present evidence suggests that the maximum temperature where molecular constituents (ATP, protein) in bacteria can function is in the range of 120°C to 130°C.

Temperature Limitations

Proteins are one of the major cellular constituents that may limit the ability of a microorganism to grow at high temperatures. Thermophiles must have heat-stable proteins, and many of the proteins in mesophiles are heat labile (unstable at high temperatures). Generally, the cytoplasmic proteins in mesophiles begin to precipitate if heated at 60°C for 8 to 10 min. The equivalent enzyme from a thermophile would generally withstand this temperature.

Growth temperature can also influence the fatty acid composition of the membranes in bacterial cells. Analysis of cytoplasmic membrane fatty acids in a microorganism grown at both a minimum and a maximum temperature confirms this. At the minimum temperature, the relative proportion of unsaturated fatty acid in the cellular lipids is high. Growth at the maximum temperature leads to a

***Table* 6.4 Tolerance of selected Eubacteria and Archaea for decreased water activity a_w**

Type	Organisms	a_w
Nonhalophiles	*Aquaspirillum* and *Caulobacter*	1.00
Marine forms	*Pseudomonads* and *Alteromonas*	0.98
Moderate halophiles	*Vibrio* species and gram-positive cocci	0.91
Extreme halophiles	*Halobacterium* and *Halococcus*	0.75

greater relative percentage of saturated fatty acids. The degree of saturation of fatty acids in the cytoplasmic membrane determines the fluidity of the membrane at a given temperature. Since the proper functioning of a cytoplasmic membrane depends on fluidity, it is not surprising that membrane composition reflects the growth temperature.

Carbon Dioxide Availability

Virtually all microorganisms require carbon dioxide (CO_2) for growth. The level necessary for optimal growth varies with the microbial species. The atmospheric level of carbon dioxide (0.03 percent by volume) satisfies the minimal requirement for virtually all heterotrophic and autotrophic microorganisms. Optimal growth of autotrophic microorganisms, however, is attained in culture when they are provided with carbonates or with air enriched in carbon dioxide. Heterotrophs cannot utilize carbon dioxide as the sole source of carbon, but they do require it for growth. Carbon dioxide is assimilated into a number of metabolic intermediates involved in synthesis and energy generation. Complete removal of carbon dioxide from the atmosphere of a growth medium often leads to cessation of growth. Some animal and human pathogens, such as *Neisseria gonorrhea,* the agent of the venereal disease gonorrhea, will not grow in laboratory culture unless they are incubated in the presence of an atmosphere containing 5 percent to 10 percent carbon dioxide.

Water Availability

All microorganisms, whether eubacterial or archaeal, require a source of water for survival. **Water availability** is a major factor in determining whether a natural environment is colonized. It also markedly influences the type of microorganisms that can inhabit an environmental niche. The term water availability differs from **presence of water** because even if it is present, solids or surfaces in the environment may absorb molecules of water, rendering it unavailable. For example, solutes such as sugar or salts that dissolve in water have a high affinity for it. If the concentration of a solute is high enough, it can withdraw water and make it unavailable to a cell. For this reason, microbes generally do not grow in solutions such as saturated salt or undiluted honey.

Water availability is expressed as **water activity (a_w),** and is defined as the vapor pressure of air over a substance or a solution divided by the vapor pressure over pure water. If a solute such as sodium chloride absorbs water, the vapor pressure in the atmosphere above that solution becomes lower. The tolerance of some Eubacteria and Archaea for decreased available water is presented in Table 6.4.

If water is placed on one side of a semipermeable membrane and concentrated salt water is placed on the other side, the water diffuses into the salt solution. This process is called **osmosis.** A bacterium placed in a culture medium is generally faced with this physical condition. If the solute concentration is higher in the medium than in the cell, water leaves the cell, and the cytoplasmic membrane collapses inward. The organism may die of dehydration, a process called **plasmolysis.** Bacteria can prevent this from occurring by increasing the solute concentration inside the cell to one greater than in the suspending medium.

A variety of solutes are used by microbes to establish this higher solute concentration, and these solutes do not interfere with normal cellular activities. Because they do not interfere, they are called **compatible solutes.** Compatible solutes may be ions pumped into the cell from the environment or they may be an organic solute synthesized by the organism. Many organisms use potassium as the solute, including those as diverse as the halophilic organisms that live in highly saline environments and *Escherichia coli,* which lives in the gastrointestinal tract of humans. The cytoplasm of halophiles ("salt lovers") is virtually saturated with potassium ions. The gram-positive cocci synthesize proline as solute. Yeasts that live in high salt or sugar concentrations synthesize polyalcohols such as sorbitol to serve as the compatible solute.

Summary

- An increase in the number of cells in a culture medium is termed **population growth.** An increase in the size of a bacterial cell is **cell growth.**

- A **viable microorganism** is one that can reproduce in the laboratory. This definition is flawed because conditions for the growth of most microorganisms present in nature have not yet been defined. Our inability to grow them results from our ignorance, not their lack of viability.

- An increase in the number of microorganisms per unit time is the **growth rate.** The **generation time** is the time required for a population to double in number.

- Addition of a small number of microorganisms to a suitable growth medium and recording the number of microorganisms present vs. time will produce a **growth curve.** The component parts of a growth curve are: **lag phase, exponential growth phase, stationary phase,** and **death phase.**

- During **balanced growth** every component of a cell culture increases by the same constant factor per unit time.

- During **exponential growth** the number of organisms present increases logarithmically, that is 1 $\longrightarrow$ 2 $\longrightarrow$ 4 $\longrightarrow$ 8 $\longrightarrow$ 16 $\longrightarrow$ 32, etc.

- The **stationary phase** is a period where the microorganism does not reproduce. It is a period of survival.

- The **wet weight** of packed cells—such as in the use of a measured cell mass for enzyme isolation or metabolic studies—is an effective method for determining biomass when survival of the microorganism is important. **Dry weight** of cell mass is more accurate but tends to destroy the microorganism. Total nitrogen is effective because it can be accomplished on a small amount of material.

- A **direct count** of eubacterial or archaeal cells in a suspension can be done with a **calibrated slide** or a **Coulter counter.**

- When microorganisms grow in a clear liquid medium they create a visible **turbidity.** This turbidity can be measured in a **spectrophotometer** and is a reliable method for determining biomass. Turbidity is evident when there are 10^7 microorganisms present per milliliter.

- **Maintenance energy** is that amount necessary to stay alive.

- The cell cycle in microorganisms cannot be measured in a single cell, as techniques for this are unavailable. Events in the cell cycle can be followed in a population that are in the same physiological state. A culture in this state is called a **synchronous culture.** A number of techniques are available for establishing synchronous cultures.

- **Growth yield** is a determination of biomass produced per mole substrate utilized.

- A microorganism growing on a nutrient under favorable conditions for a set period of time is called a **batch culture.** A culture may be maintained in the exponential growth phase for extended periods of time in **continuous culture.** Continuous culture may be accomplished in a **chemostat apparatus** or a **turbidostat.**

- Selected species of Eubacteria and Archaea can grow from a **pH** of near 0 to 11. Most soil bacteria favor a pH near neutral (7) whereas fungi generally grow better at an acidic pH (2 to 6).

- Virtually all eukaryotic organisms have a requirement for molecular oxygen in their energy generation. They are **aerobic.** Eubacteria and Archaea initially evolved in an **anaerobic** environment, and many species can thrive in the absence of O_2.

- Most microorganisms that live in the presence of air must have mechanisms for removal of toxic products of O_2 utilization. The major toxicants generated from O_2 are **superoxide, hydrogen peroxide,** and the **hydroxyl radical.**

- The most effective way to manipulate strict anaerobes in the laboratory is in a **glove box,** where all molecular O_2 can be excluded.

- The temperature ranges for growth of various microbial types: 0°–20°C, **psychrophiles;** 20°–44°C, **mesophiles,** 45°–70°C, **thermophiles;** and 70°–110°C, **hyperthermophiles.**

- **Carbon dioxide** availability is necessary for growth of virtually all microorganisms, and growth of autotrophs is stimulated by the presence of higher than atmospheric levels of CO_2.

- **Water activity** (a_w) or water availability is an important factor in growth of microorganisms. Life cannot exist without available water.

Questions for Thought and Review

1. What is the difference between cell growth and population growth? How do these fit into the cell cycle?

2. A growth curve is divided into phases designated lag, exponential growth, stationary, and death. What is happening to individual cells during each of these phases? How does this relate to microorganisms in nature?

3. Total dry weight and viable count of cells are employed in measuring cell mass. What are some of the pros and cons of each method? How can turbidity be employed in measuring microbial mass?

4. Why are some studies done with cells growing in synchronous culture? How would one obtain a synchronous culture?

5. How does a chemostat work? A turbidostat?

6. Define aerobe, anaerobe, and microaerophile. What does strict and facultative mean in reference to aerobes and anaerobes?

7. How have aerotolerant organisms adapted to tolerate the presence of toxic oxygen products?

8. What are some of the reasons why microorganisms have a fairly limited temperature range for growth?

9. Water availability is an important factor in microbial growth. How can an organism grow in high concentrations of sugar or salt?

Suggested Readings

American Waterworks Association. 1991. *Standard Methods for the Examination of Water and Waste Water.* 18th ed. Washington, DC: American Waterworks Association.

Balows, A., H. G. Truper, M. Dworkin, W. Harder, and K. H. Schleifer, eds. 1992. *The Prokaryotes.* 2nd ed. Vol. 1. New York: Springer-Verlag.

Cooper, S. 1991. *Bacterial Growth and Division: Biochemistry and Regulation of Prokaryotic and Eukaryotic Division Cycles.* New York: Academic Press.

Gerhardt, P., ed. 1993. *Methods for General and Molecular Bacteriology.* Washington, DC: American Society for Microbiology Press.

Neidhardt, F. C., J. L. Ingraham, and M. Schaechter. 1990. *Physiology of the Bacterial Cell.* Sunderland, MA. Sinauer Associates Inc.

Norris, J. R., and D. W. Ribbons. 1970. *Methods in Microbiology.* Vols. 2 and 3A. New York: Academic Press.

In order to use chemotherapy successfully we must search for substances which have an affinity to the cells of the parasites and a power of killing them greater than the damage such substances cause to the organism itself so that the destruction of the parasites will be possible without seriously hurting the organism.

Paul Ehrlich 1854–1915
Father of Chemotherapy

Chapter 7

Control of Microbial Growth

Historical Perspective
Heat Sterilization and Pasteurization
Sterilization by Radiation
Filter Sterilization
Use of Toxic Chemicals
Chemotherapeutic Agents

The title of this chapter, "Control of Microbial Growth," is essentially a contradiction in terms. We live in a world where about one-half of the total biomass is microbial, and our bodies contain more microbial cells than human cells. In reality, we coexist with microorganisms. We attempt to keep the harmful ones at bay and foster the growth of those that are helpful, but regardless of our efforts, they are always around and some can pose a serious threat. Their growth is never really under total "control."

Microbes do play an indispensable role in the cycles of matter, and many of these contributions are presented in Chapter 12. This positive role is a consequence of broad metabolic diversity among microorganisms. Microorganisms have evolved as recyclers and, as a consequence, can grow on and destroy our food, our clothing, our buildings, and even our bodies. They have a virtually unlimited ability to metabolize organic compounds, thus returning carbon dioxide to the atmosphere. There are microorganisms that are obligately parasitic, and these survive by feeding on either plants, animals, or other microbes. Others are opportunists that, under appropriate conditions, cause disease.

To counteract a constant threat of destruction by microorganisms, humans have spent much money and effort in devising preservation methods and have met with moderate success. Human populations have historically suffered from the inability to understand or gain more than limited control over microorganisms. In earlier times, whatever anyone did in thwarting the destructive microorganism was the result of an empirical observation that something worked. Thus, Nicholas Appert (in 1810) discovered that one could preserve food by heating it in a closed container (canning), American Indians dried meat, Norsemen salted fish, and many civilizations made cheese. During the last 130 years humans have come to understand the microbe, and our strategies have changed.

We have developed procedures whereby, to a limited extent, microbes can be controlled. We sterilize with heat and chemicals, we filter air or liquids, and antibacterials are now available to cure disease. We have developed immunization methods and public health measures to survive in a world of the ubiquitous microbe. This chapter discusses many of the procedures and processes that are employed to curtail the growth of microorganisms in areas where this is desirable.

Historical Perspective

The early civilizations used many methods to protect food from microbial destruction. These included the salting of meat, fish, and vegetables. Certain bacterial species—alone or combined with salt—proved effective in preserving foods such as sauerkraut. Fermented milk products such as cheese, butter, and yogurt were stable for some length of time even without refrigeration. Acidification with vinegar was also effective in preserving selected food products.

The development of effective control measures for microorganisms in general parallels the evolution of microbiology as a science (see Chapter 2). Several important early workers in medicine discovered techniques that significantly decreased the number of deaths during medical procedures. It is generally acknowledged that **Ignaz Semmelweis** (1816–1865) pioneered the use of disinfectants. He insisted that all individuals working in his hospital in Vienna wash their hands in chlorinated lime before working with patients. This brought about a marked decrease in disease and death in the obstetrics ward. **Joseph Lister** (1827–1912) was able to conduct aseptic surgery by sterilizing instruments with carbolic acid (phenol) and spraying phenol around the operating room. Robert Koch demonstrated that mercuric chloride would kill all bacteria tested, including the highly resistant endospore (see Chapter 2). Pasteur's invention of the autoclave revolutionized laboratory work, as it permitted one to sterilize media and work under aseptic conditions.

Heat Sterilization and Pasteurization

Sterilization is the destruction of all viable forms of life, including endospores. A properly packaged sterile material should be devoid of all life until opened and exposed to the outside environment. Sterilization can be attained by various procedures including fire, moist heat, dry heat, radiation, filtering, and toxic chemicals. Any procedure that disrupts an essential component of the cell—protein, nucleic acid, or cytoplasmic membrane—will lead to a loss of cell function. The more extensive the disruption, the greater the chance that the damaged microorganism will die. When you watch an egg fry you are witnessing the effectiveness of heat in denaturing (breaking down) egg protein. Some proteins are more stable than others at high temperatures, but this is only a matter of degree. Given sufficient heat, the intrinsic function of any protein will be lost. Heat, in some form, is the agent most often employed for routine sterilization.

Studies in a microbiology laboratory are dependent on the ability to sterilize media, inoculating loops, glassware, and all material involved in culturing microorganisms. Most studies are done with pure cultures (only one strain present), and this can be accomplished only when all manipulations are done under sterile conditions. Sterility is also an important requirement for surgical instruments, bandages, and in most medical procedures.

Several factors can affect any sterilization procedure. First, the presence of organic matter can protect Eubacteria or Archaea from the destructive effects of heat, chemicals, or radiation. Second, a material to be sterilized usually harbors a mixed population of microorganisms. Consequently, the regimen selected for sterilization must be designed to kill *all* of the most resistant organisms present and under the conditions that exist.

The pH of the suspending medium is also a significant factor. In an acidic medium, heat is considerably more effective in killing microorganisms than at neutrality. This has been a major factor in home canning, where acidic foods, such as fruits and tomatoes, rarely spoil. Foods that have a more neutral pH, such as corn, peas, and beans, spoil more often because on harvesting they harbor endospore-forming organisms that may survive the canning process. On rare occasions, the anaerobic spore-forming organism *Clostridium botulinum* may survive in improperly canned foods. This organism then may germinate and grow, producing botulinum toxin which causes botulism. Ingestion of minute amounts of botulinum toxin can lead to death. This was a danger in the days of extensive home canning when inefficient or improperly used pressure cookers were sometimes employed.

Technique of Heat Sterilization

Microbial death by heating occurs at an exponential rate, meaning that a fraction of the cells present in a heated solution will be killed per unit time. Consequently, a thicker suspension of microorganisms will require a longer heating time than a more dilute suspension. The effect of heat on the survival of a mesophilic bacterium is illustrated in Figure 7.1. A survival curve can be obtained by suspending bacterial cells in water and heating to a selected temperature. At intervals, an aliquot can be removed and the number of viable cells present determined by a plate count as described in Chapter 6. Boiling for 15 minutes will kill

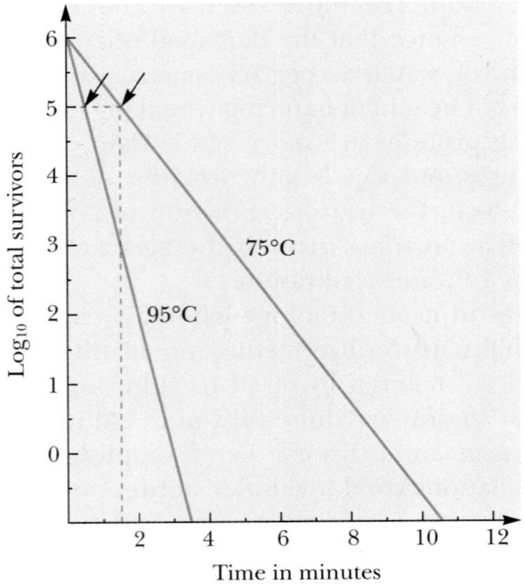

***Figure* 7.1** Effect of temperature on the survival of vegetative cells of a mesophilic bacterium. This is an exponential plot of viable cells remaining versus minutes heating at two different temperatures. The arrow points to the time (minutes) of a decimal reduction where 90 percent of the organisms initially present are killed.

Table* 7.1 Heat resistance of bacterial endospores obtained from various genera and species in the genus *Bacillus

Organism	*D.* Value[1] (minutes)
Bacillus cereus	0.8
Bacillus megaterium	2.1
Bacillus cereus var. *mycoides*	10.0
Bacillus licheniformis	24.1
Bacillus coagulans	270.0
Bacillus stearothermophilus	459.0

[1]Time required to reduce the population tenfold.

virtually all vegetative eubacterial and archaeal cells, viruses, and fungi but does not necessarily kill bacterial endospores. The length of time required to reduce a population tenfold (for example, from 10^6 to 10^5) is the **decimal reduction time,** designated *D*. Endospores of various bacterial species have differing resistance to boiling (100°C), as indicated by the *D* values listed in Table 7.1. A comparison of the decimal reduction time for a heat-sensitive microorganism and one that is relatively resistant is presented in Figure 7.2. The time required to sterilize a suspension of endospores varies over a wide range, with *Bacillus anthracis* requiring less than 5 minutes. In contrast, a suspension of *Clostridium botulinum* endospores can be heated for 5 hours or more at 100°C and some of the endospores will survive. Placed in an appropriate medium they can germinate and grow.

Moist Heat

The most effective system for sterilization and one that is generally used in the laboratory and throughout the food and pharmaceutical industries is steam under pressure. The instrument available for accomplishing this is the aforementioned **autoclave** (Figure 7.3). An autoclave is a closable metal vessel that can be filled with steam at greater than atmospheric pressure. The autoclave is designed so that the air present inside when the door is closed and sealed is driven out by steam, as illustrated in the figure. A valve closes automatically when the air passing outward reaches 100°C, an indication that steam now fills the entire inner space. Laboratory autoclaves operate at 15 psi (pounds per square inch) pressure, and at this pressure, the temperature rises to 121.6°C. At this temperature, sterilization of routine material can be achieved within 10 to 15 minutes.

Large batches of material must be autoclaved for a longer time because heat transfer to the interior can be slow. The denser the material, the slower the heat is transferred and the longer the time required for sterilization. Thus it takes longer to sterilize a flask containing 1 liter of liquid than it does the same flask when empty. The autoclave is effective for the sterilization of most laboratory growth media. Selected organic compounds such as vitamins and antibiotics may not be stable at autoclave tem-

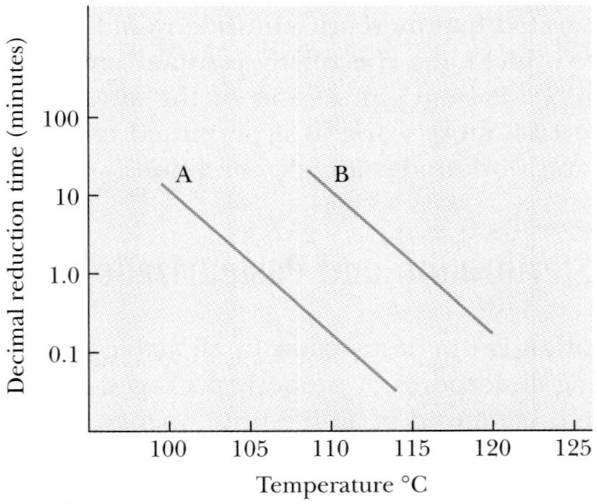

***Figure* 7.2** The decimal reduction time for an organism that is relatively heat sensitive **(A)** and another that is more heat resistant **(B).** The rate of killing increases with a rise in the temperature.

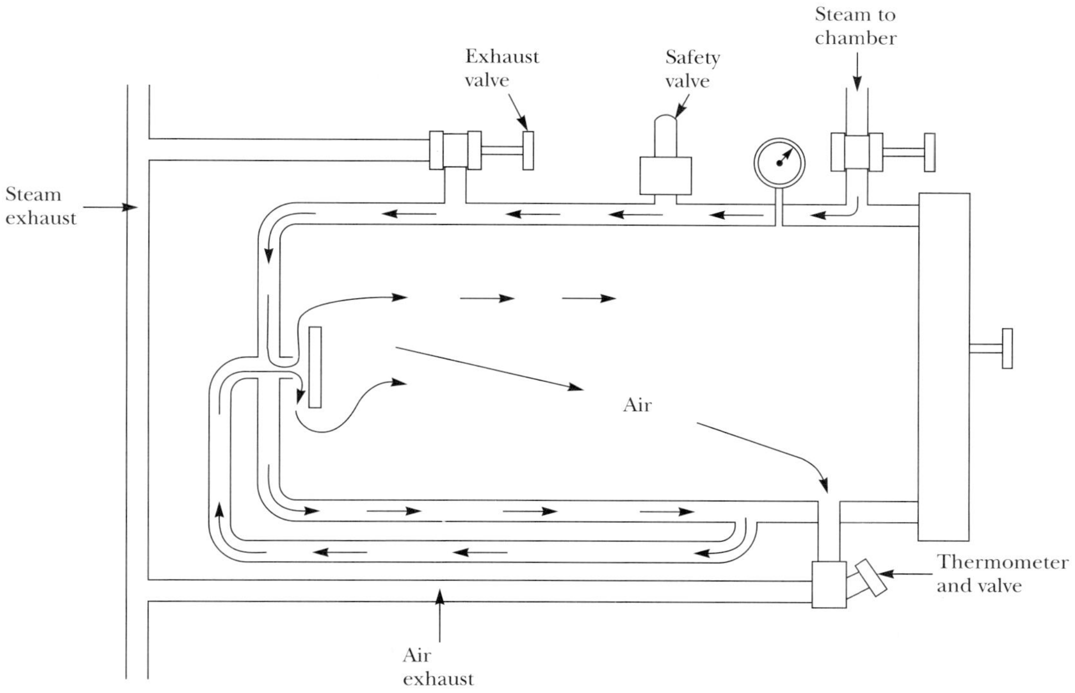

***Figure* 7.3** A laboratory autoclave in cross section.

peratures and should be sterilized by filtration or other less destructive means, as discussed later.

Dry Heat

Dry heat is effective in sterilizing glassware, pipettes, and other heat-stable, solid materials. Pipettes can be placed in metal canisters, glassware wrapped in paper, and flasks and other vessels capped with aluminum foil. This protects the objects from contamination when they are handled subsequent to sterilization. An oven that can be heated to 160°C to 180°C is used for dry heat sterilization, and the materials are retained at this temperature for 2 to 4 hours. Dry heat has been preferred for sterilizing pipettes, syringes, and other heat-stable objects because autoclaving leaves them moist. However, most laboratory autoclaves now have a vacuum cycle that will remove moisture from such samples after sterilization.

Pasteurization

Pasteurization is a process that employs low heat and is often applied to milk products and other heat-sensitive foods. It is used specifically to reduce the microbial population but does not sterilize the product. The method was devised by Louis Pasteur to control wine spoilage. Pasteurization was originally accomplished by heating to 63°C to 66°C and holding the material at this temperature for 30 minutes. This method will kill 79 percent to 99 percent of the microorganisms present in milk. A short-term "flash method" is now often employed, whereby milk is passed through coils where it is heated quickly to a temperature of 71.6°C for 15 seconds.

Pasteurization of milk was adopted as a common practice in the dairy industry because raw milk often contained *Mycobacterium tuberculosis,* the causative agent of tuberculosis, and/or *Brucella abortus,* which causes brucellosis. These organisms have now been mostly eliminated from dairy herds in the United States by mandatory testing. However, pasteurization is still employed as a precaution and to extend the shelf life of certain food products, including beer and milk. Due to the widespread consumption of dairy products, the potential spread of disease by this consumable is a constant threat.

Sterilization by Radiation

Selected wavelengths of light can cause death of living organisms, thus making **radiation** a practical method of sterilization. The entire **electromagnetic spectrum** of light is presented in Figure 7.4. The visible and infrared regions of the spectrum are important because these ranges provide the energy for photosynthetic growth (see Chapter 6). The two types of radiation that are effective **microbiocides** (literally, "microbe killers") are ultraviolet (UV) light

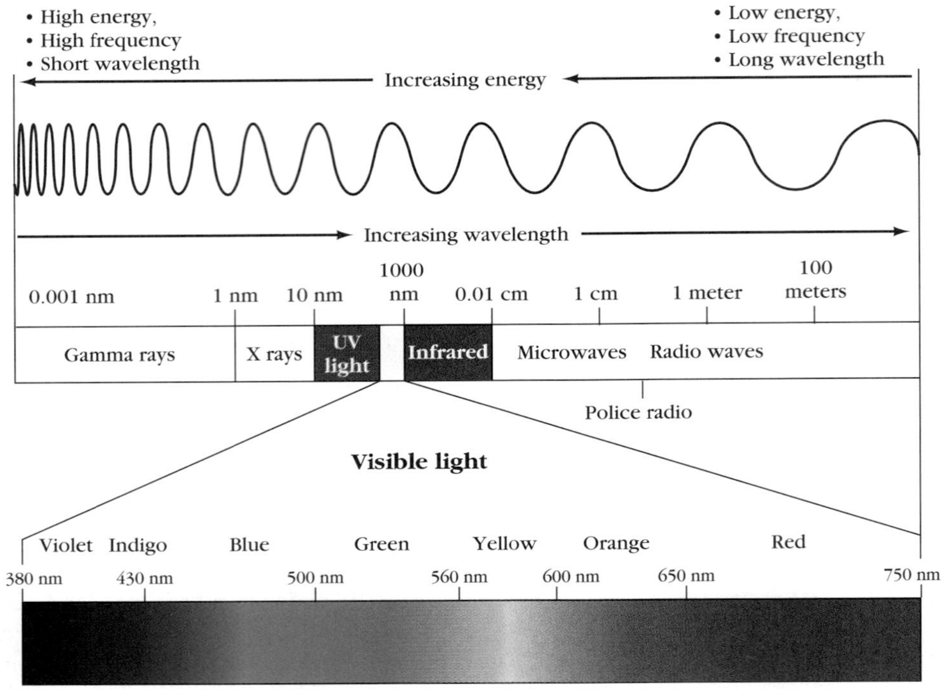

***Figure* 7.4** The electromagnetic spectrum. Humans can detect "visible" wavelengths that range from about 400 nm to 750 nm.

and ionizing radiation. Light in the UV range can disrupt cellular DNA and some other cell constituents, whereas **ionizing radiation** can cause harm to all constituents of a cell. Gamma (γ) rays, X-rays, alpha (α) and beta (β) particles, and neutrons are all forms of ionizing radiation.

Ionizing Radiation

Gamma rays, as with other forms of ionizing radiation, kill cells by causing the formation of highly reactive free radicals (see Chapter 6). These free radicals, such as the hydroxyl ion ($OH^{\cdot}$), can react with and inactivate any of the various macromolecules present in a cell. It is probable that all macromolecules in a cell are equally susceptible to damage from ionizing radiation. Since there are multiple copies of most macromolecules in a cell, damage to a limited number of the copies will not cause cell death. Generally, however, the cellular DNA will have a single copy of a specific gene, and disruption of a gene can cause cell death. Ionizing radiation is destructive to genetic material because it causes single- and double-strand breaks in the DNA double helix.

The sensitivity of species to ionizing radiation varies considerably. Endospores of *Clostridium* or *Bacillus* species are highly resistant, as are cells of *Deinococcus radiodurans. D. radiodurans* is exceedingly resistant to UV or gamma radiation and also to most mutagens. Materials that have been exposed to ionizing radiation often bear this organism as the sole survivor. The ability to repair DNA damage efficiently and rapidly is a major survival factor in radiation-resistant microorganisms.

Ionizing radiation is an effective sterilizing agent because it can penetrate deeply into objects such as containers. Consequently, sterilization can be accomplished after a product is packaged. Major sources of usable radiation for sterilization include cobalt-60, X-rays, and the neutron piles that are available at nuclear plant sites. The radiation source most often used commercially is cobalt-60. Generally, items that are heat sensitive and nonfilterable are sterilized by radiation, and among the materials sterilized in this way are plastics, antibiotics, serums, various medicinals and, in some cases, food.

Ultraviolet Light

Wavelengths of electromagnetic radiation in the UV range between 220 nm and 320 nm are particularly important biologically because these wavelengths can be absorbed by cellular constituents. More intense, shorter wavelengths of UV are emitted by the sun but are absorbed by the ozone layer in the earth's upper atmosphere. Shorter wavelength UV radiation is lethal to microorganisms and harmful to many other organisms as well. This has caused the present concern over the damage to the ozone layer over the Arctic and Antarctic areas.

Germicidal lamps and lights can be constructed to emit the shorter, damaging wavelengths. UV lamps with this potential may be placed in sterile rooms and safety cabinets to sterilize the air, but they cannot be operated when people are present. These wavelengths burn skin and may damage the eyes. UV radiation does not penetrate glass, water, or substances other than air to a signif-

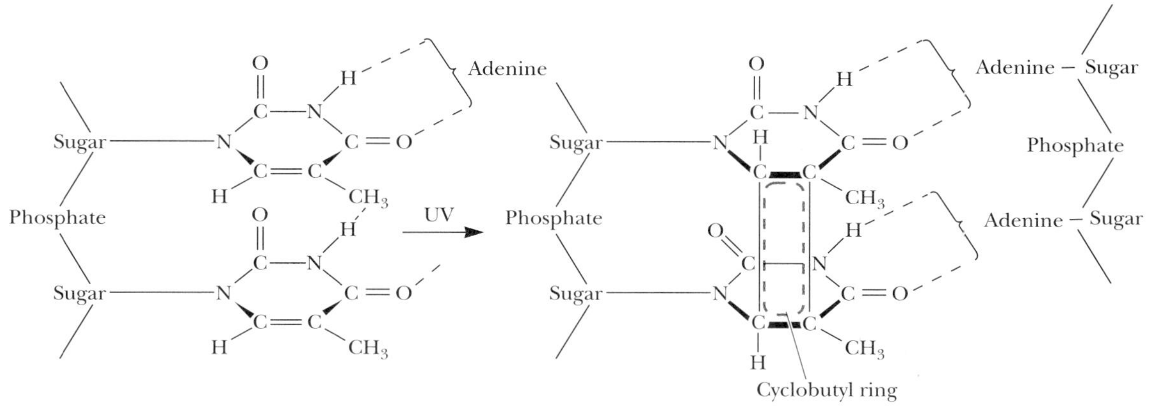

***Figure* 7.5** UV irradiation can cause dimerization of adjacent thymines on one strand of DNA. Normal base pairing during replication is disrupted by the presence of these dimers.

icant extent. This limits the effective use of UV radiation to sterilization of surface areas and air.

Cellular components that strongly absorb UV light are the purines and pyrimidines. Both are present in DNA and RNA and absorb UV maximally at 260 nm. A UV lamp will produce levels of radiation that can effectively penetrate eubacterial, archaeal, or other viable cells and cause alterations in cellular nucleic acids. The production of *thymine dimers* is probably the most important cause of damage to DNA. A dimer is formed by the linking of two adjacent thymine molecules on a single strand of DNA (Figure 7.5). The molecules linked in this fashion prevent the replication of DNA, and this can be lethal to a cell. Thymine-cytosine or cytosine-cytosine dimerization can also occur (see Chapter 13).

Microbes have DNA **repair systems** that can excise the dimers and replace them with unaltered thymine molecules. UV light is lethal if a sufficient number of dimers are created such that all of them cannot be repaired. In some cases, the enzyme repair system will make an error, which leads to mutations or death.

Filter Sterilization

The ability of **filters** to retain eubacterial and archaeal cells and allow smaller particles and liquids to pass through has been recognized for over 100 years. In 1892, Dmitrii Ivanowsky, a Russian scientist, demonstrated that tobacco mosaic virus would pass through a filter that would retain the smallest bacteria. This was confirmed and expanded by Martinus Beijerinck a few years later (see Chapter 2).

Filters have proven useful for the selective sterilization of liquids and gases. Heat-sensitive materials, such as enzymes, vitamins, and sugars, can be filtered to remove contaminating bacteria. Beer and wine are now filtered in bulk to remove bacteria that would oxidize the ethanol to acetic acid and render these drinks unacceptable (see Chapter 32). Virologists routinely filter blood serum and other culture media to prevent unwanted bacterial growth. Media for plant or animal cell culture are generally complex and unstable at high temperatures. They, too, are sterilized by filtration.

Basically, a filter is constructed such that it has a pore size that permits liquids to pass, but the pores are small enough to retain Eubacteria and Archaea. These organisms, in general, are longer than 1 μm and are greater than 0.5 μm in diameter. Effective filters must be able to retain cells of this size.

In practice a filter may be employed to separate or concentrate microbes based on their size. Bacteria can be separated from yeasts by employing a membrane with a pore diameter of about 3 μ meters. Most bacteria can pass through a filter such as this, but yeasts would be retained. Small bacteria such as *Bdellovibrio,* an organism that parasitizes other bacteria, can be concentrated by passing water or dilutions of soil through membrane filters with a pore diameter of 0.45 μ meters. The *Bdellovibrio* are only 0.25 μ meters in width and would pass through while most other microorganisms would be retained. The organisms in the filtrate can then be concentrated by centrifugation. Most early filters were composed of stacks of asbestos, sintered glass (glass heated to fuse without melting), or diatomaceous earth. These materials formed circuitous passageways that retained microbial cells. Several different types of filter are used today. A **glass fiber depth filter** is illustrated in Figure 7.6**(a)**. This random stacking of glass fibers forms a barrier impenetrable by bacteria and other particles.

The filters that are now in common use are the **membrane filters** (Figure 7.6**b**). These filters can be made of

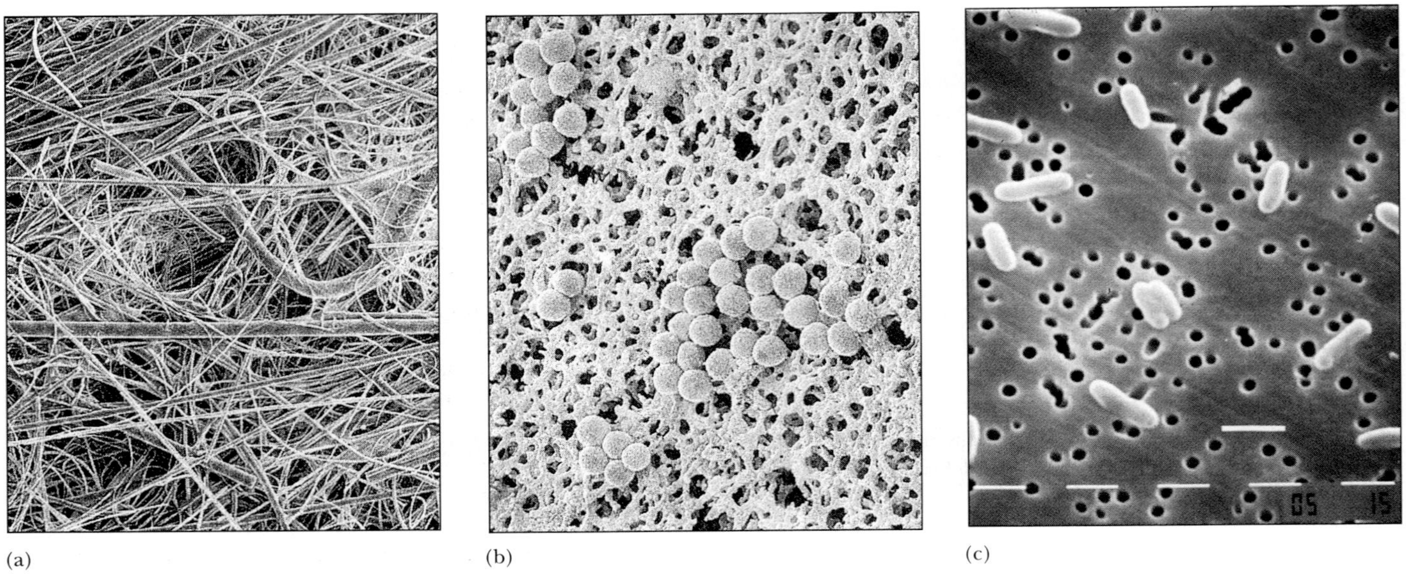

Figure **7.6** A scanning electron microscope picture of filters that can be employed to retain microorganisms: **(a)** depth filter, and **(b)** membrane filter. **(c)** A scanning electron micrograph of *Thermoleophilum album* on an Isopore® filter. The microorganism is about 1 μm in length. (**a, b,** courtesy of Millipore; **c,** courtesy of J. J. Perry)

cellulose acetate, polycarbonates, or cellulose nitrate. They are quite thin (200 μm) and become relatively transparent when wet, allowing direct microscopic examination of organisms that are trapped on the filter. A membrane filter can have a pore size ranging from 0.1 μm to 0.5 μm, with 80 percent to 85 percent of the filter being open space. This allows a rather high rate of fluid passage. With this much open space available, a filter can be placed on the upper surface of an agar medium, and nutrient will pass upward through the filter by capillary action. Following incubation and growth one can count the number of colonies that develop directly and observe their colonial morphology (see Figure 6.7).

The **Isopore® filter** is a type of membrane filter that has widespread application. Also known as the nuclepore filter, it contains clearly defined cylindrical holes that pass vertically through the membrane (Figure 7.6**c**). These filters are manufactured by treating thin sheets of polycarbonate with nuclear radiation. The radiation results in a random distribution of minute holes in the polycarbonate that can be enlarged by chemical etching. The length of time that the membrane is etched determines the size of the pores. A major advantage of the nuclepore filter is that bacteria and other objects remain on the surface of the membrane and are thus readily observable with a scanning electron microscope (Figure 7.6**c**). With other types of membrane filter, the bacteria absorb to the sides of the pores and are not so readily observable. One problem with nuclepore filters, however, is the small amount of total space on the filter that is porous. This may result in low flow rates or clogging.

Presterilized disposable filter units are now routinely employed in the laboratory (Figure 7.7). Much larger and more sophisticated filtration systems are used for large volumes in industrial settings. Large volumes of liquid are sterilized by placing the filters in stainless steel cartridges. As an alternative to centrifugation, particularly for large volumes, bacteria can also be harvested from culture fluids by using membrane filters.

Figure **7.7** A commercial filter unit. These filter units are disposable, preassembled, and presterilized. They are equipped with removable filters that can be placed on nutrient agar plates for growth and recovery of microorganisms. (Courtesy of Malgene Corporation)

Use of Toxic Chemicals

An effective method for sterilizing surgical instruments and some heat-sensitive materials is exposing them to toxic compounds in the form of gases. Sheets and bedding in a hospital are sterilized in this manner. Plastic petri dishes and other plastic laboratory containers can be sterilized by toxic gases. The filter shown in Figure 7.7 would be sterilized by gas and sealed in a closed plastic bag until used. Ethylene oxide and propylene oxide are generally the compounds of choice. Both function as alkylating agents in disrupting cellular DNA. As these gases are exceedingly toxic to humans, their use in the sterilization process must be carefully controlled. Ionizing radiation is replacing toxic gases for many applications.

Types of Antimicrobial Agents

An array of chemical agents are now available that can be used in the control of undesirable microorganisms. These chemicals are called **antimicrobial agents** (Table 7.2). Some of the agents are naturally occurring, some synthetic, and others a combination of the two. A much desired quality in an antimicrobial agent is **selective toxicity.** That is, the agent will eliminate a pest without doing significant harm to other species in the environment. Spraying plants with a chemical that kills invading bacteria or fungi but also weakens the plant is of little value. Humans cannot be cured of disease by treating them with medicines that are overly harmful to the host (see quote at beginning of this chapter).

In addition, humankind has become more and more concerned with the effects of chemical agents on the total environment. Experience has taught us that we should consider the potential harm that can come from careless application of antimicrobial substances. For example, over the years, compounds containing mercury were used in the pulping industry and other manufacturing processes to control microbial contamination. These mercury compounds eventually entered streams and rivers, resulting in considerable harm to fish and other wildlife. Stringent regulations are now in place for all potentially toxic compounds that are artificially introduced into the environment. They must be tested for safety, for biodegradability, as potential carcinogens, or for any other undesired effects on humans. Another example of a carelessly utilized agent is DDT, which was once applied widely as an insecticide. It was very effective against insects, but it accumulated in the natural environment and was very harmful to

***Table* 7.2 Common antimicrobial agents and their uses**

Compound	Use
Organic compounds	
Cresols	Disinfectant in laboratories
O-phenylphenol	Disinfectant in laboratories
Phenol	Disinfectant in laboratories
Formaldehyde	Disinfectant for instruments
Quaternary ammonium compounds	Skin antiseptic
Ethanol and isopropanol	Disinfectant for instruments
Halogens	
Chlorine gas	Disinfectant for water
Iodine solution	Skin antiseptic
Hexachlorophene	Skin antiseptic
Chlorox®	Domestic cleanser
Heavy Metals	
Mercury chloride	Disinfectant
Silver nitrate	In infant eyes to prevent opthalmic gonorrhea
Copper sulfate	Algicide
Other	
Hydrogen peroxide	Skin antiseptic
Mercurochrome	Skin antiseptic
Ozone	In drinking water and fish tanks

Table 7.3 **Categories of antimicrobial agents and their target organisms**

Antimicrobial Agent	Target Organism
Bactericide	Bacteria
Fungicide	Fungi
Algicide	Algae
Pesticide	Pests
Herbicide	Plants
Insecticide	Insects
Germicide	"Germs"

birds and other wildlife. Accumulation resulted from the inability of microbes to mineralize DDT at rates commensurate with application rates. DDT had a half-life of about 30 years in most environments, and its use has been curtailed in the United States.

A compound that kills a target group is identified by combining the name of the designated organism with the suffix *-cide,* from the Latin *cida* meaning "having power to kill." For example, a **bactericide** is a compound specifically designed to kill bacteria. Table 7.3 lists some categories of compounds that kill target groups. Other agents are made that do not kill but rather prevent or inhibit growth. These are referred to as **bacteriostatic agents** (for bacteria) or **fungistatic agents** (for fungi). Some antibacterial substances bring about **lysis,** or disintegration, of the cell and are thus termed **bacteriolytic agents.** For example, lysozyme is an enzyme that destroys the cell wall of bacteria, particularly gram-positive types. Loss of the cell wall results in uptake of water and subsequent bursting of the wall-less cell. Lysozyme would thus be considered a lytic agent. **Germicide** is a general term that describes substances that kill or inhibit microbes ("germs").

A **disinfectant** such as phenol or mercuric chloride will destroy microorganisms on contact, including those that potentially cause disease. Disinfectants are indiscriminate destroyers that, if applied to wounds, would also destroy host tissue. An **antiseptic** is a substance, such as methiolate, that kills or prevents the growth of disease-causing pathogens. An antiseptic should not do significant harm to host tissue. As a rule we use powerful disinfectants like phenol on inanimate objects and less potent antiseptics to treat wounds.

A **chemotherapeutic** agent is a chemical compound that can be applied topically, taken orally, or given by injection that will inhibit or kill microbes that cause infections. The term is also applied to compounds that are effective against viruses or tumor growth. **Antibiotics** are sometimes called chemotherapeutic agents and are widely used antimicrobials, although the term chemotherapeutic is generally reserved for products of chemical synthesis. The antibiotics will be discussed later.

Tests for Measuring Antimicrobial Activity

Robert Koch first devised a procedure for determining the effectiveness of antimicrobial compounds. Specifically, he devised a test for those with antibacterial activity. Koch dried endospores of *Bacillus anthracis* on threads and placed these in solutions of potential antibacterial agents. Endospores were selected because they are more resistant to harsh physical conditions than are other life forms. He then estimated the effectiveness of a given compound by determining how long the endospores would survive exposure to the agent. At selected time intervals, threads were removed from the test solution and placed in a nutrient medium. Growth indicated survival. The shorter the time required to kill all of the endospores, the more effective the disinfectant. Koch discovered that mercuric chloride was one of the most effective of all available compounds. Mercuric chloride has been employed as a disinfectant up to the present day. Unfortunately, use of large quantities of this mercury compound is discouraged because of the environmental problems it causes, as mentioned earlier (also see Chapter 24).

The relative strength of a germicide can be determined by serial dilution of the parent compound. The more a compound can be diluted and still inhibit the growth of a test organism, the greater its relative strength. The lowest concentration that will inhibit growth is designated the **minimum inhibitory concentration,** or **MIC** (Figure 7.8). The test organism selected, the inoculum size, the amount of organic material present, and other factors all affect the killing capacity of germicides. A strong germicide, such as ethylene oxide or formaldehyde (37 percent solution), will kill endospores and virtually all forms of life. Intermediate-strength germicides such as phenolics can kill viruses and most vegetative cells. Such agents are used for research and in hospital laboratories. Germicides with less killing power are effective against vegetative cells (fungal and bacterial) but are not very effective against endospores, fungal spores, or some viruses. These weaker agents have less toxicity to humans and can be applied directly to the skin or can be included as a component in mouthwash. The general mode of action of various germicides is presented in Table 7.4.

One standard that has been employed to compare the relative efficiency of germicides is the **phenol coefficient.** A phenol coefficient is a measure of the effectiveness of a test compound compared with the disinfecting power of phenol. For example, if a dilution of 1:100 kills a standard population of *Salmonella typhi* that is killed by a 1:50 dilution of phenol, the phenol coefficient would be 100/50, or 2. The test germicide would thus be twice as

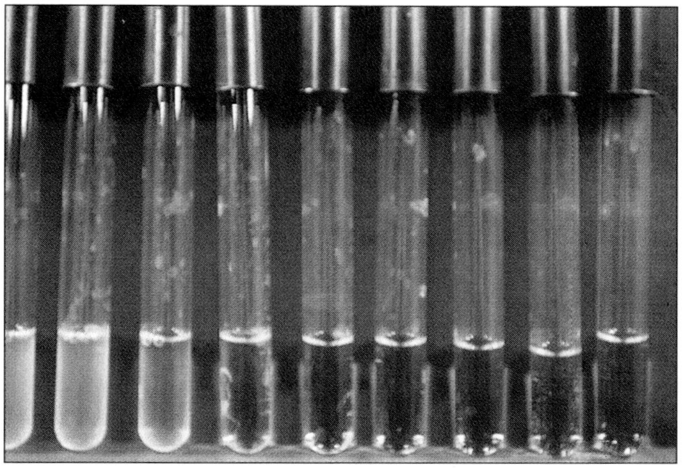

***Figure* 7.8** A tube dilution series for estimating the sensitivity of a microorganism to an antibiotic. The minimum inhibitory concentration (MIC) is the least amount of antibiotic that completely inhibits growth. The antibiotic in this test was chloramphenicol and the organism *Escherichia coli*. The level of antibiotic was 100, 50, 25, 12.5, 6.25, 3.13, 1.56, 0.78, 0.39 μg/ml. The MIC would be 3.13 μg/ml. (Courtesy of J. J. Perry)

effective as phenol in destroying the test population of *S. typhi*. However, the phenol coefficient has limited value because many factors can influence the effectiveness of disinfectants. Some are more affected by the physical environment than others. Light, oxygen, organic matter, and solid substrates can have a marked effect. Gram-positive organisms are more readily killed than gram-negatives. Members of the genus *Pseudomonas*, a gram-negative organism, are not only resistant to, but may grow on, selected disinfectants. The ability of pseudomonads to thrive in the presence of antiseptics is of particular concern for burn patients.

Chemotherapeutic Agents

The antimicrobial agents discussed in the previous section may be used as pesticides to prevent growth of undesirable organisms in the environment. They may be employed as disinfectants on laboratory benches or other solid material. They are not suitable for control of viral, bacterial, or fungal infections within the human body. They would either be too toxic or, in other cases, be inactivated by the presence of organic matter. From the time of Paul Ehrlich (see Chapter 2), scientists have sought chemicals that selectively inhibit the growth of infectious bacteria without harm to the human host. The goal has been to find compounds **selective** or **specific** for the foreign invading bacterium but with little or no harmful effect on normal body function. These investigations have been broadened in recent times to include a search for antiviral and antitumor agents.

Chemically Synthesized Chemotherapeutics

The chemotherapeutic agents called the **sulfa drugs** were discovered by Gerhard Domagk in the 1930s. Domagk was assessing the potential use of dye substances for antibacterial activity. Sulfanilamide was one of the compounds tested, and he found that it possessed antibacterial activity both *in vitro* [Latin = "in glass" (outside the body)] and *in vivo* [Latin = "living" (inside the body)]. This compound proved to be a structural analog of the vitamin *p*-aminobenzoic acid, as it was later demonstrated that the antibacterial activity of sulfanilamide could be overcome by the presence of high levels of *p*-aminobenzoic acid. Folic acid, a coenzyme in nucleic acid synthesis, is synthesized in cells by enzymatically joining a molecule of 6-methyl pterin with *p*-aminobenzoic acid and a molecule

***Table* 7.4 Mode of action of some antimicrobial agents**

Agent	Mode of Action
Alcohols	Denature proteins and dissolve membrane lipids.
Aldehydes	Combine with proteins and denature them.
Halogens	Iodine oxidizes cellular constituents and can iodinate cell proteins. Chlorine and hypochlorite oxidize cell constituents.
Heavy metals	Combine with proteins generally through sulfhydryl groups.
Phenolics	Denature proteins and disrupt cell membranes.
Quaternary ammonium compounds	Disrupt cellular membranes and can denature proteins.

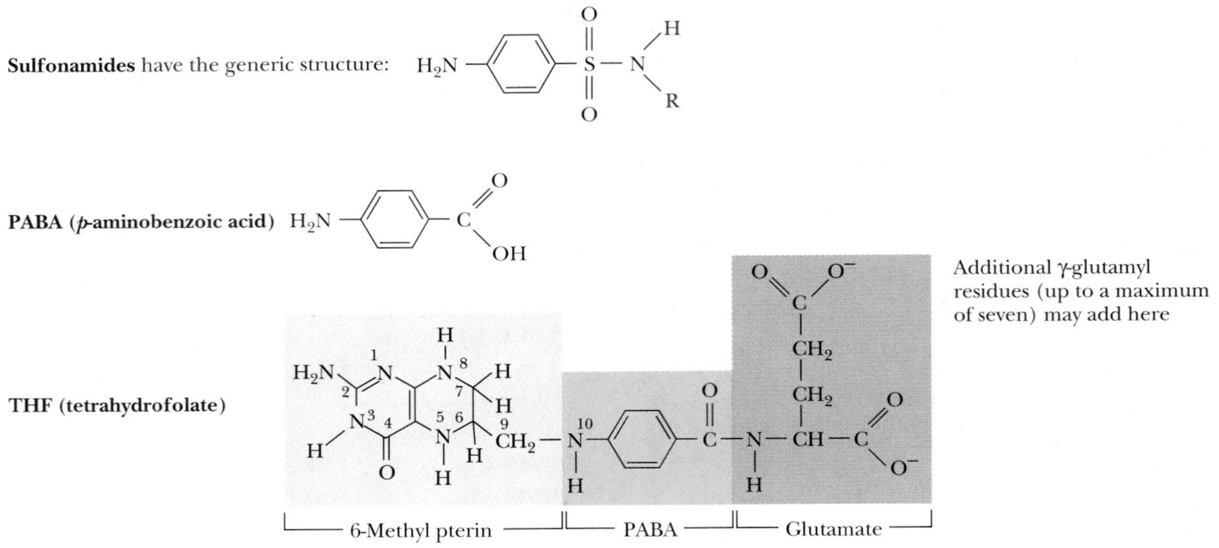

***Figure* 7.9** A model competitive inhibitor of metabolic function in a microorganism. The sulfonamide is an analogue of the vitamin *p*-aminobenzoic acid.

of glutamic acid. The structure of the coenzyme folic acid is shown in Figure 7.9. Because many species of bacteria synthesize folic acid, the presence of sulfanilamide results in the displacement of *p*-aminobenzoic acid. The product thus generated cannot form a peptide bond with glutamic acid. An inactive pteridine-sulfa conjugate results. Because folic acid is a mandatory coenzyme in the synthesis of purine and pyrimidines, an organism without functional folic acid would not survive. Animals are unaffected by sulfanilamide because they generally obtain folic acid in their diet and cannot synthesize it from the base constituents (pteridine and *p*-aminobenzoic acid).

An understanding of the selective toxicity of sulfanilamide led to a concerted search for other chemotherapeutic agents. Sulfanilamide became a model and encouraged efforts to synthesize other potentially effective antibacterial agents. Structural analogs of amino acids, purines, pyrimidines, and vitamins have been synthesized. Emphasis has been placed on synthesizing analogs of the bases in DNA and RNA (adenine, thymine, guanine, uracil, and cytosine) in an effort to find chemical compounds that have antiviral or antitumor activity. Thousands of compounds have been synthesized, but a limited number have proven effective as chemotherapeutics.

Some important antiviral agents have been developed in recent years, and these include acyclovir and AZT. **Acyclovir** is a structural analog of deoxyguanosine (Figure 7.10) and can be used in the treatment of herpes virus infections. When a herpes virus multiplies in a cell, the enzyme thymidine kinase is activated. This enzyme gratuitously phosphorylates acyclovir, resulting in the formation of an unnatural triphosphate, which blocks the DNA polymerase. As a consequence, the assembly of DNA in viral particles is stopped. **AZT** (azidothymidine) is an antiviral used in the treatment of AIDS (acquired immunodeficiency syndrome). This compound inhibits the multiplication of the retrovirus that causes AIDS. Unfortunately, neither of these antiviral agents can destroy the nonreplicating intracellular virus. Both the herpes and AIDS viruses persist for life in selected cells within the human host from the time of infection (see Chapter 28).

Antibiotics as Chemotherapeutics

The term **antibiotic,** familiar to most of us, was coined by Selman Waksman in 1953. Waksman was the discoverer of the important antibiotic streptomycin that has played a significant role in the control of tuberculosis. He defined an antibiotic as " . . . a chemical substance, produced by mi-

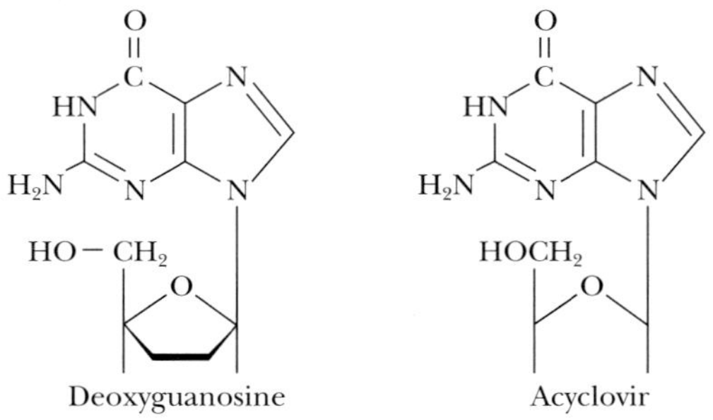

***Figure* 7.10** The antiviral agent acyclovir and the structural relationship to the nucleoside deoxyguanosine. Phosphorylation of acyclovir yields an analog of deoxyguanosine triphosphate that inhibits the virus DNA polymerase.

actinomycetes

croorganisms, which has the capacity to inhibit the growth and even to destroy bacteria and other microorganisms, in dilute solutions." Antibiotics effective in controlling disease are produced during growth by certain bacteria, actinomycetes, and fungi. In fact, the ability to produce antibiotic substances is widespread in the microbial world, and the actinomycetes are particularly adept at producing them. The antibiotics synthesized by actinomycetes have remarkable chemical diversity. The first commercially successful antibiotic used in chemotherapy was **penicillin.** This antibiotic is produced by the fungus *Penicillium notatum.* The search for antibiotic substances and their industrial production are presented in Chapter 31. The medical aspects of antibiotics are covered in Chapters 29 and 30. The present chapter briefly discusses their basic **mode of action** and the manner in which infectious organisms become resistant to them.

Antibiotics are selectively toxic to certain types of living cells and less to others. This selectivity is due to some fundamental physiological reaction in the affected cell that does not occur in the unaffected cell in quite the same way. Some antibiotics are effective against a limited range of bacteria and thus are considered to be **narrow spectrum** antibiotics. Penicillin is an example of this type because it is more effective against gram-positive organisms. A **broad spectrum** antibiotic, such as tetracycline, inhibits both gram-positive and gram-negative bacteria. Antibiotics that inhibit bacteria generally do not inhibit eukaryotes. Conversely, eukaryotes are generally not affected by antibacterial antibiotics. The Archaea are generally sensitive to few of the antibacterial antibiotics.

Mode of Action

An antibiotic interferes in some manner with the normal physiological function in a susceptible cell. The major actions of antibiotics include (1) disruption of cell wall synthesis, (2) destruction of cell membranes, and (3) interference with some aspect of protein or nucleic acid synthesis. The most useful antibiotics employed to control human infections interfere with structural or physiological reactions that are unique to the invading pathogen. This is possible because the bacteria are physiologically quite different from the eukaryotes. Consequently, antibacterial antibiotics can exploit these differences and destroy the invader without significant harm to the human host. Control of fungal, protozoan, and viral invaders is much more difficult because the physiology of these pathogens is quite similar to that of eukaryotes, including a human host. The physiological differences among the eukaryotes are simply not so distinct that these differences can readily be exploited. Thus, effective antifungal and antiprotozoan agents are generally toxic to humans. Because viral infections are intracellular, they are also difficult to control. A virus is synthesized, for the most part, by the metabolic machinery of the host, and destruction of this machinery leads to the destruction of the host.

Several antibiotics and their modes of action are listed in Table 7.5. Penicillin and related antibiotics prevent the transpeptidation reaction, which is an important step in the assembly of cell wall peptidoglycan (see Chapter 4). This results in a weakened cell wall, especially so in gram-positive organisms. Because microorganisms generally live

***Table* 7.5 The producing organisms and mode of action of several antibiotics**

Antibiotic	Producer	Mode of Action
Penicillin	*Penicillium* spp.	Blocks transpeptidation involved in cell wall synthesis
Erythromycin	*Streptomyces erythreus*	Binds to 50S ribosomal subunit and stops peptidyltransferase
Chloramphenicol	*S. venezuelae*	Binds to ribosomes blocking peptidyltransferase
Rifampicin	*S. mediterranei*	Blocks transcribing enzyme RNA polymerase
Novobiocin	*S. spheroides*	Binds to a subunit of DNA gyrase
Tyrocidine	*Bacillus brevis*	Ionophore disrupts cell membrane integrity and function
Polymyxins	*B. polymyxa*	Disrupt membrane transport and function
Streptomycin	*S. griseus*	Inhibits 30S ribosomes, blocks amino acid incorporation into peptides
Tetracycline	*S. aureofaciens*	Inhibits binding of aminoacyl RNA to ribosomes

BOX 7.1 MILESTONES

The Antibiotic Age

Fortune favors the prepared mind.
Louis Pasteur

The more I practice the luckier I get.
Lee Trevino

These quotes are a brief summation of the circumstances that led Sir Alexander Fleming (1881–1955) to the discovery of penicillin. All too often, scientific advances are attributed to chance or serendipitous good fortune, but usually they are not. More often they are the product of intuition and hard work. The discovery of penicillin resulted from Fleming's long-held and firm belief in the existence of naturally occurring substances that had useful antimicrobial properties. During World War I, Fleming worked in a field hospital in France. There he treated casualties of battle and experimented with better ways of treating deep wounds. The accepted treatment, at that time, was copious quantities of antiseptic—carbolic acid, boric acid, or peroxide. Unfortunately, in deep wounds these antiseptics were quickly neutralized by tissue, and serious infections often followed. Fleming and his colleague Sir Almroth Wright (1861–1947), a noted developer of antityphoid vaccine, observed that wounds carefully cleansed and treated with minimal antiseptic healed faster and with fewer serious infections. Stimulation of natural-defense mechanisms by encouraging the exudation of lymph into the wound site was particularly effective in promoting healing. Fleming and Wright found that bathing wounds with a saline solution was effective in stimulating the exuding of lymph. The success of this experimentation convinced Alexander Fleming that naturally produced substances could be employed successfully in the treatment of infectious disease. Fleming explained to a physician visiting his laboratory at Boulogne, "What we are looking for is some chemical substance which can be injected without danger into the blood stream for the purpose of destroying the bacilli of infection, as salvarsan destroys the spirochaetes."

Sir Alexander Fleming. (A photo by Dr. P. N. Cardew from MacFarlane, *The Man and the Myth,* Harvard University Press, 1984)

Following the war, Fleming returned to his position at St. Mary's Hospital in London. As time permitted, he

in osmotically unfavorable environments, one having a weakened cell wall may burst, or lyse. Gram-negative bacteria tend to be a bit more resistant to penicillin because their outer envelope can prevent the antibiotic from reaching the peptidoglycan layer of the cell. The Archaea are generally unaffected by antibiotics such as penicillin, which interferes with peptidoglycan synthesis. The Archaea do not have peptidoglycan in their cell wall. Some of the Archaea are sensitive to selected ionophores (bacitracin) that disrupt membranes (**Box 7.1**).

Tyrocidine and **polymyxin** are both polypeptide antibiotics that disrupt cell membranes. Both are produced by bacteria of the genus *Bacillus.* Tyrocidine is an **ionophore,** and functions by forming channels across the cytoplasmic membrane, resulting in leakage of monovalent cations. As a consequence, the organism cannot establish a proton motive force, and selective transport into and out of the cell is impaired. Polymyxin causes similar cytoplasmic membrane damage. Peptide antibiotics are not taken internally, but are applied topically to treat skin infections. A peptide antibiotic would be digested by enzymes present in the intestinal tract.

Certain antibiotics interfere directly with bacterial DNA replication. Among these is novobiocin, which binds directly to the beta subunit of the DNA gyrase responsible for unwinding supercoiled DNA during replication. Other antibiotics, such as rifampicin, block the RNA polymerase involved in transcribing the information on the DNA mol-

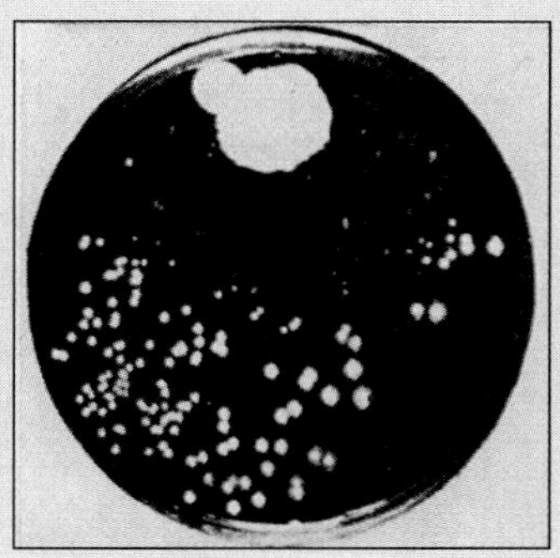

The original penicillin mold growth. (From Maurois, Andre. *Life of Sir Alexander Fleming,* E. P. Dutton Co., 1959)

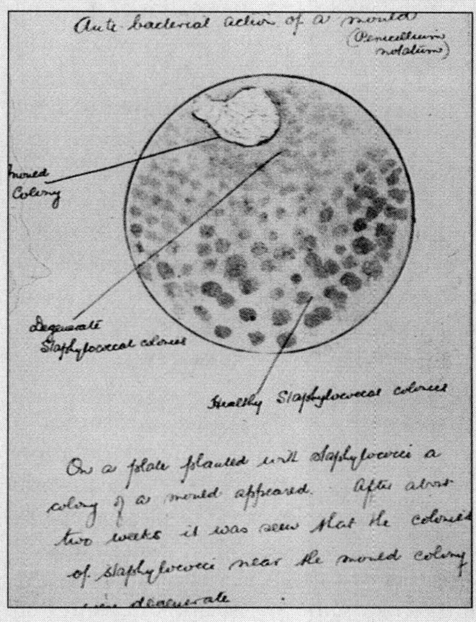

Fleming's drawing and notes. (From Maurois, Andre. *Life of Sir Alexander Fleming,* E. P. Dutton Co., 1959)

tested various naturally occurring substances for antibacterial activity. Among the materials he tested was nasal mucus, and he observed that addition of mucus to a suspension of *Staphylococcus aureus* led to a rapid clearing of the turbidity. He discovered that this clearing resulted from lysis of the bacterial cells. The factor involved was a protein called lysozyme (see Chapter 4). This discovery, made in 1921, was important scientifically, but lysozyme was not useful as a chemotherapeutic agent. It did, however, offer confirmation to Fleming that his belief in natural antibacterials was warranted.

One of Fleming's research interests was the *Staphylococcus aureus,* just mentioned, and he often had cultures growing on Petri dishes lying about the lab. One day while examining some of these old cultures he observed that one of them was contaminated with a mold and that colonies did not develop in the area adjacent to the mold growth. Fleming was intrigued and likened the phenomenon to the action of lysozyme on *S. aureus.* He then took a picture of the plate (shown above) and wrote up the observation in his research notebook. He showed the plate to a colleague with the remark, "Take a look at that, it's interesting—the kind of thing I like; it may well turn out to be important." Important, indeed: the antibacterial substance produced by the mold *(Penicillium notatum)* was penicillin. This ushered in the "antibiotic age." Fleming's monumental discovery led eventually to the conquest of such ancient scourges as pneumonia, tuberculosis, syphilis, staphylococal infections, typhus, and a host of human ills.

ecule to make messenger RNA. Messenger RNA is the template translated on ribosomes in protein synthesis (see Chapter 13). Most of these antibiotics are ineffective against the Archaea.

Many of the clinically useful antibiotics inhibit protein synthesis in Eubacteria. **Ribosomes** are the structures in all cells that are involved in the synthesis of a protein. Bacterial and eukaryotic ribosomes differ in size, protein content, and ability to bind antibiotics. Ribosomes are small, mostly spherical bodies made up of two subunits of unequal size. Bacterial ribosomes are smaller than those in eukaryotes. On the basis of sedimentation velocity during ultracentrifugation, the size of the bacterial ribosome is designated **70S** and the eukaryotic ribosome size is **80S.**

Antibacterial antibiotics that affect protein synthesis do so by binding to the bacterial 70S ribosome. This occurs with the antibiotics tetracycline, chloramphenicol, streptomycin, and erythromycin. The antifungal antibiotic cycloheximide binds to the 80S ribosome but not the 70S ribosome. As humans have 80S ribosomes, they too would be adversely affected by cycloheximide. Consequently, cycloheximide inhibits fungi but does not inhibit the growth of bacteria. A few antibiotics, including tetracycline, can bind to both 70S and 80S ribosomes. However, tetracycline can be used in humans because it does not inhibit the synthesis of protein in eukaryotic cells at the concentrations used chemotherapeutically. It does impede wound healing in humans under some circumstances.

Resistance to Antibiotics

Antibiotics have been effective in controlling many of the diseases that have been a scourge to humankind. Tuberculosis, bacterial pneumonia, syphilis, and many other infectious diseases that were once fatal can now generally be treated with antibiotics. Antibiotics have been called "wonder drugs" because they effect a dramatic cure for what had previously been incurable. But there is a problem with "wonder drugs" that became evident after widespread use. The application of penicillin as a chemotherapeutic agent led to the evolution of pathogenic bacterial strains that were unaffected by the drug. These resistant organisms retained their potent pathogenicity but were no longer controlled by the administration of penicillin. For example, when penicillin G (Figure 7.11) was first introduced, virtually all strains of *Staphlococcus aureus* were sensitive to the drug. Within a span of only ten years, essentially all staph infections acquired in hospitals were caused by strains that were resistant to penicillin.

Why do microorganisms become resistant to chemotherapeutics? Abundant microbial populations have existed throughout all of Earth's history because microbes are adaptable. For example, when oxygen appeared in the atmosphere or the climate became cooler, microbes evolved to survive in this prevalent environmental condition. Environmental stresses and constraints became natural selection processes for microbes that will survive that stress. Human control methods create a challenge to bacteria that stimulates natural selection for those organisms in the environment that can withstand the prevailing condition. Thus, the widespread use of penicillin resulted in the selection of strains that could survive in the presence of the drug. Penicillin-resistant organisms can produce an enzyme, **penicillinase,** that destroyed the antibiotic.

Microbial resistance to an antibiotic or other chemotherapeutic agent can occur for several reasons, and these are presented as examples:

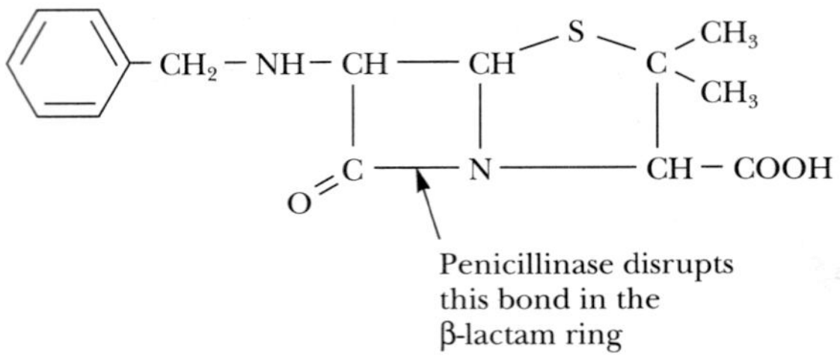

***Figure* 7.11** The source of penicillin resistance in some organisms depends on the ability to produce the enzyme penicillinase. The inhibitory properties of penicillin require that the β-lactam ring be intact.

Natural resistance

The organism may lack the structure that the antibiotic inhibits, as occurs with mycoplasma, which lacks cell walls and is thus unaffected by penicillin.

The cell wall structure or cytoplasmic membrane of an organism may be impermeable to an antibiotic.

Acquired resistance

A resistant microorganism may produce a substance that inactivates the antibiotic, as occurs with *Staphlococcus aureus* in producing the enzyme penicillinase (Figure 7.11), which disrupts the penicillin molecule.

A gradual accumulation of mutations in chromosomal DNA may result in cellular structures that will not bind the antibiotic. For example, the gene for transpeptidase synthesis in staphylococci can mutate so that the enzyme does not bind penicillin.

Another significant problem in antibiotic resistance is the acquisition of bits of extra chromosomal DNA that carry information that renders a microorganism resistant to a given antibiotic. These plasmids can be transferred cell to cell (see Chapter 15).

Summary

- The role of microorganisms in the cycles of nature is essentially destructive; human survival often depends on counteracting their activities.
- An understanding of the nature of microorganisms led to more direct means for controlling them.
- **Sterilization** is the destruction of all viable life forms. It can be attained by fire, moist heat, dry heat, radiation, filtering, or toxic chemicals.
- Microbial death by heating occurs at an **exponential rate.** The length of time required to reduce a population tenfold is termed the **decimal reduction time.**
- The most effective means for routine sterilization of media, food, and the like is steam under pressure. The common laboratory apparatus employed for this is the **autoclave.**
- Dry heat at 160°C is effective in sterilizing glassware, pipettes, and other heat-stable material.
- **Pasteurization** is a method for reducing the number of microorganisms present in milk or other products. It can destroy selected disease or spoilage-causing microorganisms.
- The wavelengths of light that can cause death of mi-

croorganisms are **UV,** which disrupts RNA/DNA, and **ionizing radiation,** which can potentially harm any constituent of a cell.

- Microorganisms have repair systems that can repair damage caused by radiation.

- Filters can be employed to sterilize liquids, and a number of types are available. Membrane filters are the most widely used.

- Toxic gases such as propylene oxide can be used to sterilize plastic petri dishes, filters, and other disposable material. Ionizing radiation is replacing toxic gas.

- **Antimicrobial agents** may be products of chemical synthesis, natural products, or a combination of the two. The favored agent has **selective toxicity** in that it destroys the target microbe but has limited activity toward other cells.

- A **disinfectant** is an agent such as phenol that destroys all living cells on contact. An **antiseptic** is less potent and prevents the growth of disease-causing organisms. Methiolate that is applied to superficial wounds is an antiseptic.

- The **minimum inhibitory concentration** (MIC) is the lowest concentration of a **germicide** that will inhibit a test organism.

- The **phenol coefficient** is a measure of the effectiveness of a test compound as compared with phenol.

- The first effective chemically synthesized chemotherapeutic agent was **sulfanilamide.** An understanding of the role of this compound in blocking vitamin synthesis in bacteria encouraged the scientific world to search for chemical agents that would be effective against viruses, protozoa, fungi, and cancer.

- By definition, an **antibiotic** is a product of microorganisms. Antibiotics have proven to be effective in controlling diseases caused by bacteria. They generally interfere with metabolic activities that occur in disease-causing Eubacteria that are not important to the host eukaryote.

- The major **modes of action** of antibiotics include: (1) disruption of cell wall synthesis, (2) destruction of cytoplasmic membranes, and (3) disruption of nucleic acid or protein synthesis.

Questions for Thought and Review

1. What is the difference between sterilization and pasteurization? What is the agent most often used in sterilization? Why are we so concerned with sterilizing things?
2. Dry heat is employed to sterilize selected materials. What are some of the materials best sterilized by dry heat?
3. What agent is most often used to sterilize air in a closed room? What would be the concern with other methods?
4. Give some advantages of sterilization by filtration. How is filtration applied in the laboratory? In industry?
5. How did Robert Koch determine the effectiveness of disinfectants? What compound did he find most effective by this method?
6. How would one calculate a phenol coefficient? Is this a useful number? What are the pros and cons?
7. Sulfanilamide is a classical example of a competitive inhibitor. How does it function?
8. What are the basic shortcomings of antiviral agents such as AZT and acyclovir?
9. Why is it so difficult to find effective antiviral and antifungal agents when antibacterials are quite plentiful?
10. Discuss the mode of action of some common antibacterials.

Suggested Readings

Block, S. S., ed. 1991. *Disinfection, Sterilization, and Preservation.* 4th ed. Philadelphia: Lea and Febiger.

Davis, B. D., and R. Dulbecco. 1990. "Sterilization and Disinfection." In Davis, B. D., et al., eds. *Microbiology.* 4th ed. Philadelphia: J. B. Lippincott Company.

Harte, J., C. Holdren, R. Schneider, and C. Shirley. 1991. *Toxics A to Z: A Guide to Everyday Pollution Hazards.* Los Angeles: University of California Press.

Levy, S. B. 1992. *The Antibiotic Paradox: How Miracle Drugs Are Destroying the Miracle.* New York: Plenum Press.

Murray, P. R., ed. 1995. *Manual of Clinical Microbiology.* 6th ed. Washington, DC: American Society for Microbiology.

Part 3

Microbial Physiology: Metabolism
A Conversation with Frederick C. Neidhardt

Professor Frederick C. Neidhardt has had a significant influence on the field of bacterial physiology, not only through his research, but also through the advanced textbooks and research treatises that he has co-authored or edited.

Professor Neidhardt's contributions to the cellular and molecular physiology of *E. coli* include the concept of catabolite repression, and many observations on the growth rate–related control of the synthesis of ribosomes and amino acid activating enzymes. He pioneered the use of temperature-sensitive mutants in exploring bacterial physiology, and discovered the heat-shock response in bacteria. For many years, he has been developing the gene-protein database of *E. coli,* a resource that links the genes of this organism to their protein products identified as protein spots following electrophoresis in two-dimensional polyacrylamide gels.

He has served the American Society for Microbiology in many posts, including being President. Currently, he holds the title of the F. G. Novy Distinguished University Professor of Microbiology and Immunology at the University of Michigan, and also serves as the Associate Vice President for Research at that institution.

JJP: What attracted you to science and more specifically, microbiology?

FCN: I've thought a lot about that question in recent years, and I've decided it's the wrong question. The question should be "*Who,* not *what,* attracted me to science and microbiology?" A series of individuals affected my life very much by helping me to discover for myself that science was for me. The first was an English teacher at Central High School in Philadelphia. Eighty percent of us were pre-meds. The teacher said he thought we should read *Arrowsmith,* and made that an assignment. I didn't find the male/female romance of the book as interesting as the tension of the pure scientist versus the physician—the tension between Gottlieb and Arrowsmith. That tension was at my core.

From Central High I went to Kenyon College, and came under the mentorship of a great teacher and friend, Professor Maxwell E. Power. He was half the biology department and he was a *Drosophila* geneticist interested in the development of the nervous system. What was important was that he recognized in me an inherent interest in what was known in those days as physiological genetics. People didn't know exactly what a gene was, but whatever it was chemically, it did something that led to either a behavioral or structural characteristic in a living organism. How that happened—what was between the gene and the behavior or structure—was what interested me. I did an honors thesis on *Drosophila* genetics.

At Harvard I met the third person who proved important to my beginnings as a scientist: Professor Boris Magasanik. He gave lectures to beginning students in Harvard's Medical Science Division and therefore gave some of the first lectures in bacteriology I'd ever heard. He quickly convinced me that what I wanted to study in Drosophila was completely

Courtesy of Fred Neidhardt

out of the question. The ideal organism to use for studies of gene function was the bacterial cell—even though there were not at that time any proper genetic tools, that is, no conjugation. But you could make mutants and you could study them biochemically. So that hooked me, and I chose Boris as my mentor.

After graduate school my postdoctoral work brought me to Professor Jacques Monod at the Pasteur Institute in Paris. I spent a year in his lab. He had a profound influence on me and supplemented what my former mentors had taught me. What they all had in common was the encouragement they gave me to be myself and to do the best science I could.

I mustn't forget to mention another teacher who influenced me greatly. This was Professor C. B. Van Niel, who taught a famous bacteriology course in the summer at Hopkins Marine Station for many years. He exposed me to a kind of bacteriology that I had not encountered at Harvard, and indeed would not encounter again for a long time—the world of the diversity of microorganisms, their role in nature and their many, many strange lifestyles. Van Niel was the great teacher.

So, what attracted me to science and to microbiology was in fact five great teachers, and, no question about it, what has continued to hold me in this endeavor have been my interactions with my colleagues and students ever since.

JJP: You have been a major contributor to our understanding of precursor metabolites and heat-shock proteins. How do these stand out in your mind?

FCN: Good question. Actually the phrase *precursor metabolites,* and the concept of the fueling reactions that produce them (as the starting materials for the biosynthetic reactions that make the building blocks of the cell), was taught to me by Professor Edwin Umbarger. So that wasn't my discovery. But you're right in a broad sense because the concept of precursor metabolites was central to what I first worked on as a graduate student with Magasanik, namely, the phenomenon that we now call catabolite repression.

It's not at all easy to see how some-

Courtesy of Fred Neidhardt

one could have worked both on precursor metabolites and on heat-shock proteins—there seems no logical connection between the two. Your question has set me thinking about the trail that led me from one subject to the other, and indeed it is a very definite trail, one called "growth of the bacterial cell." Because, you see, what I really fell in love with in bacteriology was the beauty and reproducibility of the growth curve of cultures of bacteria in steady state growth. My work with Magasanik involved what was at that time called the glucose effect. When bacterial cells like *Escherichia coli* or *Klebsiella aerogenes* are growing in a medium containing glucose and some other sugar or carbon source, the cells catabolize only the glucose, even though they have the genetic information to grow on the other compounds in the medium. Then when the glucose is used up, the cells utilize these other compounds after a growth lag. The glucose effect is an inhibition of gene action that prevents the cell from building machinery to grow on other substances. It's a beautiful piece of engineering, because glucose is a better substrate than other compounds. The cells grow fast on it, and do not waste energy and materials senselessly making enzymes to degrade the other substances in the medium. My job was to find out how glucose inhibits induced enzyme formation. I didn't solve that problem then—in the early 1950s—and forty years later it is still not completely solved. But I contributed important information, and the phrase "catabolite repression" replaced "glucose effect" because my work showed that, although glucose is the best inhibitor, it is not the only substance that blocks induced enzyme formation. Other substrates that can be metabolized faster than the cells can use the resulting precursor metabolites will also mimic glucose. I returned to catabolite repression at various times in later years, but the molecular mechanism escapes solution, except that we know that one part of the inhibition is related to decreased levels of the nucleotide, cyclic AMP, during catabolite repression.

Substances that provide the cell with the opportunity to grow rapidly cause catabolite repression. That tie-in between growth rate and regulation of gene action really interested me. As a postdoctoral fellow with Jacques Monod I continued to think a lot about growth rate, and particularly about the peculiar fact that, throughout biology, cells that grow fast have more RNA than cells that grow slowly. Now, it's easy to study growth rate with bacteria. You simply change the composition of their growth medium; the more nutrients one offers the cells the faster they grow. The exciting thing for me was to discover that the composition and size of the cells changes in these different media, but not the way you might first expect. The only thing that matters is the growth rate of the cells. Two media of very different chemical nature, if they permit the same growth rate, will produce cells of the same composition and average size. And over the wide range of possible growth rates, the composition and size of the cells varies monotonically in

So, what attracted me to science and to microbiology was in fact five great teachers, and, no question about it, what has continued to hold me in this endeavor have been my interactions with my colleagues and students ever since.

a simple and dramatic way. At their maximum rate of growth, cells are large and chock-full of ribosomes, transfer RNA and all other components of the machinery for making protein; at slow growth the cells are small and contain far fewer of these components that make proteins. What a beautiful and striking illustration of the economy of these cells! Just as in catabolite repression, the cells seem to "read" the growth medium and their physiological condition and use this information to make components that contribute optimally to rapid growth.

This gene control network, called the stringent response, may involve guanosine tetraphosphate, which, like cyclic AMP, is a nucleotide whose cellular level is related to the physiological state of the cell. Guanosine tetraphosphate plays some role, not completely understood in molecular terms, in inhibiting the synthesis of ribosomal RNA and other parts of the protein-making machinery of the cell when growth is inhibited by nutrient starvation. Like catabolite repression, the stringent control system and its role in growth rate regulation of RNA synthesis is not fully understood. Again, I have a knack for being interested in problems that defy solution!

JJP: It's very important for a new science student to appreciate how long scientific discovery takes.

FCN: Right. Well, I couldn't drop the problem of how cells adjust their ribosome synthesis to achieve a maximal growth rate, and I turned to the power of genetics to learn whether I could isolate mutants in which the elegant relationship between chemical composition of the medium, growth rate, and RNA composition was disrupted. I wanted to generate mutants in enzymes where no one had mutants before. This was in the early 1960s. You couldn't at that time isolate a mutant in an enzyme that you couldn't bypass nutritionally. You couldn't knock out, for example, an aminoacyl-tRNA synthetase, or RNA polymerase without killing the cell.

Except, along came—not from me but from phage workers—the brilliant idea of conditional mutations. That is, mutations that led to defects only under a certain condition. In this way you could theoretically obtain a mutant for any gene in the cell. The condition that proved most useful was temperature.

Temperature-sensitive mutants have a gene defect that displays itself at one temperature (usually a high temperature) but not at another (lower) temperature. Usually this occurs because the protein product of the mutant gene can't fold correctly at the higher temperature. So you isolate and grow the mutant at 30°C, and then shift the temperature to (for example) 37°C to study its behavior.

There had to be a waiting period before the field of microbial diversity could really blossom—a waiting period for new molecular techniques to come along: the powerful genetic probes that now allow us to return to the soil and water of our planet and look at the 99 percent of microbial life of which we've been ignorant. That is so exciting!

You can, therefore, ask the question in which I was interested. If you slow down the growth of a particular temperature-sensitive mutant by raising the temperature, does it still control RNA synthesis normally? The hope was that some day I would find a gene that was essential for controlling RNA synthesis. That led me into a whole decade of isolating temperature-sensitive mutants in essential enzymes.

Unfortunately, no mutant showed up with the desired characteristic of being defective in RNA control. Nevertheless, the study of temperature-sensitive mutants allowed us to make an important discovery. Here's how it came about. In the mid-1970s we had adopted the technique of two-dimensional polyacrylamide gel electrophoresis, which enables you to resolve in a two-dimensional pattern virtually all of the proteins of the cell. Through this technique you can monitor simultaneously the rate of gene expression, which is the rate of synthesis of individual proteins, of the whole cell and get a sense for what is going on under any given growth condition—or when conditions change. Well, as happens to all scientists, we asked the right question of a particular temperature-sensitive mutant, and did the right experiment—but were thrown a curve ball by nature. When we looked at the pattern of protein synthesis upon shifting the temperature from 30 to 42°C in a particular mutant which had been given to us by a colleague, we observed nothing happened of great significance. A great disappointment, but a curiosity, because the normal wild type cell had a fascinating change in protein synthesis upon the same shift in temperature. Nearly 20 proteins were greatly accelerated in synthesis, while the mutant showed almost no change in synthesis rate of these proteins. Before long it dawned on us that this was important physiology. What we had in fact discovered was that bacteria have a heat-shock response that enables them to grow after being shifted to a high temperature, and that our colleague's mutant had a defect in the molecular device that controls this response. Our discovery helped show that all living cells on this planet have a similar, evolutionarily conserved gene network response to shifts-up in temperature.

For the next decade or so I worked on heat-shock and other responses that bacteria make to stress—starvation, antibiotics and other chemical inhibitors of growth, cold, and other physical stresses. As I examined the regulation of the blocks of genes that are invoked by the cell to adapt to these conditions, I recognized the similarity to all my earlier work on growth rate and control of macromolecule synthesis, and on the gene networks called catabolite repression and the stringent response. Today I am attempting to put that all together by using two-dimensional gels to study the simultaneous operation of the cell's many global gene regulatory networks during growth under changing environmental conditions. I see now that my work has always been heading toward the goal of understanding the total adaptive gene responses of the *E. coli* cell. That's really been my vision quest.

JJP: The next question is the often-asked question: Why do people work with *E. coli* so much? What if they had chosen another organism? Why *E. coli?*

FCN: An easy question. It's not the perfect bacterium. But it has three great advantages. Its genetics is easy. Artificial transformation, transduction, and plasmid-mediated conjugation just make it a geneticist's dream. Another outstanding characteristic: it's domesticated. At least the lab strains are domesticated; meaning they have lost the ability to cause illness, and are well adapted to grow very fast in laboratory media. The third advantage is the most important, and why most of us won't switch to another organism: we're trapped. The thing going for *E. coli* now is the sheer amount of what is known about it. It's like doing a jigsaw puzzle. If you've got 60 percent done, the next part becomes easier than when you've got 4000 pieces and you're trying to find the first two that fit together. The information known about *E. coli* makes it easier to gain new bits of information. That's it. We'll solve it first.

JJP: Will we be going back to looking more at Van Niel's ideas on the diversity of microorganisms with the identification tools we now have?

FCN: I think so, but I can imagine how Van Niel would have answered that question. He'd have said, "Oh my God, I hope they don't go back to *my* ideas." But there *will* be a return to his appreciation of bacteria for their diverse forms and functions as masters of the repertoire of biochemical activities in nature. He had that appreciation and he taught it—the awe and wonder of it—to his students.

There had to be a waiting period before the field of microbial diversity could really blossom—a waiting period for new molecular techniques to come along: the powerful genetic probes that now allow us to return to the soil and water of our planet and look at the 99 percent of microbial life of which we've been ignorant. That is so exciting! It isn't going back. It's going forward and building on what Van Niel and his colleagues did. It would make him very, very happy were he alive today.

Much of our work as microbial biochemists and physiologists has been done using pure cultures—an absolute necessity. But it is being appreciated more and more that you can't understand the metabolism and regulatory mechanisms of bacteria totally until you study them back in the community in which they live in nature.

So, I see two large, new themes in my area of bacteriology: studying the organisms of which we have little knowledge, and studying how they function synergistically in the natural environment. Driving these themes—besides their intellectual excitement—is the fact that the future of the human species depends on our learning more about how the planet works. We know humans are placing stresses on the ecological balances of the biosphere, but we don't know enough about the microbial components of these equations. The growing concern about our environment will force increased attention to fundamental microbiology, and hopefully will lead to increased support of microbiological research. I see the study of microbiology just beginning—getting ready to take off.

Life is driven by nothing else but electrons, by the energy given off by these electrons while cascading down from the high level to which they have been boosted up by photons. An electron going around is a little current. What drives life is thus a little electric current, kept up by the sunshine. All the complexities of intermediary metabolism are but lacework around this basic fact.

Albert Szent-Györgyi

Chapter 8

Conservation of Chemically Generated Energy

This chapter is an introduction to the principal mechanisms whereby microorganisms can generate energy. Bear in mind that virtually all energy for life on the Earth's surface is furnished by the sun and this energy is captured during photosynthesis. Among the products of photosynthesis are ATP and energy-yielding compounds, such as sugars or starch, that support the growth of heterotrophic life forms. Animals are sustained by eating energy-rich plants and/or by devouring an animal that ate a plant. In this chapter we will discuss the mechanisms whereby chemical energy trapped in the products of photosynthesis are converted to ATP. The following chapter will present the processes whereby light energy is converted to ATP.

Basic Principles of Energy Generation

In studying the broad metabolic potentials among Eubacteria and Archaea, one should be aware that the source of energy for microorganisms is, in many cases, different from the carbonaceous substrate used for synthesis of cell material. This is a fundamental difference between these microorganisms and the animal world. The one commonality in the evolution of all viable systems is that **adenosine-5′-triphosphate (ATP),** formed from adenosine-5′-diphosphate (ADP) and inorganic phosphate (P_i), is a universal medium of biochemical energy exchange. The energy conserved in the "energy-rich" bonds of ATP can be used by the cell to do work. The oxidation-reduction reactions that are involved in the synthesis of ATP are quite varied among the prokaryotic microorganisms.

Evidence suggests that a major event in the evolution of living organisms was the advent of the ability to establish a charge separation or electrochemical potential across the cytoplasmic membrane. This charge separation may be equated to that found in a car or flashlight battery that provides a current to activate a starter or produce light. The electrochemical potential that can be established across a biological membrane can be converted to chemical energy in the form of ATP. A general representation of a charge separation across a cytoplasmic membrane is presented in Figure 8.1. Solar energy is the ultimate source of energy, and photons from the sun can be used to establish an electrochemical potential. Light energy can be used to extract electrons from water. The electrons extracted from water in the process ($HOH \longrightarrow 2H^+ + 2e^- + 1/2\ O_2$) may then be passed through membrane electron carriers to generate a charge separation. The transfer of electrons along the electron-transfer chains in the cytoplasmic membrane causes a proton to translocate to the outside of the membrane. Energy made available during photosynthesis is used to assimilate CO_2 from the atmosphere and synthesize compounds such as glucose. The energy inherent in a glucose molecule can be recovered by extracting electrons from the glucose and using these to establish a charge separation across the cytoplasmic membrane. This charge separation can be the driving force in generating ATP. The source of the electrons that bring about the extrusion of protons in microorganisms can be either organic or inorganic substrates as follows:

$$\text{Lactate} \longrightarrow \text{pyruvate} + 2\ H^+ + 2e^-$$

$$H_2 \longrightarrow 2\ H^+ + 2e^-$$

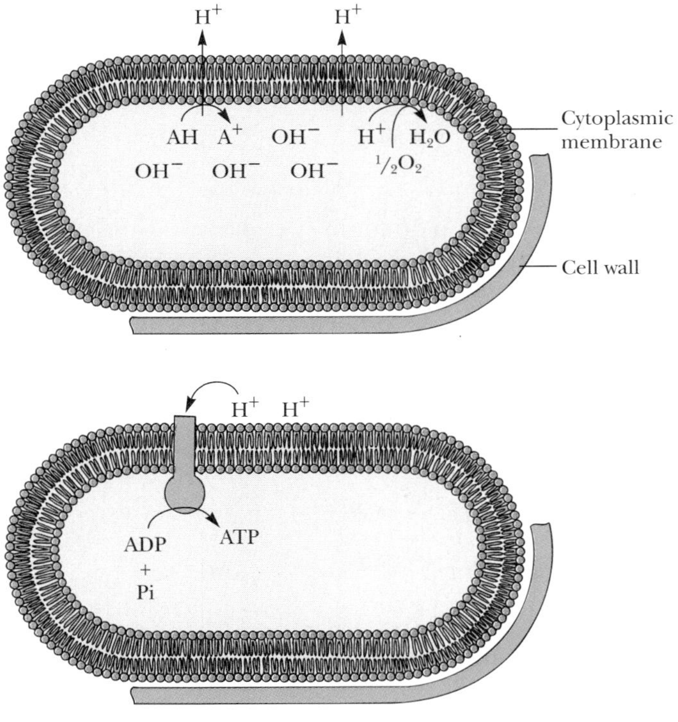

***Figure* 8.1** Establishing a charge separation across a cytoplasmic membrane. The concentration of protons (H^+) on the external surface and negative ions (OH^-) on the inside create a charge difference. This charge difference can be a source of energy to drive the synthesis of the energy-conserving compound adenosine triphosphate (ATP) and for other energy-consuming activities of the cell.

Proton accumulation on the exterior of the membrane creates a positive charge (cathode), and on the interior the charge is negative (anode). The energy inherent in the proton is captured when it enters a specific membrane channel and returns to the cytoplasm. In an aerobe the electrons involved in proton extrusion ultimately react with O_2 and protons to form H_2O (see Figure 8.1). These processes will be explained in detail later.

Eubacteria and Archaea use one or more of three general mechanisms to generate ATP: (1) respiration, (2) photophosphorylation, or (3) substrate-level phosphorylation. Respiration is the production of ATP by oxidation of reduced organic or inorganic compounds (electron donors) coupled with the reduction of inorganic or organic electron acceptors. An example of this would be the oxidation of hydrogen gas, $2H_2 + O_2 \longrightarrow 2H_2O$. This would occur in an autotrophic microorganism where hydrogen would be the reduced inorganic electron donor and O_2 would be the inorganic electron acceptor. Photophosphorylation occurs in photoautotrophic microorganisms that generate ATP by using light energy. Substrate-level phosphorylation is the synthesis of ATP by direct transfer of a high-energy phosphate from a phosphorylated organic compound to ADP to form ATP.

Aerobic oxidative phosphorylation utilizes O_2 as the terminal electron acceptor, whereas anaerobic respiration couples the electrons to other oxidized electron acceptors. Some of the compounds that can accept electrons in anaerobic respiration are NO_3^-, $SO_4^{=}$, CO_2, ferric iron, or fumarate. The products generated from these compounds would be NH_4^+, H_2S, CH_4, ferrous iron, or succinic acid, respectively.

Oxidation-Reduction Reactions

Oxidation-reduction (redox) reactions occur because electrons flow from a component of higher potential (donor)

BOX 8.1 METHODS & TECHNIQUES

Oxidation-Reduction

Redox reactions—Oxidation is the loss of electrons, and reduction is the gain of electrons. As electrons cannot exist in solution, an oxidation reaction must be coupled with a reduction. Oxidation releases energy, which can occur only if a substance is available that can be reduced. Biological reactions can involve the transfer of hydrogen atoms that consist of an electron and a proton. When the electron is removed from an atom of hydrogen one has a **proton** that will generally accompany the electron. For example:

$$2H_2 \longrightarrow 4H^+ + 4e^-$$

$$O_2 + 4H^+ + 4e^- \longrightarrow H_2O$$

Reduction potential—Is a quantitative measure of the tendency for a substance to give up electrons in biological systems. It is measured in volts and generally at pH 7, the pH of cytoplasm.

ATP—The energy generated by redox reactions is conserved in ATP, the major high-energy phosphate compound in living cells. ATP is an energy "carrier" formed during **exergonic reactions** (energy yielding), and it can drive **energonic reactions** (energy requiring). In ATP and some other phosphorylated compounds the outer two phosphate atoms are joined by an anhydride bond (see Figure 8.3). The free-energy ($\Delta G^{o\prime}$) that can be released when the high-energy bonds are hydrolyzed is − 31.8 kJ/mole. The low-energy bond of the AMP would yield only − 14.2 kJ/mole. The phosphate bond in phosphoenolpyruvate is high energy ($\Delta G^{o\prime} = -51.6$ kJ/mole) and the phosphate bond in glucose-6-phosphate is low energy ($\Delta G^{o\prime} = -13.8$ kJ/mole). The free energy available in phosphate bonds is of value to cells only when it can be coupled to a second reaction that requires energy.

In redox reactions electrons are transferred from one component to another and the acceptor is reduced. Reductions may result in the gain of one or more protons.

Cytochrome *c* undergoes $1e^-$ reduction:

$$Fe^{3+} - \text{cyt } c + 1e^- \longrightarrow Fe^{2+} - \text{cyt } c$$

NAD^+ accepts $2e^-$ and gains 1 proton

$$NAD^+ + 2e^- + H^+ \longrightarrow NADH$$

Ubiquinone accepts $2e^-$ and gains 2 protons

$$UQ + 2e + 2H^+ \longrightarrow UQH_2$$

Respiration—Is energy generation in which molecular oxygen or some other oxidant is the terminal electron acceptor. Among the latter are: nitrate, sulfate, carbon dioxide, and fumarate.

Fermentation—Is energy generation by oxidation/reduction reactions in which the substrate serves as both electron donor and electron acceptor.

Oxygenic and **anoxygenic photosynthesis**—The electron donor in **oxygenic photosynthesis** is H_2O and the overall reaction can be illustrated as follows:

$$H_2O \xrightarrow{\text{light}} 2H^+ + 2e^- + 1/2\ O_2$$

Water is the electron donor in cyanobacterial and plant type photosynthesis, and molecular oxygen is the product. Many photosynthetic bacteria are anaerobes and utilize electron donors other than H_2O. They do not produce O_2, and the process is **anoxygenic** photosynthesis. Examples of electron donors for anoxygenic photosynthesis would be:

$$H_2 \longrightarrow 2H^+ + 2e^-$$

$$H_2S \longrightarrow S = + 2H^+ + 2e^-$$

Photophosphorylation—Is formation of ATP by use of light energy.

to another of lower potential (acceptor). The donor is often termed the **reductant,** and the electron acceptor, the **oxidant.** One component in a redox reaction is oxidized and the other reduced. In chemistry we learned that oxidation is the loss of electrons and reduction the gain of electrons. The capacity for the occurrence of a redox reaction is defined as the **reduction potential** or $\mathbf{E_o}$ (see **Box 8.1**). The reduction potential of a compound is measured in volts and is arbitrarily based on the voltage required to remove an electron from H_2 under standard conditions. The hydrogen electrode has been assigned a standard value of 0.0 V (volts):

$$H_2 \longrightarrow 2H^+ + 2e^-$$

When this reaction occurs in a biological system under standard conditions, with reactants at a concentration of 1 *M* (molar) and at a pH of 7, the redox potential is actually −0.42 V. This reaction presented above would be termed a **half reaction** because electrons cannot exist alone in solution. For the reaction to occur there must be a corresponding reduction. Any half reaction must be coupled to another half reaction (a reduction) such as $1/2\ O_2 + 2e^- + 2H^+ \longrightarrow H_2O$. In a biological reaction, the amount of free energy released in a redox reaction is the difference in reduction potential between the donor and the acceptor of electrons. This energy difference is designated by $\Delta E_o'$. The term E_o' is the redox potential at pH 7 (neutrality). Thus, the free energy available from the reduction of oxygen by hydrogen is the product of two half reactions:

$$H_2 \longrightarrow 2\ H^+ + 2\ e^-\ (E_o' = -0.42\ V)$$

$$1/2\ O_2 + 2\ H^+ + 2\ e^- \longrightarrow H_2O\ (E_o' = +0.82\ V)$$

$$\text{Sum:}\quad H_2 + 1/2\ O_2 \longrightarrow H_2O$$

The total free energy change occurring in these two half reactions can be calculated by the following equation:

$$\Delta G^{o\prime} = (-nF)(\Delta E_h)$$

where $\Delta G^{o\prime}$ is the free energy under standard conditions, n is the number of electrons transferred, and F is the Faraday constant (−96.48 kJ/V).

$$\Delta E_h = E_o'\ (\text{oxidized}) - E'_o\ (\text{reduced})$$

Substituting in values for each factor gives the change in free energy as:

$$\Delta G^{o\prime} = (-2)(96.48)(-0.82 + 0.42)$$
$$= -239.27\ kJ$$

Often a redox reaction occurs in a biological system when the electron and accompanying proton are enzymatically removed from the substrate and passed to an electron **carrier** such as nicotinamide adenine dinucleotide **NAD⁺** (Figure 8.2). An electron carrier is an intermediate that **carries** an electron from a **donor** to an **acceptor.** Reduction of NAD^+ would yield $NADH + H^+$,

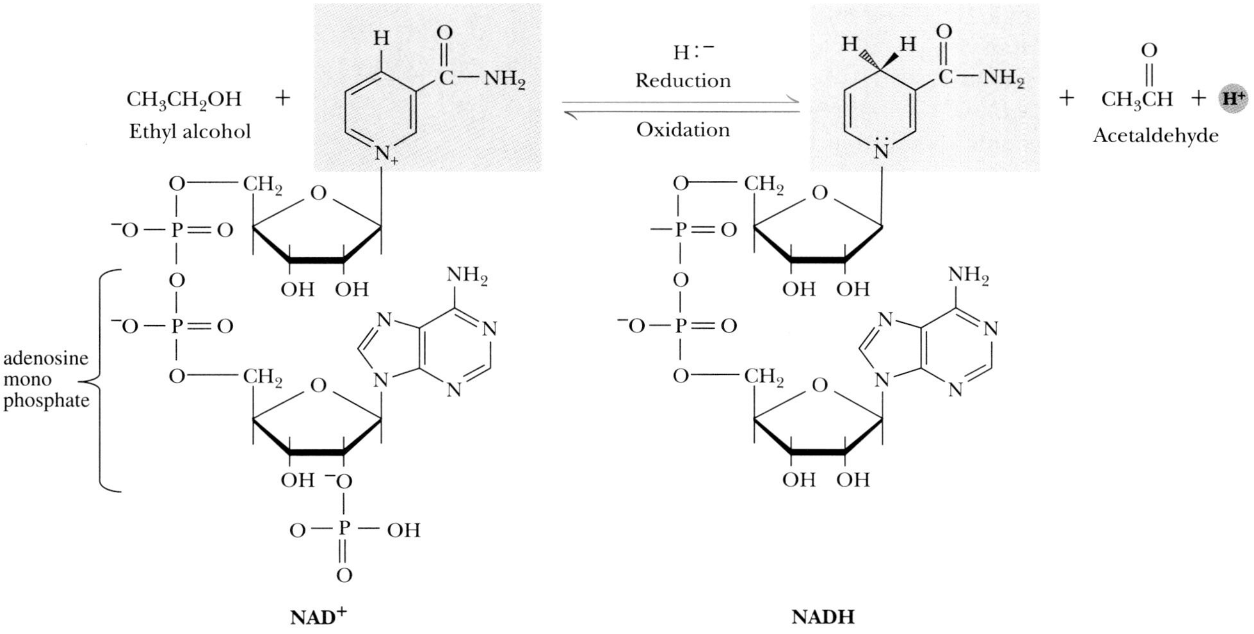

***Figure* 8.2** The structure of a major electron carrier nicotinamide adenine dinucleotide (NAD). When electrons are enzymatically removed from a donor (ethyl alcohol) they can be passed to NAD^+ to generate NAD · H. The structure of $NADP^+$ differs in the presence of a phosphate on the 2′hydroxyl of the adenosine moiety.

***Table* 8.1 Half reactions, the number of electrons transferred (*n*), and the electrode potential under standard conditions (E_o') compared to the hydrogen half cell**

Half Reaction	*n*	E_o' (V)
Ferredoxin (oxidized/reduced)	2	−0.43
$2\ H^+/H_2$	2	−0.42
$NAD(P)^+ + H^+/NAD(P)H + H^+$	2	−0.32
1,3-di-*P*-glycerate $+2H^+/$ glyceraldehyde-3-*P* + P_i	2	−0.29
Chlorophyll (P_{II})	1	−0.20
$FMN + 2H^+/FMNH_2$	2	−0.22
$FAD + 2\ H^+/FAD\ H_2$	2	−0.22
Standard half cell 2 H^+/H_2	***2***	***0.00***
Methylene blue (oxidized/reduced)	2	+0.01
Fumarate + $2H^+$/succinate	2	+0.03
Ubiquinone (oxidized/reduced)	2	+0.06
Cytochrome *b* (Fe^{3+}/Fe^{2+})	1	+0.08
Cytochrome *c* (Fe^{3+}/Fe^{2+})	1	+0.25
Chlorophyll (P_I)	1	+0.40
$NO_3^- + 2\ H^+/NO_2^- + H_2O$	2	+0.42
Fe^{3+}/Fe^{2+}	1	+0.77
$2\ H^+ + 1/2\ O_2/H_2O$	2	+0.82

which for brevity is written NADH (see **Box 8.1**). This carrier can pass the electrons derived from the donor to a series of electron carriers in a cascade of increasing E_o' (Table 8.1). The more negative E_o' half reactions tend to donate electrons and become oxidized as the more positive E_o' half reactions accept electrons and become reduced. A typical biological reaction would proceed as follows:

$$NAD^+ + 2\ H^+ + 2\ e^- \longrightarrow NADH\ (E_o' = -\ 0.32) + H^+$$

$$1/2\ O_2 + 2\ H^+ + 2e^- \longrightarrow H_2O\ (E_o' = +0.82)$$

$$\Delta G^{o\prime} = (-2)(96.48)(+0.114) = -219.97\ \text{kJ}$$

Examination of one of the half reactions shown in Table 8.1 illustrates the inefficiency of growth with a donor such as iron [Fe^{3+}/Fe^{2+} (0.77)] as the energy source and oxygen [$1/2\ O_2/H_2O(0.82)$] as the terminal electron acceptor:

$$\Delta G^{o\prime} = (-1)(96.48)(0.05) = -9.85\ \text{kJ}$$

This low value indicates that little energy can be obtained from the oxidation of one iron molecule when compared with the NADH $\longrightarrow O_2$ reaction. As previously mentioned, electrons do not exist in the free state in biological systems but are passed on to molecules that can accept them. The electron transport chains such as the cytochromes, in the membranes of microorganisms, are efficient in accepting electrons and then become donors as they pass the electrons along to an acceptor of lower redox potential. There are electron carriers that move freely in the cytoplasm and function as links between metabolic pathways and the membrane-bound electron transport chains. Those that move freely are termed **diffusible** versus the **fixed** electron carriers that are present in membranes. The major diffusible carriers are the aforementioned NAD^+, another is $NADP^+$ (NAD-phosphate), and others are flavin nucleotides, which will be discussed later.

The reduction potential for $NAD^+/NADH$ and $NADP^+/NADPH$ is equivalent at 0.32 V (see Table 8.1) and both are effective electron donors. NAD^+ is principally involved in energy-generating (catabolic) reactions, and NADP is involved mostly in biosynthesis (anabolism). NAD^+ is a **hydrogen atom** carrier and can transfer two hydrogens in metabolic reactions, such as **dehydrogenation:**

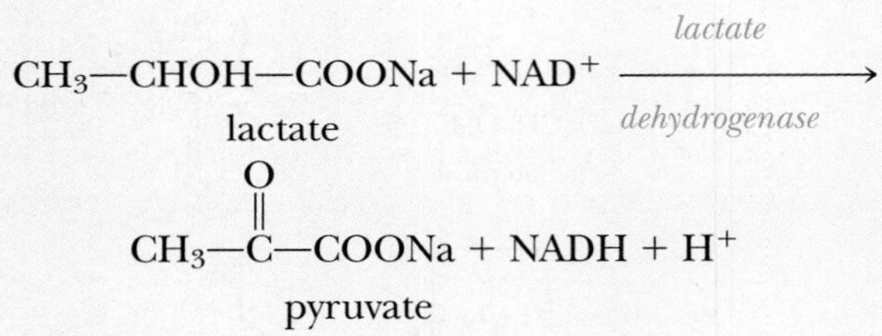

The reduction potential in NADH is delivered to the membrane-bound electron transport system, and the electron (e^-) moves through the respiratory chain and translocates protons to the **exterior** of the membrane. This creates a **charge separation** or an electrochemical potential, as mentioned previously. The proton traverses back across the membrane via a protein complex called the ATPase system, and this generates ATP. The ATPase system is a complex in the membrane that can undergo a proton-driven configuration change that covalently bonds a molecule of ADP and inorganic phosphate (P_i) to form ATP.

Energy Conservation in the Microbial World

According to the first law of thermodynamics, all forms of energy are convertible from one to another, but energy can be neither created nor destroyed. Chemical and light energy are conserved in cells via the formation of ATP. The energy available in ATP drives biosynthetic reactions and other energy-requiring processes in the living cell. All microorganisms are **dynamic systems** with steady inputs as well as outputs of both matter and energy. Every free-living microbe has the metabolic capacity to transform chemical and/or physical energy into biological energy. The major mechanisms that have evolved for ATP formation in microbial cells are redox reactions and photon-driven reactions. Redox reactions fall into two categories: **substrate-level phosphorylation** and **electron transport phosphorylation.** The mechanism whereby ATP is generated by these redox reactions is discussed here in some detail. The photon-driven mechanisms are covered in Chapter 9.

The structure of the nucleotide **adenosine triphosphate (ATP)** is presented in Figure 8.3. ATP is involved in the biosynthesis of cell constituents, transport, motility, polymerization reactions, and other activities of the cell. Two of the bonds in ATP are high-energy bonds (as noted in Figure 8.3). Hydrolysis of the high-energy phosphate bond of ATP can provide energy to various biosynthetic intermediates, rendering them more reactive. This is important in raising the free energy level of the intermediates thus activated so that they can participate in biosynthetic reactions.

In some biosynthetic reactions, other **nucleotides** that are high-energy intermediates are involved. For example, uridine triphosphate (UTP) participates in the synthesis of polysaccarides, cytidine triphosphate (CTP) is involved in lipid synthesis, and guanosine triphosphate (GTP) may be an activator in protein synthesis. These nucleotides are thermodynamically equivalent to ATP. The structures of these compounds are presented in Chapter 10.

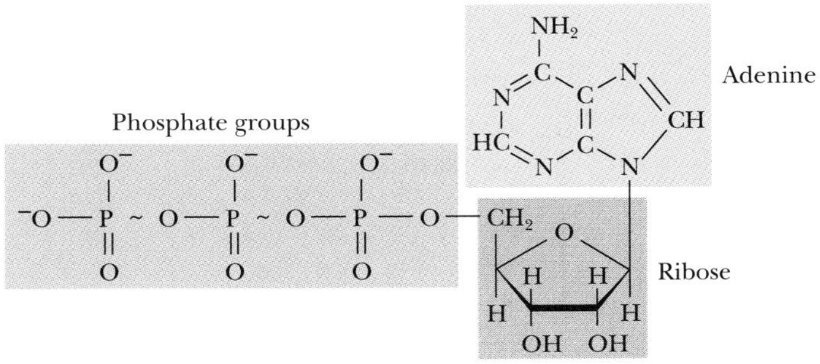

Figure **8.3** Structure of adenosine-5′-triphosphate (ATP). This nucleotide is a major carrier of chemical energy in living cells. The energy-rich phosphate bonds are designated (~).

Following is a discussion of the principal mechanisms whereby ATP is generated by both aerobic and anaerobic microorganisms.

Fermentation

Fermentation is the sum of anaerobic reactions that can provide energy for the growth of microorganisms in the absence of O_2. Fermentation is an ATP-generating metabolic process whereby one organic compound (the energy source) serves as a donor of electrons and another organic compound is the electron acceptor. The principal substrates for fermentation include carbohydrates, amino acids, purines, and pyrimidines. The most studied fermentation processes include alcoholic fermentation by yeasts, lactic acid fermentation in bacteria, and anaerobic dissimilation of amino acids by the clostridia. *Clostridium* is a genus of endospore forming strictly anaerobic bacteria. In amino acid dissimilation one amino acid is oxidized and a product (or products) of the dissimilation of a second amino acid provides an electron acceptor that permits the energy-generating reactions to continue. Fermentation is essentially a **closed system** in which the redox level of substrate and products are internally balanced. This is illustrated by the following:

$$\underset{\text{glucose}}{C_6H_{12}O_6} \longrightarrow \underset{\text{ethanol}}{2\ CH_3CH_2OH} + 2\ CO_2\ (6\ C + 12\ H + 6\ O)$$

Two examples of balanced fermentation processes that occur under strictly anaerobic conditions are **glycolysis** and the **Stickland reaction.** In both cases, ATP is generated by **substrate-level phosphorylation.** Glycolysis leads to the formation of lactic acid, whereas the Stickland reaction generates energy from redox reactions involving selected pairs of amino acids. These reactions are balanced in that the oxidation levels of the substrates and products are equivalent. The necessity for a balanced redox level in substrate and product limits the range of compounds that can serve as energy source for fermentation. Most such substrates are essentially neutral in the redox state. They cannot be highly oxidized or reduced and must provide intermediates that are sufficiently oxidized so that they can serve as electron acceptors. Examples of intermediates that can serve as electron acceptors would be pyruvate and acetaldehyde. Reduction of these compounds would produce lactic acid and ethanol. Substrate-level phosphoryla-

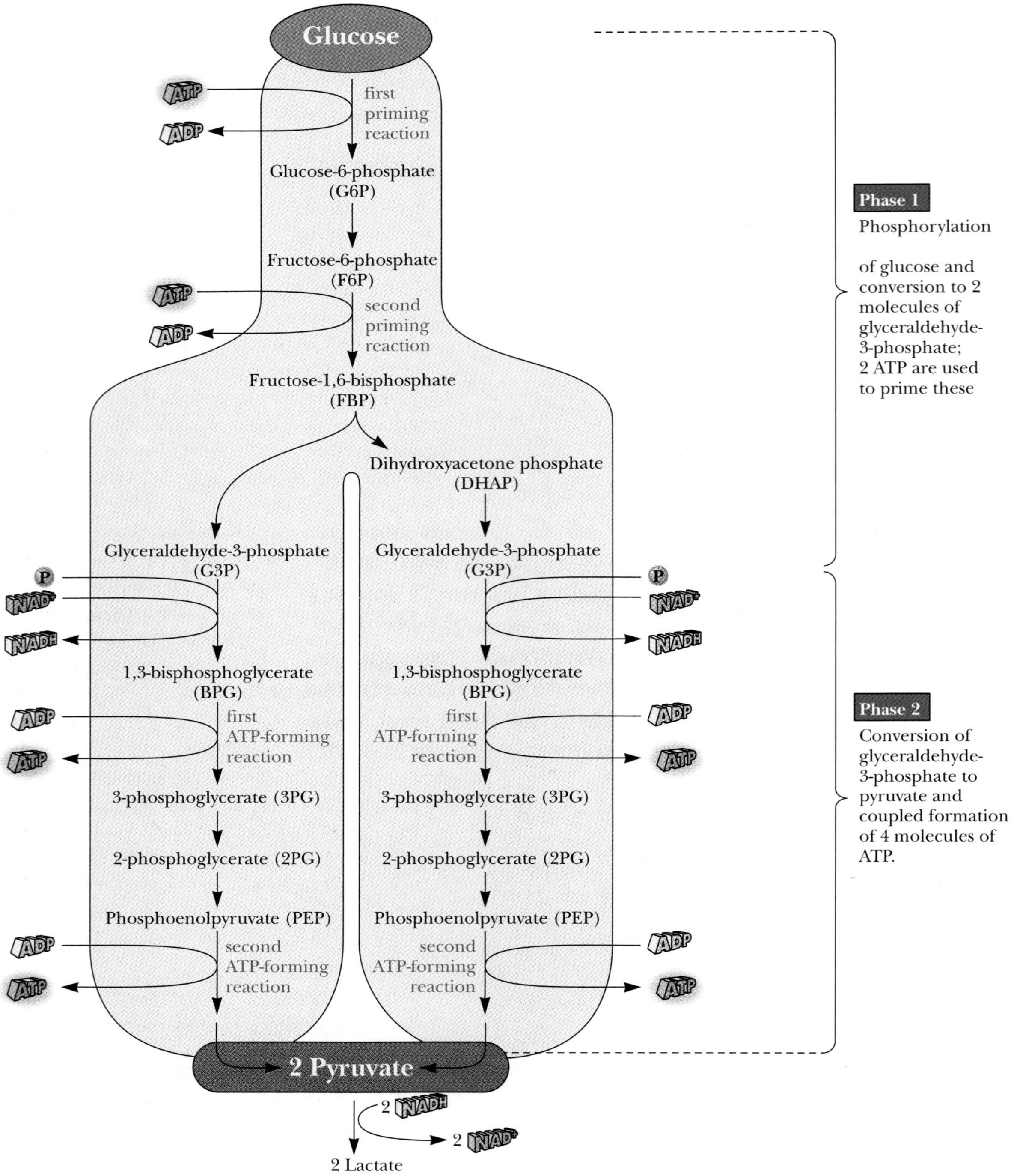

***Figure* 8.4** Glycolysis is basically an anaerobic closed system whereby a molecule of glucose can yield two molecules of lactate without involvement of molecular oxygen. There is a net yield of 2 ATP via substrate-level phosphorylation. There is no net gain or loss of electrons.

tion occurs when an **activated phosphoryl** group on a metabolic intermediate is transferred directly to ADP to form ATP. The enzymes involved in these phosphorylations are soluble and are not associated with cytoplasmic membranes; hence membrane proton gradients are not involved in this process.

One of the reactions in glycolysis that results in the stoichometric production of ATP is presented in Figure

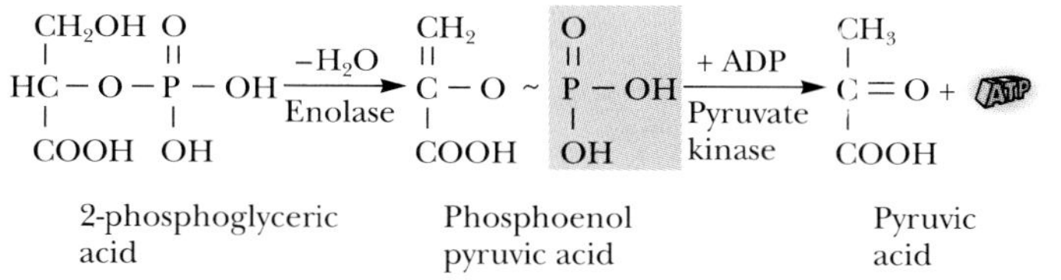

***Figure* 8.5** An example of substrate-level phosphorylation. Dehydration of 2-phosphoglyceric acid produces the anhydride with a high energy bond in phosphoenol pyruvic acid (~). This phosphate can be transferred to adenosine diphosphate to form ATP and a molecule of pyruvic acid.

8.4. One molecule of 2-phosphoglyceric acid is de ydrated by the enzyme enolase to form one molecule of phosphoenolpyruvic acid. The anhydride thus formed has a high-energy bond. The enzyme pyruvate kinase can couple this high-energy phosphate bond to one ADP to form a molecule of ATP and one pyruvic acid.

Glycolysis, also known as the Embden-Meyerhof pathway, or EMP (Figure 8.5), is the energy-producing system in many lactic acid bacteria and other microorganisms that utilize this sugar anaerobically. The pathway is also a major route of glucose catabolism in aerobic organisms. The lactic acid fermentation requires two molecules of ATP to form fructose 1,6-diphosphate, which is cleaved to two molecules of glyceraldehyde-3-phosphate. The oxidant NAD^+ in concert with glyceraldehyde-3-phosphate dehydrogenase generates a high free-energy intermediate, which undergoes a phosphate (P_i) substitution to form energy-rich 1,3-diphosphoglycerate. Further sequential reactions result in the production of four molecules of ATP from two molecules of 1,3-diphosphoglycerate. The end product of the reactions is pyruvate. The reduction of pyruvate to lactate permits the regeneration of the oxidized form of NAD^+. The net reaction is as follows:

$$\text{Glucose} + 2\ \text{ADP} + 2\ P_i \longrightarrow 2\ \text{Lactate} + 2\ \text{ATP}$$

The total amount of energy generated by glycolysis (2 ATP) is meager when compared with the 38 ATPs that can result from the complete aerobic oxidation of one molecule of glucose (see later).

The strictly anaerobic clostridia have evolved with an array of ATP-generating mechanisms that employ many different substrates. Among the substrates that can be fermented by various strains in the genus *Clostridium* are pyruvic acid, purines, pyrimidines, nicotinic acid, carbohydrates, and amino acids. Some of the metabolic capabilities of clostridia will be discussed in Chapter 32. One example of energy metabolism that occurs in various species of clostridia is presented in Figure 8.6. This is the aforementioned classic Stickland reaction, whereby one amino acid of a pair is oxidized and the other reduced. Dissimilation by this mechanism allows organisms growing in anaerobic protein-rich environments to catabolize many of the amino acid constituents of protein as a source of energy. The net result of a typical Stickland reaction is:

$$2\ \text{Glycine} + \text{Alanine} + \text{ADP} + P_i \longrightarrow 3\ \text{Acetate} + 3\ NH_3^+ + \text{ATP}$$

As in glycolysis, an internal oxidant (NAD^+) generates an activated high-energy intermediate (acetyl coenzyme A). A substrate-level phosphorylation reaction results in the uptake of one inorganic phosphate to generate acetyl phosphate with the concomitant release of coenzyme A. The high-energy phosphoryl group can then be transferred to ADP to form ATP. The structure of acetyl CoA and the location of the high-energy bond is presented in Figure 8.7. Other energy-rich compounds that can be involved in substrate-level phosphorylation are propionyl-CoA, butyryl-CoA, and butyrylphosphate.

There are a number of distinct fermentation pathways that bacterial species living under anaerobic conditions can utilize to generate energy. Generally these fermentations are categorized by the nature of the products formed. Some of the fermentations that involve glucose and the major products generated are listed in Table 8.2. There may be small amounts of other compounds produced during sugar fermentations such as alcohols, organic acids, CO_2, and molecular hydrogen (H_2). Other fermentations not involving sugars are listed in Table 8.3. One necessity is a **fermentation balance,** meaning that the oxidation/reduction state of the products is equivalent to that of the substrate.

Energy Generation via Respiration

Aerobic and anaerobic **respiration** are characterized by the transfer of reducing equivalents from a donor, such as NADH, succinate, and methanol, to a terminal electron acceptor. This transfer of electrons occurs via sequential redox reactions involving highly organized redox carriers present in the cellular membranes of all organisms that generate ATP by respiration. The passage of electrons through an electron transport chain results in the translocation of protons to the exterior of the membrane, as previously mentioned. Evidence now available affirms that ATP synthesis, during respiration, is driven by protons that accumulate on the exterior of the cytoplasmic membrane. The transfer of reducing potential from substrate to the membrane electron transport system is generally through NADH (**Box 8.2**). Considerable variation exists in the constituents of the electron transport system used by various

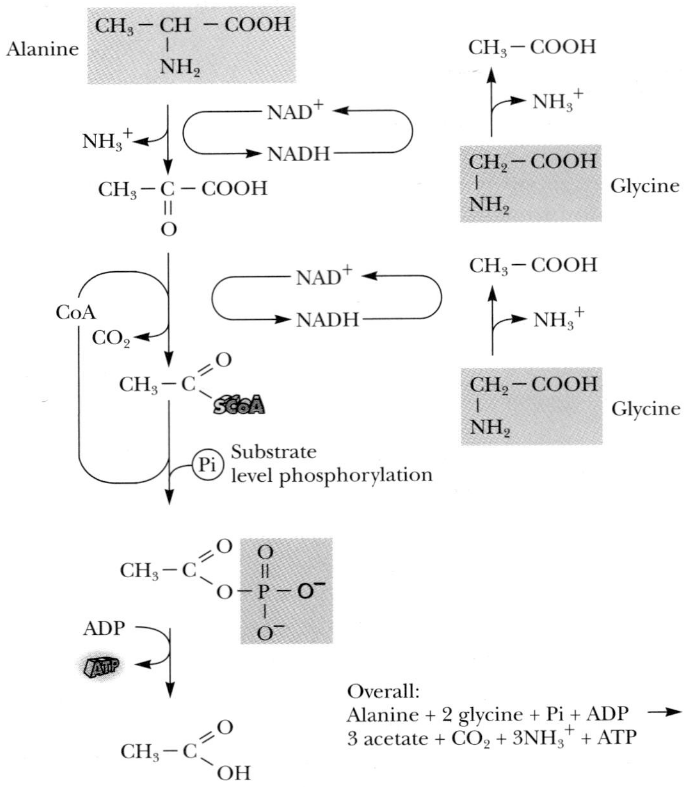

***Figure* 8.6** The Stickland reaction whereby pairs of amino acids can be utilized to generate ATP by substrate level phosphorylation. Alanine (left series of reactions) is oxidized and the two glycine molecules are reduced. The Stickland reaction occurs in anaerobic organisms such as *Clostridium sporogenes*.

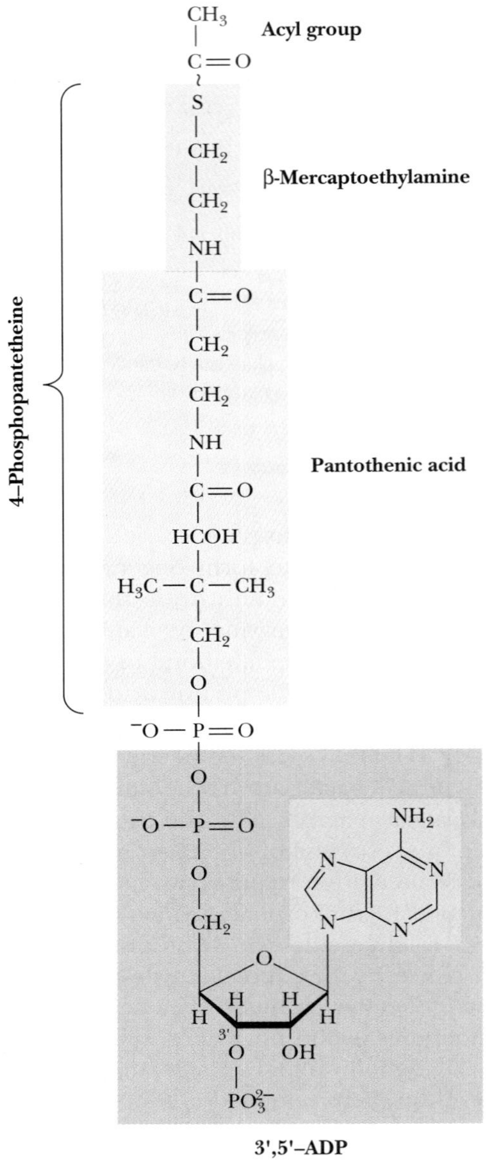

***Figure* 8.7** The structure of acetyl coenzyme A (CoA). The bond between β-mercaptoethylamine and acetic acid is energy-rich.

Eubacteria and Archaea. The total ATP yield will vary according to the efficiency of the system present. Many facultatively anaerobic and obligately anaerobic chemoheterotrophs (see Chapter 5) can replace oxygen with alternate electron acceptors to derive energy from an electrochemical gradient generated by electron transport. Among the electron acceptors that can be used during aer-

***Table* 8.2 Examples of products generated during fermentation of glucose and the organism involved**

Type	Product	Organism
Mixed acid	ethanol + acetate + lactate	*Escherichia coli*
Butanediol (neutral)	2,3-butanediol + ethanol	*Enterobacter aerogenes*
Alcoholic	ethanol	*Zymomonas mobilis*
Homolactic	lactate	*Lactobacillus acidophilus*
Heterolactic	lactate + ethanol	*Lactobacillus brevis*
Butanol/ acetone	acetone + butanol	*Clostridium butyricum*

Table 8.3 **Bacterial fermentations that utilize substrates other than sugars**

Type		
Homoacetic acid	$4H_2 + CO_2 \longrightarrow CH_3COOH$	*Clostridium aceticum*
Propionic acid	Lactate $\longrightarrow$ propionate + acetate	*Clostridium propionicum*
Acetylene	Acetylene + $H_2O \longrightarrow$ ethanol + acetate	*Pelobacter acetylenicus*
Oxalate	Oxalate $\longrightarrow$ formate + CO_2	*Oxalobacter formigenes*
Malonate	Malonate $\longrightarrow$ acetate + CO_2	*Malomonas rubra*

obic and anaerobic respiration are those listed in Table 8.4. The following section will present some of the major mechanisms for generating ATP via electron transport.

Energy-Transducing Membranes

The cytoplasmic membrane of respiring and photosynthetic prokaryotic cells, the inner membrane of the mitochondrion, and the thylakoid membrane of a chloroplast are all **energy-transducing membranes.** The nature of a typical membrane that can transduce energy is illustrated in Figure 8.8(**a**). This depicts a representative series of carriers that might be present in the membrane of a respiring bacterium. This series of reactions may also be illustrated as in 8.8(**b**). The orientation and carriers present in an electron transport chain may differ. They are dependent on the nature of the microorganism, the primary energy source, and growth conditions. The chemical

BOX 8.2 METHODS & TECHNIQUES

Efficiency in Viable Systems

The efficiency of oxidative phosphorylation can be as follows:

For the complete oxidation of 1 mole of NAD · H via the electron transport chain with oxygen as terminal electron acceptor.

$$\Delta G^{o\prime} = -nF\Delta E_h$$

n = number of electrons

F = Faraday constant

ΔE_h = redox potential difference between two half reactions at a defined pH

$NAD/NADH + H^+$ -0.32

$2H^+ + 1/2\ O_2/H_2O$ 0.82

$$\Delta G^{o\prime} = -2\ (96{,}500)(0.82 - (-0.32))$$
$$= -220\ \text{kJ/mole}$$

(theoretical)

The $\Delta G^{o\prime}$ for ATP hydrolysis = -30.5 kJ/mole

If 3 moles of ATP are synthesized/mol of NADH:

$$3 \times (-30.5) = 91.5\ \text{kJ (actual)}$$

Efficiency would be:

$$\frac{91.5}{220} = 42\%\ \text{of the energy is conserved as ATP}$$

Table 8.4 **Electron acceptors used by Eubacteria or Archaea during aerobic and anaerobic respiration. Some organisms that utilize these acceptors are presented.**

	Terminal Electron Acceptor	Product	
Aerobic			
	O_2	H_2O	*Micrococcus luteus*
Anaerobic			
	SO_4^{2-}	H_2S	*Desulfovibrio desulfuricans*
	Fe^{2+}	Fe^{3+}	*Geobacter metallireducens*
	Mn^{4+}	Mn^{2+}	*Desulfuromonas acetoxidans*
	NO_3^-	NO_2^-	*Escherichia coli*
	NO_3^-	N_2	*Thiobacillus denitrificans*
	CO_2	CH_4	*Methanosarcina barkeri*
	CO_2	CH_3COO^-	*Clostridium aceticum*
	S^0	H_2S	*Desulfuromonas acetoxidans*
	Fumarate	Succinate	*Wolinella succinogens*

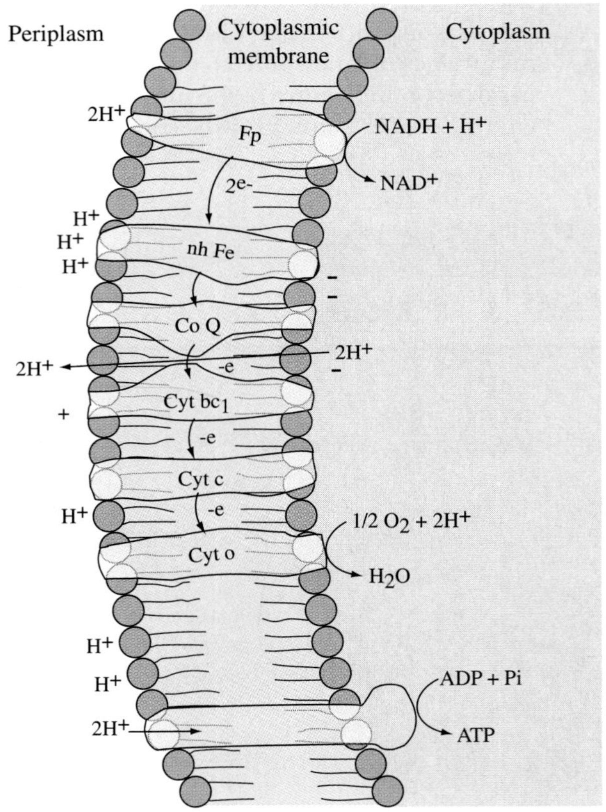

(a)

Fp	Flavoprotein
nhFe	Non-heme iron
Co Q	Coenzyme Q
Cyt bc_1	Cytochrome complex
Cyt c	Cytochrome c
Cyt o	Cytochrome o

Figure **8.8** A model **(a)** of an electron transport chain that translocates protons to the external surface of the cytoplasmic membrane. From NADH the energized electrons flow down a cascade of carriers, each step proceeding to a lower energy state and translocating protons outward. Ultimately the electrons couple with O_2 to form H_2O. **(b)** A shorthand way of presenting the components that might participate in an electron transport chain.

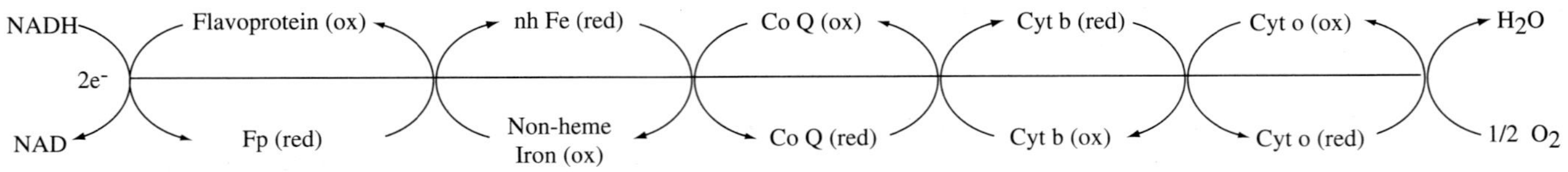

(b)

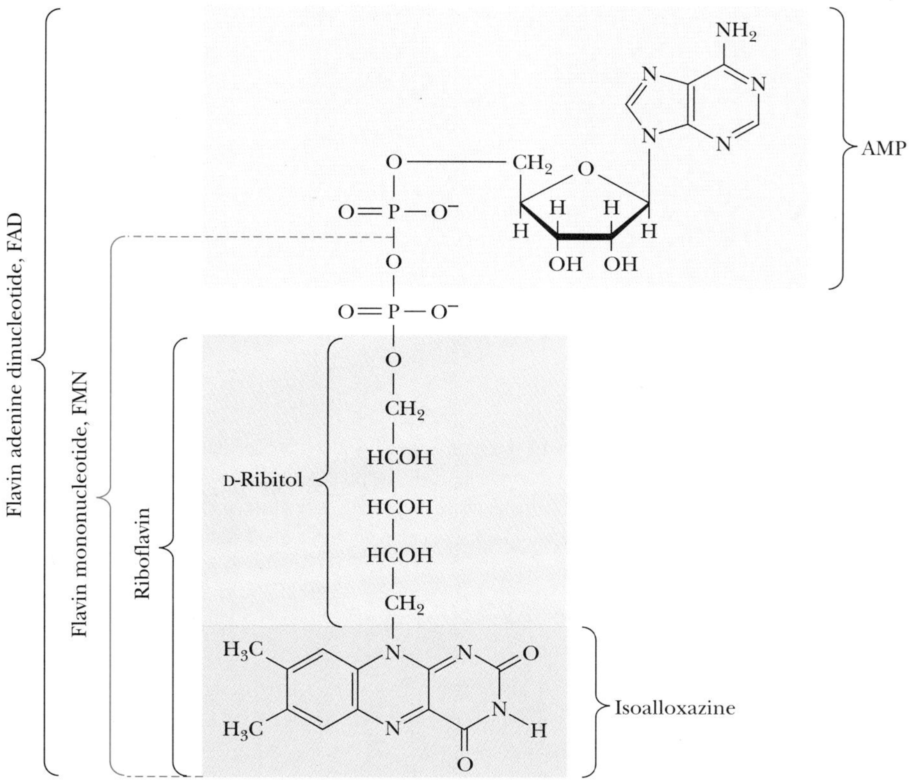

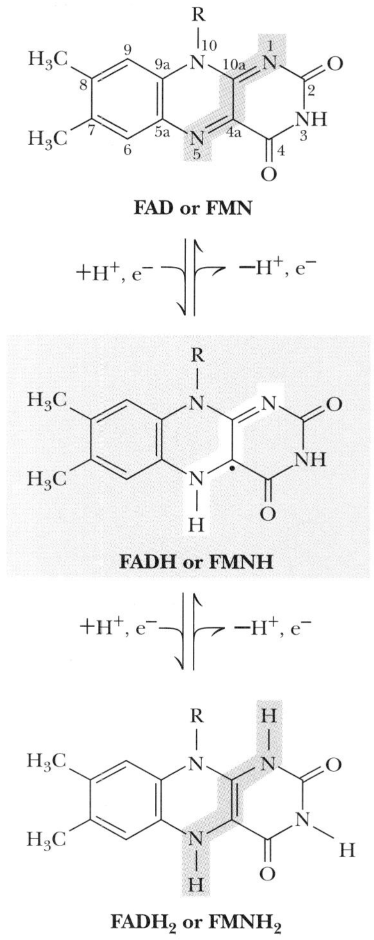

***Figure* 8.9** Structure of the flavins that are the functional part of the flavoproteins. These electron carriers are derivatives of the B-vitamin riboflavin.

reactions that establish the proton gradient are effected by enzymes that span the width of the cytoplasmic membrane. Among the common features of energy-transducing membranes are two distinct protein assemblies: (1) an ATP synthetase that catalyzes the "uphill" synthesis of ATP from ADP + P_i, and (2) a respiratory chain (as illustrated) that catalyzes the "downhill" transfer of electrons to a terminal acceptor such as O_2. Energy can also be generated by absorption of light (plant chloroplasts or photosynthetic bacteria), which drives electrons across the membrane to establish the electrochemical potential.

Generation of the Proton Motive Force

The proton motive force in respiring microorganisms is generated by the concerted actions of electron carriers (Figure 8.8**a**). We will consider variations that may occur in the types of carriers and their orientation later. In the transport system shown (Figure 8.8**b**), NADH donates two hydrogen atoms and two electrons to the flavoprotein (Figure 8.9). The reduced flavoprotein donates two electrons to the nonheme iron-sulfur protein (Figure 8.10) complex

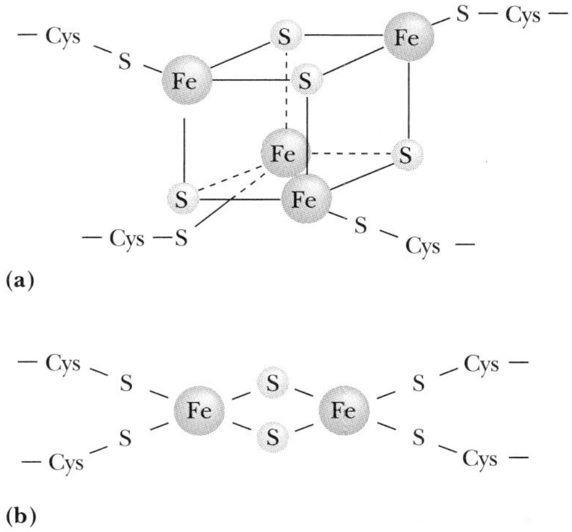

***Figure* 8.10** The arrangement of iron and sulfur in the nonheme iron-sulfur proteins, **(a)** Fe_4S_4 and **(b)** Fe_2S_2. These carriers are attached to a protein through the amino acid cysteine.

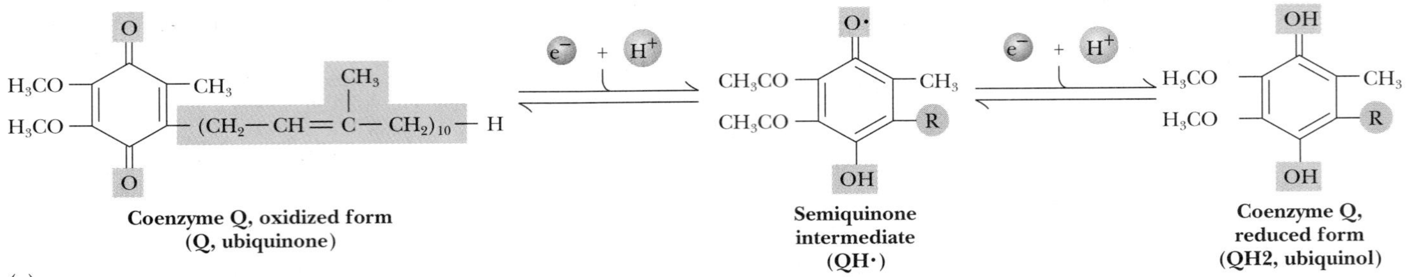

(a)

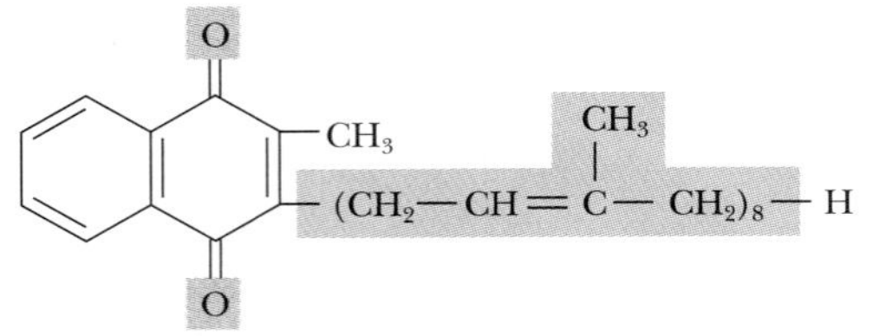

(b)

***Figure* 8.11** Coenzyme Q **(a)** is a quinone (ubiquinone) that can accept $2e^-$ and $2H^+$ to form ubiquinol. The isoprenoid side chain can vary in number of units. In microorganisms it generally varies from 6 to 10. Menaquinone (MQ) **(b)** is present in some microorganisms, particularly gram-positives.

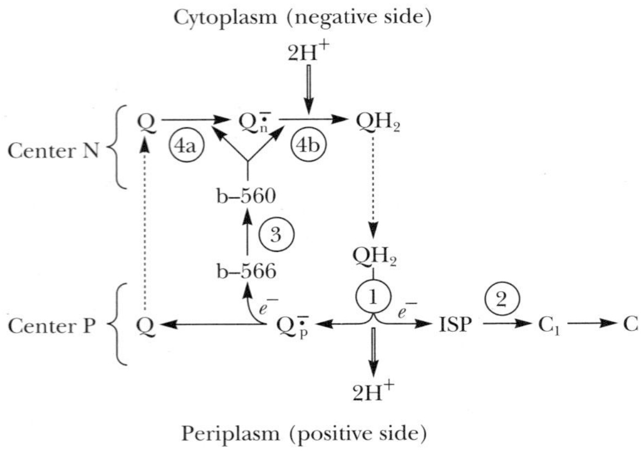

***Figure* 8.12** The "Q cycle." In this drawing Q is ubiquinone, $Q\cdot_p^-$ and $Q\cdot_n^-$ are the semiquinone, and QH_2 is ubiquinol (see Figure 8.11). Cytochrome b_{560} and cytochrome b_{566} are transmembrane proteins, and the two hemes form an electrical circuit across the membrane. Cyt b_{566} has a more negative potential and can transfer electrons to b_{560}. ISP is a 2Fe-2S iron sulfur cluster, and cyt c and cyt e are near the positive surface of the membrane. In the first half of a Q cycle a QH_2 is oxidized and 2 protons (H^+) are released at the positive surface at center P. The two electrons removed at center P diverge and follow separate paths (1). One is passed to ISP and on to cyt c_1 and cyt e (2); the other electron is transferred from $Q\cdot_p-$ to the cyt bc_1 complex (3); from b_{560} heme the electron passes on and reduces Q to $Q\cdot_n-$ (4a). In the second half of a Q cycle a second molecule of QH_2 is oxidized, transferring one electron to ISP and on to cyt c_1. A $Q\cdot_p-$ reduces b_{566} and an additional $2H^+$ is released to the outer membrane surface. These reactions are repetitive of (1), (2), (3). The heme of b_{560} reduces $Q\cdot_n-$ formed in (4a) to QH_2 (4b), taking in $2H^+$ from the inner side of the membrane. The two-step reduction of a Q at the negative side of the membrane occurs in center N. During one complete Q cycle, two QH_2 molecules are oxidized to Q and one Q is reduced to QH_2. b_{566} and b_{560} are reduced and reoxidized 2X. b_{560} reduces Q to $U\cdot_n-$ (semiquinone) during the first half of the cycle (4a) and reduces $U\cdot_n-$ to QH_2 (4b) during the second half.

Summing both halves: one QH_2 is oxidized to Q, two cyt c are reduced; two protons are consumed on the N side of the membrane and 4 protons are deposited on the P side. (Reproduced with permission from Annual Review of Biochemistry Volume 63, 1994 by Annual Reviews, Inc.)

with two protons (H^+) extruded to the periplasm. Two electrons are passed to the quinone coenzyme Q (Figure 8.11) and two protons are taken up from the cytoplasm in the reduction of CoQ. The "Q cycle" then becomes operative and transfers protons to the periplasmic (positive) side of the cytoplasmic membrane. The functioning of the Q cycle is now quite well understood, and an explanation of the transfer of protons via this cycle is presented in Figure 8.12.

Components of Respiratory Chains

There were a number of components shown in the respiratory chain that was presented in Figure 8.8. These components participate in a cascade of reactions that carry the electron from a highly activated state to the ground state where the electron reacts with the terminal acceptor. We now have a reasonable understanding of the structure/function of component links in many of the respiratory chains known to be present in bacteria, and some of these are discussed here.

Flavoprotein

Flavoproteins are molecules consisting of a protein of varied molecular mass bound to a molecule of the B-vitamin riboflavin. Flavin mononucleotide (FMN) and flavin adenine dinucleotide (FAD) were presented in Figure 8.9. The **flavins** are two electron (proton) carriers, as illustrated in the figure. FAD is generally covalently bound to the protein through a histidine group, whereas FMN is bound by its negatively charged phosphate groups. FAD and FMN have a relatively high redox potential (0.003 to 0.09) when bound to protein but have a redox potential of −0.22 V as coenzymes. The NAD-NADH couple

(−0.32 V) could transfer electrons to either the FAD/$FADH_2$ or $FMNH_2$ couple as electrons flow freely from a more negative carrier to a more positive one.

Iron-Sulfur Proteins

Iron-sulfur proteins are relatively small and contain equimolar amounts of iron and sulfur. There can be two, four, or eight atoms of each per molecule or protein. There are 2Fe/2S and 4Fe/4S iron-sulfur complexes presented in Figure 8.10, and the redox center in the 4Fe/4S complex forms a cube as shown. These centers are held in position by the amino acid cysteine. The redox potential for these membrane-bound carriers spans a broad range from −0.60 V to +0.35 V. Consequently, they can function at different stages of electron transport depending on the microbial system involved. A common iron-sulfur protein in biological systems is **ferredoxin,** which has an Fe_2S_2 configuration (see Figure 8.10).

Quinones

The terms **coenzyme Q, CoQ,** or **Q** are used to denote a family of **quinones** that are of widespread occurrence in energy-transducing membranes. Among these are the **ubiquinones,** which are the principal quinones in membranes of eukaryotes and gram-negative bacteria. Other bacteria (generally gram-positives) contain **menaquinone.** The structures of ubiquinone and menaquinone are presented in Figure 8.11. NADH can funnel electrons into the respiratory chain via specific flavoprotein/NADH dehydrogenases. Other dehydrogenases pass electrons directly to alternate flavoproteins, and these electrons enter the respiratory chain through CoQ.

Cytochrome

Cytochromes are constituents of the respiratory chain in aerobic and facultatively anaerobic microorganisms. Cytochromes are also present in some of the Archaea. Cytochromes are a heterogeneous group of compounds that are located primarily in the cellular membrane and are characterized on the basis of their spectrophotometric properties. The functional portion of a cytochrome molecule is an iron-porphyrin electron transport component, called **heme** (Figure 8.13). The heme moiety is attached to a protein. The central iron atom in the porphyrin nucleus accepts a single electron on reduction. Those cytochromes (cyt) with similar absorption characteristics are designated by lower-case letters as cyt *a*, cyt *b*, cyt *c*, and so forth. Within each of these groups there are measurable differences in spectral and biochemical properties. These are indicated by numerical subscripts (for example, cyt a_1, cyt a_2, and cyt a_3 or, in some cases, a cytochrome is identified by the wavelength at which it absorbs maximally). A spectrum of oxidized and reduced cytochrome is shown in Figure 8.14. The peaks in the reduced spectrum around 550 nm and 520 nm are called the α (alpha) and β (beta) bands.

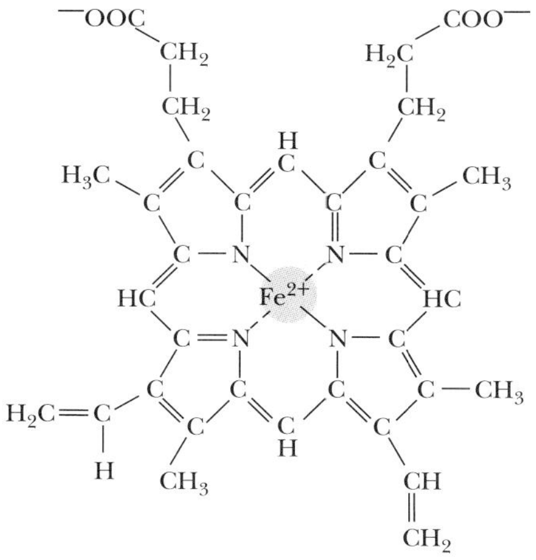

***Figure* 8.13** The heme portion of a cytochrome molecule. The tetrapyrrole is also called a porphorin. The heme is attached to a protein through the ethylene groups at the bottom of the molecule.

The redox potential of the various cytochromes ranges from −100 mV to 400 mV, their long wavelength absorption maximums are from 550 nm to 650 nm, and their molecular masses are between 12,000 and 350,000. The cytochromes located between CoQ and O_2 transport a single electron per event, but the reduction of oxygen to two molecules of water requires four electrons ($O_2 + 4\ e^- + 4\ H^+ \longrightarrow 2\ H_2O$). The cytochromes involved in the transfer of electrons to O_2 are called the cytochrome oxidases. The cytochrome oxidase in most organisms is either cytochrome a_3 or cytochrome *o*.

Much about the electron transport chain and respiratory ATP synthesis resulted from studies with chemical compounds that interfere with one or both of these processes. There are two classes of chemicals that affect

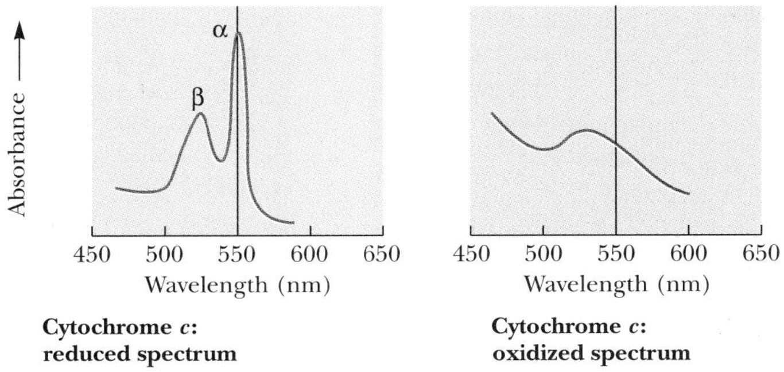

***Figure* 8.14** The absorption spectrum of reduced and oxidized cytochrome *c*. The peak at 550 nm is called α and the one at about 520 nm is β.

electron transport that were described by early workers—**uncouplers** and **inhibitors.** Uncouplers prevent the synthesis of ATP but do not interfere with electron transport. An inhibitor is a chemical that blocks both electron transport and ATP synthesis. Among the effective uncouplers are 2,4 dinitrophenol (DNP) and carbonyl cyanide-p-trifluoromethoxyphenylhydrazone (FCCP). Uncouplers permit electron transport to proceed unimpeded, but it cannot be coupled to ATP synthesis (uncoupled). This "uncoupling" occurs because DNP and FCCP are lipophilic (lipid-affinity) and can readily pass through a cytoplasmic membrane. They are also acidic and can bind protons on the outside of the membrane and carry them to the inside, thus dissipating the proton gradient. Inhibitors are compounds such as cyanide, carbon monoxide, or azide that bind to the cytochromes and block electron transport. All three of these compounds bond with the iron in heme, preventing electron flow.

Orientation in Electron Transport Chains

The orientation of carriers in the electron transport chains present in bacterial cytoplasmic membranes, in the inner membrane of eukaryotic mitochondria, and in the membranes of photosynthetic bacteria share a number of common properties. There is, however, much greater diversity in the components of the electron transport chains in the bacteria. Among the major differences between the transport chains in bacteria and eukaryotes: (1) bacteria have a greater variety of carriers, (2) many bacterial species respond to changes in growth conditions by altering their respiratory chain, and (3) the bacterial respiratory chains can be branched with alternate electron transfer pathways for different electron acceptors. This is illustrated in Figures 8.15 and 8.16, where different respiratory pathways are present depending on whether NO_3^- or O_2 serves as terminal electron acceptor. During aerobic growth, the bacterium *Paracoccus denitrificans* has a respiratory system remarkably similar to that of the mitochondrion. When *P. denitrificans* is grown anaerobically with either NO_3^- or NO_2^- as the terminal electron acceptor, there is a distinctly different electron transport chain present in the cytoplasmic membrane. The proposed scheme for electron transfer during the reduction of nitrate to gaseous N_2 is shown in Figure 8.16. The following reactions result in the conservation of energy by the accompanying synthesis of ATP:

$$NO_3^- \longrightarrow NO_2^- \longrightarrow 1/2\ N_2O \longrightarrow 1/2\ N_2$$

The electron carriers present in some organisms may be equivalent during aerobic or anaerobic growth, although synthesis of some cytochromes, such as cytochrome oxidase, are oxygen dependent.

Diversity often observed in respiratory chain components reflects the ability of bacteria to catabolize many different substrates. By using these various carriers, bacteria can extract available energy from substrates or markedly different redox potential. Electrons may flow into the respiratory chain at intermediate sites, as shown in Figure 8.17. The primary dehydrogenases such as succinate dehydrogenase can transfer electrons to CoQ at a different site than that utilized by electrons from NADH.

The orientation of the cytochrome system may be branched or linear, or, in some cases, composed of a very limited number of cytochromes depending on substrate or O_2 availability. For example, *Escherichia coli* has a linear terminal pathway during growth at high oxygen levels and a branched system when oxygen availability is limited (see Figure 8.17).

The aerobic iron-oxidizing bacterium ***Thiobacillus ferrooxidans*** has an abbreviated electron transport chain (Figure 8.18) and is an example of a bacterium that utilizes its environment to advantage in energy generation. *T. ferrooxidans* utilizes iron as energy source in environ-

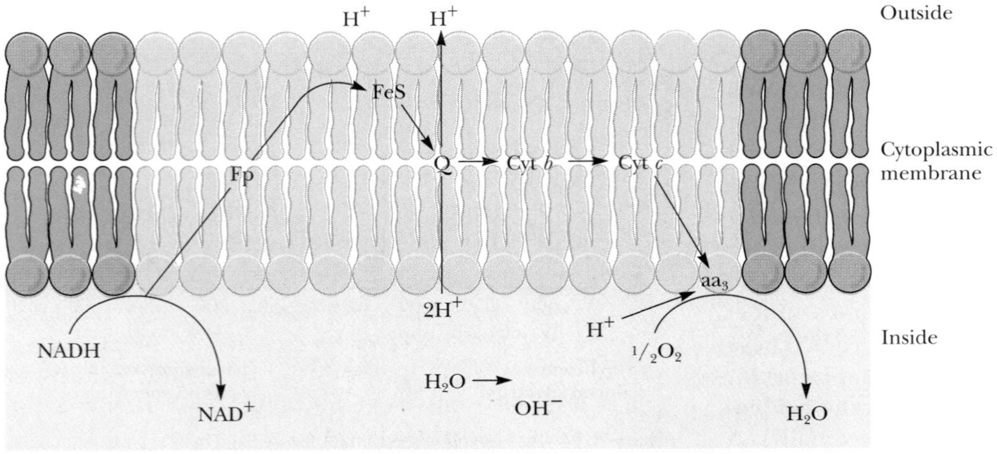

***Figure* 8.15** The aerobic electron transport chain in *Paracoccus denitrificans.*

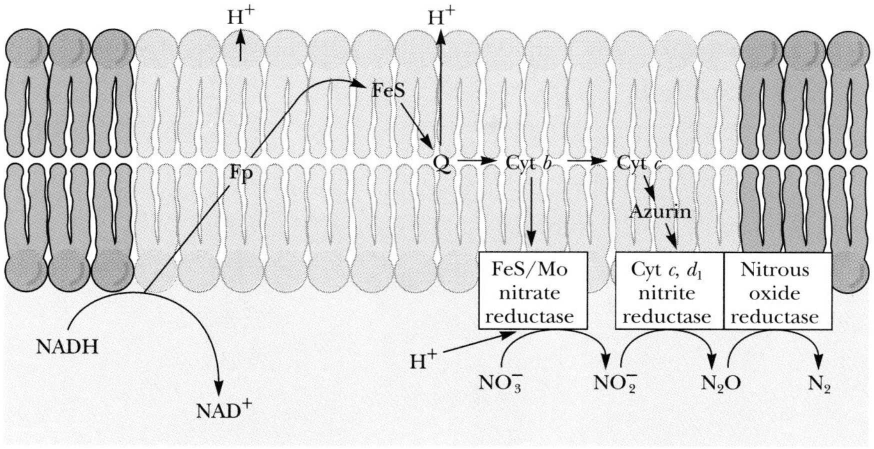

***Figure* 8.16** The electron transport chain in *Paracoccus denitrificans* during anaerobic respiration with NO_3^- as terminal electron acceptor. This microorganism can completely reduce nitrate to gaseous N_2. Azurin is a copper-containing electron transport protein.

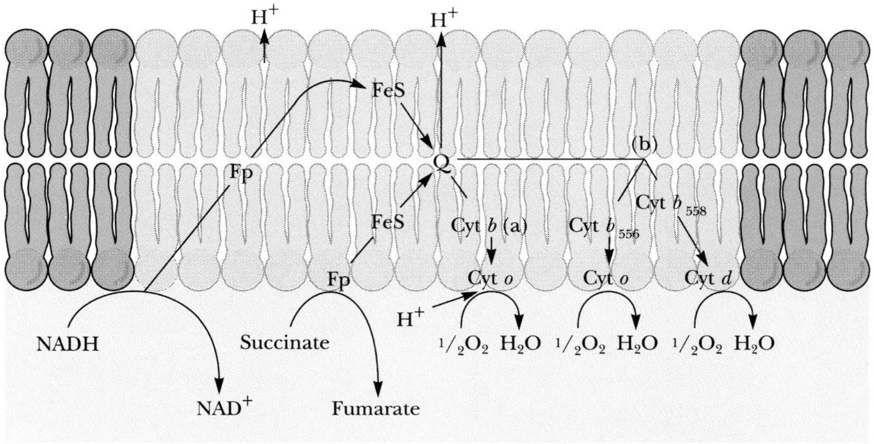

***Figure* 8.17** The electron transport chain in *Escherichia coli* following growth under highly aerobic vs. limited oxygen conditions. The high oxygen level **(a)** results in a linear chain, whereas the low oxygen level induces a branched chain **(b).** Bacteria may alter the constituents of their respiratory chain to generate the maximum amount of energy under the existing physical conditions.

ments where the external pH is about 2. The pH in the cytoplasm of the microorganism growing under these conditions would be 6 to 6.5. At pH 2 iron would be available as soluble Fe^{2+} and at the acidic pH there would be an abundance of protons (H^+) in the area surrounding the outer membrane surface. The copper-containing protein rusticyanin (see Figure 8.18) removes electrons from Fe^{2+} producing insoluble Fe^{3+}. These electrons flow from rusticyanin to cyt *c* and inward to cyt a_1 on the inner surface of the cytoplasmic membrane. The electrons are coupled to $1/2\ O_2$ with the concomitant uptake of two protons (H^+) from the cytoplasm. Uptake of protons internally by this series of reactions would result in a proton gradient across the membrane (H^+ outside OH^- inside). Protons

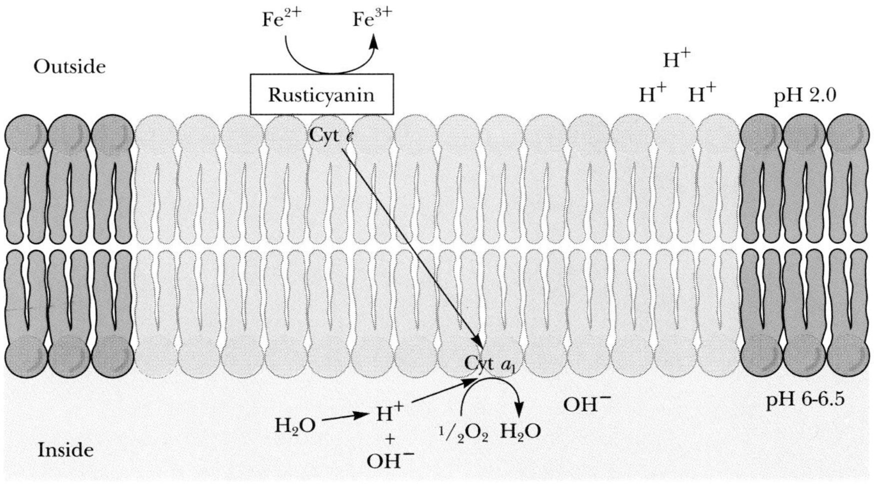

***Figure* 8.18** The electron transport chain in *Thiobacillus ferrooxidans* growing with Fe^{2+} an energy source. The major function of the chain is to pump electrons inward, and these electrons react with O_2, removing protons (H^+) from the cytoplasm. As protons are depleted inside, protons from the outer member surface are drawn through the ATPase to generate ATP.

BOX 8.3 MILESTONES

Evolution of a Concept: Chemiosmosis

Peter Mitchell was born in Surrey, England, in 1920. He attended Cambridge University, where he completed his Ph.D. in 1951. Mitchell remained at Cambridge until 1955, when he moved to the University of Edinburgh. There he directed the chemical biology unit for the next eight years. In 1961, Peter Mitchell, with his able assistant Jennifer Moyle, published a landmark paper that revolutionized thinking on cell energetics. This was the chemiosmotic hypothesis and was an outgrowth of his belief that enzymatic reactions in solution could differ from those that might occur when enzymes are embedded in a membrane. Mitchell believed that one could have a chemical reaction that could result in translocation of a chemical group. This directional (vectorial) event could cause the movement of a proton from the inner surface of the membrane to the outer surface. He affirmed that this would create an electrochemical proton gradient ($\Delta\mu H^+$) across the membrane. The relatively stable $\Delta\mu H^+$ was the central energy currency in photosynthesis respiration and respiration. The exterior of the membrane would be positively charged and the interior would be rendered negative. The protons would be drawn toward the interior and pass through a proton-translocating ATPase down the electrochemical gradient. This passage would drive ATP synthesis. This ion current would be the driving force for performance of chemical, osmotic, and, in the case of flagellar movement, mechanical work.

The scientific establishment working in the field of bioenergetics considered Mitchell's hypothesis bizarre and untenable. The central dogma of the time was that a high-energy intermediate composed of respiratory chain enzymes and an ATP-synthetase must exist. Intermediates were sought and considered but failed the ultimate test because no such factor existed. There was no "coupling" factor between respiratory chain enzymes and the ATPase system.

In 1963, ill health led to Mitchell's retirement to a small laboratory in Cornwall. He rebuilt an eighteenth-century derelict Cornish mansion called Glynn House and there he reestablished his research program. To the restful calm environs of the Glynn Research Laboratory he invited his detractors and gradually convinced all of them that the chemiosmotic hypothesis was indeed correct. In 1978 he received the Nobel Prize, and few doubted the wisdom of his theories.

Glynn House stands as an anomaly and a viable alternative to the huge collaborative efforts that receive funds from centralized funding sources. Is truly original research accomplished better in the hurly-burly of vast enterprises? Can original research always thrive best with centralized funding based on previous results? These are interesting questions in the age of big science, and Glynn House offers mute testimony that bigger is not necessarily better.

Peter Mitchell, The Nobel Laureate who proposed the chemiosmotic theory for generation of ATP. (Courtesy of *Times Union*)

Glynn House in Cornwall, England, where Peter Mitchell established his research laboratory. The Glynn Research Foundation, Ltd., now occupies Glynn House. (Courtesy of The Glynn Research Foundation)

would be drawn from the acidic outer environment inward, driving the ATPase reaction and generating ATP. The role of the abbreviated electron transport chain in *T. ferrooxidans* is to pump electrons inward, which in turn removes the protons from the cytoplasm and promotes ATP synthesis. This is a remarkable example of the versatility in the prokaryotic world.

Chemiosmotic Theory of Energy Conservation

The mechanism by which the potential energy in cytoplasmic electron carriers such as NADH can be conserved in the synthesis of ATP was an intriguing question. For decades scientists exhaustively sought (without success) an "energy-transducing intermediate" that would be involved in the following reaction:

$$\text{ADP} + \text{P}_i \longrightarrow \text{("Activated Intermediate")} \longrightarrow \text{ATP}$$

The "intermediate" is no longer of much interest because evidence now available precludes the existence of such a compound. About 20 years ago, Peter Mitchell, an English scientist, put forth the **chemiosmotic hypothesis,** which proposed that the driving force for the "uphill" synthesis of ATP is a proton gradient that can be established across a cellular membrane. The chemiosmotic theory has been the subject of much debate and trial, but it is now widely recognized as the only feasible explanation for the data available. For his contributions, Mitchell received the 1978 Nobel Prize in Chemistry (see **Box 8.3**).

The basic tenet of the chemiosmotic theory is that the electron transport chains in the respiratory membranes of mitochondria, chloroplasts, and bacteria are integrated with ATP synthesis through the establishment of the proton gradient. There are two significant events involved: (1) as discussed above, protons are translocated across the cellular energy-transducing membrane creating a proton electrochemical potential and (2) the proton potential between the inner and outer surfaces drives the ATP-hydrolyzing proton pump (ATPase) backward toward ATP synthesis. There are two distinct protein assemblies in the energy-transducing membrane:

1. The aforementioned respiratory chain catalyzes the transfer of electrons from substrates such as NADH or succinate to a terminal electron acceptor such as O_2, or, as in photosynthetic organisms, light energy drives electrons in one direction to establish a proton potential across the membrane.
2. ATPase catalyzes the synthesis of ATP from ADP and P_i.

The chemiosmotic theory is dependent on the concept of a **vectorial** (directional) **translocation.** The major role of the electron transport proteins in the respiratory chain is a **passive** one. They hold the carrier prosthetic groups that define the vectorial pathways, which are spatially oriented to promote unidirectional processes. The respiratory chain components participate in a vectorial charge separation.

During this reaction series the electron ultimately flows to the inner surface of the membrane (see Figure 8.8), where it reacts with O_2 or other terminal acceptors. The overall reaction $4e^- + O_2 + 4H^+ \longrightarrow 2H_2O$ extracts protons that are available from the cytoplasm. These protons may originate from dissociated water ($H_2O \longrightarrow H^+ + OH^-$), and removal of the protons results in excess OH^- or greater alkalinity on the inner surface of the membrane.

The energy available from an electrochemical gradient is determined by the relative pH (concentration of H^+) on the opposing side of the membrane and is called the **electrochemical potential** or **proton motive force.** This is designated as $\Delta\mu H^+$, where μ represents the electrochemical potential. The cytoplasmic membrane of bacteria and the membrane of mitochondria are impermeable to both protons (H^+) and hydroxyl (OH^-) ions and have little electrical conductivity. The proton motive force has two components: the pH gradient (ΔpH) and the electrical potential ($\Delta\psi$) between two phases separated by a membrane. This proton motive force drives the synthesis of ATP. The number of ATP molecules generated per transfer of two electrons from NADH through the respiratory chain is three in microbes having cytochrome *c* and two in those lacking this component. The coupling of the $\Delta\mu H^+$ to the production of ATP uses a second major protein complex in energy-transducing membranes, and that is the ATPase system. The function of the ATPase system is discussed next.

ATPase System

The ATPase complex is illustrated in Figure 8.19. **ATPase** is present in the energy-transducing membranes of mitochondria, chloroplasts, aerobic, and photosynthetic bacteria. It is also present in Archaea and anaerobic bacteria that do not rely on substrate-level phosphorylation. The structure of the complex is remarkably similar regardless of the source. Staining of energy-transducing membranes with a negative stain (such as phosphotungstate) reveals that knobs or mushroom-like bodies protrude from one side of the membrane (Figure 8.20). In Eubacteria, Archaea, and mitochondria, they protrude into the internal space (matrix), and in isolated thylakoid or chromatophore membranes, they project outward. Regardless of direction, these projections are functionally equivalent

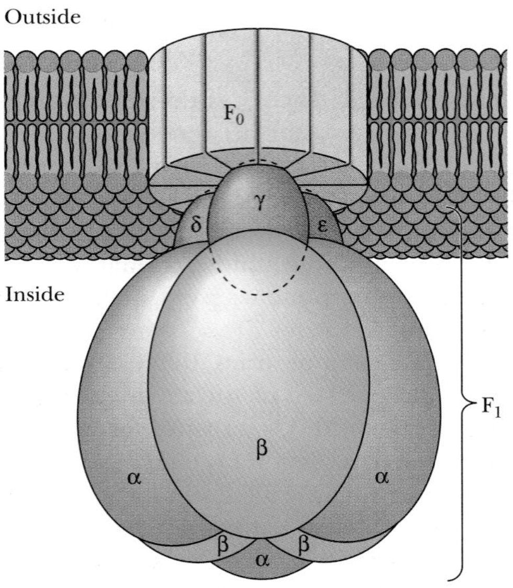

Figure **8.19** The subunit composition of ATPase. The F_0 polypeptides are integrated into the cytoplasmic membrane and channel protons to F_1. F_1 appears as knobs on the inner surface of the cytoplasmic membrane under high-resolution electron microscopy. The F_1 proteins are involved in the synthesis of ATP from ADP and inorganic phosphate.

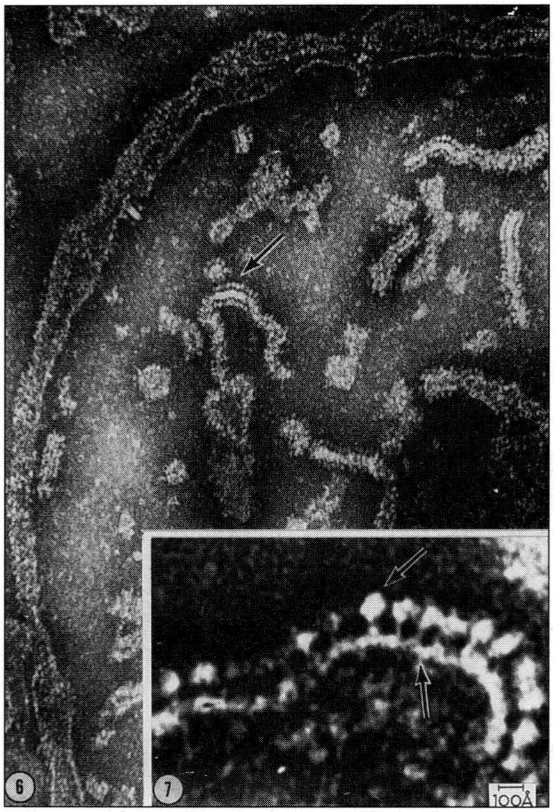

Figure **8.20** An electron micrograph of mitochondrial membrane. The round bodies (arrow) are the F_1 part of ATPase. Equivalent structures are present on the inner side of cytoplasmic membranes that have ATPase. (© Omikron/Photo Researchers)

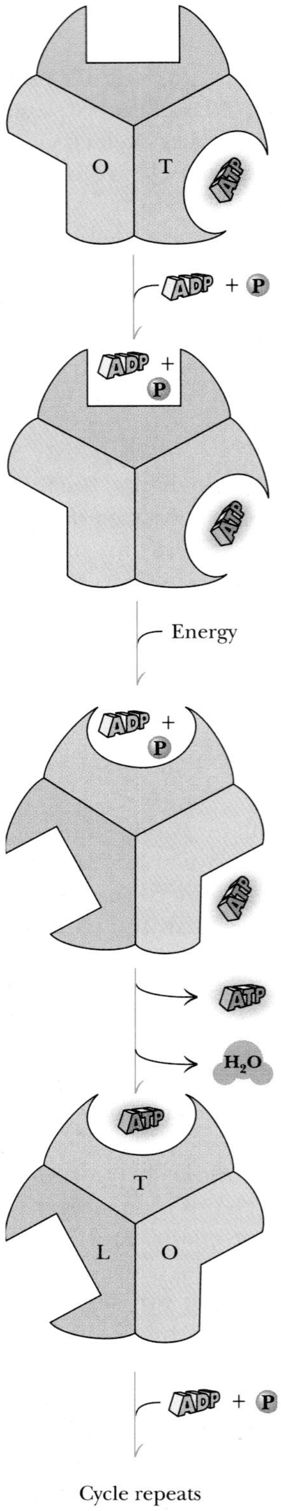

Figure **8.21** The binding change mechanism for ATP synthesis in F_1 of ATPase. This model assumes that F_1 has three interacting and conformationally distinct active sites. The open (O) has a low affinity for ligands and is inactive; the L conformation (loose affinity) takes up ADP and P_i. The tight (T) has a high affinity and effects the coupling of ADP + P_i to form ATP via a conformational change. The T site becomes an O and releases ATP. The L (with ADP + P_i) becomes a T and the cycle continues O $\longrightarrow$ L $\longrightarrow$ T $\longrightarrow$ O, etc.

because ATP is synthesized (or hydrolyzed) on the side from which the knobs project, while protons enter from the side lacking knobs.

The function of the ATPase is to utilize the $\Delta\mu H^+$ to maintain a mass-action ratio **away from equilibrium** and toward the synthesis of ATP. Remember that ATPase is an ATP-hydrolyzing enzyme. In fermentative bacteria, the complex utilizes ATP to maintain a $\Delta\mu H^+$ for active transport. Many organisms use electron donors that have a redox potential more positive than NAD^+ or $NADP^+$ (see Table 8.1). These bacteria cannot produce NADH or NADPH directly and need a source of reduced pyridine nucleotide for synthetic reactions. If they are CO_2-fixing autotrophs, they must have NADH for the reduction of 3-phosphoglycerate (they reverse glycolysis to synthesize sugars). ATP can drive a reverse $\Delta\mu H^+$ generating system that generates reduced pyridine nucleotides. There are in reality two forms of energy utilized by the bacterial cell—one ATP and the other the proton motive force. These energy forms are interconvertible by the membrane-bound F_1F_0 ATPase. The proton motive force is utilized for many purposes in the bacterial cell including: nutrient transport, membrane biogenesis, protein export, to turn flagella, and to maintain a favorable intracellular pH.

The knob on energy-transducing membranes is a series of proteins designated $\mathbf{F_1}$. This is the site of ATP synthesis and can be readily separated from the membrane (see Figure 8.19). The F_1 is composed of at least five polypeptides, and the subunit composition is quite similar regardless of the source. Isolated F_1 can catalyze an exceedingly rapid hydrolysis of ATP: One mole of F_1 can hydrolyze 10^4 moles of ATP per minute. The reverse reaction has not been observed *in vitro* (the proton motive force is required to drive the ADP + $P_i \longrightarrow$ ATP reaction). The remainder of the ATPase complex, termed $\mathbf{F_0}$, is a protein that is buried in the membrane. This protein complex is responsible for the passage of protons from the exterior of the membrane toward F_1. F_0 is a very hydrophobic ("water fearing") molecule. Intact F_0-F_1 complexes have been isolated from several bacteria by treating energy-transducing membranes with detergents.

Experimental evidence suggests that α and β subunits form the active site of F_1. Combinations of α and β from *E. coli* have considerable ATPase activity. The γ-peptide promotes the assembly of α and β into a functional unit and may serve as a gate to control the passage of H^+ to and from the active site. The major role of δ and ϵ subunits is to facilitate the binding of F_1 to F_0. There is no discernible enzymatic activity in F_0. It functions mainly to channel H^+ ion through the membrane.

The mechanism by which ATPase utilizes the proton motive force to change the equilibrium of the ATPase reaction toward ATP synthesis is not fully understood. As previously mentioned, there is no evidence that an energy-rich phosphorylated intermediate is involved. Several mechanisms have been proposed for the esterification of ADP to form ATP. One model proposes that the proton gradient induces a conformational change in the polypeptides of F_1, resulting in the synthesis of ATP by altering the relative affinities of F_1 for the substrate and the product (Figure 8.21).

In the **binding change mechanism** model it is assumed that F_1 has three conformationally different active sites. The O conformation has little affinity for ligands (small molecules that bind), L (loose) has a weak affinity for ligands, and the T (tight) has a strong affinity for ligands. Synthesis of ATP is initiated when ATP and P_i bind at an L site. Energy changes the conformation of L to a T where the esterification of P_i to ADP occurs. The T site conformation changes to O, releasing ATP, and the O site simultaneously becomes an L site.

Summary

- Virtually all energy available for biological systems is ultimately furnished by the sun.

- The universal medium of biological energy exchange is **adenosine triphosphate (ATP).** A major part of the ATP available to living organisms is generated by an **electrochemical proton gradient** across a membrane. Light from the sun can be employed to establish this gradient in photosynthetic bacteria and eukaryotes. As products of photosynthesis are the ultimate source of food for heterotrophic life, this proton gradient provides energy for viable systems.

- There are three major mechanisms for **ATP generation:** (1) respiration, (2) photophosphorylation, and (3) substrate-level phosphorylation.

- **Redox** (oxidation-reduction) reactions require an **electron donor** and an **electron acceptor.** Oxidation is the loss of electrons, and reduction is a gain of electrons. In biological redox reactions, one molecule is oxidized and another reduced. The amount of **free energy** available in a redox reaction is the difference in **reduction potential** (E_o') between the donor and acceptor of electrons. The reduction potential is the inherent ability to act as a donor or acceptor of electrons.

- In biological systems there are **electron carriers** that carry electrons enzymatically removed from a substrate to an acceptor. **Nicotinamide adenine dinucleotide** (NAD^+) is soluble and moves freely in cytoplasm. NAD^+ is a major electron carrier in cells.

- Electron carriers may also be **fixed** in membranes, and these carry electrons and accompanying protons across the cytoplasmic membrane to establish a proton gradient. **Cytochromes, iron-sulfur complexes,** and **coenzyme Q** are fixed electron carriers.

- A **fermentation** can be defined as an anaerobic reaction that provides energy when O_2 is unavailable. Fermentations are not involved with electron transport or proton gradients.

- **Substrate-level phosphorylation** is the direct formation of ATP from **adenosine diphosphate** and an activated phosphate group on a metabolic intermediate.

- **Glycolysis** is the major route of glucose catabolism in living cells. Anaerobic glycolysis is a **closed system** with products generated that are at the same redox level as glucose. For example, in the lactic acid fermentation:

$$\underset{\text{glucose}}{C_6H_{12}O_6} \longrightarrow \underset{\text{lactic acid}}{2\ C_3H_6O_3}$$

- **Respiration** occurs when reducing equivalents (e^-) from donors are passed to electron acceptors. During this process, electrons are passed from substrate to fixed electron transport chains in the membrane and these translocate protons to the exterior of the membrane. The ultimate acceptor of electrons carrying the protons outward would be oxidized molecules (O_2, NO_3^-, SO_4^{2-}, or fumarate).

- A cytoplasmic membrane that bears fixed electron carriers is called an **energy-transducing membrane.** The inner membrane of mitochondria, thylakoid membranes in chloroplasts, and the cytoplasmic membrane of respiring and photosynthetic bacteria are energy-transducing membranes.

- Among the components of respiratory chains are **flavoproteins, iron-sulfur proteins, quinones,** and **cytochromes.** There is much greater diversity in the components of prokaryotic respiratory chains than in those of eukaryotes.

- According to the **chemiosmotic theory** the proton gradient generated by electron transport is employed to drive ATPase, a series of transmembrane proteins that synthesize ATP from ADP and P_i.

- The energy available from a proton gradient depends on the pH (concentration of H^+) on the outer surface versus that on the inner surface of the membrane and is termed the **electrochemical potential.** This can occur because a cytoplasmic membrane is impermeable to protons, and protons can flow to the cytoplasm only through the ATPase system.

- The ATPase complex can function in either direction. When protons flow inward, ATP is formed. ATP can be **hydrolyzed** to ADP + P_i and establish a proton gradient.

- It is theorized that ATP is formed in the ATPase complex by binding site changes in constituent proteins of the F_1 part of ATPase. F_1 is a series of proteins that form knobs on the inner surface of bacterial membranes.

Questions for Thought and Review

1. What is the universal medium of energy exchange in viable cells? There are three major mechanisms for formation of this compound. How do they differ? What do they have in common?

2. What is a charge separation? What are terms we use to describe this charge separation?

3. A redox reaction occurs with a transfer of electrons. What are some components that might function in a redox reaction? Think in terms of fermentation and respiration.

4. If ferredoxin served as a half reaction in a redox reaction and cytochrome *c* as the other half, what would be the $\Delta G^{0\prime}$? What is a half reaction? Do half reactions actually occur in biological systems?

5. Give examples of electron carriers. How does an electron carrier function? How do diffusible and fixed electron carriers differ?

6. How do substrate-level phosphorylation and electron transport phosphorylation differ?

7. Give an example of a fermentation. What is the redox level of the substrate and products? Why? Cite examples of fermentation and how they are named.

8. What are some basic differences between fermentation and respiration? What do they have in common?

9. The electron transport chain in *Thiobacillus ferrooxidans* is unique. How does it differ functionally from that in *Escherichia coli*?

10. What is an energy-transducing membrane and why is it given this name?

11. How do electron transport chains in prokaryotes differ from those in eukaryotes? What do they have in common?

12. Outline the chemiosmotic theory.

13. Terminal electron acceptors in eukaryotes and prokaryotes differ in some respects. What are these differences and what makes the prokaryotes so unique?

14. The alternating catalytic site model has been discussed as one mechanism for ATP formation. What are the components and how do they function?

Suggested Readings

Battley, E. H. 1987. *Energetics of Microbial Growth.* New York: John Wiley and Sons.

Dawes, E. A. 1986. *Microbial Energetics.* Glasgow: Blackie and Sons, Ltd.

Harold, F. M. 1986. *The Vital Force: A Study of Bioenergetics.* New York: W. H. Freeman and Co.

Harris, D. A. 1995. *Bioenergetics at a Glance.* Cambridge, MA: Blackwell Science.

Schlegel, H. G., and B. Bowien, eds. 1989. *Autotrophic Bacteria.* New York: Springer-Verlag.

Shively, J. M., and L. L. Barton, eds. 1991. *Variations in Autotrophic Life.* New York: Academic Press.

Youvan, D. C., and F. Daldal. 1986. *Microbial Energy Transduction: Genetics, Structure, and Function of Membrane Proteins.* New York: Cold Spring Harbor Laboratory.

Life on planet Earth is completely dependent on the harvesting of the sun's light energy, and on the transmission and chemical storage of that energy by photosynthetic organisms and plants.

Peter Mitchell

Chapter 9

Conservation of Photochemically Generated Energy

Photosynthetic Organisms
Photosynthesis in Purple Nonsulfur Bacteria
Photosynthesis in Heliobacteria
Photosynthesis in Green Sulfur Bacteria
Photosynthesis in Cyanobacteria
Photosynthesis Without Chlorophyll

Photophosphorylation is the process whereby light energy is conserved as ATP. This process has much in common with oxidative phosphorylation (respiration), which was discussed in the previous chapter. The two processes are in a sense mirror images of one another. In oxidative phosphorylation, the H^+ on a reduced product of photosynthesis such as glucose is coupled with O_2 to generate energy, H_2O and CO_2. In photophosphorylation, light energy is used to extract electrons (and H^+) from H_2O (or certain other reduced substrates) producing O_2 or oxidized products. The light-activated electrons are employed to generate ATP and ultimately convert CO_2 to reduced plant or bacterial cell components. The reactions that generate ATP are considered **light reactions,** and those that convert CO_2 to organic matter are called **dark reactions.** (Light is required to drive the former but not the latter.) The importance of these processes is affirmed by the quote from Peter Mitchell. In this chapter we will discuss the diverse processes whereby light energy is conserved in photosynthetic bacteria.

Photosynthetic Organisms

Early Evolution

The evolution of complex living organisms on Earth was dependent on the development of a capacity for **biosynthesis.** Simple accretion of molecules present in the "primordial soup" would have yielded beings of very restricted capabilities. Only when sources of potential energy became available in the environment could the evolution of regenerative systems occur. The singular event that ultimately allowed for a systematic and orderly evolution of viable systems was the advent of the ability to capture the inexhaustible source of earthly energy—sunlight. Virtually all life on the surface of the earth relies directly or indirectly on photons from the sun for cellular energy. It is apparent that the orderly processes of life evolved from chaos when light-capturing chemicals evolved that coupled the energy of excitation to the production of relatively stable, "energy-rich" compounds.

The nature of primitive organisms on Earth is highly speculative, and many of their characteristics remain unknown. Examination of microorganisms now present in the environment and comparative analysis of component parts and metabolic functions lead to inevitable questions of origin and ancestry, but firm conclusions are elusive. Fermentative energy generation, akin to substrate-level phosphorylation used by present-day *Clostridium* species (discussed in Chapter 8), may have occurred in some of the earliest organisms. However, the abiotic synthesis (not produced by living organisms) of sugar under conditions that existed on the primeval Earth is questionable. Consequently, metabolism based on the fermentation of sugars is improbable.

The absence of oxidized compounds to serve as electron acceptors would have placed limitations on the effectiveness of an anaerobic proton gradient for ATP generation. Some evidence suggests that sulfate can be generated abiotically, and this compound may have served as a terminal electron acceptor for some early anaerobic microorganisms. One attractive idea is that a simple photosystem, such as that present in *Halobacterium,* could have occurred early in evolution. The generation of a proton gradient across a membrane with a simple photosensitive carotene-type compound (see later) may have been a forerunner of oxidative phosphorylation. The charge from an electrochemical gradient generated in this way could have been used to drive ATP synthesis.

A simplistic view of the energetics involved in life processes can be drawn as follows:

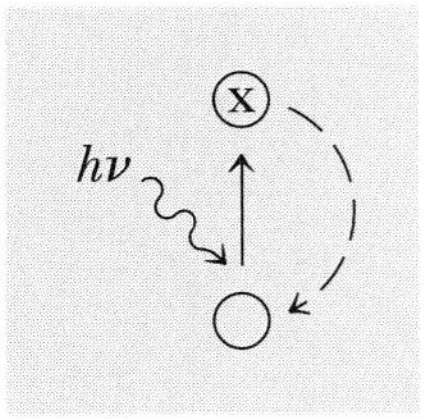

A quantum of energy from the sun, designated *hν*, drives an electron in a light-absorbing compound from the ground state (O) to a higher level of electronic excitation Ⓧ, illustrated by the vertical line. This excitation, *if untrapped,* would return to the ground state, indicated by the dashed line. Energy for "life" would be available if excitation-capturing molecules were interposed between the two circles, thereby conserving the quantum of light energy (or **exciton**) inherent in X. This can be depicted as follows:

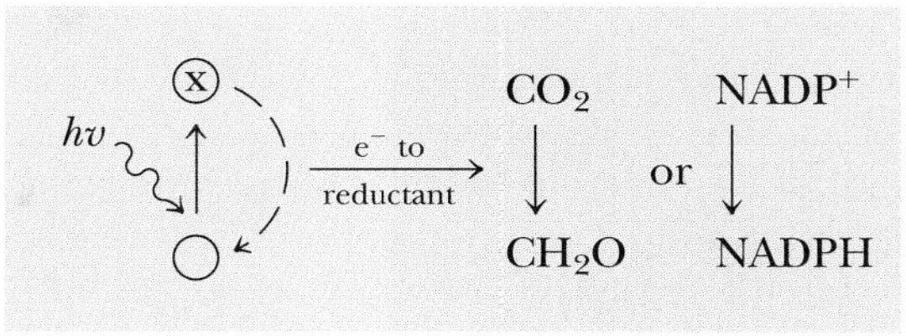

The driving force for all types of cellular processes both autotrophic and heterotrophic can be illustrated as components of this scheme:

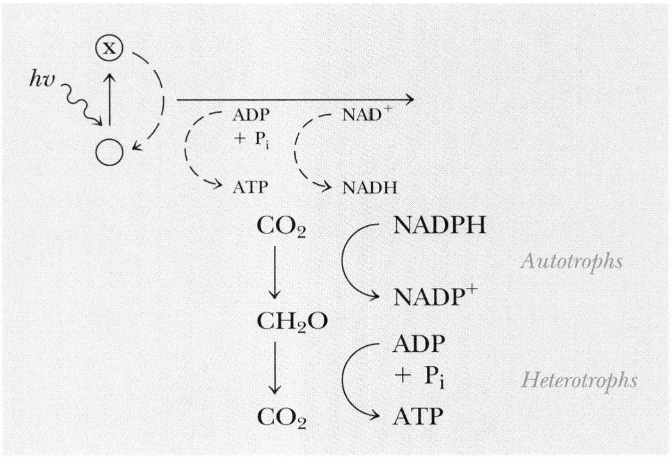

The reactions illustrated in this simple diagram present, in an abbreviated form, the reactions that support most life forms. Photophosphorylation systematically captures light energy in a chemically stable intermediate (ATP) and generates the reductant to sustain biosynthetic reactions. The reduced products serve as source of carbon and energy for heterotrophs **(Box 9.1)**.

Origin and Types of Photosynthetic Organisms

Because Eubacteria and Archaea existed on Earth for more than three billion years prior to the advent of eukaryotes it follows that light-energy generating systems originated with these microorganisms. There is speculation supported by some evidence that all primitive Eubacteria were photosynthetic. Three distinct photochemical energy-capturing systems exist in microorganisms. The first is an oxygen-generating photochemical event, the second is **anoxygenic** (that is, it does not produce oxygen), and the third uses a carotenoid compound structurally similar to animal rhodopsin (an eye pigment). The ability of green and purple photosynthetic bacteria to grow anaerobically (as well as other characteristics) suggests that they preceded the

BOX 9.1 MILESTONES

The Current of Life

A signal achievement in the evolution of viable microorganisms on Earth occurred with the creation of chemical systems that could capture light energy. The primordial soup contained compounds that could absorb light energy. Absorption of a quantum of light energy by a light-sensitive molecule could elevate a constituent electron to a higher energy level (M*) as shown in the diagram. The energy now inherent in M* has four possible fates:

I. The energy might be dissipated as **heat** as the energy is redistributed about the molecule.
II. The energy of excitation might reappear as emitted light or **fluorescence.** A photon of light is emitted as the electron returns to a lower orbital.
III. The energy of excitation may be transferred by **resonance energy transfer** to a neighboring molecule, raising an electron in the receptor molecule to a higher energy state as the photoexcited electron in the absorbing molecule returns to the ground state.
IV. The energy of excitation may change the reduction potential of the absorbing molecule to a level such that it can become an **electron donor.** Reacting this excited electron donor to a proper electron acceptor can lead to the transduction of light energy (photons) to chemical energy in the form of reducing power.

It is fate IV that permits the transduction of light energy into chemical energy, and this is the essence of photosynthesis. Virtually all life on Earth depends on this simple equation: light energy $\longrightarrow$ chemical energy. The purposeful use of this chemical energy for synthesis and reproduction led to the evolution of the first viable microbe.

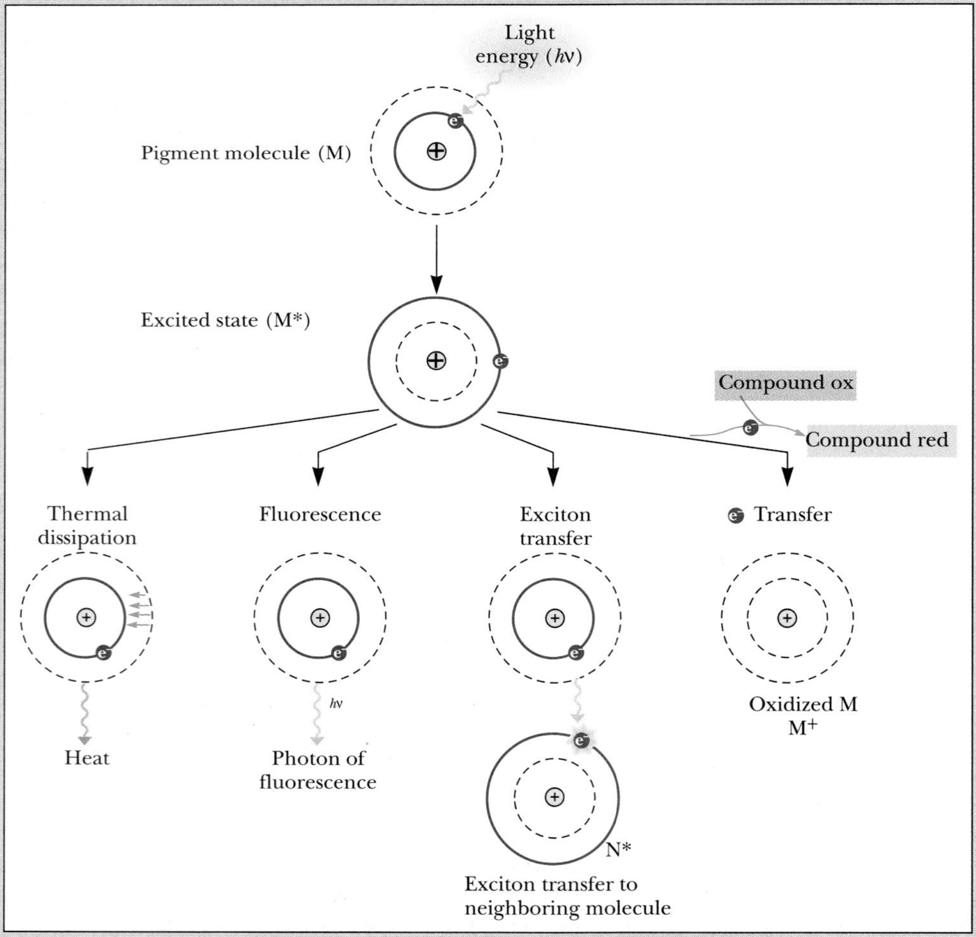

Potential events that led to the evolution of a system that harvested light energy.

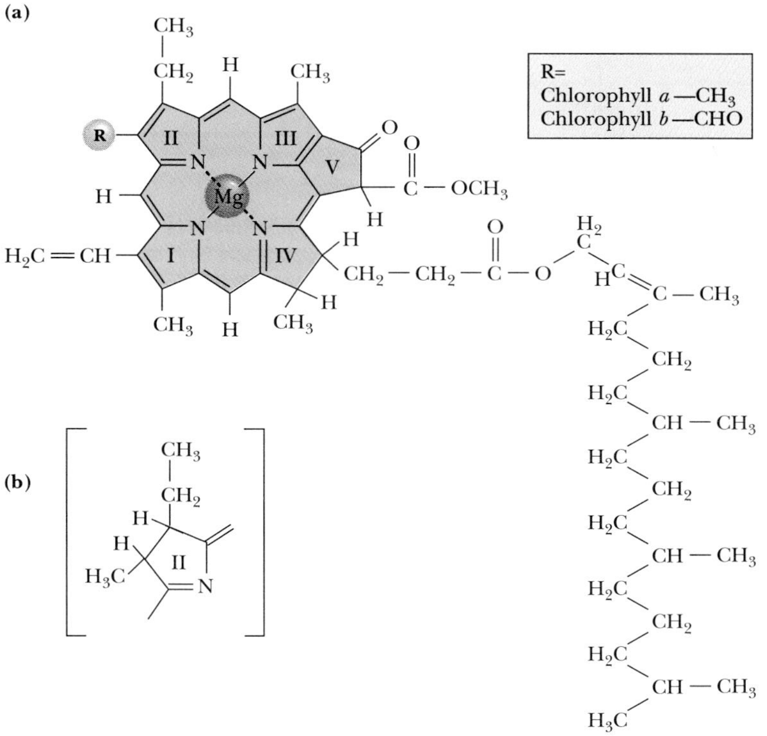

Figure **9.1** The structures of chlorophyll *a* and *b* compared with bacteriochlorophyll *a* (inset). The chlorophylls are structurally related to hemes, except Mg^{2+} replaces Fe^{2+}. The phytyl side chain on ring IV provides a hydrophobic tail that anchors the chlorophyll to membrane protein complexes.

oxygen-generating cyanobacteria in evolution. The occurrence of chlorophyll-based photosynthesis in bacteria that are phylogenetically dissimilar supports the concept that photosynthesis may have been an early event in evolution.

The three photochemical systems in microorganisms involve either a chlorophyll or bacteriorhodopsin. The chlorophylls are tetrapyrroles quite similar to the basic structure found in the heme portion of cytochromes (see Figure 8.14). The iron present in heme is replaced by a molecule of magnesium in the chlorophyll molecule. Magnesium is crucial to light capture because it is a "close shell" divalent cation that changes the electron distribution in chlorophyll pigments and produces powerful excited states. The structures of chlorophyll *a* and bacteriochlorophyll *a* are shown in Figure 9.1. There are a number of chlorophylls and they are designated by a lower-case letter—a, b, c, d, e, and g (see Figure 21.1). Bacteriorhodopsin, a carotenoid, does not contain a metal and functions in a manner markedly different from the chlorophylls. This type of photosynthesis occurs in the halophilic (salt loving) Archaea, and the characteristics of these organisms will be discussed later.

The chlorophylls are light-absorbing pigments that can participate in energy transduction. Most of the chlorophyll in a green plant or photosynthetic bacterium is involved with light gathering (Figure 9.2) and the transfer of light-generated excitation to reaction centers where the energy is used to split water ($H_2O \longrightarrow 2H^+ + 2e^- + 1/2\ O_2$) and/or drive the photophosphorylation reaction.

There are five types of photosynthetic Eubacteria: these are commonly known as the nonsulfur purples, the green sulfurs, the purple sulfurs, the cyanobacteria, and the heliobacteria. The general distinguishing properties of these microorganisms are presented in Table 9.1, and their characteristics are described in detail in Chapter 21. The various photosynthetic bacteria differ morphologically, in habitat, and physiologically. They are distinguished also by the electron donor involved in their photochemistry: the purple nonsulfur bacteria generally use H_2 or reduced organics, the green and purple sulfur bacteria use reduced sulfur compounds, the cyanobacteria use H_2O, and the heliobacteria use reduced organics.

It is reasonable to consider that the early anoxygenic photosynthetic bacterium, whether a purple or green sulfur bacterium, had a chlorophyll pigment similar to chlorophyll *a* in the cyanobacteria. Chlorophyll *a* absorbs

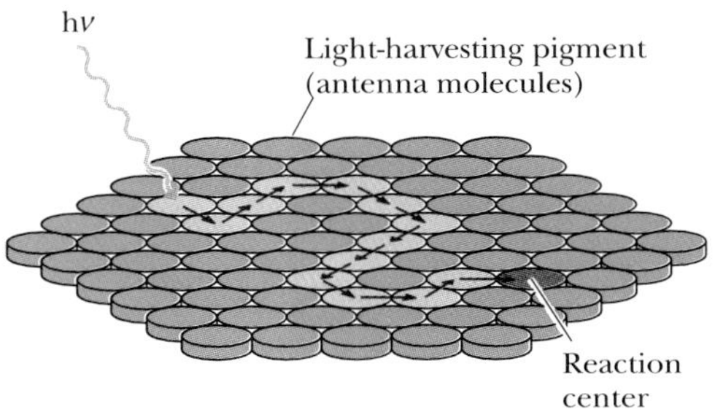

Figure **9.2** A diagram of a photosynthetic unit. The light-harvesting antenna pigments absorb and transfer light energy to the specialized chlorophyll dimer that constitutes the reaction center. The pathway of exciton transfer is shown in light green and the reaction center in orange.

***Table* 9.1 Some general properties of the various photosynthetic bacteria**

	Nonsulfur Purple Bacteria	**Purple Sulfur Bacteria**	**Green Sulfur Bacteria**	**Cyano-bacteria**	**Helio-bacteria**
Source of reducing power (e^-)	H_2, reduced organic	H_2S	H_2S	H_2O	Lactate organic
Oxidized product	Oxidized organic	SO_4^{2-}	SO_4^{2-}	O_2	Oxidized organic
Source of carbon	CO_2 or organic	CO_2	CO_2	CO_2	Lactate pyruvate
Heterotrophic growth	Common	Limited[1]	Limited[1]	Limited[1]	Required

[1]Generally limited to assimilation of low molecular weight organics during autotrophic growth.

light of a wavelength (< 700 nm) that penetrates the lower reaches of aquatic environments. The highly anaerobic photosynthetic organisms living at these depths could use these wavelengths as a source of energy. The absorption spectra for the pigments in photosynthetic microorganisms are presented in Chapter 21.

The absorption of an incident photon by a molecule of chlorophyll leads to a redistribution of electrons within the basic framework of the tetrapyrrole. This state of the molecule is known as an **excited state** and has an energy of excitation greater than the **initial** or **ground state.** This is illustrated in Figure 9.3, where the shaded portion of the porphyrin is the area of increased electron flow. Incident radiation is absorbed only if it provides the correct energy to achieve one of the allowed activated states. Photons with intermediate energies are transmitted by chlorophyll but leave it unaffected energetically.

Bacteriochlorophyll present in the green bacteria absorbs light at a wavelength of about 700 nm. This light-harvesting capacity permits growth at lower depths but does not have sufficient potential as an oxidant to extract electrons from water. Thus, evolution in the anaerobic environment yielded a photosynthetic population limited to the use of reduced molecules such as hydrogen or sulfide. Eventually (about 2.5 billion years ago) there evolved an organism with a photosystem with sufficiently strong oxidative power that it could extract electrons from H_2O, according to the following:

$$2H_2O \xrightarrow{h\nu} 4H^+ + 4e^- + O_2$$

The development of this oxygen-generating system was the fundamental event that allowed organisms to evolve beyond the need for an aquatic habitat, thus opening up the endless possibilities of life on land.

The primeval anaerobic environment on Earth bore great quantities of ferrous iron that served as a trap for O_2 generated by the first cyanobacterial-type organisms. Bands of ferric iron Fe_2O_3 called banded-iron formations (BIFs) can be seen in strata deposited during the era when oxygenic photosynthesis evolved, attesting to the probable role ferrous (Fe^{2+}) iron played. Free oxygen is a highly reactive molecule, and peroxides and other oxygen products could have been a problem had not reduced iron

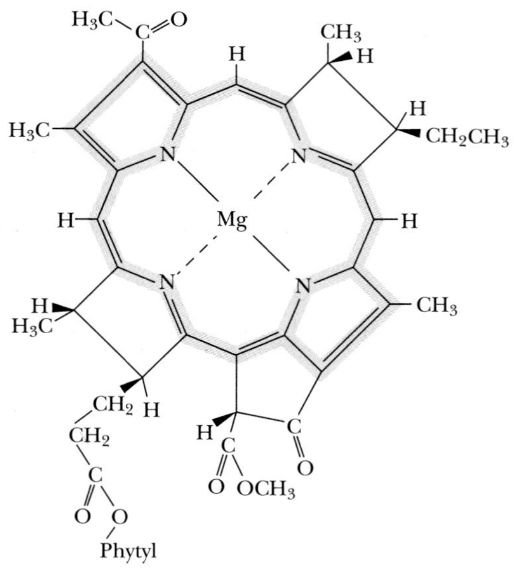

Bacteriochlorophyll

***Figure* 9.3** The shaded portion of bacteriochlorophyll *a* with nine double bonds is the site where electrons delocalize and flow about the outer area of the molecule. This provides an area for capture of photons, resulting in energy transfer and generation of an oxidant that can extract electrons from sulfides, hydrogen, or reduced organic compounds.

been present to serve as a buffer. The reduced iron absorbed the toxic molecules as the biota adapted enzymatically to handle oxygen products that were toxic. The first line of defense in the primitive O_2-generating photosynthetic organisms may have been **carotenoid pigments.** The carotenoids are long-chain hydrocarbons with extensive conjugated double bonds (Figure 9.4). The carotenoids probably evolved to provide light-harvesting capacity in the 400 nm to 550 nm range. Carotenoids also can intercept highly toxic single-state oxygen generated during photolysis of water and convert it to a less reactive state.

Ferrous iron was gradually lost through oxidation, and approximately two billion years ago O_2 began accumulating in the atmosphere. This O_2 was subjected to the following reaction in the stratosphere: O_2 + sun (UV) $\longrightarrow$ O_3. This formed an ozone (O_3) layer that screened out low wavelength ultraviolet (UV) rays that would be harmful to any organism exposed to unfiltered sunlight. Without this ozone layer, evolution would certainly have taken a different course. Ozone traps most wavelengths of sunlight below 290 nm (UV) and prevents this potentially harmful radiation from reaching the Earth's surface. This is important because UV plays havoc with compounds, such as nucleic acid, that absorb in the 260 nm range. Without this UV trap, most living creatures would have had to remain in the aquatic environment. With the ozone layer established, the lower wavelengths no longer reached the surface of the Earth in great quantity, and life on land became feasible. The diversity of living forms that were possible under the ozone layer was virtually unlimited **(Box 9.2).**

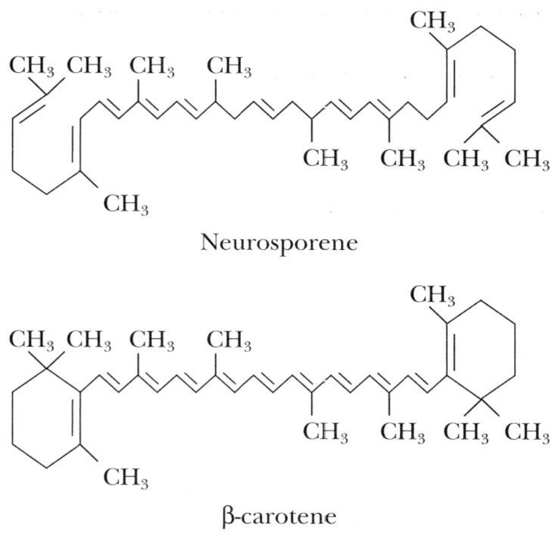

***Figure* 9.4** Structure of typical carotenoids that are present in photosynthetic microorganisms. These compounds serve a dual capacity in functioning as accessory (antenna) pigments and dissipating stray energy generated during light absorbance.

Common Features of Photosynthetic Prokaryotes

Photosynthetic microorganisms (excluding the halobacteria, which will be discussed later) have some common features but differ in the way they achieve photosynthetic growth. These bacteria use light as the source of energy, and carbon dioxide can generally be used as the source of carbon. The major energy-capturing pigment is chlorophyll (Table 9.2), and all photosynthetics have other pigments that aid them in capturing a broader range of light excitation. The sites of photochemical reactions are distinct areas called **reaction centers** and are defined as a single photochemical unit because they transfer one electron at a time. Reaction centers are protein-chlorophyll complexes within the cytoplasmic membrane of photosynthetic bacteria where quanta of light energy are converted to stable chemical energy.

The following sections cover the means whereby the photosynthetic bacterial types, purple bacteria, green sulfur bacteria, cyanobacteria, and heliobacteria, capture energy from the sun.

***Table* 9.2 The major bacteriochlorophyll in the prokaryotic photosynthetic bacteria and the primary acceptor that transfers the electrons to the energy-conserving photochemical reactions**

	Electron Donor	**Electron Acceptor**
Purple bacteria	Bacteriochlorophyll *a* and *b*	Bacteriopheophytin *a*, Q_A, Q_B
Green sulfur bacteria	Bacteriochlorophyll *c*, *d*, and *e*	Bacteriopheophytin *a* and Fe-S-protein
Cyanobacteria photosystem I	Chlorophyll *a*	Chlorophyll *a* and Fe-S-protein
Cyanobacteria photosystem II	Chlorophyll *a*	Pheophytin *a*, Qa, Qb, and plastoquinones
Heliobacteria	Bacteriochlorophyll *g*	Bacteriochlorophyll *c* Fe-S-protein

BOX 9.2 MILESTONES

Food Pyramids and the Chemistry of Sustenance

The energy that supports virtually all life forms on planet Earth comes from sunlight. Photons from the sun are captured by photosynthetic plants and bacteria, and this photon-generated energy is utilized by living cells in synthesis and reproduction. The primary producers on land are the plants, and in aquatic environments it is the phytoplankton that are at the bottom of the food pyramid. It is the primary producers that provide sustenance for the herbivores and carnivores that are so evident in the biosphere. Life as we know it is dependent on photosynthesis, and the photochemical event that converts light to chemical energy is the sustainer of life. The photosystems responsible for the bulk of the primary productivity are those that evolved in the cyanobacteria, and these bacteria are considered the progenitor of chloroplasts in green plants. Much has been learned about the structural/functional relationships of the components involved in the photophosphorylation reactions that occur in cyanobacterial-type photosynthesis.

The two photosystems PSI (NADPH producing) and PSII (O_2 generating) are presented in the accompanying figure. The NADPH is employed in the assimilation of CO_2 and in cell synthesis. Photophosphorylation and reduced pyridine nucleotide production via these reactions provides us with the oxygen we breathe and the food we eat. Humans are generally omnivores and, as such, we are sustained by primary producers and primary and secondary consumers. From an evolutionary standpoint, the role of cyanobacteria in oxygen generation and as ancestor of a highly effective photosystem is apparent.

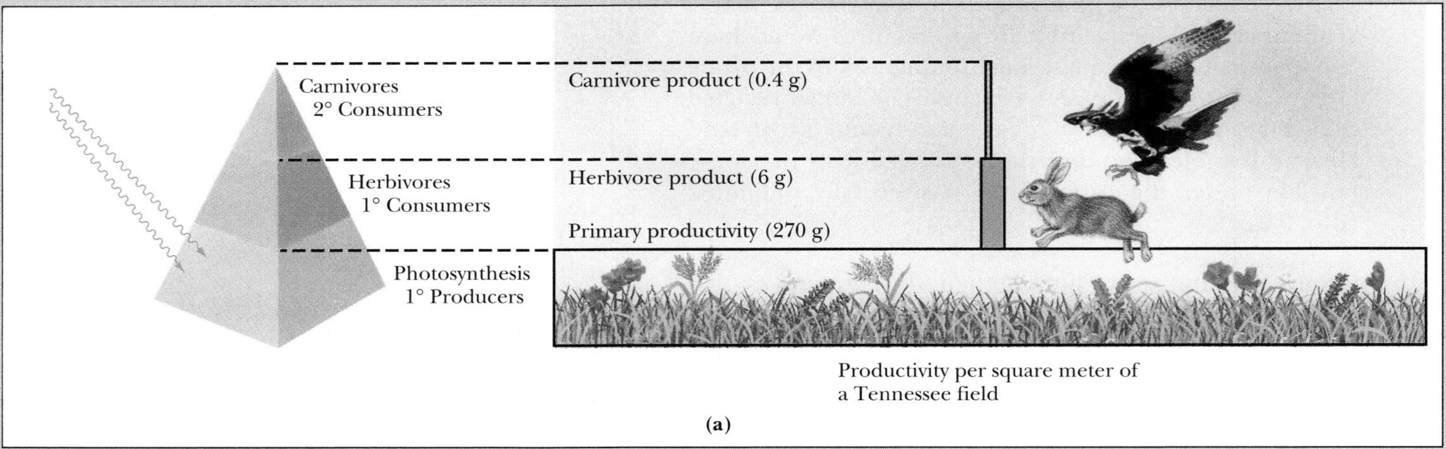

(a) The productivity in viable systems. Photosynthetic organisms convert light energy to chemical energy for use by 1° and 2° consumers.

Photosynthesis in Purple Nonsulfur Bacteria

The photosynthetic nonsulfur purple bacteria have been studied extensively, and reaction centers from *Rhodopseudomonas viridis* have been crystallized. These will be the only purple bacteria discussed in this section. These crystal structures have provided much information on the arrangement of pigments and protein subunits in reaction centers. Reaction centers have also been crystallized from *Rhodobacter sphaeroides*. The reaction centers from *Rhodopseudomonas viridis* were obtained by breaking cells by sonication (ultrasound) and differential centrifugation (centrifuging at speeds that separate material by size or weight). The reaction centers were then purified by general protein separation techniques and crystallized. These reaction center crystals yielded much data on the structural relationships between the component parts and mechanism whereby photosynthesis occurs. They have also been useful as models for studies on the photosystems in higher plants.

The composition of a reaction center in the nonsulfur purple bacterium *R. viridis* is shown in Figure 9.5. The reaction center contains four molecules of bacteri-

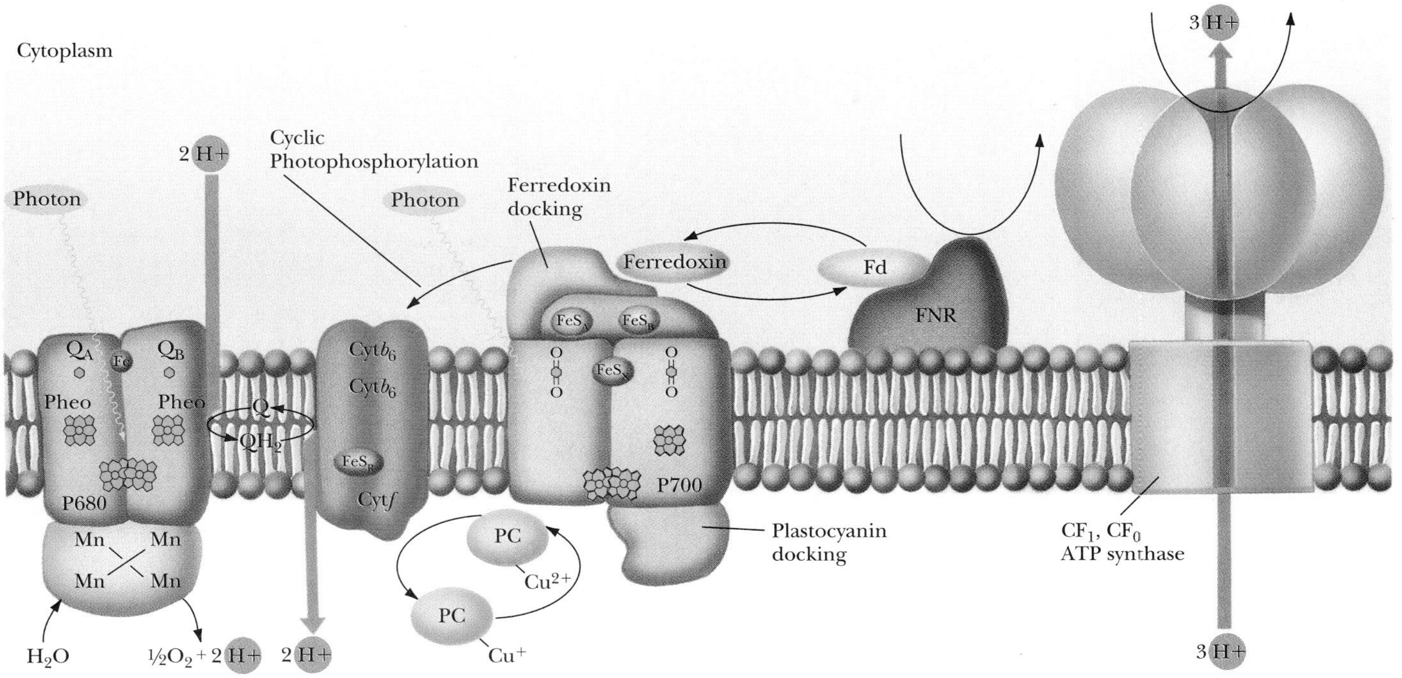

Q_ASQ_B are quinones
Pheo—pheophytin
P680 and P700 chlorophylls in PSI and PSII
Mn—manganese that is involved in cleavage of water
Cyt—cytochrome
FeS—iron-sulfur centers
FNR—ferredoxin nucleotide reductase
PC—copper-containing plastocyanin

(b)

(b) The events that occur in the membrane of a photosynthetic organism in converting photons to ATP.

ochlorophyll and two molecules of **bacteriopheophytin** (a molecule of bacteriochlorophyll *a* that lacks magnesium). A molecule of **menaquinone** and a molecule of **ubiquinone** are also present, along with a nonheme iron that is present on the cytoplasmic side of the membrane. The components are covalently linked to the light (L) and medium (M) subunits of the reaction center protein. The L and M designations are based on the sedimentation rate of these proteins on centrifugation. There is also a heavy (H) protein subunit that spans the membrane and is attached to the L-M protein complex. Two of the bacteriochlorophyll molecules are paired and share the function of a photochemical electron donor. One of the bacteriopheophytin molecules acts as the intermediary electron carrier between bacteriochlorophyll and menaquinone. The roles of the other two bacteriochlorophyll molecules and the one bacteriopheophytin molecule are not known at the present time. The electron is passed to ubiquinone and then to the cyt *b*-c_1 complex where cyclic electron flow results in the extrusion of protons.

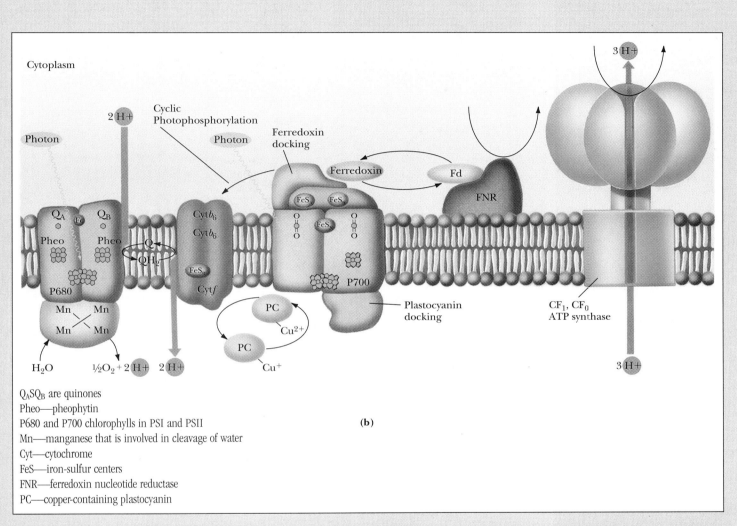

In reaction centers obtained from wild *R. sphaeroides,* one molecule of carotenoid is bound to each reaction center particle. The reaction center carotenoid is probably not involved in light gathering or excitation transport, but

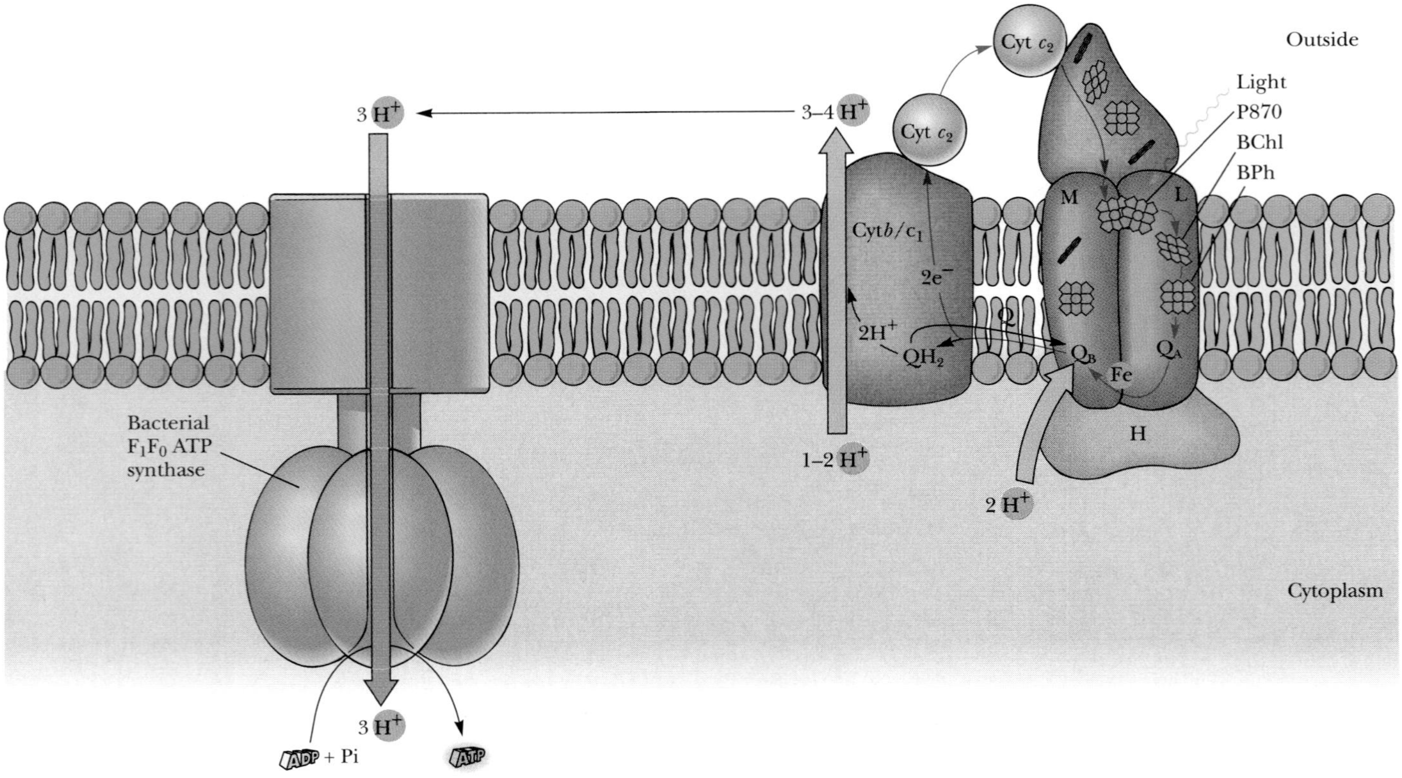

Figure **9.5** The arrangement of transport components and electron flow in the reaction center of *Rhodopseudomonas viridis.* The electron flows from the electron donor to a *c-type cytochrome* and on to a pair of bacteriochlorophyll *b* (BChl) molecules. The electron then passes to a bacteriopheophytin (Bph), to menaquinone (Q_A O_B), and to ubiquinone (Q) in the Cyt *b*-c_1 complex. The structural polypeptides are designated light (L), medium (M), and heavy (H) based on their relative size. The cyt *b*-c_1 complex drives protons to the outside of the membrane. These protons return via the ATP synthetase. By the reaction series, light energy is conserved as chemical energy.

rather it functions in carrying stray energy away from the reaction center.

Cyclic and noncyclic electron transport in photosynthetic purple nonsulfur bacteria is illustrated in Figure 9.6. Many of the purple nonsulfur bacteria are facultative and grow aerobically on organic substrates. There are three essential areas involving electron transport: I photoreduction of NAD through reversal of ATPase and generation of a $\Delta\tilde{\mu}H^+$ sufficient for this reduction; II the cyclic photophosphorylation system that generates ATP. If these microorganisms are placed in the dark on an organic substrate the energetics for nonphotosynthetic aerobic growth is as outlined in Figure 9.6. All three systems have some common components that function in either aerobic or photosynthetic growth. The major requirement for aerobic heterotrophic growth is the synthesis of the terminal oxidase system III.

Photophosphorylation occurs in the purple nonsulfur bacteria by a light-induced translocation of a proton across the energy-transducing membrane. The organization of a light-dependent transfer of electrons that would result in a $\Delta\tilde{\mu}H^+$ is shown in Figure 9.6. Cytochrome c_2 donates an electron to oxidized bacteriochlorophyll (Bchl 870). Two or more of this cytochrome type are associated with the periplasmic surface of the membrane in the vicinity of a reaction center. The specific mechanism whereby electrons flow from the substrate (photoheterotrophic growth) to the ubiquinone pool and subsequently to cytochrome c_2 has not yet been elucidated. Many reduced substrates can donate electrons, including ethanol, isobutanol, succinate, and malate. Recent evidence affirms that ferrous iron (Fe^{++}) may also serve as electron donor in the purple bacteria (see **Box 9.3**). Some of the resultant oxidized products (such as acetate) are further metabolized, while others (such as isobutyric acid) are not. The overall reaction for anaerobic photoautotrophic growth of *R. sphaeroides* can be depicted as:

$$H_2 + CO_2 \xrightarrow{h\nu} \text{cell material}$$

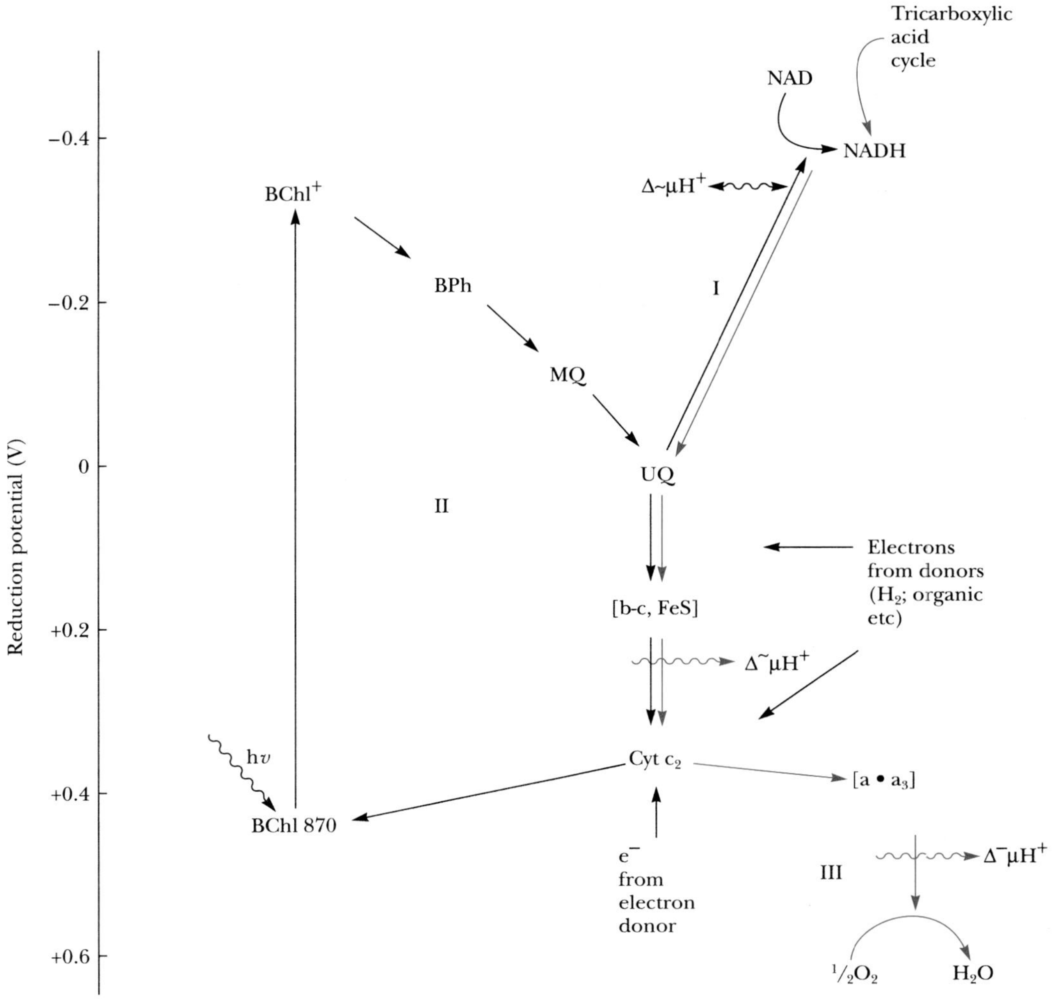

***Figure* 9.6** Pathways of electron flow in *Rhodobacter sphaeroides* during photosynthetic (anaerobic) growth and during chemoheterotrophic aerobic growth. There are a number of components that are common to both pathways. I Energy of the proton gradient ($\Delta\tilde{\mu}H^+$) that generates reduced pyridine nucleotide. II Cyclic electron flow. III Aerobic electron flow following induction of a terminal oxidase. The red arrows denote electron flow during aerobic growth.

The effect of light intensity on the synthesis of photosynthetic pigments by *R. sphaeroides* is illustrated in Figure 9.7. The rate of growth is not affected by an increase in the intensity of light. The synthesis of bacteriochlorophyll, however, is suspended temporarily, at higher light levels and resumed at a lower rate with the reestablishment of balanced growth. The level of bacteriochlorophyll per cell is lower and, on return of the culture to a weaker light intensity, there is a decreased rate of growth. The effect of oxygen on the synthesis of cellular photosynthetic pigments is similar to that observed with a change in light intensity. Cultures growing in the dark on organic substrates (facultative chemoheterotrophs) have little pigment per unit cell mass. A decrease in the oxygen supply results in pigment synthesis. The major mechanism for gaining energy anaerobically is through photophosphorylation, and a lowered oxygen supply is a signal to synthesize the light-harvesting pigments.

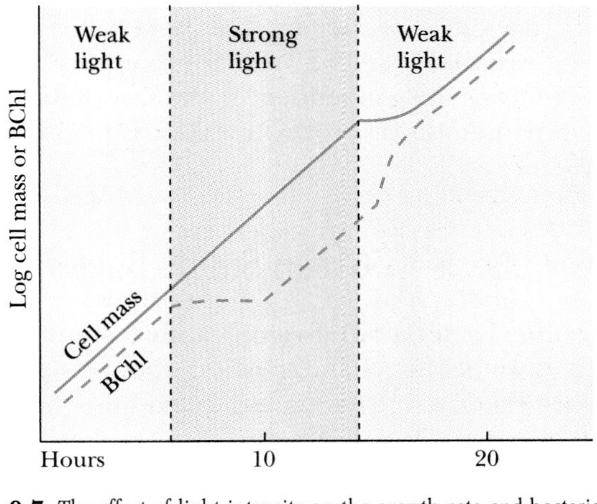

***Figure* 9.7** The effect of light intensity on the growth rate and bacteriochlorophyll synthesis in a nonsulfur purple bacterium during anaerobic growth. During balanced exponential growth the organisms synthesize sufficient bacteriochlorophyll to maintain a maximum rate of growth.

BOX 9.3 RESEARCH HIGHLIGHTS

Anaerobic Iron Oxidation

During the Archean and Proterozoic Ages (over 1.5 billion years ago), banded iron formations (BIFs) were laid down over extensive areas of planet Earth. These BIFs have been attributed to the chemical oxidation of reduced iron (Fe^{++}) by molecular oxygen generated by the cyanobacteria. The cyanobacteria evolved with the ability to utilize water as electron donor in photochemical energy generation. The resultant cleavage of water ($H_2O \rightarrow 2H^+ + 1/2\ O_2$) was the original source of atmospheric oxygen. Recent evidence suggests that anaerobic oxidation of soluble ferrous iron could also have contributed to BIFs through the activities of the anoxygenic purple bacteria. These photosynthetic bacteria can utilize reduced iron that is present in anoxic iron-rich sediments as an electron donor for photophosphorylation. As the soluble ferrous iron is oxidized to insoluble ferric state, a brown precipitate (BIF) is formed. The reaction would proceed as follows (see Figure 9.7 for details):

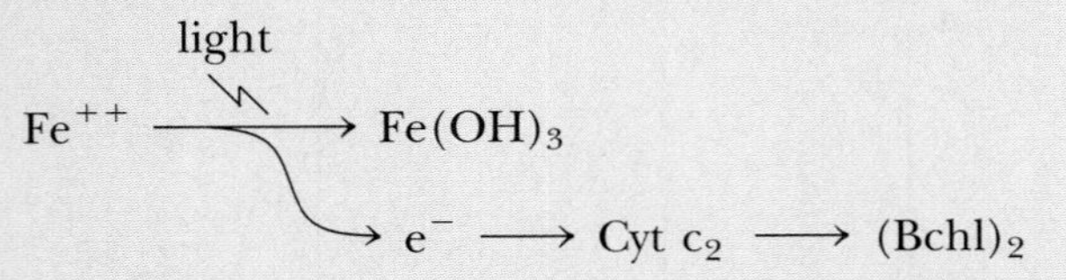

This is another example of the remarkable diversity that prevails in the microbial world.

Photosynthesis in Heliobacteria

The heliobacteria are distinctly different from other anoxygenic photosynthetic bacteria. The bacteriochlorophyll in these microorganisms is bacteriochlorophyll *g*, and the microorganisms contain no bchl *a*. Bchl *g* is unique among bacteriochlorophylls as it contains a vinyl ($—CH{=}CH_2$) constituent on ring I (see Figure 9.1) as is present in plant chl *a*. The Bchl *g* in heliobacteria absorbs maximally, at 788 nm, a wavelength not absorbed by other photosynthetic bacteria. The only sources of carbon utilized by these organisms are simple monomers such as pyruvate, acetate, and lactate. The heliobacteria cannot grow autotrophically. The reaction centers of these microorganisms are embedded in the cytoplasmic membrane, and they do not form thylakoids or chlorosome-like structures.

Photosynthesis in Green Sulfur Bacteria

Green sulfur bacteria differ significantly from purple bacteria, and in some characteristics, they resemble the cyanobacteria. These bacteria are called green sulfurs because they generally utilize reduced sulfur compounds as electron donor. The light reaction in green sulfur bacteria results in the formation of a stable reductant with a potential of approximately 540 mV, which is a much stronger reductant than that present in purple bacteria. This reductant is sufficiently negative to permit a noncyclic, light-driven reduction of NADP (Figure 9.8) without the reverse electron flow essential in purple bacteria.

The reaction centers in green bacteria have proven difficult to purify in a functional form, and thus, much remains unknown about the mechanics of photophosphorylation in these organisms. The major antenna pigments in green bacteria are two different chemical forms of bacteriochlorophyll *c* and bacteriochlorophyll *a*. The reaction center itself contains bacteriochlorophyll *a* and bacteriopheophytin *a*. The ratio of antenna bacteriochlorophyll *c* to bacteriochlorophyll *a* is 1000 to 1500 per reaction center in green bacteria, which is unique among the photosynthetic bacteria. Plants have 200 to 400 molecules of chlorophyll per center, and other photosynthetic bacteria have 50 to 100.

The origins of green sulfur photosynthetic bacteria may offer a plausible explanation for this high level of antenna molecules. These antenna pigments funnel energy of excitation to the reaction centers (see Figure 9.2). Green bacteria evidently evolved in the lower depths of aquatic environments where reduced sulfur compounds were available as electron donors in photochemical reactions. The green sulfur bacteria would therefore have existed under swarms of cyanobacteria and purple bacteria. To absorb the weak light that filtered down through the layers of cyano- and purple bacteria, the green bacteria evolved with an extensive network of antenna pigments to capture the excitation energy of the limited light available.

The way that the light-harvesting pigments are bound to the cytoplasmic membranes in green bacteria is similar

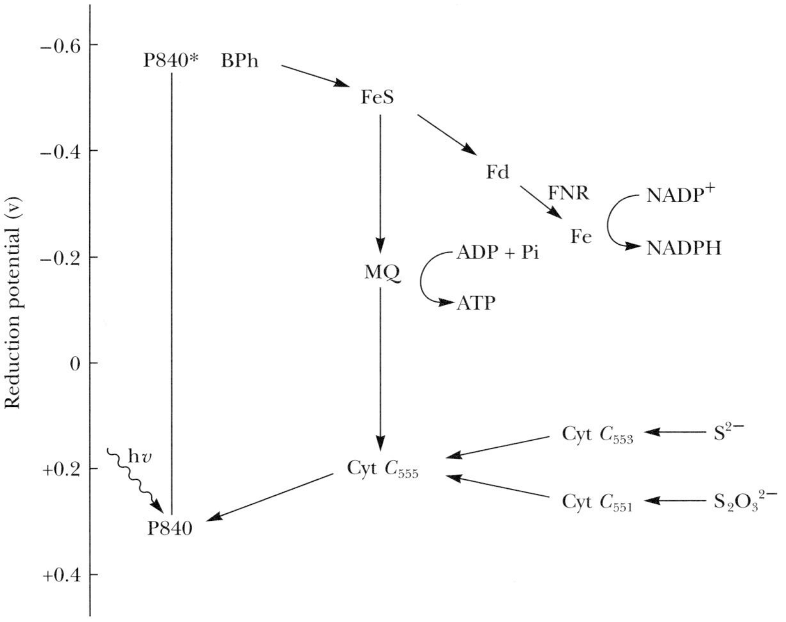

***Figure* 9.8** Electron flow during photophosphorylation in a green sulfur bacterium. These organisms have a sufficiently high reduction potential to directly reduce pyridine nucleotides. Bph is bacteriopheophytin, Fd is ferredoxin, FNR is a ferredoxin nucleotide reductase, and MQ is menaquinone. Cyt c_{553} and Cyt c_{551} transfer electrons from reduced sulfur to a pair of Cyt c_{555} in the reaction center.

to the manner in which these pigments are bound to thylakoids in the cyanobacteria. The antenna pigments in green sulfur bacteria are packed into **chlorosomes,** which are vesicles that lie along the inside of the cytoplasmic membrane, as shown in Figure 9.9 (see electron micrograph Figure 21.9). The chlorosomes are bound by a proteinaceous (nonlipid) membrane. Each vesicle contains up to 10,000 bacteriochlorophyll *c* molecules, each of which is attached to the cellular membrane by a subantenna of a bacteriochlorophyll *a*. Excitation energy, generated by light, passes through this subantenna to the reaction centers. Experimental evidence suggests that the reaction centers and probably some of the antenna bacteriochlorophyll *a* are built into a base plate (an organized set of molecules) in the cytoplasmic membrane. The high efficiency of the photosystem in green bacteria indicates that excitation energy captured by the antenna pigments in a saturated reaction center can readily be passed to another receptive center.

Electron transfer in green sulfur bacteria begins with the primary photochemical donor (designated P840) passing electrons to an acceptor molecule, considered to be an Fe-S-protein. The energy-enriched Fe-S-protein apparently transfers electrons through a soluble ferredoxin (Fd) to an Fd-NADP reductase to yield NADPH. This noncyclic pathway must be replenished with electrons, which are provided by reduced substrates such as sulfide, thiosulfate, or hydrogen. Electrons are transferred from substrates to bacteriochlorophyll in the reaction centers via specific types of cytochrome *c*, such as cytochrome c_{551} for thiosulfate and cytochrome c_{553} for sulfide (see Figure 9.8). A specific reductase for each substrate acts as an intermediate carrier of electrons from the substrate to the cytochrome *c* involved.

The overall photochemistry in green bacteria is relatively straightforward. ATP is generated through a cyclic photophosphorylation, and reduced pyridine nucleotides are produced by the noncyclic Fd-reductase system. The electrons are ultimately transformed to excitation energy,

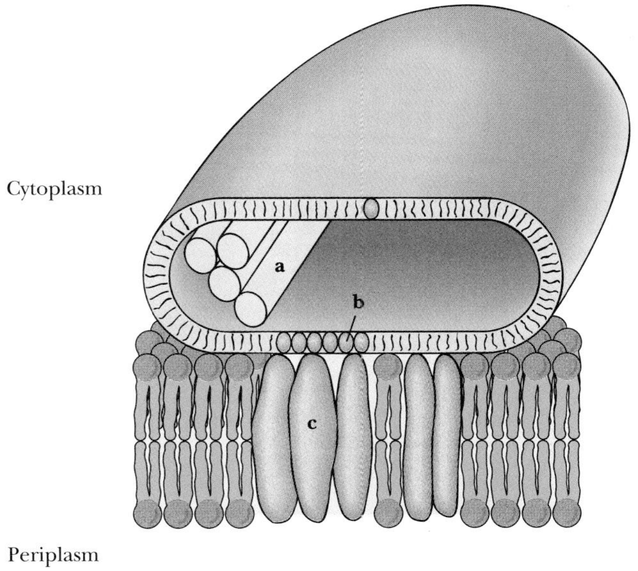

***Figure* 9.9** Model of the organization of a chlorosome. These vesicular structures are attached to the cytoplasmic membrane at the site of a reaction center. The rods **(a)** are 10 nanometers in diameter and are lipid-rich structures that contain the antenna pigments, carotenoids, and bacteriochlorophyll *c*. A light-harvesting bacteriochlorophyll *a* is present in the base place **(b)** and this is attached to the cytoplasmic membrane. The reaction centers **(c)** and other light-harvesting bacteriochlorophyll *a* is present in the cytoplasmic membrane. The vesicular structure is surrounded by a polypeptide lipid membranous matrix (not a unit membrane).

and hence, a utilizable energy-rich compound (NADH or ATP) is contributed by a reduced sulfur compound. Sulfides and other reduced sulfur compounds are of widespread occurrence in those environments in which green sulfur bacteria thrive. The primary requirements for survival of an autotrophic, anaerobic photosynthetic bacterium include a ready source of reductant for CO_2 assimilation and ATP to drive synthetic reactions. These organisms have evolved with a marvelous capacity for survival in sulfurous anaerobic environments where low levels of light are available as an energy source.

Photosynthesis in Cyanobacteria

The **cyanobacteria,** formerly called the blue-green algae or blue-green bacteria, are the only photosynthetic prokaryotes that generate O_2 during photosynthesis. The cyanobacteria are a physiologically and morphologically diverse group (see Chapter 21) that appeared on Earth an estimated 2.5 billion years ago. During their evolutionary process, they developed two independent light reactions. These are called photosystem I (PSI) and photosystem II

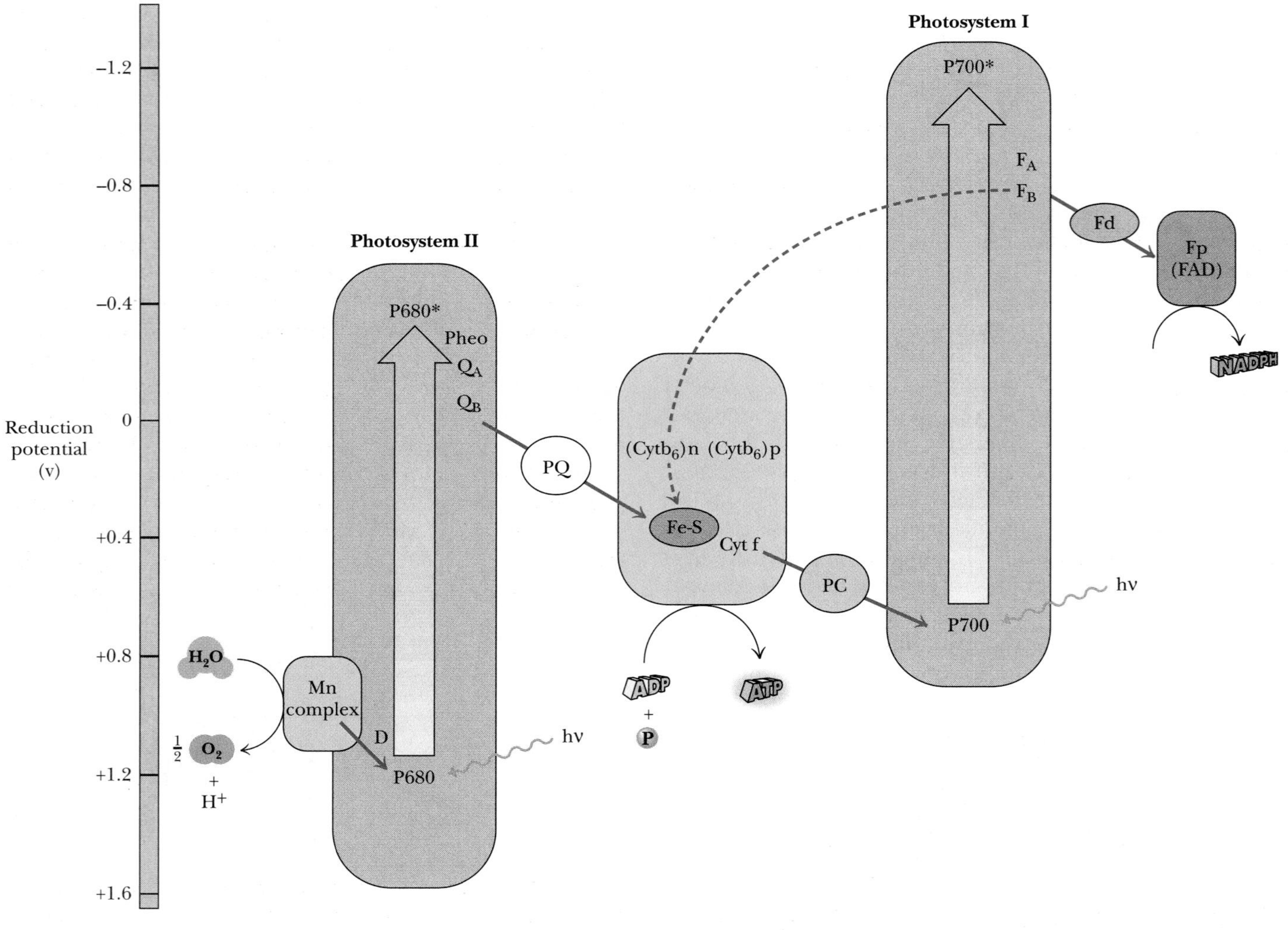

Figure **9.10** Electron flow in the reaction centers of a cyanobacterium. There are two photosystems, designated PSI and PSII. PSII is responsible for the cleavage of water and production of O_2. The chlorophyll is a dimer (P680), and the energized (P680*) electron passes from QA to QB (plastoquinones) and on to the plastoquinone pool. Cyclic electron flow can occur through cyt b_{559}. Electrons flow from the PQ pool to an iron-sulfur center cyt *f* and plastocyanin (PCu) and on to the chlorophyll P700 in photosystem I. Light drives the electron to a higher excitation state (P700*). The electrons flow to iron-sulfur centers (A and A,B) and on to ferredoxin (FD). FD is a flavoprotein enzyme ferredoxin $NADP^+$ reductase. Electrons from A may cycle back to the PQ pool.

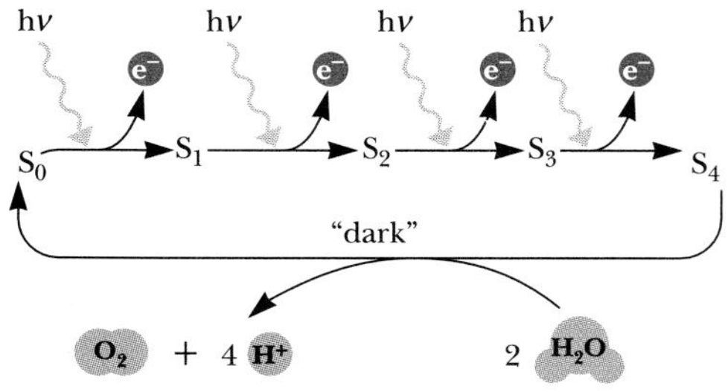

***Figure* 9.11** Oxygen evolution requires the accumulation of four oxidizing equivalents in PSII. The PSII reaction center passes through five different oxidation states when exposed to light. These states are S_0 to S_4. One e^- is removed photochemically at each oxidation state. When S_4 is reached there is a spontaneous decay to S_2 and oxidation of $2H_2O$ to O_2. $4H^+$ are concomitantly released.

(PSII), and these operate in series with a redox span from H_2O/O_2 to $NADP^+/NADPH$. The oxygen-generating photochemical systems of the cyanobacteria were instrumental in providing an atmosphere on Earth that eventually permitted the commencement of aerobic respiration.

Photosynthetic reactions in cyanobacteria are similar to and a forerunner of the reactions in green plants. The only relationship that the prokaryotic cyanobacteria have to eukaryotes is the similarity of the photosystem in these bacteria to one of the organelles in green plants—the chloroplast. The understanding of green plant photosynthesis has been enhanced by studies of the photoapparatus in the more readily manipulated cyanobacteria. These rapidly growing organisms are easily grown under laboratory conditions and offer considerable diversity; thus they have become a widely used model for study.

The cyanobacteria combine the salient features of photoreactions in both purple and green bacteria. Figure 9.10 is a diagram of electron flow in the cyanobacteria. The photosystem, designated PSI, is similar to the photosystem in green sulfur bacteria. It provides a reductant of sufficient potential to reduce pyridine nucleotide. Photosystem II is similar to the photosystem in the purple nonsulfur bacteria except that the latter lacks the water-splitting system. The light-activated system responsible for the photolysis (splitting) of water to generate oxygen and donate electrons to PSII requires four quanta of light energy. Photolysis occurs when two water molecules are bound to a light-sensitive enzyme designated the S system, which accumulates four positive charges that extract four electrons from the water. The protons are released into the medium, and O_2 is released by a series of four reactions as outlined in Figure 9.11.

One electron is released at each step, and oxygen is produced only when four separate photo acts have occurred on the S system. S_o is the enzyme at the ground state, and after the release of O_2 the enzyme returns to this state of activation. Four quanta of light are required for the photolysis of water, and a total of eight to ten quanta are necessary for the photosynthetic dissimilation of one molecule of water and assimilation of one molecule of CO_2.

A model for the organization of the pigment-protein complex involved in photosynthesis in the cyanobacteria is presented in Figure 9.12. The photosystem is present in projections termed **phycobilisomes.** A phycobilisome is a subcellular body that appears as tiny knobs that are attached to the outer surface of photosynthetic membranes. Little if any transfer of quanta occurs between photosynthetic units, and each is a discrete miniaturized power cell. The light energy absorbed in the phycobilisomes is passed preferentially through chlorophyll *a* to the reaction center of photosystem II, but some excitation energy can be transferred between photosystem I and photosystem II. The system as depicted permits a "downhill" cascade of

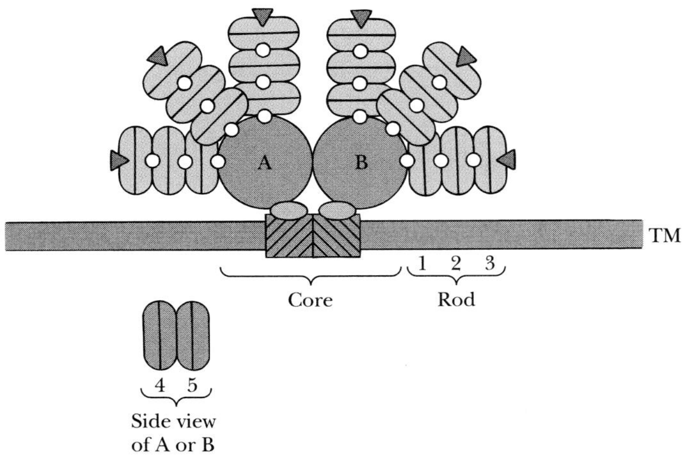

***Figure* 9.12** The arrangement of antenna pigments in the cyanobacteria. These knoblike structures, termed phycobilisomes, appear on the outer surface of the cytoplasmic membrane. The center, or core, designated A and B are allophycocyanins, which are bluish in color. The six rods radiate from the core. In *Synchococcus* sp. the rods are composed of the blue light-harvesting pigment phycocyanin. The rods are joined together by linked polypeptides (small circles). The core is joined to the thylakoid membrane (TM) by a polypeptide (dark circles) and the reaction centers are denoted as cross-hatched squares. (Reproduced with permission, Arthur Grossman, *Microbiological Reviews,* 57:725-749)

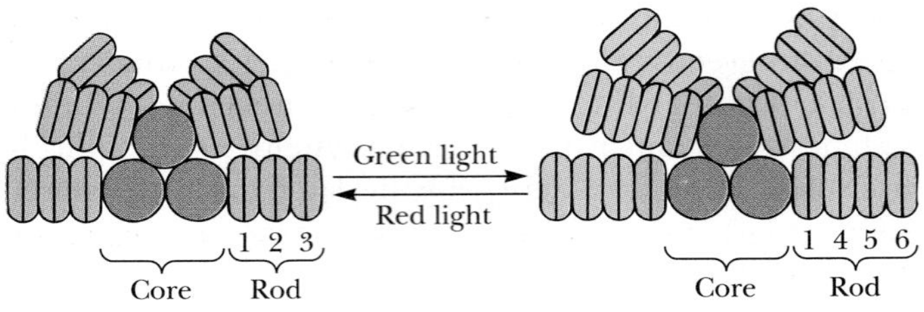

***Figure* 9.13** The composition of phycobilisomes can be altered in some cyanobacterial species by changing the wavelength of light provided for growth. This is termed chromatic adaptation and is illustrated for the organism *Fremyella diplosiphon.* When grown in green light, the rods are predominantly composed of phycoerythrin (red), and when grown in red light, the rods are predominantly made up of phycocyanin (blue). (Reproduced with permission, Arthur Grossman, *Microbiological Reviews,* 57:725-749)

excitation energy through the phycobilisome pigments to chlorophyll *a*. Growth of some cyanobacteria in light of a different wavelength will induce the synthesis of pigments that absorb that particular wavelength. For example, growth in green light results in the synthesis of phycoerythrin, which absorbs green wavelengths, whereas growth in orange light leads to increased levels of the blue pigment phycocyanin. This change in the phycobilisomes is termed **chromatic adaptation** and is illustrated in Figure 9.13.

Phycocyanin and phycoerythrin are both open-chain tetrapyrroles (Figure 9.14) coupled to proteins and collectively are called phycobiliproteins. The phycobiliproteins are light-gathering antenna pigments. Phycocyanin absorbs light at around 630 nm and phycoerythrin at around 550 nm. The chlorophylls absorb a rather narrow spectrum of light, and the accessory pigments allow the organism to capture more of the incident light that falls outside this narrow range.

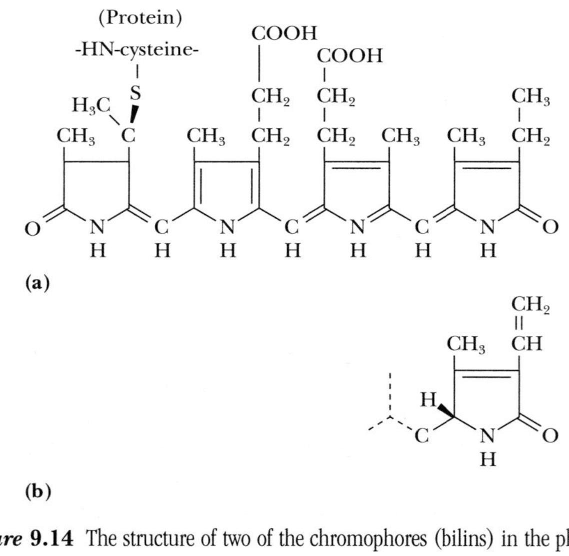

***Figure* 9.14** The structure of two of the chromophores (bilins) in the phycobilisomes of cyanobacteria: **(a)** the blue pigment phycocyanin and **(b)** the red pigment phycoerythrin differ only in the fourth ring as shown.

The primary electron acceptors in photosystem I are soluble Fe-S proteins similar in function to those in green sulfur bacteria. Electrons flow cyclically back to chlorophyll P700 via cytochromes and the copper-containing protein plastocyanin. A theoretical scheme for electron transport, ATP synthesis, and NADPH production in cyanobacteria was presented in Figure 9.10.

Photosynthesis Without Chlorophyll

The halophilic Archaea are a remarkable group of organisms that thrive in highly saline environments, as they require substantial concentrations of NaCl for growth (see Chapter 22). Of these organisms, *Halobacterium halobium* has been studied most extensively. *H. halobium* is obligately aerobic and operates a conventional membrane-bound respiratory pathway for ATP generation when sufficient oxygen is available. The solubility of oxygen in water decreases as salinity increases. Consequently the availability of oxygen beneath the surface of water with a high salt concentration is limited. When the halobacteria encounter low oxygen concentrations they synthesize purple patches within the cytoplasmic membrane. These patches can make up over one half the surface of the cellular membrane. Studies reveal that these purple patches exhibit a light-driven proton ejection system that is not inhibited by cyanide. Aerobic respiration in these organisms is inhibited by cyanide because the membrane-bound electron transport chain would be involved in generation of ATP (see Chapter 8).

The purple patches are made up of flat sheets containing a crystalline array of a single protein—**bacteriorhodopsin.** This protein has a molecular mass of 26,000, and each protein assembly is attached to a molecule of retinal through a lysyl residue in the protein (Figure 9.15**a**). Bacteriorhodopsin (vitamin A aldehyde) is similar to the visual pigments of animals. The bacteriorhodopsin absorbs light maximally at a wavelength of 570 nm and becomes bleached as it ejects a proton. The deprotonated bacteriorhodopsin then takes up a proton from the cytoplasm of the cell (Figure 9.15**b**). Bacteriorhodopsin absorbs at wavelengths at which maximum energy is available from sunlight, particularly when there is a water layer over the organism. No accessory pigments are necessary for the photochemical reaction to occur, and respiratory electron transport is not involved. The reaction is a simple light-driven proton pump.

Elucidation of this system has lent considerable credibility to the chemiosmotic theory of ATP generation. The isolated purple membrane has been reconstituted in closed vesicles, which, when exposed to light, act as proton pumps. ATPase from beef heart has been incorporated into these vesicles, where they can perform a light-dependent synthesis of ATP (see **Box 9.4**). It would be difficult to ascribe this phenomenon to any direct coupling reaction because no electron transfer is involved. Replacement of the halophile's ATPase with that from an animal confirms that no electron transfer–related phenomenon occurs in the native ATPase. Electron flow in respiration then serves only to transduce protons across the membrane.

The quantum requirement for the ejection of one proton by *H. halobium* is approximately 2 as compared with 0.5 for purple photosynthetic bacteria. Because light at a wavelength of 570 nm has a higher energy content than light at 870 nm, the overall efficiency of *H. halobium* is less than 25 percent of the efficiency of purple bacteria.

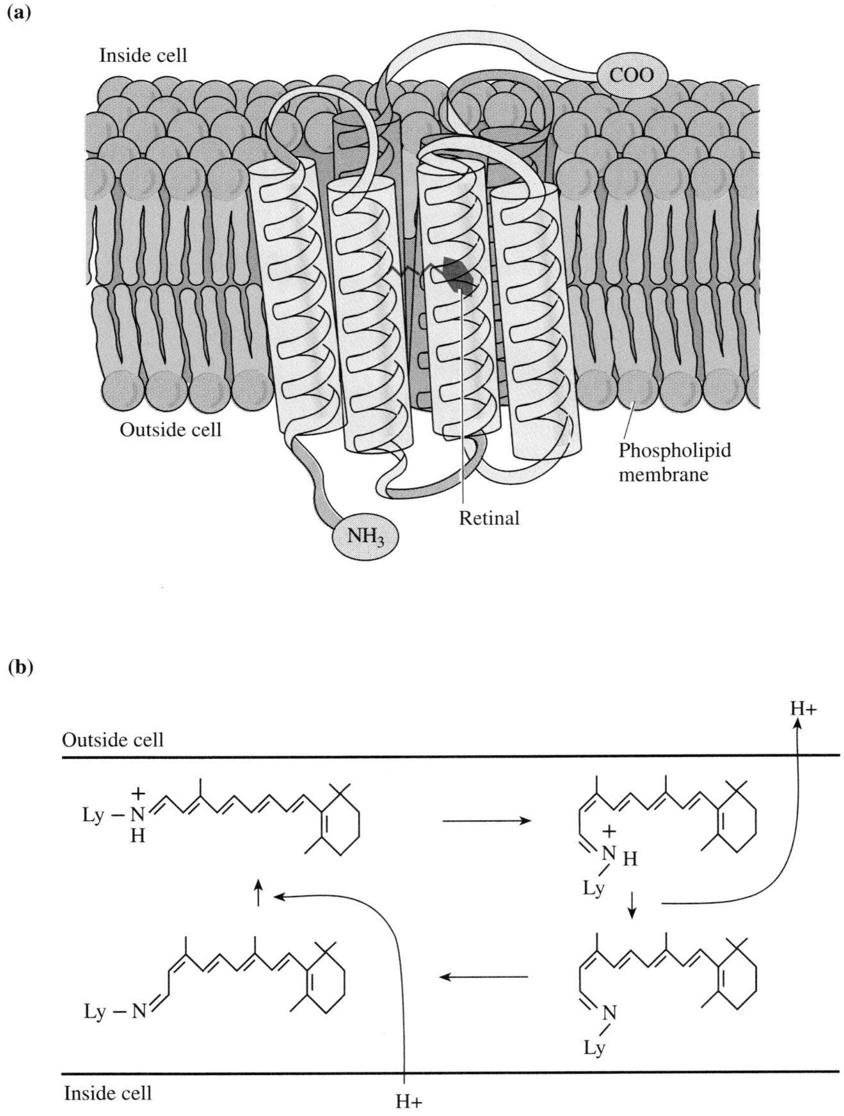

***Figure* 9.15** The light-driven proton pump present in the membranes of halophilic (salt-loving) bacteria. Bacteriorhodopsin is a pigmented protein molecule made up of seven helixes that span the membrane **(a).** The retinal molecule is attached to one of the helixes. Retinal becomes protonated when exposed to light and transfers the H^+ to a protein that undergoes a conformation change that carries the proton to the external medium. The deprotonated pigment then picks up a proton from the cytoplasm. There is no electron transport involved. The protons expelled to the periplasm can return via the ATPase system to generate ATP.

BOX 9.4 MILESTONES

Archaea, Bovines, and Chemiosmosis

In 1974, Efraim Racker, a biochemist, and Walther Stockenius, a pioneer in elucidating the role of bacteriorhodopsin in halophilic bacteria, jointly performed an experiment that confirmed Mitchell's chemiosmotic hypothesis. They reconstituted vesicles from the membranes of *Halobacterium halobium* that were inverted so that the bacteriorhodopsin would pump protons inward when exposed to light. They obtained bovine mitochondrial ATPase and reconstituted this into the vesicle as shown. When illuminated, the bacteriorhodopsin pumped protons into the vesicles and generated a proton gradient inside that drove ATP synthesis by the ATP synthase. There were two types of protein in the vesicles, one from an *Archaea* and the other from a mitochondrion. Together they synthesized ATP, as the Mitchell hypothesis predicted. There were no electrons involved in the synthesis of ATP, confirming that the role of respiratory electron transport chains is transduction of protons.

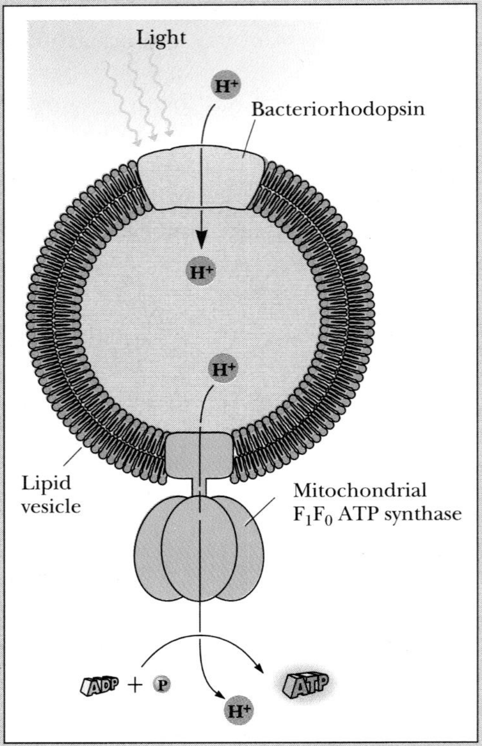

A combination of archaeal photosynthetic pigments with bovine ATPase to generate ATP. This system clearly demonstrated that the proton gradient was instrumental in ATP synthesis.

Summary

- **Photophosphorylation** is the sum of the processes that generate ATP by use of light energy.

- All life on the surface of the Earth relies directly or indirectly on energy from the sun. Photoautotrophs use photons directly while heterotrophs exist by utilizing photosynthetic products.

- **Oxygenic** photosynthesis generates O_2 as the source of electrons for this process is H_2O ($2H_2O \longrightarrow 4H^+ + O_2 + 4e^-$). **Anoxygenic** photosynthesis does not result in O_2 production. The source of electrons in anoxygenic photosynthesis is reduced organic (ethanol $\longrightarrow$ acetate) or inorganic compounds ($S_2O_3^{2-} \longrightarrow SO_4^{=}$).

- Photophosphorylation can occur in the absence of chlorophyll by employing a light-absorbing compound that can carry protons from the interior of the bacterium to the outer surface. This reaction occurs in **Archaea** that grow in highly saline environments.

- Much of the chlorophyll in photosynthetic organisms is involved in light gathering. The gathered excitons are transferred to reaction centers where the **light energy** is converted to **chemical energy.**

- There are **several chlorophylls**—a, b, c, d, e, and g—and each has a **characteristic absorption spectrum.** This permits microorganisms to grow at various depths in water because they absorb wavelengths of light that are not absorbed by those above.

- The **ozone layer** above the Earth was probably formed when microorganisms that performed oxygenic photosynthesis evolved. The O_2 produced reacted with UV in the stratosphere to form O_3. The ozone layer protected earthly creatures from harmful UV radiation and was a major contributor to evolution of life on dry land. Aquatic life was not adversely affected by UV because this wavelength does not penetrate into water.

- Reaction centers have been crystallized from purple nonsulfur photosynthetics. Much has been learned about structural arrangements in reaction centers by studying these crystals.

- Most **purple nonsulfur bacteria** grow as either a **heterotroph** or with light energy. They use most of the same respiratory system when grown by either mode.

- The reaction centers in **green sulfur bacteria** are complex and have not yet been purified. They have considerably more **antenna pigments** than are present in the purple nonsulfur reaction centers. This permits them to gather light of much lower intensity. These antenna pigments are arranged in **chlorosomes.**

- The **photosystem** in cyanobacteria and green sulfurs can reduce NADP, but that from the purple nonsulfur bacteria does not have sufficient energy to accomplish this. The latter must generate NADPH for CO_2 fixation by reverse flow of ATPase.

- The antenna pigments in cyanobacteria are arranged in projections attached to the surface of membranes termed **phycobilisomes.**

- The pigments in the phycobilisomes change with the wavelength of light available. The pigments are open-chain tetrapyrroles coupled to protein and are called **phycobiliproteins.**

- The **halophilic Archaea** are aerobic but can generate ATP anaerobically by utilizing **bacteriorhodopsin.** This compound is present in the cytoplasmic membrane and is similar to **retinal,** the eye pigment in animals. Bacteriorhodopsin can use light energy to move protons from the cytoplasm to the outer surface of the cytoplasmic membrane. The protons reenter the cytoplasm via ATPase, thus producing ATP.

Questions for Thought and Review

1. What is the basic difference between photophosphorylation and respiration? What are the common features?

2. There are basic differences in the electron donor for the three photosynthetic bacterial types—purple nonsulfurs, green sulfurs, and cyanobacteria. What are these differences? Is this related to habitat?

3. Which of the three bacterial types mentioned in question 2 would you consider the more ancient? Why? Consider the atmosphere of the primordial earth and the nature of compounds available.

4. What is the structural makeup of chlorophyll and what relationship does it have to heme in blood? Why does Mg^{2+} replace iron?

5. The evolution of the cyanobacteria with oxygenic photosynthesis was of major importance in the movement of life forms to terrestrial habitats. Explain.

6. Antenna pigments are important to photosynthetic organisms. Why? Consider all types of photosynthetic bacteria.

7. Why have the purple nonsulfur photosynthetic bacteria been a model system for clarifying the events that occur in a reaction center?

8. How does the role of electron transport differ in purple bacteria grown photosynthetically from that during aerobic/heterotrophic growth? How much actual difference is there?

9. How does a photosynthetic bacterium respond to increases in available light intensity? Consider chlorophyll and growth rate.

10. What is a chlorosome and what role does it play in green sulfur bacteria? What are the major parts?

11. How have green sulfur bacteria adapted to life at the bottom of the pond?

12. How does the photosynthetic apparatus differ among the three photosynthetic types: the cyanobacteria; the green sulfur bacteria; and the purple nonsulfur photosynthetic bacteria?

13. What is bacteriorhodopsin? How does photosynthesis with this pigment differ from that due to chlorophyll?

Suggested Readings

Balows, A., H. G. Truper, M. Dworkin, W. Harder, and K. H. Schleifer. 1992. *The Prokaryotes.* New York: Springer-Verlag.

Lawlor, D. W. 1993. *Photosynthesis: Molecular, Physiological, and Environmental Processes.* 2nd ed. Essex, UK: Longman Scientific & Technical.

Omerod, J. G. 1982. *The Phototrophic Bacteria: Anaerobic Life in the Light.* Berkeley: University of California Press.

Schlegel, H. G., and B. Bowien, eds. 1989. *Autotrophic Bacteria.* Madison, WI: Science Tech. Publishers.

Shively, J. M., and L. L. Barton, eds. 1991. *Variations in Autotrophic Life.* New York: Academic Press.

. . . a superficial survey of the biochemical field is apt to fill one with profound astonishment at the practically unlimited diversity of the chemical constituents of living organisms. . . . We have to accept the undeniable fact that all these substances have ultimately been derived from carbon dioxide and inorganic salts. . . .

A. J. Kluyver

Chapter 10

Biosynthesis of Monomers

Macromolecular Synthesis from One-Carbon Substrates
Core Reactions That Generate Biosynthetic Precursors
Glucose Metabolism and Synthesis of Precursor Metabolites
The Tricarboxylic Acid Cycle or Krebs Cycle
Nitrogen Fixation
Sulfur Assimilation
Biosynthesis of Amino Acids

The previous chapter was concerned with the strategies used in biological systems for conserving energy captured from the sun. The quanta of light that reach the Earth activate those systems that convert photons into stable, energy-rich chemicals such as adenosine triphosphate (ATP) or reduced pyridine nucleotides (NADH and NADPH). These then serve as key intermediaries in energy conservation. If ATP is considered as the medium of exchange in energy transformation, then the C—H bond might be considered a "savings bank." Generally, ATP and the electron carriers act solely as transitory go-betweens, with the C—H bond that is available in sugars, amino acids, lipids, and other constituents of cells serving as a stable, life-giving, energy source for heterotrophic growth. Thus, the formation of a C—H bond during photosynthesis provides a stable basic unit that can support the food chains on Earth. Annually, in excess of 10^{11} tons of CO_2 are fixed on Earth, principally through the mediation of roughly 40 million tons of ribulose-1,5-bisphosphate carboxylase. This key enzyme in the **Calvin cycle** (see later), the major mechanism for CO_2 fixation in autotrophs, is among the most abundant proteins in the biosphere. Light is the ultimate source of energy on the surface of the earth, and autotrophic CO_2 fixation is the ultimate sustainer of all food chains.

The combined mechanisms whereby the major constituents of the bacterial cell are synthesized from simple building blocks is the process we term **biosynthesis.** Carbon dioxide (CO_2) is the primary substrate supporting life and is incorporated into cells via an assortment of mechanisms. Among the other one-carbon compounds that are utilized by bacteria are carbon monoxide (CO), cyanide (CN^-), formic acid ($HCOO^-$), methanol (CH_3OH), and methane (CH_4). Some of these substrates are funneled through carbon dioxide and enter anabolic sequences through the classic Calvin cycle or are assimilated through other reaction sequences.

Many organisms can satisfy their nitrogen requirements by fixation of gaseous N_2 available in the atmosphere; others utilize ammonia or nitrate. Sulfur needs are met by inorganic sulfate. Phosphate for nucleic acid synthesis and generation of high-energy intermediates also can come from inorganic sources. This chapter considers the synthetic mechanisms that have evolved among the prokaryotes for the generation of the monomolecular units or **monomers** that can be polymerized to form the macromolecules of a cell. Synthesis by autotrophs is considered here as these microorganisms can construct all of their cellular components from CO_2, the most rudimentary of building blocks.

Macromolecular Synthesis from One-Carbon Substrates

Evolution ultimately led to the selection of viable organisms that grew and reproduced by using the low molecular weight substrates that were available to them. It is probable that many of these early organisms utilized light as a source of energy. The use of low molecular weight compounds in biosynthesis permitted survival without relying on any tedious random process of abiotic (or biotic) generation of building blocks. Obviously, it was essential that early in evolution there was the development of a counterpart biodegradative process that continuously replenished the low molecular weight primary substrate. Ultimately the central substrate for biosynthesis and also the end product of biodegradative activities came to be **carbon dioxide.**

A microorganism will grow and reproduce only if it can synthesize *all essential* cellular macromolecules from substrates available to it (see Chapter 5). Autotrophic organisms that use CO_2 as sole carbon source and N_2 as nitrogen source represent the ultimate in biosynthetic capacity. Research has shown clearly that the broad synthetic ability of autotrophs and other microorganisms that grow on simple low molecular weight substrates rests on the organism's effective use of a limited number of biosynthetic pathways. In this section, emphasis will be placed on anabolic sequences (biochemical processes involved in synthesis of cell constituents) as they function in Eubacteria and Archaea **(Box 10.1).** Eukaryotic organisms that can synthesize these cellular constituents do, in all likelihood, follow equivalent biosynthetic routes. Indeed, studies on the metabolic sequences in both bacteria and fungi have been of fundamental importance to our understanding of the biochemistry of animal systems.

There are four major classes of macromolecules in a eubacterial or archaeal cell: proteins, nucleic acids, polysaccharides, and lipids. These macromolecules are synthesized from a limited number of monomers. A generalized scheme of cellular synthesis is presented in Figure 10.1.

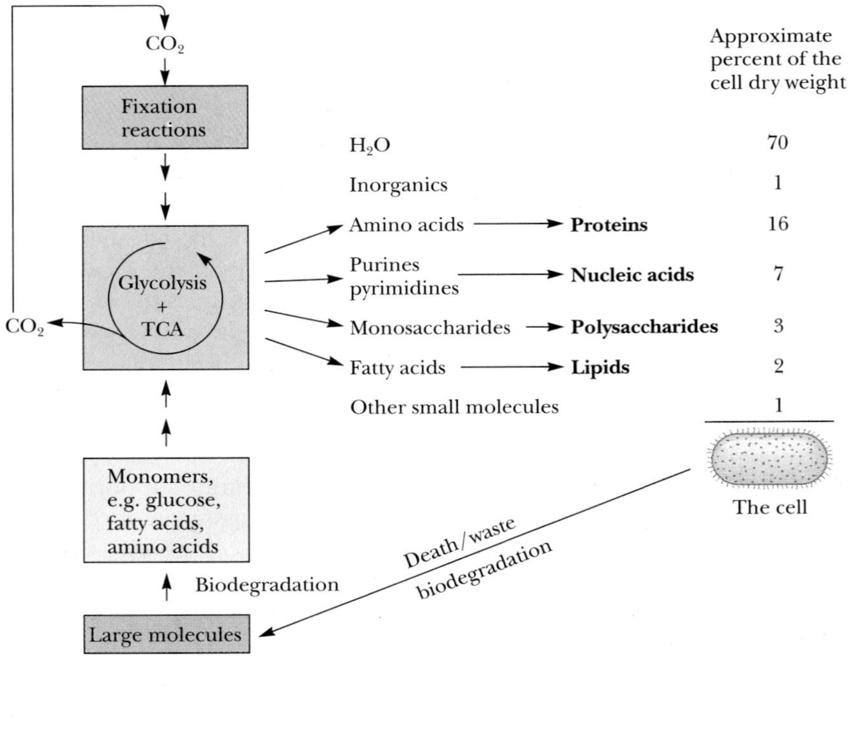

***Figure* 10.1** An overview of the reactions involved in cellular synthesis and biodegradation.

BOX 10.1 MILESTONES

The Microbe as a Paradigm

Back in 1930, the eminent Dutch microbiologist A. J. Kluyver gave a series of lectures on the metabolism of microorganisms at the University of London. His was a clear exposition on the status of the field at that time. These lectures were published in a classic treatise entitled *The Chemical Activities of Micro-Organisms* (1931, University of London Press, Ltd.). It was through the landmark studies of Kluyver and the Delft School of Microbiologists (M. Beijerinck, C. B. van Niel, and others) that bacteria became the paradigm for experimentation in biochemistry of the cell. Below is a quote from one of Kluyver's elegant lectures as it was presented to the audience in 1930.

> *However, in so far as the biochemical achievements of higher animals are concerned, it is possible that we may be dealing mainly with a rearrangement of the component units of the food, since the necessity for a complex food for this class of organisms is a well-established fact. Also in the higher, chlorophyll-containing plants, the biochemical miracle still remains obscure, since attention is focused upon the essential function of light in the metabolism of these organisms. This external energy supply enables the green cell to carry out chemical processes and it can only be estimated with difficulty how far the influence of the light on metabolism extends.*
>
> *All this becomes different when we consider the metabolism of the colourless micro-organisms. For here we find the most remarkable fact—the significance of which is often under-estimated, although known since the time of Pasteur—that a single organic compound suffices to ensure a perfectly normal development of these organisms, although they are cut off from any external energy supply. Here we find the biochemical miracle in its fullest sense, for we are bound to conclude that all the widely divergent chemical constituents of the cell have been built up from the only organic food constituent, and that without any intervention of external energy sources. The chemical conversions performed by these organisms rather resemble witchcraft than chemistry!*

In 1954 Kluyver and Van Niel presented the "John M. Prather Lectures" at Harvard University. These lectures were published in another classic entitled *The Microbe's Contribution to Biology* (1956, Cambridge: Harvard University Press). This series of lectures confirmed that, in the years between 1931 and 1954, studies on microorganisms did indeed teach us much about the biochemistry of the cell. The titles of the two books indicate the advances made in the intervening years.

According to this outline, CO_2 assimilation (fixation) reactions lead in a few steps into glycolysis and the tricarboxylic acid (TCA) cycle. The TCA cycle is also called the Krebs cycle, as it was originally proposed by Hans Krebs. Other substrates lead into these two key cycles as shown on the lower part of the diagram. All of the major monomers, such as amino acids, purines, and pyrimidines, that are polymerized to form macromolecules are readily synthesized from intermediates in glycolysis, TCA, and related pathways. (This is discussed in detail later.) The unity and simplicity (but also complexity) of biosynthetic reactions in the eubacterial or archaeal cell is illustrated in Figure 10.2. There are 75 to 100 low molecular weight compounds including sugars, the fatty acids, and amino acids derived directly or indirectly from sugars that are utilized in the processes of biosynthesis. In autotrophs these low molecular weight compounds come either directly or indirectly from CO_2 fixation.

Integration of sugar metabolism with the TCA cycle results in the generation of **precursor metabolites** at a

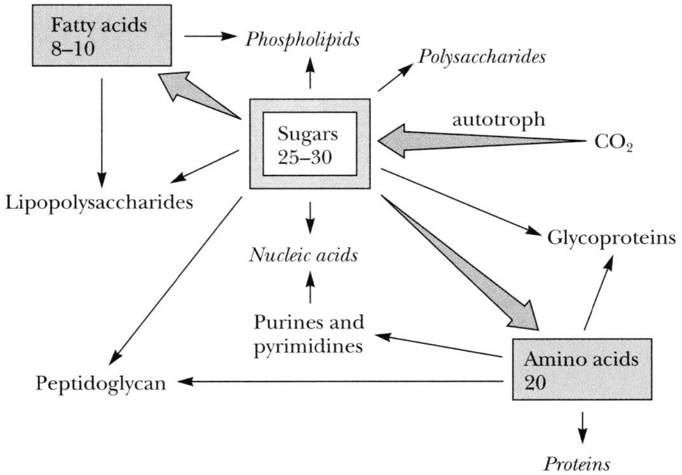

***Figure* 10.2** An outline of the central role that a limited number of compounds play in the biosynthesis of macromolecules. Sugars (for example, glucose) are among the most abundant compounds in the biosphere and a principal in providing microorganisms with the monomers that are assembled to form a viable organism.

BOX 10.2 MILESTONES

Economics of Growth

"Metabolic diversity" describes the seemingly endless capacity of microorganisms to utilize a vast array of naturally occurring chemicals (and products of chemical synthesis) as substrate for growth. They expeditiously convert these disparate substrates into the amino acids, purines, sugars, and fatty acids that make up the cell. How can a microorganism do this?

Evolution in microorganisms as evident in the processes of anabolism has proven to be exceedingly economical. Economical as defined by Webster—"marked by careful, efficient, and prudent use of resources: operating with little waste." Once Mother Nature chanced upon an effective mechanism for accomplishing a task, that mechanism became *de rigueur.* The heart and soul of anabolism then is two pathways that occur in one form or another in virtually all living cells—glycolysis and the tricarboxylic acid cycle. This is the main stream, and from this stream comes the precursors of the macromolecules that make up the cell. A simple depiction might be as follows:

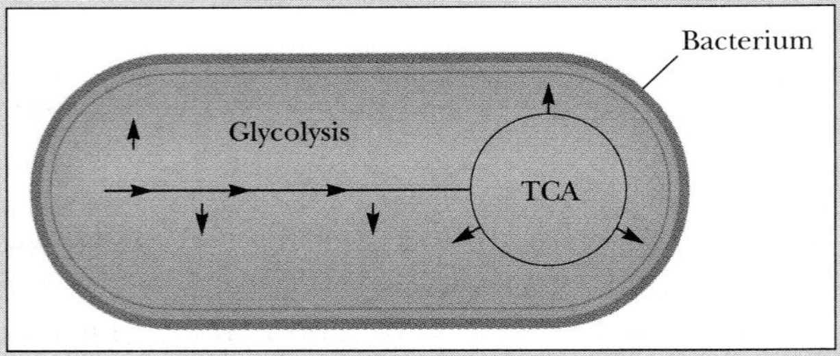

What then of "metabolic diversity"? How does the ability to utilize so many different substrates fit into the economics of evolution? Neatly—because the processes of catabolism lead directly to the main stream. All substrates that can be utilized by microorganisms as carbon source are, by a limited number of enzymatic reactions, converted to acetate, sugars, keto acids, and so forth, and these can be funneled directly into the glycolytic or TCA main stream. This would be depicted as follows:

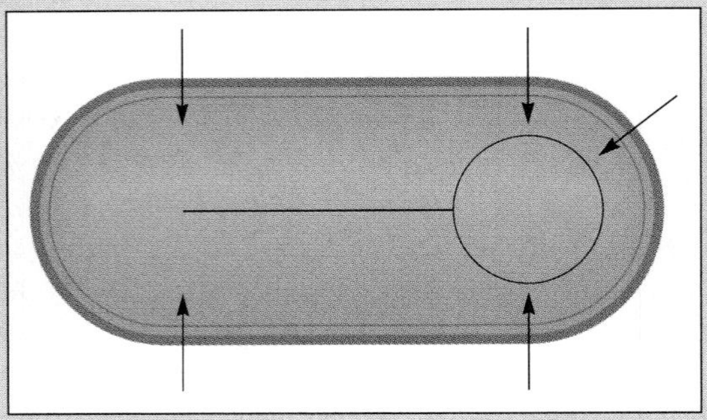

Together these catabolic and anabolic reactions generate a new cell economically and expeditiously.

The precursors that are the source of monomers for synthesis of all cell components originate in core pathways. All come from glycolysis, the tricarboxylic acid cycle, or related pathways.

All growth substrates for microorganisms enter into the core pathways after a limited number of reactions to provide the organism with the precursor metabolites.

continuous and appropriate level to promote orderly cellular reproduction **(Box 10.2).** The precursor metabolites are 12 key metabolites that are involved in the biosynthesis of all building blocks, prosthetic groups, coenzymes, and other cellular components. These metabolites and their pathways of origin are listed in Table 10.1.

The following discussion presents the metabolic reactions whereby these key precursor metabolites are generated in bacteria.

Core Reactions That Generate Biosynthetic Precursors

The key role of CO_2 as a basic building block and a product of biodegradation has been mentioned. The primary products of CO_2 fixation and of the various one-carbon assimilation mechanisms are compounds that are readily

Table 10.1 Major precursor metabolites in biosynthetic reactions

	Predominant Pathway of Origin[1]			
Metabolite	**EM**	**TCA**	**HMS**	**ED**
Glucose-6-phosphate	X		X	X
Fructose-6-phosphate	X			
Ribose-5-phosphate or ribulose-5-phosphate			X	
Erythrose-4-phosphate			X	
Glyceraldehyde-3-phosphate	X		X	X
3-Phosphoglycerate	X			
Phosphoenolpyruvate	X			
Pyruvate	X		X	X
Acetyl CoA	X		X	X
Oxaloacetate		X		
α-ketoglutarate		X		
Succinyl CoA		X		

[1]EM = Embden-Meyerhof pathway; TCA = tricarboxylic acid cycle; HMS = hexose monophosphate shunt; ED = Entner-Doudoroff pathway.

carried into the mainstream of anabolism. The precursor metabolites (Table 10.1) originate during glucose catabolism via the **Embden-Meyerhof (EM) pathway** or the **hexose monophosphate shunt,** or are products of the **tricarboxylic acid (TCA) cycle.** Some precursor metabolites are also intermediates in **Entner-Doudoroff (ED) pathway.** All free-living microorganisms obtain precursor metabolites by doing one or more of the following: (1) using these pathways, (2) operating variations of these pathways that yield equivalent products, or (3) manifesting a requirement for an anabolic product generally synthesized from a given precursor metabolite. It should be noted that a rich growth medium can provide many of the precursor metabolites and intermediates generally synthesized from them.

The three major pathways for glucose metabolism are presented in Figures 10.3, 10.4, and 10.5. The Embden-Meyerhof (EM) pathway, also termed **glycolysis,** is the major pathway for glucose catabolism and yields a net of two molecules of ATP per molecule of glucose utilized. *Two molecules of ATP* are required for the conversion of glucose to fructose-1,6-diphosphate, and two molecules of triose phosphate are formed by the aldolase cleavage of the fructose 1,6-diphosphate. Esterification of the two trioses (glyceraldehyde-3-phosphate) with inorganic phosphate yields two molecules of 1,3-diphosphoglycerate. Each phosphate group (two per each of the two molecules of 1,3-diphosphoglyceric acid) is used in the generation of a molecule of ATP (*four* total).

The Entner-Doudoroff (ED) pathway yields one ATP with the formation of two molecules of pyruvate (Figure 10.4). This pathway is of widespread occurrence in *Pseudomonas, Azotobacter,* and other gram-negative genera, but it does not occur to any great extent in gram-positive or anaerobic bacteria. The Entner-Doudoroff pathway is of greatest utility in microorganisms that use gluconate or hexuronate as substrate. Some organisms such as *E. coli* employ glycolysis for glucose metabolism, but when gluconate is provided as substrate, the key enzymes of the ED pathway are induced.

The **hexose monophosphate shunt (HMS)** is important in microorganisms during growth on hexoses and pentoses. The key reactions involved in this pathway are shown in Figure 10.5. The decarboxylation of 6-phosphogluconate provides the microorganism with pentoses that are essential for the biosynthesis of nucleotides and other cellular components. The hexose monophosphate shunt is also a ready source of NADPH, the favored nucleotide in biosynthetic reactions. Generally, the HMS is a secondary pathway operative in organisms that utilize the EMP or ED as the major dissimilatory pathway for glucose.

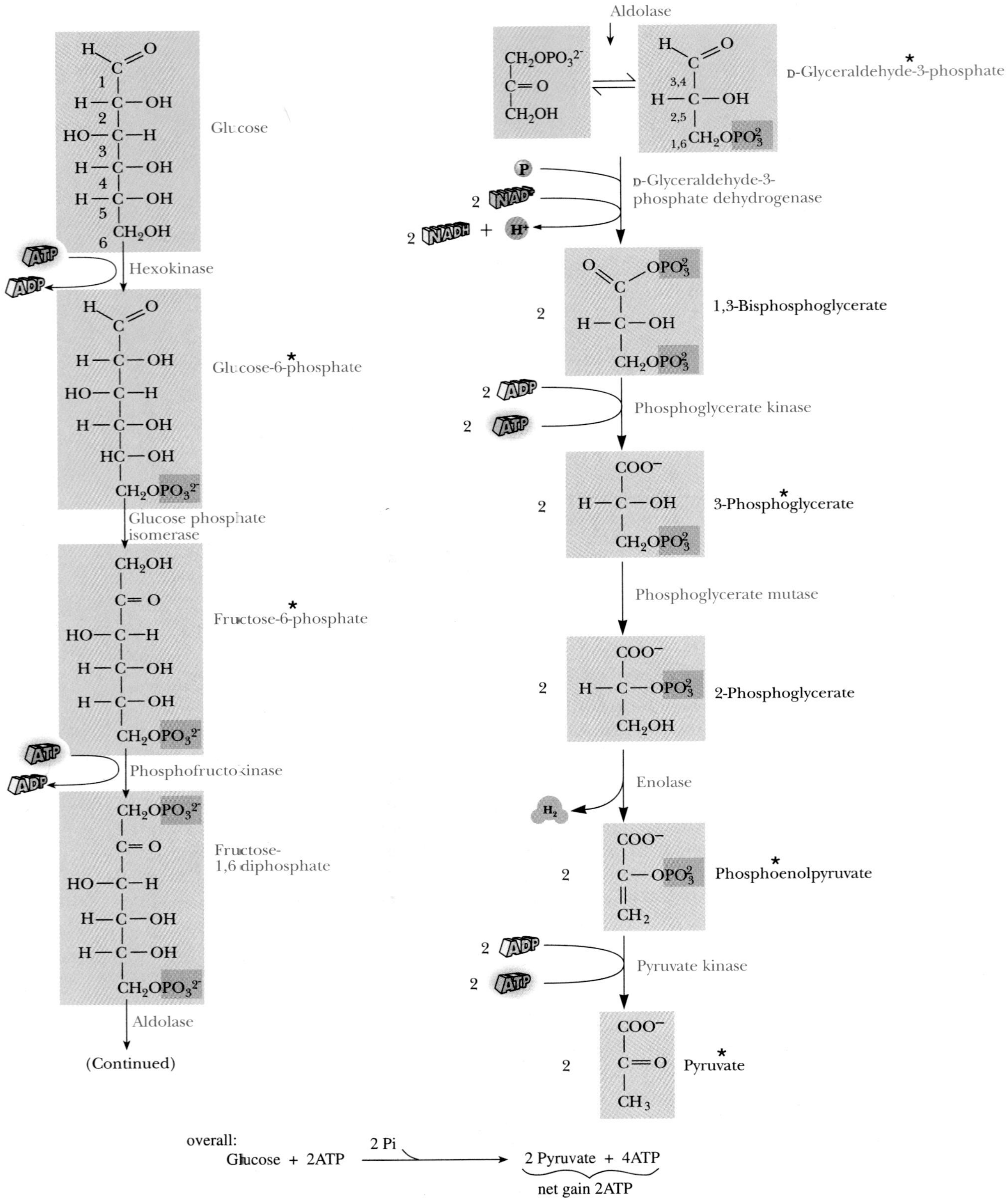

***Figure* 10.3** The Embden-Meyerhof (EM) pathway for glucose metabolism. This pathway is also termed glycolysis. The reactions from glucose to glyceraldehyde-3-phosphate are rearrangements of the molecule as a prelude to the major event—the incorporation of inorganic phosphate (substrate level phosphorylation) to form an "energy rich" anhydride bond in 1,3-diphosphoglyceric acid. The remainder of the reactions are involved in recovering all of the potential energy available. Precursor metabolites are marked by a star (*).

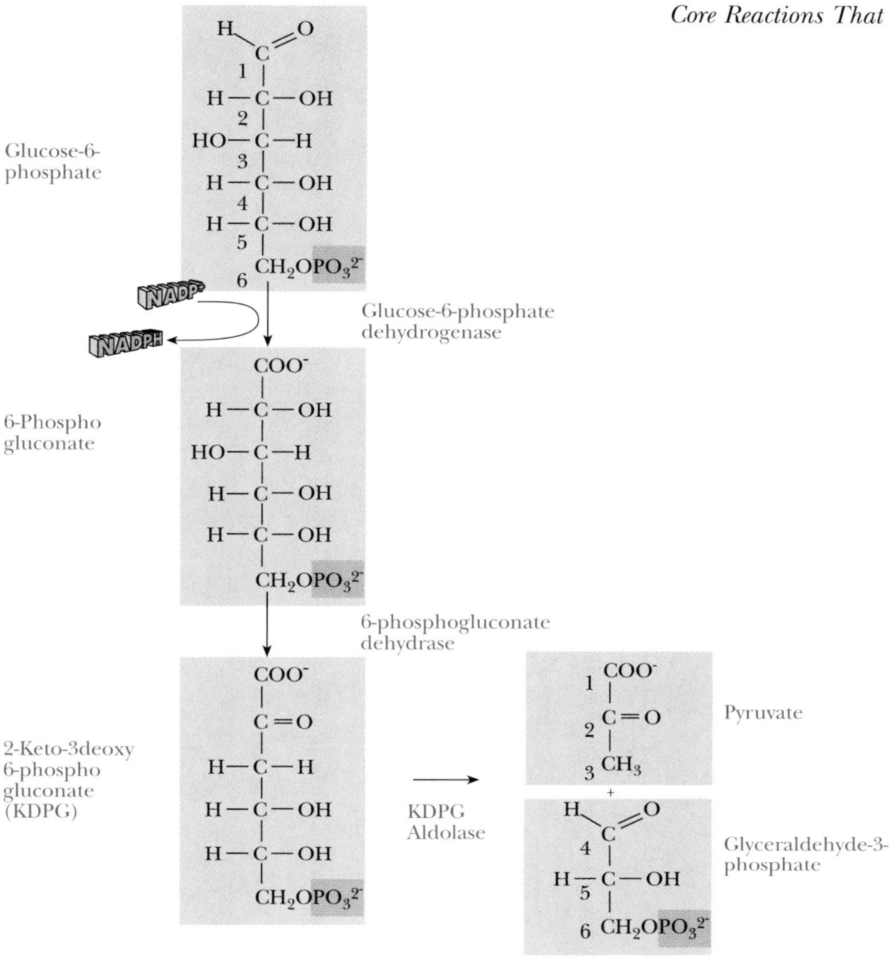

***Figure* 10.4** The Entner-Doudoroff (ED) pathway for glucose metabolism. This pathway is common among the pseudomonads and some of the other gram-negative bacteria. The steps to glucose-6-phosphate and from glyceraldehyde-3-phosphate to pyruvate are equivalent to those in glycolysis (Figure 10.3). This pathway generates only one net ATP in catabolizing one molecule of glucose to two molecules of pyruvate.

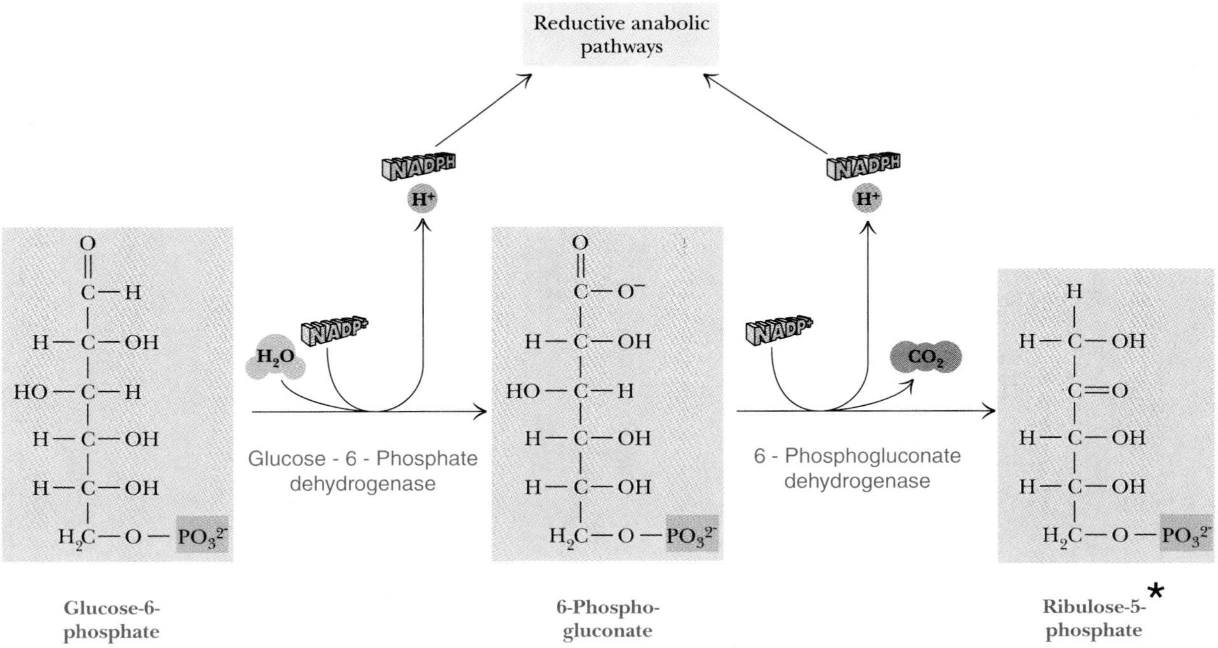

***Figure* 10.5** The hexose-monophosphate shunt (HMS), also known as the pentose phosphate pathway. This is a major source of the pentose sugars required for synthesis of nucleic acids and other cell constituents. The NADPH produced is involved in biosynthetic reactions. The ribulose-5-phosphate is a key sugar in the Calvin cycle and formation of other important intermediates (Figure 10.10).

Glucose Metabolism and Synthesis of Precursor Metabolites

The involvement of glycolysis, the hexose monophosphate shunt, and Entner-Doudoroff pathway in generating essential metabolites was outlined in Table 10.1. Pyruvate is a product of the three pathways of glucose catabolism—EM, HMS, and ED. Aerobically a microorganism can use pyruvate for the useful production of reduced pyridine nucleotides, ATP, and precursor metabolites. Anaerobically an organism utilizes pyruvate or products of pyruvate catabolism as terminal electron acceptor or a source of precursor metabolites. The balance between CO_2/precursor production in an aerobe depends on the composition of the growth medium and other factors. Aerobes gain energy by decarboxylation of pyruvate to acetyl CoA. The acetyl CoA generated by the decarboxylation of pyruvate enters the tricarboxylic acid cycle by condensing with a molecule of oxaloacetate. Aerobic organisms may completely oxidize acetate to CO_2 and water. The TCA cycle also supplies the organism with precursor metabolites—α-ketoglutarate, oxaloacetate, and succinyl CoA.

The Tricarboxylic Acid Cycle or Krebs Cycle

The reactions involved in the tricarboxylic acid cycle are presented in Figure 10.6. One turn of the cycle can result in the complete oxidation of one molecule of acetate with the concomitant generation of about 15 molecules of ATP and two of CO_2. This ATP generation occurs in aerobes that can extract the maximum amount of energy from NADH through respiration (see Chapter 8). A modified tricarboxylic acid cycle is employed by anaerobes mostly

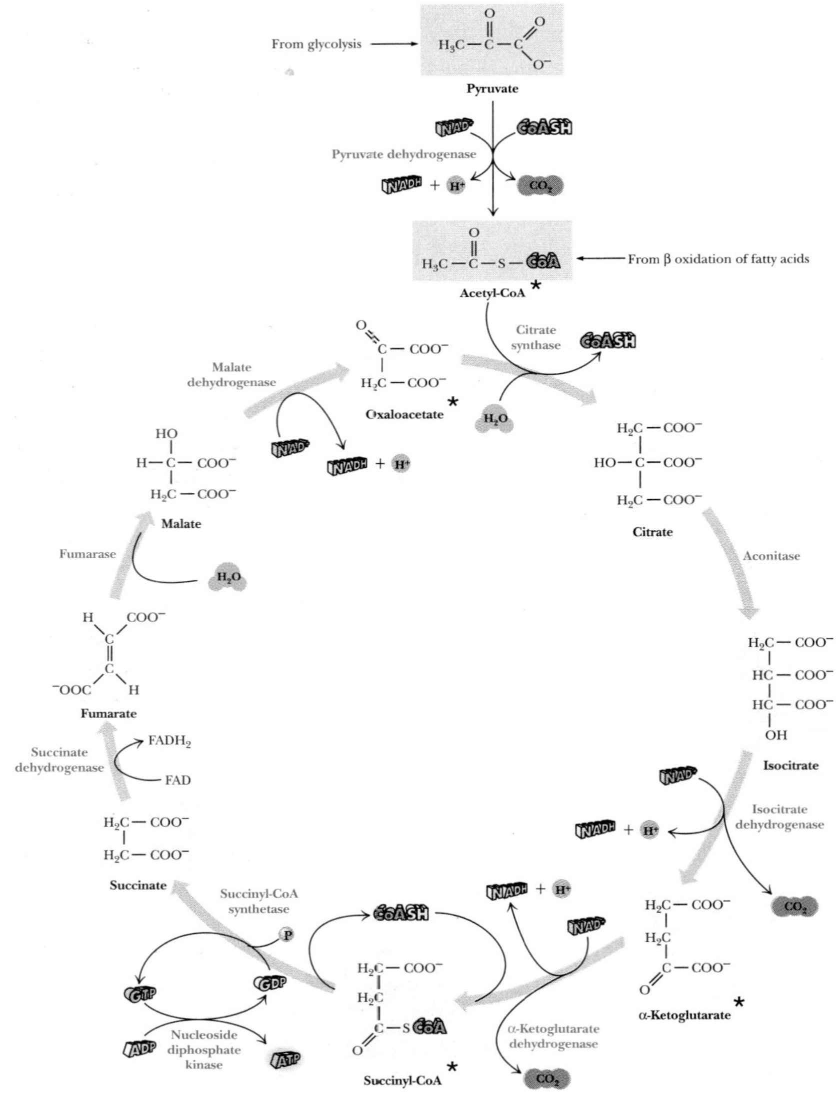

***Figure* 10.6** The Krebs cycle (tricarboxylic acid cycle or TCA). One molecule of pyruvate is metabolized aerobically to three molecules of CO_2 with a potential gain of 15 ATP. The precursor metabolites generated by the TCA cycle are starred. Microorganisms grown under anaerobic conditions generally do not synthesize the enzyme α-ketoglutarate dehydrogenase, and oxaloacetate is generated by carboxylation of phosphoenolpyruvate. Succinyl CoA is obtained by reversal of the reactions on the left-hand side of the diagram.

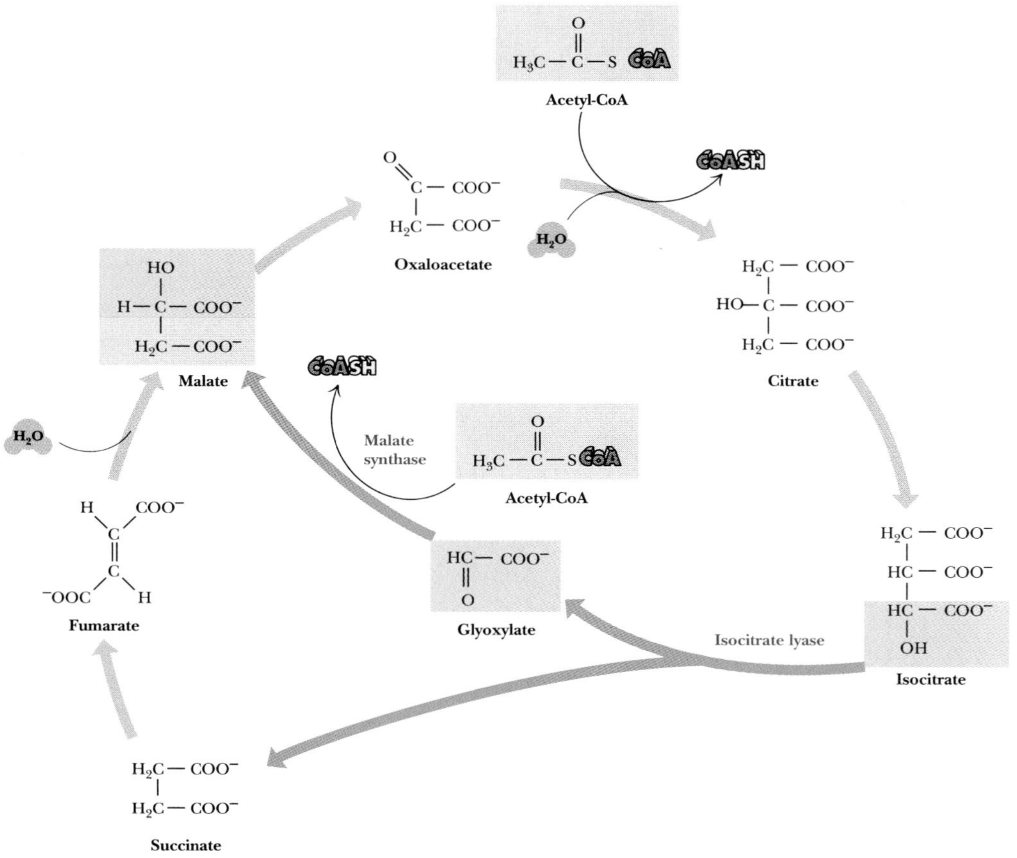

***Figure* 10.7** The glyoxylate cycle. This cycle is *induced* in microorganisms that are grown on acetate or other substrates (long-chain fatty acids) that are metabolized through a 2-carbon intermediate. The genetic information for the synthesis of an induced enzyme would be expressed only when an appropriate substrate is available. The enzyme isocitrate lyase is induced by growth on acetate. Note the number of common intermediates in this and the TCA cycle.

to provide biosynthetic intermediates. The enzyme α-ketoglutarate dehydrogenase is not generally synthesized anaerobically and, where necessary, succinyl CoA is formed by reduction of oxaloacetate. In a microorganism growing anaerobically, oxaloacetate can be synthesized by carboxylation of phosphoenolpyruvate.

Biosynthetic reactions withdraw intermediates from the TCA cycle (α-ketoglutarate to glutamate, succinyl CoA to porphyrin synthesis, etc.) and these reactions deplete the 4-carbon intermediate (oxaloacetate) that combines with acetate to form citrate. For the TCA cycle to continue the oxaloacetate must be replenished. This replenishment can be achieved in most organisms by the carboxylation of pyruvate or phosphoenolpyruvate to form a molecule of oxaloacetate. Any reaction that replenishes intermediates in a central pathway such as this is called an **anaplerotic** reaction.

The **glyoxylate shunt** is a variation of the TCA cycle and it is of considerable consequence to organisms growing on acetate, fatty acids, long-chain n-alkanes, or other substrates that are catabolized through 2-carbon intermediates. Because anabolic reactions remove intermediates from the TCA cycle, it is essential that organisms growing on 2-carbon substrate have a mechanism for replenishing substantial quantities of oxaloacetate. Obviously while growing on acetate the microorganism would not have available a 3-carbon intermediate such as phosphoenolpyruvate that can be carboxylated to form the requisite oxaloacetate. Organisms that can grow on substrates such as acetate generally have the inducible glyoxylate shunt (Figure 10.7). Isocitrate is cleaved by the inducible isocitrate lyase, yielding succinate and glyoxylate. The glyoxylate thus formed can condense with a molecule of acetate to form a molecule of malate. Thus an organism growing on acetate would have an unlimited supply of the 4-carbon intermediate to continue the cycle. Precursor metabolites such as α-ketoglutarate and succinyl CoA would be made available by reactions associated with the TCA cycle.

The Calvin Cycle–CO_2 Fixation

The major pathway for utilization of carbon dioxide as the sole source of carbon relies on a series of interacting processes designated the **Calvin cycle.** There are two major events that render an organism capable of autotrophic growth via this cycle. One is the carboxylation reaction that leads to the assimilation of a molecule of CO_2. The other is the molecular rearrangements that regenerate the CO_2 acceptor molecule. A summary of these integrated processes is presented in Figure 10.8.

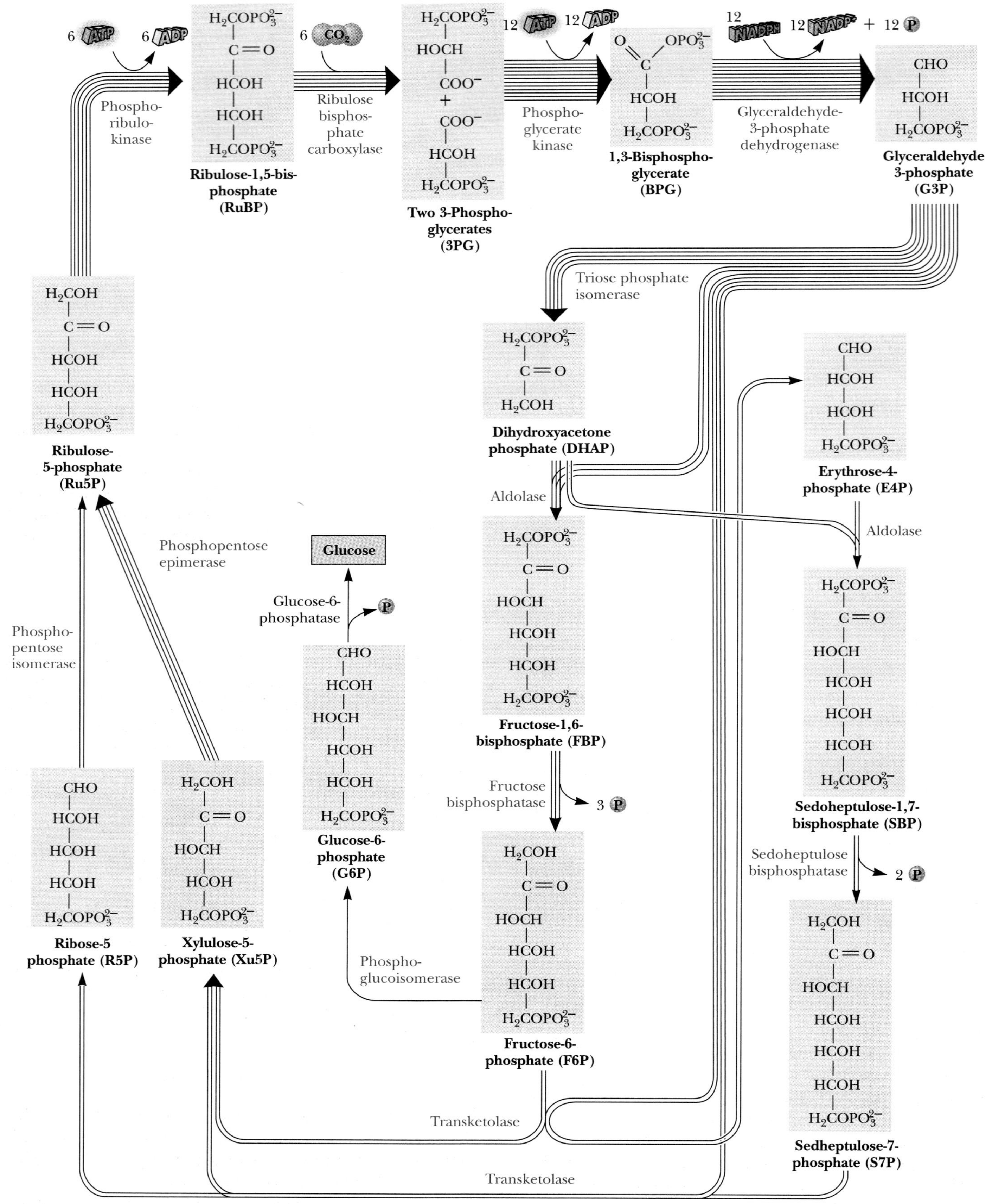
6 ATP
6 ADP
Phospho-ribulo-kinase
Ribulose-1,5-bis-phosphate (RuBP)
6 CO2
Ribulose bisphos-phate carboxylase
Two 3-Phospho-glycerates (3PG)
12 ATP
12 ADP
Phospho-glycerate kinase
1,3-Bisphospho-glycerate (BPG)
12 NADPH
12 NADP+
+ 12 P
Glyceraldehyde-3-phosphate dehydrogenase
Glyceraldehyde 3-phosphate (G3P)
Triose phosphate isomerase
Dihydroxyacetone phosphate (DHAP)
Aldolase
Fructose-1,6-bisphosphate (FBP)
Fructose bisphosphatase
3 P
Fructose-6-phosphate (F6P)
Erythrose-4-phosphate (E4P)
Aldolase
Sedoheptulose-1,7-bisphosphate (SBP)
Sedoheptulose bisphosphatase
2 P
Sedheptulose-7-phosphate (S7P)
Ribulose-5-phosphate (Ru5P)
Phosphopentose epimerase
Phospho-pentose isomerase
Glucose
Glucose-6-phosphatase
P
Glucose-6-phosphate (G6P)
Phospho-glucoisomerase
Ribose-5 phosphate (R5P)
Xylulose-5-phosphate (Xu5P)
Transketolase
Transketolase

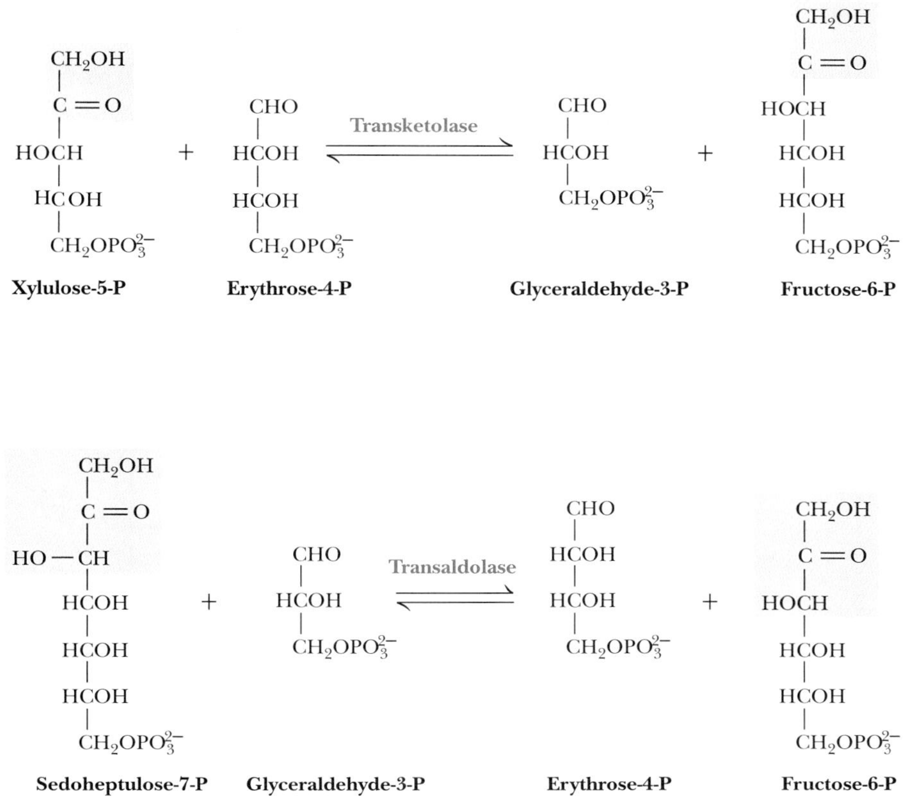

***Figure* 10.9** The involvement of transketolases and transaldolases in the interconversion of sugars. In a transketolase reaction the two carbon units from xylulose-5-phosphate can be transferred to erythrose-4-phosphate to form fructose-6-phosphate. In the transaldolase reaction the dihydroxyacetone unit from sedohepulose-7-phosphate is coupled with glyceraldehyde to form fructose-6-phosphate.

A primary reaction in the Calvin cycle is the phosphorylation of ribulose-5-phosphate to produce ribulose-1,5-bisphosphate. The enzyme involved in assimilating a molecule of CO_2 is ribulose-1,5-bisphosphate carboxylase/oxygenase (RuBisCo). The carboxylase incorporates a molecule of CO_2 into ribulose-1,5-bisphosphate with a simultaneous cleavage to form two molecules of 3-phosphoglycerate. It is noteworthy that 3-phosphoglycerate is an intermediate in glycolysis (see Figure 10.3) and further that glycolysis is a major source of precursor metabolites.

The ready regeneration of ribulose-5-phosphate would be of utmost importance to autotrophic organisms that use the Calvin cycle for CO_2 fixation. The first step in the regeneration process is the conversion of 3-phosphoglycerate to an equilibrium mixture of dihydroxyacetone phosphate and glyceraldehyde-3-phosphate, which can be condensed to form fructose-6-phosphate. The key enzymes transaldolase and transketolase then become involved, and the carbons in fructose-6-phosphate are rearranged to generate the 4-, 5-, and 7-carbon sugars, as shown in Figure 10.8. Fructose-6-phosphate can be isomerized enzymatically to glucose-6-phosphate, an important precursor metabolite in biosynthetic reactions. A key intermediate in regenerating ribulose-5-phosphate is sedoheptulose-7-phosphate because two carbons from this sugar can be added to a readily available 3-carbon compound (glyceraldehyde-3-phosphate) resulting in the formation of two molecules of 5-carbon sugar. The mechanisms whereby the enzymes transaldolase and transketolase function are outlined in Figure 10.9. These reactions also occur in heterotrophic organisms in conjunction with the hexose monophosphate shunt and supply an organism with ribulose-5-phosphate and erythrose-4-phosphate. They are also of importance in the biodegradation of various sugars (see Figure 12.3).

◀ ***Figure* 10.8** The Calvin Cycle. This CO_2 fixation pathway is predominant in plants, the cyanobacteria, and most other autotrophs. The key intermediate is ribulose-1,5-bisphosphate and the key enzyme is ribulose-1,5-bisphosphate carboxylase (RuBisCo), which incorporates a molecule of CO_2 with the cleavage of the transitory intermediate formed to two molecules of 3-phosphoglycerate. Note that 3-phosphoglycerate is an intermediate in glycolysis (Figure 10.3). The other reactions depicted are involved in regenerating ribulose-1,5-bisphosphate. The number of arrows at each step indicates the number of molecules reacting in a turn of the cycle that generates one molecule of glucose.

Alternate Pathways of 1-Carbon Assimilation

Elucidation of the Calvin cycle led some to assume that this cycle might be a universal mechanism of autotrophic 1-carbon assimilation. Experiments with green sulfur photosynthetic bacteria and other autotrophs, however, clearly showed that there must be other distinctly different pathways for CO_2 assimilation in microorganisms. When radiolabeled CO_2 was placed in the atmosphere supplied for growth of green sulfur photosynthetics, the primary products were not the expected radiolabeled sugars, but were radiolabeled fatty acids. Similar experiments with organisms that utilize methane, methanol, or other 1-carbon compounds affirmed that a number of alternate pathways for assimilation of 1-carbon compounds exist.

Anaerobic Eubacteria and Archaea that occupy niches such as swamps in which electron donors more negative than NADH are available can fix CO_2 by reductive carboxylation reactions. Among these organisms are the acetogenic anaerobes, green sulfur photosynthetic bacteria, and methanogens (methane-producing bacteria). Despite the fact that these organisms fix CO_2 via a common pathway (reductive carboxylations) they are not phylogenetically related to one another.

Chlorobium thiosulfatophilum is a green sulfur photosynthetic organism and one that utilizes a reductive carboxylation mechanism for CO_2 fixation. The pathway is outlined in Figure 10.10(**a**). This is essentially a reversal of the TCA cycle shown in Figure 10.6. Acetate is a principal intermediate in this cycle, and one complete turn essentially results in the net production of one molecule of acetyl-Co A. The acetyl-Co A can readily enter into the synthetic TCA cycle and provide the organism with the es-

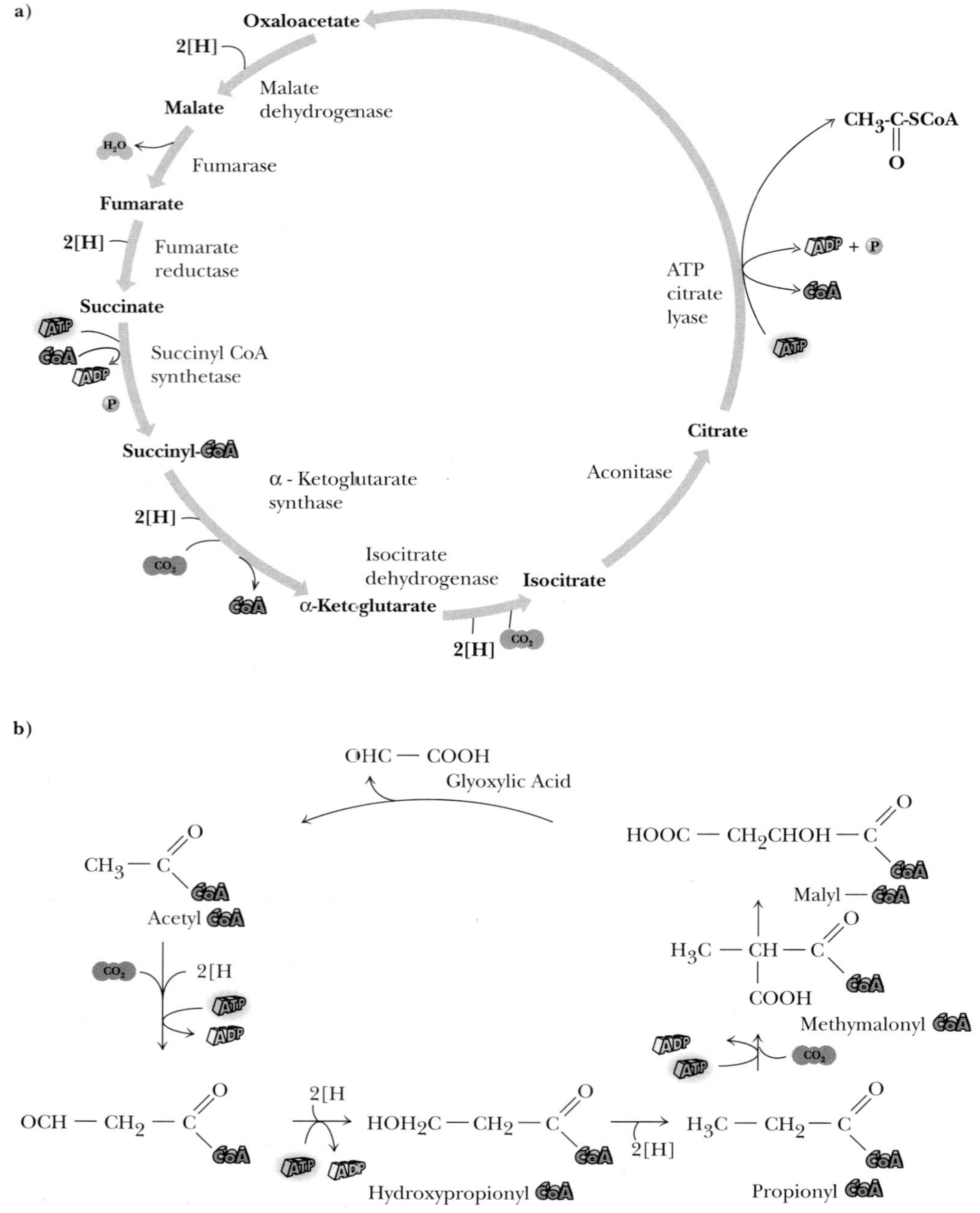

***Figure* 10.10** Two novel autotrophic pathways for CO_2 fixation. **(a)** The reductive citric acid pathway that is present in green sulfur bacteria (*Chlorobium limicola*), thermophilic hydrogen oxidizing bacteria (*Hydrogenobacter thermophilus*), and some of the sulfate-reducing bacteria (*Desulfobacter hydrogenophilus*). Note the similarity to the TCA cycle (Figure 10.6). **(b)** The hydroxypropionate pathway that is present in *Chloroflexus.* Two molecules of CO_2 are assimilated to form a molecule of glyoxylic acid.

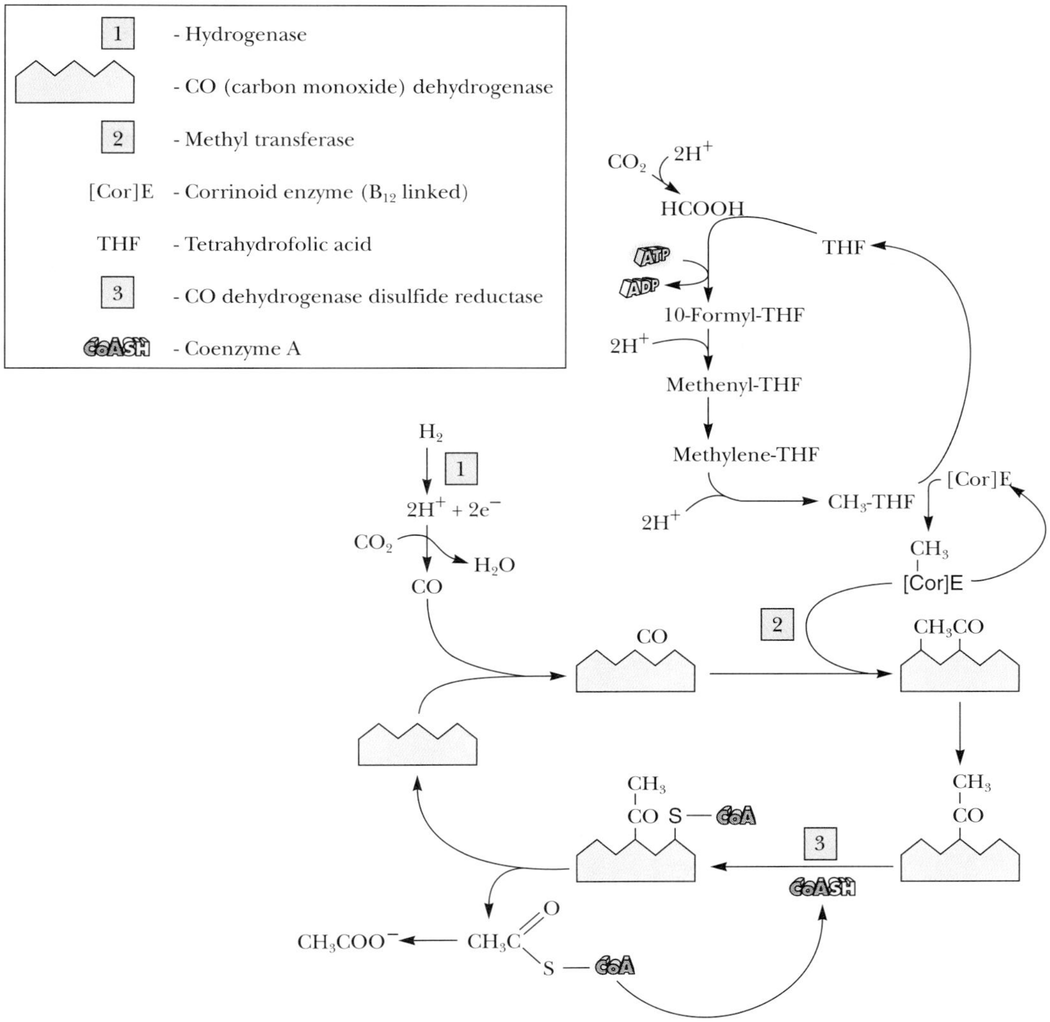

***Figure* 10.11** The reductive acetyl CoA pathway for autotrophic CO_2 fixation. This pathway is present in homoacetogenic Eubacteria (*Clostridium thermoaceticum*), most of the sulfate-reducing Eubacteria (*Desulfobacterium autotrophicum*), and selected methanogenic Archaea (*Methanosarcina barkeri*). One turn of this cycle generates a molecule of acetate from two CO_2.

sential precursor metabolites. Decarboxylation of oxaloacetate would provide the organism with pyruvate. Glycolysis leads from glucose to pyruvate, and a reversal of glycolysis can generate a molecule of glucose (two pyruvate $\longrightarrow \longrightarrow \longrightarrow$ glucose). Members of genus *Chloroflexus* have a unique pathway for CO_2 fixation. This is the hydroxypropionate pathway as illustrated in Figure 10.10(**b**). Hydroxypropionyl CoA and propionyl CoA are key intermediates, and one cycle leads to the formation of one molecule of glyoxylic acid.

A number of organisms generate acetate as a major product of anaerobic respiration. These organisms are called **acetogenic bacteria.** They produce acetate essentially by coupling two molecules of CO_2, with H_2 as their source of energy. Among these organisms is *Acetogenium kivui,* a chemolithotrophic anaerobe that obtains carbon and energy according to the following equation:

$$4\,H_2 + 2\,CO_2 \longrightarrow CH_3COO^- + 2\,H_2O$$

A general scheme for the energetics and acetate synthesis in acetogenic bacteria is presented in Figure 10.11. A key enzyme in the pathway is carbon monoxide (CO) dehydrogenase. A molecule of carbon monoxide is bound to this enzyme followed by the addition of a methyl group donated by the vitamin B_{12} corrinoid enzyme. This methyl group originates on a tetrahydrofolic acid-protein complex through the reduction of CO_2 (see Figure 10.11). The methyl group migrates and couples to the −CO attached to the CO dehydrogenase enzyme complex forming the acetyl derivative. A molecule of coenzyme A is attached to the complex, and the CO dehydrogenase catalyzes the release of acetyl CoA. The three compounds, a molecule of coenzyme A (CoA), the methyl group, and carbon monoxide are independently bound to the enzyme. The molecule of acetate formed in this series can be utilized in the synthesis of cellular components. These organisms would be considered autotrophs because they synthesize their cellular material from CO_2 and utilize H_2 as source of energy.

Many microorganisms can grow with methane as the

sole source of carbon and energy. Methanol is also utilized by bacteria and yeasts. Both methane and methanol are oxidized to the formaldehyde (HCHO) level, and this molecule can be assimilated by two distinctly different pathways:

1. Glycine + HCHO $\longrightarrow$ serine
2. Ribulose-5-phosphate + HCHO $\longrightarrow$ fructose-6-phosphate

These pathways are outlined in Figures 10.12(**a,b**). In the glycine-serine pathway, the amino acid glycine becomes the acceptor of the one-carbon intermediate. Energy is derived aerobically and occurs by the oxidation of CH_4 or CH_3OH to CO_2. In the cycle illustrated (Figure 10.12**a**), one mole of formaldehyde and one mole of CO_2 are assimilated in each turn of the cycle.

The other distinct group of methane-utilizing microorganisms employs a different mechanism for formaldehyde assimilation. This group assimilates the formaldehyde via a **ribulose monophosphate pathway,** as shown in Figure 10.12(**b**). This is somewhat similar to the ribulose-1,5-bisphosphate pathway operating in most autotrophs. Both employ a 5-carbon sugar and ultimately generate fructose-6-phosphate that can be rearranged by the enzymes transaldolase and transketolase to regenerate ribulose-5-phosphate. All of these reactions and modifications of the tricarboxylic acid cycle as occurs in anaerobes are a source of precursor metabolites for cell synthesis.

Nitrogen and Sulfur Incorporation

Nitrogen is a major component of all living cells and constitutes about 14 percent of the dry weight of a bacterium. The Eubacteria and the Archaea generally use inorganic nitrogen in the form of the ammonium ion (NH_4^+) or nitrate (NO_3^-), while some species obtain cellular nitrogen by reduction of atmospheric dinitrogen (N_2). The ammonium ion (NH_4^+) is a universally utilizable source of inorganic nitrogen for bacterial growth, while the ability to assimilate nitrate is more restricted. The number of types that can fix dinitrogen (N_2) is more limited (see Table 5.5). Both NO_3^- and N_2 enter into biosynthetic reactions through NH_4^+. Bacteria generally link NH_4^+ to organic intermediates via three major reactions. The enzymes mediating these reactions are (1) **carbamoyl-phosphate synthetase,** (2) **glutamate dehydrogenase,** and (3) **glutamate synthetase.**

Carbamoyl-phosphate is involved in the synthesis of the amino acid arginine and in the synthesis of the pyrimidine ring (see later). The synthesis of carbamoyl-phosphate occurs via the following reaction:

$$NH_4^+ + HCO_3^- + 2\,ATP \longrightarrow H_2N{-}\overset{\overset{\displaystyle O}{\|}}{C}{-}O{-}PO_3^{\,2-} + 2\,ADP + P_i + 2H^+$$

Glutamic dehydrogenase (GDH) is an enzyme that catalyzes the reductive amination of α-ketoglutarate to form glutamate (Figure 10.13**a**). The reductant for this reaction is either NADH or NADPH. **Glutamine synthetase (GS)** catalyzes an ATP dependent amidation of the γ-carboxyl of glutamate to form glutamine (Figure 10.13**b**). Glutamine is a major amine donor in the biosynthesis of major monomers in the cell including purines, pyrimidines, and a number of amino acids. Combined, the enzymes glutamic dehydrogenase and glutamine synthetase are responsible for most of the ammonium that is assimilated into cell material.

When a microorganism grows in an environment with an ample supply of NH_4^+ the following sequence of reactions would occur:

$$NH_4^+ + \alpha\text{-ketoglutarate} + NADPH \xrightarrow{GDH} \text{Glutamate} + NADP^+ + H_2O$$

$$\text{Glutamate} + NH_4^+ + ATP \xrightarrow{GS} \text{Glutamine} + ADP + P_i + H_2O$$

When NH^+ availability is limited, the glutamate dehydrogenase (GDH) reaction is not effective and glutamate synthetase (GS) becomes the sole NH_4^+ assimilation reaction. This occurs because glutamate synthetase has a much higher affinity for ammonia. Since the glutamate synthetase reaction consumes glutamate (see above) there is a need for an alternate mechanism for glutamate synthesis. The enzyme **glutamate synthase** (glutamate: oxoglutarate amino-tranferase or GOGAT) is the supplier of this glutamate through the reductive amination of α-ketoglutarate with the amide-N of gutamine as N-donor according to the following reaction:

$$NADPH + \alpha\text{-ketoglutarate} + \text{glutamine} \longrightarrow 2\ \text{glutamate} + NADP^+$$

Two molecules of glutamate are generated—one from amination of α-ketoglutarate and the other from the deamination of glutamine. Together the two enzymes GS

(a)

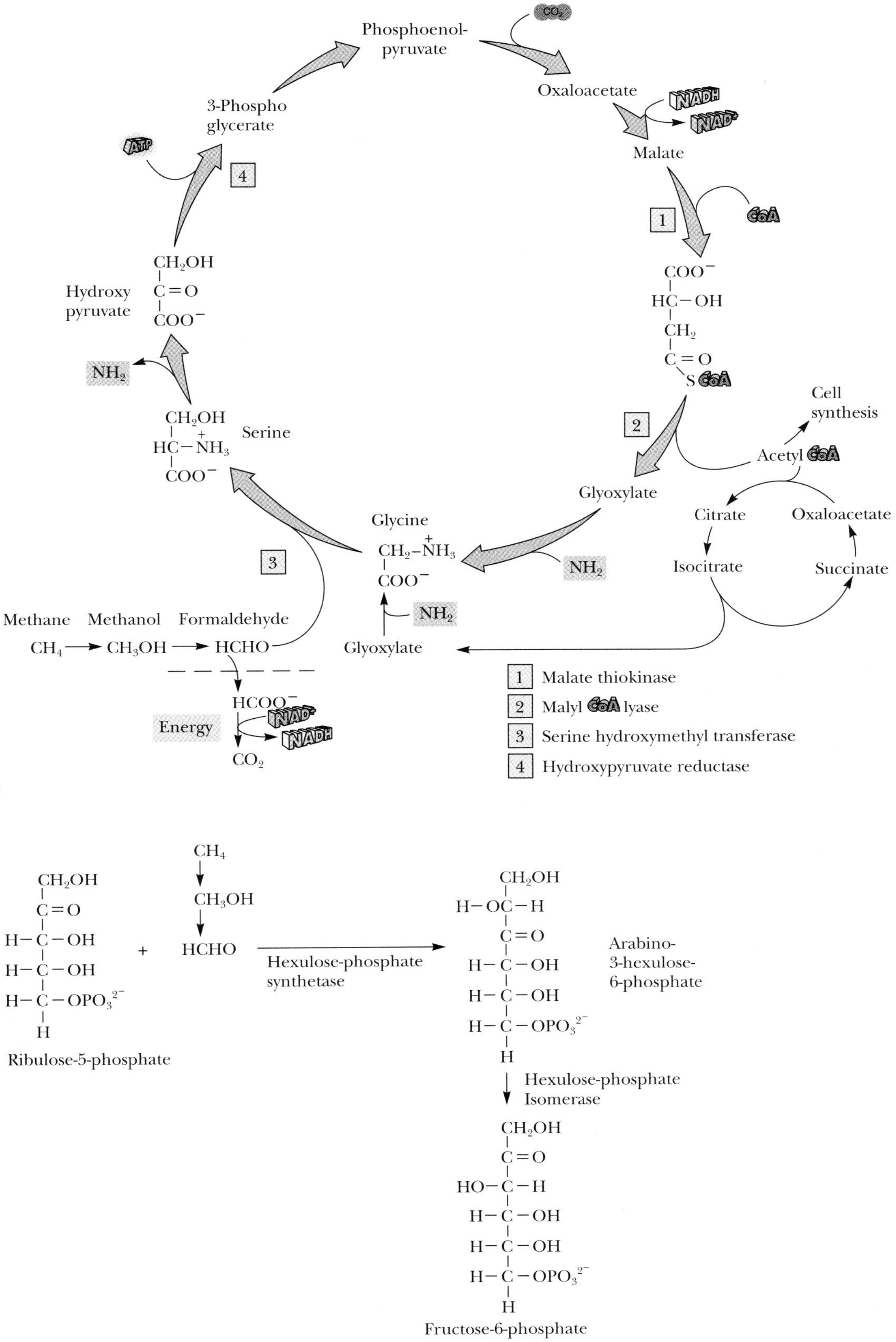

(b)

***Figure* 10.12** **(a)** The serine pathway for 1-carbon assimilation. Only structures not previously depicted are shown, and enzymes are identified that are confined to this pathway. All of the precursor metabolites can be synthesized from 3-phosphoglyceric acid (by reversal of glycolysis) or through the TCA cycle. Methane or methanol can serve as source of both carbon and energy. **(b)** The ribulose-5-phosphate pathway for assimilation of 1-carbon substrates. The ribulose-5-phosphate can be regenerated by transaldolase and transketolase reactions as outlined in Figure 10.9.

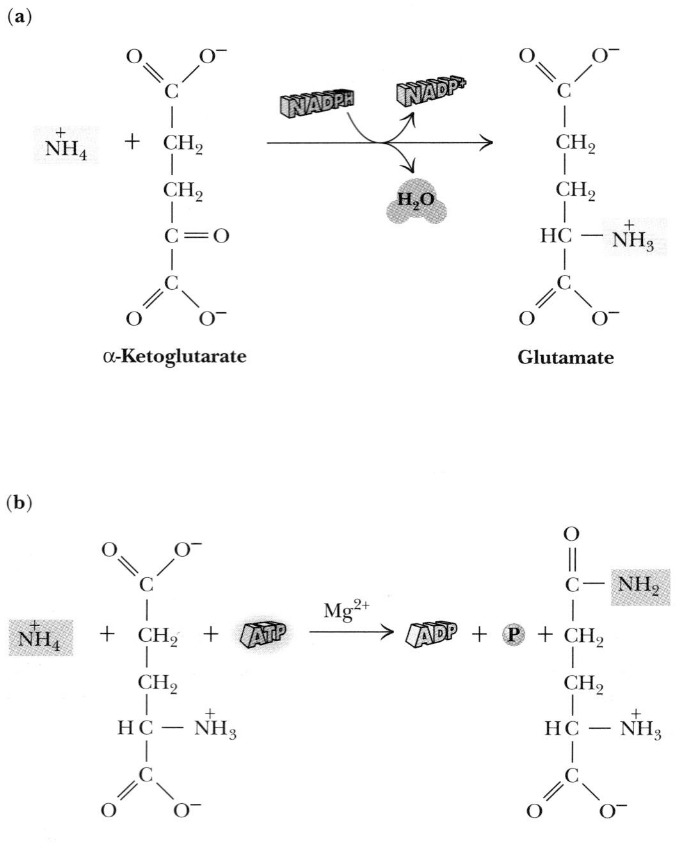

***Figure* 10.13** The glutamate dehydrogenase reaction that results in the assimilation of one molecule of ammonia **(a)** and the glutamine synthetase reaction that amidates glutamate to form glutamine **(b).**

and GOGAT constitute a major pathway of NH_4^+ assimilation and the role of GOGAT is to regenerate glutamate:

$$2\ NH_4 + 2\ ATP + 2\ \text{glutamate} \xrightarrow{GS} 2\ \text{glutamine} + 2\ ADP + 2\ P_i$$

$$NADPH + \alpha\text{-ketoglutarate} + \text{glutamine} \xrightarrow{GOGAT} 2\ \text{glutamate} + 2\ NADP^+$$

Sum:

$$2\ NH_4^+ + \alpha\text{-ketoglutarate} + NADPH + 2\ ATP \longrightarrow \text{glutamine} + NADP^+ + 2\ P_i$$

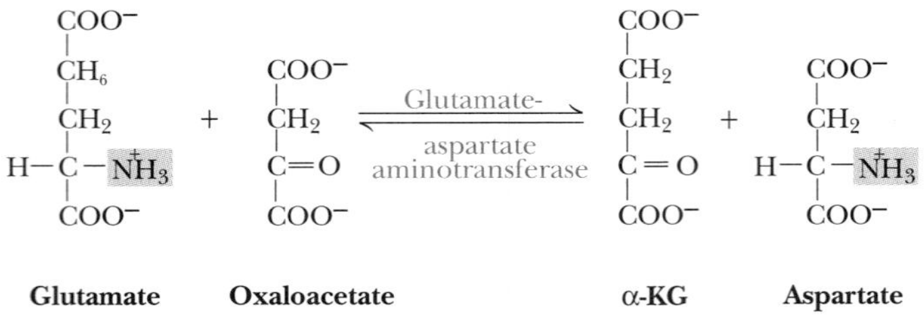

***Figure* 10.14** The glutamate-dependent transamination of α-keto acid is a fundamental mechanism for amino acid synthesis.

The amino group can be transferred from glutamic acid to other keto acids via enzymes termed **transaminases.** A typical transamination reaction is presented in Figure 10.14.

Nitrate Reduction

Two enzymatic systems are principally responsible for nitrate reduction by bacteria: (1) **dissimilatory nitrate reductase** and (2) **assimilatory nitrate reductase.** The dissimilatory pathway is induced in selected facultatively anaerobic bacteria during anaerobic growth in the presence of NO_3^-. Nitrate serves as the terminal electron acceptor, and various reduced nitrogen compounds are generated, including NH_2OH (hydroxylamine), N_2O (nitrous oxide), NO (nitric oxide), and dinitrogen (N_2). The later three compounds are gaseous. The process is termed **denitrification** and can occur in agricultural soil that becomes waterlogged and anaerobic. Denitrification is a serious problem as it can deplete the soil of nitrogen for growth of crops.

The assimilatory nitrate reductase is induced in those organisms that can utilize NO_3^- as the nitrogen source and is present only in the absence of NH_4^+. The enzymatic reaction proceeds as follows:

$$\underbrace{NO_3^- \xrightarrow{2e^-}}_{\text{nitrate reductase}}$$

$$\underbrace{NO_2^- \xrightarrow{2e^-} HNO \xrightarrow{2e^-} NH_2OH \xrightarrow{2e^-} NH_3}_{\text{nitrite reductase}}$$

The molybdenum-containing nitrate reductase system reduces the nitrogen in NO_3^-, with a valence of +5, to the ammonia level, with a valence of −3. This occurs through the joint action of the two enzymes and requires NADPH

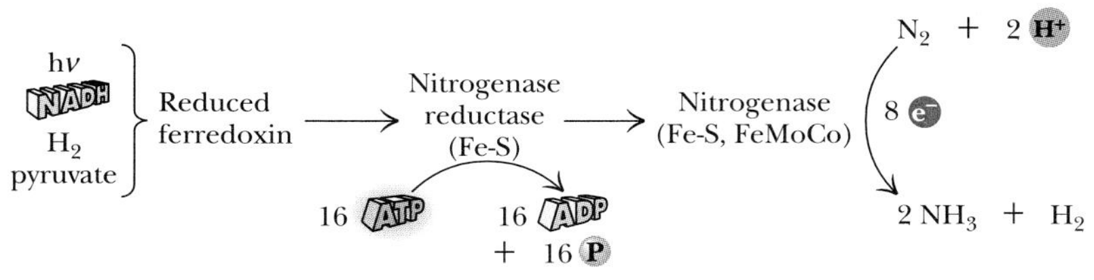

***Figure* 10.15** The nitrogenase reaction. Electrons for N_2 reduction may be generated by light, NADH, or other reactions. The primary e^- donor is reduced ferredoxin. When grown under iron-limited conditions, some bacteria can synthesize the flavoprotein flavodoxin, which is functionally equivalent to ferredoxin. Six electrons are required to reduce N_2 to 2 NH_4^+, and two more are needed for the reduction of 2 H^+ to H_2. N_2 is bound to the FeMoCo group until all requisite electrons and protons are added. There are no detectable intermediates.

and FAD; there are no free intermediates involved. Considerable energy is expended in these reactions, and consequently, NH_4^+ is preferred over nitrate as a nitrogen source in the preparation of culture media.

Nitrogen Fixation

Dinitrogen (N_2) is the major constituent of the Earth's atmosphere (80 percent), and some eubacterial and archaeal species have evolved with the capacity for **N_2 fixation** (assimilation). This N_2-fixing capacity is apparently confined to the prokaryotic world, and these reactions ultimately provide a source of nitrogenous compounds for growth of eukaryotes. The mechanisms for N_2 fixation have received considerable attention in recent years. Dinitrogen is a very unreactive molecule due to the triple bond N≡N, and because of this unreactive nature, considerable activation energy is required to reduce (fix) N_2 (i.e., convert it to NH_4^+). **Nitrogenase** (Figure 10.15) is the enzyme complex involved in N_2 fixation and is inactivated by molecular oxygen (O_2). Nitrogen fixation occurs in anaerobic, aerobic, and photosynthetic prokaryotes. All N_2 fixers that grow in the presence of air have a mechanism for keeping O_2 away from their nitrogenase system. For example, *Azotobacter sp.* are N_2-fixing aerobes but they have mechanisms for excluding O_2 from the site of fixation. Many facultative anaerobes such as *Bacillus polymyxa* fix N_2 only when growing anaerobically. Many cyanobacteria develop specialized anaerobic cells (called **heterocysts**) that become the sites of nitrogen fixation (see Chapter 20).

Nitrogenase is composed of two subunits: an iron-sulfur protein and a molybdenum-iron-sulfur protein. The overall process of nitrogen fixation is a reductive one, and the electrons required for reduction are provided by photosynthesis, respiration, or fermentation. A considerable amount of ATP (20–24 per mole N_2 fixed) is also necessary. Electrons and ~P (high-energy phosphate bonds) are passed to nitrogenase, and the reduction of N_2 occurs in the enzyme complex.

Sulfur Assimilation

Sulfur is an indispensable component of amino acids (cysteine, methionine), water-soluble vitamins (biotin, thiamine), and other key cellular constituents (Coenzyme A). Eubacteria, Archaea, and fungi can use sulfate as their source of sulfur, but it must be reduced to S^{2-} for incorporation into the constituents of cells. The key reaction in sulfur assimilation is as follows:

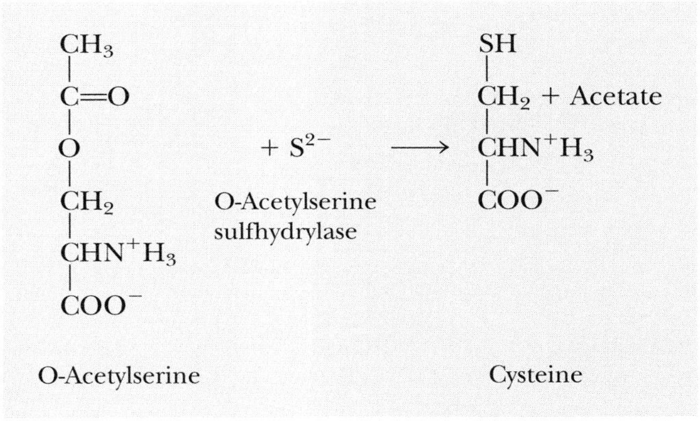

Sulfate is reduced to S^{2-} in a series of reactions that are presented in Figure 10.16. A few organisms, such as the methanogens, lack the capacity for sulfate reduction and require H_2S as a source of sulfur. There are actually limited amounts of sulfate in soil, and it is probable that microorganisms in the environment assimilate other sulfur compounds such as sulfur esters ($R—CH_2—OSO_3^-$), sulfonates ($R—CH_2SO_3^-$), sulfur-containing amino acids, and other carbon-sulfur compounds.

Biosynthesis of Amino Acids

One gram of dry bacterial cell mass contains 500 mg to 600 mg of protein. This protein is synthesized from about 20 different amino acids. An organism growing on a simple

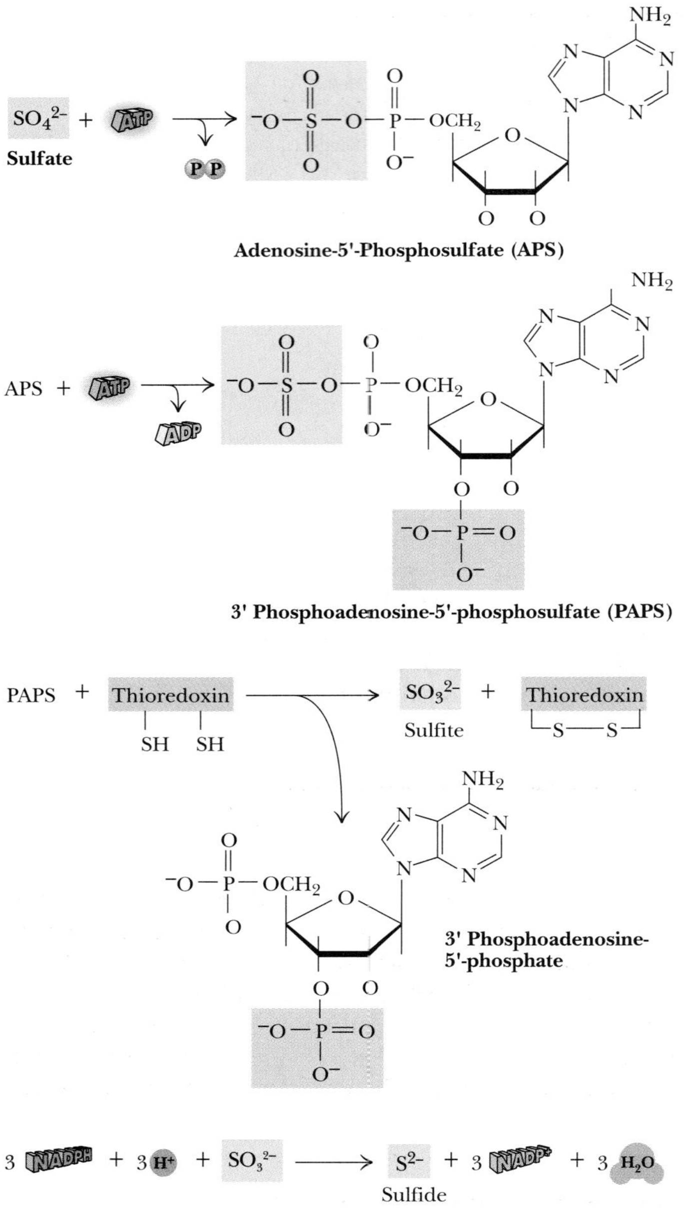

***Figure* 10.16** Sulfate assimilation through the production of sulfide (S_2^-) that is involved in the synthesis of organic sulfur-containing compounds (see Figure 10.23).

substrate, whether glucose or CO_2, expends a considerable portion of its energy and biosynthetic capacity on the synthesis of the monomeric amino acids. Much energy is also spent organizing and assembling these amino acids into functional protein units. The ability to synthesize most all of the requisite amino acids is a trait shared by virtually all of the free-living bacteria. Obviously the strict autotrophs can synthesize them all.

Amino acids are synthesized as "families" from a limited number of precursor metabolites. These families and the precursor from which they originate are presented in Table 10.2. Each of the amino acids has a one- and three-letter symbol that is used as a convenient shorthand when depicting an amino acid sequence in a peptide. These symbols are presented in Table 10.3. The biochemical reactions involved in their synthesis are presented in the discussion that follows. (The structures of the amino acids were presented in Figure 3.2.) Key intermediates are presented and a complete step-by-step depiction of the biosynthetic pathways is available in a biochemistry text, and one is listed in the suggested readings at the end of this chapter.

The reactions involved in the biosynthesis of alanine, valine, and leucine are shown in Figure 10.17. Acetolactate is formed by condensation of two molecules of pyruvate. All free-living organisms produce pyruvate via sugar metabolism, reaction sequences involved in the assimilation of 1-carbon substrates, or various other reaction sequences.

The biosynthetic reactions involved in the synthesis of the glutamate family of amino acids are presented in Figure 10.18. The amino acids are glutamate, glutamine, arginine, and proline. Hydroxyproline is present in some proteins but is synthesized from proline by hydroxylation after the proline molecule is incorporated into the protein.

The pathways leading to the large family of amino acids synthesized from oxaloacetate through aspartic acid are presented in Figure 10.19. Threonine, methionine, lysine, and isoleucine all originate from aspartate. Homoserine is a key intermediate in the synthesis of these amino acids. Lysine and diaminopimelate are dibasic amino acids that are constituents of bacterial cell walls. Lysine is present in proteins but diaminopimelate is not.

Serine, glycine, and cysteine are synthesized from the precursor metabolite 3-phosphoglycerate, as depicted in Figure 10.20. Synthesis of the aromatic amino acids phenylalanine, tyrosine, and tryptophan is more complex than the reactions involved in producing the other amino acids. The precursor metabolites are erythrose-4-phosphate and phosphoenolpyruvate (Figure 10.21). A unique pathway that yields only one amino acid, histidine, is shown in Figure 10.22. The precursor metabolite for histidine synthesis is ribose-5-phosphate.

Biosynthesis of Purine and Pyrimidine Bases

Nucleic acids are a constituent part of all living cells as they are the genetic determinant in all creatures from the simple, lowly virus to the complex mammal. Conditions on the primeval Earth included available gaseous mixtures of hydrocarbon, hydrogen, and ammonia, as well as energy from ultraviolet radiation and lightning; these combined can lead to the formation of purines and pyrimidines. Because the nucleic acid bases can be formed abiotically, they were apparently available early on in the

***Table* 10.2 Precursor of the major amino acids in protein and the family designations**

Family	Precursor Metabolite	Amino Acid
Glutamate	α-ketoglutarate	Glutamate Glutamine Proline
Alanine	Pyruvate	Alanine Valine Leucine
Serine	3-phosphoglycerate	Serine Glycine Cysteine
Aspartate	Oxaloacetate	Aspartate Asparagine Methionine Lysine Threonine Isoleucine
Aromatics	Phosphoenolpyruvate + erythrose-4-phosphate	Phenylalanine Tyrosine Tryptophan
Histidine	5-phosphoribsyl-1-pyrophosphate	Histidine

evolution of viable organisms. Polymerization of these bases created a structure that evolved as the key inheritable element in all life forms. Nucleic acid bases are also a component part of ATP, NAD^+, $NADP^+$, and some of the B-vitamins.

The structures of the bases generally present in nucleic acids are presented in Figure 10.23. The purines adenine and guanine are constituents of both RNA and DNA. The two pyrimidines that are constituents of DNA are cytosine and thymine. Uracil replaces thymine in RNA.

The origin of each component of the nine-member purine ring system is presented in Figure 10.24. The key intermediates in the synthesis of the purine bases are presented in Figure 10.25. The product of purine biosynthesis

***Table* 10.3 The one- and three-letter symbols for the common amino acids**

Alanine	A	Ala	Methionine	M	Met
Aspartic acid or Asparagine	B	Asx	Asparagine	N	Asn
Cysteine	C	Cys	Glutamine	Q	Gln
Aspartic acid	D	Asp	Arginine	R	Arg
Glutamic acid	E	Glu	Serine	S	Ser
Phenylalanine	F	Phe	Threonine	T	Thr
Glycine	G	Gly	Valine	V	Val
Histidine	H	His	Tryptophan	W	Trp
Isoleucine	I	Ile	Tyrosine	Y	Tyr
Lysine	K	Lys	Glutamine or Glutamic acid	Z	Glx
Leucine	L	Leu			

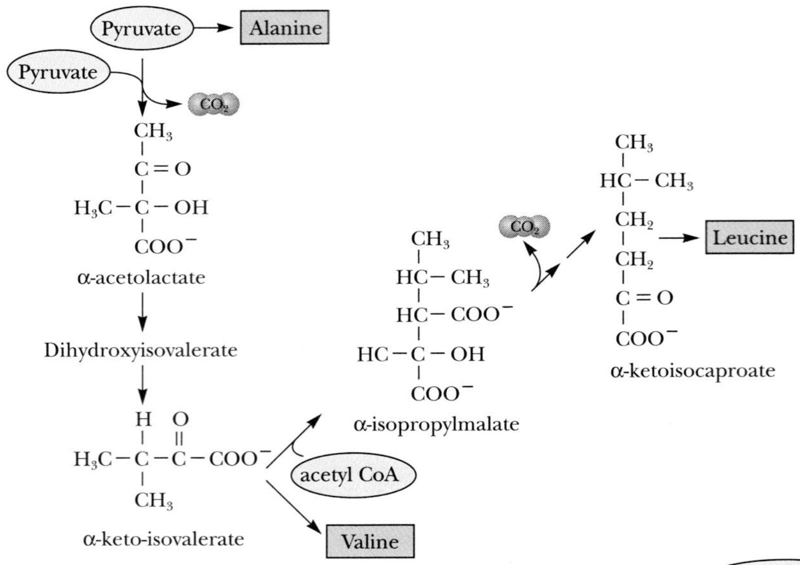

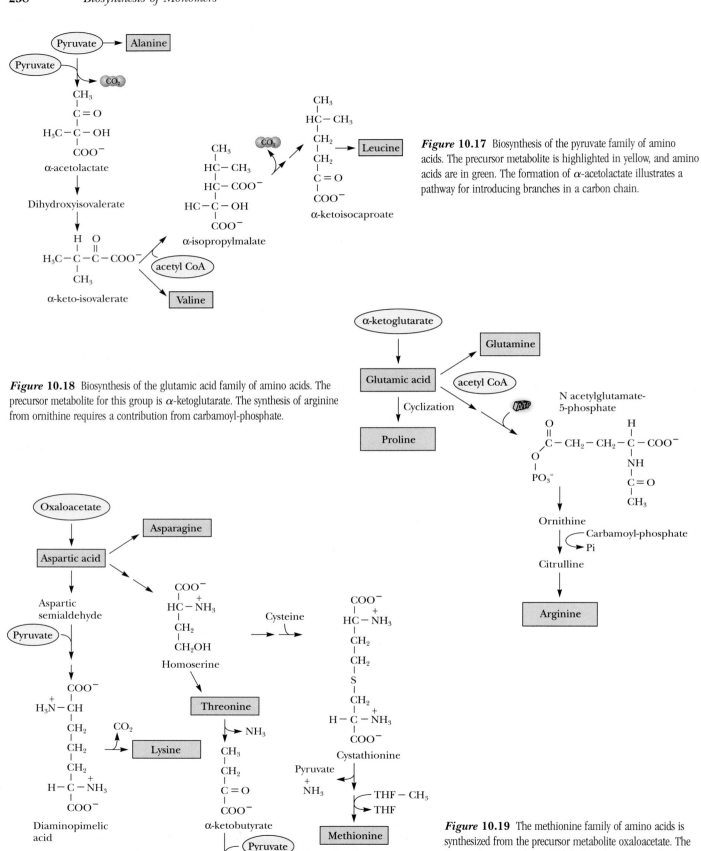

***Figure* 10.17** Biosynthesis of the pyruvate family of amino acids. The precursor metabolite is highlighted in yellow, and amino acids are in green. The formation of α-acetolactate illustrates a pathway for introducing branches in a carbon chain.

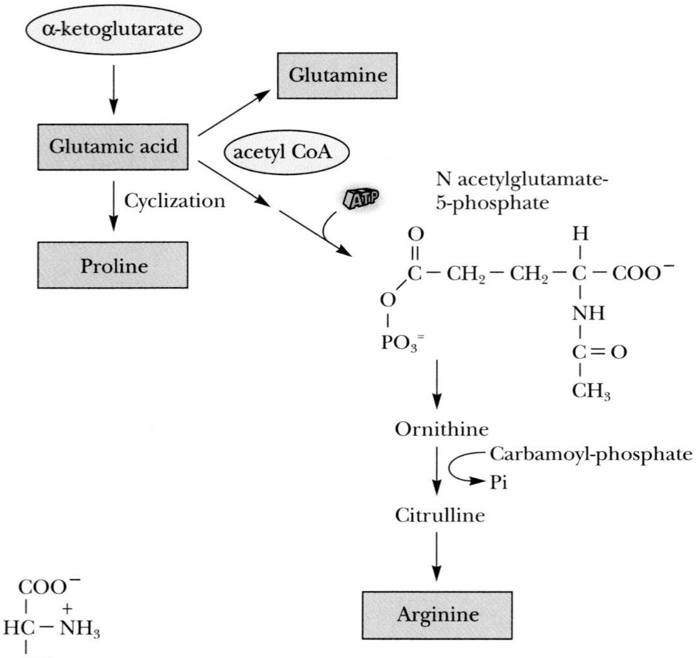

***Figure* 10.18** Biosynthesis of the glutamic acid family of amino acids. The precursor metabolite for this group is α-ketoglutarate. The synthesis of arginine from ornithine requires a contribution from carbamoyl-phosphate.

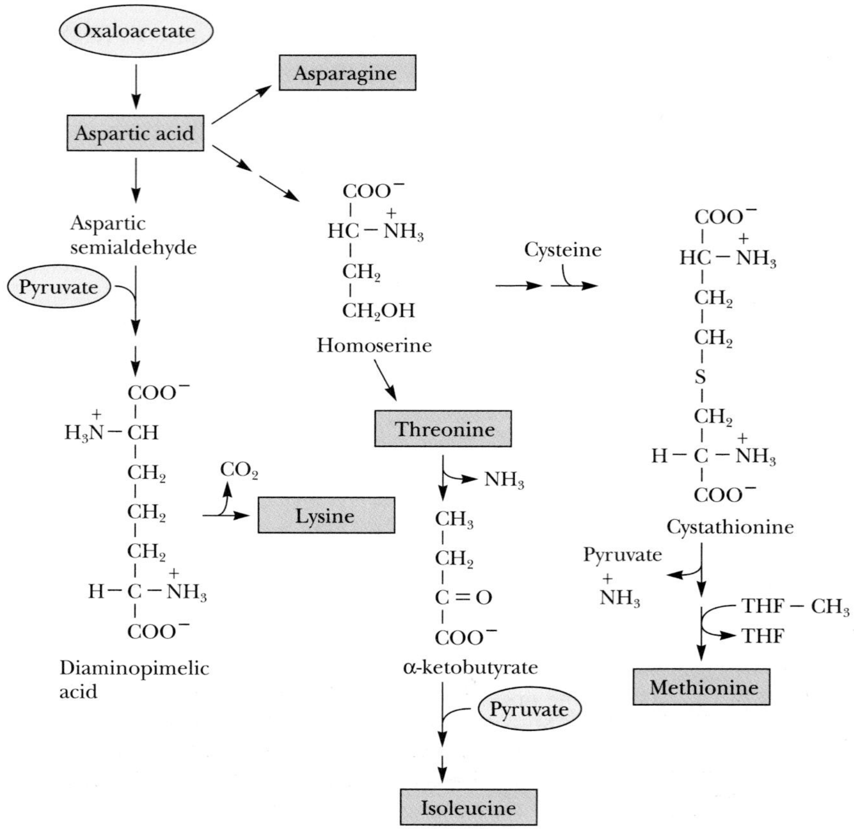

***Figure* 10.19** The methionine family of amino acids is synthesized from the precursor metabolite oxaloacetate. The intermediate aspartic-semialdehyde is the precursor for the biosynthesis of methionine, diaminopimelic acid, and lysine. Diaminopimelic acid is present in the peptidoglycan of bacterial cell walls but is not a constituent of protein.

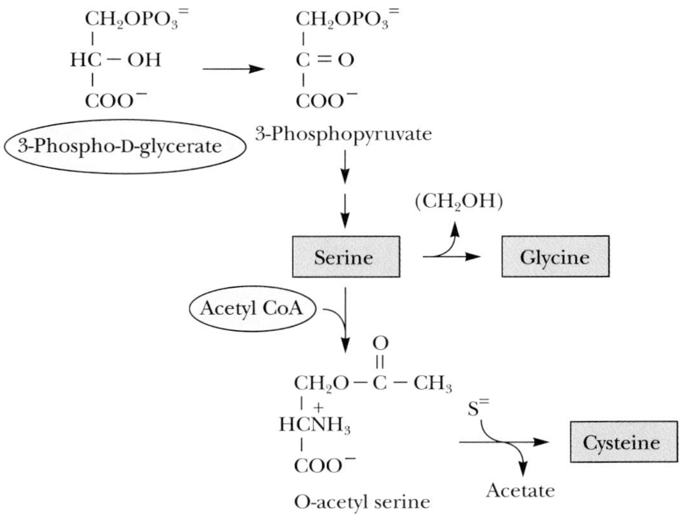

Figure **10.20** The biosynthesis of serine, glycine, and cysteine. 3-Phosphoglycerate is the precursor for these amino acids. The incorporation of sulfide ($S^=$) into cysteine is a mechanism for the incorporation of sulfur into cells.

is **inosine-5′-monophosphate (IMP).** IMP is converted to adenosine-5-monophosphate and guanosine-5′-monophosphate by the reactions shown in Figure 10.26. The remarkable aspect of purine biosynthesis is the simplicity of the components (low molecular weight) that are brought together to make up the molecule.

The biosynthesis of the pyrimidine molecule occurs by a more direct route than that for purines. Carbamyl phosphate condenses with a molecule of aspartate to form the six-member ring (Figure 10.27). The key intermediates involved in pyrimidine biosynthesis are shown in Figure 10.28. Where IMP was the key precursor of purines, **uridine monophosphate (UMP)** is the progenitor of the pyrimidines. The conversion of UMP to thymidine

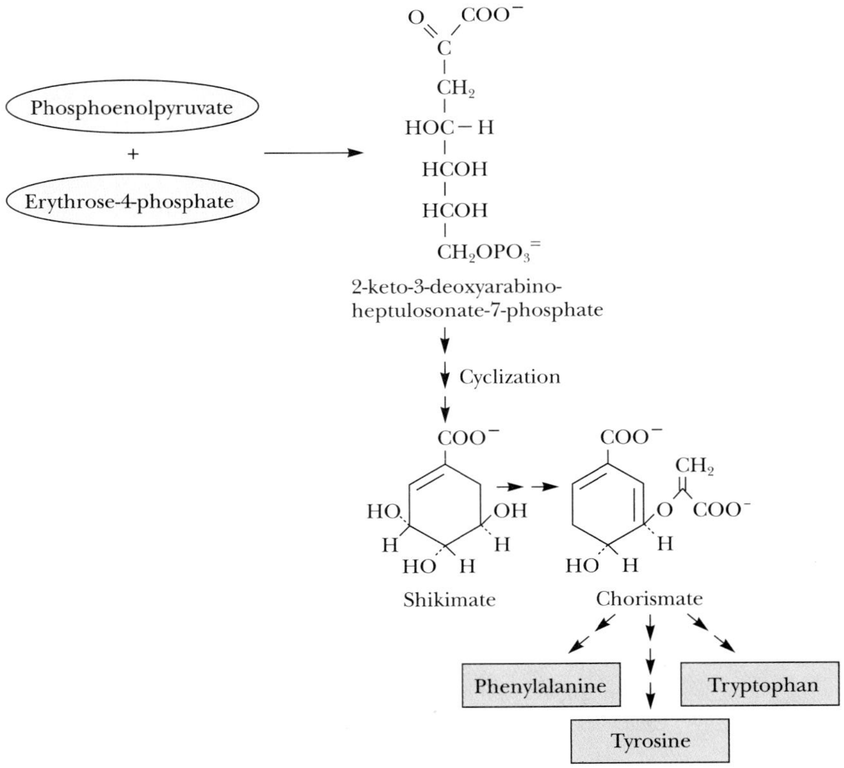

Figure **10.21** Erythrose-4-phosphate and phosphoenolpyruvate are the precursors for an array of aromatic amino acids and other cell constituents, including coenzyme Q and plastoquinone, that are involved with electron transport. Precursors of the lignin present in woody plant tissue are also synthesized from chorismate.

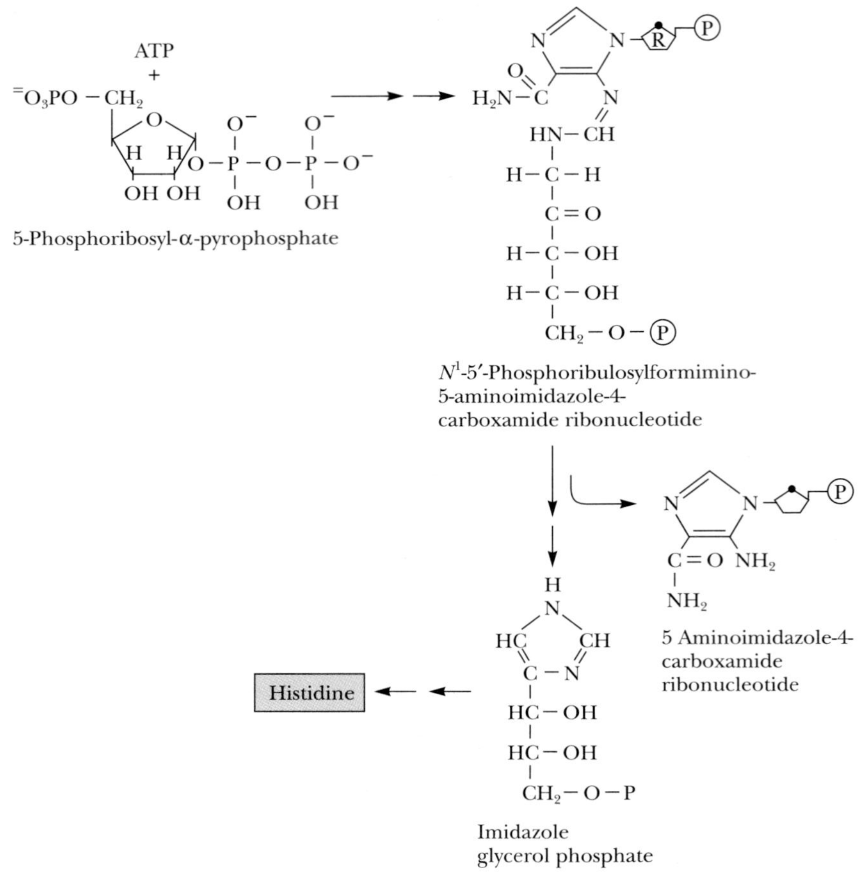

***Figure* 10.22** The amino acid histidine is synthesized from the precursor ribose-5-phosphate. This is the only biosynthetic pathway that leads to a single amino acid. The imidazole ring present in histidine is formed from parts of the ribose molecule and the six-member ring of adenine.

monophosphate and cytidine triphosphate is depicted in Figure 10.29. The pentose side chain ribose is present in RNA, and this sugar is reduced to deoxyribose in DNA by replacement of the 2′-OH group with hydrogen through the enzyme ribonucleotide reductase. The ribonucleotide reductases are generally specific for the diphosphate derivatives and reduce them to deoxydiphosphate. In some cases, the reductases function with the triphosphates. NADPH is the ultimate source of reducing equivalents for the ribonucleotide reductases, and these enzymes function through a thioredoxin reductase. Thioredoxin is a small protein that has cysteine residues that can undergo oxidation-reduction reactions.

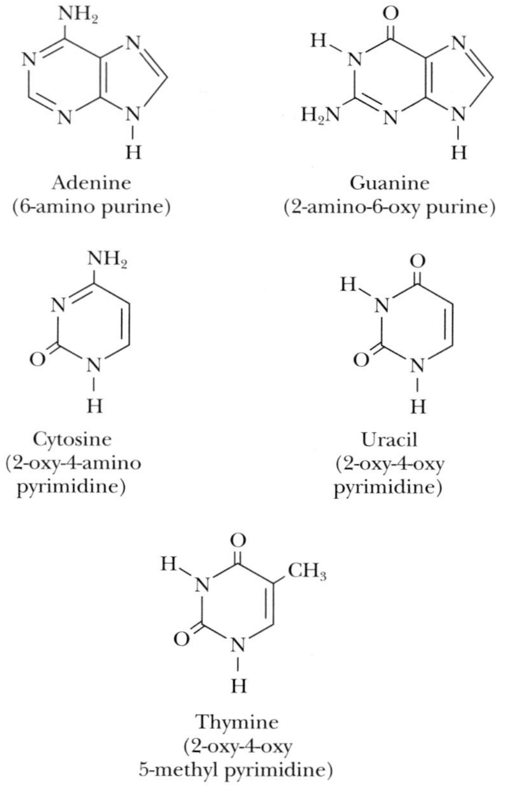

***Figure* 10.23** The structures of the major purines and pyrimidines present in DNA, RNA, and other molecules in the cell.

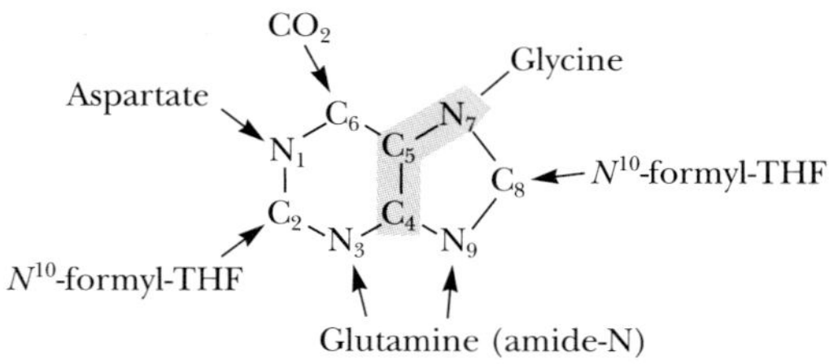

***Figure* 10.24** The origin of the nine atoms in the purine ring. The major contributor to this molecule is glycine.

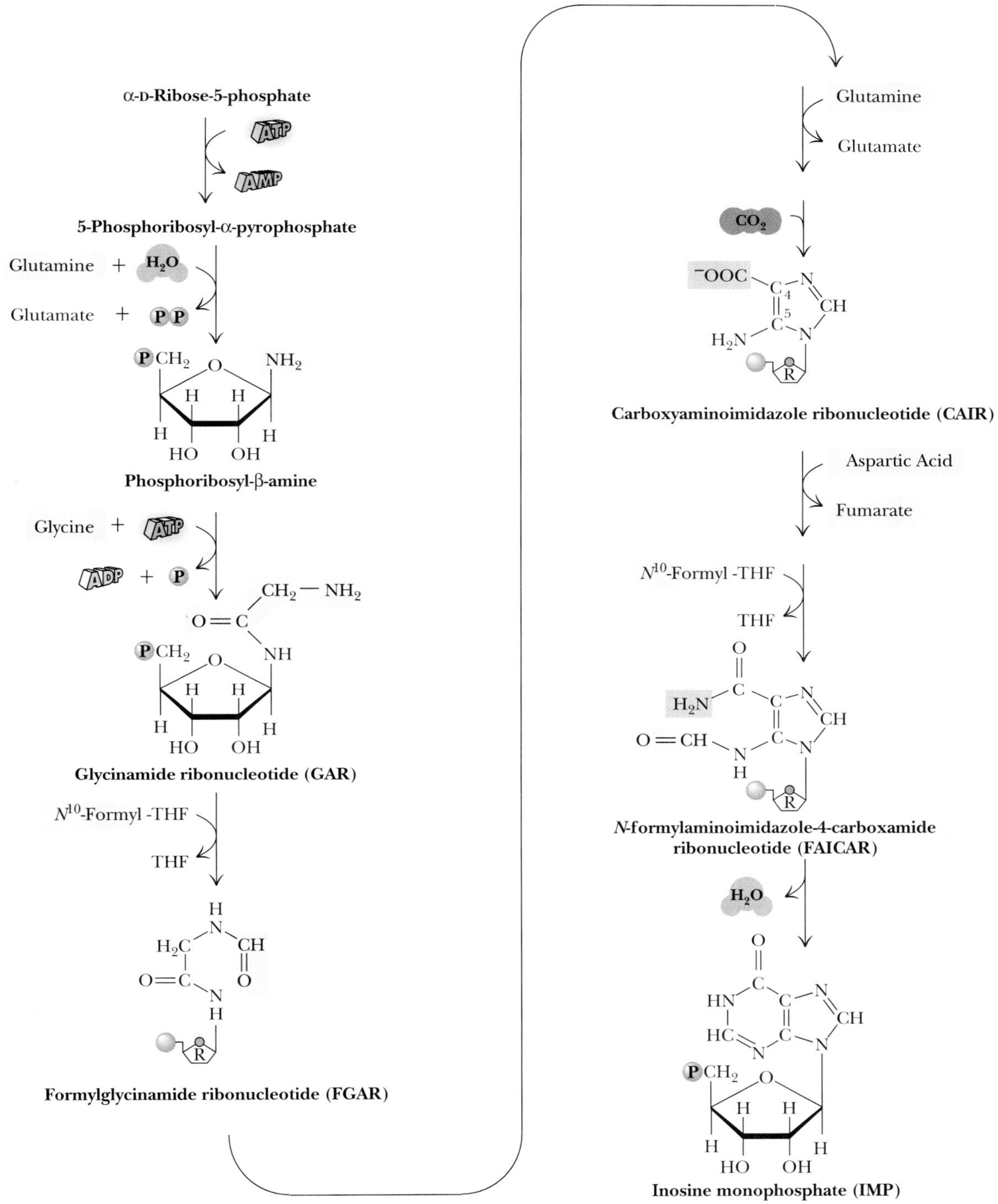

***Figure* 10.25** The major intermediates involved in the synthesis of inosine monophosphate. This figure presents the mechanism whereby each of the constituents (Figure 10.26) is incorporated into the purine nucleus.

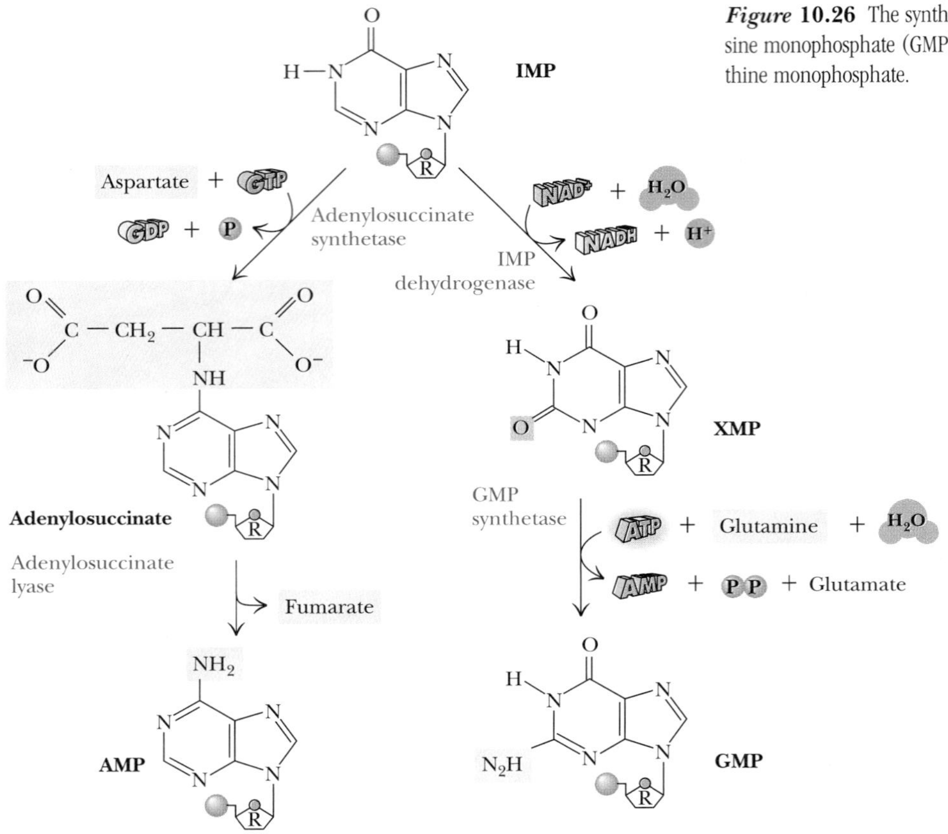

***Figure* 10.26** The synthesis of adenosine monophosphate (AMP) and guanosine monophosphate (GMP) from inosine monophosphate (IMP). XMP is xanthine monophosphate.

Synthesis of Fatty Acids from Acetate

Bacteria generally synthesize long-chain fatty acids by the condensation of a molecule of acetyl CoA with one of malonyl CoA and with NADPH as reductant. The malonyl CoA is formed by carboxylation of acetyl CoA by an ATP-driven reaction. The malonyl CoA formed reacts with the acyl carrier protein (ACP) to form the malonyl-ACP derivative. A molecule of the acetyl-ACP conjugate reacts with α,β-keto-ACP synthase (Kase) to form an acetyl Kase derivative (Figure 10.30). The acetyl Kase reacts with malonyl ACP to form an acetoacetyl-ACP. The decarboxylation of malonyl-ACP drives the condensation of the acetyl Kase and the methylene carbon of the malonyl-ACP molecule. The acetoacetyl derivative is reduced stepwise to the saturated fatty acid. There are eight independent enzyme reactions involved in fatty acid synthesis (Figure 10.30) and each involves a different protein. **Acetyl carrier protein (ACP)** is an important low molecular weight protein that is functionally equivalent to coenzyme A but activates and transfers fatty acids having chain lengths longer than the lower molecular weight groups involved with CoA.

The unsaturated fatty acids in bacteria are virtually all monounsaturated, meaning that they have one double bond. Those present in eukaryotic organisms commonly have more than one double bond. The cyanobacteria are one prokaryote that has more than a single double bond in its cellular lipids. Many bacteria can incorporate exogenously supplied polyunsaturated fatty acids into their cellular lipids, and these apparently are not harmful. Biosynthesis of unsaturated fatty acids occurs by either the aerobic or the anaerobic pathway outlined in Figure 10.31(**a,b**). A double bond introduced in the anaerobic pathway remains at the site of incorporation during the biosynthetic process. The anaerobic pathway occurs in *Escherichia coli,* lactic acid bacteria, and other anaerobes, and

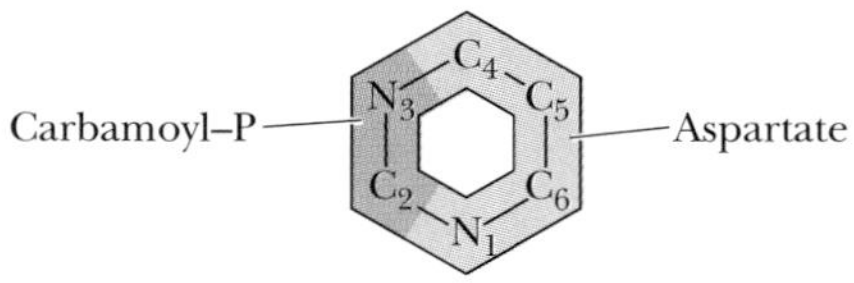

***Figure* 10.27** The origin of the six atoms in the pyrimidine ring.

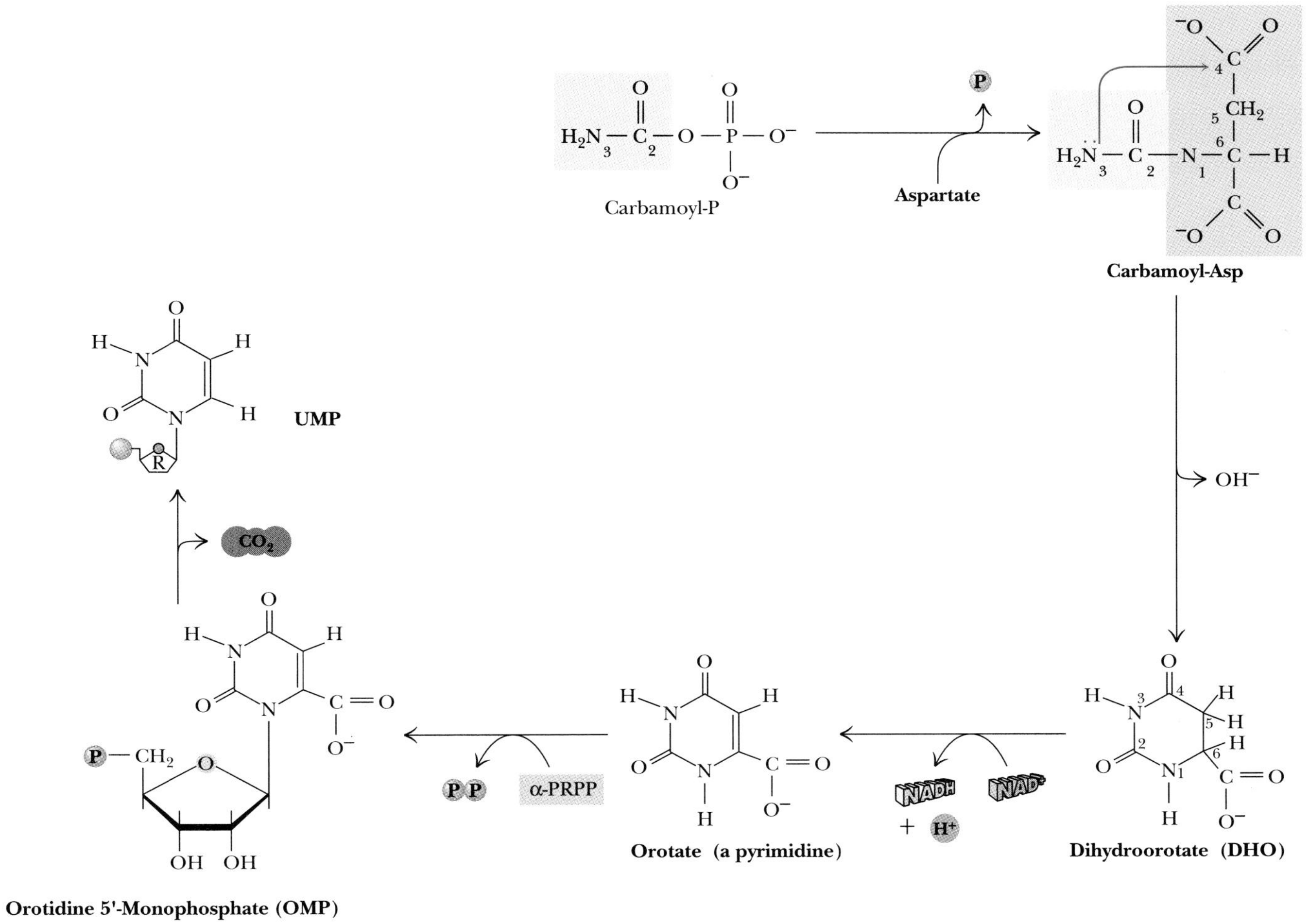

***Figure* 10.28** The biosynthesis of the pyrimidine uridine monophosphate (UMP) from carbamoyl-phosphate and aspartic acid.

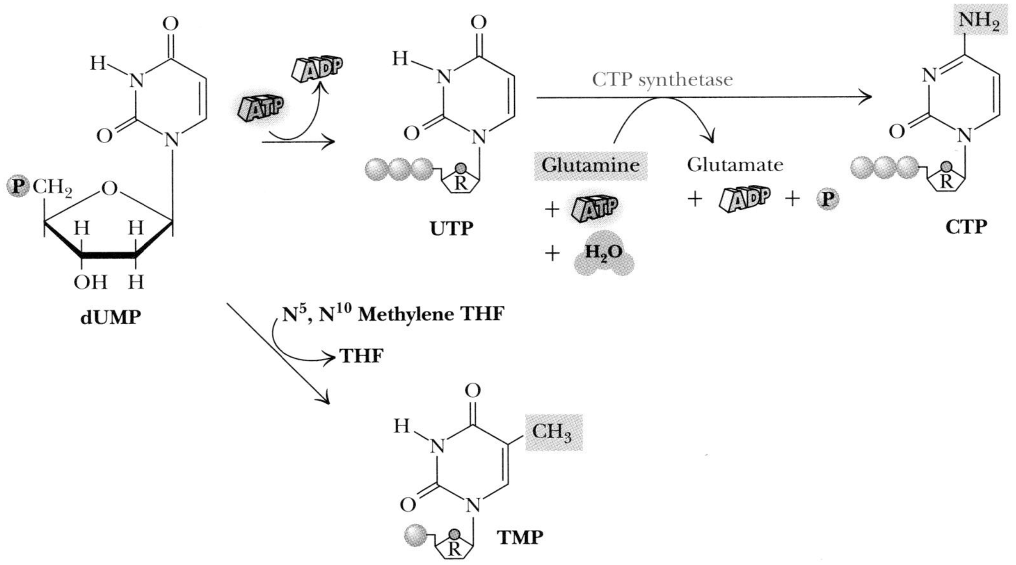

***Figure* 10.29** The synthesis of cytidine triphosphate and thymidine monophosphate from uridine monophosphate.

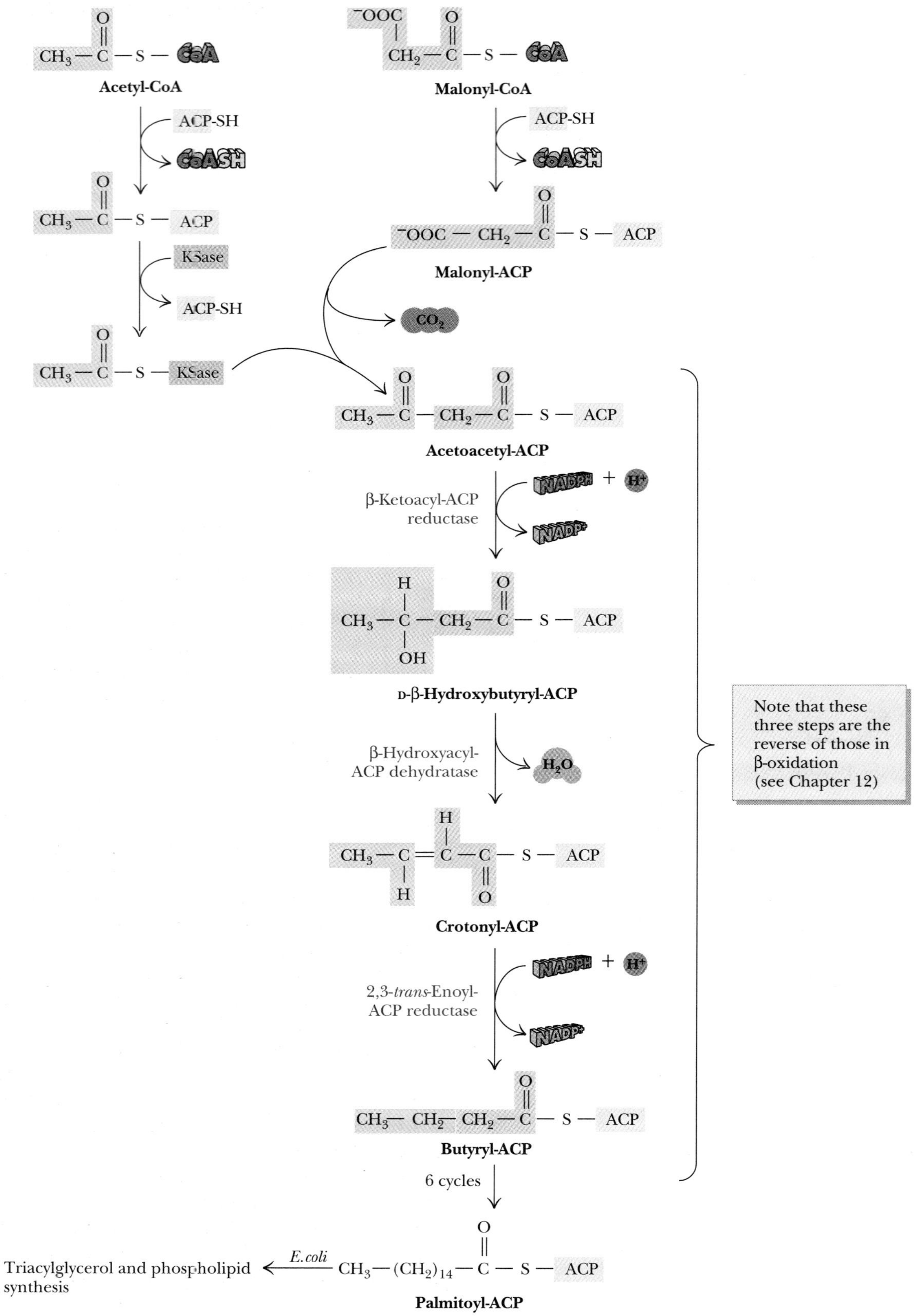

***Figure* 10.30** The synthesis of n-hexadecanoic acid (palmitic acid) from acetyl CoA and malonyl CoA. Acetyl and malonyl building blocks are introduced as conjugates of the acyl carrier protein (ACP). Decarboxylation promotes the β-keto-acyl-ACP synthase (Kase) to unite the 2-carbons derived from malonyl CoA to the growing chain. The newly synthesized fatty acids exist in cells as derivatives of ACP until they are joined to glycerophosphate moieties to form phospholipids.

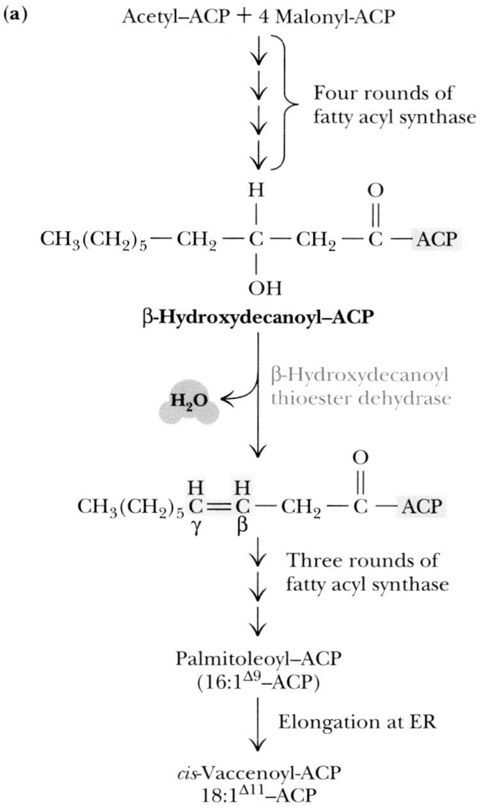

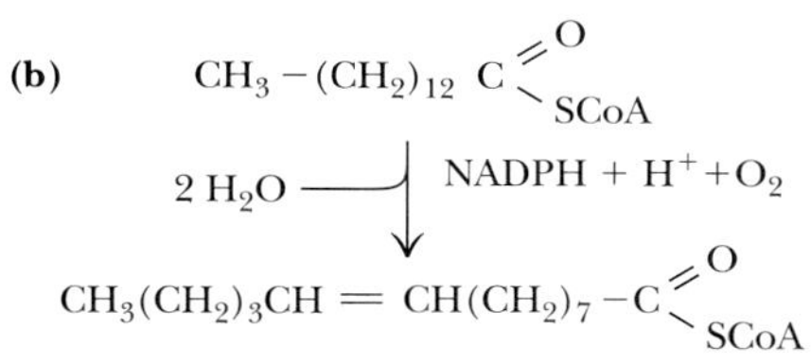

***Figure* 10.31** **(a)** The anaerobic pathway for introduction of double bonds into the growing fatty acid chain. Palmitoleoyl-ACP is synthesized by sequential additions of four 2-carbon units followed by double bond formation and additional elongation steps. Another elongation step results in the formation of the 18-carbon unsaturated fatty acid. This pathway is present in many of the gram-negative bacteria (*Escherichia coli* and *Salmonella* sp.) and anaerobic gram-positives such as the lactic acid bacteria and clostridia. The anaerobic pathway is also utilized by the cyanobacteria. **(b)** The aerobic pathway for synthesis of unsaturated fatty acids. The double bond is formed between the nine and ten carbons of the fatty acid. This pathway is utilized by aerobic bacteria and eukaryotes.

may also occur in aerobic bacteria. The aerobic pathway is utilized by micrococci, endospore-forming bacilli, some cyanobacteria, fungi, and animals. In the aerobic pathway the double bond occurs between carbons 9 and 10.

Considerable variety exists in the fatty acids present in the major lipids of Eubacteria (Figure 10.32). The chain length varies from C_{12} to C_{20}, they are straight or branched, are saturated or unsaturated, may have a cyclopropane or hydroxy constituent, and may (but rarely except in cyanobacteria) have more than one double bond. The hydroxy fatty acids that are constituents of the outer envelope of gram-negative bacteria are synthesized by the anaerobic pathway.

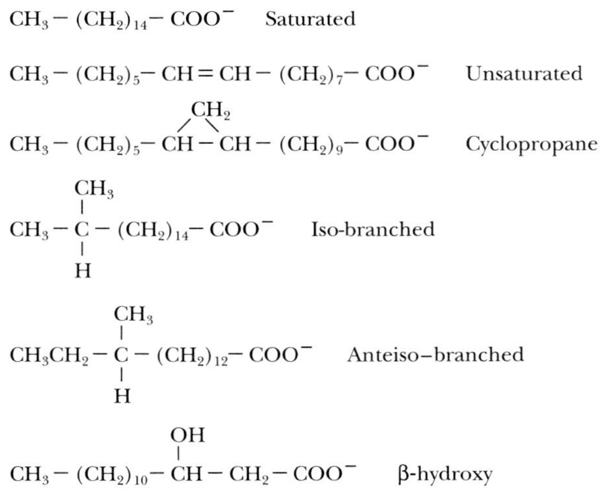

***Figure* 10.32** Fatty acids that are commonly present in the phospholipids of bacteria.

The phospholipids are assembled as outlined in Figure 10.33. The precursor metabolite, glycerol-3-phosphate, provides the backbone for polar lipid synthesis. Addition of ACP-conjugated fatty acids yields phosphatidic acid, a key intermediate in the synthesis of the membrane lipids. The pathways for the formation of phosphatylserine, phosphatidylethanolamine, and cardiolipin, commonly present in membranes of bacteria, are shown.

The carotenoids and related long-chain branched hydrocarbons are common constituents of many eubacterial and archaeal species. These compounds are composed of isoprene subunits and are often termed isoprenoids. The yellow and red pigments of photosynthetic and other bacteria and the phytol chain that anchors chlorophyll to the photosynthetic membrane are composed of isoprene units. Isoprenoid molecules strung together are present in Archaea in place of fatty acids. The synthesis of these important units is depicted in Figure 10.34.

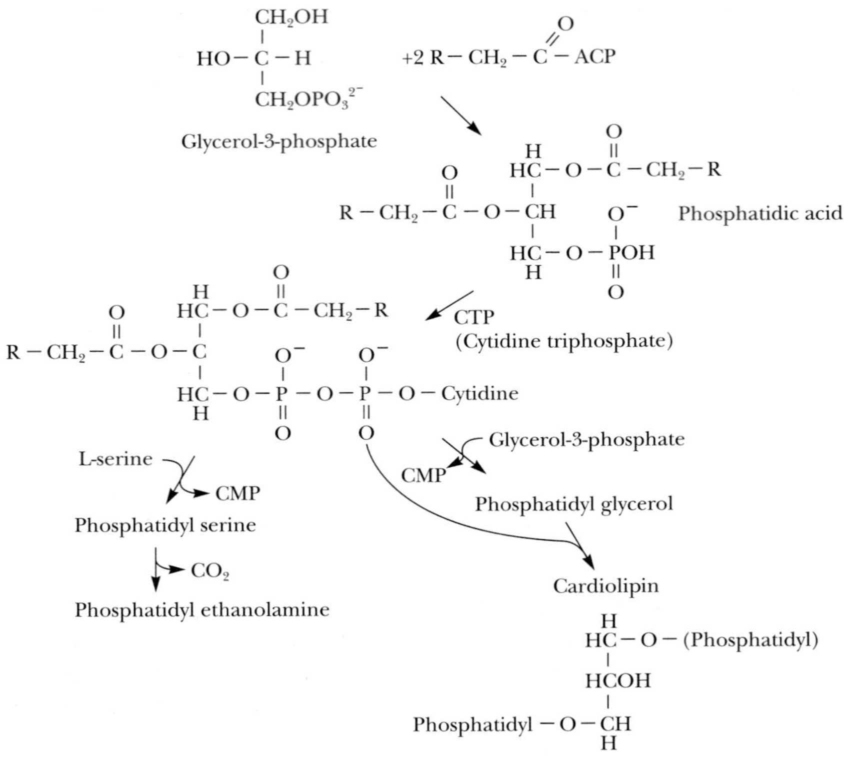

***Figure* 10.33** Synthesis of the major phospholipids from glycerol-3-phosphate and the ACP conjugate of saturated or unsaturated fatty acids. The cytidine triphosphate derivate of phosphatidic acid reacts with L serine to form phosphatidyl serine. Phosphatidyl serine is decarboxylated to yield phosphatidylethanolamine. Cardiolipin is formed by conjugating a molecule of phosphatidic acid to each end of a molecule of glycerol.

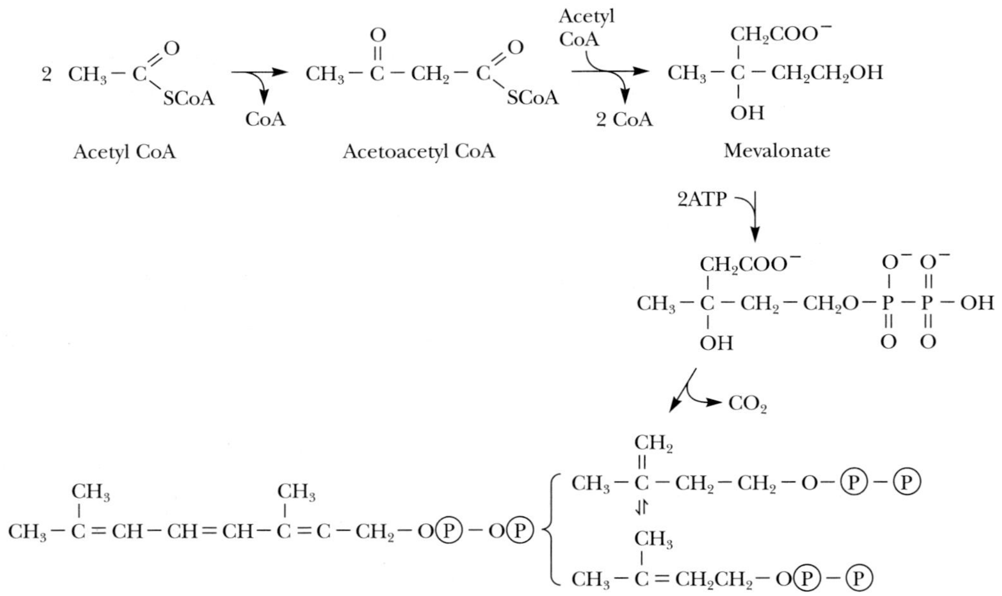

***Figure* 10.34** The branched isoprenoids are commonly present in the membrane lipids of Archaea and in the carotenoids of photosynthetic and other pigmented Eubacteria. These branched compounds are synthesized from acetyl CoA through acetoacetyl CoA and mevalonic acid.

Summary

- The **autotrophs** have the most thorough of biosynthetic systems, as they can synthesize all constituents of a living cell from CO_2. Many can obtain cellular nitrogen from atmospheric N_2.

- The **macromolecules** that make up a cell are of four types: proteins, nucleic acids, polysaccharides, and lipids.

- There are 12 monomers, termed the **precursor metabolites,** that are involved in the biosynthesis of the macromolecules in a cell. These metabolites are mostly products of **glycolysis** and the **tricarboxylic acid cycle.**

- Glucose is catabolized in bacteria by any of three pathways—**glycolysis** (Embden-Meyerhof pathway), **Entner-Doudoroff** pathway, or the **hexose monophosphate shunt.**

- The **tricarboxylic acid cycle (TCA)** may be used by aerobic organisms to completely oxidize one molecule of pyruvate to three molecules of carbon dioxide. Aerobes also utilize the TCA cycle to provide **precursor metabolites. Anaerobically** the cycle is employed predominantly to generate **precursor metabolites.**

- An **anaplerotic reaction** is one that permits cycles such as the TCA cycle to continue by replacing intermediates removed for biosynthesis.

- Carbon dioxide is generally assimilated into autotrophs via the **Calvin cycle.** In this cycle a molecule of carbon dioxide is incorporated into a 5-carbon sugar (ribulose-1,5-bisphosphate) resulting in the formation of two molecules of 3-phosphoglyceric acid. 3-phosphoglyceric acid is also an intermediate in glycolysis.

- The green sulfur photosynthetic bacteria assimilate carbon dioxide via a mechanism that is essentially a reversal of the tricarboxylic acid cycle. This is termed a **reductive carboxylation mechanism** for carbon dioxide fixation.

- **Acetogenic bacteria** can effect a back-to-back condensation of two molecules of carbon dioxide to form **acetic acid.** This is another mechanism for autotrophic carbon dioxide assimilation.

- Nitrogen is an important constituent of proteins, nucleic acids, and other important macromolecules in a cell. Ammonia (NH_4^+) is assimilated in most bacteria through amination of α-ketoglutarate to form glutamic acid. The amino group can be transferred to other compounds by **transamination.**

- Bacteria can use **nitrate** (NO_3^-) as source of nitrogen and must reduce it to the amine level. The enzymes involved are **nitrate** and **nitrite reductase**.

- **Nitrogen fixation** (use of dinitrogen as nitrogen source) is confined to the **prokaryotic world.** An array of free living, commensalistic, and photosynthetic bacteria can supply their need for nitrogen by assimilating it from the air.

- The key reaction in the incorporation of sulfur into bacteria and fungi is the formation of **cysteine** from the serine derivative **O-acetylserine.**

- Amino acids are synthesized from precursor metabolites. They are synthesized as **"families."** An example of a family of amino acids would be those synthesized from 3-phosphoglycerate—serine, glycine, and cysteine.

- The nucleic acid bases are synthesized de novo from simple substrates. The **pyrimidines** are synthesized from aspartate and carbamoyl-phosphate while the **purines** are synthesized from CO_2, glycine, two formyl groups, and an amine nitrogen from aspartic acid and two amide nitrogens from glutamine.

- Fatty acids in bacteria are synthesized from **acetyl CoA** and **malonyl CoA**. During the joining of acetyl CoA and malonyl CoA a decarboxylation occurs that drives the reaction forward.

- The **isoprenoids** such as the **carotenoids** in photosynthetic bacteria are synthesized from acetyl CoA.

Questions for Thought and Review

1. In discussing macromolecule synthesis—CO_2 was selected as the starting substrate. Why?

2. What are the four major macromolecular types that are present in a bacterial cell? What are the monomers that make up these macromolecules? Draw typical monomers that are assembled to form these macromolecules.

3. A precursor metabolite is a molecule that is directly involved in biosynthesis. Why are these monomers called precursors? What does their existence tell us about biological systems? About metabolic diversity?

4. Review the three major pathways for glucose catabolism in microorganisms. What are some of the significant differences between them? What role might each play?

5. What is an anaplerotic reaction? What is the function of these reactions? How is CO_2 involved?

6. What are the key enzymes in the Calvin cycle? How much different enzymatically is an organism utilizing CO_2 as substrate than one growing on glucose? How does the initial product of CO_2 fixation fit into the concept of the precursor metabolite?

7. The green sulfur bacteria are probably ancient-type organisms. How do they fix CO_2 and in what way is this related to aerobic metabolism?

8. Ammonia and sulfate are "funneled" into the metabolic machinery through two intermediates. Draw out these reactions.

9. What is a family of amino acids? How many families are there?

10. What low molecular weight compounds are utilized in synthesis of a purine? A pyrimidine?

Suggested Readings

Garrett, R. H., and C. M. Grisham. 1995. *Biochemistry.* Philadelphia: Saunders College Publishing.

Lawlor, David W. 1993. *Photosynthesis: Molecular, Physiological, and Environmental Processes.* 2nd ed. Essex, England: Longman Scientific and Technical.

Neidhardt, F. C., J. L. Ingraham, and M. Schaechter. 1990. *Physiology of the Bacterial Cell. A Molecular Approach.* Sunderland, MA: Sinauer Assoc., Inc.

Schlegel, H. G., and B. Bowien, eds. 1989. *Autotrophic Bacteria.* Madison, WI: Science Tech Publishers.

Usually, if nature invents something that is really good, it keeps applying the invention over and over again whenever this can help solve its problem.

Vladimir P. Skulachev

Chapter 11

Assembly of Cell Structures

The generation of **precursor metabolites** and the reactions that convert these monomers to the building blocks of a cell were discussed in Chapter 10. The precursor metabolites originate from either glycolysis, the tricarboxylic acid cycle, or allied pathways. Through a limited number of well-integrated reactions, the major monomers, including the amino acids, nucleotides, sugars, and fatty acids, are synthesized from these precursors. Rapid and orderly growth depends on the **polymerization** of monomeric building blocks to form macromolecules. Among the essential polymerization reactions are the formation of proteins from amino acids, lipids from fatty acids, and the nucleic acids from nucleotides. Macromolecules (DNA, RNA, proteins, and phospholipids) once formed are assembled to generate a cell. The entire process is illustrated in Figure 11.1. Note that proteins and RNA make up the major part of a living cell. The actual number of molecules of each macromolecular component present in an *Escherichia coli* cell is given in Table 11.1.

Replication of DNA, RNA synthesis, and the role of the nucleic acids in protein synthesis will be discussed in Chapter 13. This chapter is devoted to a brief discussion of the assembly of protein structures, the synthesis of the cell wall/envelope, and the formation of cytoplasmic membranes and inclusion bodies.

Protein Assembly, Structure, and Function

The term **protein** broadly defines molecules that are composed of one or more polypeptide chains. A **polypeptide** (not a precise term) is a chain that exceeds several dozen amino acids in length. Proteins are present in the cytoplasm, within the cytoplasmic membrane, and associated with the outer areas of a bacterium (Table 11.2). Microorganisms that can digest polysaccharides, fats, or other large molecules secrete enzymes (proteins) to the exterior of the cytoplasmic membrane that can hydrolyze large molecules to low molecular weight compounds. These low molecular weight compounds can then be transported into the cells. The outer envelopes of gram-negative bacteria contain the protein porins that are involved in passage of molecules into the periplasmic space where they can be retained by binding proteins. Transport to the

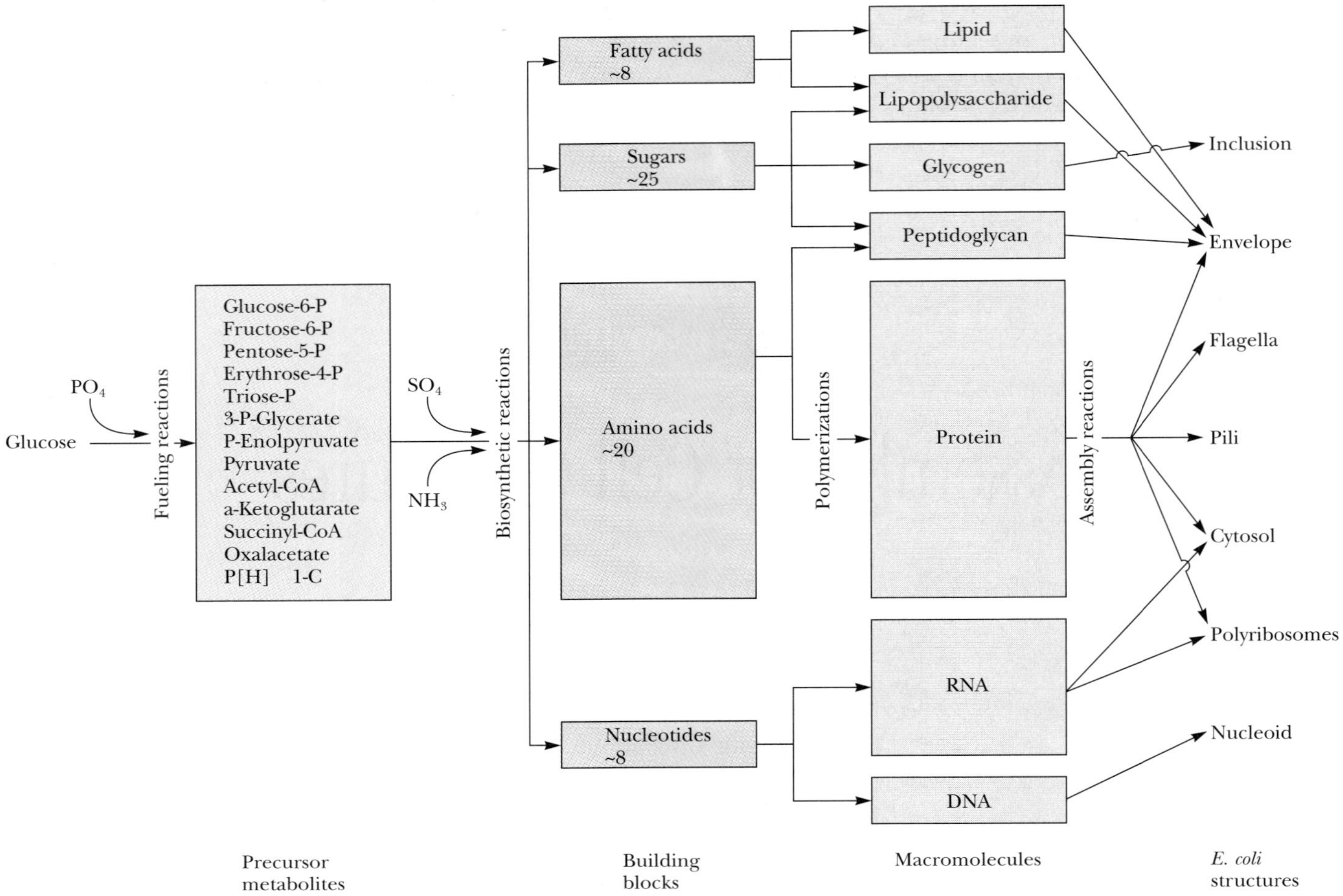

***Figure* 11.1** An overview of the anabolic reactions that lead from glucose to the structures in an *Escherichia coli* cell. The number of monomers (building blocks) needed are given in the boxes, and the size of each box is proportional to the total required to produce a cell. Polymerization reactions convert monomers to macromolecules. The macromolecules interact to form the structures that make up a cell. (Reproduced with permission, Ingraham, Maaloe, Neidhardt, *Growth of the Bacterial Cell,* Sinauer Associates, Inc. 1983)

cytoplasm occurs through the action of proteins in the cytoplasmic membrane. The electron transport systems in the microorganisms capable of respiration are located in the cytoplasmic membranes as are the proteins of the ATPase system. Proteins in the cytoplasm are involved with catabolic functions and the synthesis of DNA, RNA, and cellular components.

Proteins Are Synthesized from Amino Acids

Amino acids are joined together by **peptide bond** formation, and during this reaction a molecule of water is removed, as was depicted in Figure 3.22. A joining together of a series of amino acids results in a peptide chain referred to as the **primary structure** (Figure 11.2). The nature and character of a protein is determined to a considerable extent by the total number and sequence of amino acids in this chain. A long chain of amino acids is called a polypeptide. Functional proteins are generally folded into a three-dimensional structure, and this is the conformation that a functional polypeptide ultimately assumes. The folding pattern is determined to a considerable extent by the amino acid sequence, and the formation of a functional protein potentially can occur by unassisted self-assembly. Evidence suggests, however, that random unassisted assembly can result in a structure that is nonfunctional. There are preexisting proteins in the growing cell that act as **chaperones** to prevent the incorrect molecular interactions that would lead to nonfunctional secondary or tertiary structures. These chaperones assist in forming but are not a part of the ultimate protein product. Generally the folding of peptides to a functional secondary structure is energetically favorable; that is, it occurs without energy input.

***Table* 11.1 The overall macromolecular composition of an *Escherichia coli* cell (Adapted from Ingraham, Maaloe, and Neidhardt, 1983)**

Macromolecule	Percentage Total Dry Weight of Cell	Number of Molecules per Cell
Protein	55.0	2,360,000
RNA	20.5	
23S rRNA		18,700
16S rRNA		18,700
5S rRNA		18,700
transfer RNA		205,000
messenger RNA		1,380
DNA	3.1	2.1
Lipid	9.1	22,000,000
Lipopolysaccharide	3.4	1,200,000
Peptidoglycan	2.5	1
Glycogen	2.5	4,360
Soluble pool[1]	2.9	
Inorganic	1.0	

[1]Metabolites, vitamins.

***Table* 11.2 Location of proteins associated with a gram-negative bacterium**

Location	Proteins
Extracellular	Carbohydrases Lipases Proteases Nucleases
Outer envelope	Receptors for bacteriophage Porin proteins
Periplasm	Binding proteins involved in transport and chemotaxis Phosphatases Esterases
Cytoplasmic membrane	Electron-transport chain Proton translocating ATPase Transport proteins Lipid biosynthesis enzymes Cell wall and outer-envelope biosynthetic enzymes
Cytoplasm	Enzymes involved in catabolism of soluble substrates Enzymes involved in DNA replication
Ribosomes	Structural proteins
External surface	Fimbriae Flagella Pili

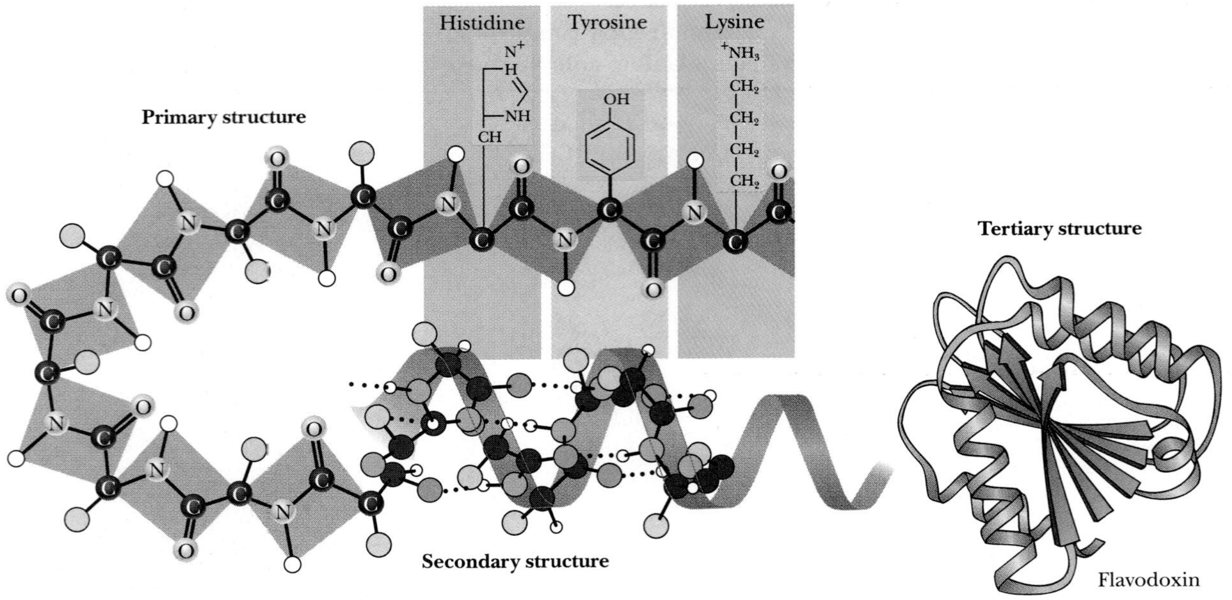

***Figure* 11.2** Formation of a peptide from single amino acids and the steps involved in the conversion of the primary peptide into a secondary structure. The secondary structure depicted is an α-helix, but a pleated sheet (Figure 11.4) may also be formed. These pleated sheets ⟶ and the α-helix are joined to form a tertiary (three-dimensional) structure. The tertiary structure is that of flavodoxin, a protein involved in the nitrogenase (N_2 fixation) reaction. Flavodoxin is composed of five parallel β-sheets and four α-helices.

Structural Arrangements of Proteins

The types of bonding that occur between domains of a polypeptide and give the protein a secondary and tertiary structure are outlined in Figure 11.3. The variability in the side chains of the 20 major amino acids (see Figure 3.21), the total number incorporated, and their sequence in a polypeptide are all factors in the biochemical characteristics of a protein. Amino acids with hydrocarbon side chains (valine, leucine, phenylalanine, and tryptophan) are hydrophobic. The hydrocarbon substituent on these amino acids is on the external surface (points outward) in pro-

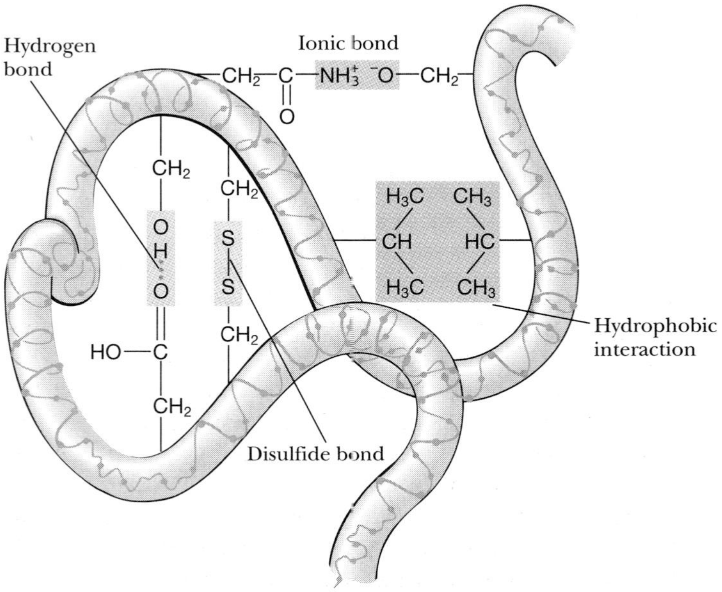

***Figure* 11.3** The major interactions that form between the domains of a polypeptide chain. These bonds determine the shape of a protein.

***Table* 11.3 Molecular mass and number of subunits in selected proteins**

	Relative Mass (M_r)	Subunits
Ribonuclease	13,700	1
Lysozyme	14,388	1
Luciferase	80,000	2
Hexokinase	100,000	2
Lactic dehydrogenase	223,000	4
Urease	483,000	5

The average molecular mass of an amino acid residue in a protein is 120 daltons (average molecular weight, mol wt, less a molecule of H_2O), and an "average" protein molecule has about 300 amino acid residues or an M_r of 36,000. A dalton is defined as 1/12th the mass of a carbon atom.

teins that are embedded in hydrophobic membrane lipids. These hydrophobic groups extend inward in most other proteins. The hydrophilic amino acids (aspartic acid, glutamic acid, lysine, arginine, and histidine) extend outward from the peptide chain and provide an advantage for proteins that function in an aqueous environment.

A single polypeptide can be a functional protein, but more often functional proteins are an aggregate of two or more similar proteins. These arrangements are called dimers, trimers, tetramers, and so forth, and the union of two or more individual proteins produces a **quaternary structure.** The quaternary structures are bound together by the noncovalent forces (Figure 11.3) that are involved in the secondary folding of a polypeptide chain. In some cases, the individual polypeptides that make up a functional protein can be different in composition and size. The **relative mass** (M_r) of a protein is the total molecular mass of the assembled subunits. Some representative proteins, their relative mass, and the number of subunits (peptide chains) that form each protein are listed in Table 11.3.

Typically, the twisting and coiling of peptide chains results in the formation of a **secondary structure** that is either helical in form, called an ***α*-helix,** or a flat arrangement designated a ***β*-sheet.** The *α*-helix is a linear polypeptide that is wound like a spiral staircase (Figure 11.4). It is held together by hydrogen bonding between the amine hydrogen of one amino acid and an oxygen from another amino acid. The hydrogen bonding occurs as the polypeptide is formed and leads to the stable helical structure. Glutamic acid, methionine, and alanine are the strongest formers of the *α*-helix.

Beta sheets (Figure 11.4) are formed when two or more extended polypeptide chains come together side by side so that regular hydrogen bonding can occur between the peptide backbone amide NH and the carbonyl C=O of an adjacent chain. Addition of more polypeptides results in a multistranded structure. The amino acids valine, tyrosine, and isoleucine are common *β*-sheet formers.

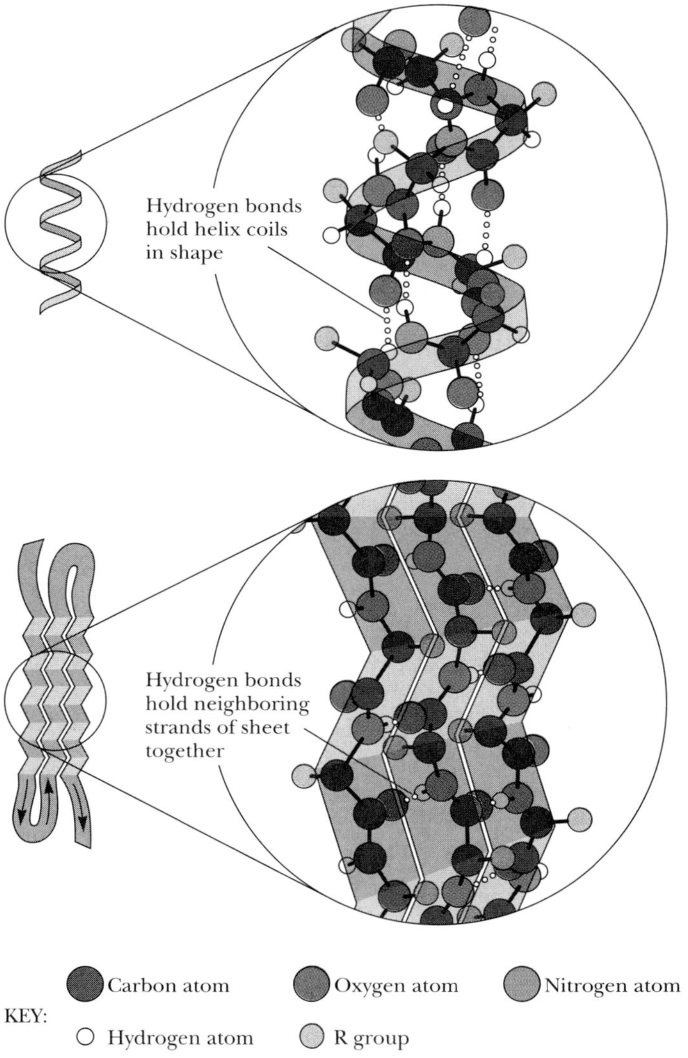

***Figure* 11.4** The bonding that retains a polypeptide chain in an *α*-helix or *β*-pleated sheet configuration. Hydrogen bonds are important determinants in secondary structures.

An example of the **tertiary structure** of a protein was presented in Figure 11.2. The protein shown is flavodoxin, an electron transport protein in sulfate-reducing bacteria. The folding of a polypeptide chain to form a discrete compact protein molecule requires considerable bending. These bends in the chain of amino acids that reverse the direction are called ***β*-bends.** Glycine and proline are frequently present in *β*-bends. A *β*-bend is a tight loop resulting when a carbonyl group of one amino acid forms a hydrogen bond with the NH_2 group of another amino acid that is three positions down the polypeptide chain. This results in the polypeptide folding back on itself.

BOX 11.1 MILESTONES

Protein Conformation

Classic experiments in the early 1960s by Christian B. Anfinsen led to the presumption that the sequence and nature of the amino acids in a polypeptide chain were the major contributing factors to the folding and final conformation of a protein. This was based on experimentation with bovine pancreatic ribonuclease, a small protein made up of 124 amino acids. The secondary structure of this protein is maintained by four disulfide bridges. Denaturation can be accomplished by cleaving these covalent disulfide linkages with reagents (urea and β-mercaptoethanol) that reduce the —S—S— bridges to —SH HS— as follows:

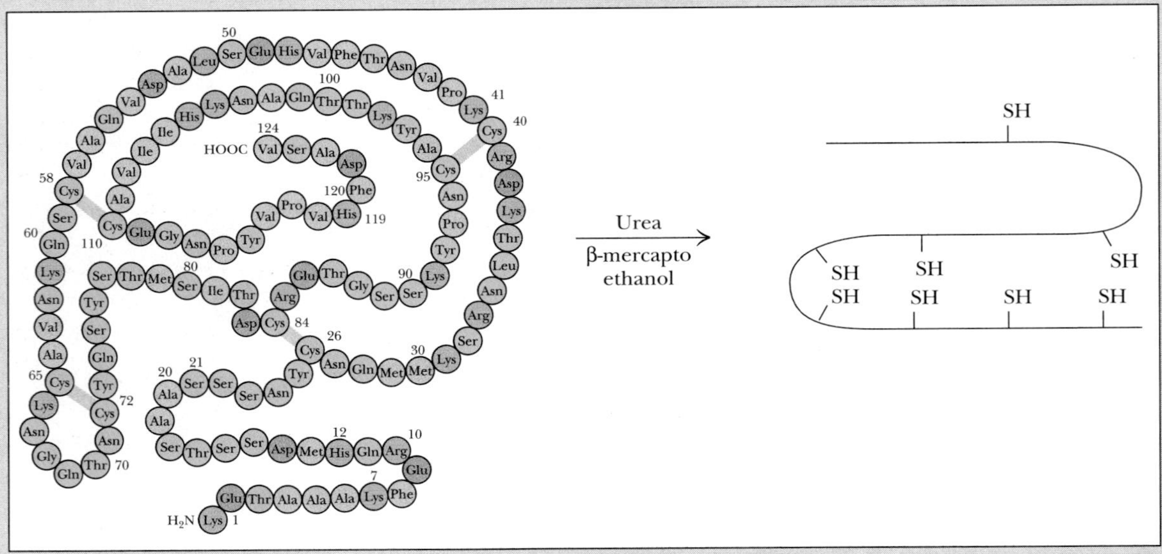

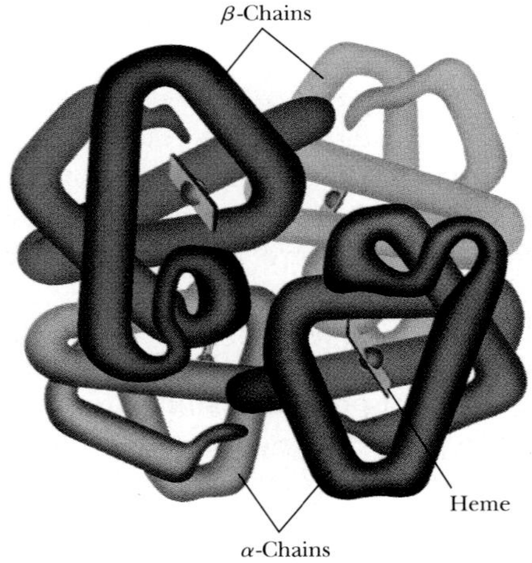

***Figure* 11.5** The quaternary structure of hemoglobin is composed of two α and two β polypeptide chains. As drawn, the β chains are the two uppermost polypeptides and the α chains are the lower half of the molecule. The darker chains in the forefront are the β_2 chain (left) and α_1 chain (right). The heme groups are represented by rectangles. Note the compact symmetry of the quaternary structure.

Many proteins are composed of more than one polypeptide chain. As mentioned previously, these subunits are adjoined by noncovalent bonding (hydrogen bonding, hydrophobic interaction, or ionic bonding). The joining of these subunits forms what is called a **quaternary structure** (Figure 11.5). A classic example of a quaternary structure is the oxygen-carrying protein hemoglobin. This protein is composed of four polypeptides—of two different kinds—and is termed an $\alpha_2\beta_2$ tetramer. The individual subunits in a quaternary structure are proteins themselves having typical secondary and tertiary structures.

Synthesis of the Cytoplasmic Membrane

One structure that is present in all Eubacteria and Archaea is the **cytoplasmic membrane,** also called the cell membrane. A cytoplasmic membrane is basically a bilayer of phospholipid interspersed with proteins. The bacterial cytoplasmic membrane is similar in its basic structure to

Disruption of the sulfide alters the conformation of the protein resulting in a loss of enzymatic activity. Removal of the denaturing agents and exposure to air results in a random reassembly and an inactive enzyme. Exposure of the denatured protein to trace levels of β-mercaptoethanol to promote the proper rearrangement of the —S—S— cross-links results in a reformation of the active enzyme. This and test tube experiments with other proteins indicated that proper folding may be an inherent property of polypeptides. Thus folding to secondary and tertiary structures would be promoted by the proper spacing of amino acids with side chains that could interact with counterpart amino acids elsewhere in the polypeptide chain. The bonds that would give the ultimate structure would be those formed by the interactions of various amino acids as illustrated in Figure 11.3. This sort of *in vitro* study apparently does not completely reflect the conditions in the cytoplasm of a cell. The levels of polypeptide required for test tube refolding are much higher than would occur *in vivo,* and the physiochemical conditions of the *in vitro* experiments would be unattainable in the living cell. In reality, studies *in vitro* that mimic *in vivo* physiological conditions result in misfolding or aggregation of the polypeptide. This misfolding is uncommon *in vivo* except with mutant proteins or protein folding at elevated temperatures.

Studies in recent years affirm that there are two classes of proteins associated with the proper folding of polypeptides in a cell. One class is the conventional enzymes that catalyze specific isomerization reactions that result in proper polypeptide conformations for proper folding. The second class of proteins are the **chaperones** that stabilize unfolded or partially folded structures and preclude the formation of inappropriate intra- or interchain interactions. A chaperone may also interact with individual proteins to promote protein-protein arrangements that yield quaternary functional structures. Among the roles that protein chaperones play are: they prevent folding of secretory proteins before translocation, are involved with assembly of bacterial viruses, and promote the assembly of pili and the assembly of the carboxysome in photosynthetic organisms. The chaperones are also important in preventing protein denaturation during environmental stress, such as elevated temperature.

◀ The secondary structure of bovine pancreatic ribonuclease is maintained by disulfide linkages. Disruption of these —S—S— bonds results in an unfolded and denatured polypeptide. The enzyme can regain activity if the disulfide bonds are properly reestablished.

other membranes—for example, those surrounding eukaryotic nuclei, mitochondria, and other organelles. The membranes of Archaea are quite different, as discussed in Chapter 22.

A basic unit in a bacterial cytoplasmic membrane is the phospholipid that was discussed in Chapter 3 and a typical one—phosphatidylethanolamine, as depicted in Figure 11.6(**a**). The polar end groups commonly present in microorganisms are ethanolamine, serine, and glycerol. Cardiolipin is a lipid that can span the membrane and is composed of two molecules of phosphatidic acid joined to the terminal hydroxyls of a molecule of glycerol (see Figure 10.40). The shorthand representation of a polar lipid is shown in Figure 11.6(**b**) and these are aligned in a lipid bilayer as shown in Figure 11.6(**c**). The circle represents the polar hydrophilic phosphate end of the molecule, and the squiggly lines represent the apolar (hydrophobic) hydrocarbon chains. A phospholipid is synthesized from two molecules of fatty acid, a glycerol phosphate, and a polar head group such as ethanolamine. The long-chain fatty acids that are generally constituents of the phospholipids in bacteria were presented in Figure 10.38.

When phospholipids are placed in aqueous solution, they assume the configurations shown in Figure 11.6(**c**). Dissolution of bacterial cytoplasmic membranes followed by careful reassembly results in membrane vesicles similar to the vesicle shown (Figure 11.6(**d**). Under proper conditions, a disrupted membrane will reassemble to form vesicles that have sidedness as present in the cell from which they were obtained. The cytoplasmic side will face inward and the periplasmic side will face outward. These vesicles have been an important tool in studies on cytoplasmic membrane function.

A vesicle can retain membrane functions such as solute transport and the ATPase system, and these functions occur without the presence of cytoplasmic proteins.

A cytoplasmic membrane in a growing cell must assimilate new membrane components and integrate these into spaces in the membrane. During integration of the newly synthesized phospholipids, the permeability and barrier functions of the membrane must be maintained.

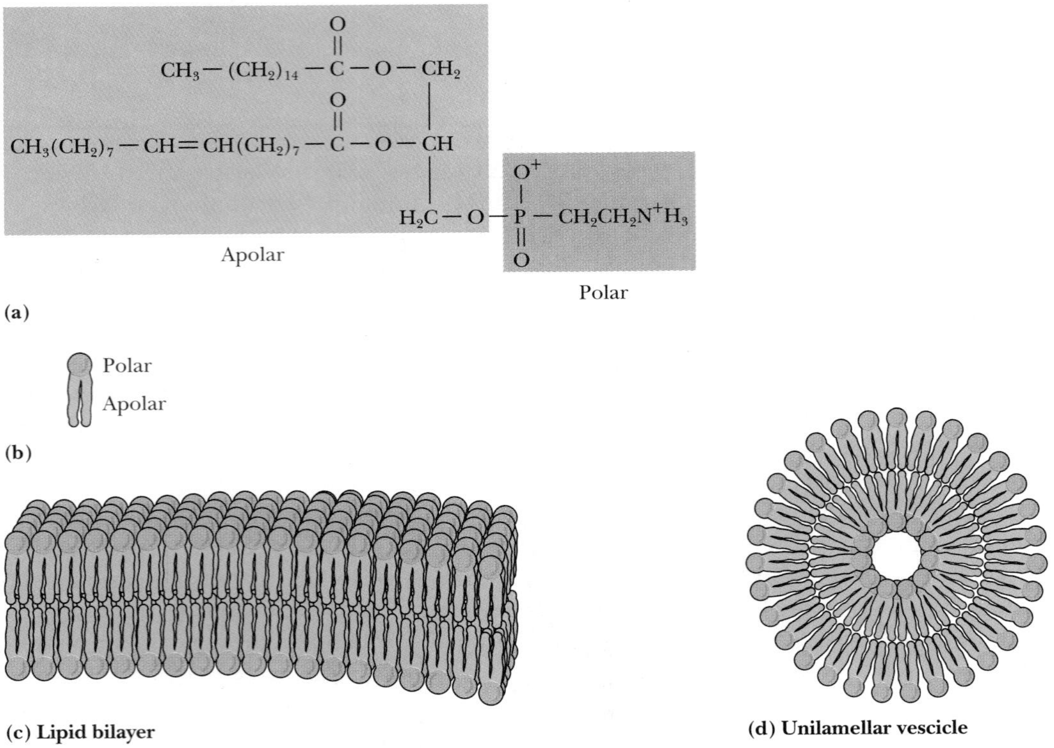

***Figure* 11.6** Components of a membrane. **(a)** Phosphatidylethanolamine with the apolar hydrophobic and the polar hydrophilic ends of the molecule indicated. The shorthand version **(b)** and the arrangement of membrane phospholipids in a lipid bilayer presented in **(c)**. A self-assembling vesicle **(d)**.

A simplistic view of a cell membrane and its components is illustrated in Figure 11.7. During growth a microorganism must generate extensive amounts of new cytoplasmic membrane. Proteins are an integral part of the membrane complex and must be added to newly synthesized membrane. The proteins that are embedded in the membrane pass completely through, extending from the cytoplasmic to the periplasmic side. Other proteins are closely associated with the inner and outer leaflet surfaces. Rarely are proteins partially embedded into the cell membrane. The peripheral proteins that are present on the outer membrane surface can be solubilized by mild detergent. The integral proteins present in the membrane matrix are released only when the membrane is dissolved.

In most microorganisms, the enzymes involved in fatty acid synthesis are located in the cytoplasm. These enzymes acylate (add fatty acid to) a glycerol-3-phosphate to form phosphatidic acid at the inner surface of the membrane. An ethanolamine, serine, or other polar molecules would then be esterified enzymatically to the phosphate group of the phosphatidic acid to form the polar lipid. These reactions occur on the inner leaflet, and expansion of the existing phospholipids bilayer provides an opening for the newly synthesized phospholipid to be integrated. This apparently occurs by some form of spontaneous self-assembly, but the exact nature of this incorporation process is not now known.

The newly synthesized phospholipid is incorporated

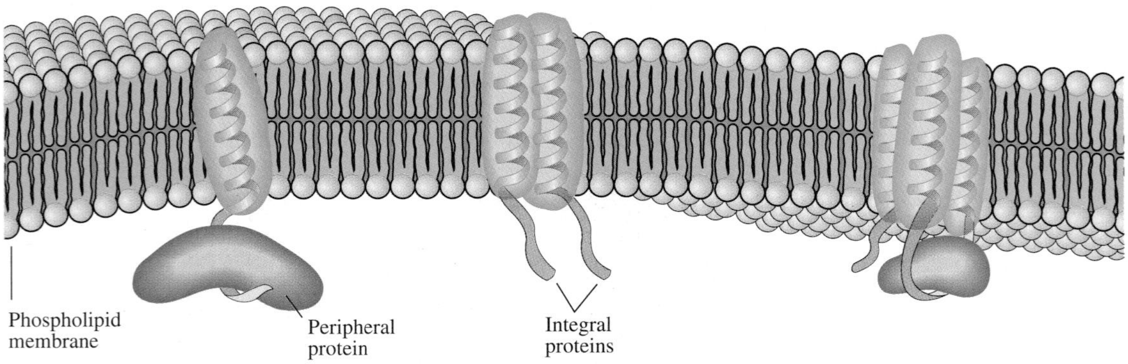

***Figure* 11.7** A typical cytoplasmic membrane with integral transmembrane proteins. The hydrophobic interior prevents the passage of most molecules. A key element in synthesis of functional membranes is the insertion of the proteins that are essential to proper function.

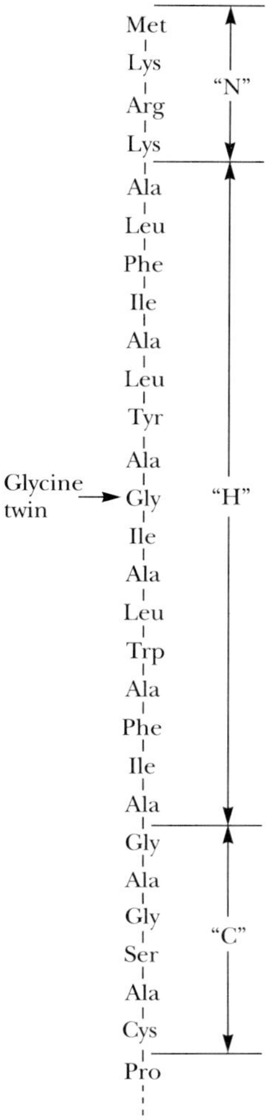

***Figure* 11.8** A typical signal peptide that might be present in a gram-negative bacterium. The "N" section is composed of amino acids that are polar and have a net positive charge, and this enables the leader to enter the cytoplasmic membrane. The "H" section is hydrophobic and inserted into the cytoplasmic membrane, and "C" is also a series of hydrophobic amino acids that are recognized by the signal peptidase that cleaves the signal peptide from the secretory protein. The "glycine turn" is the point where the signal peptide folds, as shown in Figure 11.10.

into the inner leaflet and appears in the outer leaflet in a relatively short period of time. This flip-flop from the inner to outer membrane leaflet occurs at a more rapid rate than is known for exchange of phospholipids from the inner to outer leaflet in model bilayers. This suggests that rotation from the inner to outer leaflet in a viable cell may be controlled by proteins in the membrane. The process may require ATP, but this is uncertain at the present time.

Approximately 20 percent of the polypeptides synthesized by bacteria are located in the cytoplasmic membrane or translocated across the membrane. These proteins pass into or through the cytoplasmic membrane via the general secretory pathway (GSP). Proteins transported by the GSP are called secretory proteins. In some cases polypeptides secreted by the cell are then assembled outside the cytoplasmic membrane, and these polypeptides are termed presecretory proteins. The distinguishing characteristic of secretory proteins is the presence of a **signal sequence** of at least 10 hydrophobic amine acids at a terminus of the protein.

The function of the signal sequence is to direct the secretory proteins to the cytoplasmic membrane. The leading end of the signal peptide is termed "N" and is polar with a net positive charge. A typical signal peptide is shown in Figure 11.8. The middle region of a signal peptide, termed "H" (hydrophobic) is inserted into the membrane. The "C" sequence is also hydrophobic and is recognized by a peptidase that cleaves the secretory protein from the signal peptide.

Two mechanisms exist for the passage of proteins or presecretory proteins into or through the cytoplasmic membrane. One is **posttranslational translocation,** and this occurs when a membrane protein is synthesized by polyribosomes in the cytoplasm and the complete protein is incorporated into or transmitted through the membrane. There are no cytoplasmic or membrane proteins involved in this process. The number of proteins secreted by this mechanism is very limited. A model for the posttranslational receptor-independent translocation of a secretory protein is presented in Figure 11.9.

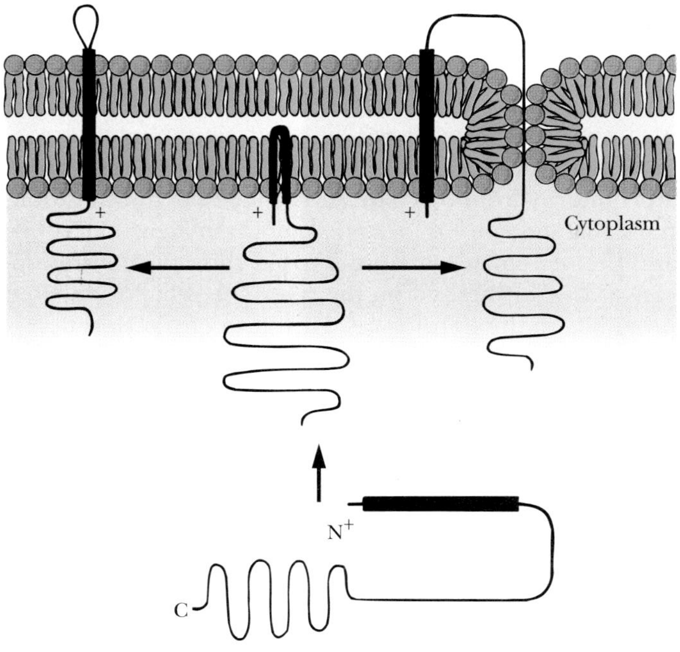

***Figure* 11.9** Model for a receptor-independent signal peptide insertion into a cytoplasmic membrane. Insertion may be followed by the formation of a transmembrane loop in which the signal peptide exports the protein or the signal peptide induces the membrane phospholipids to change conformation to provide a channel lined by the polar ends of the polar lipids. (Reproduced with permission, Anthony Pugsley, *Microbiological Reviews* 57:50-108.)

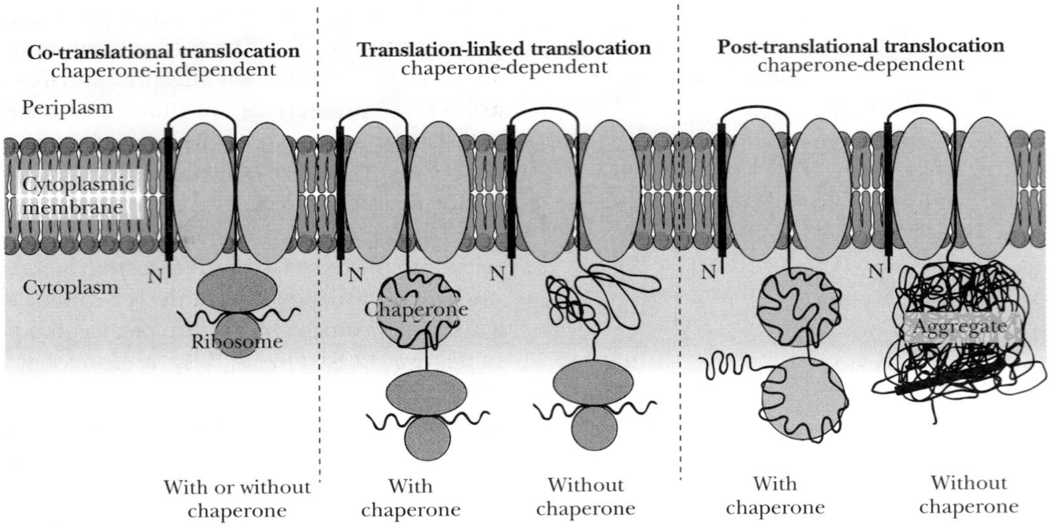

***Figure* 11.10** Model for three different modes of translation-translocation coupling of secretory proteins. **(a)** The protein is translocated as it is generated on the ribosome; **(b)** the protein is translation-linked to translocation through a chaperone to prevent the presecretory protein from folding into conformations that cannot be translocated; and **(c)** posttranslational translocation with the involvement of chaperones that prevent folding of the protein into aggregates that could not be translocated (see Figure 11.9). (Reproduced with permission from Anthony Pugsley, *Microbiological Reviews,* Vol. 57:50-108.)

The mechanisms of **cotranslational translocation** are translation-linked and generally chaperone-dependent. Posttranslational translocation mechanisms are illustrated in Figure 11.10. Most secretory proteins are translocated by the mediation of a **translocase complex.** There are an estimated 500 translocase complexes in the cytoplasmic membrane of gram-negative bacteria, and these complexes are composed of cytosolic and transmembrane proteins. The translocase proteins that span the membrane are quite hydrophobic. In cotranslational translocation, the signal sequence is located as shown in Figure 11.10(**a**), and the secretory protein is carried through the translocase as it is synthesized on the ribosome, which is in close proximity to the translocase. This form of translocation is chaperone-independent. In chaperone-dependent translation-linked translocation, the protein is synthesized away from the translocase, and the protein is associated with a chaperone following synthesis. The chaperone protein then passes the secretory protein to the translocase complex Figure 11.10(**b**). Proteins that are part of the secretory complex may act as chaperones. In the absence of a chaperone, the protein may fold prior to translocation, and secretory function would cease. In posttranslational translocation, the secretory protein must have chaperone molecules to prevent folding as illustrated in Figure 11.10(**c**). An aggregate that would occur in the absence of chaperones would not pass through the translocase complex. After the secretory protein enters the periplasmic space or is in place as a membrane protein, the signal sequence may be cleaved by a signal peptidase.

The production and secretion of fungal enzymes and other proteins is fundamentally different from that in bacteria. Fungal proteins are synthesized at the rough endoplasmic reticulum and transported in vesicles to the Golgi apparatus. Vesicles then bud from the Golgi and are transported to the cytoplasmic membrane. At the site of secretion, the membrane of the vesicle that contains protein fuses with the cytoplasmic membrane, and the protein is released to the outside.

It is known that membrane lipid synthesis and alterations can occur independently of the membrane proteins and vice versa. A constant turnover of either proteins or phospholipids (or both) occurs as organisms adjust to environmental changes. Cells increase the proportion of membrane unsaturated fatty acids as a response to a decrease in growth temperature. This occurs because a functional membrane must be fluid, and unsaturated fatty acids are more fluid at lower temperatures. Newly formed proteins may also be inserted in the membrane as an organism encounters a different substrate or other changes in growth conditions.

Biosynthesis of Peptidoglycan

Peptidoglycan (murein) is the main structural component of the eubacterial cell wall. It is a large single molecule that surrounds the cytoplasmic membrane and gives the organism its shape and rigidity. The peptidoglycan layer is quite thin in gram-negative bacteria and is multilayered in gram-positives. Murein is composed of repeating units of *N*-acetylglucosamine (NAG) and *N*-acetylmuramic acid (NAM) (Figure 11.11). NAM differs from NAG in having a lactyl group coupled to carbon three of the glucose moiety. The lactate is bonded to the glucose by an ether linkage. During biosynthesis, muramic acid has a pentapeptide attached to the carboxyl of the lactyl through a peptide bond. The tetrapeptides on adjacent muramic acid molecules are cross-linked through peptide bridges or through the dibasic amino acid in position 3 of the tetrapeptide. About 50 percent of the *N*-acetylmuramic tetrapeptides in a murein layer are cross-linked.

Biosynthesis of peptidoglycan units occurs in the cytoplasm and in the cytoplasmic membrane. The steps involved in this biosynthesis are outlined in Figure 11.12. Both *N*-acetylglucosamine and *N*-acetylmuramic acid are synthesized and coupled to bactoprenol (Figure 11.13) at the inner membrane surface. **Bactoprenol,** or undecaprenol phosphate, is a long-chain hydrocarbon that can enter cytoplasmic membranes and carry hydrophobic molecules into or through the lipid bilayer. Bactoprenol displaces the uridine phosphate (Figure 11.14), and this renders the muramyl pentapeptide sufficiently hydrophobic to allow passage through the hydrophobic membrane. During this passage, a bridge peptide such as pentaglycine may be attached to the terminal amine of lysine via a peptide bond. This would occur in gram-positive bacteria. The peptidoglycan units then enter the periplasmic space and are inserted at a growing point in the cell wall.

The cross-linkage between adjacent molecules of the pentapeptide on *N*-acetylmuramic acid is the final step in peptidoglycan synthesis. The third amino acid in the pentapeptide attached to muramic acid is a diamino acid, that is, an amino acid that contains two amine groups (diaminopimelic acid, ornithine, or lysine). The terminal amine can form a peptide bond with a carboxyl of the amino acid in position four of an adjacent chain. The divalent amino acid occupying the third position varies according to the genus of the organism. The composition of the peptide bridge can vary by species within a genus (see Chapter 4).

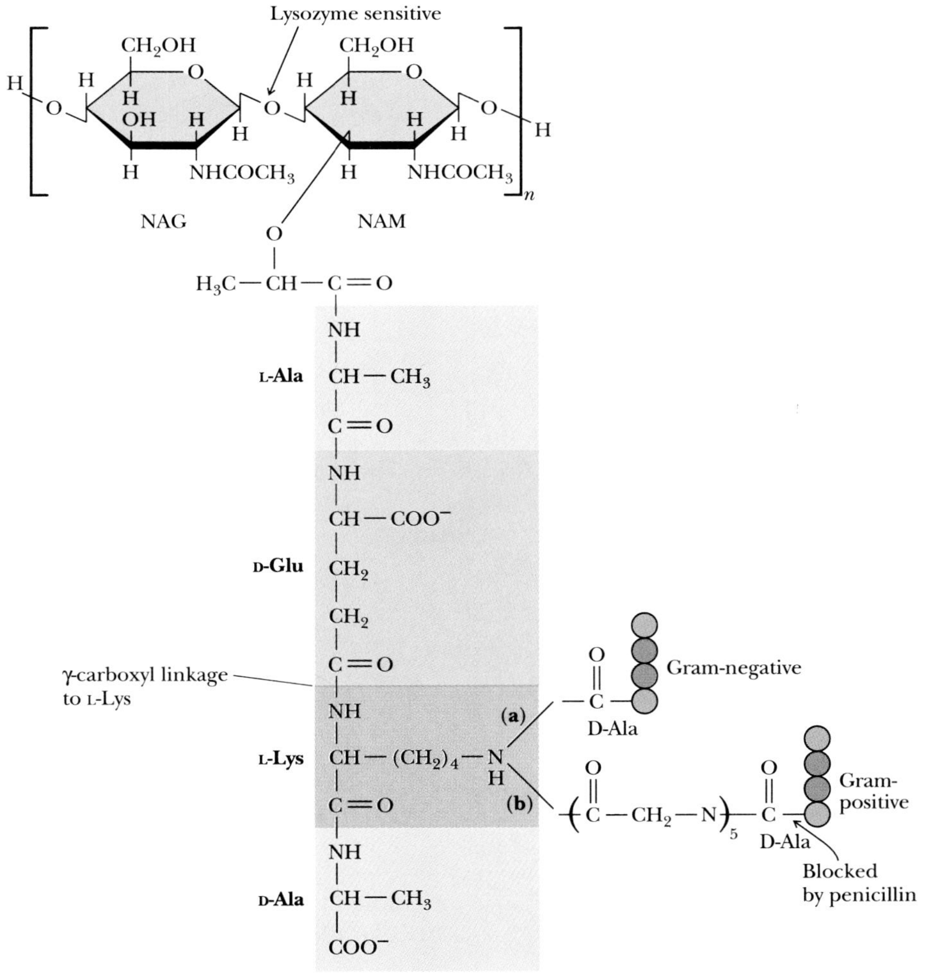

***Figure* 11.11** The basic unit of the structural peptidoglycan in the cell walls of bacteria. NAG is *N*-acetylglucosamine and NAM is *N*-acetylmuramic acid. NAM has an ether linkage to a molecule of lactic acid. The tetrapeptide is composed of L-alanine, D-glutamate, L-lysine (or other dibasic amino acid), and D-alanine. In gram-negative bacteria **(a)** the dibasic amino acid is linked directly to the terminal D-alanine molecule in an adjacent tetrapeptide. In gram-positive bacteria the linkage is through a pentaglycine (or other amino acid) bridge **(b)**. Pentaglycine is present in *Staphylococcus aureus.*

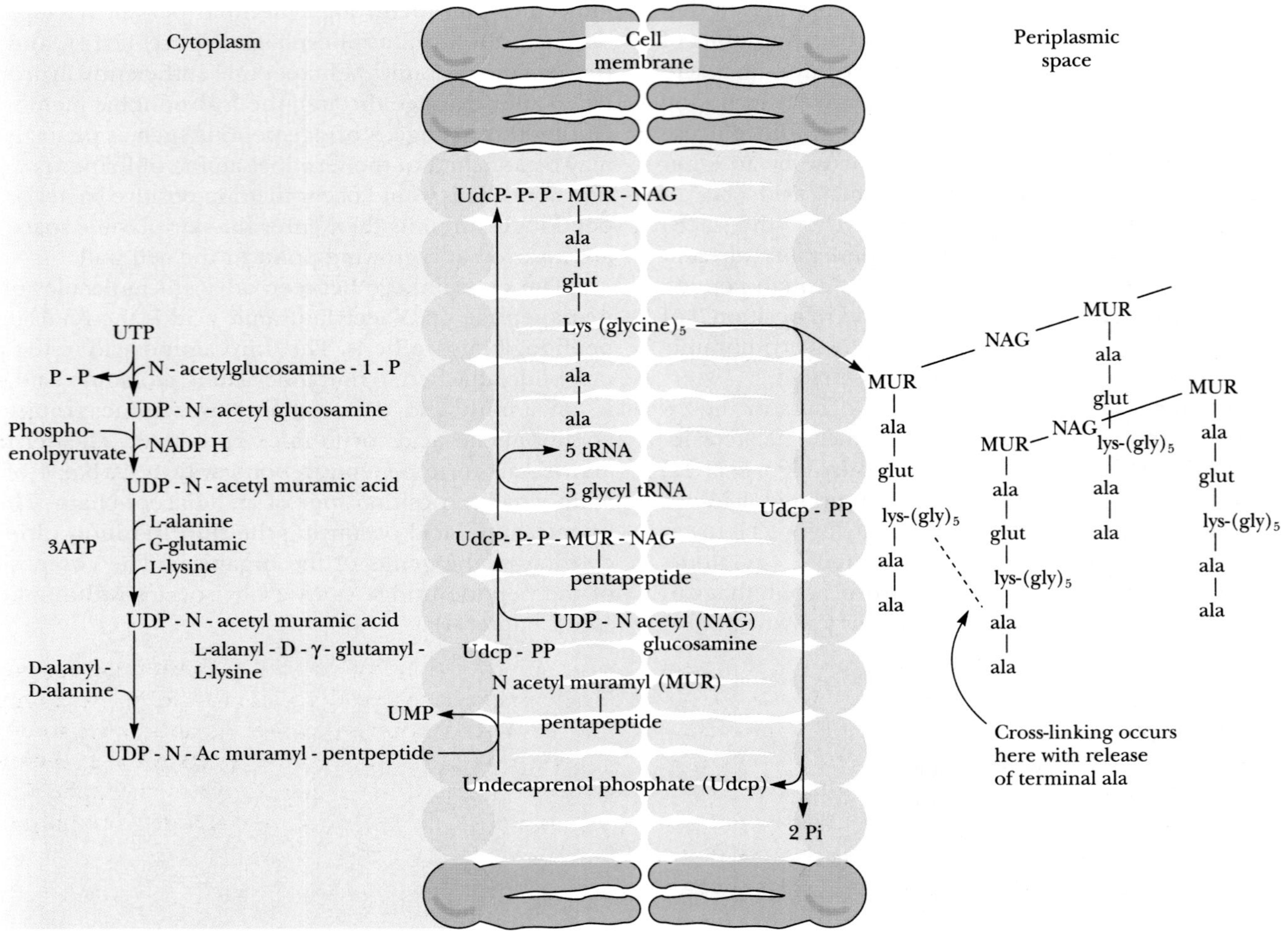

Figure **11.12** The individual steps in the synthesis of peptidoglycan. The nucleotide involved is uridine triphosphate. The uridine-*N*-acetylmuramic acid pentapeptide complex is synthesized in the cytoplasm and carried into the cytoplasmic membrane by undecaprenol. The undecaprenol pyrophosphate muramyl-*N*-acetylglucosamine moiety is assembled in the cytoplasmic membrane and integration into the cell wall occurs outside the membrane.

The bridge or dibasic amino acid attached to one muramic acid side chain forms a bond with an adjacent pentapeptide by transpeptidation. A D-alanine molecule occurs at the end of the pentapeptide (Figure 11.11). This D-amino acid is cleaved off during the transpeptidation reaction. The peptide bond between the D-alanine molecules "activates" the subterminal alanine, which promotes the reaction with the cross-linking peptide bridge. Because the reaction occurs outside the cytoplasmic membrane, there is no other source of energy for this reaction.

A battery of enzymes is involved in the covalent reactions that result in extension and cross-linking between

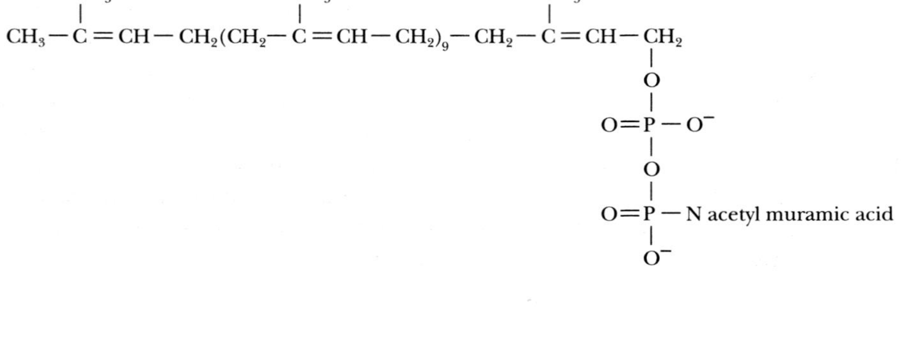

Figure **11.13** The structure of undecaprenol phosphate with a muramic acid attached.

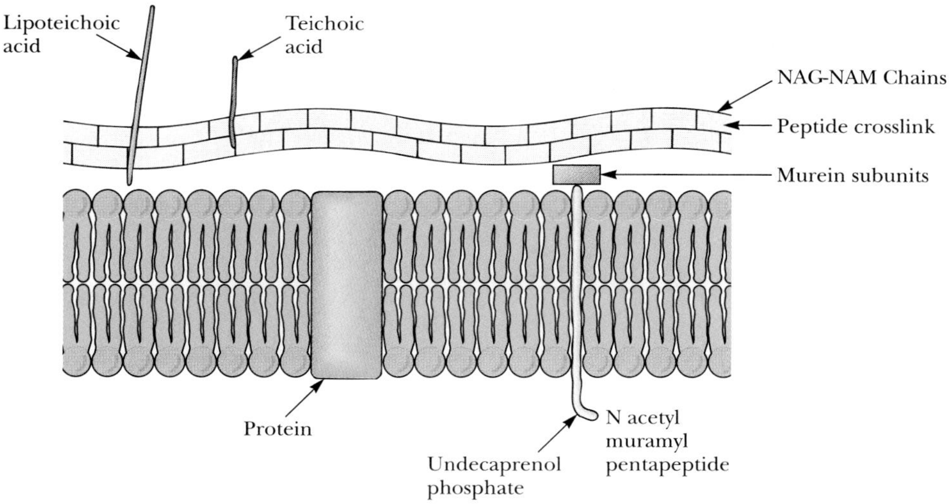

***Figure* 11.14** Structures that a gram-positive bacterium assembles outside the cytoplasmic membrane include peptidoglycan, teichoic acids, and lipoteichoic acids.

the peptidoglycan strands. The enzymes are also responsible for septation of the murein sacculus that occurs during cell division. These proteins that are involved in extension of the peptidoglycan have a unique ability to bind the antibiotic penicillin and some related antibacterials. Binding of penicillin to these proteins inhibits murein biosynthesis and can destroy the integrity of the cell. The number of penicillin-binding proteins on the surface of a bacterium varies with species. Studies indicate that the penicillin-binding proteins are involved with transglycosylation (elongation of glycan strands), transpeptidation (cross-linking), and the enzyme carboxypeptidase that cuts preexisting cross-links for new glycan insertion.

Gram-Positive Cell Walls

The structural relationship between the cytoplasmic membrane and the peptidoglycan in a gram-positive bacterium was discussed in Chapter 4. Peptidoglycan subunits are constantly synthesized and enter the space between cytoplasmic membrane and existing murein layer at points where they are linked to the existing peptidoglycan layer. Thus, a growing cell is continually adding a murein layer in the area adjacent to the cytoplasmic membrane (Figure 11.14). About 40 layers of murein surround a gram-positive cell. The layers move outward as newly synthesized murein is added to the inner layer.

Teichoic acids are an integral part of the gram-positive cell wall and are synthesized in the cytoplasm. They are assembled on a membrane carrier, probably bactoprenol, and transported across the membrane. The structures of these components was also presented in Chapter 4. The function of these sugar derivatives is not certain. It has been suggested that they regulate the autolysins that open up the peptidoglycan for expansion during growth. There is evidence that the teichoic acids present on pathogenic streptococci are involved in adherence to eukaryotic cells. The negative charge created by the teichoic acids makes the cell surface impermeable to hydrophobic compounds much as the negatively charged lipopolysaccharide does for gram-negatives.

Gram-Negative Cell Walls

The cell wall structures of gram-negative bacteria are considerably more complex than those of gram-positives (Chapter 4). Growth and expansion of the walls are consequently more elaborate. It is apparent that fatty acids, sugars, phospholipids, and proteins must move from the cytoplasmic side of the membrane to the periplasm during wall expansion in gram-negative bacteria. A typical area undergoing expansion is illustrated in Figure 11.15. Expansion occurs in minute openings in the cell wall called **Bayer's junctions.** These openings have also been termed **zones of adhesion** because the inner and outer membranes actually make direct contact. The junctions occur in the peptidoglycan layer, and all components of the outer cell envelope pass through these gaps. The components of the lipopolysaccharide present in the cell envelope are synthesized on the inner surface of the cell membrane and are carried outward with the assistance of bactoprenol. Assembly apparently occurs on the outer surface by mutual attraction. The phospholipid components are translocated by the same mechanism, and hydrophobic interactions with lipid A results in the formation of the outer layer. The structure and function of the outer envelope components in gram-negative bacteria were discussed previously (Chapter 4).

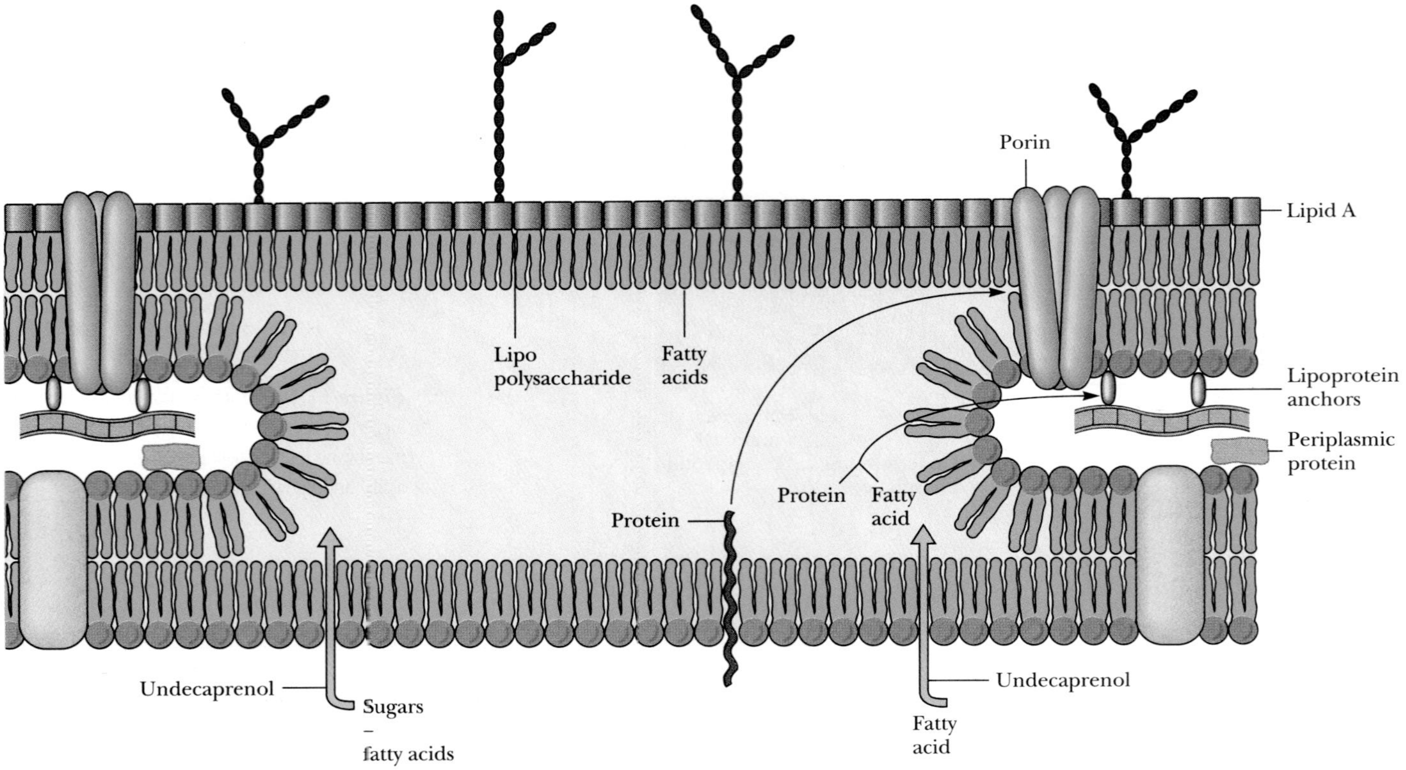

***Figure* 11.15** A Bayer's junction in the cytoplasmic membrane/peptidoglycan/outer envelope complex of a gram-negative bacterium. This is an exaggerated view for illustration and in reality would be a minute pore. The junctions would occur randomly about a cell undergoing growth. Fatty acids and sugars would be transported into the gap by undecaprenol and proteins by mechanisms described in Figure 11.11.

Proteins are secreted through the cell membrane as they are synthesized by ribosomes attached to the membrane. This process was described earlier. Many major proteins, such as porins, may self-assemble into the three-dimensional configuration shown in Figure 11.15. It is interesting to note that a mixture of lipopolysaccharide, porin proteins, and other components of the outer envelope can assemble under laboratory conditions to form a structure similar to that present in a living cell. It is apparent, however, that many of the proteins that are assembled into structures outside the cytoplasmic membrane are complexed by chaperones that promote proper assembly. For example, the pilins were previously considered to self-assemble to form pili, but it is now known that pilin proteins are complexed by proteins until they arrive at the assembly site. Pilins are assembled and added at the base of the pilus.

The lipoprotein anchors that attach the cell envelope to the peptidoglycan are formed by the interaction of a protein with the lipids present on the outer membrane.

Flagellum Assembly

The bacterial flagellar structure (Figure 11.16) is assembled by a sequential series of events. In a gram-negative bacterium the first visible structure is the M ring, which forms on the cell membrane. The S ring is added, followed by insertion of the hollow rod. The P and L rings are then added, followed by the appearance of a hook. The flagellum itself is composed of a single protein called **flagellin.** Flagellin is synthesized on ribosomes present in the cytoplasm and passes outward through the hollow rod of the growing flagellum. The flagellum is formed by self-assembly of these flagellin proteins at the distal end. The processes that control the length of flagella are not presently known. If a portion of a flagellum is broken off, it is quickly replaced by newly synthesized flagellin molecules. The gram-positive bacteria have only two rings, one in the cytoplasmic membrane and the other associated with the peptidoglycan/teichoic acid part of the wall (Chapter 4).

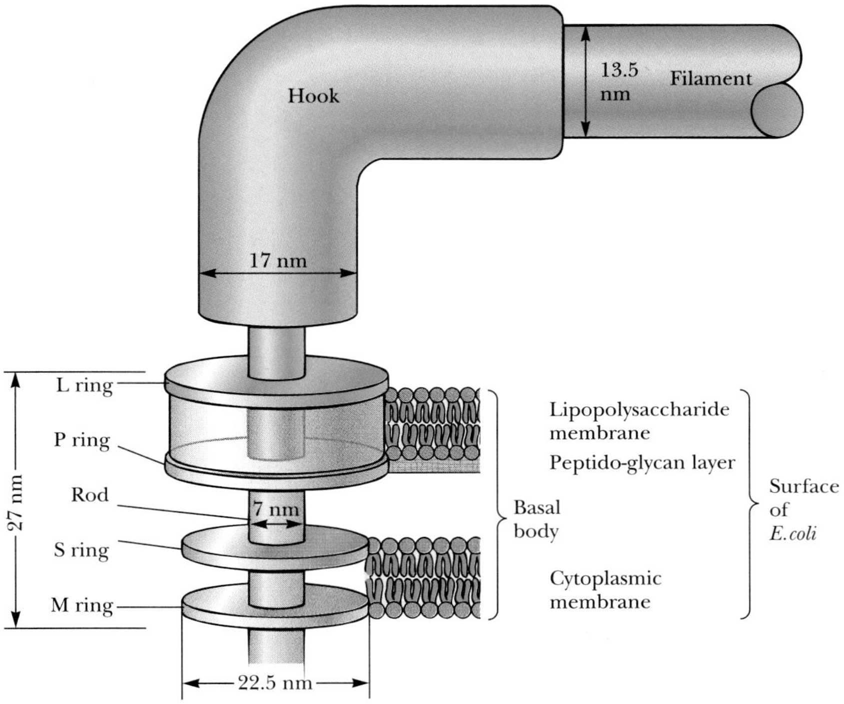

***Figure* 11.16** The parts of a flagellum that are present in a bacterium motile by flagellation. A gram-positive organism would not synthesize the L ring.

The last step in producing an active flagellum is the insertion of protein molecules near the basal body. These proteins are responsible for the rotation of the hook that results in motility.

Storage Compounds

The ability to accumulate and store organic reserve material is a common attribute among microorganisms. This accumulation generally occurs in late log phase, when sufficient nutrient is available in the growth medium (see Chapter 6). Bacteria also tend to store carbonaceous compounds when substrate is available but nitrogen and/or phosphorous levels are limited. When conditions are appropriate for active growth, the reserve polymers may disappear. Among the compounds that different species may accumulate are amino acids, fatty acids, and sugars. Virtually all stored materials are polymerized into discrete granules that appear in the cytoplasm. Polymerization is essential as it decreases acidity, basicity, or surface activity that would otherwise occur with many unpolymerized monomeric compounds. Storage granules are readily observable as electron-dense structures in thin sections of cells viewed by electron microscopy. They also may be visualized with a light microscope after staining.

Enteric bacteria, some spore formers, and certain other bacteria store glycogen. **Glycogen** is a glucose polymer similar to starch and is synthesized from adenosine diphosphate and glucose (Figure 11.17). Branching occurs by transglycosylation in which small oligosaccharides are cleaved from the end of the sugar chain by a branching enzyme, which then attaches them in a 1,6 linkage at some random point. Another common storage granule is polymerized β-hydroxybutyric acid (Figure 11.18). These granules are present in some spore formers, pseudomonads, and rhizobia. When carbonaceous substrates are unavailable, these polymers are utilized as a source of carbon and energy. Granules of polyphosphate (volutin) are stored by some microorganisms, and these granules may be bound by a protein membrane. Polymers of β-hydroxybutyric acid may also be bound by a protein membrane.

Some of the cyanobacteria store a simple polymer containing aspartic acid and arginine. The peptide polymer is as follows:

```
      |
Aspartic acid—Arginine
      |
Aspartic acid—Arginine
      |
Aspartic acid—Arginine
      |
```

Under appropriate conditions, the cyanobacteria may use this peptide as a source of both nitrogen and energy.

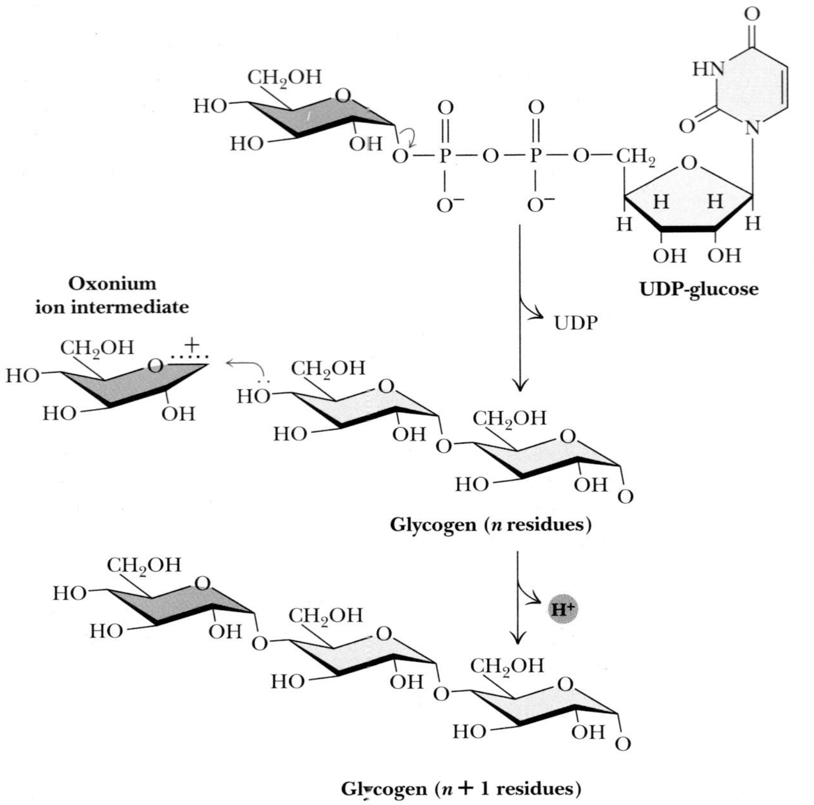

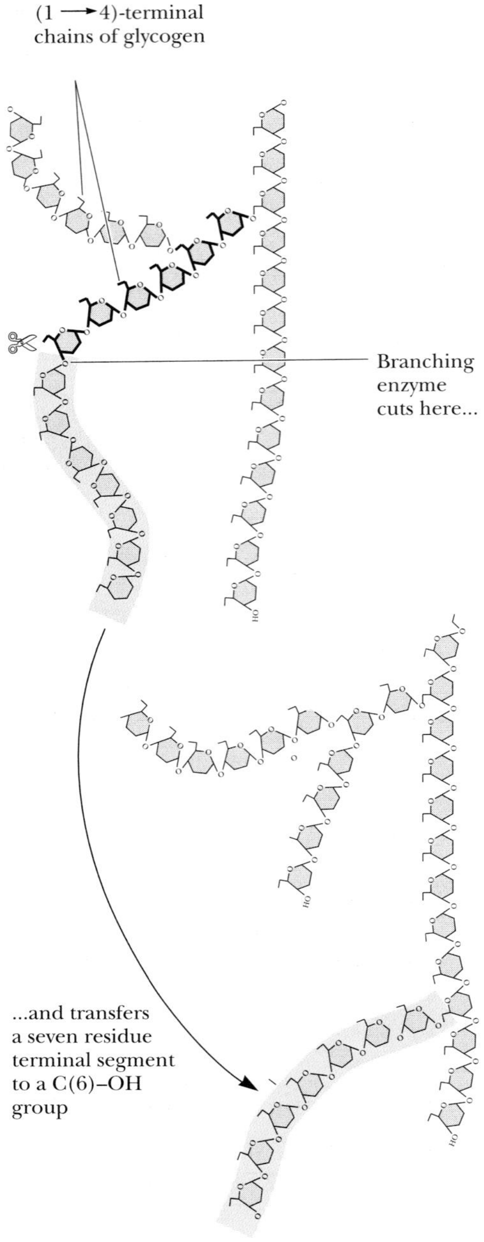

***Figure* 11.17** Steps involved in the synthesis of the branched polysaccharide glycogen. Uridine-diphosphoglucose is the nucleotide involved. Note that α1,4 linkages occur in the extended chain and that six or seven residue segments are cut off and are transferred to the C-6 hydroxyl group of a glucose residue on the same or a nearby chain.

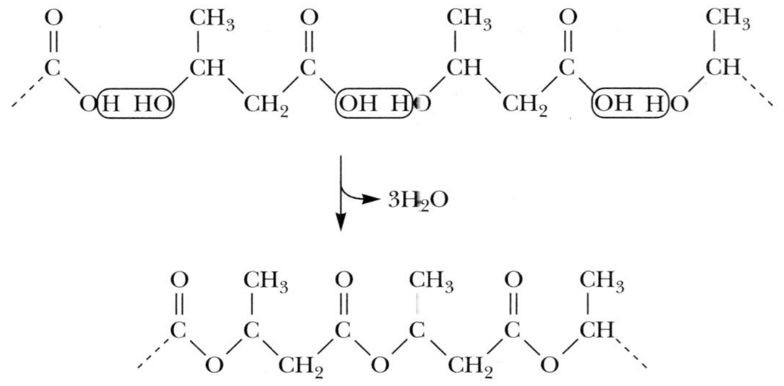

***Figure* 11.18** Granules of poly-β-hydroxybutyric acid are a common storage compound in bacteria. β-hydroxybutyric acid is formed by condensation of two molecules of acetate to form acetoacetate.

Summary

- The **monomers** that are essential for the synthesis of cell matter are **polymerized** to form macromolecules, and these are the functional components of the cell. Major macromolecules in a cell are proteins, phospholipids, DNA, RNA, and polysaccharides.

- Over one half of the dry weight of a bacterial cell is **protein.** About 20 percent is **RNA.**

- A **polypeptide** may become a functional **protein** by assuming a **tertiary configuration.** The tertiary configuration results from the joining of **secondary structures** such as the **α-helix** and **β-sheets.** Joining to two or more tertiary structures yields a **quaternary structure.**

- **Cytoplasmic membranes** are formed from bilayers of **phospholipid** interspersed with proteins.

- A **phospholipid** is made up of fatty acids, a glycerol-3-phosphate, and a polar group such as ethanolamine linked to the phosphate group.

- About 20 percent of the polypeptides synthesized by bacteria are integrated into the cytoplasmic membrane or translocated across the membrane.

- Most proteins are translocated across the membrane as they are **translated** on a ribosome. This process is termed **cotranslational translocation.**

- A **chaperone** is a protein that assists in the folding of a polypeptide chain to assume the proper functional configuration.

- The structural component of a eubacterial cell wall is **peptidoglycan.** It is composed of sugar amines that are complexed with peptides. Synthesis of peptidoglycan units occurs in the cytoplasm and within the cytoplasmic membrane. Assembly occurs outside the cytoplasmic membrane.

- **Gram-positive** microorganisms have a thick peptidoglycan layer, whereas a **gram-negative** bacterium has a thinner peptidoglycan layer. The gram-negative bacteria have an elaborate cell envelope outside the peptidoglycan layer.

- **Bacterial flagella** are based in the cytoplasmic membrane and are attached to wall peptidoglycan. The whiplike flagellum is composed of the protein **flagellin.**

- Microorganisms can accumulate excess carbonaceous material in specific **granules** inside the cell. Among the materials stored are lipids, glycogen, pyrophosphate, and peptides.

Questions for Thought and Review

1. Proteins make up over half of the dry weight of a cell. What are the diverse roles that proteins play and where are the various types located?

2. Draw a peptide bond and a peptide composed of several amino acids. Define primary, secondary, tertiary, and quaternary structure.

3. What are the two major secondary structures that a protein may assume? How do they differ and what is our shorthand for depicting them? What is the major bonding that holds polypeptides in these configurations?

4. What are some major bonds that are involved in the functional configuration of proteins?

5. Draw a polar lipid. Which is the hydrophobic and hydrophilic end of the molecule? What happens if a number of phospholipids are added to water?

6. How can a peptide pass through the hydrophobic cytoplasmic membrane? What is cotranslational insertion? What role might chaperones play in secretory protein translocation?

7. What is a Bayer's junction? How does it benefit a growing cell? Where are the major components of a gram-negative outer envelope assembled? Discuss the processes involved.

8. How does a bacterium assemble a flagellum? A pilus?

9. What are some of the major storage materials present in bacteria? Why are they generally polymerized?

Suggested Readings

Garrett, R. H., and C. M. Grisham. 1995. *Biochemistry.* Philadelphia: Saunders College Publishing.

Gerhardt, P., R. G. E. Murray, W. A. Wood, and N. R. Krieg. 1994. *Methods for General and Molecular Bacteriology.* Washington, D.C.: American Society for Microbiology.

Neidhardt, F. C., J. L. Ingraham, and M. Schaechter. 1990. *Physiology of the Bacterial Cell: A Molecular Approach.* Sunderland, MA: Sinauer Associates.

Solomon, E. P., L. R. Berg, D. W. Martin, and C. Villee. 1993. *Biology.* 3rd ed. Philadelphia: Saunders College Publishing.

All these tidal gatherings, growth and decay,
Shining and darkening, are forever
Renewed; and the whole cycle impenitently
Revolves; and all the past is future.

Robinson Jeffers

Chapter 12

Microbial Biodegradation

The previous chapters in Part 3 dealt with the major processes whereby microorganisms generate energy and the biosynthetic reactions involved in the construction of a living cell. It is axiomatic that survival, growth, and perpetuation of species depend on the endless generation of progeny. An indispensable component of survival and growth is a source of building material and the assembly of this material into viable cells. The obvious question: how long can nature generate new life without renewing the ultimate source of building blocks? Not long! The ultimate source of building blocks for cell synthesis is through the destruction of the old in an unending replenishment of the ultimate source of structural material—CO_2. **Microbial biodegradation** is nature's way of recycling the organic and inorganic components of the inert products of life. Complete biodegradation of a substrate to its constituent inorganic parts is also termed **mineralization.**

A major step in the evolution of complex biological systems was the advent, about 2.5 billion years ago, of viable organisms that could assimilate simple compounds as substrates during oxygenic (oxygen generating) photosynthesis. These microorganisms were the cyanobacteria, a group that evolved with a synthetic system that used CO_2 as carbon source and a photosystem that yielded molecular oxygen:

$$2\ CO_2 + H_2O \longrightarrow 2\ CH_2O\ \text{(cellular mass)} + O_2$$

Growth of these organisms, with the inexhaustible source of solar energy, resulted in an atmosphere containing an oxidant (O_2) suitable for the development of aerobic heterotrophs. The diversity of microbes that arose as a consequence was instrumental in the evolution of the multicelled eukaryotes.

A viable Earth can be sustained only if constituents of the cells produced by photosynthetic reactions are ultimately recycled back to CO_2 (about 0.03 percent of the gases in the atmosphere). The success in meeting this requirement is apparent from the general balance and stability evident in nature. The massive biological productivity resulting from photosynthesis, if not countered by respiration, would markedly lower the CO_2 levels in the atmosphere within about 30 years. Various microorganisms (Archaea, Eubacteria, and eukaryotes), through their metabolic activities, account for 85 percent to 90 percent of the CO_2 entering the atmosphere. Without this constant renewal of atmospheric CO_2 the plants would wither, the pH of the ocean would increase due to loss of CO_2, and much of life on Earth would be altered. Animals and plants would ultimately be doomed.

A simple depiction of the biosphere is presented in Figure 12.1. Biosynthetic reactions are those shown at or above the soil, and biodegradative reactions are those that occur below. Microbial activities are responsible for the recycling of carbon, nitrogen, and sulfur. The release of phosphate and other minerals present in living cells is also a result of microbial mineralization. The prominent role of microorganisms in perpetuating life can be summed up in two aphorisms termed **Van Niel's postulates.** These postulates describe the significant role played by microorganisms and are as follows:

1. A microorganism is present in the biosphere that can utilize every constituent part or product of the living cell as source of carbon and/or energy.
2. Microbes with this capacity are present in every environmental niche on Earth.

These postulates are a clear statement of the role that microbes play in the balance that exists between biosynthesis and biodegradation. It is also evident that as eukary-

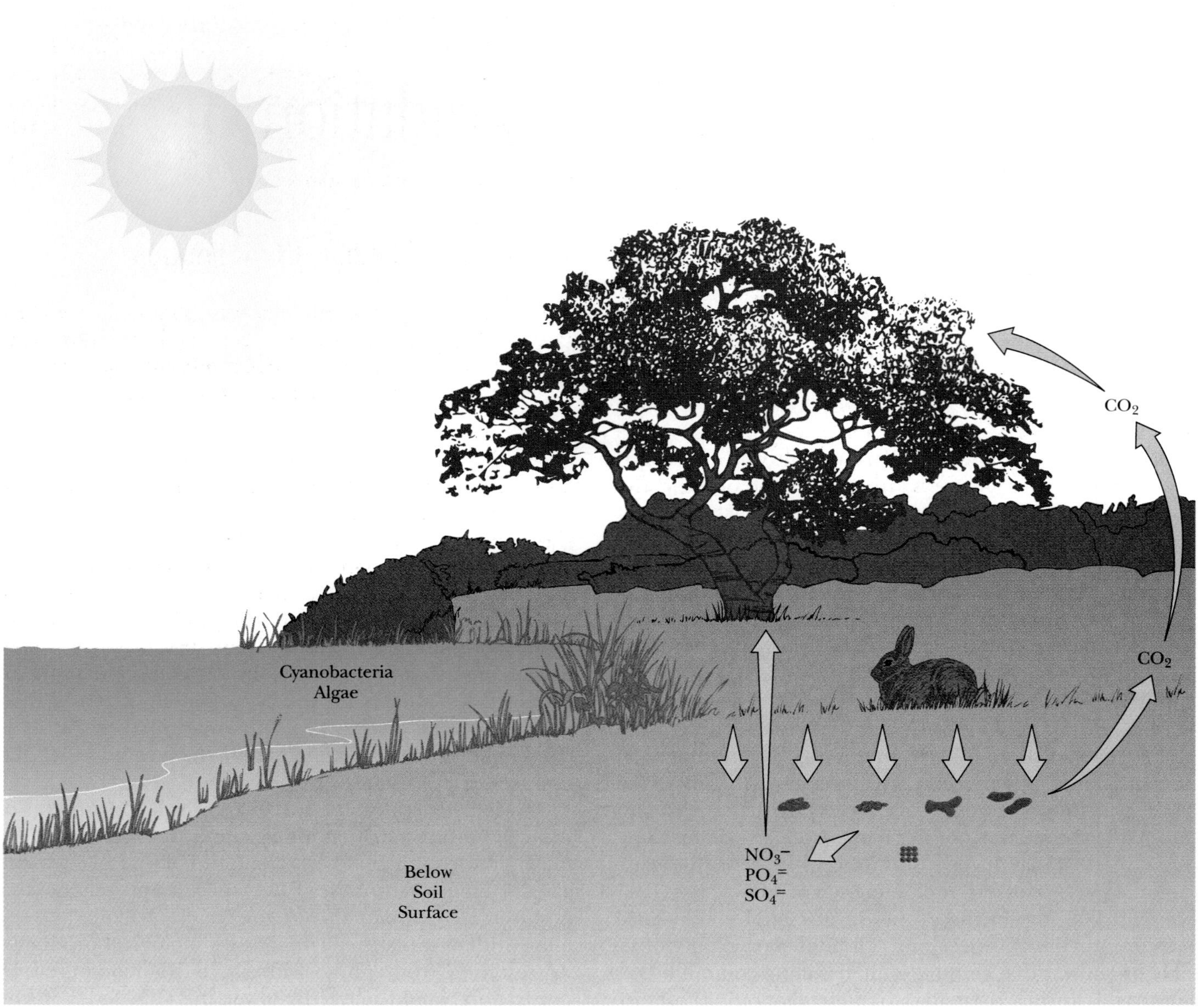

***Figure* 12.1** A generalized overview of the balance between biosynthesis and biodegradation. Biosynthesis is ultimately driven by light energy, and CO_2 is fixed into cell via photosynthesis. Biodegradation restores CO_2 to the atmosphere.

otes evolved with complex components, such as sterols or cerebrosides, a microbial population evolved in due course that mineralized these unique molecules. This chapter covers some of the mechanisms involved in the biodegradation of biological material, selected xenobiotics, and other compounds. The examples presented are but a few representatives of the vast array of biodegradative pathways that occur in nature.

Overview of the Carbon Cycle

The major pathways involved in the synthesis of cellular constituents from precursor metabolites were presented in Chapter 10. Recall that the precursor metabolites originated in glycolysis, the tricarboxylic acid cycle, and the hexose monophosphate shunt. The precursor metabolites were converted to building blocks and these (amino acids, nucleotides, etc.) were polymerized to form macromolecules. These macromolecules were then assembled to produce a cell. Eubacteria, Archaea, and eukaryotes can reverse anabolism via catabolic reactions that break down macromolecules, releasing monomers that can be utilized to generate ATP. In many instances, biodegradation retraces the pathways involved in synthesis. However, there are many notable exceptions, and an example of this is fatty acid synthesis (see Figure 10.30) and biodegradation (see later) that follow somewhat different pathways. Regardless of the route taken, macromolecules are disassembled to produce monomers quite similar to those from which they were assembled. Thus the reversal of anabolic sequences or other catabolic reactions transform the constituents of the dead cell or products of living cells to low molecular weight compounds that can be completely mineralized.

A simplistic, generalized view of aerobic biodegradation is presented in Figure 12.2. This diagram does not depict many of the intermediate reactions that may also result in the release of CO_2. The total mineralization of sugars and other metabolites is depicted in Figure 12.3. Note that each carbon in a sugar, from a three carbon sugar to one with seven carbons, can become a molecule of CO_2. Acetate, glucose, or various tricarboxylic acid cycle intermediates generated during catabolism can enter into this schema. The microorganisms that are involved in these catabolic reactions gain energy and may obtain precursor metabolites in the course of these catabolic reactions that sustain their growth. Approximately 35 percent to 45 percent of the organic carbon in a growth substrate can become a constituent part of the Eubacterium, Archaea, or fungus involved in the biodegradative process. The remaining carbons are mostly released as CO_2 or intermediates. The intermediates, in turn, are biodegraded by other microorganisms.

In some cases, biodegradative pathways are not a simple reversal of the biosynthetic route, as microorganisms can mineralize compounds that are not components of their cells. For example, plant polymers and alkaloids are not constituent parts of bacteria, but microorganisms have evolved with a potential for biodegrading them. Aromatic hydrocarbons, alkanes, ketones, and xenobiotics (products of chemical synthesis) are compounds that may be metabolized by microorganisms.

Catabolism of Selected Compounds

The number of compounds that can potentially serve as substrate for bacterial, archaeal, or fungal growth is virtually endless. The following section presents pathways involved in the catabolism of selected substrates.

Cellulose

Cellulose is considered to be the most abundant product of biosynthesis present on Earth. It is a major constituent of plants and is also synthesized by a limited number of bacterial species (*Acetobacter xylinum* and *Sarcina ventriculi*). Fungi and bacteria are the major decomposers of cellulose in nature. Fungi function mostly in acidic soils and in woody tissue where the cellulose present is protected from enzymatic attack by **lignin.** Both wood-rotting and soil

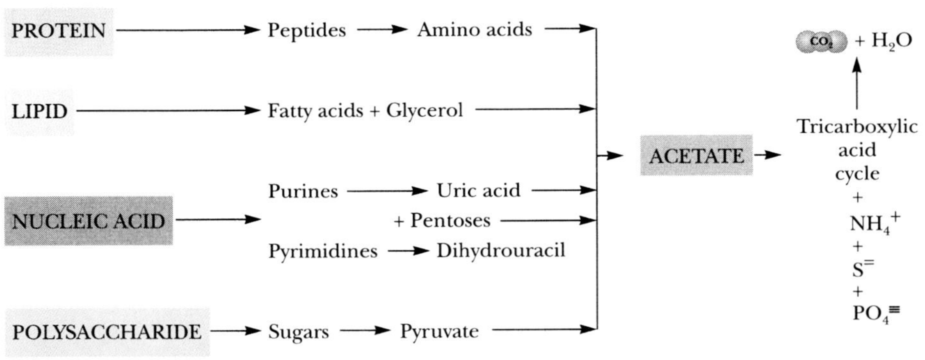

***Figure* 12.2** The fate of the major biomolecules in living cells. They are hydrolyzed by microorganisms to their constituent parts. These follow selected biodegradative routes that eventually lead to the tricarboxylic or related cycles where they are oxidized to carbon dioxide.

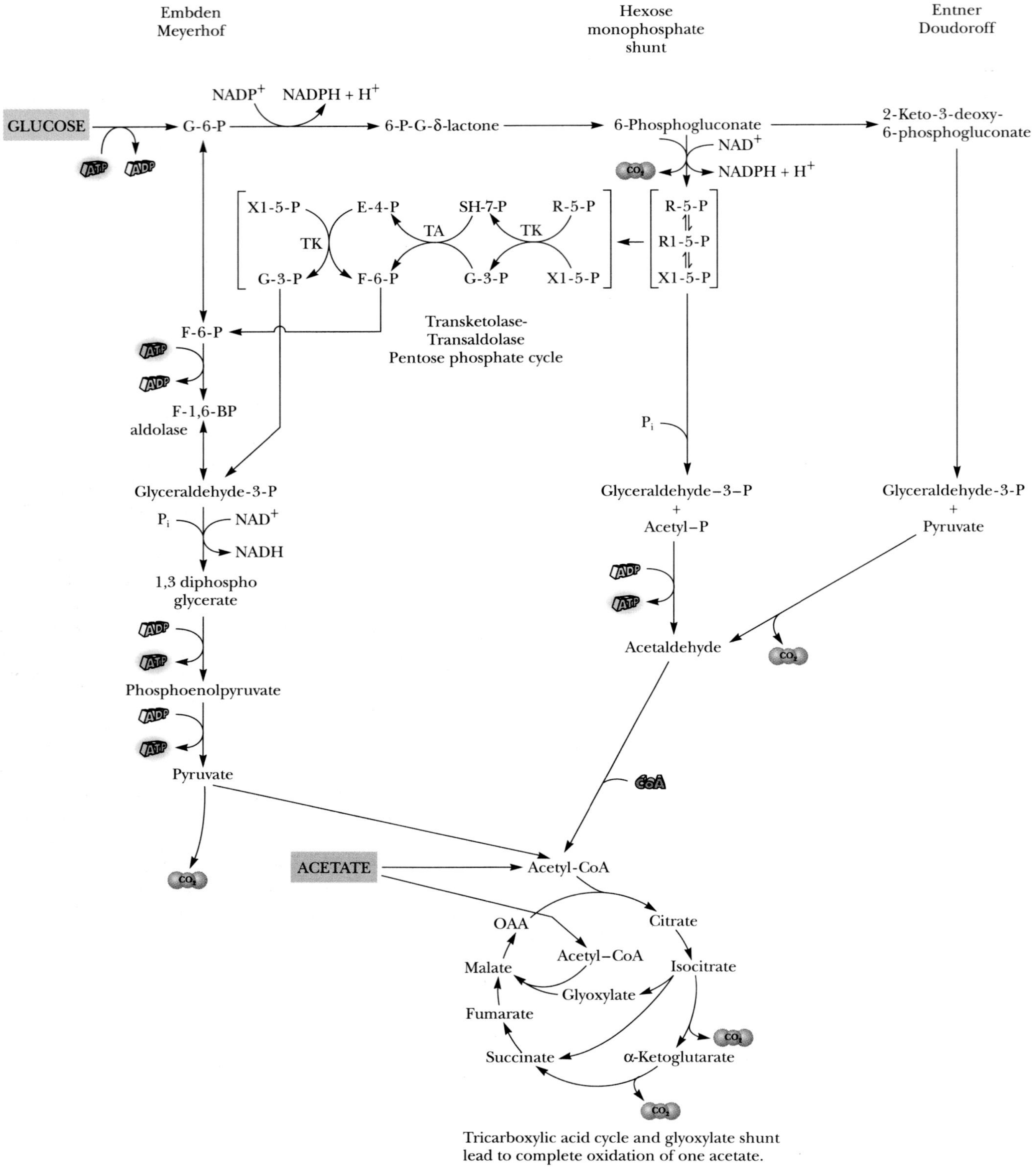

***Figure* 12.3** The reactions involved in the catabolism of various sugars and fatty acids. Transketolases and transaldolases are key enzymes in transforming the 3-, 4-, 5-, and 6-carbon sugars to intermediates that enter the tricarboxylic acid cycle. Note that ATP, NAD(P)H, and CO_2 are generated during these catabolic reactions. The abbreviations are: G-6=glucose, R=ribose, R1=ribulose, X1=xylulose, F=fructose, G-3=glyceraldehyde, E=erythrose, SH=sedoheptulose, TK=transketolase, and TA=transaldolase. Details of the Embden-Meyerhof, the hexose monophosphate shunt, and Entner-Doudoroff pathways are presented in Figures 10.3, 10.4, and 10.5.

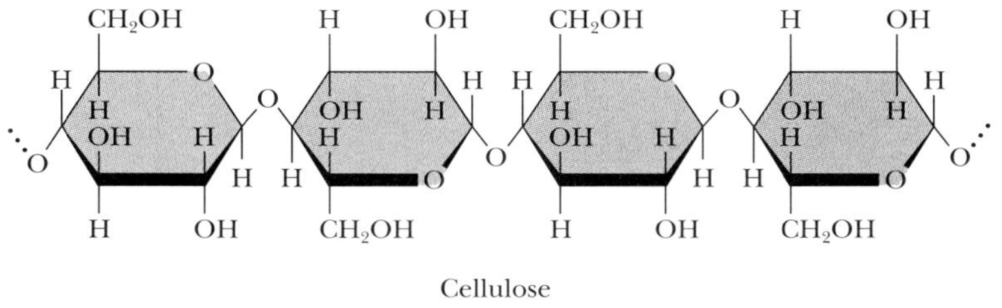

***Figure* 12.4** The structure of cellulose. The β(1-4) linkage between glucose units in this polymer is twisted 180° (note position of the —CH_2OH group). This linkage permits the polymer to grow into long fibers as are present in cotton. Starch (see Figure 12.6) has an α(1-4) linkage.

fungi are instrumental in the mineralization of the cellulose that is a constituent part of plants.

The aerobic bacteria that decompose cellulose include myxobacteria, cytophaga, sporocytophaga, and some other Eubacteria. Characteristics of these organisms will be discussed in Chapter 19. Cellulose is fermented anaerobically in soil by clostridia and in the rumen of cattle and other ruminants by a number of different anaerobic bacterial species.

Cellulose is a large polymer composed of glucose molecules that are joined by a β(1,4) linkage. An unbranched cellulose fibril can be composed of up to 14,000 glucose molecules. The structure of cellulose is presented in Figure 12.4. Decomposition of cellulose occurs by the concerted action of several enzymes:

1. Endo-β(1,4) glucanase hydrolyzes bonds in the interior of chains releasing long-chain fragments.
2. Exo-β(1,4) glucanase removes two glucose units (called *cellobiose*) from the long fragments.
3. α,β-Glucosidases hydrolyze cellobiose to glucose.

One problem in the biodegradation of cellulose is that the effectiveness of the endoglucanase is repressed (enzyme level lowered) by the presence of cellobiose or by the glucose molecules as they are released. Industrially there is an interest in large-scale conversion of cellulose to glucose, and a successful operation must overcome this repression. Production of glucose from the abundant supplies of cellulose that are available is of considerable interest as the glucose released can be fermented to ethanol and other industrial chemicals.

Starch

Starch is a major storage product in plants. There are two distinct polymers present in starch granules. One is **amylose,** which consists of unbranched *D*-glucose chains in an α(1,4)-glucosidic linkage with 200 to 500 sugar units per chain (Figure 12.5). The other polymer is **amylopectin,** composed of the same glucosidic linkages but with branching in the 1,6 position (Figure 11.17). Amylose is generally hydrolyzed by α-amylase or β-amylase. α-Amylase hydrolyzes the α(1,4) linkages to a mixture of glucose and maltose. **Maltose** is a disaccharide made up of two molecules of glucose in an α(1,4) linkage. β-Amylase hydrolyzes the α linkages of amylose at alternating positions to yield mostly maltose. These two enzymes cleave amylopectin to a mixture of glucose, maltose, and a branched core; amylopectin branching occurs every 25 to 30 sugar residues. A different enzyme designated amylo-1,6-glucosidase cleaves the 1,6 linkage responsible for branching. This enzyme in combination with α-amylase or β-amylase can lead to the complete degradation of starch to maltose or glucose. The enzyme maltase is present in starch-degrading organisms, and this enzyme can hydrolyze a molecule of maltose to two molecules of glucose. Another important industrial enzyme is glucoamylase, which splits glucose units from the nonreducing end of starch.

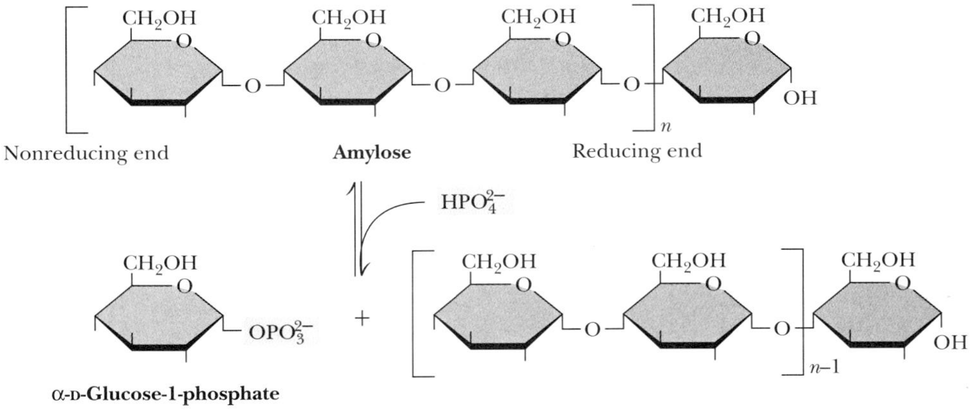

***Figure* 12.5** The structure of starch. The major difference between starch and glycogen (Figure 11.17) is in the number of branches. The α(1-4) linkage results in a bending between glucose molecules and a helical structure.

Amylases are present in many *Bacillus* species, including the obligate thermophiles *B. acidocaldarius* and *B. stearothermophilus.* The glucoamylases are produced by *Aspergillus niger, Rhizopus niveus,* and other species. Anaerobic thermophiles produce amylases that are stable at a range of temperatures and are of potential value in the production of commercial ethanol from starch. Starch digestion in the rumen of cattle and other ruminants is essential for the efficient utilization of plant material and grain. Rumen bacteria, such as *Streptococcus bovis* and *Bacteroides amylophilus,* produce α-amylase. *Entodinium caudatum,* a rumen ciliate, can hydrolyze starch to maltose and glucose.

Xylans

Xylans are heterogeneous polysaccharides, second in abundance only to cellulose among the sugar-based polymers that occur in nature. Xylans are composed xylose, a pentose sugar. A simplified structure of xylans is shown in Figure 12.6. These polymers are present in the cell walls of virtually all land plants and are also a major constituent of mature woody tissue. Xylans are degraded by bacteria and fungi by two key enzymes:

1. β(1,4) Xylanases that hydrolyze the internal bonds that link the xylose molecules yielding lower molecular weight xylooligosaccharides. An oligosaccharide contains a small number of monosaccharide units.
2. β-Xylosidases release xylosyl residues by endwise attack on the xylooligosaccharides. Pentoses such as xylose are catabolized through pathways outlined in Figure 12.3.

Glycogen

Glycogen is a carbon and energy storage product that is present in animals and bacteria. It is also called *animal starch,* and it differs from plant starch only in that its structure is more branched. Branching in glycogen (see Figure 11.17) occurs every 8 to 10 glucose residues, whereas branching in amylopectin occurs at every 25 to 30 residues. Glycogen is hydrolyzed by a glycogen phosphorylase that removes one glucose and simultaneously incorporates an inorganic phosphate to yield a molecule of glucose 1-phosphate. The enzyme glycogen phosphorylase has been isolated from the yeast *Saccharomyces cerevisiae* and the cellular slime mold *Dictyostelium discoideum.* As enteric bacteria, spore-formers, and cyanobacteria store glycogen, it is apparent that these organisms also have the glycogen phosphorylase. The glucose-1-phosphate would be metabolized by the pathways outlined in Figure 12.3.

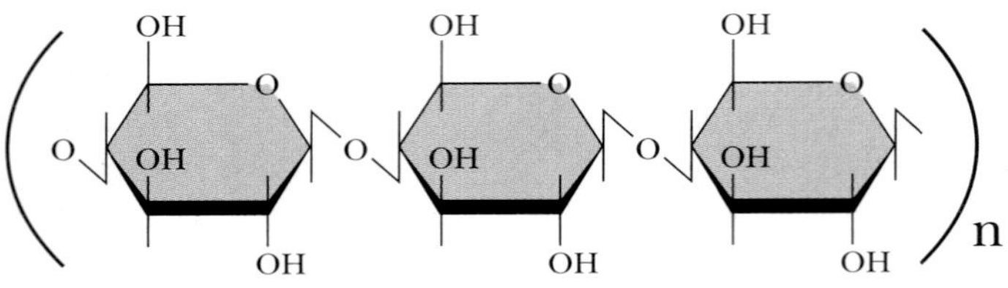

***Figure* 12.6** The structure of the common pentose polymer xylan.

Chitin

Chitin, a major constituent in the cells of plants and animals, is a polymer of *N*-acetylglucosamine units linked by β(1,4)-glucosidic bonds. Many fungi possess chitin as a cell wall component, and it is present in crustaceans (lobsters, shrimp, crabs). Chitin is the tough, outer integument of grasshoppers, beetles, cockroaches, and other insects. Stability of chitin is derived from hydrogen bonding between adjacent chains through the *N*-acetyl moiety (Figure 12.7). The widespread occurrence of chitin in soil and in aquatic environments has led to the evolution of a wide array of organisms that decompose this substance. The actinomycetes (filamentous soil bacteria; see Chapter 20) are a major utilizer of chitin. Addition of chitin as a substrate in an enrichment medium will consistently result in the isolation of diverse actinomycetes.

The enzyme chitinase cleaves chitin at numerous points in the polymer, yielding mostly the disaccharide chitobiose (*bi* = two sugars) and chitotriose (*tri* = three sugars). Another enzyme, chitobiase, then hydrolyzes these to monomers that are readily biodegraded.

Long-Chain Fatty Acid and *n*-Alkanes

Alkanes having chain lengths of 12 to 20 carbons are a minor constituent of most living cells, and longer-chain *n*-alkanes (of greater than 30 carbons) are present in many plant tissues. Paraffins in crude oil range from methane

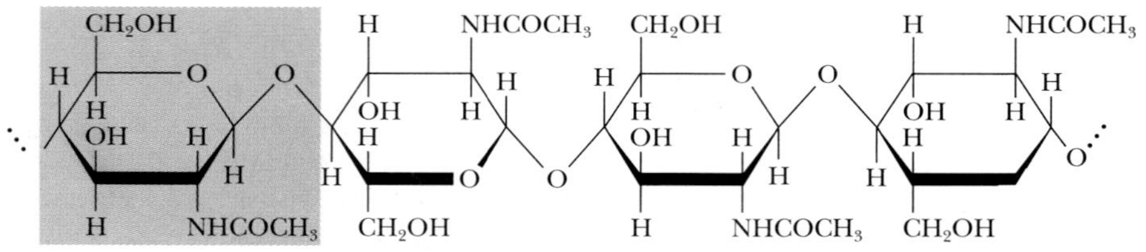

Chitin

***Figure* 12.7** The structure of chitin with β(1-4) linkages is similar to that of cellulose. It occurs as long ribbons in the cell walls of fungi and the exoskeletons of crustaceans, insects, and spiders.

to solid paraffins of considerable length. Thus alkanes are present in significant amounts in nature. Microorganisms that can utilize alkanes as growth substrate are abundant in the environment. The initial site of attack on the *n*-alkane series from *n*-butane to eicosane (C_{20}) is a terminal methyl group that ultimately yields the homologous carboxylic acid. There is substantial evidence that *n*-alkanes over 30 carbons in length may be cleaved at sites in the interior of the chain to yield more manageable, smaller molecules.

Many Eubacteria, including pseudomonads, mycobacteria, actinomycetes, and corynebacteria, can utilize *n*-alkanes as substrate. Yeasts and filamentous fungi are also capable of growth on these compounds. A general scheme for the dissimilation of a molecule of *n*-alkane to the homologous alcohol and fatty acid is presented in Figure 12.8. The acyl CoA derivative is catabolized by beta (β) oxidation as outlined in Figure 12.9.

Phospholipids

Phospholipids are present in the cytoplasmic membranes of Eubacteria and eukaryotes and account for about 10

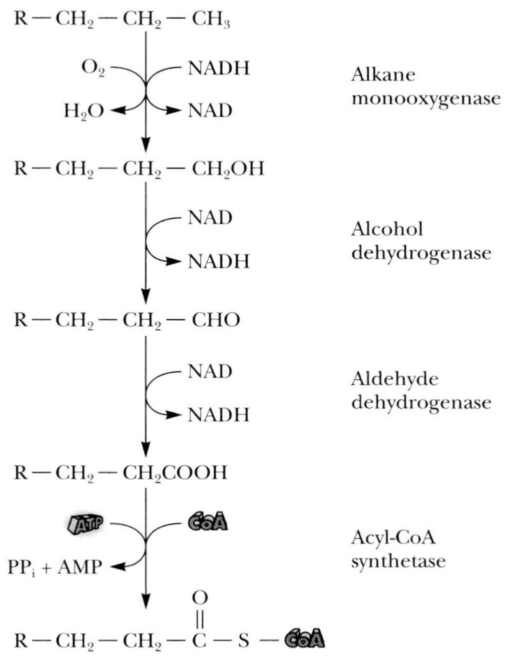

***Figure* 12.8** The stepwise oxidation of an *n*-alkane to the acyl CoA.

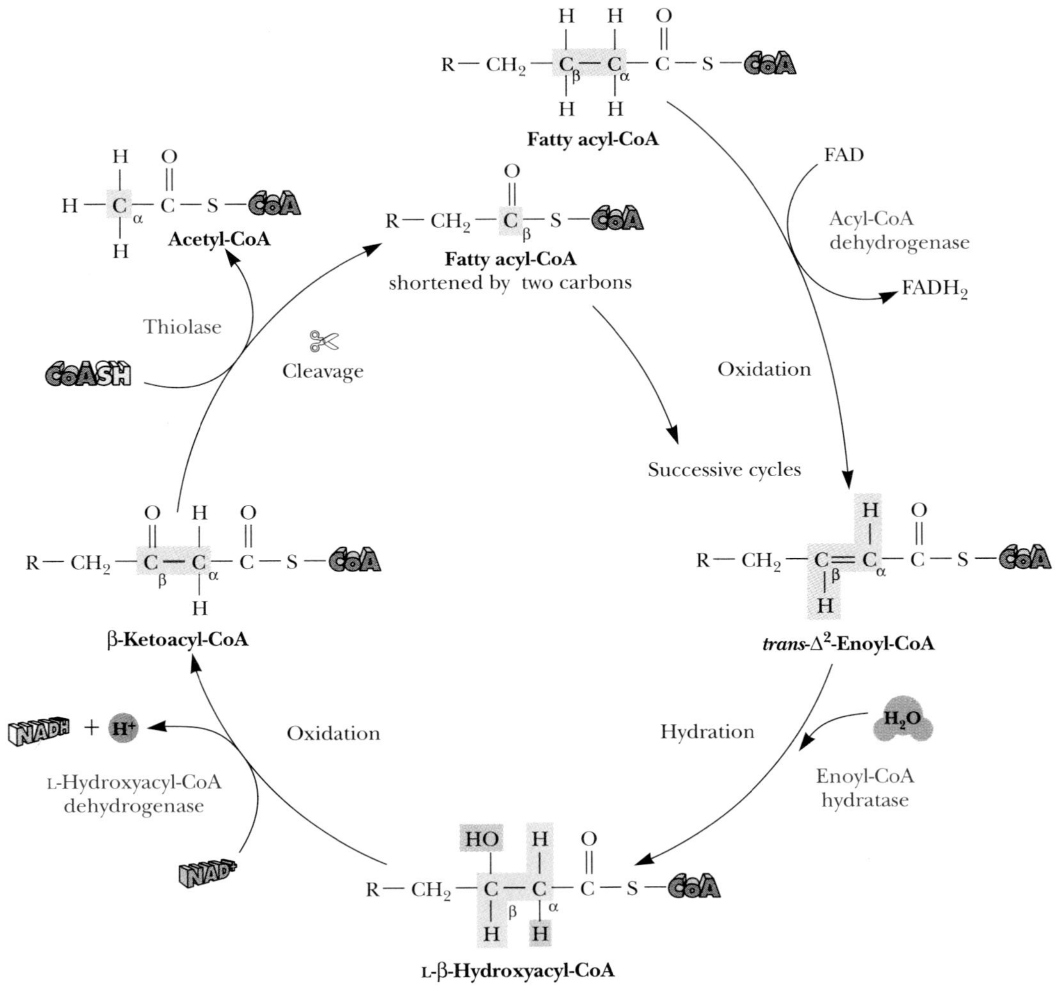

***Figure* 12.9** The pathway for β-oxidation of saturated fatty acids. Each cycle produces a single molecule of $FADH_2$, an NADH, and a molecule of acetyl-CoA. The fatty acid is shortened by two carbons. Delta (Δ) connotes a double bond, and the superscript indicates the lower-numbered carbon involved.

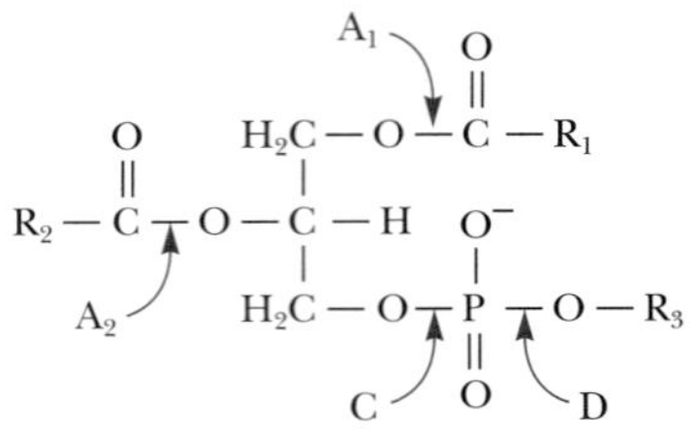

***Figure* 12.10** Phospholipases are the enzymes the cleave the polar phospholipids. There are four different hydrolytic enzymes involved termed A_1, A_2, C, and D. A complete phospholipid is illustrated in Figure 11.6.

percent of a bacterial cell dry weight. The biodegradation of a phospholipid occurs by sequential removal of components as illustrated in Figure 12.10. The fatty acid moiety is catabolized as described for *n*-alkane-derived fatty acids (Figure 12.9). Bacteria and fungi that can degrade phospholipids are commonly present in the environment.

Proteins

Proteins are a major constituent of all cells and are constantly introduced into the environment in dead cells and waste materials. Microorganisms (Eubacteria, Archaea, and fungi) that biodegrade proteins are of widespread occurrence. Microorganisms must digest these high molecular weight compounds to smaller molecules that can be transported into the cell, where they are utilized as substrate (Figure 12.11**a**). Extracellular **proteolytic enzymes (proteases)** hydrolyze proteins to polypeptides, and these in turn are cleaved by endopeptidases to oligopeptides. Endopeptidases (also called proteinases) hydrolyze peptide bonds within polypeptide chains. The endopeptidases are specific and disrupt the peptide bond only if a selected amino acid is present at the cleavage site. For example, the endopeptidase **thermolysin** from *Bacillus thermoproteolyticus* cleaves peptide linkages where leucine, phenylalanine, tryptophan, or tyrosine is one of the amino acids present.

An **exopeptidase** is an enzyme that hydrolyzes amino acids from the end of a peptide chain (Figure 12.11**b**). A **carboxypeptidase** attacks the C-terminal end of the chain, and an **aminopeptidase** removes an amino acid from the N-terminal end of a peptide chain. The enzymes involved in proteolysis act in concert and with overlapping responsibility.

The amino acids released by the hydrolysis of a protein are transported into the fungal or bacterial cells by specific transport systems. The oligopeptides may also be transported into cells, where they are hydrolyzed by peptidases. The free amino acids can be incorporated into the proteins of the cell or deaminated and catabolized to provide carbon and/or energy for growth.

Uric Acid

Birds, scaly reptiles, and carnivorous animals secrete uric acid as a product of purine catabolism. Uric acid is catabolized by selected soil microorganisms as a source of carbon and energy. Among these organisms is the unique endospore forming *Bacillus fastidiosus,* which cannot grow with any substrate except uric acid and its degradation products, allantoic acid and allantoin. The metabolic pathway for uric acid degradation is presented in Figure 12.12. Note that two molecules of CO_2 are generated via this degradative pathway with glyoxylate and urea as product.

Lignin

Biodegradation of lignin is an essential part of the carbon cycle. **Lignin** is the most abundant renewable aromatic molecule on Earth. Its natural abundance may be exceeded only by cellulose. Lignin surrounds and protects

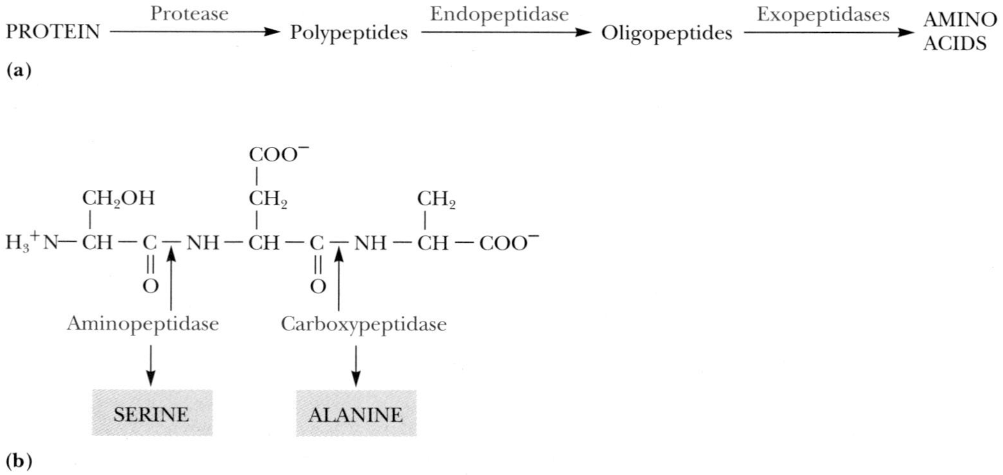

***Figure* 12.11** Proteins are biodegraded by a series of enzymes. The extracellular proteases cleave large protein molecules to polypeptides and oligopeptides (12 to 20 amino acid residues). These are cleaved by endopeptidases and exopeptidases to release amino acids. Aminopeptidases **(a)** attack the amine end of a peptide chain, and carboxypeptidases **(b)** attack the carboxyl end of a peptide. Amino acids may be deaminated, and the resultant fatty acid can be utilized as a source of carbon and energy.

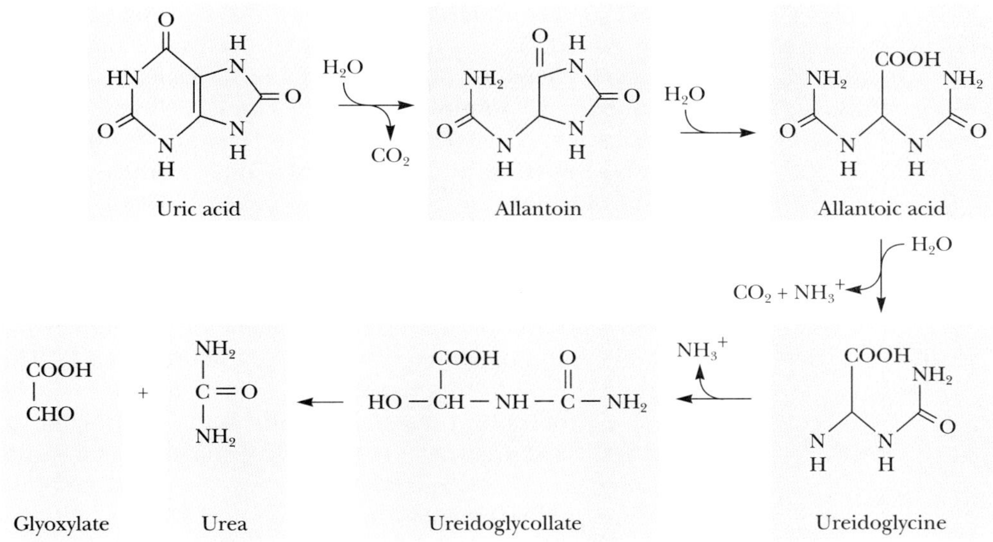

***Figure* 12.12** Steps involved in the metabolism of uric acid by *Bacillus fastidiosus*. This organism is restricted to growth on uric acid, allantoin, or allantoic acid.

the xylans and cellulose that are components of wood from destruction by microorganisms. Between 20 percent and 30 percent of wood and plant vascular tissue is composed of lignin. All higher plants, including ferns, contain lignin, but lower plants, such as mosses and liverworts, do not. Lignin is relatively indestructible, and it is probably not used by any single microbial species as sole source of carbon or energy.

Clues to the factors that render lignin so resistant to microbial destruction can be found in the reactions whereby it is synthesized. Lignin is synthesized from three precursor alcohols by free radical copolymerization (Figure 12.13**a**). These alcohols are coumaryl (*p*-hydroxycinnamyl alcohol), coniferyl (4-hydroxy-3-methoxycinnamyl alcohol), and sinapyl (3,5-dimethoxy alcohol). It is apparent from the complex structure shown in Figure 12.13(**b**) that a large polymer of lignin would present a single organism with an insurmountable task in any attempt to break it down. The molecule is a heterogeneous polymer assembled in a haphazard way that would be attacked only by enzymes with very high levels of nonspecificity. Such enzyme specificity has not evolved in nature.

Lignin is biodegraded in nature by white-rot basidiomycetes (fungi) via a mechanism befitting its synthesis—that is, a random oxidative attack. Lignin is not known to serve as growth substrate for this or any other fungus, so an alternate carbon or energy source (cosubstrate) must be available. Xylans, cellulose, and carbohydrates can serve as this cosubstrate.

Experiments with a white-rot fungus, *Phanerochaete chrysosporium*, have affirmed that lignin is degraded by enzymatic "combustion." During growth on a cosubstrate, *P. chrysosporium* synthesizes peroxidase that is indirectly involved in lignin degradation. Hydrogen peroxide is necessary for ligninase activity and is probably supplied by oxidases in the fungus that reduce O_2 to H_2O_2. Lignin is not degraded in the absence of O_2. The aromatic nuclei of the lignin are oxidized by the action of single electrons, and this produces unstable cation radicals. This leads to nonenzymatic radical-promoted reactions that disrupt the lignin molecule. This process is a nonspecific oxidation of lignin to random, low molecular weight products that are then available for mineralization by fungi or bacteria.

Aromatic Hydrocarbons

Aromatic and polyaromatic hydrocarbons are industrial chemicals that occur both as products of or intermediates in chemical syntheses. They are also a source of considerable environmental pollution. Polyaromatic hydrocarbons are formed during combustion of fossil fuels and are natural constituents of unaltered fossil fuels. Benzene, toluene, xylenes, and other low molecular weight aromatics serve as commercial solvents and are constituents of paint. Various aromatic compounds are intermediates in the synthesis of dyes, medicines, plastics, and other chemicals that have become a part of modern life. Aromatic compounds such as phenylalanine and tyrosine are constituents of living cells.

The biodegradation of aromatic hydrocarbons has been studied extensively. Two major pathways for the biodegradation of benzene are presented in Figure 12.14. The pathway used depends on the bacterial strain involved. Both result in products (pyruvate, succinate, and acetate) that can be mineralized readily via reactions outlined in Figure 12.3.

Derivatives of aromatic hydrocarbons, particularly chlorinated compounds, are widely used as pesticides, insulators, and preservatives. The accumulation of these chloroaromatics has become a serious environmental problem. For example, the cumulative annual worldwide production of **polychlorinated biphenyls (PCBs)** (Figure 12.15**a**) is presently estimated to exceed 750,000 tons. Ultimately a considerable portion of this ends up in the environment. There are 210 possible PCBs, all having the same biphenyl carbon skeleton and differing only in the number and position of chlorine atoms. PCBs are used in capacitors, transformers, plasticizers, printing ink, paint, and a host of other materials and products. **Pentachlorophenol** (Figure 12.15**a**) is another broadly applied compound that has become an environmental hazard. It is used in termite control, as a preharvest defoliant, and in wood preservation. The biodegradation of chlorinated hydrocarbons occurs via dehalogenation reactions that yield compounds that are biodegradable **(Box 12.1)**. The removal of chlorines from an organic compound may be a slow process, and cosubstrates often must be available to provide a source of energy and carbon for the dehalogenating microorganism. A pathway for mineralization of trichloroethylene is outlined in Figure 12.15(**b**).

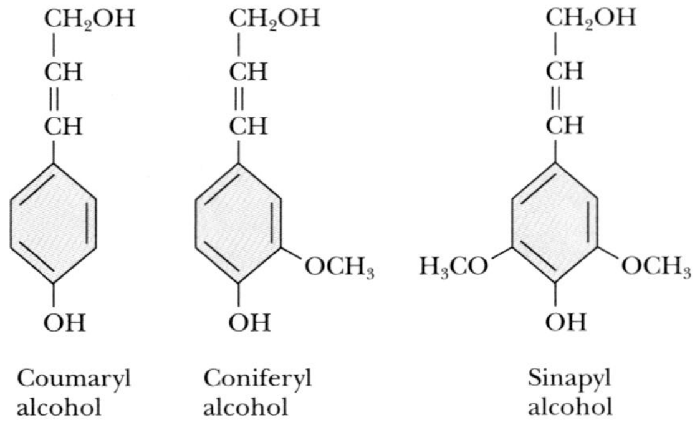

(a)

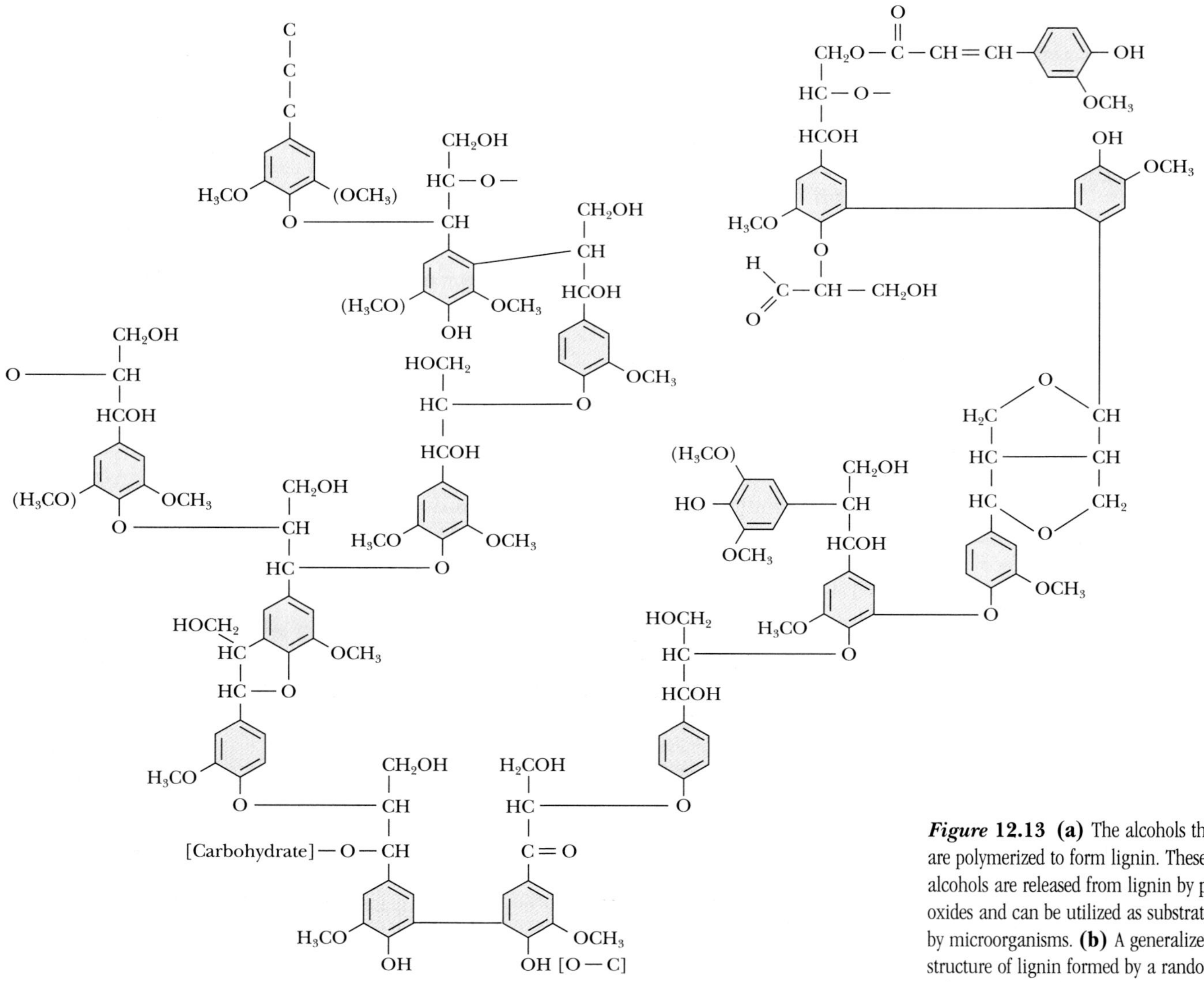

(b)

***Figure* 12.13** **(a)** The alcohols that are polymerized to form lignin. These alcohols are released from lignin by peroxides and can be utilized as substrate by microorganisms. **(b)** A generalized structure of lignin formed by a random polymerization of **(a).**

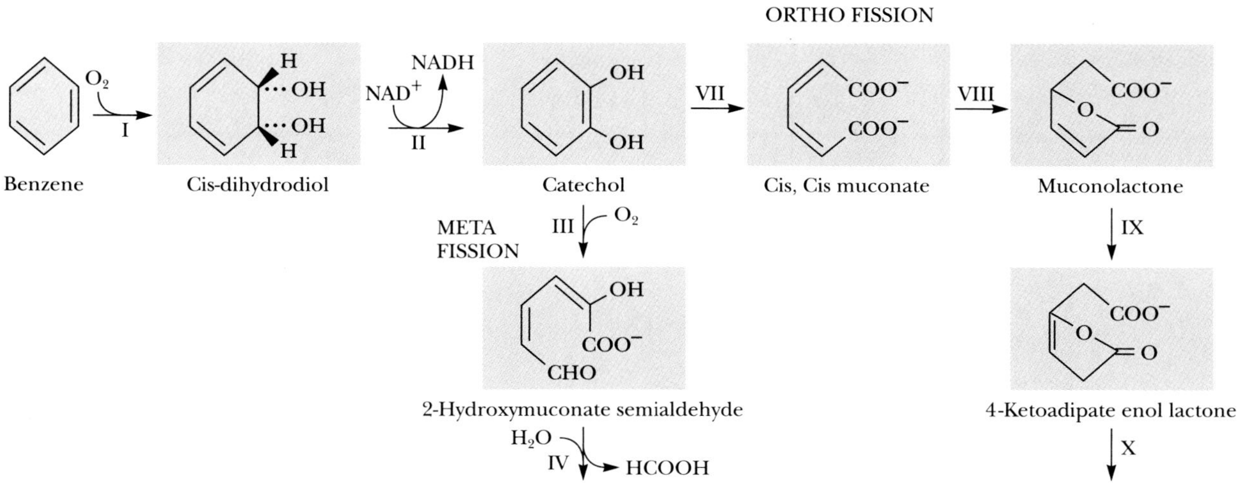

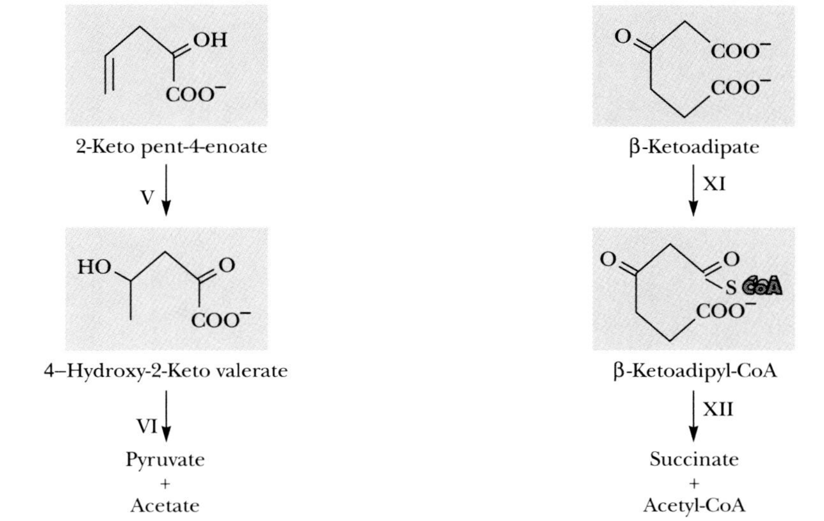

***Figure* 12.14** The two pathways for benzene degradation in bacteria. The ortho fission pathway results in a cleavage between the hydroxyl groups of catechol, and the meta pathway results in the cleavage between a hydroxyl and an adjacent carbon. The enzymes involved are: I Benzene dioxygenase; II Diol dehydrogenase; III Catechol-2,3-dioxygenase; IV 2-Hydroxymuconate semialdehyde hydrolase; V 2-Ketopentan-4-enoate hydrolase; VI 4-Hydroxy-2-ketovalerate aldolase; VII Catechol-1,2-dioxygenase; VIII Muconate lactonizing enzyme; IX Muconolacetone isomerase; X 4-Ketoadipate enolacetone hydrolase; XI 3-Ketoadipate-succinyl CoA transferase; and XII 3-Ketoadipyl-CoA thiolase.

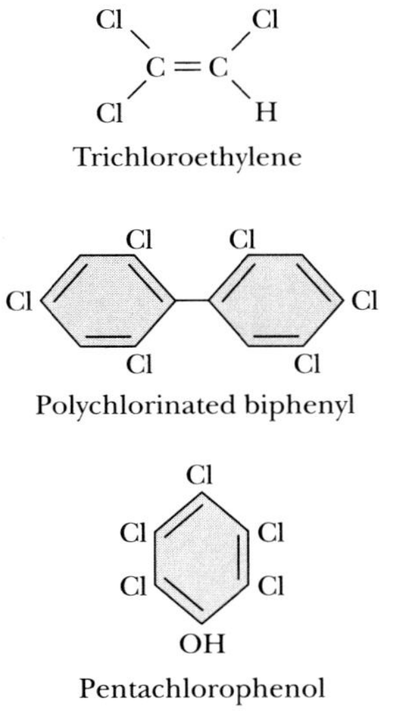

(a)

***Figure* 12.15** Typical chlorinated hydrocarbons **(a)** that are significant environmental pollutants. Trichloroethylene (TCE) is a degreasing agent, polychlorinated biphenyls (PCBs) have been employed as insulators, and pentachlorophenol (PCP) is a wood preservative. The number of chlorine molecules on a PCB varies; there are 210 possible combinations.
(b) The biodegradation of trichloroethylene by *Mycobacterium vaccae* following growth on propane. The single chlorine migrates to generate trichloroethanol followed by the sequential removal of the Cl^- ions.

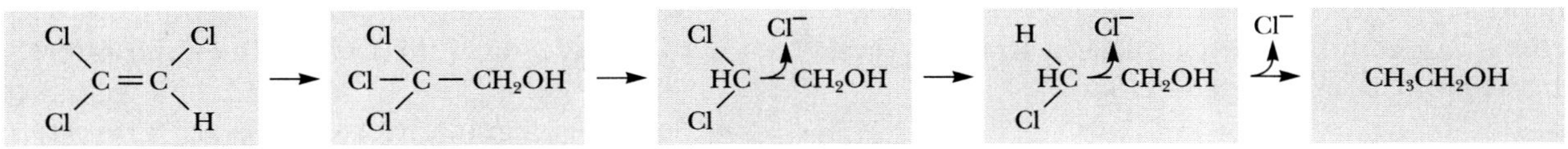

(b)

BOX 12.1 MILESTONES

Recalcitrant Molecules

***Recalcitrant**–obstinately defiant of authority or restraint; difficult to manage or operate; not responsive to treatment.*

–*Webster's New Collegiate Dictionary*

Molecular recalcitrance is a relative term that measures the inherent resistance a molecule has to microbial attack. Recalcitrance might then be an indicator of the evolution of enzyme systems capable of converting a given compound to intermediates that could flow into the mainstream of bacterial metabolism. Many of the compounds that have proven to be recalcitrant are halogenated aromatic or aliphatic compounds and are of considerable concern because they are a principal environmental pollutant. The persistence of many halogenated chemicals and their toxicity render them hazardous to environmental health. Chlorinated compounds are widely applied as pesticides, solvents, hydraulic fluids, and the like and thus eventually enter the environment. For example, pentachlorophenol (Figure 12.15) is used as a wood preservative (telephone poles, fence posts) and is an effective long-term protector against rot.

Halogenated compounds have been considered impervious to microbial assault because they are "unnatural," and microorganisms may not have evolved with enzyme systems capable of their mineralization. One may question whether the inability of microbes to dehalogenate is due to the presence of halogenated compounds in nature solely as products of chemical synthesis. Clearly this is not the case. There are over 700 naturally occurring halogenated compounds that have been identified thus far and at least seven distinct enzyme systems that can cleave the carbon-halogen bond.

Recalcitrance is not solely related to the presence of a -Cl, -Br, -F, or -I atom. The number, position, and electronegativity of the halogen-carbon(s) bonds is of fundamental importance in determining whether the dehalogenating enzymes can function. It is apparent that some halogenated compounds are not recalcitrant, while others are impervious to microbial attack. The structures and half-life of some pesticides are presented as follows:

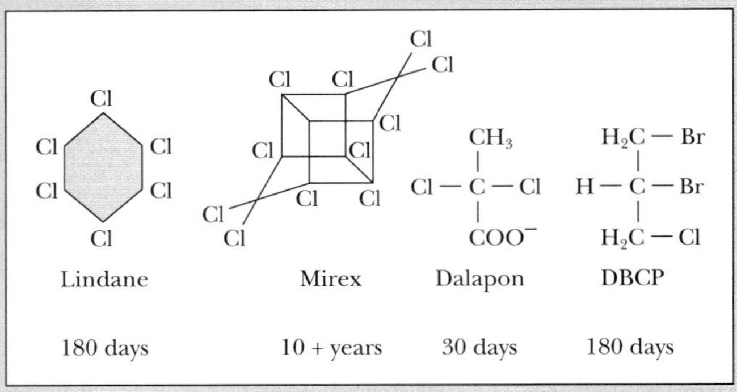

The half-life of some compounds that have been employed as pesticides. DBCP is dibromo-chloro-propane.

The half-life in soil would be for temperate regions such as the United States and would be somewhat shorter in tropical regions. Lindane and mirex are insecticides, dalapon is a herbicide, and DBCP is a nematocide. Dalapon would not accumulate in nature regardless of usage, while lindane or DBCP would gradually build up in soil with continued use. Mirex, which has been used to eradicate fire ants, is no longer applied in the United States.

Summary

- Sunlight supports biosynthetic reactions, and these reactions can continue only if photosynthate is **recycled** constantly by **biodegradation.**
- Without replenishment, the level of CO_2 in the atmosphere would support the present rate of photosynthesis on Earth for only **30 years.**
- **Microbial decomposition** is the source of 85 percent to 90 percent of the CO_2 in the atmosphere.
- Some **biodegradative pathways** are a **reverse** of **synthetic** reactions, while others follow distinctly **different** pathways.

- All products of living cells are biodegraded by some microorganism as are many products of chemical synthesis.

- **Cellulose** is the most abundant product of **biosynthesis** present on Earth. It is biodegraded by both bacteria and fungi.

- Starch is a major storage product in plants. Enzymes termed **amylases** can biodegrade starch to glucose.

- **Chitin,** the tough integument of insects and crustaceans, is mineralized by the **actinomycetes.**

- The **alkanes** are of widespread occurrence in nature and are oxidized by a **monooxygenase** that ultimately produces the homologous carboxylic acid.

- **Phospholipids** are biodegraded by the sequential removal of the fatty acids and polar group by a series of distinctly different phospholipases.

- Initial attack on large molecules such as a protein occurs by the action of **extracellular** enzymes termed **proteases.** These enzymes release lower molecular weight polypeptides that are cleaved to **oligopeptides.** There are specific enzymes **(exopeptidases)** that hydrolyze amino acids from the oligopeptides.

- **Uric acid,** a product of pyrimidine catabolism, is utilized by *Bacillus fastidiosus* as sole source of carbon and energy.

- **Lignin** is an aromatic complex that makes up 20 percent to 30 percent of woody vascular tissue. It is resistant to enzymatic attack and is depolymerized by peroxides generated by wood-rotting fungi. The products of this oxidative attack are utilized by both bacteria and fungi as a carbon source.

- **Aromatic hydrocarbons** occur widely as **environmental pollutants.** Many are degraded by bacteria and fungi.

- The chlorinated hydrocarbons are generally quite resistant to enzymatic attack. Removal of the chlorine molecules through incidental attack while metabolizing cosubstrates is a key element in ridding the environment of these compounds.

Questions for Thought and Review

1. Why is recycling important in the perpetuation of life on Earth? What event was of signal importance in the evolution of heterotrophic prokaryotes and eukaryotes?

2. What percent of atmospheric CO_2 comes from microbial degradation? How long would the present supply last without replenishment? What would be the consequence of lowered levels of CO_2 in the atmosphere?

3. What are Van Niel's postulates? How do they relate to the role of microbes on Earth?

4. Biodegradation is sometimes a reversal of biosynthesis. In some cases biodegradation follows a different pathway from that involved in biosynthesis. Explain, and consider examples of both.

5. How would a microbe attack a large molecule like cellulose? Protein? What are the similarities and differences between cellulose and protein degradation?

6. Why are amylases important to the well-being of some animals? What types of organisms produce these enzymes?

7. Where does commonality occur in the biodegradation of long-chain *n*-alkanes and long-chain fatty acids?

8. What is remarkable about lignin biosynthesis? Biodegradation?

Suggested Readings

Bitton, G., and C. P. Gerba. 1984. *Groundwater Pollution Microbiology.* New York: John Wiley & Sons.
Gibson, D. T., ed. 1984. *Microbial Degradation of Organic Compounds.* New York: Marcel Dekker, Inc.
Gottschalk, G. 1986. *Bacterial Metabolism.* 2nd ed. New York: Springer-Verlag.
Kirk, T. K., and R. L. Farrell. 1987. Enzymatic "combustion": The microbial degradation of lignin. *Annual Review of Microbiology* 41(465–505).
Young, L., and C. Cerniglia. 1995. *Microbial Transformation and Degradation of Toxic Organic Chemicals.* New York: John Wiley & Sons.
Zehnder, A. J. B., ed. 1988. *Biology of Anaerobic Bacteria.* New York: John Wiley & Sons.

Part 4

Genetics and Basic Virology A Conversation with Martha Howe

Martha Howe is a virologist and molecular geneticist whose research interests have been concerned with bacteriophage Mu. She received her AB degree from Bryn Mawr College and her Ph.D. from the Massachusetts Institute of Technology. She progressed through the professional ranks at the University of Wisconsin, where she had her first academic position. She is now Professor and Chair of the Department of Microbiology and Immunology at the University of Tennessee. Dr. Howe has received a number of honors for her research work, including the Eli Lilly Award.

JS: What led to your decision to study microbiology?

MH: While I was an undergraduate biology major at Bryn Mawr in the 1960s, I went to two seminars that I remember being very exciting—by Jim Watson and Matt Meselson. They were talking about DNA structure, making mutants, and studying regulation of genes in bacteria and bacterial viruses. I found it absolutely fascinating—to think that we could get down to the molecular level and understand how genes are controlled and how they can be reassorted by recombination. Bacteria and their viruses were the model systems of that time. They had the advantages that they were easy to grow and the generation time was very quick. You could do an experiment one day and know the result the next day. So you could design an experiment to test your ideas and very quickly learn whether they were right.

I also enjoyed the visual nature of microbiology—red and white bacterial colonies on color indicator plates, rough colonies and smooth shiny colonies, big plaques or small plaques, clear plaques or turbid plaques—each told you something about the genetic makeup of the bacterium or virus.

JS: What advice would you give to undergraduate students who are interested in pursuing a career in microbiology?

MH: The first advice I'd give is probably advice that they get from everyone, and that is to get as broad a scientific background as possible—organic chemistry, inorganic chemistry, physical chemistry, physics, math computers—so that as they continue in the biological sciences and into genetics, biochemistry, and microbiology, they really have a firm and broad foundation. Then they can take advantage of a variety of approaches to understanding and learning about microorganisms. A broad foundation is very important. It keeps doors open and lets one use a wide spectrum of approaches.

The second thing is not to be too narrow in thinking about what careers they might enjoy. In academia we tend to think of our careers as faculty members, but really the career opportunities in microbiology are much broader than that. One can go into the medical arena, with research, or into industry, into the academic arena, the environmental arena—the opportunities are very broad.

JS: What about careers for women in microbiology? Are there opportunities or hurdles? How would you characterize it?

MH: I would say there are lots of opportunities. When I got into biology at the graduate level in the late '60s the doors had just opened for women; there were a lot of women in my graduate student class. I think biology in general and microbiology in particular has been much more open to women than some of the other sciences. I'm not quite sure why, but it's been wonderful. I personally have had many opportunities, especially as a young faculty member at the University of Wisconsin, that I might not have had if I weren't a woman. There are still some pockets of prejudice and much depends on individuals and their own experiences. But I think, in general, fields in biology are open to women.

Courtesy of Martha Howe

There are now a number of women faculty climbing the academic ladder. It's quite common to expect a woman to do the same thing a man would do in terms of a career. So, I don't see any major hurdles that women have to jump that aren't the same as for men. I even think there are opportunities that women get simply because they are women, but they do have to perform; they won't get the second opportunity if they don't do well on the first.

I think the biggest hurdle is the same hurdle women face in any career, juggling career and family. They have to make decisions about their priorities in life, and it's still true that there are responsibilities at home with the family and work that has to get done. It is still taken for granted that a woman will shoulder more of the responsibility for the home. Therefore she does a day's work at work and then she does a day's work at home afterwards. I think what you'll find is that there will be some women who go the full career route, who will do just as much as any man would do. There are other women who won't want to work 24 hours a day. They want to have more time to spend with the family and therefore they will choose a slightly different career path. Both paths are valuable to society.

JS: You've told us about how you became interested in genes and genetics, but you've actually studied bacterial viruses. What led to that transition to study phages?

MH: It was actually an outgrowth of my interest in genes and genetics because phages were some of the very early model organisms that were studied, right along with bacteria. When I was a graduate student at MIT, I found that microbiology was the area that I was most interested in, and Ethan Signer, who was doing lambda genetics, had a position available in his lab. It was absolutely fascinating to me to work on lambda. The field was just learning about prophage integration and excision, the mechanisms of integration and recombination and what the different genes do.

For better or worse I worked on lambda's DNA binding protein, but at the end of four years we took however many notebooks there were, maybe six, put them on the shelf and said this just isn't going anywhere. The predominant proteins could be lost by mutation in either of two nonessential regions, both of which had a very similar effect. There was no understanding in lambda at that time how that was possible, and the two regions themselves were not at all understood. It was not at all clear that with another couple of years we would be any farther along in our understanding. My committee and I decided that we had spent enough time on it, it was not guaranteed to work and, therefore, it was not good for me to continue it.

> **I found it absolutely fascinating—to think that we could get down to the molecular level and understand how genes are controlled and how they can be reassorted by recombination. Bacteria and their viruses were the model systems of that time.**

At that point I picked up a new phage, called Mu. It was a wonderful opportunity, because there were only four people in the entire world studying Mu at that time. There had been two papers published: one paper reported Mu's discovery and demonstrated that it integrated into a number of different genes in its host, *E. coli.* The other paper described the shape and form of virus particles seen with the electron microscope. The study of Mu was wide open!

I began by isolating mutants and then characterizing what the effects were of those mutations. I actually did my Ph.D. thesis research in a year and a half. It went very quickly because of the experience I had gained in working with lambda. I quickly generated a collection of phage mutants, mapped them, and demonstrated that, although Mu could integrate into many different sites on the host chromosome, it always used unique sites in the phage chromosome. That was an important finding and it was a very exciting time.

I then did a rather unusual thing. Generally you change fields when you go from being a graduate student to being a post-doc. You want to learn something new and different to broaden your scientific expertise. Well, I had already done that as a graduate student—I had moved from one phage to another, so I decided to keep on with Mu. Along with three or four students from other labs, I had set the foundation for studying Mu. When you get onto a good thing, there is no point in giving it up. So, I have continued working on Mu ever since.

JS: What would you regard as your most important contribution to the study of Mu?

MH: I was involved in the beginning of the field—generating a genetic map for the virus, isolating mutations, assigning them to different genes, and determining the functions of those genes. Perhaps one of the most lasting contributions was that I isolated temperature-sensitive mutations in the Mu repressor. They allowed us to start with a Mu lysogen, with Mu in a dormant prophage state; then we could simply shift the temperature to denature the repressor and phage development would begin. The reason that was important is that Mu phage lysates were not terribly stable. The particles tend to adsorb to bits of cell debris that are left in the lysate. It was hard to grow and hard to keep a high titer. By using this temperature-sensitive prophage and making the host defective in the Mu receptor, we were able to get all the cells to produce Mu in a synchronous manner and reduce the binding to cell debris. These lysates allowed us to do experiments much more easily. Everyone in the field still uses those mutants.

Another thing we've done is to determine the functions of the genes: defining which ones are involved in replicating the DNA, which ones are involved in making the structural proteins for the phage particle. Mu turns out to be particularly interesting in its regulatory mechanisms. It has a DNA segment that inverts, and our lab showed that the function of this DNA segment was in fact to encode the tail fibers of Mu. Along with other labs, we demonstrated that the inversion of this DNA segment actually led to the generation of a different host range. Now we know that Mu carries genes for two sets of tail fibers within this invertable segment.

The orientation of this segment determines which set is being expressed. One set will let Mu attach to certain bacteria and the other set will let Mu attach to different bacteria. So it has a way to alternate its host range.

Over the last ten years we've focused on Mu gene regulation. We've discovered the promoters where the transcripts start. We've identified the regulatory proteins that allow the different promoters to be turned on, and now we're getting to understanding at a molecular level how this regulation works. I think we've done a number of things that have contributed to the field.

JS: How has the study of Mu been important to the field of bacterial genetics?

MH: The first impact was from Larry Taylor's original paper, where he described that Mu could cause mutations in bacteria and those mutations were associated with the prophage. That's how we came to realize that Mu could insert at many different sites. At that time it was the only phage known to do that. It causes mutations by inserting into the middle of the gene and therefore knocking out the function of that gene. So very early on it became a wonderful tool for making mutations. You simply infected a population of cells with Mu, and took the surviving cells, which were primarily lysogens, and hunted in that population for the mutant you wanted. It gave us a way to knock out genes in a specific way—it didn't alter any of the other genes of the host and the mutant gene had a tag because the prophage was there. It was a very useful alternative to chemical mutagenesis.

Secondly, Mu was one of the first transposable elements to be discovered in bacteria, along with insertion sequences, and it was one of the earliest to be studied in terms of its transposition mechanism. It transposes as it replicates, which happens a hundred times or so in a single bacterium in an hour; thus, it's a frequent process that can be studied biochemically as well as genetically. Much of the understanding of how transposition occurs has come from studies of Mu.

The third impact is the use of Mu for gene fusion technology. Malcolm

Courtesy of Martha Howe

Casadaban was making genetic constructions with Mu and discovered an unexpected product. The *lac Z* gene encoding β-galactosidase had been transferred onto Mu in such a way that when Mu inserted into a gene, β-galactosidase was expressed from the regulatory elements of that gene rather than being expressed from Mu or from the *lac* promoter. He realized that this phage would let one take a very simple enzymatic assay for β-galactosidase and use it to study regulation of a gene even though the gene product itself was not known. That became a major step forward in bacterial genetics. It is one that is used extensively now, namely, studying the elements that regulate a gene by fusing it to a good reporter gene. His Mu-*lac* phage was the beginning of that technology. The technology is also used to isolate genes whose functions are unknown but which respond to a particular environmental signal by turning on or turning off the gene. The genes are defined by where the Mu is inserted. It lets us get a handle on genes that are regulated in a common way; it's a wonderful tool.

Mu also has a number of novel features. The first one, already mentioned, is inversion of the invertable DNA segment to alternate between two host ranges. There are some other examples of DNA insertion that are now known, but it certainly is not widespread. Because Mu changes its host range, it had to evolve a way to protect its DNA from restriction as it went from one host to another. It did that by evolving a new and unusual DNA modification that protects it from cleavage by many restriction enzymes. I don't know of any other phage that has this type of modification.

Mu, like other phages, is also an excellent model system for studying gene regulation. It has revealed novel features in the mechanism of promoter activation, assembly of macromolecular complexes, and gene regulation by proteolysis of specific gene products and by processing of the primary RNA transcripts.

JS: What do you regard as the most important contribution that studies of bacterial genetics have made to microbiology in the past half century?

MH: At the beginning would be the understanding of the structure of DNA and the flow of information from genes encoded in the DNA to the messenger RNA and then to the translated protein product. This revolutionized how we think about biological systems. We now understand that DNA is the primary genetic repository and the proteins are the workers that carry out the various processes that go on in cells. Those key findings provide a foundation for all of biology.

Another was the discovery of restriction enzymes which have totally changed our approach to doing biomedical research. The discovery came from very basic research, answering an interesting and off-beat question: How do bacteria protect themselves from bacteriophages? The answer: bacteria evolved mechanisms to mark their own DNA and then use restriction enzymes to cleave any entering foreign DNA that wasn't appropriately marked. The mark turns out to be a specific modification in the recognition sequence for the restriction enzyme that prevents it from cutting the DNA.

We now use those restriction enzymes—and there are different ones from different hosts—to cut DNA at particular places so that we can manipulate genes. That has resulted in the development of recombinant DNA technology. We see the product of that technology in the very sophisticated research that we can now do, and in the production of specific protein or enzymatic products in the pharmaceutical industry. We can modify and change genes by the use of these enzymes.

A third contribution would be the discovery of plasmids. For a long time we thought of bacterial chromosomes as a single very large circle of DNA. What has become clear is that there are, in addition to that primary chromosome, smaller DNA circles which we call plasmids. They are usually extra DNA, that is, the cell can survive without them, but sometimes they give the cell a selective advantage. In bacteria that cause disease, plasmids often allow them to degrade or prevent the functioning of specific antibiotics. Often they are easily transferred from one type of bacteria to another. They are becoming an ever-increasing problem in our medical treatment of infectious diseases.

The last one is just beginning but may have equally high impact as the others I've mentioned. It is Larry Gold's work on using a genetic approach to find very small pieces of RNA or DNA that will interact with a protein and block normal activity. For example, blocking the activity of a particularly important viral protein might prevent the disease caused by that virus. He has used a novel and creative approach to taking randomly synthesized pieces of DNA, bind them to the target proteins of interest, recover the rare pieces that bind, and amplify them to generate a population that is enriched in the ability to bind to that particular protein. After multiple cycles of binding and amplification, he ends up with an oligonucleotide that reacts specifically with that particular protein. If it happens to bind in the active site or some other critical part of the protein, it may prevent that protein from functioning.

We now understand that DNA is the primary genetic repository and the proteins are the workers that carry out the various processes that go on in cells. Those key findings provide a foundation for all of biology.

So what he's doing is applying genetics, looking for something that is occurring very rarely, and asking if a therapeutic agent can be generated as a result of that. He's not to the therapeutic agents yet, but that is certainly where the research is headed. The potential there is very exciting!

A cell is DNA's way of making more DNA.

Author unknown

Chapter 13

Basic Genetics

The Nature of the Hereditary Information in Bacteria
DNA Replication
Transcription and Its Regulation
Regulation of Gene Expression
Protein Synthesis
Mutations
Mutagens
DNA Repair

All the members of a bacterial or archaeal species are remarkably similar in their structure and ability to survive in a given environment. The thousands of biochemical reactions carried out within the cell—synthesizing cellular constituents, responding to environmental stimuli—are nearly identical in all members of a species living under equivalent conditions. But how is it that this species identity is maintained as each cell moves through a predetermined life cycle and divides, producing daughter cells that are identical to the parent? During replication, a copy of instructions written in chemical code is transferred to each daughter cell. It is this genetic code, unique to each bacterial species, that safeguards the identity of the species from generation to generation.

The next four chapters will define the chemical nature of the genetic code, delineate the basic principles of informational transfer during cell division, and describe the cellular machinery responsible for translating the "words" of the code into the "actions" of the cell, such as enzymes and structures. We will also examine how the coded instructions carried by the genes can be altered, and will investigate the consequences of such alterations on traits that bacteria possess and pass on to their progeny, or in some cases, to their siblings. These mechanisms whereby a virus, with a limited amount of genetic information, can commandeer the synthetic machinery of a host cell will be examined (Chapter 14). Some practical aspects of genetic manipulations used in basic, medical, and industrial microbiology will also be discussed.

The Nature of the Hereditary Information in Bacteria

The hereditary information of all living cells is stored in blocks of deoxyribonucleic acid (DNA) called **genes;** genes are found strung together in large molecules called **chromosomes.** The complete set of an organism's genes is called its **genome.** While higher organisms carry their genes in multiple, linear chromosomes, the genome in the vast majority of bacteria is contained in one circular chromosome. A notable exception is *Borrelia burgdorferi,* the causative agent of Lyme disease, a eubacterium with a linear genome. Occasionally, bacteria contain additional, smaller, circular molecules of DNA that carry information necessary for survival in special environments. These molecules, called **plasmids,** are chemically identical to chromosomal DNA. The smallest genomes are found in viruses; viral genomes can be circular DNA (double- or single-stranded), linear DNA, or double- or single-stranded ribonucleic acid (RNA).

The flow of information within the cell may be viewed as ideas (DNA) transcribed to words (RNA) translated to actions (proteins), which necessarily include maintaining the store of DNA so that the flow can be repeated. These processes require controlled energy and are therefore regulated at many levels. The rest of this chapter reviews the mechanics that make this regulated flow possible.

Structure of DNA

Of all the reactions in a cell, the consistently accurate copying of genetic information can be considered one of the most important. Species identity is written in DNA, and reliable replication is the guardian of this identity. The key to this accuracy is found in the double-stranded structure of DNA.

Chemically, DNA is a relatively simple molecule: deoxyribonucleotides linked by phosphodiester bonds. All of these deoxyribonucleotides consist of the same sugar phosphate, 2-deoxyribose, but they differ with respect to their nitrogenous bases, as shown in Figure 13.1. Purines (adenine and guanine) and pyrimidines (cytosine and thymine) are attached to the sugar by specific *N*-glycosidic linkages. The long DNA chain is the result of links made by phosphodiester bonds between the 3′—(pronounced "three prime") and 5′—carbon atoms of each sugar. It is the order of nucleotides in a gene which carries its unique meaning, just as the order of the letters in a word conveys its meaning. Again, the accurate copying of this sequence during bacterial division allows a bacterium to transmit its specific characteristics from the parent cell to the daughter cells.

Chromosomal and plasmid DNA are always present in a double-stranded form, with hydrogen bonds holding the two strands together. Base pairing takes place between adenine and thymine and between guanine and cytosine, with a purine bonding to a pyrimidine in each case (see Figure 13.1). The number of hydrogen bonds between the bases is not identical: two hydrogens participate in a bond between adenine and thymine, while guanine and cytosine can form hydrogen bonds with three participating hydrogens. Nevertheless, this bonding maintains a distance of just under 3 Å (angstroms) between the bases of each strand, giving the DNA molecule a uniform width (see Figure 3.3).

As a result of base pairing, the sequence in each strand is **complementary,** that is, a specific sequence in one strand determines the sequence in the other. Moreover, because the two polynucleotide chains run in opposite directions with respect to their 5′—to—3′ links, they are said to be **antiparallel.** This is depicted in Figure 13.1, where the phosphodiester linkages in one strand run in the 5′—to—3′ direction while the complementary strand runs 3′—to—5′.

The linear arrangement of nucleotide bases, together with strict base pairing between specific purines and pyrimidines, puts additional constraints on the DNA molecule.

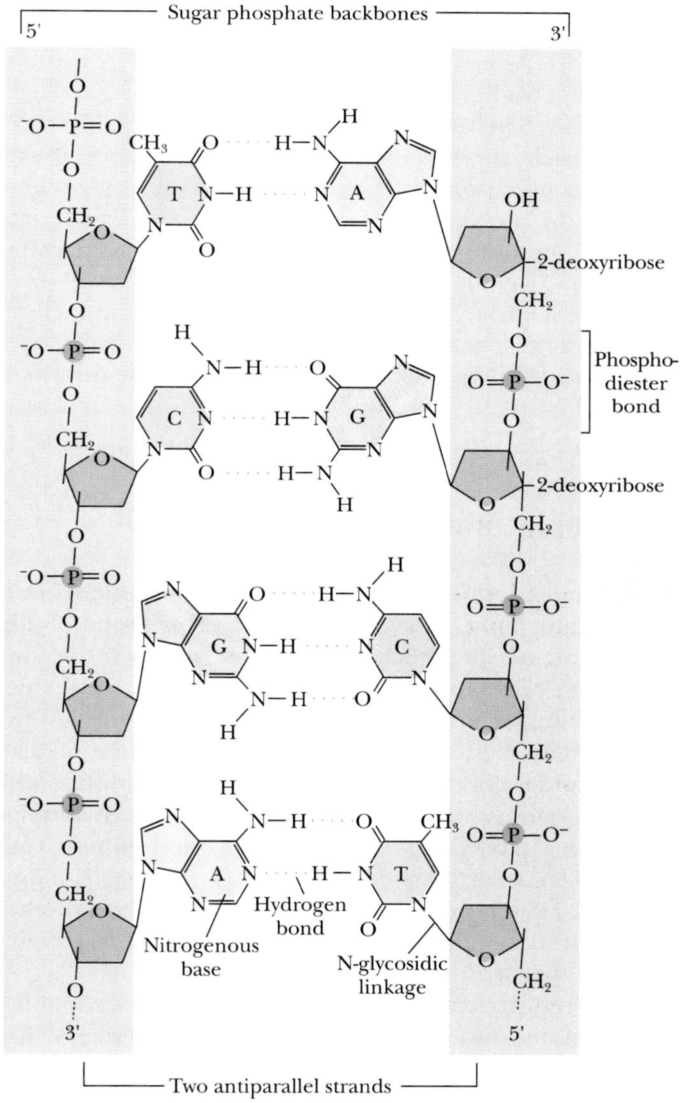

***Figure* 13.1** The arrangement of DNA duplexes in DNA. The single strands are paired A-T and G-C by hydrogen bonds. The strands are antiparallel with one strand running 3′⟶5′ and the other 5′⟶3′. The strands shown in this figure are flattened; in bacteria, DNA strands form a double helical structure.

Hydrophobic interactions between adjacent bases result in stacking, forcing the phosphate backbone into a helical configuration (see Figures 3.9 and 3.10). The predominant form of double-stranded DNA is a right-handed helix, turning in a clockwise direction, with a regular periodicity at each 10.6 base pairs, or 34 Å.

Typically, a bacterial cell contains about 5×10^6 base pairs of the four nucleotides, giving a total (unwound) bacterial chromosome length of approximately 1 mm. This single huge molecule fits into a bacterial cell, which is cylinder of approximately 0.5 μm in diameter and 1–2 μm in length, a feat comparable to a ten-meter string folding into a one-centimeter box. DNA accomplishes this by several additional higher order structural features, most important of which is **supercoiling.** Like any circular object, if a circular DNA molecule is twisted, it will cross itself. When a right-handed, clockwise helix is twisted in a left-handed direction it becomes **underwound.** DNA twisted in the same right-handed direction as the helix becomes **overwound.** The torsional stress is relieved by supercoiling, in which the double-stranded DNA twists onto itself. The condensed, supercoiled form of DNA brings together the negatively charged phosphate backbones, which are naturally repulsing. Neutralizing these charges is accomplished by a coating on the DNA that contains small proteins and a variety of cationic molecules. These factors allow packaging of the genomic DNA into a compact space without disrupting its ability to provide genetic information to the cell.

DNA Replication

Bacterial division is by an asexual process, in which a cell divides into more or less equal parts. This mode of reproduction is called **binary fission** and results in the formation of two daughter cells that are indistinguishable from the single parent cell. Constituents within the cell cytoplasm and cell envelope are divided approximately equally, and if necessary, minor differences in cytoplasmic content, such as enzymes and cofactors, can be made up by synthesis in the daughter cell immediately following cell division. The exact partition of the genetic material, however, is absolutely necessary. "Minor" errors in the genome can result in daughter cells permanently losing a genetic trait and, perhaps, in death.

The replication of the genome results in two double-helical molecules. One strand in each molecule is the parent DNA, while the other is a complementary, newly synthesized strand. The strand separation of a parent molecule, its use as a template, and incorporation into the daughter genome is termed **semiconservative** replication. The mechanism of semiconservative replication of a circular bacterial genome through two cycles is shown in Figure 13.2.

The Macro Cycle of Replication

DNA replication in bacteria is rapid and continues throughout the cycle of cell division, in contrast to replication of DNA by eukaryotic cells, which takes place during one distinct stage of the cell cycle. The replication of DNA within the circular bacterial chromosome begins at a single specific site on the bacterial chromosome, termed ***ori*C** for the **origin of replication**. The DNA is unwound to create two single-stranded templates, and synthesis proceeds from this point in opposing directions. This bidirectional replication mechanism requires the presence of two sites of synthesis of the new DNA. Each of these two sites, termed a **replication fork** (Figure 13.3), comprises a complex of molecúles that incorporate complementary nucleotides, accurately copying the template of the parent cell. When the replication forks meet at approximately the opposite side of the circular DNA duplex, each strand has been copied into a complete double-stranded helix, which can then be partitioned into daughter cells.

The replication of this enormous molecule is very rapid; replication forks proceed at a rate of 1000 base pairs (bp) per second, resulting in replication of the total 1-mm-long chromosome in about 40 minutes. This, however, is not a satisfactory speed for certain rapidly growing bacteria that can divide in 20 minutes. To take advantage of optimal growing conditions, DNA replication is reinitiated before one round of replication is completed. The genomes partitioned into the new daughter cells will already be partially replicated, enabling them to divide again more quickly. A concomitant effect is that in these rapidly dividing cells, genes near *ori* will be present in more than one copy.

The Micro Cycle of DNA Replication

Let's now examine in greater detail the events that establish the replication fork and subsequent replication. The bacterial replication apparatus is quite complex. It has been dissected with the aid of mutations in genes that specify its components and with meticulous biochemical reconstruction of the replicative events using purified enzymes. The major enzymes involved in DNA replication are presented in Table 13.1.

Replication of the chromosome of *E. coli* is initiated at *ori*C, a 245 base pair sequence that is highly conserved in gram-negative bacteria (Figure 13.4). There are four 9 base pair and three 13 base pair repeats (rich in adenine and thymine base pairs) within the *ori*C region. The replication is initiated by binding of several tetramers of an **initiator protein** termed DnaA to these 9 base pair repeats, followed by sequential addition of up to 40 DnaA monomers, until the DNA becomes wrapped around an aggregate of DnaA. The bound DnaA molecules, together with ATP, then cause separation of the DNA duplex of the adjacent 13 base repeat region, into single-strands, resulting in an "open complex." A protein designated Dna B binds to the two replication forks with the assistance of two proteins designated Dna C and Dna T. The addition of the Dna B protein completes the assembly of the **prepriming complex.** Dna B has helicase activity and, assisted by DNA gyrase, promotes unwinding of the DNA duplex. The single-strand binding protein (SSB) tetramers attach to the single strands as they arise. Unwinding exposes the base sequences in the single strands to serve as templates in replication.

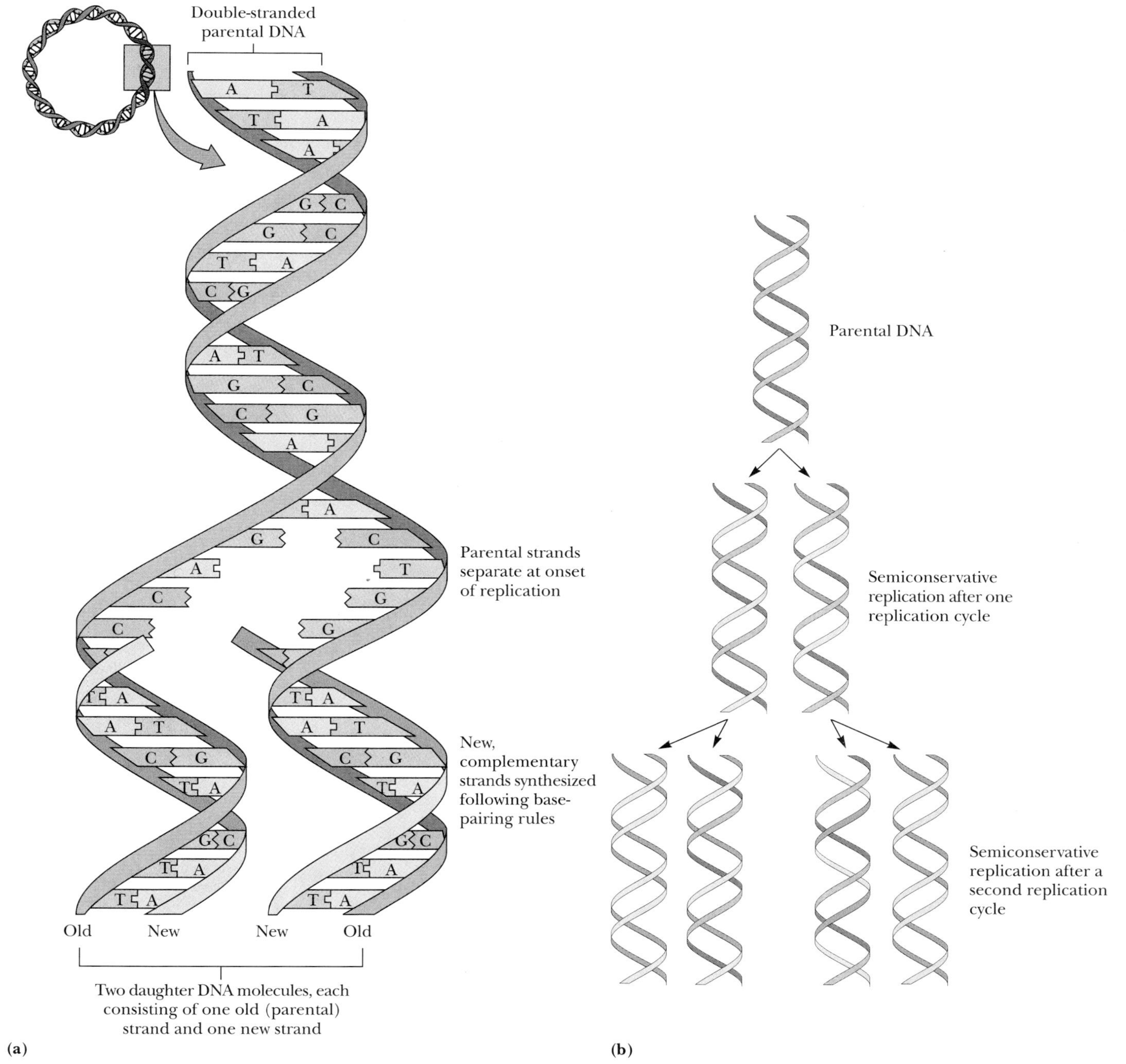

***Figure* 13.2** Semiconservative nature of DNA replication. The basic structure of DNA predicts a possible mechanism for replication, shown in **(a),** which involves copying each strand into daughter strands, each containing one new and one parental strand. Overall duplication of gentian c information by the semiconservative mechanism after two cycles of replication is shown in **(b).**

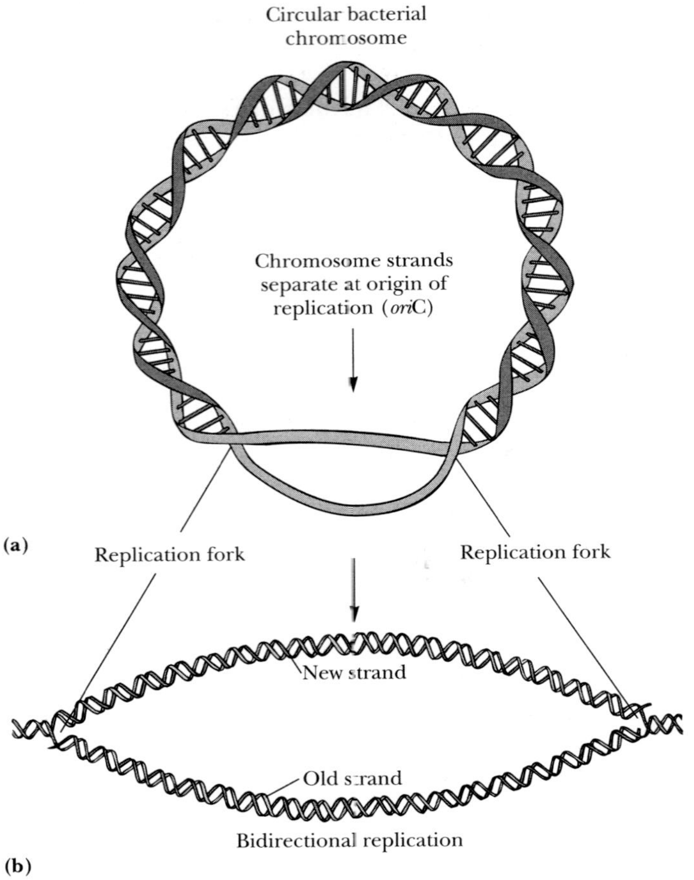

***Figure* 13.3** Overall representation of the replication of a circular bacterial chromosome. The replication is initiated at *oriC*. Two replication forks move bidirectionally until they meet, at which point the complete circular genome has been duplicated.

Further additions to the prepriming complex generates a replisome assembly that can replicate the genome directionally (Figure 13.5). The key addition is the enzyme primase that synthesizes the RNA primer necessary for DNA synthesis and elongation of the DNA strand.

DNA polymerase is the enzyme that polymerizes deoxynucleotide triphosphates into DNA. Bacteria possess three distinct DNA polymerases, DNA pol I–III. The enzyme termed DNA pol III carries out the elongation stage of DNA synthesis as the replication fork moves forward (Figure 13.6). DNA pol I in concert with DNA ligase plays a crucial role in filling in gaps and removing RNA primers, annealing short gaps, and in DNA repair. The role of DNA pol II in replication of the chromosome is not known, and mutations in the gene encoding this enzyme have no effect on bacterial cell division.

DNA pol III can add only nucleotides to a preexisting 3′—hydroxyl; consequently the template must be "primed" with a nucleotide from which the polymerase can start. For this purpose, a short RNA **primer,** complementary to the exposed template, is synthesized by an RNA polymerase called **primase.** This short RNA segment provides the necessary 3′—hydroxyl end to which single nucleotides are added. The polymerization reaction takes place in the 5′—to—3′ direction, that is, the incoming 5′—nucleotide triphosphate is linked by its α—phosphate to the 3′—hydroxyl group of the ribose on the primer. The accuracy in replication is dictated by the same hydrogen-bonding rules that hold together the nucleotide triphosphates of the original duplex: a guanine (G) is paired with a cytosine (C), and a thymine (T) with an adenine (A).

Table* 13.1 Proteins required for replication of DNA in *Escherichia coli

Protein	Function
DNA gyrase and helicase	Unwinding DNA
SSB	Single-stranded DNA binding
DnaA	Initiation factor
HU	Histone-like (DNA binding)
PriA	Primosome assembly, 3′ ⟶ 5′ helicase
PriB	Primosome assembly
PriC	Primosome assembly
DnaB	5′ ⟶ 3′ helicase (DNA unwinding)
DnaC	DnaB chaperone
DnaT	Assists DnaC in delivery of DnaB
Primase	Synthesis of RNA primer
DNA polymerase III holoenzyme	Elongation (DNA synthesis)
DNA polymerase I	Excises RNA primer, fills in with DNA
DNA ligase	Covalently links Okazaki fragments
Ter	Termination

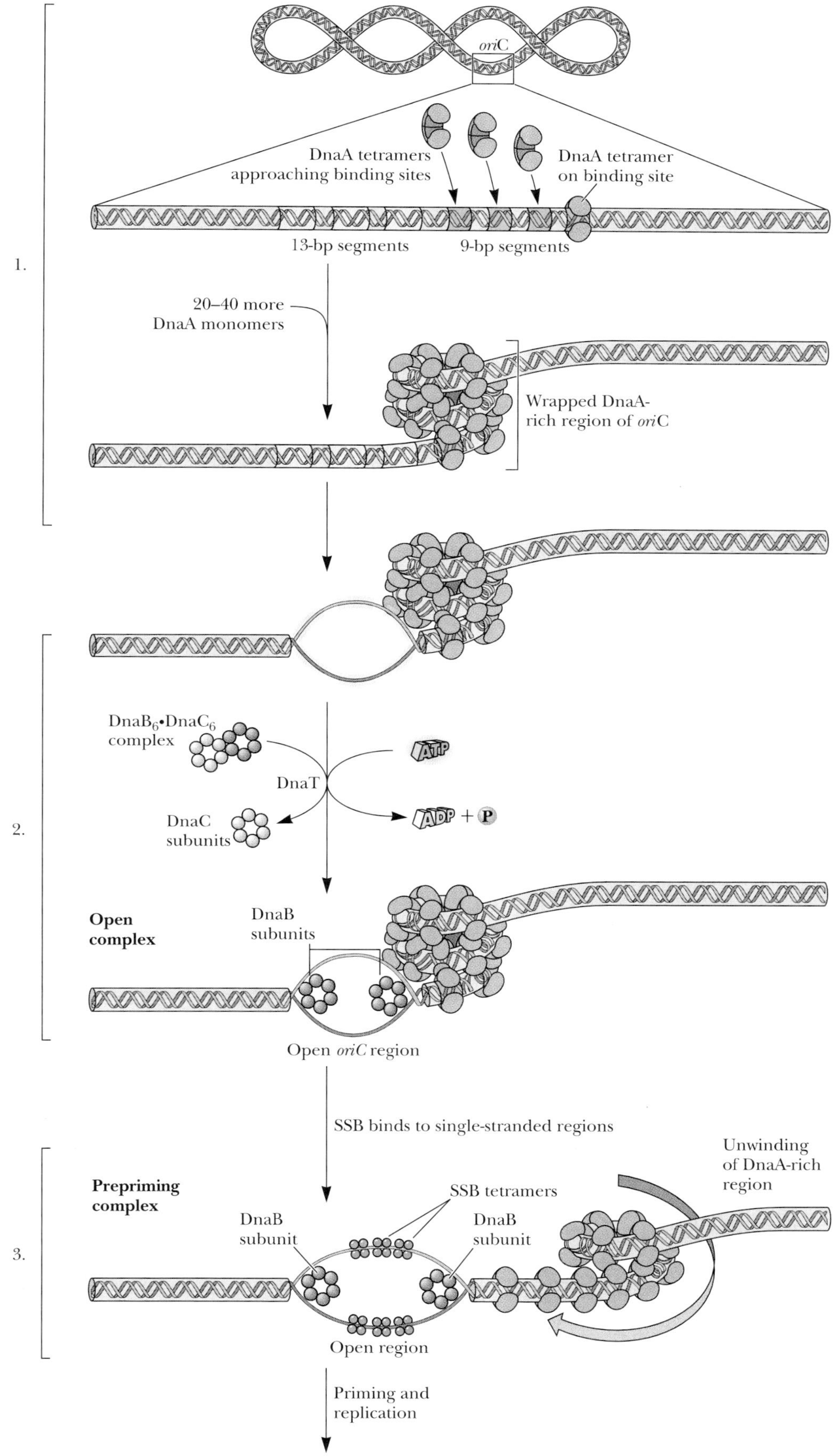

***Figure* 13.4** Events that occur during the initiation of *ori*C replication in *E. coli.* (1) DnaA protein tetramers bind at each of four 9 base pair repeats followed by the addition of 20 to 40 more DnaA molecules. (2) The adjacent 13 base pair region, rich in adenine thymine pairs, is separated to form an open complex. DnaB protein from a DnaB/DnaC complex then binds at each end of the open complex. (3) Single strand binding protein (SSB) association.

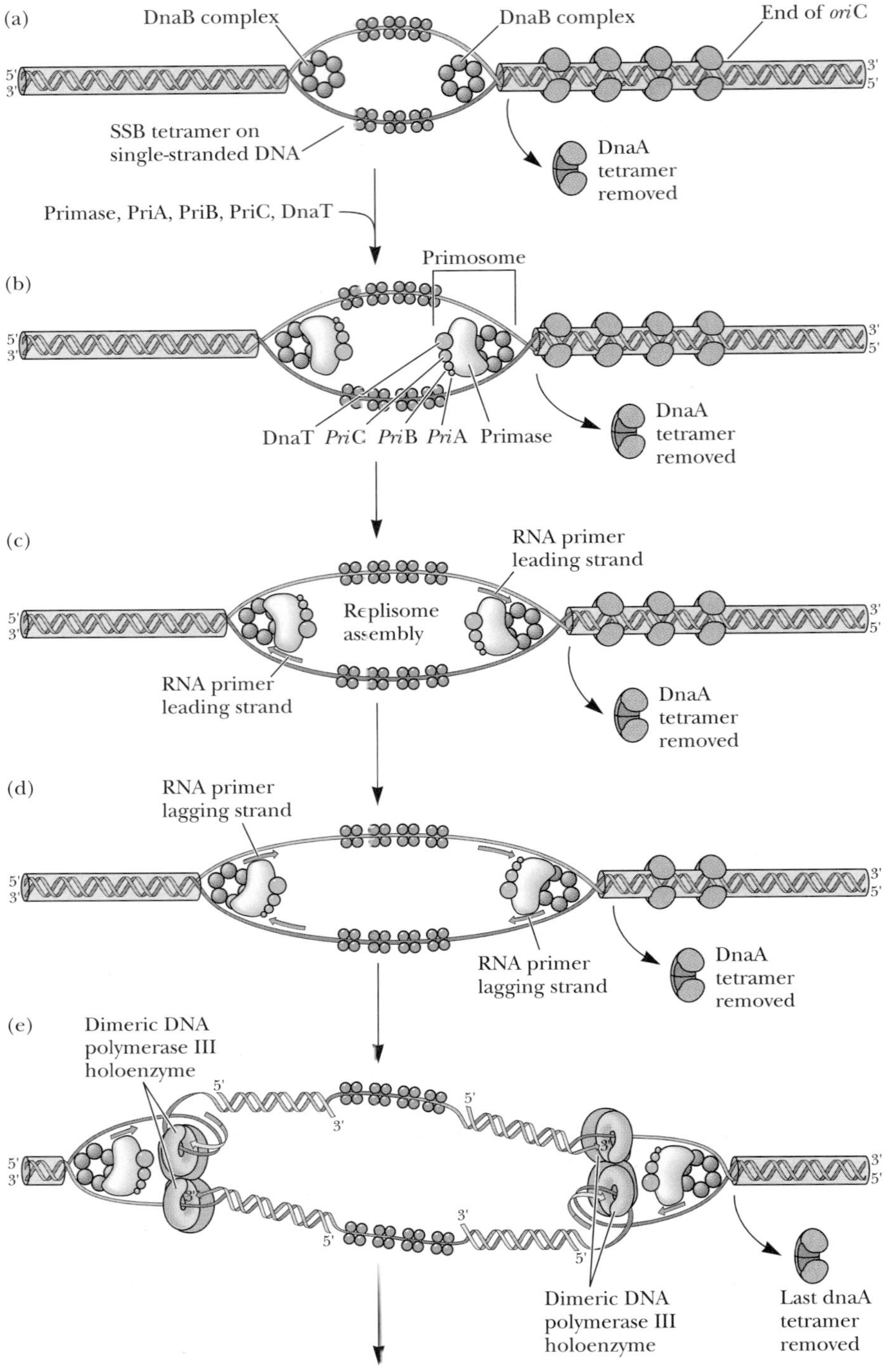

***Figure* 13.5** Formation of a replisome assembly at each fork of the separated DNA formed in the pre-priming process (Figure 13.4.) **(a)** The extra DnaA proteins are removed. **(b)** Several proteins are added at each replication fork (PriA, PriB, PriE, and DnaT) along with primase. **(c), (d)** An RNA primer is synthesized by primase for each of the single strands (leading and lagging). **(e)** A DNA polymerase III holoenzyme dimeric complex is added at each replication fork. One half of the dimer elongates by synthesizing on the leading strand and the other elongates by synthesizing on the lagging strand.

The fact that DNA pol III can only polymerize in the 5′—to—3′ direction presents a special problem regarding one of the two strands of DNA. The strand depicted above the replication fork in Figure 13.5 contains constantly available primer, and synthesis is continuous in the 5′—to—3′ direction. This so-called **leading strand** is synthesized as a single molecule complementary to the template. The same 5′—to—3′ direction of synthesis on the opposite DNA template strand, called the **lagging strand,** however, requires **discontinuous synthesis** involving several steps. First, RNA primase synthesizes a short, 10-nucleotide RNA primer about 1000 to 2000 nucleotides **upstream** (that is, in the 5′—direction) from *ori*C. This process is analogous, although not identical, to the initiation of DNA synthesis at *ori*C. The 3′—hydroxyl group then serves as a primer for polymerization of DNA from the primer back

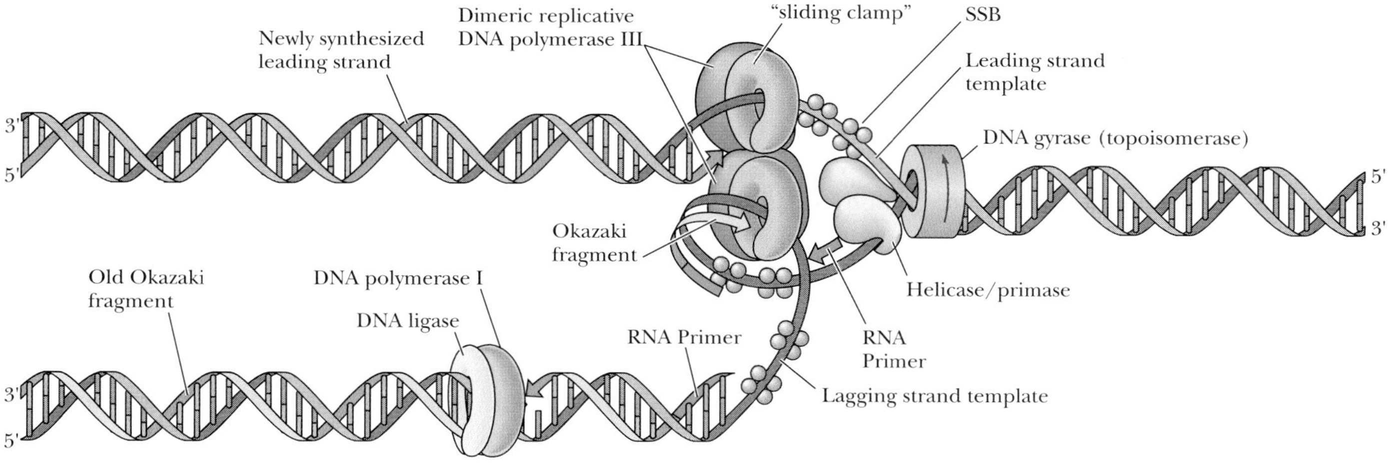

***Figure* 13.6** A complete replication form as it would appear at each of the forks. The duplex DNA is unwound by DNA gyrase and helicase and the resulting single strands are coated with ss DNA binding protein (SSB). Primers are added to the lagging strand at intervals. The dimeric DNA polymerase is bound by a "sliding clamp" and adds single nucleotides to the 3′ end of the leading strand. Okazaki fragments are added to the RNA primers on the lagging strand. DNA polymerase I and DNA ligase act downstream to remove RNA primers and replace them with the appropriate complementary nucleotide and ligate the fragments.

toward the origin of replication. The DNA segments polymerized from such an RNA primer are called **Okazaki fragments,** named for the researcher who first characterized them. In leapfrog fashion, the primase continues laying primers on the lagging strand upstream of the last Okazaki fragment as DNA synthesis proceeds. The resulting complementary strand on the lagging strand is discontinuous, consisting of a series of RNA primers linked to Okazaki fragments. The primers are removed by DNA pol I via its 5′—to—3′ exonuclease activity (discussed in the next section on DNA proofreading), and the gaps formerly occupied by RNA are filled in by polymerization of the complementary deoxynucleotide triphosphates into DNA. The final stage of synthesis is sealing the nicks between fragments by another enzyme, **DNA ligase.**

The replication of the circular bacterial chromosome terminates when two replication forks meet. Being helical in nature however, the DNA undergoes rotation to accommodate the moving replicative fork. This leads to excessive supercoiling and tension on the DNA molecule. As mentioned earlier, partial relief of this tight winding generated during synthesis, as well as the eventual untangling of the daughter strands, is directed by DNA topoisomerases. Two types of this enzyme have been isolated from bacteria. **DNA topoisomerase I** is an enzyme that creates a nick on one of the DNA strands, allowing free rotation around the phosphodiester bond opposite the nick, thus relieving the tension. This break persists only transiently and is rapidly closed. **DNA topoisomerase II** (also called **gyrase**) causes double-stranded breaks, allowing two helices that cross each other to pass through the break on one of the strands before the break is resealed. The reaction catalyzed by DNA topoisomerase II is especially important in freeing the **concatamer** (two interlocked circles) upon completion of replication, which allows free separation of each chromosome into each daughter cell during cell division.

The Mechanism of DNA Replication Proofreading

The specificity of base pairing (A:T, G:C) is the major mechanism by which reliable copying of the parent strand is guaranteed. Errors during replication take place when incorrect base is added to the growing chain, creating a **transient** mismatch. When these strands separate and are partitioned into daughter cells, the chromosome of the daughter receiving the mismatched base will be different from the parent's. This change could be deleterious to the daughter cell if the change is in an important gene. The process of DNA replication is, however, remarkably efficient and has a very low error rate. It is estimated that in *E. coli,* the frequency of replication error is less than 10^{-9} per replicated base pair. This low frequency of mismatched bases is due to the **proofreading** and repair activities associated with DNA polymerases. One system for corrections is the **mismatch repair system** present in *E. coli* that monitors newly replicated DNA for mispaired bases, removes the mismatched base, and replaces it with the correct one. Repair is by a DNA polymerase mediated local replication. To replace a mismatched base the enzyme must identify which base in the pair should be replaced.

A mismatch repair system that can be identified in the incorrect base is the **methyl directed pathway.** The parental strand in a duplex is methylated, whereas the newly formed strand is not. Methylation of bases occurs after replication. When the methyl repair system encounters a mismatched base pair it will continue along the DNA duplex until it encounters a methylated base. The system accepts the methylated strand as parental and excises and replaces all of the nucleotides on the complementary strand from this methylated base back to and including the mismatch. All three bacterial polymerases are capable of both 5′—to—3′ polymerizing and 3′—to—5′ exonucleolytic activity. The immediate repair of mismatched bases formed during replication is carried out by the 3′—to—5′ exonuclease activity of DNA polymerase III, the major polymerizing enzyme in the replicating fork. Filling in small gaps and repairing damaged DNA is accomplished by continuous degradation and synthesis in the 5′—to—3′ direction, carried out by DNA polymerase I. Additional enzymes are involved in the repair of DNA damaged by physical or chemical agents; these types of repairs are discussed later in this chapter.

Transcription and Its Regulation

There are two occasions when the cell copies its genes. In the first, **replication,** the entire genome is copied from DNA to DNA for the purpose of continuing the species. In the second, **transcription,** one or a few genes are copied from DNA to RNA for the purpose of **expressing** those genes, that is, to make proteins needed for continuing the life of the individual cell.

The genetic information encoded in the DNA almost always specifies proteins. The intermediary in the transfer of genetic information from DNA to proteins is a messenger molecule of ribonucleic acid (mRNA). Because the process of synthesizing RNA from DNA is called **transcription,** molecules of RNA are sometimes referred to as **transcripts.** Separate classes of RNA, not translated into protein, make up the protein synthesis machinery; these are ribosomal and transfer RNA (rRNA and tRNA). All types of RNA are synthesized by an identical enzymatic reaction. One significant difference between the types of transcripts is that while rRNA and tRNA are produced from larger RNA precursor molecules by enzymatic processing, bacterial mRNA is not processed after transcription. (This difference is not found in eukaryotes, where all types of RNAs are extensively processed, and the final form of the mRNA is the result of cutting and splicing from a large precursor molecule.)

Bacterial mRNA is rather labile, the average half-life being only 1.5 to 2 minutes, unlike eukaryotic mRNA, which can persist and function for hours. Short-lived mRNA enables bacteria not only to respond rapidly to environmental changes but also to conserve energy. Within minutes of a new signal from the environment, bacteria can stop putting their resources into transcription of genes that are no longer needed. Nucleotides from the degraded RNA become available for new transcripts. More importantly, the extremely energy-costly process of protein synthesis is not wasted on unnecessary proteins.

Steps in RNA Synthesis

The process of RNA synthesis is a chemical polymerization, which in principle resembles the copying of one strand of DNA during replication: initiation, elongation, and termination. Transcription, however, involves copying only a short segment of the DNA, and the product is built of ribonucleotides rather than deoxyribonucleotides. The steps in initiation of transcription are shown in Figure 13.7.

The synthesis of RNA is carried out by the enzyme RNA polymerase. Using a DNA template, it recognizes and binds to a specific sequence which indicates the beginning of a gene. This site is termed the **promoter;** the first nu-

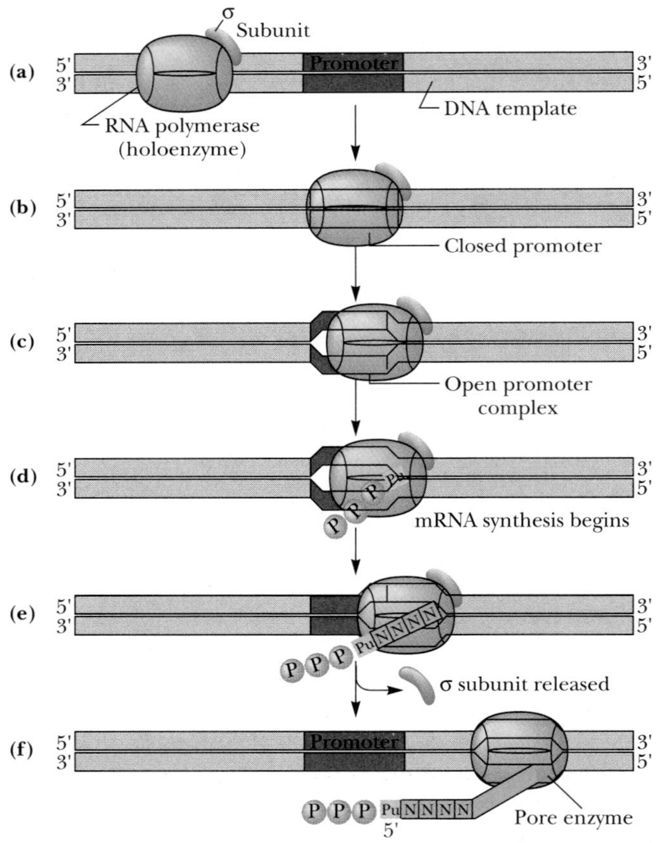

Figure **13.7** The sequence of events in initiation and elongation during transcription. **(a)** Binding of the polymerase to DNA and migration to the promoter. **(b)** Complex of RNA polymerase: closed promoter. **(c)** Unwinding of DNA with formation of an open promoter complex. **(d)** Polymerase begins mRNA synthesis by addition of ATP or GTP (purine nucleotides). **(e)** Elongation by nucleotide addition. **(f)** After addition of 5-6, nucleotides, σ is released and polymerase continues down the template.

cleotides of the DNA will be copied within 10 to 15 base pairs (bp) of the promoter. Because only one strand (called the **sense strand**) of the double-stranded helix is copied, it is necessary for the RNA polymerase to displace the complementary copy as it moves along incorporating free nucleotide triphosphates into a polymerized RNA strand.

One method used most frequently to control gene expression is to regulate how often transcription is initiated. Regulation of RNA polymerase binding allows bacteria to exert control over an entire cluster of genes specifying related functions, utilizing a common regulatory mechanism. A promoter serves as a start for transcription of a single gene, or that of an operon resulting in polycistronic mRNA. Since bacteria do not have a nucleus as eukaryotic cells do, the nascent transcript can be translated into proteins from its free end while it is yet being synthesized. In further contrast to eukaryotic transcription, bacteria frequently synthesize **polycistronic mRNA,** that is mRNA which specifies more than one gene. Bacterial genes are frequently arranged in **operons,** sets of adjacent genes that are transcribed and translated together. Genes in an operon utilize a single promoter, preceding the first gene, and transcription terminates after the last gene. Therefore, the genes within an operon are expressed together, and control of their expression involves usually the regulatory sites near the single promoter (see below). Operons generally code for enzymes that function together in a common biosynthetic or catabolic pathway. This enables the bacterial cell to coordinately regulate expression of related genes, increasing its efficiency.

The bacterial RNA polymerase is a multisubunit enzyme, consisting of 5 subunits: two Alpha (α), one Beta (β), one Beta$'$ (β'), and one sigma (σ). The subunits are assembled into a multimeric structure (called the **holoenzyme**) which binds to the promoter region. It is the sigma subunit that recognizes the promoter—the beginning of the region to be transcribed. This stable complex between the promoter and the RNA polymerase is referred to as a closed promoter complex. The subsequent steps in the initiation of transcription are melting of the double-stranded DNA and formation of the so-called open promoter complex (i.e., separating the two strands) and polymerizing the first few nucleotides, after which the sigma factor is not needed. The sigma subunit then dissociates from the holoenzyme, leaving four subunits (2α, β, and β') as a complex termed the **core enzyme** that moves along the DNA strand and polymerizes nucleotides into RNA.

Signals built into the DNA allow the RNA polymerase to identify the end of the gene or genes it is transcribing. If RNA polymerase is transcribing an operon, the termination signal is present at the end of the operon. One type of termination sequence is identified as an inverted repeat followed by a stretch of polyuridines. As an RNA copy is made of this section, same-strand pairing of the inverted repeats forms a **stem-loop structure** that causes the RNA polymerase to be released from the DNA template, terminating transcription.

Another mechanism of termination of transcription involves both a signal on the template that causes RNA polymerase to pause and a specific termination protein called rho (ρ). **Rho factor** is an ATP-dependent helicase that promotes unwinding of RNA:DNA hybrid duplexes. ρ binds to C rich regions in the RNA transcript and advances in the $5'$—$3'$ direction until it reaches the RNA polymerase. At the site where the RNA polymerase stalls at a G:C rich termination site, the ρ factor unwinds the transcript from the template and the mRNA is released.

The difference between the rho-dependent and rho-independent termination mechanism is the type of sequence information encoded at the termination site, that is, whether it leads to pausing of the RNA polymerase, or whether it leads to an RNA stem-loop. Following release of the RNA polymerase, the core enzyme associates with a sigma (σ) subunit present in the cytoplasm, forming the holoenzyme capable of recognizing another promoter and initiating transcription of another gene.

Promoters and σ Factors

Recognition of the beginning of the gene is key to determining which gene will be expressed. In bacteria there exist several species of sigma subunits that can form complexes with the RNA polymerase core enzyme. Each species of σ subunit recognizes specific promoter sequences, guiding the polymerase to specific genes.

σ^{70} (named for its molecular weight of about 70,000 daltons) recognizes the majority of genes encoding essential functions in the bacterial cell. The promoter sequences of these genes are relatively conserved, although they are not identical. These promoters often have two regions of particular similarity: a sequence TATAAT approximately 10 base pairs upstream from the site of initiation of transcription(-10 region), and a sequence TGACA, 35 base pairs upstream termed the **$-$35 region.** (The first base pair copied into RNA is referred to as $+1$, hence the TATAAT sequence is at position -10, TGACA is -35.) The TATAA is termed a Pribnow box and is named for David Pribnow, the first to recognize the importance of this sequence in transcription. These sequences were deduced by examination of a large number of different bacterial promoters. They represent what is called a consensus promoter for recognition by σ^{70}.

Bacteria contain alternative σ factors, which also form complexes with the core RNA polymerase and lead to transcription of genes that contain their cognate promoter sequences. Some of these are summarized in Table 13.2. The genes requiring alternative σ factors for transcription are involved in cellular responses to a variety of specialized conditions, including survival at elevated temperature,

Table **13.2 Some of the Alternative Sigma (σ) Factors in Bacteria and Their Cognate Promoter Sequences**

σ Factor Promoter Recognized	Genes Transcribed
σ^{70} TTGACA-17bp-TATAAT	Many and diverse
σ^{32} CNCTTGAA-14bp-CCCCATNT	Heat shock response
σ^{54} CTGGNA-7bp-TTGCA	Many and diverse
σ^{28} TAA-15bp-GCCGATAA	Chemotaxis, motility, flagellar components
σ^{29} TTNAA-17bp-CATATT	Sporulation in *B. subtilis*[1]

[1]Several different sigma factors control different genes involved in sporulation in *B. subtilis.*

synthesis of genes for nitrogen metabolism, motility, and sporulation. The alternative σ factors are a minority, and do not effectively compete with σ^{70} under normal conditions.

One of the important features of all promoters, including those that are recognized by the alternative σ factors, is that they contain specific sequences relative to +1. Therefore promoters give RNA polymerase a direction for transcription; once RNA polymerase binds to a promoter, it will move unidirectionally, as determined by the relative position of the bases in the promoter, for example, in the case of σ^{70} by the −10 and −35 sequences. Occasionally, two genes are transcribed from promoters that overlap, and the RNA is copied from opposite strands, but the position of the −10 and −35 sequence (or similar sequences for alternative σ factors) assure movement of the RNA polymerase in the correct direction for each gene.

The term **promoter strength** is used to describe the efficiency of transcriptional initiation from a given promoter. High frequency of holoenzyme binding to a promoter results in high levels of RNA synthesis, and consequently high levels of protein synthesized from a specific gene. A perfect match at both −10 and −35 would be expected to result in the highest level of gene expression. However, agreement of a specific promoter sequence is frequently not the sole determinant of the amount of a specific gene product made at any one time. In the next section, we will see that there are many ways of modulating gene expression.

Regulation of Gene Expression

Bacteria have the capacity to adapt to specific environmental conditions and they do so by altering the levels of mRNA suitable for translation. From the point of view of energetics, this is most efficiently accomplished by controlling the initiation of transcription, to prevent the unnecessary use of nucleoside triphosphates and consequent energy-costly synthesis of protein. While there are many genes expressed **constitutively,** that is, without any control of the levels of mRNA other than the strength of the promoter, there are also a large number of genes for which the ability of RNA polymerase to initiate transcription is controlled by other proteins that respond to signals from the environment of the bacterial cell.

Control at the level of transcriptional initiation can be divided into two classes, based on whether the environmental signals *facilitate* transcriptional initiation **(positive regulation)** or *interfere* with it **(negative regulation).** In certain instances, a gene or an operon can be regulated negatively under one set of conditions and positively under another. Moreover, both forms of regulation can involve small signaling molecules that regulate multiple operons scattered throughout the bacterial chromosome.

Negative Regulation

There are several modes employed to decrease the capacity of RNA polymerase to transcribe a specific gene or operon. The most common mechanism involves a regulatory protein called the **repressor.** Repressors usually bind to specific sites, termed **operators,** near the promoter region. The operator sequence is usually located between the promoter and +1 position on the DNA, and frequently overlaps the −10 region of the promoters. The binding of the repressor therefore prevents transcription by physically blocking the polymerase from either binding its operator or, if bound, from moving forward.

One of the best-studied examples of negatively regulated genes is the lactose utilization operon (*lac*) that controls catabolism of lactose by *E. coli.* The regulatory mechanism is shown in Figure 13.8. The genes of the *lac* operon are expressed only when bacteria are growing in the presence of lactose as sole carbon source. The operon includes three genes: *lacZ* (encoding the enzyme β-galactosidase), *lacY* (lactose permease), and *lacA* (transacetylase). The negative regulatory aspect of the *lac* operon comes from the ability of a repressor protein, called the lactose repressor, to tightly bind to the operator site near the promoter from which the lactose operon is transcribed, preventing binding of RNA polymerase. In the absence of lactose, none of the three genes necessary for lactose catabolism are transcribed. When *E. coli* finds itself in a medium where lactose is the only carbon and energy source, small amounts of lactose enter the cell and bind to the repressor protein. The interaction of lactose with the repressor leads to dissociation of the repressor from the operator, and RNA polymerase then can bind to the promoter of the *lac* operon. Moreover, as long as there is lactose present in the bacterial environment (and hence inside the bacterial cell), the repressor is unable to bind to the operator, and cannot interfere with the initiation of transcription. Interestingly, other synthetic molecules which resemble lactose, but are not metabolized by *E. coli,* can cause dissociation of the repressor from the operator. These small molecules and similar signaling molecules which interfere with the activities of repressor proteins, are called **inducers.** The common mechanism of all inducers is that they bind to the repressor, causing it to lose its ability to interact with the operator sequences.

While all negatively regulated systems rely on the ability of repressor proteins to block transcriptional initiation, the precise interaction with small signaling molecules varies. The inducer (lactose or its structural analogs)

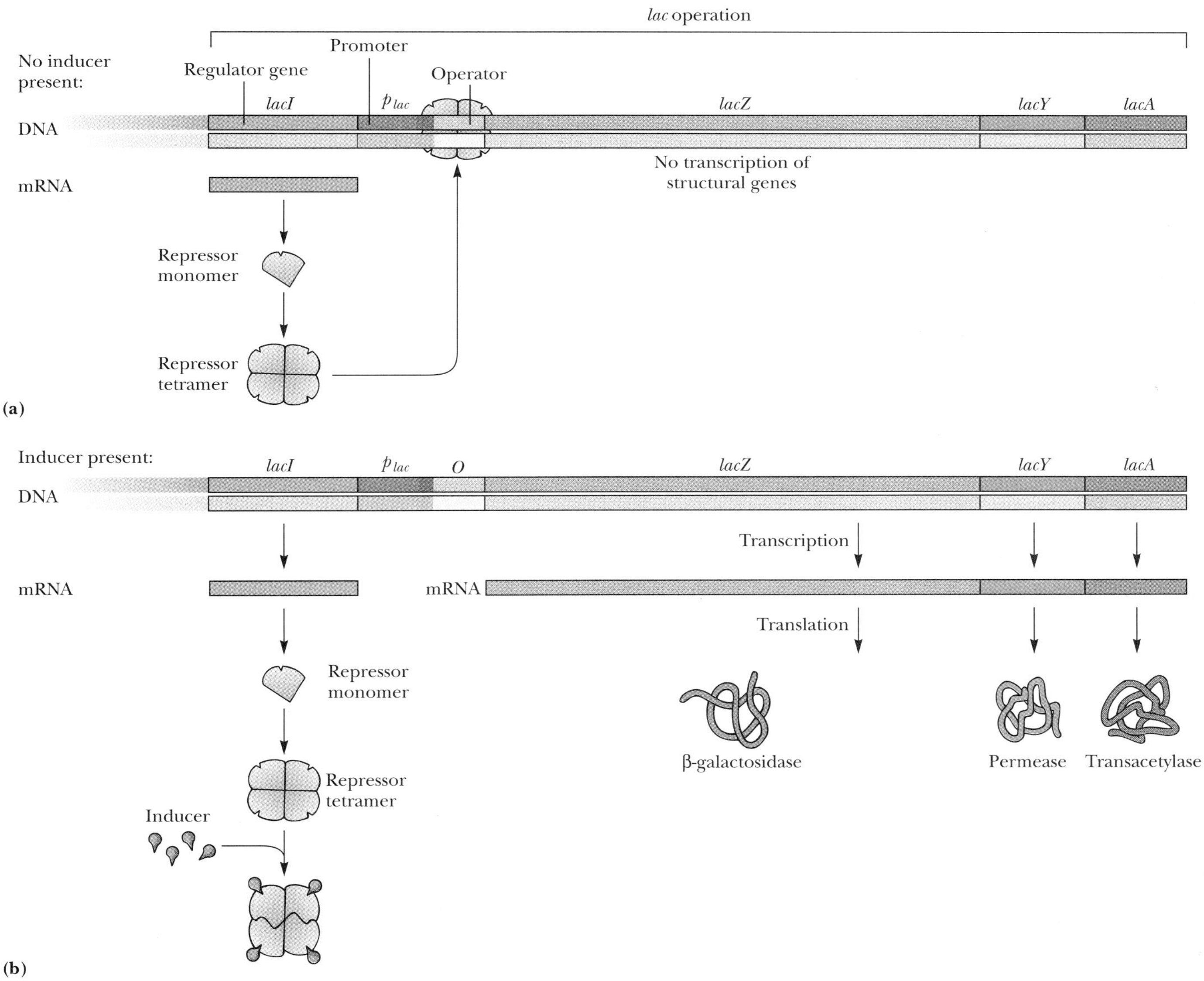

***Figure* 13.8** The mode of action of the lac operon. **(a)** The repressor binding site (the operator) slightly overlaps the promoter region, and is located between the promoter and the transcriptional start site. **(b)** If lactose becomes available, it binds to the repressor, and this complex is unable to interact with the operator sequence, leading to transcription of the operon.

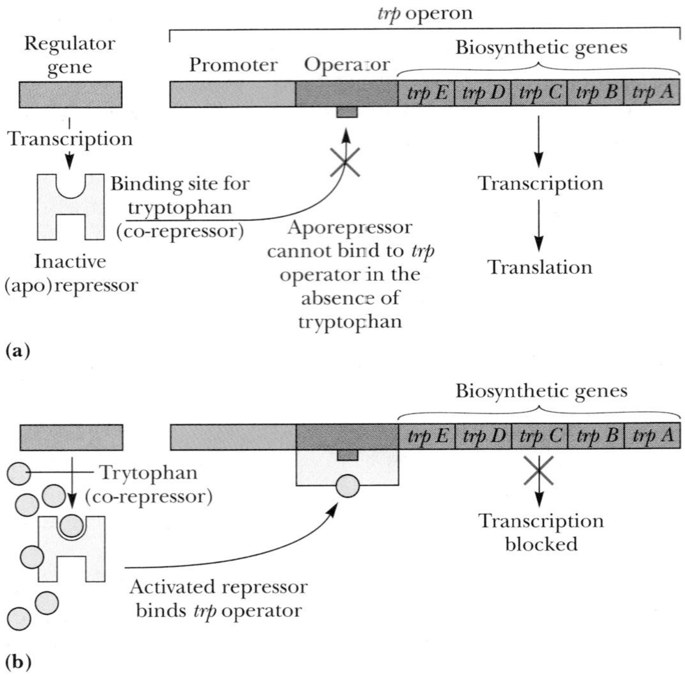

Figure **13.9** The tryptophan operon. During growth in the absence of exogenous tryptophan, the operon is read constitutively and the amino acid is synthesized. If tryptophan is present in the medium, the inactive repressor (aporepressor) interacts with the co-repressor (tryptophan) and the promoter is blocked. (See also Chapter 10, Biosynthesis of Monomers.)

causes dissociation of the repressor from the operator. Similarly, a regulatory protein (AraC), which controls genes of arabinose metabolism, acts as a true repressor, binding to a specific operator sequence, preventing the initiation of transcription in the absence of arabinose. In the presence of the inducer arabinose, the AraC-arabinose complex dissociates from the repressor, analogous to the behavior of the lactose repressor-lactose complex. The situation with the AraC protein is a little more complex, since the AraC-arabinose complex also participates in a positive regulatory mechanism (see below).

A slightly different mechanism of negative regulation involves interaction of a signaling molecule with a free, unbound repressor protein, in which this complex interacts with the operators. The most studied system of this type is the regulatory mechanism involved in transcription of the tryptophan biosynthetic genes (Figure 13.9). The "empty" repressor (called the **aporepressor**) which is not complexed with tryptophan, is unable to bind to the operator of the *trp* operon, thus tryptophan biosynthetic genes are transcribed, and bacteria synthesize their own tryptophan. When tryptophan is available in the growth medium, there is no need to synthesize the enzymes of the tryptophan biosynthetic pathway. The external tryptophan is taken up and utilized, and the *trp* operon is not transcribed. The shutoff of the *trp* operon is due to formation of an active repressor, following its binding of tryptophan. In this system, the signal molecule is called a **co-repressor,** as it actively participates with the repressor in the negative regulation of transcriptional initiation.

Positive Regulation

Positive regulation involves direct or indirect action of signaling molecules with activator proteins, stimulating transcription. While the mechanism of negative repression is obvious, that is, interference with binding of RNA polymerase, the molecular basis of activation is less clear. The activator proteins interact with sequences usually 5′ to the promoter, and these regulatory sites can be several hundred base pairs from the affected gene. The interaction of the activator protein with its cognate regulatory sequences results in a direct contact between the regulatory protein and RNA polymerase bound to the promoter. If

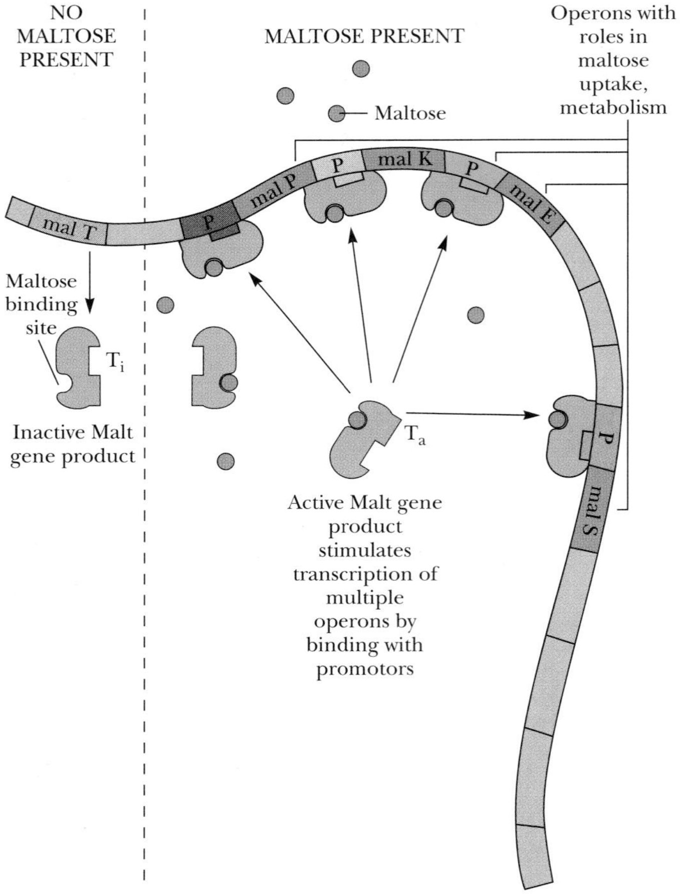

Figure **13.10** Positive control of genes of maltose utilization. The transcriptional regulator MalT, encoded by the *mal*T gene, is inactive (Ti), unless complexed with maltose. When exogenous maltose is available to the bacterium, maltose binds MalT, and this complex now functions as an activator (Ta) for transcription of genes specifying components of a maltose transport machinery and catabolism.

the binding site for the regulatory protein is far from the promoter, this requires extensive looping of the DNA between the two sequences to allow for the regulatory protein to contact RNA polymerase. Some activator proteins can recognize sequences that are present near or within promoters. In case of the promoters of positively regulated genes, which are recognized by the sigma^{70} RNA polymerase, they have −10 regions that are well conserved with other promoters, but there is a considerable variation in the sequence at their −35 regions. The most plausible explanation for the stimulatory effect of the proteins involved in the positive regulation of gene expression is that they help RNA polymerase in melting of the DNA duplex into single-stranded regions during the critical incorporation of the first few nucleotides at the initiation of transcription.

One of the most studied examples of positively regulated genes is the **maltose regulon.** The term regulon denotes a set of operons under the control of a common regulatory element. In the case of the maltose regulon, the positive regulatory protein MalT stimulates transcription of at least four different operons, involved in uptake and metabolism of maltose or related sugars (Figure 13.10). The *mal*T gene product (Ti) becomes active, that is, capable of binding to the DNA recognition sequences after complexing with maltose. All promoters of the maltose regulon contain a so-called maltose box which consists of the sequence GGAG/TGA within the −35 region.

Several negatively regulated genes are also positively controlled as well, while some genes are only regulated by activator proteins. The *lac* operon, for example, is regulated by the levels of cyclic AMP in the bacterial cell, through the activity of the catabolite activator protein (CAP), as shown in Figure 13.11. In this case, cAMP is a co-inducer, since it binds directly to the regulatory protein. In the presence of elevated levels of cyclic AMP (which takes place when preferred carbohydrate sources are depleted—another condition that allows *E. coli* to take full advantage of lactose metabolism), the cAMP-CAP complex binds to a site adjacent to the promoter, and if the operator site is unoccupied by the repressor, the transcription of the *lac* operon is enhanced several-fold above that which would result in the presence of the inducer alone.

There are several cases of positive control of initiation where the environmental signal is sensed by activator proteins through a covalent modification, which converts it into a form capable of activating transcription. The typical mechanism of conversion of an inactive regulatory protein into its active form involves autophosphorylation of a protein kinase (the **sensor**), which subsequently transfers the phosphate to the **response regulator.** The phosphorylated

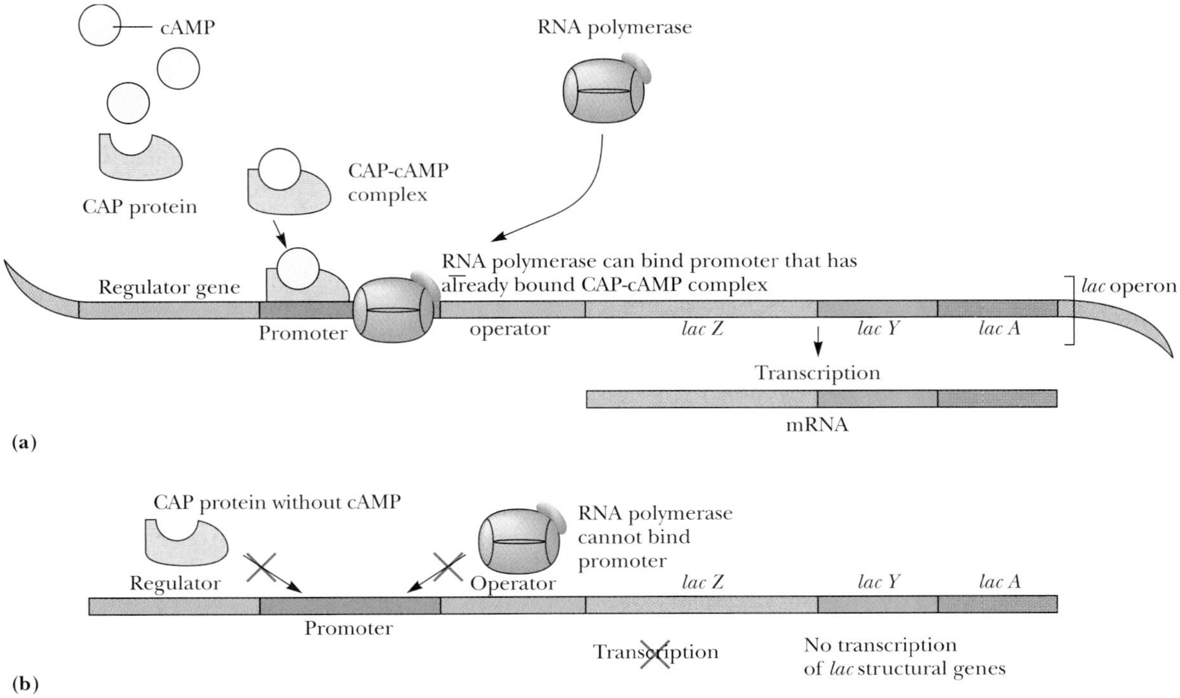

***Figure* 13.11** Catabolite repression occurs when products of catabolism regulate the rate at which structural genes are transcribed. Transcription occurs when the catabolite activator protein (CAP) complexes with cyclic AMP (cAMP) at a specific region of the promoter site **(a).** In the absence of cAMP **(b),** the CAP protein cannot bind to the gene and RNA polymerase does not bind to the promoter.

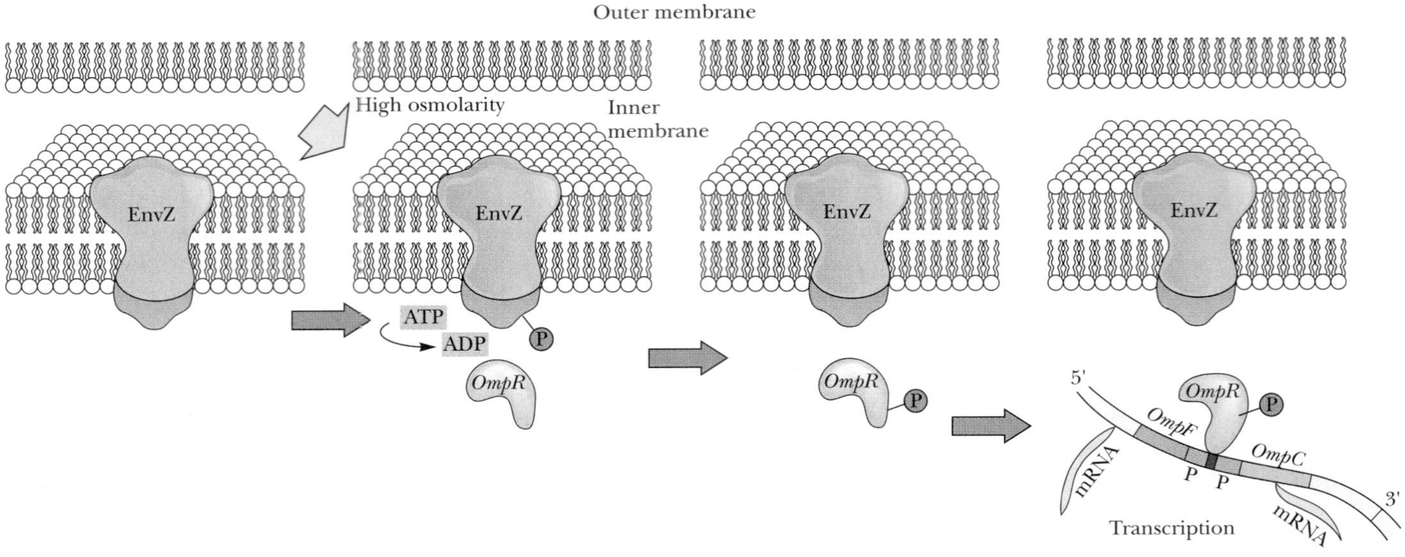

***Figure* 13.12** Osmoregulation of porin gene expression in *E. coli* controlled by a two-component regulatory system. The environmental signal, high osmolarity, is sensed by the inner membrane protein EnvZ, which responds to the signal by undergoing autophosphorylation. Phosphorylated EnvZ (EnvZ-P) transfers the phosphate to cytoplasmic protein OmpR. It is the phosphorylated form of OmpR, which functions as a transcriptional regulator, binding to promoters of porin genes *omp*F and *omp*C, controlling their expression.

regulatory molecule then binds to a specific regulatory sequence which is located on the 5′ (upstream) of the gene and facilitates transcription of a promoter from an adjacent site. This mechanism of sensing the environmental signal and converting it to transcriptional activation of specific genes always involves two components, the protein autokinase and the regulatory protein, and is called the **two-component regulatory system.** The mechanism of sensing high osmolarity by the EnvZ (sensor) OmpR (response regulator) of *E. coli* is outlined in Figure 13.12. Similar phosphorylation mechanisms have also been shown to function in controlling selective expression of genes that respond to nitrogen limitation in *Enterobacteriaceae,* nitrogen fixation in *Rhizobium,* signaling for transcription of the apparatus necessary for DNA transfer from *Agrobacterium* to plants, and biosynthesis of pili and flagella in an assortment of bacteria.

Protein Synthesis

Approximately 2000 to 3000 different polypeptides (the biochemical term for proteins) are present in a bacterial cell; the vast majority of these are the cellular structural components or enzymes. A minority are signal molecules. Chemically, proteins are relatively simple molecules, consisting of linearly linked amino acids. The order of the assembled amino acids—the total number and nature of the amino acids—distinguishes one protein from another. There are 20 different amino acids that can be used as building blocks. Amino acids are similar to one another, each consisting of a central carbon (called the α—carbon) with an amino group, a carboxyl group, a hydrogen, and a side group. Of the 20 amino acids, 19 differ only in the structure of the side groups. The twentieth amino acid, proline, differs from the others by having its side group linked to the α—carbon, resulting in a ring structure. The structures of the 20 amino acids present in proteins were presented in Figure 3.21.

Each amino acid is linked to its adjacent amino acid by a **peptide bond** (see Figure 3.22), formed when the primary amino group of one amino acid condenses with the carboxyl group of the adjacent one. Short chains of bonded amino acids are therefore called **peptides.** Because of the directional bond, one end of the peptide is the amino terminus and the other is the carboxy terminus. The sequence of linked amino acids makes up the **primary** structure of a polypeptide. Proteins, however, are rarely found in these extended configurations, but rather assume additional conformations. The **secondary structure** of a polypeptide refers to the shape of the backbone of each α—carbon and the positioning of the side groups; that is, does the backbone have a spiral shape with the side chains pointing outward, or are there regions that are relatively straight with intermittent turns? Furthermore, the polypeptide chain can fold, and this **tertiary** conformation can bring together amino acids that are distant from one another in the linear polypeptide. Moreover, independent polypeptide chains can aggregate to bring together amino acids that are present on separate polypeptide chains (see Figures 11.2, 11.3, and 11.4).

Different polypeptides consist of differing amounts of the 20 amino acids. The differences among proteins are due to the characteristics (polar, nonpolar, positively or negatively charged) and order of the amino acids in their primary structure. The distribution of the specific amino acids along the polypeptide chain is determined by the specific order of the nucleotides in the gene coding that polypeptide. The genetic information is first **transcribed** from DNA into linear RNA, called **messenger RNA** (mRNA), as it serves as the messenger of genetic information between the genes and the protein-synthesizing apparatus. Distinct triplet nucleotide sequences in mRNA, called **codons,** encode specific amino acids. The order of the codons determines the order of the amino acids in the polypeptide chain. Hence, the process of protein synthesis serves two main functions simultaneously: (1) it chemically catalyzes the linking of amino acids into a protein, and (2) it decodes the triplet codons on mRNA and converts this information to the primary structure of a protein. The process of protein synthesis is therefore called **translation.** One of the major breakthroughs in biological science took place in the early 1960s, when the genetic code was determined and the specific assignment of triplet nucleotides to each amino acid was made. There are 61 codons specifying all 20 amino acids that are components of proteins. Codon usage is relatively redundant. For example, any one of six different triplets (UUA, UUG, CUU, CUC, CUA, or CUG) in mRNA can specify leucine in the polypeptide chain. However, two amino acids have only one codon each: methionine (AUG), and tryptophan (UGG) (see Table 13.3).

In addition to codons specifying amino acids, several signaling codons determine the beginning and the end of polypeptide chains. **Initiation codons** signal the start (the amino terminus) of the polypeptide chain by coding for a methionine (AUG) that has a formyl group on the N terminus. This methionine is generally removed after the protein is synthesized. The end of the carboxyl terminus of the polypeptide is determined by **termination codons,** also called **stop codons.** Stop codons (UAA, UAG, or UGA) do not specify any amino acids; one or several of these on the mRNA will stop the RNA polymerase from continuing. The sequence of codons between the initiation and termination signals is the **reading frame.** Any one nucleotide sequence of mRNA contains three possible reading frames depending on which of three positions is used as the beginning point, as illustrated below. The one translated is determined by the location of the initiation codon upstream of this sequence:

***Table* 13.3 The genetic code: the codons (three bases) specify the amino acids incorporated into protein**

First Position (5′-end)	Second Position				Third Position (3′-end)
	U	C	A	G	
U	UUU Phe	UCU Ser	UAU Tyr	UGU Cys	U
	UUC Phe	UCC Ser	UAC Tyr	UGC Cys	C
	UUA Leu	UCA Ser	UAA Stop	UGA Stop	A
	UUG Leu	UCG Ser	UAG Stop	UGG Trp	G
C	CUU Leu	CCU Pro	CAU His	CGU Arg	U
	CUC Leu	CCC Pro	CAC His	CGC Arg	C
	CUA Leu	CCA Pro	CAA Gln	CGA Arg	A
	CUG Leu	CCG Pro	CAG Gln	CGG Arg	G
A	AUU Ile	ACU Thr	AAU Asn	AGU Ser	U
	AUC Ile	ACC Thr	AAC Asn	AGC Ser	C
	AUA Ile	ACA Thr	AAA Lys	AGA Arg	A
	AUG Met[1]	ACG Thr	AAG Lys	AGG Arg	G
G	GUU Val	GCU Ala	GAU Asp	GGU Gly	U
	GUC Val	GCC Ala	GAC Asp	GGC Gly	C
	GUA Val	GCA Ala	GAA Glu	GGA Gly	A
	GUG Val	GCG Ala	GAG Glu	GGG Gly	G

[1]AUG signals translation initiation as well as coding for Met residues.

From: Garrett and Grisham, *Biochemistry,* Figure 31.3, page 1025.

mRNA	. . . GUC	UUU	AUA	ACA	CAC	CCU	
Reading frame #1	Ala	Phe	Ile	Thr	His	Pro	
mRNA	. . . G	UCU	UUA	UAA	CAC	ACC	CU
Reading frame #2		Ser	Leu	Stop			
mRNA	. . . CU	CUU	UAU	AAC	ACA	CCC	U
Reading frame #3		Leu	Tyr	Asn	Thr	Pro	

If the desired protein contains the sequence Ala—Phe—Ile—Thr—His—Pro, the initiation codon must be located **upstream,** in multiples of three and in reading frame #1. If the initiation codon were in frame with either the second or third reading frames, proteins with totally different primary sequences would result. Note that reading frame #2 contains a stop codon that would result in a short polypeptide chain having Leu as its carboxyl terminus.

Components of the Protein Synthesis Machinery

Protein synthesis takes place on complex particles called **ribosomes.** Ribosomes, together with a number of RNA molecules and polypeptides, recognize the coded information on mRNA and translate it into a polypeptide. Ribosomes in bacterial cells are relatively large, multisubunit complexes of RNA and proteins. In its functional stage, an individual ribosome consists of two subunits, one large and one small. As mentioned in Chapter 4, the physical properties of ribosomes were originally studied by their behavior during high-speed centrifugation. Hence, the size of a ribosome and its subunits are in Svedberg units (S), the units of velocity of travel in a centrifugal field. The complete ribosome has a size of 70S, while the small subunit is 30S and the large one is 50S. A comparison of the size and the RNA/protein composition of individual subunits of the *E. coli* ribosome is presented in Table 13.4. Operons encoding ribosomal RNA genes are present in more than one copy in the bacterial chromosome. This redundancy of genes is rare in prokaryotes, and it enables bacteria to respond to sudden changes of growth conditions by synthesizing a large quantity of translational machinery.

The critical components of the translational apparatus are small RNAs that carry specific amino acids and are responsible for the conversion of genetic information into protein sequences. These **transfer RNAs** (tRNA) are the "translators" of the translation process. The tRNA molecule is about 70 nucleotides long and is folded into the configuration shown in Figure 13.13. Four extensive regions of complementarity are formed, and three open loops are present. One of these is the **anticodon loop,** containing a region precisely complementary to the codon sequence on mRNA. For example, the tRNA that is involved in adding tryptophan to the polypeptide chain contains an anticodon sequence ACC, which is complementary to the codon UGG in the mRNA. There are fewer tRNAs than there are possible codons, because incorrect base pairing in the third position of the codon can still result in incorporation of the correct amino acid into the polypeptide. This is referred to as third base wobble. The tRNA genes, as with genes for rRNA, are redundant, and tRNA is often cut out from larger precursor RNA molecules.

The attachment of the amino acid to tRNA is an enzymatic step involving recognition of a tRNA molecule by

***Table* 13.4 The structural components of *E. Coli* ribosomes**

	Ribosome	**Small Subunit**	**Large Subunit**
Sedimentation coefficient	70S	30S	50S
Mass (kD)	2520	930	1590
Major RNAs		16S = 1542 bases	23S = 2904 bases
Minor RNAs			5S = 120 bases
RNA mass (kD)	1664	560	1104
RNA proportion	66%	60%	70%
Protein number		21 polypeptides	31 polypeptides
Protein mass (kD)	857	370	487
Protein proportion	34%	40%	30%

From: Garrett and Grisham, *Biochemistry,* Table 32.1, page 1041.

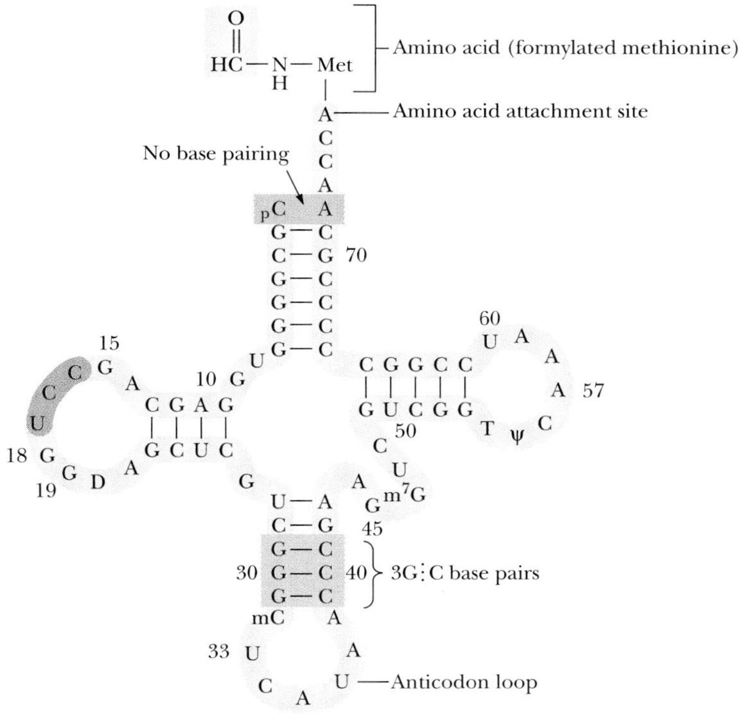

***Figure* 13.13** The structure of *E. coli* N-formyl-methionyl-tRNA^{fMet}. The base areas that differ from those in noninitiator tRNAs are highlighted. This is the initiator of peptide synthesis from mRNA.

the enzyme **aminoacyl—tRNA synthetase** and attachment of the amino acid (aa) to the end of the tRNA:

$$\text{aa} + \text{tRNA} + \text{ATP} \longrightarrow \text{tRNAaa} + \text{ADP}$$

A tRNA attached to its specific amino acid is said to be **charged.** The enzyme aminoacyl-tRNA synthetase has a specific recognition for both the tRNA and the amino acid corresponding to the codon. For example, the amino acid asparagine is added only to the tRNA containing the sequence UUA or UUG in the anticodon loop. Errors (placing the wrong amino acid onto a tRNA) are exceedingly rare. They would result in an incorrect translation of the genetic code, thereby synthesizing a polypeptide with different amino acids than those specified by the genetic information.

Mechanism of Protein Synthesis

The process of assembling a polypeptide chain is an orderly reaction consisting of the stepwise transfer of amino acids directed by an mRNA sequence. The overall scheme of protein synthesis is outlined in Figure 13.14. The process can be divided into three steps: initiation, elongation, and termination. Each of these steps is discussed here separately.

Initiation

The initiation of protein synthesis involves the interaction of ribosomes with the site on mRNA that encodes the first amino acid (the amino terminus) of the polypeptide chain. The precise placement of the N-terminal amino acid is important since the reading frame of the remaining polypeptide is determined from this point.

A series of soluble factors and the free 50S and 30S subunits are joined with mRNA and the aminoacyl-tRNA to form what is called the **initiation complex.** Initiation of protein synthesis involves binding the small ribosomal subunit to mRNA with the aid of several polypeptides called **initiation factors (IFs).** There are three IFs in the bacterial cytoplasm: IF1, IF2, and IF3. IF3 is required to stabilize the interaction of the 30S subunit with mRNA binding. IF1 is also part of the 30S—mRNA complex, and assists IF3 although its precise role in the initiation of protein synthesis is not known. IF2 functions during binding of the initiator tRNA and GTP to the ribosome.

The association of the 30S subunit with a precise site on mRNA that encodes the first amino acid of the polypeptide is accomplished by pairing a specific sequence, GGAGGU, on the mRNA to a complementary sequence, ACCUCC, on the 3′ end of the 16S ribosomal RNA. This sequence on mRNA, the so-called **ribosome binding site** (or the Shine-Delgarno sequence, named after its discoverers), is approximately 4 to 6 nucleotides from the

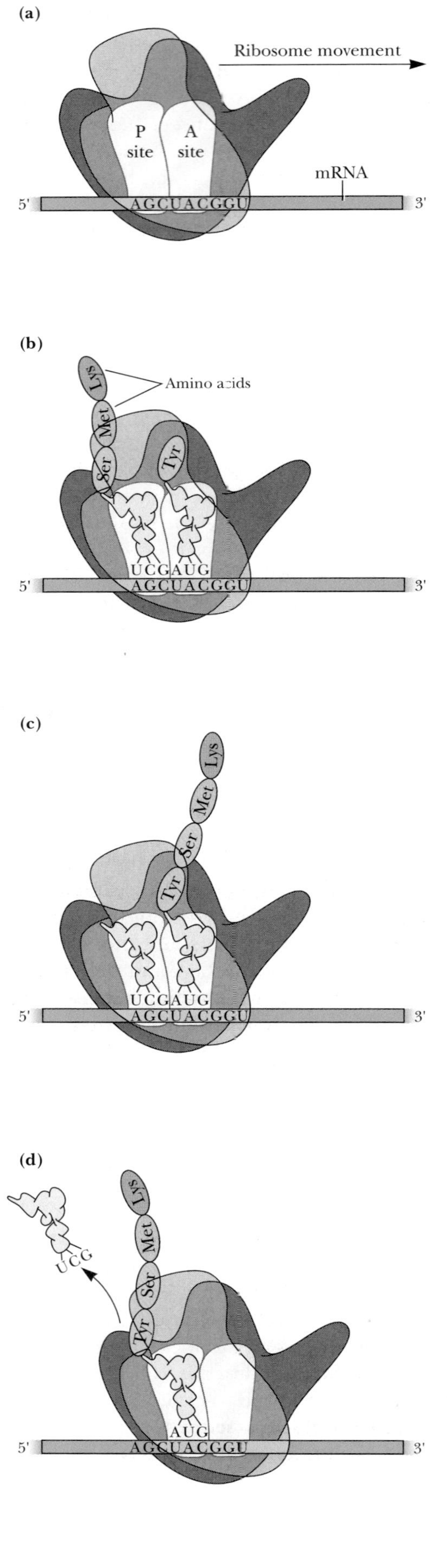

initiation codon, AUG. Thus, the interaction of the small subunits not only allows the ribosomes to find the beginning of the protein (even though there may be other AUG codons specifying methionine within a given polypeptide chain) but also determines the reading frame of the polypeptide.

Following binding of the 30S ribosomal subunit to mRNA, the first tRNA, carrying methionine, binds to this complex. In *E. coli,* the first amino acid is almost always N-formylated methionine (fMet), which is generated by first linking methionine to a specialized tRNA by amino-acyl—tRNA synthetase, and subsequently attaching a formyl group to the free amino group. The interaction of fMet—tRNA with ribosome involves interaction with initiation factor 2 (IF2) and formation of a hydrogen-paired region between the anticodon sequence UAC and the complementary codon AUG. Occasionally, initiation involves a mismatch, where the anticodon UAC pairs with GUG codon on mRNA. At the same time, a molecule of GTP binds to a specific site on the 30S subunit. The final stage in the assembly of the initiation complex involves association of the 50S subunit, hydrolysis of GTP to GDP, and release of all of the initiation factors. The ribosome is now ready to move along the mRNA and progressively synthesize a polypeptide chain.

Elongation

Following completion of the assembly of the initiation complex, the ribosome contains two sites that can bind aminoacyl-tRNA. The peptidyl site (or P-site) contains the fMet—tRNA, while the adjacent acceptor site (or A-site) is unoccupied, although it does contain the exposed triplet codon of the second amino acid in the reading frame. The formation of the first peptide bond between fMet and the second amino acid of the polypeptide chain takes place following the interaction of the amino-

***Figure* 13.14** Steps in protein synthesis. **(a)** Ribosomes have two binding sites for tRNA: the acceptor (A) site and the peptidyl (P) site. Individual codons of mRNA are also positioned within the A and P sites. This arrangement allows for precise interaction of the anticodon loop of tRNA with their cognate codons within the ribosome. **(b)** During the continuous synthesis of the polypeptide chain, the P-site is occupied by the tRNA with its nascent polypeptide chain, while the A-site contains the amino-acyl tRNA (in the example shown, Tyr-tRNA) which is in the process of being incorporated into the growing polypeptide, according to the code specified by the triplet codon of mRNA exposed in the A-site (UAC in the example). **(c)** The nascent chain is then transferred to the amino acid on the tRNA in the A-site, resulting in the extension of the polypeptide chain by one amino acid. In the example shown, Tyrosine (attached to its specific tRNA), is added to the nascent chain. **(d)** The ribosome moves one codon over, with the concomitant release of the empty tRNA. The ribosome now contains the nascent chain in its P-site leaving the next codon exposed in the A-site, which now is available for the next amino-acyl tRNA, as specified by the codon (GGU in the example shown, which will bind Gly tRNA).

acyl-tRNA charged with the second amino acid and the A-site. This is directed by the correct hydrogen-bonded, double-stranded region formed between the codon and anticodon sequence of tRNA. The binding of aminoacyl-tRNA to the A-site is a stepwise process, involving several **elongation factors (EFs)** and GTP hydrolysis. The aminoacyl-tRNA first binds to a complex of elongation factor Tu (EF-Tu) and GTP. The EF-Tu places the aminoacyl-tRNA into the A-site, with the concomitant hydrolysis of GTP to GDP. The EF-Tu—GDP complex is released. The regeneration of fresh EF-Tu—GTP complex is necessary for the next step of the elongation process; another elongation factor, EF-Ts, releases GDP from EF-Tu and reloads it with fresh GTP.

The formation of the actual peptide bond between the two amino acids within the ribosomes is accomplished by the **peptidyl transferase** activity of the ribosome. The bond formed between the amino acids results in transfer of fMet to the second aminoacyl-tRNA occupying the A-site. The final step in elongation is called **translocation,** in which the ribosome moves three nucleotides, with a concomitant release of the free tRNA from the P-site. The translocation is catalyzed by another elongation factor (EF-G) and requires hydrolysis of GTP. The translocation leaves the A-site open for the next incoming aminoacyl-tRNA, specified by the third codon of the mRNA. The subsequent elongation steps proceed in a manner entirely identical to the reaction between fMet and the second amino acid. In this process, all ribosomal elongation factors are reused and the elongation of the polypeptide requires energy in the form of GTP, one to charge EF-Tu and one to fuel the EF-G–mediated translocation. Together with the requirement of ATP to charge a tRNA, the elongation stage of polypeptide chain formation is an energy-demanding process. Various initiation and elongation steps in bacterial protein synthesis are targets of different antibiotics **(Box 13.1).**

Termination

The termination of the polypeptide chain is directed by the location of one of the termination codons immediately following the codon for the carboxy terminal amino acid. The termination codons do not code for any amino acid, so when they occupy the A-site, no tRNA has the corresponding anticodon in its loop structure. Termination

BOX 13.1 MILESTONES

Antibiotics as Inhibitors of Bacterial Protein Synthesis

The ability to synthesize proteins is essential for bacterial survival and growth in all environments. Interference with this process leads to bacterial death. A number of antibiotics, useful in antibacterial therapy for infectious disease, are directed against ribosomes and result in the inability to divide or to synthesize proteins. Because bacterial ribosomes differ substantially from eukaryotic ribosomes, these antibiotics show significant specificity, and can thus be used against infectious diseases, selectively inhibiting prokaryotic pathogens without causing any damage to protein synthesis in the human cells. A number of the commonly used antibiotics and their mechanisms of action are listed:

Antibiotic	**Mode of Action**
Streptomycin	Binds to 30S subunit; causes misreading and irreversible inhibition of protein synthesis
Other aminoglycosides, e.g., gentamicin	Similar to streptomycin although has different targets on 30S and 50S subunits
Tetracyclines	Reversibly bind to 30S subunit; inhibit binding of aminoacyl tRNA
Chloramphenicol	Reversibly binds to 50S subunit; inhibits peptidyl transferase and peptide bond formation
Erythromycin and lincomycin	Bind to 50S subunit; inhibit peptidyl transferase and translocation reactions

codons are recognized by **release factors (RFs)** that bind to the A-site. The termination codons UAA and UAG are recognized by RF1, while RF2 recognizes UAA or UGA. Release factors bound to the A-site stimulate hydrolysis of the attached chain from the last tRNA in the P-site, releasing the chain. The polypeptide is now free to fold into its native tertiary structure.

Mutations

A change in the base pair composition of the DNA within a gene is an event called a **mutation.** One consequence of having a haploid chromosome is that change in a specific gene is immediately expressed by the daughter cells since the bacterial genome lacks the second copy of each gene which, in diploid organisms, often masks the effect of mutation in one of the chromosomes. This section discusses the key experimental approach (the fluctuation test) used to demonstrate the random origin of mutations in the bacterial genome, the types of mutations that occur in bacteria, the predicted consequences of mutations, and the effects of physical or chemical mutagens on the frequency of mutations in bacteria.

Fluctuation Test

When a small amount of streptomycin (an antibiotic made by *Streptomyces*) is added to a culture of bacteria, almost all bacteria are killed within a few minutes. Similarly, bacteria can be killed by exposure to bacterial viruses (also called **bacteriophages**) that adsorb to their cell walls, inject their genomes, and, following a cycle of replication, release new viruses, with the concomitant death of the host bacterium. A small fraction, however—fewer than one in a million—can survive the exposure to potential killing agents. When these bacteria are propagated, they and their progeny are completely resistant to killing by the same agents that devastated the members of the previous sensitive population. The resistant population of bacteria owe their survival to mutations. Changes in a gene that encodes a ribosomal protein results in the ability of the streptomycin-resistant bacterium to synthesize protein even in the presence of streptomycin. Similarly, resistant bacteria that are isolated from a population of cells killed by bacteriophages have altered genes for synthesis of cell wall components that do not allow bacterial viruses to attach or enter the cell.

Since these mutant bacteria appear to arise only following exposure to the appropriate agent, it had been debated for many years whether mutations are adaptive changes, wherein a specific agent would induce the required alteration in the genetic material. An alternative hypothesis, supported by the Darwinian theory of evolution, proposed that changes occur spontaneously and continuously in a population, and it is the selective process that allows us merely to identify such bacteria that carry a specific mutation.

The controversy was resolved in the early 1940s when M. Delbruck and S. Luria devised the **fluctuation test,** which demonstrated that mutations are spontaneous and independent of a selective environment. The original test was based on the ability of *E. coli* to mutate into a form that is resistant to the killing action of bacteriophage T1. When exposed to T1, a population of 10^8 bacteria gives rise to about 10 to 50 mutants that are T1 resistant, that is, they are unaffected by the presence of this phage.

The fluctuation test is outlined in Figure 13.15. A culture of *E. coli* propagated in a flask was divided into two tubes, each containing about 10^3 bacteria. The contents of one of the tubes were further subdivided into 50 smaller tubes. All of the cultures were grown to stationary phase. From the single large test tube, as well as from the 50 individual tubes, small aliquots were spread over an agar surface, densely seeded with T1 bacteriophage that will kill susceptible bacteria, that are not resistant to T1 by virtue of a mutation. These resistant bacterial mutants can give rise to colonies even in the presence of large numbers of bacteriophage. After overnight incubation, the number of phage-resistant colonies was determined. The culture that was incubated in a single tube yielded about three to seven resistant colonies. The range of phage-resistant colonies obtained from the individually grown tubes was from none to 100 or more.

What was the reason for this variation in the number of resistant colonies? If mutations arise as a result of contact with the phage on the plate, aliquots from both large and small cultures should give rise to equivalent number of resistant colonies. If phage-resistant mutants are arising at all times, the individual cultures should have a great variation in number, depending on the time the mutation takes place during growth. That is, if mutation takes place immediately after subdivision, as many as 100 resistant bacteria will be identified in that tube following plating on a bacteriophage-containing agar plate. If mutation takes place long after subdivision or not at all, small numbers of phage-resistant bacteria will be identified. Clearly, results favor the second hypothesis because of the great fluctuation in the number of resistant bacteria that arose from the small inocula.

Types of Mutations

A mutation involving a single base pair substitution, in which a purine (G or A) is replaced by another purine, or a pyrimidine (T or C) is replaced by another pyrimidine, is called a **transition.** A **transversion** is a type of mutation wherein a purine is replaced by a pyrimidine, or a pyrim-

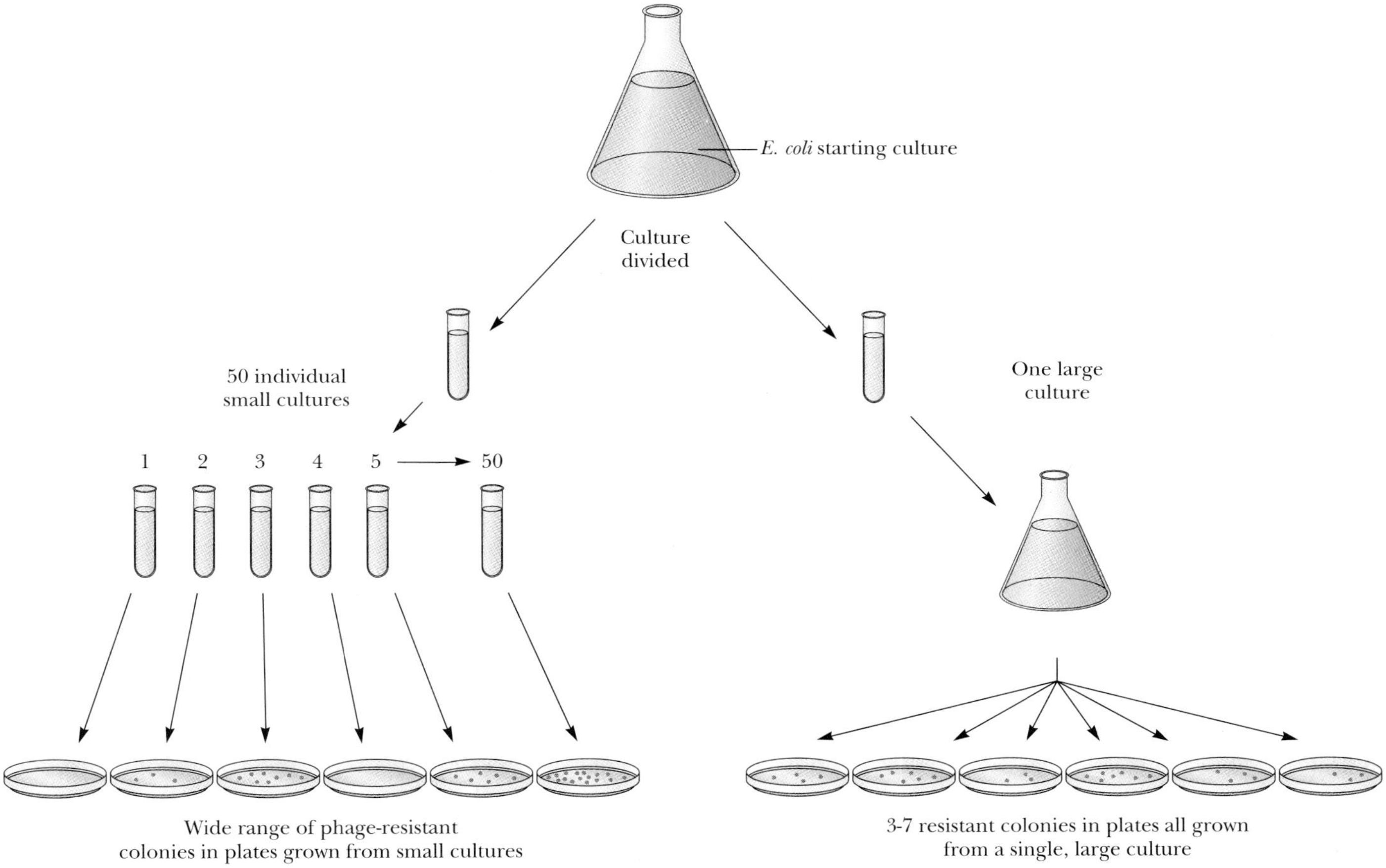

***Figure* 13.15** The fluctuation test that presented conclusive proof that mutations occurred in bacteria.

idine is replaced by a purine (Figure 10.23). These substitutions become part of a newly synthesized chain as a consequence of incorrect base pairing during the polymerization of the chain; for example, a G pairs with a T instead of a C. Deletions and insertions arise when the template or the newly synthesized strand loops out and the DNA polymerase makes a slip, resulting in duplication or elimination of the looped-out region in the newly synthesized strand. For any form of mutation to take effect, the change also has to escape the repair activity of DNA polymerase, namely, the ability to back up in the 3′—to—5′ direction, remove the incorrect base, and resynthesize the correct strand. Since base-pairing errors, loop-outs, and failure of the repair system are all very rare, mutations resulting from errors of this type arise infrequently.

Consequence of Mutations

Changes in the DNA sequence can have a number of effects in bacterial cells that inherit the mutant version of a specific gene. As an example, examine the effect of substitutions as well as deletions or insertions at the highlighted codon on the following polypeptide coding sequence: (In this example, DNA strand 1 is the template used by RNA polymerase to generate the mRNA for this protein.)

DNA strand 1	TAC	TTT	GCT	**AAT**	ACG	CTC	TAG	TTG	ACC	ATT
DNA strand 2	ATG	AAA	CGA	**TTA**	TGC	GAG	ATC	AAC	TGG	TAA
mRNA	AUG	AAA	CGA	**UUA**	UGC	GAG	AUC	AAC	UGG	UAA
Protein	Met	Lys	Arg	**Leu**	Cys	Glu	Ile	Asn	Trp	Stop

Now, consider the following consequence of several changes in the DNA encoding this hypothetical nine-amino-acid peptide, all involving substitutions that change the mRNA from specifying the codon UUA, which codes for the amino acid, leucine.

1. A transversion of the A/T base for G/C results in the codon UGA in the mRNA. This is a termination codon, resulting in synthesis of a shortened three-amino-acid peptide.
2. A transition of the T/A base pair for a G/C pair changes the codon on mRNA into UUG. This mutation has no effect on the polypeptide since UUG, like UUA, encodes leucine.
3. Another transversion of A/T to T/A alters the codon into one specifying isoleucine (AUA).

In addition to substitutions, insertions or deletions can take place. For example, insertion of one or two base pairs results in the alteration of the order of the triplet codons relative to the initiation codon. A new reading frame will originate from the site of insertion, encoding a completely different polypeptide. These so-called **frameshift mutations** often have the same result as a chain termination mutation at the same site since the new polypeptide, however long it is, usually does not specify a functional protein. Frameshift mutations terminate when ribosomes encounter a termination codon in the new reading frame. Deletion of one or two nucleotides also results in frameshift mutations.

A frameshift mutation due to insertion of a T paired with A would look like this:

DNA strand 1	TAC	TTT	GCT	**T**AA	TAC	GCT	CTA	GTT	GAC	CAT	T
DNA strand 2	ATG	AAA	CGA	**A**TT	ATG	CGA	GAT	CAA	CTG	GTA	A
mRNA	AUG	AAA	CGA	AUU	AUG	CGA	GAU	CAA	CUG	GUA	A
Protein	Met	Lys	Arg	Ile	Met	Arg	Asp	Gln	Leu	Val	

Note that as a result of the insertion, the new reading frame continues beyond the stop codon of the **wild type** (that is, original) mRNA until a new termination codon is reached. The polypeptide sequence after the site of mutation is completely different from that of the original peptide. Insertions or deletions of three nucleotides (or multiples of three nucleotides) yield true insertions or deletions of amino acids, without altering the reading frame around the site of the mutation.

Changes in a specific base sequence can have variable effects on the polypeptide that is encoded by a specific gene. Alteration of the length of the protein by a mutation to a termination codon can abolish the function of that protein. Mutations that result in substitution of one amino acid for another may have only minor effects, especially if the amino acids are similar in size, such as the example in which leucine is substituted by isoleucine. However, if the substitution is in an important part of the molecule, it can also alter the functional area of the polypeptide. Mutations that have a more dramatic effect on a protein's function or stability are those that result in substitution of amino acids having opposite charges or very different side chains. Occasionally, the change will cause altered physical properties in the polypeptide. Certain substitutions render enzymes **temperature sensitive,** such that they can function only at lower temperatures. This is presumably due to incorrect folding of the polypeptide chain, making it more susceptible to thermal denaturation. The effect of the mutation would therefore be noticeable only under **nonpermissive** conditions, that is, when the protein is denatured. Under permissive conditions, however, less than perfect folding may still result in a protein with only slightly altered activity.

Since mutations often result in the alteration of cellular metabolism, such that the cell does not function optimally, there is sufficient selective pressure for secondary compensatory mutations that repair the original defect. Such restoration by a counteractive mutation is called **suppression.** Suppression of the mutation can occur by several mechanisms, the simplest of which is exact reversal of the mutation. However, other mechanisms of suppression involve genetic changes at adjacent or distal sites on the DNA. Suppression of a base substitution mutation can be accomplished by a mutation in a different amino acid, which would nullify the disruptive effect of the first mutation in the polypeptide chain.

A specialized category of suppression deals with **nonsense mutations.** Nonsense mutations occur when a base substitution changes a codon for an amino acid into one of the chain-termination codons. **Nonsense suppressors** are mutations in genes encoding the tRNA anticodon loop corresponding to an mRNA stop codon, such that the tRNA adds an amino acid to the growing peptide rather than terminating it. Figure 13.16 presents an example of a chain-termination mutation suppressed by a mutant suppressor tRNA. Since cells still need the correct tRNA genes for normal protein synthesis, only those tRNAs that are present in more than one copy per cell can mutate to nonsense codon suppressors without killing the cell. Moreover, suppressor tRNA can interfere with normal termination of protein synthesis by a single termination codon, resulting in added carboxyterminal tails extending through the adjacent gene to the next chain termination codon. This happens relatively infrequently, since most coding sequences of polypeptides terminate with more than one in-frame termination codon.

Often, a cell function is accomplished by two proteins associating with each other, and these complexes are formed by protein-protein interactions involving distinct regions of each polypeptide called **interactive domains.** A mutation in a gene encoding the interactive domain of one such protein can be suppressed by a compensatory mutation in the second protein's interactive domain. Some mutations are more difficult to suppress, such as frameshift mutations in which the reading frame has to be corrected near the site of the insertion or deletion by insertion or deletion of one or two base pairs to restore the original triplet reading frame.

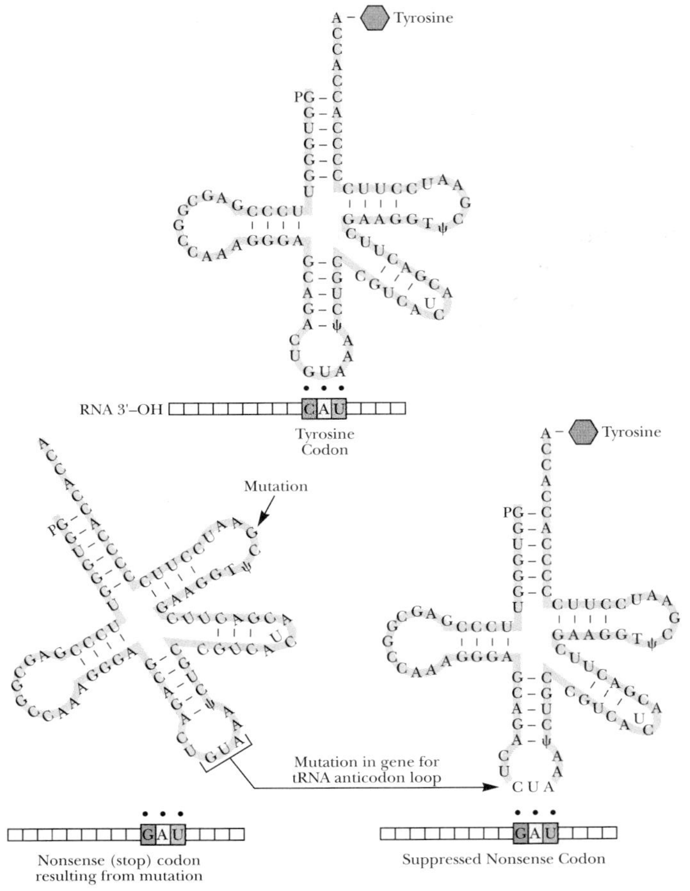

***Figure* 13.16** Suppressor mutation. In this example, codon CAU, recognized by anticodon GUA of tyrosine tRNA, mutated to GAU, a stop codon. A subsequent mutation in the gene for synthesis of tyrosine tRNA changed the anticodon sequence GAU to CUA, which now recognizes the codon and incorporates tyrosine.

Mutagens

The rate of spontaneous change in DNA is relatively low, owing to the effective repair mechanisms present in bacterial cells. The rate of mutation can be increased by exposing cells to several physical or chemical agents. Physical mutagenesis almost exclusively involves radiation damage to DNA. A variety of chemicals can also cause mutations by interacting with DNA directly or by interference with the DNA replication and repair processes. Chemical compounds affecting mutational rates are termed **mutagens,** and fall into three broad categories: structural analogs of nucleotides, chemicals that alter purine or pyrimidine bases, and agents that intercalate the DNA at the replication fork. Mutagens also affect eukaryotic DNA, including that of humans, where these chemicals are suspected to be responsible for certain types of cancer. The ability of carcinogens to be mutagenic is the basis of a simple test for environmental carcinogens **(Box 13.2).**

Mutagenesis with Base Analogues

5—Bromouracil (BU) is an example of a base analog because of its similarity to thymine. During replication, it can substitute for thymine resulting in a BU::A pair. (The double colon indicates a hydrogen bond.) However, thymine can occasionally pair with guanine in a relatively unstable transient bond. However, the BU::G bond is sufficiently stable, and during subsequent replication, the stabilized G pairs with C, resulting in a permanent transition of an AT pair for GC (Figure 13.17). Treatment with 5—bromouracil can also result in the opposite transition, in which an existing GC pair is replaced by an AT pair in daughter cells.

Mutagenesis with Agents That Modify DNA

Agents that modify DNA include a large number of compounds with nucleotide bases that readily incorporate into DNA.

1. *Nitrous acid* deaminates cytosine and adenine, converting them to uracil and hypoxanthine, respectively. During the subsequent rounds of replication, uracil pairs with adenine while hypoxanthine pairs with cytosine. The net effect of the reaction of nitrous acid with DNA is the transition of AT to GC and GC to AT.

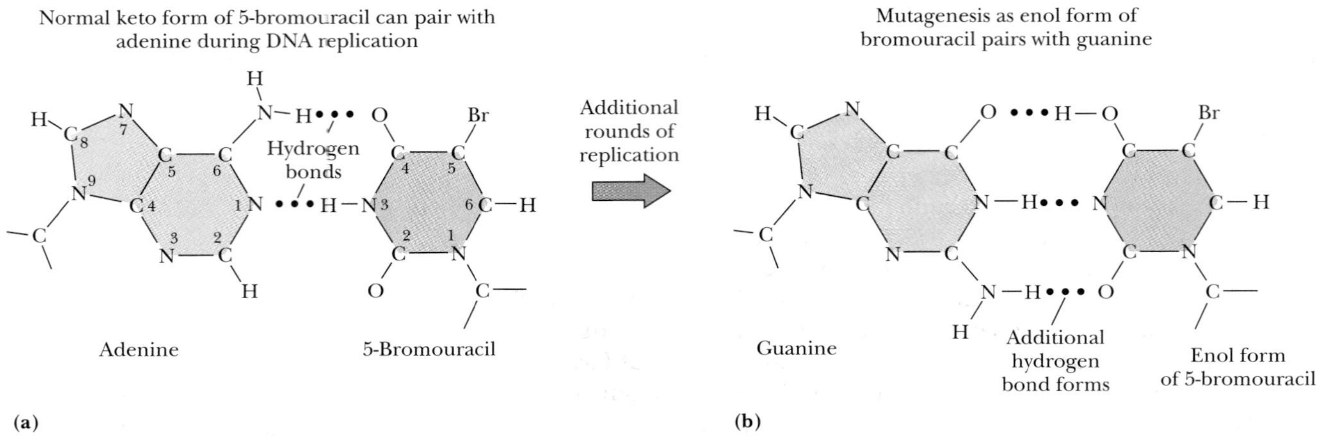

***Figure* 13.17** Mutagenic action of bromouracil. Bromouracil alternates between two chemical forms—enol and keto. **(a)** Bromouracil in the keto form pairs with adinine during replication. **(b)** During subsequent rounds of replication, the bromouracil may switch to the enol form and pair with guanine. This leads to an AT pair being replaced by G:C.

2. *Hydroxylamine and nitrous acid* specifically modify cytosine so it can pair with adenine, resulting in the transition of a GC pair to an AT.
3. *Ethylmethane sulfonate* is an alkylating agent that modifies the oxygen at position 6 of guanine by adding an alkyl group (a methyl group), converting it to O^6-methylguanine. This modified base can now weakly pair with thymine during subsequent replication. This modification leads to a GC to AT transition in one of the daughter cells. This mispairing can also lead to an AT to GC transition.
4. *Nitrosoguanidine* also alkylates guanines, resulting in a GC to AT transition. This often leads to small deletions.

Mutagens That Intercalate with DNA

Several complex organic molecules, such as acridine orange, proflavin, and ethidium bromide, interact with DNA and result in distortion of DNA. Distortion leads to DNA polymerase slippage, resulting in small insertions or deletions. Frameshift mutations are the most common types of lesions in DNA that occur as a result of interaction with intercalating compounds. The structures of all of these compounds are different, yet they all have conjugated rings of a size similar to a base pair in a DNA duplex, which they most likely insert between hydrogen-bonded adjacent bases in the duplex, resulting in the bending of DNA.

Radiation Damage

Ultraviolet (UV) radiation can cause DNA damage by covalently cross-linking adjacent pyrimidines. The most common form of cross-link is between two adjacent thymines, forming a **thymine dimer** (see Figure 7.5), although T-C and C-C dimers and 6,4 photoproducts are also often seen following UV treatment of DNA. Errors in repair may cause deletions, but pyrimidine dimers can also lead to transition and transversion of bases during subsequent replication. Ionizing radiation causes hydrolysis of phosphodiester linkages, forming the phosphate backbone of DNA; subsequent errors in repair can lead to deletions and insertions.

DNA Repair

Bacteria possess a number of DNA repair systems that can correct damaged DNA. The mutations described in the previous section thus become part of the genetic repertoire of the bacterium only if they are not repaired. Direct reversal of DNA damage is accomplished by enzymes present in many eubacterial and archaeal species. Methylated guanines, such as those caused by exposure to ethyl methyl sulfonate or other alkylating mutagens, are repaired by **methylguanine transferase,** an enzyme that recognizes the modified guanine in the DNA duplex, and transfers the methyl group onto the enzyme itself, thus restoring guanine. Adjacent pyrimidines that become cross-linked as a result of the mutagenic action of ultraviolet light are repaired by a process called **photoreactivation.** The resulting lesion is recognized by the enzyme photolyase, which, upon absorption of visible light, breaks the bond between the dimerized pyrimidines, thus restoring their original structure.

BOX 13.2 METHODS & TECHNIQUES

Ames Test

It is an established fact that many chemicals in our environment, including foods, are **carcinogens**—that is, they can induce cancers in animals and humans. Testing these chemicals is very complex, requiring exposure of experimental animals to high concentrations of compounds over prolonged time periods, and noting development of cancer. A simplified and sensitive test has been developed to examine carcinogens based on the ability of the vast majority of these compounds to induce mutations. Mutagenic ability of a chemical is tested as outlined in the figure. The test compound is exposed to a strain of *Salmonella* that has a single mutation in an enzyme of histidine biosynthesis, designated His^- (unable to grow on media without histidine). If the compound is mutagenic, it can increase the frequency of mutations in the *Salmonella* genome, and at a certain frequency will result in back mutations in the original gene responsible for the histidine requirement. Those bacteria can then synthesize their own histidine and grow on media that is not supplemented by histidine. In the Ames test, a pure culture of His^- *Salmonella* is mixed with the suspected carcinogen, and plated on minimal media lacking histidine. As a control, the same culture is plated without the carcinogen. After the plates are incubated, the number of colonies are counted. A significant increase in the number of bacteria on the test sample compared to the control, resulting from an increase in back mutations from His^- to His^+, indicates mutagenic activity of the compound. This test can be adapted for those compounds that are not mutagenic by themselves, but are metabolically converted to mutagens after ingestion. Usually, the compound is exposed to an extract from rat liver, to allow conversion to the mutagenic form.

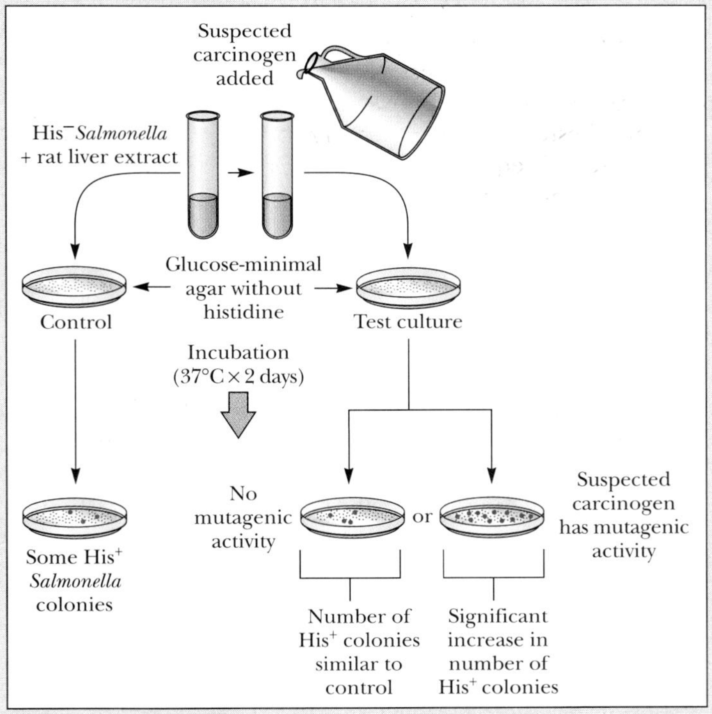

Ames test for mutagenicity/carcinogenicity. A test strain of bacteria, which requires histidine for growth because of a mutation in the histidine biosynthesis pathway, is mixed with the test compound. If the compound is mutagenic, it can cause additional mutations, reversing the histidine requirement, such that these bacteria can grow on medium lacking histidine. Since many compounds are not mutagenic in their original form, but are activated in tissues, the test also includes a homogenized liver extract.

In addition to direct reversal of mutations by enzymes, a common mechanism involves **excision repair,** in which the damaged DNA strand is enzymatically removed and the gap correctly resynthesized using the complementary strand as a template. Bulky distortions in DNA resulting from gross mispairing or ultraviolet light–induced pyrimidine dimers can be repaired by an excision mechanism that relies on the abilities of a multisubunit endonuclease

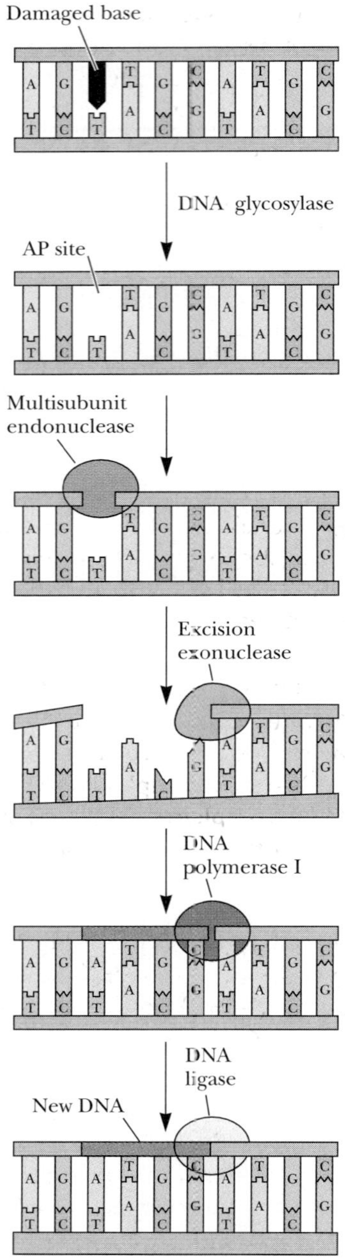

***Figure* 13.18** Excision repair. A damaged base is excised by DNA glycosylase creating an apurinic/apyrimidinic (AP) site. A multisubunit excision nuclease (consisting of UvrA, B, and C proteins) removes several nucleotides at the site and DNA polymerase I and DNA ligase repair the gap.

(in *E. coli*, encoded by genes *uvrA*, *uvrB*, and *uvrC*) to hydrolyze the phosphate backbone at two places flanking the damaged site (Figure 13.18). This results in the release of a 12-base oligonucleotide containing the damaged bases; the resulting gap is filled in by DNA polymerase I.

Another mechanism by which thymine dimers can be removed is called **recombination repair.** When DNA polymerase III reaches a thymine dimer, it skips it and continues synthesis at the other side of the dimer, leaving a gap in the newly synthesized duplex. The opposite strand is replicated normally. It is not known how DNA polymerase can initiate synthesis in the absence of primers when it skips the damaged site. The gap that is created around the damaged DNA is filled by excision of the corresponding fragment from the second strand and insertion into the gap by a mechanism similar to recombination between two related DNA molecules, as described in Chapter 15. This repair, just like recombination between DNA segments of similar sequence, is also dependent on the **RecA** protein, as well as a multisubunit nuclease (exonuclease V) that contains products of genes *recB*, *recC*, and *recD*. The newly created gap in the parental strand is then filled in by repair synthesis, since it involves copying an undamaged strand.

The ability of bacteria to sense the presence of damaged DNA involves synthesis of a number of specialized proteins, in addition to those that function in normal repair. In *E. coli*, a complex regulatory system is activated, which directs synthesis of many of these proteins. This regulatory network, called the **SOS regulon,** is negatively controlled by a repressor protein, LexA, which recognizes operator sequences for a number of genes for repair enzymes, including *umuC*, *uvrA*, *uvrB*, and *uvrC* as well as the *recA* gene. All of these genes express proteins that function at one stage or another in the above-mentioned repair systems. Clearly, it is the sensing of damaged DNA by the RecA protein, resulting in cleavage of the LexA repressor, that initiates the various DNA repair mechanisms by relieving the repression of the repair enzymes' genes. The SOS response is activated when the replication fork stalls and RecA protein binds to exposed ssDNA. The RecA-DNA complex binds LexA, and this binding induces a conformational change that causes LexA to cleave itself.

Summary

- All living cells carry hereditary information in deoxyribonucleic acid (DNA), which is organized into **genes.** In bacteria, genes are usually found in a single circular **chromosome.**

- DNA is composed of four deoxyribonucleotides: deoxyadenosine monophosphate (**dAMP**), deoxythymidine monophosphate (**dTMP**), deoxyguanosine monophosphate (**dGMP**), and deoxycytidine monophosphate (**dCMP**).

- The deoxyribonucleotides are linked together via **phosphodiester linkages** between phosphoric acid and hydroxyls on neighboring deoxyribose sugars.

- Bacterial chromosomal DNA is **double-stranded;** opposite, **antiparallel** strands are held together by hydrogen bonds formed between adenine and thymine and guanine and cytosine bases of the deoxyribonucleotides.

- DNA replicates by a **semiconservative** mode of replication, where the strands are separated and each strand is copied with the daughter genome, receiving one parental strand and a newly synthesized complementary copy.

- DNA replication begins at a single, relatively well-conserved site called the origin (*ori*). Two **replication forks** move away from the origin, as nucleotides are incorporated with the assistance of a number of proteins.

- Within each replication fork, one strand (the **leading strand**) is synthesized as single contiguous molecule, while the complementary strand (**lagging strand**) is synthesized in a discontinuous fashion, involving short RNA segments that serve as primers for DNA segments (**Okazaki fragments**). The newly synthesized DNA strand is completed by removal of RNA primers and filling in the gaps with DNA, followed by the joining of fragment into a single molecule.

- The low frequency of mistakes made during DNA replication is due to the **proofreading** and **repair** activities associated with enzymes that replicate DNA.

- If genes specify proteins, this form of RNA is called messenger RNA (**mRNA**). If a single RNA can specify several genes, it is then called a polycistronic mRNA.

- Other forms of RNA encode components of ribosomes, the ribosomal RNA (**rRNA**) or the adapter molecules for protein polymerization, called transfer RNAs (**tRNA**).

- RNA is a copy of one of the DNA strands. It consists of polymerized ribonucleotides with the same base composition as DNA, except thymine is replaced by uracil. The complementarity of the RNA copy is assured by hydrogen bonds base-pairing the nitrogenous bases during synthesis of the RNA strand.

- The beginning of the gene, a sequence called the **promoter,** where transcription initiates, is recognized by the sigma subunit. After binding, sigma dissociates from the RNA polymerase complex, leaving a form called the **core enzyme,** which continues elongating the RNA strand. The end of a gene is recognized by RNA polymerase after formation of stem-loop structure or by termination assisted by the so-called **rho** protein.

- When the product of a gene is not needed, bacteria utilize **negative regulation** to prevent its transcription. This often involves binding of a **repressor** protein to an **operator** site located adjacent to a promoter sequence. Tight binding of the repressor to a sequence that often overlaps the protein interferes with sigma-factor-directed binding of the RNA polymerase, and thus prevents transcription of the negatively regulated gene.

- Alternatively, protein factors (**activators**) stimulate transcription by binding to DNA sites upstream of promoters, and facilitate the rate-limiting incorporation of the first bases into RNA.

- The information in mRNA is converted into proteins by a process of **translation,** which involves ribosome-assisted polymerization of 20 amino acids into a sequence specified by the triplet code of mRNA, copied from its complementary DNA strand. The specific order of such codons is called the **reading frame.**

- Amino acids are incorporated into polypeptides, after they are covalently attached to tRNAs. Each tRNA recognizes a sequence of three bases on mRNA (called codons) via a complementary sequence.

- The ribosomes recognize and bind to a specific site at the beginning of the mRNA. This so-called ribosome binding site is adjacent to the codon for the first amino acid of the polypeptide, which is always methionine. Ribosomes terminate translation at specific codons (**termination,** or **stop codons**), for which there are no corresponding tRNAs.

- Changes in the base pair composition of the DNA are called **mutations.**

- Various chemicals can affect the frequency of mutations in a specific gene, either by altering the DNA or by interfering with DNA repair during replication. These are called **mutagens.**

- There are several mechanisms through which cells repair DNA and eliminate the deleterious effects of mutations.

Questions for Thought and Review

1. Contrast the basic structural differences between deoxyribonucleotides and ribonucleotides.

2. Draw the structure of deoxycytidine-deoxyguanosine (dG-dC) and deoxyadenosine-deoxythymidine (dA-dT) base pairs, as they appear in the antiparallel form in double-stranded DNA. Indicate the location of hydrogen bonds holding the strands together.

3. Distinguish between DNA synthesis on a leading and on a lagging strand, addressing the different mechanisms of initiation and elongation of the DNA strand.

4. What is the mechanism of relieving the tension in DNA caused by supercoiling during replication?

5. What would be the consequences for *E. coli* in its ability to synthesize the enzyme beta galactosidase, of the following mutations in the regulatory components of the lactose utilization operon? Explain your answer by considering bacteria that are present in medium with or without lactose.
   (a) A mutation in the operator sequence of the lac operon, such that it cannot bind the repressor protein.
   (b) Mutation in the DNA sequence of the lacI gene, such that the mutant repressor cannot bind an inducer (lactose or its structural analogue).
   (c) A two-base pair mutation in the lacI gene, changing, in mRNA, the AUG-initiating codon to UAG.
   (d) A double mutant containing both types of mutations described in (b) and (c) above.
   (e) Would a mutation in the cyclic AMP binding protein (resulting in the inability of bacteria to synthesize this protein) offset the effect of the mutation described in (b), such that the bacteria synthesize wild-type levels of beta-galactosidase?

6. Consider a polycistronic operon consisting of genes X, Y, and Z. The promoter located upstream of the gene X undergoes a mutation, such that three of the bases in the −10 region are changed. How would this affect the synthesis of proteins encoded by genes X, Y and Z? What would be the effect of a three base pair mutation in the ribosome-binding site of gene X on synthesis of proteins X, Y and Z?

7. When bacteria are exposed to antibiotics, they are killed, except for a small population of mutants that are resistant, due to their ability to undergo mutations in the target of the antibiotic. Devise a set of experiments (analogous to the fluctuation test described in the chapter) which demonstrate that mutations arose spontaneously in the bacterial population, and are not induced by exposure of the bacteria to antibiotics.

8. A bacterial culture was exposed to a mutagen, and some mutants were isolated that were unable to synthesize an essential biosynthetic enzyme. When this mutant was grown in pure culture and reexposed to the mutagen, a small fraction of bacteria regained their ability to synthesize the biosynthetic enzyme. Describe several possible mechanisms, explaining how the second exposure to a mutagen "corrected" the original mutation.

9. What are the mechanisms of removal of crosslinked thymine dimers in DNA?

10. What is the mechanism of induction of DNA repair processes, when the DNA replication fork stalls at the site of DNA damage, exposing single-stranded regions?

11. The partial DNA sequence below from the *E. coli* genome encodes a gene for a short polypeptide. The transcriptional start site is underlined.

    (a) Identify the most likely promoter for this gene. Is it recognized by the major RNA polymerase?
    (b) Write out the sequence of the mRNA transcribed from this promoter.
    (c) Translate the sequence into the polypeptide according to the genetic code.
    (d) Identify the position in the DNA sequence, where, by a single base pair substitution or a frameshift mutation, the result would be synthesis of a peptide that would have the identical sequence but half the size of the wild-type peptide.
    (e) Identify the position in the nucleotide sequence, where a mutation involving a single base pair change (insertion, deletion, or substitution) would result in synthesis of a peptide identical to the wild type throughout the sequence, but having four additional arginine residues at its carboxyl terminus.

+1

GGCTTGACACCCGGCTAGCGTAGTTGTATAATGGTCAGGCTTTGAGGAGGTCTAGGCATGAAAGCTCAACGTAAGGGGATCCGGAGTAGGAGACGGAGGTAAG
CCGAACTGTGGGCCGATCGCATCAACATATTACCAGTCCGAAACTCCTCCAGATCCGTACTTTCGACTTGCATTCCCCTAGGCCTCATCCTCTGCCTCCATTC

Suggested Readings

Birge, E. A. 1994. *Bacterial and Bacteriophage Genetics.* 3rd ed. New York: Springer-Verlag.

Joset, Francoise, and J. Guespin-Michel. 1993. *Prokaryotic Genetics.* Cambridge, MA: Blackwell Science.

Kornberg, A., and T. A. Baker. 1992. *DNA Replication.* 2nd ed. New York: W. H. Freeman and Co.

Maloy, S. R., J. E. Cronan, and D. Freifelder. 1994. *Microbial Genetics.* 2nd ed. London: Jones and Bartlett Publ.

Primrose, S. B. 1995. *Principles of Genome Analysis.* Cambridge, MA: Blackwell Science.

. . . It is, indeed, an RNA world, not a DNA world. In present-day life, DNA acts entirely by way of RNA, whereas RNA can act on its own. Viruses are a case in point.

Christian de Duve
Blueprint for a Cell:
The Nature and Origin of Life
(Reprinted with the permission of Carolina Biological Supply Company, Burlington, NC)

Chapter 14

Viruses

Self-perpetuation is the key to survival of a species and, as discussed earlier, the blueprint for "self" is present in nucleic acids (genome). Therefore a bit of nucleic acid that can generate a like nucleic acid may be considered to have some properties of life. Thus we define a **virus:** a bit of nucleic acid—RNA or DNA—surrounded by a protective coat, and this small bit of nucleic acid can regenerate by commandeering the metabolic machinery of an appropriate archaeal, bacterial, or eukaryotic host. Viral nucleic acid must gain entry into a suitable host, where it mobilizes the essential cellular synthetic capacities and compels the host to generate viral particles. Viral RNA (message) can be synthesized by polymerases available in the host or furnished by the virus. However, viruses do not contain translational enzymes, and they depend largely on the translational machinery of the host to synthesize viral proteins.

An understanding of how viruses function begins with an overview of the basic structure of the virus (Figure 14.1). At the core of the virus is the viral genome, which

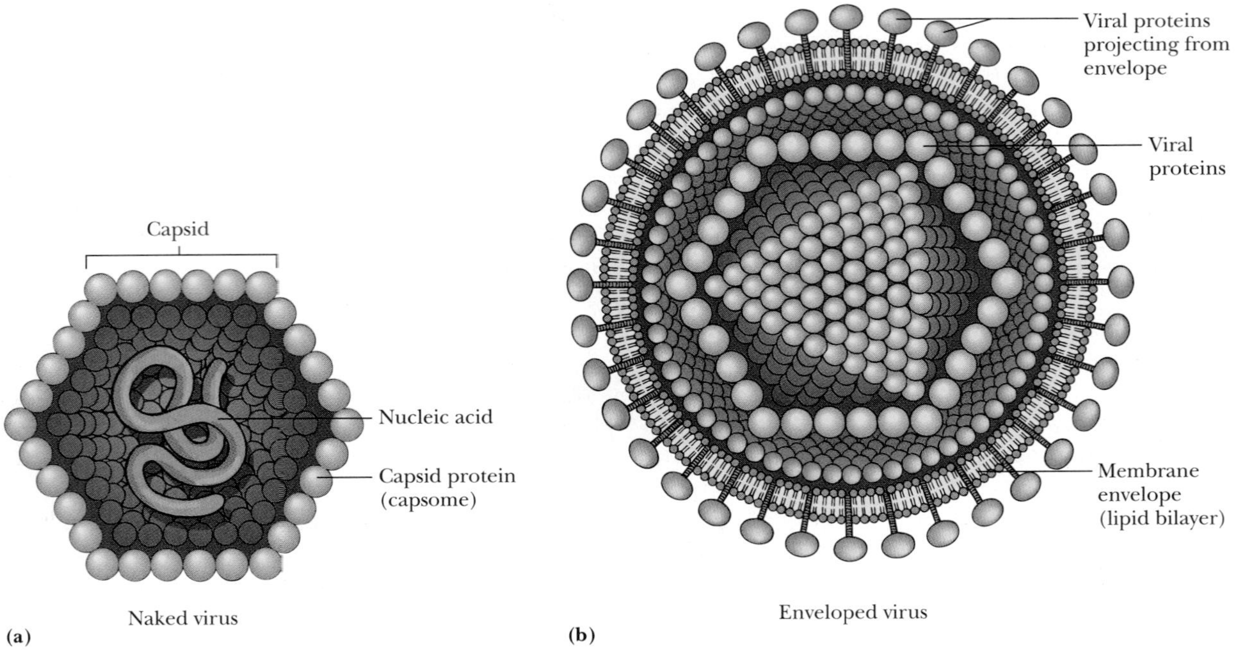

***Figure* 14.1** The general structure of a virion. The genomic nucleic acid is surrounded by a protein coat (capsid). In some viruses the protein coat is encased in an envelope composed of a lipid bilayer. Virus-specific proteins may be present in the envelope as shown. **(a)** "Naked" virus and **(b)** an enveloped virus.

can be composed of either RNA or DNA. Unlike the cells of animal, plant, or bacteria a virus rarely contains both. The nucleic acid is bounded by a protective protein coat termed a capsid. The capsid and genome together form a **nucleocapsid.** In some viruses the nucleocapsid is surrounded by a membrane envelope, and these are considered **enveloped viruses.**

All viruses exist in two states: extracellular and intracellular. In the **extracellular state** the **virion** or **virus** particle is quite inert. The virion is the form in which a virus passes from one host cell to another host. In the **intracellular state,** the viral nucleic acids induce the host cell to synthesize the virus. Such viral reproduction is called infection. A viral infection can be productive, which leads to viral progeny, or nonproductive, whereby a virus may remain dormant in the cell causing no apparent manifestations of infection. With some animal viral infections there are few, if any, cases of multiplication without overt disease symptoms (measles, for example), whereas in others the inapparent infection rate may be 100 or more cases of viral multiplication for each clinical case. This chapter is concerned with the mechanisms whereby prokaryotic and eukaryotic viruses are propagated. The first half of the chapter presents a general overview of viruses: their size; structure; how they are cultivated and quantitated; and how they reproduce. The second half considers selected viruses in some detail.

Origin of Viruses

The origin of viruses is an intriguing question and has been the subject of considerable speculation in the scientific community. There is general agreement that viruses (phage) probably infected bacteria during those millennia when Archaea and Eubacteria were the sole life form on Earth. As eukaryotes evolved, they too became subject to viral attack. There are three major theories that have been advanced for the origin of viruses: (1) viruses originated in the primordial soup and coevolved with the more complex life forms, (2) viruses evolved from free-living organisms that invaded other life forms and gradually lost functions—also known as retrograde evolution, and (3) viruses are "escaped" nucleic acid no longer under control of the cell—also termed the escaped gene theory (Figure 14.2). The great variety of viruses present in the living world affirm that they originated independently many, many times during evolution. Viruses also originated from other viruses by mutation. A brief discussion of theories of viral evolution follows.

Coevolution

The earliest self-replicating genetic system was probably composed of RNA. RNA can promote RNA polymerization, although this would be a slow process and proteins present in the primordial soup may have become promoters of RNA replication. The DNA template is much more effective and originated early in evolution. RNA then became the messenger between the DNA template and protein synthesis. Thus the genetic code came into being and permitted orderly replication.

Gradually the early replicative forms gained in complexity and became encased in a lipid sac whose purpose was to separate its metabolic machinery from the environment. This may have been the ancestor of the progenote and later the Archaea and Eubacteria. Another replicative form may have retained simplicity and may have been composed mainly of self-replicating nucleic acid surrounded by a protein coat. This entity was the forerunner of the virus and evolved to a dependence for duplication on its ability to invade and take over the genetic machinery of a host. Thus, there was coevolution of the bacterium and virus. This hypothesis has few supporters but does offer an explanation for the evolution of viruses with genomic RNA.

Retrograde Evolution

The theory that viruses originated as free-living or parasitic microorganisms is based on the assumption that a prokaryote gradually lost biosynthetic capacity and genetic information. Eventually the prokaryote evolved to nothing more than a group of genes known now as a virus. Prior to the advent of the eukaryote, bacterial and archaeal life forms were mostly free living, but predators similar to *Bdellovibrio* may also have existed. An intracellular parasitic microorganism could have become more and more dependent on the host cell for preformed metabolites, and loss of much of its genetic information could occur without dire consequence. It would need to retain only the ability to replicate nucleic acids and some mechanism for traveling cell to cell.

An equivalent occurrence in eukaryotic cells might account for the evolution of plant and animal viruses. An intracellular parasitic *Rickettsia* sp. is a bacterium that has lost much biosynthetic capacity. Intracellular parasitic bacteria such as the rickettsiae or the chlamydiae would be

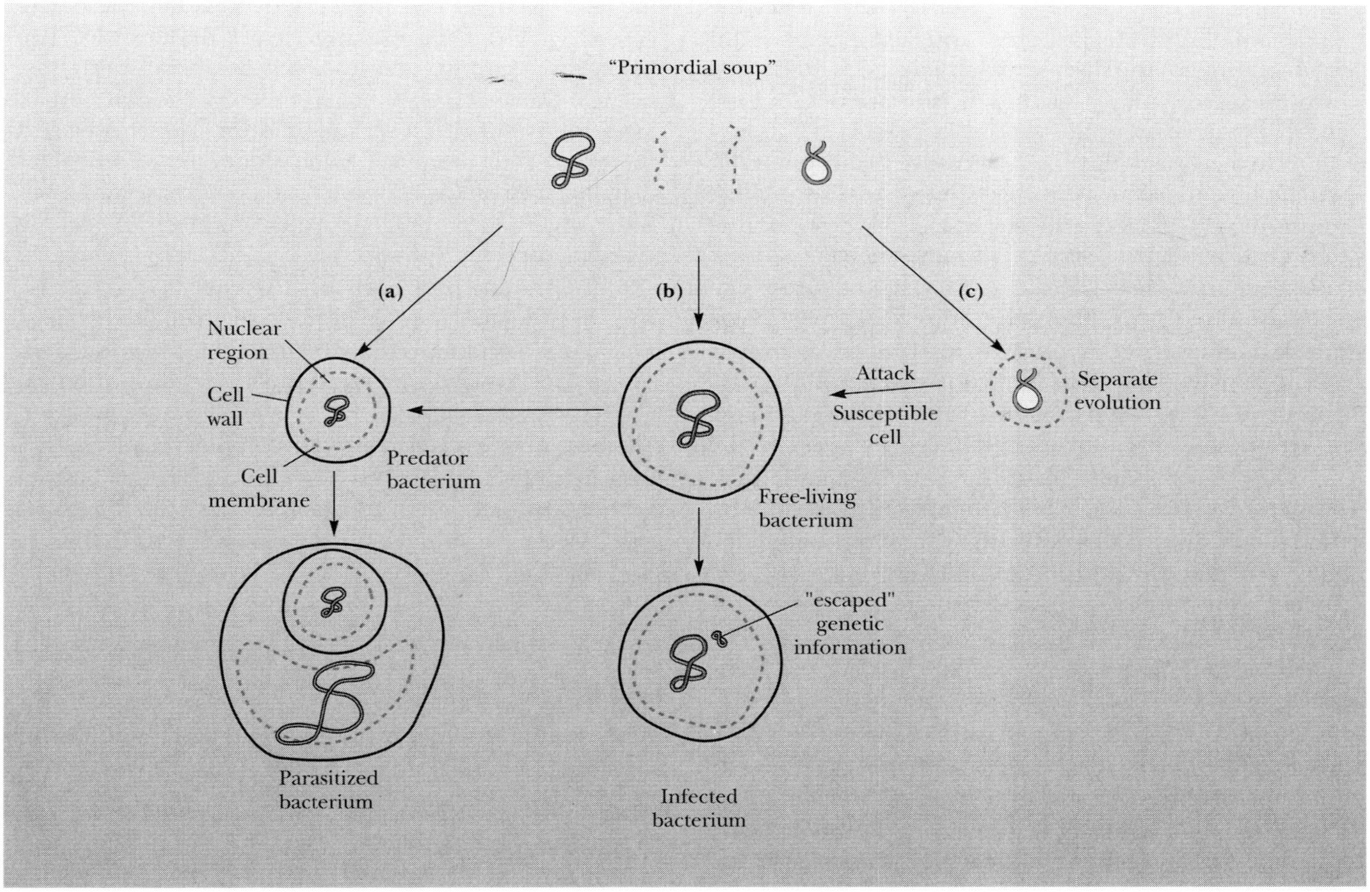

***Figure* 14.2** Theories advanced for the origin of viruses. **(a)** Coevolution of virus with primitive archaeal and eubacterial cells. **(b)** A predator that invaded a host and lost functions over time to become parasitic nucleic acid. **(c)** Genetic information that "escaped" cellular control.

BOX 14.1 RESEARCH HIGHLIGHTS

Diseases Transmitted by Protein

Several diseases of humans and other animals are attributed to *proteinaceous infectious* particles or **prions.** Prions have no nucleic acid, are heat and UV resistant, and are unaffected by agents that digest nucleic acids. They consist of about 250 amino acids and are about 1/100 the size of a small virus. The gene for prion development is apparently present in normal animals, and many believe that the protein product causes disease symptoms only if modified posttranslationally. There are theories that the protein may serve as a genetic repressor or a genetic inducer. How the defective infectious prion particles may be transmitted to an uninfected individual and actually initiate the disease process is unknown. There is no inflammatory response to prions and they are apparently too small to serve as antigens when introduced into a healthy animal. The incubation period can be years in length, hence, the name once attributed to these infective agents—slow viruses. Some of the diseases now believed to be caused by prions were once thought to be slow viruses, but not all slow virus diseases are caused by prions.

Prions cause a chronic progressive central nervous system degeneration. Collectively the diseases are called transmissable spongiform encephalopathies because they destroy neurons resulting in a spongelike appearance to brain tissue. The diseases caused by prions can be transmitted from animal to animal by injection of infected brain tissue. Among the diseases caused by prions are: kuru and Creutzfeldt-Jakob disease of humans; scrapie in sheep and goats; "mad cow" disease; and encephalopathies in mink, elk, and cats.

D. Carleton Gajdusek, of the National Institutes of Health, received the Nobel Prize in 1976 for studies on the human disease kuru. Kuru was then attributed to a "slow virus" and occurred mostly in New Guinea. The disease appeared mostly among women and small children. Gajdusek went to New Guinea and followed the course of kuru in the population. It was apparent that the disease was transmitted to women because it was their task to prepare bodies of the dead for disposal. In the ritual of preparation, the women rubbed their bodies with raw brain tissue of the deceased. Gajdusek attributed the disease to protein particles (prions) that were introduced through breaks in the skin and evidently traveled to the brain. Children were infected because they accompanied the women during the ritual. Kuru symptoms occur 1 to 15 years after inoculation and appear as a severe headache, followed by loss of coordination, and a tendency to giggle at inappropriate times. Three months after onset there is a loss of the ability to walk and swallow. Death occurs within a year.

Creutzfeldt-Jakob disease apparently passes in families due to a genetic defect. How this genetic defect causes active infection is not clearly understood. The prion involved in this disease has been transmitted by corneal transplant and through the injection of human growth hormone to dwarfed children. The cornea and the growth hormone are generally obtained from human cadavers. In recent years the growth hormone has been obtained from genetically engineered bacteria (see Chapter 32).

Prions can survive on instruments that are sterilized by storage for long periods in formaldehyde, as well as on instruments that are inadequately autoclaved. The prion involved in scrapie can survive in buried animals for years.

It has been suggested that Alzheimers and Parkinsons diseases may be caused by prions, but this has not yet been substantiated. It is clear, however, that we have much to learn about neurodegenerative diseases and the role and control of prions in animal infections.

---

examples of bacteria that potentially could regress to the viral state. The chlamydiae are bacteria that have no cell wall and cannot live outside of living cells. Bacterial origin would be more likely for complex viruses such as the poxviruses, but the genetic information in these viruses differs so markedly from that in prokaryotes that this theory has little support. The absence of any life form between intracellular bacterial pathogens and viruses is also cited as a concern, and it is difficult to explain how RNA viruses came into existence through loss of genetic information.

"Escaped" Gene Theory

This theory is more plausible than the previous two and has considerable support. It proposes that pieces of host-cell RNA or DNA gained independence from cellular control and escaped from the cell. Living organisms make duplicates of their genetic information by initiating replication at a specific site called the initiation site. The replication cycle ends when a full complement of the genome is synthesized. If initiation of replication began elsewhere on the genome, this duplication would occur independently of host control. A virus that could recognize nucleotide sequences at sites other than the start site and that carried the proper polymerase could have the capacity to produce RNA or DNA without interference from normal control mechanisms (see **Box 14.1**).

The origin of virus may have been with episomes (plasmids) or transposons, circular DNA molecules that replicate in cytoplasm and can be integrated into or excised from various sites in the host chromosome. Plasmids can also move from cell to cell carrying information such as fertility or antibiotic resistance. Transposons are bits of DNA present in both prokaryotic and eukaryotic cells that can move from one site in a chromosome to another site carrying genetic information.

There is actually one type of transposon that directs the assembly of an RNA copy of its own DNA. This transposon can use this messenger RNA to synthesize DNA via reverse transcriptase. This copy may then be inserted into the chromosome. The DNA of the transposon carries a gene for synthesis of the reverse transcriptase, and transposon elements with such properties have some similarity to retroviruses. Analysis of nucleotide sequences in viruses indicates that they are quite equivalent to specific sequences in the host cell. Evidence accumulated to date strongly supports the proposal that viruses originated from "escaped" host nucleic acid.

Size and Structure of Viruses

Electron microscopy has affirmed that viruses vary widely in size and structure. A very small virus, the polio virus, is about 20 nm (a nanometer is 1/1000 of a micrometer) in diameter. This virus actually approaches the size of a ribosome. The larger viruses, such as the poxviruses, are 400 nm long and 200 nm in width. These viruses are large enough to be visible by phase-contrast microscopy.

Electron microscopy, X-ray diffraction, biochemical analysis, and immunological techniques have provided much information on structure and composition of viruses.

Viruses are of several morphologic types including:

Icosahedral viruses. The capsids in these viruses are constructed from protein subunits termed **protomers.** In an icosahedral capsid these protomers are arranged in capsomers composed of pentamers (five sides) and hexamers (six protein subunits). The bonding between capsomers is noncovalent and weaker than the bonding within a capsomer. The hexamers form the edges and faces of the icosahedron, and pentamers are at the vertices. The 12 vertices in a polyhedron are shown in Figure 14.3(**a**), a structure that has 20 equilateral triangles as faces. The capsid of a small virus such as poliovirus would be made up from 32 capsomers, and the capsid of a larger virus (adenovirus) would be composed of 252 capsomers. Generally the capsids follow the laws of crystallography. The pentamers and hexamers may be composed of one type of protein subunit, or the proteins in the pentamers may differ from those in hexamers (Figure 14.3**b**). The icosahedral capsid is assembled and then filled with nucleic acid. Viruses have specific mechanisms to ensure that an entire genome is incorporated. There is little apparent organization of nucleic acid in the capsid.

Helical viruses. These are long thin hollow cylinders composed of distinct protomers. The helical tobacco mosaic virus (TMV) is depicted in Figure 14.4. The protein subunits are arranged in a helical spiral with the genome fitted into a groove on the inner portion of the protein. The nucleic acid is enclosed in the capsid as the protomers are assembled. The helical viruses are narrow (15–20 nm) and may be quite long (300–400 nm). They may be rigid

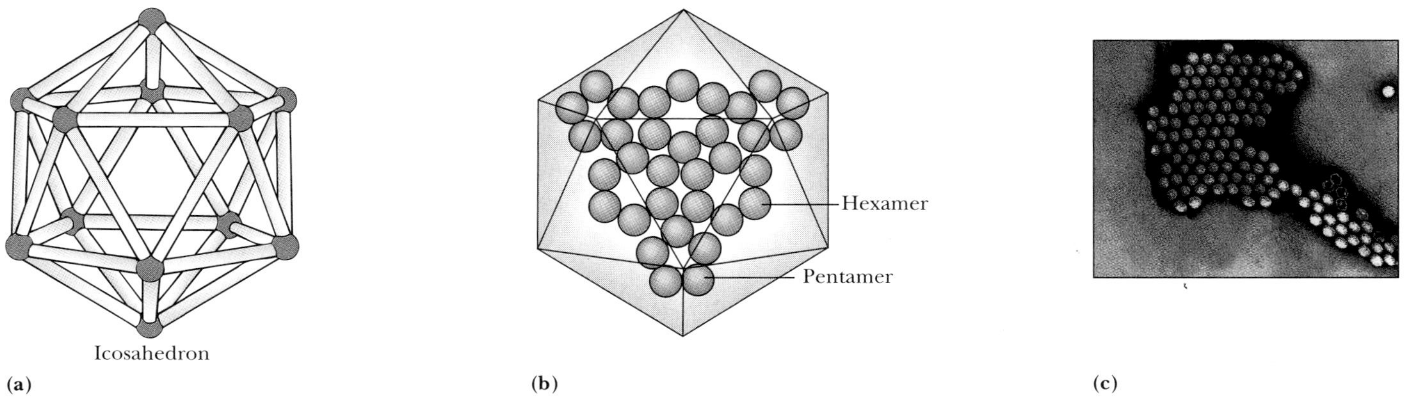

***Figure* 14.3** **(a)** A graphic presentation of an icosahedron, **(b)** the arrangement of protein pentamers at the vertices and hexamers along the edges, and **(c)** an electron micrograph of an adenovirus, an icosahedral virus. (**c,** © Oliver Meckes/E.O.S./Gelderblum/Photo Researchers, Inc.)

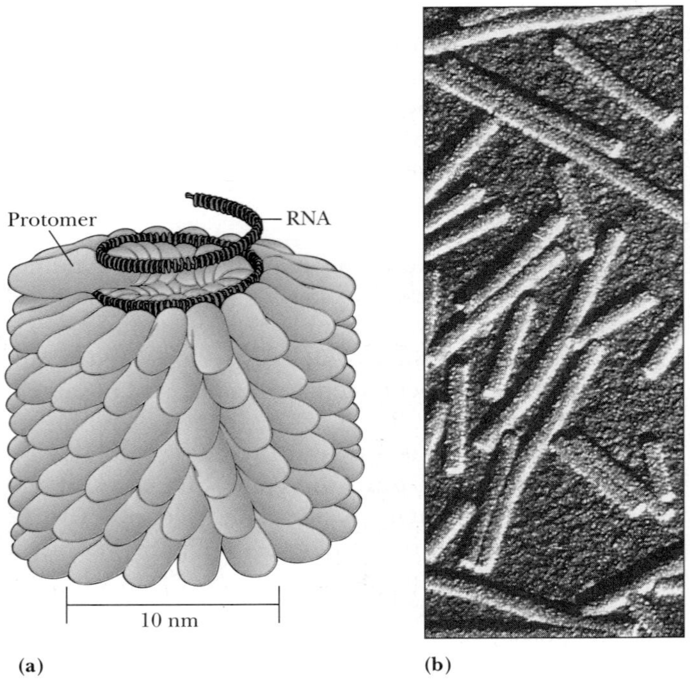

Figure **14.4** Structure of tobacco mosaic virus. **(a)** Tobacco mosaic virus is a helical virus with the protomers arranged in a hollow helix. The genomic RNA fits into a groove in the proteins. **(b)** Photomicrograph of TMV. (© Omikron Science Source)

or flexible, depending on the nature of the constituent proteins. The helix of TMV has one type of self-assembling protomer composed of 158 amino acids. There are 2130 copies of this one protein subunit in the coat of a mature virus particle. Synthesis of this size protein would be accomplished by about 500 of the 6000 nucleotides in the genome of TMV. If an organism such as TMV made several different protomers, much of the genome would be required for this purpose. There would be little genomic information to direct other necessary functions in propagation of the virus.

Enveloped viruses. These viruses have an outer membranous structure that surrounds the nucleocapsid (see Figure 14.1). The membranous envelope consists of a lipid bilayer and may have specific glycoproteins (spikes) embedded in the lipid. The glycoprotein spikes and other coat proteins are incorporated into the host membrane and are picked up by the virus as it leaves the infected cell (Figure 14.5). The glycoproteins are encoded by the virus genome and are not normal constituents of the host cell membrane. These spikes are responsible for specific attachment to host cells.

Complex viruses. These have capsid symmetry, but often the virus is assembled from parts synthesized separately (head, tail, capsomer). The *Escherichia coli* virus T4 (Figure 14.6) is an example of a complex virus. The virus is composed of a polyhedron head, a tail, whiskers, tail fibers, and other parts.

Viral Propagation

There are four rather distinct phases in viral reproduction:

1. Attachment or adsorption. There are specific sites (virus receptors) on the surface of a host cell where the virion attaches.
2. Penetration or injection. The nucleic acid of the virion gains access to the cytoplasm of the host. Bacterial and plant viruses initiating an infection must gain entry through the cell wall, while animal viruses enter through cytoplasmic membrane.
3. Gene expression and genome replication. Once inside, the viral genome is expressed and duplicated.

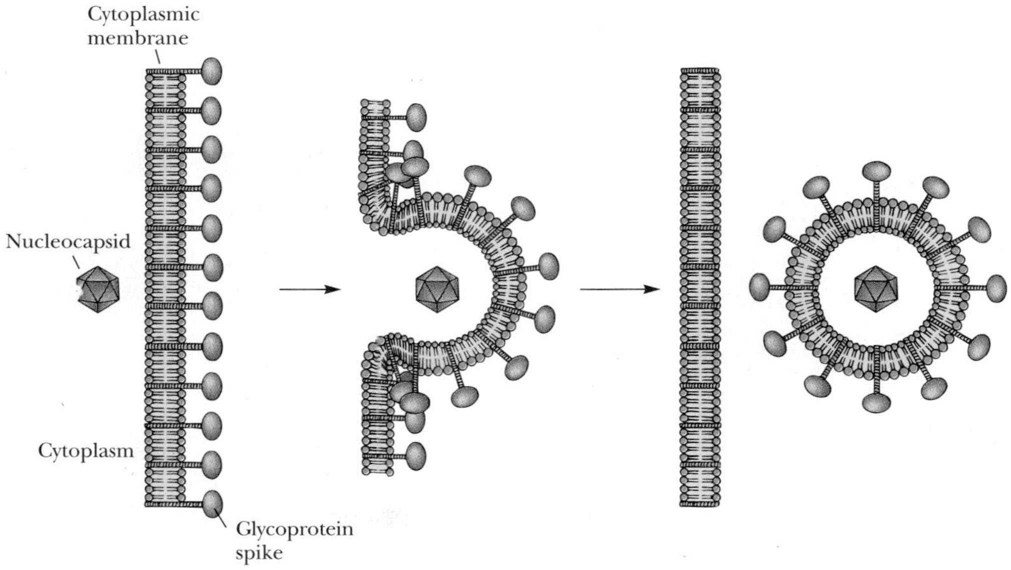

Figure **14.5** Enveloped viruses obtain the membrane coat as they exit through the nuclear membrane or the cytoplasmic membrane. The spikes are viral encoded.

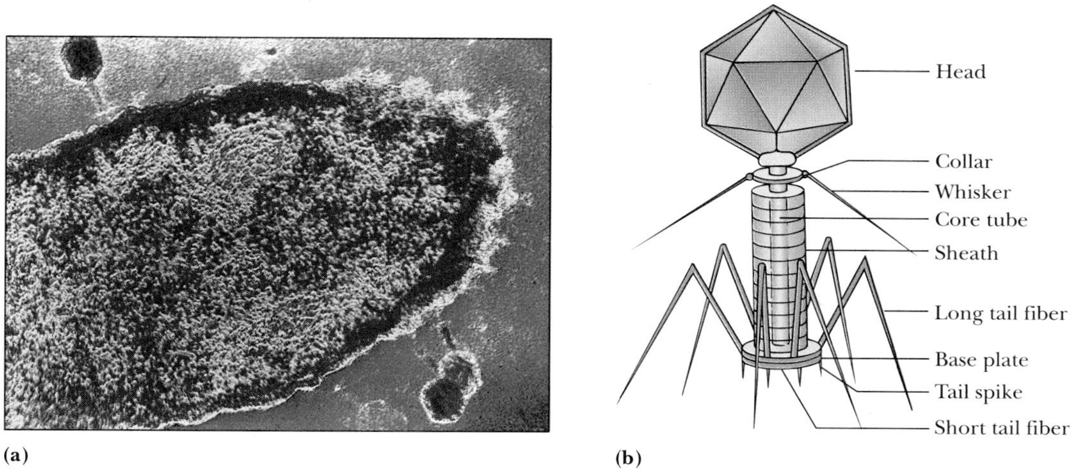

***Figure* 14.6** A complex virus. **(a)** Micrograph of T4, that attacks *Escherichia coli*. (Institut Pasteur/CNRI/Phototake) **(b)** Structure of T4.

This leads to the formation of gene products that control replication of viral nucleic acid and capsomer synthesis.

4. Assembly and release. The viral nucleic acid is packaged in the capsid and the mature virus is released. Viruses use ATP, ribosomes, and building blocks from the host; they often use host enzymes with or without modification. Generally lytic enzymes cause host cell dissolution.

These phases are illustrated in Figure 14.7. Entry of the viral genome often results in a cessation of normal host cell function and the replication of viral nucleic acid. Later in the infectious process proteins are synthesized that are part of the mature viral particle.

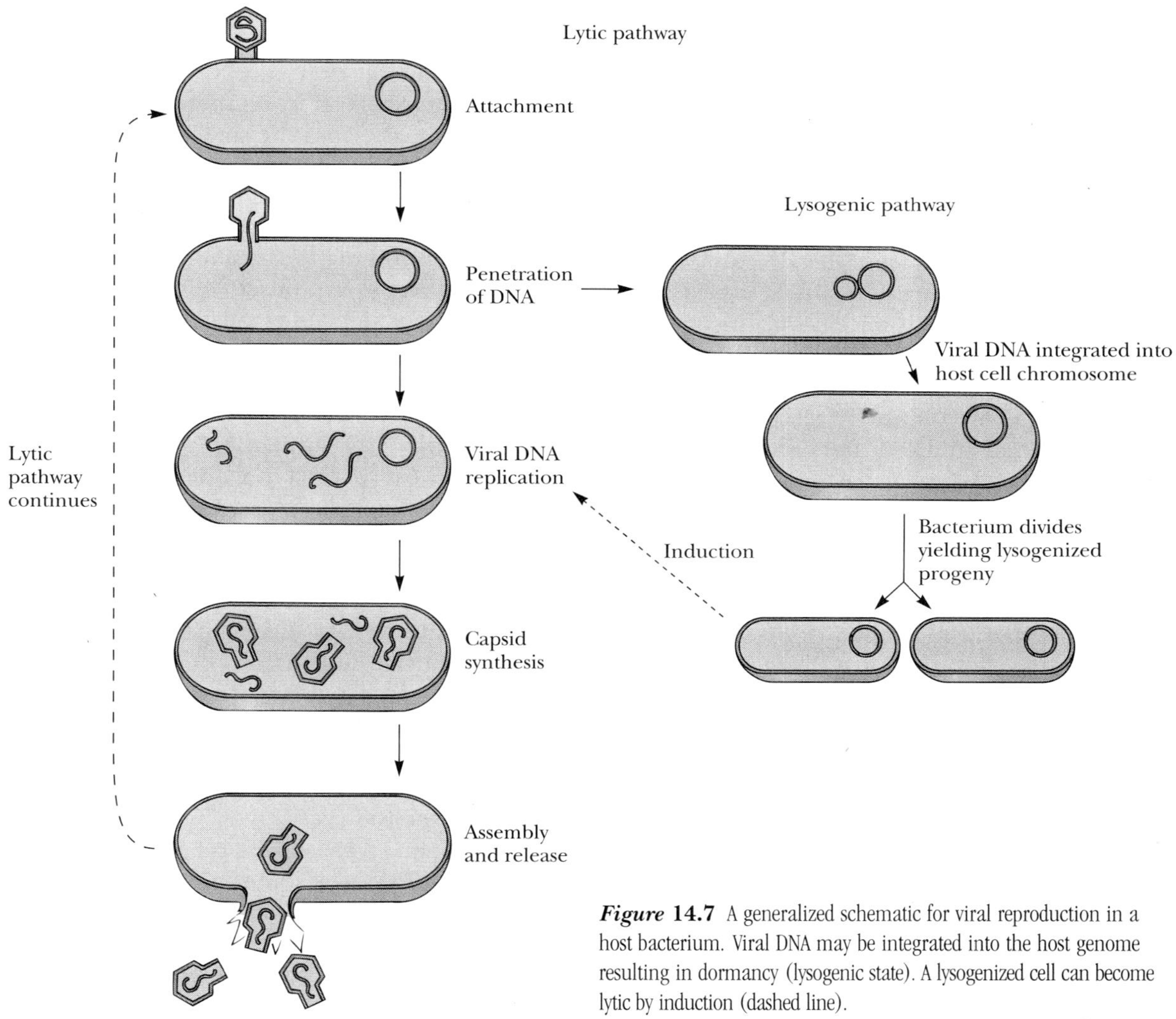

***Figure* 14.7** A generalized schematic for viral reproduction in a host bacterium. Viral DNA may be integrated into the host genome resulting in dormancy (lysogenic state). A lysogenized cell can become lytic by induction (dashed line).

Cultivation of Viruses

Viruses can be grown only within living cells. This is because a virus has no biosynthetic capacity of its own, but can manipulate a viable cell genetically to synthesize viral particles. In early experimentation, viruses were cultivated by transfer from plant to plant, animal to animal, or bacterial culture to bacterial culture. The symptoms or manifestations of disease resulting from inoculation was the principal means of identification. Animal viruses were routinely propagated in embryonated eggs. A fertile egg was incubated for 6 to 8 days at a temperature suitable for hatching and then inoculated with a viral suspension. The part of the embryo inoculated depended on the nature of the virus. Egg inoculation was often suitable for diagnostics, as the lesion produced in the embryo was characteristic of a given virus. More exacting cell-culturing techniques have been developed for use in research and diagnostic laboratories today.

Viruses that infect eukaryotic cells are now generally grown in monolayers of cells obtained from plant or animal tissue. Cells that are removed aseptically from a plant or animal can be placed in a sterile vessel made of glass or plastic. The cells may attach to the surface and, when fed properly, may form a monolayer. This monolayer is called a **cell culture;** similarly, tissue can be maintained in culture and is termed **tissue culture.** A virus inoculum can be spread over a cell monolayer or introduced into tissue culture and, if the cells in the culture are susceptible, the virus will enter cells and reproduce. After viral attachment to a monolayer, a layer of agar can be spread over the surface to limit the spread of virions to cells that are adjacent to the infected cell. In this way, areas of cell destruction can be observed and quantified.

Bacterial viruses can be cultivated by inoculating a suitable susceptible bacterial culture with a virus. Liquid cultures or cultures growing on an agar medium are suitable for this purpose.

Plant viruses can be cultivated in monolayers of plant tissue culture, in cultures of separated cells in suspension, in protoplast cultures, or in whole plants. In whole plants, the leaf may be abraded with a mixture of virus and an abrasive material, and localized necrotic lesions can be observed. Viral suspensions may be injected into a plant with a syringe thus emulating insects in nature. These methods for growing viruses in plants, animals, or bacteria can be adapted to quantify the number of viruses in a sample.

Quantitation of Viruses

To understand the nature of viruses, their reproduction and their properties, it is essential that methods be available to quantify or count the number of virus particles present in a suspension. This may be accomplished by either of two procedures: (1) directly counting the virions or (2) quantifying viruses as infectious units. A direct count can be accomplished with an electron microscope (EM). A suitable dilution of virions would be mixed with a predetermined concentration of small latex beads and the mixture examined by EM. The ratio of beads to viruses in a field would give a reasonable estimate of viral particles present. This method is effective for concentrated viruses of known morphology, but it is not practical for routine study. In addition, it does not distinguish infectious and noninfectious virus particles.

Many viruses agglutinate red blood cells (RBCs), a phenomenon termed **hemagglutination.** This is an effective method for quantitating selected viruses such as the influenza virus. Serially diluted suspensions containing virus are added to a standardized number of RBCs. The hemagglutination titer is the highest dilution of virus that causes visible aggregates. Standards for estimating the number of viruses present can be established with viral suspensions quantitated by other means. For example, a hemagglutination titer of influenza virus is equivalent to a known number of plaque-forming units from the same viral preparation. Alternatively, viruses can be quantified by measuring their effect on their host. The term **virus infectious unit** describes the smallest unit that causes a detectable effect on a specific host. The amount of a virus in a fluid can be determined by measuring the number of infectious units present. Following are some other methods for measuring virus infectious unit.

Plaque assays are an effective means for measuring the number of infectious particles in a viral suspension. This procedure is routinely applied to bacterial viruses. It can be accomplished by pouring a sterile agar medium that can support growth of the host into a petri dish. Dilutions of the virus suspension would be added to a suspension of host bacteria in molten agar and, after mixing, the virus/bacterial suspension would be poured over the surface of the aforementioned agar surface. A dilution of virus with significantly fewer virions than there are bacteria present will result in clear areas in the soft agar layer, and these are termed **plaques** (Figure 14.8). The plaque results from the destruction of a bacterium, followed by destruction of those adjacent to it. Each plaque is assumed to have resulted from an infection with one virion. Counting the number of plaques on a plate permits one to estimate the number of **plaque forming units (PFUs)** in the original suspension.

The plaque assay may also be adapted to plant or animal viruses. However, some animal viruses cause lethal cytopathic effects without resulting in discernible plaques. For these viruses degenerative changes in host cells can be quantitated by microscopic examination. Animal

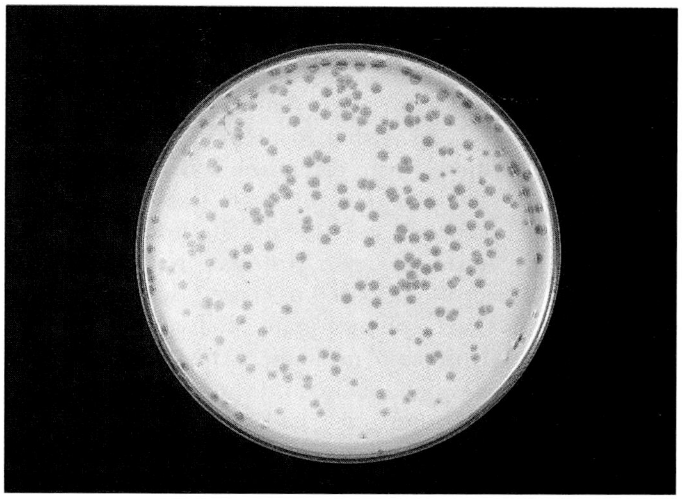

***Figure* 14.8** Plaques formed in a lawn of *Escherichia coli* by phage T4. (© Bruce Iverson)

viruses also can be quantitated by inoculating a host with serially diluted virus and determining the highest dilution that causes symptoms of disease or death. Experimental animals may be inoculated with a viral suspension, and the percent of the animals that survive can be determined. The results of this type of assay would be reported as the infectious dose (ID_{50}) or lethal dose (LD_{50}). The ID_{50} or LD_{50} is based on the occurrence of infection or lethality in 50 percent of the test subjects. This is a reasonably accurate method of quantitating animal viruses in hosts that would have broad variability in susceptibility to the agent.

Leaf assays are useful in estimating the number of plant viruses present in a suspension. A dilution of virus particles would be combined with an abrasive and rubbed into a suitable host leaf. The number of necrotic lesions evident after incubation would be an estimate of the number of viruses present in the inoculum.

The methods described can be employed effectively in propagating or quantifying plant, animal, and bacterial viruses. A determination of viral composition is practical only when the viral particles are purified from host tissue.

Purification of Viral Particles

Virus particles are larger than the macromolecules that make up a cell. They are coated with protein (capsid) either with or without an envelope. Treated as discrete large macromolecules they can be purified by techniques that are useful in the isolation of proteins. Some of the techniques employed in the isolation of virus particles are as follows:

1. Differential and density gradient centrifugation. Bacterial viruses are normally released on lysis of the host cell and no further cell disruption is required. In some cases mechanical disruption may release more of the viruses produced in an infected plant or animal. Disruption can be accomplished by alternate freezing and thawing procedures. The resultant material is centrifuged at low speed to remove particles that are larger than a virus. The suspension would then be centrifuged at high speed to precipitate viruses, and the supernatant would be discarded. The precipitate from this centrifugation would be layered on the top of a sucrose gradient (high concentration of sucrose at the bottom of the centrifuge tube and low at the top). During centrifugation a viral particle will settle in the zone where the sucrose density is equivalent to that of the virus. A larger virus will move through the gradient more rapidly since density gradient centrifugation separates by both mass and density. The concentration of sucrose in the gradient and relative speed of centrifugation would depend on the physical characteristics of the virus.
2. Precipitation of virus. Proteins can be removed from suspension by precipitation with ammonium sulfate or ammonium chloride. This procedure can be applied to virus purification by adding a concentration of the salt that precipitates most of the extraneous material, leaving the virus in suspension. The supernatant would then be treated with more concentrated salt to precipitate the virus. Where the virus might be sensitive to salt, polyethylene glycol or other materials can be employed. The salting out technique may be employed as a preliminary step prior to gradient centrifugation.
3. Denaturation techniques. Nucleic acids are tough molecular structures resistant to denaturation at a temperature or pH that tends to denature other macromolecules. Some viruses also are tolerant of organic solvents such as butanol and chloroform. Crude viral suspensions in cellular debris are mixed thoroughly with solvents and allowed to stand. The viruses will enter in the aqueous phase. The lipids will enter the solvent phase, and denatured host cell debris will be present at the interface. This is an effective method for selected viruses that are not surrounded by a membranous envelope.
4. Enzymatic digestion. A mass composed of virus and cell debris can be treated enzymatically to digest proteins and some nucleic acids. The virus must be resistant to the particular protease and

nuclease utilized. In some cases a viral suspension can be treated with ribonuclease and trypsin to digest RNA and protein.

Virus particle purification for preparation of vaccines and for other uses can be accomplished by one, or combinations, of the above methods. None of these procedures would be applicable for all viruses. When large-scale virus preparations are desired, the purification methods employed are geared to the characteristics of that particular virion.

Viral Genome and Reproduction

The genomes of eukaryotes contain solely double-stranded DNA. A virus genome differs in that it can be either DNA or RNA. Viruses have not been characterized that contain both RNA and DNA, although certain DNA viruses contain RNA primers. Some hepadna virions contain DNA-RNA hybrid molecules. The genome of viruses that infect plants, animals, or bacteria can be composed of a single strand (ss) or double strand (ds) of RNA or DNA. An exception would be the hepadna viruses that are partially double stranded and partially single stranded. A viral genome may be a single strand of RNA that is mRNA (+strand) or an RNA (−strand) that must be transcribed to produce mRNA that is then translated to make proteins. The arena viruses and phleboviruses have an RNA genome that is part + and part −strand. There are no ss RNA (−) viruses yet characterized that infect bacteria.

The pathways followed in production of mRNA from the genomic nucleic acids of various viruses are presented in Table 14.1. By convention the negative (−) strand of ds DNA is the one transcribed into mRNA. Consequently, the mRNA translated to protein is considered plus (+). The mRNA is a complementary copy of one strand of DNA. However, ss DNA Geminiviruses that infect plants make a complementary double strand and transcribe mRNA from both strands. Capsid size limits the amount of nucleic acid a virus may contain. Thus certain groups have mechanisms for producing mRNAs that lead to genetic economy not found in eukaryotes and prokaryotes. For example, ΦX174, a ss DNA bacteriophage, contains enough DNA to code for 4 to 5 proteins (1795 base triplets; 400 amino acids/protein). The virus uses overlapping reading frames, different initiation sites, and frame shifts within a reading frame to produce 11 different proteins. Parvoviruses—small ss DNA animal viruses, geminiviruses—ss DNA plant viruses, and leviviridae—a + ss RNA bacteriophage also employ overlapping reading frames. Tobamoviruses, a group of plant + ss RNA viruses including tobacco mosaic virus, create long and short polyproteins from the same genome sequences when the longer reads through the termination codon of the shorter. The promoter is a locale on double-stranded DNA where RNA polymerase attaches and is pointed in the proper direction for transcription. A ss RNA (+) strand viral genome would actually be mRNA. The RNA genome of retrovirus, such as the virus that causes AIDS, is transcribed to make ds DNA that is integrated into the host genome. From there it can direct the transcription of viral mRNA.

Some viruses produce early and late mRNA, and the early proteins resulting from translation of the early mRNA may direct the host cell to cease normal function. A series of reactions is then initiated that are directed toward production of viral messages. These result in the replication of viral nucleic acids and viral-specific enzymes. The late proteins synthesized from late mRNA are involved with the synthesis of capsid and other proteins necessary for the assembly of the mature virus. A bacterial virus may direct the production of lytic enzymes that lyse the host cell and free the mature viral particles.

DNA viruses can transcribe and replicate their genomes using cell enzymes without modification, although some code for their own enzymes and/or modify host enzymes. Replication of viral DNA parallels host DNA replication by requiring a small RNA primer to provide the 3′-OH. The ss DNA parvoviruses self-prime by using palindromes to create a hairpin at the 3′ terminus. Pox viruses also self-prime by nicking one of their ds DNA covalently closed ends to give a free 3′-OH (Figure 14.9). Eukaryotic and prokaryotic cells, with the possible exception of some plants, do not contain the RNA-dependent RNA polymerase required to transcribe and replicate the genomes of RNA viruses. The + strand RNA viruses code for the polymerase, which is made early in the infection process, while the polymerase is present in the virion of the ss stranded and ds RNA viruses. Hepadnaviruses make genomic DNA from a RNA template using a reverse tran-

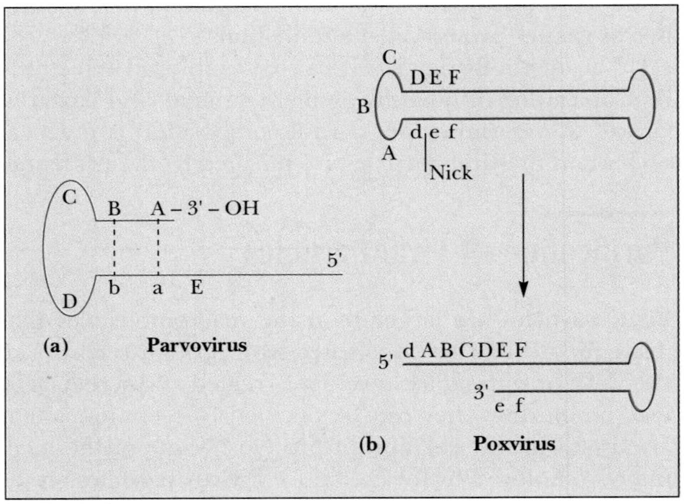

***Figure* 14.9** The generation of primers for genome replication in **(a)** parvovirus and **(b)** pox viruses.

scriptase (RNA-dependent DNA polymerase) while retroviruses employ a reverse transcriptase to replicate their RNA using a DNA intermediate.

Bacterial Viruses

The first scientist to observe and report the manifestation of a viral attack on bacteria was Frederick W. Twort, of England, in 1915. Twort noted that an agent that caused dissolution of bacterial colonies could be transferred to intact colonies and that these also were lysed. Twort also realized that the agent was active at high dilution, heat labile, and passed through a bacterial filter. He suggested that it might be a virus. In 1917, Felix D'Herelle, of the Pasteur Institute in Paris, had observed the lytic phenomenon and had called the agent a **bacteriophage,** which means "bacteria-eating." Today, the name bacteriophage can be used interchangeably with **bacterial virus,** virions that attack bacteria have essentially the same attributes as animal or plant viruses: they simply have a different host specificity.

Bacterial viruses have been widely employed as a model system for host-parasite interactions in viral pathogenesis. The viruses that attack enteric bacteria, particularly *Escherichia coli,* have been studied most. There is a broad array of viral types that attack *E. coli,* and *E. coli* is useful for study as we know much about the genetics of the organism. The value of bacteria as a model is apparent when you consider that *E. coli* has a generation time of about 20 minutes, which means that considerable cell mass can be obtained overnight. A virus that attacks *E. coli* can produce mature progeny in around 22 minutes. Thus one can follow the entire sequence of molecular events—from attachment of virus to the bacterium to lysis and release of virus—over a very short period of time.

There are viruses that attack virtually all microorganisms including thermophiles, cyanobacteria, Archaea, and Eubacteria. The number of viral types is broad and the mechanism for viral reproduction varied. Some of the viruses are complex, having a discrete head and tail. Others are quite simple. Bacterial viruses can establish a latent infection in bacteria with survival and reproduction of both virus and host. This relationship is termed **lysogeny.**

Another attribute of bacterial viruses is mutation to generate a virus with a different host range, plaque morphology, or temperature sensitivity. These are useful in genetic studies and in learning more about viral structure and function. Further, genetic recombination can occur when two viral mutants attack the same cell. The resultant recombinants are useful in mapping viral genomes and studying gene function.

Taxonomy

Bacterial viruses are generally identified by a Greek symbol, or a combination of strain number and symbol. At present there is no genus or species designation. There are over 2000 different bacterial viruses described in the literature. Unfortunately, the guidelines for describing a new virus were not established until recent years and most of those described are inadequately characterized. Relatedness among viruses is determined by host range, virion morphology, type of nucleic acid present in the genome, and other factors. Immunological identification of viral components is now employed and is of value when characterizing a specific viral group and relating it to others. A classification scheme for bacterial viruses is presented in Table 14.2.

***Table* 14.1 The pathway to mRNA followed in viruses that have differing types of nucleic acid**

The minus strand of nucleic acid is the strand transcribed to mRNA. With ds DNA the minus strand is transcribed. By convention the mRNA translated into protein is considered the plus strand and is complementary to the minus strand of DNA.

Virus Type

ds DNA ⟶ mRNA
ss DNA ⟶ ds DNA ⟶ mRNA
ss RNA (plus) acts as mRNA
ss RNA (minus)[1] ⟶ plus strand RNA is mRNA
ds RNA ⟶ minus strand transcribed ⟶ mRNA

Retroviruses

ss RNA (plus) ⟶ ss DNA (minus) ⟶ ds DNA ⟶ m RNA

[1]There are plant and animal viruses of this type, but bacterial ss RNA (minus) have not been characterized.

***Table* 14.2 The families of some major bacterial viruses**

DNA ds – linear	Size (nm)	Example and organism attack	
Siphoviridae	Head 90 Tail 200 ∞ 15	λ (E. coli)	
Myoviridae	Head 80 ∞ 110 Tail 110 ∞ 25	T4 (E. coli)	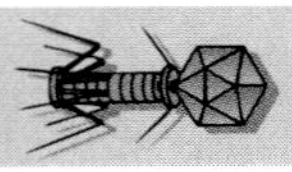
ds – circular			
Corticoviridae	60	PM2 (Pseudomonas)	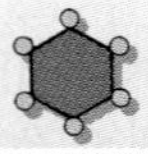
Plasmaviridae	50 to 120 (enveloped)	MV-L2 (Mycoplasma)	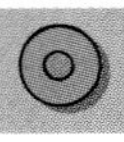
ss – circular			
Microviridae	30	ϕ ∞ 174 (E. coli)	
Inoviridae	9 ∞ 890	M13 (E. coli)	
RNA ds – linear			
Cystoviridae	85	ϕ6 (Pseudomonas phaseolicola)	
ss – linear			
Leviviridae	30	MS2 (E. coli)	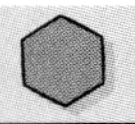

Lysogeny and Temperate Viruses

A bacterial virus may destroy the host, but bacteria and virions can also establish a mutually beneficial interaction that allows both viral propagation and survival of the host. This association is termed **lysogeny,** and a virus that can establish this relationship is called a temperate virus. The nucleic acid of a temperate virus is injected into a host, becomes integrated, and replicated along with the host genome. The progeny of a lysogenized cell are lysogenic (see Figure 14.7). In nature, bacteria may be lysogenic for one or more viruses. The lysogenic state can cause an alteration in the host outer envelope that can alter attachment sites for other bacterial viruses. Lysogeny, then, may be beneficial in that it can protect the host bacterium from attack by more virulent viruses that are related to the temperate one, and the virus benefits in self-perpetuation. Lysis of all hosts could lead to extinction of a virus, while a lysogenized dormant bacterium would retain the virus indefinitely.

In a lysogenic culture, 1/1000 or less of the infected cells become lytic and produce virions. Proteins that repress the expression of genes that would cause a lytic infection are part of the integrated prophage (provirus). It is these repressor proteins that maintain the lysogenic state. Under certain conditions the lysogenic state is lost, the virion is activated, and lysis results. This is called **induction of the lytic state.** Induction occurs at a low spontaneous rate but is accelerated in the presence of mutagens such as UV, X rays, and nitrogen mustards. Much research has been devoted to the lysogenic state because it has implications for gene-regulatory mechanisms in all organisms and for dormancy in human viruses. The integration of viral nucleic acid into the host genome is also of considerable interest in diseases such as AIDS and cancer.

Lysogenic Conversion

As mentioned above, lysogeny can alter surface components of the host bacterium and render it immune to attack by other bacteriophages. Lysogenic cells can undergo **lysogenic conversion,** thus gaining other properties that nonlysogenized counterparts do not have. These conversions result from expression of genes on the phage chromosome that are not repressed during lysogeny, and will continue to be expressed as long as the bacterium retains the prophage. Lysogenic conversion changes the phenotype of a cell and can lead to an increase in pathogenicity in some disease-causing bacteria.

Lysogenic strains of *Corynebacterium diphtheriae* produce a potent exotoxin that is the cause of diphtheria. Scarlet fever results from a toxin released by lysogenized

Staphylococcus aureus, and the botulinum toxin is also a product of prophage-bearing *Clostridium botulinum.* The epsilon phage of *Salmonella* sp. alters surface lipopolysaccharides by altering activity of enzymes involved in their synthesis. This changes the antigenic character of the surface and susceptibility of the pathogen to specific antibody. Elimination of the prophage from the aforementioned bacteria removes the virus-induced pathogenicity.

Lambda Phage

Lambda (λ) has long been a model system for understanding lysogeny and gene regulation mechanisms in its host, *Escherichia coli.* In the lysogenic state, λDNA is integrated into the host genome at a specific site (*att*B) located between the *gal* operon (for catabolism of galactose) and the *bio* operon (for biotin biosynthesis). The integration of λ phage is effected by the protein integrase (*int* gene product) that is encoded in the viral genome. The genome of λ has a region of DNA (*att*P) that is homologous to the *att*B site of the host genome. Immediately after λ infects the host its DNA is circularized. The circularized virus integrates by a single reciprocal crossover between *att*P and *att*B, leading to the lysogenic state.

Whether lysis or lysogeny occurs with λ is determined by a number of host and phage factors, but it ultimately depends on the concentration of two viral-encoded proteins, Cro and cI (the λ repressor). The *cro* and *cI* genes are adjacent in the λ genome and are transcribed in opposing directions (Figure 14.10). The adjacent promoters are designated P_{RM} ("repressor maintenance") for *cI* transcription and P_R ("rightward") for *cro* transcription. cI and Cro proteins bind to the same transcriptional operators (O_{R1}, O_{R2}, and O_{R3}) adjacent to P_{RM} and P_R. When cI binds to O_{R1} and O_{R2} it represses rightward transcription from P_R, but activates transcription from P_{RM}. Therefore, more cI repressor is made and the lysogenic state is maintained. The cI repressor prevents expression of virtually all phage genes, including those transcribed from P_L (the leftward promoter). Thus, the cI protein is both a repressor and an activator of transcription. The level of the cII protein, an activator of λ*int* gene transcription, is also important for promoting lysogeny. The cII protein is unstable and affected by the physiological state of the bacterium.

Induction of the lytic cycle depends on the relative amounts of the two repressor proteins cI and Cro. Although these repressors bind to the same operators (O_{R1}, O_{R2}, and O_{R3}) that regulate P_R and P_{RM}, they bind them in different order. The cI repressor has the highest affinity for O_{R1} and lowest for O_{R3}, whereas Cro has the highest affinity for O_{R3}. A high cI to Cro ratio will inhibit P_R and maintain lysogeny as described above. If cI levels drop (by UV-induced cleavage, for example) Cro binds first to O_{R3}, repressor synthesis from P_{RM} is inhibited, and lytic genes are transcribed from P_R and P_L. Expression of lytic genes results in excision of the prophage from the bacterial genome *att* site as a covalently closed circle.

The λ DNA replication cycle is presented in Figure 14.11. The excised, circularized genome becomes supercoiled and early replication via "theta structures" yields additional copies of the circularized λ phage genome. Later, rolling circle replication leads to synthesis of linear concatemers (chain composed of a number of repeated units of duplex genomic DNA). The concatemers are cleaved to form the individual λ phage genomes when the DNA is packaged in a separately synthesized capsid. Mature viruses then escape by lysing the host cell wall.

Mutagens and other agents that disrupt bacterial DNA and affect the integrity of the cI repressor are ultimately responsible for inducing the lytic cycle and virus release. Lysis can therefore be considered a survival mechanism for the phage.

We will now examine in some detail two other bacterial viruses that have been studied extensively and serve

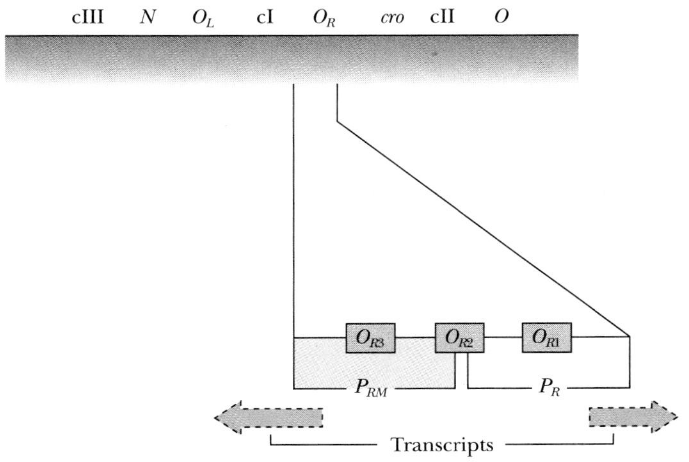

***Figure* 14.10** The critical area of the genetic map of λ phage that determines whether a lytic cycle or lysogenic state occurs.

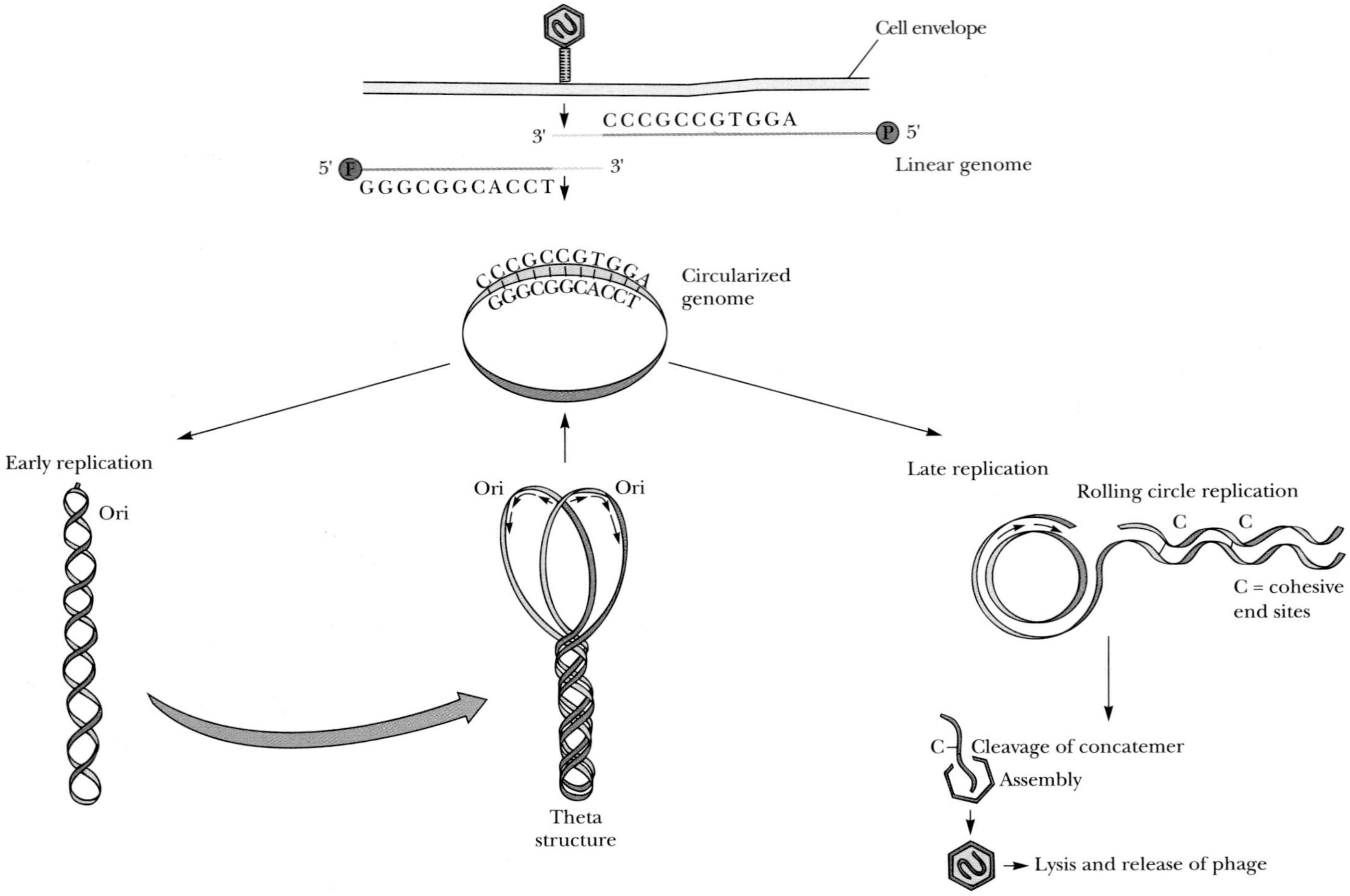

***Figure* 14.11** The replication of λ phage. On entry, the linear genome is circularized by joining the cohesive ends. In early replication the genome is circularized and the genome replicated via the theta structure. After production of a number of circularized forms, late replication occurs by rolling circle replication. The concatemer formed is cleaved at the cohesive end sites as it enters the phage capsid. The assembled virion is released by lysis.

as examples of alternative mechanisms by which bacterial viruses propagate.

M13 and T4

Bacterial viruses have proven to be excellent models for the study of viral infection and propagation. Some of the viral types and their size, structure, host, and nucleic acid components were presented in Table 14.2. Because the diversity of viral types is too large to discuss more than a limited number, only the complex virus T4 (Figure 14.6) and the filamentous M13 (Figure 14.12) will be discussed in detail.

M13

The filamentous phage M13 has an unusual property in that it can invade a cell *(Escherichia coli),* reproduce, and exit without serious harm to the host. The M13 virion is a thin filament about 895 nm in length and 9 nm in diameter (Figure 14.12). The inner diameter of the protein capsid is 2.5 nm. This encloses a single-stranded loop of DNA that extends the entire length of the filament. Stretched out, the viral genome would be nearly 2 μ meters in length. The capsid coat consists of about 2700 individual protein subunits that overlap one another. At one end of the virion are four pilot proteins that guide the virus into and out of the host. In one generation, *E. coli* infected with M13 can produce 1000 intact progeny virions. A considerable mass of virions can readily be collected (100 mg/liter growing culture) free of the host cell debris normally associated with a lytic infection.

GENOME The M13 genome is composed of 6407 nucleotides and there are 10 defined genes (Figure 14.13). The genome has an intergenic space (about 8 percent of the genome) that is the origin of DNA replication. A fila-

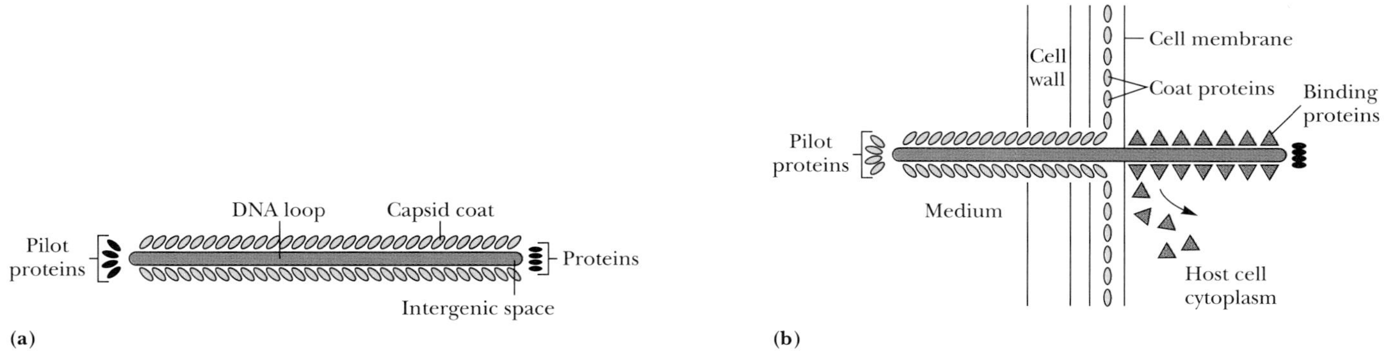

Figure **14.12** A schematic of the virus M13. **(a)** There are distinct proteins at the ends of the structure. The genome is a single strand of DNA. **(b)** The coat proteins are added as the virus exits from the cell.

mentous virus does not have the constraints on genome size, which are imposed by an icosahedral head of limited capacity and simply extend the length of the filament.

COAT The M13 coat protein is composed of subunits that contain 50 amino acid residues. The capsid protein from gene VIII accumulates on the plasma membrane of the infected cell and is assembled on the genome as the DNA exits from the cell. The binding proteins that protect the M13 genome are removed by endopeptidase as the pilot guides the genome outward.

PENETRATION AND INFECTION The DNA of M13 is not injected into the host cell as occurs with most bacterial viruses. The virion attaches at or near the F (sex factor) pilus of *E. coli.* (organisms lacking an F pilus are not susceptible to attack). Two theories have been advanced for viral penetration: (1) the virion follows a groove in the F pilus to a receptor site on the cell surface where it enters, or (2) the virion stimulates a progressive depolymerization of the pilus that retracts the tip and brings the M13 to the surface.

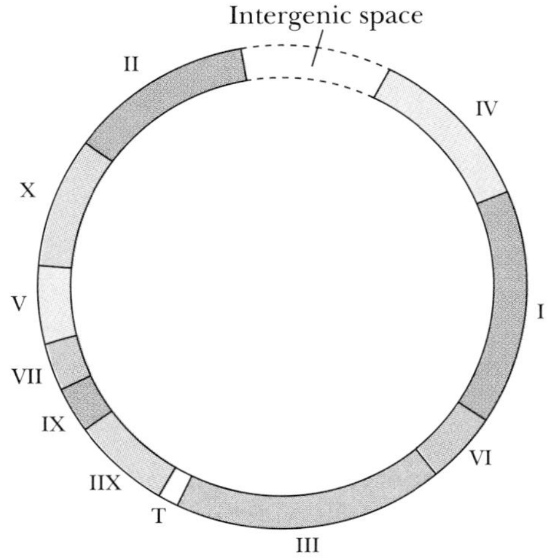

Figure **14.13** A genome map of M13. The function of the genes is as follows: T=termination point; II=production of the ds replicative form (RF) of DNA; V=binding protein that prepackages newly formed DNA; IX=minor coat protein; VIII=major coat protein; III=pilot protein guides virus into and out of host; VII, VI, I, IV are involved in morphogenesis; X is unknown; and I=intergenic space.

Entry of the intact virion is probable, and decapsidation occurs within the cell cytoplasm. Replication of the (+) single strand occurs simultaneously with coat removal. Most of the capsid protein is conserved and deposited on the plasma membrane and reappears as capsid progeny DNA. The gene 3 pilot protein present at the lead end of the virus remains bound to the genome on it is initially replicated to the duplex form. The enzymes involved in initial replication are provided by the host and are normally associated with synthesis of extrachromosomal elements. M13 does not take over the genetic apparatus of the host. The growth rate of *E. coli* rapidly producing M13 virions is reduced by about one-third.

REPLICATION The overall events involved in the replication of M13 viral DNA are presented in Figure 14.14. The process can be divided into three stages. In the first stage the viral single-stranded DNA (+) is converted to the double-stranded replicative form (RF). There are no viral-coded proteins synthesized for this step. The phage relies on the host replicative system and the gene 3 encoded A protein apparently is involved in directing this synthesis. Remember that in M13 infections the host synthetic machinery is not disrupted and the bacterial genome remains intact and functional.

In the second stage, a virus-encoded gene (2) is expressed and leads to synthesis of a replicase protein that promotes the multiplication of RFs. The negative strands of these RFs generate progeny (+) strands and serve as a template for continued production of viral genome (Figure 14.15).

In the third stage, the synthesized viral single-stranded (+) DNA is coated by binding proteins. This envelops the

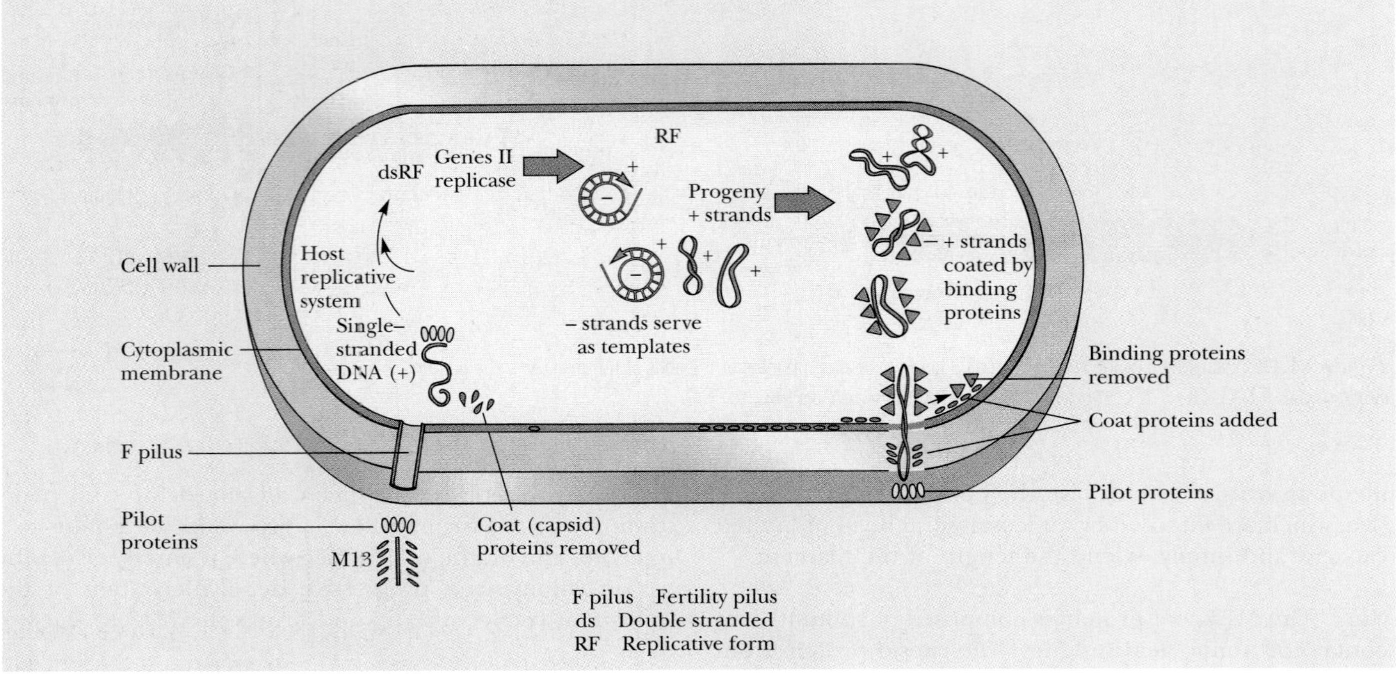

***Figure* 14.14** Summary of events that occur during replication of M13.

DNA in a form for assembly at the membrane. The function of the gene 5 binding protein is to facilitate the processes that generate (+) ss DNA. Binding also prevents the (+) strands from being employed as templates for synthesis of duplexes and protects against nucleases. As protein A guides the DNA through the cytoplasmic membrane the capsid protein replaces the binding protein. This process is not clearly understood. The mature virus is extruded through the multilayered gram-negative bacterial cell wall without disrupting the host cell.

T4

The *Escherichia coli* phage T4 is a complex phage with several distinct functional parts including a head, tail, whiskers, and tail fibers (see Figure 14.6). T4, along with T2 and T6, is one in a series of bacterial viruses called the T-even phages, which share about 85 percent DNA homology. The most studied has been T4 because it is large, complex, and has many capabilities, including the ability to propagate in a short period of time. As discussed earlier in this section, the host for T4, *E. coli,* has a doubling time of about 22 minutes, which means that in 22 minutes one cell can produce another complete and equivalent cell. The genome of *E. coli* has about 3000 genes and reproduces these in about 40 minutes. The organism can divide in 22 minutes because multiple rounds of DNA replication are ongoing in a rapidly growing cell (see previous chapter). However, when *E. coli* becomes infected with T4, the virus assumes complete control of the cell's reproductive machinery. Within about 22 minutes after infection the *E. coli* host lyses and releases up to 300 viral particles. These viral particles each contain a genome bearing 200 genes. Essentially a normally dividing *E. coli* produces about 3,000 genes in 40 minutes, while the machinery of an infected bacterium can generate 60,000 genes in an equivalent length of time. The ability to take over a cell and accomplish this feat in so short a time makes the T4 virion a prime object for study. The events that occur during a lytic infection of *E. coli* follow.

GENOME The T4 genome is among the largest of genomes relative to those present in plant, animal, or bacterial viruses. It encompasses about 200 genes and is synthesized as a **concatemer,** a continuous series of separate genomes linked in a linear fashion. As the genome is packaged in the head of the virion somewhat more than one genome (105 percent) enters. This ensures that the same gene will be present at or near the end of each genome (Figure 14.16), a phenomenon called **terminal redundancy.** The sequence of genes in each viral head is equivalent, but the genome may start and end with a different gene. Therefore, each viral particle has a complete genome plus two sets of at least one gene.

The genetic map of T4 is depicted as circular (Figure 14.17). The genome is linear as packed in a phage head and enters a bacterial host as a linear strand. All markers

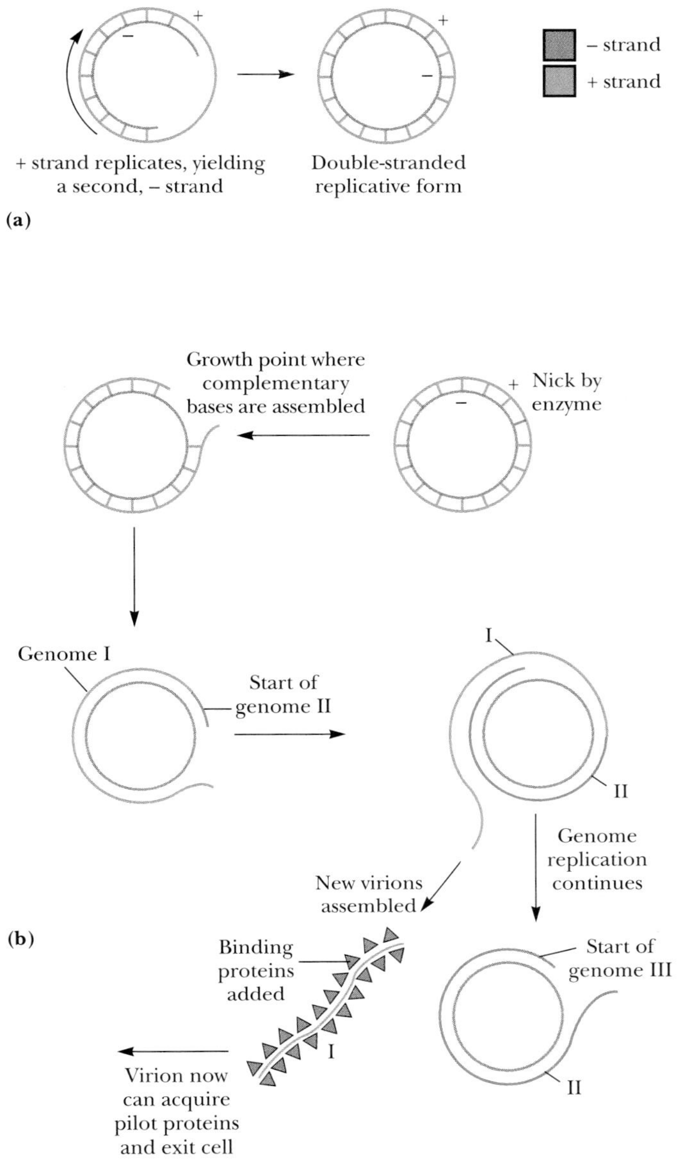

***Figure* 14.15** The viral genome of M13 is a (+) strand. **(a)** Initially, the + strand is replicated, giving rise to the (−) form. Together, + and − strands are the M13 replicative form. **(b)** The replicative form yields (−) strands for templates to generate single-stranded viral genomes. This is a rolling-coil form of genome production.

or genes in the genome are linked. Mapping indicates that each marker identified is spatially linked to other known markers. Given that the gene at the end of each genome differs, no single gene would appear at the end of the linear chain with any consistency. The genes as mapped would be linked on both sides by other known genes. Every gene would be in the middle of a linear genome with the same frequency. This phenomenon is called **circular permutation.**

The expression of T4 genes is tightly regulated. The organization of the genome is such that expressing the genes in the order they appear would follow the life cycle of the virus (Figure 14.18). Genes with related function are located in adjacent areas of the viral genome. Early genes are transcribed in a counterclockwise direction and late genes tend to be read in the opposite or clockwise direction. Genes in T4 are expressed in this way, but this is not necessarily the case in other phages.

ABSORPTION AND PENETRATION There are selected areas on the outer envelope of *Escherichia coli* that act as receptors for the fiber tips of the T4 virion. The viral particle attaches via tail fibers to the surface of the bacterium at these recognition sites. This attachment may occur at a fusion point between the inner cytoplasmic membrane and the outer cell envelope. After attachment of the base plate, the sheath reorganizes and the inner core of the tail penetrates through the cell envelope and cytoplasmic membrane. The sheath contains 24 protein rings, but during penetration the proteins slide downward to form 12 rings. The tail sheath contains ATP and the contraction has been compared with muscle activity. The core is unplugged and the DNA genome is injected into the host. The precise mechanism of injection into the host cytoplasm is not yet known.

EFFECT ON HOST Within one minute after injection of viral DNA the synthesis of host-specific macromolecules ceases. This stoppage occurs because the bacterial RNA polymerase has a strong affinity to promoters of T4 DNA, so T4 DNA transcription gets higher priority. Transcription of selected phage genes is initiated a few minutes after

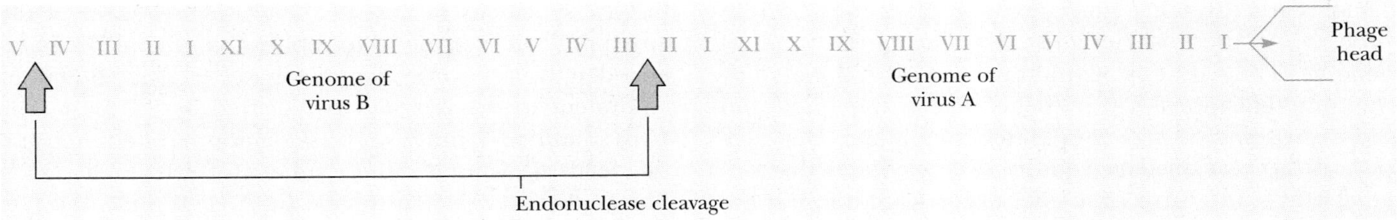

***Figure* 14.16** A concatemer of T4 as it might be cleaved. Virus A would receive duplicates of I and II.

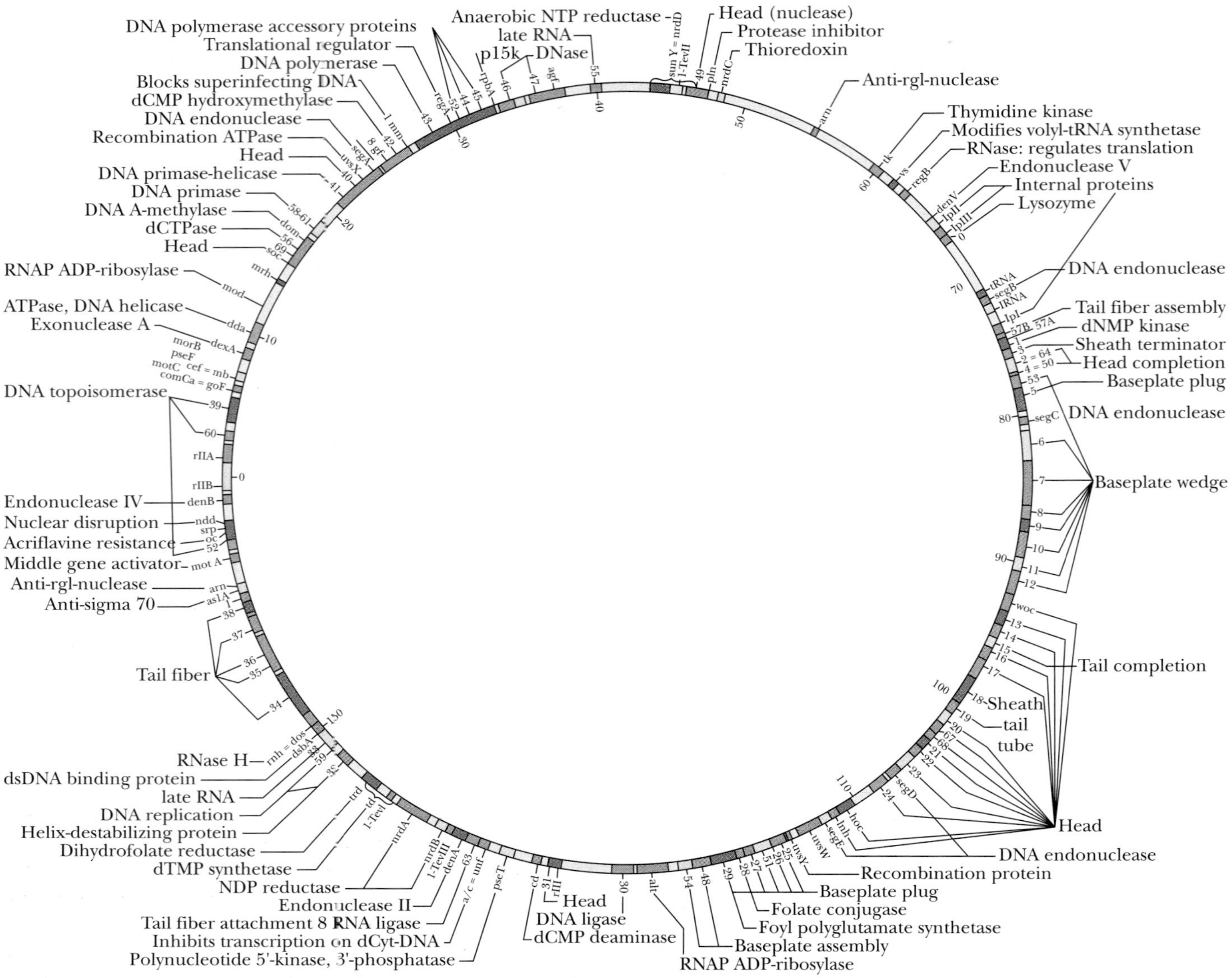

***Figure* 14.17** The genome map of T4. Although the map is drawn in a circle, in a virus particle it is linear. The open end in a given genome may be anywhere as the concatemers are cut by the endonuclease, shown in Figure 14.18. (With permission, *Molecular Biology of Bacteriophage T_4*, edited by Jim D. Karam, American Society of Microbiology, 1994)

infection. In less than 2 minutes, the host RNA polymerase begins transcribing phage mRNA. This mRNA that is translated to form early proteins is transcribed from viral DNA, and this occurs before viral specific DNA is synthesized. These early proteins and enzymes take over various processes in the host cell and effect changes in the host genome. The synthesis of viral nucleic acid is initiated by these modifications. The host genome is disrupted and moves to a site on the cytoplasmic membrane. Degradation of the host nucleus to produce nucleotides that can be assembled to form viral DNA occurs within 5 minutes after infection. In less than 10 minutes the viral attack turns the host cell into a virus-synthesizing machine.

REPLICATION To induce the host to immediately express the T4 genome the virus must take control of transcription and translation. Shortly after infection, proteins are synthesized, employing viral genome information that modifies the specificity of host RNA polymerase. Many of the enzymes actually involved in replication of viral DNA and encoded by the virus genome have catalytic functions equivalent to the enzymes synthesized by the host. The virus stimulates synthesis of larger amounts of enzyme and this leads to a more rapid synthesis of T4-specific DNA. The replication rate for phage DNA is 10 times the rate of host DNA synthesis in an uninfected, normally dividing host. Phage-induced enzymes are active in promoting this

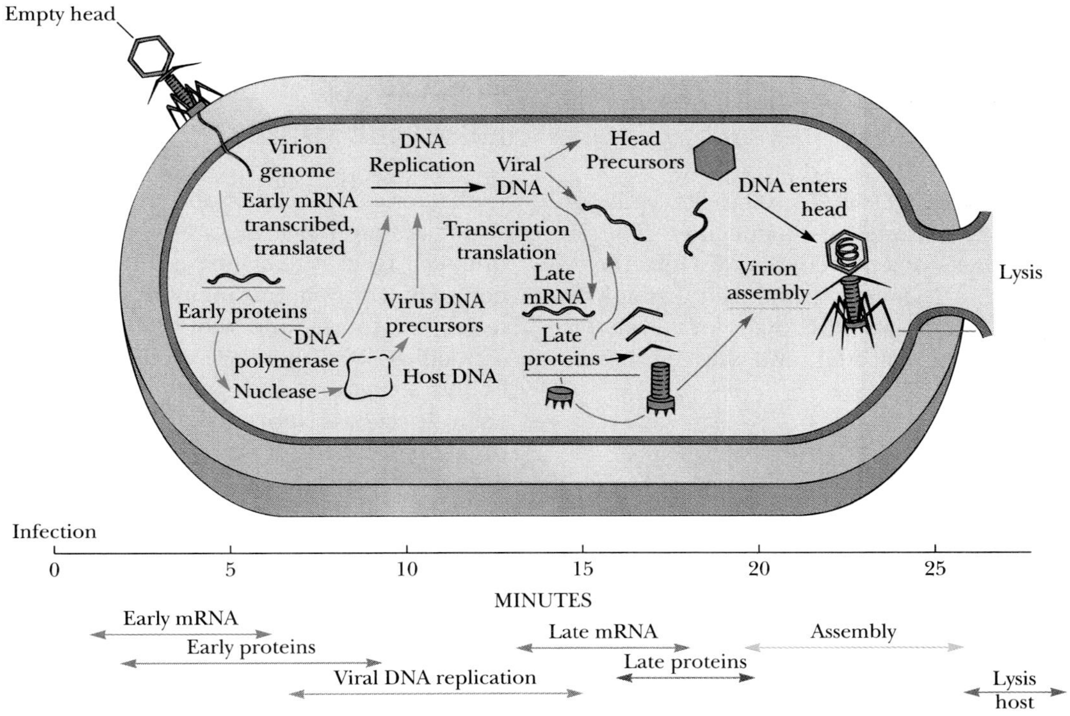

***Figure* 14.18** The order of events in a lytic infection by the *E. Coli* virus T4. The early genes produce messages for proteins that shut off host function, and delayed early genes are involved with replication of the T4 genome and nucleases that degrade the host DNA.

increase. There are at least 20 viral-encoded proteins synthesized immediately after infection, and some of these promote the synthesis of new tRNAs. These tRNAs read the T4 mRNA more efficiently and rapidly than occurs in a normal growing cell. The phage also induces the synthesis of about 30 replication proteins. These proteins are involved in nucleotide biosynthesis, DNA synthesis, and modifications of DNA.

The pyrimidine cytosine is the nucleic acid base generally paired with guanine in cellular DNA. In T4 DNA the cytosine is replaced by the unique base 5-hydroxymethyl cytosine (Figure 14.19). Endonucleases encoded by the viral genome hydrolyze the host DNA. To prevent the inclusion of unaltered deoxycytidine triphosphate (dCTP), as dCTP is released, it is converted to dCMP. The dCMP reacts with hydroxymethylate to form 5-hydroxymethyl dCMP. The 5-hydroxymethyl dCMP is then phosphorylated to form 5-hydroxymethyl dCTP that is incorporated into the viral genome. A glucosylation reaction adds a glucose molecule to the hydroxy group after the dCTP derivative is incorporated into viral DNA, and this protects the DNA of the virion from attack by host-restriction enzymes.

ASSEMBLY After the polyhedron head of T4 is completed, it is packed with viral DNA. A few low molecular weight core proteins and phage-induced enzymes are also incorporated into the T4 head. Magnesium and polyamines are also included to neutralize the genome DNA. It is remarkable that the 0.085 μ meter wide capsid of T4 can incorporate a DNA genome that is 500 μm in length.

Translation of the late mRNA results in the synthesis of tail fibers and whiskers. The base plate is assembled as a unit and the core or tube through which the DNA is

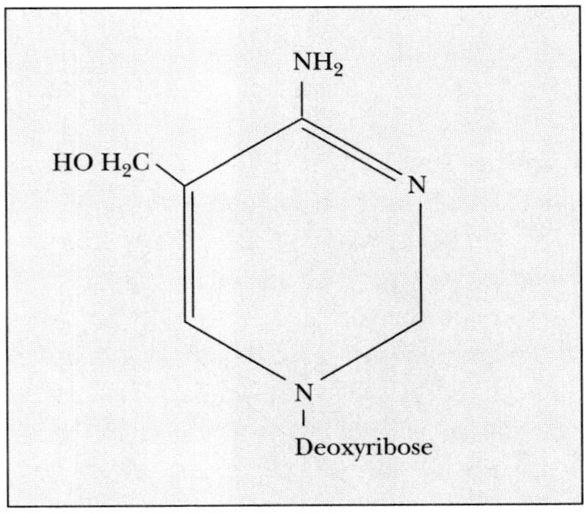

***Figure* 14.19** The unique pyrimidine present in the DNA of T4, 5-hydroxymethyl cytosine. This base often has a glucose (glycosylated) attached to the hydroxyl.

injected is assembled on this base plate. The core is capped at the distal end and encircled by the sheath proteins. The head and tail are then joined together. Whiskers are put in place, and these probably direct the placement of the long tail fibers on the base plate.

RELEASE. Lysis of *E. coli* and release of 100 to 300 viral particles occurs about 25 minutes after infection. The T4 genes direct the synthesis of a protein that disrupts the cytoplasmic membrane. A lysozyme synthesized through T4 templates moves through the disrupted membrane to breach the peptidoglycan cell wall, and the cell disintegrates releasing viral particles.

Viral Restriction and Modification

Bacteria have effective mechanisms for protection against viral invaders. The outer surface may be modified to prevent attachment, or they produce enzymes, termed restriction enzymes, that cleave DNA at one or more sites. Restriction enzymes are quite specific and only attack selected nucleotide sequences. These sequences are generally 5 to 6 base pairs in length. The host is protected against its own restriction enzymes by chemical alteration of its DNA at sites where the restriction enzyme might function. An example of chemical modification is methylation of bases at potentially susceptible sequences.

Viruses can counteract restriction enzymes by adding methyl or sugar groups (glycosylate) to the nucleic acid bases at susceptible sites. The T-even phages (T2, T4, T6) that infect *Escherichia coli* can glycosylate selected bases in DNA, rendering the genome resistant to restriction enzymes. The lambda phage methylates adenine and cytosine bases in its DNA. Methylation and glycosylation occur after replication of DNA. The host bacterium has the enzymes involved in methylation that otherwise would be employed for its own use.

Eukaryotic Viruses

Prokaryotic and eukaryotic cells differ significantly in structure, and the strategy of viral attack on these two distinct cell types reflects these differences. A prokaryote is bound by a cytoplasmic membrane containing nuclear material, ribosomes, and other cell machinery. In contrast with eukaryotes, there are no internal plasma membrane-bound structures or separate functional parts. The eukaryotic cell is more complex, with discrete internal organelles. These organelles, including the nucleus, are bound by membranes.

When a virus attacks a susceptible bacterial host, the genome of the virus must traverse the cell wall and cytoplasmic membrane. Once the genetic information enters the cytoplasm, it has the transcription/translation machinery of the cell available to it all in one compartment. In infecting a eukaryotic cell, the virus must pass through the cytoplasmic membrane and, in most DNA viruses and some RNA viruses, the viral genome must also gain access to the membrane-bound nucleus. Some representative eukaryotic viruses are depicted in Figure 14.20.

The receptor present on a virus attaches to specific sites, and this influences both the hosts it can infect and, in multicellular hosts, the specific cells that will be infected. Penetration of the viral particle into a eukaryotic cell occurs by several mechanisms and varies with the nature of the virus. Binding of unenveloped viruses to selected host membranes is irreversible, with release of the viral particle into the cytoplasm. The enveloped viruses enter the host cell by inducing fusion between the host and viral membranes or by endocytosis. A virus-encoded gly-

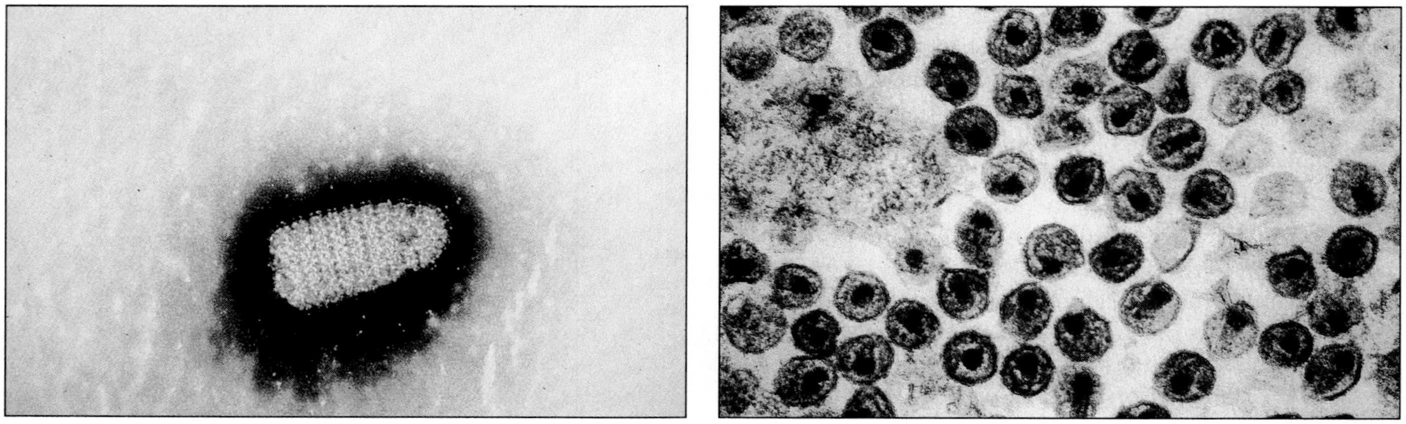

(a) (b)

***Figure* 14.20** Electron micrographs of typical animal viruses: **(a)** rabies virus × 35,000 (© CNRI/Phototake, NYC), and **(b)** human immunodeficiency virus × 200,000 (© David M. Phillips/VU).

coprotein on enveloped viruses is generally responsible for the attachment to selected host cells. Not all viruses are uncoated immediately on entry, although the nucleocapsid may be altered. For example, adenovirus virion moves to the nuclear membrane where DNA enters the nucleus and capsid proteins remain in the cytoplasm.

Transcription in eukaryotes takes place in the nucleus, and the mRNA traverses to and is translated in the cytoplasm. The eukaryotic mRNA has a polyadenylate tail consisting of 100 to 200 adenylates at the 3′ end. These assist the mRNA in exiting from the nucleus. The mRNA has a methylated guanosine at the 5′ end called the cap. This cap is required for the ribosomes to bind the mRNA.

A typical eukaryotic gene has two distinct types of sequences called the exons and introns. As described in Chapter 13, exons are protein coding and the introns are nonprotein coding. A primary transcript would contain both the exon and the intron regions. The mRNA must be processed to remove the noncoding areas before translation to protein. The polyadenylate tail and cap remain. Prokaryotic mRNA is generally polygenic, that is, it carries information from more than one gene. Eukaryotic mRNA generally carries one gene and is monogenic.

Classification, Size, and Structure

Some major animal and plant viruses are illustrated in Tables 14.3 and 14.4. These viruses, as with bacterial viral particles, are classified by the nature of the nucleic acid in the genome, although the presence/absence of an envelope is also considered. The animal viruses range from the tiny parvoviruses (20 nm) to very large poxviruses (250 × 350 nm). Many of the plant viruses are very narrow long filaments. The Potato γ virus is 10 nm across but 700 nm in length.

Types of Infection

Viruses can have different effects on the host cell. A **lytic infection** results in the destruction of the host cell. Picornaviruses such as the polio virus and the common cold virus are lytic viruses. In a **persistent infection** the virus replicates actively but the host cell retains viability, and viral production continues for an extended period of time. The mature virus can leave the host cell by budding without disrupting the integrity of the plasma membrane. A **latent infection** is one in which the virus is not actively replicating within the host. The herpesvirus that causes cold sores is an example of a virus that can be involved in a latent infection. All plant viruses cause persistent infections, and a few are involved in latent infections.

Some viral infections can cause **transformation** of the host cell. Transforming viruses can change a normal cell into a cancer cell with fewer growth factor requirements than for the normal cell. As a consequence the transformed cell reproduces more rapidly, resulting in a mass of cells called a tumor. Some tumors are self-limiting, do not spread, and are called **benign.** Others are **malignant** tumors, which spread and grow in other tissue and organs causing dysfunction or death of that tissue. Malignant tumors initiated by viruses result from genetic changes, either in expression or structure of the genes, that lead to a loss of growth regulation. Abnormal masses of cells result, and these cancers or neoplasms can grow unchecked. Phosphorylations, transcription, or enzymes involved in DNA or RNA synthesis are among the altered functions caused by tumor viruses, and these altered functions can also result from other physical factors such as mutations. Chemicals, diet, and environmental factors can cause genetic alterations that ultimately lead to cancer.

Animal Viruses

The genome of viruses that infect animals is composed of either DNA or RNA (Table 14.5). The genomes of DNA viruses are mostly double stranded but a few, such as the parvoviruses, are single stranded. The genome of RNA viruses are generally single-stranded RNA. The genome may be surrounded by a nucleocapsid or a nucleocapsid that is in turn surrounded by an envelope of varying complexity. Attachment to the host cell may involve the nucleocapsid, the envelope, or the spikes that extend out from the surface of the envelope. The unenveloped viruses enter a host cell by mechanisms that are not clearly understood. For example, the polio virus attaches to a susceptible cell, the protein capsid loses structural integrity, and the RNA protein complex is translocated into the cytoplasm. It is not clear whether the capsid enters the cytoplasm in all cases with unenveloped viruses. The mechanism of attachment and penetration of the genome of enveloped viruses into the host can occur as depicted in Figure 14.21. A limited number of representative viruses will be discussed in some detail (for further information, see the reading list at the end of the chapter). There is considerable evidence that other viruses, not yet studied in detail, will receive greater attention in the future (see **Box 14.2**).

DNA Viruses

Most DNA viruses replicate in the nucleus of the host cell. Poxviruses are an exception as they replicate in the cytoplasm. Herpesviruses are an example of a ds DNA virus that replicates in the nucleus of the host cell and is the causative agent of several human ailments including fever blisters, shingles, genital herpes, chicken pox, and mononucleosis. The herpes simplex virus is described in some detail as a typical DNA virus.

Table 14.3 Some of the Major Animal Virus Families

DNA – ds – linear	Size (nm)	Example of disease	
Poxviruses	250 ∞ 350	Smallpox, fever blisters	
Herpesviruses	180 – 200	Genital herpes	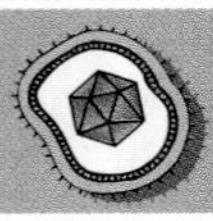
Adenoviruses	75	Conjunctivitis	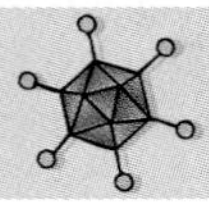
ss – linear			
Parvoviruses	20	Gastroenteritis	
ds – circular			
Papovaviruses	50	Warts (human)	
Baculoviruses	40 ∞ 400	Polyhedrosis (Lepiderm insects)	
RNA – ss – (+)			
Piconaviruses	27	Polio, colds	
Togaviruses	50	Rubella, sindbis	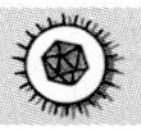
Flavivirus	40 ∞ 50	Yellow fever	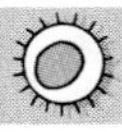
Retroviruses	80	Sarcomas, leukemia	
Coronaviruses	25	Avian bronchitis	
ss (–)			
Orthomyxoviruses	110	Influenza	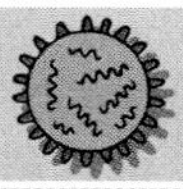
Paramyxoviruses	200	Colds, measles, mumps	
Rhabdoviruses	70 ∞ 170	Rabies	
Bunyaviruses	90	Encephalitis	
Filoviruses	50 ∞ 1000	Ebola virus	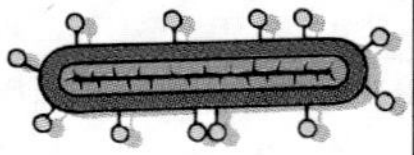
ds			
Reoviruses	65	Encephalitis, diarrhea	

***Table* 14.4 Some of the Major Plant Virus Families**

DNA ds – linear	Size (nm)	Example of disease	
Caulimovirus (pararetrovirus)	50	Cauliflower mosiac	
ss – circular			
Geminivirus (paired genome segments)	18 ∞ 30	Maize streak	
RNA ss (+) linear			
Tobamovirus	15 ∞ 300	Tobacco mosaic	
Comovirus (2 genome segments)	30	Cowpea mosaic	
Carlavirus	15 ∞ 650	Carnation latent	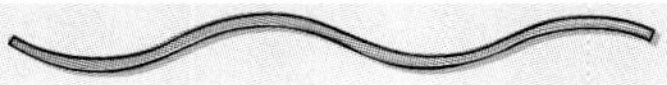
Cucumovirus (3 genome segments)	30	Cucumber mosaic	
Potyvirus	10 – 700	Potato	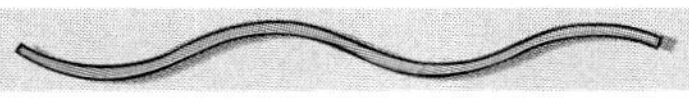
ds – linear			
Phytoreovirus (multiple genome segments)	80	Wound tumor	
ss (–)			
Tospovirus	90	Tomato spotted wilt	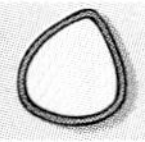
Rhabdovirus	70 ∞ 170	Lettuce necrotic yellows	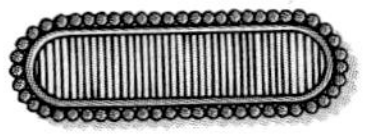

Herpesviruses

Some of the herpesviruses, such as *Herpes simplex* Type 1, can be latent for extended periods of time and become active when the host is under stress (for example, sunburn, fever). The virus remains latent in neurons of sensory ganglia and is not integrated into the host genome. The herpesvirus particle is quite complex, with an icosahedral capsid enclosed by an envelope bearing elaborate glycoprotein spikes that protrude outward (Figure 14.22). Herpesvirus is about 100 by 200 mm in size. Between the envelope and nucleocapsid is an electron-dense amor-

***Table* 14.5 The Basic Mechanisms for Production of Viral Proteins and Genomic Nucleic Acid in Animal Viruses**

Example of Disease Caused	Genome Type	Mechanism
Polio, cold	ss RNA (+)	Processing ⟶ mRNA ⟶ Proteins Replicase ⟶ (±) RNA ⟶ RNA (+)[1]
Influenza, encephalitis, mumps	ss RNA (−)	RNA dependent polymerase ⟶ mRNA ⟶ Proteins Replicase ⟶ (±) RNA ⟶ RNA (−)[1]
Encephalitis, diarrhea	ds RNA (±)	Transcriptase ⟶ mRNA ⟶ Proteins mRNA replicase ⟶ RNA (±)[1]
Retrovirus		
AIDS, Rous sarcoma	ss RNA (+)	Reverse transcriptase—DNA (−) Reverse transcriptase DNA (±) ⟶ RNA (+)[1] ⟶ mRNA

[1]Virion genome.

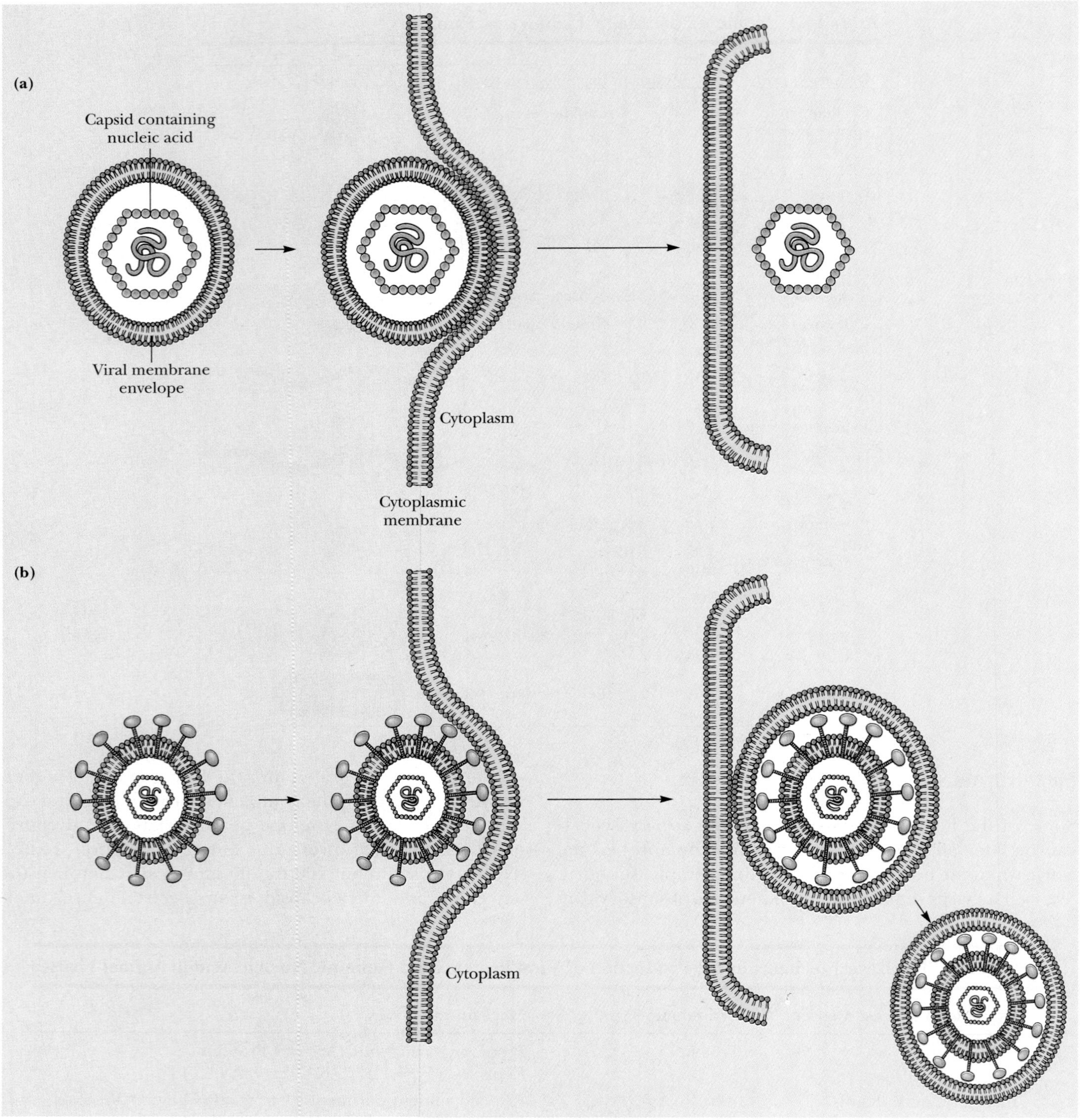

***Figure* 14.21** Attachment and penetration cf enveloped viruses into a host cell. **(a)** Membrane fusion between the host cytoplasmic membrane and the virus envelope. **(b)** Endocytosis with invagination of the host cytoplasmic membrane. The nucleocapsid is released after entry into the cytoplasm.

phous material of unknown function. The icosahedral nucleocapsid is composed of 162 capsomers made up of several distinctly different proteins. The genome is a ds DNA that carries genetic information for the synthesis of over 100 different polypeptides.

The glycoprotein spikes that protrude outward from the envelope are the means of attachment to host cell receptors. After fusion to the host cytoplasmic membrane, the nucleocapsid is extruded into the cytoplasm. The viral particle is uncoated and the viral ds DNA traverses to the host cell nucleus. Initially the viral genome directs the synthesis of mRNA that is translated into regulatory proteins and mRNA that is transcribed to generate the proteins involved in DNA replication. A later mRNA is transcribed from the viral genome and codes for the numerous proteins that are an integral part of the viral particle.

The herpesvirus DNA is synthesized in the host cell nucleus as long concatemers. These long concatemers are processed to viral genome length as they are incorporated into the nucleocapsid. The viral nucleocapsid capsomers are synthesized in the cytoplasm and the nucleocapsid is assembled in the host nucleus. The amorphous fibrous coat that surrounds the nucleocapsid and the envelope are acquired as the virus buds through the host nuclear membrane. The virus particle passes to the endoplasmic reticulum of the host cell and exits the cell from this site. In many cases the accumulation of viruses may cause disruption of the cytoplasmic membrane.

Pox Viruses

The pox viruses are complex viruses that have some characteristics of a free-living cell. One disease caused by this group, smallpox, was important historically (see Chapter 2), but it has been eradicated by worldwide vaccination. The poxvirus replicates in the cytoplasm. The synthesis of poxvirus DNA outside the nucleus is unusual, and such extranuclear DNA synthesis generally occurs only in specialized organelles such as chloroplasts or mitochondria.

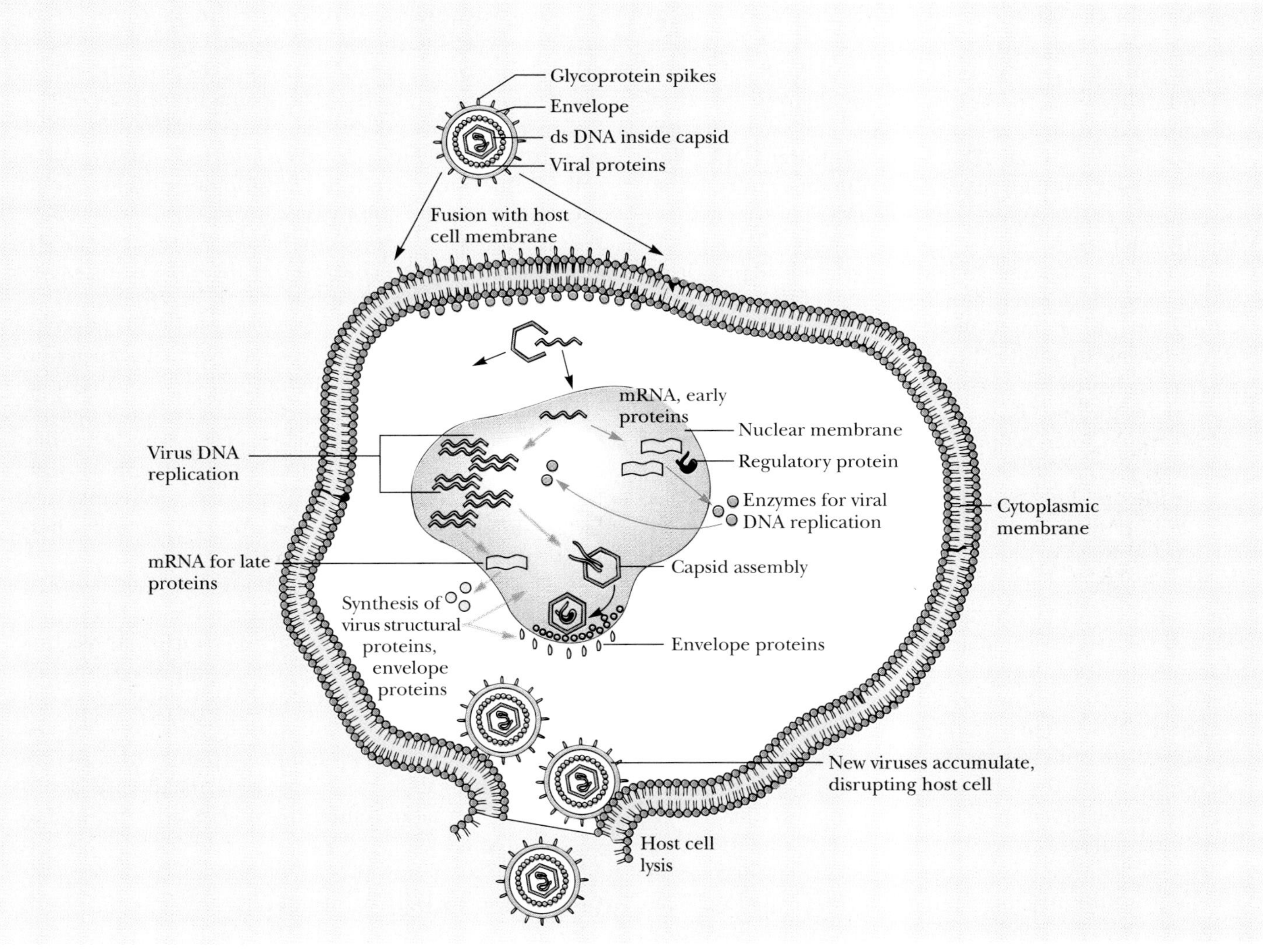

***Figure* 14.22** The replication of herpes simplex virus. The glycoprotein spikes in the viral envelope attach to host cell receptors in the cytoplasmic membrane. The nucleocapsid is uncoated, and viral DNA is transported to the nucleus. Early transcription and mRNA processing are catalyzed by host enzymes. Early proteins are involved in viral DNA replication. RNA transcripts generated in the nucleus are responsible for directing the synthesis of structural proteins. These proteins enter the nucleus and are involved in assembly of the virus. The envelope is added as the nucleocapsid leaves the nucleus. Accumulation of viruses causes disruption of the host and release of virus. The viral particle may exit from the endoplasmic reticulum, and this is not illustrated.

BOX 14.2 RESEARCH HIGHLIGHTS

Emerging Viruses

The discovery and development of antibiotics and vaccines during the late 1880s and early 1900s lead to the control of most infectious diseases, especially in developed countries. A concentrated effort by the WHO resulted in the eradication of smallpox in 1977. However, since the 1970s, 13 new viruses have been identified as major agents of infectious disease in humans, many other new viruses have been isolated, old diseases have reappeared, and established viruses have moved into new areas. Viruses that have recently appeared in humans, or are rapidly increasing in incidence, or expanding in geographic range are referred to as emerging viruses. Many have existed for years as zoonoses, often transmitted by insects or rodent urine, and have high fatality rates in their new human hosts. Most of the new viruses belong to the arena, bunya, or filo virus groups. These are all RNA viruses with an RNA polymerase that produces frequent errors that can result in mutations that support rapid adaptation to environmental changes.

Ecological changes including agricultural practices, deforestation/reforestation, dams, and climate changes; alterations in human demographics and behaviors; breakdowns in public health and/or medical procedures; and travel and commerce have contributed to the phenomenon. Bolivian hemorrhagic fever, caused by Machupo virus, reappeared in 1994 after an absence of 20 years. The disease is associated with exposure to rat urine in the process of converting grasslands to maize and in harvesting the crops. Cases of hemorrhagic fever due to Guanarito virus in Venezuela in 1989 followed clearing of forests, which brought workers into increased contact with urine from cotton rats. The construction of the Aswan dam provided new breeding sites for mosquitoes and resulted in 200,000 cases of Rift Valley Fever in 1970. Another 1,264 cases followed the activation of the Diama Dam in the Senegal River basin in 1987. The mild wet winter and spring in the four corners region of the United States in 1993 lead to increases in the mouse population and pinion nuts, which attracted humans and rodents. Fifty-eight people died from Sin Nombre, a hantavirus, which was passed to humans from the rodent urine. This virus is a close relative of the Korean hemorrhagic fever virus, identified in 1976, which affects 100,000 Chinese a year. The influence of human sexual and drug use behavior on the emergence of the HIV virus is well documented. In 1969, seventeen individuals who cared for a nurse before she died became infected with Lassa virus, which later claimed the life of one laboratory worker in the United States. The spread to the USA of the Asian tiger mosquito, *Aedes albopictus,* in tires imported from Japan has set the stage for Dengue to move in from Asia, South or Central America, and has the potential to alter the distribution of eastern equine encephalitis.

Ebola virus was first recognized in 1976 when separate outbreaks occurred in northern Zaire and southern Sudan, with case fatality rates of 90 percent and 50 percent, respectively. In 1979, a small outbreak of the Sudan type occurred. No new human infections were seen until 1995, but a monkey infected with a genetically distinct strain from the Philippines caused an outbreak in a primate colony in Reston, Virginia, in 1989. When the virus reemerged in Kikwit, Zaire, in 1995, scientists were ready to characterize the virus using molecular biology tools, and rapidly established that the virus differed from the original Zaire strain by 1.6 percent. The index case is believed to be a charcoal maker who worked in the forest, but the natural cycle of the virus and how it is transmitted to humans remains elusive. Once humans become infected, the virus is readily transmitted to other humans who come in contact with infected body fluids. The 1995 outbreak was terminated by transferring the task of preparation of bodies for burial from family members to trained volunteers.

Ecological alterations, increases in international travel, and alterations in the climate will most likely continue. The effect of global warming on insect vectors and animal reservoirs has yet to be established, but laboratory studies suggest that changing temperatures will alter the distribution of mosquitoes and thus the diseases they carry. Unless adequate public health programs are instituted and disease potential is considered in land-use decisions, emerging viruses will continue to be a problem.

Contributed by Dr. Christina L. Frazier

The poxviruses are large, and the vaccinia virus is a boxlike structure (400 × 240 × 200 nm) with an outer coat composed of protein filaments in an array that resembles a membrane. The vaccinia virus is taken up by host cells via a process that resembles phagocytosis. The plasma membrane of the host cell actually extends around the virus, resulting in viral entry. The core of vaccinia virus contains the DNA genome and several viral-encoded proteins. As the virus does not penetrate the nucleus of the host, there are no proteins generated on infection that are transcribed from host DNA. Host genes are not expressed after infection, indicating that a molecular message does enter the host nucleus to turn off the host genetic machinery. All of the proteins expressed after post infection are encoded in the vaccinia genome. Enzymes and DNA polymerase come from the virus. All of the enzymes necessary to transcribe viral DNA are brought in by the virus on infection. The mRNA transcripts of viral origin are capped and polyadenylated even before the viral coat is completely removed.

After a viral particle enters the host cytoplasm, an inclusion body is formed. If the host cell is infected by several viruses, there will be an inclusion body formed around each virus particle. It is in these inclusion bodies that viral constituents are synthesized. After synthesis, the viral parts assemble into a mature viral particle. Host nuclear function ceases on infection, and this ultimately leads to disintegration of the host cell and release of virus particles.

RNA Viruses

The genome of most RNA viruses is a single strand of RNA (ss RNA), and this strand may be the equivalent of mRNA (plus strand). The ss RNA may be complementary to mRNA (minus strand), and in this case it must be replicated to generate mRNA. Major positive-strand RNA viruses that infect humans are polio and the cold (rhinovirus). Negative-strand viruses are the causative agents of rabies, HIV, influenza, and other human ailments. Two major diseases caused by RNA viruses—polio and AIDS—will be discussed.

Poliovirus

The poliovirus has a genome composed of a single strand of RNA. Polio was a dreaded paralytic disease but, since 1954, widespread use of a vaccine has led to its virtual elimination in the United States. The polio virus multiplies in the intestinal tract of human beings, and infection is generally asymptomatic or causes mild symptoms. The paralytic form, poliomyelitis, results from passage of the virus to motor neurons in the spinal cord. Destruction of these nerve cells leads to paralysis. How a limited number of infections progress from the mild intestinal tract infection to a paralytic form is not known.

The polio virus attaches to specific receptors on cells of the host intestine and enters host cells by endocytosis. Following entry of the virus, all protein synthesis in the infected cell ceases. The RNA virus acts as messenger RNA and is about 7500 bases in length. The 5′ end of the RNA codes for coat proteins and replication proteins are coded at the 3′ end. At the 5′ terminus of the linear viral genome is a small protein attached covalently, and at the 3′ end is a polyadenylate chain. The (+) strand is copied by an RNA-dependent RNA polymerase. The (−) strand generated then serves as a template for the production of multiple copies of the (+) strand.

The viral mRNA is a monogenic code that is translated into a single, very large protein molecule. This polyprotein bears all of the proteins involved in poliovirus synthesis. After formation, the large protein is cleaved into about 20 individual proteins. Some of these proteins are structural; others are the RNA polymerase that are involved with the synthesis of (−) strand templates, and a protease that actually cleaves the polyprotein. Replication of viral RNA occurs a short time after infection by generating (+) strands from the (−) strand templates. An infected cell can contain hundreds of (−) strands that can generate up to a million viral (+) strands. The individual proteins cleaved from the polyprotein direct the synthesis of structural coat proteins, and these are assembled to form the mature viral particle.

Retroviruses

The retroviruses are defined by their ability to reverse transcribe a single-stranded RNA genome to form double-stranded (ds) DNA. The (+) single-stranded RNA genome is first copied into single-stranded (ss) DNA, and this ss DNA is replicated to form ds DNA. The viral ds DNA is an intermediate that is integrated into the host genome and resides there as a **provirus.** The ds DNA can be transcribed to generate mRNA.

The retroviruses are the causative agents of certain cancers; Rous sarcoma, a form of chicken cancer, was the first virally induced cancer described. The avian leukemia virus and murine leukemia virus are also retroviruses. The human immunodeficiency virus (HIV) is another retrovirus and the causative agent of AIDS (acquired immunodeficiency syndrome).

Retroviruses are enveloped viruses with a diploid single-stranded RNA genome. They are considered diploid because they have duplicate single strands of the RNA genome (Figure 14.23). The inner core contains the ss RNA genomes and a number of enzymes that are responsible for the early replication events. Among these enzymes are integrase, reverse transcriptase, protease, and ribonuclease. This inner core is surrounded by a protein capsid. There is a protein matrix present on the inner surface of the lipid membrane that stabilizes the viral particle. The envelope is a lipid bilayer with spikes composed of glycoprotein, and these protrude from the surface of the virion.

Infection occurs after direct exposure to the retrovirus HIV-1 through body fluids—blood, semen, or vaginal fluids. There is at least one documented case of HIV being passed in breast milk. The virus enters the bloodstream, where it may encounter a cell with a high-affinity receptor for the virus surface glycoprotein. There is a highly specific binding between the glycoprotein protruding from the exterior of the virus and a surface molecule on selected host cells. The host cell receptor is a glycoprotein

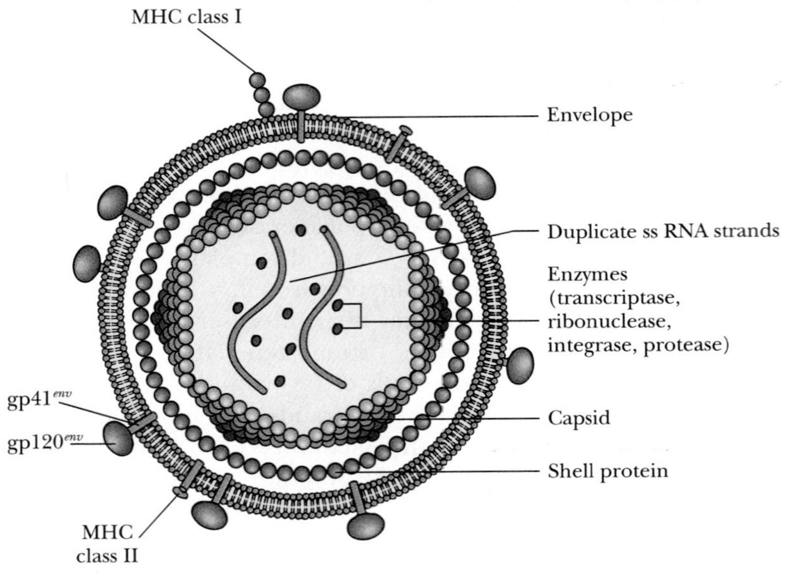

***Figure* 14.23** A cross-section of the human immunodeficiency virus, the causative agent of AIDS. The core contains several viral enzymes, including integrase, ribonuclease, reverse transcriptase, and protease. There are two copies of the ss RNA genome. The core is surrounded by a protein capsid with a second protein layer (shell protein) outside the nucleocapsid. The envelope is a lipid bilayer interspersed with glycoprotein spikes.

present in considerable quantity on the outer surface of helper T-lymphocytes, and this glycoprotein is designated CD4. The glycoprotein is also present, but in smaller amounts, on the surface of monocytes, dendritic cells, and macrophages. Any cell bearing the specific glycoprotein is termed $CD4^+$. If a cell has the CD4 glycoprotein on its surface, HIV-1 attaches and enters the cell by fusion with the host cell cytoplasmic membrane (Figure 14.24).

During HIV-1 replication the viral RNA is transcribed to DNA by the action of a specific RNA-dependent DNA polymerase (reverse transcriptase). The enzyme responsible is contained in the virus particle. The order of the genetic information in the HIV-1 genome is outlined in Table 14.6. Synthesis of DNA occurs in the cytoplasm within the first six hours of infection. After transcription of the ss RNA viral genome to ds DNA, the RNA is apparently destroyed by the viral ribonuclease. The viral reverse transcriptase that transcribes RNA into ds DNA is quite inaccurate, and there is no mechanism for correction of errors made. The error rate in HIV-1 is ten times that in other reverse transcriptases. Therefore, variants occur continu-

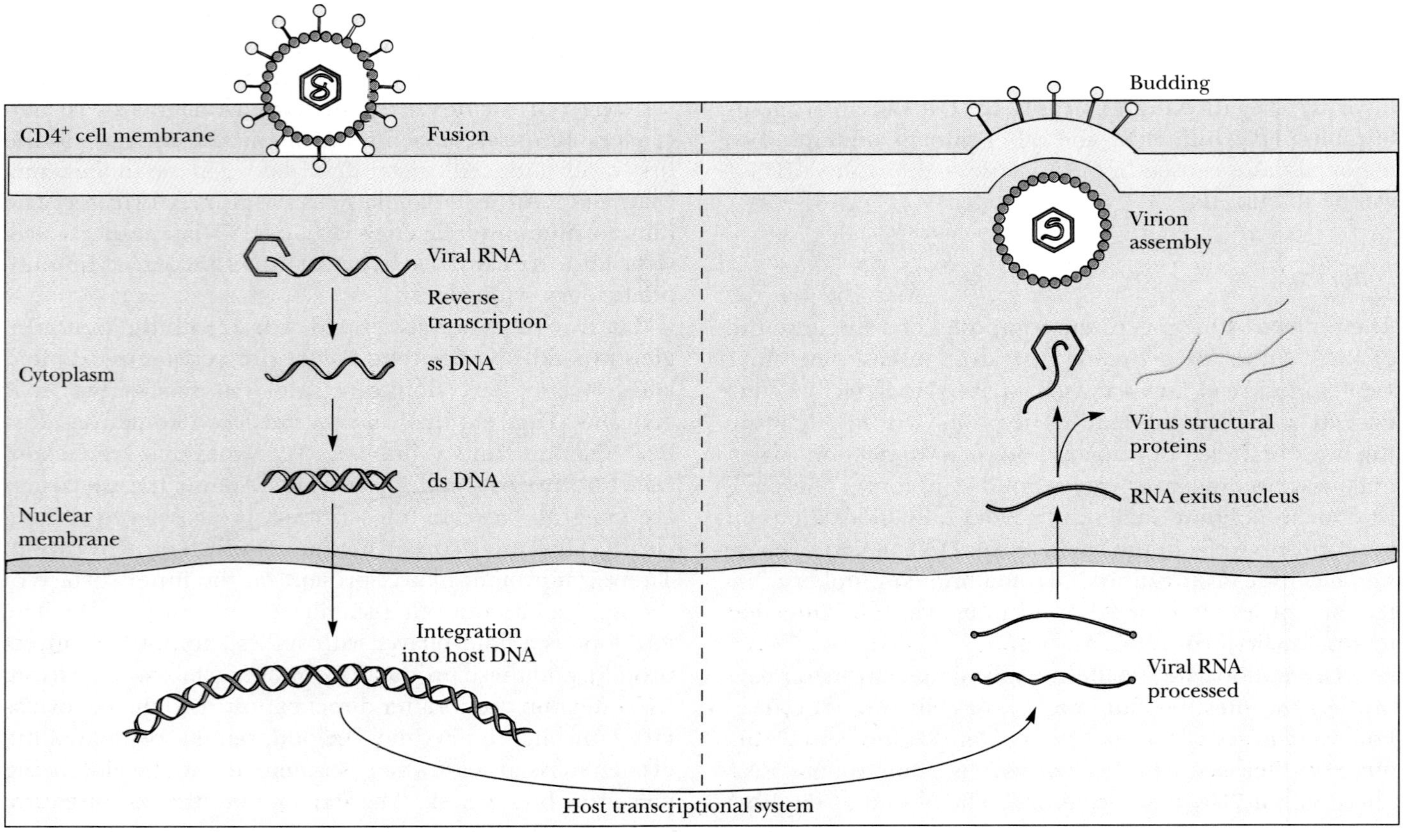

***Figure* 14.24** The replicative cycle in HIV-1. The virion attaches to $CD4^+$ host cells, and after fusion the nucleocapsid enters the host cytoplasm and is uncoated. Reverse transcription produces a ds DNA that is transferred into the host nucleus where the integrase incorporates the vital DNA into the host genome. Host transcriptional enzymes generate viral RNA. Viral RNA exits from the nucleus, viral structural proteins are synthesized, the virion is assembled, and the maturing virus buds from the host cell, picking up the viral envelope in the process. The length of time for the events depicted is highly variable, particularly that for host enzyme synthesis of viral RNA to initiate virus production.

***Table* 14.6 The Known Genes in the HIV-1 Virion and Their Functions**

gag	Nucleo-capsid proteins
pol	Enzymes—integrase, reverse transcriptase, protease, and ribonuclease
vif	Involved in infectivity
vpr	Transcription
tat	Transcription
rev	Gene expression regulator
vpw	Budding
env	Coat glycoproteins, attachment, and fusion
nef	Unknown; signals reproduction

ously, and progeny have a DNA template that varies 2 percent to 3 percent in base sequence from that of the parent virus. This variation generates progeny that may be unaffected by antibodies formed against the progenitor virus. This would give a survival advantage.

Although the genome of HIV-1 is RNA, the ultimate genetic information that generates viral progeny is ds DNA. This ds DNA is integrated into the host genome by integrase, where it remains as a provirus. The viral integrase serves three major functions: (a) it trims the ends of the viral ds DNA, (b) it cleaves host DNA, and (c) it covalently links the termini of the trimmed DNA to host DNA. The inserted viral DNA (provirus) has all of the characteristics of a cellular gene. A treatment for HIV-1 infection that might excise the provirus is improbable.

The synthesis of viral genome begins with an RNA transcript formed from the viral DNA template present in the host genome. Transcription rates are regulated by host proteins that are involved with the transcription of other host genes. Consequently, production of viral RNA transcripts results from normal activities that occur in the host. An inactive T-cell that contains the viral genome would not produce viral RNA. Macrophages that are not dividing but contain the provirus do transcribe viral RNA and produce HIV-1. These macrophages are thought to be the source of much of the virus in infected individuals.

The viral RNA transcribed from the provirus DNA serves as the progenitor of capsid proteins and enzymes needed for viral assembly. The capsid proteins are synthesized as a polyprotein. The replicative enzymes and capsid proteins are fused. These fused proteins move to the inner surface of the host cytoplasmic membrane, and viral RNA binds to the capsid protein precursor. The precursor then surrounds the replicative enzymes and viral RNA, and the spherical particle buds through the host cell membrane. The glycoprotein surface complex present on HIV-1 is formed in the endoplasmic reticulum of the host and is added to the outer envelope of the virus as it buds through the host cell surface.

It is not clear how the HIV-1 virus curtails the immune response or causes the destruction of T-cells. The $CD4^+$ T-cells are an essential part of the immunological response. Only about 1 of 400 T-cells becomes infected with HIV-1, and fewer than one in 100,000 actually produce the virus. Despite this, an HIV-1 infection will ultimately result in loss of virtually all functional T-cells in the immune system. Destruction of T-cells may occur by apoptosis (programmed cell death) initiated by cross-linking of molecules by gp120 (envelope protein) of the HIV virus and binding of antigen to the T-cell receptor. It has also been established that infected $CD4^+$ cells proliferate abnormally in the lymph nodes, ultimately causing a collapse of the lymphatic system. This would cause a reduction in number of all $CD4^+$ cells. There is recent evidence that the immune response is diminished as dendritic cells that present foreign antigens to T-cells become coated with antibody to HIV-1 virus and lose the ability to respond to the HIV antigen. It is probable that a number of factors combined are involved in the destruction of T-cells by infection with HIV-1, and no single factor is responsible for AIDS pathogenesis.

Insect Viruses

Some plant viruses are carried by and multiply in insect vectors (see following), and some of these viruses harm or kill the insect. The baculoviruses are one of the most studied of the insect viruses as they are virulent and kill susceptible insect populations. Eastern tent caterpillars and the tussock moth that defoliates Douglas fir trees are highly susceptible to baculoviruses. The East African army worm (which devastates extensive grassland areas), forest sawfly, orchard codling moth, and cotton bollworm are potentially controllable by baculovirus. However, baculoviruses sprayed on crops may not kill infected insects immediately, allowing them to mate and spread the infection to the general population. Killing can be accelerated by introducing genes for insect-specific toxins into the virus. Baculoviruses can be distributed among humans safely as the virus has a narrow host range confined to insects.

Baculoviruses are excellent vectors for carrying foreign genes into insect larval cells for the generation of useful products. Insects are an effective eukaryotic system for gene cloning through introduction of a gene into cells cultured from adult insects, or into their larvae. These insect systems have many advantages over the use of engineered mammalian cells. Genes from plants, animals, and humans expressed in insect cells generate amounts of product far in excess of that obtained from mammalian cells. The insect system also does posttranslational processing, as in animal cells, to yield functional proteins. The insect systems can be employed in the production of pharmaceutics, pesticides, and various proteins.

Plant Viruses

Although the first virus described was the tobacco mosaic virus (TMV), the plant viruses have received less attention than bacterial or animal viruses. Plant viruses may well receive increased attention in the future as the world food supply dwindles and populations grow, while many crops important in human nutrition are adversely affected by viruses. Plant viruses have proven to be rather difficult to study, but in recent years methods have been developed for growth of many of them in plant cell culture.

The plant virus structures are essentially equivalent to the bacterial and animal viruses. Many are long, flexible, very narrow helixes, but of a length approaching that of a bacterium such as *Escherichia coli*. The clostervirus, called beet yellow, is 1.3 μm in length but exceedingly narrow ($<$ 10 nanometers) and cannot be visualized in a light microscope. The capsid of a plant virus is generally composed of a single protein.

The genome of most plant viruses is composed of either single- or double-stranded RNA. There are exceptions, and the caulmovirus is composed of ds DNA and geminivirus ss DNA. Caulmovirus, the causative agent of cauliflower mosaic, is a para-retrovirus, as it forms mRNA and this message is transcribed to ds DNA by a reverse transcriptase. The geminivirus has a single (+) strand of DNA and employs host enzymes to form a ds DNA. The (−) strand produces multiple (+) strands via the rolling circle method. These (+) strands are encapsulated to form a mature virion.

A difficulty that a plant virus encounters in nature is in passing through the plant cell surface barriers. Plant viruses are disseminated by wind or vectors such as insects or nematodes. The insects that transmit viruses include aphids, leafhoppers, white flies, and mealy bugs. Viruses infect the mouth parts of these insects and are transferred to uninfected hosts during normal feeding as the insects penetrate plant cells to withdraw sap. Some plant viruses are stored in the foregut of aphids and inoculated into the plant by regurgitation during feeding. Plant viruses also enter through breaks or abrasions on plant surfaces. Some plant viruses are transmitted by seeds or pollen. Budding or grafting to nursery stock is also a concern as parent stock must be virus free to ensure propagation of healthy progeny.

A select number of plant viruses propagate in both the plant and in the insects that transmit them, and in these the plant and insect are host to the pathogen. The wound tumor viruses multiply in the tissue of leafhoppers before they move to the salivary glands. The virus is then transmitted by the bite of the insect. The RNA viruses that can reproduce in both plant and insect vectors produce viral genomes via a virus-carried RNA dependent replicase. The topovirus (tomato spotted wilt) and the rhabdovirus (lettuce necrotic yellow) are two other examples of viruses that infect the insects that transmit them.

Virus diseases of plants are classified according to apparent manifestations of disease symptoms. These manifestations include the following:

1. Mosaic diseases cause mottling of leaves, yellow spots, blotches, and necrotic lesions. These symptoms sometimes are observed on flowers. The mosaic viruses produce variegations in leaves or flowers of ornamental plants. The variegated petals observed in tulip flowers are a result of viral infection.
2. A number of viruses cause leaves to curl and/or turn yellow. Dwarfing of leaves or excessive branching can result from viral infections.
3. Wound tumors can occur on roots or stems.
4. Wilt diseases can result in the complete wilting and death of a plant.

Fungal and Algal Viruses

Plant, animal, eubacterial, and archaeal viruses are transmitted horizontally as they multiply in one cell, exit, and enter another susceptible cell. Some animal viruses such as herpesvirus or HIV may be transferred to adjacent cells by cell fusion. Fungal viruses are quite unique in that they are spread by cell-to-cell fusion, and extracellular virions have not been observed. Latent viral infections are apparently common in fungi, and the majority of *Saccha-*

romyces cerevisiae strains carry ds RNA viruses. These viruses are transferred by cytoplasmic mixing during cell fusion. There are also yeast viruses that have a genome composed of ss RNA circularized and a retrovirus that is ss RNA.

Killer strains occur among various genera of yeasts including *Saccharomyces, Hansenula,* and *Kluyveromyces.* These killer strains secrete a protein toxin that is encoded in specific regions of the genome of a latent ds RNA virus. Resistance to the toxin is also encoded in the viral genome, rendering the host immune to the toxin. The killer toxin binds to specific glucans in the cell wall of nonimmune yeasts and causes cell death by creating cation-permeable pores in the cytoplasmic membrane. The killer trait (virus) can be introduced into yeast strains employed in the fermentation industry. The presence of the killer trait protects them from the killer toxin but does not affect the fermentative ability of the yeast strain.

Filamentous fungi, such as *Penicillium chrysogenum* are also susceptible to infection by viruses. A ss RNA virus (Barnavirus) can infect the cultivated mushroom *Agaricus bisporus* and is a concern in commercial production.

Algal viruses such as phycodnavirus (ds DNA) are ubiquitous in fresh water. These viruses are host specific and attach to cell walls of susceptible unicellular eukaryotic algae. The virus causes dissolution of the cell wall at the site of attachment and viral DNA enters the algal cell. Following virus multiplication the viral particles are released by cell lysis.

It is apparent from this brief discussion that viruses have evolved that can infect all types of living cells. There is no specific chemotherapy for viral infections, and natural or induced resistance or vaccination is the only control measure now available.

Viroids

Viroids are the smallest nucleic acid–containing infectious agents known. They were first described by O. T. Diener, a plant physiologist, in 1971. Viroids have been identified only in plants and are composed solely of RNA. The RNA is made up of 270 to 380 nucleotides (Figure 14.25). This amount of RNA would be about 1/10 the genetic material present in the smallest of the viruses that have been characterized. This number of nucleotides could theoretically produce a 100 amino acid protein if every base were used in a single reading frame, or 2 proteins if a double (frame-shift) reading frame were used. However, much evidence suggests that viroids do not produce any proteins. The viroid is a circular single strand of RNA collapsed into a rod by intrastrand base pairing. Viroids are not surrounded by a protein capsid. Viroids appear in the nucleus of the plant cells and apparently do not function as mRNA. A viroid is synthesized by host RNA polymerases by employing the viroid particle as a template. The polymerase of the host cell recognizes the RNA as it would a piece of DNA.

The viroid most studied is called the potato spindle-tuber disease agent (PSTD). It is 359 nucleotides in length and may function by interfering with host gene regulation. Comparison of nucleotide sequences in viroids suggests that they are similar to the RNAs of introns. They may have evolved from introns and function by interfering with the normal splicing of introns in cells. A number of strains of PSTD have been isolated and some cause mild symptoms in plants while others cause lethal infections. The difference in virulence is due to alterations in the sequence of the nucleotides.

Viroids are the causative agent of diseases in agricultural crops including cucumbers, potatoes, and citrus fruit, and these diseases cause losses in the millions of dollars each year. Viroids may be spread by grafting in fruit trees and through the use of modern harvesting machinery. Although viroids have not been identified in animal infections, they well may be in the future. A viroid is not apparent in infected tissue without employing special techniques that involve nucleotide sequencing. Application of these techniques may lead to the identification of viroid-caused infections in humans.

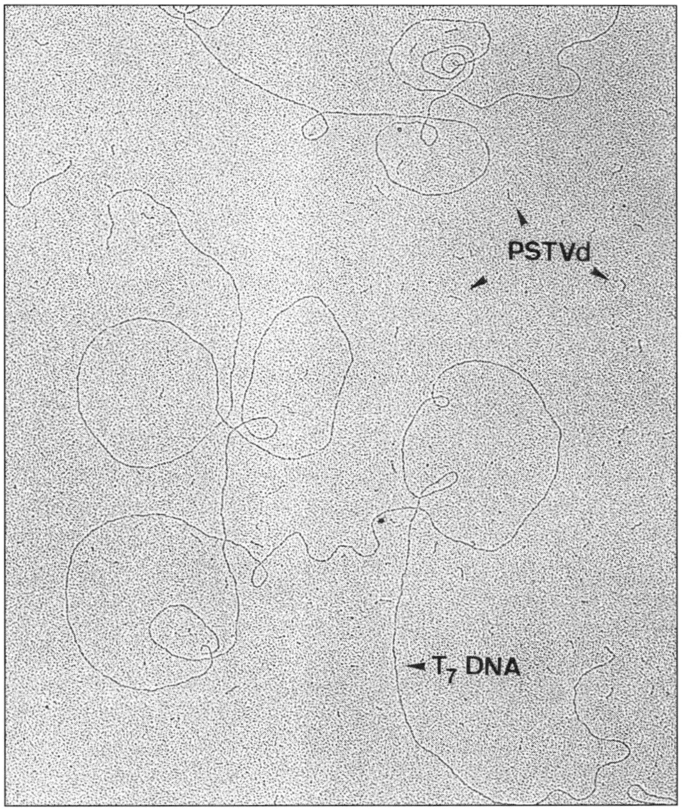

***Figure* 14.25** Comparison of the low molecular mass RNA viroid PSTVd with the genomic DNA of the *Escherichia coli* phage T7. (Courtesy of T.O. Diener, University of Maryland, College Park)

Summary

- Viruses vary in size from those that are close to the size of a ribosome (polio) to viruses that approach free-living bacteria (pox viruses) in length. Smallpox is a large virus that is visible in phase-contrast microscopy.
- A viral genome is surrounded by protein **capsid** that protects the virus when outside a host cell. Many viruses also have a membranous **envelope** that surrounds the capsid, and there may be specific glycoproteins (spikes) embedded in this envelope. These spikes are the site of attachment to host cells.
- Viruses grow only within living cells including the cells of plants, animals, Eubacteria, and Archaea. Plant and animal viruses can be grown in an appropriate cell or tissue culture.
- **Quantitation** of bacteriophage can be accomplished by placing dilutions of a viral suspension on a "lawn" of the susceptible host growing on an agar surface. The phage will lyse organisms in a confined area, forming clear areas called plaques. Plant or animal viruses can be quantitated in cell culture or by treating the host with dilutions of virus and determining the highest dilution that produces manifestation of disease.
- Viruses are generally **purified** by differential centrifugation, enzyme treatment, and other procedures.
- The viral genome is composed of either DNA or RNA. This nucleic acid may be single stranded (ss) or double stranded (ds). A viral genome may be mRNA or transcribed to produce an early message for the proteins that initiate the infectious process.
- Some **tumor viruses** and the **HIV-1 virus** that causes AIDS are ss RNA. There are also DNA tumor viruses. The ss RNA viruses are transcribed in reverse to produce ds DNA. This ds DNA is integrated into the genome of the host.
- A very elaborate bacterial virus, **T4,** infects *Escherichia coli.* This virus can enter a host cell and produce 200 to 300 viral progeny in about 25 minutes. *E. coli* can be infected by a unique virus **M13** that can enter the cell, reproduce, and exit without serious harm to the host.
- Viruses can cause a **lytic** infection that destroys the host cell, a persistent infection where the host remains viable and sheds virus for an extended time, or a **latent** infection where the virus exists in a **dormant** state.
- Viruses that infect eukaryotes transfer genetic information into the cytoplasm and, in most cases, viral nucleic acid transcription occurs in the nucleus. RNA viruses and some DNA viruses (e.g., poxviruses) are transcribed in the cytoplasm. Some eukaryotic virus particles contain viral-encoded proteins that are involved in the infectious process. Enveloped animal virus enter a cell by fusing to the cytoplasmic membrane or **endocytosis.**
- The **retrovirus** that is the causative agent of AIDS attaches to cells (macrophages, T-lymphocytes, monocytes) that have a specific glycoprotein on their surface. These cells are called **CD4$^+$** cells. The **T-cells** are ultimately destroyed by the AIDS virus resulting in a loss of immune function in the infected human.
- Herpesvirus is a typical DNA virus that can remain dormant in a host and cause an active infection when the host is under stress.
- Polio is an infection caused by an **ss RNA virus.** The genome is actually mRNA that is copied to form an RNA template that is involved in production of viral genomes.
- Plant viruses are generally transmitted by insects that suck sap from plants or enter through abrasions. Plant viruses can infect the insects that transmit them as well as plant hosts. Plants resistant to viruses can be obtained by selective plant breeding.

Questions for Thought and Review

1. What are the extremes in the size of viruses and how does this compare with sizes of bacteria?

2. What are the general morphologies of viruses? Can one predict the morphology of a virus based on the host it attacks—bacteria, animals, or plants?

3. Discuss cell culture and how it might be employed to propagate viruses.

4. Discuss techniques that might be employed to purify viral particles.

5. Why are bacterial viruses employed as a model system for the study of viral replications?

6. What are the four phases involved in viral replication?

7. How can a potential host cell protect itself from viral invasion? How can the virus overcome these barriers?

8. From an evolutionary standpoint, why would the lysogenic state be favored? Why is lambda of interest?

9. Outline the steps involved in reproduction of M13. What is the unique attribute of M13? Why is M13 an attractive virus for recombinant DNA technologies?

10. Outline the steps involved in reproduction of the virus T4.

11. Compare the mechanism whereby an animal virus and bacterial virus attach to and enter a host cell. What difference in structure of the two cell types is important in these processes?

12. Define lytic, persistent, latent, and transforming as these terms apply to animal viruses.

13. What is a retrovirus and why have these viruses been given this name?

14. Outline the life cycle of a virus like HIV. How does it attack, become integrated, and cause disease?

15. How do plant viruses spread? What difficulties do they encounter and how are these overcome?

Suggested Readings

Dimmock, N. J., and S. B. Primrose. 1994. *Introduction to Modern Virology.* Cambridge, MA: Blackwell Science, Inc.

Fenner, F., and A. Gibbs, eds. 1988. *Portraits of Viruses—A History of Virology.* Basel: Karger.

Fields, B. N., and D. M. Knipe, eds. 1995. *Field's Virology.* New York: Raven Press.

Karam, J. D. ed. 1994. *Molecular Biology of Bacteriophage T4.* Washington, DC: American Society for Microbiology.

Kornberg, A., and T. Baker. 1992. *DNA Replication.* 2nd ed. New York: W. H. Freeman and Co.

Levy, J. A., H. Fraenkel-Conrat, and R. A. Owens. 1994. *Virology.* 3rd ed. Englewood Cliffs, NJ: Prentice Hall.

Voyles, B. A. 1993. *The Biology of Viruses.* St. Louis: Mosby.

If it is ultimately proved . . . that the transforming activity . . . is actually an inherent property of the nucleic acid, one must still account on a chemical basis for the biological specificity of its action. . . . although the constituent units and general pattern of the nucleic acid molecule have been defined, there is as yet relatively little known of the possible effect that subtle differences in molecular configuration may exert on the biological specificity of these substances.

O. T. Avery, C. M. McLeod and M. McCarty
(1944, original paper on DNA as the transforming principle)[1]

[1] *Journal of Experimental Medicine* 79:137–158

Chapter 15

Genetic Exchange

Plasmids and Bacteriophages
Recombination
Genetic Exchange Among Bacteria
Mobile Genetic Elements and Transposons

Tracing the effects of a mutation introduced into the genes of higher animals and plants would take decades, as their reproductive rates are measured in years and their numbers are comparatively sparse. Countering this inability to make quick changes in their genetic make-up, animals and plants have large and complex genomes that enable them to display an array of behaviors for coping with changes, opportunities, and threats in their environment.

By contrast, the immediate effects of a mutation in a bacterium can be seen in a matter of hours in the millions of cells that arise from the original. To continue the analogy, bacteria have a limited repertoire of life skills, largely emphasizing direct metabolism and cell division. For example, their only avenue of response, if their environment changes beyond their ability to cope, is that one among them may happen to mutate in such a way that a new trait arises that meets or **adapts** to the new environment.

Random mutations in a bacterial population occur at a low but constant measurable rate. Usually, mutations are deleterious and lead to the loss of a functional gene, and this leads to the death of the cell. Even so, mutations are the main mechanism by which bacteria acquire new traits and adapt to life in changing environmental niches. However, the evolution of bacteria is greatly accelerated by their ability to acquire and express entire genes they obtain from other bacteria. This chapter discusses the mechanisms of this genetic exchange. New traits that bacteria gain by genetic exchange allow them to adapt to new environments much more rapidly than would occur by mutation alone. This adaptability is important in circumstances where newly acquired genetic information permits bacteria to survive in what previously was a potentially lethal environment. Further, the new traits may augment the metabolic versatility of the bacterium, allowing survival in an environment in which the cell previously lacked the ability to utilize available nutrients.

Plasmids and Bacteriophages

Bacterial plasmids and bacteriophages (bacterial viruses) are small genetic elements that are not part of the chromosome. They replicate in the bacterial cytoplasm and hence utilize the metabolic machinery and replicative apparatus of the host bacterial cell. Genes found on these extrachromosomal elements can specify functions that may influence the life of the bacterial host. In general, however, the genes encoded by these elements are not essential; that is, they are useful only when the bacterium finds itself in a rather specialized environment.

Plasmids

Plasmids present in bacteria are generally circular, double-stranded DNA molecules. A single bacterial cell can be completely devoid of plasmids, or it can carry many different plasmids in its cytoplasm. Plasmids are variable in size, ranging from several hundred to many thousands of base pairs. In fact, some large plasmids present in certain species of the genus *Rhizobium* can be considered to be mini-chromosomes encoding hundreds of genes. While many plasmids exist as a single copy within a bacterial cell, some plasmids are present in several to as many as 20 or 30 copies. Many, but not all, plasmids can be transferred to other bacteria by a process called **conjugation** (to be discussed later), thus bacteria that are plasmid-free can acquire plasmid-encoded genes. Usually, only a closely related member of the genus can receive plasmids, but some plasmids, called **promiscuous,** can be transferred to bacteria that are unrelated.

Plasmids Specifying Resistance to Antibiotics

Bacterial plasmids are classified on the basis of the information that is encoded in the genes of these plasmids. One very important group of plasmids are the **R-factors** (or **R-plasmids**) that encode antibiotic-resistance determinants. A bacterial species carrying one of these plasmids can be resistant to a specific antibiotic, while the same species lacking the plasmid can be killed readily by the antibiotic. A number of different mechanisms exist for plasmid-encoded antibiotic resistance, the most common of which is synthesis of enzymes that destroy the antibiotic. Another example is plasmid-encoded proteins that can alter the membrane structure such that the antibiotic cannot enter the cell and reach its target. Occasionally, a plasmid can encode a product that modifies the target of the antibiotic, rendering it resistant to the antibiotic. Table 15.1 sum-

***Table* 15.1 Common mechanisms of plasmid-encoded antibiotic resistance**

Antibiotic	Mechanism of Resistance
β-lactams	Synthesis of β-lactamases, enzymes that hydrolytically destroy the antibiotic
Chloramphenicol	Synthesis of an enzyme that acylates chloramphenicol, rendering it inactive
Aminoglycosides	Synthesis of one of several enzymes that inactivate the antibiotic by acetylation, phosphorylation, or adenylation
Tetracycline	Synthesis of a membrane protein capable of pumping the antibiotic out of the cell before it can act on the ribosomes
Erythromycin	Synthesis of an enzyme that methylates bacterial 23S ribosomal RNA; methylated ribosomes cannot bind the antibiotic
Trimethoprim	Synthesis of a mutant, trimethoprim-insensitive form of dihydrofolate reductase

marizes some common mechanisms of plasmid-encoded resistance.

The prevalence of R-factors among pathogenic (disease-causing) bacteria is of concern to physicians treating infectious disease because it can limit the range of antibiotics that can be administered to treat a particular infection. The presence of a plasmid-encoding β-lactamase in *Neisseria gonorrhoeae* requires the use of costly and less effective antibiotics for the treatment of gonorrhea, although previously susceptible strains had been killed by administration of a single large dose of penicillin to the infected patient. Many R-plasmids can move from one bacterium to another by conjugation, allowing a pathogen that is normally sensitive to therapy with an antibiotic to become resistant. A plasmid can acquire additional resistance determinants from other R-factors, therefore plasmids carrying multiple antibiotic resistance determinants are common. Figure 15.1**a** is a schematic representation of RK2, a plasmid encoding resistance to tetracycline, kanamycin, and ampicillin. This plasmid is also transferrable to other bacterial hosts, including virtually all of the gram-negative bacteria. This promiscuous behavior of a plasmid implies that the plasmid or the bacterial host carrying the plasmid has the capability of transferring it to a recipient, and once in the new host, the plasmid can stably replicate.

Virulence Plasmids

Plasmids of pathogenic bacteria may encode genes that are required for virulence. Plasmids conferring pathogenic abilities on *Escherichia coli* have been well characterized. Certain strains of *E. coli* responsible for diarrheal diseases in humans and animals carry large plasmids containing genes for two types of toxins. The same plasmid can also carry genes for **adhesins,** surface molecules that are required for mucosal colonization. A diagram of plasmid pCG86 that encodes toxin production as well as resistance determinants to three different antimicrobial agents is presented in Figure 15.1**b**. Plasmids of *E. coli* that are important in virulence can encode both hemolytic factors (that is, proteins that destroy membranes of a variety of cells, including red blood cells) as well as siderophores (molecules responsible for efficient iron uptake, a metal often limiting in infected tissues). Pathogenic gram-positive bacteria also carry plasmids that encode synthesis of toxins, such as plasmid-encoded toxin genes of *Bacillus anthracis.* A group of insecticidal toxins of *Bacillus thuringiensis* is plasmid-encoded, as are *Clostridium* neurotoxins. Specific plasmid-encoded genes are required for cell invasion in the course of diseases caused by *Shigella, Yersinia,* and *Salmonella* species.

Infection of certain plants with the pathogen *Agrobacterium tumefaciens,* which carries specialized tumor-inducing (Ti) plasmids, results in formation of crown gall tumors. The plant tumor results from the expression in the plant of genes encoded on the bacterial plasmid, a portion of which is transferred into the nucleus of the host where it becomes incorporated into the plant genome. A map of a Ti plasmid is presented in Figure 15.2. There are two regions of this plasmid important for virulence, the T-DNA (so named because it carries the genes required for tumor-induction in the host plant) and the *vir* genes (for *virulence*). The T-DNA is about 30 kilobases (kb) in length and is flanked by two 25-base-pair direct repeats. This is the region of the plasmid that becomes excised and transferred into the host genome. T-DNA genes encode enzymes responsible for the synthesis of opines, some of which are analogs of the amino acid arginine that can be used by *Agrobacteria* as a source of energy and nitrogen. In

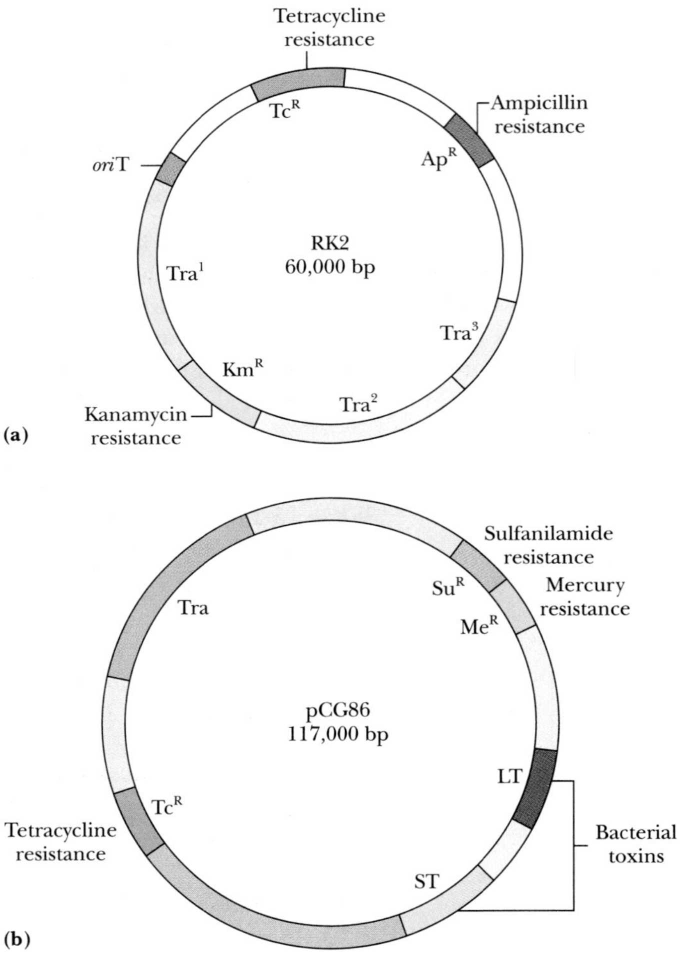

***Figure* 15.1(a)** Schematic representation of plasmid RK2. This large R-plasmid encodes resistance to tetracycline (Tc^R), kanamycin (Km^R), and β-lactam antibiotics such as ampicillin (Ap^R). Also shown are the origin of transfer (*ori*T) during conjugation and the regions encoding the various accessory transfer functions (Tra). **(b)** Plasmid pCG86 of pathogenic *E. coli.* This plasmid encodes genes for two bacterial toxins, ST and LT, as well as resistance to tetracycline (Tc^R), sulfanilamide (Su^R), and mercury (Me^R).

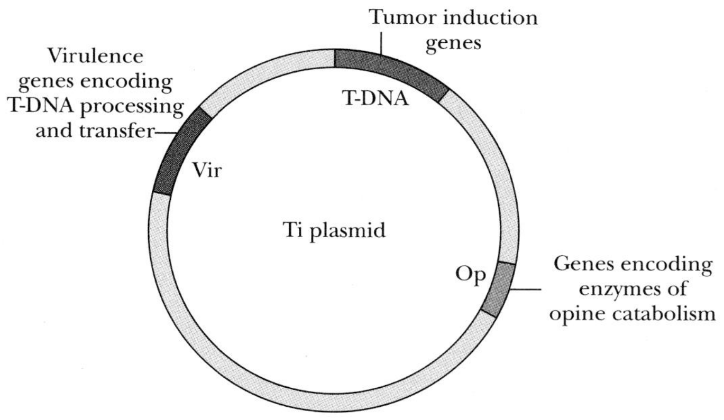

***Figure* 15.2** Ti plasmid of *Agrobacterium tumefaciens*. The important regions indicated on this drawing are the T-region, which encodes phytohormone biosynthesis enzymes responsible for induction of plant tumors; the *vir* region, which encodes functions responsible for DNA processing and transfer; and the *op* region, which codes for enzymes of opine catabolism.

the plant cell, T-DNA genes also direct synthesis of plant hormones, resulting in uncontrolled growth of the plant tissue and formation of a tumor. The *vir* genes are not transferred into the plant, but nevertheless are responsible for transfer of T-DNA. They include regulatory genes that respond to signals from the plant, as well as genes involved in processing T-DNA prior to its transfer to the plant cell (see **Box 15.1**).

Another group of plasmids can direct synthesis of proteins that are toxic to other bacteria. These bactericidal proteins, called **bacteriocins,** are encoded by specialized plasmids termed **bacteriocinogenic plasmids.** Commonly, the names of the plasmid and its toxic protein are derived from the genus or species name of the host bacterium. For example, *E. coli* can carry **colicinogenic plasmids** that encode a variety of **colicins.** Colicins kill other bacteria by a variety of mechanisms, ranging from direct damage to the cell membrane to the destruction of ribosomes. Bacteria carrying bacteriocinogenic plasmids protect themselves from the lethal bacteriocins by expressing **immunity determinants,** which are coded on the same bacteriocinogenic plasmids. The bacteriocin/immunity systems thus allow bacteria to survive while killing other bacteria that may compete with them in an environment of limited nutrients.

Plasmids Encoding Genes for Specialized Metabolism

Certain metabolic functions can also be encoded by plasmids, including the biodegradation of complex organic molecules by some species in the genus *Pseudomonas.* Large plasmids that encode catabolic pathways for aliphatic or aromatic compounds, such as toluene, naphthalene, chlorobenzoic acid, octanes, and decanes, have been described in recent years. A map of one such degradative plasmid is presented in Figure 15.3 together with the enzymes of naphthalene oxidation. This plasmid codes for the enzymes that catabolize naphthalene to pyruvate and acetylaldehyde. Such degradative plasmids allow the bacterium to utilize unusual organic compounds as sole carbon and energy source. Because many environmental

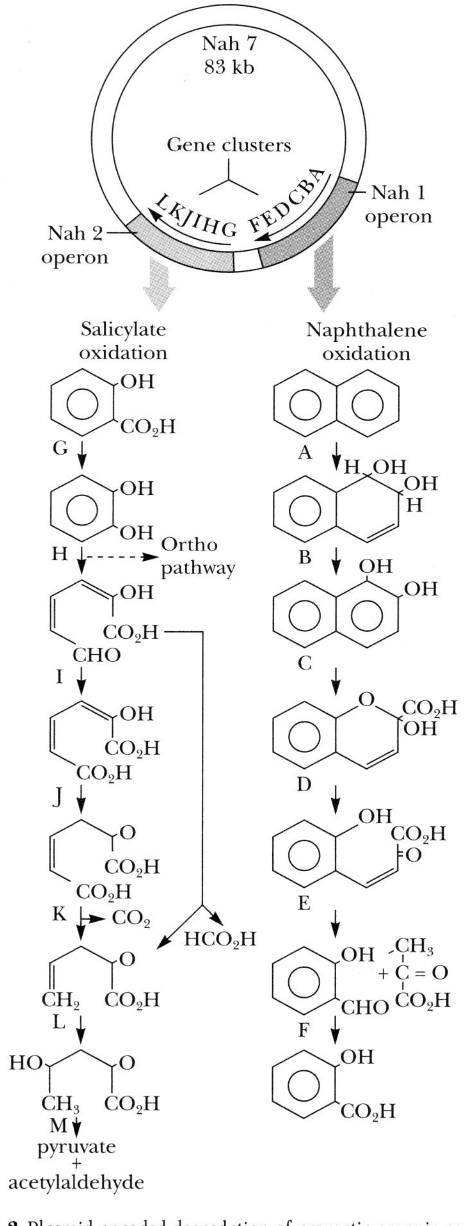

***Figure* 15.3** Plasmid-encoded degradation of aromatic organic compounds. **(a)** Plasmid NAH7 harbored by *Pseudomonas putida* contains two gene clusters (*nah*1 and *nah*2) each specifying a transcriptional unit (indicated by the arrows) encoding enzymes of naphthalene and salycilate oxidation. The *nah*1 operon encodes the enzymes (A–F) that carry out stepwise oxidation of naphthalene to salicilate, while the oxidation of silicilate to pyruvate and acetaldehyde is catalyzed by the enzymes encoded by the *nah*2 genes (G–M). **(b)** Shown are the exact intermediates of the oxidation pathways carried out by the enzymes encoded by the *nah*1 and *nah*2 operon.

BOX 15.1 MILESTONES

Sex Among Bacteria

The discovery of transformation by Avery and his colleagues, as a means of bacterial acquisition of genetic information, raised the possibility that bacteria can acquire traits by more conventional, that is, sexual, means. In 1946, Tatum and Lederberg demonstrated the exchange of genetic information among bacteria. Their simple experiment, outlined in Figure **(a)**, demonstrated transfer of genes between two strains of *E. coli*, each carrying two mutations in biosynthetic genes.

One of the strains was *E. coli* that required both biotin and methionine for growth on minimal medium, while the second was a threonine- and proline-requiring double mutant of *E. coli*. When a culture of these two mutant strains was mixed and plated onto minimal medium, a few bacteria grew that did not require any of the supplements which characterized the mutations of the parental strain. When a comparable number of each of the *E. coli* mutants were plated onto minimal medium, no wild-type bacteria were ever recovered. This was the first demonstration that recombination can occur between two bacteria, even in the absence of apparent lysis and release of DNA, since DNase present during the coincubation of the two bacterial strains failed to block recombination.

The final proof that the observed recombination requires a cell-to-cell contact was provided by Davis in 1950. His experiment, shown in Figure **(b)**, utilized a U-shaped tube, which was separated by a fritted glass filter, allowing passage of liquid, but the pores in the disk were impermeable to the bacteria. One side of the tube was inoculated with an *E. coli*–requiring threonine, leucine, and thymine, while the other side contained a methionine-requiring mutant of *E. coli*. In separate experiments, the two bacterial mutants were mixed in one of the arms of the tube. The medium was flushed back and forth, and incubated for several hours. Bacteria were recovered from the tube and plated onto minimal media.

While the tube that contained a mixture of bacteria yielded a fair number of wild-type recombinants, the tube in which bacteria were separated failed to give rise to any recombinants capable of growing on media lacking the appropriate nutritional supplements. Hence, genetic exchange by intimate contact was unambiguously demonstrated, which was subsequently shown to be mediated by plasmids mobilizing DNA from donor (male) to recipient (female) cells.

pollutants are complex organic molecules, bacteria that carry degradative plasmids are prime candidates for use in bioremediation (see Chapter 33).

Plasmid Replication

Plasmid replication is similar to that of replication of the host DNA. Starting from a specific origin of replication, the host replication machinery proceeds bi-directionally around the plasmid. In addition to the host replicative apparatus, some plasmids require expression of plasmid-encoded genes. These genes code for various enzymes that assist DNA replication, such as unwinding and separating strands, and ensuring correct partition of replicated plasmids into daughter cells.

While a bacterial cell can carry several plasmids, not all plasmids can coexist in the same cell. Plasmids that use identical or closely related replication mechanisms are incompatible in the same cell and, therefore, belong to the same **incompatibility group.** If a cell receives two plasmids of the same incompatibility group by any one of the genetic exchange mechanisms, these two plasmids will segregate during cell division such that each daughter cell will have only one of the plasmids. This property allows yet another method of classifying plasmids—on the basis of plasmid compatibility. The mechanism of incompati-

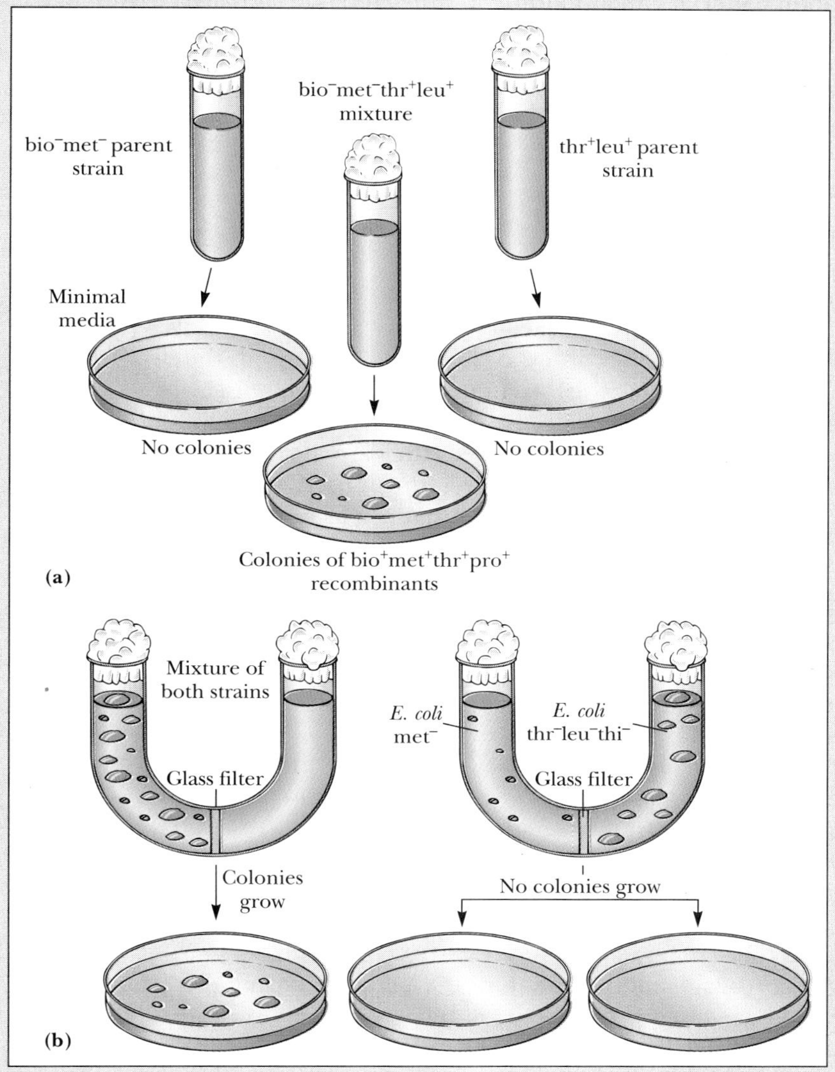

(a) Demonstration of recombination among bacteria. A mixture of *E. coli* bio^-met^- and *E. coli* thr^-pro^- double mutants were plated onto minimal medium (where they gave rise to colonies), while neither one of the parental strains was capable of growing on such medium. This implied existence of a sexual mechanism for genetic transfer.
(b) Intimate contact is required for genetic recombination to occur. A mixture of two *E. coli* mutants gave rise to wild-type recombinants only when they were incubated together. Separation by a filter prevented genetic exchange, and hence no wild-type bacteria were recovered.

bility is not well understood, but for a number of plasmids, it appears to be directed by a control system that coordinates replication with cell division.

Occasionally, plasmids can be "lost" by a bacterium, and this may occur naturally during cell division when both copies of a replicated plasmid end up in one daughter cell and the second daughter cell does not receive a plasmid. The daughter cell without a plasmid is said to be **cured.** Environmental stresses, such as exposure to extreme temperatures, nutritional limitation, or treatment with chemicals that interact with DNA, greatly stimulate curing. If the bacterium is in an environment where it depends on the expression of a plasmid-encoded gene, such as in the presence of antibiotics, the cured daughter cell will not survive.

Bacteriophages

Viruses are obligate intracellular parasites that can propagate only inside host cells. Virtually every type of living cell can be infected by one or several kinds of viruses, and Eubacteria and Archaea are no exception. While viruses can self-replicate, they do so only when inside a host cell, parasitizing virtually all of the host cell's metabolic machinery. Outside a host, they are inert molecules of nucleic acids surrounded by a protein. For these reasons, viruses are not considered living organisms, although they are generally discussed in the context of their host. Bacterial viruses utilize the metabolic and biosynthetic machinery of their host, having replication and gene regulation patterns similar to those of Eubacteria or Archaea, while viruses that infect eukaryotic cells follow the patterns of gene expression seen in eukaryotic cells. The replication of viral nucleic acid and assembly of a virus was considered in Chapter 14.

Recombination

The replication of genetic material is controlled by a set of enzymes that, by minimizing changes in the specific DNA bases and the order of blocks of genes in the chromosome, ensures the accurate transfer of genetic information to daughter cells. Errors or mutations can have potentially lethal consequences, yet they are also responsible for adaptive evolution to novel functions. A gene can acquire new properties by mutation and selection. An altered gene can arise by duplication of a related gene and extensive mutation in one of the two genes. Another very important mechanism in bacterial evolution is the acquisition of novel genetic information from another bacterium. Rearrangement of blocks of DNA within the bacterial genome and integration of newly acquired genetic information following genetic exchange is a complex process. These processes involve specialized enzymes as well as components of the DNA replication and repair machinery. Orderly rearrangement of DNA, called **genetic recombination,** is a process that occurs in both prokaryotic and eukaryotic cells.

Genetic recombination can be divided into two classes: general and site-specific. In all cases, recombination involves breakage and religation of DNA strands during the rearrangement process. General and site-specific recombination differ from one another in the requirement for homologous (that is, matching or very similar) base sequences at the site of recombination and the availability of distinct proteins that catalyze the recombination process.

General (Homologous) Recombination

General (homologous) recombination is a mechanism of genetic rearrangement involving DNA sequences that share significant sequence similarity. The key protein that controls general recombination is encoded by the *recA* gene. This type of recombination is also called **crossover recombination,** since the event appears to be a simple one-step breakage of two double-stranded DNA molecules and ligation across each strand, as outlined in Figure 15.4. At the molecular level, this process is much more complex. It can be divided into two stages: formation of a hybrid structure between two DNA molecules followed by their resolution into free DNA molecules. These steps are illustrated in Figure 15.5.

The recombination process between two double-stranded DNA molecules requires a nick, or a single-stranded gap, in one of the two DNA duplexes. The nick can be formed by physical damage or by an enzyme complex consisting of RecB, RecC, and RecD, which unwinds DNA and cuts one strand at any site containing the sequence GCTGGTCG, called the ***chi* site.** A single-stranded tail is formed, which becomes coated by a single-stranded DNA-binding proteins, including RecA. The bound form of RecA is now capable of aggregating with double-stranded DNA, searching for homology, and invading the duplex of the second DNA molecule, allowing formation of a stable duplex from the two homologous strands. The displaced "bubble" of the invaded strand is then nicked and forms a stable double-stranded structure with the other strand, resulting in a transient hybrid molecule. The nicks in both of the strands are resealed. The final stage in recombination is **isomerization,** which leads to alignment of the two remaining outside strands across each

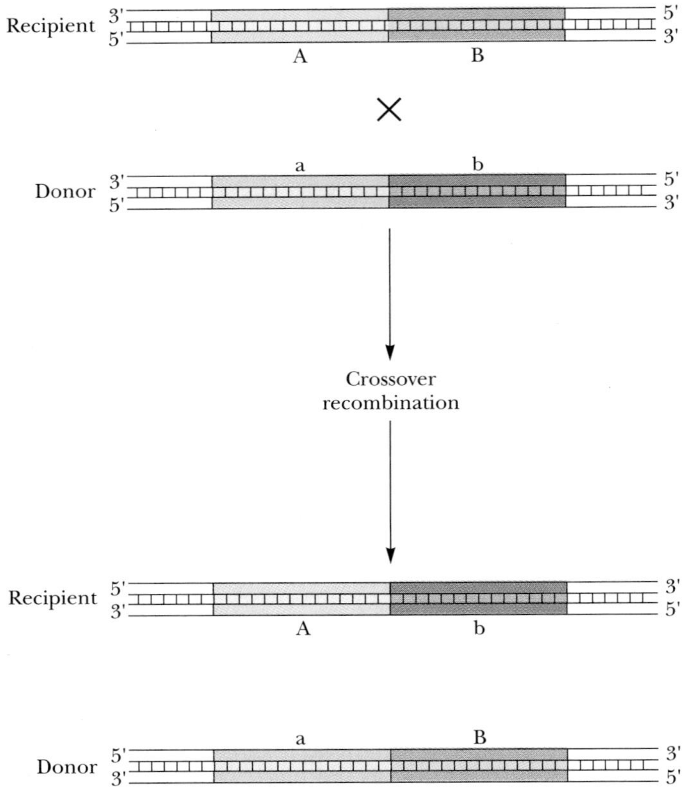

***Figure* 15.4** Exchange of two DNA segments by homologous recombination. The outcome of homologous recombination, DNA exchange between identical or highly homologous sequences, appears as a double-stranded break between recipient and donor DNA, resulting in the reciprocal exchange of genes (gene *b* for gene *B*) which are adjacent to the site of recombination.

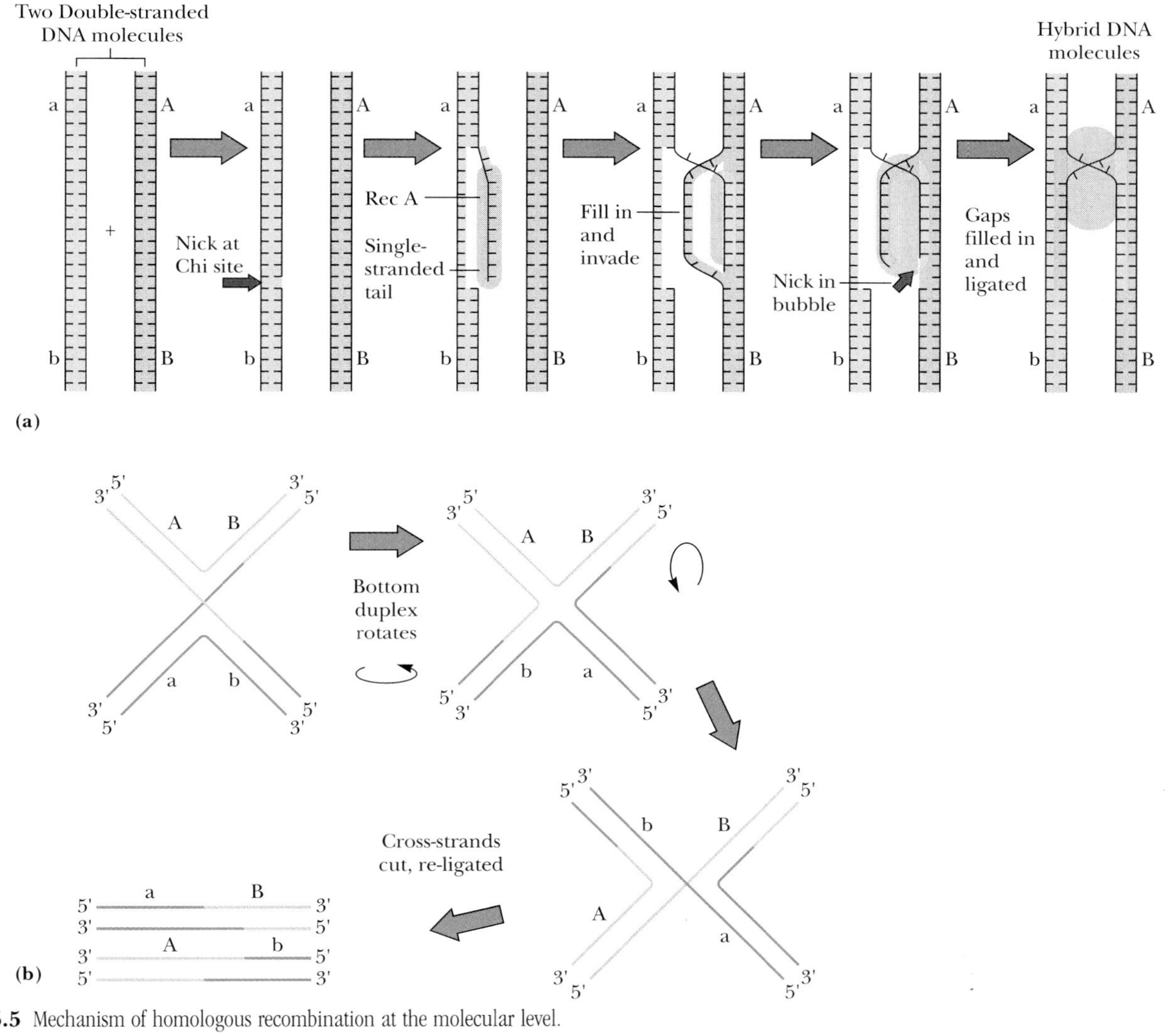

***Figure* 15.5** Mechanism of homologous recombination at the molecular level.

other. Isomerization involves simultaneous rotation of the two strands, followed by cutting and ligation of the crossed strands. This process effectively exchanges a segment of two DNA strands.

Now let's consider some of the DNA rearrangements that can result from homologous recombination. When two plasmids each containing similar DNA sequences recombine, the result is a single, large, composite plasmid (Figure 15.6**a**). A similar process involving chromosomal sequences with sequences on a plasmid leads to incorporation of the plasmid into the bacterial genome. The outcome of intrachromosomal homologous recombination depends on the relative orientation of the homologous sequences. If such sequences are in an opposite orientation (Figure 15.6**b**), the reciprocal exchange results in **inversion** of the intervening region. If the sequences are in the same orientation (Figure 15.6**c**), the result of general recombination is **excision** of a circular DNA consisting of the intervening region and potential loss of the genetic information during subsequent cell division.

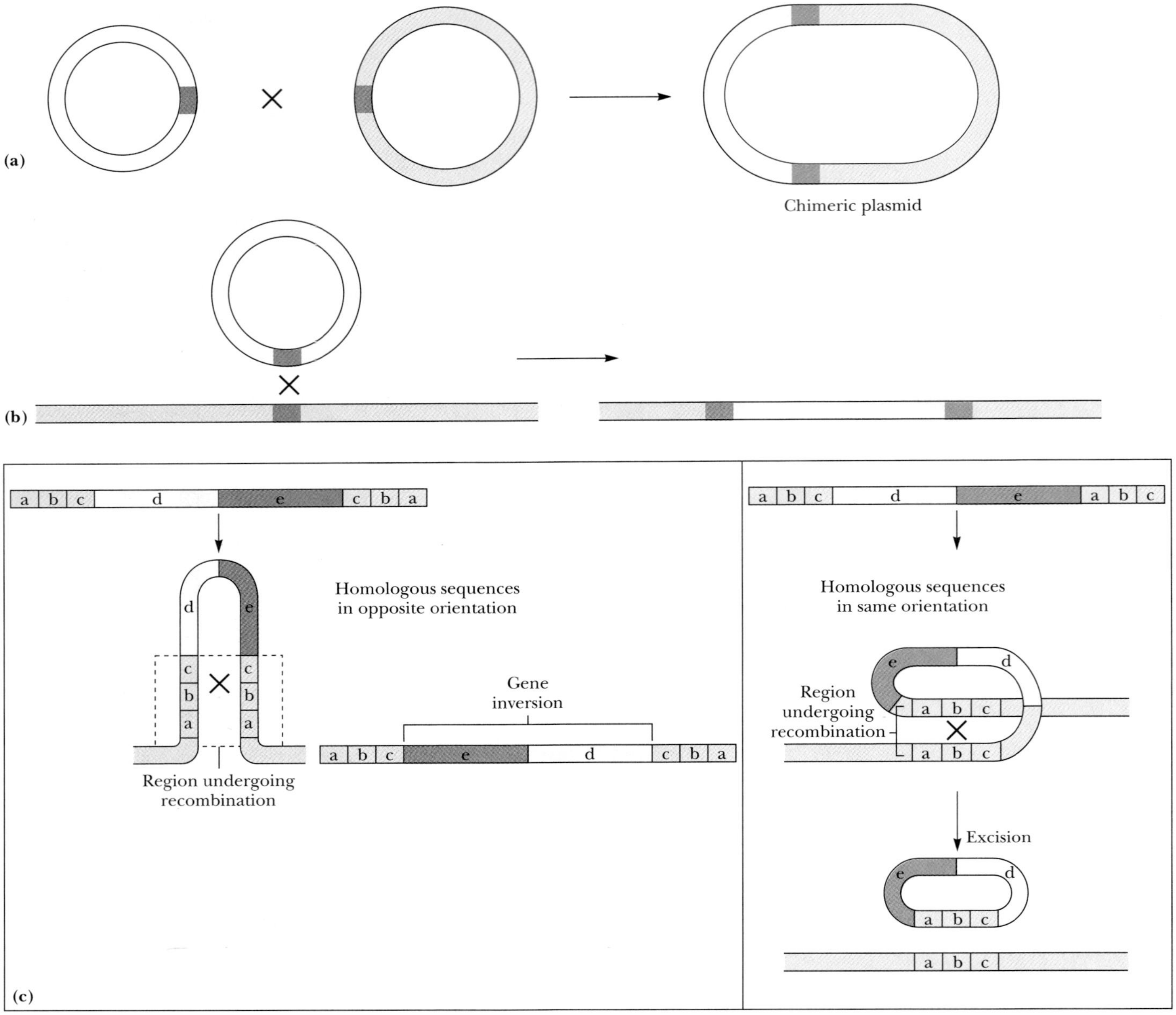

***Figure* 15.6** Homologous recombinations as they may occur in bacteria: **(a)** Homologous recombination between two plasmids, resulting in formation of a single chimeric plasmid. **(b)** Recombination between a plasmid and a chromosome containing a sequence which is also found in the plasmid. **(c)** Recombination between two segments on a chromosome containing a duplicated set of genes *a, b, c*. The consequence of this type of recombination, when the homologous sequence is in the opposite orientation (an inverted repeat of genes *a, b, c* and *c, b, a*) is inversion of the order of genes (*d, e*) that are surrounded by the region undergoing recombination. When the homologous sequence is in the same orientation (a direct repeat of genes *a, b, c* and *a, b, c*) the recombination leads to excision of the duplicated genes as well as the region between the site of recombination. The excised circular form of the DNA is usually lost during cell division.

Site-Specific Recombination

Site-specific recombination between two DNA segments does not require homology between the base pairs involved in the exchange of DNA strands. The process is called site-specific because it takes place between specific sites on the DNA, as defined by their nucleotide sequence. Moreover, site-specific recombination requires a specialized set of enzymes that recognizes these sites and mediates the recombination process. This form of recombination mediates a number of important cellular processes, including integration of viral genomes into the bacterial chromosome, insertion of specific genes by a transposition mechanism (see below), and expression of genes by inversion of promoter or regulatory sequences.

The mechanism of inversion of DNA resulting in expression, or lack of expression, of specific genes has been extensively studied in a number of microorganisms. One of the best understood examples is flagellar **phase variation** in the flagellin subunit proteins by *Salmonella typhimurium* (Figure 15.7**a**). This bacterium alternates between expression of either one of the two possible immunologically distinct flagellin proteins (H1 and H2) encoded by two different genes. This cleverly designed molecular circuit not only assures that one of the flagellin genes is expressed, but also prevents expression of the other gene. The expression of gene for H2, and repression of the gene specifying H1, is accomplished because expression of H2 is linked to the expression of another protein, a repressor that prevents expression of H1. The repressor gene *rh1* (for *r*epressor of *H1*) is a second gene of a two-gene operon, transcribed from a single promoter

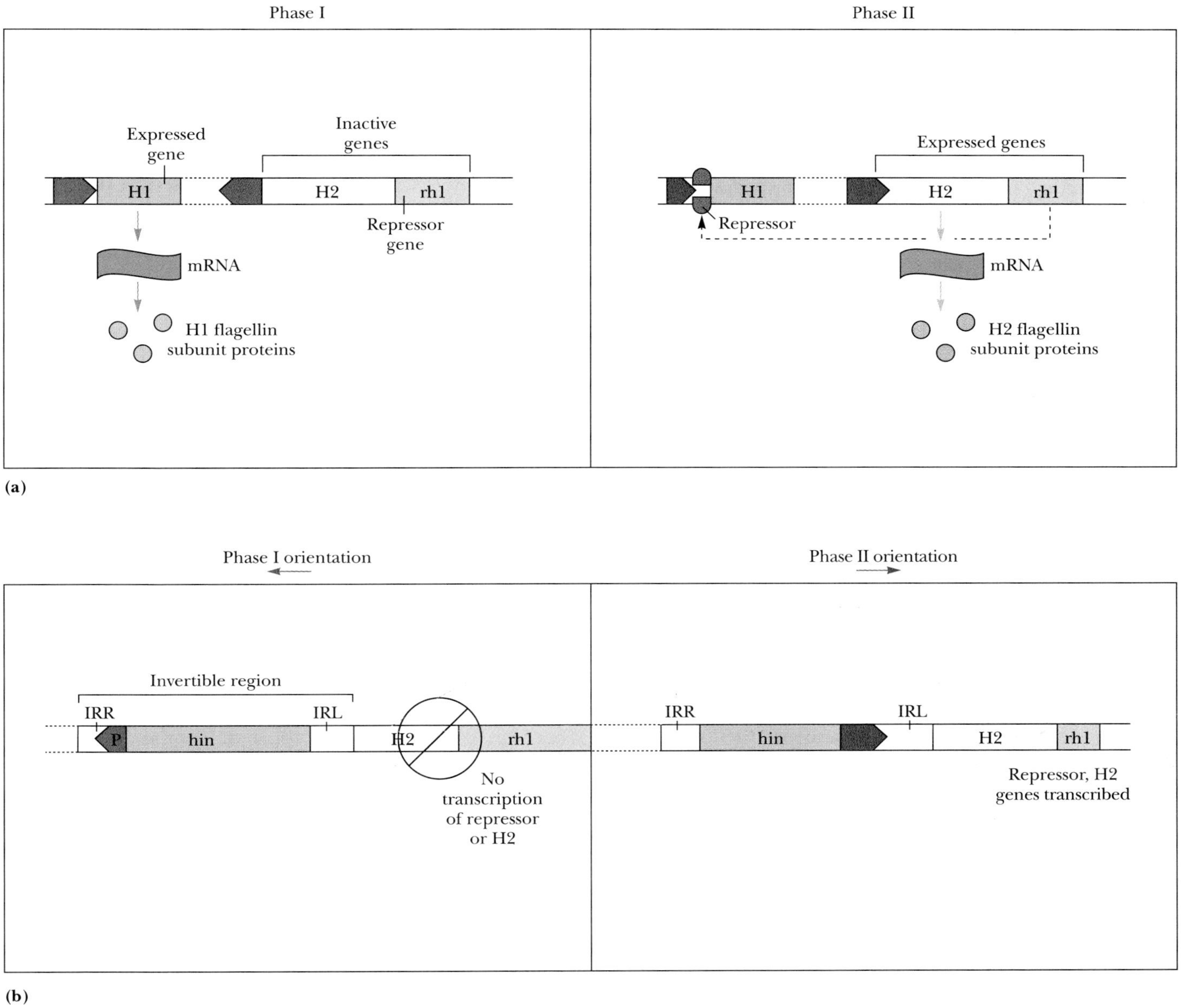

***Figure* 15.7** Phase variation during expression of *Salmonella typhimurium* flagellar proteins: **(a)** Overall scheme for the expression of H1 and H2 type flagellins and control of the H1 gene by the repressor. **(b)** Details of the inversion regions near the promoter for the operon of the genes encoding H2 and the repressor. The invertible region encodes a site-specific recombinase (Hin), which is necessary for the controlling orientation of the promoter (p) by recombination between the right and left inverted repeats (IRR and IRL).

proceeding the gene for H2. The expression of the H2-gene:*rh1* is controlled by an invertible 995 base pair (bp) DNA segment, flanked by short, 14 bp inverted repeats (IRL and IRR); the end nearest to the operon contains its promoter. At a regular frequency, the 995 bp segment is inverted, resulting in the promoter being oriented away from the genes for H2 and the repressor. The consequence of this rearrangement is concomitant loss of expression of H2 and lack of synthesis of the repressor. In the absence of the repressor protein, the gene for H1 is now expressed and these new subunits are assembled into flagella. The inversion process can be reversed (going from H1 expression to H2) by the reversal of the same rearrangement, allowing for synthesis of H2 as well as the repressor of H1.

The molecular details of this type of site-specific rearrangement have been worked out (Figure 15.7**b**). The key mediator of this process is Hin, a recombinase enzyme, whose gene *(hin)* is encoded within the 995 bp invertible segment. Hin mediates this type of site-specific recombination by binding to the ends of the invertible segment, within the IRR and IRL, where with the aid of several ancillary proteins, the proteins, together with bound DNA, assemble. Within this protein complex, the DNA is cleaved and the ends that form the opposite ends of the segments are ligated together. The consequence of inversion of the segment is a reversal of the IRR and IRL sequences, thus moving the promoter away from its site adjacent to the H2 gene. Mechanistically similar reactions are responsible for expression of genes encoding receptors for certain bacterial viruses, and those specifying pili in a number of bacterial species.

Genetic Exchange Among Bacteria

Bacterial recombination via a general or site-specific mechanism is most easily manifested following acquisition of foreign DNA by a bacterium. This is a normal process in higher eukaryotes in which sexual reproduction is initiated by fertilization of an egg that brings genetic material from two different sources. Bacteria also have the capacity to acquire DNA from other bacteria, but it is not tied to cell division. Genetic exchange in bacteria is an important vehicle for evolution and rapid adaptation to a new environment.

Bacteria can acquire novel genetic information, that is, new segments of DNA by three different mechanisms: **transformation,** in which a bacterial cell takes up free DNA from its environment; **conjugation,** which is a plasmid-mediated transfer of DNA from one bacterium to another; and **transduction,** in which a bacteriophage carries bacterial DNA from one cell to another. The ability to detect microorganisms that have received new genetic information by any one of these methods depends on expression of the gene in the new host. Often the genes become part of the bacterial genetic repertoire, and the trait thus gained is inherited by daughter cells after cell division.

Transformation

The process of giving a bacterium acquired heritable traits by exposure to a solution of DNA was a true landmark in the history of genetics, as it was the first demonstration that genes are located on DNA. In 1944, O. T. Avery, C. M. MacLeod, and M. McCarty reported that avirulent (nondisease producing) *Streptococcus pneumoniae,* lacking a protective capsule, regained the ability to synthesize a capsule and become virulent when exposed to purified DNA from a virulent strain (Figure 15.8). They further showed that the ability of the DNA preparation to transform an avirulent *S. pneumoniae* into a virulent form was lost when the DNA preparation was exposed to DNases (enzymes that destroy DNA), but not when exposed to proteinases or RNases (enzymes that destroy proteins or RNA). This proved that the transforming genetic information was carried in cellular DNA, not in proteins or RNA.

Virtually all bacteria have an ability to take up DNA from the environment, provided they are **competent. Transformation competence** is a state of a bacterial cell during which the usually rigid cell wall can transport a relatively large DNA macromolecule. This is a highly unusual process for bacteria, for they normally lack the ability to transport large macromolecules across the rigid cell wall and through the cytoplasmic membrane. Competence in some bacteria, such as *Haemophilus, Streptococcus,* or *Neisseria,* is expressed during a certain stage of cell division and most likely represents a stage when the reforming cell wall can allow passage of DNA. These bacterial genera possess **natural competence** because their cells do not require any special treatment to increase their ability to take up DNA. Not all bacteria, however, are naturally competent to take up DNA, but treatment of such cells with calcium or rubidium chloride alters their envelope and they become competent. This form of competence, termed **artificial competence,** is one of the more important techniques for introducing recombinant DNA molecules into bacteria.

What is the fate of a DNA molecule once it enters a competent cell? If the transformed DNA is a plasmid capable of replicating within that particular species of bacterium, the cell becomes plasmid-bearing and plasmid DNA replicates as an extrachromosomal element. Genes encoded on the plasmid are expressed, and synthesis of plasmid-encoded proteins allows the bacterial host to take advantage of some additional traits that its siblings without plasmids lack. For example, if the bacterium took up a plasmid-encoding gene that specified aminoglycoside antibiotic resistance, the transformant would then be resistant to an aminoglycoside, a trait certainly useful if the bacterium is in an environment containing this class of antibiotic.

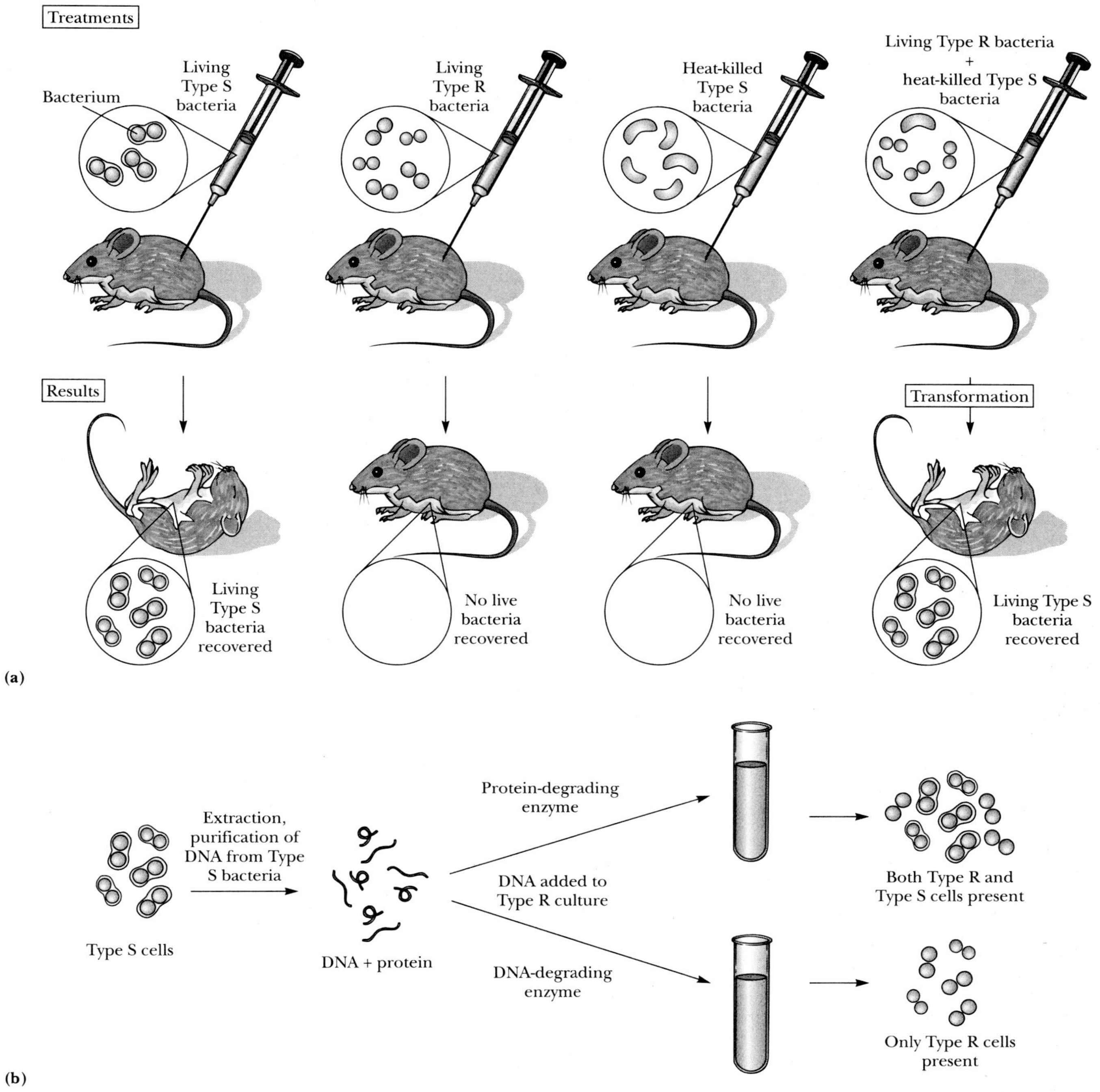

***Figure* 15.8** **(a)** Transformation of avirulent *Streptococcus pneumoniae* to virulent. While only the Type S of *S. pneumoniae* is virulent in mice, a mixture of the avirulent Type R with killed Type S can result in virulent organisms. From infected mice, live S Types are recovered. This mechanism involves an apparent acquisition, by R Type bacteria, of virulence traits from the killed S Type bacteria. **(b)** Demonstration that the transforming principle is DNA. The conversion of the avirulent R Type bacteria is due to acquisition of DNA from the S Type cells, since enzymes that degrade DNA block the ability of a fraction of the Type R bacteria to acquire the virulence trait of Type S, while enzymes that degrade proteins have allowed Type R bacteria to acquire the virulence-associated characteristics.

Transformation of bacteria with linear DNA fragments is more complex since the acquired DNA is a natural target for hydrolysis by intracellular enzymes that degrade noncircular DNA. One mechanism whereby foreign DNA can escape degradation is incorporation into the chromosome of the recipient cell via general recombination. Since incorporation of linear DNA into a bacterial chromosome by a single-crossover event would disrupt the chromosome, stable incorporation of genes requires a **double-crossover** event. As shown diagrammatically in Figure 15.9, such double recombination effectively replaces the DNA sequence from the bacterial chromosome with that of the transforming DNA.

Since general recombination requires homologous sequences, such recombination can only take place with a nearly identical DNA sequence, thus having little or no effect on the bacterial genome. Note that homology is required only around the site of crossover, and that the sequence between the two crossover sites need not be identical. Mutations can be introduced or eliminated by transforming DNA, provided the DNA participating in the crossovers flanks the DNA containing the mutation. As illustrated in Figure 15.10, a newly acquired gene can be inserted into the chromosome by transformation, given that sufficient homology exists on either side of the acquired gene to allow for efficient general recombination. Analogous events can lead to loss of the gene. The double crossover, when applied to the introduction of new or modified genetic material, is sometimes referred to as **gene replacement.** The replaced DNA fragment is generally degraded by bacterial nucleases in the cytoplasm.

It is not clear to what extent transformation contributes to natural genetic exchange. Transformation requires the presence of free DNA in the bacterial environment and there it can be readily attacked by nucleases. The availability of DNA in the environment of a bacterium would be rather limiting. However, in certain environments where bacteria live in large numbers, some bacteria die and lyse and it is not unreasonable to expect that some DNA escapes extracellular nucleases. As is the case with all mechanisms of DNA exchange, if the acquired trait gives the recipient bacterium significant survival advantage, the progeny of the rare transformant will outnumber the bacteria lacking such a trait within several generations, even if the DNA acquisition mechanism is inefficient.

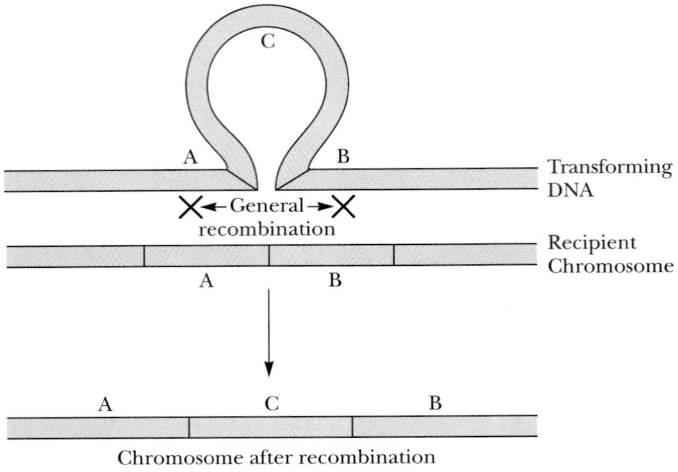

***Figure* 15.10** Introduction of new genes into bacteria by transformation. Transformation can result in the recipient bacterium acquiring new genes or sets of genes, even in the absence of homology between the transforming DNA and the chromosome of the recipient, provided there is sufficient homology in the flanking DNA to allow for recombination. Here, the transforming DNA carrying a new gene *C* is flanked by genes *A* and *B*, which allow recombination (at X), leading to insertion of the gene *C* into the bacterial chromosome.

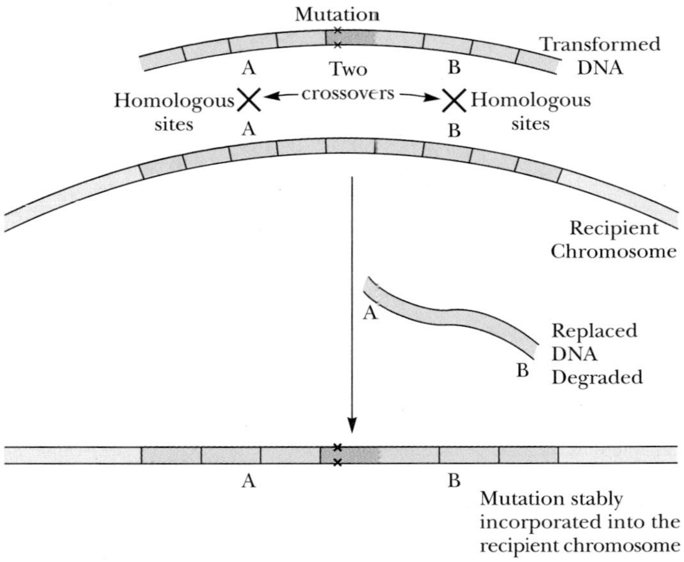

***Figure* 15.9** Introduction of a mutation into the bacterial chromosome from a piece of transformed DNA. The transformed DNA carries a mutation that incorporates into the chromosome of the recipient cells, following homologous recombination at two sites (A and B), where the DNA is similar or identical. This double crossover replaces a similar sequence in the recipient's chromosome.

Conjugation

Conjugation is a mechanism of DNA exchange mediated by plasmids. Some but not all plasmids are **transmissible,** which means they can promote their own transfer as well as mediate the transfer of other plasmids and even portions of chromosomal DNA. Bacteria that contain transmissible plasmids are called **donor** cells; those cells that receive plasmids are the **recipients.** Analogous to sexual transfer in higher eukaryotes, donor bacteria are often called male cells, while recipients are called female.

Plasmids, such as the F-plasmid of *E. coli* or the large R-plasmids of a variety of other bacteria, are circular molecules of DNA. These extrachromosomal genetic elements encode a number of genes responsible for their own replication and maintenance within the bacterial cell. A significant number of cellular genes are dedicated to the promotion of DNA transfer via conjugation. These transfer functions include several genes that encode the pilus components as well as proteins involved in the regulation of expression and biogenesis of the pilus. The pilus is an organelle formed on the surface of a plasmid-bearing donor cell that recognizes the recipient and makes the initial contact. A given plasmid can express a distinct pilus.

The process of conjugation can be divided into several steps, which are outlined in Figure 15.11. In addition to the initial contact mediated by pili, several other factors are necessary to stabilize the contact, resulting in a close association between the two mating cells and formation of a channel in the bacterial cell wall. Plasmid DNA is then processed by nicking at a specific site called the **origin of transfer,** often abbreviated ***oriT.*** The transfer from this site proceeds by a rolling circle replication mechanism, in which the 5′—region of the DNA strand is transferred to the recipient cell. The transfer process is accompanied by an immediate copying of the template strand in the donor cell in the 5′—to—3′ direction, effectively restoring the double-stranded molecule. In the recipient, the transferred strand is also converted to a double-stranded form. The process is completed by the conversion of donor and recipient plasmids to covalently closed circular forms.

The conjugative transfer described for F-plasmids or large R-plasmids requires that the plasmid contain all of the information for its own transfer. These plasmids are called **self-transmissible** since they can enter a recipient using entirely their own transfer machinery. The transfer of a self-transmissible plasmid into a recipient converts that recipient to a donor since, by inheriting a self-transmissible plasmid, it acquires the ability to conjugate with other recipients. Not all plasmids are self-transmissible. Some plasmids lack some of the genes that encode the transfer functions, rendering them incapable of transfer to another cell. These plasmids, however, need not be restricted to a permanent residence in a bacterial cell. Since a bacterium can contain more than one plasmid, these plasmids may provide various conjugal functions to other plasmids in that organism. For example, a plasmid lacking a certain transfer function, such as pili, can still transfer to a recipient cell, provided the donor cell has another plasmid that encodes pili. This "helper" plasmid thus is capable of mobilizing another plasmid for transfer. Note that, unlike the case of self-transmissible plasmids, the recipient receiving such an incomplete plasmid cannot be a donor, unless it already contains a helper plasmid.

Plasmids lacking *oriT* are not transmissible even with helper plasmids. However, these plasmids can encode transfer functions and can act as helper plasmids for other transfer-deficient plasmids. One way a plasmid lacking *oriT* or any additional transfer functions can transfer from one cell to another is by "hitchhiking" a ride on a transmissible element. This can be accomplished by having some homologous sequence on both plasmids such that a homologous recombination event fuses the two plasmids into a **chimeric** plasmid (that is, from two sources). The transfer of the chimeric plasmid is originated from *oriT* of the transmissible plasmid component.

Plasmid-Mediated Exchange of Chromosomal Genes

Transfer of genes among bacteria, mediated by self-transmissible plasmids, is not restricted to genes that are located on plasmids themselves. A number of conjugative plasmids have the ability to transfer chromosomal genes as well. This property was first discovered in the *E. coli* F-plasmid. However, it is now apparent that similar mechanisms of chromosomal transfer exist among several R-plasmids. The ability to mediate chromosomal genes transfer is referred to as the **chromosome mobilizing ability.**

The F-plasmid can mediate transfer of chromosomal genes by two related mechanisms. They both require formation of **high-frequency recombinant cells (Hfr)** in which

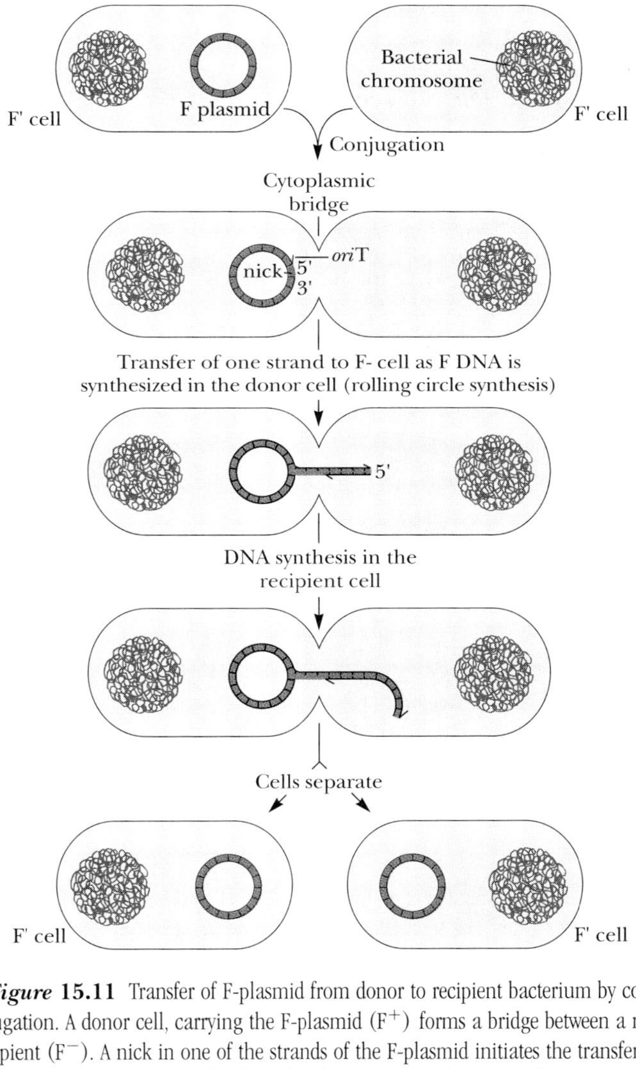

***Figure* 15.11** Transfer of F-plasmid from donor to recipient bacterium by conjugation. A donor cell, carrying the F-plasmid (F^+) forms a bridge between a recipient (F^-). A nick in one of the strands of the F-plasmid initiates the transfer of one of the strands by its leading 5′ end. During transfer, a complementary strand is synthesized in the recipient, followed by circularization of the DNA. Also during transfer, the transferred strand is resynthesized in the recipient. Once the transfer of the plasmid is completed, both cells are F^+, that is, have an intact copy of the F-plasmid and can function as donors.

the F-plasmid has integrated into the chromosome (Figure 15.12). The integrated form of the F-plasmid has no effect on the cell *per se* since it replicates along with the chromosome. The integration is not absolutely stable; in a population of cells, an equilibrium exists between bacteria with integrated F-plasmid (Hfr) and cells where the F-plasmid is not part of the bacterial chromosome. Integration is not random; a few preferred sites exist in the *E. coli* chromosome where most of the integration can take place. All of the sites of integration have a small DNA sequence in common.

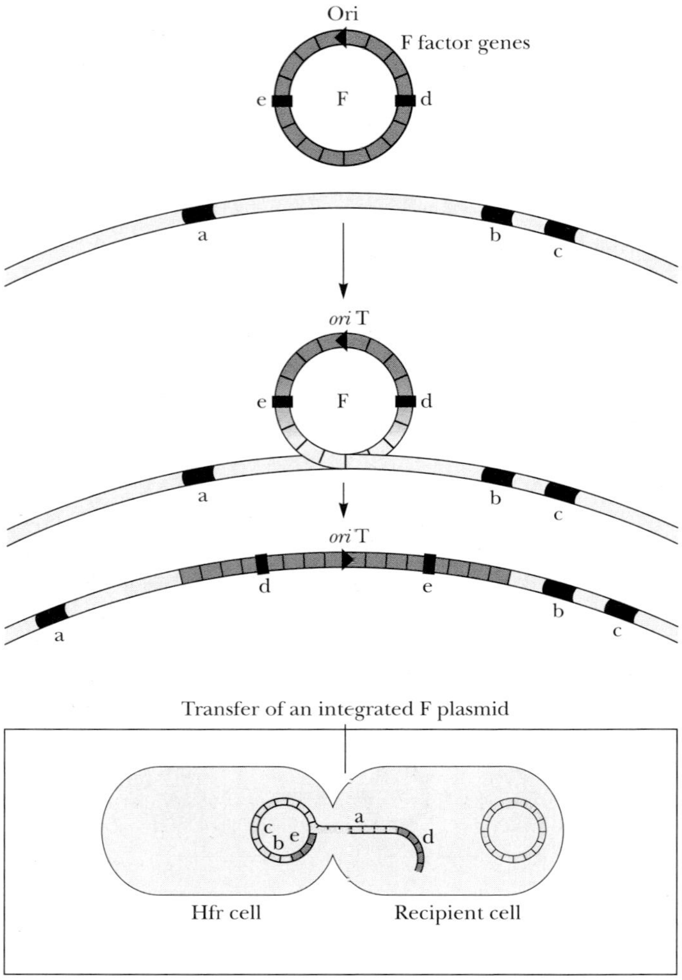

***Figure* 15.12** Formation of an Hfr and transfer of chromosomal genes into a recipient bacterium. The orientation of the origin of transfer (*ori*T), relative to the genes in the chromosome, determines the sequence of transfer and the length of time required for their transfer. In the example shown, genes *d* and *a* are transferred early, while *c*, *b*, and *e* are transferred late. Insertion of the F-plasmid in the opposite orientation allows early transfer of genes *e*, *b*, and *c*, and later transfer of *d* and *a*. This allows mapping of relative locations of genes in the bacterial chromosome, by mixing donor and recipient cells and interrupting the mating at various times by vigorous agitation to break up the conjugating pairs. After growing the cells on appropriate media, the transferred genes can be identified.

Once the F-plasmid has integrated, the cells can initiate a transfer process virtually identical to the one described for transfer of the F-plasmid itself. Following contact of an Hfr cell with a recipient, a single-stranded nick is created in the integrated F-plasmid, and transfer begins by the process analogous to the transfer of the free, nonintegrated F-plasmid. However, unlike transfer of F-plasmid, chromosomal genes are transferred in the 5′—to—3′ direction. The process continues as long as the cells remain in contact. The entire chromosome of a donor cell can be transferred into a recipient, resulting in a diploid cell containing a reformed Hfr chromosome. However, the association of donor and recipient is usually not very strong, and the amount of transferred donor chromosomal DNA is directly proportional to the duration of contact between the bacterial mating pair. Once DNA is transferred and the complementary strand is synthesized, the genes transferred by this process can undergo general recombination with the chromosome of the recipient, not unlike the fragments of DNA that a cell receives by transformation. Double crossover events are again required for integration of blocks of DNA between the sites of crossover. The recipient cell usually receives only a small portion of the chromosome of a donor, containing only a fraction of the integrated F-plasmid. Therefore, it is unable to donate DNA to another recipient. In complete matings in which the entire chromosome is transferred, however, the recipients are Hfr, that is, they are capable of further chromosomal transfer.

Bacterial conjugation via Hfr is a useful method of mapping genes in a bacterial cell. For each site of integration of F, one can mix donor and recipient cells and allow conjugation to take place in a controlled way. At various time intervals, conjugation is interrupted by shearing the suspension of bacteria, generally by disrupting contact by placing the bacterial suspension in a blender. Those recipients that received DNA from the donor and have undergone recombination are identified by their growth on specialized media. Thus, if a recipient is a mutant requiring an amino acid, recombination of the defective gene with DNA from a wild-type donor cell allows identification of such genes. The farther a gene is located from the site of integration of F, and, hence, from the origin of transfer, the lower the frequency of transfer of that particular gene to a recipient.

F-plasmids or similar R-plasmids can mediate transfer of chromosomal genes by yet another mechanism. As alluded to earlier, integration of the F-plasmid into the bacterial chromosome is not permanent, and excision does take place occasionally. When the F-plasmid excises from the chromosome by reversal of the reciprocal recombination, the site of integration is usually perfectly restored, yielding a complete F-plasmid and uninterrupted chromosomal DNA. However, occasionally imperfect excision leads to removal of a small portion of the chromosome,

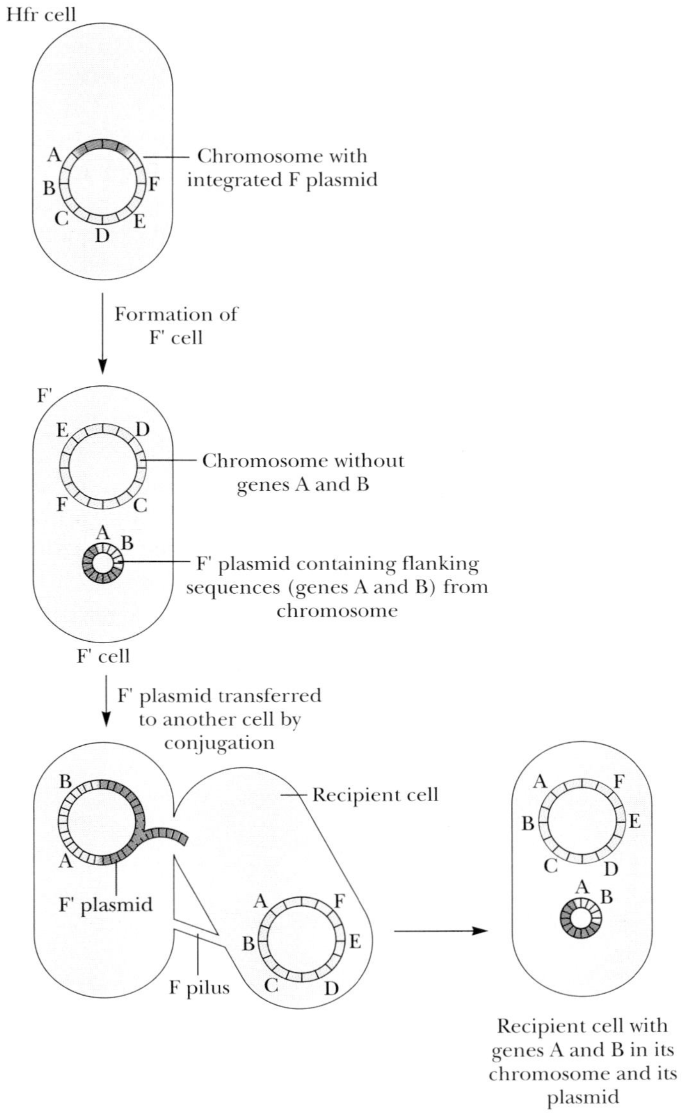

***Figure* 15.13** Formation of F′ from Hfr cell. An Hfr bacterium carries an integrated F-plasmid that can excise itself from the bacterial chromosome. If the process is imprecise, DNA sequences flanking the site of insertion can be removed from the chromosome and become incorporated in the plasmid. This modified form of the plasmid, called F′, is then capable of transferring the chromosomal sequences it carries into a new recipient by conjugation.

such that genes adjacent to the site of recombination are now part of the excised plasmid. These plasmids are called **F prime (F′),** or if the excision involved an R-plasmid, they are called **R′.** Transfer of DNA from an F′-donor (or an R′-donor) to a recipient proceeds by a mechanism identical to the transfer of the F-plasmid (Figure 15.13). In the case of plasmids that contain most of the F-plasmid's genome intact, the recipient becomes a donor. Since the recipient can contain the same genes as are on the larger F-plasmid transferred during mating with F′, the recipient may become a partial diploid (or merodiploid), that is, it contains two copies of a gene in a single cell. Moreover, regions of extensive homology exist between the chromosome and the plasmid, allowing general recombination to take place along these sequences. Insertion of an F-like plasmid into the bacterial chromosome is one way new Hfr strains can be generated, thus allowing chromosomal gene transfer from a specific location.

Transduction

The process of transferring genes between bacteria as part of bacteriophage particles is called **transduction.** Two fundamentally distinct transductional mechanisms operate among bacteriophages. One system, called **generalized transduction,** results in transfer of all bacterial genes at low but identical frequencies. During **specialized transduction,** certain genes are transferred at very high frequencies, while others are transferred at low rates or not at all.

Generalized Transduction

Generalized transduction is carried out by some but not all virulent bacteriophages. This process is outlined in Figure 15.14. Following infection of the cell, these bacteriophages undergo a complete replication cycle. During the infection process, the host chromosome is disrupted and used as a source of nucleotide—building blocks for the phage genome.

Occasionally, fairly large host DNA fragments remain in the cytoplasm at the time of viral DNA particle packaging. Infrequent packaging errors result in bacterial DNA being packaged into the capsid instead of the phage genome. Upon completion of the infection cycle, the host cell lyses, releasing the progeny virus. While most capsids contain viral DNA and are fully capable of binding to new host cells and thus repeating the infectious cycle, the rare capsid containing the host DNA will also adsorb to another host and inject its DNA into the bacterium. This interaction of a cell with a viral particle does not result in the death of the host, since the introduced DNA does not specify viral particles. Rather, the cell now contains a small fragment of the bacterial genome. This newly introduced DNA is now similar to any DNA fragment introduced to the cell by other gene exchange mechanisms, such as transformation. In the event that the transduced DNA shares homology with the bacterial chromosome, double reciprocal recombination can result in incorporation of the transduced DNA into the bacterial chromosome. This process, thus, results in transfer of DNA from one bacterium to another.

Specialized Transduction

Specialized transduction requires incorporation of the viral DNA into the bacterial chromosome, and it is there-

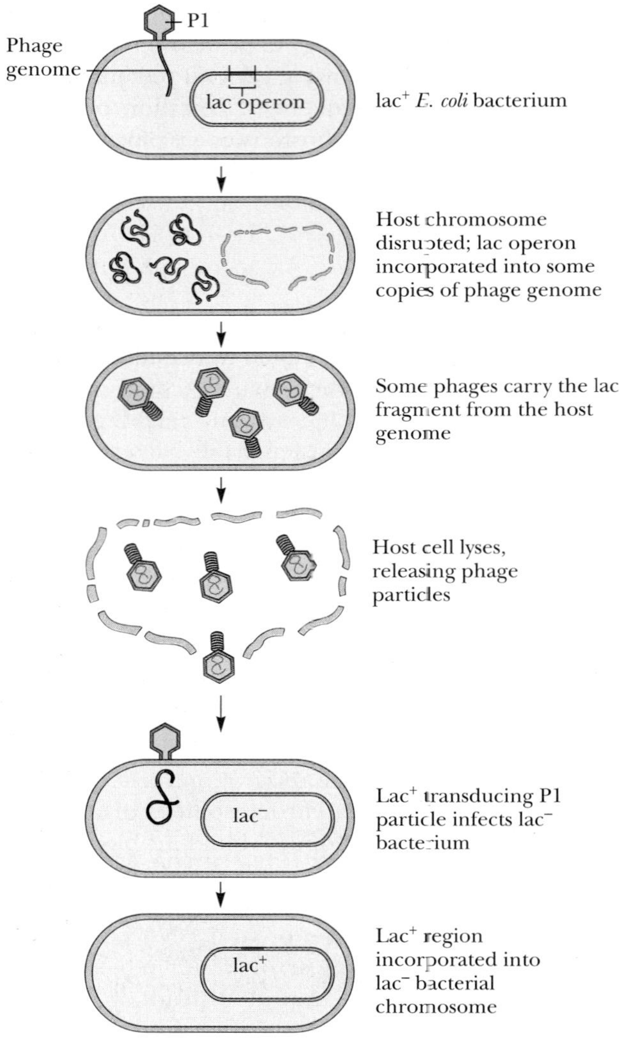

***Figure* 15.14** Generalized transduction carried out by the bacteriophage, P1. Infection of *E. coli* by bacteriophage P1 leads to production of progeny bacteriophages. Occasionally, a fragment of the bacterial chromosome is incorporated into the bacteriophage. In the example shown, a portion of the lactose operon *(lac)* is packaged into the bacteriophage, and this defective bacteriophage particle can infect another *E. coli (lac$^-$)*, which carries a mutation into the *lac* operon. The DNA fragment introduced into the cell by the bacteriophage can become incorporated into the bacterial chromosome by double crossover, resulting in restoration of the ability of the previously *lac$^-$ E. coli* to utilize lactose.

fore carried out only by lysogenic viruses. Specialized transduction occurs following formation of prophage, whereby the viral DNA integrates at a specific site in the host genome. Although the prophage replicates for many generations with the bacterial chromosome, it can occasionally excise and initiate a virulent cycle, including viral DNA replication, packaging, and lysing of the host. Occasionally, excision of the prophage is imprecise, resulting in removal of some DNA on either left, right, or both sides of the site of integration, as shown schematically in Figure 15.15. This excised DNA now is part of the viral genome; it replicates and is packaged into every viral particle. Following lysis of the cell, released virulent bacteriophage can infect other cells. Infection of a new host can lead to viral replication and continuation of the virulent cycle. However, as is the case with the original infecting phage, occasionally the transducing virus can enter into a lysogenic relationship with its new host. The transduced genes would then be part of the genome of the second bacterial cell. Since the second recipient can also contain genes homologous to those carried by the transducing phage, reciprocal recombination can result in stable incorporation of such genes into the bacterial chromosome, without the necessity of establishing a bacterial lysogen.

Specialized transduction therefore results in phage-mediated transfer of genes that are near the chromosomal attachment site of the lysogenic phage. While this process is limited to such genes, it is extremely efficient since, once the error in excision has occurred, such genes are part of the bacteriophage genome and are efficiently transferred.

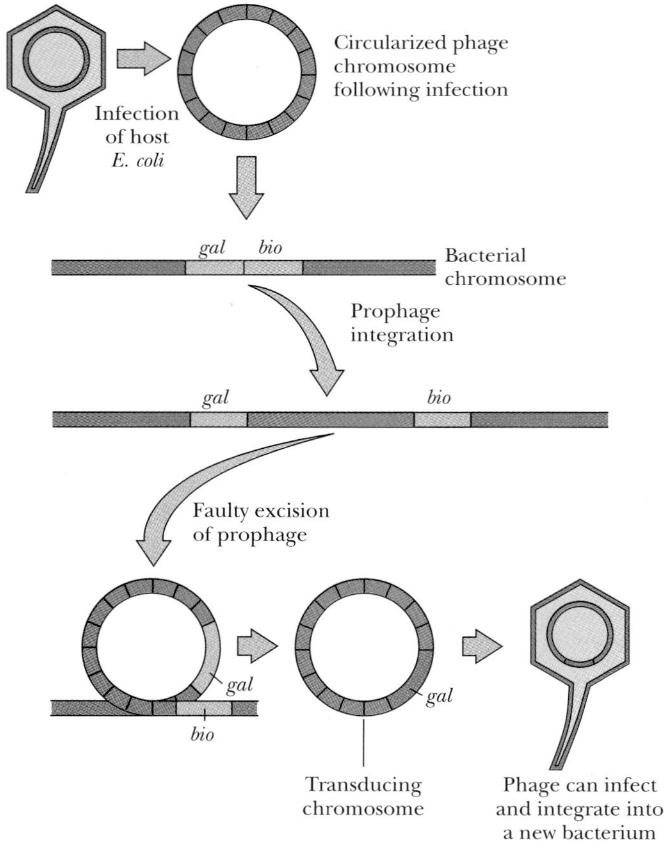

***Figure* 15.15** Specialized transduction carried out by the bacteriophage, lambda. Following infection of *E. coli* by bacteriophage lambda, the genome of the bacterium is circularized and inserted into the bacterial chromosome between the *bio* and *gal* genes. Imprecise excision of the prophage and subsequent packaging of the DNA into lambda phage results in incorporation of the *gal* gene into the viral particle. This form of lambda bacteriophage can infect other bacteria, bringing with it the *gal* gene from its previous host.

Mobile Genetic Elements and Transposons

This chapter has considered genetic exchange in terms of blocks of DNA that move from cell to cell by being integrated into defined, self-replicating units, such as viruses or plasmids. Occasionally, a gene can move from one place on the chromosome to another in that cell, or a gene can move between a chromosomal site and a plasmid site. This process is called **transposition.** Movement from one location to another by the transpositional mechanism requires recombination to take place between the transposing DNA and the site where it integrates, by distinct mechanisms that are similar to site-specific recombination because the DNA exchange leading to integration of a transposon does not require homologous sequences, as general recombination does. The site-specific recombination utilizes DNA sequences that define the site of transposition, and specialized enzymatic machinery that catalyzes the transpositional event.

The simplest form of such a mobile gene is an **insertion sequence,** abbreviated IS. A number of different insertion sequences have been characterized in bacteria, ranging in size from 700 to 5000 base pairs (bp). All IS elements have a common feature: they contain short (16–41 bp) inverted repeat sequences at their ends. The IS also encodes at least one enzyme, called **transposase,** that specifically mediates the site-specific recombination event during transposition.

Transposition of an IS element into a new site can create insertion mutations by physically disrupting a sequence that encodes a polypeptide. The presence of this large block of DNA in an **operon** (a set of genes transcribed from a single promoter) can cause polar mutations; that is, insertion in the first gene of an operon can block expression of genes downstream from the site of insertion. This is due to the inability of RNA polymerase to traverse through an IS element, presumably because of the presence of transcriptional termination signals within the element. In fact, since IS elements usually do not code for easily identifiable genes, generation of a polar mutation in an assayable gene is a common method of identifying such transposable elements.

Transposons are more complex forms of mobile genetic elements and are characterized by the presence of genes in addition to the genes needed for transposition. Most common genes found within transposons are those specifying antibiotic-resistance determinants. However, other genes, such as those encoding toxins, have also been identified on these transposable elements.

Classification of Transposons

The simplest type of transposon is a **composite transposon,** which is a mobile genetic element in which two IS elements surround a gene. Transposition then involves translocation of a copy of the entire element (two ISs and the internal genes) into a new site, using the transposase gene of one of the IS elements. Composite transposons can serve as a source of IS sequence transposition since, as long as the IS element contains functional transposase, it can transpose independently from the transposon. Examples of such transposons are shown in Figure 15.16. As can be seen in the figure, some composite transposons, such as Tn5, contain multiple drug-resistance genes within the IS boundaries. Composite transposons represent a simple way for any bacterial gene to become mobile. Two identical IS can surround a bacterial gene, and under some circumstances, that gene becomes a transposon. Every transposon requires expression of a functional transposase gene, which is generally found within at least one of the insertion sequences.

A distinct class of transposons are those modeled after **transposon Tn3,** shown in Figure 15.16(**b**). This family of transposons does not contain repeated IS elements at its ends. However, the transposon is a single DNA segment that contains small (38-bp) inverted repeats at its ends. Within the element, there are antibiotic-resistance determinants (β-lactamase in Tn3), as well as two genes involved in transposition: transposase and resolvase. Resolvase is an endonuclease.

The last class of transposons are part of several lysogenic bacteriophages. These bacteriophages insert their genomes into the bacterial chromosome by site-specific recombination and, thus, can themselves be considered moveable genetic elements. Bacteriophage Mu (Figure 15.16**c**) represents one such transposable element.

Mechanism of Transposition

Transposons are mobile and they "jump" from a **donor** site (phage, plasmid, or chromosome) into **target** sites which can be another region of the bacterial chromosome, a site on a plasmid, or a site on a phage. The target sequence can vary, depending on the transposon. Some transposons require specific sites that occur only rarely within a chromosome or plasmid. Some transposons, such as Tn10, insert into a sequence NGCTNAGCN, where N can be any base. This sequence occurs many times in a bacterial chromosome. Other transposons, such as Tn5 and Mu phage, can insert into virtually any target sequence.

It is also possible to make distinctions regarding the fate of different transposons when they move from the donor to the target sites. For some transposons, such as Tn10, this process is **conservative,** and represents true translocation, in which it is lost from the donor site after transposition to a new location. Other transposons (i.e., Tn3) move by a **replicative** mechanism: a copy of the transposon inserts into the target sequence, while a copy remains in its initial position.

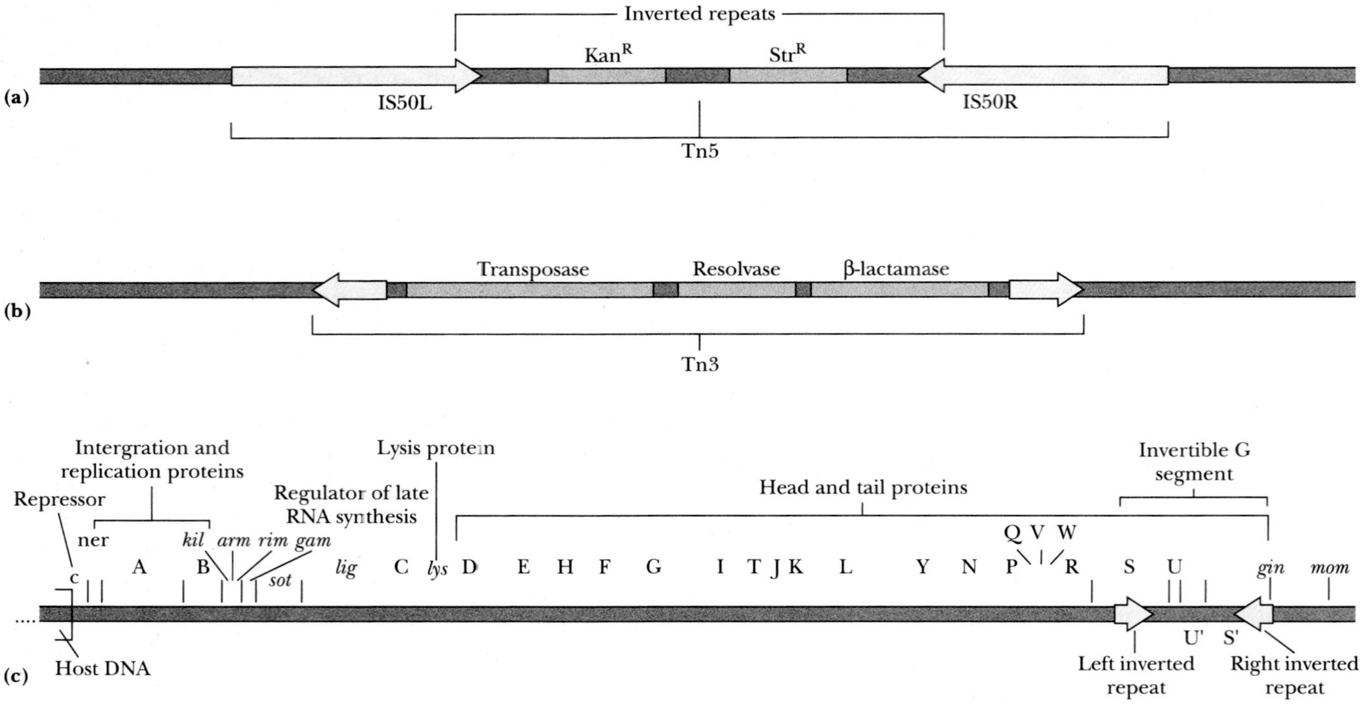

***Figure* 15.16** Structure of various bacterial transposable elements: **(a)** General structure of a composite transposon with antibiotic genes, flanked by two insertion sequences as direct or inverted repeats. Transposon Tn5 contains kanamycin- and streptomycin-resistance genes, flanked by insertion sequences IS50 in an inverted orientation. **(b)** Tn3 transposon. The open triangle indicates the short 38-base-pair inverted repeats. **(c)** Mutagenic bacteriophage Mu.

Summary

- While most of the essential genes are found on the bacterial chromosome, additional genes for a variety of cellular functions are encoded by extrachromosomal genetic elements called **plasmids.** A class of plasmids carrying antibiotic resistance genes are called **R-factors** or **R-plasmids.** A special class of plasmids, called virulence plasmids, carry genes that encode essential virulence determinants, such as **adhesins** and **toxins.**

- A plant pathogen ***Agrobacterium tumefaciens*** causes tumors in plants by transferring a portion of its plasmid (called tumor-inducing, or **Ti plasmid**) to the plant, where the Ti-encoded genes induce synthesis of plant hormones.

- **Homologous recombination** involves breakage and joining of two DNA strands along sequences of identity, or a high degree of similarity. Homologous recombination utilizes a DNA site that possesses a nick, generated by the **RecB, RecC,** and **RecD** nuclease complex, at a sequence called the ***chi* site.** Another key component, required for homologous recombination is the **RecA** protein, which binds to single-stranded DNA and identifies homologous regions in the opposite strand.

- There are three types of genetic exchange mechanisms among bacteria: (1) uptake of free DNA from the media, called **transformation,** (2) direct transfer of DNA between donor and recipient bacteria, called **conjugation,** and (3) bacteriophage-mediated transfer of DNA, called **transduction.**

- Uptake of DNA by transformation requires that the recipient bacterium be in a physiological state called **natural competence.**

- In some bacteria, chemical treatment can induce **artificial competence,** such that they can take up DNA fragments or plasmids.

- DNA exchange by conjugation requires that the donor bacterium carry a **transmissible plasmid,** which encodes a complete transfer machinery, including **conjugal (sex) pili.** Plasmid transfer by conjugation is **replicative,** that is, a copy of the plasmid is transferred from the donor to a recipient, without the loss of the plasmid in the recipient bacterium.

- *E. coli* carrying a chromosomally integrated form of a well-studied F-plasmid, called **high-frequency recombinant (Hfr),** can transfer chromosomal sequences from this integrated form, starting from the *ori*T of the F-plasmid. Upon entry into the recipient *E. coli,* the partial genome is incorporated into the chromosome. Occasionally, the copy of the entire genome can be transferred into a recipient, in which case the recipient can become a donor.

- **Specialized transduction** is the consequence of imprecise excision of the lysogenic bacteriophage from the chromosome, resulting in packaging into bacteriophage capsids, of portions of the chromosomal DNA from the region flanking the attachment site.

- **Generalized transduction** involves random packaging of chromosomal DNA fragments, instead of bacteriophage genomes, into bacteriophage capsids during the final stages of phage maturation.

- Movement of genes between various locations on the chromosomes or plasmids is referred to as **transposition.** Insertion of a transposon into a gene results in a mutation in the gene due to the disruption of its coding sequence.

- Simple mobile elements are **insertion sequences,** containing short **inverted repeats** at their ends; these encode one enzyme, **transposase.** Some genes can transpose when they are flanked by two insertion sequences. These mobile genetic elements are called **composite transposons.** Certain transposons, such as **Tn3,** lack flanking insertion sequences. However, they still contain inverted repeats as their ends; within the transposon, they encode two essential enzymes, **transposase** and **resolvase.** Depending on the type of transposon, the site of insertion—the **target site**—varies from a defined short sequence to any random DNA sequence.

Questions for Thought and Review

1. Does a bacterium bearing a self-transmissible plasmid require a functional DNA synthesizing machinery for it to transfer the plasmid into a recipient by conjugation? Does a recipient need a functional DNA synthesis for it to receive the plasmid from a donor?

2. How does the fate of a chromosomal DNA fragment differ from that of a plasmid, when introduced into bacteria by transformation?

3. When bacteria contain more than one compatible plasmid, some plasmids, which are not self-transmissible, can transfer to recipients. Describe the mechanism that makes it possible. If the same plasmid lacks an origin of transfer, can it ever transfer to a recipient by conjugation?

4. When a chromosomal gene is encoding an antibiotic-resistance determinant, and is flanked by two identical DNA sequences, homologous recombination can lead to the loss of the resistance gene, such that some bacteria become sensitive to the antibiotic. Explain.

5. Which method of DNA transfer between bacteria would not take place if the donor and recipient were separated by a filter with a pore size of 0.45 μm? Which method of transfer would be blocked by the presence of high concentrations of DNase (enzymes capable of degrading DNA)?

6. A transposon insertion into the *lac*Z gene of the lactose utilization operon results in inability of the bacterium to utilize lactose as a sole carbon and energy source. When this bacterium acquires an F′ *lac* (an F plasmid with the intact *lac*Z gene) it is still unable to utilize lactose. Provide a reasonable explanation for this phenomenon.

7. Chromosome mobilization by the integrated F-plasmid (Hfr) usually does not lead to the recipient being capable of acquiring donor functions. In contrast, F′-mediated transfer usually converts the recipient to a donor. Explain this difference in transfer mechanisms. What condition of DNA transfer results in the appearance of rare recipients that have acquired donor capabilities that are receiving DNA by the Hfr-mediated transfer?

8. How can the Hfr strain be used to map the relative order of genes?

9. A recipient *B. subtilis,* with mutations in the genes encoding enzymes of histidine and arginine biosynthesis, is transformed with DNA that came from *B. subtilis,* carrying a mutation in a ribosomal protein, rendering its resistant to streptomycin. The transformants were plated onto streptomycin, and subsequently replica plated onto minimal media containing either arginine or histidine. Growth (or lack of growth) on these supplemented minimal plates revealed that among the original streptomycin-resistant transformants, 12 percent did not require histidine anymore for growth, and 6 percent did not require arginine for growth, while the remaining 82 percent were streptomycin-resistant and retained the original requirement for both histidine and arginine. No streptomycin isolates that lost both of the arginine or histidine requirements were obtained. What is the location of the arginine and histidine biosynthetic genes, relative to the streptomycin-resistance determining ribosomal protein gene?

10. Compare and contrast specialized and generalized transduction.

11. A plasmid, carrying a kanamycin resistance gene and containing a composite transposon (two insertion sequences flanking a tetracycline-resistance gene) was introduced into a recipient by conjugation. When the bacteria were passaged for many generations in an antibiotic-free medium, a fraction of these bacteria did not contain any plasmid. When these bacteria were individually analyzed for antibiotic resistance, it was found that some were resistant to tetracycline, because they contained a chromosomal copy of the transposon. However, some were resistant to kanamycin, and the kanamycin-resistance gene was inserted into the chromosome. No isolates that contained both of the resistance genes were recovered. Explain the events in the recipient bacterial cell following introduction of the original plasmid.

12. A wild-type strain of *E. coli* is transformed by DNA from *E. coli* that has a Tn5 transposon in a gene encoding an enzyme of the tryptophan biosynthetic pathway. After plating transformants on rich medium containing kanamycin (the resistance gene carried on Tn5), the colonies are individually transferred onto minimal medium. Of 100 colonies tested, 18 required tryptophan for growth, 70 did not require any nutrients, and 12 could not grow on minimal media, even when it was supplemented with tryptophan. Provide an explanation for the three types of transformants. If the recipient was a *recA* mutant of *E. coli,* would you expect a different distribution of the three classes of transformants based on their nutritional requirements?

Suggested Readings

Hanahan, D. 1987. Mechanism of DNA transformation. In F. Neidhardt, J. Ingraham, K. B. Low, B. Magasanik, M. Schaechter, and H. E. Umbarger, eds. *Escherichia coli and Salmonella typhimurium: Cellular and Molecular Biology.* Washington DC: American Society for Microbiology.

Kleckner, N. 1990. Regulation of transposition in bacteria. *Annu. Rev. Cell. Biol.* 6:297–327.

Leach, D. R. F. 1995. *Genetic Recombination.* Cambridge, MA: Blackwell Science.

Margolin, P. 1987. Generalized transduction. In F. Neidhardt, J. Ingraham, K. B. Low, B. Magasanik, M. Schaechter, and H. E. Umbarger, eds. *Escherichia coli and Salmonella typhimurium: Cellular and Molecular Biology.* Washington DC: American Society for Microbiology.

Old, R. W., and S. B. Primrose. 1994. *Principles of Gene Manipulations.* Cambridge, MA: Blackwell Science.

Schleif, R. 1993. *Genetics and Molecular Biology,* 2nd ed. Baltimore: Johns Hopkins Press.

Smith, G. R. 1988. Homologous recombination in procaryotes. *Microbiol. Rev.* 52:1–36.

Willets, N. and R. Skurray. 1987. Structure and function of F-factor and mechanism of conjugation. In F. Neidhardt, J. Ingraham, K. B. Low, B. Magasanik, M. Schaechter, and H. E. Umbarger, eds. *Escherichia coli and Salmonella typhimurium: Cellular and Molecular Biology.* Washington DC: American Society for Microbiology.

If we should succeed in helping ourselves through applied genetics before vengefully or accidentally exterminating ourselves, then there will have to be a new definition of evolution, one that recognizes a process no longer directed by blind selection but by choice.

M. A. Edey and D. C. Johanson
Blueprints—1989
Little, Brown and Company

Chapter 16

Genetic Engineering

Previous chapters dealt with the mechanisms for gene transfer and recombination, which allow bacteria to acquire genes from other bacteria and express traits encoded by these genes. Our growing understanding of DNA replication, genetic variation, and transfer of genetic information among microorganisms, as well as of the molecular mechanisms that control gene expression, has led to the development of a new industry: **biotechnology.**

In this chapter, we will discuss the rationale, methodology, and outcome of deliberate laboratory manipulations of genes, a process called **genetic engineering.** The aim of genetic engineering, generally, is to isolate genes from eukaryotic, eubacterial, or archaeal genomes, and to introduce these genes into a different species where they will be propagated in a stable manner, resulting in a **transgenic** organism.

Basis of Genetic Engineering

Early work by bacterial geneticists with transducing phages and conjugative plasmids (see Chapter 15) that carry fragments of the bacterial chromosome suggested a possible means of isolating and manipulating individual genes. This work has promoted the study of genetic organization and expression, and has provided a mechanism for accumulating large amounts of protein for research and bulk amounts of protein for use in agriculture, industry, and medicine.

While individual genes may be isolated by classic bacterial genetic techniques (that is, using transducing phages and F′ plasmids), it took the discovery of **restriction endonucleases** to launch the field of genetic engineering. By borrowing these specialized enzymes from bacteria, researchers gained a powerful tool for manipulating genetic information. Simultaneously, increased knowledge of the replication requirements for plasmids and phages provided vehicles for isolating individual genes and propagating them in pure form—that is, **cloning** them. Finally, development of readily applied methods for determining the precise nucleotide sequence of DNA allowed studies of genetic organization, structure, and the function of gene products. Recombinant DNA technology has affected virtually every aspect of biomedical science, ranging from the production of human hormones in bacteria to the study of evolution. This chapter reviews some principles of genetic engineering and summarizes some of its more common applications.

Restriction and Modification

The ability of bacteria to exchange DNA by mechanisms, described in Chapter 15, may suggest that the bacterial genetic makeup is constantly changing. This is indeed not true; bacterial species have stable genomes. Several factors minimize genetic exchange via conjugative plasmids, transducing bacteriophages, and transforming DNA. Conjugation between different species is generally prevented by physical barriers, such as differences in the surface properties of various bacteria that preclude recognition of the recipient by sex pili—the initial step in conjugation. Similarly, transduction is limited by the types of receptors for transducing phages on the surface of the recipient. Transformation occurs at low efficiency due to the presence of natural barriers that limit uptake of "foreign" DNA by competent recipient bacteria. Transformation is further hampered by the presence of nucleases in the intercellular medium that degrade free DNA before it can be taken up by a potential recipient.

In addition to the limitations just discussed, one of the most effective barriers to free genetic exchange of DNA between unrelated bacteria is the presence of **restriction-modification** systems. These are enzymes that bacteria employ in nature to maintain the integrity of their genomes and degrade DNA that they may accidentally acquire from other bacteria by any of the genetic exchange mechanisms discussed in Chapter 15. **Restriction enzymes** are also useful in protecting bacteria from DNA viruses, by degrading the viral genetic material after it enters the bacterial cytoplasm. Restriction enzymes (or **restriction endonucleases**) introduce double-stranded breaks in foreign DNA at specific base sequences. Since these sequences are short (generally 4 to 6 base pairs), they will undoubtedly be present in the bacterial chromosome as well. So the bacteria prevent destruction of self-DNA by using **modification enzymes,** which modify the recognition sites on self-DNA for restriction enzymes, such that they are no longer degraded by these enzymes. Each bacterial species has characteristic restriction-modification enzymes that recognize unique sequences. For example, certain strains of *E. coli* contain an enzyme, *Eco*RI, which cleaves DNA whenever the sequence GAATTC occurs. The genome of *E. coli* itself has such sequences, and these are modified by methylation of the second adenine of this sequence on both strands, GAATTC, rendering the sequence resistant to *Eco*RI cleavage.

The best example of the advantage conferred by a restriction-modification system is observed in infections by bacterial viruses. The efficiency of infection of a particular host by bacterial viruses is inverse to the extent of damage to the viral DNA by the host's restriction endonucleases, following entry of the viral genome into the bacterial cell. An efficient restriction system can reduce the infection process to an exceedingly low level, such that most of the bacteria survive the infecting virus, and only 1 in 1000 bacteria infected will ever yield viral progeny. However, when these rare infective virions that infect 1 in 1000 are isolated and used to infect the same bacterial strain, the viral infection will be very successful, with hundreds of virions obtained from each host. The reason for the high level of infectivity of the second generation virions is that they survived in the first host's cytoplasm long enough to have their DNA modified by the host's modification enzymes. Thus, during the subsequent infection, this modified DNA is not recognized as foreign by the host's restriction enzymes.

This same restriction barrier is effective against the uptake of foreign DNA by transformation or conjugation. Thus, the fate of a plasmid or a DNA fragment depends on whether it is a substrate for the host's particular restriction enzymes and, if so, whether it has been protected by modification. Overall, restriction-modification systems, along with other barriers to DNA exchange, favor recombination of DNA within identical or closely related species.

Analysis of DNA Treated with Restriction Endonucleases

In practice, restriction enzymes are powerful tools for genetic engineering because they permit hydrolysis of a DNA molecule into defined fragments containing individual genes or clusters of genes. Hundreds of restriction enzymes with unique recognition sequences have been purified for this purpose, providing an array of enzymes to fragment DNA to a desired size. Restriction enzymes can cleave at a sequence by hydrolyzing a phosphodiester bond in each strand that is staggered, or they can cleave across a double-stranded DNA at the same base pair, producing a blunt end, as illustrated in Figure 16.1. The most commonly employed restriction endonucleases and their respective DNA recognition sequences are summarized in Table 16.1.

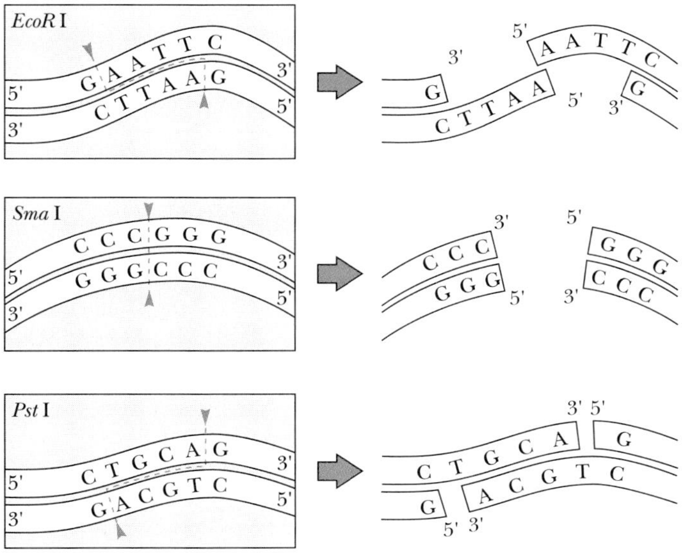

Figure **16.1** Cleavage of DNA strands by restriction enzymes. EcoR1 creates a staggered cut, leaving a single-stranded 5′ overhang. Pst1 also generates a staggered cut, with a 3′ overhang. Sma1 creates a flush blunt-ended cut. Arrows indicate the sites of hydrolysis of the phosphodiester backbone.

***Table* 16.1 Common restriction endonucleases (enzymes) and their DNA recognition sequences**

Microorganism	Restriction Endonuclease	Target Sequences, Showing Axis of Symmetry (|) and DNA Cleavage Sites (▼)	
Generates cohesive ends:			
Escherichia coli RY13	*Eco*RI	G ▼A A C T T	T T C A A▲ G
Bacillus amyloliquefaciens H	*Bam*HI	G ▼G A C C T	T C C A G▲ G
Bacillus globigii	*Bgl*II	A ▼G A T C A	T C T A G▲ A
Haemophilus aegyptius	*Hae*II	Pu▼ G C Py C G	G C Py C G▲ Pu
Haemophilus influenzae R_d	*Hind*III	A ▼A G T T C	C T T G A▲ A
Providencia stuartii	*Pst*I	C T G G ▲A C	C A▼ G G T C
Streptomyces albus G	*Sal*I	G ▼T C C A G	G A C C T▲ G
Xanthomonas badrii	*Xba*I	T ▼C T A G A	A G A T C▲ T
Thermus aquaticus	*Taq*I	T▼ A G	C G A C▲ T
Generates flush ends:			
Escherichia coli	*Eco*RV	G A T▼ C T A▲	A T C T A G
Brevibacterium albidum	*Bal*I	T G G▼ A C C▲	C C A G G T
Haemophilus aegyptius	*Hae*III	G G▼ C C▲	C C G G
Serratia marcescens	*Sma*I	C C C▼ G G G▲	G G G C C C

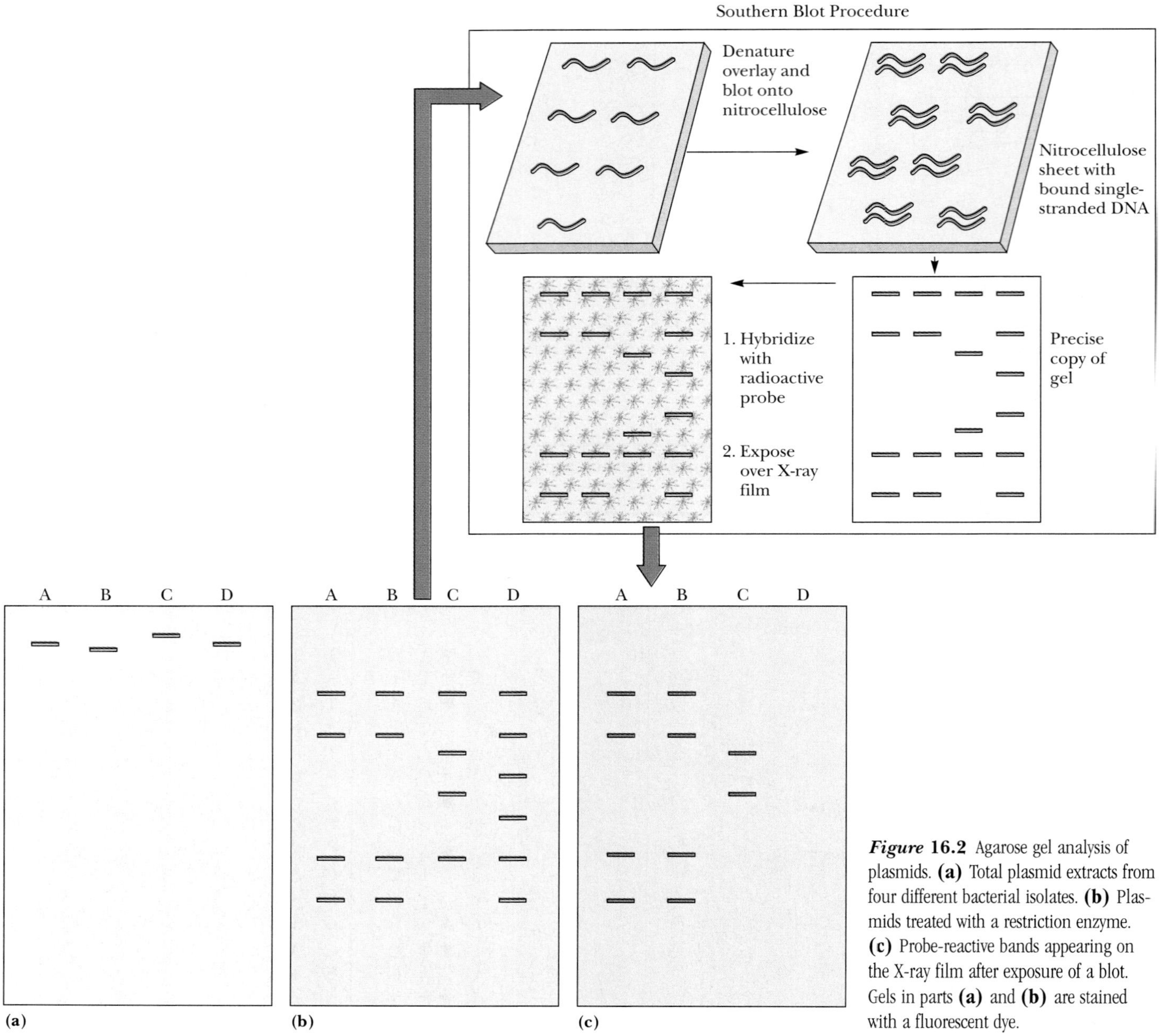

***Figure* 16.2** Agarose gel analysis of plasmids. **(a)** Total plasmid extracts from four different bacterial isolates. **(b)** Plasmids treated with a restriction enzyme. **(c)** Probe-reactive bands appearing on the X-ray film after exposure of a blot. Gels in parts **(a)** and **(b)** are stained with a fluorescent dye.

A simple method of DNA analysis is **agarose gel electrophoresis,** a procedure that separates DNA fragments by differential migration in an electrical field, applied across a hydrated porous matrix of agarose (a highly purified polysaccharide from seaweed). Among the critical parameters determining the rate of migration are the size and shape of the DNA. Agarose gel electrophoresis can thus separate plasmids based on size, and the plasmid content of independent isolates can be compared, to see whether they contain identically sized plasmids. As illustrated in Figure 16.2(**a**), different clinical isolates of the same organism can contain unique plasmids, and their respective electrophoretic mobilities can establish the size relationship among them.

The technique of comparing plasmids based on migration in an electrical field is rather unreliable, mainly because two plasmids can have similar sizes and yet be totally unrelated in sequence. Treatment of plasmids with restriction enzymes and subsequent separation by agarose gel electrophoresis allows a finer matching of sequences. If two plasmids are identical, the location of the restriction site recognition sequences is the same. Thus, following treatment with a particular restriction enzyme, an identical or similar pattern of fragments is obtained. Examples of such analysis are shown in Figure 16.2(**b**), which illustrates a case in which some of the fragments in strains A, B, C, and D are identical, although strains C and D contain a few unique fragments.

To determine the relationship among any set of sequences, a method of DNA hybridization was developed by Edward Southern in 1975. This technique, called **Southern blotting,** relies on the formation of stable duplexes (hybrids) between electrophoretically separated restriction fragments and radioactively labeled denatured DNA probes. The restriction fragments are denatured into single-stranded DNA by treating the gel with an alkali. The gel is then overlaid with a nitrocellulose sheet. Capillary action (blotting) moves the DNA onto the nitrocellulose sheet where it firmly binds. In this way, a permanent copy of the gel is obtained, with the fragments immobilized in the exact position where they migrated on the gel. This blot can then be soaked in a solution of a denatured radiolabeled probe. If the probe is one of the plasmids (for example, plasmid A), it will anneal to any complementary sequences immobilized on the solid support of the nitrocellulose. By this method, similar sequences can be radioactively "tagged" and visualized following exposure of photographic film. Results presented in Figure 16.2(**c**) indicate that plasmid A is probably identical to plasmid B, but only partially similar to plasmid C. Plasmid D is completely different from plasmid A, despite having an identical size and sharing some common-sized restriction fragments. Extra sequences found in plasmid C are probably due to recombination or transposition from other plasmids or chromosomal sites having sequences similar to plasmid A.

This powerful technique can be applied to the analysis of sequences found on specific fragments of bacterial or even eukaryotic chromosomes. DNA is extracted from cells and treated with restriction enzymes. As with plasmid DNA, fragments are separated according to size by agarose gel electrophoresis. Unlike fractionation of plasmids that yield only a handful of fragments, chromosomal DNA contains hundreds of recognition sites. Following gel electrophoresis, many bands appear, often preventing clear differentiation of one band from closely migrating bands of equivalent size. Nevertheless, following Southern blotting, a radiolabeled complementary probe can clearly identify a related sequence in one of the fragments, as shown in Figure 16.3. This method can detect proviruses in bacterial genomes (proviruses are viruses that have integrated their DNA into the host genome), as well as specific genes that encode determinants of antibiotic resistance or of bacterial pathogenicity, such as those that encode toxins.

Probes may be obtained from a variety of sources. Usually they are cloned pieces of DNA from different species that share some sequence similarity. Alternatively, a short piece of chemically synthesized DNA can be used as a

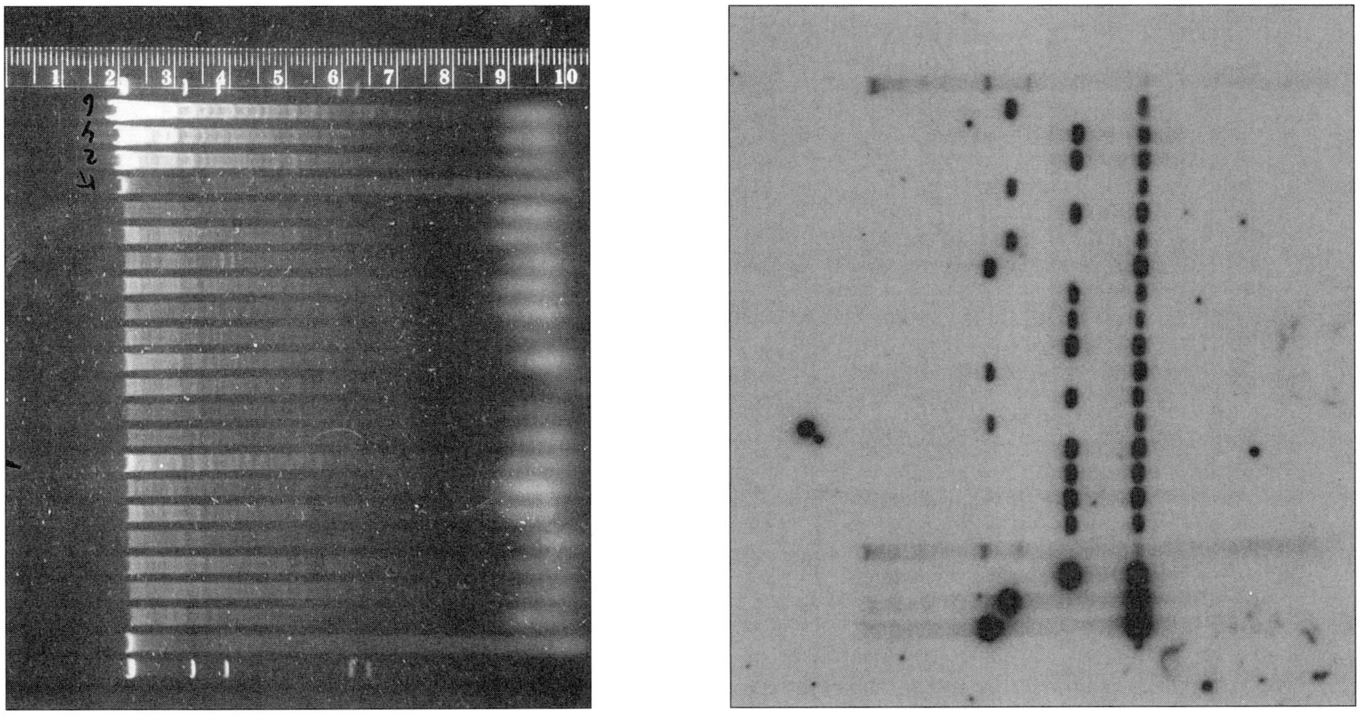

(a) Agarose gel electrophoresis (b) Southern blot

***Figure* 16.3** Southern blot analysis of chromosomal DNA treated with restriction enzymes and electrophoresed on an agarose gel. Fluorescently stained gel (a) shows that while the thousands of DNA fragments generated by the restriction enzyme treatment migrate according to their size, they appear as a continuous smear, rather than as individual "bands." Nevertheless, a DNA probe identifies the complementary sequence, as shown in the X-ray image of a blot, probed with a radioactively labeled DNA probe. (Courtesy of Steve Lory) probe. The synthetic oligonucleotide probe is made on the basis of a DNA sequence deduced from a known amino acid sequence of a protein. For example, an enzyme, X, has been purified and has an amino terminal sequence of Met · Trp · Asp · Trp. Based on the codons for these four amino acids (see Table 13.2), oligonucleotides ATGTGGGATTGG and ATGTGGGACTGG can be synthesized chemically and can be used as probes in a Southern blot hybridization to bind to fragments containing the complementary sequence. (Note that codons GAT and GAC both can encode aspartic acid, so two probes must be made.) Labeled with radioactive phosphorus, these probes hybridize with their complementary sequence on one of the DNA strands within the gene that encodes enzyme X. This probe can also be used directly to identify recombinant plasmids containing the cloned gene for enzyme X.

Construction of Recombinant Plasmids

Extrachromosomal genetic elements, such as plasmids and bacteriophages, have become the logical choice for propagation of isolated genes because they can be readily isolated in pure form from their bacterial hosts. Many different plasmids and bacteriophages are used in genetic engineering as vectors, that is, vehicles of transmission for cloning isolated genes. Most of the useful plasmid cloning vectors are small (less than 10,000 base pairs) and contain antibiotic-resistance determinants. Bacteriophage vectors contain regions that are dispensable, and these can be replaced with foreign DNA. Three commonly used plasmid vectors, pBR322 and pACYC184 and pUC18, as well as a bacteriophage vector, lambda, are shown schematically in Figure 16.4.

Construction of a recombinant plasmid begins by treating the appropriate vector with a restriction enzyme and treating the DNA to be cloned (the donor DNA) with the same enzyme. The two populations of DNA molecules are mixed, and recombinants are generated by the addition of DNA ligases, which covalently integrate the foreign DNA into the plasmid vector. DNA ligase also re-closes the vector. The mixture of ligated DNA is introduced into bacteria (usually *E. coli*) by transformation of artificially competent cells. The bacteria are plated onto selective media containing an antibiotic to which the plasmid encodes a resistance determinant. Only the cells that have taken up a plasmid will grow. Because the number of competent cells exceeds the number of **transformable** molecules, each colony that arises from plating of transformants is expected to be a clone, containing a single plasmid with a DNA insert. Individual colonies can be isolated and plasmids extracted. After treatment with a restriction enzyme, they can be analyzed by agarose gel electrophoresis. Cloning of fragments from a large plasmid into vector pUC18 is outlined in Figure 16.5.

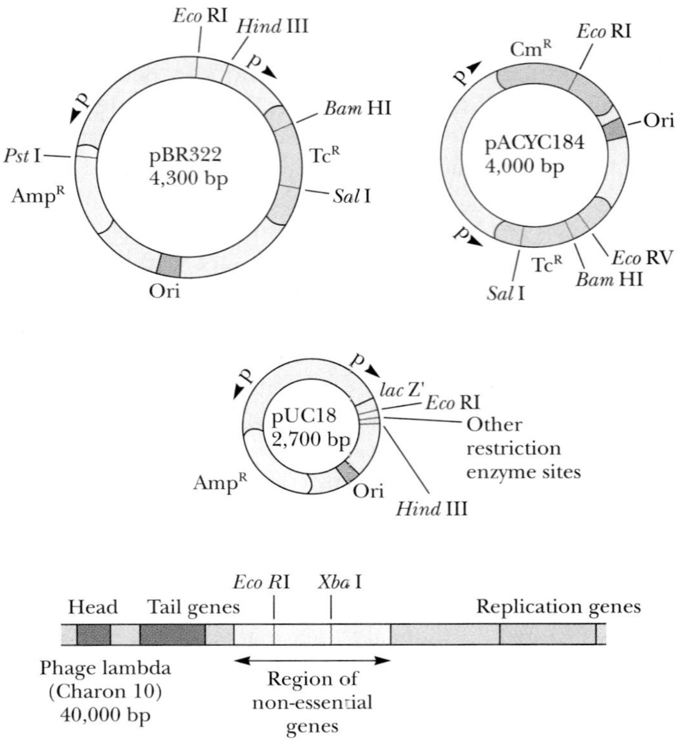

***Figure* 16.4** Vectors for construction of recombinant molecules. Plasmids pBR322, pACYC184, and pUC18 each contain several unique restriction enzyme recognition sequences (indicated by arrows) as well as resistance determinants for ampicillin (Amp^R), tetracycline (Tc^R), or chloramphenicol (Cm^R). Identification of clones carrying inserted DNA is based on disruption of antibiotic-resistance determinants in pBR322 and pACYC184, and sensitivity of transformants to a particular antibiotic. Insertion into the multiple cloning site of pUC18, which includes the recognition sequence for 13 different restriction enzymes, results in disruption of a coding sequence (*lacZ′*) for a portion of the β-galactosidase enzyme. In this case, the plasmids have to be transformed into special *E. coli* strains, which carry the gene encoding the rest of the β-galactosidase; the two portions of the protein form the active enzyme. This artificial formation of an active enzyme from two fragments (called α complementation) allows screening for insertions for colony color, containing X-gal, the chromogenic indicator of β-galactosidase. Bacterial colonies carrying the intact plasmid have a blue appearance, while colonies with plasmids that carry inserts are white. The phage lambda derivative Charon 10 contains a nonessential region needed only during the lysogenic phase of growth, which can be used to insert foreign DNA. These phages thus have only a virulent life cycle. p > indicates the locations of the various promoters of the genes contained in the vector.

The mixture of bacteria containing the various inserts is called a **library** of DNA. A simple library, such as that outlined in Figure 16.5(**a**), is represented by four different clones plus the one vector. Thus, the hundreds of possible transformants contain five populations of plasmids: four with inserts and one containing only the vector without an insert DNA. The identification of a specific clone therefore requires screening at least four randomly selected clones by agarose gel electrophoresis of the plasmid digests. A more complex library, prepared by restriction enzyme treatment of bacterial chromosomal DNA

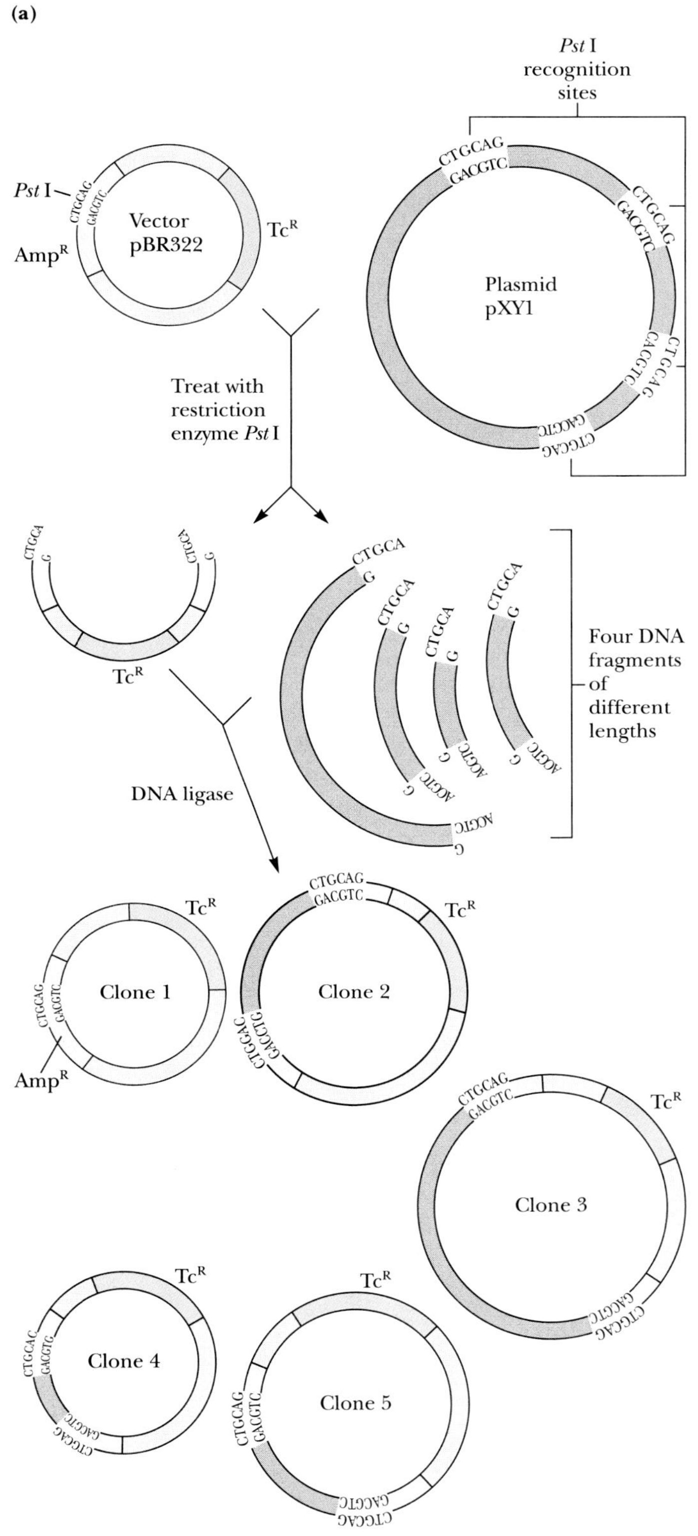

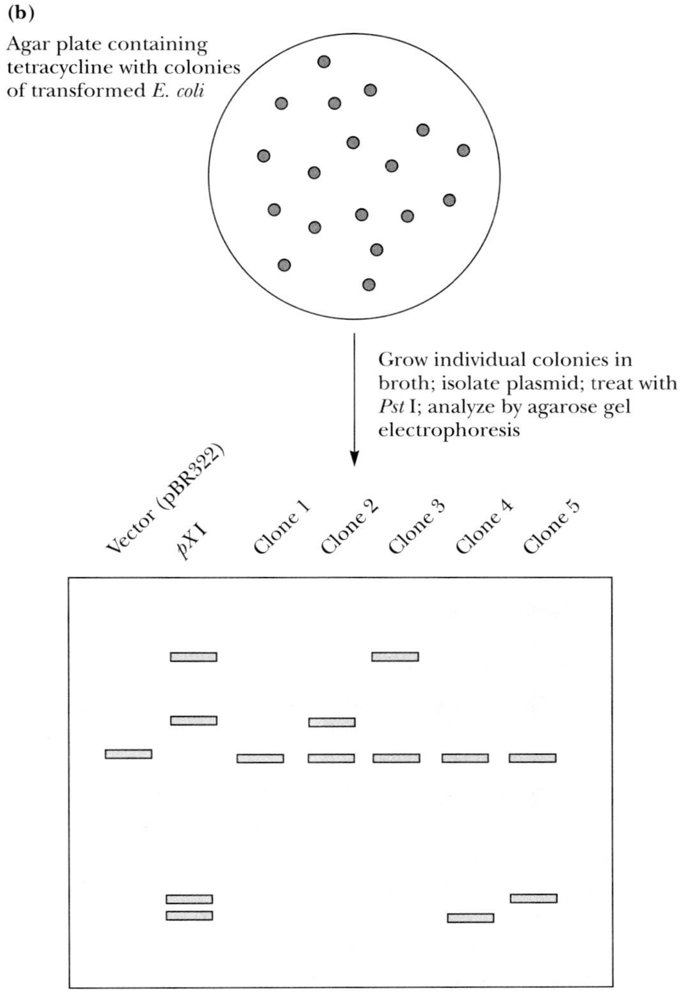

***Figure* 16.5** Cloning and analysis of recombinant plasmids. **(a)** Steps in generating a collection of bacteria carrying recombinant plasmids. Vector pBR322 and the plasmid pXYI, donor of the cloned DNA, are treated with restriction enzyme *Pst* I. Treatment of the pXYI will generate four different DNA fragments because it contains four recognition sites for *Pst* I. When the fragments from pXYI are mixed together with similarly treated pBR322, a portion of the molecules insert in the plasmid at the *Pst* I site within the region containing the coding region for the ampicillin resistance determinant, by first annealing via their complementary sequences at the 3′ overhangs. DNA ligase seals the phosphodiester backbone, resulting in a population of chimeric plasmids (that is, plasmids of mixed origin), as well as re-closed vectors. **(b)** Analysis of individual clones following transformation of the entire ligation mixture into competent *E. coli.* The transformed bacteria are plated onto media containing tetracycline which give rise to colonies that carry either re-closed plasmid or plasmids with inserted DNA from pXYI. Those bacteria that lack insert and contain an intact ampicillin-resistance determinant can be distinguished from the rest by spotting them onto plates containing ampicillin. Plasmid DNA from those colonies which contain inserted DNA—that is, they are sensitive to ampicillin killing—is extracted, treated with *Pst* I, and analyzed by agarose gel electrophoresis, comparing the gel mobility pattern to the vector and donor DNA. Clones 2, 3, 4, and 5 carry individual fragments from the donor plasmid pXYI, while clone 1 represents a vector which reclosed without an insert DNA.

ligated with a vector, contains thousands of bacteria, each carrying a recombinant plasmid with a single fragment of the chromosome. This is called a **genomic library,** and is a collection of plasmids carrying the entire bacterial chromosome as fragments in recombinant plasmids. The task of isolating genes, then, is reduced to identification of a desired clone from a library of DNA fragments in a vector.

Identification of Recombinant Clones

In construction of recombinant plasmids, one often deals with the technically more difficult problem of identifying a specific clone from the collection of recombinant plasmids. A number of different techniques are available to investigators that allow them to identify a bacterium with a specific recombinant plasmid.

Expression of a Gene in a Heterologous Host

A library of DNA fragments containing thousands of bacteria can easily be screened for a desired cloned gene, provided that gene endows the host with a unique characteristic that distinguishes its particular host from others carrying different fragments of the donor chromosome. For example, to identify a clone of a hemolysin, *E. coli* carrying a library of DNA from an organism that has a gene for a hemolysin is grown on a blood agar plate. The clone

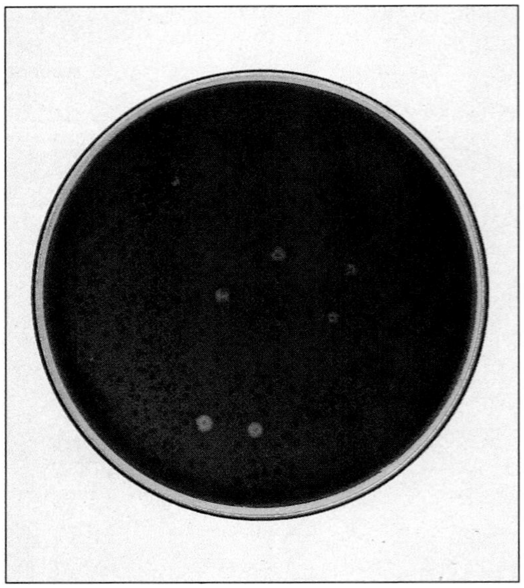

***Figure* 16.6** Identification of hemolysin-containing clones from a library of chromosomal DNA. Hemolysin is an enzyme capable of catabolizing red blood cells. DNA from a hemolytic microorganism, *Pseudomonas aeruginosa,* was treated with restriction enzyme *Bam*HI, and a library of this DNA was prepared in pBR322. The ligation mixture was plated onto blood agar plates, and those clones carrying the hemolysin gene were identified by a pale zone around the bacterial colony on the dark red plate, indicating lysis of the red blood cells in the agar. (Courtesy of Steve Lory)

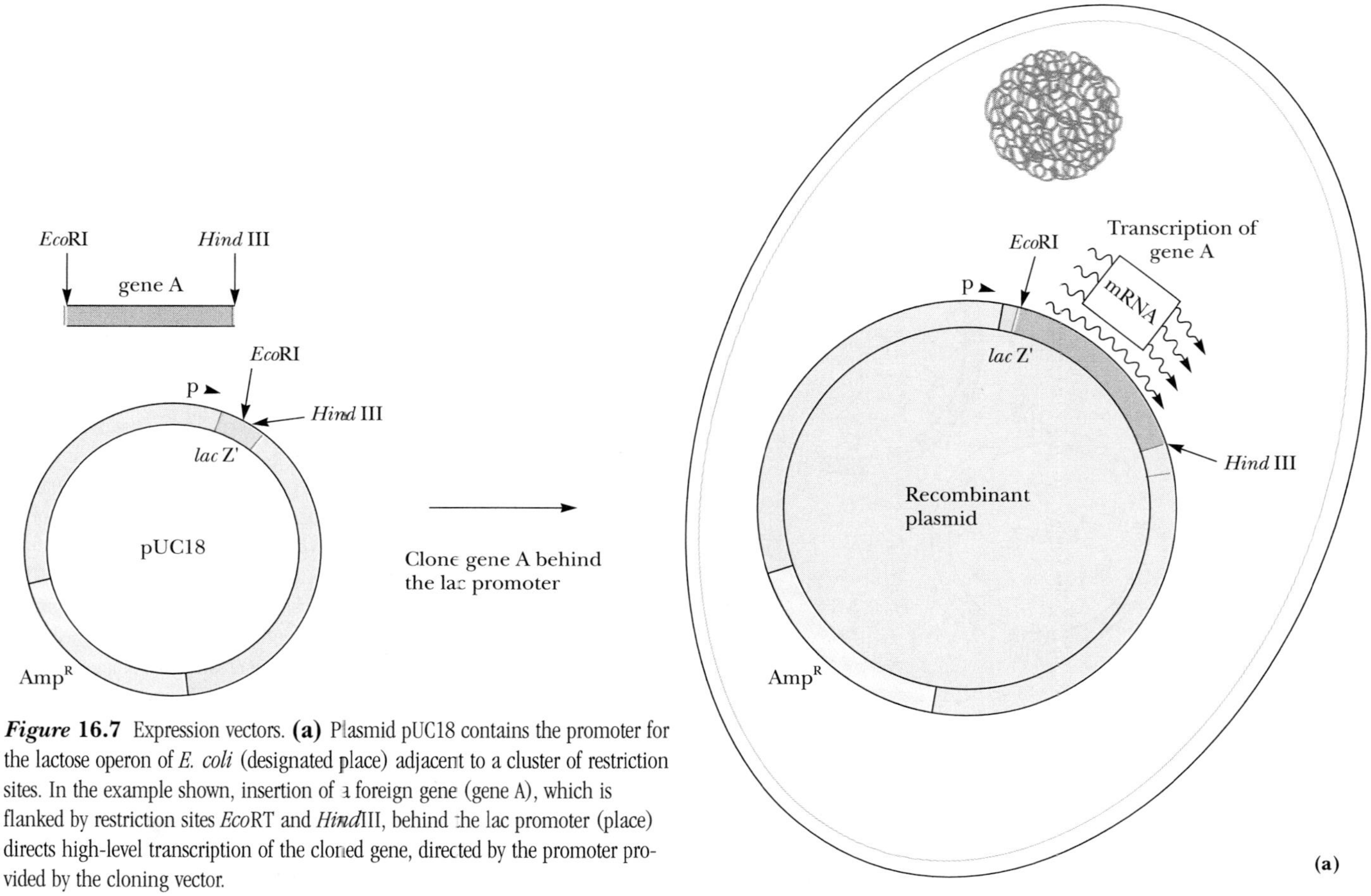

***Figure* 16.7** Expression vectors. **(a)** Plasmid pUC18 contains the promoter for the lactose operon of *E. coli* (designated place) adjacent to a cluster of restriction sites. In the example shown, insertion of a foreign gene (gene A), which is flanked by restriction sites *Eco*RT and *Hind*III, behind the lac promoter (place) directs high-level transcription of the cloned gene, directed by the promoter provided by the cloning vector.

containing the hemolysin gene can be identified by a zone of clearing around the colony, as shown in Figure 16.6. Similarly, hydrolysis of starch can be used as a test for expression of amylases. These assays rely on expression of the gene in the cloning host, and this is not always guaranteed, especially if the gene in its natural host is regulated. These regulatory signals may be absent in the host carrying the recombinant plasmid.

High-level expression of a cloned gene product can be ensured by using specialized cloning vectors containing regulatory sequences adjacent to the cloning sites. These sequences are usually promoters of highly expressed genes, and the vectors containing these promoters are called **expression vectors.** A modified strong promoter from the lactose operon and bacteriophage lambda, or T7, have been incorporated into some vectors (Figure 16.7). Expression vectors allow efficient transcription, provided the cloned gene is in the same orientation as the promoter.

Screening with DNA or Antibody Probes

Because many gene products do not have a simple assay such as screening on the agar surface of a petri plate, clones carrying these genes must be identified by DNA hybridization techniques. This requires presence of the DNA sequence in the recombinant plasmid, but does not depend on expression of the cloned gene. The method for identifying such a clone is yet another application of a localized DNA hybridization technique, **colony hybridization,** which is analogous to Southern blotting (Figure 16.8). The oligonucleotide used in the previous example to identify enzyme X can be used to screen a gene bank of DNA fragments. Bacteria growing as colonies on a plate are allowed to stick to a nylon or nitrocellulose support, which is placed on the agar surface. Once the support is removed, these bacteria, which are bound to the support as an imprint of the original plate (hence called the

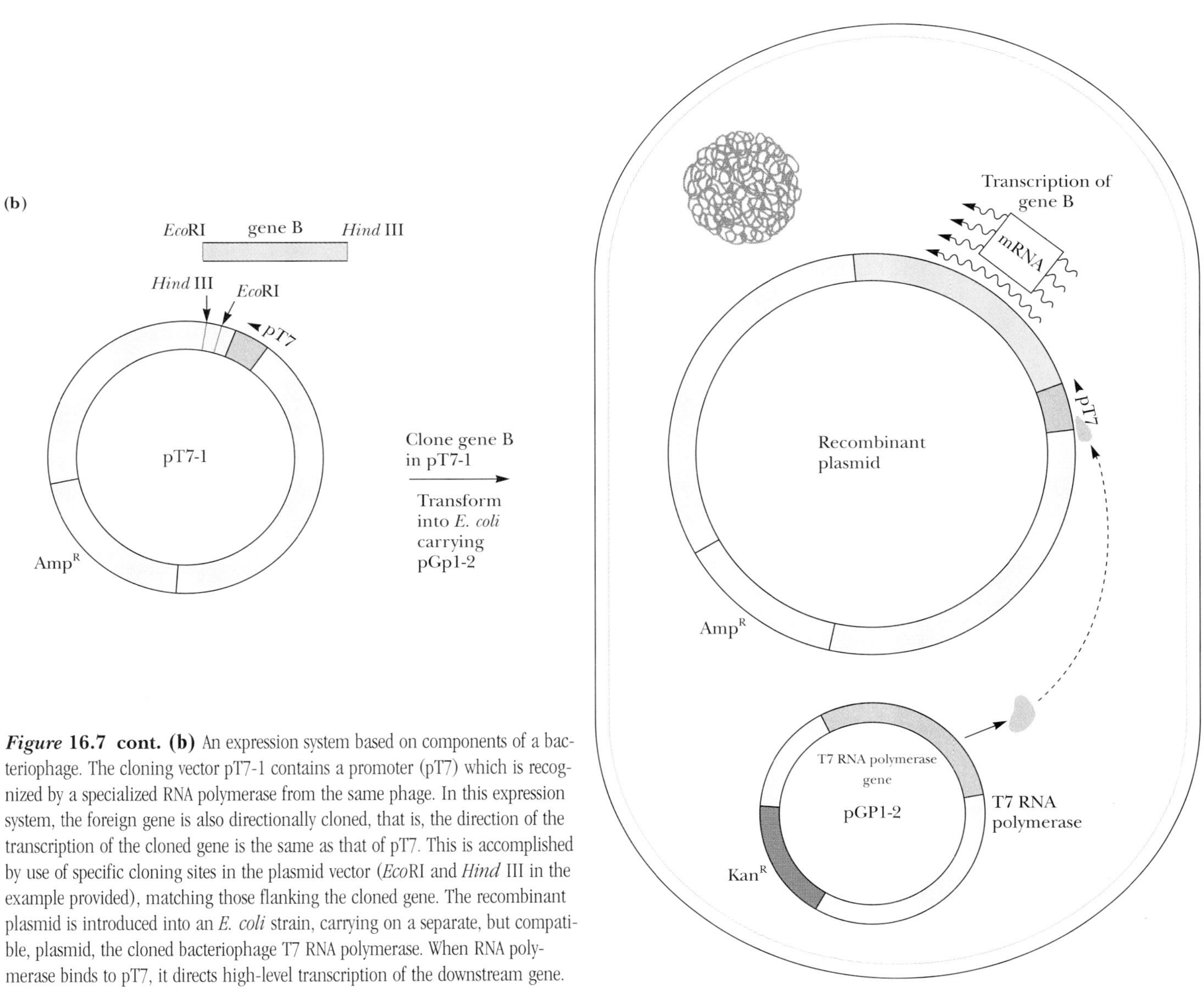

***Figure* 16.7 cont. (b)** An expression system based on components of a bacteriophage. The cloning vector pT7-1 contains a promoter (pT7) which is recognized by a specialized RNA polymerase from the same phage. In this expression system, the foreign gene is also directionally cloned, that is, the direction of the transcription of the cloned gene is the same as that of pT7. This is accomplished by use of specific cloning sites in the plasmid vector (*Eco*RI and *Hind* III in the example provided), matching those flanking the cloned gene. The recombinant plasmid is introduced into an *E. coli* strain, carrying on a separate, but compatible, plasmid, the cloned bacteriophage T7 RNA polymerase. When RNA polymerase binds to pT7, it directs high-level transcription of the downstream gene.

BOX 16.1 RESEARCH HIGHLIGHTS

Genetic Engineering of Plants Using *Agrobacterium tumefaciens* and Ti Plasmids

Crown gall disease in plants, described in Chapter 13, is caused by integration of genes from the bacterium *Agrobacterium tumefaciens* into a plant's chromosome. The genes that are transferred are T-DNA, found on the bacterium's Ti plasmid. *Agrobacterium*'s *vir* genes encode proteins that promote the transfer of T-DNA, not only into the plant cell but across the nuclear membrane and into the plant chromosome. Once integrated, the bacterial genes are expressed by the plant cell, giving the plant new properties (which in this case are beneficial to the bacterium, but harmful to the plant).

Genetic engineers have exploited this natural system by replacing the disease-causing genes with genes they believe will produce new, desirable traits in the plant. Genes "spliced" into the Ti plasmid will be carried into the plant during the process of infection, and integrated into the plant chromosome under the direction of *Agrobacterium*'s *vir* gene products. The steps involved are outlined in the accompanying figure.

A plasmid carrying T-DNA is treated with restriction enzymes to allow cloning of a foreign gene such that it is flanked by T-DNA. This plasmid is introduced into a strain of *A. tumefaciens* that contains the *vir* region, but lacks its own T-DNA. The *vir* genes and the T-DNA need not be located on the same plasmid for infection to occur. These bacteria are incubated with plant cells, which incorporate the T-DNA into their genomes. When the plant cells carrying T-DNA divide, each daughter receives a copy of the cloned DNA. This can be verified by including a marker gene among the cloned genes. For example, genes for enzymes that cause color changes in the growth medium as the cell metabolizes will identify successfully transformed colonies by their color. These cells are hormonally induced to produce roots and shoots, and grow into plants in which every cell contains a copy of the cloned DNA. Such plants will not express the new trait introduced by the T-DNA–mediated gene transfer. The foreign genes used for genetically engineering plants include those that improve growth characteristics of the plant, speed up ripening of fruit, or render the plant resistant to killing by plant insects or disease-causing bacteria.

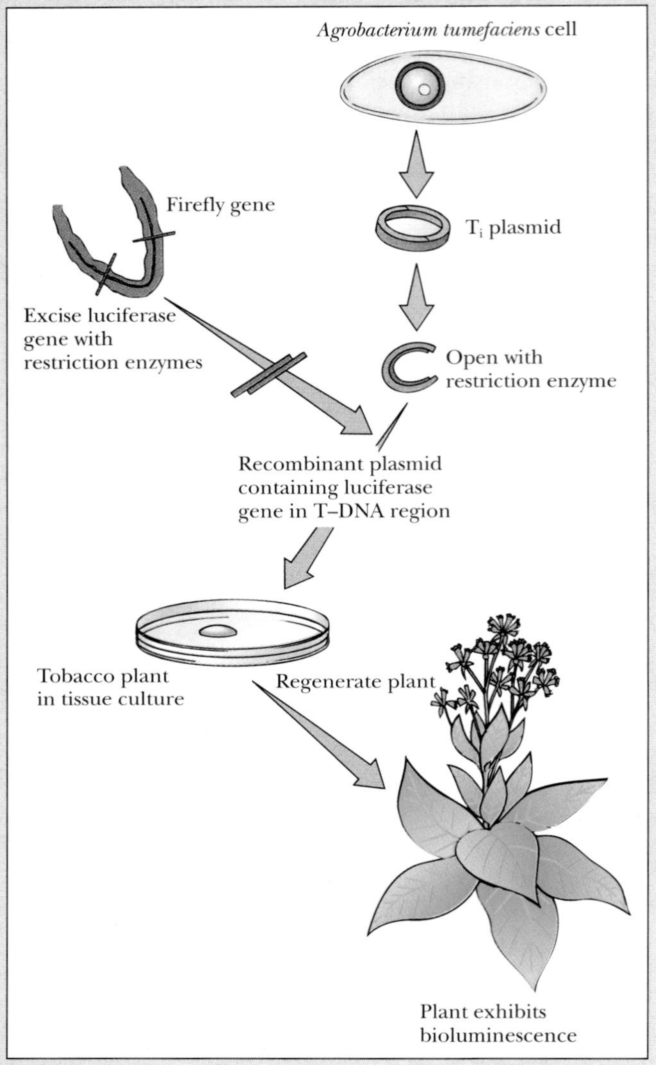

Genetic engineering of plants

replica), are lysed, and the DNA is denatured. The DNA becomes immobilized, and the radioactive probe hybridizes to the spot on the support that corresponds to the colony carrying a plasmid with the gene coding the enzyme.

A related technique based on colony hybridization uses radioactive antibody as a probe. This requires that the protein product of the cloned gene be obtained, usually from the original host. Purified protein is injected into animals (usually rabbits or mice) whose immune system responds by producing an antibody that specifically binds the foreign protein. These antibodies are harvested and

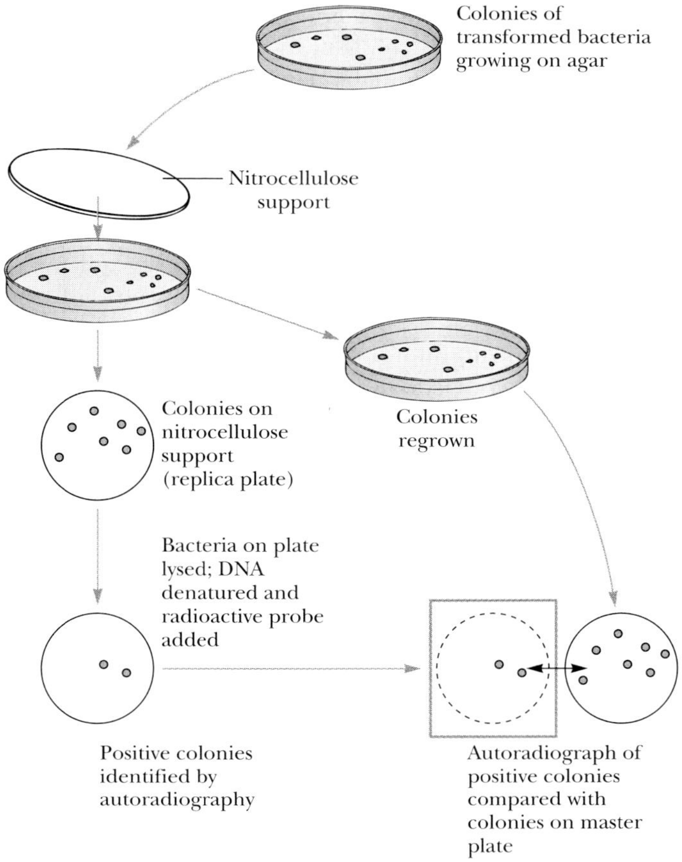

***Figure* 16.8** Colony hybridization. DNA from individual bacterial colonies on gel is immobilized on nitrocellulose. The nitrocellulose replica is then exposed to radioactive DNA probe, and positive colonies are identified.

radiolabeled. Lysed colonies, prepared as described above for colony hybridization with a radiolabeled DNA probe, can be reacted with the radioactive antibody. The probe will bind and identify only those colonies where protein is being expressed from the gene of interest.

Cloning of Eukaryotic Genes in Bacteria

The relatively simple technique of isolating prokaryotic genes in recombinant plasmids or bacteriophages can be applied to isolating genes from eukaryotic microorganisms. It is, however, likely that the cloned gene will not be expressed in bacteria, because eukaryotic genes contain intervening sequences (the final, or mature, mRNA being assembled after extensive splicing of the primary transcript). Eukaryotic genes are often cloned as copies of mRNA, called complementary DNA (cDNA). The total mRNA is first isolated from a cell, and a copy of the RNA is made using the enzyme **reverse transcriptase** (see Chapter 14). This enzyme is an RNA-dependent DNA polymerase and can be isolated from selected animal viruses. The RNA-DNA hybrid is denatured, the RNA component hydrolyzed, and a second copy of the DNA is synthesized with DNA polymerase. The final double-stranded product can be cloned into phage and plasmid vectors.

Practical Applications of Recombinant DNA Technology

Recombinant DNA technology has revolutionized biological and medical research by providing scientists with the ability to obtain large quantities of products that were previously available from their natural sources in minute quantities only. For example, rare enzymes from both bacterial and eukaryotic sources can now be obtained in gram quantities by cloning the appropriate genes in expression vectors and isolating the recombinant products from bacterial extracts. Bacteria, as the source of recombinant products, offer an additional economic advantage because they can be propagated on relatively inexpensive media and grow rapidly.

Bacterial, and more recently yeast-based, expression systems have been extensively employed in industrial-scale production of various hormones, such as insulin and human growth hormone, as well as in vaccines against hepatitis B and herpes. An added benefit of recombinant products, recently realized, is that they are often safer than their natural counterparts, being free of potentially allergenic protein contaminants.

Currently, a great deal of research is devoted to development of genetically engineered vaccines for a variety of infectious diseases. This includes introducing specific mutations into the code for proteins that make up vaccine components, such that they retain their immunogenicity but do not have harmful effects on the vaccinated patients. Genetically attenuated bacterial and viral hosts are also being developed to provide a safe and efficient way to deliver such vaccines.

An integral part of identification and study of genes is the knowledge of the precise nucleotide sequence, which allows deduction of the amino acid sequence of the protein product, using the genetic code table. Moreover, sequence information has been useful in studies of molecular evolution, by comparing sequences between common genes and thus establishing relationships among genes and organisms.

The ability to detect the presence of certain genes in bacteria has been exploited in diagnostic microbiology as well. Agarose gel electrophoresis of plasmid DNA, coupled with Southern blot analysis, often allows tracking of bacteria responsible for dissemination of antibiotic resistance genes. This type of plasmid fingerprinting allows detection of specific size categories of plasmids, as well as specific DNA sequences that reside on such plasmids as a result of transposition from the chromosome or another plasmid.

BOX 16.2 METHODS & TECHNIQUES

The Polymerase Chain Reaction

One of the more powerful techniques for isolating genes that have been developed in recent years is called the **polymerase chain reaction** (PCR). The principle of PCR is the generation of copies of two strands of DNA by a repeated sequence of denaturation of double-stranded DNA, synthesis of the complementary strands, followed by the next round of denaturation and synthesis. Because synthesis of DNA strands by DNA polymers requires primers (see Chapter 13), this technique also necessitates knowledge of limited sequence, such that the polymerization can be initiated from a short oligonucleotide which hybridizes to the ends. The PCR can start from a double-stranded DNA, or the first DNA molecule can be generated by reverse transcription of mRNA, followed by denaturation of the duplex, and hybridization of primers for the next round of polymerization. This modification of PCR is called RT-PCR. As illustrated in the figure in Box 16.1, the reaction is divided into cycles, each cycle consisting of three steps. First, the double-stranded DNA is denatured by raising the temperature of the DNA solution. Next, the temperature is lowered and complementary oligonucleotides (the primers) are allowed to anneal to the DNA. Finally, DNA polymerase is added, together with dATP, dCTP, dTTP, dGTP, and the complementary strand is synthesized. The denaturation-primer annealing-complementary strand synthesis is repeated 30 to 40 times. Since each newly synthesized complementary strand can serve as a template for the next round of PCR, the amount of DNA synthesis increases exponentially, and from a few starting DNA molecules, 20 cycles can generate over a million copies. The PCR-generated DNA can be used for many manipulations, among them DNA sequencing and preparation of DNA probes for Southern or Northern hybridizations, if they can be directly cloned into plasmid or viral vectors.

The principles of polymerase chain reaction. The three steps, denaturation of double-stranded (ds) DNA, annealing of primers to the single-stranded (ss) DNA, and extension, are shown. For RT-PCR, single-stranded DNA is generated from RNA by reverse transcriptase synthesizing a complementary DNA strand; this strand then serves as a template for the first round of annealing and extension. ▶

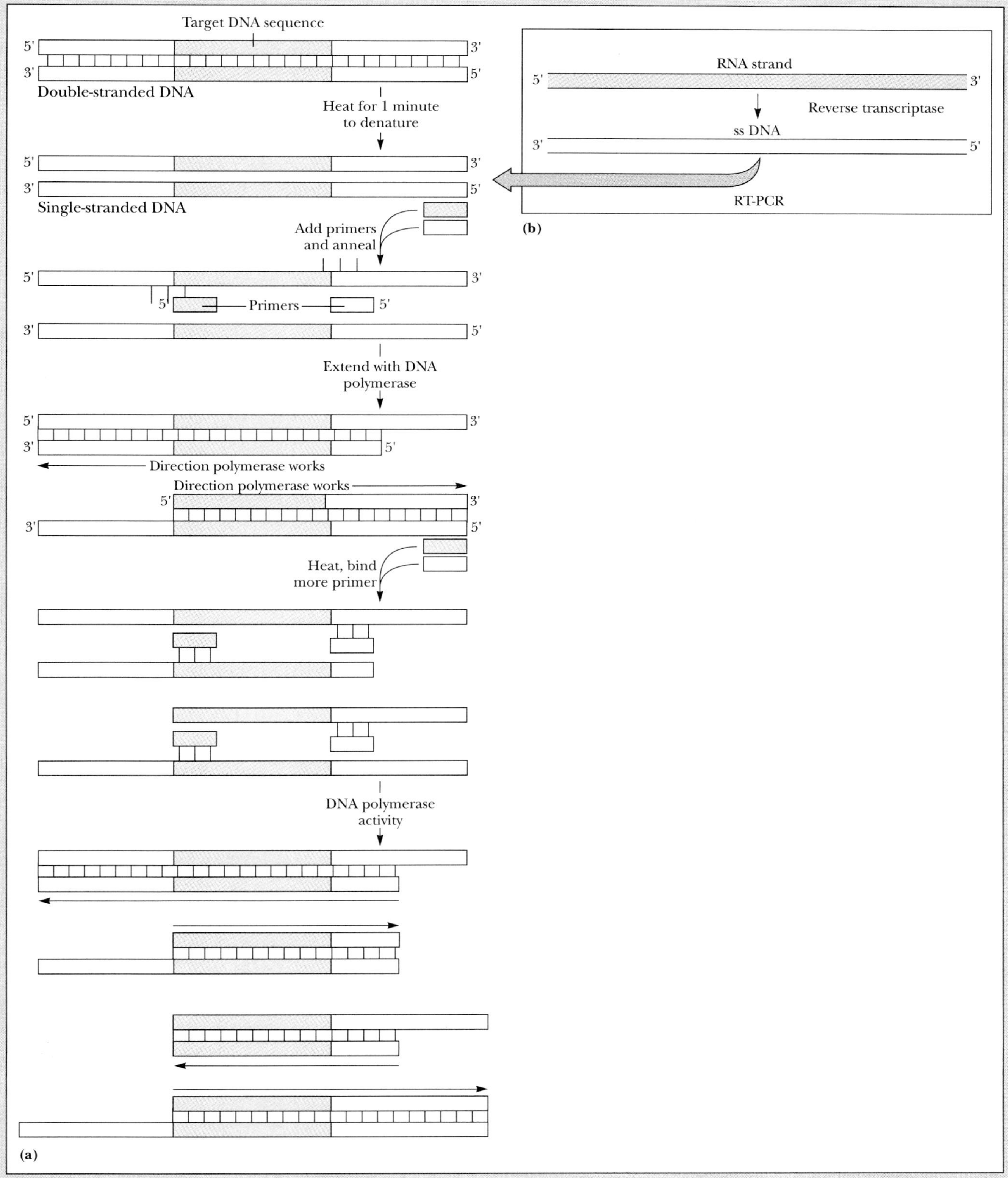
Target DNA sequence
5'
3'
3'
5'
Double-stranded DNA
Heat for 1 minute
to denature
Single-stranded DNA
Add primers
and anneal
Primers
Extend with DNA
polymerase
Direction polymerase works
Direction polymerase works
Heat, bind
more primer
DNA polymerase
activity
(a)
RNA strand
Reverse transcriptase
ss DNA
RT-PCR
(b)

BOX 16.3 METHODS & TECHNIQUES

Determination of a Specific DNA Sequence of a Gene

There are several methods available for the determination of a specific nucleotide sequence of a gene, primarily by sequencing of DNA fragments cloned in various cloning vectors. The technique most commonly employed for sequencing large fragments of DNA was devised by Frederick Sanger and colleagues in 1977. This method relies on synthesis of complementary DNA in a test tube, which contains modified nucleoside triphosphates called 2′,3′-dideoxynucleotide triphosphates (ddNTP). These dideoxynucleotides can be incorporated into the growing chain; however they lack the necessary 3′ hydroxyl group to form a phosphodiester bond with the next deoxynucleotide triphosphate (dNTP). The consequence of incorporation of ddNTP bases is termination of DNA synthesis at a precise place. DNA sequencing by this method is carried out in four reactions, each containing a mixture with a single-stranded template, containing the DNA sequence of interest and a short radioactively labeled oligonucleotide primer, which anneals to its complementary sequence and one of four ddNTPs. The location of the primer determines the starting point of the sequence analysis. The reaction is started by adding DNA polymerase I, all four dNTPs in excess, and a small amount of one ddNTP. The chain elongation reaction proceeds using dNTPs, but incorporation of a ddNTP terminates synthesis. For example, the reaction using ddATP will terminate at positions on the template containing a thymine, illustrated in the accompanying figure. The other three ddNTPs will terminate reactions corresponding to the positions of their complementary bases on the DNA. These reactions give radioactive products of different length that are size-fractionated by electrophoresis on acrylamide gels. The relative size of the fragments, and the source of the ddNTP, will allow determination of a corresponding sequence in the 5′ direction from the primer. This method allows sequence analysis of a region of DNA, up to 600 to 700 base pairs from the site of the primer hybridization. Once this sequence is known, a new primer can be chemically synthesized, radioactively labeled, and annealed to the same DNA template, or to a new template carrying an additional portion of the gene; this will serve as a starting point for the next round of sequencing.

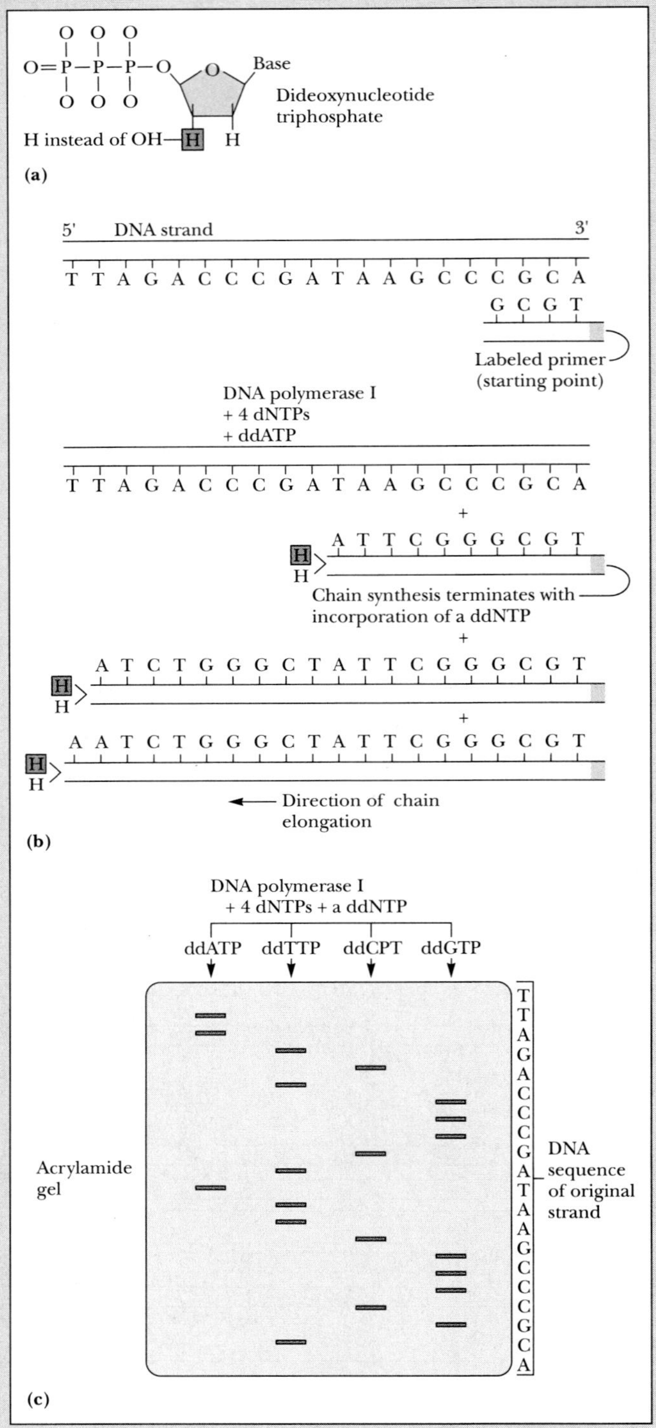

DNA sequence determination using the dideoxynucleotide triphosphate chain termination method. One of the four reactions, using ddATP, is illustrated. The top **(a)** shows the structure of dideoxynucleotide triphosphate. The analysis of the reaction products for reactions containing individual ddNTPs is shown in **(b).**

BOX 16.4 MILESTONES

How Should Genetic Engineering Be Regulated?

The relative ease of introducing foreign genes into bacteria initially created a great deal of intellectual discomfort among scientists, as well as nonscientists. Would new pathogenic bacteria be created, escape from the laboratories, and cause epidemics in an unimmunized population? Similarly, new weapons of biological warfare could be genetically engineered by introducing lethal genes into ordinarily benign bacterial hosts. These concerns led to governmental guidelines that restricted genetic engineering involving genes of virulence determinants, such as bacterial toxins. It has since become apparent that nonpathogenic bacteria cannot be made into pathogens by the expression of a single gene. Virulent phenotypes are the result of complex interactions among many traits expressed by pathogenic bacteria. While the restrictions on genetic engineering have now been greatly reduced, manipulation of genes for some highly potent toxins, such as botulinum neurotoxin, can be carried out only in specially equipped laboratories that are approved for handling highly virulent bacteria and viruses.

Summary

- Classical bacterial genetics was the foundation for recombinant DNA research, which is based on the ability of scientists to **isolate individual genes** from more complex genomes.

- Restriction endonucleases, which introduce double-stranded breaks at specific sequences, are made by bacteria. These are purified and used to fragment DNA into pieces that contain intact genes or parts of genes. Restriction endonucleases, together with their cognate **modification enzymes,** serve as barriers to unrestricted DNA exchange among different bacterial species.

- DNA fragments, generated by treatment with restriction endonucleases, are usually analyzed by **agarose gel electrophoresis,** which separates DNA fragments according to their size.

- Specific DNA sequences found within a DNA fragment fractionated by agarose gel electrophoresis can be identified by a method of in situ **hybridization** of a single-stranded DNA probe with its complementary sequence in the gel. This procedure is known as **Southern blotting.**

- Construction of recombinants involves insertion of a DNA fragment into a **vector,** which can be either **plasmid** or **virus,** followed by introduction of the plasmid into the bacterial **host** by transformation. If the vector is a bacteriophage (usually bacteriophage lambda), the recombinant DNA is packaged into the bacteriophage head and is introduced into the bacteria by infection.

- Plasmid vectors contain sites where DNA is inserted by ligation of a fragment, generated by treatment of the DNA and vector by the same restriction endonuclease. The insert DNA is joined with the plasmid, to give a chimeric circular molecule, by the action of **DNA ligase.**

- Recombinant plasmids can be propagated in pure culture as **clones.** They are identified by expression of protein product using either immunological or activity screens.

- Specialized cloning vectors, called **expression vectors,** include specific sequences to maximize expression of the cloned genes.

- Cloned genes can be altered in the laboratory, and introduced into the bacterial genomes either on plasmids or bacteriophage vectors. Even plants can be genetically engineered, using **Agrobacterium**-mediated transfer of recombinant DNA cloned within its **Ti plasmid.**

- Eukaryotic genes encoding proteins are often cloned from mRNA sequences. First, a copy of mRNA, called **cDNA,** is made by the enzyme **reverse transcriptase,** and the second DNA strand, complementary to cDNA, is made by DNA polymerase. This double-stranded form is then inserted into the appropriate plasmid or bacteriophage vectors.

- Rare sequences from complex genomes, or from a mixture of different RNA species, can be isolated by repeated denaturation-complementary strand synthesis-denaturation, called the **polymerase chain reaction (PCR).**

Questions for Thought and Review

1. Why don't bacteria that produce restriction enzymes destroy their own DNA?

2. A portion of the *E. coli* chromosome containing the *lac*Z gene (coding for β-galactosidase) is shown.

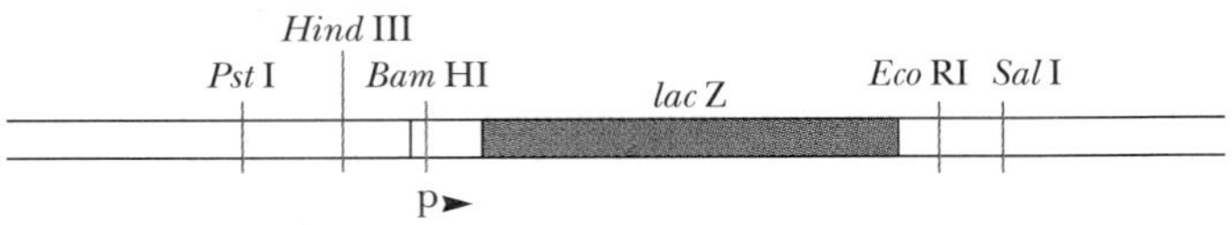

   a. Design a strategy to clone the minimal sequence, but still carrying the full coding sequence for the enzyme, into plasmid pBR322.

   b. Describe several methods to identify the recombinant *E. coli* carrying the cloned *lac*Z gene.

   c. How would your cloning strategy change if you wanted to assure that the *lac*Z gene is expressed in a host strain other than *E. coli*?

3. Some *E. coli* strains, used as cloning hosts, carry a mutation in their *rec*A gene. Why is it sometimes important to use such a mutant strain when cloning certain genes?

4. A solution of an oligonucleotide of the following sequence, 5′ CGGCCCGGGATCCGAATTCCTAGGCA 3′, is heated and mixed with the following oligonucleotide: 5′ TGGGATGGAATTCGGGATCCCGGGCCG 3′, and then allowed to cool.

   a. Identify all the sites recognized by restriction enzymes (use Table 16.1).

   b. Would a solution of the first oligonucleotide alone be recognized and cut by any restriction enzyme?

5. A patient was admitted to a hospital with a urinary tract infection, and antibiotic-sensitive *Klebsiella pneumoniae* were recovered. During hospitalization he became nonresponsive to antibiotic therapy, and the new isolates of *K. pneumoniae* were now resistant to ampicillin and kanamycin. Stool cultures of all of the patients on the same floor revealed that several of them carried *E. coli* that was also resistant to ampicillin and kanamycin. You are in charge of determining if the *Klebsiella* somehow acquired an R-plasmid from *E. coli.* What molecular tools would you use to prove the plasmid-transfer hypothesis?

6. You wish to isolate a gene from a chromosome, which is flanked by recognition sites for either *Bgl* II or *Sma*I restriction endonucleases. Is it possible to use either vector pBR322 or pACYC184, in spite of the fact that neither one of these plasmids has a recognition sequence for *Bgl* II or *Sma*I?

7. A procedure analogous to Southern blotting is called Northern blotting, where total RNA is extracted from bacteria, and size-fractionated on agarose gels. Specific mRNA sequences are identified by in situ hybridization with radiolabeled probes, which can be deduced from protein sequence. If you wish to analyze for the presence of mRNA of the gene X (encoding the protein with the N-terminal sequence Met · Trp · Asp · Trp, described on pp. 371–372), could you use, in the Northern blot, the same oligonucleotide probes (5′ ATGTGGGATTGG 3′, or 5′ ATGTGGGACTGG 3′) used to detect the gene X in the Southern blot?

8. Describe the differences in colony hybridization aimed at detecting specific DNA sequences or the related immunological techniques for the identification of clones producing a specific protein.

9. Explain the difference in the initial steps of the polymerase chain reaction (PCR) when starting from mRNA, rather than DNA.

Suggested Readings

Ausubel, F. M., R. Brent, R. E. Kingston, D. D. Moore, J. G. Seidman, J. A. Smith, and K. Struhl. 1989. *Current Protocols in Molecular Biology.* Vol 2. New York: John Wiley & Sons.

Friedberg, E. C., G. C. Walker, and W. Siede. 1994. *DNA Repair and Mutagenesis.* Washington, DC: American Society for Microbiology.

Hooykaas, P. J., and R. A. Schileperoot. 1992. *Agrobacterium and Plant Genetic Engineering. Plant Mol. Biology 19*:15–38.

Leach, D. R. F. 1995. *Genetic Recombination.* Cambridge, MA: Blackwell Science.

Old, R. W., and S. B. Primrose. 1994. *Principles of Gene Manipulation.* Cambridge, MA: Blackwell Science.

Walston, J. D., J. Tooze, and D. T. Kurty. 1983. *Recombinant DNA: A Short Course.* New York: W. H. Freeman.

Part 5

Microbial Evolution and Diversity
A Conversation with Carl Woese

Dr. Carl Woese has championed the use of ribosomal RNA for analyzing the phylogeny of organisms. Through the use of this approach he discovered that the Archaea comprise a third major Domain of life on Earth. For his work in molecular phylogeny, Dr. Woese has received many awards, including the Bergey Award and the Leeuwenhoek Medal. He is a Professor in the Department of Microbiology at the University of Illinois.

JS: One of the greatest breakthroughs in biology in this century has been the discovery that the phylogeny of organisms can be assessed by an analysis of the sequences of the subunits of conserved macromolecules. You have played a key role in this finding. Can you explain the significance of this discovery to the field of biology, and to microbiology in particular?

CW: First, as regards biology in general. What this does for biology in general is allow scientists for the first time to see the full phylogenetic tree—something that Darwin dreamed of, I think, but realized would never happen in his own lifetime. Without the understanding of the phylogenetic relationships among organisms, of course, all you had, twenty or thirty years ago, was the phylogeny of animals, and the phylogeny of plants—that turns out to be a small fraction of the overall phylogenetic tree. Now all life is unified in one phylogenetic tree that comes from one root point. And that root point represents the universal ancestor of all extant life.

As regards microbiology: In order for any (organismal) discipline to be *biological* in the full sense of that word, one has to know about the natural (phylogenetic) relationships among the various organisms and species. Whereas you could talk about the relationship among animals and among plants at least since the time of Darwin, you could never talk in meaningful terms about the phylogenetic relationships among the microorganisms, except for some trivial relationships such as between *Escherichia coli* and *Salmonella*. From the global perspective those told you almost nothing.

The example I like to use here is that of looking under the microscope at two isolates. Before molecular phylogeny was possible, a microbiologist could not tell if these two isolates were as closely related as sheep are to goats, or far more distantly related than animals are to plants. This example brings home the magnitude of the problem microbiology faced before it was possible to determine phylogenetic relationships. Now, of course, this problem has been resolved and we can determine to our heart's content the phylogenetic relationships among any microorganisms.

Now all life is unified in one phylogenetic tree that comes from one root point. And that root point represents the universal ancestor of all extant life.

Courtesy of University of Illinois at Urbana-Champaign News Bureau.

JS: Based on the phylogenetic analysis of 16S ribosomal RNA, your laboratory discovered that what was previously known as the bacteria actually comprises two major phylogenetic groups, the Eubacteria (or Bacteria) and the Archaea. And, furthermore, these two groups are separated from a third group comprising the higher microorganisms, plants, and animals. Virtually all microbiologists

have now accepted this three-domain system you have proposed. In contrast, botanists and zoologists seem more reluctant in embracing this. They still seem to believe in the five-kingdom system. Do you agree with this interpretation? And if so, why is this true, and will it change?

CW: The problem as regards botanists and to some extent zoologists, can be summed up in a phrase, "intellectual inertia." These folks were brought up to classify in terms of gross morphological similarities and differences. The idea of doing so in terms of molecular sequence is a bit hard for many of them to swallow, particularly so the botanists. Since they are suspicious of using molecules as phylogenetic measures, it is understandable that they are reluctant to accept the resultant taxonomies. You get the feeling talking to some of them that if they can't see a morphological difference, they can't intuitively accept that profound phylogenetic difference exists.

The five-kingdom scheme is essentially a mixture of taxonomic apples and oranges. You have on the one hand animals, plants, and fungi, and then protists and finally monera (or bacteria or prokaryotes or whatever you want to call them). Well the animals, plants, and fungi together make up a phylogenetic unit that is actually embedded within the protist group. So, the animal "kingdom," for example, has no higher a phylogenetic status than do any of the (many) major microbial lineages encompassed by the protists; and animals, plants, fungi, and protists together comprise a taxon equivalent to the monera. (But the monera, as you know, are not themselves a valid phylogenetic unit.) What you have with this five-kingdom classification is a set of taxa of various ranks, some even paraphyletic or polyphyletic, all arbitrarily reduced to the same rank; all are called "kingdoms." To say such an arrangement is not very useful is an understatement; it is a theory that makes no useful predictions; it is positively misleading. I don't see the five-kingdom classification persisting any longer than "intellectual inertia" will allow. So long as textbooks with the five-kingdom scheme are printed and accepted by school boards, then five kingdoms will persist. However, enlightened teachers and school boards will see how useless and misleading this taxonomy is, and begin to reject it.

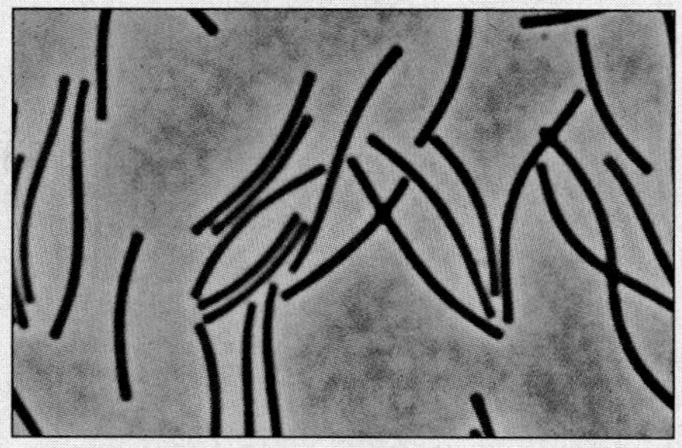

© F. Widdel/VU

JS: The Archaea, Eubacteria, and the Eucarya have a common ancestor. What do you believe were the properties of the ancestor?

CW: First of all, let me say that the question regarding the nature of that ancestor is probably one of the most important questions we biologists face today. It certainly is one of the most challenging, and working on it is going to be a lot of fun.

Regardless of what the ancestor itself was, it's clear that it gave rise to two very distinct lineages, both prokaryotic. One of the lineages, that would become the Archaea, then spun off (or contributed heavily to) a third lineage, which became the eukaryotes. Now, what was the nature of that ancestor that gave rise to these three primary lineages, Archaea, Bacteria, and eukaryotes? Was this ancestor itself a prokaryote? The answer is not the trivial "Well, if it wasn't a eukaryote, then it was a prokaryote." That's meaningless. What we are asking here is whether the latest universal ancestor was as sophisticated as modern bacteria are. What were its information processing systems like? Were they as advanced and accurate as those we know today? Did it have a metabolic capability typical of Bacteria or Archaea or eukaryotes? Were the states of the cell as many and as precisely defined as those in bacteria and Archaea? Did it possess the same general types of proteins that we see in cells today? These are the kinds of questions that need to be answered, questions that will lead us to understanding whether the universal ancestor was prokaryote-like or whether it was something more rudimentary than cells we know anything about. Finally we want to know where that ancestor itself came from: What was the succession of predecessors like, and which gave rise to it?

JS: Do you think there is any way we can get at this scientifically? The nature of this common ancestor? Are there approaches that could be taken that aren't being taken perhaps?

CW: I'm very sanguine about learning a great deal about the universal ancestor through the approaches now being taken and through some that will soon come on line. Molecular sequences can tell you an incredible amount about the history of an organism or group of organisms. It's just astounding to see how constant—how conserved—certain sequence motifs, proteins, genes have been over enormous expanses of time. You can see sequence patterns that have persisted probably for over three billion years. That's far longer than mountain ranges last, than continents retain their shape. Why, some of the familiar stars and star clusters in the heavens weren't even formed then! We think of organisms as such delicate, ephemeral things; they come and go in the wink of an eye. Yet certain patterns in their genes are among the most stable things in this universe.

JS: How do you believe the separation of the Bacteria and Archaea came about?

CW: That's actually part and parcel of the previous question. What gave rise to, what I like to call, the break-up of the universal ancestor into the primary lineages, the very first organisms that were differentiable in the evolutionary sense of the universal tree? And, no one can answer that question today. The best you can do is speculate. It may have been that certain evolutionary developments in the ancestral lineage allowed it, at some point, to set the stage for splitting into two types of organisms: one which would become the Bacteria and one which would become the common ancestor to the Archaea and the eukaryotes and then split again into those two. My own guess is that evolutionary events having to do with the deep integral parts of the cell were involved in these splits. For example, perhaps the development of chromosomal organization and the capacity to replicate the chromosome may be one such event. Certainly that organization is different in the eukaryotes than it is in the Bacteria. How it is in the Archaea is still something to be determined. Granted, the Archaea have a circular chromosome, as do almost all Bacteria. But what has not been shown is that the chromosome of the Archaea is replicated from a single starting point. For all we know the archaeal mechanism of genome replication could mimic that found in eukaryotes where there are many start points for the replication of the chromosome, more than one per chromosome. When we find that out, then we will have more facts to lay on the table and more precise speculations to make.

JS: Wolfram Zillig has proposed a model in which he suggests that a physical, perhaps some sort of geographic, separation, might have occurred which allowed for the separate evolution of two different groups, but then again it's speculation.

CW: That would have to be a kind of geographic situation as you well know, that doesn't exist now, because today Bacteria and Archaea are intermixed together in most environments. Essentially, you can go to New Zealand and find the same kinds of Archaea or Bacteria that you find in Yellowstone National Park. I don't like to invoke geographical separations to explain prokaryotic evolution. That's okay for multi-cellular eukaryotes—for animals and plants—but not for microorganisms. They don't care about geographical borders.

It's just astounding to see how constant—how conserved—certain sequence motifs, proteins, genes have been over enormous expanses of time. You can see sequence patterns that have persisted probably for over three billion years. That's far longer than mountain ranges last, than continents retain their shape.

JS: Do you believe that thermophilic prokaryotes—were the first organisms to evolve?

CW: I believe that life originated in a hot environment and was, therefore, thermophilic—yes. This idea is not too popular among some people who adhere to Oparin's ideas on the origin of life. They might say that life began at low temperatures but there was a catastrophe on Earth at some point that only thermophilic organisms could survive. But I say let's use Ockham's razor. The deepest branchings in the phylogenetic trees of both the Bacteria and the Archaea (as well as the distribution of thermophilic species throughout these trees) suggests that initially things were thermophilic. On the one hand you have the *Aquifex* and the *Thermotogales* branches in the bacteria and the branch that leads to *Chloroflexus*—those are the deepest branchings and they are all basically thermophilic. On the other hand, you've got this split in the Archaea between the Crenarcheota and the Euryarcheota, and essentially all the Crenarcheota, with a rare exception, are thermophilic; while among the Euryarcheota, again, the deeply branching things are thermophilic. The simplest and for me most plausible explanation of this is that life began at high temperatures. That leaves you with a real question, Why are there not deeply branching thermophilic eukaryotes?

JS: Do you have any plausible answers?

CW: No, but I have a recommendation: keep looking for them.

JS: Do you believe that photosynthetic organisms will be found in all major phylogenetic groups of the Bacteria and the Archaea?

CW: With regard to the Archaea, the answer is in and I think it is "No." Photochemistry among the Archaea is confined to the extreme halophiles; they are a highly derived phenotype, not an aboriginal. The halophiles somehow evolved out of one of the three subgroups of methanogens. How they then invented photosynthesis is anyone's guess.

With regard to Bacteria the answer is not so straightforward, but it is intriguing. The two most deeply branching lineages of the Bacteria, the *Aquifex* lineage and the *Thermotogales,* are not now known to be represented by any photosynthetic examples; whereas the next most deeply branching group, which includes things like *Thermomicrobium, Herpetosiphon,* and *Chloroflexus* does contain photosynthetic species, for example, *Chloroflexus.* The way I usually put it is that photosynthesis is either aboriginal or was invented early in the history of Bacteria. And, if the latter, that presents you with a wonderful and possibly solvable problem because you can hopefully see the genes from which photosynthesis arose on the lineages that branched before say, the *Chloroflexus* lineage. You will see the anlage of photosynthesis.

I've got to continue a little bit on this question: while photosynthetic representatives do not occur in all the major bacterial groups, they are indeed widespread. The alpha, beta, and gamma subdivisions of the Proteobacteria are all very much interlaced phytogenetically with photosynthetic species. The cyanobacteria and green sulfur bacteria are two more major photosynthetic groups. Over in the gram-positive bacteria, we actually came in for a big surprise. No one thought that gram-positive bacteria could be photosynthetic in the old days. Yet, here comes this wonderful discovery of Howard Gest's, called *Heliobacterium,* which is indeed phylogenetically a gram-positive bacterium, and is indeed photosynthetic. At first, few people believed this. I'm not even sure that Howard Gest, himself, believed that his find was a member of the gram-positive bacteria. But some people took it seriously and they said, "Well, if the heliobacteria are gram-positive, then they could make heat-resistant spores." And, now the method of choice for isolating *Heliobacterium* is to take a soil sample, heat it up hot enough to kill everything but spores, and then enrich it on the proper medium, and just like that you have heliobacteria.

JS: It appears that the most ancient, deepest branching bacteria involve thermophilic, hydrogen-oxidizing bacteria such as *Aquifex* and *Hydrogenobacter.* All the known extant representatives of this group require oxygen for growth,

yet it is thought that little oxygen was available prior to the evolution of cyanobacteria. Can you offer an explanation for this?

CW: Yes, at least I can tell you a lovely story about *Aquifex.* There is a debate going on among geologists, at least those who concern themselves with primitive Earth conditions, as to whether oxygen was present at low levels on the Earth's surface, early on. The consensus seems to be that there indeed was, due to processes like photolysis of water in the upper atmosphere. You know, oxygen is an absolute requirement for the hydrogen-oxidizing bacteria to live. Yet they require oxygen at low levels, very low levels. What is more, they cannot stand it at any higher levels. So some people would say, therefore, that the amounts of oxygen that these organisms require are indicative of the levels of oxygen that some geologists would postulate had to exist on a primitive Earth, in other words, very low levels.

JS: It would be interesting for students to know how you came into microbiology. If you want to say a few words about that. . . .

CW: In college, my love was math and physics and I had little or no interest in biology. But when I went to graduate school, I went as a biophysicist—and here the world of biology began to open for me. The thing that caught me more than anything was evolution. I don't know why, but I love evolution. So, by the time I was in the first decade of my career, I was focusing on the problem of the origin of the genetic code. That was at the time when the code was breaking—back in the early 1960s. I realized that the origin of the genetic code was not a cryptographic problem. It was a problem of the origin of the translation mechanism, which gave rise to the code as one facet of its evolution. I went from the code to the translation mechanism. Then I realized that you can't begin to answer any evolutionary question without having a phylogenetic framework, and lo and behold, none existed for the microorganisms. So, to understand the evolution of the genetic code, I'd first have to create a phylogeny for microorganisms—and that's as far as I ever got. But that was a pretty big step.

What's in a name? that which we call a rose,
By any other name, would smell as sweet.

William Shakespeare, *Romeo and Juliet,* Scene II.

Chapter 17

Taxonomy of Eubacteria and Archaea

Nomenclature
Classification
Identification

This part of the book discusses the variety of microorganisms that exist on earth and what is known about their characteristics and evolution. Most of the material pertains to the bacteria; however, there is one chapter on eukaryotic microorganisms. Because greater emphasis is given to bacteria in this part, this first chapter discusses how bacteria are named and classified. This will be followed by a chapter on evolution and several chapters (Chapters 19–22) concerning the properties and diversity of bacteria.

When scientists encounter a large number of related items such as the chemical elements, plants, or animals, they characterize, name, and organize them into groups. Thousands of species of plants, animals, and bacteria have been named, and many more will be named in the future as more are discovered. However, not even the most brilliant biologist knows all of the species. Organizing the species into groups of similar types aids the scientist not only to remember them but also to compare them to their closest relatives, some of which the scientist would know very well. In addition, microbiologists are interested in the evolution of bacteria. To unravel this puzzle, it is essential to understand how one species is related to another. For these reasons, the 4000 or so bacterial species have been named and, based upon their characteristics, placed within the existing framework of other known bacterial species. The branch of bacteriology that is responsible for characterizing and naming organisms and organizing them into groups is called bacterial **taxonomy** or **systematics.**

Bacterial taxonomy can be separated into three major areas of activity. One is **nomenclature,** which is the naming of bacteria. The second is **classification,** which entails the ordering of bacteria into groups based upon common properties. In **identification,** the third area, an unknown bacterium—for example, from a clinical or soil sample—is characterized to determine what species it is. This chapter covers all three of these areas.

Nomenclature

Bacteriologists throughout the world have agreed upon a set of rules for naming bacteria. These rules, called the *International Code for the Nomenclature of Bacteria* (1991), state what a scientist must do to describe a new species or other **taxon** (plural taxa), which is a unit of classification, such as a species, genus, or family. Each bacterium is placed in a genus and given a species name in the same manner as plants and animals. For example, humans are *Homo sapiens* (genus name first, followed by species), and a common intestinal bacterium is named *Escherichia coli.* This **binomial system** of names follows that proposed by the Swedish taxonomist Carl von Linné (Linnaeus; 1707–1778) for plants and animals.

In bacterial nomenclature, the root for the name of a species or other taxon can be derived from any language, but it must be given a Latin ending so that the genus and species names agree in gender. For example, consider the species name, *Staphylococcus aureus.* The first letter in the genus name is capitalized, the species name is lowercase, and they are both italicized to indicate that they are Latinized. When writing species names in longhand, as for a laboratory notebook, they should be underlined to denote they are italicized. The genus name *Staphylococcus* is derived from the Greek *Staphyl* from *staphyle,* which means a "bunch of grapes," and *coccus* from the Greek, meaning "a berry." The *o* ("oh") between the two words is a joining vowel used to connect two Greek words together. The figurative meaning of the genus name is "a cluster of cocci," which describes the overall morphology of members of the genus. The species name *aureus* is from the Latin and means "golden," the pigmentation of members of this species. The *-us* ending of the genus and species names are the Latin masculine endings for a noun (*Staphylococcus* in this case) and its adjective (*aureus*).

Questions on a point of nomenclature are published in the *International Journal of Systematic Bacteriology* (IJSB), a journal devoted to the taxonomy of bacteria that is published by the American Society for Microbiology. The question is then evaluated by the international Judicial Commission, which subsequently publishes a ruling in the journal. One recent example of a problem considered by the Judicial Commission was the question about *Yersinia pestis,* the causative agent of bubonic plague. Scientific evidence indicates that it is really just a subspecies of *Yersinia pseudotuberculosis,* a species name that has precedence over *Y. pestis* because of its earlier publication. Because of the potential confusion and possible public health issues that could arise by renaming *Y. pestis, Y. pseudotuberculosis* subspecies *pestis,* the Judicial Commission ruled against renaming the bacterium despite its scientific justification. IJSB also publishes papers that describe and name new bacteria and contains an updated listing of all new bacteria whose names have been validly published.

Classification

Classification is that part of taxonomy concerned with the grouping of bacteria into taxa based upon common characteristics. Classification systems can be either artificial or natural. **Artificial systems** of classification of bacteria are based upon *expressed characteristics* of the organisms, or the phenotype of the organism. In contrast, **natural** or **phylogenetic systems** are based upon the purported *evolution* of the organism. Until recently, all bacterial classifications were artificial because there was no meaningful basis for determining the evolution of bacteria. In contrast, plants and animals have a fairly extensive fossil record upon which to base an evolutionary classification system. Although fossils of microorganisms do exist (see Chapter 18), the simple structure of bacteria and other microorganisms does not permit their identification to species, genera, families, or even kingdoms based solely upon their morphology as revealed in fossils.

An important article published by Zuckerkandl and Pauling in 1962 suggested that the evolution of organisms might be recorded in the sequences of subunits in their macromolecules. Subsequent research in the late 1960s and 1970s has supported this concept. In particular, molecules such as ribosomal RNA and some proteins have changed at a very slow rate during evolution, and therefore their sequences provide important clues to the relatedness of the various bacterial taxa and their relatedness to higher organisms as well. Considerable success has been achieved using this approach, and the classification of prokaryotic organisms is quickly becoming phylogenetic. At this time we are in a transitional period in which much work is still underway and some controversies remain, so the phylogeny of the bacteria is still incomplete. Our current classification is a hybrid between an artificial and a phylogenetic system. This chapter first discusses the traditional system of classification and then covers what is being done to make it phylogenetic.

While considering bacterial classification, it is important to keep in mind that bacteria have been evolving on Earth for the past 3.5 to 4.0 billion years. Therefore, it should not seem surprising that two separate Domains of bacteria (prokaryotic organisms) exist—the Eubacteria and the Archaea—versus only one Domain for eukaryotes, Eucarya. And, furthermore, the Eucarya appear to be the result of fusion and symbiotic events occurring between different early prokaryotic forms of life. Because of the long period of evolution of prokaryotes, they display considerable diversity, particularly metabolic and physiological diversity, among the various groups. In contrast, the large numbers of plant and animal species, though known for their morphological diversity, fit into only one group—either the primary producer (photosynthetic) or the consumer (heterotrophic) group, respectively. The diversity of metabolic types of prokaryotes is discussed more fully in Chapter 5.

The fact that two Domains of prokaryotes exist was not at all appreciated until molecular phylogenetic studies were performed. Bacteriologists are now trying to sort out when the split occurred between these two major groups of prokaryotes and what the common or universal ancestor organism might have been. This topic is discussed in greater detail in Chapter 18.

Taxonomic Units

The basic taxonomic unit is the species, although, as mentioned earlier, some species have subspecies categories as well. The categories above the species are (sequentially) genus, family, and order (Table 17.1). Higher categories are also used such as class, division, and kingdom, but uncertainties currently exist in bacteriology about their meaning. For this reason most higher categories are simply called "groups" or "phyla" and are given vernacular names.

Each culture of an organism represents an individual **strain** or **clone** in which all of the cells are descended from one single organism. In a somewhat different sense of meaning, a strain can also refer to a mutant of a species that has changed characteristics (for example, lacks a particular gene). The strain, however, is not considered a formal taxonomic unit, and Latin names are therefore not ascribed to strains; they have only informal designations, such as *E. coli* strain O157:H7, a pathogenic strain that causes hemolytic uremic syndrome and can be lethal to children who become infected by eating contaminated food. There can also be **varieties** within species that exhibit differences. These are called **biovars.** For example, serological varieties displaying somewhat different antigens are called **serovars,** and pathogenic varieties are termed **pathovars,** ecological types, **ecovars,** and so forth. Now let us look at the individual taxa beginning with the species to see what features are typical at each taxonomic level.

Reference *Table* **17.1 • Hierarchical classification of the eubacterium, *Spirochaeta plicatilis*[1]**

Taxon	Name
Order	Spirochaetales (vernacular name: spirochetes)
Family	Spirochaetaceae
Genus	*Spirochaeta*
Species	*plicatilis*

[1]Categories higher than orders exist, but most formal names have not yet been approved and accepted.

The Species

The definition of a bacterial species differs from that of higher organisms. In higher animals, the species is defined as a group of individuals (males and females) that exhibit evident phenotypic similarities and produce fertile progeny through reproduction. In prokaryotic organisms, sexuality is uncommon and different from that of eukaryotes, so this definition is not applicable. A bacterial species comprises a group of organisms that share many phenotypic properties and a common evolutionary history and are therefore much more closely related to one another than to other species. This definition, which is very subjective, has been interpreted differently by bacteriologists in describing species. For example, at one extreme some taxonomists are called **lumpers** because they group (or "lump") fairly diverse organisms into a single species or genus even though they are quite different. An example of a lumper is F. Drouet, who has proposed reducing the number of cyanobacteria from 2,000 species to only 62! At the opposite pole, are **splitters.** These are taxonomists who consider even the slightest differences sufficient for a new species. For example, it has been proposed that the genus *Salmonella* be "split" into hundreds of different species, a separate species for each of the hundreds of different serotypes (or serovars) that are recognized based upon specific cell surface antigens of their lipopolysaccharides and flagella. However, the extreme views of lumpers and splitters illustrated here are not accepted by the majority of microbiologists.

In fact, more recently, a less arbitrary, quantitative basis has been proposed to define a bacterial species. Agreement was reached by a group of prominent bacterial taxonomists to define a bacterial species based upon genomic similarity between strains. According to this proposal, now widely accepted by bacterial taxonomists, a species is defined as follows: *two strains of the same species must have a similar mole percent guanine plus cytosine content (mol % G + C), and must exhibit 70 percent or greater DNA/DNA reassociation.* The procedures used to determine these features are described here.

MOLE PERCENT GUANINE PLUS CYTOSINE (MOL % G + C) The **mol % G + C** refers to the ratio of the amount of guanine and cytosine to total bases (guanine, cytosine, adenine, and thymine) in the DNA. Recall that since G and C are paired in the double-stranded DNA molecule by hydrogen bonds, as are A and T, they occur in equal concentrations. The formula is given as:

$$\text{mol \% G + C} = \frac{\text{moles (G + C)}}{\text{moles (G + C + A + T)}} \times 100$$

Several methods can be used to determine the mol % G + C, sometimes also called the "GC ratio," of a bacterium. All of them require that the DNA is first isolated from a bacterium and purified (Figure 17.1). Thus, it is necessary to lyse the cells to release the cytoplasmic constituents including DNA, and the DNA must then be purified to remove proteins and other cellular material. Cell lysis is typically accomplished by treatment with lysozyme

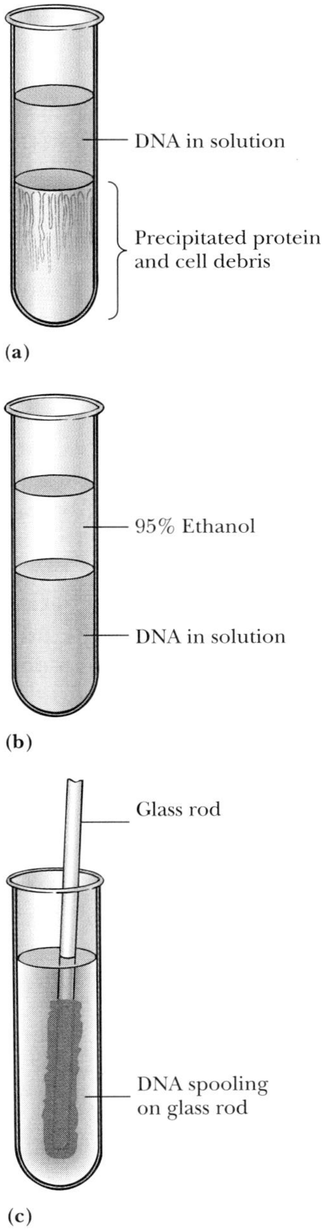

***Figure* 17.1** Typical steps involved in isolation and purification of DNA from bacteria. After cells have been grown and harvested by centrifugation, they are suspended in a saline buffer with lysozyme and a chelating agent (EDTA) and incubated at 37°C for up to several hours to disrupt the cell envelope. Detergent (sodium dodecyl sulfate) is added to enhance cell lysis by breaking up the cell membranes. Then, when the viscosity of the suspension increases, indicating lysis, sodium perchlorate and chloroform are added to precipitate the protein. The suspension is centrifuged and **(a)** the solution above the precipitated protein contains the dissolved DNA. The DNA solution is removed by pipetting into another tube; **(b)** 95% ethanol is added to precipitate the DNA which is harvested by **(c)** spooling it on a glass rod. Additional purification is achieved by repeated dissolution and reprecipitation with ethanol. RNase is added to remove RNA. Micromethods are also used for DNA isolation when small amounts are added. In this instance, harvesting is done without spooling, but by centrifugation in a microcentrifuge.

and detergents, and the DNA is precipitated with ethanol. When the DNA has been sufficiently purified it can be analyzed chemically to determine the content of each of the bases. Several different procedures can be used to determine the GC ratio of the purified DNA. We will describe three of them here.

The first method is a chemical method and can be accomplished by hydrolyzing the DNA and determining the concentration of each of the bases using an instrument called a **high pressure liquid chromatograph,** or **HPLC.**

A second procedure that is used is **buoyant density centrifugation** (see Chapter 4, Box 4.5). The density of the DNA is dependent on the amount of GC pairing versus AT pairing. GC bonds are triple hydrogen bonds (Figure 17.2), whereas AT pairs have double bonds; the greater the GC content of the DNA, the greater the density of the DNA. There is a direct relationship between the buoyant density and the mol % G + C.

The third common procedure to determine GC ratios is by **thermal denaturation.** The principle behind this method is that the hydrogen bonds between the two double strands can be broken by heating dissolved DNA. As the hydrogen bonds are broken, and the two strands separate, the absorbance of the DNA increases. This procedure, called melting the DNA, is conducted with a spec-

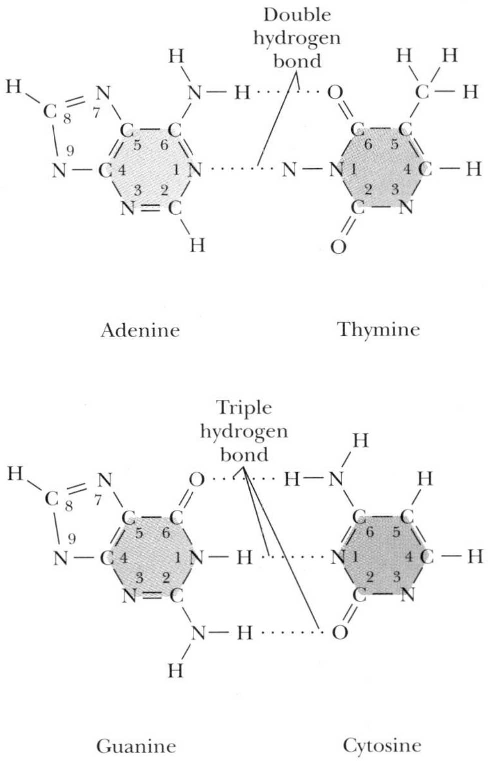

***Figure* 17.2** Two hydrogen bonds are formed in AT pairs and three in GC pairs in the DNA molecule.

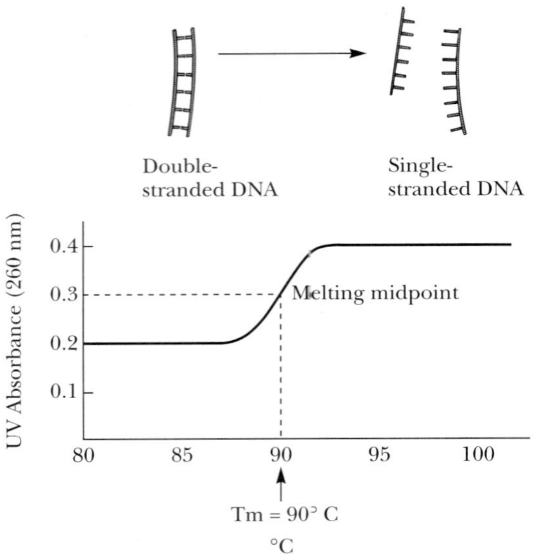

***Figure* 17.3** Melting curve for a double-stranded DNA molecule. As the temperature is raised during the experiment, the double-stranded form is converted to the single-stranded form. As this occurs the absorbance of the solution increases. The midpoint temperature, T_m, can be calculated from each curve. This process is reversible if the temperature of the solution is slowly lowered to allow the single-stranded molecules to reanneal to form double strands.

trophotometer set at 260 nm, a wavelength at which DNA absorbs strongly. The hydrogen bonding of the GC base pair is stronger than the AT pair in the double-stranded DNA molecule (Figure 17.2). Therefore, a higher temperature is required to melt DNA that has a high content of GC pairs—that is, a high GC ratio. Figure 17.3 presents a graph of the melting of a double-stranded DNA molecule. This process is accomplished by gradually increasing the temperature of a solution of the DNA in an appropriate buffer (ionic strength is important). As the temperature is raised, the melting process begins and continues until the double-stranded DNA molecule is completely converted to the single-stranded form. The absorbance increases during this melting process. The midpoint temperature (T_m) is directly related to the GC ratio of the DNA. Thus, the GC ratio can be read from a chart showing the relationship between T_m and GC content (Figure 17.4). Once the DNA has been melted, it will reanneal if the temperature is slowly lowered. Thus, the process shown in Figure 17.3 is reversible. However, if the solution is cooled rapidly, hybrid formation does not recur and the molecules of DNA are left in the single-stranded state.

Figure 17.5 shows the range of GC ratios in various groups of organisms on Earth. On this basis alone, one can see that bacteria, which have GC ratios ranging from about 20 to greater than 70, are truly a very diverse group. In contrast, higher organisms such as animals have a very restricted GC ratio range.

The GC ratio provides only the relative amount of guanine and cytosine compared to total bases in the DNA of an organism and says nothing about the inherent characteristics of the organisms or what genes are present. Indeed, *two very different organisms can have similar or even identical GC ratios.* For example, the DNA of *Streptococcus pneumoniae* and humans have the same mol % G + C content.

DNA/DNA REASSOCIATION OR HYBRIDIZATION Although the determination of GC ratio has its utility in bacterial taxonomy, it does not tell us anything about the linear arrangement of the bases in the DNA. It is the arrangement of the DNA subunits that codes for specific genes and proteins and therefore determines the features of an organism. DNA/DNA reassociation or hybridization is one method used to compare the linear order of bases in two different organisms.

DNA/DNA reassociation can be performed by a variety of different methods. In all approaches it is necessary to begin with purified DNA from the two organisms that are being compared. The DNA is then denatured by melting and DNA from the two different strains are mixed and allowed to cool together to allow reannealing to occur. This reannealing will occur both between DNA strands of the same species and between strands of the comparison species, the degree of reannealing depending upon how similar the DNAs are to one another. If two strains are very similar to one another, their DNAs will reanneal to a high degree. In contrast, if two strains are very different, then the extent of reannealing will be much less. DNA/DNA reassociation can be conducted experimentally in a number of different ways.

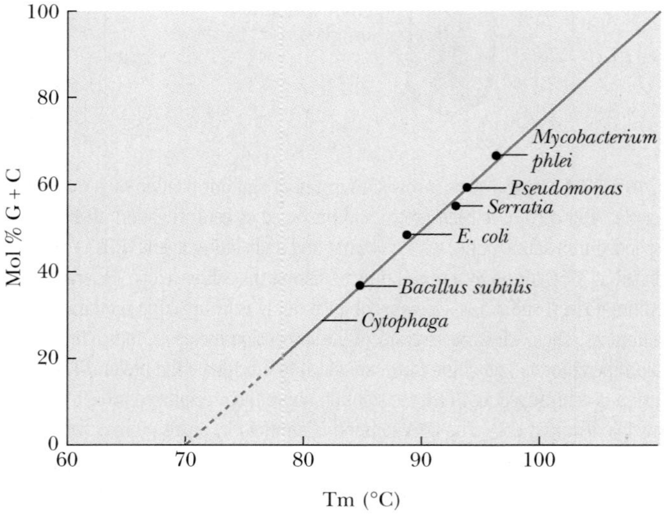

***Figure* 17.4** A graph showing the direct relationship between mol % G + C and midpoint temperature (T_m) of purified DNA in thermal denaturation experiments.

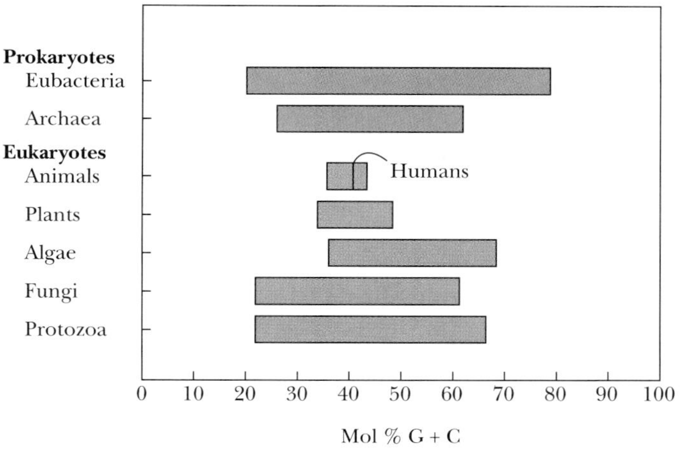

***Figure* 17.5** Range of mol % G + C content among various groups of organisms. Note the broad range of GC ratios for bacteria in comparison to plants and animals and subgroups of them.

One way to perform DNA/DNA reassociation is to radiolabel the DNA by growing the bacterium with tritiated thymidine or ^{14}C-labeled thymidine (other DNA bases can be used as well). If the bacterium takes up this labeled substrate and incorporates it into DNA, then the DNA becomes labeled. Alternatively, the DNA can be purified from the bacterium and labeled enzymatically in the laboratory.

After the DNA has been labeled and purified, it is sheared to an appropriate length by sonication. It is then ready for the hybridization experiments. First, single-stranded DNA is prepared. This is accomplished by heating the isolated DNA molecules to render them single-stranded and then cooling them rapidly to prevent reannealing.

First, let's look at the control assay for the DNA reassociation experiment (Figure 17.6). In this case a small amount of sheared radiolabeled DNA is rendered single-stranded. This is then mixed with a much larger amount of unlabeled DNA obtained from the same bacterial strain. These are heated together and cooled slowly to allow the two single-stranded groups to reanneal to form hybrid double strands. Because the amount of labeled DNA relative to the unlabeled DNA is small, there is a very low probability that it will reanneal with other labeled strands. Most of the reassociations will occur between unlabeled strands and most of the remainder will be between the labeled and unlabeled strands. The single-stranded fragments that did not reanneal are removed by enzyme digestion, and the double-stranded fragments are collected on a membrane filter or in a column. The amount of radioactivity remaining on the filter or on the column, after washing to remove low molecular weight material, represents the amount of hybrid formation between the labeled and unlabeled DNA for this identical strain. This is the *control reaction,* and the amount of radiolabel (the extent of hybridization) is considered to be 100 percent.

To determine the extent of reassociation between the strain described above and an unknown strain, similar experiments need to be performed. In this instance, unlabeled single-stranded DNA from the unknown strains is prepared and mixed with the known strain for which we

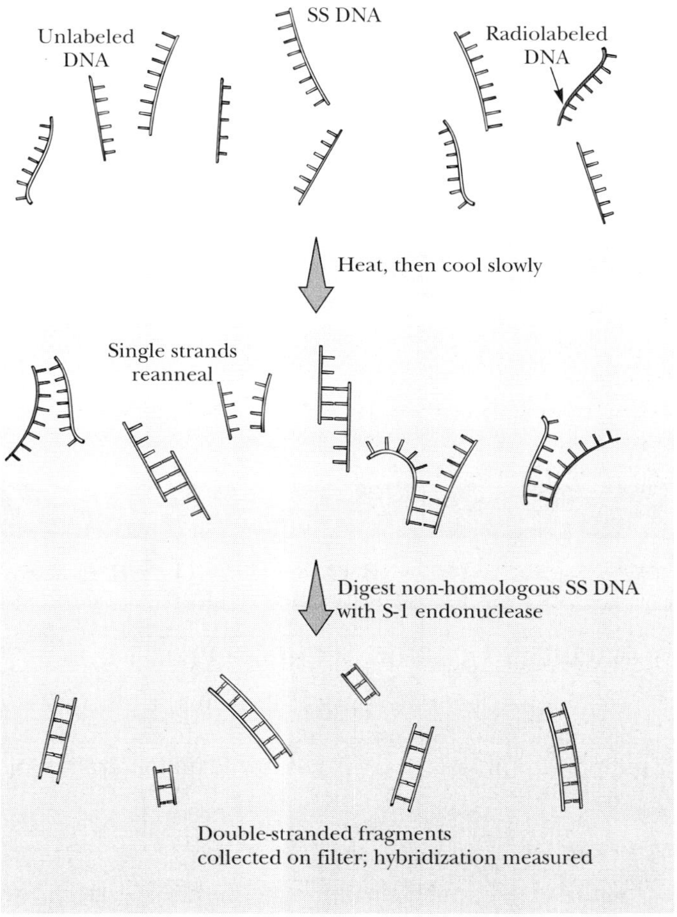

***Figure* 17.6** The process of DNA/DNA reassociation. After the DNA is radiolabeled, it is sheared and denatured to the single-stranded form. Small amounts of the labeled DNA (shown in red) are mixed with large amounts of similar single-stranded segments of nonlabeled DNA (in black) from the test organism. In the example shown, this is the same strain. The solution is heated and slowly cooled to permit double-stranded reannealing to occur. Single-stranded segments that exhibit no homology are digested with an enzyme (S-1 endonuclease), and the double-stranded segments are collected on a membrane filter. The amount of radiolabeled DNA collected on the filter represents the degree of reassociation (and hence, similarity between strains). In this example, which is the control, the degree of reassociation will be very high because it is the same strain reannealing with itself. When other strains are reannealed with the radiolabeled DNA, they will show lower degrees of reannealing (compared to the 100 percent value attributed to the control) indicative of the similarity between the strains being tested. Strains showing reannealing values of 70 percent or greater are considered to be the same species.

have labeled the DNA. As indicated earlier, those strains that show 70 percent or greater reassociation or hybrid formation with the labeled strain (determined by the amount of hybrid DNA that is radiolabeled compared with the same strain control of 100 percent shown in Figure 17.6) are considered to be the same species. Anything less is considered a different species.

The temperature at which the DNA/DNA reannealing occurs will influence the degree of reassociation between single strands. At higher temperatures only those sequences that are most similar will reanneal. These higher temperature conditions are termed more **stringent** conditions for reannealing. Scientists conducting DNA/DNA reassociation experiments typically use a reannealing temperature that is 25°C lower than the average midpoint temperature (T_m) of the DNAs being compared.

It should be noted that in DNA/DNA reassociation, the actual order of bases in the DNA is not determined, but rather the extent of reannealing between the DNAs of two different strains is assessed. Sequence analyses of entire bacterial genomes, analogous to the human genome project, have only recently been undertaken. Most of the genome of *Escherichia coli* has been sequenced, and some sequencing is also being undertaken for some methanogen genomes. However, this is a tedious, time-consuming, and expensive process, and a number of laboratories have been working for several years on *E. coli* sequencing. It is worthwhile noting that one could consider that the actual DNA base sequence of a strain is the ultimate definition of a strain—analogous to the chemical formula for a compound **(Box 17.1).**

The Genus

All species belong to a genus, the next higher taxonomic unit. When DNA/DNA reassociation is performed within a genus, some species of the genus may show little or no significant reassociation with other species. This does not prove conclusively that they are unrelated to one another, only that this technique is too specific to identify outlying members of the same genus. Therefore, DNA/DNA hybridization has limited utility for determining whether a species is a member of a known bacterial genus.

BOX 17.1 RESEARCH HIGHLIGHTS

Bacterial Genome Sequencing

The human genome project has caught the fancy of laypeople and biologists alike. It is an enormous and important project that will cost three billion dollars and take many years to complete.

Meanwhile, bacterial genome sequencing is surging ahead at a fraction of the cost because bacterial genomes are much smaller and therefore easier to sequence. Recently, a breakthrough has occurred in the strategy to carry out bacterial genome sequencing. Craig Venter, a crew of scientists at The Institute of Genomic Research (TIGR), and Hamilton Smith (a Nobel laureate) and his colleagues from Johns Hopkins have developed a new approach using random or shotgun cloning of DNA fragments of the 1.8 megabase genome of *Haemophilus influenzae*. These are then sequenced, and the overlapping areas are used to construct a map of the entire genome.

Although several groups have been working for a number of years to sequence the larger *E. coli* genome, it has not yet been completed. In contrast, the Venter and Smith groups have sequenced the entire genome of *Haemophilus influenzae* using their new approach in a relatively short time. A group in England is especially pleased with the results because they are interested in determining the sequence of some of the genes of this pathogen that are involved in pathogenesis. They have identified 18 genes that are required for toxin synthesis. Now that they have the sequences for them, they can produce mutants and see what the effect is on the disease process.

The sequences of several more bacterial genomes are now underway and will be completed in a short time. In addition to *E. coli,* the list includes two methanogens, *Methanobacterium thermoautotrophicum* and *Methanococcus jannaschii,* one thermophilic archaean, *Sulfolobus sulfataricus,* and another pathogen, *Mycoplasma genitalicum*. The information gained from analysis and comparison of sequences will aid in our understanding of evolution, genetics, and metabolism as well as in applied areas such as medicine and biotechnology. The estimated cost for sequencing a small bacterial genome is only one million dollars.

The TIGR database containing the sequence information can be obtained through the home page of *Science* (http://www.aaas.science/science.html).

The definition of most genera is based on one or more prominent phenotypic characteristics. Therefore, at this time, there is no uniform definition of what constitutes a bacterial genus. Oftentimes some striking physiological or morphological feature is present that permits it to be distinguished from other genera. For example, the genus *Nitrosomonas* is a group of rod-shaped bacteria that grow as chemoautotrophs gaining energy from the oxidation of ammonia. Another example is the genus *Caulobacter,* which are motile heterotrophic bacteria with a single polar prostheca.

Higher Taxa

As mentioned earlier, the taxonomy of prokaryotes is undergoing major changes. Although by classical taxonomy, each genus belongs to a family of similar genera, relatedness at the familial and higher levels is often uncertain for bacteria. Therefore, bacteriologists are careful about ascribing organisms to formal Latinized families and orders. Instead, vernacular group names are often used. For example, one vernacular group is the "budding and appendaged bacteria." Recent research indicates that at least two separate and unrelated groups are represented in this group, but only one Order, *Planctomycetales,* has been formally named.

As more becomes known about bacterial phylogeny, it is increasingly apparent that higher taxonomic levels do have meaning and can be distinguished from one another by comparing the sequence of subunits in certain macromolecules. This is because these macromolecules, or regions within them, have been highly conserved throughout evolution. One example of this is ribosomal RNA (rRNA). The ribosome is a very complex structure that carries out a complicated function—protein synthesis (see Chapter 11). Apparently major changes in ribosome structure have been selected against during evolution. Mutant organisms with dysfunctional ribosomes would be unable to compete with existing types and have therefore not survived. In this manner, evolution has selected against *major* changes in the ribosome. Nonetheless, some changes have occurred over the billions of years of biological evolution.

An additional advantage of using the ribosome as an evolutionary marker is that *all* organisms, from bacteria to higher plants and animals, have ribosomes. And, furthermore, the function of the ribosome as the structure responsible for protein synthesis is consistent for all classes of life. Therefore, it is possible to compare the phylogeny of all organisms with one another by analysis of a single highly conserved structure with an important cellular function. Ribosomal RNAs are not the only macromolecules that have been considered in determining relatedness at higher taxonomic levels. The proteins cytochrome *c*, ribulose bisphosphate carboxylase, and other macromolecules have also been considered, but these molecules are not present in all organisms. Since they are not universally distributed, they cannot be used to compare distantly related taxa.

Within the biological world, ribosomes share many similarities, indicating the conservative nature of the structure. Prokaryotic ribosomes contain three types of RNA: 5S, 16S, and 23S. Both 5S and 16S rRNA have been used to determine relatedness among organisms. Since the 16S molecule is larger (with about 1500 bases), it contains more information (Figure 17.7) than the smaller 5S molecule with only about 120 bases (Figure 17.8). Little work has been done to date on the 23S molecule because it is longer (about 3000 nucleotides) and is therefore more difficult to study. Thus, scientists interested in the classification and evolution of bacteria have concentrated on the 5S and 16S rRNA. The method of evaluation that provides the most information is sequence determination, especially for the 16S rRNA or its complementary DNA. It has been found that some regions of these molecules are more highly conserved than others. The more highly conserved regions permit one to compare distantly related organisms, and the more variable domains are used for examining more closely related organisms.

It appears that an analysis of 16S rRNA sequences provides important information on the evolution of prokaryotes. However, to conclude that the sequence of bases in rRNA accurately reflects the phylogeny of organisms, it is important to find totally separate and independent evolutionary markers to confirm the classification. Some work has been performed with sequencing of ATPases, elongation factors, RNA polymerases and other conserved macromolecules. The outcome of this research, which represents one of the most exciting areas of biology, is leading to the development of a complete phylogenetic classification of the Eubacteria, Archaea, and other microorganisms.

Another method of assessing similarities at the genus and higher levels is **rRNA/DNA hybridization.** This procedure is similar to DNA/DNA reassociation or hybridization (as described earlier), but in this case, the rRNA is radiolabeled. Because rDNA is more conserved than the bulk of an organism's DNA, it is possible to compare organisms that are more distantly related to one another than at the species level using this procedure.

Phylogenetic Versus Artificial Classifications

Most bacteriologists favor a phylogenetic system for the classification of bacteria, and with the advent of molecular phylogeny this hope is now being realized. However, many bacteria have not yet been isolated from nature and not all strains from culture collections have been sequenced to determine their relatedness to one another. At this time, bacterial classifications are hybrids between natural and artificial, being largely artificial, especially at

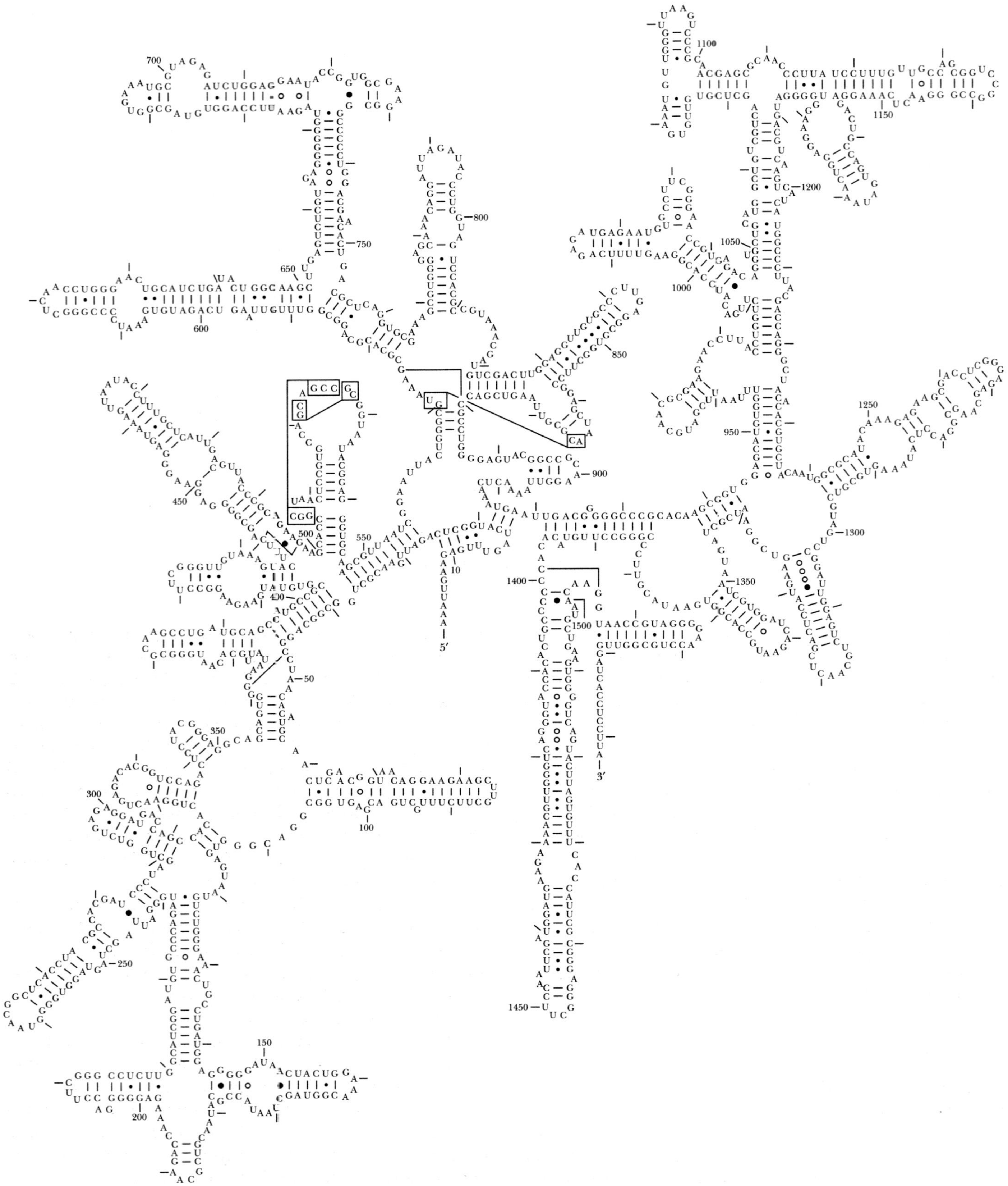

Figure 17.7 The 16S rRNA molecule from the small ribosomal RNA subunit of the bacterium, *Escherichia coli,* showing the secondary structure. Note that the bases are numbered from 1 at the 5′ end to about 1500 at the 3′ end. (Courtesy of Robin Gutell through the Ribosome Database Project by Gary Olsen and Carl Woese)

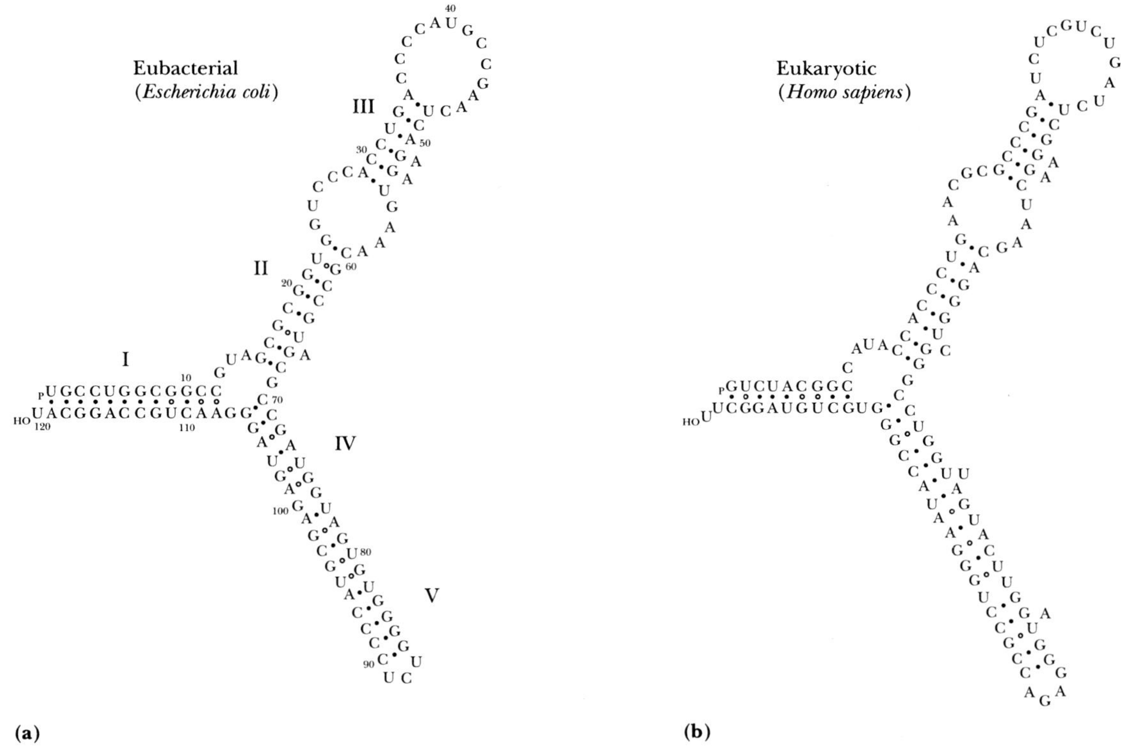

***Figure* 17.8** The 5S rRNA molecule from **(a)** *Escherichia coli* and **(b)** *Homo sapiens.*

the level of the family and genus. Although present classifications may not reflect the true phylogeny of the prokaryotes, it is extremely important to have an accepted classification because it enables scientists to readily identify known and new organisms and to communicate with one another on a common ground. The current accepted treatise that contains a complete listing of prokaryotic species and their classification is the four-volume *Bergey's Manual of Systematic Bacteriology* (1984–1989) and its more condensed edition, *Bergey's Manual of Determinative Bacteriology* (1994) (see Appendix I). In addition to containing a complete classification of prokaryotes, the more comprehensive version of *Bergey's Manual* contains a description of all known validly described bacterial species. Thus, it is the "encyclopedia" of the bacteria and has widespread acceptance among bacteriologists for the information it contains **(Box 17.2)**.

Artificial Classifications

In artificial classifications, it is important to select characteristics that clearly distinguish among organisms. Furthermore, it is important that the identifying characteristics selected are easy to determine. Two examples of simple phenotypic characteristics that have been widely used in artificial classification schemes are the Gram stain and cell shape. It turns out that each of these has some utility as an evolutionary marker, but they are rather crude indicators and by themselves yield only limited phylogenetic information. The Gram stain tells something about the nature of the cell wall. Gram-positive organisms fall into 2 of the 14 separate phylogenetic groups currently recognized: "Gram-Positive Bacteria" and "the Deinococcus group." However, the Mycoplasmas have been found to be most closely related to the gram-positives by 16S rRNA analyses (see Chapter 20) even though they stain as gram-negative bacteria. The Mycoplasmas lack cell walls, which explains why they stain as gram-negative bacteria. Therefore, they have apparently evolved from the gram-positive bacteria or vice versa.

In contrast to the gram-positive bacteria, gram-negative organisms fall into many different phylogenetic groups including peptidoglycan-containing types and non–peptidoglycan-containing types that are eubacterial as well as archaeal. Therefore, the gram-negative bacteria are very diverse phylogenetically.

At one time, some bacteriologists had proposed that the simplest, and purportedly the most stable, cell shape—the sphere—must have been the shape of the earliest bacteria. They then developed an evolutionary scheme based on this theory, in which all of the coccus-shaped bacteria were included in the same phylogenetic group. The validity of this classification has not been borne out by research. For example, there are both gram-negative as well as gram-positive cocci. Some cocci are Proteobacteria, others are nonphotosynthetic and highly resistant to ultravi-

BOX 17.2 MILESTONES

Bergey's Manual Trust

David Bergey was a professor of bacteriology at the University of Pennsylvania in the early 1900s. As a taxonomist he was a member of a committee of the Society of American Bacteriologists (SAB—now called the American Society for Microbiology) that was interested in formulating a classification of the bacteria that could be used for identification of species. In 1923, he and four others published the first edition of *Bergey's Manual of Determinative Bacteriology,* which was followed by new editions every few years. Royalties collected by the publication activities of the committee were held in SAB. When David Bergey and his coeditor, Robert Breed, requested money from the account to be used for preparation of the fifth edition, the leadership in SAB refused. After a long and bitter fight the SAB relented and turned the total proceeds (about $20,000) over to Bergey, who promptly put the money into a nonprofit trust with a board of trustees to oversee the publication of manuals on bacterial systematics. The Trust, now named in his honor as *Bergey's Manual* Trust, is responsible for the publication of *Bergey's Manual of Determinative Bacteriology,* which is now in its ninth edition, as well as other taxonomic books such as *Bergey's Manual of Systematic Bacteriology.* The Trust is headquartered at Michigan State University and has a nine-member international board of trustees as well as associate members from many countries.

Photograph of some current and past members and associate members of *Bergey's Manual* Trust attending an annual meeting in Yellowstone National Park. From left to right: Peter Sneath, Noel Krieg, John G. Holt, Karl-Heinz Schleifer, Micah Krichevsky, Jan Ursing, James Staley, Richard Castenholz, Donald Brenner, Stanley Williams, David Boone, Robert G. Murray, and Joseph Tulley. (Courtesy of David Boone)

olet light (*Deinococcus*), some are Archaea, and others Eubacteria.

However, one phylogenetic group, the spirochetes, contains all the helically shaped bacteria with axial filaments (see Table 17.1). Because of their morphology these bacteria were correctly classified with one another in the order *Spirochaetales* long before phylogenetic data confirmed the grouping. Apart from this order, overall cell shape has little meaning at higher taxonomic levels. Nonetheless, it can still be significant at the species, genus, and even family levels.

Numerical Taxonomy

When a large number of similar bacteria are being compared, computers are very useful in the analysis of the data. This aspect of taxonomy, which has been used in artificial classifications, is referred to as **numerical taxonomy.** Numerical taxonomy is most useful at the species and strain level, where phylogenetic relatedness has already been established by rRNA sequencing and DNA/DNA reassociation.

Conventional artificial taxonomy uses phenotypic tests to determine differences between strains and species. These tests are weighted so that one characteristic is considered more important than another. For example, in traditional taxonomy, the Gram stain has been given more weight in determining the classification of an organism than whether the organism uses glucose as a carbon source. In this case, all gram-positive strains would be ascribed to one family or genus, and within that group, certain species or strains would use glucose and others would not.

In contrast, in numerical taxonomy *all characteristics are given equal weight.* Thus, metabolism of a particular carbon source is considered to be as important as the Gram stain or the presence of a flagellum. In characterizing strains in this manner, a large number of characteristics are determined, and the similarity between strains is then compared by a similarity coefficient. Each strain is compared with every other strain. The similarity coefficient, S_{AB}, between two strains A and B is defined as follows:

$$S_{AB} = \frac{a}{a + b + c}$$

where a represents properties shared in common by strains A and B; b represents properties positive for A and negative for B; and c represents properties positive for B and negative for A.

Characteristics for which both strains A and B are negative are considered irrelevant because there would be many such features that would have no bearing on their similarity. For example, endospore formation is an uncommon characteristic for bacteria. It is not significant to incorporate this characteristic when comparing two species within a genus that does not produce endospores. It would, however, be of value in comparing endospore-forming organisms to closely related organisms.

In numerical taxonomy, it is best to have as many tests of characters as possible, typically at least 50 *independent* characters. Each characteristic should represent a single and separate gene. The same gene should not be assessed more than once, and therefore overlapping phenotypic tests must be avoided. S_{AB} values greater than 70 percent are expected within species and greater than 50 percent within a genus. An example of a numerical analysis is shown in **Box 17.3.**

It should be noted that the similarity coefficient can be used not only to relate phenotypic features of one organism to another, but also to relate the sequence similarity of macromolecules of different organisms. For example, in Chapter 18 the amino acid sequence of nitrogenase is compared with the 16S rDNA sequence of several strains of nitrogen-fixing bacteria (see Figure 18.5).

Phylogenetic Classification

During the 1970s a revolution occurred in bacterial taxonomy. By then data had accumulated that indicated a true natural or phylogenetic classification of bacteria was possible. What made it possible was, first of all, acceptance of the evidence that some of the macromolecules of bacteria were highly conserved, and their sequences held the key to unlocking the relatedness of bacteria to one another and to higher organisms. And, second, sequencing techniques were developed and improved so that it became easy to conduct sequencing analyses of ribosomal RNA and other macromolecules. In this section, emphasis will be given to sequencing 16S rRNA as it is the most common conserved molecule used for phylogeny in microbiology. But, before we discuss its use in phylogeny, it is first necessary to consider phylogenetic trees.

Phylogenetic Trees

Like the family tree, a phylogenetic tree contains the pattern of descendants of a biological family or group. However, whereas the family tree traces the genealogy of a family of humans, phylogenetic trees trace the lineage of a variety of different species. Thus, **phylogenetic trees** convey the purported evolutionary relationships among a group of species usually by use of some molecular attribute they have, such as the sequence of their ribosomal RNA. In this particular section we will discuss molecular phylogeny based upon a comparison of the 16S rRNA sequences of organisms.

Phylogenetic trees have two features—**branches** and **nodes** (Figure 17.9). Each node represents an individual species. External nodes (usually drawn to the extreme right of the tree) represent living species and internal nodes represent ancestors. A branch is a length which represents the distance between or time of separation of the species (nodes) from one another.

Trees may be rooted or unrooted. Figure 17.10**a** shows an unrooted tree containing three different species, A, B, and C. Unrooted trees compare one feature of a group of related organisms, such as the sequence of their 16S rRNA. There is only one shape to this particular tree with three species. In contrast, rooted trees need to have either (a) an outlying species that is very distantly related to the species being compared, or (b) an additional feature with which to compare the species—this should be a different characteristic such as the sequence of another macromolecule that is, in this example, unrelated to ribosomal RNA. A rooted tree containing three species has three possible shapes (Figure 17.10**b**).

For bacterial phylogeny, trees are constructed from information based on the sequence of subunits in macro-

BOX 17.3 METHODS & TECHNIQUES

Determination of Similarity Coefficients

In this example, eight strains, A–H, are compared to one another by ten phenotypic tests. The results are shown in the first table:

Results of phenotypic tests for eight strains, A–H

	Strains Tested							
Tests	A	B	C	D	E	F	G	H
1	+	−	+	+	+	+	−	+
2	−	+	+	−	+	−	+	+
3	+	+	−	+	−	+	+	+
4	+	−	−	−	+	+	−	+
5	+	−	−	+	+	+	−	−
6	+	+	−	+	−	−	+	−
7	−	+	+	+	+	−	−	+
8	+	−	−	+	−	+	+	−
9	+	−	+	+	+	+	−	+
10	−	+	+	−	+	−	+	+

Similarity coefficients are then determined by comparing the results of the tests for each of the strains against one another using the formula given in the text. The results are shown in the following table:

Similarity coefficients (×100 to give percent similarity) for the eight strains

Strain	A	B	C	D	E	F	G	H
A	100							
B	22	100						
C	20	38	100					
D	75	43	33	100				
E	40	33	71	40	100			
F	86	11	22	63	44	100		
G	33	67	25	33	50	22	100	
H	40	50	71	40	75	44	33	100

The information from this matrix is then used to group the strains into similar types as shown in the following matrix:

molecules. As mentioned earlier, ribosomal RNA, in particular the small subunit ribosomal RNA molecule, 16S rRNA, has been selected as the molecule of choice because of its conserved nature and length.

SEQUENCING 16S rRNA If one isolates a new bacterium and wishes to determine its phylogenetic position among the known bacteria, it is necessary to determine the sequence of its 16S rRNA. This can be accomplished in a number of ways. One of the most common ways is to use the polymerase chain reaction (PCR) to amplify the 16S rDNA from genomic DNA from the bacterium. The amplified 16S rDNA is then ligated into a cloning vector and cloned into *E. coli*. This allows a large quantity of 16S rDNA to be

BOX 17.3 METHODS & TECHNIQUES

Determination of Similarity Coefficients (Cont.)

Similarity matrix of grouped strains[1]

Strain	A	F	D	E	H	C	G	B
A	100							
F	86	100						
D	75	63	100					
E	40	40	40	100				
H	40	44	40	75	100			
C	20	22	33	71	71	100		
G	33	22	33	50	33	25	100	
B	22	11	43	33	50	38	67	100

[1]By these tests, strains A, F, and D are very similar to one another and probably comprise a single species. Likewise, E, C, and H are very similar and appear to be a separate species. Strains B and G may also be a different species, although more tests should probably be performed to substantiate this. As mentioned earlier in the chapter, phenotypic data such as this can provide an indication of relatedness at the species level, but if new species are being described, DNA/DNA reassociation tests should be performed.

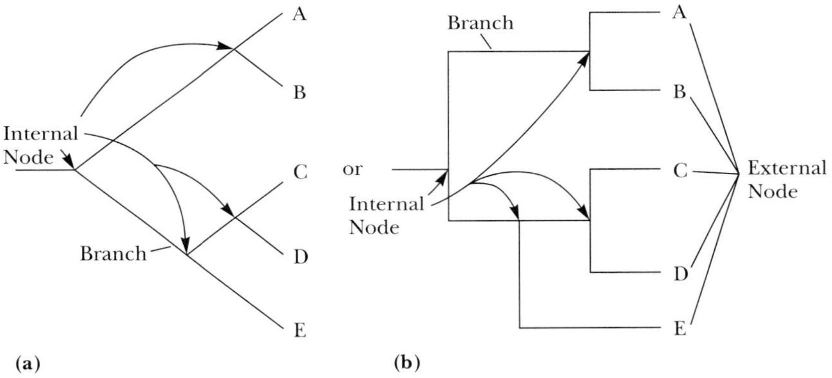

Figure **17.9** Two different formats **(a)** and **(b)** are used to show relatedness among different species A–E. External nodes, drawn on the extreme right, represent extant, known species. Internal nodes represent ancestor species. The horizontal branch length represents the time of separation or distance between the species.

produced through growth prior to sequence analysis (Chapter 16). The entire 16S rDNA can be sequenced using a standard set of oligonucleotide primers and standard sequencing techniques.

ALIGNMENT WITH KNOWN SEQUENCES The next step is to incorporate the determined linear DNA into an alignment with the sequences of other known organisms. An international database called the Ribosome Database Project, operated by Carl Woese and Gary Olsen at the University of Illinois, contains the 16S rRNA sequences of those bacteria that have been sequenced (over 4500 strains). Sequences from this database can be retrieved electronically over the Internet. Using a series of computer programs and careful manual examination, the sequence can be aligned with those retrieved from the database.

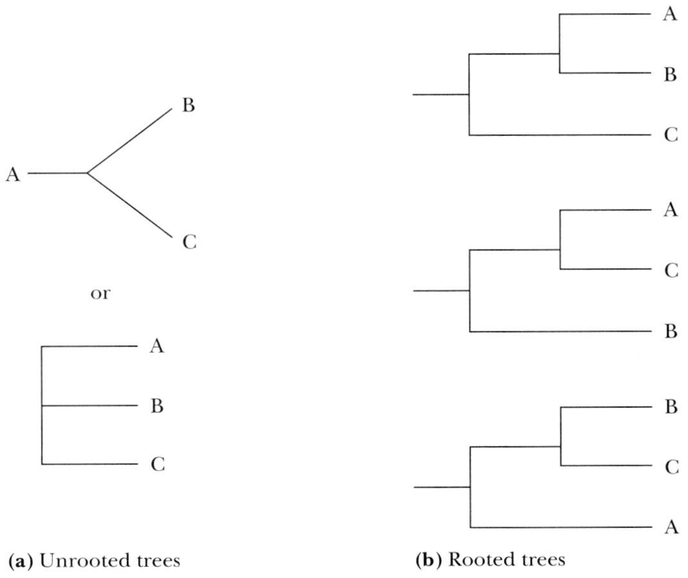

Figure **17.10** **(a)** The shape of a single unrooted tree for three different species, A, B, and C. **(b)** Three possible rooted trees for the three species.

PHYLOGENETIC ANALYSIS Having the sequence is only half of the story. It is next necessary to compare the sequence of the unknown bacterium to that of other bacteria from the database. The determination of the evolutionary relatedness among organisms can be accomplished by one of a number of phylogenetic methods. There are several types of analysis that can be used. **Distance matrix** methods are one type of approach. In distance matrix methods, the evolutionary distances, based upon the number of nucleic acid or amino acid subunits that differ in a sequence, are determined among the strains being compared. A second approach is to use **maximum parsimony** methods. In maximum parsimony the goal is to find the most simple or parsimonious phylogenetic tree that could explain the relatedness between different sequences. In both approaches, the sequence of nucleic acid subunits among different strains is compared.

Figure 17.11**a** uses a distance matrix method called the "unweighted pair group method with arithmetic mean" or UPGMA to analyze an aligned hypothetical sequence region of ribosomal RNA from four different strains. This is one of the simplest analytical methods that can be used. Figure 17.11**b**, using the same sequences, shows an analysis of the data by maximum parsimony.

In the UPGMA method a distance matrix is set up to compare the differences in sequence or "distance" between each of the four strains. There are four nucleotide differences between the sequence of organism A and the sequence of organism B, thus, d_{AB} is determined as 4. Likewise, d_{AC} is 5, d_{BC} is 5, and so forth. In this instance the shortest distance, 2, is between strains C and D. From these two strains which show the closest relationship to one another, a simple tree is constructed that shows C and D connected by a node that is half the distance between the two, that is, 1 unit. This is expressed in the actual tree (Figure 17.11**a**) as a horizontal branch length of one unit from each of the organisms to a common ancestral node.

The next step is to construct another matrix in which C and D are considered as a single composite unit and compared with A and B. Thus, A is different from (CD) by $(d_{AC} + d_{AD})/2$, or $(5 + 6)/2 = 5.5$. Likewise, B is different from (CD) by $(d_{BC} + d_{BD})/2 = 5$, and d_{AB} is only 4. From this matrix, B is closest to CD, and the length of its branch is calculated as $d_{(CD)B}/2$, or 5/2, which is equal to 2.5. From this the second branch is formed. Finally, since the distance between A and B is 4, the final branch is formed between A and B, between which is the distance, $d_{AB}/2$, which is equal to 2 units. From this reasoning a final tree (Figure 17.11**a**) is produced showing the relationship among the four different strains.

The UPGMA method is the simplest of the distance methods used. More sophisticated distance methods include transformed distance and neighbor joining methods, which will not be discussed further here.

As mentioned earlier, in maximum parsimony the goal is to identify the simplest tree that could explain the difference between two different sequences or species. This approach has its philosophical basis in **Occam's razor (named after William of Ockham),** commonly used in the sciences, which states that *the likely solution to a problem is the simplest one.* In this case, to explain the evolutionary difference between two species, one looks at the tree that has the fewest subunit changes (mutational events) that could explain their differences. This is accomplished on a computer which, at least in theory, considers all possible trees and then identifies the simplest one (the one with the fewest assumed mutational events).

In maximum parsimony it is important to recognize sites in the sequences that are being compared, termed **informative sites.** These sites are then used to determine the most parsimonious tree. For example, site 1 in the example given is not informative because all the bases are identical. Site 2 is not informative either, because three of the strains have A and one has C, suggesting that a single mutational event has occurred. Site 3 is also not informative, because trees developed from the information given at this site are distinguished from one another by two base changes—none is more parsimonious than the other. Site 5 is also not informative because trees constructed from the information at this site differ from one another by three mutations. In contrast, Site 4 is informative, because one tree (Tree #1) is more parsimonious than the other two. Sites 6 and 8 are not informative because all bases are identical, but both sites 7 and 9 are informative. Site 7 favors Tree #1 whereas Site 9 favors Tree #2. Thus, for this set of data, Tree #1 is favored two out of three times, Tree #2 is favored one out of three times, and Tree #3 is not favored at any time. Adding the changes at those three sites gives the following data: Tree #1 is the most parsimonious because a total of only four changes $(1 + 1 + 2 = 4)$ would explain its phylogeny, whereas in Tree #2, five changes are required $(2 + 2 + 1 = 5)$ and in Tree #3 six changes would be required $(2 + 2 + 2 = 6)$.

Note that the two trees that were constructed from the distance matrix and maximum parsimony methods are identical in shape, indicating that two of the strains, A and B, are closely related to one another, but not to C and D. And, likewise, C and D are closely related to one another. As you can imagine, some rather sophisticated computer programs have been developed to handle the immense amount of information inherent in longer sequences such as 16S rRNA, which contains about 1500 bases. Moreover, when such large amounts of data are being analyzed, it is often not possible to determine with 100 percent accuracy that a proposed tree is, in fact, the true tree. To help address this question, other techniques are used to test the validity of the tree. For example, in "bootstrap" analyses, random sequence positions are selected by the computer and the trees formed from them are compared with the proposed tree to see if a simpler (more parsimonious) tree can be found. In this manner, some 100 or 1000 different bootstrap comparisons might be made and provided as ev-

Organism	Site # 1	2	3	4	5	6	7	8	9
A	G	C	G	G	A	C	A	A	A
B	G	A	C	G	C	C	A	A	G
C	G	A	A	A	G	C	G	A	A
D	G	A	A	A	G	C	G	A	G

First matrix

	Organism A	B	C
B	$d_{AB} = 4$	–	–
C	$d_{AC} = 5$	$d_{BC} = 5$	–
D	$d_{AD} = 6$	$d_{BD} = 5$	$d_{CD} = 2$

Second matrix

	Organism (CD)	A
A	$d_{(CD)A} = 11/2$	–
B	$d_{(CD)B} = 10/2$	$d_{AB} = 4$

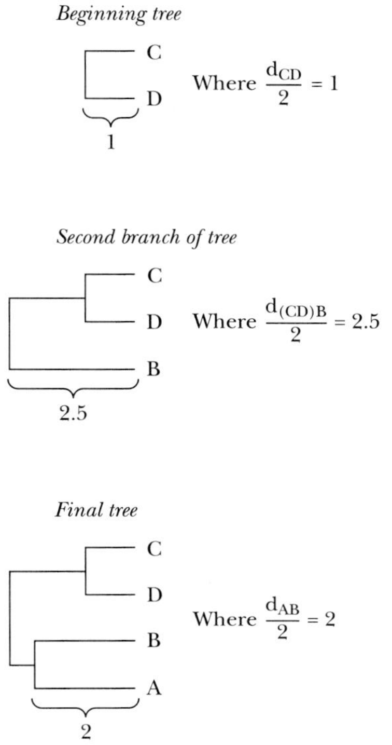

(a) The UPGMA distance method

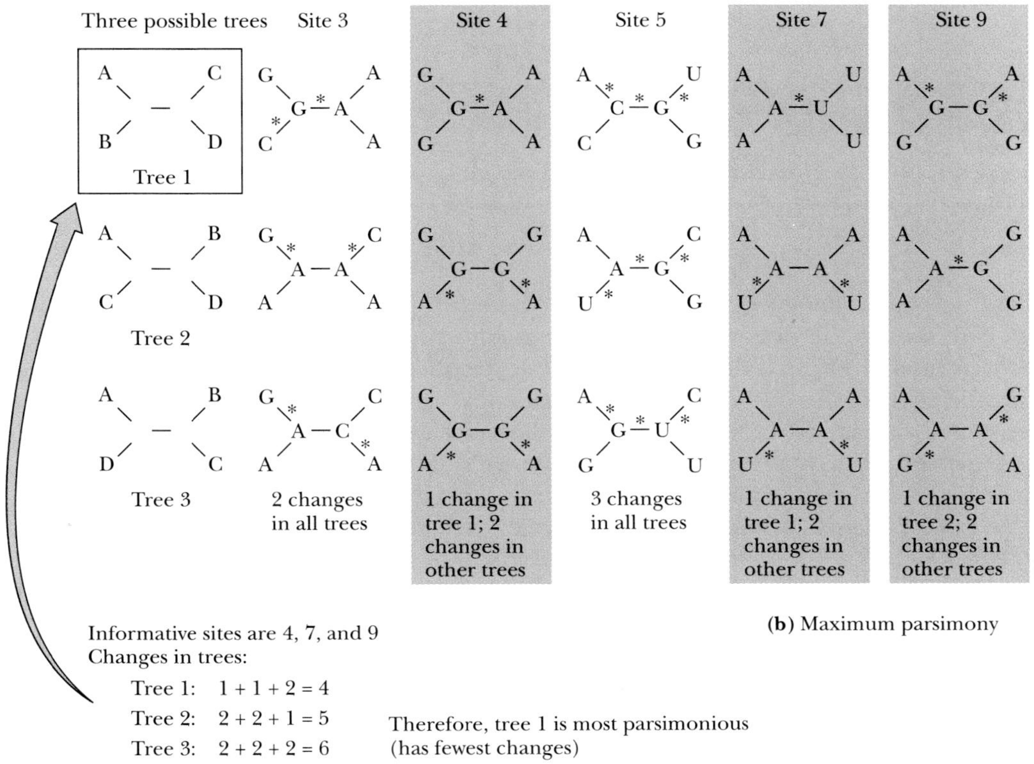

(b) Maximum parsimony

***Figure* 17.11** Four different strains, A, B, C, and D, showing a hypothetical region of their 16S rRNA that contains nine bases. **(a)** The UPGMA method of determining a phylogenetic tree. A distance matrix is constructed of all base comparisons to determine the most closely related strains. These are diagrammed first. Then, another matrix is determined to assess the distance between the first paired group (CD) and the remaining strains A and B. Since A and B are close to one another their position is determined next. Finally, the relation between CD and A and B are determined. **(b)** In maximum parsimony, it is first necessary to identify the informative sites in the sequences. Trees are constructed for sites 3, 4, 5, 7, and 9 to illustrate that only sites 4, 7, and 9 are informative because one tree is more parsimonious at those sites than the others. The most parsimonious tree is determined by an analysis of each of the informative sites (see text for details).

idence that the proposed tree is indeed the most parsimonious one.

As mentioned previously, there are other analytical methods that can be used to analyze sequence information and construct phylogenetic trees. A common one used by microbiologists is the **maximum likelihood** method, which involves selecting trees that have the greatest likelihood of accounting for the observed data. This is accomplished by assigning a probability to the mutation of any one base to any other base at each possible sequence position. From this, all possible topological trees are constructed. By integrating the probabilities for each mutation over each tree, a degree of improbability for a tree is assessed. The least improbable tree is chosen as the "true" tree.

Major Groups of Eubacteria and Archaea

Bacteriologists have begun to construct classifications using phylogenetic information from rRNA analyses. As mentioned in Chapter 1, some prokaryotes are very different from others, a revelation that came through an analysis of ribosomal RNA **(Box 17.4).** Ribosomal RNA data allow the division of all organisms on Earth, prokaryotic and eukaryotic, into three Domains: Eubacteria, Archaea, and Eucarya (eukaryotes) (Figure 17.12). These Domains can also be distinguished from one another by phenotypic tests. For example, consider the cell envelope composition of the organisms. Peptidoglycan is found only in Eubacteria, although two groups—the mycoplasmas and the *Planctomycetales*—lack it. Furthermore, the lipids of the Eubacteria and Eucarya are ester-linked, whereas they are ether-linked in the Archaea (see Chapter 4). Table 17.2 includes a summary of these and other major phenotypic characteristics that differentiate among these three Domains.

At this time 14 different phylogenetic groups, referred to here as **phyla,** are known.[1] Included are 11 phyla of Eubacteria (Figure 17.13) and 3 phyla of Archaea (Figure 17.12). Each of these phyla has specific regions or sequences in its ribosomes that are distinctive to it. These regions are therefore termed **signature sequences** (Figure 17.14; Table 17.3). A bacterium can be assigned to one of these 14 phylogenetic groups once its signature sequence is known. These 14 phyla are listed here along with a brief description of their major features. They are treated in more detail in subsequent chapters (Chapters 19 through 22), and the groups in each chapter are indicated below.

Bergey's Manual of Systematic Bacteriology has been closely followed in organizing the bacterial groups treated in this book (Table 17.4). The Eubacteria in the following 14 phylogenetic groups are treated in Chapters 19, 20, and 21.

Chapter 19. Gram-Negative Heterotrophic Eubacteria

PHYLUM 1.[1] PROTEOBACTERIA AND THE MITOCHONDRION The Proteobacteria is a very diverse group of organisms. All four of the major bacterial nutritional types are represented within this group. Some of these organisms are photosyn-

[1] The term phylum is used in this text to refer to a phylogenetic group. It is not equivalent to the taxonomic term, Phylum, used in zoology or any other official term in taxonomy.

BOX 17.4 RESEARCH HIGHLIGHTS

The Discovery of Archaea

Clearly, one of the most exciting developments in bacterial classification of the 20th century was the discovery that there are two different groups of bacteria or prokaryotic organisms. The appreciation of the difference between the true bacteria (Eubacteria) and the Archaea was the culmination of years of research by microbiologists throughout the world. However, the final piece of evidence that convinced microbiologists of this dichotomy was the discovery that these organisms had very different 16S ribosomal RNAs. This research was performed in Carl Woese's laboratory at the University of Illinois. At that time, rRNA sequencing was not done routinely in laboratories. Instead, 16S rRNA was purified and digested by a ribonuclease. The oligonucleotide fragments produced were subjected to two-dimensional electrophoresis. The pattern of spots on a two-dimensional chromatogram represented the various rRNA oligonucleotides typical of each species. Studies from Woese's laboratory demonstrated that the patterns were very different for Eubacteria and Archaea. Indeed, their studies of 18S rRNA from eukaryotic organisms indicated that the Archaea are as different from Eubacteria as they are from eukaryotes. The seminal findings of this work have changed the way microbiologists view taxonomy and phylogeny.

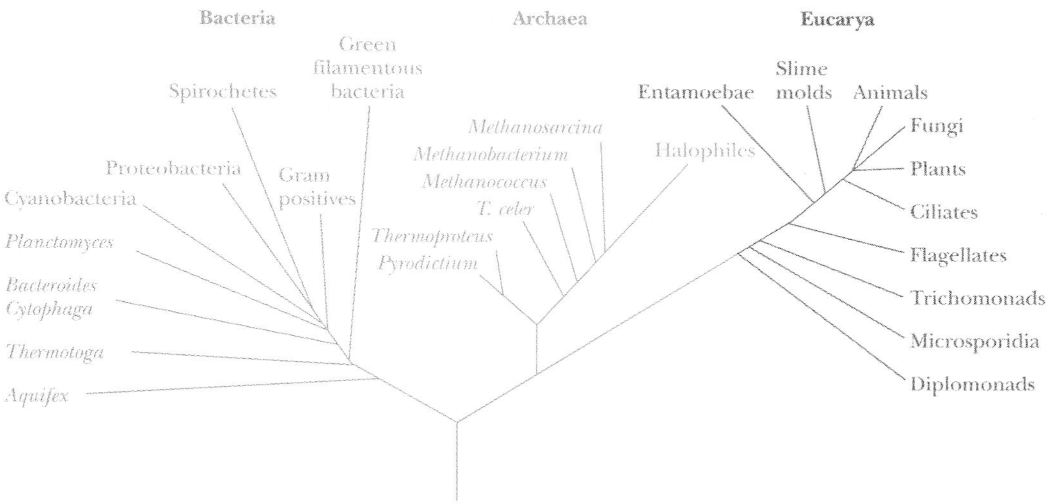

***Figure* 17.12** The three Domains of organisms—Eubacteria, Archaea, and Eucarya—based on a comparison of 16S rRNA sequences. This is a rooted tree in which the 16S rRNA sequences are compared with an elongation factor. (Adapted from Carl Woese)

thetic while some are heterotrophic and others are chemolithotrophic. The chemolithotrophic bacteria include the nitrifiers, the thiobacilli, the filamentous sulfur oxidizers (*Beggiatoa* and related genera), and many species that grow as hydrogen autotrophs. Carbon dioxide fixation, when present, is via the Calvin-Benson cycle in all of these bacteria.

This phylogenetic group contains many of the well-known gram-negative heterotrophic bacteria such as *Pseudomonas,* the enteric bacteria including *E. coli, Vibrio* and luminescent bacteria, and the more morphologically unusual bacteria such as the prosthecate bacteria. In addition, many symbiotic genera such as *Agrobacterium, Rickettsia,* and *Rhizobium* are members of this group.

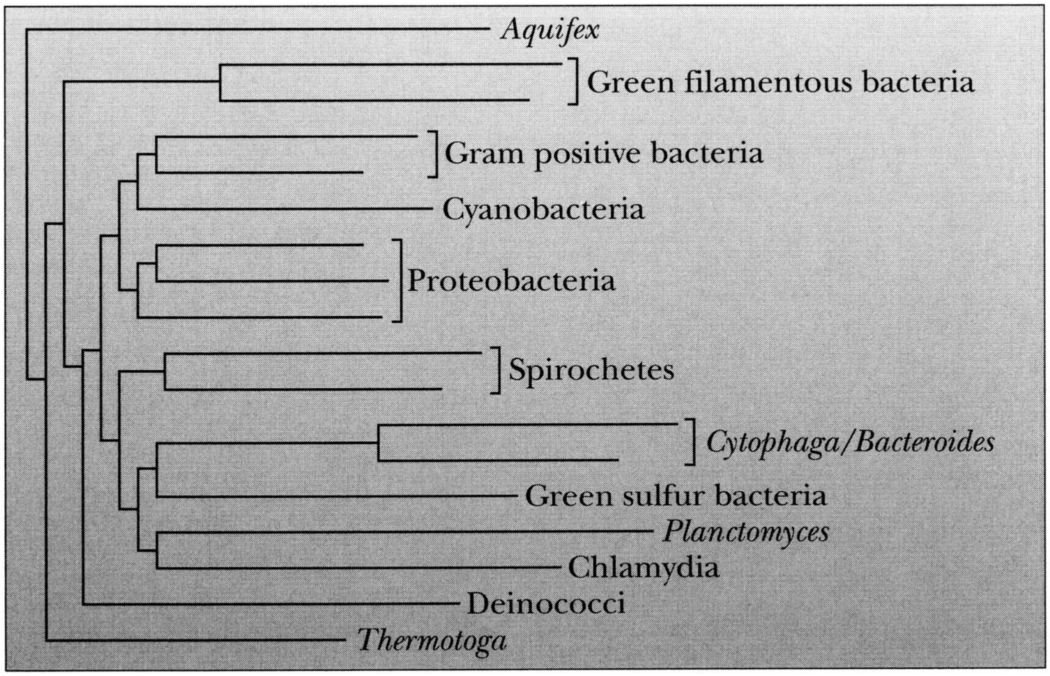

***Figure* 17.13** Twelve of the phylogenetic groups of Eubacteria as differentiated on the basis of 16S rRNA. Two other eubacterial groups are thermophiles which branch near *Thermotoga.*

Differential *Table* 17.2 • **Comparative biological properties of the three Domains**

Character	Eubacteria	Archaea	Eucarya
Cell envelope			
Peptidoglycan	+	−	−
Ester lipids	+	−	+
Ether lipids	−	+	−
Genome			
Nucleolus	−	−	+
Reverse gyrase	−	+	−
Ribonucleic acid			
Simple polymerase	+	−	−
Complex polymerase	−	+	−
Multiple polymerases	−	−	+
Polycistronic mRNA	+	+	−
70S ribosomes	+	+	−
Introns			
tRNA	−	+/−	+
rRNA	−	+/−	+
mRNA	−	−	+
Transcription			
Sensitive to antibiotics	+	−	−
Promotors			
Eubacterial type	+	−	−
Polymerase II type	−	+	+
Translation			
Diptheria toxin sens.	+	−	+
fMet tRNA	+	−	−
Antibiotic sensitivity	+	−	−
Metabolism			
Similar enzymes:			
Malate dehydrogenase	+	−	+
3-Phosphoglycerate kinase	+	−	+
Glyceraldehyde-3-phosphate dehydrogenase	+	−	+
Similar ATPase	−	+	+
Methanogenesis	−	+/−	−
Nitrogen fixation	+/−	+/−	−
Chlorophyll-based photosynthesis	+/−	−	−[1]
Energy-yielding sulfur metabolism	+/−	+/−	−
Extreme thermophily	+/−	+/−	−
Structure			
Gas vesicles	+/−	+/−	−
Single filament flagellum	+/−	+/−	−

[1]Photosynthesis occurs with prokaryotic-derived chloroplast.

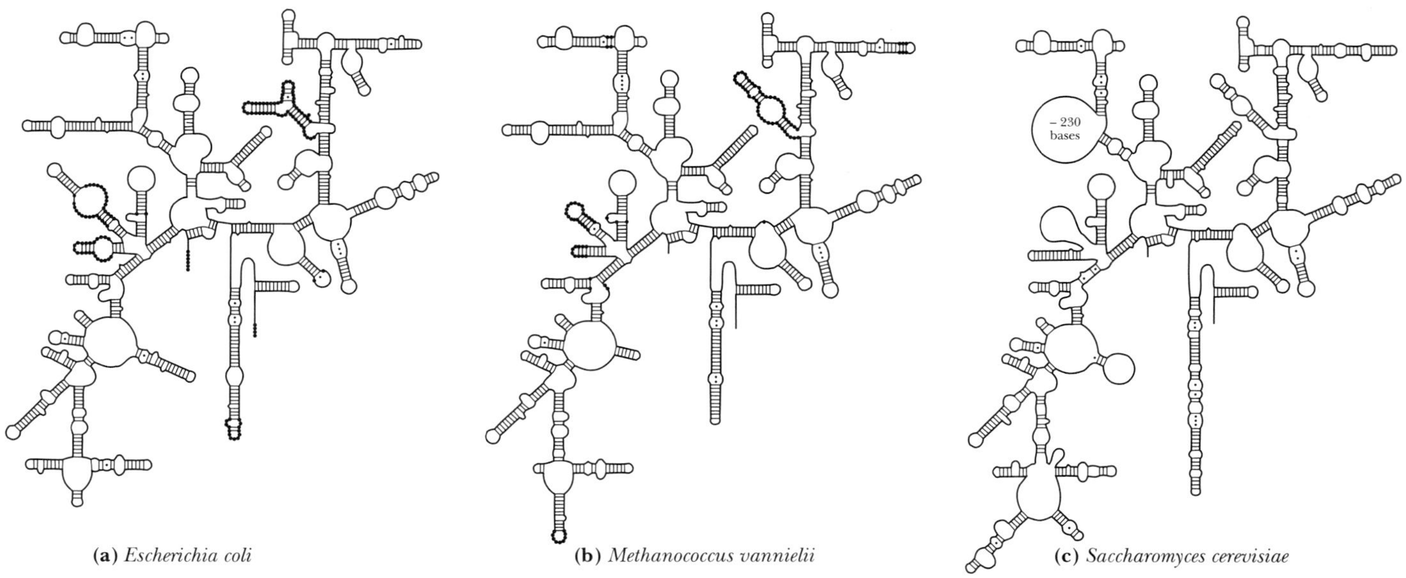

(a) *Escherichia coli* (b) *Methanococcus vannielii* (c) *Saccharomyces cerevisiae*

***Figure* 17.14** A comparison of the secondary structure of the 16S rRNA from **(a)** the Eubacterium, *E. coli,* with **(b)** the Archaean, *Methanococcus vannielii,* and the 18S rRNA from **(c)** the eukaryotic yeast, *Saccharomyces cerevisiae.* Regions with heavy lines on the prokaryotes indicate major signature sequence regions typical of Eubacteria and Archaea, respectively. (Adapted from Carl Woese)

Differential ***Table* 17.3 • Differential illustrative 16S rRNA signature sequence regions for representative Eubacteria compared with *E. coli*[1]**

Position	Consensus	*E. coli* & Proteobacteria	Spirochaetales	Flavobacteria/ Bacteroides	Planctomycetales	Gram-positive	Deinococcus	Green Sulfur	Green Filamentous	Cyanobacteria
47	C	+	U	+	G	+	+	+	+	+
48	Y[2]	+	+	+	A	+	+	+	+	+
49	U	+	+	+	+	+	+	+	+	+
50	A	+	U	+	U	+	+	+	+	+
51	A	+	+	+	+	+	+	+	+	+
52	Y	+	A	+	G	+	G	+	+	+
53	A	+	G	+	G	+	+	+	G	+
1224	U	+	+[3]	+[3]	+	+[3]	+	Y or A	G	+[3]
1225	A	+	+	+	+	+	+	+	+	+
1226	C	+	+	+	+	+	+	+	+	+
1227	A	+	+	+	+	+	+	+	+	+
1228	C	+	+	+	+	+	+	+	+	+
1229	A	+	+	+	+	G[4]	+	+	+	+
1230	C	+	+	+	+	+	+	+	+	+
1231	G	+	+	+	+	+	+	+	+	+
1232	U	+	+	+	+	+	+	+	+	+
1233	G	+	+[3]	+	+[3]	+[3]	+	+	+	A
1234	C	+	+[3]	U[5]	+	+	+	A	+	+

[1]+ means same as consensus sequence.
[2]Y is either cytosine or uracil.
[3]Other bases are found in some species.
[4]Actinomycetes (high mol % G + C Gram-positive taxa).
[5]Bacteroides group only.

Reference *Table* **17.4 • Overall organization for treatment of Eubacteria and Archaea**

Bacterial Group	This Book	*Bergey's Manual of Systematic Bacteriology*[1]
Gram-Negative Heterotrophic Eubacteria	Chapter 19	Volume 1[2]
Gram-Positive Eubacteria	Chapter 20	Volumes 2 and 4
Phototrophic, Chemolithotrophic and Methylotrophic Eubacteria	Chapter 21	Volume 3
Archaea	Chapter 22	Volume 3

[1]See Appendix for more complete treatment of organization of taxa as treated in *Bergey's Manual of Systematic Bacteriology* (First Edition).
[2]Some gram-negative heterotrophs were included in Volume 3.

Also included in this phylum are the exotic myxobacteria that form fruiting structures as well as unicellular, nongliding forms such as the bacterial predator, *Bdellovibrio*. The gram-negative sulfate reducers such as *Desulfovibrio* are also found in this group.

Finally, the mitochondrion present in almost all eukaryotes has been shown to have evolved from this group of bacteria.

PHYLUM 2. FLAVOBACTERIUM, CYTOPHAGA, BACTEROIDES, AND RELATIVES This is a diverse group containing heterotrophic aerobes and anaerobes. The nonfruiting gliding heterotrophic bacteria are filamentous and have a low DNA base composition (about 30–40 mol % G + C).

PHYLUM 3. PLANCTOMYCETES The Planctomycetes group of Eubacteria are budding, unicellular, or filamentous bacteria. These bacteria lack peptidoglycan.

PHYLUM 4. CHLAMYDIAE The Chlamydiae is a group of obligately intracellular parasites and pathogens whose closest relatives are the Planctomycetales. They also lack peptidoglycan.

PHYLUM 5. SPIROCHETES The spirochetes (Order Spirochetales) are morphologically distinct from other bacteria. Their flexible cells are helical in shape. All are motile due to a special flagellum-like structure, the axial filament, not found in other bacteria.

The remaining heterotrophic gram-negative eubacterial phyla, 6 through 9, are thermophilic bacteria that comprise the most ancient groups of the Eubacteria based upon 16S rRNA sequence analysis.

PHYLUM 6. THERMOTOGA This is a fermentative genus containing some of the most thermophilic Eubacteria known. They grow at temperatures from 55° to 90°C. Their cell lipids are unusual.

PHYLUM 7. THERMOMICROBIUM A single strain of the genus *Thermomicrobium* exists. It is a small rod-shaped thermophile which grows as a heterotroph in hot springs with an optimum temperature for growth of 70° to 75°C. The cell wall contains very low amounts of diaminopimelic acid.

PHYLUM 8. THERMOLEOPHILUM This group is represented by a single genus of thermophiles that use n-alkanes with a chain length of C13 to C20 as carbon sources for growth. The optimum temperature is 55° to 65°C.

PHYLUM 9. AQUIFICALES This order of bacteria contains the hydrogen autotrophic genera *Aquifex* and *Hydrogenobacter*. This phylogenetic group contains the most thermophilic Eubacteria known and comprises the deepest branch (earliest branch) of the Eubacteria.

Chapter 20. The Gram-Positive Bacteria

PHYLUM 10. GRAM-POSITIVE AND CELL WALL–LESS EUBACTERIA The bacteria in this group are all gram-positive, although the *Mycoplasma* group (Mollicutes) lacks a cell wall altogether and therefore stains as gram-negative. All other members of the group contain large amounts of peptidoglycan in their cell wall structure.

The gram-positive bacteria range in shape from unicellular organisms to branching, filamentous, mycelial organisms, the actinomycetes. Most gram-positive bacteria are common soil organisms; some have specialized resting stages (endospores or conidiospores) that enable them to survive during dry periods. Most members of the group are heterotrophic bacteria, although some are autotrophic. One group, the heliobacteria, is photosynthetic and produces a unique form of bacteriochlorophyll, bchl *g*.

PHYLUM 11. DEINOCOCCUS AND THERMUS This is a very small group of organisms currently represented by only three genera.

The genus *Deinococcus* contains gram-positive bacteria. However, they differ from Phylum 10 bacteria in showing strong resistance to gamma radiation and ultraviolet light. The second genus, *Thermus,* contains thermophilic, rod-shaped bacteria. Ornithine is the diamino acid in the cell walls of both *Thermus* and *Deinococcus.*

Chapter 21. Phototrophic, Chemolithotrophic, and Methylotrophic Eubacteria

PHYLUM 1. PROTEOBACTERIA The chemolithotrophic and photosynthetic members of the Proteobacteria are included in this chapter.

PHYLUM 12. GREEN SULFUR BACTERIA The green sulfur bacteria are anoxygenic photosynthetic bacteria. Some are unicellular forms and others produce networks of cells. None are motile by flagella or gliding motility. Some have gas vacuoles. They use the reductive tricarboxylic acid (TCA) cycle rather than the Calvin-Benson cycle to fix carbon dioxide.

PHYLUM 13. GREEN NONSULFUR BACTERIA AND RELATIVES This group contains the genus *Chloroflexus,* a green gliding bacterium that is very metabolically versatile. Members can grow as heterotrophs or photosynthetically. Carbon dioxide is not fixed by either the Calvin-Benson cycle or the reductive TCA cycle but by a special pathway known only for this group of bacteria.

PHYLUM 14. CYANOBACTERIA, PROCHLORALES, AND THE CHLOROPLAST The group contains the only bacteria that carry out oxygenic photosynthesis. This is a diverse group of bacteria ranging from unicellular to multicellular filamentous and colonial types. Some grow in association with higher plants and animals. All cyanobacteria use the Calvin-Benson cycle for carbon dioxide fixation. The chloroplast found in all eukaryotic photosynthetic organisms evolved from this group of bacteria.

Chapter 22. The Archaea

The Archaea are divided into the following three phylogenetic groups:

PHYLUM 1. METHANOGENS The methanogens (methane-producing) are noted for their ability to produce methane gas from simple carbon sources. Some use carbon dioxide and hydrogen gas, whereas others use methanol or acetic acid. These bacteria are extreme anaerobes that grow at the lowest redox potentials (-0.35 V) of all bacteria. Some of these bacteria fix carbon dioxide, but they use neither the Calvin-Benson cycle nor the reductive TCA cycle.

PHYLUM 2. EXTREME HALOPHILES The extremely halophilic bacteria grow only in saturated salt brine solutions. They lyse if they are placed in distilled water.

It should be noted that there is an overlap between the extreme halophiles and methanogens. Thus, some recent isolates of methanogens grow in high salt environments.

PHYLUM 3. EXTREME THERMOPHILES Some of these organisms grow at temperatures above the boiling point. All can grow at 70°C or higher. Most rely on sulfur metabolism either as an energy source or as an electron sink. For example, some oxidize reduced sulfur compounds aerobically to produce sulfuric acid. Others reduce elemental sulfur and use it as an electron acceptor to form hydrogen sulfide.

Identification

The final area of taxonomy is identification. Bacteriologists are often confronted with determining the species of a newly isolated organism. For example, clinical microbiologists need to know whether a specific pathogenic bacterium is present so they can properly diagnose a disease. Food microbiologists need to determine whether *Salmonella, Listeria,* or other potentially pathogenic bacteria are present in foods. Dairy microbiologists need to keep their important lactic acid bacteria in culture in order to produce a uniform variety of cheese. Brewers and wine makers need to keep their cultures pure in order to inoculate the proper strains for quality control of their fermentations. Analysts at water treatment plants need to make sure that their chlorination treatment is effective in killing coliform bacteria in the treated water and distribution systems. Microbial ecologists need to identify bacteria that are responsible for important processes such as pesticide breakdown and nitrogen fixation.

The process of identification first assumes that the bacterium of interest is one that has already been described and named. This is the usual case for most clinical specimens or specimens from known fermentations. However, microbial ecologists often find that the organism they are interested in is new. It is estimated that less than 1 percent or so of the total numbers of prokaryotic species have been isolated, studied in the laboratory, and named (Chapter 24). Therefore, it is not always possible to identify a bacterium that has been isolated from an environmental sample.

Phenotypic Tests

Phenotypic tests based on readily determined characteristics are often used to identify a species. Most are simple to perform and inexpensive. Furthermore, the amount of time and equipment required for conducting genotypic tests such as DNA/DNA reassociation precludes their use in routine diagnosis. By performing a battery of some 10 to 20 simple phenotypic tests, it is often possible to determine the genus, and perhaps even the species, of a clinically important bacterium, although some taxa are much more difficult to identify than others.

Traditional methods for identification require growing the organism in question in pure culture and performing a number of phenotypic tests. For example, if one wishes to identify a rod-shaped bacterium, the first test would be a Gram stain. If the organism is a gram-negative rod, the next questions to ask include the following: is it motile, is it an obligate aerobe, is it fermentative, does it have catalase, can it grow using acetate as a sole carbon source? The answers to these questions will direct the investigator toward the next step in identification. On the other hand, if it is a gram-positive rod, then a completely different set of tests would need to be performed—such as a test for endospore formation. Regardless of the group, it is also very helpful to determine whole cell fatty acid compositions (see below).

These tests take time to perform, and the appropriate tests for one genus of bacteria differ from those for others. Thus, the results of one set of tests will determine which tests will need to be performed for further clarification of the taxon. Sometimes several weeks might be required to conduct all the tests needed to identify a strain. Rapid tests are extremely helpful, especially in the medical, food, and water-testing areas. Fortunately, there are routine tests that can be performed for most clinically important bacteria that allow for their rapid identification. Several companies have now produced commercial kits that are helpful in assisting the identification of unknowns.

An increasingly popular approach to identification of eubacterial unknowns involves characterization of their fatty acids. The fatty acids are found in membrane lipids and are readily extracted from the cells and analyzed. Different species of Eubacteria produce different types and

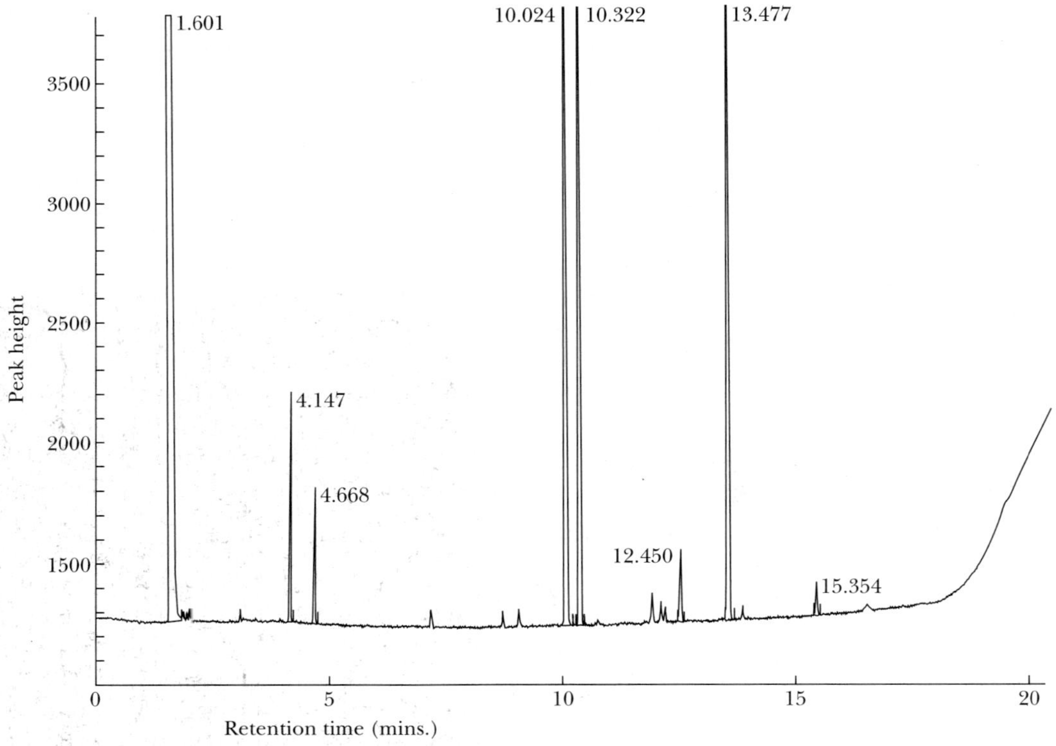

***Figure* 17.15** A fatty acid methyl ester (FAME) chromatogram of an unknown species showing chromatographic column retention times (rt) and peak heights. Major peaks are from left to right: solvent (1.601 rt); 10 : 0 (4.147 rt, a ten-carbon saturated fatty acid); 12 : 0 (4.668 rt, a 12-carbon saturated fatty acid); 16 : 1 omega 7 cis (10.024 rt, a 16-carbon fatty acid that is unsaturated at the omega 7 position), that is, the double bond is between the 7th and 8th carbon atoms from the hydrocarbon (not the carboxyl) end. Cis indicates the hydrogens are in the cis positions at the unsaturated site; 16 : 0 (10.322 rt, a saturated 16-carbon fatty acid); 16 : 0 2OH (12.450 rt, a 16-carbon saturated fatty acid with an hydroxyl group at the second carbon from the omega or hydrocarbon end); 18 : 1 omega 7 cis, omega 9 trans, and omega 12 trans (13.477 rt, this is one or a mixture of three fatty acids with an unsaturated bond at the three positions given, the chromatographic column does not separate these three fatty acids); 19 : 0 cyclo omega 8 (15.354 rt, this is a 19-carbon fatty acid with cyclo-carbon atom at the 8th position from the hydrocarbon end). (Courtesy of MIDI)

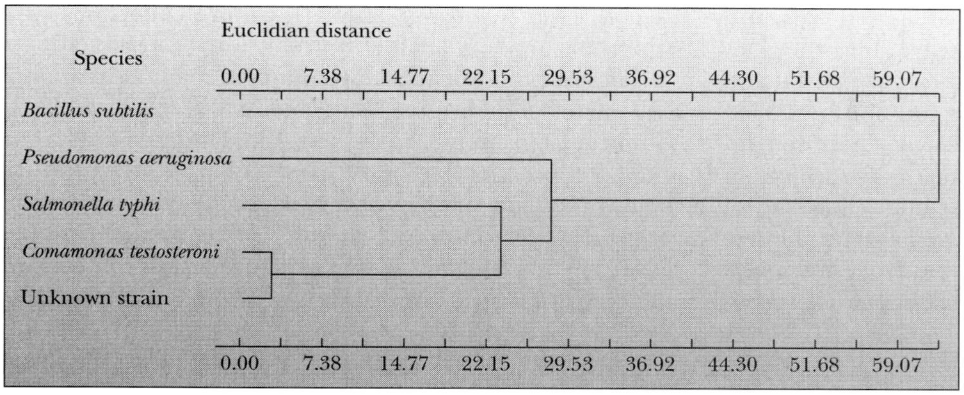

***Figure* 17.16** A dendrogram constructed using a computer program based on the FAME composition of several bacterial species along with the unknown from Figure 17.15. Differences are recorded as Euclidean distances. Note that the unknown is very similar to *Comamonas testosteroni.* The unknown is sufficiently closely related to the library strains of *Comamonas testoteroni* to be considered a member of the same species. Therefore, it is possible to identify many unknown bacteria on the basis of their FAME composition. (With permission from Microbial ID, Inc., Newark, Delaware)

ratios of fatty acids. The Archaea, of course, do not produce fatty acids, so this procedure is not of value for their identification. However, they are not important in disease or in most commercial applications.

The fatty acid analysis procedure involves hydrolyzing a small quantity of cell material (about 40 mg is all that is needed) and saponifying it in sodium hydroxide. This is acidified with HCl in methanol so that the fatty acids can be methylated to form methyl esters. The fatty acid methylated esters (FAME) are then extracted with an organic solvent and injected in a gas chromatograph. The resulting chromatograph (Figure 17.15) can be used to identify the fatty acids that are indicative of a species. Commercial firms have developed databases of fatty acid profiles that can be used for the identification of species (Figure 17.16). The advantage of this procedure is that many samples can be analyzed quickly and without great effort. However, all organisms must be grown under controlled conditions of temperature and length of incubation, and on an equivalent medium.

Nucleic Acid Probes and Fluorescent Antiserum

One exciting area of current research and commercial application involves the development of DNA or RNA "probes" that are specific for the signature sequences of rRNA (Table 17.4) or some other appropriate gene such as an enzyme characteristic of a species of interest. The probes are labeled in some manner so that the hybridization can be visualized. This is accomplished either by making them radioactive or by tagging them to a fluorescent dye or an enzyme that gives a colorimetric reaction. An illustration of this process is shown in Figure 17.17. By the proper selection of probes it is possible to identify an organism to a kingdom or genus or species by demonstrating specific hybridization to the probe.

Potential applications for probe technology are considerable. A number of commercial firms are already marketing probes to identify pathogenic bacteria from clinical samples. A goal of this technology is to enable the rapid identification of organisms directly from clinical or environmental samples without actually growing them in culture first.

Fluorescent antiserum tests are also useful. For example, *Legionella* spp., the causative agents of Legionnaire's disease, are very difficult to cultivate. However, good fluorescent antisera are available for the identification of these species directly from clinical samples or even from environmental samples.

Culture Collections

Unlike higher organisms, most bacteria can be easily grown in pure culture and preserved by **freeze-drying**

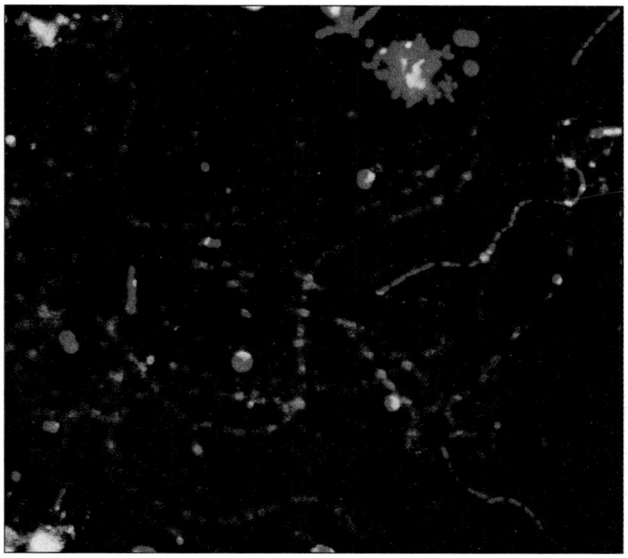

***Figure* 17.17** A light photomicrograph of an activated sludge sample showing a diverse mixture of bacteria. Cells of the beta-Proteobacteria have been stained with a fluorescein-labeled ribosomal RNA probe specific for that group and appear green. Cells that appear red are of the gamma-Proteobacteria, which are distinguished by another ribosomal RNA probe specific for that group that is labeled with rhodamine. (Courtesy of M. Wagner and Karl Schleifer)

(lyophilization), handled in small test tubes and vials, and readily sent anywhere in the world. As lyophils, they can remain viable for 20 years or more and can be revived and studied by anyone anywhere. Cultures can also be frozen at −80°C in vials containing 15% glycerol. These remain viable for many years. Thus, unlike plants for which an herbarium is used to preserve the specimens collected of an original species, bacteria are preserved as **type cultures** that are subclones of the original viable **type strain** of a species. *The type strain of a species is the one on which the species definition has been based.* Through **culture collections** type strains are made available to professional microbiologists throughout the world. Many countries maintain national collections of microorganisms such as the American Type Culture Collection (ATCC) in the United States. For example, a microbiologist from India who wishes to determine if he or she has a new species, can obtain the original type strain from a culture collection and conduct DNA/DNA reassociation assays and other tests to compare the type strain with his or her own isolates.

Summary

- Bacterial **taxonomy** or **systematics** consists of three areas: nomenclature, classification, and identification. **Nomenclature** is the naming of bacteria. An *International Code for the Nomenclature of Bacteria* has been published containing the rules for naming Eubacteria and Archaea.

- **Classification** is the organization of bacteria into groups of similar species. Bacteria are classified in increasing hierarchical rank from species, genus, family, to order. Although higher taxonomic ranks such as class and division exist, they have little meaning at this time.

- **Artificial classifications** are based not on the evolution of organisms but on other features such as cell shape. **Phylogenetic classifications** are based on the evolution of a group of organisms. Bacterial phylogeny is now being developed based on the sequence information from the highly conserved macromolecule, 16S rRNA.

- A **bacterial species** is a group of similar strains that show at least 70 percent DNA/DNA hybridization. Organisms of the same species will have similar if not identical mol % G + C content. Organisms that have the same GC ratio are not necessarily similar.

- **Phylogenetic trees,** with branches and roots, can be constructed based upon the sequence of macromolecules such as rRNA. The length of the branch represents the inferred difference (number of changes) between organisms. **External nodes** represent extant species, whereas **internal nodes** represent ancestor species. **Rooted** trees are based on a comparison of (a) related species and an outgroup species or (b) the sequence of one macromolecule such as 16S rRNA versus an entirely unrelated conserved macromolecule such as ATPase.

- **Identification** is the process whereby unknown cultures can be compared to existing species to determine if they are similar enough to be members of the same species.

- **Type strains** of all species must be deposited in one or more **type culture collections,** repositories where strains are preserved by lyophilization and deep freezing. The culture collections provide cultures to microbiologists from all over the world so they can compare unidentified strains to the official type strains.

Questions for Thought and Review

1. Why is it important to name and classify bacteria?

2. What procedure(s) are necessary to identify a bacterial isolate as a species?

3. Differentiate between an artificial and a phylogenetic classification.

4. In what ways does the classification of bacteria differ from that of eukaryotic organisms?

5. How do the Archaea differ from Eubacteria? From eukaryotes?

6. How is DNA melted and reannealed, and why is this useful in bacterial taxonomy?

7. How would you go about identifying a bacterium that you isolated from a soil habitat?

8. Why is morphology of little use in bacterial classification? Is it of any use?

9. What is *weighting,* and should phenotypic features be weighted in a bacterial classification scheme?

10. Distinguish between lumpers and splitters.

11. If you were working in a clinical laboratory, outline the types of procedures you would use to identify isolates. Why do you recommend using the procedures you suggest?

12. Why is ribosomal RNA of use in bacterial classification?

13. Compare the information obtained from determining the DNA base composition (GC ratio) with that obtained by DNA reassociation experiments.

Suggested Readings

Bergey's Manual of Determinative Bacteriology. 1994. J. G. Holt, Editor-in-Chief. N. R. Krieg, P. H. A. Sheath, J. T. Staley, S. T. Williams (eds). Baltimore: Williams and Wilkins Co.

Gerhardt, P. ed. 1993. *Methods for General and Molecular Microbiology.* Washington, DC: American Society for Microbiology.

Li, W. H., and D. Grauer. 1992. *Fundamentals of Molecular Evolution.* Sunderland, MA: Sinauer Associates, Inc.

Computer Internet Resources

American Type Culture Collection
World Wide Web at: http://www.atcc.org/
This has a listing of all the bacterial strains deposited in the American Type Culture Collection.

Ribosome Database Project (RDP), University of Illinois
World Wide Web at: http://rdp.life.uiuc.edu
This has information on the 16S rRNA sequences of over 4000 bacterial species. The database allows one to conduct phylogenetic analyses of unknown strains whose 16S rRNA sequence has been determined.

Hence without parents, by spontaneous birth,
Rise the first specks of animated earth. . . .
Organic life beneath the shoreless waves
Was born and nursed in ocean's pearly caves;
First, forms minute, unseen by spheric glass,
Move in the mud or pierce the watery mass;
These, as successive generations bloom,
New powers acquire, and large limbs assume;
Whence countless groups of vegetation spring,
And breathing realms of fin and feet and wing.

Erasmus Darwin,[1] from "The Temple of Nature," 1802

Chapter 18

Evolution

Origin of Earth and Life
Microbial Evolution and Biogeochemical Cycles
Evolution of Plants and Animals
Sequence of Major Events During Biological Evolution

Our solar system originated through physical and chemical processes. After Earth formed, organic compounds were produced abiotically, and organisms subsequently originated and evolved. These first organisms were microorganisms, and they had a profound impact on Earth and the formation of its biosphere. Certain bacterial groups played especially crucial roles early on in Earth's development. For example, geochemical and fossil evidence indicates that the production of oxygen in the atmosphere was due to the photosynthetic activity of cyanobacteria. This is of monumental significance because oxygen is essential to all higher plant and animal life forms: they could not have evolved unless microorganisms evolved first. To understand how this happened, we begin with a discussion of the Earth's origin and the evolution of life.

Origin of Earth and Life

The origin of the Earth and the evolution of life on our planet has been a long, slow process. The universe, which is estimated to be 18 billion years old, began with a "Big Bang" that produced two principal elements, hydrogen (^{1}H) and helium (^{4}He) with smaller amounts of ^{2}H, ^{3}He, and ^{3}Li. Following the Big Bang, the universe expanded and continues to do so today. At its periphery the original low atomic weight elements condensed to form clouds of gases of the lighter elements and dust of the heavier elements that evolved from the original lighter gases. Ultimately by gravitational contraction, the dusts accreted to

[1]Erasmus Darwin was the grandfather of Charles Darwin.

form our solar system with the Sun and its planets. Thus, Earth was formed.

Earth and its solar system were formed about 4.5 billion years (4.5 Giga annums, Ga) ago. This date was determined by use of slowly decaying radioactive isotopes whose decay occurs at a constant rate independent of temperature and pressure. The isotope most relied on for dating such ancient events is potassium (^{40}K), which decays to argon (^{40}Ar), with a half-life of 1.26 billion years. Radioisotopic methods are also used for dating strata in sedimentary rocks in which fossils occur as a means of dating the period during which the fossilized life forms lived.

Fossil Evidence of Microorganisms

By the nineteenth century it was known that fossils were the remains or impressions of plants and animals that had been preserved in sedimentary rocks. At that time, accurate dating methods were not available, so the estimated dates were only guesses. It is now known that some of the organisms that became fossilized, such as the extinct dinosaurs, lived millions of years ago on Earth. By examining rocks, paleontologists came to several conclusions regarding the evolution of life. They noted that fossils nearest the surface—that is, in the most recently deposited sedimentary rocks—were structurally more complex than those in deeper layers, indicating that those nearer the surface were more advanced forms of life. Fossils found in deeper strata were increasingly more simple in structure. The deepest, and hence, oldest sedimentary rocks contained fossils of very simple animals such as trilobites. The gradation of simple organisms in the most ancient rocks to more complex forms in the more recent sedimentary rocks argued for an evolutionary process in which more complex forms of plants and animals arose from simple creatures. Geologists and paleontologists worked hand in hand to develop time scales for sedimentary rock deposits and named the various time periods of Earth's history based upon fossil records (Table 18.1).

At the time of Charles Darwin (1809–1882) the fossil record extended back to a period that has now been dated at 600 million years before present. (Rocks older than that contained no plant and animal fossils.) This later became known as the Precambrian era (Table 18.1). In the 1950s, two American scientists, Stanley Tyler and Elso Barghoorn, made a startling discovery. They reported to the scientific world that they had found fossils of microorganisms in sedimentary rocks dated to the Precambrian period (Table 18.2). This important discovery provided the first convincing evidence that the earliest life forms on Earth were microorganisms **(Box 18.1).**

Microbial fossils were discovered in special undulating sedimentary rocks called **stromatolites** (Figure 18.1**a**). Fossil stromatolites are multilayered structures containing calcium carbonate along with the fossils of filamentous microorganisms (Figure 18.1**b**). Living stromatolites still exist on Earth today. The classic columnar-shaped stromatolites occur in intertidal marine areas such as Shark Bay, Western Australia (Figure 18.2). These living stromatolites contain microorganisms that deposit calcium carbonate and

***Table* 18.1 Geological timetable of Earth**

Era	Period	Epoch	Years Before Present (millions)
	Quaternary	Holocene	0.01
		Pleistocene	2.5
Cenozoic	Tertiary	Pliocene	7
		Miocene	25
		Oligocene	38
		Eocene	54
		Paleocene	65
Mesozoic	Cretaceous		145
	Jurassic		190
	Triassic		225
Paleozoic	Permian		280
	Carboniferous		345
	Devonian		395
	Silurian		430
	Ordovician		500
	Cambrian		570
Precambrian			600–4500

Table 18.2 **Precambrian geological timetable of Earth**

Geological Era	Characteristics	Years Before Present
Cambrian	Fossils of higher organisms	0.6 Ga–present
Precambrian	Before higher life forms	
Proterozoic	Onset of oxygen production	2.6–0.6 Ga
Archaean	First sedimentary rocks formed; first stromatolites	3.8–2.6 Ga
Hadean	No sedimentary rocks	4.5–3.8 Ga

other minerals and are therefore responsible for the formation of the successive layers of the stromatolite structure. Other precursors of fossil stromatolites are microbial **mat communities** that occur extensively in intertidal marine environments throughout the world (Figure 18.3**a**). Photosynthetic microorganisms, including cyanobacteria as well as photosynthetic bacteria and heterotrophic microorganisms, are found residing in their respective layers in living stromatolite structures (Figure 18.3**b**). These mat communities produce a flatter, less columnar-shaped struc-

BOX 18.1 MILESTONES

The Discovery of Microbial Fossils

In the early 20th century an American geologist, Charles Doolittle Walcott, was studying Precambrian sedimentary rocks in Glacier Park in Northwestern Montana. He noted that some had curious undulating wavelike structures to them (these are now called stromatolites) and postulated that they were fossilized forms of Precambrian reefs. His theory was doubted by contemporary scientists and remained untested for many years.

American micropaleontologists, Stanley Tyler from the University of Wisconsin and Elso Barghoorn from Harvard University, were the first to test Walcott's hypothesis. They were studying stromatolites taken from one- to two-billion-year-old Precambrian Gunflint Chert deposits from the Great Shield area in the Great Lakes vicinity of North America. When they examined sections of these stromatolites using the light microscope they discovered microbial fossils.

Microbiologists were incredulous when Tyler and Barghoorn first reported their seminal observations in the 1950s to 1960s, as most microbiologists did not believe microbial fossils existed. However, their photomicrographic evidence was so clearcut they convinced a whole generation of skeptical microbiologists. More recently micropaleontologists have discovered microbial fossils in stromatolites that are over 3.5 Ga old, less than one Ga after the origin of Earth!

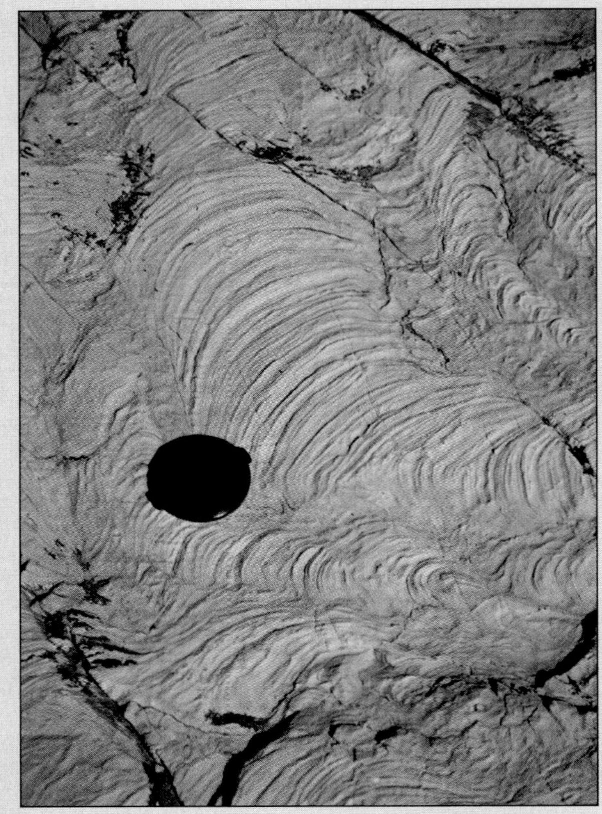

The undulating layers of this sedimentary rock from Glacier National Park are stromatolites that contain fossil microorganisms. Note lens cap as size scale marker. (Courtesy of Beverly Pierson)

(a)

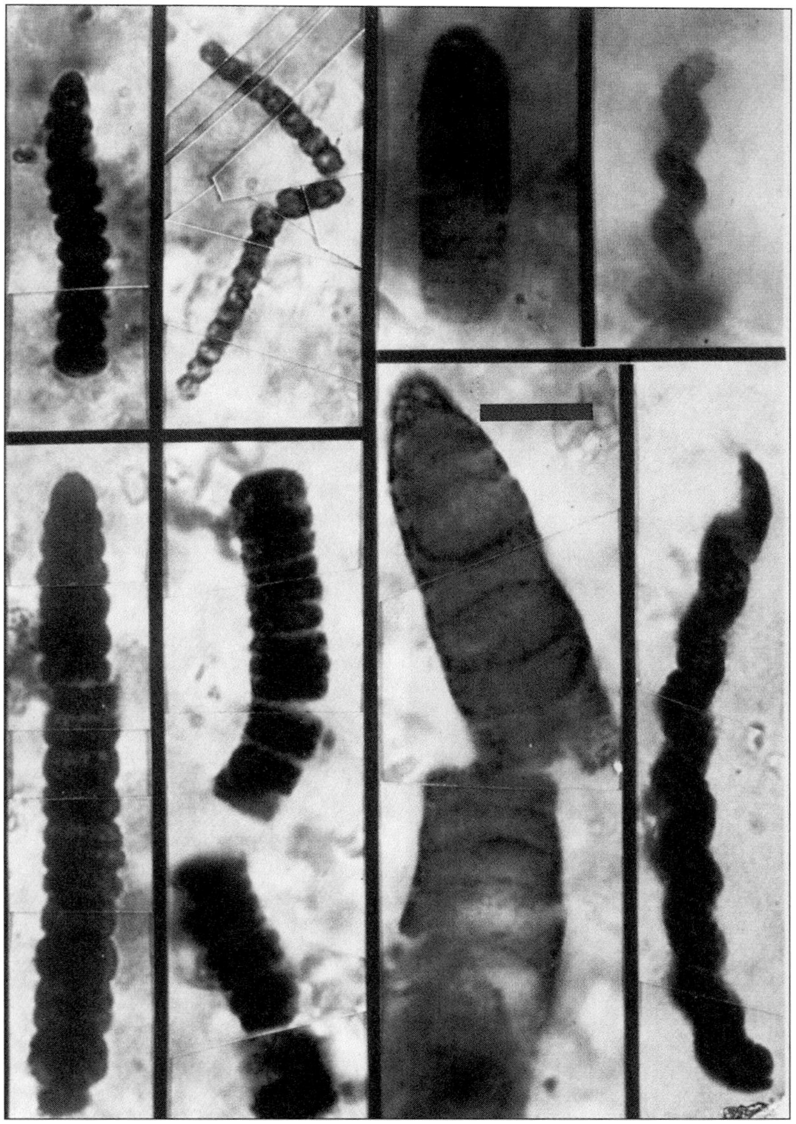

(b)

***Figure* 18.1** **(a)** Fossil columnar stromatolites from Glacier National Park shown in cross section. A U.S. quarter is shown for size comparison. (Courtesy of Beverly Pierson) **(b)** Filamentous microbial fossils observed in sections through 860 million-year-old stromatolites from the Bitter Springs formation in central Australia. Bar is 10 μm. (Courtesy of William Schopf)

***Figure* 18.2** Living columnar stromatolites from Shark Bay, Australia. The size of the largest ones is about one meter in diameter. (Courtesy Beverly Pierson)

(a) (b) (c) (d)

***Figure* 18.3** Microbial mat communities. Although it is not apparent from a distance, this marine intertidal community **(a)** in Massachusetts, called Sippewisset Marsh, contains a microbial mat community. Some areas are sectioned off by ribbons for research purposes. **(b)** Though photosynthetic bacteria are not obvious from examination of the mat community at a distance, the community is easily seen just beneath the surface by cutting through the upper layers of the sand using a razor blade, which is shown to provide a size scale. **(c)** A photograph of a vertical section taken of the mat showing the four layers of photosynthetic microorganisms. The thumb serves as a size marker. The green top layer contains cyanobacteria, the underlying purple layer is of purple sulfur bacteria, the third orange layer is of a different purple sulfur bacterium, and the bottom olive-green layer contains green sulfur bacteria. Each layer is about one mm thick. (**a, b, c** Courtesy of Beverly Pierson). Although the Sippewisset Marsh mat occurs during the summer months, winter storms disrupt it, and it reforms the next summer season. In contrast, some mat communities remain stable for many years such as this one **(d)** from Laguna Mormona (Laguna Figueroa), Baja California del Norte, Mexico, which shows multiple years of buildup and is several cm deep. The green surface layer contains living cyanobacteria. (Courtesy of William Schopf)

ture than the classical stromatolites, but they are produced in similar environments by similar microorganisms. Evidently, during some major geological events these living stromatolites became fossilized and preserved in the sedimentary deposits.

Fossil stromatolites formed by their characteristic, now fossilized microorganisms have been dated at greater than 3.5 Ga before present, and therefore are found in some of the earliest sedimentary deposits on Earth. It therefore appears that microorganisms originated on Earth within a billion years of its formation. In fact, during the two-billion-year period between 3.5 and 1.5 Ga ago, living mat and stromatolite communities were the dominant form of life, which covered vast areas of intertidal zones on the planet.

Although mat communities are common in intertidal areas on Earth today, the classical columnar stromatolites are much more rare. Presumably the evolution of higher life forms led to the development of organisms that could ingest the microorganisms in stromatolite communities, and eventually led to their demise in many areas on Earth.

So, except in special environments such as Shark Bay, which is hypersaline, columnar stromatolites have disappeared.

Origin of Life

Early fossils provide evidence that microbial life existed on Earth within a billion years of its formation, but we still know very little about this early period. Questions such as how did life originate? what were the first forms of life? and what were the conditions on Earth that permitted the origin of life? are important and intriguing, but are very difficult to answer due to a lack of evidence. However, from what we know of life and the early history of the planet, the process can be partially reconstructed. For example, we know that life cannot exist without liquid water. This means that, at the time life originated, the temperature someplace on Earth, must have been between 0°C and 100°C (at atmospheric pressure). Furthermore, we know that the atmosphere was anaerobic. Oxygen could not have formed chemically in any great amount, and certainly not in sufficient quantities to account for its 20 percent makeup of the atmosphere today.

Another precondition for the origin of life is the need for organic compounds. It is inconceivable that cells could have originated *de novo* in the absence of organic compounds, which are part and parcel of all living organisms and biological processes. Thus, an important question is: can organic compounds be produced abiotically, that is, in the absence of organisms? The first experiments to address this question were conducted by Stanley Miller in 1953. He constructed an apparatus involving a flask in which a mixture of gases typifying those thought to be present in the early Earth's atmosphere could interact. The experimental device mimicked the prebiotic biosphere itself (Figure 18.4). The flask contained 500 ml of water representing the "ocean" and an "atmosphere" consisting of a gas mixture of methane, hydrogen, and ammonia. The water was boiled, producing steam, which rose into the atmosphere and mixed with the gases. A condenser was subsequently used to cool the gases again. Miller included as a source of energy a 60,000-volt spark discharge that represented lightning in the atmosphere. The gases and water were recirculated and the process was run continuously. In a matter of a few days of operation of his apparatus that simulated early conditions on Earth, a dark, tarry liquid was produced. This material was analyzed and found to contain, in addition to tarry hydrocarbons, a variety of other organic compounds such as glycine, alanine, lactate, glycolate, acetate, and formate, as well as smaller amounts of other organic compounds. Thus, at least some organic materials could be formed under anaerobic conditions resembling those found in the early history of Earth. This discovery provided support for the so-called **primordial soup** theory of evolution, which holds that organic materials, derived from inorganic precursors, were formed in aquatic areas on earth before organisms originated.

Since Miller's experiments, similar ones have been conducted by other investigators using gas mixtures whose compositions were more like those of the gases that are now thought to more closely resemble the atmosphere of early Earth. These are gases that are now released from the Earth's mantle by volcanoes and are called fumarolic gases. In addition to the gases and water that Stanley Miller used, the fumarolic gases include large amounts of carbon dioxide, nitrogen, sulfur dioxide, and hydrogen sulfide. Ultraviolet light has been successfully used as an alternate energy source for the spark discharge. In addition, the Earth was much warmer then, and the heat from within Earth's crust would have influenced many of these early reactions. In all cases in which conditions were anaerobic, as they were on early Earth, organic compounds similar to those found by Miller were synthesized.

The overall results of these Miller-type experiments indicate that organic compounds can be readily synthesized from inorganic compounds under conditions that resembled Earth's prebiotic environment. Thus, a large variety of organic compounds would have been present early on in this Precambrian primordial soup. Furthermore, these organic compounds would have been present in sufficiently high concentrations that the primordial soup would have resembled a thick bouillon (Figure 18.5). Not all compounds produced in the soup were suitable for early life forms. Over time these would have degraded and diluted as biological activity began to dominate in the production of more compatible biological compounds.

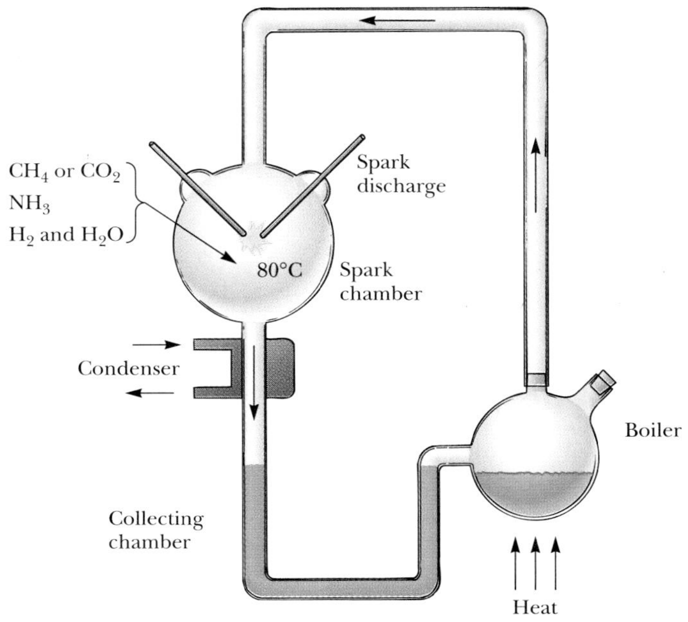

***Figure* 18.4** A diagram of the apparatus Stanley Miller used to produce organic compounds from inorganic sources. The gas mixture contained hydrogen, methane, and ammonia.

***Figure* 18.5** An illustration showing possible scene of early Precambrian Earth. The atmosphere lacked oxygen but contained gases produced by volcanic eruptions. Conditions would have been favorable for abiotic synthesis of organic compounds.

The most difficult questions that still remain unanswered are, how is it that the first biological entity originated, and what was it? We can only speculate on the answers to these questions. A number of scientists are interested in this problem, and perhaps someday it will be resolved. However, it is beyond the scope of this book to delve into the various possibilities that have been proposed. This much we do know. The first biological entity must have been a progenitor of microbial life. One popular current theory is that the first protoorganisms were based on RNA. A number of arguments can be made in support of an "RNA World," as discussed in **Box 18.2.**

BOX 18.2 RESEARCH HIGHLIGHTS

RNA World

Proponents of an "RNA World" believe that RNA was the first macromolecule to evolve. A number of arguments favor an early form of "life" based on RNA. For example, RNA, like DNA, can store genetic information as it does in RNA viruses. Furthermore, unlike DNA, some RNA molecules called **ribozymes** carry out biochemical reactions like enzymes. For example, the enzyme RNase P contains an active site that has an RNA moiety which actually cleaves transfer RNA. In addition, it is the ribosomal RNA, not protein, that is responsible for the formation of peptide bonds in protein synthesis. Indeed, RNA is essential to protein synthesis and therefore probably existed before protein.

Proponents of an RNA world visualize a simple protoorganism that could carry out primitive metabolic activities. It contained RNA in its cytoplasm that was capable of catalytic and templating (reproductive) activities. A simple, probably inorganic, barrier separated it from the environment. Later on, proteins evolved that were synthesized from the RNA. Still later, lipid membranes evolved. Ultimately DNA itself evolved as a means of storing genetic information. Increasingly, the concept of an RNA world is being accepted by biologists as a logical step toward the evolution of cellular organisms.

Microbial Evolution and Biogeochemical Cycles

Although we do not know which organisms were the first biological entities on Earth, various theories have been presented. The Russian evolutionist A. I. Oparin argued that it was likely a simple heterotrophic bacterium. He reasoned that autotrophic organisms are inherently more complex and therefore would not have evolved first. Although autotrophs can live on simple nutrients, he believed they were more complex from an evolutionary standpoint—reasoning that they would not only need to have pathways for generation of energy, but, in addition, they would need pathways to carry out carbon dioxide fixation. In contrast, simple fermentative heterotrophs require only a few enzymes for ATP generation. Therefore, they could have used organic compounds that were formed abiotically, early in Earth's history.

Some contemporary scientists such as Carl Woese, at the University of Illinois, have argued that the first forms of life were probably anaerobic and thermophilic because the deepest branches in the Eubacteria and the Archaea are thermophilic prokaryotes (see conversation with Carl Woese). Furthermore, strong evidence indicates the early Earth was not only anaerobic, but very warm and volcanically active. Woese calls the progenitor of microbial life the Progenote, the prototype precellular "organism" that gave rise to both the Eubacteria and the Archaea, and ultimately the Eucarya as well. It is likely that the Progenote would have had an ability to concentrate essential chemicals and to carry out simple reactions, and it perhaps contained RNA but no proteins or DNA.

Wolfram Zillig, a German biochemist, has proposed that the Progenote populations must have separated physically, possibly geographically, into two communities early in Earth's history and that this separation led to the evolution of the two main lines of descent—the Eubacteria with ester-linked lipids in their cell membranes and the Archaea with ether-linked lipids. (For further discussion of Zillig's ideas, see later and Figure 18.7.) However, all theories on the origin of life, and microorganisms in particular, are speculative and will not be addressed in great detail here.

Possible Early Eubacteria and Archaea

An example of a possible early type of heterotrophic metabolism is that of the lactic acid bacteria. These Eubacteria metabolize anaerobically, have no requirement for oxygen, and generate energy from the glycolytic pathway, which requires only a few enzymes (see Chapters 8 and 20). They do not require electron transport systems with ATPases for energy generation, but instead obtain energy entirely from substrate-level phosphorylation (Chapter 8). Likewise, they are nonmotile and therefore would not need to produce flagella or to generate proton gradients through chemiosmosis. Of course, they require enzymes for some anabolic processes, and they need a cell membrane. It should be emphasized here that present-day lactic acid bacteria are much more complex than their early ancestors. It is likely that the metabolic characteristics of their metabolism would have been harbored in some primitive, noncellular entity that lacked DNA and an ester-type cell membrane.

Another possible early life form could have been an Archaea whose metabolism resembled that of the methanogens. These organisms have simple nutritional requirements. Some grow autotrophically, generating energy from the oxidation of hydrogen gas and using carbon dioxide as a sole source of carbon (Chapter 22). Furthermore, they are all anaerobic and could have existed in an environment like that of the early Earth.

Other early autotrophic forms might have been the hydrogen bacteria that obtain energy from the oxidation of hydrogen gas. Both eubacterial and archaeal hydrogen bacteria are known. Interestingly, all contemporary thermophilic hydrogen bacteria require oxygen but apparently only in very low concentrations.

Although photosynthetic bacteria may not have been the first organisms, it is believed that they evolved early. It is widely accepted that the ability to carry out photosynthesis using chlorophyll-type compounds evolved shortly after the split between the Eubacteria and the Archaea. Only one group of the Archaea, the extreme halophiles, have photosynthetic species, and their type of photosynthesis uses a rhodopsin-type compound rather than a chlorophyll. The first photosynthetic organisms may have resembled the eubacterial purple or green sulfur groups in their metabolism. These bacteria carry out photosynthesis anaerobically using hydrogen sulfide as a reducing source for carbon dioxide fixation (see Chapters 9 and 21) by these overall unbalanced reactions:

$$\text{(a)}\quad CO_2 + H_2S \longrightarrow (CH_2O)_n + S^0$$

$$\text{(b)}\quad CO_2 + S^0 \longrightarrow (CH_2O)_n + SO_4^{=}$$

where $(CH_2O)_n$ represents organic material.

The volcanic atmosphere of early Earth would have been ideal for these organisms in that it provided abundant quantities of carbon dioxide and hydrogen sulfide, the essential ingredients for their photosynthesis. This type of photosynthesis is termed **anoxygenic photosynthesis** because it proceeds in an anaerobic environment without the production of oxygen.

There is considerable evidence that microorganisms have had major impacts on Earth's biosphere since its origin and the origin of life. The best evidence of this comes from information about the production of oxygen on Earth.

Cyanobacteria and the Production of Oxygen

To understand the importance of cyanobacteria in the production of oxygen, we must first review their metabolism. The metabolism of the cyanobacteria is similar to that of the anoxygenic photosynthetic bacteria discussed previously (see Reaction (a) above). However, there is one major difference: cyanobacteria use water in place of hydrogen sulfide as the hydrogen donor. Thus, the overall formula for their type of photosynthesis is as follows:

$$\text{(c)}\ CO_2 + H_2O \longrightarrow (CH_2O)n + O_2$$

This type of photosynthesis is termed **oxygenic photosynthesis** because oxygen is produced in the process. C. B. van Niel, an American microbiologist who studied photosynthetic bacteria, noted that the O_2 produced in Reaction (c) must be derived from the water molecule rather than from the carbon dioxide in the substrates of Reaction (c). He concluded this based upon analogy to Reaction (a) above for the anoxygenic photosynthetic bacteria. His theory was confirmed when scientists used radiolabeled water, $H_2{}^{18}O$, to show that the label ended up in the oxygen produced ($^{18}O_2$). Thus, the oxygen comes from a reaction referred to as the "water splitting" reaction. This reaction, which is the key reaction of oxygenic photosynthesis, is found in all algae and in plants as well.

It is thought that the cyanobacteria evolved at least 2.5 to 3.0 Ga ago. At this time the Earth's atmosphere still lacked oxygen. But by 2.5 to 1.5 Ga ago, oxygen had been produced. This is known because it is at that time that the **banded iron formations** were formed in the oceans. The bands on this formation are alternating millimeter-thick layers of quartz and iron oxides. These formations contain partially oxidized forms of iron (FeO and Fe_2O_3) that could only have been produced in the presence of oxygen. However, banded iron formations are not formed today because the concentration of the oxygen in the atmosphere and in the oceans is too high. Instead, contemporary iron deposits are called **red beds,** so named because they contain hematites (Fe_3O_4), a more highly oxidized form of iron which gives them their red color. Therefore, it is thought that the banded iron formations were produced during the period in which oxygen was first formed but before significant concentrations were free in the atmosphere, that is, between about 2.5 and 1.7 Ga. The source of the oxygen needed to oxidize the iron is believed to have come from oxygenic photosynthesis first carried out by cyanobacteria.

Initially, when the cyanobacteria first began producing oxygen on Earth, its concentration would have remained very low. This is because it is highly reactive chemically and would have combined with the large amounts of highly reduced compounds that existed on Earth at the time. These reduced compounds, such as ferrous iron and sulfides, would have reacted with free oxygen, thereby preventing oxygen from accumulating rapidly in the atmosphere. Therefore, the oxygen concentration in the atmosphere increased very gradually over the last 2 to 3 billion years to attain its present-day level of about 20 percent.

One of the recent exciting discoveries about cyanobacteria is that some of them can also carry out anoxygenic photosynthesis according to Reaction (a) (see also Chapter 21). This finding suggests that the cyanobacteria may have evolved from anoxygenic photosynthetic bacteria similar to purple or green sulfur bacteria. However, some variant must have evolved over millions of years that could use water in place of hydrogen sulfide as a reductant in photosynthesis and therefore could split water and carry out oxygenic photosynthesis (Reaction (c)). This important process may have evolved by natural selection as a means of replacing scarce supplies of hydrogen sulfide with abundant supplies of water for photosynthesis.

The oxygen produced by cyanobacteria would have been toxic to early life forms. Fortunately, the environment was very highly reduced so that free oxygen would not have been available in the atmosphere for many millions of years after the first oxygenic cyanobacterium produced it. This lengthy time provided favorable conditions for the selection and evolution of peroxidases which would protect sensitive bacteria from the oxidizing effects.

Precambrian Nitrogen Cycle

Another important biogeochemical event was the evolution of the nitrogen cycle. Biological nitrogen fixation is regarded as an early process in the evolution of the Earth's biosphere. There are several reasons why it is thought to have evolved early. First of all, it is a process that is unique to prokaryotic organisms, including species of both the Eubacteria and Archaea, which are believed to have preceded eukaryotes. Although only prokaryotes carry out this process, some nitrogen-fixing bacteria and cyanobacteria grow in association with higher plants such as the legumes (Chapter 25).

Another reason it is thought that nitrogen fixation was an early process is that it is commonly found in Eubacteria and Archaea that are thought to have evolved early. For example, virtually all purple and green sulfur bacteria are able to fix atmospheric nitrogen. Likewise, many methanogens and cyanobacteria are able to carry out this process. In addition, nitrogen fixation occurs best anaerobically, a condition of the early environment on Earth. It seems likely that it could have evolved as a means of providing nitrogen to organisms when fixed forms of nitrogen such as nitrate and ammonia were absent or available only in limiting concentrations. The atmosphere has always contained nitrogen gas in large quantities.

It is well known that bacteria can readily transfer genes from one species to another through various processes such as transformation, conjugation, and transduction (see Chapter 15). Such transfer between different species is referred to as horizontal gene transfer, and many examples of this type of transfer exist, such as the transfer of antibiotic resistance genes through plasmids. Some scientists have argued that the genes involved in nitrogen fixation (*nif* genes) evolved early, and therefore passed to descendants through vertical gene transfer. In contrast, others believe they evolved recently and have been transferred from one species to another quickly through horizontal gene transfer. To test this hypothesis, experiments were performed to examine the *nif* H (nitrogenase) protein from various Eubacteria with their 16S rRNA sequence. The results indicate that the same pattern of evolution exists for both features (Figure 18.6). This evidence supports an early evolution of *nif* H. If it had evolved only more recently and been transferred from one organism to another by horizontal gene transfer, the *nif* H proteins from different species should be very similar to one another and therefore show a different pattern of evolution from the 16S rRNA sequence. However, this is not the case. These data support an independent evolution of this feature in each of the descendants that can be traced to an ancient common ancestor. Indeed, nitrogenase probably evolved prior to the separation of the Protoeubacterial and Protoarchaeal lineages.

Prior to oxygen production by cyanobacteria, oxidized forms of nitrogen were rare. All nitrifying bacteria require oxygen for the oxidation of ammonia and nitrite, and therefore had probably not evolved in the anaerobic Precambrian period. Thus, it is doubtful that the process of nitrification existed in the early history of the Earth. Furthermore, without nitrate formed by nitrifiers, denitrification could not occur. Therefore, the nitrogen cycle may have been very abbreviated in the Precambrian (Figure 18.7), initially consisting primarily of the process of conversion of amino-nitrogen of organisms to ammonia by ammonification and the reverse. Subsequently, the process of nitrogen fixation may have evolved due to the unavailability of adequate amounts of ammonia for the increasing biomass present on Earth. The ample supplies of nitrogen gas in the atmosphere since prebiotic times could have provided conditions for the selection and evolution of nitrogen-fixing bacteria that could use this previously unused source of nitrogen. Even after nitrogen fixation evolved, the hypothetical, truncated Precambrian nitrogen cycle, based on these considerations, would have been considerably simpler than the contemporary nitrogen cycle (Chapter 24).

Molecular Evolution of Eubacteria and Archaea

As discussed in Chapter 17, molecular approaches are now used to study the evolution of prokaryotic organisms as well as eukaryotic organisms. In particular, sequencing of the 16S rRNA has become the favored method for reconstructing the evolution of bacteria. The various trees that have been constructed by sequence analyses provide some clues as to what the early forms of bacterial life were like (see Figure 17.13). As more organisms are isolated in pure culture and analyzed, a better picture of the early forms of life will be constructed. But based upon previous

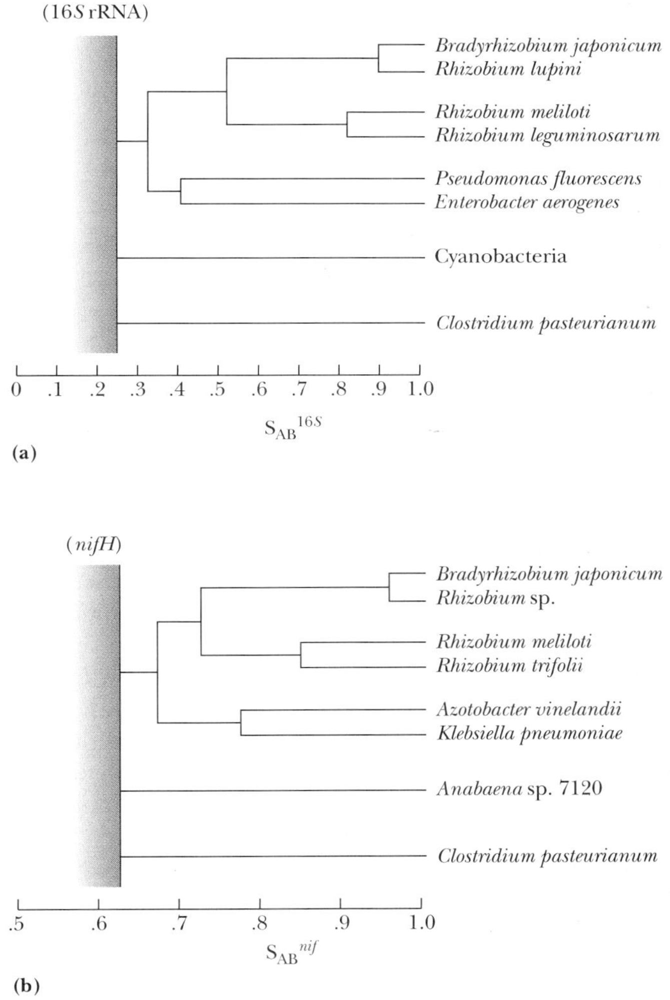

***Figure* 18.6** Trees constructed from similarity coefficients of **(a)** 16S rRNA sequences of various nitrogen-fixing species, and **(b)** amino acid sequences of the *nif* H gene from species from these same phylogenetic groups. Note that the patterns produced are superimposable on one another. (From H. Hennecke et al., *Archives of Microbiology* 142:342–348 [1985])

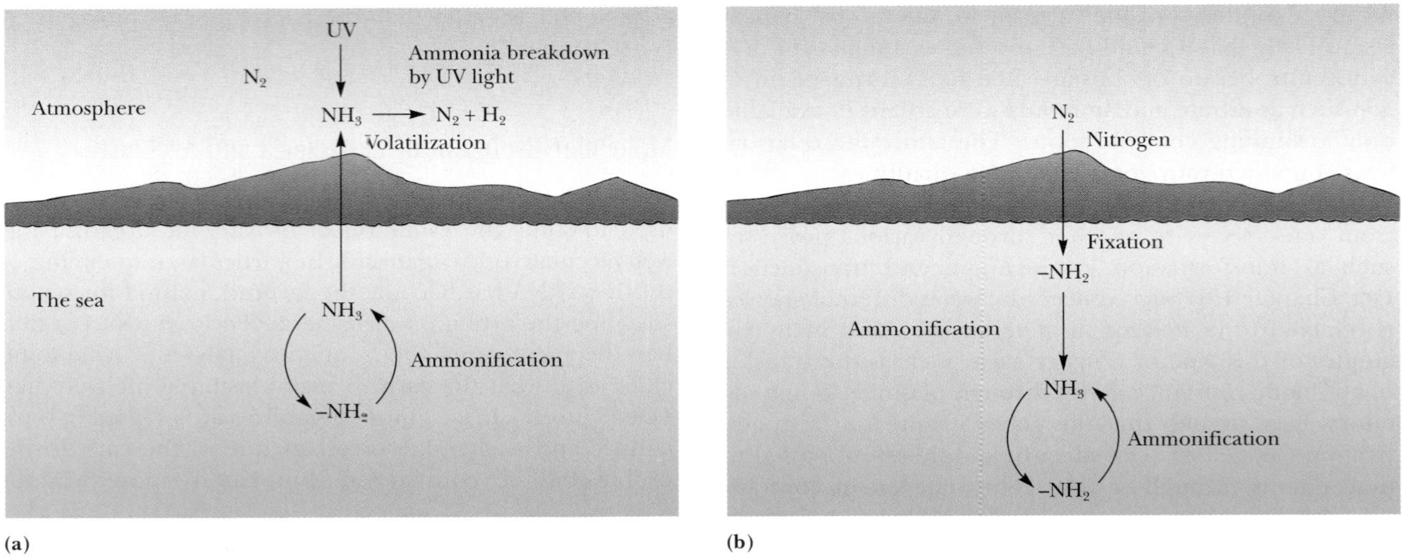

***Figure* 18.7** A simplified nitrogen cycle that may have been characteristic of the early Precambrian. Initially the predominant forms of nitrogen **(a)** would have been nitrogen gas (N_2) and ammonia, both from volcanic emissions from Earth. Ammonia entering the atmosphere would have been broken down to hydrogen and N_2 by ultraviolet radiation. Later on **(b)** nitrogen fixation would have arisen to provide additional nitrogen when ammonia became scarce. Compare these cycles with the contemporary one shown in Chapter 24.

discussions, it seems probable that early forms of life were thermophilic and anaerobic.

Evolution of Plants and Animals

The actual origin of eukaryotic organisms is obscure at this time. The most popular hypotheses about eukaryotic evolution stem from the ideas of scientists such as Wolfram Zillig. He has proposed that eukaryotic organisms, the Eucarya, evolved through a fusion event that occurred between an ancestor of the Eubacteria and an ancestor of the Archaea (Figure 18.8). According to this theory the Progenote had a permeable cell membrane (possibly protein) that allowed genetic communication with other species. As evolution proceeded, some event such as geographic isolation led to the separation of two different populations—the Preeubacteria and the Prearchaea. During this period of separation, the two groups of organisms developed different metabolic patterns and genetic systems. Also at this time, the organisms evolved their own characteristic lipid cell membranes, ester-linked fatty acid lipids for the Preeubacteria, and ether-linked isopranyl lipids for the Prearchaea. According to Zillig, it was at about this time that the fusion event occurred between these two cell types, giving rise to the eukaryotic organisms.

Similarities that Eucarya share with Eubacteria and Archaea are cited as evidence supporting this hypothesis. For example, the cell membranes of Eucarya and Eubacteria contain fatty acids linked to glycerol by ester linkages. In contrast, Archaea do not produce long-chain fatty acids and use ether linkages in their isopranyl cell membranes (see Chapter 11, Figures 11.6 and 11.7, and Chapter 22). This evidence suggests that, if a fusion event occurred, it occurred at the time the eubacterial ancestor had its cell membrane and therefore this was incorporated into the Eucarya.

Other evidence supporting this hypothesis comes from comparisons of metabolic and genetic features (see Table 17.3). Thus, the DNA sequence of cytoplasmic enzymes of Eucarya such as glyceraldehyde-3-phosphate dehydrogenase (GAPDH), L-malate dehydrogenase, and 3-phosphogylcerate kinase are very similar to those of Eubacteria, and differ considerably from those of the Archaea. ATP synthase (ATPase) sequences of Archaea closely resemble those of Eucarya. Furthermore, the genetic machinery of Archaea more closely resembles that of the Eucarya. For example, the two RNA polymerases pol2 and pol3 of Archaea are quite similar to those of the Eucarya. In addition, the amino acid sequences of ribosomal proteins, standard promoters, and translation factors all show a strong resemblance to those of Eucarya.

Although much of this is speculative, it is interesting to note that as scientists further dissect organisms at the molecular level, they are beginning to infer likely, if not actual, evolutionary events that occurred billions of years ago.

Impact of Oxygen on the Evolution of Plants and Animals

The evolution of oxygenic photosynthetic organisms had a profound impact, not only on the chemistry of the Earth, but on the evolution of all higher forms of animal and plant life. Virtually all plants and animals use oxygen and carry out aerobic respiration, a process that occurs in the subcellular organelle, the mitochondrion. Mitochondria are not found in Eubacteria or Archaea. However, *the mitochondrion is eubacterial.* By this we mean that: (a) it has its own DNA, but its DNA is not bound by a nuclear membrane, and (b) it has ribosomes, but the ribosomes are not the larger 80S ribosomes of eukaryotes, but the smaller ribosomes of prokaryotes. Indeed, its very size and structure are reminiscent of a gram-negative bacterium without peptidoglycan. Furthermore, sequence analyses of mitochondrial 16S rRNA place its ancestor within the Eubacteria in the Proteobacterial line of descent (Figures 18.8 and 18.9).

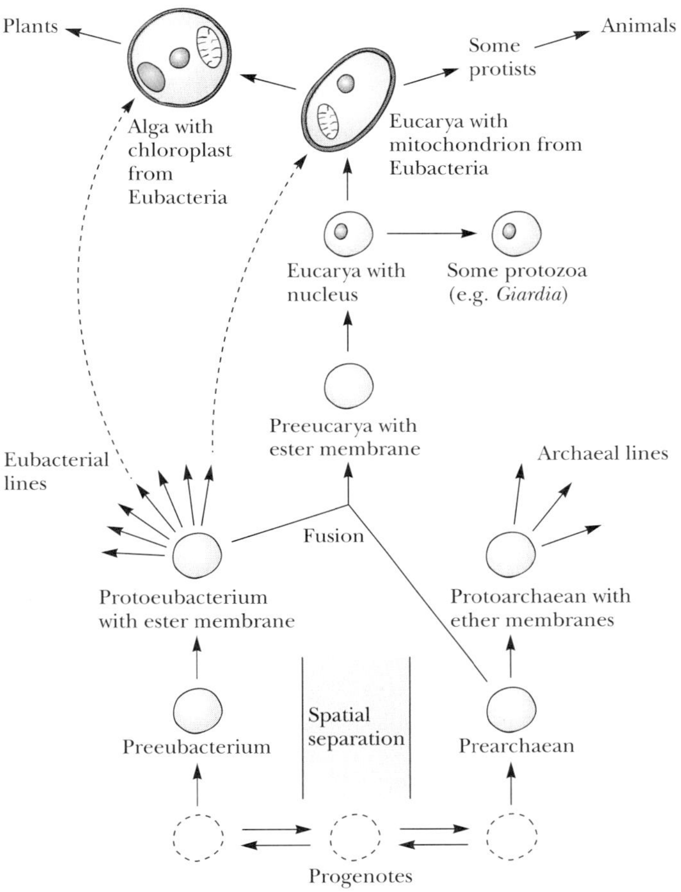

***Figure* 18.8** Evolution of the main lines of descent. This diagram illustrates the possible origin of prokaryotic and eukaryotic organisms from a Progenote. Early on the Progenotes, possibly precellular RNA forms without organic cell membranes, were separated probably geographically from one another. This separation led to the independent evolution of Preeubacteria and Prearchaea with their characteristic cell membrane structures. As suggested by Wolfram Zillig, a fusion event between a Prearchaean and a Preeubacterium could have given rise to a Preeukaryote. Subsequently, the mitochondrion would have developed a symbiosis between some Eubacteria and early protist in the evolutionary line toward plants and animals. However, some protozoal lineages such as *Giardia* spp. which lack mitochondria now, probably evolved before this symbiosis occurred. The chloroplast, also from Eubacteria, probably evolved through a mutualistic symbiosis with a eukaryote to give rise to algae and later on higher plants.

Similarly, the chloroplast, the photosynthetic organelle of algae and higher plants, bears a striking resemblance to that of another prokaryotic group, the cyanobacteria. Thus, the chloroplast, like the mitochondrion, is prokaryotic. It, too, has DNA but no nuclear membrane. And, it, too, has 70S ribosomes. And finally, like the mitochondrion, it lacks peptidoglycan though it is descended from one of the eubacterial phyla. Sequence analysis of the 16S ribosomal RNA places the chloroplast within the cyanobacteria branch of the Eubacteria (Figures 18.8 and 18.9).

Endosymbiotic Evolution

The theory of **endosymbiotic evolution** has been developed to explain the origin of mitochondria and chloroplasts. This theory states that early in the evolution of eukaryotic organisms, certain prokaryotic organisms (the premitochondrion and prechloroplast-type organisms) developed intracellular symbioses with eukaryotic cells. As time proceeded, the partners in these symbioses became more and more interdependent and certain bacterial structures and functions (such as the peptidoglycan layer) were lost.

It is also interesting to note that some Eucarya, such as the protozoan, *Giardia lamblia,* lack mitochondria—suggesting that they evolved before the mitochondrion was incorporated into eukaryotic organisms (Figure 18.8). Nonetheless, the lipid membranes of *Giardia* and its glyceraldehyde-3 phosphate dehydrogenase (GAPDH) are both very similar to that of Eubacteria, suggesting that, if the fusion hypothesis of Zillig is correct, the fusion event occurred before the mitochondrion was incorporated into eukaryotes.

By 16S rRNA sequencing of mitochondrial ribosomal DNA, it has been possible to construct the evolution of the mitochondrion (Figure 18.8). Regardless of the eukaryotic source that has been analyzed, all are descended from the alpha line of Proteobacteria.

Further support for the endosymbiotic theory is found in intracellular bacterial associations with protozoa. For example, certain protozoa harbor bacterial cells as "parasites" that provide unique features to their hosts (Figure 18.10). One type of association is the so-called "killer paramecium." This strain of *Paramecium* contains a eubacterium called a kappa particle. The kappa particle lives inside the protozoan and is responsible for the production of an organic compound that kills other paramecia that do not harbor the kappa particle. These kappa particles still contain their peptidoglycan layer. Because of this the symbiosis is not as highly evolved as that of the mitochondrion.

Another example occurs in certain flagellate and amoeboid protozoans that have photosynthetic organelles called **cyanellae.** The cyanellae appear to be cyanobacte-

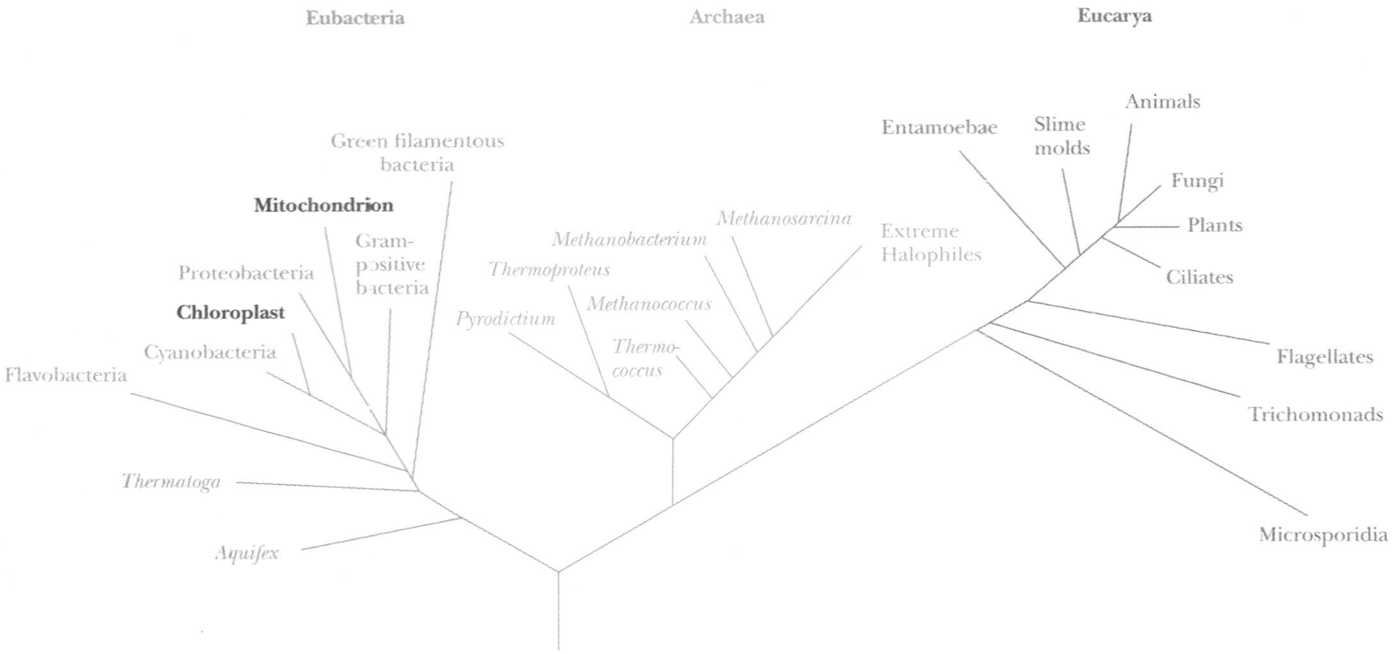

***Figure* 18.9** Diagram showing the evolution of various biological groups based on 16S rRNA sequence analysis and showing also the ancestral relationships of mitochondria and chloroplasts.

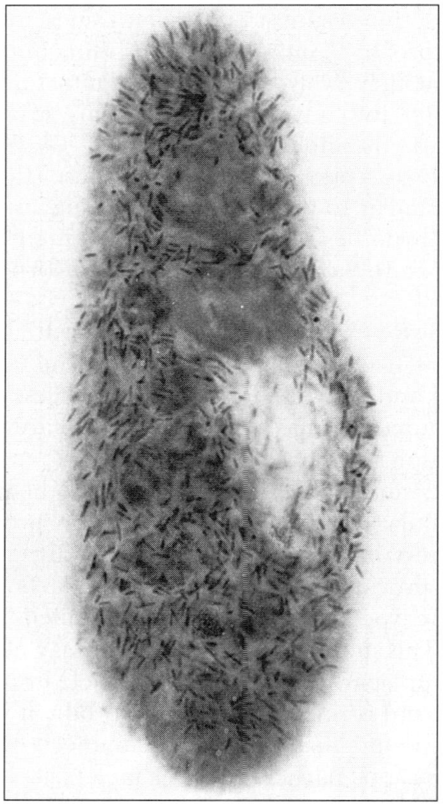

***Figure* 18.10** A *Paramecium* containing numerous bacteria called "lambda particles" as endosymbionts. (Courtesy of John Preer Jr., Louise Preer, and Artur Jurand, Indiana University)

ria replete with all their eubacterial features, including peptidoglycan (see Chapter 23).

Clearly plants and animals, through evolution, have obtained many of their genes from microorganisms. In this manner genes have been available from a vast pool of different types of organisms. The macroorganism can therefore be viewed, at least in part, as a chimera of different genes, some coming directly from microorganisms and others having been modified through mutation and selection.

Origin of the Ozone Layer

Crucial to the evolution of higher life forms was the development of the ozone layer in Earth's stratosphere. Ozone is produced by a photochemical oxidation reaction of oxygen in the upper stratosphere. Thus, it is believed to have first formed as a consequence of oxygen production by early cyanobacterial photosynthesis. This layer is important in that it strongly absorbs ultraviolet light, which is a very active oxidizing agent and mutagen. Some aquatic forms of life could have evolved and lived largely unaffected by UV radiation prior to the formation of an ozone layer because water also strongly absorbs UV radiation. However, higher forms of terrestrial life would have been unprotected from solar UV and could not have evolved until the ozone shield was established.

It is clear that oxygen and the development of the ozone layer had a profound impact on the evolution of plants and animals. What is especially intriguing is the diversity and complexity of multicellular eukaryotic life forms that have all evolved during the past 600 million years, particularly in terrestrial environments. In contrast, prokaryotes, which have had about 3.5 Ga to evolve, have rather simple structures. Despite their simple and largely unicellular morphologies, the genetic diversity of prokaryotes is vast as illustrated by the broad range in mole % G + C of their DNA (see Figure 17.5). Whereas plants and animals have ranges of about 20 percent in their G + C content, the bacterial range is greater than 50 percent, extending from a low of about 22 percent to a high of about 75%. The large diversity of prokaryotes has occurred with respect to their metabolic processes, which encompass a variety of anaerobic, autotrophic, aerobic, and heterotrophic activities, few of which occur in higher living organisms. In contrast, eukaryotes, despite their great morphological diversity, have rather limited metabolic capabilities.

Sequence of Major Events During Biological Evolution

The timetable shown in Figure 18.11 portrays, to the best of our current knowledge, the probable major sequential events during biological evolution. The initial composition of the atmosphere was determined in large part by volcanic emissions. From these anaerobic gases and water, ultraviolet light led to the production of organic compounds. Within a billion years of the Earth's formation, the first microorganisms appeared. These were likely thermophilic, anaerobic bacteria that were able to use organic compounds as energy sources.

Anoxygenic photosynthetic bacteria had likely evolved by 3.0 to 3.5 Ga ago. These bacteria may have been the first organisms to fix nitrogen gas. These communities of organisms are thought to have colonized all aquatic habi-

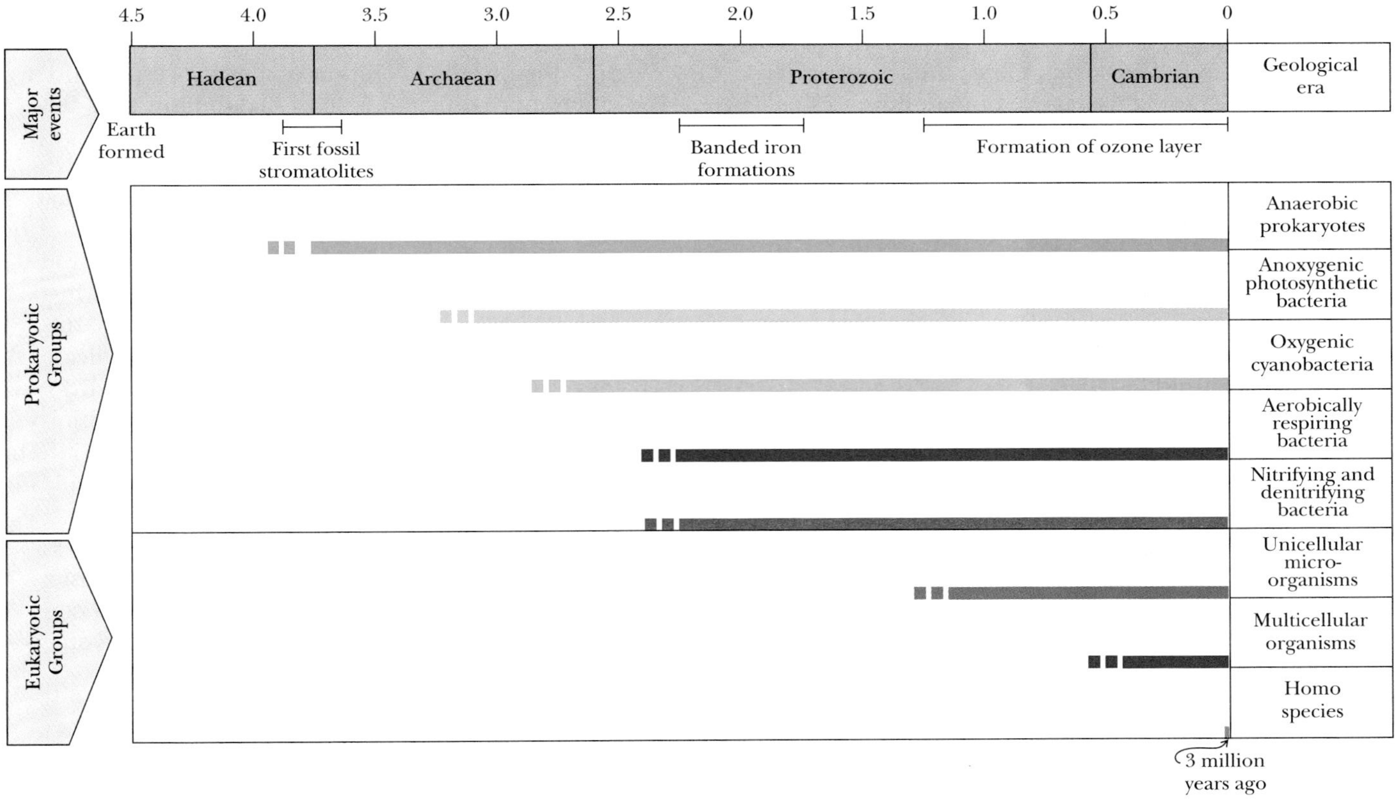

***Figure* 18.11** A timetable showing the major geological events, beginning with the formation of Earth 4.5 Ga before present, the first evidence of stromatolites, the period of banded iron formation, and the development of the ozone layer following the appearance of oxygen in the atmosphere. The evolution of various eubacterial, archaeal, and eukaryotic groups is also shown. Note that human species have occupied Earth only a minute fraction of Earth's geological time, namely 3 million years, which is less than a thousandth of the time since cellular life forms arose.

tats favorable for life. Photosynthesis would have had a major impact on the biosphere of the Earth. By producing large amounts of organic carbon, the growth of heterotrophic organisms would have been greatly enhanced. At the same time, methanogenic Archaea must have evolved using carbon dioxide as a carbon source and releasing large amounts of methane into the atmosphere.

The cyanobacterial branch of photosynthetic bacteria probably first appeared between 2.5 and 3.0 Ga ago. The production of oxygen would have had a major impact on many different processes in the biosphere. First, it would have caused the oxidation of reduced inorganic compounds such as iron sulfides, which were present in enormous amounts early on. In addition, it would have been a poison to the anaerobic bacteria that had already evolved. Furthermore, it would have adversely impacted processes such as nitrogen fixation, which occur best under reducing conditions. Moreover, it would have led to the evolution of aerobic, respiratory bacteria, which have a more efficient mode of metabolism. And, of course, it would have provided the key conditions for the evolution of higher forms of life that had obtained their mitochondria from Proteobacteria and chloroplasts from the cyanobacteria.

Another consequence of oxygenic photosynthesis was the development of an ozone layer, which was conducive to the evolution of land plants and animals because it protects from ultraviolet radiation. Aquatic organisms would have been less impacted by ultraviolet light because water strongly absorbs its radiation.

It is interesting, from the standpoint of human evolution, that the genus *Homo* dates back to only about 3 million years ago, and the species *Homo sapiens,* a mere 100,000 years or so ago. Humans are therefore very recent participants in the biosphere, and very dependent on the other processes and organisms that occur on Earth.

Summary

- The formation of Earth and its solar system occurred by physical and chemical processes about 4.5 Ga ago. Microbial fossils were first reported by micropaleontologists only recently (1950s). The oldest microbial fossils are found in stromatolites that have been dated to 3.5 to 3.8 Ga ago. **Stromatolites** are fossils of laminated sedimentary rock produced by microorganisms. Intertidal mat communities are the most common type of living stromatolites currently found on Earth.

- Many organic molecules can be formed abiotically from inorganic compounds in an anaerobic environment if there is an energy source such as ultraviolet light or lightning.

- The **RNA world** theory holds that RNA was the first macromolecule and served both as a biochemical catalyst and as a template for reproduction in primitive "organisms."

- The cyanobacteria are thought to be the first oxygen-producing **(oxygenic) photosynthetic** organisms. The oxygen produced by cyanobacteria resulted in the formation of the banded iron formations as well as the protective ozone layer, and led to the high concentration of oxygen in the atmosphere. Through natural selection, oxygen in the atmosphere led to the evolution of aerobic respiration, a much more efficient type of metabolism than fermentation and anaerobic respiration.

- The theory of **endosymbiotic evolution** holds that mitochondria and chloroplasts and possibly other organelles of higher animals and plants are derived from prokaryotic organisms that developed symbioses with higher organisms. Evidence from 16S rRNA sequence analysis of mitochondria indicates they evolved from the alpha Proteobacteria. Evidence from 16S rRNA sequence analysis of chloroplasts indicates they evolved from cyanobacteria. The **cyanelle** is a cyanobacterium symbiont that resides in some protozoa.

- Higher organisms may have evolved from a fusion event between Preeubacteria and Prearchaea coupled with endosymbioses events with prokaryotic organisms.

Questions for Thought and Review

1. Describe what you believe to have been the first true bacterium on Earth. What were its properties and why?
2. What are the strengths and weaknesses of the prebiotic soup theory of evolution?
3. Compare and contrast oxygenic and anoxygenic photosynthesis.
4. How do you explain that some eukaryotes do not have mitochondria?

5. As discussed in earlier chapters, D-amino acids occur only rarely in nature. Some are found in the cell walls and capsules of Eubacteria. What does this mean in terms of evolution?

6. Are humans more like Eubacteria or Archaea? Why?

7. Do you believe it would be possible to place chloroplasts in human skin cells? Would humans become photosynthetic? Would they make us become "little green people"?

8. Describe the biosphere on Earth before the evolution of cyanobacteria.

9. How much information can be obtained by the discovery and dating of microbial fossils? Can the type and metabolic activity of the microorganism be determined?

10. Is there life on Mars? On the Moon? Explain your answer.

Suggested Readings

Butcher, S. S., R. J. Charlson, G. H. Orians, and G. V. Wolfe, eds. 1992. *Global Geochemical Cycles.* New York: Academic Press.

Edey, M. A., and D. C. Johanson. 1989. *Blueprints. Solving the Mystery of Evolution.* Boston: Little, Brown, and Co.

Schopf, J. W. 1983. *Earth's Earliest Biosphere.* Princeton, NJ: Princeton University Press.

Schopf, J. W. 1978. "Evolution of Life." *Scientific American.* 239:110–138.

Schleifer, K. H., and E. Stackebrandt. 1985. *Evolution of Prokaryotes.* Orlando, FL: Academic Press.

Woese, C. R. 1987. "Bacterial Evolution." *Microbiological Reviews* 51:221–271.

The extraordinary diversity...among the bacteria is shown by a simple enumeration of the properties which they can possess. They can be unicellular, multicellular, or coenocytic; permanently immotile, or motile by any one of three distinct mechanisms; able to reproduce by binary fission, by budding, or by the formation of special reproductive cells, such as the conidia of actinomycetes; photosynthetic or non-photosynthetic.

R. Y. Stanier, *The Bacteria,* 1964

Chapter 19

Gram-Negative Heterotrophic Eubacteria

Proteobacteria (Heterotrophic Members)
Other Heterotrophic Eubacterial Groups

Several different phylogenetic groups of gram-negative Eubacteria have been identified by 16S rRNA analysis, each representing a separate evolutionary line (see Figure 17.13). The heterotrophic members of each of these phylogenetic groups will be discussed individually in this chapter (Table 19.1). One of these groups, the Proteobacteria, also contains chemoautotrophic and photosynthetic taxa that will be discussed in Chapter 21.

There is considerable morphological and physiological diversity among the heterotrophic gram-negative Eubacteria. Most are unicellular rods, vibrios, and cocci that divide by binary transverse fission. However, some have life cycles with two or more types of cells that divide by budding, and one group, the myxobacteria, have a complex developmental cycle culminating in the formation of an elaborate fruiting structure. Many are flagellated, but others move by gliding motility or gas vacuoles. Stalks are produced by some. Some are obligate aerobes, but others are facultative or obligate anaerobes. Each of the major phylogenetic groups is treated individually in the sections below.

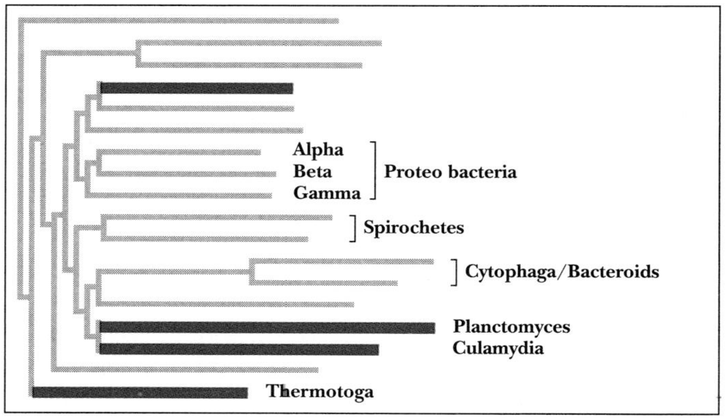

Proteobacteria (Heterotrophic Members)

The *Proteobacteria,* also called the "purple bacteria" because of its photosynthetic species, comprises the largest and

Reference *Table* **19.1 • Major gram-negative eubacterial phylogenetic groups treated in Chapter 19**

Groups and Subgroups	Page Number
Proteobacteria (heterotrophic members)[1]	
Fermentative Rods and Vibrios	(p.431)
Enteric bacteria	(p.432)
Vibrio and related genera	(p.437)
Pasteurella and *Hemophilus*	(p.438)
Zymomonas	(p.439)
Chromobacterium	(p.439)
Oxidative Rods and Cocci	
Pseudomonads	(p.439)
Azotobacter and other free-living, nitrogen-fixing bacteria	(p.442)
Rhizobium and related genera	(p.444)
Agrobacterium	(p.445)
Vinegar Bacteria: *Acetobacter* and *Gluconobacter*	(p.446)
Legionella	(p.447)
Neisseria and related genera	(p.447)
Rickettsia	(p.448)
Sheathed bacteria	(p.449)
Prosthecate bacteria	(p.450)
Bdellovibrio: The bacterial predator	(p.459)
Spirilla and *Magnetospirillum*	(p.460)
Myxobacteria: Fruiting, gliding bacteria	(p.461)
Sulfate- and sulfur-reducing bacteria	(p.464)
Other Heterotrophic Eubacterial Groups	
Cytophaga* and *Bacteroides	
Cytophaga group	(p.466)
Other heterotrophic gliding bacteria	(p.467)
Bacteroides and related taxa	(p.468)
Planctomycetales	(p.469)
Chlamydia	(p.473)
Spirochetes	(p.473)
Thermophilic Gram-Negative Eubacteria	(p.475)

[1]See Chapter 20 for chemolithotrophic and photosynthetic members.

most heterogeneous phylogenetic group of Eubacteria. Photosynthetic, chemoautotrophic, and heterotrophic bacteria are all found within this single phylum, indicating a complex and extensive evolutionary history. The current evolutionary chart, based on 16S rRNA analyses, indicates there are five major subgroups of Proteobacteria, designated as the **alpha, beta, gamma, delta,** and **epsilon** subdivisions. The most important genera of each of the subdivisions of Proteobacteria are listed in the Appendix (Appendix 2). Many species of common, well-known heterotrophic bacteria are included in these groups. As mentioned previously, *all of the autotrophic members of the Proteobacteria, both chemoautotrophic and photosynthetic, will be treated in Chapter 21.* In the next sections, the most important genera of the heterotrophic Proteobacteria are discussed.

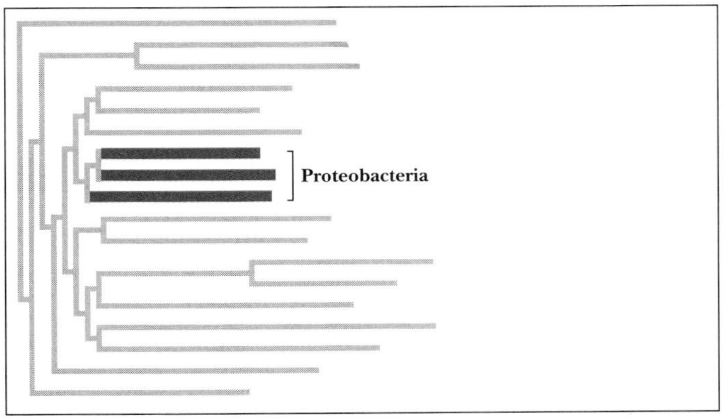

Fermentative Rods and Vibrios

Many of the most important members of the Proteobacteria are fermentative. None of these is obligately anaerobic, but instead grow as facultative aerobes. Several important groups are known, all of which are members of the gamma group of the Proteobacteria.

Enteric Bacteria

The enteric bacteria or enterobacteria are small, nonsporeforming rods typically about 0.5 μm in width and from 1 μm to 5 μm in length. In general, they have simple nutritional requirements. Some, however, require vitamins and/or amino acids for growth. Most of the bacteria in this group are motile, and if so, they have peritrichous flagella. Enteric bacteria are facultative aerobes that, under anaerobic conditions, ferment glucose to form an array of end products. They are catalase positive, oxidase negative, and most reduce nitrate to nitrite when oxygen availability is limited. The DNA from the various species exhibit considerable homology based on hybridization tests (Figure 19.1). These organisms are metabolically and genetically similar but demonstrate considerable diversity in ecology and pathogenic potential for humans and other vertebrate animals (cold and warm blooded), insects, and plants. The intimate association of many enterics with higher eukaryotes suggests a long evolutionary relationship. Major genera in the enteric group and their habitats are shown in Table 19.2.

The enteric bacteria have had considerable influence on human history (see Chapter 2). Their medical importance is evident, as members of this group are the causative agents of diseases such as plague (*Yersinia pestis*), typhoid fever (*Salmonella typhi*), and bacillary dysentery (*Shigella dysenteriae*). Those that are not pathogens can cause disease under appropriate conditions and are therefore called **opportunistic** or **secondary pathogens.** Because they are so firmly associated with humans, a huge number of enterobacteria have been isolated and characterized. For example, 1464 different serogroups have been defined as serovar subspecies in the genus *Salmonella.* The considerable medical and epidemiological importance of the enterics is discussed in Chapters 29 and 31.

The enteric bacteria are separated into two physiological groups based on the products they generate during glucose fermentation. The **mixed acid fermenters** (Figure 19.2) produce significant amounts of organic acids. In contrast, the **2,3-butanediol fermenters** (Figure 19.3), produce mostly neutral compounds. *Escherichia coli* is a mixed acid fermenter. It metabolizes glucose via the Embden-Meyerhof pathway to pyruvate. The reduced NADH generated by glyceraldehyde 3-phosphate dehydrogenase is reoxidized with the formation of three major acids: acetic, lactic, and succinic acids (Figure 19.3). Formic acid is catabolized by the enzyme formic hydrogenlyase to CO_2 and H_2 in a ratio of one to one (Table 19.3). Fermentation of 100 mol of glucose by *E. coli* leads to the formation of about 128 mol of organic acids.

Butanediol fermentation results in the formation of neutral products and considerable quantities of CO_2 (see Table 19.3). *Enterobacter aerogenes* is a typical butanediol-fermenting enteric bacterium. Formic acid is cleaved to form $H_2 + CO_2$, but the significantly higher proportion of CO_2/H_2 occurs because other reactions produce CO_2 but not hydrogen. Only about 20 mol of acidic product are generated from 100 mol of glucose. A distinctive and therefore diagnostically important intermediate called **acetoin** is formed in this fermentation (Figure 19.3).

The precise identification of enterics is of considerable importance in public health microbiology and epidemiology. Consequently, a variety of diagnostic tests (see

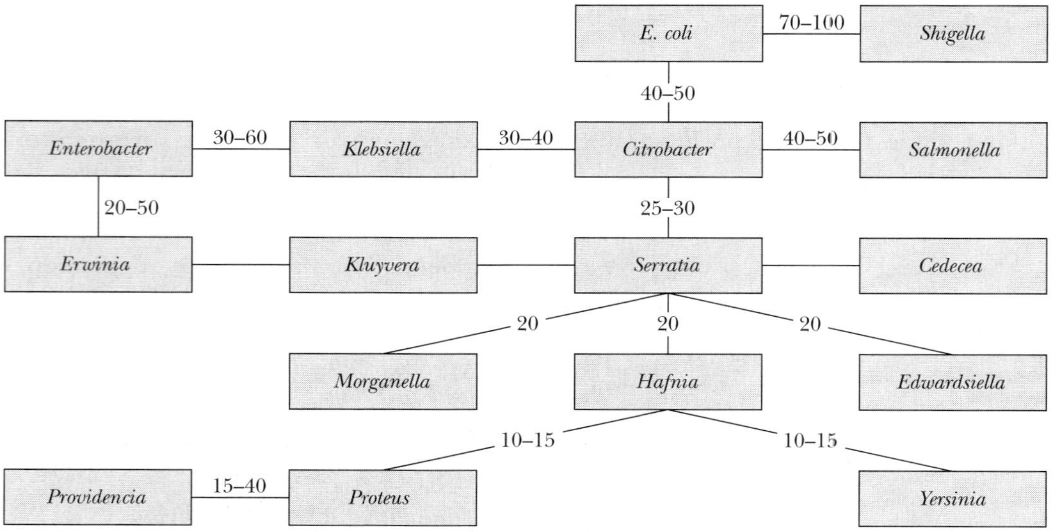

***Figure* 19.1** Relatedness among the enteric bacteria as shown by DNA/DNA reassociation.

Table **19.2 Normal habitat of the enteric bacteria**

Genus	Habitat
Escherichia	Normal intestinal tract of warm-blooded animals
Salmonella, *Shigella*, and *Providencia*	Intestinal pathogens of humans and other primates
Hafnia	Feces of humans, other animals and birds
Edwardsiella	Cold-blooded animals; may be pathogenic for eels and catfish
Proteus	Intestinal tract of humans and other animals; also soil and polluted water
Morganella	Mammal and reptile feces
Yersinia	Humans, rats, and other animals
Klebsiella, *Citrobacter*, *Enterobacter*, and *Serratia*	Human intestine; also soil and water
Erwinia	Plants as pathogenic, saprophytic, or epiphytic microflora

Table 19.4 for examples) have been devised to identify genera in this group (Table 19.5, on page 436). Identification of a strain from a clinical specimen entails a large variety of diagnostic tests. Typically, the clinical microbiologist performs these tests routinely, along with whole-cell fatty acid methyl ester analysis (see Chapter 17) using prepackaged automated or semi-automated formats that are available from commercial manufacturing firms. The results of these tests can be subjected to computer analysis to find the best-fit species identification for the unknown strain.

Escherichia coli, the best known species of the enteric bacteria and arguably the most widely known of all the

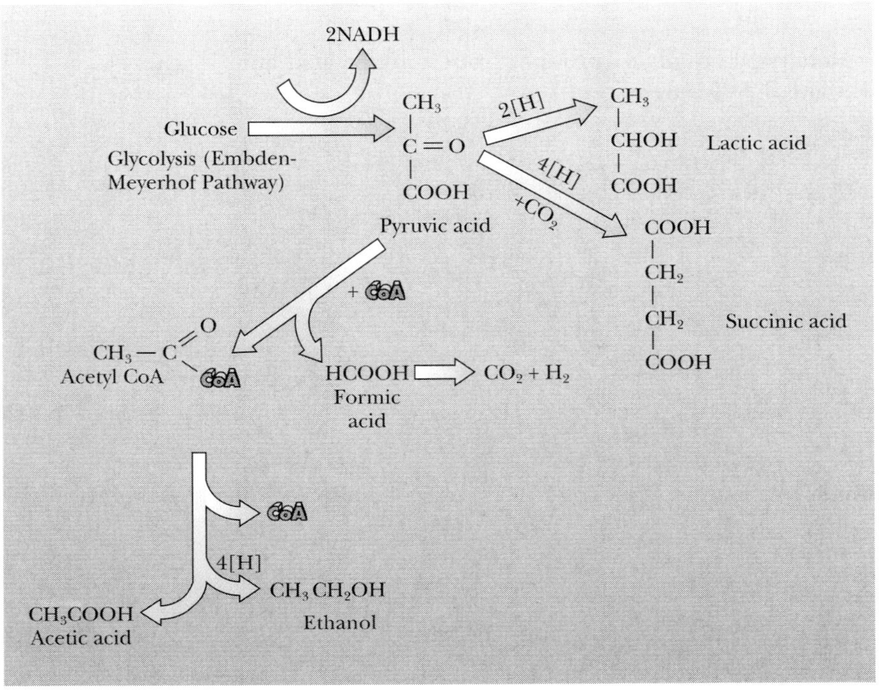

***Figure* 19.2** Mixed acid fermentation of glucose by *Escherichia coli.* Glucose is catabolized by the Embden Meyerhof pathway of glycolysis to yield pyruvate. Pyruvic acid is converted to lactic acid, succinic acid, acetic acid, ethanol, formic acid, carbon dioxide, and hydrogen gas.

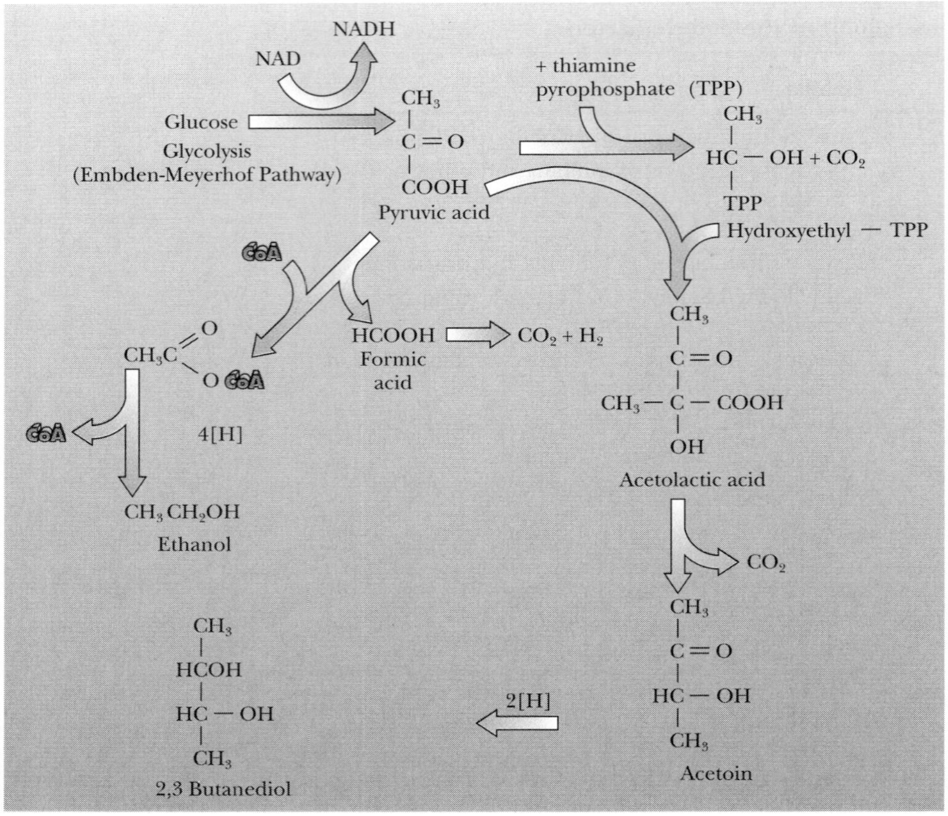

***Figure* 19.3** Butanediol variation of the mixed acid fermentation in which more neutral products are formed as well as a higher ratio of CO_2 to H_2.

bacteria, has been more thoroughly studied than any other living organism. This common eubacterium is a normal inhabitant of the intestinal tract of humans and most other warm-blooded animals. As a facultative aerobe of the intestinal tract, it is well equipped to survive anaerobically and consume any oxygen that enters this environmental niche. This activity is essential for maintaining anaerobic conditions in the large intestine. Although it is an important species in the human intestinal tract, it is not the dominant one. Many other, but not all, bacteria in this group

Differential ***Table* 19.3 • Relative amounts of product from a mixed-acid and a neutral (Butanediol) fermentation**

	Moles of Product per 100 Moles of Glucose Fermented	
	Mixed Acid Fermentation *(Escherichia coli)*	**Neutral Fermentation** *(Enterobacter aerogenes)*
Acetic acid	36	0.5
2, 3-Butanediol	0	66
Ethanol	50	70
Lactic acid	79	3
Succinic acid	11	0
Formic acid	2.5	17
H_2	75	35
CO_2	88	172
Total moles acid	128.5	20.5
Ratio CO_2/H_2 produced	1.2	4.9

Descriptive ***Table*** **19.4 • Types of tests used to differentiate enteric bacteria from one another**

Test	Description
Urease	Streak the organism on urea agar base. Urea agar contains high levels of urea (2.0%) and phenol red indicator. Organism producing urease liberates ammonia that turns indicator red near streak.
Indole	Grow organism in a peptone medium with high tryptophan content. Test for indole production after growth.
Motility	Tubes contain tryptose in a soft agar medium (0.5% agar). Inoculate by stabbing down the center of the agar. Diffuse growth out from the inoculum indicates motility.
Methyl red	A buffered glucose-peptone medium containing methyl red indicator, turns red if large amounts of acid are produced as in mixed acid fermentation.
Acetoin (Voges-Proskauer)	A liquid medium containing glucose is inoculated and after growth, assayed for acetoin.
Citrate	Organism is tested for its ability to grow with citrate as sole carbon source.
H_2S production	Inoculate agar medium containing ferrous sulfate by stabbing. Blackening of the agar indicates sulfide production.
Phenylalanine deaminase	After growth on a medium containing phenylalanine (0.1%), assay for production of phenylpyruvic acid by adding ferric chloride. A green color indicates deaminase activity.
Ornithine decarboxylase	Measured manometerically by following CO_2 release.
Gas from glucose	Gas production from glucose or other sugars is measured by placing a Durham tube in a broth containing sugar. A Durham tube is a small test tube that is placed, inverted, in the broth before autoclaving. Autoclaving causes the tube to be filled. Gas production is indicated by a gas bubble within the Durham tube.
β-galactosidase	Growing organism on a lactose medium. Add a disc containing O-nitrophenyl-β-D-galactoside to the surface. After 15–20 min the yellow O-nitrophenol is released if the enzyme is present.

also live in the intestinal tracts of animals, hence accounting for the general name of "enteric bacteria" for these organisms.

Because *Escherichia coli* is found in the intestinal tract of warm-blooded animals, it is an important test organism for fecal contamination of food and drinking water. As *E. coli* does not grow or survive for long periods in food, water, or soil, its presence in these environments is indicative that they have been contaminated by fairly recent fecal material and therefore, should not be consumed. Thus, *E. coli* is termed an **indicator bacterium** whose presence in the environment indicates fecal contamination from warm-blooded animals (see Chapter 33).

Shigella dysenteriae is the causative agent of **bacillary dysentery,** a severe type of gastroenteritis. The species of this genus are so closely related to *Escherichia coli,* as determined by DNA/DNA reassociation, that they could be considered the same species. However, for historical reasons and because of their distinctive pathogenesis, a separate genus has been maintained.

Differential ***Table*** **19.5 • Characteristics used to distinguish the common genera of enteric bacteria**[1]

Genera	Methyl-Red	Voges-Proskauer	Citrate Utilization	Urease	Indole Production	Motility	H_2S Production	Gas from Glucose	β-Galactosidase	Ornithine Decarboxylase	Phenylalanine Deaminase	mol % G + C
Mixed-Acid Fermenters												
Escherichia	+	−	−	−	+	+	−	+	+	+	−	48–52
Salmonella	+	−	V	−	−	+	+	+	V	+	−	50–53
Shigella	+	−	−	−	V	−	−	−	V	V	−	49–53
Edwardsiella	+	−	−	−	+	+	−	+	−	+	−	53–59
Citrobacter	+	−	+	+	V	+	−	+	+	+	−	50–52
Proteus	+	V	V	+	V	+	V	+	−	V	+	38–40
Morganella	+	−	−	+	+	+	−	+	−	+	+	50
Providencia	+	−	+	V	+	+	−	+	−	−	+	39–42
Yersinia	+	−	−	+	−	−	−	−	+	V	−	46–50
2, 3-Butanediol Producers												
Klebsiella	+	+	+	V	V	−	−	+	+	−	−	56–58
Enterobacter	V	+	+	−	−	+	−	+	+	+	−	52–60
Serratia	V	+	+	−	−	+	−	V	+	−	+	52–60
Erwinia	+	+	−	−	−	+	+	−	V	−	−	50–58
Hafnia	V	V	−	−	−	+	−	+	+	+	−	48–49

[1]Symbols: + = positive for most strains; − = negative for most strains; V = variable within genus.

Salmonella typhi causes **typhoid fever** and gastroenteritis. As mentioned earlier, there are hundreds of serovars in this genus. The antigens for the serovars are determined from three types of surface polymers including the outer cell membrane lipopolysaccharides ("O" antigens), the flagella ("H" antigens), and the outer layer polysaccharides ("Vi" antigens). These serovars are helpful in tracing the source of organisms during an outbreak of typhoid fever.

Another important pathogenic genus is *Yersinia,* which contains the dreaded agent of **bubonic plague,** *Yersinia pestis.* During the fourteenth century, in a scourge called the black death, bubonic plague spread throughout Europe resulting in the death of over 25 percent of the population, a proportion much greater than caused by any war.

Klebsiella pneumoniae can cause a type of bacterial pneumonia as well as urinary tract infections, but it is not normally pathogenic. As a soil and freshwater group, this genus is noted for its lack of motility, its formation of a capsule, and the ability of many strains to fix nitrogen. In other respects it is very similar to the common soil bacterium, *Enterobacter aerogenes.*

The genus *Proteus* is well known for its active **urease,** which breaks down urea to ammonia and carbon dioxide.

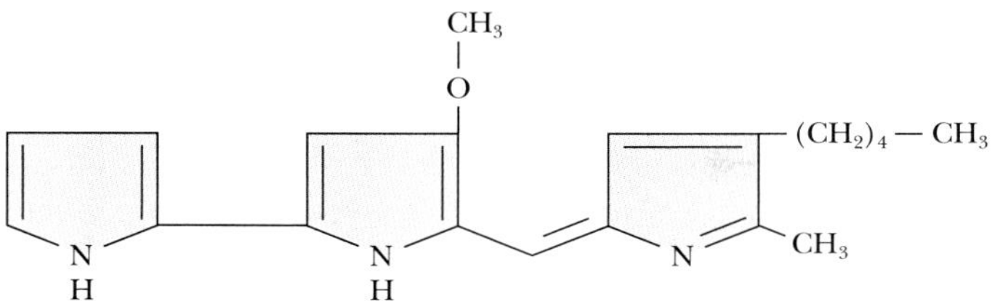

***Figure* 19.4** Chemical structure of prodigiosin, the bright red pigment from *Serratia marcescens.*

It is also noted for its active motility, called **swarming,** which results in colonies rapidly moving over the surface of an agar plate.

Serratia marcescens is a distinctive enteric bacterium because of its red pigment. The pigment of this bacterium is a bright red, pyrrole-based compound called **prodigiosin.** Prodigiosin is a tripyrrole (Figure 19.4) akin to the tetrapyrroles of chlorophyll, cytochrome, and heme. The function of prodigiosin in *S. marcescens* is still unknown.

One of the most recent isolates of the enteric group is *Xenorhabdus,* which grows in association with nematodes. Some strains can actually produce light, a phenomenon referred to as **bioluminescence.** This fascinating capability is more widely found in the *Vibrio-Photobacterium* group discussed below. Bioluminescent bacteria are usually found in symbiotic associations and they are therefore discussed in more detail in Chapter 25.

Vibrio and Related Genera

Vibrio, Photobacterium, Aeromonas, and *Plesiomonas* are gram-negative facultatively anaerobic straight or curved rods (Table 19.6). All are motile via polar flagella. All are capable of respiration or fermentation. None can use nitrate as the terminal electron acceptor in respiration. Most are oxidase positive and utilize glucose as the sole source of carbon and energy. These four genera are primarily aquatic organisms; those that live in seawater require 2 percent to 3 percent NaCl or a seawater-based medium for optimum growth. Some species are pathogenic to humans and other vertebrates. Some are pathogenic to fish, frogs, and certain invertebrates.

Certain species of *Photobacterium* and *Vibrio* are bioluminescent and emit a blue-green light. It is interesting to note that this activity is of no known value to the bacterium except through its symbiosis with fish—in which the bacteria live as symbionts in their light organs, special areas on the fish where these bacteria are maintained in high concentrations (see Chapter 25).

The polar flagellum of *Vibrio* species is enclosed in a sheath. The sheath is continuous with the outer membrane of this gram-negative bacterium. Some species also produce lateral flagella when they are grown on solid media. These lack a sheath and have a shorter wavelength than the polar flagellum. Such flagellation, which is unique to this group, is termed **mixed flagellation.** Apparently, synthesis of the lateral flagella occurs when the microenvironment is particularly viscous, such as on an

Differential ***Table* 19.6 • Characteristics of genera of the *Vibrio* group**

Genus	Morphology (size μm)	Luminescence	Mol % G + C	Extracellular Enzymes	Habitat
Vibrio	Straight or curved rods (0.5–0.8 × 1.4–2.6)	+ or −	38–51	+	Aquatic, intestine of marine animals
Photobacterium	Rods (0.8–1.3 × 1.8–2.4)	+	40–44	+	Marine environment as symbionts in luminous fish organs
Aeromonas	Rods (0.3–1.0 × 1.0–3.5)	−	58–62	+	Freshwater, marine, sewage, and animal pathogens
Plesiomonas	Rods (0.8–1.0 × 3.0)	−	51	−	Fish, aquatic animals, some mammals

agar surface. In this manner the additional flagella aid in the cell's movement.

Vibrio species live primarily in marine and estuarine environments **(Box 19.1)**. Many live in the intestines or on the outer surfaces of marine animals. Almost all grow best with added sodium ions in the medium and therefore are typically grown in media with marine salts. Most species grow on a variety of organic compounds (sugars, amino acids, and organic acids) and do not require growth factors. Extracellular hydrolases are produced by many species and include amylase, lipase, chitinase, and alginase. Also some are agar digesters, that produce depressions or cavities on the surface of agar media where their colonies grow. One species, *Vibrio natriegens,* has the shortest doubling time reported for any bacterium (9.8 min at 37°C).

Vibrio cholerae is the most thoroughly studied species in the group because it is the causative agent of **cholera.** Cholera is an epidemic human disease that occurs in India and more recently in Peru and other underdeveloped countries. It is transmitted by fecal contamination of water and food. Effective water treatment has eliminated the disease from most developed countries.

Photobacterium species, like *Vibrio,* are widespread in marine habitats and are also found associated with fish. The luminescent species, *P. phosphorum,* can be isolated by incubating marine fish, squid, or octopus partially submerged in seawater at 10°C to 15°C. After 15 h to 20 h, luminescent areas are observed and streaked on plates for purification.

Aeromonas species live in both fresh and marine waters as well as sewage. *Aeromonas salmonicida* is a strict parasite that causes severe diseases of salmon and trout and can also infect humans. It occurs in the blood and kidneys of fish.

Plesiomonas shigelloides causes a disease similar to shigellosis, a type of bacterial dysentery. Although it has been isolated from fish and land mammals, it is not a constituent of the normal flora of humans. The disease is apparently spread through contaminated water.

Pasteurella *and* Haemophilus

Pasteurella and *Haemophilus* are parasites of invertebrates and are often pathogens for both mammals and birds. All are nonmotile rods that are oxidase positive. They are generally fastidious microbes requiring organic nitrogen sources, B-vitamins, amino acids, and hematin for growth. Characteristics of this group are shown in Table 19.7.

Pasteurella are the causative agents of diseases in cattle and of a fowl cholera in poultry. They are parasites of the mucous membranes of the respiratory tract of mammals and birds. Some species are found in the digestive tracts of animals, but they rarely cause human disease.

BOX 19.1 RESEARCH HIGHLIGHTS

Are Marine Bacteria Unique?

The oceans began forming very early in the evolution of Earth. Their saltiness increased as land was eroded by rainfall and the rivers carried the dissolved salts into the sea. The total concentration of salts now equals a salinity of about 3.5 percent or, as biological oceanographers state, 35‰ (parts per thousand). Thus, marine bacteria must be able to grow in a saline environment with a lower water activity than their freshwater relatives.

Some microbiologists have defined marine bacteria as those that grow optimally when the salinity of the medium, or growth environment, is about 3.5 percent. To test this, special media are prepared that are amended with the appropriate concentrations of sea salts; this is referred to as artificial sea water, or sterile seawater itself may be used.

In addition to the preference to grow best at 3.5 percent salinity, some microbiologists have proposed that true marine bacteria must also have an absolute requirement for sodium ions. This can be tested by preparing a medium in which the sodium ions are replaced entirely by potassium ions. Using this definition, if the organism cannot grow under these conditions, it is not regarded as a marine organism.

Recently it has been discovered that strains of *Caulobacter* isolated from marine environments differ phylogenetically from those isolated from freshwater. This is consistent with the notion that, millions of years ago, an independent evolution of at least some groups of marine bacteria occurred, so that they have now separated phylogenetically from their freshwater relatives, in a manner analogous to that found for marine fish and mammals.

Differential ***Table*** **19.7 • Characteristics of the *Pasteurella* group**

Genus	Morphology	Mol % G + C
Pasteurella	Ovoid or rod (0.3–1.0 × 1.0–2.0 μm)	40–45
Hemophilus	Coccobacillus (0.3–0.5 × 0.5–3.0 μm)	37–44

Haemophilus is a genus of obligate parasites of the mucous membranes of humans and other animals. Blood or blood derivatives are required for growth. *Haemophilus influenzae* was originally isolated from patients with viral influenza and was thought to be the causative agent of the disease, hence the name. The organism is present in the nasopharynx of healthy individuals and is probably a secondary invader in infections of the respiratory tract. This is the first organism whose entire genome has been sequenced (see **Box 17.1**).

Zymomonas

Zymomonas is a gram-negative nonmotile rod (1.0–1.4 × 2–6 μm) that is facultatively anaerobic and oxidase negative. The mol % G + C is 47 to 50. Strains are occasionally microaerophilic and some are strictly anaerobic. *Zymomonas* is often a spoiler of beer and cider in which it produces a heavy turbidity and unpleasant odor due to formation of acetaldehyde and sulfide. The organism ferments glucose anaerobically via the Entner-Duodoroff pathway (see Chapter 8):

$$\text{Glucose} \longrightarrow \text{Ethanol} + \text{Lactate} + CO_2$$

Small amounts of acetaldehyde and acetyl methyl carbinol are also produced during this fermentation. The palm wines of the Far East and Africa use *Zymomonas mobilis* as the fermenting agent. The organism is also involved in transforming the sugary sap of various agaves to pulque, a fermented beverage produced in Mexico. *Zymomonas* is found on honey bees and in ripening honey because it tolerates high concentrations of sugars.

Chromobacterium

Chromobacterium violaceum is one of the most striking of all bacteria because it produces deep violet colonies on a solid medium. The pigment, called **violacein** (Figure 19.5), is produced in significant amounts during growth on tryptophan, and it bears a structural relationship to this amino acid. Members of the genus are motile by polar flagella and are rod-shaped (0.6–0.9 × 1.5–3.5 μm). *C. violaceum* occurs primarily in soil and water and is especially common in tropical soils.

Oxidative Rods and Cocci

Many of the Proteobacteria are obligately aerobic, nonfermentative rods and cocci that obtain their energy through aerobic respiration. They use a large variety of organic compounds as energy sources, depending on the genus. Some species of *Pseudomonas* can use more than 100 different organic monomers including amino acids, sugars, and organic acids. Others in this group are nitrogen fixers. Some are widespread in aquatic and soil environments, whereas others are obligate parasites of humans and other animals.

Pseudomonads

The pseudomonads are a group of strict aerobes that utilize oxygen as a terminal electron acceptor. Some species also utilize nitrate as an alternate electron acceptor in anaerobic respiration, and therefore can grow anaerobically. The pseudomonads, however, do not ferment, which is a hallmark of the enteric and *Vibrio* groups. The pseudomonads are gram-negative rods (generally 0.5 × 1–4 μm), and virtually all species are motile by one or

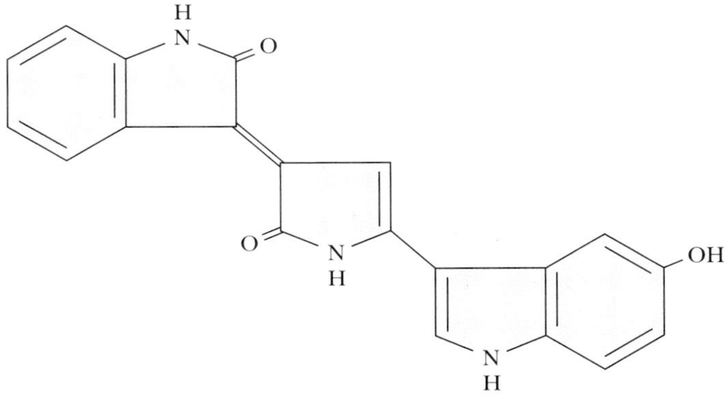

***Figure* 19.5** Chemical structure of violacein, the purple pigment of *Chromobacterium violaceum.*

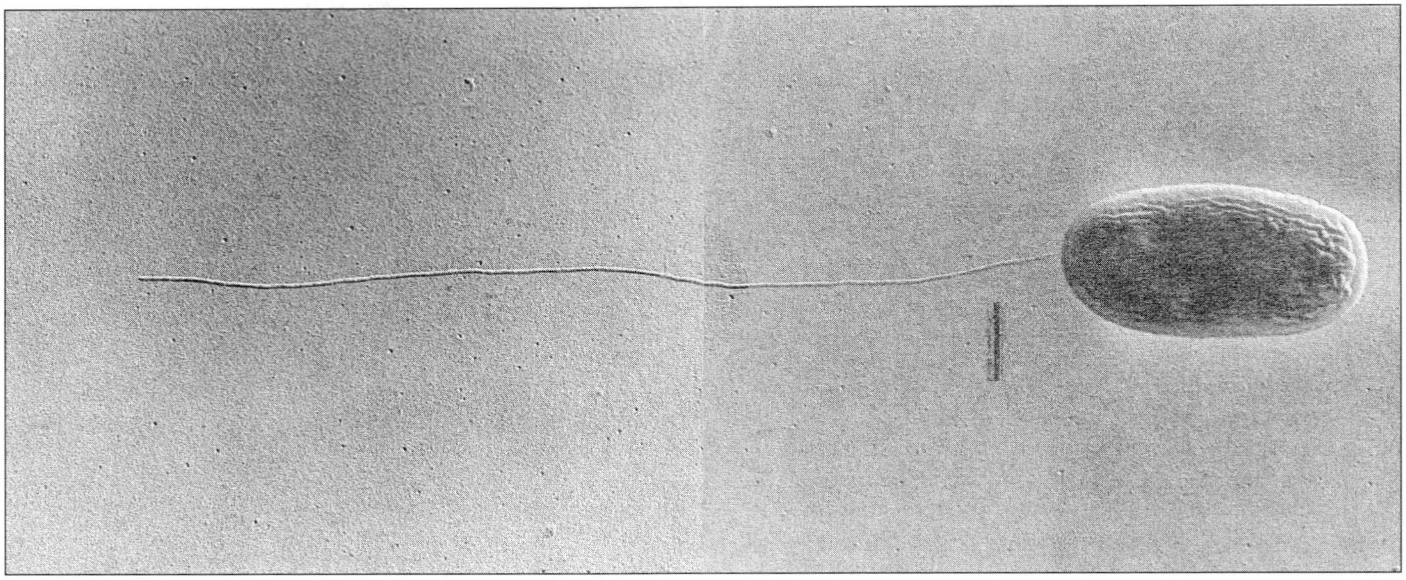

***Figure* 19.6** Electron micrograph showing a typical *Pseudomonas* cell with a polar flagellum. Bar is 0.5 μm. (Courtesy of Judith Bland)

more polar flagella (Figure 19.6). The pseudomonads are oxidase and catalase positive. **Oxidase positive** organisms produce large amounts of cytochrome oxidase. The pseudomonads can be differentiated from the *Vibrio* and enteric groups by a few simple diagnostic tests (Table 19.8). The pseudomonad group consists of several genera in the gamma Proteobacteria, including *Pseudomonas, Xanthomonas, Zoogloea* as well as others in the alpha and beta groups. These are differentiated from one another by the tests outlined in Table 19.9.

The genus *Pseudomonas* is the most thoroughly studied and characterized of the group. It is of widespread occurrence in nature. Some species reside in soil, some in water, and some on surfaces such as human skin. Species are defined on the basis of physiological characteristics. Some species produce water-soluble fluorescent pigments. These are yellow-green pigments called **pyocyanin** and **pyoverdin** that diffuse into the medium and fluoresce under ultraviolet light. The pyocyanins are blue-colored phenazines (Figure 19.7). The chemical structure of the pyoverdins is not completely known because they are quite unstable. However, they are believed to be **siderophores,** which are iron binding compounds that bring ferric ion from the external environment into the cytoplasm.

Differential ***Table* 19.8 • Differentiation of the pseudomonads from the *Vibrio* and enteric bacteria**

Characteristic	Pseudomonads	Vibrios	Enterics
Flagella	Polar	Polar or mixed	Peritrichous
Oxidase[1]	+[2]	+	−
Glucose fermentation	−	+	+

[1]The oxidase test is a measure of the oxidative capacity of an organism. It is determined by picking up a colony of the test organism with a platinum loop and placing a smear directly on filter paper containing two or three drops of a 1% solution of tetramethyl-p-phenylenediamine dihydrochloride If the smear turns violet in less than 10 sec, the organism is considered oxidase positive (contains cytochrome oxidase).

[2]Some *Xanthomonas* strains are oxidase negative.

Differential *Table* **19.9 • Characteristics used to distinguish the common genera of pseudomonads**[1]

Subphylum / *Genus*	*Growth factor requirement*	*Hydrogen autotroph*	*Floc formation*	*Marine*	*Plant pathogen*
Gamma Proteobacteria					
Pseudomonas	–	+/–	–	+/–	V
Xanthomonas	+	–	–	–	+
Stenotrophomonas	+	–	–	–	–
Zoogloea	+	–	+	–	–
Alpha and Beta Proteobacteria					
Comamonas	–	–	–	–	–
Deleya/Halomonas	–	–	–	+	–
Hydrogenophaga	–	+	–	–	–

[1]Symbols: + = positive for most strains; – = negative for most strains; V = variable by strain within a genus.

Members of the genus *Pseudomonas* exhibit remarkable nutritional versatility. Most can grow on 50 or more different substrates, and some can use over 100 different organic compounds. Because of this metabolic versatility, *Pseudomonas* species are very important in the degradation of organic compounds in soil and aquatic environments. Among the substrates utilized are sugars, fatty acids, di- and tricarboxylic acids, alcohols, aliphatic hydrocarbons, aromatic hydrocarbons, amino acids, various amines, and other naturally occurring compounds as well as many humanly produced chemicals such as chlorinated hydrocarbons. Glucose is catabolized by many of the pseudomonads via the Entner-Duodoroff pathway (see Chapter 8). *Pseudomonas* species have a limited capacity to hydrolyze polymeric compounds.

Pseudomonas species are used extensively as biochemical tools in the elucidation of catabolic pathways. They have been particularly useful in ascertaining pathways in the catabolism of aromatic compounds (see Figure 12.15). The genetics of many fluorescent strains has been clarified. Both conjugational and transduction systems have been established. The genetic information for synthesis of the enzymes involved in catabolism of many uncommon

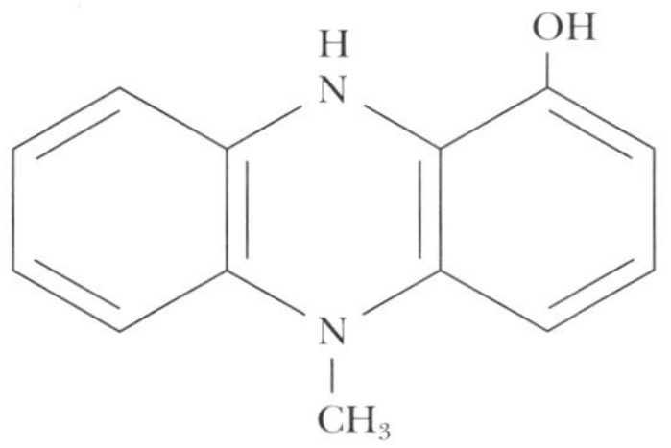

***Figure* 19.7** Chemical structure of pyocyanin, the phenazine pigment of *Pseudomonas aeruginosa.*

organics occurs on transferable plasmids. Among these are genes for catabolism of salicylate, camphor, octane, and naphthalene. It should be noted that the genus *Pseudomonas* has changed greatly since its taxonomy was evaluated by rRNA analysis. The original genus has now been split into several additional genera including *Comamonas* (an alpha subdivision genus that does not produce fluorescent pigments), *Deleya* and *Halomonas* (which are marine), *Stenotrophomonas* (with its single species, *S. maltophilia*), *Hydrogenophaga* (H_2-oxidizing facultative chemolithotrophs), and *Burkholderia.*

Pseudomonas aeruginosa is a human pathogen. It is especially important in humans as an opportunistic pathogen of compromised hosts. For example, it causes serious skin infections of burn victims and grows in the lungs of patients with cystic fibrosis. However, it can also cause urinary tract and lung infections of normal individuals. Normally it is found in soils and is an important denitrifying genus. Certain other *Pseudomonas* species are animal pathogens such as *P. mallei,* which causes glanders in horses.

A number of species of *Pseudomonas* including *Pseudomonas syringae* are plant pathogens. *P. syringae* owes its invasiveness to its ability to form ice crystals that damage the plant tissue **(Box 19.2).**

Xanthomonas is a genus of yellow-pigmented plant pathogens. The yellow pigment is a brominated aryl polyene of unknown function (Figure 19.8) called **xanthomonadin.** Disease symptoms caused by *Xanthomonas* vary with plant species and the strain involved. Stem wilt, leaf necrosis, and other manifestations of disease are observed. These are ascribed to toxins, enzymes, ice nucleation activity, and other metabolic products that accumulate during the growth of the bacterium in its host tissue.

BOX 19.2 RESEARCH HIGHLIGHTS

Ice Nucleation

Certain gram-negative plant pathogenic bacteria, including some species of *Pseudomonas, Xanthomonas,* and *Erwinia,* cause ice nucleation. These bacteria normally colonize plant leaves in the spring and summer. In the fall, when the temperature begins to dip toward freezing, they cause frost damage by forming ice crystals at temperatures somewhat higher than normal freezing temperatures. The ice crystals damage the plant leaf cells, which then exude nutrients that these bacteria utilize. Ice nucleation activity is caused by a protein located in the outer cell membrane of these bacteria. This protein has a distinctive repeating amino acid region, which is the site of ice nucleation.

Pseudomonas syringae, one of the ice-nucleating species, is also used to prevent ice nucleation damage to crops. To accomplish this, a mutant strain having a deletion of the ice-nucleation gene, an "ice-minus" strain, is sprayed on crop plants early in the spring. The concept is to enable the mutant bacterium to colonize the leaves before a wild-type, "ice-plus" strain reaches the leaves. Then, later on, when a pathogenic, "ice-plus" strain arrives at the plant leaf, it will be prevented from colonizing because the ice-minus mutant is already there. This ecological concept is termed "preempting the niche," the result of which is that the ice-plus phytopathogen cannot colonize and subsequently cause damage to the plant.

The ice-nucleating bacterium, *Pseudomonas syringae,* is also used commercially in snow making. The strain is added to water in snow making machines used in ski areas to raise the temperature at which the water freezes to make snow. In this manner, less energy is required to cool the water to sufficiently low temperatures.

Zoogloea is another pseudomonad genus. It forms masses of cells, called **zoogloea**—for which the genus was named—enclosed in a polysaccharide gelatinous matrix (Figure 19.9). *Zoogloea* occur in organically polluted freshwater and in wastewaters such as activated sludge in sewage treatment plants.

Azotobacter *and Other Free-Living Nitrogen-Fixing Bacteria*

The ability to utilize atmospheric N_2 as a sole source of cellular nitrogen is widespread among prokaryotic genera. Several otherwise unrelated aerobic gram-negative species are characterized in Table 19.10. Although these organisms are genetically distinct, they share a number of characteristics and are thus considered here as a group. All are free-living bacteria found in soil and waters. Most strains are motile. These aerobic organisms have an oxygen-sensitive, N_2-fixing nitrogenase system (see Chapter 10). Under favorable conditions, they fix about 10 mg of N_2 per gram of carbohydrate consumed. Because nitrogenase is sensitive to oxygen, it must be protected from excessive O_2 in order for the fixation process to be induced and function. All of the organisms can utilize ammonia as their nitrogen source, and nitrogenase is induced as a response to the depletion of ammonia. All species produce copious quantities of a slimy, mucoid capsular material when growing with atmospheric N_2 as nitrogen source. In nature, this mucoid layer may limit the amount of oxygen that reaches the cell or it might harbor aerobic bacteria that consume O_2 in the immediate vicinity.

Azotobacter was originally isolated by Martinus Beijerinck at Delft University and is the most thoroughly studied of the free-living nitrogen fixers. The cells are large,

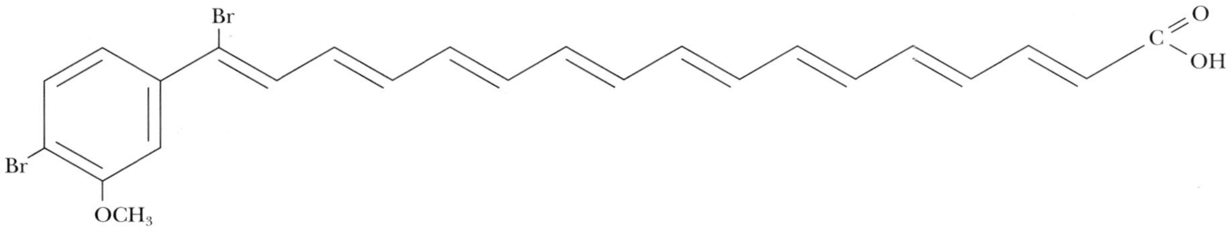

***Figure* 19.8** Chemical structure of the brominated pigment from *Xanthomonas* spp., xanthomonadin.

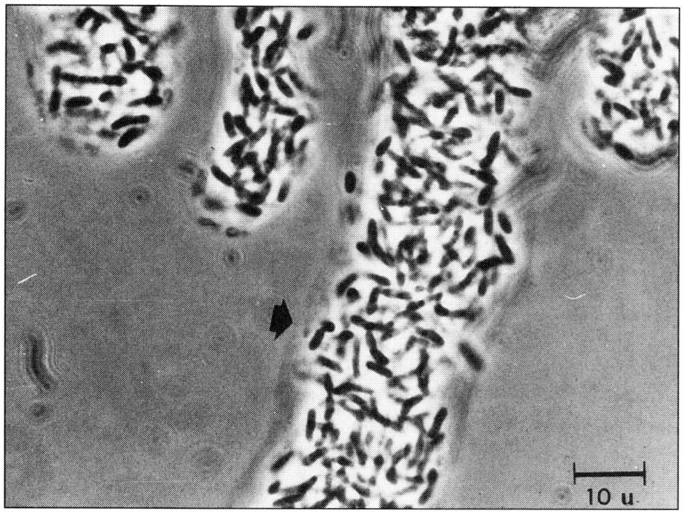

***Figure* 19.9** A zoogloeal cell mass. Each finger-like projection contains numerous cells embedded in an extracellular slime matrix which appears transparent. (Courtesy of R. Unz)

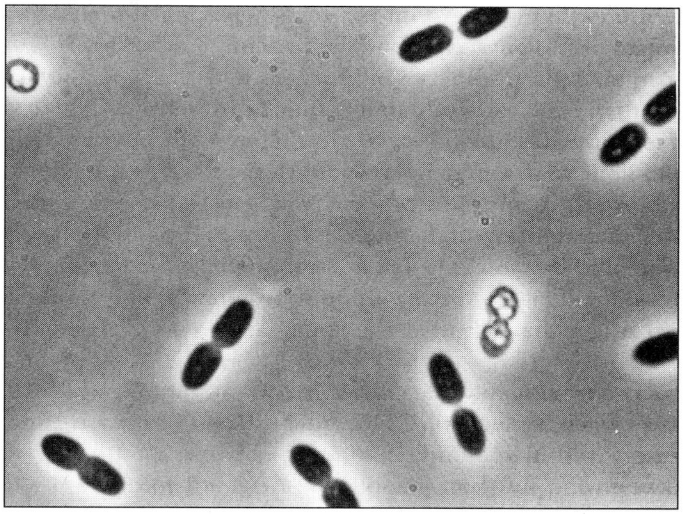

***Figure* 19.10** Cells of an *Azotobacter* sp. The cell that is rounded and refractile is a microcyst. The other refractile cell pair contains PHB granules. Bar equals 10 μm. (Courtesy of J. T. Staley)

pleomorphic ovoid rods, 1.2 μm to 2.0 μm in diameter (Figure 19.10). *Azotobacter* has an exceedingly high respiration rate (O_2 uptake in the presence of substrate) which may aid in retaining low O_2 levels in the cell. As mentioned, nitrogenase is not functional in the presence of molecular oxygen.

Members of the genus form cysts that are resting bodies resistant to drying and radiation but not to heat. The cyst coat surrounds a cell that in other respects is quite similar to a typical vegetative cell. The cysts are not completely dormant because they can oxidize exogenous energy sources. The cyst-forming process can be triggered by supplying the organism with 1-butanol. Removal of the cyst coat by mechanical disintegration or metal chelators yields a viable dividing cell.

Azotobacter spp. are generally found in rich soil that is slightly acid to alkaline in pH and high in phosphorus. Sugars, organic acids, and fatty acids serve as substrates for their growth. The species most prevalent in soil is *Azotobacter chroococcum*. *A. vinelandii* is the most studied and characterized from a biochemical and genetic standpoint. The more efficient nitrogenase in *Azotobacter* species contains molybdenum, but recently, a vanadium-containing nitrogenase has been characterized. This enzyme may

Differential ***Table*** **19.10 • Free-living nitrogen-fixing bacteria**

Genus	Form Cysts or Resting Bodies	Autotrophic Growth	Mol % G + C	Habitat
Azotobacter	+	−	63–67	Rich soil, high in phosphate, slight acid to alkaline pH
Beijerinckia	+	−	55–61	Tropical soil; acid pH
Azomonas	−	−	52–59	Soil and water
Azospirillum	+	+/−	69–71	Free-living in soil or with roots
Derxia	−	+	69–73	Tropical soil
Xanthobacter	−	+	65–70	Wet soil and water

serve as a back-up in soils containing low levels of molybdenum.

Beijerinckia is the genus of nitrogen fixers found in acidic soils, particularly of tropical regions. *Beijerinckia* spp. form straight or curved rods with rounded ends. The cells are often misshapen, with polar lipid bodies surrounded by a membrane. The lipid bodies are composed of poly β-hydroxybutyric acid. Nitrogen-fixing colonies form copious quantities of tenacious elastic slime. Fixation is enhanced by reduced O_2 tension. Cysts containing one cell may occur in *Beijerinckia;* however, during growth on a nitrogen-free medium, several individual cells will be enclosed in a common capsule. *Beijerinckia* spp. use glucose, fructose, and sucrose as the sole source of carbon, but grow poorly on glutamate.

Several other genera are also important nitrogen fixers including *Azomonas,* which resembles *Azotobacter,* and *Derxia,* which is common in acidic tropical soils. *Derxia* spp. can grow on sugars, alcohols, and organic acids. They can be distinguished from *Beijerinckia* by their ability to grow as a hydrogen-utilizing chemoautotroph. *Xanthobacter* species can also grow as facultative hydrogen-utilizing chemoautotrophs. *Xanthobacter* also use a variety of carbon sources including 1-butanol, 1-propanol, and tricarboxylic acid cycle intermediates. *Xanthobacter* occurs in wet soil and water containing decaying organic matter.

Azospirillum is unique in that it is a nitrogen fixer that grows in association with the roots of grains such as corn. These are plump curved rods that grow either as free-living bacteria in soils or may be associated with the roots of monocotyledonous plants including cereal grains, grasses, and tuberous plants. They do not produce root nodules. *Azospirillum* spp. form distinct capsules around the cell that give some resistance to dessication. Sugars and organic acids such as malate are used as substrates; some strains are also hydrogen chemoautotrophs. Nitrogen is fixed only under microaerophilic conditions.

Rhizobium *and* Bradyrhizobium: *Symbiotic Nitrogen-Fixing Bacteria*

Rhizobia may be free-living or they may develop a symbiotic nitrogen-fixing association with leguminous plants **(Box 19.3).** Legumes are plants that bear seeds in pods and include soybeans, clover, alfalfa, peas, vetch, lupines, beans, and peanuts. The bacteria penetrate the roots, or in some cases, stems of the plant and form a nodule. The bacteria reside in this nodule, and under appropriate conditions, fix atmospheric nitrogen providing a source of nitrogen for plants growing in soil deficient in this nutrient. The symbiotic association of these bacteria and legumes is discussed more fully in Chapter 25.

Some of the characteristics of *Rhizobium* and *Bradyrhizobium* are given in Table 19.11. *Rhizobium* is generally involved in nodule formation in plants in the temperate zone. It is faster growing (with a generation time of 4 h) and produces colonies 2 mm to 4 mm in diameter within 3 to 5 days on a yeast-mannitol-mineral salts agar medium. *Bradyrhizobium* nodulates some plants in the temperate zone but is also effective in nodulation of tropical leguminous plants. *Bradyrhizobium* grows slowly with a generation time of 8 h. It produces colonies on the yeast-

BOX 19.3 MILESTONES

History of Symbiotic Nitrogen Fixation

Root nodules were observed on leguminous plants by early scientists. Marcello Malpighi (1628–1694), an Italian anatomist, drew elaborate pictures of leguminous plants depicting root nodules. During the early nineteenth century, scientists realized that the amount of combined nitrogen (nitrate and ammonium concentration) in soils controlled the production of cereal grains. In contrast, the growth of legumes appeared independent of the nitrogen content. It was also noted that the total combined nitrogen measured in soils after the growth of legumes was greater than that which was present before. For this reason, it became common practice to grow legumes in nitrogen-poor soil and plow the resultant crop back in. The soil was then fit for growth of cereal grains. This alternation of legumes and grains was the beginning of the routine practice of crop rotation.

In 1888, Martinus Beijerinck isolated and cultivated bacteria from root nodules. He demonstrated that these bacteria could cause nodule formation in legumes grown in sterile soil. This confirmed that a relationship existed among the nodule, the bacterium, and nitrogen replenishment in soil. The potential utility in taking advantage of this for growing food crops without addition of nitrogen fertilizer was evident to agriculturalists. Since that time, research on the symbiosis between legumes and rhizobia has been an important and exciting area.

Descriptive ***Table* 19.11 • Characteristics of free-living *Rhizobium* and *Bradyrhizobium* species**

Gram-negative rods (0.5–0.9 × 1.2–3.0 μm)
Aerobic
Motile
Produce copious amounts of extracellular slime
Mol % G + C = 59–64
Metabolize glucose via Entner-Duodoroff pathway
Fix N_2 under special microaerophilic conditions
Some species can grow as H_2 chemoautotrophs

mannitol medium that do not exceed 1 mm in diameter in 5 to 7 days. There is a specificity in the species of legume nodulated by organisms within these genera. The leguminous plants infected by a particular strain are considered to be a cross-inoculation group (Table 19.12). The three described species of *Rhizobium* are *R. leguminosarum*, *R. meliloti*, and *R. loti*. The three major strains of *R. leguminosarum* are *trifolii*, *phaseoli*, and *viceae*.

Agrobacterium

Members of the genus *Agrobacterium* induce tumors in dicotyledonous, but not monocotyledonous plants. The disease is commonly called **crown gall** as it frequently occurs at the soil-stem interface, the crown of the plant (Figure 19.11). Other varieties of the disease are hairy root and cane gall. The major species involved in gall formation is *Agrobacterium tumefaciens*, whereas *A. rhizogenes* causes hairy root disease. The genus *Agrobacterium* is closely related to *Rhizobium*. They are gram-negative rods (0.6–1.0 μm × 1.5–3 μm) that are motile by peritrichous flagellation. They are aerobes with a mol % G + C of 57 to 63. Agrobacteria utilize a variety of carbohydrates, organic acids, and amino acids as carbon sources. Some species use ammonia or nitrate as nitrogen sources, while others require amino acids and other growth factors. *Agrobacterium* spp. are common in soil and are often abundant in the rhizosphere, or root zone, of plants. In the past, *A. tumefaciens* received considerable attention as a model of tumor induction with possible implications for tumor formation in humans. Now it is of commercial importance as a mechanism for introducing genetic information into plants.

Agrobacterium tumefaciens generally attacks plants at a wound site at the root stem interface. It is believed that the bacterium attaches at the wound site by one of two mechanisms: (1) specific components of the bacterial envelope, such as glucans, interact with plant cells, or (2) fimbriae are formed by the *A. tumefaciens* strains that anchor the bacterium to the plant cells. The entire bacterium does not enter the plant. Instead, all oncogenic (tumor-pro-

Descriptive ***Table* 19.12 • Cross-inoculation groups nodulated by *Rhizobium* and *Bradyrhizobium* species[1]**

Species	Plant Genus	Common Name
Rhizobium leguminosarum		
	Pisum	Field pea
	Lathyrus	Pea
	Vicia	Vetch
	Lens	Lentil
	Phaseolus	Bean
	Trifolium	Clover
Rhizobium meliloti		
	Melilotus	Sweet clover
	Medicago	Alfalfa
Rhizobium loti		
	Lotus	Trefoil
	Lupinus	Lupines
	Mimosa	Mimosa
Bradyrhizobium japonicum		
	Glycine	Soybean
	Arachis	Peanut

[1]Species listed normally cause root nodules on some, but not necessarily all, species of legume listed.

***Figure* 19.11** Photograph of crown galls on a plant. This plant has been artificially wounded by cutting with a knife and inoculated with *Agrobacterium tumefaciens.* (Courtesy of E. W. Nester)

ducing) strains have a large conjugative plasmid, designated T_i for tumor inducing, that is released into the site of infection. This plasmid carries several pieces of information:

1. Virulence genes that induce tumor formation
2. Genes for substances that regulate the production of plant growth hormones, auxins, and cytokinins
3. Information that directs the plant to synthesize **opines** (Figure 19.12), special amino acids that serve as specific substrates for *Agrobacterium* species in their soil environment
4. Information that permits bacterial catabolism of opines

Only a specific part of the T_i plasmid enters the plant cell. This portion, designated T-DNA (where T stands for transforming) is a series of genes residing in the T_i plasmid between two 23-base pair direct repeat sequences. The T-DNA carries the information listed above and is inserted into the plant chromosome at various sites and is maintained in the plant cell nucleus. Once a plant cell is transformed, its progeny continue to exhibit the tumor-producing characteristic in the absence of bacteria. There is considerable evidence that T-DNA gene expression is controlled by plant regulatory elements, that is, the plant actually assists in the disease process.

The T_i plasmid is an excellent vector for introducing genetic information into a plant. A gene has been placed in a tobacco plant by genetic engineering (see **Box 16.1**). Genetic engineering of plants by this method is restricted so far to dicotyledenous plants because these are susceptible to the infection. These include potatoes, tobacco, tomatoes, and some 640 other plant species that can be infected by strains of the bacterium. Some traits that would be beneficial to establish in crop plants include herbicide resistance, resistance to plant pathogens, resistance to adverse conditions (drought, heat, and salinity) and improved nutritional value. Considerable research is now devoted to attempts to introduce the T_i plasmid into monocotyledenous plants such as corn, wheat, rice, and other cereal grains that are not susceptible to the bacterium (see conversation with E. W. Nester).

Vinegar Bacteria: Acetobacter *and* Gluconobacter

Vinegar has been in use as a preservative and flavoring agent throughout human history. As early as 6000 B.C., the Babylonians depicted methods for converting ethanol from beer to vinegar (acetic acid). Two genera of bacteria, *Acetobacter* and *Gluconobacter,* are the major microbes responsible for vinegar production (Table 19.13). Both of

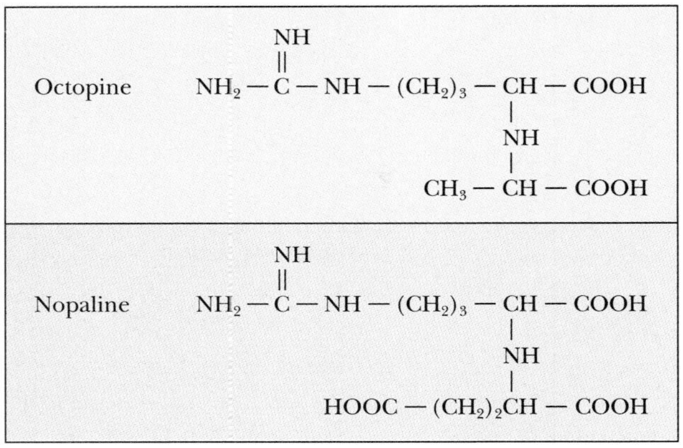

***Figure* 19.12** Formulas for two unusual amino acids, octopine and nopaline, called opines which are produced by plants infected with *Agrobacterium tumefaciens.* It is interesting to note that the genes for producing these opines are transferred from the T_i plasmid to the plant genome. The plant then produces and excretes the opines which can be degraded by the same *Agrobacterium* strain living in the soil. This suggests one reason as to how the *Agrobacterium* strains benefit from the crown gall association.

Differential *Table* **19.13 • Differences between *Acetobacter* and *Gluconobacter***

Characteristic	*Acetobacter*	*Gluconobacter*
Cellulose production	+ or −	−
Complete tricarboxylic acid cycle	+	−
Flagellation	Peritrichous	Polar
Mol % G + C	56–60	56–64

these genera are natural inhabitants of flowers, fruit, sake, grape wine, brewers yeast, honey, and cider. *Gluconobacter* is found on the beechwood shavings of vinegar generators (see Chapter 32).

Gluconobacter and *Acetobacter* can be differentiated from other gram-negative genera by their ability to oxidize ethanol to acetic acid at low pH (4.0–5.0). Both genera also cause pink disease of pineapple and rotting of apples and pears. They are pests in the brewing and wine industry because they convert ethanol to acetic acid. Besides acetification, they cause ropiness, tubidity, and off-flavors in beverage products, including soft drinks.

Gluconobacter cells are also ellipsoidal to rod-shaped and often form enlarged, irregular forms. Some strains are motile by polar flagella. Extensive analysis of about 100 strains indicates there is only one species in this genus, designated *Gluconobacter oxydans*. *Gluconobacter* is a strict aerobe, but it lacks a complete tricarboxylic acid cycle and is therefore called an **underoxidizer.** As a result, it produces acetic acid as a terminal product. It is **ketogenic,** meaning that it oxidizes glucose to 1-ketogluconate, and some strains produce 5-ketogluconic acid. *Gluconobacter oxydans* also can produce intermediates such as dihydroxyacetone from glycerol.

Acetobacter cells are also ellipsoidal to rod-shaped gram-negative organisms but they are motile by peritrichous flagellation. These are obligately aerobic bacteria that prefer an acid pH for growth. DNA/rRNA hybridization studies confirm that *Acetobacter* belongs in a different genus from *Gluconobacter. Acetobacter* is called an **overoxidizer** because it can oxidize acetic acid all the way to carbon dioxide as a result of its complete tricarboxylic acid cycle. *Acetobacter* is also ketogenic. It converts sorbitol to 2-keto sorbitol and glucose to 2-keto- and 5-ketogluconic acid. For this reason this genus is used commercially for the production of ketonic acids such as ascorbic acid (vitamin C).

Some strains of *Acetobacter* form extracellular cellulose microfibrils that surround the dividing cell mass. Therefore the cells become embedded in a large mass of cellulose microfibrils. The cellulose is formed internally and secreted through pores in the lipopolysaccharide outer envelope. In static culture, cellulose-synthesizing cells are favored, and in shake culture, cellulose-free mutants predominate. Cellulose production is rare among bacteria, and its purpose is unknown. It has been postulated that cellulose production aids in maintaining the organism on the surface of liquids so they have oxygen available for growth. However, many aerobic organisms survive in equivalent environments without such an elaborate mechanism for flotation.

Legionella

DNA/DNA hybridization and 16S rRNA sequencing confirm that *Legionella* species and related organisms form a distinct group among the aerobic, gram-negative rods. *Legionella pneumophila,* which is the causative agent of Legionnaires' disease, is the type species. All species in the genus have been implicated in human respiratory disease. Morphologically, they are relatively small bacteria (0.5 × 2 μm). All strains are motile by polar or lateral flagella. Members of this genus have complex nutritional requirements, and L-cysteine and iron salts must be provided for growth. All are strict aerobes that utilize amino acids as their source of carbon and energy. They cannot oxidize or ferment carbohydrates.

Strains of *Legionella* and associated genera are commonly found in ponds, lakes, and wet soil. They also thrive in the warm water associated with evaporative cooling towers. It is probable that aerosols from this source caused an outbreak of the severe respiratory infection among attendees at an American Legion Convention in Philadelphia in 1976. Several fatalities resulted from this outbreak. Difficulties in culturing the organism and assessing its nature are a problem in prevention and treatment of the disease. It is not transferred from one human to another by contact. Sensitive immunofluorescence techniques indicate that *Legionella* strains are of widespread occurrence in the environment. Thus, they must either be of low pathogenicity or be transferred to humans only rarely.

Neisseria *and Related Genera*

The *Neisseria* group are gram-negative, aerobic, immotile bacteria of varying morphology (Table 19.14). *Neisseria* is the major genus of the group. The best known species of *Neisseria* is *N. gonorrhoeae,* the causative agent of the com-

***Differential Table* 19.14 • *Neisseria* group of gram-negative bacteria**

Genus	Cell Shape (size)	Mol % G + C	Oxidase Reaction	Habitat
Neisseria	Cocci (0.6–1.0 μm)	47–54	+	Mucous membrane of mammals
Moraxella	Plump rods (1.0–1.5 × 1.5–2.5 μm)	40–48	+	Mucous membrane of warm-blooded animals
Kingella	Rods (1 × 2–3 μm)	47–55	+	Mucous membrane of humans
Acinetobacter	Rods (0.9–1.6 × 1.5–2.5 μm)	38–47	−	Soil, water, sewage

mon venereal disease, **gonorrhea** (see Chapter 29). Another pathogenic member of this genus is *N. meningitidis,* a causative agent of **cerebrospinal meningitis.** Other species are typically found in the nasopharynx and respiratory passages of warm-blooded animals. These rarely cause disease and are considered part of the normal microflora. *Neisseria* is best cultivated on chocholate-blood agar at 37°C in a 3% to 10% atmosphere of CO_2.

Moraxella is found in the eyes and upper respiratory tract of humans and other warm-blooded animals. Some species cause conjunctivitis (eye inflammation) in humans and bovines. *M. catarrhalis* is a normal inhabitant of the nasal cavity of humans, but it is an opportunistic pathogen associated with disease in unhealthy patients.

Most strains of *Acinetobacter* have simple nutritional requirements, and as the genus name implies, are immotile. *Acinetobacter* is a group of nutritionally diverse organisms that can utilize a wide array of substrates including selected hydrocarbons. They cannot use hexoses as the sole source of carbon and energy, although they can oxidize aldose sugars to the corresponding sugar acid. For example, glucose + O_2 $\longrightarrow$ gluconic acid. *Acinetobacter calcoaceticus* is a prominent member of this genus and is readily isolated from soil and water.

Kingella, the final genus of this group, is a common organism found in the mucous membrane of humans as part of the normal flora.

Rickettsia

Rickettsia and *Chlamydia* are genera that are unable to grow outside a living host cell. Instead, they live as intracellular parasites of animals and are therefore called **obligate intracellular parasites.** Major differences between the two genera are outlined in Table 19.15. *Chlamydia* spp. do not have muramic acid, and hence, have no peptidoglycan. Curiously, they are susceptible to penicillin, suggesting they may have a modified form of peptidoglycan. 16S rRNA analyses indicate *Chlamydia* are not members of the Proteobacteria and are not at all closely related to *Rickettsia,* which are members of the alpha Proteobacteria. The *Chlamydia* comprise their own completely separate phylum of Eubacteria and are discussed later in the chapter.

Rickettsia are parasitic in humans and other vertebrates. Members of this genus are the causative agents of typhus (*R. prowazekii*), Rocky Mountain spotted fever (*R. rickettsii*), and scrub typhus (*R. tsutsugamushi*). Rickettsias are often associated with arthropods, usually intracellularly. Arthropods such as ticks, lice, and fleas are vectors that introduce the *Rickettsia* spp. into humans and other mammals through a bite (the medical aspects are considered in more detail in Chapter 29).

Rickettsia spp. are short rods (0.3–0.5 × 0.8–2.0 μm) that have not yet been cultivated away from host tissues (Figure 19.13). They multiply readily in embryonated eggs

Differential Table* 19.15 • Differences between *Chlamydia* and *Rickettsia

Characteristics	*Chlamydia*	*Rickettsia*
Peptidoglycan	−	+
Developmental cycle of alternating special cell types	+	−
ATP generation via oxidation	−	+
Mol % G + C	41–44	29–33

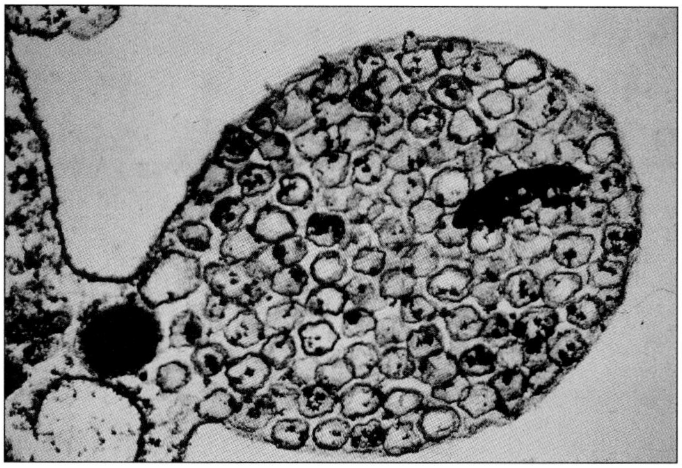

***Figure* 19.13** A photomicrograph showing *Rickettsia* cells infecting animal cells. (© Science VU-PIADC Visuals Unlimited)

or metazoan cell culture, provided that the host cells are viable. They have a generation time in vivo of 8 to 9 hours. Their cell walls are typical of gram-negative Eubacteria with muramic acid and diaminopimelic acid in their peptidoglycan.

The rickettsias derive energy by oxidation of glutamic acid via the tricarboxylic acid cycle. The generation of NADH is coupled to an electron transport system for ATP synthesis. They require a source of AMP and other nucleoside monophosphates. *Rickettsia* spp. have limited synthetic capacity but can synthesize low molecular weight proteins and lipids. Monomers such as amino acids must be provided by the host cell.

Rickettsia prowazekii multiplies to a high density in the cytoplasm of a host causing disruption of host cells. The released bacteria move on and infect other cells. *R. rickettsii* produces small numbers of cells in the cytoplasm of infected cells that escape to extracellular spaces. These then go on to infect other host cells.

Sheathed Bacteria

Some bacteria produce **sheaths,** which are distinctive layers formed external to the cell wall (Figure 19.14). The three major genera in this group are *Sphaerotilus, Leptothrix,* and *Crenothrix.* They are differentiated from one another on the basis of their deposition of ferric hydroxide and manganese oxide as well as other features (Table 19.16).

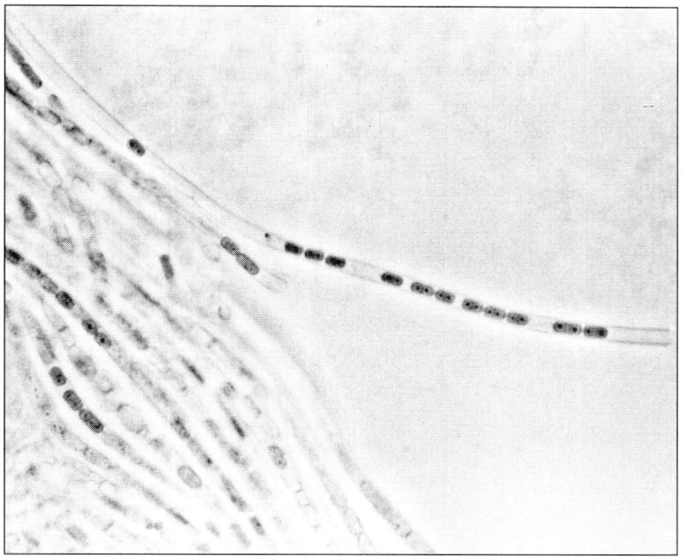

***Figure* 19.14** *Sphaerotilus* filaments showing the cells in chains within a sheath. The sheath appears as a transparent tube about 1.5 μm in diameter. (Courtesy of J. T. Staley)

Sphaerotilus and *Leptothrix* are large rod-shaped bacteria (1–2 × ~ 10 μm) that are motile by polar flagella. They live in flowing aquatic habitats where they attach to inanimate materials such as rocks and sticks. They undergo a characteristic life cycle (Figure 19.15). Single motile cells have a holdfast structure that allows them to attach to inanimate substrata such as sand grains. When they are attached, they become sessile. These cells then reproduce by binary transverse fission to form a chain of cells enclosed in the sheath. As the chain grows and elongates, motile cells are produced and released from the unattached end. These motile swarmer cells repeat the life cycle and provide the organism with a means of dispersal to other habitats.

Differential ***Table* 19.16 • Differences among *Sphaerotilus*, *Leptothrix*, and *Crenothrix***

Genus	Mol % G + C	Flagellation	Distinguishing Features
Sphaerotilus	70	Polar tuft	May deposit iron oxide
Leptothrix	69–71	Single polar	Deposits iron oxide; oxidizes Mn^{+2}
Crenothrix	Unknown	Unknown	Tapering sheath

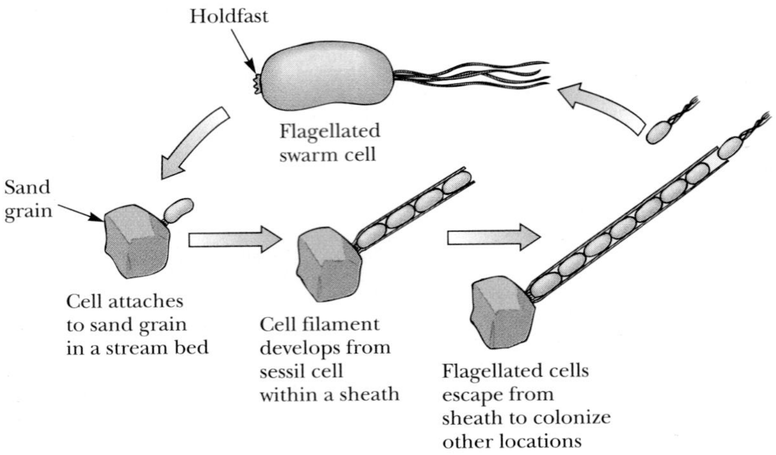

***Figure* 19.15** A diagram of the life cycle of *Sphaerotilus natans.*

These bacteria are aerobic heterotrophs that use a variety of organic substrates for growth, including sugars, alcohols, and organic acids. All species known require vitamin B_{12} for growth. They are common in flowing aquatic habitats where they utilize the nutrients that pass by them in the water. They are particularly common in the Pacific Northwest where they grow downstream of pulp mills and sewage treatment facilities. They can be a nuisance because, when they grow profusely in enriched habitats, clumps of them can break off and clog fishermen's nets and water treatment inlets. Their growth has been controlled considerably since pulp mills have been required by the Environmental Protection Agency to treat pulping plant effluent prior to discharge into receiving streams.

The sheaths of *Sphaerotilus natans* have been analyzed chemically. They have a complex composition similar to that of the outer membrane of gram-negative bacteria consisting of protein, carbohydrate, and lipid.

Leptothrix species appear dark brown or black due to the deposition of iron and manganese oxides in their sheaths (Figure 19.16). These organisms are commonly found in iron springs. Samples from such springs are filled with encrusted sheaths that are devoid of cells. Conditions away from the mouth of the spring are not conducive to the growth of these bacteria, and the cells leave the sheaths or are lysed. However, samples taken close to the mouth of the spring where conditions are more reduced contain cells within the sheath.

Some controversy exists over the ability of these bacteria to obtain energy from the oxidation of reduced iron and manganese ions. As yet there has been no conclusive demonstration of chemolithotrophy using these inorganic substrates. The pH of the environment in which these organisms grow is near neutral, and virtually all iron would be already oxidized to the ferric state. However, manganese is available in part in the Mn^{2+} state and could be used as a source of energy, as suggested early on by Winogradsky. However, although enzymes have been found that enable the oxidation of Mn^{2+} to the Mn^{4+} state in *Leptothrix,* as yet there has been no demonstration of ATP production in this process. Many other bacteria are known to be able to oxidize and/or deposit manganese and iron, but with the exception of *Thiobacillus ferrooxidans* and *Gallionella ferruginea* (see Chapter 21), none are known to be able to derive energy from this process.

Crenothrix is a genus that has been known for over a century, but has not yet been successfully grown in pure culture. Although it resembles the other sheathed bacteria, it has a tapered sheath resembling a cornucopia (Figure 19.17) and produces small cells from the larger unattached end of the sheath.

Prosthecate Bacteria

Prosthecate bacteria are unicellular Eubacteria that have appendages extending from their cells that give them re-

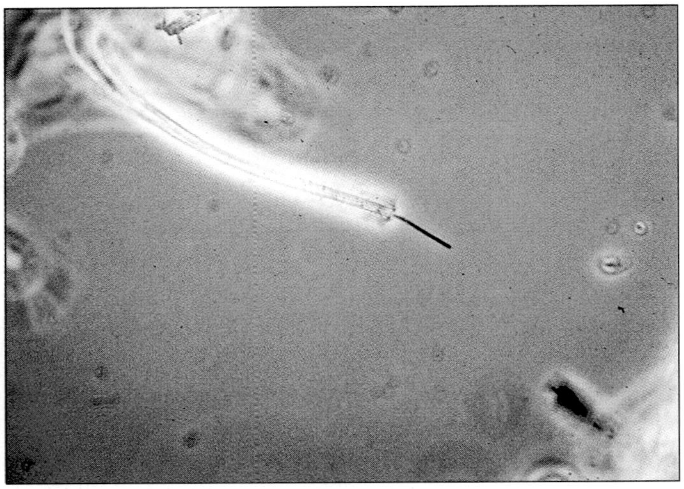

***Figure* 19.16** A phase photomicrograph of a filament of a *Leptothrix* sp. from a natural sample showing that the FeO(OH) and MnO_2 that have accumulated in its sheath have imparted a brownish color. The sheath is about 1.5 μm in diameter. (Courtesy of J. T. Staley)

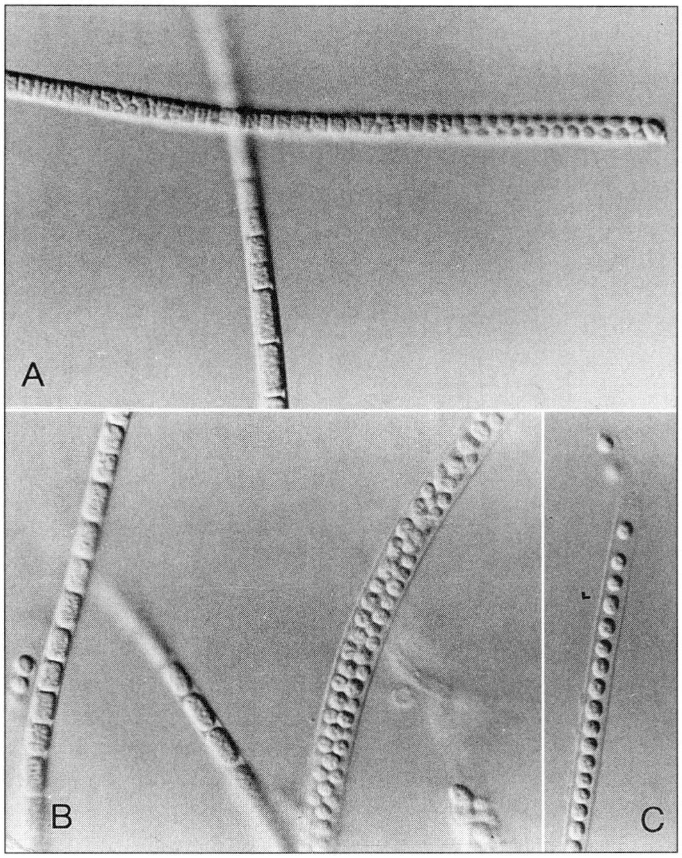

***Figure* 19.17** *Crenothrix polyspora* from a freshwater source. Note the slightly tapered sheaths with the smallest cells near the tips in **(A)** and **(B)**. Cells are being released from the sheath in **(C)**. Nomarski interference photomicrograph. (Courtesy of P. Hirsch)

markable and distinctive cell shapes (Figure 19.18**a**). The appendages, called **prosthecae** are, by definition, **cellular appendages,** *or extensions of the cell that contain cytoplasm* (Figure 19.18**b**). There are two important consequences of having these appendages. First, they increase the surface area of the cell, resulting in a higher surface area to volume ratio thereby allowing the cells increased access to the nutrients in the environment. Since these organisms live in aquatic environments that have low nutrient concentrations, the prosthecae provide greater exposure to the nutrients they require for growth. Studies have shown that these organisms have very high affinity nutrient uptake systems, which enable them to take up nutrients at very low concentrations. They are therefore examples of **oligotrophic** bacteria. Second, the increased surface area is also important for these planktonic organisms because it provides greater friction or drag and therefore slows their settling out from the plankton.

There are three major groups of prosthecate bacteria: the **caulobacters,** the **hyphomicrobia,** and the **polyprosthecate** bacteria (Table 19.17). Almost all heterotrophic

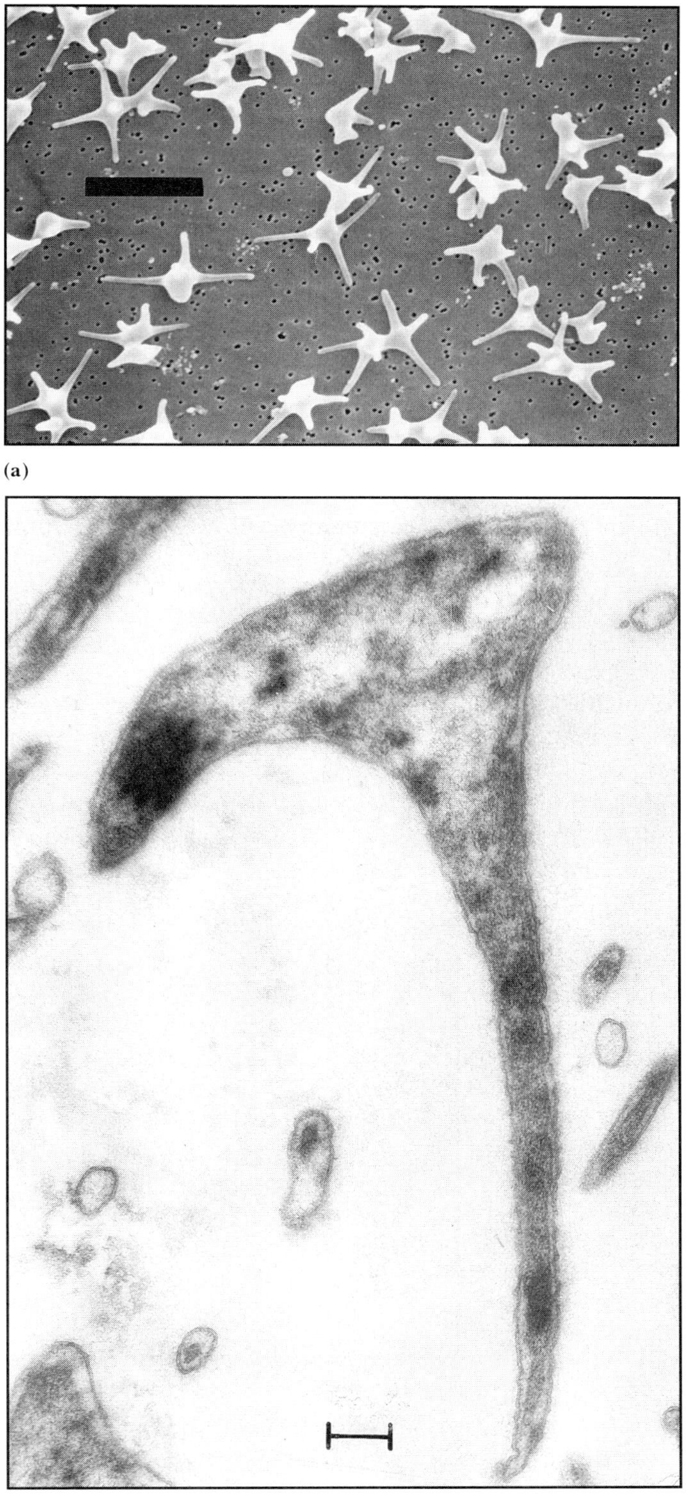

(a)

(b)

***Figure* 19.18** **(a)** Scanning electron micrograph showing the prosthecate bacterium, *Ancalomicrobium adetum.* Each cell has several long prosthecae extending from it. Bar is 5.0 μm. (Courtesy of A. van Neerven and J. T. Staley) **(b)** Thin section through *Ancalomicrobium adetum* cell and prostheca. Note that the cell wall and cell membrane extend around the appendage that contains cytoplasm. Bar is 0.2 μm. (Courtesy of J. T. Staley)

***Differential Table* 19.17 • Groups of heterotrophic prosthecate bacteria**

Group	Budding	Prosthecae (no. per cell; location)	Reproductive Role of Prostheca
Caulobacters	–	One (rarely two); polar or subpolar	–
Hyphomicrobia	+	One to several; polar[1]	+
Polyprosthecate bacteria	+	Several to many	–

[1]The genus *Pedomicrobium* can produce prosthecae at other locations on the cell surface.

prosthecate bacteria are members of the alpha group of the Proteobacteria, but not all alpha Proteobacteria have prosthecae (Appendix 2). Each of the prosthecate groups is considered individually below.

Caulobacters

There are three genera of caulobacters (Table 19.18). The best studied genus is *Caulobacter* with a single polar prostheca, which in this genus is commonly called a "stalk" (Figure 19.19). *Caulobacter* spp., and certain other prosthecate and or budding bacteria, have a life cycle (Figure 19.20) consisting of two stages: **(a)** a motile stage and **(b)** a sessile, prosthecate stage. In the motile stage the cell is called a **swarmer cell,** which has a single polar flagellum and, at the same site, bears a **holdfast,** a special adhesive organelle that mediates attachment. The holdfast allows the cell to attach to detritus or other particulate material in the environment. Before the swarmer cell divides, it develops a polar prostheca at the same site as the flagellum and loses motility. Normally by this time in its development, the cell would be attached to a detritus particle or a dead algal cell in the environment by virtue of its sticky holdfast. The *Caulobacter* cell then elongates and produces a daughter swarmer cell at the nonprosthecate pole. Next, the daughter cell becomes motile and separates from the mother cell to become free in the planktonic environment and ready to repeat the cycle. The mother cell continues to produce daughter cells from the same position on its cell surface as long as conditions are favorable for its growth. A life cycle such as this with two separate morphological forms (a flagellated nonprosthecate cell and a nonflagellated prosthecate cell) is termed a **dimorphic life cycle.** Because of its life cycle and simple prokaryotic nature, *Caulobacter* spp. have become model organisms in the study of **cellular differentiation** or **morphogenesis (Box 19.4).** (See also Conversation with Lucy Shapiro)

In pure cultures the holdfasts of *Caulobacter* enable several cells to attach together to form a rosette structure (Figure 19.19**c**), something which is not found in the natural environment.

These organisms also produce structures called **crossbands** in their prosthecae (Figures 19.19**d** and 19.21). Studies have shown that these are produced at the time of cell division. Therefore, it is possible to determine the "age" of a mother cell by counting the number of crossbands (hence, the number of times it has divided), much like counting tree rings.

***Differential Table* 19.18 • Genera of caulobacters**

Genus	Motility	Mol % G + C	Special Features
Caulobacter	+	62–67	Polar prostheca with crossbands
Asticcacaulis	+	55–61	Subpolar prostheca with crossbands
Prosthecobacter	–	54–60	Polar prostheca; no crossbands

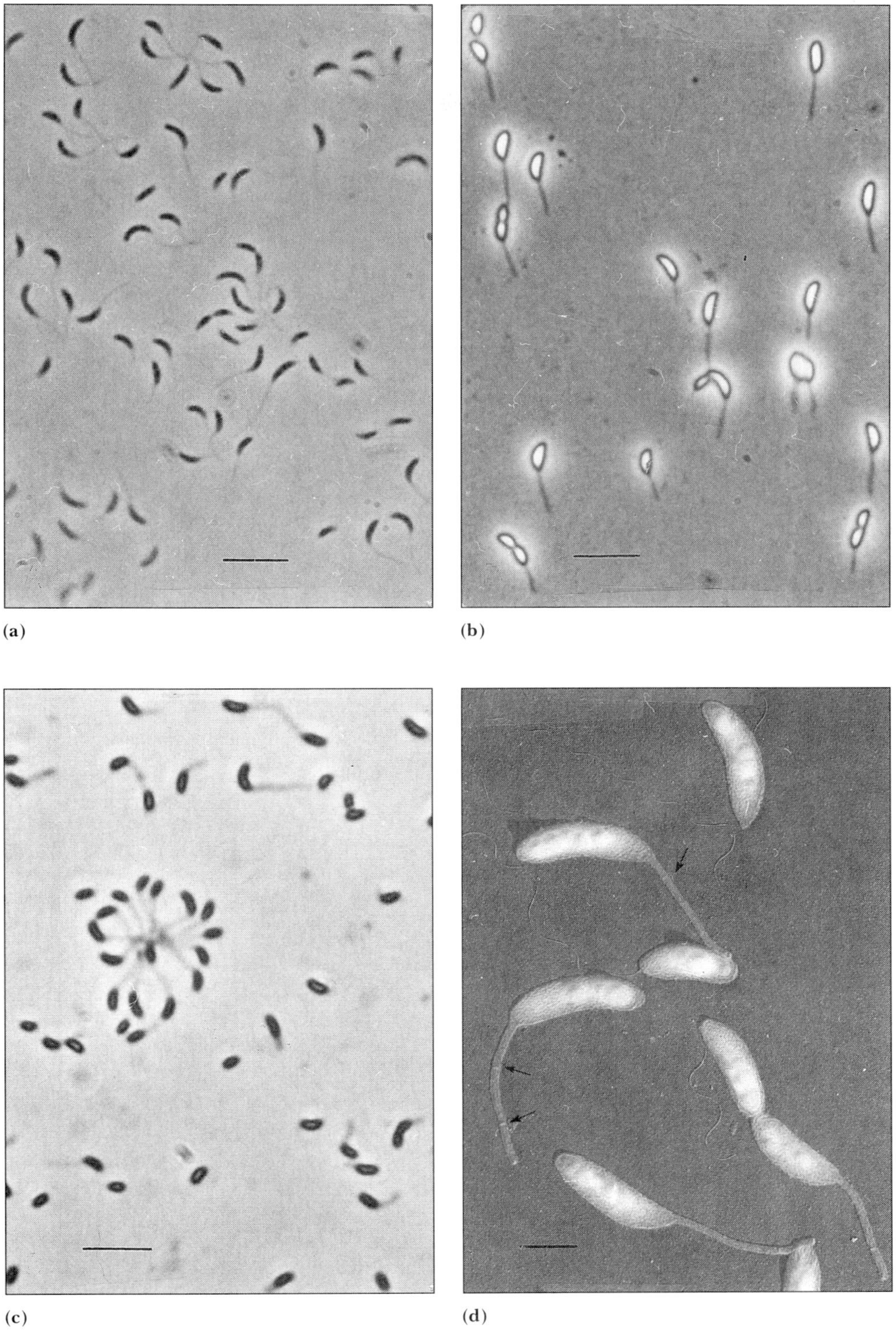

***Figure* 19.19** **(a)**, **(b)**, and **(c)** Phase photomicrographs of *Caulobacter crescentus* showing polar prostheca. Refractile cells in **(b)** are due to accumulation of poly-β-hydroxybutyric acid (Bar, 5.0 μm). **(d)** Electron micrograph of shadowed preparation of *C. crescentus.* Bar is 1.0 μm. Note stalk crossbands (arrows). (Courtesy of J. Poindexter)

These bacteria are metabolically similar to *Pseudomonas* species. They utilize a wide variety of soluble organic carbon sources, including sugars, amino acids, and organic acids, which are found in low concentrations in aquatic habitats (see Chapter 24). They use the Entner-Duodoroff pathway and TCA cycle and are active in aerobic respiration. They occur in both freshwater and marine habitats as well as in soils.

The genus *Asticcacaulis* is less frequently encountered in most habitats. This genus is very similar to *Caulobacter* in all respects except that its flagellum and prostheca are in a subpolar position on the cell surface and one species

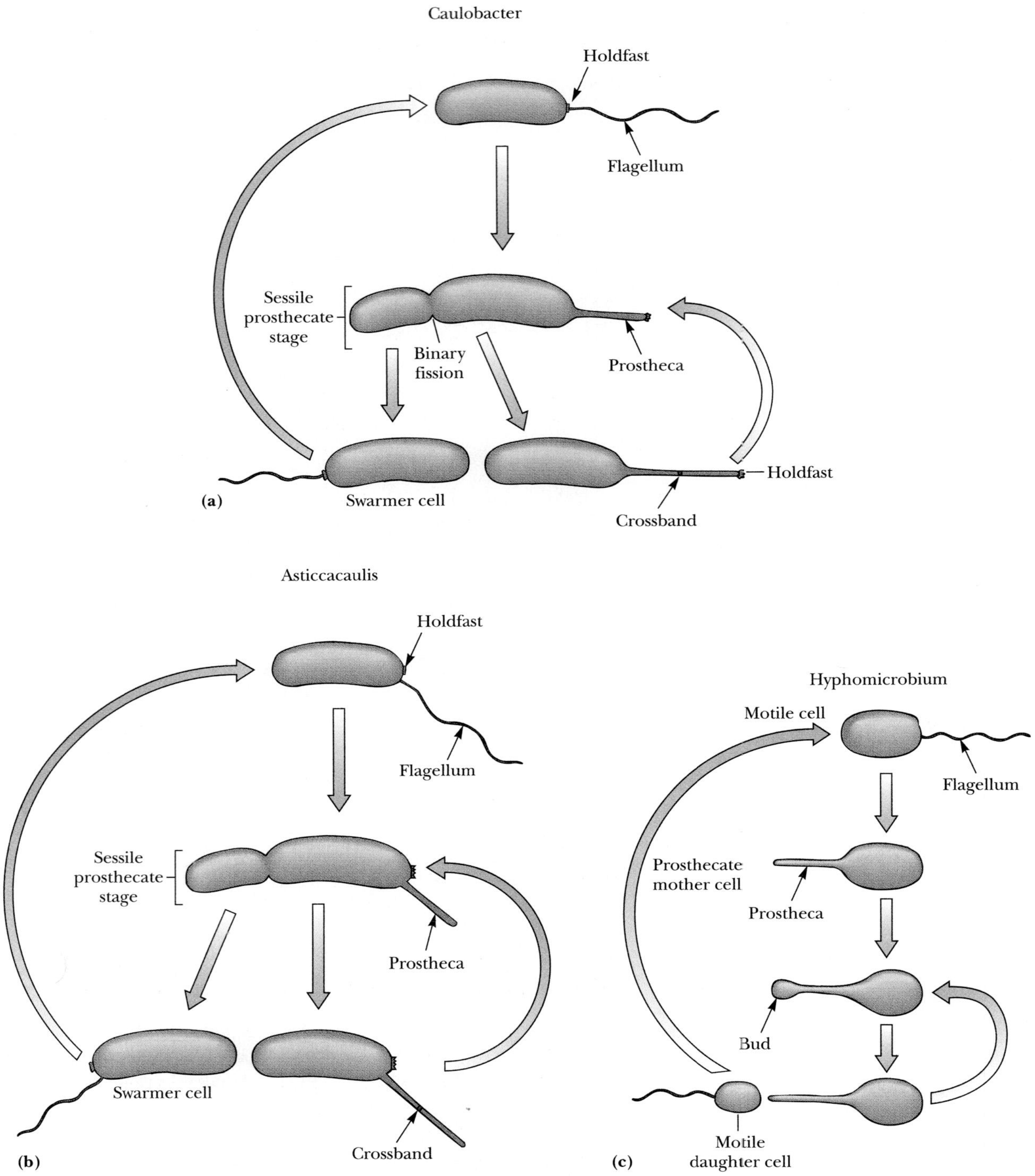

***Figure* 19.20** Diagram of life cycle of **(a)** *Caulobacter,* **(b)** *Asticcacaulis* and **(c)** *Hyphomicrobium.* Note that all produce a flagellated swarmer cell. *Hyphomicrobium* species form buds off their prosthecae.

produces two lateral prosthecae. However, the holdfast remains in a polar position. Thus, rosettes formed by pure cultures hold the cells together so that the prosthecae extend away from the center.

The genus *Prosthecobacter* is quite different from either *Caulobacter* or *Asticcacaulis.* First, it has no motile stage and second, it has no crossbands in the appendage. However, like *Caulobacter* spp. it has a single polar prostheca with a distal holdfast (Figure 19.22). Recent evidence indicates that this genus is in a different phylogenetic group of the Eubacteria, not sufficiently studied to be discussed further in this text. Therefore, it is not closely related to *Caulobacter.* This may be an example of **convergent evolution** in which the aquatic habitat has allowed for the selection and evo-

BOX 19.4 RESEARCH HIGHLIGHTS

Caulobacter crescentus, a Model Prokaryote for Studying Cell Morphogenesis or Differentiation

Several laboratories have been studying the dimorphic life cycle of *Caulobacter crescentus*. This organism is ideally suited for such analysis because it undergoes a transformation from a flagellated swarmer cell to an immotile, prosthecate cell during its life cycle. In the cell cycle, the cell must be programmed to "switch on" and "switch off" various biosynthetic processes at specific times. Furthermore, the cell has the capability of localizing these events at one pole of the cell or the other.

Let us consider initially the flagellated swarmer cell. Not only does this stage of the life cycle have a polar flagellum and is active chemotactically, but the flagellated pole of the cell also contains the holdfast, a specific pilus structure, and DNA phage receptor sites. Thus, there is strong polar orientation of functions. Then, at some time during the differentiation process, the swarmer cell loses its flagellum and pili, and begins to develop a prostheca (called a "stalk") and a holdfast at the same pole of the cell.

Only after the stalk forms does the cell switch on DNA replication. The stalked cell elongates, DNA replication occurs, and the polar differentiation process whereby the new swarmer cell is formed occurs.

Not surprisingly, the length of time it takes for a swarmer cell to undergo cell division is longer (typically an additional 30 min) than it takes a cell with a stalk to divide (90 min). This is due to the length of time required for the swarmer cell differentiation process. At this time, the molecular biological events during the life cycle of *C. crescentus* are being analyzed genetically and biochemically in several laboratories (see Conversation with Lucy Shapiro).

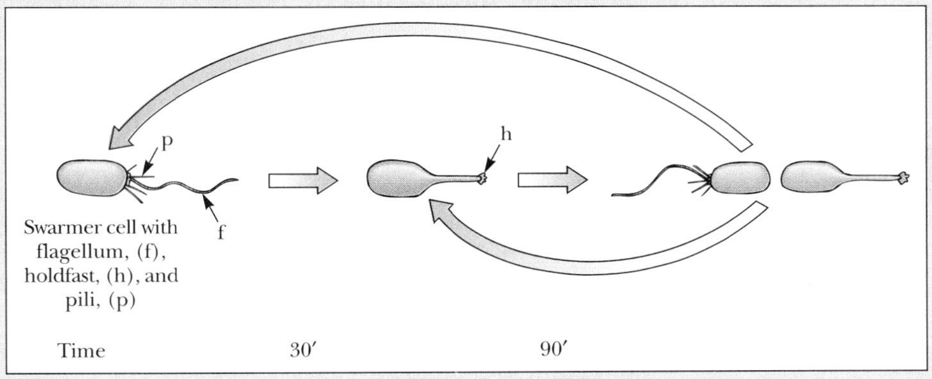

Life cycle of *Caulobacter crescentus*.

lution of two similar morphological types from two separate phylogenetic groups. Alternatively, they may have had an ancient, common ancestor. Even though it is not a member of the Proteobacteria, this genus is treated here for simplicity. The niche of *Prosthecobacter* species differs from *Caulobacter* in that they prefer certain disaccharides as growth substrates rather than monosaccharides and amino acids, which are preferred by *Caulobacter*.

Caulobacters are widespread in aquatic habitats. Studies indicate they are found in lakes of all trophic states from oligotrophic to eutrophic. They comprise a significant proportion of the total heterotrophs in oligotrophic and mesotrophic habitats, making up as much as 50 percent of the viable heterotrophic count during their periods of greatest abundance (see Chapter 24). Although the natural function of the caulobacter prostheca is not yet well understood, it is known that it becomes longer during conditions of phosphate limitation. Since phosphate is a common limiting nutrient in aquatic habitats, especially during summer algal blooms, this appendage may enhance its uptake and improve their ability to survive when it becomes limiting during the summer periods.

Hyphomicrobia

Like the caulobacters, hyphomicrobia comprise a group of several genera with similar morphological traits (Table 19.19). They all have polar prosthecae and divide by forming a bud at the tip of it (Figure 19.23). Buds are usually motile by a single flagellum; some species produce a holdfast for attachment. In this group, the holdfast is located on the surface of the cell, not at the tip of the prostheca.

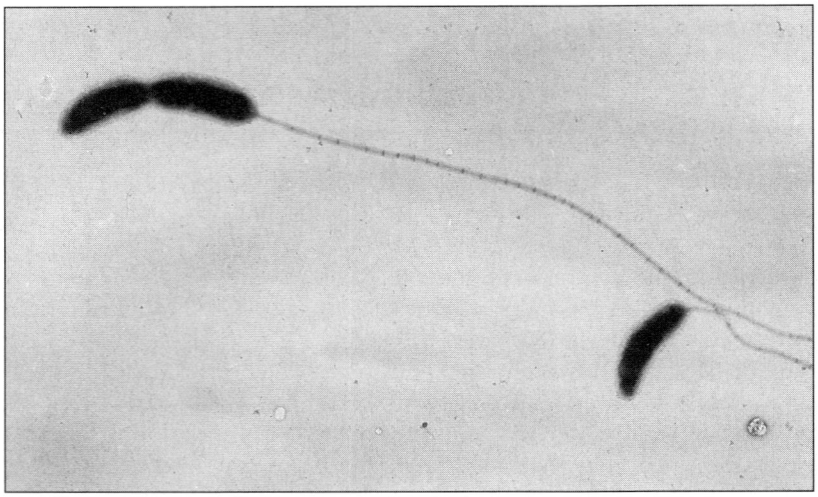

***Figure* 19.21** Electron micrograph of a *Caulobacter* cell showing 31 crossbands in stalk. (Courtesy of J. T. Staley and T. Jordan)

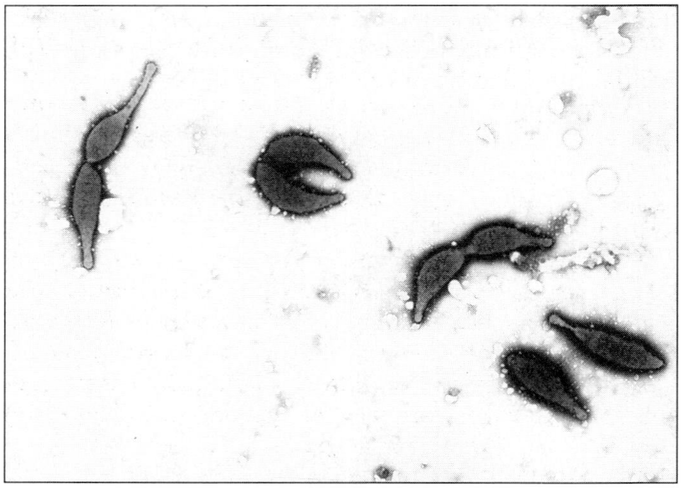

***Figure* 19.22** Electron micrograph of *Prosthecobacter fusiformis* showing several cells in the process of division. Note the absence of flagella and crossbands in the prosthecae. (Courtesy of J. T. Staley)

These organisms also have a dimorphic life cycle consisting of a nonprosthecate motile daughter cell and a prosthecate mother cell (Figure 19.20**c**). The motile daughter cell must first form a prostheca. It then divides by a process in which a bud develops from the tip of the prostheca. The daughter cell enlarges, becomes motile, separates from the mother cell, and repeats the cycle. Meanwhile, the appendaged mother cell develops a new bud either at the same site as the earlier bud or on a newly formed polar appendage. Apparently, if conditions are favorable for growth, an unlimited number of buds can be produced from the mother cell, in the same manner as caulobacters.

Hyphomicrobium spp. are primarily methanol utilizers, although they can also use some other one-carbon compounds such as methylamine as their carbon source. They use the serine pathway for methanol assimilation (Chapter 10). Some carry out denitrification using nitrate as an electron acceptor in anaerobic respiration and oxidizing methanol as the energy source.

Hyphomonas and *Hirschia* look identical to *Hyphomicrobium,* but they do not use methanol as a carbon source. Instead, they use amino acids and organic acids as carbon

Differential ***Table* 19.19 • Genera of hyphomicrobia**

Genus	Mol % G + C	Carbon Source	Special Features and Habitat
Hyphomicrobium	59–65	Methanol	Polar prosthecae; soil, freshwater
Hyphomonas	57–62	Amino acids	Polar prosthecae; marine
Hirschia	45–47	Amino acids, organic acids	Polar prosthecae; marine
Pedomicrobium	62–67	Acetate, pyruvate	Lateral and polar prosthecae; soil, lakes

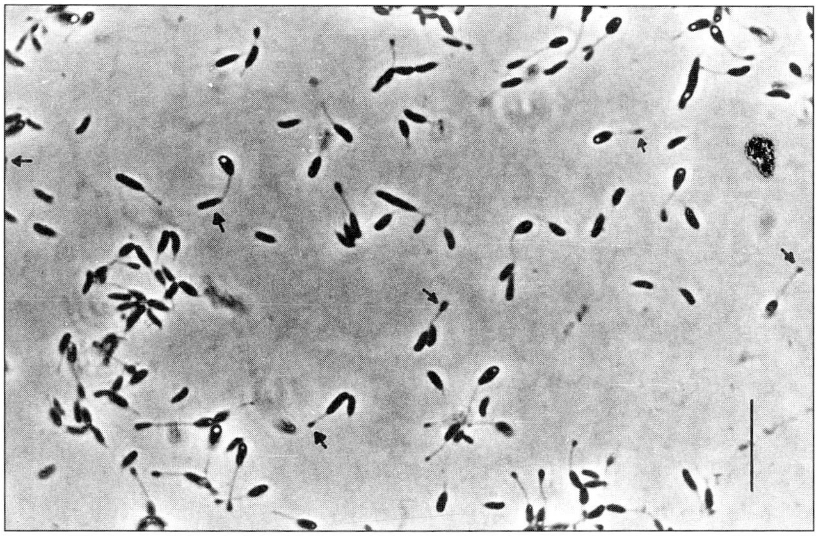

(a)

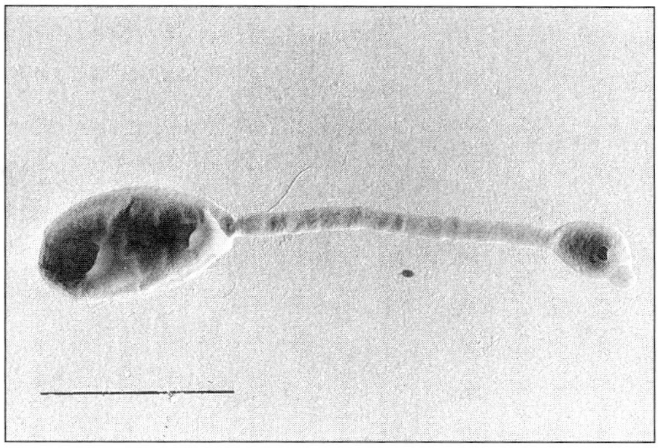

(b)

***Figure* 19.23** **(a)** Cells of *Hyphomicrobium zavarzinii* showing bud formation at the tips of the prosthecae (arrows). Bright area in some cells is poly-β-hydroxybutyric acid. Bar is 5 μm. **(b)** Electron micrograph of a budding cell of *H. facilis*. Bar is 1 μm. (Courtesy of P. Hirsch)

sources. Amino acids are required. The life cycle of these bacteria is similar to that of *Hyphomicrobium* despite their metabolic differences.

Pedomicrobium spp. are quite different from the other hyphomicrobia. They are soil and aquatic bacteria that grow on acetate and pyruvate as well as more complex organic compounds such as fulvic acid. They are best known for their ability to oxidize reduced iron and manganese compounds and form deposits of the oxides on the cell surface. Some have even been implicated in the natural formation of placer gold (gold dust) because biological structures resembling their cell colony morphology can be seen in certain gold dust deposits. Thus, they are of interest to geologists because this implies that gold dust formation may be due to a biological phenomenon, as are many other geochemical processes (see Chapter 24).

Their life cycle is similar to that of the other hyphomicrobia. However, they commonly produce more than one prostheca per cell, and some of these are in nonpolar positions (Figure 19.24). The buds are more characteristically rod-shaped rather than coccoidal like the other hyphomicrobia.

The hyphomicrobia are widespread in soils and aquatic environments. *Hyphomicrobium* are commonly found in freshwater and soil habitats. Most *Hyphomonas* and *Hirschia* spp. have been isolated from marine habitats. *Pedomicrobium* was first reported in the soils, but they have also been obtained from freshwater lakes and water pipelines where they form a coating of manganic oxides if the water contains reduced manganese ions (Mn^{2+}).

Polyprosthecate Bacteria

Polyprosthecate bacteria are a diverse group of bacteria represented by several genera (Table 19.20). Unlike other prosthecate bacteria, they have several prosthecae per cell, which gives their cells starlike shapes. In fact, the genus *Stella* produces six prosthecae all in one plane, so they resemble a perfect six-pointed star (Figure 19.25).

The genus *Prosthecomicrobium* is the most common member of this group. They are found in freshwater, marine, and soil environments. Some species are motile by a single subpolar flagellum, whereas others are immotile or produce gas vacuoles (Figure 19.26).

Unlike all other heterotrophic prosthecate bacteria, the genus *Ancalomicrobium* (Figure 19.18) is a fermentative facultative aerobe. Indeed, it is a mixed-acid fermenter whose products are identical to that of *Escherichia coli* (see Table 19.3). However, unlike *E. coli*, it produces prosthecae and gas vacuoles and is well adapted to live and com-

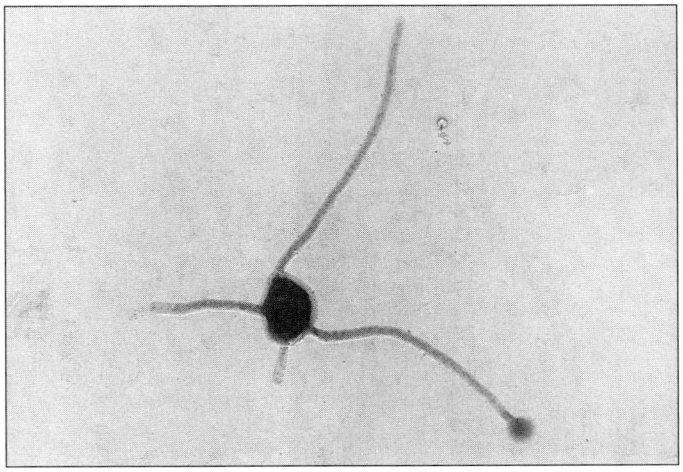

***Figure* 19.24** Electron micrograph of a *Pedomicrobium* cell showing a bud at the tip of one prostheca. (Courtesy of R. Gebers)

***Differential Table* 19.20 • Genera of polyprosthecate bacteria**

Genus	Mol % G + C	Special Features
Prosthecomicrobium	64–70	> 10 Short prosthecae around cell; aerobic
Ancalomicrobium	70–71	< 10 Long prosthecae; facultative aerobe; fermentative
Stella	69–74	Star shaped; prosthecae in one plane

pete successfully in aquatic habitats where nutrients are in much lower concentrations than in the intestinal tract of animals. *Ancalomicrobium adetum* uses a variety of sugars and other organic carbon sources for growth and requires one or more B-vitamins such as thiamine, biotin, riboflavin, or B_{12}.

Like the caulobacters, these bacteria live in aquatic habitats and soils, particularly those that are oligotrophic to mesotrophic. Their appendages enhance the uptake of nutrients in habitats where nutrients occur in low concentrations. Each genus and species specializes in the uptake of a group of low molecular weight dissolved organic nutrients, such as sugars and organic acids, and are therefore responsible for keeping these nutrients in low concentrations in oligotrophic environments. In this manner, they maintain a competitive edge against eutrophic bacteria that require organic nutrients in much higher concentrations and are therefore not competitive in oligotrophic environments.

Conversely, the low maximum growth rates of the pros-

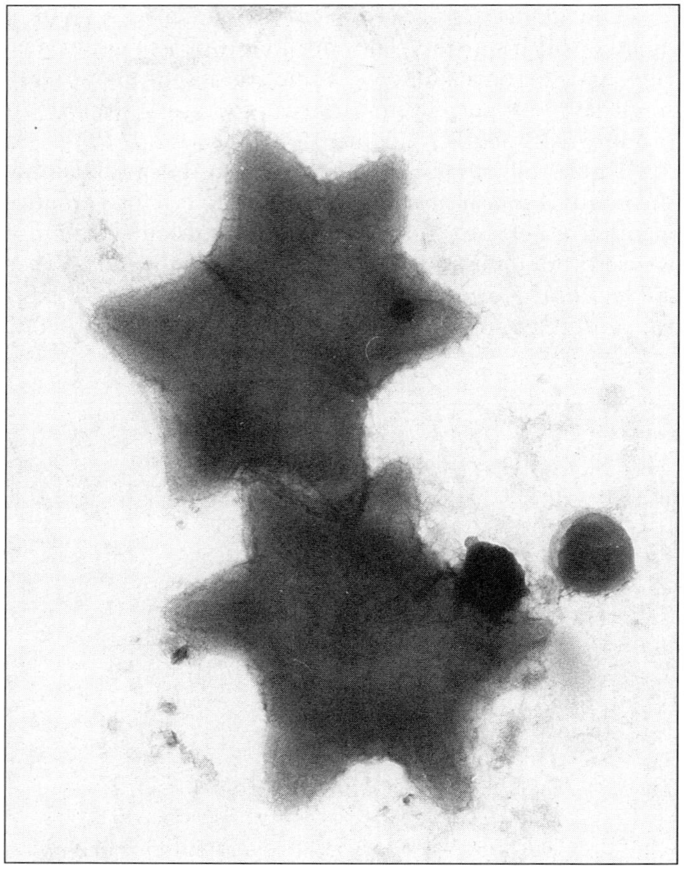

***Figure* 19.25** Two cells of *Stella humosa,* a six-pointed, star-shaped budding bacterium. (Courtesy of J. T. Staley)

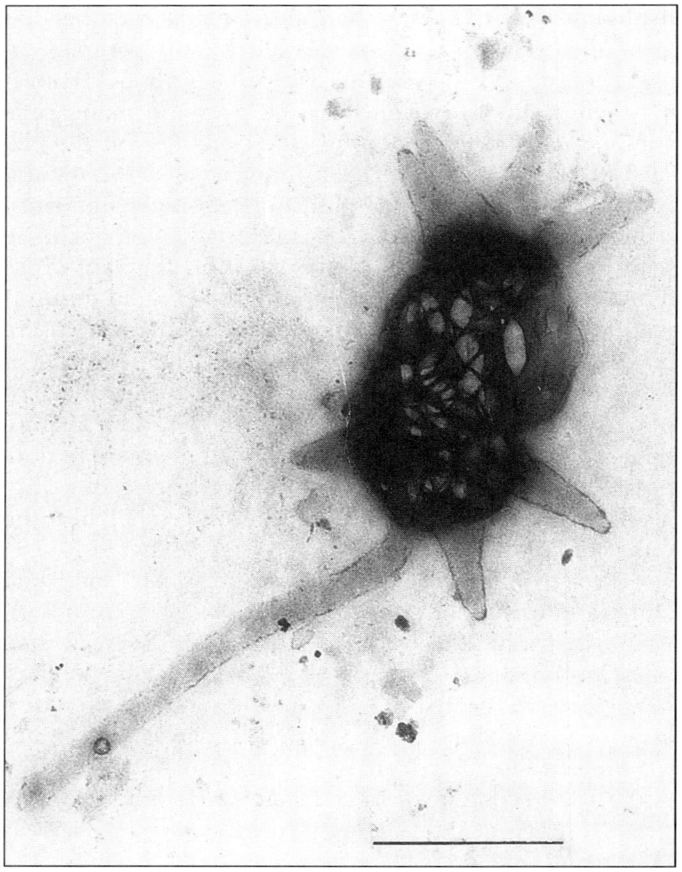

***Figure* 19.26** Electron micrograph of a gas vacuolate cell of *Prosthecomicrobium pneumaticum.* Bar is 1 μm. (Courtesy of J. T. Staley)

thecate bacteria (generation times are longer than 2 hours in laboratory culture) mean that they cannot outgrow fast-growing eutrophic bacteria such as the enteric bacteria in more enriched habitats, for example, the intestinal tract, which has much higher concentrations of organic compounds.

Bdellovibrio: The Bacterial Predator

The genus *Bdellovibrio* contains unique gram-negative bacteria that actually parasitize other gram-negative bacteria. They are small vibrios, only about 0.25 μm in diameter. The *Bdellovibrio* cell has a polar flagellum that rapidly propels the organism through the environment until it collides with an appropriate host cell. It hits the host cell with such force that the host is moved several micrometers. The *Bdellovibrio* cell then attaches to the gram-negative host at its nonflagellated pole and rotates rapidly at the site of attachment (Figure 19.27). After a few minutes, during which it produces enzymes that are responsible for the breakdown of the cell wall, it enters the cell envelope of the host and moves into the periplasmic space. This process of penetration requires the production of enzymes such as proteases, lipases, and a type of lysozyme, all of which enable it to penetrate the outer lipopolysaccharide and peptidoglycan cell wall layers so that it can enter the cell. This entire process takes several minutes. The *Bdellovibrio* cell does not break the cell membrane and lyse the cell but instead resides within the periplasmic space during its parasitism. The host cell becomes rounded during this process due to the destruction of its cell wall, and is called a **bdelloplast.** Once the *Bdellovibrio* cell enters the periplasm, it loses its flagellum and becomes immotile.

While in the periplasm, the *Bdellovibrio* cell produces enzymes that degrade the host cell macromolecules including its DNA, RNA, and protein. The degradation products are then used by *Bdellovibrio* in making its own protein, DNA, and RNA. The *Bdellovibrio* cell enlarges as growth proceeds to form a long, helical cell during the next 4 hr or so (Figure 19.28). When mature, this cell divides by multiple fission to form up to six motile daughter cells in a typical unicellular gram-negative bacterium such as *E. coli* or *Pseudomonas* spp. The *Bdellovibrio* prog-

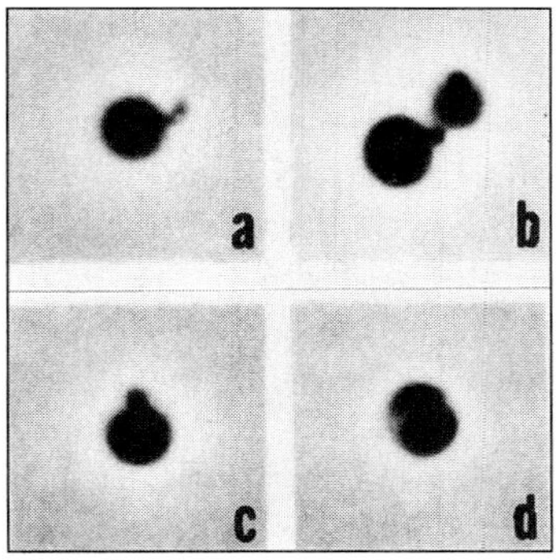

***Figure* 19.27** Life cycle of *Bdellovibrio bacteriovorus.* **(a)** Small *Bdellovibrio* cell has attached to an *E. coli* cell. The *E. coli* cell has begun to form a spheroplast. **(b)** and **(c)** *Bdellovibrio* has begun to enter the *E. coli* cell. **(d)** *Bdellovibrio* cell appears as a small vibrio in periplasm of the *E. coli* host cell. (Courtesy of R. J. Seidler)

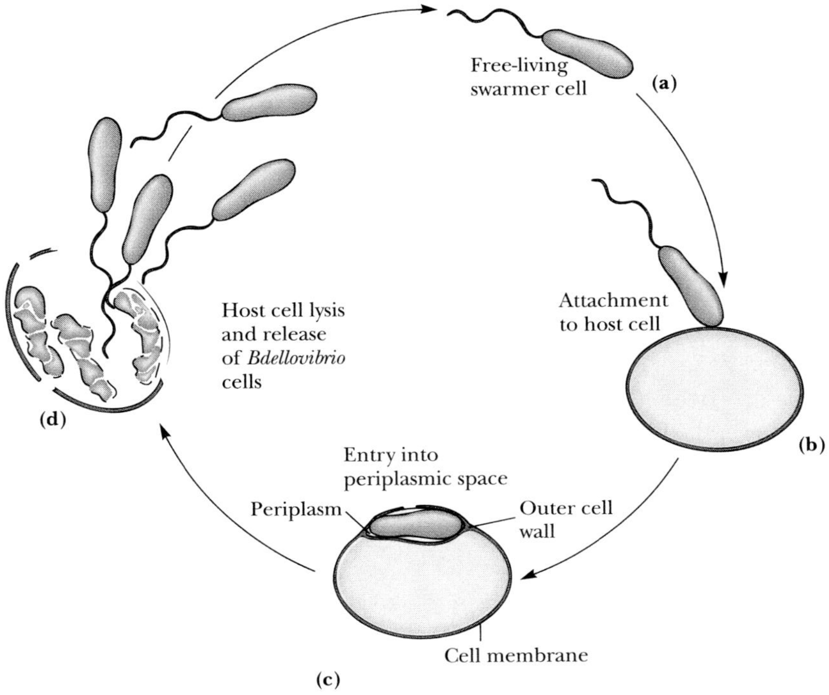

***Figure* 19.28** Life cycle of *Bdellovibrio bacteriovorus.* **(a)** Free-living stage consists of a motile swarmer cell. This swarmer cell **(b)** attaches to a gram-negative host cell, **(c)** invades its periplasmic space, where it grows. **(d)** When the nutrients are exhausted, the four to six daughter cells lyse the host cell and leave.

eny then swim away from the dead host cell and are free to repeat the life cycle.

One of the remarkable features of this organism is that it actually produces plaques such as phage plaques on plates of host bacteria. It was this feature that enabled the German microbiologist, Heinz Stolp, to discover the bacterium when he was isolating phage from soil and sewage samples. The plaques form after most phage plaques have already developed, and they keep on enlarging, making them unusual and suspicious. When wet mount preparations from such plaques were examined in the light microscope, the small bacterium was observed quickly darting about among the cells of the host species.

Bdellovibrio cells are obligately aerobic with a tricarboxylic acid cycle. They do not use sugars but rather degrade proteins and use the amino acids as their organic carbon source. Studies of the physiology and metabolism of this bacterium have been conducted with some mutant strains, the so-called host-independent strains, that can be grown in the absence of host cells.

Bdellovibrio spp. are widespread in aquatic environments as well as in soils and sewage. Their major role appears to be analogous to that of virulent bacteriophages in controlling the numbers of gram-negative bacteria. However, they have a broader host range than typical phages and carry out a much more active type of parasitism. It is curious that they have not yet been found to be chemotactic, an ability that would seem to be of great significance in locating prey in the dilute environments where they reside.

Spirilla and *Magnetospirillum*

The spirillum is one of the more common bacterial cell shapes. Like the rod and coccus, different phylogenetic groups contain spirilla. Many of the heterotrophic spirilla are members of the Proteobacteria (Table 19.21), but spirilla are also found in other phyla, too, such as the cyanobacteria (genus *Spirulina*) and even in the Archaea (methanogen genus *Methanospirillum*).

Previously in this chapter we discussed *Azospirillum*, a nitrogen-fixing bacterium associated with monocotyledonous plants. It is a member of the alpha Proteobacteria as is the unusual magnetite-containing genus, *Magnetospirillum*. The genera *Spirillum, Aquaspirillum,* and *Comamonas* also contain spirilla-shaped bacteria, but they are all in the beta Proteobacteria. Finally, the gamma Proteobacteria also contains spirilla such as the genus *Oceanospirillum* and the photosynthetic genus, *Thiospirillum*, which will be discussed in Chapter 21.

The heterotrophic Proteobacterial spirilla are primarily microaerophilic and grow by the oxidation of organic acids. All are motile with either a single polar flagellum at each pole, or with polar flagellar tufts (lophotrichous flagellation). The genus *Spirillum* contains the largest species, *S. volutans* (see Figure 4.57), which move by a flagellar tuft at each pole, and cells that are healthy are in constant motion. These bacteria live in natural aquatic habitats and are particularly successful at growing when nutrients are in low concentration.

Magnetospirillum magnetotacticum is one of the most fascinating bacteria known. Each cell contains small crystals of magnetite (Fe_3O_4) that are aligned along the long axis of the cell (Figure 19.29). The cells actually orient themselves and move in a magnetic gradient. Therefore, if a magnet is placed against a test tube containing the cells, they will move toward and accumulate where the magnet is located. The magnetite crystals are located inside a membrane called the **magnetosome.** It is unclear whether this membrane is in direct contact with the cell membrane.

These bacteria, which are microaerophilic, prefer to live in the sediment of freshwater and marine habitats. The magnetic forces of the earth provide a direction for the cell to move down and toward the sediments where the

Differential ***Table*** **19.21 • Common genera of spirilla**

Genus	Mol % G + C	O_2 Requirement	Other Features
Spirillum	36–38	Microaerophilic	Large cells, freshwater
Aquaspirillum	49–66	Aerobic, microaerophilic	Freshwaters
Magnetospirillum	65	Microaerophilic	Magnetotactic, freshwater
Oceanospirillum	42–51	Aerobic	Marine
Ancylobacter	66–69	Aerobic	Forms rings, freshwater
Campylobacter	30–38	Microaerophilic	Intestinal tract
Helicobacter	35–44	Microaerophilic	Upper digestive tract

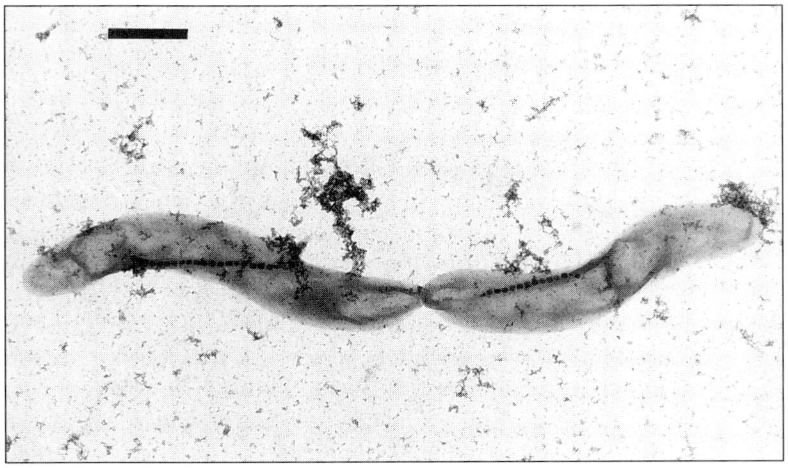

(a)

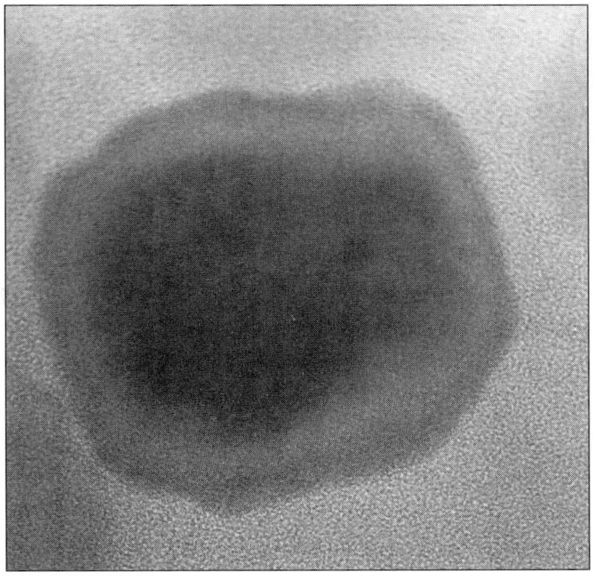

(b)

***Figure* 19.29 (a)** *Aquaspirillum magnetotacticum* cell showing its magnetosome and magnetite particles. (Courtesy of N. Pellerin and J. T. Staley) **(b)** A high resolution electron micrograph showing the structure of the magnetite crystal from *Aquaspirillum magnetotacticum.* (Courtesy of M. Sarikaya)

concentration of oxygen is low and iron is readily available to the cell. This type of movement is termed **magnetotaxis.** The magnets allow the cell to orient themselves toward the North Pole if they are in the northern hemisphere or toward the South Pole if they are in the southern hemisphere. It should be noted that the magnetic force of Earth is not sufficient to "pull" the organism, but only to orient the cell. The flagella provide the force that permits cell movement.

Materials engineers and scientists are very interested in knowing how bacteria can make such small and perfect magnets. For this reason, understanding the molecular biology of this process is one of the exciting areas of research for the future.

It is interesting to note that many other species of bacteria produce magnetite, but very few others have been isolated in pure culture. Some of them, like *Magnetospirillum,* are members of the Proteobacteria, but recent analyses of natural samples indicates that other eubacterial phyla also have magnetotactic members.

Ancylobacter (formerly *Microcyclus*) is unusual in that it forms ring shapes (see Figure 4.28). Most strains are gas vacuolate and are quite versatile physiologically. In addition to growing as an ordinary heterotroph using sugars and amino acids as carbon sources, it is able to degrade methanol and also grows as a hydrogen autotroph. This is a common genus found in freshwater environments.

Campylobacter and *Helicobacter* are curved microaerophilic bacteria that can cause disease in humans and other animals. Both are members of the epsilon subphylum of the Proteobacteria. *Campylobacter fetus* causes enteritis in humans, and *Helicobacter pylori* is implicated as the causative agent of peptic ulcers.

Myxobacteria: The Fruiting, Gliding Bacteria

The **myxobacteria** (slime bacteria) are named for the production of extracellular polysaccharides and other material they excrete as they glide on surfaces. Myxobacteria are gram-negative, rod-shaped bacteria (Figure 19.30) that have complex life cycles. The cells live together in close association with one another as they grow. Most species require complex nutrients such as amino acids and protein for growth and are therefore commonly found on animal dung in soil environments. They produce extracellular hydrolytic enzymes to degrade protein, nucleic acids, and lipids. Using these enzymes, they are able to lyse and degrade bacteria and other microorganisms as their princi-

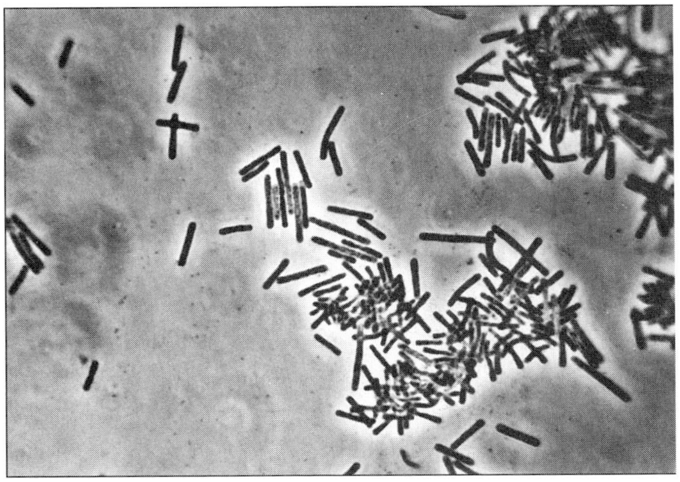

***Figure* 19.30** Typical rod-shaped vegetative cells of the myxobacterium, *Chondromyces crocatus.* (Courtesy of H. McCurdy)

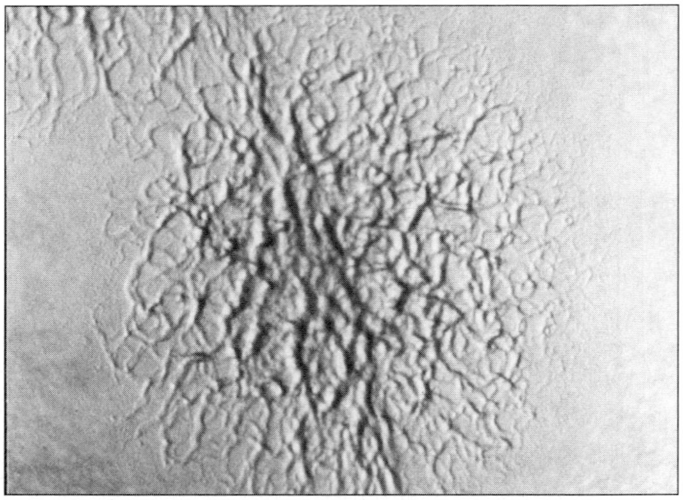

***Figure* 19.31** Swarms of the myxobacterium, *Myxococcus fulvus,* moving across an agar surface. Each finger-like projection contains hundreds of cells. (Courtesy of H. Reichenbach)

pal nutrient source. These bacteriolytic species do not use sugars as carbon sources for growth. Although they have an active tricarboxylic acid cycle, they do not have a complete Embden-Meyerhof pathway, which explains their inability to use sugars.

One genus of myxobacteria, *Polyangium,* lives on tree bark and degrades cellulose as its organic energy and carbon source and otherwise uses inorganic nutrients for growth. Therefore, it does not require the complex organic nitrogen sources needed by others. All myxobacteria, including *Polyangium,* are aerobic respiratory bacteria.

The most remarkable aspect of the developmental process of these bacteria is their ability to aggregate together to form a **swarm** (Figure 19.31) and then, as nutrients or water become depleted, to develop a complex **fruiting structure** (Figure 19.32), a dormant stage in which reproductive structures called **microcysts** (Figure 19.33) are produced. Each cell of the swarm is converted into a

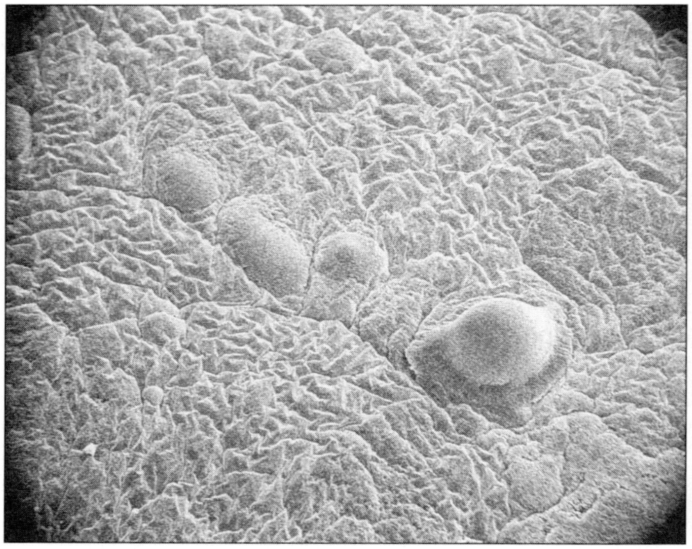

(a)

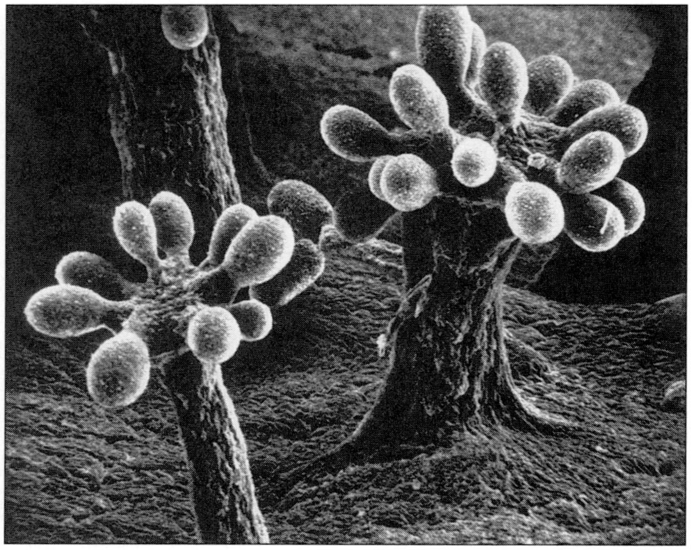

(c)

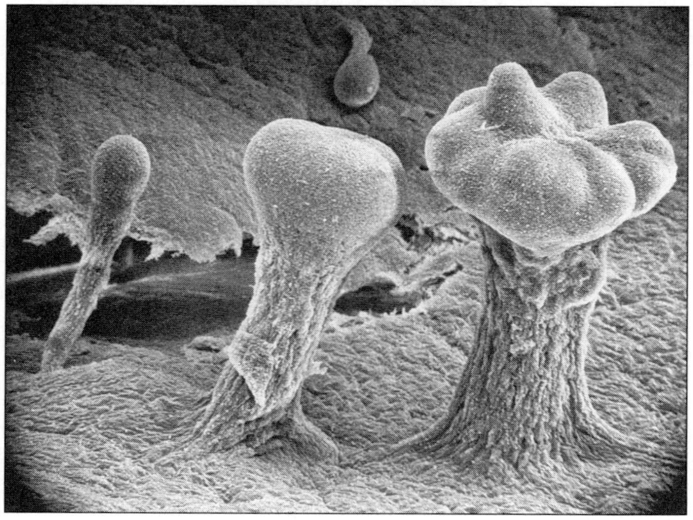

(b)

***Figure* 19.32** Scanning electron micrographs showing the development of the *Chondromyces crocatus* fruiting structure. **(a)** Cells have aggregated and just begun to rise to form the structure. **(b)** The rising fruiting structures are shown at various stages of differentiation. **(c)** The mature fruiting structure is about 1 mm in height. Each of the "fruits" of the fruiting structure is a sporangium that is filled with microcysts. (Courtesy of Patricia Grilione and Jack Pangborn)

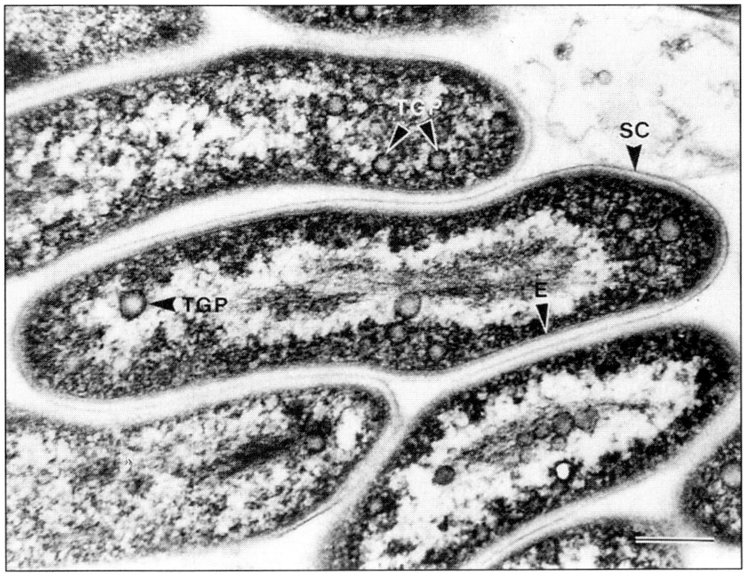

***Figure* 19.33** A thin section through microcysts of *Cystobacter fuscus* showing the cell membrane (E) and spore coat (SC). Bar is 0.5 μm. (Courtesy of H. McCurdy)

microcyst (also called **myxospore**) in the fruiting structure. The microcysts are resistant to dessication, thereby enabling the organism to survive until conditions are again favorable for growth. The microcysts can be dispersed much like the spores of fungi that form analogous fruiting structures for reproduction and dispersal. In some species, the microcysts are formed in larger structures called **sporangia.** These sporangia may be formed within the cellular mass or borne on **stalks.** Stalked varieties such as *Chondromyces crocatus* have an elaborate morphology (Figure 19.32). Table 19.22 lists the important genera of these bacteria. Because they are simple, one-celled organisms with complex life cycles, these bacteria have become model organisms for the study of developmental processes. The species that has been most thoroughly studied is *Myxococcus xanthus.*

Differential ***Table* 19.22 • Important genera of myxobacteria**

Genus	Mol % G + C	Distinguishing Features
I. Cells tapered		
Myxococcus	68–71	Spherical microcysts; no sporangia
Archangium	67–69	Rod-shaped microcysts; no sporangia
Cystobacter	68–69	Rod-shaped microcysts in nonstalked sporangia
Melittangium	Unknown	Rod-shaped microcysts in sporangium on single stalk
Stigmatella	68–69	Rod-shaped microcysts in sporangium borne singly or in clusters on stalk that may be branched
II. Cells have blunt, rounded ends; microcysts resemble vegetative cells		
Polyangium	69	Rod-shaped cells; form nonstalked sporangium; may degrade cellulose
Nannocystis	70–72	Coccoidal cells; form nonstalked sporangium
Chondromyces	69–70	Rod-shaped cells; sporangia formed singly or in clusters on stalks

Gliding motility in these bacteria is poorly understood. As with other gliding bacteria, these organisms must be in contact with a surface to be able to move. They produce extracellular slime as they glide, leaving a trail of slime as well as an indentation or actual groove, called a slime track or trail, if they are on agar (Figure 4.40). Once a track has been formed by one cell, others follow it in preference to making their own. The cells maintain a "herd" or "pack" instinct while gliding. Only rarely do they venture away from the pack except at the advancing edge of growth, and even then it is only momentary. In this period of growth, myxobacteria are actively involved in the degradation of particulate organic material that includes other microorganisms. They elaborate several lytic enzymes as they move, and if a host bacterium is approached by the pack, its cells are lysed even before physical contact is made. Therefore, although they resemble *Bdellovibrio* spp. in that they parasitize other bacteria, they do not actually enter individual host cells. Furthermore, they have a broader host range and can parasitize gram-positive as well as gram-negative bacteria.

The myxobacteria cells come together to aggregate at the onset of drying or nutrient depletion. Apparently, some as yet unknown chemical signal draws the cells together. They pile on top of one another to form a mound, which eventually becomes the fruiting structure. The fruiting structure develops into a distinctive shape characteristic for each species. These fruiting structures as well as the vegetative cell masses are pigmented in vivid yellow, orange, or red colors, depending upon the species.

In the simplest of the fruiting structures, such as the one formed by *Myxococcus xanthus,* the cells aggregate together to form a small, raised mound about 1 mm in diameter and height. Each of the cells in the fruiting structure is converted into a microcyst. The microcyst then is a modified cell which shows greater resistance to both drying and ultraviolet light exposure than does the vegetative cell. Although the microcysts also exhibit somewhat greater resistance to heating, they are not similar morphologically or physiologically to the more inert, heat-resistant endospores of the gram-positive bacteria.

The myxobacteria with the most elaborate fruiting structures form their microcysts in a special structure called a **sporangium.** Sporangia may contain hundreds of microcysts. In some species, the sporangium lies on the same surface as the vegetative growth. However, in the more exotic forms, the cells aggregate, climb on top of one another, and produce a stalk. The stalk consists of hardened, extracellular slime. The sporangia then develop on top of the stalk (Figure 19.32). Clearly, these are among the most complex prokaryotic organisms.

Most myxobacteria are bacteriolytic and require complex organic materials for growth. They obtain these nutrients by killing and lysing other bacteria. To perform this activity, the myxobacteria produce lysozyme and some of their own antibiotics, most of which have not yet been characterized. Once they lyse the host bacterial cells, they degrade their proteins, lipids, and nucleic acids. Most myxobacteria live in animal dung, which contains other bacteria in high concentrations. However, *Polyangium* are cellulolytic and grow by degrading wood cellulose. Thus, they are found on logs and bark in forested areas.

Sulfate- and Sulfur-Reducing Eubacteria

The sulfate- and sulfur-reducing Eubacteria are a physiologically and ecologically related group of bacteria. They are strict anaerobes that generate energy by anaerobic respiration of a variety of organic compounds. Elemental sulfur (S°) or oxidized sulfur compounds such as sulfate act as terminal electron acceptors that are converted to H_2S in the process of dissimilatory sulfur reduction. Some of the common genera of the sulfate and sulfur reducers and their characteristics are given in Table 19.23.

Not all members of this group are closely related to one another phylogenetically. Most eubacterial sulfate reducers, however, such as the common genus *Desulfovibrio,* are members of the delta subphylum of the Proteobacteria. Although *Desulfonema* differs from the others in that it is filamentous, moves by gliding, and stains as a gram-positive bacterium (Figure 19.34), phylogenetic analysis indicates it is also a member of the delta Proteobacteria. Its closest relative is *Desulfosarcina variabilis,* a packet-forming (sarcina) coccus.

In contrast, *Desulfotomaculum* is not a Proteobacterium. It is a genus of endospore-forming bacteria with a broad range of DNA base composition (37–50 mol % G + C). 16S rRNA sequence analysis confirms that this endospore-forming genus does not belong to the Proteobacteria, but instead is closely related to *Clostridium* and other gram-positive bacteria. Nonetheless, it is treated here because of its physiological and ecological similarity.

Sulfur- and sulfate-reducers are abundant in anaerobic soil, as well as freshwater, estuarine, and marine sediments. They are especially common in estuarine and marine sediments which are in constant contact with seawater that is rich in sulfate. Aquatic and terrestrial habitats that become anaerobic through decomposition of organic matter are rich in sulfate and sulfur reducers. Sulfate-reducing bacteria also live in the intestinal tract of some animals.

Sulfur- and sulfate-reduction is not limited to these eubacterial genera considered here. Some Archaea (see Chapter 22) also carry out the reduction of sulfur compounds. In addition, it should be noted that other ordinary heterotrophic bacteria, such as certain enteric bacteria and pseudomonads, can reduce sulfate to sulfide; however, these organisms cannot use sulfate as the sole terminal electron acceptor in anaerobic respiration, the hallmark of the dissimilatory sulfate reducers.

Differential *Table* **19.23 • Representative genera of sulfate- and sulfur-reducing bacteria[1]**

Genus	Cell Shape	Carbon Source	Electron Acceptor
Desulfovibrio	Vibrio	Lactate, ethanol	SO_4^{2-}; SO_3^{2-}; $S_2O_3^{2-}$
Desulfococcus	Coccus	Lactate, ethanol	SO_4^{2-}; SO_3^{2-}; $S_2O_3^{2-}$
Desulfosarcina	Ovoid; aggregates	Lactate, benzoate	SO_4^{2-}; SO_3^{2-}; $S_2O_3^{2-}$
Desulfobacter	Rod; vibrio	Acetate	SO_4^{2-}; SO_3^{2-}; $S_2O_3^{2-}$
Desulfobulbus	Oval	Lactate, acetate	SO_4^{2-}
Desulfomonile	Rod	Pyruvate, benzoate	SO_4^{2-}; SO_3^{2-}; $S_2O_3^{2-}$
Desulfomicrobium	Rod	Lactate	SO_4^{2-}; SO_3^{2-}; $S_2O_3^{2-}$
Desulfobotulus	Vibrio	Higher fatty acids	SO_4^{2-}; SO_3^{2-}; $S_2O_3^{2-}$
Desulfobacterium	Rod; vibrio	Lactate; acetate, Higher fatty acids	SO_4^{2-}; SO_3^{2-}; $S_2O_3^{2-}$
Thermodesulfobacterium	Thermophilic rod	Lactate	SO_4^{2-}
Desulfurella	Rod	Acetate	S^0
Desulfuromonas	Oval, curved rod	Acetate, propionate	S^0
Desulfotomaculum[2]	Gram-positive spore-forming rod	Lactate, fatty acids	SO_4^{2-}

[1]It should be noted that many thermophilic Archaea are also dissimilatory sulfur- and sulfate-reducing organisms (see Chapter 22).

[2]Although *Desulfotomaculum* spp. stain as gram-negative bacteria, they are spore-forming bacteria that are related to the gram-positive bacteria.

The organic energy sources used by sulfur- and sulfate-reducing bacteria (such as acetate, lactate, ethanol, pyruvate, and butyrate) are end products of fermentation produced by other bacteria. As noted in Table 19.23, many sulfur reducers cannot catabolize acetate because they have an incomplete tricarboxylic acid cycle. Those with a complete tricarboxylic acid cycle oxidize acetate completely to CO_2. In addition to the respiratory generation of ATP with sulfate as the terminal electron acceptor, some genera carry out fermentations. Thus, *Desulfococcus, Desulfosarcina,* and *Desulfobulbus* ferment pyruvate or lactate to propionate and acetate. This provides an electron flow that can be utilized in ATP generation.

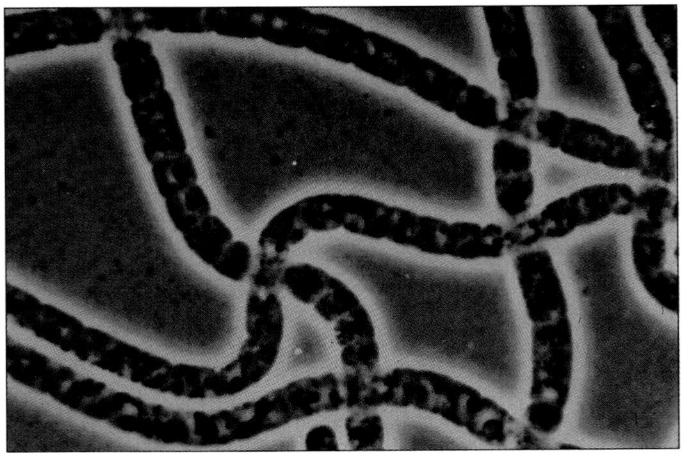

***Figure* 19.34** *Desulfonema,* a genus of gliding filamentous sulfate-reducing bacteria. (© F. Widdle/VU)

Other Heterotrophic Eubacterial Groups

As can be seen from the material presented heretofore, the Proteobacteria are a very diverse group of Eubacteria. Several other heterotrophic eubacterial groups also exist and are discussed below. Unlike the Proteobacteria, however, none of them are yet known to have chemoautotrophic or photosynthetic members.

Cytophaga and *Bacteroides*

This diverse collection of bacteria contains heterotrophic, nonphotosynthetic organisms. Some are aerobic or facul-

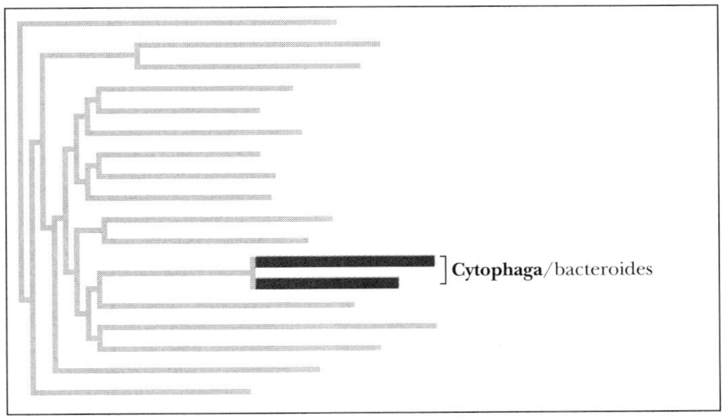

tative aerobic and move by gliding motility as represented by the cytophaga group, whereas others are anaerobic nonmotile fermenters represented by the *Bacteroides* group. These bacteria and their associates are discussed here individually.

Cytophaga *Group*

These are gliding, rod-shaped to filamentous bacteria (Figure 19.35), some of which are over 100 μm in length, depending upon the genus (Table 19.24). Unlike the fruiting myxobacteria, which are also gliding heterotrophs, these bacteria have very low DNA base compositions (mol % G + C of about 30–45). Most are obligately aerobic, but some are fermentative. Because of their gliding motility their colonies are thin and spreading, almost always with a yellow to orange (or, rarely, red) pigmentation. The pigments are cell-bound carotenoids and **flexirubins** (Figure 19.36), some of which are unusual biologically because they are chlorinated. One diagnostic feature of these bacteria is the "flexirubin reaction" in which the color of the colony changes from yellow to purple to red-brown when it is flooded with alkali (20% KOH).

These bacteria are aerobic respirers or facultative aerobic fermenters. Therefore, they are thought to have an Embden-Meyerhof glycolytic pathway and a TCA cycle. Electron transport systems contain menaquinones as the respiratory quinone.

A typical fermentation of glucose by *Cytophaga succinicans* produces succinate, acetate, and formate (Figure 19.37). *C. succinicans* requires carbon dioxide for anaerobic growth and it is thought that they produce succinate by condensation of phosphoenolpyruvate with carbon dioxide, followed by a reduction to succinate.

The genera *Cytophaga, Flexibacter* (Figure 19.38), and *Microscilla* are best known for their ability to degrade macromolecules. Indeed, Sergei Winogradsky described the first member of *Cytophaga* as a cellulose-degrading organism. Cellulose is degraded while these bacteria grow and glide in direct contact with the cellulose fibers. These organisms produce a cell membrane–bound endoglucanase as well as soluble periplasmic exoglucanases that are responsible for cellulose degradation. Some species of marine origin can degrade agar, and others—from marine, freshwater, and soil habitats—degrade chitin. In addition, almost all species are able to degrade starch, and

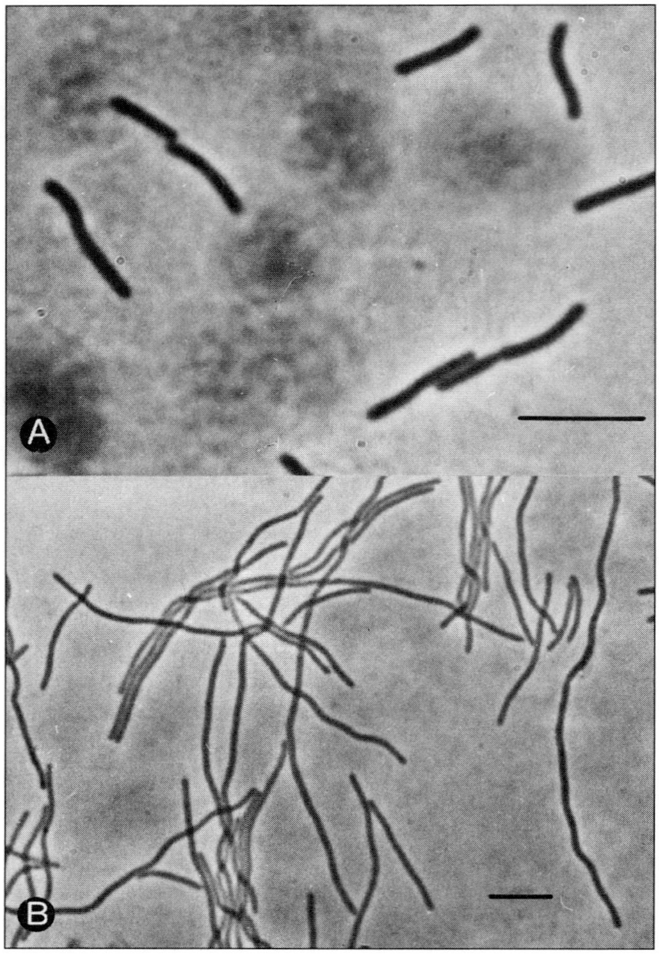

***Figure* 19.35** A chitin-degrading marine *Cytophaga, C. latercula* grown at **(A)** 21° C and **(B)** 30° C. Bar is 5 μm. (Courtesy of Hans Reichenbach)

***Differential Table* 19.24 • Principal genera of the *Cytophaga* group**

Genus	Mol % G + C	Cell Length	Special Features
Cytophaga	30–45	1.5–15 μm	Some degrade cellulose, agar, or chitin
Capnocytophaga	33–41	2.5–6 μm	Oral cavity
Flexibacter	37–47	15–50 μm	Some degrade chitin or starch; freshwater
Microscilla	37–44	10– > 100 μm	Like *Flexibacter,* but marine, and do not degrade chitin

A new set of flagella appear at the middle of a cell, a septum is laid down, and the two cells separate. Some strains are multicellular and grow to 250 μm or longer.

This free-living genus is the best understood of all the spirochetes. The most thorough studies on the metabolism, physiology, and motility have been conducted with members of this genus.

Cristispira

This genus was assigned the name *Cristispira* because members of the genus form a bundle of 100 or more periplasmic flagella. When these flagella are intertwined on the protoplasmic cylinder, they distend the outer sheath, forming a ridge or crest. *Crista* is the Latin word for "crest." *Cristispira* divide by binary transverse fission. The genus is widely distributed in both marine and freshwater molluscs (clams, oysters, and mussels). They inhabit the crystalline style that is a part of the digestive tract of these invertebrates. They can also be found in other organs in molluscs. The *Cristispira* are not pathogenic and occur in healthy univalve and bivalve molluscs. None, however, has ever been isolated in pure culture.

Treponema

Members of the genus *Treponema* live in the mouth, digestive tract, and genital areas of humans and other animals. One of the ancient scourges of humans is due to a member of this genus, *Treponema pallidum,* the causative agent of syphilis. *T. pallidum,* subspecies *petunae,* causes yaws, a contagious disease in tropical countries (see Chapter 29). The treponemes are strictly anaerobic or microaerophilic. The human pathogens are probably microaerophiles but have not been grown on artificial media. The treponemes that have been cultured are anaerobic and ferment carbohydrates or amino acids. Some require long-chain fatty acids for growth. Others require the short-chain volatile fatty acids present in rumen fluid.

Borrelia

Some species of *Borrelia* have been grown in culture. All are microaerophilic with complex nutritional requirements. Two noted pathogenic types occur in the genus. *B. burgdorferi* is the causative agent of Lyme disease, which is transmitted to humans by the deer tick *Ixodes dammini.* Other species are tick-borne pathogens of relapsing fever. Both diseases are discussed more fully in Chapter 29.

Leptospira

Leptospira is the simplest of the spirochetes. Species in this genus have only two fibrils in their axial filament. The cells are tightly coiled and the fibrils of the axial filament rarely overlap. Frequently the cells have a hook at both ends. Most species are free-living organisms that occur in soil, freshwater, and marine environments. Some are parasites of humans and other animals, while others are pathogenic. *Leptospira* spp. are obligate aerobes with a DNA base composition of 35 to 41 mol % G + C. The major cell wall diamino acid is diaminopimelic acid.

The source of carbon and energy for members of this genus is long-chain fatty acids (15 carbons or more) or long-chain fatty alcohols. They cannot synthesize fatty acids and directly incorporate substrates into cellular lipids. Based on DNA/DNA hybridization considerations, at least seven species of *Leptospira* are known, but only two are recognized. These are the free-living *Leptospira biflexa* and the pathogen, *L. interrogens.* The latter is the causative agent of leptospirosis, a disease known worldwide. The natural host for leptospirosis is rodents and other animals (see Chapter 29).

Spirochetes in Termites

Microscopic examination of the microbiota in the hindgut of termites and wood-eating cockroaches indicates that spirochetes are part of the normal flora. Some appear to be free-living in the gut fluids, whereas others are **ectosymbionts** attached to the surface of protozoa that also inhabit these areas. These bacteria were first observed in 1877 by J. Leidy and were regarded as vibrios or spirilla. Electron microscopic examination, however, confirmed that they are spirochetes. Many morphological types occur in the hindgut of each termite species. For example, one *Pterotermes occidentis* specimen examined had at least 15 types based on size of the cell body, wavelength and amplitude of primary coils, and number of periplasmic flagella. The spirochetes vary in size from 0.2×3 μm to as large as 1×100 μm. The number of periplasmic flagella ranges from few to over 100. Little is known of the physiology of these spirochetes, as they have not yet been grown in culture. They are apparently anaerobic because they lose motility when exposed to air. The presence of the spirochetes appears to be beneficial to the hosts because their removal results in a shortening of the host's life span.

Thermophilic Heterotrophic Gram-Negative Eubacteria

Prior to 1965, a few species of the gram-positive, endospore-forming genera *Bacillus* and *Clostridium* were considered to be the major thermophilic bacteria. During the mid-1960s, Thomas Brock initiated a study of the microbial populations of thermal springs of Yellowstone National Park in Wyoming. He and his co-workers discovered an array of gram-negative bacteria in these hot springs and other thermal environments. These investigations led to extensive studies worldwide and the resultant discovery of a diverse and fascinating array of thermophilic species. Some of the organisms isolated were Eubacteria, whereas others are Archaea (see Chapter 22).

Some of the genera of eubacterial thermophiles that have been cultured since that time from various hot springs and other geothermal environments are listed in

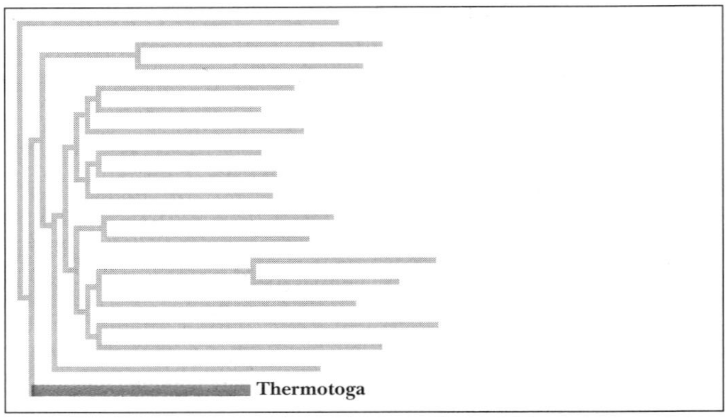

Table 19.28. All are heterotrophic gram-negative rods of various sizes with optimum growth temperatures in the range of 40°C to 100°C. Most would be considered moderate thermophiles in contrast to some of the extreme thermophilic Archaea that grow optimally at 100°C or higher. Some of the genera are aerobes and others are anaerobes. It is interesting to consider that several genera—*Thermomicrobium, Thermotoga, Thermoleophilum,* and *Aquifex*—represent rather deep branches in the eubacterial lineage and are not closely related to any other known mesophilic eubacterial groups. *Thermus* is grouped with the radiation-resistant genus, *Deinococcus.*

Thermomicrobium

Thermomicrobium roseum is an aerobic, pink-pigmented, nonmotile bacterium originally isolated from Toadstool Spring in Yellowstone National Park. The temperature at this site was 74°C. *Thermomicrobium* cells are short pleomorphic rods and dumbbell shapes. They grow best on a medium using dilute concentrations (0.5 percent) each of yeast extract and peptone.

Thermus

This is a genus of aerobic nonmotile bacteria that forms filaments up to 200 μm in length. Colonies are often pigmented yellow to red due to the formation of carotenoids. Many strains form rotund bodies, 10 μm to 20 μm in diameter, that are produced by the association of individual cells. The role of these bodies is unknown. *Thermus* is of widespread occurrence, and strains have been isolated from thermal environments throughout the world. They grow in hot springs and in other thermal environments such as hot water heaters. One species, *Thermus aquaticus,* is noted in molecular biology for its heat-stable DNA polymerase, the Taq polymerase, that is used in polymerase chain reaction technology. This genus is a member of the *Deinococcus* division, which is discussed with the gram-positive bacteria in Chapter 20.

Thermoleophilum

These are very small gram-negative rods (see Figure 7.6). Colonies are devoid of pigments, and the cells are nonmotile. The organism has been isolated from various hot springs and also from environments that are considered nonthermal. The minimum growth temperature for *Thermoleophilum* strains is 45°C, a temperature found even in some soils exposed to direct sunlight. The unique feature of this genus is its inability to grow on any substrate except for *n*-alkanes of 13 to 20 carbons in length.

***Differential Table* 19.28 • Characteristics of thermophilic Eubacteria**

Genus	O_2 Requirement	Temperature Range (°C)	Mol % G + C	Habitat (Substrate)
Thermus	Aerobe	40–85	60–67	Hot water heaters (glucose; acetate)
Thermomicrobium	Aerobe	45–75	64	Hot springs (tryptone; yeast extract)
Thermoleophilum	Aerobe	45–70	68–71	Hot springs (use alkanes)
Thermotoga	Anaerobe	55–90	46–51	Marine thermal vents (sugars)
Fervidobacterium	Anaerobe	40–80	33–40	Hot springs (sugars)
Thermosipho	Anaerobe	35–77	60–63	Marine thermal vents (yeast extract; peptone)
Aquifex[1]	Aerobe	60–95	40	Hot springs H_2 autotroph
Hydrogenobacter[1]	Aerobe	60–80	38–44	Hot springs H_2 autotroph

[1]These two genera will be treated in Chapter 21.

Thermotoga, Fervidobacterium, and *Thermosipho*

These genera are closely related anaerobic organisms isolated from submarine thermal environments. These bacteria represent one of the deepest and most slowly evolving branches of any known lineage of Eubacteria. *Thermotoga* can ferment sugars such as glucose to lactate, acetate, CO_2, and H_2. *Thermosipho* grows on rich media such as yeast extract and requires the amino acid cysteine. Cells of these three genera are surrounded by a proteinaceous sheathlike structure, a so-called **toga** that balloons over the ends of the rods. A number of rods in a chain may be surrounded by one sheath; the function of the sheath is unknown. Some strains of *Thermotoga* are motile by monotrichous flagellation.

Aquifex and *Hydrogenobacter*

The *Aquifex* group comprises the most thermophilic species of Eubacteria known. The maximum growth temperatures of some species exceed 95°C. They are obligate hydrogen autotrophs that do not grow on organic compounds and therefore are treated in Chapter 21.

Summary

- The **Proteobacteria** is the most diverse phylum of Eubacteria, containing heterotrophic, chemoautotrophic, and photosynthetic bacteria. The Proteobacteria include the alpha, beta, gamma, delta, and epsilon subgroups. The **enteric bacteria** including *Escherichia coli, Enterobacter, Erwinia, Salmonella,* and *Shigella,* are all facultative aerobes that ferment sugars and belong to the gamma subgroup. Some enteric bacteria cause human disease—for example, **typhoid fever** (*Salmonella typhi*), **plague** (*Yersinia pestis*), and **bacterial dysentery** (*Shigella dysenteriae*).

- *Vibrio* species are typically marine or estuarine members of the gamma Proteobacteria; some are pathogenic to fish and humans. *Vibrio cholerae* is the causative agent of **human cholera.** The *Vibrio* group are facultative aerobic bacteria that ferment sugars. Some *Vibrio* and *Photobacterium* spp. produce light by **bioluminescence** using bacterial luciferase; many luminescent bacteria live as symbionts in the light organs of fish.

- **Pseudomonads,** also Proteobacteria, include the genera *Pseudomonas, Xanthomonas,* and *Zoogloea,* and are obligately aerobic, oxidase positive, gram-negative rods that are motile by polar flagella. *Pseudomonas* species may be pathogens of humans (*P. aeruginosa*) or plants (*P. syringae*). *Pseudomonas* spp. are very versatile in their degradative abilities; many can use over 100 different sugars, organic acids, amino acids, and other monomers as sole carbon sources for growth. Some *Pseudomonas* spp. can degrade toxic compounds such as toluene and naphthalene, and the genes for this degradative ability are borne on plasmids. Some *Pseudomonas* and *Xanthomonas* spp. are plant pathogens that cause ice nucleation, which damages plant cells.

- *Azotobacter, Beijerinckia, Derxia,* and *Azomonas* are free-living soil bacteria capable of **nitrogen fixation.** *Rhizobium* and *Bradyrhizobium* are **symbiotic nitrogen fixing** bacteria that are associated with legumes where they form root nodules.

- *Agrobacterium tumefaciens* causes **crown gall** disease in dicotyledinous plants. *Agrobacterium tumefaciens* carries a plasmid, called the **T_i (tumor inducing) plasmid,** the DNA of which is incorporated into the plant host DNA upon infection; normally this plasmid contains genes that cause gall formation and plant disease.

- The Proteobacteria contain many other important genera. Included are *Acetobacter* and *Gluconobacter,* which are aerobes that produce acetic acid and are therefore called the **vinegar bacteria.** *Legionella pneumophila* is the pathogen that causes **Legionnaire's disease.** *Neisseria* is a genus of gram-negative cocci that cause **gonorrhea** (*N. gonorrhoeae*) and **bacterial meningitis** (*N. meningitidis*). *Rickettsia* and *Chlamydia* are two genera of obligate intracellular parasites. **Sheathed bacteria,** including *Sphaerotilus* and *Leptothrix,* live in flowing freshwater habitats; some deposit iron oxides and manganese oxides in their sheaths.

- **Prosthecae** are cellular appendages produced by some mostly aquatic bacteria. Three groups of prosthecate bacteria exist: the caulobacters, the hyphomicrobia, and the polyprosthecate bacteria.

- **Bdellovibrio** is a bacterium that is a predator of other gram-negative bacteria. Spirilla are typically microaerophilic, heterotrophic bacteria that are members of the alpha, beta, and gamma Proteobacteria. *Magnetospirillum magnetotacticum* is a **magnetotactic bacterium** that has magnetite crystals that allow it to orient in a magnetic field.

- The **myxobacteria** are gliding bacteria that have a complex life cycle. Cells move in "packs" when nutrients are plentiful; when nutrients are depleted, the cells aggregate and undergo a complex fruiting process in which they are converted to a more resistant stage called a **microcyst,** which is formed in a **fruiting structure;** microcysts can germinate when conditions for growth again become favorable.

- **Sulfate- and sulfur-reducers** oxidize organic compounds (some can also use hydrogen gas) and pass the electrons on to sulfate and sulfur in an anaerobic respiration; the end product is H_2S. Sulfate reducers are particularly common in marine and estuarine environments where sulfate is plentiful. With the exception of *Desulfotomaculum,* all eubacterial sulfate reducers are members of the Proteobacteria.

- The *Cytophaga–Bacteroides* phylum contains aerobic gliding bacteria and anaerobic fermenters, respectively. *Cytophaga* are aerobic or facultative aerobic; most species are noted for degradation of polymeric substances. They are frequently pigmented yellow to orange in color.

- *Bacteroides* spp. are common fermentative bacteria that are found in the digestive tract of animals, where they occur in very high concentrations.

- The **Planctomycetales** comprise a separate phylogenetic group containing budding bacteria that are widespread in aquatic habitats and soils. *Chlamydia,* which are the closest relatives of the Planctomycetes, are termed **intracellular energy parasites** because they require ATP from their host cells.

- The **spirochetes** comprise their own phylogenetic group of helical bacteria with unique **periplasmic flagella,** called **axial filaments,** that impart a characteristic motility which is especially well adapted to highly viscous environments. Several spirochetes are pathogenic including *Treponema pallidum* (syphilis), *Borrelia burgdorferi* (Lyme disease), and *Leptospira* (leptospirosis).

- The deepest phylogenetic branches of the Eubacteria contain **thermophiles** including *Thermotoga, Thermoleophilum,* and *Aquifex*—some of which grow at temperatures as high as 95°C. *Thermoleophilum* is an aerobic genus that uses alkanes as a carbon source for growth. The *Thermotoga* group (*Thermotoga, Thermosipho,* and *Fervidobacterium*) produce a characteristic external sheath. *Aquifex* and *Hydrogenobacter* form the **deepest eubacterial branch** and are aerobic, **obligate hydrogen autotrophs.**

Questions for Thought and Review

1. The Proteobacteria constitute an extremely diverse group of bacteria. Cite examples of their diversity. How could such great diversity occur in a single bacterial group?
2. Some bacteria move by gliding, some have flagella, and others are nonmotile. What are the advantages or disadvantages of each type of existence?
3. How is it that some gliding bacteria are found in the Proteobacteria and others are found in a completely different evolutionary group?
4. What is the role(s) of the prostheca? Why do some bacteria have more than one?
5. When you enter a winery you smell vinegar. Why?
6. What are possible desirable features that could be engineered into plants by use of the *Agrobacterium* T_i plasmid? Could undesirable features be engineered, too?
7. Do any Eubacteria lack peptidoglycan? Would you expect them to be related to an ancient group of the Eubacteria or a rapidly evolving one?
8. Of what advantage to an organism is the sheath?
9. Compare the motility of spirochetes to that of flagellated and gliding bacteria.
10. Why is it that only prokaryotic organisms can carry out nitrogen fixation?
11. Why are *Legionella* species difficult to grow?
12. How is it possible that some bacteria can grow at high temperatures?

Suggested Readings

Balows, A., H. G. Trüper, M. Dworkin, W. Harder, and K. H. Schleifer, eds. 1992. *The Prokaryotes.* 2nd ed. Berlin: Springer-Verlag.

Bergey's Manual of Determinative Bacteriology. 1994. 9th ed. J. G. Holt, Editor-in-Chief. Baltimore: Williams and Wilkins.

Bergey's Manual of Systematic Bacteriology. 1984–89. 1st ed., Vol I (N. Krieg and J. G. Holt, eds) and III (J. T. Staley, N. Pfennig, M. Bryant, and J. G. Holt, eds). Baltimore: Williams and Wilkins.

The most important discoveries of the laws, methods and progress of Nature have nearly always sprung from the examination of the smallest objects which she contains.

J. B. Lamarck, *Philosophie Zoologique,* 1809

Chapter 20

Gram-Positive Eubacteria

Common Gram-Positive Eubacteria and *Mollicutes*
Deinococcus and *Thermus*

The common gram-positive Eubacteria comprise a spectrum of morphological types ranging from unicellular organisms to branching, filamentous, multicellular organisms. Most are heterotrophs that grow as saprophytes existing off nonliving organic materials. Gram-positive bacteria are particularly abundant in soil and sediment environments. Some are important plant and animal pathogens. This phylogenetic group contains some of the simplest heterotrophic organisms from a metabolic standpoint, such as the lactic acid bacteria. A few gram-positive bacteria are hydrogen autotrophs that make acetic acid from carbon dioxide and H_2. Only one photosynthetic group, the heliobacteria, has been described. In addition, *Deinococcus* and *Thermus,* which form a separate phylogenetic group, will also be discussed in this chapter because the deinococci are gram-positive.

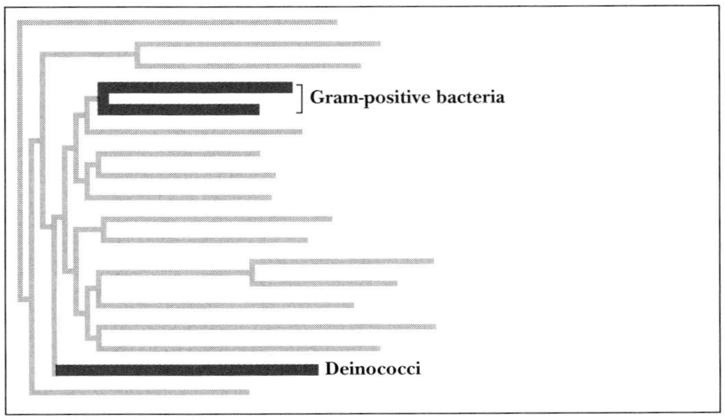

Common Gram-Positive Eubacteria and *Mollicutes*

Gram-positive Eubacteria contain significant amounts of peptidoglycan in their cell walls (see Chapter 4). Their cell walls, however, lack lipopolysaccharide, a characteristic component of the cell walls of gram-negative Eubacteria. They also lack a periplasmic space and therefore have a simpler cell wall architecture. However, one subgroup, the *Mollicutes* or **mycoplasmas,** lack a cell wall altogether. Nonetheless, they are close relatives of the gram-positive bacteria based on 16S rRNA sequence analyses.

This phylogenetic group contains both low as well as high mol % G + C subgroups. Most of the unicellular species have a low mol % G + C content whereas the **mycelial** species, which have a characteristic filamentous,

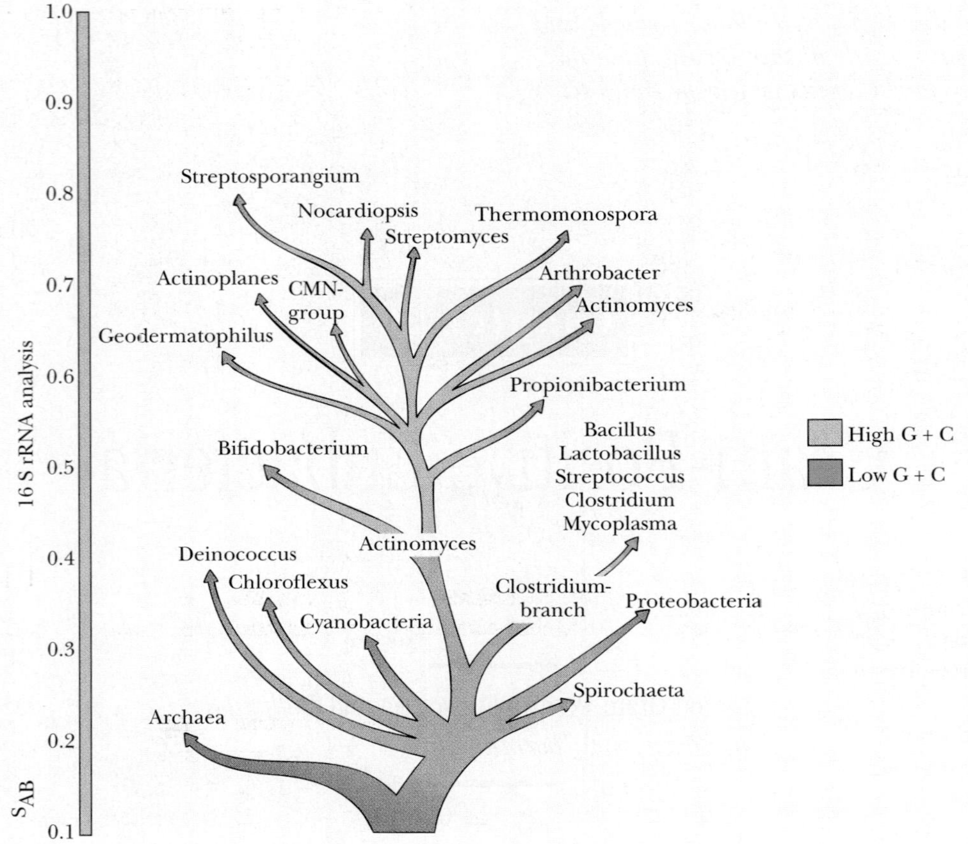

***Figure* 20.1** A diagram showing the relatedness of gram-positive bacteria to other phylogenetic groups by use of 16S rRNA analyses. Note that the unicellular, low mol % G+C lactic acid bacteria and endospore-formers are found in one branch, whereas the mycelial high mol % G+C actinomycetes are found in a separate branch. (Adapted from E. Stackebrandt)

branching morphology, all have a high mol % G + C content. The phylogenetic relatedness of the gram-positive bacteria to other bacterial phyla and themselves is shown in Figure 20.1. Note that the low GC group branches off separately from the high GC group. Each subgroup of the phylum is discussed individually below beginning with the unicellular, low mol % G + C taxa.

Lactic Acid Bacteria

As their name implies, the **lactic acid bacteria** are noted primarily for their ability to ferment sugars to produce lactic acid. This family contains several important genera whose activities in human health and the environment are familiar to all of us. Most of us have had sore throats, called "Strep" throat, caused by *Streptococcus* species. The genus *Streptococcus* contains several important pathogens, such as *Streptococcus pyogenes,* which causes scarlet fever and rheumatic fever.

The lactic acid bacteria are also important in food and dairy microbiology as well as in agriculture. The fermentation of vegetable materials is used in the manufacture of pickles and sauerkraut. Milk fermentation by lactic acid

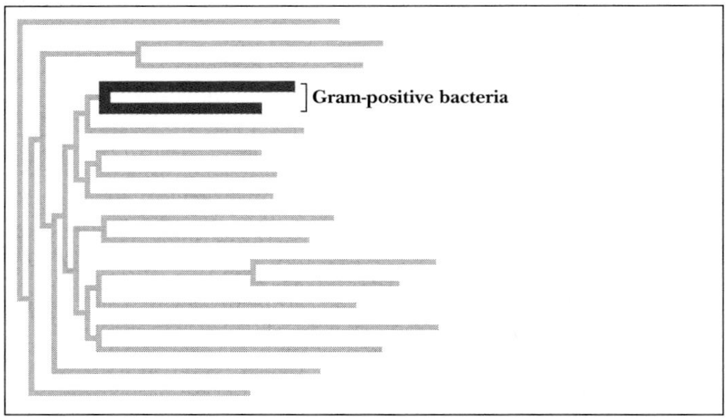

bacteria leads to the manufacture of commercial products such as yogurt, buttermilk, and cheeses.

All lactic acid bacteria are **aerotolerant anaerobes:** that is, they are facultative aerobes that grow in the presence of oxygen but do not use it in respiration. Instead, they produce energy by fermentation of sugars. In addition, most of them have complex nutritional requirements. Most require amino acids and vitamins, and some even need purines and pyrimidines to grow. Not surprisingly, they are found in organic-rich environments such as de-

Differential *Table* **20.1 • Principal genera of lactic acid bacteria**

Genus	Morphology	Fermentation Type	Lactic Acid Form
Streptococcus	Cocci in chains	Homofermentative	L-
Leuconostoc	Cocci in chains	Heterofermentative	D-
Pediococcus	Cocci in tetrads	Homofermentative	DL-
Lactobacillus	Rods	Homo- or Heterofermentative	Varies with species

caying plant and animal materials. Table 20.1 lists the major genera of lactic acid bacteria, as well as some of their common features.

Physiology and Metabolism

Although all lactic acid bacteria carry out the lactic acid fermentation, they can be separated into two different groups based on the type of fermentation process. The **homofermentative** bacteria carry out a simple fermentation in which lactic acid is the sole product from sugar fermentations:

$$\text{Glucose} \longrightarrow 2 \text{ Lactic acid}$$

The Embden-Meyerhof or glycolytic pathway is used in this fermentation. The pyruvic acid formed is reduced by the enzyme lactic acid dehydrogenase to produce the characteristic end product, lactic acid (Figure 20.2).

In contrast, the **heterofermentative** lactic acid bacteria produce ethanol and carbon dioxide as well as lactic acid:

$$\text{Glucose} \longrightarrow \text{Lactic acid} + \text{Ethanol} + CO_2$$

The heterofermentative pathway (Figure 20.3) lacks the key enzyme aldolase present in the glycolytic pathway of the homofermentative lactics. Although homofermenters obtain 2 moles of ATP for each mole of glucose fermented, the heterofermenters obtain only 1 mol, as can be seen by examining their pathway. This pathway involves an initial oxidation of glucose to 6-phosphogluconic acid, which is decarboxylated to form CO_2 and ribose 5-phosphate. The pentose formed is converted to (a) the three-carbon intermediate, glyceraldehyde 3-phosphate, which in turn gives rise to one molecule of ATP by substrate-level phosphorylation in the formation of lactic acid, and (b) acetyl phosphate, which is reduced to acetaldehyde and then to ethanol.

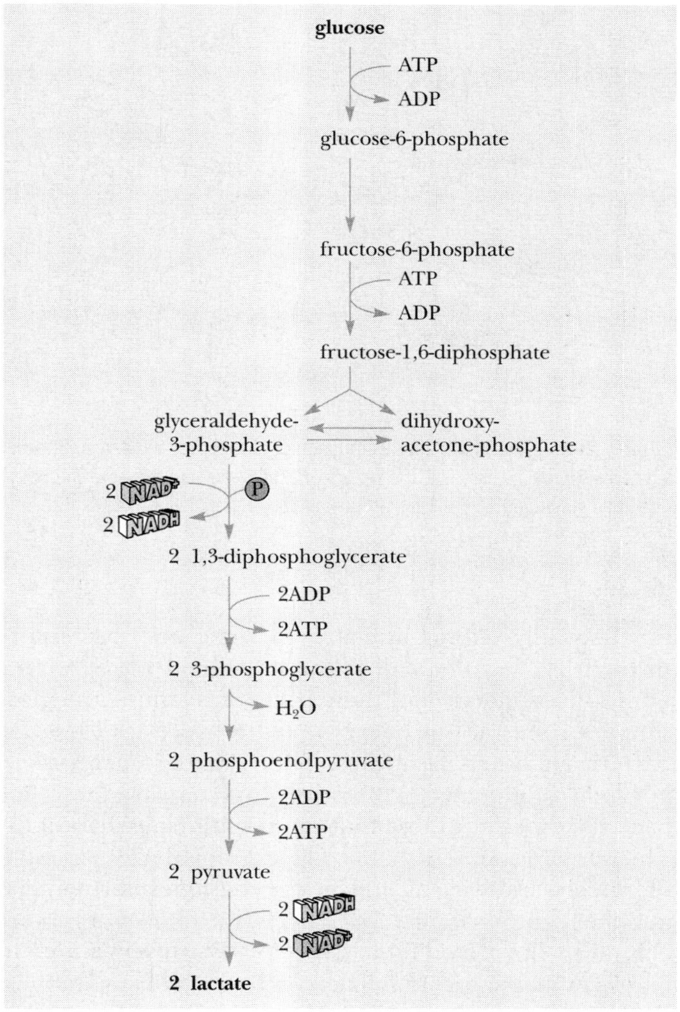

***Figure* 20.2** The pathway for dissimilation of glucose by homofermentative lactic acid bacteria. These bacteria use the glycolytic pathway for formation of pyruvate and lactic acid dehydrogenase to produce lactic acid as the final end product.

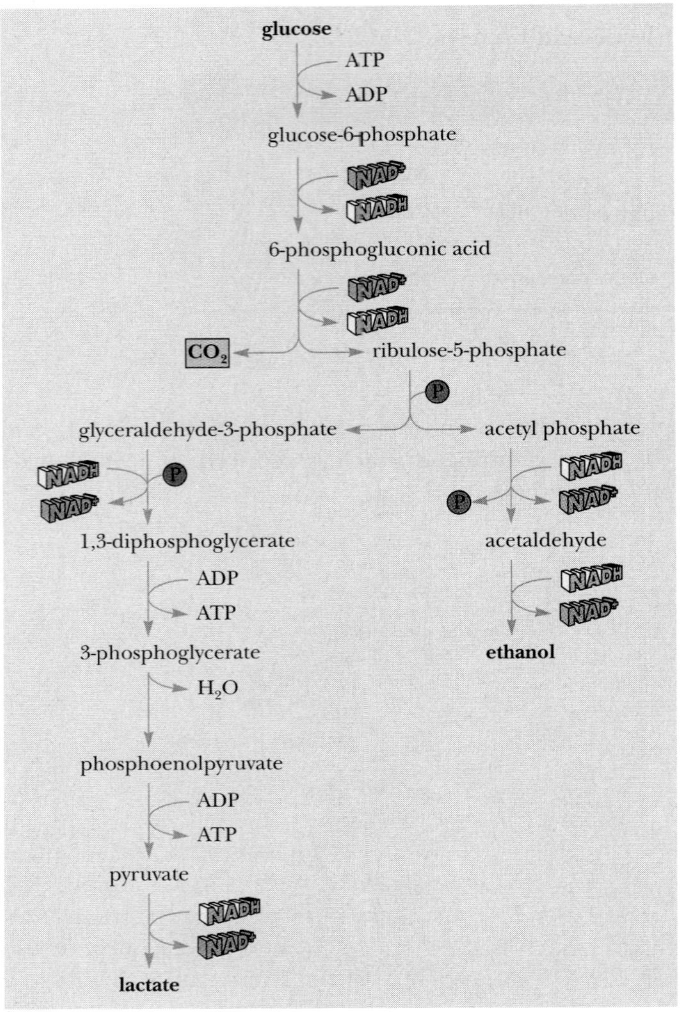

***Figure* 20.3** The pathway for dissimilation of glucose by heterofermentative lactic acid bacteria which lack the key enzyme, aldolase, and therefore do not use the Embden-Meyerhof fermentation (see text for details).

The only energy available to lactic acid bacteria is through ATP generated by substrate-level phosphorylation. The metabolism of these bacteria is among the most simple of the various energy-yielding processes found in bacteria or other organisms. The resulting energy yield per mole of glucose (Y_{ATP}) is very low compared to other bacteria, especially those that can respire (see Chapter 8). The homofermenters that produce 2 mol of ATP per mole of glucose are twice as efficient as the heterofermenters, and they capture about 23 percent of the total energy available in this process (Table 20.2). Although very little energy is available for these bacteria by such fermentations, they have been remarkably successful in establishing themselves in important niches in the environment, as discussed later.

The optical form of lactic acid produced by lactic acid bacteria varies with the genus. Some produce one stereoisomer, the D-form, which rotates light toward the right (*dextro* rotary) whereas others produce the L-form (left or *levo* rotary). The reason for these differences lies in the stereospecificities of the lactic dehydrogenase enzyme itself. Some species of lactic acid bacteria produce enzymes that make both D- and L-forms, resulting in a racemic mixture of the two.

Lactic acid bacteria can grow in the presence of oxygen but are unable to use it metabolically. They lack cytochrome enzymes (which have iron-containing heme groups they cannot synthesize) and an electron transport system with which to generate ATP by electron transport phosphorylation. Lactic acid bacteria do have flavoproteins, however, and when they are exposed to oxygen they produce hydrogen peroxide which can be toxic to the cells.

It is interesting to note that most lactic acid bacteria are **catalase negative;** that is, they are unable to make the enzyme catalase which, like cytochromes, consists of iron-containing heme-proteins that degrade peroxides in the following manner:

$$2\ H_2O_2 \longrightarrow 2\ H_2O + O_2$$

A few lactic acid bacteria make a special manganese-containing catalase and are therefore phenotypically catalase positive. Those species lacking catalase have other enzymes, **peroxidases,** that degrade hydrogen peroxide through organic compound-mediated reductions. In this manner they are protected from the toxic effects of hydrogen peroxide.

Another toxic form of oxygen is the superoxide anion, O_2^-, which is formed in various physical and biochemical ways by single electron reductions of O_2. The lactic acid bacteria, as well as most aerobic and facultative aerobic bacteria, protect themselves from this strong oxidizing agent by producing the enzyme **superoxide dismutase,** which decomposes superoxide as follows:

$$2\ O_2^- + 2\ H^+ \longrightarrow O_2 + H_2O_2$$

Note that hydrogen peroxide is formed by the enzyme. The concerted action of superoxide dismutase and either

Descriptive *Table* **20.2 • Comparison of the efficiency of the homofermentative versus the heterofermentative lactic acid bacteria**[1]

	Energy Available (kjoules)
Glucose (one mol)	2,870
Pyruvate (1,326 kjoules/mol × 2 mols)	2,652
Theoretical energy available in oxidation of glucose to pyruvate	218
Energy captured by ATP synthesis (30 kjoules per mol ATP):	
a. Homofermenters (2 ATP/mol glucose)	60
b. Heterofermenters (1 ATP/mol glucose)	30
Efficiency of process	
a. Homofermenters (60/218 × 100%)	27%
b. Heterofermenters (30/218 × 100%)	13%

[1]Because pyruvate is more oxidized than glucose, the two moles of it formed from glucose contain 218 kjoules less energy than the initial mol of glucose. Two molecules of ATP are synthesized in the homofermentative process and each of these is equivalent to 30 kjoules energy per mol of glucose. Thus, 60 kjoules of energy have been captured in the formation of ATP by homofermenters representing an efficiency of 27%. In contrast, heterofermentative lactic acid bacteria are only half as efficient because they produce only one molecule of ATP per molecule of glucose fermented. Also note that some 92% of the total energy available in the glucose molecule is still available in the two molecules of pyruvic acid formed (2652/2870 × 100 = 92%), indicating that much of the original energy still remains in pyruvate.

catalases or peroxidases results in the destruction of these toxic oxidizing agents. In contrast to the lactic acid bacteria, many obligate anaerobes do not produce these enzymes, so lethal cellular oxidations can occur when they are exposed to the atmosphere.

Nutrition and Ecology

The lactic acid bacteria are regarded as **fastidious** organisms because almost all species have complex nutritional requirements. With the exception of *Streptococcus bovis* from the intestine of cattle, virtually all require preformed amino acids and vitamins, and some even require purines and pyrimidines. Thus, although they do not live as intracellular pathogens of other organisms, they live in habitats with other microorganisms, animal or plant tissues, or decaying organic substances that serve as sources of these materials. Because they require these compounds presynthesized, they conserve some energy that would be required for their synthesis.

Despite their rather limited metabolic capabilities and their complex nutritional requirements, the lactic acid bacteria have survived well in selected environments due to their specialization in sugar fermentation. The principal habitat groups of the lactic acid bacteria are shown in Table 20.3.

One of the principal habitats of lactic acid bacteria is decomposing plant materials. These bacteria ferment the hexoses and pentoses that predominate as decomposition products. The lactic acid formed prevents the growth of many other organisms, especially if the pH is lowered to 5.0 or so. For example, in pickle fermentations, the cucumbers are placed in a vat with a small amount of salt and seasonings. They are covered with water and incubated at or about room temperature. Initially, aerobic and facultative aerobes grow and through aerobic respiration remove the free O_2 that is dissolved in the water in the vat. In a short time, the vat becomes anaerobic and the plant sugars begin fermenting. The initial fermentation is accomplished by the genus *Streptococcus* because it grows at higher pH values than the genus *Lactobacillus.* The pH is lowered by this group until it reaches about 5.5 to 6.0 at which point *Lactobacillus* spp. begin to grow. They ferment the remaining sugars and lower the pH even further, to about 4.5. As long as conditions remain anaerobic, the plant materials are preserved very well due to the low pH of the lactic acid. Therefore, the pickles keep many weeks, or, if refrigerated or canned, for many months.

The same general protocol is followed for the manufacture of sauerkraut and kim chee, the Korean bok choy product. The characteristic flavor of sourdough bread is also due to the fermentation of lactic acid bacteria. Likewise, in agriculture, silage is made by placing hay of the right moisture content in silos or, more recently, in large plastic bags to prevent the access of air and to allow the fermentation of plant materials. The resulting silage can be used during the winter months when fresh hay is not available.

Lactic acid bacteria are of importance in the dairy industry as well. These bacteria are the natural souring agents of milk. They gain access to the milk from plant materials such as the feed or, in some cases, from the cat-

***Differential* *Table* 20.3 • Habitats of lactic acid bacteria**

Habitat	Predominant Group	Activity or Product
Decomposing plant material	*Streptococcus* spp. and *Lactobacillus plantarum*	Pickles, kim chee, silage, and sauerkraut
Dairy	*Streptococcus lactis, Lactobacillus casei, L. acidophilus, L. delbrueckii, Leuconostoc mesenteroides, L. lactis*	Cheeses, yogurt, etc.
Gastrointestinal tract of animals (oral)	*Streptococcus salivarius, S. mutans,* and *Lactobacillus salivarius*	Normal flora, dental caries
(intestinal)	*Streptococcus faecalis*	Intestine; some urinary tract pathogens
Mammal vagina	*Streptococcus* spp. and *Lactobacillus* spp.	Normal flora

tle themselves. If milk is permitted to sour naturally, these bacteria become established, with first the *Streptococcus* group followed by the *Lactobacillus* group, as discussed in pickle fermentation. They ferment milk sugar, lactose, and form lactic acid in the process. When the pH drops sufficiently, the principal milk protein, casein, is precipitated. Thus, the **curd,** or the precipitated casein, is separated from the **whey,** the remaining fluid. It is this same process that is used for the commercial manufacture of cheeses. However, special strains called **starter cultures** of lactic acid bacteria are used by dairies for the manufacture of cottage cheese, yogurt, acidophilus milk, butter, buttermilk, and various cheeses. The distinctive flavor and aroma of butter is due to diacetyl formed spontaneously from acetoin produced by some lactic acid bacteria in butter production.

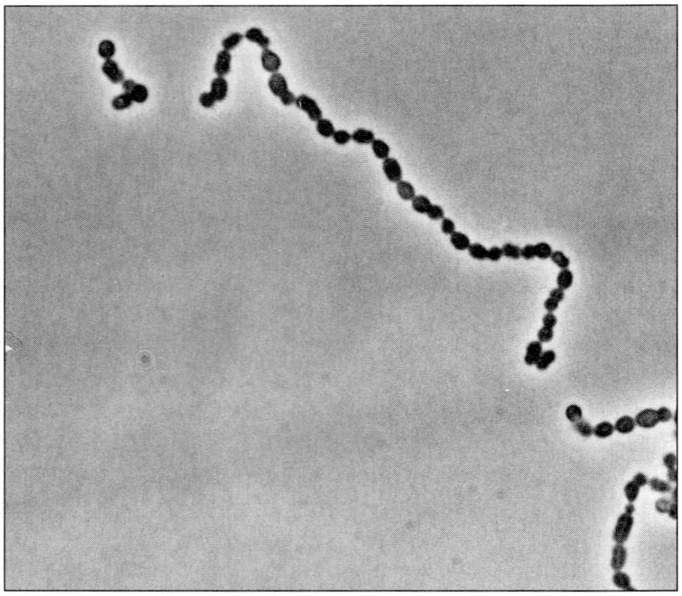

***Figure* 20.4** A typical appearance of a *Streptococcus* sp. in chains of cocci as seen in the phase microscope. (Courtesy of J. T. Staley)

Another habitat of the lactic acid bacteria is the gastrointestinal tract of animals. They grow in the oral cavity where they ferment sugars such as sucrose with the formation of lactic acid. Both *Streptococcus* and *Lactobacillus* species occur in the mouth, where the lactic acid they produce can cause decay of the enamel, resulting in cavities in teeth. These bacteria also are indigenous in the small intestine of humans and other animals. The predominant intestinal species in humans is *Streptococcus faecalis.*

A final habitat of significance is the vagina of female mammals. Both *Streptococcus* and *Lactobacillus* reside there, where they ferment glucose from the glycogen normally secreted by the vagina of adult females. This resulting lactic acid serves as a barrier to vaginal infection.

Some of the more common genera of lactic acid bacteria are discussed individually below.

Streptococcus

This genus of cocci (Figure 20.4) contains several important species (Table 20.4). They all grow poorly on media and form only small, pinpoint colonies during growth. *Streptococcus lactis* is a common dairy organism found in milk and other dairy products. It does not cause disease but is of industrial significance. Note that it is incapable of causing the hemolysis of red blood cells (Table 20.4). Therefore, it lacks the ability to produce **hemolysins,** special enzymes that are responsible for this activity. Hemol-

Differential *Table* **20.4 • Differential characteristics of important species of the genus *Streptococcus***

Species	Group Name	Hemolysis of Red Blood Cells	
		Alpha (greening)	Beta (clearing)
S. lactis	Dairy group	–	–
S. salivarius	Viridans group	+	–
S. faecalis	Enterococcus	Variable	–
S. pyogenes	Pyogenic	–	+
S. pneumoniae	Pneumococcus	+	–

ysis is regarded as a **virulence factor:** that is, it is a characteristic that enhances the pathogenicity of a species.

Streptococcus salivarius is one of many bacteria that grow in the upper part of the gastro-intestinal tract, the oral cavity of humans. This organism lives off the sugars supplied in the diet or produced from starches by amylases secreted by the salivary glands. A close relative of this species, *Streptococcus mutans,* produces a capsule in the presence of disaccharides that enables it to attach to enamel surfaces and cause tooth decay. Members of this genus, the so-called **viridans group** (from Greek meaning "green") cause a partial clearing or greening effect on blood agar plates (Figure 20.5**a**). This phenomenon is referred to as **alpha-hemolysis,** but in fact red blood cells are not actually lysed in this process. In alpha-hemolysis, hemoglobin is converted to methemoglobin *in vivo,* and this form of hemoglobin has a reduced capacity to carry oxygen. Most of the members of the viridans group are not pathogenic but are implicated in tooth decay.

Another group of the streptococci, appropriately informally named the **"enterococci,"** inhabit the lower part of the human digestive tract. *Streptococcus faecalis* is a normal and therefore nonpathogenic inhabitant of the human intestine—particularly abundant in the large intestine. This environment is anaerobic and contains ample nutrients, making it an ideal habitat for growth. Some of the enterococci are pathogenic, causing infections of the urinary tract and sometimes endocarditis.

The **pyogenic group** of *Streptococcus* characteristically produce hemolysins that completely lyse red blood cells, resulting in a clearing on a blood agar plate prepared with mammalian blood. This type of hemolysis is termed **beta-hemolysis** (Figure 20.5**b**). The pyogenes group of streptococci are serious human pathogens. In addition to hemolysins, they produce other virulence factors that are important for pathogenicity, such as fibrinolysin, an enzyme that dissolves blood clots, an erythrogenic toxin that causes the characteristic rash of scarlet fever, and leuco-

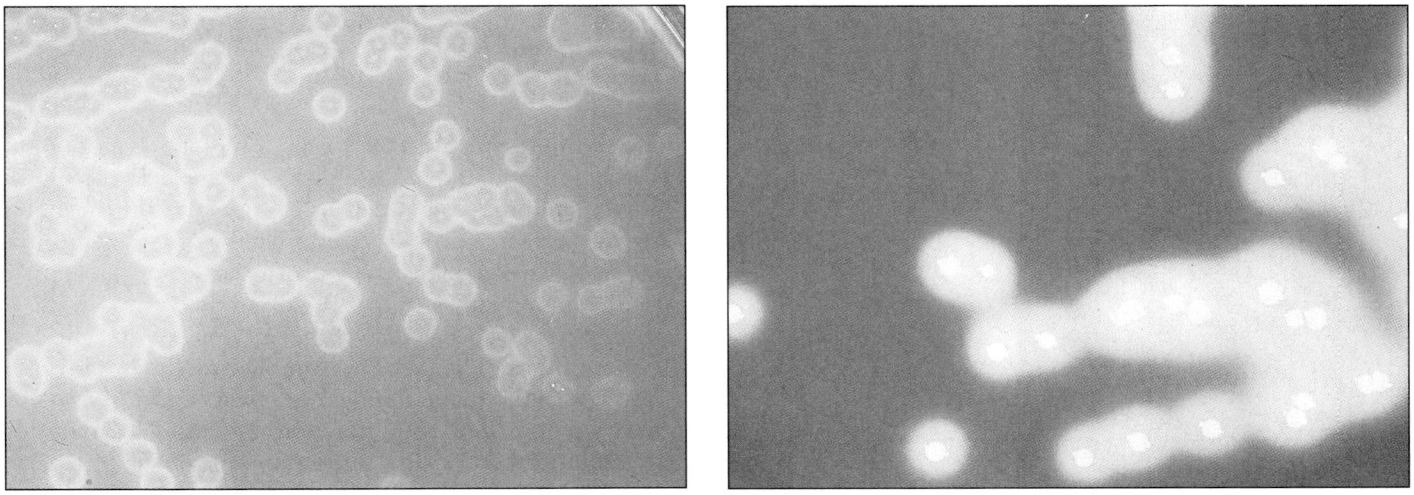

(a) **(b)**

***Figure* 20.5 (a)** The appearance of alpha-hemolysis or "greening" hemolysis on a plate of red blood cells. Note that the colonies of this viridans group *Streptococcus* sp. are very small, typical of the pinpoint colonies of this group and **(b)** The appearance of β-hemolysis caused by *Streptococcus pyogenes.* The blood cells are completely lysed in the vicinity of the colony, making the blood agar transparent and colorless. (Courtesy of Dale Parkhurst and J. T. Staley) cidin, which destroys white blood cells. Furthermore, they produce the enzyme hyaluronidase, which attacks hyaluronic acid in connective tissue. These organisms are ideally adapted for growth in warm-blooded animals because their growth temperature range is between 10°C and 45°C. *Streptococcus pyogenes* is the causative agent of several diseases including scarlet fever, puerperal fever of post-delivery mothers, and endocarditis, an inflammation of heart tissue.

Streptococcus pneumoniae comprises its own group and is an important causative agent of pneumonia; it is therefore informally referred to as the **"pneumococcus."** Unlike the other members of the genus, this species does not typically form long chains of cells under most conditions of cultivation, but rather grows as pairs of cocci, so it is also sometimes called the **"diplococcus."** Virulent members of this species have a capsule that is important for their pathogenicity (Figure 20.6). If the capsule is removed, the cells are much more readily phagocytized. Only strains with capsules can cause pneumonia. Encapsulated strains form colonies that are described as "smooth" in texture and are distinguished from unencapsulated strains that form "rough" colonies. As discussed in Chapter 15 these bacteria served as important experimental organisms in Avery and McCleod's demonstration that genetic material can be passed from one organism to another even if the donor organism is dead. This was the first demonstration of the phenomenon of genetic transformation.

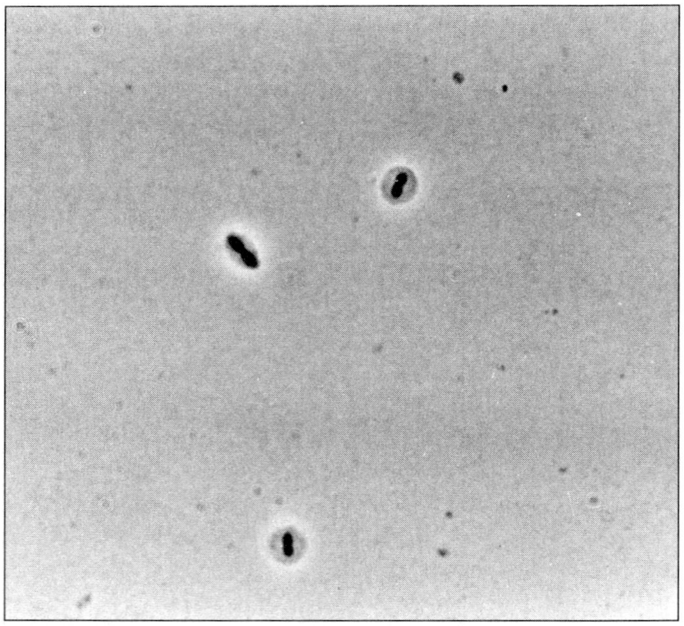

***Figure* 20.6** These cells of *Streptococcus pneumoniae* have been stained in the presence of antiserum, which reacts with the capsule so that it can be visualized. This is the "Quellung" reaction, which is German for "swelling," because the apparent size of the organism is increased by staining the capsule. Note that the capsule extends a considerable distance away from the cells. (Courtesy of J. T. Staley)

Streptococcus is also known for the devastating tissue damage caused by some strains infecting humans. This rare disease has a rapid onset and, if not treated quickly, may result in loss of life.

Lactobacillus

This genus of rod-shaped lactic acid bacteria is also widespread in the same habitats in which the genus *Streptococcus* resides, including plant materials and the oral and genital tracts of humans. However, unlike *Streptococcus,* the genus *Lactobacillus* does not harbor any pathogenic strains.

These bacteria can grow at lower pH values than the streptococci. As mentioned previously they do not grow during the initial stages of fermentations, but only when the pH is lowered to 5.5 to 6.0. In turn they continue to produce lactic acid and lower the pH of the environment to values below 5.0. This effectively eliminates the genus *Streptococcus.* This is a bacterial example of an ecological succession.

Other Important Genera

Two other genera of importance are *Leuconostoc* and *Pediococcus. Leuconostoc* spp. are common in dairy products as well as in the oral cavity. They are especially well known as producers of **dextrans** (alpha-1,6-glucans), which they form as capsular materials in the fermentation of sucrose, a disaccharide consisting of glucose and fructose. *Leuconostoc mesenteroides* produces an extracellular enzyme called dextran sucrase that converts the glucose of the sucrose molecule to dextran and releases the fructose into the environment.

$$\text{Sucrose} \longrightarrow \text{Dextran} + \text{Fructose}$$

Therefore, the result is the formation of a polymer of glucose subunits linked in the alpha 1-6 position. Actually, the glucose subunits are added to the glucose on the initial sucrose molecule primer. Thus, each dextran molecule has one subunit of fructose as its terminal sugar. Dextran is used commercially as a blood plasma extender.

Pediococcus species also carry out important lactic acid fermentations. They can be troublesome in the brewing industry, as the name *Pediococcus damnosus* implies, in that they can interfere with the normal yeast alcoholic fermentation by producing undesirable products of the lactic acid fermentation.

Listeria

Listeria is important because it is a food-borne pathogen that causes **listeriosis,** a type of gastrointestinal disease. Al-

Differential *Table* **20.5 • Other gram-positive cocci and rods**

Genus	Shape	Oxygen Requirement	Habitat (mol % G + C)
Staphylococcus	Cocci in clusters	Facultative aerobe; fermentative	Warm-blooded animals, food (30–35)
Micrococcus	Cocci in clusters	Obligate aerobe	Soil; airborne dust (66–72)
Listeria	Rod	Facultative aerobe	Widely found; animal pathogen (36–38)
Renibacterium	Rod	Aerobic	Fish pathogen, salmonids (53)
Geodermatophilus	Filamentous	Aerobic	Soil (73–75)
Frankia	Filamentous	Aerobic	Alder—N_2 fixing symbiont (66–71)

though usually not fatal, it can cause death in infants and debilitated individuals. Like *Lactobacillus, Listeria* is a gram-positive rod (Table 20.5) that is catalase positive and oxidase negative. It also ferments sugars to form L-lactic acid. However, unlike *Lactobacillus,* it is motile (peritrichous flagella), produces cytochromes, and grows aerobically by respiration. In addition, it grows at low temperatures and therefore multiplies in the refrigerator. For this reason it has been a problem with refrigerated foodstuffs such as cheese and fish.

Renibacterium

Renibacterium is a slow-growing, rod-shaped bacterium noted as the causative agent of bacterial kidney disease (BKD) in salmonid fish (Table 20.5). It has been cultivated in pure culture only recently. It is an obligate aerobe that uses sugars as carbon sources for growth and requires cysteine.

Staphylococcus

Like the lactic acid bacteria, members of the genus *Staphylococcus* grow anaerobically and produce lactic acid from sugar fermentation. These bacteria are also gram-positive cocci, but they occur in grapelike clusters of cells (Figure 20.7), rather than chains (Table 20.5). The genus differs from the lactic acid bacteria in nutrition and physiology. For example, staphylococci do not have the complex nutritional requirements typical of lactic acid bacteria. Furthermore, they produce heme pigments and carry out aerobic respiration with an electron transport system containing cytochromes.

Staphylococci are associated with the skin and mucosal membranes of animals. Indeed, all humans carry *Staphylococcus epidermidis,* a nonpigmented coccus, as a normal inhabitant on the surface of their skin. *Staphylococcus aureus* is a human pathogen capable of causing a variety of health problems. Unlike *S. epidermidis, S. aureus* is pigmented a gold color. *S. aureus* also produces several virulence factors that enhance its ability to cause disease. One distinctive factor called **coagulase** is an enzyme that causes

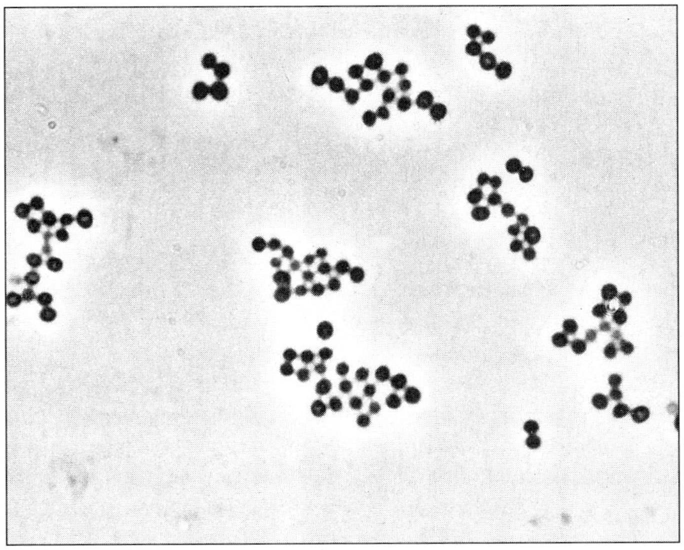

Figure **20.7** Typical grapelike clusters of *Staphylococcus aureus.* (Courtesy of J. T. Staley)

fibrin to clot. Another factor, leukocidin, attacks leukocytes. Several other factors can be produced by pathogenic strains including β-hemolysin, lipase, fibrinolysin, hyaluronidase, deoxyribonuclease, and ribonuclease.

Staphylococcus aureus is a normal inhabitant of the nasopharynx region in humans. Some humans harbor a specific strain for many years and are therefore called **carriers.** Infants come into contact with the organism during their first week of life. The strain they receive may come from the mother, other close relatives, or hospital personnel. Normally these strains do not result in any infection. However, an unhealthy infant is susceptible to the colonizing strain, and serious skin infections can result. Treatment of such infections can be difficult because many *S. aureus* strains are antibiotic resistant.

Staphylococcus aureus can also cause a common type of food poisoning. Such strains produce an **enterotoxin,** a type of **exotoxin** (exotoxins are to be distinguished from endotoxin produced by gram-negative bacteria, which is their lipopolysaccharide; see Chapter 29). Foods that have been prepared and not refrigerated for extended periods of time allow for the growth of the toxin producer. Problem foods include cooked meats, potato salads, and cream desserts such as eclairs. During summer months when the weather is warm and families have picnics, conditions are ideal for potential problems. Foods that are not refrigerated properly and are served on subsequent days are especially hazardous. The symptoms of *Staphylococcus* food poisoning are acute and memorable. The toxin that is produced while the bacterium grows in the food takes effect within a few hours of ingestion and causes headache and dizzyness, fever, diarrhea and nausea that persists for 24 hours or so. During this period the inflicted individual feels extremely uncomfortable and is totally incapacitated. Fortunately the condition is self-limiting and most healthy individuals recover fully after 24 to 48 hours.

Toxic shock syndrome (TSS) is a well-publicized malady caused by *Staphylococcus aureus.* Women who use tampons during their menstrual periods are susceptible to this. *S. aureus* grows and produces toxins in the vagina. If unchecked, TSS can be fatal.

Endospore Formers

The bacterial endospore is a very distinctive structure produced by relatively few genera. The two principal genera are the aerobic to facultatively anaerobic genus *Bacillus* (Figure 4.8**a**) and the obligately anaerobic genus *Clostridium.* Three additional genera of ecological importance include *Desulfotomaculum,* a group of anaerobic sulfate-reducing bacteria discussed in Chapter 19, *Sporosarcina,* a soil coccus, and *Thermoactinomyces,* a thermophile (Table 20.6).

With the exception of *Sporosarcina,* which is a coccus, all endospore formers are rod-shaped bacteria that have the typical gram-positive cell wall ultrastructure and stain gram-positive to variable. Many species are motile with peritrichous flagella. The endospore has been the subject of many years of study inasmuch as it is an important example of cellular differentiation or morphogenesis.

Endospore Characteristics and Development

The endospore, as the name implies, is a spore or resting stage unique to bacteria that is formed within the cell. It confers to the bacterium that produces it an ability to survive during periods of dessication and high temperature, two conditions often encountered in soil and rock environments where these organisms are most commonly found. They are more resistant not only to heat and dessication than typical bacteria, but also to ultraviolet radiation and disinfection. Experiments have shown that the endospores of some species can survive storage in the ambient environment at least 50 years.

Endospores appear highly refractile when observed by phase contrast microscopy (Figure 4.8**a**). This high refractility is due to the low water content of the spore. Not only is the endospore dehydrated, but it is rich in calcium ions and a unique compound, dipicolinic acid (Figure 20.8**a**). These are believed to form a chelate complex be-

Differential *Table* **20.6 • Important genera of endospore-forming bacteria**

Genus	Oxygen Requirement	Mol % G + C	Nutrition/Physiology
Bacillus	Aerobe or facultative aerobe	25–69	Aerobic respiration or fermentation
Clostridium	Obligate anaerobe	22–55	Fermentation
Desulfotomaculum[1]	Obligate anaerobe	37–50	Anaerobic respiration (sulfate reduction)
Thermoactinomyces	Obligate aerobe	52–55	Aerobic thermophile; compost
Sporosarcina	Obligate aerobe	40–42	Some degrade urea

[1]Discussed with dissimilatory sulfate reducers in Chapter 19.

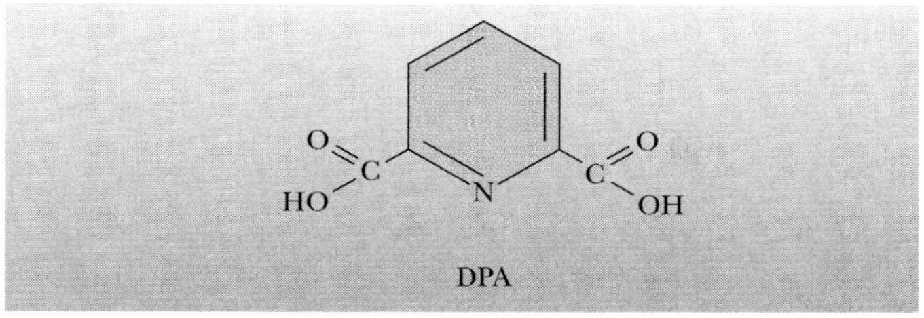

(a)

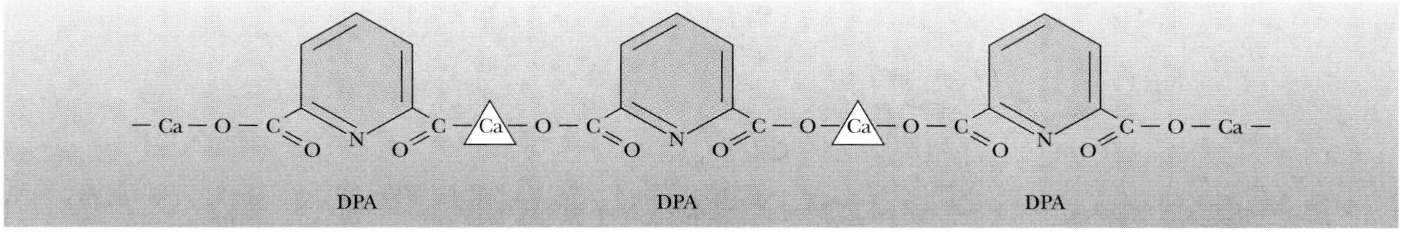

(b)

***Figure* 20.8 (a)** Chemical formula of dipicolinic acid (DPA). **(b)** Proposed structure of the calcium-dipicolinic acid chelate.

cause of their equal molar concentrations and chemistry (Figure 20.8**b**) and are thought to be responsible for the heat resistance of the endospore and other unique properties.

Endospores are very difficult to stain using ordinary staining procedures. For example, in the Gram stain, the endospore does not take up either the primary stain or the counterstain and therefore appears colorless upon completion of the staining process. Endospores, however, can be stained by special staining procedures. One common procedure involves steaming the smear with the dye malachite green. After this is completed the smear is counterstained with safranin. In this procedure the cell appears red while the endospore appears green.

Endospore-forming bacteria have two phases of growth, **vegetative growth,** which is the normal period of growth and reproduction, and **sporulation,** the period of spore formation (Figure 20.9). The endospore is produced at the end of the exponential phase of growth. A number of stages are noted in sporulation (Figure 20.10). After the period of vegetative growth ceases, two resulting molecules of DNA are present in the cell. These first coalesce and are then separated from one another by a cell membrane formed during a special differentiation process. An engulfment stage ensues in which the membrane from one incipient cell grows around the membrane of the other until it completely surrounds it. The engulfed incipient daughter cell, termed the **forespore,** eventually becomes the mature endospore. The other "cell" that contains the endospore is called the **sporangium.** At this time the process of sporulation is irreversible.

During sporulation the spore becomes increasingly refractile and develops into four distinct layers, a **spore coat,** the **core wall** with its modified peptidoglycan, the **cortex,** and the **core** wherein the calcium ions and dipicolinic acid are found along with the nuclear material. Thin sections through endospores show each of these features (Figure 20.10; 20.15). These layers are bound by a membrane called the **exosporium.** Eventually the original cell lyses to release the free endospore.

The endospore is a resting stage that can survive long periods of time in the environment. However, it remains viable and under appropriate conditions can undergo the process of **germination,** whereby it begins to develop into a vegetative cell again (Figure 20.9). In this process the endospore takes up water, swells, and breaks the spore coat and exosporium, releases its calcium dipicolinate, and begins multiplying as in normal vegetative growth.

Some endospores are known to be able to survive boiling temperatures for long periods of time. These were the

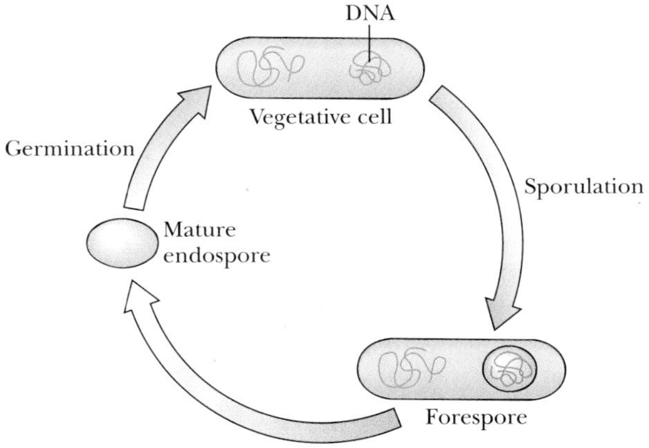

***Figure* 20.9** Diagram of the life cycle of an endospore-forming bacterium. The vegetative cell forms an endospore when nutrients become depleted in the process of sporulation. Subsequently, when conditions for growth are favorable the endospore can germinate to form a vegetative cell again.

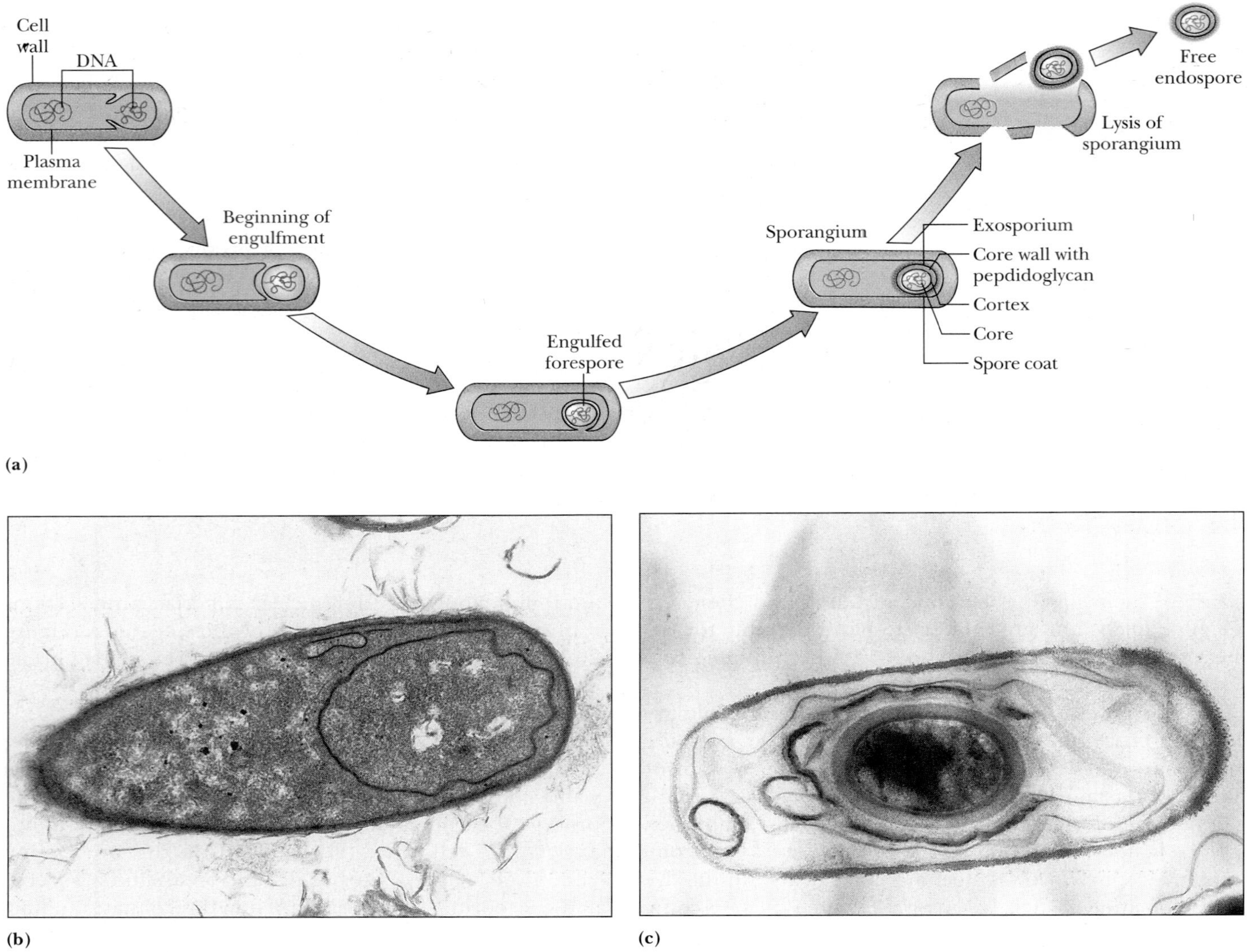

***Figure* 20.10 (a)** Illustrations of the stages of sporulation as determined from electron micrographs. Thin sections through an endospore forming bacterium at **(b)** early stage of spore formation and **(c)** the mature spore. (Courtesy of S. Pankratz)

heat-resistant structures that caused so many problems for the proponents and opponents of the theory of spontaneous generation discussed in Chapter 2.

The location of the endospore in the cell and its shape are important features in classification. Most species have oval-shaped endospores that are located in a central to subterminal location within the cell or sporangium (see Figure 4.8**a**). At the other extreme, some species, such as *Clostridium tetani,* produce a terminal endospore that is larger than the cell diameter, giving the sporangium a club-shaped appearance (Figure 20.11).

Habitats of Endospore-Forming Bacteria

The endospore-forming bacteria are found primarily in soils and aquatic sediments and muds. Like other gram-positive bacteria, they are not normally found in planktonic habitats, at least in low-nutrient aquatic habitats such as oligotrophic lakes and the ocean. The endospore is an ideal survival device for soil bacteria. Organisms that can convert to resting stages during dry periods in soil environments can survive periods of prolonged dessication. These bacteria are also found on rock weathering surfaces in deserts where hot, dry periods are common.

Endospore-forming bacteria can be selectively isolated from habitats by first pasteurizing the soil or mud sample at 80°C for 10 min. Although this kills normal soil bacteria, the heat-resistant endospores of most of these bacteria will not be affected. In fact, some endospore formers do not germinate unless they are first heat shocked, that is, exposed to high temperature first.

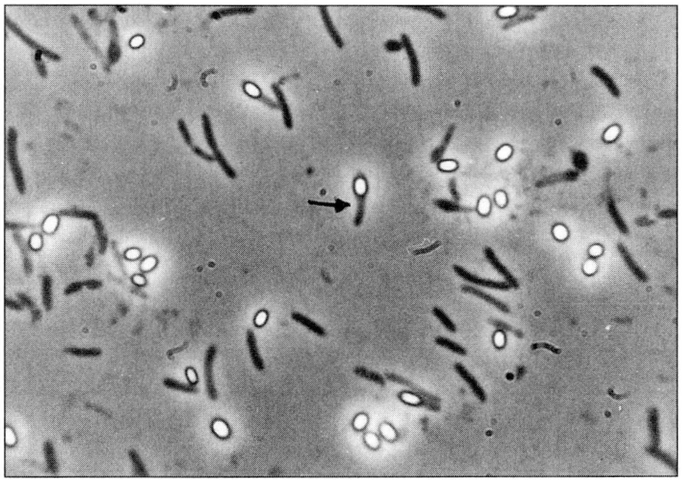

***Figure* 20.11** A phase photomicrograph of *Clostridium tetani.* Note that the endospores are larger than the cell, giving the cells a club-shaped appearance. (Bar is 5 μm.) (Courtesy of J. T. Staley)

Bacillus

This genus contains aerobic to facultative aerobes. Unlike the strictly anaerobic genus *Clostridium, Bacillus* species produce catalase, which may explain the difference in the survival of these two genera in the presence of oxygen. The various species of this genus are classified into groups based on their cell morphology, using in particular the shape of the endospore and its location. Table 20.7 provides a grouping of some of the major species based upon cell morphology.

Bacillus subtilis is a small, obligately aerobic spore former that is a common soil inhabitant. It is noted for its ability to degrade plant polysaccharides and pectin, and some strains are even known to produce a rot in potato tubers. It can cause a condition known as "ropy bread" due to polymer production during growth after baking contaminated bread.

Bacillus licheniformis resembles *B. subtilis,* but it is a facultative aerobe that ferments sugars. It carries out a characteristic fermentation:

$$3\ \text{Glucose} \longrightarrow 2\ \text{Glycerol} + 2\ \text{2,3-butanediol} + 4\ CO_2$$

Most strains of *B. licheniformis* carry out denitrification, which is a relatively unusual attribute within this genus.

Bacillus megaterium (from the Greek *megaterium* meaning "big beast"), as the species name implies, is very large. Because it is large it is used by bacteriologists as a teach-

***Table* 20.7 Some major species of *Bacillus* and their properties**

Group	Major Species	Distinctive Properties
I. Oval spores		
A. Sporangium not swollen		
	B. subtilis	Common soil form, aerobic gramicidin-producer
	B. licheniformis	Denitrifier, fermentative; bacitracin producer
	B. megaterium	Large rod
	B. cereus	Common soil form, fermentative
	B. mycoides	Like *B. cereus,* but distinctive hairlike colonies
	B. anthracis	Like *B. cereus,* but causes anthrax
	B. thuringiensis	Insect pathogen
B. Sporangium swollen		
	B. stearothermophilus	Thermophile
	B. circulans	Colony rotates in circles
	B. polymyxa	Nitrogen fixer, antibiotic (polymyxin) producer
	B. popilliae	Insect pathogen
II. Spherical spores, nonfermentative		
	B. pasteurii	Degrades urea, grows at high pH

ing organism in introductory laboratories for Gram staining as well as in research laboratories interested in morphology and ultrastructure. Like *B. subtilis,* this species in an obligate aerobe with very simple nutritional requirements (ammonium or nitrate can serve as the sole nitrogen source).

Bacillus cereus and *B. mycoides* are also quite large and are among the most common soil bacteria. They grow anaerobically by sugar fermentation and require amino acids for growth. *B. mycoides* is a common air contaminant and is often found on plates in microbiology laboratories.

Closely related to *Bacillus cereus* are two important pathogens. *B. anthracis* causes anthrax of animals as well as humans. Unlike *B. cereus,* it is immotile. The source of infection of humans is largely through domestic farm animals, although many other animals including horses, mink, dogs, and even birds, fish, and reptiles can acquire the disease. In humans, the disease occurs most commonly in sheep and cattle tanners and meat workers. The organism normally invades through open cuts in the skin but can also infect through the lungs. The organism produces a gummy D-glutamic acid polypeptide capsule, an important virulence factor during invasion. The disease symptoms are produced by a protein exotoxin formed during growth in the host tissues. The toxin acts by causing physiological shock and ultimately kidney failure. The toxin is heat labile. The spores survive well in the environment, so animals that have died of the disease present a problem for disposal; they are usually buried. *B. anthracis* is the only member of the genus that is pathogenic to humans. This organism has been studied as a prospective agent in biological warfare (see **Box 20.1**).

BOX 20.1 MILESTONES

Biological (or Germ) Warfare

Pathogenic microorganisms have been studied as agents of germ warfare by many technologically advanced countries. *Bacillus anthracis* is one of the pathogens that has been studied most because of the hardiness of the endospore, the ability to infect populations using aerial dispersal, and the deadliness of pulmonary anthrax.

At the International Biological Weapons Convention in 1972 many nations, including the United States and the Soviet Union, signed an agreement to ban research on the development and use of biological weapons. However, some countries continue to study biological warfare even today.

In 1979 a tragic incident occurred in Sverdlovsk, a city of 1.2 million inhabitants located 1400 km east of Moscow in the former Soviet Union. This city was the site of a biological weapons research facility. Apparently, due to an accidental release from the military research plant, 77 individuals were infected with anthrax and 66 subsequently died. This epidemic occurred during the cold war between the Soviet Union and the United States and after they were signatories to the 1972 Convention. Official information released by the Soviet Union regarding the outbreak stated that individuals developed *gastrointestinal,* not pulmonary, anthrax after *eating* contaminated meat. In this manner the Soviet Union covered up its continued activities in biological warfare after the 1972 ban.

Following the collapse of the Soviet Union, the real cause of the epidemic was determined in a study led by a U.S. scientist, Matthew Meselson (*Science,* 266: 1202–1208, 1994). This group of scientists discovered that, in fact, most individuals died of *pulmonary* anthrax, indicating that the outbreak was due to airborne dispersal and inhalation of spores, not from ingestion of meat as officially claimed. This tragic and accidental epidemic provides some insight into the deadliness of *B. anthracis* as a biological warfare agent.

In the past, the United States has also studied the offensive and defensive aspects of biological weapons. In the 1960s, an aerosol of *Serratia marcescens* was released by airplanes over the San Francisco metropolitan area, with the intent of monitoring the effectiveness of its aerial dispersal, an important aspect of bacterial dissemination. Because of its distinctive red pigment, prodigiosin (see Chapter 19), it is easy to determine the concentration of this species when air samples are plated on agar media. At that time *S. marcescens* was not regarded as a pathogen. However, several individuals who were being treated in hospitals in the Bay Area acquired previously unknown *S. marcescens* lung infections due to the aerial spraying, resulting in one death.

Movement is now afoot to put more "teeth" in a treaty on the ban of biological weapons. Russia, the United Kingdom, and the United States have already agreed to limited inspections of biological weapons facilities whose purpose, at least ostensibly, is for defensive reasons only. Hopefully, in the near future all nations will agree to a complete ban.

The best known insect pathogen in this genus is *Bacillus thuringiensis.* This bacterium produces a large amount of a crystalline protein during the sporulation process. The protein crystal can be seen with both the light microscope and the electron microscope (Figure 20.12). It is formed in the sporangium along with the spore and is referred to as a parasporal body. This protein crystal is toxic to insects. Vegetative cells of *B. thuringiensis* and its spores are found on plant leaves and are ingested by insects that feed on the leaves. The most common hosts are the larval stages of moths and butterflies (Class *Lepidoptera*). These insects have an alkaline gut that dissolves the toxin. The toxin is a neurotoxin that causes paralysis and death of the caterpillar. Presumably the bacterium benefits from its host by growing on the dead carcass (see **Box 20.2**).

Bacillus stearothermophilus is a thermophile with a temperature growth range between 45°C and 65°C. This particular thermophile has been known for many years and has been found in hot springs, deserts, and soils.

Bacillus polymyxa is a facultative anaerobe that ferments sugars with the production of 2,3-butanediol, ethanol, acetic acid, and CO_2. It is also distinctive in being one of the few members of the genus to carry out nitrogen fixation, which it does when growing anaerobically.

Bacillus circulans is similar to *B. polymyxa* but is peculiar in its formation of motile colonies. The colonies of this organism actually rotate and can move across a petri dish due to the concerted action of the flagella.

Another member of the genus that is particularly noteworthy is *Bacillus pasteurii.* This bacterium is especially suited for growth on urea. For example, it does not grow on ordinary nutrient broth unless urea is added to it. As the urea is degraded, ammonia is produced and the pH rises:

$$CO(NH_2)_2 + H_2O \longrightarrow CO_2 + 3\ NH_3$$

The organism can tolerate pH values between 8.0 and 9.5, much higher than typical bacteria. Because of its ureolytic ability, it is commonly found in urinals and can be easily isolated from these sources or from soils by the addition

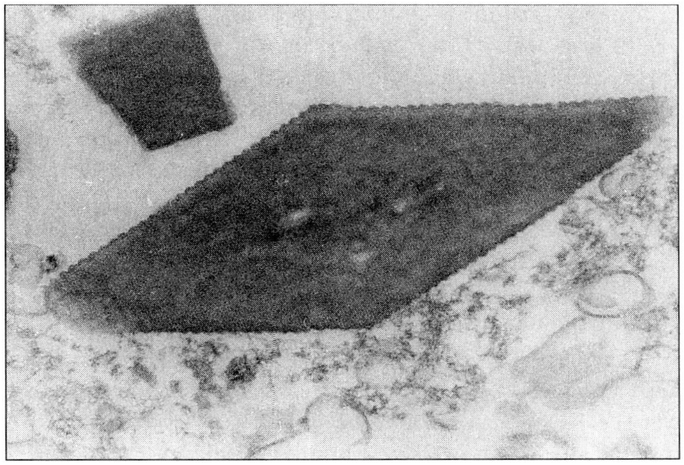

***Figure* 20.12** A thin section electron micrograph of the Bt protein crystal. (Courtesy of S. Pankratz)

BOX 20.2 RESEARCH HIGHLIGHTS

Biological Pesticides

Bacillus thuringiensis toxin, often referred to simply as **Bt,** is currently being used as a biological insecticide. Commercial preparations are available that are used to eradicate tussock moths and other troublesome insects. The advantage of using biological agents against such pests is that other animals and plants are not affected because of the high specificity of its action, and, since the toxin is of biological origin, it is readily degraded in the environment. In contrast, DDT, which was used for control of these same pests, had devastating effects on other members of the food chain. Furthermore, it undergoes biomagnification (see Chapter 33), and persists in the environment for a long time. It was for this reason that DDT was banned in the United States and many other countries.

Some strains of *Bacillus thuringiensis* as well as other species of *Bacillus, B. popillae* and *B. lentimorbis,* produce toxins effective against other insects. The commercial development of the mosquito toxin promises to be useful in the control of malaria.

In a quite different approach aimed at controlling plant insect infestation, these toxin genes are now being incorporated into plants by genetic engineering using *Agrobacterium tumefaciens.* Thus, when insects eat plant leaves, they will ingest the toxin and be killed by the genetically engineered bioinsecticide.

of urea to nutrient broth. Although other bacteria can degrade urea, they cannot do so at the high pH values of this bacterium. Thus, other bacteria are eliminated from environments where this is a major activity because the resulting high pH is so selective. Some *Bacillus* species, such as *Bacillus alcalophilus,* grow at even higher pH values. These are typically found in alkaline soils such as in deserts. Some have been reported to grow at pH 10 and above.

Clostridium

This genus of anaerobic spore formers is even more diverse than *Bacillus* and contains some important environmental species as well as human pathogens. *Clostridium* spp. are grouped according to their fermentative abilities into several major subgroups (Table 20.8).

These bacteria are obligate anaerobes that lack cytochromes and an electron transport system. Thus, they rely solely on the formation of ATP by substrate level phosphorylations during fermentation of various carbon sources. Although these organisms are obligately anaerobic, they are not as difficult to grow as methanogenic bacteria. Unlike methanogens, they can be transferred in the air and incubated easily in ordinary anaerobe jars in which oxygen has been removed by reaction with H_2. Also, most species can be grown in liquid media in tubes that have been supplemented with sodium thioglycollate. Thus, they can be maintained in tubes on the lab bench without additional precautions.

The genus is metabolically diverse. Two species, *Clostridium cellobioparum* and *C. thermocellum,* carry out an anaerobic fermentation of cellulose to form the disaccharide cellobiose, which is ultimately fermented to produce acetic and lactic acids, ethanol, H_2, and CO_2 as the major end products. *C. cellobioparum* is found in the rumen of cattle and sheep and enables these higher animals to utilize cellulose in their diet (see Chapter 25). Interestingly, this organism is inhibited when too much H_2 accumulates in the fermentation. These bacteria live in close association with other ruminant microbes, the methanogenic bacteria, which remove the hydrogen in the formation of methane gas. This interaction involving two different groups is referred to as an **interspecies hydrogen transfer** and is an example of synergism (see Chapter 25). *Clostridium thermocellum* is common in decaying soils containing cellulose.

A major group of the clostridia ferment sugars and occasionally starch and pectin to form butyric acid, acetic acid, CO_2, and H_2 as the principal end products. The pathway for this fermentation is shown in Figure 20.13. In this **butyric acid fermentation,** glucose is fermented to pyruvic acid via the Embden-Meyerhof pathway. The pyruvate is split to carbon dioxide and hydrogen gas in the formation of acetyl coenzyme A. Some of the acetyl CoA is used for ATP generation in the formation of acetic acid via acetyl phosphate. Also, the acetyl CoA can be condensed with another molecule of acetyl CoA to form acetoacetyl CoA, which is the precursor of butyric acid.

Some of the butyric acid bacteria produce fewer acids and more neutral products in prolonged fermentations. These are the so-called acetone-butanol fermenters. Butanol is formed from butyryl CoA via butyrylaldehyde (see Figure 20.13) and acetone and isopropanol are formed from acetoacetyl CoA by decarboxylation and subsequent reduction, respectively.

Differential ***Table* 20.8 • Major groups and representatives of *Clostridium***

Group or Substrate	Fermentation Type	Representative Species	Distinctive Products or Features
Cellulolytic	Acetate, lactate ethanol, H_2, CO_2	*C. cellobioparum*	Rumen organism
		C. thermocellum	Soil and sewage
Saccharolytic	Butyrate (or butanol-acetone); proteolytic	*C. butyricum*	
		C. pasteurianum	Nitrogen fixer
		C. acetobutylicum	
		C. perfringens	Wound infections (gas gangrene)
Amino acid pairs	Stickland reaction	*C. sporogenes*	
		C. tetani	Causes tetanus
		C. botulinum	Causes botulism
Purines	Uric acid and other purines fermented to acetic acid, NH_3, CO_2	*C. acidurici*	
		C. fastidiosus	
H_2 and CO_2	Acetogen	*C. aceticum*	Acetic acid formed

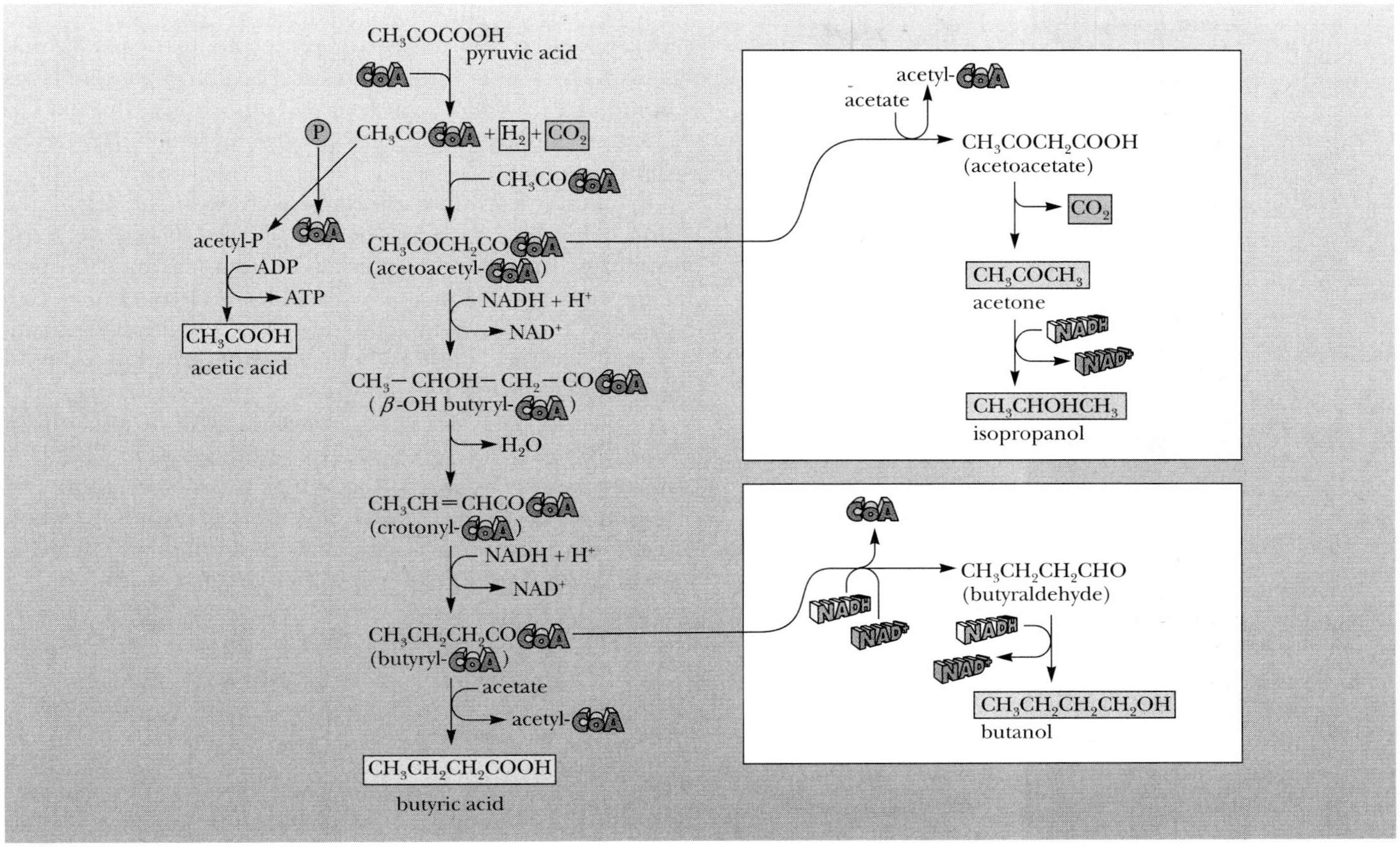

***Figure* 20.13** The butyric acid fermentation (see text for details).

These so-called **butyric acid bacteria** include *C. pasteurianum,* which is also able to fix nitrogen, a property shared by some other species in this group. As the acids accumulate in the butyric acid fermentation, some species, such as *C. acetobutylicum,* begin to produce more neutral compounds including butanol and acetone. The acetone-butanol fermentation has been used commercially. It was especially important during World War I because of the need for acetone in munitions manufacture. In this regard, it is interesting to note that recently clostridia have been shown to be degraders of munitions, such as the explosive TNT (trinitrotoluene) (see Chapter 33).

Pectin-fermenting clostridia play a role in "retting," which is used in making Irish linen from flax. In this process, which was empirically determined centuries ago, natural plant stems of flax are bundled together and immersed in water. Conditions become anaerobic, and the pectin that cements the plant cells together is degraded by clostridia and other organisms, thereby freeing the fibers for linen manufacture.

Many butyric acid species of *Clostridium* are proteolytic, that is, they carry out the anaerobic hydrolysis of proteins to form amino acids. The amino acids can then be fermented with the production of ATP. Figure 20.14 shows an example of a typical fermentation for glutamic acid. The end products of this particular fermentation are the same as the butyric acid fermentation, except that ammonia is also formed.

Some of the proteolytic clostridia carry out unique fermentation reactions, called Stickland reactions, in which two amino acids are catabolized. For example, L-alanine can be oxidized and L-glycine reduced by some species, leading to the formation of the final products: acetic acid, CO_2, and NH_3, and ATP (see Figure 8.6).

Some disease-producing amino acid fermenters include *Clostridium tetani,* the causative agent of tetanus, and *C. botulinum,* which is responsible for botulism. *C. tetani* can grow in wounds exposed to soil, which is the normal environment of the organism. The bacterium does not need to grow very much in the tissue because the toxin it produces is extremely potent. The disease is best handled by preventive medicine: children are given DPT [diphtheria, pertussis (whooping cough), and tetanus] immunizations to build up a natural antiserum immunity to the organism's toxin before any exposure to the bacterium through wound infections. However, if the skin of an im-

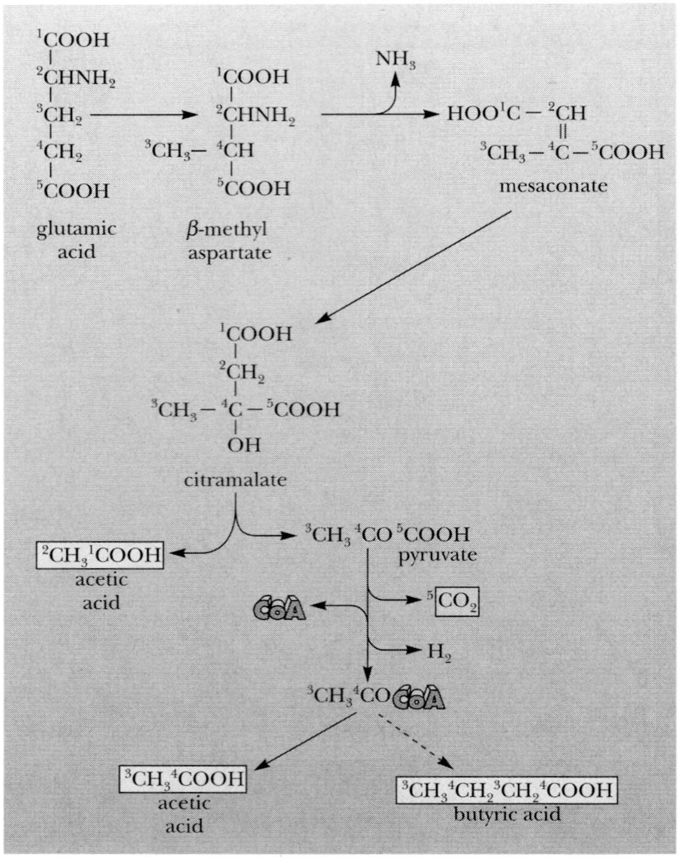

***Figure* 20.14** Many clostridia ferment glutamic acid. The carbon atoms of the glutamate are numbered here so that the enzymatic activity can be followed more closely. The initial reaction involves a rearrangement of the molecule to form β-methyl aspartate. This is then deaminated to produce mesaconate, which is oxidized to form citramalate. The citramalate is cleaved to produce acetic acid and pyruvic acid. The pyruvate is cleaved to form carbon dioxide and hydrogen gas, ultimately resulting in the formation of butyrate and more acetate.

munized person is cut or punctured and exposed to dirt or soil, it is recommended that the individual receive both a booster shot of DPT and antitoxin.

Clostridium botulinum is another common soil bacterium and is also a pathogen. It causes a fatal food poisoning called **botulism.** Botulinum toxin is a protein exotoxin that is excreted from the bacterium as it grows. It affects the normal release of acetylcholine from motor nerve junctions, thereby causing paralysis. It is one of the most potent toxins known (see **Box 20.3**).

Clostridium perfringens is another important pathogen and food-poisoning organism. Like *Staphylococcus aureus* this species produces an enterotoxin. The *C. perfringens* enterotoxin is formed during sporulation and can be seen as a protein crystal in the sporangium (Figure 20.15). This bacterium is widely distributed in soils and commonly contaminates foods. In addition, it is particularly troublesome during times of war to foot soldiers. Soldiers' wounds that are exposed to soil can become infected by the bacterium. It grows profusely in the tissues and produces gas that clogs blood vessels, which can lead to compromised blood circulation and eventually "gas gangrene," which may necessitate amputation.

Other *Clostridium* spp. are involved in the fermentation of purines such as uric acid *(C. fastidiosus),* with the resulting formation of acetic acid, ammonia, and CO_2 (see Figure 12.12). Uric acid is excreted as a nitrogenous waste product of birds, analogous to urine excretion by mammals. Therefore, these bacteria are widespread in soils and rookeries.

Clostridium aceticum is a representative of a group of clostridia that can grow as chemolithotrophs. This bacterium is a hydrogen autotroph that gains its energy by the oxidation of hydrogen. This is an example of an **acetogenic** bacterium in which carbon dioxide is reduced to form acetic acid in the following overall reaction:

$$2\ CO_2 + 4\ H_2 \longrightarrow CH_3COOH + 2\ H_2O$$

A number of non–spore forming gram-positive bacteria are also acetogenic. These include the genera *Acetobacterium* and *Acetogenium* (Table 20.9). All of these bacteria fix carbon dioxide by the acetyl coenzyme A pathway (Figure 10.11), a much different mechanism of carbon dioxide fixation from the Calvin-Benson cycle. These bacteria can also grow as ordinary heterotrophs by utilization of sugars in fermentations.

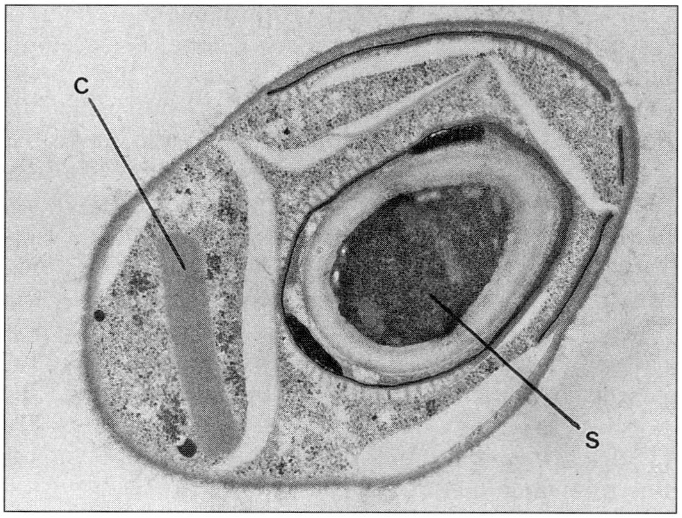

***Figure* 20.15** A thin section showing the spore (s) and enterotoxin crystal (c) of *Clostridium perfringens.* (Courtesy of Judith Bland)

BOX 20.3 MILESTONES

Food to Die For!

Because of the toxin it produces, *Clostridium botulinum* is a deadly pathogen to humans. Most incidents of botulism occur from eating improperly sterilized home-canned foods, such as vegetables that are contaminated with soil containing the bacterium. If the bacteria are not killed by canning, they can grow in the canned food and release the toxin into the contents of the can. Because the toxin can be readily denatured by boiling, it is primarily a problem in foods such as peas that are often used cold in salads without prior cooking.

When botulism is detected in commercially canned food, it can have disastrous financial impact on manufacturers. A few years ago, a handful of individuals from around the United States were diagnosed with botulism. By examining the consumption of infected individuals, epidemiologists quickly identified the source of the problem as canned salmon. Ultimately, the cans were traced to a defective canning machine in one plant in Alaska. When the news broke, sales of canned salmon plummeted, and the canned salmon market took years to recover, despite assurances by canners that the problem had been corrected.

Desulfotomaculum

Although this is also a genus of anaerobic spore formers, it differs from the genus *Clostridium* in two important ways. First, it carries out an anaerobic respiration involving sulfate and is therefore a sulfate-reducing bacterium (see Chapter 19 for more information on sulfate reducers). Second, it contains some cytochrome enzymes (cytochrome *b* but not c_3) and therefore has a limited electron transport system, which is needed as a means of passing the electrons generated from the anaerobic oxidation of organic compounds onto sulfate in the formation of hydrogen sulfide. However, ATP is not generated by chemiosmotic processes, but only by substrate-level phosphorylation. Species using lactic or pyruvic acid as carbon sources oxidize these incompletely to produce acetic acid and carbon dioxide. Species using acetic acid oxidize it completely to carbon dioxide.

Desulfotomaculum is found in soils, geothermal regions, sediments, and anaerobic muds. It is also known to produce off-flavors in the canned food industry referred to as "sulfur stinker." One species, *D. nigrificans,* is thermophilic and grows at temperatures from 45°C to 70°C, whereas the others are mesophilic.

Sporosarcina

This is the only group of endospore-producing bacteria that have a coccus shape. They divide to form groups of cells in tetrads and packets of eight. They are obligately aerobic and motile. The two species are *Sporosarcina ureae,* a urea degrader, and *S. halophila,* from salt marshes. *S. ureae* is found in urban settings, particularly in soils frequented by dogs. Media supplemented with 3% urea select against most other bacteria due to the resulting high pH produced by the growth of this species.

Differential ***Table* 20.9 • Acetogenic bacteria**

Taxon	Special Features
Clostridium aceticum	Endospore-former
Acetobacterium spp.	Non–spore forming rods; mesophilic; also ferments sugars to produce acetate
Acetogenium kivui[1]	Non–spore forming rod; thermophilic (opt. temp. 66°C); also ferments sugars to form acetate

[1]Stain as gram-negative, but have a gram-positive cell wall type.

Thermoactinomyces

As the name implies, *Thermoactinomyces* is a genus of thermophilic bacteria. Like *Bacillus,* this genus forms endospores and is aerobic. Likewise, studies of its 16S rRNA indicate it is most closely related to the other endospore-forming bacteria, although some taxonomists classify it with the actinomycetes. The organism is a septated, filamentous bacterium. Many endospores are formed within the same filament. The organism has been found in decaying compost piles, where temperatures can become so hot from decomposition that they begin spontaneous burning.

Mollicutes, Cell Wall-Less Bacteria

The *Mollicutes* are a major taxonomic group of cell wall-less bacteria that stain as gram-negative. However, studies of their 16S rRNA have clarified their relatedness to other bacteria and indicate they are most closely allied with gram-positive bacteria in the genus *Clostridium.*

Because they completely lack a cell wall, the organisms have unusual shapes (Figure 20.16). They are also among the smallest of the bacteria and the simplest in structure. Some have a diameter of about 0.25 μm, which approaches the theoretical minimum for the size of an organism, in other words, a structure large enough to contain the DNA, ribosomes, and necessary enzymes with which to perform the functions of life. Their genomes are also small, having an approximate molecular weight of 5×10^8 which is comparable to that of the obligately parasitic chlamydia (see Chapter 19).

Their lack of a cell wall makes the *Mollicutes* particularly fragile osmotically. Many species must be grown on complex media and some require sterols to stabilize their plasma membranes (Table 20.10). Because they cannot synthesize sterols, they must obtain them from the medium in the form of cholesterol often added as serum. However, some species do not require sterols.

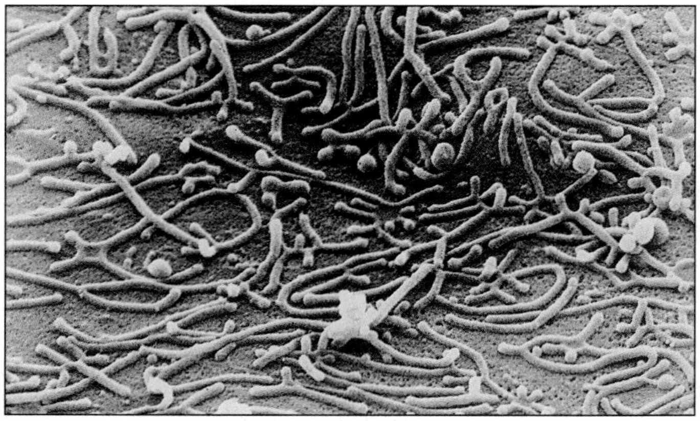

Figure **20.16** The typical "fried egg" appearance of colonies of *Mycoplasma pneumoniae.* (© CNRI / Phototake NYC)

Mycoplasma

One important genus of the Mollicutes is *Mycoplasma.* Colonies of *Mycoplasma* species appear as "fried eggs" with dense, granular centers and more transparent outer region. Some members of this genus obtain energy by fermentation of sugars through the Embden-Meyerhof pathway to form lactic, pyruvic, and acetic acids. Others degrade arginine as an energy source or derive energy from acetyl coenzyme A by phosphoacetyl transferase and acetate kinase.

Many species of this genus are parasitic or pathogenic to animals. They are found particularly associated with mucoid epithelial tissues. *Mycoplasma pneumoniae* causes a type of bacterial pneumonia. Unlike pneumococcal pneumonia, however, it is not possible to use penicillin as an antibiotic in its treatment because *M. pneumoniae* lacks peptidoglycan. However, mycoplasmas are inhibited by tetracycline and chloramphenicol.

Other Important Genera

Ureaplasma spp. obtain energy through the degradation of urea to ammonium ions and the subsequent conversion of ammonium to ammonia plus protons from which they produce ATP chemiosmotically. These are parasites in animal genitourinary tracts and respiratory tracts.

Acholeplasma spp. lack a requirement for sterol. They produce **lipoglycan** which may help stabilize their cell membranes. Unlike the lipopolysaccharide of gram-negative bacteria, lipoglycan is not linked to a lipid A backbone. They resemble *Mycoplasma* species in that they obtain energy from sugar fermentation. None are known to be pathogens, but they are common parasites of animals.

Spiroplasma spp. have a characteristic helical shape (Figure 20.17). They obtain energy by sugar fermentation. They are found primarily as parasites of plants and insects. Insects feeding on infected plants can obtain the bacterium from plant phloem and carry it to another plant.

Anaeroplasma are strict anaerobic organisms found in the rumen of cattle and sheep. They ferment starch and other carbohydrates to produce acetic, formic, lactic, and propionic acids, as well as ethanol and carbon dioxide. They lyse other bacteria as well, but are not known to be pathogenic to their hosts.

Other Single-Celled or Filamentous Gram-Positive Genera

The following three genera are gram-positive Eubacteria commonly found in the environment. *Micrococcus* and *Geodermatophilus* are common soil genera, and the third genus, *Frankia,* is a plant symbiont (Table 20.5).

Differential ***Table* 20.10 • Important genera of *Mollicutes***

Genus	Mol % G + C	Sterol Requirement	Habitat	Distinctive Properties
Mycoplasma	23–40	+	Animals	
Ureaplasma	27–30	+	Animals	Degrade urea
Acholeplasma	27–36	−	Animals	
Spiroplasma	25–31	+	Plants; insects	Helical shape
Anaeroplasma	29–34	+	Cattle rumen	Obligate anaerobes

Micrococcus

Although *Staphylococcus* and *Micrococcus* resemble one another superficially in both being nonmotile, gram-positive cocci, they are considerably different. The genus *Micrococcus* is a group of obligately aerobic organisms with a high DNA base composition (66 to 72 mol % G + C), whereas the genus *Staphylococcus* is a group of facultative aerobes having a low DNA base ratio (30 to 35 mol % G + C).

The genus *Micrococcus* is a common soil organism that is dispersed in air. Some species also reside on the skin of humans and other mammals. Most strains produce carotenoid pigments that give their colonies a yellow to red pigmentation. The pigments may protect these organisms from ultraviolet light during their dispersal in air by absorbing lethal ultraviolet radiation. Some of them produce packets of cells that give them a characteristic appearance (Figure 20.18**a**). Some *Micrococcus* spp. produce capsules giving colonies a characteristic gummy appearance (Figure 20.18**b**).

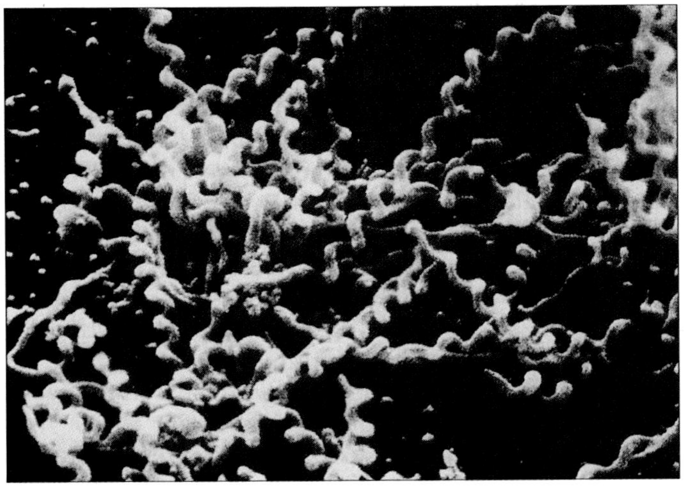

***Figure* 20.17** The cells of the cell wall-less genus, *Spiroplasma,* has a characteristic helical shape. (© David M. Phillips / VU)

Geodermatophilus

This unusual genus, with the aptly named species, *Geodermatophilus obscurus,* is a common inhabitant of desert soils and rocks. Like *Micrococcus,* it is an obligate aerobe that utilizes a variety of sugars as carbon sources for growth. Colonies produce a greenish-black pigment, probably a melanin. The organism grows as masses of cocci to form a gummy colony that rises above the agar surface. Motile cells are produced by some strains.

Frankia

This genus is an important plant symbiont. It produces root nodules similar to that formed by *Rhizobium;* however, it associates with a variety of nonleguminous plants such as *Alnus* (alder), *Ceanothus* (wild lilac), and *Casuarina* (Australian pine or she-wood). Like *Rhizobium, Frankia* carries out nitrogen fixation while growing in the plant as a symbiont, thereby benefiting the plant.

Frankia is very difficult to cultivate away from its host plant. The organism is microaerophilic and produces an aerial as well as substrate mycelium. The aerial mycelium develops a sac or **sporangium** that is referred to as multilocular because it is compartmentalized into many individual spores (Figures 25.16–25.19).

Coryneform Bacteria

The coryneform bacteria are a group of organisms, primarily soil forms, that show characteristic branching cells with the formation of Y, V, or Chinese character shapes (Figure 20.19). The Y shapes are due to rudimentary branch formation, a common characteristic of the filamentous gram-positive bacteria, the actinomycetes (except for some of the cyanobacteria, gram-negative bacteria do not produce true branches in their filaments). The characteristic V shapes occur after cell division. Cells have an outer cell wall layer not shared by other bacteria. When the cells have completed division, turgor pressure is exerted on the space between the dividing cells, and they

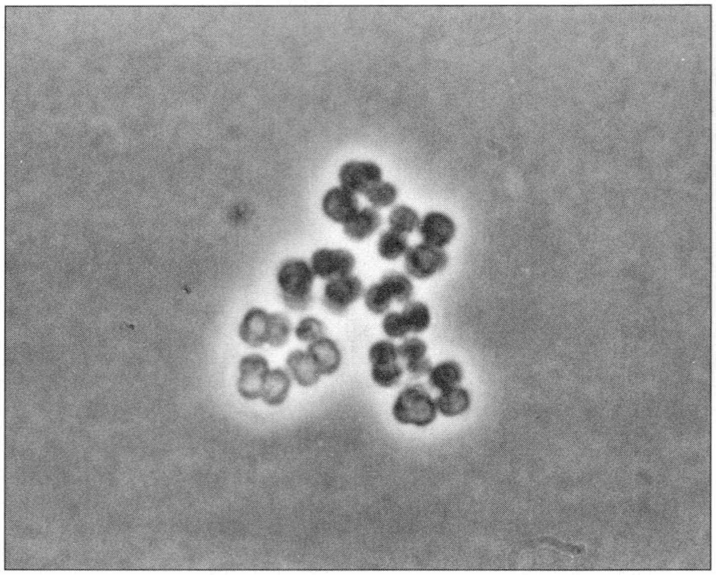

(a)

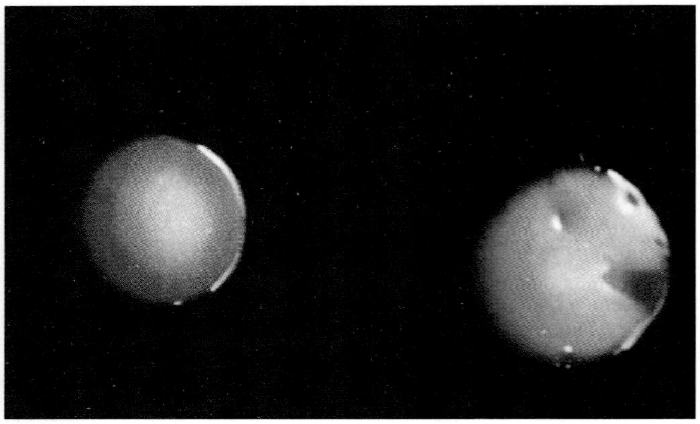

(b)

***Figure* 20.18** **(a)** *Micrococcus luteus* as it appears in the phase microscope. Note the occurrence of cells in packets. (Courtesy of J. T. Staley) **(b)** Colonies of *Micrococcus roseus* without (left) and with (right) capsules. Note the larger, gummy appearance of the encapsulated colony as well as a mutant sector that does not produce the capsule. (Courtesy of Wesley Kloos)

separate incompletely. This is called a post-fission snapping movement. Because the two cells do not separate completely from one another, a V configuration is produced.

The coryneform bacteria appear to be intermediate morphological forms between lactobacilli and true mycelial actinomycetes. Indeed, studies of the 16S rRNA places this group in such a position (see Figure 20.1). Four principal genera in this group include *Propionibacterium, Corynebacterium, Arthrobacter,* and *Bifidobacterium* (Table 20.11).

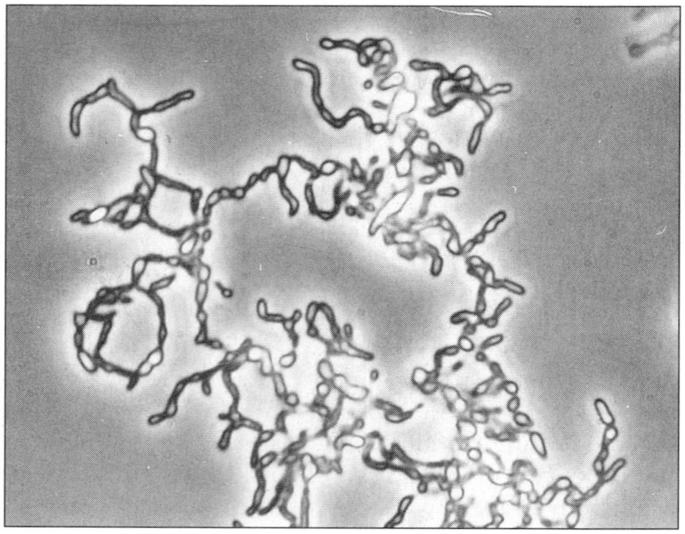

***Figure* 20.19** A typical Chinese character arrangement of cells of the coryneform bacterium, *Corynebacterium diphtheriae.* (Courtesy of J. T. Staley)

Propionibacterium

This genus is so named because they produce propionic acid as a principal product of their fermentation. They are gram-positive aerotolerant fermentative bacteria that are found in two different habitats. One habitat occupied by the classic type of *Propionibacterium* is the intestinal tract of animals. An allied habitat is cheese. Indeed, the characteristics of Swiss (Ementhaler) cheese are due to the growth of propionibacteria. These organisms take the lactic acid formed by the lactic acid bacteria in cheese fermentation and further metabolize it to propionic and acetic acids as well as CO_2 (Figure 20.20). These reactions are actually more complex, comprising a series of steps beginning with the oxidation of lactic acid to pyruvate and then transformation of the pyruvate (Figure 20.21). The gas production results in the characteristic holes, called "eyes" of Swiss cheese (if gas is not produced in adequate amounts, holes are not formed and the cheese is said to be "blind"). Although cheese was the initial source of these bacteria, their true natural habitat has been traced to the rumen of cattle and other animals. Rennin, an enzyme used for the curdling of milk in the production of cheeses, is obtained from the stomach of calves. Therefore, rennin contaminated with propionibacteria was the likely original source of these organisms in the cheese-making industry. Sugars are also fermented by propionibacteria. They use the Embden-Meyerhof pathway to product pyruvate which is then fermented as shown in Figure 20.20.

The other major habitat of propionibacteria is the skin of mammals. One species, *Propionibacterium acnes,* is found on the skin of all humans. It grows in the sebaceous gland (*not* sweat glands) where it produces propionic acid in

***Differential* *Table* 20.11 • Important genera of coryneform bacteria**

Genus	Shape	Mol % G + C	O_2 Requirement	Habitat
Arthrobacter	Rod-coccus	59–66	Obligate aerobe	Soil
Corynebacterium	Irregular rods, V-shapes	51–60	Facultative aerobe	Soil; animal pathogen
Propionibacterium	Club-shaped rods	57–67	Facultative aerobe	Animal intestine; cheese
Bifidobacterium	Rods	55–67	Anaerobic	Animal intestine

abundance. It ferments the lactic acid produced by *Staphylococcus epidermidis* to form propionic and acetic acid as previously discussed. Humans fall into two groups based upon the numbers of *P. acnes* they harbor on their skin. Some have over 1 million per cm^2 of skin surface, whereas others have fewer than 10,000 per cm^2. Since propionic and acetic acids are volatile fatty acids with distinctive smells, they give animals, including humans, a characteristic natural scent.

Corynebacterium

This genus is a group of common aerobic soil organisms. The first isolate of the genus was *Corynebacterium diphtheriae,* which is a normal inhabitant of the oral cavity of animals, where it lives as a parasite. Pathogenic strains of this species carry a piece of DNA they have received from a phage by lysogenic conversion. This genetic material is responsible for production of the protein exotoxin, the principal virulence factor for diphtheria. The disease is rare now, having been controlled largely through vaccination.

Soil "diphtheroids" which appear as club-shaped rods (Figure 20.22) are similar nutritionally and metabolically to the animal parasites. All are immotile and most are facultative aerobes that produce propionic acid during fermentation of sugars. Some species cause diseases of plants.

Arthrobacter

Arthrobacter spp. are also common soil inhabitants. Sergei Winogradsky was the first to note that small coccoid cells were abundant in soils. Actually, the coccoid cells are one of two cell types exhibited by this genus. When cells are actively growing, they grow as irregular rods (Figure 20.23**a**). However, as the cells enter stationary phase they become shorter and rounded in appearance (Figure 20.23**b**). In fact, some studies have shown that the length of the rod is directly related to the growth rate of the bacterium. If they are growing slowly, they are shorter than if they are growing more rapidly. These bacteria are gram-positive, but some may stain as gram-negative cells. Some strains produce motile cells.

Arthrobacter spp. are obligate aerobes that use a variety of sugars as carbon sources. Many grow on a simple medium with ammonium as a nitrogen source, although some strains require biotin or other vitamins.

Bifidobacterium

Bifidobacterium is a genus of anaerobic irregular rod-shaped bacteria that ferment sugars to acetic and lactic acids (Fig-

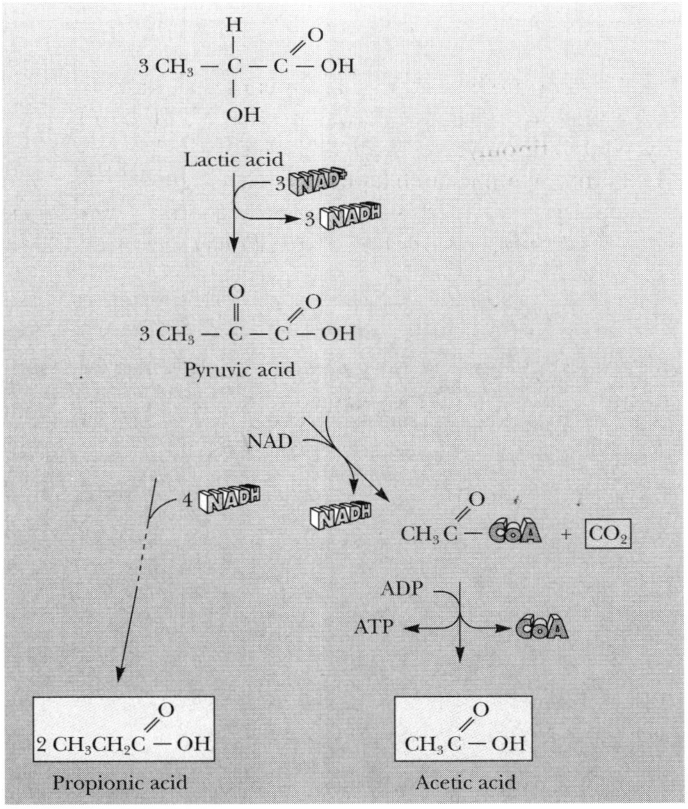

***Figure* 20.20** Propionic acid fermentation of lactic acid by *Propionibacterium* spp. First, 3 mol of lactic acid are oxidized to pyruvic acid, which is the source of 2 mol of propionic acid and 1 mol each of acetic acid and carbon dioxide. ATP is generated from acetyl phosphate. The route to propionic acid from pyruvate is shown in Figure 20.21.

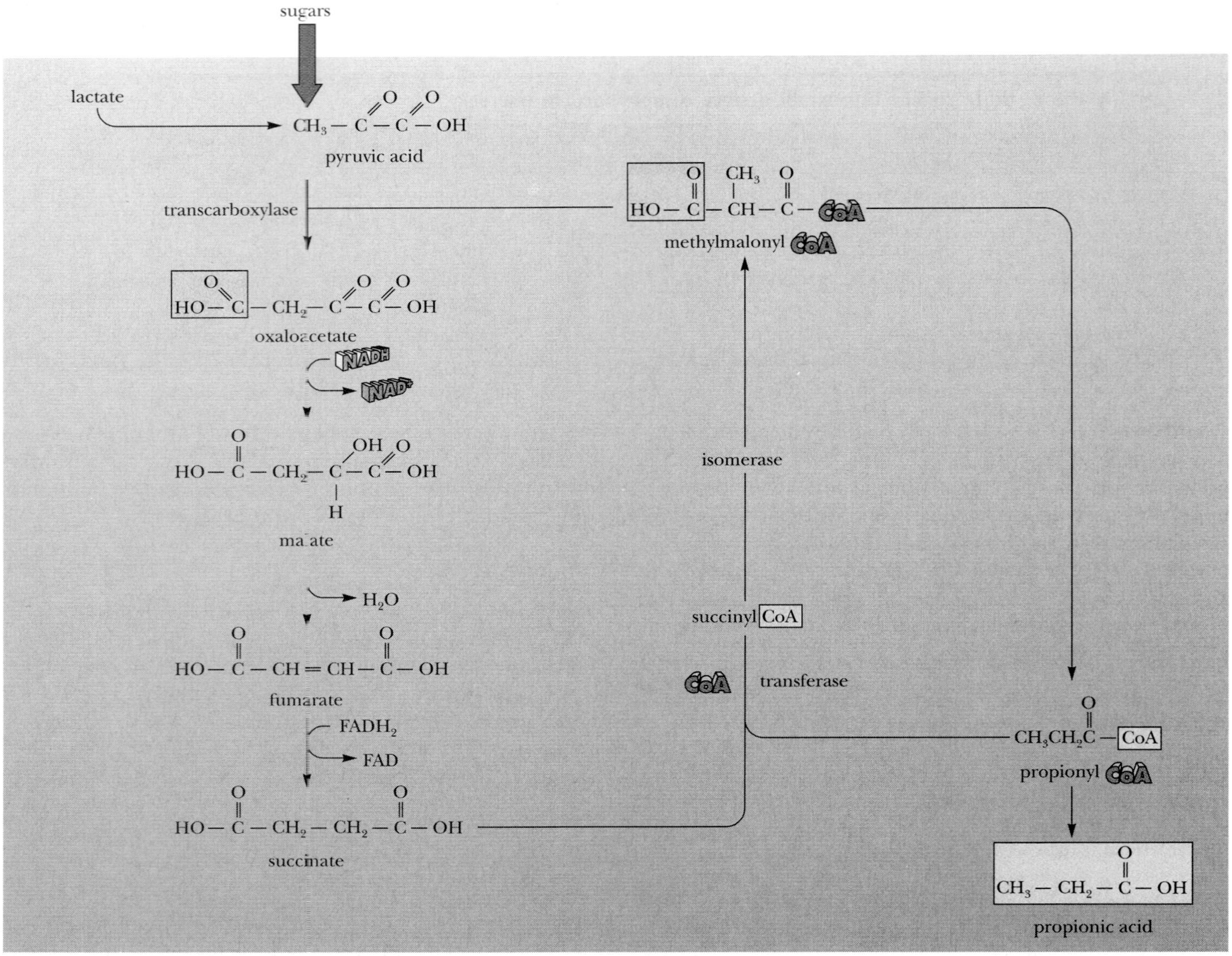

***Figure* 20.21** In the propionic acid fermentation, either sugars or lactic acid can be a substrate. Sugars are broken down by the Embden-Meyerhof scheme to produce pyruvate, and ATP is generated by glycolysis. If lactic acid is the substrate, it is first oxidized (see Figure 20.20) to form pyruvic acid. The pyruvate is converted to oxaloacetate by receiving a carboxyl group from methymalonyl-coenzyme A via a transcarboxylase. Propionyl CoA is the other product of this reaction and it transfers its CoA to succinate to form the end-product propionic acid. Succinate is formed from the oxaloacetate via a reversal of the TCA cycle.

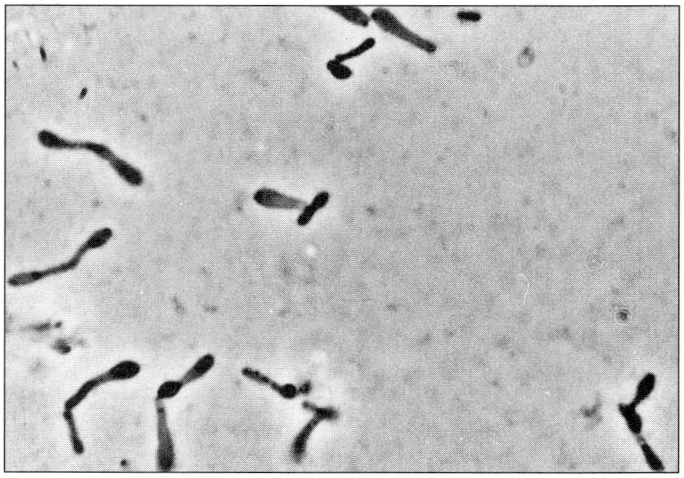

***Figure* 20.22** Club-shaped diphtheroids of *Corynebacterium diphtheriae*. Phase micrograph. (Courtesy of J. T. Staley)

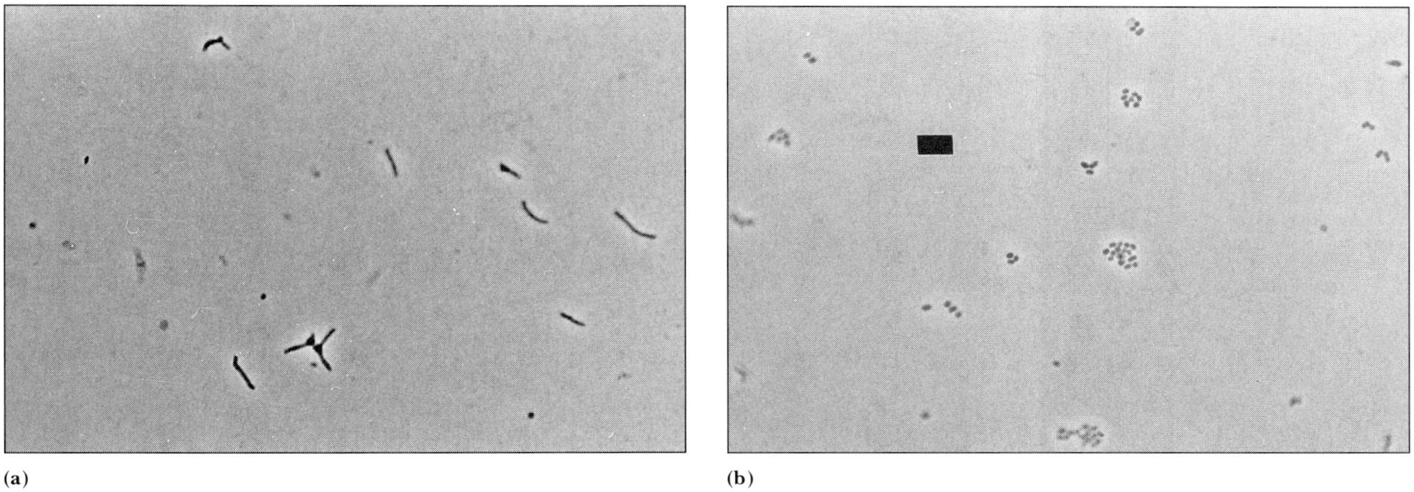

Figure 20.23 Cells of an *Arthrobacter* species **(a)** during active growth and **(b)** during stationary phase growth. Bar is 5 μm. (Courtesy of Fred Palmer and J. T. Staley)

ure 20.24). They are found primarily in the intestinal tract of animals. One species, *B. bifidus,* is commonly found in the intestines of infants that are breast-fed and is therefore a pioneer colonizer of the human intestinal tract. It is particularly well adapted to growing on human breast milk, which contains an amino sugar disaccharide not found in cow's milk. This species is unusual in that it requires amino sugars for growth.

Non-Mycelial Actinomycetes

The "true" actinomycetes are mycelial organisms, that is, the cells produce branches as well as filaments. Therefore, they have a mycelial growth habit resembling that of the true fungi. One of the first organisms studied in this group was the genus *Actinomyces,* after which the group is named. Some species produce only a **substrate mycelium** in which the growth is within or on the surface of the agar or other growth medium. The substrate mycelium is coenocytic, lacking cell septa. Some genera produce an **aerial mycelium** as well as a substrate mycelium (Figure 20.25). The aerial mycelium can form special reproductive spores called **conidia** or conidiospores. Conidia are more resistant to ultraviolet light, can survive well under dry conditions, and are disseminated by the wind as a means of dispersing the organism in soil environments. However, unlike endospores, they are not particularly resistant to high temperatures. Almost all of the actinomycetes are nonmotile; however, some types produce flagellated spores that permit dispersal in aquatic habitats.

As mentioned earlier, there are many genera in the gram-positive bacteria that are transitional between unicellular forms and mycelial forms. One such group already discussed in the coryneform group, which show evidences of true branching in their Y forms. Two other genera that appear to be intermediate are still largely unicellular, but are considered true actinomycetes because they can form true branches. These are the genera *Mycobacterium* and *Nocardia.* These will be discussed here along with the more highly differentiated mycelial forms that include the genera *Actinomyces, Micromonospora, Streptomyces,* and *Actinoplanes.*

Mycobacterium

The mycobacteria are nonmotile rod-shaped bacteria that may show true branching and typically bundle together to form cordlike groups (Figure 20.26). They are aerobic rods that occur naturally as saprophytes in soils. Unlike other bacteria, they produce a distinctive group of waxy substances called **mycolic acids** (Figure 20.27) that are covalently linked to the peptidoglycan. In addition, they have a polymer of arabinose and galactose (called an *arabinogalactan*) bound to the peptidoglycan. The mycolic acids make the organisms difficult to stain using ordinary simple staining procedures; nonetheless they are regarded as being gram-positive. These same mycolic acids make these bacteria **acid-alcohol fast,** which refers to the ability of these organisms, once stained with a solution of basic fuchsin in phenol, to withstand decolorization with acidified ethanol during a staining procedure called the Ziehl-Neelsen stain. The basic fuchsin binds strongly to the mycolic acids. This acid-alcohol fast property is not found in any other bacterial group, making it an excellent differential property for distinguishing these from all other bacteria.

Mycobacteria grow on simple inorganic media with ammonium as the sole nitrogen source and glycerol or acetate as a carbon source. Incorporating lipids into growth media tends to enhance growth of some species. Their slow

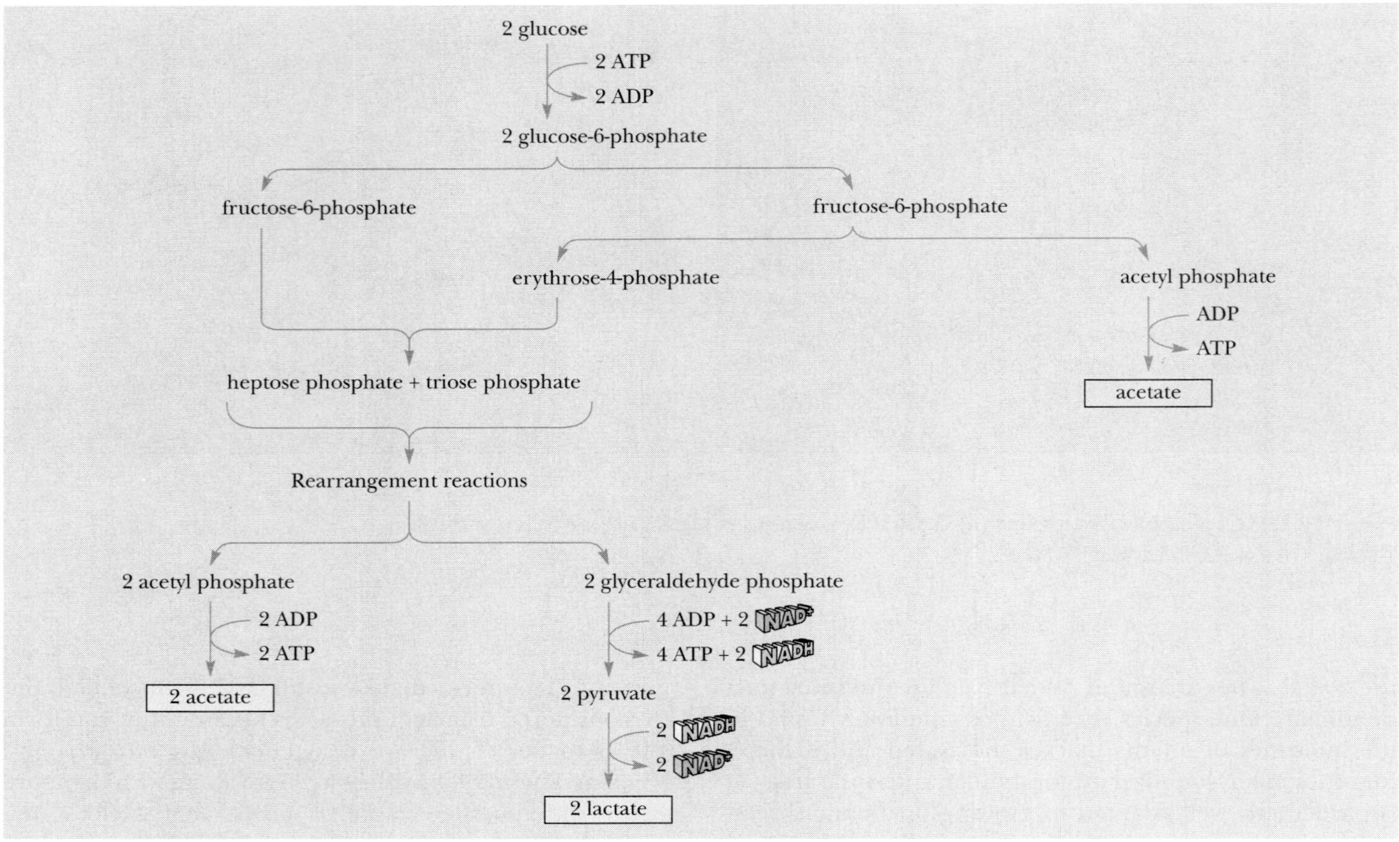

***Figure* 20.24** The bifidobacteria carry out an unusual lactic acid fermentation that results in the conversion of 2 mol of glucose to 3 mol of acetate and 2 mol of lactic acid. Glucose is first phosphorylated as in the glycolytic pathway. It is next converted to fructose 6-phosphate, which is cleaved and produces two phosphorylated compounds, erythrose 4-phosphate and acetylphosphate. Acetylphosphate generates ATP and acetic acid. The erythrose 4-phosphate and a molecule of fructose 6-phosphate are then used to form a seven-carbon and a three-carbon sugar intermediate, which after rearrangement result in the production of acetic acid and lactic acid.

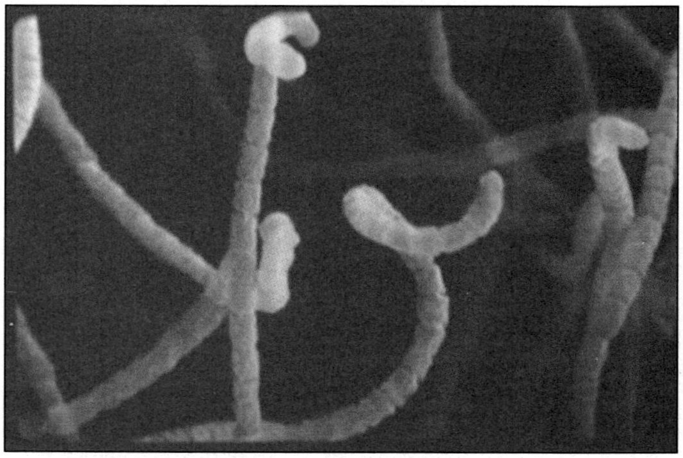

***Figure* 20.25** The aerial mycelium of a *Streptomyces* sp. (© Frederick P. Mertz / Science VU)

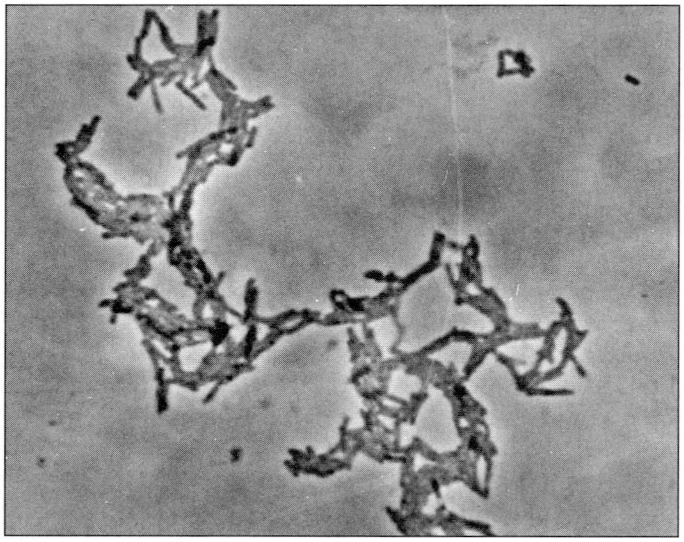

***Figure* 20.26** Cells of *Mycobacterium tuberculosis* showing its tendency to aggregate into cordlike structures. (Courtesy of J. T. Staley)

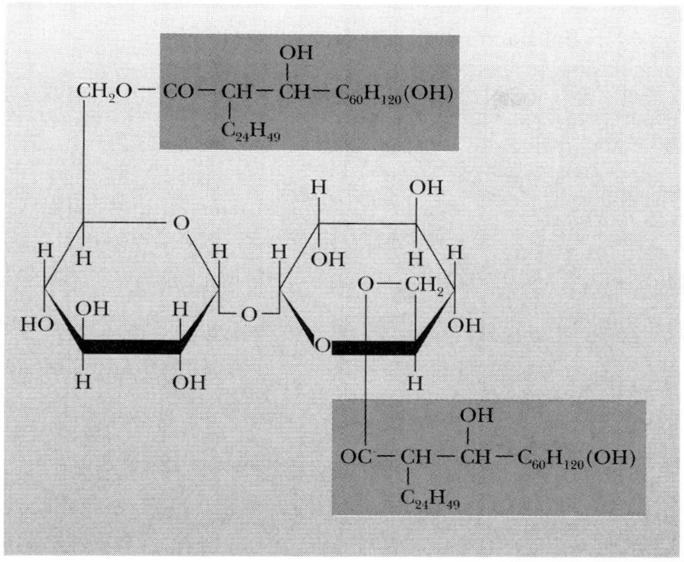

***Figure* 20.27** General formula for mycolic acids.

growth is in part due to the hydrophobicity of the organisms, which prevents rapid uptake of dissolved nutrients.

Two species are important pathogens of humans. *Mycobacterium tuberculosis* is the causative agent of tuberculosis, and *M. leprae* causes leprosy. See Chapter 29 for more detail on these diseases.

Nocardia

The nocardiae are separated into two groups. One group is acid-fast and therefore similar to the mycobacteria, whereas the other group is nonacid-fast. These nonmotile, rod-shaped to filamentous organisms show somewhat greater branching than the mycobacteria. Some strains even show extensive mycelial development (Figure 20.28) and may even produce an aerial mycelium. One feature characteristic of this group is that the mycelium, if formed, tends to fragment into small units during stationary growth phase.

Nocardia are common soil bacteria. They are obligately aerobic heterotrophs whose metabolism has not been well studied. As a group and individually, they utilize a wide variety of sugars and organic acids as carbon sources for growth.

Rhodococcus

The genus *Rhodococcus* is widely distributed in soils and aquatic environments. It is particularly well known for its ability to degrade hydrocarbon compounds. It produces capsular materials that are excellent dispersants and emulsifiers of hydrocarbon compounds. The cells grow at the interface between oil and water and are particularly adept at degrading aliphatic hydrocarbons. Some species are pathogenic to animals.

Mycelial Actinomycetes

The more highly branched actinomycetes are common soil organisms that are prokaryotic counterparts of the fungi (Table 20.12). These organisms are not acid-fast. Some of them produce an aerial mycelium that has a fruiting structure bearing spores called conidiospores on conidia. These

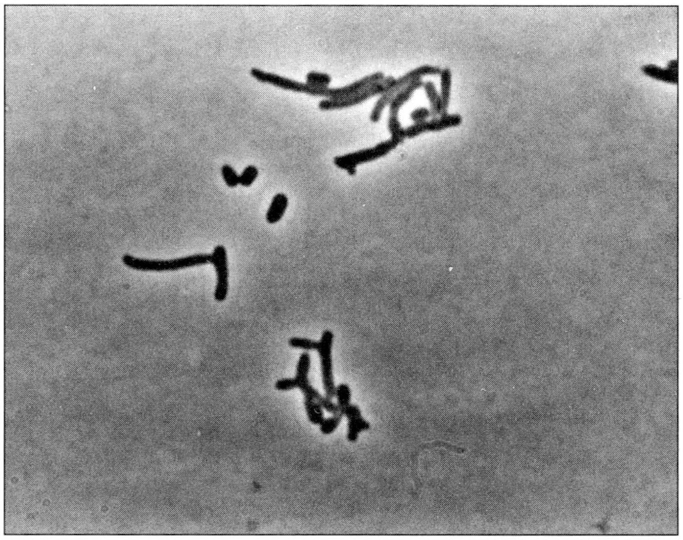

***Figure* 20.28** Cells of *Nocardia* sp. showing branching. (Courtesy of J. T. Staley)

Differential *Table* **20.12 • Important genera of branching and Mycelial bacteria (Actinomycetes)**

Genus	Shape	Mol % G + C	Habitat	Special Features
Mycobacterium	Rods; some branching	62–70	Animal pathogen	Mycolic acid; acid-fast
Nocardia	Slight to extensive branching	64–69	Soil	Mycolic acid; some are acid-fast
Rhodococcus	Coccus or filamentous mycelium	60–69	Soil; some pathogens	Hydrocarbon degraders
Actinomyces	Rods; some branching	58–63	Oral cavity; some pathogens	Only facultative aerobes in group
Micromonospora	Substrate and aerial mycelium	71–73	Soil	Single conidium
Streptomyces	Substrate and aerial mycelium	69–78	Soil	Many conidia
Actinoplanes	Substrate and aerial mycelium	72–73	Soil	Motile sporangiospores

are asexual spores, sometimes referred to as exospores to differentiate them from endospores.

Actinomyces

This genus, for which the entire group was named, is atypical of other mycelial members, primarily because it is a group of anaerobic or facultatively aerobic bacteria. They ferment sugars such as glucose to produce formic, acetic, lactic, and succinic acids and do not carry out aerobic respiration. Organic nitrogen compounds are required for growth, and supplemental carbon dioxide greatly enhances it. Though they are mycelial, they do not produce an aerial mycelium. They grow in the oral cavity of animals and can cause serious infections, such as lumpy jaw by *Actinomyces bovis.*

Streptomyces

The foremost genus of the mycelial actinomycetes is *Streptomyces.* This is a diverse group of soil bacteria that produce aerial as well as substrate mycelia. The **hyphae,** or filaments, of the aerial mycelia differentiate to form asexual conidiospores. The spores of this genus are formed in chains at the tips of aerial hyphae (see Figure 20.25**b**). Species differences are based partially on the morphology of the spores. For example, some have a warty or spiny appearance, whereas others are smooth. Physiological differences, especially antibiotic production, are also important.

Members of this genus grow on simple inorganic media supplemented with a variety of organic carbon sources including glucose or glycerol; vitamins are not required. They metabolize by aerobic respiration. In addition to simple organic carbon sources, some can use polysaccharides such as pectin, chitin, and even latex. If a culture is started with conidiospores, they first germinate to produce the vegetative or substrate mycelium. After this, the aerial mycelium is formed. If the medium has a sufficiently high carbon-to-nitrogen ratio, the aerial mycelium differentiates to produce conidiospores when the nutrients have been depleted. The conidia are actually formed within the outer wall of the filament; however, they do not at all resemble endospores. The conidia become pigmented blue, gray, green, red, violet, or yellow colors, but the color can be influenced by medium composition. The substrate mycelium can also be pigmented.

Streptomyces occur in viable concentrations of 10^6 to 10^7 per gram in soil environments. They are responsible for imparting the characteristic "earthy" odor to soil. This is due to the **geosmins,** a group of volatile organic compounds they produce during growth. Geosmins cause odors and flavors in drinking water supplies as well. However, cyanobacteria also produce geosmins and are more likely to be responsible for this problem in water supplies.

This group is noted for the commercially important antibiotics they produce including streptomycin, chloramphenicol, and tetracycline (see Table 7.5). Approximately half of the commercially produced antibiotics are derived from this genus (see Chapters 7 and 32 for more detail). In addition, they are sources of anticancer drugs used in chemotherapy (see Table 32.9).

Although this is an extremely important genus of soil microorganisms, little is known of their ecological roles. It is not even known for certain that they produce the antibiotics while growing in their natural habitat. This is a plausible hypothesis because they could use them to inhibit competitors. The proliferation of great numbers of species in this genus (over 500 species have been proposed) in part reflects the need for pharmaceutical firms who study these bacteria to propose a new species when they wish to patent and manufacture a new antibiotic.

Other genera of mycelial actinomycetes are differentiated from this genus primarily by the cell wall composition and the morphology of the conidiospore-bearing structure.

Actinoplanes

This is a genus of actinomycetes that produces a flagellated spore. Like *Streptomyces*, this organism has both a substrate and an aerial mycelium. However, the spores are produced within a sac or sporangium, a stage that can survive dessication. When conditions in the environment are moist and favorable for germination and growth, the sporangium ruptures and releases the motile spores.

Heliobacteria

The **heliobacteria** are one of the most recently discovered bacterial groups. These bacteria are the only phototrophic gram-positive bacteria known. They are photoheterotrophs that require organic compounds as carbon sources and use light for energy generation. They have a unique type of bacteriochlorophyll called bacteriochlorophyll *g* (see Chapter 9). Although they stain as gram-negative bacteria, an analysis of their 16S rRNA places them with the gram-positive bacteria. Furthermore, some members of the group produce heat-resistant bacterial endospores similar to those of *Bacillus* spp.

To date, only two different genera have been discovered in this group: *Heliobacterium* and *Heliobacillus*. *Heliobacterium* spp. are gliding bacteria, whereas members of *Heliobacillus* are motile by peritrichous flagella. The heliobacteria grow as obligate anaerobes carrying out anoxygenic photosynthesis using organic compounds such as pyruvate as carbon sources.

The finding of photosynthetic bacteria in the gram-positive phylum is consistent with photosynthesis being a primitive characteristic and shared by many of the other eubacterial groups.

Deinococcus and *Thermus*

Deinococcus

Members of the genus *Deinococcus* look superficially like *Micrococcus* spp. in their morphology, in that most species are nonmotile cocci (Figure 20.29**a**). However, they differ from the micrococci in their strong resistance to gamma

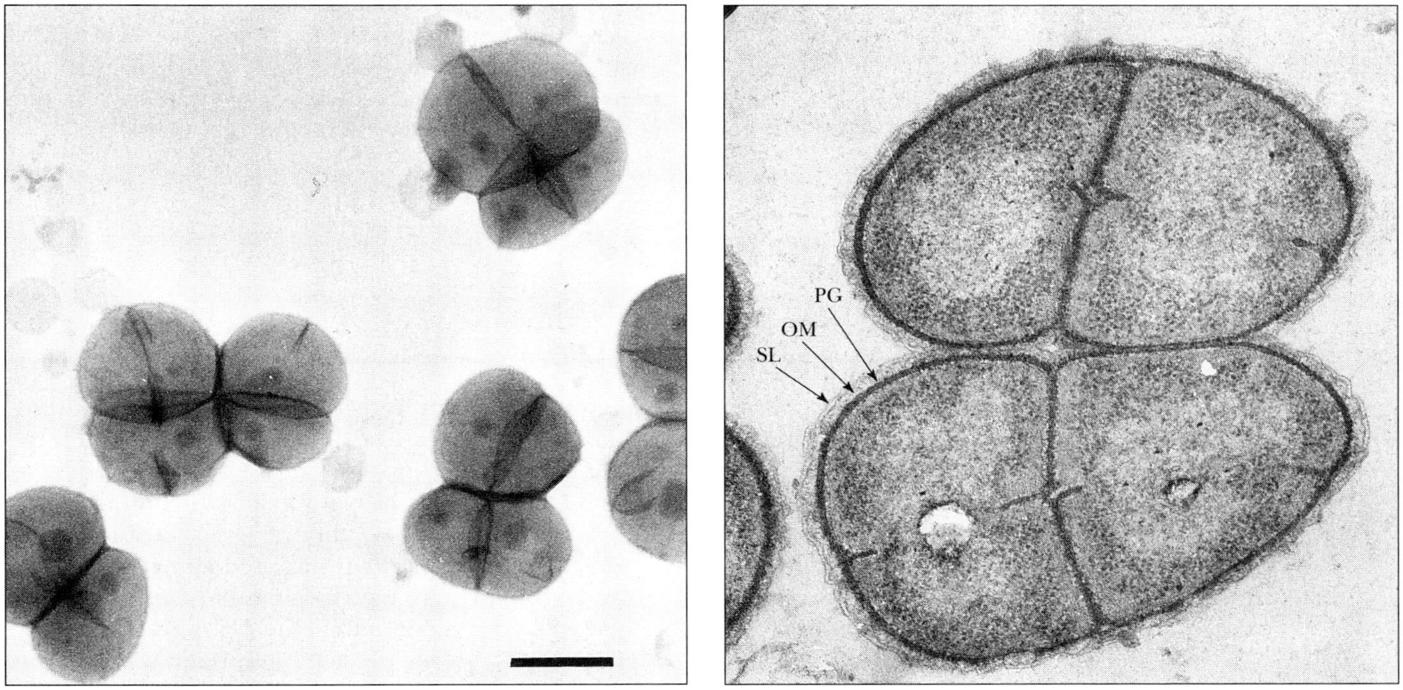

(a) (b)

***Figure* 20.29** **(a)** Electron micrograph of actively growing cells of *Deinococcus radiodurans* showing their tendency to grow as four-celled packets or tetrads. Note that the septa are forming as "curtains" rather than as iris diaphragms. Bar is 1 μm. **(b)** Thin section through *Deinococcus radiodurans* showing its multilaminated cell wall structure consisting of an outer single S-layer (SL), an outer membrane (OM), and a thick peptidoglycan layer (PG). (Courtesy of R. G. E. Murray)

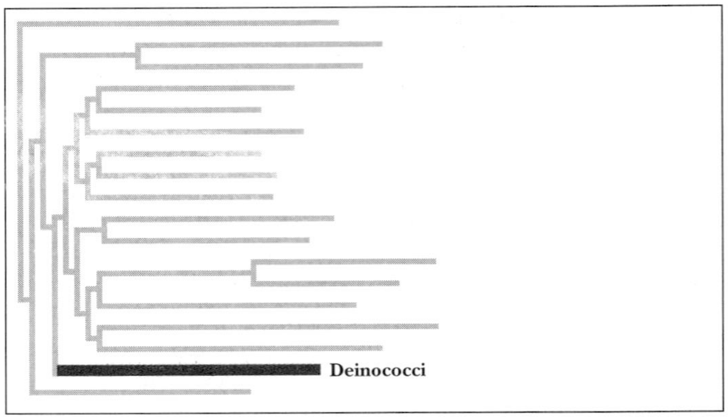

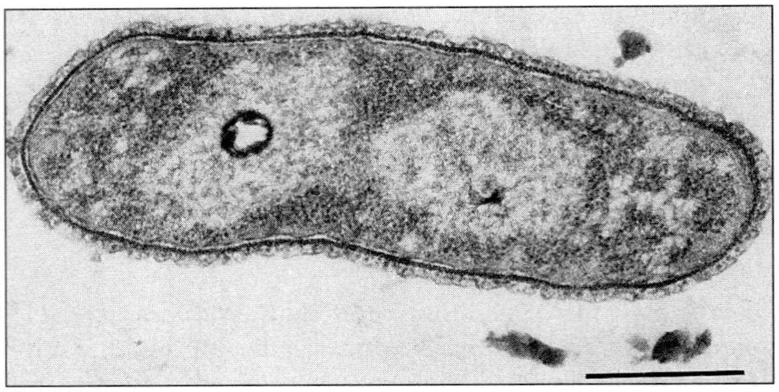

***Figure* 20.30** Thin section through *Deinobacter grandis,* a rod-shaped member of the deinococci group. The outer membrane appears as a looped layer inside of which is the darker peptidoglycan layer. Bar is 0.5 μm. (Courtesy of R. G. E. Murray)

radiation and ultraviolet light as well as their complex cell wall structure (Figure 20.29**b**). The genus also differs because the diamino-amino acid of its peptidoglycan is L-ornithine and it lacks phosphatidylglycerol in its membrane. *Deinococcus radiodurans* was first isolated from foods that had been sterilized by gamma irradiation, indicating its importance in food microbiology. The genus is also noted for its pink to red pigments, which are carotenoids. The optimum temperature for growth is between 25°C and 35°C. Some members of the deinococci, such as *Deinobacter grandis,* are rod-shaped (Figure 20.30).

Thermus

T. aquaticus is quite different from *Deinococcus.* It is a nonmotile, rod-shaped to filamentous organism ranging from 5 to 100 μm or more in length (see also Chapter 19, where other thermophilic Eubacteria are treated). Unlike *Deinococcus,* it stains like a gram-negative bacterium, but like *Deinococcus* has ornithine as the diamino acid in its cell wall peptidoglycan. Colonies are pigmented yellow, orange, or red. Its optimum growth temperature ranges from 70°C to 75°C. It is an obligate aerobe that grows as a heterotroph with ammonium salts as the nitrogen source and sugars as carbon source.

Thermus aquaticus is one of the most common thermophilic bacteria known. It grows in most neutral pH hot springs, but is also found in ordinary home water heaters. The usual way to isolate it is to inoculate a broth medium from hot tap water and incubate at 70°C. This bacterium is currently of interest in biotechnology because its DNA polymerase (called *Taq* polymerase) is heat resistant and survives multiple heating and cooling cycles used in polymerase chain reaction technology.

Summary

- Lactic acid bacteria ferment sugars to produce lactic acid either by the **homofermentative** or **heterofermentative** pathway. All energy generated by the lactic acid bacteria is via substrate level phosphorylation. More energy is available to the homofermentative (2 ATP/mol glucose) compared to the heterofermentative (1 ATP/mol glucose) lactic acid bacteria.

- Lactic acid bacteria are **aerotolerant anaerobes** that are indifferent to the presence of oxygen in their growth environment. Most lactic acid bacteria are nutritionally **fastidious,** requiring complex nutrients for growth including vitamins, amino acids, and purines or pyrimidines.

- Several groups of lactic acid bacteria exist including those associated with plants, dairy products, and in the mouth, intestine, and vagina of animals. In natural fermentations of plant materials and dairy products, the fermentation undergoes a succession in which the *Streptococcus* group predominates first and lowers the pH to 5.5 or so, then the *Lactobacillus* group succeeds and lowers the pH even further, to about 4.5. The low pH acts as a preservative for the foodstuff.

- Some *Streptococcus* spp. cause dental caries, others cause scarlet fever, puerperal fever, pneumonia, tissue necrosis, and endocarditis. *Streptococcus* spp. may cause **alpha-** or **beta-hemolysis** of red blood cells.

- *Staphylococcus epidermidis* and *Propionibacterium acnes* are members of the normal microbiota of human skin; all humans are colonized by these two bacterial species. *Staphylococcus aureus* is a normal resident of the nasopharynx of humans, but can also be a human pathogen; it is noted for causing skin infections, food poisoning, and toxic shock syndrome.

- *Bacillus* spp. are aerobic to facultatively aerobic, endospore-forming rods commonly found in soils. *Clostridium* spp. are obligately anaerobic spore-forming rods. **Endospores** are specially modified cells that are heat-resistant resting stages of various genera including *Bacillus, Clostridium, Sporosarcina, Desulfotomaculum,* and *Thermactinomyces.*

- Endospores are formed when the nutrients are depleted in the environment; in cultures this occurs at the beginning of stationary phase growth. Endospore formation involves producing a spore within a cell. After DNA replication and cell membrane separation, one of the resulting units engulfs the other, which will become the endospore; several stages are involved. Endospores have a low water content and high concentrations of calcium and **dipicolonic acid,** a unique compound not found elsewhere.

- Some *Bacillus* spp. such as *B. thuringiensis,* are insect pathogens that produce a characteristic protein crystal that is toxic to moths. *Bacillus anthracis* is the causative agent of anthrax in cattle and humans.

- *Clostridium* spp. ferment cellulose, sugars, amino acids, or purines. In the **Stickland fermentation** two amino acids are catabolized: one amino acid is fermented, the other is oxidized. One species, *Clostridium aceticum,* is an acetogen that obtains energy from oxidation of H_2 and fixes CO_2 via the acetyl coenzyme A pathway. *Clostridium botulinum* is the causative agent of botulism; *Clostridium tetani* is the causative agent of tetanus.

- The **mycoplasmas** or Mollicutes, lack a cell wall, but are close relatives of *Clostridium.* Many species, such as *Mycoplasma pneumoniae,* are pathogenic to humans.

- *Micrococcus* spp. differ from *Staphylococcus* in that they are obligate aerobes that are common soil inhabitants and have a higher mol % G + C content.

- *Propionibacterium* is noted for its production of propionic acid from lactic acid; it is responsible for Swiss cheese production. *Bifidobacterium bifidus,* which ferments sugars to acetic and lactic acids, is part of the normal microbiota of breast-fed infants.

- *Corynebacterium* is the formally named genus of the coryneform bacteria that are common soil organisms. *Corynebacterium diphtheriae* lives as a parasite of the oral cavity of humans. Strains that cause diphtheria have received a piece of DNA from a phage that encodes for the diphtheria protein exotoxin.

- The **actinomycetes** are a group of branching filamentous organisms, some of which produce both a **substrate** and an **aerial mycelium.** *Mycobacterium* spp. are acid-fast and produce mycolic acids; two species, *M. tuberculosis* and *M. leprae* (leprosy), are well-known pathogens.

- *Streptomyces* is a common soil actinomycete noted for its production of antibiotics such as tetracyclines, chloramphenicol, streptomycin, and erythromycin. *Streptomyces* impart an "earthy" smell to soil due to the production of **geosmins.**

- *Heliobacterium* and *Heliobacillus* are photoheterotrophic bacteria that use organic compounds such as pyruvate as carbon sources for growth and derive energy from sunlight.

- *Deinococcus* is a genus that superficially resembles *Micrococcus,* but it differs from *Micrococcus* in being highly resistant to UV and gamma radiation and falls into its own separate phylum along with *Thermus,* a thermophilic heterotrophic bacterium.

Questions for Thought and Review

1. What evidence indicates that the Gram stain has phylogenetic significance? Name groups that are exceptions.
2. The primitive Earth was anaerobic and contained anaerobic bacteria. The modern world is aerobic (with anaerobic sediments), but it has highly evolved exotic aerobic organisms, such as humans. Do you believe there is a parallel to this in the evolution of gram-positive bacteria?
3. What bacterial groups do we ingest in large numbers with certain foods?
4. What is the significance of the bacterial endospore?
5. In what ways are gram-positive bacteria known to be commercially important?

6. List several genera of coccus-shaped gram-positive bacteria. How do they differ from one another?

7. What are the photosynthetic members of the gram-positive bacteria?

8. Compare the genus *Bacillus* to the genus *Clostridium*.

9. Identify pathogenic species of the gram-positive bacteria. What is the basis for their pathogenecity?

10. Identify agriculturally useful gram-positive bacteria and discuss their value.

11. What are some examples of fermentations carried out by gram-positive bacteria? Are they commercially important?

12. Why are some bacteria acid-fast?

Suggested Readings

Bergey's Manual of Determinative Bacteriology. 9th ed. 1994 (J. G. Holt, Editor-in-Chief). Baltimore: Williams and Wilkins.

Bergey's Manual of Systematic Bacteriology. 1st ed. 1984–89. Volumes II (P. Sneath and J. G. Holt, eds.) and IV (S. Williams and J. G. Holt, eds.). Baltimore: Williams and Wilkins.

Lederberg, J., ed. 1992. *Encyclopedia of Microbiology.* San Diego, CA: Academic Press.

The Prokaryotes, 2nd ed. 1992. (A. Balows, H. G. Trüper, M. Dworkin, W. Harder, and K. H. Schleifer, eds.). Berlin: Springer-Verlag.

The sulfur bacteria . . . life processes are played out according to a much simpler scheme; all their activities are maintained by a purely inorganic chemical process, that of sulfur oxidation.

S. Winogradsky, 1887

Chapter 21

Phototrophic, Chemolithotrophic, and Methylotrophic Eubacteria

This chapter discusses those bacteria that obtain energy from (a) light (phototrophic metabolism), from (b) the oxidation of inorganic compounds (chemolithotrophic metabolism), or from (c) the oxidation of one-carbon compounds such as methane and methanol (methylotrophic metabolism). Some of these bacteria are *autotrophic,* that is, they are able to use carbon dioxide as their sole source of carbon for growth. However, a few of the photosynthetic bacteria require organic compounds and are therefore photoheterotrophic. Furthermore, the methane- and methanol-oxidizing bacteria do not fix carbon dioxide, but have special pathways whereby they use methane and methanol as both their energy source and their carbon source for growth. Because they are unique in using these one-carbon compounds as energy sources and are closely related phylogenetically to some of the chemolithotrophic bacteria, they are included in this chapter. The various groups that are treated here and their properties are listed in Table 21.1. We begin by considering the photosynthetic Eubacteria.

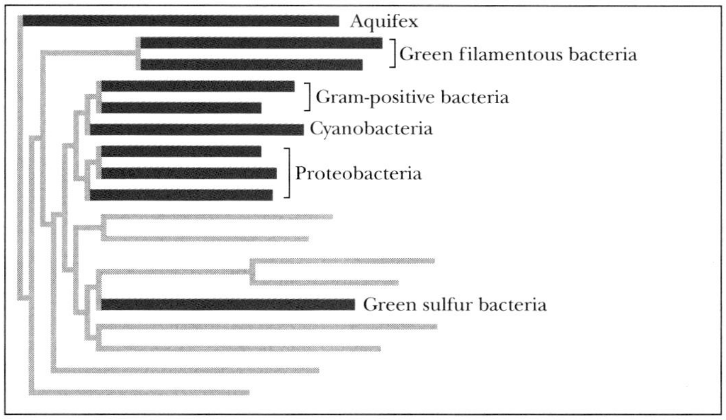

Phototrophic Eubacteria

All photosynthetic organisms from phototrophic bacteria to higher plants obtain energy from sunlight. The reactions

Differential *Table* **21.1 • Phylogenetic groups and properties of photosynthetic, chemolithotrophic, and methylotrophic Eubacteria**

Group Subgroup	Energy Source	Carbon Source; Carbon Metabolism
I. Phototrophs		
Proteobacteria		
Purple sulfur	Light	CO_2; Calvin-Benson
Purple nonsulfur	Light	CO_2 and organic; Calvin-Benson
Green sulfur	Light	CO_2; reductive TCA cycle
Green filamentous	Light[1]	CO_2 or organic; hydroxypropionate pathway
Cyanobacteria and prochlorales	Light	CO_2; Calvin-Benson
Heliobacteria	Light	Organics; photoheterotrophic
II. Chemolithotrophs		
Proteobacteria		
Nitrifiers	NH_3 or NO_2^-	CO_2; Calvin-Benson
Sulfur oxidizers	$S^=$, S^0, or $S_2O_3^=$	CO_2, Calvin-Benson or heterotrophic
Gallionella	Fe^{2+}	CO_2; Calvin-Benson
H_2 Bacteria[2]	H_2	CO_2; variable
III. Methylotrophs		
Proteobacteria		
Methanotrophs		
Type I	CH_4	CH_4 or CH_3OH[3]; ribulose monophosphate pathway
Type II	CH_4	CH_4; serine pathway
Methylotrophs	CH_3OH[3]	CH_3OH[3]; serine pathway

[1]The green filamentous bacteria are very versatile metabolically. See text for details.
[2]Hydrogen bacteria are found in a number of different phylogenetic groups and therefore use a variety of different carbon dioxide fixation pathways. See text for more details.
[3]Methanol is one example of one- or two-carbon compounds that can be used by these bacteria. See text for details.

used to capture the energy from sunlight and transform it into chemical energy are referred to as the **light reactions** of photosynthesis (Chapter 9). The energy generated from photosynthesis as well as the reducing power, produced as NADPH or NADH, are required for the carbon dioxide fixation reactions. Because the actual carbon dioxide fixation reactions do not directly involve sunlight, they are referred to as the **dark reactions.** Before discussing the various photosynthetic Eubacteria, the light and dark reactions that were treated in Chapter 9 are reviewed briefly below.

Light and Dark Reactions

The light reactions initially involve absorption of radiant energy by pigments in the cells. The pigments fall into three groups: **reaction center chlorophyll** pigments located in the photosynthetic membranes are chlorophyll molecules that play a direct role in generating ATP by photophosphorylation. The type of reaction center chlorophyll varies from one group of organisms to another (Table 21.2). **Antenna** or **light-harvesting chlorophyll** molecules are associated with the reaction center and are involved in harvesting light radiation for the reaction center chlorophyll. Finally, photosynthetic microorganisms also have **accessory pigments,** such as carotenoids, that collect light energy and transfer it to the chlorophyll molecules by fluorescence. The nature of these accessory pigments varies from one bacterial group to another (Table 21.2).

Photosynthetic bacteria produce a variety of chlorophyll pigments (Table 21.2; Figure 21.1). Bacteriochlorophyll *a* (Bchl *a*) is a common reaction center chlorophyll used by both the purple and green bacteria. In contrast,

Differential *Table* **21.2 • Characteristics of photosynthetic Eubacteria**

Group	CO_2 Fixation Pathway	Reaction Center Chlorophyll/ Oxygen Relation	Antenna (and Accessory Pigments)
Proteobacteria	Calvin-Benson	Bchl *a*; anoxygenic	Bchl *b*; (carotenoids)
Green sulfur	Reductive TCA	Bchl *a* anoxygenic	Bchl *c*; Bchl *d* Bchl *e*; (carotenoids)
Green filamentous	Hydroxypropionate	Bchl *a*; facultative	Bchl *c*; (carotenoids)
Heliobacteria	Heterotrophic	Bchl *g*	Carotenoids
Cyanobacteria (and prochlorophytes)	Calvin-Benson	Chl *a*; oxygenic	Phycobilins or Chl *b*

Bchl *b* is produced only by purple bacteria, whereas Bchl *c*, Bchl *d*, and Bchl *e* are produced by various green bacteria, and Bchl *g* is produced by the heliobacteria. Unlike other prokaryotes, the cyanobacteria produce chlorophyll *a* as their reaction center chlorophyll. It is interesting to note that the bacteriochlorophylls absorb very long wavelength light, some of which are in the infrared region (wavelength > 800 nm) not detected by human eyesight (Figure 21.1). In contrast, chlorophyll *a* and *b* absorb much shorter wavelength light in the visible range. Because each photosynthetic group has its own characteristic absorption spectrum for light, these groups can coexist in the same habitat without directly competing for available light.

ATP is synthesized by a proton gradient established across the photosynthetic membranes associated with the reaction center where the electron transfer reactions occur (Chapter 9). Thus, like typical heterotrophic bacteria, photosynthetic bacteria generate ATP by use of proton pumps and their associated ATPases. The unique feature of photosynthesis is the generation of electron gradients using pigments that harvest sunlight.

A common biochemical pathway used by phototrophic bacteria for carbon dioxide fixation is the Calvin-Benson cycle (Figure 10.8). This pathway, however, is not used by all photosynthetic Eubacteria. The green sulfur and green filamentous bacteria use different pathways for carbon dioxide fixation (see below).

Phototrophic Groups

The phototrophic Eubacteria fall into five major groups: the photosynthetic Proteobacteria, the Green Sulfur bacteria, the Green Filamentous bacteria, the Heliobacteria, and the Cyanobacteria which are differentiated from one another on the basis of phylogeny, carbon dioxide fixation pathways and pigments (Table 21.2). Each of the photosynthetic eubacterial groups will be discussed in greater detail below.

Proteobacteria

The photosynthetic Proteobacteria, commonly called the purple bacteria because of their reddish or purplish coloration, comprise two separate subgroups, the **purple sulfur bacteria** and the **purple nonsulfur bacteria.** Like all Proteobacteria that fix carbon dioxide, both of these subgroups use the Calvin-Benson cycle, although they utilize different compounds for reducing power for the dark reactions. The purple sulfur bacteria use reduced sulfur compounds such as sulfide or elemental sulfur as their source of electrons. In contrast, the purple nonsulfur bac-

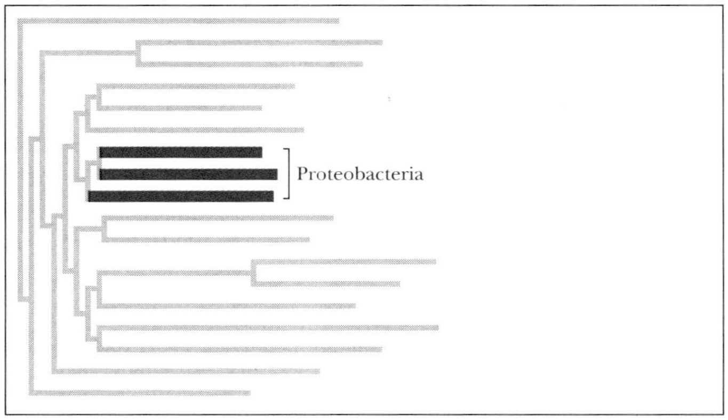

(text continued)

Figure **21.1** **(a)** General chemical structure of the chlorophyll molecule. The R-groups vary from one type of chlorophyll to another as shown in **(b).**

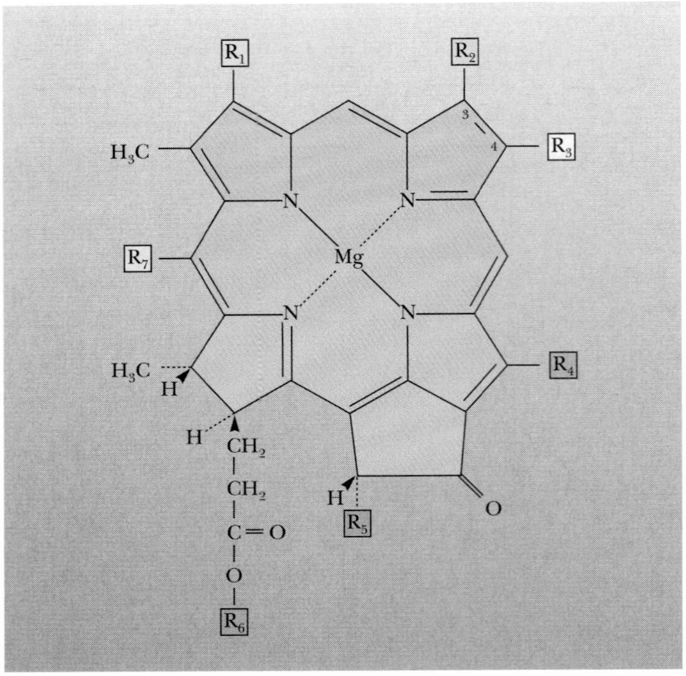

(b) Structure of R-groups of chlorophyll molecules from prokaryotes as well as the absorption maximum of each chlorophyll type. P refers to the phytyl ester ($C_{20}H_{39}O-$); G refers to the geranylgeraniol ester ($C_{10}H_{17}O$); and F refers to the farnesyl ester ($C_{15}H_{25}O$). Where two or more R-groups are indicated, mixtures occur in the native chlorophyll structure. (Adapted from A. Gloe, N. Pfennig, H. Brockmann, and W. Trowitsch, 1975 *Archives of Microbiology, 102:* 103–109 and A. Gloe and N. Risch, 1978, *Archives of Microbiology, 118:* 153–156)

Chlorophyll	R1	R2	R3	R4	R5	R6	R7	Absorption maximum of cells(nm)
Bacterio-chlorophyll *a*	$-C(=O)-CH_3$	$-CH_3$	$-CH_2-CH_3$	$-CH_3$	$-C(=O)-O-CH_3$	P/G	$-H$	805 830–890
Bacterio-chlorophyll *b*	$-C(=O)-CH_3$	$-CH_3$	$=C(H)-CH_3$	$-CH_3$	$-C(=O)-O-CH_3$	P	$-H$	835–850 1020–1040
Bacterio-chlorophyll *c*	$-C(H)(OH)-CH_3$	$-CH_3$	$-C_2H_5$ $-C_3H_7$ $-C_4H_9$	$-C_2H_5$ $-CH_3$	$-H$	F	$-CH_3$	745–755
Bacterio-chlorophyll *d*	$-C(H)(OH)-CH_3$	$-CH_3$	$-C_2H_5$ $-C_3H_7$ $-C_4H_9$	$-C_2H_5$ $-CH_3$	$-H$	F	$-H$	705–740
Bacterio-chlorophyll *e*	$-C(H)(OH)-CH_3$	$-C(=O)-H$	$-C_2H_5$ $-C_3H_7$ $-C_4H_9$	$-C_2H_5$	$-H$	F	$-CH_3$	719–726
Bacterio-chlorophyll *g*	$-C(H)=CH_2$	$-CH_3$	$-C_2H_5$	$-CH_3$	$-C(=O)-O-CH_3$	F	$-H$	670,788
Chlorophyll *a*	$-C(H)=CH_2$	$-CH_3$	$-C_2H_5$	$-CH_3$	$-C(=O)-O-CH_3$	P	$-H$	440,680
Chlorophyll *b*	$-C(H)-CH_2$	$-C(=O)-H$	$-C_2H_5$	$-CH_3$	$-C(=O)-O-CH_3$	P	$-H$	645

teria typically use organic compounds, especially organic acids and alcohols, as their source of reducing power. Photosynthesis in purple sulfur bacteria is represented by the following two reactions (reactions are not balanced):

$$H_2S + CO_2 \longrightarrow (CH_2O)_n + S^0 \qquad [21.1]$$

$$S^0 + CO_2 + H_2O \longrightarrow (CH_2O)_n + H_2SO_4 \qquad [21.2]$$
where $(CH_2O)_n$ represents organic carbohydrate carbon

Photosynthesis in purple nonsulfur bacteria is represented by:

$$H_2A + CO_2 \longrightarrow (CH_2O)_n + A \qquad [21.3]$$
where H_2A = organic compound such as ethanol or an organic acid

As can be seen by the above three reactions for carbon dioxide fixation, oxygen is *not* formed during photosynthesis by these bacteria. This type of photosynthesis is therefore referred to as **anoxygenic photosynthesis.** As discussed in Chapter 9, the electron flow in anoxygenic photosynthesis is noncyclic involving only one photosystem, Photosystem I, with Bchl *a*. Also, a characteristic of this type of photosynthesis is that it occurs at a relatively high redox potential, about −0.15 volts. Thus, NADPH cannot be directly formed from the electron flow; instead energy must be expended by reverse electron flow to obtain the NADPH needed for CO_2 reduction (Figure 21.2; also see Chapter 9 for more detail).

The two purple Proteobacterial photosynthetic groups are discussed individually below.

PURPLE SULFUR BACTERIA The purple sulfur bacteria are large unicellular bacteria (Figure 21.3) that may reach over 6.0 μm in diameter. Many are motile with polar flagella. Some have gas vacuoles, and all can utilize hydrogen sulfide in photosynthesis and form elemental sulfur granules inside their cells (Figure 21.3).

The photosynthetic Proteobacteria have extensive intracellular membrane systems that contain the photopigments and photoreaction centers. These membranes are extensions of the cytoplasmic membrane and may be vesicular, lamellar, or even tubular (Figure 21.4). The photosynthetic pigments are all bacteriochlorophylls with Bchl *a* found in all but one species, *Thiocapsa pfennigii,* which contains Bchl *b*. They also contain carotenoid pigments (Figure 21.5). These carotenoid pigments actually mask the green to blue bacterial chlorophyll pigments thereby giving the organisms their characteristic red, orange, and

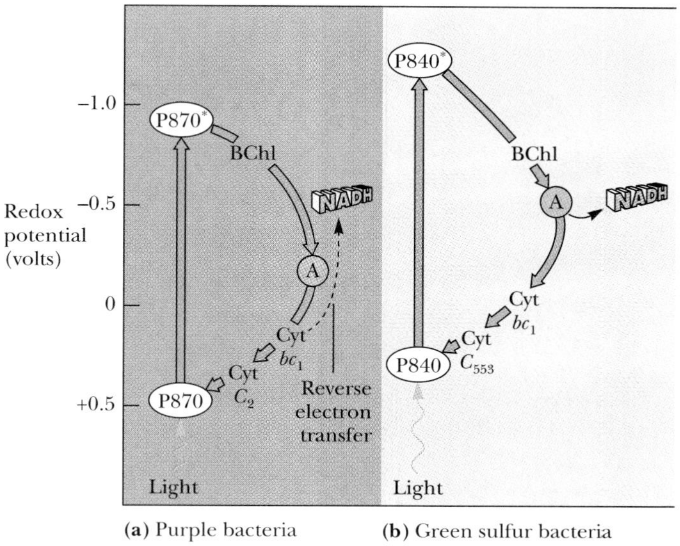

***Figure* 21.2** Diagram illustrating the redox potential for electron transfer reactions in photosynthesis by **(a)** purple and **(b)** green sulfur bacteria. Note that the redox potential of the green sulfur bacteria is sufficiently low to produce NADH directly from the electron acceptor (A) whereas reversed electron transport is required by purple bacteria. Bchl = Bacteriochlorophyll; P840 and P870 are the reaction centers of green and purple bacteria, respectively, and contain Bchl *a*. Asterisk indicates activated photosystem.

purple colors (Figure 21.6). A variety of carotenoid pigments may be present, depending on the species. The pigmentation of an organism is related to the amount of the various carotenoid pigments produced. For example, lycopene, which is brown in color, is the biochemical precursor of spirilloxanthin, which is purple. The amounts of these two pigments and their intermediates, and hence,

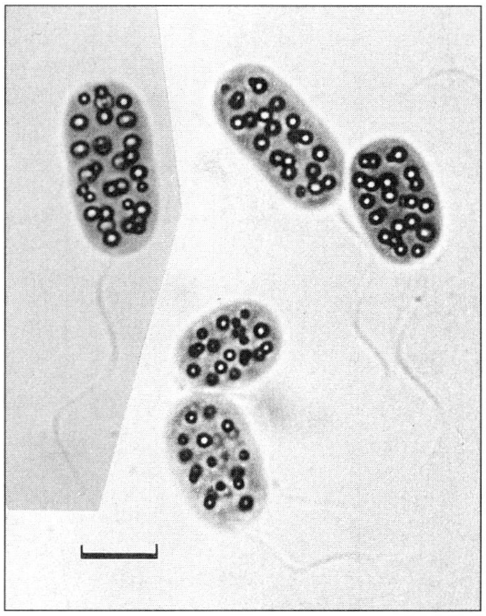

***Figure* 21.3** A phase photomicrograph of *Chromatium okenii*. Note the polar flagellum (which is actually a tuft) as well as the internal sulfur granules. Bar is 5 μm.(Courtesy of N. Pfennig)

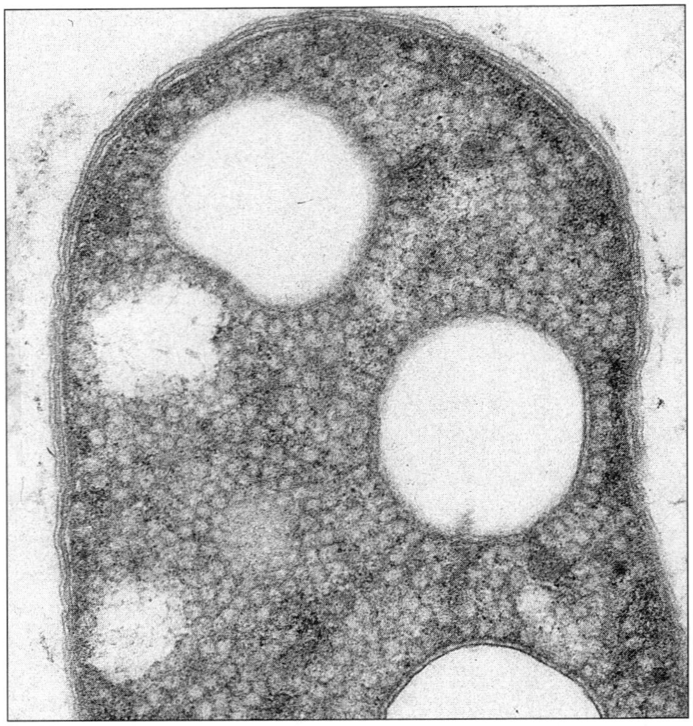

(a)

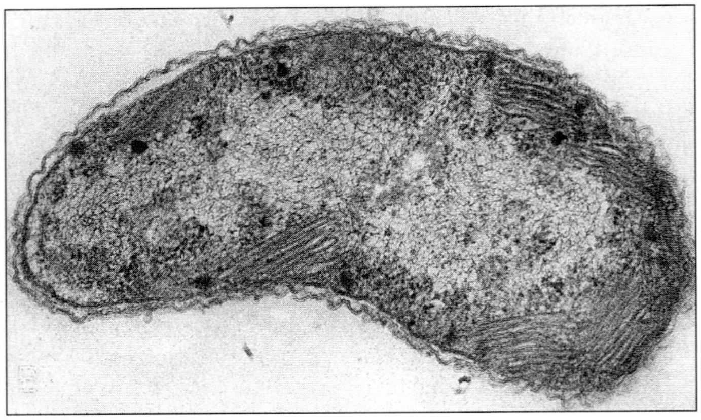

(b)

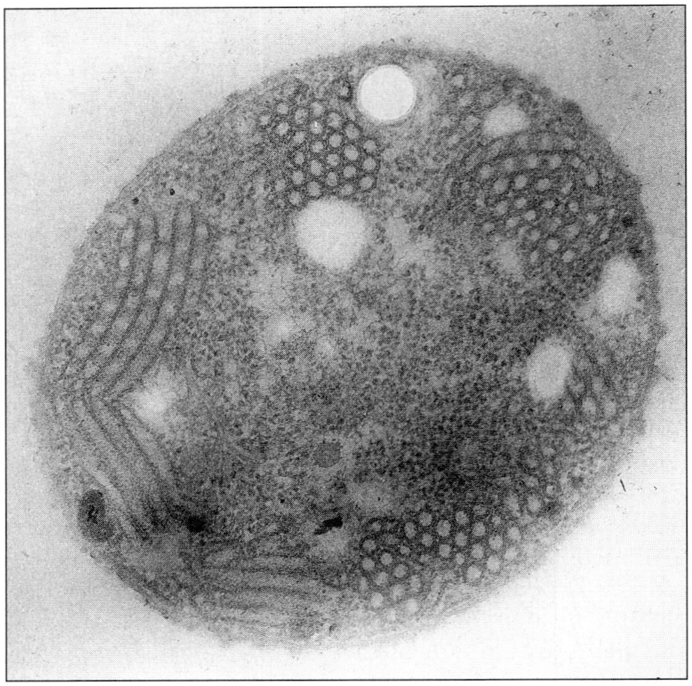

(c)

***Figure* 21.4** Thin sections showing examples of the various types of intracellular photosynthetic membranes of purple bacteria. **(a)** Vesicular type from *Chromatium vinosum;* empty areas in cell are sulfur granules; **(b)** lamellar type from *Rhodospirillum molischianum*, and **(c)** tubular type from *Thiocapsa pfennigii*. (**a** and **c** courtesy of S. Watson, J. Waterbury, and C. Remsen; **b** courtesy of G. Drews)

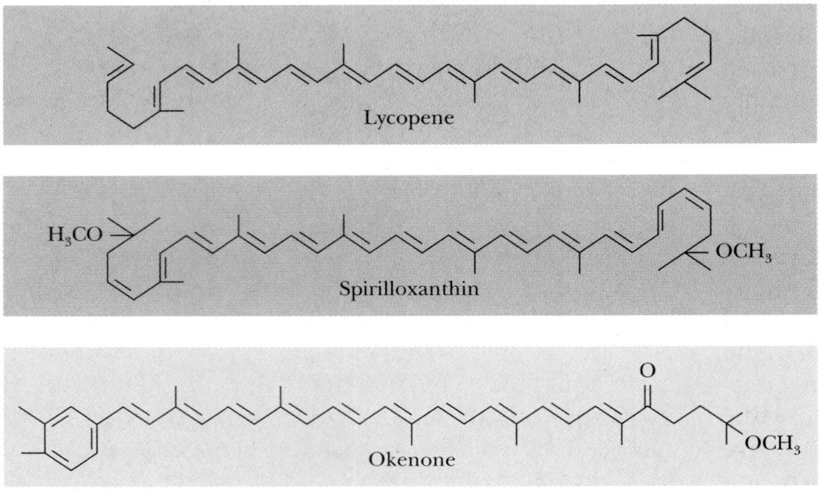

***Figure* 21.5** Structures and color of common carotenoid pigments of purple bacteria.

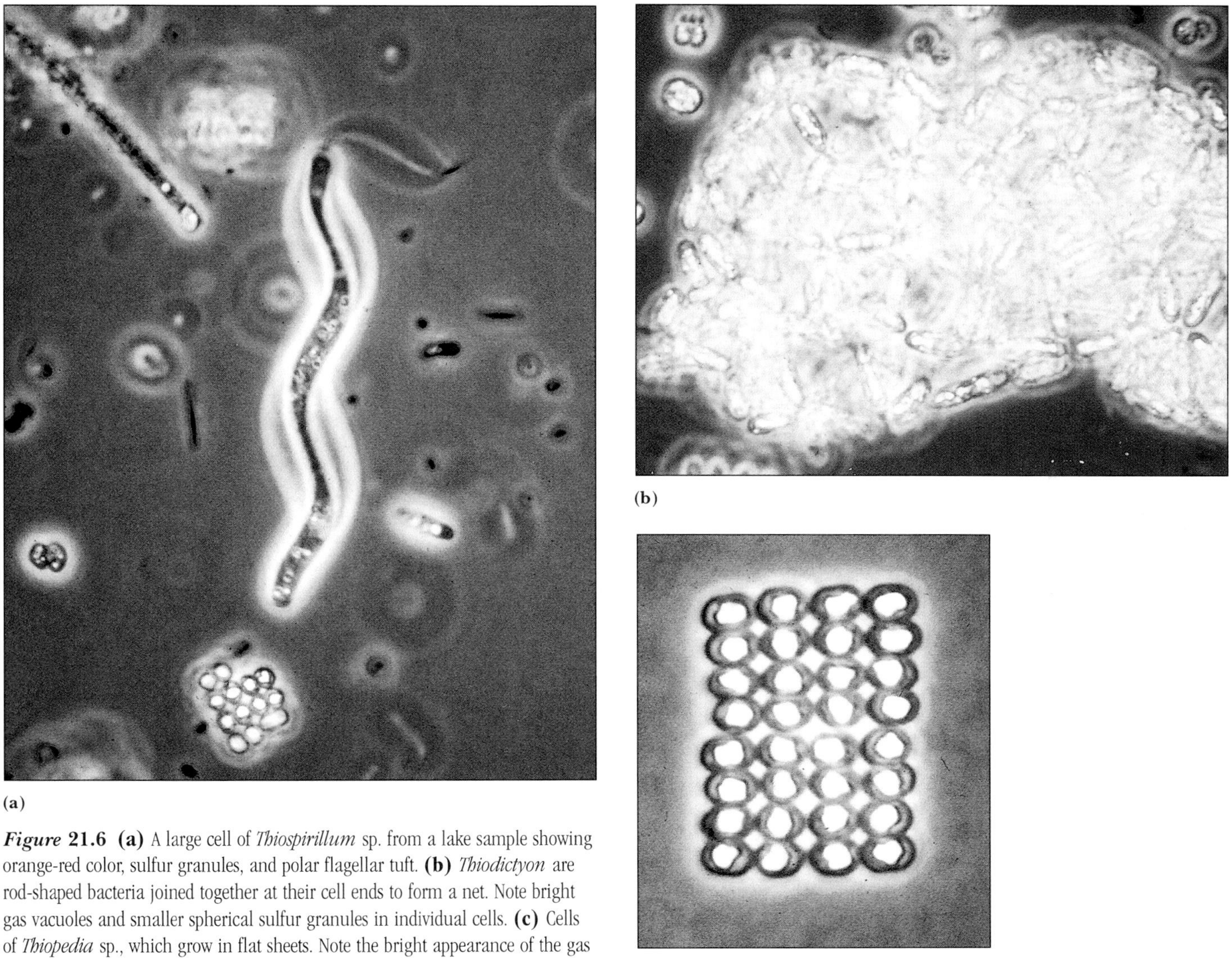

(a) (b) (c)

***Figure* 21.6 (a)** A large cell of *Thiospirillum* sp. from a lake sample showing orange-red color, sulfur granules, and polar flagellar tuft. **(b)** *Thiodictyon* are rod-shaped bacteria joined together at their cell ends to form a net. Note bright gas vacuoles and smaller spherical sulfur granules in individual cells. **(c)** Cells of *Thiopedia* sp., which grow in flat sheets. Note the bright appearance of the gas vacuoles in each cell. (Courtesy of J. T. Staley)

the color of an organism will vary depending on the environment and growth state of the organism.

Table 21.3 lists important genera and their characteristics. *Chromatium* is one of the most common genera. This is a genus of large, polarly flagellated rods (Figure 21.3). The genus *Thiospirillum,* as the name implies, contains helical bacteria with polar flagellar tufts (Figure 21.6**a**). *Thiodictyon* is a genus of net-forming rods (Figure 21.6**b**) and *Thiopedia* species form flat plates of coccoid cells with gas vacuoles (Figure 21.6**c**). All purple sulfur bacteria are members of the gamma group of the Proteobacteria.

Under anaerobic conditions in light, purple sulfur bacteria grow as photolithoautotrophs using sulfide or elemental sulfur as an electron donor and carbon dioxide as carbon source. They are best known for this physiological activity. In addition, many species can also use organic compounds, such as ethanol or pyruvate as shown in Reaction 21.3 above, which is typical of purple nonsulfur bacteria. Under such growth conditions, these bacteria still require sulfide, however, because they cannot carry out assimilatory sulfate reduction. Many species can also use hydrogen gas as a reducing agent for photosynthesis under anaerobic conditions. Virtually all species are nitrogen fixers. Although most are obligate anaerobes, some can grow under microaerobic to aerobic conditions in the dark. When doing this they obtain energy from the oxidation of inorganic compounds such as hydrogen gas or sulfide or even organic carbon sources in the same manner as some chemolithotrophs and heterotrophs, respectively.

The purple sulfur bacteria are most commonly found in anaerobic environments where sulfide is abundant and light is available (see Chapter 24). Such conditions occur in the anaerobic hypolimnion of many eutrophic lakes, at the surface of marine and freshwater muds, or in sulfur springs (Figure 21.7). A characteristic vertical layering of phototrophic bacteria occurs in such environments. The

Differential ***Table*** **21.3 • Characteristics of genera of purple sulfur bacteria**

Genus	Flagella	Gas Vacuoles	Mol % G+C	Morphology
Chromatium	+	−	48–70	Rod
Thiocystis	+	−	61–68	Coccus
Thiospirillum	+	−	45–46	Spirillum
Thiocapsa	−	−	63–70	Coccus
Lamprobacter	+	+	64	Ovoid cells
Lamprocystis	+	+	64	Coccus
Thiodictyon	−	+	65–67	Rod; forms network of cells
Amoebobacter	−	+	64–66	Coccus singly or clusters
Thiopedia	−	+	62–64	Cocci in one plane

cyanobacteria are found nearest the surface, the purple sulfur bacteria are beneath them, and the green sulfur bacteria occur at the lowest depth. This pattern of vertical stratification is identical to that found in microbial mat communities as discussed earlier in Chapter 18.

The genus *Ectothiorhodospira* is closely related to the purple sulfur bacteria. It resembles the purple sulfur bacteria in its metabolism, pigmentation, and internal membrane systems. Like the purple sulfur bacteria, *Ectothiorhodospira* spp. are members of the gamma Proteobacteria. Their cells are small, motile, and vibrioid to rod-shaped. Depending on the species, they live in marine habitats, alkaline soda lakes, or hypersaline lakes. Some species are gas vacuolate. Unlike the purple sulfur bacteria, however, the sulfur granules they produce are deposited *outside* of the cell (hence the name *ectothio-*, which means *extracellular sulfur*).

PURPLE NONSULFUR BACTERIA The purple nonsulfur bacteria share characteristics of both the heterotrophic proteobacteria and the purple sulfur photosynthetic bacteria. Like many bacteria they can grow aerobically (in the absence of light) as heterotrophs using certain organic substrates. However, if they are grown anaerobically in the light, they carry out photosynthesis, much like the purple sulfur bacteria. They differ from purple sulfur bacteria in

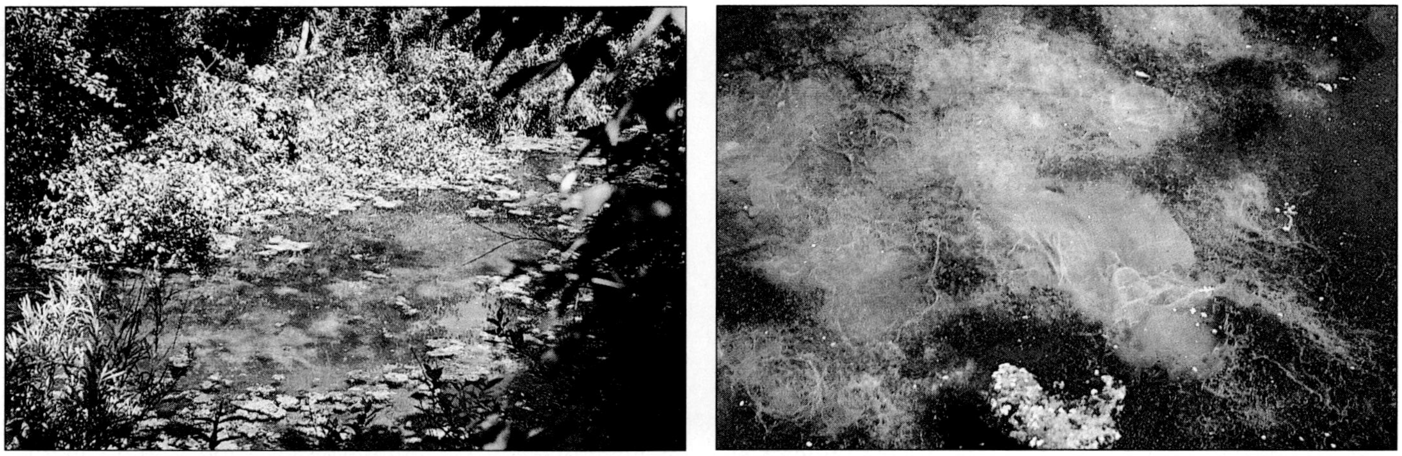

(a) (b)

***Figure* 21.7 (a)** A sulfur spring. Yankee Springs in central Michigan. The darker green growth that occurs on the top in some areas is due to cyanobacteria. **(b)** Close-up view of the purple "veils" from Yankee Springs containing several different species of purple sulfur bacteria. (Courtesy of Peter Hirsch)

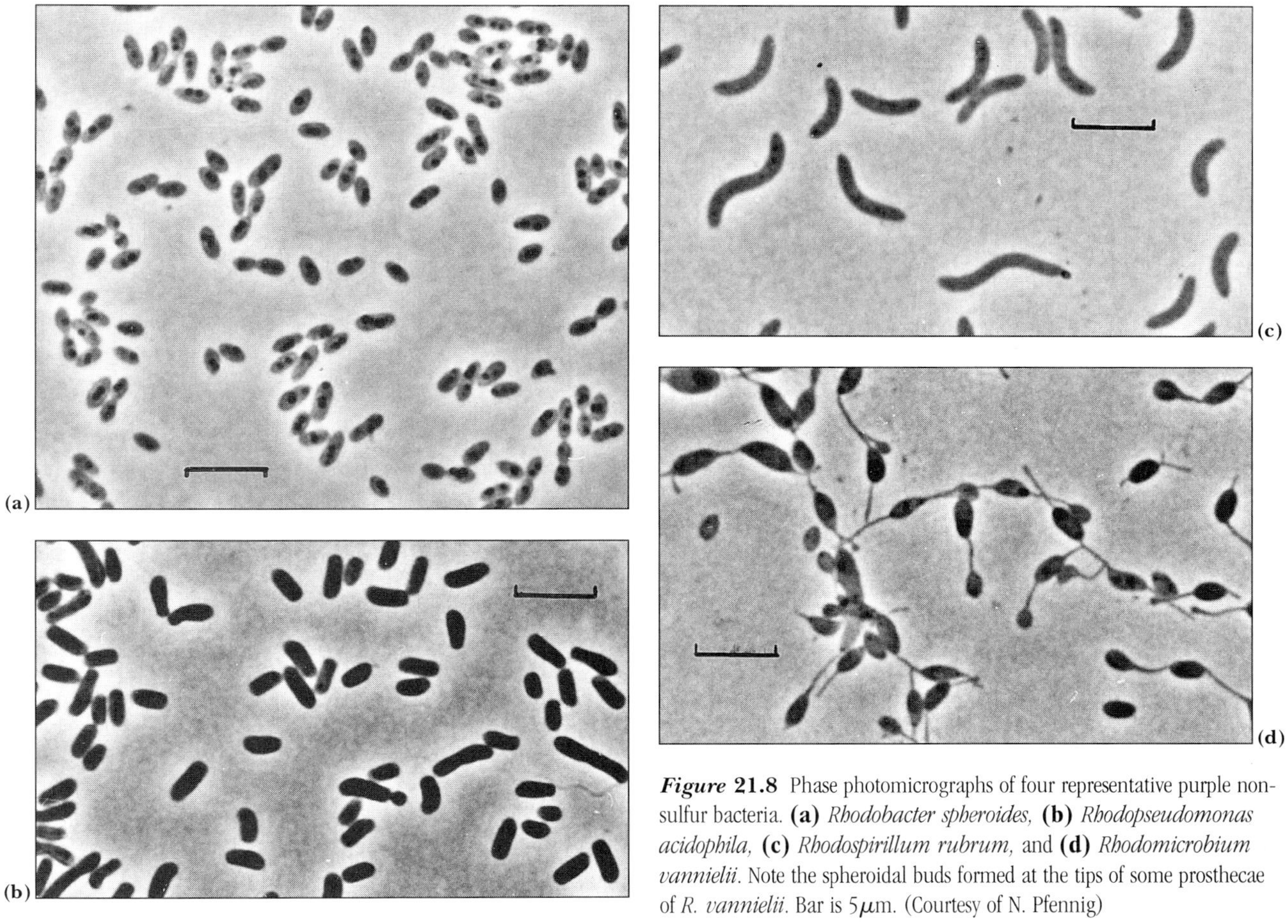

***Figure* 21.8** Phase photomicrographs of four representative purple nonsulfur bacteria. **(a)** *Rhodobacter spheroides,* **(b)** *Rhodopseudomonas acidophila,* **(c)** *Rhodospirillum rubrum,* and **(d)** *Rhodomicrobium vannielii.* Note the spheroidal buds formed at the tips of some prosthecae of *R. vannielii.* Bar is 5μm. (Courtesy of N. Pfennig)

that organic carbon sources, such as organic acids (for example, acetate, pyruvate, or lactate) or ethanol, rather than sulfide, are preferred as electron donors for carbon dioxide fixation. Some can also grow as photolithoautotrophs, using hydrogen gas or sulfide as reducing agents in the same manner as the purple sulfur bacteria. However, they do not tolerate high concentrations of sulfide and for this reason are called nonsulfur purple bacteria. When sulfide is available in low concentrations, those that use it as a reductant for photosynthesis do not form sulfur granules inside the cells. Either elemental sulfur is formed outside of the cell or the sulfide is oxidized all the way to sulfate.

The cells of most purple nonsulfur bacteria are smaller than those of the purple sulfur bacteria. Representatives of several common genera, *Rhodobacter, Rhodopseudomonas,* and *Rhodospirillum* are shown in Figure 21.8. They also differ from the purple sulfur bacteria in that individual cells do not appear pigmented when observed with the light microscope. This is because their smaller cells are not thick enough to contain sufficient pigments to reveal their true color.

Like the purple sulfur bacteria they, too, produce intracytoplasmic photosynthetic membranes when growing photosynthetically in the light under anaerobic conditions. These membranes are extensions of the cell membrane and may be either vesicles or lamellae. They produce more of these membranes and photosynthetic pigments under lower light intensity as a means of compensating for the decrease in available light. These organisms also contain the same carotenoid pigments as produced by the purple sulfur bacteria, with characteristic types for each species (Figure 21.5). Many of the species in this group are motile by flagella. None is known to produce gas vacuoles. There are eight genera (Table 21.4).

Most of the nonsulfur purple bacteria are members of the alpha purple group of Proteobacteria, and are therefore phylogenetically related to the prosthecate, budding bacteria. Indeed, *Rhodomicrobium* spp. (Figure 21.8**d**) and some *Rhodopseudomonas* spp. produce prosthecae and divide by budding—confirming their close phylogenetic relatedness to the heterotrophic prosthecate bacteria. Three genera, *Rhodocyclus, Rhodovivax,* and *Rhodoferax,* are members of the beta-Proteobacteria. Unlike the purple sulfur bacteria, none are members of the gamma Proteobacteria.

While discussing the purple bacteria, it is important to note that several new genera, producing bacteriochlorophyll *a*, that superficially resemble purple nonsulfur

Differential ***Table* 21.4 • Genera of purple nonsulfur bacteria**

Genus	Mol % G+C	Cell Morphology and Features
Alpha Proteobacteria		
Rhodospirillum	60–66	Helical cells
Rhodobacter	64–73	Ovoid to rod-shaped; neutrophilic
Rhodopila	66	Ovoid to rod-shaped; acidophilic
Rhodopseudomonas	61–72	Ovoid to rod-shaped; budding
Rhodomicrobium	61–63	Prosthecate, bacterium
Beta Proteobacteria		
Rhodocyclus	64–73	Curved rod to ring-shaped
Rhodoferax	59–61	Curved rod
Rhodovivax	70–72	Curved rod

bacteria have been discovered recently. However, unlike the purple nonsulfur bacteria, these bacteria produce Bchl *a* while growing under *aerobic* conditions. The role of these bacteria in nature is not yet understood **(Box 21.1).**

Green Sulfur Bacteria

The green sulfur bacteria comprise a separate group of photosynthetic bacteria with an independent phylogeny (Chapter 17). They are so named because most species are green due to the presence of characteristic bacteriochlorophylls (Figure 21.1), predominantly Bchl *c*, Bchl *d*, or Bchl *e*. Others appear brown in color due principally to carotenoid pigments (Figure 21.9). In addition, all strains contain small amounts of Bchl *a* as their reaction center chlorophyll. They are termed "sulfur" bacteria because they carry out anoxygenic photosynthesis using H_2S as the reductant for carbon dioxide fixation in the same manner as the purple sulfur bacteria. However, they grow at a reduced redox potential and, because of their electron transport chains, are able to reduce NADPH directly without resorting to reverse electron transport, thereby saving energy (see Figure 21.2).

The green sulfur bacteria and the green filamentous bacteria resemble one another in that they both have the **chlorosome** as their intracellular membrane system for photosynthesis. The chlorosome is a cigar-shaped struc-

BOX 21.1 RESEARCH HIGHLIGHTS

Aerobic Bacteria Producing Bacteriochlorophyll *a*

Purple sulfur, purple nonsulfur, and green sulfur photosynthetic bacteria produce bacteriochlorophyll *a* only under anaerobic conditions. In contrast, several genera of aquatic bacteria have been recently discovered that produce Bchl *a* only under *aerobic* conditions. Many of these bacteria can make ATP by photophosphorylation, but all studied so far require organic compounds as carbon sources. Some of these bacteria have recently been isolated from marine habitats including the genera, *Erythrobacter* and *Roseobacter*, whereas others such as *Porphyrobacter*, are from freshwater habitats. All are members of the alpha Proteobacteria as determined by 16S rRNA sequence analyses.

The role of these bacteria in natural habitats is uncertain. However, their common occurrence and widespread distribution suggest that they may comprise an important part of the normal microflora of aquatic habitats. Clearly these bacteria will be an area of considerable research interest in the future.

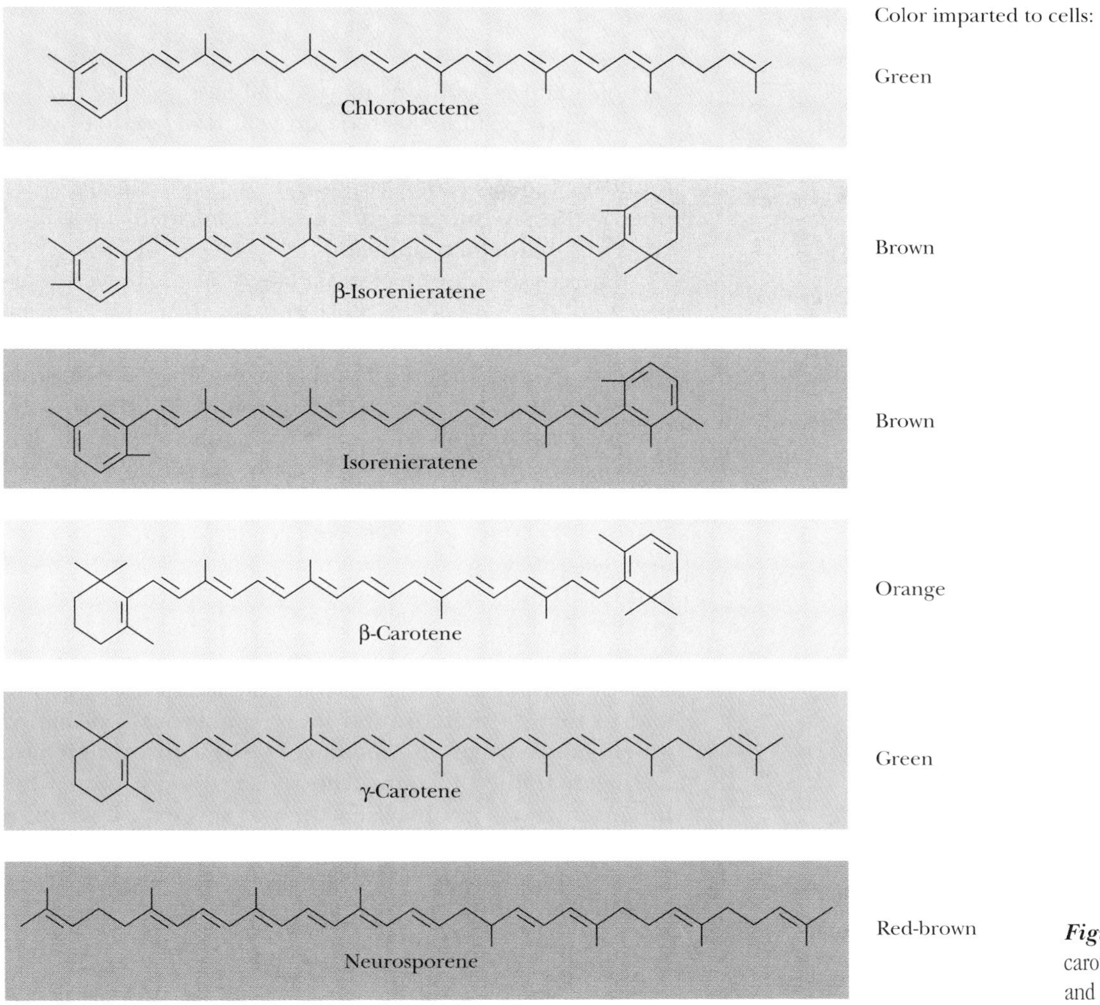

***Figure* 21.9** Chemical structures of carotenoid pigments from green bacteria and heliobacteria.

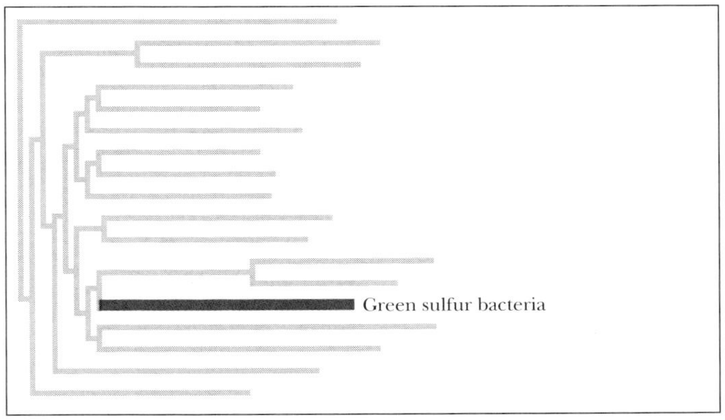

ture situated next to the cell membrane (Figure 21.10). It is here and in the cytoplasmic membrane that the photopigments and photoreaction center of these bacteria is located. These two groups of bacteria are considered individually here and in the following section because they differ in their phylogeny, are found in different habitats, and have their own carbon dioxide fixation reactions.

The green sulfur bacteria are small organisms, much smaller than the purple sulfur bacteria. Like typical Eubacteria, their cell diameter is about 0.5 to 1.0 μm. None are motile by flagella, although many have gas vacuoles and use them to regulate their vertical position in stratified habitats (see Chapters 4 and 24). One genus, *Chloroherpeton,* moves by gliding motility.

Like the purple sulfur bacteria, the green sulfur bacteria use sulfide as an electron donor in photosynthesis. They, too, form elemental sulfur as an intermediate in this process, but the sulfur granules are deposited *outside* the cells in the same manner as in *Ectothiorhodospira* (Figure 21.11). They can continue to oxidize the sulfur granules, however, and ultimately produce sulfate. Therefore, the overall photosynthetic reactions are identical to that of the purple sulfur bacteria shown above (Reactions 21.1 and 21.2). However, these bacteria do not use the Calvin-Benson cycle for carbon dioxide fixation. Instead, they use the reductive tricarboxylic acid cycle (see Figure 10.10**a**). Also, unlike the purple bacteria, the green bacteria are strict anaerobes. Although most use carbon dioxide as their sole carbon source, some can also use simple organic acids in the presence of sulfide and carbon dioxide. Again, as with the purple sulfur bacteria, these bacteria must have sulfide when using organic compounds, because they cannot carry out assimilatory sulfate reduction.

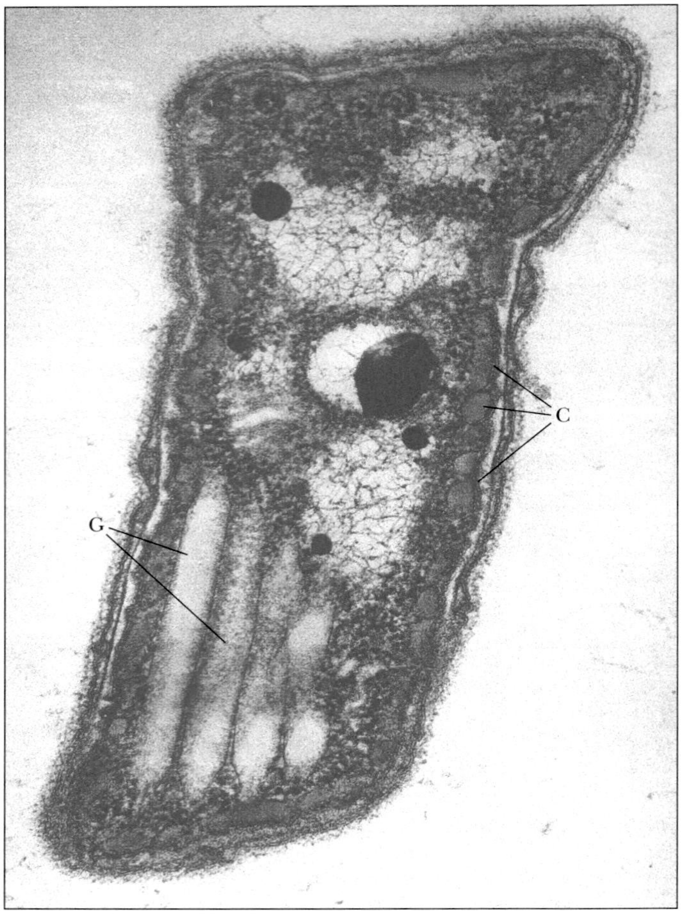

***Figure* 21.10** Thin section through a green sulfur bacterium showing the oval chlorosomes (C) located just inside the cell membrane. Note also the large transparent gas vesicles (G). (Courtesy of G. Cohen-Bazire)

Several genera of green sulfur bacteria are known, each with its own characteristic morphology (Table 21.5). *Chlorobium* is a common genus of unicellular rods whose species are found in many anaerobic, sulfide-rich habitats (Figure 21.11). *Pelodictyon* is also very common. It is a genus of rod-shaped, gas vacuolate bacteria whose cells divide to produce a netlike structure similar to the purple sulfur genus, *Thiodictyon.* Some species appear green in color (Figure 21.12**a**) whereas others appear brown (Figure 21.12**b**) due to differences in pigment composition. *Ancalochloris* spp. which have not yet been isolated in pure culture, resemble *Pelodictyon* in that they form netlike microcolonies, but also produce prosthecate cells (Figures 21.12**c** and **d**). These bacteria are found in the plankton of the anaerobic hypolimnion of lakes and in freshwater and marine muds and marine or saline anoxic habitats. *Prosthecochloris* also is prosthecate and forms nets and is therefore similar to *Ancalochloris,* but it is found in saline habitats and does not produce gas vacuoles. The genus *Chloroherpeton* contains rod-shaped gliding bacteria with gas vacuoles and is found in marine habitats (Figure 21.13).

The green sulfur bacteria grow at lower redox potentials than the purple sulfur bacteria; therefore they can be found in very low light conditions beneath all the other photosynthetic microorganisms in thermally stratified lakes and mat communities.

The consortium species, such as *"Chlorochromatium aggregatum,"* are one of the most unusual forms of bacterial life. They are found in the same anaerobic sulfide-rich habitats as the green sulfur bacteria. They are unique among the phototrophic bacteria in that they consist of a symbiotic association between two separate organisms. For this reason they cannot be officially named as a species, but instead quotation marks are used to designate an association between two different species. They appear as multicellular aggregates in the anoxic, sulfide-rich lower depths of lakes (Figure 21.14**a**). At the center of the aggregate, or microcolony, is a large, nonpigmented motile rod, usually seen in the process of division, which propels the consortium (Figure 21.14**b**). Attached to the large rod and forming a shell around it are numerous rod-shaped green sulfur bacteria. Members of the consortium have not been studied in pure culture, so it is not clear what the interactions are between the two species. However, it appears that the large rod is an anaerobic heterotroph that obtains organic nutrients excreted from the green sulfur bacteria, and the green sulfur bacteria gain motility from the large rod (Figure 24.14**c**).

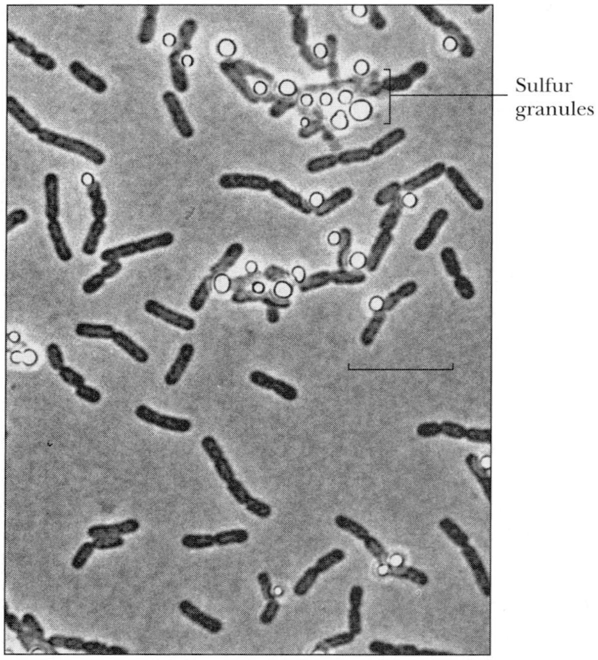

***Figure* 21.11** A pure culture of a *Chlorobium limicola* showing external deposition of sulfur granules, which appear as bright spheres. Phase contrast microscopy. Bar is 5 μm. (Courtesy of N. Pfennig)

Differential ***Table*** **21.5 • Important genera of green bacteria**

Group Genus	Mol % G+C	Gas Vacuoles	Morphology
Green Sulfur Genera			
Chlorobium	49–58	−	Single cells, rods
Prosthecochloris	50–56	−	Prosthecate; marine or saline habitats
Pelodictyon	48–58	+	Rods in netlike clusters
Ancalochloris	Unknown	+	Prosthecate; freshwater
Chloroherpeton	45–48	+	Long rods; gliding
Consortium species	Unknown	+/−	Microcolonial aggregates of two separate species
Green Filamentous Genera			
Chloroflexus	55	−	Gliding; moderate thermophile
Heliothrix	Unknown	−	Gliding; moderate thermophile
Chloronema	Unknown	+	Gliding; mesophile

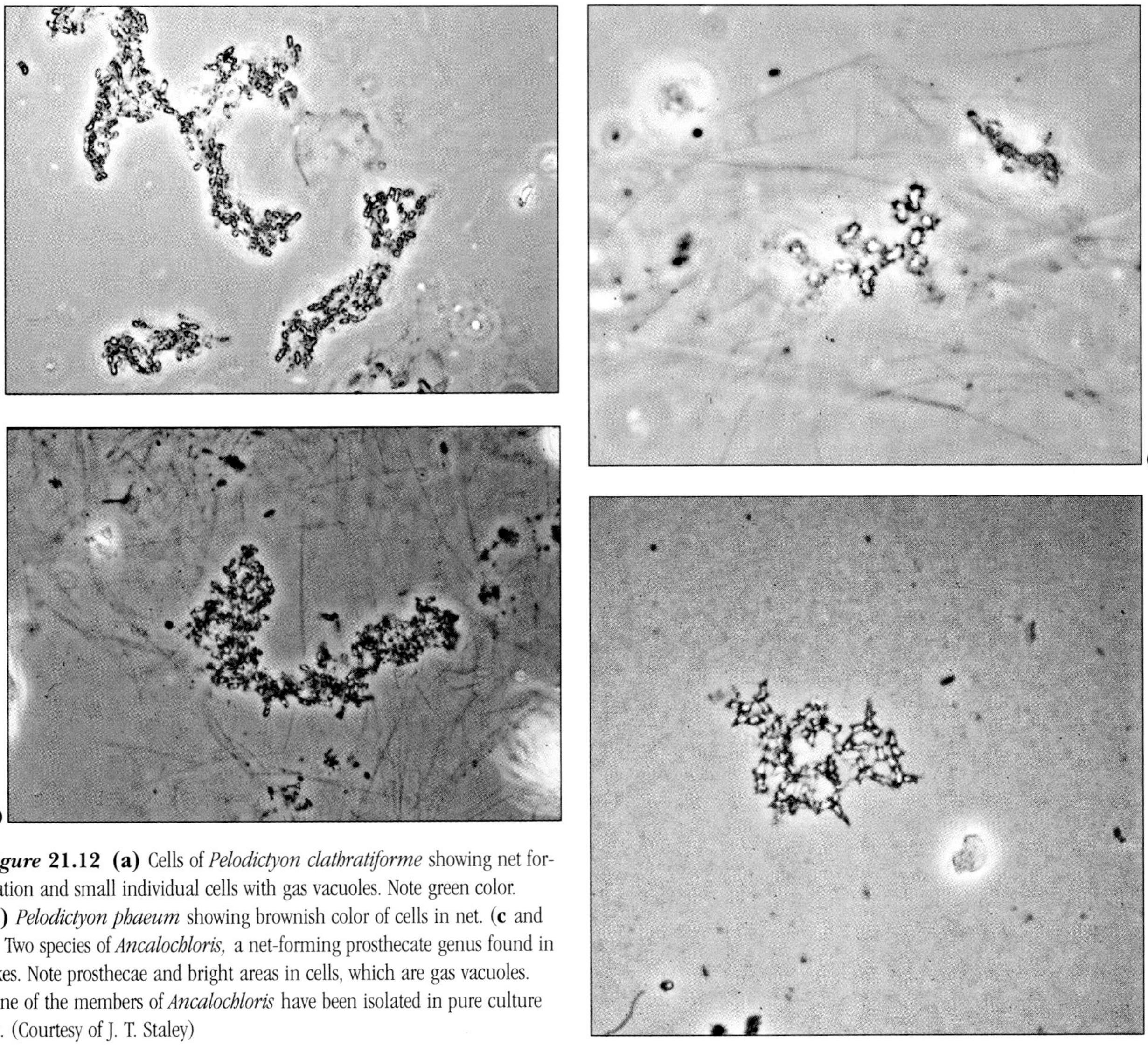

(a) (b) (c) (d)

Figure **21.12 (a)** Cells of *Pelodictyon clathratiforme* showing net formation and small individual cells with gas vacuoles. Note green color. **(b)** *Pelodictyon phaeum* showing brownish color of cells in net. (**c** and **d**) Two species of *Ancalochloris,* a net-forming prosthecate genus found in lakes. Note prosthecae and bright areas in cells, which are gas vacuoles. None of the members of *Ancalochloris* have been isolated in pure culture yet. (Courtesy of J. T. Staley)

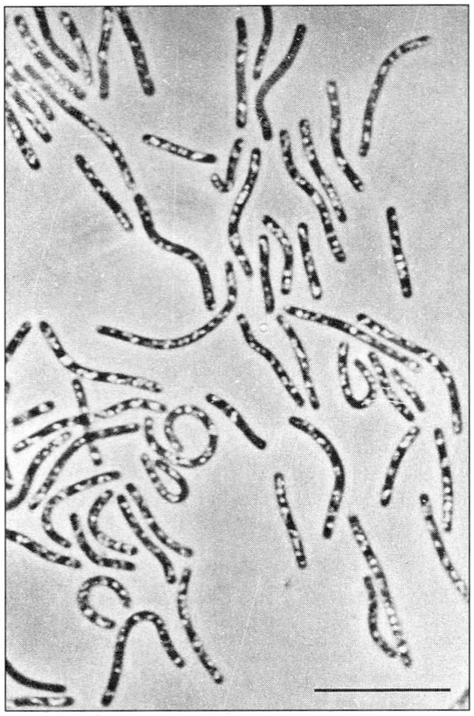

***Figure* 21.13** Phase photomicrograph of *Chloroherpeton* cells. Note gas vacuoles. Bar is 10 μm. (Courtesy of Jane Gibson and John Waterbury)

Green Filamentous Bacteria

The green filamentous bacteria resemble the green sulfur bacteria only in that they contain chlorosomes and are phototrophic. They resemble the green sulfur genus, *Chloroherpeton,* most closely in that they move by gliding motility. However they are filamentous and multicellular organisms that evolved as a separate phylogenetic group. Three genera are currently known including *Chloroflexus, Heliothrix,* and *Chloronema.* The latter two genera have not been isolated in pure culture as yet. In addition, this phylogenetic group contains a genus of heterotrophic gliding bacteria, *Herpetosiphon.*

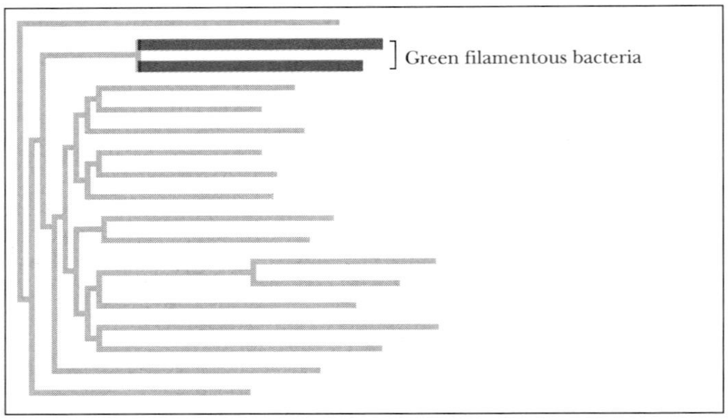

The most thoroughly studied genus is *Chloroflexus.* This long, thin (0.5–1.0 μm diameter) filamentous gliding bacterium (Figure 21.15) was first isolated from hot springs where it forms distinct layers in mats; however, some strains have also been found in mesophilic habitats. The temperature range for growth of the thermophilic strains is 45°C to 70°C.

The nutrition of *Chloroflexus aurantiacus* is complex. It grows best as a photoheterotroph, anaerobically in the light. It utilizes a variety of organic carbon sources including sugars, amino acids, and organic acids during photoheterotrophic growth. *C. aurantiacus* can also grow as a chemoheterotroph aerobically in the dark. Finally, it grows

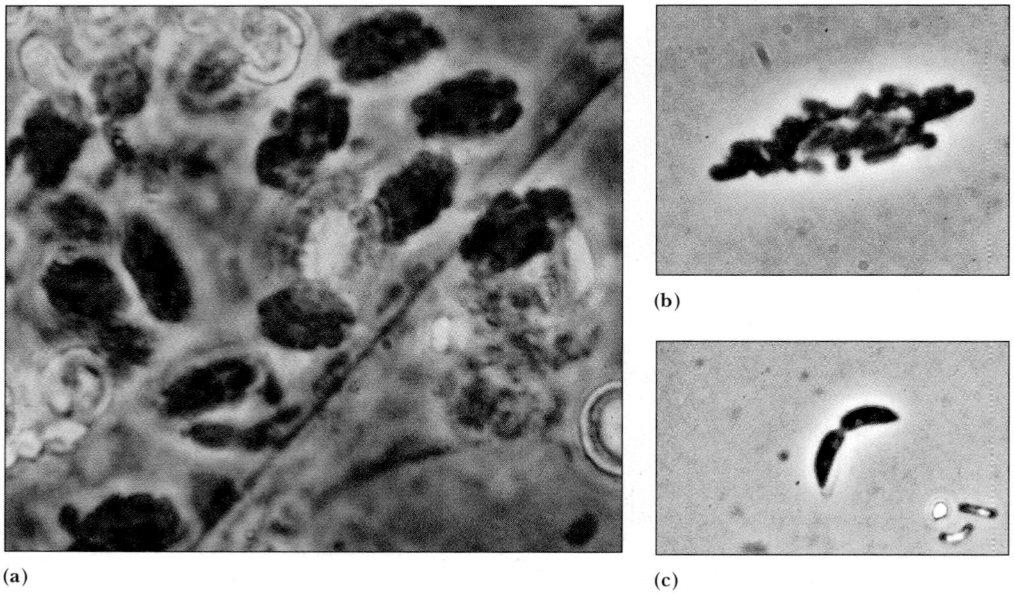

***Figure* 21.14 (a)** Several microcolonies of the consortium species "Chlorochromatium aggregatum" are shown by phase microscopy in this lake sample. **(b)** One microcolony has been partially disrupted by rubbing the coverslip against the slide. Note the large internal rod-shaped cell. **(c)** After further treatment the microcolony is completely disaggregated to reveal the large internal rod-shaped organism undergoing division. Note the flagellar tuft at one pole. (Courtesy of J. T. Staley)

as a photoautotroph anaerobically in the light using sulfide or hydrogen gas an as electron donor.

It is interesting to note that *Chloroflexus aurantiacus* uses neither the Calvin-Benson cycle nor the reductive tricarboxylic acid cycle for carbon dioxide fixation. The best evidence available suggests that instead it uses a unique pathway, the hydroxypropionate pathway, for carbon dioxide fixation (see Figure 10.10**b**).

Like *Chloroflexus, Heliothrix* is a genus of filamentous gliding organisms that grow in alkaline hot springs at temperatures from 35°C to 56°C. It uses Bchl *a* as its primary photosynthetic pigment but is colored orange due to the large amount of carotenoid pigments produced. It has not been isolated in pure culture. *Chloronema* has not been isolated in pure culture either. It is a filamentous, gas vacuolate genus with chlorosomes and a sheath and is found in sulfide-rich lake habitats.

Herpetosiphon is a little studied genus of filamentous gliding heterotrophic bacteria that produce a sheath, and are only mentioned here because they are shown to be closely related to *Cloroflexus* by 16S rRNA sequence analysis. Thus, as in many bacterial phylogenetic groups, both phototrophic and heterotrophic members occur. *Herpetosiphon* spp. are aerobic heterotrophs that live on a variety of organic carbon sources. Some species degrade cellulose and chitin. Both marine and freshwater species occur.

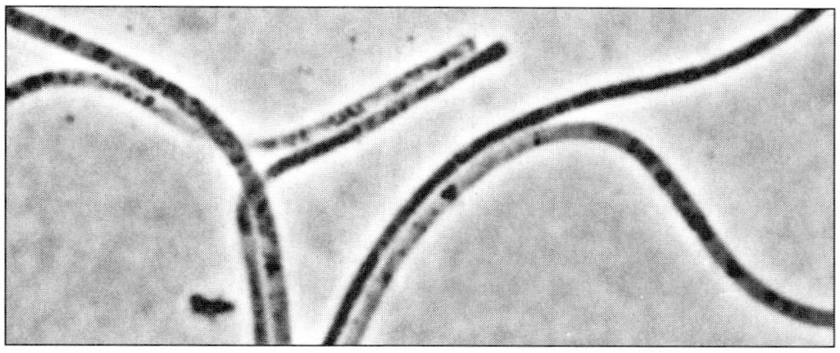

(a)

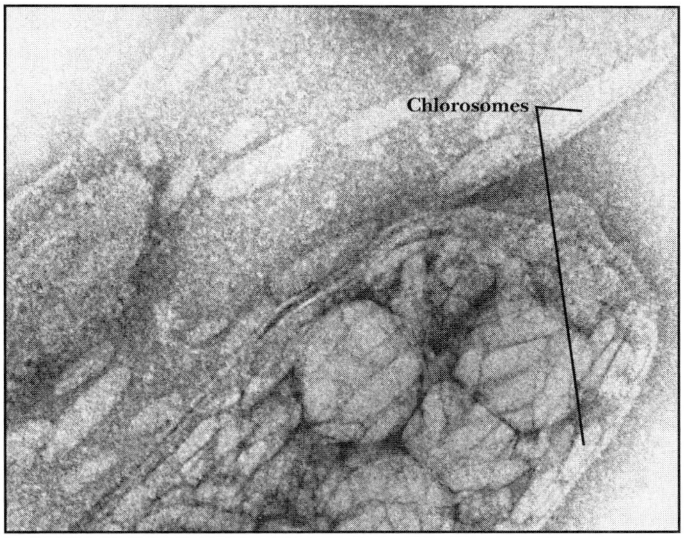

(b)

***Figure* 21.15** Filaments of *Chloroflexus aurantiacus* **(a)** as seen by light microscopy and **(b)** by electron microscopy (bar is 0.5 μm). Note the chlorosomes in the electron micrograph. (**a** courtesy of R. Castenholz and **b** courtesy of M. Broch-Due)

Heliobacteria

The heliobacteria are the most recently discovered photosynthetic bacteria. They are unique because they comprise the first group of gram-positive phototrophic prokaryotes and are therefore treated in Chapter 20. There are currently two genera, *Heliobacterium* and *Heliobacillus.* Both organisms appear to be closely related to *Clostridium,* and some strains even produce true endospores that are rich in calcium and dipicolinic acid. They are all photoheterotrophic, requiring an organic carbon source for growth, and in this respect resemble the photoheterotrophic purple nonsulfur bacteria. Acceptable carbon sources include organic acids such as acetate or pyruvate.

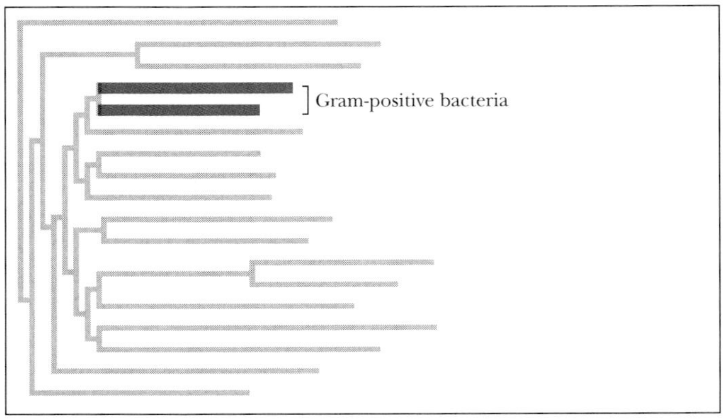

They are also unique in that their bacteriochlorophyll, Bchl *g*, is not found in other organisms and most closely resembles chlorophyll *a* (Figure 21.1**b**). Photosynthesis in the heliobacteria is reminiscent of that in the green sulfur bacteria in that the primary reduced electron acceptor is sufficiently reduced (-0.5 V) that NAD^+ can be reduced directly without the use of reverse electron flow, as required by the purple bacteria and green filamentous bacteria. Cells are red-brown in color due to their characteristic carotenoid pigment, neurosporene (Figure 21.9).

The heliobacteria are found in alkaline soil environments. Their distribution and function in soil habitats is not yet well understood.

Cyanobacteria

This group of photosynthetic bacteria differs from those discussed previously in that they carry out **oxygenic photosynthesis** in which oxygen is evolved:

$$CO_2 + H_2O \longrightarrow (CH_2O)n + O_2 \qquad [21.4]$$

Oxygen is derived from water by "water splitting," a process characteristic of all oxygenic photosynthetic organisms including algae and higher plants. All oxygenic photosynthetic organisms have photosystem II and chlorophyll *a* (see Chapter 9). It is in photosystem II that the water splitting reaction occurs, resulting in oxygen formation.

The cyanobacteria were traditionally classified with the true algae as the blue-green "algae" because they have chlorophyll *a* and carry out oxygenic photosynthesis. However, since it was discovered by electron microscopy that their cell structure is truly prokaryotic (Figure 21.16), they have been claimed by bacteriologists and are now named by the bacteriological code and classified accordingly. Furthermore, phylogenetic analyses of their 16S rRNA indicates they are a separate line of descent of the Eubacteria. At this time there is a dual system of classification of cyanobacteria, because they are also still named and classified by botanists using the botanical code.

STRUCTURE AND PHYSIOLOGY The fine structure of cyanobacteria is typical of other gram-negative bacteria in that they have a multilayered cell wall containing peptidoglycan and an outer membrane. However, not only do they have these layers, but their cell walls usually contain other layers as well (Figure 21.17).

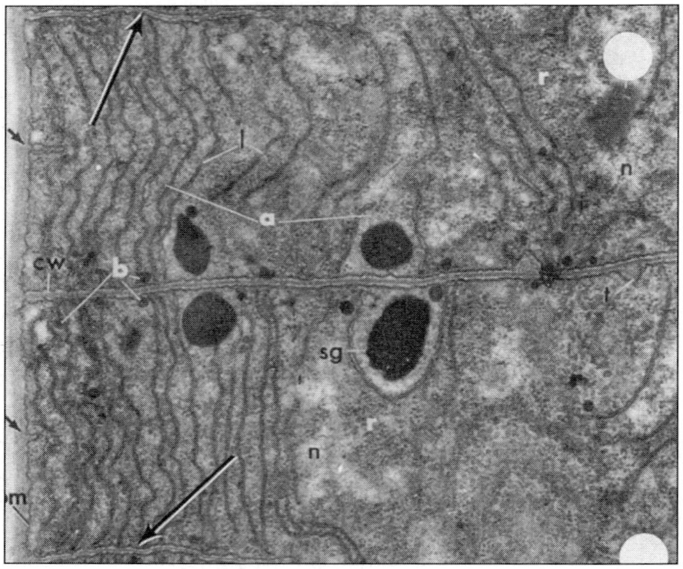

***Figure* 21.16** Thin section through a filamentous cyanobacterium, *Symploca muscorum,* an oscillatorian species, showing the prokaryotic nature of the cell with nuclear area (n). Note also the thylakoid lamellar membranes (l) throughout the cells, the plasma membrane (pm), the cell wall (cw), septa at various stages of ingrowth (large and small arrows), glycogen granules (a), cyanophycin granules (sg) and beta-granules (b). Bar is 0.4 μm. (Courtesy of H. S. Pankratz and C. C. Bowen)

In addition to chlorophyll *a* and cytochromes, the cyanobacteria have characteristic, unique pigments called **phycobilins,** including **allophycocyanin, phycocyanin,** and **phycoerythrin** (see Figure 9.14). The phycocyanins give the cyanobacteria their characteristic blue-green color.

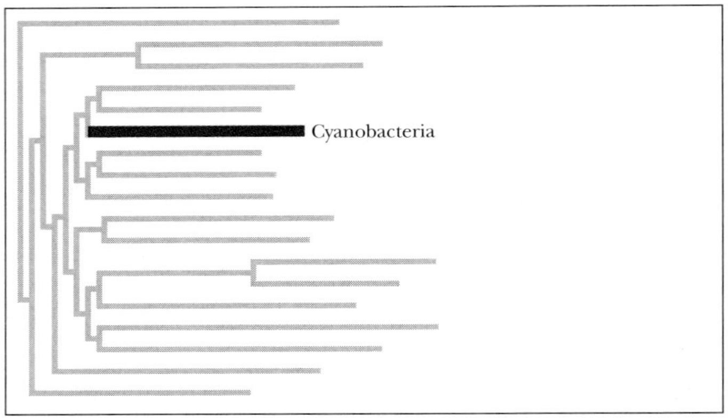

However, not all cyanobacteria are blue green in color. Some are red in color due to another phycobilin pigment called phycoerythrin. These pigments are covalently bonded to proteins in complexes that are called **phycobiliproteins**—phycocyanobilins for the blue pigments or phycoerthyrobilins for the red pigments.

The photosynthetic pigments of cyanobacteria are located in lamellar membranes called **thylakoids** (Figures 21.16 and 21.17). These intracellular membranes are studded with small structures called **phycobilisomes** which contain the phycobiliproteins (see Figures 9.12 and 9.13). The thylakoid complex serves as the photosynthetic reaction center for photosynthesis in cyanobacteria. Thus, both the physical process of light harvesting as well as the chemical process of electron transfer occur here.

The cyanobacteria fix carbon dioxide via the Calvin-Benson cycle. One of the key enzymes in this cycle is ribulose-bisphosphate carboxylase (RuBisCo). Some cyanobacteria store this enzyme in structures in the cell referred to as **carboxysomes**, a feature shared by some of the thiobacilli that will be discussed later in this chapter. Most cyanobacteria do not require vitamins and do not utilize organic compounds, and are therefore excellent examples of photolithoautotrophs. Some species, however, can photoassimilate simple organic compounds such as acetate. They do not use these as energy sources, but simply as carbon sources for growth, thereby relying on photophosphorylation as their sole means of formation of ATP.

One of the recent exciting discoveries regarding cyanobacteria is that some of them can perform anoxygenic photosynthesis as well as oxygenic photosynthesis. When they carry out anoxygenic photosynthesis, sulfide, but not sulfur, serves as the electron donor. Sulfur granules are formed outside the cells. This discovery has ma-

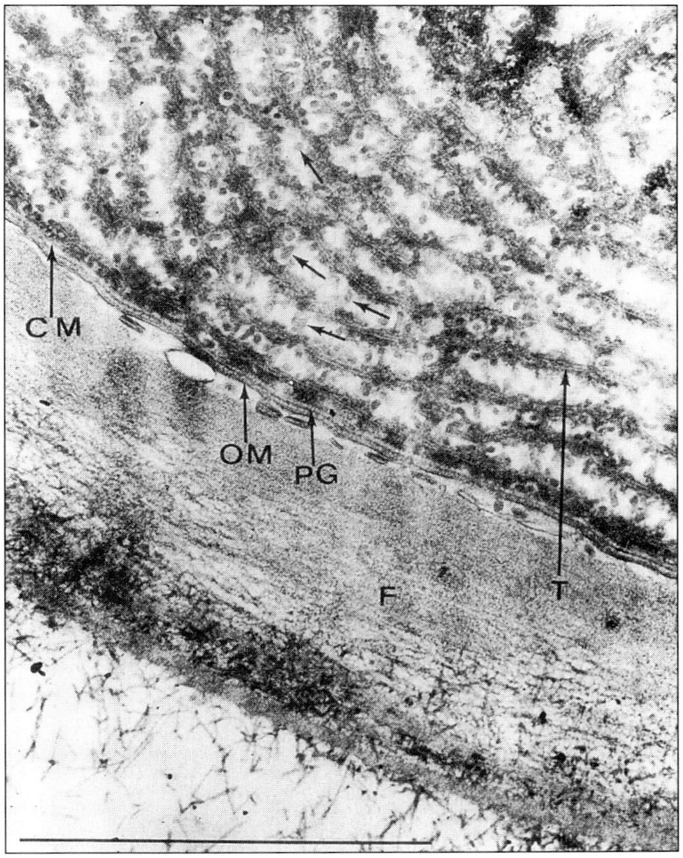

***Figure* 21.17** Thin section showing the cell envelope layer of a cyanobacterium, a *Dermocarpa* sp. In addition to the plasma or cell membrane (cm), peptidoglycan layer (pg), and outer membrane (om) typical of gram-negative bacteria, there is an additional outlying layer of thick fibrous material (F). Note also the thylakoid membranes (T) and glycogen granules (arrows). Bar is 1.0 μm. (Courtesy of J. Waterbury and R. Stanier)

jor implications for evolution suggesting that the cyanobacteria may have evolved early on from anoxygenic photosynthetic bacteria and may therefore have had a common ancestor with the green sulfur and/or purple sulfur bacteria (see Chapter 18).

One of the unique features of cyanobacteria is that they can store a compound called **cyanophycin** (see Figure 21.16). This is unique in that it is a polymer of aspartic acid with each residue of aspartic acid containing a side-group of arginine:

```
-asp-asp-asp-asp-
  |   |   |   |
 arg arg arg arg
```

Cyanophycin is unique in that it is the only nitrogen-containing storage granule of prokaryotic organisms. Since many environments contain low concentrations of nitrogen, it is often a limiting nutrient; thus, cyanophycin is a useful storage product. Also, cyanophycin can be used as an energy source during breakdown. Arginine dihydrolase leads to the synthesis of ATP:

$$\text{arginine} + \text{ADP} + \text{P}_i + \text{H}_2\text{O} \longrightarrow \text{ornithine} + \text{ATP} + \text{NH}_3 + \text{CO}_2$$

ECOLOGICAL AND ENVIRONMENTAL SIGNIFICANCE OF CYANOBACTERIA The cyanobacteria are very important ecologically. Their major significance is that they are important contributors to carbon dioxide fixation, and hence, primary production in aquatic environments. Whereas the anoxygenic purple and green photosynthetic bacteria contribute at best about 10 percent of the annual primary production of the habitats in which they occur, the cyanobacteria and prochlorophytes are responsible for up to half. Moreover, the cyanobacteria are very common in all aquatic habitats, whereas the anoxygenic photosynthetic bacteria are restricted to special anaerobic habitats rich in sulfide. Thus, in the oceans where algae and cyanobacteria are the dominant primary producers, up to almost half of the primary production is due to cyanobacteria and prochlorophytes. Similarly high rates of primary production are attributable to them in freshwater habitats.

Cyanobacteria are also important members of contemporary mat communities. They form the uppermost layer in mats where they carry out oxygenic photosynthesis (**Box 21.2**). The mat communities in which they dominate are common in thermal habitats, polar lakes, intertidal marine habitats and hypersaline environments throughout Earth. Some species also grow well on the surface of soil.

Because cyanobacteria require sunlight and oxygen for their metabolism, they grow at the surface of the aquatic habitats and mat communities where they reside. This position is especially high in solar radiation. In order to protect their cells from mutating ultraviolet radiation, many species produce pigments that absorb ultraviolet light. One such pigment is called scytonemin (see **Box 21.3**).

Many cyanobacteria are nitrogen fixers. Since nitrogen is a common limiting nutrient in many marine and some freshwater environments, their contribution to the nitrogen budget can be extremely important.

It should be noted that some cyanobacteria carry out undesirable activities in the environment. For example, certain gas vacuolate species are responsible for **nuisance blooms** in freshwater habitats (Figure 21.18). Nuisance blooms occur in the midsummer or fall. These gas vacuolate cyanobacteria float to the surface of the lake and are carried onshore by winds. They accumulate and subsequently decay onshore where they cause odors. Under nor-

BOX 21.2 RESEARCH HIGHLIGHTS

Cellular Absorption Spectra of Photosynthetic Eubacterial Groups

As mentioned in the text, the various photosynthetic eubacterial groups occupy the same aquatic habitats—such as mat communities in springs and intertidal zones or the water columns of lakes, ponds, and marine ecosystems. The pattern in the vertical distribution of these photosynthetic groups is always the same: the cyanobacteria reside at the surface, nearest the oxygen, the purple sulfur bacteria lie beneath them, and the green sulfur bacteria occupy the deepest layer, which is farthest from the oxygen and closest to the sulfide. Each of these groups has its own characteristic bacteriochlorophyll or chlorophyll pigments and accessory pigments such as the phycobiliproteins and carotenoids, and each of these pigments has its own characteristic absorption spectrum (see figure). Since each group absorbs light of a different wavelength, each is able to obtain some of the light radiation from the sun because the overlying groups filter out only some of the wavelengths.

It should be noted that the photosynthetic bacteria that reside below the cyanobacteria in these vertically stratified communities receive less energetically favorable light for photosynthesis because of the light's longer wavelength. Furthermore, if these photosynthetic groups are water column communities, little light remains available for the purple and green sulfur bacteria that live in the deepest zones because much of the light is absorbed by the overlying water.

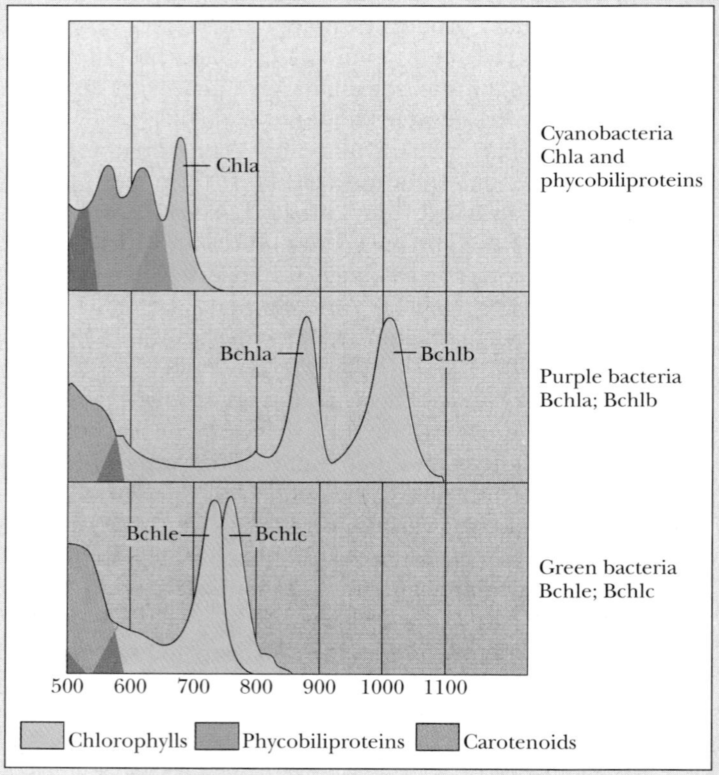

Absorption spectra of bacteriochlorophyll and chlorophyll from various photosynthetic eubacterial groups shown in the vertical stratification in which they occur in natural communities. The cyanobacteria are shown at the surface, the purple bacteria underneath, and the green bacteria occupy the lowest layer.

mal conditions, the gas vacuoles allow the cells to float at or near the surface of lakes and ponds where they grow at a location where photosynthesis is most active.

Some species, such as *Microcystis aeruginosa,* produce neurotoxin compounds that can actually kill animals, including cattle and dogs that eat them. Fortunately they do not normally accumulate in such large quantities, so this problem is sporadic in occurrence.

Cyanobacteria can cause odor and off-flavor tastes in drinking waters as well. The typical cause is the produc-

BOX 21.3 RESEARCH HIGHLIGHTS

Scytonemin, an Ancient Prokaryotic Sunscreen Compound

All organisms are susceptible to mutation caused by ultraviolet light. Humans can develop skin cancer if exposed to excess solar radiation. Animals produce melanin pigments to absorb ultraviolet radiation, and many plants produce flavonoid compounds. However, to further retard incoming radiation for humans, a variety of chemicals are commercially produced that are used as effective sunscreens to absorb damaging ultraviolet radiation. For example, *p*-aminobenzoic acid (PABA), a precursor in the synthesis of the vitamin folic acid (see Figure 7.9), is commonly used in sunscreen and sunblocking lotions to absorb ultraviolet radiation. The reason for selecting these compounds is that PABA and other aromatic compounds are excellent absorbers of ultraviolet light.

Many cyanobacteria produce their own remarkable sunscreen compound called **scytonemin.** Although this substance was named about 150 years ago, only recently has its chemical formula been determined by the cyanobacteria expert, Richard Castenholz, and his colleagues at the University of Oregon.

Considering its complex aromatic structure, it is not surprising that this compound is an excellent UV blocking agent. Scytonemin is especially effective at absorbing UV-A (325–425 nm) but is also effective at absorbing UV-B (280–325 nm) and UV-C (maximum absorption at 250 nm).

Scytonemin is deposited in the extracellular sheath of the cyanobacteria that produce it. There it can intercept the UV light before it reaches the cytoplasm and DNA of the cell. It has been estimated that at the normal concentrations of scytonemin found in a trichome, 85 to 90 percent of the incident UV will be absorbed by this biological sunscreen before it enters the cell.

Because sheathed cyanobacteria are found in early fossil stromatolites, it is likely that these cyanobacteria evolved scytonemin early on. The UV radiation would have been much more intense early in Earth's history than it is now.

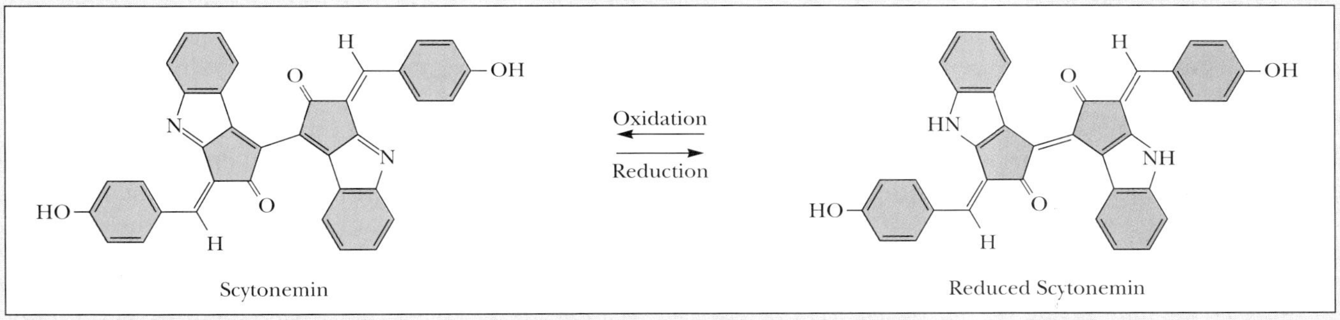

The chemical structure of scytonemin in the oxidized and reduced forms.

tion of geosmin compounds which are similar chemically to the geosmins produced by actinomycetes. This problem typically occurs in the late summer or fall when these cyanobacteria are common in lakes and reservoirs that may be used for the water supply in municipalities.

MAJOR TAXA OF CYANOBACTERIA Several orders of cyanobacteria exist classified on the basis of their morphological characteristics (Table 21.6). They range in morphology from unicellular rods and cocci to filamentous types, some of which show branching division. Many species have not yet been isolated in pure culture. This is due to the difficulty that microbiologists have in cultivating them. Part of the problem is that many cyanobacteria produce an external capsule or sheath that bacteria attach to, making isolation especially difficult. Consequently, the cultivation of these bacteria in pure culture will be a major area of research for bacteriologists well into the next century. A brief description of each order and representative species follows:

***Figure* 21.18** Phase photomicrograph of *Microcystis aeruginosa*. Bright areas in cells are gas vacuoles. (Courtesy of R. W. Castenholz)

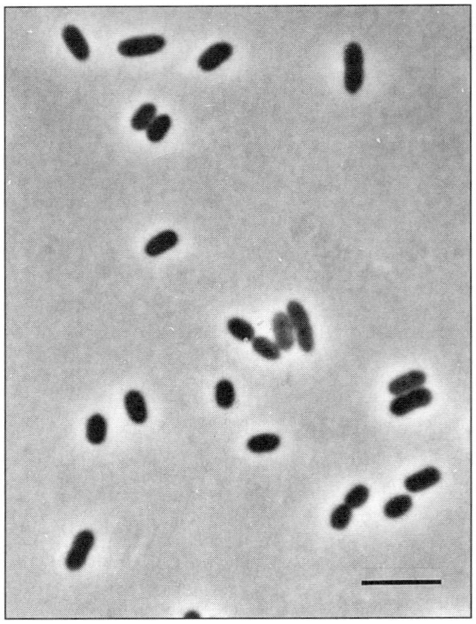

***Figure* 21.19** Cells of *Synechococcus,* a common marine cyanobacterium. Bar is 5 μm. (Courtesy of J. Waterbury and R. Rippka)

Chroococcales are common cyanobacteria that are widely distributed in fresh and marine waters. Table 21.7 provides a listing of the important genera of the *Chroococcales,* some of which are discussed below.

Synechococcus spp. (Figure 21.19) which are unicellular coccoid to rod-shaped cyanobacteria, are among the most important photosynthetic organisms in marine habitats. It has been estimated that they account for approximately 25 percent of the primary production that occurs in typical marine habitats (see Chapter 24).

The genus *Gloeothece* is remarkable in being able to fix atmospheric nitrogen in the presence of oxygen. Perhaps the capsular material of this organism (Figure 21.20) is somehow involved in keeping the oxygen concentration low within the cells. However, it is known also that some unicellular cyanobacteria carry out nitrogen fixation during the dark period when oxygen is not formed by photosynthesis, and this may account for the ability of this genus to carry out this process.

Another notable genus in this order is *Microcystis.* Members of this genus are gas vacuolate and are one of the principal genera responsible for nuisance water blooms (see Figure 21.18). Furthermore, these organisms may produce a toxin that can be deadly to animals that ingest them. *Chaemosiphon,* another genus in this order, is one of the few unicellular cyanobacteria that divide by budding.

Pleurocapsales have an unusual morphology. They grow as unicellular organisms that enlarge in size and undergo internal cell divisions to form **baeocytes,** small spherical reproductive cells, that are released when the division cycle is complete. The baeocytes then develop into larger cells and repeat the division cycle. Some members of this order form multicellular filaments, but in these, too, baeocyte formation occurs.

Oscillatoriales are common filamentous gliding bacteria. Because the cells in the filaments are in very close contact to one another and act in a concerted fashion to ef-

Differential ***Table* 21.6 • Cyanobacterial orders**

Order	Distinctive Features
Chroococcales	Unicellular or multicellular, nonfilamentous
Pleurocapsales	Baeocytes formed
Oscillatoriales	Straight filaments without specialized cells
Nostocales	Straight filaments with heterocysts
Stigonematales	Branching filaments

Differential *Table* 21.7 • **Important genera of *Chroococcales***

Genus	Mol % G + C	Characteristics
Chaemosiphon	46–47	Divide by budding
Synechococcus	47–56	Rods of freshwater, marine, and hot spring origin
Gloeothece	40–43	Cocci in common sheath; nitrogen fixer
Microcystis	45	Cocci; gas vacuolate; may cause toxic water blooms

fect motility, they are regarded as being truly multicellular. The name for this gliding multicellular filament, which may contain 50 to 100 or more cells, is the **trichome** (Figure 21.21**a, b,** and **c**). The trichome may divide to form groupings of 5 to 15 cells called **hormogonia** which are short, gliding filaments used for dispersal and reproduction. These are commonly formed during asexual reproduction in all types of trichome-producing cyanobacteria. A list of the common genera of this order is given in Table 21.8.

Oscillatoria spp. (Figure 21.21**a** and **b**) are common in freshwater habitats and hot springs and can be found growing on the surface of soils. They vary in pigmentation from light green to almost black. Some are even red due to phycoerythrin pigments. Lake forms may be gas vacuolate (Figure 21.21**b**). Their marine counterpart is the genus *Trichodesmium.* These bacteria are gas vacuolate and grow in clusters of filaments. *Trichodesmium* spp. are able to carry out nitrogen fixation even though they do not have specialized cells, that is, heterocysts (see *Nostocales* below). *Lyngbya* spp. are similar to *Oscillatoria.* However, in addition, they also have a prominent sheath (Figure 21.21**c**).

Spirulina spp. are unusual in that they have a helical trichome (Figure 21.21**d**). They are also used as a food source by Africans in Chad, and more recently *Spirulina* is sold as a dietary supplement in health food stores.

Nostocales comprise an important group of nitrogen-fixing filamentous bacteria (Table 21.9). *Anabaena* spp. (Figure 21.22) are widely distributed in freshwater and saline habitats. They, like other members of the order *Nostocales,* have specialized cells called **heterocysts** that are the sites of nitrogen fixation (Figure 21.22). Heterocysts are formed in the filaments from ordinary cells. Their location within the filament is distinctive for the species. Some are located at the end of the filament, whereas others are located within it. When ammonia or other forms of fixed nitrogen are depleted from the environment, heterocyst formation is induced. During the process in which the cell is converted to a heterocyst, the synthesis of phycobilins, the antennae pigments for photosynthesis, is stopped. Therefore, the heterocyst is no longer able to photosynthesize, so the production of oxygen ceases. Since oxygen is inhibitory to nitrogen fixation, this enables the heterocyst to carry out fixation when nitrogenase is synthesized. A considerable amount is known about heterocyst differentiation and the process of nitrogen fixation in *Anabaena* spp. (see **Box 21.4**).

Some species produce **akinetes,** which are special cyst-like resting cells. These are usually somewhat larger than the normal vegetative cells in filaments (Figure 21.22) and are ordinarily formed after the heterocysts have been produced. Typically they do not have gas vesicles even though normal vegetative cells may have them. They can survive better under conditions of starvation and dessication than the vegetative cells.

One species of *Anabaena, A. azollae,* grows in a symbiotic association with the small water fern, *Azolla.* The bacterium is harbored in a special sac of the fern (see Figure 25.27). The water fern obtains fixed nitrogen from *A. azollae,* and in return the bacterium has a protected place where it can grow. Water ferns are widely distributed in aquatic habitats such as rice paddies, and this association is important in ensuring the fertility of such habitats.

Aphanizomenon flos-aquae as well as *Anabaena flos-aquae* are common gas vacuolate organisms that cause water

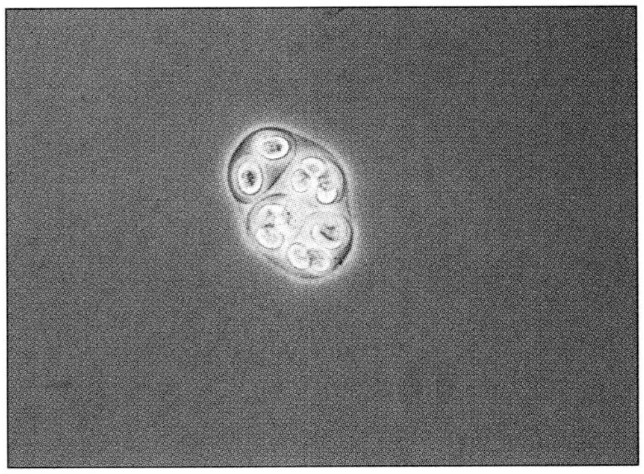

***Figure* 21.20** Phase photograph of a *Gloeothece* sp. showing the cells within a sheath. (Courtesy of J. T. Staley)

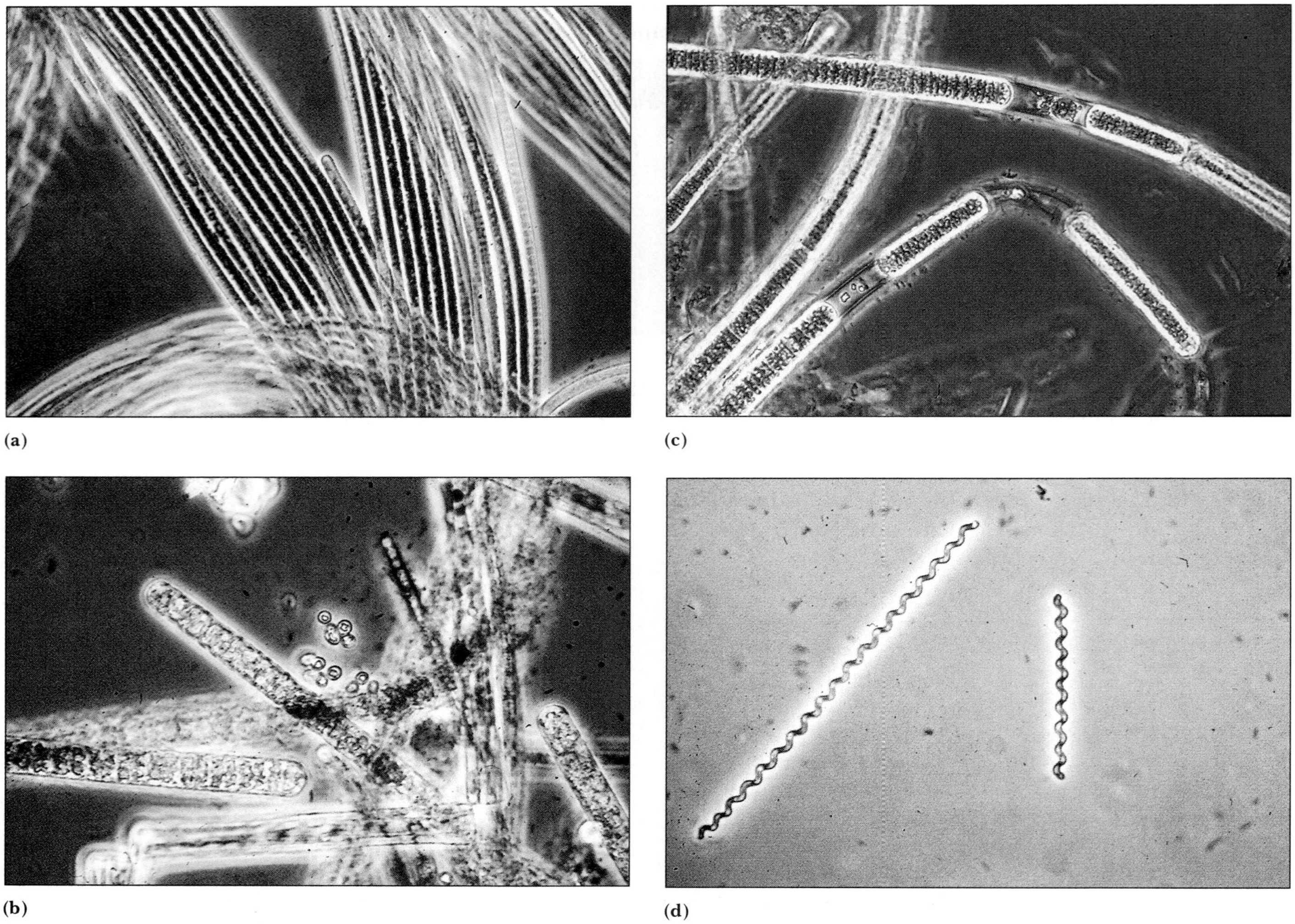

***Figure* 21.21** **(a)** *Oscillatoria formosa,* a typical oscillatorian species showing filamentous growth habit. Each of the multicellular filaments shown is a trichome. **(b)** A gas vacuolate oscillatorian from a lake sample. **(c)** An example of a *Lyngbya* species with its dark sheath. **(d)** *Spirulina,* a helical, filamentous cyanobacterium. (Courtesy of J. T. Staley)

blooms. *Aphanizomenon flos-aquae* occurs in characteristic colonies or "rafts" of trichomes that glide back and forth against one another.

Nostoc spp. are also common in aquatic habitats. They closely resemble *Anabaena* spp.; however their filaments are enclosed within a sheath. Indeed, the sheathed structures can be as large as golf balls or even baseballs in habitats where they occur (Figure 21.23). Like *Anabaena* spp. they have heterocysts and therefore carry out nitrogen fixation.

Differential ***Table* 21.8 • Common genera of *Oscillatoriales***

Genus	Mol % G + C	Description
Oscillatoria	40–50	Straight trichomes; some gas vacuolate; freshwater or hot springs
Trichodesmium	Unknown	Resemble *Oscillatoria,* but marine origin; fix nitrogen
Lyngbya	42–49	Sheathed; resembles *Oscillatoria*
Spirulina	54	Helical trichomes

Differential *Table* **21.9 • Important genera of *Nostocales***

Genus	Mol % G + C	Distinctive Features
Anabaena	35–47	Untapered trichomes with heterocysts; most gliding; many gas vacuolate
Aphanizomenon	Unknown	Like *Anabaena,* however, trichomes aggregate to form colonies; gas vacuolate
Nostoc	39–45	Like *Anabaena,* but sheathed
Rivularia	Unknown	Tapering filaments with polar heterocysts
Gleotrichia	Unknown	Tapering filaments with polar heterocysts in microcolonies

The *Rivularia* group differs from the other heterocystous filamentous genera in that they have tapering filaments. In addition, as shown by the genus *Gloeotrichia* (Figure 21.24), they also have heterocysts which are located at the base of the filament and are responsible for nitrogen fixation.

Stigonematales contain branching clusters of filamentous organisms. They have the most complex morphology of any of the cyanobacteria. Some produce heterocysts as well as hormogonia and akinetes.

Prochlorophytes

Prochloron and *Prochlorothrix* differ from cyanobacteria in that they possess chlorophyll *b* as well as chlorophyll *a,* and lack phycobilins. At one time they were postulated as a possible evolutionary intermediate between cyanobacteria and some of the true algae. However, 16S rRNA phylogenetic analyses indicate they are members of the cyanobacteria. The genus *Prochloron* grows in symbiotic association with marine animals called didemnids and have not been cultivated in pure culture yet. *Prochlorothrix*

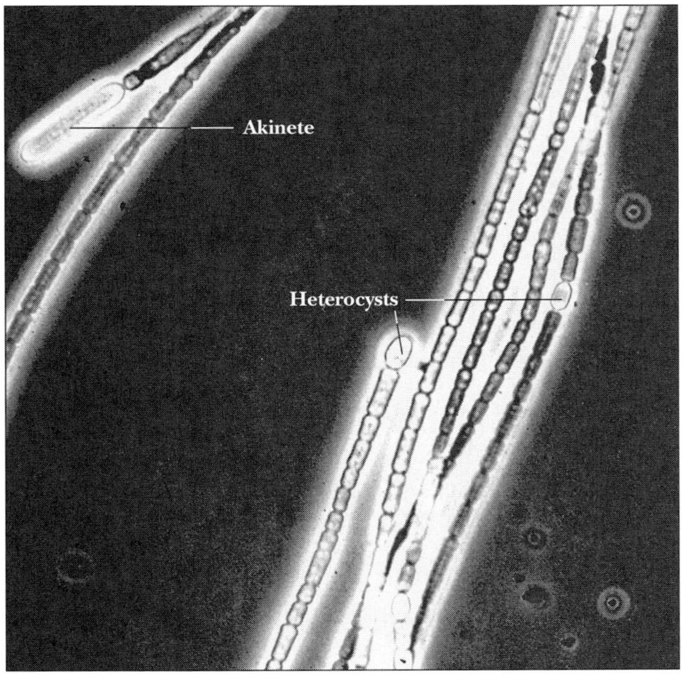

***Figure* 21.22** An *Anabaena* species showing typical filaments with heterocysts and akinetes. (Courtesy of J. T. Staley)

***Figure* 21.23** A large colony of *Nostoc pruniforme* from Mare's Egg Spring in Klamath County, Oregon. Note the smaller new colonies forming on its sides. (Courtesy of W. K. Dodds and R. W. Castenholz)

BOX 21.4 RESEARCH HIGHLIGHTS

The Cyanobacterial Heterocyst

The heterocyst is a specialized cell produced by some filamentous cyanobacteria. The heterocyst is formed from a typical vegetative cell through a developmental process induced when the concentration of fixed nitrogen forms such as ammonia and nitrate are depleted from the environment. Once the heterocyst has formed, it is unable to undergo cell division and therefore resembles a resting cell, such as a cyst. A differentiation process occurs whereby the heterocyst develops a complex coat of three predominant layers—an outer fibrous layer (F), a homogeneous layer (H), and an inner laminated layer (L)—which are thought to restrict the passage of oxygen and other gases including nitrogen into the heterocyst. Nitrogen, which is needed by nitrogenase, enters the heterocysts through the ends of the heterocysts, called the microplasmodesmata.

The pre-heterocyst cell loses its ability to produce phycobilins, the antennae pigments for photosynthesis. This shuts off Photosystem II, and therefore the production of oxygen, as well as fixation of carbon dioxide cease. Photosystem I, however, remains intact, allowing the heterocyst to produce ATP by photophosphorylation. This ATP is needed by nitrogenase for nitrogen fixation. The reducing power needed for nitrogen fixation is produced through metabolism of carbohydrates that are produced by adjacent photosynthesizing cells.

The decrease in oxygen concentration coupled with the decrease in fixed nitrogen available to the organism are signals for the induction of nitrogenase and the process of nitrogen fixation. The ammonia formed by nitrogenase is carried to adjoining cells through the formation of glutamine. The heterocyst has a high level of the enzyme glutamine synthetase (GS), which carries out the following reaction:

$$\text{glutamate} + NH_3 \longrightarrow \text{glutamine}$$

The glutamine (gln) formed from the heterocyst is converted in the adjoining vegetative cells to glutamate (glu) by an enzyme called glutamine oxoglutarate amido transferase (GOGAT), which couples glutamine to alpha-ketoglutarate to form two molecules of glutamate. Part of the glutamate is recycled to the heterocyst (see figure), and part is transferred to adjacent cells as a form of utilizable fixed nitrogen. In this manner, organic, fixed nitrogen is produced that can be used by the vegetative cells in anabolic processes.

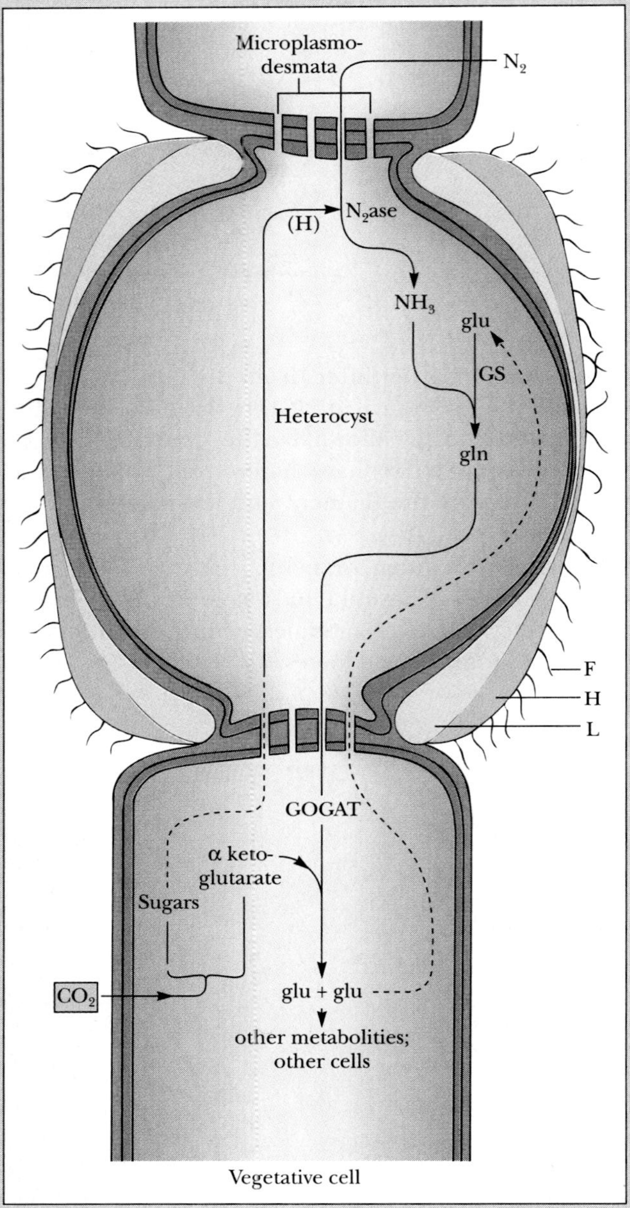

Diagram of the heterocyst showing its structure and the manner in which fixed nitrogen is formed and transferred from the heterocyst.

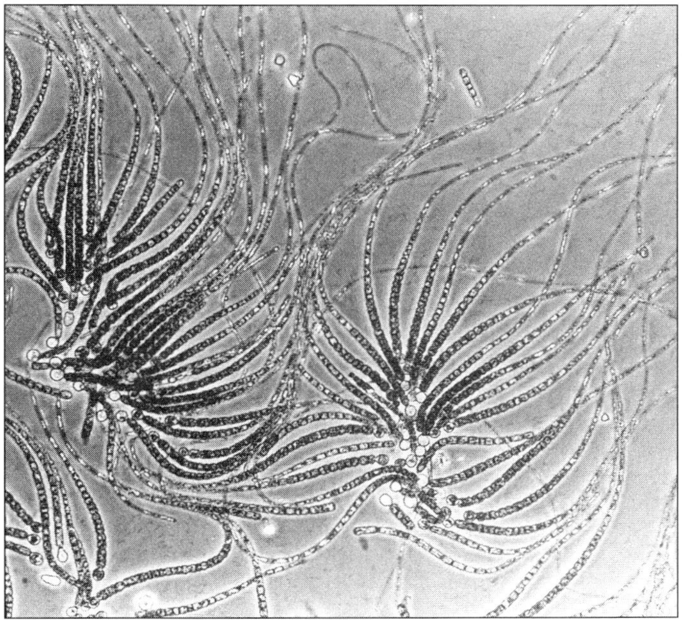

Figure **21.24** Tapering filaments are typical of rivularian cyanobacteria such as *Gloeotrichia echinulata*. Note the gas vacuoles in cells and the heterocysts at the base of each tapering filament (see cover photo). (Courtesy of J. T. Staley)

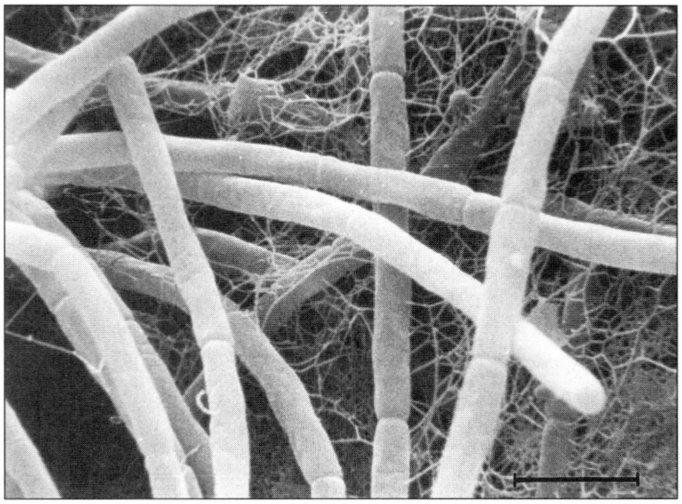

Figure **21.25** A scanning electron micrograph of *Prochlorothix,* a marine prochlorophyte. Bar is 3 μm. (Courtesy of W. Krumbein)

(Figure 21.25) is a free-living marine bacterium. Like the cyanobacteria, prochlorophytes carry out oxygenic photosynthesis and appear to be significant contributors to the process of carbon dioxide fixation in the open oceans.

The Chloroplast

Studies of 16S rRNA isolated from the chloroplasts of plants and algae indicate they have a common origin. Furthermore, their ancestors all came from the cyanobacteria branch of the Eubacteria, and they would be placed with this group of the prokaryotes if they were free-living microorganisms. These results strongly support the endosymbiotic theory of evolution and argue that the progenitor of the chloroplast was a cyanobacterium that evolved into a close association with eukaryotic cells early in evolution (see Chapters 18 and 25).

Chemolithotrophic Eubacteria

The **chemolithotrophic bacteria** obtain energy from the oxidation of reduced inorganic compounds. Several different groups exist and are differentiated from one another based on the inorganic compounds they oxidize as energy sources (Table 21.10). **Nitrifying bacteria** oxidize reduced nitrogen compounds, **sulfur oxidizers** oxidize reduced sulfur compounds, **iron bacteria** oxidize reduced iron, and **hydrogen bacteria** oxidize hydrogen gas. One special group of hydrogen bacteria produce acetic acid

Differential *Table* **21.10 • Groups of chemolithotrophic and methanotrophic bacteria**

Group	Energy Source(s)	Product(s)
Nitrifiers	NH_4^+ or NO^{2-}	NO_2^- or NO_3^-
Sulfur oxidizers	H_2S, S^0, $S_2O_3^{2-}$	SO_4^{2-}
Iron oxidizers	Fe^{2+}	Fe^{3+}
Hydrogen bacteria	H_2	H_2O
Carboxydobacteria	$H_2 + CO$	$H_2O + CO_2$
Acetogenic bacteria	H_2	CH_3COOH
Methane oxidizers[1]	CH_4	$CO_2 + H_2O$

[1]Methane may be regarded as the simplest organic compound. However, the metabolism of methane oxidizers is very similar to that of the nitrifiers; therefore they are included here.

from carbon dioxide and are therefore called **acetogens.** Another special group of hydrogen bacteria, the **carboxydobacteria,** oxidize carbon monoxide and hydrogen.

By and large, these chemoautotrophic activities are uniquely associated with bacteria. Only rarely have higher organisms been found that can carry out any of these processes, and when they have been found, as for example nitrification by some fungi, their activity is not considered to be significant in the environment. In contrast, these processes carried out by bacteria are well known and considered to be extremely important in the biogeochemical cycles of nature (see Chapter 24). Each of the chemoautotrophic groups is discussed individually below in greater detail.

Nitrifying Bacteria

As their name implies, the nitrifying bacteria use reduced inorganic nitrogen compounds, ammonia and nitrite, as

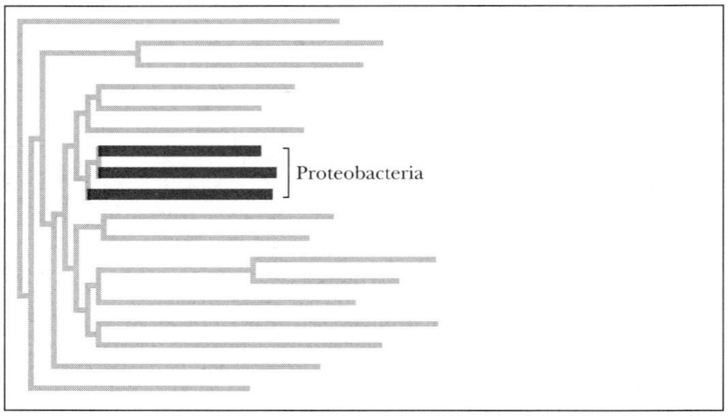

their sources of energy for growth. Ammonia is oxidized to nitrite by one group of nitrifiers, called the **ammonia oxidizers:**

$$2\ NH_3 + 3\ 1/2\ O_2 \longrightarrow 2\ NO_2^- + 3\ H_2O \quad [21.5]$$

Nitrite is oxidized to nitrate by another group called the **nitrite oxidizers:**

$$NO_2^- + 1/2\ O_2 \longrightarrow NO_3^- \quad [21.6]$$

No single nitrifying bacterium is able to oxidize ammonia all the way to nitrate, although pure cultures of some fungi have been reported to carry out this process. These fungi, however, are not considered important in the environment. Normally the two groups of nitrifying bacteria grow in close association in the environment to carry out the two-step sequential oxidation. The two processes appear to be tightly coupled so that nitrite, which can be toxic to higher organisms, does not accumulate in high concentration.

Nitrifying bacteria are found in all soil and aquatic habitats and are especially common in alkaline or neutral pH environments where ammonia rather than ammonium ion is abundant, and where ammonia is the form of nitrogen used as the energy source. Also, more carbonate is available for CO_2 fixation in moderately alkaline environments.

Nitrifiers are superb examples of chemolithotrophs. They grow well in pure culture or completely inorganic media. However, most species are able to use organic carbon sources such as acetate for growth and thus are facultative chemolithotrophs. They grow poorly in culture even under the best of conditions, having generation times of 24 hours or more. Rarely do they grow to high enough cell densities to produce turbidity in media even though their activities in the medium can be readily demonstrated. The reason for their poor growth in pure culture is not well understood; however, it may be due in part to the toxicity of nitrite. We first discuss the ammonia oxidizers, which are sometimes also called the nitrosofying bacteria.

Ammonia Oxidizers

The oxidation of ammonia to nitrous acid involves two major steps with hydroxylamine, NH_2OH, as an intermediate (see also Chapter 8):

$$NH_3 + O_2 + XH_2 \xrightarrow{AMO} NH_2OH + H_2O + X \quad [21.7]$$

$$NH_2OH \xrightarrow{HAO} HNO + H_2O \xrightarrow{HAO} HNO_2 + 2\ H^+ + 2\ e^- \quad [21.8]$$

where AMO = ammonia monooxygenase
HAO = hydroxylamine oxidoreductase
X = a hydrogen carrier such as NADH
$\Delta G°'$ (reactions 7 and 8 combined) = − 275 kJ/mol

Several enzymes are involved in the oxidation pathway to form nitrous acid. The first enzyme in this pathway is the key enzyme, ammonia monooxygenase, AMO, which carries out the oxidation of ammonia to hydroxylamine. Ammonia monooxygenase shows rather broad specificity. Thus, it can also oxidize methane, making these bacteria resemble methane-oxidizing bacteria, which have a methane monooxygenase of similar broad specificity (see methanotrophic bacteria section below). A variety of inhibitors affect the ammonia monooxygenase including acetylene, which is used as a blocking inhibitor in ecological studies, as well as 2-chloro-6-trichloromethylpyridine (nitrapyrin).

Energy is generated in reaction 8, in which hydroxylamine is oxidized to HNO and subsequently to nitrous acid. This reaction is catalyzed by the enzyme hydroxylamine oxidoreductase (see reaction 8). HNO spontaneously degrades to produce N_2O, nitrous oxide (laughing gas). Electrons from the oxidations are passed through an extensive membrane-bound system of cytochromes (giving the cultures a red appearance). The free energy of this reaction is sufficiently high to generate ATP. Acid produced by oxidation of the inorganic nitrogen compounds results in the formation of protons that are passed through the membrane, and ATP is formed by a cell membrane ATPase. Reducing power generated during the oxidation process is used for the carbon dioxide fixation reactions. Most ammonia oxidizers are members of the alpha group of Proteobacteria, but some are gamma Proteobacteria.

Nitrite Oxidizers

Nitrite oxidation is a one-step oxidation process carried out by the enzyme nitrite oxidoreductase. Water is the actual electron donor for the oxidation process:

$$NO_2^- + H_2O \xrightarrow{NOR} NO_3^- + 2\ e^- + 2\ H^+ \qquad [21.9]$$

where NOR = nitrite oxidoreductase
$\Delta G^{\circ\prime} = -76$ kJ/mol

As shown in reaction 9, protons are present in the periplasmic space, thereby forming a proton motive force across the cell membrane. The electrons from reaction 9 are passed onto oxygen through the cytochrome system, resulting in the formation of water in the cytoplasm (see Chapter 8). Under standard conditions this reaction provides less energy per mole of substrate oxidized than ammonia oxidation. However, as with ammonia oxidizers there is sufficient energy available in this process to generate ATP, and this is accomplished by chemiosmotic means. However, unlike the ammonia oxidizers, the nitrite oxidizers must produce NADPH by reverse electron transport (see also Chapter 9).

Table 21.11 lists some of the genera of ammonia and nitrite-oxidizing bacteria, and representatives of selected genera are illustrated in Figure 21.26. Most nitrifying bacteria have a complex internal membrane system that contains cytochromes (Figure 21.26). These intracytoplasmic membranes are thought to contain the cytochrome systems that act as the site of ammonia or nitrite oxidation as well as the site of generation of NADH. *Nitrobacter* is a member of the alpha Proteobacteria, whereas others are members of the beta Proteobacteria.

Ecological Importance of Nitrifiers

Although nitrifiers occur in relatively low concentrations in habitats, they are very active metabolically. Nitrite does not accumulate in most environments because of the tight coupling between ammonia oxidation and nitrite oxidation. Therefore, nitrite, which is toxic (and mutagenic) to plants and animals, typically occurs in very low concentrations in environments because of its rapid oxidization to nitrate.

The activities of nitrifying bacteria are especially important in soil environments because of the role they play in nitrogen cycling (see Chapter 25). Ammonium nitrogen is a preferred source of nitrogen as a crop fertilizer.

Differential *Table* **21.11 • Important genera of nitrifying bacteria**

Group/Genus	Mol % G + C	Special Characteristics
I. Ammonia oxidizers		
Nitrosomonas	45–54	Common rod
Nitrosospira	53–55	Tightly coiled helix
Nitrosococcus	48–51	Coccus
Nitrosolobus	53–56	Multilobed coccus
II. Nitrite oxidizers		
Nitrobacter	60–62	Common rod, divides by budding
Nitrospina	58	Thin rod
Nitrococcus	51	Marine genus
Nitrospira	50	Tightly coiled helix; common marine genus

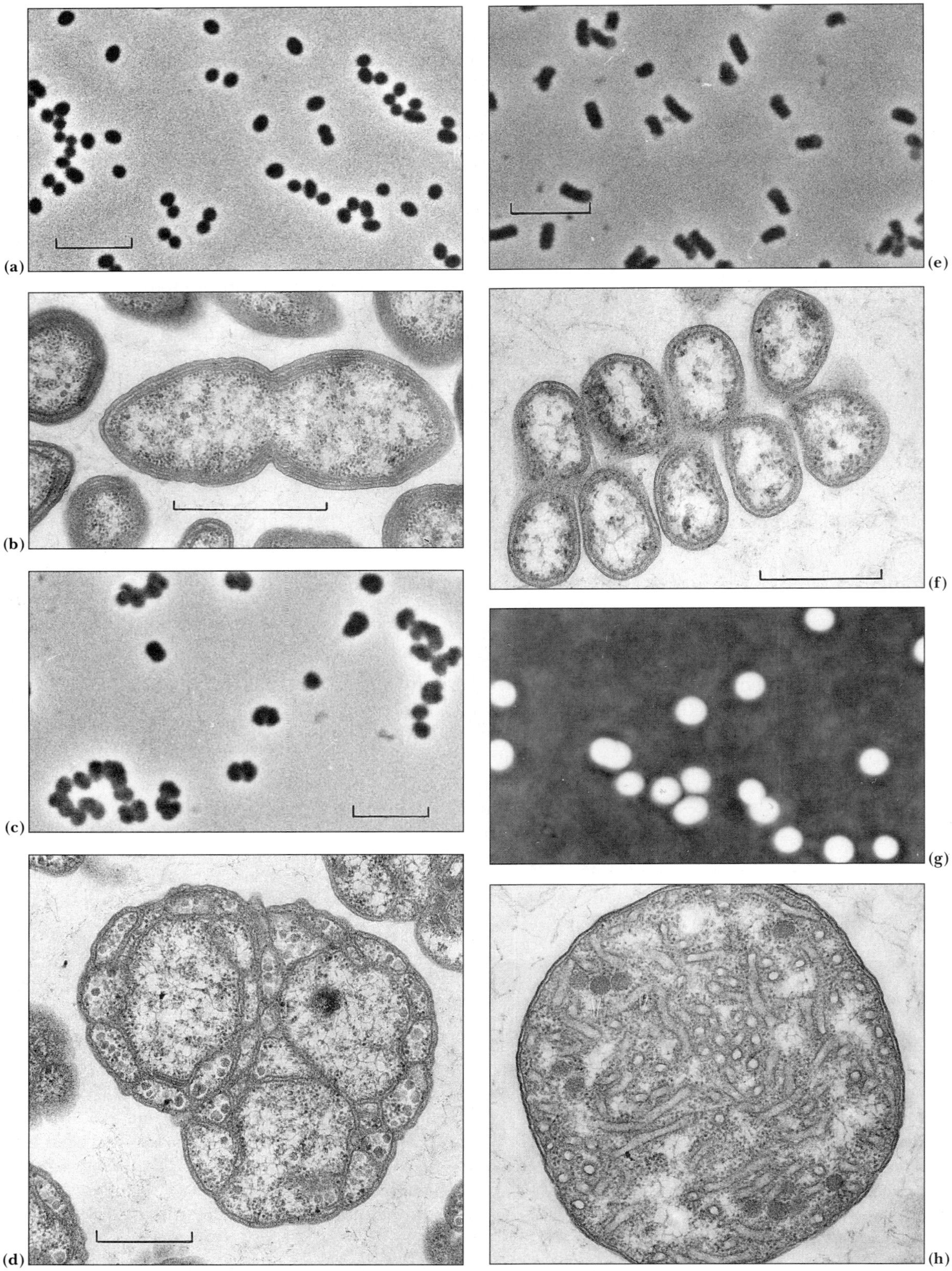

***Figure* 21.26** Phase photomicrographs and thin sections, respectively, of nitrifying bacteria. *Nitrosomonas europaea* (**a** and **b**), *Nitrosolobus multiformis* (**c** and **d**), *Nitrosospira briensis* (**e** and **f**) and *Nitrococcus mobilis* (**g** and **h**). *Nitrosolobus* cells are unusual in that they are compartmentalized as shown in **d.** Also, note that *Nitrosospira* is so tightly coiled that it appears as a rod in the phase photo, **e.** Bar is μm for phase photomicrographs and 1, 0.5, 0.5, and 0.1 μm, respectively, for thin sections. (Courtesy of S. Watson)

The principal reason for this is that ammonium is more readily retained in soils because it is positively charged. Nitrate, on the other hand, is highly soluble and readily leached from soils. Moreover, nitrate formed by nitrifiers can be converted to nitrogen gas by denitrifiers, and therefore lost from soils. For these reasons the inhibitor nitrapyrin (see above) is produced commercially as a "fertilizer" to inhibit nitrification and thereby prevent nitrogen loss from soils.

It is also noteworthy that N_2O, an intermediate formed in nitrification as well as denitrification, is important environmentally in that it can react photochemically with ozone in the upper atmosphere (see Chapter 24) and destroy it.

Sewage contains high concentrations of ammonia due to the decomposition of proteins and amino acids (see Chapter 33). As a result, ammonia is discharged in large quantities in sewage outfalls. Like organic compounds, ammonia has a high biochemical oxygen demand (BOD, see Chapter 33), because the process of nitrification requires oxygen. Therefore, even treated sewage, which has low organic carbon concentrations, can cause low oxygen concentrations that reduce the quality of receiving waters because of nitrification. For this reason, modern wastewater treatment systems are designed to encourage the growth of nitrifying bacteria *within* the treatment plant so that downstream waters will be unaffected by ammonia (see Chapter 33). Otherwise, receiving waters could become anaerobic by the activity of these bacteria even though it contains little organic material.

Sulfur and Iron Oxidizers

Several different groups of lithotrophic bacteria obtain energy by the oxidation of reduced inorganic sulfur compounds. Two of these are Eubacteria—the "unicellular sulfur oxidizers" and the "filamentous sulfur oxidizers," both groups of which are members of the Proteobacteria. In addition, one proteobacterium that is a sulfur oxidizer, *Thiobacillus ferrooxidans,* can also carry out the oxidation of pyrite, FeS_2, with the formation of ferric oxide and sulfate and can obtain energy from both iron oxidation as well as the oxidation of sulfur. Thus, it will also be considered in this section. Other sulfur oxidizers, such as the family *Sulfolobaceae,* are Archaea. The eubacterial groups will be discussed here, and the *Sulfolobaceae* will be discussed with the Archaea in Chapter 22.

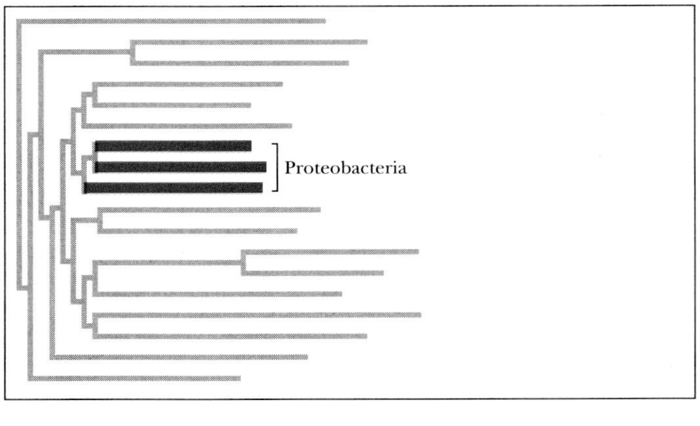

Some eubacterial sulfur oxidizers use sulfide as an energy source. The sulfide can be supplied as hydrogen sulfide, or as a metal sulfide such as iron or copper sulfide. Some use elemental sulfur, S^0, and some use thiosulfate, $S_2O_3^{2-}$. Some can use all three forms of sulfur. The final product formed in all cases is sulfuric acid. The overall reactions for the complete oxidation of these various sulfur forms are shown below:

$$S^{2-} + 4\ O_2 \longrightarrow 2\ SO_4^{2-} \qquad [21.10]$$

$$2\ S^0 + 3\ O_2 + 2\ H_2O \longrightarrow 2\ H_2SO_4 \qquad [21.11]$$

$$S_2O_3^{2-} + 2\ O_2 + H_2O \longrightarrow 2\ SO_4^{2-} + 2\ H^+ \qquad [21.12]$$

The production of sulfuric acid (as sulfate or sulfuric acid) leads to the lowering of pH during the growth of these bacteria. Indeed, the acidophilic members of the sulfur oxidizers, as exemplified by *Thiobacillus thiooxidans,* can produce enough acid to lower the pH to 1.0, equivalent to 0.1 N H_2SO_4, and still remain viable. Clearly, these are the ultimate proton pumpers! These oxidations yield energy for the cells as shown in Figure 21.27.

The unicellular sulfur and iron oxidizers will be treated in the next section, and this will be followed by a discussion of the filamentous sulfur oxidizers.

Unicellular Sulfur Oxidizers

There are three major genera of unicellular sulfur oxidizing Eubacteria: *Thiobacillus, Thiomicrospira,* and *Thermothrix. Thiobacillus* is a genus of polarly flagellated rods (Figure 21.28) that is widespread in soil and aquatic habitats. *Thiomicrospira* is a genus of small, polarly flagellated vibrios that are common in marine habitats. The final genus, *Thermothrix,* resembles *Thiobacillus* spp. in being a polarly flagellated rod. However, it grows at temperatures from 40°C to 80°C as a filamentous organism under the low oxygen conditions found in thermophilic hot springs where it lives. Table 21.12 lists important species from each of these genera and describes some of their salient features.

The genus *Thiobacillus* is especially diverse metabolically. At one extreme are obligate chemolithotrophs, such as *Thiobacillus thiooxidans,* which do not use organic compounds at all. These obligate chemolithotrophs are also called chemolithoautotrophs because they use (a) chemicals as energy source (hence, *chemo-*), (b) inorganic

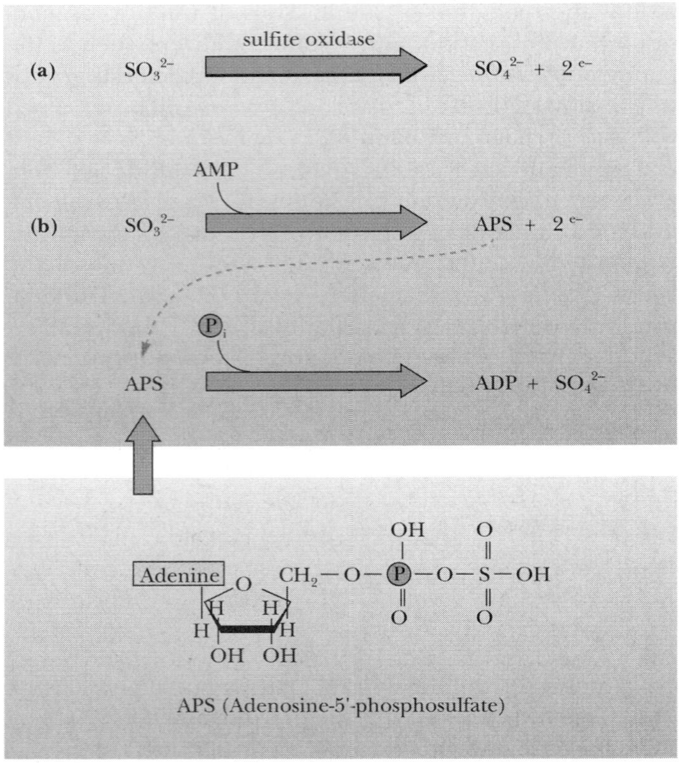

***Figure* 21.27** Pathways for oxidation of reduced sulfur compounds and generation of energy by *Thiobacillus* spp. Sulfide, elemental sulfur, and thiosulfate are reduced sulfur forms that can serve as substrates for oxidation by the enzyme sulfide oxidase to form sulfite. The sulfite formed is metabolized by one of two different pathways. **(a)** The principal pathway involves the oxidation of sulfite to sulfate by the enzyme sulfite oxidase. ATP is generated by proton pumping. **(b)** Alternatively the sulfite is oxidized to sulfate, which is transferred to AMP to form adenosine phosphosulfate (APS). In this latter pathway, ATP is formed by a proton gradient and ADP by substrate level phosphorylation.

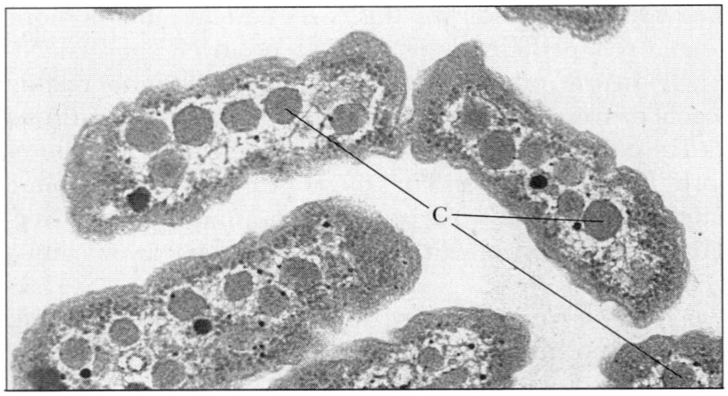

***Figure* 21.28** A thin section of rod-shaped cells of *Thiobacillus neapolitanus* showing the typical gram-negative cell structure. Note the carboxysomes (c) in the cells. Cells are about 0.5 μm in diameter. (Courtesy of J. Shively)

sources of electrons (that is, *litho-* meaning rock), and (c) carbon dioxide as a sole carbon source (hence, *autotrophic*). Many autotrophic species produce carboxysomes in their cells (Figure 21.28). *Thiobacillus novellus* is an intermediate group of so-called facultative chemolithotrophs, or mixotrophs, that fix carbon dioxide but can also assimilate organic carbon sources for anabolic purposes. Finally, at the other extreme, are some ordinary heterotrophs including some *Pseudomonas* species that oxidize reduced sulfur compounds but do not obtain energy from this process and cannot utilize carbon dioxide as a carbon source.

Filamentous Sulfur Oxidizers

The filamentous sulfur oxidizers are found at the interface of sulfide-containing muds and the aerobic environment. For example, they are common on the surface of marine and freshwater sediments and in sulfur springs. These types of habitats, which have an active sulfur cycle,

Differential ***Table* 21.12 • Characteristics of unicellular sulfur-oxidizing bacteria**

Genus/Species	Nutritional Group	pH Optimum	S Source	Mol % G + C
Thiobacillus thiooxidans	Chemolitho-autotroph	2–4	S^0, $S_2O_3^{2-}$	51–53
T. ferrooxidans	Chemolitho-autotroph	2–4	S^0, S^{2-} (incl. metal sulfides)	53–65
T. thioparus	Chemolitho-autotroph	6–8	S^0, S^{2-}, $S_2O_3^{2-}$	63–66
T. acidophilus	Mixotroph	2–4	S^0, $S_2O_3^{2-}$	61–64
T. novellus	Mixotroph	6–8	$S_2O_3^{2-}$	67–68
T. denitrificans	Chemolitho-autotroph	6–8	S^{2-}, S^0, $S_2O_3^{2-}$	63–68
Thiomicrospira pelophila	Chemolitho-autotroph	6.5–7.5	S^{2-}, S^0, $S_2O_3^{2-}$	44
Thermothrix thiopara	Mixotroph	6–8	S^{2-}, $S_2O_3^{2-}$	Unknown

Differential *Table* **21.13 • Filamentous sulfur-oxidizing bacteria**

Genus	Mol % G + C	Distinctive Features
Beggiatoa	37–51	Long gliding filaments with sulfur granules
Thiothrix	52	Rosette-forming; gliding gonidia
Thioploca	Unknown	Gliding filaments in sheath

are called **sulfureta.** There are several important genera as presented in Table 21.13.

The genus *Beggiatoa* is by far the best-known member of the filamentous sulfur bacteria (Figure 21.29). Studies of this organism by Sergei Winogradsky led to his proposal of the concept of chemolithotrophy (see **Box 21.5**). Although Winogradsky did not study *Beggiatoa* in pure culture, several strains of *Beggiatoa* have now been isolated. All isolated strains produce sulfur granules and are known to oxidize sulfides, but some are facultative chemoautotrophs that use acetate as a carbon source.

The classical metabolic type of *Beggiatoa,* as proposed by Winogradsky, is a strict chemolithotroph that uses sulfide as an energy source and oxidizes it completely to sulfuric acid. Furthermore, it is also grows chemoautotrophically, using carbon dioxide as a sole source of carbon. Strains of this type have been isolated from the marine environment, and they use the Calvin-Benson cycle for carbon dioxide fixation.

However, heterotrophic types are currently known and have been studied in pure culture. The heterotrophic strains are common in freshwater habitats and sulfur springs. They oxidize sulfide only to sulfur, not sulfate, and they use acetate as a carbon source and obtain energy by its oxidation. These strains may also be able to fix carbon dioxide by the Calvin-Benson cycle as low levels of ribulose bisphosphate carboxylase (RuBisCo), a key enzyme in this pathway, have been detected.

Thiothrix species are found not only in sulfur springs, but also in sewage treatment facilities as a bacterium that causes bulking (see Chapter 33). The filaments contain sulfur granules and closely resemble *Beggiatoa* spp.; however, they produce holdfasts, form rosettes (Figure 21.30**a**), and divide by generating gliding gonidia in a manner analogous to the genus *Leucothrix.* They also produce a thin sheath (Figure 21.30**b**). Little is yet known about the physiology and metabolism of this group although some strains have been cultivated in pure culture.

(a)
(b)
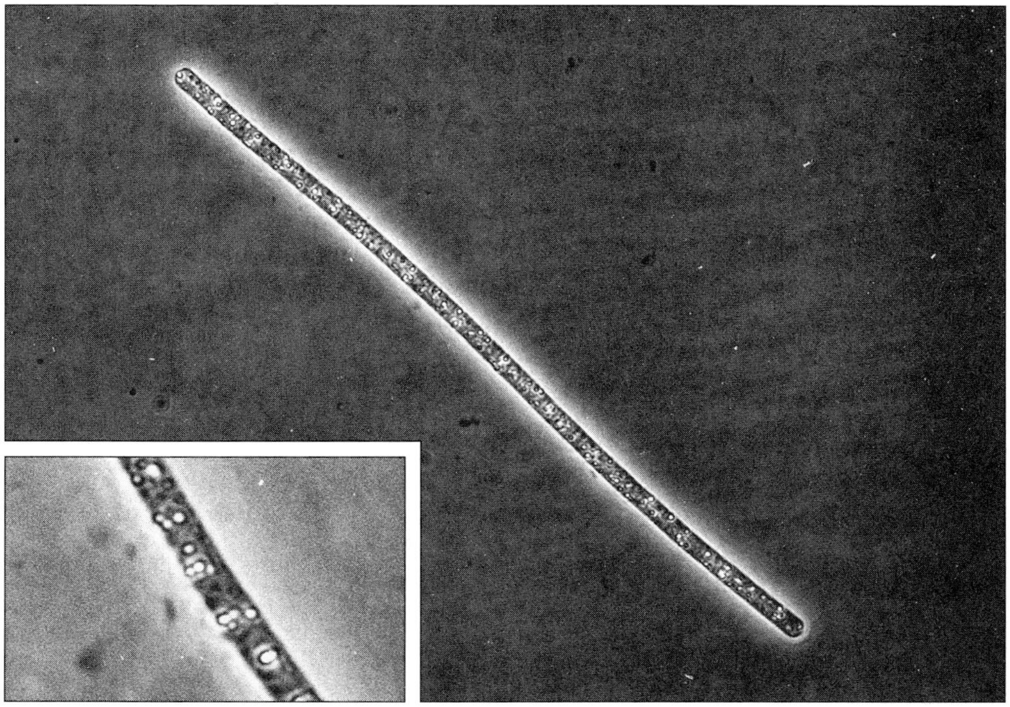

***Figure* 21.29** (**a** and **b**) Phase photomicrographs of the filamentous, gliding sulfur-oxidizing bacterium, *Beggiatoa*. **(b)** Enlargement shows intracellular sulfur granules that appear as bright areas within cells. (Courtesy of J. T. Staley)

BOX 21.5 MILESTONES

Winogradsky and the Concept of Chemolithotrophy

Sergei Winogradsky, a scientist who was born and raised in Czarist Russia, became one of the most remarkable figures in microbiology (see Chapter 2). His most important contributions relate to the phototrophic bacteria and the chemolithotrophic bacteria. For example, he described by direct microscopic observation several genera of the purple sulfur bacteria. During the past century, many microbiologists have worked on these bacteria, and Winogradsky's classification of them still holds largely unchanged today.

Sergei Winogradsky was also the first to grow nitrifying bacteria successfully. He cultivated them on a silica gel medium because they do not form colonies on agar media. Most isolates of nitrifiers are now obtained by dilution of enrichment samples to extinction using liquid inorganic media. To accomplish this, active enrichments containing the bacteria are diluted serially with many replicates. By chance, some of the highest dilution tubes with growth will receive only one cell of a nitrifier of interest, and from this a pure culture can be obtained. In this tedious manner, representatives of several new genera have been isolated in pure culture. Many of these strains were first isolated in pure culture using this technique in the late Stanley Watson's laboratory at Woods Hole Oceanographic Institute. Most pure cultures must be maintained in liquid media to keep them viable.

But Winogradsky's most important contribution was to our understanding of chemolithotrophy, a discovery he made while he was a young man. In 1885, Sergei Winogradsky traveled from Russia to Strasbourg to study with the eminent scientist Anton De Bary. At that time Dr. Bary was fascinated by the observation that the filamentous bacterium, *Beggiatoa,* stored sulfur granules in its cells. He likened the storage of sulfur in this bacterium to that of starch in plant cells and wondered about the role of sulfur in the metabolism of these bacteria. De Bary conveyed his enthusiasm with this phenomenon to Winogradsky, who immediately initiated studies on the growth of *Beggiatoa* and the role of sulfur granules in the life cycle of the microorganism. Others had also observed sulfur granules in these bacteria, and the prevailing view was that they were a product of sulfate reduction. The elemental sulfur granules were considered intermediates in the generation of sulfides in swamps and other aquatic habitats. This seemed reasonable inasmuch as *Beggiatoa* was common in sulfide-rich environments. And, furthermore, the growth of this bacterium was stimulated by the addition of sulfate to crude laboratory cultures.

Winogradsky collected filaments of *Beggiatoa* from sulfur springs in the nearby Alps. He noted that the sulfur granules disappeared when he observed filaments in the light microscope over a period of time, even though sulfate was added to the preparations. However, when he bubbled hydrogen sulfide through the filaments, the granules quickly reappeared. Therefore, he reasoned that the sulfur storage was due not to sulfate reduction, but to *sulfide oxidation.* And, moreover, in order to explain the disappearance of the sulfur granules from the cells, he reasoned that they might be further oxidized to sulfate.

To verify that the sulfur granules were being oxidized to sulfate, he placed a barium chloride solution in with the sulfur-containing filaments, and noted that barium sulfate crystals were formed. This was clear evidence that the elemental sulfur granules were oxidized to form sulfuric acid.

From this study, Winogradsky proposed the novel concept that *Beggiatoa* actually obtained energy for growth by the oxidation of inorganic sulfur. He suggested further that the organism utilized carbon dioxide as a source of carbon for growth.

Sergei Winogradsky's studies were a monumental contribution to biology not only because they filled in a missing link of the sulfur cycle, but because they led to our understanding of chemolithotrophy.

Another important filamentous sulfur bacterium is the genus *Thioploca.* It is found in lake sediments and in the sediments of certain marine intertidal zones, such as in the Gulf of Mexico and off the coast of Chile, where it serves as an important food source for marine animals. Pure cultures have not yet been obtained of these bacteria. They resemble *Beggiatoa* spp. in that they produce filaments with sulfur granules; however, in addition, they produce a sheath in which one or more filaments reside (Figure 21.31). In the natural habitat it is not possible to determine whether a given filament from one sheath actually originated there or simply glided there from another

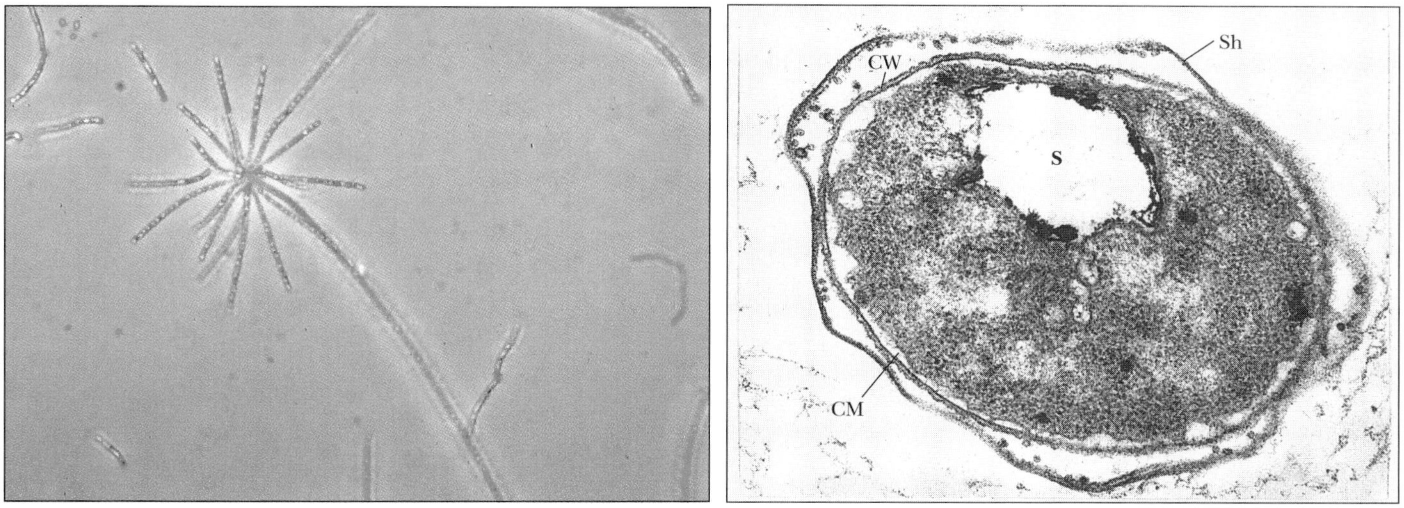

(a) (b)

***Figure* 21.30** **(a)** A rosette of *Thiothrix* from a sulfur spring. Bar = 10 μm. (Courtesy of J. Bland and J. T. Staley) **(b)** An electron micrograph cross-section through a *Thiothrix* cell showing its outer sheath (Sh) as well as the typical gram-negative cell structure including outer cell wall (CW) and cell membrane (CM). Note also the sulfur granules (S), which are bound by a membrane. Bar is 0.5 μm. (Courtesy of J. Bland and J. T. Staley)

adjacent one. One of the unique attributes of this genus is that the cells appear "empty" when they have been examined by thin section in the electron microscope (Figure 21.31**c**). The reason for this is unclear; however, it has been proposed that the organism compensates metabolically for having such large filament diameters by producing vacuoles inside the cells that force the cytoplasm against the outlying membrane. These vacuoles appear as empty spaces inside the cytoplasm of cells observed in thin section. If this is borne out by further experimentation,

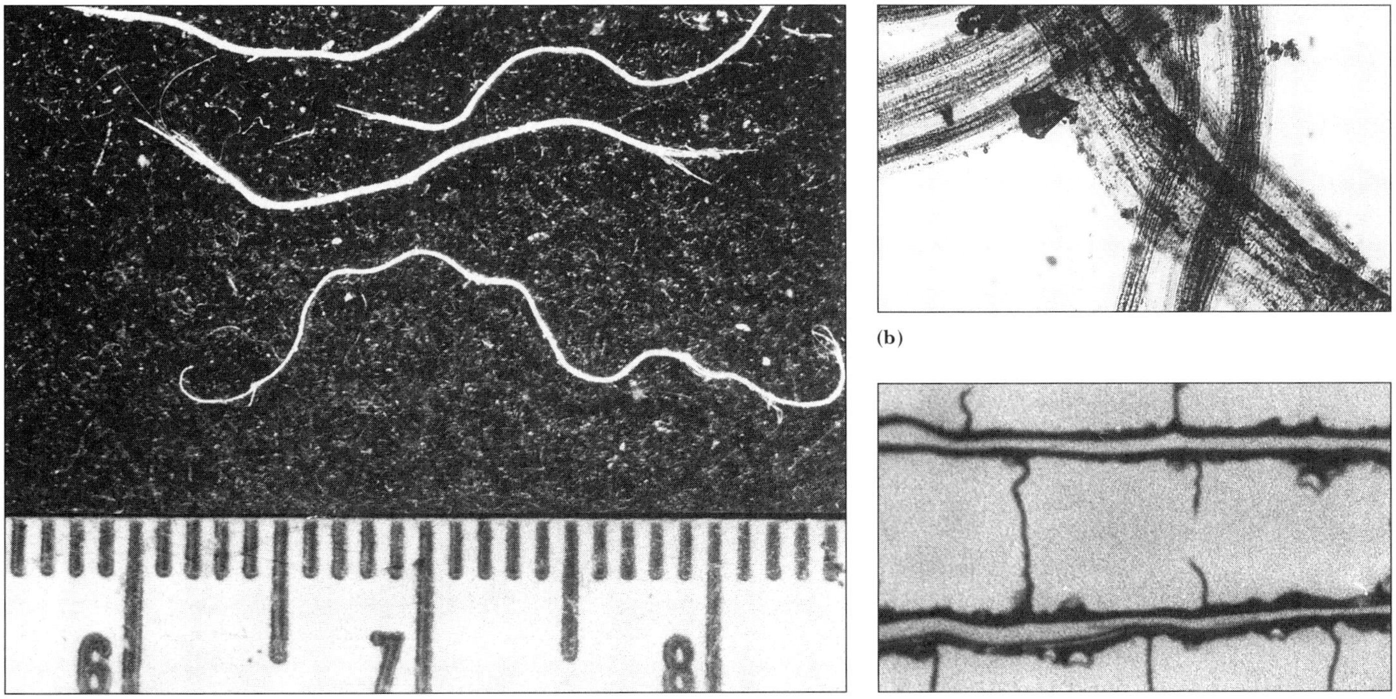

(a) (b) (c)

***Figure* 21.31** **(a)** Sheathed filaments of *Thioploca* washed from marine intertidal sediments. Compare the size of the filaments with the mm scale shown in the foreground. These filaments are large enough to be visible to the unaided eye. **(b)** Ensheathed filaments of *Thioploca* as they appear by brightfield light microscopy. Note that each sheath contains several filaments. Bar is 0.2 mm. (Courtesy of S. Maier) **(c)** Thin section through *Thioploca* filaments showing their empty appearance. Apparently the cytoplasm is held against the cell membrane by internal vacuoles. Bar is 10 μm. (Courtesy of S. Maier)

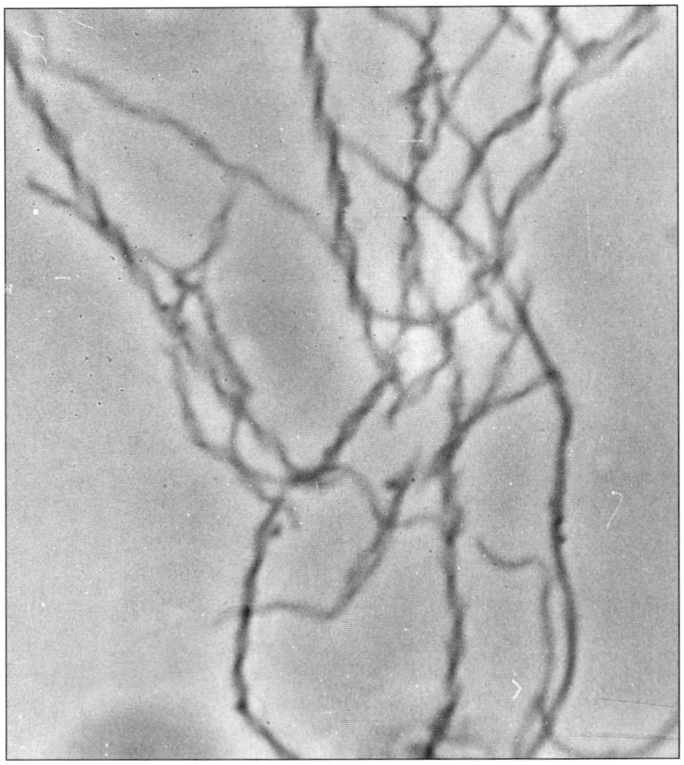

(a)

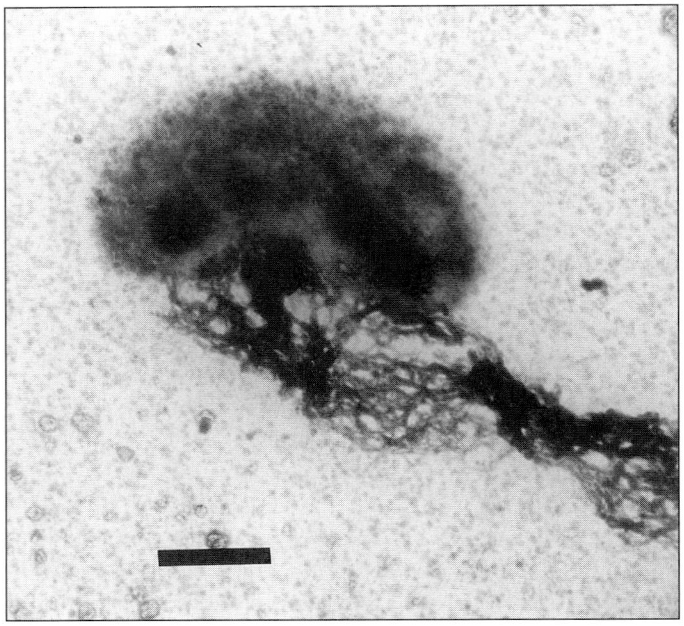

(b)

***Figure* 21.32 (a)** Twisted stalks of *Gallionella ferruginea* as they appear when obtained from natural spring sources. **(b)** Electron micrograph of *Gallionella ferruginea* showing vibrioid appearance of cell and stalk formation from concave side of cell. Bar is 0.1 μm. (Courtesy of H. Hanert)

then it may be a unique mechanism by which an organism enhances the exchange of nutrients in the dilute environments in which it lives.

Iron Oxidizers

Very few known bacteria are able to obtain energy by the oxidation of ferrous iron. One reason is that known iron oxidizers require oxygen for the process. However, since iron oxidation at typical physiological pHs—that is, near neutrality (pH 4 to 8)—occurs spontaneously in the presence of oxygen, the bacteria that carry out this process either live at very low pH conditions or at very low oxygen concentrations where ferrous ion remains reduced. Two different bacteria are known to have one or the other of these characteristics.

Iron oxidation by *Thiobacillus ferrooxidans* occurs at low pH. These bacteria grow in association with acidophilic sulfur oxidizers and are capable of acidophilic sulfur oxidation themselves. In this environment ferrous ion is abundant, and these bacteria have been shown to be able to use it as an energy source (see Figure 8.18).

In contrast, *Gallionella ferruginea* grows at neutral pH in an environment very low in oxygen. This bacterium lives in iron springs where it appears as masses of twisted filaments that look reddish brown from the oxidized iron, FeO(OH). These are the so-called stalks of the bacterium (Figure 21.32**a**). The cell appears as a small vibrio that produces the inorganic stalk of ferric hydroxide from the concave side of the cell (Figure 21.32**b**). As the organism grows, the stalk elongates and twists to form a helix. When the cells divide, the stalk bifurcates. Meanwhile, the cells remain small and almost indiscernible but continue to produce more stalk material as they grow. This iron-oxidizing bacterium is better known for its stalk than for itself because when one observes it in the microscope one primarily sees the massive amounts of iron oxide–encrusted stalks.

Gallionella ferruginea is a chemoautotrophic iron oxidizer that belongs to the alpha Proteobacteria. Although it has been difficult to obtain *Gallionella ferrugenea* in pure culture, some strains have been isolated. Results of physiological studies indicate that this bacterium is a true chemoautotroph because it has the enzymes of the Calvin-Benson cycle.

Hydrogen Bacteria

Hydrogen gas (H_2) is an excellent energy source for bacterial growth. Among the Eubacteria, several types of hydrogen utilizers are found. One group are mesophilic, facultative hydrogen chemolithotrophs that can also use organic carbon sources for growth. Another group, the Aquificales, are thermophilic Eubacteria. A third group also is involved in the oxidation of carbon monoxide. Still another eubacterial group are anaerobic bacteria called acetogenic bacteria because they produce acetic acid from carbon dioxide. Each of these four groups is discussed individually below.

Mesophilic Hydrogen Bacteria

The ability to use hydrogen gas as an energy source is widespread among Eubacteria. At one time all hydrogen bacteria were placed in the genus *"Hydrogenomonas."* However, many of these bacteria were found to be closely related to other heterotrophic bacteria from a variety of different genera (Table 21.14). Furthermore, with the exception of the thermophilic genera *Hydrogenobacter* and *Aquifex,* all aerobic hydrogen bacteria are facultative hydrogen bacteria that can use organic compounds as energy sources as well as hydrogen gas. Therefore, most are facultative chemolithotrophs or mixotrophs, not obligate chemolithotrophs.

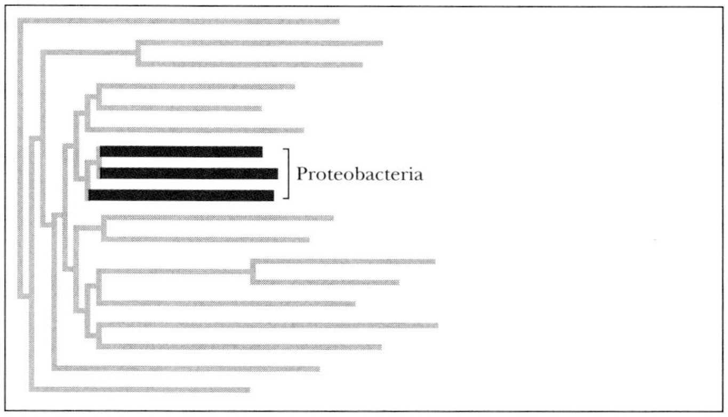

Typical hydrogen oxidizers obtain energy from the oxidation of hydrogen gas using a membrane-bound hydrogenase:

$$H_2 \longrightarrow 2\,H^+ + 2\,e^- \qquad [21.13]$$

The electrons generated are passed through an electron transport chain, and ATP is generated by proton pumping and membrane-bound ATPases. Only a few genes are needed for the hydrogen oxidation process. In some species these have been found on plasmids. This suggests that the ability to generate energy from hydrogen can be genetically transferred from one species to another. We call this ability lateral gene transfer. However, in addition to having a hydrogenase and an electron transport system for energy generation, these bacteria also need to have the enzymes for carbon dioxide fixation in order to grow autotrophically. All of the hydrogen-oxidizing Eubacteria studied so far that can grow on carbon dioxide use the Calvin-Benson cycle for carbon dioxide fixation. One of the most thoroughly studied members of this group is *Alcaligenes eutrophus.*

Carboxydobacteria

One special group of hydrogen bacteria obtains energy from the oxidation of hydrogen and carbon monoxide to form carbon dioxide. These are called **carboxydobacteria.** The biochemistry of the pathway has not been studied

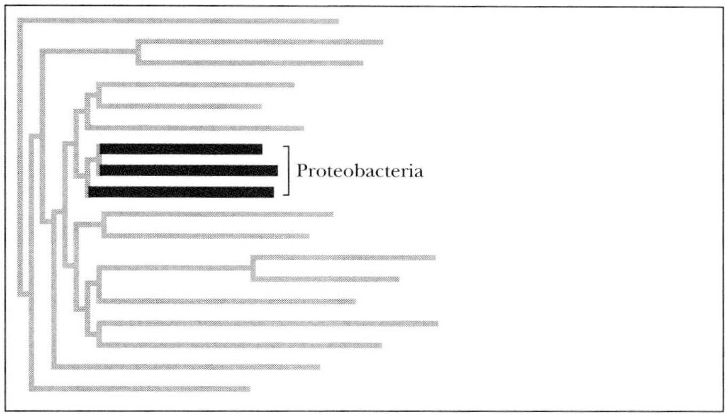

***Table* 21.14 Genera of heterotrophic Eubacteria in which hydrogen autotrophy has been reported**

Gram-Positive Genera	Gram-Negative Genera
Arthrobacter	*Alcaligenes*
Bacillus	*Ancylobacter* (formerly *Microcyclus*)
Mycobacterium	*Aquaspirillum*
Nocardia	*Derxia*
	Flavobacterium
	Hydrogenobacter
	Paracoccus
	Pseudomonas
	Rhizobium
	Spirillum

well, although it is thought that an oxidoreductase is responsible for the following overall reaction:

$$CO + H_2 + 1/2\ O_2 \longrightarrow CO_2 + 2\ H^+ + 2\ e^- \quad [21.14]$$

Some hydrogen-oxidizing bacteria including *Alcaligenes eutrophus,* as well as members of the genera *Azotobacter* and *Hydrogenophaga,* carry out this process. However, not all hydrogen bacteria are capable of carbon monoxide oxidation. *Carboxydomonas* has been described as a distinct genus of carbon monoxide degraders. It is a thermophilic bacterium isolated from hot springs.

Carbon monoxide is produced through the incomplete combustion of wood, coal, and oil and therefore is common in the environment. CO is lethal to humans and other animals because it reacts with hemoglobin irreversibly. Fortunately for animals this diverse group of bacteria appears to be widely distributed in soils and aquatic habitats and is very efficient in degrading CO in the environment. As a result, CO is found in extremely low concentrations in the atmosphere.

Acetogenic Bacteria

Acetogenic bacteria are anaerobic bacteria that can use hydrogen as an energy source and carbon dioxide as a terminal electron acceptor and produce acetic acid as an end product in the following overall reaction:

$$4\ H_2 + 2\ CO_2 \longrightarrow CH_3COOH + 2\ H_2O \quad [21.15]$$

In all acetogens so far studied, carbon dioxide is fixed by the acetyl-CoA pathway (see Chapter 10). Some of the acetate produced when grown autotrophically is then used as a carbon source for the cell.

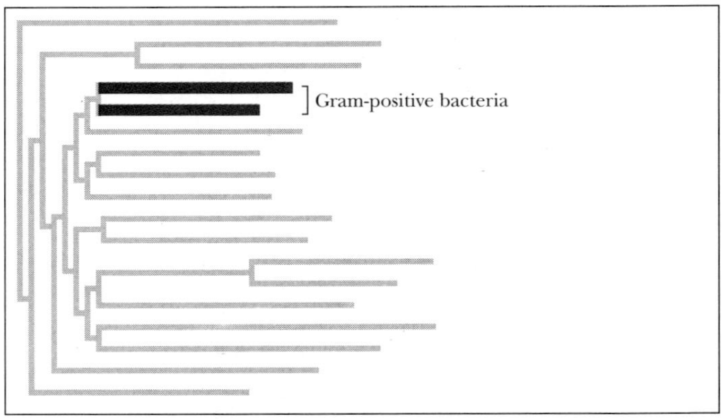

In the acetyl-CoA pathway, one of the carbon dioxide precursor molecules is reduced to carbon monoxide by carbon monoxide dehydrogenase, using one molecule of H_2. This is then covalently complexed with a methyl vitamin B_{12} group derived originally from another carbon dioxide that is reduced initially to a formyl group and then to a methyl group by a tetrahydrofolate (THF)-containing enzyme in the presence of H_2. ATP is required initially in this process. Some ATP is synthesized in the last reaction step in which acetyl-CoA forms acetate (see Figure 10.11 for details).

The bacteria, which are gram-positive, include *Acetobacterium woodii, Acetogenium kivui,* and *Clostridium aceticum.* All of them can also grow heterotrophically and produce acetate through sugar fermentations using the glycolytic pathway. In this process, pyruvate is first produced as the precursor in the formation of acetic acid and carbon dioxide.

Thermophilic Hydrogen Bacteria

This group comprises the order Aquificales and contains the thermophilic genera *Hydrogenobacter* and *Aquifex,* which are thermophilic bacteria that grow in hot springs. *Hydrogenobacter* strains grow with an optimum temperature of 70°C to 75°C. Some strains of *Aquifex* grow at temperatures as high as 95°C, the highest temperature so far known for Eubacteria. These bacteria form their own branch, which comprises the deepest branch of the Eubacteria.

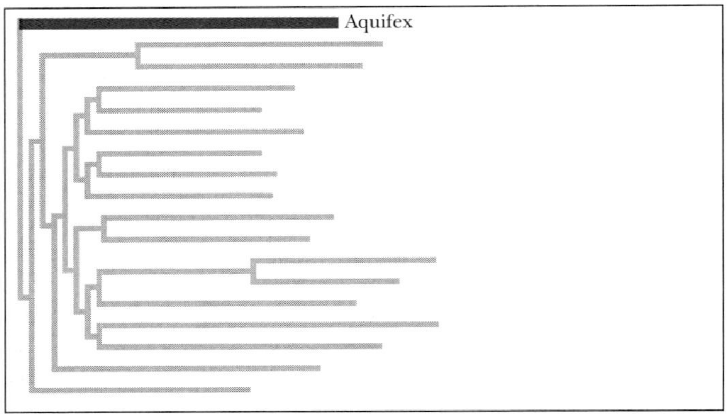

None of the species in this group grow on organic carbon sources. So, unlike all previously described hydrogen bacteria, these are the only ones that can be regarded as obligatory hydrogen chemolithotrophs. All known species require oxygen. This is surprising in the sense that these are the most deeply branched of the Eubacteria, and yet they require oxygen, which was presumably absent in the early biosphere of Earth (see Chapter 18). Part of the explanation may lie in the fact that these bacteria require oxygen in extremely low concentrations (see "A Conversation with Carl Woese"). Carbon dioxide is fixed via the reductive tricarboxylic acid cycle, not the Calvin-Benson cycle.

Methylotrophic Eubacteria

Methane and methanol are considered by some to be two of the simplest organic compounds. However, it could be

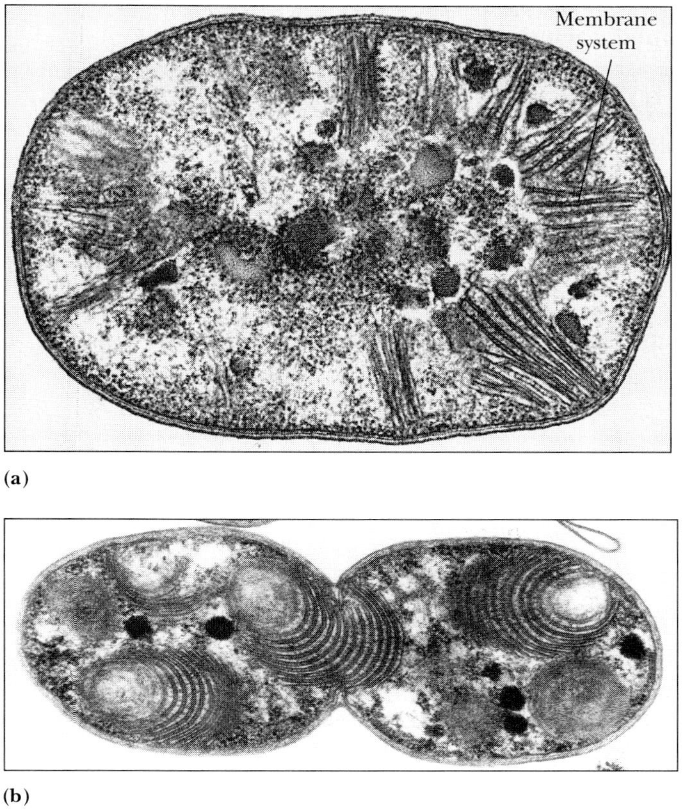

***Figure* 21.33** Thin sections through Type I methanotrophs, **(a)** *Methylomonas agile* and **(b)** *Methylomonas methanica* showing the typical transverse membrane systems. (Courtesy of C. Murrell)

argued that true organic compounds have carbon–carbon bonds. Of course, microorganisms are not mindful of human definitions. The methylotrophic bacteria treat methane and methanol from a purely physical and biochemical standpoint—as substrates that can provide energy for growth. **Methanotrophic bacteria** use methane as a carbon source and oxidize it to carbon dioxide. These bacteria can also use methanol and a variety of other one-carbon and a few two-carbon compounds as carbon sources. Another separate group of bacteria use methanol, but not methane gas, as a carbon source for growth. Therefore, these are not methanotrophic bacteria, but are instead called **methylotrophic bacteria.** It should be noted that methanotrophic bacteria are also methylotrophic because they can also use methanol and other one-carbon compounds for growth. For this reason this section is entitled the methylotrophic bacteria.

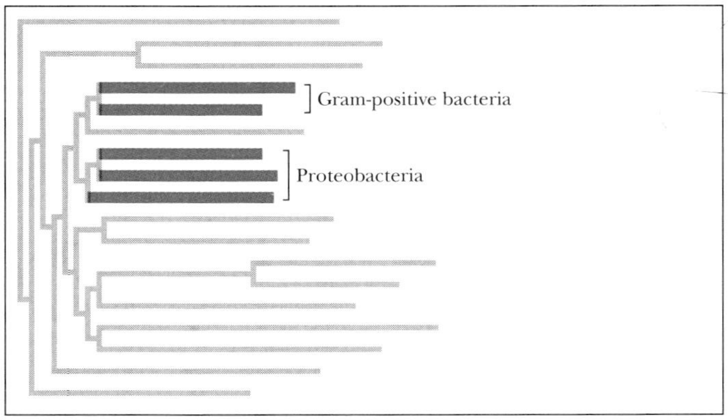

Methanotrophs

There are two groups of methane-oxidizing bacteria, both of which have internal membrane systems similar to those of the nitrifying bacteria. The **Type I methanotrophs** have a system of internal membranes that lie perpendicular to the long axis of the cell (Figure 21.33), whereas the **Type II methanotrophs** have membranes that lie parallel to the cell membrane (Figure 21.34). The genera that have been described are listed in Table 21.15. The Type I methanotrophs are members of the gamma Proteobacteria, whereas the Type II methanotrophs belong to the alpha Proteobacteria.

Some of the methanotrophs produce a resting stage, called cysts or exospores. The cysts are analogous to those produced by *Azotobacter* spp. in that they have complex cell wall layers and may store poly β-hydroxybutyrate. They are formed during conditions of nutrient depletion and can germinate during periods of nutrient sufficiency.

Energy Generation from Methane

The methanotrophs are all gram-negative obligate aerobes that derive energy from the oxidation of methane to carbon dioxide. The first reaction is unique to these bacteria

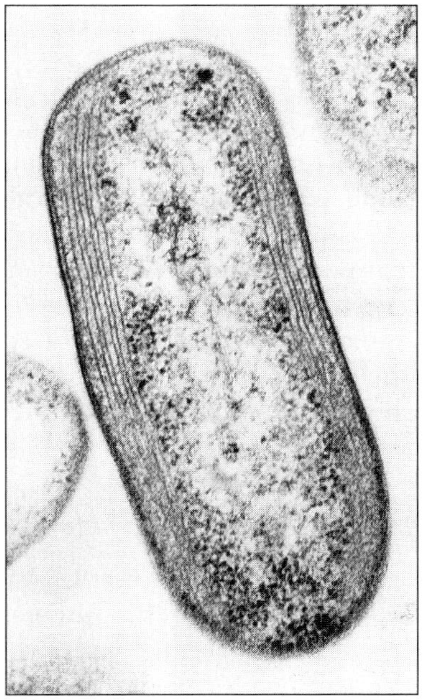

***Figure* 21.34** Thin section of a Type II methanotroph, *Methylocystis parvus,* showing the membranes lying along the periphery of the cell. (Courtesy of C. Murrell)

Differential *Table* **21.15 • Genera of methanotrophic bacteria**

Group	Genus	Mol % G + C	Special Features
Type I Methanotrophs			
	Methylobacter	50–54	Rod, polar flagellum, cyst
	Methylococcus	62–64	Coccus, nonmotile, cyst
	Methylomonas	50–54	Rod, polar flagellum, cyst
Type II Methanotrophs			
	Methylocystis	62–63	Rod or vibrio, nonmotile
	Methylosinus	62–63	Curved rod, polar flagellum, exospore

and involves an enzyme called methane monooxygenase (MMO), which converts methane to methanol as shown below:

$$CH_4 + O_2 + XH_2 \longrightarrow CH_3OH + H_2O + X \quad [21.16]$$

where X = hydrogen carrier (cytochrome carrying electrons). The MMO is a complex copper-containing enzyme that can be either membrane-bound or free in the cytoplasm. The next reaction involves a methanol dehydrogenase that converts the methanol to formaldehyde as follows:

$$CH_3OH \longrightarrow HCHO + 2\ e^- + 2\ H^+ \quad [21.17]$$

The final reactions involved in carbon dissimilation for energy generation include either an oxidation via formate to carbon dioxide (see Chapter 10), or a more complex route to carbon dioxide called the ribulose monophosphate (RMP) pathway (Chapter 10). Many methanotrophs use the ribulose monophosphate pathway for energy generation but some of these can also use the serine pathway. Some of the species using the serine pathway have a distinctive quinone, pyrrolo-quinoline quinone (PQQ), as a hydrogen acceptor.

Carbon Assimilation by Methanotrophic Bacteria

To assimilate carbon, the Type I methanotrophs use the RMP pathway (see Figure 10.12**b**), whereas the Type II methanotrophs use the serine pathway (see Figure 10.12**a**). In both pathways formaldehyde serves as the substrate for organic carbon synthesis. Thus, they are not autotrophs in that they cannot use carbon dioxide as their sole source of carbon.

Ecological and Environmental Importance of Methanotrophic Bacteria

The methanotrophic bacteria are very important environmentally. They are responsible for the oxidation of methane produced by methanogenesis and natural gas seeps in the biosphere. Methane is a greenhouse gas that has been increasing in the atmosphere at a rate of about 1 percent per year during the past decade. If it were not for the activity of these bacteria, the gas would increase at much more rapid rates, with deleterious environmental effects. These bacteria reside in soil and aquatic habitats where oxygen, as well as methane, is available. Therefore they live at the interface between the reduced zones of the environment in which methane is produced and oxygenated zones exposed to air. Although methane gas escapes in part from shallow ponds and marshes, in deeper water bodies it is largely oxidized by the activity of methanotrophs before it reaches the surface of the water column. Methane also reaches the atmosphere through the activities of termites as well as ruminant animals that harbor methanogenic bacteria.

The MMO of methanotrophs can also remove halogen ions from toxic halogenated compounds. This process is called dehalogenation, or dechlorination if the compound is chlorinated. Brominated compounds can be similarly dehalogenated. The enzyme is most effective at dehalogenating low molecular weight, one- and two-carbon, aliphatic compounds such as chloroform and trichloroethene. The methanotrophs do not derive energy from these dechlorination reactions. This is an example of **cometabolism** (see Chapter 24) in that methane must be provided in order for the methanotrophs to carry out the dechlorination process. They cannot obtain energy from the dehalogenation reactions. These bacteria are important agents in dehalogenation of wastes containing low molecular weight halogenated compounds and are therefore of considerable importance in bioremediation (see Chapter 33).

Table 21.16 **Nonmethanotrophic methylotrophic eubacterial genera**

Gram-Positive Genera	Gram-Negative Genera
Arthrobacter	*Acinetobacter*
Bacillus	*Alcaligenes*
Mycobacterium	*Ancylobacter* (formerly *Microcyclus*)
Streptomyces	*Hyphomicrobium*
	Klebsiella
	Paracoccus
	Pseudomonas
	Rhodopseudomonas

The methanotrophic bacteria share similarities with ammonia-oxidizing bacteria. Members of both groups have been shown to be able to carry out the oxidation of each others' substrates, ammonia and methane, using their characteristic monooxygenases. In addition, both are members of the Proteobacteria, and therefore phylogenetically related to one another. However, neither group obtains energy from the oxidation of the other's substrate, and therefore are not considered to be ecologically important in the other's niche.

Other Methylotrophs

As mentioned heretofore, the term **methylotroph** refers to a microorganism that can utilize one-carbon compounds such as methane, methanol, or methylamine, as sole carbon and energy sources for growth. This is a general term that encompasses methanotrophs discussed above as well as bacteria such as *Hyphomicrobium* spp. that can grow on methanol as sole carbon source. A list of heterotrophic genera that are known to be methanol or methylamine users is provided in Table 21.16. All of these bacteria that have been studied use the serine pathway for carbon assimilation and grow by aerobic or nitrate respiration. It should be emphasized that none of them are known to use methane gas and none of them produce the enzyme methane monooxygenase. Also, all of them are facultative methylotrophs that can use other organic compounds for growth including sugars, amino acids, organic acids, and the like, depending on the taxon.

One interesting newly discovered habitat for methylotrophic bacteria is on the surface of leaves. Methanol is derived from the degradation of methoxylated compounds produced by the plant. Strains especially adapted to growth on the plant leaf surface use the volatile methanol that is released. These strains are all pink in pigmentation.

Summary

- There are several major phylogenetic groups of photosynthetic Eubacteria including the **purple bacteria,** the **green bacteria,** the **heliobacteria,** and the **cyanobacteria.**

- The **purple photosynthetic bacteria,** which comprise two groups, the **purple sulfur** bacteria and the **purple nonsulfur** bacteria, carry out **anoxygenic photosynthesis,** in which oxygen is not produced and either reduced sulfur compounds or organic compounds are used as electron donors for carbon dioxide fixation, which occurs anaerobically. The purple bacteria, like other **Proteobacteria,** fix carbon dioxide by the Calvin-Benson cycle. The coloration of the purple bacteria is determined by **carotenoid pigments** which mask the **bacteriochlorophyll *a*** that is found in the reaction center where **photophosphorylation** occurs.

- The **green photosynthetic bacteria** have a special structure called the **chlorosome,** which contains the photosynthetic pigments and reaction center. In addition to Bchl *a*, they produce other green or brown bacteriochlorophylls, Bchl *c*, Bchl *d*, and Bchl *e*. The green sulfur bacteria use the **reductive TCA cycle** for carbon dioxide fixation. The **consortium species** consist of two separate organisms, a green sulfur bacterium and a heterotroph, living in close symbiotic association. *Chloroflexus aurantiacus,* a green filamentous bacterium, has its own pathway for carbon dioxide fixation, termed the **hydroxypropionate pathway.**

- The **heliobacteria** are members of the gram-positive bacteria and are photoheterotrophic. Some form endospores.

- There are two groups of oxygenic photosynthetic bacteria, the **cyanobacteria** and the **prochlorophytes.** The cyanobacteria have Photosystem I and Photosystem II and carry out the water-splitting reaction to form oxygen from water. The accessory photopigments of the cyanobacteria are phycobilins, **phycocyanin** and **phycoerythrin.** These are located in phycobilisomes, which are structures on the photosynthetic lamellae termed **thylakoids**. Some cyanobacteria store an amino acid polymer called **cyanophycin**, which is composed of repeating units of aspartic acid and arginine.

- The multicellular structure typical of many filamentous cyanobacteria is termed a **trichome.** Most cyanobacteria that are nitrogen-fixing carry out the process in a specialized modified vegetative cell called the **heterocyst.**

- Several groups of chemolithotrophic bacteria exist including the **nitrifiers,** the **sulfur oxidizers,** the **iron oxidizers,** and the **hydrogen bacteria.** The **nitrifying bacteria** comprise two separate groups, the **ammonia oxidizers** and the **nitrite oxidizers.** The **sulfur-oxidizing Eubacteria** also consist of two groups, the **unicellular** species, such as *Thiobacillus* and *Thiomicrospira,* and the **filamentous** species, such as *Beggiatoa.* Many sulfur-oxidizing bacteria grow as **facultative chemolithotrophs** or **mixotrophs** by requiring or using various organic carbon sources.

- Only two genera of **hydrogen bacteria** are known to grow as chemolithotrophs. These are the thermophilic genera, *Hydrogenobacter* and *Aquifex,* which comprise a deeply rooted phylogenetic group. **Acetogenic bacteria** are gram-positive anaerobes that can use hydrogen as an energy source and fix carbon dioxide by the acetyl-CoA pathway. **Carboxydobacteria** are a special group of hydrogen bacteria that can oxidize hydrogen and carbon monoxide.

- **Methylotrophic bacteria** use one-carbon compounds such as methanol or methane as a sole carbon and energy source for growth. **Methanotrophic bacteria** can use methane as a sole source of carbon; they comprise two groups, the Type I and Type II methanotrophs, which have internal membranes where the enzyme, methane monooxygenase, resides. Type I methanotrophs use the ribulose monophosphate pathway for carbon assimilation. Type II methanotrophs use the serine pathway for carbon assimilation. Methane-oxidizing bacteria are all aerobic and oxidize methane produced by methanogens in natural environments.

Questions for Thought and Review

1. Distinguish between the terms chemoautotroph, chemolithoautotroph, photoautotroph, and mixotroph.
2. What are the various carbon dioxide fixation pathways used by autotrophic Eubacteria? Which organisms use which pathways?
3. What distinguishes anoxygenic from oxygenic photosynthetic bacteria? What groups fall into each category?
4. What is the evidence that prochlorophytes are or are not intermediates in the evolution of higher algae?
5. Compare the metabolism of *Escherichia coli, Methylococcus* spp., *Beggiatoa* spp., and *Chromatium okenii.* How do you explain that they are all members of the gamma Proteobacteria?
6. Identify the major groups of chemolithotrophic bacteria, indicating their phylogenetic group, their source of energy, their principal products of metabolism, and their carbon dioxide fixation pathway.
7. Compare the morphological and metabolic diversity of the various cyanobacterial orders and the Prochlorales.
8. Can you identify a bacterium discussed in this chapter that can grow as either a typical heterotroph, a hydrogen bacterium, or a methylotrophic bacterium? Why do you think it would have such diverse capabilities?
9. What is an acetogenic bacterium? A hydrogen bacterium?
10. How are autotrophic bacteria more diverse metabolically than higher autotrophic organisms? Why?

Suggested Readings

Bergey's Manual of Determinative Bacteriology. 9th ed. 1993 (J. G. Holt, Editor-in-Chief). Baltimore: Williams and Wilkins.

Bergey's Manual of Systematic Bacteriology. 1st ed. 1989. Volume III (J. T. Staley, N. Pfennig, and M. Bryant, eds.). Baltimore: Williams and Wilkins.

The Prokaryotes, 2nd ed. 1992. (A. Balows, H. G. Trüper, M. Dworkin, W. Harder, and K. H. Schleifer, eds.). Berlin: Springer-Verlag.

The two ultimate lines of bacterial descent are no more closely related to one another than either is to the cytoplasmic aspect of the eukaryotic cell.

Science 209 (1980): 457–463.

Chapter 22

Archaea

When Carl Woese and his collaborators at the University of Illinois began to study the phylogenetic relationships among bacteria during the mid-1970s, they discovered that one group of bacteria was very different from all other bacteria. By analyzing RNA sequences, they found that the 16S rRNAs of these bacteria were no more closely related to the eubacterial rRNA than to the 18S rRNAs of eukaryotes. They call these different bacteria Archaea because they appeared to resemble the primitive bacteria believed to exist in Archaean times 3 to 4 billion years ago. This discovery piqued the interest of microbiologists around the world. After much debate, microbiologists now generally agree that the Archaea are very different from other organisms. In recognition of this fact, they currently are classified as **Archaea,** one of the three primary Domains of organisms. Their discovery is of great importance to the study of microbiology.

Why Study Archaea?

Microbiologists studying the Archaea contribute to our basic knowledge in three areas. First, Archaea offer a valuable model system because some of their features closely resemble those of the eukaryotes. Their simpler cell structure and smaller size may prove to be a valuable research tool in understanding the function of eukaryotic-type RNA polymerase, RNA splicing, and other processes common to both domains.

Second, many Archaea are extremophiles, and they flourish under conditions that are lethal to most Eubacteria and eukaryotes. Thus, studies involving the extremely thermophilic and extremely halophilic Archaea may lead to an understanding of mechanisms whereby microorganisms can grow in these harsh environments. A question of special interest is how the macromolecules of these

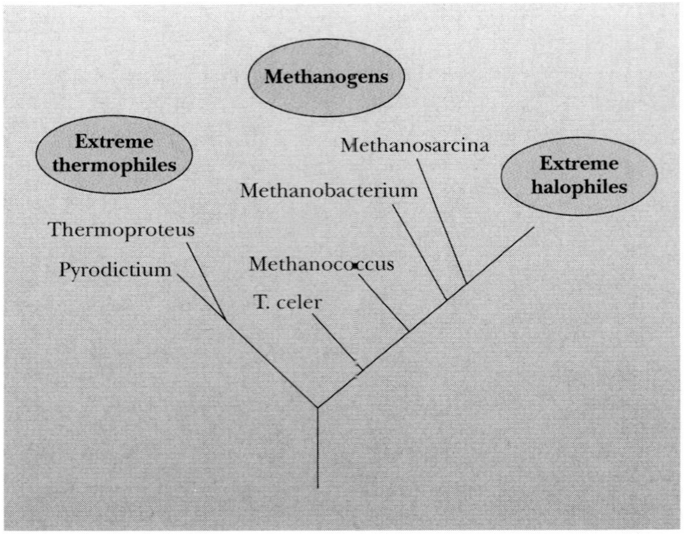

extremophiles maintain their structure and activity under conditions that denature the macromolecules of most organisms. For instance, the optimal growth temperature of many extremely thermophilic Archaea is above the melting temperature of their DNA. The mechanism by which these organisms maintain a functional chromosome is not fully understood. This question is also of practical importance for the design of commercially valuable thermostable enzymes.

Third, the discovery and subsequent study of Archaea offer valuable insights into the nature of the earliest organisms. Although most biochemical features of organisms are absent in the fossil record, we can deduce these properties from modern organisms such as Archaea. By studying modern organisms, we also know that form changes rapidly throughout evolution but that basic biochemical processes are inherited by even very remote ancestors. There still is much debate over whether Archaea are "primitive" organisms or whether they are as "evolved" as modern Eubacteria, whose line of descent separated early. In any case, Archaea figure prominently in models on the origin of life.

Models on the Origin of Life

The importance of the discovery of the Archaea to the study of the origin of life may be illustrated by the following example. First, consider the theory that all modern organisms evolved from two lines of descent, a proposition that we know to be false (Figure 22.1, model A). Properties that are the same in both lines of descent, such as the genetic code, were probably inherited from the common ancestor; however, for other properties, such as the structure of the ribosome, we may conclude very little. For instance, did the 80S eukaryotic ribosome evolve by the addition of complexity to the simpler 70S eubacterial ribosome, or did the 70S ribosome evolve by subtraction from, and a more efficient construction of, the 80S ribosome? Either scenario is plausible and this problem cannot be solved without additional evidence.

Now, consider that all living organisms evolved from three lines of descent that arose very early in life's history, a proposition that many scientists believe to be true. The location of the origin of life is not known in this scenario, but it is probably close to the point where all three lines intersect and on one of the three branches (that is, not at the center). This scenario may be represented by the three models B, C, and D (Figure 22.1). Only one of these can be correct, but the correct one has not been determined with certainty. Again, properties that are common to all three lines of descent, such as the genetic code, were probably inherited from the common ancestor. But for properties that are different, scientists can draw firmer conclusions. For example, both the Archaea and the Eubacteria possess 70S ribosomes, and eukaryotes possess 80S ribosomes. (The chloroplastic and mitochondrial ribosomes are of relatively recent origin and are not included in this discussion; see Chapter 23.) Did the universal ancestor contain 70S or 80S ribosomes? If models B and C are correct, then it is very likely that the univer-

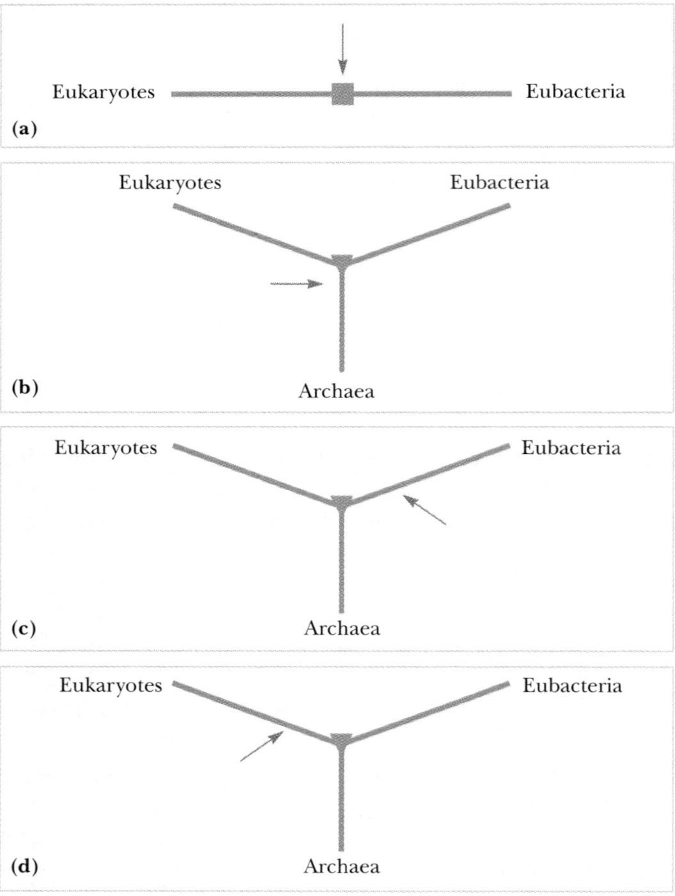

***Figure* 22.1** Possible models for ancient evolution. The arrows indicate the origin of life in each model.

sal ancestor contained a 70S ribosome. If model D is correct, then it is likely that the ancestor of the Eubacteria and the Archaea contained a 70S ribosome. However, the properties of the common ancestor with the eukaryotes have not yet been determined.

Variability of Archaea

In addition to being very different from other organisms, the Archaea are very different from one another. In fact, the variability within the Archaea is probably on the same order of magnitude as that found in the Eubacteria, even though many fewer Archaea have been identified. Two explanations for this current distribution seem plausible. One, until recently, few investigators have systematically tried to isolate Archaea. Because little effort has been expended in this area, few have been found. Two, the Archaea appear to be limited to extreme environments or to niches for which there is little direct competition with the Eubacteria. It is possible that the Archaea are unable to compete successfully in the temperate environments extant on most of the present-day Earth. Their numbers may be small because they may have been restricted to a few habitats. However, their variability may be large because they are an ancient line of descent.

Biochemical Differences and Similarities to Other Organisms

The Archaea are composed of three major phenotypic groups. These include the methane-producing bacteria, the extreme halophiles, and the extreme thermophiles. Although the physiology of these bacteria differs, some

Differential *Table* **22.1 Comparison of Archaea to Eubacteria and Eukaryotes**

	Archaea	Eubacteria	Eukaryotes
Typical Organisms	Methane-producing bacteria, halobacteria, extreme thermophiles	Enteric bacteria, cyanobacteria, *Bacillus,* etc.	Fungi, plants, animals, algae, protozoa
Typical Size (diameter)	0.5–4 μm	0.5–4 μm	>5 μm
Physiological Features	Aerobic and anaerobic, extremely thermophilic and halophilic	Aerobic and anaerobic, photosynthetic	Largely aerobic, photosynthetic
Genetic Material	Small circular chromosome, plasmids and viruses, histone-like proteins present	Small circular chromosome, plasmids and viruses, histones absent	Complex nucleus with more than one large linear chromosome, viruses, histones present
Differentiation	Frequently unicellular, cellular differentiation infrequent	Frequently unicellular, cellular differentiation infrequent	Unicellular and multicellular, cellular differentiation common
Cell Wall	Protein, glycoprotein, pseudomurein, wall-less	Murein and LPS, protein and wall-less forms rare	Great variety, peptidoglycan absent
Cytoplasmic Membrane	Glycerol ethers of isoprenoids, site of energy biosynthesis	Glycerol esters of fatty acids, site of energy biosynthesis	Glycerol esters of fatty acids, sterols common
Intracytoplasmic Membranes	Generally absent	Generally absent; when present they frequently contain large amounts of protein	Common in organelles like mitochondria and chloroplasts, nucleus, Golgi apparatus, endoplasmic reticulum and vacuoles, site of energy biosynthesis
Protein Synthesis	70S ribosome, insensitive to chloramphenicol and cycloheximide, diphthamide present in elongation factor	70S ribosome sensitive to chloramphenicol, insensitive to cycloheximide, diphthamide absent in elongation factor	80S and 70S (organelle) ribosomes, insensitive to chloramphenicol (80S), sensitive to cycloheximide (80S), diphthamide present in elongation factor
Locomotion	Simple flagella, gas vesicles	Simple flagella, gliding, gas vesicles	Complex flagella, cilia, legs, fins, wings
RNA Polymerase	Complex	Simple	Complex

features are common to most if not all Archaea (Table 22.1). The cell membranes are composed of isoprenoid-based glycerol lipids. Murein is absent in their cell walls, and it is usually replaced by a protein envelope. The enzyme DNA-dependent RNA polymerase, which copies DNA to form messenger RNA, is different from the eubacterial type. Lastly, Archaea are insensitive to many common antibiotics that are potent inhibitors of Eubacteria or eukaryotes.

Archaeal Membranes

Like the membrane lipids of the Eubacteria, the archaeal lipids are composed of hydrophobic side-chains linked to glycerol. However, many of the details are different (Figure 22.2). In place of fatty acids joined by an ester linkage to glycerol, most archaeal lipids are composed of isoprenoid side-chains joined by an ether linkage to glycerol. Isoprenoids are branched-chain alkyl polymers based upon a five-carbon unit synthesized from mevalonate. In Eubacteria and eukaryotes, they are found in the side-chains of quinones and chlorophyll, intermediates in the biosynthesis of sterols like cholesterol, and in rubber. They are never major components of glycerol lipids. Although branched-chain fatty acids are sometimes present in eubacterial lipids, they are not synthesized from mevalonate. Instead, they are generally synthesized from the branched-chain amino acids or volatile fatty acids. In Archaea, the ether linkage of the isoprenoid side-chains to glycerol is also distinctive. The substituted glycerol contains one asymmetrical carbon. Therefore, it exists in two stereoisomeric forms (Figure 22.3). In the Archaea, only the stereoisomer called 2,3-*sn* glycerol is present. The glycerol lipids of the Eubacteria and the eukaryotes contain only the other stereoisomer of glycerol, 1,2-*sn* glycerol. Although a few Eubacteria contain ether-linked fatty acids, they have the typical eubacterial stereochemistry. Therefore, the reactions involved in the biosynthesis of these ether linkages may be very different in the Eubacteria and the Archaea.

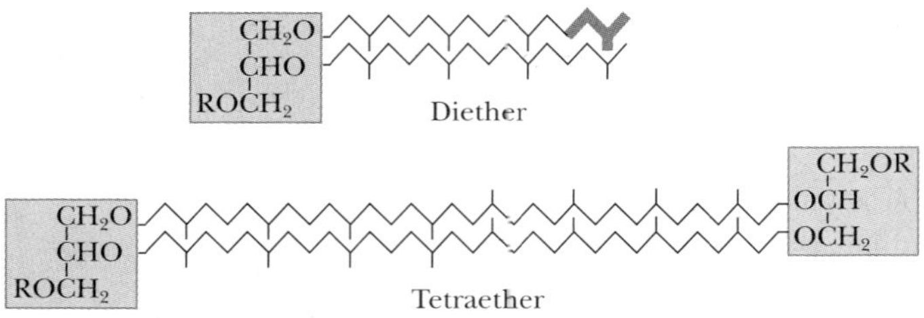

***Figure* 22.2** Typical archaeal glycerol lipids contain isoprenoid side-chains linked to glycerol by an ether bond. A single C-5 isoprenoid unit is shaded in the diether. The tetraether is believed to span the membrane. One glycerol moiety is on the cytoplasmic side of the membrane and the second is on the periplasmic side.

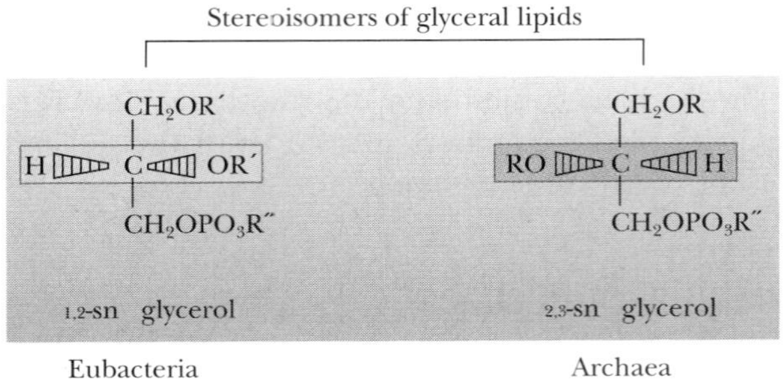

***Figure* 22.3** Stereochemistry of the glycerol lipids in the Eubacteria (1,2-*sn* glycerol) and the Archaea (2,3-*sn* glycerol). R = ether-linked isoprenoid side-chains, R′ = ester-linked fatty acids, and R″ = hydrophilic amino acids or sugars.

Archaeal Cell Walls

The archaeal cell walls differ from those in the gram-negative and gram-positive Eubacteria. Cell walls in Archaea that are composed of murein have not been observed. Instead, the cell walls are generally composed of S-layers: protein subunits arranged in a regular array on the cell surface. Frequently, polysaccharides are also associated with the envelope, although in most cases their structures have not yet been elucidated. The S-layers from many Archaea are very sensitive to detergents such as SDS (sodium dodecyl sulfate), which solubilizes proteins. The cells of these Archaea also lyse rapidly in solutions containing low concentrations of detergent. Although S-layers are common in Eubacteria, they are usually a minor cell wall component found outside the murein cell wall. Some exceptions to this generality are the budding Eubacteria of the genera *Planctomyces* and the thermophilic aerobe *Thermomicrobium roseum*. In these Eubacteria, the S-layer is the major cell wall component.

The cell envelopes of some Archaea are similar in composition to the cell envelopes of certain Eubacteria. *Thermoplasma* is an archaeum that lacks a cell wall, as do the eubacterial mycoplasmas. Some methanogens, such as *Methanobacterium*, contain a cell wall polymer called pseudomurein, which is strikingly similar to murein, the peptidoglycan in Eubacteria. Pseudomurein is composed of polysaccharides crosslinked by amino acids much like murein (Figure 22.4). However, the polysaccharide is composed of N-acetylglucosamine (or N-acetylgalactosamine) and N-acetyltalosaminuronic acid. Muramic acid, the common saccharide component of murein, is not present, nor are D-amino acids, which are common in eubacterial peptidoglycan. In other respects, many of the chemical and physical properties of pseudomurein and murein are equivalent. Both are resistant to proteases, or enzymes that hydrolyze peptide bonds. Both provide the cell with a rigid sacculus. In addition, many of the Archaea that contain pseudomurein also stain gram-positive.

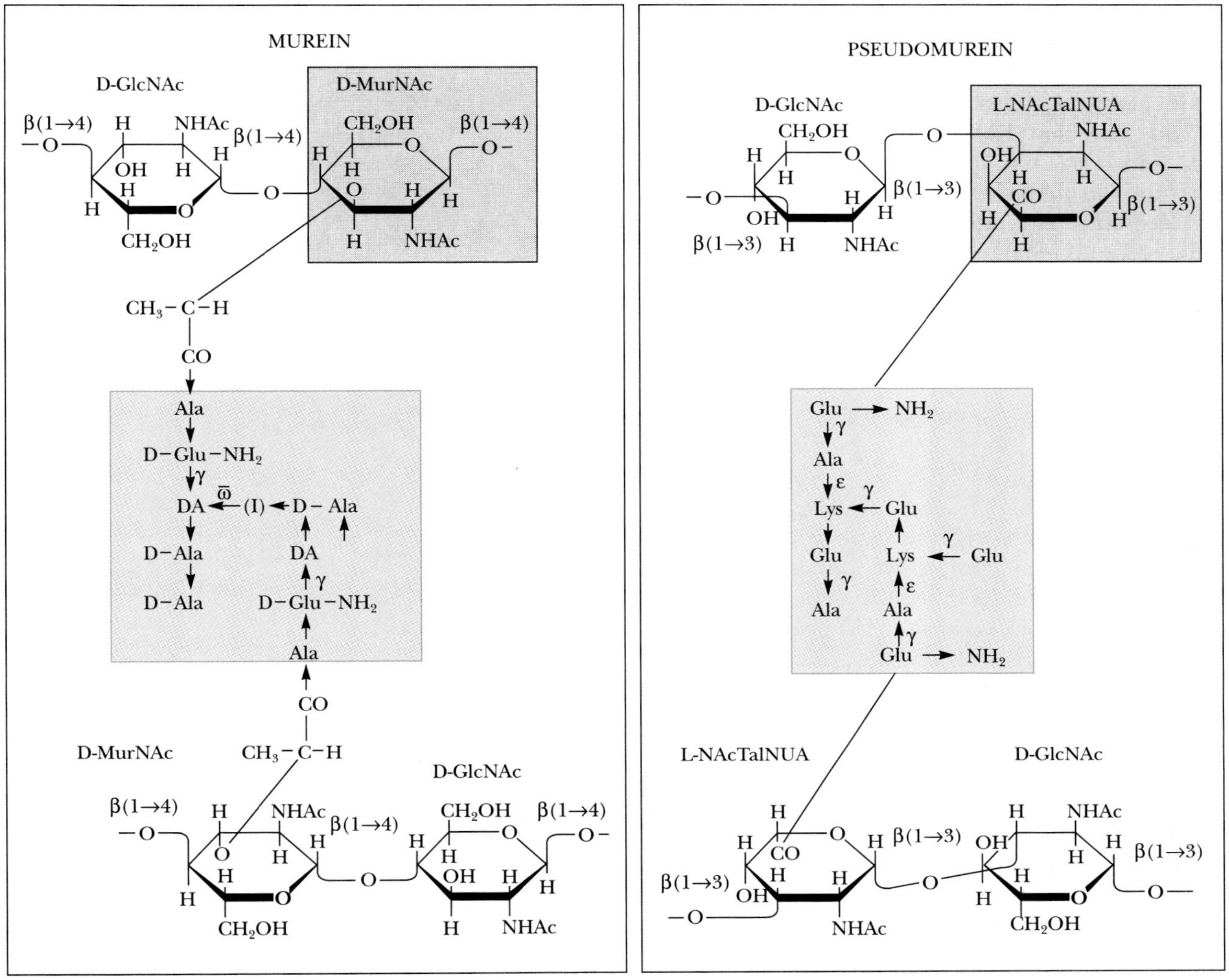

***Figure* 22.4** Comparison of the structure of the peptidoglycan (murein) in Eubacteria to the structure of pseudomurein in the Archaeum *Methanobacterium*. Major differences include the substitution of *N*-acetyl-*L*-talosaminuronic acid (*L*-NAc TalNUA) for *N*-acetyl-*D*-muramic acid (D-MurNAc) (yellow shading) and *L*-amino acids for *D*-amino acids (pink shading). Note also the differences in the bonds in the polysaccharide chain. *N*-acetyl-*D*-glucosamine (D-GlcNAc) is present in both polymers. DA is a diamino acid like diaminopimelic acid, lysine, or ornithine. (Modified from O. Kandler and H. König, 1985, *Cell Envelopes of Archaeobacteria.* In C. R. Woese and R. S. Wolfe, ed., *The Bacteria,* Vol. 8, pp. 413–457.)

Archaeal RNA Polymerase

The enzyme DNA-dependent RNA polymerase synthesizes messenger RNA, which is complementary to the DNA sequence of genes. Because this function is essential to all living cells, this enzyme is considered to be quite ancient. The eubacterial enzyme is relatively simple in structure, and consists of a core enzyme with three subunits, β, β', and α. Binding of the core enzyme to DNA also requires an additional subunit, called the sigma factor (σ) (see Chapter 13). In contrast, the eukaryotic polymerase is much more complex and contains 9 to 12 subunits. The archaeal enzyme more closely resembles the eukaryotic polymerase than the eubacterial enzyme. It has a complex subunit structure, and the amino acid sequence of some of the subunits closely resembles that of the eukaryotic enzyme.

Antibiotic Sensitivity of Archaea

Because the biochemistry of the Archaea is unlike that of other organisms (Table 22.1), it might be expected that the sensitivity to antibiotics would also be different. This is true. For instance, because the Archaea do not contain murein, they are not sensitive to most antibiotics that inhibit eubacterial cell wall synthesis. Thus, most Archaea are not affected by very high concentrations of penicillin, cycloserine, vancomycin, or cephalosporin, all of which are inhibitors of murein synthesis. Because of this selectivity, these antibiotics are also useful additions to enrichment cultures for archaeal strains. Likewise, the DNA-dependent RNA polymerase from the Archaea is not inhibited by rifampicin, which inhibits the eubacterial enzyme at low concentrations. Protein synthesis is also not affected by the common antibiotics chloramphenicol, cycloheximide, and streptomycin, although neomycin is inhibitory at high concentrations. Tetracycline is also a weak inhibitor even though it inhibits protein synthesis in both Eubacteria and eukaryotes. Thus the structure of the archaeal ribosome is quite different from the eubacterial and eukaryotic ribosome.

Similarities to Other Microorganisms

Equally interesting are the similarities between the Archaea and other microorganisms. For instance, even though the difference is more pronounced between the archaeal and eubacterial 16S rRNAs than between any two eubacterial 16S RNAs, archaeal and eubacterial 16S rRNAs are still identical at 60 percent of their positions. Similarly, the genetic code is essentially the same in all organisms, many of the major anabolic pathways are equivalent, and many of the same coenzymes and vitamins are present in all organisms.

In conclusion, the divergencies found in the Archaea are a matter of degree, and a major challenge facing microbiologists is to understand the significance and evolution of this diversity. This understanding will enable us to address the significance of "horizontal" evolution, or gene transfer between distantly related organisms, in bacterial evolution. In eukaryotes, the mitochondrion and chloroplast are two well-documented cases of horizontal evolution. Did similar events occur in bacterial evolution? Currently, the evidence is mixed. Certain properties such as methanogenesis and chlorophyll-based photosynthesis exist solely in the archaeal and the eubacterial Domains, respectively. For these properties, horizontal evolution probably did not occur between Domains. In contrast, nitrogen fixation is found in both Domains, and it is possible that either the Archaea or Eubacteria acquired the genetic information for nitrogen fixation late in evolution. A detailed knowledge of the biochemistry and molecular biology of both archaeal and eubacterial types will be necessary to answer questions such as this.

Major Groups of Archaea

As mentioned previously, the Archaea contain three major phenotypic groups: the methane-producing bacteria or the methanogens, the extreme halophiles, and the extreme thermophiles. The phylogenetic relationships among these bacteria are complex (Figure 22.5). The Archaea can be divided into two groups on the basis of their rRNA structure. One group includes only the extreme thermophiles. The second group contains all the methanogenic bacteria. The extreme halophiles are closely related to each other and to one branch of the methanogens within this group. Some of the extreme thermophiles also appear as deep branches of this second group. Thus, both the methanogens and the extreme halophiles are related phylogenetically as well as phenotypically. In contrast, the extreme thermophiles are not a closely knit phylogenetic group.

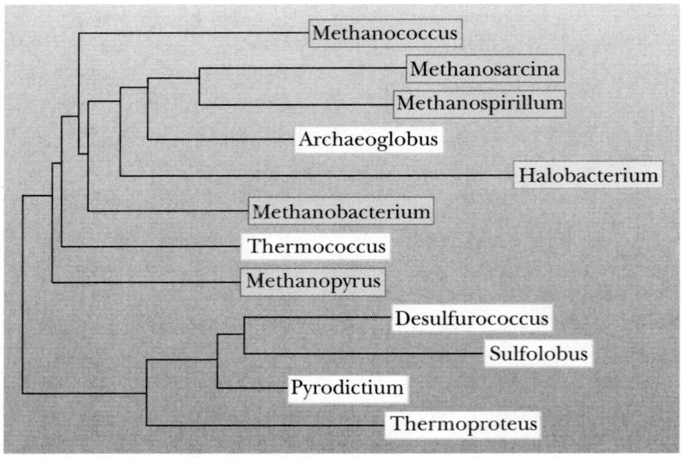

***Figure* 22.5** Phylogeny of the major genera of the Archaea based upon the sequences of the 16S rRNA. The major groups of the methanogens are closely related. The extreme halophile *Halobacterium* and *Archaeoglobus* are also related to the methanogens in the orders *Methanosarcinales* and *Methanomicrobiales*. In contrast, the extreme thermophiles are found in both of the two major lines of descent in the Archaea.

Methane-Producing Bacteria or Methanogens

Methanogens are the only Archaea that are truly cosmopolitan (see **Box 22.1**). By this we mean that these strict anaerobes are present in most anaerobic environments on Earth, including waterlogged soils, rice paddies, lake sediments, marshes, marine sediments, and the gastrointestinal tracts of animals. Generally they grow in association with anaerobic Eubacteria and eukaryotes, and they participate in anaerobic food chains involved in degradation of complex organic polymers to CH_4 and CO_2.

Methanogenesis is a significant component of the carbon cycle on Earth. The methane-producing bacteria release 3–7 × 10^{14} g of methane into the Earth's atmosphere each year. Because methane is a greenhouse gas, it contributes to the warming of the planet. The present atmospheric concentration of methane is about 1.9 ppm. However, this value is more than twice the concentration found in air frozen in ice cores before the Industrial Rev-

BOX 22.1 RESEARCH HIGHLIGHTS

Exploring the Global Distribution of Archaea

The Archaea have apparently been far less successful than the Eubacteria in evolving species that could successfully colonize the predominant soil and aquatic ecosystems on Earth. Until recently the only group deemed successful in this regard is the methanogens, which are present in aquatic sediments and the intestinal tracts of animals. However, Ed DeLong's laboratory at the University of California, Santa Barbara, has obtained evidence that an archaeal line related to the hyperthermophilic Archaea occurs in marine habitats. Using shotgun cloning of 16S rDNA from various marine habitats, including some from Antarctica, his laboratory discovered the characteristic signature of the hyperthermophiles. More recent evidence suggests that these Archaea may be symbionts of sponges.

olution. Currently, the concentration of atmospheric methane is increasing at a rate of about 1 percent per year. Thus, its contribution to greenhouse warming is expected to increase.

A summary of the properties of the major genera of methane-producing bacteria is given in Table 22.2. All methane-producing bacteria have the ability to obtain their energy for growth from the process of methane biosynthesis. To date, no methanogens have been identified that can grow using any other energy source. Thus, these bacteria are obligate methane producers. In bacterial nomenclature, the names of genera of methane-producing bacteria contain the prefix "methano-." This prefix distinguishes them from an unrelated group of Eubacteria, the methylotrophic bacteria, which consume methane. The names of genera of methylotrophic bacteria contain the prefix "methylo-."

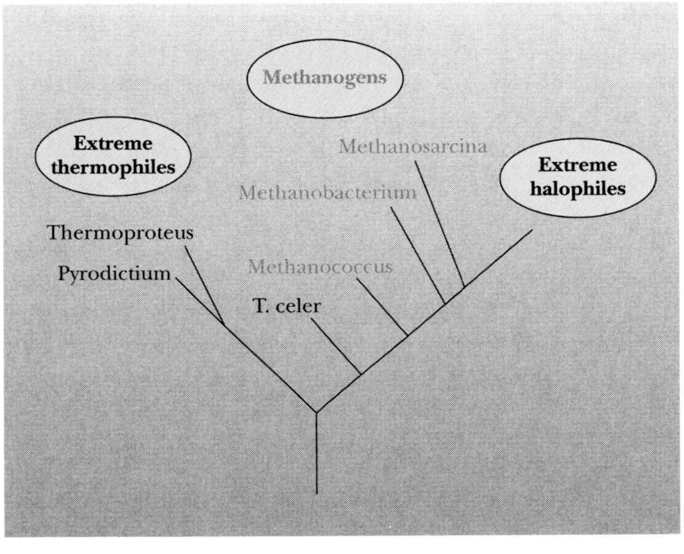

The substrates for methane synthesis are limited to a few types of compounds (Table 22.3). The first type is used by the methanogens that reduce CO_2 to methane. The major electron donors for this reduction are H_2 and formate. In addition, some methanogens can use alcohols like 2-propanol, 2-butanol, and ethanol as electron donors. In the oxidation of 2-propanol, acetone is the product. Because eight electrons are required to reduce CO_2 to methane, four molecules of H_2, formate, or 2-propanol are consumed. Even though formate is a reduced C-1 compound, it is oxidized to CO_2 before CO_2 is reduced to methane. The second type of substrate for methanogenesis includes C-1 compounds containing a methyl carbon bonded to O, N, or S. Compounds of this type include methanol, monomethylamine, dimethylamine, trimethylamine, and dimethylsulfide. The methyl group is reduced to methane. The electrons for this reduction are obtained from the oxidation of an additional methyl group to CO_2. Because six electrons can be obtained from this oxidation and only two are required to reduce a methyl group to methane, the stoichiometry of this reaction is three molecules of methane are formed for every molecule of CO_2 formed. An exception to this rule is found in *Methanosphaera stadtmaniae,* which can use only H_2 to reduce methanol to methane. Apparently, this bacterium lacks the ability to oxidize methanol. The third type of substrate is acetate. In this reaction, the methyl (C-2) carbon of acetate is reduced to methane using electrons obtained from the oxidation of the carboxyl (C-1) carbon of acetate. This reaction is called the aceticlastic reaction because it results in the splitting of acetate into methane and CO_2.

Biochemistry of Methanogenesis

Elucidation of the pathway of methane synthesis has been a major challenge to microbial physiologists. Despite simple substrates, the complexity of the pathway is compara-

Descriptive *Table* **22.2 Summary of properties of the methane-producing Archaea**

	Morphology	Major Energy Substrates[1]	Temperature Optimum (°C)	Cell Wall[2]
Order *Methanobacteriales*				
Family *Methanobacteriaceae*				
Genus *Methanobacterium*	Rod	H_2, (formate)	35–40	Pseudomurein
Methanothermobacter	Rod	H_2, (formate)	55–70	Pseudomurein
Methanobrevibacter	Short rod	H_2, formate	30–38	Pseudomurein
Methanosphaera	Coccus	H_2 + methanol	36–40	Pseudomurein
Family *Methanothermaceae*				
Genus *Methanothermus*	Rod	H_2	83–88	Pseudomurein + protein
Order *Methanococcales*				
Family *Methanococcaceae*				
Genus *Methanococcus*	Coccus	H_2, formate	35–40	Protein
Methanothermococcus	Coccus	H_2, formate	65	Protein
Family *Methanocaldococcaceae*				
Genus *Methanocaldococcus*	Coccus	H_2, (formate)	85	Protein
Methanoignis	Coccus	H_2	88	Protein
Order *Methanomicrobiales*				
Family *Methanomicrobiaceae*				
Genus *Methanomicrobium*	Rod	H_2, formate	40	Protein
Methanogenium	Coccus	H_2, formate	30–57	Protein
Methanoplanus	Plant or disc	H_2, formate	32–40	Glycoprotein
Methanolacinia	Rod	H_2	40	Glycoprotein
Methanoculleus	Coccus	H_2, formate	37–60	Glycoprotein
Methanofollis	Coccus	H_2, formate	37–40	Glycoprotein
Family *Methanospirillaceae*				
Methanospirillum	Spirillum	H_2, formate	30–40	Protein + sheath
Family *Methanocorpusculaceae*				
Genus *Methanocorpusculum*	Small coccus	H_2, formate	30–37	Glycoprotein
Order *Methanosarcinales*				
Family *Methanosarcinaceae*				
Genus *Methanosarcina*	Coccus, packets	(H_2), $MeNH_2$, Ac	35–50	Protein + HPS
Methanococcoides	Coccus	$MeNH_2$	23–35	Protein
Methanolobus	Coccus	$MeNH_2$	30–40	Glycoprotein
Methanohalophilus	Coccus	$MeNH_2$	26–36	Protein
Methanohalobium	Flat polygons	$MeNH_2$	50	ND
Methanosalsus	Coccus	$MeNH_2$	45	ND
Family *Methanosaetaceae*				
Genus *Methanosaeta* (*Methanothrix*)	Rod	Ac	35–60	Protein + sheath
Order *Methanopyrales*				
Family *Methanopyraceae*				
Genus *Methanopyrus*	Rod	H_2	98	Pseudomurein

[1]Major energy substrates for methane synthesis. $MeNH_2$ is methylamines, Ac is acetate. Parentheses means utilized by some but not all species or strains.

[2]Cell wall components include HPS for heteropolysaccharide. ND is not determined.

Reference *Table* **22.3 Free energies for typical methanogenic reactions**

Reaction	$\Delta G^{\circ\prime}$ (kJ/mol of CH_4)
Type 1	
$CO_2 + 4\ H_2 \longrightarrow CH_4 + 2\ H_2O$	−130
$4HCOOH \longrightarrow CH_4 + 3\ CO_2 + 2\ H_2O$	−120
CO_2 + 4 (isopropanol) $\longrightarrow CH_4$ + 4 (acetone) + 2 H_2O	−37
Type 2	
$CH_3OH + H_2 \longrightarrow CH_4 + H_2O$	−113
$4\ CH_3OH \longrightarrow 3\ CH_4 + CO_2 + 2\ H_2O$	−103
$4\ CH_3NH_3Cl + 2\ H_2O \longrightarrow 3\ CH_4 + CO_2 + 4\ NH_4Cl$	−74
$2\ (CH_3)_2S + 2\ H_2O \longrightarrow 3\ CH_4 + CO_2 + 2\ H_2S$	−49
Type 3	
$CH_3COOH \longrightarrow CH_4 + CO_2$	−33

ble to bacterial photosynthesis. In addition, many of the enzymes of methanogenesis are extremely sensitive to oxygen, and their study requires specialized equipment and techniques. Even after extensive research, many features of bacterial methanogenesis are still not well understood.

The reduction of CO_2 to methane requires five novel coenzymes or vitamins (Figure 22.6). Elucidation of the structure and function of these coenzymes was achieved primarily in the laboratories of R. S. Wolfe at the University of Illinois, G. D. Vogels at the University of Nijmegen, and R. K. Thauer at the Phillips University in Marburg, Germany. The structures of these coenzymes represent a fascinating blend of novel and familiar themes in biochemistry. Methanofuran is an unusual molecule containing an aminomethylfuran moiety. During methanogenesis, the aminomethylfuran serves as a C-1 acceptor, forming formylmethanofuran. The use of a furan moiety as a reactive center is unique. In contrast, methanopterin is very similar to folate in structure and function. In both molecules, the reactive center is the tetrahydro-form and composed of a pterin and *p*-aminobenzoic acid (PABA). However, the side-chains are strikingly different. Interestingly, unusual pterins are common in other Archaea as well. Coenzyme F_{430} is also familiar. Named because it absorbs light at 430 nm, it is a tetrapyrrole with Ni at the active center. Although the ring structure is more reduced than other common tetrapyrroles containing Fe (heme), Mg (chlorophyll), and Co (cobamide), the biosynthesis and function of coenzyme F_{430} are probably similar. Coenzyme M and 7-mercaptothreonine phosphate are two thiol-containing coenzymes unique to methanogens. Although thiols are common functional groups in other cellular components such as coenzyme A and cysteine, the coenzymes from methanogens have some very unusual features. 7-Mercaptoheptanoylthreonine phosphate contains a threonyl phosphate residue, which has previously been known to be present only in phosphorylated proteins. A phosphorylated amino acid in a coenzyme is extremely rare. Similarly, coenzyme M is a sulfonate while most highly polar coenzymes are phosphate derivatives.

The requirement for a large number of unusual coenzymes in methanogenesis has profound implications. Although the biosynthetic pathways for the coenzymes have not been clarified completely, they are probably unique because these coenzymes are not present in other living cells. Therefore the ability to utilize methanogenesis as an energy source requires a large amount of genetic information for coenzyme biosynthesis in addition to methanogenesis. For this reason, methanogenesis probably evolved only once and all methanogens are related.

Methanogenesis from CO_2 Reduction Occurs Stepwise

CO_2 is bound to carriers and reduced successively through the formyl, methylene and methyl oxidation-reduction levels to methane (Figure 22.7). Initially, CO_2 binds to methanofuran (MFR) and is reduced to the formyl level. The mechanism whereby this initial binding occurs is not known. The formyl group is then transferred to tetrahydromethanopterin (H_4MPT), which is the C-1 carrier for the next two reductions. Thus, after dehydration of formyl-H_4MPT to methenyl-H_4MPT, the C-1 moiety is reduced to methylene-H_4MPT and then to methyl-H_4MPT. The methyl group is then transferred to coenzyme M. The last reaction forms methane from the reduction of methyl coenzyme M (CH_3-S-CoM). The methylreductase, the en-

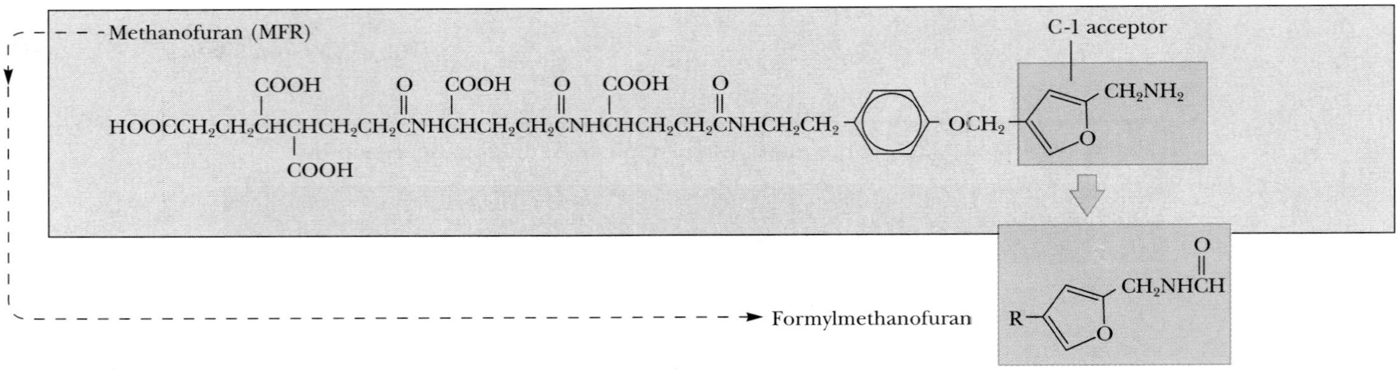

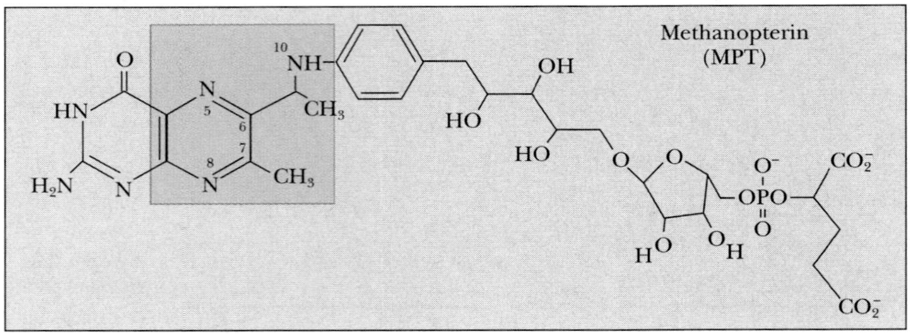

Tetrahydromethanopterin
(H_4MPT)

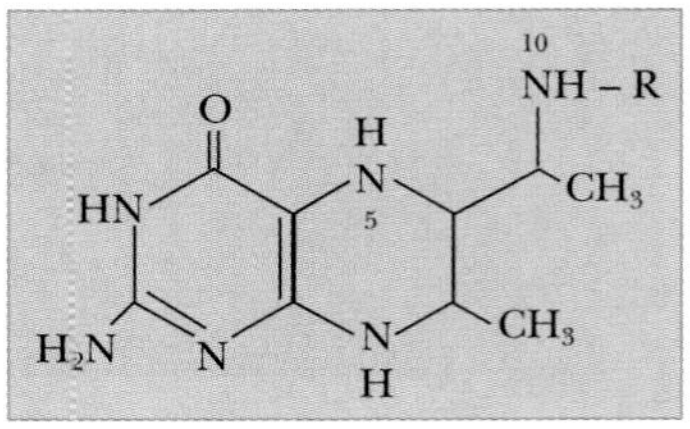

5-formyltetrahydromethanopterin
($HCO-H_4MPT$)

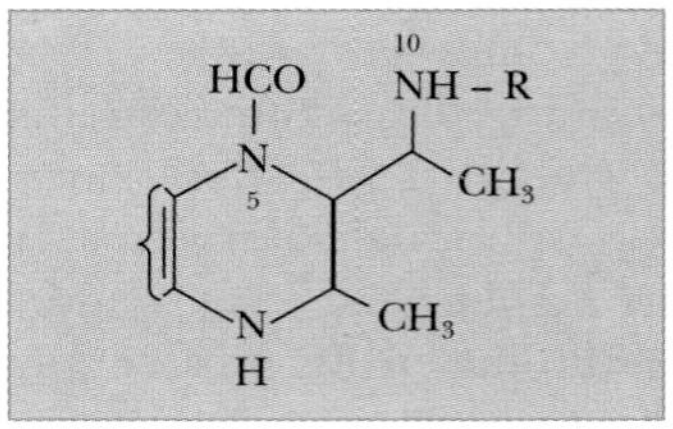

5,10-methenyltetrahydromethanopterin
[5,10-(=CH-)H_4MPT]

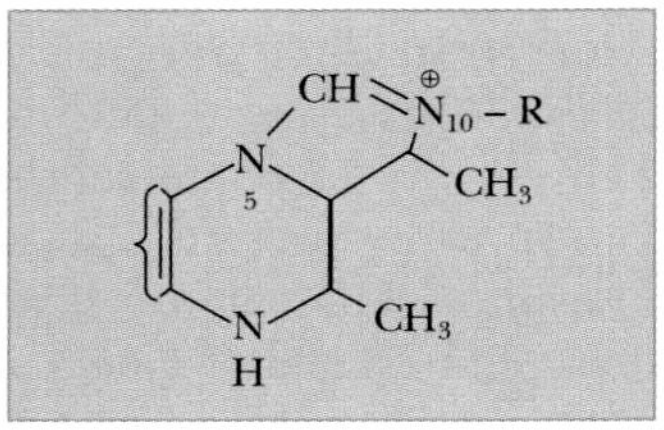

5,10-methylenetetrahydromethanopterin
($CH_2=H_4MPT$)

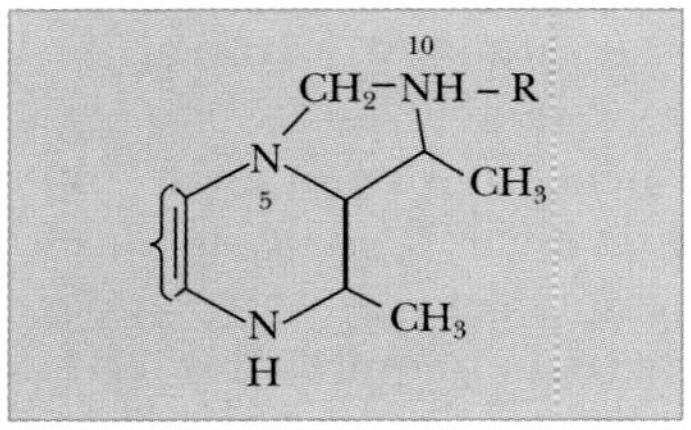

5-methyltetrahydromethanopterin
(CH_3-H_4-MPT)

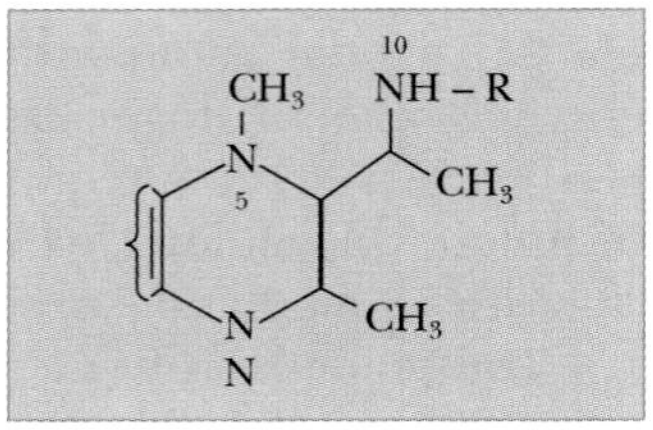

Coenzyme F_{420}

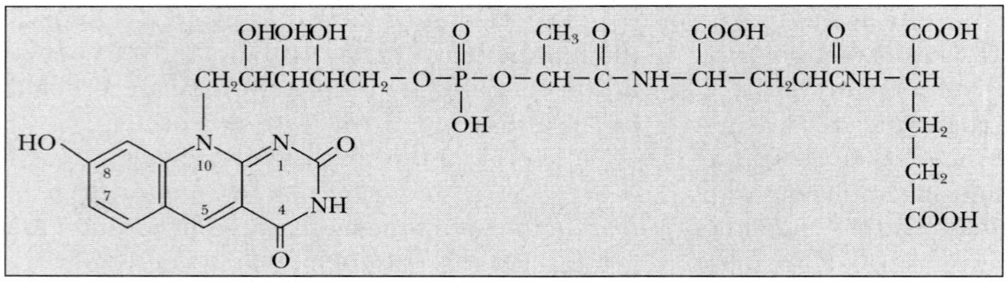

Coenzyme M

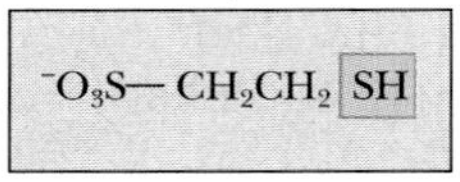

Methyl coenzyme M

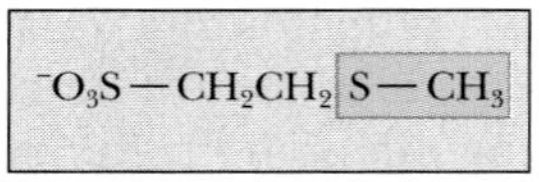

Coenzyme F_{430}

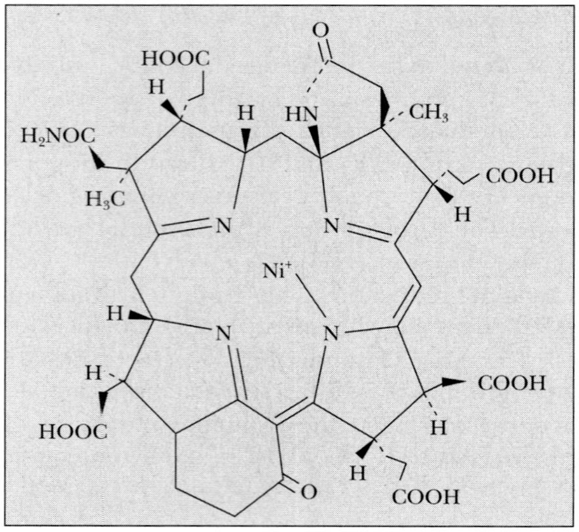

7-mercaptoheptanoylthreonine
phosphate (HS-HTP)

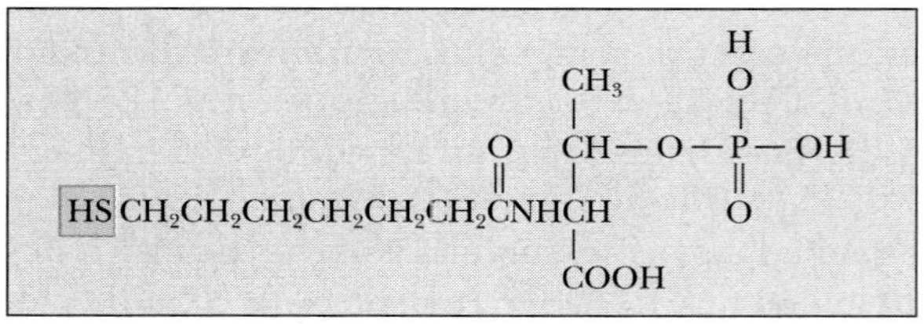

◀ ***Figure* 22.6** Coenzymes of methanogenesis. Except for coenzyme F_{420}, these coenzymes have only been found in the methane-producing bacteria and closely related Archaea.

zyme that catalyzes this reaction, requires two additional coenzymes, coenzyme F_{430} and 7-mercaptoheptanoylthreonine phosphate (HS-HTP). HS-HTP reduces CH_3-S-CoM to form methane plus the heterodisulfide of coenzyme M and HS-HTP (Figure 22.7). This mixed disulfide is then reduced to regenerate the thiols.

The electrons for the reduction of CO_2 to methane are obtained from H_2, formate, or alcohols. For some of these reductions, the electron carriers have not been fully characterized. However, for at least the reduction of methenyl-H_4MPT to methylene-H_4MPT and then methyl H_4MPT, the electron carrier coenzyme F_{420} is utilized. Coenzyme F_{420} is named because of its absorption maximum at 420 nm (see Figure 22.6 for the chemical structure of F_{420}). Although it was discovered in methanogens, coenzyme F_{420} is present in other organisms; for example, in *Streptomyces,* it functions in antibiotic synthesis. In many Eubacteria and eukaryotes, coenzyme F_{420} is essential for photoreactivation of UV-damaged DNA. In methanogens, coenzyme F_{420} is present in very high levels, and it is an electron carrier in many biosynthetic reactions in addition to methanogenesis.

Methanogenesis from Other Substrates

Methyl coenzyme M is a central intermediate in the pathway of methane synthesis from methyl compounds and acetate. For methyl compounds, the C-1 group is first transferred to a cobamide-containing protein and then to coenzyme M. The methyl group is then either reduced to

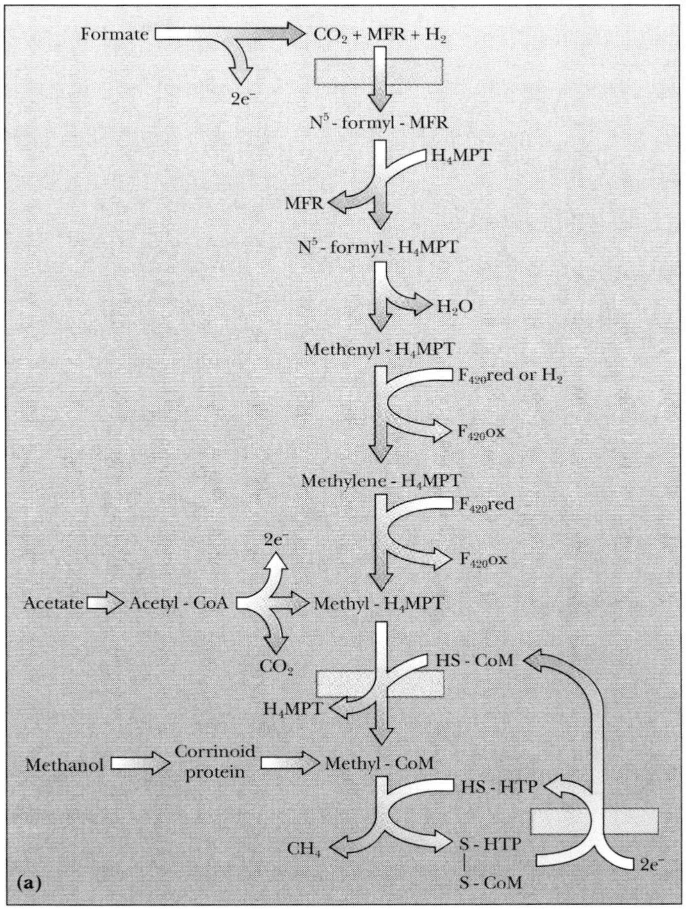

***Figure* 22.7** Pathways of methanogenesis. **(a)** Methyl coenzyme M (methyl-S-CoM) is a central intermediate in methanogenesis from CO_2, acetate, and methanol or methylamines. For some of the reactions, either F_{420} or H_2 is the electron donor. In cases designated by "2e," the immediate electron donor is not known. Shaded reactions indicate potential points of coupling with the proton motive force. MFR = methanofuran, H_4MPT = tetrahydromethanopterin, HS-HTP = 7-mercaptoheptanoylthreonine phosphate. **(b)** The aceticlastic reaction for biosynthesis of methyl-H_4MPT from acetate. Fd_{ox} and Fd_{red} are the oxidized and reduced forms of ferredoxin, respectively.

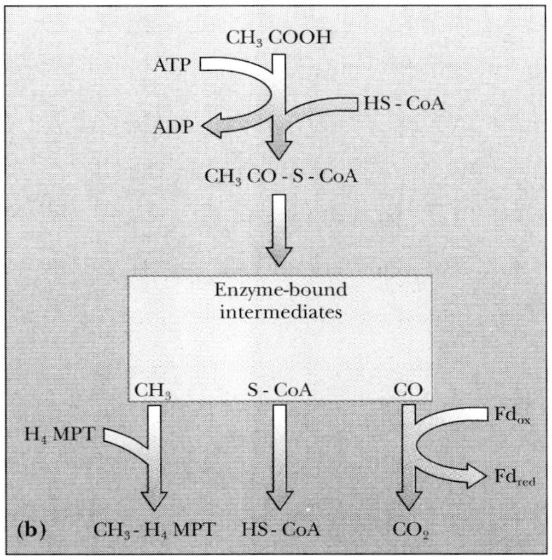

methane or oxidized by reversing of the pathway of CO_2 reduction.

Acetate must first be activated with ATP to form acetyl-CoA (Figure 22.7**b**). Acetyl-CoA is then cleaved to form an enzyme-bound methyl group, an enzyme-bound carbon monoxide (CO), and HS-CoA. The methyl group is transferred to coenzyme M via tetrahydromethanopterin. The enzyme-bound CO is oxidized to CO_2, and the electron carrier ferredoxin is reduced. The ferredoxin is then used to reduce CH_3-S-CoM to methane. The enzyme complex that oxidizes acetyl-CoA is also called CO dehydrogenase because it oxidizes CO to CO_2. A similar enzyme is found in the homoacetogenic clostridia and many anaerobic autotrophs including the hydrogenotrophic methanogens. In these autotrophic bacteria, acetyl-CoA biosynthesis is the product of CO_2 fixation. All the organic molecules in the cell are derived from acetyl-CoA.

Bioenergetics of Methanogenesis

Methanogenesis is a type of anaerobic respiration and serves as the primary means of energy generation in these bacteria. Three coupling sites to the proton motive force have been identified: the reduction of CO_2 to formylmethanofuran, the methyl transfer from methyl-H_4MPT to HS-CoM, and the reduction of the heterodisulfide to HS-HPT and HS-CoM (see Figure 22.7). The first of these reactions is endergonic in the direction of methane synthesis, and the proton motive force is required to drive the formation of formylmethanofuran. At the remaining sites, the reactions are exergonic, and either a sodium or proton motive force is generated. These ion gradients may then be used for ATP biosynthesis, CO_2 reduction, or other energy-requiring reactions in the cell.

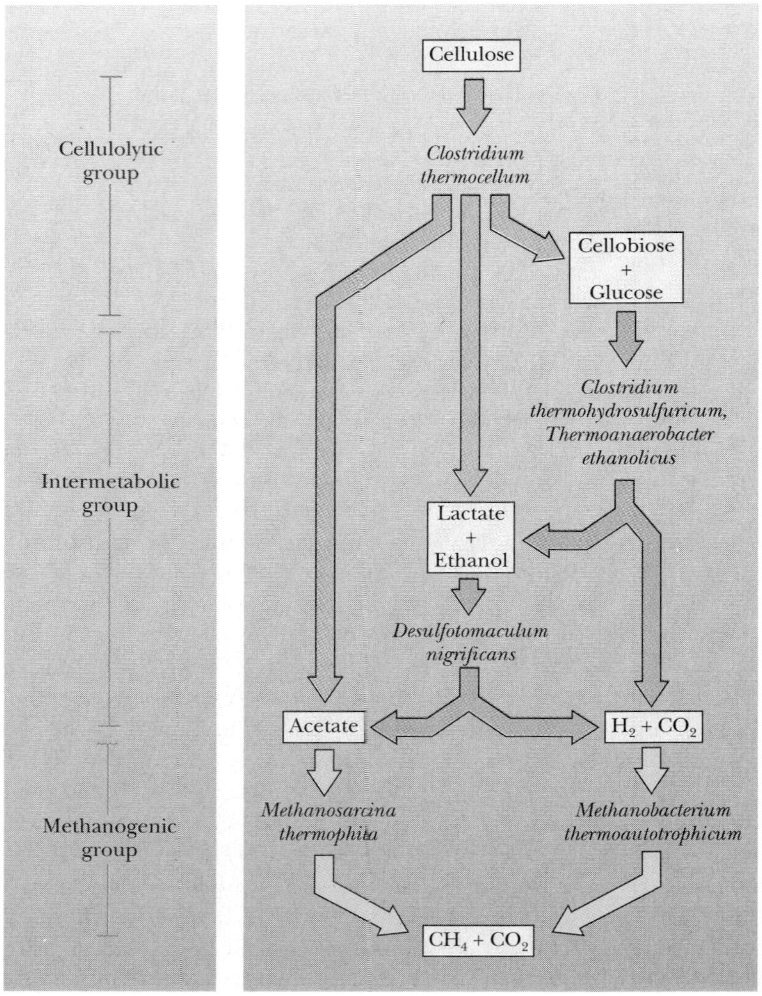

***Figure* 22.8** Typical anaerobic food chain found in a thermophilic bioreactor. Cellulose is converted to methane and CO_2 by the action of the cellulolytic, intermetabolic, and methanogenic groups of microorganisms. In this fermentation, about 95 percent of the combustion energy of the cellulose is released as methane. (Modified from J. Wiegel and L. G. Ljungdahl, 1986. The importance of thermophilic bacteria in biotechnology, *CRC Critical Reviews in Biotechnology* 3:39–108.)

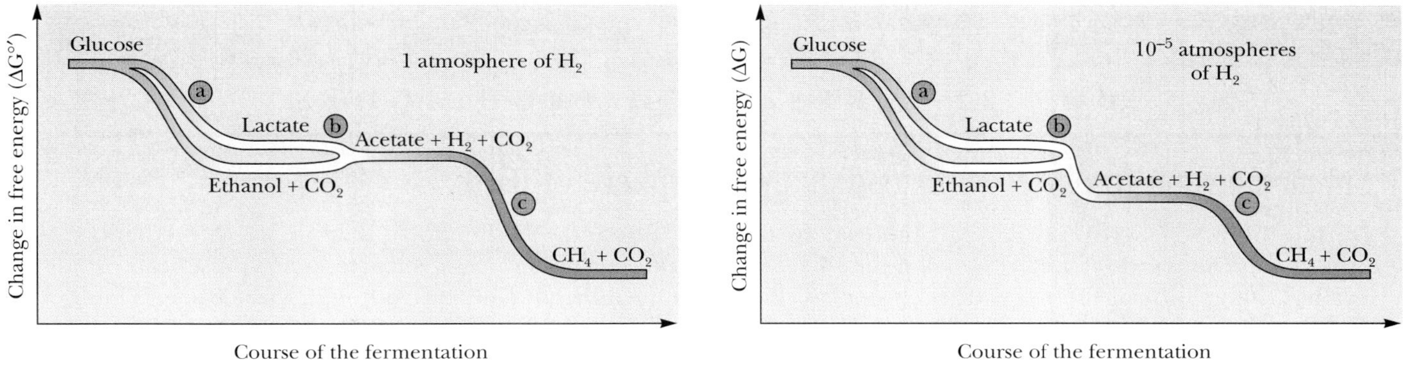

***Figure* 22.9** Changes in free energy (ΔG) during a typical fermentation of glucose to methane under standard conditions (1 atmosphere of H_2) and low partial pressures of H_2 (10^{-5} atmospheres of H_2). The reactions catalyzed by each of the three groups of microorganisms required for the fermentation are **(a)** the saccharolytic intermetabolic group, **(b)** the syntrophic intermetabolic group, and **(c)** the methanogen group. Under high partial pressures of H_2, the thermodynamics of the fermentation of sugars to lactate and ethanol are very favorable. The microorganisms that can catalyze these reactions grow and these products are formed. However, the thermodynamics of the fermentations of lactate and ethanol are unfavorable or only slightly favorable. Therefore, the bacteria that catalyze these reactions cannot grow, these compounds accumulate, and the fermentation ceases. In contrast, under low concentrations of H_2, the fermentations of lactate and ethanol are favorable, and the intermetabolic group of organisms grow well. Therefore, lactate and ethanol are rapidly converted to acetate, H_2, and CO_2. Notice that the free energy now available for the growth of the methanogens is diminished, and growth of the methanogens may limit the rate of the complete fermentation of glucose to methane.

Ecology of Methanogens

The methane-producing bacteria flourish in anaerobic environments where sulfate, oxidized metals, and nitrate are absent. In these environments, the substrates for methanogenesis are readily available as the fermentation products of Eubacteria and eukaryotes. The methanogens catalyze the terminal step in the anaerobic food chain where complex polymers are converted to methane and CO_2. These principles are illustrated in a typical methanogenic food chain for a thermophilic bioreactor (Figure 22.8). Polymers are first degraded by specialized microorganisms, such as the cellulolytic bacteria. The major products are simple sugars such as glucose and the disaccharide cellobiose; lactate, and volatile fatty acids (VFAs) such as acetate; and alcohols such as ethanol. These products are further metabolized by the intermetabolic group. These microorganisms convert simple sugars to VFAs and alcohols. They also convert VFAs and alcohols to acetate and H_2, which are major substrates for the methanogenic bacteria.

INTERSPECIES HYDROGEN TRANSFER Hydrogen is a key intermediate in this process. Under standard conditions, when the H_2 partial pressure is 1 atmosphere, the fermentations of VFAs and alcohols to acetate and H_2 are thermodynamically unfavorable (Figure 22.9). Therefore, the microorganisms that catalyze these reactions cannot grow. The VFAs then accumulate to very high concentrations that are toxic to the cellulolytic and intermetabolic groups of microorganisms, and the fermentation of cellulose ceases. However, if the methanogenic bacteria are present, H_2 is rapidly metabolized, and its partial pressure is maintained below 10^{-3}–10^{-4} atmospheres. Under these conditions, the fermentations of VFAs and alcohols are thermodynamically favorable (Figure 22.9). Because these compounds are rapidly metabolized, their concentrations are maintained below the toxic levels. This interaction between the H_2-producing intermetabolic organisms and the H_2-consuming methanogens is called interspecies hydrogen transfer. In many anaerobic environments, it is a key regulatory mechanism. Because the H_2-consuming methanogens play a critical role, they are said to "pull" the fermentation of complex organic polymers to methane and CO_2.

Interspecies hydrogen transfer is not limited to methanogenic food chains. In environments rich in sulfate or nitrate, anaerobic Eubacteria oxidize H_2, VFAs, and alcohols. When sulfate is present, this activity is catalyzed by the sulfate-reducing bacteria. Because the oxidation of H_2 with sulfate as the electron acceptor is thermodynamically more favorable than when CO_2 is the electron acceptor (as in methanogenesis), the sulfate-reducing bacteria outcompete the methanogens for H_2. For similar reasons, the sulfate-reducing bacteria also compete successfully with the methanogens for other important

BOX 22.2 RESEARCH HIGHLIGHTS

Subterranean Microbiology: Extending the Biosphere of Earth

During the past few years it has become evident that life extends thousands of meters beneath the surface of Earth. Only microbial life exists at such depths. Both Eubacteria and Archaea thrive in these environments. Among the more interesting subterranean habitats are the thermal oil reservoirs that are at temperatures from 50°C to $>$ 100°C. Thermophilic Archaea have been isolated from such areas in the North Sea, Alaska, and an oil-bearing area near Paris called the Paris Basin. These bacteria are sulfur-reducing thermophilic Archaea. It is not yet clear what the primary energy sources of these bacteria might be.

An intriguing recent discovery is the finding by Todd Stevens's group at Batelle Laboratories Northwest (Richland, Washington) that archaeal methanogens and sulfate-reducing Eubacteria are present in subterranean basalt aquifers. Although these bacteria are not thermophilic, they are interesting in that they appear to obtain their energy from the hydrogen that is generated by chemical–physical, basalt/water reactions. If this is confirmed, it would indicate that these autotrophic bacteria have developed a food chain that is independent of photosynthesis. This is important as it suggests that the primitive autotrophic organisms that evolved on Earth may have been hydrogen-oxidizing bacteria.

substrates such as acetate and formate. Therefore, methanogenesis is greatly limited in marine sediments that are rich in sulfate. The oxidation of H_2 with nitrate, Fe^{3+}, and Mn^{4+} as electron acceptors is also thermodynamically more favorable than methanogenesis. The denitrifying and iron- and manganese-reducing bacteria also are more successful than the methanogens when these electron acceptors are present. In conclusion, methanogenesis only dominates in habitats where CO_2 is the only abundant electron acceptor for anaerobic respiration.

METHANOGENIC HABITATS Just as aerobic microorganisms rapidly deplete the O_2 in environments rich in organic matter to establish anaerobic conditions, sulfate-reducing bacteria, iron- and manganese-reducing bacteria, and denitrifying bacteria frequently consume all the sulfate, Fe^{3+}, Mn^{+4}, and nitrate in anaerobic environments and rapidly establish the conditions for methanogenesis. In these environments, CO_2 is seldom limiting because it is also a major fermentation product. Thus, methanogenesis is the dominant process in many anaerobic environments that

Differential *Table* **22.4 Summary of properties of extremely halophilic Archaea**

	Morphology	Substrates[1]	pH Optimum	Optimum NaCl (M)
Family *Halobacteriaceae*				
Genus *Halobacterium*	Long rod	Amino acids	5–8	3.5–4.5
Haloarcula	Short pleomorphic rods	Amino acids, CH_2O	5–8	2.0–3.0
Haloferax	Pleomorphic	Amino acids, CH_2O	5–8	2.0–3.0
Halococcus	Coccus	Amino acids	5–8	3.5–4.5
Natronobacterium	Rod	CH_2O, organic acids	8.5–11	3.5–4.5
Natronococcus	Coccus	CH_2O	8.5–11	3.0–4.0

[1]CH_2O is carbohydrate. Oxygen is the electron acceptor for all the extremely halophilic Archaea. Some species may also grow anaerobically by fermentation or denitrification.

contain large amounts of easily degradable organic matter. Especially important environments of this type include freshwater sediments found in lakes, ponds, marshes, and rice paddies. Here methane synthesis can be readily demonstrated by a simple experiment first performed by the Italian physicist Alessandro Volta in 1776. In a shallow pond or lake, select a site where litter or other organic debris has accumulated. When the debris is stirred with a stick, gas bubbles will escape from the sediment to the surface. Collect these bubbles in an inverted funnel closed at the top with a short piece of tubing and a pinch clamp. Collect about four liters of gas. To demonstrate that the gas is methane, ignite it. First, the funnel is partially submerged to pressurize the gas. When the clamp is opened slowly and a lit match is touched to the escaping gas, the gas burns with a blue flame. Care must be taken to keep your eyebrows and face clear of the burning gas.

Methanogenesis also occurs in other habitats. Methane is formed by the methane-producing bacteria in the anaerobic microflora of the large bowel. About one-third of healthy adult humans excrete methane gas. Some methane is also absorbed in the blood and excreted from the lungs. The predominant methanogen in humans is *Methanobrevibacter smithii,* which is a gram-positive coccobacillus that utilizes H_2 or formate to reduce CO_2 to methane. In individuals that excrete methane, *M. smithii* is present at 10^7–10^{10} bacteria per gram dry weight of feces, or between 0.001 and 12 percent of the total number of viable anaerobic bacteria. Why the numbers of *M. smithii* fluctuate so greatly in apparently healthy individuals remains a mystery. *Methanosphaera stadtmaniae* is also present in the human large bowel, but in much lower numbers. This interesting methanogen will grow only by the reduction of methanol with H_2. It cannot reduce CO_2 to methane or utilize acetate and methylamine. Like *M. smithii,* it is also gram-positive and contains pseudomurein in its cell walls.

The rumen is another major habitat for methanogens, and about 10 percent to 20 percent of the total methane emitted to the Earth's atmosphere originates in the rumen of cows, sheep, and other mammals. In the rumen, complex polymers from grass and other forages are degraded to volatile fatty acids, H_2, and CO_2 by the cellulolytic and intermetabolic groups of bacteria, fungi, and protozoans. The VFAs are absorbed by the animal and are a major energy source. Thus, little methane is produced from acetate. The H_2 is used to reduce CO_2 to methane, which is emitted. Methanogenesis represents a significant energy loss to the cow, and 10 percent of the caloric content of the feed may be lost as methane. *Methanobrevibacter ruminantium* is the predominant methanogen in the bovine rumen. It is a coccobacillus that utilizes H_2 and formate. Some strains require coenzyme M in addition to volatile fatty acids for growth. Because growth is proportional to the concentration of coenzyme M in the medium, these strains have been used in a bioassay for coenzyme M.

In the rumen, in freshwater sapropel, and in marine sediments, many of the anaerobic protozoans are associated with methanogenic symbionts. For rumen ciliates, methanogenic bacteria are attached to the cell surface. For some species, up to 100 percent of the individual ciliates are associated with methanogens. In aquatic sludge or sapropel, many protozoans contain endosymbiotic methanogens. These include both ciliates and amoebae. The methanogens are present in high numbers and distributed throughout the cytoplasm. The marine ciliate *Metopus contortus* also contains numerous endosymbiotic methanogens of the species *Methanoplanus endosymbiosus.* This disc-shaped methanogen has a protein cell wall and lyses rapidly in 0.001% sodium dodecyl sulfate. It is located in parallel rows on the cytoplasmic side of the pellicle and on the surface of the nuclear membrane, and it is present in very high numbers, greater than 10^{10} methanogens per ml of cytoplasm. *M. endosymbiosus* is probably associated with the hydrogenosomes, which are microbodies that convert pyruvate to acetate, H_2 and CO_2. The methanogen is then the electron sink for H_2. Conceivably, the ciliate may use this mechanism to divert reducing equivalents away from the sulfate-reducing bacteria and limit the production of H_2S, which is very toxic.

Although methanogens are most frequently found at the bottom of anaerobic food chains associated with the intermetabolic microorganisms, in some ecosystems they are the primary consumers of geochemically produced H_2 and CO_2 (see **Box 22.2**). The submarine hydrothermal vents found on the ocean floor expel large volumes of very hot water containing H_2 and H_2S. As the water cools from several hundred degrees Celsius to the temperature of the ocean, zones suitable for the growth of thermophilic methanogens are established. *Methanocaldococcus jannaschii* is found in such an environment. It has an optimum growth temperature of 85°C and a protein cell wall. Methane produced by this bacterium escapes into the surrounding water where it is utilized by symbiotic methylotrophic Eubacteria living in the gills of the clam *Calyptogena,* the body cavity of the tube worm *Riftia,* and the mussel *Bathymodiolus.* In this ecosystem, the methanogen is at the top of the food chain.

Extreme Halophiles

The extremely halophilic Archaea are all closely related, and they comprise only six genera (Table 22.4). They all grow only at concentrations of NaCl above 1.8 M. Most species will also grow at concentrations greater than 4 M, which is close to the saturation point for NaCl in water. In addition, two genera are alkalophilic, and these bacteria grow only at basic pHs.

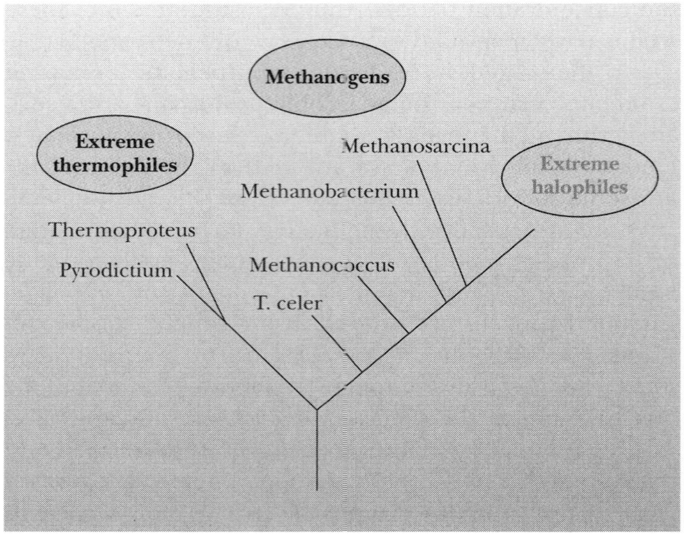

Although the trivial name halobacteria generally refers to the extremely halophilic Archaea, some Eubacteria, algae, and fungi can also grow in very high concentrations of NaCl. Examples of extremely halophilic Eubacteria include the purple photosynthetic bacterium *Ectothiorhodospira* and the actinomycete *Actinolyspora*. In addition, *Halomonas* is a haloterant Eubacterium that grows optimally near 1 M NaCl, but it also grows slowly at 4 M NaCl. Although these Eubacteria share with the halobacteria the ability to grow in high concentrations of salt, they are unrelated to the halobacteria, and their biochemistry and physiology are quite different.

The halobacteria are obligate or facultative aerobes. Most utilize amino acids, carbohydrates, or organic acids as their principal energy sources. *Halobacterium* oxidizes glucose by a modification of the Entner-Doudoroff pathway (Figure 22.10). In this modification, 2-keto-3-deoxygluconate-6-phosphate is the first phosphorylated intermediate, and glucose-6-phosphate is not involved. A further modification in which none of the intermediates are phosphorylated is also present in the extremely thermophilic Archaea (Figure 22.10). Carbohydrates are further oxidized by pyruvate oxidoreductase and the complete tricarboxylic acid cycle. NADH generated by the TCA cycle is oxidized by an electron transport chain very similar to that found in the Eubacteria. In the absence of oxygen, *Haloferax denitrificans* is also capable of growth by anaerobic respiration with nitrate as terminal electron acceptor. *Halobacterium salinarium* can ferment arginine in the absence of oxygen or nitrate.

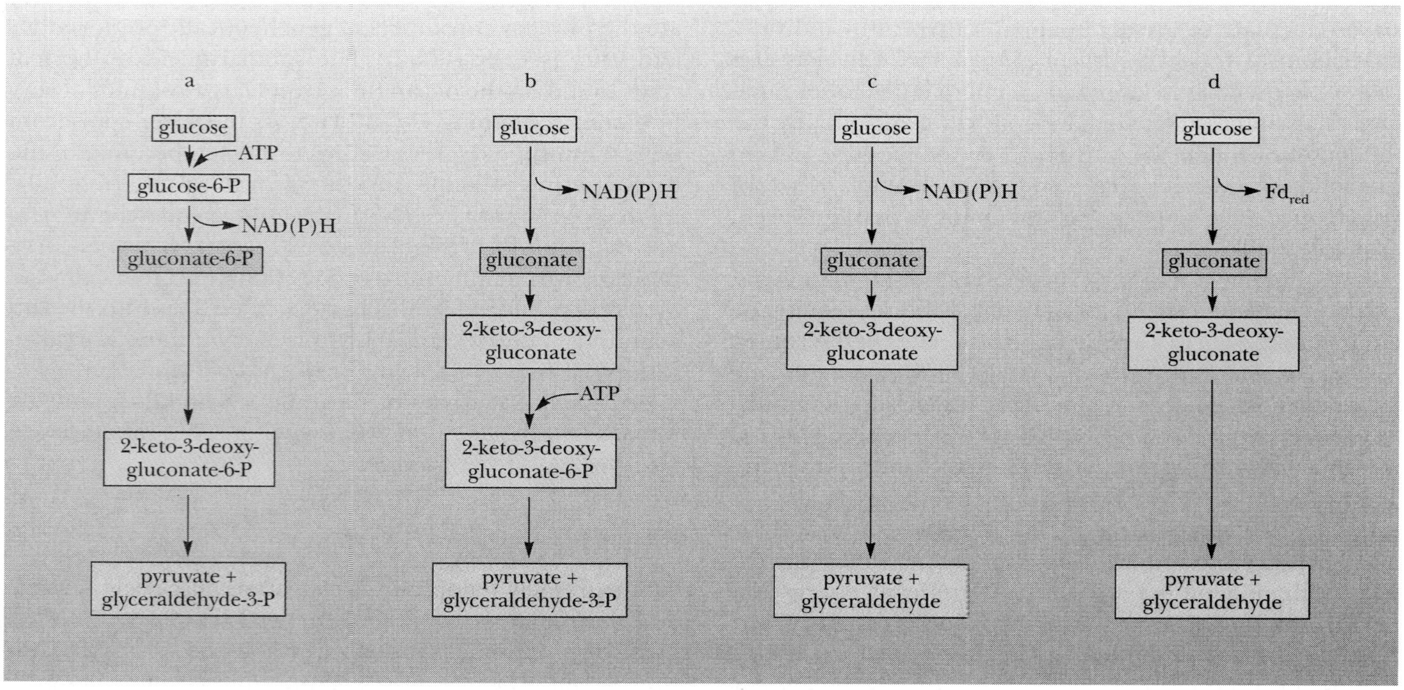

***Figure* 22.10** Modifications of the Entner-Doudoroff pathway in the Archaea. **(a)** Pathway common in *Pseudomonas* and other Eubacteria. **(b)** Pathway found in *Halobacterium* and *Clostridium aceticum*. **(c)** Pathway found in the extremely thermophilic Archaea *Sulfolobus* and *Thermoplasma*. **(d)** Pyroglycolytic pathway found in the hyperthermophilic Archaea *Pyrococcus*. Fd is ferredoxin.

Many halobacteria are also capable of a unique type of photosynthesis. Under low partial pressures of oxygen, they insert large amounts of a protein called *bacteriorhodopsin* into their cytoplasmic membrane. This protein forms patches that are called purple membrane. The color is due to the presence of the pigment retinal, which is also the common visual pigment in eyes. The retinal is covalently bonded to a lysyl residue of bacteriorhodopsin via a Schiff base (see Figure 9.15). In the dark, the retinal is in the all-*trans* configuration and the Schiff base is protonated. Upon the absorption of light, the retinal isomerizes to the 13-*cis* configuration, and the Schiff base is deprotonated. A proton is also expelled outside the cell. The Schiff base then returns to the protonated form by removing a proton from the cytoplasm, and the retinal reverts to the all-*trans* isomer. In this fashion, bacteriorhodopsin functions as a light-driven proton pump. The proton motive force it generates is then used for ATP synthesis and transport. The halobacteria have at least two additional retinal-based photosystems. One is a chloride pump called halorhodopsin. The second is important in phototaxis and is called "slow rhodopsin."

Although retinal is the visual pigment in the eye, rhodopsin functions very differently from bacteriorhodopsin. In rhodopsin, the retinal is not covalently bonded to the protein opsin. In the dark, retinal is in the 11-*cis* configuration (Figure 9.15). Exposure to light causes an isomerization to the all-*trans* configuration. Proton movement does not occur. Instead, the change in retinal causes a conformational change in the protein, which leads to the activation of a second messenger in the visual response. Because the mechanisms of the two retinal-based photosystems are fundamentally different, they may be an example of convergent evolution.

Photosynthesis in the halobacteria is also very different from the chlorophyll-based system found in the Eubacteria and chloroplasts. The electron transport chain does not participate in retinal-based photosynthesis, and it is not possible to directly reduce $NADP^+$ from electron donors that have unfavorable redox potentials, like water or elemental sulfur. Moreover, extensive light-trapping pigments like the light-harvesting chlorophylls, which are elaborated by photosynthetic Eubacteria in dim light, have not been described in the halobacteria. Therefore, the retinal-based system is more suitable for growth at high light intensities.

The halobacteria are adapted specifically for growth in high concentrations of salt. The water activity (a_w) of the cytoplasm of all microorganisms is equal to or less than the a_w of their medium; this means that the concentration of solutes in the cytoplasm equals or exceeds the concentration outside the cell. For halophilic microbes, the concentration of intracellular solutes must be very high. In the halobacteria, the intracellular solute is KCl, which is often found at concentrations greater than 5 M. High concentrations of salts also tend to weaken the ionic bonds necessary to maintain the conformation of proteins and the activity of enzymes. To maintain their structure, the proteins from the halobacteria have an increased number of acidic and hydrophobic residues, and their enzymes frequently require high concentrations of salts for activity. In the absence of salts, many of their enzymes denature. Likewise, below NaCl concentrations of 1 M, the protein cell walls of many halobacteria lose their structural integrity, and the cells lyse.

The halobacteria are common in salt lakes and salterns, or ponds used to prepare solar salt by the evaporation of sea water. Frequently, they are abundant and impart a red color to the brine. The red color is due to the presence of C_{50} carotenoids called bacterioruberins in the cell envelope. Although nonpathogenic, the halobacteria can spoil fish and animal hides treated with solar salt.

In salt lakes, the total concentration of salts is generally between 300 and 400 grams per liter, but the specific ions present may vary greatly. For instance, the Dead Sea between Jordan and Israel contains abundant chloride, magnesium, and sodium ions. In the Great Salt Lake in Utah, chloride, sodium, and sulfate ions predominate. The pH of these lakes is near neutrality. In contrast, the Wadi Natrum in Egypt contains high concentrations of bicarbonate and carbonate ions, which maintain the pH near 11. The alkalophilic halobacteria, *Natronobacterium* and *Natronococcus,* have been found in these lakes. In the salt lakes, the halobacteria mineralize organic carbon to CO_2. The organic carbon is produced by other halophilic microorganisms including the green algae *Dunaliella,* cyanobacteria, and *Ectothiorhodospira.*

Extreme Thermophiles

Although somewhat of a misnomer, the remaining Archaea are referred to as the extreme thermophiles. This name encompasses a disparate group of Archaea that are not particularly related, either phylogenetically or physiologically. While most species are extremely thermophilic, some are only moderate thermophiles. Moreover, some methanogens are extremely thermophilic but are not included in this group.

The name is justified because extreme thermophily is almost exclusively a property of Archaea. Although many more thermophilic Eubacteria have been described, only two genera, *Thermotoga* and *Aquifex,* have optimal temperatures for growth equal to or greater than 80°C. In contrast, many Archaea have optimal temperatures greater than 80°C, and *Pyrodictium,* the most extreme thermophile yet characterized, has an optimal temperature of 105°C. Organisms such as *Pyrodictium,* that can grow at 90°C or above, are also called hyperthermophiles. Interestingly, extremely thermophilic Archaea are abundant just above the temperature where thermophilic Eubacteria become rare

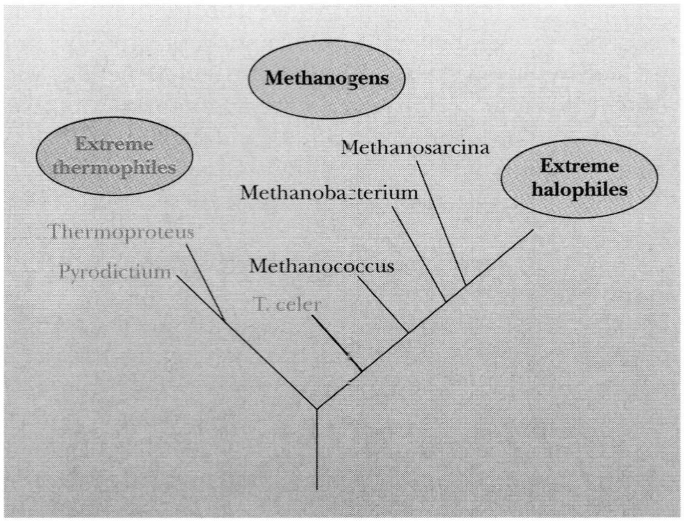

(Figure 22.11). The Archaea with optimal temperatures below 80°C are either methanogenic, extremely halophilic, or both moderately thermophilic and acidophilic. Thus, they occupy niches in which Eubacteria are absent or rare. A possible explanation for this distribution is that Eubacteria exclude the Archaea from temperate environments and Archaea are abundant only in habitats where Eubacteria grow poorly, such as high temperature or salt, or generate energy via mechanisms not utilized by the Eubacteria, such as methanogenesis.

The extremely thermophilic Archaea may be further subdivided into the obligate and facultative aerobes and the obligate anaerobes (Table 22.5). The aerobes are all acidophilic, and have pH optima close to 2.0. The obligate anaerobes are mostly neutrophilic, and have pH optima greater than 5.0.

Importantly, the extremely thermophilic Archaea are a recent discovery. More than 90 percent of the species that have been characterized were isolated after 1980, largely due to the pioneering work of Karl Stetter at the University of Regensburg, Wolfram Zillig at the Max Planck-Institut in Martinsried, and their colleagues in Germany. Because of their recent discovery, many aspects of their biology, physiology, and biochemistry are not yet understood. However, even at this early stage in their study, the extremely thermophilic Archaea have challenged many of the basic tenets of modern biology.

Habitats and Physiology

The extremely thermophilic archaeal microorganisms grow above the temperature limit for photosynthesis. Therefore, they are generally restricted to environments where geothermal energy is available. These areas include hot springs, solfatara fields, geothermally heated marine

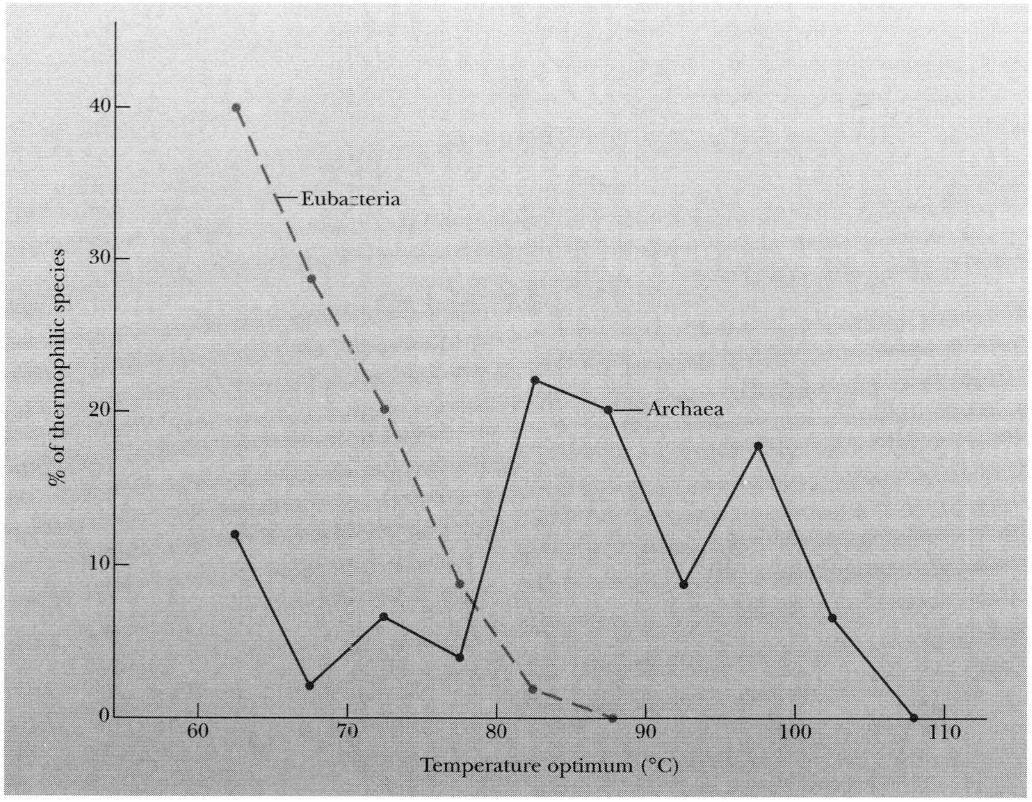

***Figure* 22.11** Optimum temperature for growth of thermophilic Archaea and Eubacteria. The percentage of 45 thermophilic species from each domain is given. (Modified from T. D. Brock, 1985, Life at high temperatures, *Science* 230:132–138.)

Differential ***Table*** **22.5 Summary of properties of the extremely thermophilic Archaea**

	Morphology	Electron Donor[1]	Electron Acceptor	Temperature Optimum (°C)	pH Optimum
Genus *Thermoplasma*[2]	Coccus	Organic	O_2, S°	59	2.0
Order *Sulfolobales*					
Family *Sulfolobaceae*					
Genus *Sulfolobus*	Irregular coccus	Organic, S°, H_2S, Fe^{2+}	O_2, MoO_4^{2-}, Fe^{3+}	70–87	2.5–3.0
Acidianus	Irregular coccus	H_2 S°, organic	S°, O_2	70–90	2.0
Metallosphaera	Coccus	S°, organic	O_2	75	3.0
Desulfurolobus	Irregular coccus	H_2, S°	S°, O_2	80	2.5
Stygiolobus	Irregular coccus	H_2	S°	80	2.5–3.0
Order *Thermococcales*					
Family *Thermococcaceae*					
Genus *Thermococcus*	Coccus	Organic	S°	88	5.8–6.5
Pyrococcus	Coccus	Organic	S°	100	6.2–7.0
Order *Thermoproteales*					
Family *Thermoprotaceae*					
Genus *Thermoproteus*	Thin rod	Organic, H_2	S°	85–88	5.0–6.8
Pyrobaculum	Thin rod	H_2	S°	100	6.0
Thermofilum	Thin rod	Organic	S°	85–90	5.0–6.0
Family *Desulfurococcaceae*					
Genus *Desulfurococcus*	Coccus	Organic	S°	85	6.0–6.5
Thermodiscus	Disk	Organic	S°	87	5.5
Pyrodictium	Disk, filaments	H_2	S°	105	5.5–6.5
Staphylothermus	Coccus	Organic	S°	92	6.5
Genus *Hyperthermus*[2]	Irregular coccus	Peptides, H_2	S°	100	7.0
Order *Archaeoglobales*					
Family *Archaeglobaceae*					
Genus *Archaeoglobus*	Coccus	H_2, lactate, organic	SO_4, S_2O_3	75–80	6.0–7.0

[1]Organic substrates include amino acids, sugars or yeast extract.

[2]Higher taxa (order and family) not named.

sediments, and submarine hydrothermal vents. In these environments, H_2, H_2S, and S° are abundant energy sources, and many of these microorganisms are chemolithotrophic and capable of some form of sulfur metabolism. *Acidianus infernus* is a particularly interesting example. It is an obligate autotroph and uses CO_2 as its sole carbon source. During aerobic growth, S° is oxidized to H_2SO_4 (Figure 22.12). Acid production maintains the pH at very low levels, and this organism is therefore an acidophile. Under anaerobic conditions, H_2 is oxidized and S° is concomitantly reduced to H_2S. Thus, S° functions as either the electron donor in the presence of O_2 or the electron acceptor in the presence of H_2. Other extremely thermophilic Archaea are capable of either one of these two modes of sulfur metabolism. The aerobe *Sulfolobus* oxidizes S° and organic compounds. It is also an acidophile.

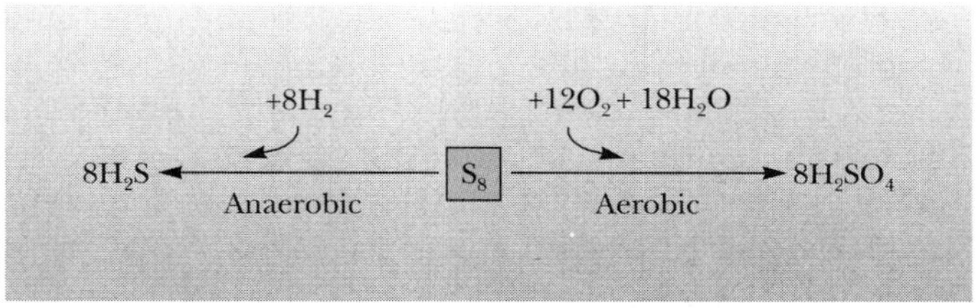

***Figure* 22.12** Sulfur metabolism of the facultative aerobe *Acidianus infernus*. During aerobic growth, elemental sulfur (S°) is oxidized to sulfuric acid. During anaerobic growth, elemental sulfur is reduced to hydrogen sulfide. The aerobic metabolism of *Acidianus* is typical of the obligately aerobic extreme thermophiles. The anaerobic metabolism is typical of the obligately anaerobic extreme thermophiles.

The anaerobes in the order *Thermoproteales* oxidize H_2, sugars, or amino acids using S^o as an electron acceptor. In addition, growth by fermentation or other types of anaerobic respiration may also occur.

Thermoplasma is an unusual Archaeum that was first isolated from burning coal refuse piles. It has since been found in hot springs as well, and two species have been identified, *Thermoplasma acidophilum* and *T. volcanium*. *Thermoplasma* is a facultative aerobe and grows on the peptide components of yeast extract. It has not been grown in defined medium. During aerobic growth, it utilizes an electron transport chain consisting of cytochrome *b* and a quinone. Nevertheless, its bioenergetics are poorly understood, and ATP is probably synthesized by substrate level phosphorylation. *Thermoplasma* lacks a cell wall, yet it is osmotically stable and contains glycoprotein and an unusual lipopolysaccharide in its cell membrane. Interestingly, *Thermoplasma* also contains a basic DNA-binding protein that is functionally very similar to the histones found in eukaryotes. However, a comparison of its amino acid sequence reveals that it is more closely related to the eubacterial DNA-binding proteins than to the histones.

Archaeoglobus is another exceptional extreme thermophile. It is the only Archaeum capable of sulfate reduction. Phylogenetically, it is a close relative of the methanogens and contains tetrahydromethanopterin, methanofuran, and coenzyme F_{420}. However, coenzyme F_{430} and coenzyme M are absent. During growth on lactate, *Archaeoglobus* utilizes an aceticlastic system similar to that found in the acetotrophic methanogens to oxidize acetyl-CoA (Figure 22.13). Methyltetrahydromethanopterin is then oxidized by the reverse of the pathway of CO_2 reduction.

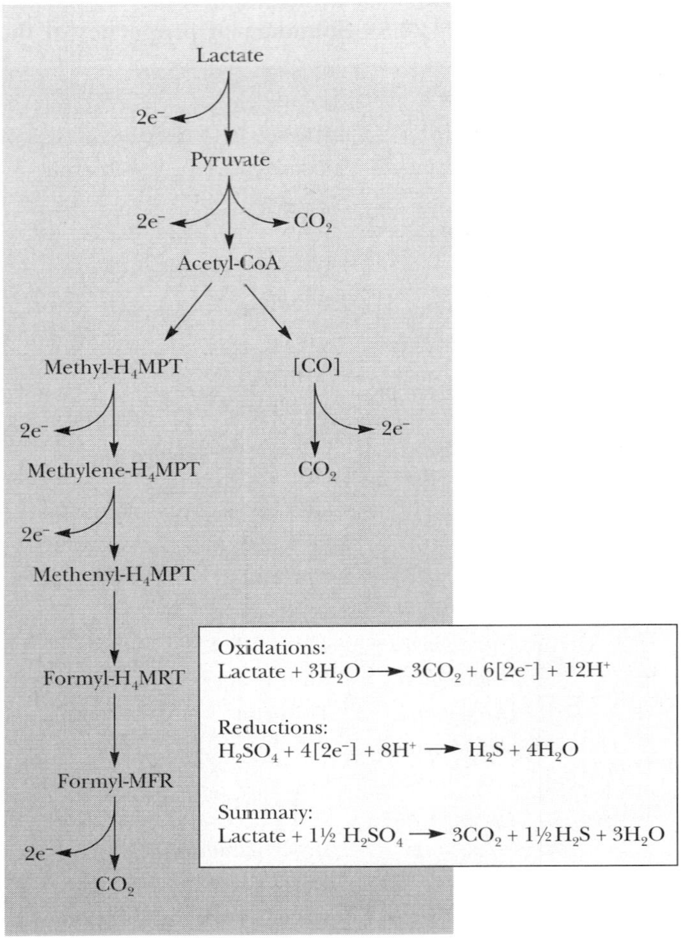

***Figure* 22.13** Lactate oxidation by the sulfate-reducing archaebacterium *Archaeoglobus*. MFR = methanofuran, H_4MPT = tetrahydromethanopterin.

Perspectives and Applications

The extremely thermophilic archaeal species are of scientific interest for several reasons. First, these bacteria grow at very high temperatures, putatively the upper temperature limit for life. The precise value for the upper limit is significant because it sets one boundary on the biosphere. Thus, Earth scientists studying geothermally heated aquifers, hot springs, deep vents, and similar geological formations need to know this boundary to determine where biological transformations can occur. Additionally, in the search for life on other planets, the upper temperature limit determines where life as we know it can thrive. Currently, the upper temperature limit for life appears to be about 110°C, which is the temperature maximum for growth of *Pyrodictium* and *Methanopyrus*. Studying the growth conditions and physiology of these organisms as well as isolating new species from geothermal environments may be of benefit in understanding the key events in geothermal environments that permit life at high temperature.

THERMOADAPTION A second area of interest is the mechanisms utilized by these bacteria to cope with high temperature. Most proteins and nucleic acids from mesophilic organisms denature at temperatures well below 100°C. Although a few mechanisms for growing at high temperatures have been elucidated, the overall adaptation to growth under extreme condition is still poorly understood.

One important mechanism of thermoadaption is the biosynthesis of specialized proteins and enzymes that are both thermostable and maximally active at high temperatures. For example, many enzymes from *Pyrococcus* are most active above 95°C and stable for several days at that temperature. These enzymes are virtually inactive at room temperature. While the biochemical basis for this thermoadaption is only partly understood, it probably requires an increase in the number of ionic and hydrophobic bonds that determine the tertiary structure.

However, not all biomolecules from extreme thermophiles are stable at high temperatures. Double-stranded DNA and the secondary structure of ribosomal RNA denatures well below the temperature optima of many hyperthermophiles. In some bacteria, special DNA-binding proteins enhance the thermostability of doubled-stranded DNA in vitro. Presumably, these proteins protect the chromosome in the living cell. Some small molecules are also thermolabile. NADH and NADPH are very unstable above 90°C. During growth on sugars, *Pyrococcus* utilizes the pyroglycolytic pathway, where glucose is oxidized by a ferredoxin-dependent glucose oxidoreductase instead of the conventional $NAD(P)^+$-dependent glucose dehydrogenase (see Figure 22.10). *Pyrococcus* also contains a ferredoxin-dependent aldehyde oxidoreductase instead of the conventional glyceraldehyde dehydrogenase. Presumably, this ferredoxin-dependent pathway is less temperature sensitive than the $NAD(P)^+$-dependent pathway.

Some extreme thermophiles also contain high concentrations of intracellular anions that may contribute to thermostability. Two compounds, called "thermoprotectants," that may have this role are di-inositol-1,1′-phosphate from *Pyrococcus* and cyclic 2,3-diphosphoglycerate (or cDPG) from some methanogens. However, thermoprotectants have not been identified in all extreme thermophiles, so this mechanism may not be general.

INDUSTRIAL APPLICATIONS Because of their high temperature optimum and thermostability, enzymes from extreme thermophiles may have important commercial applications. Activity at high temperature is important because most industrial processes operate between 50°C and 100°C. Enzymes with a temperature optimum in this range are cheaper to use because less protein is required for the same amount of activity. Thermostability is important because enzymes are expensive. Thermostability is also correlated with greater stability at mesophilic temperatures and resistance to denaturing chemicals. One of the major uses of moderately thermophilic enzymes is alkaline proteases in laundry detergent. These enzymes have high activity at 50°C, the temperature of wash water. They are very stable at room temperature, which extends their shelf-life in the supermarket. They are also resistant to the other components of laundry detergent that can denature enzymes, like surfactants and metal chelators. While proteases from extreme thermophiles have not yet been used in laundry detergents, many are being tested in this capacity.

An example of extremely thermophilic enzymes currently used industrially is the DNA polymerase from *Pyrococcus* and *Thermococcus.* The biotechnology industry uses these enzymes for DNA sequencing and the polymerase chain reaction. In addition to their high activity and thermostability, these enzymes are especially attractive because they have a very low error rate.

ORIGIN OF LIFE The extreme thermophiles represent very deep branches of the Archaea. Similarly, the extreme thermophiles, *Thermotoga* and *Aquifex,* represent the deepest branches of the Eubacteria. Based upon these observations, it has been proposed that the earliest organisms were extreme thermophiles. If true, extant extreme thermophiles may have retained some ancient characteristics that were lost in the evolution of mesophiles. Thus, the unusual enzymes and pathways common in extreme thermophiles may be remnants of the earliest organism. These microorganisms have been described as "living fossils" for this reason. Even if this hypothesis is false, it is a refreshing perspective on the origin of life.

Summary

- The **Archaea** are a diverse group of prokaryotes distantly related to the familiar **Eubacteria.**
- Unusual biochemical features of the Archaea are the presence of **isoprenoid-based glycerol lipids,** protein cell walls, a eukaryotic type of RNA polymerase, and insensitivity to many common antibiotics.
- The major physiological groups of the Archaea are the **methanogens,** the **extreme halophiles,** and the **extreme thermophiles.**
- The Archaea are widely distributed in aerobic and anaerobic, mesophilic and thermophilic, and freshwater and extremely halophilic environments.
- Methanogensis is a form of anaerobic respiration and requires a complex biochemical pathway that contains five unique or otherwise rare coenzymes: **methanofuran, methanopterin, coenzyme M, coenzyme F_{420},** and **7-mercaptoheptanoylthreonine phosphate.**
- Methanogens catalyze the last step in the anaerobic food chain in which organic matter is degraded to CH_4 and CO_2. This food chain is the basis of many **symbiotic associations** of the methanogens with eukaryotes and Eubacteria.
- The extreme halophiles among the Archaea contain **bacteriorhodopsin,** a light-driven proton pump that contains **retinal.** This type of photosynthesis differs fundamentally from chlorophyll-based photosynthesis, common in plants and Eubacteria, because it is not coupled to an electron transport chain.

- The enzymes from the extreme halophiles are unusually salt tolerant and many also require salt for activity and stability.
- Organisms with a growth optimum greater than 80°C are called **extreme thermophiles.** Organisms that grow above 90°C are called **hyperthermophiles.** Most extreme thermophiles and hyperthermophiles are Archaea.
- The extreme thermophiles are common in geothermal habitats where H_2 and reduced sulfur compounds are abundant.
- The extreme thermophiles are a rich source of thermostable enzymes, which may have valuable applications in biotechnology.

Questions for Thought and Review

1. What was the basis for establishing the Archaea as a domain? On what is the three-domain concept based?
2. Give major reasons for studying the Archaea. What can we learn from Archaea that is not available from the Eubacteria?
3. What are some of the distinguishing features of the Archaea? How do these affect antibiotic sensitivity?
4. Where would one find methanogens in nature? How do they obtain energy?
5. Methanogens have some unusual coenzymes: to what eubacterial coenzymes are these related and what are their functions? Some coenzymes are unique to methanogens; give an example.
6. How does the presence of nitrate or sulfate affect methanogenesis? Why?
7. Are the halobacteria autotrophs or heterotrophs? Give reasons for your answer.
8. What is bacteriorhodopsin? How does photosynthesis with this compound differ from chlorophyll-based photosynthesis?
9. How have halobacteria adapted to growth at high salt concentration? Solutes? Proteins?
10. Define an extreme thermophile. What habitats do they occupy?
11. What can we learn from the hyperthermophiles? How might they contribute to industry?
12. Which of the archaeal groups is found in typical environments living in association with many other organisms? Why?
13. Recently a marine sponge has been found to harbor a symbiotic bacterium that is most closely related to the thermophilic Archaea. Explain the significance of this.
14. Which group of Eubacteria do the hyperthermophiles most closely resemble physiologically?

Suggested Readings

Balows, A., H. G. Truper, M. Dworkin, W. Harder, and K. H. Schleifer, eds. 1992. *The Prokaryotes.* 2nd ed. New York: Springer-Verlag.

Ferry, J. G. 1993. *Methanogenesis, Ecology, Physiology, Biochemistry & Genetics.* New York: Chapman & Hall.

Jones, W. J., D. P. Nagle, Jr., and W. B. Whitman. 1987. "Methanogens and the Diversity of Archaebacteria". *Microbiological Reviews* 51: 135–177.

Kates, M., D. J. Kushner, and A. T. Matheson. 1993. *The Biochemistry of Archaea (Archaebacteria).* New York: Elsevier.

Stetter, K. O. 1986. *Diversity of Extremely Thermophilic Archaebacteria.* In T. D. Brock (ed.) *Thermophiles: General, Molecular, and Applied Microbiology.* New York: John Wiley & Sons, Inc.

Wiegel, J., and L. G. Ljungdahl. 1986. "The Importance of Thermophilic Bacteria in Biotechnology". *CRC Critical Reviews in Biotechnology* 3: 39–107.

Woese, C. R., and R. S. Wolfe. 1985. *Archaebacteria.* New York: Academic Press, Inc.

And in the summer, when I feel disposed to look at all manner of little animals, I just take the water that has been standing a few days in the leaden gutter up on my roof, or the water out of stagnant shallow ditches: and in this I discover marvellous creatures.

From letter of Antony van Leeuwenhoek to Constantijn Huygens, Dec. 26, 1678

Chapter 23

Eukaryotic Microorganisms

Characteristics of Eukaryotic Cells
The Protistan Groups—Algae and Protozoa
The Fungi
Eukaryotes and Multicellularity

Comparisons of rRNA sequences have demonstrated that there are three primary divisions in the living world. Members of two of these groups, the Archaea and Eubacteria, are prokaryotic. That is, they lack a membrane-bound nucleus which contains the cell's DNA. All known members of the third group are **eukaryotic,** which means that they contain a true nucleus. Presumably, members of this third group did not arise as eukaryotes, and were originally structurally simpler forms. Whatever has become of these structurally simpler forms is not known. It may be that some members of the eukaryotic lineage, which lack a nucleus, still exist somewhere in the world, but no such organisms have been found yet.

One of the most apparent features of eukaryotic cells is their knack for multicellularity, their capacity to cooperate to form single organisms containing millions of cells. There are no prokaryotic equivalents of oak trees, whales, or mushrooms. All of the large organisms in the world are eukaryotic, which makes it possible to overlook the existence of eukaryotic microorganisms. Nonetheless, such unicellular forms exist, and they form the bulk of the eukaryotic world. In age and genetic diversity, members of this group, the protists, dwarf the three multicellular eukaryotic kingdoms traditionally recognized—the green plants, animals, and fungi. Not all protists are unicellular; multicellularity has arisen many times among the eukaryotes.

After a discussion of general eukaryotic characteristics, the major protistan and fungal groups will be described in this chapter.

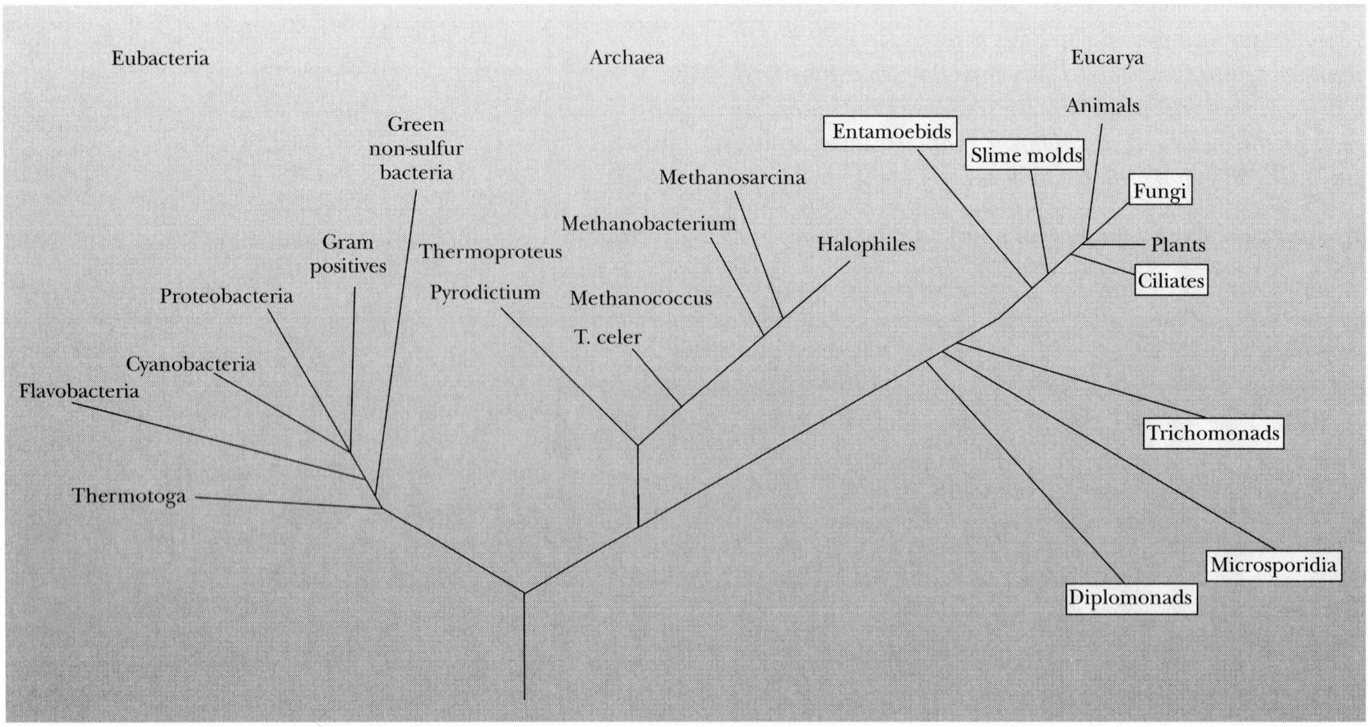

Characteristics of Eukaryotic Cells

There are many major differences between eukaryotes and both groups of prokaryotes (see Figures 1.6 and 1.7). These differences include the presence of a nucleus, mitosis, sexual recombination, internal membrane systems, mitochondria and chloroplasts, the way cell shape is maintained, and the way cells move (Table 23.1). A discussion of these differences follows.

The Nucleus

The presence of a nucleus in eukaryotic cells is only one manifestation of the differences that exist between prokaryotes and eukaryotes in the way that DNA is stored in cells and distributed between cells at the time of division. Most eukaryotes have histones associated with their DNA. These are basic proteins around which the DNA is wrapped. Histones have been found in only one prokaryote.

Eukaryotes tend to have much more DNA than prokaryotes do. This may be due to the greater physical complexity of eukaryotes; it may take many more genes to make a eukaryotic cell. Whatever the reason, the difference in the amount of DNA can be dramatic. The length of DNA in an *E. coli* genome is 1.3 millimeters. The length of DNA in a human cell nucleus is 1.8 meters! Eukaryotes are no longer able to divide up their DNA in the simple way that bacteria employ during cell division, and they have had to develop more elaborate mechanisms to ensure the proper segregation of chromosomes to daughter cells.

Eukaryotic nuclei contain many chromosomes instead of having a single chromosome as prokaryotes do. Breaking the DNA up into smaller pieces presumably makes it easier to move the DNA around in the cell at the time of division and easier to keep strands from being caught in the fission furrow that will constrict the parent cell into two new daughter cells. Humans have 46 chromosomes, and even the smallest of these contains more DNA than the entire *E. coli* chromosome does. Some eukaryotic microorganisms have hundreds of chromosomes.

Before division occurs, the chromosomes are duplicated. In prokaryotes, these replicated chromosomes are attached to the cell membrane. The points of attachment, which are initially right next to each other, are gradually spread apart by the growth of the cell and the expansion of the membrane. A transverse extension of the cell wall forming between these sites divides the cell in two and segregates the chromosomes into the daughter cells.

Mitosis

Instead of having to split only two chromosomes apart from each other, eukaryotes have a large number to separate. A complex three-dimensional scaffold called a spindle is required to place all the pairs of chromosomes into the proper orientation for separation from each other and also to effect the separation (see Figure 1.8).

Differential *Table* **23.1 Major differences between eukaryotes and prokaryotes**

Prokaryotes	Eukaryotes
No membrane around the cell's DNA	Nucleus present
No mitosis	Mitosis
No meiosis	Meiosis
No histones (one exception)	Most have histones
Few internal membrane systems	Many internal membrane systems
No tubulin	Microtubules composed of tubulin present
No actin	Actin present
No internal cytoskeleton	Internal cytoskeleton
Flagella with flagellin	Flagella with microtubules and dynein
No motility system based on actin or tubulin	Motility systems based on actin or tubulin
No ingestion of particles	Ingestion of particles (phagotrophy) widespread
No mitochondria	Mitochondria usually present
No chloroplasts	Chloroplasts sometimes present
Multicellularity rare	Strong tendency toward multicellularity; multicellular forms common
70s ribosomes	80S ribosomes with more proteins and larger ribosomal RNAs than in 70s ribosomes
No separate 5.8S rRNA	5.8 S rRNA almost always present

Some of the spindle fibers from opposite poles overlap each other. Some only go partway across and are connected to the chromosomes. The fibers attached to the chromosomes shorten at the same time that the longer fibers slide along each other to lengthen the spindle. The combined effect of these actions is to spread apart the chromosomes. After the chromosomes have moved toward the poles of the spindle and are divided into two groups, the spindle, which just formed before division, starts to break down again. As the spindle disappears, the two groups of chromosomes differentiate into two new daughter nuclei, and the cell divides between them.

In most eukaryotes, the individual chromosomes are visible only during nuclear division. During most of the life of a cell, the chromosomes exist as long extended strands of DNA and associated proteins too thin to be seen in a light microscope. Before division occurs, the chromosomes condense into tightly packed masses of DNA and protein that are thick and short enough to be seen with a light microscope. The actual distance that the chromosomes are moved apart by the spindle is usually quite small, so this condensation may be necessary for separation to occur. Without compaction, the chromosomes might not be adequately separated by the spindle, just as the strands in a plateful of spaghetti would not be separated from each other by sticking a fork into the mass and moving it one centimeter this way or that.

Normally, nuclear division immediately precedes cell division; the two processes are coupled. However, this is not always true. In some eukaryotic microorganisms, growth is accompanied by nuclear division but not by cell division, so that a very large cell with many nuclei is formed. This shows that nuclear and cell division are separate processes, and since it is sometimes desirable to talk about only one or the other, these two types of division have their own names. Nuclear division is called mitosis, and cell division is called cytokinesis. Production of daughter cells through mitosis and cytokinesis is called asexual reproduction.

The details of mitosis vary a great deal. In some organisms, the spindle is formed inside the nucleus, and the nuclear envelope remains intact while the chromosomes move to opposite sides. The old nucleus is eventually pinched in half to form two new ones. This is called closed mitosis. In many organisms, including humans, the spindle forms outside of the nucleus and chromosomes become attached to the spindle after the nuclear membrane breaks down. Membranes form again at the end of mitosis around each of the two clumps of chromosomes to form two new nuclei. This is called open mitosis. In humans, the spindle fibers grow out of two specific cytoplasmic areas that have small bodies called centrioles in their centers. Centrioles are structurally like the kinetosomes found at the bases of cilia and flagella. The nucleating centers from which spindles grow in other eukaryotes do not all have centrioles. Other structures, or no visible structure at all, may be present in protists. The chromosomes are condensed to varying degrees in different groups. In some cases, discrete chromosomes are never seen. In other cases, chromosomes are permanently condensed, and may be seen whether the nucleus is dividing or not. All of these little details are useful in taxonomy, and in deciding who is evolutionarily related to whom. Some of these features will be mentioned in the discussion of individual groups. Even given the fact that eukaryotes have decided to fragment their genome and move the pieces around on a spindle, there is a great deal of variety possible in how this can be accomplished, and protists seem to have tried out many of the variations.

Sexual Recombination

This process of cell division only allows mutations to build up slowly over evolutionary time. A set of mutations, which together may be quite advantageous, could only appear in an evolutionary lineage in the length of time it takes for all of them to independently appear. Evolution could proceed more rapidly if genes were shuffled between organisms in a manner that bypasses the limitations of mitosis, in which the genome is passed from generation to generation whole and uncontaminated with DNA from elsewhere.

There is such a process built into the life cycles of many eukaryotes, and it is called sexual recombination. It is frequently referred to as sexual reproduction, because in the familiar variant seen in multicellular organisms, this recombination accompanies the production of new individuals. Two gametic cells, one from each parent, fuse to form a zygote, which will grow into a new individual. However, in unicellular eukaryotes, in which a gamete is by necessity the whole organism, the fusion of cells actually results in the net *loss* of individuals (perhaps in this case, it would be more correct to speak of sexual deproduction!). It is necessary to separate the concept of genetic recombination occurring through sex from the idea of the multiplication of individuals in order to have an accurate impression of the role of sex in the lives of protists.

Genetic recombination also occurs in prokaryotes, but the way in which it is incorporated into the life cycle constitutes a major difference between the prokaryotes and eukaryotes. In the case of bacteria, recombination involving the direct passage of chromosomal DNA from one cell to another occurs only sporadically. It is effected by the temporary union of two bacterial cells, and the passage of a strand of DNA from the donor cell to the recipient cell (see Chapter 15). The recipient cell is then left to integrate the new DNA into its own genome through the physical recombination of the two pieces of DNA by a process of snipping and gluing. Leftover pieces are discarded, so that at the end of the process, the recipient cell still has just a single chromosome. The donor cell does not lose any chromosomal genes, since the transferred segment is copied before being transferred. Chromosomal recombination in bacteria is a haphazard process in which variable amounts of DNA are passed from the donor to the recipient cell. Bacteria may also pass plasmids to each other, and DNA may also be exchanged by viruses moving between cells.

Sexual recombination in eukaryotes is both more uniform and more complete. Instead of the temporary union of cells seen in bacteria, there is usually a complete fusion of cells. This is followed by the fusion of the two nuclei, so that a cell is produced that contains the complete genomes of both parent cells. Both genomes are retained, and genes from both genomes are used in making RNA. Novel combinations of mutations are brought together in the same cell by combining genomes from different organisms in this way. Fusing together nuclei in each generation would result in organisms having enormous amounts of DNA, and could not proceed indefinitely. Eukaryotes have invented a way of periodically halving the DNA in a cell. This is accomplished by a modified version of mitosis called meiosis (see Figures 1.8 and 1.12). In life cycles of eukaryotes that have a sexual phase, zygote formation is balanced by meiosis. In mitosis, chromosomes are replicated once and then separated once by the action of the spindle into the two daughter cells. In meiosis, chromosomes are replicated once, but then separated twice on spindles into four daughter cells. To see how this works, consider the fate of a gene present in a single copy in each gamete. The copies may or may not be identical, depending on whether or not one of the gametes carries a mutation. When a zygote is formed, it will have two copies of the gene, one from each parent gamete. When the zygote undergoes mitosis, each of its chromosomes is first replicated, so that it briefly has four copies of the gene, two copies from each parent. The spindle will separate these so that two copies wind up in each of the two daughter cells formed after cytokinesis. Furthermore, mitosis always occurs in such a way that these two copies represent one copy from each parent. When meiosis occurs, the initial division separates the four copies into two, and then these sets of two are split apart in the second meiotic division. Ultimately, two cells will each have a single copy of the gene that came from one gametic ancestor and the other two cells will have single copies of the gene that was originally present in the other gametic ancestor.

Although the most visible events of the sexual process in eukaryotes consist of this process of mixing and sorting of whole chromosomes, a process that has no counterpart in the prokaryotic world, the mechanical events seen in bacterial recombination at the DNA level are also occurring. During meiosis, homologous chromosomes (those chromosomes containing the same sets of genes) line up in physical contact with each other. Individual chromosomes sometimes break at this time. They may rejoin, reforming the same original chromosome, or they may recombine with homologues to form a new chromosome consisting of a mixture of pieces from the two different homologues (see Figure 1.12). Thus the processes occurring in prokaryotes remain in eukaryotes although they have been overshadowed by the novel and more visible method of shuffling around discrete chunks of DNA to produce new combinations of genes.

Terms have been created to describe the **ploidy** level, the number of copies of a genome present in a cell. The gametes described above, with one copy of each chromosome, are said to be **haploid.** The zygote they produce through their fusion, in which there are two copies of each different chromosome, is **diploid.** Shortly before its nucleus divides, it would be **tetraploid,** since the chromosomes would replicate, and there would be four copies of each different chromosome.

The organisms most familiar to us—humans and other vertebrates—spend most of their lives as diploids. The haploid state, seen in our gametes, is a very transient state. Haploid cells do not lead an independent existence in our life cycle. Eukaryotic microorganisms, on the other hand, vary tremendously in how meiosis and zygote formation are incorporated into the life cycle. Which ploidy state is dominant in the life cycle varies from group to group. The green algae like *Chlamydomonas* (Figure 23.1) are haploid organisms throughout most of their life. It is the haploid cells that swim around, photosynthesize, and divide to form more individuals asexually. Some of these cells occasionally fuse with each other to form a zygote. The zygote quickly undergoes meiosis to form more haploid cells. It is the diploid state that has a transient existence in the life cycle of a green alga. In other forms, both haploid and diploid stages are long-lived forms that lead an independent existence. Both cell types can multiply. This is true of some red algae, for example. Both the haploid and diploid stages undergo mitosis to produce multicellular "plants." These sometimes look so different from each other that they were regarded as different species until the life cycle was determined (Figure 23.2). These differences in ploidy level and life cycles are only seen in eukaryotes. They are not found in prokaryotes and only exist as a consequence of the type of genetic recombination that the eukaryotes have developed.

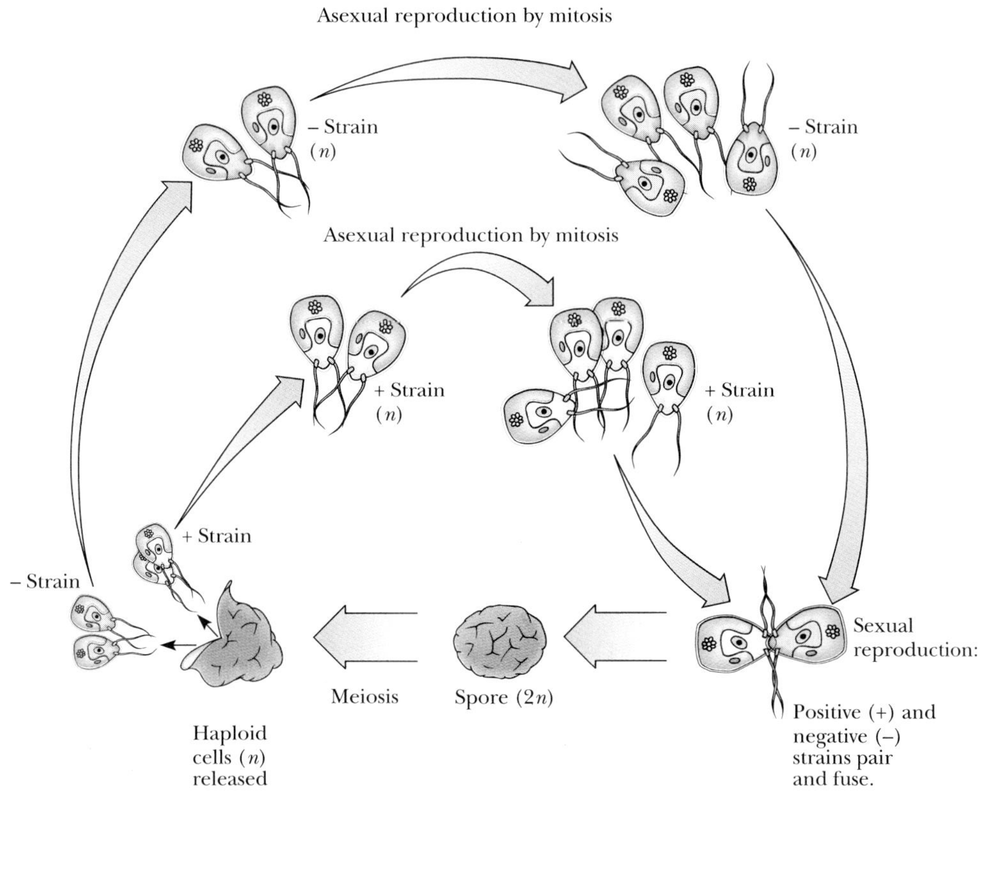

***Figure* 23.1** The *Chlamydomonas* life cycle. The biflagellated cells that swim around and multiply asexually by mitosis are haploid cells. They are of two different mating types, "+" and "−," and are sometimes induced to fuse to form diploid zygotes. The zygotes can undergo a period of dormancy before undergoing meiosis and producing more active cells. The actively swimming, multiplying cells in the *Chlamydomonas* life cycle are always haploid.

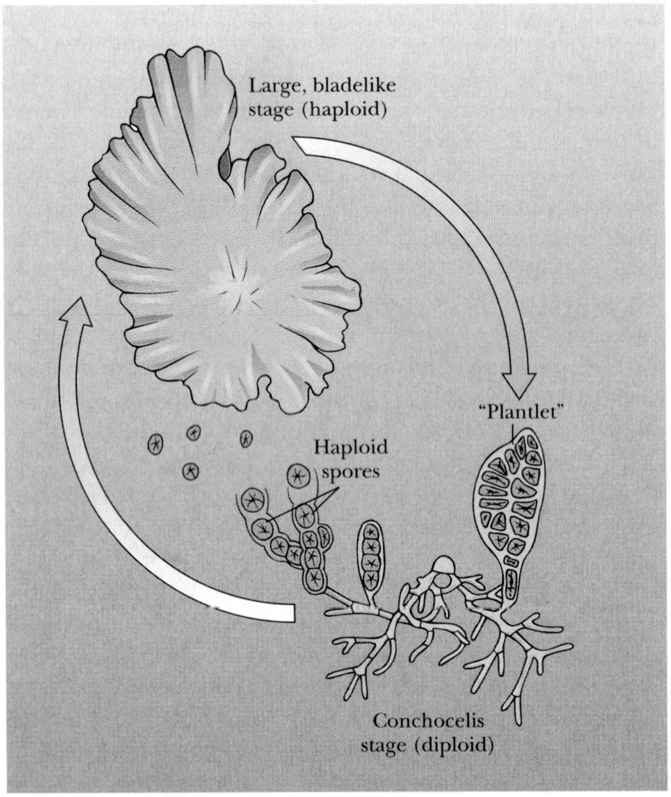

***Figure* 23.2** The *Porphyra* life cycle. The haploid stage is a large leaflike structure. The diploid conchocelis stage is a tiny inconspicuous filamentous structure that grows in mussel shells. The two stages were originally described as different genera; *Conchocelis* was the genus name assigned to the small filamentous form.

Internal Membrane Systems

The separation of chromosomes from the cytoplasm by a nuclear envelope is just one manifestation of the much greater use of internal membranes by eukaryotic cells than by prokaryotic cells. Most prokaryotes do not have well-developed internal membrane systems, whereas all eukaryotes have at least a nuclear envelope, and usually much more than that.

The nucleus is actually bounded by two closely spaced membranes. These two membranes come together at a series of points around the surface of the nucleus. Channels which connect the inside of the nucleus to the cytoplasm on the outside are present at these junctions. The space between the two membranes is continuous with the lumen of the endoplasmic reticulum, a system of channels and vesicles that extend throughout the cell and which are lined with an extension of the outer nuclear membrane. The endoplasmic reticulum is a site of fatty acid synthesis and metabolism. It is also where proteins that will be secreted from the cell are synthesized. They are made on ribosomes that line the endoplasmic reticulum and they accumulate in the lumen as they are made. These secretory proteins are further processed in the Golgi apparatus. This organelle consists of a series of flattened, closely spaced vesicles that in cross-section look somewhat like a stack of hollow plates or bowls. Proteins enter one side of the Golgi apparatus from the endoplasmic reticulum and leave the other side in small vesicles that bud off from the surface. The Golgi apparatus is a polarized organelle; material passes through in a one-way flow as on a factory assembly line. In the Golgi, carbohydrates or phosphates can be added to proteins, or an inactive protein can be cleaved by enzymes to form a biologically active molecule.

Lysosomes also bud off from the Golgi. These are small, membrane-bound organelles that contain digestive enzymes. They fuse with vacuoles in the cell and release their contents into it, to digest the material within. The vacuolar contents may be old organelles that need to be broken down and replaced with new organelles. They may also be things that were deliberately brought into the cell from outside, such as an item of food, as in the case of a eukaryote feeding on bacteria.

Eukaryotic microorganisms also frequently contain contractile vacuoles, membranous structures that collect and expel water from the cell (Figure 23.3**a**). Protists are permeable to water, and when the solute concentration is higher inside the cell than outside, water is drawn into the cell by osmosis. Without a way to remove this water, the cells would swell and burst. Water collects inside the contractile vacuole, which periodically opens up to the outside of the cell and expels its contents (Figure 23.3**b**).

These membranous systems are all connected with each other by an ongoing traffic of small vesicles that are constantly fusing with and budding from the larger membranous organelles. The membrane systems in eukaryotic cells are very active structures in a constant state of flux.

Mitochondria and Chloroplasts

Many eukaryotes have two other membrane-bound organelles, quite unlike those described above. These are the mitochondria and chloroplasts. They are actually the remnants of Eubacteria taken into ancestral eukaryotic cells eons ago. Rather than being digested, they established a symbiotic relationship with their host cell. The host cells fed them and moved them around while the eubacterial metabolic pathways provided their hosts with much more efficient ways of generating ATP, and with a way of using sunlight as an energy source. The pathways involved and the nucleotide sequences of genes in the organelles indicate that mitochondria are descendants of proteobacteria, and that chloroplasts are highly modified cyanobacteria. Mitochondria generate ATP through the oxidation of organic molecules and through the agency of a cytochrome electron transport chain. Photosynthesis takes place in chloroplasts in the same way that it occurs in cyanobacte-

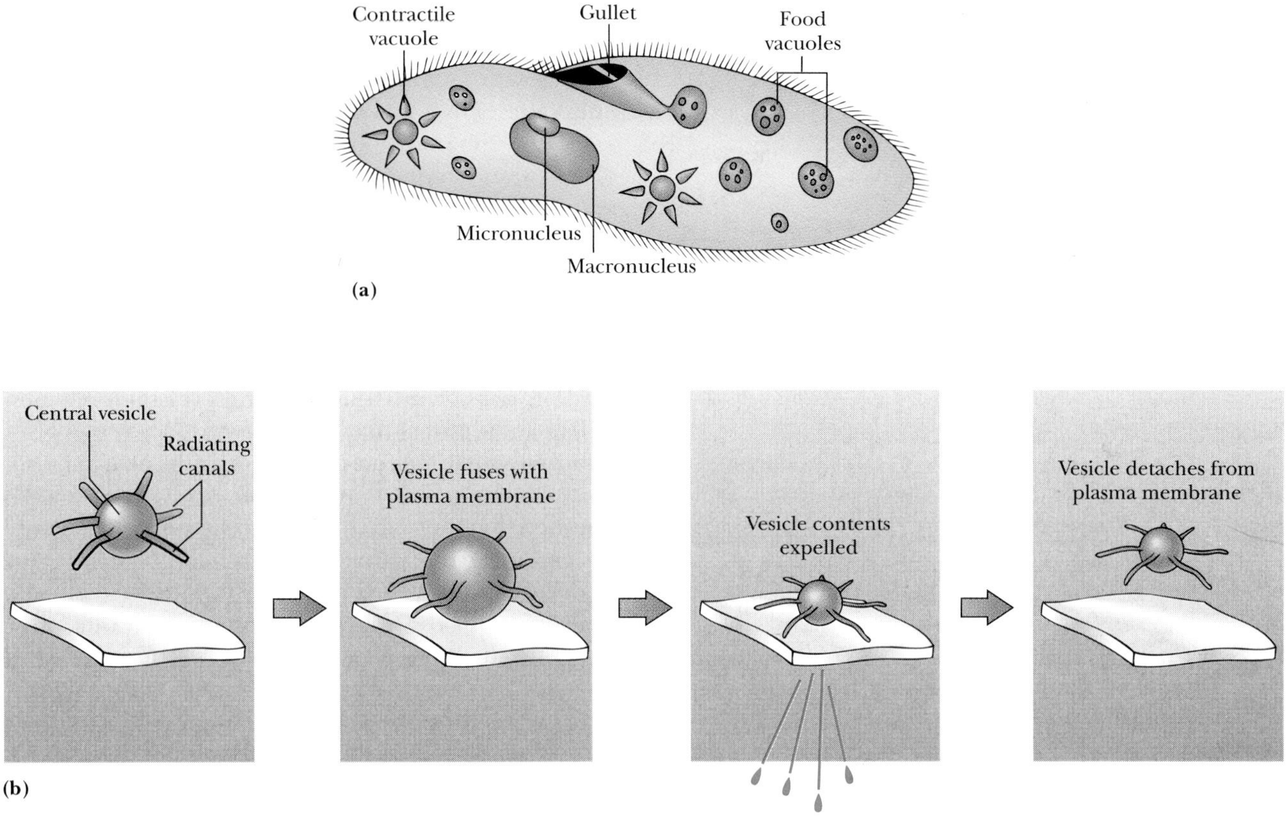

***Figure* 23.3** The contractile vacuole system of *Paramecium caudatum. P. caudatum* **(a)** is a ciliated protozoan in which the activity of the large contractile vacuole system **(b)** can easily be watched with the light microscope. The cell also contains a macronucleus, from which mRNA is transcribed; a micronucleus, which is only used in the sexual phase of the life cycle; a gallet, down which bacteria are swept into food vacuoles for digestion and extrusive organelles called trichocysts, which line the cell surface. Radiating canals collect water from the cytoplasm and send it into a central vesicle. The central vesicle periodically fuses with the plasma membrane and expels the water from the cell.

ria. The bacteria have degenerated to such an extent that an independent existence is no longer possible for them. All that remains of their bacterial past is a couple of biochemical pathways and a handful of genes.

Mitochondria have an outer and an inner membrane. The inner membrane runs parallel to the outer membrane in part, but is also drawn into a series of folds which project inward into the mitochondrion. These projections of the inner membrane are called cristae (see Figure 1.9). Cristae may be tubular, lamellar, or discoid. The crista type is characteristic of a given group of eukaryotes, but what physiological significance these physical differences have, if any, isn't known. Not all eukaryotes have mitochondria, but most do.

A smaller number of eukaryotes have chloroplasts (Figure 23.4**a–c**). Protists that have chloroplasts are called **algae.** All eukaryotes that have chloroplasts also have mitochondria. Inasmuch as the acquisition of chloroplasts and mitochondria were independent events, there is no obvious reason why a eukaryote that lacked mitochondria could not have acquired chloroplasts, but this has apparently never happened. It may be that protists lacking mitochondria have a sensitivity to oxygen, which would prevent them from harboring an oxygen-generating organelle. Mitochondria-bearing protists require oxygen; mitochondria won't work without it. Any anaerobic protists originally sensitive to oxygen would have had to adapt to the presence of oxygen for mitochondria to become established in their cytoplasm. They would then be able to acquire chloroplasts without being damaged by them.

Chloroplasts also have internal membranes. The arrangement of membranes varies between algal groups and is of taxonomic significance. So are the types of pigments found in the chloroplasts. Pigments are the organic molecules that absorb the light used in photosynthesis. Chlorophyll is universally present in chloroplasts. They all contain chlorophyll *a* and may contain chlorophyll *b* or *c* as well (see Figure 9.1). These different chlorophylls differ from each other in only minor ways, involving the nature of the side groups attached to the main ring system.

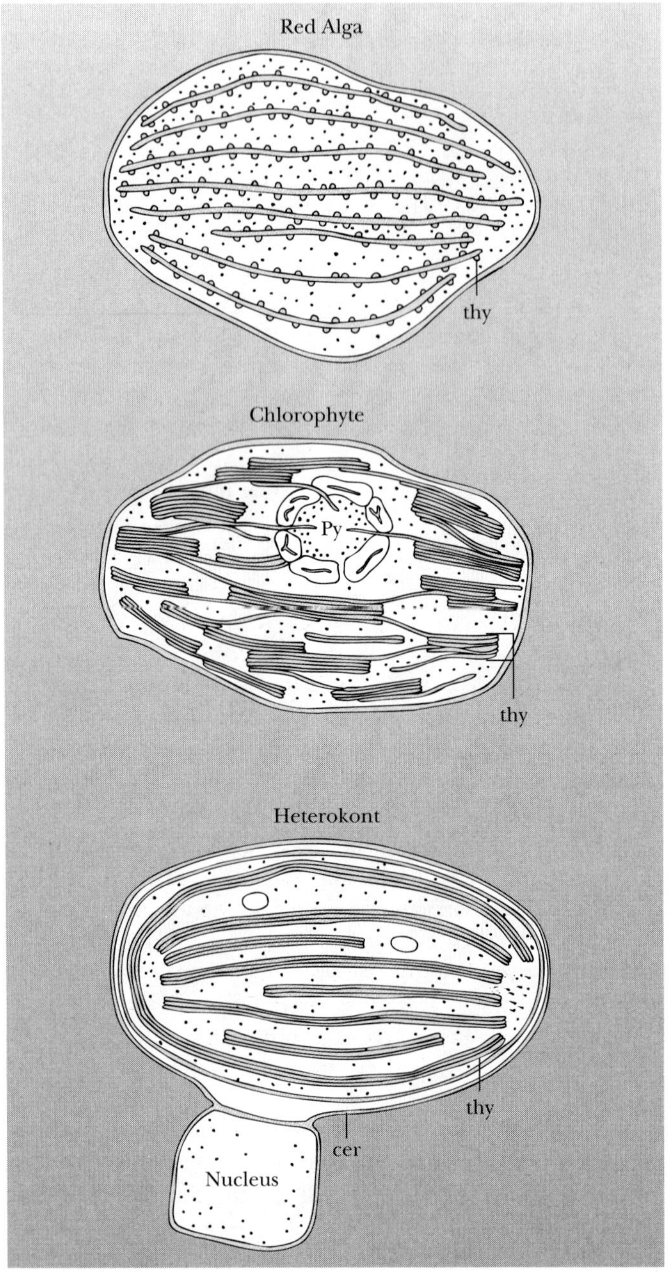

***Figure* 23.4** Some of the basic types of chloroplasts seen in protists. In red algal chloroplasts **(a),** the internal chloroplast membranes, the thylakoids (thy), exist singly and are covered with small proteinaceous grains, the phycobilisomes, which assist in light harvesting. The thylakoids in chlorophyte chloroplasts **(b)** are arranged in stacks. Starch synthesis takes place on a structure called a pyrenoid (py). The thylakoids in heterokont chloroplasts **(c)** are set in groups of three. There is an outer chloroplast membrane (cer) which is part of the endoplasmic reticulum system and is continuous with the nuclear membrane.

The Cytoskeleton and Movement

Prokaryotic and eukaryotic cells differ in the way that cell shape is maintained. Most prokaryotes have a stiff external envelope, the cell wall, but nothing inside the cell for maintaining its shape. Eukaryotes usually have an internal **cytoskeleton,** a system composed of fibrous elements. Some (algae, plants, and fungi) have also secondarily acquired an external cell wall of some type.

Three general types of fibrous elements having a wide distribution among eukaryotes are recognized. These are microtubules, microfilaments, and intermediate filaments. **Microtubules** are long hollow fibers 24 nanometers in diameter. They are made up of molecules of the protein tubulin. **Microfilaments** are about 7 nanometers wide and they consist of two strings of actin monomers wound around each other. **Intermediate filaments** are intermediate in size. Protists have all three of these elements as well as a few others that don't fit into these categories. Microtubules may extend in parallel underneath the surface of a cell like so many microscopic ribs (Figure 23.5) or be grouped together into bundles to support extensions from the cell surface (Figure 23.6**a–b**). Microfilaments running underneath the cell surface in an ill-defined sheath give the surface of the cell some strength and rigidity. These structures do not necessarily have a fixed length. They can be made as long or short as necessary, and some are constructed temporarily for a particular task. This gives eukaryotic cells a plasticity in shape not ordinarily seen in prokaryotes.

By attaching accessory proteins to these stiff elements, they can be made to slide past each other. This provides a way of moving things around inside the cell. The spin-

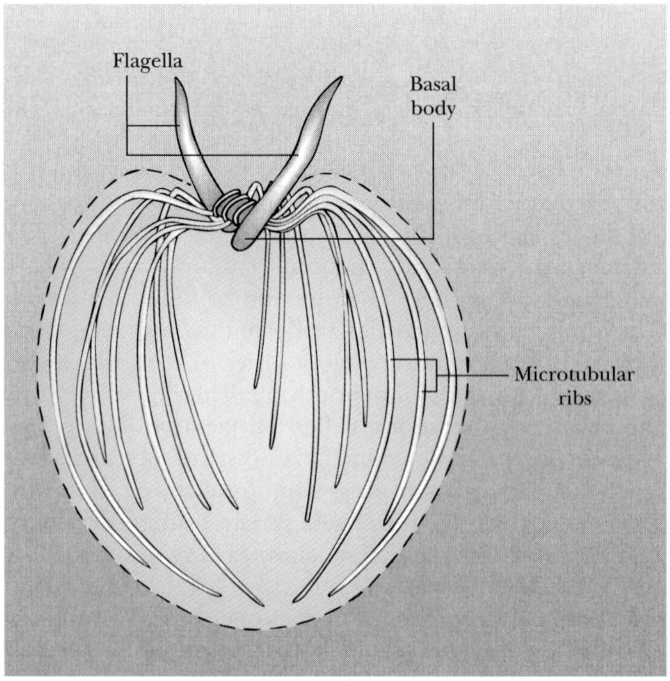

***Figure* 23.5** Microtubular ribs radiating out from the flagellar basal bodies in *Chlamydomonas* run underneath the surface of the cell. The outline of the cell is indicated by a dotted line.

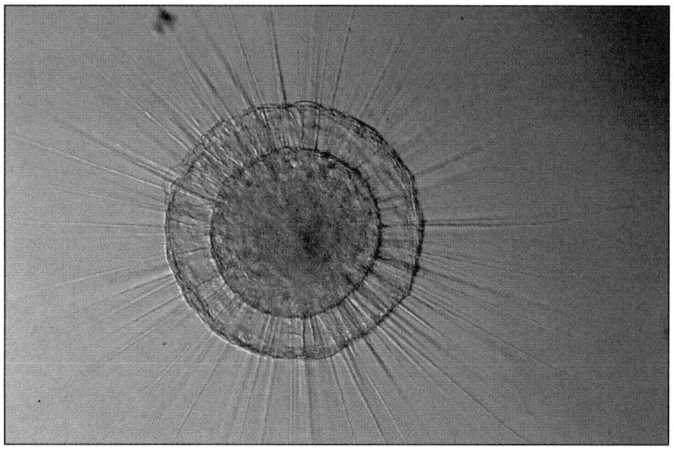

(a)

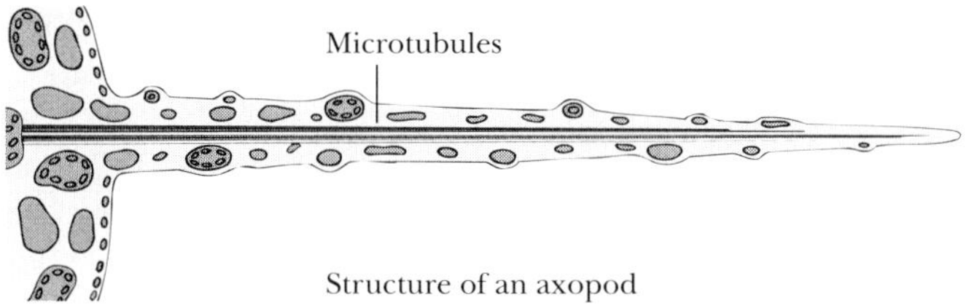

(b)

***Figure* 23.6** Heliozoans **(a)** use long thin projections from the body called axopods for capturing prey. Small organisms stick to the axopods after bumping into them, and are then drawn down to the body of the cell and digested. (© T. E. Adams/VU) Each axopod **(b)** is supported by a bundle of microtubules.

dle used in nuclear division is such a microtubular structure; the microtubules slide along each other to lengthen the spindle. Microtubules can also be used to move the cell itself, or the water around the cell. The hairlike microtubule-containing entities which project from the cell and carry out these tasks are the cilia and flagella (see Figure 1.10). They are ultrastructurally alike and operate by the same basic mechanism. However, flagella tend to be longer, show a greater variety of movement patterns, and tend not to be used in large synchronously beating groups, as cilia often are.

A basal body or kinetosome (two names for the same object) is present in the cytoplasm immediately underneath the cilium or flagellum; they are an outgrowth of a basal body. A basal body is ultrastructurally identical to a centriole, which also in some more indirect way organizes the growth of microtubules. The basal body consists of nine groups of microtubules set near the perimeter of the organelle. Each group contains three microtubules. The cilium above it has two central microtubules and nine groups of two microtubules set around the periphery. A series of short arms extend along the length of one microtubule in each group. These are composed of the protein dynein. The dynein arms walk along the nearest microtubule of the neighboring pair. If the microtubules were free, they would slide along each other as the microtubules in a spindle do. But the microtubules are attached at the base of the cilium, so the walking action of the dynein arms causes the tubules to bend. This happens rapidly and repeatedly, so that the cilia are said to be beating. A field of cilia beating this way generates sufficient force to propel the cell forward. Even a single flagellum can propel a small cell through the water. The bases of cilia and flagella are anchored in the cytoplasm by a complex array of microtubules and microfibrils that radiate away from the kinetosomes. The composition and arrangement of these accessory elements is characteristic for a given group of protists. Consequently, the ultrastructure of these elements (which together with the kinetosome and flagellum or cilium constitute the **kinetid**) has been used to determine which group of protists enigmatic organisms belong to, and to infer evolutionary relationships between groups.

Actin is used in the muscles of animals, where actin and myosin molecules slide along each other in a manner similar to that seen in microtubular movement, although the details of the processes are very different. Actin is also part of the contractile system responsible for pinching a eukaryotic cell in two during division. Actin is responsible for the movement of amoeboid organisms as well. Amoebae are highly deformable cells (Figure 23.7) that move through temporary extensions of the body called *pseudopods*. The body flows into an advancing pseudopod. Pseudopods can also be used to capture food particles. The way in which actin is used during amoeboid movement is by no means as clear as the way it is used in muscle contraction. Actin cross-linking and unlinking seems to be involved in the alternation between the fluid and

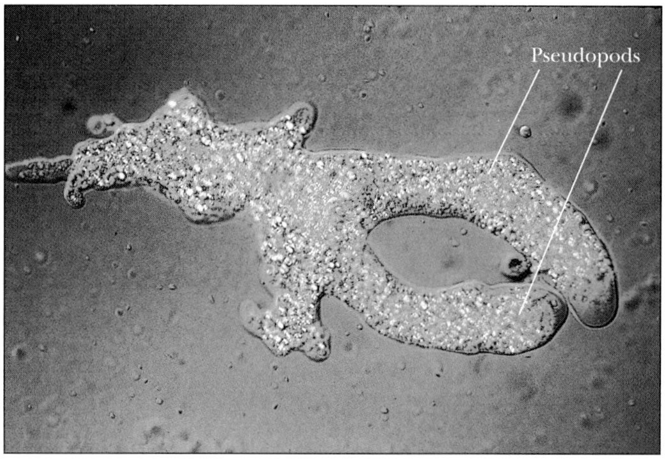

***Figure* 23.7** Amoebae move and trap food by using temporary extensions of the body called pseudopods. The vacuole containing the prey item is called a food vacuole. Lysosomes, which contain digestive enzymes, fuse with the vacuole. Undigestible remains are expelled. (© M. Abbey/VU)

more rigid states that allows pseudopod formation to occur.

The absence of a thick, stiff outer cell covering as a permanent feature of the eukaryotic cell together with the possibility of localized shape changes allow eukaryotes to engulf particles, a process not possible in prokaryotes. This process, called phagocytosis (Figure 23.8), is a basic method of feeding in eukaryotes. Large food particles are brought into the cell inside of vacuoles which form at the cell surface where food is being trapped. Lysosomes then fuse with the vacuoles, releasing enzymes that digest the food. In addition to phagocytosis, protists acquire food through **saprotrophy** (the absorption of dissolved organic material) and **phototrophy** (photosynthesis by chloroplasts).

Protists can be easily sorted out by the way they move, a readily visible characteristic. Flagellates have flagella, ciliates have cilia, amoebae have pseudopods, and some parasitic forms don't move much and lack specialized structures for movement. These divisions used to be incorporated into classification schemes at a high taxonomic level, where they were regarded as fundamental differences. Now the names are used only in a descriptive sense, only as a matter of convenience. The same organisms can have flagella and pseudopods at different stages in the life cycle. Conversely, rRNA comparisons have shown that organisms having the same motility system can be extremely distantly related to each other. Even humans illustrate the problem in using the type of motility system present to create the major divisions in a classification system. We have flagellated gametes, ciliated tracheal epithelium, and amoeboid macrophages. All are present simultaneously in the same organism, encoded by the same genome. These are merely common cell types, alternate states that the same cytoplasm can be in, not markers of basic phylogenetic significance.

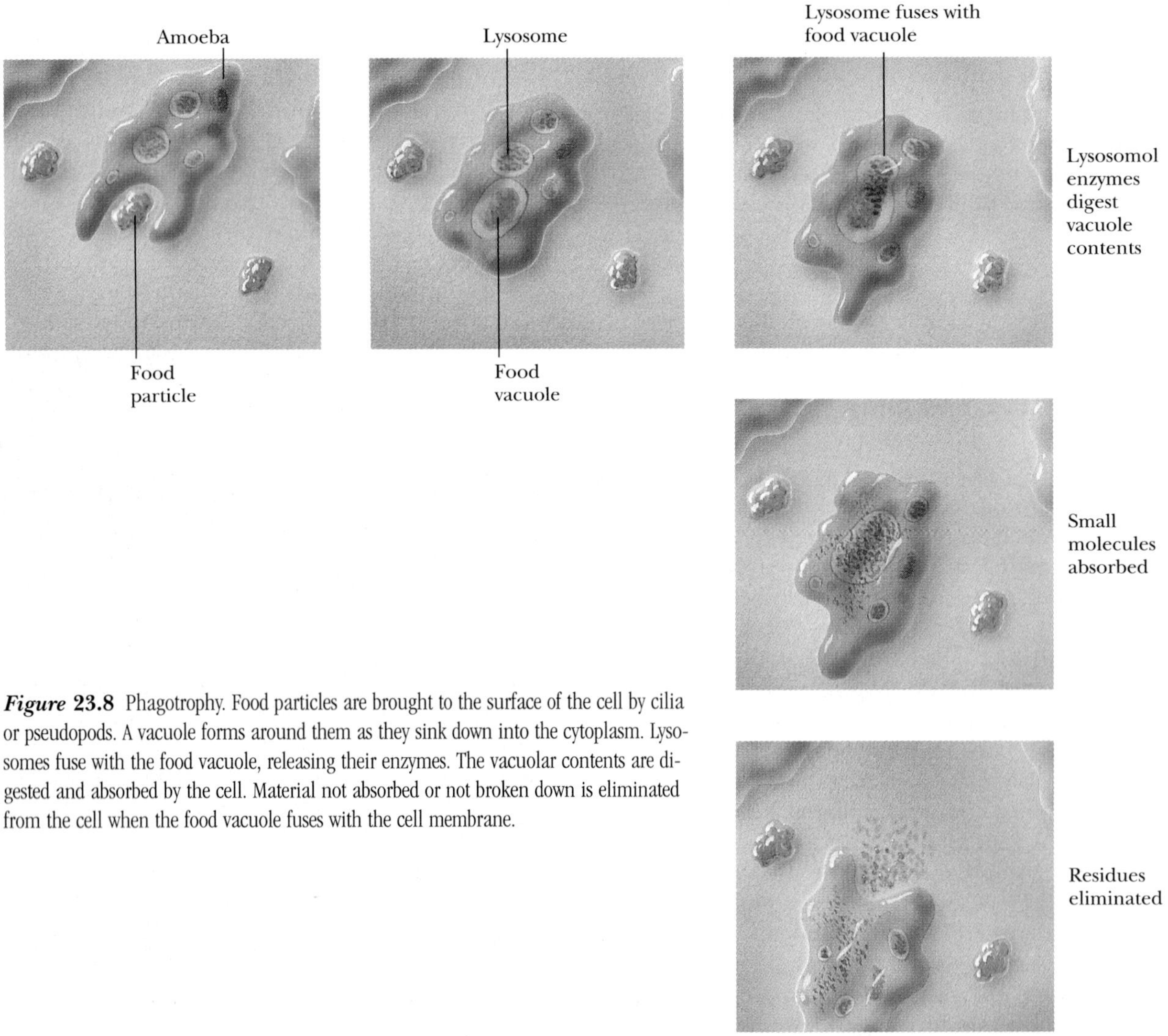

***Figure* 23.8** Phagotrophy. Food particles are brought to the surface of the cell by cilia or pseudopods. A vacuole forms around them as they sink down into the cytoplasm. Lysosomes fuse with the food vacuole, releasing their enzymes. The vacuolar contents are digested and absorbed by the cell. Material not absorbed or not broken down is eliminated from the cell when the food vacuole fuses with the cell membrane.

The Protistan Groups—Algae and Protozoa

These two words are frequently seen in discussions of protistan classification, where they are used to divide the protists in half—into plant-like and animal-like groups. The word "algae" is used to denote photosynthetic protists, while the word "protozoan" is used to refer to most of the colorless, phagotrophic forms. Apart from the practical problem of protists that refuse to be either plant or animal, and are both phagotrophic and photosynthetic, there is another problem. Both ultrastructural and molecular studies have shown that neither the algae nor the protozoa, no matter how they are defined, represent separate monophyletic units. Algae have arisen independently on several occasions through the independent acquisition of chloroplasts, and such groups are completely mixed together with colorless, or "protozoan," groups. This, in turn, means that the taxa "Algae" and "Protozoa" should not be used in a classification scheme that seeks to represent as accurately as possible the evolutionary history of the organisms being considered. The words can still be used as terms of convenience, to denote a set of organisms sharing some characteristics, in the same way that words like "predator" or "commensal" are used. The words "algae" and "protozoa" no longer have the taxonomic significance that they once did.

Organization of Groups in this Chapter

For the purpose of organizing a general review of the entire group, protists can be divided up into smaller groups based on the type of mitochondrion that they have—whether they have mitochondria at all, and if they do, which of the three major mitochondrial types it is. Although mitochondrial type is not a completely reliable indicator of relationship, a comparison with other types of evidence shows that it provides a reasonable first approximation, and is much better than dividing the protists into algae and protozoa.

In this chapter, three major groups of amitochondrial protists, which may have separated from other lineages before the acquisition of mitochondria, will be described. These are the microsporans, diplomonads, and parabasalians.

The groups with discoid cristae are the englenoids, kinetoplastids, and the amoeboflagellates.

Groups mentioned later that have tubular cristae include the slime molds, alveolates (dinoflagellates, ciliates, and apicomplexans), heterokonts (chrysophytes, xanthophytes, brown algae, diatoms, and oomycetes), foraminifera, radiolaria, and some heliozoans.

Groups with lamellar cristae discussed below include the red algae, green algae, cryptomonads, choanoflagellates, and chytrids, as well as the three generally recognized multicellular kingdoms—plants, animals, and fungi.

As in the case of prokaryotes, ribosomal RNA has been of great use in figuring out relationships. A tree representing the current best guess as to how protists evolved, based on RNA sequence comparisons, is shown in Figure 23.9. Organisms having the same type of mitochondrial cristae tend to be grouped together in an rRNA tree.

Protists Without Mitochondria

If chloroplasts and mitochondria are the remains of bacterial endosymbionts, it stands to reason that the first eukaryotes were probably colorless and anaerobic. If chloroplasts and mitochondria are subtracted from a eukaryotic

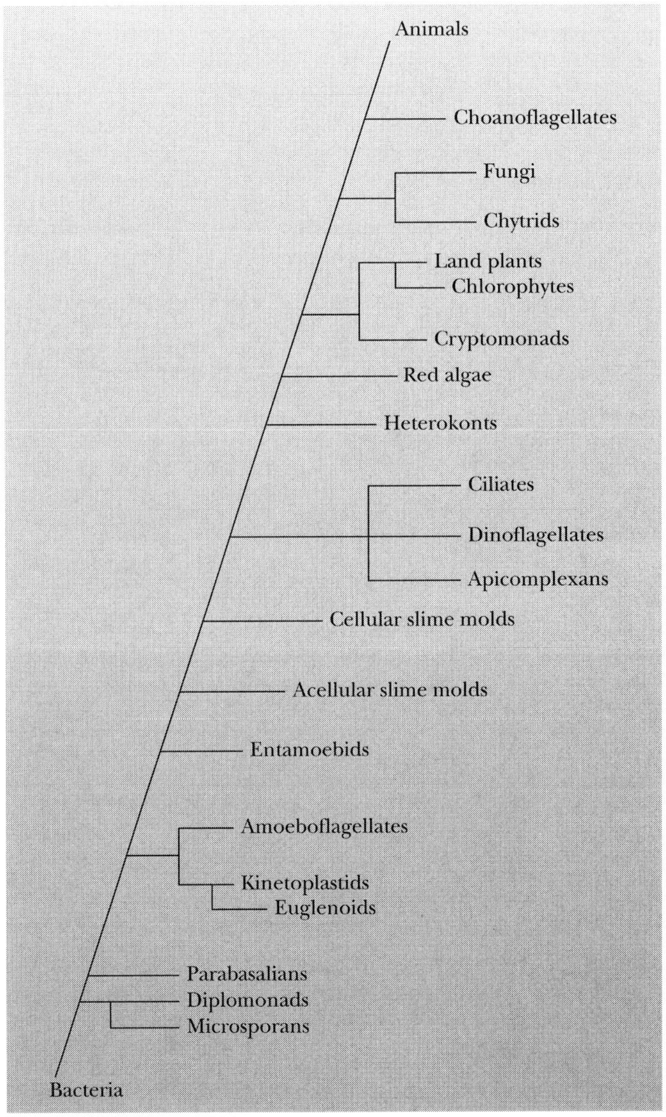

***Figure* 23.9** Tentative phylogeny of the eukaryotes based on ribosomal RNA sequence comparisons.

cell, what is left behind is a colorless anaerobic bag of cytoplasm. rRNA comparisons show that, indeed, such forms are in the deepest branches of the eukaryotic world. Most of these protists now live inside of animals, especially in the intestinal tract. These are probably relict populations left over from a time before cyanobacteria had oxygenated the Earth's atmosphere. They are what remains of a premitochondrial protistan world, now forced out of most habitats by oxygen and replaced by species with mitochondria. Anaerobic habitats suitable for protists are not very common now, but one very good habitat that remains is the gut of an animal. It is filled with food in the form of bacteria and dissolved organic material, and the oxygen concentration is vanishingly low.

Three lineages have been identified as emerging from near the base of the eukaryotic world. Their current representatives are the microsporans, diplomonads, and parabasalians. A discussion of these lineages follows.

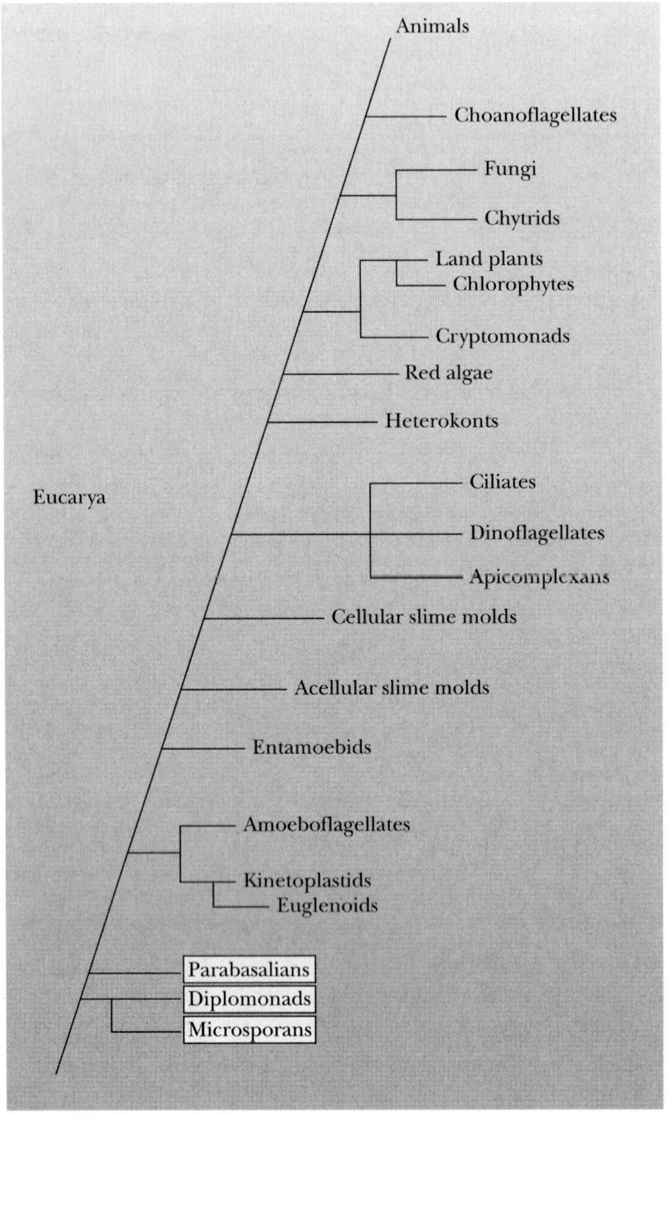

The Microsporans

The **microsporans** (Figure 23.10) are a group of about 800 species of protists that live as intracellular parasites in tissues of animals and other protists. The infective stage is transmitted from host to host in a resistant spore, which is typically about 4 μm to 5 μm in length (Figure 23.10**a**). Most spores contain a coiled hollow filament which is everted and can penetrate a host cell after the spore is ingested. The microsporan cytoplasm then travels down the tube into the host cell where it begins growing and dividing (Figure 23.10**b–d**). Ultimately, the microsporan goes through a division sequence in which a new generation of spores are produced. They may be passed from the body in feces or urine, or retained until the death of the host, depending on the location in which sporulation occurs.

All microsporans undergo rounds of division leading to spore production (sporogony) (Figure 23.10**f–i**). Most are also known to undergo at least one earlier round of multiplication (merogony) (Figure 23.10**e**) leading to the production of more growth stages in the host.

There are no flagella or centrioles at any stage of the life cycle. Dense plaques on the nuclear membrane serve as microtubular organizing centers in mitosis, which is closed. Meiosis has been seen, but how sexual recombination fits into the life cycle is mostly unknown.

Microsporans apparently lack lysosomes and may also lack a Golgi apparatus. Structures regarded as Golgi exist, but they do not have the morphology typically seen in eukaryotes. Microsporans also have unusual ribosomal RNAs. In addition to being prokaryotic in size, the 5′ end of the large rRNA is not split off to form a 5.8S piece as it is in other eukaryotes. It is tempting to regard these peculiarities as primitive eukaryotic traits remaining in the microsporan lineage, but we do not in fact know which traits are primitive and which have been secondarily acquired. There were presumably free-living members of this lineage at one time, inasmuch as their current hosts appeared later in evolutionary history. They could not have been parasites without organisms to parasitize. Free-living forms might exist, but none have been found.

The Diplomonads

The **diplomonads** are a group of amitochondrial flagellates. Many exist as symbionts in the intestines of both vertebrates and invertebrates, but a few are free-living and are found in sewage treatment plants and organically polluted water. *Giardia intestinalis* (= *G. lamblia*) (Figure 23.11), which Leeuwenhoek found in his own stools, was the first diplomonad discovered. It is the only member of the group that has been intensively studied. The organism is somewhat pear-shaped as viewed from above, with the rounded end being anterior. It is dorsoventrally flattened, and the bottom surface bears a stiff bilobed disk. There

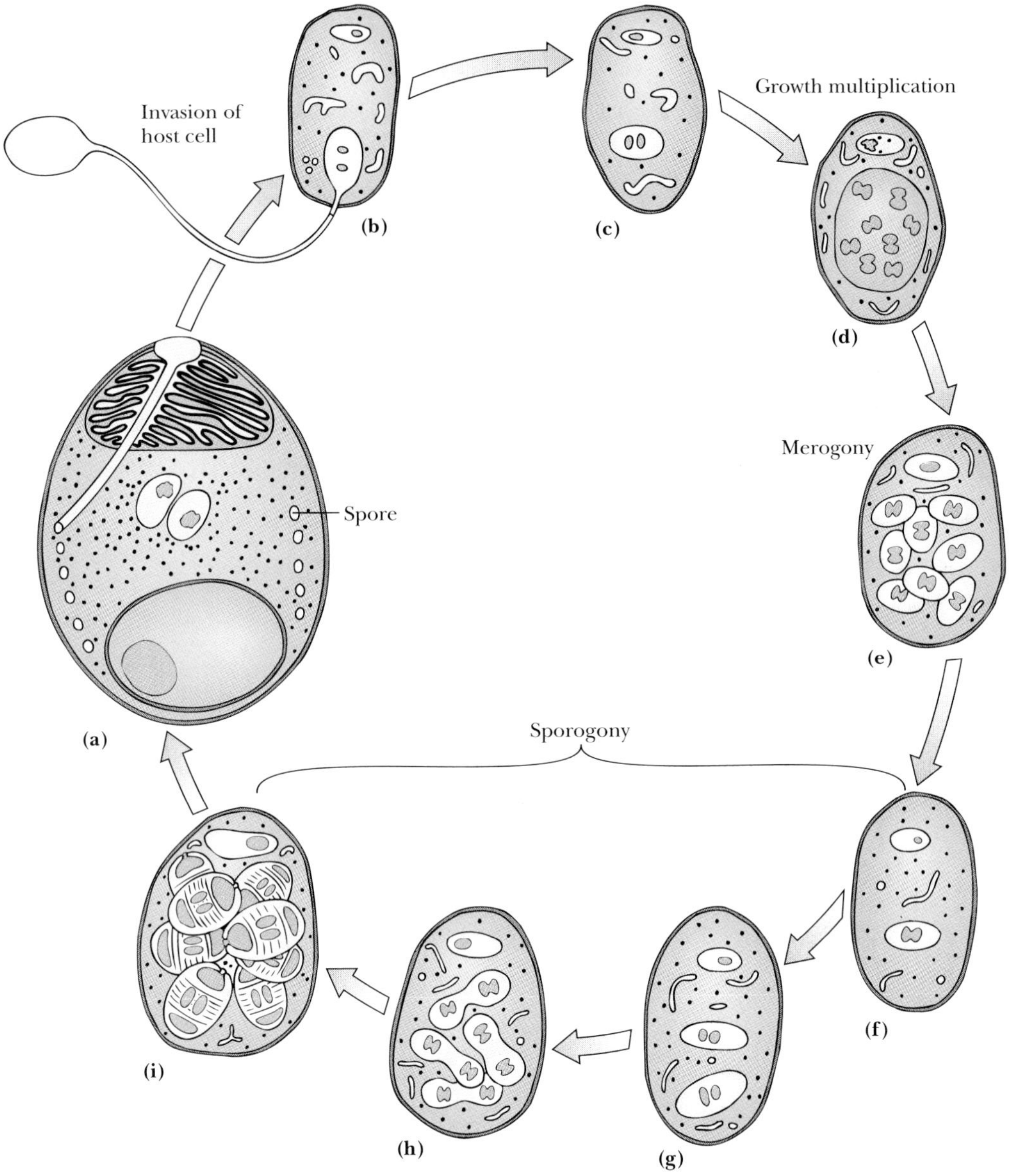

***Figure* 23.10** A microsporan lie cycle. The mature spore **(a)** contains a coiled up filament that everts and through which the cytoplasm of the spore travels to infect a cell **(b).** Growth and multiplication ensue. Merogony produces more invasive stages that will infect other cells in the same host **(e, c).** At some point, the cells start to undergo sporogony **(f–i);** the multiplication products differentiate into spores.

are two nuclei, each of which is associated with four flagella. This appearance of being a double organism is the source of the group's name (diplo, meaning "double").

G. intestinalis lives and multiplies in the small intestine of many vertebrates, where it absorbs dissolved nutrients. It is able to adhere to the intestinal surface by using the ventral flagella and disk. The upper small intestine is completely coated with these parasites in heavy infections. This interferes with the absorption of fats and leads to some of the pathology associated with *Giardia* infections. Humans can experience severe diarrhea, flatulence, weight loss, and abdominal pain, or remain asymptomatic. *Giardia* infections are frequently acquired by campers and hikers in mountainous areas in the western United States who drink directly from streams, or contaminate their hands and utensils by washing them in streams. Wild animals are the source of *Giardia* in such situations.

Giardia cells dislodged from the small intestine and swept further back into the intestinal tract are induced to encyst when they enter the colon with feces and begin to dehydrate. These resistant cysts are then passed out in the feces. Cysts are not found in diarrheic stools, since dehydration, the stimulus for encystment, does not occur. Active flagellates are passed out, but these cannot survive outside of the body. This means that it is people with few or no symptoms who spread *Giardia* most efficiently. Drinking water contaminated with cysts is the most common way of becoming infected with *Giardia,* but any means by which viable cysts can be passed from person to person serves to transmit the disease.

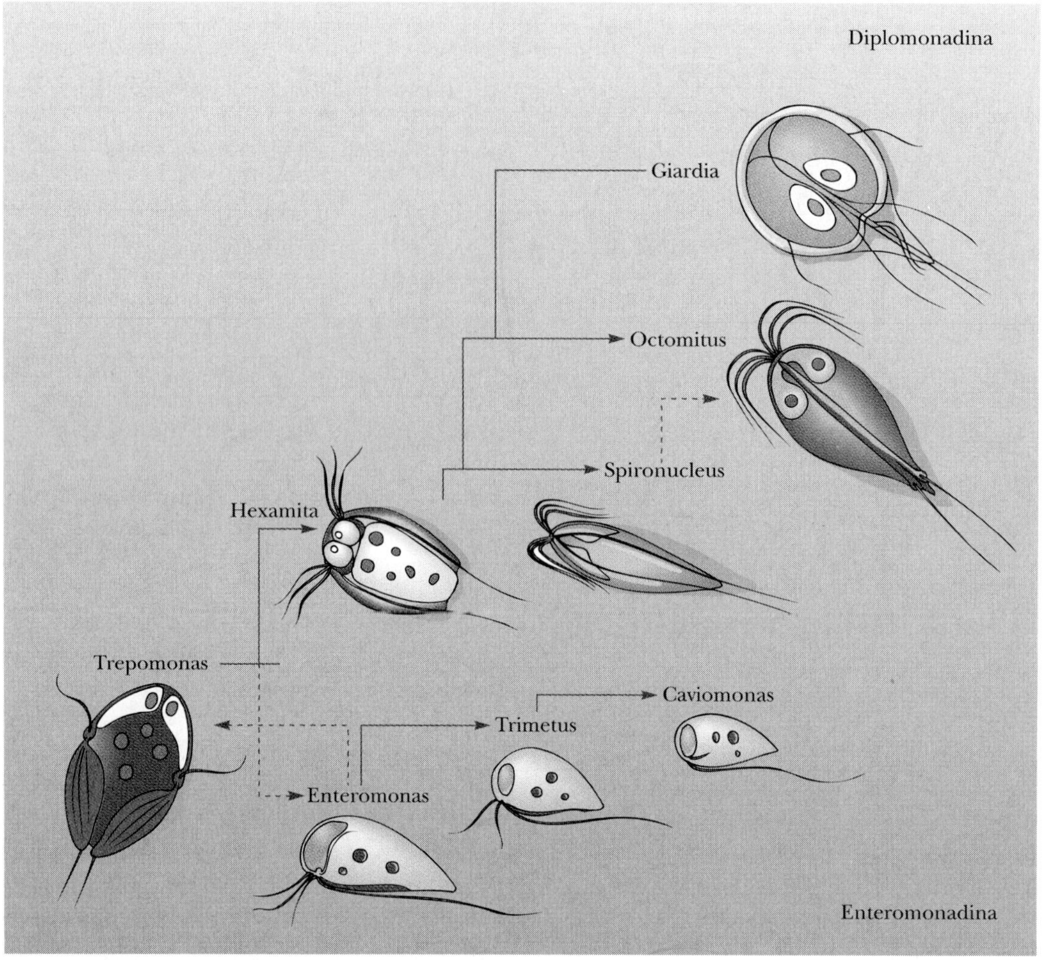

***Figure* 23.11** A diagram that shows the diversity of diplomonad genera and how they might be related to each other. Dashed lines are used when there are two possible alternatives for the origin of a genus.

Ultrastructural studies have revealed the probable phylogeny of diplomonads (Figure 23.11) and shown that a series of uninucleate intestinal symbionts, the enteromonads, also belong to this group. Enteromonads double their cell organelles before undergoing division, indicating how the morphological state displayed by forms like *Giardia* could have arisen from simpler forms through an arrest of cell division. Only asexual reproduction occurs; meiosis and cell fusion have never been observed in the diplomonads.

Although a Golgi apparatus is frequently said to be lacking, a membrane system possibly recognizable as a rudimentary Golgi system appears transiently during encystment, while large quantities of material are being secreted from the cell to produce a cyst. In contrast to microsporans, kinetosomes and flagella are present, lysosomes exist, and a 5.8S rRNA has become separated from the large rRNA.

The Parabasalians

The **parabasalians** are predominantly symbiotic, although there are a few free-living forms. The symbiotic forms typically live in the digestive tract of arthropods and vertebrates, where they consume the bacteria present in the same habitat. All the species living in the intestine are harmless, and those living in termites actually benefit their host. A few parabasalians, such as *Trichomonas vaginalis* of humans and *Tritrichomonas foetus* of cattle, live in the vertebrate reproductive system. These can cause disease.

In contrast to the two protistan groups described above, the Golgi apparatus is highly developed in the parabasalians. It is large enough to be easily seen with a light microscope, although its structural details can't be made out, and was given the name "parabasal body" before its identity was revealed by electron microscopy. The presence of this "parabasal body" (meaning "body beside

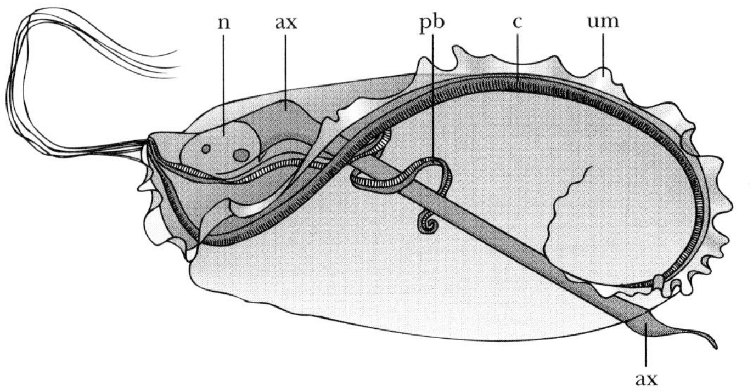

Figure **23.12** A trichomonad. A nucleus (n) is located at the anterior end of the cell. There are four anterior flagella, and a posterior flagellum running along the body, attached to it by a cytoplasmic flap, forming an undulating membrane (um). The undulating membrane is supported by a fibrous cytoskeletal element in the cytoplasm called a costa (c). The microtubular axostyle (ax) runs from the nucleus all the way through the posterior end of the cell. This species has a long parabasal body (pb).

the basal body") is one of the diagnostic features of the group.

The structurally simplest parabasalians, the **trichomonads** (Figure 23.12), are uninucleate. Each nucleus is typically associated with 4 to 6 flagella. One of the flagella is recurrent while the rest are directed anteriorly. The recurrent flagellum is usually attached for much of its length to the cell surface, forming an undulating membrane. A pipelike array of microtubules called an axostyle extends posteriorly from the basal bodies and nucleus at the anterior end of the cell and often passes all the way through the cell. The possession of an axostyle is another diagnostic feature of the group that is visible through light microscopy.

Parabasalians have an unusual mode of mitosis in which the nuclear membrane remains intact, but the spindle is extranuclear. Chromosomes and the spindle fibers attached to them connect with each other at the nuclear surface. Meiosis also occurs in some species. It has been best studied in genera found in the hindgut of the wood-eating roach *Cryptocercus.* In a classic series of papers, L. R. Cleveland showed that the life cycles of these flagellates were controlled by the molting cycle of their insect host.

Parabasalians are also characterized by the possession of organelles called hydrogenosomes. It was originally thought that they were mitochondria, but after they were isolated and studied, it became apparent that they were quite unlike mitochondria. They are double-membraned, mitochondrial-sized structures in which pyruvate is converted to acetate and H_2 under anaerobic conditions through the activity of iron-sulfur proteins. The processes carried out in hydrogenosomes are entirely unlike the oxidative processes that occur in mitochondria, but are similar to processes found in some anaerobic bacteria. It was suggested that hydrogenosomes evolved from anaerobic bacteria in a manner parallel to the development of mitochondria from proteobacteria in aerobic hosts. However, hydrogenosomes also occur in a few other protists occupying anaerobic habitats, and a series of transitional forms that suggest the derivation of hydrogenosomes from mitochondria has recently been found in ciliates. If hydrogenosomes are modified mitochondria, the source of their enzyme complement becomes an interesting problem.

The largest and most impressive parabasalians (Figure 23.13) live in the hind guts of termites and wood-eating roaches where they are responsible for the digestion of cellulose. The protists ingest wood particles floating around with them in the hindgut and may become quite distorted by the large pieces they have taken in. They break down the cellulose anaerobically into small soluble molecules, mostly acetate, which they then excrete. The termite can absorb these and digest them further aerobically. This means that even though termites ingest wood, they actually live on dilute vinegar! This is a true mutualistic association; neither the flagellates nor the termites can live without each other. At first glance these large flagellates do not resemble their smaller trichomonad ancestors, since they may have hundreds of nuclei and thousands of flagella. However, their derivation from simpler parabasalians through two evolutionary routes, one leading to the **calonymphids** and one to the **hypermastigids,** can be understood by considering what happens to the organelles while the cells are increasing in size and complexity.

The large calonymphids may be regarded as a parabasalian in which the organelles of a trichomonad have been multiplied. A species like *Coronympha octonaria* (Figure 23.13**a**) with 8 nuclei also has 8 axostyles, 8 parabasal bodies, and 32 flagella. It is essentially 8 parabasalian bodies contained within one cell. Electron microscopy has shown that in a cell such as *Calonympha grassii* (Figure 23.13**b**), which is covered by rings of nuclei and hundreds of flagella, each nucleus is associated with a small Golgi, four flagella, and an axostyle. It is essentially hundreds of parabasalian bodies stuck together in the same cell. The large axostyle observed to pass through the center of the body in such forms is actually a compound structure made up of all the individual axostyles originating near each nucleus. In the hypermastigids (Figure 23.14), only the flagella are multiplied. There are only one or several nuclei per cell, but each is associated with a field of hundreds or thousands of flagella rather than just four. Calonymphids are polymerized trichomonads, while in hypermastigids only the flagella have been multiplied.

These flagellate symbionts of roaches and termites have, in turn, their own bacterial endo- and ectosymbionts. In most cases the nature of the association is unknown.

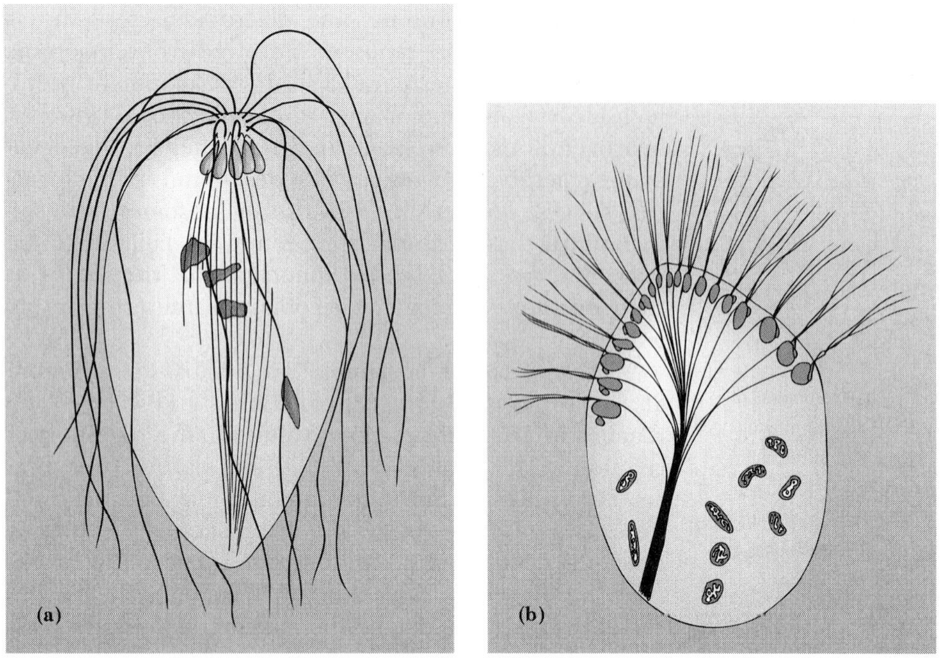

***Figure* 23.13** **(a)** The calonymphid *Coronympha octonaria.* **(b)** The calonymphid *Calonympha grassi.* The entire top of the cell is covered with nuclei, but they are only pictured in one plane of the cell.

However, in some cases, motile bacteria coat the external surface of the flagellates and are actually responsible for propelling the protist through the gut contents. Some of the flagellates have developed special structures to which the bacteria are attached (Figure 25.29). Margulis has suggested that the cilia and flagella of eukaryotes arose through this type of motility symbiosis. There isn't yet any compelling evidence for this idea, and how cilia and flagella originated isn't known.

There are two other amitochondrial flagellate groups that may be related to the diplomonads or parabasalians.

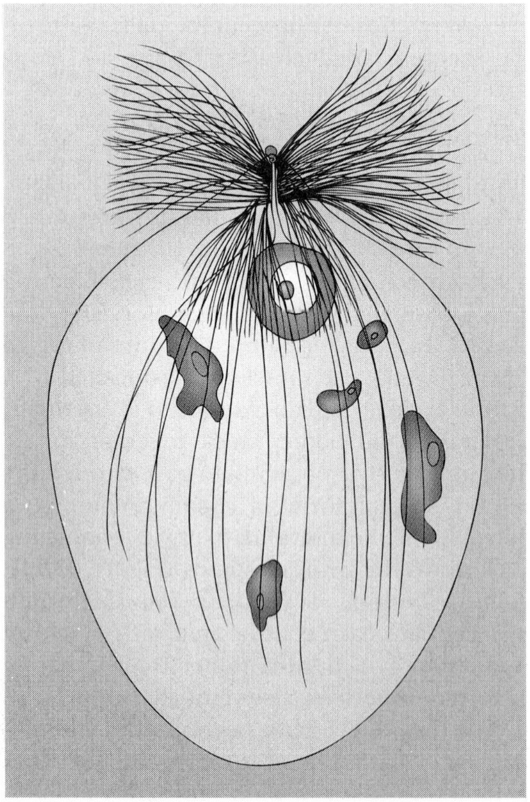

***Figure* 23.14** The hypermastigid *Barbulanympha* from the intestine of the wood-eating roach *Cryptocercus.*

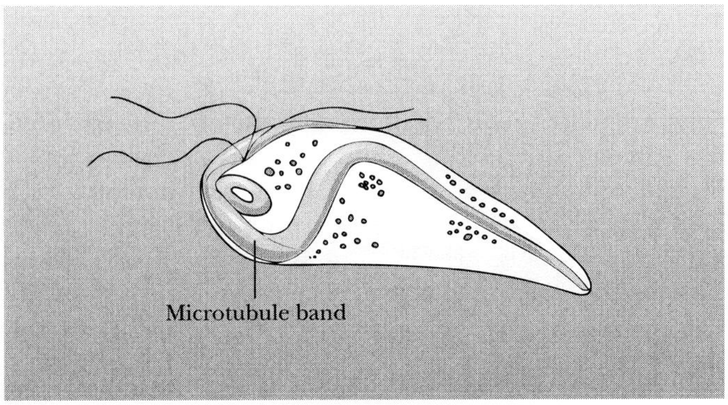

***Figure* 23.15** An oxymonad. These are harmless symbionts that live in the intestines of insects and vertebrates. They have four flagella that emerge from the anterior end of the cell, and a wide band of microtubules that curves through the body.

No rRNA sequences from these members of these groups are available yet, so where they fit into a protistan phylogenetic scheme is not known. The oxymonads (Figure 23.15) and retortamonads (Figure 23.16**a–b**) are intestinal symbionts of arthropods and vertebrates. No free-living forms are known. They have neither mitochondria nor a Golgi apparatus.

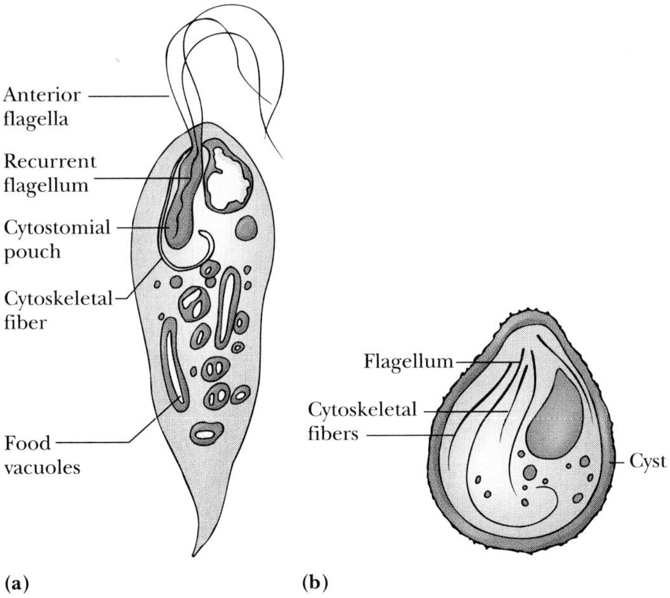

***Figure* 23.16 (a)** The retortamonad *Chilomastix*. Retortamonads live in the intestine of insects and vertebrates, where they eat bacteria. There are three anterior flagella and a recurrent flagellum, which is located inside of a permanent cytostomial pouch. The recurrent flagellum is used for drawing bacteria into the cytostome, where they are phagocytized. The cytoplasm is usually full of food vacuoles. There is a cytoskeletal fiber which supports the cytostomial area. Members of this genus are transmitted from host to host by characteristic pear-shaped cysts **(b)** passed out in the feces.

Discocristate Protists

The aerobic, mitochondria-possessing, eukaryotes found in branches closest to the base of the phylogenetic tree are the protists that contain mitochondria with discoid or ping-pong-paddle-shaped cristae at some stage in the life cycle (trypanosomes have discoid and tubular cristae at different stages). Two flagellate groups, the euglenoids and kinetoplastids, belong here together with the more distantly related amoeboflagellates.

THE EUGLENOIDS The euglenoids are particularly common in organically enriched freshwater, but may be found in virtually any aquatic environment. A few are parasitic. They usually have two flagella which emerge from a pear-shaped

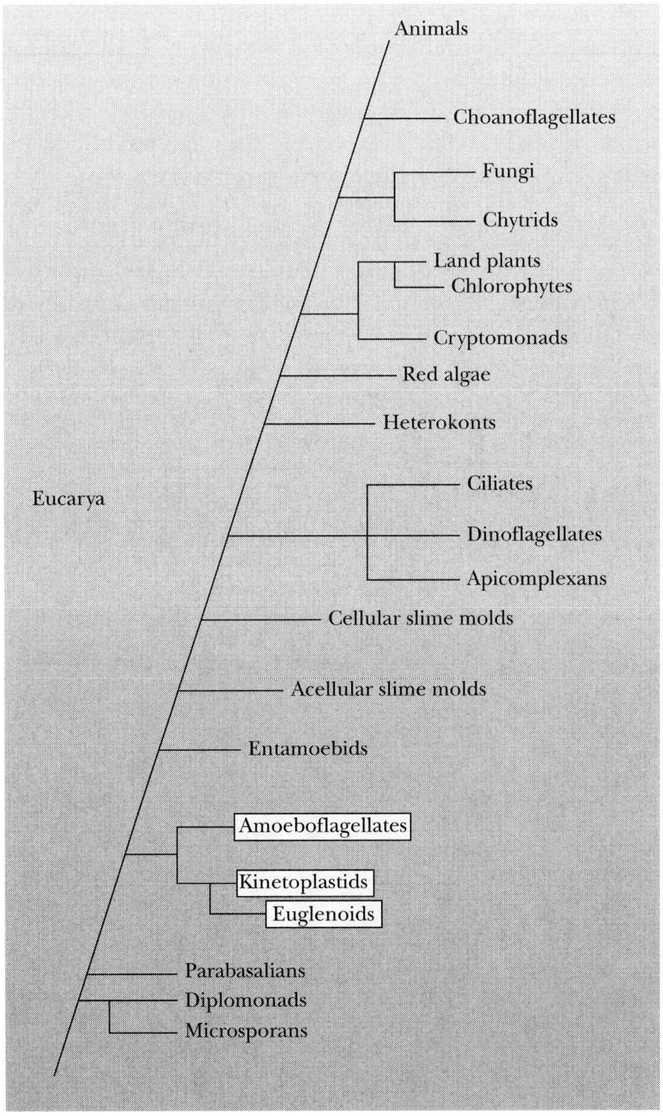

cavity at the anterior end of the cell. In some species, such as *Euglena gracilis* (Figure 23.17), one of the flagella is absent or reduced. Small hairs, called mastigonemes, project laterally along one side of emergent flagella. An unusual internal rod (paraflagellar rod) runs the length of the flagellum. The cell membrane is underlain by proteinaceous strips (Figure 23.17, insert). Euglenoids may be phototrophic, phagotrophic, or saprotrophic. Some combine saprotrophy with either phototrophy or phagotrophy. Species of *Euglena* can photosynthesize in the light or grow in the dark by taking up dissolved organic material. The trophic divisions that separate the animals (phagotrophy), land plants (phototrophy), and fungi (saprotrophy) are blurred together in the euglenoids.

Phototrophic species have chloroplasts containing chlorophylls *a* and *b*. The source of the euglenoid chloroplast is unknown, but it was probably acquired through the ingestion of a eukaryotic alga rather than through the ingestion of a cyanobacterium, since there are too many membranes around the chloroplast to be accounted for by a simple internalization of a bacterium. Phototrophic euglenoids also have an orange carotenoid-containing eyespot in the cytoplasm near the base of one of the flagella, which is swollen. The eyespot and flagellar swelling interact in some way to produce the phototactic response characteristic of the photosynthetic species. They prefer subdued light, avoiding both bright light and darkness.

It is easy to produce permanently colorless strains of photosynthetic euglenoids by treatment with a variety of chemical agents, and it appears that chloroplast loss also occurs in nature. There are colorless euglenoids morphologically identical to *Euglena* species, with the exception that chloroplasts are absent.

Only about a third of the euglenoids are phototrophic. The others are colorless phagotrophs or saprotrophs. Phagotrophic species have an ingestion apparatus consisting of several bundles of microtubules which operate like hooks to draw prey into the body (Figure 23.18**a, b**). The ingestion apparatus is a separate structure from the reservoir; food does not enter the body through the reservoir.

THE KINETOPLASTIDS The kinetoplastids are the second flagellate group found among the discocristate forms. The flagellar apparatus of kinetoplastids is similar to that of euglenoids because it has (primitively) two flagella that emerge from a depression; paraxial rods in the flagella; and, sometimes, mastigonemes on the flagella. The microtubular elements radiating from the flagellar basal bodies in the cytoplasm also have the same arrangement in the two groups. The euglenoids and kinetoplastids are sometimes classed together as the Euglenozoa.

The kinetoplastids acquired their name from the kinetoplast, an object near the base of the flagellum that took up stains used by early protozoologists (Figure 23.19). It is now known to be a mass of mitochondrial DNA, very different in organization from the mitochondrial DNA of other organisms. In most eukaryotes, the mitochondrial genome is contained in a small circular chromosome. There are several copies of this in each mitochondrion,

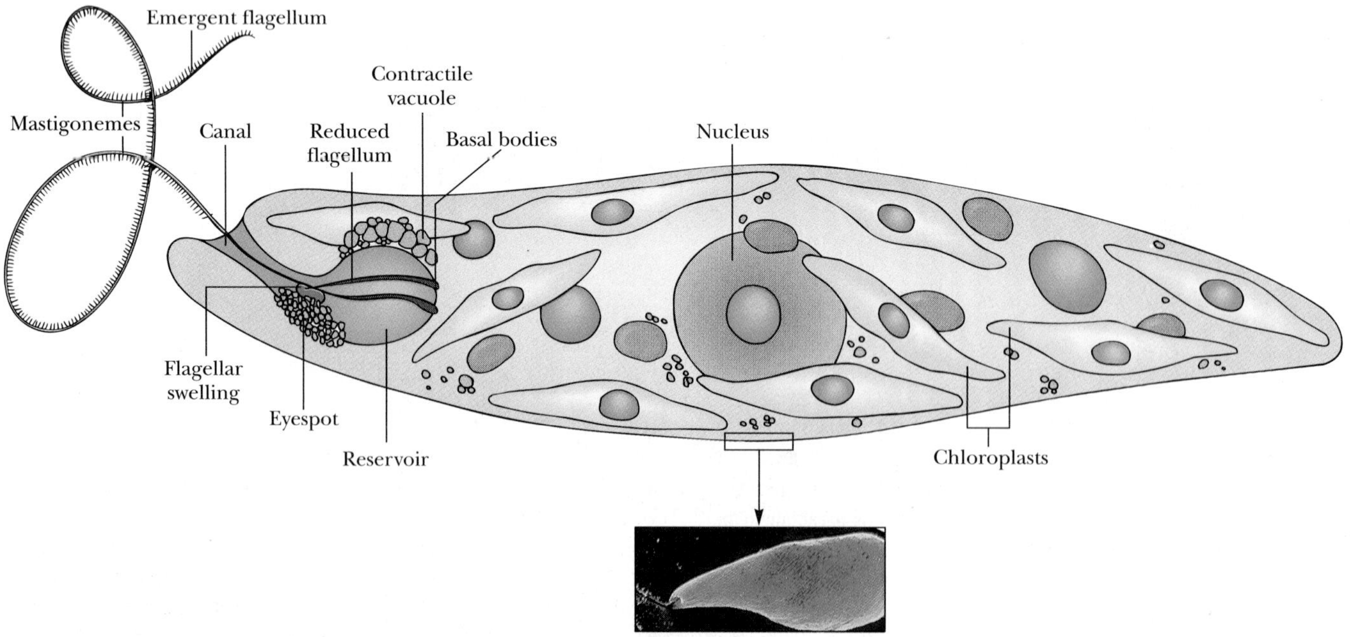

***Figure* 23.17** The phototrophic euglenoid *Euglena gracilis.* (Insert: © David M. Phillips/VU)

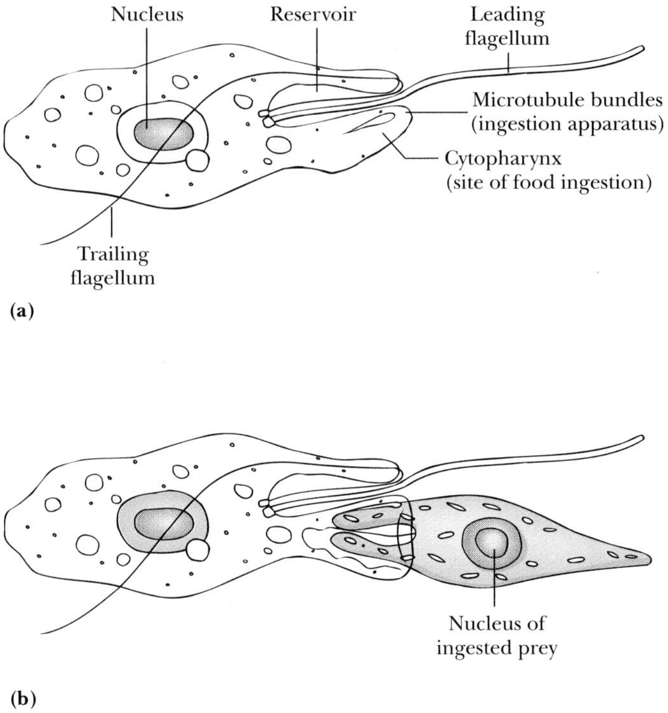

***Figure* 23.18** The colorless phagotrophic euglenoid *Peranema.* **(a)** general appearance. **(b)** feeding.

but not so many that they can be stained and seen with a light microscope. In the few kinetoplastid species whose mitochondria have been well-studied, the mitochondrial genome is a network of thousands of interlocked circular chromosomes. A small number of large chromosomes ("maxicircles") encode the genes commonly found in mitochondria, while thousands of much smaller "minicircles" exist that do not encode any complete genes. The coding regions are also unusual in that many of the transcripts produced from them do not encode functional proteins. These transcripts cannot be translated into normal mitochondrial proteins until U (uracil) residues are post-transcriptionally inserted into the proper places in mRNA. The reason for this type of organization is entirely unknown. The mitochondrial genomes of kinetoplastids are being intensively investigated by scientists who are interested in finding out how such a strange genetic system arose and how it works.

There are two groups of kinetoplastids. One of these groups, the bodonids, contains many free-living forms that are very common in both soil and water. These ingest bacteria through a microtubule-lined cytopharynx (meaning "cell throat") (Figure 23.20). Traces of such a pharynx can be found in some euglenoids when they are examined ultrastructurally, which further supports the belief that

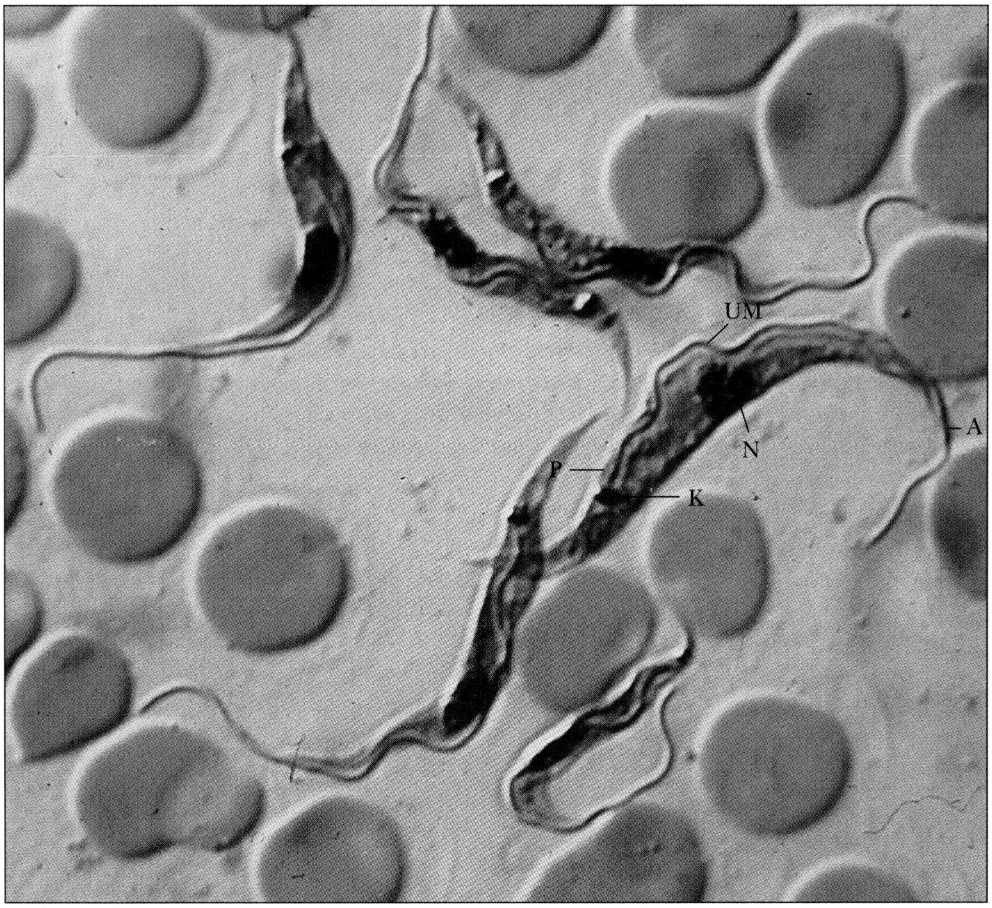

***Figure* 23.19** *Trypanosoma lewisi* from the blood of a rat. The nucleus (N), kinetoplast (K), and undulating membrane (UM) can be seen. The flagellum of the undulating membrane emerges at the posterior (P) end of the cell and extends beyond the anterior end of the cell (A) as a free flagellum (F). (© D. S. Snyder/VU)

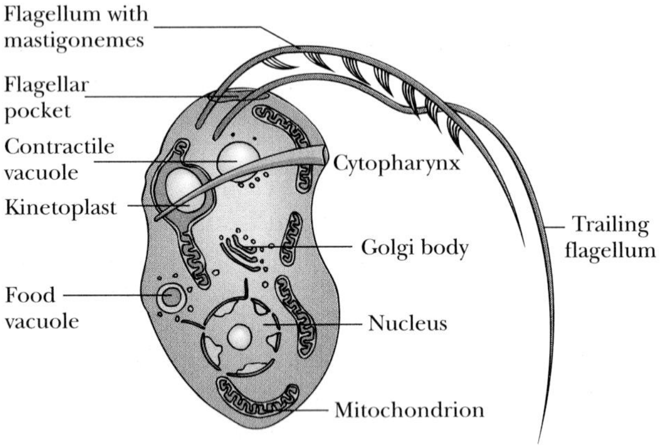

***Figure* 23.20** Diagram of a bodonid, showing details visible with an electron microscope.

euglenoids and kinetoplastids are related to each other. There are also a few parasitic species, which are found primarily in aquatic organisms.

The second group of kinetoplastids, the trypanosomatids (see Figure 23.19), evolved from bodonids. They lack a cytopharynx and feed on dissolved organic material. All have only a single flagellum, but may have a second, barren kinetosome. They are especially common in the intestines of insects, where they probably originated. They feed saprotrophically on the food present in the intestine of their insect hosts. Insects feeding on animal and plant juices have introduced them into other organisms, and they have gradually adapted to life in these secondary hosts. Members of the genera *Leishmania, Trypanosoma,* and *Schizotrypanum* include some important human parasites found in tropical and subtropical regions.

Leishmania species are transmitted to humans via the bites of sandflies. They cause a spectrum of diseases in which the pathology that develops is related to a great extent to a particular *Leishmania* strain's ability to spread through the body from the site of infection. The most benign disease caused by members of this genus is tropical sore. A small ulcer, in which the *Leishmania* cells are multiplying, forms around the region of the bite. It is a self-limiting infection which lasts only a few months to a year before disappearing. It leaves behind a scar in the skin. A more serious disease, mucocutaneous leishmaniasis, can be produced when the *Leishmania* strain is capable of spreading away from the region of the bite and causing secondary lesions elsewhere in the body, away from the point where they were introduced. Some strains can spread to the nose and mouth, where they cause the destruction of the soft tissues and the cartilage. This can lead to severe disfigurement. Finally, *Leishmania donovani* causes visceral leishmaniasis. *L. donovani* preferentially replicates throughout the body in the cells of the reticuloendothelial system, specifically in the macrophages, which they subsequently destroy. Because macrophages ingest and destroy microbes invading the body, their destruction leaves the body without one of its main immune system components. People infected with *L. donovani* frequently die of other diseases against which they are no longer able to defend themselves.

Species in the genus *Trypanosoma,* transmitted by tsetse flies, are responsible for African sleeping sickness and serious livestock diseases. Trypanosomes are actually quite common in all classes of vertebrates. Those found in fish are transmitted by leeches rather than insects.

Schizotrypanum cruzi, transmitted by reduviid bugs ("kissing bugs"), causes Chagas' disease in Central and South America. This is commonly a long, chronic disease. In the human body, *S. cruzi* cells preferentially multiply in nerve and muscle cells, and gradually cause degenerative changes in the heart or intestinal tract. A high percentage of people in such areas who die of heart failure at an early age are actually succumbing to the effects of an *S. cruzi* infection.

Kinetoplastids, which have been introduced into plants by their insect hosts, are found in the genus *Phytomonas.* These species are transmitted to a variety of plants by hemipterans. Most live in the latex of euphorbs and asclepiads, but they occur in other plants as well. Some species damage palm trees and cause a wilting disease in coffee plants.

THE AMOEBOFLAGELLATES The third group of discocristate organisms is the amoeboflagellates. They live in soil, fresh water, and sea water. As the name implies, many have both an amoeba and a flagellate stage in the life cycle (Figure 23.21**a–c**). The production of these different stages is not connected with sexual recombination or ploidy level changes. Both amoebae and flagellates reproduce asexually, if they reproduce at all. Transformation of amoebae into flagellates is triggered by dilution of the medium. In nature, this could occur after a rain. It may be advantageous for a soil amoeba to detach from the substrate and rise up above it so that it can be carried into a new habitat by the rainwater flowing over a particular patch of ground. The stability of the flagellate stage varies. In *Naegleria,* the flagellates are incapable of either feeding or dividing, and revert to amoebae after several hours. In other genera such as *Tetramitus* and *Percolomonas,* flagellates feed and divide indefinitely. The flagellate stages from different species are remarkably unlike each other; nothing visible through light microscopy would suggest their close relationship.

The manner of pseudopod formation by the amoebae is distinctive, and characteristic of the group. In contrast to many other amoebae that send out exploratory pseudopodia in several directions at once, the amoeboflagellates tend to move through one rapidly advancing pseudopod at a time. This gives them a sluglike outline from above, which is responsible for their description in protistological literature as limax amoebae ("limax" means "slug" in Latin and is the genus name for the common

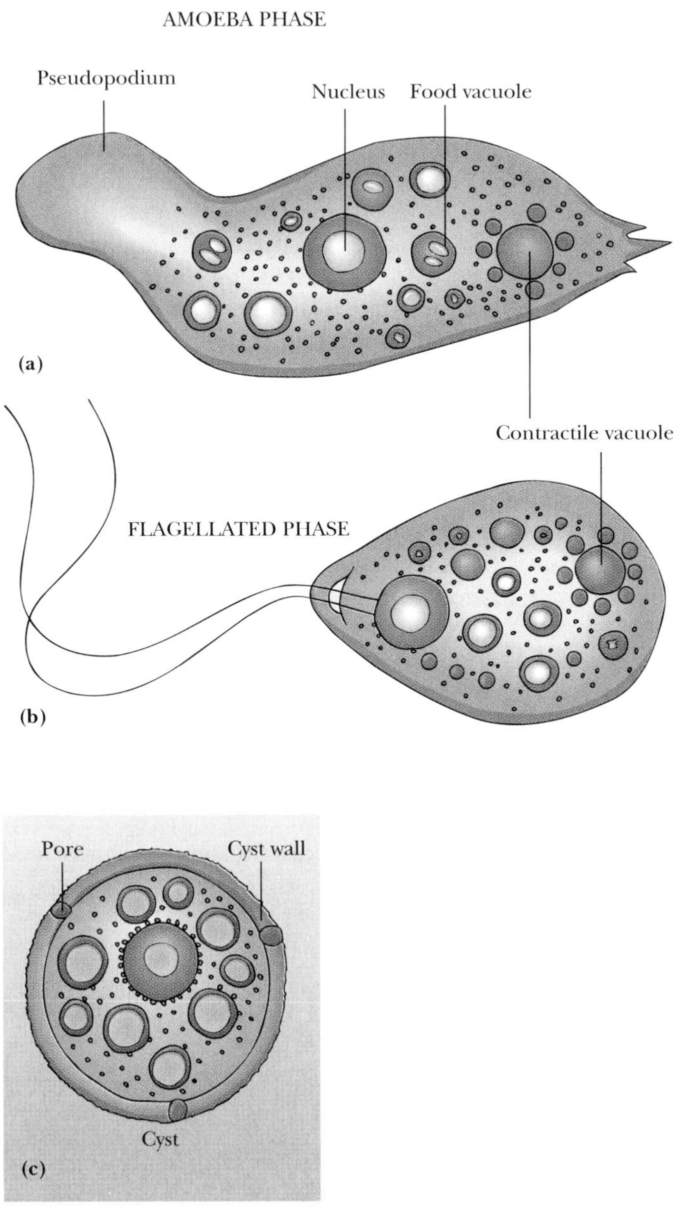

***Figure* 23.21** Stages in the life cycle of the amoeboflagellate *Naegleria gruberi,* which lives in soil and freshwater. The amoeba **(a)** is the dominant phase in the life cycle, and is the only stage that feeds and multiplies. It can change into a flagellate **(b),** which swims around for a few hours before reverting to an amoeba. If its food supply is exhausted or it is in a habitat which is drying out, the amoeba encysts **(c).** When conditions are favorable again, the amoeba excysts, managing to squeeze through a pore in the cyst wall considerably smaller in diameter than the amoeba is.

garden slugs of Europe and North America). Pseudopod formation is eruptive, with hyaline cytoplasm flowing out ahead of the body in what looks like a series of spills. A cyst is also present in the life cycle, and is formed by the amoebae whenever environmental conditions become unfavorable for further reproduction. Only the amoebae form cysts and emerge from cysts; the flagellates are not capable of cyst formation.

Amoeboflagellates are bactivorous forms that multiply very rapidly under suitable conditions, some having a generation time of less than two hours. This might not be much compared to a prokaryote, but it is unusually fast for a eukaryote. Most species are completely innocuous free-living forms, but a thermophilic species, *Naegleria fowleri,* is capable of infecting humans, in whom it causes a fatal meningoencephalitis. Although it can live in any thermally elevated body of fresh water, people most commonly become infected in hot springs and warm, unclean swimming pools. The amoebae normally live on bacteria present in the water, but if they are accidentally drawn into the nose of a swimmer, they can invade the nasal mucosa and migrate up the olfactory nerves to the brain. There they cause the tissue destruction that results in death. There is no good treatment.

Tubulocristate Protists

Protists in the next region up in the tree from the discocristate organisms have mitochondria with tubular

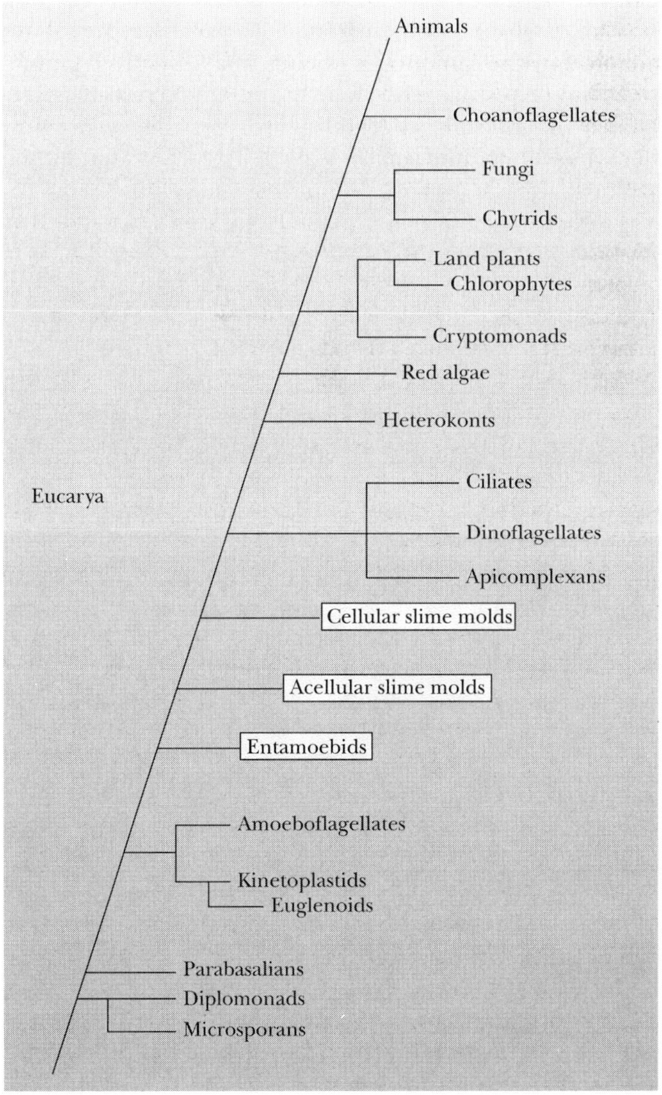

cristae, if they have mitochondria at all (not all do). Three groups of amoeboid organisms appear first in this set of organisms in an rRNA-based protistan phylogeny. These are the cellular slime molds, such as *Dictyostelium,* acellular slime molds such as *Physarum,* and parasitic amoebae in the genus *Entamoeba.* The branches leading to these forms are separate enough from each other to indicate that they are not directly related, although which protistan groups might connect them have not yet been identified. The non-amoeboid alveolates and heterokonts are also tubulocristate protists.

Cellular and Acellular Slime Molds

Slime molds were long regarded as being a type of fungus (hence the name "mold") because they produce spore-bearing, multicellular structures that extend up from the substrate into the air, as do true fungi. However, they are completely unrelated to fungi. Their possession of similar structures is an example of **convergent evolution,** which is the development of similar characteristics by unrelated organisms because they have adapted in the same way to a similar environmental challenge. In this case, they have adapted in the same way to being small, relatively immobile, and terrestrial. The slime molds are terrestrial amoebae, living in soil, in leaf litter or on the surfaces of decaying wood and plant leaves. They are not motile enough to transport themselves into new habitats. Aquatic protists are carried around by the movement of water. But stumps and forest soil don't move much; how are the amoebae that live there effectively dispersed? In the same way as are the fungi—by raising resistant spores up above the soil, where they can be moved around by wind, rainwater, or animals in whose fur they have been trapped. Such a strategy also encourages the development of multicellularity, since a single protistan cell cannot raise itself very far up off the substrate. A multicellular structure can be much larger. There are a few unicellular slime molds, the protostelids, but most are multicellular. The two slime mold groups have very different life cycles, so it is likely that they are not closely related. The two kinds of slime molds and the fungi developed the same elevated spore-bearing structures independently.

The cellular slime molds (see Figure 1.31**a–f**) exist as uninucleate soil amoebae while feeding. They can be stimulated by starvation to aggregate and form multicellular masses called slugs which rise from the substrate and differentiate into a stalk that supports a mass of spores. Under suitable conditions single amoebae emerge from each spore to begin feeding and multiplying again.

The acellular slime molds (Figure 23.22) form similar resistant terrestrial structures, although their life cycles are different. Haploid amoebae emerge from the spores. They either fuse with one another or are transiently transformed

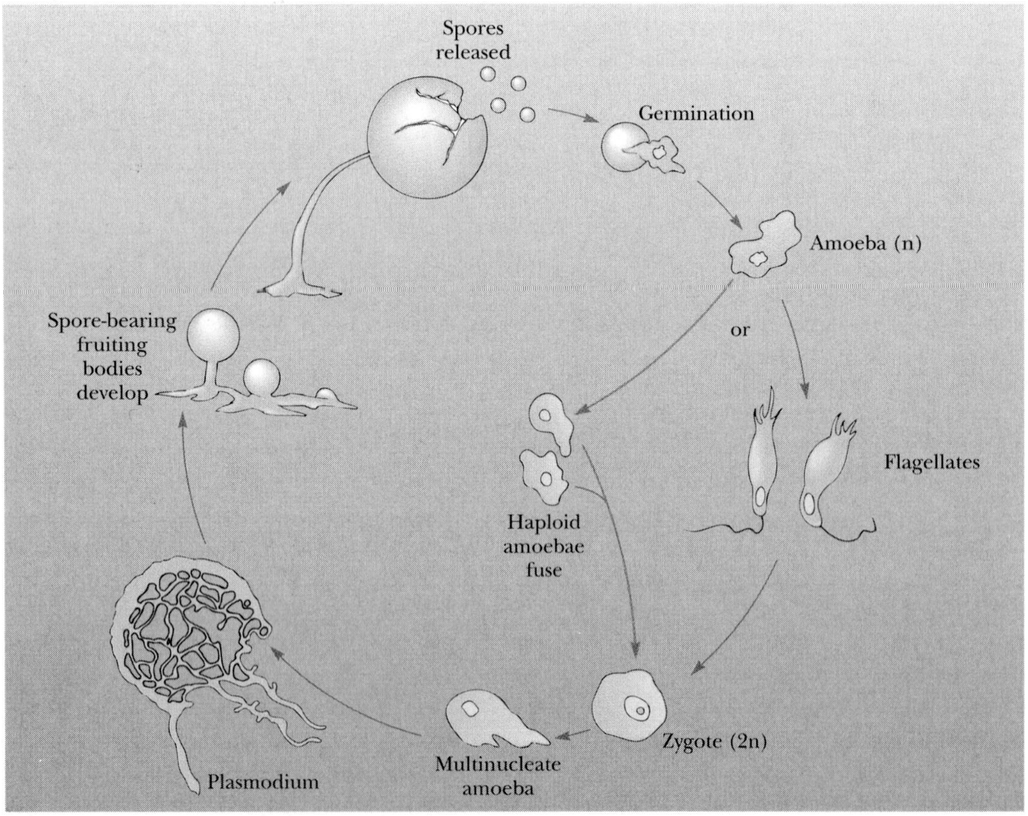

***Figure* 23.22** The life cycle of an acellular slime mold. It is superficially similar to that of a cellular slime mold. However, the stage producing the fruiting bodies is not a multicellular structure formed by the aggregation of amoeba. Rather, it is one single, gigantic, multinucleate amoeba that grew without dividing from a small uninucleate amoeba.

into flagellates that fuse with one another before reverting to amoebae. The end product in either case is a diploid amoeba that grows tremendously in size while consuming bacteria on the substrate on which it crawls. Mitosis occurs repeatedly so that the cytoplasm contains thousands of nuclei. This plasmodial stage is much larger than the slugs produced by cellular slime molds. They can frequently be seen in the woods as orange or yellow patches on decaying wood. Eventually the plasmodium is stimulated to form spore-bearing structures similar to those of the cellular slime molds. Meiosis occurs during spore formation.

Entamoebae

Many species of *Entamoeba* exist. Found in the intestines of vertebrate and invertebrate hosts, where most of them consume bacteria, they lack mitochondria. Several species live in humans. One of these, *Entamoeba histolytica,* causes amoebic dysentery. It alone, among the intestinal amoebae living in humans, is capable of secreting proteolytic enzymes that digest away cells lining the intestine, forming large ulcers. *E. histolytica* is transmitted by cysts in the same manner as *Giardia.* Several hundred million people worldwide are infected with this organism, which makes it an important human parasite. Fortunately, most people infected are asymptomatic. The factors that cause the amoebae to do very little harm sometimes and a great deal of harm at other times aren't known.

The Alveolates

A cluster containing the dinoflagellates, apicomplexans, and ciliates, appears next on the tree. These three groups share a unique type of cell surface consisting of a typical unit membrane underlain by a layer of vesicles developed to different extents in the three groups. In ciliates, the vesicles, called **alveoli,** are usually well-developed and form a pellicle somewhat resembling bubble packing material. In dinoflagellates, these alveoli frequently contain cellulose plates, forming a cell wall just beneath the surface membrane. In apicomplexans there is usually no space between the membranes, so the cell is effectively surrounded by three closely spaced unit membranes, one right on top of the next.

THE DINOFLAGELLATES The dinoflagellates are an extremely diverse group of organisms, both morphologically and ecologically. Most have two flagella oriented at right angles to each other, which emerge halfway down the side of the cell (Figure 23.23). One flagellum is wrapped around the circumference of the cell in a groove, while the other extends posteriorly. Those species that have a covering of cellulose plates are called armored dinoflagellates. The most common dinoflagellates lack histones and have an extranuclear spindle during division, but this is not true of

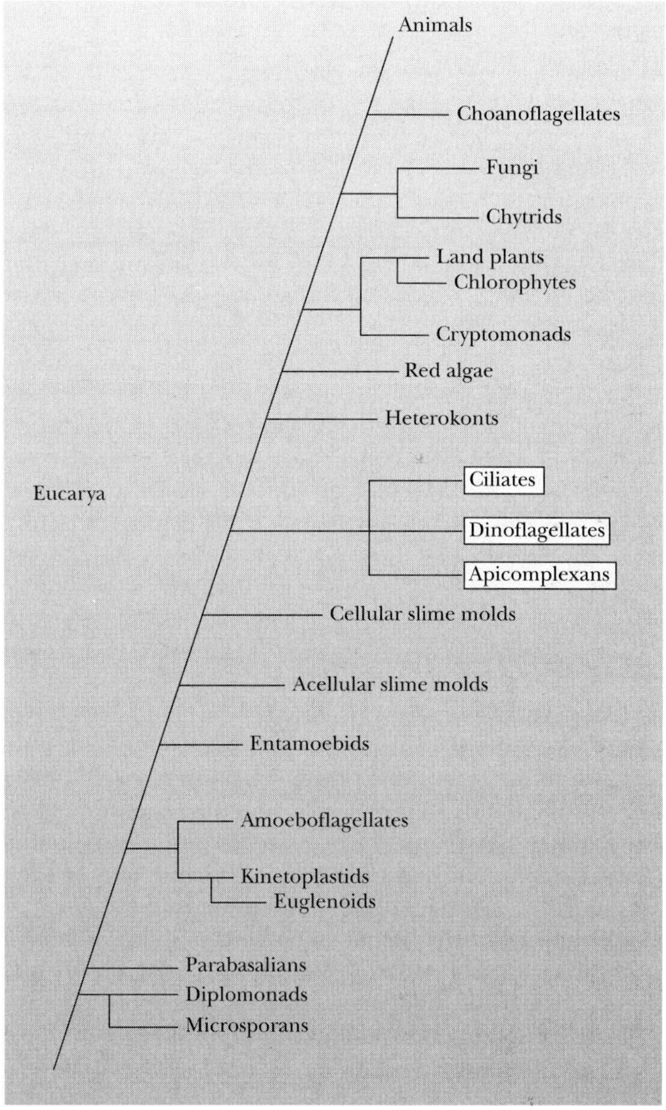

all species; these are secondary modifications of a more standard nuclear configuration. Most dinoflagellates whose life cycle is known are haploid organisms with a zygotic meiosis, like the green alga *Chlamydomonas.*

Most are free-living photosynthetic forms whose chloroplasts contain chlorophylls *a* and *c.* Colorless saprotrophs and phagotrophs exist, and many photosynthetic forms are also phagotrophic. Symbiotic photosynthetic forms ("zooxanthellae") live in the tissues of a wide variety of marine invertebrates, especially in tropical waters. They use carbon dioxide and ammonia produced by their hosts and in turn provide their hosts with sugars and amino acids. Zooxanthellae are abundant in reef-building corals and must be present in the coral for reef building to occur.

Some free-living photosynthetic forms produce toxic compounds that accumulate in the water during "red tides" (temporary blooms of dinoflagellates that occur when conditions for growth are optimal). These toxins are

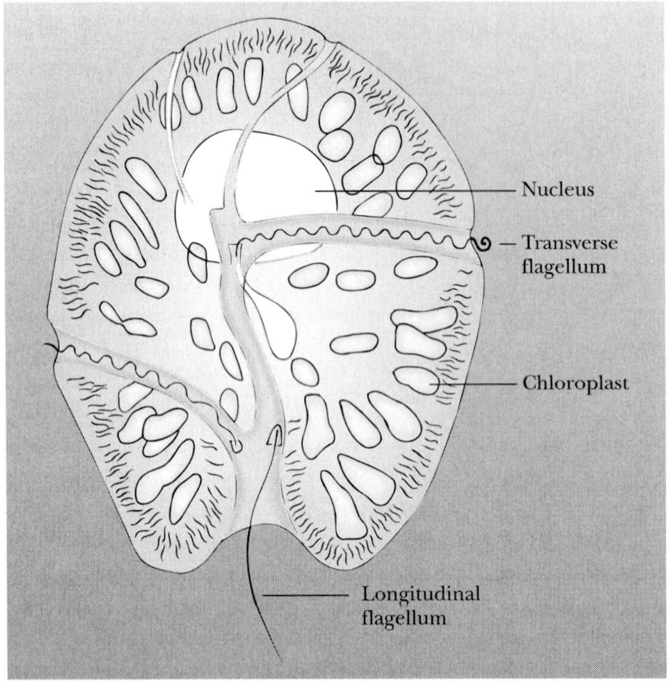

***Figure* 23.23** The dinoflagellate *Gyrodinium*.

responsible for massive fish kills. Still others produce toxins that are passed up the food chain after the dinoflagellates are eaten, becoming more concentrated at each trophic level. Although the toxins are harmless to fish and shellfish, they can be lethal to humans. Paralytic shellfish poisoning and ciguatera poisoning both result from human ingestion of dinoflagellate toxins carried in the tissues of other organisms. Why dinoflagellates synthesize these compounds is unknown.

THE APICOMPLEXANS Apicomplexans are parasites of animals. They probably arose from parasitic dinoflagellates living in the intestines of marine invertebrates. Most species still live in or on cells lining the intestine, and are transmitted by resistant spores passed out with feces. A few have penetrated more deeply into the tissues of their hosts, and must rely on other means of being passed from host to host. Members of this group are united by features of the life cycle and ultrastructural details rather than by readily visible gross morphological features; members of the different groups can look very much unlike one another at the light microscopic level. Most live intracellularly during at least a part of the life cycle. They all have a sexual phase in the life cycle, after which they all produce a distinctive stage known as a sporozoite. Sporozoites (Figure 23.24) contain an apical complex which usually consists of a polar ring, conoids, rhoptries, and micronemes, although in some species one or more of these components may be missing. These structures are used in penetrating the cells of their hosts. Possession of this complex gives the group its name. All stages except the zygote are haploid, as in their dinoflagellate relatives.

Perhaps the best-known apicomplexans are the *Plasmodium* species, which are responsible for causing malaria in humans and other animals (Figure 23.25). This is one of the genera containing species that have migrated more deeply into host tissues. In this case, completion of the life cycle depends on a mosquito picking up parasites from the blood of an infected vertebrate. Zygote formation occurs in the mosquito's intestine. The sporozoites formed afterwards migrate to the salivary glands of the mosquito, to be injected when it takes its next blood meal. In humans, the *Plasmodium* cells initially begin multiplying in liver cells, but eventually move into blood cells. For unknown reasons, multiplication of all the individuals in the blood is synchronous, so that blood cells are lysed and toxins are released in bursts. This causes the periodic chills and fever that a malaria victim experiences.

THE CILIATES The third alveolate group is the ciliates (Figure 23.26). They have a body with cilia in at least one stage of the life cycle, two types of nuclei (see Figure 23.3), and a peculiar method of sexual recombination called conjugation. Ciliates typically have two sets of cilia, somatic and oral cilia, as shown in Figure 23.26**a**. In *Tetrahymena* **(Box 23.1)**, somatic cilia are organized into longitudinal rows called kineties which run the length of the cell. The somatic cilia move the ciliate through the water. The oral cilia are set in groups on each side of the cytostome (meaning "cell mouth"). On one side, the cilia are aligned in a single closely set row, like the slats in a picket fence, to form an undulat-

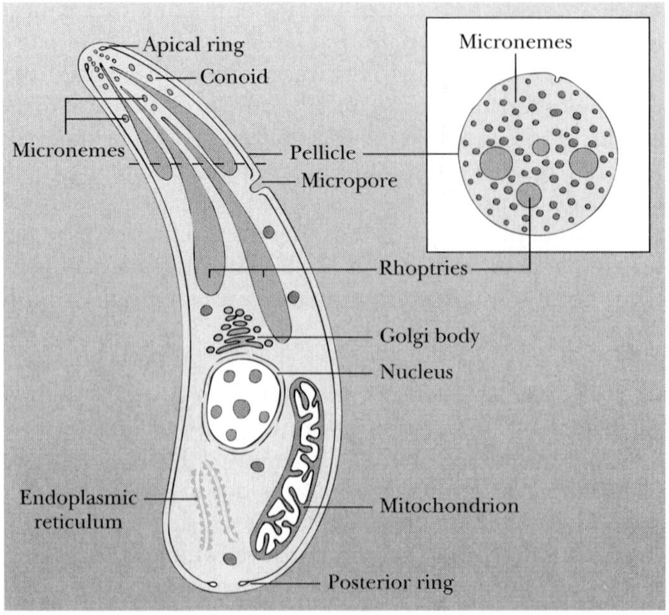

***Figure* 23.24** Diagram of a sporozoite showing features visible with an electron microscope.

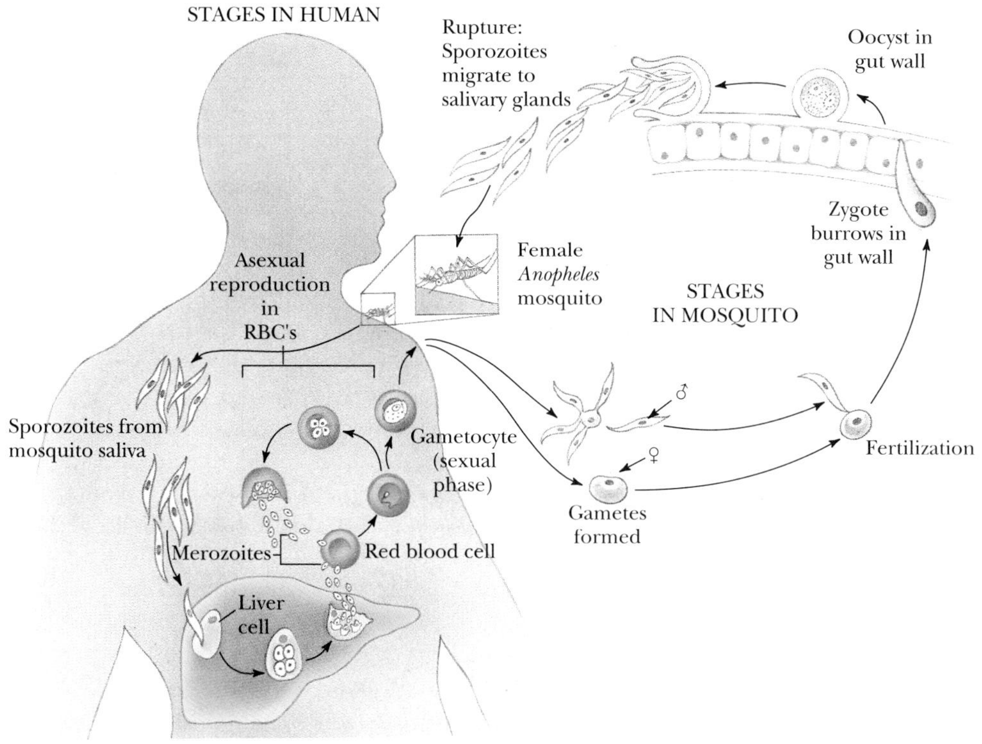

***Figure* 23.25** The life cycle of *Plasmodium vivax,* one of the species that cause malaria in man. Sporozoites are injected into a person when a mosquito takes a blood meal. Multiplication occurs first in the liver and then in blood cells. Repeated cycles of asexual reproduction occur in blood cells. Gametes in blood cells fuse to form a zygote when the blood is taken into a mosquito. The zygote penetrates the intestinal epithelium of a mosquito and grows on the outside surface. It divides to form sporozoites, which migrate to the salivary gland of the mosquito to repeat the cycle.

ing membrane. This is structurally different from the undulating membrane of parabasalians, but they both have the same name because they looked alike to early microscopists. On the other side of the mouth, the cilia are arranged in three rectangular clumps, like three housepainter's brushes. These are called membranelles. The oral cilia are used in filtering bacteria from the water and sending them into the cytostome. This basic organization has been modified in the different lines of ciliate evolution, and differences in ciliary patterns served as the basis for early taxonomic schemes. Somatic cilia may be reduced or absent, and the oral cilia may be reduced or lengthened in comparison to what is seen in *Tetrahymena.* The great majority of ciliates are phagotrophs, and many of these filter-feed on bacteria or small flagellates.

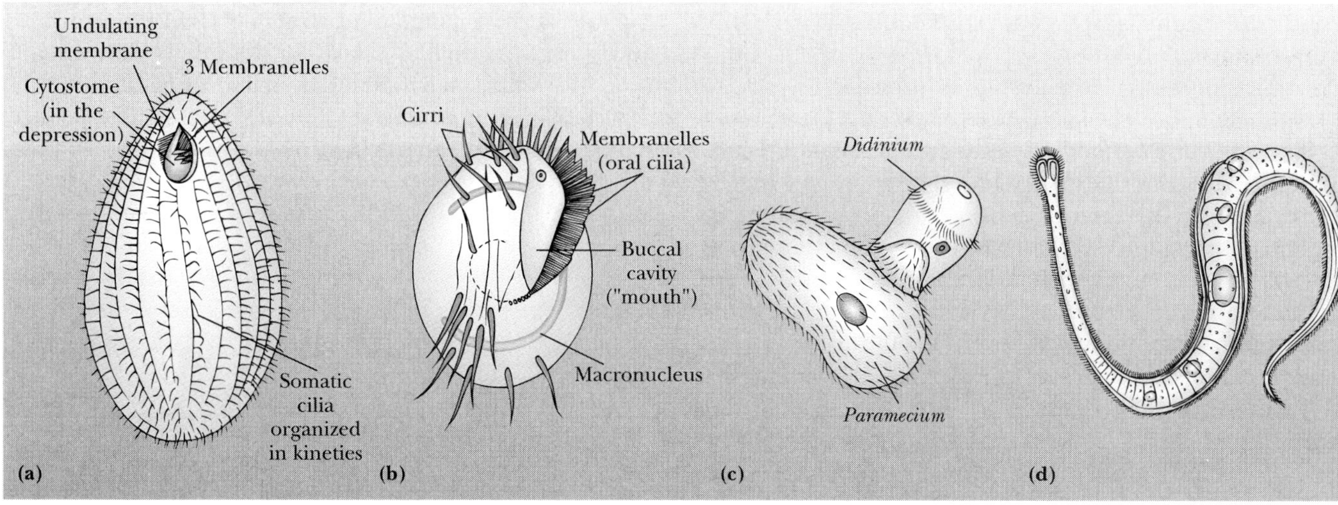

***Figure* 23.26 (a)** The ciliate *Tetrahymena.* The rows of somatic cilia can be seen, as well as the specialized oral cilia around the mouth. **(b)** *Euplotes* is an example of a hypotrich. They use their oral cilia to filter bacteria and small protists out of the water. The oral cilia are arranged in the same way as in *Tetrahymena,* but cover a larger area. The somatic cilia are grouped together into structures called cirri. They are used as little legs in walking across surfaces. **(c)** *Didinium* feeding on *Paramecium. Didinium* attaches to and immobilizes *Paramecium* by firing harpoon-like organelles into it. **(d)** *Tracheloraphis,* a karyorelictid ciliate that lives in the sand of seashores.

BOX 23.1 MILESTONES

A Protistan *E. coli*

Experimental biology has always relied on a small number of organisms that are particularly amenable for use in certain types of investigations. These "model systems," as they are called, provide the means of elucidating phenomena that are believed to be widespread if not universal, but which cannot be studied conveniently in most organisms. As science advances and the questions that need answering change, so do the organisms that are used. Thus, the development of classical genetics depended on the fruit fly, but as genetics became molecular, fruit flies were displaced by *E. coli* and its phages (and ironically, as developmental biology became molecular, fruit flies replaced other metazoans!).

Earlier in this century, the development of biochemistry and cell biology required colorless, animal-like cells capable of jumping through the experimental hoops that scientists were holding up at the time. Growth factors were being identified, and it was desirable to see how universally these were needed, and if they played the same roles in all cell types. Metabolic pathways were being elucidated, and it was desirable to see how universal they were. It was becoming possible to study some basic eukaryotic processes, such as cell division and phagocytosis, and cells that could be induced to divide synchronously and phagocytize on command were needed. A eukaryotic cell as easy to grow and as obedient as *E. coli* was required in order to study these things.

The necessary organism was provided by Andre Lwoff, who in 1923 succeeded in getting *Tetrahymena* (then known as *Glaucoma*) (see Figure 23.26a) into pure culture. This was the first ciliate to be grown **axenically** (meaning "in the absence of other organisms"). Getting it into pure culture was a remarkable achievement, as it was done in an age when antibiotics were not available to prevent the growth of bacteria in a rich organic medium, and nothing was known about the nutritional requirements of ciliates. Even more remarkable was the fact that Andre Lwoff was a 21-year-old student at the time. Both Andre Lwoff and *Tetrahymena* went on to greatness; Andre Lwoff gave up ciliates and received the Nobel Prize in 1965 for his work on lysogeny, and *Tetrahymena* became for a time *the* model eukaryotic cell.

The number of papers published on *Tetrahymena* has increased exponentially in the years since it was first put into pure culture. Although the first studies were nutritional, it has now been experimentally poked and prodded in almost every way that a eukaryotic cell can be—with one exception. It is more difficult to use for modern genetic studies than some other eukaryotic cells are.

However, one very important genetic discovery was made by using *Tetrahymena*. The Central Dogma (DNA makes RNA which makes protein) contains an implicit evolutionary conundrum. If the formulation is true, how could it all have started? If proteins are all encoded for in DNA, where did the proteins required to make the first string of DNA come from, and how could they have been preserved in the coding system? Thomas Cech found the answer unexpectedly, while studying an intron in *Tetrahymena* rRNA. He found that the intron removed itself from the rRNA, without the involvement of any enzymes. rRNA itself displayed enzymatic activity. Subsequent studies on RNA showed that it is capable of carrying out a wide variety of activities. RNA is capable of producing larger strings of RNA on its own, and is even capable of catalyzing peptide bond formation, the critical step in building up a protein. It is likely that the first genetic systems were RNA-based; RNA contained both the inherited sequence information and the catalytic capacities required for its own replication and inheritance. DNA was probably incorporated later, as a more stable repository of genetic information. Proteins were probably added in to increase catalytic versatility. *Tetrahymena* provided the first glimpse into the primeval RNA world, which preceded the genetic system that all life is now based on.

The basic set of oral cilia seen in *Tetrahymena* can be enlarged in such forms to increase the filtering capacity, as is the case in hypotrichs (Figure 23.26**b**). The undulating membrane in hypotrichs is much larger, and the three membranelles have become many membranelles. Hypotrichs are very efficient filter feeders.

Some ciliates feed on very large particles, such as other ciliates, and cannot rely on ciliary filters for trapping their prey. In these forms, the oral cilia are frequently reduced or absent. *Didinium,* which feeds on the ciliate *Paramecium,* is an example of such a species (Figure 23.26**c**). *Didinium* has harpoon-like organelles called toxicysts, which are fired into *Paramecium* on contact and serve to hold predator and prey together while *Paramecium* is slowly swallowed.

Ciliates have two types of nuclei. One is large and called the macronucleus while the other is small and called a micronucleus. The micronucleus is diploid and transcriptionally inactive. The macronucleus is large and usually polyploid. Many copies of each gene are present. Almost all the messenger RNA present in a ciliate's cytoplasm comes from the macronucleus. The micronucleus is used only during the sexual phase of the life cycle.

In most eukaryotes, sexual recombination occurs when two gametic cells fuse. This is followed by nuclear fusion and the development of the zygote into a new individual. However, in ciliates, there is usually no permanent cell fusion. Rather, two ciliates fuse temporarily and exchange haploid nuclei with each other. They then separate from each other. During conjugation, the micronucleus undergoes meiosis, but there is no cell division. The consequence is that the ciliate contains a number of haploid nuclei. The macronucleus and all of the haploid micronuclei except two disintegrate. It is one of these nuclei that is passed to the partner. Each partner winds up with one of its own haploid nuclei and one from its partner. These two nuclei then fuse, forming a diploid zygotic nucleus containing genes from both parents. At some point after the exchange, the cells separate. The first mitotic divisions after separation are not accompanied by cytokinesis, so that the number of nuclei present in the cells increases. After generation of the correct number of nuclei, differentiation into macro- and micronuclei occurs. Then cell division accompanies nuclear division once again, so that the number of nuclei remain constant in the cells through succeeding generations.

The karyorelictids (Figure 23.26**d**) are the group of ciliates thought to have nuclei most like those of the ancestral ciliates. Their macronuclei are diploid and incapable of division. Their number must be maintained by the differentiation of micronuclei each time cell division occurs. This state supposedly represents an earlier stage in the evolution of macronuclei than that displayed by other ciliates. These forms are characteristically found crawling among sand grains on the seashore. They are long, thin, flexible forms adapted to live among the interstices of such environments. They have well-developed longitudinal rows of somatic cilia.

The Heterokonts

The next major group to emerge in an rRNA tree is made up of the heterokonts and their relatives. A common morphological type among members of this group is a small flagellate with two flagella. One flagellum is directed anteriorly and has small hairs projecting laterally from it on both sides. The second flagellum is naked and projects posteriorly. "Heterokont" refers to this difference in flagella. Many species also have a peculiar helical structure at the point where the basal body and flagellum come together.

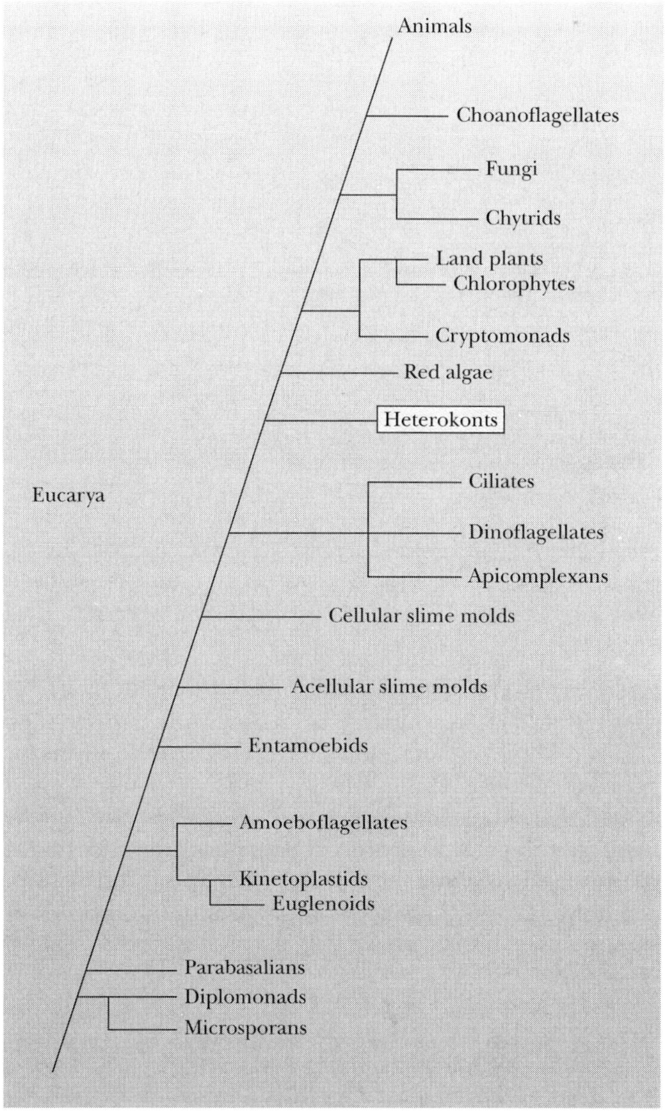

Ochromonas danica (Figure 23.27) is a well-studied chrysophyte, or golden-brown alga, and illustrates the basic chrysophyte features. The chloroplasts contain chlorophylls *a* and *c*. The outer chloroplast membrane is continuous with the outer nuclear membrane. There is a Golgi apparatus between the flagellar bases and the nucleus. Food vacuoles are frequently present; many chrysophytes are phagotrophic as well as photosynthetic. Chrysophyte chloroplasts frequently contain an orange eyespot, located near the base of the posterior flagellum. As in the case of the euglenoids, the flagellar base associated with the eyespot is swollen. There are also multicellular filamentous forms, and a few amoeboid species as well. Chrysophytes tend to be freshwater species with a preference for slightly

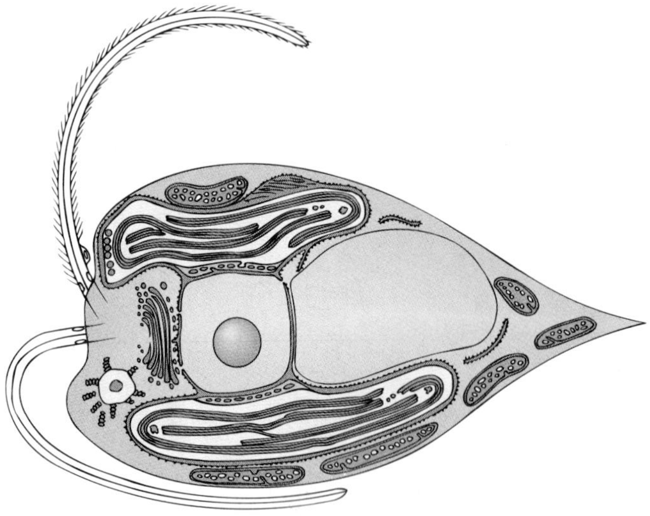

***Figure* 23.27** A diagram of the heterokont *Ochromonas.* The two kinds of flagella can be seen. The endoplasmic reticulum extending from the nucleus surrounds the chloroplast. This organization is characteristic of heterokonts.

acidic water. Populations tend to reach maximum size in cool water, such as when ice is melting in the spring.

Another group of heterokonts is the xanthophytes, or grass-green algae. Chlorophylls *a* and *c* are both present in the chloroplasts, but chlorophyll *c* is present only in very small amounts. Most live in freshwater. The majority of species are nonmotile coccoid or filamentous forms with cell walls (Figure 23.28**a–b**). Many produce zoospores of a chrysophyte structure, which demonstrates their relationship to the golden-brown algae.

The diatoms, or bacillariophytes (Figure 23.29), are also part of the heterokont assemblage. This is the largest group of heterokonts; more than 10,000 species have been described. They are abundant in most aquatic habitats, and are responsible for an estimated 20 percent of the primary productivity of the Earth. At first glance, they appear to have little in common with the chrysophytes and xanthophytes. The body is enclosed in two siliceous valves that fit together, and no flagella can be seen. However, the chloroplasts are ultrastructurally the same as those of chrysophytes, and the pigment composition is also the same. Furthermore, some species produce male gametes having flagella with lateral hairs. Other flagellated heterokont algae produce siliceous scales and spines that adorn the body; diatoms merely carry this ability to an extreme.

The brown algae, the kelps and wracks, also belong to this group (Figure 23.30). Although the most obvious stages in the life cycle are large multicellular leaflike structures, they reproduce by means of chloroplast-bearing biflagellated unicells having the appearance of chrysophytes. At all stages, the chloroplasts are chrysophyte-like.

There are several colorless heterokonts, such as the oomycetes (Figure 23.31). Oomycetes are aquatic organisms similar in appearance to fungi; they form masses of white threads on objects decaying in the water such as dead fish, or even minute objects such as pollen grains. They

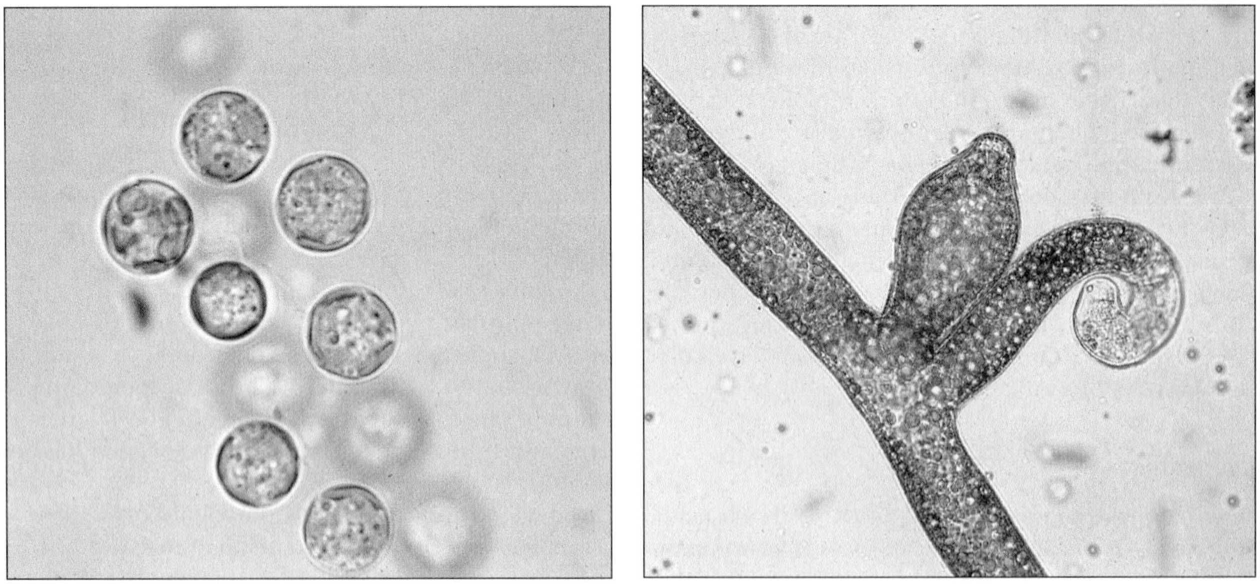

(a) (b)

***Figure* 23.28** Xanthophytes. **(a)** *Botrydiopsis,* a coccoid form and **(b)** *Vaucheria,* a filamentous form. (© Dr. Phillip Sze)

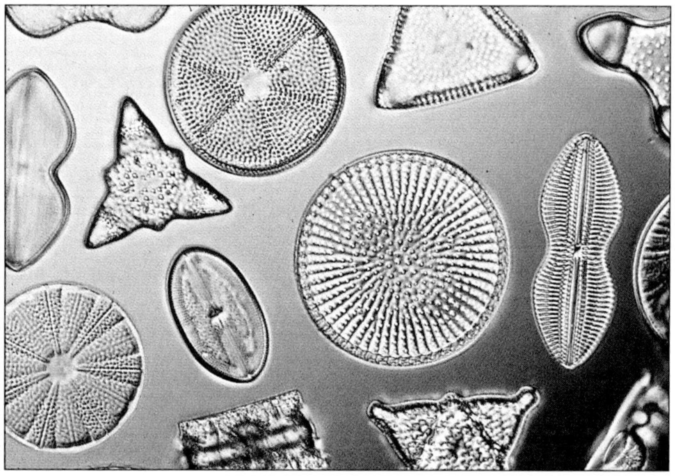

***Figure* 23.29** Marine planktonic diatoms. (© Manfred Kage/Peter Arnold, Inc.)

form motile zoospores having a typical oomycetes structure, although without chloroplasts. Long regarded as a connecting link to the true fungi, oomycetes are shown by rRNA sequences to be quite unrelated to the true fungi. They have acquired their superficially similar life cycle and morphology through convergent evolution; they occupy a niche in aquatic environments comparable to that occupied by the higher fungi in terrestrial environments. rRNA sequences indicate that the first heterokonts were colorless, with chloroplasts being acquired later.

Lamellocristate Protists

All of the aerobic forms mentioned so far except for the discocristate species have mitochondria with tubular cristae. This is the most common type of cristae existing among the protists. A few groups have lamellar cristae. This is of interest because so do the animals, fungi, and land plants. Most of the forms with lamellar cristae cluster together in rRNA-based phylogenetic trees. This means that the ancestors of all three major multicellular eukaryotic groups probably belonged to this small pool of protistan groups. Protists having mitochondria with lamellar cristae include the rhodophytes, or red algae; the chlorophytes, or green algae; the cryptophytes (cryptomonads); the choanoflagellates; the chytrids (or chytridiomycetes); and a few amoebae.

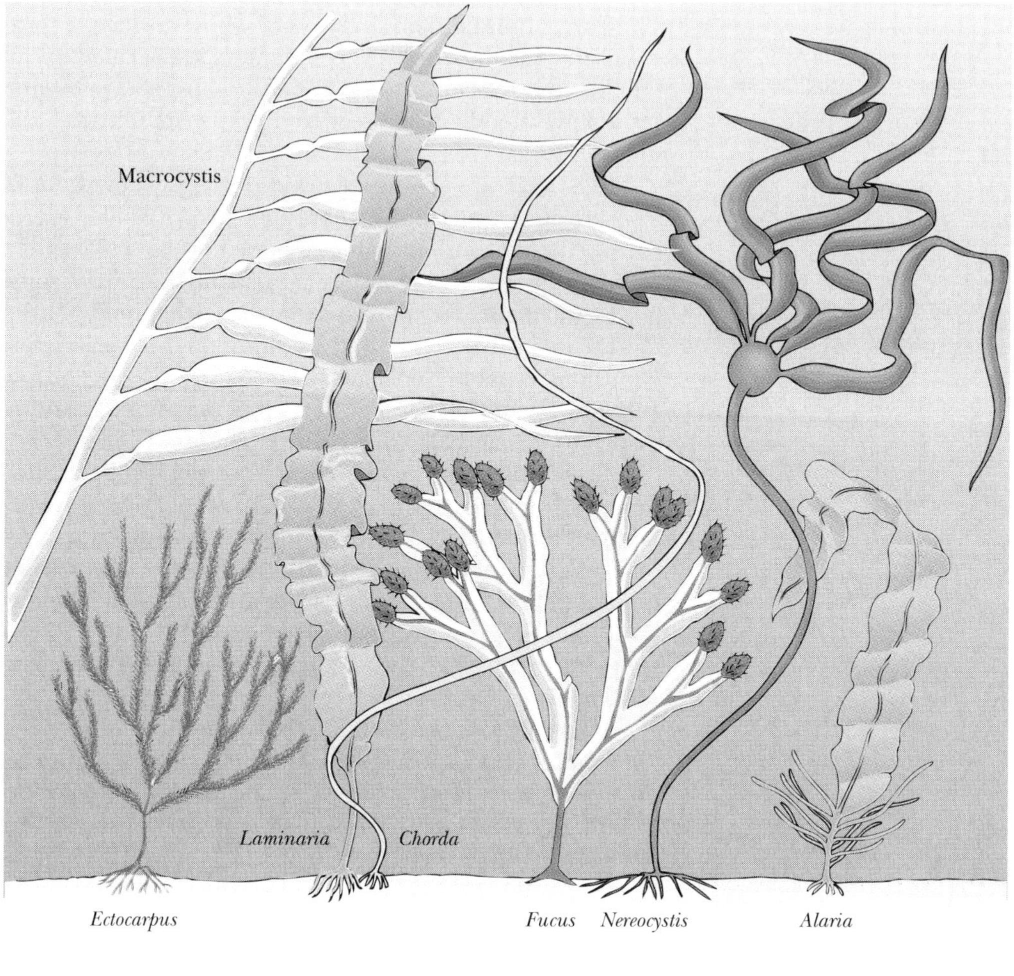

***Figure* 23.30** Representative brown algal genera.

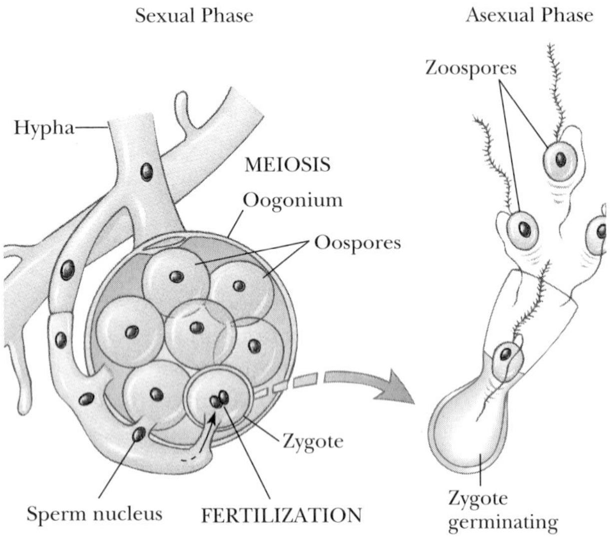

***Figure* 23.31** Hyphae and oospore-containing oogonia of the oomycete *Saprolegnia.* The oospores are produced in the sexual phase of the life cycle. Oomycetes also produce flagellated zoospores asexually.

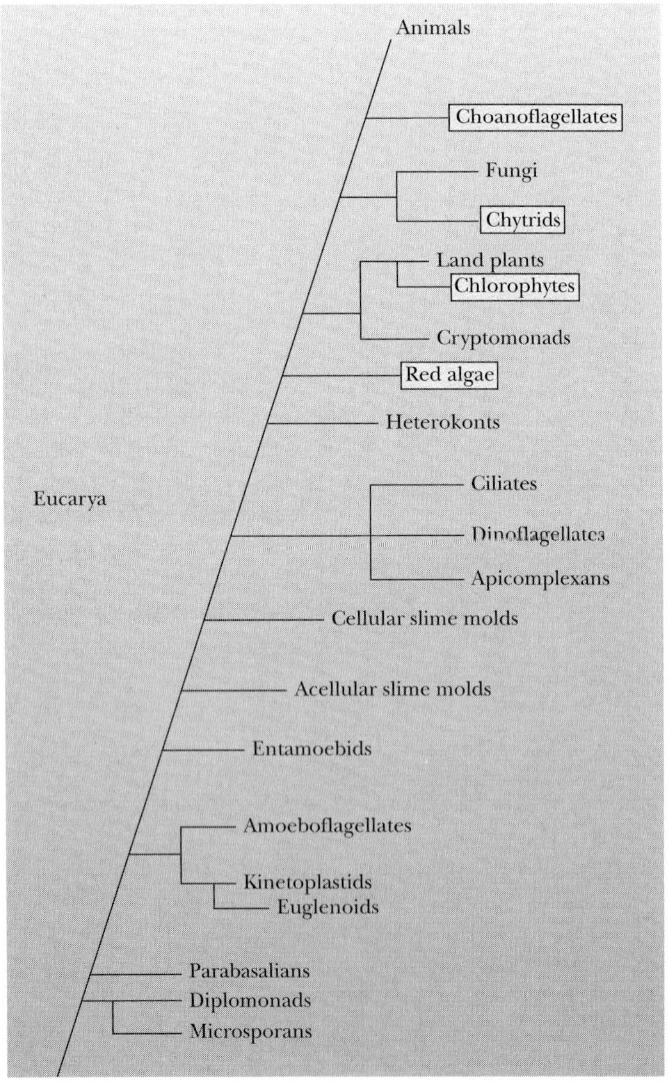

The Red Algae

The rhodophytes, or red algae (see Figure 23.2), are predominantly photosynthetic forms, ranging from unicellular to large multicellular bladelike forms. Most are multicellular. In contrast with other major algal groups, there are no free-swimming, flagellated unicellular forms. No such forms exist as independent species or even as a phase in the life history of a multicellular form. Red algae lack flagella, basal bodies, and centrioles at all stages of the life cycle. The life histories themselves can be quite complex, with different morphologically distinct stages associated with different ploidy levels. Surprisingly, forms displaying simple asexual reproduction can also have life cycles composed of morphologically distinct generations.

Another feature unique to the red algae, but not present in all of the species, is the so-called pit-connections existing between cells in multicellular forms. The term is inappropriate inasmuch as these are plugs rather than connections. During cell division, the cell walls formed between daughter cells grow inward and stop before the daughter cells are completely compartmentalized. The remaining holes are plugged with proteinaceous material—the pit "connection" (Figure 23.32).

Most red algae live in tropical marine environments. They are abundant in coral reefs. Some, referred to as corallines, produce an extracellular coating of calcium carbonate, as corals do. They can be the most abundant and productive members of a reef community, helping to build the reef itself.

Red algae are economically important. They produce a series of unusual polysaccharides, two of which are of considerable economic value. Agar is used in making bacteriological media, and also as a thickening agent in some foods. Carrageenan is also used as a thickening agent in foods such as ice cream and pudding, and for stabilizing emulsions in paint. Red algae are also used directly as food. *Porphyra* is eaten in the Far East (it is the "nori" found in sushi), and is actually farmed in Japan and China.

Cryptomonads (Figure 23.33) are small biflagellated unicells found in both freshwater and marine environments. They tend to be more common in cool water. Most of them are photosynthetic. Both flagella, which have mastigonemes, emerge subapically near a gullet. The gullet is lined with unique extrusive organelles called ejectosomes. In the body they look like very short rods, but unwind like a short roll of crepe paper when fired.

The chloroplasts contain chlorophylls *a* and *c*. The chloroplast structure is unusual in that a set of membranes outside of the plastid itself contain eukaryotic-sized ribosomes and a nucleomorph, which appears to be a degenerate nucleus. This suggests that the "chloroplast" might represent the remnants of a eukaryotic alga ingested by some ancestral cryptomonad. rRNA sequencing supports this idea; small-subunit rRNA genes from the nucleo-

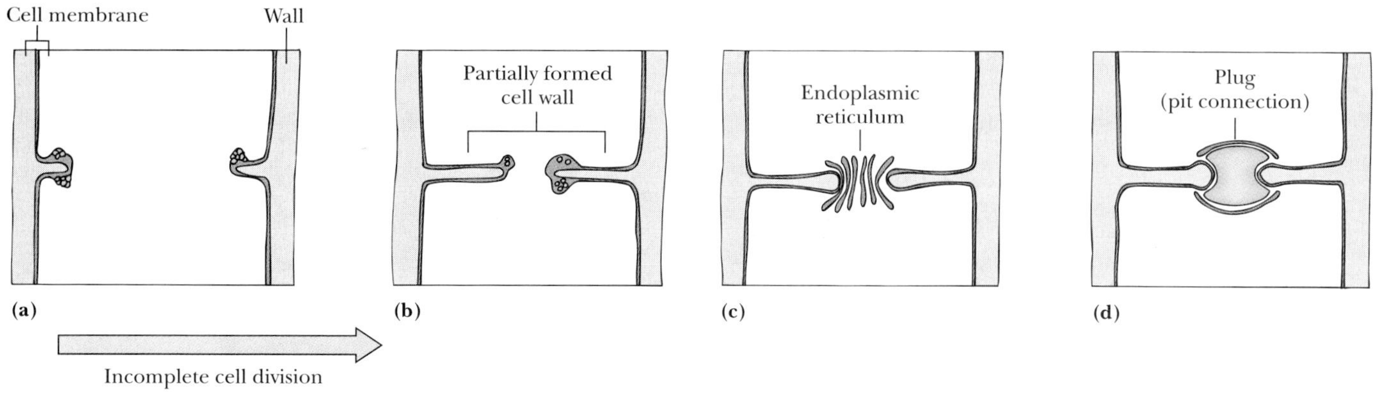

***Figure* 23.32** Diagram showing the formation of a pit connection between two red algal cells. **(a, b)** Division of the cell begins, but remains incomplete. **(c)** Material is deposited in the opening in the incompletely formed cell wall. **(d)** It eventually becomes a plug.

morph are eukaryotic in nature and group with the red algae in rRNA trees, while the nuclear cryptomonad sequences are quite different. The chloroplasts of cryptomonads are apparently the remnants of ingested red algal cells. Instead of being digested, they remained in the cell and lost structures that were no longer needed. This process parallels the development of chloroplasts from cyanobacteria in other algae.

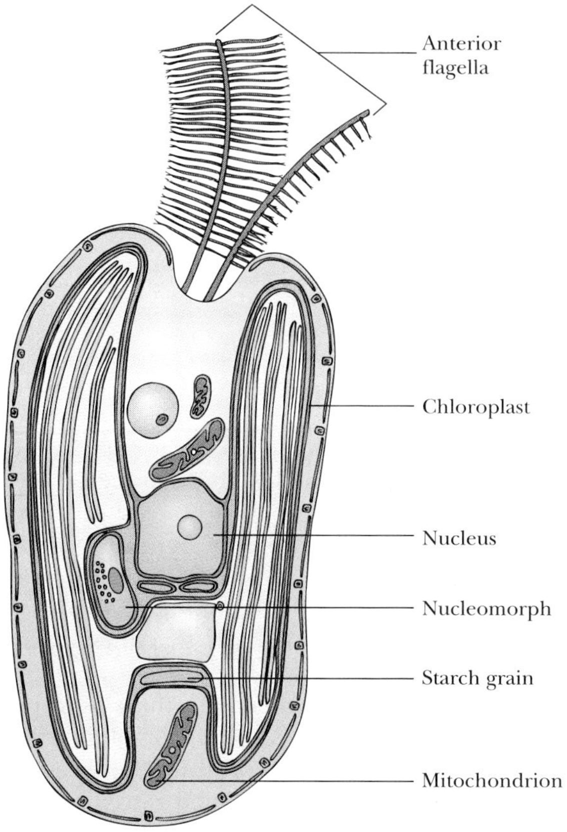

***Figure* 23.33** Diagram of a cryptomonad showing the two anterior flagella, the nucleus, the chloroplast, the pyrenoid, the nucleomorph, a starch grain, and mitochondria.

The Green Algae

The large and diverse group of green algae, the chlorophytes, branch off near the top of the tree, near the point at which the fungi, higher plants and animals emerge. The land plants undoubtedly evolved from a chlorophyte ancestor. The morphology of the simplest forms is demonstrated by *Chlamydomonas* (Figure 23.34). Two naked flagella emerge from the anterior end of the cell. There is a single cup-shaped chloroplast containing chlorophylls *a* and *b*. It may also contain an eyespot. There is a cell wall, which is composed of glycoprotein. Many different lineages have produced multicellular forms, which may be filamentous, leaflike, or even spherical in shape, as in *Volvox* (see Figure 1.21). *Volvox* colonies are hollow spheres whose surfaces consist of hundreds or thousands of *Chlamydomonas*-like cells held together. Cell differentiation occurs, so that a subset of cells in the colony produce gametes.

Choanoflagellates (Figure 23.35) are small, primarily marine, phagotrophic organisms. They have a single flagellum that beats in such a way as to draw a stream of water toward the cell body. As the water flows past the body, it is drawn through a filter consisting of a ring of tentacles running around the top part of the cell. Bacteria are trapped by these tentacles and consumed by the cell. Choanoflagellates frequently produce a delicate lorica, an external basket or shell, in which the flagellate sits.

In phylogenetic trees based on RNA sequences, choanoflagellates are the protistans that are closest to the animals, so animals probably arose from them or an unknown protistan group quite a bit like them. Cells having a similar collar of tentacles, and which feed in the same way, are found in sponges.

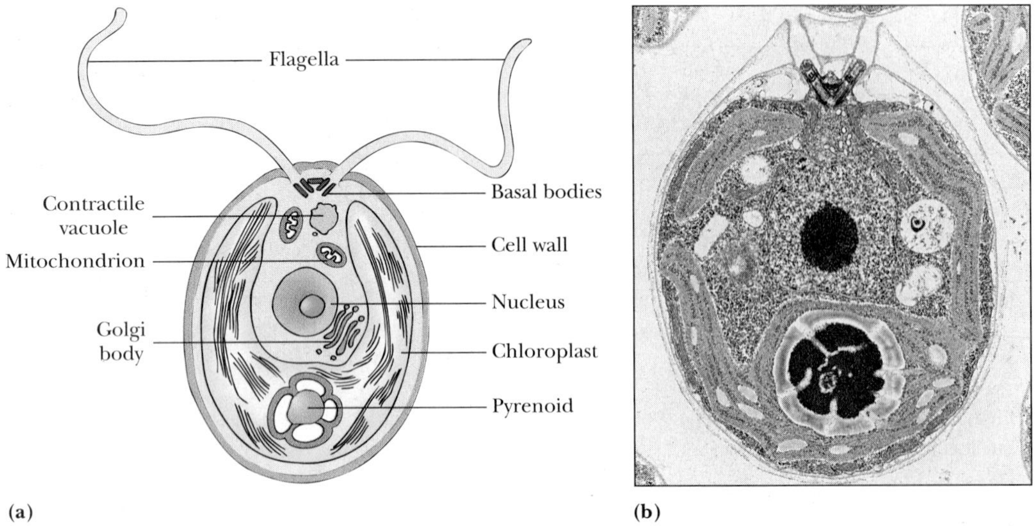

***Figure* 23.34** **(a)** A diagram of *Chlamydomonas* showing features visible with an electron microscope **(b)**, including the flagella and basal bodies, the contractile vacuole, mitochondria, the chloroplast, the pyrenoid and starch grains, the Golgi body, the nucleus, and the cell wall. (© 1991 W. L. Dentler, The University of Kansas/Biological Photo Services)

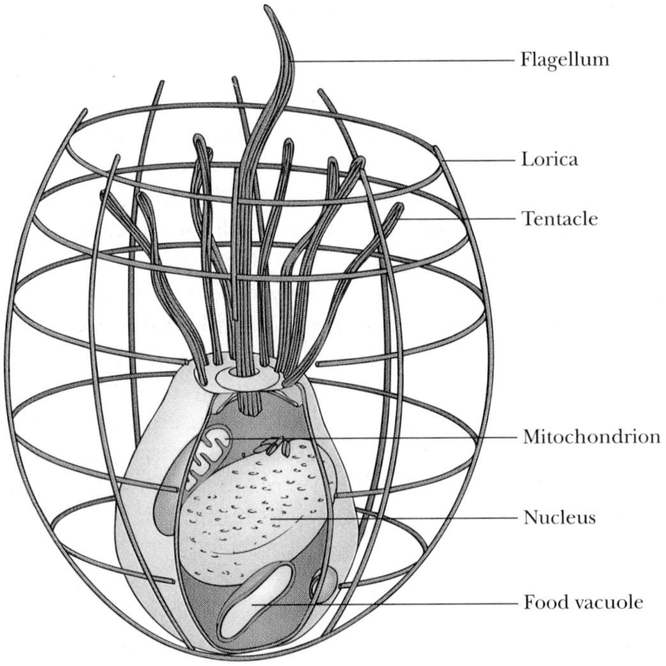

***Figure* 23.35** A choanoflagellate with a lorica. An apical flagellum is surrounded by a set of tentacles that trap food. There are food vacuoles in the cell as well as a nucleus and mitochondria with lamellar cristae. The cell sits inside of a delicate lorica.

Amoeboid forms other than the entamoebae and slime molds are scattered throughout an rRNA tree, indicating that amoebae arose from flagellated cells on many occasions. Large amoeboid groups, containing many species, include the foraminifera (Figure 23.36), heliozoa (see Figure 23.6), and radiolaria (see Figure 1.29). The foraminifera are marine amoebae with tests of calcium carbonate or an organic matrix in which sand grains are incorporated. Most are benthic forms that spread out an anastomosing net of pseudopods with which they trap food. The heliozoans and radiolarians extend long thin rays from the body (heliozoan means "sun animal") composed of bundles of microtubules covered by a thin layer of cytoplasm. Small food organisms are trapped on these rays.

The Fungi

The different types of true fungi are clearly related to each other and form a monophyletic unit. How they arose from protists is a little clearer now than in the past. In the past, organisms such as slime molds and oomycetes were presumed to connect the fungi to the protists; in addition to their protistan features, they produced fruiting bodies and this allowed one to imagine how fungal life cycles may have arisen. Studies on ultrastructure and life cycles, as well as rRNA comparisons, show, in fact, that these forms are quite unrelated to fungi.

The chytridiomycetes, which are the closest relatives of higher fungi in rRNA trees, are the one link remaining between the true fungi and the protists. They produce sporangia like fungi, but have flagellated zoospores (see Figure 1.24). The zoospores usually have two basal bodies, only one of which has a flagellum. The flagellum emerges at the posterior end of the cell. The life cycle is extremely variable. Sexual reproduction may or may not occur, and

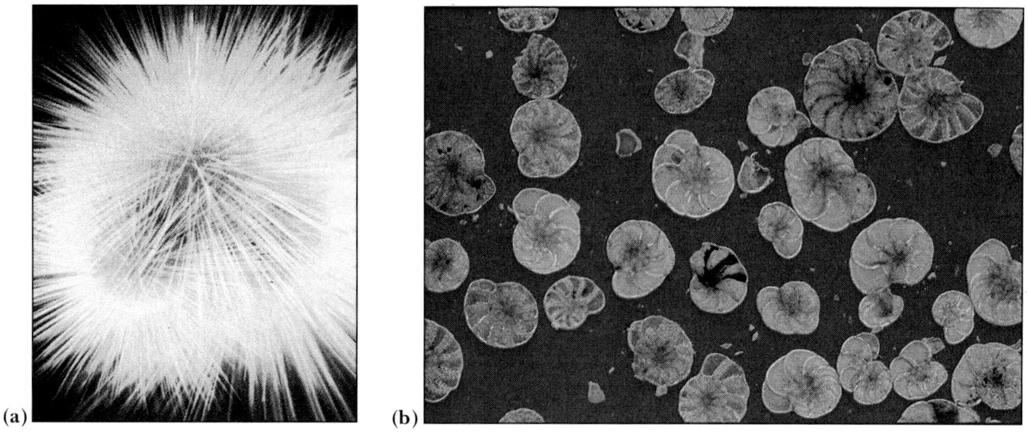

***Figure* 23.36** Foraminiferans live in a shell from which they extend a net of long pseudopods, which operate like small marine spider webs. The foraminiferan feeds on the organisms that get stuck to the pseudopods. **(a)** The central test is hard to see in this living individual. (**a,** © John D. Cunningham/Visuals Unlimited). **(b)** Foraminiferal tests; it is easy to see why foraminifera were regarded as minute molluscs when they were first discovered. (**b**, © M. I. Walker/Photo Researchers, Inc.)

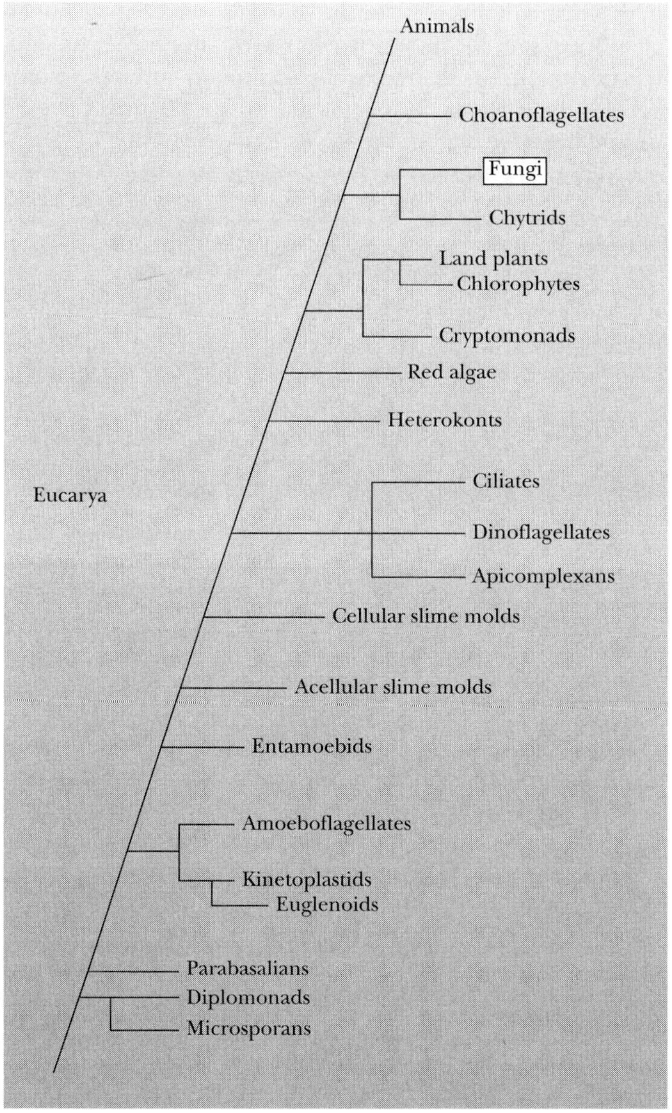

when it does occur can involve gametes or the fusion of rhizoids. The entire organism may consist either of a single cell or a multinucleate branching mass resembling the mycelia of fungi, albeit much smaller. Most live in water or soil, but there are some multiflagellated chytridiomycetes that live in the rumen of cattle. A few chytridiomycetes are parasitic. One of these, *Synchytrium endobioticum,* causes black wart disease in potatoes.

The true fungi are colorless, saprotrophic, predominantly terrestrial organisms that reproduce through the production of spores. Their cell walls are chitinous. Flagella do not occur at any stage of the life cycle. Vegetative growth occurs by the elongation of threadlike filaments called hyphae. Growth may be so rapid that it can be seen through a microscope. The basic structure of a growing fungus is hyphal. All of the large macroscopic structures produced by fungi, such as toadstools and puffballs, are masses of hyphae and the spores produced by them. The collection of hyphae forming one organism is called a mycelium.

During growth, division of the nuclei may be accompanied by the formation of septa, marking the boundaries of one cell. However, septa formation is usually incomplete, leaving a pore connecting compartments, and sometimes doesn't occur at all. This means that the contents of a mycelium are a single connected multinucleate mass of cytoplasm.

Hyphae grow over and through the organic material sustaining them, frequently secreting enzymes to break the substrate into smaller organic molecules that the hyphae can then absorb. Fungi are important decomposers in terrestrial ecosystems, digesting grass, leaves, dead trees, and dead animals. Some attack living plants and animals; ringworm and athlete's foot are two common human diseases caused by fungi. Fungi may reproduce either sexually or asexually, and in either case, the special propagule produced is a spore.

The true fungi are divided into four major groups. These are the zygomycetes, ascomycetes, basidiomycetes, and the deuteromycetes (also called the Fungi Imperfecti). Three of these are natural groups whose members are united by a common evolutionary history. The deuteromycetes are an artificial group made up of members whose incomplete life cycles (incomplete in comparison with the other fungi) prevent determination of their true taxonomic placement.

The Zygomycetes

Rhizopus stolonifer (Figure 23.37), a bread mold, is an example of the zygomycetes. The cycle begins when a spore lands on a piece of bread and germinates. Grayish hyphae grow rapidly across the bread. Some grow down into it and absorb nutrients. Others grow up into the air and produce a generation of asexually produced spores. The tips of the hyphae growing up into the air swell, and nuclei collect in these swellings. The tip eventually becomes a spherical structure called a sporangium. The protoplasm in it breaks up into spherical uninucleate masses, each of which is ultimately covered by a resistant cell wall; these are the spores. The sporangium wall breaks and the spores are released. If they settle on a piece of bread, the spores germinate and the cycle is repeated. The asexual production of new generations can continue indefinitely in this manner. However, sexual reproduction can also occur. This necessitates the presence of two mycelia of compatible mating types on the bread.

When hyphal tips from two such mycelia contact each other, a septum is formed in each hypha at some distance behind the contact point. The wall between the tips is then broken down so that cytoplasm and nuclei from the two parental hyphae mix in this chamber. Eventually the two types of nuclei fuse in pairs, forming zygotes. The wall of the chamber becomes a thick black structure called a zygospore. After several months, the zygospore germinates, at which time the zygotes undergo meiosis, and a sporangium containing the division products emerges. The spores from this sporangium germinate and grow into hyphae if they land on a suitable piece of bread. They may reproduce either sexually or asexually. The organism is haploid throughout its entire life cycle except at the zygospore stage. Most zygomycetes are fairly inconspicuous terrestrial fungi. A few are parasitic in animals. Some species have practical uses; one of these, *R. oryzae,* is used in making sake.

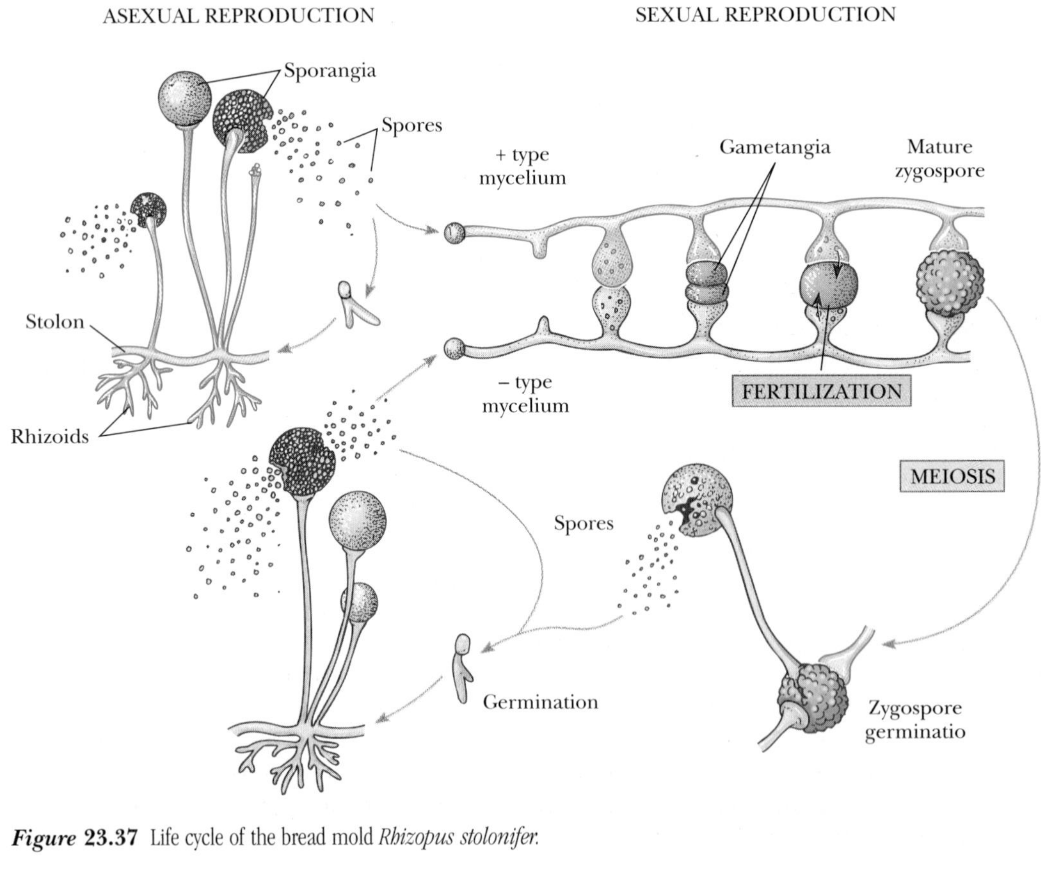

***Figure* 23.37** Life cycle of the bread mold *Rhizopus stolonifer.*

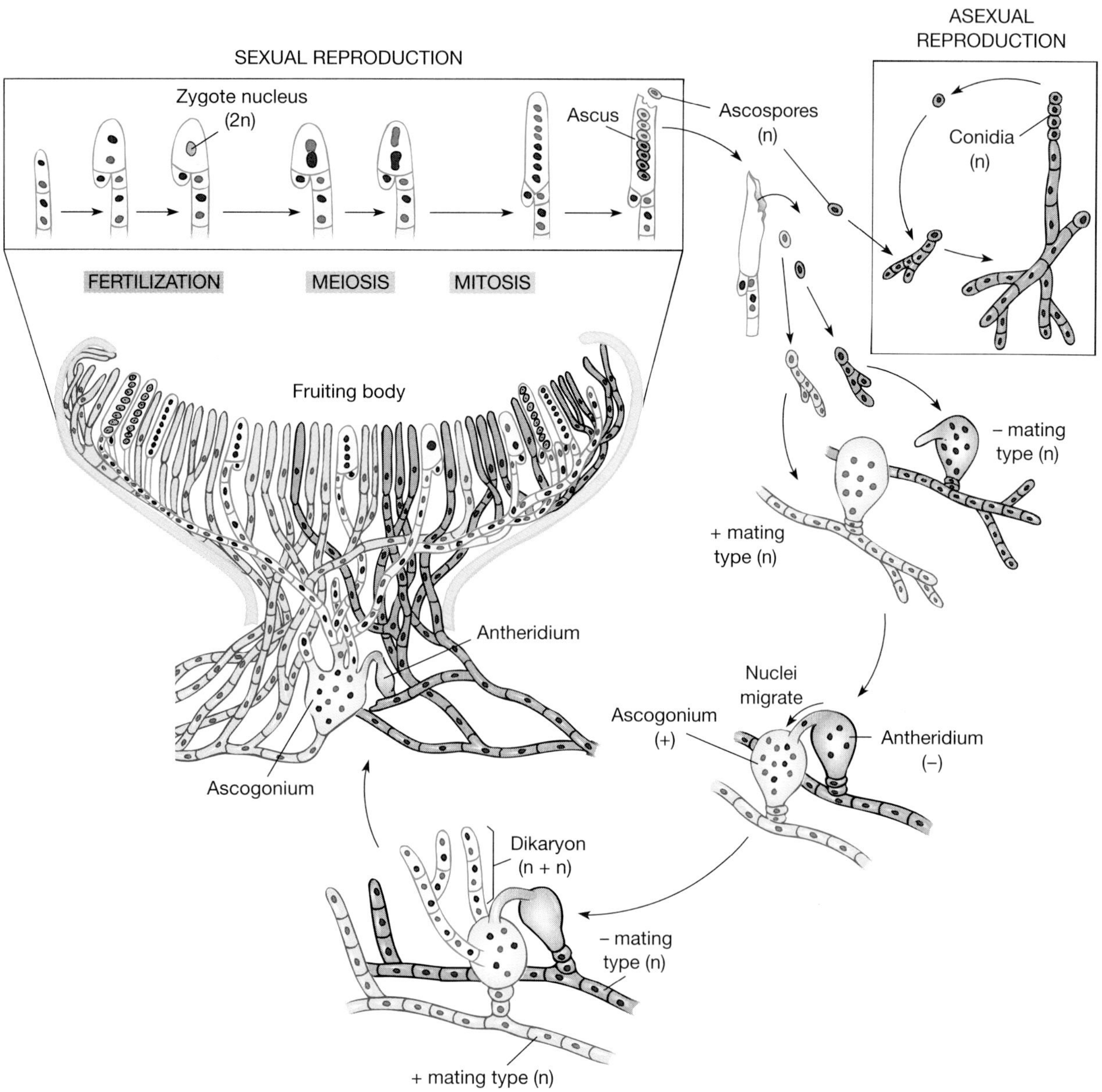

***Figure* 23.38** Diagram of an ascomycete. This cup fungus is formed from a mass of hyphae that grew out of two different spores. At the point where they met, one hypha donated nuclei to the other hypha. The region containing the donated nuclei grew into a larger structure called an ascogonium. Hyphae emerging from the ascogonium are made up of cells containing nuclei from both parents. Such a cell is called dikaryon. The two nuclei fuse, immediately undergo meiosis, and produce spores in an ascus. This is shown diagrammatically in the hyphal tips in the top of the cup fungus. Spores can also be produced asexually at the tips of hyphae. These spores are called conidia.

The Ascomycetes

The ascomycetes may produce spores asexually as do the zygomycetes, although not all ascomycetes are capable of asexual reproduction. As in the zygomycetes, spores are produced at the ends of raised hyphae. However, no sporangium encloses the spores. The spores, called conidia, are arranged in chains extending outward from the tip of the hyphae. The feature uniting members of the group is the formation of an ascus (a little bag containing the products of meiosis) during sexual reproduction (Figure 23.38).

When haploid hyphae of two different mating types meet, the nuclei of one are transferred into the body of the other. The hypha receiving the nuclei is regarded as female and called an ascogonium. Somatic hyphae around the fertilized ascogonium are in some manner stimulated to form a protective covering for it. New hyphae grow out from the ascogonium, containing a mixture of nuclei from the two-parent hyphae. During the growth of these ascogenous hyphae, septa are produced that compartmentalize the cytoplasm. Each cell contains two nuclei, one from each parent. The penultimate cell of each ascogenous hypha grows into an ascus. The two nuclei fuse and then undergo meiosis to produce four daughter nuclei. These may undergo mitosis, in which case the developing ascus will contain eight nuclei. Eventually each nucleus and the cytoplasm around it differentiates into a spore. These will germinate when they reach a suitable food source and give rise to the next mycelial generation.

Superficially, the organisms in this group are very dissimilar, but all are united by the production of asci during sexual reproduction. Ascomycetes include cup fungi, powdery mildews, and even yeasts. Yeasts do not form mycelia; they are unicellular forms that reproduce by splitting in half or budding. Sexual reproduction also occurs when two haploid yeast cells of complementary mating types fuse. A diploid zygote is formed that can multiply by dividing repeatedly, producing many separate daughter cells. At some point, meiosis is induced, and four haploid cells come to lie within a single mother cell. This is the equivalent of an ascus.

Many ascomycetes are important to humans. In addition to yeasts, which are used in baking as well as brewing beer and wine, the group includes truffles and a series of pathogens of crop plants. *Ceratocystis ulmi* attacks American elms. *Endothia parasitica* has wiped out American chestnut trees. Powdery mildews on fruit can cause enormous losses. *Claviceps purpurea,* the ergot fungus, is a historically important member of the group. It is a parasite of rye which causes potentially fatal illness in animals, including humans who eat the grain or bread made from it. It induces psychotic delusions and convulsions. Epidemics of ergotism occurred frequently in the Middle Ages, when it was known as St. Anthony's Fire.

The Basidiomycetes

Basidiomycetes include puffballs, smuts, mushrooms, and bracket fungi. The above ground, fruiting portion that humans see is a small fraction of the total mycelial mass. Formation of the macroscopic fruiting bodies requires the union of two mycelia of different mating types (see Figure 1.26). Mycelia that grow out from a spore in the soil are incompletely septate. Pores in the septa allow cytoplasm and nuclei to pass throughout the mycelial mass. Nonetheless, there is one nucleus in each compartment. When the growing hyphae meet hyphae of a compatible mating type, nuclei are exchanged. The new nuclei multiply very rapidly and migrate throughout the mycelium, so that the compartments become binucleate. When these binucleate hyphae grow, nuclear division is synchronous, so that new cells remain binucleate. At some point, the characteristic macroscopic fruiting body for the species is formed and spores are produced; the tip cell is transformed into a basidium, on which spores will be borne. The two nuclei in a developing basidium fuse, forming a diploid zygote. This then undergoes meiosis, so that four haploid nuclei are produced. At the same time, four protuberances appear on the basidium. The nuclei migrate to the tips of these, and, together with the surrounding cytoplasm, are transformed into spores. These are disseminated by wind, rain, and the movement of animals and give rise to the next generation. Some basidiomycetes also produce conidia, but spores produced in basidia are the primary means of reproduction. Mushrooms produce enormous numbers of basidia on their gills, and can produce billions of basidiospores.

The rusts can have more complex life cycles, in which two hosts are required. An example is provided by *Puccinia graminis,* which causes black stem rust in grains and also grows on barberry. The primary mycelium, formed when a spore germinates on a barberry plant, grows throughout the tissues of the plant. The cells have one type of nucleus. The hyphae do not simply touch and fuse as in the mushrooms. Rather, the hyphae differentiate into male and female structures, defined on the basis of which hyphae donate nuclei and which accept nuclei. The male structures secrete nectar that attracts insects. While feeding and crawling over the plant, insects carry spermatia to the receptive female hyphae. Through convergent evolution, these fungi have developed a method of fertilization much like that employed by flowering plants. However, this does not lead directly to the production of a large secondary mycelium containing cells with two nuclei, as in the mushrooms. Rather, binucleate spores are produced. These do not develop further until they land on a wheat plant or related grain. After germination occurs, binucleate hyphae are produced that penetrate the plant and grow between the cells. These mycelia can asexually produce masses of spores that germinate on contact with other wheat plants and produce mycelia capable of producing still more spores. These rounds of asexual reproduction by secondary binucleate mycelia correspond to asexual reproduction through the production of conidia in other fungi.

At some point, a second type of binucleate spore is produced in which the two nuclei fuse. These are resistant spores that overwinter in a resting state. When they germinate, a very short hypha emerges, which will be the rust's basidium. Meiosis takes place, and the four nuclei migrate into the hypha. Septa are produced between them, and each cell produces a spore.

The life cycle of the rust is, then, a more complicated version of a typical basidiomycete life cycle. The primary mycelium exists on one plant, the secondary mycelium on another. The transition from primary to secondary mycelium is aided by the production of special reproductive structures, and the passage from one plant to another is accomplished through the insertion of spore production into the life cycle at the time transfer must be accomplished. Building up a large mass of secondary mycelia depends on spreading from plant to plant, which is again accomplished through spore production. Rather than being produced on a large structure like a mushroom, the basidiospores are produced on tiny dispersed structures. Both mushrooms and rusts display the basic features of a basidiomycete life cycle.

The Deuteromycetes (Fungi Imperfecti)

The Fungi Imperfecti or deuteromycetes ("secondary fungi") do not produce asci or basidia but have conidia like those of the ascomycetes or basidiomycetes. They are believed to be ascomycetes or basidiomycetes that have lost the sexual part of the life cycle. Many species in this group are economically important. The fungus that produces penicillin belongs here, as do the fungi used in the production of several cheeses. Among the pathogens are those that cause celery leaf spot, celery blight, and potato blight.

Eukaryotes and Multicellularity

All three of the main multicellular groups emerge from approximately the same place in an rRNA tree. There was no reason to believe that they would be so closely related, inasmuch as multicellularity has arisen in many protistan lineages. The red algae, heterokonts, chlorophytes, and ciliates all show a series of forms ranging from uni- to multicellular. The cellular slime molds are all multicellular. Genuine differentiation of cells occurs in these groups; the algae have cells specialized for reproduction, while in slime molds multicellularity and differentiation are necessary to produce the fruiting bodies. In these forms, cells remain together after dividing. The situation seen in the parabasalians, in which some cell types represent, in essence, a multiplication of cell parts, is another reflection of the basic protistan evolutionary drive to escape the limitations of unicellularity. Too many words are wasted in contriving definitions of multicellularity meant to differentiate animals, plants, and fungi in a qualitative way from the protists. It would be better to recognize the widespread tendency to multicellularity among the protists and regard these other three groups as small, recent offshoots of the protistan world that reflect this basic tendency.

Summary

- **Eukaryotes** constitute the third major division in the living world. They are organisms whose DNA is separated from the cytoplasm by a **nuclear** membrane. They may be unicellular or multicellular. Eukaryotes other than plants, animals, or fungi are called **protists.** Protists, even in the same group, may be unicellular or multicellular.

- Eukaryotic cells contain more DNA than prokaryotic cells, and transmission of DNA to daughter cells at the time of division requires a special mechanism called **mitosis.** Mitosis normally immediately precedes cell division, or **cytokinesis.** Genetic recombination in eukaryotes involves combining the chromosomal complements of two gametes by their fusion to form a zygote. This doubling of DNA must be balanced at some other point in the life cycle by a halving of the DNA. This is accomplished by **meiosis,** which also randomly sorts homologous chromosomes into the gametes.

- Other membrane-bound organelles found in eukaryotic cells include the **endoplasmic reticulum,** the **Golgi apparatus, lysosomes, contractile vacuoles, mitochondria,** and **chloroplasts.** Mitochondria and chloroplasts are the remains of endosymbiotic Eubacteria. Chloroplast-bearing protists are called algae. They feed **phototrophically. Phagotrophic,** or particle-eating, protists are called **protozoa.** Some members of both groups also feed **saprotrophically,** by absorbing dissolved nutrients.

- Protists contain structural elements collectively called the **cytoskeleton,** which are responsible for maintaining the cell shape. Three cytoskeletal components found among most protists are **microtubules, microfilaments,** and **intermediate filaments.** Locomotion is by means of **pseudopods, cilia,** or **flagella.** Cilia and flagella grow out of **basal bodies** or **kinetosomes,** which are ultrastructurally identical to the **centrioles** found at the poles of mitotic spindles.

- rRNA sequences are useful in revealing relationships among eukaryotes, just as they have been for prokaryotes. The deepest branches in rRNA trees lead to three groups of protists which lack mitochondria—**microsporans, diplomonads,** and **parabasalians.** Microsporans are all parasitic. They lack organelles of locomotion and live intracellularly. Diplomonads and parabasalians are flagellated cells.

Most live in the intestinal tract of animals. A few are free-living. Many parabasalians live in termites, where they are responsible for cellulose digestion.

- The aerobic protists found in the deepest branches of rRNA trees are those whose mitochondria contain discoid cristae—the **euglenoids, kinetoplastids,** and **amoeboflagellates.** Euglenoids and kinetoplastids are flagellates. Kinetoplastids have unusual mitochondria containing such large amounts of DNA that the mass can be seen with a light microscope. Several kinetoplastids cause serious diseases in humans and domestic animals. Amoeboflagellates frequently have both flagellate and amoeboid stages in the life cycle.

- The next aerobic group is composed of protists having mitochondria with tubular cristae. Most protists belong here. Included are most groups of **amoebae**, the cellular and acellular **slime molds,** the **alveolates,** and the **heterokonts.**

- The alveolates include **dinoflagellates, ciliates,** and **apicomplexans.** Dinoflagellates have two flagella set at right angles to each other. They live in a wide variety of habitats and may be either free-living or symbiotic. Ciliates have a covering of cilia, organelles of motility structurally identical to flagella but more numerous. They have two types of nuclei—**macronuclei,** from which the cell's mRNA is transcribed, and **micronuclei,** which are only used in the sexual phase of the life cycle. Ciliates do not form zygotes as part of the sexual process. Rather, two cells fuse temporarily, exchange haploid nuclei, and then separate. Apicomplexans are all parasitic and include such species as those that cause malaria. All apicomplexans produce an invasive stage called a **sporozoite.** The sporozoite contains an apical complex of organelles involved in the penetration of host cells.

- Heterokonts include **chrysophytes, diatoms, brown algae, xanthophytes,** and **oomycetes.** These groups produce, in at least one stage of their life cycle, a small biflagellated cell with an anteriorly directed flagellum bearing rows of small hairs and a trailing naked flagellum. Many species have chloroplasts. If chloroplasts are present, they contain chlorophylls *a* and *c*. The outer chloroplast membrane is continuous with the outer membrane of the nucleus.

- A small number of protistan groups contain mitochondria having lamellar cristae, as do the plants, animals, and fungi. These are the protists most closely related to the multicellular groups. The **red algae** are photosynthetic forms which lack flagella and can have complex life cycles. Some forms are involved in reef-building, and others are of economic importance. **Cryptomonads** are small flagellates which contain a chloroplast that represents the remnants of a red algal cell ingested by an ancestral cryptomonad long ago. **Green algae** may be unicellular or multicellular. They contain chlorophylls *a* and *b*, as do the terrestrial plants, to which they are related. **Choanoflagellates** are small, colorless, filter-feeding organisms, that trap food in a collar of tentacles surrounding their single flagellum. They are structurally similar to some cells found in sponges. **Chytrids** are related to fungi. They produce sporangia as fungi do, but have flagellated zoospores.

- The four groups of true fungi are very similar to each other and are closely related. The basic structure of vegetative growth is the **hypha,** which is a multinucleate threadlike structure. The mass of hyphae constituting a single fungal individual is a **mycelium.** Asexual reproduction can occur through spore formation. Spores are also produced as part of sexual reproduction, and the differences in the sexual cycle provide some of the distinguishing features of the main groups. In the **zygomycetes,** zygotes are formed at the point of contact between hyphae of two different mating types. In the **ascomycetes** and **basidiomycetes,** the nuclei from two hyphae are mixed together in a growing mycelial mass for a much more extended period. In the ascomycetes, zygote formation and meiosis leading to spore formation occur in a sack called an **ascus.** In basidiomycetes, zygote formation and meiosis occur in a **basidium,** on which the spores are eventually borne. The fourth group of fungi, the **deuteromycetes,** contain fungi that have lost the sexual phase of the life cycle.

Questions for Thought and Review

1. What are some of the ways in which eukaryotes differ from the two prokaryotic groups?
2. How does mitosis vary among the protistan groups?
3. What are some of the different ways in which sexual recombination is incorporated into protistan life cycles?
4. Which protistans move by using flagella? Cilia? Pseudopods? More than one type of organelle?

5. Which protistan groups contain parasitic forms?

6. Which protists have chloroplasts?

7. List some organelles unique to a given group of protists, by which they may be identified.

8. How are the different groups of fungi distinguished?

9. What are some of the ways fungi affect human life?

Suggested Readings

Anderson, O. R. 1988. *Comparative Protozoology.* New York: Springer-Verlag.

Bold, H. C., and M. J. Wynne. 1985. *Introduction to the Algae.* 2nd ed. Englewood Cliffs, NJ: Prentice-Hall, Inc.

Kendrick, B. 1992. *The Fifth Kingdom.* 2nd ed. Waterloo, Ontario: Mycologue Publications.

Lee, J. J., G. F. Leedale, D. Patterson, and P. Bradbury. 1996. *Illustrated Guide to the Protozoa.* 2nd ed. Lawrence, KS: Society of Protozoologists.

Lee, R. E. 1989. *Phycology.* 2nd ed. Cambridge: Cambridge University Press.

Margulis, L. 1993. *Symbiosis in Cell Evolution.* 2nd ed. San Francisco: W. H. Freeman and Co.

Margulis, L., J. O. Corliss, M. Melkonian, and D. J. Chapman. 1990. *Handbook of Protoctista.* Boston: Jones and Bartlett.

Margulis, L., and K. Schwartz. 1988. *Five Kingdoms: An Illustrated Guide to the Phyla of Life on Earth.* 2nd ed. New York: W. H. Freeman.

Patterson, D. J., and S. Hedley. 1992. *Free-Living Freshwater Protozoa: A Color Guide.* Boca Raton, FL: CRC Press, Inc.

Sleigh, M. 1989. *Protozoa and Other Protists.* London: Edward Arnold.

Sze, P. 1986. *A Biology of the Algae.* Dubuque, IA: Wm. C. Brown.

Part 6

Microbial Ecology
A Conversation with Eugene Nester

Dr. Eugene Nester is a bacterial geneticist who studies the interaction between the plant pathogenic bacterium, *Agrobacterium tumefaciens,* and its host plants. His laboratory discovered that DNA from the Ti plasmid of the bacterium becomes incorporated into the host plant's DNA. This breakthrough led to the use of *Agrobacterium* to genetically engineer dicotyledonous plants and his receipt of the first Australia Prize in Science. Dr. Nester served as Chair of the Department of Microbiology at the University of Washington for many years.

JS: What led to your decision to study microbiology?

EN: To be very honest, it was probably something that seems rather trivial now. I read several books, in junior high school, that related to microbiology and scientific discovery. The one I remember the best was *Arrowsmith,* by Sinclair Lewis, the story of a scientist at Rockefeller who discovered bacterial viruses and how they might be used to cure bacterial diseases of the intestine. I found this story extremely exciting, and the idea of being able to work on the unknown and discover something that had not been observed before seemed like a very interesting occupation. After that I read *The Microbe Hunters,* by Paul de Kruif—a renowned microbiologist, who turned to a career in writing. Microbiology seemed like the area to study. I focused on microbiology in my thinking during high school. I went to Cornell and majored in bacteriology in the College of Agriculture. Today I tell students to keep an open mind as long as possible before deciding on a career choice—something I did not do myself. However, I have never been sorry.

JS: What advice would you give to undergraduates who are interested in pursuing a career in microbiology?

EN: Well, I think it is important to get a really firm foundation in microbiology, biochemistry, genetics, physical sciences, informatics—now. Be aware that there will be an increasing need to look at things in a quantitative fashion.

Courtesy of Eugene Nester

JS: Your laboratory was the first to demonstrate that DNA from a bacterium could be incorporated into the nucleus of plants. What was your reaction when you first saw the evidence for this?

EN: Tremendous satisfaction! We had looked for this for about five years. We realized that if we were going to take the next step in understanding the mechanism by which *Agrobacterium* causes plant tumors, we were going to have to determine whether or not bacterial DNA was in fact in the plant cell nucleus. A group of us worked pretty darn hard at it and had many frustrating moments. When we finally showed it, we realized great satisfaction. Although we personally had expected to find bacterial DNA in the transformed plant, many prominent scientists were very skeptical of this possibility. They felt it was impossible for bacterial DNA both to become part of a eukaryotic genome, and then be expressed.

JS: It was really quite revolutionary when you discovered it.

EN: It really was and even today it remains the only documented case of prokaryotic DNA becoming integrated into a eukaryotic genome. It was ironic that, after we published our data, the great debate on recombinant DNA was going on. One argument against carrying out recombinant DNA experiments was that these laboratory experiments would never occur in nature. *Agrobacterium* is nature's genetic engineer!

JS: What exciting areas do you see developing during the next decade in basic research studies of plant-bacterial interaction?

EN: Future studies will probably emphasize the plant's role in the whole process. In the case of *Agrobacterium* and *Rhizobium,* the bacteria are relatively easy to study. Now it is going to be important to understand the role of the plant in the pathogenic and symbiotic process. What are the plant receptors that interact with the bacterium? What role do various plant structures play in the transfer of the DNA in the case of *Agrobacterium* into plant cells? In the case of *Rhizobium,* investigators will look primarily at the plant side and ask such questions as: What is the receptor for the nod factors that turn on plant genes, and what genes do they turn on? How do these gene products result in the development of nodule formation?

JS: Because of the economic importance of plants in genetic engineering, a number of biotechnology companies have been established. What do you believe will be the major discoveries of these companies in the next 10 to 20 years?

EN: Many new genetically engineered plant products will be developed. Your guess is as good as mine as to what some of these will be. Plants will certainly be used as the means to produce a variety of high-value products, such as new oils and substitutes for plastics. Experiments are underway to develop plants to produce vaccines against a variety of infectious diseases. By genetically engineering trees to produce enzymes that degrade toxic compounds, planting trees in a contaminated area may be a way to decontaminate the soil. Also, plants are being made resistant to a variety of bacteria, to viruses, to herbicides, to insect pests; and then, the quality of plants, specifically tomatoes, is being improved, in terms of longer shelf life.

© John D. Cunningham/VU

JS: Has that met any consumer resistance?

EN: Not much. Certainly from what you read in the papers . . . people like Jeremy Rifkin want to boycott it, feeling that the FDA made a terrible mistake. You hear of chefs saying they will never use genetically engineered products in their restaurants. However, the public seems to be accepting these products quite well.

JS: Is this concern related to people not understanding what genetic engineering is, and being worried that it occurred in a lab setting?

EN: I think so. A report from the National Research Council of the National Academy of Sciences came out a few years ago stating that the end product should be considered only as to what it is and not as to how it was reached. If you reach insect resistance by genetic engineering and then you reach it through plant breeding by traditional means, the end product is the same. You shouldn't consider one product acceptable and the other not acceptable. Farmers have been breeding quite willfully (and successfully of course) many of the properties into plants that are now being achieved in the laboratory in much shorter periods of time.

JS: Can you explain why it is that genetically engineered plants are now allowed to be grown in the environment, but the government does not permit the release of genetically engineered bacteria? Do you think this policy will or could be changed and why?

EN: Bacteria have a bad reputation in people's minds. They cause disease and they're harmful. Of course, you can't see them. Once they're out there, you don't know that they are there. And you can't call them back if they escape. You don't know where they're going, whereas with plants you can see them. I think government policy on genetically engineered bacteria will probably change, but it may take a longer time than it did for plants. There certainly is resistance. The fact that you can exchange DNA between microorganisms quite readily also makes it difficult to convince people that it won't be harmful to release certain organisms into the environment.

. . . we were going to have to determine whether or not bacterial DNA was in fact in the plant cell nucleus. A group of us worked pretty darn hard at it and had many frustrating moments. When we finally showed it, we realized great satisfaction.

JS: Do you believe that what has been learned from the study of bacterial-plant interactions is applicable to bacterial-animal associations or other symbiotic interactions? If yes, in what ways?

EN: I think so. I think basically they are very, very similar. What we know about the initial steps in the interaction are the same in animal-bacterial and plant-bacterial interactions: the attachment process, the signaling between the host and the bacterium, the particular mechanism by which signals are recognized in animal-bacterial and plant-bacterial interactions. One important concept common to both systems is that many of the genes involved in the interaction are active only when the interaction occurs.

JS: However, is it true that there are no other known systems that are like the *Agrobacterium* system, in which bacterial (or plasmid) DNA is incorporated into the host genome?

EN: That is true so far. I wonder if we just haven't looked hard enough.

JS: One of the themes in microbiology is that of comparative biology; that is, all organisms are similar to one another in physiology, metabolism and genetics. Furthermore, since microorganisms, especially bacteria, are simple one-celled organisms that are very easy to grow and handle in the lab, they serve as ideal subjects with which to study metabolic processes and genetics. There has been a heyday in using microorganisms to study basic aspects of biology. Now, there is less justification for using microorganisms for the study of higher organisms. However, there still remains in microbiology that aspect of their tremendous diversity. We now know, in point of fact, that comparative biology is true only to a certain extent. There is a diversity of other microorganisms out there that do things differently.

EN: I think the comparative biology approach worked out very well, because nothing was known. When other systems were so difficult to study, microbes were the obvious choice and they served as wonderful models. And now, we know so much about those aspects. It is time to look at the unique nature of microbes.

JS: That brings us to the next question. Bacteria are so surprising because of their metabolic diversity. Can you provide one or two examples of a time when you learned something new, that could not have been foreseen, that really surprised you? What permitted the development as you understand it?

EN: I suppose the very common features of gene transfer, whether it be from bacteria to bacteria or bacteria to plants. It is not surprising that the ability of bacteria to conjugate with plants could not be foreseen. When I was studying *Bacillus subtilis* many years ago, we could achieve DNA-mediated transformation only if we used the same strain as donor and recipient. In fact, that was one reason that some people suggested it was foolish to consider the possibility that DNA could be transferred from a bacterium to a plant.

JS: You did a post-doc with Joshua Lederberg. At that time, bacterial genetics was in a much more primitive state of understanding, and surely there have been major advances since that time.

EN: The Lederberg lab had discovered gene transfer by conjugation and by transduction. DNA-mediated transformation was known in other systems. Restriction enzymes were unknown at the time. Just the descriptive aspects of the gene transfer mechanisms were known, but not the details. In fact, we know as much about the way *Agrobacterium* transfers its DNA into plant cells as we know about how a bacterium transfers its DNA through conjugation into another bacterium. It's still a black box.

JS: For many years you studied the genetics of *Bacillus* and bacterial transformation. What led you to change your research direction?

EN: We were studying the synthesis of small molecules, aromatic amino acids, and their regulation in *Bacillus subtilis.* I was also studying competence mechanisms; how do bacteria achieve the ability to take up and integrate into their genome high molecular weight DNA? At that time, in the early 70s, there wasn't much interest in how small molecules were synthesized. The vogue at that time was nucleic acids and proteins—macromolecules. Everybody wanted to know how DNA was transcribed and replicated. Graduate students really didn't want to work in the lab on how tryptophan and phenylalanine were synthesized and how their synthesis was regulated. So it was clear that it would be difficult to attract graduate students to study this subject. About this time a number of things occurred that changed the direction of my research program. First of all, I guess I was sort of looking around to see if there were new research areas to study. I became aware of the *Agrobacterium* system through a colleague at Northwestern who was working on *Agrobacterium.* It sounded like a very intriguing system.

> **One argument against carrying out recombinant DNA experiments was that these laboratory experiments would never occur in nature. *Agrobacterium* is nature's genetic engineer!**

> **Farmers have been breeding quite willfully (and successfully of course) many of the properties into plants that are now being achieved in the laboratory in much shorter periods of time.**

Here was this bacterium that interacts with a wounded plant and, even though you killed the bacterium, a crown gall tumor still developed. And, you could grow the tumor with its altered properties indefinitely in the absence of the bacterium. We also knew the transformed cells grew in the absence of auxin, and auxin is derived from tryptophan or from indole, so this tied into what we had been doing in aromatic amino acid biosynthesis.

In addition, a biochemist colleague of mine, Milt Gordon, who was working on tobacco mosaic virus at the time, was also looking for new challenges. He knew a great deal about plants, had a wonderful plant growth facility, and is a fine scientist and a good colleague. At about the same time, Dr. Mary-Dell Chilton was looking for a part-time post-doc position, and she was very skilled in nucleic acid technology. This was our original research team. We were fortunate to have two talented graduate students, Tom Currier and Bruce Watson.

JS: What has been done with respect to genetically engineering monocotyledonous plants?

EN: That's certainly a very interesting area of research for a number of reasons. *Agrobacterium* infects a wide variety of plants, but apparently not monocots. The grasses, such as wheat, corn, rice, barley, and oats are the most important plants. So other techniques, such as shooting pellets coated with DNA, are being used as a means of delivering DNA into plant cells. However, this technique is labor-intensive, expensive, and fraught with other problems. From a basic science standpoint, we'd like to understand why *Agrobacterium* does not transform these plants efficiently. Because if we could understand this, we could modify *Agrobacterium* to make a much more useful tool in plant genetic engineering. We've been working on rice as a model system. Rice is relatively easy to work with in terms of tissue culture, but it's still fairly tricky. In 1994 a group of scientists in Japan reported an efficient *Agrobacterium* transformation system for rice. It gives us great satisfaction to know that it can be done. Although we always believed that *Agrobacterium* could transfer its DNA into monocots, some respected scientists had stated very openly that it would be a waste of time to consider using *Agrobacterium* to transform monocots. The Japanese report has proved them wrong.

JS: Are there any special features that one would want to put in monocots?

EN: In the near future, probably some of the same kinds of genes that are being put into dicots, such as resistance to viruses and resistance to insect pests.

The ultimate aim of ecology is to understand the relationships of all organisms to their environment.

R. E. Hungate. *The Bacteria.* 1962.

Chapter 24

Microorganisms and Ecosystems

The Microenvironment
Counting Microorganisms in the Environment
Autecology
Habitats
Dispersal, Colonization, and Succession
Biogeography of Bacteria
Synecology, Part I: Biogeochemical Cycles
Synecology, Part II: Assessment of Microbial Biomass and Activities

Microorganisms are found in all **ecosystems** that are major geographic entities, such as a lake, that contain both biotic and the associated abiotic components of the environment. Indeed, microorganisms even occupy some of their own special ecosystems, **microbial ecosystems,** in areas that are too inhospitable for macroorganisms, including thermal, hypersaline, acidic, alkaline, and anoxic habitats. Furthermore, microorganisms live in the intestinal tracts and on the surface of virtually all animals, form close associations with the plant rhizosphere, and are therefore found wherever animals and plants are located. For these reasons, the **microbial biosphere** is considerably more extensive than the biosphere of macroorganisms.

The chemical transformations that are carried out by microorganisms are often unapparent. For example, consider the tree leaves that fall on the soil in autumn. By late spring they have disappeared. But what happened to them? Actually, they were the foodstuff of a myriad of microorganisms and some higher organisms. The microorganisms colonized the leaves in the fall and began breaking down the leaf tissues with their cellulases, amylases, and other enzymes. To the casual observer there was no obvious indication that bacteria and fungi were present, let alone that they were largely responsible for the degradation that occurred. Other microbial processes are even less apparent. For example, the process of nitrification and the bacteria that carry it out are not detectable by our unaided senses, so it is natural that most people are simply unaware this important process even occurs. In point of fact, these and many other microbial processes that occur in nature cannot be readily differentiated from chemical processes, certainly not without careful scientific investigation.

Although it is true that the study of the ecology of macroorganisms and microorganisms share common principles and ecosystems, the small size of microorganisms, their rapid growth rates, the many unique geochemical transformations they perform, and the necessity for studying them in pure culture make microbial ecology a special area of investigation. Their small sizes means that they create their own microscale environment as will be discussed next.

The Microenvironment

The typical bacterial cell is about one millionth the length of a human being. Because microorganisms are so small, the natural environment, or **microenvironment,** in which they live and carry out their activities is correspondingly small. The microorganism carries out important biochemical reactions that influence the physical and chemical conditions of the microenvironment immediately around its cells. As a result, the concentration of substrates and products is different in the vicinity of the cells as compared to the bulk environment that is measured with ordinary electrodes and chemical analyses. Therefore, in order to study what is happening in the microenvironment, it is necessary to amplify these activities with appropriate instruments, a challenging aspect for scientists. For example, microscopes are needed to magnify the cells so they can be visualized. Figure 24.1 shows typical microcolonies of bacteria that have grown on an electron microscope grid attached to a glass slide and immersed in a lake. Note that two of the cells can be identified as a *Caulobacter* sp. because of their distinctive prosthecae.

The small size of the microenvironment requires that special consideration be given to designing instruments to measure conditions in the vicinity of the organisms. **Microelectrodes** for the measurement of oxygen or pH are examples of such instrumentation. These fine probes come equipped with tips as small as 5 μm to 10 μm diameter. With these instruments microbial ecologists have demonstrated that the oxygen concentration and pH near colonized areas can be quite different in a span of a few micrometers (see **Box 24.1**).

For these reasons, microbial ecologists need to be constantly aware that the growth and activities of microorganisms occur in the environment at a microscale. Nonetheless, the concerted action of enormous numbers of microorganisms are responsible for producing metabolic products and chemical gradients that occur at macroscopic scales in soils, in aquatic environments, and in the atmosphere.

Two basic issues that microbial ecologists need to address concern the abundance and location of microorganisms of interest in the natural environment. The census of microorganisms is essential for an understanding of their importance in an ecosystem. Therefore we will initially discuss techniques that are used for identifying and counting microorganisms.

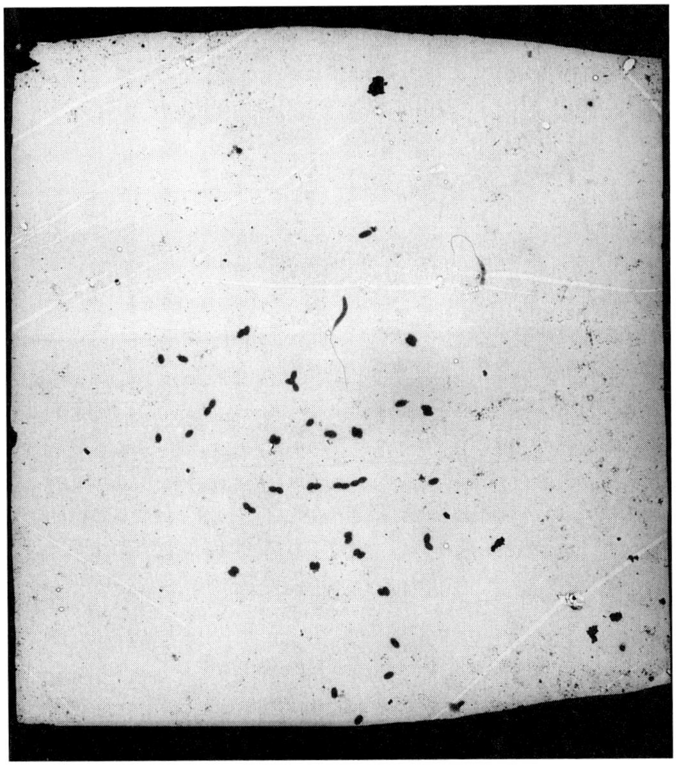

***Figure* 24.1** A microcolony of rod-shaped bacteria growing on an electron microscope grid attached to a glass slide that was immersed in a lake. Note also the two attached *Caulobacter* cells, with their prosthecae. (Courtesy of J. T. Staley)

Counting Microorganisms in the Environment

In this section we will concentrate on counting or enumerating bacteria—in particular heterotrophic bacteria—in natural habitats. When enumerating bacteria in natural environments, it is important to recognize that not all small organisms in the habitat are heterotrophic bacteria. Depending on the environment, large numbers of cyanobacteria, small algae, or autotrophic bacteria may also be present. As we will discuss, these other microorganisms can usually be readily distinguished from heterotrophic bacteria.

The fundamental problem of counting bacteria in natural environments is challenging for three primary reasons. First, bacteria must be visualized with a microscope because they are so small. Second, unlike plants and animals, when bacteria are viewed in the microscope, most appear as nondescript rods, spirilla, or cocci. It is therefore generally impossible to distinguish one species from another. Third, even if we can see and count them, determining whether they are dead or alive is not a simple matter. Initially we will address how bacteria are counted and then consider how it is possible to determine whether or not they are alive or metabolically active.

BOX 24.1 RESEARCH HIGHLIGHTS

Thinking Small: A Microbial World Within a Marine Snow Particle

Marine snow is found in the water column of marine habitats. It is called marine snow because, as these particles sediment in the water column, scuba divers note that they bear a strong resemblance to falling snowflakes. The marine snow particles range in size from 1 to about 10 mm in diameter and may contain photosynthetic algae and cyanobacteria as well as heterotrophic bacteria and protozoa.

Microprobes, with tips that are only 5 μm to 10 μm in diameter, were used by A. L. Alleridge and Y. Cohen to measure the pH and oxygen concentration at the surface and at various points in the interior of marine snow particles. Microgradients of oxygen and pH were found from the surface to the interior of the particles. For example, the oxygen concentration in the interior of the particle can be half that of the bulk water in which the marine snow particle is found. It is lower inside the particle because of aerobic respiration, which depletes the oxygen more quickly than it can be replenished by diffusion.

Total Microscopic Count

One of the best ways to determine the total number of bacteria in a sample is to count them by staining them with a fluorescent dye such as acridine orange. Acridine orange binds to the RNA and DNA of a cell. In this staining procedure, the sample containing the bacteria is fixed, immediately after collection, with formaldehyde or glutaraldehyde to preserve it. After fixation, samples can normally be stored in the refrigerator for several months, if necessary, before they are counted. They are then stained with acridine orange and filtered onto a black, nonfluorescent filter to collect and concentrate the bacteria. Cells that have taken up the acridine orange appear green or orange when observed by fluorescence microscopy. This is one of the best procedures known for accurate counting of bacterial cells (Figure 24.2). Furthermore, because this is an epifluorescence procedure that uses incident light (see Chapter 4), it is possible to observe cells attached to soil particles and other opaque materials. Other nucleic acid-binding fluorescent dyes, such as DAPI (4′,6-diamidino-2-phenylindole), can also be used for total microscopic counting.

Cyanobacteria that might be present in the sample can be distinguished from other bacteria because the cyanobacteria have chlorophyll *a*, which produces a natural red fluorescence without acridine orange staining. Thus, control samples are typically prepared to determine whether cyanobacteria and algae containing chlorophyll *a* are present in a sample. It should be noted that it is not possible to distinguish living from dead cells by this technique—*this is a total count of both living and dead cells.* Furthermore, it is not possible to distinguish lithotrophic bacteria from heterotrophic bacteria, although in most environments chemolithotrophs would be expected to occur in low numbers relative to the heterotrophs.

An alternative to microscopic counting is the use of **cell sorters** or **flow cytometers,** instruments developed for separation of blood cells in medicine. With minor modifications these instruments can be used to separate microbial cells from one another, based upon size. Furthermore, with special detectors it is possible to separate fluorescent particles (such as acridine orange–stained cells) from other particles, and to examine that group individually. This automated procedure is being increasingly used in microbial ecology.

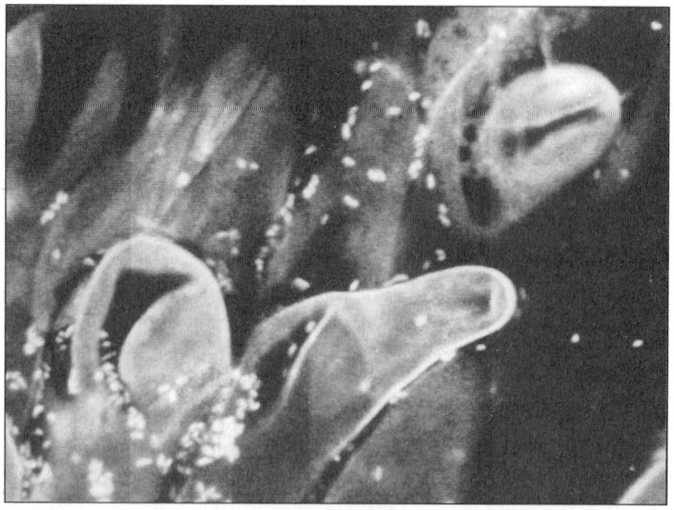

***Figure* 24.2** A photomicrograph of acridine orange–stained bacteria growing in the rhizosphere of a legume. Taken with a confocal laser photomicroscope. (Courtesy of Frank Dazzo)

Viable Counts

If one wishes to know the numbers of bacteria that are alive in a sample, it is necessary to use a viable counting procedure in which the bacteria are actually grown. The traditional method used for this is the spread plating procedure, in which a medium prepared to grow the heterotrophic bacteria is first poured into the petri dish (see Figure 6.6), and a dilution of bacteria is spread over the surface of the plate with a sterile glass rod. This procedure is superior to pour plating because the high temperature required to maintain the agar in a molten state while mixing with the bacteria kills many psychrophilic and mesophilic bacteria. An alternative to plate counts is the use of liquid media. For example, samples can be quantitatively diluted and inoculated into a liquid growth medium for viable enumeration, such as the most probable number (MPN) procedure (see Figure 6.8). This procedure may provide slightly higher counts of viable bacteria; however, there are serious drawbacks to all viable counting procedures, as will be discussed later.

The Bacterial Enumeration Anomaly

Major discrepancies are found between the total acridine orange counts and viable counts of bacteria from many habitats. For example, in **oligotrophic** (low concentrations of nutrients) or **mesotrophic** (moderate concentrations of nutrients) aquatic habitats, less than 1 percent of the total acridine orange–staining bacteria can be grown on the best of media. In contrast, in **eutrophic** environments (rich in nutrients) such as waste waters, the recovery of bacteria can approach 100 percent of the total count (Table 24.1). The inability to recover viable bacteria from oligotrophic and mesotrophic environments is called the **bacterial enumeration anomaly** or the "Great Plate Count" anomaly. There are three possible explanations for this result. First, there is no such thing as a "universal medium" that will permit the growth of all heterotrophic bacteria, so not all viable bacteria are capable of growth on the medium employed. Second, it is possible that many of these bacteria do not grow well enough on artificial media, under laboratory conditions, to form colonies. And finally some, perhaps most, of these bacteria are dead. To address the enumeration anomaly issue, three alternative microscopic procedures have been developed to identify metabolically active ("living") bacteria from natural samples: microautoradiography, the INT reduction technique, and the nalidixic acid cell enlargement technique. Each of these procedures is described below.

Microautoradiography

Microautoradiography entails the use of a radiolabelled substrate such as tritiated acetate or tritiated thymidine. If a microorganism can use acetate as substrate or assimilate thymidine, it will incorporate the radiolabelled material into its cells as it would a normal nonradioactive substrate. When an organism takes up the radiolabelled substrate, the cells become radioactive and can then be identified by autoradiography. In using this approach, the proportion of metabolically active organisms (that is, those that incorporate the label into their cells) to total cells can be readily determined.

Several steps are used to conduct microautoradiography. First, a radiolabelled substrate (or a mixture of them) is added to a sample from the habitat and incubated under the same conditions as the habitat (Figure 24.3**a**). After an appropriate length of time, usually 15 to 60 min, the reaction is stopped, the sample is fixed, and an autoradiogram is prepared (Figure 24.3**b–c**) to determine what proportion of the organisms incorporated the radiolabelled substrate into their cells and are now radioac-

***Table* 24.1 Bacterial enumeration anomaly illustrated for a mesotrophic lake, Lake Washington, compared with a eutrophic pulp mill aeration lagoon**

Trophic Status	Bacterial Cells/ml		% Recovery[1]
	Total Count	*Viable Count*	
Mesotrophic			
Lake Washington	3×10^6	2×10^3	0.067
Eutrophic			
Pulp mill oxidation lagoon	2.1×10^7	3.1×10^7	~100[2]

[1]This refers to the ability to cultivate the bacteria on a plating medium; it is the ratio of viable count divided by the total microscopic count.

[2]In this particular instance the viable count was actually somewhat higher than the total count due to minor uncertainties in measurement. Therefore, the recovery is considered to be 100%.

tive, indicating they are metabolically active. After the autoradiogram is developed (Figure 24.3**d**), it is examined under a microscope (Figure 24.3**e**). Metabolically active cells (those that are now radioactive) have exposed silver grains in the emulsion above them (Figure 24.4). Thus, the radioactive cells can be easily counted. Total cell counts are determined by acridine orange counting, so the percent that are active is determined by dividing the radioactive cells by the total cell count.

INT Reduction Procedure

In the **INT reduction procedure** a tetrazolium dye (INT) is used as the substrate. It is commonly metabolized by aerobic respiring bacteria if they are alive and metabolically active. They use it as an electron acceptor and reduce it to form a precipitate called formazan, which appears as a black deposit in the cell. When the INT reduction procedure is combined with acridine orange total counting, the organisms that are metabolically active can be distinguished from those that are not. More recently a **fluorogenic** dye, CTC, has been developed that produces a red fluorescent formazan. A fluorogenic compound, while not fluorescent itself, can be metabolized to produce a fluorescent compound. This dye should provide a better assay inasmuch as fluorescent deposits are easier to detect in cells than nonfluorescent ones.

Nalidixic Acid Cell Enlargement Procedure

In the **nalidixic acid cell enlargement procedure,** a small amount of yeast extract is added to a sample along with nalidixic acid, which inhibits DNA replication. Metabolically active cells that are sensitive to this antibiotic cannot divide, but their cells continue to enlarge in size as they grow. Therefore the proportion of cells from a sample that become very large compared to those that have not en-

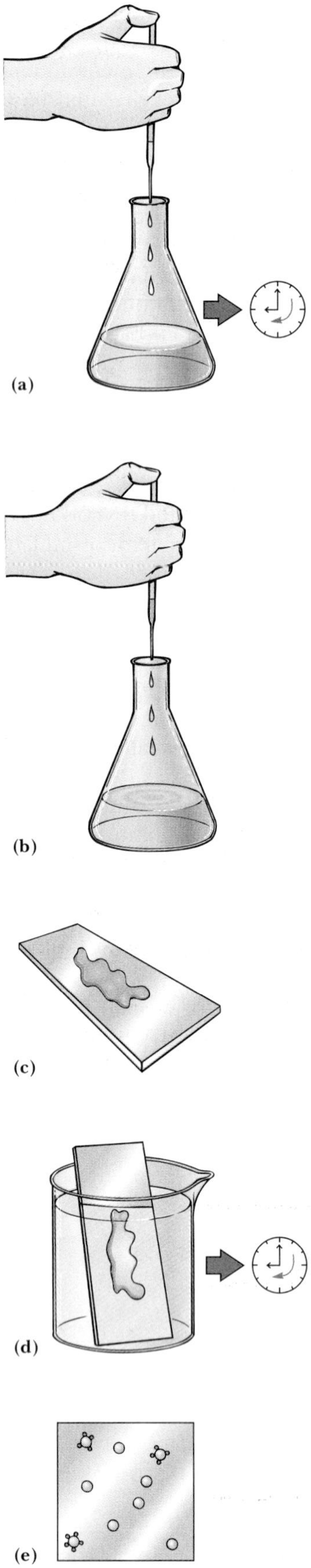

(a)
^{3}H organic substrate added to sample and sample incubated under natural conditions
(b)
Lugol's iodine added to kill cells; cells harvested by centrifugation
(c)
Cells placed on glass slide and fixed to glass
(d)
Slide coated with liquid photographic emulsion in the darkroom; the preparation is incubated in the dark to expose silver grains
(e)
Slide is developed by photographic procedures and cells are located by microscopy. Cells that have taken up the substrate are marked by exposed silver grains.

◀ ***Figure* 24.3** Preparation of a microautoradiogram. **(a)** Samples are collected in a container and immediately exposed to the radiolabelled substrate. **(b)** After about 15 to 60 minutes the reaction is stopped by killing the cells with Lugol's iodine, a preservative. The cells are harvested by filtration or centrifugation and placed on a slide. **(c)** Cells are fixed to the slide by gentle heating and the excess (unincorporated) radioactivity of the substrate is removed by washing. **(d)** The smear is taken into a darkroom where it is covered with a special liquid photographic emulsion whose silver grains become exposed during radioactive decay. After a period of incubation for about a week in the dark (the emulsion, like photographic film, is sensitive to light as well) the radioactive decay exposes silver grains on the emulsion of the preparation. **(e)** The preparation is then developed like photographic film, and individual organisms are then viewed under the microscope to determine if they are metabolically active by looking for silver grains exposed in the vicinity of the cells (see Figure 24.4).

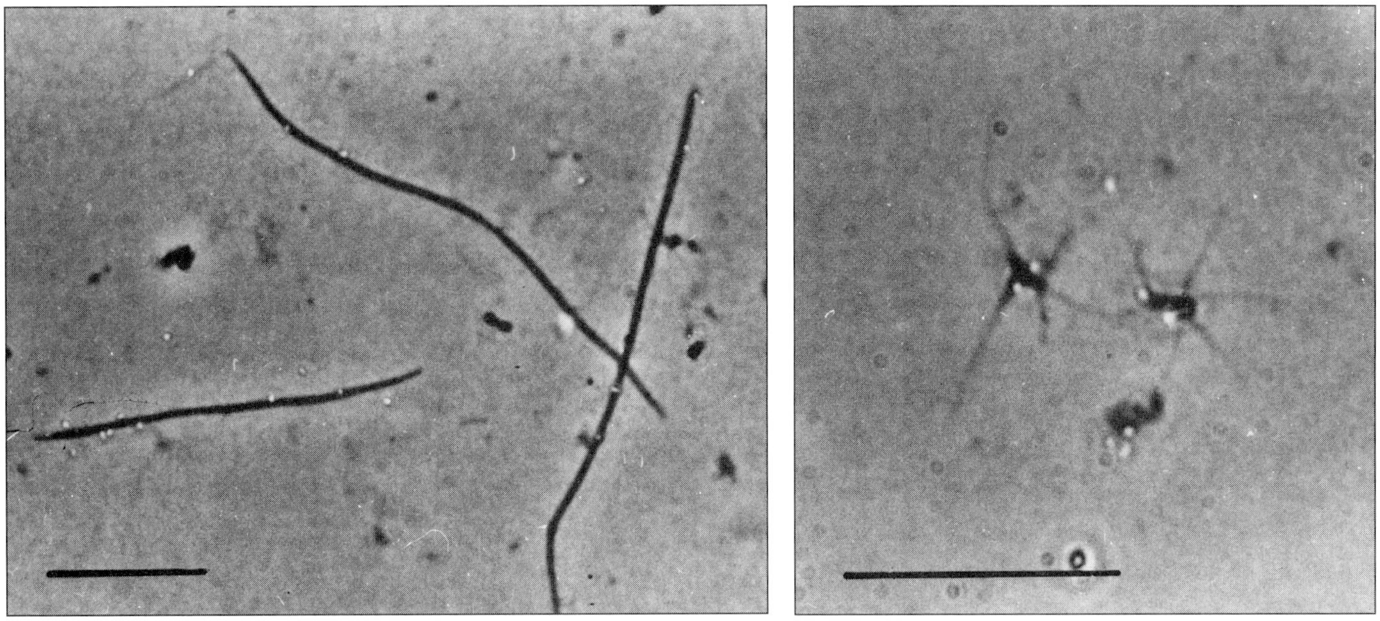

(a) (b)

Figure **24.4** An autoradiogram of **(a)** a filamentous bacterium and **(b)** *Ancalomicrobium adetum* from a pulp mill treatment lagoon. The exposed silver grains (arrows) show clearly in the emulsion overlying the cells that are metabolically active. In this case the substrate was tritiated acetate. (Courtesy of P. Stanley and J. T. Staley)

larged provides a measure of the percentage of metabolically active cells.

All of these procedures have their own pitfalls and drawbacks; however, they provide similar results when used to assess viability in typical mesotrophic and oligotrophic environments (Table 24.2). Unlike viable plating, in which less than 1 percent of the cells can be recovered, about half of the cells from most oligotrophic and mesotrophic habitats are metabolically active. Thus, either these organisms are different physiologically from known bacteria and therefore do not grow on typical plating media, or the cells are too debilitated to actually multiply on plating medium, or a combination of both.

Although there are drawbacks to using plating and other traditional viable counting procedures to determine total viable counts of heterotrophs, viable counting procedures work well if one is enumerating a specific group of organisms that are known to grow under certain laboratory conditions. For example, as mentioned previously, it is possible to determine the seasonal distribution of the genus *Caulobacter* in lakes using viable counting procedures (Figure 24.5).

Table **24.2 Comparison of acridine orange counts with microautoradiographic counts using a mixture of tritiated amino acids in Chesapeake Bay[1] samples**

	Bacterial Cells/ml × 10^6		
	Acridine orange	*^{3}H-Amino acids*	*% Active*
August 5	10	6	60
September 21	15	14	93
February 9	4	1.5	38

[1]8.5 m depth (1979–1981).

Source: Data from P. S. Tabor and R. A. Neihof, *Applied and Environmental Microbiology* 48:1012–1019, 1984.

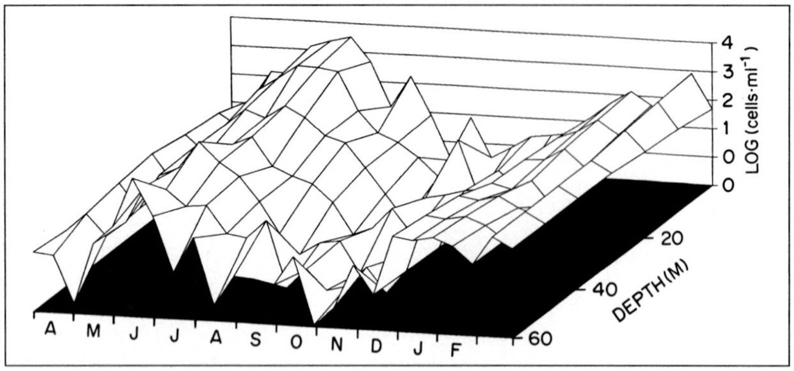

***Figure* 24.5** The seasonal and spatial distribution of *Caulobacter* species in Lake Washington. The concentration of cells is shown on the "y" axis, the time of the year on the "x" axis (months are listed by first letter), and the depth of the lake on the "z" axis. Note that the number of *Caulobacter* cells increases during the early spring period, especially in the surface water of the lake. However, there is a precipitous drop in cell numbers during the summer months. (Courtesy of J. T. Staley, A. E. Konopka, and J. Dalmasso)

Autecology

Autecology is the study of the ecology of a species. It considers the individual microorganism in the microenvironment, as well as population studies of the species in the natural habitat in which the organism lives. Because microorganisms are very small, such studies are challenging and often painstaking, but they are possible and can be very informative to the microbial ecologist.

Identification of Species in Natural Environments

As mentioned previously, it is not usually possible to identify a bacterium to the level of species, genus, or perhaps even Domain by observing it in the microscope. This is because most bacteria have very simple shapes, at least as can be discerned by light microscopy. Thus, members of the genus *Nitrosomonas* appear very similar to *Escherichia coli,* although the former is a genus of chemolithotrophic ammonia-oxidizing bacteria, whereas the latter is an ordinary heterotroph.

Fluorescent Antibody Approach

The problem of *in situ* identification of microorganisms can be solved by the use of **fluorescent antibodies** that can be used microscopically to identify a species. A fluorescent dye, such as fluorescein, is covalently tagged to a whole cell antiserum or to a monoclonal antiserum prepared from the organism of interest. An environmental sample that contains the organism can be stained with the fluorescent antiserum and examined microscopically using an ultraviolet or halogen light source. Cells of the organism of interest will fluoresce because they have been labelled by the specific antiserum. In fact, this is the preferred procedure for detection and identification of *Legionella* species in natural samples, because these bacteria are so difficult to grow.

Nucleic Acid Probes

More recently, considerable interest has been generated using **nucleic acid probes** (either RNA or DNA) for the identification of microorganisms from the environment. Although these techniques are not yet perfected, they hold considerable promise. In one application of this procedure, an oligonucleotide that is specific for a species or higher taxonomic group can be tagged with a fluorescent dye. Then the hybridization reaction is prepared directly on a glass microscope slide with cells from a natural sample and the preparation observed using a fluorescent microscope.

Biomarkers

Biomarkers have also been used to assess the presence and concentrations of specific bacterial groups. A **biomarker** is a chemical substance that is uniquely produced by a specific organism or group of organisms. Thus, if a natural sample is analyzed for a specific biomarker and it is found in the sample, we can conclude that cells of this particular type of organism are present in the sample. For example, environmental samples that have Eubacteria with peptidoglycan in their cell walls can be analyzed for the presence of peptidoglycan or for muramic acid, a unique amino sugar constituent of it. When analyzing for muramic acid, the identification is at a rather broad taxonomic level that would include almost all of the Eubacteria. Ideally, the biomarker selected should be one readily determinable; otherwise one might as well use other approaches, such as the fluorescent antiserum procedure for specific groups.

A popular group of biomarkers currently used for bacteria is phospholipids. Different groups of bacteria have different phospholipid types and therefore can be differentiated from one another in natural communities by assessing the type of phospholipid present.

It should be noted that, although biomarkers have also been used as biomass indicators for certain groups, biomarkers may survive in the environment even though the bacterium that produced them is no longer alive. Therefore, caution should be exercised in assessing the significance of finding a biomarker from a natural sample.

Biomarkers have useful applications. For example, some *Bacteroides* species that grow in the human intestine produce coprastanol, a product of cholesterol degradation. Because of its association with the human intestinal tract, coprastanol has been used as an indicator of human sewage contamination in natural environments. Therefore, if a water sample contains this biomarker, it is suggestive evidence that human fecal material has contaminated the water supply. However, a recent study has shown that this particular biomarker assay has some limitations. Coprastanol was found in marine sediments near Antarctica far from human habitation and influence. Subsequent research determined that the source was certain marine mammals, including whales and some seals! So, coprastanol is not as much a marker for human feces as it is for the bacteria that live in the intestinal tracts of mammals that transform cholesterol to coprastanol.

Sampling Habitats and Assessing the Distribution of Species

By combining a sampling program of a habitat with a means of identification of a bacterium, it is possible to conduct distributional studies of species in habitats. The first concern facing the microbial ecologist is, "How can I obtain samples from the environment?"

Sampling the Habitat

Sampling habitats is not a trivial problem. Imagine the difference between sampling for the bacteria that occur in the intestine of an animal compared to sampling for marine planktonic bacteria from the Sargasso Sea. Completely different approaches are used to obtain samples of microorganisms from each habitat. For intestinal samples, the microbiologist has some difficulty in obtaining samples—stool samples could be used for lower intestinal bacteria, but more elaborate techniques would be required to sample the small intestine. Furthermore, the microbiologist needs to maintain anaerobic, aseptic conditions during the collection process. The marine microbiologist needs to find an appropriate location to obtain samples and employs entirely different collection gear (Figure 24.6). This text cannot describe all of the techniques used

(a) (b) (c)

***Figure* 24.6** Sampling gear used by marine microbiologists. **(a)** A plankton ▶ net containing *Sargassum* from the Sargasso Sea is being examined by microbial ecologist J. E. Hobbie, **(b)** a bathythermograph for measuring temperature and depth (hydrostatic pressure), and **(c)** a Nansen bottle for collection of water samples for chemical analyses. (Courtesy of J. T. Staley)

Table 24.3 **Guidelines for microbial sample collection and processing**[1]

Purpose of Sampling	Preferred Handling Technique
Total count (direct microscopic count)	Fix sample immediately after collection with formaldehyde or glutaraldehyde. Store at refrigerator temperatures in the dark. Aseptic collection is preferred, but is not always essential as long as vessel is clean and microbial counts are high.
Viable count	Collect aseptically. Enumerate as soon as possible after collection. Chill sample and keep under conditions of the habitat until processed. Incubate at or near the in situ conditions of the environment (i.e., pH, temperature, redox potential, etc.).
Activity measurements	Perform assay as soon as possible. Perform in situ if at all possible. May have to use a container for certain assays (e.g., radioisotopes), but these should be incubated under conditions of the environment, if not right in the environment.
Chemical analyses	Collection container should be chemically inert. Analyses should be performed soon after collection, especially for reactive compounds such as hydrogen sulfide, dissolved oxygen, etc. Special chemical fixation techniques are used for more reactive chemicals.

[1]As soon as a sample of microorganisms is taken from its natural habitat, conditions will change. Part of the change will be caused by the microbes themselves as they metabolize. However, physical and chemical changes also occur. It is important to minimize changes, as they can affect the results of a study. The actual sampling procedure will vary from one habitat to another.

for sampling for microorganisms; however, some general guidelines are provided in Table 24.3.

In addition, it is desirable to measure certain physical and chemical properties of the habitat at the time of sample collection. For example, in both habitats described above, it is important to measure the temperature at the time of collection. In the marine habitat it is important to record the geographic location from which the sample was taken, the time of sample collection, the hydrostatic pressure, and the temperature—using special instruments such as the bathythermograph (Figure 24.6**b**). These parameters, as well as others, provide valuable information that is useful in interpreting what is happening in the environment.

Assessing the Distribution of a Bacterial Species

Autecological studies may be designed to assess seasonal changes in the distribution of a microbial species as well as its spatial location in the habitat. For example, nitrifying bacteria have been enumerated in marine environments and in soils by use of fluorescent antibody techniques. This procedure is quite species-specific, at least for this group, so it provides a means of assessing the temporal and spatial distributions of particular nitrifying species.

Morphologically distinctive microorganisms such as many of the cyanobacteria and anoxygenic photosynthetic bacteria can be enumerated microscopically. Thus, photosynthetic bacteria can be counted in lakes by identification using the light microscope (Figure 24.7). As with all microorganisms, the distribution of these groups is confined to certain strata in the lakes in which they reside. The lower depth of each photosynthetic bacterial group is determined by its need for sulfide obtained from the sediments of the lake, whereas its upper depth is determined by its need for light obtained from the surface of the lake.

Viable counting procedures can also be used. For example, as mentioned previously, the seasonal and spatial distribution of *Caulobacter* spp. has been determined in freshwater lakes (Figure 24.5). Note that this bacterium is most abundant in the surface waters of the lake, and furthermore, that it is most numerous in the spring, when algal blooms occur in the lake.

Evaluation of the Activities of a Species

To assess the activity of a species in the environment, microbiologists may follow the "ecological" Koch's postulates approach. The approach can be summarized by paraphrasing the postulates used for studies of pathogenesis. However, in this instance the microbiologist is not attempting to verify that the bacterium causes a disease, rather that it is responsible for a particular process or chemical transformation. These modified Koch's postulates for microbial ecology are as follows:

1. The species believed responsible for a transformation must be found in environments in which the process is occurring.
2. The species must be isolated from the environment where the process occurs.
3. The species must carry out the process in the laboratory in pure culture.
4. The species that carries out the process in the laboratory must be shown to carry out the process in the environment in which the process occurs.

Of course, unlike infections that are typically mediated by a single, specific pathogen or parasite, ecological processes in the environment are frequently mediated by groups of organisms growing in concert with one another. Such mixed groups containing two or more species are referred to as a **consortium.** An example of a consortium-mediated environmental process is the process of formation of stromatolites, which are produced by a variety of organisms acting together (see Chapter 18). This structure could not be formed in the laboratory using a single pure culture of an alga or cyanobacterium. But it is formed in nature by the concerted action of several different species. Thus, it may not always be possible to use Koch's postulates with pure cultures to show that a specific process is caused by a specific microorganism. However, what has been said of the species above would also be true of the consortium. So, the word "consortium" can be substituted for "species" in the environmental Koch's postulates stated above. A microbial ecologist studying such a consortium should attempt to identify the members of the consortium, and ideally should also determine what each member of the consortium contributes to the overall activity.

Generally speaking, testing the first three postulates is straightforward: An organism suspected of causing a transformation can be found in various environments in which the process occurs and can be isolated in pure culture. Once in the laboratory it can be shown to be able to carry out the process. The final step is much more challenging, however, because it is incumbent upon the microbiologist to demonstrate that the species or consortium is carrying out this process in the environment. One approach in assessing this is to combine microautoradiography with an identification procedure such as the fluorescent antibody procedure. For example, if one wonders whether a specific heterotrophic bacterium is using acetate in the environment, then the species can be identified by fluorescent antibody, and the uptake of radiolabelled acetate associated with the chemical transformation can be noted by the exposed silver grains on the same autoradiogram.

Microautoradiography can also be used with morphologically identifiable bacteria, such as *Ancalomicrobium adetum* (Figure 24.4**b**). Indeed, quantitative autoradiography can be achieved in some cases as with *A. adetum.* This species was found to comprise a minor component of the acetate-utilizing community in this study, in comparison with a filamentous organism (Figure 24.4**a**—data not shown). This does not mean that *A. adetum* is insignificant in this habitat. It is more likely that it has a different niche. For example, from pure culture studies in the laboratory it is known that *A. adetum* preferentially uses sugars as carbon sources. Sugars are also abundant in pulp mill effluents and are probably being utilized as carbon sources by this species.

As discussed previously, for those microorganisms that cannot be identified by their morphology, their identity can be made using either (a) fluorescent antibody, or (b) a fluorescently labelled nucleic acid probe to the 16S rRNA of the organism (see Chapter 17) or to some specific gene found in the organism. Microautoradiography can be used in this approach to assess activity. Or, another probe could be prepared to hybridize with the specific mRNA of an enzyme that is of interest to study. Since mRNA is a transitory compound in bacteria that typically lasts 15 to 30 minutes or, in some cases, up to two or three hours, showing that it is present in a cell is strong evidence that the cell is producing that enzyme. Studies using nucleic acid probes are still in their infancy; however, they show great promise for future ecological work.

Finally, it is very important that the organism that one has shown to be responsible for a process be kept in pure

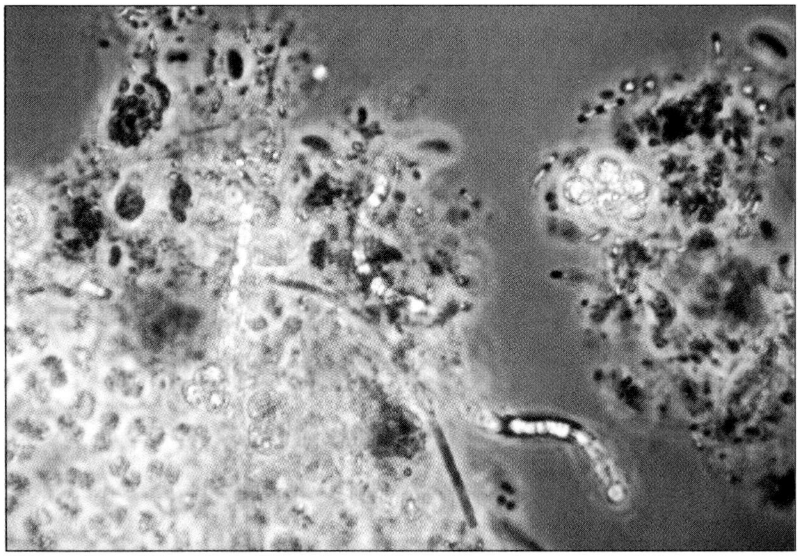

***Figure* 24.7** A sample from the anaerobic hypolimnion of a lake showing a variety of photosynthetic bacteria identifiable by their characteristic morphology. (Courtesy of J. T. Staley)

culture and deposited in a culture collection. If it comprises a new species, then it should also be described and named.

Habitats

As previously mentioned, bacteria grow in some of the most unusual and extreme habitats on Earth. In this chapter, the more common natural habitats are considered–including freshwater, marine, and terrestrial environments. The most unusual habitats are occupied by special bacteria and are not considered here. Instead they are discussed in Chapters 19 through 22, where bacterial diversity is treated, as well as in Chapter 25, which considers symbiotic associations.

Freshwater Habitats

Aquatic habitats comprise some of the most important habitats for microorganisms. Even waters of the most pristine lake or areas in the open ocean contain hundreds of thousands or millions of bacteria per ml. Bacteria can be easily concentrated on filters of samples from aquatic habitats and therefore counted without interference from such debris as is found in the soil environment.

Freshwater habitats include lakes, ponds, rivers, and streams. Most of these inland waters have very low concentrations of salt. However, some, such as the Great Salt Lake and the Dead Sea, are very salty because they do not have outlets to rivers or the sea. The saltiness occurs because the water evaporates from the lake, leaving behind the salt, which becomes more and more concentrated over time. After hundreds or thousands of years the salt concentration increases, as it has in the oceans. Sometimes these lakes become saltier than the sea. Table salt is obtained from salt lakes or from marine habitats using special evaporation ponds (see Figure 1.1**c**). The high salinity in such habitats favors the growth of halophilic bacteria including *Halobacterium* spp., and algae such as *Dunaliella.* These salt lakes will not be considered further here.

Lakes

Many freshwater lakes of sufficient depth that are located in temperate zones become thermally stratified during the summer months. In such lakes the surface water becomes warmer than the underlying water. This occurs because the surface is increasingly heated by the sun's radiation during the spring and summer months. Summer winds are not strong enough to mix the lighter warm surface layers with the more dense cold water that lies beneath. Therefore, the warm surface water layer becomes separated from the colder bottom water, resulting in **thermal stratification** of the lake. In contrast to deep, temperate-zone lakes, shallow lakes and ponds do not stratify because the winds keep them intermittently mixed.

In lakes where thermal stratification occurs, the surface layer is called the **epilimnion,** and the lower body of water is called the **hypolimnion** (Figure 24.8). The zone of transition in temperature between the surface and the lower depths is called the **metalimnion** or **thermocline.**

The period of thermal stratification persists throughout the summer season and into early fall. However, as the air temperature cools in the fall, the surface water temperature decreases and the water becomes more dense. At some point the phenomenon of **fall turnover** occurs, in which the density of the surface water and the underlying water are almost identical. Winds can then readily mix the lake thoroughly from top to bottom. This period of mixing continues throughout the wintertime, unless the temperatures become so cold that the lake actually freezes over. If freezing occurs, then the spring warming must first melt the ice and warm the melted surface water to 4°C (water is most dense at 4°C) before it will mix with the underlying water. In either case, at least one period of mixing occurs during the winter months in typical temperate-zone lakes.

These physical features of lakes make them ideally suited for spring and summer blooms of algae and cyanobacteria and the resultant growth of bacteria. In the early spring months the nutrients in the lake water are high in concentration because they have been brought up to the surface from the lower depths and sediments during the mixing period. At this time, the concentration of dissolved nutrients such as phosphate, ammonium, and nitrate is constant at all depths. Then, as the daylight period lengthens in the spring it provides both increased light and warmer temperatures—conditions favorable for algal growth and other biological processes. Thus, spring and summer blooms of algae and cyanobacteria occur. At this same time the increased warmth of the surface waters results in a decrease in surface water density and, combined with wind, enables the separation of the epilimnion from the hypolimnion.

The algae and cyanobacteria, called **primary producers,** serve as a source of food for **primary consumers,** such as protozoa and small crustaceans—for example, water fleas (*Daphnia* spp.). These primary consumers in turn serve as food for larger animals called **secondary consumers** (larger crustaceans and small fish), which feed on them, and they in turn provide food for **tertiary consumers** (larger fish, ducks, turtles, and so on) and higher members of the food chain called **quaternary consumers** (for example, raccoons and humans). Thus, a **food chain** or food web is established in which the carbon and energy flow from one group of organisms to another (see Figure 1.17).

Heterotrophic bacteria degrade the excess organic material excreted by algae and decompose the remains of dead algae and other organisms that inhabit the lake.

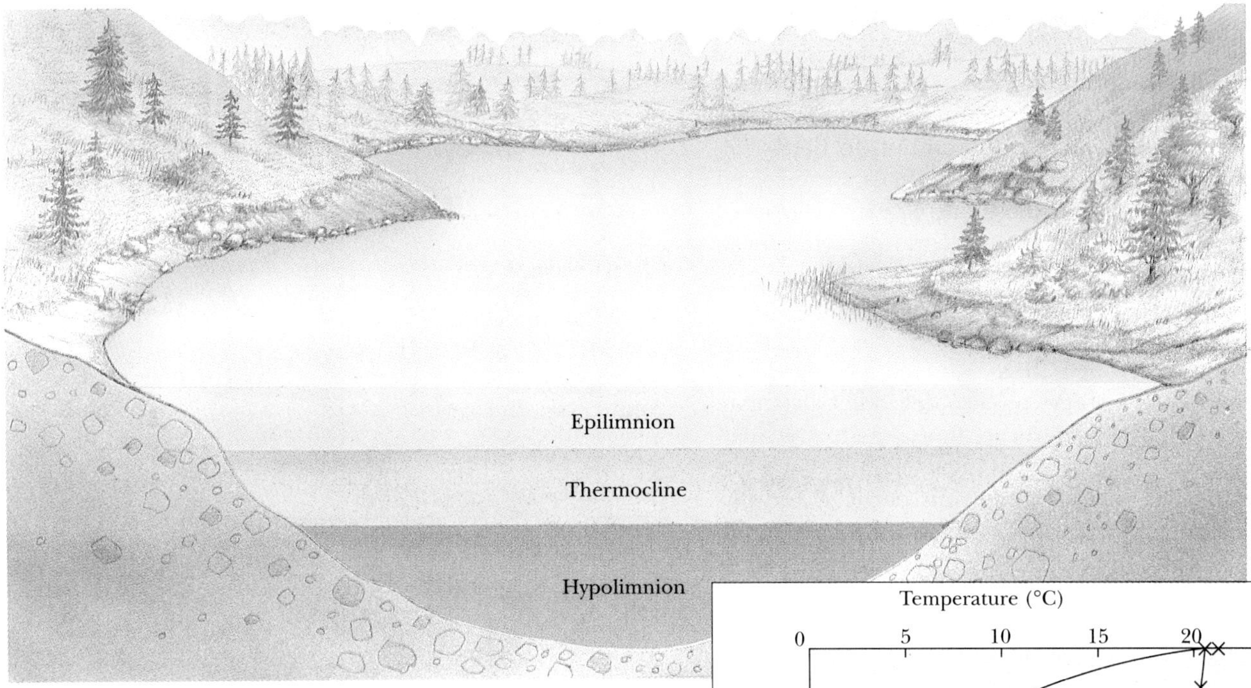

(a)

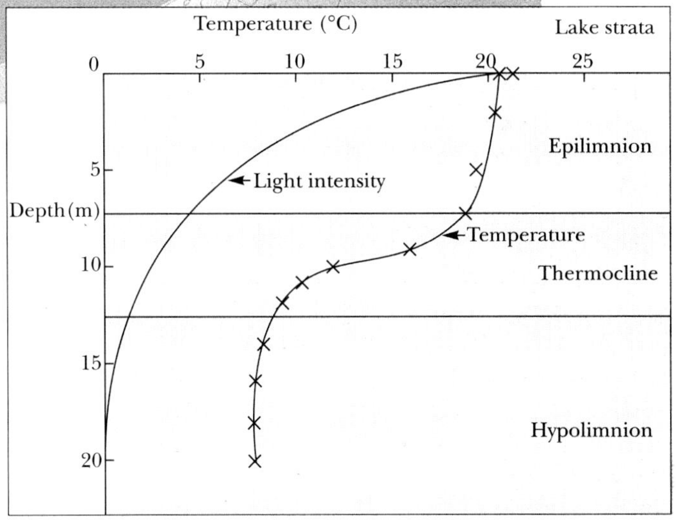

(b)

Figure **24.8** **(a)** A cross-section through a thermally stratified lake showing the positions of the epilimnion, thermocline, and hypolimnion. **(b)** A limnological graph showing the depth of the lake on the "y" axis and the concentration or intensity of a parameter on the "x" axis. In this graph, temperature and light intensity are shown at different depths in a typical thermally stratified lake during the summer months.

These heterotrophic bacteria are called **mineralizers** or **decomposers** because they convert the organic material back into inorganic material so that it can be recycled (see Chapter 12). Most heterotrophic bacteria that are indigenous to planktonic aquatic habitats are referred to as **oligotrophs,** because they can grow using very low concentrations of organic nutrients. They are to be distinguished from **eutrophs** (also called copiotrophs), which grow in organic-rich environments such as the intestinal tract of animals.

Oligotrophs exhibit low growth rates and have high affinity uptake systems in their cell membranes that allow for the uptake of the carbon sources occurring in low concentrations in planktonic habitats. By their activities, the concentrations of soluble organic substrates are kept at very low levels ($\mu g/\ell$ concentrations) in mesotrophic and oligotrophic environments.

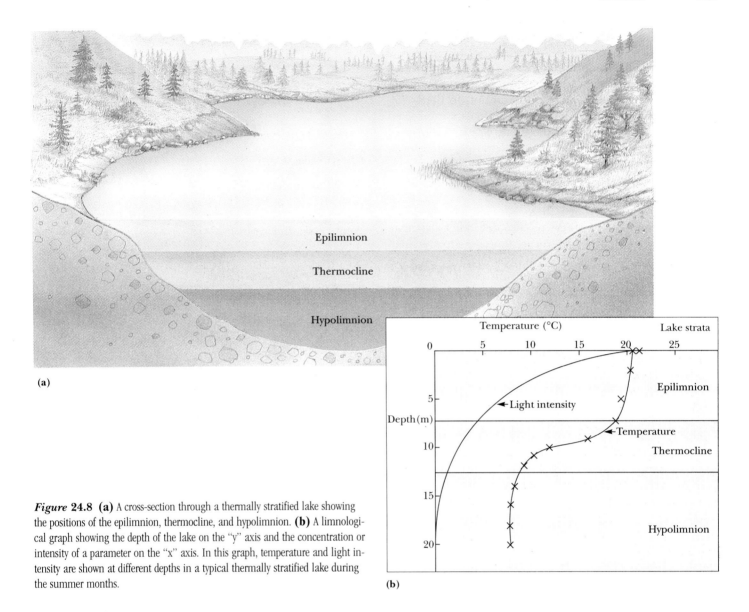

In contrast, eutrophic bacteria have rapid growth rates when they are provided with high concentrations of nutrients. In lake water, nutrients are available only in low concentrations, so the eutrophic bacteria are not competitive with oligotrophic species from these natural communities. However, the intestinal tracts of all of the consumer animals in aquatic food chains contain eutrophic heterotrophs, which play a major role in the decomposition of food stuffs eaten by the consumers. These foods are converted into amino acids, vitamins, organic acids, and the like that are absorbed in the intestinal tracts of the animals and used as their principal energy sources and sources of organic building materials. It is important to note that almost as much of the food consumed by the animal goes to produce microorganisms in the intestine that are ultimately defecated, as to that which goes to the animal's nutrition and energetics. *Therefore, symbiotic microor-*

ganisms of the animal intestine play a major, often overlooked, role in all food chains and food webs.

The **microbial loop** model best explains what occurs in aquatic habitats (Figure 24.9) to account for the mineralization of organic material. In addition to the bacteria, protozoa and small invertebrate animals are involved in the degradation of organic materials. The heterotrophic bacteria utilize dissolved organic material excreted by algae and cyanobacteria. Because they have rigid cell walls and high affinity uptake systems for organic compounds, the bacteria grow principally on dissolved organic compounds (although through their extracellular cellulases, chitinases, and so forth they can also degrade particulate organic materials). The bacteria are then ingested by small protozoa, which also ingest small algae. The bacteria are commonly attached to organic **detritus** particles (nonliving organic particulate material), which they are degrading (Table 24.4). Unlike bacteria, protozoa and other invertebrates have mouths and therefore ingest the smaller bacterial cells and small detritus particles by **phagotrophic** nutrition. These protozoa, which are microbial **grazers** (consumers that ingest bacteria, algae, and bacterially coated particles), are in turn eaten by larger protozoa, and so on.

Although grazers do not degrade all the particulate material during its passage through their digestive tract, after defecation, their fecal pellets can be degraded by bacteria or even reingested again by another grazer, so the process is repeated until the particulate is ultimately completely degraded. Therefore, it is the combined efforts of bacteria and these small grazing animals that result in the ultimate decomposition of organic materials in aquatic ecosystems.

Eutrophic lakes that are sufficiently enriched with nutrients may develop an anaerobic hypolimnion in the summer months during the period of thermal stratification.

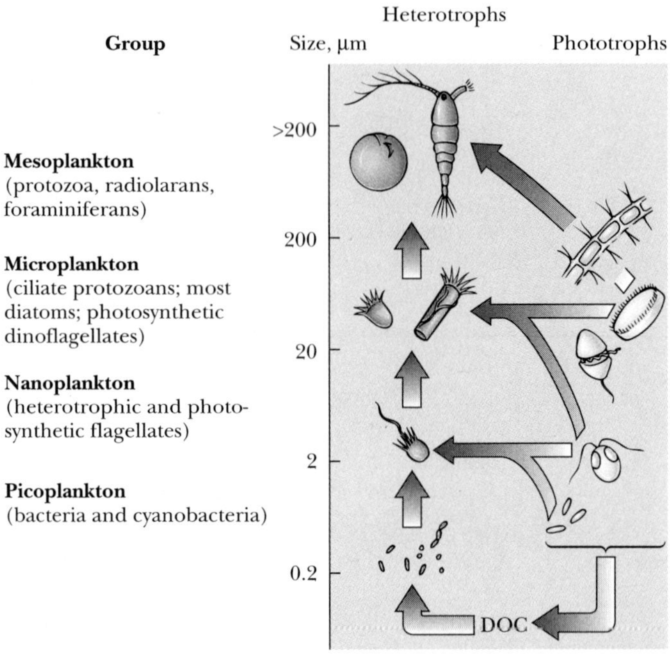

***Figure* 24.9** The microbial loop as found in a marine habitat. This model, based on the size scale of the microorganisms, is used to explain the mineralization of organic material. The bacteria use dissolved organic carbon (DOC) excreted by the phototrophs as their primary carbon source. The bacteria and cyanobacteria, which comprise the picoplankton, are ingested by flagellate protozoa (heterotrophic nanoplankton), which are in turn ingested by microplanktonic ciliate protozoans. These ciliates also ingest larger phytoplanktonic species including dinoflagellates and larger diatoms. Copepods, radiolarians, and foraminiferans serve as the highest-level consumers in the microbial loop.

This occurs because the hypolimnion is physically separated from the surface layers of the lake during stratification, so the amount of oxygen available in the hypolimnion is limited to that which was initially available at the time of thermal stratification and the small amount that dif-

***Table* 24.4 Composition of a natural water sample**

Constituent	Characteristics or Description
I. Dissolved material	Obtained by either centrifugation or filtration through a 0.2 μm filter to remove particulates
A. Organic components	
1. DOC	Dissolved organic carbon
2. DON	Dissolved organic nitrogen
B. Inorganic materials	
II. Particulate materials	Obtained as centrifuged material or collected on 0.2 μm pore-size filter
A. Living material	Living cellular material is called biomass
B. Nonliving materials	
1. organic	Nonliving organic material is called detritus
2. inorganic	

fuses to it from the epilimnion. Anaerobic conditions develop in the hypolimnion in the following manner. The high concentration of organic nutrients in the eutrophic lake will be degraded in the hypolimnion, initially by aerobes, because aerobic respiration is such an energetically advantageous process. This results in oxygen depletion and the hypolimnion becomes anaerobic.

Anaerobic conditions in the hypolimnion permit the growth of many other bacterial groups during the summer months. For example, if light penetrates to the hypolimnion and sufficient hydrogen sulfide is produced in the sediments, then photosynthetic bacteria grow and may produce hypolimnetic blooms. Common photosynthetic bacteria that may produce anaerobic blooms in this habitat include the purple sulfur bacteria and the green sulfur bacteria (Figure 24.7). Other anaerobic processes such as fermentation, denitrification, sulfate reduction, and methanogenesis occur at successively deeper depths in the hypolimnion or in the sediments of the lake.

Gas vacuolate organisms are especially common in thermally stratified lakes (Figure 24.7). These organisms can readily stratify in the lake at a light intensity or nutrient concentration level that is satisfactory for them. Thus, the gas vacuolate aerobic photosynthetic cyanobacteria grow at the surface of the lake where high light intensities occur, while the gas vacuolate purple and green sulfur bacteria are present at greater depths in the lake where there is both sulfide and at least some light.

Rivers and Streams

Flowing water habitats are quite different from lakes. They are more shallow and usually remain aerobic except in nonflowing backwaters and in their sediments. Frequently, microorganisms that inhabit these environments, like *Sphaerotilus* spp., attach to rocks and sediments so that they can take advantage of the nutrients that flow by them. Streams in natural areas receive large amounts of leaf fall and other dead organic material in the autumn and winter, and are therefore important habitats for leaf and litter decomposition.

Flowing habitats receive runoff from agricultural and urban areas, and are greatly influenced by the nature of the material they receive. In urban areas, rivers frequently are used for sewage disposal and may receive varying loads of organic material and bacteria, depending on the degree of treatment of the sewage (see Chapter 33). Industrial effluents are also extremely important in larger cities and commercial waterways. In agricultural areas, excess fertilizers and pesticides reach the rivers through runoff and may pose serious pollution problems.

Marine Habitats

Marine environments are extremely important because about two thirds of the Earth's surface is covered by oceans. The predominant primary producers in the marine environment are planktonic algae (called phytoplankton) that serve as the base of the food chain. The consumers in the ocean that are highest on the food chain are larger fish, sharks, toothed whales, and some other mammals. It is interesting to note that the krill-eating baleen whales, the largest mammals, occupy a rather low position in the food chain.

Because so much of the Earth is covered by oceans, the role of marine microorganisms in primary production and other biogeochemical activities greatly surpasses that of microorganisms in inland lakes and rivers. Biological oceanographers study the biogeochemical processes that are occurring, some of which have important implications on global climate.

Measuring primary productivity in the ocean is no small feat. Because of its vast size and seasonal and temporal variability, a few measurements here and there do not adequately assess activities. More recently, satellite mapping procedures have been used to assess global productivity in the oceans. These procedures rely on analyzing photos that show chlorophyll *a* and associated pigments in the phytoplankton (Figure 24.10). Although the quantitative information of these photos is limited due to the inability to observe phytoplankton beneath the surface, they provide valuable information on the extent and intensity of blooms over large scales heretofore impossible to study.

Other marine microbial activities are also important. For example, it has recently been discovered that dimethyl sulfide (DMS) is generated as a product of metabolism by some phytoplankton. DMS is volatile and enters the atmosphere, where it reacts photochemically with oxygen to form dimethyl sulfoxide (DMSO) and then sulfate. The sulfate in turn serves as a raindrop nucleator, which results in cloud formation. Cloud formation affects the intensity of solar radiation on an area in the ocean thereby causing a decrease in primary production and a decrease in DMS production. Thus, there is an internal feedback mechanism regulating primary production, temperature, and even weather (Figure 24.11).

The anaerobic processes that occur in marine sediments are similar to those that occur in freshwater sediments. From the sediment surface going downward, the following are sequentially encountered: aerobic respiration, ferric oxide respiration, denitrification, sulfate reduction, and finally methanogenesis (Figure 24.12). However, there are some major differences. For example, in marine systems, there is much more sulfate than in freshwater habitats, so sulfate reduction is a much more important process in this habitat.

Typical marine bacteria are different from freshwater bacteria. The higher salt concentration (about 3.5 percent) makes marine bacteria moderate halophiles (see **Box 19.1**). Thus, one of the characteristics of a typical marine bacterium is that it will have an optimum salinity of about 3.5 percent, whereas a freshwater bacterium exhibits

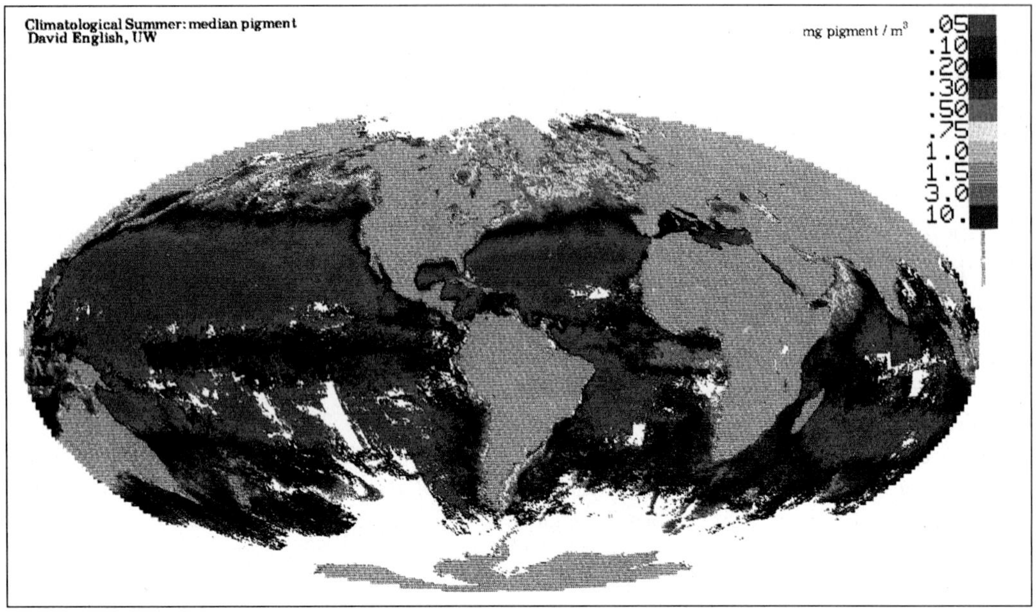

***Figure* 24.10** A global satellite spectral image showing the absorption of chlorophyll-like pigments of Earth's oceans during summer months. Note the greater abundance (yellow and orange colors) of photosynthetic pigments off the coastal regions in the Northern Hemisphere. The Coastal Zone Color Scanner Project of NASA's Goddard Space Flight Center produced and distributed the original data. (This image was processed by David English at the School of Oceanography, University of Washington using the University of Miami DSP system developed by Otis Brown and Robert Evans. Courtesy of David English.)

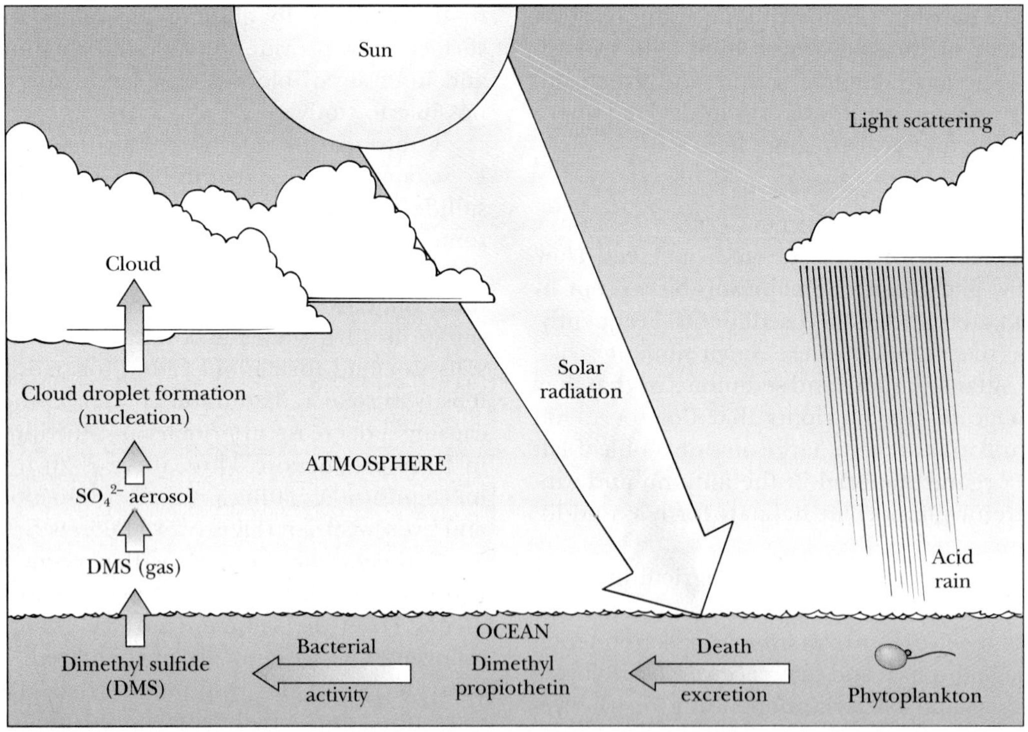

***Figure* 24.11** The proposed pathway for formation of clouds in the marine environment. Certain species of algae produce dimethyl propiothetin, which is thought to serve as an osmolyte in these algae. This compound is released from the algae when they die or through excretion. Dimethyl propiothetin is hydrolyzed by some bacteria to form dimethyl sulfide (DMS) and acrylic acid. The DMS is volatile, so some escapes into the atmosphere where it reacts chemically to produce SO_2 and then sulfate, SO_4^{2-}. The SO_4^{2-} serves as a water droplet nucleating agent, thereby producing clouds. This process has an internal feedback mechanism to control it: the greater the cloud cover, the less solar radiation for growth of phytoplankton and therefore less DMS production with lower sulfate production and fewer clouds. It is also interesting to note that rainfall is acidic because of the sulfate from the DMS. Therefore, this is an example of a process that causes a type of natural acid rainfall.

	Electron acceptor	Process	Approximate depth in sediment (cm)
Water column			
Sediment	O_2	Aerobic respiration	0 – 0.5
	Fe O(OH)	Iron respiration	0 – 0.5
	NO_3^-	Denitrification	0.5 – 5
	SO_4^{2-}	Sulfate reduction	5 – 100
	CO_2	Methanogenesis	> 100

***Figure* 24.12** Several electron acceptors are used in the biodegradation of organic substances in marine sediments. In the surface of the sediment, aerobic respiration occurs in which organic materials are oxidized by oxygen. At slightly lower depths where oxygen is depleted, particulate iron oxides are used as an electron acceptor with the formation of reduced ferrous ion. At even lower depths, nitrate (denitrification), then sulfate (sulfate reduction), and finally carbon dioxide (methanogenesis) are used as electron acceptors in anaerobic respirations, resulting in the breakdown of organic materials. The actual depths will vary from one location in the sediments to another depending on the amount of organic material, temperature, and other factors.

a lower optimum concentration. Also, most marine bacteria have an absolute requirement for Na^+, a feature not found in typical freshwater or terrestrial bacteria.

One special microbial environment found in oceanic polar regions of the Earth is the sea ice microbial community (SIMCO). This habitat develops during the early spring when the sunlight reaches the polar region. The sea ice, which can be over 2 meters thick, is colonized by algae (mostly diatoms), bacteria, and protozoa (Figure 24.13**a, b**). They grow in the lower 10 to 20 cm of the ice just above the seawater (see Figure 1.1**b**). Thus, these organisms grow at temperatures below the freezing point of

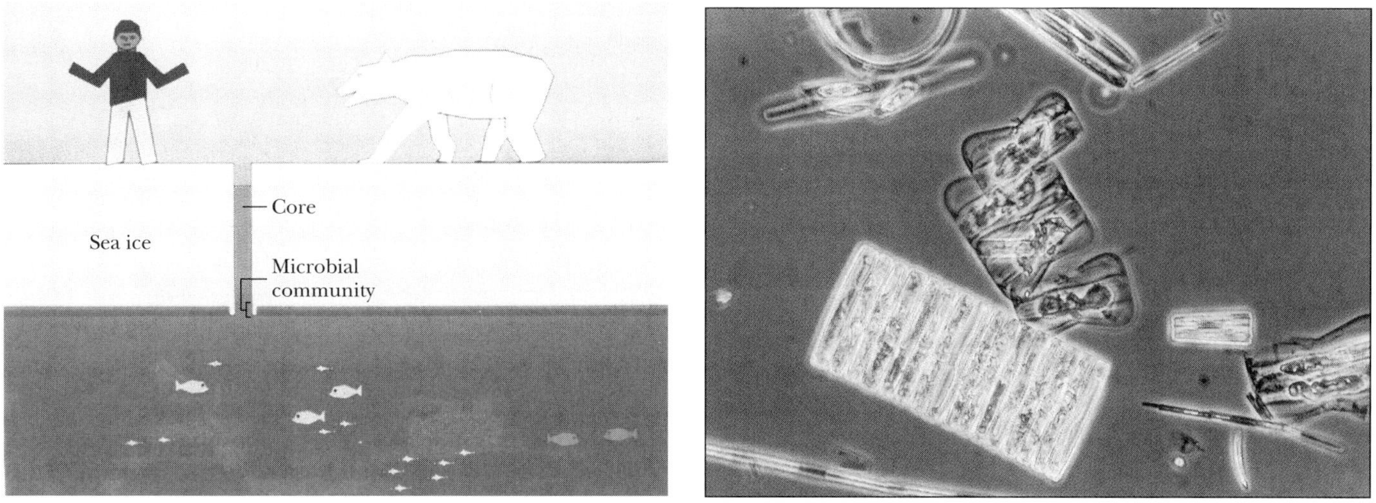

(a) **(b)**

***Figure* 24.13** **(a)** A diagram showing the sea ice overlying the water column at a North Polar location. The ice is approximately two meters thick and the sea ice microbial community is shown as a brown layer at the interface between the ice and the water column. A core has been cut from the sea ice. (Courtesy of John Gosink) **(b)** Diatoms predominate in the sea ice community. (Courtesy of J. T. Staley) seawater (−1.8°C). They do this by growing in the brine pockets between ice crystals.

Terrestrial Environments

Terrestrial environments are extremely important for human civilization. Modern agriculture relies on good soils for both plant and animal productivity. Such soils contain a mineral component that is derived from rock weathering, as well as an organic component that is derived primarily from organisms that previously resided in the habitat. A discussion of both soil microbiology and rock microbiology follows.

Soils

The soil environment is very complex. It is made up of inorganic minerals and the weathered remains of rocks as well as organic material, called **humus.** Humus comprises the partial decomposition products of plant organic matter and as such contains many of the less degradable or **refractory** organic plant substances such as lignin and humic acids. Soils vary greatly in composition and fertility depending upon climate, geology, and vegetation. Depending on the composition of the soil, the climate, and the types of vegetation, soils are classified into various groups. Soil texture, one important feature of classifications, is determined by the relative proportion of clay, silt, and sand particles that the soil contains (Figure 24.14).

The soil profile is another important feature used in classification of soils (Figure 24.15). The profile of a soil is determined by making a vertical cross section or cut into the soil. Each layer, called a **horizon,** has a characteristic composition and color due to the activities that occur there. For example, the A horizon at the surface of the soil is dark in color due to the large amounts of organic matter (humus) that accumulate there. Below this is the E (for eluviation or "wash out") horizon, which is lighter in color and contains resistant minerals such as quartz. The deeper B horizon is a zone of accumulation (illuviation or "wash into") that contains materials such as Fe, Al, and silicate minerals. Each soil type has its own characteristic soil profile.

Unlike aquatic environments, water is not always available in soils. Nonetheless, water is absolutely required for the growth of all organisms. Thus, microorganisms that occur in soils grow intermittently, that is, only when there is sufficient soil moisture available for them. Microorganisms have made special adaptations to withstand these intermittent periods when water is unavailable. Some produce cysts (*Azotobacter* and myxobacteria); endospores (*Clostridium* and *Bacillus*); or conidiospores (the actinomycetes), specialized cells that survive periods of dessication. Then, when moisture is available through rain or dew, these germinate to produce metabolically active vegetative cells.

In soils, the predominant primary producers are plants, not algae, although algae and cyanobacteria may grow on the surface of soils. Plants synthesize complex organic constituents such as cellulose, hemicellulose, and lignin. These types of organic material are insoluble in water and generally more refractory to decomposition than algal polysaccharides and starch. Nonetheless, heterotrophic bacteria and fungi are known to degrade these substances. The rate and extent of degradation of these substances varies. For example, peat bogs are low pH environments that have complex organic constituents such as humic acids and lignin that degrade more slowly than they accumulate. Therefore, over many centuries the refractory organic substances may be converted to coal or petroleum deposits under the appropriate geological conditions. However, almost all of the organic material that is synthesized by plants is degraded biologically and goes back into the soil and air and is thereby made available for recycling.

The "health" of soils, called **soil tilth,** is extremely important for agriculture. Microorganisms play major roles in soil nutrient cycles, making nutrients available on the one hand and removing nutrients on the other hand, depending on the process. For example, nutrients are made available through the recycling of organic compounds, as

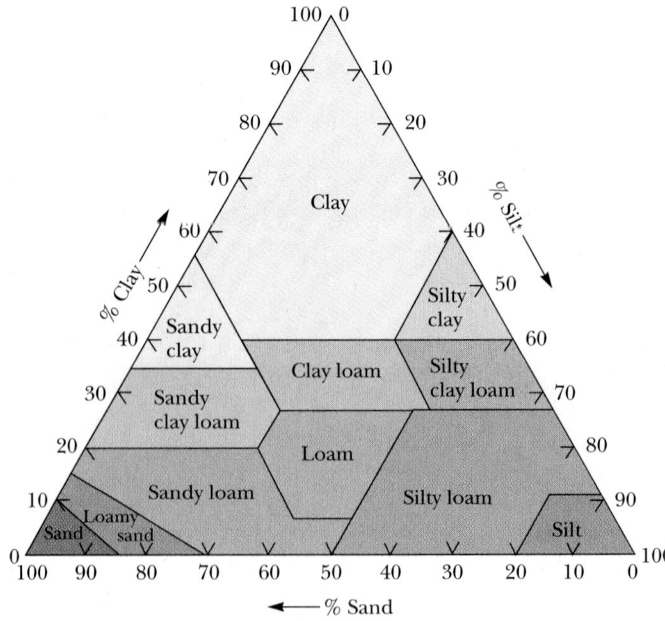

***Figure* 24.14** A soil textural triangle illustrating how soils are described based on the proportion of sand, silt, and clay particles.

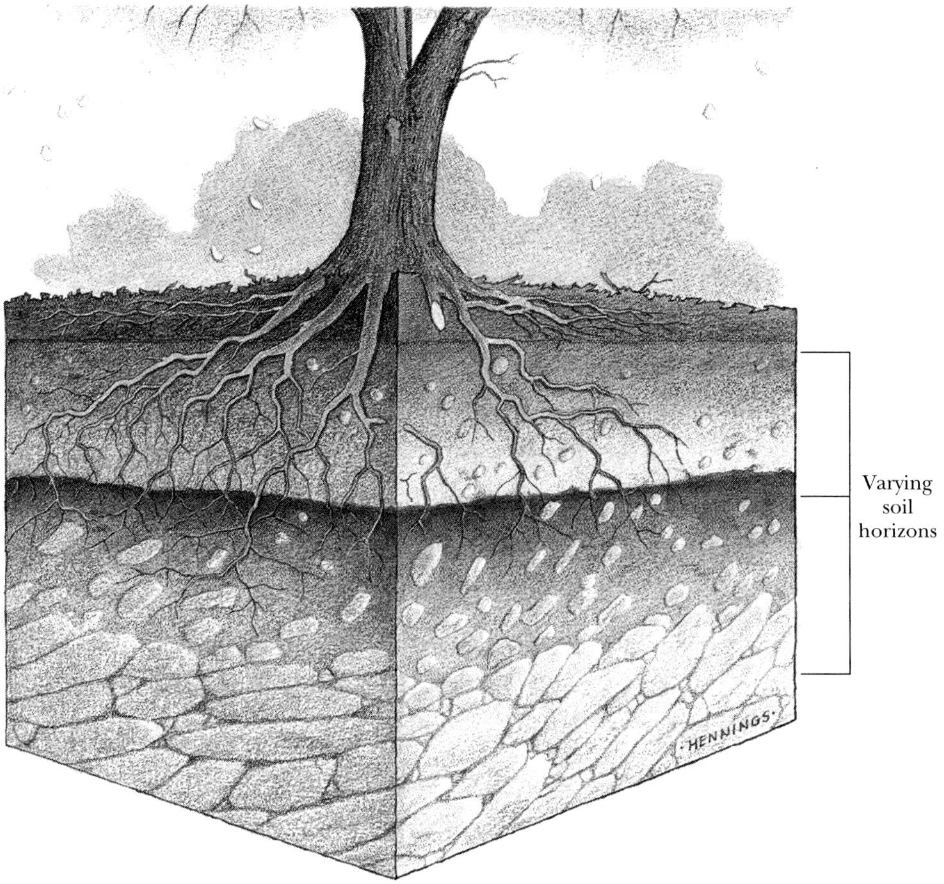

***Figure* 24.15** Diagram of a soil profile. Each profile is separated into layers called horizons that are distinguished by different chemical composition, colors, and microbial activities.

previously discussed. Likewise, important transformations in the nitrogen cycle occur in soils (see the upcoming discussion on Nitrogen Cycle). Among the nitrogen transformations that are beneficial include nitrogen fixation, both symbiotically and asymbiotically, as well as ammonification, the removal of ammonia from organic sources. Less obviously beneficial is the process of nitrification that converts ammonia (normally a good source of nitrogen for plants) to nitrite and nitrate, which is an energetically less favorable form of nitrogen for plants. The process of denitrification is actually harmful to soil fertility because it results in a conversion of nitrate nitrogen back into atmospheric nitrogen gas, in effect reversing the nitrogen fixation process.

Rocks

Although rocks are not normally thought of as an environment for living organisms, they are, in fact, common habitats for microorganisms. Even the most inhospitable rock in the driest desert is likely to harbor microorganisms. It is the growth of microorganisms on rocks in conjunction with physical and chemical processes (such as freezing of water, wind action, and acid rainfall) that cause rock weathering. As mentioned previously, weathering processes result in soil formation.

Algae and lichens are common primary producers on rocks (Figure 24.16**a**). They may grow on the surface of rocks **(epilithic)** (Figure 24.16**b**), in cracks **(chasmolithic)**, or even on the underneath surface **(hypolithic)** if the rocks are translucent. Lichens are particularly well suited for growth on rocks because the fungal component that dominates in biomass can withstand severe dessication and provide inorganic nutrients for the algal component. In turn, the algae produce organic nutrients for the growth of the fungus symbiont (see Chapter 25). Some fungi can also grow as microcolonies on rock surfaces in desert and arid areas in environments too severe for the growth of lichens (Figure 4.14). Apparently they obtain their nutrients from windblown dust.

One of the most remarkable findings has been that some lichens and algae can actually grow *inside* of rocks (Figure 1.1**d**). Such growth is called **endolithic.** Indeed, endolithic microorganisms that grow in rocks in Antarctica are among the most extreme forms of terrestrial life on Earth (see **Box 24.2**).

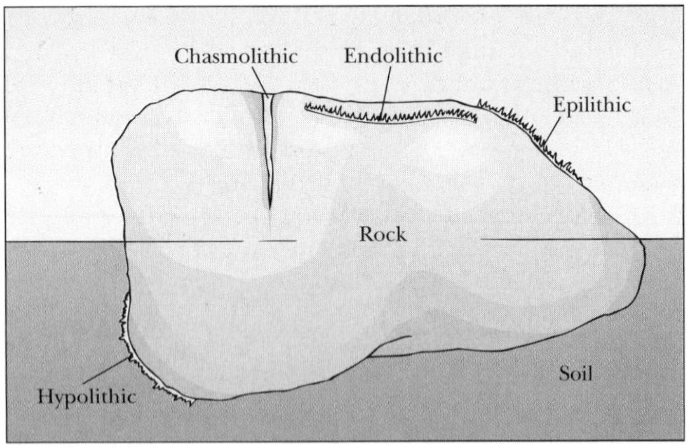

(a)

(b)

***Figure* 24.16** Microbial growth on and in rocks. **(a)** Epilithic organisms grow on the surface, chasmolithic in cracks, endolithic inside the rock, and hypolithic on the buried side of the rock. **(b)** Epilithic lichens colonizing rocks in Antarctica. The green "grassy"-appearing growth below the rocks is moss. (Courtesy of J. T. Staley)

BOX 24.2 RESEARCH HIGHLIGHTS

Endolithic Microorganisms of Antarctica

The Victoria Land Dry Valleys near the U.S. Antarctic base at McMurdo Sound comprise one of the most extreme environments on Earth. The two principal features that limit biological growth and activity are the low temperature and low water activity (humidity). As a result, "soils" taken from these environments appear sterile, having extremely low counts of bacteria. For this reason, the Dry Valleys have been compared to the surface of Mars.

However, some microorganisms do live in the terrestrial environments of Antarctica. They reside *inside* the rocks in the Dry Valleys. The rocks are porous sandstones that obtain some moisture from snowfall. During brief periods of the midsummer, conditions allow for melting. During these periods the endolithic lichens obtain water and are able to photosynthesize and grow. For the remainder of the year, they exist in a natural freeze-dried or lyophilized state. Carbon-dating by E. I. Friedmann, the microbial ecologist who discovered this remarkable community of microbial life, indicates that these lichens are thousands of years old.

(a)

(a) A view in the Victoria Land Dry Valleys. Rocks in foreground with a reddish color contain endolithic lichens.

Dispersal, Colonization, and Succession

In order to disseminate from one location to another, microorganisms must survive during transit under nongrowing conditions. Many bacteria and higher microorganisms have special dispersal and survival stages such as endospores, cysts, or conidiospores that allow them to maintain their viability for many days, weeks, or even years. Other organisms can survive for long periods of time in the dirt of birds claws, or in the digestive tract of a bird or on plant and animal materials eaten by a bird. These birds can serve as vectors to carry a variety of microorganisms from one continent to another. Similar mechanisms occur with other animals, although their migration patterns are more limited in scale.

Perhaps the microorganisms that are best known for their abilities to traverse long distances are the fungi. Under favorable conditions for dispersal, their airborne spores can be carried many miles. Some of these microorganisms cause plant diseases and their long-distance aerial dissemination is therefore a matter of concern to agriculture. Others are important allergens that affect the pulmonary systems of sensitive individuals.

A different illustration of dispersal and colonization is that of the ruminant animal (see Chapter 25). Ruminant animals maintain a dense community of anaerobic bacteria, including methanogens which are very sensitive to oxygen, in their forestomach. However, newborn calves have a sterile intestinal tract. After they are weaned and begin eating plant materials, they develop the typical anaerobic microflora of adult cattle. Because of the sensitivity of this community to oxygen, long-distance dissemination of the organisms is highly unlikely. The anaerobic microbiota is most likely transmitted from the mother through her saliva, which contains small numbers of these bacteria that

(b)

(c)

(b) A photo of the rocks being studied by microbial ecologists from the late J. Robie Vestal's laboratory; **(c)** a sandstone rock has been broken to show the endolithic lichen community, which appears as a dark band beneath the surface. See also Figure 1.1**d** for a close-up view of the algal and fungal layers of this community. (Courtesy of J. R. Vestal)

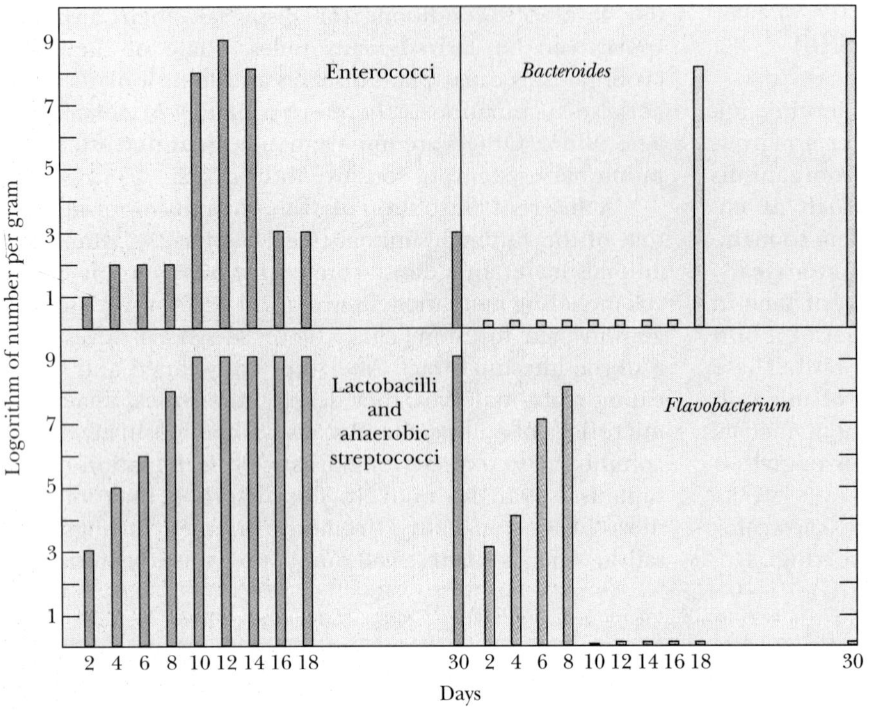

***Figure* 24.17** Colonization of a newborn mouse intestinal tract. Note that not all early colonizers are successful. As the intestinal tract becomes anaerobic, obligate aerobes, such as *Flavobacterium* spp., can no longer survive. (Reproduced from R. W. Schaedler, R. Dubos, and R. Costello *The Journal of Experimental Medicine,* 1965, 122:59–66.)

are brought to the mouth through regurgitation and cud chewing.

Of course, newly exposed habitats, such as a new volcanic land mass, will not be colonized immediately. The process of colonization takes time. Furthermore, the colonizing or **pioneer** species may not necessarily be one of the species that survives in the new habitat. Colonization leads to the phenomenon of **succession,** in which the species composition of the habitat changes over time. An example of a microbial succession is one that occurs in the colonization of the intestinal tract of newborn mice (Figure 24.17). In this example, the initial colonizing organisms include a *Flavobacterium* species, a *Lactobacillus* species, and enterococci. Over a few days' time, the succession proceeds and colonization by the obligately anaerobic genus *Bacteroides* occurs. Eventually the *Flavobacterium,* one of the pioneer species and an obligate aerobe, disappears. Thus, changes have occurred in the intestinal tract as it matures, changes that are brought about in part by the colonizing organisms including the *Flavobacterium.* These changes have resulted in creating a habitat with new features (among them, anaerobiosis) that no longer allow the *Flavobacterium* to survive. In contrast, the others, which are facultative anaerobes, remain, although the concentrations of all are affected by the changed conditions they helped bring about. Eventually a **climax state** is attained in which the types and concentrations of microorganisms attain an equilibrium in the intestine. This equilibrium can be disrupted by intestinal infections or antibiotic therapy.

Biogeography of Bacteria

L. M. G. Baas-Becking, a Dutch microbiologist, stated that "Everything is everywhere, the environment selects." By this he meant, for example, that if a particular soil type in Europe is very similar to a soil type in North America with similar climatic conditions, the same species of free-living bacteria, such as *Bacillus cereus,* is expected to live in both places. Thus, this hypothesis argues for most bacteria having a **cosmopolitan,** or worldwide distribution. The basis for this hypothesis is that microorganisms are readily dispersed throughout Earth: they are carried by the wind across the ocean as well as by animals—especially birds and humans, which travel long distances. When they find a favorable habitat, they will colonize it and grow. Therefore, over relatively short periods of time a bacterium from Europe could be brought to North America or vice versa. Although this hypothesis is widely accepted by microbiologists, it has not been carefully tested. Recently, however, microbiologists are using molecular techniques to analyze samples from various habitats such as soil and sea ice to determine the validity of the hypothesis. Data to date raise questions. Recent data taken from diverse geographic regions suggest that, at least, some bacteria are **endemic** to specific locales, that is, no identical bacteria can be found in widely separated habitats that are similar. However, it is difficult to preclude the possibility that the same species might be found in much lower concentrations in other habitats. The answer to this question is important because it will help provide information on the total number of bacterial species on Earth.

Synecology, Part I: Biogeochemical Cycles

Synecology is the study of natural communities of organisms. At this level of study the scientist is not concerned with the activities or distribution of a single species but rather in the processes of a trophic group of species or the entire microbial community of the habitat. For example, whereas the autecologist might be interested in the distribution and activity of a single species of thiobacilli such as *Thiobacillus thioparus* in lakes, the synecologist would be interested in the whole community of sulfur-oxidizing species in the lake and the rates at which this process is occurring. In this section, one aspect of synecology is considered, namely the important roles that microorganisms play in biogeochemical cycles. Later on, the actual methods used to assess microbial activities in biogeochemical cycles will be discussed.

Because microorganisms, particularly bacteria, evolved billions of years before higher organisms, they were largely responsible for the first biosphere on Earth and the origin and evolution of the biogeochemical cycles. They continue to play major roles in these transformations. Of course, the cycles have continued to change since the evolution of plants and animals and are influenced also by human industrial and military activities.

Atmospheric Effects of Nutrient Cycles

From a global perspective, the effects of the nutrient cycles are most keenly felt in the atmosphere. This is because all terrestrial living forms exist bathed in the atmosphere. The animals are dependent on oxygen that is provided by primary producers, and through respiration they expel carbon dioxide, an essential nutrient for plants. Because of their reliance on air, animals and plants are profoundly affected by noxious airborne pollutants. Furthermore, the chemical composition of the atmosphere affects global temperatures and incident ultraviolet light radiation. Two major current concerns are the buildup of greenhouse gases that cause global warming and the release of anthropogenic chemicals such as the chlorofluorocarbons that destroy the ozone layer (which protects all living organisms from UV radiation).

Microorganisms are involved in many of the transformations that affect the atmosphere. For example, nitrous oxide, N_2O, an intermediate product formed in two different bacterial processes, nitrification and denitrification, reacts photochemically with ozone, O_3, in the atmosphere thereby depleting it (see Nitrogen Cycle, below).

Both methane and carbon dioxide, other important constituents of the atmosphere, are greenhouse gases produced in part through microbial activities. Methane is produced largely through the activities of methanogenic bacteria, although it is also released from natural gas seeps and by oil drilling operations. Its concentration has been increasing in the atmosphere at a rate of about 1 percent per year for the past 15 to 20 years. The biological sources of methane include ruminant animals, termites, and bogs and swampy areas, all of which contain methanogenic bacteria. In the northern hemisphere an important nonbiological source is oil drilling operations, which release lighter hydrocarbons such as methane into the atmosphere.

Carbon Cycle

Microorganisms play major roles in the carbon cycle (Figure 24.18) an element which is a key constituent of living organisms. In the example shown, there are four separate **reservoirs** for carbon, the ocean, the atmosphere, the lithosphere, and the terrestrial biosphere. The amount of carbon in each reservoir, termed the **mass,** is given in PgC (where $1P = 10^{15}g$). Thus, the atmosphere of Earth contains 725 Pg carbon dioxide, 3 Pg methane, and 0.2 Pg carbon monoxide. Arrows designate the **flux** or rate of movement of carbon from one reservoir to another in terms of PgC/year. Therefore, in the ocean, a balance occurs between the flux of carbon to and from the atmosphere (80 PgC/year). The mass of the marine biota in the surface water of the Ocean (3 PgC) is responsible for the processes of carbon **assimilation** (largely due to carbon dioxide fixation by primary producers), **respiration,** and **decomposition.**

Please note that the fluxes to and from a box are balanced. Thus, in this example, 50 PgC/year are assimilated by the marine biota, 45 PgC/year are respired, and the final 5 PgC/year are particulate materials (largely nonliving phytoplankton and other organisms) that sediment into the deeper layers. The deeper water contains large amounts of inorganic carbon in the form of carbonate, bicarbonate, and carbonic acid. It also contains substantial amounts of dissolved organic carbon (DOC) and particulate organic carbon (POC). Only a small portion of the carbon, 0.2 Pg/year, reaches the sediments.

This model also accounts for inputs of carbon to the atmosphere through volcanic eruptions (methane, carbon dioxide, and carbon monoxide) as well as fossil fuel burning and deforestation all of which occur in the terrestrial compartment. Although this model is not very detailed, it provides important information about the major reservoirs of carbon and the most important fluxes that occur between the reservoirs. More and more complex models can be generated if more detailed experimental data are available.

Since this text is primarily focused on the microbial aspects of biogeochemical cycles, particular emphasis will be given to primary production, decomposition, and methane formation and oxidation.

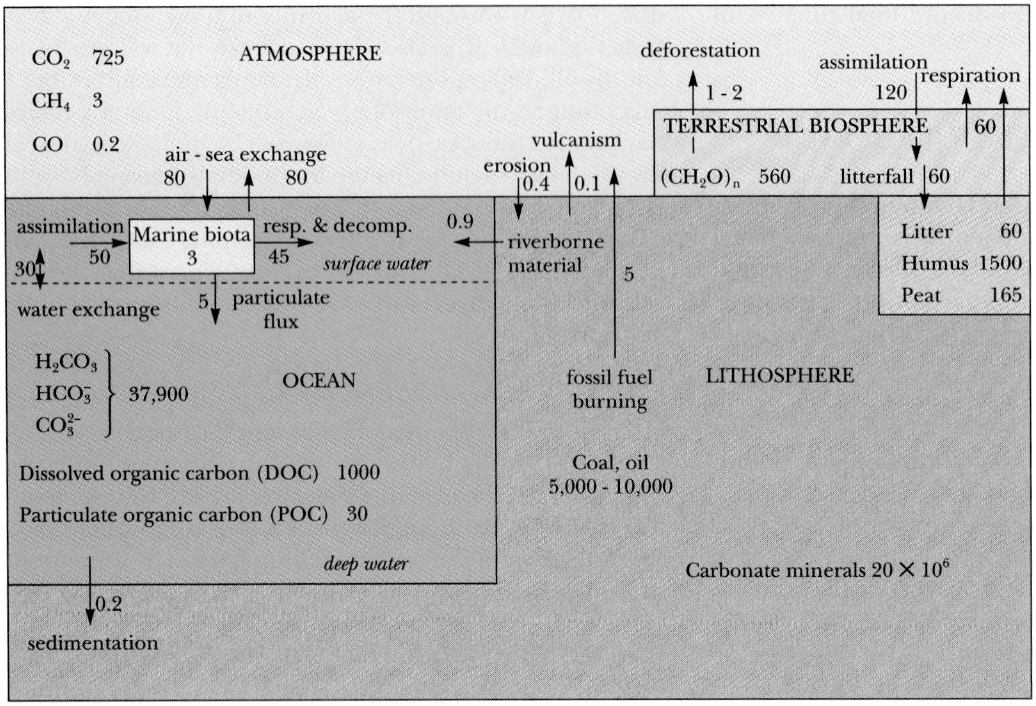

***Figure* 24.18** The global carbon cycle. (Adapted from K. Holmén from *Global Biogeochemical Cycles* Academic Press.) Reservoir masses are in PgC, and fluxes are in PgC/year.

Primary Production

Primary producers fix carbon dioxide and convert it to organic material (Table 24.5). Strictly speaking, only the photosynthetic organisms are considered to be truly primary producers because they derive their energy directly from sunlight. The chemolithotrophic bacteria require chemical sources of energy (reduced inorganic compounds such as hydrogen sulfide, ammonia, or hydrogen), and these are generally provided through geochemical activities or other biological processes that depend ultimately on sunlight. For example, chemolithotrophic sulfur oxidizers require reduced forms of sulfur derived either from geochemical sources or from reduced sulfur sources provided by sulfate reducers who obtain their energy from organic materials ultimately produced by photosynthesis.

The principal terrestrial primary producers are higher plants that are the major producers of organic material. In freshwater and marine habitats, which account for about two thirds of the surface area of Earth, the algal and cyanobacterial phytoplankton are the principal primary producers. In contrast, anoxygenic photosynthetic prokaryotes and chemolithotrophic bacteria play much more restricted roles in primary production except in specialized habitats.

Decomposition of Organic Material

Organic material derived ultimately from primary producers resides in living organisms and the nonliving organic material derived from them. Plants and animals and most microorganisms carry out **respiration** to form carbon dioxide and water. For animals and aerobic, heterotrophic microorganisms, respiration is the mechanism by which they obtain ATP for their metabolism.

Bacteria and fungi are the ultimate recyclers of nonliving organic material. They live as saprophytes on organic material from dead plants and animals as well as other microorganisms. They are aided in this process by higher animals that ingest particulate organic materials (herbivores and carnivores) that have bacteria residing in their intestinal tract. The process is chemically analogous to respiration, but involves the degradation of nonliving organic material to obtain energy for growth. This process is called organic decomposition or degradation. If the organic compound is degraded completely to inorganic products such as carbon dioxide, ammonia, and water, the process is called mineralization.

Bacteria and fungi are especially well suited for the degradation of polymeric organic compounds that are refractory to higher organisms. Cellulose, chitin, and lignin are examples of this. Cellulose, derived from plants, and chitin, derived largely from crustaceans, insects, and some fungi, are degraded by many bacteria and fungi. The white rot fungi are the principal known degraders of lignin, a process that has only recently been chemically characterized. A wider variety of microorganisms can degrade soluble organic compounds including organic acids, amino acids, and sugars.

Almost all organic material produced on Earth is degraded by microbial activities. However, small amounts accumulate in sediments as shown in Figure 24.18. Over time, these accumulations have resulted in the formation of coal and petroleum deposits.

Methanogenesis and Methane Oxidation

Major anaerobic processes in the carbon cycle result in the fermentation of organic compounds to organic acids, and generate gases such as hydrogen and carbon dioxide. Additional degradation by methanogens results in the formation of methane gas in highly reduced sediments and the digestive tracts of ruminant animals and termites. Methane-oxidizing bacteria and certain yeasts degrade methane in the biosphere, but some escapes to the atmosphere where it becomes a greenhouse gas.

Nitrogen Cycle

All living organisms require nitrogen, since it is an essential element in protein and nucleic acids. Animals require organic nitrogen sources which they obtain through digestion of plant or animal tissues. Plants use inorganic nitrogen sources such as ammonia or nitrate. Most bacteria can use ammonia or nitrate as nitrogen for growth but some, such as the lactic acid bacteria, may require one or more of the amino acids in their diet.

Microorganisms play several important roles in the nitrogen cycle (Figure 24.19). They are responsible for several processes not carried out by other organisms. A discussion of the major processes follows.

Nitrogen Fixation

Only prokaryotic organisms carry out nitrogen fixation (see Chapter 10). Some Eubacteria and Archaea produce the enzyme nitrogenase, which is responsible for this process. Nitrogenase is found in both photosynthetic prokaryotes as well as in heterotrophic prokaryotes (Table 24.6). Since nitrogen is often a limiting nutrient in terrestrial and aquatic habitats, this process is important because it recycles nitrogen back into the biosphere.

Nitrogen is fixed into organic nitrogen (as the amino group in the amino acid glutamine). This can be transformed into other amino acids using various enzymes available in the organisms that fix the nitrogen or it can be made available to the organism's symbiotic partner (if it is a symbiotic fixer). In this manner, N_2, a rather inert gas, is made available to organisms directly from the atmosphere.

Ammonification

Organisms release nitrogen from their cells or tissues in the form of ammonia, a common decomposition product. This process, called **ammonification,** is hastened by the

***Table* 24.5 Primary production (carbon dioxide fixation) among the Eubacteria and Archaea[1]**

	Examples of Microorganisms
I. Photosynthetic Primary Producers[2]	
A. Eubacteria	Cyanobacteria
	Purple bacteria
	Green sulfur bacteria
	Green filamentous bacteria
B. Archaea	None
II. Chemosynthetic Primary Producers	
A. Eubacteria	Nitrifiers
	Sulfur oxidizers
	Iron oxidizers *(T. ferrooxidans)*
	Aquificales (hydrogen oxidizers)
	Acetogens
B. Archaea	Methanogens
	Sulfur oxidizers

[1]Not all members of each group are primary producers.

[2]Some photosynthetic groups are not included because, although they use light for photophosphorylation, they cannot fix carbon dioxide; these include the heliobacteria, the purple nonsulfur bacteria, and the halobacteria.

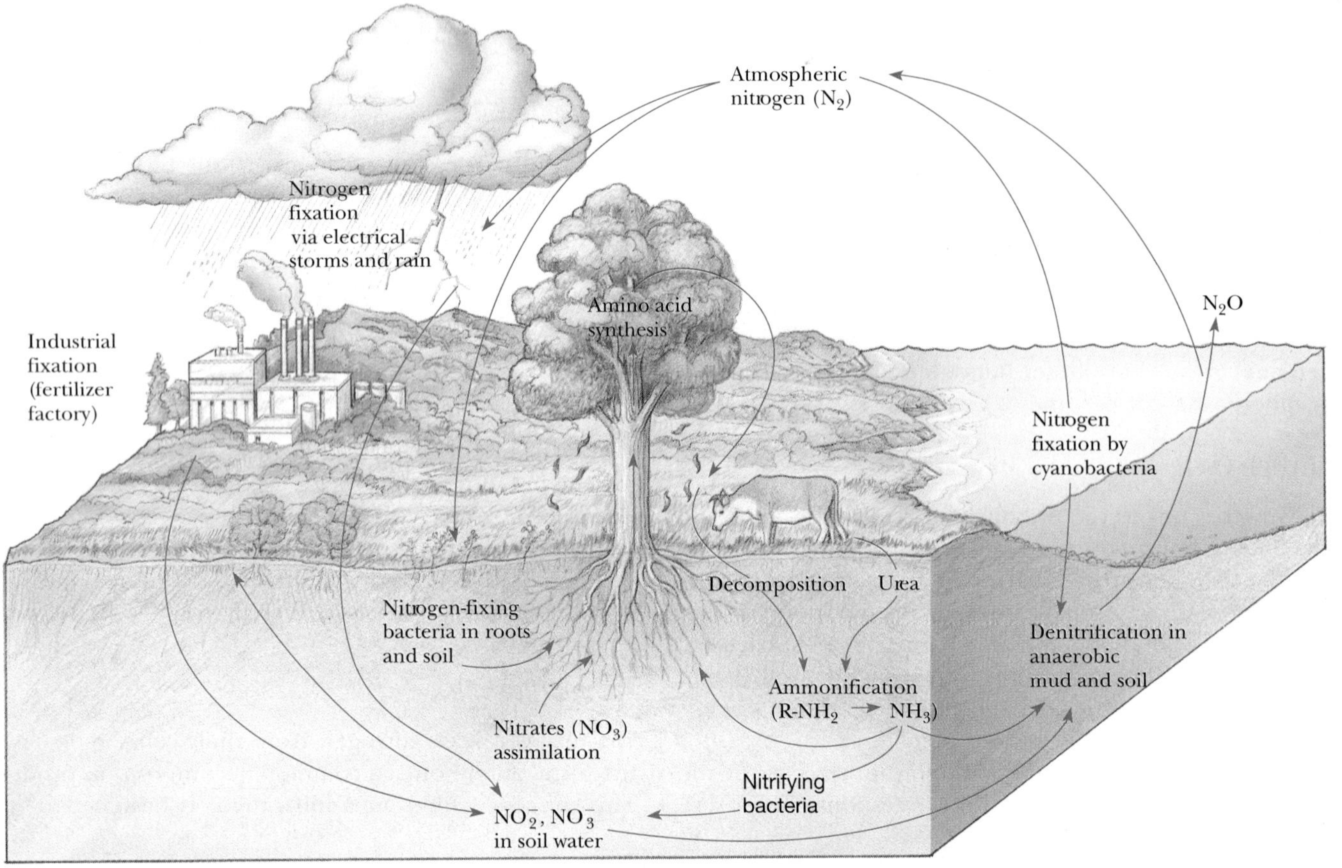

***Figure* 24.19** The nitrogen oxidation-reduction cycle. Atmospheric nitrogen, N_2, can be fixed by some prokaryotic organisms, resulting in the formation of R-NH_2, which represents organic nitrogen as in the amino acid form. Ammonia is produced by ammonification of organic nitrogen obtained from plants, animals, and microorganisms. Ammonia formed from ammonification can be used as a nitrogen source by many plants and microorganisms. Ammonia may also be oxidized by nitrifying bacteria first to nitrite and then nitrate. As with ammonia, nitrate can serve as a nitrogen source for many plants and bacteria by a process referred to as assimilatory nitrate reduction. Animals require organic nitrogen and must obtain this nitrogen from other animals or plants. Denitrifying bacteria use nitrate as an electron acceptor in anaerobic respiration and produce nitrous oxide and nitrogen gas, which are released to the atmosphere.

activity of some bacteria having deaminases that remove amino groups from organic nitrogenous compounds to form ammonia.

Nitrification

Ammonia produced by ammonification can be used directly by many plants as a source of nitrogen for synthesis of amino acids and other nitrogen-containing organic compounds.

However, ammonia can be oxidized by a special group of bacteria called nitrifiers. These bacterial chemolithotrophs obtain energy from this two-step process (Chapter 21) in which ammonia is first oxidized to nitrite, and the nitrite is subsequently oxidized to nitrate. Like nitrogen fixation, this process is uniquely associated with bacteria; during it some nitrous oxide (laughing gas), N_2O, is produced. This is an important gas because it reacts photochemically with ozone according to the following reactions.

$$N_2O + h\upsilon \longrightarrow N_2 + O$$
$$N_2O + O \longrightarrow NO$$
$$NO + O_3 \longrightarrow NO_2 + O_2$$

where $h\upsilon$ represents a photon.

The result of these reactions is that the ozone, O_3, in the protective ozone layer becomes depleted.

Nitrate is much more readily leached from soils than is ammonia. If excessive amounts of nitrate are leached from soils it can accumulate in runoff water and in wells. When the concentrations become high enough the water becomes unfit as a drinking source for humans. This oc-

***Table* 24.6 Representative prokaryotic groups containing species that carry out nitrogen fixation[1]**

	Group
I. Eubacteria	
A. Eubacteria (nonsymbiotic)	
1. Photosynthetic	Purple sulfur bacteria
	Green sulfur bacteria
	Purple nonsulfur bacteria
	Cyanobacteria
2. Heterotrophic	*Azotobacter*
	Clostridium
B. Eubacterial symbionts	*Rhizobium* (legumes)
	Frankia (alder trees)
	Cyanobacteria (lichens; *Azolla*)
II. Archaea	Methanogens

[1]This list is illustrative and does not contain all taxa that are capable of nitrogen fixation; not all members in each group are capable of nitrogen fixation.

curs because the nitrite formed in the intestinal tract by nitrate-reducing bacteria can have an extremely adverse effect by interacting with hemoglobin in the bloodstream to produce methemoglobin. This causes **methemoglobanemia** and, if it is not properly diagnosed, may result in the death of infants because of its effect on respiration. The color of infants becomes blue (hence the vernacular name for the ailment, "blue babies"). This malady occurs primarily in agricultural areas that receive excess nitrogen fertilizer.

Denitrification

A number of bacteria carry out nitrate respiration. In this process, which occurs preferentially in an anaerobic environment, nitrate is ultimately converted into nitrogen gas. This process is called **denitrification.** It occurs predominantly in waterlogged areas that have become anaerobic. From an agricultural perspective, this is an undesirable process because it results in a loss of fixed nitrogen back to the atmosphere. The intermediates of this process are similar to those of nitrification, including the formation of N_2O. Some of the bacteria that carry out this process include various *Pseudomonas* species, *Thiobacillus denitrificans,* and *Paracoccus denitrificans.*

Nitrate Fermentation

In some anaerobic environments, such as cattle rumen, nitrate is not converted to nitrogen gas by denitrification, but is fermented to ammonia by the resident bacteria. This process is called **nitrate fermentation.**

Sulfur Cycle

The **sulfur cycle** is also an important biogeochemical cycle. Sulfur, as with carbon and nitrogen, is needed by living organisms, where it is a component of proteins and other cell constituents. However, it is not required in such large amounts as carbon and nitrogen and has not been reported as a limiting nutrient in environments. Higher plants usually obtain sulfur in the form of sulfate, whereas animals obtain it through amino acids (cysteine, cystine, and methionine) in their diet, either from plant protein or from the protein of other animals or microorganisms (as in the ruminant animals). In contrast, microorganisms use sulfur compounds in other ways. For example, some use sulfur-containing substances as energy sources, others use sulfur compounds as electron acceptors in anaerobic respiration, and some as hydrogen donors for photosynthesis. Because of these diverse activities, the sulfur cycle is dominated by microbial processes and is one of the most fascinating of the elemental cycles (Figure 24.20). The oxidation state of inorganic sulfur ranges in its cycling from -2 in sulfide to $+6$ in sulfate.

Sulfur Oxidation

Reduced inorganic forms of sulfur including not only sulfide and elemental sulfur, but also thiosulfate and other ions as well, can be oxidized by several different groups of organisms, as has been discussed previously (Table 24.7). This process is called **sulfur oxidation.** Photosynthetic bacteria use hydrogen sulfide produced by sulfate reducers in anaerobic environments as electron acceptors for the

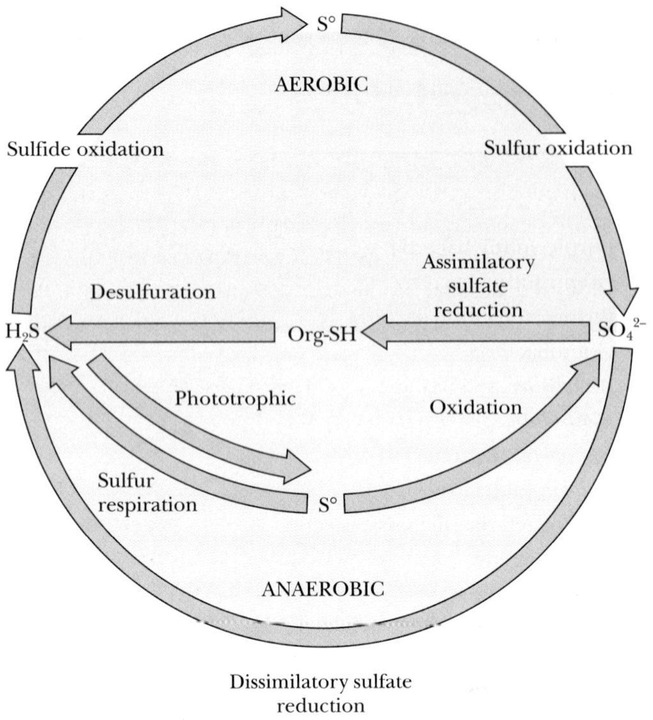

***Figure* 24.20** The sulfur oxidation-reduction cycle. Org-SH refers to organic sulfhydryl groups that occur in amino acids and other organic compounds. H_2S is used as an energy source for unicellular and filamentous sulfur-oxidizing bacteria, which can oxidize the sulfide first to elemental sulfur, S°, and then to sulfate. Sulfate is used by many plants and microorganisms as a sulfur source for formation of sulhydryl groups in amino acids. This process is referred to as assimilatory sulfate reduction. Organic sulfur is converted to H_2S by desulfuration. Phototrophic purple and green sulfur bacteria use sulfide as an electron donor anaerobically in photosynthesis and produce elemental sulfur and sulfate from this process. Sulfate is used by sulfate-reducing bacteria as an electron acceptor in dissimilatory sulfate reduction. Elemental sulfur can also be used by some bacteria in anaerobic respiration.

ultimate reduction of carbon dioxide for the synthesis of organic materials. Included in this group of organisms are the purple sulfur and green sulfur bacteria. These organisms oxidize the sulfide to elemental sulfur and ultimately to sulfate (see Chapter 21).

In addition, certain nonphotosynthetic bacteria also oxidize reduced sulfur compounds. Some of these are chemosynthetic and therefore use the reduced sulfur compounds as energy sources and inorganic carbon as a carbon source for growth. Others are heterotrophic. For the

***Table* 24.7 Bacterial groups responsible for the oxidation of reduced sulfur compounds**

Eubacteria	Oxidative Activity
Photosynthetic bacteria	
Purple sulfur	$H_2S \rightarrow S^o \rightarrow SO_4^{2-}$
Green sulfur	$H_2S \rightarrow S^o \rightarrow SO_4^{2-}$
Selected cyanobacteria	$H_2S \rightarrow S^o$
Chemosynthetic bacteria	
Filamentous sulfur oxidizers (e.g., *Beggiatoa*)	$H_2S \rightarrow S^o \rightarrow SO_4^{2-}$
Unicellular sulfur oxidizers (e.g., *Thiobacillus, Microspira*)	$H_2S \rightarrow S^o \rightarrow SO_4^{2-}$
Heterotrophic bacteria	
Filamentous sulfur oxidizers (e.g., *Beggiatoa*)	$H_2S \rightarrow S^o \rightarrow SO_4^{2-}$
Unicellular sulfur oxidizers (e.g., some *Pseudomonas* spp.)	$H_2S \rightarrow S^o \rightarrow SO_4^{2-}$
Archaea	
e.g., *Acidianus, Sulfolobus*	$H_2S \rightarrow S^o \rightarrow SO_4^{2-}$

most part these are obligately aerobic organisms, although a few can use nitrate as an electron acceptor in nitrate respiration. Both filamentous and unicellular Eubacteria are involved in this process as well as some acidophilic, thermophilic Archaea.

Sulfur Reduction

The groups of sulfur-reducing bacteria are less well studied. The best-studied group are sulfate-reducing Eubacteria, which obtain carbon from organic compounds and use sulfate as an electron acceptor in sulfate respiration. Some of these also use hydrogen gas as an energy source and grow autotrophically by carbon dioxide fixation. This process of sulfate reduction is referred to as **dissimilatory sulfate reduction** to distinguish it from **assimilatory sulfate reduction**, the process by which plants, algae, and many aerobic bacteria obtain sulfur for synthesis of amino acids. The dissimilatory sulfate reduction process (see Chapter 19) requires large amounts of sulfate and occurs in anaerobic muds. It is the sulfate-reducing bacteria that produce the sulfide smell typical of black muds from lake and marine sediments.

A number of extreme thermophilic Archaea such as *Pyrodictium* spp. carry out anaerobic respiration in which sulfur, not sulfate, is used as an electron acceptor.

Other Cycles

Microorganisms play important roles in many other elemental cycles. For example, all of the metallic elements have their own cycle, and bacteria are involved in all of them. More common microbial transformations occur with iron, manganese, calcium, mercury, zinc, cobalt, and the other metals. Ionized forms of the heavy metals precipitate protein and other macromolecules, explaining their toxicity. Many bacteria carry plasmids having enzymes that carry out the oxidation/reduction or other transformations of these heavy metal ions, thereby detoxifying them. However, other bacteria can mobilize heavy metals. For example, some methanogens, in the presence of mercury, produce dimethyl mercury, which is volatile. In general, however, the metal cycles do not have an atmospheric stage, and are not nearly as mobile as nitrogen and sulfur.

Iron and manganese oxides can serve as electron acceptors in anaerobic systems. These elements are accordingly reduced to their Fe^{2+} and Mn^{2+} states, which are soluble and therefore more mobile than their oxidized forms. These transformations are especially important in sediments. Because iron is needed by almost all organisms, its availability is restricted due to its insolubility under most aerobic conditions. Thus, bacteria have developed special iron transport compounds termed siderophores to bring Fe^{3+} into the cells (see Figure 5.7).

Synecology, Part II: Assessment of Microbial Biomass and Activities

One of the most important areas of microbial ecology is measuring the rates of microbial processes. Knowing this allows scientists to assess the significance of the contribution of the microorganisms of interest. However, first of all, it is essential to know whether or not a process is carried out by microorganisms. Therefore, following a discussion of biomass, this topic will be treated and it will be followed by a discussion of the measurement of typical processes in the environment.

Microbial Biomass

In the same sense that autecologists are interested in knowing the numbers or quantities of a particular species in the habitat, the synecologist is interested in knowing the quantity of all microbial cytoplasm. This is the **microbial biomass,** or *concentration of living microbial cytoplasm* in the environment. A variety of techniques have been used for this, all of which have limitations. Because of its widespread use, one of the procedures, the **ATP biomass procedure,** is discussed here. ATP is found in the cytoplasm of all living organisms but not in dead organisms. Thus, the approach used in this procedure is to concentrate organisms from a specific area or volume of the environment and extract the ATP, the amount of which is indicative of the quantity of living cytoplasm. Therefore, if the microorganisms, or the bacteria in particular, can be separated from other organisms in a natural sample, the ATP concentration can be determined, and the microbial or bacterial biomass, respectively, can be estimated. To determine bacterial biomass, the larger microorganisms such as phytoplankton and protozoa are removed from a water sample by plankton nets or filtration and the bacteria are captured on small pore-size (0.2 μm diameter) filters. ATP is extracted from the harvested cells on the filter and the ATP content is determined with an ATP Photometer. The firefly lantern luciferin/luciferase reaction is used to quantify the ATP by measuring the light emitted by the reaction, the amount of which is directly proportional to the ATP content and, therefore, the amount of bacterial biomass in the sample.

Although this is a straightforward analysis, note that it offers only an approximation of the number of living cells. Different species have different ATP contents, and, furthermore, cells of a given species will contain different concentrations of ATP depending upon how metabolically active the cells are. Thus, as with viable counting and other procedures, there are limitations. Nonetheless, this assay can provide important information to the microbial ecologist (see **Box 24.3**).

BOX 24.3 RESEARCH HIGHLIGHTS

Effect of Mount Saint Helens' Eruption on Lakes in the Blast Zone

On May 18, 1980, Mount Saint Helens, a volcano in Washington State, erupted. The cataclysmic event devastated the forest in the blast zone area north of the mountain. All of the trees and other vegetation in the blast zone were killed due to the heat and force of the blast, the ensuing mud slides, and the fall of ashes and other pyroclastic materials.

Ecologists set out to determine the effect of the eruption on lakes in the blast zone on the north side of the volcano. Although it was difficult to obtain samples immediately following the eruption, after a short time helicopters were flown in and small-volume samples were collected that could be used for analysis of total and viable counts of bacteria, as well as analysis of chlorophyll *a* and ATP. Fortunately, some samples had been obtained from Spirit Lake prior to the eruption so that the chlorophyll *a* and ATP levels could be compared with those collected after the eruption (Table 1).

Curiously, those results indicated that, following the eruption, there was a decrease in chlorophyll *a*, but an increase in ATP. This was surprising because, generally, the most significant biomass in a lake is due to algae, and yet, in this instance, the concentration of chlorophyll *a* actually decreased. Thus, the increase in ATP could not be explained by an increase in phytoplankton growth.

The explanation for the increase in biomass, as indicated by the increase in ATP concentration, was that there was tremendous growth of bacteria. This result was first indicated because the initial plates used for viable counting of the bacteria showed there were much higher concentrations of bacteria in the blast zone lakes north of the volcano than were found in control lakes located south of the volcano (Merrill Lake and Blue Lake), which were not affected by the blast (Figure c). In fact, there were so many bacteria that higher dilutions were needed to accurately count them. The concentrations of viable heterotrophic bacteria in blast zone lakes north of the volcano (for example, Coldwater Lake), where the effects of the eruption were most severe, increased dramatically to over 10^6/ml in the first summer, and this pattern continued into the next year after the eruption (Table 2). In contrast, the control lakes south of the mountain (Merrill Lake) were unaffected by the eruption with a concentration of about 10^3 viable cells per ml (compare these numbers with Lake Washington and the pulp mill oxidation lagoon in Table 24.1). It was also interesting to note that not only did the viable counts of bacteria increase in the blast zone lakes, but they approached the total microscopic cell count (acridine orange counts), indicating that the plating procedures for viable counts were able to recover most of the bacteria in the blast zone lakes. As mentioned in the text, the phenomenon of higher recoveries of bacteria ($>$ 1 percent of total microscopic count) is found in lakes that are eutrophic (high in nutrient concentration) but not oligotrophic (low in nutrients). This indicated that the lakes in the Mount Saint Helens blast zone, which had previously been oligotrophic, became eutrophic due to the effects of the eruption. Indeed, the lakes were converted from olig-

***Table* 1. Characteristics of Spirit Lake prior to and following 18 May 1980 volcanic eruption**

Feature	4 April 1980	15 July 1980	Enrichment
Chlorophyll *a* (μg/l)	2.5	0.3	33-fold
ATP (μg/l)	0.25	4.3	17-fold
DOC (μg/l)	0.83	39.9	48-fold
Dissolved Oxygen (mg/l)	Unknown	2.35[1]	
Temperature	15°C	22–24°C[2]	

[1]Although oxygen was not measured on 4 April, it would have been expected to be saturated at 4°C and therefore greater than 12 mg/l.

[2]A temperature of 35°C was actually measured in Spirit Lake on 19 May, the day after the eruption. On 15 July the temperature was still much higher than it would have been normally at this time of year due to hot water entering the lake from the volcano. (Data from Wissmar, R. C., A. H. Devol, J. T. Staley, and J. R. Sedell. 1982. Biological Responses of Lakes in the Mount St. Helens Blast Zone. *Science,* 216:178–181.)

Table 2. **Bacterial concentrations (per ml) and recovery from a blast zone lake compared to a control lake following Mount Saint Helens' eruption**[1]

Date (mo/day)	Control Lake			Blast Zone Lake		
1980–1981	Merrill		%	Coldwater		%
	Viable	*Total*	*Recovery*	*Viable*	*Total*	*Recovery*
6/30/80	1.0×10^3	5.4×10^5	0.19	$> 10^4$	7.4×10^6	—
9/11/80	2.0×10^2	1.4×10^6	0.01	2.5×10^6	6.4×10^6	39.1
4/30/81	n. d.[2]			2.2×10^6	2.7×10^6	81.5
6/29/81	n. d.			1.8×10^6	2.5×10^6	72.0

[1]Viable counts were on CPS (caseinate, peptone, starch) plates, and total counts were determined by acridine orange epifluorescence counting.

[2]Not determined.

Data from J. T. Staley, L. G. Lehmicke, F. E. Palmer, R. W. Peet, and R. C. Wissmar [*Appl. Environ. Microbiol.* 43:664–670 (1982)].

otrophic mountain lakes to eutrophic lakes following the eruption (Table 2).

So, why were there such high concentrations of bacteria in the lakes? The dramatic increase in the concentration of heterotrophic bacteria in the blast zone lakes occurred because of the tremendous increase in nonliving organic matter (dead vegetation and animals) that became available when the plants and animals were killed following the eruption. The organic matter leached from the watersheds into the rivers and lakes making the lakes naturally eutrophic and the bacteria thrived.

(a)

(b)

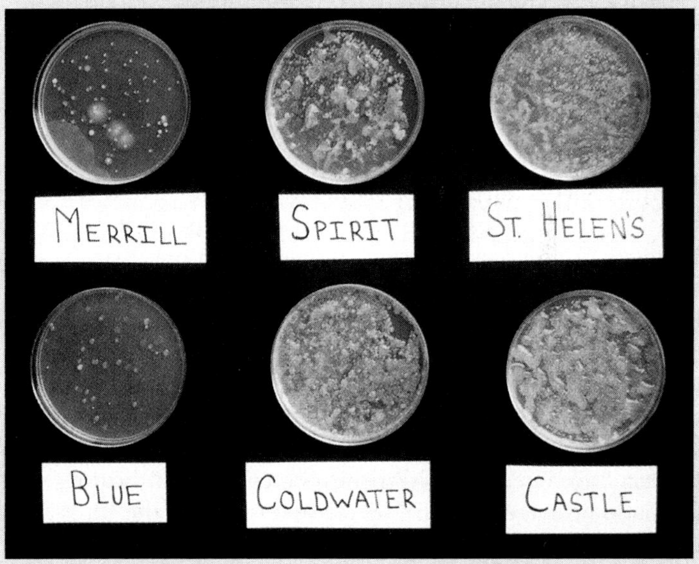

(c)

Photographs of Spirit Lake north of Mount Saint Helens **(a)** before and **(b)** after the eruption of the volcano. Note the dramatic impact the eruption had on the biological communities. (**a,** Courtesy R. C. Wissmar **b,** J. T. Staley). **(c)** All plates received 0.1 ml water samples taken from lakes near Mount Saint Helens shortly after the eruption. Note that all lakes in the blast zone showed overgrowth of bacteria (Spirit, Saint Helens, Coldwater, and Castle) compared to the control lakes (Blue and Merrill). (**c**, Courtesy of J. T. Staley)

Is It a Microbial or Chemical Process?

When examining a transformation in nature, the first question the microbiologist asks is, "Is this transformation a chemical or biological process?" It is not always clear whether a particular activity is performed by bacteria and other microorganisms or by a chemical mechanism. Consider, for example, the weathering of rocks, buildings, and monuments. It is well known that a variety of microorganisms including algae, bacteria, fungi, and lichens grow on the surfaces of rocks and can be involved in weathering of the surfaces. They do this by producing organic acids and chelating agents that gradually dissolve away minerals. However, the chemical effects of acid rain and the physical effects of freezing and thawing are examples of nonbiological processes that also cause weathering. For this reason, it is not always easy to determine if a specific rock is being eroded by microorganisms or by physical and chemical processes. Furthermore, both chemical and biological processes often work in concert with one another in effecting a transformation. And because these processes are so slow, they may be difficult to study in the field.

An early example in which a process was identified as a biological process is nitrification, which was studied by Schloesing and Müntz (Chapter 21). They added soil to a cylinder with a drain and trickled sewage effluent containing nitrogen in the form of ammonia through the column. In a few days the effluent from the column contained nitrate, not ammonia. This indicated to them that the process of nitrification was occurring in the soil in the column (see previous discussion on nitrification). To determine whether this process was a biological process, they first added boiling water to the soil column. They noted that the process abruptly ceased. They reasoned that this must therefore be a biological process because, if it were chemical, the increased heat would speed up the process, not stop it. They then treated another nitrifying column with chloroform, which is toxic to most organisms. Again the process stopped. Since chloroform would not be expected to inhibit most chemical processes, they concluded that this result also supported the hypothesis that this was a biologically mediated process. Subsequently, nitrifying bacteria have been isolated from such columns indicating that biological agents are indeed responsible for the transformation.

As discussed previously, Koch's postulates have been used in microbial ecology to verify that a microorganism is responsible for the process. This and similar approaches have been used for many years to establish whether natural processes are mediated by microorganisms or are chemical transformations. It is increasingly found that microorganisms are responsible for many reactions that were previously thought to occur only by chemical means. However, for processes such as some types of rock weathering that proceed very slowly in nature, it is not always easy to demonstrate whether the process is caused by biological or chemical and physical agents. Nonetheless, it has been well established that microorganisms play an important role in rock weathering processes.

Stable isotopes can also provide important information about whether a process is biological because of isotope discrimination. For example, consider the sulfur cycle. There are nine isotopes of sulfur, four of which are stable (nonradioactive): the normal isotope ^{32}S, and three ^{33}S, ^{34}S, and ^{36}S; the natural abundance of each in Earth's crust is respectively, 95.0, 0.76, 4.22, and 0.014 percent. Unlike chemical processes, biological processes exhibit **isotope discrimination,** that is, organisms preferentially use lighter isotopes. For example, sulfate-reducing bacteria preferentially reduce $^{32}SO_4^{-2}$ rather than the heavier isotopic form of the ion, $^{34}SO_4^{-2}$. As a result, they produce a sulfide that is lighter (contains more ^{32}S than ^{34}S) than the substrate. From this information, it has been concluded that elemental S (S^o) deposits on Earth that have a lighter sulfur isotope content than that of Earth's crust are derived from biological sources. Presumably they are the consequence of an active sulfur cycle in which some sulfur-producing bacteria such as the photosynthetic sulfur bacteria used biogenic sulfide formed by sulfate reducers to produce a lighter sulfur deposit than would be expected if a chemical process were involved.

Likewise, the sulfur cycle itself has been dated to about 3.0 to 3.5 Ga ago on the basis of observing lighter than expected sulfur compounds in sedimentary deposits. In sedimentary deposits older than 3.0 to 3.5 Ga, however, the isotope ratios ($^{34}S/^{32}S$) of the sulfur-containing minerals are the same as that of Earth's crust, indicating that the complete sulfur cycle was not yet functioning on a global scale prior to that time.

Rate Measurements

Ecologists are interested in determining how quickly transformations occur in nature, and to do this, they must measure the rates of the reactions. This may be accomplished in several ways, such as direct chemical measurements of the substrates and products, or by more sophisticated procedures using isotopes. The following section describes some of the various techniques for rate measurements and their applications.

Direct Chemical Analyses

In some cases, rate measurements are rather straightforward and can be determined by simply analyzing for the changing concentrations of chemicals (substrates and products). An example of this is the measurement of primary productivity by measurement of the oxygen evolved during photosynthesis. As discussed earlier, the overall reaction for primary production follows:

(a)

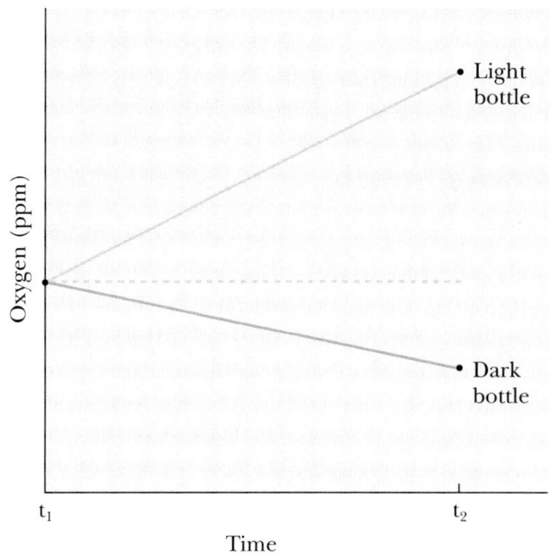

(b)

***Figure* 24.21** The measurement of primary production in aquatic habitats by oxygen production. **(a)** Water samples are collected for each depth in which photosynthesis is being measured. These are placed into three glass-stoppered bottles which are typically returned to the water column for incubation. One bottle is the experimental bottle and is incubated in the light unamended. The other is fixed by use of Lugol's iodine solution to serve as a control. The third bottle is incubated in the dark to control against nonphotosynthetic primary production. These are then incubated for 24 hours. Oxygen is measured at zero time and after the incubation period. **(b)** This graph illustrates the change in concentration of oxygen from time one, t_1, to a subsequent time, t_2, usually 24 hours later to measure primary production. (See text for details.) The measurement of primary production using the $^{14}CO_2$ tracer procedure uses the same types of bottles as for the oxygen procedure. However, prior to incubation a trace amount of radiolabelled $NaH^{14}CO_3$ is added to initiate the reaction. For the analysis of primary production, the amount of label taken into the cells is measured using a scintillation counter. Therefore, after incubation, the cells must be concentrated on a filter before placing them in the scintillation vial. Primary production is determined as the amount of carbon dioxide taken up by the algae in the experimental bottle compared with that taken up in the dark bottle.

$$6\ CO_2 + 6\ H_2O \longrightarrow (C_6H_{12}O_6)_n + 6\ O_2$$

where $(C_6H_{12}O_6)_n$ = organic carbon formed.

Thus, the amount of carbon dioxide fixed (that is, the primary production) is directly related to the oxygen evolved on a mole to mole basis. Therefore the rate of primary production can be determined simply by measuring the amount of oxygen produced for a 24-hour period of time during photosynthesis. In aquatic habitats, this is accomplished in glass-stoppered bottles containing water collected from a lake or marine habitat. Three bottles are used at each depth (Figure 24.21**a**). One bottle is called the "dark bottle" because it is made opaque to prevent light from entering it. Although photosynthesis cannot occur in this bottle, respiration, which is the reverse of the photosynthesis reaction, does occur. The second bottle, called the "fixed control" is fixed with formalin at zero time so that no biological activity can occur in it. The third bottle, called the "light bottle" or experimental bottle, is allowed to incubate and photosynthesize as it would in nature except that it is now occurring in an enclosed container. These bottles are incubated usually by submerging at the depth of collection preferably for an entire 24-hour period. The reaction is then stopped by addition of formalin.

Primary production is determined from measurements of the concentration of oxygen in the different bottles (Figure 24.21**b**). At zero time the oxygen concentration would be expected to be the same in all bottles. After a 24-hour incubation period, the oxygen concentration would be expected to be highest in the experimental bottle in which photosynthesis occurs. The difference in oxygen concentration in this bottle during the time of incubation is a measure of the **net primary production** (NP) that occurred in the volume of water tested and is calculated as follows:

$$\text{Net primary production/day} = [O_2]_{L2} - [O_2]_{L1}$$

where $[O_2]_{L1}$ is the moles of oxygen per liter at zero time in the light bottle and $[O_2]_{L2}$ is the moles of oxygen per liter after the incubation period in the light bottle.

The **gross primary production** (GP) is directly related to the net primary production plus that amount of oxygen that was used in respiration (Figure 24.21**b**). This latter value can be calculated as the amount of oxygen used during incubation in the dark bottle where only respiration could occur. So, the gross photosynthesis is:

$$\text{Gross primary production/day} = \text{net primary production} - [O_2]_{D2}$$

The oxygen concentration in the fixed control should not have changed during the incubation. If it did, this would indicate either a leak in the bottle or some other problem with the assay that would indicate to the researchers that the experiment was flawed.

Use of Chemical Isotopes

Chemical isotopes can also be used for rate measurements (see Chapter 3). The major advantage of using them is the extreme sensitivity that they provide for rate analysis. Two types of isotopes are available, depending upon the substrate. As mentioned above, stable isotopes are not radioactive, but their atomic weight differs from that of the common form of the element. **Radioisotopes** exhibit radioactive decay. These isotopes occur normally in nature, but they are found in reduced amounts compared to the normal or common isotope. Both stable and radioactive isotopes are used to determine the rates of ecological transformations. Whether a stable or radioactive isotope is used is determined by several factors such as the equipment available for analysis and the half-life (the length of time required for half of the radioactivity to be lost or decay) of the radioisotope available. For example, for carbon isotope analysis, most scientists determining rates of processes use carbon-14 (^{14}C) because it can be simply quantitated using a scintillation counter, and it has a long half-life (over 5000 years). On the other hand, ^{13}C, a stable isotope, requires that a mass spectrometer be available for analysis, and this is much more expensive than a scintillation counter and the analysis is much more labor intensive. However, with nitrogen, the choice is simple: ^{15}N is used even though a mass spectrometer is required, because the half-life of ^{13}N is only a few minutes, making it impracticable for virtually all ecological experiments.

PRIMARY PRODUCTIVITY The most common way to measure primary productivity is by the use of isotopic procedures, namely using ^{14}C-labelled sodium bicarbonate. The utility of this can be seen by referring to the reaction given previously for primary productivity. So, rather than measuring oxygen production, which is an indirect measure of primary production, one measures carbon dioxide fixation into biomass using ^{14}C-labelled bicarbonate. As mentioned earlier, this radioisotopic method is much more sensitive than ordinary chemical methods and is therefore the preferred technique.

The experiment is set up in the same manner as the oxygen experiment—that is, with both light and dark bottles used as well as a fixed control (Figure 24.21). These are then incubated, this time with a small amount **(trace)** of the ^{14}C-labelled bicarbonate. The amount of bicarbonate added is small relative to the total amount already in the water, which is measured chemically. Therefore, the concentration of the substrate in the sample is unaffected by the addition. This is called the **tracer approach,** because the **pool size** or natural substrate concentration is not affected by the amount of radiolabel added because so little is added relative to the total amount of natural substrate. By following the fate of the radiolabel, the biological transformation can be assessed.

In this procedure, the amount of carbon dioxide fixed is determined by the amount of radiolabel taken into the cells during photosynthesis. Thus, upon completion of the incubation, the algae are harvested from the experimental and control bottles by filtration and the amount of radiolabel CO_2 incorporated into their cells is determined by scintillation counting of the filters. The net photosynthesis is the difference between the amount of radiolabel taken up in the experimental bottle less the amount taken up in the dark bottle. The fixed control is used to assess whether chemical effects such as adsorption might have caused artifacts in the experiment.

The photosynthetic rate is calculated as follows:

$$\text{Photosynthetic rate (mg c}/\ell/\text{h)} = \frac{\text{cpm}_L - \text{cpm}_D}{\text{CPM}_{\text{added}}} \times \frac{\text{Total } ^{12}\text{C(mg}/\ell) \times 1.06}{\text{Incubation time (hours)}}$$

where cpm_L and cpm_D equal the counts per minute found on filters in the light bottle and dark bottle, respectively, and CPM_{added} is the amount of radiolabel added to each bottle in the form of ^{14}C-bicarbonate; total ^{12}C is the amount of carbon in the bicarbonate pool in the habitat; 1.06 is the isotope discrimination factor.

A major assumption of these tracer techniques is that the labelled substrate acts just as the normal substrate would for carbon dioxide fixation. In the case of carbon dioxide, the ^{14}C label is not only radioactive, but it is heavier than the naturally occurring ^{12}C. This heavier ^{14}C is discriminated against by phytoplankton by a factor of 6 percent. Therefore the factor 1.06 is multiplied by the entire photosynthetic rate to determine the actual rate and correct for this factor. Isotope discrimination is largely a problem with small molecules and ions such as carbon dioxide and sulfate and some of the low molecular weight organic compounds. As discussed previously, isotope discrimination occurs with both stable as well as radioactive isotopes whose masses are different from the normal isotope.

HETEROTROPHIC ACTIVITIES The measurement of heterotrophic activities is more complicated than that of primary production. This is because in primary production, a single substrate, carbon dioxide, is being metabolized by primary producers, whereas for heterotrophic activities, there are hundreds of different organic substances that serve as substrates in the environment. Furthermore, some of these are soluble (DOC) and others are insoluble (POC). Nonetheless, radioisotopes are available for many DOC compounds such as sugars, amino acids, and organic acids, and either tritium (^3H)-labelled or ^{14}C-labelled forms can be used.

The general reaction that occurs aerobically in respiration is shown as follows:

$$(CH_2O)_n + O_2 \longrightarrow CO_2 + H_2O$$

where $(CH_2O)_n$ represents organic carbon.

Thus, overall, this is the reverse reaction of photosynthesis. One advantage of using ^{14}C-labelled organic compounds is that it is possible to collect the carbon dioxide that is respired and quantitate it. This is not possible if the tritiated form of the organic substrate is used. Furthermore, the radioactive water formed with tritium is difficult to separate from the soluble organic substrate and products produced in the reaction. Therefore, ^{14}C-labelled substrates are most commonly used except for microautoradiography (see Figures 24.3 and 24.4).

Another major advantage of using isotopes is that it is possible to determine rates in systems in which a **steady state** or **approximate steady state** occurs. This is a condition in which the concentrations of substrates and products remain constant from one time to the next. This commonly occurs for extended periods of time in nature because the rate of formation of a product is equivalent to the rate at which it is removed or metabolized. Therefore, it is meaningless to measure the chemical concentration of the substrate and product at time zero and some time later, because they will be approximately the same, and if they are somewhat different it is due to the combined activity of formation and utilization. Nonetheless, utilization is occurring in the natural habitat. By adding a radiolabelled substrate to the natural substrate pool as a tracer, it is possible to then identify (label) that part of the substrate pool and follow it in the formation of the product.

It is this tracer approach that is most commonly used in measuring the rate of the heterotrophic activity or metabolism of dissolved organic compounds in environments. This is usually accomplished with short-term incubations using sealed flasks (Figure 24.22). If the actual concentration of the substrate in the natural habitat can be measured, then the reaction can be quantitated by labelling the pool with a tracer label of that organic compound. By knowing the natural concentrations of the substrate, (C_o), the amount that is radiolabelled at zero time, (cpm_s), and the amount of radiolabel in the product (or that which is taken up by the metabolizing bacteria plus the amount respired to form CO_2), cpm_p, it is possible to calculate the total amount of substrate that has been incorporated into the product, X, during a certain period of time (μg/ℓ/hr) by a simple proportion:

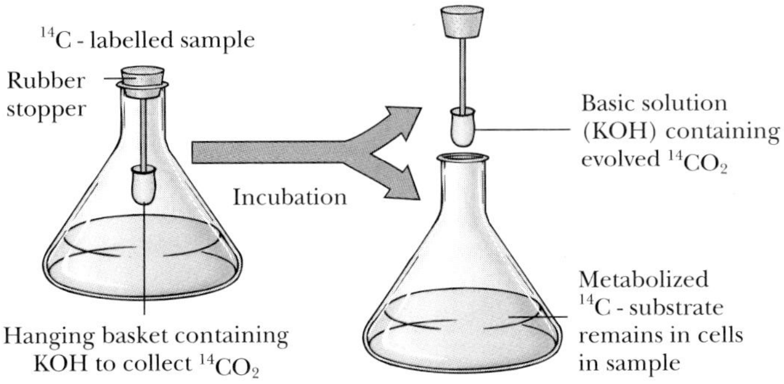

***Figure* 24.22** An incubation vessel used for assessing rates of dissolved organic carbon utilization by microorganisms in aquatic habitats using the tracer label approach. Suspended from the rubber stopper is a basket containing a basic solution (KOH). To start the reaction, ^{14}C-labelled organic compound is added to the bottle. After incubation, ideally for only 15 to 30 minutes, the $^{14}CO_2$ evolved by mineralization can be collected in the basket in a basic solution. Sulfuric acid is added to drive the labelled CO_2 into the basket. Most of the other metabolized ^{14}C-substrate, which can be collected on a membrane filter and assessed by scintillation counting, will be found in the cells.

$$C_o/cpm_s = X/cpm_p$$

where C_o is the concentration of the substrate in μg/ℓ/; cpm_s is the counts per minute of substrate; cpm_p is the counts per minute of the product; and X is the amount of substrate taken up by metabolizing cells plus the amount of CO_2 respired during the time of incubation.

In these experiments, some of the substrate will be converted into carbon dioxide, which is collected in a separate vial in the experimental vessel (Figure 24.22). Most of the remaining radiolabel in the product will have been incorporated into cells. This is determined by harvesting cell material on a membrane filter and determining total radioactivity along with that of the mineralized carbon dioxide using a scintillation counter. It is the sum of the ^{14}C–CO_2 and ^{14}C-cellular material that is determined to provide the value for cpm_p.

In addition to the aspect of the decay of radioisotopes, there is the question of **specific activity** of the radiolabelled compound. This refers to how radioactive (cpm/μg) or "hot" the substrate is. Because dissolved organic compounds occur at extremely low concentrations in natural environments, if we are to use the tracer approach, it is important to add small amounts of the radiolabelled substrate to the pool so that the substrate concentration is unaffected by the added radiolabel. This requires that the radiolabelled substrate received from the manufacturer contain as many labelled molecules as possible. Otherwise, the natural substrate will be diluted too much to provide a sufficiently accurate and sensitive assay.

Radioisotopic methods can also be complemented with autoradiography, as discussed earlier in this chapter. This permits the scientist to associate the uptake of a specific compound with a specific microorganism. Even though the organism might not be identifiable morphologically, it can be rendered identifiable by labelling it with a fluorescent antiserum or labelled nucleic acid probe and then examining it by both immunofluorescence and by light microscopy.

Alternate Substrates

For some activities it is possible to use an alternate substrate in place of the natural substrate for determining the reaction rate. Perhaps the best example of this is the use of acetylene in place of nitrogen for the measurement of nitrogen fixation. The overall reaction for nitrogen fixation involves three reductive steps in which N_2 ($N \equiv N$) is reduced first to $HN = NH$, then to $H_2N–NH_2$, and then ultimately to 2 NH_3. Acetylene, which is also a triple-bonded compound, occupies the active site of nitrogenase and is reduced to ethylene as follows:

$$HC \equiv CH \longrightarrow H_2C = CH_2$$

Remarkably, acetylene is actually preferred to N_2 by nitrogenase, so the reaction can be carried out even in the presence of the natural substrate, N_2. The main advantage in measuring nitrogen fixation by this method is that both acetylene and ethylene can be easily quantitated using a gas chromatograph, whereas the only nitrogen isotope that is usable is the stable isotope $^{15}N_2$ and its analysis is tedious and requires a mass spectrometer.

However, there are drawbacks to all of these types of assays. The disadvantage here is that the actual rate of the reaction is influenced by the substrate provided. Furthermore, the acetylene is reduced by only one reductive step rather than the three required for complete nitrogen fixation. For these reasons it is not possible to directly determine in situ rates by these procedures. Nonetheless, they provide excellent relative measures of nitrogen fixation in systems where nitrogen fixation is known to occur. Furthermore, assays can be standardized using $^{15}N_2$.

Inhibitors

Inhibitors are sometimes used to measure the rates of reactions. If the inhibitor specifically stops the process of interest and has no other effects, then the rate of the process can be determined by the rate at which the substrate accumulates. For example, molybdate ion is a known inhibitor of sulfate reduction. Therefore, it can be incorporated into assays to inhibit sulfate reducers. One example where this has been used is in the determination of how effectively sulfate reducers compete with methanogens for hydrogen H_2 in sediment communities. The sulfate reducers are inhibited with molybdate to assess uptake of hydrogen by methanogens in the environment. However, the use of such inhibitors may cause other effects on the community that may adversely alter the results, so they must be used with caution.

Summary

- **Microbial ecology** is the study of the interactions between microorganisms and the **abiotic** and **biotic** components of their environment. The **habitat** of the organism is where the organism lives (its "address"), and the **niche** of the organism is its role (its "profession") in the ecosystem. **Ecosystems** are physical areas on Earth, such as lakes, where organisms live. The microbial biosphere extends to areas on Earth not occupied by other organisms.

- **Total counts** of bacteria, including dead and living organisms, are usually conducted by using fluorescent dyes and a fluorescence microscope. Viable counts are made using growth media. The **bacterial enumeration anomaly** refers to the fact that the recovery of culturable bacteria from typical mesotrophic and oligotrophic aquatic and soil habitats is typically less than 1 percent of the total count.

- **Autecology** is the study of an individual species in its environment. Species can be identified in natural samples by use of fluorescent antibody techniques or by nucleic acid probes. Using these procedures it is possible to determine the spatial and temporal distribution of a species in its environment. The activity of an individual microorganism can be assessed in the environment by microautoradiography.

- **Biomarkers** are substances that are uniquely associated with a species or other group of organisms and can therefore be used for their identification.

- Microorganisms inhabit all habitats known including freshwaters, marine waters, soils and rocks. These may be aerobic or anaerobic, acidic or alkaline, cold or hot. The primary role of microorganisms is that of **decomposition** or **mineralization,** which is the breakdown of organic materials to simpler organics or inorganic materials. However, some microorganisms, such as the algae and cyanobacteria, are **primary producers** and others, primarily the protozoa, are **consumers.**

- Microorganisms are readily dispersed from one habitat on Earth to another. Therefore, it has been postulated that all species are **cosmopolitan** in their **biogeography** or distribution. However, this hypothesis has not been rigorously tested.

- **Synecology** is the study of natural communities of organisms. Microorganisms participate in all **biogeochemical cycles** where they often play unique roles. For example, nitrogen fixation, nitrification, and aerobic and anaerobic sulfide oxidation are examples of unique microbial processes.

- The **biomass** of microorganisms is the sum total of all living microbial protoplasm in a habitat at a given time. ATP measurements are often used to assess biomass in microbial communities.

- The rates of microbial processes can be measured by direct chemical analyses or by use of radioisotopes. Radioisotopic measurements of rates typically use the **tracer approach,** in which the substrate **pool** is labelled by small amounts of the radioisotope. After a short period of time, the amount of label that leaves the substrate pool or enters into the product pool is measured to provide an assessment of the rate of activity.

Questions for Thought and Review

1. What is meant by biomass and biomarkers, and how are they used in microbial ecology?
2. How would you go about determining the total and viable numbers of nitrifying bacteria in a lake? Do you believe probes would be of any use?
3. Discuss how the tracer approach is used to measure rates of microbial processes in nature.
4. Why does the hypolimnion of a eutrophic lake often become anaerobic during the summer months? Does this affect the temperature of the hypolimnion?
5. If you were counting only the total and viable heterotrophic bacteria in a mesotrophic lake and saw that the viable numbers increased a hundredfold in a matter of two weeks, what would you believe was the cause?
6. Define succession and provide an example of it. Some successions are seasonal and are repeated each year. Why?
7. In what ways are microorganisms important to the atmosphere? To the greenhouse gases? To the ozone layer?
8. How can microorganisms live on rocks?
9. Compare the sulfur cycle and the nitrogen cycle. Which do you believe is more important to the biosphere? Why?
10. What is the acetylene reduction technique, and what is its use?

Suggested Readings

Atlas, R. M., and R. Bartha. 1993. *Microbial Ecology.* 3rd ed. Redwood City, CA: The Benjamin Cummings Publishing Co., Inc.

Global Biogeochemical Cycles. 1992. (S. S. Butcher, R. J. Charlson, G. H. Orians, and G. V. Wolfe, eds.) London: Academic Press, Ltd.

Microorganisms in Action: Concepts and Applications in Microbial Ecology. 1988. (J. M. Lynch and J. E. Hobbie, eds.). Oxford: Blackwell Scientific Publications.

In symbiosis two dissimilar organisms become intimately associated. This type of vital relationship was apparently first observed by Reinke in 1872, who named the phenomenon consortism. In 1879, De Bary observed and recorded the same phenomenon, and introduced the term symbiosis, which has since been adopted into common usage.

I. E. Wallin, *Symbionticism and the Origin of the Species*, 1927

Chapter 25

Beneficial Symbiotic Associations

Functions of Symbiotic Relationships
Establishment of Symbioses
Types of Symbioses
Chemosynthetic Symbioses
Plant–Microbe Symbioses
Animal Symbioses
Syntrophy
Commensalism

Evolution of eukaryotic life forms began some one billion years ago in a biosphere dominated by the Eubacteria and Archaea. Throughout evolution, the prokaryotes had competed with one another, and as eukaryotes emerged these newcomers had to compete for space and food and to cope with microbial predation; survivors were those with the capacity to coexist in the environment. Selective pressure obviously favored those species that adapted most effectively. Not only did the evolving prokaryotes and eukaryotes find a way to survive in the presence of other species but many formed beneficial interspecies alliances. These alliances are a biological phenomenon termed **symbiosis**, which is the living together of two dissimilar organisms in a more or less intimate association (see **Box 25.1**). These interactions are outlined in Table 25.1.

This chapter will be devoted to the symbiotic associations involving microorganisms where the interaction is mostly beneficial to one or both of the participants. Parasitism, antagonism (amensalism), and competition will not be discussed. Harmful interactions between humans and other species will be discussed in Chapters 26, 29, and 30. Generally, the smaller organism in a symbiotic association is considered the **symbiont**, the larger the **host**. Many of the symbiotic interactions are of critical importance to life on Earth. For example, the microbiota in the digestive system of mammals plays an indispensable role in the health and survival of the animal. The gut flora of termites and other insects is instrumental in the digestion of complex food sources. Some plant/microbe interactions result in fixation of atmospheric nitrogen, while others provide nutrients for both partners. And a distinct biosphere of

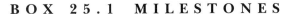

BOX 25.1 MILESTONES

The Eukaryotic Cell—A Sophisticated Symbiotic Association

There is considerable speculation regarding the origins of the composite eukaryotic cell. The progenitor eukaryotic cell arose some one billion years ago in a world teeming with prokaryotes of endless diversity that included aerobes, anaerobes, thermophiles, autotrophs, heterotrophs, photosynthetics, and so forth. About the only thing that these prokaryotes had in common was small size, a cytoplasmic membrane, ribosomes, and DNA that floats free in the cytoplasm.

How did evolution proceed from the boundless diversity among the prokaryotes to animal-type cells that are all functionally much the same? The animal-type cell requires preformed organic food, it respires for energy generation, and DNA reproduces by mitosis. Compared with bacterial cells these cells are dull and unexciting. What event(s) in the origins of the animal-type cell resulted in such conformity?

The signal event in the evolution of the progenitor eukaryotic cell occurred when an aerobic bacterium took up residence within the confines of a **thermoplasma**-like bacterium. Whether this entry was by predation or engulfment is not known. This captive organism was undoubtedly the progenitor of the mitochondrion. Mitochondria in eukaryotic cells actually retain many of the characteristics of their former free-living status. They have their own genetic apparatus, and enclosed in the mitochondrion membrane is DNA, mRNA, tRNA, and ribosomes similar to those in bacteria. As with bacterial DNA the mitochondrion DNA is not arranged in chromosomes nor is it associated with histones. Mitochondria divide by pinching off to form two mitochondria, and genetic transfer in the DNA is similar to that in bacteria. What the putative aerobic bacterium ⟶ mitochondria did give to eukaryotes is their remarkably similar metabolism. Virtually all eukaryotes respire, and their metabolism is geared to accommodate this fact. In this symbiotic association, the host cell provides nutrient, and the mitochondrion symbiont provides energy—resulting in the survival of both. The primitive eukaryotic cell further evolved with more internal structure, and some 700 million years ago the typical animal cell evolved.

How and when did the prototype plant cell emerge? The characteristic plant cell came into being about 460 million years ago. The ancestor of the chloroplast present in plants was probably a prochloron. Prochlorons are photosynthetic prokaryotes that possess chlorophyll *a* and *b* as do green plants. The thylakoids in chloroplasts are similar to those in the prochlorons. Cyanobacteria contain chlorophyll *a* and phycobilin in place of chlorophyll *b*. The pigments present in prochlorons and their arrangement is a major factor in considering these organisms rather than cyanobacteria as the progenitor of chloroplasts.

What is a prochloron? Prochlorons are a group of unicellular or filamentous microbes that perform oxygenic photosynthesis. They are either free living or associated with invertebrate hosts. Attempts to culture these organisms have not generally been successful.

There are a number of theories on the evolution of the eukaryotic cell, and many have merit. As we learn more about cells—both prokaryotic and eukaryotic—our ability to define the evolutionary events that led to the present-day living world will also move forward.

Table 25.1 **Interactions that can occur between species in a symbiotic association**

	Species Interacting	
	A	*B*
Mutualism	+	+
Syntrophy	+	+
Commensalism	+	0
Parasitism	+	−
Antagonism (amensalism)	−	0
Competition	−	−

+ organism gains; 0 not affected; − organism harmed.

Table 25.2 **Some defined beneficial symbiotic relationships between microorganisms and eukaryotes**

Plant/Microbe	Symbiotic Partners
Rhizosphere	Bacteria–plant roots
Mycorrhizae	Fungi–plant roots
Actinorrhizae	Actinomycetes–plant roots
Lichens	Fungi–cyanobacteria or algae
Symbiotic N-fixers	Rhizobium–legume
	Azolla (waterfern)–cyanobacteria
Animal/Microbe	**Symbiotic Partners**
Forage-eating animals (cattle, sheep, deer)	Animal–rumen bacteria and protozoa
Wood-eating animals	Termites–bacteria and protozoa
Birds	Leaf eaters–bacteria
	Wax eaters–bacteria

interacting organisms, dependent on energy emanating from the Earth rather than from the Sun, exists in the depth of the ocean.

In this chapter we discuss some of the described beneficial relationships between species such as chemosynthetic symbioses, plant-microbe symbioses, animal symbioses, syntrophy, and commensalism (Table 25.2). This discussion is by no means complete. While our understanding of the activities of individual components of ecosystems is considerable, our understanding of species interactions is still in its infancy.

Functions of Symbiotic Relationships

Every symbiotic association is unique, and the benefits or harm that may accrue to each of the symbionts is likely as varied as there are such relationships. Symbionts often provide for the partner some physical or chemical element that a **free living** organism would normally gain from its environment. To ascertain the role of members in a symbiotic relationship it is helpful if one separates the partners and determines the needs of each. This is impractical in many cases because some symbionts do not survive without the presence of a partner(s). Organisms in these associations are termed **obligate symbionts.** Among those that can be separated it is not certain that, when cultivated in axenic culture, the symbiont and host perform as they would in a symbiotic association.

There are four functions generally attributed to beneficial symbiotic associations and these are:

1. Protection
2. Provision of a favorable position
3. A recognition device
4. Nutritional benefits

This chapter will present evidence that in many symbiotic associations there can be considerable overlap between these functions.

Protection Microbes in nature are quite often subject to adverse physical conditions. Availability of water, pH, and temperature extremes are among the physical conditions microbes encounter. An endosymbiont residing within a cell can be protected from adverse environmental conditions. The host cell can prevent desiccation and variances in osmotic pressure, prevent predation, and in a warm-blooded animal, extremes of temperature. These symbionts in turn might protect the host from invasion by pathogens (see Chapter 26).

Favorable position Photosynthetic organisms need access to light, and there are a number of examples that could be cited where a symbiotic association fills this need. For example, a lichen is composed of a fungus and a cyanobacterium or eukaryotic alga. The fungus adheres to stable structures such as rocks, and the photosynthetic symbiont entrapped by the fungus is thereby favorably exposed to the sun.

Recognition devices Many marine invertebrates and fish have bioluminescent bacteria on their surface or as endosymbionts in special organs. These bioluminescent bacteria emit light that is involved in schooling, mating, or attraction of prey.

Nutrition The involvement of one symbiont or partner in providing nutrient for the other partner is common in favorable symbiotic associations. For example, the symbiotic nitrogen-fixing bacteria that infect alder, legumes, and other plants provide the eukaryote with nitrogen in exchange for a place to grow and reproduce.

Establishment of Symbioses

The evolution of symbiotic relationships occurred through a progressively greater interdependence of two different species. As the interdependence became pronounced it was essential that mechanisms evolve that would ensure continuity of both partners from generation to generation. There are two mechanisms for ensuring that each generation obtains the requisite symbionts: (1) the host transmits symbionts directly to progeny at each generation, or (2) each new generation must establish interactions within the environment.

Direct transmission occurs in many endosymbiotic associations. For example, the algal symbionts that are associated with protozoa would have no invasive ability and would have difficulty in gaining access to the protozoan host. To ensure that each of the protozoan progeny receives its algal symbiont, both the host and symbiont divide at about the same rate. The host cell becomes host to two symbionts and on division of the protozoa the daughter cells each receive an algal cell. The daughter cell can control division of the symbiont and, when two are present, the daughter cell may then divide. This assures perpetuation of this favorable symbiotic interaction.

Sexually reproducing animals can transmit symbionts by infection of the egg cytoplasm. Some insects carry microbial symbionts in specialized cells called **mycetocytes.** These mycetocytes are maintained in organs called **mycetomes.** Symbionts released from mycetomes are passively transported to ovaries by the lymph system and are taken up by the egg. In some insects the intestine is linked through a duct to the vagina of the female. Intestinal symbionts move through this duct and coat eggs as they pass through. When larvae hatch they feed on the eggshell, thus infecting themselves with the intestinal symbionts.

The normal flora of mammals is generally obtained through close association with adults. In most mammals the intestinal tract is sterile at birth and gains its microbiota by ingestion.

Types of Symbioses

Symbiotic interactions are generally beneficial for at least one of the partners. In fact, the terms "mutualism" and "symbiosis" are sometimes thought to be synonymous, but there are two general types of symbioses: **mutualism** in which both partners benefit from the association; and **commensalism** in which one partner benefits from the association and the other partner is neither harmed nor benefited. **Parasitism** is an association where one partner benefits and the other is harmed, and these interactions will not be discussed. As previously mentioned, the smaller organism in a symbiotic association is considered the **symbiont** and the larger one the **host.** An **ectosymbiont** lives outside but in close proximity to the host, whereas an **endosymbiont** exists within the cells of the host. A nutritional cooperation where two organisms combine to synthesize a required growth factor or to catabolize a substrate unavailable to either alone is termed **syntrophy.** A **consortium** is a stable culture composed of two or more component organisms. Generally, consortia are loose-knit associations less defined than symbiotic or syntrophic interactions.

Chemosynthetic Symbioses

Chemosynthetic symbioses occur in environments that are dependent on an energy source other than sunlight. In these environments, energy sources are provided from the depths of the Earth. One such environment has been discovered in the deep sea. Light does not penetrate deeper than 100 meters into the ocean, so that most of the ocean is permanently dark. We know that more than 75 percent of our planet's biosphere lies below 1000 m of seawater. At this depth it is obvious that no photosynthesis can occur. The temperature ranges from 2° to 4°C. The photosynthetic organic material produced by phytoplankton in the oceans' surface waters is largely recycled in the upper 300 m layer, and only a fraction, about 5 percent, reaches deeper waters in the form of sinking particulate matter. About 1 percent is estimated to reach the deep-sea floor at 3000 to 4000 m. This limited food supply controls the scarce animal populations and their life strategies in this desert-like deep-sea environment.

It was, therefore, very surprising when in 1977, during oceanographic studies on plate tectonics and related volcanic activities on the sea floor, dense and thriving populations of novel marine invertebrates were discovered at depths from 1000 to 3700 m with biomass orders of magnitude higher than possibly supported by a photosynthetic food supply. How was this possible? These complex animal communities were found near "hydrothermal vents," which emit hot and highly reduced water. Some of its major chemical constituents, H_2S, H_2, and CH_4, serve as reductants for chemosynthetic growth of microorganisms. A variety of feeding mechanisms among the various planktonic or sessile animals make use of this unusual microbial base of the food chain at these ecosystems. However, the largest portion of the biomass appears to be produced by a new form of symbiosis between chemolithoautotrophic bacteria and certain marine invertebrates. Although detailed studies on living specimens of these

***Figure* 25.1** The mussels *(Bathymodiolus thermophilus)* that are present at hydrothermal vent sites of the 9° N East Pacific Rise at a depth of 2520 meters (1.58 miles). (Courtesy of H. W. Jannasch)

***Figure* 25.2** Population of the "giant" white clams *(Calyptogena magnifica)* 20 to 32 cm in length, located within cracks between boulders of "pillow lava" at the 21° N East Pacific Rise hydrothermal vent site at a depth of 2610 m. (Courtesy of H. W. Jannasch)

invertebrates have not yet been done, the physical characteristics of the animals and tissues from them indicate that they metabolize by a remarkable symbiotic association.

Symbiotic Invertebrates

Three predominant species of symbiotic invertebrates constitute the bulk of these deep-sea hydrothermal vent ecosystems: mussels *(Bathymodiolus thermophilus)*, Figure 25.1; the "giant" white clam *(Calyptogena magnifica)* specimens are up to 32 cm in length, Figure 25.2; and the vestimentiferan tube worms *(Riftia pachyptila)*, Figure 25.3. While some of the mussels are not symbiotic and appear to feed directly on bacterial suspensions within the immediate plume of warm vents (at about 25°C), the symbiotic animals also occur outside the immediate vent emissions in clear water where the concentration of particulate organic matter is much too low to serve as a food source. Subsequently, it was found via electron microscopy and enzymatic studies that the gill tissue of the bivalves contain prokaryotic endosymbionts that are involved in the production of ATP through the oxidation of sufficiently reduced inorganic sulfur compounds (hydrogen sulfide and thiosulfate) and in the reduction of CO_2 to organic carbon. Specifically indicative of microbial metabolism (Figure 25.4) is the presence of ribulose bisphosphate carboxylase (RuBisCo) and adenosine 5′-phosphosulfate reductase (APSR). While microbial symbionts may be absent from mussels or occur in variable quantity, the white clams appear to depend on their presence.

The same is true for the vestimentiferan tube worms that reach lengths of more than 2 m and can occur at densities that appear to exceed the highest concentration of biomass per area known anywhere in the biosphere. Detailed anatomical studies revealed the absence of a mouth, stomach, and intestinal tract—in short, the absence of any ingestive and digestive system. Instead, the animals contain the so-called trophosome (Figure 25.5) in their body cavity, which is a tissue consisting of coccoid prokaryotic cells interspersed with the animals' blood vessels.

A "chemoautotrophic potential" in these worms was described by listing the enzymes catalyzing the synthesis of ATP via sulfur oxidation (rhodanese, APSR, and ATP sulfurylase) as well as the Calvin cycle enzymes RuBisCo and ribulose 5′-phosphate kinase (APK). ADP sulfurylase (ADPS) and phosphenolpyruvate carboxylase (PEPC) were also present. None of these enzymes were detected

***Figure* 25.3** A population of vestimentiferan tube worms *(Riftia pachyptila)*, about 40 cm to 1 meter long, at the 21° N East Pacific Rise hydrothermal vent site at a depth of 2620 m. (Courtesy of H. W. Jannasch)

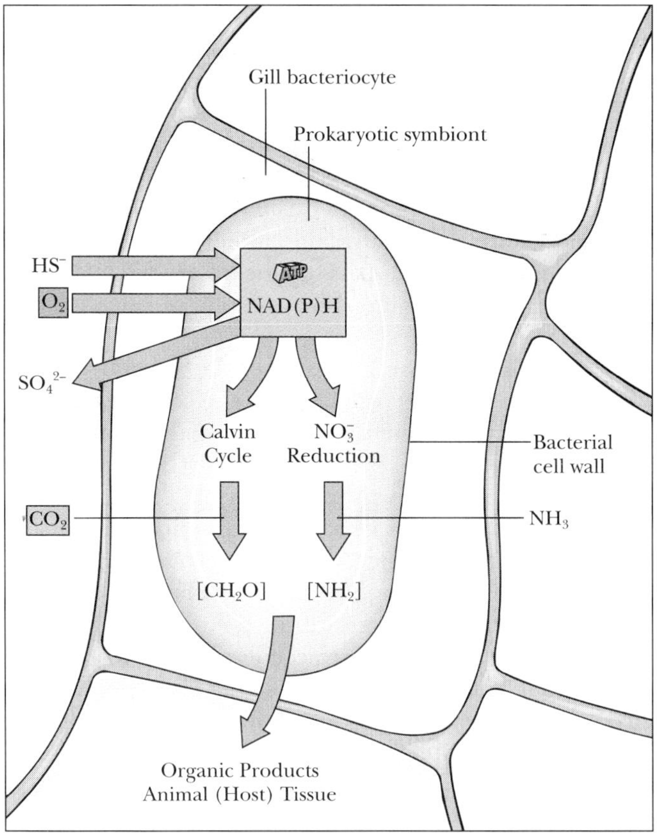

***Figure* 25.4** Schematic of metabolic processes in prokaryotic symbionts within gill "bacteriocytes" of vent bivalves; the presence of enzymes catalyzing ATP production, the Calvin cycle, and nitrate reduction have been demonstrated. (Adapted from Felbeck et al., Chapter 2 from *The Mollusca,* Academic Press 1983)

within the worm tissue proper. The trophosome may represent more than half of these worms' wet weight. The necessary simultaneous transport of oxygen and hydrogen sulfide from the retractable plume of gill tissue to the trophosome is carried out by an extracellular hemoglobin of the annelid-type blood system (Figure 25.6). The spontaneous oxidation of H_2S and the respiratory poisoning appears to be prevented by the presence of a sulfide-binding protein.

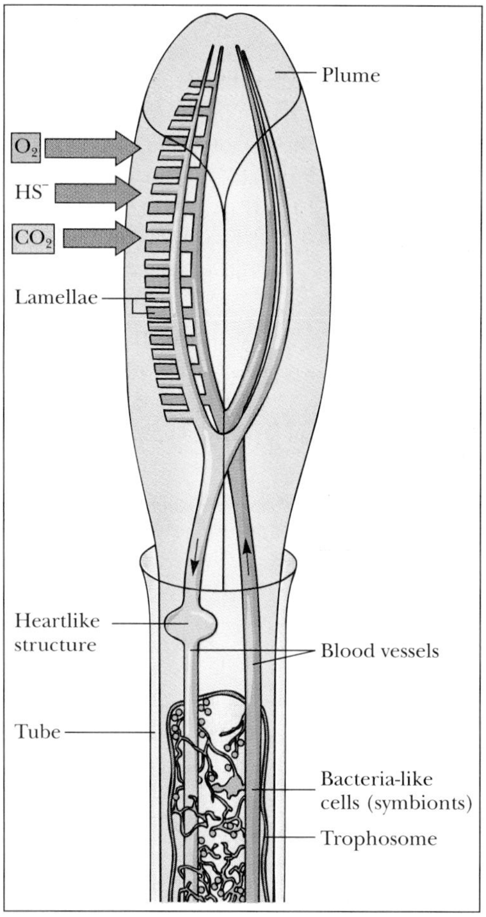

***Figure* 25.6** Scheme of the dissolved oxygen transport system in *Riftia pachyptila* from the gill "plume" to the symbiotic trophosome tissue. (Courtesy of H. W. Jannasch)

No chemolithotrophic symbiont has so far been isolated from any of the above-mentioned vent invertebrates.

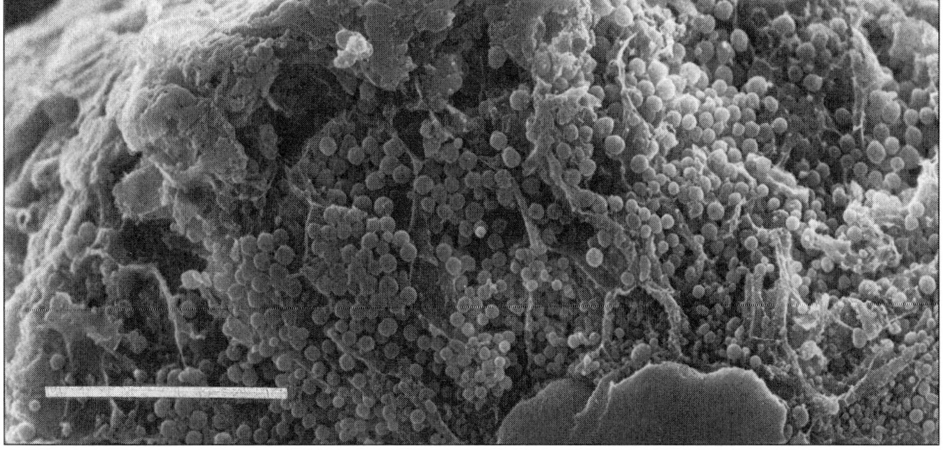

***Figure* 25.5** Trophosome tissue of *Riftia pachyptila* consisting of coccoid prokaryotic cells, 3 to 5 μm in diameter interspersed among blood vessels of the eukaryotic animal tissue. (Courtesy of H. W. Jannasch)

A study on concentrated and purified cell suspensions obtained from mussel gill tissue and from tube worm trophosomes by density gradient centrifugation revealed that the symbionts of these two types of animals are distinctly different. While the symbionts of the mussel *Bathymodiolus thermophilus* use thiosulfate as the only electron donor and are clearly psychrophilic—that is, show an optimal CO_2-uptake activity at 16°C with a maximum just beyond 22°C—the trophosome symbionts of *Riftia pachyptila* oxidize hydrogen sulfide and are active at temperatures up to 35°C (Figure 25.7). In addition to differences in cell size, the mussel symbionts use free oxygen while the worm symbionts depend on its uptake from oxidized hemoglobin.

From the analyses of 5S and 16S rRNA nucleotide sequences it became apparent that the prokaryotic symbionts of all chemolithoautotrophically existing marine invertebrates so far investigated belong in different albeit closely related groups. A predominantly symbiotic chemosynthesis, based upon the oxidations of H_2S as well as CH_4, has now also been found in anoxic marine sediments of nontectonic origin in shallow waters. The symbiotic methanotrophy has been recently found in mussels similar to *Bathymodiolus* at shallow hydrocarbon leakages in the Gulf of Mexico as well as at so-called cold seeps at the bottom of the West Florida continental slope at a depth of 3200 m. Since it cannot be defined as a chemolithotrophic symbiosis, it is mentioned here only in passing. Generally speaking, the detour via the discovery of deep-sea vents was apparently necessary to detect this new metabolic way of life. No symbiotic nitrification has yet been demonstrated.

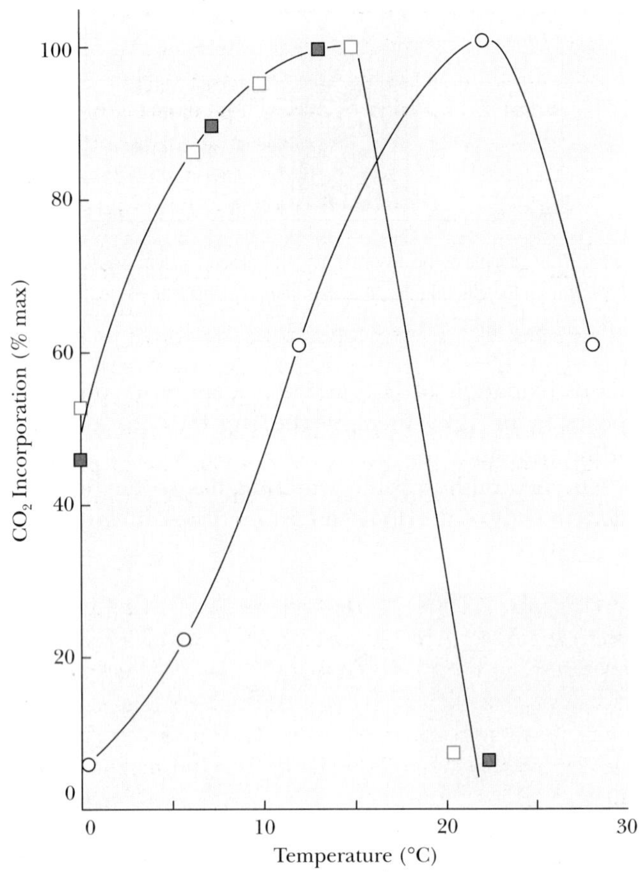

***Figure* 25.7** Incorporation of ^{14}C radiolabelled bicarbonate by homogenates of mussel *(Bathymodiolus thermophilus)* gill trophosome at various temperatures. Open squares: gill homogenate incubated with 200 μm $Na_2S_2O_3$; Filled squares: purified fraction of gill homogenate incubated with 400 μm $Na_2S_2O_3$; and Dots: trophosome homogenate incubated with 600 μm Na_2S. Maximum incorporation rates in these three experiments were 6.0, 1.0, and 3.3 nmol CO_2 mg^{-1} protein h^{-1}, respectively. (Adapted from Belkin et al., *Biological Bulletin,* 170 : 110–112. 1986)

The most surprising characteristic of the chemolithotrophic-symbiotic sustenance of the copious deep-sea communities is its efficiency. Considering the point source of the geothermally provided energy for chemoautotrophic bacterial growth in the normal food chain, filter-feeding on bacterial cells from the quickly dispersing vent plumes appears highly wasteful. This problem was overcome by transferring the chemosynthetic production of organic carbon to a site within the animal where the electron donor as well as the acceptor are made available with the aid of the respiratory system. This symbiotic association combined the metabolic versatility of the prokaryotes with the genetic and differentiative capabilities of the eukaryotes and takes advantage of a most direct and efficient transfer of the chemosynthate to the animal tissue. These transfer processes and the biochemical interactions between the microbial and animal metabolisms are presently the focus of research in this area.

In a geobiochemical context, the term "primary production" of organic carbon can be used for the aerobic chemolithotrophic fixation of CO_2 in the deep sea, keeping in mind that the necessary free oxygen is a product of photosynthesis (Figure 25.8). This qualification does not apply in the case of anaerobic chemosynthesis—for instance, methanogenesis from H_2 and CO_2. The same is true when contrasting the geothermal or "terrestrial" source of energy for the support of life at deep-sea vents with the "solar" energy for the support of life at the continental and ocean surface.

Plant–Microbe Symbioses

Various symbiotic relationships exist between plants and microbes. Symbionts may be present on a leaf surface (the phyllosphere) or in the soil surrounding the roots (rhizosphere). Microbes in such a relationship are considered ectosymbionts. Microbes such as fungi may invade roots and form a relationship known as mycorrhizas. Bacteria and actinomycetes can invade roots and grow there with the formation of nitrogen-fixing nodules.

Rhizosphere

The **rhizosphere** is the thin layer of soil remaining on plant roots after taking the plant from its environment and shak-

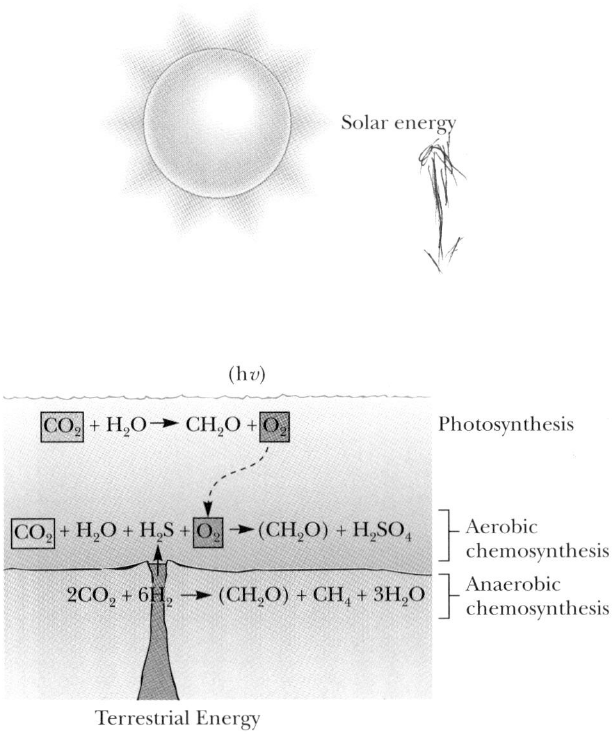

***Figure* 25.8** Scheme contrasting solar versus terrestrial generation of energy for photo- and chemosynthetically sustained ecosystems at the sea surface and in the deep sea. (Courtesy of H. W. Jannasch)

ing it. The microbial population in the rhizosphere is generally one to two orders of magnitude higher than in surrounding, root-free soil. The number of microorganisms in the rhizosphere quite frequently reaches 10^9/gram of soil. The root system of plants can be quite extensive in area. For example, a typical cereal grain root system can be 160 to 225 meters in length with an average diameter of 0.1 mm. The total root surface area would be 62.8 cm^2. This provides ample opportunity for a significant microorganism/root interaction.

Ectorhizosphere organisms are those that grow in the soil immediately surrounding the root, while *endorhizosphere* organisms penetrate the root itself where they feed directly on the plant. The *ectorhizosphere* dwellers consume root exudates, lysates, and sloughed off root cells. The total carbon released through the root system (including CO_2) can be as much as 40 percent of the total plant photosynthate. Root exudates include sugars, amino acids, vitamins, alkaloids, and phosphatides. The release of root exudates is frequently stimulated by the presence of selected microorganisms.

The rhizosphere has a higher proportion of gram-negative, nonsporulating, rod-shaped bacteria than would be present in adjacent soil. Gram-positive organisms are scarce in the rhizosphere. Ammonifying and denitrifying bacteria are present, and many of these require growth factors such as B-vitamins and amino acids, which are supplied by root exudates.

Among the direct plant/microbe interactions that have been described are the following:

1. The organic fraction of wheat and barley exudate stimulates the growth of *Azotobacter chroococcum.* The bacterium can fix nitrogen that can be utilized by the plant. Many other species of *Azotobacter* are associated with root surfaces or are present in the rhizosphere and provide plants with nitrogenous compounds.

2. Anaerobic clostridia occupy the *rhizosphere* of submerged seawater plants *Zostera* (eel grass) and *Thallasia* (turtle grass). The root exudate released by these plants supplies clostridia with a carbon and energy source, and the bacteria are nitrogen fixers.

3. *Azospirillum* sp. inhabit the cortical layer of tropical grasses and feed on plant carbon. The bacteria in turn fix abundant quantities of nitrogen for the growth of both.

4. *Desulfovibrio* are abundant around the roots of rice, cattails, and other swamp-dwelling plants. The sulfate-reducing *Desulfovibrio* generate H_2S in quantities that would harm developing plants. The H_2S-oxidizing bacterium *Beggiatoa* is also present in the *rhizosphere* of swamp-dwelling plants. Oxidation of sulfide by *Beggiatoa* under these oxygen-limiting conditions results in the production of potentially toxic levels of H_2O_2. *Beggiatoa* does not produce catalase, the enzyme that decomposes H_2O_2. A catalase-like activity surrounds the root tips and decomposes the peroxide produced by the bacterium to water and oxygen (Figure 25.9). Thus, the *Beggiatoa* removes sulfide that could limit plant growth, and the plant provides catalase to prevent peroxide autointoxication of the bacterium.

Mycorrhizae

There are fungal species that are present in nature only in association with roots. These root/fungal interactions are termed *mycorrhizae,* and most fungi involved in these associations are obligate symbionts. An estimated 80 percent of vascular plants have some type of mycorrhizal involvement and over 5000 species of fungi form mycorrhizal associations. The dramatic effect of *mycorrhizae* on the growth of citrus seedlings is illustrated in Figure 25.10. The fungi obtain nutrients from the plant and, in turn, produce plant growth substances that induce morphological alterations in roots. The affected roots are short and dichotomously branched. Mycorrhizal fungi can be grown in culture and mixed with rooting soil resulting in better growth of the plant. Two types of mycorrhizal associations have been observed: ***ectomycorrhizae*** and ***endomycorrhizae.***

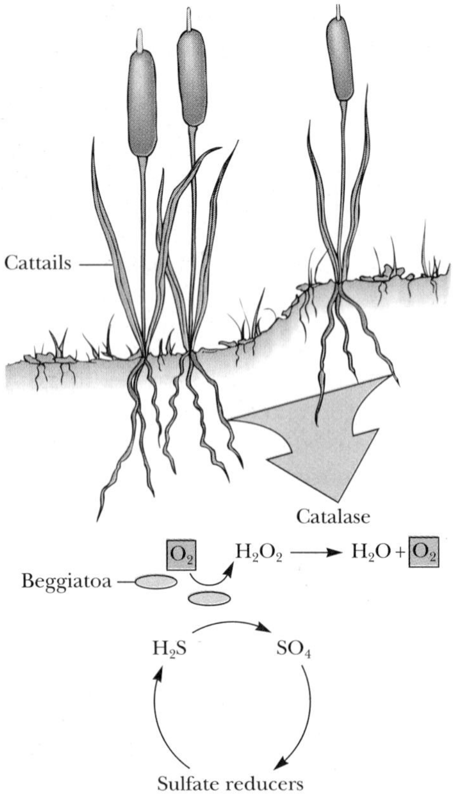

***Figure* 25.9** Growth of cattails *(Typha latifolia)* in an anaerobic swamp environment sustained by a symbiotic association with the sulfur-oxidizing bacterium *Beggiatoa* sp. In the low oxygen environment the bacterium produces toxic H_2O_2 but lacks catalase to remove the peroxide. This enzyme is supplied by the plant. Sulfide that would be toxic to the plant is converted to sulfate by the bacterium.

Ectomycorrhizae

These fungi grow as an external sheath about 40 nm thick around the root tip, and are present mainly on roots of forest trees, such as conifers and oaks. The fungi penetrate intercellular spaces of the epidermis and cortical re-

***Figure* 25.10** The effect of mycorrhizal organisms on the growth of citrus seedlings. The plants on the bottom were not inoculated with mycorrhizal organisms, whereas the two plants at the top were inoculated. (Courtesy of L. F. Grand)

***Figure* 25.11** The effect of mycorrhizal fungi on the growth of loblolly pine seedlings. The seedlings at the top **(a)** were not inoculated with mycorrhizal fungi and those on the bottom **(b)** were inoculated. (Courtesy of L. F. Grand)

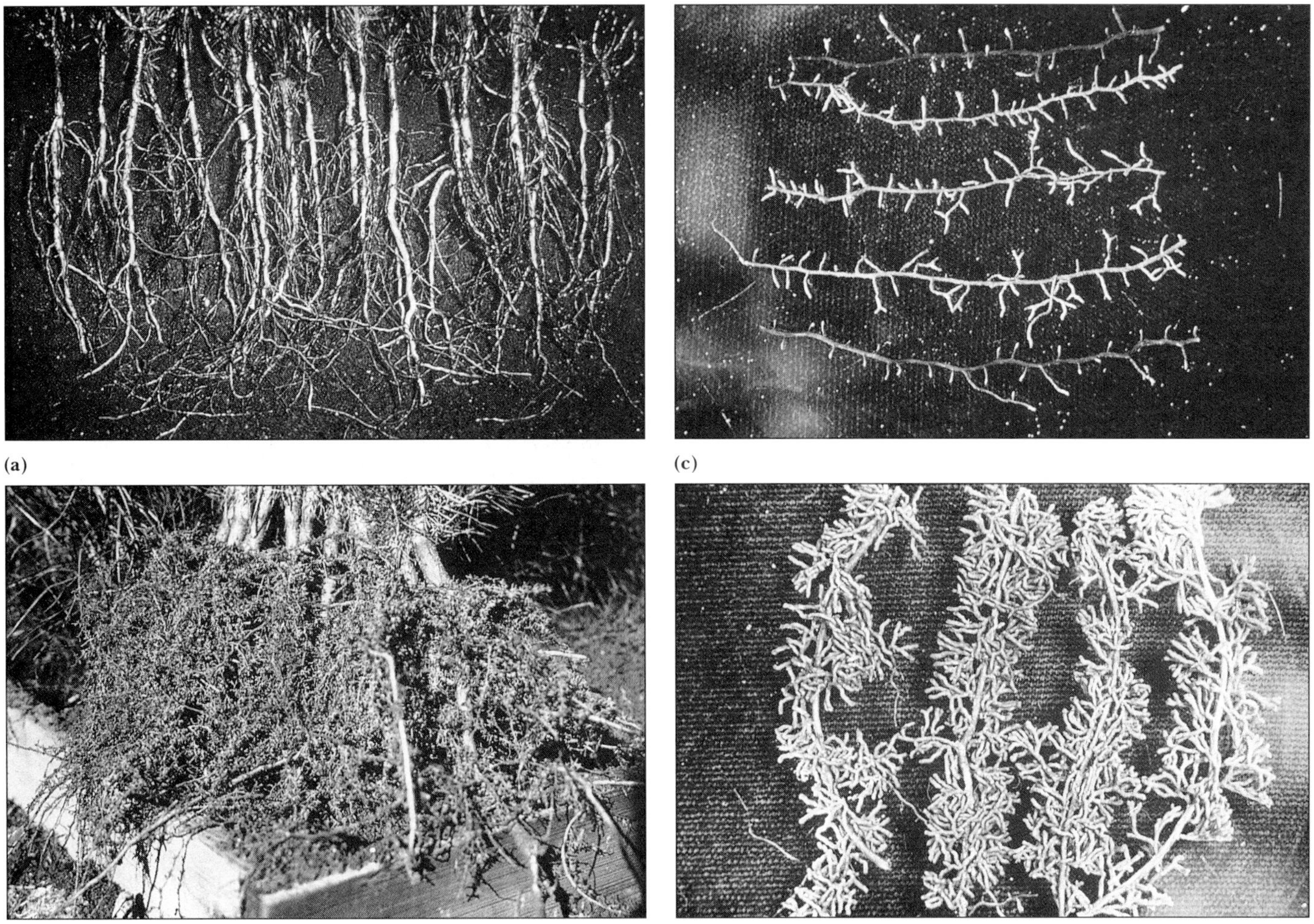

(a) (b) (c) (d)

Figure **25.12** The effect of mycorrhizal fungi on the development of loblolly pine roots: **(a)** uninoculated and **(b)** roots of an inoculated seedling. A closeup view: **(c)** uninoculated and **(d)** inoculated seedlings. (Courtesy of L. F. Grand)

gions but not into root cells. The relative growth of loblolly pine seedlings uninoculated and inoculated with the fungus *Pisolithus tinctoris* is illustrated in Figure 25.11**a,b.** The appearance of the roots of these plants is shown in Figure 25.12**a,b,c,d.** A cross section of a root tip is shown in Figure 25.13.

Endomycorrhizae

In this symbiosis there is significant invasion of cortical cells with some fungal hyphae extending outside the root. This fungal/plant relationship is present in plants such as orchids. In fact, an orchid cannot grow without these fungal associations.

Other symbiotic relationships are found mostly in wheat, corn, beans, and pasture and rangeland grasses where the fungi form **intracellular** structures called vesicles and arbuscules (Figure 25.14). This type of association is designated a vesicular-arbuscular mycorrhiza. The fungi involved in this relationship are difficult to grow separately from the plant.

A mycorrhizal relationship is beneficial to the plant in many ways. It provides longevity to feeder roots; improved nutrient absorption; selective absorption of ions; and, in some cases, resistance to plant pathogens. Plants growing in wet environments have endomycorrhizal relationships

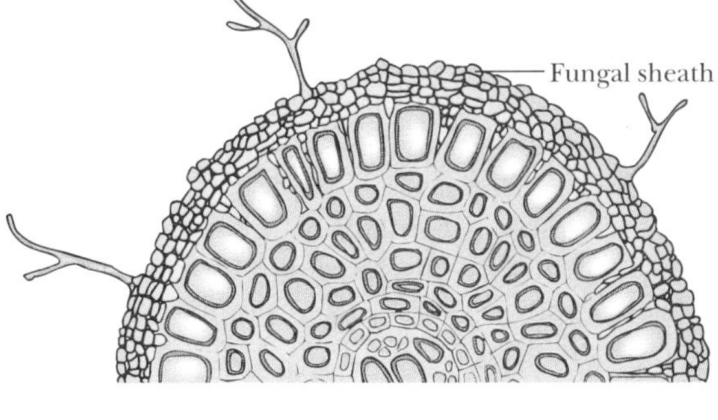

Figure **25.13** Cross-section of a root with an ectomycorrhizal relationship. The fungal hyphi appear on the outer root surface and intercellular spaces.

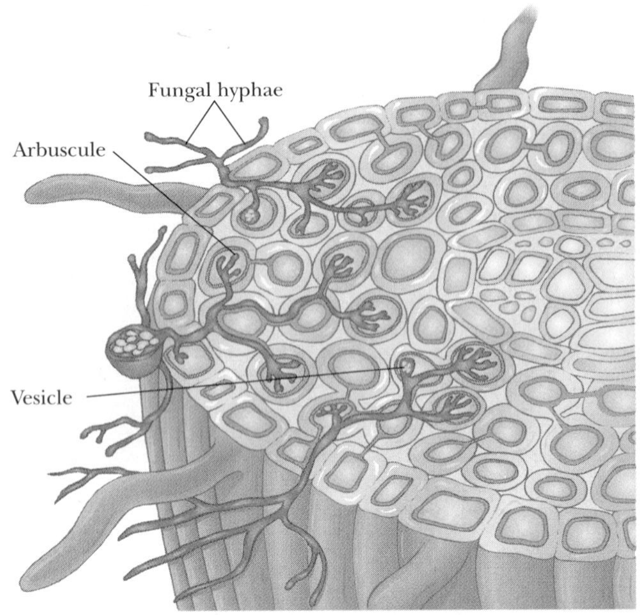

***Figure* 25.14** Cross-section of endomycorrhizal penetration of a root. Intracellular growth is evident along with "tree-like" and "vesicle-like" internal structures.

that increase availability of nutrients such as phosphorus. In arid environments the fungi aid in water uptake. As a result, mycorrhizal plants often thrive in poor soil where nonmycorrhizal plants cannot.

Actinomycetes

Pioneer plants, such as alder (Alnus) and bayberry (Myrica), grow in moist, nitrogen-poor environments. These woody plants and shrubs often grow in bogs, dredge spoil, and abandoned open pit mines. The ability to establish a symbiotic relationship with nitrogen-fixing bacteria is in part the reason these plants survive in such an inhospitable environment. The nitrogen fixers in these pioneering plants are filamentous actinomycetes in the genus *Frankia.* The symbiotic relationship between plant and *Frankia* is functionally equivalent to that present in the legume/*Rhizobium* symbiosis (see later). The striking difference is that *Rhizobium* sp. will only interact with one family, the legumes. *Frankia* will colonize the roots of plants that are quite phylogenetically unrelated. Symbiotic associations are known to occur between *Frankia* and 7 different plant orders, 8 families, and 14 genera. More will undoubtedly be discovered.

The endosymbiotic relationship between plant and *Frankia* apparently occurs through root hairs. Development of nodules leads to densely packed, coral-like branching roots that cease growth (Figure 25.15**a**). The nodules contain leghemoglobin, and they can be quite large (Figure 25.15**b**). It is only in recent years that *Frankia* strains have been isolated in culture. The organism grows

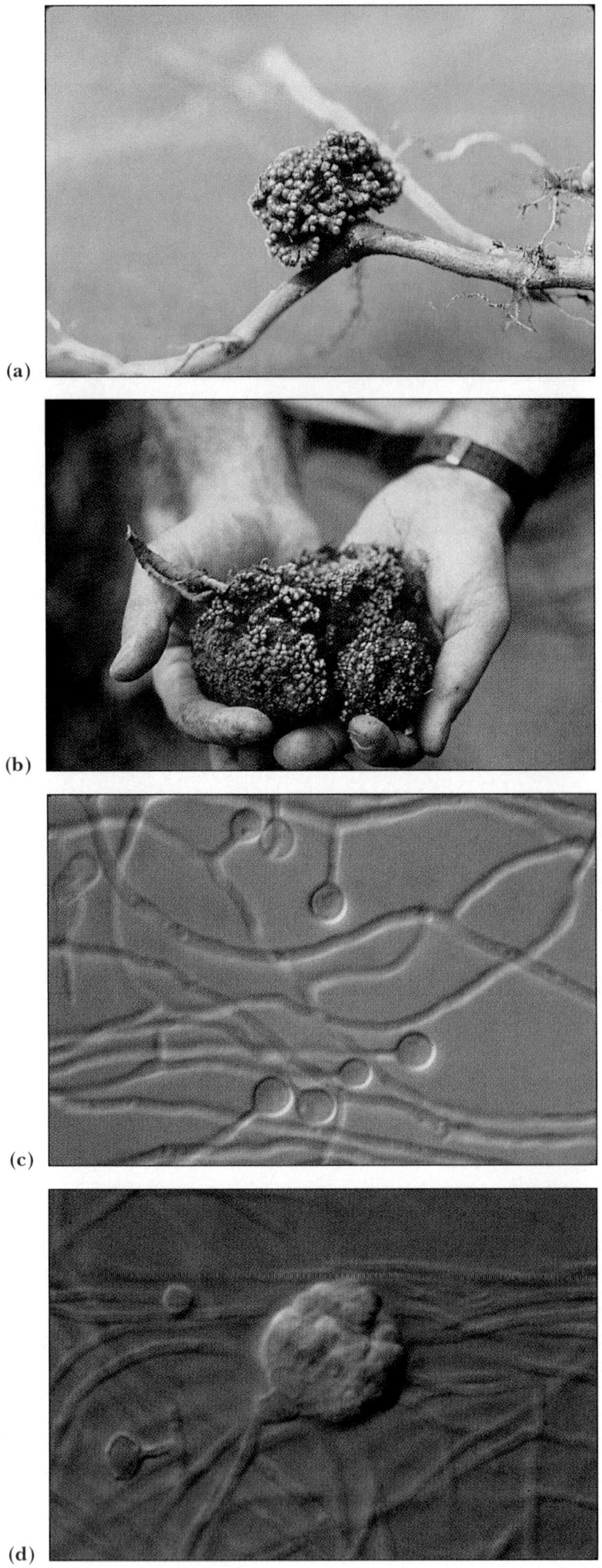

***Figure* 25.15** **(a)** A large nodule on the root of Alder *(Alnus)* caused by the symbiotic nitrogen-fixing actinomycete *Frankia*. (Courtesy of J. M. Ligon) **(b)** A large nodule containing *Frankia*. (Courtesy of J. M. Ligon) **(c)** A light microscope picture of thick-walled terminal bodies of *Frankia*. These bodies are involved in N_2 fixation. The vesicle is about 5 μm in diameter. (Courtesy of M. D. Stowers) **(d)** A reproductive body of the genus *Frankia*. It is considerably larger than a terminal body. The sporangium is about 30 μm in diameter. (Courtesy of M. D. Stowers)

slowly, and intact cells can fix nitrogen under atmospheric levels of oxygen. Cell extracts from *Frankia* are sensitive to oxygen. The development of nitrogenase activity coincides with a differentiation of terminal areas of the filaments. These thick-walled terminal bodies are designated vesicles (Figure 25.15**c**). Purified fractions from these vesicles can fix nitrogen. The thick walls of the vesicles apparently are barriers to atmospheric oxygen and protect the oxygen-sensitive nitrogenase. The reproductive spores of *Frankia* are different from, and much larger than, the vesicles (Figure 25.15**d**).

Rhizobium–Legume

Two major eubacterial genera that establish symbiotic nitrogen-fixing associations with leguminous plants are *Rhizobium* and *Bradyrhizobium.* The general properties of these organisms were presented in Chapter 19. Nitrogen fixation by these microbes occurs within nodules that develop on roots of leguminous plants (Figure 25.16**a**). The consequence of a well-established symbiotic association in a peanut plant is evident in Figure 25.16**b**. This symbiotic association is beneficial both economically and ecologically. Commercial production of ammonia for fertilizer consumes considerable quantities of natural gas. Run-off of fertilizer added to soil is a major cause of eutrophication (nutrient enrichment) of natural waters. Therefore, growing food crops without the necessity of adding nitrogen fertilizers is clearly advantageous.

(a)

(b)

***Figure* 25.16** **(a)** Nitrogen-fixing nodules resulting from invasion of a peanut plant *(Arachis hypogala)* by a member of the genus *Bradyrhizobium*. (Courtesy of T. J. Schneeweis) **(b)** The appearance of peanut plants that have been inoculated with root-nodulating rhizobial strains compared with plants that were uninoculated (light - colored in middle of picture). (Courtesy of T. J. Schneeweis)

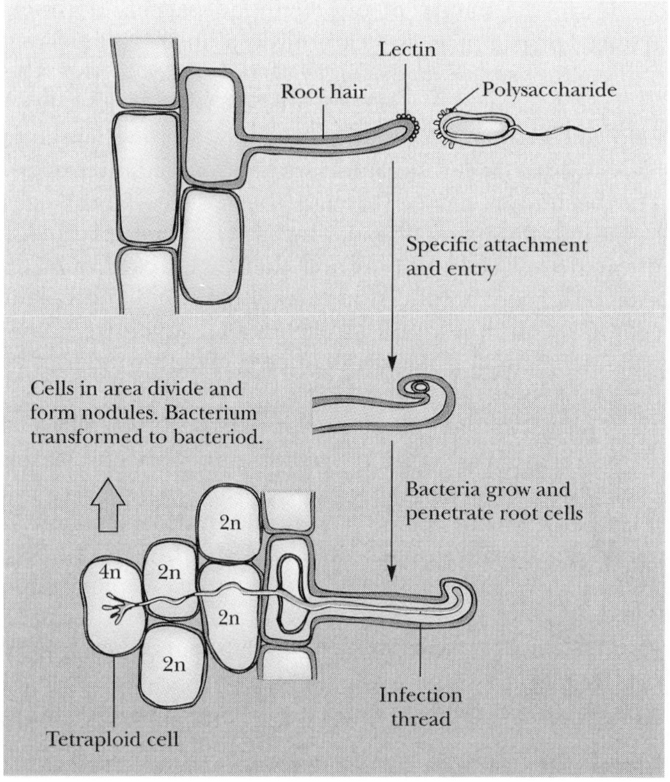

***Figure* 25.17** The attachment and invasion process involved in the development of a root nodule. The rhizobia reproduce in the tetraploid cell and can establish a mutually beneficial symbiotic association.

The development of the symbiotic association in legumes has been studied extensively, and the events leading to successful nodulation are now quite well understood. The *Rhizobium* or *Bradyrhizobium* species moves to, and specifically attaches to, the root hair of a legume within a **cross-inoculation** group (Figure 25.17). The rhizobia are specific in the plant species they can infect. The narrow range of plant species attacked by a rhizobium strain is called a cross-inoculation group. The root hair surface has a specific protein, a **lectin,** that interacts with a specific polysaccharide on the outer layer of the bacterial cell wall. Cross-inoculation group specificity apparently resides in the nature of the lectin/polysaccharide interaction. For example, the carbohydrate in the outer layer of the *R. leguminosarum* strain *trifolii* is 2-deoxyglucose. The lectin in the clover that can be infected has a marked specificity for this sugar. Following binding, the root hair curls back and the bacterium enters via an invagination process. The bacterium moves through the cortex of root cells forming an **infection thread.** The bacterium passes directly through root cells, not around them. The bacterium divides and forms an infection thread through diploid plant cells and continues until it reaches a tetraploid root cell. A limited number of tetraploid cells occur at random in the plant root.

The tetraploid cell along with adjacent diploid cells are stimulated to divide repeatedly, probably due to the production of cytokinins (plant hormones) by the bacterium. This plant cell division results in the formation of a nodule. Bacteria divide in vacuoles within the nodule and are surrounded by portions of the root cell membrane (peribacteroid membrane). Most of the bacterial population is transformed to branched, club-shaped, or spherical **bacteroids** (Figure 25.18). The irregularly shaped bacteroids have 10 to 12 times the volume of the free-living bacteria.

The bacteroids are incapable of cell division and ultimately lyse and are absorbed by the plant. When a plant dies, the nodule disintegrates and the bacteroids lyse and are mineralized by soil microbes. There are a limited number of dormant rod-shaped bacteria present in nodules, and these are the survivors that reproduce in the soil, where they can infect other legume roots in the vicinity or are available to infect plants at a later date.

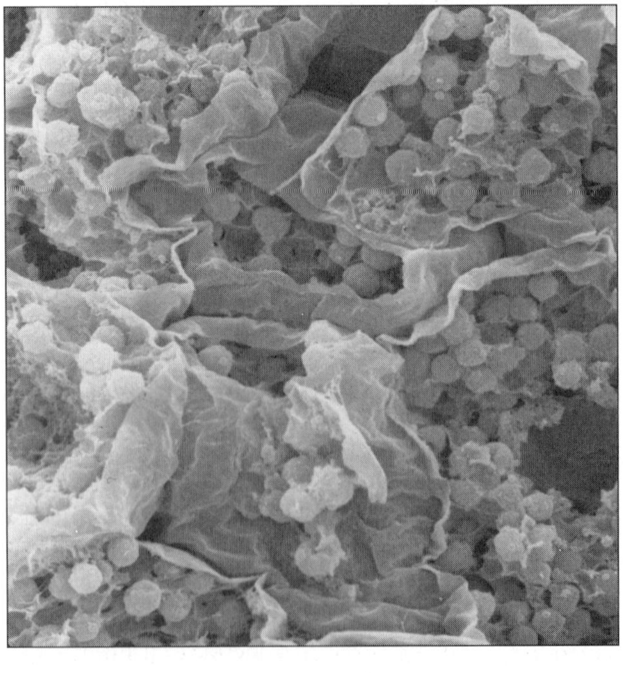

***Figure* 25.18** A scanning electron micrograph of the bacteroids present inside a nodule. (Courtesy of T. J. Schneeweis/R. Steece)

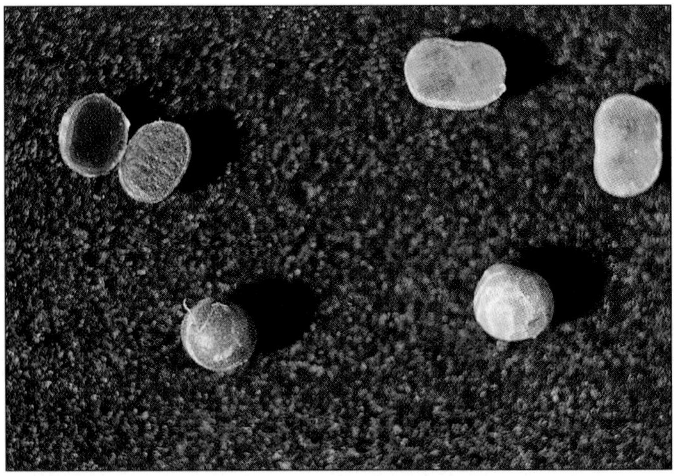

***Figure* 25.19** The appearance of an effective nitrogen-fixing nodule (right) with reddish coloration resulting from the presence of the O_2-carrying leghemoglobin. The ineffective nodule (left) has no pigment. (Courtesy of T. J. Schneeweis)

There are a number of biochemical events that occur during the infection process. Over 20 polypeptides are synthesized by either the bacterium, plant, or both, which play a role in the development of the symbiotic association. The nitrogenase and associated factors are contributed by the bacterium, as under certain conditions the free-living microorganism can fix nitrogen. The bacteria have large plasmids, bearing information that is transcribed to produce an effective nitrogen-fixing nodule. **Nodulins** are synthesized by the plant only after infection and are involved in the generation of compounds necessary for nodule formation. Leghemoglobin is synthesized and concentrated in plant cell cytoplasm, surrounding vacuoles that enclose the bacteroids. The protein portion, globin, is coded by plant DNA and the protoheme portion is coded by bacterial DNA. Leghemoglobin is an iron-containing heme quite similar to animal hemoglobin. Hemoglobins have a high affinity for oxygen and bind it tightly but reversibly. Leghemoglobin binds oxygen in the nodule and maintains an oxygen tension sufficient to allow the aerobic bacteroids to respire and generate ATP. It restricts the oxygen available to a level that does not inactivate the oxygen-sensitive nitrogenase system. The ratio of oxygen bound to leghemoglobin to free oxygen in the nodule is in the order of 10,000/1. Nodules that are effective in nitrogen fixation are red (contain leghemoglobin) in the interior (Figure 25.19). The bacteroids receive photosynthate from the plant, in the form of tricarboxylic acid cycle intermediates, which they oxidize to generate ATP. Nitrogen fixation is an energy-expensive process, as it requires 15 to 20 moles of ATP hydrolyzed per mole of nitrogen fixed.

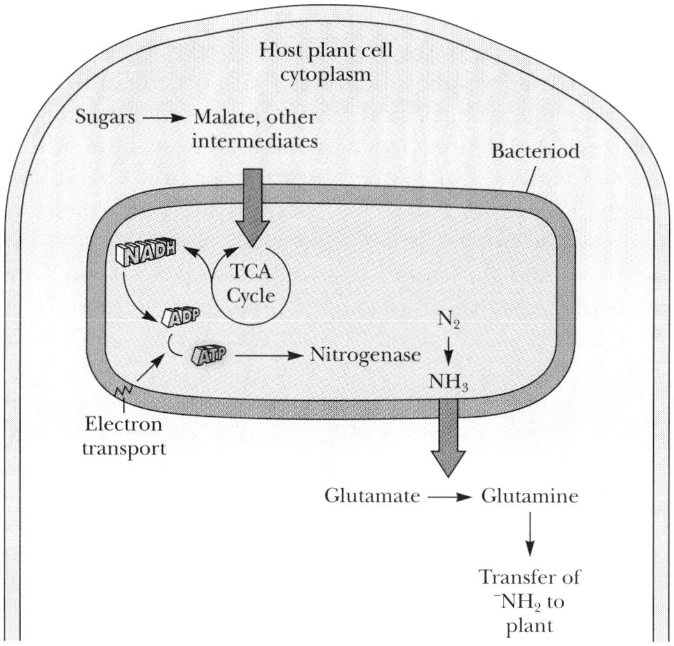

***Figure* 25.20** An overview of the biochemical events that occur in a successful symbiotic relationship. The plant provides intermediates that can be oxidized by the bacterium, providing reducing power (NADH) and ATP to drive the N_2-fixing reactions. The bacteroid provides NH_3 for growth of the plant. (See also Figure 10.13 for more detail.)

About 95 percent of the nitrogen fixed is delivered to the cytoplasm of the host cell. The overall process is diagrammed in Figure 25.20.

The number of nodules formed in an agricultural crop such as soybeans is proportional to the number of rhizobia in soil. The number of rhizobia necessary to achieve the desired level of nodulation is about 10^4 bacterial cells/gram of soil. It is a common agricultural practice to add the appropriate nodule-forming culture to seeds during planting. Nodulation can occur without the establishment of an effective symbiotic nitrogen-fixing relationship. There is considerable research underway to find bacterial strains to apply to seed to gain the maximum yield of legume.

Azorhizobium–Legume

Sesbania rostrata is a leguminous plant that grows in moist areas of Senegal and Madagascar. This plant has the potential for development of a symbiotic association with root- and/or stem-nodulating bacteria. Isolation and characterization of the rhizobial strains from nodules suggest that the stem nodulators should be placed in a separate genus *Azorhizobium*.

Under microaerophilic conditions, *Azorhizobium* can grow with gaseous nitrogen as a sole source of nitrogen. Leghemoglobin has not been observed in *Azorhizobium*. The establishment of an *Azorhizobium* nitrogen-fixing system in stems occurs by a sequence similar to that described for root nodulation. The root-nodulating strains found in *Sesbania* are of the genus *Rhizobium*, and are closely related to *Rhizobium meliloti*.

Cyanobacteria–Plant

Azolla is an aquatic fern that grows on the surface of still waters in temperate and tropical regions. It typically grows on rice paddies after the harvesting of the rice crop. An ancient practice in Southeast Asia was to allow the fern to grow on the surface of a paddy after harvest, thereby ensuring a nitrogen supply for the subsequent growing season. The success of this practice results from the ability of nitrogen-fixing cyanobacteria to grow within the tissue of the fern leaves. The cyanobacterium involved is *Anabaena azollae*, and during symbiotic growth 15 to 20 percent of the cells in a trichome are converted to heterocysts, (Figure 25.21) the anaerobic nitrogen-fixing cell. Free-living *Anabaena* generally have 5 percent of the total trichome as heterocysts. On the leaf surface of *Azolla* there are mucilage-containing cavities that, during development, are open to the outside. These later close, and the plant cells entrap filaments of *Anabaena* within the leaf structure. *Azolla* growing without symbionts require a source of fixed

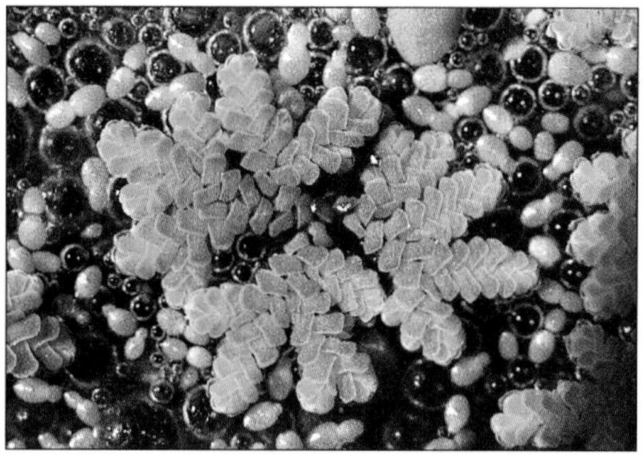

(a)

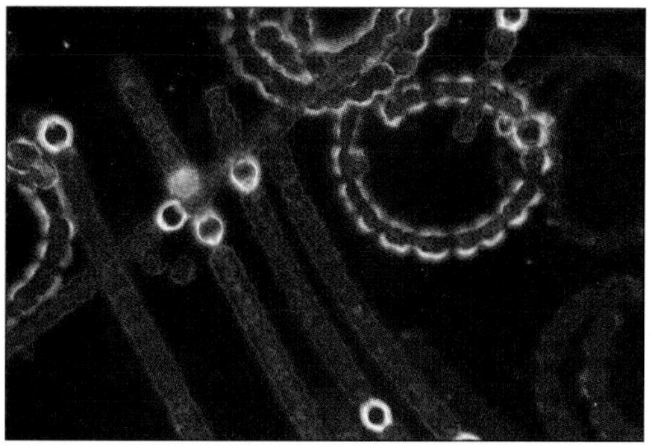

(b)

***Figure* 25.21** Fronds of the aquatic fern *Azolla* that can harbor nitrogen-fixing cyanobacteria **(a)** Trichomes of *Anabaena azollae* with round heterocysts in which nitrogen fixation occurs **(b)** (**a,** © William S. Omerod/VU and **b,** © Dennis Drenner) nitrogen. Another plant that has a symbiotic association with cyanobacteria is the tropical shrub *Gunnera macrophylla.* The symbiont in this case is *Nostoc punctiforme. N. punctiforme* occupies special glands at leaf nodes present in the lower stem. This strain of *Nostoc* also forms heterocysts. There are similar symbioses involving liverworts and cyanobacteria, and the symbionts are also of the genus *Nostoc.*

Bacteria–Plant Leaf

Bacterial populations associated with leaf surfaces (phyllosphere) often reach one hundred million per gram of fresh leaf material. Under the humid conditions of the tropics the phyllosphere is constantly moist, and this provides a habitat for a diverse population of bacteria. The species present on leaves are significantly different from those that inhabit the soil in areas adjacent to the plant. The source of nutrients for phyllosphere organisms are the organic acids and sugars that leach from the leaf. Some of the organisms present are nitrogen fixers of the following genera: *Klebsiella, Beijerinckia,* and *Azotobacter.* There is evidence that these nitrogen fixers provide the host plant with nitrogenous compounds, but the interactions involved are not yet clear.

Cyanobacteria–Fungal

Lichens are a symbiotic association readily observed on rocks, tree bark, and soil. A lichen is composed of a fungus and a cyanobacterium or eukaryotic alga. Typical lichens are depicted in Figure 25.22. The fungus and cyanobacterium or alga are so closely associated that they are considered a unitary vegetative body. The composite is classified as a distinct biological entity. The morphology and metabolic relationship in any particular lichen is so constant and reproducible that it can be assigned to both a genus and a species. There are over 20,000 species of lichen described. Most of the fungi in lichens are ascomycetes, with some basidiomycetes from tropical regions. One fungal species can have as partners many different algal types, and each would be a separate species differing in morphology and the metabolic interrelationship. It is possible to separate and grow the fungal and phototrophic partners in an appropriate medium. The phototrophic partner is often present as a free-living organism in nature and grows readily when separated. The fungal component grows slowly in axenic culture and generally requires complex carbohydrates for growth. These fungi do not produce the same type of fruiting bodies in the absence of phototrophs.

The general structure of a lichen was presented (see Figure 1.33). The fungal mycelium may penetrate into the outer layer of the algal cell without disrupting its cytoplasmic membrane. This close association is termed a fungal **haustorium.** The lower cortex forms a rootlike extension that serves as a hold-fast, allowing the lichen to inhabit surfaces such as rocks and tree bark. The photosynthetic partner supplies the fungus with organic nutrients. The fungus scavenges minerals for the association and synthesizes sugar alcohols, such as mannitol or sorbitol, which are compatible solutes that absorb water from the atmosphere. The composite lichen permits the fungus and its photosymbiont to inhabit extreme ecosystems where neither could survive on its own. Many grow at low temperatures in high altitudes and in polar environments. The so-called reindeer moss in the tundra of polar regions is actually a lichen.

Lichens are an effective measure of air pollution, as they are very sensitive to sulfur dioxide, ozone, and toxic metals. Apparently, air pollutants are absorbed by the

(a)

(b)

***Figure* 25.22** Lichens vary in their shape, color, and appearance. **(a)** Crustose lichen commonly grows on rocks and tree trunks. (© Walter H. Hodge/Peter Arnold, Inc.) **(b)** A foliose lichen appears leaflike. (© L. West 1993/Photo Researchers, Inc.)

lichen and the toxic component destroys the chlorophyll in the photosynthetic partner. Cities with air pollution problems are devoid of lichens. A lichen can be white, black, yellow, green, or various shades of red or orange. The pigments have long been extracted and employed as textile dyes. The dyes that give the distinct color in Harris tweed, for example, have traditionally come from lichens. Other compounds produced by lichens are litmus, the acid-base indicator employed in chemistry, and some of the essential oils used in perfumery.

Animal Symbioses

As we learn more about the microbiota of healthy animals we recognize that the microbial populations that are present play more than a passive role. For example, the microbes normally present in the intestinal tract of animals are indispensable to the host's well-being. Some of the beneficial roles that symbiotic microbes play in animal health have been defined and will be discussed in this section.

Autotrophic Bacteria–Protozoa

Sediments at the bottom of stagnant waters are rich in organic matter. These anaerobic sediments support the life of various ciliates, amoebae, and flagellates. These heterotrophic protozoa generate energy by oxidation of organic compounds and must have an electron sink to rid themselves of H^+. They accomplish this through the endosymbiotic methanogens that are spread throughout their cytoplasm. Under anaerobic conditions, an amoeba such as *Pelomyxa* excretes measurable quantities of CH_4. One amoeba can play host to more than 10,000 methanogens. Electron micrographs of protozoa from the surface of sediments indicate that those that harbor methanogens may lack mitochondria. Mitochondria are present in free-living protozoa and are involved in electron transfer to oxygen. As the protozoa adapted to anaerobiosis they lost this organelle and are able to transfer electrons to the endosymbiotic methanogens. Much of the methane generated in the upper layer of anaerobic sediments originates from the methanogens that inhabit protozoa, and not from free-living methanogens.

Ruminant–Symbioses

Ruminants are herbivorus animals that have a complex digestive system composed mainly of a large specialized fermentation chamber called the **rumen.** Among the grazing animals that have a rumen are domestic animals such as cattle and sheep, as well as giraffes, buffalo, and elk. The reticulum or forestomach of these animals is the first chamber of the multicompartmental digestive system (Figure 25.23).

Ruminant animals feed on grasses and other plant materials that are rich in cellulose, hemicellulose, and starch. Although these animals are not themselves capable of degrading these somewhat refractory carbon sources, their rumen harbors a complex community of anaerobic bacteria and protozoan microorganisms that can degrade them.

The cattle chew the plant materials thoroughly. They have a special small compartment called the reticulum, located in the anterior of the rumen, that is used for regurgitation. This activity is commonly referred to as "chewing the cud."

The rumen is a strictly anaerobic environment, so the polysaccharide substrates are depolymerized to the constituent sugars, and these are fermented to the final principal products: the volatile fatty acids (VFAs) including acetic, propionic, and butyric acids (Figure 25.24). These acidic products are neutralized by sodium bicarbonate produced in the saliva. The rumen has redox potential (-0.35 volt) that is low enough to permit the growth of methanogenic bacteria. These methanogens produce methane gas from the hydrogen and carbon dioxide derived from the sugar fermentations. It is estimated that a mature cow belches about 100 liters of methane gas per day (the rumen holds about 40 liters).

Due to this symbiotic relationship, the animal is able to utilize plant materials that, by themselves, are indigestible. The volatile fatty acids are absorbed through the epithelial wall of the rumen and (to a lesser extent) the omasum, which have absorptive tissues. The VFAs are taken to the liver via the blood system where they are oxidized and used as energy and as a carbon source. Ruminant animals can survive on very simple nitrogen sources

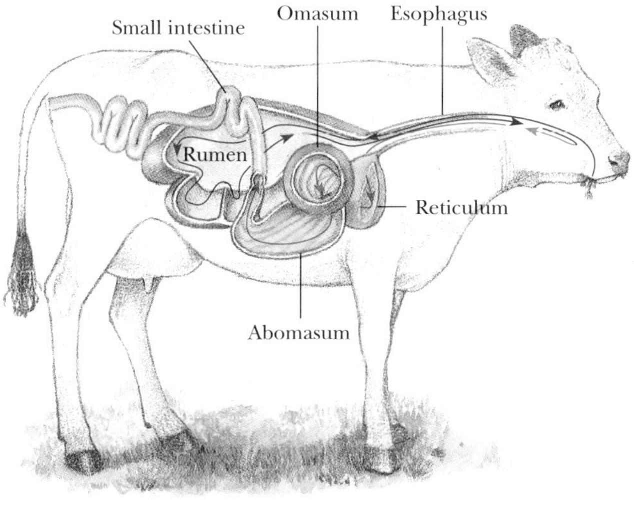

***Figure* 25.23** Diagram of the rumen and gastrointestinal system of a ruminant. The location and relative size of the rumen in a cow. The arrows indicate the direction of foodstuff flow.

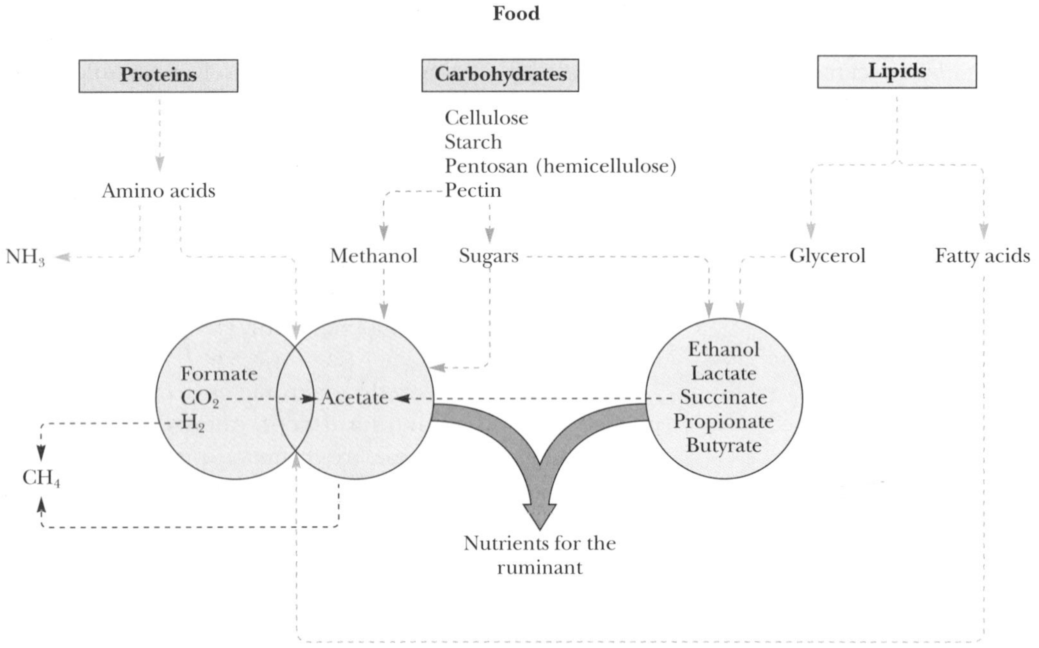

***Figure* 25.24** The major biochemical reactions involved in production of the fatty acids that essentially "feed" the ruminant. Methane serves as a hydrogen and is removed from the animal by belching.

such as ammonium salts or even urea. Microbes in the rumen use these inorganic nitrogen sources and incorporate the amine nitrogen into substrates from which they make their cellular protein and nucleic acids. Thus, although ruminant animals cannot make protein from ammonia and fatty acids, they obtain protein and amino acids from the degradation of the microbes that reside in the rumen. This process occurs in the animal's true acid stomach, the abomasum, the next compartment in the digestive system, as well as in the small and large intestines.

From a microbial ecology perspective the rumen fermentation has been a classic source of information on microbial interactions. For example, the first methanogenic bacteria that were cultivated were obtained from the rumen. Moreover, it was the study of ruminant metabolism that led to an understanding of the tight coupling in metabolism that can occur between different species. The concept of "interspecies hydrogen transfer," in which one species (a fermentative bacterium) produces hydrogen gas and another species (a methanogen) uses it quickly as a substrate for methane gas production, originated from a study of rumen metabolism. The combined activity of these two species maintains an extremely low concentration of hydrogen gas. If hydrogen were to accumulate, it would inhibit the activity of the fermentor, and the fermentation could not proceed. Even though hydrogen concentrations are low, the rate of turnover is extremely high. This same tight coupling, or interspecies hydrogen transfer, occurs in anaerobic sediments, where methanogens also proliferate.

Bacteria–Bird

Two fascinating cases of beneficial symbiotic associations occur in specialized species of birds. One is the beeswax-eating honey guide, and the other is the leaf-eating hoatzin. The honey guides (genus *Indicator*) (Figure 25.25)

***Figure* 25.25** A black-throated honey guide. These birds can utilize beeswax as a food source because they are host to bacteria that can cleave to the wax and form fatty acids. (© Nigel Dennis/Photo Researchers, Inc.)

are small, sparrow-sized birds that live in India and on the African continent. Honey guides seek out cracks and crevices in trees and cliffs where swarms of honey bees have established residence. The birds attract the attention of sweet-toothed honey badgers (or humans), who are aware that the presence of these birds affirms that honeycombs can be found nearby, and the honey guide instinctively knows that these animals will rip open the bees' habitation to obtain honey. The bird can then feed on the honeycomb that remains.

The primary source of food for the honey guide is beeswax. The digestive tract of the bird does not produce esterases and other enzymes necessary for the utilization of the wax esters. However, the intestinal tract of the honey guide has two inhabitants that can digest beeswax: *Micrococcus cerolyticus* (cero-wax lyticus-splitting) and a fatty-acid-cleaving yeast, *Candida albicans*. The micrococcus requires a growth factor, and this is available in the digestive tract of the honey guide. Thus we have a remarkable symbiotic association between a bird species and microbes.

The hoatzin *(Opisthocomus hoazin)* is a leaf-eating (folivorous) bird for whom leaves are the sole source of food. The hoatzin is a large bird, about 750 grams, whose habitat ranges from the Guianas to Brazil (Figure 25.26). It is found mostly in riverine swamps, forests, and oxbow lakes.

Cellulosic components of leaves are the major carbon source for hoatzins. They have, therefore, evolved with a rumen-like crop. The crop and esophagus are the major digestive organs in this species and the pH and physical environment in these organs supports the growth of bacteria at concentrations equivalent to those in the ruminant. Volatile fatty acids are produced by the bacteria present in the crop and esophagus including acetic, propionic, butyric, and isobutyric acid. The crop and esophagus make up about 75 percent of the total digestive system. The volatile fatty acids are absorbed by the small intestine. As one might expect, the gizzard is very small in this species.

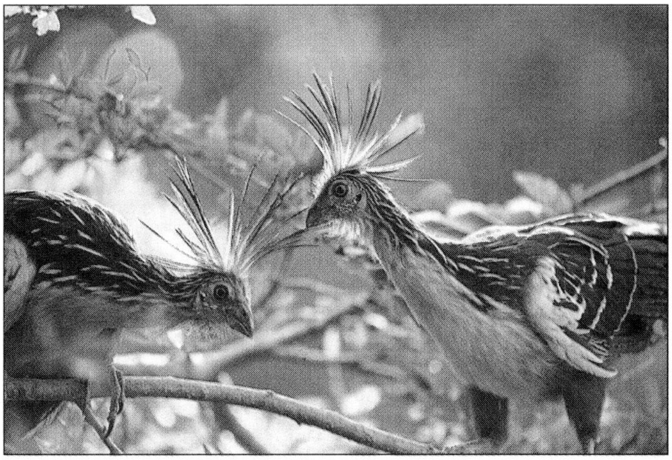

***Figure* 25.26** The foliage-eating hoatzin. The hoatzin is the smallest animal that has a rumen-like system and the only bird species with this capability. (© Roland Seitre/Peter Arnold, Inc.)

The hoatzin is by far the smallest warm-blooded animal with a foregut that functions as a fermentation vat (rumen). All other animals that do this are mammals. There are some significant anatomical and behavioral modifications in the hoatzin to make room for the relatively large crop. They have a reduced sternum area for flight muscle attachment, and the bird is a poor flyer. The young take more than 60 days before they are able to fly, and survival mandates that the juveniles have some means for protection against predation. Hoatzins have wing claws that are utilized in a rapid crawling movement for escape from predators. They move to and dive into water when threatened.

Termites–Microbes

The termite is an interesting social insect that relies on symbionts of one type or another for survival. The basic wood diet of termites is not available as nutrient for the insect unless it is predigested by their symbiotic associates. The enzymes essential for disassembly of complex cellulosic compounds can be supplied by either external or internal microbial populations. In some cases, the cellulases are produced by microbes present in the termite gut; in other cases, it acquires digestive enzymes by eating externally grown fungi. Termites may also harbor nitrogen-fixing organisms that supply nitrogenous compounds to supplement a diet low in available organic nitrogen.

The "higher" termites are those that obtain digestive enzymes by ingesting fungi that are grown in the termite nest. The fungi are members of the genus *Termitomyces*, a basidiomycete. Termites moving from one habitat to another carry fungal spores to the site where they establish a new fungal garden. These fungal populations are cultivated much in the manner of fungus-gardening ants. Growth of fungi on the cellulosic material they are fed by termites ensures that the mycelium will contain cellulases. These fungi can digest lignin, which usually protects cellulose from microbial attack. How this occurs is not now known. The termites that cultivate fungi do not have internal cellulase-producing protozoa or bacteria.

The gut of "lower" termites has a population of protozoa that digest cellulose and release monomers and various metabolites (Figure 25.27). These serve as the source of nutrients for the termite. The gut protozoa are anaerobic and ferment cellulose to acetate + CO_2 + H_2. The acetate is absorbed through the hind gut of the termite and is oxidized aerobically to provide the termite with ATP and cell carbon. The termites also harbor bacterial popu-

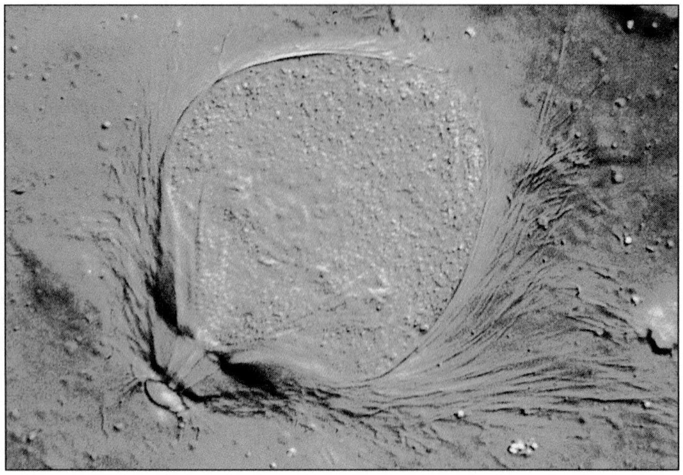

***Figure* 25.27** An example of mutualism of a worker termite (left) of the wood-eating genus *Reticulitermes*. (© William J. Weber/VU) The gut of these termites is populated by protozoa (right), of the genus *Trichonympha,* which has the ability to depolymerize cellulose to provide nutrient for the termite. (© M. Abbey/VU)

lations capable of nitrogen fixation. *Enterobacter agglomerans* is the nitrogen fixer in some species and is particularly important during insect development. Various spirochetes are also associated with protozoan populations in the termite gut. These spirochetes attach to the protozoan cell and provide locomotion much as cilia do in other protozoa (Figure 25.28). The protozoa benefit from a decreased energy requirement necessary for motility, and the spirochetes obtain nutrients from the termite.

Methanogens are also present in protozoan cells as endosymbionts. These bacteria can be identified readily, as they contain large amounts of unique cofactors (F_{420} deazoflavin and F_{450}, a pterin) that autofluoresce when illuminated with blue-green or blue light. The methanogens are the recipients of an interspecies transfer of H_2 providing them with a source of energy. They utilize CO_2 as electron acceptor with CH_4 as product. Both CO_2 and H_2 are products of the protozoan fermentation of cellulose, as mentioned above.

Some termites also harbor homoacetogenic bacteria that generate acetate by the following reaction (see Figure 10.11):

$$2\ CO_2 + 4\ H_2 \longrightarrow CH_3COOH + 2\ H_2O$$

The reduction of CO_2 to acetate in the hindgut can provide over one third of the energy requirements of the termite.

Bioluminescent Bacteria–Animal

The capacity for light emission (bioluminescence) is rather widespread in the biological world. Bioluminescence has been observed in bacteria, fishes, insects, molluscs, and annelids. In some instances, the light is produced by the animal's tissues (as in fireflies) and in others by symbiotic luminescent bacteria living in or on the

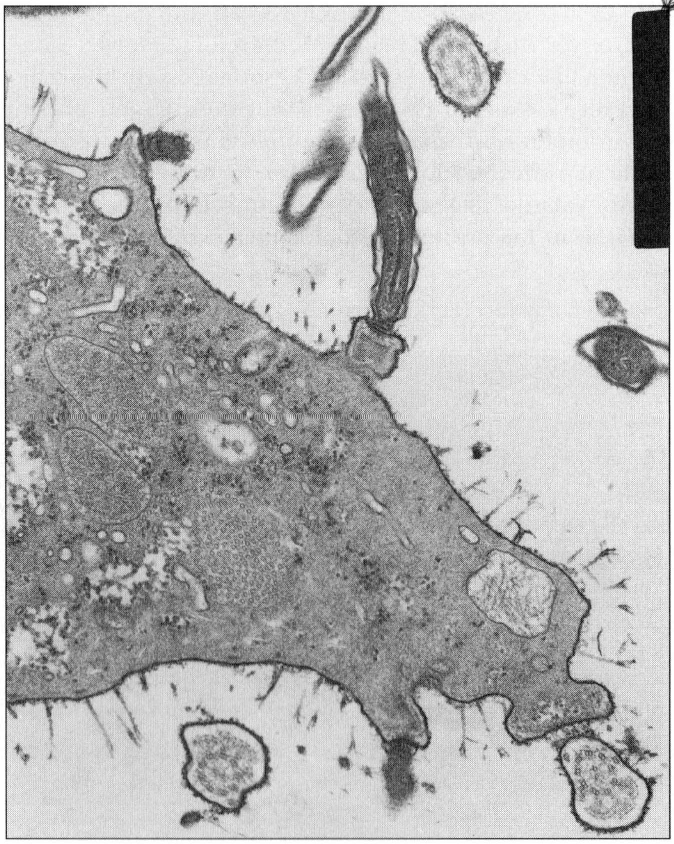

***Figure* 25.28** Spirochetes attached to the surface of the protozoan, *Personympha*. The motility of the protozoan is dependent on the presence of these bacteria. (× 50,000) (© J. A. Breznak and H. S. Pankratz)

animal. Luminescence is employed by the animal host as a recognition device and may be involved in mating, schooling, for prey attraction, or in predator avoidance.

A number of species in the genus *Photobacterium* emit light, as do some members of the genus *Vibrio.* Both genera are abundant in marine environments. These bacteria are often associated with marine fish and live as saprophytes on dead fish. Luminescent bacteria can readily be observed by partly immersing a marine fish in a shallow pan containing seawater. The pan can be placed in a cold room at 10° to 15°C and, after a few days, luminescent bacteria can be isolated from the surface of the fish. Luminescent bacteria can also be isolated by filtering seawater onto a membrane filter and placing the membrane on a rich organic medium in a petri plate. The luminescent colonies can be visualized by taking the plate, following a few days of incubation, from a brightly lit room to one that is dark.

It is difficult to assign a role to bioluminescence in marine bacteria. They emit light only when grown to a dense suspension. The enzyme luciferase, a flavin (FMN); O_2; and a long-chain aldehyde (RCHO) are necessary for luminescence to occur. This reaction is as follows:

$$FMNH_2 + O_2 + RCHO \xrightarrow{\text{luciferase}} FMN + RCOOH + H_2O + LIGHT$$

Luciferase is induced in *Vibrio fischeri* only when a metabolic product, the autoinducer (N-β-ketocaproyl homoserine lactone), accumulates to a critical level in the growth medium. This critical level is attainable in a rich growth medium but would not occur among free-living organisms in the marine environment.

Some marine fish have luminous glands that are an integral part of the animal. These glands are actually pouches that contain cultures of bioluminescent bacteria. For instance, the ponyfish *(Gazza minuta)* has tubular structures within this pouch that contain *Photobacterium leiognathi* at densities up to 10^{11} per ml of fluid. Another bioluminescent fish, the flashlight fish, has special pouches under the eye (Figure 25.29) where *P. leiognathi* grows as an ectosymbiont. The fish can control the emission of light by drawing a fold of dark tissue over the gland. Some fish have luminescent bacteria in open glands where they are nourished directly by the fish. The squid *(Doryteuthis kensaki)* has luminous glands containing *P. leiognathi* embedded in the ink sac, and these are partly enclosed in tissue. It is believed that in all these animals the bacterial symbionts are a result of infection and that they do not enter through egg passage.

Syntrophy

The term "syntrophy" literally means "feeding together." A syntrophic relationship occurs when two or more species live in an association that is mutually beneficial. They may exchange growth factors or nutrients. An example of syntrophy would be the exchange of vitamin precursors by the filamentous fungus *Mucor ramannianus* and a strain of the yeast *Rhodotorula.* Thiamine (vitamin B_1) is required by all free-living organisms, but neither the yeast nor the fungus can synthesize the complete molecule. The fungus

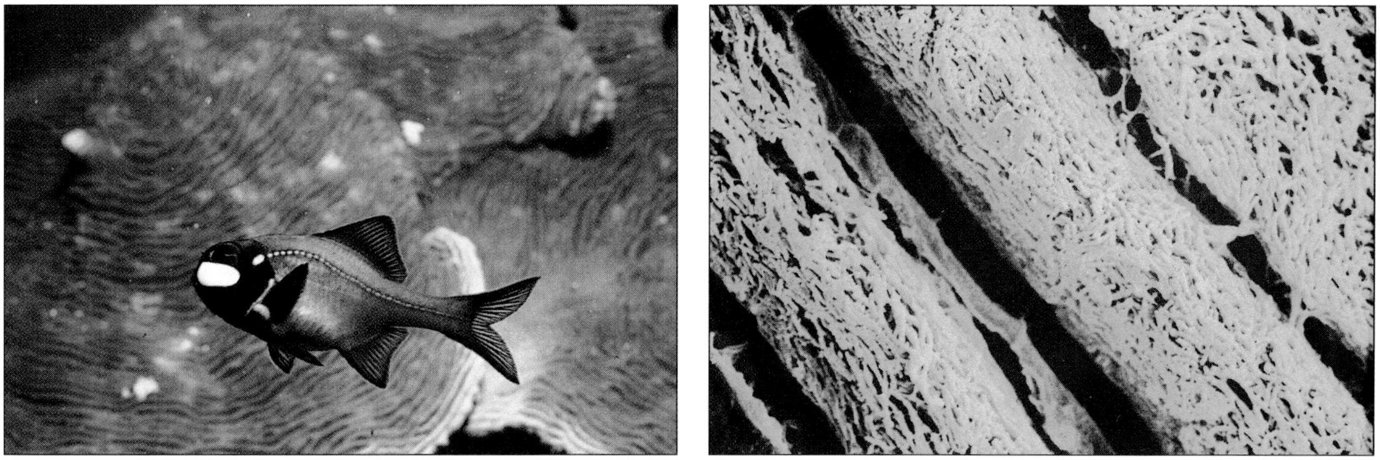

(a) **(b)**

***Figure* 25.29** Bioluminescence in marine fish. **(a)** The Atlantic flashlight fish *Kryptophanaron alfredi* with a luminous patch under the eye. (© Fred McConnaughey 1989/Photo Researchers, Inc.) **(b)** Scanning electron micrograph of bacteria inside the luminous organ. (© James G. Morin/VU) can synthesize the pyrimidine precursor, but not the thiazole; and the yeast synthesize the thiazole but not the pyrimidine.

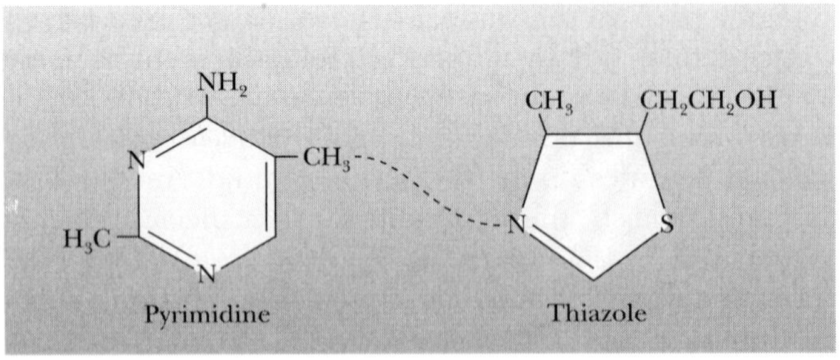

Living in close association, the organisms exchange precursors and survive in environments that lack the essential vitamin (dotted line is bond that is present in thiamine).

A syntrophic relationship also occurs in anaerobic methanogenic ecosystems. *Syntrophomonas* and *Syntrophobacter* have been isolated from these environments, where they grow in association with methanogens or sulfate reducers. *Syntrophomonas* metabolizes saturated fatty acids ranging from butyrate to octanoate, via β-oxidation, to acetate (even-number carbon chain substrate) or acetate + propionate (odd chain). For example, n-pentanoic acid would be metabolized as follows:

$$\text{n-pentanoic acid} \longrightarrow \text{acetate} + \text{propionate} + H_2$$

Syntrophobacter oxidizes only propionate as follows:

$$\text{propionate} \longrightarrow \text{acetate} + CO_2 + H_2$$

Both *Syntrophomonas* and *Syntrophobacter* utilize protons ($H^+ + H^+ \longrightarrow H_2$) as terminal electron acceptor. These organisms can gain sufficient energy for growth only if the H_2 generated is consumed. The immediate removal of the H_2 is necessary to drive the reaction and to provide ATP to the organisms. They can survive only in syntrophic associations with *Desulfovibrio* or *Methanospirillum*, which can remove the H_2 produced.

$$SO_4^{=} + H_2 \longrightarrow H_2S \quad \text{or} \quad CO_2 + H_2 \longrightarrow CH_4$$

Associations of this type are apparently of widespread occurrence in nature but have received little attention. It is probable that the reason why so few of the organisms visible microscopically in natural environments have been cultured is our inability to reproduce them in such conditions as they exist in nature.

Commensalism

Commensalism is a one-sided biological situation where one organism provides sustenance to another. The provider neither is harmed nor does it receive apparent benefit. A prime example would be the microbial production of extracellular depolymerases. The monomers released provide substrate for the organism that produces these and also for other organisms in that environmental niche. Polysaccharides, proteins, and nucleic acids are some of the macromolecules that are degraded in this way. The removal of oxygen from a confined environment by a facultative anaerobe would be another example. The resultant anaerobic environment would sustain growth of strict anaerobes. *Escherichia coli* in the human gut removes O_2 ingested with food, thus providing a favorable environment for the anaerobic bacteroids. The bacteroids neither benefit nor harm the *E. coli*.

The biodegradation of cycloalkanes in the environment also occurs mostly by commensalism. Cycloalkanes are a significant part of crude oil; the gasoline fraction can be 7 to 10 percent cyclohexane. Despite the large amounts of oil that have seeped to the Earth's surface, the soil is devoid of microorganisms that can degrade cycloalkanes. Gaseous alkanes, such as propane or butane and n-alkanes of any length, are utilized readily by a host of microbes in virtually all soils.

However, cyclohexane disappears from fertile soil because it is readily oxygenated by enzymes called monooxygenases. Microorganisms in soil that grow on methane or other reduced substrates rely on monooxygenases. While growing on a substrate such as methane, any cyclohexane in contact with the cell would simultaneously be oxidized. This oxidation of a nongrowth substrate would be considered **cooxidation.** Recall that most methanotrophs are obligate for methane as substrate and can utilize no other substrate for growth. The product of this enzymatic reaction, cyclohexanone, is readily metabolized by resident microorganisms in soil or water. This sequence of events is illustrated in Figure 25.30. Commensalistic relationships of this nature may be of widespread occurrence in nature.

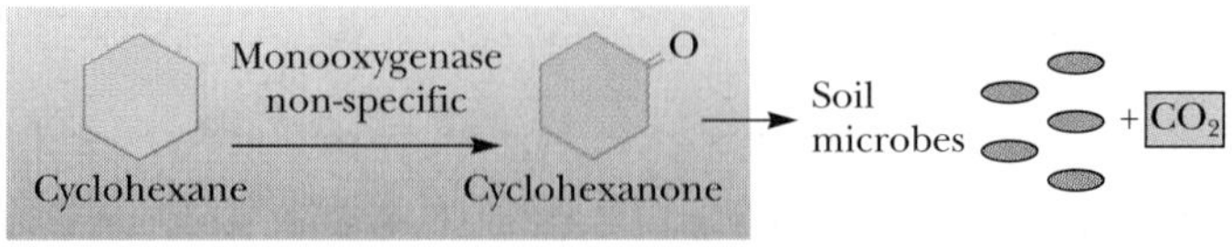

***Figure* 25.30** Example of biodegradation through commensalism. The cyclohexane is gratuitously oxidized by microorganisms that have a mono-oxygenase that nonspecifically oxidizes cyclohexane. Soil organisms that can grow on the cyclohexanone are abundant in soil.

Summary

- The eukaryote evolved in the prokaryotic world. It was essential that the eukaryote evolve with the capacity to co-exist and cope with the ubiquitous prokaryote. Alliances between the eukaryotes and prokaryotes are of common occurrence.

- Two or more dissimilar organisms may form an alliance that is termed a **symbiosis.**

- Symbiotic associations can benefit the **symbionts;** among the advantages: protection, favorable position in the biosphere, recognition, or nutritional benefits.

- Symbiotic relationships are often beneficial to at least one partner. **Mutualism** is an interrelationship where all partners benefit. **Commensalism** is an association where one partner benefits and others are neither harmed nor benefited.

- A **consortium** is a relatively stable but loose-knit association of organisms in an environment. Nutritional cooperation, where organisms exchange growth factors, is termed **syntrophy.**

- Animals survive at depths greater than 1000 m in the ocean because they have developed beneficial symbiotic associations with **chemosynthetic** organisms that live within their bodies. Among those animals are clams, mussels, and tube worms.

- **Plant–fungal** and **plant–bacterial** associations are common in nature. Fungi develop in or on roots, and these are termed **mycorrhizal relationships.** Bacteria may invade roots or stems of selected plants and form **nodules** that are involved with **nitrogen fixation.**

- Bacterial associations with plants that result in the formation of nitrogen-fixing nodules generally occur with **legumes** (peas, clover, etc.). The microorganisms involved are members of the genus ***Rhizobium*** or ***Bradyrhizobium.*** Formation of nitrogen-fixing nodules on roots of alder, a woody plant, is a result of invasion by ***Frankia*** sp., a filamentous actinomycete.

- A **lichen** is a symbiotic association between a **fungus** and a **cyanobacterium** or **eukaryotic alga.**

- **Heterotrophic protozoa** survive in anaerobic sediments because they have **methane-producing** endosymbionts that serve as an electron sink. Many of these protozoa lack mitochondria, and the bacteria serve the general function of this organelle.

- **Herbivorous** animals survive on vegetation because they have a bacterial population in their digestive tract that converts plant material to utilizable foodstuff. This occurs in a large internal **fermentation vat** termed the **rumen.** The electron sink in the rumen is **methane formation** by the bacterial population.

- There are symbiotic relationships in the avian world that permit birds to utilize beeswax or foliage as foodstuff. The **honey guide** is a beeswax eater and is host to bacteria that cleave the wax to utilizable fatty acids. The **hoatzin** can survive on foliage because it has a rumen, much like that in mammals (cows and deer).

- Termites depend on bacterial populations in their gut to digest the wood they consume.

- **Bioluminescence** is common in the biological world. Generally animals in marine environments that emit light do so because they are host to **luminescent bacteria.**

- **Syntrophic** associations are mutually beneficial and generally are involved with synthesis or exchange of growth factors or other metabolic products.

- **Commensalism** occurs when a microorganism alters a compound in such a way as to render it utilizable by another organism in the environment. The alteration is of no benefit to the organism that is responsible.

Questions for Thought and Review

1. What is symbiosis? What are the major functions of symbiotic associations? How does mutualism or commensalism differ from parasitism? Could many of these associations have originated through parasitism?
2. How does a symbiotic association get established?
3. Define: ectosymbiont, endosymbiont, syntrophy, and consortium. What are some examples of each?

4. How does a deep-sea chemosynthetic ecosystem differ from the ecosystems on the surface of the Earth? Which animal systems on the Earth's surface resemble those in the deep sea?

5. Outline the "feeding" of a tube worm. What are some of the unique features of a tube worm?

6. Microbiologically, what is a rhizosphere and how does this environment differ from surrounding soil? What are benefits that the plant and microorganism receive in symbiotic associations of this type?

7. How do ectomycorrhizae and endomycorrhizae differ? Think of some of the ways that these relationships benefit a plant. What type of plant is involved in each of these?

8. What is *Frankia?* Where is it found? What does it do?

9. Outline the steps involved in the establishment of root nodules on a leguminous plant. What is the function of leghemoglobin?

10. Microbiologically, how do "higher" and "lower" termites differ?

11. What function is served by methanogens present in the protozoa that reside in anaerobic environments? Is this similar to ruminants? How?

12. What symbiotic associations occur in birds? Discuss the role of the bacterial symbionts in nutrition of the animal.

13. Outline the microbes' functions in a ruminant. What are the benefits to the animal? To the bacterium?

14. What is the chemical reaction involved in bacterial luminescence? Why have animals established relationships with luminescent bacteria? How would you isolate some of these organisms?

15. Give an example of a syntrophic relationship and explain how it differs from a common symbiotic association.

Suggested Readings

Childress, J. J., H. Felbeck, and G. N. Somero. 1987. Symbiosis in the Deep Sea. *Scientific American* 256: 115–120.

Felbeck, H., J. J. Childress, and G. N. Somero. 1983. "Biochemical Interactions Between Molluscs and Their Algal and Bacterial Symbionts." In: *The Mollusca,* Vol. 2, *Environmental Biochemistry and Physiology* (331–358). New York: Academic Press.

Harley, J. L., and S. E. Smith, eds. 1984. *Mycorrhizal Symbiosis.* New York: Academic Press.

Jannasch, H. W. 1985. *The Chemosynthetic Support of Life and the Microbial Diversity at Deep Sea Hydrothermal Vents.* London: Proceedings of the Royal Society. B225: 277–297.

Jannasch, H. W. 1989. Chemosynthetically Sustained Ecosystems in the Deep Sea. In: *Biology of Autotrophic Bacteria.* H. G. Schlegel and B. Bowien, eds., pp. 147–166. Madison, WI: Science Tech. Publ.

Jones, L. J., ed. 1985. Hydrothermal Vents of the Eastern Pacific: an Overview. *Bulletin of the Biological Society of Washington.* No. 6, 545 pp. Vienna, VA: INFAX Co.

Margulis, L., and R. Fester, eds. 1991. *Symbiosis as a Source of Evolutionary Innovation: Specialization & Morphogenesis.* Cambridge, MA: MIT Press.

Margulis, L., and D. Sagan. 1986. *Microcosmos. Four Billion Years of Evolution from Our Microbial Ancestors.* New York: Summit Books.

Perry, Nicolette. 1990. *Symbiosis: Nature in Partnership.* New York: Sterling Pub. Co.

Healing is a matter of time, but it is sometimes also a matter of opportunity.

Hippocrates, 460–377 B.C.

Chapter 26

Host–Microbe Interactions

The Normal Human Microbiota
Nonspecific Host Defenses
Pathogenesis Mechanisms
Toxins

Billions of bacterial cells are present in and on the human body. Most are innocuous, and many play a beneficial role and are directly involved in our well-being. Every healthy human plays host to a distinct microbial population that is termed the **"normal microbiota."** Bacteria are the major component of this microbiota, with a few fungi also present. The alliance between the human **host** and the normal microbiota is generally **mutualistic** and in some cases **commensalistic** (see Chapter 25). Some of the normal microbiota may be **opportunists** and cause infection if the host defenses are lowered. Humans also have a **transient** microbiota composed of microbes that generally do not reproduce extensively or establish a long-term residence.

Some of the microorganisms that humans encounter in daily life are **parasitic,** and their presence can cause harm to the host. Mostly these parasites do not colonize; they are transient and cause no apparent injury. In some cases a parasite can become established and bring injury to the host and is termed a **pathogen.** The amount of damage that a parasite may cause depends on its **pathogenicity.** Pathogenicity is a relative term and is dependent on the capacity of the parasite to inflict damage and the level of resistance in the host. **Virulence** is a quantitative term that measures the capacity of a population of parasitic organisms to inflict damage (see **Box 26.1**). An **infection** is a situation where the parasite actually grows in or on the host and inflicts damage.

This chapter is concerned with three aspects of microbe–human relationships: first a description of the normal microbiota of the human host, second a description of the host defenses against pathogens, and third some of the mechanisms employed by pathogens to combat the normal host defenses.

The Normal Human Microbiota

Prior to birth, the human fetus is virtually free of microorganisms. During birth and immediately thereafter the newborn is invaded by bacteria. The respiratory system and gastrointestinal tract are quickly colonized by microorganisms that circumvent the normal host defenses. These colonizers are picked up from the environment and are generally beneficial. Microorganisms readily establish an "ongoing" residence in most of those areas that are exposed to the external environment—the skin, the oral cavity, the respiratory tract, the gastrointestinal tract, and the genitourinary system. However, microorganisms are not normally present in organs, internal tissues, blood, or the

BOX 26.1 MILESTONES

Evolution of Virulence

English rabbits were introduced into Australia in 1859 by British sportsmen interested in game hunting. There were no carnivorous animals on the Australian continent that were natural enemies of the rabbit, so the populations increased at an incredible rate. The agricultural interests in Australia became alarmed at the loss of grazing lands to the rapidly growing rabbit community, and they pressured the government to find a solution to this problem.

In 1950, a myoxma virus that was prevalent in South American rabbits was introduced into the Australian rabbit populations. This virus had low virulence in the South American rabbit populations but a much higher virulence for the European rabbit strains. The myoxma virus spread rapidly in Australia and within a few years killed over 99 percent of the rabbit population. Death rates were highest in the summer, as the virus was spread by mosquito vectors. This daring experiment in pest control was followed closely by the scientific community in Australia.

There were two interesting phenomena that became evident over time—one in the rabbit population, the other in the virus itself. Controlled laboratory studies confirmed that over a six-year span the virus had become less virulent than the original strain. The rabbit became more resistant to the virus and less susceptible to ravages of the disease. Apparently, survivors of the initial epidemic were selected in some way for resistance to infection. The rabbit population in Australia now stays at about 20 percent of the previous high number. Experimental evidence indicated that resistance in rabbits was related not to an improved immunological defense system but to a physiologically altered rabbit. This study indicates that both the susceptible host and the virulent virus evolved in a way that allowed both to survive. It is axiomatic that any pathogenic agent that kills all of its hosts will quickly become extinct. Evolution then favors both the pathogen and the host.

lymphatic system. A **diseased state** may be indicated if microorganisms are present at these sites.

Human health is dependent on the ability of the body to establish and maintain a consistent "normal" indigenous microbiota. Yet factors, such as prolonged antibiotic therapy, can alter or suppress the normal microbiota and lead to serious health problems. Use of oral antibiotics can alter the normal gastrointestinal tract microbiota, increasing the risk of developing a disease such as pseudomembranous colitis. This disease is caused by the rapid proliferation of *Clostridium difficile,* an organism normally present in very low numbers. Broad-spectrum antibiotics taken orally may also reduce the resident microbiota of the mucous membranes in the oral cavity and respiratory system, resulting in yeast or other fungal infections. There are many examples of the adverse effects resulting from the destruction of the normal microbiota, confirming that the indigenous microbiota of humans and animals in general is a primary defense against invasion by harmful pathogens.

Some areas of the human body provide a more favorable environment for bacterial growth than others. The physical conditions such as pH, moisture, osmotic pressure, and temperature at a given site are constant—with some variability from site to site. The nutrient available and other factors combined result in a distinct microbiota associated with each area of the body. Many microorganisms can survive on human skin, but internal mucous membranes have a larger and more varied microbiota. There are an estimated 10^{13} (100 trillion) bacteria associated with an adult human, and this number is far greater than the total number of animal cells present. A general picture of the microorganisms present in various parts of a human is presented in Figure 26.1.

Skin

The intact skin or epidermis is impermeable to microbes and is considered the first line of defense against invasion. As mentioned, there is a normal continuous microbiota on human skin and these inhabitants must have a source of nutrient. There are two key sources of secretions on human skin, the sweat glands and sebaceous glands (Figure 26.2). Both secrete compounds that may retard growth of some bacteria but also are compounds that can be utilized by others as a source of nutrient. The sweat glands secrete lysozyme, an enzyme that destroys the peptidoglycan layer of microorganisms, and lactic acid that lowers the pH to a range of 3 to 5, an acidity that discourages excessive colonization. The sebaceous glands secrete lipid materials

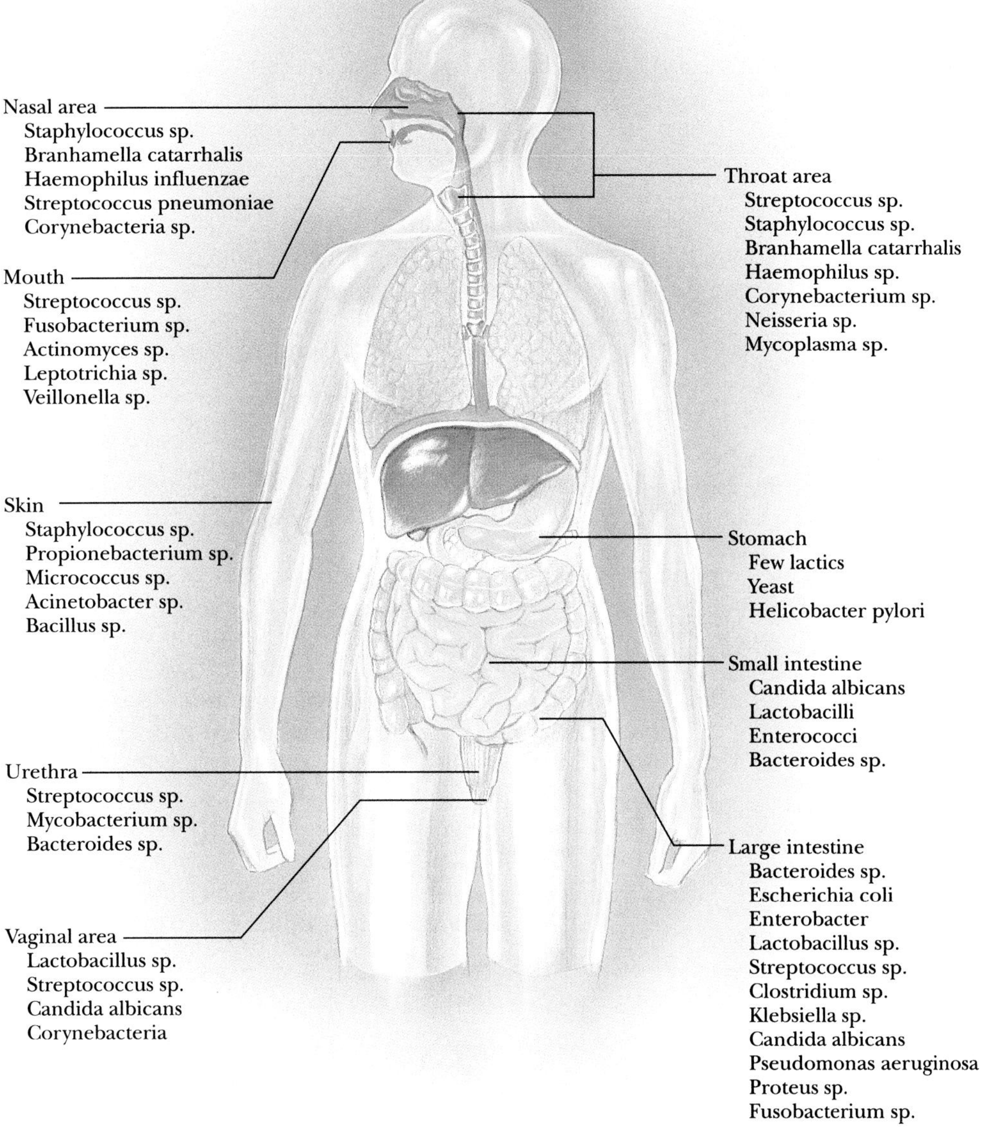

***Figure* 26.1** Microorganisms that have been identified and present as "normal microbiota" on/in the human body.

that prevent overwetting, overdrying, and abrupt changes in temperature of the epidermis layer. These lipids also prevent absorption of foreign material. Sebum, the secretion of sebaceous glands, contains fatty acids that are inhibitory to many microorganisms. Catabolism of the lipids present in sebum by propionic acid bacteria releases fatty acids such as oleic acid that can be inhibitory to others. The presence of inhibitory compounds and dryness of some areas of the skin results in a modest bacterial population of 10^2 to 10^4 per square centimeter.

The moist areas of skin including the armpits, scalp, and feet are a more favorable environment for microbial growth. The area around hair follicles has a relatively high concentration of sebum, lipids, free fatty acids, wax alcohols, glycerol, and hydrocarbons, which are products of the secretory glands. These compounds provide substrate for a population of about 10^6 bacteria per square centimeter. The sweat glands also secrete urea, amino acids, lactic acid, and salts that favor bacterial proliferation. These nutrients support the growth of a varied constant population of mostly gram-positive bacteria including *Staphylococcus* sp., *Micrococcus* sp., *Corynebacterium* sp., and *Propionibacterium* sp. (see **Box 26.2**). The bacterial population changes at puberty, with the appearance of greater numbers of *Propionibacterium acne* and selected species of staphylococci such as *S. capitis* and *S. auricularis.* Diet has

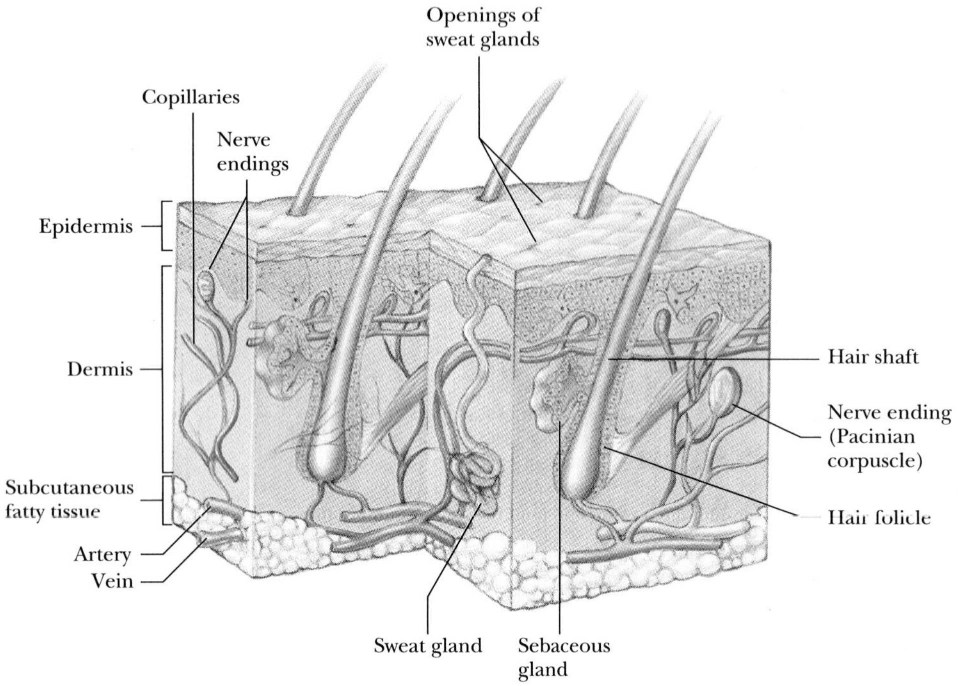

***Figure* 26.2** A cross-section of human skin. The sweat and sebaceous glands secrete material that sustains bacterial populations on the skin.

little measurable influence of the microbiota, but antibiotic therapy can alter the skin populations.

Respiratory Tract

The respiratory tract consists of the mouth, nasopharynx, throat, trachea, bronchia, and lungs. All of these locales are moist and a potential area for microbial colonization. The **mouth** is a favorable environment for bacterial growth because it is relatively nutrient rich and the pH is favorable, as are other physical conditions. There are about 10^8 bacteria per ml fluid in the mouth. These microorganisms originate from colonies that adhere to surfaces of the tongue, gums, and teeth. Constant saliva production and swallowing tend to remove bacteria that are unable to stick to a surface. Lysozyme and lactoperoxidase are present in saliva and have some inhibitory effect on bacterial growth. Saliva contains about 0.5% dissolved solids, consisting of proteins—salivary enzymes and mucoproteins. *Streptococcus salivarius* is commonly present and proliferates by adhering to the surface of the tongue. **Teeth** are a calcium phosphate crystalline material that readily becomes colonized. In the newborn, before development of teeth, lactobacilli and streptococci predominate in the oral cavity. As teeth develop, anaerobes such as *Streptococcus mutans* invade and attach to the enamel surface. Oral streptococci produce glucans (polysaccharides) that bind bacteria together on the tooth surface. These aggregates of bacteria and organic matter are known as *plaques* (Figure 26.3). *Streptococcus sanguis* and anaerobic actinomyces are among the microbes in these aggregations. The microbes in the plaque can catabolize sucrose to lactic and other organic acids that etch teeth, causing dental caries.

The **nasopharynx** and **nose** are areas with adequate moisture and are constantly inoculated with microorganisms as one breathes. The **nostrils** have a population similar to that of facial skin with *Staphylococcus* sp. and other gram-positive bacteria predominating. The nasopharynx (above the soft palate) is colonized by avirulent strains of *Streptococcus pneumoniae, Neisseria meningiditis, Branhamella* sp., and *Hemophilus* sp. The **oropharynx** (between the soft

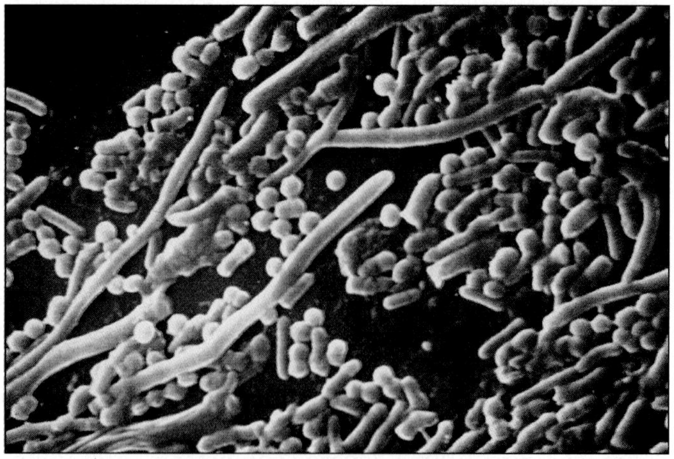

***Figure* 26.3** A scanning electron micrograph of dental plaque showing the abundant growth of various bacteria. The bacteria are cemented to the tooth by polymers they secrete. (© Fred E. Hossler/VU)

BOX 26.2 RESEARCH HIGHLIGHTS

Staphylococci Colonizers

Staphylococci are one of the major bacterial colonizers of human skin. Of the bacteria present on the body surface and the anterior nares, usually over 50 percent are staphylococci. *Staphylococcus aureus* was the first member of this genus isolated and was described by F. J. Rosenbach in 1884. He isolated the organism from an infected wound. The second species described was *S. epidermidis* and was isolated by C. E. A. Winslow in 1908. Both *S. aureus* and *S. epidermidis* are gram-positive cocci and are facultative anaerobes. They are generally differentiated from one another by the coagulase test. *S. aureus* generally produces coagulase (a blood-clotting enzyme) and *S. epidermidis* does not. The common habitat for staphylococci is humans and other animals, and their occurrence in nature is generally attributed to human or animal sources. *S. aureus* is a serious, opportunistic pathogen in humans and the causative agent of boils, impetigo, pneumonia, osteomyelitis, endocarditis, meningitis, toxic shock syndrome, and several other infections. *S. epidermidis* strains are a clinical problem in patients with prosthetic heart valves or other implanted devices. Antibiotic resistant strains of the staphylococci are a persistent and acute problem, particularly in nosocomial (hospital acquired) infections. Despite the early recognition that staphylococci are a common organism on the skin of humans, there was relatively little research done on the characterization and infectivity of these human parasites. This changed markedly in the 1970s.

In the early 1970s, Wesley Kloos of the United States, in collaboration with Karl-Heinz Schleifer of Germany, initiated a systematic study of the staphylococci. Over the ensuing years they and others determined that there were at least 12 species of the genus *Staphylococcus* that inhabit the human skin (see figure). *Staphylococcus* are the dominant organism on skin with members of the genus *Micrococcus* second to fifth in abundance. There are at least two to three species of micrococci present on most humans, although on some people there may be as many as seven species present. Coryneforms, *Acinetobacter* sp., *Enterobacteria* sp., *Bacillus* sp., and *Streptomyces* sp. may also be present in significant numbers.

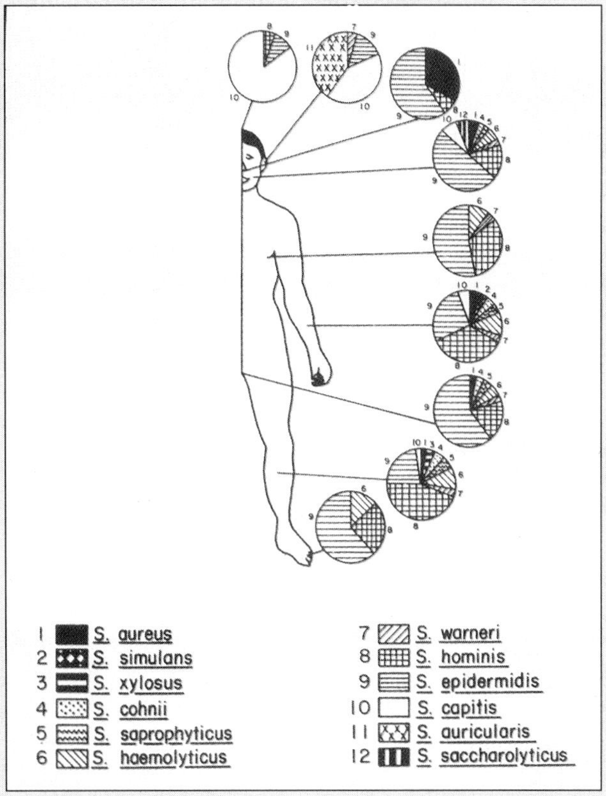

The distribution of various species of *Staphylococcus* on external areas of humans. (Courtesy of Wesley Koos)

These studies affirm that an array of staphylococci are common colonizers of the human skin. Other species are adapted to life on other animals: *S. intermedius,* the domestic dog and other carnivores; *S. hyicus,* the pig and other ungulates; *S. caseolyticus,* the cow, hoofed mammals, and the whale; *S. felis,* the cat; and *S. caprae,* humans and goats. There are now 32 named species of *Staphylococcus* peculiarly adapted to life on animal hosts. Despite the potential pathogenicity of many of these parasitic organisms, the host and parasite have evolved with a reasonable tolerance for one another. Unfortunately, this stand-off is broken at times, and staphylococci can be a serious and deadly enemy.

palate and larynx) has a microbial population that is similar to the population present in the nasopharynx but includes in addition some micrococci and anaerobes of the genera *Prevotella* and *Porphyromonas.*

The **lower respiratory tract,** which includes the trachea, bronchi, and lungs, is generally devoid of bacteria. (This would apply to a healthy individual.) Dust and other particles that are inhaled mostly settle out on mucous membranes of the upper respiratory tract before reaching the lower regions. Particles bearing microorganisms that reach the lower respiratory tract are removed by the muco-ciliary escalator. The mucus lining the area traps particles, and the beating cilia push the mucus layer upward where the material enters the throat and is subconsciously swallowed.

Gastrointestinal Tract

The **stomach** is constantly exposed to bacteria through swallowing of saliva and mucus or by ingestion of food. Few survive: there are fewer than 10 microbes per ml of stomach fluid. The low pH, due to HCl and digestive enzymes, destroys most of the organisms that reach the stomach. Some acid-tolerant lactobacilli and yeasts are present. The stomach of some humans is colonized by the microaerophilic bacterium *Helicobacter pylori,* an organism implicated in the development of peptic ulcers (see Chapter 29).

The **small intestine** is divided into three sections—the upper part is the **duodenum,** the middle the **jejunum,** and the lower part the **ileum.** The number of bacteria present increases in concert with digestion of food as it moves downward through the small intestine. The duodenum has a limited bacterial population (fewer than 10^3 per ml) due to stomach acids, bile secreted by the gallbladder, and pancreatic secretions. The microbiota is mainly gram-positive cocci and bacilli. The jejunum population consists of enterococci, lactobacilli, and corynebacteria. The yeast *Candida albicans* is also commonly found in this area. The ileum microbiota resembles that of the large intestine with large numbers of ***Bacteroides*** sp. and facultative anaerobes such as *Escherichia coli.* The facultative organisms function to remove O_2.

The **large intestine** is also called the **colon.** This area is actually a fermentation vat populated by masses of anaerobic bacteria. Ingested food is the basic substrate for these organisms. There are about 10^{10} to 10^{11} bacterial cells per gram of mass in the colon, and more than 300 different species are present that have been identified. There are several hundred times as many strict anaerobes as there are facultative microorganisms and there are no strict aerobes. An adult excretes 3×10^{13} bacteria per day, as 25 to 35 percent of feces is bacterial mass. The microbiota of the digestive tract supplies a human with several vitamins including B_{12}, biotin, riboflavin, and vitamin K. The healthy adult colon microbiota consists of gram negatives including bacteroides (*Bacteroides fragilis, B. melaninogenicus,* and *B. oralis*) and *Fusobacterium* sp. Among the gram positives are members of the following genera—*Bifidobacterium, Eubacterium, Lactobacillus,* and *Clostridium.* Less than 1 percent of the microbiota consists of *Escherichia coli, Proteus* sp., *Klebsiella* sp., and *Enterobacter* sp. There may also be a few harmless protozoa such as *Trichomonas hominis* present in healthy human adults.

The newborn infant has a sterile gut that quickly becomes populated by bacteria. With breast-feeding, the bacteria are mostly of the genus *Bifidobacterium* obtained from the skin of the mother. Human milk contains an amino disaccharide that is required by this bacterium. Bottle-fed infants have a complex colon population with *Lactobacillus* species among the dominant organisms. As a child moves to a solid diet, the microbial population of the gut gradually changes to resemble that of the adult from whom the typical microbiota is obtained.

Genitourinary Tract

The kidney, ureter, and urinary bladder of the healthy adult are free of microorganisms. The upper part of the urethra near the bladder is sterile because of mechanical flushing, and there may be some antibacterial activity exerted by urethral mucous membranes. Bacteria are present in the lower part of the urethra in both males and females. The adult female genital tract has a complex normal microbiota. The character of this microbiota changes during the menstrual cycle. The adult vagina is colonized by acid-tolerant lactobacilli that convert the glycogen produced by vaginal epithelia to lactic acid. This maintains the pH at 4.4 to 4.6. Microbes that can tolerate this pH—the lactic acid bacteria, enterobacteria, coryneforms, the yeast *Candida albicans,* and various other anaerobic bacteria—are commonly present.

Nonspecific Host Defenses

The normal microbiota is important to our daily well-being as the natural populations crowd out and interfere with colonization by less desirables. Every day a few of these microbes that occur as normal microbiota may gain access to the bloodstream in inoffensive ways. A scratch, an abrasion, a minor trauma that may occur when brushing one's teeth, or chewing food permit bacteria to enter tissue. Once in the tissue, they enter the bloodstream. Generally, these organisms are of little consequence as they are removed quickly by specialized circulating cells, called **phagocytes,** that move to foreign material and engulf it. The phagocytes are also a part of the immune system that protects us against foreign invasion. (For a detailed discussion of the immune system, see Chapter 27.)

We begin this section with a discussion of the barriers that keep foreign invaders out of our system and specific host defenses that immediately attack any invader that does pass through the skin or mucous membranes. A human has two types of defense should a potentially harmful agent pass through the barriers erected against invasion. These are the specific defenses mediated by the immune system (humoral) and the nonspecific defenses (cellular) that are continuously functioning in a healthy human.

Physical Barriers

We possess natural physical barriers that protect us from the potential invaders that we encounter in our daily lives. These include skin and mucous membranes, which, under normal conditions, are virtually impenetrable to microbes. Membranes in the eyes, lungs, intestines, and urinary tract are constantly washed by fluids that move over the surface and remove foreign matter that is not tightly attached. Another barrier is the cilia of the respiratory passages that continually wave and push foreign material outward.

The acidity of the stomach (pH 2) destroys most ingested microorganisms leaving the natural microbiota of the intestines unchallenged. The moderately acidic pH of skin also inhibits microbial growth by all that are not adapted to grow there. In addition, the sebaceous glands in the skin release oils and waxes that are converted to organic acids, and these lower the skin pH. The acidity of the vagina precludes growth of organisms other than the normal lactic acid bacterial microbiota or other acid-tolerant species. Because pathogenic microorganisms are generally quite tissue specific and attach and colonize particular cell types, they are readily removed from nontarget areas. For example, washing your hands will remove *Salmonella* and other intestinal pathogens. *Shigella* or *Vibrio cholerae* are also readily washed away, as they have no mechanism for attachment to skin. However, these organisms are a potential problem if ingested since they may tolerate the stomach acidity and colonize the intestinal tract.

Chemical Defenses

The body maintains an array of chemical defenses against invasion by microorganisms. Among the major chemical defenses is lysozyme, an enzyme that is present in blood, sweat, tears, saliva, nasal secretions, and other body fluids. Lysozyme is an effective lytic agent for gram-positive bacteria. The enzyme cleaves the $\beta 1 \rightarrow 4$ glycosidic bond between N-acetylglucosamine and N-acetylmuramic acid in peptidoglycan. Gram-negative bacteria are more resistant to lysozyme since their peptidoglycan layer is buried beneath the outer cell envelope.

Lactoperoxidase is an enzyme present in saliva and milk. Lactoperoxidase effects a reaction between chloride ions and H_2O_2 resulting in the generation of toxic singlet oxygen that can kill bacteria. In addition a basic polypeptide, termed β-lysin, is released from blood platelets that move to cuts and abrasions and is effective in killing gram-positive bacteria. The natural barriers, nonspecific enzymes, and other bactericidal agents keep us disease free for much of our lives. If a virulent pathogen does breach these defenses the diseased state is a distinct possibility. We do become diseased at times because humans have little natural resistance to selected infectious agents. Exposure to viruses such as the AIDS virus or herpes will lead to infection. Many of the sexually transmitted bacterial diseases (gonorrhea, syphilis, chlamydial urethritis) are readily transmissable. We have little resistance to these, and infection generally follows exposure.

The Inflammatory Process

An effective nonspecific reaction to tissue insult and invasion by a bacterium or a foreign object is the **inflammatory response.** This response is comprised of an ordered series of events that localizes and generally eliminates an invader (Figure 26.4). The processes involved also remove damaged tissue and heal the area by restoring the normal condition. Injury or bacterial invasion initiates the inflammation response with an immediate dilation of capillaries and an increased blood flow into the area. This is apparent as a localized reddening and swelling. Disruption of cells by the injury releases cell contents that lower pH in the intracellular area. This activates kallikrein (an enzyme) at the site of injury that splits **bradykinin** from a precursor molecule. Bradykinin attaches to sites on the capillary walls and to mast cells that underlie the skin and surround the blood vessels. The attachment causes openings to occur in the capillaries through which leukocytes and other phagocytic cells can move into the infected area. The reaction between bradykinin and the mast cells leads to the release of histamine and serotonin. The histamine acts by dilating capillaries and makes intercellular junctions in capillary walls wider bringing more blood into the area.

The binding of bradykinin to capillaries stimulates the release of prostaglandins (PGE_2 and $PGF_{2\alpha}$). These prostaglandins cause tissue swelling in the area of inflammation. Phospholipases release arachidonic acid, which is an essential fatty acid for humans and serves as a precursor for other prostaglandins (E_2 and $F_{2\alpha}$), thromboxane A_2, and leukotrienes (Figure 26.5). Pain results when prostaglandins bind to nerve endings, causing them to produce a pain impulse.

Phagocytes respond to inflammation and chemotactically move to the site of injury. The first phagocytic cells that enter the injured area are the **neutrophils.** These cells are closely followed by macrophages. There is a local increase in temperature that may stimulate the inflammatory response and accelerate the rate of phagocytosis.

The flow of fluids into the trauma area brings in quantities of red blood cells, causing a restriction in blood flow. Greater numbers of leukocytes, polymorphonuclear neu-

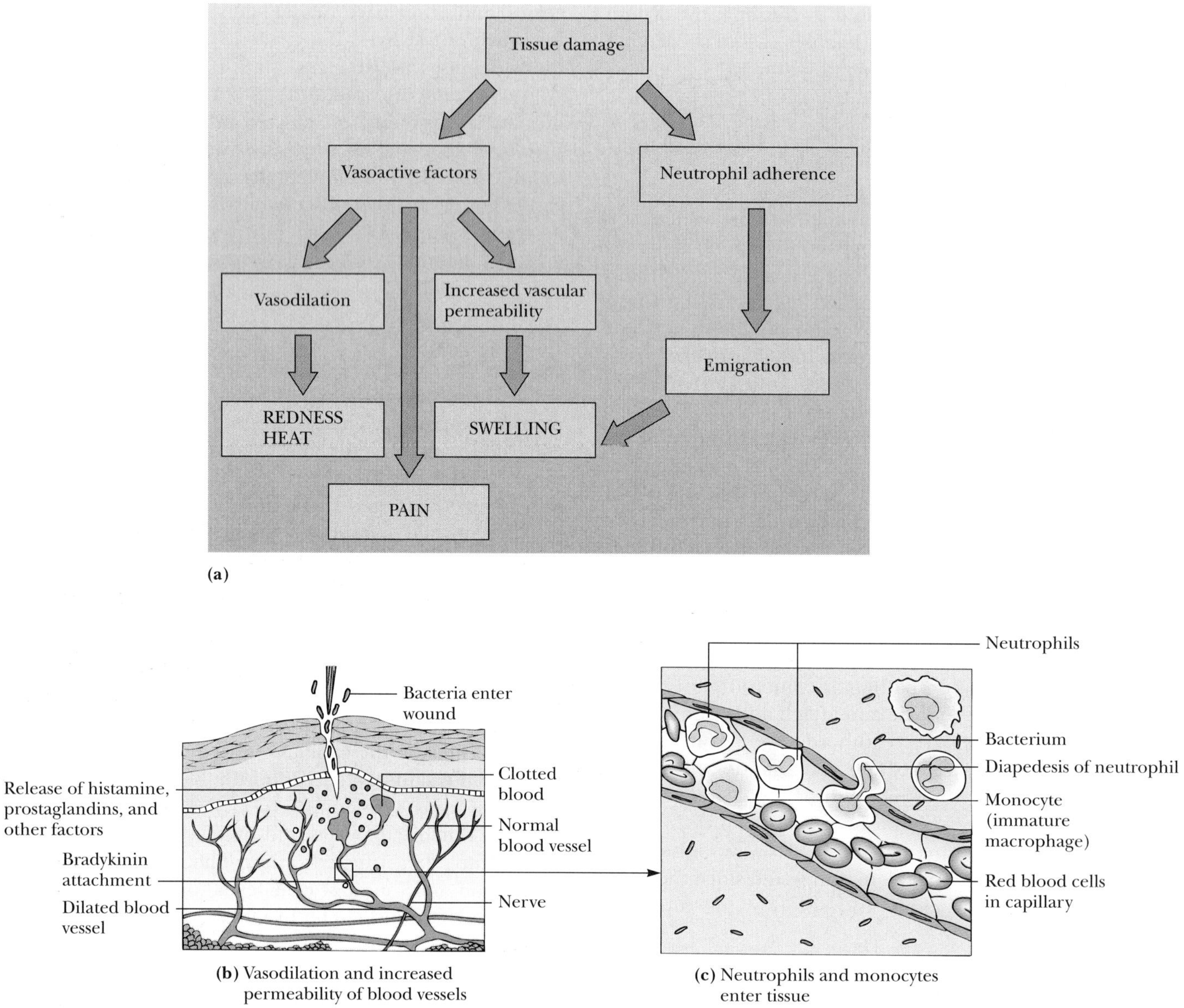

***Figure* 26.4** A diagram of the nonspecific inflammatory response to the presence of foreign material such as bacteria. These nonspecific defenders can eliminate the invader.

trophils (PMNs), and monocytes enter the injured area and chemotactically move to any bacterial invaders. If the injured area is large, a fibrinous blood clot forms, which can prevent the spread of invaders further into the affected area.

After the invader is removed, the wound is repaired by a proliferation of epithelial and endothelial cells. There is no single event that results in the removal of an invading pathogen. The actions of a number of responses—dilation of capillaries, clot formation, release of chemotactic factors, and action of phagocytes—combine to eliminate the foreign invader.

Phagocytes and Phagocytosis

Human blood contains an array of distinct cell types (Table 26.1). Erythrocytes carry oxygen, platelets are the cellular component of the blood clotting system, lymphocytes stimulate, regulate, and generate antibodies, and phagocytes ingest foreign matter. All of the blood cell types listed in Table 26.1 are derived from a multipotent stem cell produced in bone marrow (Figure 26.6).

Phagocytic cells are of considerable importance in maintaining the healthy state. The presence of a specific antibody alone will not prevent infection. We know this because humans with defects in the phagocytic response

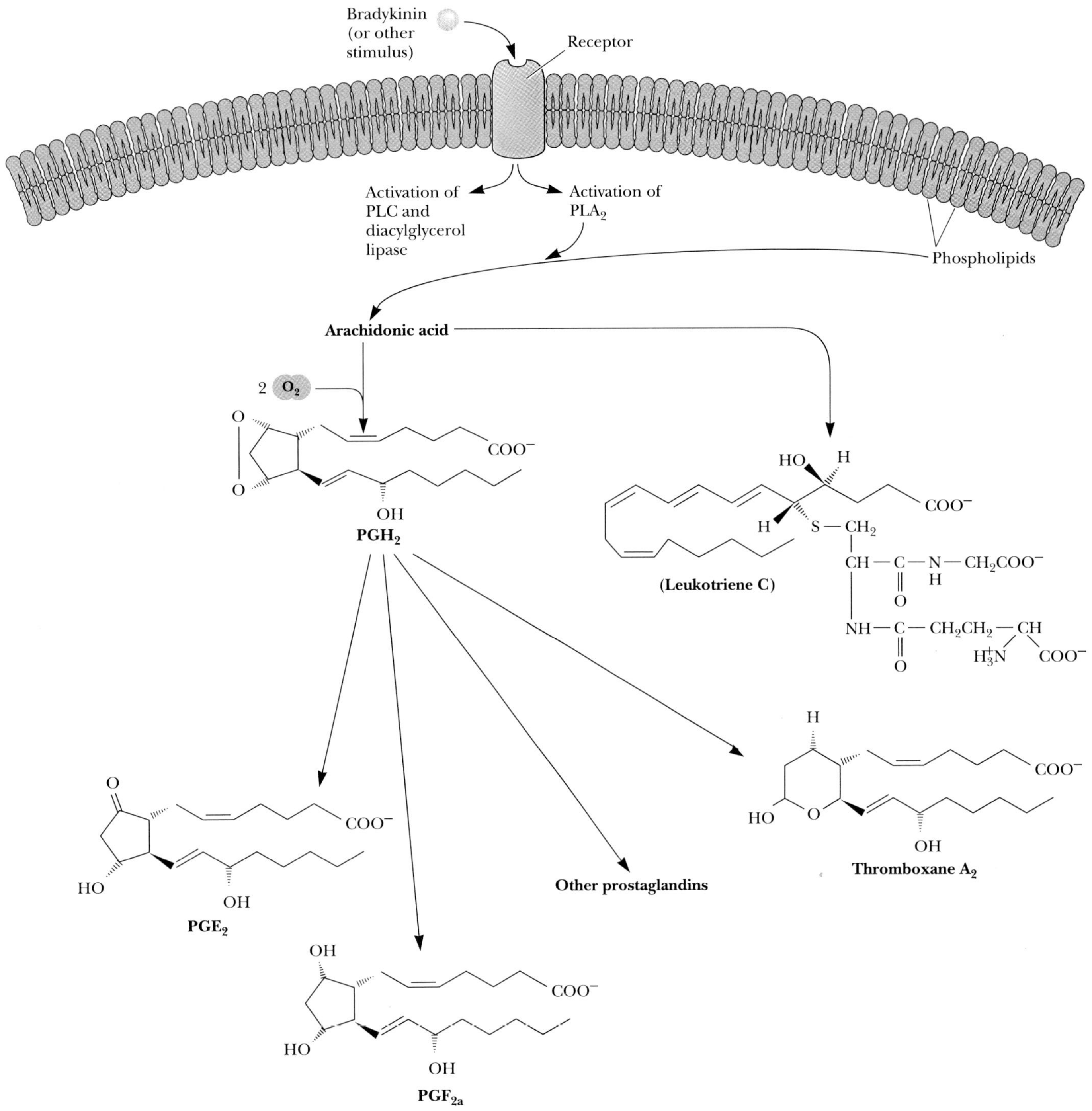

***Figure* 26.5** Synthesis of a prostaglandin and a leukotriene from arachidonic acid. These are examples of chemical compounds that are released during the inflammatory response.

are subject to frequent episodes of infectious disease. Phagocytes pass from the blood into tissue and are strategically placed throughout the body where they can quickly respond.

The two major phagocytic cell types are the granulated polymorphonuclear leukocytes (PMNs) and monocytes (macrophages). A description of these phagocytic cells follows.

Polymorphonuclear Leukocytes (PMNs)

The **PMNs** are circulating leukocytes (white blood cells) that remove debris, including bacteria, by engulfment.

***Table* 26.1 The cellular components of normal human blood**

	Range	Function
Red blood cells (Erythrocytes)	Male: 4.2–5.4 million/μl Female: 3.6–5.0 million/μl	Oxygen transport; carbon dioxide transport
Platelets	150,000–400,000/μl	Essential for clotting
White blood cells (Leukocytes)	5,000–10,000/μl	
Neutrophils	About 60% of WBCs	Phagocytosis
Eosinophils	1–3% of WBCs	Role in allergic response; destroy parasites
Basophils	1% of WBCs	Mast cell progenitors
Lymphocytes	25–35% of WBCS	Produce antibodies; destroy foreign cells
Monocytes	6% of WBCs	Differentiate to form macrophages

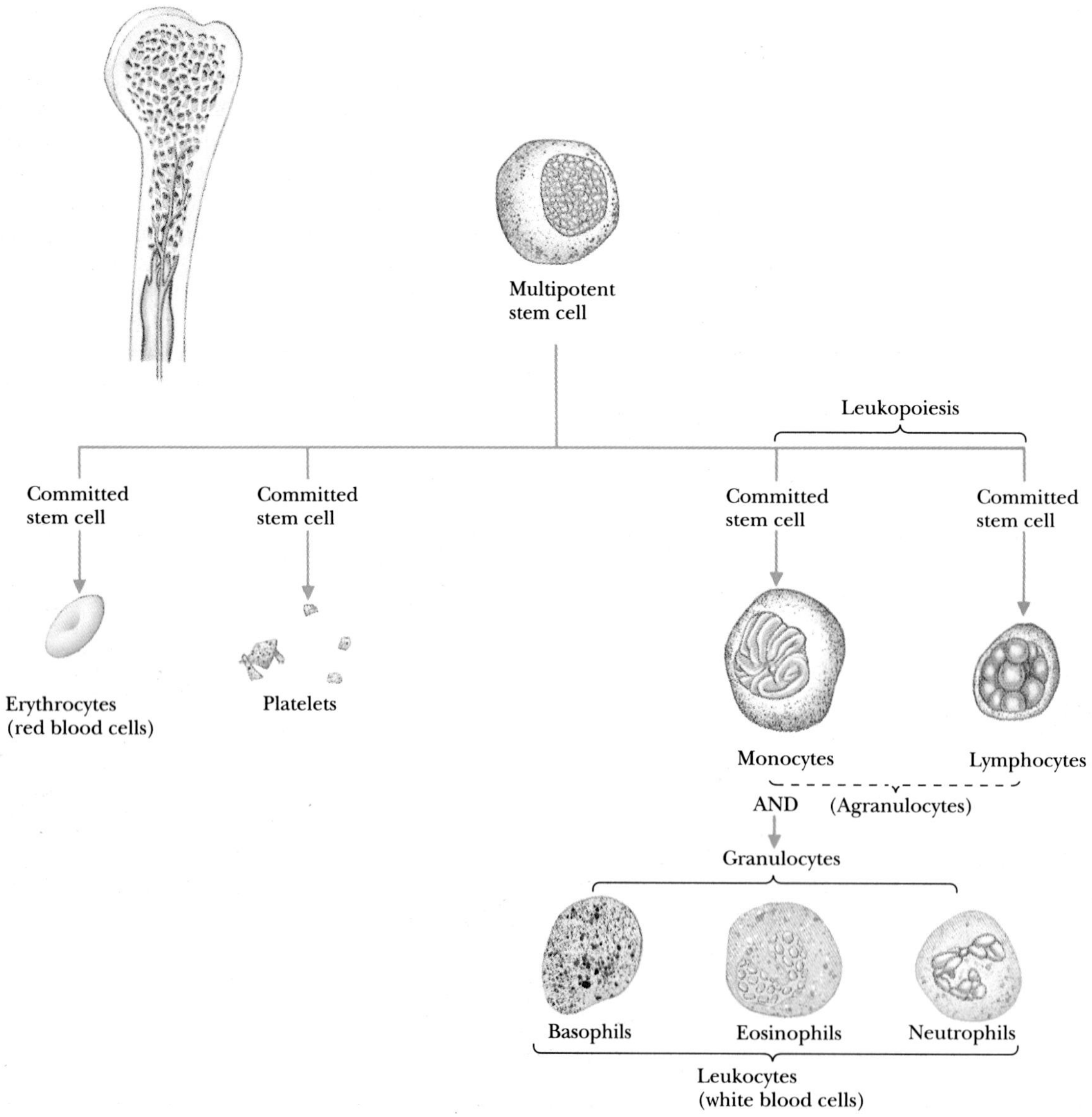

***Figure* 26.6** The origin of various cells that are present in human blood. The multipotent stem cell is produced in bone marrow and is differentiated into the blood cells.

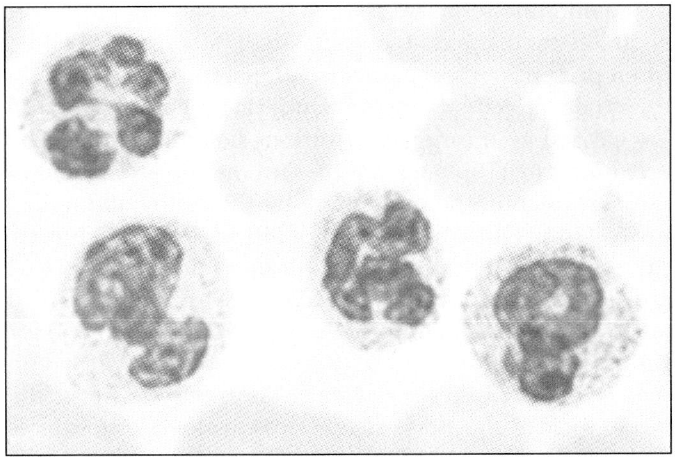

***Figure* 26.7** An electron micrograph of a neutrophilic granulocyte (a polymorphonuclear leukocyte or PMN) from rabbit bone marrow, showing two lobes of its dense nucleus and cytoplasmic granules. (© Manfred Kage/Peter Arnold, Inc.)

The PMNs move rapidly (40 μ meters/min) and are the first phagocytic cells to arrive at a site of inflammation. They can enter tissue by squeezing between the cells that line blood vessels and capillaries, a process called **diapedesis** (see Figure 26.4). PMNs are short-lived cells (half-life, 6 to 8 hours) that are present as a response to a variety of infections. Their increased presence is due to a stimulated rate of release from bone marrow, a response termed **leukocytosis.** Because they are short-lived, a relatively high number of PMNs in the blood is indicative of an active infection. In a chronic infection, PMNs are replaced by long-lived macrophages.

The nucleus of a PMN is divided into a number of segments (Figure 26.7)—hence, the name polymorphonuclear. PMNs also contain numerous granules that stain with neutral dyes (origin of the names granulocyte and neutrophil). These granules are **lysosomes,** small bags of enzymes that are a constituent of all animal cells. Lysosomes serve a number of functions. In the PMN, they contain hydrolytic enzymes that can disrupt and hydrolyze major constituents of a bacterial cell.

The sequence of events that occur during phagocytosis is depicted in Figure 26.8. The foreign object (here a bacterium) is engulfed and immediately enclosed in a vacuole called a phagosome, which is formed by invagination of the phagocyte plasma membrane. The phagosome separates from the plasma membrane. Immediately after phagocytosis and separation, a lysosome fuses with a phagosome releasing hydrolytic enzymes into the fused cell, called a **phagolysosome.** More than 60 distinct enzyme systems are released into the phagolysosome. Enzymes released by the lysosome include proteases, lipases,

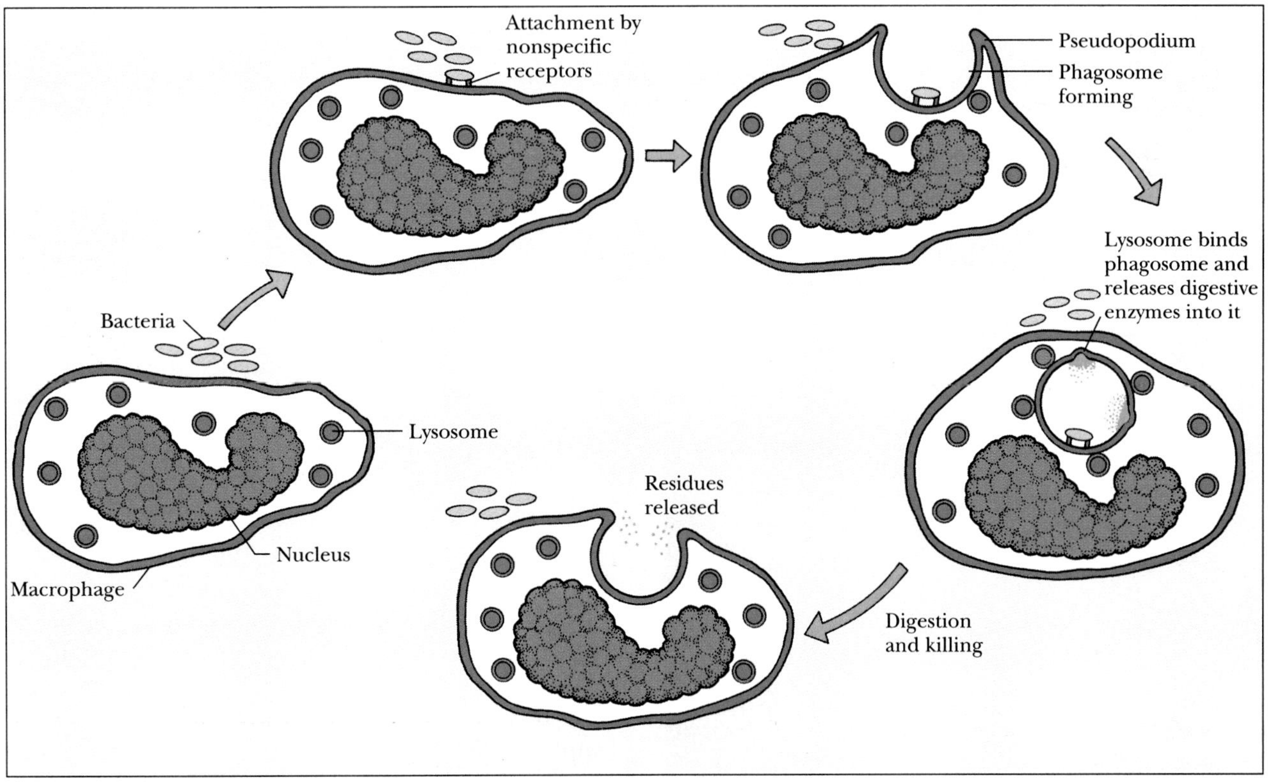

***Figure* 26.8** Events that occur during phagocytosis. The macrophage engulfs the bacterium, which leads to killing and digestion of the invader.

carbohydrases, RNAase, DNAase, acid phosphatase, peroxidase, and a myeloperoxidase system.

Following phagolysosome formation there is a burst of metabolic activity and stores of polyglucose (glycogen) are metabolized (respiratory bursts). Glucose released from glycogen is catabolized anaerobically via both glycolysis and the hexosemonophosphate shunt, resulting in the formation of reduced pyridine nucleotides (NADH and NADPH). The reduced pyridine nucleotides are utilized in production of peroxide and superoxide. These reactions are summarized as follows:

Peroxide formation $\longrightarrow H_2O_2$
$$NADPH + O_2 + H^+ \longrightarrow H_2O_2$$

Superoxide formation $\longrightarrow O_2^-$
$$NADPH + 2O_2 \longrightarrow 2O_2^- + H^+ + NADP^+$$

Myeloperoxidase – halide – peroxide system $\longrightarrow$ Hypochlorite
$$Cl^- + H_2O_2 \longrightarrow ClO^- + H_2O$$

The enzymes and oxidants (ClO^-, H_2O_2, O_2^-) kill and digest the microbial cell, and the debris is excreted from the PMN. Normal body functions then remove the relatively harmless debris from the blood.

Monocytes (Macrophages)

Macrophages are large cells (Figure 26.9) that play an immediate role in preventing invasion and also are involved in the immune response (see Chapter 27). Monocytes that circulate in the blood pass through blood vessels and, when present in tissue, are termed macrophages. Some of these become circulating or **wandering** macrophages and are carried by the lymph to various tissue locations. The wandering macrophages are present in lung alveoli and in the peritoneum, the membrane that lines the abdominal cavity. Others have a limited mobility and are **fixed** to tissue. Macrophages are present on epithelial surfaces that are exposed to the external environment. Fixed macrophages also line vessels through which blood or lymph flows and are present in liver, spleen, and lymph nodes. Macrophages sometimes are referred to as the mononuclear phagocyte system or **reticuloendothelial system.** The macrophages are actually a second line of defense since they appear at a site of trauma after the PMNs respond in the inflammatory response.

The macrophages are long-lived cells that are important in acute and chronic infections. They have fewer granules than PMNs, but the fixed macrophages continuously synthesize lysosomal enzymes. Macrophages engulf foreign invaders, form phagosomes, and fuse with lysosomes to generate a phagolysosome. They produce peroxide and contain the various hydrolytic enzymes. Macrophages apparently lack some of the activities of the PMN, including the hypochloride-generating myeloperoxidase system.

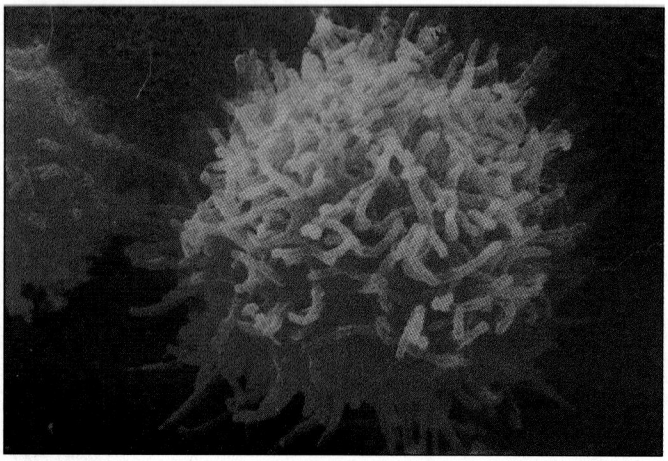

***Figure* 26.9** A macrophage. Note the finger-like projections that can seek out and capture foreign material. (© E. Shelton/D. Fawcett/VU)

The Lymphatic System

The **lymphatic** system functions in conjunction with the cardiovascular system (Figure 26.10). It consists of a network of lymph-carrying vessels, ducts, nodes, and other lymphatic tissues including the spleen (Figure 26.11). The lymphatic system is a significant part of the nonspecific host defense. The system carries lymph, a fluid that lacks red blood cells and most serum proteins, through tissue and back to the circulatory system. Lymph carries macrophages through tissue and picks up waste and debris and transfers it to the lymphatic capillaries. The lymphatic capillaries are small tubes with a single layer of endothelial cells that are permeable to interstitial constituents. The interstitium is the fluid-filled space between tissue cells. Low levels of the interstitial fluid enter the lymphatic capillaries and flow into the larger lymphatic vessels. The network flows through lymph nodes and eventually drains into the cardiovascular system.

The lymphatic system is a nonspecific system that monitors the health of tissue and is also the site of the specific immunological **(humoral)** response. The lymphatic system is the juncture where the cellular (nonspecific) and specific (humoral) responses to invasion come together. The lymphatic vessels have valves at certain points that permit only a forward flow. The lymph nodes are lined with macrophages that can engulf and digest foreign material carried by the lymph from interstitial spaces. The spleen

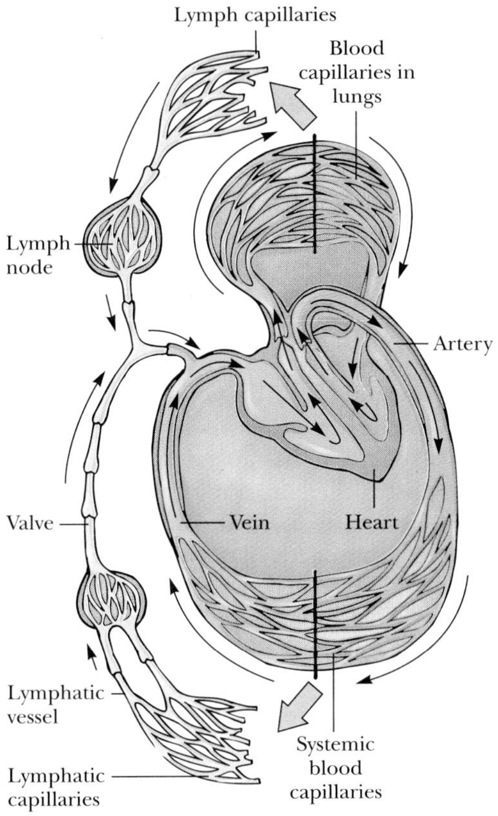

***Figure* 26.10** The interaction between the cardiovascular system and the lymphatic system. The red denotes oxygenated blood, and the blue oxygen-depleted blood.

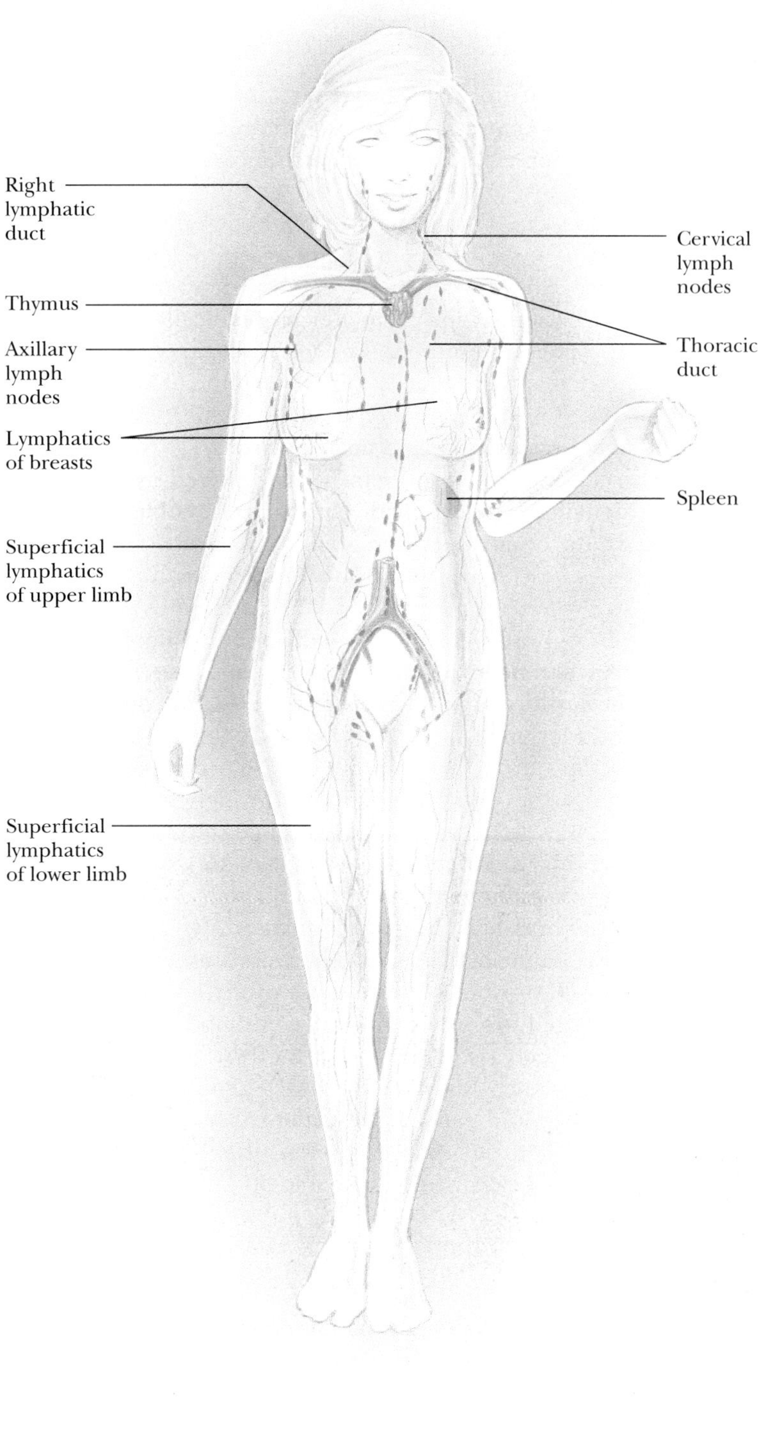

***Figure* 26.11** The lymphatic system in a human. The lymph-carrying capillaries are widely distributed while the lymph nodes occur in a limited area.

is the largest lymphoid organ and has a vascular network containing macrophages. The macrophages of the spleen phagocytize aged erythrocytes and other waste particles and remove them from the system. The normal host defenses are successful in maintaining most humans in a reasonable state of health. As previously mentioned these defenses are not ideal and are ineffective against many pathogens. Pathogens do have effective mechanisms that can overcome those of the host, and these will now be considered.

Pathogenesis Mechanisms

Despite the presence of natural barriers and the various nonspecific host defenses, diseases do occur. Our natural defenses can be breached by abrasions or cuts, allowing entry to pathogens, and they can initiate the infectious process. Diseases can be caused by any pathogen that penetrates the skin, mucous membranes, or epithelial surfaces, and successfully combats the natural defenses. An organism may have the capacity to attach specifically to selected membranes or epithelia and produce toxic substances that damage host cells. There is little defense against many of these pathogens, and infection is inevitable if the organism reaches a specific site.

Disease-causing organisms can be categorized as **obligate, accidental,** or **opportunistic** pathogens. An **obligate pathogen** would be an organism such as *Neisseria gonorrhoeae* or *Streptococcus pyogenes* that does not survive outside a living host. Survival then depends on the pathogen's ability to move from host to host and to **adhere** and **colonize.** An **accidental pathogen** would be an organism such as *Clostridium tetani* that is ubiquitous in nature and causes disease only under unusual circumstances. Tetanus caused by this organism can be fatal, but the disease occurs when *C. tetani* enters the body accidentally through a deep wound. The ability to cause a pathological condition does not play a significant role in the survival of this species. An **opportunistic pathogen** is an organism that does not affect a healthy host but can cause disease under certain circumstances. Some normal microbiota may cause disease if tissue is injured or when the resistance of the host is decreased. Prolonged use of antibiotics or immunosuppressive agents can decrease resistance of the host to opportunistic pathogens. Diseases, such as AIDS, suppress the immune response and render the patient susceptible to an array of opportunistic microorganisms that would not cause disease in a healthy human (Table 30.4).

Adherence and Penetration

As mentioned earlier, infectious organisms rarely have the ability to penetrate intact skin, mucous membranes, or external epithelia. Rather, they depend on various **adherence factors** that permit them to attach to selected tissue where they colonize and initiate the infectious process. There are instances where bacteria cause disease without specific attachment. Minor breaks in the skin or mucous membrane can lead to infection by *Staphylococcus aureus* or *Streptococcus pyogenes,* as these organisms are commonly present at these sites. Organisms can also enter wounds or burn areas by passive transmission on airborne dust particles or water droplets. *Pseudomonas aeruginosa* is a constant threat to burn patients, as this opportunist may be carried passively on airborne dust particles.

Most acute infections originate on mucosal surfaces of the respiratory, gastrointestinal, or genitourinary tracts. The ability of the pathogen to initiate the infectious process is frequently related to its capacity to cling to specific cells. The organisms that initiate infections on surfaces generally have factors that aid them in this process (Figure 26.12). These adherence factors are quite specific and do not interact with epithelial cells indiscriminately. Those that attach to throat mucosal cells would probably not attach to intestinal epithelia and *vice versa.* Some of the adherence factors that are involved in attachment of pathogens to host cells are listed in Table 26.2.

An organism that can cause disease in selected areas of the body may have distinct fimbriae or other surface factors that are involved with attachment. Different types of fimbriae are present on strains that attach to different tissues. For example, *Escherichia coli* strains produce fimbriae with differing lectin specificities, and these effect adherence to different host tissues and cause distinctly different diseases. A lectin is a protein on the bacterial cell surface that can bind to specific areas on the glycoprotein surface of another cell. Human pathogenic *E. coli* strains that cause urinary tract infections have P (pyelonephritic) type fimbriae, whereas S type fimbriae are present on *E.*

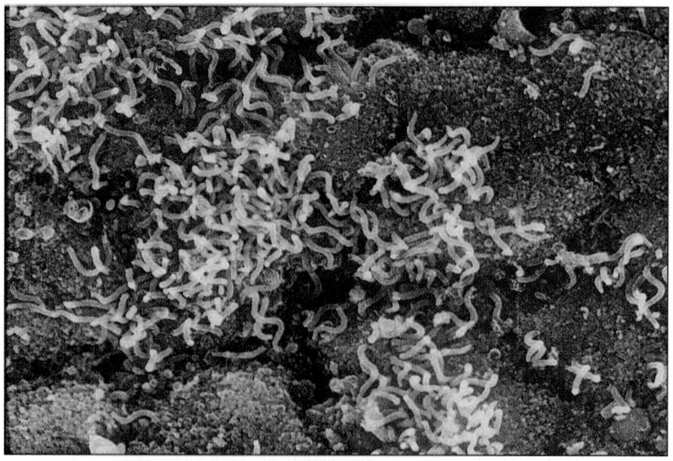

***Figure* 26.12** Specific adherence of microorganisms to epithelia. (© Veronika Burmeister/VU)

***Table* 26.2 Adherence factors involved in attachment of organisms to host cells**

Adherence Factor	Example
Fimbriae (adhesion proteins)	*Proteus mirabilis*–urinary tract infections
	Neisseria gonorrhoeae–attach to urinary epithelia
	Salmonella–attach to intestinal epithelia
	Streptococcus pyogenes–M protein attaches to epithelia
Capsule (glycocalyx)	*Streptococcus mutans*–dextrans attach to teeth
	Streptococcus salivarius and *S. sanguis*–attach to tongue epithelia
Teichoic acids Lipotechoic acids	*Staphylococcus aureus*–attach to nasal epithelia

coli strains that cause intestinal infections in the neonate. Genetic alteration that results in the inability to synthesize specific fimbriae leads to a loss of virulence in these organisms. *Neisseria gonorrhoeae* fimbriae attach to sites on the surface of buccal mucosa or cervical epithelial cells in the genitourinary tract. Strains that do not have adherence factors (fimbriae) do not cause disease. These organisms are swept away by the action of cilia or flushing actions.

Diseases that affect one animal species generally do not affect animals of other species. Different animal species have chemically distinct surface components on their epithelial cells. If the pathogen cannot adhere, it cannot infect and colonize.

Pathogens often remain localized on mucosal surfaces, where they reproduce and cause damage through the production of toxic substances. Whooping cough *(Bordetella pertussis)*, diphtheria *(Corynebacterium diphtheriae)*, and cholera *(Vibrio cholerae)* are examples of such diseases. Other microbes cause infection because they penetrate the epithelia and grow in the submucosa area. Once a pathogenic organism penetrates the natural barriers it may be carried by the lymphatic system or bloodstream to other areas of the body. *Neisseria meningitidis*, the causative agent of meningitis, is one such disease. A localized infection permits the organism to pass into the lymphatic system where it is carried to the bloodstream. It eventually may be transported to the meninges, where it adheres and reproduces, resulting in meningitis.

The ability of microbes to adhere is not necessarily detrimental to the host. Our natural flora also possess adherence and survival mechanisms. The presence of these organisms at various sites precludes invasion by potential pathogens that would endanger our health.

Colonization and Virulence Factors

To initiate the infectious process, pathogens must be able to establish a **growing** colony. If an organism possessed the ability to invade but could not increase in number, the natural defenses of the host would generally prevail. Nutrients are generally unavailable on surfaces as the normal microbiota utilizes them as they are produced. We know that a pathogenic organism cannot grow unless there is a source of soluble growth substrates such as sugars, amino acids, or fatty acids. Some minerals are also essential. Such essential nutrients are not present in the intercellular space. Therefore, host cell disruption is frequently necessary to obtain nutrients.

Host cell disruption can occur through activities of enzymes produced by pathogenic bacteria. Some of these enzymes that are released by pathogens are presented in Table 26.3. Collagenase, elastase, hyaluronidase, and lecithinase are enzymes that disrupt host cells by breaking down collagen, hydrolyzing the intercellular cement, or disrupting phosphatidylcholine, a component of the host cytoplasmic membrane. Once the host cell has been disrupted, its soluble cellular contents are then available to the pathogen. Coagulase is an enzyme produced by *Staphylococcus aureus* that promotes the formation of clots that prevent phagocytic cells from reaching and destroying the invader. *Staphylococcus aureus* and *Streptococcus pyogenes* also can synthesize an enzyme (streptokinase) that dissolves

***Table* 26.3 Some enzymes produced by pathogenic bacteria that promote invasion of the host**

Enzyme	Organism	Function
Collagenase	*Clostridium*	Breaks down collagen in connective tissue
Coagulase	*Staphylococcus aureus*	Clot formation around point of entry protects from host defenses
Elastase	*Pseudomonas aeruginosa*	Disrupts membranes
Hyaluronidase	*Streptococcus* *Staphylococcus* *Clostridium*	Hydrolyzes hyaluronic acid-intercellular cement
Lecithinase	*Clostridium*	Disrupts phosphatidylcholine in membranes
Streptokinase	*Staphylococcus* *Streptococcus*	Digests fibrin clots

clots, permitting the pathogen to move further into tissue.

Virulence, as previously discussed, refers to the ability of a pathogen to cause disease. There are several virulence factors, and these include the ability to bind iron, phase variation, and plasmid encoded virulence factors. A highly virulent organism is one that, given the opportunity, will cause an acute infection. It does not necessarily produce significant amounts of toxic material but then it must be highly invasive. Conversely, a pathogen may have little invasive ability but does synthesize potent toxins. *Streptococcus pneumoniae* has considerable invasive ability but does not produce a significant toxin. Yet, before the advent of antibiotics this organism was a leading cause of death. Organisms such as *Clostridium tetani* and *Clostridium botulinum* have virtually no invasive ability but produce highly potent toxins. Introduction of *C. tetani* into a wound can be fatal. Ingestion of botulinum toxin, even in small quantity, can also be fatal.

The virulence of a pathogenic organism can be measured by determining the number of microorganisms of a particular strain that must be injected or given to an animal to kill 50 percent of the population within a fixed period of time. This is called an LD_{50} (lethal dose–50% killing). Mice are often employed in LD_{50} tests. The more virulent pathogen will kill a test animal with exceedingly low doses. This is illustrated in Figure 26.13, where strain A would be highly virulent and strain B less virulent.

Iron is essential for the synthesis of cytochromes, the electron carriers that are involved in energy generation. Vertebrates retain iron in their system in a soluble state by binding reduced iron to high-affinity glycoproteins: lactoferrin in milk, tears, saliva, mucus, and intestinal fluids; and transferrin, the iron transport protein in plasma. The level of free iron in body fluids is less than 10^{-8} M. Unless an organism that requires iron can take it from lactoferrin or transferrin of the host it will have little chance of survival. Some bacteria can synthesize low molecular weight compounds (catechols or hydroxamates) that have a **higher affinity** for iron than lactoferrin or transferrin. These compounds are called **siderophores,** and there are receptors on cytoplasmic membranes of bacteria for the siderophore-iron complex. This complex can transfer the iron pulled away from the host to the growing bacterium. The genetic loss of the ability to synthesize siderophores can lead to a loss of virulence. Feeding soluble iron to in-

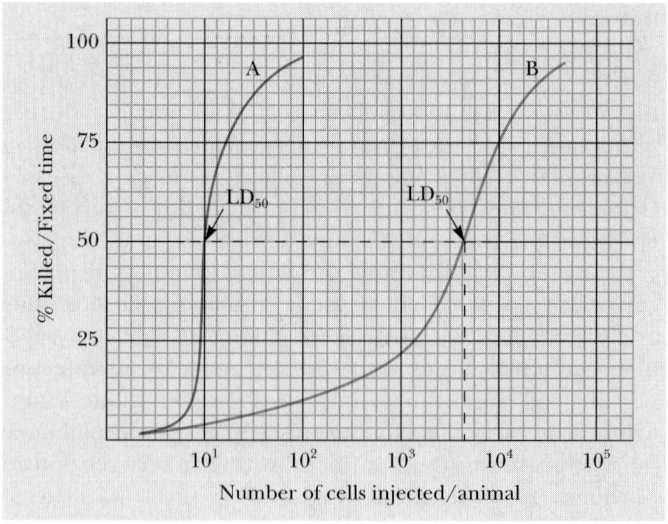

***Figure* 26.13** The LD_{50} for two strains of pathogenic bacteria. Strain A is more virulent, as fewer organisms kill 50 percent of the population in a shorter time.

Table 26.4 Virulence factors that are generally encoded in plasmids

Organism	Factor	Disease
Escherichia coli	Enterotoxin	Diarrhea
Clostridium tetani	Neurotoxin	Tetanus
Staphylococcus aureus	Coagulase enterotoxin	Boils/skin infections, food poisoning
Streptococcus mutans	Dextransucrase	Tooth decay
Agrobacterium tumefaciens	Tumor	Crown gall
Staphylococcus sp. *Streptococcus* sp.	Antibiotic resistance	Various infections

fected animals can markedly increase the severity of an infection and can lower the LD_{50}.

Another virulence factor is **phase variation.** Bacterial fimbriae are effective antigens, and host antibodies are often directed at the proteins in fimbriae. We previously discussed the role of fimbriae in adherence to host mucosa. The chromosomes of some pathogens have a complete gene for fimbriae synthesis that is routinely expressed. Other sites on the bacterial chromosome have unexpressed fimbriae genes that code for fimbriae of a different amino acid sequence. Recombination between fimbriae-coding genes can change part or all of the gene that directs fimbriae synthesis. This generates redesigned fimbriae that are not affected by antibodies that have been elicited by the protein in the original fimbriae. A virulent bacterium with altered fimbriae would therefore have an advantage against host defenses.

Virulence factors may be encoded in the plasmids of pathogens. This permits a rapid passage of genetic information from a limited number of virulent pathogenic microorganisms to a population with lower virulence. The organisms in an infection that have **plasmid encoded virulence factors,** such as antibiotic resistance, would survive and pass the information to others, increasing the level of antibiotic resistance in a large proportion of the population. Some plasmid-mediated virulence factors are listed in Table 26.4.

Interaction of Pathogens and Phagocytic Cells

A pathogenic organism can gain entrance into a host and establish residence only by overcoming host defenses. As mentioned earlier, one major defense system that a pathogen may encounter is the phagocyte. As a result, many successful pathogens have developed mechanisms for combating phagocytic cells.

To counteract phagocytes, some pathogens produce antiphagocytic factors. Some of these antiphagocytic factors are listed in Table 26.5. **Leukocidins,** produced by pneumococci, streptococci, and staphylococci, attach to

Table 26.5 Antiphagocytic factors produced by bacteria and their mode of action

Factor	Action
Leukocidins	Specific lytic agent for leukocytes including phagocytes.
Hemolysins	Form pores in host cells including macrophages. Streptolysin O affects sterols in membranes. Streptolysin S is a phospholipase.
Capsules (glycocalyx)	Long polymers of carbohydrate—physically prevents engulfment.
Fimbriae	(1) Bind to surface components of phagocytes, prevent close contact and phagocytosis may not occur. (2) Phase variation—a change in the antigenic composition.

the leukocyte membrane and promote disruption of the granulous lysosomes. The subsequent release of the hydrolytic enzymes from the lysosome destroys the integrity of the phagocytic cell. **Hemolysins** act by destroying the plasma membrane of host cells, including leukocytes. A primary function of hemolysins in invasion is to release nutrients from the cell for use by the pathogen.

Capsular material surrounding pneumococci and meningococci protect these organisms from phagocytosis. A pneumococcus without a capsule is not very effective in invasion as it will be readily captured and destroyed by phagocytes. *Borrelia recurrentis* has multiple genes for cell surface proteins. Alterations in cell surface proteins negate the actions of the macrophage/antibody defense enabling *B. recurrentis* to survive in infected individuals for long periods of time.

Brucella abortus, the causative agent of undulant fever in humans, has cell envelope components that inhibit phagocytic digestion after the bacterium is engulfed. By an unknown mechanism, the organism prevents degranulation (phagolysosome formation). *B. abortus* can multiply in phagosomes and survive in the human body for extended periods.

Legionella pneumophila forms microcolonies inside monocytes and is thereby resistant to normal host defenses. Also, virulent strains of *Mycobacterium tuberculosis* can grow in macrophages, and the intracellular bacteria accumulate to form a nodule. In an active infection, the nodules are surrounded by connective tissue forming the **tubercle** that is characteristic of tuberculosis. Chest X rays are employed to detect the presence of these tubercles in lungs. *Salmonella typhi,* the causative agent of typhoid fever, survives and multiplies in macrophages. When growing in macrophages they are protected from circulating antibodies.

The rickettsia that cause diseases such as Rocky Mountain Spotted Fever and typhus are intracellular pathogens. These rickettsia can grow in the cytoplasm or in the nucleus. Apparently, the organism growing in cytoplasm can pass through the cytoplasmic membrane without disrupting the membrane. This allows the organism to infect adjacent cells. When growing in the nucleus, the rickettsia do not pass through the nuclear membrane to the cytoplasmic area. This passage would destroy the host cell.

Toxins

Adherence, virulence factors, and a pathogen's defenses against phagocytosis are critical to a microorganism in gaining entry into a human host. The role of these factors has been discussed. The presence of bacteria is important in the infectious process but their mere presence rarely is the cause of disease symptoms. These symptoms are the result of **toxic agents** formed by pathogens and released into the body of the host. The toxins produced by bacteria are of two basic types—endotoxins and exotoxins. The major differences between an endotoxin and an exotoxin are outlined in Table 26.6. A discussion of these toxins follows.

Endotoxins

An endotoxin is actually part of the lipopolysaccharide complex that forms the outer envelope of gram-negative bacteria. Endotoxin is released upon lysis of the organism, or in some instances, during cell division. The lipopolysaccharide complex is composed of Lipid A, core polysaccharides, and the O-polysaccharide side chain (see Figure 11.15). The polysaccharide component has little apparent toxicity but is important because it tends to solubilize the complex. Lipid A exhibits virtually all of the toxic properties of the intact lipopolysaccharide complex. Endotoxins are quite similar regardless of the organism from

***Table* 26.6 Characteristics of endotoxins and exotoxins**

Endotoxins	Exotoxins
Heat stable	Heat labile 60°–80°C
Weakly immunogenic	Immunogenic
Cause fever	Cause no fever
Toxic at high doses	Can be lethal at low concentrations
Similar regardless of source	Different genera produce different toxins
Released on lysis of bacterium	Released by live bacterium
Not generally harmed by chemicals that affect proteins	Inactivated by chemicals that affect proteins

which they originate. They are rather stable to heat and are weak inducers of an antibody response. Consequently, it is difficult to establish an effective immunity to endotoxins.

Endotoxins are not toxic at low levels and are not highly specific as to the site they affect. The release of endotoxins into the human body at high levels can induce acute inflammatory responses that can lead to serious illness and/or death. Endotoxins damage the lining of blood vessels (endothelium) leading to the release of the Hageman factor and interleukin 1. Interleukin 1 is associated with events in the immunological response and will be discussed in Chapter 27. The Hageman factor is blood-clotting factor XII and can initiate the blood-clotting cascade. A cascade is a series of reactions that are "triggered" by a single event. Clotting of blood results from a cascade that ultimately leads to cross-linking of a fibrin network. The activation of the Hageman factor results in the major enzyme cascades outlined in Figure 26.14. The clotting cascade leads to the activation of thromboplastin and the development of blood clots. In the absence of adequate control mechanisms, thrombosis (blockage of blood vessels) can occur. This leads to widespread coagulation of blood in the vascular system. The clotting removes platelets from the blood system at rates that exceed their replacement and may cause hemorrhaging elsewhere in the body. This hemorrhaging can lead to the failure of essential organs such as the kidney.

The Hageman factor concurrently activates the cascade that leads to fibrinolysis (digestion of fibrin deposits). This may cause further hemorrhaging, with a decrease in blood pressure, depressed respiration, and loss of consciousness. The presence of high levels of endotoxin can bring about uncontrolled clotting/fibrinolytic cascades that cause circulatory collapse and potentially death. Activation of the kallikrein system activates the full inflammatory response discussed earlier. This response would promote vasodilation and drainage of blood from vessels, and this would further decrease blood pressure.

An exceedingly sensitive test for trace amounts of endotoxin is the Limulus amoebocyte lysate assay. This test employs amoebocytes of the horseshoe crab, *Limulus polyphemus.* An endotoxin, even in trace amounts, reacts with the clot protein from these circulating amoebocytes and causes clotting. This clotting can be measured spectrophotometrically and is an effective sensitive specific assay for endotoxins.

Exotoxins

An exotoxin is a heat-labile toxic protein released into the surrounding medium by a growing microorganism. If the bacterium releases an exotoxin within the human body it may travel from the focus of infection to other areas. The exotoxin produced by a species is generally unique, differing in both structure and function from other exotox-

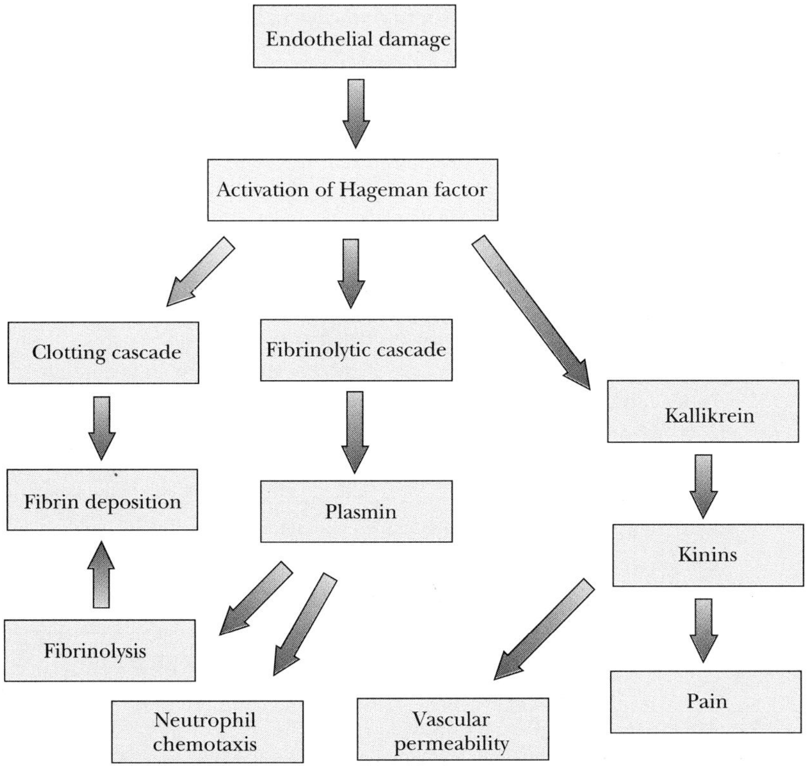

***Figure* 26.14** The reactions that follow endothelial damage by endotoxins. Both the cascade of reactions that lead to clotting and a series of reactions that lead to destruction of clots (fibrinolysis) are initiated by activation of the Hageman factor. The inflammatory response is also activated through kallikrein.

ins. The effects of exotoxins on tissues can be classified by site of action as follows:

Enterotoxins—cause dysentery and other intestinal distress. An example would be the cholera toxin.

Neurotoxins—affect nerve impulse transmission. Tetanus and botulinum toxin are neurotoxins.

Cytotoxins—destroy cells by inhibiting the synthesis of proteins. A cytotoxin can harm the heart, liver, and other organs. Diphtheria toxin is an example.

Pyrogenic toxins—stimulate the release of cytokines, leading to fever and shock. An example is the exotoxin produced by *Staphylococcus aureus.*

The exotoxins stimulate an effective antibody response, so infants are routinely inoculated against some of the diseases caused by exotoxin-producing microorganisms. Among the diseases that can be prevented by vaccination are diphtheria, tetanus, and whooping cough. A number of exotoxins, along with producing organisms and disease symptoms are outlined in Table 26.7. The mode of action of specific exotoxins will be discussed in Chapter 29.

Toxins and Fever

Pyrogens are substances that cause an increase in the temperature in the human body. This increase is generally referred to as a **fever.** The normal temperature in humans is 37°C (98.6°F) and varies little in a healthy individual. The body temperature is controlled by the hypothalamus, a part of the brain. Body temperature generally increases with strenuous exercise and decreases during sleep. A major factor in increased body temperature during an infection would be the presence of endotoxins. Selected

***Table* 26.7 Some exotoxins produced by bacteria**

Exotoxin	Producing Organism	Disease	Effect
Diphtheria toxin	*Corynebacterium diphtheriae*	Diphtheria	Inhibits protein synthesis; affects heart, nerve tissue, liver
Botulism toxin	*Clostridium botulinum*	Botulism	Neurotoxin; flaccid paralysis
Perfringens toxin	*Clostridium perfringens*	Gas gangrene	Hemolysin, collagenase, phospholipase
Erythrogenic toxin	*Streptococcus pyogenes*	Scarlet fever	Capillary destruction
Pyrogenic toxin	*Staphylococcus aureus*	Toxic shock syndrome	Fever, shock
Exfoliative toxin	*Staphylococcus aureus*	Scalded skin syndrome	Massive skin peeling
Exotoxin A	*Pseudomonas aeruginosa*	—	Inhibits protein synthesis
Pertussis toxin	*Bordetella pertussis*	Whooping cough	Stimulates adenyl cyclase
Anthrax toxin	*Bacillus anthracis*	Anthrax	Pustules; blood poisoning
Enterotoxin	*Escherichia coli*	Diarrhea	Water and electrolyte loss
Enterotoxin	*Vibrio cholerae*	Cholera	Water and electrolyte loss
Enterotoxin	*Staphylococcus aureus*	"Staph" food poisoning	Diarrhea, nausea
Enterotoxin	*Clostridium perfringens*	Food poisoning	Permeability of intestinal epithelia
Neurotoxin	*Clostridium tetani*	Tetanus	Rigid paralysis

microbes, some immunological processes, and some exotoxins can also cause a temperature increase. Localized heat may also accompany an inflammatory response.

A temperature increase engendered by gram-negative endotoxins results from the interaction of the toxin with phagocytes and other leukocytes. During phagocytosis, **endogenous pyrogens** are released by the phagocytes that travel to the hypothalamus and interfere with the thermostat that controls body temperature. A rise in temperature can occur within 20 minutes of this pyrogen/hypothalamus interaction. The factor released from phagocytes that acts as an endogenous pyrogen is the cytokine interleukin-1. Interleukin-1 is a protein secreted by macrophages that are involved with nonspecific host defenses. Lymphotoxin, a cytotoxic factor released by lymphocytes is also an endogenous pyrogen.

An increase in temperature may be a positive response, as it can stimulate T-cells, increase the rate of phagocytosis, and promote immunological processes in general. There is some evidence that an increase in temperature may decrease the amount of iron available in the system. This would interfere with microbial growth. As many pathogens grow optimally at 37°C an increase of 2° to 3°C could curtail microbial growth. "Chills" that can occur with a fever is an effort by the body to increase muscular activity to drive the temperature upward. A temperature much above 41°C (105.8°F) can lead to systemic damage and be fatal.

Summary

- Most of the microorganisms normally present in or on humans have an alliance that is either **mutualistic** or **commensalistic.** In some cases the microorganism is a **parasite.**

- A **parasite** that causes measurable injury to the host is a **pathogen. Virulence** measures the ability of a pathogen to inflict damage. An **infection** is a situation where the pathogen grows in the host and inflicts damage.

- A healthy human has a **natural microbiota** that is beneficial.

- The skin has a natural bacterial population that is mostly **gram-positive bacteria.** The number on drier areas is 10^2 to 10^4 per square centimeter, and there are about 10^6 per square centimeter in the moist areas. Natural skin secretions include amino acids, fatty acids, urea, lactic acids, lipids, and mineral salts. These provide nutrient for the normal microbiota.

- Fluid in the mouth has about 10^8 bacteria per ml. The basic bacterial population is not dislodged by swallowing as they adhere to the teeth, tongue, and other surfaces.

- The **nasopharynx** has adequate moisture and nutrient for growth of bacteria.

- The **lower respiratory tract** is mostly devoid of bacteria. Dust and other particles that enter this region are moved upward to the throat by action of **cilia** and swallowed.

- The number of organisms in the stomach is limited to some acid-tolerant lactobacilli and yeasts. There are fewer than 10 microbes per ml of stomach fluid. Colonization of the stomach by *Helicobacter pylori* can cause peptic ulcers.

- The small intestine has an increasing number of bacteria as it proceeds downward.

- The large intestine is a fermentation vat with 10^{10} to 10^{11} bacterial cells per gram mass, and the species diversity is broad.

- The **genitourinary** tract of a healthy human is relatively free of bacteria but the lower part of the urethra has a bacterial population.

- The most important defense against infection is barriers such as skin, membranes, and **cilia** that ward off invaders. **Acidity** in the stomach and vagina are also effective in protection against pathogens. **Colonization** by nonpathogens can crowd out potential invaders.

- The **inflammatory response** is an important nonspecific defense against entry of pathogens. This response brings **phagocytic** and other protective cells to the site of injury.

- **Phagocytes** are body defense cells that engulf and destroy invaders. Phagocytic cells have **lysosomes,** compartments containing an array of destructive enzymes. Polymorphonuclear leukocytes and macrophages are important phagocytic cells.

- The **lymphocytes** are a nonspecific system that constantly monitors health of tissue.

- **Adherence** to specific tissue is a major aspect of **pathogenesis.** A pathogen generally has to adhere and produce **toxic** substances to cause disease.

- Pathogens can be divided into three general categories: those that live only in and infect a host are **obligate** pathogens, those that can cause infection if given access by trauma are **accidental pathogens,** and those that infect individuals debilitated by such factors as lowered immunity are **opportunistic pathogens.**

- Virulence can be measured by the number of microorganisms that must be administered to a host population to kill 50 percent in a fixed period of time. This is called an LD_{50} (lethal dose).

- Some pathogens can survive in phagocytes, and this disseminates the pathogen within the host.

- **Disease** is generally caused by toxic substances produced—not by the presence of microorganisms.

- The toxins produced by microorganisms are the **endotoxins** that are part of the bacterial cell and the **exotoxins** that are released into the surrounding medium.

Questions for Thought and Review

1. How do virulence and pathogenicity differ? How would one measure virulence?

2. Define: host, commensalism, infection, and normal microbiota.

3. What are some factors on human skin that promote growth of bacterial inhabitants? Some that inhibit or limit bacterial colonization?

4. Why is brushing one's teeth important in preventing dental caries? How do microorganisms contribute to caries?

5. How do the bacterial populations of the stomach and large intestine differ qualitatively and quantitatively? What role do the facultative aerobes play in the human gastrointestinal tract?

6. What is the source of the acidic pH in the genitourinary tract of females? How does the bacterial flora in this area prevent invasion by pathogens?

7. Is inflammation a positive or a negative in defending against infection? Give reasons for your answer.

8. What are phagocytes? How do PMNs differ from macrophages? What does a high PMN count in human blood signify?

9. The lymphatic system is important in removing bacteria and other debris from tissue. How does this occur? What is the role of the lymphatic system? How does a humoral response differ from a cellular response? How do these relate to the lymphatic system?

10. How does an obligate pathogen differ from an accidental pathogen? What is an opportunist? Cite examples.

11. Adherence is important in infection. Why? Cite diseases where adherence is not essential.

12. What are some of the virulence factors and how do they function in promoting the survival of pathogens? What are some of the antiphagocytic factors?

13. How do endotoxins and exotoxins differ chemically, in toxicity, heat stability, and origin?

14. How do endotoxins and exotoxins differ in their mode of action? What are the reasons for this? How would you determine whether an intravenous fluid might contain endotoxin?

Suggested Readings

Ayoub, E. M. 1990. *Microbial Determinants of Virulence and Host Response.* Washington, DC: American Society for Microbiology.

Iglewski, B., and V. L. Clark, eds. 1990. *Molecular Basis of Bacterial Pathogenesis.* New York: Academic Press.

Kreier, J. P., and R. F. Mortensen. 1990. *Infection, Resistance and Immunity.* New York: Harper & Row.

Roth, J. A., ed. 1988. *Virulence Mechanisms of Bacterial Pathogens.* Washington, DC: American Society for Microbiology.

Salyers, A. A., and D. D. Whitt. 1994. *Bacterial Pathogenesis.* Washington, DC: American Society for Microbiology Press.

Schaechter, M., G. Medoff, and D. Schlessinger, eds. 1989. *Mechanisms of Microbial Disease.* Baltimore: Williams and Wilkens.

Sherris, J. C., ed. 1990. *Medical Microbiology, an Introduction to Infectious Diseases.* 2nd ed. New York: Elsevier Science Publishing.

Part 7

Immunology and Medical Microbiology
A Conversation with Stanley Falkow

Professor Stanley Falkow from Stanford University is an internationally renowned medical bacteriologist. He is well known for his studies of the genetics of bacterial pathogenesis. In particular, he has investigated genes involved in pathogenesis that are located on plasmids and more recently "pathogenesis islands." His career has been especially remarkable because his laboratory has covered, in-depth, such a broad array of bacterial pathogens. Dr. Falkow is the recipient of many scientific awards. He was recently elected President of the American Society for Microbiology.

JS: What led to your decision to study microbiology?

SF: When I was 11 years old, I went to the library and happened to see a book called *The Microbe Hunters*. I took it home and I read it. I decided there and then that I would be a microbiologist. Prior to that, I thought I might be an airplane mechanic. But my mother said to me, "Whoever heard of a Jewish airplane mechanic?" From then on, when anyone asked me what I wanted to do, I said I wanted to be a bacteriologist.

In subsequent years, I've talked to a number of people who were influenced by the same book, but most wanted to be physicians. I had no desire to be a physician. I was more attracted to the description of life in the laboratory and experimentation. I was so taken by microscopy. I was fascinated by Koch's description of looking in the microscope and discovering the anthrax bacillus and watching its growth. So I worked. I actually bartered my services at a toy store. I put bicycles together (poorly) for which I was repaid, not with money, but with a genuine Gilbert Hall of Science microscope. I made a hay infusorian, which means I threw grass in water and prayed. By accident, I happened to find a rotifer. Have you ever seen a rotifer? Rotifers are remarkable. I saw this transparent animal with pulsating cilia. I could make out an internal structure. And I thought, my god, you can really understand things by looking through a microscope! I was hooked at this point.

JS: So this was when you were in high school?

SF: This occurred in grammar school. But I was very undisciplined. I still may be. I was a poor student. This didn't change until my junior year in high school. I had been told by my high school counselor that I should save my parents and myself a whole lot of money and join the army, that I wasn't going to amount to anything. I might have followed his advice except that the Korean War broke out. I decided to get all A's.

JS: Tell us about your undergraduate experience.

SF: I went to the University of Maine at Orono. When I arrived at Maine, I was the beneficiary of some remarkable teachers. The head of the department was E. R. Hitchner and under him was a cadre of professors who taught bacteriology—Charles Buck, Frank Dalton, and a biochemistry professor, Charles Radke, and in biology, a parasitologist, Marvin Meyer. They could not have been nicer. They nurtured me, for want of a better word. They were interested in teaching me to do research. I now understand how fortunate I was. I had these people virtually to myself.

(Courtesy of Dr. Stanley Falkow)

I knew from the outset that I wanted to be a medical bacteriologist. At the end of my freshman year at Maine, I wrote a letter to the head of the hospital laboratory in Newport, Rhode Island, the city where I was raised. I wrote, very dramatically, that I would do anything if they would let me work in the lab and teach me medical bacteriology. He wrote me back and (unfortunately) took me at my word. He said, "We can't pay you, but if you'll agree to be an assistant in the autopsy room, we will teach you bacteriology." It was 1952 and I was 18 years old, and suddenly I was in an autopsy room. At the same time I was paired with a bacteriologist named Alice Schaeffer Sauzette. She had a classical microbiology training and a masters in bacteriology. On my first day, I sat there surrounded by all these jars of different kinds of stools and other effluvia. Suddenly, I was learning medical bacteriology. I did this every summer and vacation during college. During the academic year, I was learning the basic information that one needed on the biology of microorganisms.

JS: So you went to Michigan to graduate school?

SF: I went to the University of Michigan because Hitchner said to me he had never had a student accepted at Michigan. On that very intelligent basis, I chose a graduate school. In retrospect, I would encourage students to take some time off in their education to find out about the reality of their career choice if they possibly can, because I think that it is useful to know what your career choice actually entails. And then, go learn. But I didn't have much of a choice in that decision. I actually left my studies at Michigan and went back to the hospital laboratory. For two years I did nothing but full-time hospital laboratory work including autopsies. I was in a hospital lab without many research opportunities. I misidentified the *Citrobacter* as a *Salmonella*. One doesn't mean much clinically and the other, of course, is very important. Of course, when it was sent out for confirmation, I was mortified to learn I had made a mistake. I'm sure the patient was as well.

So, I tried to find out how to better distinguish between these two organisms. I was referred to a Scandinavian journal, which discussed a very difficult biochemical test that distinguished the two. I began to do a research project to come up with a simpler way to make the distinction. In the two years I was in the hospital lab, I actually published two papers. My research budget was something like $12.87. But the point is that I learned you could do satisfying research with very few tools as long as you had a question to ask and an experimental way of addressing the question. You could obtain answers and those answers

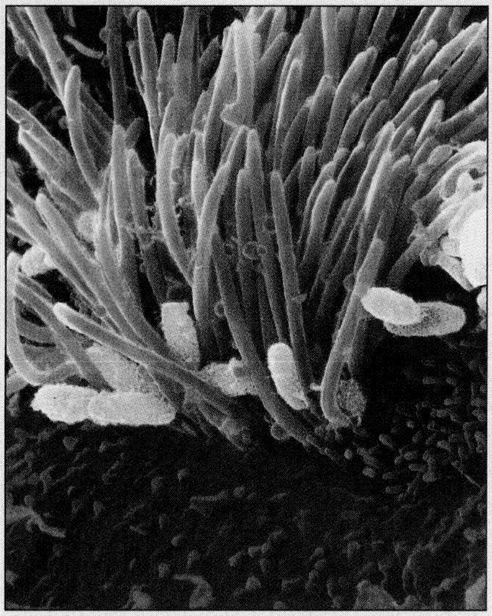

Bordetella pertussis. (© NIBSC/Science Photo Library/Photo Researchers, Inc.)

could be satisfying and could be of enough significance that you could communicate the results to other people by publishing a scientific article. That was a very important lesson for me.

All through my early career, I was the beneficiary of people who cared about me. I have a real concern now that this kind of mentoring is not as prevalent as it used to be. I found these people to be quite selfless. They considered it part of their career. In my career, I have trained well over 100 people in my laboratory. If I have a legacy, it is these people. It is not anything I have ever published or anything of that nature. It is the people whom I have trained.

JS: Your laboratory has been a major contributor to our understanding of pathogenic bacteria, particularly in your seminal studies on plasmid forms and features. Of your major contributions in the area of pathogenesis, which ones stand out in your mind and why?

SF: Well, it's hard to say. I don't think I've personally made a seminal discovery. There are a few things that we did in the laboratory that were done by my students rather than by me. For example, while Allison Weiss was a graduate student at the University of Washington, she was able to establish a genetic system for *Bordetella pertussis*. A vaccine was available and the disease seemed to be curtailed, so there wasn't a lot of interest in looking at it. Then the vaccine was found to be not as good as everyone thought. We needed a new vaccine and everyone suddenly realized that no one knew anything about the biology of this organism. How do you approach it? Here was a young woman who was not even 30 years old. She pioneered the genetic analysis of *Bordetella pertussis* and identified the bacterial genes associated with pathogenicity. Allison's work established the foundation for a new generation of pertussis vaccines. Her work was seminal in the literal sense.

Shortly after I arrived at Stanford, a postdoctoral fellow, Ralph Isberg, was able to use a clever method to identify genes associated with the bacterial invasion of animal cells. This work, I think, had a major impact on the field of medical bacteriology. There were other research findings. Many were serendipitous.

You know when most people think of medical bacteriology, they think of the disease not the microbe. . . . I often say that disease is a distraction from understanding the biology of the host-parasite interaction.

You know when most people think of medical bacteriology, they think of the disease not the microbe. Koch was interested in tuberculosis and its cure. I often say that disease is a distraction from understanding the biology of the host-parasite interaction. Actually, 95% of the people infected with the tubercle bacillus don't go on to develop clinical disease. But they do continue to carry the organism for the remainder of their life. To me the interesting aspect of tuberculosis is not the disease but how the organism manages to persist. What genes are turned on by the bacteria to activate our host defense mechanisms. We now have the technology, I think, to go after questions like this.

JS: What exciting areas do you see developing in the next decade or so in basic research aspects of pathogens?

SF: Even though bacterial chromosomes are being sequenced, we understand, at best, the function of only one-third of them. And our understanding is limited to the catalytic action of the enzyme. How an enzyme choreographs its interaction with the multitude of other enzymes in its cell to permit a microorganism to live in a particular niche remains a future challenge.

Understanding the evolution of how microorganisms become pathogenic is going to be an important area of research. It is a bit like Jurassic Park. We try to work backwards using DNA sequences to understand the steps that took place in what must have been ancient interactions between predator and prey. It is important to recognize that amoeba and nematodes and other small creatures depend on bacteria as a food source. Bacteria had to learn to resist and that was the beginning of the early predator-prey relationship. Later the bacteria not only resisted, but turned the tables on their predators. The evolution of many pathogenic traits probably can be traced back to the interactions of an amoeba and other organisms. I think we will find that the bacterial genes important in plant pathogenicity were the progenitors to what we see as human pathogens in bacteria.

JS: That comes to another question—the recent work of Ausubel. It's quite interesting that he's found that the same genes of *Pseudomonas aeruginosa* are involved in both plant and animal pathogenesis.

SF: Right. *Pseudomonas* is not really a very good animal pathogen.

JS: Can you elaborate?

SF: One of the things I have always found fascinating is that *Pseudomonas aeruginosa* has an admirable array of toxins and other pathogenic traits, but is harmless to normal humans and normal animals. It's only devastating to those animals that are immunocompromised. Why does the organism come armed this way? Well, Ausubel's data indicates it must be plants. And, why not? When you look at the structural basis for pathogenicity in *Pseudomonas* and you look at the structural basis for the other animal pathogens, say *Salmonella,* they have a lot of striking similarities. Indeed, homologous genes indicate homologous strategies that have evolved including contact-dependent secretion of virulence factors.

JS: There always seems to be another bacterial pathogen that is responsible for a new disease or malady that has not been previously attributed to a bacterium. Recent examples include *Helicobacter pylori* for ulcers, *Borrelia burgdorfii* for Lyme disease and the suggestion of *Chlamydia pneumoniae* for some types of myocardial infarction. Do you believe there will be more surprises?

SF: Host-parasite relationships are dynamic. We've recognized for only a little over a hundred years that microorganisms can cause a disease. Historically, the most important determinant of human survival has been infectious diseases. There's no reason to believe that this fact won't continue to be the case. However, what have been called "new diseases" in many cases are what I call "diseases of human progress" in the sense that we have shifted the balance of predator-prey relationships. Lyme disease has appeared because we have interfered with a healthy relationship that existed between a tick, a mouse, and a deer. Humans suddenly entered into the equation. In the initial interaction, the tick doesn't get very sick from the *Borrelia burgdorferi,* nor does the mouse, nor does the deer, although they are very good at transmitting it to one another. To be a pathogen, an organism has to replicate at the expense of the cellular integrity of something else, but not necessarily kill it or even make it clinically sick. The two animals that get the most ill from *Borrelia burgdorferi* are humans and their faithful companion, the dog. They are recent parts of this cycle.

> **I'm no less excited about things going on in my lab now than I was when I was young. In many ways it's much more exciting than it was when I was just beginning. Young people worry that there will be nothing left for them to discover. But the best is yet to be.**

There are subtle changes in human behavior we don't even think about. For example, ask contemporary students, "How many of you take a bath on a regular basis?" Very few people do. Most Americans and Western Europeans take showers. People don't think of that as a major behavioral change. Yet, 50 or less years ago, more people took baths rather than showers. For the transmission of certain diseases, that's an enormous difference. We don't just immerse ourselves in water now, we actually enter a fine aerosol that contains organisms in the environment that get by our purification systems. In Legionnaire's disease, the causative microbe evolved to live in a fresh water environment possibly inside amoeba. People taking a shower now deliver an organism like *Legionella* by aerosol into their lungs. This may not be a problem to "normal" people, but there are plenty of people who are immunocompromised. In susceptible individuals, *Legionella* grows as well in an alveolar macrophage as it does in amoeba.

JS: Bacteria are so surprising because of their great metabolic diversity. Can you provide one or two examples of a time when you learned something new that could not have been foreseen earlier but really surprised you?

SF: Most every day. I am constantly amazed at the way microorganisms do their thing. In their interactions with host cells, there was a view that pathogens were these brutish things, spewing poisons and raising havoc with our body, almost as if they were evil. The definition of pathogen was any microorganism capable of causing disease. I don't view pathogens any differently than I view any other microbial specialist. I don't think it's appreciated that a pathogen is just a microorganism whose survival is dependent on causing some degree of cellular damage, because to live, it has to get to a place where other microbes can't. The bacterial pathogens and the organisms that we harbor, those that live on us, understand our biology better than we do—at least in a microcosm. I argue that when we study a pathogenic organism, we not only learn about the human medical disorders and how pathogens work but we also learn about human biology. You always learn about the host biology when you learn about a pathogenic microbe's biology.

What surprises me is that we know so little about the organisms we carry. Numerically, we carry more bacterial cells than we do our own cells.

JS: What advice would you give to undergraduate students who are interested in pursuing a career in microbiology?

SF: Microbiologists will have the luxury of having the first sequences of the whole organism. There is a number of people who think that if you have the DNA sequence, you know about the biology of that organism. That's nonsense. But it does remove much of the tedium of research and it provides us with important clues about the organization of chromosomes and the structure and function of bacterial genes. Having a sequence also gives us more time to develop experiments and to think about the biology. The laboratory will expand from studies of isolated bacterial populations in the confines of a test tube into an explanation of the actual environment in which the organisms live. I think we will have probes and the ability to monitor the interactions of organisms in the real world. Microbiology will be at the forefront. Microbiology always has been at the forefront of most biological sciences because of the relative simplicity of the organisms and the tools that have been available. When you work with the simplest free living things in the world, then you always have that kind of advantage. I think it will be possible for microbiologists—as we were the pioneers for the development of genetics and molecular biology in this century—to be the pioneers in understanding the biology of a lot of microorganisms, the actual biological interactions and complexities. We will remain the pioneers of the life sciences for the foreseeable future.

I'm no less excited about things going on in my lab now than I was when I was young. In many ways it's much more exciting than it was when I was just beginning. Young people worry that there will be nothing left for them to discover. But the best is yet to be.

Stimulate the phagocytes. Drugs are a delusion. Find the germ of the disease; prepare from it a suitable antitoxin; inject it three times a day, quarter of an hour before meals; and what is the result? The phagocytes are stimulated; they devour the disease; and the patient recovers—unless, of course, he's too far gone.

Bernard Shaw—*The Doctor's Dilemma*

Chapter 27

Defense of the Body: Basic Features of the Immune System

The living animal body contains all the components necessary to sustain life. It is warm, moist, and full of nutrients. As a result, animal tissues are highly attractive to a vast array of microorganisms. A striking example of the power of the body's defenses is demonstrated by the speed at which an animal body begins to decompose after death. A live animal can prevent microbial invasion for many years, but once dead, it can be destroyed within hours. The tissues of living, healthy animals are resistant to microbial invasion because they have a vast array of defense mechanisms, and the survival of an animal depends on the successful defense of the body against microbial invasion. This defense is encompassed by the discipline of immunology and is the subject of this chapter.

Because of the essential nature of the defense of the body, it is critical that the body effectively exclude all invaders that may cause disease or reduce an animal's ability to survive. As might be expected, therefore, the body does not rely on a single mechanism for this defense. It must have multiple, overlapping defense systems available.

Some are broadly effective against a range of invaders; others are specific for individual infectious agents. Some will act at the body surface to exclude invaders; others will act deep within the body to destroy organisms that have broken through the outer defenses. Some will be optimized to defend against bacterial invaders, some against viruses that live inside cells, and even some against large invading organisms such as the parasitic worms. Clearly the defenses of the body form a complex system of overlapping and interlinked defense mechanisms that together will be able to destroy or control almost all invaders. A failure in these defenses either because the immune system is destroyed (as occurs in AIDS) or because the invading organisms can overcome or evade the defenses (as in a disease such as rabies) will inevitably result in death. The immune system is not simply a useful system to have around. It is essential to life itself.

The Levels of Host Defense

Because the successful exclusion of microbial invaders is essential for an animal's survival, it is not surprising that animals use a great variety of defense strategies. Indeed, the body has multiple layers of defense (Figure 27.1). The first and most obvious defenses are the physical barriers to invasion. Thus the skin provides an effective barrier. If it is damaged, the wound healing process ensures that it is repaired very rapidly. On the other body surfaces, simple physical defenses include the "self-cleaning" processes—coughing, sneezing, and mucus flow in the respiratory tract, vomiting and diarrhea in the gastrointestinal tract, and urine flow in the urinary system. The presence of an established normal microbiota also serves to exclude many potential invaders.

Physical barriers, although very helpful, cannot be entirely effective in themselves. Given time and persistence, an invader will eventually overcome mere physical obstacles. The second layer of defenses, therefore, consists of responsive defense mechanisms. These are specific defense mechanisms that can be focused on threatened areas to temporarily enhance local defenses. As was discussed in Chapter 26, in the body this focused defense response occurs in the form of inflammation. In inflammation, local changes or damage in tissues brought about by microbial invasion result in increased blood flow and the accumulation of cells that attack and destroy the invaders. These cells, called neutrophils and monocytes, can usually destroy most invading organisms and prevent their spread to uninfected areas of the body. The body also uses some enzymes that are triggered by the presence of invaders to cause microbial destruction—the complement system. Some of the cells involved in inflammation may also repair damaged tissues and return them to normal once the invaders have been destroyed.

These focused processes contribute significantly to the defense of the body. Animals that have defective inflammatory responses will die as a result of overwhelming microbial invasion. Nevertheless, these processes cannot offer the ultimate solution to the defense of the body. What is really needed is a defense system that can stop invaders, destroy them, and then learn from the process. This would be a system that can learn to recognize the invaders when it encounters them again and that can respond even more rapidly and effectively. This type of **adaptive response** is the function of the **immune system** and the topic of this chapter. As an animal mounts specific immune responses against an invader, the chances of successful invasion by that organism will eventually be reduced to very low levels. The immune system, therefore, provides the ultimate

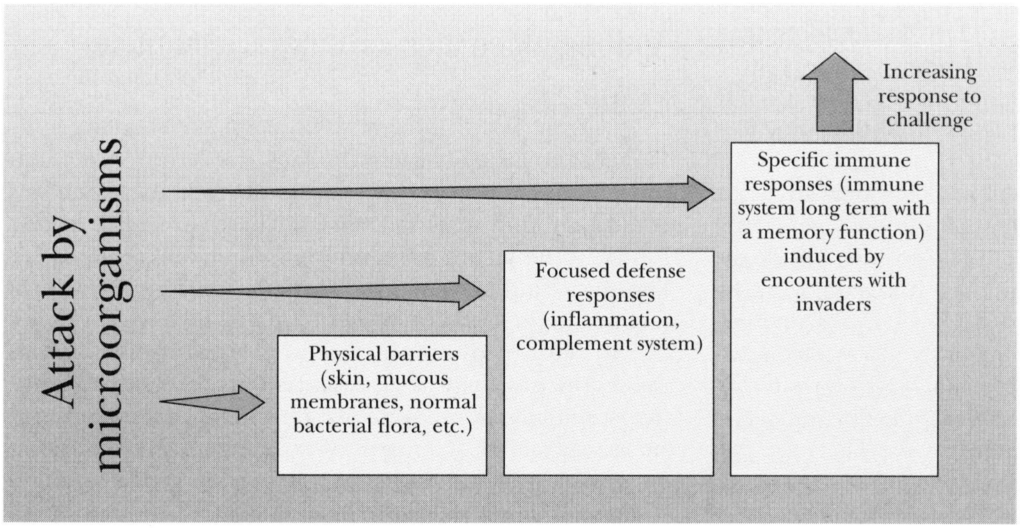

***Figure* 27.1** There are three forms of barriers to invading organisms: first, the purely physical barriers to invasion; second, the focused responses that serve to limit microbial spread; and third, the specific immune responses that respond to invasion by increasing in potency and, as a result, form an impenetrable barrier.

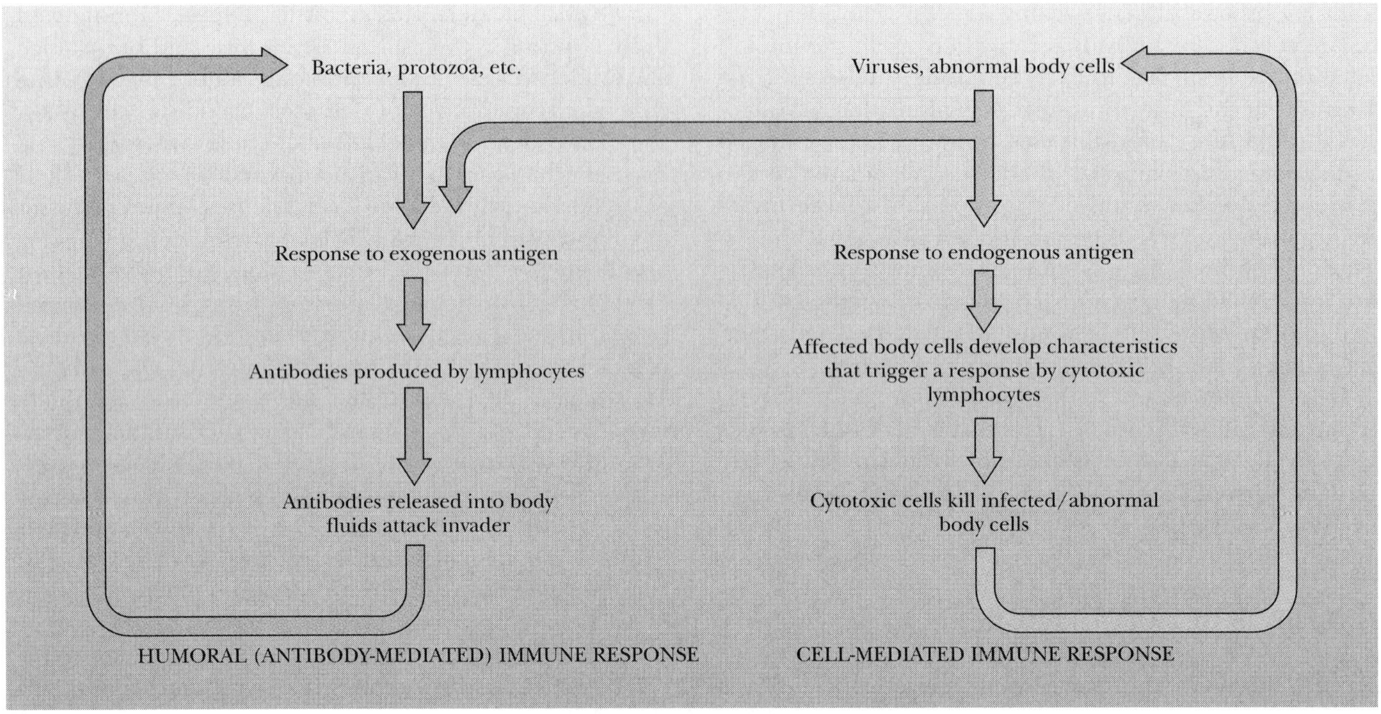

***Figure* 27.2** The two major forms of the immune response and their role in the body.

defense of the body. Its importance is readily seen when it is destroyed. Thus in AIDS patients, loss of the immune defenses of the body leads inevitably to death as a result of invasion by a whole host of microorganisms.

As indicated, the immune system is a remarkably effective defense system. It can recognize foreign invaders, destroy them, and retain the memory of the encounter. If the animal encounters the same organism a second time, the immune system responds more rapidly and more effectively. Such a sophisticated system must of necessity be complex. One reason for this complexity is the wide variety of microorganisms that must be defended against. Thus the immune system, to be effective, has to be able to combat this whole array of invaders. In addition, it is the task of the immune system to ensure that the cells of the body function normally. If abnormal cells arise, such as, for example, cancer cells, it is essential that they too be rapidly eliminated.

If we examine the list of potential invaders that the immune system must counter, they fall into two major categories. One category consists of organisms that invade directly from outside the body, including, for example, most bacteria, many protozoa, newly arriving viruses and invading helminths. The other category consists of organisms that originate or live inside the body's own cells. These include viruses formed in infected cells, intracellular bacteria or protozoa, and cancer cells. The immune system falls into two major branches based upon resistance to the two categories of invaders (Figure 27.2). Thus, one branch of the immune system is primarily directed against the extracellular or exogenous invaders. These invaders are mainly destroyed by proteins called **antibodies.** This type of immune response is sometimes called the **humoral immune response** since antibodies are found in body fluids (or humors). The other major branch of the immune system is directed against the intracellular or endogenous invaders, which cause cellular abnormalities. These abnormal cells are destroyed by specialized cytotoxic cells. This type of response is therefore called the **cell-mediated immune response.**

Antibody-Mediated (Humoral) Immune Responses

Early in the history of immunology it was recognized that the substances that provide immunity to many organisms could be found in blood serum. These protective factors found in the serum of an immunized animal are proteins known as antibodies. For example, antibodies against tetanus toxin are not present in the serum of normal horses but are produced following exposure to tetanus toxin. Tetanus toxin is an example of a foreign substance that stimulates an immune response. The general term for such a substance is **antigen.** If an antigen is injected into an animal, then antibodies will be produced that can bind to that antigen and ensure its destruction. Antibodies will bind only to the specific antigen that stimulates their production. For example, the antibodies produced following exposure to tetanus toxin bind only to tetanus toxin. If serum containing these antibodies is mixed with a solution of tetanus toxin, then a visible precipitate will develop as a result of the combination of antibody with the toxin. In addition, the antibody neutralizes the toxin so that it is no longer toxic for animals. In this way, antibodies protect animals against the lethal effects of the toxin.

The time-course of the antibody response to tetanus toxin can be followed by taking blood samples from a horse at intervals after injection of the toxin (or injection of the chemically detoxified toxin—called tetanus toxoid, a much safer procedure) (Figure 27.3). The blood is allowed to clot and the clear serum removed. The amount of antibody in the serum may be estimated either by measuring the amount of precipitate formed on adding toxin or, alternatively, by measuring the ability of the serum to neutralize a fixed amount of toxin. Both methods yield approximately the same result. Following a single injection of toxin into a horse that has never been previously exposed to it, no antibody is detectable for several days. This lag period lasts for about one week after the injection. When antibodies appear in serum, their level climbs to reach a peak by 10 to 14 days before declining and disappearing within a few weeks. The amount of antibody formed, and therefore the amount of protection conferred, during this first or **primary response** is relatively small.

If, sometime later, a second dose of toxin or toxoid is injected into the same horse and the antibody response again followed, then the lag period lasts for no more than two or three days. The amount of antibody in serum then rises rapidly to a high level before declining slowly. Antibodies may be detected for many months or years after

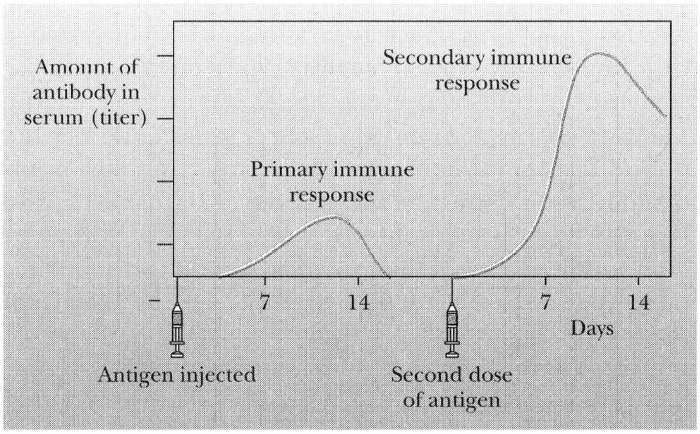

Figure **27.3** The time course of immune responses mediated by antibodies. Following first exposure to an antigen, the primary immune response is slow and weak. Second or subsequent exposures trigger a rapid response characterized by production of a high level of antibodies.

this injection. A third dose of the antigen given to the same animal results in an immune response characterized by an even shorter lag period and a still higher and more prolonged antibody response. As will be described later in this chapter, the antibodies produced after repeated injections are better able to bind and neutralize the toxin than those produced early in the immune response. The stimulation of the immune responses to infectious agents by repeated injections of antigen forms the basis of vaccination.

As we have seen, the response of an animal to a second dose of antigen is very different from the first in that it occurs much more quickly, antibodies reach very much higher levels, and it lasts for much longer. This **secondary response** is very specific in that it can be provoked only by a second dose of the antigen. A secondary response may be provoked many months or years after the first injection of antigen, although its size tends to decline as time passes. A secondary response can also be induced even though the response of the animal to the first injection of antigen was so weak as to be undetectable. These features of the secondary response indicate that the antibody forming system possesses the ability to "remember" previous exposure to an antigen. It should be noted, however, that repeated injections of antigen do not lead indefinitely to greater and greater immune responses. The level of antibodies in serum is controlled, so that they eventually stop rising, even after multiple doses of antigen or exposure to many different antigens.

Cell-Mediated Immune Responses

If a piece of living tissue such as skin is surgically removed from one individual and grafted onto another, it usually survives only for a few days before being destroyed by the recipient. This process of **graft rejection** is significant because it demonstrates the existence of a mechanism whereby foreign cells, differing only slightly from an animal's own normal cells, are rapidly recognized and eliminated. Even cells with minor structural abnormalities may be recognized as foreign by the immune system and eliminated though they are otherwise apparently healthy. These abnormal cells include aged cells, virus-infected cells, and cancer cells. The immune response to foreign cells as shown by graft rejection demonstrates the existence of a surveillance system that identifies and removes abnormal cells.

If a piece of skin is transplanted from one human to a second, unrelated human, it will survive for about ten days (Figure 27.4). The grafted skin will initially appear to be healthy, and blood vessels will develop between the graft and its host. By one week, however, these new blood vessels will begin to degenerate, the blood supply to the graft will be cut off, and the graft will eventually die and be shed. If a second graft is taken from the original donor

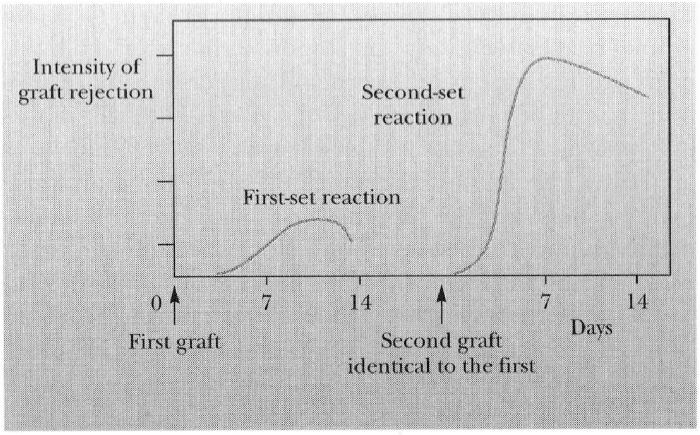

***Figure* 27.4** The time course of graft rejection, a cell-mediated immune response. Rejection of a first graft is relatively slow. Rejection of a second graft from the same donor is rapid and intense. This secondary immune response may occur months or even years after sensitization.

and placed on the same recipient, then that second graft will survive for no longer than one or two days before being rejected. Thus the rejection of a first graft is relatively weak and slow and analogous to the primary antibody response, whereas a second graft stimulates very rapid and powerful rejection similar in many ways to the secondary antibody response. Graft rejection, like antibody formation, is specific, in that a secondary response occurs only if the second graft is from the same donor as the first. Like antibody formation, the graft rejection process also possesses a memory, since a second graft may be rapidly rejected many months or years after loss of the first.

However, the graft rejection process is not entirely identical to the process involved in protection against tetanus toxin since it cannot be transferred from a sensitized to a normal animal by means of serum antibodies. The ability to mount a secondary reaction to a graft can only be transferred between animals by means of living cells. The cells that can do this are called **lymphocytes** and are found in the spleen, lymph nodes, or peripheral blood. The process of graft rejection is mediated primarily by lymphocytes and not by serum antibodies. It is a good example of a cell-mediated immune response.

Antigens

Since the function of the immune system is to defend the body against invading microorganisms, it is essential that these organisms be recognized as soon as they invade the body. The body must be able to recognize that these are foreign if they are to stimulate an immune response. In fact, the immune system can recognize and respond to specific components of these foreign organisms called antigens.

Microbial Antigens

The four major components of the bacterial surface include the cell wall, the capsule, the pili, and the flagella (Chapter 4). The cell wall of gram-positive organisms is largely composed of peptidoglycans (chains of alternating N-acetyl glucosamine and N-acetyl muramic acid cross-linked by short peptide side chains). The cell wall in gram-negative organisms is a polysaccharide-lipid-protein structure. Most of the antigenicity of gram-negative bacteria is associated with the polysaccharide component. This consists of an oligosaccharide attached to a lipid (lipid A) and to a series of repeating trisaccharides. The structure of these trisaccharides determines the antigenicity of the organism. Many bacteria are classified according to this antigenic structure. For example, the genus *Salmonella* has been classified into about 2000 species on this basis. Their polysaccharide antigens are called O antigens. These cell wall lipopolysaccharides of gram-negative bacteria are toxic and thus are also called **endotoxins.**

Bacterial capsules may be either polysaccharides or proteins. The polysaccharides are usually good antigens, possibly because they contain hexosamines, which seem to be essential for their antigenicity. These capsules protect bacteria against phagocytosis, and anticapsular antibodies therefore protect an infected animal. Capsular antigens are collectively known as K antigens. Pili, or fimbriae, are short projections that cover the surfaces of some gram-negative bacteria. They are classified as F or K antigens. Pili attach the bacteria to cells and may play a role in bacterial conjugation. Antibodies to pili may have an important protective function since they can prevent bacteria from sticking to body surfaces. Bacterial flagella consist of a single protein called flagellin. Flagellar antigens are collectively known as H antigens.

Other significant bacterial antigens include the porins, the heat-shock proteins, and exotoxins. The **porins** are proteins that form the pores on the surface of gram-negative organisms. Heat-shock proteins are generated in large amounts in stressed bacteria. The exotoxins are toxic proteins secreted by bacteria or released when they die. They are highly immunogenic and promote the production of antibodies known as **antitoxins.** Exotoxins, when treated with a mild denaturing agent like formaldehyde, lose their toxicity but retain their antigenicity. Toxins modified in this way are called **toxoids.** Toxoids may be used to prevent disease caused by bacteria such as *Clostridium tetani.*

Viruses are very small, obligate intracellular parasites that consist of **virions,** a nucleic acid core surrounded by a layer of protein subunits (Chapter 14). The protein layer is termed the capsid, and the subunits are called capsomeres. The capsid proteins are good antigens, well capable of provoking antibody formation. Some viruses may also be surrounded by an envelope containing lipoproteins and glycoproteins. When a virus infects an animal, the proteins in the virions act as antigens and trigger an immune response. Viruses however, are not always found free in the circulation but live within cells, where they are protected from the unwelcome attentions of antibodies. Indeed, virus nucleic acid can be integrated into a cell's genome. In this situation, the virus genes code for the production of new proteins, which may be carried to the surface of infected cells. These proteins, although they are synthesized within an animal's cells, are recognized as foreign and can provoke immune responses. These newly synthesized foreign proteins are called **endogenous antigens** to distinguish them from the foreign antigens that enter from the outside and are called **exogenous antigens.**

Nonmicrobial Antigens

Invading microorganisms are not the only source of foreign material encountered by the body. Food may contain many foreign molecules that, under some circumstances, may trigger an immune response and cause an allergic reaction. Likewise inhaled dusts can contain foreign particles such as fungal spores or pollen grains so that antigens from these may enter the body through the respiratory system. Foreign molecules may be injected directly into the body as in a snake (or mosquito) bite, or deliberately by a physician when administering a vaccine or a blood transfusion. Foreign proteins may be injected into animals for experimental purposes. Organ grafts are a very effective way of administering a large amount of foreign material to an animal.

Molecules vary in their ability to act as antigens and stimulate an immune response. In general, foreign proteins make the best antigens. Almost all proteins with a molecular weight greater than 1000 daltons are antigenic. Many of the major antigens of microorganisms—such as the clostridial toxins, bacterial flagella, virus capsids, and protozoan cell membranes—are proteins. Other important antigenic proteins include components of snake venoms, serum proteins, cell surface proteins, milk and food proteins, hormones, and even antibody molecules themselves.

Clearly not all foreign molecules are capable of stimulating an immune response. Stainless steel pins and plastic heart valves, for example, are commonly implanted in humans without triggering an immune response. The lack of antigenicity in the large organic polymers, such as the plastics, is due not only to their molecular uniformity but also to their inertness. They cannot be degraded and processed by cells to a form suitable for triggering an immune response. Conversely, since immune responses are antigen driven, foreign molecules that are unstable and destroyed very rapidly may not persist for a sufficient time to stimulate an immune response.

The cells that respond to antigens **(antigen-sensitive cells)** do not normally respond to molecules originating within an animal **(self-antigens).** They will respond, however, to foreign molecules that differ even in minor respects from those normally found within the body. This lack of reactivity of the immune system to normal body components occurs because any self-reactive cells are selectively destroyed, leading to a state of "self-tolerance." This destruction occurs as the cells are developing. Those cells that respond inappropriately to self-antigens are triggered to die in a process called **negative selection.** If antigen-sensitive cells do not encounter a self-antigen when immature, tolerance to that antigen will not develop. Thus the cells of the mature immune system may still be able to respond to self-antigens that have been hidden within tissues or cells. For example, the sperm-forming cells in the testes develop only after puberty and then are separated from the rest of the body by a tissue barrier. As a result, the cells of the immune system do not normally encounter sperm antigens and, therefore, retain the ability to respond to them. If the testes are injured, sperm antigen may escape and reach the bloodstream, where the antigen-sensitive cells will regard it as foreign and mount an immune response against it. The development of antisperm antibodies is a common sequel to vasectomy in men.

Epitopes

Foreign particles, such as bacteria, nucleated cells, and red blood cells, are a complex mixture of proteins, glycoproteins, polysaccharides, lipopolysaccharides, lipids, and nucleoproteins. The immune response against such a foreign particle is, therefore, a mixture of many simultaneous immune responses against each of the foreign molecules.

A single large molecule such as a protein can also be shown to stimulate multiple immune responses. Large molecules have regions, against which immune responses are directed. These regions, usually on the surface of the molecule, are called **epitopes,** or **antigenic determinants** (Figure 27.5). In general, the number of epitopes on a

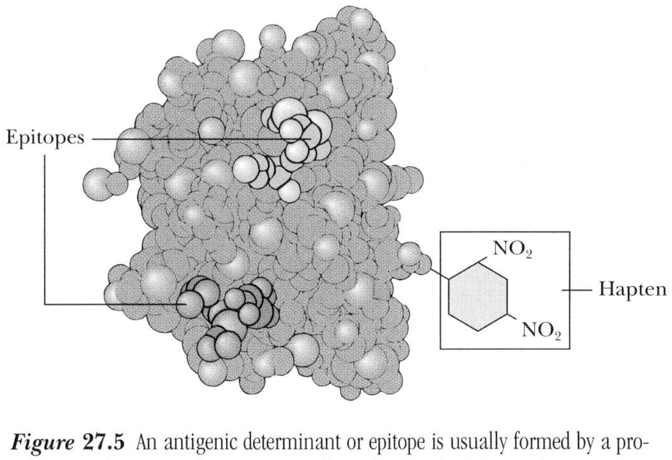

***Figure* 27.5** An antigenic determinant or epitope is usually formed by a projection or other unique area on the surface of a protein. A hapten is simply a new epitope formed by conjugating a new chemical group to the antigen surface.

molecule is directly related to its size. When we describe a molecule as foreign, therefore, we are implying that it contains epitopes that are not found on self-antigens. The immune system recognizes and responds to such foreign epitopes.

Cells and Tissues of the Immune System

As with other body systems, the immune system is composed of many different types of cells. These include cells that trap foreign antigens, cells that process these antigens into a form that can be recognized, and cells that recognize the foreign antigens and respond by mounting an immune response.

Neutrophils

The first line of defense involves a population of blood cells called **neutrophils** (Figure 27.6). Neutrophils attack and eat invading bacteria in a process called phagocytosis in which the organisms are taken up into the neutrophil, killed, and then destroyed. Obviously neutrophils play a critical role in host defenses. Their functions were discussed in Chapter 26.

Macrophages

A second population of phagocytic cells are the **macrophages.** These too play a key role in the initial defenses against invasion and were discussed in detail in Chapter 26. Unlike neutrophils, however, macrophages play a key role in the initiation of immune responses (Figure 27.7). Foreign particles that enter the blood and are not destroyed by neutrophils are phagocytosed by macrophages in the spleen and liver. Other macrophages within tissues trap and process antigens that invade those

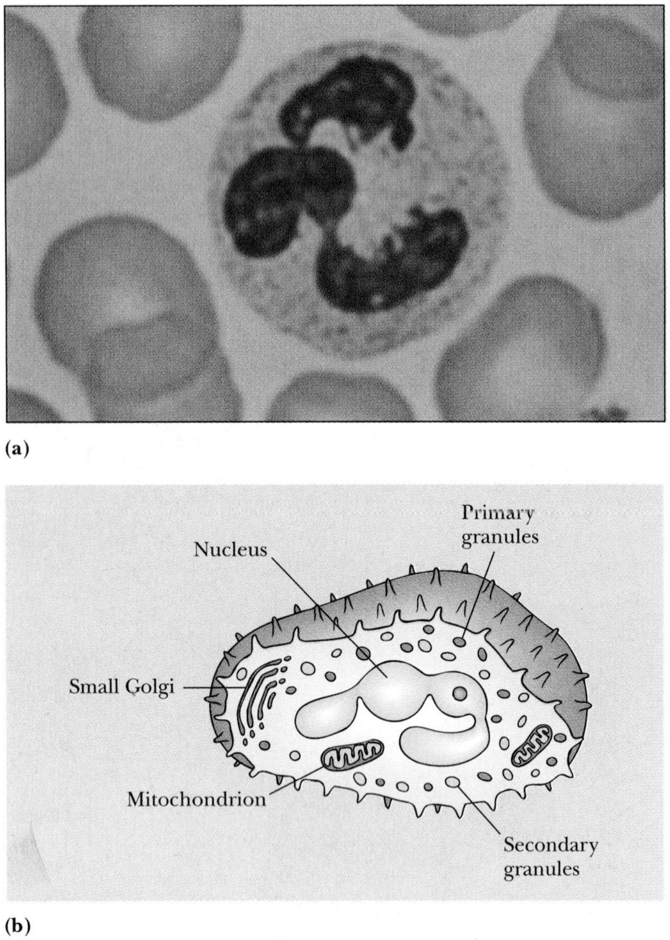

***Figure* 27.6** The structure of neutrophils. **(a)** A light micrograph of a neutrophil from human blood. (© Manfred Kage/Peter Arnold, Inc.) **(b)** A schematic diagram showing the structure of a neutrophil.

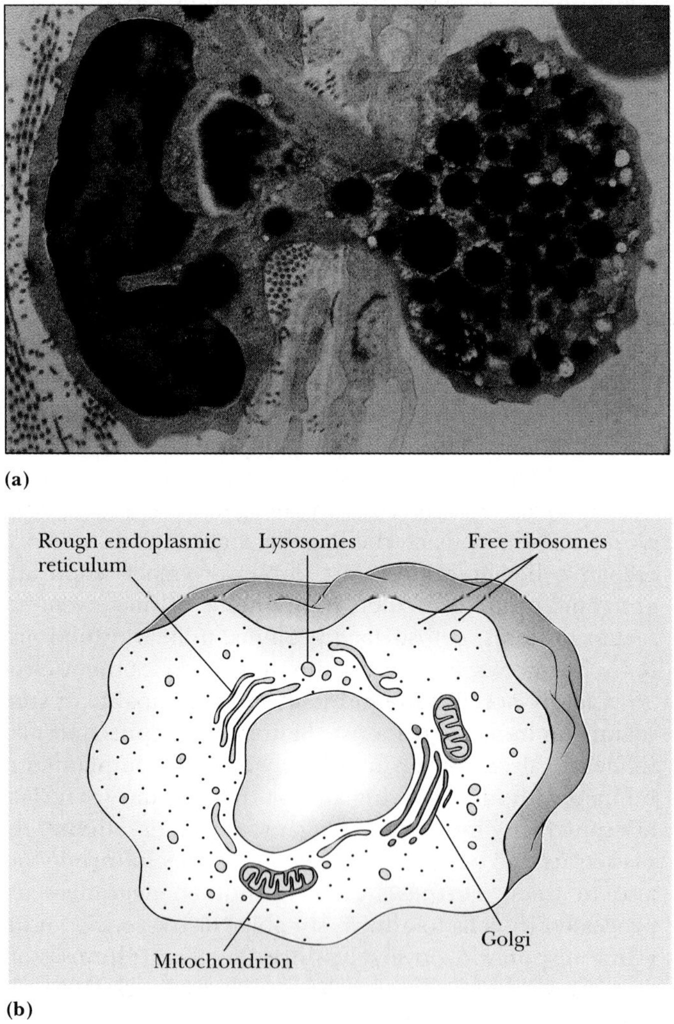

***Figure* 27.7** The structure of macrophages. **(a)** Macrophage moving from blood into tissue. (© David M. Phillips/VU) **(b)** A schematic diagram showing the structure of a macrophage.

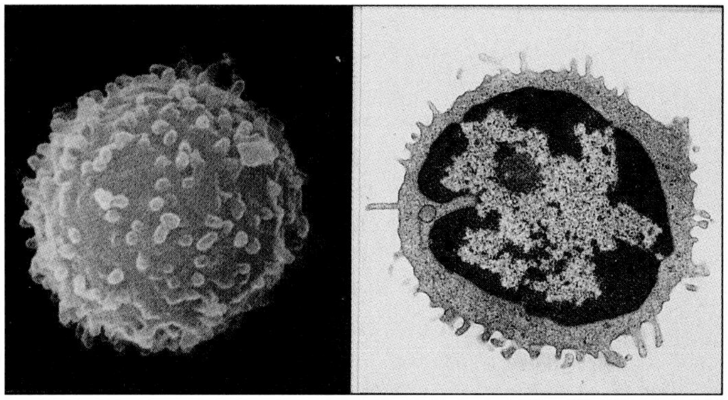

(a)

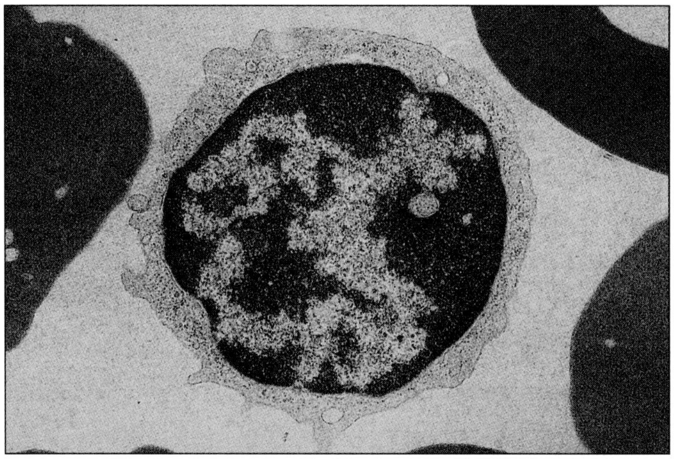

(b)

***Figure* 27.8** The structure of lymphocytes. **(a)** Comparative SEM and TEM views of a T-cell. (© Don Fawcett/VU) **(b)** An electron micrograph of a B-cell (× 15,000). (© David M. Phillips/VU)

tissues. Macrophages phagocytose particles in a manner similar to neutrophils, but they are able to do so repeatedly and they live much longer than do neutrophils.

Although most of the foreign material phagocytosed by macrophages is destroyed within these cells, some is not destroyed but is instead processed for initiation of an immune response. Fragments of the phagocytosed proteins are linked to a specialized cell surface receptor molecule called a MHC molecule. The antigen-MHC molecule complex is then carried to the macrophage surface where it is presented to lymphocytes. Antigen fragments bound to MHC molecules are recognized by a population of lymphocytes called **helper T-cells.**

In addition to their ability to phagocytose and process antigens, macrophages secrete proteins called **cytokines.** Cytokines are required for helper T-cells to respond optimally to processed antigen. Only when a helper T-cell is exposed to a processed epitope and the correct cytokines at the same time will it respond by initiating an immune response.

Macrophages are not the only cells that can process antigen and present it to lymphocytes. Other antigen-presenting cells include dendritic cells, a population of specialized macrophage-like cells found in lymph nodes and the skin, and B-cells, a population of lymphocytes also involved in antibody production.

Lymphocytes

The cells that recognize and respond to antigens are called **lymphocytes.** They are small cells with a large rounded nucleus and little cytoplasm (Figure 27.8). Each lymphocyte has receptors that bind specifically to a single epitope (one lymphocyte-one epitope). Lymphocytes that are first produced in the bone marrow of young animals are a mixture of cells with antigen receptors directed against a random assortment of different epitopes (Figure 27.9). Because some of these lymphocytes have receptors that can bind to self-antigens and so have the potential to cause damage, the cells carrying these receptors must be eliminated before they establish themselves in the body. The

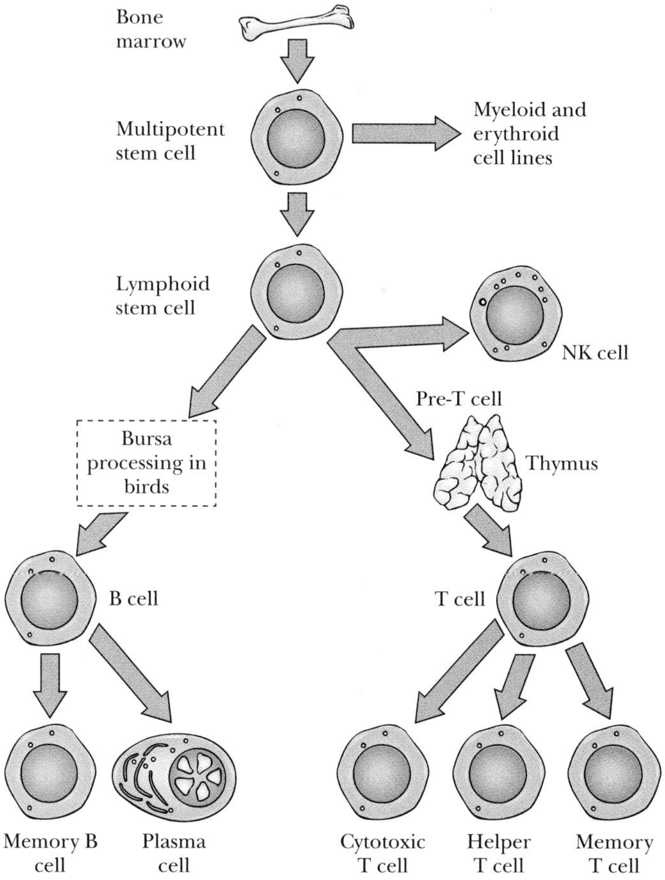

***Figure* 27.9** Lymphocytes originate in the bone marrow. Their subsequent development is determined by the environment in which they mature. B-cells develop in the bone marrow of most mammals and in the bursa of Fabricius in birds. T-cells develop within the thymus. NK (natural killer) cells appear to develop from prethymic lymphocytes.

organs in which lymphocytes develop and in which those with self-reactive receptors selectively are eliminated include the **thymus,** the **bone marrow,** and some Peyer's patches in the intestine. (In birds, a special organ called the **bursa** of Fabricius serves this function.) Lymphocytes that undergo development and selection in the thymus are called **T-cells.** Some T-cells act as helper cells, promoting the immune responses, while others mediate the cell-mediated immune responses. Lymphocytes that undergo development in the bursa of Fabricius or equivalent tissue are called **B-cells** and are responsible for antibody production.

The importance of organs such as the thymus in the development of the immune response becomes clear when they are surgically excised. Removal of the thymus from a newborn mouse prevents the production of functional T-cells. As a result, a neonatally thymectomized animal cannot reject foreign tissue grafts—a cell-mediated immune response. Removal of some Peyer's patches in mammals or the bursa of Fabricius in birds prevents B-cell production and so prevents antibody formation.

The organs in which mature lymphocytes reside and where they encounter and respond to foreign antigens include the **lymph nodes,** the **spleen,** the **bone marrow,** and the **tonsils,** as well as lymphoid tissues in the intestine, the lungs, and other body surfaces. B-cells and T-cells are found in all of these organs. The B-cells tend to remain stationary, which means that antigens are brought to them by the flow of lymph or blood. In contrast, the T-cells circulate between the blood, lymph, and secondary lymphoid organs (presumably looking for abnormal cells). T-cells account for about 70 percent of blood lymphocytes. The structure of the lymphoid organs provides an optimal environment for cell interaction and the development of immune responses.

The two major lymphocyte populations, T-cells and B-cells, cannot be distinguished by any structural features. However, they do possess distinctly different receptors for antigens. The B-cell receptors **(BCR)** for antigens are membrane-bound antibody molecules. As a result, B-cells can be identified by the presence of antibody molecules (also called immunoglobulins) on their surface. B-cells account for about 15 percent of blood lymphocytes. The T-cell antigen receptors **(TCR)** also consist of characteristic protein molecules on the cell surface. These antigen receptors are associated with a set of other surface proteins used to transmit signals to the cell. The most important of these signaling proteins, called CD proteins are, CD3, CD4, and CD8. All mature T-cells possess CD3, about two thirds carry CD4, while about one third carry CD8. The remaining 15 percent of blood lymphocytes are neither T- nor B-cells. These cells can kill some foreign cells even in an unimmunized animal. For this reason they are called **natural killer** (or **NK**) **cells.** NK cells probably play a key role in defense against cancer.

Antigen-Binding Molecules

The cells of the immune system will respond to foreign antigens only if they can first recognize them. In order to do this they need specific antigen-binding surface receptors. Several different antigen-binding receptors have been identified.

Proteins are constructed by assembling together several modules or **domains**—small subassemblies containing about 100 amino acids. For example, in proteins that bind to cell surfaces one domain contains hydrophobic amino acids so that it penetrates the lipid bilayer on the cell surface. Other domains may be responsible for the structural stability of a protein, or for its biological activities. In antibody molecules one domain is used to bind antigen while other domains are responsible for other biological functions. The presence of similar domains in proteins of dissimilar function suggests that they have a common origin. Proteins can be assigned to families or superfamilies based on their domain structure.

One protein superfamily that plays a key role in the immune system is the **immunoglobulin superfamily** (Figure 27.10). The members of this superfamily all contain at least one immunoglobulin domain. In a typical immunoglobulin domain, the peptide chains weave back and forth to form a pleated sheet that folds into a sandwich-like structure. Two forms of immunoglobulin domain exist, one called the variable domain, the other called the constant domain. These are discussed later in this chapter. Immunoglobulin domains were first identified in antibody molecules (otherwise known as immunoglobulins). They have since been found in many other proteins that play a key role in the functioning of the immune system. They in-

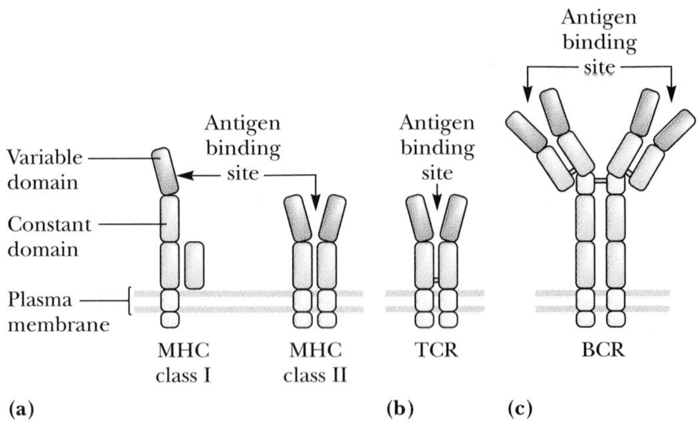

***Figure* 27.10** The four different types of antigen receptors that belong to the immunoglobulin superfamily. **(a)** Histocompatibility antigens are antigen receptors of broad specificity found on antigen-processing cells. **(b)** The T-cell antigen receptor (TCR) is a heterodimer found only on T-cells. **(c)** The B-cell antigen receptor (BCR) is an immunoglobulin molecule found both on B-cell surfaces and free in serum, where it serves as an antibody.

clude the B-cell antigen receptors (BCR), the T-cell antigen receptors (TCR), and the **MHC class I and II molecules.** Most are found on cell surfaces and none have enzymatic activity. In many cases, cellular interactions are mediated by binding two different members of the superfamily as, for example, between TCR and MHC molecules.

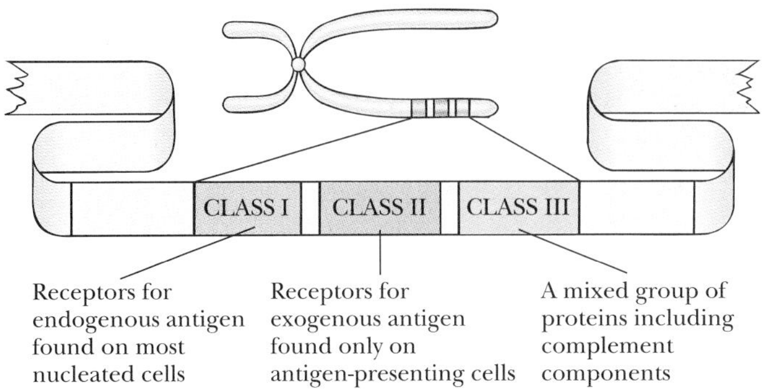

***Figure* 27.11** The structure of the major histocompatibility complex.

MHC Molecules

In order to trigger an immune response, antigen processing requires not only the fragmentation of antigen molecules inside cells, but also the binding of these fragments to appropriate antigen-presenting receptors on the cell surface. These antigen-presenting receptors are called **MHC molecules.** They are specialized glycoproteins coded for by genes located in a gene complex called the **major histocompatibility complex (MHC)** (Figure 27.11). The absolute requirement for antigens to be bound to MHC molecules in order to trigger an immune response is called **MHC restriction.** The genes in the MHC, by coding for MHC molecules, determine which antigens are processed and presented. The MHC therefore provides the major genetic component of infectious or autoimmune disease resistance or susceptibility.

All vertebrates from cartilaginous fish to mammals possess MHC genes, and they are maintained in the genome as a linked set. This gene complex can be very large. For example, the human MHC contains about four megabases, which is two to three times larger than the mouse MHC and about the same size as the total genome of *E. coli.*

Each MHC contains three classes of gene. Class I genes code for receptors found on the surface of most nucleated cells (and on red cells in some species). These class I genes can be divided into those that are highly polymorphic **(class Ia genes),** and those that show very little polymorphism **(class Ib genes).** (Polymorphism refers to inherited structural differences between proteins.) Some genes related to the class Ib genes are found outside the MHC on a different chromosome.

A second class of genes, called **class II genes,** code for MHC molecules found on the surface of antigen-presenting cells (macrophages, dendritic cells, and B-cells). Because both class I and class II MHC molecules are receptors that bind antigen fragments and present antigen to lymphocytes, they effectively regulate the immune responses.

Class III genes are also found within the MHC. They do not code for antigen receptors but rather for many different proteins many of which are not directly linked to antigen presentation. Some of these class III genes code for complement proteins. The complement system is a set of serum proteins responsible for protection against microbial invasion (page 720).

Each MHC probably contains all three classes of genes, although their number and arrangement vary. The collective name given to the proteins produced by the MHC also varies between species. In humans these molecules are called HLA (Human Leukocyte Antigens).

MHC Class Ia Proteins

Class Ia proteins are antigen receptors found on the surface of most nucleated cells (Figure 27.12). They consist of two peptide chains. One chain called α is linked to a much smaller chain called β2-microglobulin (β2M). The α chain is inserted in the lipid bilayer of the cell surface and forms the antigen binding site on these molecules. Class Ia α chains show great polymorphism. That is, their amino acid sequences are very variable. X-ray crystallography has demonstrated that the α chain is folded so that it forms an open-ended groove. This groove binds antigenic peptides provided by antigen processing within the cell. The variable regions form the walls of the groove and determine its shape. The shape of the groove determines which antigen peptides can be bound and thus determines the ability of an animal to respond to a specific epitope.

MHC Class Ib Proteins

Mammals possess many nonpolymorphic MHC class Ib genes. They code for receptors on the surface of regulatory and immature lymphocytes and on hematopoietic cells. These receptors each consist of a membrane-bound α chain associated with β2 microglobulin. Their shape is similar to that of class Ia molecules and they do have an antigen-binding groove. Unlike MHC class Ia molecules that present peptide fragments to antigen-sensitive cells, class Ib molecules present common peptides or lipids from bacteria.

MHC Class II Proteins

MHC class II molecules are receptors found on the professional antigen-processing cells (B-cells, dendritic cells, or macrophages). They are glycoproteins consisting of two

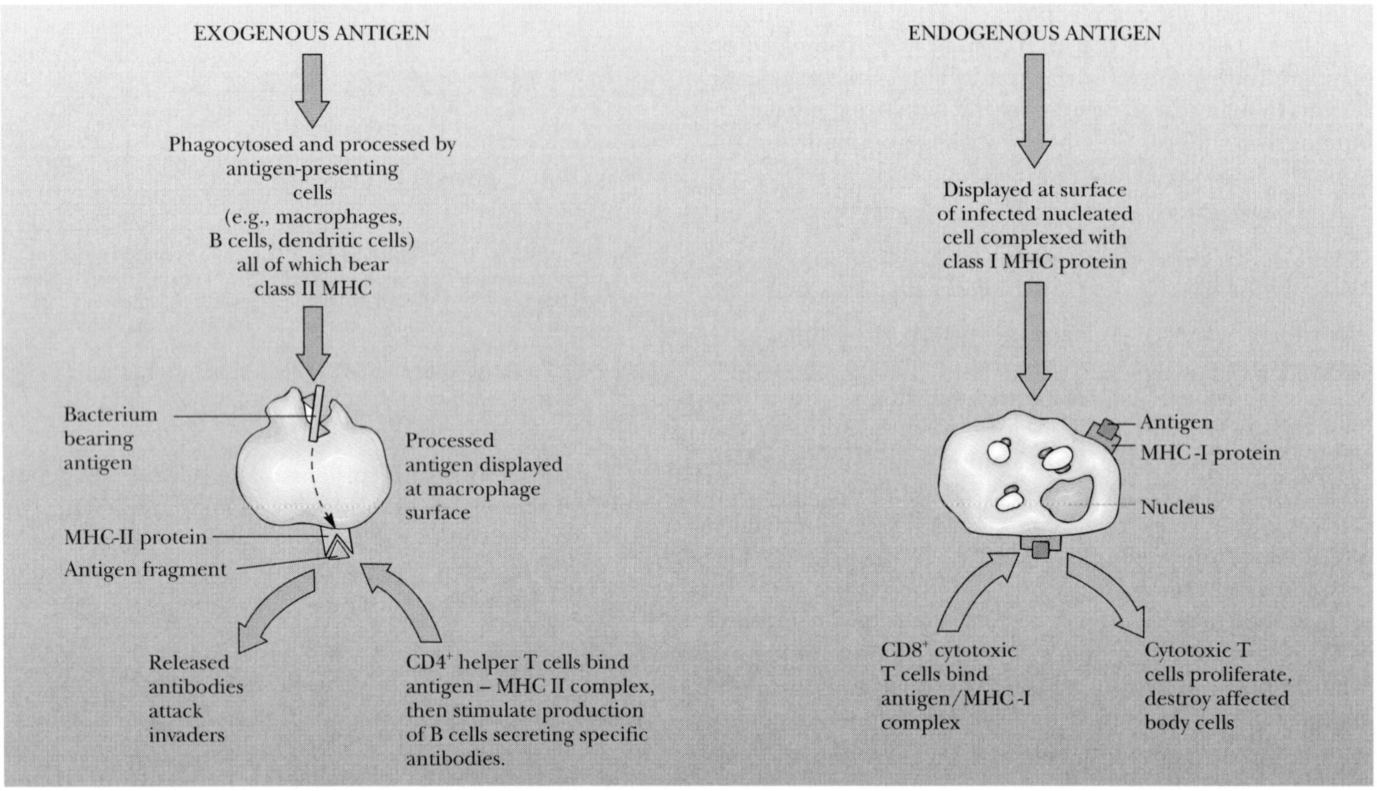

***Figure* 27.12** Two different classes of histocompatibility antigen act as antigen receptors. Class I receptors are found on most nucleated cells. They present antigen to CD8 positive cytotoxic T-cells. Class II receptors are found on antigen-presenting cells and present antigen to CD4 positive helper T-cells.

linked polypeptide chains called α and β. They have an antigen-binding groove formed jointly by these α and β chains. The groove has a structure similar to that seen in MHC class I molecules. Polymorphism results from variations in the amino acid sequences in the peptide chains lining the groove.

MHC Polymorphism

The antigen-binding site of an MHC molecule is fairly nonspecific in that the groove can actually bind many different peptides. Nevertheless, one MHC molecule cannot possibly bind all peptides made available to it. As a result, it is likely that only one or two peptides from an average protein will bind to any given MHC molecule. The efficiency of antigen binding by MHC molecules determines resistance to infectious diseases. Each individual usually has a limited number of MHC molecules available to bind antigen. However, because of polymorphism, these differ widely between individuals. Extensive polymorphism protects the population as a whole from disease. Because of MHC polymorphism, most individuals in a population will have a unique set of class Ia molecules and these will be able to bind a great diversity of epitopes.

When a completely new infectious disease strikes, it is likely that at least some individuals will have receptors that can bind and present the new microbial epitopes. These individuals will be able to respond to the agent and so become immune. No single MHC molecule confers significant advantages on any individual. This reflects the futility of the host attempting to match invading organisms in antigenic variability. A microorganism will always be able to mutate to evade the immune response very much faster than a mammalian population can develop resistance. Any mutations in the MHC, while they may increase resistance to one organism, may, at the same time, decrease resistance to another. It is more advantageous for each member of a population to have a different set of MHC molecules as antigen receptors so that an organism spreading through a population will have to adapt anew to each host.

The T-Cell Antigen Receptor

T-cell antigen receptors or TCRs are found only on the T-cell surface. Each T-cell possesses about 10,000 to 20,000 identical TCRs (Figure 27.13). Because all the TCRs on one cell are identical, each T-cell will only bind and respond to a single epitope. Each TCR consists of two antigen-binding peptide chains associated with a large number of other glycoproteins. These associated glycoproteins amplify the signal generated by the binding of an antigen to the TCR and transmit it to the T-cell. Two types of TCR have been identified. One consists of two chains called γ and δ (γ/δ). The other consists of two different chains called α and β (α/β). In humans, as well as mice, between 1 and 10 percent of T-cells carry γ/δ receptors. The remaining 90 to 99 percent of T-cells carry α/β receptors.

The four TCR peptide chains α, β, γ, and δ are similar in structure. Each TCR chain is divided into four domains. The domain at the N-terminus contains about 100 amino acids whose sequence is highly variable. This is therefore called the variable domain (V-domain). The next domain is about 150 amino acids in length and has a constant amino acid sequence; this is called the constant domain (C-domain). A third, very small domain passes through the T-cell membrane, while the C-terminal domain is located within the cytoplasm of the T-cell. The two chains are joined together by a disulfide bond to form a stable heterodimer. Because each TCR consists of paired peptide chains, the V-domains of each chain together form a groove that can bind antigenic peptides. The shape of this antigen binding groove varies between different TCR dimers because of the variable amino acid sequences in the V-domains. The specificity of the binding between a TCR and an antigenic peptide is determined by the receptor shape generated by the paired V-domains.

When the variability in the V-domains is examined in detail, it is found that within each V-domain there is an area of the chain where the amino acid sequence is very highly variable indeed (Figure 27.14). This is the region that actually comes into contact with the antigen. For this reason it is called the complementarity determining region (CDR). The antigen-binding site of the TCR is formed by the paired CDRs that line a groove on the molecule's surface. The rest of the V-domain has a fairly constant sequence and is called the framework region.

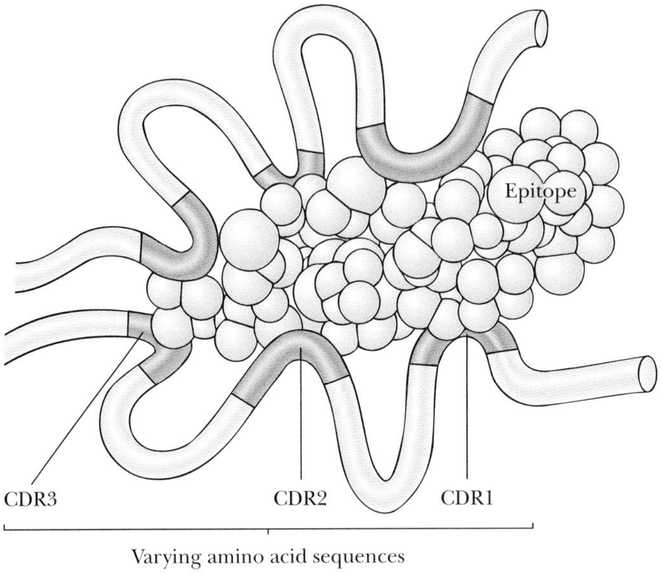

***Figure* 27.14** The binding of antigen by immunoglobulins or TCR peptide chains. The precise conformation of these chains is determined by the amino acid sequence in this region of the molecule. Changes in the amino acid sequence alter the shape of the binding site and hence regulate their abilities to bind to antigens.

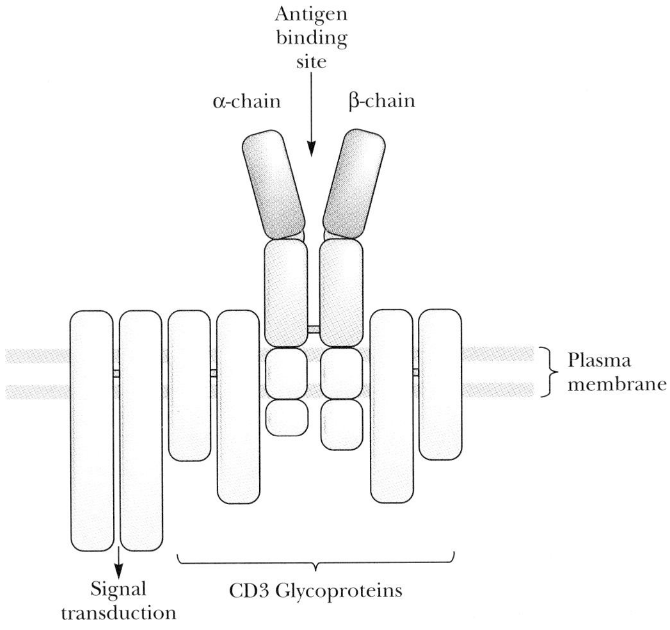

***Figure* 27.13** The structure of the TCR.

The variety of T-cells and hence TCRs available to an animal is carefully regulated (Figure 27.15). An animal must possess T-cells that can respond to as many different foreign antigens as possible while, at the same time, not bind self-antigens. At least four different mechanisms are involved in regulating this. First, the basic TCR repertoire is determined by the number of different TCR germline genes available. Second, those T-cells whose TCRs bind to self-antigens are selectively destroyed before they can develop fully (the cells commit suicide in a process termed apoptosis). Third, the surviving T-cell population proliferates following interaction between TCRs and self MHC antigens. Thus T-cells with receptors that do not bind self-antigens are stimulated to multiply. Finally, specific subpopulations (clones) of T-cells are increased when their TCR binds appropriately presented foreign antigen.

The binding of the MHC antigen complex to the TCR must send a signal to the T-cell in order to trigger a response. The antigen-binding chains of the TCR are linked to a set of glycoproteins called CD3 and the CD3 complex

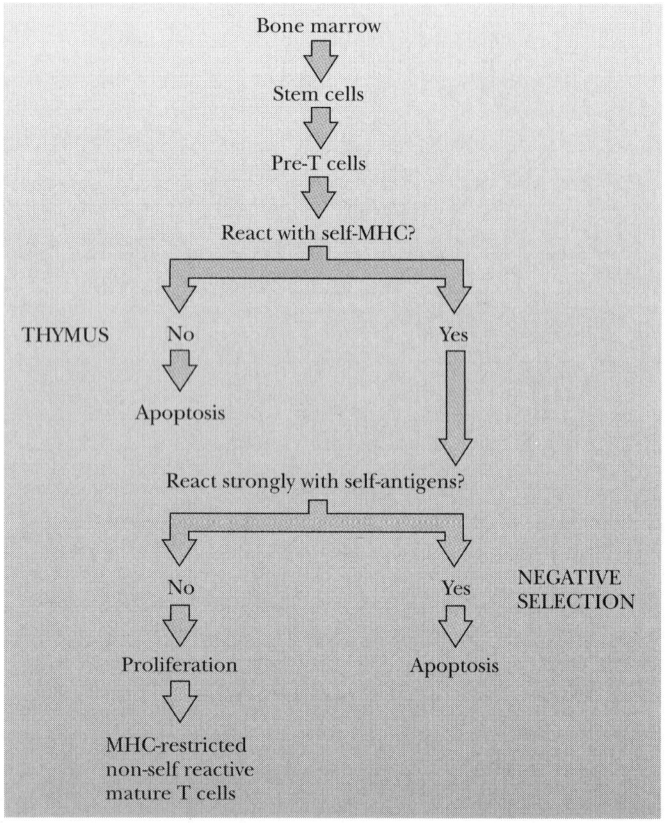

***Figure* 27.15** The regulation of TCR generation. This primarily occurs within the thymus, where self-reactive cells commit suicide in a process called apoptosis, while cells that can respond to foreign antigens bound to MHC class II molecules are stimulated to proliferate.

transfers a signal from the TCR to the cell. One of two other proteins are also associated with the TCR, CD4, or CD8. The presence of CD4 or CD8 determines which class of MHC molecule is recognized by a T-cell. Thus CD4 binds to MHC class II molecules on antigen-presenting cells and helps trigger a helper response. CD8, in contrast, binds to MHC class I molecules on antigen-presenting cells and triggers a cytotoxic response.

The B-Cell Antigen Receptor

Each B-cell is covered with about 200,000 to 500,000 identical antigen receptors (BCR) (Figure 27.16). The BCR belongs to the same class of receptors as the TCR—the immunoglobulin superfamily. Like the TCR, the BCR can be divided into two functionally different components. One component binds antigen, and the other component sends a signal to the B-cell. Unlike the TCR, however, the BCR may be released from the B-cell surface and bind antigens when in solution. Antibodies are simply soluble forms of BCR. They all belong to a class of proteins called immunoglobulins. In addition to binding to specific antigen and signaling the B-cell to respond appropriately, the BCR also triggers phagocytosis of antigen, leading to its processing and the expression of peptide fragments on MHC class II molecules. B-cells thus play two roles. They respond to antigen by making antibodies, while at the same time they can act as antigen-processing cells.

The antigen-binding structure of the BCR or of soluble immunoglobulin consists of four peptide chains. Two of these chains are identical, are about 50 kDa, and are called **heavy chains.** The other two chains are also identical, are about 25 kDa, and are called **light chains.** The light chains are only half the length of the heavy chains and are linked by disulfide bonds to the N-terminal half of the heavy chains so that the complete molecule is shaped like the letter Y (Figure 27.17). The tail of the "Y" (called the Fc region) is inserted into the B-cell surface membrane. The arms of the "Y" (called the Fab regions) bind antigen. The antigen-binding sites are formed by the groove between a light and heavy chain. Thus, unlike the TCR, the BCR has two antigen-binding sites.

Light chains consist of two domains. The amino-acid sequences in the C-terminal domain of each light chain are identical. In contrast, the sequences in the N-terminal domain are different in each light chain examined. For this reason, these are referred to as the constant (C_L) and variable (V_L) domains, respectively.

The heavy chains of a typical immunoglobulin consist of four or five domains. The amino-acid sequence of the N-terminal domain is highly variable and it is therefore called the variable (V_H) domain. The remaining domains show few sequence differences and so form a constant (C_H) region.

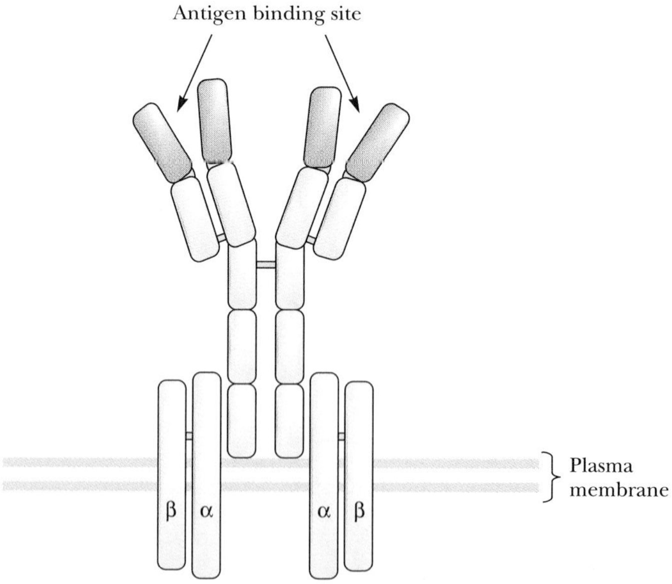

***Figure* 27.16** The structure of the BCR. The antigen-binding component has only a very small transmembrane domain. Signal transduction is therefore the responsibility of the Ig-α and Ig-β heterodimers.

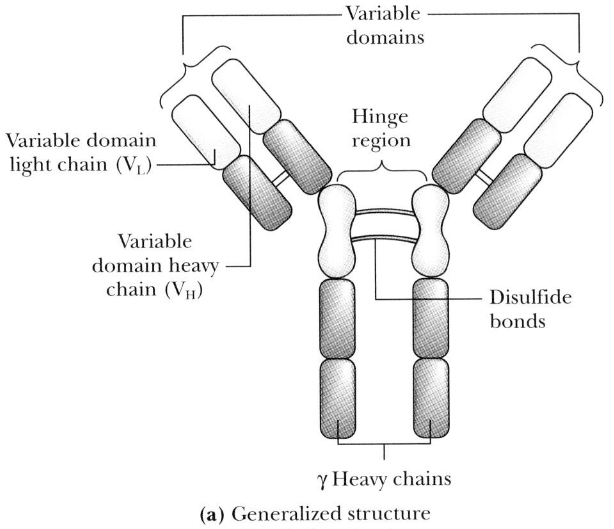

(a) Generalized structure of an IgG antibody

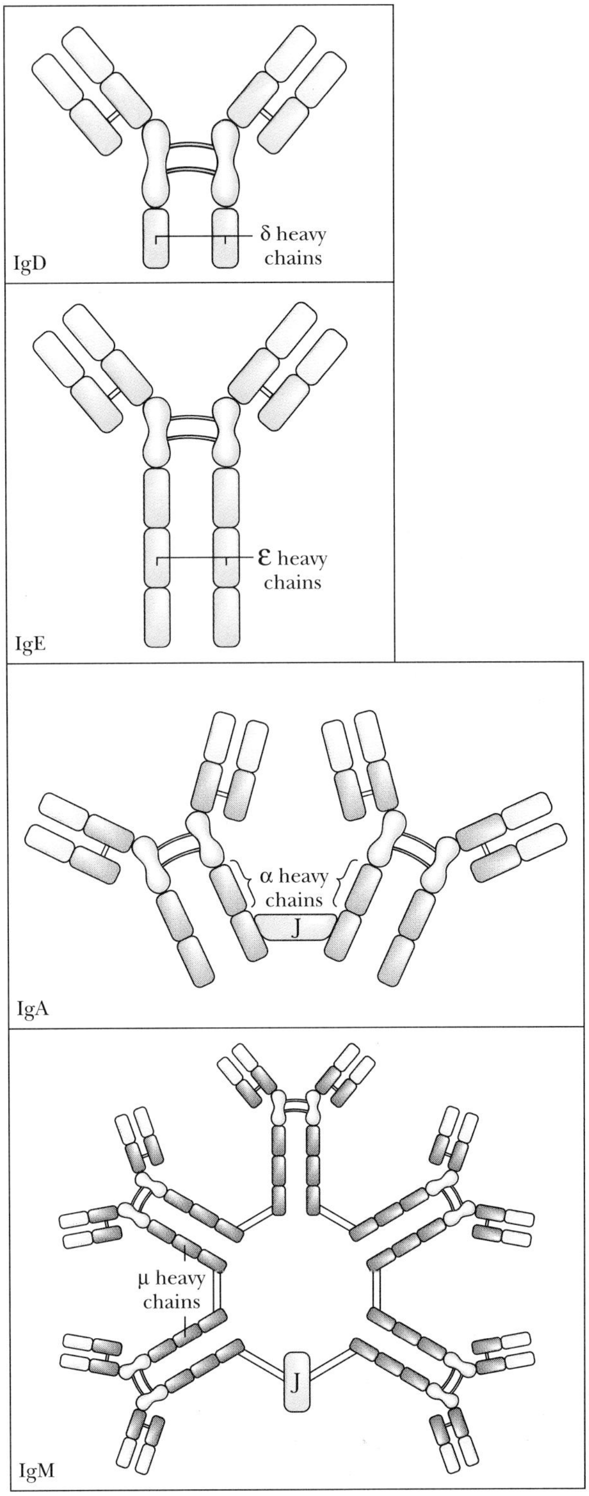

(b)

***Figure* 27.17** The structure of the different immunoglobulin classes.

There are five distinct types of immunoglobulin heavy chains called α, γ, δ, ϵ and μ that determine the immunoglobulin class or isotype (Figure 27.17). Immunoglobulins with two α heavy chains are called immunoglobulin A (IgA), those with two γ chains are called IgG, and so forth. There are two classes of immunoglobulin that act as BCRs on immature B-cells: IgM, which has two μ heavy chains, and IgD, which has two δ heavy chains. When an immune response is triggered and B-cells begin to respond, the cell changes the heavy chains in its BCR. Thus, in the spleen and lymph nodes, a stimulated B-cell switches from making BCRs using μ chains and instead uses γ heavy chains to form IgG. On body surfaces, the B-cells use α heavy chains to make IgA or ϵ heavy chains to make IgE.

When the amino-acid sequences of immunoglobulin V-domains from both light and heavy chains are examined in detail, their similarities to the TCR are apparent. Thus sequence variation is largely restricted to three smaller complementarity-determining regions (CDRs) within the variable domain. Between these hypervariable regions there are relatively constant framework regions. As in the TCR, the hypervariable regions determine the shape of the antigen-binding site, and they determine just to which antigens an antibody can bind.

Since heavy chains are paired, domains come together in pairs to form specialized regions within the BCR/immunoglobulin molecule. These paired domains provide a structure by which antibody molecules can exert their biological functions. Thus, V_H and V_L form a paired domain that serves to bind antigen. Other paired domains contain a site for the activation of the complement system and a site that binds to Fc receptors on phagocytic cells.

B-cell receptor immunoglobulins cannot send signals directly to the cell since their cytoplasmic domains contain only three amino acids. However their C-terminal and transmembrane domains are linked to two other proteins called Ig-α and Ig-β, which act as signal transducers.

Binding of specific antigen to a TCR or BCR results in activation of several different genes and transcription factors within the lymphocyte. As will be discussed later, more than one signal is required for T- and B-cells to be induced to proliferate, since linking to antigen receptors alone is not sufficient. Antigen only provides a signal that activates resting T- and B-cells and tells them to enter the mitotic cycle, while other signals from soluble proteins and other receptors are required to induce the activated cells to actually divide and differentiate.

BCRs and Antibodies

The function of the BCR parallels in many respects the function of the TCR, but with certain key differences. Most notably, both the BCR and antibody molecules can bind to intact antigen molecules in solution whereas the TCR can only bind and respond to antigen fragments presented on an MHC molecule. This difference in the antigen-recognizing ability of B- and T-cells is significant in that B-cells can respond to a much greater variety of antigens than can T-cells. Likewise, antibodies are directed not against breakdown products of antigens, but against intact antigen molecules.

We know a lot about antibodies since they can be readily isolated from body fluids such as blood or milk. Antibody molecules are glycoproteins called immunoglobulins (which may be abbreviated to Ig). The term **immunoglobulin** is used to describe all soluble BCRs. Immunoglobulins fall into five different classes (or isotypes) based on their use of specific heavy chains. The immunoglobulin class found in highest concentrations in serum is called immunoglobulin G (abbreviated to IgG) (Figure 27.18). The class with the second highest concentration in the serum of most mammals is immunoglobulin M(IgM). The third highest concentration in most mammals is immunoglobulin A (IgA). IgA is, however, the predominant immunoglobulin in secretions such as saliva, milk, or intestinal fluid. Immunoglobulin D (IgD) is primarily a BCR and is rarely encountered in body fluids. It may not be present in all mammals. Immunoglobulin E (IgE) is found in very low concentrations in serum and mediates allergic reactions.

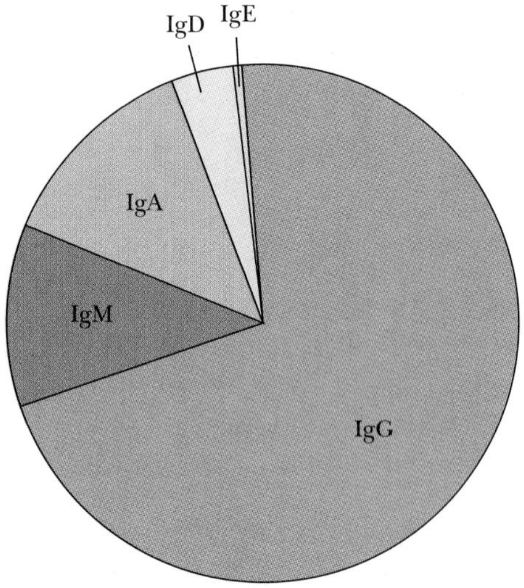

***Figure* 27.18** The relative proportions of different immunoglobulin classes in human blood.

Immunoglobulin Classes

Immunoglobulin G is made and secreted by cells found in the spleen, lymph nodes, and bone marrow. These cells are called plasma cells and are simply fully differentiated B-cells. IgG is the immunoglobulin found in highest concentration in blood and plays the major role in antibody-mediated defense mechanisms. It has a typical BCR structure, with two light chains, two γ heavy chains, and a molecular weight of 180 kDa. Because it is the smallest of the immunoglobulin classes, IgG can escape from blood vessels more easily than can the others. This is especially important in inflamed tissues where the increase in vascular permeability permits IgG to participate in the defense of tissue fluid and body surfaces. IgG will readily bind to foreign organisms and promote their destruction by phagocytic cells. (In the absence of antibodies, phagocytosis of bacteria is fairly slow and inefficient. If, however, the organisms are coated with antibody, the antibody binds to receptors on the phagocytic cell and greatly enhances the process. This phenomenon is called opsonization.) IgG antibodies can also activate the complement system but only if sufficient molecules have accumulated in a correct configuration on the antigen surface.

Immunoglobulin M is also made and secreted by plasma cells in the spleen, lymph nodes, and bone marrow. It is found in the second highest concentration after IgG in most mammalian serum. While on the B-cell surface and acting as a BCR, IgM is a single 180 kDa immunoglobulin monomer. However the secreted form of IgM is a polymer consisting of five subunits linked by disulfide bonds in a circular fashion. A small peptide called the J chain binds two of the units to complete the circle. IgM is the major immunoglobulin class produced during a primary immune response. It is also produced in a secondary response, but this tends to be masked by the predominance of IgG. Although produced in small amounts, IgM is considerably more efficient (on a molar basis) than IgG at complement activation, opsonization, neutralization of viruses, and agglutination. Because of their very large size, IgM molecules are usually confined to the bloodstream and are probably of little importance in conferring protection in tissue fluids or body secretions, even at sites of acute inflammation.

Immunoglobulin A is secreted by plasma cells mainly located in the tissues just under body surfaces. Thus it is made in the walls of the intestinal tract, respiratory tract, urinary system, skin, and mammary gland. Its serum concentration in most mammals is usually lower than that of IgM. Individual IgA molecules are normally secreted as dimers joined by a J chain. Each IgA molecule has a typical four-chain structure consisting of paired light chains and two α heavy chains. IgA produced in body surfaces may either pass through epithelial cells into external secretions or diffuse into the bloodstream. Thus most of the IgA made in the intestine is carried into the intestinal lumen. The IgA is transported through intestinal epithelial cells bound to a receptor called the polymeric immunoglobulin receptor (pIgR) or secretory component. Secretory component binds to IgA dimers to form a complex called secretory IgA (SIgA) and in this way protects the IgA from digestion by intestinal proteases (Figure 27.19). Secretory IgA is the major immunoglobulin in the external secretions. As such, it is of critical importance in protecting the intestinal, respiratory, and urogenital tracts, the mammary gland, and the eyes against microbial invasion. IgA does not activate the complement system nor can it act as an opsonin. It can, however, agglutinate particulate antigen and neutralize viruses. The major mode of action of IgA is to prevent the adherence of antigens to the body surfaces. IgA is also unique in that it can act inside cells.

Immunoglobulin E, like IgA, is made predominantly by plasma cells located beneath body surfaces. It is a typical Y-shaped, four-chain immunoglobulin. IgE is found in extremely low concentrations in serum. Because of these low concentrations, it cannot act simply by binding and coating antigens as the other immunoglobulins do. IgE acts as a signal-transducing molecule. Thus IgE molecules are normally found bound to receptors on mast cells and basophils. When antigen binds to this IgE it triggers the rapid release of inflammatory molecules from these cells. The resulting acute inflammation will enhance local defenses and help eliminate the invader. IgE mediates Type I hypersensitivity reactions (Chapter 28) and is largely responsible for immunity to invading parasitic worms.

Immunoglobulin D is primarily a BCR, and only small amounts of soluble IgD are secreted by B-cells. IgD molecules consist of two δ heavy chains and two light chains. IgD has a long hinge region and is unusually susceptible to proteases generated when blood clots. As a result, IgD is not present in serum.

Other Structural Features

In the arms of the "Y"-shaped immunoglobulin there is a groove located between the two variable domains, V_H and V_L. The peptides of the CDRs line this groove and, as a result, the surface of the groove shows great variation in

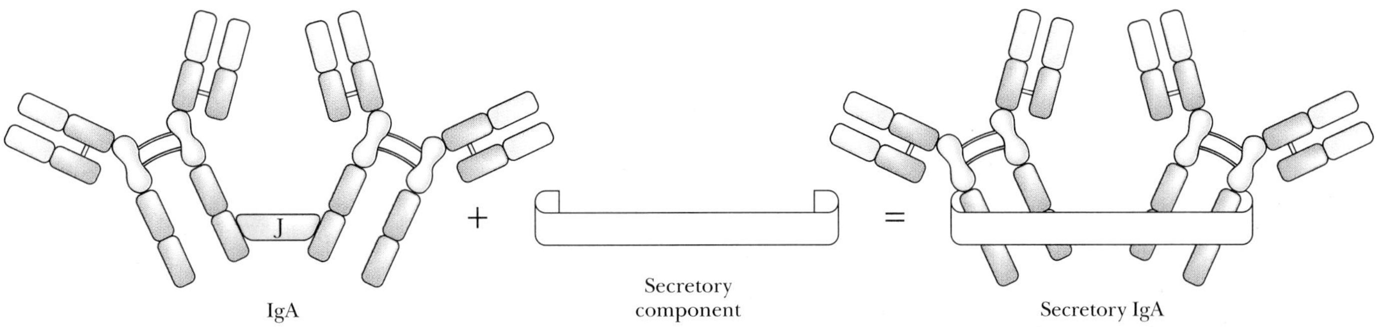

***Figure* 27.19** The structure of secretory IgA.

BOX 27.1 METHODS & TECHNIQUES

Myelomas and Hybridomas

If a single B-cell becomes cancerous it may give rise to a plasma cell tumor known as a **myeloma.** Because tumors arise from a single precursor cell, myelomas secrete a single homogeneous immunoglobulin product called a myeloma protein. Myeloma proteins may belong to any immunoglobulin class. Because myeloma cells also break down bone, the presence of myelomas growing within bones may lead to multiple fractures. The increased viscosity of the blood as a result of its very high immunoglobulin content can lead to heart failure. Individuals with myelomas are profoundly immunosuppressed because of the overwhelming commitment of the body's immune resources to the production of cancerous plasma cells. Thus, these patients commonly suffer from severe bacterial infections.

Unfortunately, since plasma cells become cancerous in an entirely random manner, the myeloma proteins that they produce are not usually directed against any antigen of practical importance. It would be highly desirable to have a method to obtain large quantities of absolutely pure, specific immunoglobulin directed against an antigen of interest. This can be done by fusing a normal plasma cell, making the specific antibody, with a myeloma cell that can grow in tissue culture. The resulting mixed cell is called a **hybridoma.**

The first stage in making a hybridoma is to generate antibody-producing plasma cells. This is done by immunizing a mouse against the antigen of interest. Two to four days after injecting antigen, its spleen is removed and broken up to form a cell suspension. These spleen cells are mixed in culture medium with cells of a special mouse myeloma-cell line. It is usual to use myeloma cells that do not secrete immunoglobulins since this simplifies purification later on. Polyethylene glycol is added to the mixture. This compound induces many of the cells to fuse together. If the fused cell mixture is cultured for several days, any unfused spleen cells will die. The myeloma cells are eliminated by using a culture medium that will not support their growth. Only hybrid cells can survive. On average, about 300 to 500 different hybridomas can be isolated from a mouse spleen, although not all will make antibodies of interest. After culturing for two to four weeks, the growing cells can be seen and the culture fluid can be screened for the presence of antibodies. Clones that produce the desired antibody are grown in mass culture.

Within recent years, **monoclonal antibodies** produced by hybridomas have become the preferred source of antibodies for much immunological research. They are absolutely specific for single epitopes and are available in large amounts. Because of their purity they can function as standard chemical reagents. They are rapidly being incorporated into clinical diagnostic techniques where large quantities of antibody of consistent quality are required.

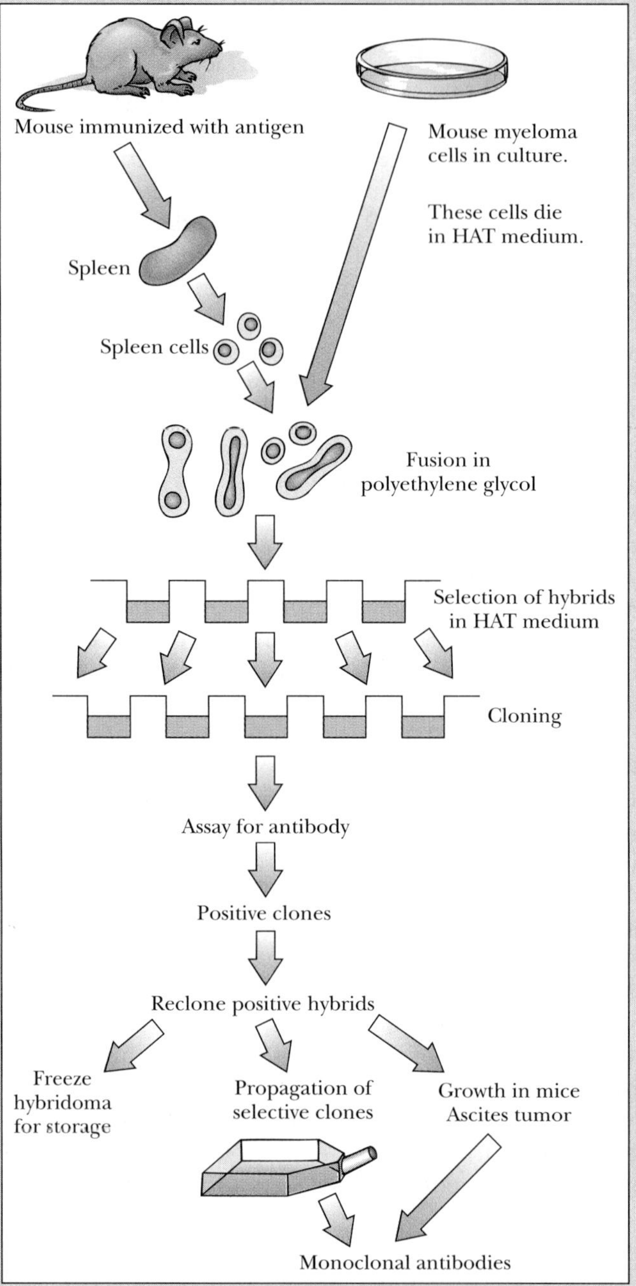

The production of monoclonal antibodies from cloned hybridoma cells. The hybrids are generated by fusing spleen cells with myeloma cells in the presence of polyethylene glycol. The fused cells are allowed to grow and clones producing the antibodies of interest are selected for further culture. The antibodies are harvested from the culture fluid.

its shape. This groove forms the antigen-binding site. The CDRs from both light and heavy chains contribute to the binding of antigen. Because immunoglobulins are bilaterally identical, the CDRs on each of the Fab regions are also identical. As a result, the molecule has two identical antigen-binding sites, and is able to bind two identical epitopes together. The presence of a hinge region in the middle of each heavy chain gives immunoglobulin molecules great flexibility. Thus bacteria may be clumped together by antibody molecules in a process called agglutination. If sufficient soluble protein molecules are cross-linked by antibody, they may come out of solution and so form a precipitate (**Box 27.1**).

Regulatory Molecules

Cells of the immune system secrete a complex mixture of proteins that regulate the immune responses by signaling between cells. The generic term for these regulatory proteins is **cytokine.** Cytokines differ from conventional hormones in several important respects. For example, they have effects on a wide variety of cells and tissues, unlike classical hormones, which tend to affect a single target organ. Second, cells rarely secrete only one cytokine at a time. For example, macrophages secrete at least four, IL-1, IL-6, IL-12, and TNF-α. Third, cytokines appear to be "redundant" in their biological activities in that many different cytokines may have similar effects. For example, IL-1, TNF-α, TNF-β, IL-6, and MIP-1α all cause a fever. This has given rise to the concept of a **cytokine network,** a complex web of interactions between all the cell types of the immune system mediated by many different cytokines. (The nomenclature of cytokines is not based on any systematic relationship between these proteins. They were originally named after their cell of origin or the bioassay used to identify them. Thus **lymphokines** came from lymphocytes and **monokines** came from macrophages or "monocytes." This nomenclature is not, however, satisfactory since characteristically these molecules come from many different cell types. The term cytokine is therefore preferred.)

Cytokines

Interleukins are cytokines that regulate the interactions between lymphocytes and other leukocytes. They are numbered sequentially in the order of their discovery and at least seventeen have been identified to date. Thus typical interleukins are interleukin 1 (IL-1), secreted by macrophages, and interleukin 2 (IL-2), secreted by T-cells. Because their definition is so broad, the interleukins are a very heterogeneous mixture of proteins with very little in common except their name.

Interferons are glycoproteins synthesized in response to virus infection, to immune stimulation, or to a variety of chemical stimulators. They inhibit virus growth by interfering with viral RNA and protein synthesis. Five types of interferon are recognized: interferon alpha (IFN-α), interferon beta (IFN-β), interferon gamma (IFN-γ), interferon omega (IFN-ω), and interferon tau (IFN-τ).

Tumor necrosis factors are two related cytokines, one derived from macrophages (TNF-α) and one derived from T-cells (TNF-β). As their name suggests, they both have the ability to kill some tumor cells, although this is not their primary function.

Many cytokines act as **growth factors** and stimulate cells to grow. They are thus very important in ensuring that the body is supplied with sufficient cells to defend itself.

Chemokines are a family of small proteins that play an important role in inflammation. A typical example of a chemokine is interleukin 8 (IL-8). They may have some antiviral activity.

Functions of Cytokines

Cytokines act on a variety of cellular targets. They may, for example, bind to receptors on the cell that produced them and so have an **autocrine effect.** They may bind to receptors on cells in very close proximity to the cell of origin—a **paracrine effect.** Or they may spread throughout the body and affect cells in distant locations—an **endocrine effect.**

When cytokines bind to receptors on target cells they can have a variety of effects on the behavior of the cells. They may, for example, induce the target cell to divide or differentiate, or they may stimulate the production of new receptors or of other protein products. Alternatively they may inhibit these effects—inhibiting division, differentiation, or new protein synthesis. Most cytokines act on a large number of different target cells, perhaps inducing different responses in each one, which is called **pleiotropy.** Many different cytokines may act on a single cellular target (redundancy). Thus B-cell function is affected by IL-2, IL-4, IL-5, and IL-6. Some cytokines work best in association with other cytokines (synergy). This synergy can occur when one cytokines act on a target cell at the same time. Thus the combination of IL-4 and IL-5 stimulates B-cell switching to IgE production. Synergy can also occur in sequence when one cytokine induces the receptor for another cytokine. For example, IL-1 acts on B-cells and induces expression of the IL-2 receptor. Once this receptor is expressed, then the B-cell becomes responsive to IL-2. Finally some cytokines may antagonize the effects of others. The best example of this is the mutual antagonism of IL-4 and IFN-γ.

Defensive Molecules

Although antibodies can bind to invading organisms they cannot, by themselves, kill anything. That function is reserved either for cells such as neutrophils or macrophages, or for a complex enzyme system called the complement system.

The Complement System

The **complement system** is an enzyme system whose activation eventually results in the destruction of cells or invading microorganisms. Complement is an essential part of the body's defenses but it must be carefully regulated, since uncontrolled activation may lead to massive cell or tissue destruction.

The 19 serum proteins that form the complement system are either labeled numerically with the prefix C—for example, C1, C2, C3; or designated by letters of the alphabet—B, D, P, and so forth. They are synthesized at various sites throughout the body. Most C3, C6, C8, and B are made in the liver, while C2, C3, C4, C5, B, D, P, and I are made by macrophages. Neutrophils store large quantities of C6 and C7. As a result, these components are readily available for defense at sites of inflammation where macrophages and neutrophils accumulate. The genes for B, C4, and C2 are located in the MHC class III region.

Classical Complement Pathway

The complement system can be activated through three different pathways. The most important of these is called the classical complement pathway (Figure 27.20). The first component of the classical complement pathway is a protease called C1, which is activated when IgG or IgM binds to an antigen. Free immunoglobulin molecules cannot activate C1 since only when an immunoglobulin binds to an antigen is the activating site exposed.

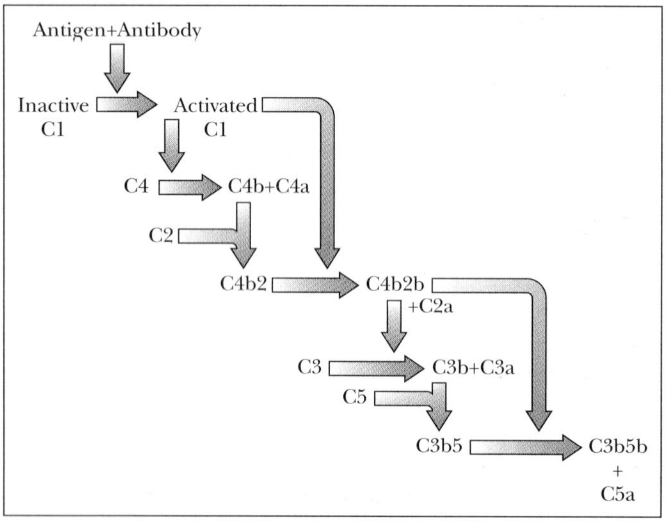

***Figure* 27.20** The classical complement pathway.

C1 can only be activated by binding to two closely spaced activating sites. These are formed by single molecules of IgM or paired molecules of IgG bound to antigen. The polymeric IgM structure readily provides two such activating sites. Two IgG molecules, in contrast, must be located very close together in order to have the same effect. As a result, IgG is much less active than IgM in activating complement. When it binds to these sites, C1 changes from an inactive to an enzymatically active form. C1 may also be activated directly by the surface proteins of some viruses, or by bacteria such as *E. coli* and *Klebsiella pneumoniae.*

Although the complement components are numbered, they do not act in numerical order! Thus activated C1 acts on two proteins, C4 and C2. When C4 is activated, an exposed thioester bond binds the C4b to nearby cell surfaces. Thus if a bacterium activates the complement system, the C4 will bind to the bacterial surface. C2 binds to C4 to form a complex protease whose substrate is the third component of complement C3 and is called "classical C3 convertase." C3 is the complement component in highest concentration in serum. Classical C3 convertase acts on C3 to split off a small fragment called C3a. The remaining biologically active portion of the molecule is called C3b. These active C3b molecules also bind to nearby surfaces.

Alternative Complement Pathway

In addition to the classical complement pathway as described above, C3 may be activated by a second mechanism called the alternative complement pathway (Figure 27.21).

Native C3 breaks down slowly but spontaneously in plasma, and, as a result, small amounts of C3a and C3b are continuously generated in normal serum. The newly formed C3b binds to cell surfaces. Under normal circumstances, this cell-bound C3b is rapidly inactivated by proteases called factors H and I. Whether these work or not depends on the nature of the cell surface. When factor H interacts with sialic acid and other neutral or anionic polysaccharides on cell surfaces, its binding to C3b is enhanced, factor I is activated, and the C3b is destroyed. On the other hand, on surfaces deficient in sialic acid, factor H does not bind to C3b, factor I is inactivated, and the C3b persists. Most mammalian cell surface glycoproteins are heavily sialylated and thus do not trigger the alternative complement pathway. Thus in a normal, stable situation, factors H and I destroy C3b as fast as it is generated and, as a result, the alternative pathway for C3 activation remains inactive. In contrast, surfaces that permit activation of C3b include bacterial cell walls, bacterial lipopolysaccharides, viruses, and aggregated immunoglobulins. In the presence of an activating surface, the binding of factor H to C3b is inhibited and the C3b

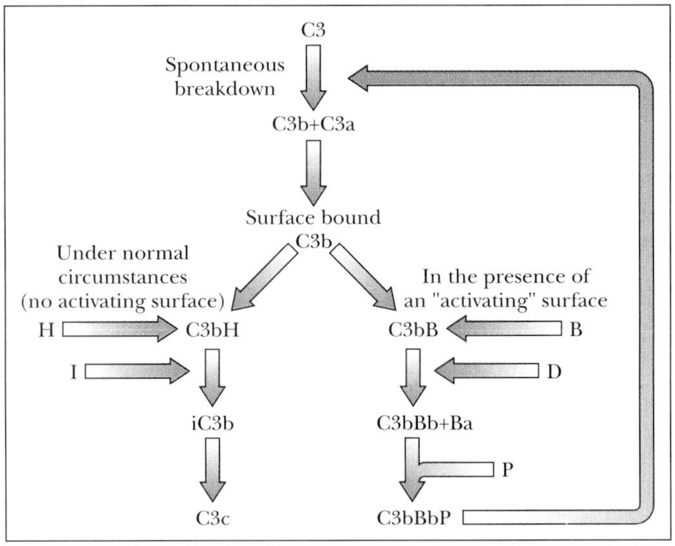

Figure **27.21** The alternative complement pathway.

then binds to proteins called factors B and P to form the alternate C3 convertase. This can split C3 and so generate active C3b.

Terminal Complement Pathway

The terminal pathway is initiated by the binding of C5 to surface-bound C3b. When bound to C3b on a cell surface C5 becomes susceptible to the C3 convertases. They split off a small peptide called C5a while the part of the molecule remaining attached to C3b is called C5b. This cleavage exposes a site on the C5b that in turns binds C6, C7, C8, and C9. Twelve to 18 C9 molecules come together with the C5b678 complex to form a tubular transmembrane pore called the membrane attack complex (MAC). The MAC forms a large doughnut-shaped structure that inserts itself into a cell membrane and forms a transmembrane channel (in other words, a "hole"). If sufficient MACs are formed on the cell membrane, lysis of the target occurs. Thus complement can effectively punch holes in and kill bacteria. Even if the terminal pathway is not activated, the presence of components such as C3 on the microbial surface will promote opsonization.

Mannose-Binding Proteins

A third pathway of activating the complement system involves mannose-binding proteins. When macrophages ingest bacteria or other foreign material they are stimulated to secrete the cytokines IL-1, IL-6, and TNF-α. These cytokines act on hepatocytes, stimulating them to secrete mannose-binding proteins. Mannose is a major component of bacterial cell wall glycoproteins. The mannose-binding protein thus coats bacteria in the bloodstream and enhances complement activation by the alternative pathway. Mannose-binding protein has structural similarities to C1q and in association with a serum protease may also activate the classical pathway.

The Immune Responses

Each of the key components described above plays a role in the immune response. Processed antigen is presented to helper T-cells (Figure 27.22). These cells in turn secrete cytokines that permit B-cells to respond to antigen and trigger antibody production. Antibodies in body fluids mark invading organisms for destruction by cells such as neutrophils or macrophages or by the complement system. Other T-cells act to destroy virus-infected or otherwise abnormal cells.

Helper T-Cell Responses

As mentioned earlier, helper T-cells require more than exposure to antigen in order to respond to antigen. The binding of a peptide to its specific TCR and CD4, even when associated with a MHC class II molecule, is insufficient by itself to trigger a response. Indeed, the binding of antigen alone may "turn off" a T-cell and lead to tolerance. In addition to antigen and MHC class II, two other signals are needed to induce a T-cell response (Figure 27.23). One signal comes from the cytokines interleukin 1 (IL-1) and interleukin 12 (IL-12) secreted by the antigen-presenting cells. The other signal is generated by binding between T-cells and antigen-presenting cells and thus involves cell surface adhesion molecules.

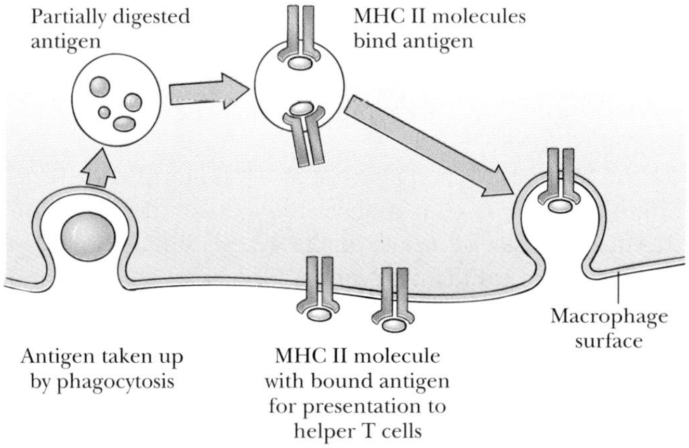

Figure **27.22** The processing of a foreign antigen within macrophages. Antigen fragments are bound to class II histocompatibility antigen molecules and presented on the macrophage surface. In order to trigger a T-cell response, the presence of interleukin 1 is also required.

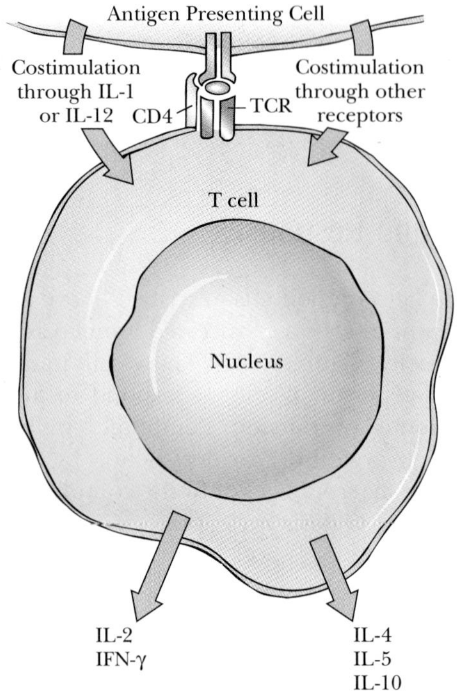

***Figure* 27.23** The helper T-cell response. Helper T-cells, appropriately triggered by antigen MHC class II and interleukin 1, synthesize helper factors such as interferon-γ, interleukin 2, or interleukin 4. These two interleukins are synthesized by different helper cell subsets and stimulate different types of antibody response.

Costimulation

Helper T-cells are stimulated by either IL-1 or IL-12 secreted by antigen-presenting cells. The secretion of these cytokines is triggered by many different stimuli, including phagocytes of bacteria, parasites, or other foreign particles. Cytokine synthesis can also be induced in the antigen-presenting cell by T-cell contact through signals transmitted through the TCR. IL-1 may either be secreted into the tissue fluid or remain bound to the cell surface. Membrane-bound IL-1, together with properly presented antigen, stimulates division of adherent T-cells. IL-12 is secreted into the surrounding fluid where it binds to receptors on the T-cells.

In addition to the effects of antigen and IL-1 or IL-12, optimal helper T-cell responses require that the T-cell and the antigen-presenting cell bind together and interact. Several different interacting pairs of molecules participate in this process. For example, CD2 binds to CD58, CD40 binds to CD40L, CD28 binds to CD80, and CD11a/18 binds to CD54. These interactions are of two types. Some of these interactions simply tie both cells together so that other signals can be efficiently transmitted between them. CD2-CD58 and CD11a/18-CD54 binding are of this type. The other interactions actually transmit signals to the T-cell. While by far the most important of these is the signal from the interaction between the TCR and MHC-bound antigen, other key signals are provided by the binding of CD40 to CD40L and of CD28 to CD80. The stimulation of a T-cell by appropriately presented antigen is not an all-or-nothing affair. Because multiple signals are involved, it is clear that T-cells may become partially activated so that they express some, but not all, functions.

Superantigens

When T-cells are exposed to a processed foreign antigen, usually only a very small proportion of the population, perhaps less than one in 10,000 cells, have TCRs that can bind the antigen and hence respond to it. However, some microbial proteins are unique in that they can stimulate as many as one in five T-cells to divide. It was originally thought that these proteins acted as nonspecific mitogens. This is not the case. They only stimulate cells whose TCRs contain certain specific V-domains in their β chains. These powerful antigens have been called **superantigens** (Figure 27.24). All superantigens come from microbial sources such as streptococci, staphylococci, and mycoplasma, and from viruses such as mouse mammary tumor virus, rabies virus, and possibly human immunodeficiency virus (HIV). The responses to superantigens are not MHC restricted (that is, they don't depend on specific MHC receptors), but the presence of MHC antigens is required for an effective response. Superantigens do not bind to the antigen-binding groove of the MHC class II molecule but attach elsewhere on its surface. Superantigens bind to both

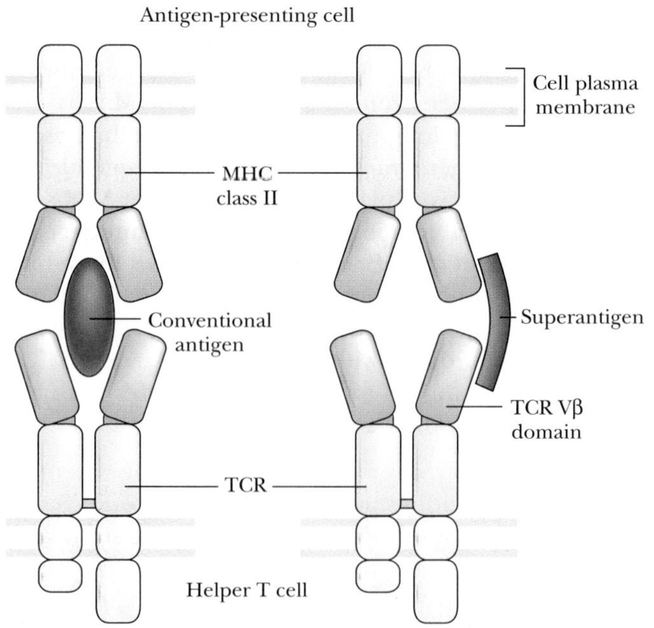

***Figure* 27.24** Superantigens link the TCR strongly to MHC II molecules by binding to sites on the MHC and on the TCR Vβ domain.

BOX 27.2 RESEARCH HIGHLIGHTS

Superantigens Are Toxic

Toxic shock syndrome is a disease that usually affects menstruating women. It is characterized by development of a fever, low blood pressure leading to vascular collapse, the development of a skin rash resembling scarlet fever, conjunctivitis, and damage to organs such as the kidney, liver, and intestine. The disease is associated with the use of vaginal tampons that permit overgrowth of certain strains of *Staphylococcus aureus.* These bacteria grow in the tampon and produce a toxin called TSST-1. TSST-1 is a potent superantigen binding specifically to the Vβ2 regions on the TCR. As a result it acts as a T-cell mitogen and as many as 20 percent of normal T-cells in human blood may respond to it. In order to act as a superantigen and stimulate these T-cells, TSST must first bind to MHC class II molecules on antigen-presenting cells such as macrophages. The toxin-MHC class II complex then binds to the Vβ2 positive T-cell and stimulates it to respond.

The clinical signs of toxic shock syndrome are therefore a result (at least in part) of the massive release of cytokines, especially TNF-α and IL-1, by stimulated T-cells or by macrophages activated by these T-cells.

a TCR Vβ domain and to an MHC class II molecule on the antigen-presenting cell. As a result, they tightly crosslink the T-cell and the antigen-presenting cell. Because of this strong binding, a very powerful T-cell response is triggered. Superantigens can stimulate the secretion of large amounts of cytokines to produce a disease called toxic shock syndrome (**Box 27.2**) (see Chapter 29).

Helper T-Cell Subpopulations

There are two subsets of helper T-cells distinguishable by the mixture of cytokines that they secrete. They are called helper 1 (Th1) and helper 2 (Th2) T-cells (Figure 27.25). These subsets respond to antigen and costimulators provided by different antigen-presenting cells. For example, Th2 cells respond optimally to antigen presented by macrophages and less well to antigen presented by B-cells. Th1 cells, in contrast, respond optimally to antigen presented by B-cells. The relative proportions of these antigen-presenting cells determine the nature of the helper T-cell response.

Th1 cells characteristically secrete IL-2, and interferon-γ (IFN-γ) after stimulation by antigen and interleukin 12. These cells primarily act as helper cells for cell-mediated immune responses such as the delayed hypersensitivity reaction (see Chapter 28) and macrophage activation. They thus promote resistance to intracellular organisms such as *Mycobacteria.* Th1 cells lack IL-1 receptors and will not therefore respond to IL-1.

Th2 cells secrete IL-4, IL-5, IL-10, and IL-13 after exposure to antigen and interleukin 1. This mixture of cytokines stimulates B cell proliferation and immunoglobulin secretion. A Th2 response is associated with enhanced immunity to some helminth parasites such as *Toxocara canis,* but with decreased resistance to Mycobacteria and other intracellular organisms. Th2 cells have IL-1 receptors and require IL-1 from macrophages of dendritic cells as a costimulator.

A third helper T-cell population has also been described that secretes a cytokine mixture that is representative of both Th1 and Th2. These cells, called Th0 cells,

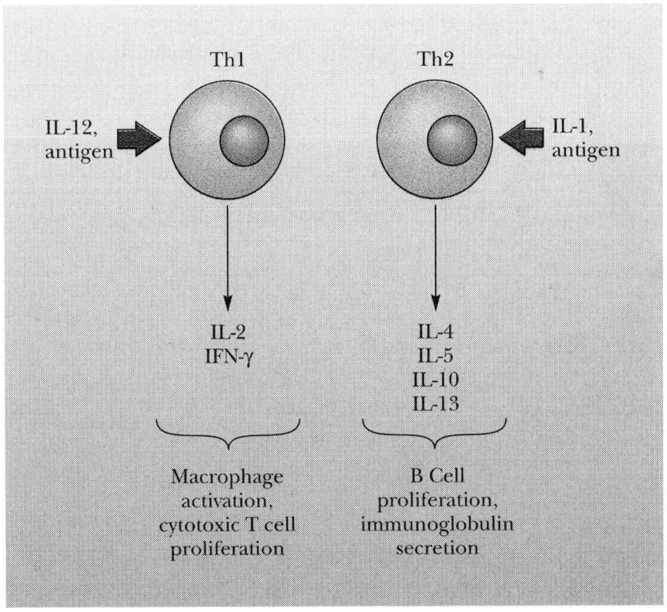

***Figure* 27.25** Helper cell subpopulations in the mouse secrete distinctly different sets of cytokines.

may be precursors of Th1 and Th2 or cells that are in transition between the two populations. When cultured in the presence of IL-4, Th0 cells become Th2 cells. When cultured in the presence of IL-12, they become Th1 cells.

B-Cell Responses

Binding of antigen to the BCR triggers activation of B-cells (Figure 27.26). As with T-cells, however, more than one signal is required for B-cells to respond fully since binding of antigen receptors alone is not sufficient. Additional signals such as binding by IL-2 or IL-4 as well as helper T-cell contact are required to fully activate B-cells. It is suggested that the antigen provides an initial signal that activates resting B-cells, while a "progression" signal such as IL-2 or IL-4 induces activated cells to actually divide.

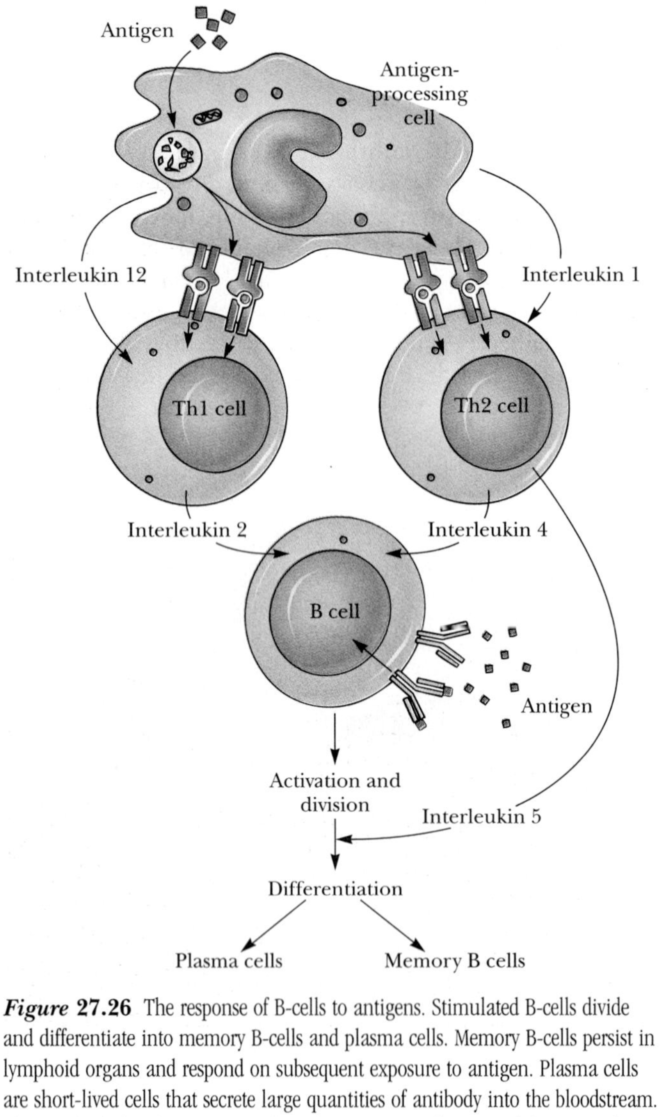

***Figure* 27.26** The response of B-cells to antigens. Stimulated B-cells divide and differentiate into memory B-cells and plasma cells. Memory B-cells persist in lymphoid organs and respond on subsequent exposure to antigen. Plasma cells are short-lived cells that secrete large quantities of antibody into the bloodstream. This antibody has the same specificity for antigen as the BCR of the original responding B-cell.

It is appropriate to reemphasize at this stage that the B-cell receptor for antigen can bind to free intact antigenic molecules in solution. This is very different from the TCR that can only bind and respond to processed antigen fragments presented on an MHC molecule. This difference in the antigen-recognizing ability of B- and T-cells is significant in that B-cells can respond to a much greater variety of antigenic stimuli than T-cells. Likewise, antibodies are directed, not against breakdown products of antigens, but against intact antigen molecules. As a result, antigen-antibody interactions usually depend on maintenance of the three-dimensional conformation of the antigen. A good example of this is seen with tetanus toxoid, as described at the beginning of this chapter.

Antigen Processing by B-Cells

B-cells are effective antigen-processing cells because they can specifically bind antigen through the BCR. B-cells ingest and process this antigen prior to linking it to MHC class II molecules and presenting it to helper T-cells, while at the same time secreting IL-1 and IL-12. Each B-cell will, of course, bind only one antigen. This makes them much more efficient than macrophages that must ingest any foreign material that comes their way. This is especially true in a primed animal, where many B-cells may be available to bind a specific antigen. B-cells can thus activate helper T-cells with 1/1000 of the antigen concentration of nonspecific antigen-processing cells such as macrophages.

Helper T-Cells and B-Cells

B-cells require physical contact with a helper T-cell in order to initiate their responses to antigen. This contact probably occurs through both MHC class II molecules and through antigen. B-cells ingest, process, and present antigen to helper T-cells through MHC class II molecules on the B-cell and the TCR complex on the Th-cell. This interaction not only triggers the Th-cell into activity but also activates the B-cell.

Activation by Cytokines

Th1 cells secrete IL-2 and IFN-γ that, although they promote B-cell proliferation and polyclonal immunoglobulin secretion, do not stimulate specific antibody formation. Th2-cells, in contrast, secrete IL-4, IL-5, IL-6, IL-10, and IL-13 that strongly enhance B-cell production of IgG1, IgA, and IgE. IL-2 is secreted by Th1-cells in response to appropriately presented antigen and IL-12. Although resting B-cells express very few IL-2 receptors (IL-2R), the presence of IL-2 stimulates the appearance of more IL-2R. The IL-2 then acts with antigen and T-cells to trigger a B-cell response. These three stimuli together make the B-cells enlarge and begin to divide. IL-4 secreted by activated Th2-cells then stimulates the growth and differentiation of activated B-cells. IL-5 secreted by Th2-cells acts on activated B-cells to enhance their differentiation into plasma cells.

Response of B-Cells to Antigen

When a mature B-cell encounters antigen that binds to its surface receptors in association with the correct stimuli from helper T-cells, it responds by division and differentiation. The responding B-cells lose their IgD BCR and secrete IgM specific for the inducing antigen. As their response to antigen continues, these immunoglobulin-producing B-cells switch their heavy-chain constant-region production from μ to γ, ϵ, or α while retaining their original variable regions. As a result, they can produce immunoglobulins of different classes without altering their antigen specificity.

An appropriately stimulated B-cell will enlarge and divide repeatedly. Some of its progeny cells develop a rough endoplasmic reticulum, increase their rate of synthesis, and start to secrete large quantities of immunoglobulins. These antibody-secreting cells are called plasma cells (Figure 27.27). Within a few days, the cell switches to synthesizing another immunoglobulin class. This switch results in a change from IgM to IgG, IgA, or IgE production. Class switching is controlled by IL-4, IFN-γ, and TGF-β. Thus IL-4 directs mouse B-cells to produce IgG1 and IgE while it directs human B-cells to produce IgG4 and IgE. However, IL-4 alone is insufficient for class switching, and additional signals are required to complete the process. IFN-γ stimulates a switch to IgG2a and IgG3 in mouse B-cells and effectively suppresses the effects of IL-4. As an immune response progresses, there is a gradual increase in antibody affinity for antigen. This increase is due to progressive somatic mutation and selection within responding B-cell populations **(Box 27.3).**

Cytotoxic T-Cell Responses

Antibodies are designed to bind to free antigen in the circulation and hasten its elimination. Not all foreign material, however, is found in extracellular locations. All viruses and many bacteria can grow inside cells in sites inaccessible to antibody. Antibodies are therefore of limited usefulness in defending the body against these organisms. Viruses and other intracellular organisms are eliminated by two processes. In one process, the infected cell is killed rapidly so that the invader has no time to grow. In the second process, the infected cell is altered or activated so that it develops the ability to destroy the intracellular organism. Both processes are mediated directly or indirectly by T-cells. The antigens that trigger these responses arise from intracellular locations and are therefore called endogenous antigens.

Endogenous Antigens

Every time a cell makes a protein, a sample is processed and peptides carried to the cell surface bound to MHC class I molecules. (Remember that MHC class I molecules are expressed on all nucleated cells) (Figure 27.28). If the peptide is not recognized by T-cells then no response is triggered. If, however, the peptide-MHC complex binds to a TCR then an immune response occurs. The T-cells that respond to endogenous antigens characteristically express CD8 on their surface since CD8 is the specific ligand for MHC class I molecules.

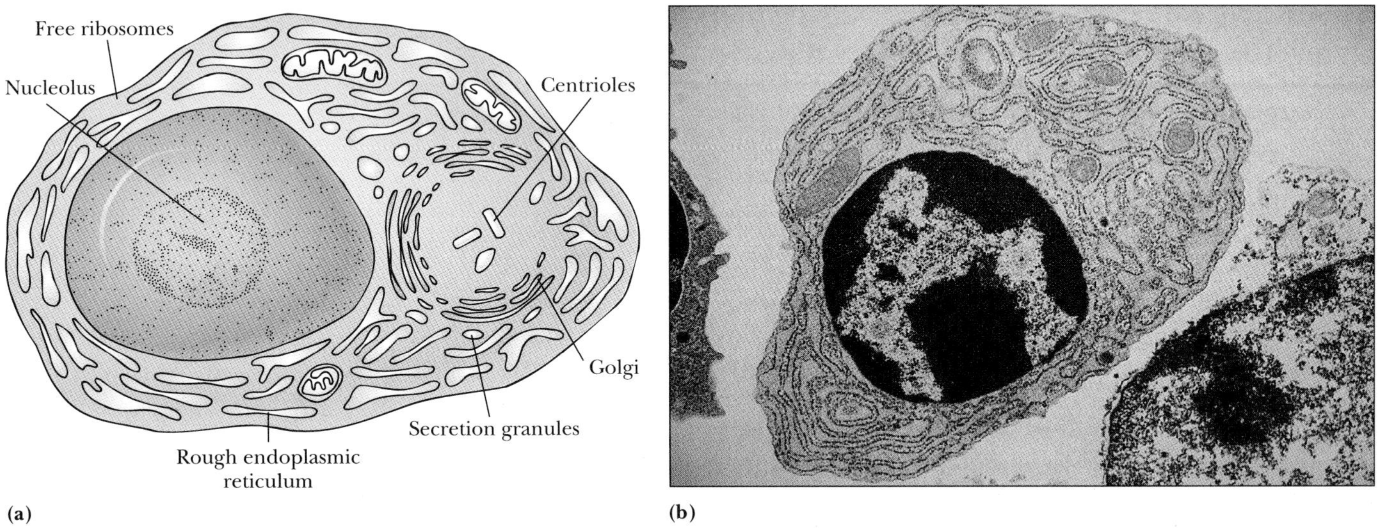

***Figure* 27.27** A typical plasma cell. **(a)** The cytoplasm is densely packed with rough endoplasmic reticulum. As a result, this cell can synthesize and secrete very large quantities of immunoglobulin. Plasma cells live for only a few days. **(b)** A TEM of a plasma cell (× 17,500). (© K. G. Murti/VU)

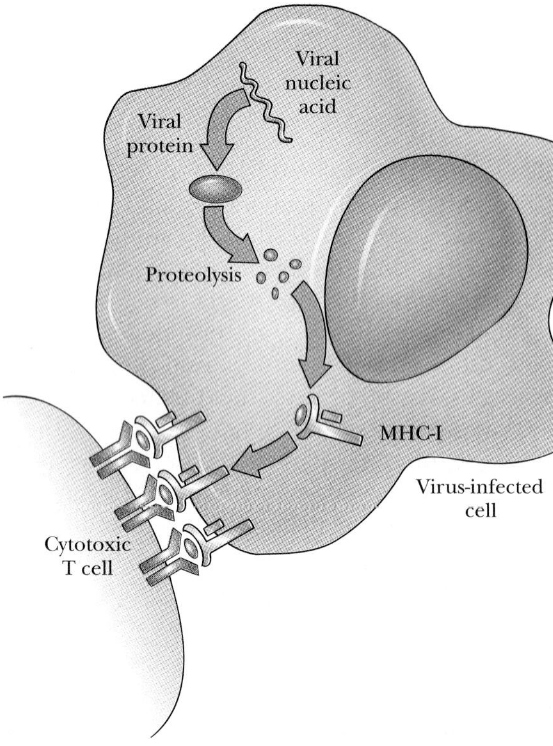

***Figure* 27.28** Viruses that invade cells replicate within the nucleus or cytoplasm, as the cell is obliged to synthesize viral proteins. Some of these proteins are fragmented, bind to class I histocompatibility antigens, and are carried to the cell surface. Here they are recognized by CD8 positive T-cells. These T-cells then destroy the target cells.

Programmed Cell Death

Any multicellular organism must be able to ensure that the production of new cells is approximately equal to the destruction of old cells in order to maintain a constant size. Likewise an animal must be able to remove old, damaged, or abnormal cells that would otherwise interfere with normal tissue functions. Thus cell death is a physiological process that is necessary for normal body function. This physiological cell death is not random but is a well-defined active process known as programmed cell death. It is commonly reflected by a set of characteristic changes in the structure of a cell called **apoptosis.** Apoptosis has distinctive features that are very different from pathological cell death or **necrosis** (Figure 27.29). For example, the cells' nuclear chromatin condenses against the nuclear membrane. The cells shrink as a result of water loss and detach from the surrounding cells. Eventually, nuclear fragmentation and cytoplasmic budding occur to form characteristic fragments that are rapidly phagocytosed and destroyed by nearby macrophages. The fragmentation of the cells does not lead to the release of cellular contents, and the phagocytosis of the apoptotic bodies does not lead to inflammation. As a result, cell death can occur without damage to adjacent cells or tissues.

In immunology, the most important example of programmed cell death is in T-cell-mediated cytotoxicity. When a cell is recognized by a CD8+ T-cell as expressing endogenous antigen, it triggers a response that results in the death of the antigen-presenting cell. This apoptosis is mediated by **cytotoxic T-cells.** These T-cells defend the body by destroying foreign or altered cells such as virus-infected cells, cancer cells, and foreign organ grafts.

Although most cells can be triggered to undergo apoptosis by certain, very specific signals, the cytotoxic T-cell is able to induce apoptosis in any cell it recognizes. This is a well-controlled process since at least two and possibly three interacting cells are required for an effective cytotoxic T-cell response to occur. These include a cell expressing endogenous antigen on its surface (the target cell), a CD8+ effector T-cell (the cell that will actually trigger apoptosis), and to a lesser extent, a Th1-cell.

The effector T-cell receives two key signals. The first is IL-2 secreted by Th1-cells. This binds to receptors on CD8+ T-cells, triggering them to make more IL-2 receptors. The second signal comes from the endogenous antigen/MHC class I complex on the surface of the abnormal cell. This binds to the TCR-CD8 complex of the T-cells. The combination of signals initiates cell division and activates the cytotoxic process. Responding T-cells, as they divide, eventually differentiate into memory-cell and cytotoxic-cell populations.

Once fully activated, cytotoxic T-cells trigger apoptosis in their targets through two pathways (Figure 27.30). One pathway involves the secretion of proteins called per-

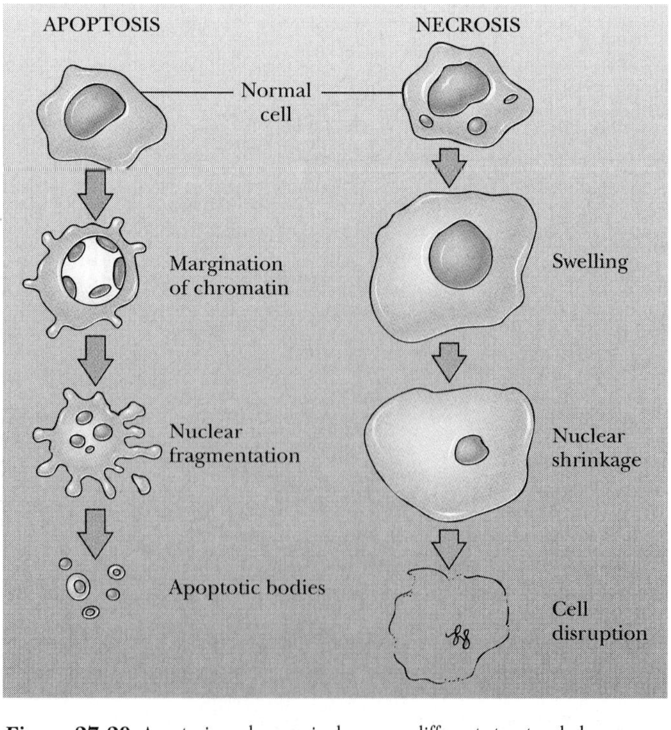

***Figure* 27.29** Apoptosis and necrosis show very different structural changes.

BOX 27.3 RESEARCH HIGHLIGHTS

Somatic Mutation

Although it was known for many years that the secondary antibody response was characterized by increased levels of antibody production, it was only relatively recently that it was shown that these antibodies were much more effective in binding antigen than those produced in a primary immune response. Thus, as an immune response proceeds, the antibodies produced improve both quantitatively and qualitatively.

The improvement in antigen binding seen as the response proceeds is a result of two processes; cell selection and somatic mutation. These processes cause gradual changes to occur in the structure of antibody antigen-binding sites so that their ability to bind antigen is enhanced.

An antibody response is initiated when antigen binds to a BCR. This stimulates the B-cell to divide and differentiate. The intensity of this stimulus is directly related to the strength of binding of the antigen with the receptor. Antigen binding weakly to the receptor will stimulate a weak immune response; antigen binding strongly will stimulate a much greater response. Thus those B-cells whose receptors bind the antigen best will be preferentially stimulated to respond. Since the BCR is eventually secreted in the form of antibodies, it is clear that the antibodies produced during an immune response will be derived from the B-cells that bind and respond most strongly to antigen.

The results of this cell selection are however amplified by the process of somatic mutation. This process, occurring in the genes of responding B-cells results in random changes occurring in the structure of the BCR antigen binding sites. Many of these changes will reduce the ability of the BCR to bind antigen. In these cases the B-cell will no longer be stimulated by antigen and will cease to participate in the response. If however, the changes induced by somatic mutation enhance antigen binding by a BCR, then such B-cells will be preferentially stimulated. The population of responding B-cells will gradually be dominated by those cells that bind antigen most strongly. As the response progresses therefore, the combination of somatic mutation and selective stimulation of B-cells with the strongest antigen-binding ability will lead to production of antibodies that bind antigen most strongly. As a result, invading organisms will be confronted by an antibody response of gradually increasing effectiveness and so will be progressively less able to invade the body.

forins (the perforin pathway). The other pathway is mediated through a cell surface receptor called CD95 or *fas* (the CD95 pathway). Both pathways require physical contact between the cytotoxic T-cell and its target.

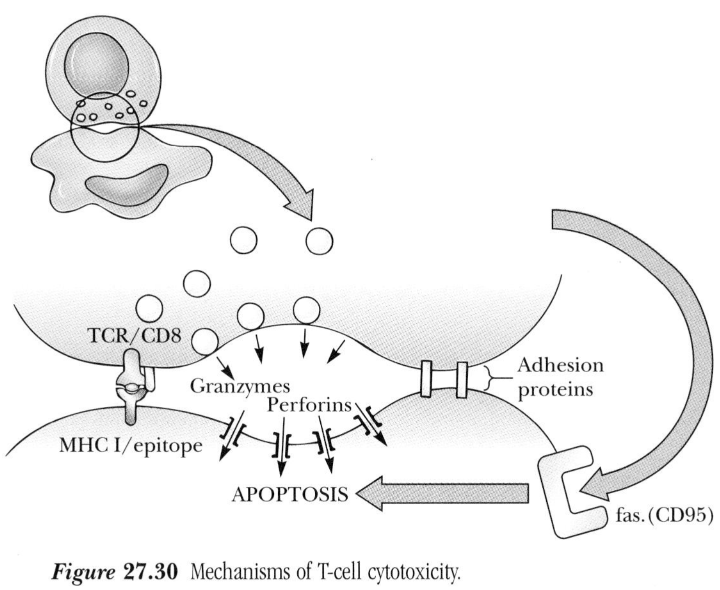

***Figure* 27.30** Mechanisms of T-cell cytotoxicity.

The Perforin Pathway

Cytotoxic T-cells adhere to their targets through the TCR binding to a peptide bound to an MHC class I molecule and CD8 binding to the MHC molecule. The first step in the lethal hit process occurs within a few minutes of binding. The T-cell granules are released at the region of cell contact. The granules contain perforins, which polymerize in the target cell membrane to form circular transmembrane pores. Proteases called granzymes are then injected by the T-cell into the target, where they trigger apoptosis. Within seconds after contact between a T-cell and its target, the organelles and the nucleus of the target cell show irreversible apoptotic changes. The membrane attack complexes formed by polymerized perforins have a similar structure and function to the complexes formed by C9 of the terminal complement pathway. The T-cell can then disengage itself and move on to find another target.

The CD95 Pathway

A transmembrane glycoprotein called CD95 (or fas) also plays a key role in T-cell-mediated cytotoxicity. CD95 is a receptor found on many nucleated cells. When bound by its appropriate ligand, CD95 signals a cell to undergo programmed cell death. Its ligand, CD95L, is thus a killing factor, while CD95 is its receptor. Cytotoxic CD8+ T-cells express high levels of CD95L and can induce apoptosis when they come into contact with CD95+ targets. Expression of CD95 on target cells is an essential prerequisite for the destruction of some targets by cytotoxic T-cells.

Immunologic Memory

A characteristic feature of both the B- and T-cell-mediated immune responses is memory—the ability to mount an enhanced immune response on second or subsequent exposure to antigen. Memory is mediated by long-lived memory cells.

B-Cell Memory

Memory cells form a reserve of antigen-sensitive cells to be called upon on subsequent exposure to an antigen. If a second dose of antigen is given to a primed animal, it will encounter a large number of memory B-cells, which respond in the manner described previously for antigen-sensitive B-cells. As a result, a secondary immune response is much greater than a primary immune response. The lag period is shorter, since more antibody is produced and it is detected earlier. IgG is also produced in preference to the IgM that is characteristic of the primary response. In the case of the B-cell response, one reason the primary immune response terminates is because many responding B-cells and plasma cells are simply removed by apoptosis. If all these cells died, however, immunological memory could not develop. Clearly some B-cells must survive to become memory cells. The survival of memory cells is under the control of a gene called *bcl*-2. Activation of the *bcl*-2 gene and expression of its product allow a cell to avoid death by apoptosis and to differentiate into a memory cell. Thus *bcl*-2 is a "survival gene" that ensures the prolonged survival of these cells by preventing apoptosis.

Germinal centers within lymphoid tissues are the sites of antigen-driven cell proliferation, somatic mutation, and positive and negative selection of B-cell populations (Figure 27.31). About six days after the response begins, B-cells stimulated by antigen and helper T-cells migrate to a germinal center, where they divide rapidly. During this stage of rapid B-cell division, the BCR V region genes mutate at a rate of about one mutation per division. This somatic mutation generates large numbers of B-cells whose BCRs differ from those of the parent cell. Once these cells have been clonally expanded they migrate to the periphery of the germinal center where they encounter antigen on dendritic cells. Antigen is trapped by dendritic follicular cells in germinal centers. Because of somatic mutation, some of the germinal center B-cells will bind this antigen with greater affinity, but many, perhaps the great majority, will bind the antigen less strongly. If somatic mutation has resulted in greater affinity of the BCR for the antigen, then B-cells with these receptors will divide further and leave the center to form either plasma cells or memory cells. The majority of mutated BCRs will, however, show reduced antigen binding. The cells with these receptors will undergo apoptosis, and their remains are removed by macrophages. Thus the B-cells that emerge from a germinal center are a very different population from the population of cells that entered it.

T-Cell Memory

When T-cells respond to antigen by the production of cytotoxic cells, memory T-cells are also generated. Memory T-cells can be distinguished from naïve T-cells by expressing different cell surface antigens, by secreting a different mixture of cytokines, and by having different activation requirements. The life span of mature T-cells is very long, as shown, for example, by the slow decline in their numbers after adult thymectomy. At the end of a primary immune response, when antigen becomes limiting, many effector T-cells are eliminated. This may be a device to prevent the repertoire of naïve T- and B-cells from being swamped with primed cells.

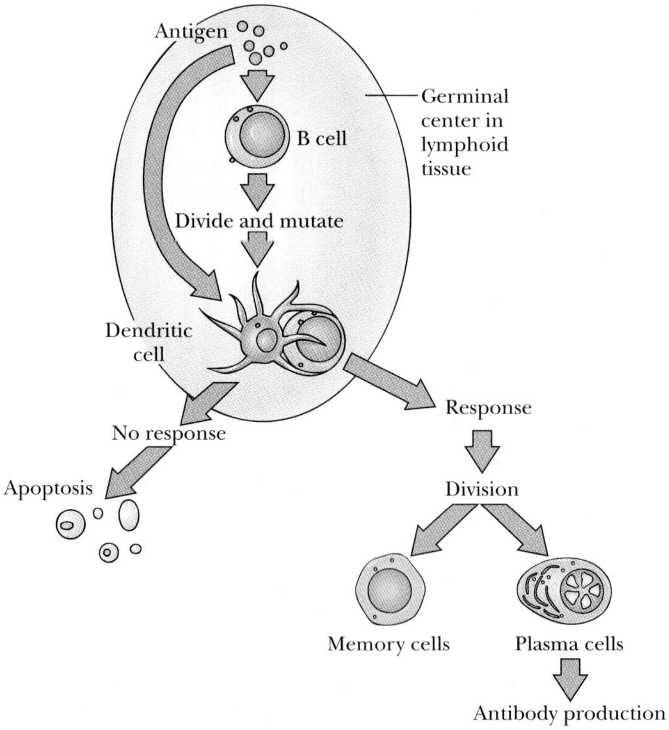

***Figure* 27.31** The function of germinal centers.

Regulation of the Immune Response and Tolerance

The immune system can recognize and respond to foreign invaders and can learn from the experience so that the body responds faster and more effectively when exposed to the invader a second time. However, there is a risk associated with this—the risk of damaging normal body components. One of the main reasons that the immune system is so complex is that considerable effort must be put into ensuring that it will attack only foreign or abnormal tissues and will be tolerant of normal healthy tissues. As might be anticipated, several different mechanisms ensure that self-tolerance is established. In addition, the immune responses must be carefully regulated to ensure that they are appropriate with respect to both quality and quantity.

Since both T and B lymphocytes generate antigen-binding receptors at random, it is clear that the production of self-reactive cells cannot be prevented. An animal cannot control the amino acid sequences and hence the antigen specificity of these receptors. Tolerance induction therefore requires the selective elimination or suppression of those lymphocytes that develop receptors that can bind self-antigens.

The ability of a T-cell to respond to an antigen depends on having the correct TCR. Tolerance to an antigen will result if an animal lacks T-cells with receptors specific for that antigen. Several processes are involved in eliminating T-cells with self-reactive TCRs. First, the mechanisms used to generate TCR diversity will inevitably result in the production of nonfunctional TCRs. For example, two thirds of possible gene arrangements will be out of frame. Presumably cells bearing these nonfunctional TCRs undergo apoptosis very early in life. Second, there is positive selection of T-cells carrying receptors that can bind the animal's own MHC molecules. Third, there is negative selection (or clonal deletion) of T-cells with receptors for self-antigens that may be potentially able to mount damaging autoimmune responses. This negative selection leads to self-tolerance.

Unlike the T-cell receptor repertoire, antibody diversity is generated within the B-cell population in two phases. The first occurs by gene segment rearrangement in the primary lymphoid organs; the second phase involves random somatic mutation in germinal centers in secondary lymphoid organs. Thus B-cells have several opportunities to become reactive to self-antigens, and self-reactive B-cells can be readily found in lymphoid organs. These cells do not necessarily make autoantibodies. If all the components necessary for a B-cell response to antigen are not present (i.e., antigen-presenting cells and helper T-cells), or if regulatory T-cells are active, then the B-cell may be unable to respond effectively to antigen and so appear to be nonfunctional. Since their response depends on helper T-cells, the failure of B-cells to produce autoantibodies may simply result from the absence of self-reactive helper T-cells rather than any change in the B-cells themselves. This is not, however, a foolproof method of preventing self-reactivity. In situations in which T-cells but not B-cells are tolerant, antibody production may be stimulated by using either cross-reacting epitopes or a foreign carrier molecule to stimulate nontolerant helper T-cells. Thus, other mechanisms must also be available to ensure B-cell tolerance. These mechanisms include destruction of self-reactive clones, loss of B-cell function, B-cell exhaustion, and blockage of B-cell receptors.

Summary

- The immune system is a key component of the body's defenses against invading microorganisms.

- These invaders fall into two groups, those that invade the body directly from the exterior and those that actually arise within the body. Examples of the former include most bacteria and newly arrived viruses. The latter group include viruses growing within infected cells and intracellular bacteria.

- Two types of immune response have evolved to combat these invaders. Bacteria and other **exogenous antigens** tend to stimulate an **antibody-mediated immune response.** These antibodies in body fluids are able to assist in the destruction of extracellular organisms. Viruses and other **endogenous antigens,** in contrast, tend to stimulate a **cell-mediated immune response.** In this type of response, cytotoxic T-cells destroy the infected cells.

- Both forms of immune response require that foreign antigens first be processed. Thus exogenous antigens are taken up by phagocytic cells such as **macrophages,** B-cells, or **dendritic cells,** broken down into small peptides and inserted in a groove on receptors called **MHC class II molecules.** Endogenous antigens are produced within infected cells. They, too, are broken down into small fragments and inserted in a groove on receptors called **MHC class I molecules.** This processed antigen is then presented to a population of lymphocytes with specific antigen receptors.

- Exogenous antigen on a MHC class II molecule is presented to **helper T-cells** that can recognize the antigen while at the same time cross-linking the MHC molecule to a receptor called **CD4** on the T-cell surface. This binding, in the presence of proteins secreted by the antigen-processing cell called **interleukin 1** and **interleukin 12,** is sufficient to trigger the helper T-cell to respond and secrete its own cytokines.

- At the same time as a helper T-cell is responding to processed antigen, another population of **lymphocytes,** called **B-cells,** can recognize free, unprocessed antigen but cannot respond by itself. The **cytokines** secreted by the stimulated helper cells will enable B-cells to respond to antigen by dividing and secreting their receptor molecules into the extracellular fluid. These soluble, antigen-binding proteins are called **antibodies.**

- Antibodies, or **immunoglobulins,** can bind to foreign antigen and mark it for destruction. Antibody-coated organisms may be destroyed either by the activities of phagocytic cells or by activation of a group of enzymes called the **complement system.**

- Endogenous antigen synthesized by an infected cell is recognized by members of another T-cell population—**cytotoxic T-cells.** These T-cells link to infected cells through their antigen receptor as well as by binding between their surface molecule CD8 and the MHC class I molecule on their target. This binding will trigger the T-cell to kill the infected cell within a few minutes.

- In both forms of immune response, not only do lymphocytes respond to foreign antigen by attacking it in some way, but some of the responding cells alter their characteristics and become long-lived **memory cells.** These memory cells are therefore available to respond on subsequent exposure to the same antigen. Thus the immune response characteristically is enhanced, each time it encounters an antigen, and the efficiency of the process becomes progressively greater.

- In addition to the main protective pathways, foreign invaders can be attacked by other processes. Thus T-cells may secrete cytokines that activate macrophages and greatly enhance their ability to kill invading organism.

- A third population of lymphocytes called **natural killer** (NK) cells also has cytotoxic properties and will kill both virus-infected cells and tumor cells.

Questions for Thought and Review

1. What are the two major cell types that phagocytose antigen? In what ways do the functions of these cells differ? Describe what might happen to an individual whose phagocytic cells are defective.

2. Which cell types usually process foreign (exogenous) antigen in order to trigger an immune response? How is this processing accomplished?

3. Class I and class II histocompatibility molecules play a key role in antigen presentation. How do they do this? What are the key differences between the functions of these molecules?

4. Outline the key differences between T- and B-cells with respect to origin, function, and fate.

5. What are cytokines? List the important T-cell-derived cytokines and outline their function. Which cytokines are secreted by Th1-cells and Th2-cells?

6. Draw a labeled diagram showing the structure of an IgG molecule. Indicate clearly how this molecule binds to an epitope.

7. What is the role of IgA in the defense of the body? How does this molecule function within the intestinal lumen without being digested?

8. Why might the body require both cytotoxic T-cells and NK cells?

Suggested Readings

Abbas, A. K., A. H. Lichtman, and J. S. Pober. 1991. *Cellular and Molecular Immunology.* Philadelphia: W. B. Saunders Co.

Burton, D. R. 1990. "Antibody, the flexible adaptor molecule." *Trends Biochem Sci.,* **15:**64–69.

Brown, E. J. 1995. "Phagocytosis." *Bioessays,* **17:**109–117.

Callard, R., and A. Gearing. 1994. *The Cytokine Facts Book.* London: Academic Press.

Coffman R. L. 1989. "T-helper heterogeneity and immune response patterns." *Hosp Pract.,* **24:**101–133.

Cohen, J. J. 1993. "Apoptosis." *Immunol Today,* **14:**126–130.

Goodenough, U. W. 1991. "Deception by pathogens." *American Scientist,* **79:**344–355.

Johnston, R. B. 1988. "Current concepts: Immunology, monocytes and macrophages." *N. Engl J. Med,* **318:**747–752.

Keegan, A. D., and W. E. Paul. 1992. "Multichain immune recognition receptors: Similarities in structure and signaling pathways." *Immunology Today,* **13:**63–68.
Monaco, J. J. 1992. "A molecular model of MHC class I-restricted antigen processing." *Immunol Today,* **13:**173–179.
Silverstein, Am. M. A. 1989. *A History of Immunology.* San Diego, CA: Academic Press.
Tizard, I. R. 1994. *Immunology: An Introduction,* 4th ed. Philadelphia: Saunders College Publishing.
Vitteta, E. S., R. Fernandez-Botran, C. D. Myers, and V. M. Sanders. 1989. "Cellular interactions in the humoral immune response." *Adv Immunol.,* **45:**1–105.

Throughout nature infection without disease is the rule rather than the exception.

Rene Dubos, in *Man Adapting*

Chapter 28

The Defense of the Body: Immunity to Infectious Agents

Immunity to Foreign Organisms
Immunity to Bacteria
Immunity to Viruses
The Elimination of Foreign Cells
Immunity on Body Surfaces
Vaccines and Vaccination
Defects in the Immune System
Excessive Immune Function

The immune system evolved in order to defend the body against microbial invasion. A functioning immune system is necessary for survival. In the absence of an effective immune system, the body will be invaded and destroyed by a host of microorganisms. Because the defense of the body is critical, it is not surprising that an animal's immune system acts through many different mechanisms. Some defenses are on the body surface, some are deep in the body, some are directed against bacteria, and others are directed against viruses. In no case is the defense of the body entrusted to a single defense mechanism. Survival is much too important to rely on only one line of defense.

Immunity to Foreign Organisms

Resistance to infectious agents is mediated by several different immune processes. In general, antibodies are the primary defense mechanisms against extracellular organisms—such as most bacteria and many protozoa. In contrast, cell-mediated immune responses are the primary defense mechanisms against intracellular organisms such as viruses (see Figures 27.2 and 27.12). It is of interest to note that this is confirmed by observations on individuals who have a defective immune system. For example, individuals who cannot make antibodies usually die as a result of overwhelming bacterial infections. In contrast, those who have a T-cell deficiency frequently die as a result of infections with viruses, intracellular bacteria, or protozoa.

Immunity to Bacteria

There are four basic mechanisms by which the specific immune responses combat bacterial infections (Figure 28.1). These are the neutralization of toxins or enzymes by antibody; the killing of bacteria by antibodies, complement, and lysozyme; the opsonization and phagocytosis of bacteria by neutrophils and normal macrophages; and the phagocytosis and destruction of intracellular bacteria by activated macrophages. The relative importance of each of these processes depends upon the organisms involved and the mechanisms by which they cause disease.

Immunity to Toxigenic Bacteria

In diseases caused by bacteria that produce **exotoxins** such as the clostridia or *Bacillus anthracis,* the immune system must not only eliminate invading organisms but also neutralize their toxins. Unfortunately, destruction of these bacteria may be difficult, especially if they are embedded in a mass of dead tissue. Nevertheless, antibodies can readily neutralize bacterial toxins by blocking the combination of the toxin with its receptor on the host cell. IgG is almost totally responsible for toxin neutralization since IgM molecules do not usually bind sufficiently strongly. If IgM binds to the toxin, it is easily dissociated and the toxin will remain active. The neutralization process involves competition between a receptor and antibody for the toxin molecule. Once the toxin has combined with the receptor, antibody will be relatively ineffective in reversing this combination. This is seen in practice when the dose of tetanus immune globulin required to produce clinical improvement is much greater than that required to prevent the onset of disease.

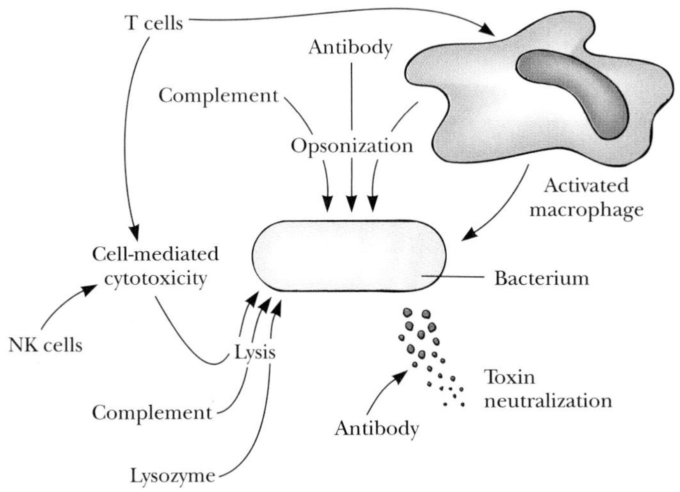

Figure **28.1** The mechanisms by which the immune responses can protect the body against bacterial invasion. Antibodies effectively bind to extracellular bacteria and mediate either opsonization or complement-mediated lysis. Recent studies have suggested that T-cells and NK-cells may have a direct bactericidal effect.

Immunity to Invasive Bacteria

Protection against invasive bacteria is mediated by antibodies directed against antigens found on the microbial surface. For example, while normal bacteria are very inefficiently phagocytosed, antibody-coated bacteria are readily phagocytosed. Some bacteria resist phagocytosis by having a thick protective capsule. Antibodies directed against antigens on the capsule neutralize its antiphagocytic properties, thus opsonizing the bacteria and permitting their destruction by phagocytic cells. In organisms lacking capsules, antibodies directed against cell wall antigens serve a similar function. The fate of many bacteria depends on the immunoglobulin class produced and whether complement is activated. Thus, IgG is a more effective opsonin that IgM in the absence of complement. IgM, however, is more effective at activating complement than IgG and thus is the most effective opsonin in the presence of complement. During a primary immune response, the low level of the IgM response is compensated for by its quality, so ensuring early and efficient protection.

While some bacteria are phagocytosed by neutrophils or macrophages, others may be killed while they are free in the circulation. This bactericidal activity is mediated by antibodies, complement, and lysozyme. The outer cell wall, composed of complex carbohydrates, may first be damaged by lysozyme. Complement can then act on the exposed inner cell membrane and so kill the bacteria. Complement may also act in the absence of antibody. Gram-negative organisms activate the alternate complement pathway by inhibiting factor-H activity, or directly activate the classical pathway through mannose-binding protein.

The Heat Shock Protein Response

Many new proteins are produced by cells stressed by heat, starvation, exposure to oxygen radicals, toxins such as heavy metals, protein synthesis inhibitors, or viral infections. The heat shock proteins (HSP) are the best understood of these new proteins. HSP are present in all organisms at very low levels at normal temperatures. Mild heat treatment such as a low-grade fever will induce HSP in cells and increase their levels significantly. For example, HSP climb from 1.5 percent to 15 percent of the total protein in stressed *Escherichia coli.* These proteins increase thermotolerance so that the cells or organisms can function normally at a higher temperature. There are three major heat shock proteins, HSP 90, HSP 70, and HSP 60.

When a bacterium is exposed to the respiratory burst within a neutrophil, large quantities of HSP 60 are generated by the stressed organism. HSP 60 is the immunodominant antigen induced by Mycobacteria, *Coxiella burnetti,* Legionella, Treponema, and Borrelia infections. These HSP are highly antigenic, and it is suggested that an anti-HSP response may be a major defense against pathogens.

Modification of Bacterial Disease by Immunity

The development of an antibacterial immune response will clearly influence the course of an infection. At its best, it will result in a cure. In the absence of a cure, however, the disease may be modified. This depends on whether a cell-mediated or antibody response is generated. Thus the type of helper T-cell that develops during an immune response will profoundly affect the course of disease. As described in Chapter 27, cell-mediated responses are usually required to control diseases caused by intracellular bacteria. Only activated macrophages can prevent the growth of these organisms. Macrophage activation requires that Th1-cells produce IL-2 and IFN-γ. When the macrophages are activated, they can localize or cure these infections. If, in contrast, Th2 responses are stimulated, then macrophages are not activated, and chronic progressive disease results. An example of this is seen in leprosy (Figure 28.2).

Leprosy occurs in two distinct clinical forms called tuberculoid and lepromatous leprosy. Tuberculoid leprosy, found in about 80 percent of cases, is associated with resistance to the disease as a result of strong cell-mediated immune responses to the leprosy bacillus. Patients with tuberculoid leprosy have few organisms in their tissues and therefore few lesions. The lesions are infiltrated with macrophages and the patients give a positive delayed hypersensitivity reaction to intradermally injected bacterial extract (called lepromin) (this is a cell-mediated response, page 737). Most of these patients do not have antibodies to *Mycobacterium leprae.*

In contrast, lepromatous leprosy, found in about 20 percent of cases, is associated with low resistance to the disease. Patients with lepromatous leprosy do not mount a detectable cell-mediated response but instead make high levels of antibodies against *M. leprae.* They may also develop immune-complex lesions (Type III hypersensitivity, page 754) and autoantibodies to normal tissues. The lesions in these patients are densely infiltrated with enormous numbers of leprosy bacilli and contain very few macrophages. The prognosis of lepromatous leprosy is much poorer than for tuberculoid leprosy. Recent studies have clarified the mechanisms involved in the development of lepromatous leprosy. These patients mount an immune response preferentially utilizing Th2-cells. As a result, large amounts of IL-4 and IL-10 are produced. The IL-10 has a direct inhibitory effect on macrophage cytokine synthesis. It reduces the production of IL-12. This

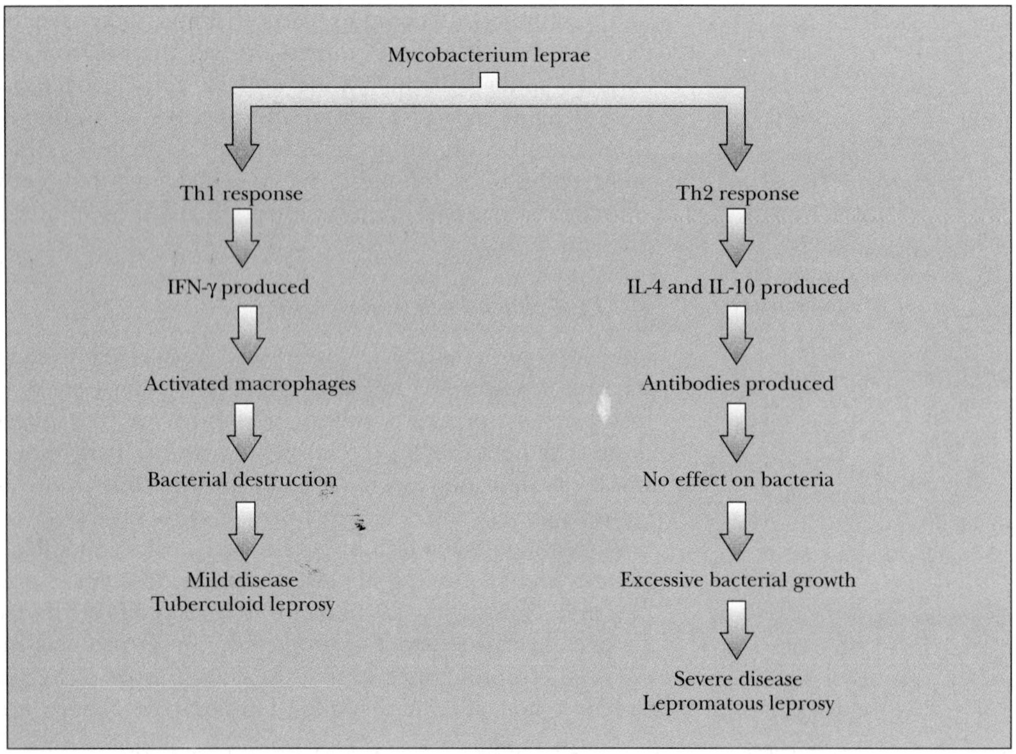

***Figure* 28.2** The two clinical forms of leprosy. Most people develop the mild disease as a result of mounting a Th1 response leading to the development of cell-mediated immunity. An unfortunate minority mount a Th2 response and produce ineffective antibodies. As a result they develop severe disease.

reduction in IL-12 production will result in a decrease in Th1-derived IFN-γ. Since IFN-γ is necessary for macrophage activation, this will directly reduce a patient's ability to control an intracellular organism such as *M. leprae.* In contrast, patients who develop tuberculoid leprosy mount a response dominated by Th1-cells. Their lesions contain IFN-γ, which triggers significant macrophage activation.

Evasion of Host Defenses by Bacteria

Bacteria, like most organisms, are not well served either by the death of their host or by their own elimination. They have therefore evolved mechanisms to evade the immune responses. Some organisms such as *E. coli, Staphylococcus aureus, Mycobacterium tuberculosis,* and *Pseudomonas aeruginosa* secrete molecules that can depress phagocytosis by neutrophils. For example, *Pasteurella haemolytica,* a major cause of pneumonia in cattle, secretes a protein that kills alveolar macrophages. Another bacterium, *Pseudomonas aeruginosa,* secretes proteases that can destroy IL-2.

Many bacteria can defend themselves against the oxygen radicals produced by phagocytic cells by producing enzymes such as superoxide dismutase and catalase. The pigments responsible for the yellow color of *Staphylococcus aureus* can quench singlet oxygen and so enable the bacteria to survive the respiratory burst. Strains of *E. coli* and salmonellae may carry plasmids that confer resistance to complement-mediated lysis. Other protective factors secreted by pathogenic bacteria include the production of proteases specific for IgA by *Neisseria gonorrhoeae, Streptococcus pneumoniae, Neisseria meningitidis, Haemophilus influenzae,* and *Streptococcus sanguis.*

The ability to grow within cells enables an organism to avoid destruction by antibodies. In addition, intracellular bacteria such as *L. monocytogenes* can travel from cell to cell without exposing themselves to antibodies in the extracellular fluid. This strategy is not, however, free of risk. Endocytosed organisms may be transported to acidic lysosomes and destroyed by lysosomal enzymes. Thus these intracellular bacteria have to take steps to avoid being killed by the cell. They do this in three major ways. One way is to remain inside the phagosome but ensure that lysosomal enzymes do not enter. Thus *Chlamydia trachomatis* can inhibit both acidification and fusion between the bacteria-containing phagosome and lysosomes. This has the additional benefit that chlamydial antigens are not processed and presented on MHC class I molecules, thus preventing attack by cytotoxic T-cells. Other organisms that follow this intraphagosomal approach include *Legionella pneumophila* and *Toxoplasma gondii.* Some organisms, such as *Mycobacterium tuberculosis,* follow a second route. They can survive in the phagolysosome because they are resistant to destruction by lysosomal enzymes. A third survival technique practiced by intracellular bacteria is to escape from the phagolysosome into the host cell's cytoplasm. An organism that does this is *L. monocytogenes.*

Immunity to Viruses

In contrast to the immune response against bacteria, antiviral immunity is, in many cases, long-lasting. This may be related to virus persistence within cells, perhaps in a slowly replicating or a nonreplicating form as typified by the herpesviruses. The persistent virus may periodically boost the immune response of the infected animal and in this way, generate prolonged immunity. This type of infection can only be eliminated by an immune response directed against infected cells, and even this may be ineffective if the viral antigens are not presented on the cell surface. The immune responses in these cases, although not able to eliminate the virus, may prevent the development of disease and therefore be protective.

Interferons

Interferons were briefly introduced in Chapter 27 as antiviral and regulatory cytokines. Some are produced by virus-infected cells within a few hours after viral invasion, and high concentrations of interferon may be achieved within a few days *in vivo,* at a time when the primary immune response is still relatively ineffective. For example, in individuals exposed to influenza virus, peak serum interferon levels may be reached one to two days later and then decline, although they are still detectable at seven days (Figure 28.3). In contrast, antibodies are usually not detectable in serum until five to six days after virus administration.

Interferons are glycoproteins with molecular weights between 20 and 34 kDa. Five major classes are recognized: interferon α (IFN-α), a family of at least twenty different molecules derived from virus-infected leukocytes; inter-

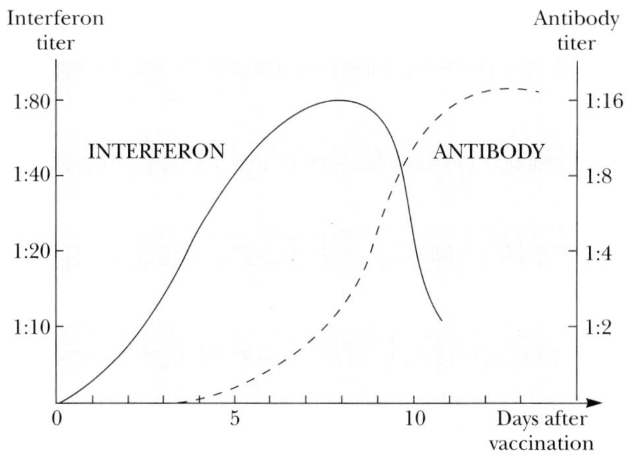

***Figure* 28.3** The sequential production of interferon and antibody.

feron β (IFN-β), derived from virus-infected fibroblasts; interferon γ (IFN-γ), a lymphokine derived from antigen-stimulated T-cells; and interferons ω and τ, proteins secreted by the embryonic placenta.

Production of interferons is brought about by the association between viral nucleic acid and host cell ribosomes, resulting in derepression of target-cell genes coding for interferon production. Interferon is secreted by infected cells and binds to receptors on other nearby cells. IFN-α and IFN-β share the same receptor while IFN-γ uses a different one. Binding to the receptor results in the development of resistance to virus infection within a few minutes and peaks five to eight hours later. Several different mechanisms are involved in this antiviral activity.

Interferons stimulate the production of many new proteins in target cells. Some of these proteins have antiviral activity. One of these is an RNAase that cleaves viral mRNA and so inhibits viral protein synthesis. Another protein induced by IFN-γ, especially in activated macrophages, is nitric oxide synthase. The nitric oxide produced by this enzyme has antiviral activity and so can prevent virus growth in interferon-activated macrophages. A third antiviral protein induced is a protein kinase that causes inhibition of viral protein synthesis by preventing the elongation of viral double-stranded RNA. A fourth new protein is called Mx. Mx protein can inhibit the transcription and translation of influenza virus mRNA.

Specific Immunity

The body employs several strategies to defend itself against viruses (Figure 28.4). Antibodies protect against free viruses in body fluids. Cell-mediated immune responses, being designed to destroy intracellular organisms are, however, of greatest importance in antiviral immunity. Individuals who are deficient in their cell-mediated immune responses will commonly die from uncontrolled viral infections.

Destruction by Antibodies

The capsids of viruses consist of antigenic glycoproteins, and it is against these and the viral envelope proteins that antiviral antibodies are directed. Antibodies may either destroy free viruses, prevent virus infection of cells, or destroy virus-infected cells. Since viruses first must bind to receptors on cells in order to infect them, antibodies may neutralize viruses by blocking their attachment to these receptors. The efficiency of this form of virus neutralization varies with the virus. The T-even phages (T2, T4, and so on), which have only one critical attachment site, can be neutralized by blocking that site with a single immunoglobulin molecule. On the other hand, other viruses, such as influenza, have multiple binding sites. Neutralization of influenza will, therefore, occur only when all of these are blocked and will thus require much higher immunoglobulin levels. Small virus particles may behave as if their entire surface is critical.

The combination of antibody with virus is not lethal in itself, since splitting of virus-antibody complexes can release fully infectious virus. Nevertheless, antiviral antibodies may act as opsonins, promoting the phagocytosis of virions by macrophages; they may clump virions, so reducing the number of infectious units available for cell invasion; or they may initiate complement-mediated virus neutralization. The complement cascade may be activated by antibody bound to virus surfaces, by activation through the alternate pathway, or by direct activation of C1 by viral glycoproteins. For viruses that have a membrane-like envelope, the complement pathway may cause virolysis. Other viruses, however, may be neutralized by C4a and C3b. C3b probably neutralizes the virus by blocking critical sites in the same way antibodies do.

Antibodies and complement are active, not only against free virions, but also against viral antigens expressed on the surface of infected cells. These infected cells may be destroyed by antibody-mediated opsonization or by complement-mediated lysis. Antibody binding to the surface of a virus-infected cell does not always destroy it. It may strip off viral antigens or, alternatively, suppress their production. As a result, infected cells may cease to carry viral antigen on their surfaces and become persistently infected.

Antibody-dependent cellular cytotoxicity (ADCC) is probably of major importance as a mechanism of destruction of virus-infected cells. In this process, cytotoxic cells with Fc receptors bind and lyse antibody-coated tar-

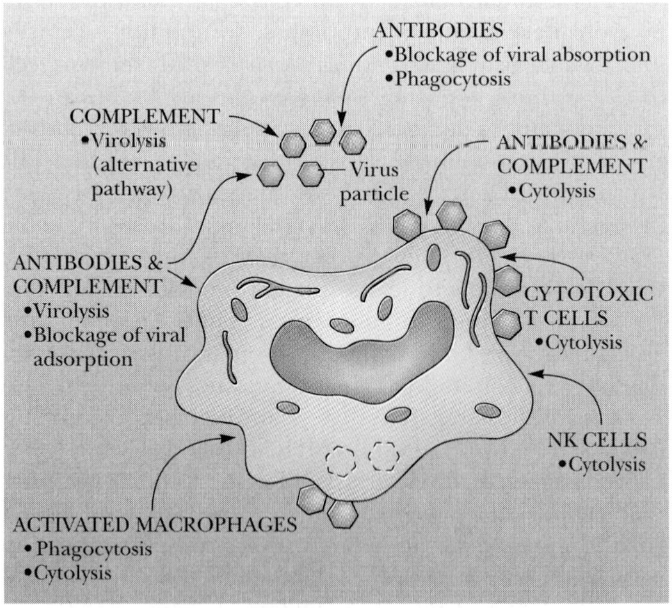

***Figure* 28.4** The ways in which the immune system can protect the body against viruses.

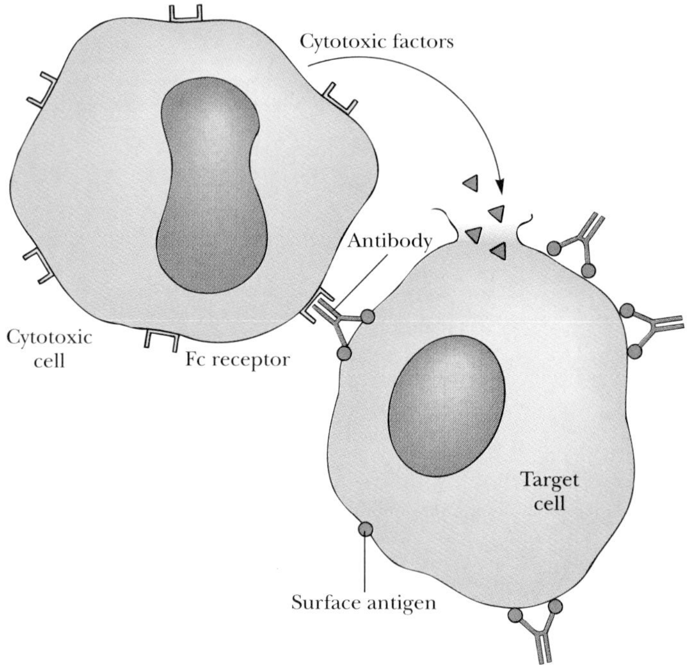

***Figure* 28.5** The mechanism of ADCC.

get cells (Figure 28.5). ADCC is mediated by concentrations of antibody several hundredfold less than those required to produce complement-dependent lysis of the same cells. The membrane lesions that develop on ADCC targets are similar to, but larger than, complement lesions and complement does potentiate ADCC. The cells that have been implicated in ADCC include macrophages, neutrophils, eosinophils, and NK cells.

The immunoglobulins involved in virus neutralization include IgG and IgM in serum and IgA in secretions. IgE may also play a protective role, since humans with a selective IgE deficiency show increased susceptibility to respiratory infections.

Cell-Mediated Immunity

Cell-mediated immunity directed against virus-infected cells is by far the most important mechanism of resistance to viruses. Virus-infected cells commonly develop new antigens on their surface. These new antigens include virus proteins exposed when viruses bud from the surface of the cell, other viral antigens expressed on the cell surface, or antigens induced by the virus on the host cells. These viral antigens are usually proteins or glycoproteins. Cell-mediated destruction of virus-infected cells is mediated through the actions of cytotoxic T-cells or NK cells. (See Chapter 27.) The pathogenesis of some viral diseases may therefore stem from direct damage to virus-infected cells by T-cell-mediated cytotoxicity rather than from any damaging effect of the virus.

Modification of Viral Disease by Immunity

In some cases, the immune responses directed against viruses may enhance, rather than prevent, the expression of disease. A good example is lymphocytic choriomeningitis (LCM). In this disease, transmission of the virus from female mice to their offspring during pregnancy causes an asymptomatic infection. The virus is not cytotoxic, and infected cells are not destroyed because of neonatal tolerance. Transmission between immunologically competent adult mice, on the other hand, leads to severe neurologic disease since cytotoxic lymphocytes attack virus-infected cells within the brain. Immunosuppressive treatment will therefore prevent the development of clinical LCM although such treated animals will remain persistently infected.

Evasion of Host Defenses by Viruses

The relationship between a host and a virus must be established on the basis of mutual accommodation so that its long-term continuation is ensured. Failure to establish such a relationship results in elimination of either host or virus, an undesirable result for one or both parties. One aspect of this adaptation involves the virus avoiding destruction by the immune system.

This type of evasive process has been observed in measles-infected cells in humans and in LCM-infected cells in mice. Antibody against measles virus will normally kill measles-infected cells. If, however, there is insufficient antibody present to be immediately lethal, then an infected cell may respond by removing the measles antigens from its surface. Once these antigens are lost, infected cells are not attacked. Removal of the measles antibody permits reexpression of measles antigen.

In some virus infections immunosuppression of persistently infected animals may permit disease to occur. The association between stress and the development of some virus diseases is well recognized (for example, "cold sores" caused by herpes simplex virus), and it is possible that the increased steroid production in stressful situations may be sufficiently immunosuppressive to permit activation of latent viruses or infection by exogenous ones.

Evasion of the immune response may also result from antigenic variation. This is seen among the influenza viruses (Figure 28.6). These viruses possess surface antigens, of which the hemagglutinins and neuraminidases are the most important. There are 13 different hemagglutinins and 9 neuraminidases among the type A influenza viruses, and they are identified according to a system recommended by the World Health Organization. Influenza viruses in a population change their antigenic structure as mutation and selection gradually change the amino acid sequences of their hemagglutinins and neuraminidases in a random fashion. As a result, there is a

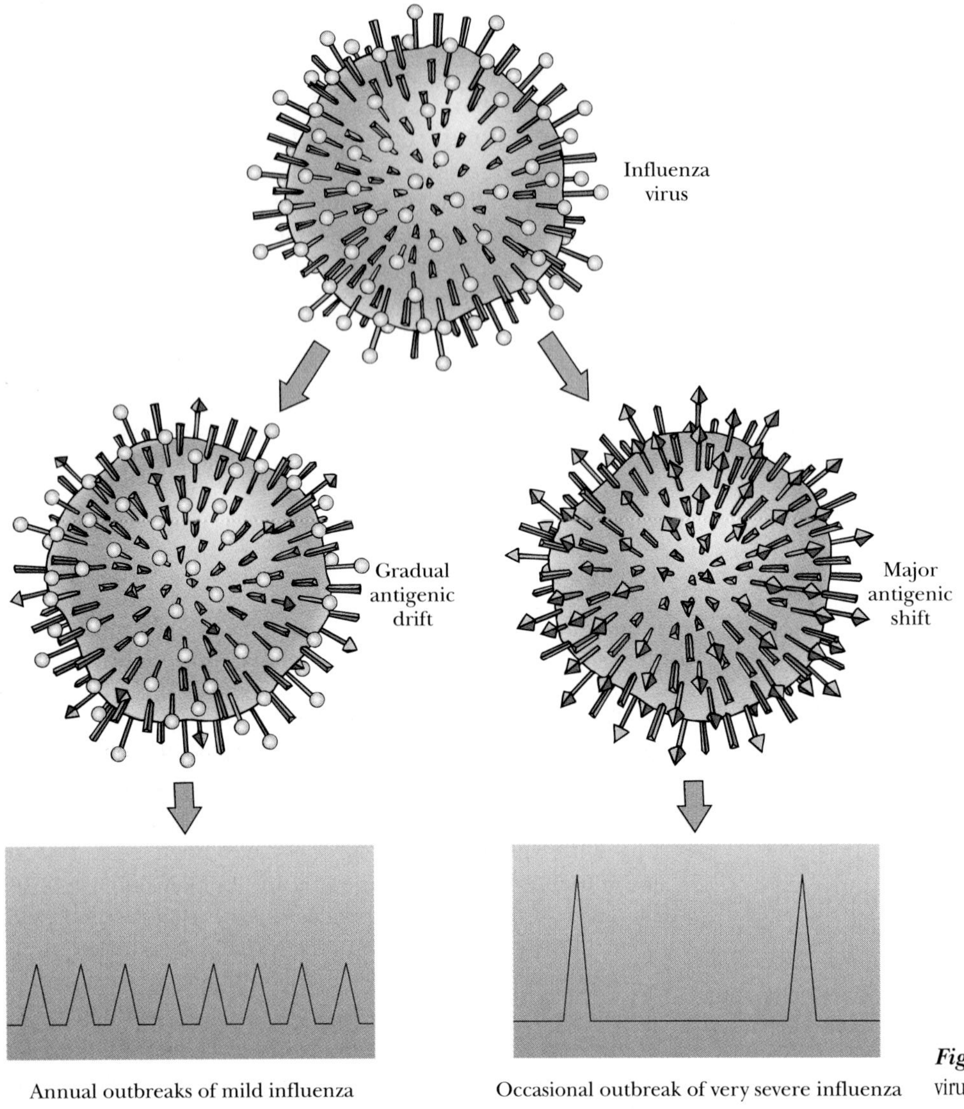

***Figure* 28.6** Antigenic variation in influenza viruses.

gradual change in the antigenicity of each subtype. This antigenic drift permits the virus to persist in a population for many years. Influenza viruses also sporadically undergo a major change in antigenic structure (antigenic shift), in which a new strain develops whose hemagglutinins are very different from those of previous strains. Such a major change is due to recombination between two different strains of virus. It is the development of these influenza viruses with a completely new antigenic structure that accounts for the periodic major epidemics of this disease.

Of special interest to immunologists are the virus diseases in which profound immunosuppression occurs. In many cases, the immunosuppression may be due to virus-induced destruction of lymphoid tissues. In some of these, only the primary lymphoid organs are involved. For example, mice may be infected by a herpesvirus that causes massive necrosis of the thymic cortex. In poultry, the virus of infectious bursal disease destroys the bursa of Fabricius. Some viruses damage secondary lymphoid organs. For example, measles infection is associated with depressed delayed hypersensitivity reactions and reduced lymphocyte responses to mitogens. This immunosuppression persists for several weeks after the infected individual has recovered clinically. Its causes may include not only lymphocyte destruction and arrest of T-cell division but also the development of suppressor cells. The most significant of these immunosuppressive viral diseases is acquired immune deficiency syndrome (AIDS) in humans, caused by human immunodeficiency virus (HIV-1). (See Chapters 14 and 30.)

A more complex interaction between lymphocytes and virus is seen in infectious mononucleosis. The cause, Epstein-Barr virus (EBV), infects only B-cells, most of which possess specific receptors for the virus. EBV activates B-cells, causing them to proliferate. These abnormal, activated B-cells, are, however, recognized as foreign by T-cells. As a result, cytotoxic T-cells attack and destroy the infected

B-cells. It is these cytotoxic T-cells that are the atypical mononuclear cells found in the blood of patients with infectious mononucleosis. Sometimes, the T-cells fail to prevent the abnormal B-cell proliferation induced by EBV. In this case, two alternative syndromes may develop: fatal infectious mononucleosis or an uncontrolled leukemia-like syndrome called Burkitt's lymphoma.

Certain genetic backgrounds predispose individuals to fatal mononucleosis, and once infection occurs the patients lose most of their blood lymphocytes (lymphopenia). Burkitt's lymphoma is a tumor of B-cells. It is generally restricted to certain areas of the tropics because malaria infection predisposes its development. EBV may also be implicated in the development of lymphomas in patients treated with the immunosuppressive drug cyclosporine.

The Elimination of Foreign Cells

The cell-mediated immune responses to endogenous antigens probably evolved in order to control the invasion of the body by viruses. However, the mechanisms involved are such that they will also ensure that any cell synthesizing proteins that can be recognized by T-cells will stimulate a cytotoxic T-cell response. Thus a T-cell-mediated cytotoxic response will occur when cells from a genetically different individual (an allograft) are grafted into the body. These cells, bearing foreign MHC molecules and proteins, effectively trigger their destruction. Similarly, some cancer cells, if they are sufficiently different from normal, will also trigger their own destruction.

Graft Rejection

Once an organ is surgically transplanted into a patient, its cells will be visited by the host's T-cells. They will respond to the foreign cells. When these cells enter the graft, they recognize its MHC class I molecules (or possibly peptides associated with these class I molecules) as foreign, and bind and destroy the vascular endothelium and other accessible cells by direct cytotoxicity (Figure 28.7). As a result, hemorrhage, platelet aggregation, thrombosis, and stoppage of blood flow occur. The grafted organ dies because of the failure of its blood supply. When T-cells are stimulated by antigen, they produce, among other factors, interferon-γ. IFN-γ causes the cells of the graft to express increased quantities of MHC molecules. During allograft rejection, MHC class I and especially MHC class II expression is increased in transplanted tissues. As a result, the graft becomes an even more attractive target for the host's cytotoxic T-cells. Although cytotoxic T-cells are most important in destroying foreign grafted tissues, antibodies also play a significant role in graft rejection. Antibodies directed against graft MHC class I molecules act with complement and neutrophils, or through antibody-dependent cytotoxic cell activity, to cause vascular endothelial cell destruction.

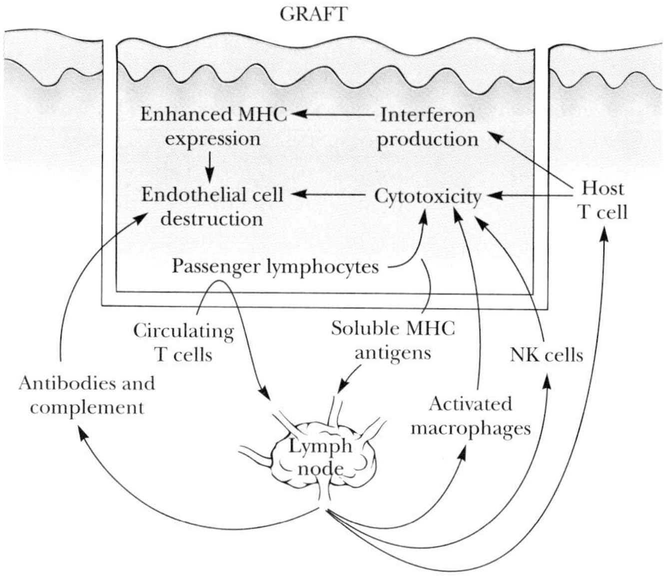

***Figure* 28.7** Some of the mechanisms involved in the rejection of an allograft.

Destruction of Cancer Cells

Since the function of the cell-mediated immune system is the elimination of abnormal cells, it follows that it may play a role in the elimination of cancer cells. This will be especially true if the cancer cell has proteins on its surface that are different from those found on normal cells (Figure 28.8). These new or altered proteins may be recognized by the cells of the immune system and so trigger immunological attack.

When organ transplantation became a common and widespread procedure as a result of the use of potent immunosuppressive drugs, it was observed that patients with

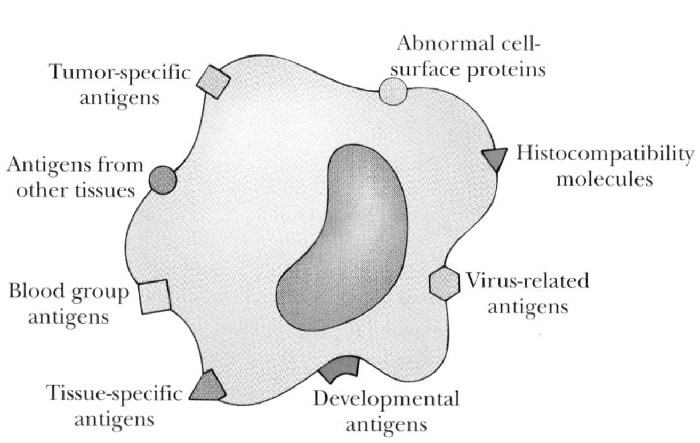

***Figure* 28.8** Some of the great variety of new antigens that may appear on the surface of tumor cells and provoke an immune response.

prolonged graft survival were many times more likely to develop cancer than were normal individuals. It has also been observed that patients with congenital or acquired immunodeficiency syndromes show an increased tendency to develop malignant tumors. (For example, AIDS patients may develop Kaposi's sarcoma.) It was therefore suggested that the immune system was responsible for the prevention of cancer. It was from this suggestion that the concept of a surveillance function for the cell-mediated immune system was developed.

Evidence has shown, however, that the view of the T-cell system as a system that identifies and destroys tumor cells alone is no longer tenable. One of the most important pieces of evidence to discredit this theory is the observation that T-cell-deficient mice are no more susceptible than normal mice to chemically induced or spontaneous tumors. The natural resistance of T-cell-deficient mice, and indeed of all normal animals, to tumors is probably dependent on NK cells **(Box 28.1).** Although the original surveillance hypothesis has therefore had to be greatly modified, there is good evidence that antitumor defense mechanisms do exist and may be enhanced in order to protect an individual against cancer. However, there is a great difference between the strong and effective cell-mediated immune response triggered by allografts and the quantitatively much poorer responses to the very weak antigens associated with tumor cells.

Tumor cells that are functionally different from their normal precursors may also be antigenically different, in that they gain or lose cell-membrane proteins. Even when tumor cells develop completely new antigens, the body may not be able to recognize or respond to them. Tumors induced by oncogenic viruses tend to gain new antigens characteristic of the inducing virus. These antigens, although coded for by the viral genome, are not part of the virus particle. An example of this type of antigen is the FOCMA antigen found on the neoplastic lymphoid cells of cats infected with feline leukemia virus.

If tumor cells are sufficiently different from normal, they will be regarded as foreign and stimulate an immune response. The major mechanisms of tumor cell destruction involve NK cells and cytotoxic T-cells, although activated macrophages and antibodies may also participate in this process (Figure 28.9).

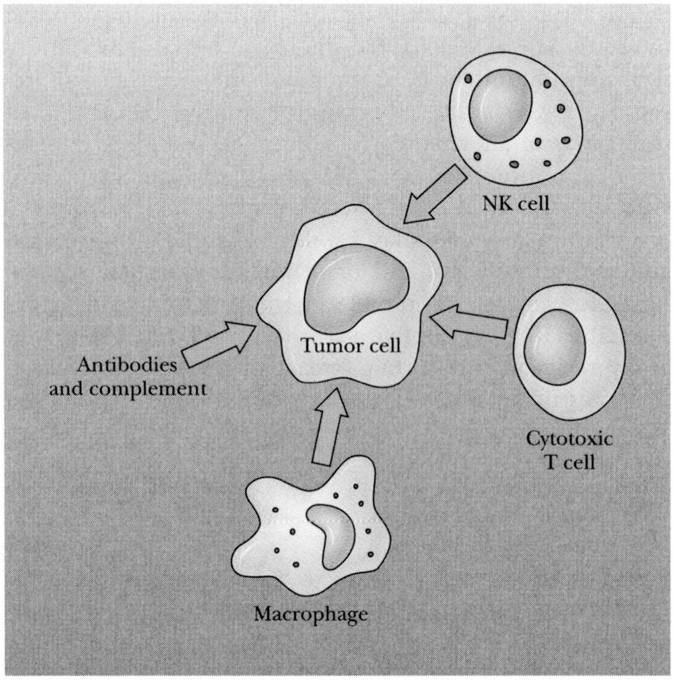

***Figure* 28.9** The mechanisms of tumor cell destruction.

Immunity on Body Surfaces

Although mammals possess an extensive array of defense mechanisms within tissues, it is at the body surfaces that invading microorganisms are first encountered and largely repelled or destroyed (Figure 28.10). The protective systems at body surfaces achieve this by establishing local environmental conditions suitable for only the most adapted microorganisms. These surfaces are populated by an extensive microbiota that, because it is well adapted, has low pathogenicity and effectively prevents the establishment of other, more poorly adapted and potentially pathogenic organisms. This environmental defense system is supplemented by immunological mechanisms in areas where the physical barriers to invasion are relatively weak. (These nonspecific defenses were described in Chapter 26.)

One major function of the skin is to act as a barrier to invading microorganisms. The skin carries a stable resident bacterial flora. If the skin environment is altered, then the skin microbiota is disturbed, its protective properties are reduced, and microbial invasion may result. For example, skin infections tend to occur in areas such as the axilla or groin, where both pH and humidity are relatively high. The importance of the resident flora is also seen in the digestive tract, where it is essential not only for the control of potential pathogens but also for the digestion of some foods, such as cellulose in the diet of herbivores. In addition, the natural development of the immune system depends upon the continuous antigenic stimulation provided by intestinal flora.

If the normal microbiota of the intestine is eliminated or its composition drastically altered (by aggressive antibiotic treatment, for example), then dietary disturbances result and the overgrowth of potential pathogens may occur. In the stomach of some animals, the gastric pH may be sufficiently low to have some bactericidal and viricidal effect, although this effect varies greatly between species and between meals, and the pH in the center of a mass of ingested food may not necessarily drop to low levels. Some foods, such as milk, are potent buffers.

Farther down the intestine, the resident bacterial flora ensures that the pH is kept low and the contents anaerobic. The intestinal flora is also influenced by the diet; for instance, the intestine of breast-fed infants tends to be colonized largely by lactobacilli, which produce large quantities of bacteriostatic lactic and butyric acids. These acids inhibit colonization by potential pathogens, such as *Escherichia coli,* so that infants suckled naturally tend to have fewer digestive upsets than infants weaned early in life. In the large intestine, the bacterial flora is mainly composed of strict anaerobes.

Lysozyme, the antibacterial and antiviral enzyme, is synthesized in the gastric mucosa and in macrophages within the intestinal mucosa. As a result, it is found in large quantities in intestinal fluid.

In the urinary system, the flushing action and low pH of urine generally provide adequate protection; however, when urinary stasis occurs, urethritis resulting from the unhindered ascent of pathogenic bacteria is not uncommon. In adult women, the vagina is lined by a squamous epithelium composed of cells rich in glycogen. When these cells desquamate, they provide a substrate for lactobacilli that, in turn, generate large quantities of lactic acid, which protects the vagina against invasion. Glycogen storage in the vaginal epithelial cells is stimulated by estrogens and

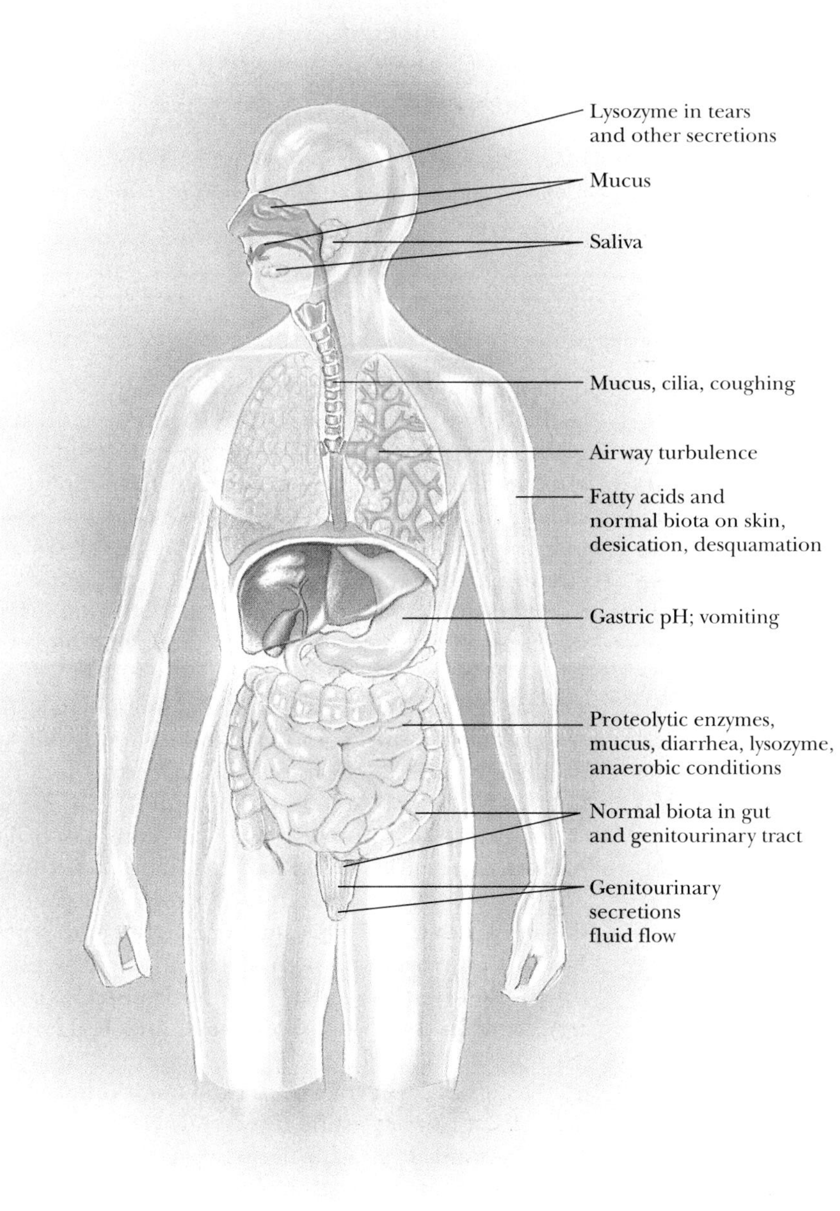

***Figure* 28.10** The nonimmunological surface protective mechanisms.

BOX 28.1 RESEARCH HIGHLIGHTS

Natural Killer Cells

About 15% of the lymphocytes found in blood are neither T- nor B-cells but form a distinct third population of cytotoxic lymphocytes called natural killer (NK) cells. These are characterized by their ability to kill tumor, virus-infected and some normal cells in the absence of previous sensitization. NK-cells in most mammals are large, granular, lymphocytes. NK-cells and T-cells probably arise from a common precursor in the bone marrow. T-cells, however, depend on the thymus for their development while NK-cells do not.

NK-cells do not express CD3, CD4, BCR, or TCR and are thus readily distinguished from T- and B-cells. Because they do not express either the TCR or the BCR, NK-cells cannot recognize antigen through any known antigen receptor. Nevertheless, they have several different ways of recognizing target T-cells. For example, they selectively kill target T-cells that fail to express MHC class I molecules. Thus loss of MHC class I expression even by otherwise normal cells will make them susceptible to NK-mediated lysis. Many viruses use reduction of MHC class I expression as a strategy to hide within infected cells and tumor cells often do not express MHC class I. Thus the use of NK-cells to monitor class I expression on cells makes sense. It appears that NK-cells are programmed to kill by signals found on most nucleated cells. However the MHC class I-antigen complex provides a negative signal sufficient to block this process. In the absence of MHC class I, killing is automatic. This NK-cell-mediated killing is mediated, like T-cells, through perforins and cytotoxins (TNF-α and TNF-β) as well as through CD95L.

NK-cells are active against a wide range of targets including foreign cells, cancer cells and virus-infected cells. They also have activity against bacteria such as *Staphylococcus aureus* and *Salmonella typhimurium* as well as some fungi.

thus occurs only in sexually mature individuals. Because of this, vaginal infections tend to be commonest prior to puberty and after the menopause.

The respiratory tract differs from the other body surfaces in that it is in intimate connection with the interior of the body and it is required to allow unhindered access of air to the alveoli. Particles in air entering the respiratory tract are removed by turbulence that directs them onto its mucus-covered walls, where they adhere. The turbulence is brought about by the shape of the turbinate bones, the trachea, and the bronchi. This "turbulence filter" removes particles as small as 5 μm before they reach the alveoli.

The walls of the upper respiratory tract are covered by a layer of mucus and provided with "antiseptic" properties through its content of lysozyme and IgA. This mucus layer is in continuous flow, being carried from the bronchioles up the bronchi and trachea to the pharynx. Here the "dirty" mucus is swallowed and digested in the intestinal tract. Particles smaller than 5 μm that can reach the alveoli are phagocytosed by alveolar macrophages.

Immunoglobulin A

In addition to the environmental and chemical factors that protect body surfaces (Chapter 27), one major immunoglobulin is found in secretions such as saliva, intestinal fluid, nasal and tracheal secretions, tears, milk, colostrum, urine, and the secretions of the urogenital tract. Immunoglobulin A appears to have evolved specifically for the purpose of protecting body surfaces. In the intestine, it is produced in amounts that exceed all other classes combined.

The IgA monomer is a typical four-chain, Y-shaped structure. It is usually found as a dimer with one molecule bound by a J chain to the other molecule. IgA is made by plasma cells located in the region immediately beneath the epithelial cells in response to local antigenic stimulation (Figure 28.11). Some of this IgA binds to a receptor for polymeric immunoglobulins (pIgR) on the submucosal surface of epithelial cells. The complex of IgA and pIgR is then endocytosed and transported across the epithelial cell. When it reaches the exterior surface, the endosome fuses with the plasma membrane and exposes the IgA to the intestinal lumen. The extracellular portion of the pIgR is then cleaved by proteolytic enzymes so that the IgA, with the receptor peptide still attached, is released. This peptide is called secretory component (SC).

Because IgA is transported through intestinal epithelial cells, it can act inside these cells (Figure 28.12). Thus IgA can bind to viral proteins inside epithelial cells and interrupt their replication. In this way, the IgA can pre-

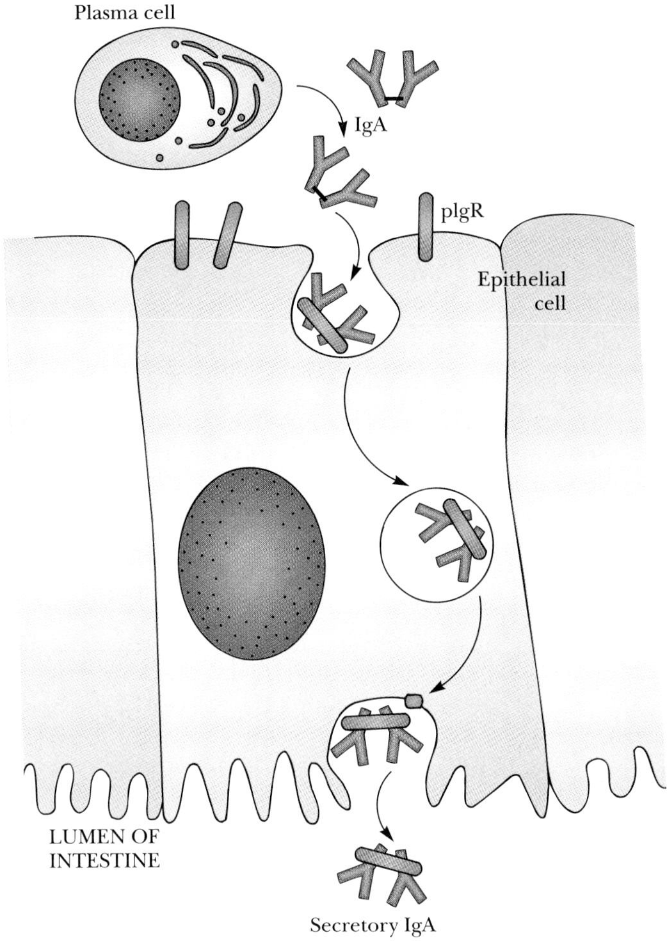

***Figure* 28.11** IgA is secreted by mucosal plasma cells and binds to receptors (pIgR) on the interior surface of intestinal enterocytes. The bound IgA is taken into the enterocytes and passed in vesicles to the cell surface. Once in the intestinal lumen, the IgA receptor is cleaved from the cell and remains bound to the IgA. In this state, it is called secretory component and serves to protect the IgA from degradation.

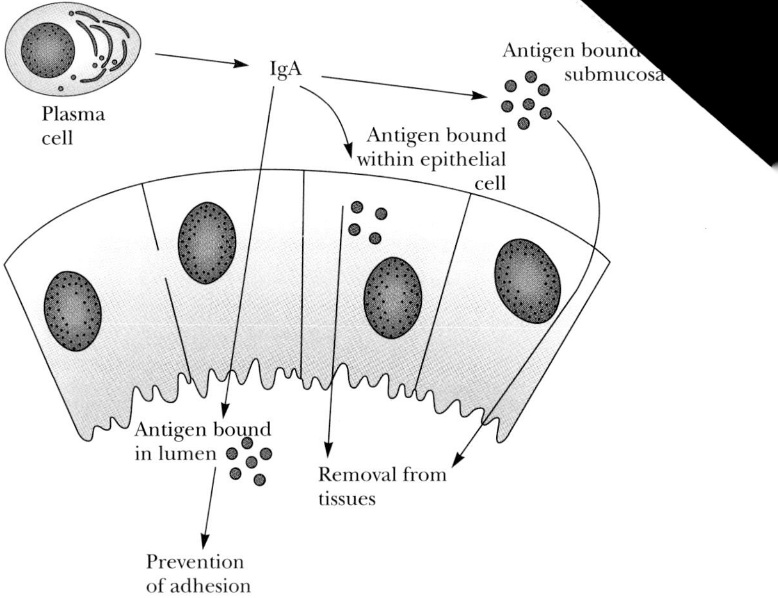

***Figure* 28.12** IgA is unique in that it can act in three locations. It can bind antigen in tissue fluid or in enterocytes, as well as in the intestinal lumen. The bound antigen in tissues or enterocytes is carried to the intestinal lumen.

vent viral growth before the integrity of the epithelium is damaged. This is a unique example of an antibody acting in an intracellular location. The second unique function of intracellular IgA is to excrete foreign antigens. Thus IgA can bind to antigens that have penetrated to the submucosa. Once bound, the immune complexes will bind to pIgR and be actively transported across the mucosal epithelial cells into the intestinal lumen. IgA can therefore act at three different levels to exclude foreign antigens; within the submucosa, within epithelial cells, and within the mucosal secretions.

The B-cells in the intestinal wall, upon encountering antigen, divide and some differentiate into plasma cells. Some of these responding B-cells also migrate to the bloodstream. These cells leave the bloodstream in the intestinal tract, the respiratory tract, the urogenital tract, and the mammary gland (Figure 28.13). Thus stimulation by antigen at one surface site permits antibodies to be synthesized and secondary responses to occur at distant surfaces.

The movement of IgA secreting B-cells from the intestine to the mammary gland is of major importance, since it provides a route by which intestinal immunity can be transferred to the newborn through milk. Oral administration of antigen to a pregnant animal will thus result in the appearance of IgA antibodies in its milk. In this way, the intestine of the newborn animal will be flooded by antibodies directed against intestinal pathogens.

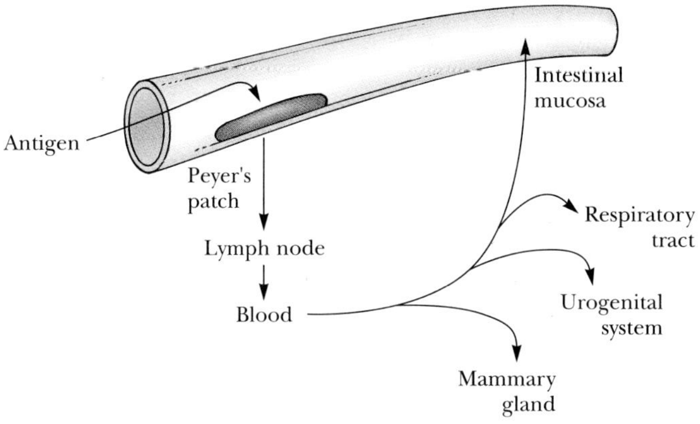

***Figure* 28.13** When a local immune response is triggered in the intestinal wall, IgA-producing B-cells leave the site of antigen invasion and migrate to other surfaces. As a result, IgA in mother's milk may protect the intestine of an infant against diarrhea.

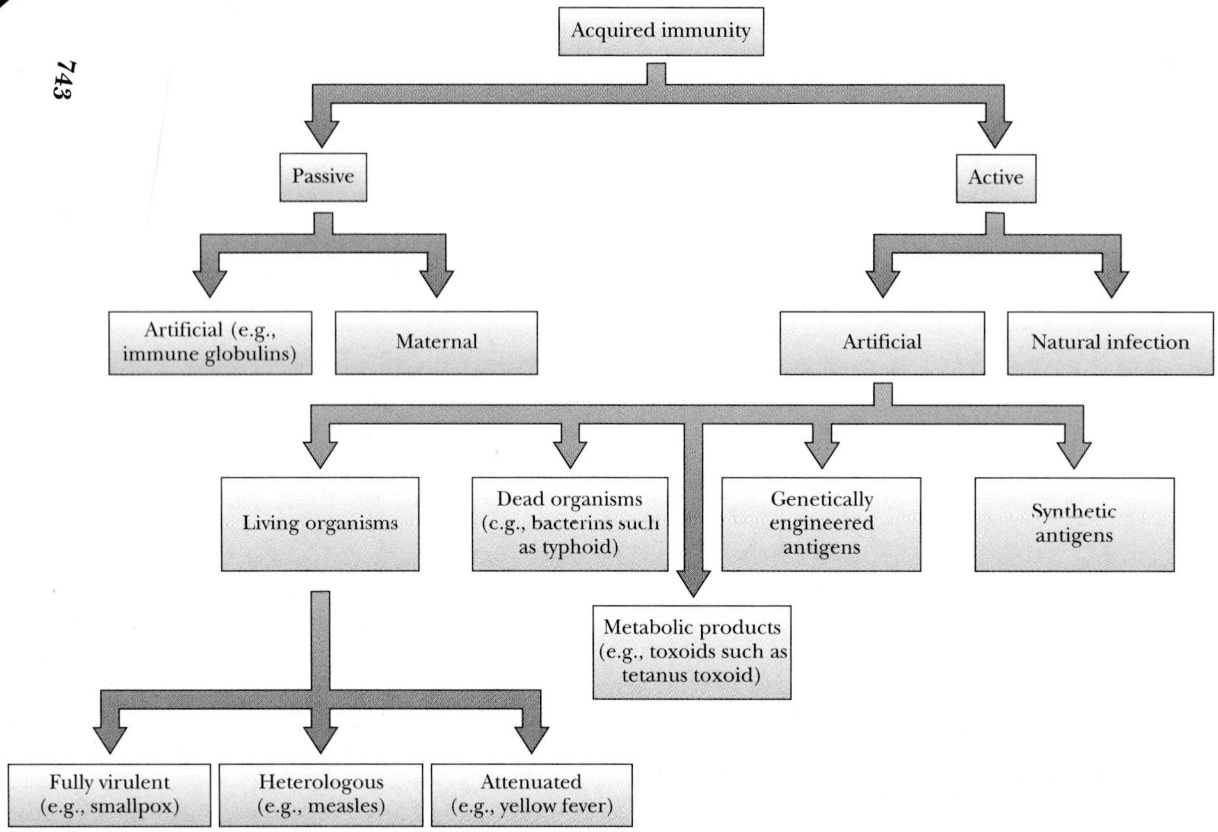

***Figure* 28.14** A classification of the different types of acquired immunity and of the methods employed to induce protection.

Vaccines and Vaccination

The use of vaccines to control infectious diseases such as smallpox, rabies, tetanus, anthrax, cholera, and diphtheria has been one of the great successes of modern medicine. In the United States, the administration of effective vaccines has reduced the number of reported cases of diphtheria, measles, mumps, pertussis, poliomyelitis, rubella, and tetanus by at least 97 percent. No other form of disease control has had such an effect on the reduction of mortality. Indeed, vaccination has been responsible, in large part, for the recent phenomenal increase in the world's population.

Types of Immunization Procedures

Passive immunization and active immunization are the two methods by which an individual can be made resistant to an infectious agent (Figure 28.14). Passive immunization produces temporary resistance by transferring antibodies from a resistant to a susceptible individual. These antibodies give immediate protection, but since they are gradually removed, the protection wanes and the recipient eventually becomes susceptible to infection once again.

Active immunization involves administering antigen to individuals so that they respond by mounting a protective immune response. It has several advantages over passive immunization. Reimmunization or exposure to the infectious agent results in a secondary immune response. The major disadvantage of active immunization is that the protection is not immediate (Figure 28.15). Its advantage is that it is long-lasting and capable of restimulation.

Passive Immunization

Passive immunization requires that antibodies produced in a donor animal by active immunization be given to susceptible animals in order to confer immediate protection. These antibodies may be raised in animals of any species and against a wide variety of pathogenic organisms. One of the most important passive immunization procedures has been the production of antibodies against tetanus in horses. The antibodies, called immune globulins, are produced in young horses by a series of immunizing injections. The toxins of the clostridia are proteins, and they can be made nontoxic by treatment with formaldehyde. Toxins treated in this way are called toxoids. Initially, the horses are inoculated with toxoids, but once antibodies are produced, subsequent injections contain purified

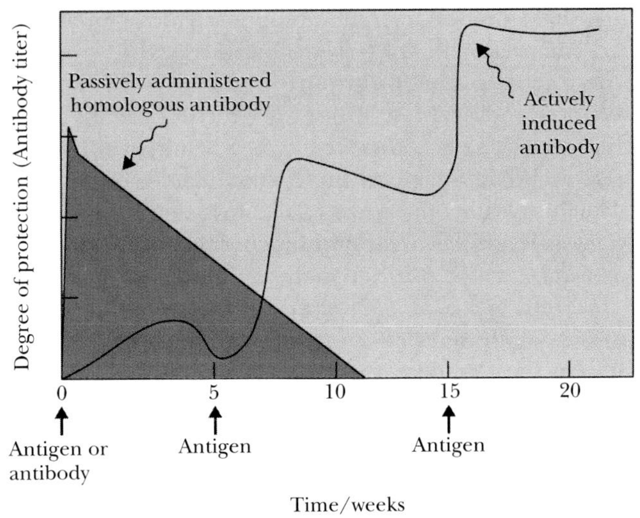

***Figure* 28.15** The levels of serum antibody (and hence the degree of protection) conferred by active and passive methods of immunization.

toxin. The responses of the horses are monitored, and once their antibody levels are sufficiently high, the horses are bled. Plasma is separated from the horse blood and treated in order to concentrate and purify the globulin fraction. This purified antibody preparation is then dialyzed, filtered, titrated, and dispensed.

Because of the problems encountered when horse antiserum is used in humans, it is preferable to use antiserum of human origin. This human immune serum globulin (ISG) is obtained from volunteers who have recently been immunized against tetanus. Other common antisera used in humans include those against measles, hepatitis A and B, rubella, varicella, rabies, diphtheria, group B streptococci, pseudomonas, cytomegalovirus, tetanus, and botulism. Thus measles ISG is given to immunodeficient patients who may not be able to mount an active immune response. Rabies ISG is used to treat individuals bitten by rabid animals.

Active Immunization

The most important advantages of active immunization as compared with passive immunization are the prolonged duration of protection and the recall and boosting of this protective response by repeated injections of antigen. The ideal vaccine to be used for active immunization should be cheap, stable, and adaptable to mass vaccination. It should give prolonged strong immunity without adverse side effects, and ideally should stimulate an immune response distinguishable from that due to natural infection so that vaccination does not interfere with diagnosis. It is important that vaccines be free of adverse side effects. This clearly precludes the use of virulent living organisms. Thus, if living organisms must be used in a vaccine, they must be treated in such a way that they lose their disease-producing ability. The term used for this loss of virulence is **attenuation.**

Commonly used methods of attenuation involve adapting organisms to unusual environmental conditions so that they lose the ability to replicate uncontrollably in their usual host. For example, the bacillus Calmette-Guérin (BCG) strain of *Mycobacterium bovis* was rendered avirulent by growing it for 13 years on a bile-saturated medium. At the end of that time, the organisms had lost the ability to grow and cause disease in humans. BCG is used as a vaccine against tuberculosis.

Viruses can also be attenuated by growth in abnormal culture conditions. Prolonged tissue culture, especially in cells of a species that the virus does not normally infect, has the effect of reducing the virulence of viruses. Thus attenuated living poliovirus vaccine contains viruses grown in monkey-tissue culture. Rubella virus vaccine may contain virus prepared in duck-embryo cells. As an alternative to tissue culture, some viruses may be attenuated by growth in eggs. These include the vaccine strain of yellow fever and strains of influenza virus.

Attenuation of organisms for use in vaccines has both advantages and disadvantages (Figure 28.16). In general, vaccines containing living organisms are effective and give prolonged, strong immunity. Because, in effect, they cause transient infection, only a few inoculating doses are required and adjuvant need not be employed. As a result, there is less chance of provoking adverse hypersensitivity reactions. In addition, live virus vaccines may provoke a rapid protective response through stimulation of interferon production.

On the other hand, live vaccines may be difficult and expensive to produce. There exists the possibility that they may contain dangerous extraneous organisms and, more important, may cause disease as a result either of residual virulence (i.e., they may not be fully attenuated) or of reversion to a more virulent form. In general, live vaccines cannot be used in patients suffering from an immunodeficiency such as AIDS.

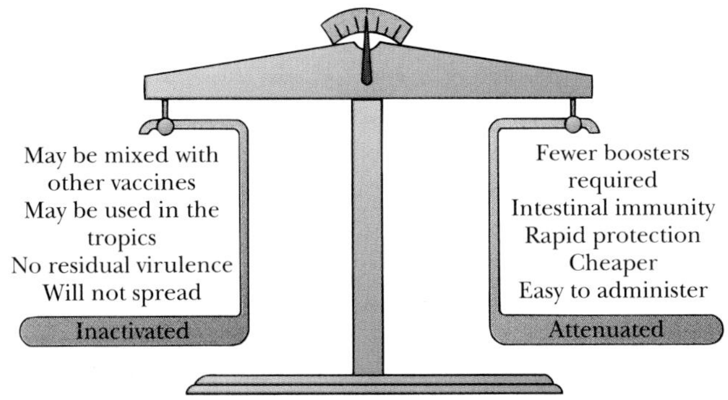

***Figure* 28.16** Advantages and disadvantages of live and dead vaccines.

To avoid these disadvantages, it is common to use organisms that have been killed or inactivated. In such cases, the "dead" organisms should be as antigenically similar to the living organisms as possible. Therefore, a crude method of killing microorganisms, such as heating, which causes extensive protein denaturation, tends to be unsatisfactory. If chemical inactivation is to be used, the chemicals must produce little change in the antigens that are responsible for protective immunity.

The relative advantages of vaccines containing modified live or "dead" organisms can be seen in poliomyelitis vaccines. Live poliovirus vaccines contain attenuated poliovirus originally grown in epithelial cells from monkey kidney. The vaccine is administered orally to children as a liquid dropped onto a sugar cube. As a result, the vaccine strains colonize the intestine and provoke an immune response similar to that induced by natural infection by poliovirus. This response includes the production of secretory antibodies in the intestine as well as serum antibodies. Consequently, the vaccine produces a long-lasting humoral and intestinal immunity, thus reducing the need for repeated boosters. The administration of the vaccine is simple and does not require trained personnel, and the vaccine itself is relatively cheap and can be grown on cultures of human cells, thus reducing the need to use increasingly rare and precious monkeys. On the other hand, since oral polio vaccination results in infection of recipients, the vaccine virus may cause occasional cases of paralytic polio. It has been estimated that the rate of paralytic disease in polio vaccine recipients is about one case in four million doses of vaccine distributed, although there is about one case of paralytic poliomyelitis in two million doses of vaccine distributed among individuals in contact with vaccinees. Although this may be considered a very low risk, consider that paralytic polio, if not lethal, is a severely crippling disease. Thus the few individuals who develop polio as a result of exposure to vaccine may have their lives ruined. Society surely owes it to these individuals to compensate them for this; it is a small price to pay for keeping the rest of us free from this disease.

In the search for effective vaccines, unwanted contaminants can increase the risk of side effects without adding anything to the efficacy of the vaccine. For example, vaccines prepared from viruses or bacteria grown in culture commonly contain culture-fluid components. Likewise, many organisms carry antigens that are not protective but which may be toxic. Thus it is usual to purify vaccines as much as possible before use. The end point of this process is the isolation of pure, protective antigens. The most obvious examples of such "pure" antigens are the vaccines against diphtheria and tetanus, which utilize not the whole bacterium but a purified preparation of exotoxin detoxified by treatment with formaldehyde. These bacterial toxoids have proved to be excellent and effective immunizing agents. Another example of this type of vaccine is that against bacterial meningitis, which contains purified bacterial cell-wall polysaccharides against two types of *Neisseria meningitidis*—groups A and C. Whole cell *Bordetella pertussis* vaccines, although effective, are associated with several adverse side effects. As a result, acellular pertussis vaccines have been produced. These contain purified bacterial subcomponents and cause significantly fewer adverse side effects. Antipneumococcal vaccine now contains a mixture of purified polysaccharides from each of 23 different types of *Streptococcus pneumoniae*. These are known to cause 85 to 90 percent of pneumococcal pneumonia in the United States.

Other Approaches to Vaccine Production

Although conventional vaccines have been successful in controlling infectious diseases, there is always a need for improvement. Several new approaches are being studied in attempts to make vaccines more effective, cheaper, and safer (Figure 28.17). For example, it has become possible to modify the genes of organisms deliberately so that they become irreversibly attenuated. A vaccine is now available against a herpesvirus that causes the disease in pigs called pseudorabies. The thymidine kinase (TK) gene has been removed from this virus. Thymidine kinase is necessary for herpesviruses to replicate in nondividing cells such as neurons. Viruses from which the TK gene has been removed are able to infect nerve cells but cannot replicate and cannot therefore cause disease. As a result, this vaccine not only confers effective protection, but by blocking cell invasion by virulent pseudorabies viruses also prevents the development of a persistent carrier state. Alternatively, other genes may be deleted such as in *Vibrio cholerae* where part of the toxin gene may be removed. This deliberate modification of virulence is a much more effective and

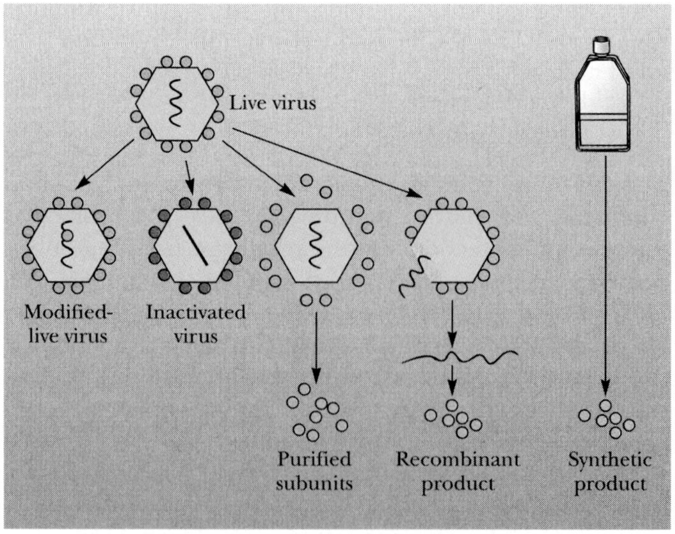

***Figure* 28.17** Other approaches to vaccine production.

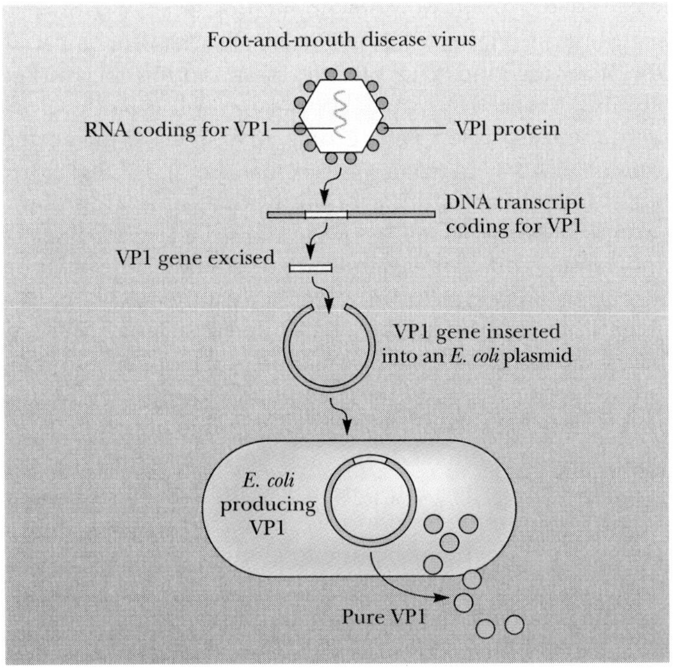

Figure **28.18** The production of a recombinant viral protein, in this case foot-and-mouth disease virus VP1, for use in a vaccine.

logical method of reducing virulence than the previous hit-or-miss methods of attenuation.

Recombinant DNA techniques can be used to isolate DNA or RNA coding for a protein antigen of interest. This can then be placed in a bacterium, yeast, or other cell and permitted to code for that protein. The first attempt to use gene cloning to prepare a vaccine was with foot-and-mouth disease virus. This virus is extremely simple, and the protective antigen is well characterized. The RNA of the foot-and-mouth disease virus was isolated and duplicated into DNA by means of the enzyme reverse transcriptase. The DNA was then carefully cut by restriction endonucleases so that it only contained the gene for the protective antigen. This DNA was then inserted into an *E. coli* plasmid, the plasmid inserted into *E. coli,* and the *E. coli* grown (Figure 28.18). These recombinant bacteria synthesize large quantities of antigen that may be harvested, purified, and incorporated into a vaccine. The process can be highly efficient since 4×10^7 doses of foot and mouth vaccine can be obtained from 10 liters of *E. coli* grown to 10^{12} organisms/ml.

A very successful recombinant vaccine against hepatitis B has been in use in the United States since 1984. It is especially effective when administered together with immune globulin to newborn children from high-risk mothers.

Recombinant DNA techniques are useful in any situation where protein antigens need to be synthesized in large and pure quantities. Unfortunately, pure proteins such as these are often poor antigens because they are not effectively delivered to antigen-sensitive cells and cannot therefore provoke a sufficiently intense response. An alternative method is to clone the genes of interest into a living carrier organism.

Genes coding for specific protein antigens may be cloned directly into an organism such as vaccinia virus. Vaccinia virus is easy to administer by dermal scratching. It has a large genome that makes it relatively easy to insert a new gene, and it can express high levels of the new antigen. Moreover, the proteins undergo appropriate processing. Individuals vaccinated with genetically modified vaccinia virus make high levels of antibodies against the introduced antigen.

Use of Vaccines

Most vaccines are designed to be administered by intramuscular or subcutaneous injection. While, ideally, they should provide protective immunity after a single dose, it is usually necessary to administer several "booster" injections in order to induce a high level of immunity. Injections, however, are relatively time-consuming and painful and bear a risk of carrying unwanted organisms into an individual. Thus, when large numbers of people or animals must be vaccinated under less than ideal conditions, other methods of vaccination are employed. For example, poliomyelitis or typhoid vaccines may be given orally. Polio vaccine is given as a solution containing live virus, while the typhoid vaccine is administered in the form of an enteric-coated capsule containing live-attenuated organisms.

In vaccination, it is usually desirable to enhance the normal immune response by administering an **adjuvant** with the antigen. Many different compounds have been employed as adjuvants. Some adjuvants act by slowing the release of antigen into the body, but in many cases their mode of action is unknown. The immune system, being antigen driven, responds to the presence of antigen and terminates that response once antigen is eliminated. It is possible to slow the rate of antigen release into the tissues by first mixing it with an insoluble adjuvant. Injected into a patient, this forms a focus, or "depot." Examples of depot-forming adjuvants include insoluble aluminum salts, such as aluminum hydroxide, aluminum phosphate, and aluminum potassium sulfate (alum). When antigen is mixed with one of these salts and injected into an animal, a macrophage-rich granuloma forms in the tissues. The antigen within this granuloma slowly leaks into the body and so provides a prolonged antigenic stimulus. Antigens that normally persist for only a few days may be retained in the body for several weeks by means of this technique.

An alternative method of forming a depot is to incorporate the antigen in a water-in-oil emulsion called Freund's incomplete adjuvant. The oil stimulates a local, chronic, inflammatory response, and the antigen is slowly leached from the emulsion. If killed tubercle bacilli

(Mycobacterium tuberculosis) are incorporated into the water-in-oil emulsion, the mixture, called Freund's complete adjuvant, is extremely potent. This adjuvant cannot be used in humans.

Vaccination Schedules

A simple immunization schedule can be followed as children grow up (Figure 28.19). It may be modified as appropriate if vaccination is begun at a later age than normal. In addition, certain legal requirements, such as the requirement that children must be vaccinated against measles before admission to school, will also influence the timing of vaccination schedules.

Since newborn animals are passively protected by maternal antibodies, it is sometimes difficult to successfully vaccinate them in early life. Thus, if measles vaccine is given to children before 12 months of age, a second dose must be given at about 15 months in order to ensure adequate protection.

The interval between booster doses of vaccine varies, but, as a general rule, inactivated vaccines generally produce a weak immunity that requires frequent boosters, perhaps as often as every six months (as in the case of older cholera vaccines). On the other hand, live vaccines produce a much more persistent immunity. For example, BCG vaccine does not require boosting, and yellow fever vaccine requires a booster dose only every ten years.

Vaccination Failures

The immune response, like other biological phenomena, never confers absolute protection and is never equal in all members of a vaccinated population. The range of responses in a large random population follows a normal distribution (Figure 28.20). This means that while most individuals respond to a vaccine by mounting an average immune response, a small proportion will mount a poor immune response. This group may not be protected against disease in spite of vaccination. It is, therefore, highly improbable that every individual in a large population will be protected by vaccination. The significance of this varies. Thus, for any individual, the lack of protection is serious. However, from a public health viewpoint, less than 100% protection may be quite satisfactory if it prevents the spread of disease within the population. This phenomenon is called **herd immunity;** its effect is due to the reduced probability of a susceptible individual encountering an infected one and so contracting disease.

A second type of apparent vaccine failure can occur when the immune response is depressed. As mentioned earlier, this effect may be due to clinical immunosuppression as in AIDS, or it may be due to a high stress situation. Extremes of cold and heat or malnourishment may all inhibit the immune response. Other reasons for vaccine failure include the possibility that the individual was incubating the disease prior to vaccination, or to the excessive use of alcohol while swabbing the skin, a procedure that may inactivate a virus vaccine.

Hazards of Vaccination

Residual virulence, toxicity, and allergic reactions are the three most important problems associated with the use of vaccines. The most important contraindications to vaccination involve the administration of live vaccines to immunodeficient patients. Thus, it is critical that no live vaccine be given to patients with immunodeficiency diseases

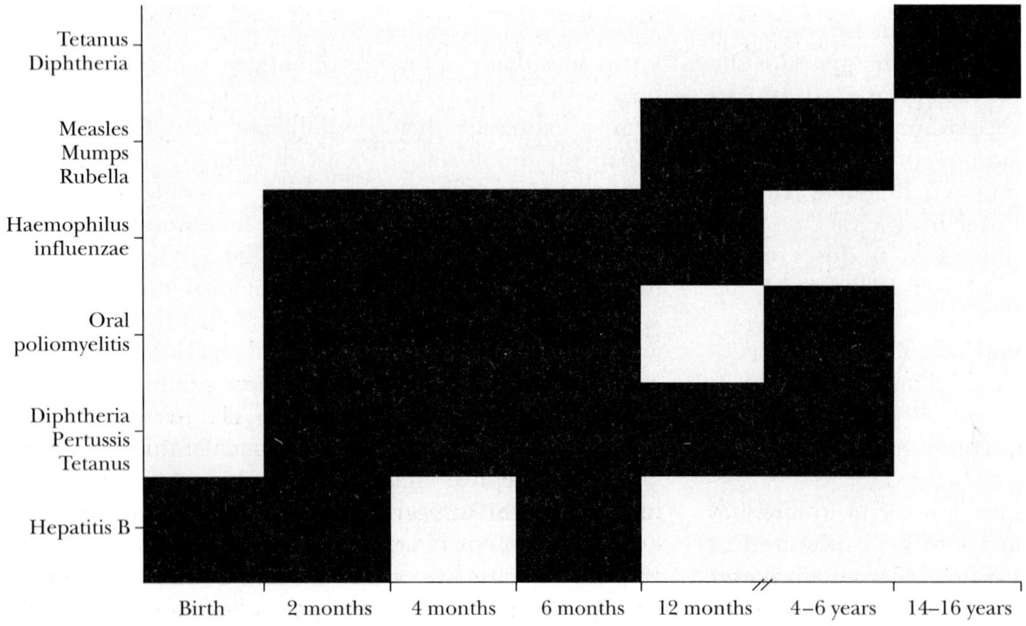

***Figure* 28.19** A recommended vaccination schedule for children.

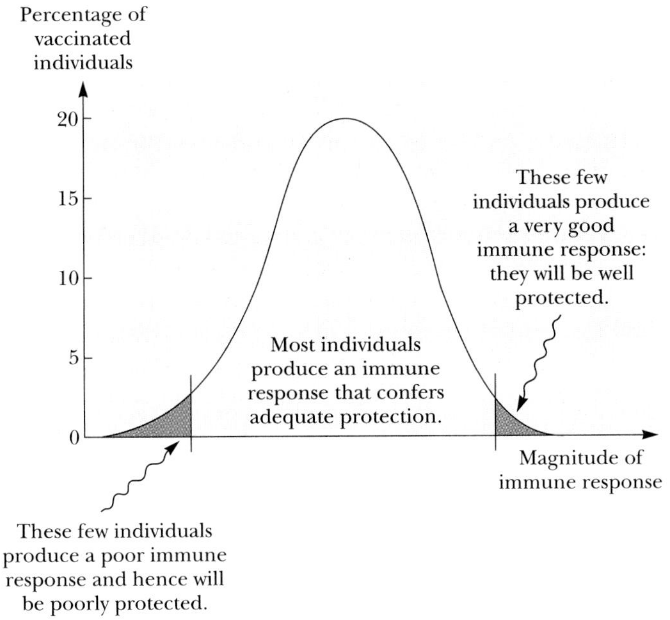
Percentage of vaccinated individuals
20
15
10
5
0
These few individuals produce a very good immune response: they will be well protected.
Most individuals produce an immune response that confers adequate protection.
Magnitude of immune response
These few individuals produce a poor immune response and hence will be poorly protected.

***Figure* 28.20** The normal distribution of protective immune responses in a population of vaccinated individuals.

(such as AIDS), patients with cancers that adversely affect immunity (such as leukemias and lymphomas), and patients undergoing immunosuppressive therapy.

While inactivated vaccines and toxoids may be given to pregnant women, live viral vaccines, especially measles and rubella, should not be given because of the potential risk to the developing fetus. On the other hand, under conditions where the risk of disease is high, as in yellow fever or poliomyelitis, the theoretical risk may be less important and the vaccine given to pregnant women.

Vaccines containing whole killed gram-negative bacteria, such as the cholera, whooping cough, and typhoid vaccines, may be toxic owing to the presence of endotoxins. These endotoxins may cause soreness, fever, malaise, and muscle pain.

Like any antigen, vaccines can provoke hypersensitivity reactions. (See final section of this chapter.) An immediate hypersensitivity can occur not only in response to the organisms but also to contaminating antigens, such as egg proteins in measles, mumps, and influenza vaccines. Patients with a history of allergies to eggs should not be given these vaccines. Some vaccines may contain trace amounts of antibiotics, for example, neomycin, and should not be given to patients with a history of allergy to this drug.

Benefits of Vaccination

Many studies have been made of the actual benefits conferred by vaccination of children. Thus, for example, in one year (1983) it has been estimated that, without measles vaccination in the United States, 3,325,000 cases of measles would have occurred instead of 2,872 actual cases. Since measles vaccine was introduced into the United States, it has been estimated that it prevented 52 million cases of measles, 5,200 deaths, and 17,400 cases of mental retardation and produced a net savings of $5.1 billion to society. The overall savings in costs to society as a result of measles, mumps, and rubella vaccination in 1983 were calculated to be $1.3 billion, with a benefit-cost ratio of 14:1. There is no doubt that routine childhood vaccination confers huge benefits on society as a whole and has been largely responsible for the control of viral diseases in our society. Few other scientific disciplines can claim to have had such an impact. Of major significance on a global scale is the very successful Expanded Program on Immunization, organized by the World Health Organization with help from UNICEF, the World Bank, Rotary International, and developing countries themselves. This program aims to provide immunization against seven major diseases to at least 80 percent of the children in underdeveloped countries. These diseases are tetanus, poliomyelitis, tuberculosis, diphtheria, measles, and whooping cough. Hepatitis B has recently been added to this list.

Defects in the Immune System

Any defect in the immune system that reduces the ability of an individual to mount an immune response or interferes with the ability of their phagocytic cells to ingest and destroy bacteria or process antigen will result in infection. These defects are of two general types. Some defects are inherited as a result of a genetic mutation. Thus they have no obvious predisposing cause and are called **primary immunodeficiencies** (Figure 28.21). Because they are inherited, their effects are seen in infants and young children. The children suffer from recurrent, severe infections that will kill them before they get very old. The second major type of defect in the immune system is one caused by a known agent such as a drug or virus that destroys lymphocytes. This type of immunodeficiency is called a **secondary immunodeficiency.** Secondary immunodeficiencies commonly occur in adults or in the aged. Like immunodeficiencies in children, they are recognized clinically by either recurrent severe, intractable infections, or by an increased susceptibility to cancer.

Defects in Phagocytosis and Antigen Processing

Neutrophils ingest and destroy invading foreign material, especially bacteria. If neutrophils are absent or ineffective, recurrent, severe, bacterial infections will result. The numbers of neutrophils in the blood may be reduced by drugs, radiation, overwhelming infections, or some autoimmune diseases. Cytotoxic drugs and radiation cause a loss of neutrophils because they are selectively toxic for rapidly dividing cells. Since the neutrophil precursors in the bone

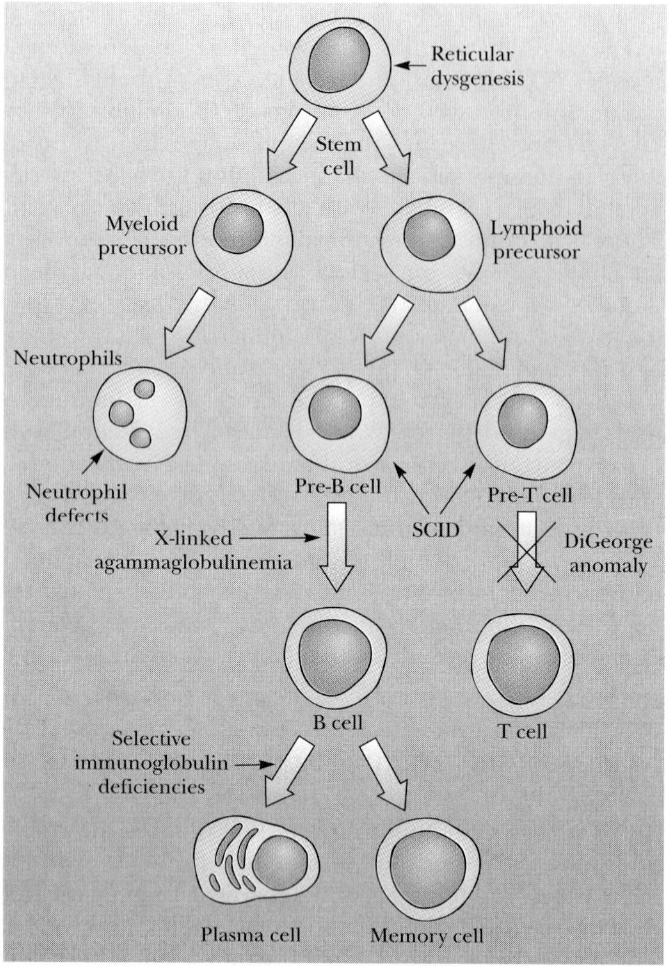

***Figure* 28.21** The points in the immune system where development blocks may lead to immune deficiencies. A failure in stem cell development results in a combined immunodeficiency. A T-cell defect leads to a loss of cell-mediated immunity while a B-cell defect causes a failure of antibody formation.

marrow are rapidly dividing, they are very susceptible to destruction.

Both neutrophil deficiencies and defective neutrophil function have been identified. The final clinical effect is similar: recurrent, uncontrolled bacterial infections. Neutrophils can fail to function in many different ways. Inherited defects have been described involving each of the major stages in phagocytosis: chemotaxis, opsonization, attachment, ingestion, or digestion. The most important group of neutrophil defects is collectively called **chronic granulomatous disease.** The most common form of this is due to a defect in a gene located on the X chromosome; as a result, it only occurs in boys. (In girls, the presence of a second X chromosome ensures that a normal gene is present, and the disease does not occur.) Chronic granulomatous disease usually develops in early childhood when the children begin to suffer from recurrent infections (pneumonia, dermatitis, lymphadenitis, osteomyelitis, sepsis) and multiple abscess formation caused by organisms such as *Staphylococcus aureus, Serratia marcescens,* **Salmonella, Pseudomonas, Nocardia,** and **Aspergillus** species. The lesions in chronic granulomatous disease are associated with an NADPH oxidase defect that results in a failure to mount a respiratory burst.

Another inherited defect is called **leukocyte adherence deficiency,** which results from an inherited abnormality of the proteins that allow neutrophils to adhere to blood vessel walls at sites of inflammation (Figure 28.22). In deficient individuals, neutrophils cannot respond to chemotactic agents or bind to vascular endothelial cells. As a result, they cannot enter tissues to destroy invading organisms or participate in extravascular inflammation. Affected individuals suffer from recurrent infections and fail to make pus, while, at the same time, they have large numbers of neutrophils in their bloodstream.

Primary Immunodeficiencies

Inherited mutations may lead to defects in lymphoid cell function. As a general rule, a defect in the production of lymphoid stem cells will result in a severe deficiency of both the antibody-mediated and the cell-mediated immune responses. A defect that occurs only in the T-cell development pathway will be reflected by an inability to mount cell-mediated immune responses, although antibody production may also be affected as a result of loss of

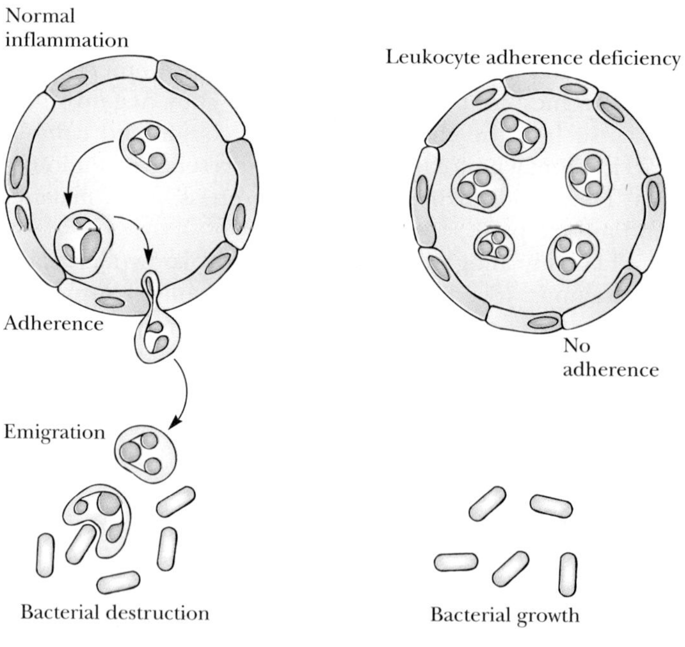

***Figure* 28.22** Integrins are required to bind neutrophils firmly to blood vessel walls. This permits the neutrophils to emigrate to sites of bacterial invasion. In the absence of integrins, neutrophil emigration fails to occur. As a result, invading bacteria can grow unmolested.

helper cell function. Similarly, a lesion restricted to the B-cell system will result in an absence of antibody-mediated immune responses.

Combined Deficiency Diseases

In the combined immune deficiencies, neither T-cell nor B-cell systems develop and no immune responses can be mounted. These combined immune deficiencies generally present as severe recurrent infections during the first weeks of life. Affected infants commonly develop oral candidiasis, pneumonias that are caused by low-grade pathogens such as *Pneumocystis carinii,* and chronic diarrhea. Unless successfully treated, these children will die by the age of two.

Combined immunodeficiencies may be diagnosed by an absence of lymphocytes in peripheral blood, reduced or absent serum immunoglobulins, and an inability to mount both cell-mediated and antibody-mediated immune responses. In some cases the nature of the lesion has been identified. These include abnormal T-cell antigen receptors, abnormal cytokine receptors, failure in gene rearrangements, defective signal transduction, and defective expression of MHC class II molecules. One of the most significant autosomal recessive forms of CID is due to an inherited deficiency of the enzyme adenosine deaminase (ADA). As a result, toxic purine metabolites accumulate within T-cells and kill them. The net effect of an ADA deficiency is the selective destruction of T-cells and a loss of cell-mediated immune responses. T-helper-cell activity is lost so antibody production is also defective.

Deficiencies of the T-Cell System

If the thymus fails to develop, the resulting disease, called the DiGeorge anomaly, is characterized by an absence of T-cell function (Figure 28.23). Because of this, individuals suffering from the DiGeorge anomaly fail to thrive and develop infections with yeasts, chronic pneumonia, and diarrhea, especially as a result of virus infection. They have few circulating lymphocytes and none with the characteristics of T-cells. Immunoglobulin levels and antibody responses are relatively normal. Possible treatments include the use of thymic hormones or transplantation of thymic epithelial cells, obtained either from an early fetus or from a cell culture.

Deficiencies of the B-Cell System

There are several different types of B-cell defect. When children are born, they possess IgG acquired by transplacental passage from their mother. This passively acquired maternal IgG provides protection against infection for several months. As this maternal antibody wanes, children normally develop their own antibodies. In B-cell deficient children, however, these antibodies fail to develop. As a result, the children gradually begin to suffer from severe, recurrent infections with such organisms as *S. pneumoniae, S. pyogenes,* and *Haemophilus influenzae,* although they show normal resistance to viruses such as vaccinia, rubella, or mumps. They usually have no B-cells or plasma cells, and their lymphoid organs lack cells in the B-dependent areas. Treatment is by repeated administration of immune serum globulin in an attempt to passively immunize the children against infection.

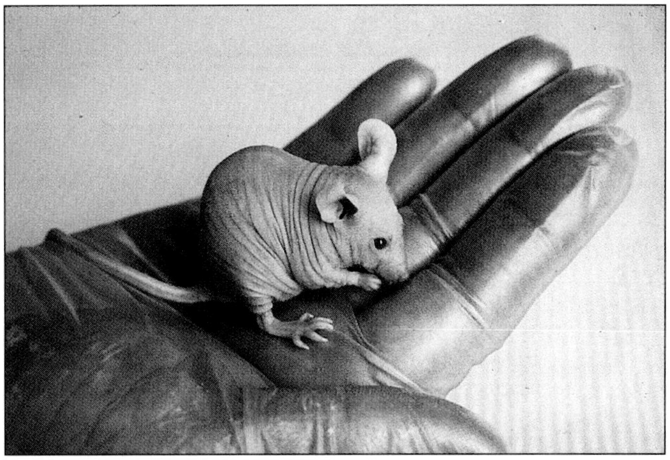

Figure 28.23 "Nude" mouse. This animal is T-cell deficient since it fails to make a thymus. (© NIH/Science Source/Photo Researchers, Inc.)

There are a variety of other congenital immunodeficiency syndromes in which there is a selective deficiency of some but not all of the major immunoglobulin classes. These deficiencies may be due either to the absence of a specific class or to production of nonfunctional antibodies. The most common primary immunodeficiency, occurring in about one in 600 Europeans, is a selective IgA deficiency. This disease commonly leads to recurrent respiratory and digestive tract infections. IgA deficiency is also associated with severe allergies, since excessive quantities of IgE are commonly produced. In the absence of IgA, antigens may be able to penetrate mucosal surfaces and reach and stimulate IgE-producing B-cells.

Secondary Immunodeficiencies

Acquired immunodeficiencies result when the immune system is damaged as a result of malnutrition, virus infections, or exposure to toxic chemicals or drugs. The most significant of these result from infection with a virus called human immunodeficiency virus (HIV). HIV causes AIDS-acquired immune deficiency syndrome.

The concept that viruses could destroy lymphocytes and so induce a profound immunodeficiency is not a new one. It had long been recognized that diseases such as measles in man or distemper in dogs were accompanied

BOX 28.2 RESEARCH HIGHLIGHTS

AIDS Vaccines

There is, at the present time, no effective vaccine available against AIDS. There are two main reasons why such a vaccine has not yet been developed. First, HIV mutates very rapidly, producing a mixture of antigenically distinct viruses even within one infected individual. Second, HIV is transmitted both as cell-free virions and by HIV-infected cells, and different types of response may be required to counteract this. In addition, there is a question of usage. Should the vaccine be employed to prevent infection of healthy individuals or, alternatively, to slow progression of the disease in those already infected with HIV? In experimental primate models, vaccines against HIV have been protective for only a short period and only when the vaccine strain is identical to the infecting virus. Likewise, it is clear that natural infection with HIV does not confer significant protective immunity.

Traditional approaches to vaccine development have involved immunization with either inactivated virus, or recombinant HIV proteins using an adjuvant. Unfortunately, the immunity induced is brief and does not stimulate the production of cytotoxic T-cells. An alternative approach, the use of a modified live virus in the vaccine, has a major drawback: the mutation rate of the virus is such that it could easily become virulent—an unacceptable result. Because of these problems, considerable attention has been paid to newer approaches to the development of an AIDS vaccine. Synthetic HIV protein epitopes have been ineffective since they generally contain too few epitopes. Live recombinant vaccines using HIV recombinants in vaccinia, mycobacteria, or enteric bacteria have not stimulated a significant protective response in animals, nor have combined approaches using live recombinant products together with recombinant vaccines been any more successful. Some potential HIV vaccines are currently being tested for safety in humans. These include vaccines containing recombinant envelope glycoproteins, synthetic peptides, and recombinant pox viruses. Unfortunately, none has been highly protective in animal models.

by a significant immunosuppression and, as a result, infected individuals were more susceptible than normal to other, secondary, infections. Nevertheless, none of these viruses induced an immunodeficiency of the severity of that induced by the human immunodeficiency virus (HIV).

Beginning in 1979, physicians began to notice a remarkable increase in cases of two unusual conditions among active male homosexuals. These conditions were a rare tumor called Kaposi's sarcoma and a severe pneumonia caused by a fungus called *Pneumocystis carinii.* Affected individuals were found to be profoundly immunosuppressed as a result of an almost total loss of their T-cells. Evidence suggested that the disease was a form of immunodeficiency transmitted by an infectious agent, so it was called the acquired immune deficiency syndrome. (See Chapters 14 and 30 and **Box 28.2** for a coverage of AIDS.)

Excessive Immune Function

Under some circumstances, immune responses may cause inflammation or tissue damage in their efforts to clear the body of foreign material. This form of reaction is called **hypersensitivity** and can have several different mechanisms.

Hypersensitivities

Type I Hypersensitivity

Allergies (or Type I hypersensitivity reactions) occur as a result of the release of pharmacologically active molecules from mast cells (Figure 28.24). These molecules include a mixture of highly irritant chemicals such as histamine, serotonin, kinins, prostaglandins, and leukotrienes. This release from mast cells is caused by the combination of antigen with IgE bound to the mast cell surface (Figure 28.25). Thus, if an individual is allergic to ragweed pollen, he or she will make IgE antibodies against the antigens in that pollen. These antibodies will bind to mast cells throughout the body.

Once a sensitized individual is exposed to ragweed pollen, the pollen grains stick to body surfaces and their soluble antigens diffuse into the tissues. When the antigens bind to the IgE on the mast cells, the mast cells release their cytoplasmic granules. Once the granules are

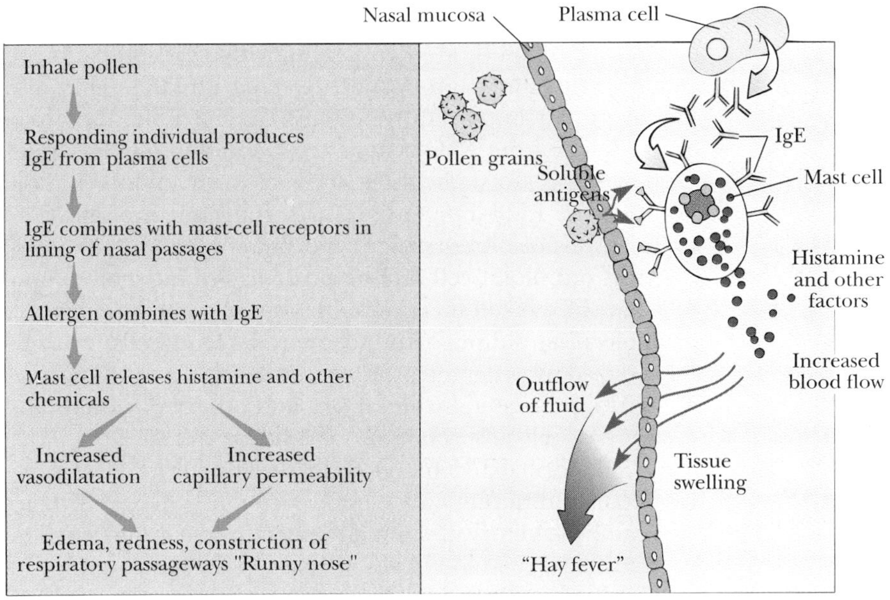

***Figure* 28.24** The pathogenesis of hay fever. Inhaled ragweed pollen attaches to IgE on mast cells in the nasal mucosa and triggers mast cell degranulation. The clinical signs of hay fever, itching, pain, inflammation, and fluid release result from the release of histamine, kinins, and other pharmacological agents from these mast cells.

exposed to extracellular fluid, they release their chemical contents. If the antigen is inhaled the contents cause nasal irritation, fluid exudation, and all the signs associated with hay fever. Allergies mediated in this way are characterized by a rapid time course, inflammation, fluid accumulation in tissues, and infiltration of the inflammatory site by eosinophils attracted by chemotactic agents released from mast cell granules. If antigens are inhaled and gain access to the bronchi in the lungs, local mast cell degranulation and histamine release causes the contraction of bronchial smooth muscle resulting in severe difficulty in breathing. This condition is called **asthma.** If antigens gain access to the bloodstream (for example, through an insect sting), large numbers of mast cells in many tissues may degranulate rapidly and cause a lethal shock syndrome called **anaphylaxis.** Thus, some allergic individuals, when stung by insects, may collapse and suffocate as their bronchi contract.

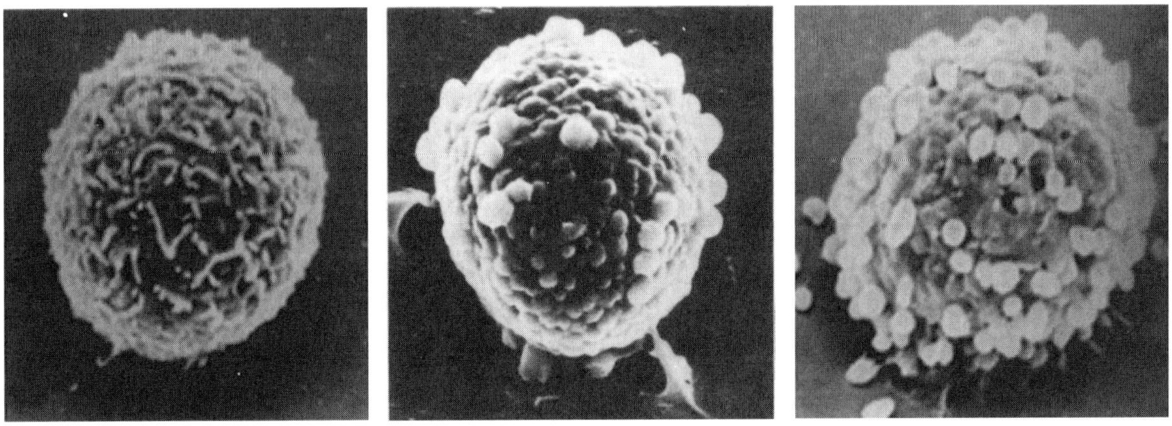

(a) **(b)** **(c)**

***Figure* 28.25** Scanning electron micrographs of a normal rat mast cell (A), a sensitized mast cell fixed five seconds after exposure to antigen (B), and a sensitized mast cell fixed 60 seconds after exposure to antigen (C) (× 3000). (From Tizard I. R. and Holmes W. L., 1974. *Int. Arch. Allergy Appl. Immunol.* 46, 867–879. Used with permission from the publisher, S. Karger, Basel.)

Type II Hypersensitivity

Type II hypersensitivity reactions are those in which tissue or cell damage is the direct result of the actions of antibody and complement. Antibodies directed against antigens on the surface of nucleated cells or red blood cells will cause their lysis in the presence of complement (Figure 28.26). If the immunoglobulin is not one that activates complement, then the target cell may still be destroyed by phagocytosis or ADCC. Thus this type of hypersensitivity is a purely destructive process. The best example of type II hypersensitivity is the destruction of foreign red blood cells following an incompatible blood transfusion.

Red blood cells, like nucleated cells, possess characteristic cell-surface proteins. Unlike the MHC molecules of nucleated cells, however, red blood cell surface proteins or blood groups do not determine an animal's ability to mount an immune response, although they do influence graft rejection (grafts between individuals incompatible in the major blood groups are rejected rapidly). They also have other important functions. For example, the glycoproteins of the ABO system are anion and glucose transporter proteins, the Kell antigen is an endopeptidase, while the Kidd antigens are urea transport proteins. The functions of others such as the Rh antigen are unknown. Most red blood cell-surface proteins are integral components of the cell membrane. However, some blood-group molecules, although found on red blood cells, are synthesized at other sites within the body. These molecules are found free in serum, saliva, and other body fluids and are passively adsorbed onto red blood cell surfaces.

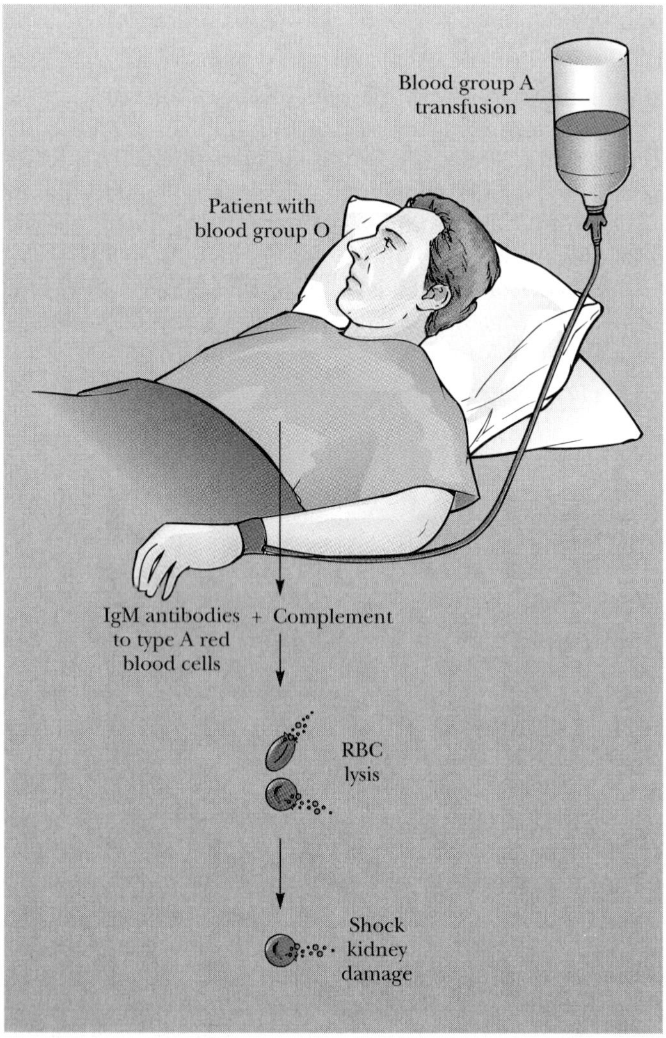

***Figure* 28.26** Lysis of foreign red cells by antibodies and complement is an example of Type II hypersensitivity.

If normal red blood cells are administered to a genetically different (allogenic) recipient, their cell membrane proteins will stimulate an immune response. The transfused red blood cells will be rapidly destroyed in the bloodstream by antibody and complement and through opsonization and phagocytosis by the cells of the mononuclear-phagocytic system. This type of cell destruction is classified as a Type II hypersensitivity reaction.

Blood Groups and Disease

It has already been pointed out how certain histocompatibility antigens are closely associated with disease prevalence or susceptibility. Since blood-group antigens are not necessarily linked to the MHC and have no antigen-presenting ability, there is no *a priori* reason why they should be linked to specific disease conditions. Nevertheless, there is a clear association between stomach ulceration and nonsecretors of blood group O. (This has recently been shown to result from the fact that stomach ulcers are caused by a bacterium called *Helicobacter pylori.* This organism preferentially binds to blood group antigens that have a terminal blood group O fucose. These molecules are located on the surface of gastric epithelial cells.)

Type III Hypersensitivity

The formation of immune complexes through the combination of antibody with antigen is the first step in many immunological processes. One of the most significant of these processes is the **complement cascade.** When complement-activating immune complexes are deposited in tissues, chemotactic factors are produced and lead to a local accumulation of neutrophils. These neutrophils release their lysosomal enzymes and oxidizing radicals, and these in turn cause local tissue destruction. Lesions generated in this fashion are called Type III or **immune-complex mediated hypersensitivity reactions.**

The severity and significance of Type III hypersensitivity reactions depends, as might be expected, upon the amount and site of deposition of immune complexes. Two major types of reactions are recognized. One is a local one

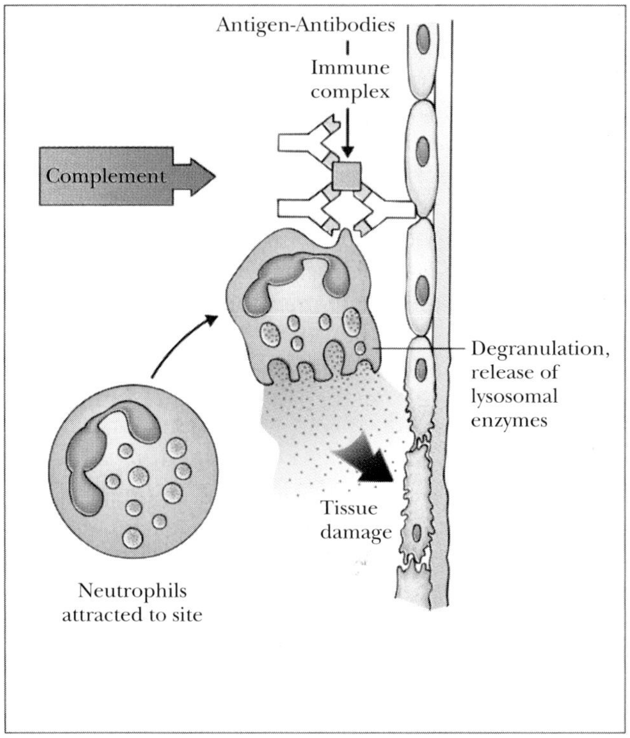

***Figure* 28.27** The mechanisms involved in the Arthus reaction.

called the **Arthus reaction,** named after the biologist who first described it. The Arthus reaction occurs when immune complexes are deposited in tissues (Figure 28.27). Arthus reactions may be induced in any tissue into which antigen can be deposited.

A second form of Type III hypersensitivity reaction results when large quantities of immune complexes are formed in the bloodstream. This occurs when antigen is administered intravenously to a hyperimmune recipient. Complexes generated in this way are deposited in the walls of blood vessels. Local activation of complement then leads to neutrophil accumulation and the development of inflammation in blood vessel walls (vasculitis). Circulating immune complexes are also deposited in the glomeruli of the kidney. Therefore, the occurrence of inflammatory lesions in glomeruli (glomerulopathy) is also characteristic of this type of hypersensitivity.

Several important clinical syndromes are associated with the development of immune complex-mediated lesions in humans and animals. All are associated with the persistence of antigen in the bloodstream in the presence of antibodies.

One interesting form of Type III hypersensitivity is called **hypersensitivity pneumonitis** (Figure 28.28). Repeated inhalation of very small spores from molds, fungi, or actinomycetes will stimulate an immune response. As a result, individuals continuously exposed to these spores will develop high-titered precipitating serum antibodies to spore antigens. If exposure to these spores continues, spore antigens will encounter antibody within alveolar walls, immune complexes will be deposited, complement will be activated, and a local inflammatory response will develop. This inflammation is characterized by a massive neutrophil accumulation and tissue damage. Fluid accumulates in the alveoli. Clinically, this type of hypersensitivity is associated with difficulty in breathing (dyspnea) occurring five to ten hours after exposure to the antigen. Once recognized, it can be prevented from recurring by removal of the source of the antigen. Many forms of hypersensitivity pneumonitis in humans are mediated in this way. They are usually named after the source of the offending antigen. Thus, farmer's lung develops as a result of chronic exposure to actinomycete spores released when working with moldy hay. Pigeon-breeder's lung arises following exposure to the dust from pigeon feces, mushroom-grower's disease is due to hypersensitivity to inhaled spores from actinomycetes in the soil used for growing mushrooms, and librarian's lung results from inhalation of dusts from old books, and so forth.

Type IV Hypersensitivity

Type IV hypersensitivity reactions result from a T-cell-mediated response to antigen. Because of the need for T-cells to migrate to the site of antigen deposition, they usually take more than 24 hours to develop and are therefore called **delayed hypersensitivity reactions.** The best example of a Type IV reaction is the tuberculin reaction. This

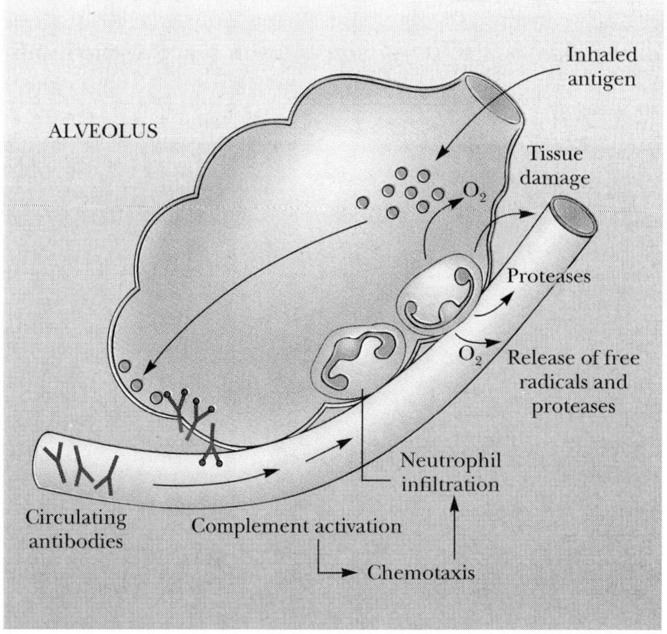

***Figure* 28.28** The mechanism of hypersensitivity pneumonitis.

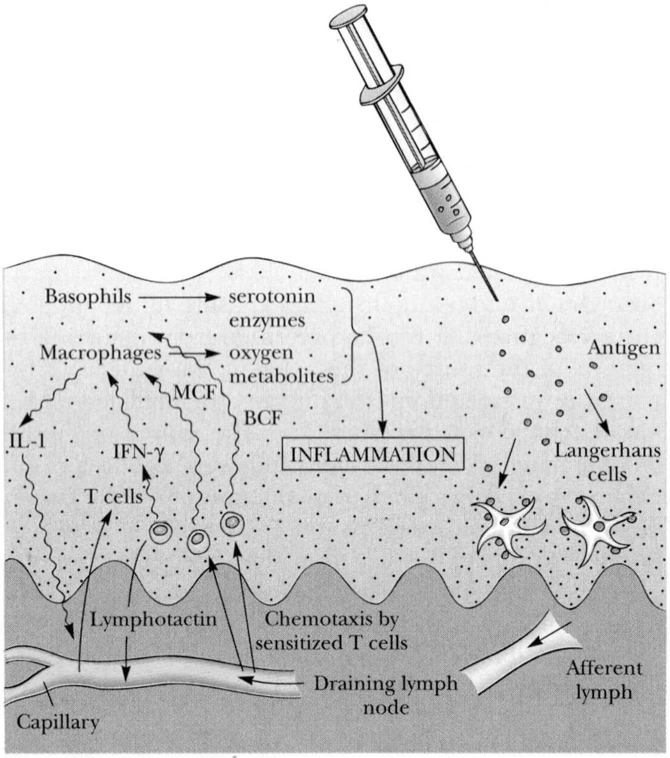

***Figure* 28.29** Schematic diagram depicting the pathogenesis of the delayed hypersensitivity reaction. Langerhan's cells are antigen processing cells found in the skin. MCF, macrophage chemotactic factor; BCF, basophil chemotactic factor.

is an inflammatory response produced in the skin in response to intradermal inoculation of an extract of the *Mycobacterium tuberculosis* (Figure 28.29). It is used in the diagnosis of tuberculosis. Type IV reactions may also occur in the skin as a result of contact with reactive chemicals. A good example of this is the skin's reaction to the resins of the poison ivy plant. In this case the reaction is directed against chemically modified skin cells. The mechanisms of Type IV hypersensitivity are complex but involve T-cells, macrophages, mast cells, and basophils.

Although the intradermal tuberculin reaction is artificial because antigen is administered by injection, a similar host response occurs if living tubercle bacilli lodge in tissues. *M. tuberculosis* is resistant to intracellular destruction until a cell-mediated immune response has developed. As a result, the hypersensitivity reaction to whole organisms is prolonged, and, consequently, macrophages accumulate in large numbers. Many of these macrophages attempt to ingest the bacteria and die in the process, whereas others fuse to form multinucleated giant cells. The lesion that develops around invading tubercle bacilli, therefore, consists of a mass of dead tissue containing both living and dead bacteria; the lesion is surrounded by a layer of macrophages, which in this location are known as epithelioid cells. The entire lesion is called a tubercle (Figure 28.30). Persistent tubercles may become relatively well organized, and collagen may be laid down, resulting in the formation of a granuloma. (Interleukin 1 stimulates collagen production by fibroblasts and hence contributes to this process.) Granuloma formation is a frequent result of local persistent inflammation.

Autoimmunity

Although the immune system is specifically designed to recognize and to respond to foreign antigens, on occasion the body mounts an immune response against normal body components. This type of response is called an **autoimmune response.** Not all autoimmune responses cause disease. For example, aged cells are naturally destroyed by autoantibodies. Autoimmunity is not a rare event. In fact,

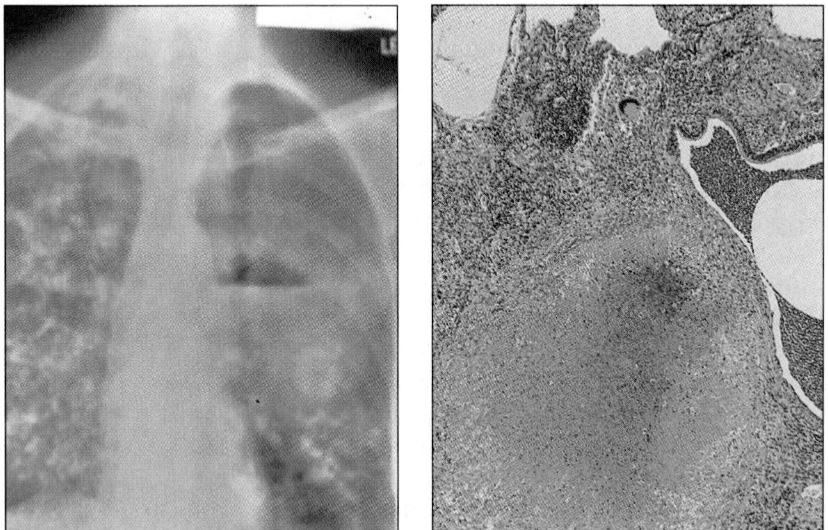

(a) **(b)**

***Figure* 28.30** **(a)** A chest radiograph showing the presence of many large opaque spots in the lungs. These are tubercles–granulomatous reactions to the presence of Mycobacterium tuberculosis in tissues (© Science VU/Visuals Unlimited). **(b)** A histological section of a tubercle in the lung. The dark central mass is a mixture of dead cells and bacteria. It is surrounded by layers of macrophages (© John D. Cunningham/Visuals Unlimited).

all normal individuals possess T-cells and B-cells able to react to self-antigens. These lymphocytes are normally suppressed by control mechanisms. Autoimmune responses come to our attention most commonly, however, when they cause disease. Autoimmune diseases have many contributing factors, and it is difficult to identify a specific cause (Figure 28.31). They may result from a loss of regulatory cells, from exposure of hidden epitopes to which the body is not tolerant, from crossreactions between microbial and body antigens (molecular mimicry), or as a result of some viral infections. The development and severity of autoimmune diseases is determined, in large part, by the histocompatibility antigens present in the affected individual. Autoimmune diseases occur as a result of the tissue destruction and inflammation brought about by each of the four types of hypersensitivity mechanisms. Some autoimmune diseases such as autoimmune thyroiditis, encephalitis, and diabetes mellitus are specific for a single organ or cell (thyroid, brain, islet cells, respectively). Other autoimmune diseases involve many organs. An example of such a disease is systemic lupus erythematosus, in which autoantibodies are made against a wide variety of normal tissues including skin, blood cells, muscle, and nucleic acids.

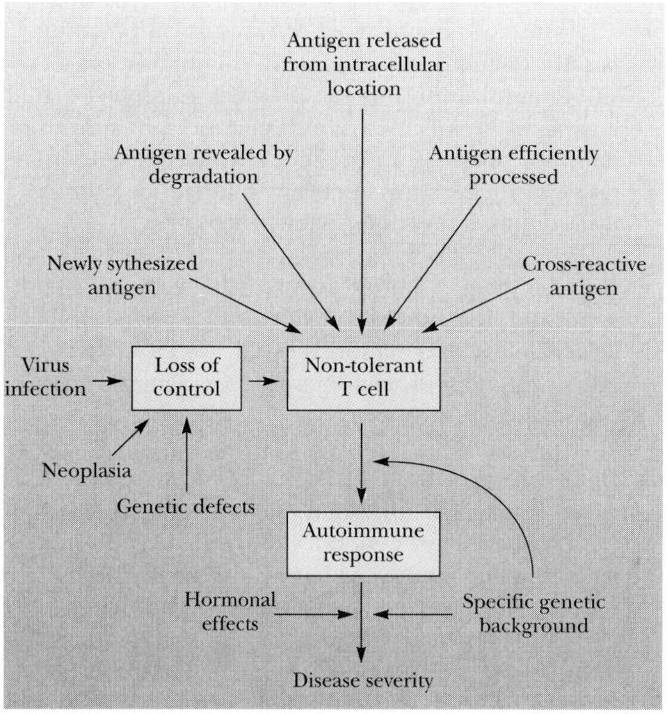

***Figure* 28.31.** A simplified scheme for the pathogenesis of autoimmune disease.

Summary

- The prime role of the immune system is to combat invasion by infectious agents.
- Extracellular bacteria are attacked through antibody-mediated mechanisms. These include **opsonization, complement-mediated lysis** and toxin neutralization.
- Intracellular bacteria are attacked by activated macrophages.
- The course and severity of bacterial disease can be greatly modified by the immune responses. Thus Th2 responses are ineffective in combating intracellular organisms.
- Bacteria and viruses employ many different strategies to ensure their survival within the animal body.
- Because viruses are obligate intracellular organisms they are largely combated by cell-mediated mechanisms involving destruction of infected cells by **cytotoxic T-cells** and **NK cells.** Antibodies may, however, play an important role in the destruction of extracellular virus particles. Destruction of virus-infected cells may contribute to the severity of virus diseases.
- The elimination of abnormal cells by the immune system gives rise to the process of **graft rejection.** It is also believed that the immune system is able to identify and destroy many cancer cells. The most important mechanism of anticancer defenses involves NK cells.
- Since invading microorganisms must gain access to the body through its surfaces, it follows that many of the host defense mechanisms must be located on these surfaces. Thus, body surfaces possess an array of **nonspecific defense mechanisms** including self-cleaning processes and the presence of a dense established normal flora.
- The immunoglobulin class that is largely responsible for the defense of surfaces is **IgA.** Secretory IgA is resistant to proteolytic digestion. It prevents microorganisms from binding and invading body surfaces. IgA can also work within both cells and tissues.
- Antibodies in mother's milk are produced by cells whose precursors arose in the mother's intestine.

- There are two major types of vaccination procedures, **passive immunization** and **active immunization.** Passive immunization provides immediate protection but wanes rapidly. Active immunization, in contrast, does not protect immediately, but when it develops it provides long-lasting immunity that can be boosted by additional injections of vaccine.

- Because vaccination induces a biological response, it cannot protect all the individuals in a population. Nevertheless, vaccination is by far the most cost-effective way to protect a population against infectious disease.

- Defects in the immune system are of two major types. Primary or inherited defects result from genetic causes. They generally affect the very young. These defects can occur in any arm of the immune system. Thus there can be defects in phagocytosis and antigen processing, defects in T-cell function, defects in B-cell function and combined immunodeficiencies where neither B- nor T-cell function develops. As a general rule, B-cell defects result in increased susceptibility to bacterial disease while T-cell defects result in increased susceptibility to virus infections and cancer.

- Excessive or inappropriate immune function may be expressed in the form of **hypersensitivities** or **autoimmunity.** Hypersensitivity reactions—inflammation or tissue damage resulting from immune processes—may occur through four major pathways. Thus in type I hypersensitivity, antigen binding to mast cell–bound IgE causes these cells to degranulate, releasing many pharmacologically active molecules into the tissues. This may cause local reactions, as in hay fever or asthma. If severe, it may cause systemic anaphylactic shock.

- On occasion, antibodies and complement may cause cell death directly. Thus, if an incompatible blood transfusion is given to an individual it will trigger an immune response. As a result, antibodies and complement will destroy the transfused red cells. This transfusion reaction is a form of Type II hypersensitivity.

- Immune complexes are normally rapidly removed from the circulation by phagocytic cells as they are formed. If, however, excessive immune complexes are formed, they may be deposited in tissues. As a result of local complement activation these immune complexes may cause acute inflammation—Type III hypersensitivity.

- The fourth type of hypersensitivity differs from the others in that it is cell-mediated. A typical example of Type IV hypersensitivity is the tuberculin reaction. The delayed hypersensitivity reaction, mediated by T-cells, occurs at the site of injection of tuberculin, a protein derived from the tubercle bacillus. It only develops in individuals sensitized by exposure to this organism.

- Autoimmune responses occur when the immune system attacks normal body components as a result of a loss of self-tolerance. Autoimmune diseases can affect any organ in the body. Some autoimmune diseases affect single organs such as the thyroid gland or the skin. Other autoimmune diseases affect multiple organ systems.

Questions for Thought and Review

1. Outline the different ways in which the immune system defends the body against invasion by bacteria. Are there differences in the defense mechanisms engaged against intracellular and extracellular organisms?

2. In what ways does the complement system defend the body against bacteria and viruses? Speculate on the consequences of a deficiency of such complement components as factor H, C3, and C9.

3. What are the roles of Th1- and Th2-cells in resistance to invading organisms? How do these cells influence the course of a disease such as leprosy?

4. What is the role of IgA in the defense of the body? How does this molecule function within the intestinal lumen without being digested?

5. What role might IgE and Type I hypersensitivity play in the defense of body surfaces?

6. Outline the predicted consequences of a loss of T-cell function and a loss of B-cell function. What differences may be observed between the clinical consequences of these defects?

7. The immune system may occasionally cause disease. Classify the ways in which this may occur.

8. To what extent is it likely that hypersensitivity reactions may in fact increase the severity of bacterial diseases?

Suggested Readings

Ada, G. 1990. "The immunological principles of vaccination." *Lancet.,* **335:**523–526.

Agre, P. C., and J. P. Cartron, eds. 1992. "Protein blood group antigens of the human red cell. Structure, function and clinical significance." Baltimore: Johns Hopkins University Press.

Baringa, M. 1992. "Viruses launch their own 'Star Wars.'" *Science,* **258:**1730–1731.

Brandzaeg, P. 1995. "Molecular and cellular aspects of the secretory immunoglobulin system." *APMIS,* **103:**1–19.

Fearon, D. T. 1988. "Complement, C receptors and immune complex disease." *Hospital Practice,* **23:**63–72.

Goodenough, U. W. 1991. "Deception by pathogens." *Amer. Sci.,* **79:**344–355.

Gupta, S. 1990. "New concepts in immunodeficiency diseases." *Immunol. Today,* **11:**344–346.

Lichtenstein, L. M. 1993. "Allergy and the immune system." *Sci. Amer.,* **269:**117–124.

Mazanec, M. B., C. S. Kaetzel, M. E. Lamm, et al. 1992. "Intracellular neutralization of virus by immunoglobulin A antibodies." *Proc. Natl. Acad. Sci., USA.,* **89:**6901–6905.

Oldstone, M. A. 1990. "Viral persistence and immune dysfunction." *Hosp. Pract.,* **25:**81–98.

Paul, W. E. 1993. "Infectious diseases and the immune system." *Sci. Amer.,* **269:**91–97.

Tizard, I. R. 1994. *Immunology: An Introduction,* 4th ed. Philadelphia: Saunders College Publishing.

Moreover he will bring upon thee all the diseases of Egypt, which thou was afraid of; and they shall cleave unto thee.

Deuteronomy

Chapter 29

Major Microbial Diseases of Humans

Skin Infections
Respiratory Diseases
Fungal Diseases
Bacterial Diseases of the Gastrointestinal Tract
Urinary Tract Infections
Sexually Transmitted Diseases
Vectorborne Diseases
Protozoa as Agents of Disease
Zoonoses

The symptoms and ravages of infectious disease have been recorded throughout history. Biblical and ancient writings describe the crippling and suffering caused by various scourges; the words *plague* and *disease* have 17 citations in the concordance of the King James Bible. Ancient drawings show humans scarred by smallpox or crippled by polio. The epidemic of syphilis that spread through Europe in the latter part of the 15th century killed thousands. Although earlier, this epidemic was ascribed to the crews of Christopher Columbus in the belief that the disease came from Indians in the New World, it is now established that there was no syphilis in the pre-Columbian New World and that syphilis probably came to Europe with migrations from northern Africa. The devastations of "Black Death," the plagues that periodically ravaged Europe, are well recorded (see **Box 2.1** and discussion in Chapter 2). Typhus was often the final arbiter in war: the disease destroyed Napoleon at Moscow and killed three million Russian soldiers during World War I.

An understanding of the relationship between a pathogenic microbe and disease symptoms has increased markedly since the pioneering studies of Robert Koch (Chapter 2). The minor limitation imposed by one of Koch's postulates (a microbe must be culturable outside the animal) has been overcome by improved isolation and culturing techniques. This permitted a broader understanding of viral and bacterial infectious agents that cannot be cultured outside a living host.

This chapter will discuss some of the significant infectious diseases and manifestations of infections. A limited number of fungal and protozoan diseases will also be described. The presentation will be by route of entry or site of infection such as respiratory, gastrointestinal tract, and so on. Presentation in this order is for convenience but in general is satisfactory because most infectious microbes are selective in the portal of entry. Major exceptions are *Staphylococcus aureus* and *Streptococcus pyogenes,* organisms that can cause disease at various sites in or on the human body.

Skin Infections

Skin and external infections often observed in humans include impetigo, wound infections, acne, ringworm, and athlete's foot (Table 29.1). **Impetigo,** for example, is a skin infection that occurs frequently in child care centers and schools where youngsters are exposed to one another under confined conditions. Impetigo is caused by *Streptococcus pyogenes* or *Staphylococcus aureus* and is transmitted by scratching or other interpersonal contacts. Streptococcal impetigo is characterized by the appearance on the skin of vesicles that are eventually covered by a reddish crust (Figure 29.1). Staphylococcal impetigo generally occurs around the nose and spreads about the face. Impetigo can be controlled by topical application of antibiotic ointments.

Another common skin infection is acne. **Acne** occurs during adolescence, when the endocrine system is most active. Hormonal activity stimulates sebaceous glands to overproduce sebum. The sebum can accumulate within the gland and become infected by a normal skin inhabitant, *Propionibacterium acnes.* The consequence is an inflammatory response with swelling and reddening in the area. Tetracycline is commonly given to patients suffering from chronic acne.

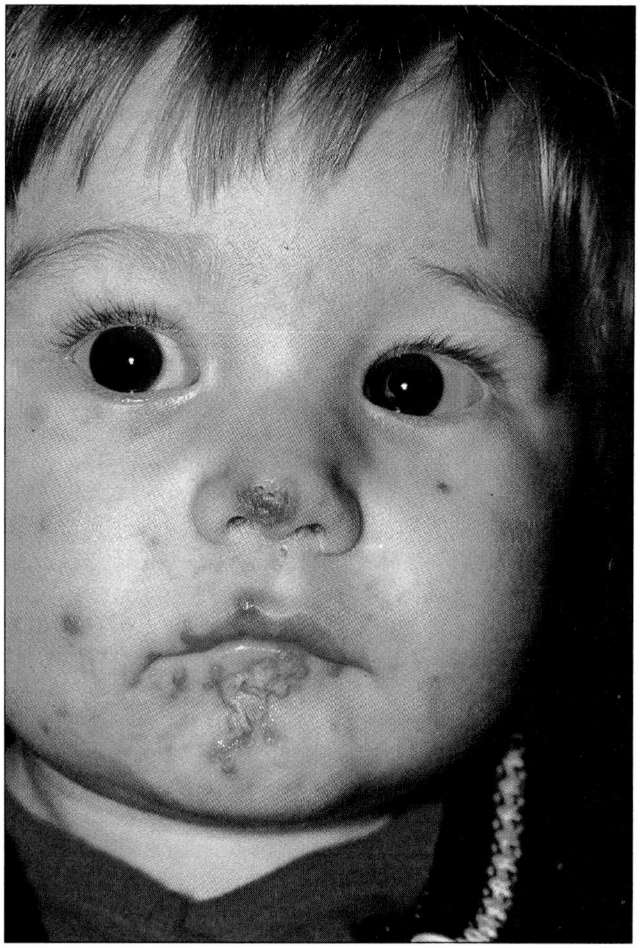

***Figure* 29.1** Impetigo on the face of a child. Note the inflammatory response (redness and swelling) around the lesions. Impetigo is caused by *Staphylococcus aureus, Streptococcus pyogenes,* or a combination of these microorganisms. (© Dr. P. Marazzi/Science Photo Library)

***Table* 29.1 Major diseases of the skin[1]**

Disease	Organism	Symptoms
Impetigo	*Staphylococcus aureus* *Streptococcus pyogenes*	Skin lesions
Wound infections (burns)	*Pseudomonas aeruginosa*	Infection followed by systemic invasion; produces toxins
Acne	*Propionibacterium acnes*	Redness and swelling of sebaceous glands
Ringworm (*Tinea corporis*)	*Trichophyton mentagrophytes*	Fungal invasion of skin; ring of inflammation
Scalp ringworm	*Trichophyton tonsurans*	Hair follicle infection and loss of hair
Athlete's foot (*Tinea pedis*)	*Trichophyton rubrum* *Trichophyton mentagrophytes* *Epidermophyton floccosum*	Peeling and cracking of skin between toes
Groin ringworm	*Epidermophyton floccosum*	Fungal infection of the groin

[1]The list is not exhaustive and other diseases occur. References to these are presented at the end of the chapter.

Pseudomonas aeruginosa is commonly present in air and water and is an opportunistic pathogen in large open wounds such as extensive burns. This organism infects over 25 percent of burn patients and can produce toxins and proteases that contribute to pathogenicity. *P. aeruginosa* is naturally resistant to many antibiotics and is difficult to control.

Fungal diseases of the skin are of widespread occurrence in humans. Ringworm and athlete's foot are familiar to most of us through personal experience and products advertised on TV. The level of this advertising is indicative of the incidence of these infections in human populations. These fungi invade the keratinized part of skin and collectively are called **dermatomycoses** (Figure 29.2). It is estimated that up to 70 percent of the population in the United States is subclinically infected with athlete's foot. The causative agents of these diseases are presented in Table 29.1. The infections are mostly caused by three genera—*Epidermophyton, Microsporum,* and *Trichophyton*—and can be treated successfully with topical applications of imidazole drugs. Severe fungal infections of the skin are best treated by a physician.

Staphylococcus aureus is probably the most versatile pathogen that afflicts human beings. Infections by *S. aureus* are difficult to prevent as virulent strains of the organism are carried in the nasopharynx by up to 50 percent of the human population. These pathogens normally inhabit the skin, intestine, and vagina. The organism is readily spread from asymptomatic carriers by touching, sneezing, coughing, or passage on inanimate materials. Overall, there are at least 14 species of *Staphylococcus* that colonize humans, with coagulase positive *S. aureus* being most often associated with disease. *S. epidermidis* and *S. saprophyticus* can be infectious if resistance of the host is compromised. *S. epidermidis* is the most prevalent of the staphylococci that infect in the vicinity of devices such as prosthetic joints, artificial heart valves, and catheters. The organism produces a polysaccharide capsule that adheres tightly to artificial materials. *S. saprophyticus* frequently causes cystitis, a urinary tract infection in women. Cystitis is an inflammation of the urinary bladder resulting from microorganisms traveling from the urethral opening. It occurs less frequently in males as their urethra is longer and there is less chance that microbes will travel to the bladder.

Infections caused by *S. aureus* are listed in Table 29.2. Generally, these infections result from a breakdown in the natural defense system. Serious staphylococcal infections occur most often in individuals with a decreased ability to phagocytize and destroy the organism. This is frequently the case in the neonate, in surgical and burn patients, in individuals on immunosuppressants, or in individuals lacking a normal immune response. Scratches and blocked pores can lead to pimples, impetigo, and various superficial or deep-seated skin infections. Deep infections, such as carbuncles, can lead to infections of the lymph nodes.

Most *S. aureus* strains have a cell wall component, protein A, that binds to immunoglobulins and interferes with the normal phagocytic response. Infected lymph nodes and internal foci of infections can result in rapid multiplication and dissemination of the organism into the bloodstream. The organism produces a number of toxins and enzymes that aid in the invasive process (Chapter 26). One enzyme, coagulase, is particularly effective in inducing the production of a fibrin clot that surrounds the bacterial cell, preventing the normal host defenses from phagocytizing the invader. *S. aureus* also produces leukocidin, an anti-phagocytic factor that destroys leukocytes.

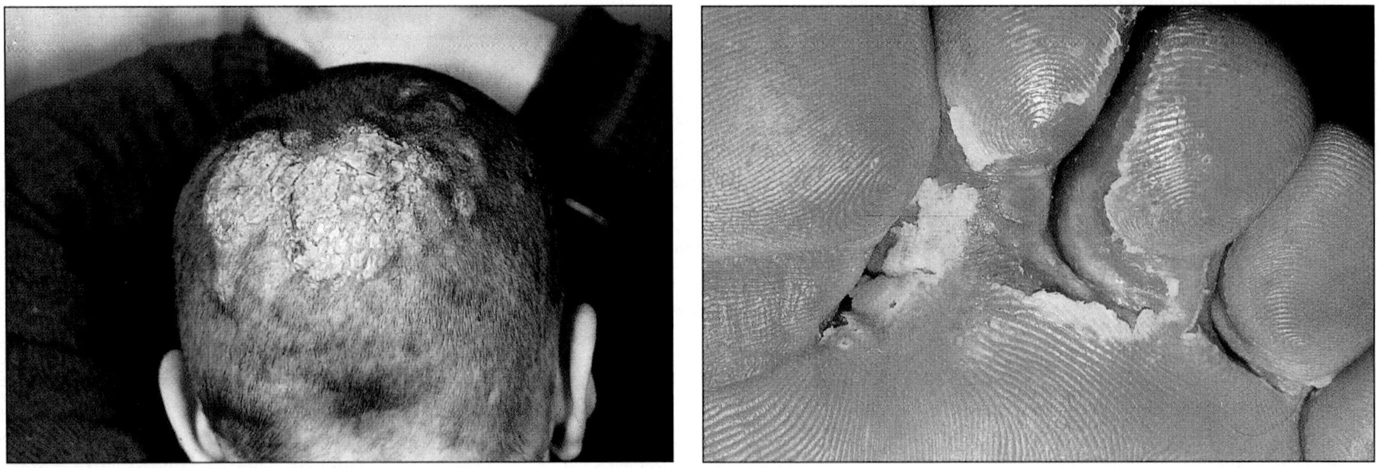

(a) (b)

***Figure* 29.2** Fungal infection of the skin. **(a)** Ringworm on the head is caused by *Tinea capitis* (© Everett S. Beneke /VU), and **(b)** ringworm (athlete's foot) in the toe area is caused by *Tinea pedis*. (© Science Photo Library)

Table 29.2 **Some infections caused by *Staphyloccus aureus***

Disease	Site
Pimples	Anywhere on the body
Impetigo	Generally on face
Boils	Anywhere
Carbuncles	Anywhere
Lymph nodes	Near site of infection
Blood (septicemia)	Carried from site of infection
Osteomyelitis	Bones
Endocarditis	Heart
Meningitis	Brain
Enteritis (food poisoning)	Gastrointestinal tract
Nephritis	Kidneys
Toxic shock syndrome	Hypotension
Staphylococcal scalded skin syndrome	Skin peeling
Wound infections	Skin
Inner ear infections	Ear
Respiratory infections	Pharynx Laryngitis Bronchitis Pneumonia

Production of leukocidin and coagulase aid the organism in overcoming host defenses. Viral diseases can also predispose an individual to invasion by staphylococci. Influenza may be followed by a staphylococcal pneumonia that may be fatal. It is probable that many deaths ascribed to viral influenza can be ascribed to staphylococci. Other viral infections of the respiratory system can cause tissue trauma, creating an opening for invasion by *S. aureus.* Meningitis and laryngitis are some of the infections that result. Staphylococcal infections can be treated with cephalosporin, penicillin, cloxacillin, and other antibiotics.

Toxic shock syndrome, first described in 1978, is a syndrome that occurs most frequently in menstruating women who use tampons. The disease symptoms are a sudden fever, diarrhea, vomiting, red skin rash, and low blood pressure. The drop in blood pressure can lead to an irreversible state of shock and death. The death rate in confirmed cases is between 5 percent and 12 percent. The disease also occurs in surgical patients of both sexes where there is an instance of internal colonization by *S. aureus.*

Respiratory Diseases

Infectious diseases of the mouth and respiratory system are a concern because they are readily communicated from human to human via air droplets created by sneezing, coughing, talking, or by kissing and other human contacts. They are in the air we breathe, and there is no possible way to avoid exposure. Some potential sites of respiratory disease are presented in Figure 29.3. Although microorganisms do not survive well in outdoor air, the atmosphere is constantly inoculated by organisms from soil, plants, humans, and animals. Indoor air has higher numbers of microbes, and these are generally disseminated by the residents occupying the space. If large numbers of humans are crowded into a room, the air will be populated by microbes common in the respiratory tract. The gram-positive organisms such as *Micrococcus, Staphylococcus,* and *Streptococcus* can survive in air, as they are quite resistant to drying and relatively resistant to ultraviolet light.

Sneezing and coughing are significant mechanisms for expelling infectious droplets into the air. A sneeze droplet can move at 100 meters per second for short distances, and a single sneeze can expel 10,000 to 100,000 bacteria into the air. A list of infections transmitted via the respiratory route is presented in Table 29.3. Viruses can also be transferred via this route and will be discussed in Chapter 30. Immunization has reduced the scourge of many respiratorily transmitted diseases, and antibiotics are effective against others. Diphtheria, pneumonia, and tuberculosis, although no longer the peril that were a few generations ago, remain a concern.

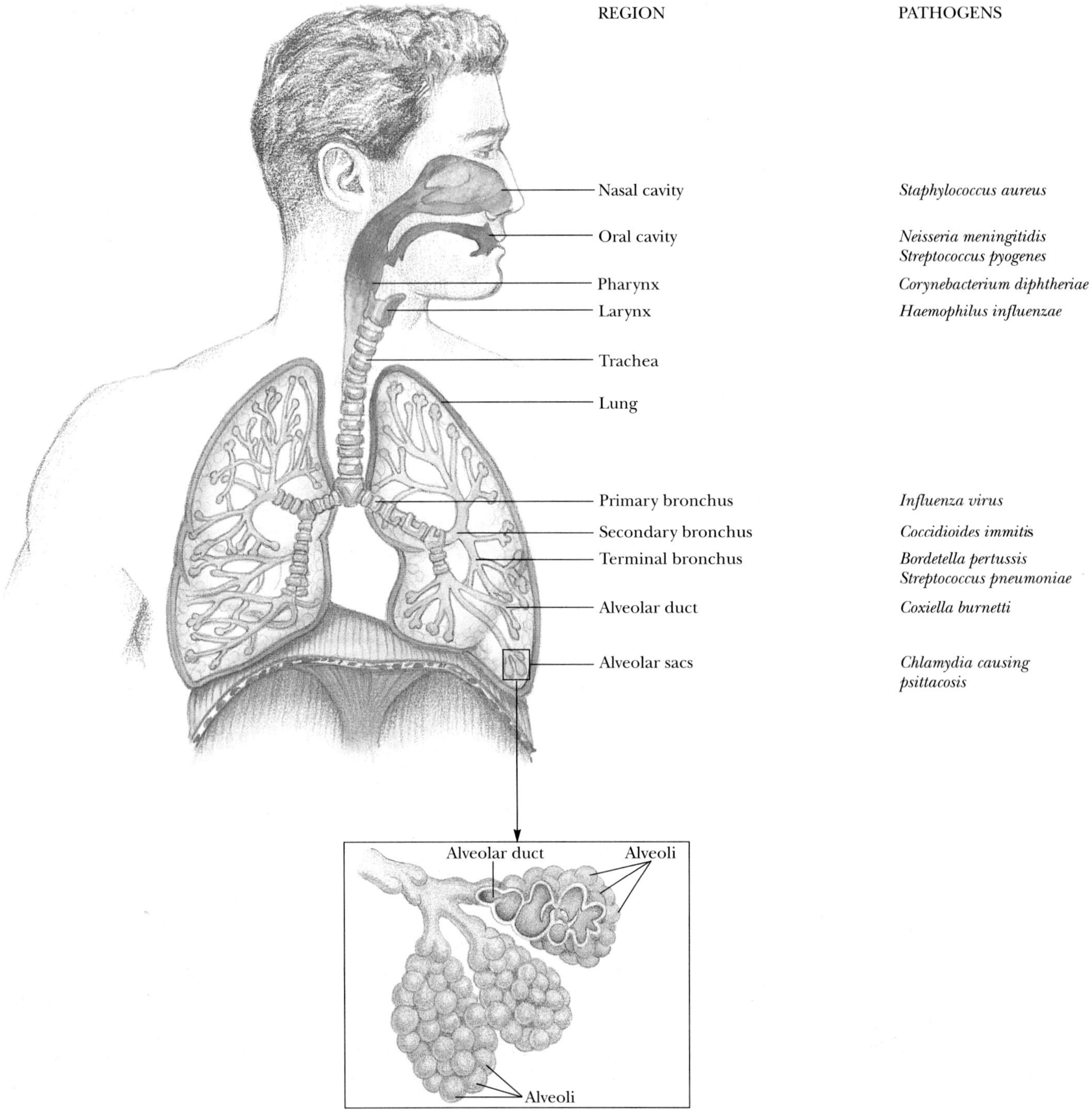

***Figure* 29.3** The human respiratory system and sites where various pathogens may localize and cause infections.

Diphtheria

Diphtheria is caused by *Corynebacterium diphtheriae,* a gram-positive, nonmotile bacillus that is pleomorphic and may appear club shaped. Diphtheria was once a leading cause of death in children, but immunization has reduced the number of cases in the United States to fewer than five per year. Diphtheria is still a concern in economically deprived urban areas of the world.

C. diphtheriae enters the body through the respiratory route and adheres to tissue in the throat. Lysogenic strains excrete an exotoxin (cytotoxin), and inflammation is the immediate response. The cytotoxin kills host cells, and these are intermixed with leukocytes and bacteria form-

Table **29.3 Diseases transmitted predominantly via the respiratory tract and the organisms involved**

Disease	Organism
Diphtheria	*Corynebacterium diphtheriae*
Whooping cough	*Bordetella pertussis*
Pneumonia	*Streptococcus pneumoniae* *Staphylococcus aureus* *Legionella pneumophila* *Mycoplasma pneumoniae* *Klebsiella pneumoniae* *Pneumocystis carinii*
Streptococcal infections	*Streptococcus pyogenes*
Meningitis	*Haemophilus influenzae* *Neisseria meningiditis* *Streptococcus pneumoniae*
Tuberculosis	*Mycobacterium tuberculosis*
Psittacosis	*Chlamydia psittaci*
Leprosy	*Mycobacterium leprae*
Coccidioidomycosis	*Coccidioides immitis*
Aspergillosis	*Aspergillus fumigatus* *Aspergillus flavus*

ing a dull gray pseudomembrane on the throat mucosa (Figure 29.4). The disease is diagnosed by presence of this membranous material and by culturing of the microorganism. The pseudomembrane can extend over the tracheal opening, blocking the passage of air into the lungs and can result in suffocation and death. Other symptoms of diphtheria are caused by transport of the cytotoxin to other areas, resulting in lesions in the kidney and heart.

The diphtheria toxin has been studied extensively, and its mode of action is quite well understood. It was the first extracellular exotoxin described. Two French scientists, Pierre Roux and Alexandre Yersin, discovered that culture filtrates of the bacterium *C. diphtheriae* were lethal to test animals. This affirmed that the lethality in diphtheria infections was due to toxic material released by the bacterium.

The toxin is produced by strains of the bacterium lysogenized by bacteriophage B. Toxin is produced at low iron concentrations because the repressor that prevents expression of the *tox*$^+$ gene in phage is iron containing. If the iron concentration is too low or unavailable, the repressor is not formed and the *tox*$^+$ gene is transcribed. Diphtheria toxin has enzymatic activity that cleaves nicotinamide from NAD and catalyzes the attachment of the

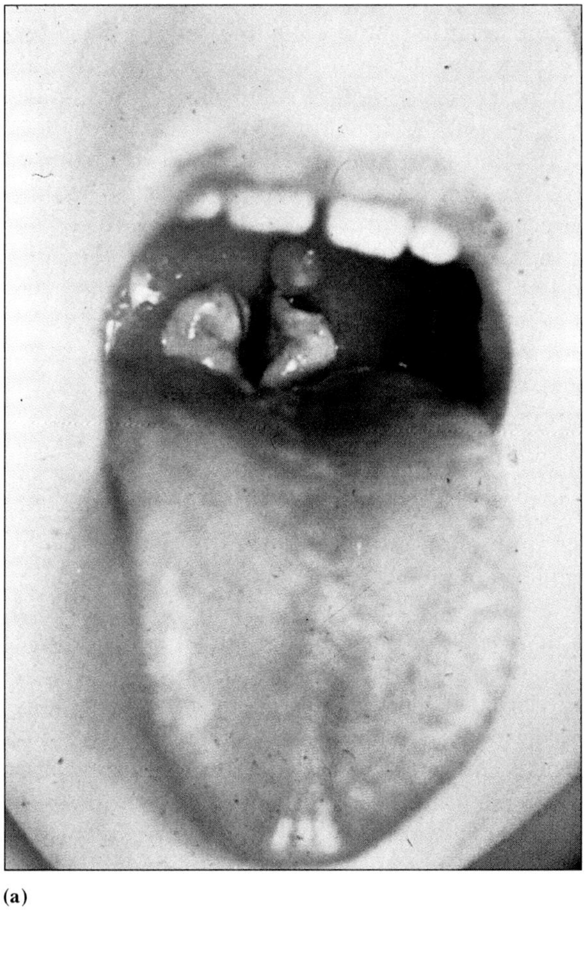

(a)

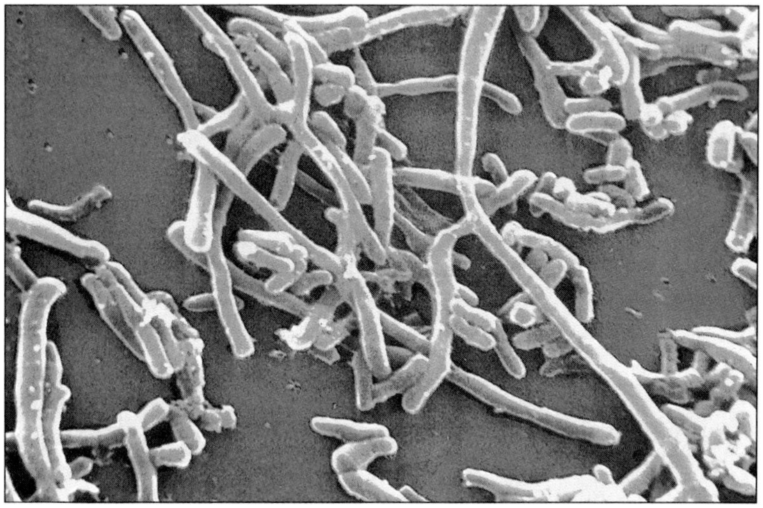

(b)

***Figure* 29.4** Diphtheria results from attachment and colonization of the oropharynx by *Corynebacterium diphtheriae.* **(a)** The grayish pseudomembrane that develops in the throat (© Courtesy of Centers for Disease Control), and **(b)** stained cells of the diphtheriod. (© CNRI/Science Photo Library)

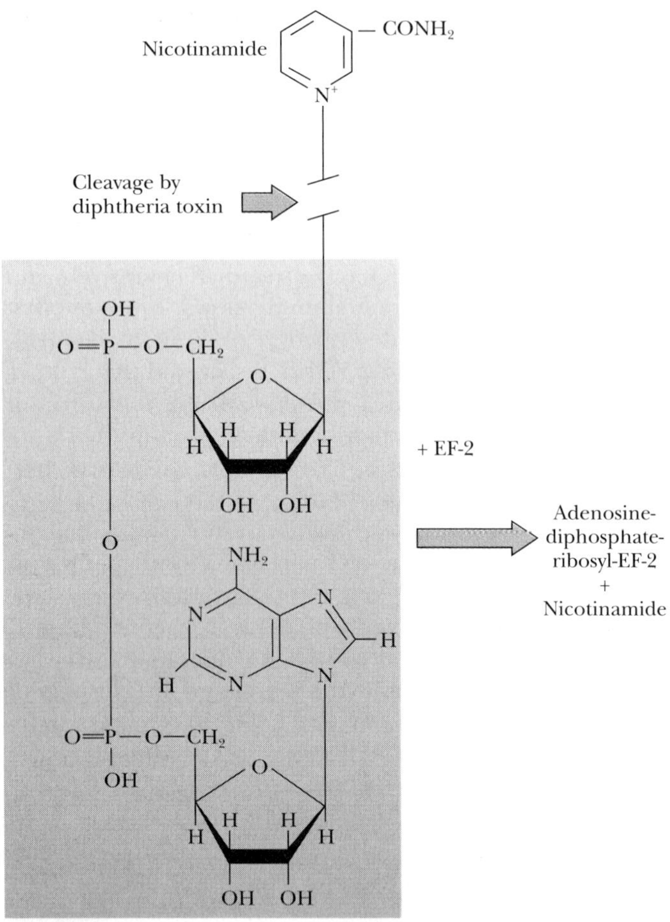

***Figure* 29.5** The mode of action of the diphtheria toxin. The toxin cleaves the bond in NAD and covalently links the adenosine-diphosphate-ribosyl portion to the elongation factor (EF-2).

resultant ADP-ribose to elongation factor-2 (Figure 29.5). Elongation factor-2 is normally involved with the growth of the polypeptide chain in protein synthesis. The addition of the ADP-ribose to EF-2 completely inactivates the factor and halts protein synthesis. Diphtheria toxin is relatively potent since a few molecules will kill a single host cell. It is interesting to note that *Pseudomonas aeruginosa,* a common organism in water, soil, and food, secretes a functionally similar toxin. *P. aeruginosa* is not a major health problem as it does not have the virulence factors necessary to sustain an infection, but it is a concern in burn patients.

Antibiotics, such as penicillin, are effective against *C. diphtheriae,* but the antitoxin should also be given, as the antibiotic does not prevent the action of the cytotoxin if it has been produced.

Whooping Cough

Prior to the introduction of an effective vaccine, **whooping cough** afflicted over 95 percent of children in the United States and resulted in about 4,000 deaths per year. The etiologic agent is *Bordetella pertussis,* a fragile gram-negative aerobic coccobacillus, that is nonmotile and often encapsulated. There are now fewer than 2,000 cases per year in the United States and fewer than ten deaths. However, worldwide, whooping cough is responsible for the deaths of over 500,000 young children each year.

Whooping cough is highly contagious in young children and is communicated by respiratory discharges from infected individuals. The incubation period is 7 to 14 days. The organism binds to ciliated epithelial cells of the bronchi and trachea. The adhesion that binds to cilia is termed the filamentous hemagglutin because the purified adhesion will agglutinate erythrocytes. After binding to cilia, *B. pertussis* produces several virulence factors, including a pertussis toxin, an extracytoplasmic adenyl cyclase, the filamentous hemagglutin, and a tracheal cytotoxin. The pertussis toxin increases tissue sensitivity to histamine and serotonin that are related to increased levels of adenyl cyclase. An increased level of cyclic AMP induced by the adenyl cyclase inhibits the action of phagocytic cells. The tracheal cytotoxin is a *B. pertussis* cell wall peptidoglycan precursor. The cytotoxin destroys the ciliated respiratory epithelial cells, and the destruction of these cilia in bronchi brings about an accumulation of mucus, bacteria, and host cell debris in the lungs.

Infection with *B. pertussis* begins with a catarrhal stage, which involves the inflammation of a mucous membrane, sneezing, and relatively mild coughing. After one to two weeks, the paroxysmal stage is reached and is marked by violent coughing. These violent coughs are the body's effort to rid the lungs of accumulated debris. The violent cough is followed by a "whoop," the sound of incoming air; hence, the name. Coughing can be so violent that the victim suffers **cyanosis** (turns blue from O_2 insufficiency), vomiting, and convulsions. The catarrhal and coughing stage can last six weeks, with the convalescent period lasting even longer.

Whooping cough is diagnosed by plating throat swabs on Bordet-Gengou agar and staining the resultant growth with a fluorescent antibody. Colonies of *B. pertussis* have a typical identifiable appearance on Bordet-Gengou agar. Smears may also be stained directly with the fluorescent antibody. The disease can be treated with erythromycin and other antibiotics.

Vaccination of infants is recommended at two months of age, as the disease can occur in infants and has a high mortality in infants less than one year old. A killed cell vaccine is employed in immunization against pertussis, and the vaccine itself can cause severe systemic toxic reactions. In the last ten years, there have been about fifty cases of brain damage and ten to twenty deaths due to reactions to the vaccine. The legal liabilities associated with these unfortunate reactions resulted in the curtailment of vaccine production. Congress has passed a no-fault compensation bill to provide monetary rewards in cases where brain damage occurs. This compensation comes from a surcharge on pertussis vaccine. Vaccines that are composed of purified component parts of organisms are under trial. These are made from formalin-treated pertussis toxin and the adhesion hemagglutinin. The first vaccines were heat-killed, whole-cell preparations that contained adenyl cyclase, pertussis toxin, the hemagglutin, and structural components of the cell. Some of these components, while not essential for an immune response, may have caused the serious reactions to the vaccine.

Pneumonia

Pneumonia is an inflammatory reaction in the alveolar region of the lungs. The infection can be caused by viruses, mycoplasma, Eubacteria, or fungi. The most frequent infectious agent is *Streptococcus pneumoniae. S. pneumoniae* is the etiologic agent of around 70 percent of pneumonia cases in the United States. Formerly called *Diplococcus pneumoniae, S. pneumoniae* is a gram-positive coccus that grows in pairs (Figure 29.6). Cases of pneumonia generally result from strains that are present as normal host flora. Most often pneumonia is a secondary infection that follows a viral infection, a respiratory tract injury, debilitating injuries, or chronic illness. There are over 300,000 pneumonia cases per year in the United States, with a death rate between 10 percent and 20 percent.

S. pneumoniae has low invasive ability, but does produce a cytolytic toxin that binds to cell membranes. The actual function of this toxin in pathogenesis is not known, and the capsule is considered to be more important in the development of the disease. The capsule protects the organism from phagocytosis. Organisms without capsules are readily removed by phagocytes. The bacterium grows rapidly in alveolar spaces, and the alveoli become filled with blood, bacteria, and phagocytic cells. Acute lung inflammation is a consequence of this fluid buildup. Symptoms of pneumonia are a sudden onset of chills, labored breathing, and pleural pain. Untreated, the bacterium may escape from the lungs and cause a bacteremia (presence in blood) which, if carried to the middle ear and sinuses, can cause serious infections. The organism can travel to the heart and cause endocarditis (inflammation of the heart valves) or pericarditis (inflammation of the pericardium, a membrane surrounding the heart), both of which can have long-lasting effects on health. Penicillin is effective against *S. pneumoniae,* but resistant strains have evolved, and these cases are treated with antibiotics such as erythromycin. Diagnosis of pneumonia is by chest X-ray or culturing of sputum (expectorated matter).

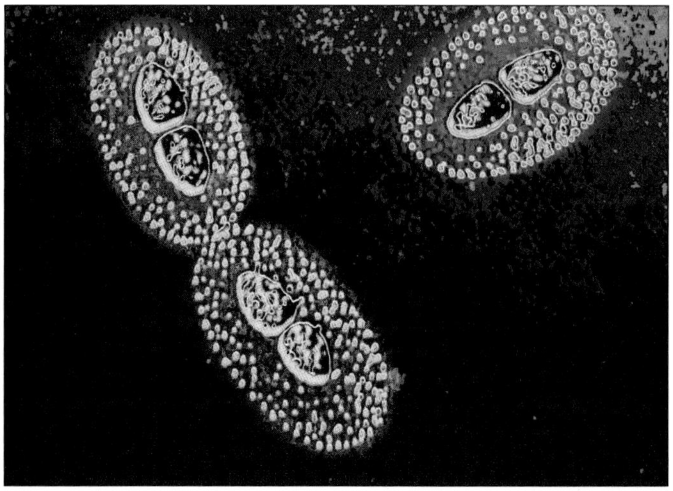

***Figure* 29.6** The diplococcus arrangement of *Streptococcus pneumoniae.* The virulent stains of the bacterium are surrounded by a thick polysaccharide capsule. (© Alfred Pasieka/Peter Arnold, Inc.)

S. aureus pneumonia may occur as a secondary infection following an initial infection with the influenza virus. Staphylococcal pneumonia is a serious infection as this organism has considerable invasive ability. The symptoms are similar to those of *S. pneumoniae.* Individuals with cystic fibrosis are quite susceptible to staphylococcal pneumonia. Treatment would be with penicillin derivatives or other antibiotics that are effective against gram-positive organisms.

Another pneumonia that is of relatively recent description is **Legionellosis (Legionnaire's disease),** so named because fatal cases of the disease occurred in attendees at a 1976 American Legion Convention in Philadelphia. At that time, the disease could not be ascribed to any known organism. Since then, *Legionella pneumophila* has been identified as the causative organism. Isolation and characterization have confirmed that *L. pneumophila* is unrelated to any of the other organisms that cause respiratory disease. It is a gram-negative pleomorphic rod that stains poorly and does not grow on media employed in diagnosing pneumonia caused by staphylococci and streptococci. Hence, earlier isolation protocols did not recover *L. pneumophila,* and the disease may have erroneously been attributed to a virus.

Legionellosis is spread by inhaling mist from air conditioning cooling towers, breathing contaminated soil, or being in contact with water from areas where epidemics have occurred. There is no evidence that the organism is spread by human contact. Despite the fact that the organism has complex nutritional requirements, including a need for high iron availability, it apparently survives in cooling water towers, on vegetable misters, and on shower heads. *L. pneumophila* can infect amoebae such as *Hartmanella verminoformis,* and this may be a source of virulent organisms in nature.

Legionellosis occurs most often in elderly and otherwise debilitated individuals. Symptoms include fever, headache, chest pains, muscle aches, and bronchopneumonia. The organism produces a hemolysin, an endotoxin, and a cytotoxin, but the exact role of these invasive agents is unknown. The organism can survive and multiply in the phagosomes of macrophages, and it can cause tissue destruction by the production of a cytotoxic exoprotease. *L. pneumophila* has been implicated in skin abcesses and inflammation of internal organs such as the heart and kidneys. Diagnosis is generally by fluorescent antibody or by the presence of antibody in the serum of the host. Treatment is with erythromycin or rifampin.

Mycoplasma pneumoniae is the etiological agent for a disease called primary atypical pneumonia. It spreads from person to person among school children, those in military service, and others living in close quarters. The symptoms include chills, fever, and a general malaise, but these symptoms are often mild and virtually unnoticed. The mild nature of the disease is the reason for calling it atypical. Diagnosis is by chest X-ray. *M. pneumoniae* can be isolated from the sputum of patients and identified by staining with fluorescent antibodies. Treatment is with tetracycline, erythromycin, or rifampin.

Pneumonia can be caused by *Klebsiella pneumoniae,* a bacterium that is present as part of the respiratory tract microbiota in about 5 percent of the population. It is a gram-negative nonmotile rod. The organism can cause severe pneumonia that may result in chronic ulcerative lesions in the lungs. Fewer than 5 percent of pneumonia cases can be ascribed to *K. pneumoniae.* The infection can be treated with gentamycin.

Another form of pneumonia is caused by a fungus that was previously considered a protozoan. The organism is *Pneumocystis carinii* and occurs in a wide variety of animals, including humans. Serological examination suggests that up to 90 percent of healthy children have had a mild infection with this organism. It can produce a severe pneumonia in undernourished infants. In recent years, *P. carinii* has been identified frequently in cases of fatal pneumonia in immunocompromised individuals. *P. carinii* occurs in over three fourths of AIDS patients and results in a fatal infection in about one half of those infected. It is diagnosed by fluorescent antibody staining. Treatment is with sulfamethoxazole and trimethoprim given intravenously. Recovery rate in immunocompromised individuals is in the 50 percent range.

Streptococcal Infections

Streptococcus pyogenes is the etiologic agent for diseases commonly known as **streptococcal infections.** It is a gram-positive coccus that divides in one plane, and the individual cocci tend to remain together and form long chains (Figure 29.7). The organism is the causative agent of skin infections (impetigo) as described previously. *S. pyogenes* also is involved with other severe infections that are of considerable consequence to human hosts.

The streptococci are **aerotolerant fermenters,** so called because most grow in the presence of air but obtain energy by a lactic acid fermentation of sugar. Generally, systemic infections by *S. pyogenes* are the result of a weakened host defense. Pathogenic strains of *S. pyogenes* secrete an active hemolysin that can lyse erythrocytes. The streaking of some strains of streptococci on blood-agar plates results in a greenish discoloration around colonies, indicating that there has been an incomplete destruction of red blood cells (α-hemolysis). The β-hemolytic streptococci effect a complete lysis of red blood cells with a clear zone around the colony.

The major human pathogens are β-hemolytic and have an invasin (virulence factor that promotes invasion) called the M protein at the cell surface. The M protein is antiphagocytic and can maintain an infection in the absence of adequate treatment or specific antibody.

Many strains of *S. pyogenes* are surrounded by a capsule composed of hyaluronic acid. As hyaluronic acid is a

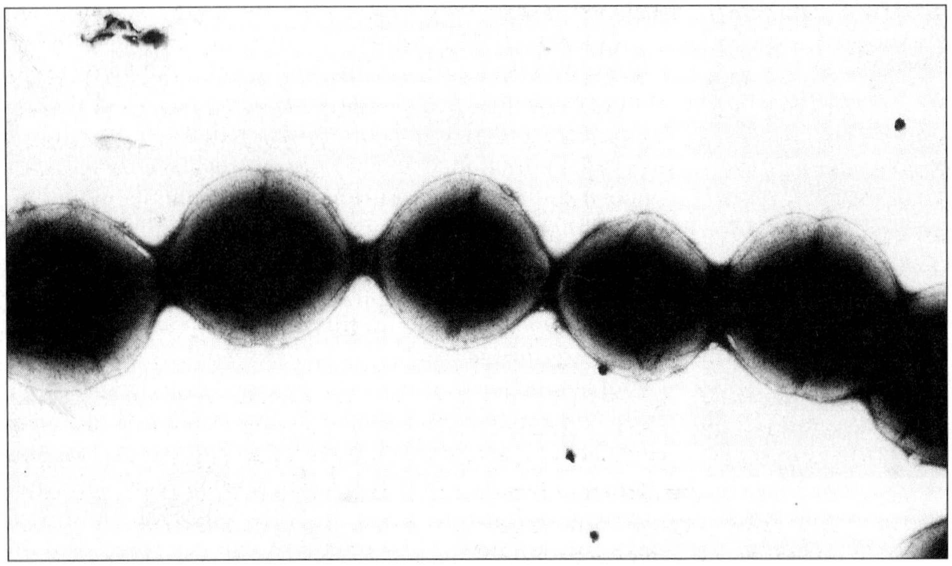

***Figure* 29.7** A long chain of *Streptococcus pyogenes* results from division and adherence of the cells at the point of division. (© Biological Photo Service)

normal host cell constituent, this gives the bacterium protection from phagocytosis. Other invasive molecules produced by streptococci were discussed in Chapter 26.

S. pyogenes is the causative agent of several familiar acute infections, including streptococcal pharyngitis, scarlet fever, rheumatic fever, and glomerulonephritis. **Streptococcal pharyngitis** is the most common infection caused by *S. pyogenes.* It can be an acute infection, with reddened tonsils and with a swollen pharynx and the production of purulent exudate. Often the local lymph nodes become enlarged and there is a high fever. The infection is commonly known as "strep throat" and may spread to the middle ear, where an acute infection can cause a loss of hearing. On rare occasions, the infection breaks through the throat epithelia and is carried by the bloodstream to the meninges, causing meningitis. Epidemics of strep throat are the result of personal contact with infected persons or carriers. Control with antibiotics is essential as streptococcal sore throat can be followed by scarlet or rheumatic fever and glomerulonephritis.

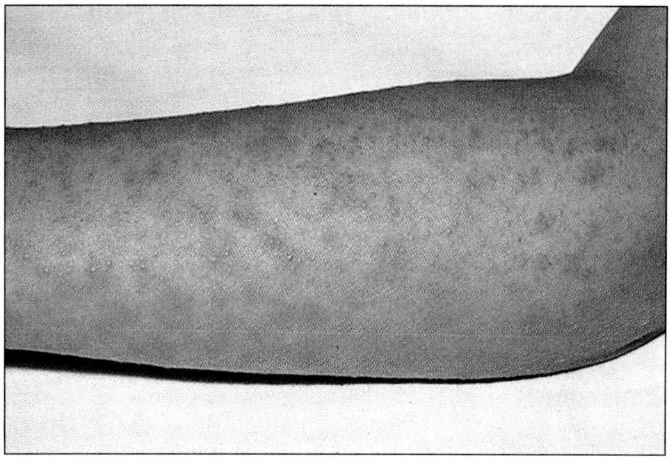

***Figure* 29.8** The rash associated with scarlet fever. The erythrogenic toxin produced by *Streptococcus pyogenes* causes a hypersensitivity that results in the pink-red rash. (© Science Photo Library)

Scarlet fever is caused by lysogenic strains of *S. pyogenes.* The phage carries the information for the production of a pyogenic exotoxin (erythrogenic toxin). The erythrogenic toxin can diffuse throughout the body and induce the skin rash that is typical of scarlet fever (Figure 29.8). The pyogenic toxin is a low molecular protein with two functional parts. One part is a heat labile portion that induces fever and suppresses the immune system. The other part is heat stable and increases the pyrogenicity and lethality of the toxin. The rash is a hypersensitive reaction (see Chapter 27) to the heat stable part of the molecule. Scarlet fever can be controlled with antibiotics. It is very important to halt the infection at an early stage for, untreated, the infection can lead to rheumatic fever.

Rheumatic fever follows a streptococcal sore throat caused by certain strains of *S. pyogenes.* This disease occurs in 3 percent to 5 percent of pharyngitis patients that fail to receive proper treatment. The availability of antibiotics has led to a marked decrease in the incidence of rheumatic fever in the Western world, although it remains a significant problem in developing countries. Rheumatic fever can result in permanent damage to the heart. There are two major theories that have been advanced to explain this heart damage. One is that antigens, possibly the M protein, get disseminated about the body and are deposited in joints and in heart tissue. When antibodies produced by the host interact with the antigens adhering to heart tissue, some damage occurs. Another theory holds that some strains of *S. pyogenes* have an antigen that induces antibodies that cross-react with a similar antigen in the heart, causing heart damage. The latter is similar to autoimmune reactions where a person's own antibodies react with his or her own tissue.

Glomerulonephritis, often a consequence of streptococcal infection, is an inflammation of glomeruli in the kidney. It may be an autoimmune response, as the streptococcal toxin and kidney share a common antigen. Antibodies to the toxin cross-react and harm kidney tissue. The incidence of this disease is low, and most patients undergo a slow healing of the damaged glomeruli. Others (under 10 percent) may have a chronic kidney disease.

Meningitis

Meningitis is an inflammation of the meninges. The meninges are any of three membranous layers that envelope the brain and spinal cord. These layers comprise the dura matter, located next to the bone; the middle arachnoid layer; and the pia matter, located next to the nerve tissue. The space between the arachnoid and the pia is filled with cerebrospinal fluid. When meningitis is suspected, the cerebrospinal fluid is tapped and examined for the presence of microorganisms.

A number of organisms involved in respiratory infections may be responsible for meningitis. These include *Haemophilus influenzae, Neisseria meningitidis,* or *Streptococcus pneumoniae,* all of which can be transmitted through the bloodstream to the meninges. It is estimated that 43 percent of the cases of meningitis are caused by *H. influenzae,* 27 percent by *N. meningitidis,* and 11 percent by *S. pneumoniae.*

Meningitis symptoms are sudden fever, stiffness in the neck, headaches, and often delirium, followed by convulsions and coma. *N. meningitidis* is called the meningococcus and is a gram-negative diplococcus. The initial infection resulting from *N. meningitidis* invasion is in the nasopharyngeal area. In a limited number of cases, however, the organism enters the bloodstream and causes a systemic infection. This can be a rapidly fatal stage of the disease as it may cross the blood-brain barrier and infect the meninges. Treatment of meningococcal infections i with penicillin. Erythromycin and chloramphenicol are also effective. Meningitis is such a severe, fatal disease that anyone in contact with a victim is generally treated with antibiotics as a precautionary measure.

H. influenzae meningitis is also an exceedingly dangerous disease, and survivors can suffer neurologic disorders. Ampicillin has been an effective therapy, but ampicillin-resistant strains of *H. influenzae* are increasing in number, requiring the use of chloramphenicol as an alternative. Meningitis caused by *S. pneumoniae* can also be treated with chloramphenicol.

Tuberculosis

The etiologic agent of **tuberculosis** is *Mycobacterium tuberculosis,* commonly referred to as the tubercle bacillus or TB. It was isolated in 1882 by Robert Koch, who confirmed the relationship between the organism and the disease. Historically and currently, tuberculosis is a devastating disease that continues as a leading cause of human suffering and death. Worldwide there are 10 million active new cases of tuberculosis each year and 3 million deaths. In the United States alone, there are over 20,000 active new cases per year and 12,000 deaths. Prior to the availability of effective antibiotics against *M. tuberculosis,* victims of the disease were practically bedridden, as rest was the only known treatment. Every state had special sanitaria where tuberculosis patients were isolated from the general population.

M. tuberculosis is a thin rod, sometimes bent and club-shaped. The organism is acid-fast, a staining property where cells stained with hot carbofuchsin are not decolorized by acid alcohol, and this can be used as a diagnostic test. *M. tuberculosis* is aerobic with relatively simple nutritional requirements. The *organism* has a high lipid content (over 40 percent cell dry weight), and the lipid contains unique fatty acids called mycolic acids (Figure 29.9). Mycolic acids vary somewhat in size and in amount of hydroxylation, and are frequently linked to sugars to form glycolipids called mycosides. A mycoside associated with virulent mycobacterial strains is depicted in Figure 29.10. Here two molecules of mycolic acid are attached to the disaccharide trehalose, and this compound is referred to as cord factor. Cord factor is believed to be responsible for the severe weight loss (cachexia) observed in tuberculosis patients.

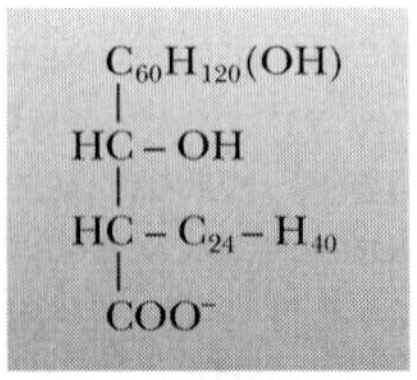

MYCOLIC ACID

ı The structure of a typical mycolic acid, a long-chain waxy fatty 'cid-fast microorganisms such as *Mycobacterium tuberculosis.*

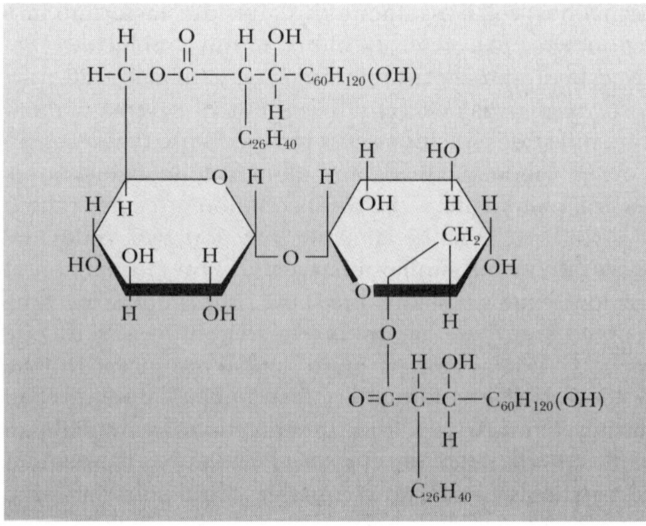

***Figure* 29.10** Structure of a **cord factor,** a mycoside that is present in virulent strains of *Mycobacterium tuberculosis.* The appearance of bacteria in colonies with this factor is a side by side cord.

Tuberculosis is transmitted respiratorily via small droplets introduced into the air by infected individuals. The organism is highly infectious, and limited epidemics can result from active infections in individuals who are in contact with large numbers of people. *M. tuberculosis* can also be transmitted by ingestion of contaminated food. Milk was once a major vehicle for transmission but was controlled when pasteurization conditions were established that killed the heat-sensitive bacterium. Inhalation of droplets containing the tubercle bacillus can lead to propagation within the lungs. In some people the disease is brief, and in others it is progressive, leading to a loss of virtually all functional lung tissue. The severity depends on the invader's virulence and the host's resistance.

Growth of *M. tuberculosis* in the lungs results in inflammation and the development of lesions. The bacteria reproduce in phagocytic cells and can be disseminated by lymph and blood to other body organs and tissue. As the bacteria propagate in lesions, they produce a mass consisting of bacteria, lymphocytes, and macrophages. The lesion becomes enmeshed in fibrous connective tissue forming a nodule called a tubercle. A tubercle contains viable bacteria that can be dormant indefinitely. During dormancy the infected individual is asymptomatic, and this would be considered a primary infection. The dormant primary infection may be activated by malnutrition, stress, hormonal imbalance, or decrease in immune function. Symptoms of the disease—fatigue, weight loss, and fever—are obvious only after many extensive lesions are formed.

The formation of tubercles is mostly a delayed type hypersensitivity reaction (see Chapter 27). Consequently, individuals in the primary stages of the disease are hypersensitive to protein fractions from *M. tuberculosis*. The **tuberculin test** given to children and to others during physical examinations is based on this hypersensitivity. A small amount of protein from *M. tuberculosis* is placed intradermally (beneath the skin) and observed for redness and swelling. Reddening after a few days (tuberculin positive) indicates that the individual has been exposed to the organism or has had an inapparent infection.

Diagnosis of tuberculosis is by the presence of *M. tuberculosis* in sputum. Presence of lesions results in the expectoration of bloody sputum laden with *M. tuberculosis*. X-ray examination of the chest is employed to reveal lung damage and tubercles (see Figure 29.11). Treatment formerly was with streptomycin, but this has been supplanted with a combination therapy of rifampin, isoniazid (INH), pyrazinamide (PZA), and ethambutol (EMB) or streptomycin. The regimen depends on the susceptibility of the infecting organisms to the drugs. The total duration of treatment is six months to ensure that all intracellular and intranodular organisms are destroyed.

In recent years, strains of *M. tuberculosis* have evolved that are resistant to antibiotics commonly employed to treat tuberculosis. These resistant strains are prevalent among AIDS patients and will ultimately spread to the general population. Poverty and inadequate efforts at eradication worldwide also contribute to the potential pool of resistant strains.

Leprosy

Leprosy is a dreaded disease that afflicts about 14 million people worldwide, mostly in tropical countries. A few hundred new cases of leprosy are diagnosed in the United States each year. The organism responsible for leprosy is *Mycobacterium leprae,* which is rod-shaped and acid-fast. *M. leprae* can be obtained from lesions but has not been grown in axenic culture. The organism will grow in the foot pads of mice and causes a systemic infection in the armadillo. Attempts to infect human volunteers have not been successful.

The epidemiology of leprosy is somewhat a mystery. Most humans apparently are not susceptible to infection, although it afflicts twice as many men as women. Children are more susceptible than adults. Transfer by human contact is possible, as individuals with lepromatous leprosy have over 10^8 *M. leprae* in their nasal discharge. This may be the mechanism whereby the organism is transmitted to susceptible individuals, but epidemics do not occur. There are two major forms of leprosy in humans: tuberculoid leprosy or lepromatous leprosy.

Tuberculoid leprosy is often a mild self-limiting disease. It is nonprogressive and occurs as a delayed hyper-

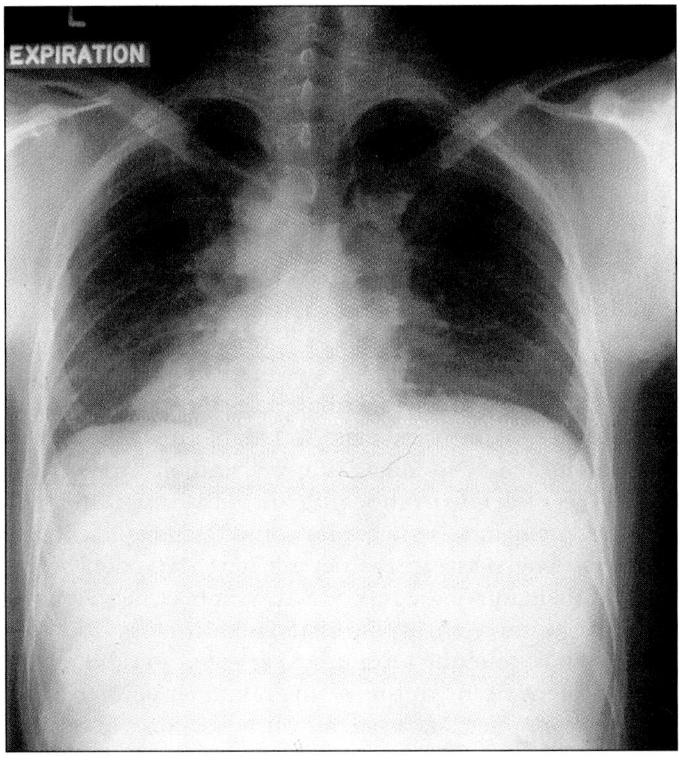

(a)

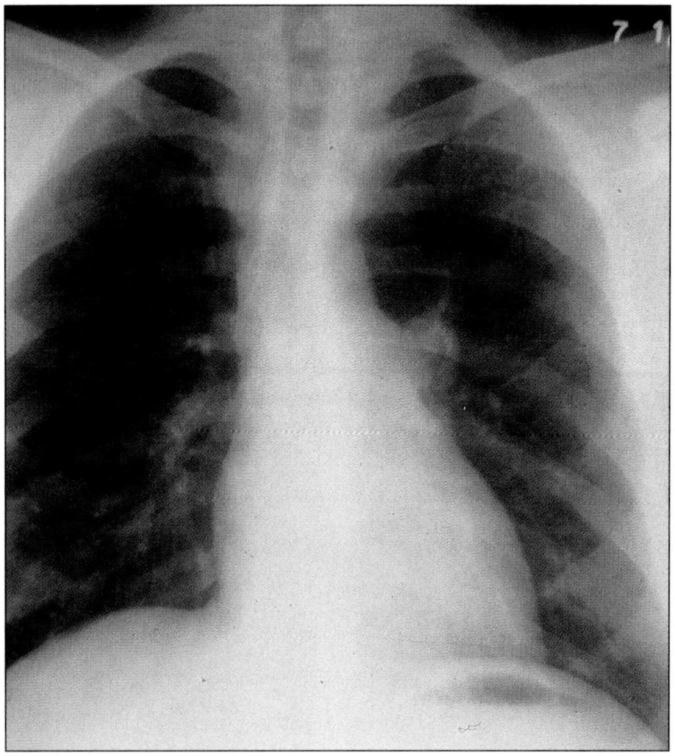

(b)

***Figure* 29.11** The chest X-ray on the left **(a)** is that of a normal lung (© 1994 SIU Biomed Comm.), the X-ray on the right **(b)** is an advanced case of pulmonary tuberculosis (© SUI/VU). The white patches in the diseased lungs are due to lesions caused by bacterial propagation.

sensitivity reaction to proteins in *M. leprae* or a normal cellular immune response. Damaged nerves and loss of sensation occur in areas where lesions exist. Rarely are intact organisms isolated from material taken from these lesions. Spontaneous recovery from tuberculoid leprosy frequently occurs.

Lepromatous leprosy, on the other hand, is a progressive, nasty disease that ultimately leads to death. *M. leprae* is found in virtually every organ of the body. Skin lesions are common, often pigmented, and organisms can be obtained for examination by skin scraping. Skin nodules infected with *M. leprae* appear about the body. Nerves are damaged, and mucosal lesions are abundant. Lesions in the mucous membranes of the nose lead to the destruction of cartilage and nasal deformities. The eye also can be infected, and this can lead to blindness. Loss of fingers or toes is also common (Figure 29.12).

Diagnosis of leprosy in the early stages is difficult. The occurrence of numbness concurrent with the isolation of the acid-fast bacterium are the major diagnostic features. It has been suggested that a material (lepromin) purified from infected armadillo tissue might be employed as a diagnostic tool. This would be analogous to the use of the tuberculin test for tuberculosis. In practice it has not been very effective.

Treatment of leprosy is difficult. The incubation period for tuberculoid leprosy is three to six years and for lepromatous three to ten years. Without a reliable diagnostic tool, treatment is delayed until the full onset of visible symptoms. Dapsone (4,4′-sulfonylbisbenzamine) has been given to patients but is quite toxic. A variant of the drug, diacetyl dapsone, is less toxic and appears to be more effective. Unfortunately, *M. leprae* resistant to the dapsones have emerged. Vaccines that might be employed in tropical areas where leprosy is a concern are under development.

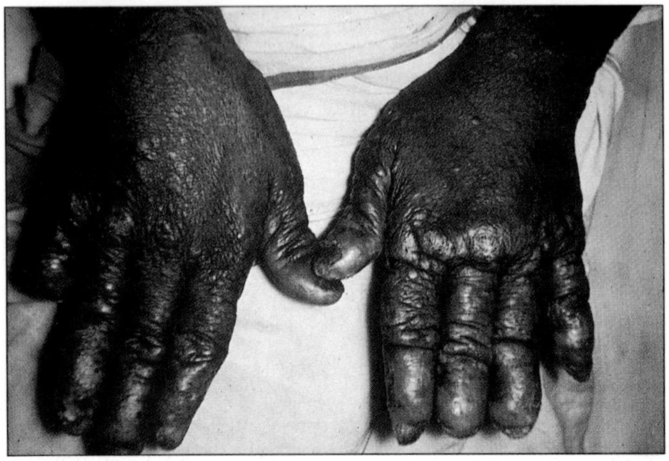

***Figure* 29.12** A case of human leprosy. The infection leads from a loss of sensation in skin to deformity and lesions that result in the loss of fingers and/or toes. (© Biological Photo Service)

Psittacosis (Ornithosis)

Psittacosis is an avian disease that afflicts many species of birds. The causative agent is *Chlamydia psittaci,* an obligate intracellular prokaryotic parasite related to gram-negative bacteria. The disease is generally latent and asymptomatic in birds. Crowding and unsanitary conditions, such as those encountered in transporting birds, can lead to active infections. The organism is present in virtually every organ of an infected bird, and considerable numbers are excreted in the feces. The consequent inhalation of dried feces by bird handlers can cause human infections. The inhaled organism travels to the liver, spleen, and lungs. Lungs become inflamed, hemorrhagic, and with symptoms of pneumonia. The pneumonia can be fatal.

A fatal meningitis can also occur in a limited number of psittacosis patients. The disease can be diagnosed both by isolation of *C. psittaci* from blood or sputum and serologically. Treatment with tetracycline is effective.

Fungal Diseases

There are a number of human respiratory diseases caused by fungi. In normal respiration, fungal spores are inhaled and in some cases these fungi can propagate in the lungs. Among these diseases are histoplasmosis, coccidioidomycosis, and aspergillosis. A brief description of some significant fungal respiratory diseases follows.

Histoplasmosis

Histoplasma capsulatum is the etiologic agent of **histoplasmosis,** the most common fungal disease in humans. An estimated 40 million Americans have been infected with histoplasmosis, and 250,000 new cases occur each year. *H. capsulatum* is widely distributed in soil and is generally associated with birds and bats. Abundant numbers are present in chicken coop litter, bat caves, and bird roosts. The disease is endemic in the Ohio and Mississippi River valleys. *H. capsulatum* is dimorphic, producing hyphae in soil and budding yeasts in the disease state. Infection occurs upon inhalation of the reproductive spores (microcondia) that are present on hyphal filaments. In most cases, the normal host immune response is sufficient to rid the body of the infectious organism. A small number of infected individuals may develop a systemic histoplasmosis of variable severity. The symptoms range from none to coughing, fever, and pain in joints. Lesions may appear in the lungs and become calcified. Generally the disease is resolved by host defenses but may become chronic, with bouts of lung infection occurring periodically. Diagnosis is by examination of sputum or by a test similar to the tuberculin test described earlier employing a fungal protein. The disease can be treated with amphotericin B or ketoconazole, but relapses may occur.

Coccidioidomycosis

The etiologic agent of **coccidioidomycosis** is *Coccidioides immitis,* a filamentous fungus present in arid soils. Most cases in the United States occur in the Southwest. Infection of the lungs occurs by inhalation of the reproductive arthrospores. Over half of primary infections are asymptomatic, and most of the remainder produce a mild to a severe pulmonary disease. Chronic or acute disseminated infections occur in fewer than 0.5 percent of those infected. Symptoms of the disease are a mild cough, fever, chest pains, and headache. Chronic cases occur where lung tissue may develop localized cavities filled with the spherules (cylindrical bodies) of *C. immitis.* Sputum from patients will generally have spherules of the organism. Amphotericin B is the treatment of choice for coccidioidomycosis.

Aspergillosis

Aspergillosis occurs in individuals frequently exposed to conidiospores of the filamentous fungi in the genus *Aspergillus.* The disease occurs more often in immunosuppressed individuals. Conidiospores are inhaled into the lungs and cause a delayed type hypersensitivity reaction that appears as a typical asthma attack. Inhaled conidiospores can form colonies in the lungs. These colonies may remain small and confined or can spread to other organs. In patients with severely limited immune function the mycelia can completely fill the lungs. Treatment of aspergillosis is difficult, as the fungus is resistant to most antifungal agents.

Candidiasis

Candida albicans is a part of the normal microbiota of the gastrointestinal tract, mouth, and vaginal area. The organism does not usually cause disease because the bacterial populations in these areas suppress the growth of *C. albicans.* In the absence of the normal microbiota the fungus can proliferate and cause disease involving the skin or mucous membranes. It has become an important nosocomial pathogen. Oral candidiasis (thrush) infects neonates born to mothers with *Candida* infections of the vagina. In cases of candidal vaginitis, the infection can be transmitted to the male during sexual intercourse. The resultant infection of the penis is termed balanitis. Thrush also occurs in immune compromised individuals. Systemic candidiasis can occur in AIDS patients. Individuals whose hands are immersed in water for long periods, such an cannery workers, often develop candidiasis infections of the hands or around the nails. Candidiasis is best treated by relieving the underlying conditions that result in infection.

Cryptococcosis

Cryptococcosis is a yeast infection caused by *Cryptococcus neoformans,* a common soil inhabitant. Most human infections result from inhalation of the organism on dried pigeon feces. *C. neoformans* does not infect pigeons but grows prolifically on the droppings. The primary infection is in the lungs, and most cases are asymptomatic or undiagnosed and cure is spontaneous. Crytococcosis is a serious problem in AIDS victims, causing infections in the brain and meninges. Cryptococci are present in the spinal fluid of 10 percent to 15 percent of AIDS patients. Cryptococcosis can be diagnosed with latex beads coated with rabbit antibodies to the polysaccharide capsule of the yeast. Presence of *C. neoformans* is indicated by aggregation of the beads. Cryptococcosis is fatal in immunocompromised individuals if untreated. Amphotericin and 5-fluorocytosine are the drugs of choice for treatment.

Bacterial Diseases of the Gastrointestinal Tract

Diseases of the gastrointestinal tract occur frequently and range from a mild upset stomach to fatal cases of botulism. Proper sanitation, clean water, refrigeration, and public health measures have been effective in lowering the incidence of these diseases in the Western world. Because many gastrointestinal diseases are spread by fecal contamination of food and water, they continue to be a serious concern in developing countries. Typhoid fever and cholera are endemic in many parts of Southeast Asia and South America due to a lack of proper sewage treatment.

The intestinal tract diseases are of two major types:

1. Food poisoning results from ingestion of food on which bacteria have grown. In such cases, bacteria release toxins that cause the disease symptoms. These toxins are then ingested along with the food.
2. Food-borne infections occur if one ingests food or drinks liquids that are contaminated with disease-causing bacteria (see **Box 29.3**). The bacteria propagate in the intestinal tract. In some cases, the infection occurs directly from hand to mouth after touching fecal-contaminated objects.

Food poisoning symptoms usually occur within a few hours after ingestion of contaminated food (Table 29.4). The duration is relatively short, and recovery can occur rapidly. The exception, of course, is botulism, which can be fatal. The incubation time for food-borne infections is

Table 29.4 Gastrointestinal diseases caused by food poisoning bacteria[1]

Organism	Time to Onset of Symptoms in Hours	Symptoms
Staphylococcus aureus	1–6	Nausea, vomiting, diarrhea
Clostridium botulinum	18–36	Dizziness, double vision, swallowing and breathing problems
Bacillus cereus	1–6	Vomiting
	10–12	Abdominal pain, diarrhea, nausea

[1]These organisms grow and produce toxins in the food prior to ingestion, except *B. cereus* may also grow for a short time and produce enterotoxin *in vivo.*

generally longer than for food poisoning cases. The duration of disease symptoms can also be considerably longer. The major route of transmission for food infections is directly or indirectly related to fecal contamination. Most of the organisms listed in Table 29.5 are inhabitants of the intestinal tract of various animals and do not survive in nature for any extended period of time. Consequently, they survive by passing from one animal to another through fecal contamination of food or water.

Recent evidence indicates that the presence of the bacterium *Helicobacter pylori* in the stomach may cause stomach ulcers. This section presents a discussion of ulcers and some of the major food poisoning and food infection agents.

Ulcers

Peptic ulcers occur in the stomach and duodenum of humans. These ulcers result from erosion of the highly alkaline mucus that forms a protective layer over the gastric epithelia. This mucus protects the epithelial cells from the acid and pepsin secretions that regularly enter the stomach. Medically, ulcers have been ascribed to several factors including genetic predisposition, excess secretion of acids and pepsin, and diet. It has long been the prevailing dogma that acidity and other factors render the stomach an inhospitable environment and that it is essentially free of bacterial colonization. In 1977, a microaerophilic bacterium, later named *Helicobacter pylori,* was discovered as a

Table 29.5 Food-borne and water-borne infections of the gastrointestinal tract

Organism[1]	Infection	Source of Infection
Salmonella typhimurium	Salmonellosis	Poultry, eggs, animals
Salmonella typhi	Typhoid fever	Water, food
Vibrio cholerae	Cholera	Water, food
Shigella dysenteriae	Shigellosis	Water, food
Campylobacter jejuni	Campylobacteriosis	Poultry, shellfish
Escherichia coli (enteropathogenic)	Travelers diarrhea	Uncooked vegetables, salads, water
Yersinia enterocolitica	Enterocolitis	Milk
Listeria monocytogenes	Listeriosis	Milk, cheese
Vibrio parahaemolyticus	Hypersecretion	Shellfish
Clostridium perfringens	Hypersecretion	Meats, gravy, stews
Clostridium difficile	Pseudomembranous colitis	Natural inhabitant of GI tract

[1]These organisms survive in the intestinal tract and produce toxins.

frequent colonizer of the human stomach. Studies conducted since 1985 have confirmed that this bacterium is a causative agent of gastric inflammation, peptic ulcers, and gastric cancer.

Infections with *H. pylori* occur throughout the world and are more prevalent in developing countries than those already developed. The organism synthesizes an adhesion that is involved in attachment to stomach epithelial cells and synthesizes a urease that generates ammonia that may protect against acidity. Many strains produce a cytotoxin that promotes infection. The organism can be eliminated from infected patients by combinations of tetracycline, metronidazole, and bismuth.

Food Poisoning

Staphylococcus aureus is considered the agent most frequently encountered in cases of **food poisoning.** *S. aureus* is present in the nasopharynx of 10 percent to 50 percent of adult populations with a higher incidence in the nasal passages of children. It is inevitable that through negligence, unsanitary practice, or accident some *S. aureus* cells will get into food during preparation. Two conditions must be met, however, before food poisoning can occur: the contaminated food must be suitable for growth and toxin production by the bacterium; and the food must stand at a temperature that will permit growth of the bacterium. For *S. aureus* this would be room temperature or warmer. *S. aureus* effectively produces enterotoxin during growth on such foods as ham or chicken salad, cream-filled pastry, custards, salad dressing, and mayonnaise. To prevent enterotoxin production, food should be refrigerated before preparation, kept cold during preparation, and refrigerated immediately afterward. Any foods containing creams or mayonnaise that are exposed to room temperature for any length of time should be discarded. It is foolhardy to take foods such as those listed above on a picnic and permit them to sit in the sun. Refrigeration is an absolute necessity.

Staphylococcal food poisoning generally occurs within six hours after ingestion of toxin-containing food. The toxin is heat and acid stable. Severe nausea, vomiting, and diarrhea are the symptoms, and they rarely last longer than 24 hours. But, food poisoning can be a serious problem for individuals weakened by other health problems. Treatment in severe cases is by administration of intravenous fluids to prevent dehydration. Antibiotic therapy is of no value, because the causative agent is the toxin.

Bacillus cereus is an endospore-forming, gram-positive bacillary organism that is a significant source of food poisoning. *B. cereus* is commonly present in soil and can be introduced into food on dust particles. Food poisoning by this organism occurs with various foods but particularly with rice. Rice is often cooked gently and left at room temperature. The heating induces endospores to germinate and, while sitting at room temperature, the organism grows extensively and produces enterotoxin. Ingestion of large numbers of bacteria may also result in growth of the *B. cereus* in the gut, and there it produces enterotoxin. This might also be considered a food infection. Creamy foods and meat have also been implicated in *B. cereus* food poisoning. Diagnosis is by symptoms and examination of suspected food. A count of 10^5 *B. cereus* cells per gram of food is a strong indicator that *this* organism is involved. *B. cereus* produces two different enterotoxins. Vomiting occurs in one to six hours after eating contaminated food and is caused by a cholera-like enterotoxin that stimulates adenyl cyclase. Abdominal pain and profuse diarrhea 4 to 16 hours after ingestion of contaminated food is indicative of yet another enterotoxin, but the action of the toxin is not clear.

Clostridium botulinum is the etiological agent of the dreaded disease, **botulism.** *C. botulinum* is widely distributed in soil, on lake bottoms, and in decaying vegetation. The endospores can contaminate vegetables, meat, and fish. Historically, botulism occurred following the ingestion of home-canned, low-acid vegetables such as peas, beans, corn, mushrooms, and meats. The endospores of *C. botulinum* are quite heat resistant and survive unless the processing is complete (120°C for 15 minutes). Smoked meats, salted fish, spiced meat, and other foods have been implicated in botulism. Botulinum produces a series of toxins that are among the most potent toxins known: one ml of culture fluid from a botulinum culture could kill 2 million mice. The lethal dose for a human is estimated to be in the range of 1 μg toxin. Some of the botulinum toxins are coded on the genes of lysogenic bacteriophages that attack only this organism. Botulinum toxin is a neurotoxin that attacks nerve cells, and this leads to paralysis. The toxin binds to a receptor on nerve synapses and enters the nerve cell. The release of acetylcholine from the nerve cell is blocked, muscles cannot contract, resulting in a flaccid paralysis (Figure 29.13). The toxin is quite heat labile and is destroyed by boiling foods for 10 minutes. The organism has no invasive ability and cannot grow and produce toxin in the adult intestinal tract. Botulism occurs infrequently in wounds. Infant botulism (occurring at 5 to 35 weeks of age) occurs by ingesting *C. botulinum* endospores that may be present in baby food. These endospores germinate and grow in the intestine of an infant.

There are an estimated 50 to 60 cases of botulism in the United States annually. The symptoms of botulism appear 18 to 36 hours after ingestion and include dizziness, weakness, double vision, difficulty in swallowing, and labored breathing. Death results from flaccid paralysis of respiratory muscles. Diagnosis is by noting symptoms and by testing for the presence of the toxin in the bloodstream. Antitoxins can be given to victims of botulinum poisoning to remove unbound toxin from the blood. Animals are also susceptible to botulism, and many die from eating fer-

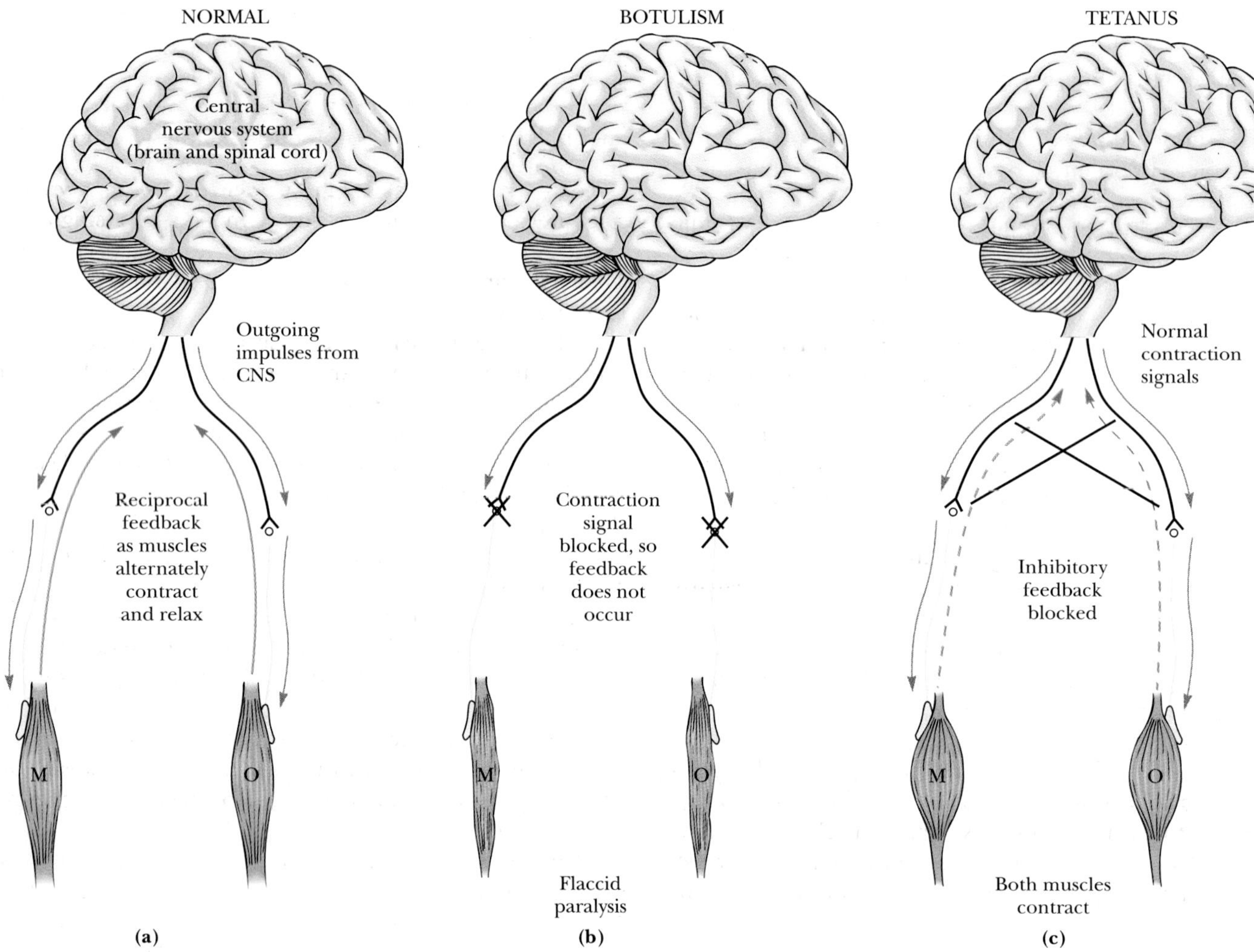

***Figure* 29.13** The mechanisms whereby botulism and tetanus toxin affect skeletal muscle activity. Skeletal muscles work in antagonistic pairs. Normally, movement results when one muscle contracts while its antagonist relaxes. Neurologically, when one nerve impulse from the central nervous system (CNS) signals muscle A to contract, another impulse blocks the neural message from the antagonist muscle B, which would otherwise prevent muscle A from contracting. Essentially, when muscle A contracts, muscle B relaxes, and vice versa. Botulism toxin blocks CNS commands for contraction; without contraction of its opposing partner, the antagonist also remains inactive and relaxed. The result is flaccid paralysis. By contrast, tetanus toxin blocks the reciprocal inhibitory action of muscle pairs, so both muscle A and muscle B contract resulting in painful, often excruciating, and sometimes fatal muscle cramping.

mented grain. Ducks that eat vegetation from the bottom of ponds often get botulism, resulting in neck muscle flaccidity. The disease in fowl is called "limber neck."

Tetanus is caused by an exotoxin produced by *Clostridium tetani.* The organism is present in soil and in the feces of many animals. It has no invasive ability and enters tissue through punctures, gunshot, or other wounds.

The tetanus toxin is composed of two peptide chains linked by a disulfide bond, and the larger of the chains has a specific receptor that binds to neuronal gangliosides. The tetanus toxin blocks the function of a nerve factor that normally allows muscles to relax. In some unknown way, tetanus toxin prevents the release of inhibitory transmitters (glycine and gamma aminobutyric acid) responsible for muscle relaxation. As a result, opposing muscles that permit movement by alternating contraction/relaxation are in a constant state of contraction (Figure 29.13). This results in a painful spastic paralysis. The spasms can involve muscles of the jaws and, consequently, the disease has been termed "lockjaw." The toxin can also cause spasms in the respiratory muscles, resulting in suffocation.

Food- and Water-Borne Infections

Salmonellosis is an infection fostered by microorganisms in the genus *Salmonella.* The species most commonly associated with the human infection is *S. typhimurium.* There are over 2000 serological types within this genus. These serological types differ by one or more cellular (O) or flagellar (H) antigens.

Various species of salmonellae are present in the intestinal tract of animals and birds. As a result, poultry, eggs, and beef can be contaminated with the organism. People who handle these foods, as well as asymptomatic carriers, are a source of food contamination. Hands contaminated by handling an infected animal such as a dog or cat can be a source of disease. Some years ago small pet turtles were available in variety stores and pet shops. These turtles were *Salmonella* carriers and responsible for an estimated 300,000 cases of human salmonellosis per year. Import and sale of these animals is now banned. Water polluted with human or animal waste is another source of the disease.

There are several million cases of salmonellosis in the United States each year. Most cases of salmonellosis follow ingestion of uncooked egg products such as custards, meringues, or eggnog. Undercooked meat, particularly chicken, is also a source of infection.

Enterocolitis is the most often observed manifestation of salmonellosis. The symptoms occur 10 to 30 hours after infection and include abdominal pain, nausea, vomiting, and diarrhea. The symptoms last from two to five days but may last longer. The organism multiplies in the intestine and produces an enterotoxin with cholera-like activity and a cytotoxin that destroys epithelial cells. Feces from a salmonellosis patient can contain one billion salmonellae per gram. Diagnosis is by isolation and identification of the infectious agent. Salmonellosis can induce a loss of fluid that may be a serious complication in children, the elderly, and debilitated individuals. Treatment is by replacement of lost electrolytes through intravenous fluids.

Another salmonella species, *Salmonella typhi,* is the etiological agent of **typhoid fever.** There are about 500 cases of typhoid fever in the United States each year and an estimated 2000 carriers. Typhoid fever epidemics have occurred throughout human history (see **Box 31.3**). The only reservoirs for *S. typhi* infections are humans, so carriers and improperly treated sewage are the sources of infection. *S. typhi* is passed from feces to hands to food, and public health measures have been established to eliminate food handlers from being carriers. Typhoid infections are caused by ingesting contaminated food, by putting contaminated objects in one's mouth, or by drinking contaminated water. The onset of symptoms occurs after a 10- to 14-day incubation period, and the disease is marked by headaches, abdominal pain, a high temperature, and rose-colored spots on the abdomen. *S. typhi* cells can penetrate the small intestine and spread to the lymphoid system. During the active stages of the disease, *S. typhi* is present in the blood and is disseminated to the spleen, liver, bone marrow, and gall bladder. Victims shed the bacterium for three months or more, while carriers often have an infected gall bladder that constantly sheds *S. typhi* into the bile duct and on into the intestines. Diagnosis is by isolation of the organism from blood, urine or feces, and through serological identification. Patients are treated with chloramphenicol, which must be continued for at least two weeks to clear the organism from the gall bladder.

The etiological agent for **cholera** is *Vibrio cholerae,* a gram-negative, comma-shaped organism with a single polar flagellum. The organism adheres to epithelia in the small and large intestine and secrets colleragen, the cholera enterotoxin.

Cholera is a fearful disease spread by fecal-oral transmission through water and food. It is inevitable that cholera will occur in areas that have breakdowns in sewage disposal systems or human waste pollutes water supplies (Figure 29.14). Cholera is endemic in India, Pakistan, Bangladesh, and other Asian countries where sewage disposal is inadequate. Serious outbreaks of cholera occurred in Peru and Brazil in recent years, killing over 10,000 people.

The incubation time from ingestion to disease symptoms is from a span of hours to three or more days and is dose dependent. One must ingest 10^8 to 10^9 organisms in water to acquire an infection because the stomach acidity can wipe out smaller doses. Ingestion of food contaminated with *V. cholerae* can result in the disease caused by fewer organisms, as food tends to protect the organism from stomach acidity.

Cholera has one major symptom, an acute diarrhea with excretion of eight to fifteen liters of liquid per day. This liquid contains vibrio cells, epithelial cells, and mucus. Untreated, 60 percent of the patients succumb to dehydration. Treatment by intravenous electrolyte and liquid replacement lowers the death rate to less than

***Figure* 29.14** Cholera is a widespread occurrence in Southeast Asia, as humans swim and bathe in rivers that are contaminated with human waste. (© Biological Photo Service)

1 percent. Cholera is diagnosed symptomatically and by the isolation of the organism from feces. Serology is employed to determine the strain of organism involved. Treatment with streptomycin or other antibiotics may shorten the course of infection. Cholera can be prevented by implementing adequate sewage treatment procedures and by purifying drinking water.

The cholera toxin consists of three or more polypeptide chains. There is an A_1 and A_2 chain with five B subunits arranged in a circular fashion. The B subunit(s) has one function and that is to bind to a specific ganglioside glycolipid on the epithelia of the ileum and large intestine. The A chains enter the cell, and the A subunit activates adenyl cyclase as follows:

$$\text{ATP} \xrightarrow{A_1} \text{cyclic AMP}$$

The increased levels of cyclic AMP effect a release of chloride ion and bicarbonate from mucosal cells that line the intestine. These inorganic ions accumulate in the lumen (inner space) of the intestine, and the normal passage of sodium ion into the mucosal cells is blocked. This creates an osmotic imbalance so that water passes through the intestinal epithelia into the lumen (Figure 29.15). This accumulation of water causes the severe diarrhea associated with the disease.

Shigella dysenteriae is another organism responsible for intestinal disease. It is the causative agent of **shigellosis,** a form of dysentery. Humans are the only known hosts for *S. dysenteriae.* The infectious organism is transmitted via the fecal-oral route. Food and water are the major routes of transmission, while direct feces-to-mouth transmission is probably responsible for the prevalence of this disease in preschool children. Most of the disease is found among children in day care centers. *S. dysenteriae* remains localized in intestinal epithelial cells and effects a loss of fluid and ulceration of the colon wall. The shigella toxin kills absorptive epithelial cells, and diarrhea results from this interference with liquid absorption. Isolation and identification of *S. dysenteriae* in feces is one means of diagnosis, but a direct swab of lesions in the colon and subsequent identification of the organism serologically is better. Treatment is by fluid replacement and with the antibacterials sulfamethoxazole and trimethoprim.

Campylobacter jejuni is the etiological agent for a major diarrheal disease termed **campylobacteriosis.** *C. jejuni* is a gram-negative curved cell with a single polar flagellum. It is microaerophilic. *C. jejuni* is an inhabitant of the intestinal tracts of wild and domestic animals. The organism can cause sterility and abortion in cattle and sheep. It is transmitted to humans via the fecal-oral route in contaminated food and water. Underprocessed chicken and raw milk are

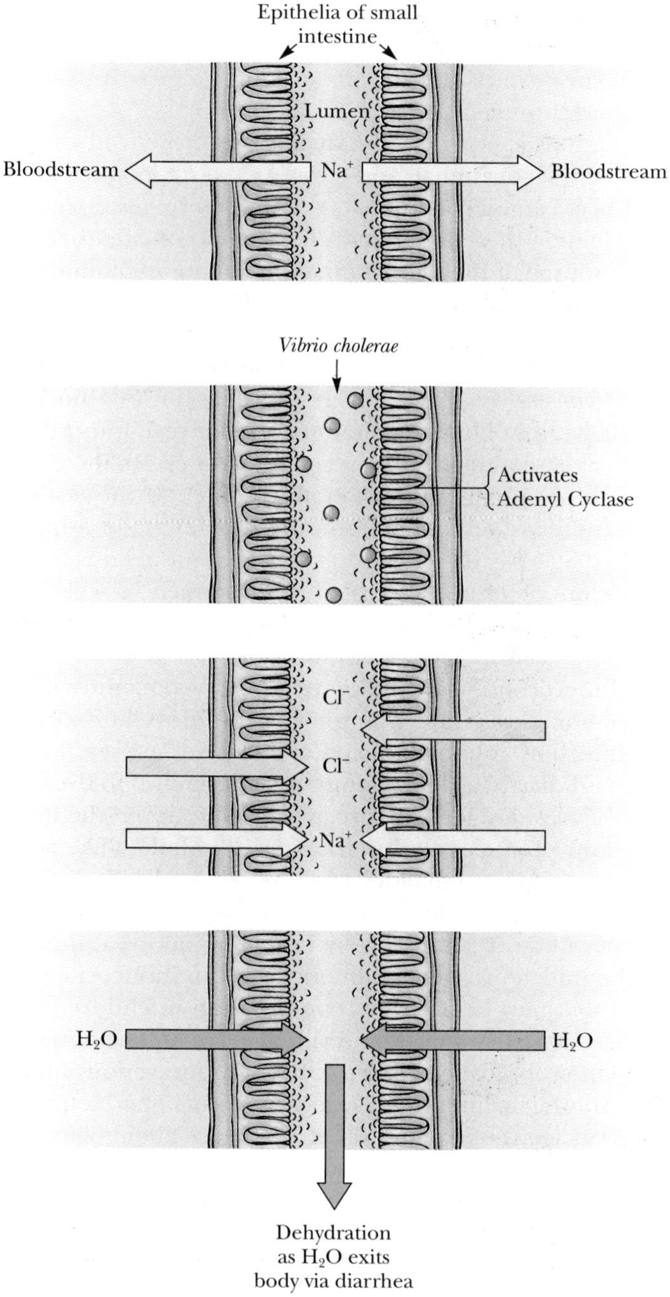

***Figure* 29.15** The mode of action of cholera toxin. The enterotoxin activates adenyl cyclase, resulting in increased cyclic AMP in the epithelial cells of the small intestine. This activates the movement of Cl^- ion into the lumen, which is followed by Na^+ ion. The osmotic imbalance from the increased NaCl draws water from tissue into the lumen, resulting in dehydration of tissue and diarrhea.

common sources of campylobacteriosis. Children can get the infection from domestic dogs. The incubation period after ingestion (fewer than ten organisms can cause infection) is two to four days. Symptoms include a high fever, nausea, abdominal cramps, and watery, bloody feces. The enterotoxin responsible for campylobacteriosis is similar in activity to the cholera toxin. Another toxin produced

BOX 29.1 MILESTONES

A Danger Lurking

There were hundreds of food poisoning cases in the northwestern United States during the winter of 1992–93. The public became acutely aware of the problem when four people, including three small children, died. The infections that led to the fatalities were traced to the consumption of undercooked hamburgers at Jack-in-the-Box restaurants. The organism involved was a strain of *Escherichia coli* designated O157:H7. Since 1982, at least 15 food poisoning outbreaks have been attributed to this organism, and generally the food involved was hamburger.

E. coli O157:H7 is present in the intestines of cattle and may contaminate meat during slaughter. Surface organisms in meat are thoroughly inoculated during grinding. Strain O157:H7 can grow slowly at temperatures as low as 44°F (7°C), and unless meat is cooked quickly, considerable growth could occur after the grinding process. If the interior of a hamburger patty does not reach 155–160°F (68–71°C) during cooking, the infectious pathogen can survive. Ingestion may lead to infection. The symptoms of O157:H7 infection are a severe hemorrhagic colitis, abdominal cramps, and bloody diarrhea. In 10 percent of the cases blood clots are formed in the kidneys, there is destruction of erythrocytes, and there is the possibility of kidney failure. The elderly and the very young are the most likely victims.

There are 6.5 million food-related illnesses per year in the United States and an estimated 7000 deaths. Illness from *E. coli* O157:H7 is probably underdiagnosed, and perhaps 20,000 cases per year do occur. Over the years our animal-raising practices and food handling have changed considerably. Regulations for inspecting fresh meat, poultry, and seafood are mostly sensory evaluations; however, smell, consistency, and appearance can detect only flagrant contamination.

The Jack-in-the-Box restaurants followed cooking protocols established years ago. Should we be more diligent and current in our regulations related to food handling? The fatalities in the Northwest indicate that we should.

by *C. jejuni* is a cytotoxin (toxic to cells) that causes colitis. Diagnosis is by isolation of the organism from feces and serological identification. The disease can be treated with erythromycin.

Escherichia coli is a normal inhabitant of the human digestive tract. There are strains of *E. coli* that cause diarrheal disease, particularly evident in individuals traveling between countries—hence the name "travelers diarrhea" (as well as many other names). These pathogenic strains have virulence factors, carried by plasmids. The K antigen is one of the surface antigens that permit attachment and invasion of mucosal cells. Pathogenic *E. coli* strains attach to the intestinal mucosa by pili or fimbriae and produce an enterotoxin. In 1993, an outbreak of a food-borne *E. coli* infection killed four people in the Seattle area and alerted public health officials to the dangers of *E. coli* O157:H7. This pathogenic strain can be transmitted in meat products, particularly ground beef (see **Box 29.1**).

Yersinia enterocolitica is the etiologic agent of an **enterocolitis** in humans. The organism is present in the intestinal tract of cats, dogs, rodents, and domestic farm animals. It can also be isolated from lake and well water. As with many of the previously mentioned enteric diseases, enterocolitis is a fecal-oral transmitted infection. *Y. enterocolitica* adheres to and invades intestinal epithelial cells. The symptoms include fever, diarrhea, and abdominal pain. A form of arthritis can occur several days after onset of acute enteritis. Antibiotics such as gentamicin may shorten the duration of the infection.

Listeria monocytogenes causes an infection called **listeriosis.** *L. monocytogenes* is a small, gram-positive diphtheroid organism. The organism is unusual in that it has a complex lipopolysaccharide on the cell surface similar to the toxins associated with gram-negative bacteria. *L. monocytogenes* is mainly an animal pathogen and is transmitted to humans through dairy products. Milk that has been improperly pasteurized and cheese manufactured from such milk are the major causes of infection. *L. monocytogenes* infects pregnant women, and the organism can pass through the placenta to the fetus. In the newborn, the disease is generally pneumonia at birth or shortly thereafter. About two to four weeks after birth, the listeriosis infection appears as meningitis. This disease occurs most often in immunosuppressed individuals (such as AIDS patients) than in otherwise healthy people. Listeriosis can be treated with penicillin or tetracycline.

Vibrio parahaemolyticus is a marine bacterium that grows only in the presence of moderate salt concentrations. It is the etiologic agent of a food poisoning that follows ingestion of uncooked seafood.

Clostridium perfringens is an endospore-forming anaerobic bacterium. Ingestion of food contaminated with 10^8 or more *C. perfringens* cells per gram will cause abdominal pain and diarrhea. When vegetative cells are present in the small intestine they sporulate and release an enterotoxin. The mode of action of *C. perfringens* enterotoxin is similar to that of the cholera toxin. This microorganism can grow rapidly in warm beef, poultry, gravies, or stew, and outbreaks attributed to *C. perfringens* result from ingestion of one of these food products.

Clostridium difficile is an anaerobic microorganism that forms endospores. *C. difficile* is present in the intestinal tract of some adults, but the number of organisms is quite low and it causes no apparent infection. If antibiotics such as clindamycin or ampicillin are administered to an individual in an amount sufficient to destroy the normal flora, *C. difficile* can be a problem. Overgrowth by the organism can occur, and a severe necrotizing (death of tissue) process is initiated. The infection is termed pseudomembranous colitis. It is interesting to note that 50 percent to 75 percent of neonates are colonized by *C. difficile* but are completely asymptomatic, but can serve as a reservoir.

Urinary Tract Infections

Infections of the urinary tract are second only to respiratory infections in number of cases annually. Urinary tract infections are by far the prevalent nosocomial infection, causing about 40 percent of all hospital-acquired infections. These infections can occur from the kidneys *(pyelonephritis),* to the bladder *(cystitis),* to an inflammation of the urethra *(urethritis).* Infection of the kidneys generally will lead to infection of the entire urinary tract, but more often infections spread by ascending from the urethra. Proper treatment of urethritis with antibiotics (amoxicillin or trimethoprim) will prevent the spread of the infectious agent to the bladder and kidneys. As the urethra in females is much shorter than that in males, the incidence of urinary tract infections in females is much higher.

The organism most often associated with urinary tract infections is *Escherichia coli,* the causative agent in 80 percent of the cases. Other organisms involved are *Proteus mirabilis, Pseudomonas aeruginosa,* and other facultatively aerobic gram-negative rods and cocci. Usually a single pathogen is involved in an infection, and diagnosis is by determining the number of organisms present in urine. The bladder is generally sterile, but epithelial cells that line the lower end of the urethra are colonized by bacteria. Urine commonly has 10,000 to 100,000 organisms per ml. Greater numbers or presence of specific types can indicate infection. However, in some cases of kidney infection, the number of organisms present in urine is relatively low. Diagnosis depends on the presence of irritation and pain, especially during urination. Catheterization of hospital patients is a leading cause of infection by opportunistic pathogens present in the urethra or on the skin.

Sexually Transmitted Diseases

Sexually transmitted diseases (STDs) are a growing health problem throughout the world. These diseases affect an alarming portion of the young adult population and may be transmitted to the newborn by transfer during fetal development or at birth. The etiological agents for STDs are bacteria and viruses. The major bacterial diseases are gonorrhea, syphilis, and chlamydial infections. All three bacterial STDs are eminently curable and have been for the last 40 years. Genital herpes and acquired immune deficiency syndrome are the leading viral infections (see Chapter 28). STDs have become as much a social problem as a medical concern. The greatest incidence of STDs is in young people (15 to 35), the age group that is least likely to seek treatment. Increased sexual activity and multiple partners make control of the diseases particularly difficult.

Gonorrhea

The etiological agent for **gonorrhea** is *Neisseria gonorrhoeae,* a gram-negative, generally diploccocal bacterium. The organism is readily killed by drying, sunlight, and UV light. *N. gonorrhoeae* is transmitted in virtually every case by direct contact. Sexual activity between infected and uninfected individuals leads to dissemination of the pathogen. The pili on pathogenic strains of *N. gonorrhoeae* attach to microvilli of mucosal cells of the genitourinary tract (Figure 29.16). The microorganism can be transmitted to mu-

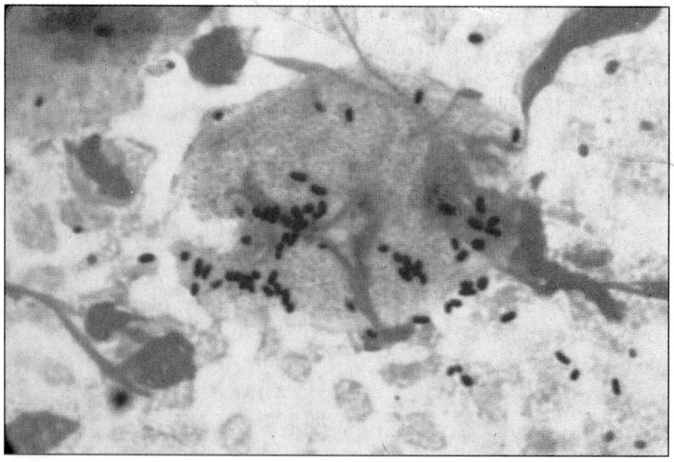

***Figure* 29.16** Presence of *Neisseria gonorrhoeae* in a urethral smear from an infected female (× 1,000) (© Martin Rotker/Phototake NYC)

cosa of the eye, throat, and rectum. The bacterium is phagocytosed by cells on the mucosal surface and transported to intercellular spaces and submucosal tissue. Phagocytosed gonococci are protected from host defenses. The bacterium can infect the eyes of newborns during birth. Untreated, this infection may lead to serious eye infections and is responsible for blindness in much of the developing world. In the United States, tetracycline, erythromycin, or silver nitrate is applied to the eyes of neonates at birth as a preventive measure. This has essentially eliminated eye infections and subsequent blindness.

The incidence of gonorrhea in the United States is alarming, with perhaps as many as one million new cases occurring annually. This figure would indicate that 1 of every 250 Americans is infected each year, and most are in the 15 to 35 age group. Considering that individuals in this group represent about one third of the population, the incidence within the group would be about 1 in 80, a public health concern. Over 50 percent of uninfected females will contract gonorrhea with one exposure to an infected male. An uninfected male will become infected about 20 percent of the time if exposed to an infected female. The incidence of gonorrhea increased over fivefold in the period following the first widespread use of oral contraceptives. A higher level of sexual activity and decreased use of condoms in the period from 1975–78 is considered responsible for this marked increase. The AIDS scare has now led to further awareness of the dangers of "unsafe" sex. This has caused some lowering of the incidence of gonorrhea in the last few years.

Infection with *N. gonorrhoeae* does not confer immunity, and many of the cases observed are reinfections. There is some evidence that chronic gonorrhea in females can induce antibody formation, but with the large number of gonococcus serotypes present in the population, reinfection with another strain is likely. The high incidence of gonorrhea is primarily due to the absence of distinct symptoms in most females. These females are unaware of the infection, and the mild vaginitis and minor discharge is attributed to other causes. Such chronic carriers serve as a significant reservoir of infection. The gonococcus can infect the urethra, vagina, cervix, and fallopian tubes of females before it is detected, and it can cause pelvic inflammatory disease, which can lead to sterility.

Males are rarely asymptomatic. Two to eight days after exposure they experience a painful urethritis and discharge of pus that contains numerous leukocytes. These leukocytes bear intracellular gonococci, and a microscopic examination of stained exudate is an effective diagnostic tool.

A major problem in treating gonorrhea is the increasing incidence of antibiotic resistance. Penicillin has been the treatment of choice, but penicillinase-producing resistant organisms have evolved. Tetracycline and erythromycin have been employed as alternatives. Recently, penicillin-resistant strains have appeared having a chromosomal-mediated resistance to penicillin that also confers some resistance to tetracycline and erythromycin. At the present time gonorrhea can be treated by use of alternative antibiotics. Unless public health measures or individual responsibility lowers the incidence of this disease, it is likely that antibiotic-resistant strains will become more of a threat in the future.

Syphilis

Treponema pallidum is the causative agent of the historically significant disease, **syphilis,** which reached epidemic proportions in Europe in the 16th century. It spread rapidly and caused much suffering. Apparently, *T. pallidum* was considerably more contagious in earlier times than it is today. About one in ten people who have a onetime exposure to syphilis actually will contract the disease. There are 25,000 to 35,000 new cases diagnosed each year in the United States. The lower incidence as compared with gonorrhea is attributed to the absence of asymptomatic female carriers. The highest incidence of syphilis is among homosexual males.

Syphilis is transmitted by sexual contact. The organism enters through mucous membranes or through minor breaks in the skin during sex. About one case in 10 occurs in the oral region. Pregnant women can pass the spirochete *in utero* to the fetus—a disease termed congenital syphilis.

There are three recognizable stages in a syphilitic infection:

Primary stage Within ten days to three weeks after exposure, a small painless reddened ulcer occurs at the site of infection. This ulcer is called a chancre and may appear on the penis or in the vaginal or oral regions (Figure 29.17). The chancre contains spirochetes, and at this stage the infection is readily transmissible through intimate contact. In some cases, the

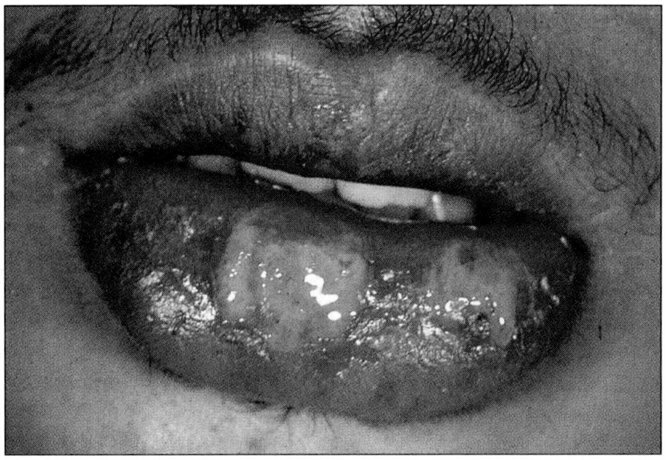

***Figure* 29.17** A primary lesion from a syphilis infection. (© 1991 Steven J. Nassenblatt/Custom Medical Stock Photo)

symptoms disappear and the disease progresses no further. In others, the spirochete enters the bloodstream and is distributed throughout the body. The primary stage lasts from two to ten weeks. The disease then enters the secondary stage.

Secondary stage After the chancre(s) heal(s), skin lesions occur about the body, including the soles of the feet and palms of the hand. They also appear on mucous membranes. The patient is generally feverish. Individuals in this stage of the infection are all serologically positive. Infected individuals can transmit the disease. After four to eight weeks, the disease enters a latent period and is not normally infectious except maternally to the fetus. At this stage, about one-fourth of the cases are essentially cured, one-fourth remain latent with relapses to the secondary stage, and about one-half of these enter the tertiary stage.

Tertiary stage Untreated tertiary syphilis is a degenerative disease. Lesions occur on skin, in bones, and in nervous tissue. Lesions often occur in the central nervous system, and this leads to mental retardation, blindness, and insanity. Paralysis often occurs, and those afflicted with tertiary syphilis walk with a stumbling gait. Lesions in the cardiovascular system can be fatal. There are actually few spirochetes in the lesions, and most of the damage is attributed to hypersensitivity to the organism and its products.

Diagnosis can be made when chancrous lesions are examined by darkfield microscopy of the exudate (Figure 29.18). Serological tests are widely employed by diagnosis and depend on the presence of antitreponemal antibodies to indicate the presence of *T. pallidum.* Penicillin is an effective treatment and is particularly so in primary stages of the disease. More drugs and prolonged treatment are necessary in the later stages. There is no cure for symptoms of tertiary syphilis.

Chlamydiae

Chlamydia trachomatis is an obligate intracellular parasite that causes several different infections in humans. *Chlamydia* are small bacteria that reproduce only in host cells. They are obligate parasites and have an exceedingly small genome reflecting their limited synthetic ability. *Chlamydia* reproduce when an elementary body attaches to a cell surface and is phagocytosed into the cell (Figure 29.19). The elementary body then forms an initial body and these form inclusions containing reticulate bodies. The reticulate bodies reform to elementary bodies that are infectious. These elementary bodies are released on host cell lysis. The cycle from elementary body to reticulate body and release of infectious elementary bodies requires 48 to 72 hours. The incidence of these venereal infections greatly outnumbers gonorrhea, and there are an estimated 3 to 10 million new cases in the United States each year. *C. trachomatis* is also the etiological agent of **trachoma,** an eye infection that occurs worldwide. Over 10 million cases of blindness, principally in developing countries, are attributed to this disease. The organism is passed to the newborn during birth and causes conjunctivitis and respiratory distress. Between 8 percent and 12 percent of pregnant women in the United States have chlamydial infections of the cervix, and one half of babies born to these infected individuals have conjunctivitis. Application of tetracycline to the eyes prevents development of the disease. Much of the pneumonia in infants is caused by this organism.

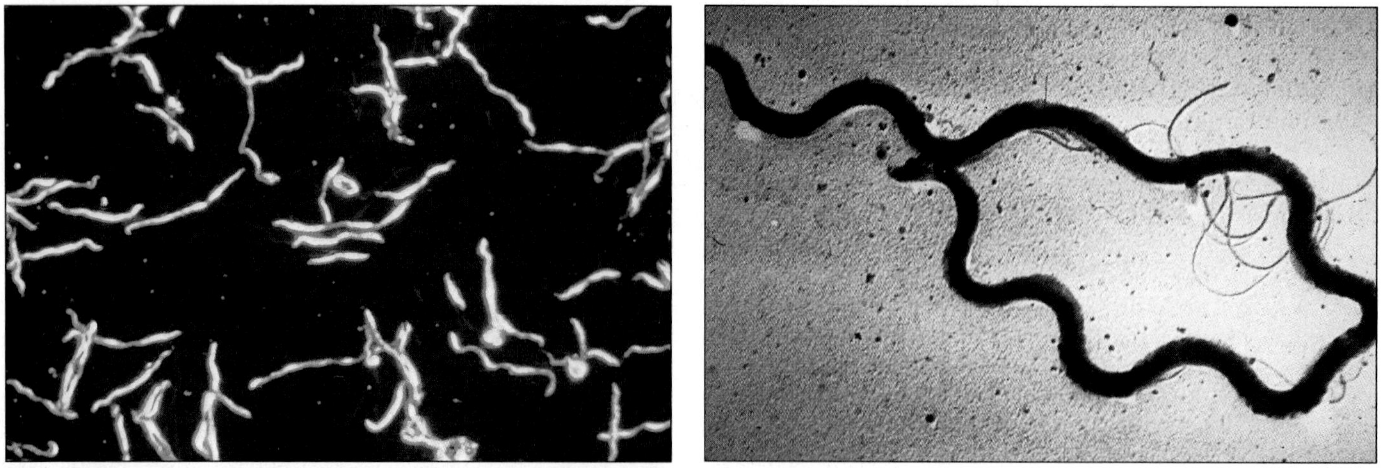

(a) (b)

***Figure* 29.18** The spirochete that is the causative agent of syphilis. **(a)** Darkfield microscopy of exudate from a chancre. (© George J. Wilder/VU) **(b)** Electron micrograph of *Treponema pallidum.* (© Science VU/Visuals Unlimited)

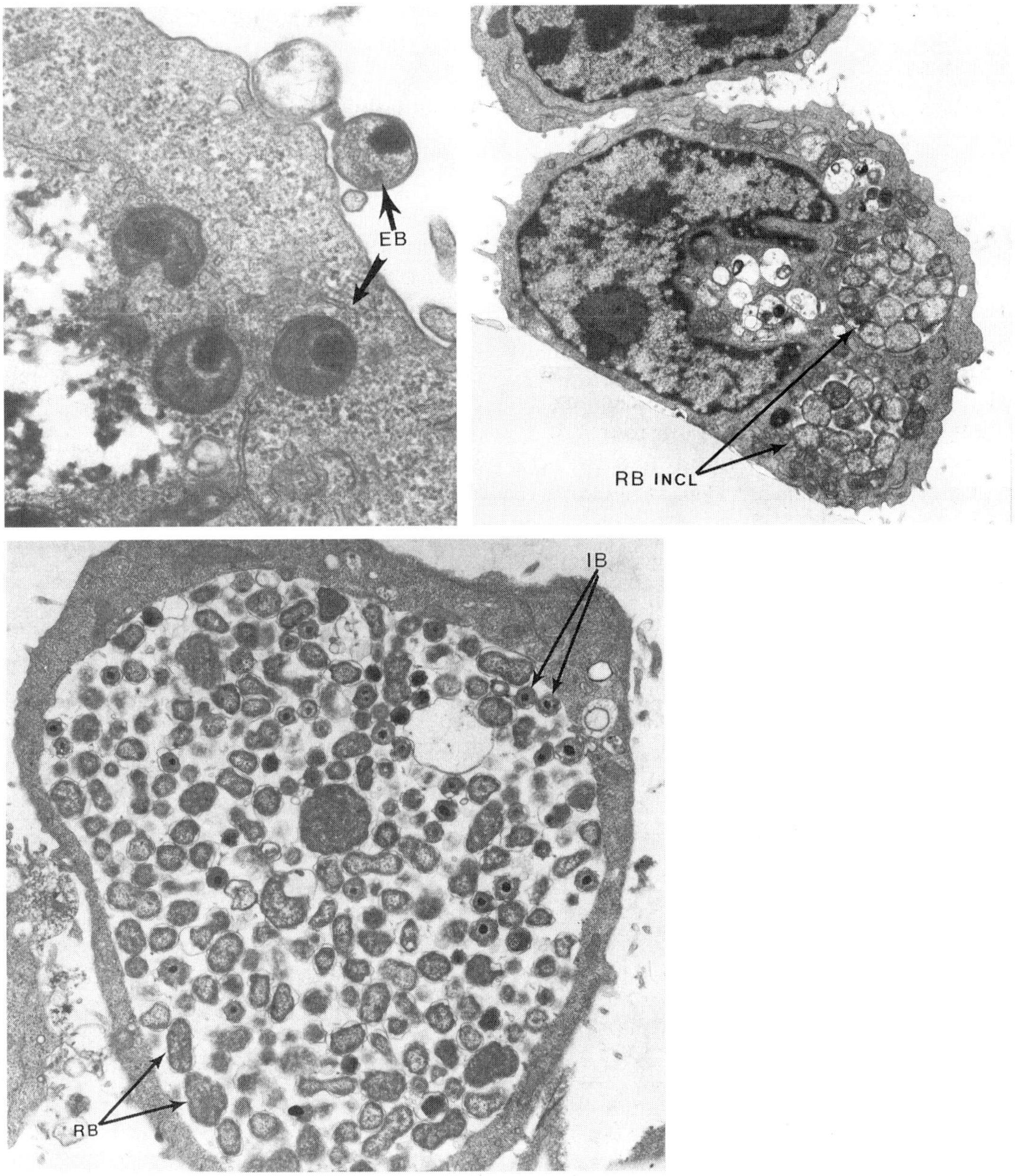

***Figure* 29.19** The life cycle of *Chlamydia trachomatis*. The elementary body (EB) is the infectious particle. After entering the host cell the initial body (IB) propagates as reticulate bodies (RB) in inclusions (INCL). On cell lysis the newly formed EBs are released. (Courtesy of Priscilla Wyrick, UNC–Chapel Hill)

Chlamydial nongonococcal urethritis (NGU) is probably the most often acquired sexually transmitted disease. In males, *C. trachomatis* causes urethritis, and in females it causes urethritis, cervicitis, and pelvic inflammatory disease. An estimated 50,000 women become sterile each year from NGU. Inapparent infections are common in sexually active adults. Diagnosis is difficult and is made by isolating the organism. Generally, identification is by staining material, taken by swab, with fluorescent antibody. Treatment with tetracycline and erythromycin is effective. Penicillin is ineffective, as *C. trachomatis* lacks a peptidoglycan cell wall. As a significant number of gonorrhea patients are also infected with chlamydiae, it is essential that both diseases be treated.

Another strain of *C. trachomatis* is responsible for a disease in males called ***Lymphogranuloma venereum.*** This is a sexually transmitted infection that causes swelling of lymph nodes about the groin. It is diagnosed by the presence of swollen lymph nodes and treated with tetracycline or sulfonamides.

Vectorborne Diseases

Vectorborne diseases are transmitted from one infected animal to another by means of an intermediary host. This intermediary host is termed a **vector** and is most often a biting insect (arthropod or louse) that transmits the pathogen while feeding on an uninfected host. The animal that is the constant source of the infectious agent is called the **reservoir.** Most reservoirs are mammals that may or may not be adversely affected by the infectious organism. Some tick and mite vectors also serve as reservoirs. This occurs in cases where the infectious bacterium is transovarially (by infected eggs) transferred from parent to insect progeny.

The bacteria involved in vectorborne diseases are obligate parasites. They generally cannot reproduce or survive in nature outside a living host. Those that cause a fatal infection in the reservoir animal must be carried by the vector to another living host. A list of vectorborne diseases is presented in Table 29.6. The etiological agents for many of these diseases are in the order *Rickettsiales,* and in the genera *Rickettsia, Rochalimaea,* and *Coxiella.*

Tick-borne diseases can be prevented by proper sanitation and by ridding the human environment of rats and other vermin. To avoid diseases such as Lyme disease and Rocky Mountain spotted fever one should dress properly when entering areas infested with ticks. Wear long-sleeved shirts and long pants; pants should be tucked in socks, and shoes should be worn. Use of insect repellants is of value, and one should check for ticks after leaving an area that is potentially infested. These are summertime diseases of high incidence when ticks are feeding.

Plague

The etiological agent for **plague** is *Yersinia pestis,* a gram-negative nonmotile coccobacillus. Historically, the plague has killed more humans than any other bacterial disease. In the 6th century more than 100 million died from the disease, and in the 14th century a plague epidemic killed more than one fourth of the people in Europe (see **Box 29.2**).

Y. pestis normally afflicts rodents and is endemic in the Southwestern United States in prairie dogs, ground squirrels, wood rats, chipmunks, and mice. These animals are relatively resistant to plague, and the disease resides in them during interepidemic periods. If domestic rats interact with these reservoir animals, the domestic animals become infected (Figure 29.20). Fleas feeding on these infected animals can cause an epidemic of plague in the domestic rat populations. Plague is a fatal disease in domestic rats. When the rats die, the fleas travel to other rats. If none are available, they move to and feed on humans,

***Table* 29.6 Vectorborne bacterial infections of humans**

Disease	Etiological Agent	Vector	Reservoir	Transmission
Rocky Mountain spotted fever	*Rickettsia rickettsii*	Wood tick (western U.S.) dog tick (eastern U.S.)	Mammals	Tick to human
Q fever	*Coxiella burnetii*	Animal tick	Ticks, cattle	Tick to human, carcass to human
Tularemia	*Francisella tularensis*	Wood tick	Ticks, rabbits, other mammals	Tick to human, animal to human
Relapsing fever	*Borellia recurrentis*	Body louse	Humans	Human to louse to human
Lyme disease	*Borrelia burgdorferi*	Deer tick	Deer, mice, voles	Flea to human
Typhus (epidemic)	*Rickettsia prowazekii*	Human louse	Humans, flying squirrels	Louse to human
Typhus (endemic)	*Rickettsia typhi*	Rat flea	Rats, ground squirrels	Flea to human
Plague	*Yersinia pestis*	Rat flea	Rats	Flea to human, human to human
Scrub typhus	*Rickettsia tsutsugamushi*	Mite	Mites	Mite to human
Trench fever	*Rochalimaea quintana*	Body louse	Lice, humans	Louse to human

BOX 29.2 MILESTONES

A Sad Game

There is a children's game where the youngsters stand in a circle holding hands and sing:

Ring-a-ring o'rosie
A pocketful of posie
A-tishoo! A-tishoo!
We all fall down.

When the last line is recited, the children drop to the ground. There are a number of versions of this poem, and it may have originated in the days of the Great Plague of London (1665). Thus, this pleasant little diversion may well have had a macabre origin, as evidenced by the following:

"Ring-o-rosie"—referred to the rosy rash associated with the symptoms of plague.
"Posies of herbs"—were carried during epidemics to ward off the plague.
"A-tishoo"—referred to sneezing, which was considered a final fatal symptom.
"All fall down"—meant that everyone dropped over dead.

Ring-a-ring-a-roses,
A pocket full of posies;
Hush! hush! hush! hush!
We're all tumbled down.

An illustration of the game Ring-a-ring o'rosie from *Mother Goose,* by Kate Greenaway (1881). (Courtesy of The Granger Collection, New York)

thereby transmitting the disease. Killing rats after the local population of fleas is infected can drive the fleas to humans and increase the incidence of plague.

The blood of a domestic rat infected with plague contains large numbers of the *Y. pestis* organism. The flea ingests this blood, and the bacterial coagulase induces a clot in the proventriculus of the flea. This prevents food from passing on to the stomach of the flea. The flea becomes hungry and feeds ravenously but regurgitates due to the intestinal block. The regurgitated blood contains large numbers of infectious bacteria that enter the bite site, infecting the host. Feces deposited by the flea in the area of the bite also can contain infectious bacteria that can enter the wound and cause infection.

The plague bacillus moves from the entrance site to regional lymph nodes, which become enlarged and tender due to growth of the bacterium. The enlarged node is called a bubo and is the origin of the name bubonic

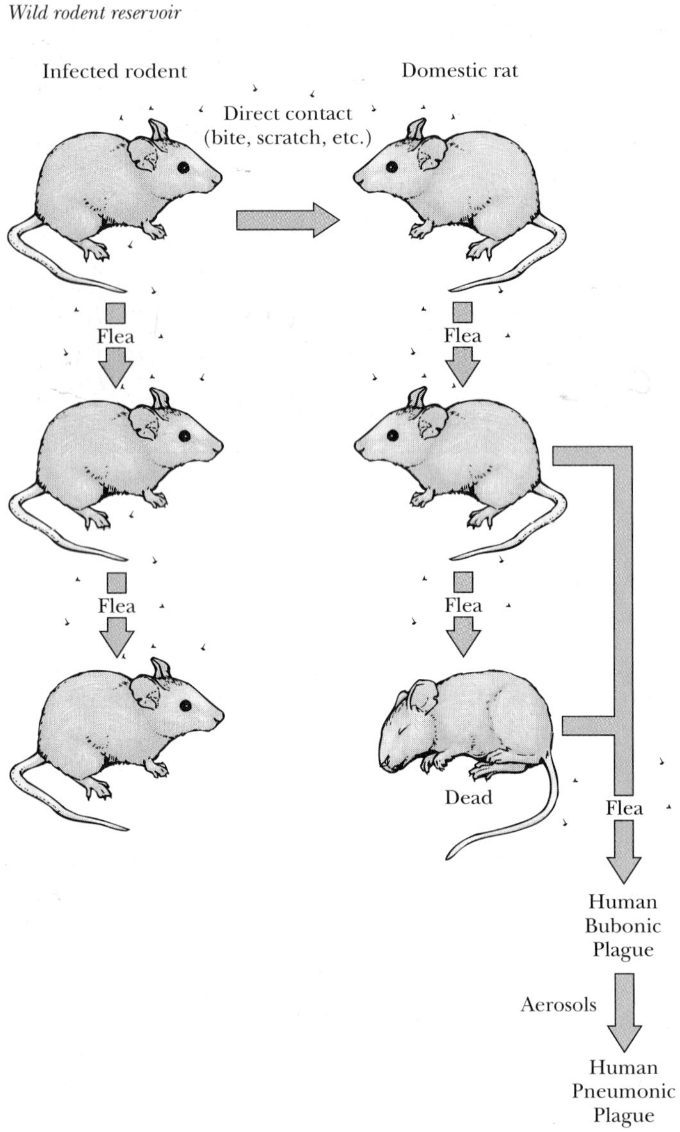

***Figure* 29.20** The transmission of *Yersinia pestis* from the reservoir wild rat population to domestic rats and humans. The microorganism can be transmitted from human to human in aerosols.

plague (Figure 29.21). The organism can be engulfed by polymorphonuclear leukocytes and destroyed while those phagocytosed by macrophages are not. The organism in the macrophage is maintained at a favorable temperature (37°C) and in a low Ca^{++} environment. Low Ca^{++} promotes the expression of virulence factors. The bacilli thrive in macrophages and travel to regional lymph nodes. They are carried via the bloodstream to the spleen, liver, and other organs, and they cause a subcutaneous (below the skin) hemorrhage that appears as dark areas on the body surface. These darkened areas and the high level of mortality are the reason for calling plague "Black Death." Movement of the bacilli to the lungs results in pneumonic plague, a disease that is transmitted via the respiratory route. Untreated, pneumonic plague has 100 percent fatality, and bubonic plague has a fatality rate of 75 percent.

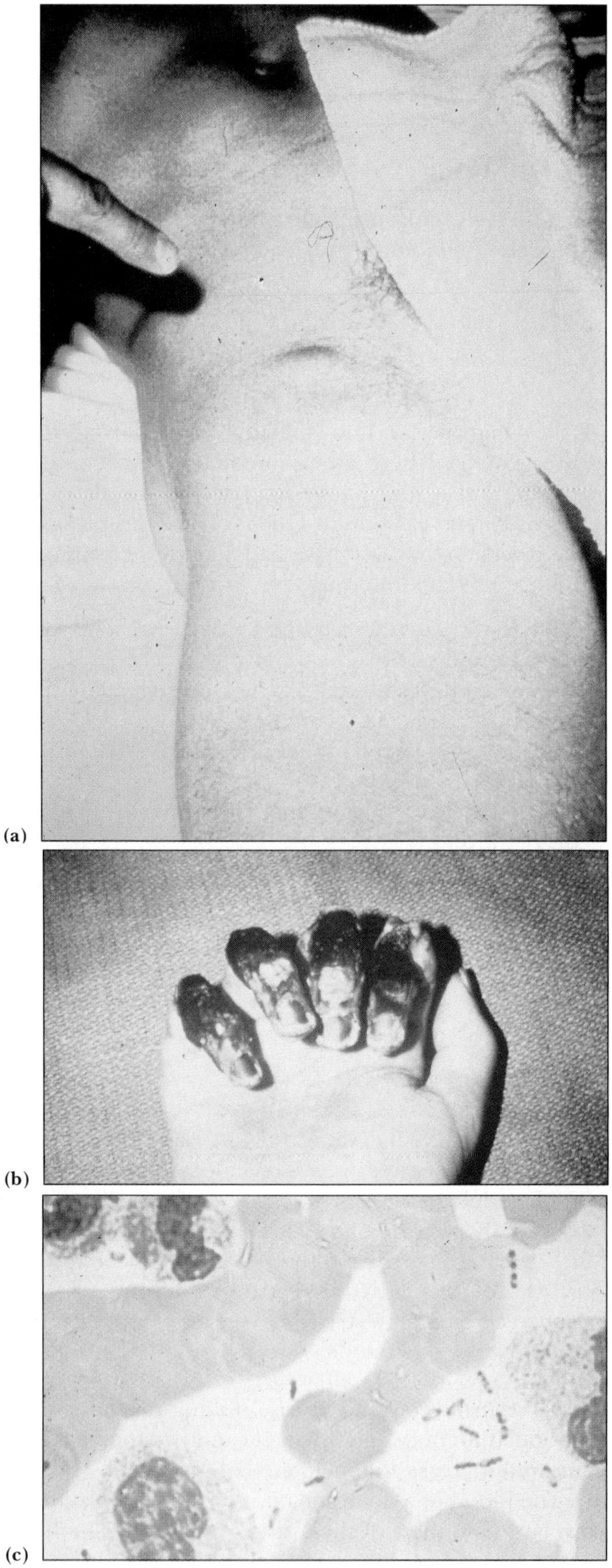

***Figure* 29.21** The manifestations of plague infection in humans. **(a)** A bubo (swelling) in the armpit area. **(b)** Gangrene in hand tissue. **(c)** Cells of *Yersinia pestis* are visible as small cells in the lung tissue of a victim of pneumonic plague. (Courtesy of Centers for Disease Control)

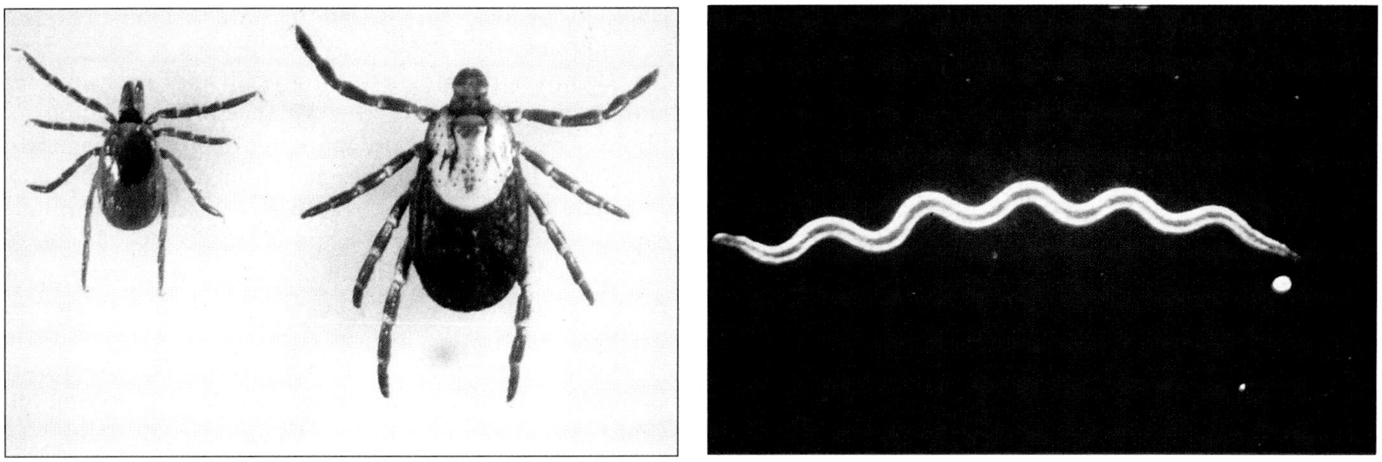

(a) (b)

***Figure* 29.22** **(a)** The deer tick *(Ixodes dammini),* the major vector of Lyme disease in eastern United States. The adult male is on the left and the female on the right. The live tick is about one-seventh the size shown. (© Pfizer Research) **(b)** The spirochete *Borrelia burgdorferi,* the etiological agent of Lyme disease. (© Charles W. Stratton/Science VU)

The plague bacillus produces a number of toxins, and among these are a protein/lipoprotein complex that prevents phagocytosis. There is an exotoxin termed **murine toxin** produced by virulent strains of *Y. pestis* that blocks respiration in laboratory mice and probably has a similar activity in human plague.

Plague is diagnosed by symptoms and isolation of the organism from aspirates of the buboes. Sputum can be the source of isolates in pneumonic plague or the organism can be identified serologically. Treatment with streptomycin, chloramphenicol, or tetracycline is effective if given early in the infectious process.

Lyme Disease

Lyme disease results from infection by the spirochete, *Borrelia burgdorferi.* The tick-borne disease was first recognized in individuals in Lyme, Connecticut—hence its name. The disease first appeared in 1975 as cases of arthritis, and in 1982 the etiologic agent was identified. Lyme disease is the most rapidly growing of the tick-borne diseases, with over 8000 cases occurring each year. Although Lyme disease is primarily observed in the northeastern and middle atlantic states it has now spread to virtually every state in the United States. The major reservoirs for *B. burgdorferi* are the deer and white-footed mouse. The tick involved is mainly the deer tick, *Ixodes dammini* (Figure 29.22**a**) although other ticks may also be infected with the spirochete (Figure 29.22**b**). The deer tick feeds on birds and other animals, rendering them potential sources of Lyme disease. The tick that harbors and transmits Lyme disease may feed on white-footed mice and pick up the bacterium that causes human ehrlichiosis (*Ehrlichia* sp.) and/or the protozoan that causes human babesiosis (*Babesia microti*). A bite by a tick carrying these pathogens can transmit from one to all of these diseases. A higher percentage of deer ticks are infected with *B. burgdorferi* than occurs in ticks of other tick-borne diseases. The spirochete is transferred transovarially so the deer tick can be a reservoir as well as the vector. The deer tick is quite small and not readily seen on human skin. The first manifestation of disease is a skin lesion at the site of the insect bite, which spreads to form a large rash area (Figure 29.23). At this time, the disease is readily treated with tetracycline. Disease symptoms include headaches, chills, backache, and general malaise. Untreated Lyme disease progresses to a chronic stage, and the symptoms at this stage are arthritis and a numbness in limbs. The organism may attack the central nervous system, affecting vision and causing paralysis. The best preventive measure is to avoid tick habitats. But if you are bitten and develop a rash, headache, or chills, see a physician immediately.

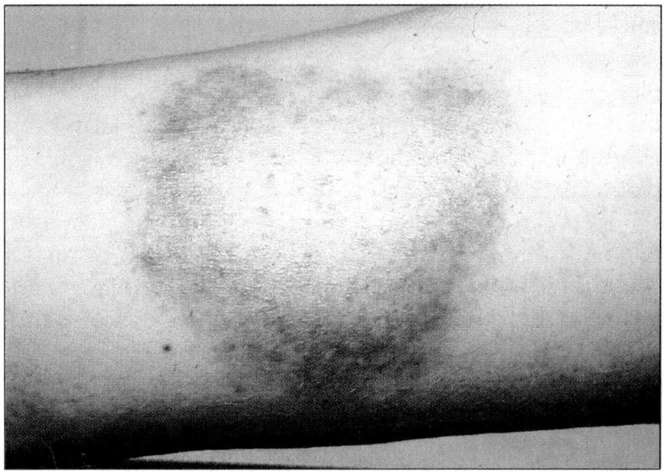

***Figure* 29.23** The characteristic rash of Lyme disease at the site of the tick bite. The rash area increases over several days, and effective treatment should start at this stage. (© Kenneth E. Greer/VU)

Rocky Mountain Spotted Fever

The etiological agent for **Rocky Mountain spotted fever (RMSF)** is *Rickettsia rickettsii,* a disease first recognized in Bitterroot Valley, Montana. The organism is named for H. T. Ricketts, who died of typhus while investigating that rickettsial disease. RMSF is caused by the bite of an infected tick and is acquired in humans from feces deposited by the tick in and about the bite site. In the western United States the rickettsia is carried by the wood tick *(Dermacentor andersoni)* and in the East by the dog tick *(D. variabilis). R. rickettsii* is passed from parent to tick progeny by transovarial passage. *Dermacentor andersoni* is a slow-feeding, hard-shelled tick quite distinct from the fast-feeding ticks that are the vectors of Lyme disease. Consequently, removal of the RMSF tick up to three hours after it initiates feeding can prevent infection. There are about 1000 cases of RMSF in the United States each year, with more cases in North Carolina and adjacent states than in the Rockies. Dogs are readily infected with *R. rickettsii,* and organisms are present in the blood for two weeks after infection. A dog may serve as a reservoir.

The symptoms of RMSF include a fever, severe headache, and a rash that appears first on the palms and soles of the feet. The organism proliferates in the endothelial linings of small vessels and capillaries. The resultant hemorrhage and necrosis is responsible for the observed rash. Gastrointestinal problems can also occur. RMSF is generally diagnosed via symptoms and by establishing a history of tick bite(s). Tetracycline is an effective treatment, provided it is administered early on in the infection.

Typhus

There are two types of the disease typhus, and they are caused by different rickettsial species that share common antigens. The two types are epidemic and endemic typhus. The etiological agent for epidemic typhus is *Rickettsia prowazekii,* and that for endemic typhus is *Rickettsia typhi.*

Epidemic typhus occurs in areas of substandard living conditions and poor sanitation. The major reservoir for typhus is the human, and the vector is the louse. Flying squirrels can be infested with lice carrying the rickettsia. Passage to humans from the flying squirrel has occurred when these animals move into the attic of a home. There is no transovarial transfer of *R. prowazekii* in the louse, and the infected insect will die within several weeks.

Infection of humans occurs while the louse is feeding as the infectious organisms are present in feces or regurgitated blood of the insect. The disease is marked by a general bacteremia with the rickettsia growing in endothelial cells of blood vessels. Symptoms include a high fever (104°F) with headaches, chills, fever, and delirium. During the fifth or sixth day of illness, a rash develops. Recovery or death occurs two to three weeks after the onset of symptoms.

Typhus is diagnosed during epidemics by symptoms and, in sporadic cases, by the presence of a circulating antibody. The Weil-Felix test can be employed in diagnosis. This test is based on the fact that strains of the genus *Proteus* share common antigens with *R. prowazekii.* If antibody is present in serum, it will agglutinate *Proteus* cells. Tetracycline and chloramphenicol are effective in treating typhus. Vaccines are also available for use in areas where typhus occurs.

Endemic typhus, also called **murine typhus,** occurs sporadically in areas throughout the world, including the Gulf Coast states, and rarely in epidemics. Immunity to epidemic typhus *(R. prowazekii)* will confer immunity to endemic typhus *(R. typhi).* The primary reservoir for endemic typhus is the rat or ground squirrel. Neither the rat nor the rat louse is adversely affected by the infection. The disease symptoms, diagnosis, and treatment are the same as for epidemic typhus.

Scrub Typhus

Scrub typhus is also a rickettsial disease, and the etiological agent is *Rickettsia tsutsugamushi.* This disease was of some consequence during World War II and the Vietnamese war. The reservoir is the rodent, and the vector is a mite. The mite transmits the disease from rodent to rodent. It is transferred transovarially (female mite to her eggs), and the mites lay eggs in soil that hatch into larvae (chiggers) that feed on animals, including humans. The disease symptoms are similar to those encountered with typhus. A primary lesion, called an eschar, occurs in this disease at the location of the bite. Diagnosis is by the Weil-Felix reaction or by means of serological tests. Tetracyclines and chloramphenicol are effective, but treatment should be prolonged because relapses do occur. A successful vaccine has not been developed for scrub typhus.

Q Fever

Q fever received the name Q (Query) because the etiological agent for the disease was long unknown. We now attribute the disease to *Coxiella burnettii,* a rickettsial organism remarkable for its ability to survive outside a host. Arthropods are both a reservoir and vector of Q fever and transmit it from animal to animal. The organism forms an endospore-like body that permits survival on animal carcasses, hides, and in feces. Outbreaks in humans occur among animal handlers and through consumption of contaminated milk that has not been pasteurized. Q fever causes an influenza-like illness with prolonged fever, headaches, and chills. Pneumonia and chest pains are also symptoms. The disease is insidious in that an endocarditis (inflammation of the heart) can occur months or years after infection. Immunological tests are employed in diagnosing Q fever, and treatment is with tetracycline.

Relapsing Fever

Humans are the reservoir for *Borrelia recurrentis,* the causative agent of **relapsing fever.** *B. recurrentis* is a spirochete that forms coarse irregular coils that are 0.5 μm wide and over 20 μm in length. The disease is transferred from human to human by the body louse. A bite will transfer the spirochete into the bloodstream, through which it passes to the spleen, liver, kidneys, and gastrointestinal tract. Multiple lesions occur in these organs. After a fever and four to five days of illness, the disease symptoms disappear. In most cases, a relapse (hence, relapsing fever) occurs after about ten days with a return of the manifestations of disease. These bouts of wellness and relapse occur three to ten times before full recovery occurs. The relapse is attributed to the development of strains with different surface proteins on the outer membrane that are unaffected by the original immune response. A period of time is necessary before the host can generate antibodies to combat this new form of organism.

The disease can be diagnosed by clinical symptoms and by the presence of *B. recurrentis* spirochetes in the blood. Dark-field microscopy is effective in visualizing spirochetes. The treatment of choice is penicillin.

Tularemia

The etiological agent for **tularemia** is *Francisella tularensis,* an extremely pleomorphic, small, gram-negative bacterium. *F. tularensis* is a facultative anaerobe that grows better in an aerobic atmosphere.

The wood tick is both a reservoir and vector of *F. tularensis* in the Northwest and Midwest, but tularemia is transmitted in the Southwest by a deer fly. The organism is transferred transovarially from one tick generation to the next. *F. tularensis* has been isolated from mammals, birds, fish, and invertebrates. Domestic cats that eat infected animals can transmit the disease to humans. It occurs worldwide mostly from handling of infected animals, particularly rabbits.

F. tularensis is an exceedingly virulent organism but the reason is unclear. The organism grows intracellularly in PMNs. There are three major manifestations of tularemia:

Ulceroglandular Occurs from the bite of an infected tick or from contact with an infected animal through a scratch. A lesion occurs at the point of entry, and in seven to eight days an ulcer develops. The organism moves to lymph nodes and through the bloodstream to the liver, spleen, and other organs. Ulcers develop in these organs.

Pneumonic Generally spread by the respiratory system during the skinning or handling of infected animals. This form of the disease can spread from human to human through droplets.

Typhoidal Occurs after ingestion of contaminated food or water. Symptoms are a high fever and gastrointestinal distress.

Tularemia can be diagnosed by microscopy of smears obtained from lesions and visualized with fluorescent antibodies. Tetracycline is effective in the treatment of tularemia.

Trench Fever

Trench fever is a disease that apparently occurs mostly during the unsanitary conditions associated with war. Trench fever first occurred in 1915 (World War I), when over one million military personnel contracted the disease. The etiological agent is *Rochalimaea quintana.* The primary reservoir is the human, and body lice are the vectors. The rickettsia does not kill the louse nor is it transferred to progeny. Disease symptoms include chills, fever, headache, and muscle pain. These symptoms last for five days; hence, the name *quintana.* Patients are incapacitated for up to two months. Relapses often occur, and antibiotics such as chloramphenicol must be given for an extended time to eliminate the disease.

Cat Scratch Disease

There are over 24,000 cases of cat scratch disease in the United States each year. A bite or scratch by a cat can result in cat scratch fever and occurs most often in young children. Papules (red swellings) are evident 7–12 days after injury, and lymph nodes in the vicinity become tender. As the disease progresses to the lymph nodes they swell and become pus filled. Enlarged nodes may require surgical drainage, but the disease symptoms are generally mild and localized. Recovery requires a few weeks. The etiological agent is *Bartonella henselae,* a small curved gram-negative rod that can be isolated on blood agar under a CO_2 atmosphere. There may be other causes of this human disease, but no other agent has been isolated.

Rat Bite Fever

There are two etiological agents for rat bite fever—*Spirillum minus* and *Streptobacillus moniliformis.* Both are apparently inhabitants of the oral cavity of rats. *S. minus* is a gram-negative rod that can be isolated *in vitro.* Infections with *S. minus* cause a relapsing fever and rash. *S. moniliformis* causes a fever, rash, and arthritis. Rat bite fever is rare in the United States and can be treated with antibiotics such as tetracycline.

BOX 29.3 RESEARCH HIGHLIGHTS

The Water We Drink—Safe or Not

A severe outbreak of a diarrheal disease occurred in Milwaukee, Wisconsin, in April of 1993. This outbreak affected one-fourth of the population of Milwaukee and served as a warning to all Americans that:

1. Our water supplies are under great stress.
2. Disease is lurking and can create havoc with the least breakdown in our defenses.
3. We must not take our water supplies for granted.

The pathogen involved in Milwaukee was a protozoan of the genus *Cryptosporidium*. *Cryptosporidium* is a common protozoan in mammals and birds that should be filtered from drinking water during routine water treatment. Ingestion of *Cryptosporidium* or cysts *(Cryptosporidia)* in drinking water gave 370,000 citizens of Milwaukee acute diarrhea, nausea, and stomach cramps. Undoubtedly there were fatalities from dehydration and other complications, but the exact number of these will never be known.

Why did this happen? The epidemic occurred in early spring, and runoff from the spring thaw may have overstressed the water treatment plant. Dairy farms abound around Milwaukee, and cattle harbor the protozoan. Water from feedlots may have carried the parasites into catchments that supply the city with potable water.

In routine treatment of water, coagulants are added, prior to sand filtration, that remove particles from the water. Alum (aluminum potassium sulfate dodecahydrate) was the coagulant employed in Milwaukee until the fall of 1992, when it was replaced with polyaluminum hydroxychloride. This replacement was due to an EPA concern that alum can leach lead from water pipes, and Milwaukee water supplies contained unacceptable levels of lead.

During March of 1993, employees at one of the water treatment facilities noted that the treated water coming through the plant was unusually turbid. The spring thaw runoff water was evidently not clarified by the replacement coagulant employed to settle particulates. Chlorination does not kill the *Cryptosporidia* cysts or those of other pathogenic protozoa. The water treatment plant in question was closed down, and the city returned to alum as coagulant.

A report on drinking water quality was released in April of 1993 by the General Accounting Office. The report concluded that water quality assurance programs across the United States are "flawed and underfunded." The conclusion was that 198,000 water supplies are inadequately monitored. This is a problem we cannot ignore.

Protozoa as Agents of Disease

Protozoa are the causative agents of many diseases worldwide. Tropical areas near the equator are particularly subject to protozoan diseases, and malaria is considered to be among the world's leading health problems. Malaria infects 150 million humans each year and is responsible for over 2 million deaths. Although malaria has been essentially eradicated in the United States, conditions exist for the reintroduction of the disease. Amebiasis and giardiasis are two protozoan diseases that are a concern in the United States, and these and malaria will be discussed briefly. Another pathogenic protozoan, *Cryptosporidium*, recently caused a health crisis in Milwaukee, Wisconsin **(Box 29.3)**.

Amebiasis

Entamoeba histolytica is the etiologic agent of amebiasis, also known as amebic dysentery. Worldwide there are an estimated 400 million active cases that cause over 100 thousand deaths each year. There are 4000 to 5000 new cases in the United States annually. The disease is most prevalent in warmer climates with inadequate water treatment and sanitation. The infection occurs when the mature cysts of *E. histolytica* are ingested. The cyst moves through the stomach, and the cyst coat is removed after arriving in the small intestine. The amoeboid cell called a trophozoite moves to the large intestine and penetrates the intestinal mucosa. In acute amebiasis, lesions are formed that lead to severe ulceration. The trophozoites can multiply and invade other intestinal mucosa while feeding on erythrocytes and bacteria. The lesions become infected with bacteria. *E. histolytica* may penetrate the large intestine and produce lesions in the liver, lungs, or brain.

The trophozoites may establish a mild or asymptomatic form of the disease called **chronic amebiasis.** Individuals with this form of the disease become carriers and constantly shed cysts into the environment. Victims of chronic amebiasis are a concern because they serve as a constant reservoir for infective cysts. Amebiasis is a health concern in homosexual males.

The symptoms of amebiasis are severe dysentery or colitis. The diarrhea is accompanied by blood and mucus shed from the extensive intestinal lesions. Amebiasis can be diagnosed by the presence of trophozoites in the feces filled with ingested red blood cells. Treatment is different for carriers than those with active infections. Diiodohydroxyquin is the drug of choice for asymptomatic carriers, and aralen phosphate is given in acute amebiasis. The best preventive measure is quality water. Unfortunately chlorination of water does not destroy the cysts of *E. histolytica.*

Giardiasis

Giardia lamblia is the agent of an intestinal disease called **giardiasis.** *G. lamblia* is an ancient eukaryote that lacks a mitochondrion, and its rRNA has many bacteria-like features. Many domains in *G. lamblia* rRNA resemble that in the Archaea. Giardiasis is the most prevalent waterborne diarrheal disease in the United States. It occurs most often in young children, and there are about 30,000 cases per year. Giardiasis is endemic in child day care centers.

About 7 percent of the population in the United States are healthy carriers of giardia and shed the cysts in their feces. Transmission of the cysts to humans generally occurs through ingestion of contaminated food or water. Outbreaks often occur in campers or back-packers in wilderness areas when they drink seemingly clear fresh water. Water can be contaminated by domestic or wild animals including beavers, rodents, or deer.

Acute cases of giardiasis are marked by severe diarrhea, cramps, flatulence, and dehydration. The asymptomatic carriers may have intermittent mild cases of diarrhea. Ingestion of the cysts is followed by uncoating and release of trophozoites. These trophozoites inhabit the lower small intestine where they attach to the mucosal wall through small discs (Figure 29.24). They apparently feed on mucosal material. Diagnosis of giardiasis is by examining for cysts in mucosal material shed in feces. The mucus sheath in the intestinal tract is shed normally, and the trophozites or cysts would be present in this material. A commercial ELISA test (see Chapter 31) is available for detection of the *G. lamblia* antigens in fecal material. Treatment is with atabrine or metronidazole. As with *E. histolytica,* the cysts of *G. lamblia* are resistant to chlorination. Campers should boil water to prevent infection. The disease can be prevented in larger populations by proper water treatment.

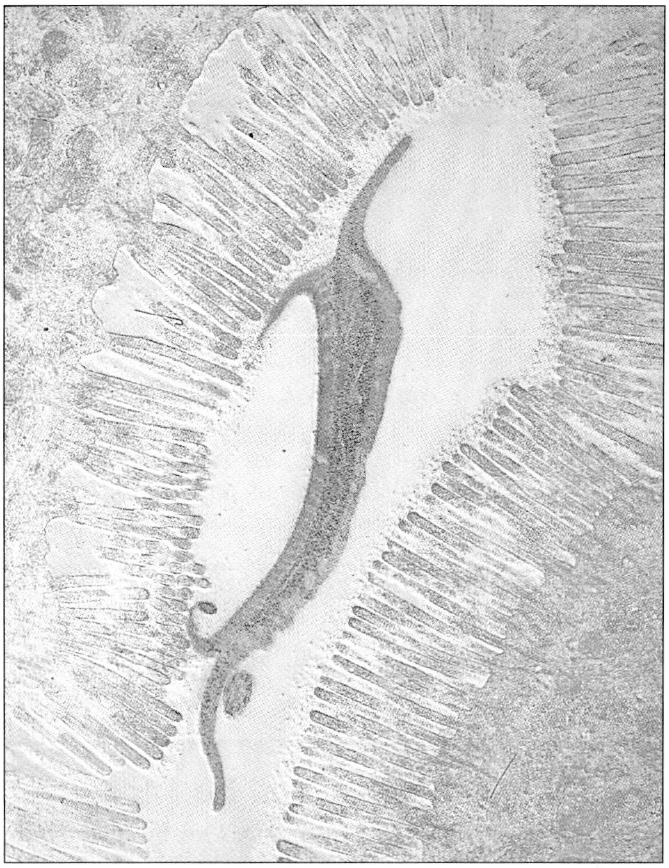

***Figure* 29.24** Scanning electron micrograph of the attachment of *Giardia lamblia* to epithelia of the small intestine. When the organisms become detached, they leave a circular impression on the surface. (© CNRI/Science Photo Library)

Malaria

The etiologic agents of malaria are several species of protozoa in the genus *Plasmodium,* and the disease is prevalent in countries near the equator. Nearly all adults in the middle regions of Africa and in India have been infected. While the disease has essentially been eradicated in the United States since 1945, it has made a comeback through infected immigrants. There are about 1000 cases per year recorded in the United States.

Malaria results from the bite of a female mosquito of the genus *Anopheles.* The female injects anticoagulants while sucking blood, and the saliva that carries the anticoagulants transmits the malarial parasite into the human bloodstream. Male mosquitoes do not suck blood and do not carry the protozoan.

The two phases in the life cycle of the protozoan *Plasmodium vivax* are illustrated in Figure 23.28. When *P.*

***Table* 29.7 Zoonoses that are transmitted to humans**

Disease	Organism	Transmitted by	Host	Symptoms
Anthrax	*Bacillus anthracis*	Endospore inhalation or animal contact	Cattle, goat, sheep	Cutaneous lesions
Bovine tuberculosis	*Mycobacterium bovis*	Milk	Cattle	Lesions in bone marrow of hip, knee
Brucellosis	*Brucella abortus*	Milk, skin/eye contact, or bite	Cattle	Fever, headache
	B. melitensis		Goat	
Leptospirosis	*Leptospira interrogans*	Urine, contaminated water, skin contact	Rodent, dogs, cattle	Meningitis, headaches, fever
Listeriosis	*Listeria monocytogenes*	Raw milk	Cattle	Meningitis
Yaws	*Treponema pallidum* (subsp. *pertenue*)	Bite	Fly	Skin ulcers, lesions

vivax enters the bloodstream it passes to the liver where the sporozoites undergo an asexual fission; in about seven days there is the release of merozoites. These merozoites invade red blood cells where they reproduce. The red blood cells rupture and release merozoites, and these infect other red blood cells. *P. vivax* forms microgametocytes in the infected human red blood cells, and there can be taken into the stomach of a female mosquito while feeding on an infected human. Sexual reproduction in the mid gut of the mosquito leads to the production of sporozoites that travel to the salivary glands of the mosquito. The sporozoites and anticoagulants are then injected into another victim during feeding.

The symptoms of malaria—chills followed by fever and sweating—are caused by the presence of merozoites and debris from the synchronized rupture of blood cells. A victim may suffer several attacks, and this is followed by remission that can last for days to months. Relapses occur as dormant parasites become activated and emerge from the liver. Malaria cannot be spread by human to human contact, as it requires the mosquito vector for transfer. This is somewhat unique in infectious diseases.

Malaria is diagnosed by finding the parasites in red cells. Serological tests are also available. Treatment is with chloroquine or other quinine derivatives. The only preventive measure is to destroy the breeding grounds for *Anopheles* mosquitoes. DDT was employed extensively for this purpose during the period from 1946 to the 1960s, but use of this pesticide has been curtailed.

Zoonoses

Zoonoses are infections that normally occur in animals but can infect humans through contact with infected animals, animal products, or bites. A number of vectorborne diseases have already been discussed. There are a number of others, and some of these are presented in Table 29.7. Many of these diseases occur rarely in the United States, but outbreaks of diseases such as listeriosis do occur when milk is improperly handled. Brucellosis has been a concern, but widespread testing of cattle has eliminated this disease from dairy herds in the United States.

Summary

- Humans are subject to an array of skin infections, of both bacterial and fungal origin. Common cutaneous infections caused by bacteria are **staphylococcal** infections and by fungi, **athletes foot** or **ringworm.**

- Respiratory infections are difficult to control because they are generally transmitted in minute air droplets.

- **Diphtheria** was once a leading cause of death in children but has been virtually eliminated in the United States by a massive immunization program. **Whooping cough** is another disease that has been controlled in the United States by vaccination. Worldwide, both diseases are a concern.

- **Pneumonia** is an inflammation of the lungs that results from infection by viruses, mycoplasma, Eubacteria, or fungi. The leading etiologic agent ***Streptococcus pneumoniae*** is often a secondary infection that follows a viral infection. ***Streptococcus pyogenes*** is a potent pathogen and can infect the **meninges** or cause **scarlet** or **rheumatic fever.**

- **Meningitis** is an **inflammation** of the meninges that can be caused by ***S. pyogenes, Haemophilus influenzae,*** or ***Neisseria meningitidis.*** Meningitis is a severe, often fatal, disease.

- **Tuberculosis** is an infectious disease, generally occurring in the lungs. It can be cured with antibacterial agents, but new antibiotic resistant strains are developing at an alarming rate, especially among AIDS patients. An estimated 10 million new cases of tuberculosis occur worldwide each year.

- **Leprosy** is an ancient disease that has not yet been eliminated, but treatment is possible.

- **Fungal diseases** are a problem—in part, because there are few compounds that will cure infections and those available are quite toxic. **Histoplasmosis, coccidioidomycosis,** and **aspergillosis** are lung diseases caused by fungi.

- Gastrointestinal tract diseases are of two major types—food poisoning from ingestion of food on which toxin-producing microorganisms have grown, and food infections from drinking liquids or ingesting food contaminated with disease-causing bacteria.

- Recent evidence indicates that **peptic ulcers** in humans are caused by growth of a bacterium *Helicobacter pylori* in the stomach.

- **Botulism** is caused by ingesting food on which *Clostridium botulinum* has grown. This microorganism produces one of the most potent toxins known.

- A common food infection is **salmonellosis,** caused by various strains of the genus *Salmonella. S. typhi* is the etiological agent of typhoid, a serious human infection. **Cholera** is a disease caused by drinking water or eating food that has been contaminated with fecal waste.

- The incidence of urinary tract infections is second only to respiratory infections in the United States in total cases. These are of frequent occurrence in hospital patients.

- **Sexually transmitted diseases (STDs)** are a major health problem. These diseases are preventable, many curable, and are a societal problem.

- **Gonorrhea** is an STD that occurs most often in the 15 to 35 age group and is of considerable concern because the organism responsible for the disease is gaining resistance to the antibiotics that are employed in treatment.

- **Syphilis** is an STD that occurs less frequently than gonorrhea but is more of a menace because of the severity of the disease.

- **Vectorborne** diseases are transmitted from one infected animal to another via an intermediate host. Most of these diseases are caused by obligate parasites such as the **rickettsia. Lyme disease** and **Rocky Mountain spotted fever** are vectorborne diseases that are rapidly increasing in incidence. Among the vectorborne diseases are **typus, Q fever, relapsing fever, tularemia,** and **trench fever.**

- **Plague** has historically been a major killer of humans. It is carried to humans by rat fleas.

- The protozoan disease **malaria** is considered to be the leading health problem in the world. The disease infects 150 million humans each year and is responsible for 2 million deaths. Malaria is passed to humans by the bite of the **Anopheles mosquito.**

- **Amebiasis** and **giardiasis** are protozoan diseases of the intestinal tract. Inadequate treatment of water is responsible for the transmission of these diseases.

- **Zoonoses** are infections that are passed to humans through contact with infected animals.

Questions for Thought and Review

1. Staphylococcus is a versatile pathogen. What are some of the infections caused by this organism? How can they be controlled?

2. Why are respiratory diseases so difficult to control? What influence did this have on the development of vaccines to prevent these diseases?

3. What is pneumonia? Which organisms cause this disease?

4. *Streptococcus pyogenes* is the etiological agent for a number of infections. Cite some of these and the symptoms. What factors give this organism invasive ability?

5. What is meningitis? As it is caused by several organisms, which is the most dangerous?

6. How does tuberculosis affect an individual? How is it diagnosed? Why is it making a comeback?

7. Fungi are the causative agents of respiratory diseases in humans. Cite examples.

8. What factors are important in lowering the incidence of gastrointestinal tract infections?

9. How does "food poisoning" differ from "food infection"? Give examples. How can either of these be prevented?

10. What are the major sexually transmitted diseases (STDs) that are caused by bacteria? Are they curable? Are they more a societal problem or a medical problem? Give reasons for your answer. What is a major health concern with gonorrhea?

11. Define vector and reservoir. Cite examples. Can an organism be both?

12. Where did Lyme disease get its name? Why is this disease a danger?

13. What are the symptoms of Rocky Mountain spotted fever? If one is bitten by a dog tick and later has such symptoms, what course should be followed?

14. Protozoan pathogens are a problem in many areas of the world. Where is amebiasis of greatest concern? Giardiasis?

15. Malaria is the world's greatest health problem. Why? How can it be prevented?

Suggested Readings

Benenson, A. S., ed. 1990. *Control of Communicable Diseases in Man.* 15th ed. Washington, DC: American Public Health Association.

Mandell, G. L., R. G. Douglas, Jr., and J. E. Bennett. 1990. *Principles and Practice of Infectious Disease.* 3rd ed. New York: John Wiley & Sons.

Miller, V. L., ed. 1994. *Molecular Genetics of Bacterial Pathogenesis.* Washington, DC: American Society for Microbiology.

Roth, J. A., C. A. Bolin, K. A. Brogden, F. C. Minion, and M. J. Wannemuehler. 1995. *Virulence Mechanisms of Bacterial Pathogens.* Washington, DC: American Society for Microbiology.

Sherris, J. C., ed. 1990. *Medical Microbiology, an Introduction to Infectious Disease.* 2nd ed. New York: Elsevier Science Publishers.

Shulman, S. T., J. P. Phair, and H. M. Sommers. 1992. *The Biologic and Clinical Basis of Infectious Disease.* 4th ed. Philadelphia: W. B. Saunders.

Stine, G. J. 1992. *The Biology of Sexually Transmitted Diseases.* Dubuque, IA: William C. Brown Communications, Inc.

Volk, W. A., D. C. Benjamin, R. J. Kadner, and J. T. Parsons. 1991. *Essentials of Medical Microbiology.* 4th ed. Philadelphia: J. B. Lippincott Co.

And hell itself breathes out contagions to this world

Shakespeare, *Hamlet*

Chapter 30

Major Viral Diseases of Humans

Viral Infections Transmitted via Respiration
Viral Pathogens with Human Reservoirs
Viral Diseases with Nonhuman Reservoirs

A virus is a piece of DNA or RNA that can enter and take over the normal functions of a specific host cell. The virus selfishly supersedes the host cell genomic DNA, forcing the synthetic machinery of the host to function solely in the synthesis of viral particles. The affected cell can be bacterial, plant, or animal. The viral genetic material originated in the susceptible organism (see Chapter 14) and is attuned to the synthetic/genetic machinery of the host it infects. Viral assault on a specific host cell generally leads to disintegration of that host. However, there are temperate viruses that can lysogenize prokaryotes or establish a latent infection in eukaryotic cells (Chapter 14). An active infection in virtually all instances results in cell destruction. This is the mechanism that has evolved to disseminate viruses, and even lysogenic or latent viruses must have an infectious cycle for survival.

Chemotherapeutic agents that can cure viral infections in humans do not now exist. The nature of viral replication is such that administering an agent to rid the body of a viral infection, as we do routinely for bacterial infections, is not feasible. The physiological explanation for this rests on the marked disparity between the components of a bacterium and a eukaryotic cell in contrast with the similarities between a virus and its host. Bacteria generally have peptidoglycan in their cell wall, and an antibiotic (penicillin, vancomycin) can interfere with cell wall synthesis without harm to the eukaryotic host as it lacks peptidoglycan. Differences between the protein-synthesizing machinery in prokaryotes and eukaryotes can be exploited, and inhibition of protein synthesis in a bacterial invader with antibiotics (tetracycline, streptomycin) will not so adversely affect the host. Thus antibiotics are effective in destroying selected bacterial invaders and in "curing" disease.

Viruses, on the other hand, present a different scenario as they are considered to have originated from the type of cell they infect. A virion can attach to, invade, and commandeer the synthetic functions of that "ancestor" host cell. It is very difficult to find exploitable differences in the synthetic machinery of a host that is producing viral genome versus activities involved in reproducing itself. A chemotherapeutic agent that would interfere with virus replication would likely have a comparable adverse effect on essentially equivalent reactions in a healthy cell. Experience has affirmed that this is essentially the case. There are a limited number of compounds that are administered for specific viral infections—adamantine for influenza A, acyclovir for herpesvirus infections, azidothymidine for HIV—but none are actually a cure. The available antiviral compounds block penetration by the virus or attack vulnerable areas in synthesis of virus-specific nucleic acids. They will be discussed in appropriate sections of the text.

This chapter is devoted to a discussion of major viral infections in humans. The transmission, entry, and symptoms of viral infections will be discussed along with mechanisms for prevention and/or control of the viral pathogens.

Viral Infections Transmitted via Respiration

Diseases transmitted from person to person via airborne droplets are difficult to control. Normal, everyday human contacts can lead to a broad distribution of infectious particles. Highly contagious diseases such as influenza, which are spread via the respiratory route, can be transmitted rapidly through a susceptible population and can reach localized epidemic proportions among human populations that have not previously been exposed.

Because many of the infections transmitted via respiration are so contagious, much emphasis has been placed on developing immunization programs for them, especially since quarantine or isolation of the ill has not proven effective. In some cases (smallpox), immunization has been totally effective, and in other cases (common cold), it has been a virtual failure. In the case of the common cold and influenza viruses, vaccination has been particularly difficult because these viruses are quite mutable and there are many antigenic types. Consequently, development of an immunity to the strain currently passing through the population does not confer immunity to an epidemic that might occur in the future.

Viral influenza and/or the common cold affect 85 to 90 percent of the American people each year. All other infectious diseases **combined** affect fewer than 20 percent of the population. This section presents some of the major viral infections that are spread via the respiratory route (Table 30.1).

Chicken Pox

Chicken pox is caused by a member of the **herpesvirus** group termed the **varicella-zoster virus.** Chicken pox is a highly contagious infection that is mainly a childhood disease. Prior to the development of an effective vaccine there were over 2.5 million cases of chicken pox in the United States annually. The infection begins, after an incubation period of about 14 days, in the mucosa of the upper respiratory tract and replicates at the site of entry. The virus is carried to the bloodstream and lymphatics where replication continues. Chicken pox is characterized by a fever and small reddish vesicles that erupt on the skin (Figure 30.1**a**). The rash may occur over much of the body but is more severe on the trunk and scalp than on the extremities. The contagious period begins a few days before the appearance of lesions and subsides a few days after fever ends. The vesicles are painful and itchy and occur over two to four days as the virus replicates. Scratching by young victims of the disease can cause infection and minor scarring. The virus can move to the respiratory and gastrointestinal tracts with more severe consequences. Damage to lungs and blood vessels in essential organs can be fatal.

A primary infection confers a lifetime immunity to chicken pox but can lead to the presence of dormant virus in the nuclei of sensory nerve roots. These dormant viruses can be activated by physiological or mental stress or by immunosuppressants, and this activation leads to an active infection commonly called **shingles** (Figure 30.1**b**). When activated, the virus moves to sensory nerve axons where it replicates and damages sensory nerves. Skin eruptions occur in the area of nerve damage and are quite painful and itchy. Generally these lesions appear about the trunk hence the name *zoster* (Greek for girdle).

***Table* 30.1 Droplet or airborne viral diseases of humans**

Infectious Disease	Viral Agent	Control by Vaccination	Type[1]
Chicken pox	Herpesvirus	+	E ds DNA
Rubella	Togavirus	+	E ss RNA
Measles	Paramyxovirus	+	E ss RNA
Smallpox	Vacciniavirus	Eradicated	E ss RNA
Mumps	Paramyxovirus	+	E ss RNA
Influenza	Orthomyxovirus	±	E ss RNA
Common cold[2]	Rhinovirus	–	N ss RNA
	Coronavirus	–	E ss RNA

[1]Type E = enveloped, N = naked capsid.

[2]Also spread by contact and formites.

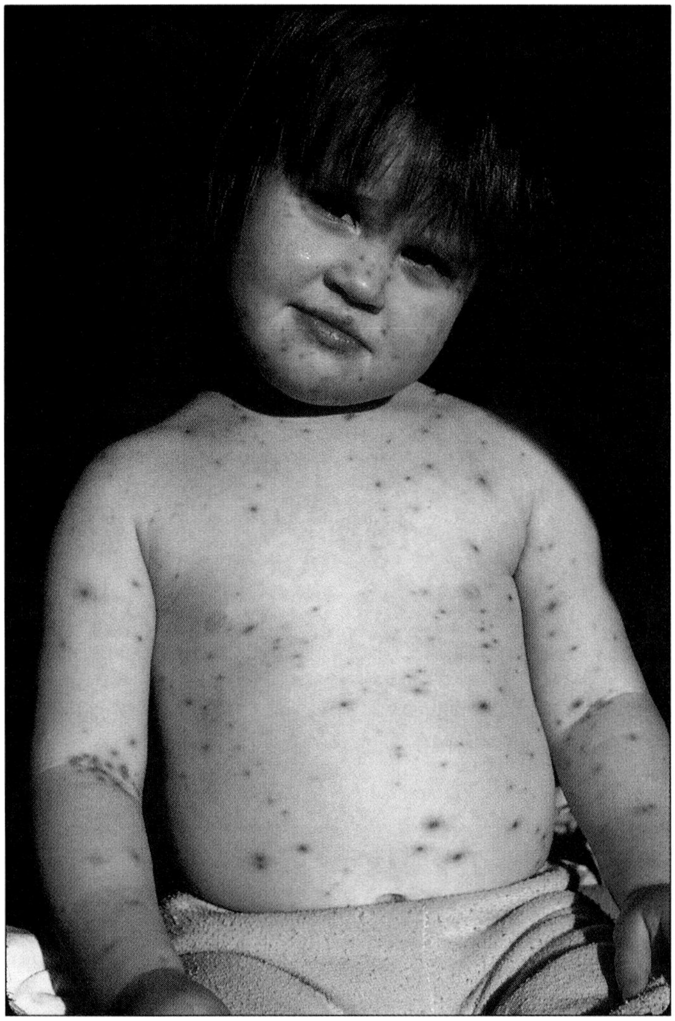

(a)

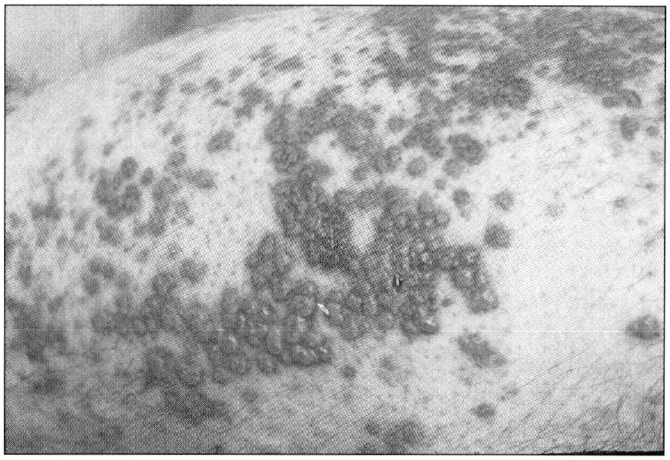

(b)

***Figure* 30.1** Symptoms of herpesvirus infection: **(a)** vesicular lesions associated with chicken pox (varicella) (©1990 Edward Ely/Biological Photo Service), and **(b)** shingles (herpes zoster lesions). (©1991 SIU Biomed Comm./Custom Medical Stock Photo)

A vaccine has been employed in Japan for many years, and in 1995 a vaccine was approved for distribution in the United States. The chicken pox vaccine is a live attenuated varicella virus and is administered to children on the same schedule as the MMR (measles, mumps, rubella) series.

Rubella

The infectious agent for **rubella (German measles)** is a **togavirus** disseminated by respiratory secretions from infected individuals. It is moderately contagious, and infection results in a rash of red spots (Figure 30.2) and a mild fever. The illness is of short duration, and complete recovery occurs within three to four days.

The rubella virus infects the upper respiratory tract and spreads to local lymph nodes. It moves throughout the body via the bloodstream. The incubation period

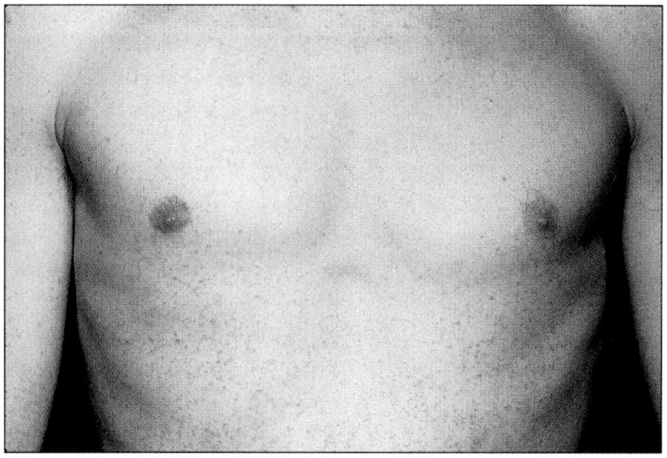

***Figure* 30.2** A rash of red spots caused by rubella (German measles). The spots are not raised. (© Barts Medical Library/Phototake NYC)

varies from 2 to 3 weeks, but the virus is present in the upper respiratory tract for more than a week before the onset of the rash. Transmittable virus is present in the respiratory tract for a period of at least 7 days after the disappearance of the rash. This is another case where an infected individual can transmit a disease prior to as well as after the onset of symptoms.

Rubella virus can cause tragic symptoms in pregnant females who become infected with the virus. The virus can cross the placenta, and infect the fetus during the first few months of pregnancy. This fetal infection may result in congenital defects in the newborn. These defects include hearing loss, microcephaly, cerebral palsy, heart defects, and other abnormalities. Fetal death or premature birth may also occur. A neonate that has been exposed to rubella *in utero* can shed (transmit) the virus for one to two years after birth, and an estimated 20 percent of infected babies do not survive the first year.

A live attenuated virus vaccine is administered in the MMR series to preschool children in the United States. It is about 95 percent effective, and widespread use may develop a herd immunity that will ultimately eliminate the disease. This would substantially reduce the possibility of rubella infections in at-risk pregnant women. Herd immunity occurs when a high percentage of the population is immune to an infectious agent and there are so few carriers that the disease is virtually nonexistent.

Measles

Measles, also known as **rubeola,** is a highly contagious viral disease caused by a **paramyxovirus.** It is an acute systemic infection marked by nasal discharge, cough, delirium, eye pain, and high fever. A rash (Figure 30.3) can appear over the entire body, and lesions in the oral cavity are rather common. The measles virus enters through the respiratory tract or conjunctiva (eye). The incubation period is about 11 days after exposure, and the rash appears three days after the onset of fever, cough, and other symptoms. These symptoms persist for 3 to 5 days and are followed by the rash, which spreads from the head downward over a subsequent 3- to 4-day period.

Measles is a severe disease and is frequently followed by secondary infections. Pneumonia, inner ear infections, and, infrequently, encephalitis may occur. This encephalitis has nearly a 25 percent mortality rate and can cause neurological damage in survivors.

Today there are few cases of common measles in the United States due to a comprehensive vaccination program (MMR) that was initiated in the mid-1960s. The vaccine employed is an attenuated live virus and induces immunity in over 95 percent of recipients. Infants from 12 to 18 months old are vaccinated, a mandatory practice in most states. There have been some outbreaks in recent years among poor inner-city children, immigrants, and illegal aliens who have not been immunized. Each year there are more than 50 million cases of measles worldwide causing the death of about one million people. Occasional outbreaks occur on college campuses and are a concern because adults generally have a more severe infection than that observed in children.

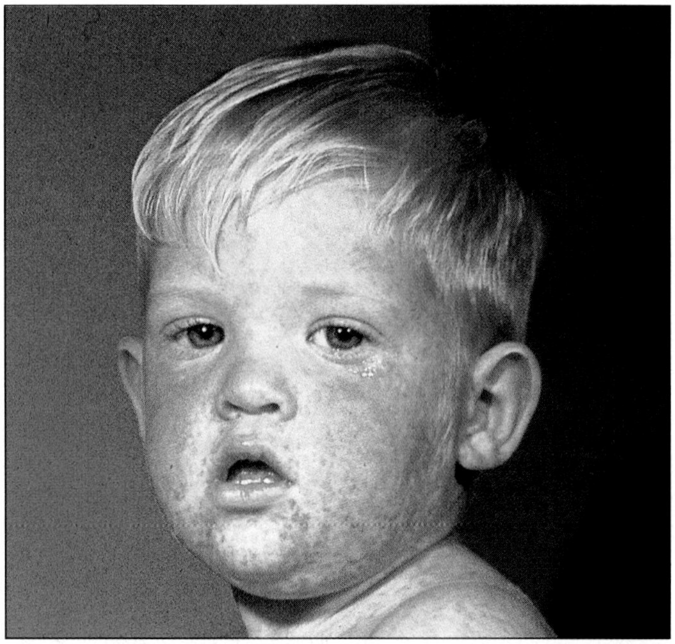

***Figure* 30.3** The characteristic rash associated with rubeola (measles) in a young child. The rash begins on the face and moves downward to the trunk. (© Science VU/Visuals Unlimited)

Smallpox

Smallpox has been a deadly disease that has caused untold misery and death throughout recorded human history (see **Box 30.1**). The smallpox virus entered a potential victim through the respiratory tract and had an incubation period of 12 to 16 days. Regional lymph nodes became infected. The virus then entered the bloodstream (viremia) and was disseminated throughout the body. The first symptoms were fever, chills, headache, and exhaustion. The virus continued multiplying in mucous membranes, in internal organs, and on the skin. Lesions occurred over the entire body and were filled with fluid. The lesions often erupted and became infected. These lesions were a considerable concern under the unsanitary conditions that existed in much of the world.

There were two strains of the smallpox virus: (a) variola major, which produced a severe infection, and (b) variola minor, which caused a mild infection. The death rate from variola major infections approached 50 percent; whereas, variola minor had less than a 1 percent mortality rate. Infection with variola minor resulted in a lifelong immunity to both strains. In fact, during the years when

(a)

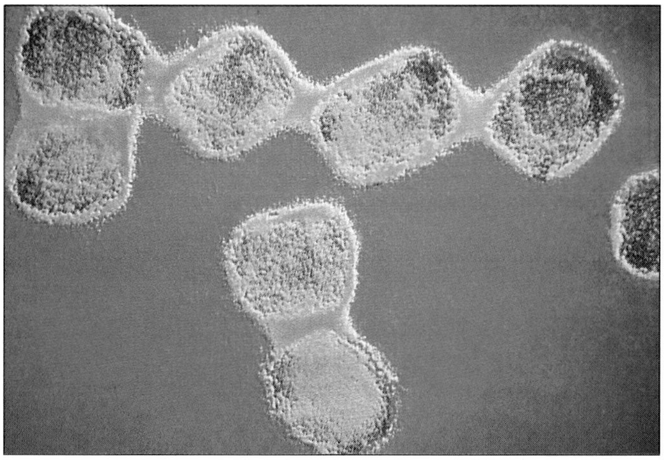

(b)

***Figure* 30.4** The last known victim of naturally acquired smallpox was Ali Maow Maolin, shown in **(a)**. The pustules are characteristic of smallpox and this infection occurred October 1977 in Somalia. (© Courtesy of WHO/J. Wickett) Maolin received treatment and survived. **(b)** An electronmicrograph of the smallpox virus. (© Institut Pasteur/CNRI/Phototake)

smallpox was rampant, individuals exposed themselves to patients infected with the mild form in an effort to gain immunity to the more deadly strain.

The World Health Organization initiated a worldwide program for smallpox immunization in 1966. The effort was funded by the United States, Russia, and other industrialized nations. All humans were inoculated. Because humans are the sole host for the virus, the virus was eradicated and the program was a great success. The last known case of smallpox occurred in 1977 in Somalia (Figure 30.4**a**; the virus is depicted in 30.4**b**). The individual received treatment and survived. The disease was declared eradicated in 1979 (see **Box 30.2**).

Mumps

Mumps was a common disease in children prior to the development of an effective vaccine. Mumps is caused by a **paramyxovirus** and is spread via the respiratory route. Disease symptoms occur 18 to 20 days after exposure. The mumps virus multiplies in the upper respiratory tract and progresses to local lymph nodes. The virus moves from the local lymph nodes to the bloodstream, and the resultant viremia disseminates the infection throughout the body. The obvious symptom of the disease is inflammation and swelling of the parotid glands on one or both sides of the neck (Figure 30.5). The parotids are salivary glands located below and in front of the ear. Enlargement of these glands can lead to difficulty in swallowing. Infection can develop in the meninges, pancreas, ovaries, testes, or heart. Infection of the pancreas may lead to juvenile diabetes. In males who have reached puberty the viral infection can become localized in the testes, a painful condition that in some cases results in sterility. Encephalitis can also occur as an aftermath of mumps infections, but is quite rare.

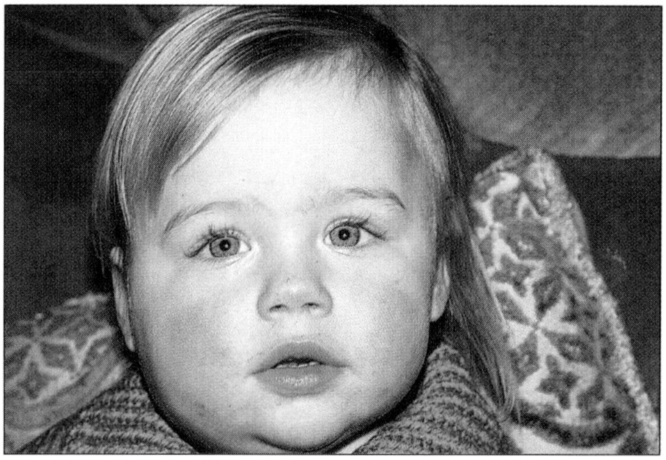

***Figure* 30.5** Enlargement of the parotid gland that is typically present in mumps patients. (© Science Photo Library)

BOX 30.1 MILESTONES

Smallpox Epidemics in the New World

Infectious diseases have played a major role in shaping the course of human evolution. Viral and bacterial infections have destroyed armies and devastated entire civilian populations. In reality, prior to large-scale vaccination and the use of antibiotics, disease played a major role in control of human populations, and many of these deadly diseases were due to viruses. One such virus was **smallpox,** a constant threat to Old World inhabitants. Smallpox became a threat to New World inhabitants, also, when explorers, fortune seekers, and immigrants brought the disease to the Americas. Because native populations had no previous exposure to smallpox, they had no resistance to the infection. As a consequence, it was exceedingly deadly to the natives of the New World.

The effect of smallpox epidemics on the Native American populations is well documented by historical records. Two examples are presented to illustrate this point. One was the purposeful employment of the smallpox virus against Native Americans and is a cruel story of "germ warfare" to destroy a perceived enemy. Another is the effect of an accidental exposure of the Aztecs, and their inability shortly thereafter to defend themselves against the conquistadors.

During prerevolutionary days Sir Jeffery Amherst was the Commander-in-Chief of the British forces in North America. At the time, General Amherst was preoccupied with containing a coalition of native tribes that had been harassing the western frontiers of Pennsylvania, Maryland, and Virginia. This coalition was under the leadership of Chief Pontiac of the Ottawa tribe. The natives, under Chief Pontiac, had captured several forts along the western frontier in defiance of the colonists who wanted to move westward. To solve this problem, Sir Jeffery Amherst proposed the use of "germ warfare." In a 1763 letter to Colonel Henry Bouquet, commander of the western forces, he offered the following advice:

> *could it not be contrived to send the smallpox among those disaffected tribes of Indians? We must on this occasion use every strategm in our power to reduce them.*

During July of that year, Colonel Bouquet replied to Amherst,

> *I will try to inoculate the ——— with some blankets that may fall in their hands, and take care not to get the disease myself.*

Drawn by Elizabeth Perry.

The results of their diabolical plan are unknown, but it is a disgraceful episode in our historical dealings with Native Americans.

On another occasion a Spanish conquistador inadvertantly brought smallpox to Mexico. In June of 1520, Hernando Cortes (1485–1547) and his army marched into the Tenochititlan, which later became Mexico City. The Aztec emperor was Montezuma, an ineffective and a weak leader. Cortes arrested Montezuma and demanded that he deliver treasures to the Spaniards to ensure their safety. Montezuma subsequently collected the treasures to satisfy Cortes's demands. Due to intrigues and treachery by Cortes's men, the Aztecs rallied around Montezuma's brother, Cuitlahuac, to drive out the Spaniards. Cortes attempted to gather his loot and escape the island with his men.

As they were attempting their escape, the Aztecs attacked and blocked the Spaniards' passage to the mainland. Many Aztecs and Spaniards were killed. Unbeknownst to the Aztecs, one Spaniard killed in battle was infected by the smallpox virus. The Aztecs looted the dead, including the victim with smallpox. Within two weeks, smallpox infected the Aztecs and killed over one fourth of the native population. Because Aztecs had never been exposed to smallpox prior to arrival of the Spaniards, they had no resistance to it. Many in the Aztec army succumbed, including the leader Cuitlahuac.

A few months later, Cortes returned and easily defeated the weak and demoralized Aztec army, looting the fortress city of Tenochititlan.

Drawn by Elizabeth Perry.

BOX 30.2 MILESTONES

To Preserve or to Destroy?

In 1798 Edward Jenner, an English physician, developed an effective vaccination procedure to protect against smallpox (see Chapter 2). Despite this discovery, smallpox continued to kill humans throughout the 19th and well into the 20th century. A worldwide immunization program initiated in 1966 led to the eradication of the disease by 1977. There has not been a documented case of smallpox in the last 20 years.

Smallpox virus survives at two sites in the world: the Centers for Disease Control in Atlanta and the Institute for Viral Preparations in Moscow, Russia. There are three major strains of the smallpox virus, and the Russian scientists were to establish a genetic map of two strains and the United States the other. All strains were to be destroyed when this task was accomplished. Scientists are divided on the issue of destroying all of the viral strains now maintained. Some favor destruction, others preservation. Those for preservation point out that genetic maps have not revealed some of the information that only the intact virus can reveal, so questions remain. Among these questions: how did smallpox kill people? Can the virus yield clues about other diseases? Are there similar diseases that could potentially emerge? Has the virus been fully exploited?

The CDC maintains about 400 vials of smallpox virus and the Russians about 200 vials. These vials are maintained in liquid nitrogen and guarded closely. The fear of biological warfare has led to smallpox vaccination of the Russian, Canadian, Israeli, and United States armies. The rest of the world is essentially unprotected.

Should all remaining cultures of the virus be destroyed? This would put an end to the goal of complete eradication of the dread disease. Should it be maintained for scientific reasons? Once destroyed it would not be available to the scientific world or to potential terrorists. The World Health Organization (WHO) in Geneva has recommended that the smallpox stocks should be eradicated by June 30, 1999. The final decision to get rid of the virus will probably be made by WHO during 1996.

There is only one antigenic type of the mumps virus. Once the virus is contracted, immunity is apparently lifelong. A live, attenuated virus vaccine was developed in 1967 that has proven to be at least 95 percent effective. This vaccine is part of the MMR series and has reduced the number of cases in the United States to fewer than 3000 per year. A mumps vaccination prior to entering school is mandatory in most states. There have been mumps outbreaks in states that do not have a mandatory vaccination program.

Influenza

Viral influenza is caused by an **orthomyxovirus.** It is an enveloped virus with an RNA genome surrounded by a matrix protein, a lipid bilayer, and glycoprotein (Figure 30.6). Influenza epidemics have occurred throughout recorded history, and pandemics (worldwide epidemics) have occurred with some frequency. The 1918–1919 pandemic, which was caused by a particularly severe and deadly strain of the influenza virus, killed at least 20 million people. The 1957 pandemic, which was the most recent, originated in central China and was called the Asian flu. The virus involved with this pandemic was a mutant of considerable virulence. From its origination in the interior of China in February 1957, the viral strain moved to Hong Kong in April. Air and naval traffic carried the infectious agent worldwide, as shown in Figure 30.7. During two weeks in October, a peak of 22 million cases were reported.

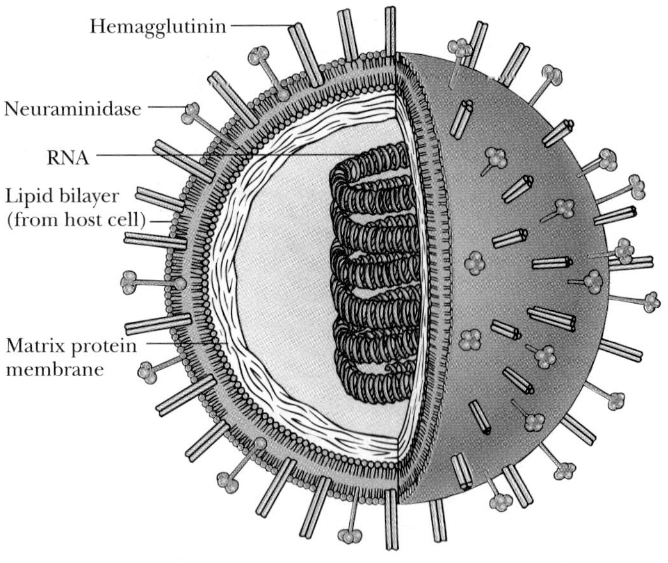

***Figure* 30.6** The influenza virus, depicting the location of the hemagglutinin and neuraminidase that are involved in invasion.

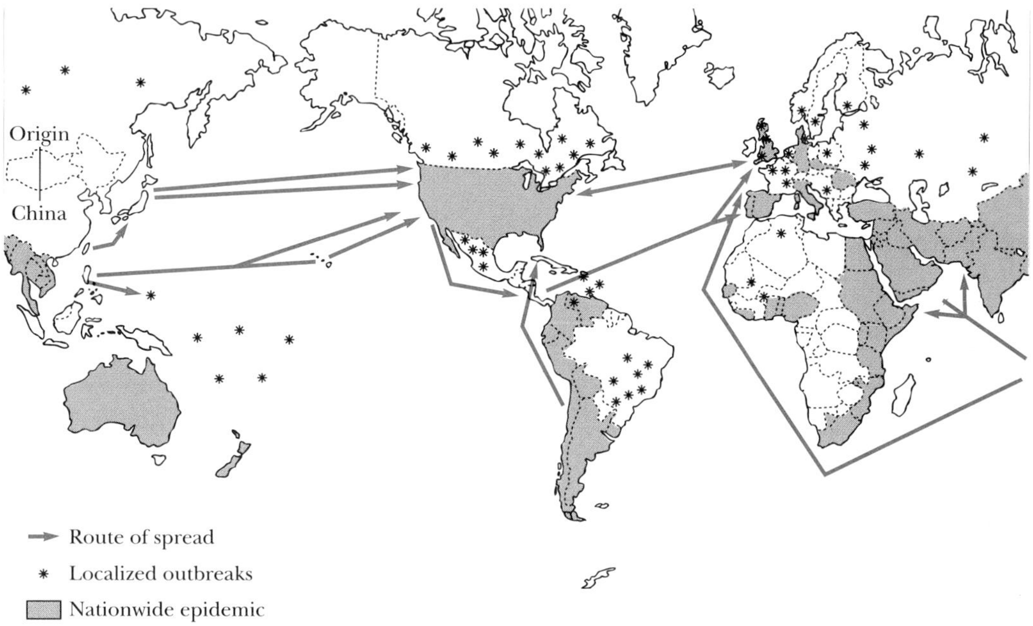

***Figure* 30.7** Route followed by the influenza pandemic that originated in the interior of China in 1957 and spread throughout the world.

The influenza virus that infects humans does not affect any other host. It is mainly transmitted through the air on droplets expelled by sneezing or coughing (Figure 30.8). The flu symptoms (chills, fever, headache, malaise, general muscular aches) appear after a 1- to 3-day incubation period and last for 3 to 7 days.

Once inhaled, the influenza virus infects the mucous membranes of the upper respiratory tract. The neuraminidase structure on the viral envelope hydrolyzes the mucus that lines the epithelial cells in the respiratory system. This allows the virus to reach the epithelial cell surface where the hemagglutinin spikes attach to the cell. Penetration occurs by endocytosis. The virus moves from the upper to the lower respiratory tract, destroying epithelial cells. The destruction of cells in the respiratory system can lead to secondary infections, such as bacterial pneumonia, caused by *Staphylococcus aureus, Streptococcus pneumoniae,* or *Haemophilus influenzae.* These are a significant problem in infants, the elderly, and otherwise debilitated individuals. Secondary infections are the leading cause of death during influenza epidemics, and prior to availability of antibiotics these deaths generally occurred from pneumonia.

The genome of the influenza virus is segmented into eight distinct fragments. When more than one virus infects a cell there can be a reassortment of genes, which yields progeny with recombinant genomes. These progeny may then produce an altered-surface glycoprotein. This altered glycoprotein can be unaffected by the antibodies that are present from a previous bout with the influenza virus. This change in the antigenic character of the viral particle is called **antigenic drift.** It is the major reason why influenza virus epidemics can occur repeatedly. There is some concern that the human influenza virus and a pig influenza virus might exchange genetic information that would generate a virulent form of this virus. Such a virus could cause a pandemic because there would be no resistance to this viral strain.

Vaccination against influenza is difficult because of the antigenic drift that yields novel strains. Polyvalent vaccines that are composed of a number of viral strains obtained from previous epidemics are generally administered.

***Figure* 30.8** A high-speed photograph of a sneeze. (© Matt Meadows/Peter Arnold, Inc.)

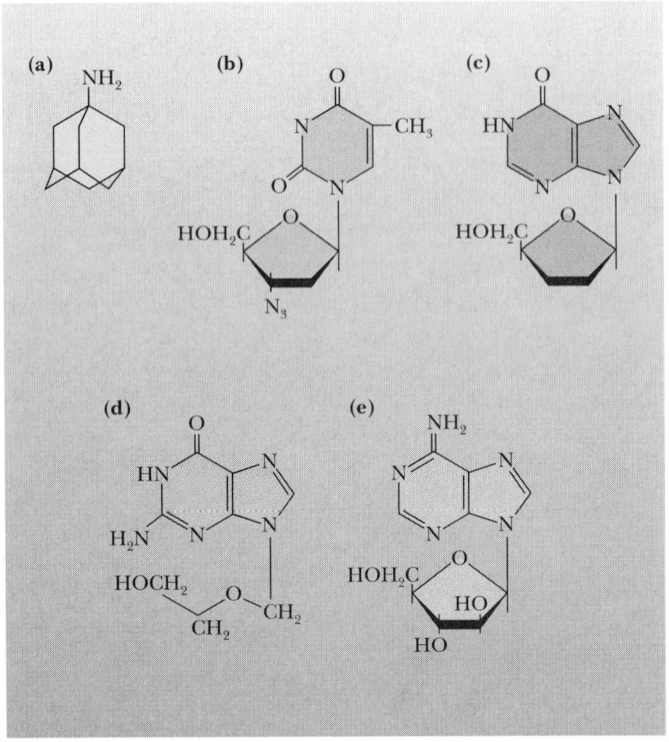

***Figure* 30.9** The structures of chemotherapeutic agents that have been of value in prevention or treatment of viral infections: **(a)** amantadine (influenza), **(b)** azindothymidine (AIDS), **(c)** dideoxyinosine (AIDS), **(d)** acyclovir (herpes), and **(e)** adenine arabinoside.

Presently, worldwide outbreaks are monitored so that the virus involved can be isolated and vaccines to new strains prepared and made available. It is the goal of these efforts to forestall a pandemic. A drug, amantadine (Figure 30.9**a**) is effective in treating influenza A if administered after exposure or early in the infection. Amantadine, given before symptoms appear, can reduce the incidence of infections by 50 to 75 percent. If given early in an infectious process amantadine may shorten the course of the infection. The drug blocks penetration by and uncoating of influenza virus particles. It is generally given to high-risk individuals.

Reyes syndrome is an acute encephalitis that can occur in children following some viral infections. Chicken pox and influenza infections treated with salicyclates, such as aspirin, increase the incidence of this syndrome. For this reason, children experiencing flu-like symptoms should not be given aspirin or related medicinals. Reyes syndrome is marked by brain swelling that results in injury to neuronal mitochondria. Liver damage can also occur. Reyes syndrome has about a 40 percent fatality rate, and mental deficiency can occur in survivors.

Guillain-Barre syndrome is a delayed reaction to influenza infection that generally occurs within eight weeks. Guillain-Barre can be a reaction to the virus or to a vaccine. This disease damages the cells that myelinate peripheral nerves. This demyelination results in limb weakness and sensory loss. Recovery is complete in virtually every case.

Colds

The common cold affects 40 percent to 45 percent of Americans each year and with varying severity. It is estimated that Americans miss more than 200 million work or school days annually due to symptoms of the common cold. Rhinoviruses are involved in about 75 percent of all cases and are the most common etiological agent (Figure 30.10). Colds caused by the rhinoviruses generally occur in spring and fall.

Another group of viruses associated with the common cold are the coronaviruses (Figure 30.11). These viruses generally are involved with colds that occur in mid-winter. The coronavirus envelope has projections that resemble a solar corona, hence the name. Coronaviruses are involved in about 15 percent of all cases. About 10 percent of all colds are caused by various other viruses.

The cold viruses may be transmitted in airborne droplets or by direct contact with infected individuals. Airborne transmission in volunteers has been difficult to demonstrate. The cold viruses can survive on inanimate objects for hours, and fomites are considered to be involved in transmission. Touching inanimate objects such as doorknobs that are contaminated with viral particles as a result of handling by an infected person is considered a major route of transmission.

Cold symptoms include inflammation of the mucous membranes, nasal stuffiness, "runny" nose, sneezing, and

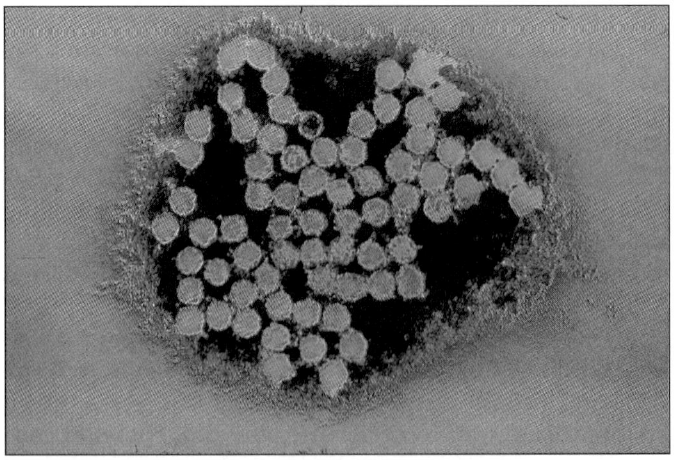

***Figure* 30.10** An electronmicrograph of a rhinovirus, an etiologic agent of colds (© 1992 Science Photo Library/Custom Medical Stock Photo)

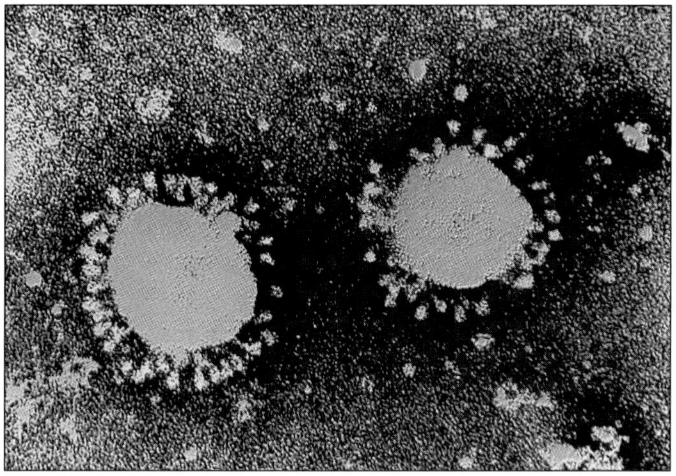

***Figure* 30.11** An electronmicrograph of a coronavirus, another causative agent of human colds. Note the projections from the viral envelope that give the virus its name. (© 1990 Custom Medical Stock Photo)

a scratchy throat. The symptoms can last for about one week. Clinical symptoms develop in only about one half of infected individuals. Individuals with no apparent symptoms are a constant source of infection to the general population. Treatment includes aspirin, antihistamines, nasal sprays, and plenty of fluids.

There are over 110 known serotypes of the rhinovirus and, as with the influenza virus, the cold virus is subject to antigenic drift. Effective antibodies are not generally produced, and any immunity developed is short-lived. There is little hope for a successful immunization program for the common cold as there are too many distinctly different viruses (rhinovirus, coronavirus, adenovirus, and so on) that cause colds and too many serotypes. Diagnosis of colds is by symptoms.

Viral Pathogens with Human Reservoirs

There are a number of highly contagious viral diseases that are transmitted directly from human to human (Table 30.2) and are difficult to control. The human serves as **reservoir,** and the viruses are transmitted in various ways including contaminated blood, by sexual contact, or intrauterine to a fetus. The viruses are also transmitted through kissing, through touching, or via the fecal-oral route from contaminated water. Ingestion of raw or poorly cooked shellfish that has been exposed to raw sewage is a danger. Swimming in septic water can also promote the transmission of selected viruses. Common warts are caused by a virus that is passed from human to human by direct contact, not by touching frogs.

This section presents information on the major viral diseases that are transmitted via human contact. These are the major diseases in the United States and, along with others, they are prevalent in other areas of the world.

Acute T-Cell Lymphocytic Leukemia

There is a growing awareness that viruses can transform mammalian cells, and this can lead to sarcoma, leukemia, lymphoma, or other cancers. One virus that can transform cells is the **human T-lymphotropic virus (HTLV),** a retrovirus (see Chapter 14).

There are two types of HTLV: HTLV-1 and HTLV-2. These viruses are transmitted through transfusion with contaminated blood, between IV drug users who exchange needles, or via sexual contact. The oncogene-bearing viruses may also be transmitted by breast feeding. The HTLV-1 infects the CD-4$^+$ lymphocytes and is not cytolytic.

The leukemia virus (HTLV-1) causes cancer after long latent periods by promoting outgrowth of infected cells in

***Table* 30.2 Viral diseases with humans as reservoir that spread directly or indirectly through human contact**

Disease	Mechanism of Transmission
Human T-cell leukemia Serum hepatitis AIDS	Contaminated blood, exchange of needles between IV drug abusers, sexual contact
Genital herpes Venereal warts	Sexual contact
Infectious mononucleosis Herpes simplex	Mouth-to-mouth contact
Infectious hepatitis A Gastroenteritis	Fecal-to-oral route, raw shellfish
Polio	Water, food, swimming pools
Warts	Direct contact

two ways. The virus may integrate near host gene sequences that are involved in growth control and activate their expression or activate promoters that stimulate overgrowth of cells. Uncontrolled cell growth can transform the lymphocyte and induce the outgrowth of a clone of aberrant T-cells. This clone of HTLV growth-stimulated lymphocytes reproduces with the accumulation of the aberrant leukemia cells. HTLV-1 infection is generally asymptomatic but can progress to acute T-cell leukemia in about 5 percent of those infected. The leukemic cells are pleomorphic and have lobulated nuclei. The disease is generally fatal within a year after diagnosis, and there is no treatment.

Human oncoviruses may have exceedingly long latency between infection and development of the disease. This latency can be 30 or more years. They will not transform cells *in vitro,* which makes experimental investigation difficult. The presence of the virus can be detected by the identification of virus-specific antibodies in the patient's serum. There is no cure, and prevention is by preventing transmission of the viral particle. At the present time, human blood supplies are not examined for viruses such as HTLV.

There are a number of viruses that can cause tumors in humans, a number of which are listed in Table 30.3.

Acquired Immune Deficiency Syndrome (AIDS)

A brief discussion of the clinical effects of the AIDS infection follows. For a discussion of the effect of HIV, the causative agent of AIDS on the immune system, see Chapter 28; for a discussion on transmission and development of the virus, see Chapter 14. Transmission of the HIV virus occurs by direct contact with body fluids (blood, semen, vaginal secretions) of an infected individual. Passage of contaminated body fluids to the bloodstream of an uninfected individual inevitably will lead to infection. The populations most at risk of acquiring HIV are homosexual and bisexual males, IV drug users, and heterosexuals who have sexual intercourse with IV drug users, prostitutes, or bisexual males. HIV can also be transmitted by blood transfusions or from infected mothers to the newborn.

There is a mild form of the disease in which antibodies to the human immunodeficiency virus (HIV) are present in blood, but symptoms are minimal and generally go unnoticed. The antibodies do not eliminate the virus because the virus becomes part of the genome of CD-4^+ T lymphocytes in an infected individual. During this period, diagnosis is accomplished by the presence of antibodies against HIV virus, and the viral infection is called ARC (AIDS-related complex). Symptoms of ARC include fever, malaise, headache, and weight loss. The symptoms generally occur within weeks of infection and last for 1 to 3 weeks. The ARC form of infection can develop into full-blown AIDS at any time. There are an estimated 1 million people in the United States who are HIV positive. The Centers for Disease Control and Prevention (CDCP) estimates that over 215,000 individuals died of AIDS from the beginning of 1993 to the end of 1994 with 60,000 to 70,000 new cases anticipated per year through the year 2000.

Exposure to HIV can lead to the inapparent response mentioned above or to immediate symptoms. Symptoms of AIDS occur when the bulk of one's CD-4^+ T lymphocytes are destroyed by the presence of the HIV virus. The lymphocytes that are destroyed are a major part of the normal immune response, and, consequently, individuals with AIDS cannot generate an effective immune response. An AIDS patient is vulnerable to attack by opportunistic infections. Opportunistic infections (Figure 30.12) are those not generally observed in immunologically competent individuals (those with a normal immune response) but are opportunists in that they take advantage of a reduced host immune system (Table 30.4). Infections in AIDS patients generally originate from eukaryotic microbes, and these infections by eukaryotes are difficult to control. For an active case of AIDS the diagnosis is often made by identification of an array of secondary infections including *Pneumocystis carinii* pneumonia. Pneumocystis pneumonia is a fungal disease that can be diagnosed by finding the organism in sputum of infected individuals.

Another symptom of AIDS is Kaposi's sarcoma, which causes purplish blotches on the skin of legs and feet. In immunocompetent individuals these lesions cause no difficulty and are self-healing. In AIDS-infected people the cancer can spread to the lungs, lymph nodes, and brain. Other malignancies are Burkitt's lymphoma, primary brain lymphoma, or immunoblastic lymphoma. Viral infections occur as secondary invaders and are also a problem, particularly the cytomegalovirus that can cause retinal damage and blindness. **Herpes** simplex can cause chronic ulcers or bronchitis. Other viral infections can be carried to the brain tissue by macrophages causing overproduction of nerve cells that surround the neurons. This results in a loss of mobility and brain function. *Candida*

***Table* 30.3 Some viruses known to cause tumors in humans**

Virus	Cancer
Hepatitis B virus	Liver cancer
Human T-cell lymphotrophic virus	T-cell leukemia
Herpes simplex type 2	Cervical cancer
Epstein-Barr virus	Burkitt's lymphoma
	Nasopharyngeal carcinoma
Papilloma virus	Cervical cancer

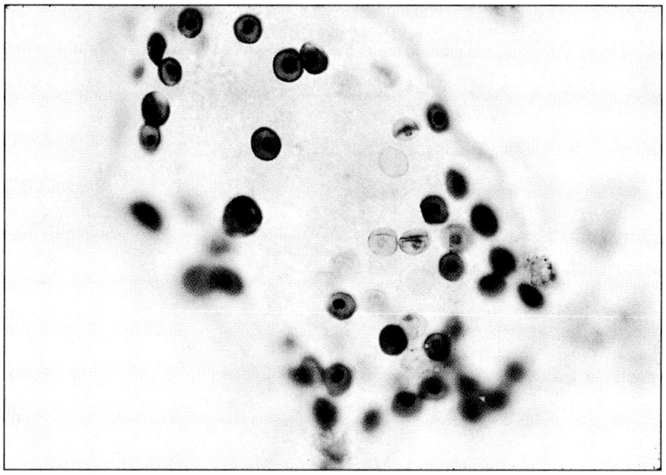

(a)

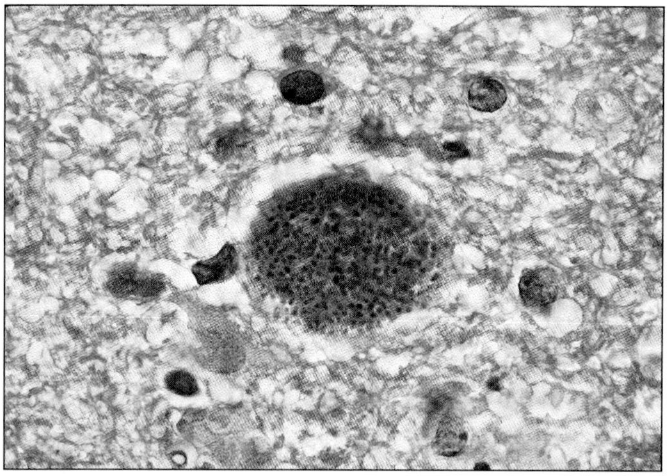

(b)

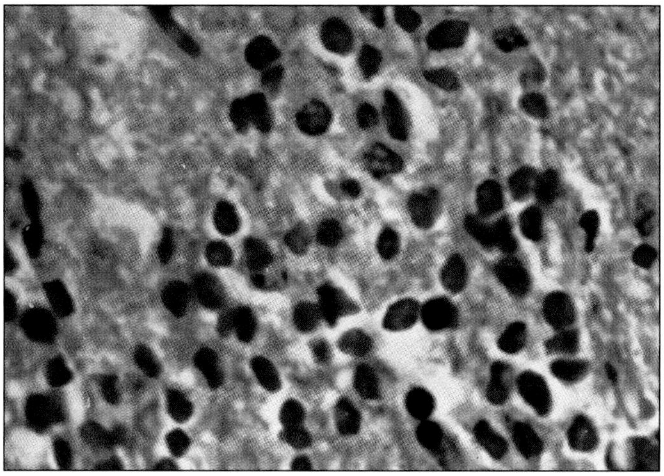

(c)

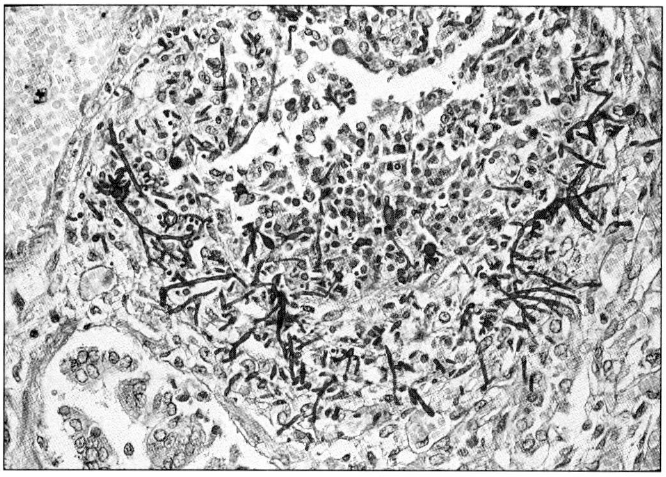

(d)

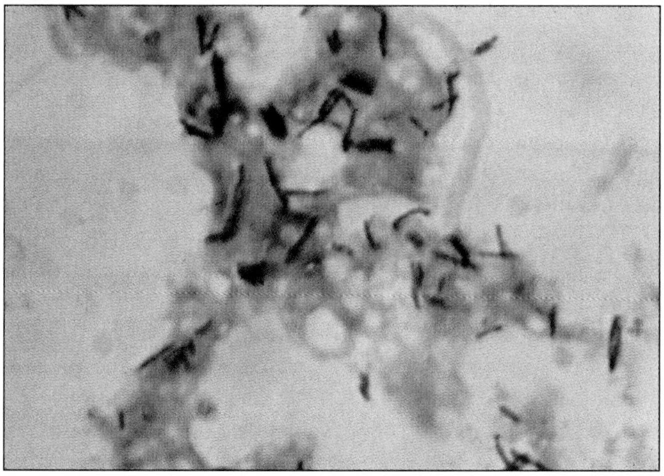

(e)

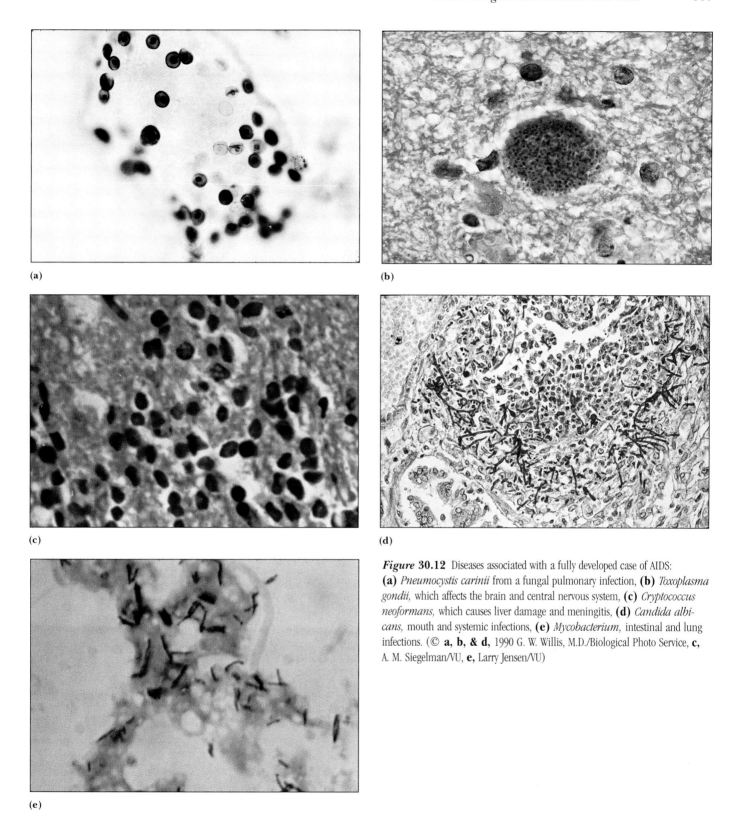

***Figure* 30.12** Diseases associated with a fully developed case of AIDS: **(a)** *Pneumocystis carinii* from a fungal pulmonary infection, **(b)** *Toxoplasma gondii,* which affects the brain and central nervous system, **(c)** *Cryptococcus neoformans,* which causes liver damage and meningitis, **(d)** *Candida albicans,* mouth and systemic infections, **(e)** *Mycobacterium,* intestinal and lung infections. (© **a, b, & d,** 1990 G. W. Willis, M.D./Biological Photo Service, **c,** A. M. Siegelman/VU, **e,** Larry Jensen/VU)

albicans is present in the mouth and alimentary tract of healthy individuals. In the immune-compromised AIDS patient it can cause severe mouth infections. Another yeast infection, cryptococcosis (caused by *Cryptococcus neoformans*) can cause a damaging infection in meninges and brain. Toxoplasmosis (*Toxoplasma gondii*) and cryptosporidiosis (*Cryptosporidium*) are protozoan diseases that afflict AIDS patients. The former causes brain damage and the latter a severe form of long-lasting diarrhea. Non-Hodgkins lymphoma and other malignant lesions can

Table 30.4 Major opportunistic pathogens associated with active cases of AIDS

Organism	Symptoms
Eukaryotic	
Protozoan	
Cryptosporidium	Intestinal infections
Toxoplasma gondii	Brain, central nervous system
Fungi	
Pneumocystis carinii	Pulmonary pneumocystosis
Cryptococcus neoformans	Meningitis, liver damage
Candida albicans	Oral thrush, systemic infections
Histoplasma capsulatum	Systemic infections
Coccidiodes immitis	Pneumonia
Helminths	
Strongyloides stercoralis	Enteric infections
Viral	
Herpes	Ulcerative lesions
Cytomegalovirus	Pneumonia, blurred vision
Prokaryotic	
Mycobacterium tuberculosis	Tuberculosis
Mycobacterium sp	Lymph node infections
Salmonella sp	Septicemia
Streptococcus pyogenes	Pneumonia

occur in the AIDS-infected, and these are cancers that are normally removed by immune host defenses.

There is a concerted effort to control AIDS by developing vaccines or effective chemotherapeutics. (See **Box 28.1**, AIDS vaccines). There are vulnerable parts of the AIDS virus, including the surface glycoproteins that might induce an effective immune response under appropriate conditions. Much effort is being expended to develop a vaccine with these or other viral proteins as antigen. Azidothymidine (AZT) is the current drug of choice in AIDS treatment. It is not a cure but can prolong the life of the patient. AZT is a synthetic nucleotide-base analogue (Figure 30.9**b**). During viral replication, AZT is incorporated into the site normally occupied by thymidine. The next nucleotide cannot be put in place because the N_3 blocks addition of another nucleotide. The AZT molecule serves as a chain terminator and is particularly effective in blocking the polymerase involved in reverse transcriptase. However, there is evidence that resistance to AZT is occurring in strains of the AIDS virus. Other chemotherapeutic agents are under development. Combined therapy with AZT and other chemotherapeutics may be of value. Dideoxyinosine (ddI) (Figure 30.9**c**) is a drug that can block reverse transcriptase and has been administered for AIDS infections.

Cold Sores (Fever Blisters)

Herpes simplex type 1 (HSV-1) is the causative agent of **cold sores** or **fever blisters.** This virus is a member of the **herpesvirus,** a group that includes common animal pathogens. The herpes viruses are noted for their ability to cause latent infections that may be activated days or years after the initial infection. Zoster, the etiological agent of shingles, is one such virus and was discussed previously. Herpes simplex is a double-stranded DNA virus with a genome that is surrounded by an icosahedral nucleocapsid. The capsid is enclosed in a membrane envelope, with glycoproteins inserted in this envelope.

HSV-1 infection results in blisters around the mouth, lips, and face (Figure 30.13). The blister results from host and virus-mediated tissue destruction. The lesions heal in one to three weeks without treatment. It is estimated that up to 90 percent of adults in the United States possess antibodies against HSV-1. During an active infection the virus is present in oropharyngeal secretions and is spread by salivary exchange. The incubation period is three to five days. After a primary infection the virus moves to trigeminal nerve ganglia and remains in the latent state. A viral attack can occur sporadically throughout the life of an infected individual. Clinical systems are triggered by stresses such as excessive sunlight, emotional upset, fever, trauma, immune suppression, or hormonal changes. When the latent virus is activated it travels down peripheral nerves to the lips, mouth, or facial epidermis, resulting in a recurrence of the fever blister(s). **Acyclovir** (see Figure 30.9**d**), a guanine analogue, is effective in treating herpes infections. When acyclovir is phosphorylated it resembles deoxyguanine triphosphate and inhibits the viral DNA polymerase. The thymidine kinase is a viral-directed enzyme coded by the virus. Acyclovir would be activated selectively in a virus-infected cell. It does not affect dormant virus and is not a cure. **Adenine arabinoside** is another chemotherapeutic that disrupts DNA polymerase in herpes virus (Figure 30.9**e**).

Genital Herpes

Genital herpes is generally caused by **herpes simplex** virus type 2 (HSV-2), with about 10 percent of clinical cases the result of HSV-1 infection. Genital herpes is associated with the anogenital region and is transmitted by direct sexual contact. A primary infection occurs after an incubation period of one week, but many primary infections are asymptomatic. The virus causes painful blisters on the penis of males. In females, the lesions appear on the cervix, vulva, and vagina. The blisters are a result of cell lysis and a local inflammatory response. They contain fluid and infectious viral particles. During this stage the virus is readily transmitted to unprotected sex partners. The blisters resulting from a primary infection heal spontaneously in two to four weeks. Recurrent genital herpes infections are generally shorter and less severe than the initial episode. These lesions that occur after primary infection heal in seven to eight days. As with HSV-1, the HSV-2 virus travels to nerve cells and remains latent. Genital herpes can recur every few weeks, a few times per year, or not at all. An active case can be activated by sunlight, sexual activity, or fever. An estimated 50 percent of those infected with HSV-2 never have symptoms and are unaware of a primary infection. These individuals may shed the virus and can be major transmitters of the disease.

Herpes virus has been associated with cervical cancer in females, and the virus can be transmitted to an infant during birth. There are over 2000 babies born in the United States annually with HSV-2. The symptoms in the neonate vary from a latent inapparent infection to brain damage and death. To avoid infecting the newborn, cesarean section is advised if there are virus-infected lesions in the genital area. Unfortunately, the virus can be present in genital secretions without overt symptoms, thereby exposing a vaginally delivered infant.

There is no known cure for herpes virus infections. Oral and topical administration of acyclovir can lessen the symptoms of the disease. Genital herpes has become a significant sexually transmitted disease, with an estimated 20 million infected individuals and about 500,000 new cases occurring in the United States each year.

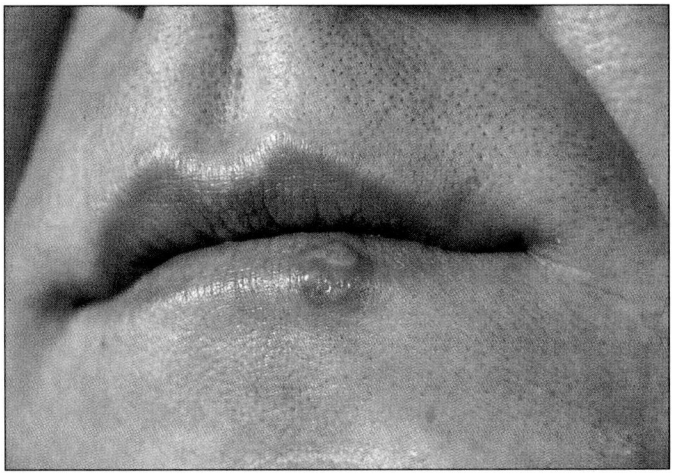

***Figure* 30.13** Cold sores are also known as fever blisters. They are caused by Herpes simplex type 1. (© CDC/Science Source/Photo Researchers, Inc.)

Infectious Mononucleosis

Infectious mononucleosis is caused by a human herpes virus called Epstein-Barr virus (EBV). The virus is named for its codiscoverers and was originally isolated when T. Epstein and Y. Barr were seeking the etiologic agent for a malignancy of lymph nodes called Burkitt's lymphoma.

The EBV can be present in the mouth-throat area and is commonly called "kissing disease"; the infection occurs most frequently among those of college age. The symptoms include enlarged lymph nodes and spleen, sore throat, nausea, and general malaise. These symptoms may be accompanied by a mild fever. Fatigue is a common complaint of mononucleosis patients. The symptoms last for one to six weeks and the only treatment is rest.

Once contracted, EBV enters lymphatics where the virus multiplies and infects B-cells. The B-cells proliferate rapidly and are altered in appearance. These atypical lymphocytes are the earliest indication of a mononucleosis infection. The presence of antibody against EBV is another and a more reliable indicator of the disease.

Burkitt's lymphoma is a malignancy that affects children in Uganda and other central African countries. The EBV virus has also been implicated in nasopharyngeal carcinoma, which occurs more frequently among the Chinese than people in the Western Hemisphere. There is evidence that another agent acts as cocarcinogen in a malignancy such as Burkitt's lymphoma. Areas where the malignancy occurs have a high incidence of malaria, a disease known

BOX 30.3 MILESTONES

Conquering Polio

During the first half of the 20th century, paralytic polio was among the most dreaded diseases in the industrialized world. It struck randomly in human populations and the epidemiology was poorly understood. We now know that improved sanitation conditions in the more advanced countries was a significant factor in the spread of paralytic polio. The virus normally survived in the intestinal tract of infants and conferred immunity. Exposure at a very early age was the rule in populations where sanitation was less favorable. If a child was not exposed at an early age but randomly contracted the disease at an older age, paralysis was more likely. Consequently, the chance of contracting paralytic polio increased with populations with an improved standard of living. During the period when polio was rampant, the "Mothers March of Dimes" was organized. This was a rather fashionable organization that collected dimes to support research on this feared disease.

The development of viral vaccines was stymied by the inability to obtain quantities of viral particles. Less virulent virus, such as that used in smallpox, or one altered to incite an antibody without a full-blown infection was a must. Until the middle of the 20th century, there was no available technique for propagating virus outside a living host. It was then (1946) that John Enders, Thomas Weller, and Frederich Robbins began a collaboration at the Infectious Disease Laboratory at Boston Children's Hospital. Together they developed techniques for propagating poliomyelitis virus in tissue culture. This was the essential step in the development of the polio vaccine and a major breakthrough in the study of viruses in the laboratory. These three scientists worked together from 1946 to 1952, and the tissue culture methods they developed could be used for selection of strains with altered pathogenicity—a prerequisite for developing effective vaccines for any viral disease. Polio, measles, and mumps vaccines resulted from the techniques developed in John Enders's laboratory.

Enders, Weller, and Robbins received the Nobel Prize for their contribution in 1954.

to suppress the immune system. Suppression of the immune system would permit EBV to proliferate in B-cells. Individuals who have AIDS and consequently are immunosuppressed are often victims of an EBV infection.

Poliomyelitis

Poliomyelitis, also known as **polio** or **infantile paralysis,** is caused by a picornavirus that may survive in food or water for lengthy periods of time. The virus is transmitted by ingestion of contaminated food or water. Public swimming pool water was considered a source of polio infection when the disease was of widespread occurrence in the United States prior to the mid-1950s **(Box 30.3).** The incubation period is generally 7 to 14 days. The polio virus can multiply in the oropharynx and may be transmitted through respiratory droplets during the early stages of infection. Multiplication occurs in tonsils and intestinal mucosa, and the virus moves from the intestinal tract to lymph nodes. It then invades through the bloodstream and is disseminated throughout the body.

There are three forms of clinical illness that can result from invasion by the polio virus. The most common form is an **abortive infection.** This infection is asymptomatic or causes some fever, sore throat, nausea, and vomiting. Recovery is rapid, and the symptoms are rarely attributed to the polio virus. **Aseptic** infection results in similar symptoms except that some of the virus enters the central nervous system, resulting in a stiff neck and back. In these infections, full recovery occurs in about one week. **Paralytic polio** is marked by destruction of motor neurons in the anterior horn of the spinal cord. Because these cells transmit impulses to the motor fibers of the peripheral nerves, their destruction results in paralysis. If the infection involves neurons in the medulla, bulbar poliomyelitis or loss of respiratory function results. Destruction of other neurons can result in loss of motor function and muscle paralysis.

The Salk vaccine for polio was introduced in 1954 and was an inactivated virus administered by injection. An attenuated viral vaccine was developed by Albert Sabin in 1962 that is given orally. The oral vaccine is more effective than the inactivated virus because it induces an infectious cycle that establishes antibody protection in the intestinal mucosa. As a consequence, vaccinated individuals cannot serve as a reservoir of infectious particles. It is now feared that the polio virus might become established among immigrants and poor children in the inner city because many of these youngsters are not properly immunized. As we depend on herd immunity to prevent the

Table 30.5 **Characteristics of hepatitis viruses**

Hepatitis Virus	Common Name	Virus Type	Envelope	Transmission	Incubation Period (days)
Hepatitis A	Infectious	Picornavirus (RNA)	No	Fecal-oral	15–50
Hepatitis B	Serum	Hepadnavirus (DNA)	Yes	Blood Sexually	45–160
Hepatitis C	Post-transfusion Non-A, Non-B	Flavivirus (RNA)	Yes	Blood Sexually	14–180
Hepatitis D	Delta	Unknown (RNA)	Yes	Parenteral Sexually	15–64
Hepatitis E	Enteric Non-A, Non-B	Calicivirus (RNA)	No	Fecal-oral	15–50

spread of many diseases, it is important that all individuals be immunized. No vaccine is effective in 100 percent of the individuals inoculated. It is the goal to have 90 to 95 percent of the potential victims immune, and this will remove reservoirs. Without a sufficient number of reservoirs the disease will disappear. In the absence of herd immunity a new outbreak of polio could occur and have decidedly disastrous effects on those infected.

Hepatitis

Hepatitis means inflammation of the liver, and five distinct viruses have been implicated in hepatitis infections. It is probable that other viral agents of hepatitis are yet to be discovered. The characteristics of the viruses now known to be involved in hepatitis are presented in Table 30.5. These diseases are generally transmitted by exchange of blood/body fluids or via the oral-fecal route.

Hepatitis A infections are acquired by ingesting raw oysters, clams, or mussels harvested from fecally contaminated water or handled by an infected carrier. Person-to-person transmission can occur in day care centers and other institutions where sanitary conditions are difficult to maintain. The infectious hepatitis virus is exceedingly stable and can tolerate heating at 56°C for 30 minutes. It is also unaffected by many disinfectants. An estimated 40 percent of all acute cases of hepatitis are caused by Hepatitis A.

Hepatitis A has an incubation period of 15 to 50 days. The virus multiplies in the intestinal epithelia and moves via the bloodstream to the liver. It often causes a mild, self-limiting infection. In severe infections, nausea, vomiting, and fever are the symptoms. Hepatitis A causes a degenerative infection in the liver that can lead to enlargement and possible blockage of biliary excretions. This situation leads to jaundice (release of bile pigments, giving a yellowish color to the skin). Damage to the liver is immune-mediated in that antibodies to the virus attack selected hepatic cells. Recovery takes about 12 weeks, and permanent liver damage is possible. Infections in children are often asymptomatic. Hepatitis A virus appears in the feces of infected individuals 10 days before onset of symptoms and for several weeks after recovery. An active case results in immunity, and about 40 percent of the people in the United States have antibodies to infectious hepatitis. As symptoms can be more severe in adults than in children, a prophylactic treatment is available to individuals exposed to the virus. Immune serum globulin administered during the incubation period will prevent symptoms in 80 to 90 percent of those exposed.

Hepatitis B virus (HBV) is a more serious threat to human populations than is hepatitis A. An estimated 300 million individuals worldwide are carriers of the virus. Of these, 40 percent will probably succumb to liver disease. There are about 300,000 humans in the United States infected each year with HBV, with about 5,000 deaths annually. Individuals infected with hepatitis B have three different antigenic particles present in their serum (Figure 30.14). The Dane particle is the infective form of the virus. The two incomplete particles (spherical and tubular) have the hepatitis B surface antigen as part of the particle and are present in greater quantity than is the infective Dane virus. The presence of these incomplete antigenic particles in serum is an effective indicator of an infection. Their presence is also important in screening blood for hepatitis B that might be used in transfusions.

The hepatitis B virus is transmitted by needle sharing, acupuncture, tattooing, and potentially by ear piercing. The virus can also be transferred during blood transfusions and in semen, and it is present in saliva and sweat. The virus traverses the placenta, and infants born to infected mothers generally become carriers.

The incubation period for hepatitis B virus varies from 45 to 160 days. The first symptoms include fever, anorexia, and general malaise. This is followed by nausea, vomiting,

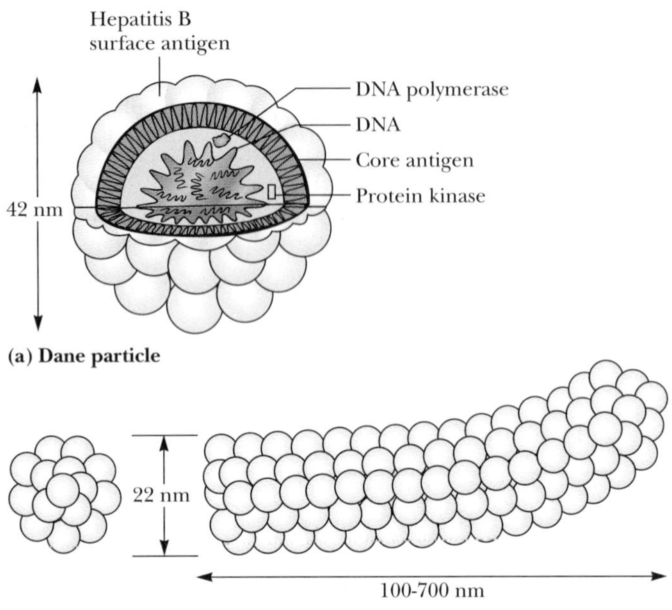

(a) Dane particle

(b) Incomplete hepatitis B viruses

***Figure* 30.14** The antigenic particles present in the serum of hepatitis B patients. The Dane particle is the virion. The others are incomplete viral particles and actually exceed the infectious virions in number.

abdominal pain, and chills. Hepatitis B viral infections can cause extensive liver degeneration, resulting in jaundice and release of enzymes into the bloodstream that are otherwise present in the liver. The hepatitis B virus is associated with primary hepatocellular carcinoma, and this disease accounts for many of the deaths that occur worldwide.

There is no cure for hepatitis B infections, but vaccines for prevention of the disease have been developed. These vaccines are recommended for administration to infants at birth with follow-up doses at 2, 4, and 6 months of age. Hospital workers and others who work in high-risk situations might also benefit from vaccination. Passive immunity with HBV immunoglobulin can be of value in individuals exposed to HBV, but it must be administered within a week of exposure.

The **hepatitis C** virus was formerly called Non-A, Non-B because all known cases not caused by A or B resulted from infection by this virus. Hepatitis C virus is a concern in blood transfusion because 5 to 10 percent of transfusion recipients are victims of this form of hepatitis. IV drug users also have a high incidence of hepatitis C infections. Generally the epidemiology of hepatitis C is similar to that of hepatitis B virus, and the symptoms are equivalent to those of A or B but generally milder.

The **hepatitis D** virus is a defective satellite virus that was formerly known as the **delta agent.** Hepatitis D virus infections occur only in individuals infected with the hepatitis B virus. It can be cotransmitted with HBV or acquired after infection with the hepatitis B virus. Hepatitis B virus must be actively replicating for the D virus to generate complete virions. A coinfection with hepatitis D and HBV leads to more severe liver damage and a higher mortality rate than occurs with an HBV infection. The coinfection occurs mostly in high-risk individuals such as drug abusers.

The **hepatitis E** virus is a problem in developing countries, and the symptoms and course of infection are similar to that for hepatitis A virus. However, the mortality rate for hepatitis E is about 10 times higher than that for hepatitis A viral infections. The mortality rate in pregnant women infected with hepatitis E virus is about 20 percent, and the reason for this high rate of mortality is unclear.

Warts

Papilloma viruses cause a variety of **skin warts** and epithelial lesions in humans. Most common warts are benign and disappear or are readily removed by minor surgery or a chemical agent. Some warts, however, may progress to a malignant condition. One such wart is the **genital wart.** Transmission of genital warts among human populations is a growing concern. It is estimated that between 10 and 30 million individuals in the United States may be infected with genital papilloma virus or carry the virus in a latent state. The number of new cases each year is estimated to be in excess of 15 million, with a high prevalence among teenagers.

The human papilloma viruses are nonenveloped, double-stranded DNA viruses. The nucleocapsid is composed of protein (Figure 30.15). Warts are transmitted by direct contact and occur mainly on skin and mucous membranes. They are not highly contagious. Virtually everyone has had or seen a wart. They are spread by scratching and most often occur in children or young adults. The projections from the skin are a result of skin cell proliferation and/or a longer life span for infected skin cells that accumulate. Common warts can be treated by applying dry ice or liquid nitrogen to the wart or by laser. Application of podophyllin over an extended time is effective in removing most warts. Podophyllin is a resin extracted from the dried roots of the May apple.

A growing health concern is the spread of **anogenital warts.** These warts are generally transmitted sexually and have an incubation period of one to six months. They appear singly or in dense growth on the squamous epithelia of the external genitalia and rectal area. A related papilloma virus that is also transmitted sexually can cause cervical cancer in females. The virus infects the genital tract and progresses to cervical carcinoma. Vaccination with inactivated papilloma virus has been employed with limited success in anogenital warts. Vaccines and other control measures will become more important if a closer relationship between the papilloma viruses and carcinoma is established.

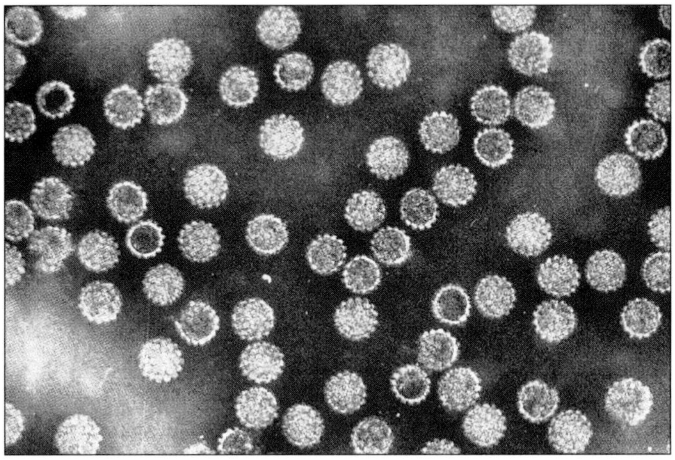

***Figure* 30.15** The human papilloma virus. Note the orderly array of protein capsomers on the surface of these wart-causing viruses. (© Science VU/Visuals Unlimited)

Viral Diseases with Nonhuman Reservoirs

There are a considerable number of viruses that infect humans that reside primarily in nonhuman reservoirs. Most are vectorborne and transmitted by the bite of an arthropod, such as a tick or mosquito. The arthropod carrier that bites the infected primary reservoir animal and then proceeds to bite a human host will transfer the infectious virus to humans. Some viruses that reside in nonhuman reservoirs are contracted directly by animal bites, from dust, or through skin abrasions. A list of natural hosts and vectors is presented in Table 30.6.

It is apparent from the range of primary hosts involved that prevention of these diseases through the control of the reservoir would be difficult. Obviously reservoirs such as squirrels, rodents, and birds cannot be exterminated, and controlling the infectious agents in a wild animal population is highly unlikely. The total population of many of the vectors, such as the mosquito, might be lowered, but complete elimination is virtually impossible. The practical solution to this problem would be vaccination of domesticated animals and/or human hosts. Unfortunately, effective vaccines are not available for many of these infectious diseases.

This section discusses some of the many viral infections transmitted to humans from nonhuman reservoirs.

Colorado Tick Fever

Colorado tick fever is one of the most common tick-borne diseases in the United States. The disease is caused by *Dermacentor andersoni,* and occurs mostly in the western and northwestern United States and in western Canada.

Symptoms include chills, headache, muscle aches, and fever. The symptoms appear after a three- to six-day incubation period and last for seven to ten days. The virus infects red blood cells and may persist for 20 weeks after symptoms subside. Fatalities are rare but can occur in young children. The virus apparently traverses the placenta in pregnant females and can cause abnormalities or death of a fetus. A viral vaccine has been developed but is impractical for general use.

The primary reservoir for Colorado tick fever is rodents. The virus is passed transovarially, and the tick can be both reservoir and vector. It is a reservoir because it passes the virus to other ticks and a vector when it passes the virus to humans. The disease occurs most often among campers in tick-infested areas. Prevention is by avoiding areas where ticks might be present and by using tick repellants.

***Table* 30.6 Viral infections with nonhuman primary hosts and transmitted to humans by vectors**

Disease	Viral Agent	Natural Host(s)	Vector
Colorado tick fever	Arbovirus	Squirrels, chipmunks	Tick bite (*Dermacentor andersoni*)
Encephalitis			
California	Arbovirus	Rats, squirrels, horses, deer	Mosquitos (*Aedes* sp)
St. Louis	Arbovirus	Birds, cattle, horses	Mosquitos (*Culex* sp)
Eastern equine	Arbovirus	Birds, fowl	Mosquitos (*Aedes*)
Venezuelan	Arbovirus	Rodents	Mosquitos (*Aedes* sp, *Culex* sp)
Western equine	Arbovirus	Birds, snakes	Mosquitos (*Culex* sp)
Lymphocytic choriomeningitis	Arbovirus	Mice, rats, dogs	Dust, food
Rabies	Rhabdovirus	Skunks, raccoons, bats	Domestic dogs and cats
Yellow fever	Togavirus	Monkeys, lemurs	Mosquitos (*Aedes* sp)

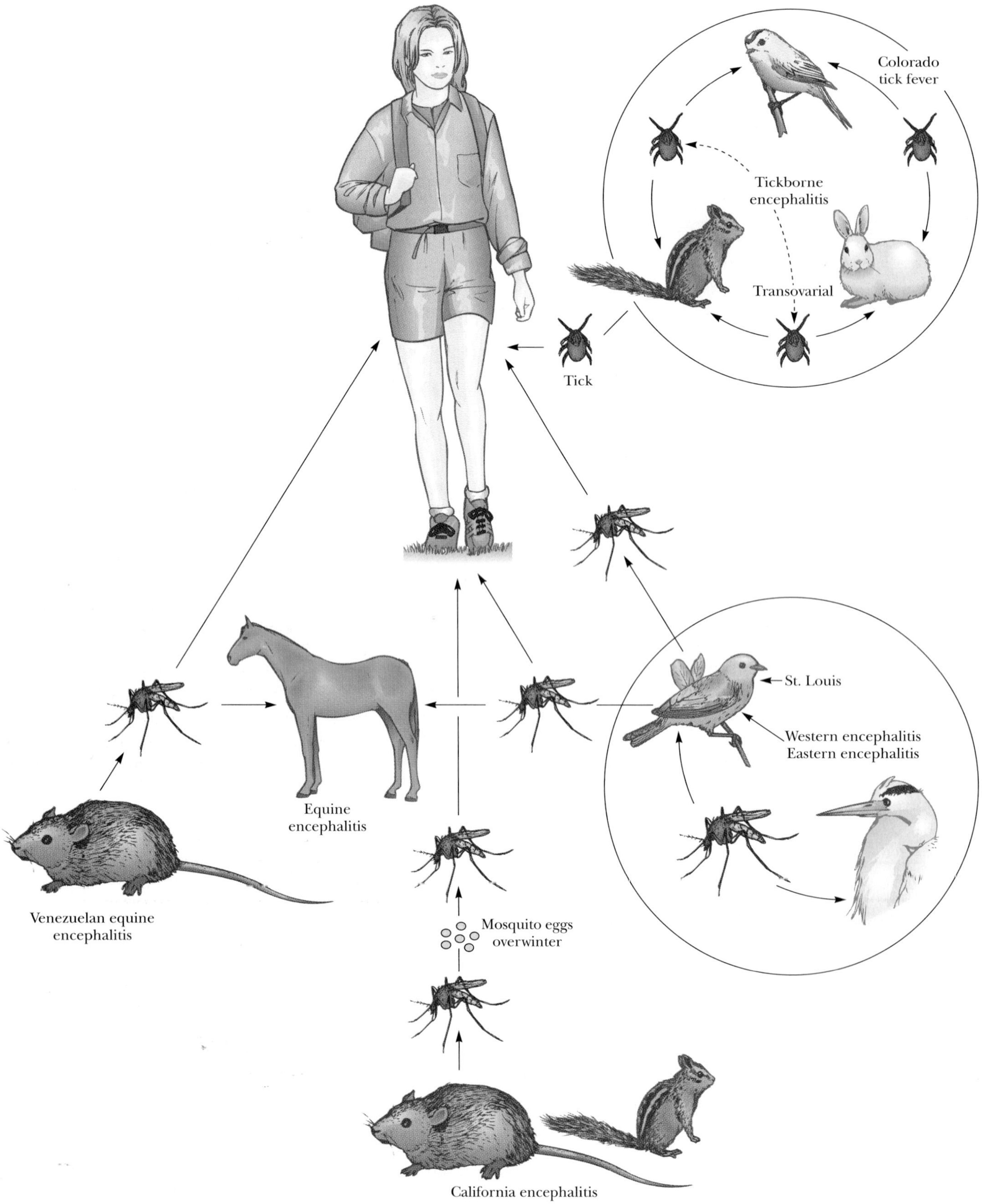

***Figure* 30.16** Epidemiology of the arboviruses that are the etiologic agent of human encephalitis infections.

Encephalitis

Encephalitis is a term that describes several diseases characterized by inflammation of the brain, which can lead to neurological damage or fatalities. The major etiological agents of encephalitis in the United States are **arboviruses,** which are transmitted by arthropod vectors. Transmission of the virus to humans is illustrated in Figue 30.16.

St. Louis encephalitis is caused by a **flavivirus** and epidemics of this disease occur sporadically in the south and southeast. Birds are the most common reservoirs. Western, eastern, and Venezuelan equine encephalitis are caused by an **alphavirus** and are transmitted to humans and horses by mosquitoes, although the horse is not an important reservoir. Symptoms of infection occur three to seven days after exposure. The first bout of infection results in a systemic infection with fever, chills, headache, and flulike symptoms. This is followed by a second phase where the virus moves to the central nervous system and multiplies in the brain. Eastern equine encephalitis can be a severe infection in humans and has a high mortality rate (over 50 percent). Neurological disorders are a potential consequence for survivors of the disease.

The elimination of breeding sites for vectors might be an effective control measure for these diseases. Vaccines are available for individuals working with the eastern and western equine encephalitis virus. A live vaccine against the Venezuelan equine virus is available for use in domestic animals.

Lymphocytic Choriomeningitis

Lymphocytic choriomeningitis is an infectious disease transmitted to humans from companion animals. The etiological agent is an **arenavirus** carried by dogs, guinea pigs, rats, and mice. After a 10- to 14-day incubation period, the arenavirus infects macrophages, and the macrophages then release mediators that may cause damage to meninges or choroid membranes. Generally the meninges are not involved and symptoms are confined to a fever, muscle aches, and nausea. There is the possibility that encephalitis may result, and about 25 percent of infected individuals do have central nervous system damage.

Hamsters and mice can be chronically infected with the lymphocytic choriomeningitis virus and remain asymptomatic. These animals shed the virus in saliva, urine, and feces. Infection of humans occurs by aerosols, contaminated food, or inhalation of contaminated dust particles. Prevention is by avoiding hamsters or thorough washing of hands after handling these animals.

Rabies

The causative agent of rabies is a bullet-shaped virus of the rhabdovirus group (Figure 30.17). The virus is maintained in nature in carnivorous animals. Skunks and raccoons are the principal reservoir, with significant occurrences in bat and fox populations (Table 30.7). Feral dogs and cats or wild animals carry the disease to domestic animals. Humans are generally exposed through bites or via contact with diseased vectors. The rabies virus proliferates in saliva; consequently aerosols in bat caves, where rabid bats may nest, are a potential source of infection.

When transmitted in aerosols, the virus infects the epithelial tissue that lines the upper respiratory system. Generally, rabies is passed among animal reservoirs through a bite or exposure to virus-laden saliva. Although a young skunk infected with rabies will eventually succumb to the disease, it may survive to reproductive age. The progeny

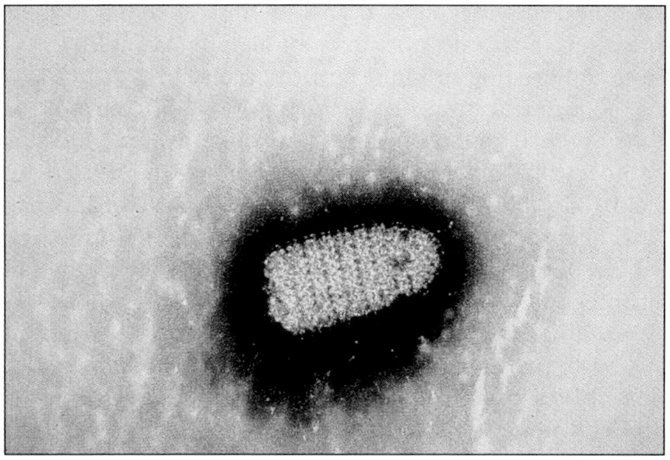

***Figure* 30.17** An electronmicrograph of the rabies virus. The virus is asymmetrical. (© CNRI/Phototake NYC)

***Table* 30.7 Major sources of the rabies virus**

Animal Group	Percent
Wild Animals	
Skunk	43.0%
Raccoon	27.7%
Bat	13.3%
Fox	2.5%
Others	1.7%
Domestic Animals	
Dog	3.6%
Cat	3.5%
Cattle	3.7%
Others	1.0%

then become infected and pass the disease along to their offspring. Foxes do not survive long after a rabies infection and are not an important reservoir.

Rabies is transmitted to humans through aerosols or entry of virus-laden saliva into an open wound, and this can occur by a bite or during handling of an infected animal. The virus multiplies in tissue at the site of inoculation and can remain localized for days or months. The length of time required for the virus to move from the site of infection to the brain depends on several factors: (a) the size of the inoculum, (b) the proximity of the wound to the brain, and (c) the host's age and immune status. The rabies virus travels from the wound site to the peripheral nerve system, further to dorsal ganglia, and on through the spinal cord to the brain. Infection occurs in the spinal cord, brainstem, and cerebellum. The virus can move from the brain to the eye, salivary glands, and other organs. Viral replication in the brain results in the presence of cytoplasmic inclusions in the affected neurons. These inclusions are called **Negri bodies** and have been a principal diagnostic test for the disease. Diagnosis is now based on a fluorescent antibody test on nervous tissue from a suspected victim.

Symptoms of rabies do not appear until the virus reaches the brain. The initial symptoms include fever, malaise, headache, gastrointestinal upset, and anorexia. After 2 to 10 days, neurological symptoms associated with rabies appear. Hydrophobia occurs in 25 to 50 percent of patients and is characterized by jerky contractions of the diaphragm when the victim attempts to swallow water. Generalized seizures and hallucinations follow, ending in coma and death.

Humans bitten or otherwise exposed to potential rabies carriers are given post-exposure prophylactic treatment. Unless it can be confirmed that the suspected animal carrier does not have a rabies infection, treatment is routinely initiated. The victim is immunized with a vaccine plus a dose of equine antirabies serum or human rabies immunoglobulin. This passive immunization provides protective antibody until antibodies are produced in response to the vaccine.

The control of human rabies depends on the effective elimination of the disease in the wild animal populations that serve as reservoirs. Domestic dogs and cats are routinely vaccinated against the disease, however vaccination of wild animals by capture/release is out of the question. Administration of an oral vaccine by placing it on bait is one possible mechanism for widespread inoculation of wild animal populations. An experiment of this type is now underway on a secluded island off the East Coast of the United States. The test is confined to an island because there is a remote chance that the attenuated virus placed on bait might revert to virulence when given to a large animal population. This experiment in mass inoculation is ongoing and, if successful, it might be of value in eliminating rabies from animal reservoirs.

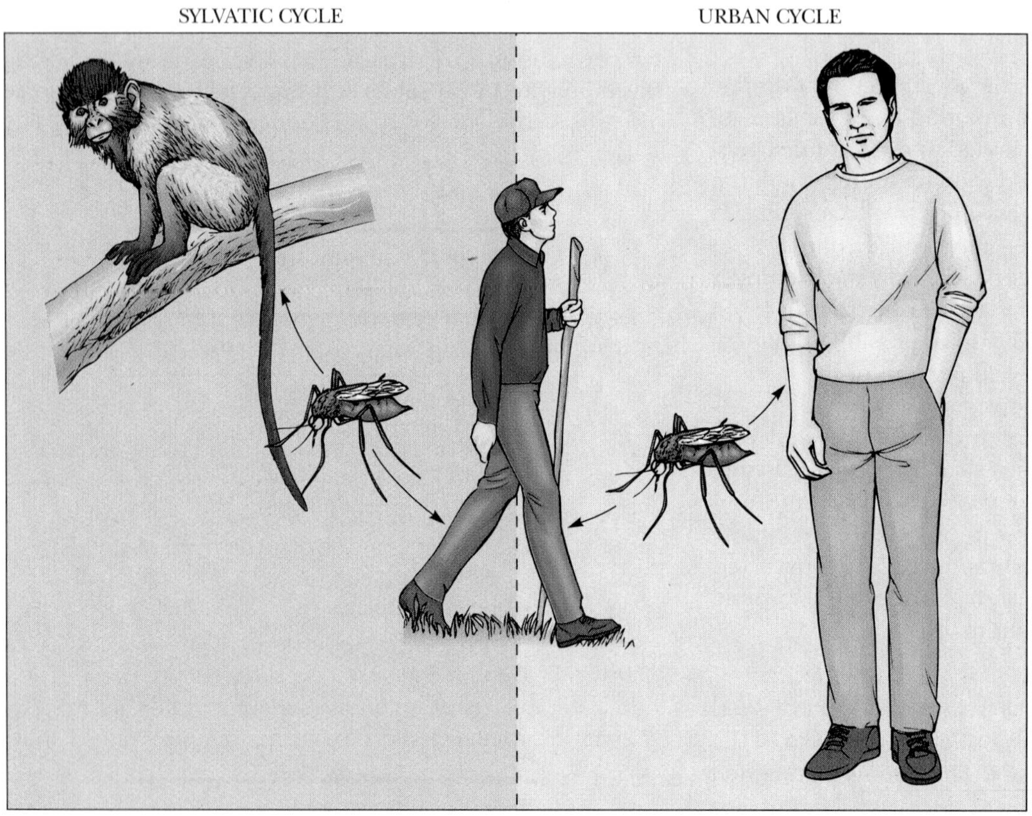

***Figure* 30.18** The urban cycle is involved in the spread of yellow fever from human to mosquito to human. In the sylvatic cycle the monkey serves as reservoir.

Yellow Fever

Yellow fever, caused by an **arbovirus,** was the first human infection shown to be caused by a virus. It is also the first viral infection demonstrated to be vector borne. Yellow fever is transmitted by a mosquito called ***Aedes aegypti.*** The last epidemic in the United States occurred in 1878 and killed over 13,000 people. There has not been a reported case in the United States in the past 70 years, but the potential for reintroduction is real because the vector is present in the southern United States. The reservoir for the disease in Central and South America is the monkey.

The yellow fever virus enters the human body through a mosquito bite and goes to local lymph nodes. The virus multiplies in lymph nodes and the spleen and results in a systemic infection. Damage to liver, kidneys, heart, and hemorrhaging of blood vessels follows the systemic infection. The name yellow fever originated because the liver damage results in jaundice. Fever, chills, headache, and nausea are also symptoms of the disease.

Where yellow fever is endemic it is spread by distinct epidemiological patterns. The urban cycle is human to human, with the mosquito serving as vector, and sylvatic cycle, where mosquitoes transmit the virus between monkeys and occasionally to humans (Figure 30.18).

Elimination of breeding grounds for the vector is an important and effective control measure. A vaccine is available that, when administered intradermally, leads to a lifetime immunity.

Emerging Viruses

The human immunodeficiency virus (HIV), the etiological agent of AIDS, was first described in 1981. Over the last 15 years this newly emerged virus has had a devastating effect on humans worldwide. In addition, there are other viral infections that have emerged in the last 20 years that may be a serious threat to human populations. A hantavirus outbreak in the Southwest in the 1990s caused over 30 fatalities and has a 70 to 75 percent death rate among victims. The virus is apparently disseminated in dust containing rat feces; whether it may gain other mechanisms of transmission is unknown. Other viral diseases that have emerged include Ebola virus and mosquito-borne infections such as dengue fever and chikungunya. Whether these viruses will become a danger to humans is difficult to assess (see Box 14.2).

Summary

- **Viruses** can infect bacteria, plants, and animals. Infections by viruses are more difficult to control than infections caused by bacteria because the metabolic machinery set in motion by the virus is quite equivalent to that in healthy cells. **Inhibition** of **viral synthesis** is harmful to the host.

- Infections spread in **airborne droplets** such as colds and influenza are difficult to control because the daily life of humans involves person-to-person contact.

- **Chicken pox** is a highly contagious disease that generally affects children. It can be a serious infection, and a vaccine has now been developed to control the disease. **Shingles** is the result of a dormant viral infection and can occur in later life.

- **Rubella (German measles)** is a viral disease that, prior to the discovery of a vaccine, was particularly dangerous if infections occurred in pregnant women, as the virus caused considerable harm to the fetus. Rubella can be controlled by **vaccination** of children. Another measles **rubeola** was a concern because it caused permanent neurological damage to some children. It also can be prevented by vaccinating infants.

- A deadly disease that has been eliminated throughout the world is **smallpox.**

- **Mumps** was a common childhood disease that caused inflammation of the **parotid glands** and spread to other areas of the body with unfortunate consequences. This disease can be controlled by vaccinating young children before they enter school.

- **Influenza** can occur in worldwide epidemics (**pandemics**). It is a respiratory disease that can lead to pneumonia. Vaccines lower the risk and/or severity of the disease.

- **Common colds** affect 40 percent to 45 percent of Americans each year and result in an estimated 200 million lost school/work days annually. They can neither be prevented nor cured, and treatment is by over-the-counter drugs that relieve symptoms.

- Viruses can transform normal human cells to cancerous cells. **Human T-cell leukemia** virus is one such and is transmitted human to human in contaminated blood, exchange of IV needles by drug users, or sexual contact.

- **Acquired immune deficiency syndrome (AIDS)** has become a leading killer of young adults. AIDS transmission can occur by transfusions with contaminated blood but most often is transmitted by sexual contact. Once contracted, the virus can be unnoticed, but it gradually develops into an active infection that destroys the immune response. The patient is subject to attack by an array of "opportunistic" microbes including viruses, fungi, and bacteria. There is no cure or prevention except "**safe sex**."

- **Herpes simplex type 1** virus is the etiological agent of **cold sores** that appear on lips and mouth of host individuals during their lifetime. Another type herpes (type 2) causes **genital herpes,** a growing problem in the United States. **Acyclovir** is a chemotherapeutic agent that can relieve symptoms of herpes infections, but it is not a cure.

- **Infectious mononucleosis** is a disease caused by herpesvirus that occurs often in college-age individuals. It causes fatigue, and rest is the only cure.

- Another virus disease that can be prevented by vaccination is **poliomyelitis.** It has not been eliminated because worldwide use of the vaccine has not been attained. **Polio** can cause paralysis by destruction of motor neurons.

- **Warts** are caused by viruses, and some infections can have dire consequences. **Anogenital warts** transmitted by sexual contact can cause **cervical cancer** in females.

- There are a number of **encephalitis** diseases that are characterized by inflammation of the brain. They are transmitted to humans by vectors that include arthropods and mosquitoes. Elimination of vector breeding sites is the only preventive measure now available.

- **Rabies** is generally spread by the bite of a domestic animal and wild animals such as skunks and raccoons that serve as **reservoir.** Feral dogs and cats can carry the virus from the reservoir to domestic animals. Treatment is with antiserum and vaccination of exposed individuals.

- **Yellow fever** is caused by an arbovirus carried by a mosquito *(Aedes aegypti)*. The disease causes severe liver damage but has been eliminated in the United States at the present time.

Questions for Thought and Review

1. Why are viral infections so difficult to control? How might this relate to the origin of virus? In considering this, would one assume that fungal or protozoan infections might also be a problem?

2. Why were the viral diseases that are transmitted via the respiratory route of such interest to the vaccine producers?

3. What are the symptoms of "shingles," and how does one get this disease?

4. How was smallpox eradicated? Why was it possible from an economic standpoint?

5. What is "herd" immunity? How does this relate to mandatory vaccination? What are some dangers when preschool children do not receive adequate immunization programs?

6. Why is the influenza virus such a problem? Periodically there are epidemics and pandemics of this disease. Why?

7. Why would an individual, once infected with HIV, have it for life despite chemotherapy? Elimination of a retrovirus from an infected individual is not feasible. Why?

8. AIDS patients have a major problem with infections by viruses and eukaryotic opportunists. What are some of these, and why is an AIDS victim so vulnerable to these?

9. Herpesviruses cause latent infections. Discuss.

10. Genital herpes is a disease that is increasing in frequency. What are the dangers?

11. The Epstein-Barr virus is of some concern in the United States. Why? Why is it more of a problem in other countries?

12. Polio may be eliminated from the world as was smallpox. What are some problems in achieving this goal?

13. Can warts cause cancer?

14. What are some of the major viral diseases in humans that have nonhuman reservoirs? How can these diseases be controlled?

15. What animals serve as major reservoirs for rabies?

Suggested Readings

Alcamo, I. E. 1993. *AIDS: The Biological Basis.* Dubuque, IA: Wm. C. Brown.

Joklik, W. K., H. P. Willett, D. B. Amos, and C. M. Wilfert. 1992. *Zinsser Microbiology.* 20th ed. Norwalk, CT: Appleton & Lange.

Mandell, G. L., R. G. Douglas, Jr., and J. E. Bennett. 1990. *Principles and Practice of Infectious Disease.* 3rd ed. New York: John Wiley & Sons.

Murray, P. R., G. S. Kobayashi, M. A. Pfaller, and K. S. Rosenthal. 1994. *Medical Microbiology.* 2nd ed. St. Louis: Mosby–Year Book, Inc.

Sherris, J. C., ed. 1990. *Medical Microbiology, an Introduction to Infectious Disease.* 2nd ed. New York: Elsevier Science Publishers.

Shulman, S. T., J. P. Phair, and H. M. Sommers. 1992. *The Biologic and Clinical Basis of Infectious Disease.* 4th ed. Philadelphia: W. B. Saunders.

Volk, W. A., D. C. Benjamin, R. J. Kadner, and J. T. Parsons. 1991. *Essentials of Medical Microbiology.* 4th ed. Philadelphia: J. B. Lippincott Co.

White, D., and F. Fenner. 1994. *Medical Virology.* 45th ed. San Diego: Academic Press.

Alas! regardless of their doom
The little victims play;
No sense have they of ills to come
Nor care beyond today.
Thomas Gray, 1716–1771

Chapter 31

Epidemiology and Clinical Microbiology

The health and well-being of human populations depend, in large measure, on the control of communicable infectious diseases. Human history is a record of devastation by infectious diseases, and up to the early years of the 20th century they were the major cause of death worldwide (**Box 31.1**). Infectious diseases continue to take a toll in the developing nations of Asia, Africa, and South America, where they account for nearly 50 percent of all deaths. In the developed countries, including the United States, Canada, and Western Europe, fewer than 8 percent of all deaths are from infectious disease. The decreased death rate in the developed nations was brought about by controlling both individual and community-wide factors that would otherwise contribute to the spread of disease.

Worldwide, very few diseases have been combated successfully, but one major affliction that caused much human misery has been eradicated, and that is smallpox. There is still a great opportunity to relieve human misery and improve the health of humans in both the developing and developed nations. The periodic spread of infec-

BOX 31.1 MILESTONES

The Major Causes of Death in the 20th Century, United States

In 1900, there were 318 deaths per 100,000 population caused by selected infectious diseases and 215 due to heart disease and cancer. Public health measures instituted in the first half of the twentieth-century lowered the death rate from these same infectious diseases dramatically and the death rate in 1992 is even lower. Public health measures, antibiotics, and immunization have combined to protect people in the United States from the scourge of infectious disease. Heart disease and cancer continue as leading causes of death.

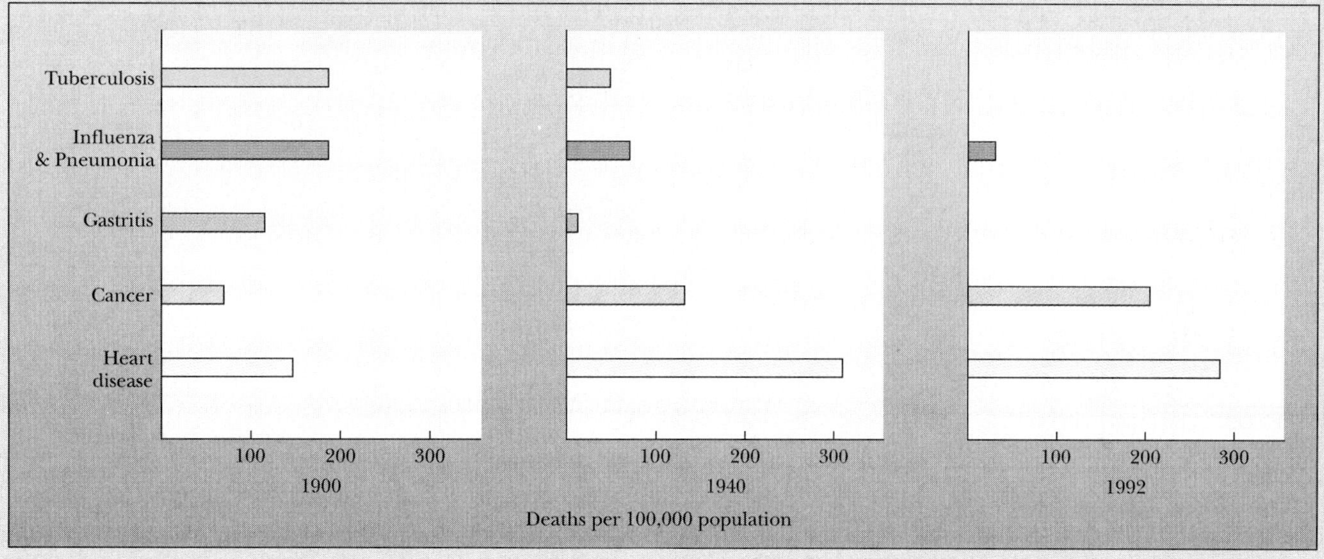

The leading causes of death in the United States in 1900, in 1940 before antibiotics were widely used, and in 1992 after antibiotics were available.

tious diseases such as influenza to virtually every nation is a reminder that communicable diseases must be a concern. The current catastrophic pandemic of the human immunodeficiency virus (HIV) confirms that uncontrolled highly communicable diseases can still leave their mark on society.

Two major aspects of infectious disease control are: (1) an accurate determination of the etiological agent involved in infections and (2) the incidence of this infectious disease in the population. This is where epidemiology and the clinical laboratory come together. The modern clinical laboratory has diagnostic capability that can determine precisely the strain of an agent involved in a disease outbreak and accomplish this in a short period of time. This permits the epidemiologist to pinpoint the origin and course of an epidemic and to take measures that might curtail the spread of the infection.

Epidemiology

The science concerned with the distribution and prevalence of communicable disease in populations is called **epidemiology.** Epidemiologists work worldwide to monitor infectious diseases in order to institute measures for their potential control. The role of epidemiologists in pinpointing the outbreak of infectious disease is illustrated by the discovery of AIDS. In June of 1981 the Centers for Disease Control and Prevention in Atlanta compiled data indicating that there was an unusually large number of deaths among immunodeficient males in the Los Angeles area. These deaths were caused by opportunistic pathogens and involved homosexual males. This discovery was announced a full two years before the retrovirus responsible for the disease was isolated and identified. This is a

BOX 31.2 MILESTONES

The Great Plague in London

The Great Plague in London (1665) was a reasonably well-documented epidemic. However, the number of deaths reported is a conservative figure, as many went unrecorded. The numbers for this drawing are mostly from Burial Registers of Churches. Many, including Quakers, were buried in their gardens, and massive numbers of others were buried without record. The population of London at the time of the Great Plague was estimated to be about 400,000 inhabitants. The number that succumbed in the plague epidemic was somewhere between 25 percent and 40 percent of the population.

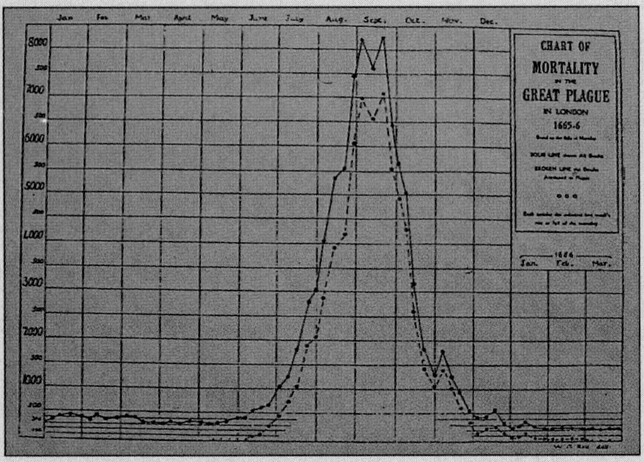

The dramatic death rate that occurred in London during the Plague epidemic of 1665–66. (From Bell, *The Great Plague in London*–1665, p. 238)

prime example of the effectiveness of epidemiologists in determining that the frequency of a disease of unknown etiology is increasing. Since the identification of HIV, the virus has spread rapidly and is now a concern in all of the continents that are inhabited by humans.

Terminology

The word "epidemic" appears ominous because it suggests that a disease may be spreading that will affect a massive number of people. In reality the word **"epidemic"** means that a disease is occurring in the population at a higher than normal frequency (see **Box 31.2**). Most infectious diseases occur at a low, but constant frequency in the population and are considered **endemic.** This is illustrated in Figure 31.1: **(a)** an endemic level, **(b)** an epidemic in western United States and in eastern Australia, and **(c)** a **pandemic,** when the frequency of a disease increases throughout the world. Major pandemics caused by the influenza virus occurred in 1918 and 1957, and HIV is now considered to have reached pandemic proportions.

Epidemics fall into three categories: sporadic, seasonal, or contact (Figure 31.2). A **sporadic** disease affects a certain percent of the population throughout the year. Endemic diseases occur sporadically. The incidence may increase slightly during the year but overall the number of cases is quite constant. A **seasonal** disease is one such as Lyme disease or Rocky Mountain spotted fever that is transmitted only in those months when the arthropod vectors are active. A **contact** disease is one such as colds, which are spread via contact with an infected individual.

The **incidence** of an infectious disease is defined as the number of newly reported cases per a given time in a population. Generally the incidence of a disease is calculated by the following formula:

$$\text{Incidence} = \frac{\text{number of newly reported cases per a given time}}{\text{100,000 people at risk}}$$

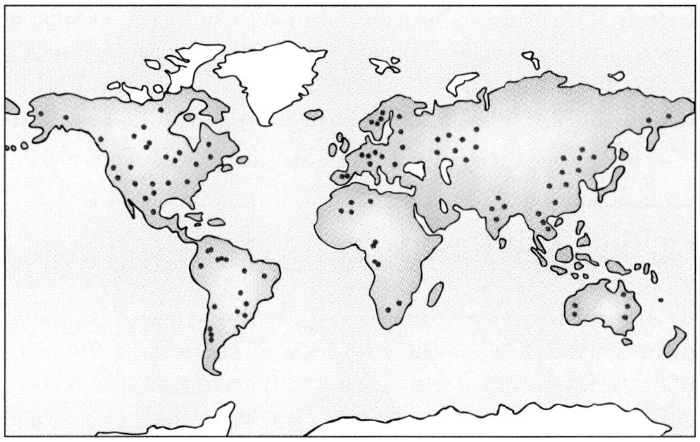

(a)

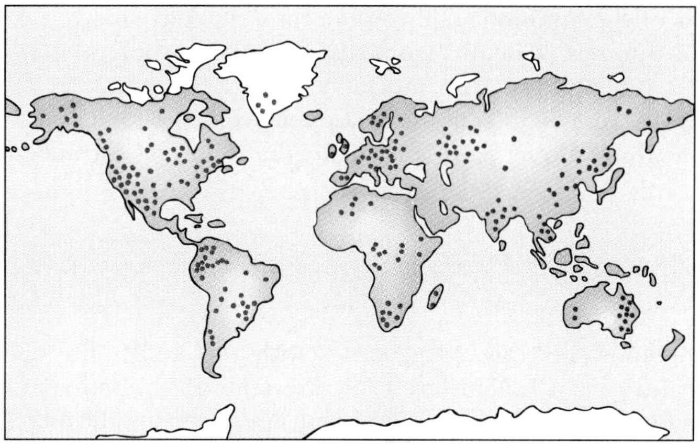

(c)

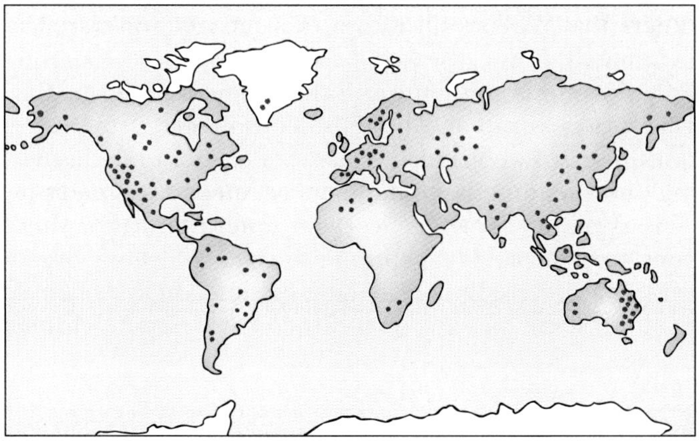

(b)

***Figure* 31.1** The occurrence of infectious diseases in human populations. Each dot would represent multiple cases of a specific disease: **(a)** endemic, **(b)** an epidemic in western United States and eastern Australia, and **(c)** a pandemic with increased incidence in locales worldwide.

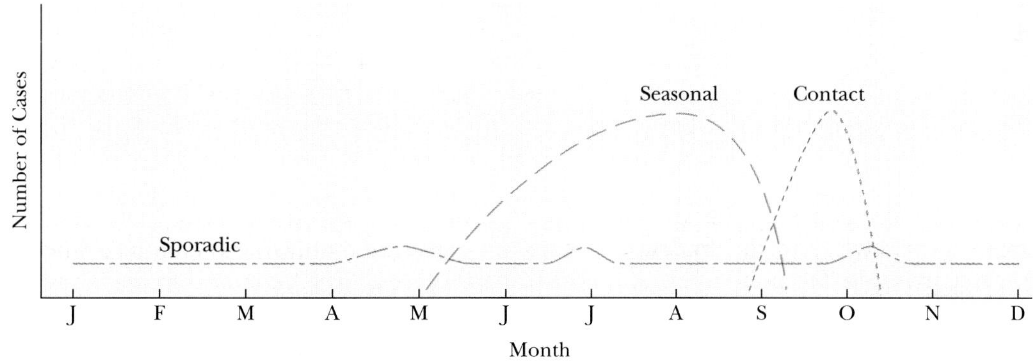

***Figure* 31.2** The incidence of disease in a population: sporadic shown in orange, seasonal (summer) shown in green, and contact shown in purple.

The **prevalence** of an infectious disease is the percent of the total population that has the disease at a given time. For example, a seasonal disease is more prevalent during a limited number of months of the year. The term **outbreak** is also used and indicates that there is a relatively large number of cases in a limited area.

Mortality and Morbidity

There are infectious diseases that cause a limited number of fatalities and others that are fatal to a considerable percentage of those infected. Botulism is generally fatal, but few succumb to staphylococcal food poisoning. The coming of the antibiotic age resulted in a marked change in the number of deaths from diseases such as pneumonia and tuberculosis. The **mortality rate** is the number of deaths from an infectious disease relative to the total number who contract it. It is given by the following formula:

$$\text{Mortality rate} = \frac{\text{deaths due to a disease}}{\text{total infected}}$$

The mortality rate often varies with the availability of health care. Cholera has a low mortality rate when adequate medical care is available but increases significantly if intravenous fluids and other health care measures are unavailable.

Morbidity is the incidence of infectious disease, both fatal and nonfatal, in a population. It is calculated as follows:

$$\text{Morbidity rate} = \frac{\text{new cases in a selected period}}{\text{total population}}$$

The morbidity rate is a definitive figure and serves as an accurate measure of the general health of a population. In the developed nations, the illness rate is considerably higher than the fatality rate, although both are relatively low for most diseases. The low fatality rate results from better health care, and the availability of antibiotics and other chemotherapeutics. Fatality figures, in general, are not an accurate measure of health.

Carriers and Reservoirs

A significant problem in controlling communicable infectious diseases is the presence of **carriers** in the population. Carriers are individuals with asymptomatic or subclinical infections that are not serious enough to require treatment or curtail one's activities. Often carriers continue normal duties while "toughing it out" with their disease symptoms, and they tend to deny to themselves and others that they are ill with a transmissable disease. Unfortunately these diseased people can expose everyone they encounter with the infectious microorganism they bear. Carriers can also be those convalescing from infectious diseases who return to work or school while harboring and shedding large numbers of infectious organisms. This is particularly a concern in enteric diseases where convalescent patients continue to shed infectious organisms for a considerable length of time. In the case of typhoid, some patients (2 to 5 percent) become **chronic**

BOX 31.3 MILESTONES

Typhoid Mary

Typhoid Mary, Mary Mallon, has attained the dubious honor of having much of her life story discussed in virtually every microbiology text. Hers is a classical case of the chronic carrier wreaking havoc wherever she traveled. Tracking her down as the source of typhoid outbreak tested the epidemiological investigatory abilities of the Health Department in New York City in the early 1900s. Yet, the Health Department's handling of the case parallels issues we face today in the AIDS dilemma, where the rights of individuals are pitted against the perceived welfare of the populace.

Mary Mallon was a Swiss immigrant who, in 1901, had a serious case of typhoid fever that resulted in a gall bladder permanently infected with *Salmonella typhi.* The infection generated large numbers of the typhoid bacillus that entered the gastrointestinal tract and were shed in her feces. Unfortunately, Ms. Mallon was a cook and housekeeper in New York City and worked in several homes. As she moved from position to position she left behind 28 cases of typhoid fever. The New York Health Department, headed by Dr. George Soper, tracked down Mary Mallon and had her arrested. The authorities offered to remove her gall bladder to effect a cure. She refused and was released after three years of imprisonment on a pledge never to cook or handle food and to report periodically to the Health Department. Mary Mallon immediately disappeared, changed her name, and became a cook in hotels, hospitals, and sanitaria. She apparently recognized that the disease was caused by her presence, because she quit her job when an outbreak occurred and moved on in order to evade authorities. After five years, Mary Mallon was intercepted during a typhoid outbreak at a hospital. She spent her remaining 23 years on North Brother Island in New York City's East River, where she died in 1938. Typhoid Mary was the source of an estimated 200 cases of typhoid fever, much suffering, and several deaths.

BOX 31.4 MILESTONES

John Snow and Cholera

John Snow (1813–1858), a British physician, was the first to recognize that humans were a reservoir for cholera. Snow realized that the feces of cholera patients were highly infectious, and he surmised correctly that human waste in drinking water could transmit the disease. Snow followed the incidence of cholera from 1853 to 1855 in a wide area of London. At that time, there were two major water systems supplying water to homes in that area: the Southwark & Vauxhall Company and the Lambeth Company. John Snow followed the cholera epidemic in homes of equivalent living standards but supplied by one or the other of these companies. It was obvious that inhabitants of houses supplied by Southwark & Vauxhall had a markedly higher incidence of cholera than those supplied by the Lambeth Company.

In the first seven weeks of the epidemic, 315 people per 10,000 had died in houses whose water was supplied by the Southwark & Vauxhall Company compared to 37 per 10,000 houses of those whose water was supplied by the Lambeth Company. Snow looked at the water supply for the two companies and found that the Lambeth Company obtained its water from the Thames above the city, whereas the Southwark & Vauxhall Company drew water from the Thames in an area where untreated sewage from London entered the river. He concluded that the source of the cholera epidemic was the sewage-contaminated water.

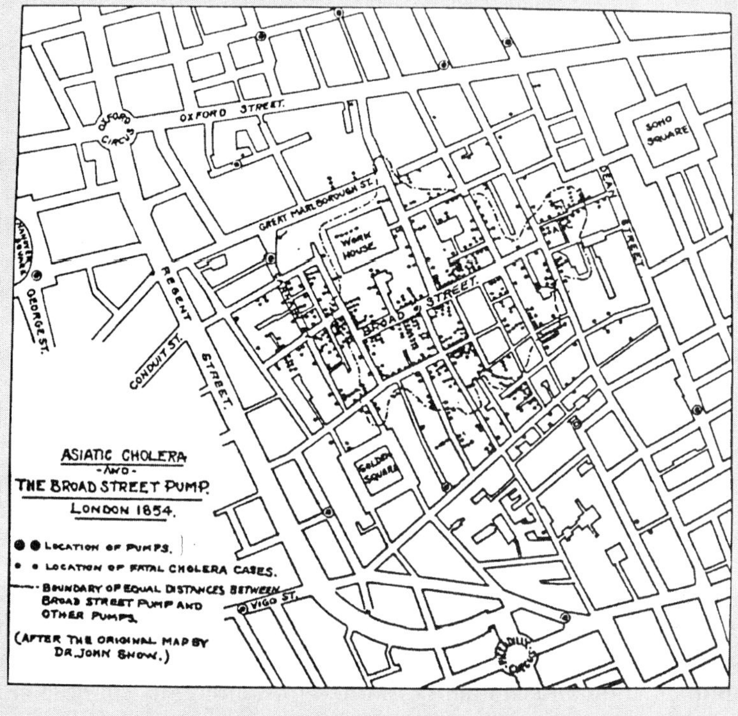

The incidence of cholera in the vicinity of the Broad Street pump among a population that drew water from the Thames in an area that was contaminated by human waste. (From Cosgrove, *History of Sanitation*, p. 92)

A child drinking from the can of an official water carrier, previously used by an infected person. "There is death in the cup." The lack of proper disinfection resulted in many fatalities. (Courtesy of The Bettman Archive)

carriers. A classical case of a chronic typhoid carrier was Mary Mallon, known as Typhoid Mary (see **Box 31.3**).

A **reservoir** is the animate or inanimate site where an infectious disease is maintained in nature between outbreaks. For a communicable infectious disease to survive it must meet certain criteria. If it survives only in living hosts, it must have a means of escaping from one host and of traveling and gaining entry into another potential host. A reservoir for diseases transmitted through human waste is food or drinking water, which is amply illustrated in **Box 31.4.** An infectious agent generally does not kill all available hosts. The goal of the parasite in the ideal relationship is for both the host and parasite to survive. If the infectious agent does cause a fatal disease, survival requires that some of the organisms move to another host before the host's death.

There are a number of diseases termed zoonoses that cause either human or animal infections. The animal is the reservoir, and a number of these diseases are discussed in Chapter 29.

Modes of Transmission of Infectious Diseases

Survival of infectious diseases depends on their transmission from host to host. In most cases, disease-causing organisms do not propagate outside a host or a reservoir. As stated previously, many infectious diseases are present in the general population at an endemic level. Under appropriate conditions, infectious agents such as the influenza virus may cause an epidemic or the evolvement of virulent strains that can be transmitted worldwide, resulting in a pandemic. There are two general types of localized epidemics, as outlined in Figure 31.3. A **common-source** epidemic would occur if a considerable number of humans were to eat from a large batch of contaminated food. A **host-to-host** epidemic can occur when susceptible children are brought together during the first days of school. Historically, childhood diseases, such as measles, were epidemic in a primary school during September of each year. A number of the major bacterial diseases that occur in humans are presented in Table 31.1, along with potential measures for their control.

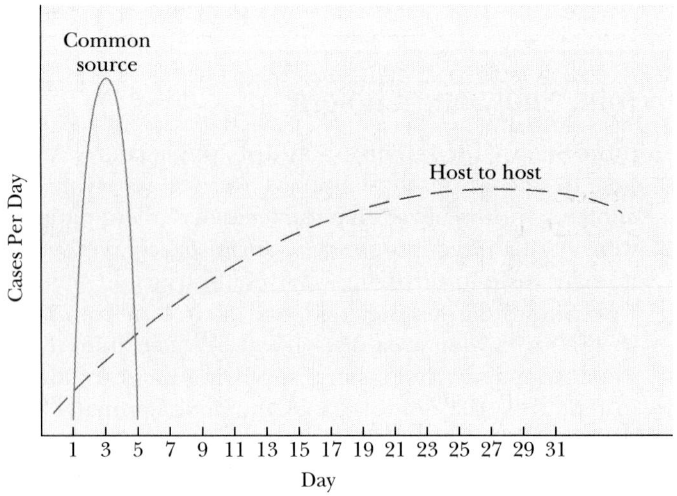

***Figure* 31.3** An epidemic contracted from a common source contrasted with a host-to-host epidemic. Common source (blue) occurs when a group of individuals drink or ingest contaminated food and host-to-host (red) occurs when an infection such as viral influenza spreads by contact.

Human Contact and Respiratorily Contracted Diseases

Human contact and respiratory diseases are passed directly by human interaction. Transfer may occur by touching, kissing, coughing, sneezing, or, in some cases, on dust particles. Streptococcal and staphylococcal skin infections, such as impetigo, are often transferred among children by scratching. An unimpeded sneeze expels infectious droplets that are about 10 μm in diameter and contain one to several bacteria (see Figure 30.8). The number of droplets per sneeze can be in the thousands. A hearty sneeze can travel at 100 meters/second and contain up to 100,000 bacteria or virus particles. Tuberculosis is a disease transmitted in respiratory droplets and is a constant threat. There are about 20,000 new cases of tuberculosis each year in the United States, and the mortality rate is 10 to 15 percent. Tuberculosis is generally a slowly progressing, chronic disease, and infected individuals can be asymptomatic for months or even years. These individuals are carriers and may transmit the disease to those they encounter. The tuberculin test is broadly employed to identify potential carriers. Leprosy is a poorly transmitted contact disease, and one must be predisposed to contract it. There are 200 to 300 new cases annually. Treatment of infected individuals is the only known method for control. Vaccination is an effective control for respiratorily transmitted bacterial diseases such as whooping cough and diphtheria. It is also effective against viruses such as mumps, as well as chicken pox, rubeola, and measles, which are respiratorily transmitted (see Chapter 30).

Water-, Food-, and Soil-Borne Infections

Many of the water- and food-borne diseases are caused by pathogenic microorganisms that survive solely in the human gastrointestinal tract. Humans are the natural host for bacteria that cause cholera, typhoid fever, and shigellosis. Because these pathogens can survive for a limited time in raw sewage, they can be transmitted by contaminated water supplies. Proper sewage disposal and care in water treatment are important in control of these diseases. Cases of cholera and typhoid fever are rare in the United States, but shigellosis, a form of dysentery, is quite common and can be transmitted from infant to infant in child care centers. Outbreaks also occur in elementary schools. Proper sanitation is the best mechanism for control of shigellosis. Pasteurization of milk, proper cooking practices, washing of hands, and immunization of cattle are important in preventing the transmission of food-borne diseases. As a result of such measures, brucellosis in cattle and bovine tuberculosis are no longer a problem in developed nations. Legionnaire's disease is associated with mists created by cooling towers for air conditioning systems and vegetable misters in grocery stores, and the causative agent may also be present in various aquatic environments. The control of *Legionella pneumophila* is difficult, and the disease occurs sporadically across the United States. Proper treatment of infected individuals is proba-

Table **31.1 Major bacterial diseases of humans, sources of infection, and potential control**

Disease	Primary Reservoir	Potential Means for Control
Human Contact and Respiratorily Contracted		
Streptococcal infections	Humans	Antibiotics, vaccine for pneumonia
Staphylococcal infections	Humans	Antibiotics, antiseptics
Meningitis	Humans	Specific antibiotics
Tuberculosis	Humans	Test and treat infected persons
Whooping cough	Humans	Vaccinate infants
Diphtheria	Humans	Vaccinate infants
Leprosy	Humans	Obtain proper treatment, vaccinate in endemic areas
Pneumonic plague	Humans	Eliminate rats and fleas
Water-, Food-, and Soil-borne		
Cholera	Humans	Treat sewage and water, observe proper sanitation
Typhoid fever	Humans	Pasteurize milk, treat sewage properly, inspect food handlers
Shigellosis (dysentery)	Humans	Observe proper sanitation
Salmonellosis	Beef, poultry	Cook meat and eggs properly
Campylobacter	Animals, poultry, and water	Pasteurize milk, cook food thoroughly
Tetanus	Soil	Vaccinate
Brucellosis	Cattle	Immunize cattle and pasteurize milk
Botulism	Soil	Properly can and cook food
Staph food poisoning	Humans	Refrigerate food
Legionnaire's disease	Aquatic environments	Clean misting equipment or do not use
Pseudomonas infections	Dust	Ensure clean air in burn wards
Sexually Transmitted		
Gonorrhea	Humans	Eliminate carriers, practice safe sex
Syphilis	Humans	Eliminate carriers, practice safe sex
Chlamydia	Humans	Eliminate carriers, practice safe sex
Louse-borne, Human to Human		
Trench fever	Humans	Proper sanitation, control lice
Relapsing fever	Humans	Control ticks and lice
Typhus (epidemic)	Humans	Proper sanitation, vaccinate
Vectorborne		
Rocky Mountain spotted fever	Mammals, birds	Wear protective clothing and examine body for ticks
Tularemia	Rodents, rabbits	Observe proper care when cleaning wild rabbits
Lyme disease	Deer	Wear protective clothing
Bubonic plague	Rats	Control rats, proper sanitation
Typhus (endemic)	Rodents	Control rats, vaccinate
Scrub typhus	Mites	Control mites
Animal Contact		
Leptospira	Vertebrates	Control rodents, vaccinate domestic animals
Anthrax	Soil	Sterilize wool, hair, other animal products
Psittacosis	Birds	Control bird imports
Q fever	Cattle	Vaccinate animal handlers

bly the best way to combat this disease. Anthrax has been an important soil-borne disease because the causative organism (*Bacillus anthracis*) can form endospores that survive in soil. Cattle killed by anthrax have been a source of the endospores, but improved health practices in the animal industry have curtailed the spread of this disease. Probably the greatest threat from anthrax is use by a mad despot as a germ warfare agent.

Sexually Transmitted Diseases

The sexually transmitted bacterial diseases listed in Table 31.1 occur only in humans. They are a significant societal problem that can be solved only when people assume responsibility for their behavior. All can be treated and cured, yet millions of new cases of gonorrhea and chlamydial infection occur each year. An increasing incidence of antibiotic-resistant strains, particularly with the gonococcus, poses a threat to public health. These diseases can be controlled only by public education, public responsibility, and a greater awareness of the problem. Because immunization to prevent the sexually transmitted diseases is not on the horizon, the use of condoms and other safe sex practices are the only control now available.

Three sexually transmitted viral infections are epidemic in the United States. These are genital herpes (HSV-2), the human immunodeficiency virus (HIV), and genital warts. There are over 20 million individuals in the United States infected with HSV-2 and about 500,000 new cases occur each year. Herpes in the neonate is one of the most common life-threatening infections, as there are about 2,000 babies born each year afflicted with this disease. HIV has reached pandemic proportions, with one million HIV positive individuals in the United States and millions infected worldwide (Figure 31.4). The incidence of genital warts is increasing at an alarming rate and is now one of the most prevalent sexually transmitted diseases, particularly in promiscuous young adults.

Louse-Borne, Human-to-Human Diseases

Louse-borne diseases are those that are passed among human populations by lice bites; they are best controlled by proper sanitation. Epidemic typhus and trench fever are louse-borne diseases and are generally associated with substandard living conditions. Sanitation and elimination of lice can effectively control the spread of these diseases. Relapsing fever is a louse-borne disease that occurs mainly among campers who spend time in areas infested with rodents. Campsites should be selected with care in locales where this disease occurs.

Vector-Borne Infections

Two vector-borne infections that are a concern in the United States are Lyme disease and Rocky Mountain spotted fever. The number of cases of both are increasing, and the sole control measure is avoidance of the ticks that serve as vectors. Protective clothing and insect repellent are essential when entering areas potentially occupied by ticks. One should check for ticks on one's body after leaving such areas. Rat control has been effective in the control

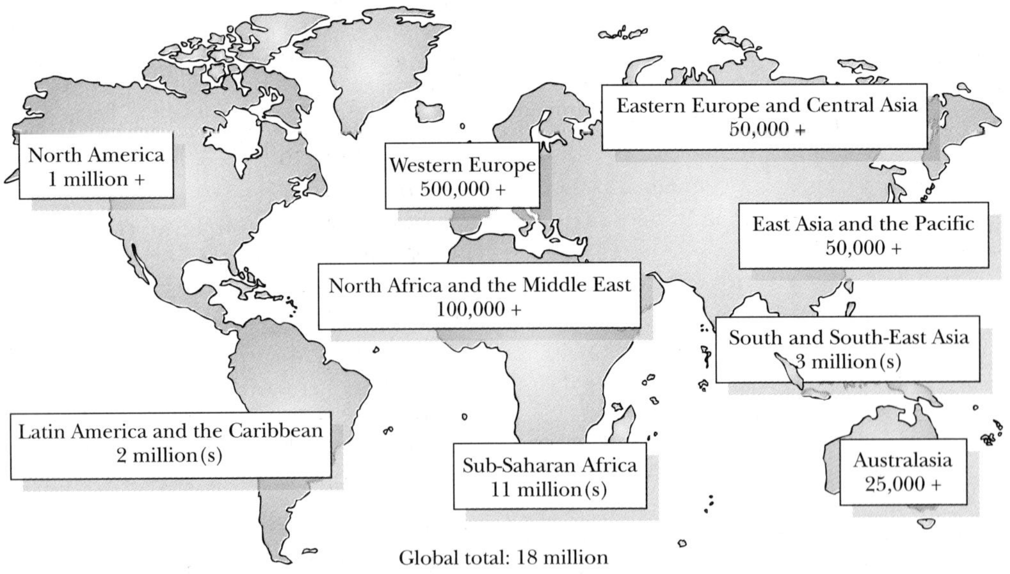

***Figure* 31.4** The estimated distribution of adult HIV infections through late 1994. (Reproduced by permission of WHO, from: *Weekly Epidemiological Record,* 70(2): 7 (1995))

of vector-borne diseases such as plague and endemic typhus. Tularemia cannot be controlled in wild animal populations. Humans can avoid infection by this organism by not dressing freshly killed wild rabbits; or, if doing so, one should wear rubber gloves and proceed with care.

Animal Contact Diseases

Animal contact diseases, such as brucellosis, are found sporadically in the United States. The spread of brucellosis is controlled by vaccination of cattle and pasteurization of milk. Anthrax, a disease found among workers exposed to wool and goat hair, is difficult to control because the causative agent, *Bacillus anthracis,* is an endospore former that can survive in soil. Sterilization of wool and goat hair is the only known control measure. Leptospirosis can be controlled by eliminating rats and by vaccinating domestic animals. Animal handlers can and should be vaccinated against Q fever. Psittacosis is an ever-present danger in domestic birds and is particularly hazardous for workers in poultry slaughterhouses. Pigeons are also a reservoir for the disease.

Nosocomial (Hospital-Acquired) Infections

To effectively invade a host, the pathogen must counteract the host's defenses, which normally are quite effective in removing invading organisms. Unfortunately, in hospital environments, patients' normal defenses are often compromised, and infections by opportunistic or accidental invaders do occur. These are called **nosocomial infections.**

Nosocomial infections occur for a number of reasons. The most prevalent include the following:

1. Illness or chemotherapeutics may compromise the immune system.
2. Abrasions or openings in epithelial barriers caused by surgery, catheters, syringes, respirators, or other tools employed by the physician to examine the inner body offer the opportunistic pathogen a site to establish infection.
3. Exposed tissues of burn or wound patients offer access to airborne microorganisms.
4. Cross-infection from patient to patient in crowded wards as well as from hospital workers or physicians to patient is always a concern.
5. The overuse of antibiotics in hospitals has resulted in a hospital environment in which drug-resistant microorganisms are ubiquitous.
6. The hospital environment selects for pathogens. As few hospitals have isolation wards, a virulent organism can find a reservoir in patients, and this pathogen can be transmitted to other patients.
7. Hospital pathogens may bear plasmids that carry information for multiple drug resistance. As a result, drug resistance can spread rapidly among a population of pathogenic microorganisms.

***Table* 31.2 The relative frequency of nosocomial infections by body site**

Site of Infection	Percent of Total Infections
Urinary tract	40%–42%
Respiratory	16%–18%
Surgical wound	17%–20%
Bacteremias	6%–7%
Skin infections	6%–7%
Other	12%

Major Nosocomial Infections

It is estimated that two million patients become infected each year during hospitalization and these infections cause fatalities in over 100,000 patients. The nosocomial infections are frequently caused by bacteria that are part of the normal human flora. In a healthy nonhospitalized individual, these microorganisms would not be invasive. The most prevalent organism in nosocomial infections is *Escherichia coli.*

Infections of the urinary tract are the nosocomial infection most often encountered in the hospital environment (Table 31.2). Of the total urinary tract infections, one third are caused by *E. coli* and another third by *Pseudomonas aeruginosa, Enterococcus faecalis,* or *Staphylococcus epidermis.* Most of the remainder are caused by other gram-negative bacteria. Infection is generally a consequence of urinary catheterization of immobilized patients. Respiratory infections that appear as a form of pneumonia are often encountered, and these may be caused by *Pseudomonas aeruginosa, Staphylococcus aureus,* or *Klebsiella* sp. These nosocomial infections, which can cause death, result from respiratory devices and the inability of the ill patient to clear his or her lungs. *Staphylococcus aureus* and pyogenic Streptococci are a major concern in surgical patients. An estimated 7 to 12 percent of all surgical patients have post-operative infection problems. When the gastrointestinal or genitourinary tracts are involved in the surgery, the number of postoperative infections is 2 to 3 times higher.

Public Health Measures

The general health of people in the United States has improved dramatically since the time of our Founding

BOX 31.5 MILESTONES

Public Health Measures and Human Well-Being

The effectiveness of public health measures on the curtailment of infectious diseases in New York City is apparent from this chart. Cholera, a disease transmitted by fecal contamination, was a considerable cause of death prior to the establishment of the Board of Health and Health Department in 1866. Smallpox, yellow fever, and typhus were also a significant health problem. Chlorination of water was initiated in 1910 and pasteurization of milk in 1912, and these measures resulted in better control of infectious disease. Note the number of deaths from the influenza pandemic in 1918.

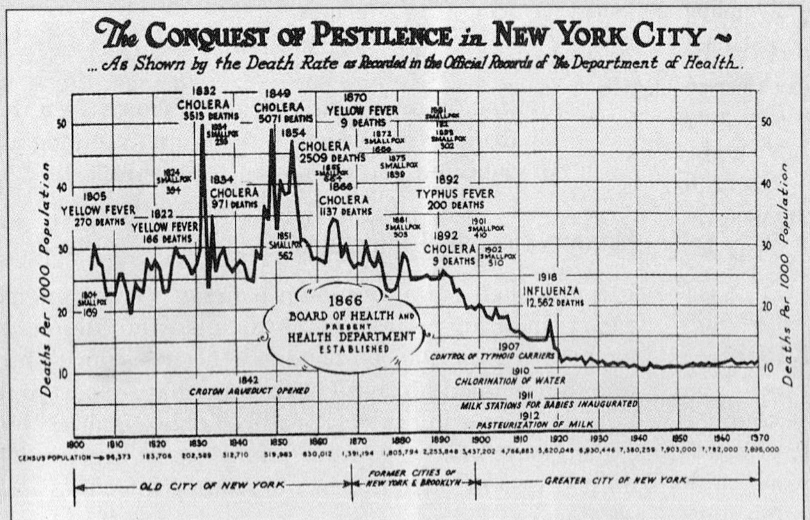

Figure 1.1 Crude death rate per 1,000 population: New York City, 1804–1970 (courtesy of the Department of Health, City of New York)

The effect of public health measures on the incidence of disease. (Courtesy of the Department of Health, New York City)

Fathers. Life expectancy for a baby born today is about double that in the late eighteenth century. Nutrition has improved markedly, with fresh fruits and vegetables available year around. Housing has also improved, and working conditions are generally less stressful. We have eliminated the threat of many deadly diseases such as yellow fever, and antibiotics have alleviated dreaded diseases such as tuberculosis. A combination of these factors has tended to make our lives longer and healthier. How has this happened? By diversified efforts termed **public health,** which refers to the overall health of populations and the efforts of local, state, and the federal public health officials to maintain reasonable health standards **(Box 31.5).**

Each state in the United States has a publicly funded organization generally known as the State Health Department. The name may vary but the responsibilities are much the same, and a major function of this organization is monitoring the incidence of infectious diseases in the population. In many states, the department provides a diagnostic laboratory for infectious organisms unidentifiable by local clinical laboratories. A major role of the State Health Agency is to follow the incidence of disease in the state and to communicate this information to a national agency, which is the Centers for Disease Control and Prevention (CDCP) in Atlanta. The CDCP is a subunit of the United States Public Health Service (USPHS). The USPHS is a branch of the Department of Health and Human Services.

The epidemiology unit of the State Health Department has a requirement that licensed physicians report cases of selected communicable diseases that occur in patients under their care. For example, listed in Table 31.3 are the communicable diseases reported to the state epidemiologists in North Carolina. Other states would have similar requirements. Many of these diseases are then reported to CDCP. It is also required that restaurants and other food or drink establishments report outbreaks of food-borne disease in employees and customers. Many dis-

***Table* 31.3 Communicable diseases reported to the epidemiology division of a state health department (North Carolina) and to the Centers for Disease Control and Prevention**

AIDS	Hepatitis A	Q fever[1]
Amebiasis	Hepatitis B	Rabies, human
Anthrax	Hepatitis non A or B	Reyes syndrome
Blastomycosis[1]	HIV infection	Rocky Mountain spotted fever
Botulism	Kawasaki syndrome[1]	Rubella
Brucellosis	Legionellosis	Salmonellosis
Campylobacter infection	Leprosy	Shigellosis
Chancroid	Leptospirosis	Syphilis
Chlamydia	Lyme disease	Tetanus
Cholera	Lymphogranuloma venereum	Toxic shock syndrome
Dengue[1]	Malaria	Trichinosis
Diphtheria	Measles (rubeola)	Tuberculosis
Encephalitis[1]	Meningitis, pneumococcal[1]	Tularemia
Foodborne diseases[1]	Meningitis, viral	Typhoid, acute
Clostridium perfringens	Meningococcal disease	Typhoid, carrier[1]
Staphylococcal	Mumps	Typhus, epidemic[1]
Other/unknown	Plague	Whooping cough
Gonorrhea	Polio, paralytic	Yellow fever
Granuloma inguinale	Psittacosis	
Haemophilus influenzae (invasive)[1]		

[1]Diseases not reported to the CDCP.

Source: Information provided by the State of North Carolina Health Services.

eases, such as cholera, plague, and hepatitis, must be reported within 24 hours so that immediate control measures can be implemented.

The effectiveness of our public health institutions is very evident. Potential epidemics are tracked and, in many cases, spread of disease curtailed. An example of this was the potential for measles (rubeola) outbreaks on college campuses during recent years. The vaccine employed during the infancy of most college students at that time did not confer a long-lasting immunity. There was the risk of an epidemic in this population, and, after a few cases were documented, it was made mandatory that all students lacking proof of the proper immunization report for vaccination. This prevented a potential harmful outbreak, as rubeola can be a serious infection in young adults.

Control Measures for Communicable Diseases

As we examine diseases and their reservoirs (see Table 31.1), it is evident that some diseases are more readily controlled than others. Diseases that can be prevented by vaccination during infancy are controllable, but only when adequate vaccination programs are followed. The same applies to most waterborne infectious diseases where adequate water treatment can virtually eliminate the disease. Diseases transmitted in milk can be controlled by maintaining healthy dairy herds and by the pasteurization of milk before consumption or prior to conversion to cheese or other milk products. On the other hand, respiratory infections and those passed orally are quite difficult to control. A viral influenza outbreak can potentially spread unchecked around the world. Humans are gregarious, and infectious droplets resulting from talking, coughing, or a partly stifled sneeze are a hazard to everyone who comes in contact with the infected individual. The influenza virus itself can be exceedingly virulent and constantly changes such that immunity to past epidemics does not ensure protection against the next.

Reservoir Control

Control of the reservoirs that sustain infectious microorganisms is effective in some cases but not in others. Domestic cattle have been a reservoir for human infections, but this source has been substantially eliminated by rigorous control measures. *Brucella abortus* causes spontaneous abortion in cattle, and vaccination prevents the spread of the disease among animals and to humans. Cattle are also immunized for bovine tuberculosis to prevent transmission of this infectious microorganism to humans through milk. There are strict rules on the extermination of cattle that may become infected with lumpy jaw, a disease caused by *Actinomyces bovis.* The elimination of diseased sources is a practical mechanism for preventing the spread of an infectious disease. It is, however, essential that *all infected sources* in the potential reservoir be eliminated, otherwise the disease can erupt and travel rapidly through a nonimmune, susceptible population.

The reservoir for a number of significant diseases is rodents, so it is mandatory that domestic rats be destroyed or driven from populated areas to control diseases spread by these rodents. Rats are one wild animal that has evolved with the unique ability to live among humans. Their destructive habits, appetite for grain, and disease-causing potential are all a threat to human welfare. During 1993, a viral infection caused a number of deaths in the Southwestern United States that were traced to a hantavirus that was spread by inhalation of aerosols of rat urine and feces. The living conditions of the victims were a primary contributor to the infection.

Quarantine

Animals are frequently quarantined before they are transported from region to region or country to country. Great Britain has eliminated rabies from the British Isles, and has imposed strict regulations on the import of animals to ensure that they will not bring in the disease. The United States enforces a quarantine on cattle and other animals before import to ensure that they are free from infectious diseases.

Quarantine was once commonly practiced in the United States at the local level. If a family had a member with scarlet fever or certain other transmissible diseases, a notice would be posted outside the door of their home. The notice warned visitors that an infected individual resided there and they should not visit lest they carry the disease elsewhere. In cases where the patient was a youngster, all siblings were barred from attending school. Prior to the availability of antibiotics, tuberculosis patients were confined to a state-controlled sanitarium. Hawaii and other tropical areas had leprosy colonies where individuals infected with the disease were confined. Quarantine of humans in the United States, except in isolation wards of hospitals, has been discontinued. By international agreement, some diseases are quarantinable. These diseases include cholera, plague, yellow fever, typhoid, and relapsing fever. Those infected with one of these diseases may be barred from moving from country to country.

Food and Water Measures

Food inspection is a common practice, and among the facilities subject to government inspection are abattoirs and meat processing plants. Milk pasteurization follows mandated procedures and has curtailed the spread of brucellosis, bovine tuberculosis, typhoid, listeriosis, and other diseases. Breakdowns in the pasteurization process have led to common-source epidemics. In recent years, there were two listeriosis (*Listeria monocytogenes*) epidemics: one in Massachusetts (1983) resulting from ingestion of insufficiently pasteurized milk and the other in California (1985) from ingestion of Mexican-style cheese, which was obviously made from contaminated milk. There were 104 deaths in these two episodes. Cases such as this are testimony to the need for strict quality control in the food we consume. Foods sold to the general public rightfully must meet reasonable public health standards for quality and sanitation. Despite these precautions we continue to have periodic incidents that cause considerable human misery.

Water treatment and purification (in the developed nations) have virtually eliminated the threat of cholera, typhoid, and other waterborne diseases. Sewage treatment plants in our municipalities have been effective in processing domestic sewage and in preventing the spread of diseases from this source. A major improvement has been to separate storm runoff drainage from the domestic sewage systems. In years past, raw sewage and storm water bypassed the sewage disposal plant during heavy storms, which resulted in considerable local pollution problems.

Human and Animal Vaccination

Vaccination has been successfully employed in controlling a number of fearful diseases. There are mandatory laws among the various states that require proof of vaccination before a child can enter school. The basic immunization program is initiated in an infant at about two months of age, and children in the United States are immunized against an array of diseases (Table 31.4). There is a growing concern that immunization programs are not begun

***Table* 31.4 The recommended immunization schedule for infants and young children in the United States**

Age	Vaccine Employed
Birth	Hepatitis B
2 months	Diphtheria; pertussis; tetanus (DPT) Hemophilus b (Hib) Poliomyelitis (OPV)
4 months	DPT; OPV; Hib Hepatitis B
6 months	Hepatitis B DPT; OPV; Hib
12–15 months	DPT; Hib; chicken pox, measles, mumps, rubella (MMR)
4–6 years	OPV; DPT; MMR

at an early enough age, particularly among the disadvantaged. It is a concern that this delay might establish a significant pool of children up to five years of age who are at risk of contracting and disseminating communicable diseases. Some adults are inadequately immunized and potentially can contract childhood diseases. Extensive immunization of children has been successful to date in virtually eliminating many childhood diseases such as smallpox, diphtheria, pertussis, and poliomyelitis, and this may be giving the adult population a false sense of security. Outbreaks in uninoculated preschoolers could cause limited outbreaks among adults.

Herd immunity is the immunity engendered to a communicable infectious disease when a major segment of the total population is immune to that disease. The greater the percent of a total population immune to an infectious agent the greater the chance that the entire population will be protected. It is not necessary that 100 percent of a population be immune for a disease to be eliminated. In practice, few vaccines confer immunity in 100 percent of the individuals vaccinated. Generally, herd immunity is conferred if 70 percent or more of the population is immune. For highly contagious diseases, such as viral influenza, the percentage of the total susceptible population that must be immune to confer herd immunity is close to 95 percent.

Vaccination of pets for rabies has been effective in controlling this disease in domestic animals. Unfortunately, there is a large reservoir of the disease in skunks, raccoons, and other carnivores. Rabid animals among these species are a threat to humans. An oral vaccine is under trial in France and, in the United States, on an island off Virginia in an effort to develop immunity in wild animal populations.

Antibiotic Resistance

The development of antibiotic resistance in pathogenic organisms is a major public health concern. One somewhat controversial practice in this regard is feeding antibiotics to cattle to increase their growth rate. About one half of the annual production of antibiotics in the United States is incorporated into feed for poultry, swine, and cattle. Tetracycline and neomycin are two antibiotics that may be added to animal feed. As the animals consuming these antibiotics generally harbor potential pathogens, such as *Salmonella* sp, the indiscriminate use of antibiotics in feed could select for antibiotic-resistant strains. The animal and pharmaceutical industries do not consider this a danger and it is not restricted. One antibiotic that was added to animal feed in the past—penicillin—is no longer used for this purpose.

The Centers for Disease Control and Prevention has estimated that 70 percent of the salmonellosis outbreaks that occurred in the years 1971 and 1983 involved resistant strains that came from animals. In a documented case, a unique strain of *Salmonella newport*, identified by plasmid profile, was involved. Eighteen cases of salmonellosis were traced to one farm in South Dakota where the antibiotic-resistant strain apparently originated.

World Health

The World Health Organization (WHO) was established in 1948 through the United Nations and is now headquartered in Geneva, Switzerland. The WHO is committed to the control of diseases and promoting health for the people in over 100 member countries. The organization is involved with population control, availability of food, and efforts to curtail disease through education. The WHO provides information on developing safe drinking water sources. Worldwide immunization programs against such diseases as diphtheria, poliomyelitis, and tuberculosis are the goal of the WHO, and it is the intention of the WHO that means be found to bring these and other diseases such as malaria and leprosy under control. The WHO was instrumental in the eradication of smallpox (**Box 31.6**), and it strives for equal success against other diseases.

Problems in Developing Nations

Diseases that were once prevalent in the Western World remain a concern in developing countries. As previously mentioned, infectious diseases cause nearly 50 percent of the deaths in developing nations, but fewer than 8 percent of deaths in developed countries. This higher death rate in the developing countries is due to inadequate sanitation, nutritional deficiencies, and substandard housing. The lack of medical care and immunization programs also contribute to a high mortality rate. Diseases listed in Table 31.5 are a concern in developing nations, and some of these result from a lack of adequate supplies of clean water. Cholera, a serious problem in Southeast Asia and a threat in some South American countries, could be eliminated by providing populations with safe drinking water. Typhoid fever is another infectious disease that would be virtually eradicated by the availability of safe water supplies. It is estimated that up to 10 percent of the population in Latin America are typhoid carriers. Inadequate treatment of sewage is a concern, as this is a potential source of typhoid infection from the ever-present carrier reservoir.

Problems in Developed Nations

The developed countries are not without health problems caused by infectious organisms. In both developed and developing countries, HIV is increasing at an alarming rate and is now considered a pandemic (see Figure 31.4). Gonorrhea continues to be a problem despite the availability of an effective cure for the disease; an estimated 1 of 35

BOX 31.6 MILESTONES

A Success Story: The Eradication of Smallpox

Smallpox is a dreaded disease that was eradicated by a successful worldwide immunization program. Smallpox is an ancient disease, and the millions of deaths caused by it are engraved in the 3000-year record of human history. Through the first quarter of the twentieth century, thousands of cases occurred annually in the United States. The disease was eliminated in the United States by about 1960 through a long-term extensive vaccination program in preschool children. WHO initiated a worldwide eradication program in 1966. As the disease was eliminated in the developed nations, the campaign's thrust turned to India, Africa, South America, and other developing countries. By 1977, the world appeared free of smallpox. In 1980, the WHO declared the disease eradicated. As a result, children are no longer vaccinated for this disease.

There are several reasons why the eradication of smallpox was so successful and that equivalent programs for other diseases might be less so. Among the advantages of the smallpox vaccine: it was available at low cost, the vaccine raised a high level of immunity, the vaccine did not require refrigeration, and inoculation was by pin pricks with no need for syringes. Thus, individuals in remote villages would be vaccinated. Not all immunization programs are so amenable.

young adults will contract gonorrhea each year. Legionnaire's disease and shigellosis are endemic, and shigellosis is particularly contagious among young children in child care centers. Lyme disease and Rocky Mountain spotted fever are increasing in incidence with no adequate control measures in sight. Vigilance is essential, and the epidemiology programs in the various states and CDCP are doing an excellent job in monitoring communicable diseases in the United States.

Problems for Travelers

When people travel from developed to developing countries they can encounter serious health problems. Immunizations are required or recommended for travel by United States citizens to many areas of the world (Table 31.6). Cholera vaccination is required when entering areas of southeast Asia and central Africa, where the disease is endemic. Yellow fever is still a concern in many Central and South American countries, and vaccination is often required for those traveling to these areas. Vaccination is recommended for typhoid if one is traveling to some Central and South American countries and parts of Africa and Asia. Diseases for which there are no immunizations, such as dengue fever, malaria, and typhus, are also prevalent in some developing countries. Travelers should check with health authorities before journeying to countries where communicable diseases not encountered in the United States are a threat.

***Table* 31.5 A number of the diseases that are a major concern in developing nations**

Diseases	Country in Which Found
Cholera	Southeast Asia, Central Africa
Salmonellosis	Most
Shigellosis	Most
Yellow fever	Central and South America
Encephalitis	Most
Typhus	Middle East
Typhoid fever	Middle East, Central and South America
Tuberculosis	Most

Clinical and Diagnostic Methods

Early identification of the microorganism that is causing an infectious disease is of utmost importance. Epidemics can be tracked effectively only where the etiological agent is correctly identified. Therapeutic agents, such as antibiotics, can be chosen more efficiently when the infectious agent involved is known and antibiotic resistance/susceptibility have been determined. Identifying etiological agents can minimize the severity of a disease and shorten the recovery time when the proper treatment is immediate and appropriate. In many epidemiological studies, the characterization of the microbe must be carried beyond species to the strain involved. Recent technological advances have provided the clinical laboratory with techniques that effectively shorten the time required for accurate identification of pathogens.

***Table* 31.6 Immunization recommended or required for travel to developing countries**

Disease	Traveling to
Vaccination Required	
Cholera	Southeast Asia, Albania, Malta, Central Africa, South Korea
Yellow fever	Central and South America, African countries
Vaccination Recommended	
Plague	Rural areas of Africa, Asia, and South America
Serum hepatitis	Africa, Indochina, Russia, Central and South America
Typhoid fever	Africa, Asia, Central and South America (Specific areas)
Diphtheria; polio; tetanus; measles; mumps; rubella (most U.S. citizens already immunized)	All

Searching for Pathogens

There are a number of factors that predispose an individual to the diseased state. Significant among these are the general health of the host, previous contact with the microbe, past medical history, exposure to toxic agents or chemicals, and traumatic or other insults not of microbial origin. These and other elements have considerable bearing on whether an individual will contract a particular disease.

The term "pathogen" can be applied to few organisms if one considers a pathogen to be an organism that **always** causes clinical symptoms. The indigenous microbiota of a human is a varied population, and many of these organisms may cause symptoms of disease under appropriate conditions. Lowered resistance can result in a clinical infection, and it is often difficult to determine which of many organisms present in the diseased state is the one responsible. Clinical microbiologists are trained to sort through the organisms present in a specimen and make decisions on the most likely source of clinical symptoms.

Obtaining Clinical Specimens

The accuracy of a diagnosis is quite dependent on the quality of the specimen delivered to the clinical laboratory. The specimens generally obtained from a patient would be one or more of the following: blood, urine, pus/exudate from a wound, biopsy tissue, feces, sputum, or cerebrospinal fluid. Throat or nasal swabs and fluid aspired from an abscess are regularly submitted to the clinic for diagnosis. Care must be taken to prevent contamination of a specimen by extraneous microorganisms after the specimen is taken. It is essential that the specimen obtained be sufficiently large that all desired tests can be accomplished on that single specimen.

Blood

Presence of bacteria in blood (bacteremia) is generally an indication that there is a significant focus of infection somewhere in the patient's body. If a few bacteria enter the bloodstream the organism will probably be cleared quickly by natural host defenses. However, when the number of microorganisms present in blood is significant, an acute focus of infection is probable, with the shedding of microorganisms into the bloodstream. This causes a general septicemia (rapid propagation of pathogens in the blood) that results in fever, chills, and shock. Blood samples for analysis would always be taken aseptically with a sterile syringe, and the blood should be delivered into a bottle containing a suitable medium and anticoagulant to prevent clots. (Clots may entrap bacteria and make isolation of the microbe difficult.) One part of the blood sample would be incubated aerobically, the other anaerobically. The inoculated culture bottles are placed in an incubator that automatically monitors CO_2 production. Should growth occur either aerobically or anaerobically, the microorganism must be isolated in axenic culture and identified. The organisms most commonly associated with blood infections are *Staphylococcus aureus, Streptococcus pyogenes, Pseudomonas aeruginosa,* and enterics.

Urinary Tract

Urinary tract infections are frequently caused by gram-negative bacteria similar to those that are part of the natural human microbiota. As urine itself can support the growth of bacteria, care must be taken to ensure that once a specimen is taken it does not stand for any length of time at room temperature before analysis. If not analyzed immediately, the sample should be refrigerated. The bacteria generally involved in urinary tract infections are gram-negatives such as *Escherichia coli* and species of *Klebsiella, Proteus,* or *Enterobacter.* Urine samples are analyzed by direct count of organisms present and by spreading an aliquot over the surface of a MacConkey agar and a blood agar plate employing a calibrated loop. The number of colonies that develop on the blood agar plate is a direct measure of the number of organisms in the urine specimen. If there are 10^5 or more organisms per milliliter of urine, a urinary tract infection is indicated. MacConkey agar is a selective medium for gram-negative bacteria. It contains bile salts and crystal violet that inhibit gram-positives, the sugar lactose, and neutral red as a dye indicator. *Enterobacter* and *Escherichia* ferment lactose, and the colonies take up neutral red, imparting a reddish color. *Proteus, Salmonella,* and *Shigella* do not ferment lactose, and they form white or clear colonies. Several media that can be employed in characterizing gram-negative bacteria encountered in the clinical laboratory are presented in Table 31.7. The blood agar plate should be examined for the presence of staphylococci and streptococci. These organisms are less frequently present than the gram-negatives in urinary tract infections.

Other procedures are followed when the infection might be caused by a sexually transmitted infectious microorganism. Gonorrhea is one of the most common infectious diseases among young adults, and the etiologic agent is *Neisseria gonorrhoeae.* A Gram stain of the purulent urethral discharge in males can be a rapid and reasonably accurate diagnostic procedure. If the symptoms are less obvious, the clinical specimen from males and females must be cultured on selective and nonselective media. A primary medium is the Thayer-Martin medium, which contains the antibiotics vancomycin, nystatin, and colistin. These antibiotics inhibit the growth of many microorganisms commonly present in such specimens but not the growth of *N. gonorrhoeae.* Some pathogenic strains of *N. gonorrhoeae* may be inhibited by vancomycin, so nonselective chocolate agar should also be inoculated. Chocolate agar is prepared from heated blood and is a source of growth factors for *N. gonorrhoeae.* It also absorbs toxic material that may be present in a rich medium that would otherwise inhibit the growth of the gonococcus. The inoculated plates should be incubated in an atmosphere of 5 percent CO_2. Commercial probes have been developed for *N. gonorrhoeae* and are a rapid diagnostic tool.

Wound Exudate

Specimens obtained in suspected cases of syphilis would be the exudates from open lesions or material from lymph nodes in the affected regions. Dark-field microscopy will often reveal the presence of spirochetes. Staining is not generally effective, as *Treponema pallidum* stains poorly and is about 0.2 μm in diameter—near the limit of light resolution. The organism cannot be cultured on laboratory media. Identification can be by fluorescein-labeled, anti-treponemal antibodies or a slide flocculation test. The flocculation tests are based on the presence of antibody to a specific cardiolipin antigen in the serum of syphilis patients. *Chlamydia trachomatis* can be identified by inoculating exudate into cell culture and monitoring for growth of the intracellular pathogens.

Table 31.7 Some agar-base media that can be employed to differentiate clinically significant gram-negative bacteria by colonial appearance

Medium	*Enterobacter aerogenes*	*Escherichia coli*	*Salmonella* sp.	*Shigella* sp.	*Proteus* sp.	Gram +
Bismuth sulfate	Mucoid silver sheen	Little growth	Black with metallic sheen	Inhibited to brown	Green	No growth
Eosin methylene blue	Pink	Purple with black centers	Colorless	Colorless	Colorless	No growth
Salmonella–shigella	Cream to pink	No growth	Colorless	Colorless	Colorless	No growth
Sodium azide agar	No growth	No growth	No growth	No growth	No growth	Growth
MacConkey agar	Pink to red	Pink to red	Colorless	Colorless	Colorless	No growth

Tissue/Abscess

Tissues from biopsies and material from skin lesions or wounds should be streaked on blood agar and other rich media. Incubation would be both anaerobic and aerobic. Care must be taken in obtaining and handling such specimens, as infections of this type are often caused by strict anaerobes and would be adversely affected by contact with atmospheric oxygen. Exudate from infected areas can be collected in a syringe by aspiration and taken directly to the clinic for diagnosis. Such samples may be examined by microscopy after staining with fluorescent antibody. The type and the appearance of the lesions as well as the history of the patient may suggest possible etiological agents and indicate which fluorescent antibody should be applied. Anthrax, plague buboes, and tularemia lesions have distinct characteristics, and confirmation can be made quickly by serological techniques.

Fecal

The pathogens most often associated with fecal specimens are food or waterborne microorganisms, and among these are *Vibrio cholerae, Campylobacter jejuni,* and species of *Salmonella* and *Shigella.* Careful handling of fecal specimens is essential, for left at room temperature for a brief period of time, the specimens can quickly become acidic. This acidity will kill pathogens such as *Salmonella* or *Shigella* as they are sensitive to a low pH. To overcome this, fecal samples are generally collected in a buffered medium and delivered quickly to the clinic for diagnosis. The selective medium of choice for fecal specimens is MacConkey or eosin-methylene-blue agar (Table 31.7). Samples should also be streaked on blood agar to determine whether gram-positive pathogens, such as staphylococcus or streptococcus, might also be present.

Sputum

Sputum and material obtained from the upper respiratory tract should be streaked on blood or chocolate agar. Blood agar is useful for growing *Streptococcus pyogenes, Streptococcus pneumoniae,* and *Staphylococcus aureus. Neisseria meningitidis* and *Haemophilus influenzae* can be detected on chocolate agar. Acid-fast staining of smears and serological tests can be applied where *Mycobacterium tuberculosis* might be involved in the infection.

Cerebrospinal Fluid

Cerebrospinal fluid specimens are cultured on rich media such as blood and chocolate agar. Organisms that are involved in meningitis (*Neisseria meningiditis, Streptococcus pneumoniae,* and *Haemophilus influenzae*) grow on these rich media. Since meningitis is a potentially fatal infection, an immediate diagnosis is mandatory. ELISA tests (see later) are a major tool in determining the nature of the pathogen that might be involved in infections of this type because this permits an immediate diagnosis and treatment.

Identifying Pathogens

The major goal of a clinical microbiology laboratory is a prompt, precise identification and characterization of the pathogen involved in an infection. Antibiotic resistance/sensitivity of the causative agent of infections is an essential element in selecting a proper treatment. If the pathogen responsible for an infection is resistant to the antibiotic generally employed in such cases, successful treatment depends on administering an alternative antibiotic.

Specimens delivered to the clinical laboratory for microbiological analysis may follow one or more of the following routes:

1. The specimen may be examined microscopically following staining or examined directly with a dark-field or phase-contrast microscope.
2. A specimen can be streaked or cultured on an enrichment, selective, or differential medium (Table 31.8).
3. The specimen may be subjected directly to serological, immunofluorescence, ELISA, or other diagnostic procedures.

Direct microscopic examination of stained clinical material can be of value with selected specimens. This can give a preliminary diagnosis that can then be confirmed by isolation and identification of the pathogen. In suspected cases of tuberculosis, sputum is subjected to the acid-fast stain, a diagnostic tool for bacteria that have a waxy coat. Cervical scrapings from females and urethral discharge from males may be Gram stained to visualize the typical gonococcus involved in gonorrhea. There may also be observable polymorphonuclear leukocytes that contain the characteristic diplococcal *Neisseria gonorrhoeae* cells. In the case of leprosy, *Mycobacterium leprae* is identified by direct observation of acid-fast stained specimens from leprous lesions as the organism cannot be grown outside the host. A preliminary diagnosis of syphilis is also possible by dark-field microscopy. Examinations for animal parasites in blood, feces, and so on, are generally done by direct microscopic examination of properly collected specimens.

Staining and viewing some specimens microscopically as they are received in the clinical laboratory is not necessarily productive. The specimen from a given source will contain an array of microorganisms, and microscopic examination will not indicate which is the agent of infection. In such cases, samples are inoculated directly into or on appropriate growth media. The medium employed will depend on the source of the specimen. After growth on selective or differential media (Table 31.8) an experienced clinical microbiologist can select the best pathway to follow in identification. This would be influenced by microscopic examination, colonial morphology, and specimen source. Microorganisms present in colonies from primary-enrichment, selective, or differential media can be subjected to growth-dependent tests for identification (Tables 31.9 and 31.10). These organisms would also be tested for antibiotic sensitivity/resistance.

Table 31.8 Basic types of media that would be employed in the clinical microbiology for isolation of bacterial cultures from specimens

Characteristics Media—Test bacteria for specific metabolic activities, enzymes, or growth characteristics.

- **Citrate agar**—Has the ability to utilize sodium citrate as sole carbon source.
- **TSI** (Triple-sugar-iron) **agar.** Three sugars are lactose, glucose, and sucrose together with sulfates and a pH indicator. Used in slants to determine relative use of each sugar and generation of sulfide.
- **SIM** (Sulfide, indole, motility) **agar.** Is a test for production of sulfide from sulfate, indole from tryptophan, and motility.

Differential and **Selective Media**—Contain chemicals.

- **Brilliant green agar**—The dye inhibits gram-positive bacteria selecting for gram-negatives.
- **Sodium tetrathionate broth**—Inhibits normal inhabitants of intestinal tract, favors *Salmonella* and *Shigella* species.
- MacConkey agar—Contains bile salts and crystal violet that inhibits gram-positives and many fastidious gram-negative organisms; favors enterobacteria.
- **Mannitol salt agar**—Contains 7.5 percent NaCl that is inhibitory to most organisms and favors growth of staphylococci.
- **Eosin methylene blue**—Partly inhibits gram-positives. Eosin gives *Escherichia coli* a metallic greenish sheen. *Enterobacter aerogenes* colonies are pink. Other species are less pigmented.

Enrichment Media—Contains blood, serum, meat extract, or other nutrients that favor the growth of fastidious bacteria, particularly when present in low numbers. Often employed for clinical specimens such as cerebrospinal fluid.

Table 31.9 Growth and colonial characteristics of some commonly isolated organisms on differential agar media

	EMB		Mannitol Salt	
Organisms	**Growth[1]**	**Color of Colony**	**Growth**	**Color of Colony**
Enterobacter aerogenes	++	Pink no sheen	Inhibited	
Escherichia coli	+++	Purple-green metallic sheen	Inhibited	
Klebsiella pneumoniae	++	Green metallic sheen, mucoid colony	Inhibited	
Salmonella typhimurium	++	Colorless	Inhibited	
Staphylococcus aureus	Inhibited		+++	Yellow
Staphylococcus epidermidis	Inhibited		++	Red

[1] ++ good growth +++ excellent growth

***Table* 31.10 Major diagnostic tests employed to differentiate pathogenic bacteria in a clinical laboratory**

Test	Purpose	Potential Application
Acid-fast stain	Tests for organisms with high wax (mycolic acid) content in cell surface that stain with hot carbolfuchsin and cannot be decolorized with acid alcohol.	*Mycobacterium tuberculosis* is acid-fast
Catalase	Enzyme decomposes hydrogen peroxide $H_2O_2 \rightarrow H_2O + O_2$.	*Staphylococcus* from *Streptococcus*
Citrate	Transports and utilization as sole carbon source.	Classification of enteric bacteria
Coagulase	Causes clotting of plasma.	*Staphylococcus aureus* from saprophytic staphs
Decarboxylases	Tests for ability to decarboxylate amino acids such as lysine, ornithine, or arginine.	Classification of enteric bacteria
Esculin	Tests for cleavage of a glycoside.	Separate streptococci
β-galactosidase	Demonstrates an enzyme that cleaves lactose→glucose + galactose.	Separate enterics and identify pseudomonads
Gelatin liquefaction	Enzymatic hydrolysis of gelatin.	Identify clostridia and others
Gram stain	Used as a first and primary differential test.	
Hemolysis	Hemolysis of red blood cells; α-hemolysis, indistinct zone of hemolysis, some greenish to brownish discoloration of medium; β-hemolysis, clear, colorless zone around colonies.	Pathogenic streptococci
Hydrogen sulfide	Demonstrates H_2S formation by sulfate reduction or from sulfur-containing amino acids.	Identify enterics
Indole	Determines hydrolysis tryptophan to indole.	Separate enterics
KCN growth	Tests for ability of microorganisms to grow in the presence of cyanide (inhibits electron transport).	Aerobes/anaerobes
Lipase	Presence of the enzyme that cleaves ester bonds of fats yielding fatty acids + glycerol.	Separate clostridia
Methyl red	Checks for acid production from glucose via mixed acid fermentation.	Separate enterics

continued

***Table* 31.10** (*Continued*)

Test	Purpose	Potential Application
Motility	By microscopic examination or diffusions through soft agar, shows the ability to move.	Motile strains
Nitrate reduction	Reduction of $NO_3 \rightarrow NO_2$ (nitrate employed as terminal electron acceptor).	Enterics and others
Oxidase test	Demonstrates presence of cytochrome c which oxidizes an artificial electron acceptor.	Enterics from pseudomonads
Phenylalanine deamination	Deaminates the amino acid to phenylpyruvic acid.	Proteus group
Protease	Tests for ability to digest casein.	Separate *Bacillus* species
Sugar utilization	Growth on pentoses, hexoses, or disaccharides, producing acid and gas.	Many
Urease	Detects an enzyme that splits urea to $NH_3 + CO_2$.	Separate enterics
Voges-Proskauer	Detects acetoin as product of glucose fermentation and indicates neutral fermentation.	Differentiate bacillus species and enterics

Growth-Dependent Identification

Identification of various potential pathogens can be achieved by growth-dependent methods, and the procedure followed is generally based on preliminary identification of the microorganism involved.

Commercial diagnostic packets containing various media and selected differential tests have been developed. Among the manual rapid test systems is the API 20E system for the identification of enteric bacteria. These systems (Figure 31.5) are compact and simple to inoculate, and many clinical isolates can be analyzed in a short period of time. The sugars and diagnostic tests in these manual systems can be varied and used to identify microorganisms based on their presumptive identification. Compact test kits have replaced many of the individual tube tests (Table 31.11) that were routine in clinical laboratories in years past and are still a part of the laboratory section of a general microbiology course.

Serologic Identification

Culturing of microorganisms, including viruses and selected bacteria, is not always possible or practical in the clinical laboratory. There are commercial test kits that can detect presence of an antigen or antibody, and these can be employed to provide a rapid diagnosis of many infectious agents. These immunologic systems can identify pathogens in clinical specimens rapidly and accurately without culturing of the infectious agent. Rapid immunologic test kits are available to detect: *Haemophilus influenzae* type b, *Neisseria meningitidis* from cerebrospinal fluid, *Streptococcus pneumoniae, Helicobacter pylori, Bacteriodes fragilis,* respiratory syncytial virus, herpes type 1 and 2, HIV antigens, and other suspected pathogens.

In some cases, diagnosis may rely on the presence of antibody in the serum of a patient. Detection of antibody does not necessarily distinguish between an active and a previous infection. There must also be sufficient time between the onset of the disease and application of serological tests to permit the formation of circulating antibody, a period of 3 to 7 days.

Automated Identification Systems

Fully automated systems have been developed for the identification of infectious microorganisms. These systems provide a constant monitoring of growth-dependent reactions and the susceptibility of the microorganism involved to antimicrobial agents. Such systems have a number of advantages over the manual manipulations that have been employed for many decades. Among these are:

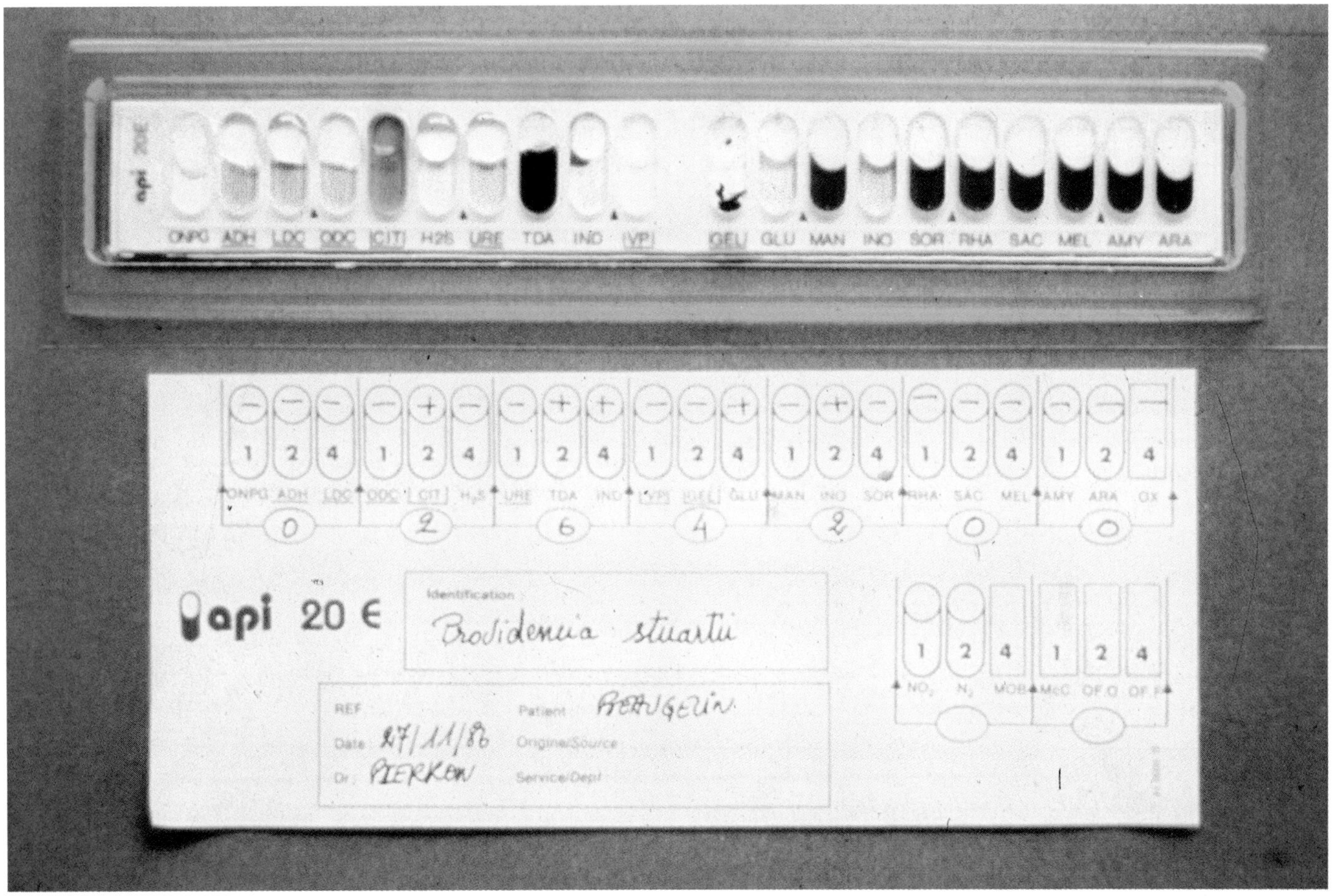

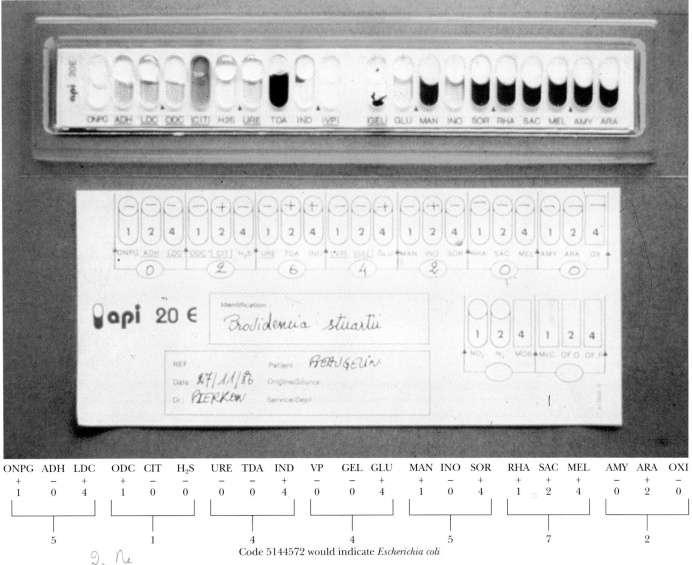

ONPG	ADH	LDC	ODC	CIT	H_2S	URE	TDA	IND	VP	GEL	GLU	MAN	INO	SOR	RHA	SAC	MEL	AMY	ARA	OXI
+	−	+	+	−	−	−	−	+	−	−	+	+	−	+	+	+	+	−	+	−
1	0	4	1	0	0	0	0	4	0	0	4	1	0	4	1	2	4	0	2	0
	5			1			4			4			5			7			2	

Code 5144572 would indicate *Escherichia coli*

***Figure* 31.5** The API 20E system for the rapid identification of enteric organisms. The profile number can be analyzed by computer or a manual entitled *API Profile Index*. The numbers for each test are weighed and the seven-digit number (code) will identify most of the enteric bacteria. The coded results at the bottom are for the test-strip shown at the top. The abbreviations indicate the following: ONPG–β-galactosidase; ADH–arginine dihydrolose; LDC–lysine decarboxylase; ODC–ornithine decarboxylase; CIT–citrate utilization; H_2S–hydrogen sulfide production; URE–urease; TDA–tryptophan deaminase; IND–indole production; VP–acetoin production; GEL–gelatin liquefaction; GLU–fermentation of glucose, MAN–mannitol, INO–inositol, SOR–sorbitol, RHA–rhamnose, SAC–sucrose, MEL–melibiose, AMY–amygdalin, ARA–arabinose; and OXI–oxidase test (done separately). (Reproduced with permission of VITEK)

1. Sample handling is minimized.
2. There is a decrease in the chance of human error, as there are fewer manipulations.
3. The systems deliver results continuously, as each test specimen is read and recorded separately. There are no batch readings so results are available at the earliest time possible.
4. These systems have broad applicability and can accommodate updating as new infectious agents, chemotherapeutics, and technical capabilities emerge.
5. They require very little space.

One such system is depicted in Figure 31.6**a**. There are a number of systems available, and this one has been selected as an example. The basic identification module for the system is a clear plastic card (Figure 31.6**b**) that has

***Table* 31.11 Growth-dependent tests that differentiate members of the enterobacteriaceae**

Tests	*Citrobacter freundii*	*Edwardsiella tarda*	*Enterobacter aerogenes*	*Escherichia coli*	*Klebsiella pneumoniae*	*Proteus vulgaris*	*Providencia alcalifaciens*	*Salmonella paratyphi*	*Salmonella typhi*	*Serratia marcescens*	*Shigella dysenteriae*	*Yersinia pestis*
Indole	−	+	−	+	−	+	+	−	−	−	±	−
Methyl red	+	+	−	+	−	+	+	+	+	−	+	+
Voges-Proskauer	−	−	+	−	+	−	−	−	−	+	−	−
Ornithine decarboxylase	−	+	+	±[1]	−	−	−	+	−	+	−	−
Motility	+	+	+	+	−	+	+	+	+	+	−	−
Gelatin liquefaction	−	−	−	−	−	+	−	−	−	+	−	−
KCN (growth in)	+	−	+	−	+	+	+	−	−	+	−	−
Glucose acid	+	+	+	+	+	+	+	+	+	+	+	+
Glucose gas	+	+	+	+	+	±	±	+	−	±	−	−
Lipase	−	−	−	−	−	+	−	−	−	+	−	−
$NO_3^- \rightarrow NO_2^-$	+	+	+	+	+	+	+	+	+	+	+	+
Lactose utilization	±	+	±	+	+	−	−	−	−	−	−	−
H_2S on TSI	±	−	−	−	−	+	−	−	+	−	−	−
Citrate	+	−	+	−	+	−	+	−	−	+	−	−

[1] ±most strains positive.

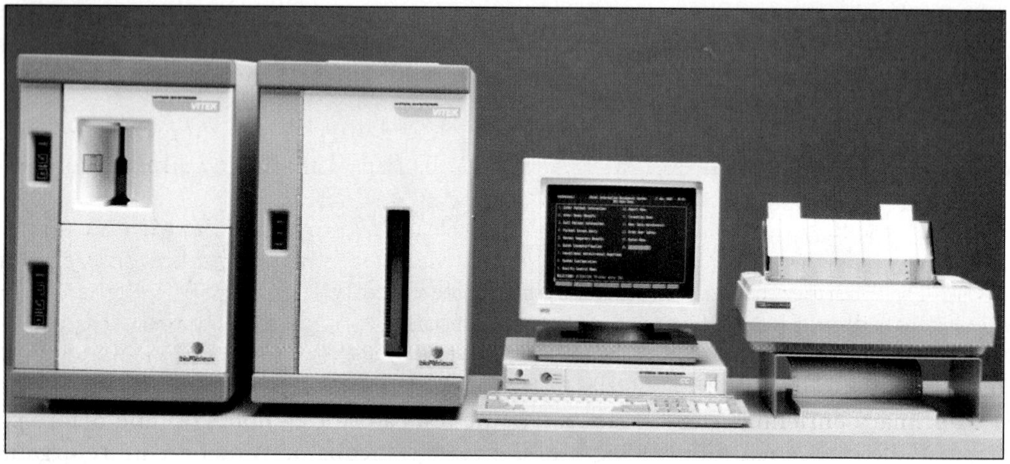

(a)

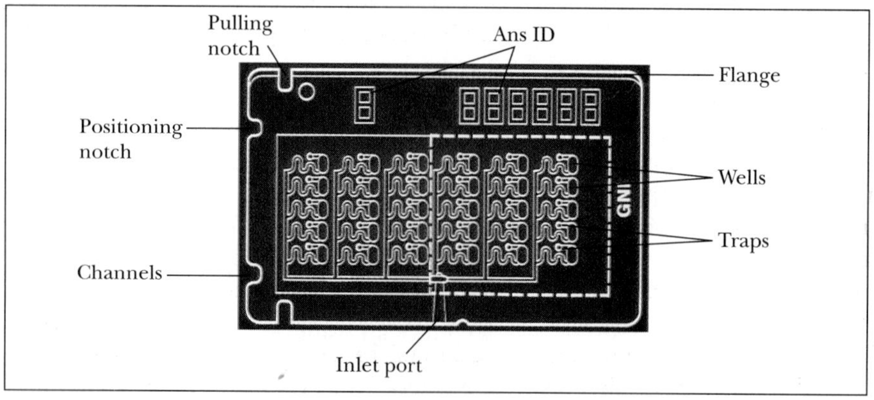

(b)

***Figure* 31.6 (a)** The automated Vitek[R] system for the identification of clinical isolates. The system includes a monitor to record visual results and a printer to record chartable patient reports. **(b)** A test card that would be placed in the automated Vitek[R] system for identification of a clinical isolate. The test card is clear plastic measuring 2.25 × 3.5 inches and has 30 individual wells. Each well contains a selected dried medium and/or constituents for a specific test such as sulfide production. The bacterial suspension is introduced through the inlet port. The notches hold the card in place during incubation and manipulations involved in reading the tests. The wells are all connected to channels from the inlet port where the inoculum enters. Traps are the sites where fermentation gases collect. (Courtesy of bioMerieux VITEK, Inc.)

***Table* 31.12 Identification cards available that will selectively identify commonly encountered organisms**

Test Card	Range	Genera Detected (speciation within the genus is generally determined)
GNI	*Enterobacteriaceae, Vibrionaceae,* Glucose nonfermenters	*Achromobacter, Acinetobacter, Citrobacter, Eikenella, Enterobacter, Escherichia, Flavobacterium, Hafnia, Klebsiella, Pasteurella, Proteus, Providencia, Pseudomonas, Salmonella, Serratia, Shigella, Vibrio, Yersinia*
GPI	Coagulase positive and negative Staphylococci, Enterococcus, β-hemolytic streptococci, *Corynebacterium, Listeria, Erysipelothrix*	*Corynebacterium, Enterococcus, Erysipelothrix, Listeria, Staphylococcus, Streptococcus* (viridans group)
YBC	Clinically significant yeasts	*Candida, Crytococcus, Geotrichum, Hansenula, Pichia, Prototheca, Rodotorula, Saccharomyces, Sporobolomyces, Trichosporon*
ANI	Anaerobic bacteria	*Actinomyces, Bacteroides, Bifidobacterium, Capnocytophaga, Clostridium, Eubacterium, Fusobacterium*
NHI	Fastidious bacteria	*Actinobacillus, Branhamella, Cardiobacterium, Eikenella, Garderella, Haemophilus, Kingella, Moraxella, Neisseria*
UID	Detect, identify, and enumerate most common urinary tract pathogens	*Citrobacter, Escherichia, Enterococcus, Klebsiella/Enterobacter, Proteus, Serratia, Staphylococcus, Pseudomonas,* yeast
EPS	Enterics	*Salmonella, Shigella, Yersinia*

30 wells containing dried medium and/or constituents for differential tests. There are a number of test cards available and they differ in the substrates that are present in the wells. The selection of a test card would be based on the presumptive identification of the primary isolate. Seven different cards and their uses are described in Table 31.12. A test card would be inoculated as illustrated in Figure 31.7**a**. Ten test cards can be inoculated in about 3 minutes by drawing in a suspension of an organism that has been **isolated** in **pure culture** by primary enrichment or on a selective/differential medium. Following inoculation the test card fits into a carousel that can hold up to 30 test cards (Figure 31.7**b**). The carousel containing the inoculated test cards is then placed in the reader incubator module at a selected incubation temperature.

Each carousel is automatically read once every hour. To accomplish this the test card is automatically removed from the slot and passed through a photometric reader. Turbidity or color changes are recorded and transmitted to a computer system that analyzes the data points. The data points that are collected each hour are electronically compared with a known data base. When sufficient data are available, the system will print out a most likely (primary) and next most likely identification of the microorganism present in the original inoculum. The probability for each is given. The data may also be displayed on a computer screen for reference.

The time required to complete the identification of a clinical isolate varies. Analyzing for the enterobacteria takes from 4 to 6 hours, depending on species. The glucose nonfermenting gram-negative bacteria require 4 to 18 hours. The gram-positives such as staphylococci require 4 to 15 hours. Identification of yeasts is accomplished by incubating the test card for 24 hours off line before placing it in the reader/incubator. A blood culture organism can be identified in 4 to 6 hours, and susceptibility to antimicrobials can be run simultaneously with the other tests. Generally an appropriate treatment can be initiated in fewer than 8 hours.

Susceptibility cards are available to determine the antimicrobial sensitivity of a microorganism isolated in primary culture. Test cards contain a standard growth medium with graded levels of different antimicrobials. The instrument is programmed to provide information on the antimicrobial that would be most effective. The effectiveness of the antimicrobial would be based on the identification of the organism, origin of the specimen, and other considerations. The effective minimal inhibitory concen-

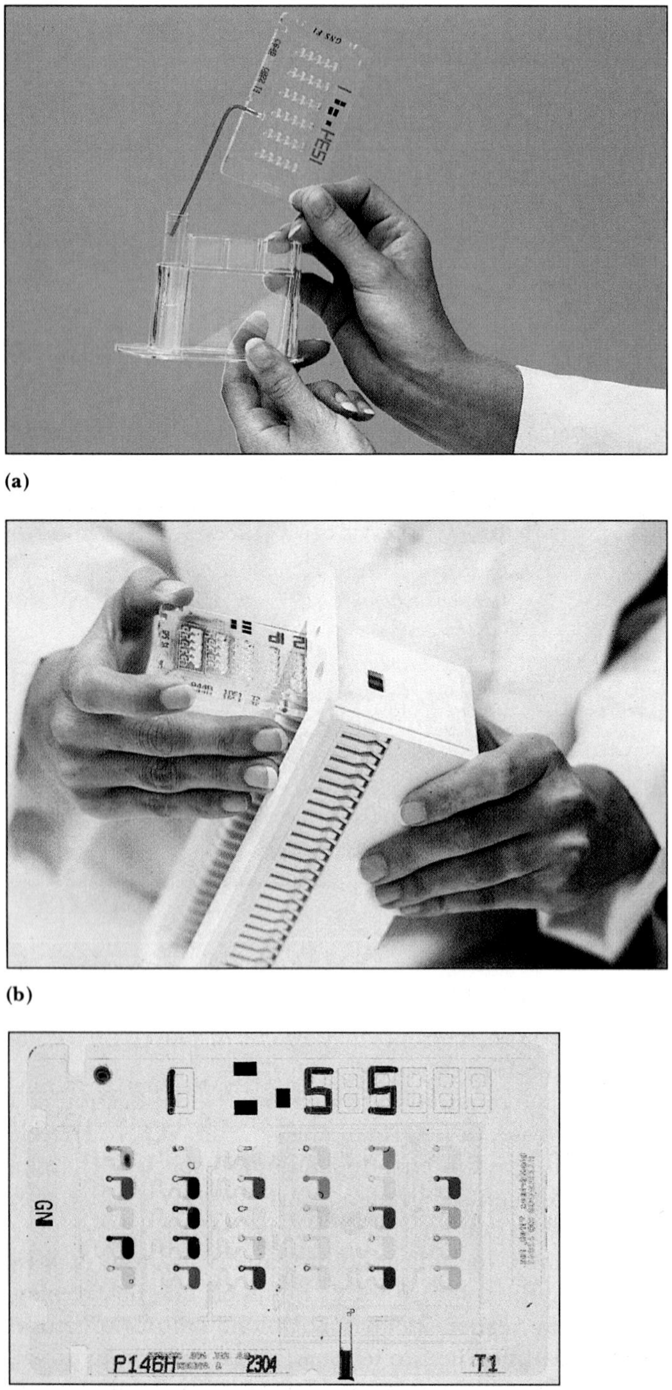
(a)
(b)
(c)
GN
P146H
2304
T1

***Figure* 31.7 (a)** The bacterial suspension is introduced under aseptic conditions. A disposable tube is used to draw the suspension of a prescribed dilution into the test card. **(b)** Placing the inoculated test card in the incubation chamber. **(c)** A test card module. (Courtesy of bioMerieux VITEK, Inc.)

tration (MIC) of an appropriate antimicrobial would be printed out in 6 hours or less. There are at present nine different standard test cards for gram-negative bacteria, two for gram-positive, and four for general antibiotic susceptibility. Each test card has 11 to 12 antimicrobials. A manual method for determining the sensitivity of clinical isolates is the Kirby-Bauer disk diffusion assay, which is discussed later.

Antigen/Antibody Techniques

Virtually all microbial species and even a strain of a pathogenic microorganism will have at least one antigen that is unique. Analytical techniques are now available for the recovery and purification of these unique antigens. A purified antigen can be employed to generate a monoclonal antibody that will react specifically with that antigen (see Chapter 27). Monoclonal antibodies to specific antigens are an effective and highly specific diagnostic tool.

One of the diagnostic tests that employs monoclonal antibodies is the **direct ELISA** test (Figure 31.8). ELISA is

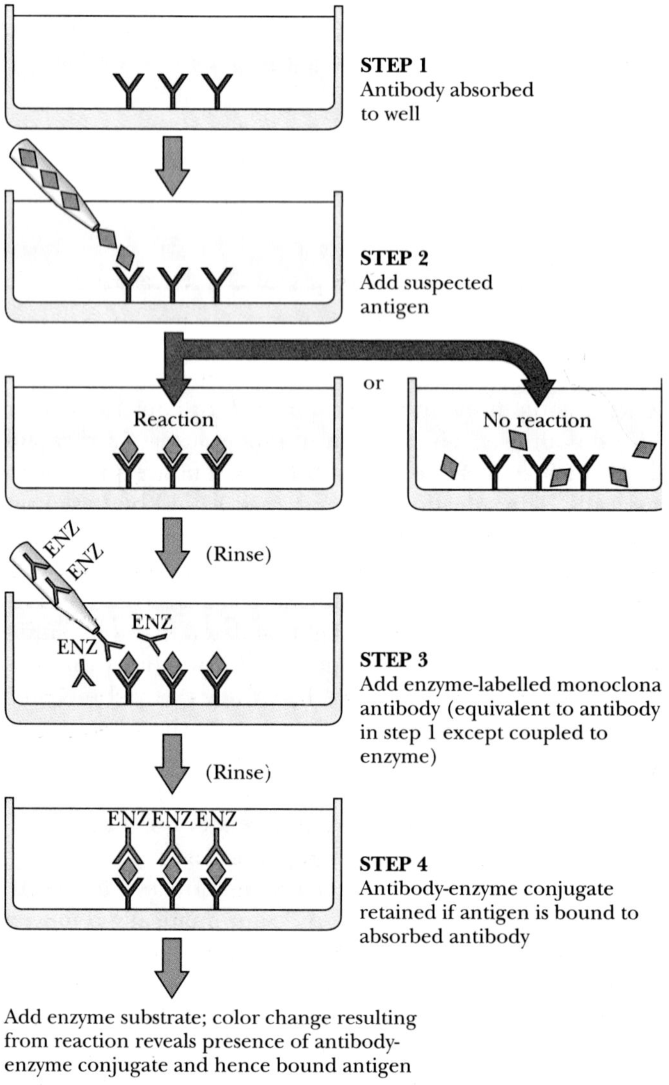
STEP 1
Antibody absorbed to well
STEP 2
Add suspected antigen
or
Reaction
No reaction
ENZ
(Rinse)
STEP 3
Add enzyme-labelled monoclona antibody (equivalent to antibody in step 1 except coupled to enzyme)
(Rinse)
ENZENZENZ
STEP 4
Antibody-enzyme conjugate retained if antigen is bound to absorbed antibody
Add enzyme substrate; color change resulting from reaction reveals presence of antibody-enzyme conjugate and hence bound antigen

***Figure* 31.8** A direct enzyme-linked immuno sorbent assay (ELISA) employing monoclonally produced specific antibodies for detection of an antigen.

an acronym for enzyme-linked immuno sorbent assay. The direct test requires that a specific monoclonal antibody be available and be absorbed onto the walls of a well in a microtiter plate (Step 1). A suspension of serum or fluid that one suspects may contain a specific antigen is added. If the antigen is present it will react with the antibody (Step 2) that is absorbed to the wall of the well. Binding of the antigen would be specific and strong enough to withstand rinsing that would remove unreacted antigen.

A suspension of an enzyme-labeled monoclonal antibody would then be added (Step 3), and this antibody is equivalent to the antibody absorbed to the well in Step 1 (it would react with the same antigen). This antibody has one modification in that a reporter enzyme is conjugated to it without affecting the capacity of the antibody to react with the specific antigen. Enzymes that are conjugated to antibodies are generally those that will give a discernible visual color when the enzyme reacts with substrate.

If the antigen is retained by the antibody at Step 2, the antibody-enzyme conjugate (Step 4) will also be retained specifically and not removed by rinsing. Addition of a substrate for the conjugated enzyme will then result in a readily observed color change. No color change indicates that no antigen was absorbed at Step 2 and the specimen did not have an antigen present that could react with the antibody—a negative test.

The **indirect ELISA** test is one that determines whether a specific antibody is present in a specimen such as serum (Figure 31.9). One infectious disease that might be diagnosed by the indirect ELISA assay is an HIV (human immunodeficiency virus) infection. Because blood transfer is a potential means for spreading the HIV virus, it is essential that the possible presence of the virus be determined in material such as blood for transfusion. HIV infection induces circulating antibody formation to specific antigens on the HIV particle. These antibodies are formed during the early stages of HIV infection before the immune system is compromised. Detection of these antibodies in the serum of a patient confirms exposure to the HIV virus.

The procedure for an indirect ELISA test is outlined in Figure 31.9. In this case, the antigen is absorbed to the wall of a microtiter plate (Step 1). Serum that might contain antibody against this antigen is added to the microtiter well (Step 2). Rinsing removes any antibodies that do not specifically attach to the antigen absorbed to the well. However if antibody does attach to the antigen it can be detected at Step 3 by addition of an antibody-enzyme conjugate suspension. The antibody in this case is one that specifically reacts with human immunoglobulins (generally anti IgG). Presence of the antibody (Step 2) to the antigen results in the formation of a complex (Step 4). Addition of a substrate for the enzyme gives a visual indication that the antibody against the original antigen was present in the serum added at Step 2. The indirect ELISA can be employed in diagnosing infections such as salmonellosis, plague, brucellosis, syphilis, tuberculosis, leprosy, and Rocky Mountain spotted fever.

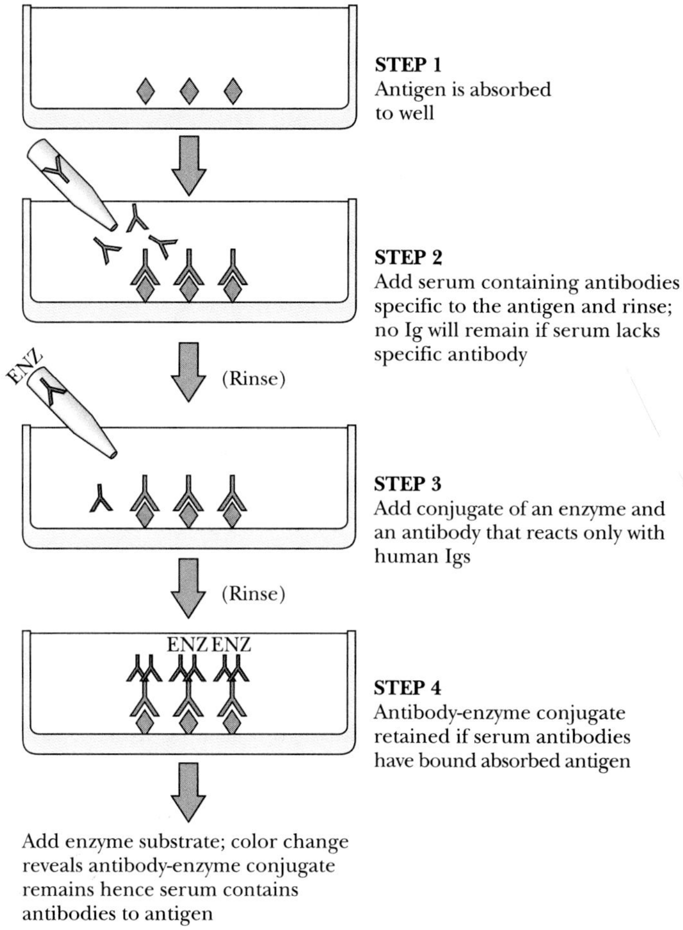

STEP 1
Antigen is absorbed to well
STEP 2
Add serum containing antibodies specific to the antigen and rinse; no Ig will remain if serum lacks specific antibody
ENZ
(Rinse)
STEP 3
Add conjugate of an enzyme and an antibody that reacts only with human Igs
(Rinse)
ENZ ENZ
STEP 4
Antibody-enzyme conjugate retained if serum antibodies have bound absorbed antigen
Add enzyme substrate; color change reveals antibody-enzyme conjugate remains hence serum contains antibodies to antigen

***Figure* 31.9** The indirect ELISA test to determine the presence of a specific antibody in the serum of a patient.

Detection by Immunofluorescence

Antibodies tagged by chemically bonding fluorescent dyes to them are often employed in a clinical diagnostic laboratory. A fluorescent dye is one that absorbs light of one wavelength, becomes excited, and emits light at another, generally longer, wavelength. This technique can be employed, for example, in diagnosing the genital tract infection caused by *Chlamydia trachomatis.* To accomplish this, a smear of cervix scrapings is placed on a slide and a specific antichlamydial antibody is applied to the smear. The antibody employed in detection can be polyclonal or monoclonal and obtained from a selected animal species by employing elementary bodies of *Chlamydia trachomatis* as antigen. The antibody is coupled to a fluorescent dye. Unreacted antibody-dye conjugate is removed by rinsing. The slide is viewed with a fluorescence microscope, and presence of the pathogen is indicated by a green fluorescence against the dark background of counterstained cells. In place of microscopy, a fluorometer can be employed that measures the total amount of light emitted at the appropriate wavelength.

A problem in identification of pathogens by immunofluorescence is that of shared antigens between species. This is particularly the case with enteric bacteria. Employing monoclonal antibodies that interact solely with an antigen unique to a pathogen lessens the problem with cross-reacting antibodies. Immunofluorescence can be a useful technique in identifying pathogens that grow slowly or are difficult to isolate. Mycobacterial and chlamydial infections and brucellosis are diseases where isolation of the organism is time consuming but early treatment is essential. Other specimens that might be examined by immunofluorescence procedures are gonorrheal exudates, feces for cholera, smears from bubonic plague buboes, tularemia lesions, and legionellosis aspirate. This technique is applicable where clinical symptoms indicate that these pathogens may be responsible for the observed symptoms.

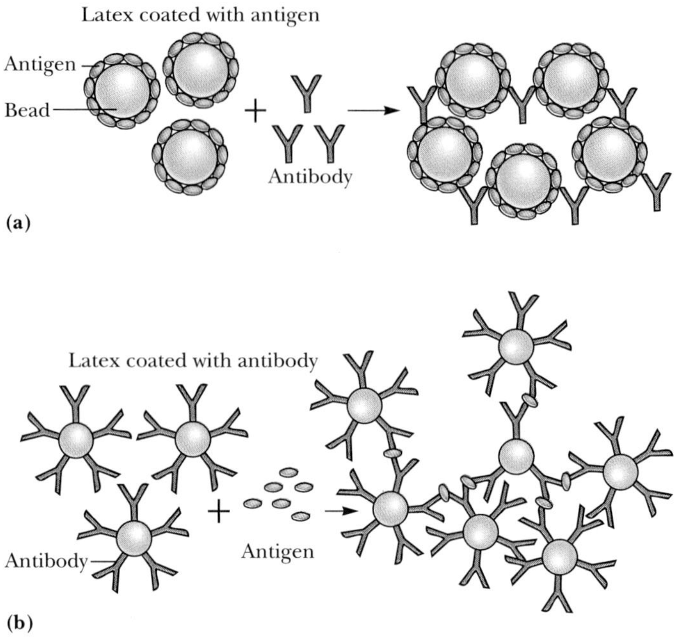

Figure **31.10** The latex bead agglutination procedure for rapid detection of **(a)** antibodies or **(b)** antigens.

Agglutination

When antigens and antibodies interact, a complex develops that is, quite often, large enough to be visible with the naked eye. This clumping of antigen/antibody is called **agglutination.** Recall (Chapter 27) that both antibodies and antigens have a valence greater than one and that a network of antigen/antibody is probable. Agglutination has been adapted to the clinical laboratory for the rapid detection of either antigens or antibodies. It is a relatively inexpensive and specific test. The agglutination test is augmented by employing antigen- or antibody-coated latex beads (Figure 31.10). The beads are spherical and about 0.8 μ of a meter in diameter. Proteins (antigen or antibody) adhere tightly to the polystyrene surface of the bead. In some procedures, particles of charcoal coated with the appropriate antigen or antibody may be employed. When an antigen or antibody is fixed to the bead and a complementary antibody or antigen is added, the agglutination reaction occurs in 30 seconds or less (if positive). A drop of urine, serum, spinal fluid, or other specimen in suspension can be added to determine the presence therein of a specific complementary antigen/antibody. The test has been employed to identify *Staphylococcus aureus, Neisseria gonorrhoeae,* and *Haemophilus influenzae.* Yeast infections, caused by organisms such as *Cryptococcus neoformans* can be diagnosed by agglutination tests.

Plasmid Fingerprinting

A plasmid is extrachromosomal DNA that replicates autonomously in a bacterial cell. Plasmids are present in many bacterial genera, including many of the pathogenic microorganisms. Bacterial **strains** that are related generally contain the same number of plasmids and are of equivalent molecular weight. Determining the number and size of plasmids in strains of a species can be a measure of relatedness. To analyze for plasmids, a cell mass must be lysed to free up intracellular components. The plasmid DNA is then separated from chromosomal DNA by density gradient centrifugation, and the plasmid DNA is applied to an agarose gel (Figure 31.11) and subjected to electrophoresis. The migration rate in agarose is inversely proportional to the molecular weight. The DNA can be visualized by staining with ethidium bromide. The greater the number of bands with equivalent migration rates in the gel the more accurate the identification. The bands present are called a **plasmid profile** or **fingerprint.** Plasmid fingerprinting is not a positive test in that one strain of a species may have several plasmids and another strain may have few or none. However, if they have several plasmids in common, it is strong presumptive evidence that strains are related.

Plasmid profiles are of value in tracing the origin of specific strains that may be involved in a local epidemic such as a nosocomial infection. If one has a series of *Staphylococcus* infections, a plasmid profile of isolated organisms can give confirmation of a potential common origin. If the organisms involved have different profiles, it would provide evidence that the infections do not have a common origin. Food poisoning cases caused by enterotoxin from coagulase-positive staphylococci may be traced by this technique. The *Staphylococcus aureus* strain in the nasal passage of a food handler can be subjected to the profile test. If a strain of *S. aureus* present on food eaten by an individual stricken with food poisoning has an identical plasmid profile to that from the food handler this would provide evi-

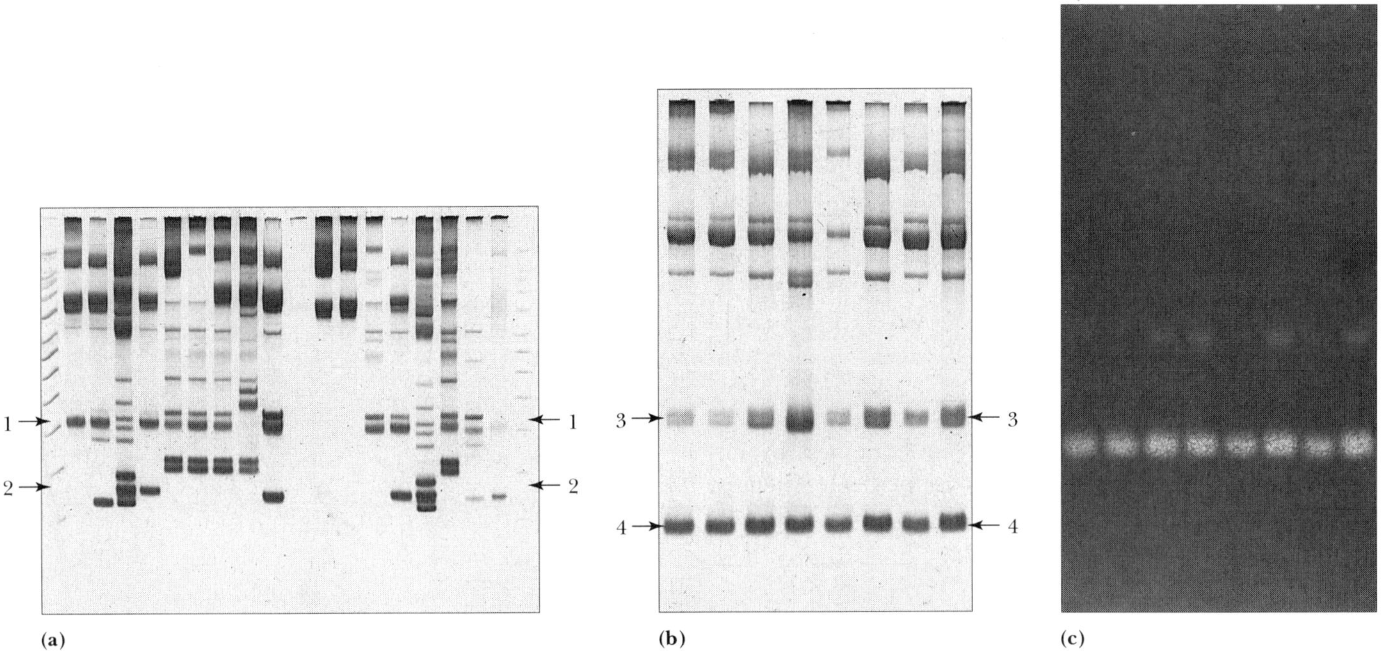

***Figure* 31.11** **(a)** Plasmid profiles for different strains of *Staphylococcus epidermidis* isolated from various body regions of a patient receiving erythromycin therapy for a streptococcal infection. *1* is the tet A plasmid for tetracycline resistance. *2* is the erm C plasmid for erythromycin resistance. **(b)** Plasmid profile for one strain of *Staphylococcus epidermidis* isolated from different body regions of a patient receiving clindamycin therapy for acne. The open circular form is *3* and the covalently closed-circular form is *4* of the erm C plasmid **(c)** is a Southern blot of **(b)** with an erm C gene probe. (Courtesy of W. E. Kloos)

dence that the food handler was a source of the staphylococcal infection.

Phage Typing

Bacterial viruses (bacteriophages) are quite specific in the species or, in many cases, the bacterial strain they can infect. This specificity is based on the presence of viral receptors on the surface of the bacterium where the virus attaches to initiate the infectious process. Virus susceptibility can be employed as an indicator of genus, species, and possibly a strain.

Virus susceptibility can be determined by spreading a selected bacterium over an agar surface. The plate is squared off by marking with a wax pencil on the bottom of the plate. A drop of different viral suspensions would be applied to each square. After suitable incubation the plate would be examined for plaques indicating which virus was capable of attacking that bacterium. Phage typing is useful in tracing localized epidemics as just described under plasmid profiles.

Nucleic Acid Probes

Techniques are now available through advances in molecular biology that can be applied to the rapid and precise identification of infectious agents. These techniques take advantage of the fact that the nucleotide sequences in one genus or species differ in a measurable way from those in another. A pathogenic microorganism that produces toxins, enzymes involved in invasion, or other virulence factors has genetic information that is unique. The information for these virulence factors resides in the DNA of the producing organism. DNA diagnostics depend on the identification of unique nucleotide sequences that are present in a given pathogen but not present in other microorganisms.

Clinical application of DNA technology relies on the tight binding that occurs between complementary single strands of DNA. A single strand of DNA of known sequence that is complementary to a single strand in the DNA of a clinical isolate is termed a probe. If a clinical isolate has DNA sequences that are complementary to those in a probe, they can hybridize and these hybrids are detectable. It is essential that the double-stranded DNA of the clinical isolate be made single stranded by heat or alkali to permit hybridization. If hybridization does occur it can be detected by incorporating a reporter molecule into the probe (Figure 31.12). The reporter that is part of the probe may be a fluorescent dye, a radioisotope, or an enzyme. These reporters are effective because they can be detected at an exceedingly low concentration. Radiolabeled reporters are the most sensitive of all.

Nucleic acid probes are superior to immunological procedures in many ways. Antibodies are sensitive to temperature, pH, and other physical conditions. Harsh conditions can be applied to clinical specimens to remove interfering material, resulting in a mixture that would not be suitable for antibody reactions. Nucleic acid probes are more stable and functional under these conditions.

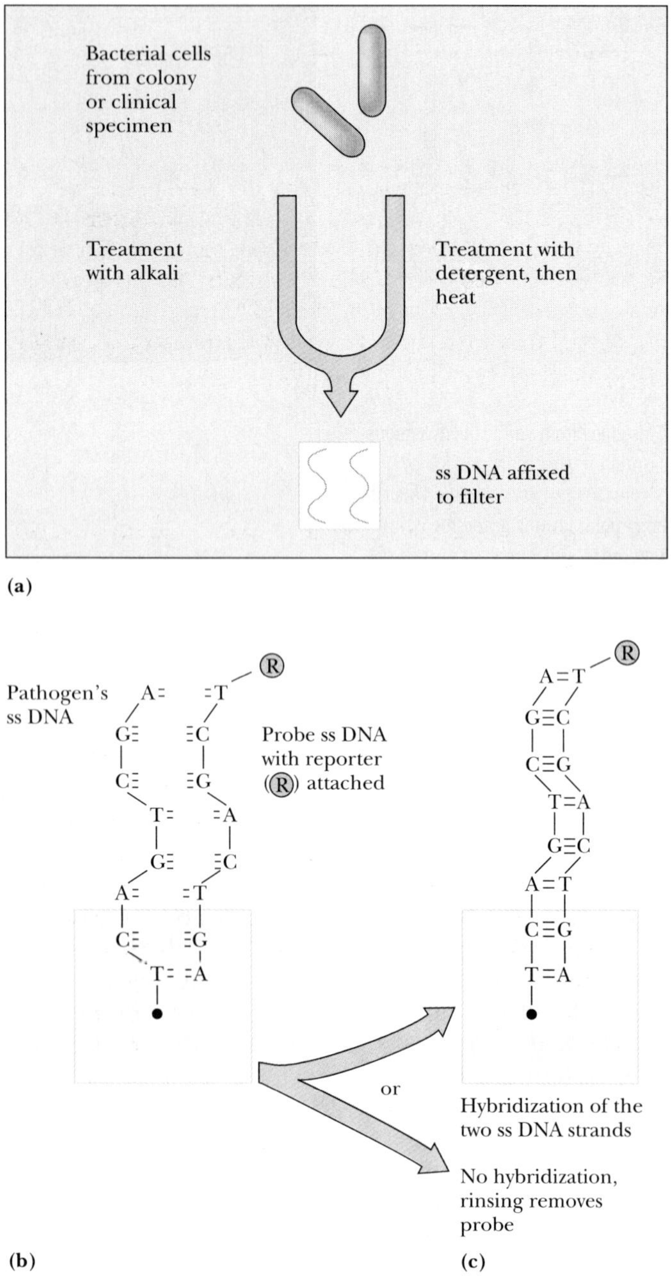

***Figure* 31.12** A DNA probe for detection of a specific microorganism in a clinic specimen. First, DNA is extracted from cells and rendered single stranded (ss DNA). The single strands then are affixed to a filter. **(a)** Next, a specific probe obtained by chemical synthesis with an attached reporter (enzyme) is added. **(b)** If hybridization occurs, the probe and reporter will remain after rinsing. **(c)** Substrate for the enzyme is added and a color change will indicate a positive test. The reporter may also be a radiolabel or fluorescent dye.

Probe Diagnosis

A probe assay is outlined in Figure 31.12. The DNA from a microorganism obtained from a specimen can be released by treatment with strong alkali and the single-stranded (ss DNA) affixed to a filter. The source of ss DNA may also be a bacterial colony treated with detergent to release the cellular DNA (ds DNA), and this can be rendered single stranded by heating. The reporter molecule can be a fluorescent dye, an enzyme, or radiolabel. Radiolabel is the most sensitive, and this technique can detect less than 1 μ gram of DNA. There are available DNA probes that will bind to complementary strands of ribosomal RNA (r RNA). A hybrid of rRNA:DNA is more sensitive than the DNA:DNA hybrids, and fewer microorganisms are required for an accurate determination. DNA:DNA probes have been effective in diagnosis of rickettsial and meningitis infections among others. They are also effective in identifying infectious viruses.

A dipstick probe has been developed for the identification of infectious agents in food, as well as crude clinical specimens. The dipstick method is outlined in Figure 31.13. In this rapid and specific method, a clinical specimen is treated with alkali to release DNA and generate single strands from any microorganism present. The nucleic acid probe is added and, after a suitable incubation under conditions that promote hybridization, the probe can be recovered with a capture probe. The nucleic acid probe would have a nucleotide sequence attached that permits hybridization to a series of nucleotides on the capture probe. The sequence on the probe would be poly T or poly A and the corresponding bases on the capture probe would be poly A or poly T. The reporter on this type probe differs from that on the nucleic acid probes mentioned above. A low molecular weight molecule such as dioxygenin would be chemically linked to the nucleic acid probe (D in Figure 31.13). Following incubation of the alkali-treated clinical specimen and the nucleic acid probe one would add nucleases to destroy any single-stranded probe that is unhybridized. A dipstick that bears the capture probe base sequence would be inserted into the hybridization solution. The capture probe with any hybridized DNA would be exposed to antibodies to the dioxygenin reporter, and these antibodies would be coupled to a fluorescent dye or a detectable enzyme. Whether hybridization occurred would be determined by observing an enzyme reaction or fluorescence by microscopy.

Probes are of particular value in diagnosing *Chlamydia trachomatis* and organisms that are involved in respira-

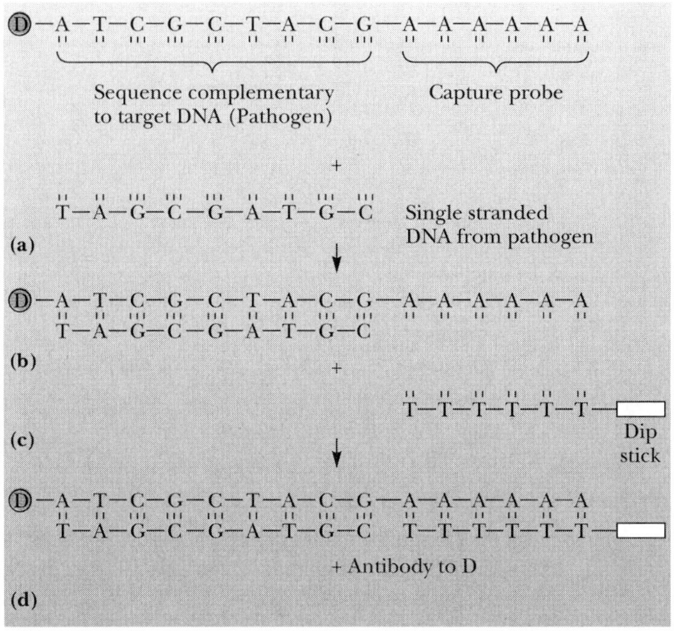

***Figure* 31.13** A dipstick assay procedure for recovery of ss DNA. A clinical specimen is treated with alkali to release ss DNA. A molecule of dioxygenin (D) is attached to the nucleic acid probe **(a).** If hybridization occurs, it would appear as in **(b)** excess unreacted probe is destroyed by nucleases. The dipstick **(c)** will hybridize with the capture probe after the hybridized DNA is recovered. Antibody to dioxygenin that has an enzyme or fluorescent dye attached would be employed to determine the presence of nucleotides from the pathogen.

tory infections. Where crude specimens are to be analyzed (sputum, feces, discharges), a probe is an effective technique because isolation of the organism is not essential. It is also a method for rapid diagnosis for viruses and other organisms that may not be grown in culture.

Polymerase Chain Reaction (PCR)

Probes can detect small amounts of DNA. It does, however, require the DNA from about 10^5 to 10^6 virus particles and is a more reliable method with the availability of several hundred bacteria. Sensitivity of nucleic acid diagnostics can be increased markedly by increasing the amount of DNA present in a sample via the polymerase chain reaction (see Chapter 16). The DNA released from one or a few bacterial cells is cut with endonuclease and rendered single stranded. The single-stranded DNA can be amplified via the addition of a primer and DNA polymerase. The primers are short pieces of single-stranded DNA (20 to 30 bases) and are produced in a DNA synthesizer. One-million-fold increases in the amount of DNA can be achieved in about 1 hour. Analysis of the DNA can be accomplished with a probe or by identification of selected DNA sequences on agarose gels. Many analyses, including this, have been automated in the clinical laboratory.

Antibiotic Sensitivity

Antibiotics or chemotherapeutics are the treatment available for most all bacterial and mycotic infections. As previously mentioned, resistance to these inhibitory compounds is a major problem in the treatment of many infectious diseases. Resistance to one or more antibiotics is now encountered with many infectious microorganisms and is a considerable concern in nosocomial infections. The earlier a patient is treated with an appropriate antibiotic the better.

The susceptibility of a clinical isolate to an antibacterial agent can be determined by the automated system mentioned previously or a disc diffusion assay. The disc assay widely employed was developed by W. Kirby and A.W. Bauer and is termed the Kirby-Bauer method. In this procedure a filter disc is impregnated with a measured amount of antibiotic and the disc is placed on an agar surface. The disc will absorb moisture from the agar, effecting the diffusion of antibiotic into the agar medium. The distance that the antibiotic moves outward from the disc depends on the solubility, concentration, and other characteristics of the antibiotic. As the antibiotic diffuses outward, a concentration gradient is established with the highest level in the immediate area of the disc.

In practice, one would inoculate a liquid growth medium by touching an inoculating loop to a colony growing on a primary isolation plate and transferring the sample to the liquid medium. The inoculated liquid medium would be incubated at an appropriate temperature. When turbidity is evident, an aliquot is used to inoculate an agar medium. This is accomplished by dipping a sterile cotton swab into the liquid and spreading the organism evenly over the agar surface. The antibiotic-containing discs are placed on the surface and the plate is incubated. After a

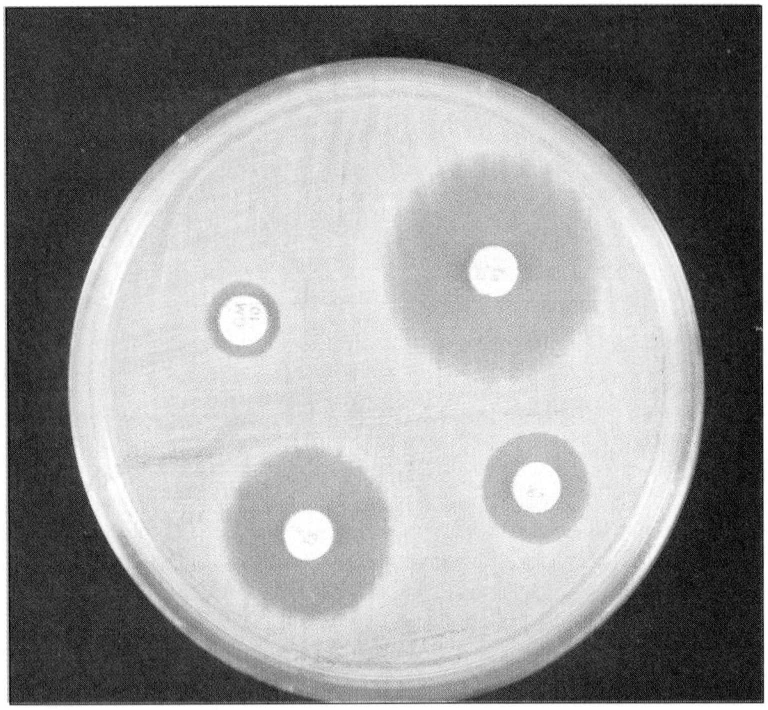

***Figure* 31.14** The Kirby-Bauer procedure for determining the sensitivity of a clinical isolate to antibiotics. A specific dilution of a broth culture is spread over an agar surface with a cotton swab. Paper discs containing a known amount of antibiotic are placed on the surface. After suitable incubation, the size of the zone of inhibition is determined. The susceptibility of the clinical isolate is estimated by consulting a chart (Table 31.13). (Courtesy of J. J. Perry)

12- to 18-hour incubation period the zones of inhibition can be measured (Figure 31.14).

The susceptibility of an organism to an antibiotic is directly related to the diameter of the zone of inhibition. The zone diameters can be interpreted by employing a table prepared by Kirby-Bauer (Table 31.13). This table was developed by determining minimum inhibitory concentration (MIC) values for many pathogenic bacteria and relating the MIC to the diameter of an inhibition zone that would be obtained with that concentration of antibiotic.

Minimum inhibitory concentration (MIC) values are determined as follows: A series of tubes containing an appropriate medium and graded levels of antibiotic are inoculated with equivalent amounts of a suspension of the clinical isolate. The tubes are examined for growth after a suitable incubation period. The lowest concentration of antibiotic that inhibits growth completely is considered the MIC (see Figure 7.8).

Clinical diagnostics has made remarkable advances in the past few years. Automation and rapid specific tests permit accurate identification of an infectious agent in short periods of time. Early treatment saves lives, shortens the illness and convalescence time, and saves money. The epidemiologist and clinical laboratory in concert contribute much to our well-being.

***Table* 31.13 Diameter of the zone of inhibition as a measure of susceptibility to selected antibacterial agents**

Antibacterial	Amount on Disc (μg)	Zone Diameter (mm) Resistant	Sensitive
Ampicillin	10	28 or less	29 or more
Erythromycin	15	13 or less	18 or more
Gentamycin	10	12 or less	15 or more
Tetracycline	30	14 or less	19 or more

Summary

- **Infectious diseases** were the major cause of death worldwide before antibiotics were available. They now are responsible for about 8 percent of the deaths in the developed countries, but 50 percent of the deaths in developing countries are due to infectious diseases.

- The science concerned with the prevalence and distribution of diseases in the population is **epidemiology.**

- An **epidemic** occurs when a disease occurs in the population at a higher than normal frequency. A constant low frequency of disease is called the **endemic** level and an epidemic worldwide is a **pandemic.**

- The **incidence** of a disease is the number of cases in a susceptible population of 100,000. **Mortality** rate is deaths per total infected, and **morbidity** is defined as the number of new cases during a period divided by the total population.

- A **carrier** is an infected individual who transmits a disease to others. A **chronic carrier** is an individual permanently infected and shedding an infectious disease. A **reservoir** is the site where an infectious disease–causing microorganism is maintained between outbreaks.

- Human contact and respiratory diseases are passed by human interaction such as touching, coughing, or sneezing. Among these would be **colds, influenza,** and **tuberculosis.**

- **Water-** and **food-borne** diseases are generally caused by organisms that infect the human gastrointestinal tract. They are transmitted through fecal contamination.

- Of major public health concern are the **sexually transmitted diseases. HIV, chlamydial infections,** and **herpes** are all on the increase as are **genital warts.**

- Epidemics of **vectorborne** diseases such as **Rocky Mountain spotted fever** and **Lyme disease** occur during summer months, and incidence of Lyme disease in the United States is increasing at an alarming rate.

- An infection acquired in a hospital is termed a **nosocomial infection.** Nosocomial infections affect an estimated two million patients each year. The prevalent organism is ***Escherichia coli.*** **Staphylococcal** and **streptococcal infections** are also a problem.

- Disease incidence is followed in most states by requiring that physicians report about 50 different diseases to the State Health Department. They in turn report many of these to the Communicable Disease Center in Atlanta.

- Many diseases are controlled in the United States by **vaccination** of babies and young children. Among the vaccinations are **mumps, measles, rubeola, diphtheria, pertussis (whooping cough), tetanus, polio, haemophilus influenza b, chicken pox,** and **hepatitis B. Smallpox** was eradicated worldwide by vaccination.

- **Reservoirs** that sustain infectious agents are a problem in epidemiology, and their elimination is important in controlling disease. This is not possible in all cases.

- **Food inspection** and **water sanitation** are important public health measures.

- A real problem in controlling infectious diseases is development of **drug resistance.**

- Rapid identification of pathogens involved in an infection leads to better treatment. A **clinical laboratory** is charged with this responsibility. Much of the identification has been automated.

- A **specimen** is the material delivered to the clinic. It must be handled carefully and may consist of blood, urine, sputum, etc.

- **Pathogens** can be tentatively identified by microscopic examination, but complete identity generally requires isolation and growth in culture.

- Automated procedures are now available that can identify clinical isolates in less than 20 hours and also determine resistance/sensitivity to antibiotics.

- An ELISA test (enzyme-linked immuno sorbent assay) is a diagnostic procedure that determines whether a specific antigen or antibody is present in a suspension.

- Antibodies labeled with **fluorescent** dyes are an important diagnostic tool. A specimen is placed on a slide and dye-labeled antibody is applied. Unreacted antibody is rinsed away and the slide is viewed with a fluorescence microscope. **Agglutination** tests with antibody or antigen-coated beads, **plasmid profiles,** and **phage typing** are other available diagnostic tools.

- **Nucleic acid probes** are a technique for detecting pathogens. The methods take advantage of the complementarity of the nucleotides in a probe and equivalent unique base sequences in the genome of a pathogen. Probes are 60 to 80 nucleotides long and generally a product of chemical synthesis.

- When small numbers of a pathogen might be present in a clinical specimen the **polymerase chain reaction** can be applied to increase the level of DNA present to readily detectable levels.

- A determination of **antibiotic sensitivity/resistance** in an infectious microorganism is necessary in designing a proper treatment.

Questions for Thought and Review

1. Define epidemic, endemic, and pandemic.
2. How does a sporadic disease differ from one that is seasonal? Give examples of seasonal diseases.
3. Two terms employed in disease reporting are mortality and morbidity. In fact the Centers for Disease Control publishes the *Morbidity and Mortality Weekly Report.* What is the significance of these two terms?
4. What is a disease reservoir? Why are diseases spread from animal reservoirs such a problem?
5. How are human-contact diseases spread? Give examples of such diseases. How can they be controlled?
6. How can sexually transmitted diseases be controlled? Which are treatable?
7. Why are nosocomial diseases such a concern? Can they be readily controlled?
8. We have lessened the incidence of the childhood diseases by vaccination. What are some of these diseases? Why are they still a concern?
9. What is the agency that is responsible for control of disease worldwide? What major success has it had?
10. What are some major disease problems in developed countries? Developing countries?
11. Why is it important that a disease-causing microbe be isolated and identified?
12. What types of specimens are sent to the clinical laboratory? How are some of these treated once there? What are identification procedures generally employed?
13. What is an ELISA test? How is it performed?
14. Nucleic acid probes are of growing importance in the clinical lab. How are they used?
15. What is PCR? How is it employed in virus identification and why is it necessary?

Suggested Readings

Baron, E. J., L. R. Peterson, and S. M. Finegold. 1994. *Bailey and Scott's Diagnostic Microbiology.* 9th ed. St. Louis: The C. V. Mosby Co.

Beaglehole, R., R. Bonita, and T. Kjellstrom. 1993. *Basic Epidemiology.* Geneva: World Health Organization.

Centers for Disease Control. *Morbidity and Mortality Weekly Reports.* Atlanta: Centers for Disease Control.

Centers for Disease Control. *Surveillance.* (Annual). Atlanta: Centers for Disease Control.

Conte, J. E., Jr., and S. L. Barriere. 1992. *Manual of Antibiotics and Infectious Diseases.* Philadelphia: Lea and Febiger.

Mandell, G. L., J. E. Bennett, and R. Dolin. 1995. *Principles and Practice of Infectious Diseases.* 4th ed. New York: Churchill Livingstone.

Murray, P. R. (Editor-in-Chief). 1995. *Manual of Clinical Microbiology.* 6th ed. Washington, DC: American Society for Microbiology.

Part 8

Applied Microbiology
A Conversation with Arnold L. Demain

Professor Arnold L. Demain, Professor of Industrial Microbiology in the Department of Biology at the Massachusetts Institute of Technology (MIT) in Cambridge, Massachusetts, has been in the forefront of the field of biotechnology for more than 40 years. He is best known for his research in the elucidation and regulation of the biosynthetic pathways leading to penicillins and cephalosporins. His ability to integrate basic studies and industrial applications is his greatest strength.

Dr. Demain has trained the people who now direct many of the world's leading pharmaceutical and chemical companies as well as departments of industrial microbiology and biochemical engineering. Although his accomplishments are many, he considers his most important honors to be the awards as "Best Teacher of the Year" from the Nutrition and Food Science Department at MIT and "Teacher of the Year" from the MIT graduate students.

Interview conducted by Edith Brady, Executive Editor at Saunders College Publishing.

EB: Dr. Demain, what led to your decision to pursue microbiology as a career?

AD: Well the answer involves my family. My grandfather was a pickler and so was my father. I was interested in science, but I was more interested in art, commercial art. My father took me at sixteen years of age to Michigan State to meet a professor of pickle fermentation, Professor F. W. Fabian, and put me in his care. One thing led to another and I went into microbiology. At first I didn't find it that exciting, but as years went by, I found it more and more exciting. Once I got turned onto the idea of a career in this field, there was virtually nothing stopping me.

EB: You mentioned fermentation and that you started out in the pickle business. Was it food microbiology and the process of fermentation that originally interested you?

AD: Yes, I started out in the food microbiology lab of Professor Fabian at Michigan State, which mainly dealt with pickle fermentations and pickle packing. But then a mentor of mine convinced me to go to Berkeley for my Ph.D. He was an earlier student of the same laboratory and the same professor. This mentor was Prof. John Etchells of North Carolina State University and the U.S. Department of Agriculture.

Courtesy of Arnold Demain.

At Berkeley and later at Davis, I became exposed to this great field of biochemistry and metabolism. I became interested in enzymes which led me to my mentors at the University of California, Professor Herman Phaff and the head of our department, Professor Emil Mrak. Professor Mrak obtained some interviews for me. Although I technically graduated from the Microbiology Department, I actually did my work in the Department of Food Science. And so I was sent two places for interviews. One was General Foods in Hoboken, New Jersey, and the second was Merck and Co. in Rahway, New Jersey. I received offers from both, but the Merck offer came first and I accepted it and that put me into the area of pharmaceutical fermentations. So I went to a penicillin factory in Danville, Pennsylvania for my first job and after a year and a half I moved up to the main laboratory in Rahway, New Jersey. I learned that there was a lot more to microbiology than pickles and antibiotics, that microorganisms could be used for many things—produce commercial enzymes, produce steroids that could be used for arthritis and rheumatism, make amino acids for the food and nutrition field, and many types of medicinals other than antibiotics.

Then came along the world of biotechnology. Biotechnology is as old as industrial microbiology going back to pre-Biblical days. But what we call the new biotechnology was born in California in 1973 with the discovery of recombinant DNA technology. This opened up the whole field of microbiology to include many things that are not microbial at all, like making interferon in an animal cell and things like that. I am a person who is involved in biotechnology but I stress the application of microorganisms in that field. For example, right now in the laboratory, we're working on immunosuppressing agents, cholesterol-lowering agents and antibiotics such as cephalosporins. We're studying secondary metabolism in microgravity for possible application in space. We're doing that for NASA. We're working on processes to make liquid fuel from waste materials. Anything that deals with the application of microbial and other forms of cells is within our realm of research.

EB: As described, you began your professional career as an industrial microbiologist. When you left industry you moved to the university where you continued your interests in antibiotic research. What led you to go to the university and continue your productive studies on antibiotic synthesis?

AD: I certainly never expected to be a teacher. Never in my life had I even thought about it. I loved research. I loved research more than I loved becoming part of top management in industry. So even if I had stayed in industry at Merck, I would not have gone further up the corporate ladder. I would not have wanted to go further than department head, which I became. The next step would have been some sort of a vice-president level, where the people who would report to me would not be researchers. As department head, I could still be involved in research.

Then, I was approached at an ASM meeting in Detroit in 1968 as to whether I would consider coming to MIT. At the time, I was 41 years old, and knowing that MIT was a great institution, I certainly was interested. I consulted with one mentor of mine at Merck, Dr. David Hendlin, and his advice was to go. I could continue my research career for many years. That was one reason that I left industry. By the way, I enjoyed industry very much and it was with great mixed emotions that I left.

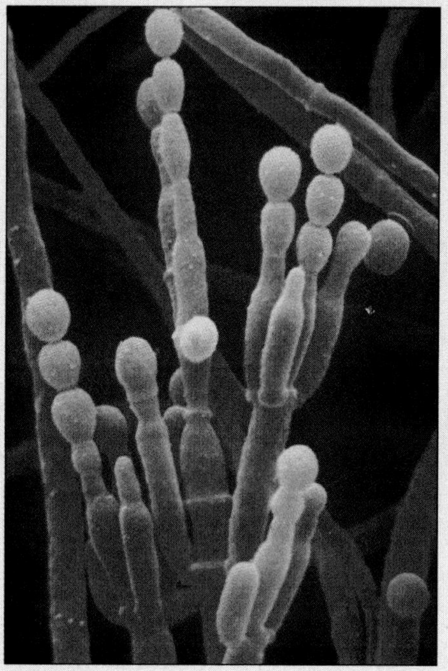
SEM of penicillin spores. (© G. Shih-R. Kessel/VU.)

EB: Antibiotics have given us a degree of control over some diseases such as pneumonia and tuberculosis. Resistance to antibiotics is spreading in previously susceptible microbial populations. What can be done? More screening for producing microbes? More chemical synthesis? Both?

AD: All of the above. One reads in the popular press today that scientists think we had won the war against the microbes—the infectious microbes. That is a terribly erroneous statement. All of us in the world of antibiotics, whether academics or industrial people, realize that this is a continual and never-ending war. The organisms will always mutate to resistance. New diseases are evolving all the time. AIDS certainly wasn't around in the early years. This is a never ending struggle between our minds, our technical abilities, and the organisms in nature. As far as antibiotic-resistant organisms are concerned, we must continue. We must continue. There is no question of stopping the fight despite the fact that some of our industrial companies have given up in the area of antibacterials. If it's not the companies, then it must be the government. There is no way we can allow the pathogenic microorganisms the free course into our bodies and allow them to maim us and to kill us.

Now how do we do that? We have lots of new techniques and lots of old techniques that still work well. Yes, natural screening, exploring biodiversity whether it is on land or in sea, whether it's microbial, plants, or even animals that make antibiotics. All of these natural sources are needed. Then we have the new combinatorial chemistry libraries: Chemists have learned to build a complicated mix of compounds similar to what we would find in natural organisms in nature; these combinatorial libraries have to be searched also. Of course, we still do have the chemical companies, the pharmaceutical companies with thousands of compounds on their shelves. These have to be screened. The question is how do we screen them? Well, we have an all new development in high throughput screening. There are companies that make the machines, all involving robotics, miniaturization, and automation. One can go through one million samples a year today where we used to be satisfied with 50,000 samples per year. The innovativeness comes from the ability to devise new screens that target our major problems, whether they be infectious diseases, which we're talking about now, cancer or heart disease, or various other non-infectious diseases. The battle must go on. Companies will go in and out of the battles. But the most successful companies will be those

> **I am a person who is involved in biotechnology but I stress the application of microorganisms in that field. . . . Anything that deals with the application of microbial and other forms of cells is within our realm of research.**

that hang in for the long run.

EB: What is the future for industrial microbiology? Is it declining or growing?

AD: We have to add the term "biotechnology" today, because so much of what we do today involves organisms other than microorganisms such as animal cells in culture. An industrial microbiologist of old would today be doing cell biology. So, industrial microbiology is part of a greater scheme which we call biotechnology and both of these are growing rapidly. It's an area that is very exciting with the new techniques being added. Ever since the discovery of recombinant DNA, our potential to discover and to make things that we could never make before has increased. This is a growing area. Unfortunately most American students are not interested in this field. It's a field that attracts the excitement of bright students from the Orient, Latin America, and Europe, but unfortunately not from the United States. We have a problem here that has to be addressed or else all of the industrial labs will end up being manned by scientists from other nations, and American students will be on the street looking for jobs.

EB: What are the major contributions made to human welfare through research on microorganisms since 1950? Do you have any ideas on models for metabolic studies?

AD: Of course the antibiotics are the major contribution. Many antibiotics came out after 1950. Penicillin only became a realistic product during the Second World War which ended in 1945. And streptomycin had been developed by the late 1940s. But many of the important antibiotics came along during and following the 1950s. So we have to attribute these years to years of discovery and commercialization of many antibiotics. Additionally, many antibiotics are used for things other than infectious disease. We have antibiotics being used as herbicides or other types of pesticides. We have antibiotics being used to lower cholesterol in human beings. We have antibiotics being used as immunosuppressive agents for transplanting hearts, and livers, and kidneys. So the broadening of the use of antibiotics or secondary metabolites in general has been quite a development.

Then, of course, we have the developments in molecular biology which came out of the study of *Escherichia coli* and led to the recombinant DNA era in 1973. Since 1973, we have some 10 or 20 new products on the market which are mainly pharmaceutical proteins, but some are in the food area and some in the agricultural area. We also have the area of diagnosis which includes monoclonal antibodies, polyclonal antibodies, and DNA probes. And, let's not forget the discovery of the polymerase chain reaction (PCR) at CETUS which has so revolutionized the world of forensic medicine and diagnosis and also research itself. These are the main discoveries. We are moving slowly on cancer. Many of our antibiotics are used as the chemotherapeutic agents of cancer. I could just go on and on. The enzyme business involves the use of microbial enzymes and exploits the discovery of organisms that grow under very unusual conditions such as above the temperature of boiling water and at almost freezing temperatures. From these weird organisms, particular enzymes are being obtained and are being used in commerce. I'd better stop because I could just go on and on and on.

> **In a way, many of our students leave the university knowing *E. coli* and nothing more when it comes to microorganisms. They totally miss the area of biodiversity, which is so important to the future of the pharmaceutical field.**

There is the whole area of the environment that is being impacted in a major way now by microbiology. I refer to the use of natural organisms, the use of microbial communities, and the use of engineered organisms to degrade pollutants, purify our water supplies, and eventually supply the motor fuel after we run out of cheap petroleum in the next century. We will have to have some way of keeping automobiles going and it's not going to be petroleum. It's going to be ethanol from microbial fermentation.

EB: In the early 1950s, Bacteriology Departments became Microbiology Departments, and in the 1970s and 1980s, many then became part of Cell or Molecular Biology Departments. What are your thoughts on microbiology as a separate discipline?

AD: This is a tough question because I can see both sides of the issue. It was correct for bacteriology departments to become microbiology departments because one cannot really study bacteria alone without thinking of yeast and filamentous fungi. These are respectable microorganisms. Any microbiologist, anyone who even calls himself a bacteriologist, has to consider the importance of higher forms such as yeasts and molds. But the major change has been the recent elimination of microbiology departments and replacement with molecular biology or cell biology, or as in MIT, a biology department.

Microbiology survives quite well in these departments. For example, we have at MIT a large biology department with 50 to 60 different lab groups and professors, and of those, a number are dealing mainly with microbial biology. You cannot have successful molecular biology and genetics without involving yourself with microorganisms; it's impossible. They're simple; they're models. Maybe that's the answer to the second part of the question. Microorganisms are great models with which to teach biology; great models to understand metabolism, even of higher forms of life. We can certainly study the same types of metabolism in microorganisms.

The problem that I worry about is the disappearance of the botany departments, the zoology departments, the horticulture departments. These are disciplines in which it is very difficult to get training. You can still get them in some of the state universities that emphasize agriculture. But when they disappear where are you going to get trained in mycology? How is anybody going to know anything about a filamentous fungus or a tree or a plant, if everything is molecular biology? So, there's a balance here that is necessary to preserve. Our knowledge of the plant kingdom, the animal kingdom, and the microbial kingdom cannot be lost by everything moving into the area of molecular biology. It's scary.

Think of it: Taxol is an anticancer agent obtained from trees. We still have people who know botany and who helped in that particular case, but if botany disappears from our universities, we're going to be out of luck. The move from the whole organism to the cell and from the cell to the particular elements of the cell, whether they be nucleic acid, protein, what have you, is again a scary thing. A cell is just not a bag of these things. A cell is an integrated, "intelligent," evolved structure that allows the many cellular reactions to work together. We can never forget about the whole cell.

In a way, many of our students leave the university knowing *E. coli* and nothing more when it comes to microorganisms. They totally miss the area of biodiversity, which is so important to the future of the pharmaceutical field.

So, as I say, I view this with mixed emotions. I think that when any one department changes completely into molecular biology, it is the responsibility of that department to maintain training in some of the more traditional fields. Biodiversity and microbiology are absolutely essential for the development of science students.

EB: How would you advise a young college student interested in entering the field (microbiology) today? What courses, areas of research, etc?

AD: My advice has always been for students to get as broad a training as possible. Try not to specialize as an undergraduate. This is my advice to a potential microbiology student. Emphasize chemistry, biochemistry, genetics and expose yourself to a chemical engineering course. These are things you should be getting in addition to your microbiology courses. Try to avoid universities that have too many or too few microbiology courses. Try to find those in which microbiology is put together in a few basic courses to allow you enough time to take chemistry, biochemistry, genetics, and chemical engineering along the way. Specialize when you get to graduate school. Avoid taking a master's degree unless you have to. Move as fast as you can into your Ph.D. degree, then you can specialize, but make sure that at the same time you are specializing in grad school, you take advantage of some of the great professors who may be in other departments. Take those courses whether you take them for credit or just sit in on them. There are great resources at most universities. You don't want to miss the opportunity to be exposed to great educators and researchers.

There is no field of human endeavor, whether it be in industry or in agriculture, whether it be in the preparation of foodstuffs or in connection with problems of shelter and clothing, whether it be in the conservation of human and animal health and the combating of disease, where the microbe does not play an important and often a dominant part.

Selman A. Waksman, 1943

Chapter 32

Industrial Microbiology

The Microbe
Major Industrial Products
Foods, Flavoring Agents and Food Supplements, Vitamins, and Beverages
Organic Acids
Enzymes, Microbial Transformations, and Inocula
Inhibitors
Genetically Engineered Microorganisms

Industrial microbiology is an ancient practice. It is reasonable to assume that this technology was born when ancient civilizations took advantage of the fermentative capacity of microorganisms to produce alcoholic beverages, as well as leavened bread and cheese. There is ample evidence that the early Sumerian City States (4th millenium B.C.) fermented barley to produce beer, and grapes to produce wine. In addition to producing foodstuff for trade, ancient civilizations unwittingly rendered perishable food more stable through the activities of microorganisms. Cheese and fermented or salted meat are transportable, and these staples enabled early civilizations to move about more freely. Until the Frenchman Francois Appert developed canning methods in 1809, salting or fermenting was the major means for preserving food.

During the first half of the 20th century, wine, beer, and vinegar manufacture, and bread yeast production moved from the realm of an ancient art to an established science. Large-scale microbial processes for the manufacture of citric and lactic acid were also developed. Acetone, butanol, and ethanol were produced in commercial quantity by fermentation, but these are now mostly products of chemical synthesis. The second half of the 20th century has been an era of dramatic change in the fermentation industry. Much of this change originated with the discovery that antibiotics are an effective weapon against disease. The search for novel antibacterial, antifungal, antitumor, and antiviral agents has become a major industry, with Japan, the United States, and Western Europe leading the way. There are other products generated by microbes that have received considerable attention. Among these are vitamins, sterols, organic acids, amino acids, flavoring agents, enzymes, and fermented foods/beverages.

There are two standard procedures one can employ to improve microbial strains that can generate useful metabolites. These two techniques are **mutation** and **selection.** Utilizing these techniques, scientists subject a producer strain to mutagenic agents and select the most prolific of the progeny. Through repeated and rigorous application of such regimes, phenomenal increases in yield of useful metabolite have been realized. For example, the original strains of *Penicillium chrysogenum* that yielded 1.2 milligrams of penicillin per liter have been genetically altered to produce 50 grams (50,000 milligrams) per liter.

Knowledge gained over the past 15 years has markedly broadened the potential for production of useful compounds by microorganisms. This new field of endeavor is broadly known as **biotechnology.** Biotechnology is the use of living systems (usually microbial) to solve technical problems or achieve technological goals. In practice, one may manipulate the genetic information in a microorganism in such a way that it can gain new functions. For example, genes for the biosynthesis of human insulin (for diabetes treatment) have been introduced into bacteria, enabling the bacterium to synthesize this hormone. With newly gained genetic information a microorganism can develop the potential for synthesis of an array of commercially useful products. Among the products now obtained from microorganisms are human growth factor, blood proteins, interferon, and vaccines.

Microorganisms can have an impact on the economic strength of a country and this is illustrated by the development of the fermentation industry in Japan. An estimated 5 percent of the gross national product in that country rests on the broad capabilities of the microbe.

The Microbe

The **fermentation industry** is the collective term applied to the commercial processes that exploit the capacity of microbes to make useful products. Every microorganism employed in the fermentation industry is unique and carefully selected for the ability to perform a desired task. When newly isolated from the environment or selected from stock cultures, the potential source of useful products is not significantly different from related species. It requires careful manipulation and selection of strains that result in the production of profitable quantities of commercial product.

It is not uncommon for a microorganism present in the environment to synthesize more of a metabolite than is required for its own growth and maintenance. Metabolites such as amino acids, B vitamins, and nucleic acid bases may be overproduced by a microorganism and excreted into the proximate area in the microenvironment. Other microorganisms in the environmental niche can then assimilate these compounds. One strain may overproduce one metabolite but require for its growth a metabolite released by another strain. Together the organisms can survive. Such harmonious associations occur widely in nature, and microorganisms isolated from various niches commonly have requirements for one or more simple metabolites. Those microorganisms that tend to overproduce in nature may, when isolated in pure culture and subjected to the rigors of laboratory growth, produce even greater quantities of a given metabolite. Unbalanced growth in the laboratory may also lead to the expression of inherent genetic information and the production of identifiable quantities of metabolites such as antibiotics.

Recycling of organic matter in nature depends on the ability of microbes to synthesize enzymes that disassemble macromolecules. Among these are lipases, amylases, and proteinases. Several of these enzymes that disrupt macromolecules are of commercial interest.

Microbial strains that are valued for the synthesis of commercial products are highly selected specialists and would not survive outside the laboratory setting. Chosen for their genetic stability in the fermentation process and consistency in production, they are fast-growing organisms that produce commercial products in a short period of time. Such strains generally must be harmless to humans and to the environment because containment of industrial microbes involved in large-scale fermentations is virtually impossible.

Primary and Secondary Metabolites

Some metabolites are synthesized and appear in the medium during active growth, and others appear as the organism completes the growth cycle and enters the stationary phase. **Primary metabolites** are those produced during active growth, and those synthesized after the growth phase nears completion are designated **secondary metabolites** (Figure 32.1).

A primary metabolite is generally a consequence of energy metabolism and necessary for the continued growth of the microorganism. Recall the discussion of anaerobic metabolism (Chapter 8) where a product of glycolysis serves as an electron acceptor for the reduced pyridine nucleotide generated by glyceraldehyde-3-phosphate dehydrogenase (Figure 32.2). In the absence of oxygen, the terminal electron acceptor could be pyruvate, and the product of anaerobic energy generation would be lactic acid. The microorganism involved would not generate ATP (or grow) without the concomitant production of the primary metabolite, lactic acid.

The production of secondary metabolites is a result of complex reactions that occur during the latter stages of primary growth. In a microorganism that produces sec-

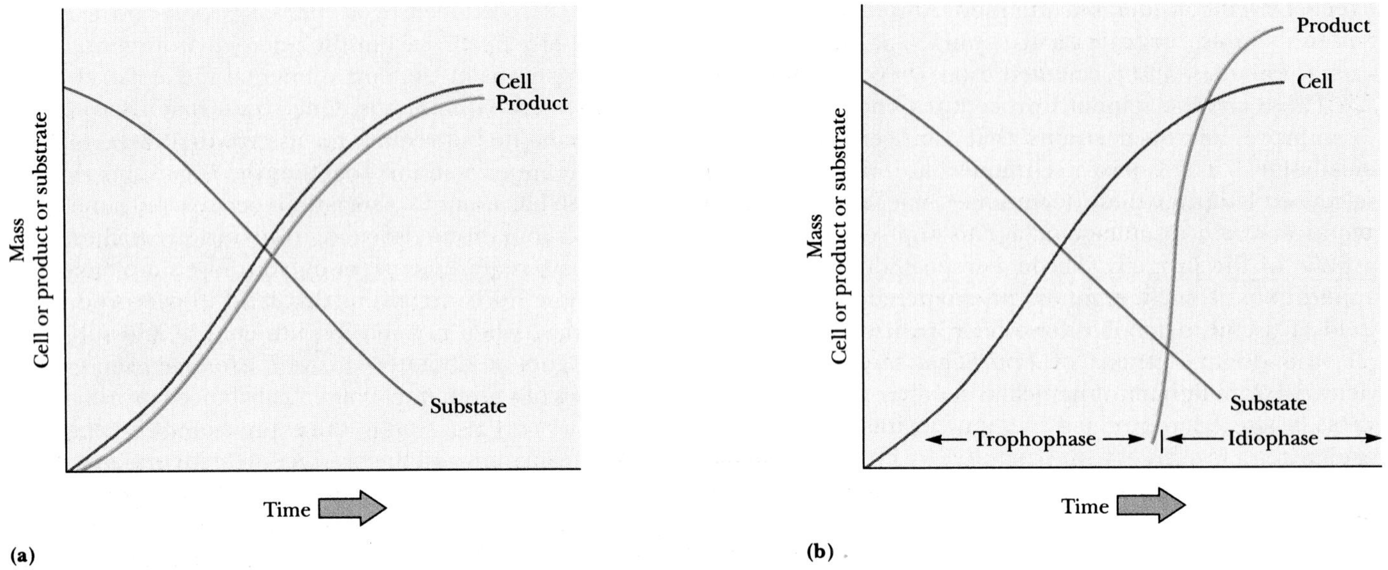

***Figure* 32.1** Time course for substrate utilization, increase in cell mass, and production of **(a)** primary metabolites and **(b)** secondary metabolites.

ondary metabolites, the growth stage is called the **trophophase.** The phase involved in production of secondary metabolites is the **idiophase.** In some cases, enzymes involved in trophophase growth shift during the idiophase to the production of metabolite. Citric acid fermentation by *Aspergillus niger* is an example of a product resulting from a metabolic shift. During the trophophase, a substrate such as glucose is mostly used for synthesis and energy generation. In generating energy, much of the substrate is oxidized to CO_2. During idiophase, glucose is mostly converted to citric acid with little respiration (CO_2 production). In the production of secondary metabolites, such as antibiotics, a different scenario prevails. An actinomycete that produces an antibiotic will synthesize a battery of enzymes during the latter stages of exponential growth and early on in the stationary phase that are involved in the synthesis of the antibiotic. The synthesis of these enzymes is a considerable undertaking as, for example, there are over 300 genes involved in the synthesis of chlorotetracycline by *Streptomyces aureofaciens.* There are at least 72 intermediates formed and 27 of these have been isolated and characterized. The antibiotic originates from

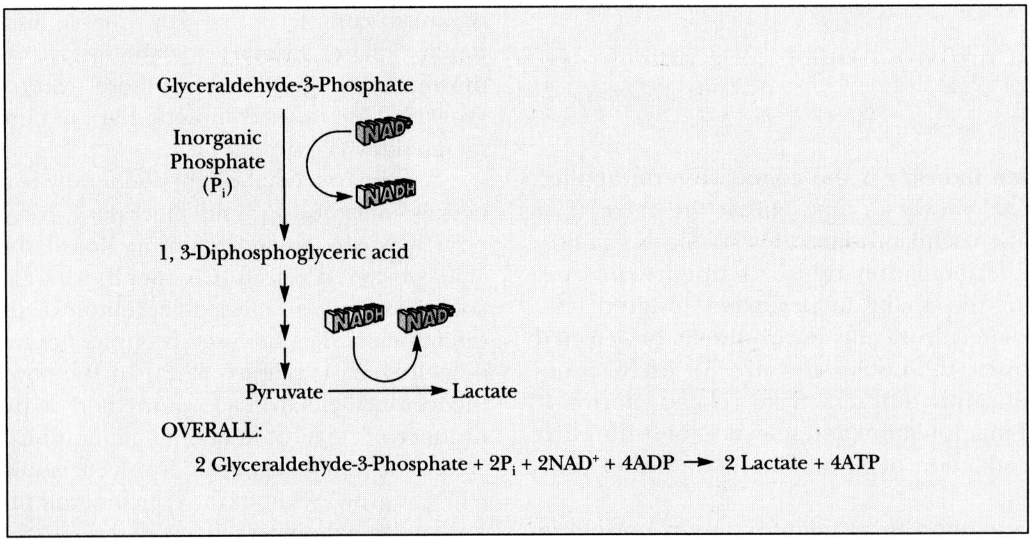

***Figure* 32.2** The production of lactate as a primary metabolite. Glyceraldehyde-3-phosphate dehydrogenase generates NADH, which is reoxidized by reduction of pyruvate to lactate.

precursor metabolites and very few of the enzymes involved in chlorotetracycline synthesis are required during cell growth. For a discussion of antibiotic function, see Chapter 7.

While there are thousands of secondary metabolites synthesized by Eubacteria of the genus *Streptomyces,* each is produced by one or a few species. The secondary metabolites are produced as a family of closely related chemical structures. The total amount of antibiotic produced by an altered actinomycete depends on the composition of the growth medium. The role of antibiotics in the life cycle of organisms that produce them is poorly understood. It is difficult to project what the role of many of these substances might be if present in miniscule amounts under the restricted growth conditions encountered in nature. There have been suggestions that antibiotics act as auto-inhibitors to prevent germination of the actinomycete spores in the absence of adequate nutrient. However, there is very little evidence that this explanation is appropriate for many species. Antibiotics and other metabolites are produced in measurable quantity only under unusual laboratory or commercial growth conditions.

Major Industrial Products

The diverse industrial products generated by microorganisms can be assigned to categories as outlined in Table 32.1. It is evident from a reading of this list that these products affect our lives with some regularity. Bread, cheese, pickles, olives, and mushroom soup are common foodstuffs in our diet. Each of these is a direct consequence of microbial activity. Although the role of microbes in the production of cheese and beer is widely known, few are aware that virtually all soft drinks contain citric acid, a product of microbial synthesis.

Flavor enhancers, such as monosodium glutamate (MSG) and inosinic acid, are products of microbial metabolism. Vinegar is a commonly used flavoring agent and preservative. Many of the B vitamins can be produced in significant quantity by selected microbial cultures, but only riboflavin and B_{12} are produced by microbial fermentation. Chemical synthesis of the others (pantothenic acid, biotin, thiamin, etc.) is less costly and the favored process for bulk synthesis.

***Table* 32.1 Major commercial products obtained from microbes**

I. Foods, Flavoring Agents and Food Supplements, and Beverages
- Foods
  - Fermented meat
  - Cheeses and milk products
  - Edible mushrooms
  - Leavened bread-baker's yeast
  - Coffee
  - Pickles, olives, sauerkraut
  - Single-cell protein
- Flavoring agents and food supplements
  - Vinegar
  - Nucleotides
  - Amino acids
  - Vitamins
- Beverages
  - Wines
  - Beer, ale
  - Whiskey

II. Organic acids
- Citric acids
- Itaconic acid

III. Enzymes and microbial transformations
- Commercial enzymes
- Sterol conversions

IV. Inhibitors
- Biocides
- Antibiotics

V. Products of Genetically Engineered Microbes
- Insulin
- Human growth factor

The conversion of low-cost biologically inactive sterols to the activated form can be accomplished by microorganisms. Antibiotics synthesized by microbes have had a profound effect on human health and longevity. Microbes can be genetically engineered to produce a variety of proteins and peptides. One such protein is insulin, which is indispensable for human diabetics. Other products of recombinant DNA are human growth factor, interferon, blood clotting factors, and vaccines.

Introduction of diverse genetic information into rapidly growing microorganisms that will induce them to synthesize valuable products is a fertile field for future development. Such manipulations are the basis for the rapidly expanding field of **biotechnology.**

The above was a brief outline of the scope of industrial microbiology. Some of the various major commercial products generated by microorganisms will be presented in the following sections.

Foods, Flavoring Agents and Food Supplements, Vitamins, and Beverages

There are many facets to the application of microbial processes in the food industry. In some cases the microorganism itself may be a potential food source. Generally, however, they are employed as agents to convert foods such as milk or meat to a favorable product. Microorganisms can synthesize flavoring agents or enhancers, and these can be important in the quality and acceptance of foods. Food supplements from fermentation can serve as preservative agents or improve the nutritional value of a food. The alcoholic beverage industry is economically the most significant of all commercial processes that involve microorganisms. The microorganisms responsible for these diverse food-related products differ, and the following is a discussion of a number of these industrial processes.

Foods

The use of bulk microbial cell mass as a source of human or animal food has fascinated researchers for many years. This potential food source is termed **single cell protein** or **SCP.** SCP could be of considerable value to nations where climatic conditions cause agricultural crops to fail or where populations are constantly threatened by starvation. For people with insufficient agricultural resources, an alternative source of food is therefore important. A potential source of food could be the bulk harvesting of bacterial and/or yeast cells grown on substrates that in themselves would not sustain humans **(Box 32.1).**

Bacteria and yeast cells are attractive food sources for

BOX 32.1 MILESTONES

Bacteria as Food

The consumption of microbes as food has generally been limited to mushrooms (filamentous fungi) and yeasts. Yeasts are eaten as single-cell protein or as a food supplement (see text). Bacteria—as food—have been collected and eaten since ancient times by tribes in the central African republic of Chad. The organism is a cyanobacterium, *Spirulina platensis,* and is collected from the bottoms of seasonally dried-up ponds and shallow waters around Lake Chad. The cyanobacterial mats are dried in the sun and cut into small cakes, called **Dihe.** Dihe is rich in protein (62 percent to 68 percent of dry matter) and an important supplement to diets of the desert nomads. *Spirulina* has been grown in France, and the potential yield of 10 tons of protein per acre dwarfs the yield obtained from wheat (0.16 tons/acre) and beef (0.016 tons/acre). *Spirulina* is now sold as a food supplement in health food stores in the United States. It is marketed as dried cakes or powdered product.

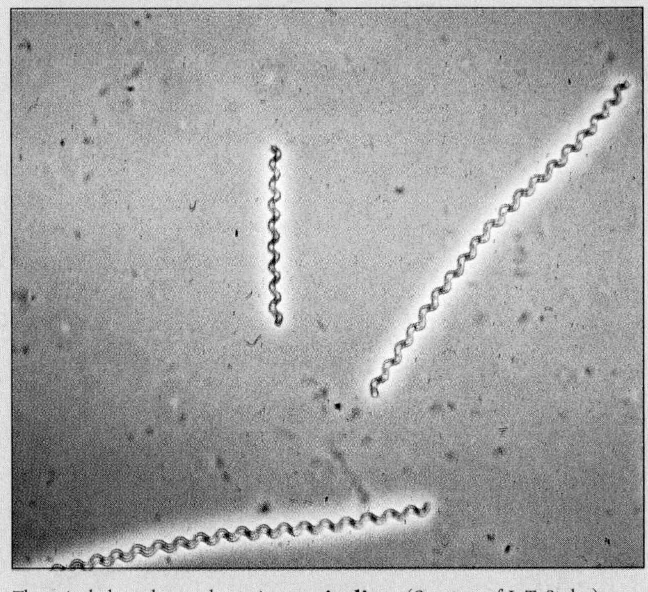

The spiral-shaped cyanobacterium, **spirulina.** (Courtesy of J. T. Staley)

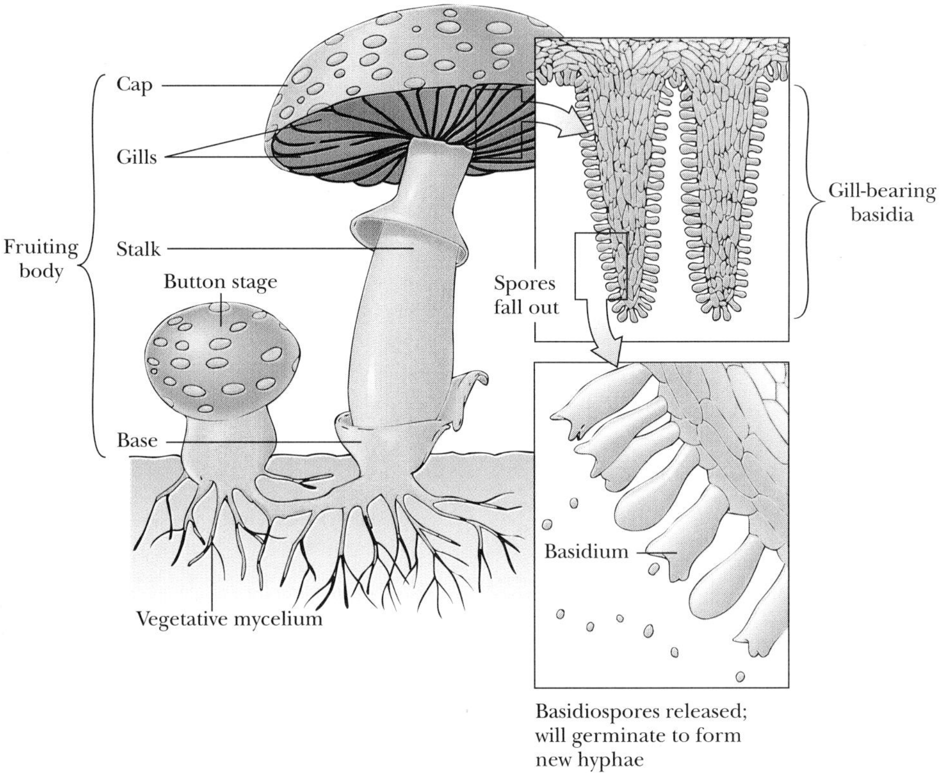

***Figure* 32.3** The developmental cycle of a mushroom. The vegetative mycelium that produces the fruiting structure is all underground.

a number of reasons. Microorganisms grow rapidly on low-cost substrates such as methane, methanol, grain husks, or other waste plant material. Mineral requirements can be met by inorganic sources. They are readily harvested, and generally have a high nutrient content. Moreover, the content of the microbial cell can be manipulated by growth conditions. For example, a yeast cell grown on a low-nitrogen, high-carbon substrate will be higher in lipid, and increased levels of nitrogen will lead to a higher protein content. Yeast cells are rich in essential B vitamins. However, yeast protein is generally low in sulfur-containing amino acids and is not sufficiently balanced to provide for human or animal needs. Yeast cells also have a high nucleic acid content, and if consumed in quantity, may cause gout or other ailments. Bulk yeast cells can be added to wheat or corn flour to provide B vitamins and selected amino acids resulting in a nutritionally balanced product.

Mushrooms are filamentous fungi, and several species are important as human food. The annual consumption of mushrooms in the United States exceeds 300,000 tons. A mushroom is the fruiting body of a mycelial fungus that generally grows on decaying organic matter in soil or wood. The mycelium grows beneath the surface and under favorable conditions will develop the familiar stalked structure. The stalk and cap are actually compact fungal hyphae, and the gills are the site of reproductive spore formation (Figure 32.3). *Agaricus bisporus* and *Lentinus edulus* are mushrooms grown commercially in the United States and commonly sold in grocery stores (Figure 32.4). A generalized flow chart for growth of *A. bisporus* is presented in Figure 32.5. Shiitake mushrooms are an edible mushroom that is favored in Japan and is gaining popularity in the United States. The shiitake fungus, *L. edulus*, is a cellulose decomposer and grows best on waste lumber from hardwood trees. Logs or waste lumber are soaked in water and inoculated by placing spawn in holes drilled in the wood. Mushrooms have little nutrient value and are eaten for their distinct flavor and texture. They are low in protein, are essentially fat free, and provide only low levels of B vitamins. For economic reasons, attempts have been made to grow bulk mushrooms in aerobic fermentors, but the product lacks flavor, texture, and desirability.

***Figure* 32.4** Two mushroom species that are produced commercially: *Agaricus bispora* on the left and *Lentinus edulus* on the right. (Courtesy of J. T. Staley)

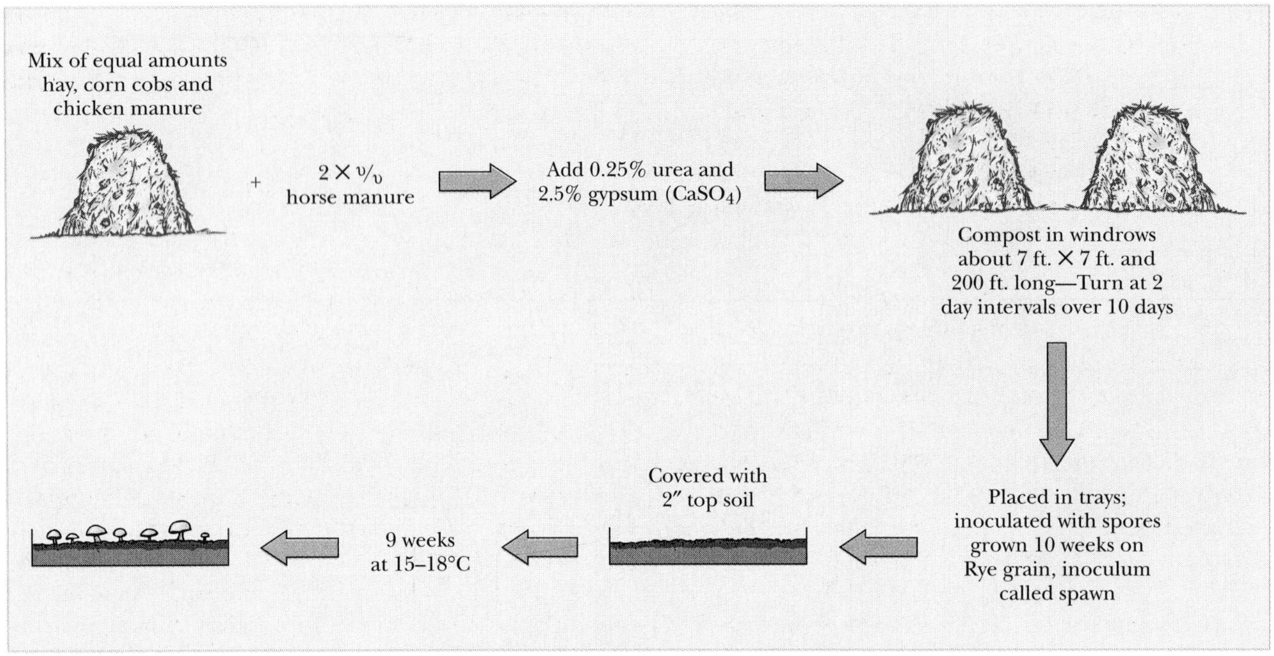

***Figure* 32.5** Flow chart for the commercial production of the edible mushroom *Agaricus bisporus.* The mushrooms are harvested and the compost on which they grew is pasteurized and recycled.

Cheese

Cheese is an ancient food whose use, in all likelihood, precedes recorded history. The art of cheese making originated in the Middle East. From this point it traveled to Rome and on to Northern Europe and England, and from there it was spread throughout the world. There are well over 400 varieties of cheese known, although many are similar, differ essentially in name only, and are generally grouped according to texture (Table 32.2).

The manufacture of cheese can be divided into four phases: (a) coagulation of milk to form a curd, (b) separation of the curd from the whey, (c) shaping of the curd to permit acid production and proper moisture content, and (d) ripening to achieve flavor and texture. Handling of the material at each of these steps and the nature of the milk employed have a marked influence on the finished product. Most cheese is made from cow's milk, but many varieties around the world are produced from the milk of goats, sheep, mares, camels, and water buffalo.

The first step in cheese production is the coagulation of the milk proteins to form a semisolid curd. This curd entraps fat globules and many water-soluble materials. Coagulation is effected by adjusting the pH of the milk to the isoelectric point of casein and/or adding milk-clotting enzymes. The clotting enzymes are acid proteases such as rennin, an enzyme originally obtained from calves' stomachs, but now produced by bioengineered microorganisms. Acidification of milk can be accomplished by adding selected cultures of bacteria. Bacteria that ferment lactose to lactic acid are effective acidifiers, and organisms generally are strains of *Streptococcus lactis, Lactobacillus cremoris, Lactobacillus bulgaricus,* or *Streptococcus bulgaricus.* A portion of the bacteria present at the time of curd formation is trapped and remains inside the cheese. These organisms remain and are involved in the aging/ripening process.

After the curd is separated from liquid, the pH level, salting, and shaping of the curd all contribute to the characteristics of the finished cheese. The ripening process, with or without added microbes, also affects the properties of the cheese product. Fungi or bacteria may be added to the ripening curd, and these add flavors or consistency not contributed by the microorganisms that were present initially. For example, the cheese commonly termed blue cheese is inoculated with spores of *Penicillium roquefortii.* If

***Table* 32.2 The major types of cheese**

Soft	Semisoft	Hard	Very Hard
Unripened	Blue	Cheddar	Parmesan
Cottage	Roquefort	Gouda	
Mozzarella	Limburger	Swiss	
Cream	Muenster		
Ripened			
Brie			
Coulommiers			
Camembert			

one examines roquefort cheese in cross section, the "blue veining" where the fungal spores were inoculated are quite evident. *Penicillium camemberti* is added to the surface of a cheese known as Camembert. The fungus produces proteases that transform the cheese to a smooth creamy consistency. Soft cheeses are ripened for a few weeks, semisoft for several months, and a hard cheese like Parmesan is ripened for a year or more. Cheese making remains an art with some science mixed in.

Yogurt and other fermented milk products are common foodstuffs consumed throughout the world. Commercial yogurt is made from pasteurized low-fat milk thickened by the removal of water or addition of powdered milk. The thickened milk is inoculated with a mixture of *Streptococcus thermophilus* and *Lactobacillus bulgaricus* and incubated at 45°C for about 12 hours. Growth of *S. thermophilus* produces lactic acid, and *L. bulgaricus* contributes lactic acid and aromatics that give the product a distinct taste.

Yeast

Yeast cells are a source of single-cell protein and a potential dietary supplement. Numerous commercial biochemicals are obtained from yeast, including enzymes, nucleosides, and natural vitamins. A common component of media for microbial growth is yeast extract, the water-soluble fraction of autolyzed yeast. It is a source of B vitamins, amino acids, and other growth factors.

The production of bakers' yeast annually in the United States exceeds 100,000 tons. The most common use of yeast cells is in the baking industry. Most leavened breads that are marketed throughout the world are products of yeast metabolism **(Box 32.2).** Yeast is added to dough to cause "rising" or leavening prior to baking. During kneading, the dough is manipulated to aid in the yeast fermentation. As yeast ferments, it converts sugar to ethanol and CO_2. During baking, the CO_2 and ethanol are driven off, leaving the odd-shaped holes in the baked product. The escaping gas (CO_2) that is trapped in wheat gluten results in an expansion of dough generally referred to as the "dough rises."

Bulk yeast used in baking or as a food supplement is produced in large fermentors. A flow chart for yeast growth is presented in Figure 32.6. A small inoculum is transferred batch-wise to larger and larger tanks as shown. Before inoculation into the largest fermentor (S-3) the yeast cells are concentrated and added as a thick slurry.

BOX 32.2 MILESTONES

The Staff of Life

The baking of leavened bread is clearly depicted in reliefs from Egyptian tombs that are over 4500 years old. The number of individuals and the scale depicted in the relief suggest that it may have been a commercial operation. The Old Testament of the Bible affirms that leavened bread was favored:

> *Seven days shall ye eat unleavened bread . . .*
>
> Exodus 12:5

Here the consumption of unleavened bread was a sacrifice to be made by the people of Israel. The presence in leavened dough, prior to baking, of an ingredient that was transferable to fresh dough and would transform it to leavened dough was also a part of Biblical lore:

> *. . . Know ye not that a little leaven leaveneth the whole lump?*
>
> I Corinthians 5:6

It was not until the middle of the 19th century that people understood that the leavening agent was actually a microbe. The early transferable leaven was probably a yeast, a heterofermentative lactic acid bacterium, or a combination of the two.

A more recent example of the employment of a "little leaven" is in the legends of the sourdough in Alaska. The gold seekers who went to Alaska in the late 19th century became known as sourdoughs because they carried a dough containing a mixture of lactic acid bacteria and a yeast. The "sour" flavor in sourdough bread and some rye breads is due to the presence of lactic acid generated by the bacteria.

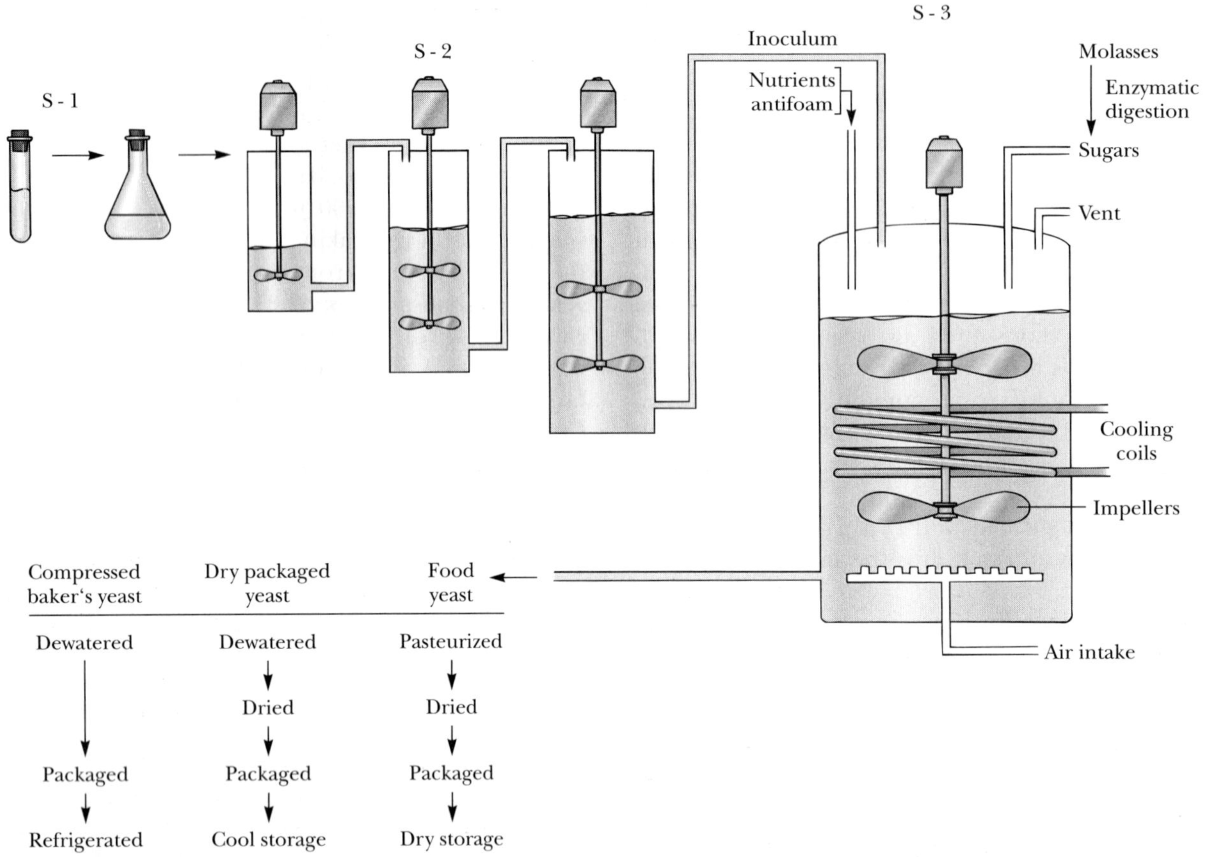

***Figure* 32.6** A flow chart for the production of baker's and food yeast. The organism is *Saccharomyces cerevisiae.* S-1 is growth of the inoculum. S-2 is growth of the organism in increasingly larger tanks, and at S-3 the fermentation vessel can be 200,000 liters or larger. Molasses is enzymatically digested to sugar monomers to provide a substrate for yeast growth.

Beet or cane sugar is the source of carbon and energy, and the fermentor is intensely aerated for rapid growth. The addition of the sugar is carefully monitored to ensure that it is used up immediately, otherwise ethanol will accumulate at the expense of more yeast cells. High aeration also prevents ethanol accumulation. Yeast cells are sold for baking as a compressed block, as a cake of moist yeast, or as activated dry yeast. Yeast cake is about 70 percent moisture and must be refrigerated to maintain activity. Dry yeast is produced by vacuuming away most of the water, resulting in long shelf life.

Fermented Meat

There are specialty meat products that are manufactured by a bacterial fermentation. Among these are summer sausage, Lebanon bologna, salami, and cervalat. The organisms generally involved in meat fermentations are strains of *Pediococcus cerevisiae* and *Lactobacillus plantarum.* Country cured hams are also a food that gains flavor and texture from fungi that are natural surface contaminants. These fungi are members of the genus *Penicillium* and *Aspergillus.*

Other Food Products

Other foods that are products of microbial fermentation are listed in Table 32.3. Acidification of cabbage by a lactic acid fermentation is an age-old method for long-term storage; the product is commonly called **sauerkraut.** Traditionally, it was prepared by placing shredded cabbage in a crock or other deep container where anaerobiosis precludes growth of spoilage organisms. Salt (NaCl) was added at 2.2 percent to 2.8 percent to restrict growth of most gram-negative bacteria. The natural lactic acid flora of the cabbage predominated under these conditions and brought about the fermentation. In the food industry, *Lactobacillus mesenteroides/Lactobacillus plantarum* are used as inoculum in making sauerkraut. The lactic acid produced lowers the pH to a level at which little decomposition occurs and the finished product is stable for extended periods of time. A typical fermentation requires about 30 days.

Pickles are produced by placing small cucumbers in a salt brine along with dill or other flavoring agents. The salt (NaCl) concentration, initially about 5 percent, is increased to 15.9 percent over a six- to nine-week period. The high salt concentration extracts sugar from the cu-

***Table* 32.3 Foods produced by microbial fermentation**

Product	Fermenting Microorganism
Cocoa beans	*Candida* sp., *Geotrichum* sp., *Leuconostoc mesenteroides*, others
Coffee beans	*Erwinia* sp., *Saccharomyces*
Sauerkraut	*Lactobacillus mesenteroides, Lactobacillus plantarum*
Soy sauce	*Rhizopus oligosporus, Rhizopus oryzae*
Olives	*Leuconostoc mesenteroides, Lactobacillus plantarum*
Pickles	*Pediococcus cerevisiae, Lactobacillus plantarum*

cumbers, and these sugars are fermented to lactic acid. Home production of pickles is a result of fermentation by the indigenous lactic acid flora of cucumbers. Commercial production is accomplished by sterilizing the pickles and adding cultures of *Pediococcus cerevisiae* and *L. plantarum* to ensure a uniform fermentation and product.

Coffee beans develop as berries with an outer pulpy and mucilagenous envelope that must be removed before roasting. The pulp can be removed mechanically, but the pectin that contains the mucilage is removed best by a pectinolytic bacterium *Erwinia dissolvens.* Fungi are also involved in this process. Apparently microbes do not contribute to the flavor or aroma of the finished coffee bean.

Chocolate is derived from cacao, a bean that is enclosed in the fruit of a plant that grows in parts of Asia, Africa, and South America. The beans are removed from the fruit and fermented in piles or boxes for several days. The fermentation occurs in two phases: (a) the sugars from the pulp are converted to alcohol and (b) the alcohol is oxidized to acetic acid. The yeasts involved in producing alcohol during the first phase are essential to the development of the chocolate flavor and aroma.

Fermentation of **olives** occurs after the olives have been treated with 1.6 percent to 2.0 percent lye at 24°C for four to seven hours. This treatment removes the bitter principal (oleuropein) from the green olive. The lye is removed by soaking, and the olives are placed in oak barrels and 7.0 percent salt brine is added. The brined olives are inoculated with *Lactobacillus plantarum* and fermented for eight to ten months.

Soy sauce is prepared by inoculating soy beans with *Aspergillus oryzae* or *A. soyae,* followed by a three-day incubation. The beans release fermentable sugars, peptides, and amino acids. The fungi-coated beans are placed in liquid, and salt (sodium chloride) is gradually added to a final concentration of 18 percent. The total incubation period is one year at room temperature. The addition of salt leads to a replacement of the filamentous fungi by lactic acid bacteria and yeast. The lactic acid bacteria are predominantly strains of *Lactobacillus delbrueckii,* and the yeast is *Saccharomyces rouxii.* The liquid resulting from the long-term fermentation is soy sauce.

Flavoring Agents and Food Supplements

Under appropriate conditions, selected bacteria and fungi can produce organic compounds that give a desirable flavor and/or aroma to food. Many of these originated as preservatives and were retained as flavoring agents despite the advent of refrigeration and other better means of preservation. Sauerkraut, mentioned previously, was originally prepared to preserve cabbage for consumption over the long winter. It is now prepared mostly because of the flavor imparted by the bacterial fermentation. This discussion is concerned with flavoring agents, flavor enhancers, and supplements such as selected amino acids that are added to food to enhance their nutritional value.

Vinegar

Vinegar making is at least 10,000 years old and came about originally with the production of wine. It is essentially sour (acetic acid) or spoiled wine, and the name vinegar comes from the French *vin* (wine) and *aigre* (sour). Vinegar has been valued as a flavoring agent, preservative, medicine, and consumable beverage. The Romans and Greeks favored dilute vinegar as a beverage and produced it by introducing air into wine casks. An early method of vinegar production was to put an air inlet in the top of a wine cask and draw out vinegar as needed. Fresh wine was then added to replace the vinegar removed. F. T. Kutzing (see Chapter 2) theorized in 1837 that microbes were involved in vinegar production, and this was confirmed in 1868 by Louis Pasteur.

The production of vinegar in the United States amounts to about 160,000,000 gallons per year. Two thirds of the vinegar produced is used in commercial products

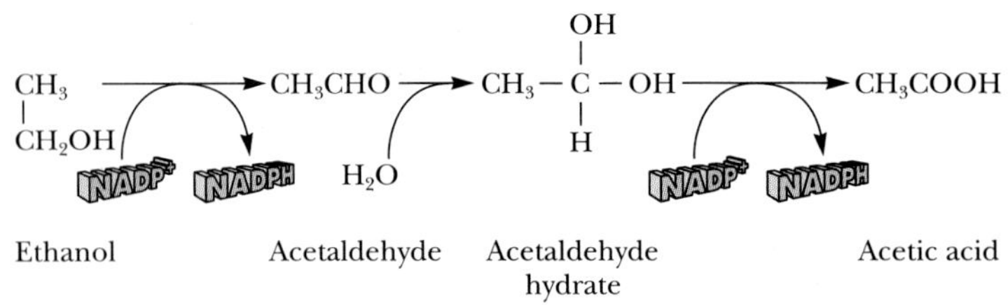

such as sauces and dressings, and in the manufacture of pickles and tomato products. The remainder is used for domestic purposes. There are two major systems for the commercial production of vinegar, the trickling generator and submerged fermentation. The basic reaction in vinegar formation is shown above.

The organisms employed in commercial vinegar production are *Acetobacter aceti* and *Gluconobacter oxydans.* There are hundreds of bacterial subspecies or strains that produce acetic acid, and mixed cultures with these other acetic acid bacteria are present during vinegar production.

One of the commercial systems for producing vinegar is the Frings-type trickling generator (Figure 32.7). The generator is constructed of cypress or redwood and packed with curled beechwood shavings. The collection chamber at the bottom holds about 3500 gallons. A pump circulates the ethanol-water-acetic acid from the collection chamber and sparges it evenly over the upper surface. The temperature is controlled by flow rate at 29°C at the top and 35°C at the bottom. The system is aerated by a blower at the bottom, and the aeration rate is carefully controlled to prevent evaporation of the acetic acid. Ethanol must be fed constantly to sustain the bacteria, as otherwise they will perish. The maximum acetic acid in solution is 13 percent to 14 percent. The life of a well-packed and maintained generator is 20 years. This is an example of a **continuous fermentation.**

The other major system for vinegar production is the Frings acetator, which is a submerged, **batchwise fermentation** process (Figure 32.8). The Frings acetator is composed of a stainless steel tank with a high-speed mixer that constantly stirs the microbes, air, ethanol, and nutrients so as to provide a favorable environment for bacterial growth. Small air bubbles are introduced at the bottom to provide O_2. Aeration can cause foaming, and this is controlled by a foam destroyer at the top of the tank. The tank is operated at 30°C, and this temperature is maintained by circulation of cooling water. Ethanol and nutrients containing the bacterial inoculum are added to the acetator in a batch. An instrument monitors the ethanol content. When the alcohol level decreases to 0.2 percent by volume, about one third of the finished product is removed and fresh nutrients and ethanol are added. To make 12 percent vinegar takes about 35 hours. The acetator is more efficient than the trickling generator.

Several types of vinegar are manufactured and are classified by the origin of the ethanol that the bacterium oxidizes to acetic acid. The basic flavor component in vine-

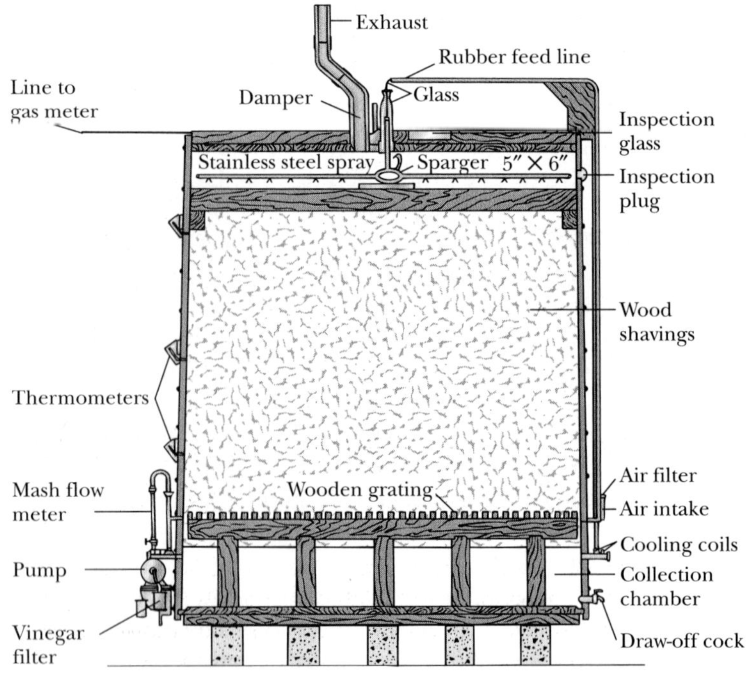

***Figure* 32.7** A Frings-type trickling generator employed in the commercial production of vinegar. The shavings are of air-dried beechwood.

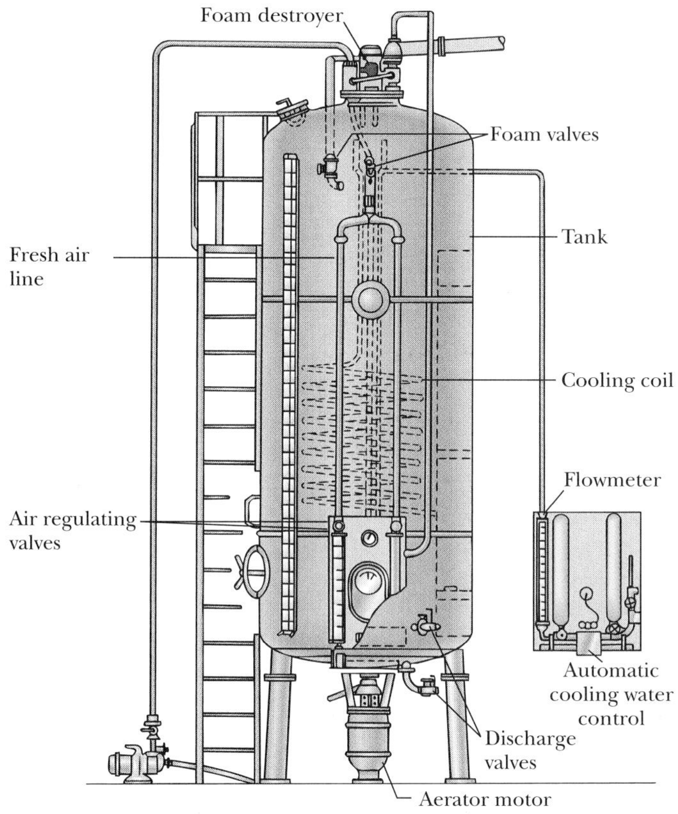

***Figure* 32.8** A submerged system for vinegar production. The Frings acetator is a batchwise fermentation.

gar is acetic acid, but distinct flavors are contributed by the original source of the ethanol. The most widely used types of vinegar include the following:

Distilled white vinegar—made from distilled ethanol. The origin of the ethanol may be fermentation ethanol or it may be obtained by chemical synthesis.
Cider vinegar—made from the fermented juice of apples.
Wine vinegar—made from a low-quality wine that has been subjected to aerobic oxidation. Either white or red wine can be converted to vinegar.
Malt vinegar—produced from the alcohol obtained by fermenting corn or barley starches that have been treated with enzymes to release sugars for fermentation.

Nucleotides

For centuries, the Japanese have used dried seaweed, dried bonita, or other fish as flavoring agents. Two of the major ingredients in these dried products that enhance the flavor of the food to which they were added are monosodium glutamate and the histidine salt of inosinic acid. Inosinic acid is a nucleotide, and the production of this and other nucleotides will be discussed here. Monosodium glutamate fermentations will be discussed in the section with other amino acids. As **flavor enhancers** the descending order of effectiveness among the nucleotides is guanylic acid, inosinic acid, and xanthylic acid. Nucleotides are added to enhance the flavor of soups and sauces and are effective at very low concentrations (0.005 percent to 0.01 percent). Production of 5′ IMP and 5′ GMP is about 4000 tons/year. Nucleosides and nucleotides produced by microorganisms are also chemically altered and employed for chemotherapeutic purposes.

The nucleotides were originally produced by enzymatic hydrolysis of yeast RNA. *Candida utilis* is rich in RNA and was one of the sources. Over half the commercial production of nucleotides is now by fermentation. The production of 5′ IMP is by direct fermentation or through the microbial production of inosine that can be chemically phosphorylated. Strains of *Brevibacterium ammoniagenes* that are auxotrophs for adenine and guanine will produce 30 grams/liter inosine. Some microorganisms can produce 5′ IMP by direct fermentation but must have an altered membrane permeability. Nucleosides (no phosphoryl group) will traverse a normal cytoplasmic membrane but nucleotides (phosphorylated) will not. Selected permeability mutants of *B. ammoniagens* can produce about 27 grams/liter 5′ IMP.

The production of 5′ GMP is generally by a combination of fermentation and chemical synthesis. A purine auxotroph of *Bacillus megaterium* is employed that accumulates the intermediate in the purine biosynthesis, aminoimidazole carboxyamide ribose (AICAR). The *B. megaterium* strain will produce 16 grams/liter AICAR (Figure 32.9). Guanosine-excreting mutants are also employed in commercial production of 5′ GMP.

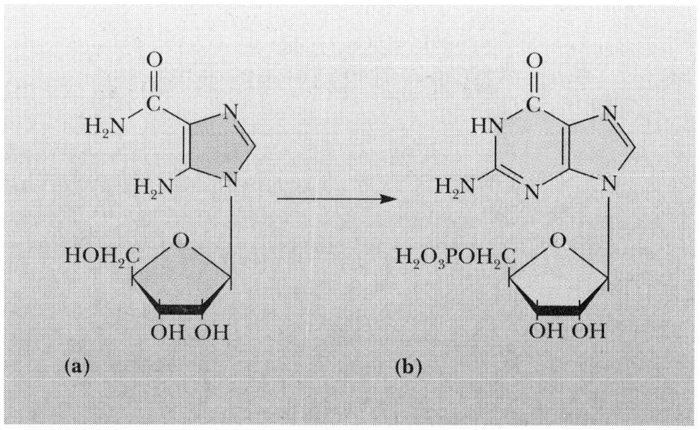

***Figure* 32.9** An example of a combined microbiological/chemical process in generating a commercial product. AICAR **(a)** aminoimidazole carboxamide ribose—an intermediate in purine biosynthesis—is produced microbiologically and transformed chemically to 5′ guanosine monophosphate **(b).**

Table **32.4 Uses of amino acids**

Sodium Glutamate	Flavor Enhancer
L-lysine L-methionine L-threonine L-tryptophan	Added to plant-derived foods to improve nutritional quality
L-tryptophan L-histidine	Antioxidant in powdered milk
Aspartate Alanine	Improve flavor of fruit juices
L-aspartyl-l-phenylalanine Methyl ester (aspartame)	Low-caloric sweetener
Mixed amino acids	Infusion solutions for surgery patients

Amino Acids

Amino acids are a major industrial product with an annual worldwide production of over 400,000 tons. The manufacturing processes that are involved in amino acid production generally originated in Japan. Amino acids are used as food additives, in medicine, or as a starting material in chemical synthesis. The major amino acid produced is glutamic acid at 80 percent with lysine second in volume at 10 percent of the total. Some of the applications for amino acids are presented in Table 32.4. The biologically active form of amino acids is the L-form and is the isomer produced by fermentation. Chemical synthesis yields a mixture of the D and L isomers.

Amino acids are synthesized in families based on the precursor metabolite that serves as the starting material (see Chapter 10). The precursor metabolite is an intermediate in the tricarboxylic acid cycle, produced during glycolysis or the pentose phosphate pathway. The synthesis of individual amino acids is tightly regulated, and the amount made corresponds to the amount necessary to synthesize cell protein. The amount of any amino acid synthesized during growth is generally regulated by end product inhibition (see Chapter 13) of enzyme activity or repression of enzyme synthesis. The first step in the biosynthesis of a particular amino acid may be inhibited by the presence of that amino acid. Another control mechanism that regulates amino acid biosynthesis is repression. Repression is an interaction of the product (amino acid) with a repressor protein and this, in turn, interacts with the operon. The genetic information not transcribed because the operon is blocked is responsible for the production of m-RNA involved in synthesis of that amino acid. If one wishes to obtain a culture that will produce excessive amounts of an amino acid, then both feedback inhibition and/or repression must be overcome. Otherwise, presence of modest levels of the amino acid will inactivate the enzymatic machinery involved in its production. Through mutation and selection, microorganisms are available that are efficient producers of most of the common amino acids.

Some of the strains that have been commercially developed are presented in Table 32.5. The major glutamic

Table **32.5 Commercially produced amino acids and the microbes involved in their synthesis**

Amino Acid	Organism	Approximate Yield Grams/Liter	Carbon Source
L-glutamic	*Corynebacterium glutamicum*	> 100	Glucose
L-lysine	*Corynebacterium*	39	Glucose
L-lysine	*Brevibacterium flavum*	75	Acetate
L-threonine	*Escherichia coli* K_{12}	55	Sucrose

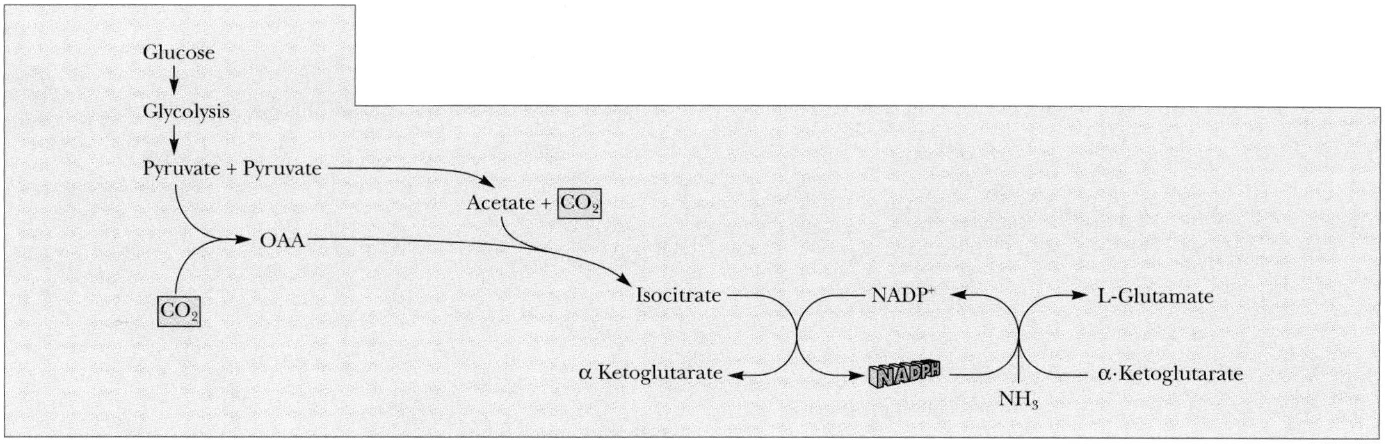

***Figure* 32.10** The basic reactions involved in the synthesis of glutamic acid.

acid–producing organisms have a requirement for biotin and lack significant levels of α-ketoglutarate dehydrogenase. When provided with low levels of biotin, the organism synthesizes leaky and inadequate cytoplasmic membranes (biotin is involved in synthesis of fatty acid). These leaky membranes permit the overproduced glutamic acid to be excreted into the medium. The anaplerotic reaction with pyruvate + CO_2 produces malate to replenish the 4-carbon intermediate involved in the tricarboxylic acid cycle. This 4-carbon intermediate is normally generated by α-ketoglutarate dehydrogenase in aerobic bacteria. The basic reactions involved in glutamic acid synthesis are presented in Figure 32.10.

L-lysine can also be produced in high yield by direct fermentation. The organism responsible is a mutant of *Brevibacterium flavum* that requires homoserine and leucine. Lysine is excreted into the medium by active transport.

Vitamins

Mutation and selection can lead to microbial strains that overproduce any of the B vitamins. Only two of the B vitamins, riboflavin and Vitamin B_{12}, are produced by fermentation, as chemical synthesis of the others is less costly. The structures of Vitamin B_{12} and riboflavin are presented in Figure 32.11. Vitamin B_{12} is a complex molecule that, interestingly, is only synthesized by procaryotic microorganisms. It is required by humans and supplied in food or is produced by the normal intestinal microbiota. Pernicious anemia occurs in humans that obtain insufficient levels of dietary B_{12}.

The annual current production of Vitamin B_{12} is about 10,000 kg. Unpurified B_{12} is added to swine and poultry feed at about 12 mg/ton, and this figure affirms that B_{12} is active at exceedingly low supplement levels. Vitamin B_{12} is synthesized in bacteria from intermediates common to heme or chlorophyll synthesis (see Chapter

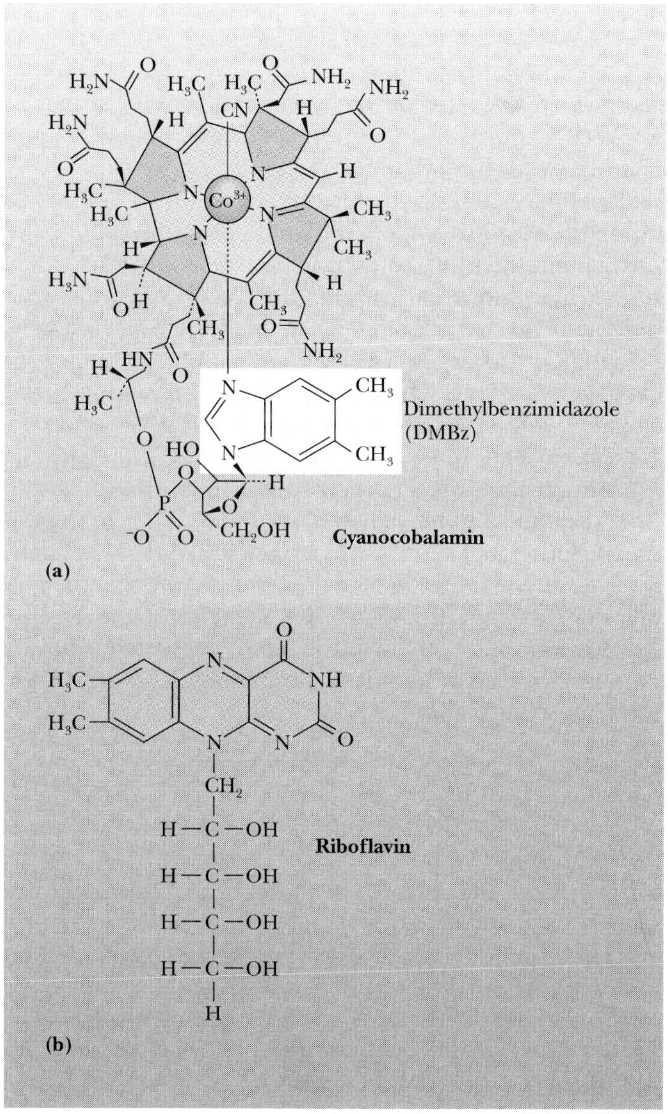

***Figure* 32.11** The structures of **(a)** Vitamin B_{12} and **(b)** riboflavin, two members of the vitamin B complex produced commercially by fermentation.

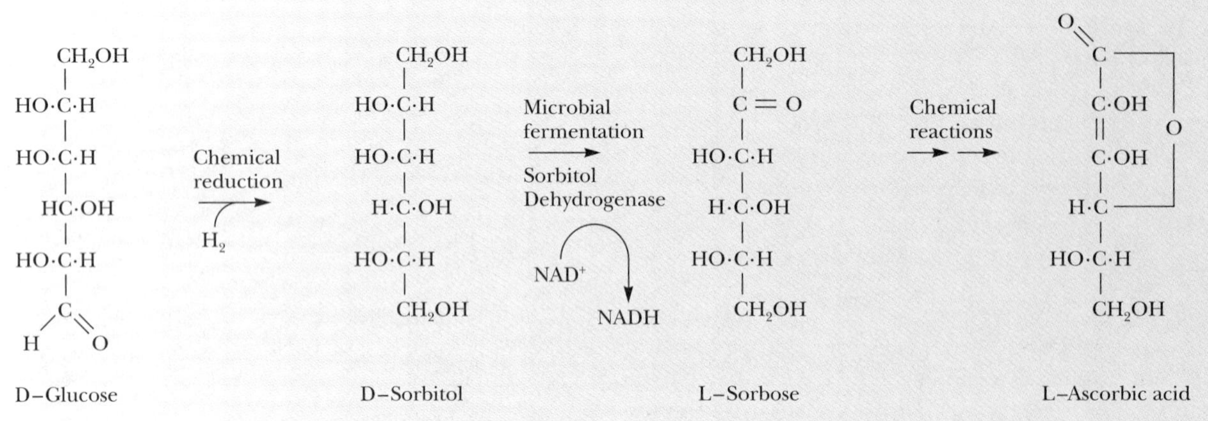

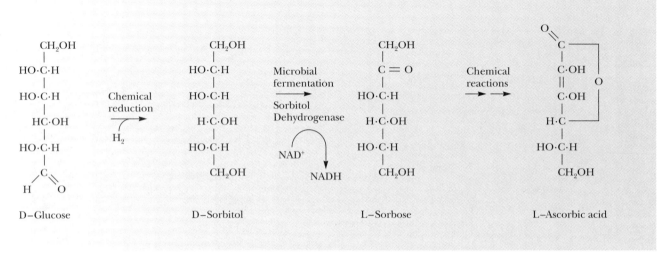

***Figure* 32.12** Production of Vitamin C (ascorbic acid) by a combination of fermentation and chemical reactions.

10). The organisms involved in Vitamin B_{12} production are *Propionibacterium shermanii* and *Pseudomonas denitrificans*. The latter produces about 60 mg/liter under appropriate growth conditions.

Riboflavin is present in milk, eggs, meat, and other food. It is added as a supplement to bread and milk to improve their nutritional quality. Yeasts such as *Ashbya gossypi* can be employed in the commercial production of riboflavin, and the yield can be in excess of 7 grams/liter.

Vitamin C (ascorbic acid) is commonly taken by itself or in vitamin preparations. It is also employed as an antioxidant in food manufacture. World production by a combination of fermentation and chemical synthesis exceeds 35,000 tons per year. The starting substrate, D-glucose is reduced chemically to D-sorbitol. A fermentation step leads from D-sorbitol to L-sorbose, and this is then chemically converted to ascorbic acid (Figure 32.12). The microbe involved in the dehydrogenation is *Acetobacter suboxydans*.

Beverages Containing Alcohol

Consumption of **alcohol-containing beverages** such as beer and wine is as old as civilization itself. Fermented alcoholic beverages are produced worldwide from a variety of plant products that contain utilizable carbohydrates. Sugars from fruit or starch and polysaccharides from grain are major sources of fermentable sugar. A list of alcoholic beverages consumed in various areas of the world is presented in Table 32.6. Wine, beer, and distilled beverages are the major alcoholic beverages produced commercially. Yeasts of the genus *Saccharomyces* are the major organisms involved in alcoholic fermentations.

Honey was gathered by the earliest recorded civilizations, and it is safe to assume that some of it became fermented when diluted with water. This fermented honey is considered to have been the first consumable alcoholic beverage. Mead is the name of the alcoholic product of honey fermentation, and the word "mead" is the oldest word associated with drinking. It has been traced to the early Sanskrit language.

Wine

There are many types of **wine** (grape, peach, pear, dandelion, etc.), and each is a product of an ancient art. Grape wines are the most widely produced and are divided into red and white varieties. Other wines are made by variations and refinements in the production of these two wine types. Wine is manufactured in many areas of the world, but Europe is the largest producer with well over half the worldwide production. Italy, Spain, Germany, France, and Portugal are major producers because the climatic conditions there are favorable for the growth of quality grapes. Over 20 percent of the worldwide wine production occurs in North and South America. New York, California, and

***Table* 32.6 Major alcoholic beverages and sources of sugars for fermentation**

Wine	Fruit sugars
Beer	Barley (sometimes rice or corn)
Whiskey	Rye, corn, barley
Tequila	Agave cactus
Rum	Sugar cane molasses
Mead	Honey
Sake	Rice
Gin	Corn, other grain
Vodka	Corn, potatoes

the state of Washington are favorable North American areas for growth of grapes and wine production. The remainder is produced in Australia, Asia, and Africa.

White wine is made from white grapes or can be made from red grapes provided the skins are removed from the **must** (must is crushed grapes ready for fermentation). Red wines gain the red pigmentation when they are partly fermented (three to ten days) in the presence of the red skins. During this period the alcohol formed extracts and solubilizes anthrocyanin pigments present in the grape skin. Rose (pink) wines are made from pink grapes or more often from red grapes, but the time of exposure to the skins is for 12 to 36 hours, and much less of the anthrocyanin pigments is solubilized. A dry wine is one in which most of the sugar has been fermented to ethanol. A sweet wine has more residual sugar. A fortified wine such as sherry or port contains higher levels of ethanol, and this is attained by adding ethanol distilled from other wine or from fermented grain.

Yeasts are part of the natural microbiota of grapes. Crushed grapes can undergo a natural fermentation process that eventually produces wine. Natural fermentations are not favored commercially because the natural microbiota can be inconsistent, and, as a result, the wine produced may not be potable. In a commercial process, the grapes are crushed and the must is treated with sulfur dioxide (SO_2) or sometimes pasteurized to destroy the natural microbiota. The must is then inoculated with a proprietary culture of *Saccharomyces ellipsoides* to bring about the desired fermentation. Laboratory-bred yeasts are favored because they have been selected to tolerate 12 percent to 15 percent ethanol, whereas the natural microbiota will cease fermenting when the ethanol content reaches 3.5 percent to 4 percent. Quality wine contains 8 percent to 12 percent ethanol.

Flow charts for red and white wine fermentations are presented in Figure 32.13. White wine fermentations are quite direct. The grapes are crushed, destemmed, and skins and solids (pomace) removed. The must is treated with SO_2 and yeast is added. After fermentation the wine is aged in casks, tanks, or bottles. During a process called racking, red wine is drawn from the primary fermentation vat, placed in casks, and stored at a low temperature. During the lengthy aging process for some of the quality red wine a second fermentation occurs. This is termed the malolactic fermentation. Residual malic acid in the wine is fermented by lactic acid bacteria and converted to lactic acid and CO_2. This lowers the pH of the wine and improves the quality. After aging, wines are clarified in a process called fining. Fining can be accomplished by filtering through casein, diatomaceous earth, or bentonite. A few wines are clarified by centrifugation.

Wine fermentations may be accomplished in tanks holding from 50 to 50,000 gallons, but it is essential that temperature of the fermentation be controlled. Heat generated during metabolism can raise temperatures above the tolerance level of the yeast. Temperature should be maintained at 20 to 24°C to produce quality wine. The aging process is usually much longer for red wine than for white wine. This aging is done at lowered temperature to improve flavor and aroma. Sparkling wines such as cham-

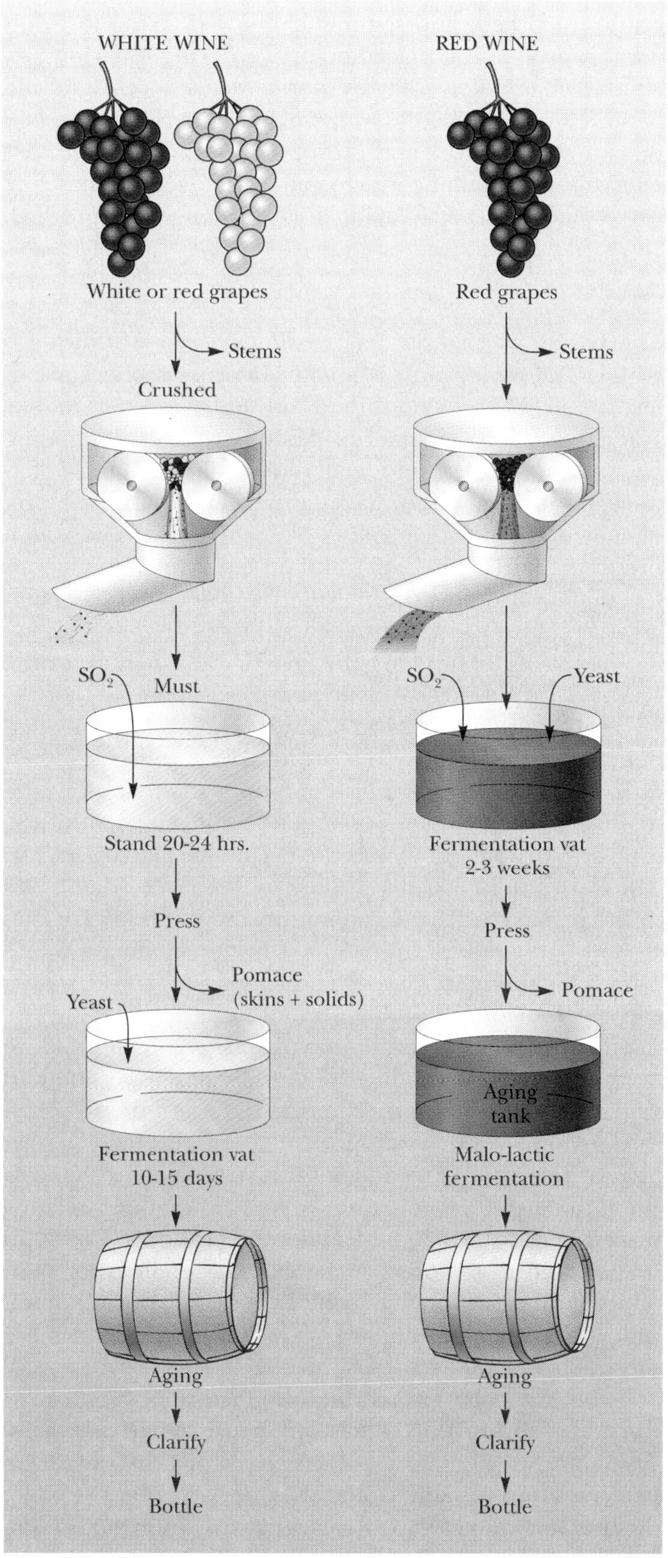

Figure 32.13 Flow charts for the production of red and white wines.

pagne are aged in bottles with a secondary yeast fermentation that generates the carbonation (bubbles). Some champagnes are carbonated by CO_2 injection after the fermentation. Champagne is aged in bottles that are slanted so the cork faces down, allowing sediments to gravitate to the neck of the bottle. Periodically the neck is frozen, the bottle uncorked, and sediment removed. Wine lost during this process is replaced before recorking. All wines, including champagne, that are stoppered with a cork should be stored with the bottles lying on their side. This will retain moisture in the cork and prevent air from entering. Presence of air would encourage growth of bacteria that convert ethanol to acetic acid, thus "spoiling" the product.

Beer

The basic raw material for brewing of beer worldwide is barley. The barley must be malted, a process whereby barley starch is converted to a substrate (sugar) that is amenable to fermentation. Malt beverages—including beer, stout, porter, and malt liquor—are made by fermenting malted barley. Sake, an alcoholic beverage produced in Japan, is a product of malted rice. The starch present in barley, rice, and other grains is not available for fermentation unless the polymer is hydrolyzed to generate low molecular weight sugars. Ethanol-producing microorganisms do not have the genetic information for the synthesis of polysaccharolytic enzymes.

The malting process is a series of manipulations that depolymerize starch to free sugars. These sugars are then fermented under anaerobic conditions to ethanol. In the malting process, barley is steeped in cold water until the grains absorb sufficient water to germinate (Figure 32.14). The wet grain is placed in rolling tubs or large open boxes in the presence of air. Temperature and humidity are controlled to initiate germination. During germination, enzymes, including amylases and proteinases, are activated, causing the release of sugars and amino acids from the malting grain. The germination process is halted before sprouts develop by drying the germinated grain in a kiln with a flow of warm air. The dried grain is then broken into small pieces in a grinding mill. Adjuncts such as corn grits or rice are often added at this stage and subjected to the milling process to expose starch. The milled grain is placed in the mashing tank and water is added. The mash is subjected to periods of warming and cooling to room temperature to extract all soluble material. During mashing, the proteinases and starch-hydrolyzing enzymes are most active. Following mashing, the solids are removed, and the liquid portion is the fermentable portion called wort. The wort is cooked in a brew kettle to halt enzymatic action and to destroy any microorganisms that might be present. Hops are added, and the wort is heated to extract tannins and other flavoring agents from the hops. The wort is sterilized by live steam and filtered to remove hops

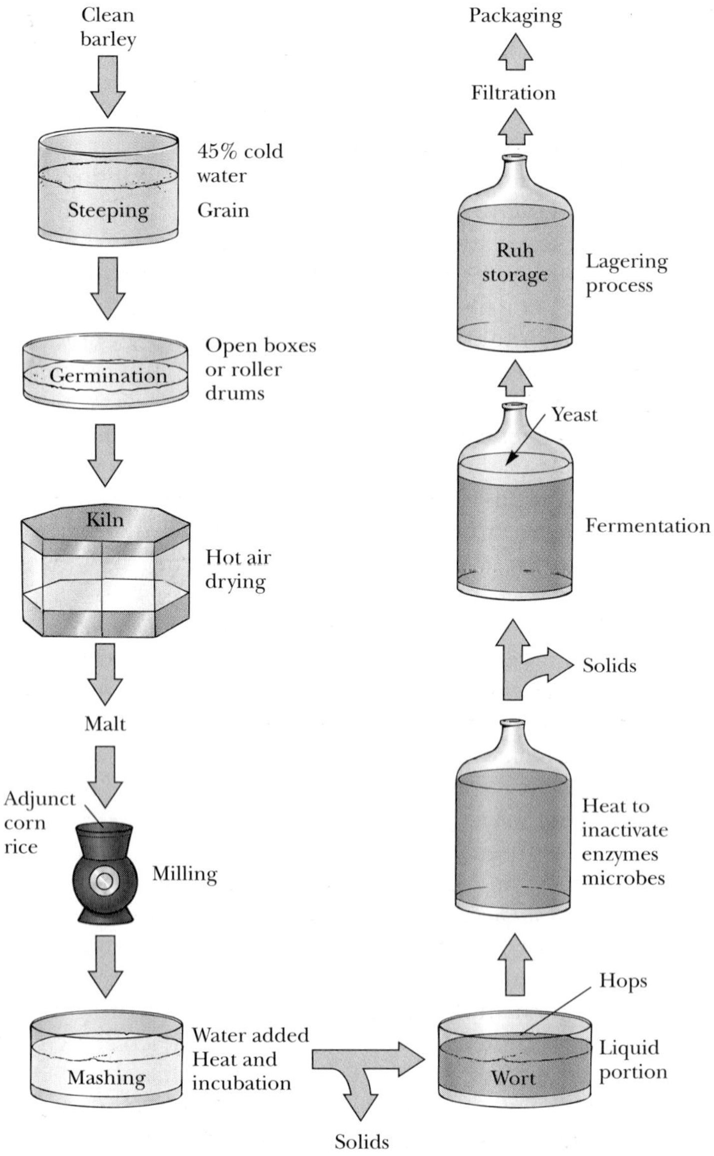

***Figure* 32.14** The steps involved in the manufacture of lager beer.

and any precipitates. The wort is then transferred to the fermentation vessel.

Quality beer production depends on careful attention to the details involved in the preparation of wort. During these processes, pH, length of heating/cooling cycles, removal of precipitates, and other factors are important in producing a desirable product. Another major concern in beer manufacture is the quality of added water. It must be low in carbonates and calcium and properly balanced in other minerals. The brewing industry developed historically in areas where quality water was available, such as Pilsen, Czechoslovakia; Munich, Germany; and Dublin, Ireland.

After the wort has been cooled and filtered, it is diluted with water and brewery yeast is then added. The fermentation may be carried out by a top-fermenting yeast (*Saccharomyces cerevisiae*) or a bottom-fermenting yeast (*Saccharomyces carlsbergensis*). The top fermenter is carried to the top (distributed throughout) by the CO_2 produced, whereas the bottom fermenter tends to settle to the bottom. The top fermentation is carried out at 14 to 23°C for five to seven days, and the product is ale. Bottom fermentation is carried out at 6 to 12°C and produces beer.

Following the fermentation, beer is pumped into a large tank for maturation. Yeast is added at about one million cells per ml, and the beer is allowed to "rest" for several weeks at 0.1°C. This process is called the *ruh* (German, meaning "rest") or lagering. Most European and American beer is lager beer. Ale is stored for aging at 4 to 8°C.

The finished beer is clarified by centrifugation or filtration, CO_2 is added, and the beer is packaged. Draft beer (kegs, bottles, or cans) is generally passed through membrane filters to remove all microorganisms. Prior to the availability of effective filter systems, bottled beer was pasteurized to prevent spoilage by acetic acid bacteria, and this pasteurization produced off-flavors. Draft beer was superior because it was unpasteurized but had to be kept cold to prevent spoilage.

Distilled Beverages

Distilled spirits are an extension of the brewing process. The fermentation liquid is heated in a container attached to cooling coils, and the volatiles are condensed and collected. Since ethanol has a boiling point of 78.5°C it can be readily concentrated from the fermentation medium by distillation. This yields a product with a higher alcohol content than can be achieved by fermentation.

There are a number of distilled alcoholic beverages, including gin, vodka, whiskey, rum, and brandy. Gin and vodka are essentially distilled grain alcohol. Gin is produced when the alcohol is refluxed over juniper berries to extract their distinct aroma and flavor. Rum is manufactured by distilling fermented cane sugar molasses, and brandies are made from distilled wine. Whiskey is defined by the grain employed in the fermentation. Rye whiskey must have 51 percent or more rye grain in the malting process, and bourbon must have at least 51 percent corn.

Scotch and some bourbons are made by a **sour mash** fermentation. A sour mash is produced by introducing a homolactic bacterium into the mash (see above), and growth of the lactic acid bacteria lowers the pH to about 4.0. The grain employed in producing Scotch is mostly barley. The distillate from any malted grain fermentation is colorless. It does contain volatile products other than ethanol that are generated during the mashing process or fermentation. These volatile products are responsible for the differing flavors among the various whiskeys. The brownish color of the final product results from the aging process. Whiskey is aged in wooden barrels that have been charred on the inside. During aging, desirable flavors are enhanced, and the level of some less desirable substances such as fusel oils may be decreased. Fusel oils are mixtures of propanol, 2-butanediol, amyl alcohol, and other alcohols. Whiskey manufacture is an ancient art and involves little science. The manufacture of whiskey (spelled whisky in Scotland) apparently originated among the Highland Scots and moved from there to other areas of the world. Most Scotch whisky sold in the United States is a blend of the product of several different fermentations. A single-malt Scotch is produced in one fermentation. It is interesting to note that there are over 300 different single-malt Scotches produced in Scotland.

Organic Acids

Citric acid is an important **organic acid** synthesized by a microbial fermentation. Over 130,000 tons are produced worldwide each year. Originally, citric acid was extracted from citrus fruit (a lemon is 7 percent to 9 percent citric acid). The market for citric acid obtained by extraction was controlled by cartels established in citrus-growing areas, and the price was high. In 1923 a microbial fermentation was developed that yielded citric acid, and the price plummeted as availability increased. About two thirds of the citric acid produced commercially is used in foods and beverages. It is present in soft drinks, desserts, candies, wines, frozen fruit, jams/jellies, and ice cream. In the pharmaceutical industry, iron citrate is a source of dietary iron. Citric acid is also used as a preservative for stored blood, tablets, and ointments. Citrate is also replacing polyphosphates in detergents due to the banning of phosphates as an environmental hazard.

Citric acid was first produced microbially by a surface fermentation. The nutrient medium was placed in open pans, inoculated with the appropriate fungus, and the mycelial mat floated on the surface. Today, most citric acid comes from a submerged fermentation in large, stainless steel fermenters. The organisms responsible for production are strains of *Aspergillus niger*. Sucrose is generally the substrate, and the citric acid production occurs during the idiophase. Citric acid is a secondary metabolite, and the fermentation would follow the pattern presented in Figure 32.1. During the trophophase, mycelium is produced and CO_2 is released. During the idiophase, glucose and fructose are metabolized directly to citric acid; little CO_2 is produced. During a properly operated fermentation about 70 percent of the sugar is converted to citric acid. A critical factor in the fermentation is the level of iron and other minerals. If copper and iron are present in other than low levels, the production of citric acid is adversely affected. The fermenter should be stainless steel or glass lined, and copper fittings must be avoided.

There are a number of organic acids synthesized in small quantity by microbes and, in fact, fungi will produce an array of these and other metabolites from sugar. The problem in producing these by fermentation is twofold: no real market exists for these products, and chemical synthesis is less costly. One compound that is produced by fermentation is **itaconic acid,** which is used in methacrylate resins as a copolymer. Adding 5 percent of this product to printing ink increases the ink's ability to adhere to paper. It also has potential as a detergent. The organism involved in production of itaconic acid is a strain of *Aspergillus terreus.* In submerged fermentation, the yield from beet molasses is 85 percent theoretical. The chemistry of production is illustrated in Figure 32.15.

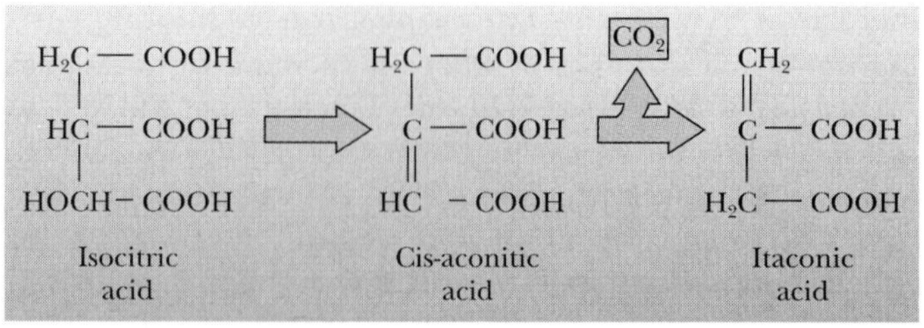

***Figure* 32.15** The biochemical reactions involved in the synthesis of itaconic acid.

Enzymes, Microbial Transformations, and Inocula

In a previous section we discussed the enzyme rennin, employed in cheese manufacture. This is but one of the many applications of commercially produced enzymes. Macromolecules are composed of monomers that are of potential value, and selective hydrolytic enzymes are available that cleave them from macromolecules. Microorganisms can also transform large molecules of little value to ones of considerable value and at low cost.

Enzymes

Enzymes are a valuable industrial product, and most commercial enzymes are obtained from microorganisms. In nature, microorganisms utilize enzymes to disassemble proteins, starch, pectin, lipids, and other insoluble large molecules to produce monomers that can serve as a carbon and/or energy source (see Chapter 12). These hydrolytic enzymes are generally extracellular or at the cell surface and are produced by both bacteria and fungi. Extracellular enzymes can be recovered from the growth medium and have found many applications in commercial processes. Several of these are outlined in Table 32.7. An advantage of microbial enzymes over chemical agents is specificity and their activity at modest temperatures and pH. Chemical hydrolysis of glycosidic or peptide bonds requires much more drastic conditions. The production of enzymes by microorganisms can be enhanced by genetic manipulation, and strains have been obtained where 25 percent of the total cell protein is a specific enzyme.

The production of bacterial proteases is a leading industrial process in both bulk and value. Over 500 tons of these enzymes are produced each year and utilized in the manufacture of detergents and cheese (rennin). The en-

***Table* 32.7 Microbial enzymes and their commercial application**

Enzyme	Genus of Producer	Use
Bacterial proteases	*Bacillus* *Streptomyces*	Detergents
Asparaginase	*Escherichia* *Serratia*	Antitumor agent
Glucoamylase	*Aspergillus*	Fructose syrup production
Bacterial amylases	*Bacillus*	Starch liquefaction, brewing, baking, feed, detergents
Glucose isomerase	*Bacillus* *Streptomyces*	Sweeteners
Rennin	*Alcaligenes* *Aspergillus* *Candida*	Cheese manufacture
Pectinase	*Aspergillus*	Fruit juice clarification
Lipases	*Micrococcus*	Cheese production
Penicillin acylase	*Escherichia*	Semisynthetic penicillins

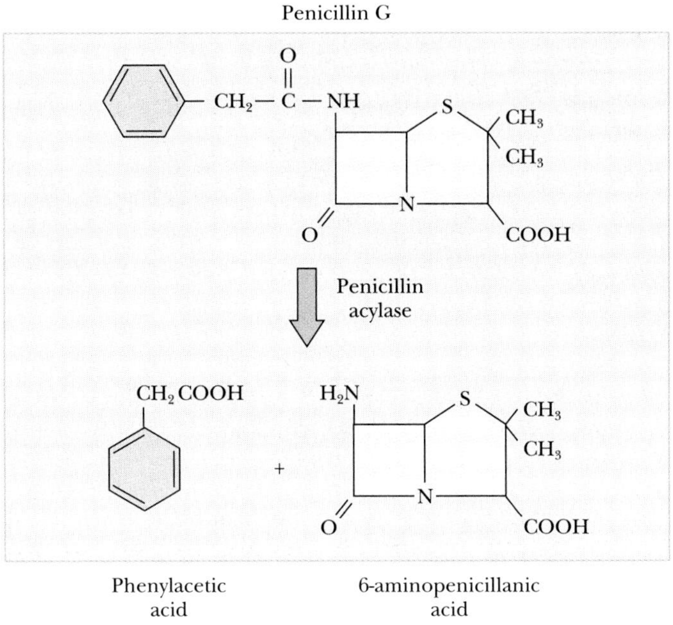

***Figure* 32.16** Cleavage of penicillin G generated by fermentation to phenylacetic acid and 6-aminopenicillanic acid. The latter can be chemically modified to produce effective antibacterials.

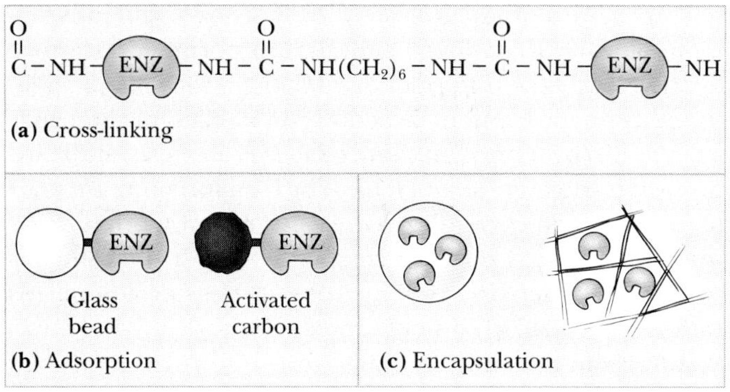

***Figure* 32.17** Immobilization of enzymes: **(a)** cross-linking to glutaraldehyde, **(b)** adsorption to inert particles, and **(c)** encapsulation by trapping in a matrix.

zymes used in detergents are mostly proteases from selected strains of *Bacillus amyloliquefaciens.* The proteases are added to detergents to promote the solubilization of stains present in laundry. A significant problem with the addition of proteases to detergents has been the development of allergic responses among users. This has been overcome by microencapsulation and other techniques that produce dustless protease preparations.

The amylases are enzymes that hydrolyze starch. There are a number of amylases, and they hydrolyze starch in different ways to short-chain polymers (dextrins) and maltose. Other enzymes hydrolyze the dextrins and maltose to glucose. The α-amylases are enzymes that split internal α-1,4 glycosidic bonds. These enzymes are produced commercially by members of the endospore-forming genus *Bacillus.* Fungi such as *Aspergillus* sp. also produce these enzymes. Glucoamylases split glucose from the nonreducing end of starch. These enzymes are in commercial demand for the production of fructose syrups. Fructose is sweeter than glucose and is the favored sugar in soft drinks and in syrups. *Aspergillus niger* is a producer of glucoamylase. Conversion of glucose to fructose can be accomplished with a glucose **isomerase** from bacteria. The most important of the glucose isomerase producers are *Streptomyces* and *Bacillus coagulans.*

Asparaginase is employed medically in the treatment of some leukemias and lymphomas. The metabolism of these tumor cells requires L-asparagine, and the enzyme asparaginase cleaves asparagine to aspartic acid and ammonia. This starves the tumor cell, thus leading to its destruction. *Penicillin acylases* are produced by bacteria and fungi. The enzymes cleave penicillin to 6-aminopenicillanic acid and phenylacetic acid (Figure 32.16). Penicillin produced by fermentation can be cleaved to obtain the 6-aminopenicillanic acid moiety, and substitution of this molecule yields the various semisynthetic penicillins now in widespread use (see later discussion of antibiotics).

Enzyme preparations have a relatively short shelf life at room temperature and must be stabilized if they are stored for any period of time. A soluble enzyme can be stabilized in several ways. Adding substrate will stabilize some enzymes; adding low levels of organic solvent such as acetone will stabilize others. Cation additions restrict some enzymes to a tertiary configuration and increase their stability. In general, the most effective means of stabilization is through immobilization. There are three major ways that an enzyme can be immobilized: (1) cross-linking, (2) chemically bonding or adsorption to an inorganic or organic carrier, or (3) encapsulation in a gel or polymer (Figure 32.17).

Cross-linking of an enzyme involves the linkage of functional groups of the enzyme with a cross-linking reagent. A common reagent is glutaraldehyde, and this reacts through the amino groups on the enzyme. Enzymes may be adsorbed on or bonded to glass, activated carbon, starch, or other carriers. They may also be bonded to ionic exchangers. Microencapsulation places the free enzyme (or free cells) inside a semipermeable membrane. Cellulose triacetate can be employed as a fibrous network to entrap an enzyme.

Microbial Transformations

Microorganisms may have relatively nonspecific enzymes at or near their outer surface that will alter compounds they contact even though the microorganism cannot utilize the altered product. Monooxygenases are typical of this type of enzyme. An example of a **biotransformation** would be the hydroxylation of benzene to phenol by microorganisms that have a monooxygenase; generally, microorganisms that effect this biotransformation cannot further catabolize the phenol.

A dramatic event in the course of medicinal chemistry was the discovery that microorganisms can successfully hydroxylate the steroid nucleus. The anti-inflammatory activity of a sterol depends on the presence of an oxygen or hydroxyl in the 11-position (Figure 32.18). Prior to the discovery that microbes could hydroxylate the steroid nucleus with a virtual stoichiometric yield, the hydroxylation was accomplished chemically. This simple, low-cost microbial hydroxylation had a marked effect on the cost and availability of biologically active sterols. A complete process for the manufacture of biologically active sterols is outlined in Figure 32.19. The starting material, diosgenin, is a constituent of a yam that is grown in Mexico.

The versatility of microorganisms is apparent from data presented in Figure 32.18, as species are available that can hydroxylate the sterol nucleus in any of the positions as shown. Naturally occurring hormones such as progesterone, testosterone, estrone, and those from the adrenal cortex differ in number, type, and location of substituent groups or double bonds. Synthetic hormones that differ from the naturally secreted hormones can readily be synthesized for medical uses. Many derivatives have been synthesized, and the uses of synthetic sterol products are presented in Table 32.8.

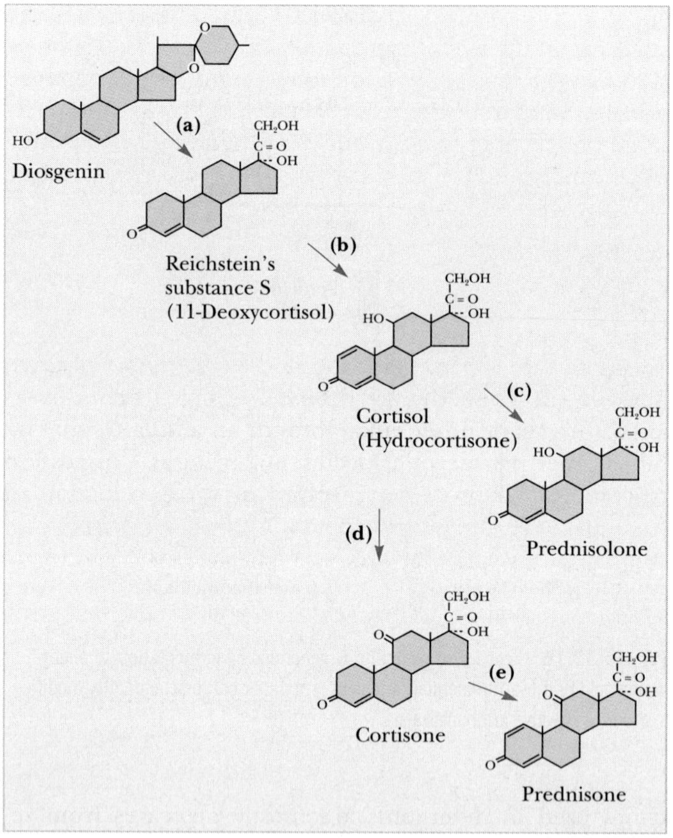

Figure **32.19** The conversion of diosgenin, a plant sterol, to medically useful anti-inflammatory agents: **(a)** chemical conversions, **(b)** 11 β hydroxylation by *Curvularia lunata,* **(c)** dehydration by *Corynebacterium simplex,* **(d)** a chemical oxidation, and **(e)** dehydration with *C. simplex.*

Inocula

Bulk microbial cell mass is an industrial product that can be employed as inocula in a number of processes. Lactic acid bacteria are grown in bulk by fermentation companies and sold to cheese makers and meat companies. As previously mentioned, pasteurized milk is the starting material for cheese production. Use of pasteurized milk is essential in preventing the spread of milk-borne diseases.

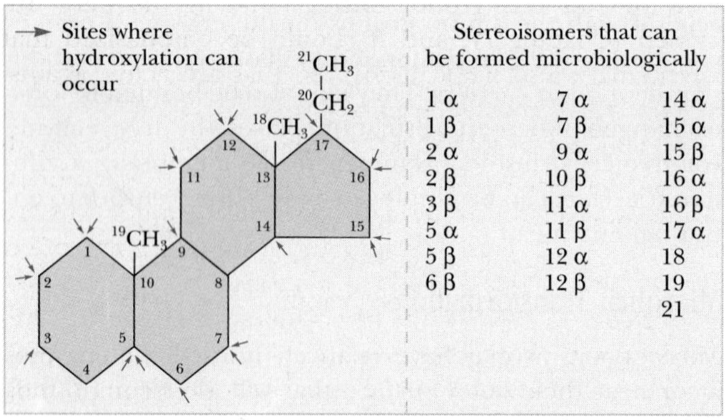

Figure **32.18** Hydroxylations of the sterol nucleus that can be accomplished by selected microorganisms. There are two sterioisomers possible in many of the positions, and these are designated α or β.

Pasteurization also lowers the natural populations that may be less desirable, permitting the more desirable species in the inoculum to become dominant. Manufacture of summer sausage and other fermented meats can also be achieved best by inoculating with the appropriate bacterium.

Soils in which leguminous crops are planted often have a low population of symbiotic nitrogen-fixing bacteria. Therefore planting leguminous crops in these soils will

Table **32.8 Uses of synthetic sterols**

Anti-inflammatory agents
Sedatives
Antitumor agents
Dermatology
Ocular disease
Cardiovascular therapy
Oral contraceptives

not lead to significant symbiotic nitrogen fixation. Inoculation of seeds with a compatible strain of *Rhizobium* or *Bradyrhizobium* prior to planting can significantly enhance root nodulation and nitrogen fixation. The symbiotic nitrogen fixers are grown commercially and sold as inocula. Experimental evidence has proven that this is a worthwhile practice.

Inhibitors

Microbes can produce catabolites that are antagonistic to other biota. These may be a **biocide** (toxin) that kills selected insects or other eukaryotes or an **antibiotic** that inhibits other prokaryotic species. Following is a discussion of commercial processes involved in the production of some of these antagonistic compounds.

Biocides

The most successful commercial product generated by a microorganism that is employed in insect control is a protein synthesized by *Bacillus thuringiensis.* This protein appears as a parasporal crystal during sporulation and is released along with the spore on disintegration of the progenitor (Figure 32.20). The vegetative cell of *B. thuringiensis* has little toxicity, but the protein crystal is effective in killing lepidopteran insects. It is most effective in insects that have an alkaline pH in their midgut, as the protein is solubilized at a high pH and the toxin is released. The proteins from ingested spores are cleaved by gut proteases, and these polypeptide toxins destroy gut epithelial cells. Gut contents are then released into the blood of the insect. This results in paralysis and death of the insect. The toxin is harmless to any life form that has a gut with a neutral or acid pH.

Bacillus thuringiensis grows well in submerged fermentations and, under appropriate conditions, will produce spores and the parasporal body in 30 hours. The spore/crystal is incorporated into an inert carrier and marketed as an insecticide to be dusted on plants. The active compound is naturally occurring and therefore degraded by soil microbes. Unlike many chemical insecticides, it is not an environmental hazard.

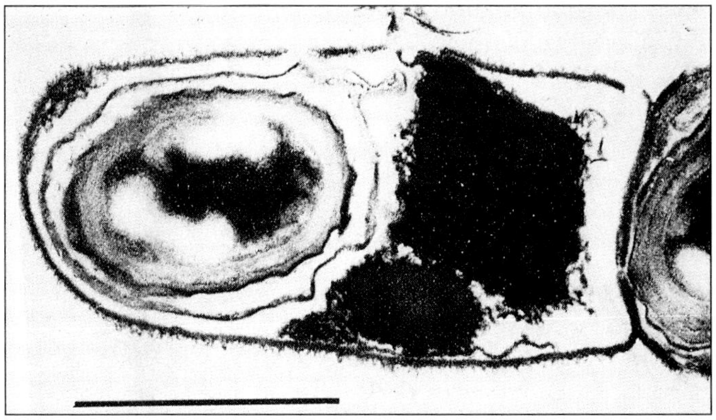

***Figure* 32.20** The spore and parasporal crystal produced by *Bacillus thuringiensis.* The crystal is an effective insecticide. (Courtesy of Art Aronson)

***Table* 32.9 Antibiotics that are effective against eukaryotic cells**

Producing Organism	Substance Produced
Antifungals	
Streptomyces griseus	Cycloheximide
Streptomyces noursei	Nystatin
Penicillium griseofulvum	Griseofulvin
Streptomyces nodosus	Amphotericin B
Antitumor	
Streptomyces peucetius	Daunorubicin
Streptomyces antibioticus	Actinomycin C
Streptomyces caespitosus	Mitomycin C
Streptomyces verticillus	Bleomycin
Antihelminth	
Streptomyces avermitilis	Avermectin

Antibiotics

The use of antibiotics in controlling microbial growth has been discussed (Chapter 7). Approximately 10,000 different **antibiotics** have been characterized, but fewer than 100 of these are of value and in commercial production. Antibiotics are produced mainly by fungi and Eubacteria. The useful antibiotics are mostly produced by a large group of filamentous soil bacteria known collectively as the actinomycetes. **Streptomyces** is the genus among the actinomycetes that has proven to be the most prolific producer, and the search for commercially valuable antibiotics has centered on this group. Antibacterial agents were discussed (Chapter 7), and it should be emphasized that agents have also been discovered that are active against some fungi, tumors, and parasites (Table 32.9).

There are many antibiotics now available for the medical profession, but the search for antibiotics continues. Antibiotics against bacteria, fungi, viruses, and tumors are actively sought. This search includes antimicrobials effective against disease-causing microbes now seemingly under control. One of the key problems in the use of antibiotics and chemotherapeutics is the pathogen's development of resistance to the agent. Antibiotic-producing organisms generally bear genetic information that renders them resistant to the antibiotic they produce. This genetic information (an antibiotic resistance gene) is thus available in nature and can pass laterally through various microbes and, ultimately, to disease-causing organisms. Resistance to antimicrobial agents can also occur through mutation and selection. Hence, antibiotics are actively sought that are not cross-resistant to those in current use. Cross-resistance is the phenomenon that occurs when a bacterium becomes resistant to one antibiotic that spontaneously gives resistance to another. It is also obvious that new agents that are effective against bacteria, fungi, viruses, or tumors would be of great value.

Following is a discussion of the industrial approach to the discovery of new and useful antibiotics. Much of the methodology could be applied with some adaptation to a search for biocides, growth factors, or other useful compounds.

Search for New Antibiotics

The indispensable factor in any search for novel compounds is the development of an **assay** system. The search is termed a screen, and the detection system is the assay. The assay must give accurate, immediate results at a low cost. Whether the screen is for inhibitors, novel enzymes, or organic or amino acids, an efficient assay is indispensable. The overall procedure employed in the search for antibiotics is as follows:

1. Identify a source of potential producers (the screen) of the desired product (antibiotic). In searching for antibiotics, all past experience affirms that actinomycetes, particularly the *Streptomycetes,* are the best microbe to screen. They produce the greatest variety of effective antibiotics, and these antibiotics are most often of low toxicity and therefore of commercial potential. The best sources of actinomycetes are neutral to slightly alkaline soils, composts, and peat. Samples from such areas would be taken to the laboratory to initiate the screening procedure.
2. Isolation of organisms can be accomplished by spreading suitable dilutions of the soil on the surface of an agar-containing medium. Petri plates with a medium made up of hydrolyzed casein (animal), soytone (plant), and yeast extract (microbe) are suitable. Sugar content should be low in order to limit the overgrowth of the slower growing actinomycetes by rapidly growing pseudomonads or bacilli. The leathery, compact dry colonies of the actinomycetes are readily apparent to an experienced technician. Colonies of the selected organisms are selected, transferred to a suitable growth medium, and retained for further testing.
3. The capacity of the isolated microorganism to produce an antibiotic substance can be ascertained by using a plate test method. The actinomycete is inoculated across an agar plate as depicted in Figure 32.21. *Streptomyces* and other actinomycetes form a compact continuous colony at the streak across the plate. After a few days the test organisms are streaked perpendicular to the actinomycete and can be varied depending on the type of inhibitory agent sought. Usually a gram-positive and gram-negative bacterium are employed to select for potential agents against infectious bacteria of this type. A mycobacterium would select for potential agents against tuberculosis and a yeast for antifungals. Rarely are antibiotics effective against all four unless the compound is cytotoxic. If a significant zone of inhibition against any of the test organisms is apparent, the actinomycete passes to another phase.
4. Assay for novel activity. It is evident that, with the vast number of described antibiotics, the newly discovered activity might be due to an antibiotic already described. It is of utmost importance that producers of known antibiotics be eliminated from the screen quickly. This is accomplished by following several possible protocols. First, one grows the selected actinomycete in liquid culture and subject it to a paper disc plate assay to ensure that antimicrobial activity is repeatable (Figure 32.22). Various identification procedures then might be applied to the growth medium from the initial fermentation,

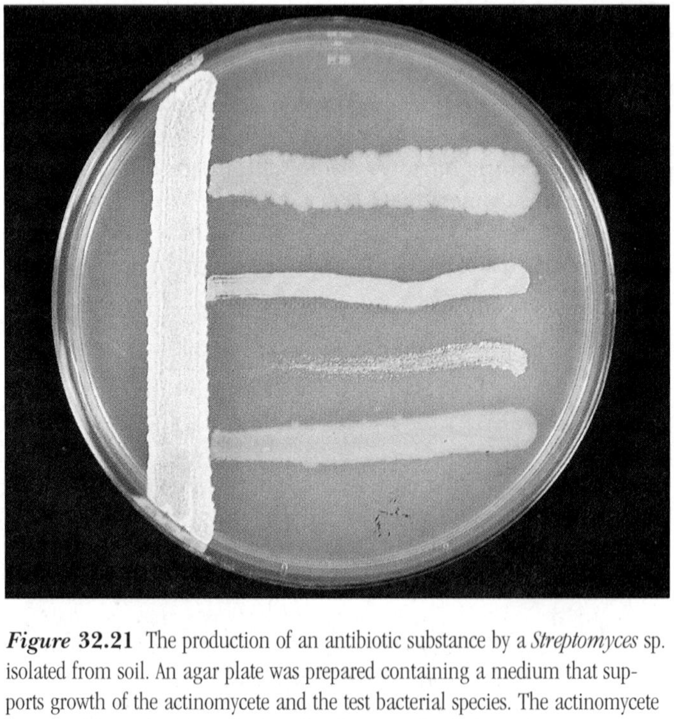

***Figure* 32.21** The production of an antibiotic substance by a *Streptomyces* sp. isolated from soil. An agar plate was prepared containing a medium that supports growth of the actinomycete and the test bacterial species. The actinomycete inoculated is on the left, and after a few days growth, the bacteria were cross-streaked. These are (top to bottom), *Bacillus megaterium, Micrococcus luteus, Staphylococcus epidermidis,* and *Escherichia coli.* The actinomycete produced an antibiotic that inhibited *S. epidermidis.* (Courtesy of J. J. Perry)

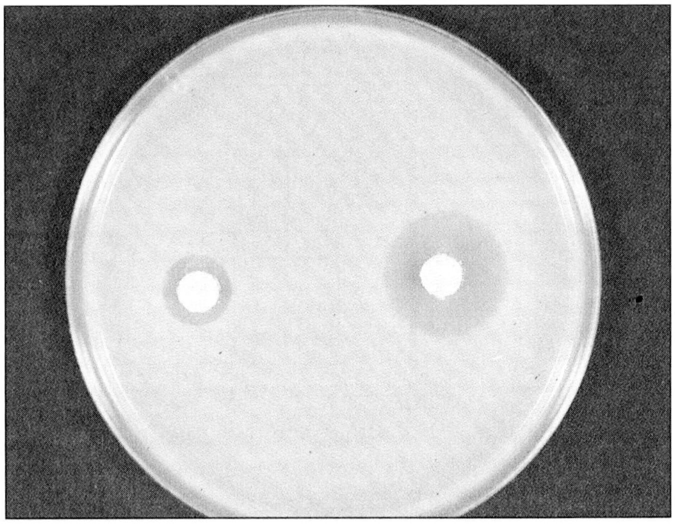

***Figure* 32.22** The detection of antibiotic activity by an agar plate assay. Graded amounts of fermentation broth are added to the filter paper disc and placed on the surface of an inoculated agar plate. The size of the zone is proportional to the amount of antibiotic present. (Courtesy of J. J. Perry)

including paper chromatography, electrophoresis, pigment production, absorption spectra, antimicrobial spectrum, and others. Activity against microbes known to be resistant to antibiotics now in use can also be determined. If the antimicrobial substance appears novel and interesting, it would then be promoted to a more extensive characterization.

5. Employ advanced culture screening. At this stage the isolated organism is subjected to a series of procedures to increase the amount produced to levels that permit isolation and characterization. Experience tells us that newly isolated organisms produce only a few micrograms of antibiotic. Often they synthesize a series of chemically related antimicrobials, so no single one of these would be present in significant amounts. Various parameters are tested for their positive effect on antibiotic production: pH, aeration, growth substrate, length of fermentation, and other conditions. These efforts to increase the amount of antibiotic produced are monitored by assay techniques to ensure that the antimicrobial is the same as that from the original isolate. Favorable results will lead to production in laboratory-scale fermentors. Chemists are then recruited in efforts to chemically isolate the antimicrobial substance. If chemical isolation of the compound proves successful and it appears novel, animal testing would begin.
6. Initiate animal testing. Mouse tests tell whether the compound is toxic or destroyed by mammalian enzymes. Such tests determine whether the antimicrobial is effective *in vivo*. If all goes well, the antibiotic proceeds to higher animal tests, human tests, and, in the rare case, ultimately to the pharmacy. The clinician would then have a new weapon against disease.

Commercial Production

The commercial production of antibiotics is accomplished in huge fermentors (Figure 32.23). These fermentors vary in size from less than 40,000 to over 500,000 liters. Size is limited by the ability to provide sufficient aeration for maximum growth. The heat generated during rapid growth is removed by cooling coils. The optimum temperature for a typical *Streptomyces* fermentation is 26 to 28°C. Aeration

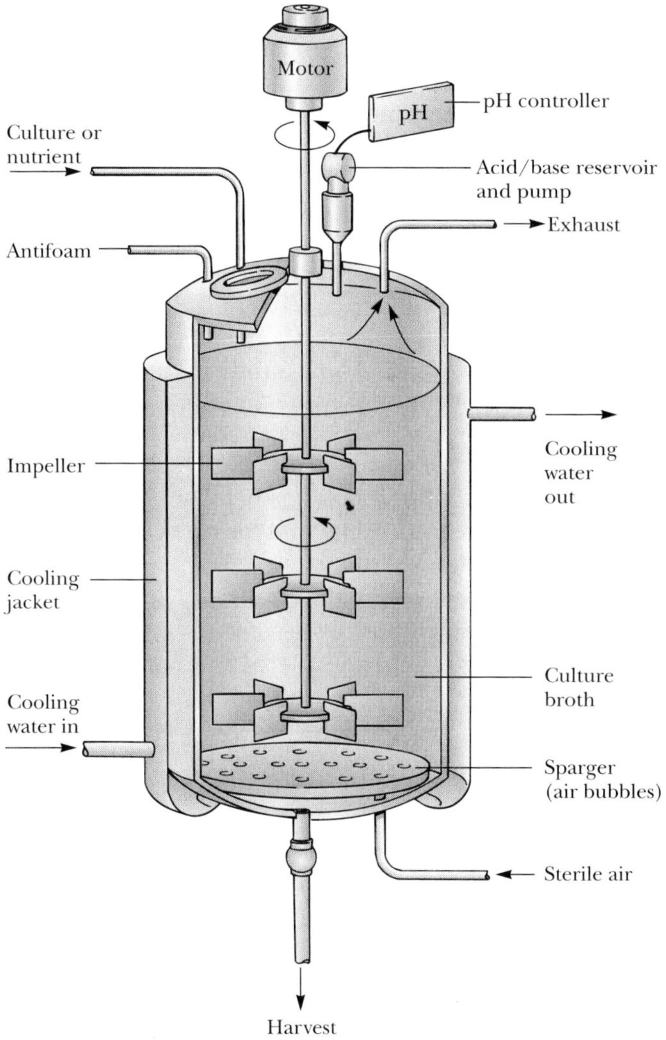

***Figure* 32.23** The basic components of a fermentor that would be employed in commercial production of antibiotics. Fermentors range from a few liters in size to 500,000 liters.

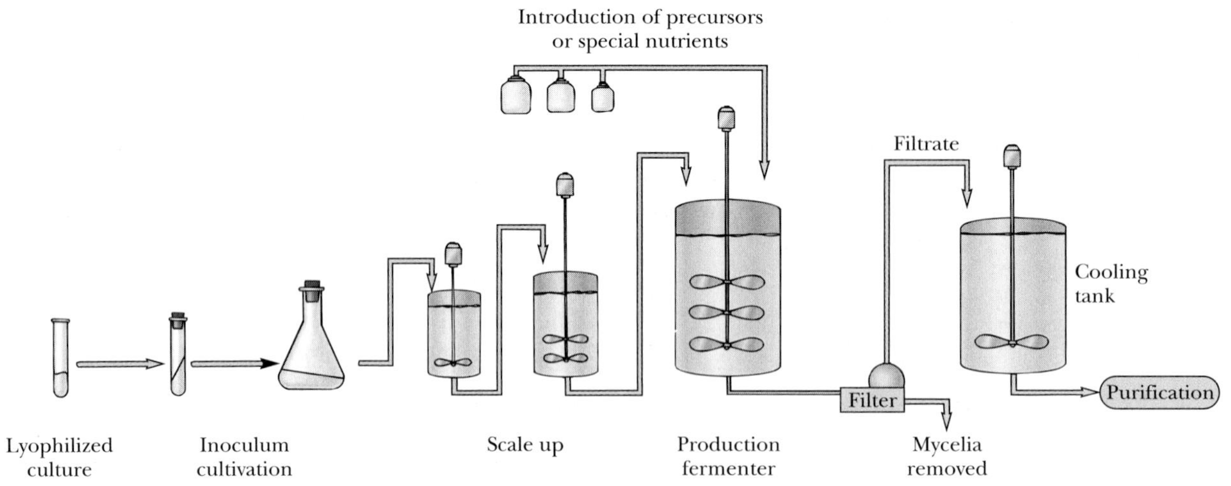

***Figure* 32.24** Flow chart for the commercial production of antibiotics.

is a significant factor and varies somewhat, but generally requirements are for 0.5 to 1.5 volumes of air/volume of the fermentor/minute. This requires that huge amounts of sterile air be introduced into the fermentor through spargers. Oxygen is poorly soluble in water, and the impellers must turn at over 100 rpm to ensure adequate air saturation of the growth medium. The pH is monitored to maintain the optimum pH for antibiotic production. Antifoam agents are added to control foaming under these high aeration/mixing conditions. The impellers are driven by powerful motors, and the shaft of the impeller passes from the outside to the inside of the fermentor. Contamination of the growth medium is a considerable concern and can cause serious economic loss. The fermentor must be constructed so that it can be sterilized completely, usually with live steam passed through the cooling coils, and every outlet must be constructed to preclude contamination.

The medium in a fermentor varies but generally consists of corn steep liquor, a molasses-like substrate, or various combinations of sugar/yeast extract/soytone. The addition of substrates as the fermentation progresses is sometimes desirable. Precursors of antibiotics obtained by chemical synthesis may also be added as the fermentation progresses. The elapsed time for a fermentation run is about four days.

The fermentation process is outlined in Figure 32.24. The antibiotic-producing stock cultures are carefully maintained either by lyophilization, under liquid nitrogen, or a storage condition that has proven to result in a stable culture. To initiate the fermentation process, the stable organism is cultivated in a series of fermentation steps, or scale-ups. At each step the culture is monitored for purity. The final scale-up inoculum is sizable to ensure that the final fermentation occurs quickly and without a lag phase. After growth (trophophase) and antibiotic formation (idiophase) the fermentation medium is filtered to remove mycelium, and the antibiotic is then recovered.

Antibiotics can be classified according to their antimicrobial spectrum, mechanism of action, producer strain, manner of biosynthesis, or chemical structure. None of these classification schemes is entirely satisfactory

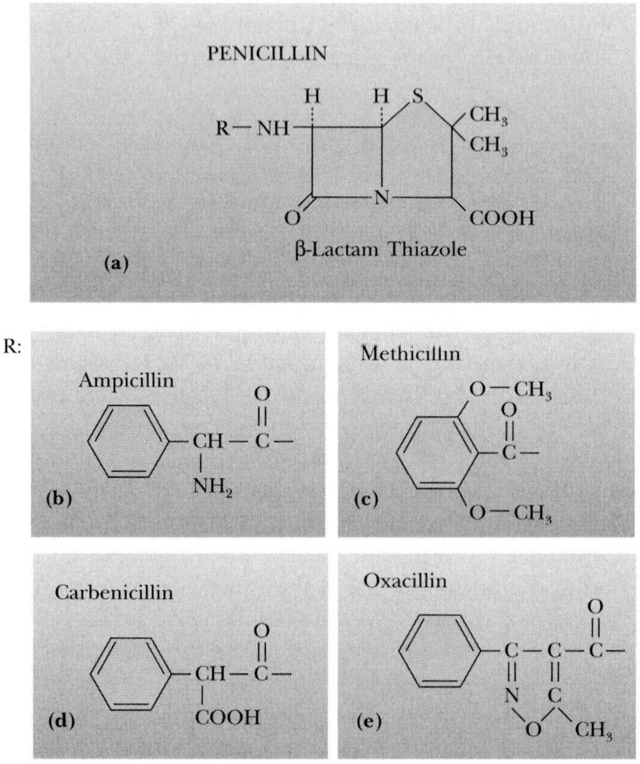

***Figure* 32.25** Chemical derivatives of the 6-aminopenicillanic acid: **(a)** moiety of penicillin, **(b)** ampicillin, **(c)** methicillin, **(d)** carbenicillin, and **(e)** oxacillin.

and no classification scheme is presented here. A discussion follows on some of the major antibiotics now employed in the clinic.

PENICILLIN Penicillin is called a β-lactam because it contains the unusual β-lactam ring. This antibiotic has been in use for about 50 years, and microbes resistant to it have become a serious concern. Semisynthetic penicillins are more effective against microbes resistant to the natural product and are antibiotics of choice clinically. The role of microbial acylase in generating 6-aminopenicillinoic acid was shown in Figure 32.16. The structures that can be added to this moiety to form effective antibacterials are illustrated in Figure 32.25.

CEPHALOSPORIN C

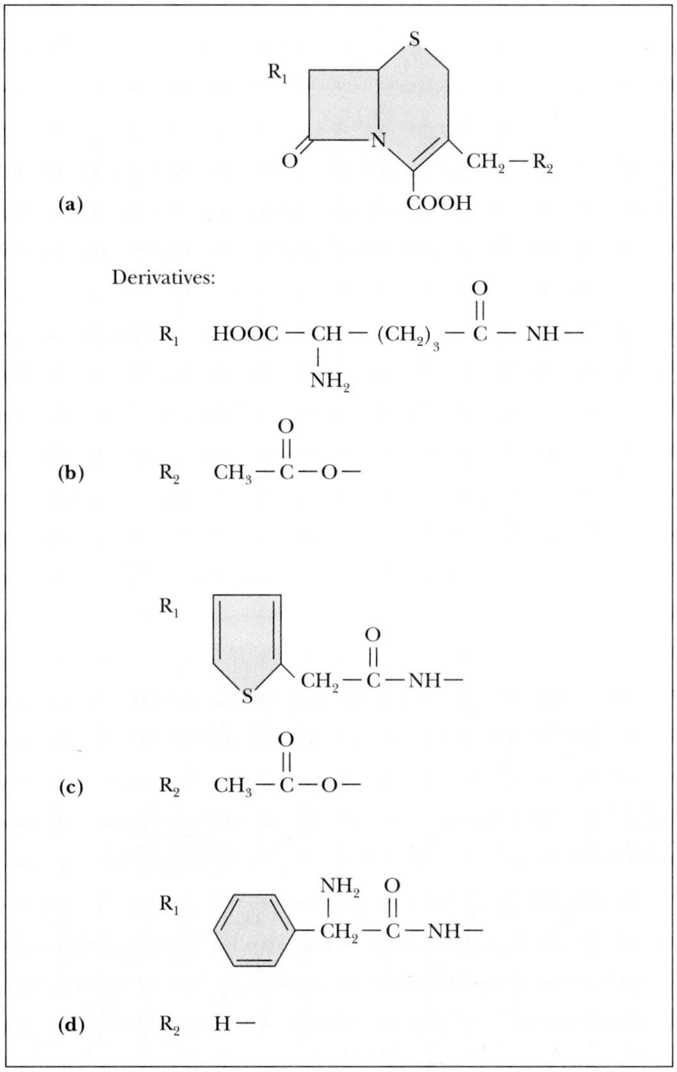

***Figure* 32.26** Cephalosporin c and some semisynthetic derivatives employed in treatment of infections: **(a)** β-lactam ring of cephalosporin, **(b)** cephalosporin c, **(c)** cephalotin, and **(d)** cephalexin.

CEPHALOSPORIN This is also a β-lactam antibiotic and produced by the fungus *Cephalosporium acremonium.* The cephalosporins have lower toxicity than penicillin and a somewhat broader antimicrobial spectrum. They are also resistant to the penicillinase produced by penicillin-resistant microbes. The cephalosporins are produced by fermentation, and some semisynthetics are produced by chemical modification. The basic structure of cephalosporin is illustrated in Figure 32.26 and semisynthetic modifications that lead to other commercial products are shown.

STREPTOMYCIN Many antibiotics are derivatives of sugars, and a major antibiotic in this group is streptomycin (Figure 32.27). These antibiotics contain amino sugars linked through glycosidic bonds to other sugars. The discovery of streptomycin was of great medical importance because it was the first effective drug against the scourge of tuberculosis. Resistance to streptomycin has been a serious problem, and other drugs are now commonly used. Because resistance to the various antituberculosis drugs is a concern, multiple drugs are administered in adequate doses and over a period of time. The antibiotic now employed along with synthetic drugs in tuberculosis treatment is rifampin.

MACROLIDES Rifampin is a semisynthetic antibiotic synthesized from rifamycin. Rifamycin is produced by *Nocardia*

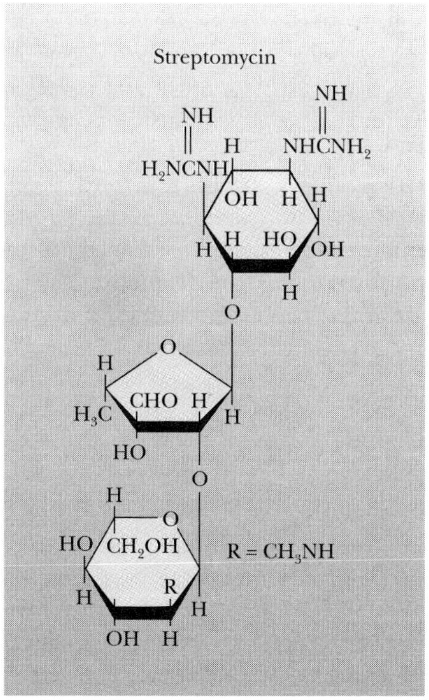

***Figure* 32.27** The structure of streptomycin, an aminoglycoside antibiotic.

mediterranei and is a *macrolide* antibiotic. Macrolides are antibiotics that have large lactone rings bonded to sugars. The structure of rifampin is presented in Figure 32.28. Rifampin is a specific inhibitor of bacterial DNA–dependent RNA polymerase. Other synthetic compounds employed in tuberculosis treatment are isoniazid and pyrazinamide.

TETRACYCLINES The tetracyclines are a major group of antibiotics effective against both gram-positive and gram-negative bacteria. They are also effective against *Rickettsia, Mycoplasma, Leptospira, Spirochetes,* and *Chlamydia.* Some semisynthetic derivations have also been developed to counteract the bacterial drug-resistance problem. The tetracyclines are all a naphthacene core with various added constituents (Figure 32.29).

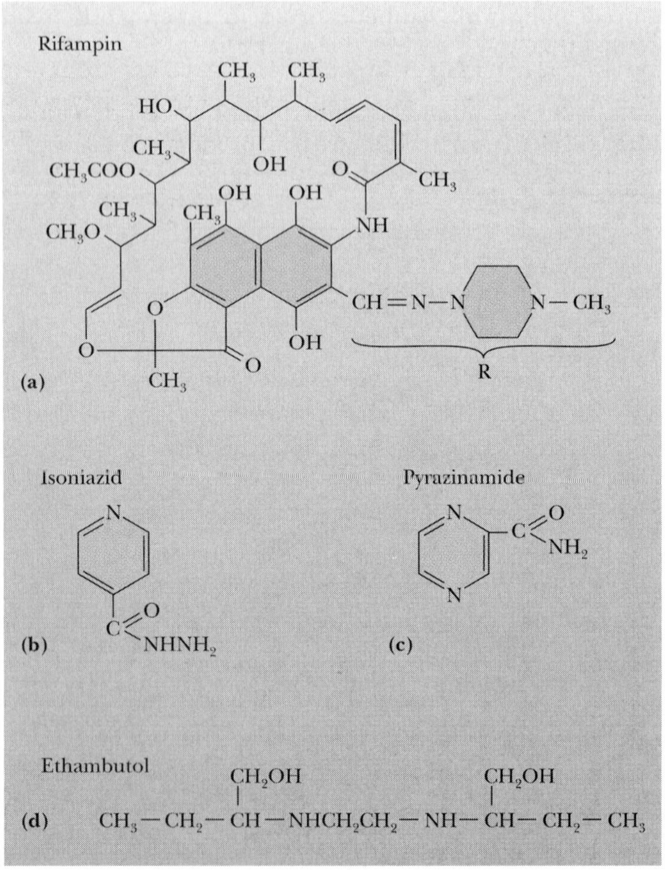

***Figure* 32.28** Structure of **(a)** rifampin, which is synthesized from the antibiotic rifamycin by addition of 1-amino-4-methyl piperazine (R). Also shown are **(b)** isoniazid, **(c)** pyrazinamide, and **(d)** ethambutol, which are administered in combination with rifampin or streptomycin in tuberculosis treatment.

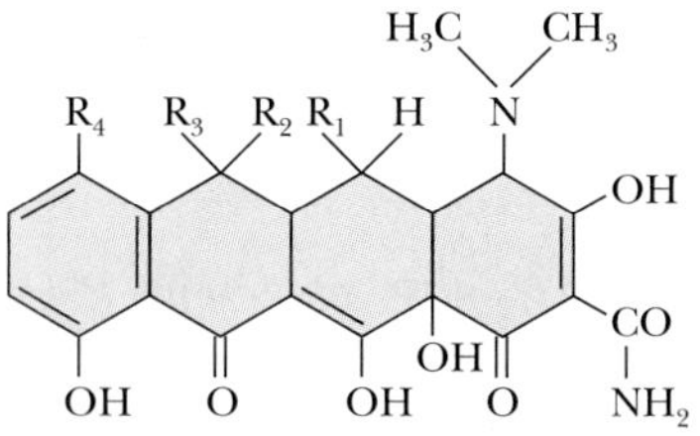

Tetracyclines	R_1	R_2	R_3	R_4	Production strain
Tetracycline	H	OH	CH_3	H	Streptomyces aureofaciens (in chloride-free medium) or through chemical modification of chlortetracycline
7-Chlortetracycline (Aureomycin)	H	OH	CH_3	Cl	S. aureofaciens
5-Oxytetracycline (Terramycin)	OH	OH	CH_3	H	S. rimosus
6-Demethyl-7-chlortetracycline (Declomycin)	H	OH	H	Cl	S. aureofaciens (+ inhibitor)
6-Deoxy-5-hydroxy-tetracycline (Doxycylcine)	OH	H	CH_3	H	Semisynthetic
7-Dimethylamino-6-demethyltetracycline (Minocyclin)	H	H	H	$N(CH_3)_2$	Semisynthetic
6-Deoxy-6-demethyl-6-demethyl-6-methylene 5-hydroxytetracycline (Methacycline)	OH	$=CH_2$		H	Semisynthetic

***Figure* 32.29** Structures of the tetracyclines. Some are natural products of *Streptomyces* strains and others are semisynthetic chemical derivatives.

Genetically Engineered Microorganisms

Genetically engineered microorganisms may have genes incorporated in them that direct the biosynthesis of compounds not associated with normal growth. Techniques for mobilizing genetic information and transferring it between prokaryote species or prokaryotes and eukaryotes are now established (see discussion of recombinant DNA techniques, Chapter 16). These techniques induce a microorganism to produce many compounds that are useful in treating human ailments.

There is considerable interest in genetic engineering in developing microorganisms that can produce human-related proteins. Proteins are difficult to synthesize, and extracting clinically useful proteins from natural material is a costly and inefficient process. Blood proteins and hormones are generally present in an animal in limited amounts, and purification of proteins from tissue, glands, or blood is expensive and difficult. Engineered bacteria grow rapidly on simple nutrient and potentially produce blood proteins, hormones, and the like in unlimited amounts. As with any endeavor, there are pitfalls in producing pharmaceuticals through genetic engineering. Getting the gene into a foreign cell can be difficult, and, once it is in the cell, protecting the gene from destruction by host nucleases or having the organism express the gene is not always a certainty. Getting the product from the cell and purifying it can also present problems. The regulations that must be met to gain approval for human use can be tedious, time consuming, and expensive. Proving efficacy through animal and human trials and safety assurance can take years. The Food and Drug Administration must approve drugs before they can be released for clinical use, and the FDA tends to be appropriately deliberate and cautious.

Human Insulin

Insulin is a small-protein (Figure 32.30) composed of two peptide chains: peptide A, consisting of 21 amino acids, and peptide B, with 30 amino acids. Until recent years, the insulin used in the treatment of human diabetes was extracted from the pancreas of slaughtered animals. Porcine- and bovine-derived insulin had two significant shortcomings. They were not precisely the same as human insulin structurally and hence not as effective, and they also carried animal proteins that caused allergic responses in some diabetics. In 1974, research was initiated on incorporating the human insulin gene in *Escherichia coli.* By 1984, commercial production of human insulin by *E. coli* was feasible.

To engineer a bacterium to produce insulin, a synthetic gene was made for each of these polypeptide chains. The gene for each was incorporated separately via a plasmid vector. The gene was linked to the end of the gene-encoding β-galactosidase. This is done so that the insulin polypeptide can be coproduced and secreted along with the enzyme. The two gene-engineered strains, one producing chain A and the other chain B, are grown separately. The peptide produced is separated from β-galactosidase and chemically joined to form human insulin. About one half of the insulin used for diabetes treatment is now obtained from engineered bacteria, and the other half is recovered from slaughtered animals.

Somatotropin

Somatotropin, a human growth hormone used to treat pituitary deficiencies that result in dwarfism, was previously obtained from the pituitary glands of cadavers. But obtaining enough hormone was both difficult and costly, which severely limited its use. Growth hormone from animals was tried but was not effective. As a result, efforts were made to put the genetic information for synthesis of

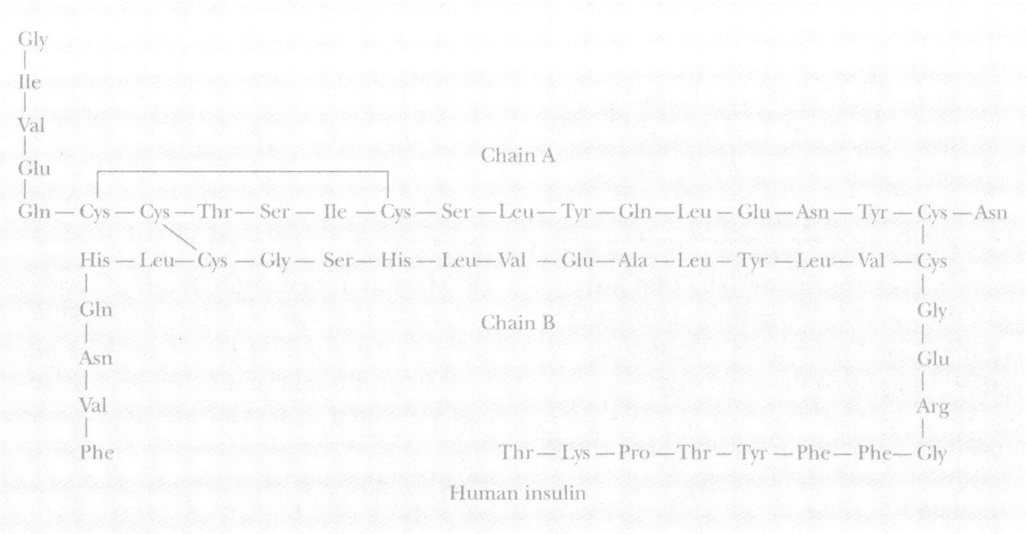

***Figure* 32.30** The amino acid sequences in chains A and B of human insulin. These two chains are synthesized separately by genetic information cloned into microorganisms and joined chemically to form the active hormone.

somatotrophin into bacteria. These efforts have been successful, and the hormone is now obtained from genetically engineered strains of *E. coli*.

Vaccines

The incorporation of genes for antigen synthesis is now possible due to advances in biotechnology. Antigens thus synthesized have many advantages over antigens extracted from pathogenic bacteria or viruses. They are generally superior to attenuated bacterial or viral preparations. Antigens generated by bacteria are less costly, more readily purified, and free from contamination by other proteins. One vaccine now obtained from a genetically engineered yeast is that for hepatitis B. Hepatitis B is often transferred person to person in contaminated blood and is a problem for IV drug users, dialysis patients, and individuals who require multiple transfusions. The hepatitis B virus cannot be grown except in blood of intentionally infected primates. The blood of individuals chronically ill with hepatitis B virus contains a protein particle called HBs Ag. The particle is not harmful and can be employed as an effective inducer of anti-hepatitis B antibody. The protein has been cloned into *Saccharomyces cerevisiae* and is now the approved source of the antigen for human immunization.

There is promise in this technique for obtaining antigens related to other infectious agents, particularly where propagation of the causative agent *in vitro* is difficult.

Other Cloned Genetic Systems

The potential for developing cloned genetic systems that will serve as a ready source of medically important compounds is endless. A number of these engineered systems are now available or in the development stage. Discussion of some of these systems follows.

Interferons

Interferons are proteins, normally synthesized by cells, that interfere with viral propagation (see Chapter 14). These may also be effective as anticancer agents.

Blood Proteins

Gene engineering for the production of a number of important blood proteins is under development. **Tissue plasminogen** activator is involved in dissolving blood clots, especially during wound-healing processes. This protein is useful in dissolving blood clots in the heart or embolisms in other areas of the body. Blood-clotting factors are also essential for hemophiliac patients. These factors are now obtained from pooled blood. With the constant worry about AIDS and other viruses in pooled blood, a gene-engineered source would be most welcome.

Bone Growth Factor

This is used for the treatment of osteoporosis patients and those with fractured bones. Bone growth protein is a hormone that occurs in intracellular materials of bones. It induces undifferentiated cells to differentiate into bone-forming cells. As the level of this protein declines with age, it is necessary to replace it in some individuals with hormones produced in gene-engineered microorganisms.

Epidermal Growth Factor

This protein stimulates wound healing. A ready source of this factor from genetically engineered bacteria would be a boon to burn and surgery patients.

Potential for Future Developments

The future for construction of genetically engineered microorganisms capable of generating a wide array of useful substances seems limitless. The purity of proteins, essential in human well-being is considerably higher than in those same proteins extracted from animal material. Fear of contamination by known or unknown human viruses is also eliminated. The cost of production is minimized, as microbes can be grown on low-cost nutrient and in unlimited amounts. The future for this area of industrial microbiology is indeed bright.

Summary

- **Industrial fermentations** are the source of both the antibiotics that cure disease, and yeast employed in bread making, vitamins, flavoring agents, organic acids, enzymes, and fermented foods/beverages.

- Microorganisms that generate useful industrial products are derived from naturally occurring populations by mutation and selection.

- **Biotechnology** is a field of research and development that is based on studies with microorganisms.

- Metabolites generated during active growth are considered **primary metabolites,** while those produced after exponential growth are **secondary metabolites.** The **trophophase** is active growth and the **idiophase** is involved in production of secondary metabolites.

- Microorganisms may be consumed as food or, more importantly, may convert foods such as meat, milk, cabbage, cucumbers, or olives to a favored product. Mushrooms are actually microbes (fungi) that are consumed directly.

- Bulk yeast is used in the baking industry and as a food supplement.
- A lactic acid fermentation of meat produces Lebanon bologna; fermentation of cabbage yields sauerkraut. Chocolate and soy sauce are products of fermentations, as are olives and pickles.
- Vinegar production in the United States amounts to about 160,000,000 gallons per year. Much of this is used in commercial products (sauces, dressings, etc.). Vinegar is a product of ethanol oxidation to acetic acid.
- **Flavor enhancers** are compounds that add little flavor to food but enhance the flavor of the foodstuff to which they are added. Monosodium glutamate and nucleotides are flavor enhancers. These are products of industrial fermentation.
- Riboflavin and B_{12} are vitamins produced commercially by microorganisms. Other B vitamins are obtained from chemical synthesis. Some steps in the commercial conversion of glucose to Vitamin C are carried out by microorganisms.
- Wine, beer, and other alcoholic beverages are products of fermentation. The starting material in wine fermentation is grapes, and for beer and other alcoholic beverages it is grain. The starch in grain must be depolymerized to yield sugars that are then fermented by yeasts such as *Saccharomyces cerevisiae.*
- **Distilled beverages** (gin, vodka, whiskey) are essentially distillate from grain alcohol. The flavor is imparted by the volatiles that distill over with the alcohol.
- **Citric acid,** a secondary metabolite, produced by *Aspergillus* sp. is added to soft drinks, candy, and other foods. It is also replacing polyphosphates in detergents.
- **Proteases** produced by bacteria such as *Bacillus amyloliquefaciens* are an additive in detergents as a sustainer. **Amylases** are employed commercially to convert starch to free sugars.
- **Enzymes** have a short shelf life and are stabilized by **immobilization.** Cross-linking functional groups to reagents, bonding to inert material, or encapsulation are effective in prolonging the shelf life of enzymes.
- **Biotransformation** of low-cost inactive plant sterols to active forms can be accomplished with microorganisms. Sterols are important medicinally as anti-inflammatory agents and for other purposes.
- A **natural biocide** from a bacterium is the protein crystal formed during sporulation of *Bacillus thuringiensuis.* The product is sold as an **insecticide.**
- About 10,000 **antibiotics** have been characterized, but there are fewer than 100 in commercial production. Most useful antibiotics are produced by **actinomycetes.** Among the many antibiotics are the **penicillins, cephalosporins, streptomycins,** and **tetracyclines.**
- Microorganisms have been genetically engineered to produce **insulin, human growth hormone, vaccines,** and other clinically useful compounds. This is a major area for future development.

Questions for Thought and Review

1. What events occurred in the second half of the 20th century that had a marked effect on the fermentation industry?
2. Define primary and secondary metabolites and cite examples of each. What do trophophase and idiophase mean in terms of industrial products?
3. What are some foods that are products of microbial fermentations? The nature of the microbe involved?
4. There are four major steps involved in cheese manufacture. What is the influence at these phases on the final product? How does one cheese differ from another?
5. How are microbes utilized in production of: (a) fermented meat, (b) sauerkraut, (c) pickles, (d) coffee beans, (e) chocolate, (f) olives, and (g) soy sauce?
6. Outline the major processes in vinegar manufacture. How do the major types of vinegar differ?
7. What are some of the major uses of amino acids?
8. How does red wine differ from white wine?
9. In beer manufacture, what are: (a) organisms involved, (b) malting, (c) wort, (d) top fermentation, (e) bottom fermentation, and (f) lagering?
10. What are distilled spirits? How does bourbon differ from Scotch and rye? How does whiskey differ from whisky?
11. Give some uses of citric acid. What organisms are involved in the citric acid fermentation?

12. Name three industrial-type enzymes. What are their uses? Why do microorganisms produce these enzymes?

13. What is a biotransformation? How does it apply to steroid manufacture?

14. Outline the steps in the search for a new antibiotic.

15. Genetically engineered organisms are involved in production of several medical products. What are some of these and their uses?

Suggested Readings

Bollon, A. P. 1984. *Recombinant DNA Products: Insulin, Interferon and Growth Hormone.* Boca Raton, FL: CRC Press, Inc.

Crueger, W., and Crueger, A. 1990. *Biotechnology: A Textbook of Industrial Microbiology.* 2nd ed. Sunderland, MA: Sinauer Associates.

Demain, A. L., and N. A. Solomon. 1986. *Manual of Industrial Microbiology and Biotechnology.* Washington, DC: American Society for Microbiology.

Hershberger, C. L., S. W. Qullner, and G. Hegerman. 1989. *Genetics and Molecular Biology of Industrial Microorganisms.* Washington, DC: American Society for Microbiology.

Jay, James M. 1991. *Modern Food Microbiology.* 4th ed. New York: Van Nostrand Reinhold Co.

Yauchinski, S. 1985. *Setting Genes to Work: The Industrial Era of Biotechnology.* New York: Viking.

. . . We live in the Age of Bacteria (as it was in the beginning, is now and ever shall be, until the world ends)

Stephen Jay Gould, *Nature,* 1993

Chapter 33

Environmental Microbiology

Sewage (Wastewater) Treatment
Drinking Water Treatment
Landfills and Composting
Pesticides
Bioremediation
Acid Mine Drainage

Heterotrophic microorganisms are ideally suited for the disposal of organic waste materials. Organic materials serve as their energy and carbon sources and therefore are not considered waste materials to them. Furthermore, because of the great diversity among heterotrophs, some grow under almost all environmental conditions of temperature, pH, nutrient concentration, and oxygen concentration. Therefore, microorganisms can utilize these materials almost anywhere they exist in the environment. Moreover, from society's point of view, it costs virtually nothing to use microorganisms to dispose of waste organic materials, whereas chemical treatment is costly and sometimes ineffective. And, in addition, biological processes occur at low temperatures and usually do not produce toxic end products. Indeed, it is difficult to stop microorganisms from degrading wastes because it is a major role they play in the recycling of organic materials on Earth. For these reasons, environmental engineers have designed bioreactors to utilize microorganisms for the treatment of waste materials.

Environmental microbiologists are also concerned about pathogenic bacteria and their potential impact on the public health. Public health microbiologists want to ensure that the drinking water supplies are safe. In addition, shellfish and other food products obtained from environmental sources should be free of contaminating bacteria. The introduction of pathogenic bacteria into the environment can occur through contamination by sewage, incomplete treatment of sewage, improper handling of garbage, or by other means.

Sewage (Wastewater) Treatment

As in natural environments where fecal material from animals is degraded primarily by microbial activities, modern municipal sewage treatment systems utilize microbial degradation as the principal means to degrade these organic materials. However, until the 1900s, cities did not treat sewage. They simply collected the **raw** or **untreated sewage** (also termed **wastewater**) using a sewer system, and discharged the raw sewage or wastewater into a receiving body of water (either a river or the marine environment, depending on the location of the city). However, it soon became apparent that raw sewage discharge in this manner was inadequate for two major reasons. First, the organic materials in the sewage produce undesirable ecological impacts on the receiving water. Second, among the large numbers of bacteria and other microorganisms discharged in the raw sewage, some might be pathogenic. Thus, the receiving water could not be regarded as safe for water contact sports such as swimming and waterskiing or for drinking water. A detailed discussion of these issues follows.

Ecological Impact of Raw Sewage on Receiving Water

Imagine a hypothetical river within a city called City A that is located upstream from another city, City B (Figure 33.1). City A discharges its raw sewage directly into the river, introducing fecal material containing large numbers of bacteria as well as large amounts of organic material directly into the river. The organic effluent serves as a substrate for the growth of heterotrophic bacteria in the river. Their aerobic respiration activities remove oxygen dissolved in the water. Indeed, if the organic load in the waste is sufficiently great, the resultant bacterial growth removes all of the oxygen, causing anaerobic conditions in the water. Anaerobic conditions pose grave dangers to the animals that live in the river. For example, "fish kills" could result, because fish breathe the dissolved oxygen in the water through their gills. If there is no oxygen, they, and other animals living in the river, will die. In addition to the loss of animal and aquatic plant life, the river becomes a less desirable recreational area for fishermen, boaters, and others.

The extent of the impact varies depending on the size of City A, the volume and flow rate of the river, the temperature, and a host of other factors. Despite all this, rivers do have a natural capacity to restore themselves. For ex-

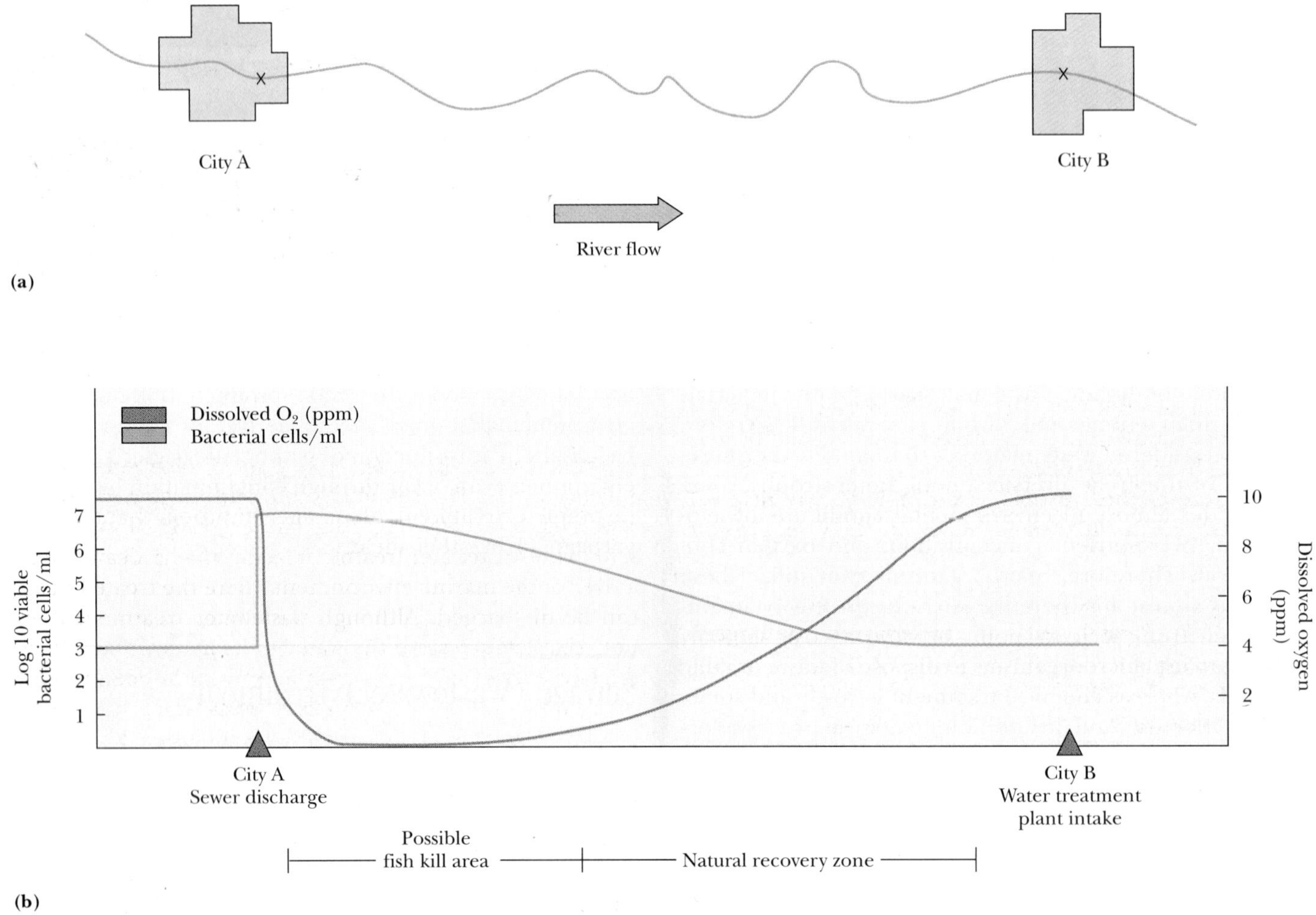

***Figure* 33.1** **(a)** A diagram showing a hypothetical river with City A discharging raw sewage upstream from City B. **(b)** The raw sewage discharge downstream of City A results in an increase in bacterial numbers (from the sewage) and a decrease in oxygen concentration (from growth of respiring bacteria). Note that further downstream the numbers of bacteria decrease and the dissolved oxygen concentration increases due to natural recovery processes.

ample, oxygen is reintroduced into the water by two activities. First, the normal process of algal photosynthesis restores some of the oxygen. Second, the turbulence of the water in the river increases the diffusion of oxygen from the overlying air. In addition, the organic material from the raw sewage is eventually degraded by the heterotrophic bacteria in the river and, when it is depleted, the numbers of bacteria will decline to approach levels comparable to those encountered above City A's wastewater discharge (Figure 33.1). The result of these processes is that the stream undergoes a natural recovery process—it purifies itself; however, this process may not occur for many miles downstream, perhaps several days' transit from the point of introduction of the wastewater.

Public Health Impact of Raw Sewage Discharge

When very large cities are located on rivers, their effluent can be so significant that normal recovery processes do not occur prior to the point at which the river reaches the next city. As a result, cities located downstream on the river can be greatly affected by the discharges in a number of ways. Of primary importance, the river water could not be safely used as a source of drinking water because it might carry waterborne infectious disease agents discharged from the feces of infected individuals in the city upstream (Table 33.1). Furthermore, beaches and other recreational areas affected by fecal effluent cannot be used for contact sports because of the possibility of disease. Commercial marine shellfish fisheries and private shellfish harvesting (Figure 33.2) will be adversely impacted, as well, because clams and oysters siphon particulate materials, including bacteria, resulting in high concentrations of this matter in their digestive tracts. Although the shellfish themselves may not be affected by the bacteria, the shellfish become contaminated by bacteria and cannot be marketed for human consumption.

Because of these problems with raw sewage disposal, municipalities in the United States are now required to treat their wastewaters *before* they are discharged into receiving waters. The sewage is collected using a system of sewer pipes, usually operating under gravity flow, and is sent to the wastewater treatment facility that is located near a river or the marine environment where the treated waste can be discharged. Although wastewater treatment does not completely restore the water to its natural state, it reduces the levels of organic matter and the concentrations of bacteria. The three degrees of wastewater treatment, primary, secondary, and tertiary, are discussed below.

***Figure* 33.2** A sign at a Seattle, Washington, city beach near the discharge point of effluent from a sewage treatment plant. The sign indicates that fish and shellfish may be contaminated with coliform bacteria. (Courtesy of J. T. Staley)

***Table* 33.1 Partial list of waterborne pathogenic microorganisms**

Microorganism	Disease
Bacteria	
Salmonella typhi	Typhoid fever
Vibrio cholerae	Cholera
Shigella dysenteriae	Bacterial dysentery
Protozoa	
Entamoeba histolytica	Amoebic dysentery
Giardia lamblia	Giardiasis
Viruses	
Infectious hepatitis	Hepatitis

Primary Wastewater Treatment

The simplest sewage treatment process is termed **primary wastewater treatment** because it involves a single step, namely, the settling of the particulate materials and their separation from the dissolved material.

In primary treatment, the sewage coming into the plant is first passed through a screen to remove sticks, plastic bags, and other large pieces of material that might

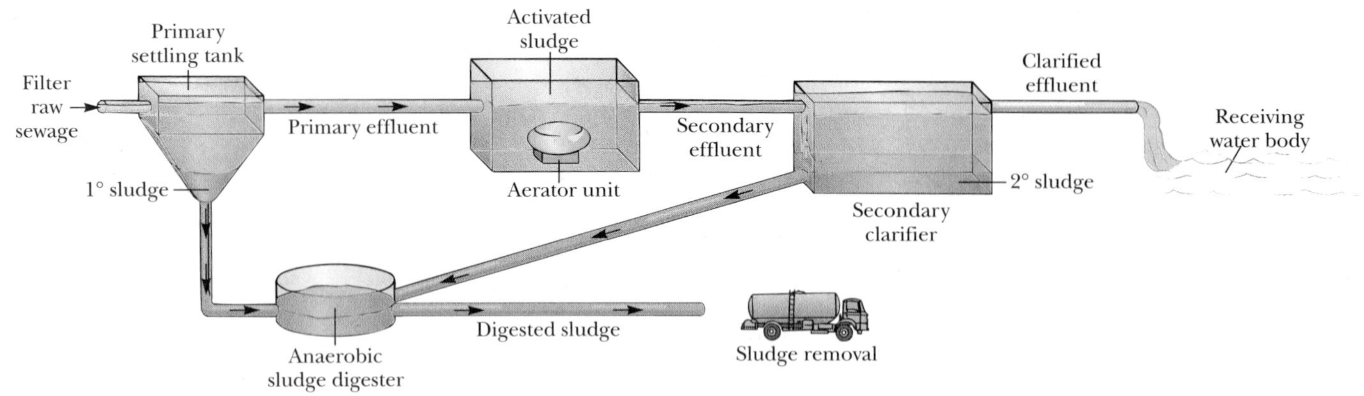

***Figure* 33.3** Diagram of a typical secondary wastewater treatment plant. Raw sewage entering the plant is first screened to remove debris, then is settled in the primary (1°) tank. The primary sludge passes to the anaerobic digestor and the liquid effluent passes to the secondary (2°) activated sludge tank. The effluent from this tank is settled in the secondary clarifier, with the sludge passing to the anaerobic digestor. Part of the sludge is recirculated to maintain the active bacterial degrading population in the activated sludge tank. The liquid effluent is chlorinated before discharge.

interfere with treatment. The wastewater then flows into the primary treatment tank—a large tank that acts as a settling basin to allow the heavier particulate material (the **primary sludge**) to settle to the bottom (Figure 33.3). Oils that float on the surface of this tank are removed by a skimming bar. Thus, primary treatment is simply a physical process whereby the liquid portion of the wastewater (which comprises over 99 percent of the waste) is separated from the solids or sludge.

Secondary Wastewater Treatment

The typical treatment systems used in the United States are **secondary wastewater treatment** systems. They are called secondary treatment systems because two sequential processes are used in wastewater treatment: primary treatment followed by secondary treatment (Figure 33.3).

The **secondary treatment** process is a microbiological process. The liquid portion from the primary tank is passed into another tank (the secondary wastewater tank) and is then aerated. Different types of aerators have been used. An older type is referred to as a **trickling filter.** In this system, the fluid is sprinkled over a bed of rocks that are about the size of a fist. The wastewater is aerated while it passes from the sprinkler and percolates through the bed of rocks. Bacteria grow on the surface of the rocks and degrade the organic materials that are in the wastewater. In actual fact, the bacteria grow as a layer on the surface of the rocks. Those near the surface are aerobic, but those underlying the surface layer are anaerobic fermentative bacteria.

Another type of secondary treatment process is referred to as the **activated sludge** process. In this type of treatment, the effluent from the primary treatment tank is aerated by bubbling air through it in the secondary treatment tank. Aerobic bacteria grow in colonial groups called **flocs** and degrade the organic material in the waste. Among the bacteria that occur here is the genus *Zoogloea,* a bacterial species that forms flocullant colonies.

The next step in the treatment consists of settling out the particulate materials from the secondary treatment tank in an additional tank, called the **secondary clarifier.** The liquid portion of this material has much lower levels of organic material in it and can be satisfactorily discharged into a receiving stream or other body of water. However, before final discharge, the fluid is chlorinated to kill many of the potentially pathogenic microorganisms that might be present.

There is also sludge from this secondary clarifier. As in the primary sedimentation tank, this settles out and must also be treated. In typical treatment plants, the primary and secondary sludges are combined and treated as a whole in the **anaerobic sludge digestor.** This is a large-capacity tank in which anaerobic bacteria carry out the final steps of anaerobic degradation of organic materials. These tanks contain fermentative bacteria that produce organic acids, alcohols, carbon dioxide, and hydrogen gases. Methanogenic bacteria grow on the acetic acid and on the hydrogen and carbon dioxide in these reactors and produce methane gas. The methane produced can be reclaimed and used as an energy source for heating in the treatment plant or elsewhere.

Anaerobic sludge digestion is a slow process and does not result in a complete conversion of sludge to gases. The undigested sludge that remains after treatment is rich in nutrients and must be disposed of elsewhere. It has been used as a fertilizer in agriculture for the growth of plants that are not used for human consumption since there is concern about viruses and other microbes that might have survived the treatment process.

The organic material in wastewater imparts what sanitary engineers refer to as a **biochemical oxygen demand** or **BOD** to the waste. The BOD test is used to measure the amount of oxygen demand, or the amount of oxygen required, to decompose the organic material in four days at 20°C in a wastewater sample. Initially, raw sewage has a very high BOD, but if treatment has been successful, the BOD level is reduced significantly. For example, trickling filters working satisfactorily can reduce the BOD content of raw sewage by about 75 percent. Activated sludge treatment units are even more effective when working properly and can reduce the BOD content by about 85 percent.

Thus, secondary treatment removes much of the organic material from wastewater. In fact, the actual BOD reduction is effected by the bacteria and other microorganisms that grow in the secondary treatment units. These microorganisms are doing what they would normally do in nature, but, by doing it in a wastewater treatment plant, the process can be monitored and controlled. The microbes carry out these processes efficiently and with little cost to society.

Large numbers of bacteria are found in secondary effluent. Most of these bacteria are organisms that have grown on the organic materials of the wastewater. However, some of them are coliform bacteria, such as *Escherichia coli,* which come from the human feces being treated in the sewage. Because some of the coliform bacteria are waterborne pathogens, it is desirable to kill them. Thus, the effluent from the secondary clarifier is disinfected, usually with chlorine, before it is finally discharged into the receiving water. Although this process is not 100 percent effective in disinfection of the effluent, it does dramatically reduce the numbers of coliform bacteria that leave the wastewater treatment plant.

Microbial Treatment Problems

The normal degradative activities occurring in the secondary treatment system depend on the types and status of the microorganisms growing in the reactors. One microbiological problem that may occur in the secondary clarifier units is called **bulking.** Bulking refers to the poor settleability of sludge. Ideally, the particulate organic material should be completely removed prior to discharge into a receiving stream or other outlet. Thus, it is important that bacterial cells and other debris settle to the bottom of the tank where they can be removed and transferred to the sludge digestor. Some species of bacteria are undesirable in secondary clarifiers because they interfere with settling. Filamentous organisms such as *Sphaerotilus natans* and *Thiothrix nivea* are examples of bacteria that interfere with normal settling. Other bacteria such as some *Nocardia* spp. float on the surface and do not settle. When these bulking organisms are too numerous, treatment conditions need to be modified to remove them, and sometimes this is difficult to accomplish.

Anaerobic digestors may also encounter problems. Under some circumstances they "go sour"—they become acidic and cannot support the normal populations of fermenters and methanogens. When this occurs they need to be reseeded with inocula from other digestors.

Tertiary Wastewater Treatment

Although secondary treatment of wastewaters is effective at removing organic matter from the wastes, it does not remove the inorganic by-products of their activity. As can be seen from the overall treatment formula below, inorganic substances such as ammonia and phosphate are produced.

$$\text{Organic material} \longrightarrow CO_2 + NH_3 + SO_4^{-2} + PO_4^{-3} + \text{trace elements}$$

The inorganic products of organic degradation are excellent nutrients for the growth of algae. As a result, secondary effluent can result in excessive algal growth and lead to eutrophication of receiving waters (see **Box 33.1**). Thus, it is possible to have successfully treated a wastewater by secondary treatment only to have it cause enrichment in the receiving water.

The major nutrients in effluents from secondary sewage treatment systems that are of concern in enriching receiving waters are phosphate and ammonia. These nutrients can be removed from secondary effluent by treatment processes referred to as **tertiary wastewater treatment.** Both chemical and microbiological tertiary treatment processes have been developed; however, chemical processes are quite expensive and are beyond the scope of this discussion.

Ammonia removal is effected biologically by a two-step process. The first step is nitrification, the aerobic oxidation of ammonia to nitrite and nitrate, by nitrifying bacteria (see Chapter 21). Although these are chemolithotrophic bacteria, they grow well in properly treated secondary effluent. During this first step, the ammonia is converted to nitrate. In order to remove the nitrate, it is necessary to carry out denitrification. Denitrification results in the anaerobic conversion of nitrate to form N_2O and N_2 gases, which will dissipate into the air. Therefore, nitrogen is removed from the water as a gas. It should be noted that denitrification by *Pseudomonas* spp. requires organic carbon. An effective and inexpensive carbon source, such as methanol, is usually added to these systems for this purpose.

This two-step process for nitrogen removal can be carried out in normal secondary wastewater treatment

BOX 33.1 RESEARCH HIGHLIGHTS

Eutrophication of Lake Washington

A classical case of lake eutrophication occurred in Lake Washington in Seattle in the 1960s. Due to the inflow of secondary effluent from sewage treatment plants for municipalities located around the lake, the lake was becoming increasingly enriched. Thus, even though the organic material of the sewage was being satisfactorily removed during secondary wastewater treatment at the sewage treatment plants, inorganic nutrients such as phosphate were being released into the lake. Dr. W. T. Edmondson, a limnologist at the University of Washington, noted that phosphate was normally limiting to algal growth in the lake, and his research showed that increased levels from treated wastewater were responsible for the increased growth of algae and resultant eutrophication.

Based upon Dr. Edmondson's recommendations, a metropolitan agency was formed and a decision was made to divert all of the wastewater effluent away from the lake. It is now discharged into a much larger body of water, Puget Sound. Since the diversion occurred in the late 1960s, Lake Washington has returned to its normal mesotrophic status.

systems if they are closely monitored. A period of aeration to enhance nitrification is followed by a period of anaerobic incubation in the secondary clarifier that favors denitrification.

Phosphate removal occurs by using bacteria that carry out **luxurious uptake** of phosphate. *Acinetobacter* spp. are used for this process. These bacteria can accumulate polyphosphate granules in excess of normal metabolic needs during periods of active growth. They can then be settled out along with the phosphate in a settling tank.

In practice, very few plants in the United States currently use tertiary treatment. However, this may change as the need for protecting our receiving waters increases.

Drinking Water Treatment

As discussed earlier, if a city receives its drinking water from a river, the water may be contaminated with effluent from wastewater treatment plants from cities upstream. In fact, even if there is no city upstream, it is possible that the water can be contaminated from wild or domestic animals that may harbor infectious waterborne microbial agents such as *Giardia lamblia,* enteric viruses, or bacteria. In addition, septic tanks from rural areas may also discharge wastewater into the river. Therefore, it is essential for the city to treat its drinking water before it is distributed to its citizens.

Depending on the source of water, the treatment may consist only of disinfection by chlorination, or the treatment may be much more extensive. Consider the case of a large city located on a major river such as the Mississippi. Furthermore, consider that the city is downstream of many other cities. In this case, it is important that the city use an extensive treatment system with coagulation, sedimentation, filtration, and disinfection (Figure 33.4). These steps are accomplished in a drinking water treatment plant or facility.

Coagulation is performed by adding alum (a salt of aluminum sulfate) to the water. When the alum is added to the water it forms aluminum hydroxide flocs that adsorb particulate materials. The flocs are then settled out by **sedimentation** in a holding tank. The next step is **filtration.** The water is passed through a sand filter system to remove the bacteria. Either rapid or slow sand filters are used, and both are effective at removing particulate material, including bacterial cells. If the treatment plant is operating effectively, about 99 percent of the bacteria present in the raw, untreated water are removed after sand filtration. The final step is **disinfection** of the water, usually accomplished by chlorine. The dissolved gas forms a hypochlorite solution (as in household bleach), which effectively disrupts the cell membranes of bacteria, thereby killing them. Disinfection does not kill all of the bacteria, however, so small numbers of bacteria remain in the water.

The drinking water must then be passed through a distribution system before it is consumed. In order to be assured that the water is safe for drinking (potable), it needs to be tested. Two approaches have been taken to ensure that drinking water is safe or potable. The first approach is to analyze for the **residual chlorine** level. If chlorine is still present in the water at the tap, at a level of at least 1 ppm, then it may be considered potable. However, as we have stated, even if there is some residual chlorine, bacteria may still be present. Therefore, microbiological tests are routinely performed to determine water potability.

Current microbiological testing uses an **indicator bacterium.** The standard indicator bacterium used in the United States is *Escherichia coli.* Because *E. coli* is found in the intestinal tract of all humans and some other warm-blooded animals, its presence in a drinking water is *indicative* of fecal contamination. Why use an indicator bacterium? Why not just look for a particular pathogen? The reason is that, if a pathogen is present, it would be expected to occur in very low numbers relative to *E. coli.* This is because the pathogen may be from only one infected individual or carrier out of hundreds or thousands of people who live in the city upstream. Therefore, it is extremely difficult to detect the pathogenic bacterium, because so few of them would be released into the sewer system in relation to the numbers of *E. coli* since *E. coli* is carried by all individuals. Furthermore, even if a test could be devised for one pathogenic bacterium, what about all the other pathogens? Because there are many different pathogens, separate tests would have to be performed for each one of them. In contrast, everyone harbors *E. coli* in his or her intestinal tract. Therefore, if *E. coli* is found in a drinking water sample, it indicates that the water has been contaminated with fecal material, and this finding alone indicates that it is unsafe for drinking.

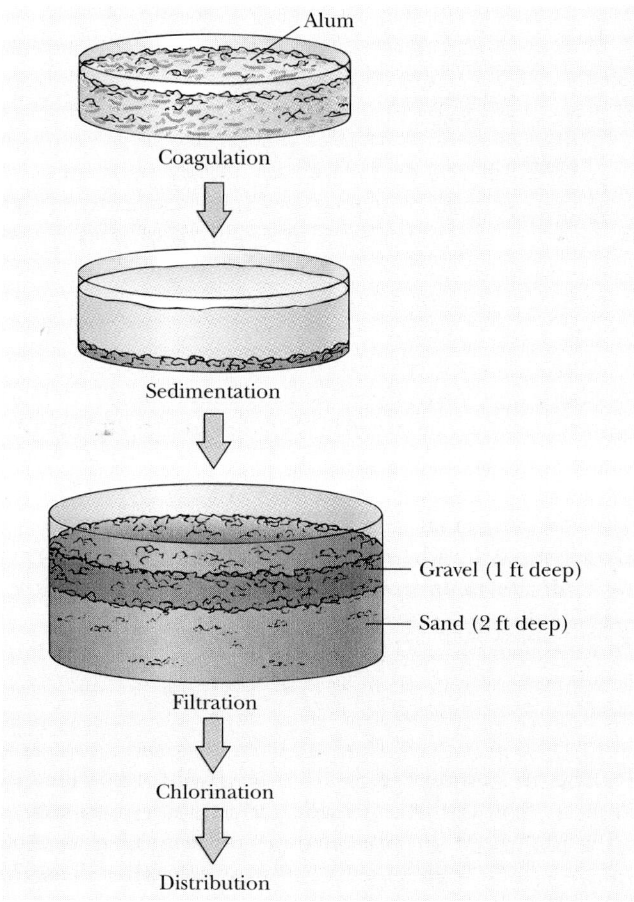

***Figure* 33.4** Diagram of a drinking water treatment system showing how the raw water is first flocculated with alum, settled in a settling tank, passed through a sand filter, and then disinfected with chlorine before it is distributed in the water system.

Although *E. coli* is the indicator bacterium of choice, it is not simple to identify it in natural samples because many other bacteria closely resemble it. It belongs to a group of enteric bacteria called the **coliform bacteria.** *Coliform bacteria are defined as nonspore-forming gram-negative rods that ferment lactose to form acid and gas in 24 to 48 hours at 35°C.* Thus, coliform bacteria are defined by experimental conditions. Indeed, several different types of tests are used to identify coliform bacteria. A brief description of some of the more common ones follows.

Total Coliform Bacteria Tests

One group of tests, the **total coliform bacteria tests,** analyze for the number of these bacteria in a drinking water sample. There are several tests for total coliform bacteria, including the most probable number and membrane filter tests.

Most Probable Number (MPN) Test

This is the oldest test for total coliform bacteria and is still used as a standard. The test is performed in three different stages. The first test, called the **presumptive test,** is designed to determine whether or not coliform organisms might be present in a sample. In this test, a series of lactose broth tubes (lauryl tryptose medium) are inoculated with the drinking water sample to be tested (Figure 33.5**a**). Typically, fivefold replicates at each of several dilutions are used. The first set of five tubes, the lowest dilution, are inoculated with 10 ml portions of the water sample to be tested. Because 10 ml is such a large volume of inoculum, 10 ml of the medium is made up double strength, so that it ends up as single strength after addition of the 10 ml inoculum. In the next dilution, the five replicate single-strength tubes are inoculated with 1 ml inocula. Higher dilutions are attained using dilution blanks. The tubes contain lactose, a pH indicator, bromothymol blue, and an inverted vial to detect gas production. They are incubated at 35°C for 24 hours and then read for acid and gas production. If they are positive for gas, they are regarded as presumptive positive. They are incubated an additional 24 hours, and any additional positive tubes are also considered positive presumptive.

The MPN test is quantifiable. Each positive tube is scored at each dilution. Then, the number of positive tubes at each dilution can be used to determine the quantities of presumptive total coliform bacteria. For example, assume all five tubes from the first two dilutions were positive, three from the third were positive, and only one of

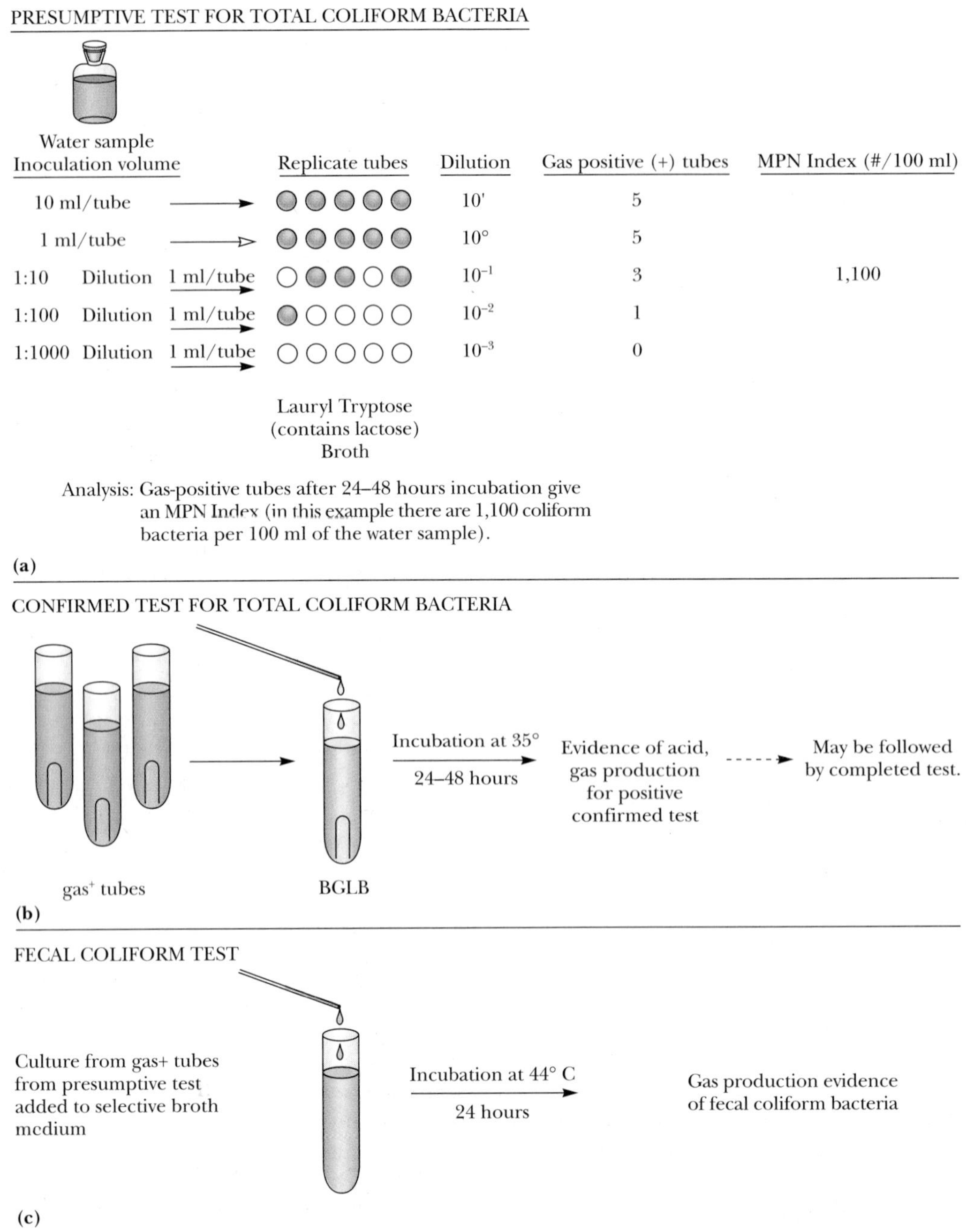

***Figure* 33.5** Diagram showing the steps used for total and fecal coliform analysis by the most probable number test. **(a)** Presumptive test, **(b)** Confirmed test, and **(c)** Fecal coliform test.

the fourth positive. The results are recorded as 5, 3, 1 for the final three dilutions. These results can be converted into a quantifiable number taken from a statistical most probable number table (see American Public Health Association's Standard Methods for the Examination of Water and Wastewater). In this particular case, the number of total coliform bacteria is 1100 per 100 ml of the original sample. Note that this concentration is given as *per 100 ml* and is called the **coliform index.**

The presumptive test is followed by the **confirmed test** (Figure 33.5**b**), which provides additional supportive information about the presence of coliform bacteria. At 24 and 48 hours, each positive presumptive tube is used to inoculate a tube of another broth medium, brilliant green lactose bile broth (BGLB). These BGLB tubes are incubated at 35°C for another 24 to 48 hours to *confirm* acid and gas production. If these are positive, they are considered to be confirmed. At this point, it is still not possible to say that coliform bacteria or *E. coli* are present in the sample, only that the presence of coliform bacteria has been confirmed.

However, a confirmed positive is considered adequate for concern about the potability of the water. For drinking water analysis, even a single positive confirmed test is considered serious. Additional testing is required to verify what the bacterium is. This is accomplished by the **completed test.** In the completed test, a loopful of culture from a positive confirmed BGLB tube is streaked on eosin methylene blue (EMB) plates. If coliform bacteria are present, they will produce colonies with a typical metallic green sheen, following incubation at 35°C for 24 hours. Cells from these colonies must be gram-stained to determine if they are gram-negative, nonsporeforming bacteria. If gram-negative, nonsporeforming bacteria are found, then all the criteria have been met for fulfilling the definition of a coliform bacterium.

It is noteworthy that there are many coliform bacteria besides *E. coli.* Thus, even though this lengthy procedure has been undertaken, it is still not possible to state that *E. coli* is present in the sample. Further tests are needed to verify this. Moreover, some of these coliform bacteria, such as *Enterobacter aerogenes,* are common soil bacteria and do not reside in the intestinal tract of warm-blooded animals. Therefore, even a positive completed coliform test would not necessarily mean that fecal coliform bacteria are present in a water sample.

Membrane Filter Tests

Membrane filter tests have also been developed for the identification of total coliform bacteria. These tests entail utilizing a sterile 0.45 μm pore size membrane filter to collect cells from the water sample to be tested. These filters are then placed on an absorbant pad that contains nutrients for the growth of bacteria. They are incubated for 24 hours at 35°C, and the colonies of total coliform bacteria are identified by their characteristic metallic sheen.

Fecal Coliform Bacteria Tests

One disadvantage of the total coliform bacteria tests is that they select for a variety of bacteria, some of which, such as *Enterobacter aerogenes,* are not indigenous to the intestinal tract of animals. Research has shown that *E. coli* strains can grow at elevated temperatures when compared to soil species such as *E. aerogenes.* Therefore, elevated temperature tests have been developed for identifying **fecal coliform bacteria,** strains of which are more likely to be *E. coli.*

Most Probable Number Test

For this test for fecal coliform bacteria, positive tubes from the presumptive MPN total coliform bacteria test described previously are used to inoculate a selective broth medium, which is then incubated at 44°C for 24 hours (Figure 33.5**c**). Cultures that grow on this medium and produce gas are regarded as fecal coliform bacteria, and most of them, if identified, turn out to be *E. coli.* As with MPN total coliform bacteria, the actual concentrations of fecal coliform bacteria in a sample can be quantified by this technique.

Membrane Filter Test

A membrane filter test has also been developed for fecal coliform bacteria. This test is usually performed directly on water samples. Therefore, the cells are exposed to 44°C and a selective medium immediately after filtration, when the membrane filter is placed on a selective *E. coli* (EC) medium and incubated at 44°C. Colonies of fecal coliform bacteria produce a characteristic blue-colored colony on this medium.

All viable counting procedures for coliform testing have limitations. Among the most serious of these is the fact that the tests take so long to perform. If a city is really concerned about an outbreak of some intestinal disease, it would be at least 2 to 4 days before it would be confirmed that coliform bacteria were present. And, as we have already mentioned, even though coliform bacteria are found, since they cannot be identified as *E. coli* using any of the above tests described, it is really not known whether the water sample has fecal contamination. Thus, new research is aimed at developing more rapid and specific tests for *Escherichia coli* and other enteric bacteria.

Other Techniques

Several different approaches are being taken to develop alternative procedures for testing the potability of drinking waters. One new approach is the presence/absence test.

Presence/Absence Test

The presence/absence test or P/A test looks for total and fecal coliform bacteria using a single large volume of the drinking water sample (100 ml) as an inoculum rather than using a series of tubes or membrane filters. In this test, only a positive or negative is recorded for a given sample. Thus, it is not possible to quantify the numbers of total or fecal coliform bacteria, but only to say they are present or absent in a particular 100 ml sample.

This type of test is very simple to perform and thereby enables small municipalities serving from 100 to 10,000 citizens to analyze their own water.

Colorimetric and Fluorogenic Tests

Another approach taken to simplify the analysis of total and fecal coliform bacteria in drinking waters uses **colorimetric** or **fluorogenic compounds.** These compounds,

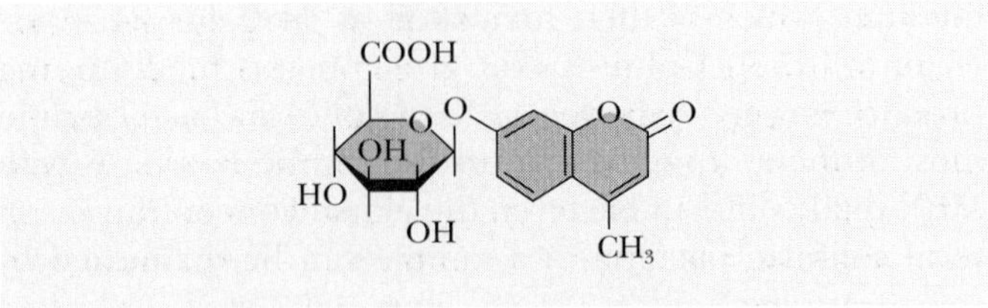

MUG
(4-methylumbelliferyl-β-D-glucuronide)

***Figure* 33.6** The chemical structure of 4-methyl-β-D-umbelliferyl glucuronide (MUG). Although this fluorogenic compound is not fluorescent, if cleaved by glucuronidase, the fluorescent product, methylumbelliferyl, is released.

when cleaved enzymatically, release either colored end products that can be seen visually (colorimetrically) or fluorescent end products that can be detected when illuminated with short wavelength radiation (ultraviolet radiation), respectively. Rapid tests have been developed for total and fecal coliform bacteria using these compounds. For example, one colorimetric test for total coliform bacteria uses orthonitrophenyl-β-D-galactoside (ONPG) as a substrate for the lactose-splitting enzyme, β-galactosidase, found in *E. coli* and other enteric bacteria that ferment lactose. When this enzyme is present, it cleaves the galactoside, producing orthonitrophenol, which has a characteristic yellow color. As a result, a yellow color is indicative of the presence of coliform bacteria. An example of the use of a fluorogenic substrate is 4-methyl-β-D-umbelliferyl glucuronide or MUG (Figure 33.6). Almost all strains of *E. coli,* but few other enteric bacteria, produce a glucuronidase enzyme that cleaves this substrate to release the methyl umbelliferyl compound that fluoresces when illuminated with ultraviolet radiation. By use of these two tests in combination, total coliform and fecal coliform tests have been developed.

The use of colorimetric and fluorogenic compounds appears to be gaining favor for the analysis of water samples. As with more traditional tests, these require that the bacteria grow in the vials and that they be metabolically active. However, the newer tests do not require lengthy incubations and follow-up inoculations.

Perhaps the future for detection of *E. coli* and waterborne pathogens lies in the development of nucleic acid probe procedures to look for specific organisms directly in natural samples. Some preliminary results suggest that polymerase chain reaction (PCR) procedures might be developed that could greatly accelerate the detection of fecal contamination.

Landfills and Composting

Microbes are also involved in the degradation of garbage and in the process of composting. This is not surprising because these contain large amounts of organic material, and are therefore good substrates for heterotrophic microorganisms.

Landfills

All cities collect garbage, called **solid waste,** from private homes and take this to a location away from the city where it is placed in the ground or **landfill.** This is then covered with earth and allowed to decay naturally (Figure 33.7).

Landfills contain not only potato peels, tin cans, and discarded cardboard boxes, but also plastic bags, styrofoam packing materials, broken appliances, and hazardous

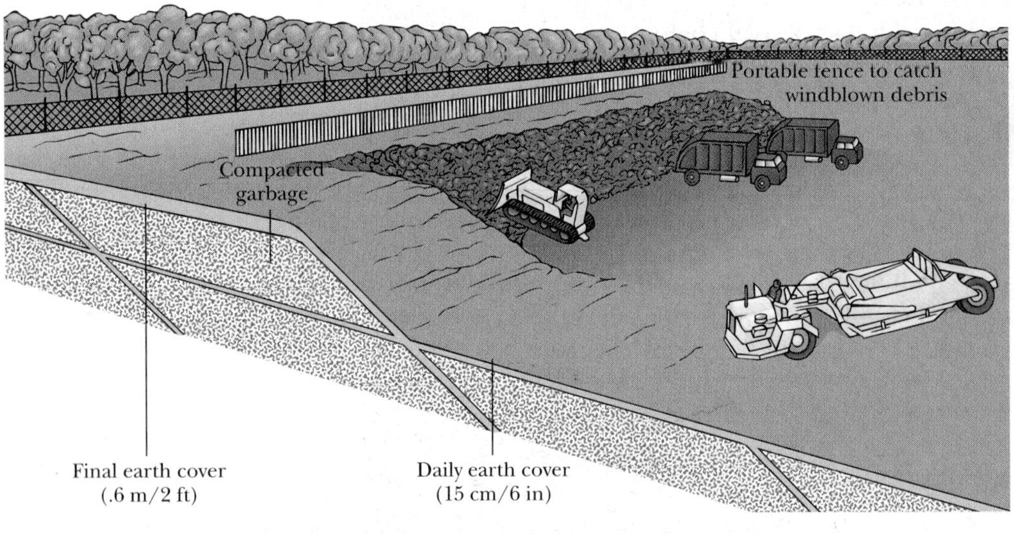

***Figure* 33.7** Sketch showing a typical landfill. Solid waste (garbage) is unloaded into the open earth pit. After the pit is full, soil is used to cover it.

chemicals. As a result, during the 1960s and 1970s, most old-fashioned garbage dumps became contaminated sites.

More recently, new regulations have been imposed to improve the treatment of solid waste materials. Recycling is practiced widely in the United States. Thus, cardboard, aluminum, and glass are usually treated separately from garbage. Likewise, hazardous materials are also handled in a separate way. So, landfills are beginning to look more like old-fashioned garbage dumps again.

Microbial degradation is the major way in which the organic materials in landfills are broken down. Because the landfilled material is covered with soil, conditions become anaerobic. Therefore, fermentations are very important. Anaerobic bacteria such as methanogens and those involved in cellulose decomposition, including *Clostridium* spp., grow in such systems. As a result of the activity of the anaerobic microorganisms, methane and carbon dioxide are major gases released. In addition, organic acids and alcohols are produced. These, and heavy metals derived from corroding materials, may be leached from the landfill and have deleterious effects on receiving streams (Figure 33.8). These heavy metals tend to accumulate in the environment and have adverse effects on plant and animal life. Thus, it is important to consider the effluents from landfills to see that they do not interfere with the environment of the watershed.

(a)

(b)

***Figure* 33.8** **(a)** The small stream contains leachate from a landfill. **(b)** The gills of this salmon fingerling from a hatchery downstream of the leachate show heavy bacterial and fungal growth due to the enrichment effects of the effluent. (Courtesy of James Staley and James Huff)

Composting

Composting is the process whereby plant materials are decomposed. Since plant materials (leaves and stems) are very high in organic content but low in nitrogen, they are not as readily degraded as most garbage. In composting, the material is placed in a pile or windrow in which the composting process is allowed to occur (Figure 33.9). This may be in a container or may simply be in a compost heap on the ground.

Composting typically occurs over a period of weeks. The most important microbial groups involved are the bacteria, in particular the actinomycetes, as well as fungi. The process undergoes several stages, including a period of heat generation in which mildly thermophilic microorganisms are selected. Eventually much of the material being composted is degraded so that the volume is depleted. This material can be used as "mulch," an organic amendment to soils used for gardening. However, it is important to recognize that heavy metals and pesticides or other hazardous materials used in gardening may be concentrated by the mulching process.

The practice of commercial composting is becoming more popular. In this practice, grass clippings, leaves, tree trimmings, and other yard waste are collected and mulched in large windrows under controlled conditions. This practice has certain advantages over landfills in that the material can be sold back to the public as a mulch.

***Figure* 33.9** A compost heap. (Courtesy of G. Buttner/Naturbild/OKAPIA/Photo Researchers, Inc.)

Pesticides

Pesticides are chemical substances manufactured by the chemical industry. Such chemically synthesized organic compounds, not previously present on Earth, are termed **xenobiotic compounds** (the term **xenobiotic** is derived from Greek and means literally "stranger (*xeno-*) to the biotic environment." Pesticides are used for control of insects (insecticides) or weeds (herbicides).

DDT is perhaps the best known pesticide (Figure 33.10**a**). It interferes with synthesis of chitin, the tough polymeric material composed of β 1-4 glucosamine, found in the exoskeleton of insects and crustaceans. Because it is such an effective insecticide, DDT has been used all over the world. DDT is an example of a persistent pesticide. Unlike natural organic materials that are readily degraded by microorganisms in the carbon cycle, some xenobiotic compounds, such as DDT, are not. Therefore, DDT persists in the environment for long periods of time. It is a recalcitrant molecule (see Chapter 12).

Furthermore, DDT is concentrated in the food chain in a process called **biomagnification** (Figure 33.10**b**). While it is usually found in relatively low concentrations in soils and aquatic habitats, it is concentrated in the fatty tissues of animals. Thus, as it passes through the food chain it accumulates in the higher animals. The highest consumers in the food chain—fish, birds, humans, and other mammals—can accumulate large concentrations in their tissues. Indeed, some humans have been found with levels greater than 50 ppm (parts per million) in their tissues, much higher than the concentration in animals, such as beef (5.0 ppm), that are permitted for human consumption.

DDT has proven to be especially harmful to birds who eat insects or fish that have high concentrations of the pesticide. The high levels of DDT in birds interfere with normal eggshell formation. The weak eggs are frequently broken in the nest prior to hatching, thereby affecting the bird reproduction rates. In the United States, eagles, ospreys, and other birds of prey have been particularly adversely impacted. For this reason, DDT has been banned for use as a pesticide in this country. Banning of DDT has resulted in increased eagle populations in the United States.

Another example of a herbicide that is quite persistent in the environment and has also been banned in the United States is 2,4,5-T. Although it contains only one additional chlorine atom, it is much more resistant (2–3 years to degrade) in the soil compared with 2,4-D (3 months to degrade) (Figure 33.11). 2,4-D is also referred to as "agent orange," which was used on a large scale as a defoliant in the Vietnam War.

Because of the problems associated with some persistent xenobiotic compounds, only readily degraded pesti-

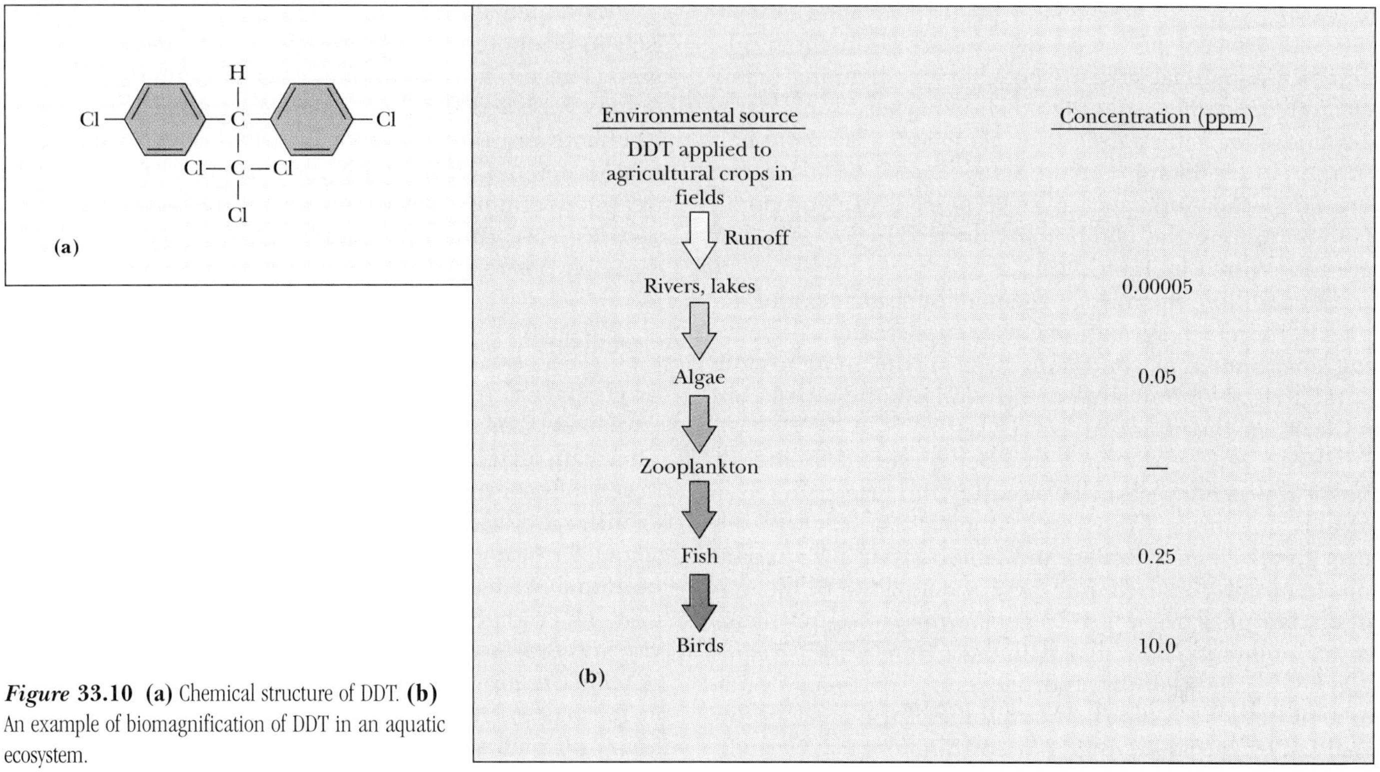

***Figure* 33.10** **(a)** Chemical structure of DDT. **(b)** An example of biomagnification of DDT in an aquatic ecosystem.

cides are now permitted for use in the United States. Most of these break down through microbial activity in a period of days or weeks following application.

As mentioned above, readily degraded pesticides have now replaced persistent pesticides. However, another completely different approach has also been taken: the use of **biological insecticides.** The best example of this is the use of the **Bt protein** produced by *Bacillus thuringiensis* (see Chapters 20 and 32). These compounds are produced normally by microorganisms in the environment and therefore can be readily degraded. Furthermore, they are highly specific in their activity compared with DDT. Thus, specific larval stages of insects are affected, without any effect on birds and higher animals. It is manufactured by growing bacteria in industrial fermentors and is used not only for the control of plant-eating insects, but for other insects as well. For example, some species of *Bacillus* produce proteins that destroy mosquitos, and they can be used for the control of Anopheles mosquitos that carry malaria.

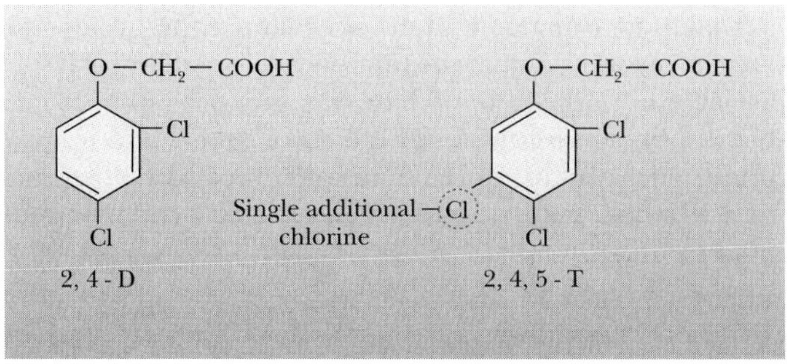

***Figure* 33.11** Comparison of the structure of 2,4-Diphenoxyacetic acid (2,4-D) and 2,4,5-T.

Bioremediation

The rapid growth of the chemical industry during the last 100 years has significantly altered our lifestyle and improved our standard of living. The "chemical age" has also created problems, including monumental environmental pollution. Today, in the United States alone, over 50,000 hazardous waste sites have been identified. Of these, 1200 have been designated by the Environmental Protection Agency (EPA) as Superfund sites. These particular sites are so hazardous that the federal government has set aside billions of dollars for their cleanup. About 40 million Americans live within four miles of a Superfund site.

Furthermore, it has been estimated that 15 percent of the five to seven million underground storage tanks in the United States are leaking. Most of these tanks, such as at gasoline stations or for home heating oil, contain petroleum products, but some contain considerably more hazardous chemicals.

Moreover, groundwaters in the United States are often dangerously polluted with an array of chemicals, chiefly commercial solvents used for dry cleaning or for cleaning machinery and high technology electronic parts. Groundwater is the source of drinking water for about half of the people in the United States, and it serves as the drinking water supply for over 95 percent of rural populations. Among the compounds present in groundwater are those listed in Table 33.2.

In addition, the mismanagement of pesticides and fertilizers, largely in agricultural areas, has created threatening environmental conditions even in the most remote rural areas. Not only are our fish and wildlife at risk, but hazards exist for humans as well.

What can be done? A number of chemical and/or physical procedures can be used for the removal of hazardous chemicals from environments. These chemicals can then be concentrated and stored safely. They could be chemically oxidized and thereby detoxified by incineration, but most are not so treated as this causes problems in air pollution. Potential pollutants from waste streams can be concentrated by absorption on a solid phase, such as activated charcoal. These are some of the mechanisms that can be employed to rid our environment of hazardous chemicals. Certain other combinations of absorption and extraction methodologies are also available. However, these processes are usually very expensive and not always applicable, particularly for contamination problems in soils and other natural environments. A more practical method for remediating environments stressed by hazardous waste is by **bioremediation.**

***Table* 33.2 Organic compounds that have been detected in drinking water wells**

Compound	Highest Level Reported mg/liter
Trichloroethylene	27.30
Toluene	6.40
1,1,1-Trichloroethane	5.44
Acetone	3.00
Methylene chloride	3.00
Dioxane	2.10
Ethyl benzene	2.00
Tetrachloroethylene	1.50
Cyclohexane	0.54
Chloroform	0.49
Di-n-butyl-phthalate	0.47
Carbon tetrachloride	0.40
Benzene	0.33
1,2-dichloroethylene	0.32

Bioremediation is defined as the use of living organisms to promote the degradation of environmental pollutants. The applications of bioremediation methodology, in many ways, are extensions of the technology that has been so effective in the treatment of urban sewage and industrial wastewater. Bioremediation has been applied to the treatment of unusual contaminations including munitions such as TNT **(Box 33.2)** and even the treatment of radioactive wastes **(Box 33.3).**

Organisms

Evolution has resulted in a vast array of microorganisms that have a broad and flexible biodegradative capacity. They can survive and destroy or detoxify chemicals in a variety of environmental habitats (hot, cold, low pH, with or without oxygen, and so on). Often the organisms best suited for bioremediation are the species that are indigenous to a particular polluted habitat or one that is similar to it. The indigenous microbe has proven that it can survive and grow in that particular environment. Also, mixed populations are superior, in many cases, to axenic cultures in the biodegradation of chlorinated aromatic hydrocarbons and pollutants.

Genetically altered organisms tend to be less stable than unaltered native populations. This, of course, is not always the case, and genetically selected microbes have been employed successfully in bioremediation. However, it is illegal to introduce genetically *engineered* bacteria into natural environments at this time because of concern about the unknown effects on natural populations. Therefore, genetically engineered bacteria cannot be used for *in situ* bioremediation—that is, they cannot be inoculated into the polluted environment to degrade the toxicant compounds (see Conversation with E. W. Nester).

The genes for many catabolic enzymes involved in biodegradation are carried on plasmids. Therefore, enhancing the number of these plasmids in a microbe selected for bioremediation is advantageous.

Hazardous environments are often formidable in that they generally contain more than one type of toxic substance. For example, it is not uncommon that a site will contain heavy metals such as cadmium, lead, or mercury, as well as chlorinated and other organic compounds. It is therefore essential that the microbes selected for such sites are able not only to remediate one group of compounds, but also to survive the toxic effects of others and to grow under such unfavorable conditions. Increased tolerance to acid, base, temperature, salinity, or heavy metals is often essential.

Advantages of Bioremediation

Bioremediation has a number of advantages over other processes for remediation, such as broad applicability and low cost. Incineration requires that the toxicant be in a burnable form, and combustion itself may produce toxic smoke. This hazardous smoke problem can be eliminated by applying rigorous conditions during incineration, which often is expensive. Another advantage to bioremediation is the low risk of exposure to hazardous chemicals during cleanup. Chemical/physical excavation methods of remediation often require the handling of the hazardous material. Ideally, sediments in riverbeds, harbors, and lakes should be remediated without disturbing the sediments. When sediments are disturbed, this increases the chance of enlarging the area polluted or of transporting them downstream or to adjacent environments. When bioremediation is successful, there is minimal danger to the environment, and the products generated, such as CO_2, H_2O, and fatty acids, are innocuous.

Problems Associated with Bioremediation

The character of the polluted site and the nature and concentration of the contaminant(s) are significant factors in bioremediation. Hazardous waste sites can contain crude petroleum, refined fuels, solvents, complex mixtures of organic compounds, metals, acids, bases, brines, or various combinations of these. Many pollutants mimic naturally occurring cellular constituents and are degraded by enzymes present in the biodegrading microbes. For example, xenobiotic compounds are hazardous chemicals specifically selected for by the organic chemical companies that synthesize them, for their resistance to biological attack.

Most chemically synthesized compounds are biodegradable, although many are biodegraded at an unacceptably slow rate. Hence, they remain in the environment for long periods of time. However, as with domestic sewage, the real problem is in bringing the pollutant in contact with the proper microbe under appropriate conditions of O_2, pH, and nutrients. Soils and water can be depleted of nitrogen, phosphorous, or other minerals required for microbial growth. In many cases an alternate energy source is essential for rapid biodegradation to occur.

Mineralization to CO_2 is the goal in bioremediation. This does not always occur, and often the pollutant is transformed to a secondary product. This is acceptable when the secondary product is less toxic than the parent compound and is utilized by other organisms present in the environment. It is unacceptable when the product is more toxic than the parent pollutant and is not readily biodegraded. Generally, catabolism of an environmental toxicant yields a more soluble and, hence, more mobile product. Hazardous chemicals are generally reduced organic compounds, and introduction of a hydroxyl group leads to the more soluble product.

Methodology of Bioremediation

Bioremediation can be accomplished *in situ* (without excavation and recovery) or by methods that require recovery and aboveground treatment. *In situ* treatment of contaminated soil involves adding nutrients or microbes to the otherwise undisturbed soil. Nutrients include sources of nitrogen, phosphorous, and/or an alternate energy source. This augmentation encourages the growth of the indigenous microbes that can catabolize the target pollutant(s). It may also be necessary to sparge air into the polluted area to promote growth of aerobic microorganisms. Microbes that can catabolize selected contaminants may also be inoculated into the environment for *in situ* bioremediation processes.

There are a number of bioremedial strategies that involve removing or recovering the polluted environment aboveground, followed by remediation and replacement after restoration. Some of these methodologies of environmental restoration are as follows.

Pump and Treat

These procedures can be employed effectively in bioremediation of polluted water such as groundwater. The groundwater is pumped to the surface, nutrients are added, and the water is reinjected into the contaminated zone. For a groundwater pollutant such as trichloroethylene (TCE) the nutrients added would be methane and O_2, inasmuch as the principal organisms responsible for degradation of this compound are the methanotrophic bacteria. The biodegradation of the major groundwater pollutant, trichloroethylene (TCE), a common dry cleaning and degreasing solvent, is carried out by these methanotrophic bacteria (Figure 33.12). Although the methanotrophs do not utilize TCE as a carbon and energy source, the methane monooxygenase they produce is able to dechlorinate and degrade this compound. This is an example of **cometabolism.** Thus, the methanotrophs need to be supplied with methane to accomplish this process. The process of treatment may take months or even years, depending on the degree of contamination of the site.

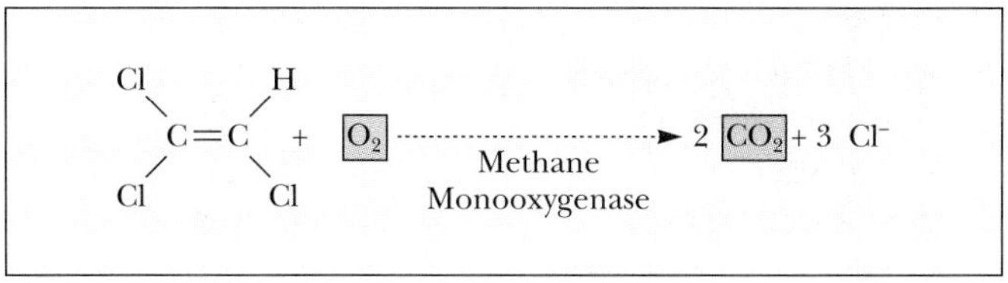

***Figure* 33.12** Overall action of methane-utilizing bacteria in trichloroethylene degradation. Intermediates are not shown. In the mixed populations of the groundwater environment, a variety of other heterotrophic bacteria would also be involved.

BOX 33.2 RESEARCH HIGHLIGHTS

Some Bacteria Get a Charge from TNT!

Military sites are often contaminated by a variety of toxic compounds. Among the most problematic are munitions such as TNT (trinitrotoluene). This well-known explosive is a common contaminant of soils near munitions storage areas. In some instances contaminated soils may contain more than 5 percent TNT!

A team of microbiologists at the University of Idaho led by Ron Crawford has recently discovered that TNT is readily biodegradable by *Clostridium bifermentans* (Figure **a**). All that is needed is to add starch to the contaminated soils and the TNT is degraded anaerobically to harmless compounds. This is an example of cometabolic activity—the bacteria cannot use TNT as an energy source, but they can degrade it if starch is available as a carbon source for their growth (Figure **b**). Researchers at the University of Idaho have developed a patented treatment process in which contaminated soils are mixed in large reactor containers with starch and an inoculum of the bacterium. This anaerobic treatment is now commercially available and is used successfully for the removal of TNT from contaminated sites (Figure **c**).

(a)

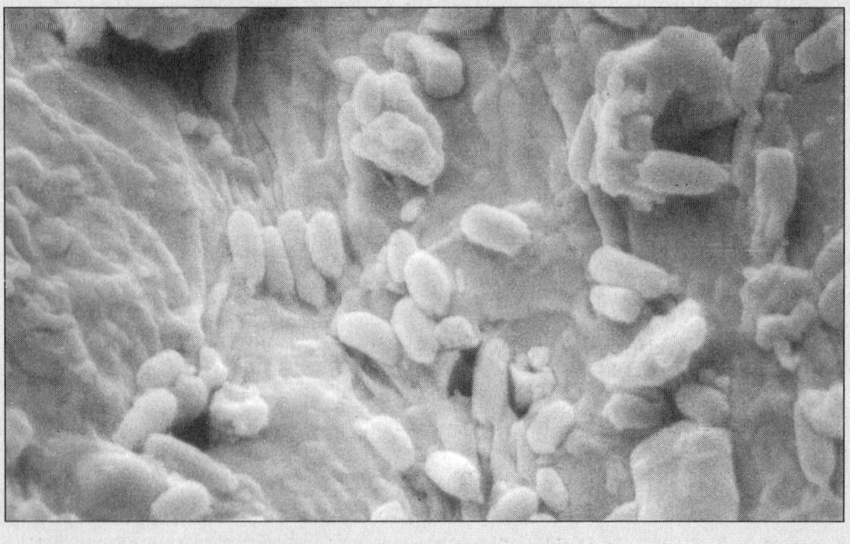

(a) An SEM of the *Clostridium bifermentans* that is responsible for degradation of TNT. It is adsorbed onto peat, which is used to inoculate soil containing TNT. (Courtesy of S. Sembries and R. Crawford)

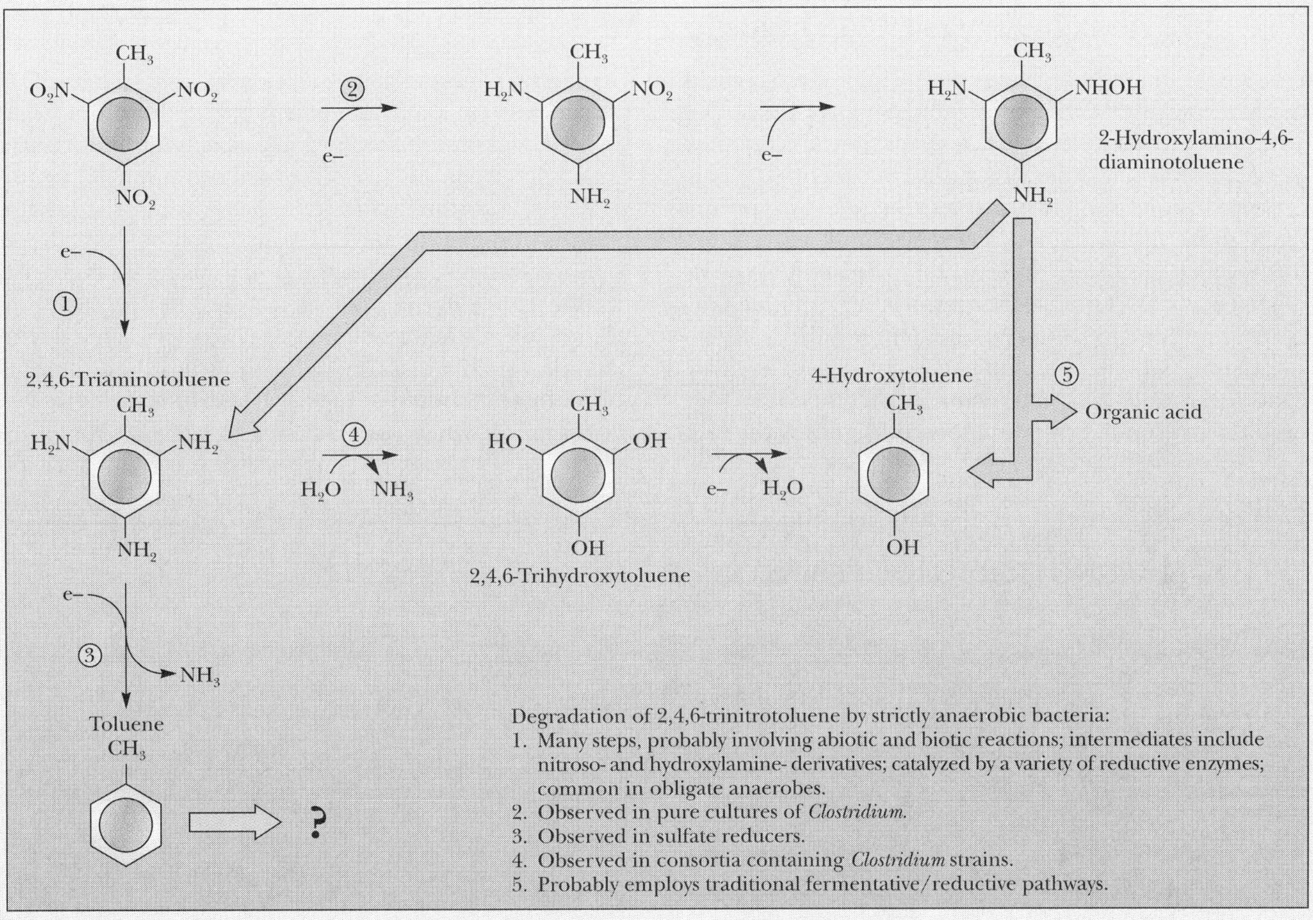
CH3
O2N
NO2
NO2
②
e–
H2N
NH2
NHOH
2-Hydroxylamino-4,6-diaminotoluene
①
2,4,6-Triaminotoluene
④
H2O
NH3
HO
OH
2,4,6-Trihydroxytoluene
H2O
4-Hydroxytoluene
⑤
Organic acid
③
Toluene
?
Degradation of 2,4,6-trinitrotoluene by strictly anaerobic bacteria:
1. Many steps, probably involving abiotic and biotic reactions; intermediates include nitroso- and hydroxylamine- derivatives; catalyzed by a variety of reductive enzymes; common in obligate anaerobes.
2. Observed in pure cultures of Clostridium.
3. Observed in sulfate reducers.
4. Observed in consortia containing Clostridium strains.
5. Probably employs traditional fermentative/reductive pathways.

(b) The probable pathway by which TNT is degraded by *Clostridium bifermentans.* (Courtesy of R. Crawford)

(c) The bioreactor used for anaerobic bioremediation of TNT-contaminated soils. (Courtesy of R. Crawford and D. Crawford)

BOX 33.3 RESEARCH HIGHLIGHTS

Role of Bacteria in Concentration of Radioactive Waste

One of the incredible things about microorganisms is the fact that they can grow in the presence of moderate levels of radioactivity. Of course, they are mutated by radiation as are other organisms, but their large concentrations and the effects of natural selection permit many to survive and grow. Therefore, they may, in the future, play important roles in the management of radioactive wastes. One interesting positive example concerns the reduction of uranium compounds.

Often, radioactive chemicals occur in very low concentrations and are mixed with other chemicals. Microbiologist Derek Lovley and his colleagues at the U.S. Geological Survey have developed and patented a process for concentration of radioactive uranium using bacteria. They use metal-reducing bacteria that oxidize organic compounds while reducing uranium from the U(VI) state to the U(IV) state. Whereas U(VI) is soluble, the reduced form, U(IV), is insoluble and therefore precipitates out from the other wastes. This process thereby concentrates and removes the radioactive compounds from the waste material (Figure **a**). One of the bacteria capable of carrying out uranium reduction is the common sulfate-reducing bacterium, *Desulfovibrio desulfuricans,* which uses lactate as a carbon source.

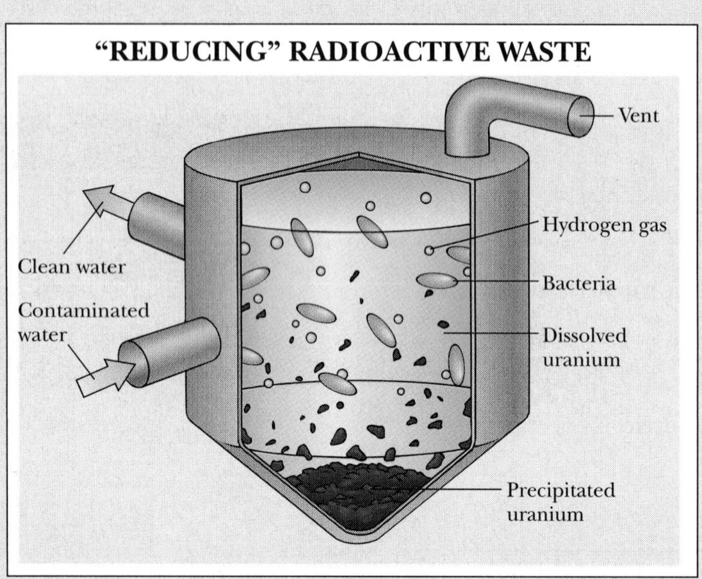

A diagram showing a reactor vessel in which radioactive wastes are concentrated by reduction and precipitation. (Courtesy of D. Lovley)

Bioreactors

Bioreactors are often employed to bring together the pollutant and the biodegrading microbe. A bioreactor follows the general principles involved in an industrial fermentor (see Chapter 32). A slurry of soil or groundwater is placed in the fermentor and an inoculum is added. This inoculum may be activated sludge from a sewage disposal plant, an appropriate pure culture or a mixed microbial culture, or an inoculum from a contaminated site. The pure or mixed cultures may be added as a suspension or on a solid support. Supports employed are activated carbon, plastic spheres, glass beads, or diatomaceous earth. The attachment of bacteria to solid support systems was discussed in Chapter 32. Bioreactors can be operated as a continuous culture system where about 80 percent of the material is removed periodically. The remaining 20 percent serves as inoculum for the added contaminated material.

Batch reactors may be placed in sequence so that the microbial population in each has the capacity to biodegrade selected chemicals in the slurry or water. For example, bacteria such as *Pseudomonas putida* are able to degrade chlorinated phenolic compounds and use them as a carbon source for growth. However, this species is not able to degrade TCE. The preferred group of organisms for degradation of this compound is the methanotrophs, as discussed previously. Furthermore, the conditions necessary for the growth of methanotrophs (methane gas as energy substrate) are not necessary for *P. putida*. Thus, a sequence of two reactors, the first containing *P. putida* with conditions ideal for its growth, followed by a separate reactor for methanotrophs would result in the complete remediation of both chlorinated phenols and TCE.

The sequential reactor arrangement is particularly important when the contaminated material contains a chemical that is generally toxic to microbes. Removal of this chemical by a selected microbe renders the rest more susceptible to bioremediation. Sequential bioreactors also permit the operation of one anaerobically and the following one aerobically. This latter arrangement is particularly useful when highly chlorinated compounds, such as PCBs, are present. It is known that the most highly chlorinated forms of the PCBs are degraded only anaerobically **(Box 33.4)**. Thus, if the first stage reactor is anaerobic, the more highly chlorinated compounds will be partially dehalogenated, and the less chlorinated products from that reactor can be transferred to an aerobic reactor where they can be degraded completely.

Mound or Heap

In this procedure, soil is placed in mounds on a plastic liner. Water, microbes, and nutrient are trickled over the soil. The biodegradation of many pollutants, including chlorinated aromatics, is augmented by also adding an energy source that is readily utilizable. The effluent from the process can be collected and recycled and the process operated until the pollutants are degraded.

Land Farming

Selected pollutants can be removed from waste streams by running the waste into a confined soil basin. Nutrients may be added to the soil and the soil can be tilled to increase mixing and aeration. This method is effective where fertile soil lies over a firm, relatively impermeable clay base. The clay base impedes penetration of the hazardous waste into the groundwater.

An Example of Bioremediation

Soil bioremediation was successfully conducted at an oil tank farm facility in California. The soil remediated came

BOX 33.4 RESEARCH HIGHLIGHTS

Degradation of Polychlorinated Biphenyl Compounds (PCBs)

PCBs are chlorinated double-ringed aromatic compounds (Figure **a**). Depending on the position and number of chlorine atoms, over 200 different varieties of PCB, called congeners, exist (Figure **b**). Like DDT, which is similar chemically, PCBs accumulate in the fatty tissues of animals that consume them. Until recently, the most highly chlorinated congeners of the PCBs were not known to be degraded by microorganisms.

Some of the chemical manufacturers that produced PCBs are located on the Hudson River, which became contaminated by these compounds. In the 1980s it was discovered that the contaminated sediments had lower concentrations of the higher chlorinated PCBs than expected, suggesting they were being degraded in the environment. Subsequent laboratory studies at Jim Tiedje's laboratory at Michigan State University confirmed that PCBs were being degraded. However, this breakdown occurred only under *anaerobic* conditions. This was a truly exciting discovery because it indicated that these compounds could be degraded by bacteria and therefore could be removed from the environment. Furthermore, it indicated that anaerobic conditions were necessary for degradation of these and probably other highly chlorinated compounds.

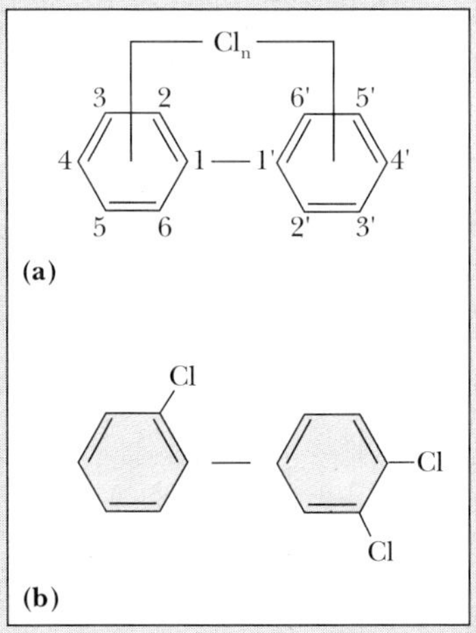

(a) General structure of PCB compounds. Depending on the location of the chlorine atoms, some 200 or so different congeners are possible. **(b)** One of the possible PCB congeners, 2, 3′, 4′ trichlorobiphenyl.

from under concrete-lined tanks, which had been used to store fuel oil for ships. The soil pollutants consisted of hydrocarbons that were from 10 to 35 carbons in length. The contaminated soil varied in total petroleum hydrocarbon (TPH) content from one place to another at the site, but specific areas had up to 30,000 ppm total petroleum hydrocarbons. Soil treatment units were set up so that a unit of excavated soil for bioremediation had 5,000 to 6,000 ppm total petroleum hydrocarbon.

The nutrients added were phosphate, ammonia, and nitrate. The disappearance of hydrocarbon and the relative number of the predominant heterotrophic bacteria present were monitored over a period of several weeks (Figure 33.13). Data presented indicates that the total ppm petroleum hydrocarbon decreased from 5000 ppm initially to about 600 ppm in a period of 14 weeks. The bacterial count suggested that over the course of study the total number of a specific orange-pigmented bacterial species increased about threefold (Figure 33.14). This organism was tested in the laboratory for the ability to utilize petroleum hydrocarbons. The organism grew on pentadecane (C-15), octadecane (C-18), docosane (C-20), hexacosane (C-26), and the branched hydrocarbon pristane. These hydrocarbons could serve as sole source of carbon and energy, indicating that the orange-pigmented bacterium was probably involved in the bioremediation process.

The factor that limited this natural population of bacteria from utilizing the petroleum hydrocarbons at this site was that the levels of required nutrients (nitrogen and phosphorus) in the soil were too low. In the absence of these nutrients, the petroleum hydrocarbons could not be degraded. Addition of nutrient and oxygen resulted in rapid bioremediation.

It is also noteworthy that the workers involved in this bioremediation effort added commercial preparations of hydrocarbon-utilizing bacteria. They wanted to determine whether such preparations would increase the rate of petroleum degradation. They found that the commercial preparations did not alter the rate of petroleum hydrocarbon disappearance. This illustrates an important principle—that the microbes present in an environment are best able to thrive there. As hydrocarbons are natural constituents of living cells, all environments, including those polluted with hydrocarbons, contain microorganisms capable of utilizing them. The commercial microbe preparation probably had little effect because it contains a population that was not adapted to survive well in that environment.

The Future of Bioremediation

Bioremediation has a promising future, although it is not the sole answer to our huge problem of hazardous waste in the United States. The low cost and broad applicability to many hazardous waste problems would make bioremediation a method of choice. However, much work and funding will be required to apply bioremediation to those situations where it can be used successfully.

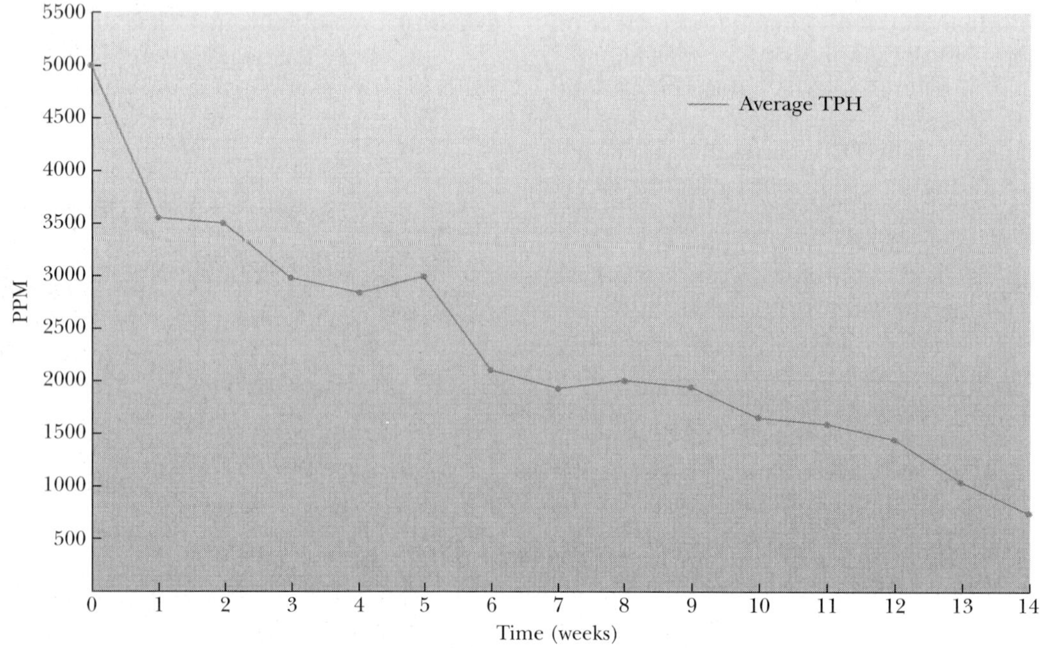

***Figure* 33.13** The rate of disappearance of total petroleum hydrocarbons (TPH) from a treatment site in California (adapted from G. S. Sayler, R. Fox, and J. W. Blackburn, eds. 1991. *Full-scale Bioremediation of Contaminated Soil and Water:* New York: Plenum).

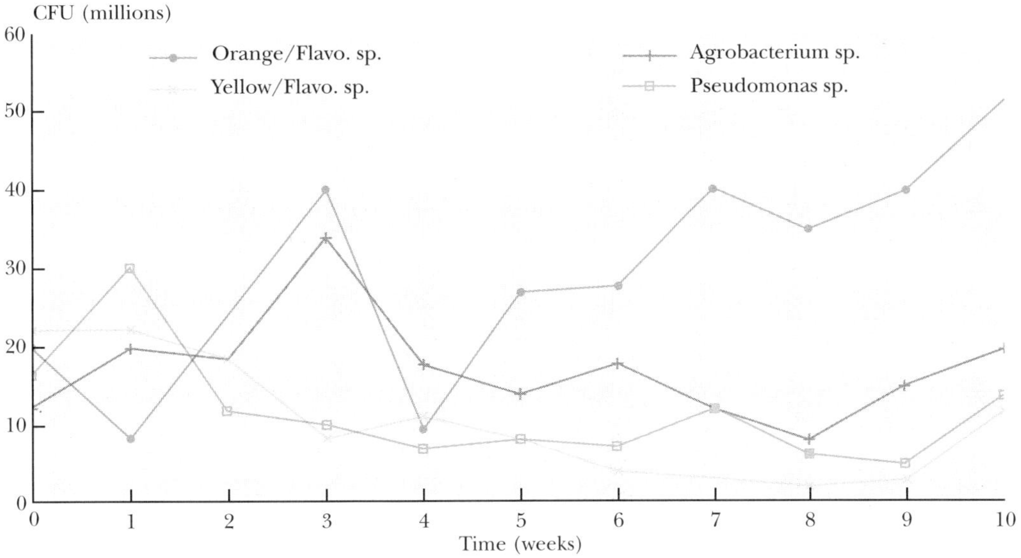

***Figure* 33.14** Analysis of the dominant heterotrophic bacterial species during petroleum hydrocarbon degradation (adapted from G. S. Sayler, R. Fox, and J. W. Blackburn, eds.).

Acid Mine Drainage

Strip mining practices, in which the surface soil is removed to expose the minerals of interest for recovery, may result in the problem of acid mine drainage. This problem occurs when the underlying minerals contain significant amounts of sulfide minerals such as pyrite, FeS_2. Strip mining exposes these reduced sulfides to oxygen and rainwater. As a result, thiobacilli flourish and oxidize the sulfur in pyrite to sulfuric acid. The iron can be oxidized by thiobacilli to form iron oxides.

In areas in Appalachia and in certain midwestern states such as Indiana, coal deposits contain large quantities of pyrite. Strip mining has exposed the sulfides and iron to oxidative activity by the thiobacilli. Runoff waters leached from these sites can have pH values of as low as 2.5 to 4.5—so low that fish are killed, as are the aquatic plants that live in the stream or receiving waters (Figure 33.15).

There is no simple way to remedy acid mine drainage. Therefore, EPA now requires that strip mining areas are reburied with the overlying soil when the mining operation is completed. In this manner, the sulfides are again removed from exposure and oxidation by thiobacilli.

Coal and oil deposits that are high in sulfur pose problems for the environment. If they are burned, they can cause acid rain. Thus, it is desirable to remove the sulfur from them before burning takes place. One way in which this can be accomplished is by use of thiobacilli to oxidize the sulfur and sulfides to sulfate, which can be removed in solution before the fuel is burned.

(a)

(b)

***Figure* 33.15** Acid mine drainage area showing **(a)** exposed sulfur ("yellow boy") and **(b)** dead plants next to a lake of pH 2.5. (Courtesy of A. E. Konopka)

Summary

- Because of their diverse capabilities of degrading organic materials, microorganisms, particularly bacteria, are used in the treatment of **wastewater** (sewage). Typical sewage treatment plants are called **secondary wastewater treatment** facilities because they entail two successive treatment systems, **primary treatment,** involving the physical separation of particulate material **(sludge)** from the fluid portion of sewage, and **secondary treatment,** in which organic material is degraded by microbial communities. Most secondary wastewater treatment facilities use the **activated sludge process,** an aerobic process, for reduction in organic materials, a major component of the biochemical oxygen demand (BOD) in wastewaters. **Anaerobic sludge digestors,** in which methanogenesis occurs, are used to reduce the organic content of sludge. **Biological tertiary treatment** is used to further treat secondary effluent in some facilities to remove nitrogen (by nitrification followed by denitrification) and phosphorus by *Acinetobacter* spp.

- **Drinking water treatment** is necessary to ensure that the public's health is protected from pathogenic bacteria that may contaminate drinking water supplies. Drinking water is treated at filtration plants that use several process steps including **coagulation** with alum addition, **sedimentation** to remove particulate materials, including bacteria, adsorbed to the alum, and **filtration** through sand filters. The final treatment is **disinfection,** usually by the addition of chlorine gas which forms hypochlorite solution, to disinfect the water before it is distributed to households and other consumers in the drinking water supply system.

- The safety of drinking water is tested by assaying for **total** and **fecal coliform** bacteria. *Escherichia coli,* which is both a total and fecal coliform bacterium, is used as the **indicator** organism, whose presence indicates the contamination of a water supply by mammalian feces. Unfortunately, positive total and fecal coliform tests can be caused by other bacteria, such as *Enterobacter aerogenes,* which is a soil organism, so additional tests are needed to validate the presence of fecal contamination by *E. coli.*

- **Landfills** (garbage dumps) are used to treat **solid waste** materials. Anaerobic bacteria degrade the organic materials placed in landfills. **Composting** involves the degradation of organic materials, such as leaves and grass, using microorganisms.

- **Bioremediation** is the process whereby toxic materials are degraded by microorganisms that are capable of degrading almost all organic substances—including halogenated compounds such as PCBs, almost all pesticides, aromatic and aliphatic hydrocarbons, and even munitions. Bioremediation processes are being used to clean up **EPA Superfund sites** scattered about the United States.

- **Acid mine drainage** occurs when mining operations expose reduced sulfur compounds such as pyrite to air and water. **Thiobacilli** oxidize the sulfides to produce sulfuric acid and may lower the pH to as low as 2.5 to 4.5. This acid mine drainage kills fish and plants in the watershed. The only practical solution to this problem is to cover up the mine after the minerals have been removed so that air and water do not reach the sulfides.

Questions for Thought and Review

1. Differentiate between primary and secondary wastewater treatment.
2. Compare the trickling filter with activated sludge treatment.
3. What is the microbiology involved in tertiary treatment for nitrogen removal? For phosphate removal?
4. What do you believe should be done to or with the undigested sludge from anaerobic digestors?
5. Why is it so difficult to identify *Escherichia coli* in water samples?
6. Is *Escherichia coli* a pathogen?
7. What would be the ideal microbiological test for water potability?
8. Explain the phenomenon of biomagnification.
9. Provide an example of cometabolism.
10. What evidence is there that selection of naturally occurring bacteria happens following contamination of a site by hydrocarbon compounds?

Suggested Readings

American Public Health Association–American Water Works Association. 1990. *Water Pollution Control Federation: Standard Methods for the Examination of Water and Wastewater.* 17th ed. Washington, DC: APHA.

Britton, G., and C. P. Gerba, eds. 1984. *Groundwater Pollution Microbiology.* New York: John Wiley & Sons, Inc.

Chaudry, G. R., ed. 1994. *Biological Degradation and Bioremediation of Toxic Chemicals.* Portland, OR: Dioscorides Press.

Guthrie, F. E., and J. J. Perry, eds. 1980. *Environmental Toxicology.* New York: Elsevier North Holland Inc.

Mitchell, R. 1992. *Environmental Microbiology.* New York: Wiley-Liss, Inc.

Omenn, G. S., ed. 1988. *Environmental Biotechnology: Reducing Risks from Environmental Chemicals through Biotechnology.* New York: Plenum.

Sayler, G. S., R. Fox, and J. W. Blackburn, eds. 1991. *Environmental Biotechnology for Waste Treatment.* New York: Plenum.

Appendix A

Classification of Bacterial Genera from *Bergey's Manual of Systematic Bacteriology*[1]

VOLUME I

Selected Gram-Negative Heterotrophic Eubacteria

THE SPIROCHETES

Genus **Borrelia**
Genus **Brachyspira**
Genus **Cristispira**
Genus **Leptonema**
Genus **Leptospira**
Genus **Serpulina**
Genus **Spirochaeta**
Genus **Treponema**
Genera of Insect Gut Spirochetes

AEROBIC/MICROAEROPHILIC, MOTILE, HELICAL/VIBRIOID GRAM-NEGATIVE BACTERIA

Genus **Alteromonas**
Genus **Aquaspirillum**
Genus **Azospirillum**
Genus **Bdellovibrio**
Genus **Campylobacter**
Genus **Cellvibrio**
Genus **Halovibrio**
Genus **Helicobacter**
Genus **Herbaspirillum**
Genus **Marinomonas**
Genus **Micavibrio**
Genus **Oceanospirillum**
Genus **Spirillum**
Genus **"Sporospirillum"**
Genus **Vampirovibrio**
Genus **Wolinella**[2]

NONMOTILE (OR RARELY MOTILE), GRAM-NEGATIVE CURVED BACTERIA

Genus **Ancylobacter**
Genus **"Brachyarcus"**
Genus **Cyclobacterium**
Genus **Flectobacillus**
Genus **Meniscus**
Genus **"Pelosigma"**
Genus **Runella**
Genus **Spirosoma**

GRAM-NEGATIVE AEROBIC/MICROAEROPHILIC RODS AND COCCI

Subgroup 4A

Genus **Acetobacter**
Genus **Acidiphilium**
Genus **Acidomonas**
Genus **Acidothermus**
Genus **Acidovorax**
Genus **Acinetobacter**
Genus **Afipia**
Genus **Agrobacterium**
Genus **Agromonas**
Genus **Alcaligenes**
Genus **Alteromonas**
Genus **Aminobacter**
Genus **Aquaspirillum**
Genus **Azomonas**
Genus **Azorhizobium**
Genus **Azotobacter**
Genus **Beijerinckia**
Genus **Bordetella**
Genus **Bradyrhizobium**
Genus **Brucella**
Genus **Chromohalobacter**
Genus **Chryseomonas**
Genus **Comamonas**
Genus **Cupriavidus**
Genus **Deleya**
Genus **Derxia**
Genus **Ensifer**
Genus **Erythrobacter**
Genus **Flavimonas**
Genus **Flavobacterium**
Genus **Francisella**
Genus **Frateuria**
Genus **Gluconobacter**
Genus **Halomonas**
Genus **Hydrogenophaga**
Genus **Janthinobacterium**
Genus **Kingella**
Genus **Lampropedia**
Genus **Legionella**
Genus **Marinobacter**
Genus **Marinomonas**
Genus **Mesophilobacter**
Genus **Methylobacillus**
Genus **Methylobacterium**
Genus **Methylococcus**
Genus **Methylomonas**
Genus **Methylophaga**
Genus **Methylophilus**
Genus **Methylovorus**
Genus **Moraxella**
Genus **Morococcus**
Genus **Neisseria**
Genus **Oceanospirillum**
Genus **Ochrobactrum**
Genus **Oligella**
Genus **Paracoccus**
Genus **Phenylobacterium**
Genus **Phyllobacterium**
Genus **Pseudomonas**
Genus **Psychrobacter**
Genus **Rhizobacter**
Genus **Rhizobium**
Genus **Rhizomonas**
Species **Rochalimaea henselae**
Genus **Roseobacter**
Genus **Rugamonas**
Genus **Serpens**
Genus **Sinorhizobium**
Genus **Sphingobacterium**
Genus **Thermoleophilum**
Genus **Thermomicrobium**
Genus **Thermus**
Genus **Variovorax**
Genus **Volcaniella**
Genus **Weeksella**
Genus **Xanthobacter**
Genus **Xanthomonas**
Genus **Xylella**
Genus **Xylophilus**
Genus **Zoogloea**

Genus **Bacteroides**
Genus **Taylorella**
Genus **Wolinella**[2]

[1]This classification has been updated using material from *Bergey's Manual of Determinative Bacteriology*, Ninth Edition (1994). Reprinted by permission Williams and Wilkins, Baltimore, MD.
[2]Note that some genera are listed in more than one place.

FACULTATIVELY ANAEROBIC GRAM-NEGATIVE RODS

Subgroup 1 Family Enterobacteriaceae

Genus **Arsenophonus**
Genus **Budvicia**
Genus **Buttiauxella**
Genus **Cedecea**
Genus **Citrobacter**
Genus **Edwardsiella**
Genus **Enterobacter**
Genus **Erwinia**
Genus **Escherichia**
Genus **Ewingella**
Genus **Hafnia**
Genus **Klebsiella**
Genus **Kluyvera**
Genus **Leclercia**
Genus **Leminorella**
Genus **Moellerella**
Genus **Morganella**
Genus **Obesumbacterium**
Genus **Pantoea**
Genus **Pragia**
Genus **Proteus**
Genus **Providencia**
Genus **Rahnella**
Genus **Salmonella**
Genus **Serratia**
Genus **Shigella**
Genus **Tatumella**
Genus **Xenorhabdus**
Genus **Yersinia**
Genus **Yokenella**

Subgroup 2 Family Vibrionaceae

Genus **Aeromonas**
Genus **Enhydrobacter**
Genus **Photobacterium**
Genus **Plesiomonas**
Genus **Vibrio**

Subgroup 3 Family Pasteurellaceae

Genus **Actinobacillus**
Genus **Haemophilus**
Genus **Pasteurella**

Subgroup 4 Other Genera

Genus **Calymmatobacterium**
Genus **Cardiobacterium**
Genus **Chromobacterium**
Genus **Eikenella**
Genus **Gardnerella**
Genus **Streptobacillus**
Genus **Zymomonas**

GRAM-NEGATIVE, ANAEROBIC, STRAIGHT, CURVED, AND HELICAL BACTERIA

Genus **Acetivibrio**
Genus **Acetoanaerobium**
Genus **Acetofilamentum**
Genus **Acetogenium**
Genus **Acetomicrobium**
Genus **Acetothermus**
Genus **Acidaminobacter**
Genus **Anaerobiospirillum**
Genus **Anaerorhabdus**
Genus **Anaerovibrio**
Genus **Bacteroides**
Genus **Butyrivibrio**
Genus **Centipeda**
Genus **Fervidobacterium**
Genus **Fibrobacter**
Genus **Fusobacterium**
Genus **Haloanaerobium**
Genus **Halobacteroides**
Genus **Ilyobacter**
Genus **Lachnospira**
Genus **Leptotrichia**
Genus **Malonomonas**
Genus **Megamonas**
Genus **Mitsuokella**
Genus **Oxalobacter**
Genus **Pectinatus**
Genus **Pelobacter**
Genus **Porphyromonas**
Genus **Prevotella**
Genus **Propionigenium**
Genus **Propionispira**
Genus **Rikenella**
Genus **Roseburia**
Genus **Ruminobacter**
Genus **Sebaldella**
Genus **Selenomonas**
Genus **Sporomusa**
Genus **Succinimonas**
Genus **Succinivibrio**
Genus **Syntrophobacter**
Genus **Syntrophomonas**
Genus **Thermobacteroides**
Genus **Thermosipho**
Genus **Thermotoga**
Genus **Tissierella**
Genus **Wolinella**[2]
Genus **Zymophilus**

DISSIMILATORY SULFATE- OR SULFUR-REDUCING BACTERIA

Subgroup 1

Genus **Desulfotomaculum**

Subgroup 2

Genus **Desulfobulbus**
Genus **Desulfomicrobium**
Genus **Desulfomonas**
Genus **Desulfovibrio**
Genus **Thermodesulfobacterium**

Subgroup 3

Genus **Desulfobacter**
Genus **Desulfobacterium**
Genus **Desulfococcus**
Genus **Desulfomonile**
Genus **Desulfonema**
Genus **Desulfosarcina**

Subgroup 4

Genus **Desulfurella**
Genus **Desulfuromonas**

ANAEROBIC GRAM-NEGATIVE COCCI

Genus **Acidaminococcus**
Genus **Megasphaera**
Genus **Syntrophococcus**
Genus **Veillonella**

THE RICKETTSIAS AND CHLAMYDIAS

I. Rickettsias
II. Chlamydias

Endosymbionts
Protozoa Endosymbionts
  Ciliates
  Flagellates
  Amoebae
  Taxa of Protozoal Endosymbionts
    Genus I. **Holospora**
    Genus II. **Caedibacter**
    Genus III. **Pseudocaedobacter**
    Genus IV. **Lyticum**
    Genus V. **Tectibacter**
Endosymbionts of Insects
  Blood-sucking insects
  Plant sap-sucking insects
  Cellulose and stored grain feeders
  Insects feeding on complex diets
  Taxon of Insect Endosymbionts
    Genus Blattabacterium
Endosymbionts of Fungi and Invertebrates other than Arthropods
  Fungi
  Sponges
  Coelenterates
  Helminthes
  Annelids
  Marine worms and mollusks

VOLUMES II AND IV
Gram-Positive Bacteria

GRAM-POSITIVE COCCI

Genus **Aerococcus**
Genus **Coprococcus**
Genus **Deinobacter**
Genus **Deinococcus**
Genus **Enterococcus**
Genus **Gemella**
Genus **Lactococcus**
Genus **Leuconostoc**
Genus **Marinococcus**
Genus **Melissococcus**
Genus **Micrococcus**
Genus **Pediococcus**
Genus **Peptococcus**
Genus **Peptostreptococcus**
Genus **Planococcus**
Genus **Ruminococcus**
Genus **Saccharococcus**
Genus **Salinicoccus**
Genus **Sarcina**
Genus **Staphylococcus**
Genus **Stomatococcus**
Genus **Streptococcus**
Genus **Trichococcus**
Genus **Vagococcus**

ENDOSPORE-FORMING GRAM-POSITIVE RODS AND COCCI

Genus **Amphibacillus**
Genus **Bacillus**
Genus **Clostridium**
Genus **Desulfotomaculum**
Genus **Oscillospira**
Genus **Sporohalobacter**
Genus **Sporolactobacillus**
Genus **Sporosarcina**
Genus **Sulfidobacillus**
Genus **Syntrophospora**

REGULAR, NONSPORING GRAM-POSITIVE RODS

Genus **Brochothrix**
Genus **Carnobacterium**
Genus **Caryophanon**
Genus **Erysipelothrix**
Genus **Kurthia**
Genus **Lactobacillus**
Genus **Listeria**
Genus **Renibacterium**

IRREGULAR, NONSPORING GRAM-POSITIVE RODS

Genus **Acetobacterium**
Genus **Acetogenium**
Genus **Actinomyces**
Genus **Aeromicrobium**
Genus **Agromyces**
Genus **Arachnia**
Genus **Arcanobacterium**
Genus **Arthrobacter**
Genus **Aureobacterium**
Genus **Bifidobacterium**
Genus **Brachybacterium**
Genus **Brevibacterium**
Genus **Butyrivibrio**
Genus **Caseobacter**
Genus **Cellulomonas**
Genus **Clavibacter**
Genus **Coriobacterium**
Genus **Corynebacterium**
Genus **Curtobacterium**
Genus **Dermabacter**
Genus **Eubacterium**
Genus **Exiguobacterium**
Genus **Falcivibrio**
Genus **Gardnerella**
Genus **Jonesia**[2]
Genus **Lachnospira**
Genus **Microbacterium**
Genus **Mobiluncus**
Genus **Pimelobacter**
Genus **Propionibacterium**
Genus **Rarobacter**
Genus **Rothia**
Genus **Rubrobacter**
Genus **Sphaerobacter**
Genus **Terrabacter**
Genus **Thermoanaerobacter**

THE MYCOBACTERIA

Genus **Mycobacterium**

THE ACTINOMYCETES

NOCARDIOFORM ACTINOMYCETES

Subgroup 1 Mycolic Acid-Containing Bacteria

Genus **Gordona**
Genus **Nocardia**
Genus **Rhodococcus**
Genus **Tsukamurella**

Subgroup 2 Pseudonocardia and Related Genera

Genus **Actinobispora**
Genus **Actinokineospora**
Genus **Actinopolyspora**
Genus **Amycolata**
Genus **Amycolatopsis**
Genus **Kibdelosporangium**
Genus **Pseudoamycolata**
Genus **Pseudonocardia**
Genus **Saccharomonospora**
Genus **Saccharopolyspora**

Subgroup 3 Nocardioides and Terrabacter

Genus **Nocardioides**
Genus **Terrabacter**

Subgroup 4 Promicromonospora and Related Genera

Genus **Jonesia**[2]
Genus **Oerskovia**
Genus **Promicromonospora**

GENERA WITH MULTILOCULAR SPORANGIA

Genus **Dermatophilus**
Genus **Frankia**
Genus **Geodermatophilus**

ACTINOPLANETES

Genus **Actinoplanes**
Genus **Ampullariella**
Genus **Catellatospora**
Genus **Dactylosporangium**
Genus **Micromonospora**
Genus **Pilimelia**

STREPTOMYCETES AND RELATED GENERA

Genus **Intrasporangium**
Genus **Kineosporia**
Genus **Sporichthya**
Genus **Streptomyces**
Genus **Streptoverticillium**

MADUROMYCETES

Subgroup 1 Streptosporangium and Related Taxa

Genus **Microbispora**
Genus **Microtetraspora**
Genus **Planobispora**
Genus **Planomonospora**
Genus **Spirillospora**
Genus **Streptosporangium**

Subgroup 2 Actinomadura

Genus **Actinomadura**

THERMOMONOSPORA AND RELATED GENERA

Genus **Actinosynnema**
Genus **Nocardiopsis**
Genus **Streptoalloteichus**
Genus **Thermomonospora**

THERMOACTINOMYCETES

Genus **Thermoactinomyces**

OTHER GENERA

Genus **Glycomyces**
Genus **Kitasatospora**
Genus **Saccharothrix**

THE MYCOPLASMAS (OR MOLLICUTES): CELL WALL-LESS BACTERIA

Genus **Acholeplasma**
Genus **Anaeroplasma**
Genus **Asteroleplasma**
Genus **Mycoplasma**
Genus **Spiroplasma**
Genus **Ureaplasma**

VOLUME III

Phototrophic, Chemolithotrophic, and Selected Heterotrophic Eubacteria and Archaea

---

ANOXYGENIC PHOTOTROPHIC BACTERIA

Subgroup 1

Genus **Amoebobacter**
Genus **Chromatium**
Genus **Lamprobacter**
Genus **Lamprocystis**
Genus **Thiocapsa**
Genus **Thiocystis**
Genus **Thiodictyon**
Genus **Thiopedia**
Genus **Thiospirillum**

Subgroup 2

Genus **Ectothiorhodospira**

Subgroup 3

Genus **Rhodobacter**
Genus **Rhodocyclus**
Genus **Rhodomicrobium**
Genus **Rhodopila**
Genus **Rhodopseudomonas**
Genus **Rhodospirillum**

Subgroup 4

Genus **Heliobacillus**
Genus **Heliobacterium**

Subgroup 5

Genus **Ancalochloris**
Genus **Chlorobium**
Genus **Chloroherpeton**
Genus **Pelodictyon**
Genus **Prosthecochloris**
Consortia, symbiotic aggregates

Subgroup 6

Genus **Chloroflexus**
Genus **Chloronema**
Genus **Heliothrix**
Genus **Oscillochloris**

Subgroup 7

Genus **Erythrobacter**

OXYGENIC PHOTOTROPHIC BACTERIA

I. THE CYANOBACTERIA

Subgroup 1 (= Order: Chroococcales)

Genus **Chamesiphon**
Genus **Cyanothece**
Genus **Gloeobacter**
Genus **Gloeocapsa**
Genus **Gloeothece**
Genus **Microcystis**
Genus **Myxobaktron**
Genus **Synechococcus**
Genus **Synechocystis**

Subgroup 2 (= Order: Pleurocapsales)

Genus **Chroococcidiopsis**
Genus **Dermocarpa**
Genus **Dermocarpella**
Genus **Myxosarcina**
Genus **Pleurocapsa**
Genus **Xenococcus**

Subgroup 3 (= Order: Oscillatoriales)

Genus **Arthrospira**
Genus **Crinalium**
Genus **Lyngbya**
Genus **Microcoleus**
Genus **Oscillatoria**
Genus **Pseudanabaena**
Genus **Spirulina**
Genus **Starria**
Genus **Trichodesmium**

Subgroup 4 (= Order: Nostocales)

SECTION A (FAMILY: NOSTOCACEAE)

Genus **Anabaena**
Genus **Aphanizomenon**
Genus **Cylindrospermum**
Genus **Nodularia**
Genus **Nostoc**

SECTION B (FAMILY: SCYTONEMATACEAE)

Genus **Scytonema**

SECTION C (FAMILY: RIVULARIACEAE)

Genus **Calothrix**

Subgroup 5 (= Order: Stigonematales)

Genus **Chlorogloeopsis**
Genus **Fischerella**
Genus **Geitleria**
Genus **Stigonema**

II. The Prochlorophytes (Order Prochlorales)

Genus **Prochloron**
Genus **Prochlorothrix**

AEROBIC CHEMOLITHOTROPHIC BACTERIA AND ASSOCIATED ORGANISMS

Subgroup 1 Colorless Sulfur-Oxidizing Bacteria

SECTION A MORPHOLOGICALLY CONSPICUOUS BACTERIA

Genus **Macromonas**
Genus **Thiobacterium**
Genus **"Thiodendron"**
Genus **Thiospira**
Genus **Thiovulum**

SECTION B LESS CONSPICUOUS SULFUR-OXIDIZING BACTERIA

Genus **Thermothrix**
Genus **Thiobacillus**
Genus **Thiomicrospira**
Genus **Thiosphaera**

SECTION C RELATED BACTERIA

Genus **Acidiphilium**

Subgroup 2 Iron- and Manganese-Oxidizing and/or -Depositing Bacteria

Genus **Gallionella**
Genus **"Leptospirillum"**
The Magnetotactic Bacteria
Genus **"Metallogenium"**
Family **"Siderocapsaceae"**
Genus **"Siderocapsa"**
Genus **"Naumanniella"**
Genus **"Ochrobium"**
Genus **"Siderococcus"**
Genus **Sulfobacillus**

Subgroup 3 Nitrifying Bacteria

SECTION A NITRITE-OXIDIZING BACTERIA

Genus **Nitrobacter**
Genus **Nitrospina**
Genus **Nitrococcus**
Genus **Nitrospira**

SECTION B AMMONIA-OXIDIZING BACTERIA

Genus **Nitrosomonas**
Genus **Nitrosococcus**
Genus **Nitrosospira**
Genus **Nitrosolobus**
Genus **"Nitrosovibrio"**

BUDDING AND/OR APPENDAGED BACTERIA

Subgroup 1 Prosthecate Bacteria

Genus **Ancalomicrobium**
Genus **Asticcacaulis**
Genus **Caulobacter**
Genus **Dichotomicrobium**
Genus **Filomicrobium**
Genus **Hirschia**
Genus **Hyphomicrobium**
Genus **Hyphomonas**

Genus **Labrys**
Genus **Pedomicrobium**
Genus **Prosthecobacter**
Genus **Prosthecomicrobium**
Genus **Stella**
Genus **Verrucomicrobium**

Subgroup 2 Order Planctomycetales

Genus **Gemmata**
Genus **"Isosphaera"**
Genus **Pirellula**
Genus **Planctomyces**

Subgroup 3 Other Budding and/or Appendaged Bacteria

Genus **Angulomicrobium**
Genus **Blastobacter**
Genus **Ensifer**
Genus **Gallionella**
Genus **Gemmiger**
Genus **Nevskia**
Genus **Seliberia**

SHEATHED BACTERIA

Genus **"Clonothrix"**
Genus **Crenothrix**
Genus **Haliscomenobacter**
Genus **Leptothrix**
Genus **"Lieskeela"**
Genus **"Phragmidiothrix"**
Genus **Sphaerotilus**

NONPHOTOSYNTHETIC, NONFRUITING GLIDING BACTERIA

Subgroup 1 Single Celled, Rod-Shaped Gliding Bacteria

Genus **Capnocytophaga**
Genus **Chitinophaga**
Genus **Cytophaga**
Genus **Flexibacter**
Genus **Flexithrix**
Genus **Lysobacter**
Genus **Microscilla**
Genus **Sporocytophaga**
Genus **Thermonema**

Subgroup 2 Flattened, Filamentous Gliding Bacteria

Genus **Alysiella**
Genus **Simonsiella**

Subgroup 3 Sulfur-Oxidizing, Gliding Bacteria

Genus **Achromatium**
Genus **Beggiatoa**
Genus **Thioploca**
Genus **"Thiospirillopsis"**
Genus **Thiothrix**

Subgroup 4 The Pelonemas

Genus **"Achroonema"**
Genus **"Desmanthos"**
Genus **"Pelonema"**
Genus **"Peloploca"**

Other Genera

Genus **Agitococcus**
Genus **Desulfonema**
Genus **Herpetosiphon**
Genus **"Isosphaera"**
Genus **Leucothrix**
Genus **Saprospira**
Genus **Toxothrix**
Genus **Vitreoscilla**

THE FRUITING, GLIDING BACTERIA: THE MYXOBACTERIA

Genus **Angiococcus**
Genus **Archangium**
Genus **Chondromyces**
Genus **"Corallococcus"**
Genus **Cystobacter**
Genus **"Haploangium"**
Genus **Melittangium**
Genus **Myxococcus**
Genus **Nannocystis**
Genus **Polyangium**
Genus **"Sorangium"**
Genus **Stigmatella**

ARCHAEA: THE METHANOGENS

Subgroup 1

Genus **Methanobacterium**
Genus **Methanobrevibacter**
Genus **Methanosphaera**
Genus **Methanothermus**

Subgroup 2

Genus **Methanococcus**
Genus **Methanocorpusculum**
Genus **Methanoculleus**
Genus **Methanogenium**
Genus **Methanolacinia**
Genus **Methanomicrobium**
Genus **Methanoplanus**
Genus **Methanospirillum**

Subgroup 3

Genus **Methanococcoides**
Genus **Methanohalobium**
Genus **Methanohalophilus**
Genus **Methanolobus**
Genus **Methanosarcina**
Genus **Methanothrix**

ARCHAEAL SULFATE REDUCERS

Genus **Archaeoglobus**

EXTREMELY HALOPHILIC, AEROBIC ARCHAEA (HALOBACTERIA)

Genus **Haloarcula**
Genus **Halobacterium**
Genus **Halococcus**
Genus **Haloferax**
Genus **Natronobacterium**
Genus **Natronococcus**

CELL WALL-LESS ARCHAEOBACTERIA

Genus **Thermoplasma**

EXTREMELY THERMOPHILIC AND HYPERTHERMOPHILIC S^0-METABOLIZERS

Subgroup 1

Genus **Acidianus**
Genus **Desulfurolobus**
Genus **Metallosphaera**
Genus **Sulfolobus**

Subgroup 2

Genus **Pyrobaculum**
Genus **Thermofilum**
Genus **Thermoproteus**

Subgroup 3

Genus **Desulfurococcus**
Genus **Hyperthermus**
Genus **Pyrococcus**
Genus **Pyrodictium**
Genus **Staphylothermus**
Genus **Thermococcus**
Genus **Thermodiscus**

Appendix B

Genera in Subgroups of Proteobacteria

Heterotroph		Chemoautotroph/Methanotroph	Phototroph
Alpha Proteobacteria			
Acetobacter	*Hyphomicrobium*	*Nitrobacter*	*Rhodobacter*
Acidiphilium	*Hyphomonas*		*Rhodomicrobium*
Agrobacterium	*Mycoplana*		
Rhodopseudomonas			
Ancalomicrobium	*Paracoccus*		*Rhodopila*
Ancylobacter	*Pedomicrobium*		*Rhodospirillum*
*Aquaspirillum**	*Phyllobacterium*		
Azospirillum	*Phenylbacterium*		
Beijerinckia	*Prosthecomicrobium*		
Blastobacter	*Rhizobium*		
Bradyrhizobium	*Rochalimaea*		
Brucella	*Stella*		
Caulobacter	*Xanthobacter*		
Gemmobacter	*Zymomonas*		
Gluconobacter			
Beta Proteobacteria			
Achromobacter	*Kingella*	*Methylomonas*	*Rhodocyclus*
Alcaligenes	*Leptothrix*	*Methanolomonas*	
*Aquaspirillum**	*Neisseria*	*Methanomonas*	
Bordetella	*Oligella*	*Nitrosococcus*	
Chromobacterium	*Simonsiella*	*Nitrosolobus*	
Comamonas	*Spirillum*	*Nitrosospira*	
Derxia	*Taylorella*	*Nitrosovibrio*	
Eikenella	*Vitreoscilla*	*Thiobacillus**	
Janthinobacterium	*Xylophilus*		
Gamma Proteobacteria			
Acinetobacter	*Leucothrix*	*Beggiatoa*	*Ectothiorhodospira*
Aeromonas	*Lysobacter*	*Thiomicrospira*	*Amoebobacter*
Alteromonas	*Marinomonas*	*Thiothrix*	*Chromatium*
Alysiella	*Moraxella*		*Lamprocystis*
Azomonas	*Oceanospirillum*		*Thiocapsa*
Azotobacter	*Pasteurella*		*Thiodictyon*
Branhamella	*Plesiomonas*		*Thiospirillum*
Deleya	*Pseudomonas*		
Enteric bacteria family including *Escherichia, Salmonella, Shigella, Proteus, Citrobacter, Klebsiella, Serratia, Yersinia*	*Ruminobacter*		
	Serpens		
	Vibrio group including *Vibrio, Listonella, Photobacterium, Shewanella*		
Frateuria	*Xanthomonas*		
Halomonas	*Xylella*		
Legionella			
Delta Proteobacteria			
Bdellovibrio	*Desulfuromonas*		
Desulfobacter	Myxobacteria including *Chondromyces, Cystobacter, Myxococcus, Nannocystis, Sorangium, Stigmatella*		
Desulfobulbus	*Pelobacter*		
Desulfococcus			
Desulfonema			
Desulfovibrio			

*indicates genera that have species in more than one subgroup

Glossary

Accessory pigment Pigments that trap light energy and transfer it to reaction center chlorophyll molecules in reaction centers. Carotenoids and phycobiliproteins are examples.

Accidental pathogen A microorganism that does not generally cause disease in its normal life cycle. Can cause disease if introduced in an unusual way.

Acetyl CoA pathway An autotrophic CO_2 fixation pathway utilized by selected anaerobic bacteria. The assimilation of two molecules of CO_2 results in formation of acetyl CoA.

Acid-fastness A property of some organisms such as mycobacteria that have a high lipid content in their outer wall. These organisms retain hot carbol-fuchsin stain when rinsed with acid-alcohol.

Acidophile An organism that preferentially grows at a pH below 5.4. Sulfolobus is one example.

Acquired immunity A specific immunity acquired by exposure to an antigen or by transfer of immune antibodies to a non-immune individual.

Activation energy Energy introduced that renders molecules capable of chemical interactions.

Active immunity An immune state generated by challenge with a specific antigen.

Active site The specific area of an enzyme where substrate is bound forming an enzyme-substrate complex.

Active transport Movement of ions or molecules across the cell membrane at an expenditure of energy.

Acute stage The stage in the infection cycle where symptoms are most pronounced.

Adenosine-5´-triphosphate (ATP) The energy carrier in all living cells.

Aerobe An organism that utilizes molecular oxygen as terminal electron acceptor in aerobic respiration.

Aerosol Liquid droplets suspended in air.

Aerotolerant An organism that does not utilize molecular oxygen as terminal electron acceptor but is not harmed by O_2.

Agar A sulfur containing polysaccharide of marine algal origin that is used as a solidifying agent in culture media.

Agglutination Sticking together of microbes, blood cells, or antigen-antibody causing an observable clump.

AIDS Acquired immune deficiency syndrome. An infection caused by HIV the human immunodeficiency virus.

Akinetes Resting bodies formed by some cyanobacteria.

Alcoholic fermentation Anaerobic metabolism where alcohol (ethanol, butanol, etc.) and CO_2 are a major product.

Algae Photosynthetic eukaryotic aquatic organisms.

Alkaliphiles Organisms that inhabit alkaline lakes and grow at a pH up to 11.5.

Allergy Inappropriate immune response to an antigen (allergen).

Allostery Change in the conformation of a protein resulting from the attachment of a compound at a site other than the reactive site. Generally a reversible inactivation of a protein, such as an enzyme.

Ames test A procedure for determining the mutagenic potential of a chemical.

Anabolism Series of reactions involved in the synthesis of macromolecules from monomers. Energy is generally required.

Anaerobe An organism that does not employ oxygen as terminal electron acceptor.

Anaerobic respiration Electron transport oxidation where sulfate, sulfur, nitrate, CO_2, or other oxidized compounds are utilized as terminal electron acceptor.

Anamnestic response Rapid immune response to antigens to which the host has previously been exposed.

Anaphylactic shock A destructive reaction between an antigen and antibody. Results in smooth muscle contraction.

Anaplerotic reactions Replacement reactions that permit the tricarboxylic acid cycle to continue. The replenishing of oxaloacetic acid by carboxylation of pyruvate is a key anaplerotic reaction.

Anion A negatively charged atom.

Anoxygenic photosynthesis Type where O_2 is not produced. Employs electron donors other than H_2O. H_2S, H_2, and reduced organic compounds would donate electrons in anoxygenic photosynthesis.

Antibiotic A metabolite produced by a microorganism that inhibits or destroys other microorganisms.

Antibody A glycoprotein present in body fluids that can bind specifically to an antigen. An immunoglobulin.

Anticodon The three base sequence on tRNA that is complementary to the three base codon of mRNA.

Antigen A substance recognized by the body as non-self and can elicit the immune response. An immunogen.

Antigen presenting cell (APC) Phagocytic cells that process antigens and present them to T-cells in the immune response.

Antigenic determinants Areas on an antigen that elicit an immune response. Also termed epitopes.

Antigenic drift Antigenic variation resulting from alterations in genes producing an altered antigen that can evade specific antibody.

Antigenic shift A change in antigenic character resulting from movement of segments of DNA in the genome to other sites.

Antimetabolite Compound that inhibits metabolism in microorganisms. General structure resembles a natural metabolite and competes for active sites on enzymes.

Antimicrobial agent Anything that inhibits or kills microbial cells.

Antiparallel The strands in double stranded DNA are oriented in opposing directions—one $3' \rightarrow 5'$ the other $5' \rightarrow 3'$.

Antiseptic A chemical generally applied externally that destroys microbes without serious harm to the tissue.

Antiserum A serum that contains specific antibodies.

Antitoxin An antibody specific for a toxic substance.

Apoptosis Elimination of aberrant cells by programmed cell death.

ARC *A*ids *R*elated *C*omplex is a series of symptoms associated with an HIV infection and may lead to an active case of AIDS.

Archaea One of the three Domains of living organisms formerly termed Archaebacteria.

Aseptic technique The manipulation of microorganisms such that contamination by undesirable organisms is prevented.

Assimilation Uptake of nutrient and conversion to cellular components.

Attenuation Weakening of a pathogen to be employed in immunization or a genetic regulation mechanism that terminates transcription.

Autecology Relationship of individual organisms or species and their environment.

Autochthonous Indigenous microbial populations.

Autoimmune Disorder in which individual becomes immune to self.

Autoimmune disease State where antibodies are formed against self-antigens.

Autolysin Enzymes in bacteria that form breaks in peptidoglycan permitting incorporation of newly formed wall units during growth.

Autolysis Process where a peptidoglycan is disrupted causing cell lysis caused by autolysins.

Autotroph An organism that can utilize CO_2 as sole source of carbon.

Auxotroph A mutant that has lost the ability to synthesize a metabolite normally synthesized by the microbial strain.

Axenic culture A culture free of contaminating organisms. A pure culture.

B cell (lymphocyte) An antibody bearing cell that can differentiate during the immune response to secrete antibody (plasma cell) or to form a memory cell.

Bacillus (bacillary) An oblong bacterium also called a rod.

Bacteremia Presence of bacteria in the blood stream; generally transient.

Bactericide Agent that kills bacteria. Mercuric chloride is a bactericide.

Bacteriocin Protein produced by a bacterial strain that kills related strains.

Bacteriophage Virus that infects Eubacteria or Archaea.

Bacteriostatic Inhibitor of bacterial reproduction or growth but does not kill.

Bacteroid An osmotically fragile bacterium that lives within protected areas such as nodules or the intestinal tract.

Balanced growth condition When all metabolic intermediates required for growth are available at the appropriate level.

Barophiles Organisms that thrive at high hydrostatic pressure.

Barotolerant Can grow under high pressure but generally grows better at normal pressure.

Base composition Relative percent of the total cellular DNA that is guanine-cytosine (G+C). Remainder is adenine-thymine (A+T).

Batch culture Growth of a microorganism in a closed vessel, on a suitable medium, at an appropriate temperature, and for a selected time.

Binary fission Simple division of a bacterium to yield two cells.

Bioconversion (biotransformation) Biologically induced modification of a substrate to generate a more useful product. The substrate converted is not utilized for growth.

Biochemical oxygen demand (BOD) Oxygen required by microorganisms to catabolize organic matter present in water.

Biodegradation Destruction of organic compounds by microorganisms.

Bioluminescence Production of light by living organisms.

Biomagnification Accumulation of substances, generally deleterious, by consumer organisms.

Biomarkers A chemical substance produced by an organism that serves as a signature for the presence or activity of that organism or group of organisms.

Biomass Total living cellular material in an environment.

Bioremediation Removal of toxic or undesirable environmental contaminants by microorganisms.

Biosynthesis (anabolism) Synthesis of macromolecules from monomers during cell growth.

Biotechnology Use of living organisms to generate useful industrial products. Usually involves genetic manipulation.

Broad-spectrum antibiotic Antimicrobial that is effective against both Gram-positive and Gram-negative bacteria.

Budding Asexual reproduction where progeny arise from protuberances on the surface of the parent cell.

Calvin cycle The major pathway for CO_2 fixation during photosynthesis in plants, algae, cyanobacteria, and many photosynthetic bacteria. Some chemolithotrophs also utilize this pathway.

Capsid The protein coat that surrounds viral nucleic acid.

Capsomers Protein subunits that together form an icosahedral capsid.

Capsule An organized layer of biosynthetic origin that surrounds the outer wall or envelope of a bacterium.

Carboxysomes Inclusion bodies present in autotrophs comprised of ribulose-1,5-bisphosphate carboxylase. This enzyme is involved in CO_2 fixation via the Calvin cycle.

Carcinogen An agent, generally a mutagen, that causes cancer.

Carotenoid Red to yellow pigments found in many bacteria. Can serve as accessory pigments in photosynthetic organisms. Also a pigment in many non-photosynthetic microorganisms.

Carrier An infected individual that transmits disease to others. Usually a carrier has a subclinical infection.

Catabolism Reactions involved in reduction of larger molecules to smaller molecules generally with the release of energy.

Catabolite repression Curtailment of the synthesis of selected enzymes by the availability of glucose or other metabolites as substrate.

Catalyst A compound that increases the reaction rate without itself being altered.

Cationic Having a positive charge.

CD4 Specific antigen on selected cells such as T-cells. $CD4^+$ cells are host for HIV.

Cell Basic unit of living matter.

Cell cycle Sequence involved in growth/division of a cell.

Cell-mediated immunity process Whereby T-cells destroy foreign or infected cells.

Cell wall A tough structure surrounding the bacterial cytoplasmic membrane that gives the cell shape and protection.

Chaperones Specific proteins that are involved in the folding and assembly of other proteins.

Chemical oxygen demand (COD) Amount of chemical oxidant necessary to oxidize organics in water to CO_2.

Chemiosmosis Development of a chemical potential across a cytoplasmic membrane that can drive ATP synthesis. Protons are driven outward by electron carriers or light.

Chemolithotroph (chemoautotroph) A microorganism that obtains its energy by oxidation of inorganics and utilizes CO_2 as carbon source.

Chemoorganotroph (heterotroph) A microorganism that obtains its energy by oxidation of organic compounds and utilizes carbonaceous substrates for growth other than CO_2.

Chemoreceptors Proteins in the cytoplasmic membrane or periplasm that bind chemicals involved in chemotaxis.

Chemostat A continuous culture apparatus that feeds medium, with one substrate at limiting concentration, into a culture vessel and permits removal of cells at a rate that maintains steady state growth.

Chemotaxis Movement of a microorganism toward a chemical attractant or away from a repellant chemical.

Chemotherapy Treatment of disease with chemicals that destroy the agent without serious harm to the host.

Chlorophyll A tetrapyrrole with a molecule of magnesium in the center that captures light energy during photosynthesis.

Chloroplast The chlorophyll containing organelle in photosynthetic eukaryotes.

Chlorosome An oblong protein-bound structure in green sulfur bacteria that contains the light gathering pigments. It is located on the inner surface of the cytoplasmic membrane.

Chromosome The site of cellular DNA. The Archaea and Eubacteria generally have one circular chromosome. One microorganism (*Borrelia burgdorferi*) is known that has a linear chromosome. Eukarya are multichromosomal.

Chronic infection Long term infection.

Classification Placing organisms in groups based on phylogenetic relatedness.

Clonal selection Theory that clones of specific B or T-cells are stimulated to reproduce when an appropriate antigen binds to their surface.

Clone Genetically identical progeny of a single parent.

Cloning vector A bit of replicative DNA that can transport inserted foreign DNA into a recipient cell.

Coccus (coccoid) A spherical bacterium.

Codon A sequence of three bases (purines or pyrimidines) in messenger RNA that codes for a specific amino acid.

Colicin Plasmid encoded proteins produced by enterics that harm other enteric bacteria.

Coliform Gram-negative facultative aerobic bacteria that ferment lactose with formation of gas within 48 hours at 35°C.

Colony A visible assemblage of microorganisms growing on a solid surface. Generally from reproduction of a single cell.

Colony-forming unit (CFU) Single cell that can form a colony on a solid medium.

Cometabolism Transformation of a non-growth substrate by microorganisms grown or growing on a utilizable substrate.

Commensalism A symbiotic association where one organism benefits and the other(s) is unaffected.

Common-source epidemic One where all victims are infected from a single source. Food poisoning is an example.

Community A mixed population of microorganisms in a natural habitat or microcosm.

Compatible solute Organic or inorganic source that increases the internal osmotic pressure inside a bacterium to the level outside.

Competent A bacterial cell that can take up DNA fragments and be transformed.

Competitive inhibitor A chemical analogue that replaces a natural substrate and competes for active sites.

Complement A series of proteins in the blood that act sequentially (cascade) and are involved in removal of antigen/antibody complexes.

Complex medium (undefined) A culture medium in which the chemical composition of the ingredients is not precisely determined.

Concatemer Viral genomic material in which individual genomes are linked end-to-end. The linear structure is split to individual genomes on viral assembly.

Congenital Disease contracted maternally during gestation or birth.

Conjugation Transfer of genetic information by cell-to-cell contact.

Consortium As assembly of several bacteria in which all benefit to some degree from the association.

Constitutive enzyme An enzyme not subject to regulation, always expressed during growth.

Co-repressor A low molecular weight compound that functions in repression of an enzyme.

Cortex Dense area between the endospore coat and core in endospore forming bacteria.

Covalent bond A chemical bond resulting from sharing of electrons.

Cross-resistance The development of resistance to one antimicrobial agent effects resistance to another.

Crossing-over Where adjacent DNA strands are exchanged.

Crustose Form of a lichen that is compact and grows close to a substrate.

Culture A strain growing on a laboratory medium.

Culture medium A liquid or solid nutrient on which microorganisms can be grown.

Cyanobacteria The Eubacteria that can grow photosynthetically via oxygenic photophosphorylation. Morphologically diverse.

Cyclic photophosphorylation Cyclical movement of electrons in photosystem I.

Cytochrome Heme proteins involved in electron transport.

Cytoplasm The liquid contents of a cell. Surrounded by a cytoplasmic membrane.

Cytoplasmic inclusion An intracellular storage granule in a bacterium.

Cytotoxin A toxic material that destroys cells. Some pathogens produce these compounds.

D value Decimal reduction time. Time required to reduce a population to one-tenth the original.

Dark-field microscopy A process for indirect illumination such that the specimen is illuminated against a dark background.

Death phase The phase in batch culture where the viable population is in decline.

Decomposition The break up of complex material to constituent parts.

Defined medium One in which the composition and quantity of all components are known.

Deletion Excision of a part of a gene causing mutation.

Denaturation Change in a protein or other macromolecule that destroys activity.

Dendritic cell An antigen presenting cell present in lymph nodes and spleen.

Denitrification Reduction of oxidized forms of nitrogen to nitrogen gas via anaerobic respiration.

Deoxyribonucleic acid (DNA) A nucleotide polymer composed of deoxyribonucleotides joined by phosphodiester bonds. The genome of an organism is composed of DNA.

Diauxie Biphasic growth that can occur when two substrates are available in a culture medium.

Differential media Culture media that permit growth of one bacterial type and inhibit others or permit a microorganism to demonstrate specific biological properties.

Differentiation Changes in the structure of a microorganism during the growth cycle.

Dimer Compound formed by joining of two monomers.

Disease Disturbance of normal structural or functional capacity of an organism.

Disinfect A way to destroy all microorganisms.

DNA library (gene library) Cloned DNA fragments that represent the genes in the entire genome of an organism.

Domain The highest level of classification of all life based on rRNA analysis. The three domains are Eubacteria, Archaea, and Eucaryota. A domain is also used to describe an area of a macromolecule.

Doubling time Time required for a population to double in number.

Early message Messenger RNA formed immediately after infection that is translated to catalytic proteins that disrupt host cell function.

Ecosystem Total organismic community and associated abiotic components of a selected environment.

Ectomycorrhizae Association between fungi and root tip where the fungi form a sheath about the root.

Electron acceptor The component that accepts electrons during an oxidation-reduction reaction.

Electron donor The component that donates electrons during an oxidation-reduction reaction.

Electron transport chain Series of electron acceptors and donors that transfer electrons from an electron carrier such as NADH to a terminal acceptor such as O_2. Also function during photophosphorylation.

Electrophoresis Separation of charged molecules such as protein or DNA in an electrical field.

Electroporation Use of an electrical pulse to alter the cytoplasmic membrane promoting uptake of DNA fragments.

Embden-Meyerhof (glycolysis) An anaerobic pathway that catabolizes glucose to two pyruvic acid molecules and generates two molecules of ATP.

Endemic disease One that is present at a constant low level in a population.

Endergonic reaction One that requires energy input to proceed.

Endocytosis Uptake of a particle, such as a virus, by enclosing via membrane extension and pinching off the vesicle formed.

Endogenous pyrogen A protein that induces fever in a host.

Endomycorrhizae A fungus-plant root association in which the fungus penetrates into the root cells.

Endophyte A microorganism, often a cyanobacterium, that lives within a plant.

Endospore A heat resistant dormant cell that forms within the cell of selected bacteria.

Endosymbiosis Growth of a microorganism within another organism generally beneficial to both.

Endosymbiotic theory The theory that the mitochondrion, chloroplast, and other organelles in eukaryotes arose through an endosymbiotic association of bacteria with eukaryotic ancestors.

Endotoxin The lipopolysaccharide portion of the outer envelope of Gram negative bacteria released by cell lysis or during growth and is toxic to animal hosts.

Energy Capacity to do work.

Enrichment culture Technique that involves addition of a selected substrate to a culture medium to isolate an organism that can grow on that substrate. Physical conditions may also be adjusted to obtain a specific type organism.

Enteric Intestinal. Often employed to describe microorganisms associated with the intestinal tract.

Enterotoxin A toxin that adversely affects the intestinal tract.

Entner-Doudoroff A non glycolytic pathway of glucose catabolism that produces pyruvate and glyceraldehyde-3-PO_4.

Envelope The outer structure in Gram negative bacteria or the membranous layer that surrounds the capsid of some viruses.

Enzyme A highly specific protein catalyst.

Enzyme-linked immunosorbent assay (ELISA) A technique that is employed to detect specific antibodies employing enzymes as the detector system.

Epidemic A marked increase in the number of individuals affected by an infectious disease during a given period of time.

Epidemiology The study of prevalence, incidence, and transmission of infectious diseases.

Epilimnion The aerobic warmer layer of water in a stratified lake that lies above the thermocline.

Episome A plasmid that can integrate into the host genome or function separately.

Epitope A specific area on an antigen that elicits an antibody response.

Etiologic agent The specific cause of a disease.

Eubacteria One of the Domains along with Archaea and Eukarya.

Eukarya One of the Domains. Composed of organisms that have a membrane bound nucleus and division of DNA by mitosis.

Eutrophic An environment, usually aquatic, overly rich in nutrient.

Exergonic A reaction that liberates energy.

Exon Sequences in a split gene that code for messenger RNA.

Exotoxin A toxin released by cells generally during growth.

Exponential growth Balanced growth where each microbial cell is dividing during a fixed period of time.

Extracellular enzyme One excreted by a microorganism to cleave large molecules to a size transportable into the cell.

Extreme Employed with thermophiles or halophiles to indicate that they grow at the highest temperature or salt concentration of microorganisms presently described.

Facilitated diffusion Carrier mediated transport across a cytoplasmic membrane.

Facultative Term applied to aerobes and autotrophs to indicate that the organism can grow either aerobically/anaerobically or autotrophically/heterotrophically under appropriate conditions.

Fastidious Complex growth requirements generally due to the inability of a microorganism to synthesize requisite monomers.

Feedback inhibition Condition where the end product of a biosynthetic pathway can curtail activity of enzymes involved earlier in the sequence of synthetic reactions leading to that end product.

Fermentation Production of ATP via catabolic sequences in which organic compounds serve as electron donor and organic intermediates as electron acceptor.

Filamentous A microorganism that in normal growth forms long strands.

Fimbriae Filamentous structures on bacteria that play a role in adherence or in formation of pellicles or masses of cells.

Flagella Structures involved in motility of bacteria. They are mainly composed of a protein called flagellin.

Flavoproteins Riboflavin derivatives that are electron carriers.

Fluid mosaic model The accepted structure of cytoplasmic membranes with the phospholipid bilayers forming a double track. Proteins are embedded through the membrane or peripherally at either surface.

Fluorescence Emission of light of distinct wavelength after activation with light of a different wavelength.

Fluorogenic compound Is a chemical substance that becomes fluorescent when cleaved enzymatically.

Foliose Lichens that have a leaf like appearance.

Fomites Inanimate objects that can harbor pathogens and transmit them to hosts.

Food infection An infection caused by ingestion of food contaminated with disease causing pathogens.

Food poisoning Illness caused by toxins present in food due to growth of microorganisms on the foodstuff prior to ingestion.

Food web Interlinked food chains.

Fragmentation Reproduction in actinomycetes where filaments break into individual cells.

Frameshift mutation Mutations that arise when the RNA polymerase begins producing mRNA at a base other than the proper codon. One base off results in a protein of different amino acid sequence.

Free energy Total energy available to do useful work.

Fruiting body A reproductive structure in some bacteria such as myxobacteria or actinomycetes. Fungi commonly produce fruiting structures.

Fruticose A lichen that is bush or tree-like.

Fungi Heterotrophic eukaryotes that have rigid walls.

Fungicide An agent that kills fungi.

G+C The DNA of an organism has guanine paired with cytosine and adenine with thymine. G+C refers to the percent of the total DNA that is guanine-cytosine pairs.

Gas vesicles Organelles with protein membranes that fill with gas in aquatic bacteria; serve as flotation devices.

Gastroenteritis Inflammation of the stomach lining or intestines often caused by food poisoning or infections.

Gene A segment of the genome that codes for a specific polypeptide, protein, tRNA, or rRNA.

Generalized transduction Transfer of any part of a bacterial genome when packaged randomly in a phage capsid.

Generation time Time required for a population to double in number.

Genetic code Triplet nucleotide sequences that specify a specific amino acid in a protein chain.

Genetic engineering Modification of the genome of an organism.

Genetic map The precise sequence of genes in the genome.

Genome The complete genetic information in a cell or virus.

Genotype Heritable genetic information in a cell.

Genus A grouping of organisms that are closely related phylogenetically.

Geosmin Organic molecules produced by actinomycetes and some cyanobacteria that give soil or water a distinct aroma.

Germicide An agent that kills bacteria.

Germination Loss of dormancy in an endospore.

Glider An organism that is motile without aid of flagella. Gliders move only when on a semisolid surface.

Glycocalyx Equivalent to capsule, the layer outside the cell envelope or wall.

Glycogen A branched polysaccharide composed of glucose used as a storage granule.

Glycolysis Anaerobic conversion of glucose to pyruvate via the Embden-Meyerhof pathway generating ATP.

Glycosidic bonds Covalent bonds between sugars in polysaccharides.

Glyoxylate cycle A modification of the tricarboxylic acid cycle where isocitrate is cleaved to form succinate and glyoxylic acid. The latter condenses with acetate to form malate. Functional in organisms growing on two-carbon compounds such as acetate.

Gram stain A differential stain that separates bacteria on retention of the dye crystal violet. Those with an outer envelope are decolorized by an ethanol wash (Gram-negative) while those without an outer envelope retain the dye (Gram-positive).

Growth factor Low molecular weight compounds that must be added to growth media for selected organisms because they cannot synthesize them.

Growth rate The rate at which bacteria reproduce.

Growth yield A measure of cellular mass generated per ATP produced from a substrate.

Halophile An organism that requires very high salt (sodium chloride) for growth.

Hapten A molecule (generally low molecular weight) that cannot elicit an antibody response against self unless coupled with a larger molecule.

Heat shock proteins Proteins produced under stress, particularly heat that protects the cell.

Helix A helical structure present in DNA and some proteins formed by hydrogen bonding.

Helper T-cell A T lymphocyte that cooperates with a B-cell in initiating the antibody response.

Hemagglutination Coagulation of red blood cells.

Hemolysins Bacterial toxins that can disrupt cytoplasmic membranes of cells. Generally hemolysins are assayed by use of red blood cells. Important in tissue invasion to release substrates for growth of pathogens.

Herd immunity Immunity to a pathogen in the majority of potential hosts results in the inability of the pathogen to cause disease.

Heterocyst Specialized cells in cyanobacteria that are sites of nitrogen fixation.

Heterofermentation Fermentation of sugars to a mixture of products.

Heterotroph An organism that utilizes preformed organic substrates as major source of carbon.

High energy compound One that yields free energy on hydrolysis.

Holdfast Material produced by some sessile microorganisms that aids in attachment to solid surfaces.

Homofermentation (homolactic) Fermentation of sugars, mostly to lactic acid.

Host An organism on which a parasite can grow.

Human immunodeficiency virus (HIV) The retrovirus that is the etiological agent of AIDS.

Humoral immunity An immune response involving antibodies.

Hybridoma Fusion of a cancer cell to a specific B-cell to generate monoclonal antibodies.

Hydrogen bond A weak, but important, bond between a hydrogen atom and an electronegative atom such as oxygen or nitrogen. Important in helix formation and other macromolecular interactions.

Hydrolysis Cleavage of a structure by addition of water.

Hydrophilic Affinity for and solubility in water.

Hydrophobic Lacking affinity for water. Mostly insoluble in water.

Hypersensitivity Harmful, exaggerated immune response.

Hyperthermophile An Archaeum or Eubacterium that has an optimum growth temperature above 80°C.

Hypolimnion The layer in a stratified lake that has a uniform cold temperature and a low level of oxygen.

Icosahedral A virus capsid having 20 equilateral triangular faces and 12 corners.

Idiophase End of the exponential growth phase when secondary metabolites are synthesized in selected organisms such as antibiotic producing actinomycetes.

Immobilized enzyme One attached to a solid support. May effectively convert substrate to product.

Immune Resistance to infectious agents generally due to specific antibodies.

Immune response Bodily response to the presence of an antigen.

Immunization Eliciting of an immune response by introduction of specific antigen.

Immunoblot Electrophoretic transfer of immobilized proteins to nitrocellulose sheets to detect complementary proteins with specific antibody.

Immunodeficiency Inability to generate normal antibody responses to specific antigens.

Immunogen (antigen) Substance that will elicit an immune response.

Immunoglobulin (Ig) Blood protein fraction that is composed of antibodies.

Incidence Number of cases of a disease in a subset of the general population.

Indicator organism A bacterial species that is employed to determine whether an environment is contaminated by human waste. An organism that does not survive for extended periods but generally outlives enteric pathogens would be an effective indicator.

Inducible enzyme Enzyme that is synthesized in the presence of a specific substrate (inducer).

Infection Presence and growth of an organism within a host.

Infection thread A minute tunnel through the roots of a legume in which *Rhizobium* cells migrate to reach a tetraploid cell in which they establish a nodule.

Inflammation A localized response to tissue injury. Characterized by pain, swelling, and redness. The host's mechanism for getting leukocytes and other protective cells to the site.

Initiation site Area of the genome where replication originates.

Insertion Placing of a piece of DNA into the interior of a gene.

Integration Incorporation of a DNA sequence into the genome.

Interference Resistance in a lysogenicized bacterial cell to invasion by another virus.

Interferons Glycoproteins that stimulate virus infected cells to produce antiviral proteins that inhibit viral nucleic acid synthesis.

Intron Noncoding intervening sequences in a split gene. Does not appear in the ultimate RNA product.

Invasiveness Innate ability of a pathogen to produce substances that aid in dissemination of the organism within the host.

In vitro Outside a living organism, such as a test tube experiment.

In vivo In a living organism.

Ionizing radiation High energy radiation that effects loss of electrons from atoms.

Ionophore A compound that disrupts cytoplasmic membranes resulting in leakage of cytoplasm.

Isotopes Form of an element with an increased number of neutrons, but normal electron and proton complement.

Joule A unit of energy.

Kilobase pairs 1,000 base pairs in a fragment of DNA.

Kirby-Bauer An antibiotic disk diffusion assay for susceptibility of a clinical isolate to antibiotics or chemotherapeutics.

Koch's postulates Expression of rules for proving the relationship between a specific microorganism and disease.

Lactic acid fermentation Generally glycolytic with lactic acid as major product.

Lag phase Period after inoculation where there is no increase in population.

Late message Messenger RNA produced sometime after infection coding for proteins that are involved in virion synthesis.

Latent virus One present in a host without causing detectable symptoms.

Leaching Release of minerals from ore by microbial action.

Lectin Surface protein of a plant cell where microorganisms, such as nitrogen fixation bacteria, can attach.

Legume Plant that can develop nodules for nitrogen fixation.

Lethal dose 50 (LD_{50}) Number of microorganisms or level of toxin that will kill 50 percent of a test population within a fixed time.

Leukocidin A microbial toxin that will destroy phagocytes.

Leukocyte A white blood cell.

Lichen A beneficial symbiotic association between a fungus and a cyanobacterium or alga.

Lipopolysaccharide (LPS) Complex structure containing fatty acids and sugars present on the outer envelope of Gram-negative bacteria.

Log phase Exponential phase in growth curve.

Lophotrichous Tuft of flagella at one or both poles of a rod shaped bacterium.

Lymph Clear yellowish fluid that bathes tissue and flows through the lymphatic system.

Lymphocyte A leukocyte involved in antibody formation or in cellular immune responses.

Lyophilization Rapid dehydration of frozen material in a vacuum.

Lyphokine Protein secreted by lymphocytes such as activated T-cells. Mediators in the immune response that transmit signals cell to cell.

Lysis Physical disintegration of a cell.

Lysogenicized cell An archaeal or eubacterial cell bearing a prophage.

Lysosome An organelle in eukaryotic cells that contains digestive enzymes.

Lysozyme Enzyme that disrupts the β(1-4) bond in peptidoglycan causing cell disruption.

Lytic cycle Life cycle of a virus that effects lysis of the host cell.

Macromolecule A large molecule formed by polymerization of small molecules.

Macrophage Phagocytic cell present in blood, lymph, and tissue. These cells destroy pathogens and some are involved in the immune response.

Magnetosomes Magnetite (Fe_3O_4) particles that are present in selected bacteria. They are tiny magnets that align the bacterium in a magnetic field.

Magnetotactic Bacteria that orient themselves in a magnetic field.

Maintenance energy Energy required by a cell for remaining viable.

Major histocompatability complex (MHC) Cell surface antigens present in all individuals that are the unique marker of self. They are encoded by a family of genes.

Mast cell Cells that line blood vessels and secrete histamine and other active products in the inflammatory response and hypersensitivity.

Medium The nutrient employed in the growth of microorganisms.

Meiosis Division of a diploid cell to form two haploid daughter cells.

Memory cell Differentiated specific B-cells formed during the immune response that can convert to plasma cells when activated by the presence of a specific antigen.

Mesophile Microorganisms that grow optimally at temperatures between 18 and 40°C.

Messenger RNA (mRNA) Single stranded RNA complementary to template DNA formed via transcription. mRNA generally carries information for one polypeptide and is translated by ribosomes.

Mesosome Cytoplasmic membrane invagination in bacteria.

Metabolism The sum of all biochemical events in a cell.

Metachromatic granules Polyphosphate polymers stored by some bacteria.

Methanogen Archaea that generate methane in anaerobic environments.

Methylotroph Aerobic Eubacteria that utilize methane, methanol, and other one carbon compounds as carbon and energy source.

Microaerophile Microorganisms that require O_2 but at lower levels than atmospheric pressure.

Microbiota Microscopic organisms collectively present in an environmental niche.

Microcyst Resting spherical structures formed by some Cytophaga and Azotobacteria.

Mineralization Complete biodegradation of organic matter to CO_2 and inorganics.

Minimum inhibitory concentration (MIC) Smallest amount of an antimicrobial agent that inhibits growth of microorganisms.

Mitosis The division of nuclear material in a eukaryote yielding two nuclei of identical chromosomal composition.

Mixed acid fermentation One that generates mixed organic acids (lactic, acetic, etc.) as products.

Mixotroph A microorganism that assimilates organic carbon sources while using inorganic energy sources.

Monoclonal antibody An antibody of a specific type produced by a clone of identical plasma cells.

Monomer A low molecular weight intermediate utilized in the synthesis of cellular macromolecules.

Monotrichous A bacterium with a single flagellum.

Morbidity rate Incidence of a particular disease in a population during a specified time period.

Mordant A substance that increases the affinity of a cell for a dye.

Mortality rate Ratio of deaths from a particular disease relative to the total number infected.

Most probable number (MPN) Measure of the number of microorganisms by dilution. End point is highest dilution yielding growth.

Motility Purposeful movement of a microorganism.

Mutagen A physical agent (radiation) or chemical that induces mutation.

Mutation Heritable change in the genetic make up of a species. A mutant.

Mutualism A symbiotic association where the partners will gain.

Mycorrhiza Fungal-plant root associations.

Natural killer cells (NK) Lymphocytes present normally (nonimmune) that nonspecifically destroy aberrant host cells or invaders.

Neurotoxin One that harms nerve tissue.

Nitrification Oxidation of ammonia to nitrate in the environment.

Nitrogen fixation Reduction of atmospheric nitrogen to ammonia. A property present in some Eubacteria and Archaea. The enzyme involved is nitrogenase.

Nonsense codon One that does not code for an amino acid but as a signal to terminate protein synthesis.

Northern blot A technique of hybridization of single stranded RNA or DNA to nucleic acid fragments attached to a matrix.

Nosocomial infection An infection acquired in a health care facility or hospital.

Nucleic acid probe A labeled single strand of nucleic acid that can hybridize with a complementary strand in a crude mixture. Employed in pathogen identification.

Nucleocapsid The basic unit of a virion, nucleic acid surrounded by protein capsid.

Nucleotide A monomeric unit that can polymerize to form nucleic acid. Composed of sugar (ribose or deoxyribose), phosphate, and a purine or pyrimidine base.

Nutrient Any substance that is assimilated by a microorganism during growth.

Obligate A term employed by microbiologists to indicate an absolute requirement. Obligate aerobe, obligate autotroph, etc.

Okazaki fragments Short fragments of DNA involved in discontinuous replication of DNA on the lagging strand.

Oligotroph A microorganism that can live under low nutrient conditions.

Oncogene A gene that when expressed can convert a normal cell to a tumor cell.

Operator A specific segment of DNA at the start of a gene where a protein (repressor) can bind to control mRNA synthesis.

Operon A sequence in DNA that is controlled by an operator and contains one or more structural genes.

Opines Derivatives of the amino acid arginine or sugars that are synthesized in plants transformed by *Agrobacterium tumefaciens.* Utilization of these compounds is restricted to the crown gall inducing pathogenic agrobacteria.

Opportunist A microorganism that is generally harmless but can cause disease under certain conditions or in an immuno-compromised host.

Opsonins Proteins that promote phagocytosis.

Organelle A membrane bound functional structure in a cell.

Osmophiles Microorganisms that grow best in media of high solute concentration.

Outbreak Sudden high incidence of disease in a given population.

Oxidation Donation of electrons.

Oxidation-reduction (redox) reaction Coupled reaction where one partner donates electrons and the other accepts them. One is oxidized, the other reduced.

Oxidative phosphorylation Employment of the electron transport system to generate a proton motive force that provides energy for ATP synthesis.

Oxygenic photosynthesis Cyanobacterial or plant type photosynthesis where water serves as electron donor resulting in oxygen production.

Pandemic An epidemic worldwide. HIV is now a pandemic.

Parasite A symbiotic association where one organism benefits and the other is harmed.
Passive diffusion Movement of a molecule from an area of high concentration to one of lower concentration.
Passive immunity A short lived immunity gained by transferring specific immune antibodies to a non-immune individual.
Pasteurization Heating a fluid to temperatures that destroy spoilage or disease causing organisms.
Pathogen A disease causing organism.
Pathogenicity A relative term indicating the disease causing potential of a microorganism.
Pentose phosphate pathway A pathway that oxidizes glucose-6-phosphate to ribulose-5-phosphate. A source of 5-carbon sugars for DNA and RNA synthesis.
Peptide bond A covalent bond formed between the carboxyl of one amino acid and the amine of another formed by dehydration.
Peptidoglycan (murein) The polymeric cell wall structure present in most Eubacteria. It is formed by alternating units of N-acetylglucosamine and N-acetylmuramic acid. The N-acetylmuramic acids are cross-linked by amino acids.
Peptones Proteins that are enzymatically digested that are used in preparation of culture media.
Periplasm (periplasmic space) A space between the cytoplasmic membrane and peptidoglycan layer (Gram-positives) and the outer envelope (Gram-negative).
Peritrichous Having flagella distributed over the surface of a bacterium.
Permease A protein in the cytoplasmic membrane that transports material inward.
Phagocyte A blood cell that can capture and digest foreign material.
Phagocytosis Engulfment of a foreign particle or bacterial cell.
Phagosome A membrane-bound vacuole formed in phagocytes by invagination of the cell membrane about a foreign particle.
Phenol coefficient Relative strength of a disinfectant compared with phenol.
Phenotype Expressed properties of a microorganism.
Phospholipid The component part of a cytoplasmic membrane composed of fatty acids, glycerol phosphate, and generally with a polar molecule linked to the phosphate.
Photoautotroph An organism that can utilize light as energy source and CO_2 as carbon source.
Photoheterotroph A microorganism that utilizes light energy while assimilating organic compounds as carbon source.
Photophosphorylation Generation of a proton motive force by use of light energy. This energy drives ATP synthesis from ADP and inorganic phosphate in ATPase.
Photosynthesis Use of light energy to generate chemical energy for cell maintenance and CO_2 assimilation.
Phototaxis Purposeful movement of an organism toward favorable wavelengths of light.
Phototroph Organism that utilizes light as source of energy.
Phycobiliproteins Phycoerythrin (red) and phycocyanin (blue) pigments that trap light in phycobilisomes.
Phycobilisomes Specialized structures on cyanobacterial membranes involved in photophosphorylation.
Phylogeny The evolutionary history and genetic relationships between organisms.
Pili Protein filaments present on bacterial cells that are involved in conjugation.
Pinocytosis A process in eukaryotes for uptake where the cytoplasmic membrane encloses material and transports it inward.
Plaque A clear area in a lawn of cells on an agar surface resulting from lysis of cells by a virus, or a microbial colony attached to and growing on a tooth.
Plasma cell A short lived differentiated B-lymphocyte that synthesizes and secretes quantities of specific antibody.
Plasmid A double strand of circularized DNA that exists and replicates independently in a cell. May integrate into the chromosome. Can carry information for specialized catabolic enzymes or drug resistance.
Plasmolysis A bacterial cell placed in high solute may lose water causing shrinkage of the cytoplasmic membrane and potential death.
Pleomorphic Bacteria that are variable in shape.
Plus strand Viral nucleic acid that is of a base sequence that can serve as mRNA.
Point mutation Affecting a base pair in a specific location of a genome.
Polar flagellum One at the end or both ends of a bacillary bacterium.
Poly-β hydroxybutyrate (PHB) A bacterial storage product consisting of a linear polymer of β-hydroxybutyric acid.
Polymerase chain reaction (PCR) Amplification of DNA *in vitro* by synthesis of specific nucleotide sequences from a small amount of DNA. This technique employs oligonucleotide primers complementary to sequences in the DNA and heat-stable DNA polymerases.
Polymorphonuclear leukocyte (PMN) Motile white blood cells that specialize in phagocytosis.
Porin Proteins that form channels in the outer envelope of Gram-negative bacteria.
Porters Membrane proteins involved in transport, both inward and outward.
Posttranslational modification Processing of the RNA transcript after formation to generate mRNA.

Precursor metabolites Twelve intermediates that originate in glycolysis, pentose phosphate pathway, or tricarboxylic acid cycle from which all cellular constituents are synthesized.

Prevalence Total percent of the population infected with a disease at a given time.

Pribnow box A base sequence located about 10 base pairs upstream from the transcription start site. It is the binding site for RNA polymerase.

Primary metabolites Products secreted during the growth phase. Lactic acid or ethanol would be examples.

Primary producer Autotrophic organisms that fix atmospheric CO_2 thus providing sustenance for the ecosystem.

Primer A polynucleotide to which the DNA polymerase attaches during DNA replication.

Prion An infectious particle in which no nucleic acid has been detected.

Prokaryote A name previously applied to all microorganisms that lacked a nuclear membrane. Now replaced by Archaea and Eubacteria.

Promoter The region on DNA at the start of a gene where RNA polymerase binds to initiate transcription.

Prophage State where a temperate viral genome is integrated into the host and replicates in concert with the host genome.

Prostheca Extension of the wall and cytoplasmic membrane to form hyphae, stalks, or unusual shaped bacteria.

Protease An enzyme that cleaves amino acids from a protein.

Protomer Subunit of a viral capsid.

Proton motive force (PMF) An energized state of a cytoplasmic membrane created when the electron transport system extrudes protons. The positivity on the exterior and negativity on the interior creates a charge separation that can drive ATP synthesis.

Protoplast An osmotically sensitive bacterial or fungal cell resulting from removal of the cell wall. The cytoplasmic membrane remains intact.

Prototroph (wild type) Parent organism from which an auxotrophic mutant has been derived.

Pseudomurein Modified peptidoglycan that is present in the cell walls of some Archaea.

Psychrophile A cold loving organism that can grow optimally as 12–15°C and at temperatures below 0°C.

Psychrotroph An organism that grows slowly at 0°C but optimally at around 20°C.

Public Health The general health of the population monitored by state and federal agencies.

Pure culture (axenic) A culture containing a single strain.

Purple membrane Areas in cytoplasmic membranes of halophilic bacteria containing bacteriorhodopsin.

Pyogenic Pus forming infection.

Pyrogen A fever-inducing molecule.

Quarantine Restriction of movement of individuals to prevent spread of a contagious disease. Also to isolate animals to ensure that they do not have a disease.

Quellung reaction Enlargement of a capsulated microorganism in the presence of antibodies to capsular antigen.

Racking Removal of sediments formed in wine bottles during fermentation.

Radioimmunoassay A technique that employs radioisotope-labeled antibody or antigen to detect presence of specific material in body fluids.

Radioisotope An element with a surplus of neutrons that spontaneously decays with emission of detectable radioactive particles.

Reaction center Complex containing multiple bacteriochlorophyll or chlorophyll molecules where photophosphorylation occurs.

Recalcitrance A relative term indicating the resistance of a molecule to microbial attack. Generally measurable by the half-life of the compound in the environment.

Recombinant DNA technology Genetic engineering involving the introduction of a gene into a vector such as a plasmid and introducing this into another species to yield a recombinant molecule.

Recombination Combination of genetic material from two separate genomes.

Reduction potential Tendency of a reduced molecule to donate electrons or an oxidized molecule to accept electrons.

Refraction Bending of light as it passes from one medium to another.

Regulation Sum of the cellular processes that control enzyme function to ensure balanced growth.

Replication Copying of genomic DNA to generate a duplicate copy.

Replication fork A Y-shaped structure where double stranded DNA separates permitting each of the single strands to be replicated.

Repressible enzymes The synthesis of these enzymes can be curtailed by the presence of an external intermediate termed the repressor, a specific protein.

Repressor protein A protein that can bind to an operator to prevent transcription.

Reservoir The site or host in nature where pathogenic microorganisms reside and act as a source of infection for humans or other species.

Respiration Energy-yielding catabolic reactions that utilize organic or inorganic compounds as electron donor and organic or inorganic electron acceptors.

Restriction endonucleases Enzymes that cleave DNA sequences at specific sites. Probably originated to protect cells against viruses. Now employed in genetic engineering.

Retrovirus Viruses that have a single stranded RNA genome that is replicated by reverse transcription to form a single stranded complementary DNA copy that is duplicated to form double stranded DNA. This double stranded DNA can be integrated into the genome of the host.

Reverse transcriptase An RNA dependent DNA polymerase that generates DNA from RNA.

Rhizosphere The area surrounding a plant root.

Rho A protein that functions to dissociate RNA polymerase after transcription is complete.

Ribonucleic acid (RNA) A polymer composed of ribonucleotides joined by phosphodiester bonds. The bases in RNA are uracil, adenine, guanine, and cytosine.

Ribosomal RNA (rRNA) The RNA present in a ribosome and involved in protein synthesis.

Ribosome The cellular organelle where protein synthesis occurs.

Ribozyme RNA that can function as an enzyme.

RNA polymerase The enzyme that synthesizes mRNA from the DNA template.

Root nodule Enlarged structure on a leguminous plant root where nitrogen-fixing endosymbionts live.

R plasmids Plasmids that carry antibiotic resistance genes.

Rumen The large vat in a ruminant (herbivore) where cellulosic material is digested anaerobically generating organic acids that sustain the animal.

Saprophyte An organism that generally grows on decaying organic matter.

Scale-up Gradual increase in size of industrial fermentation to large fermentors for production of useful chemicals.

Secondary metabolite Products that are synthesized near the end of the exponential growth phase and during early stationary phase.

Selection Establishing conditions where organisms with a predetermined desired trait are favored.

Selective medium One that favors growth of selected microorganism(s). The constituents may inhibit others.

Sense strand The strand of DNA that RNA polymerase copies to produce mRNA, rRNA, or tRNA.

Septicemia Infection of the blood stream with microorganisms or bacterial toxic products.

Serum Fluid portion of blood resulting from removal of blood cells and fibrinogen.

Sheath A tube-like structure that encloses chains of bacterial cells.

Shine-Dalgarno sequences A series of nucleotides on bacterial mRNA that binds to a specific sequence on 16S rRNA in order to properly orient the mRNA on the ribosome.

Siderophore A low molecular weight compound that complexes with ferric iron and makes it available for transport into a bacterium.

Sigma A protein that aids RNA polymerase in recognizing the promoter at the start site of a gene.

Signal sequence Hydrophobic amino acids on the lead end of proteins that move through cytoplasmic membranes to carry proteins outward that function in the periplasmic space.

Single cell protein Microbial cell biomass that is primarily a source of protein in animal or human nutrition.

S-layer A structured layer composed of protein or glycoprotein that covers the surface of selected bacterial species.

Slime layer Diffuse polymeric material exterior to the cell wall of microorganisms.

Solid waste The portion of sewage or industrial waste that settles out from the liquid phase.

SOS response A repair system that is induced in microorganisms that have been damaged.

Southern blot Hybridization of single strands of DNA or RNA to single stranded DNA immobilized on a matrix.

Species Closely related strains that differ from all other strains.

Spheroplast Removal of the rigid cell wall peptidoglycan from a Gram-negative bacterium yields an osmotically fragile cytoplasmic enclosed body. A spheroplast may contain the outer membrane.

Spike Projection from the outer envelope of a virus.

Spontaneous generation The long disproved hypothesis that living organisms can arise from inanimate matter.

Stalk An elongated structure that emanates from a bacterial cell often used in attaching the cell to a solid surface.

Starter culture An inoculum consisting of favored microorganisms employed in initiating industrial fermentations.

Stationary phase Phase in batch culture when growth ceases.

Sterile Free of all living things.

Sterilization Destruction of all living things.

Strickland reaction An ATP generating reaction employed by some clostridial species where one amino acid is oxidized and another serves as electron acceptor. The products are fatty acids and ammonia.

Stock culture A carefully maintained, stored culture from which working cultures are obtained.

Strain Descendants of a single organism.

Streak plate Isolation of colonies by spreading a culture over an agar surface with an inoculating loop.

Stromalolite A fossilized microbial mat often of cyanobacterial origin. Can be a rounded microbial mat of photosynthetic microorganisms.

Substrate The substance on which an enzyme acts. Also, the nutrient(s) on which a micoorganism grows.

Substrate level phosphorylation Generation of a high energy phosphate bond by reacting an inorganic phosphate with an activated organic compound.

Supercoiled DNA Twisted double strand of circular DNA.

Suppressor mutation A mutation that overcomes the effects of another mutation.

Svedberg unit A measure of the sedimentation rate of a macromolecule on centrifugation. Dependent on mass and density.

Symbiosis Living together or interaction of dissimilar organisms.

Synecology A branch of ecology that is involved with the development, structure, and distribution of ecological communities.

Synthetic medium A defined medium. One of known composition; qualitatively and quantitatively.

Synthrophy A species interaction where metabolic capabilities are shared to permit both to thrive.

T-lymphocyte (T-cell) A lymphocyte that matures in the thymus. T-cells are involved in many of the cell-mediated immune responses and in activating specific B-cells in the immune response.

Taxonomy Science of identification, classifiation, and nomenclature.

Teichoic acids Glycerol or ribitol polymers joined by phosphates present in cell walls of Gram positive bacteria.

Temperate virus One that can infect a host without effecting a lytic cycle. The viral genome may be integrated into the host and be replicated along with the host genome.

Template A strand of nucleic acid (DNA or RNA) that can specify the base sequence in a complementary strand.

Thermocline The layer of water in a thermally stratified lake where the temperature changes markedly with depths. The thermocline lies between the warmer epilimnion and colder hypolimnion.

Thermophile A microorganism that grows optimally at temperatures above 50°C with an upper limit of 75–80°C.

Thylakoid A series of flattened photosynthetic membranes that are impermeable to ions and whose function it is to pump out protons to establish a proton gradient essential for ATP synthesis. These membranes contain chlorophyll, electron transport chains, and specific proteins required for photosynthesis.

T*i* plasmid A plasmid present in the gall forming *Agrobacterium tumefaciens.* The plasmid has been used as a vector in transferring genes to plants.

Titer The highest dilution of antiserum that reacts visibly with an antigen.

Tolerance Generally refers to the nonresponse of antibodies to self.

Toxemia Effects of toxins present in the blood stream.

Toxic shock syndrome A potentially fatal massive inflammatory response to the presence in blood of the exotoxin of *Staphylococcus aureus.*

Toxigenicity The relative ability of a microorganism to produce toxin.

Toxin A potentially injurious substance, generally protein or lipopolysaccharide, produced by microorganisms.

Toxoid An exotoxin, such as tetanus toxin, that has been modified so that it does not cause damage but will elicit an antibody response.

Tracer A substrate that is radiolabelled in order to follow its incorporation into a living organism.

Transamination Transfer of an amine from one intermediate to a keto group of another.

Transcription Synthesis of a complementary strand of RNA from a single strand of double stranded DNA. Enzyme involved is a DNA dependent RNA polymerase.

Transduction Transfer of genetic information from one bacterium to another by a bacterial virus.

Transfer RNA (tRNA) A small RNA molecule that binds to a specific amino acid and delivers it to a ribosome during translation.

Transformation Gene transfer in microorganisms whereby free DNA is taken up by a cell and selected portions are integrated into the genome.

Transgenic Transfer of foreign genetic information to plants or animals by recombinant DNA.

Translation Synthesis of a protein with the aid of a ribosome using the information delivered by mRNA.

Transpeptidation Formation of the peptide bridge between the peptides that extend from N-acetyl muramic acid of peptidoglycan.

Transposable element Genetic material that can move from one genomic site to another.

Tricarboxylic acid cycle (Kreb's cycle, citric acid cycle) A series of reactions whereby a molecule of acetate is completely oxidized to CO_2 generating ATP. Intermediates in the cycle are some of the precursor metabolites involved in biosynthesis.

Trichome A series of bacterial cells in a filament that have close contact to one another. The cells lie parallel lengthwise.

T-suppressor (TS cell) A cell involved in turning off the antibody production.

Ultra-violet radiation Wavelengths of light (170–397 nm) that disrupt DNA nucleic acid sequences at the 5′ side of a DNA or RNA molecule.

Vaccine A material that can elicit beneficial antibodies to immunize against disease. Employed in vaccination.

Vector Any of various arthropods, animals that carry pathogenic organisms from one host to another, often from reservoir to humans.

Vehicle (formite) An inanimate object involved in transmission of infectious organisms.

Vesicle A membrane bound body. The membrane may be protein or lipid bilayer.

Viable Alive, but in microbiological terms, able to reproduce.

Viable count Determination of total viable cells in a sample.

Virion A mature virus particle that is the extracellular form of the virus.

Viroid An infectious agent of plants that is a very small single stranded RNA. Not associated with protein nor is it transcribed to generate protein.

Virulence A relative measure of the invasiveness and pathogenicity of a microorganism. Often measured by LD_{50}.

Virus An acellular infectious agent composed of DNA or RNA and able to commandeer the synthetic machinery of a host to generate virus.

Wastewater The liquid portion of sewage or output from a manufacturing process.

Water activity (a_w) A quantitative measure of water availability in an environment.

Wildtype Parental type as isolated from nature.

Winogradsky column A glass cylinder packed with mud bearing photosynthetic bacteria. The lower area is anaerobic and upper relatively aerobic. Placed in light all types of photosynthetic Eubacteria can thrive.

Xenobiotic A synthetic chemical.

Xerophile A microorganism that grows optimally under low a_w conditions.

Yeast A unicellular fungus.

Zoonoses A disease generally associated with animals other than humans but can be transmitted to humans.

Index

Page numbers in **boldface** indicate a figure; page numbers in *italics* indicate tabular material.